Point Group Symmetry Applications

Methods and Tables

Point Group Symmetry Applications

Methods and Tables

Philip H. Butler
University of Canterbury
Christchurch, New Zealand

PLENUM PRESS • NEW YORK AND LONDON

Library of Congress Cataloging in Publication Data

Butler, Philip H
Point group symmetry applications.

Bibliography: p.
Includes index.
1. Symmetry groups. 2. Point defects. 3. Crystallography, Mathematical. I. Title.

QD911.B89 530.1'2 80-17947
ISBN 0-306-40523-7

A Division of Plenum Publishing Corporation
233 Spring Street, New York, N. Y. 10013

Printed in the United States of America

Preface

The mathematical apparatus of group theory is a means of exploring and exploiting physical and algebraic structure in physical and chemical problems. The existence of structure in the physical processes leads to structure in the solutions. For group theory to be useful this structure need not be an exact symmetry, although as examples of exact symmetries we have that the identity of electrons leads to permutation symmetries in many-electron wave functions, the spatial structure of crystals leads to the Bloch theory of crystal eigenfunctions, and the rotational invariance of the hydrogenic Hamiltonian leads to its factorization into angular and radial parts.

In the 1930's Wigner extended what is known to mathematicians as the theory of group representations and the theory of group algebras to study the coupling coefficients of angular momentum, relating various properties of the coefficients to the properties of the abstract group of rotations in 3-space. In 1949 Racah, in a paper on rare earth spectra, showed that similar coefficients occur in other situations. Immediately a number of studies of the coefficients were begun, notably by Jahn, with his applications in nuclear physics. In the years since then a large number of physicists and chemists have added to the development of a general theory of the coefficients, or have produced specialized tables for a specific application. Applications now range from high-energy physics to biology.

Two unfortunate consequences of the wide range of applications coupled with the simultaneous development of the appropriate mathematics are that separate notations and terminologies have been introduced in the different areas, and that the most powerful techniques have not been available to everyone. The theory of coupling coefficients for compact groups has now advanced to a stage that I have felt it timely to produce a complete set of coupling coefficients for all the point groups—those groups that are subgroups of the group of orthogonal transformations in 3-space, $\mathbf{O}_3$. I have taken the opportunity to include a modern introduction to the relevant group theory, together with a summary of the properties of the coupling coefficients.

Group theory produces quantitative results. It does not produce only selection rules. The key theorem is the Wigner–Eckart theorem. Although the proof is rather short, the Wigner–Eckart theorem is a very significant part of the book, for it is usually the key to the applications. The theorem provides numerical relations between matrix elements; it is the link between group theory as a piece of mathematics and group theory as a simple computational tool. Some atomic and molecular applications are included, but, needless to say, no attempt has been made to make the book complete in this respect.

A serious barrier to the fruitful application of the Wigner–Eckart theorem has been the existence of conflicts in the notations and phase choices of previous workers. My notation and my phase choices are closely based on the one book known in all subject areas, *The 3j and 6j Symbols*, by Rotenberg, Bivins, Metropolis, and Wooten. That book gives a large tabulation for the group $\mathbf{SO}_3$ in the $\mathbf{SO}_2$ (JM) basis. This book complements the above in that we give as complete a set of tables for all other bases of $\mathbf{O}_3$, namely the bases formed by using any of the 32 crystallographic point groups or the noncrystallographic point groups (the icosahedral group $\mathbf{K}$, $\mathbf{D}_\infty$, and $\mathbf{D}_n$ and $\mathbf{C}_n$, n finite).

It is a pleasure to thank many colleagues for the advice and assistance given to me during the preparation of this book. Brian Wybourne, Mike Reid, Ric Haase, Clare Churcher, Paul Bickerstaff, Alec Ford, Geoff Stedman, and Susan Piepho deserve special mention. I wish to thank also: Janet Warburton, who carefully and cheerfully typed and retyped a difficult manuscript; Alan Wilkinson and others of the University Computer Centre who gave their computer expertise; Cliff Strange, of Printset Processes, who taught me the basic typesetting skills; Betty Bruhns, of Plenum, who shared the problems of table layout; and my wife Pamela, who has given years of patient encouragement.

Philip Butler

Contents

1. Introduction 1

1.1. A Guide to the Use of the Text 2

1.2. Notation and Phase Conventions 4

2. Basic Concepts 7

2.1. Irrep Spaces 8

2.2. Group–Subgroup Bases 18

2.3. Coupling Coefficients 23

2.4. Character Theory 28

2.5. Complex Conjugation 36

2.6. Spin Irreps and Labeling Irreps 38

3. The *jm* Factors and *j* Symbols 43

3.1. The *jm* Factors 44

3.2. The *j* Symbols 48

3.3. Properties of *jm* Factors and *j* Symbols 55

3.4. Computational Methods 68

3.5. Phase and Multiplicity Choices 71

4. The Wigner–Eckart Theorem 83

4.1. Basis Operators 84

4.2. The Wigner–Eckart Theorem 86

4.3. Coupled Tensors 89

4.4. The Standard Racah Tensors of $\mathbf{SO}_3$ 94

5. $\mathbf{O}_3$ and Its Subgroups 99

5.1. Direct Product Groups 100

5.2. Isomorphic Subgroups 102
5.3. Specification of x, y, z Axes and Bases 107

6. Properties of the Dihedral Groups 129
6.1. The Structure of the Dihedral Groups 130
6.2. The Chain $\mathbf{SO}_3 \supset \mathbf{D}_\infty \supset \mathbf{SO}_2$ 132
6.3. Finite Subgroups of $\mathbf{D}_\infty$ 136

7. Fractional Parentage Coefficients 139
7.1. The Parentage Concept 140
7.2. Continuous Matrix Groups 142
7.3. Complete Parentage Schemes 146
7.4. Strong-Field Parentage Schemes 149

8. Time Reversal 153
8.1. Time Reversal of States 153
8.2. Time Reversal and Selection Rules 160

9. Applications 171
9.1. The Free-Ion Hamiltonian 171
9.2. The Zeeman Interaction 176
9.3. Ligand Fields 182

10. Programming Notes 187

11. Group Information Tables 189
11.1. The Cyclic Groups 191
11.2. The Dihedral Groups 195
11.3. The Tetrahedral Groups 200
11.4. The Octahedral Groups 202
11.5. The Icosahedral Groups 205

12. Branching Rule Tables 207
Introduction 207
Tables 210

13. *jm* Factor Tables 217
Introduction 217
Tables 220

14. $3j$ and $6j$ Symbol Tables 429
Introduction 429
Tables 431

15. $9j$ Symbols 463
Introduction 463
Tables 464

16. Bases in Terms of Spherical Harmonics 513
16.1. Rotation Matrices in the JM Basis 514
16.2. Spherical Harmonics in Rectangular Coordinates 518
Tables 522

REFERENCES 553
SUBJECT INDEX 557
TABLE INDEX 566

1

Introduction

The immediate purpose of this book is to present tables of vector coupling coefficients for angular momentum, where the basis vectors are chosen to have any one of the many possible point group symmetries. The quantum theory of angular momentum is a major algebraic tool in Condon and Shortley's 1935 book, *The Theory of Atomic Spectra*. Wigner and Racah (see Biedenharn and van Dam 1965) simplified the algebra of angular momentum by applying various theorems of group theory, especially Schur's two lemmas, to the rotation group in three dimensions. These developments contributed to many advances in the theory of atomic and nuclear spectra in the 1950s and 1960s (see especially Judd 1963 and Vanagas 1972). Generalizations to coupling coefficients of other groups were developed as part of this mathematical development toward obtaining solutions to these many-body quantum mechanical problems. Much of this early development of Wigner's and Racah's work was in extending Racah's group-theoretic treatment of the coefficients of fractional parentage and their treatment of tensor operators.

Griffith (1962) led the way with respect to the topic of this book. He calculated tables of coupling and recoupling coefficients for various point groups, and showed how the familiar techniques of angular momentum could be carried over to point group symmetries. In 1965 the Racah–Wigner calculus, as these methods are sometimes called, was shown to hold (with appropriate modifications) for any group which has finite-dimensional unitary representations (Derome 1966, Derome and Sharp 1965). These papers were rather incomplete. They said that if certain free phases are appropriately chosen, then certain simplifications occur, but they did not say how to make such choices. Also, like Wigner's preprint (1940), these papers did not discuss the question of tensor operators, and tensor operators are an essential part of the Wigner–Eckart theorem.

In recent years many small tables of coupling coefficients for the point groups have been published, but unfortunately few have used Derome and Sharp's work. Phases have not been chosen such that a consistent definition of $3jm$ factors can be made. A $3jm$ factor is defined here as a symmetrized coupling factor (isoscalar factor). A $3jm$ is defined so as to have much the same symmetry properties as Wigner's $3j$ symbol. Indeed, we shall see that Wigner's $3j$ symbol is a special case, being a $3jm$ factor for the $\mathbf{SO}_3 \supset \mathbf{SO}_2$ group chain.

The textual material of this book should be sufficiently detailed to enable a graduate student in physics or chemistry to learn and to be able to use the essential group theory behind coupling coefficients. At the same time the book has been arranged so that an experimentalist experienced in the quantum theory of angular momentum can use the tables to apply the methods of tensor operators in any point group basis with a minimum of reading of the mathematical background.

In Section 1.1 we discuss in detail the layout of the various topics within the book. In the other section of this introductory chapter various phase conventions and notational choices are discussed. These two topics arise in various places within the book, but it is useful to give a self-contained description of these matters. The guiding principle behind our notations and conventions has been to treat Rotenberg, Bivins, Metropolis, and Wooten (1959) as the standard for $\mathbf{SO}_3 \supset \mathbf{SO}_2$ choices.

The diagram techniques of angular momentum theory (Jucys, Levinson, and Vanagas 1960) have been extended to the general case by Stedman (1975, 1976). He has shown that these techniques merge with the techniques of Feynman diagrams and he has chosen his notation for his diagrams accordingly. Readers who prefer diagrammatic expressions to algebraic expressions should find his papers readable.

1.1 A Guide to the Use of the Text

The text of this book has a single object, to enable the reader to make the best use of the tables. The text starts with the assumption that the reader knows little or no group theory but is familiar with elementary quantum mechanics, for example, he has seen the standard derivation of the solution of the hydrogenic Schrödinger equation. At the other end of the scale, we cover a great deal more group theory and coupling coefficient theory than is needed by someone interested only in intelligent use of the tables. A number of sections scattered throughout the book describe properties of the coefficients that are relevant only to the reader who wants a full understanding of how the tables were computed.

The chapters containing tables (Chapter 11 onward) also contain very brief summaries of the essential properties of the $3jm$ factors and $6j$ symbols. For someone who has regularly used the $3j$ and $6j$ tables of Rotenberg, Bivins, Metropolis, and Wooten (1959) or the $9j$ tables of Jucys (1968) these summaries may be sufficient to convert their calculations from the SO_3–SO_2 basis of standard angular momentum theory to the point group basis of their choice. Because one tends to use particular types of tables at a time and most tables are rather small, their order has been chosen so that character tables appear together, then the branching rules, then the $3jm$ factors, and so on.

Those with less experience, but interested in a quantum mechanical energy level calculation, would need to start at the beginning of the book. Chapter 2 begins with the vector space theory and group theory necessary to define an irreducible representation space (irrep space) from first principles. By Section 2.3 one has a simple introduction to angular momentum theory which the author believes has several advantages over the usual infinitesimal operator approach, the most important advantage being that the results are seen to be true even if the JM basis is not used. Sections 3.1, 3.2, and 4.1 define jm factors, j symbols, and basis operators, respectively. The former are the coefficients that are tabulated and the latter concept leads to a very simple derivation of the Wigner–Eckart theorem, again true in any basis.

We view the Wigner–Eckart theorem as the single result of group theory which is most important in applications. The Wigner–Eckart theorem applied to a group **G** factors a matrix element into a $3jm$ symbol for **G** and a G-reduced matrix element. The $3jm$ symbol contains all the structure derivable from the structure of **G**. The group **G** need not be an invariance group of the problem; it need only be a group under which the problem has a known structure. A very important and novel aspect of our tables is that the $3jm$ symbols are obtained as a product of $3jm$ factors corresponding to the subgroup structure of the various point groups. This factorization allows a perturbation treatment to proceed according to a descent in symmetry. For example, if a perturbation contains three terms which are scalar under $\mathbf{SO}_3$, the octahedral group **O**, and the group $\mathbf{D}_3$, respectively, the calculation can proceed in an SO_3–O–D_3 basis. The group structure of each operator can be used simultaneously. The tables for the SO_3–O $3jm$ and SO_3–K $3jm$ factors are our largest, covering all possibilities up to and including $J=8$.

Explicit calculations are performed in Section 5.3 and in Chapter 9. The reader may find it instructive to work through the calculations done there. The algebraic expressions are written out in full, and the appropriate numbers are found in the tables and substituted in the formulas.

1.2 Notation and Phase Conventions

A complete theory of coupling coefficients has been developed rather slowly over the past fifty years. Biedenharn and van Dam (1965) give a history of the development of what they call the quantum theory of angular momentum (see also Smorodinskii and Shelepin 1972 and Karasev and Shelepin 1973). As Biedenharn and van Dam emphasize, the mathematical tools were usually developed for specific applications and much less effort has been put into organizing the general results. In the recent physics and chemistry literature one can find many papers which compute a set of coupling coefficients for a specific purpose. Many of these papers introduce a notation of their own. Sometimes the notation is chosen to relate to the problem in hand, sometimes it is chosen so that the coefficients will not be confused with a similar set computed by another author, and sometimes it is chosen in apparent ignorance of the other notations.

Considerable care has been taken in choosing notations and phases for this book. With few exceptions, the notation and terminology have been chosen to correspond as closely as possible to the notation and terminology used by Rotenberg, Bivins, Metropolis, and Wooten (1959). Most texts written since then have used this same notation when discussing angular momentum in the SO_3–SO_2 basis. However, few choose the same notation for the same mathematics with different bases. Many revert to Racah's original notation for $3jm$, $6j$, and $9j$ symbols, namely V, W, and X coefficients, while others invent new names, e.g., Γ symbols.

One of the reasons for using the V, W, and X type notation has been that this notation was chosen before all the symmetry properties of the coefficients were known, and many recent authors have not been aware that general results on the symmetries of $3jm$, $6j$, and $9j$ symbols are available (see especially Derome and Sharp 1965).

All the equations we give are true for all point groups; indeed they are true for most representations of most groups used in quantum mechanics. The notation currently in use was developed mainly by Wigner (1940), who studied simply reducible groups, namely groups for which all irreps are real and for which there is no product multiplicity. Most of the generalizations of Wigner's notation have been in use for some years. This notation is such that Wigner's results and notation are recovered by omitting multiplicity indices (which we denote by Latin letters) and omitting complex conjugate signs (the asterisk, *).

One change in notation from Rotenberg *et al.* is that we distinguish between *jm* factors (or *jm* symbols) and *j* symbols. Coupling coefficients and coupling (or isoscalar) factors are related to *jm* symbols and *jm* factors, while the various reordering and recoupling coefficients and phases are simply related to *j* symbols. With this nomenclature, the *j* symbols contain information about one group and one group only. For example, the 6*j* symbols for $\mathbf{SO}_3$ tabulated by Rotenberg *et al.* may be used unchanged in any problem in which $\mathbf{SO}_3$ occurs. On the other hand, the 3*jm* factors tabulated by the same authors (and called 3*j* symbols by them) are specific to problems in which the *JM* (or $\mathbf{SO}_3 \supset \mathbf{SO}_2$) basis is used. Thus, when computing the matrix elements of the spin–orbit interaction in a crystal field of approximate octahedral symmetry, in which the basis states are labeled by the irrep labels of $\mathbf{SO}_3 \supset \mathbf{O} \supset \mathbf{D}_3 \supset \mathbf{C}_3$, one will use the 6*j* symbols of $\mathbf{SO}_3$, and the 3*jm* factors for the three imbeddings $\mathbf{SO}_3 \supset \mathbf{O}$, $\mathbf{O} \supset \mathbf{D}_3$, and $\mathbf{D}_3 \supset \mathbf{C}_3$. One has recourse to both the tables of this book and to the tables of Rotenberg *et al.*

We make some minor further changes in the notation of Wigner's preprint (1940) as published in Biedenharn and van Dam (1965). In particular, Wigner's 1*j* symbol becomes a 2*jm* factor with a simplified notation because we have now shown 2*jm* factors to be real. We introduce the names 2*j* and 3*j* phases for some reordering phases. Wigner used no name because for simply reducible groups they have simple algebraic values. We tabulate 3*j* phases at the head of the 6*j* symbol tables.

Our phases for groups other than $\mathbf{SO}_3$ are new. Nobody else has a full set of fully symmetrized coefficients. For the *JM* basis we use the phase conventions of Rotenberg *et al.* These are also used by many authors, in particular by Messiah (1965), Judd (1963), Wybourne (1965), and Piepho and Schatz (1981) (who will also use our tables).

The phases for 3*jm* factors and 6*j* symbols are partially fixed in Section 3.3 by the symmetrization requirements. Remaining phases and multiplicities are fixed in an ad hoc manner in Section 3.5, namely the first relevant coefficient computed is chosen positive.

It is occasionally necessary for us to use explicit rotation matrices in terms of Euler angles and to distinguish active and passive rotations. It is a topic in which it is easy to confuse conventions. Wolf (1969) has a good discussion of the conflicts which have arisen between otherwise standard texts. For the purpose of constructing examples we need a convention. We follow Messiah (1965), whose rotation operator $\exp(-i\theta J_z)$ rotates the system through an angle θ about the z axis (an active rotation). Likewise, we follow his convention regarding the order of rotations for Euler angles. We include explicit rotation matrices for the JM basis in Chapter 16.

Chapter 8 gives a careful treatment of time reversal, a topic prone to as many misunderstandings of conventions as the above (see Judd and Runciman 1976).

A somewhat trivial but nevertheless important aspect of notation is the question of which typeface and alphabet is used. We use boldface exclusively for sets of objects, a group **G** is a set of elements R or operators O_R; a matrix $\boldsymbol{\lambda}(R)$ is a set of matrix elements $\lambda(R)_{l'l}$; a matrix irrep $\boldsymbol{\lambda}$ or $\boldsymbol{\lambda}(\mathbf{G})$ is a set of matrices one for each $R \in \mathbf{G}$, $\boldsymbol{\lambda}(\mathbf{G}) = \{\boldsymbol{\lambda}(R)\colon R \in \mathbf{G}\}$. The letter λ is merely a label, as in the ket partner $|\lambda l\rangle$. Greek letters label irreps, Latin letters are used for all other objects. However, the letters J and M label the $\mathbf{SO}_3$ and $\mathbf{SO}_2$ irreps **J** and **M**. One has that $\mathbf{J}(\mathbf{SO}_3)$ is the set of matrices $\mathbf{J}(\alpha, \beta, \gamma)$, where (α, β, γ) is a typical element of $\mathbf{SO}_3$, consisting of a rotation through Euler angles α, β, and γ. The matrix elements are written $J(\alpha, \beta, \gamma)_{\mu'\mu}$, a notation similar to $\mathfrak{D}^{(J)}_{\mu'\mu}(\alpha, \beta, \gamma)$, which is used by many authors. If we know J or λ, say $J = \frac{7}{2}$ or $\lambda = 3$, we write $\frac{\mathbf{7}}{\mathbf{2}}(\mathbf{SO}_3)$ for the set of all rotation matrices in $\mathbf{SO}_3$, and $\mathbf{3}(\mathbf{G})$ for the set of all rotation matrices of **G**.

2

Basic Concepts

To apply the tools of "group theory" to physical problems, one does not need much of what a mathematician would mean by the term. Rather, one needs a good grasp of a few linear algebra ideas, in particular the ideas relating to a basis of a vector space. The operations of the group map the vector space to itself and this leads to the concept of minimal invariant subspaces (irrep spaces) of a vector (ket) space.

Group theory, in the mathematical meaning of the phrase, leads to the tools of character theory, from which one important theorem is required, namely that the transformation properties of a ket space may be deduced from the characteristics (the traces) of the operators. Group theory itself provides us with Schur's lemmas. These lemmas are proved in the first section, after we have defined irrep spaces and given a few examples.

Section 2.2 is devoted to an analysis of possible basis choices for ket-vector spaces. We see that the symmetry properties of vectors go a long way toward fixing bases—such symmetrized bases are most suitable in physical and chemical applications. Basis transformation coefficients defined here describe the change of basis from one symmetry scheme to another. Section 2.3 introduces a second kind of unitary transformation of basis, namely the construction of kets of given symmetry as linear combinations of products of kets with known symmetries. The coefficients are known as coupling coefficients and some of their simpler properties are discussed in this section, but the main development is left until Chapter 3.

The tools of character theory are deduced in Section 2.4. For those only interested in applications this section can be omitted, for we use it solely as a means of deducing which irreps occur. All necessary tables are presented in Chapters 11 and 12.

2.1 Irrep Spaces

A *ket space* $\mathbf{V}$ is a set of kets (or vectors) $|i\rangle$ such that any complex linear combination of kets in the space is also in the space, and where also an *inner product* $\langle i|j\rangle$ exists on the space. An inner product for any pair of kets $|i\rangle, |j\rangle$ in $\mathbf{V}$ is a complex number and $\langle i|i\rangle$ is a positive real number (zero only if $|i\rangle=0$).

Various sets of Schrödinger wave functions clearly form such inner product spaces; the inner product $\langle i|j\rangle$ of two wave functions ψ_i, ψ_j is formed as $\int \psi_i^* \psi_j \, d\tau$, where the integral is over 3-space. When we present typical functions throughout this book we use the spinors α, β and the coordinate functions x, y, z. If $\mathbf{r}$ is the position of a particle, then $x(\mathbf{r})$ or just x is the x coordinate of that particle. The three functions x, y, z form a basis for a ket space we call $\mathbf{V}_r$. An equivalent space is the space spanned by the three p orbitals $\mathbf{V}_p$, with basis functions the p_x, p_y, and p_z orbitals.

Ket spaces are particular cases of vector spaces. The mathematical concept of a vector space requires linearity properties such as those obeyed by position vectors, wave functions, electric fields, etc. Many such examples have the inner product (length and orthogonality) properties as well, but the space of linear operators introduced in Chapter 4 is an example of a vector space without an inner product. Corresponding to every vector space one can construct a unique *dual space* of operators called *functionals*; for example, one has corresponding to position coordinate x_i the differential $\partial/\partial x_i$, and one has corresponding to wave function ψ_j the "scalar product operator" $\langle j| = \int d\tau \, \psi_j^*$ which, when acting on $\psi_i = |i\rangle$, gives

$$(\langle j|)|i\rangle = \int d\tau \, \psi_j^* \psi_i = \langle j|i\rangle$$

Such correspondences led Dirac (1930) to call the wave functions (vectors) of quantum mechanics *kets* and the "scalar product operators" (functionals) *bras*. They combine to give $\langle j|i\rangle$, a "bracket."

A *group* $\mathbf{G}$ is a set of abstract entities $R, S, T, \ldots$, known as the elements of $\mathbf{G}$, that may be combined together by the *group multiplication* RS in such a way that the following conditions hold:

closure: $RS \in \mathbf{G}$ (2.1.1a)

association: $(RS)T = R(ST)$ (2.1.1b)

identity: there exists E, such that $\forall R$, $ER = RE = R$ (2.1.1c)

inverse: $\forall R$ there exists R^{-1} such that $RR^{-1} = R^{-1}R = E$ (2.1.1d)

The mathematical symbol $\forall$ means "for all."

The physically interesting and indeed mathematically interesting groups are the *non-Abelian groups*, that is, groups for which it is not true for all elements that $RS=SR$. The rotation group in three dimensions provides the prime example of this, in that, for example, finite rotations about different axes do not in general commute.

A representation space $\mathbf{V}_\lambda$ *of a group* $\mathbf{G}$ is a vector space of dimension $|\lambda|$ together with a linear operator O_R for each $R\in\mathbf{G}$, such that the operators map the space to itself and are homomorphic to the abstract group. That is, if $R,S\in\mathbf{G}$, then there exist unique operators O_R, O_S, O_{RS} such that

$$O_R\colon\quad V_\lambda\to V_\lambda\qquad \forall R\in\mathbf{G} \tag{2.1.2}$$

$$O_R O_S = O_{RS}\quad \forall R,S\in\mathbf{G} \tag{2.1.3}$$

Expressed in terms of a set of basis kets (numbered from either zero or one)

$$\text{basis}\,(V_\lambda)=\{|\lambda l\rangle\colon\quad l=1,2,\ldots,|\lambda|\} \tag{2.1.4}$$

the first condition becomes

$$O_R|\lambda l\rangle=\sum_{l'}|\lambda l'\rangle\langle\lambda l'|O_R|\lambda l\rangle=\sum_{l'}|\lambda l'\rangle\lambda(R)_{l'l} \tag{2.1.5}$$

namely that the image ket belongs to $\mathbf{V}_\lambda$ and can therefore be written in terms of the basis. The coefficients $\langle\lambda l'|O_R|\lambda l\rangle=\lambda(R)_{l'l}$ form a matrix $\boldsymbol{\lambda}(R)$. [Some authors would write the matrix as $\mathbf{D}^\lambda(R)$, $\mathbf{D}^{(\lambda)}(R)$, or $\boldsymbol{\Gamma}^\lambda(R)$.] The second condition is

$$O_R O_S|\lambda l\rangle=\sum_{l''l'}|\lambda l''\rangle\lambda(R)_{l''l'}\lambda(S)_{l'l} \tag{2.1.6}$$

$$=O_{RS}|\lambda l\rangle=\sum_{l''}|\lambda l''\rangle\lambda(RS)_{l''l} \tag{2.1.7}$$

Note that we write operators on the left and matrix transformations on the right so as to preserve the order of the subscripts. This is automatic in the Dirac bra–ket notation, for $\Sigma_{l'}|\lambda l'\rangle\langle\lambda l'|$ is the unit operator in (2.1.5). One multiplies on the left by the unit operator, then reinterprets the symbols $(|\lambda l'\rangle\langle\lambda l'|)O_R|\lambda l\rangle$ as a ket times a matrix element, $|\lambda l'\rangle(\langle\lambda l'|O_R|\lambda l\rangle)$.

EXAMPLE 2.1. Consider the Abelian abstract group $\mathbf{C}_3$ consisting of three elements E, A, B, where $AA=B$, $AB=BA=E$. We may form a realization of this abstract group by taking the following physical operators: O_E, the identity operator; O_A, a rotation by $2\pi/3$ about the z axis; and O_B, a rotation of $4\pi/3$ about the same axis. These operators act on our $\mathbf{V}_r$ basis kets $|r1\rangle=x$, $|r2\rangle=y$, and $|r3\rangle=z$ so that, for example,

$$O_A|r1\rangle=|r1\rangle(-1/2)+|r2\rangle(\sqrt{3}/2)$$

$$O_A|r2\rangle=|r1\rangle(-\sqrt{3}/2)+|r2\rangle(-1/2)$$

$$O_A|r3\rangle=|r3\rangle$$

giving, for this three-dimensional representation r, that the representation matrix $\mathbf{r}(A)$ is

$$\mathbf{r}(A)=\begin{pmatrix} -1/2 & \sqrt{3}/2 & 0 \\ -\sqrt{3}/2 & -1/2 & 0 \\ 0 & 0 & 1 \end{pmatrix}$$

One can use the group properties to prove many properties of the representation matrices. Throughout this book we assume $\mathbf{G}$ is a compact group, perhaps finite. Most importantly, one may then choose the basis kets so that all representation matrices are finite-dimensional unitary matrices. (A Lie group is said to be compact if its parametrization consists of a finite number of bounded parameter domains, or, more generally, a compact metric space is one in which every infinite sequence has a limit point.)

A *reducible* representation space is a space that the group splits into two or more orthogonal representation subspaces. For example, it is obvious that under $\mathbf{SO}_2$, the group of all rotations about the z axis, the three-dimensional space $\mathbf{V}_r$ spanned by $\{x, y, z\}$ splits into the two subspaces $\mathbf{V}_{xy}$ and $\mathbf{V}_z$ spanned by $\{x, y\}$ and $\{z\}$. It is less obvious that the two-dimensional subspace $\mathbf{V}_{xy}$ is further reducible into the two $\mathbf{V}_1$ and $\mathbf{V}_{-1}$ spanned by the one-element sets $\{x+iy\}, \{x-iy\}$.

EXAMPLE 2.1 CONTINUED. The three-dimensional space $\mathbf{V}_r$ of $\mathbf{C}_3$ has the alternative basis $|r00\rangle = z$, $|r10\rangle = -(1/\sqrt{2})(x+iy)$, and $|r-10\rangle = (1/\sqrt{2})(x-iy)$ and we have

$$O_A|r00\rangle = |r00\rangle$$

$$O_A|r10\rangle = |r10\rangle \exp(-\tfrac{2}{3}\pi i)$$

$$O_A|r-10\rangle = |r-10\rangle \exp(\tfrac{2}{3}\pi i)$$

with similar results for O_B, showing that the three one-dimensional spaces $\mathbf{V}_0, \mathbf{V}_1, \mathbf{V}_{-1}$ with bases $\{|r00\rangle\}, \{|r10\rangle\}, \{|r-10\rangle\}$ are each irreducible representation spaces of $\mathbf{C}_3$.

Check that this is also true for $\mathbf{SO}_2$ by using the fact that if $R_z(\alpha)$ is a rotation about the z axis by angle α, then we have

$$R_z(\alpha)(x, y) = (x\cos\alpha + y\sin\alpha, -x\sin\alpha + y\cos\alpha) \tag{2.1.8}$$

(Note that we have active operations, rotating the particle, not the coordinate system.)

Example 2.1 illustrates the fact that the action of a group $\mathbf{G}$ on any space will lead to a natural basis (a mathematician would say a canonical basis). The basis is in terms of *irreducible representation subspaces* (henceforth abbreviated *irrep spaces*) $\mathbf{V}_\lambda$. A *matrix irrep* $\boldsymbol{\lambda}(\mathbf{G})$ is the corresponding set of all matrices $\boldsymbol{\lambda}(R)$ for all operations R of the group $\mathbf{G}$,

$$\boldsymbol{\lambda}(\mathbf{G}) = \{\boldsymbol{\lambda}(R): \quad R \in \mathbf{G}\} \tag{2.1.9a}$$

where for all $R, S \in \mathbf{G}$

$$\boldsymbol{\lambda}(R)\boldsymbol{\lambda}(S) = \boldsymbol{\lambda}(RS) \tag{2.1.9b}$$

The irreps are all one dimensional only in the case of Abelian groups (groups for which all operators commute). A familiar example of multidimensional irreps is for the group of rotations in 3-space, where possible basis functions are the spherical harmonics and the irrep labeled J is of dimension $|J| = 2J+1$.

The basis kets of an irrep space are often called the *partners* of the irrep. We use this descriptive name.

The irreducibility says nothing about possible basis choices within the irrep $\boldsymbol{\lambda}$, nor will it distinguish between irrep spaces which transform the same way, *similar irreps*. Schur's lemmas (proved later) do show, however, that the decomposition of a representation space into irrep subspaces is essentially unique.

Consider any linear combination of the partners of two similar irreps $\mathbf{V}_{a\lambda}$ and $\mathbf{V}_{b\lambda}$. For example, take the two one-dimensional irreps of $\mathbf{SO}_2$ with bases $\{z^2\}$ and $\{x^2+y^2\}$, which transform the same way under rotations about the z axis (in this case they are absolutely invariant). The linear combination $\{pz^2+q(x^2+y^2)\}$ for any p,q is also absolutely invariant. Inspection of (2.1.5) shows this to be true in the general case if we satisfy a few requirements. Let $|a\lambda l\rangle$ be a typical partner of $\mathbf{V}_{a\lambda}$ and $|b\lambda l\rangle$ that of $\mathbf{V}_{b\lambda}$. The statement that $\mathbf{V}_{a\lambda}$ transforms similarly to $\mathbf{V}_{b\lambda}$ means that with a certain basis choice the transformation properties are exactly the same. Namely the partners of $\mathbf{V}_{b\lambda}$ may be chosen so that if

$$O_R|a\lambda l\rangle=\sum_{l'}|a\lambda l'\rangle\lambda(R)_{l'l} \tag{2.1.10}$$

then

$$O_R|b\lambda l\rangle=\sum_{l'}|b\lambda l'\rangle\lambda(R)_{l'l} \tag{2.1.11}$$

so that $\boldsymbol{\lambda}(R)$ depends only on the transformation properties of $\mathbf{V}_{x\lambda}$ ($x=a$, b, etc.).

We have the three kinds of labels (which may themselves be sets of labels) x, λ, l for partners of an irrep. The label λ gives the transformation properties (labels the irrep), l is a label distinguishing the various partners of the irrep, and x distinguishes other properties, e.g., whether it is a z^2 or x^2+y^2 function or some mixture of the two in the above example. Do not confuse the use of label x, used to denote non-group-theoretic properties of the kets, with the ket (coordinate function) x used in examples. If the group is $\mathbf{SO}_3$, the irrep label λ is the angular momentum label written l, j, s, L, J, or S, depending on context (we shall usually use J). The label x may contain radial information or any other information not related to $\mathbf{G}$, e.g., as to whether it is a one-, two-, three-,… particle function, whether the particles are molecules, atoms, electrons, nucleons, quarks, etc. Our label l contains the specific information on the basis. In applications it can always be (a set of) labels giving transformation properties under the action of subgroups of $\mathbf{G}$. For example, M is used if this remaining information is the transformation property of the basis ket about the z axis,

$$R(\theta, z)|xJM\rangle = |xJM\rangle e^{-i\theta M} \tag{2.1.12}$$

With this choice of labels we have that the transformation properties of a basis ket $|x\lambda l\rangle$ are completely independent of the label x, label λ contains general information on the irrep—its dimension and character in particular—and the label l contains all the particular information that distinguishes the partners of the basis choice.

For example, take the space $\mathbf{V}(p)$ of one-electron p states and the space $\mathbf{V}(p^2P)$ of two-electron P states (where spin is ignored completely). We could choose various bases of $\mathbf{V}(p)$; commonly one would take either

$$p_{+1} = \frac{-1}{\sqrt{2}}(p_x + ip_y), \qquad p_0 = p_z, \qquad p_{-1} = \frac{1}{\sqrt{2}}(p_x - ip_y)$$

or

$$p_x, p_y, p_z$$

as the partners of $\mathbf{V}(p)$. However, we must fix our choice and then use the same kind of choice for the two-electron partners, for otherwise the transformation equation

$$O_R|xPl\rangle = \sum_{l'} |xPl'\rangle P(R)_{l'l} \tag{2.1.13}$$

would not give matrix elements $P(R)_{l'l}$ which were independent of x. Lemma II of Schur will say precisely how much freedom we have for the partners of the second space.

The above statements on similar irreps are crucial to the whole development, so we reiterate our procedure. Given two irrep spaces $\mathbf{V}_{x_1\lambda_1}$ and $\mathbf{V}_{x_2\lambda_2}$ of a group $\mathbf{G}$ and a set of partners $|x_1\lambda_1 l_1\rangle$ of $\mathbf{V}_{x_1\lambda_1}$, there may exist a set of partners $|x_2\lambda_2 l_2\rangle$ of $\mathbf{V}_{x_2\lambda_2}$ such that the matrices $\lambda_1(R)$ and $\lambda_2(R)$ are equal for every R. If such a basis exists, the irrep spaces are said to be *equivalent* or *similar*. (A basis transformation of a vector space is a similarity transformation.) For every set of equivalent irrep spaces the partners are to be chosen in some standard fashion to give matrices $\lambda(R)$

independent of the labels x. If, on the other hand, the irreps are inequivalent, then no relationship between the respective partners is needed. However, the next section shows that systematic choices of partners are available using subgroup transformation properties.

Having said what we mean by a partner, we are in a position to look at the irreducible aspects of an irrep space. An irrep space is a representation space which has the additional property that it cannot be reduced into subspaces which transform independently (are invariant) under the action of the group. Consider the definition of *invariant spaces* first. We say that a space $\mathbf{V}_\lambda$ is invariant under a group if, taking any ket $|\lambda l\rangle$ in $\mathbf{V}_\lambda$, the kets $O_R|\lambda l\rangle$ all belong to $\mathbf{V}_\lambda$. The two-dimensional subspace $\mathbf{V}_{xy}$ of the space $\mathbf{V}_r$ of Example 2.1 is invariant under rotations about the z axis, whereas $\mathbf{V}_{xz}$ with basis $\{x, z\}$ is not. The kets $O_R z$ are all in $\mathbf{V}_{xz}$ (they are all just z), but the kets $O_R x$ contain in general some y component. The terms "invariant space under $\mathbf{G}$" and "representation space of $\mathbf{G}$" are interchangeable. Now add the definition of irreducibility, namely that an irrep space cannot be split into representation subspaces. The statement that $\mathbf{V}_\lambda$ is an irrep space under the group $\mathbf{G}$ is thus equivalent to the statement that taking any ket $|\lambda l\rangle$ whatsoever from $\mathbf{V}_\lambda$, the kets $O_R|\lambda l\rangle$ for all $R\in\mathbf{G}$ span $\mathbf{V}_\lambda$. To see this, consider the space $\mathbf{V}_{xy}$ above, which is reducible. Now, $\{O_R x\colon R\in\mathbf{C}_3\}$ spans $\mathbf{V}_{xy}$. Every vector in $\mathbf{V}_{xy}$ can be written in at least one way as a linear combination of the three vectors $O_R x$, $R=E, A, B$ (e.g., $y=Ax-Bx$). However, $\{O_R(x+iy)\colon R\in\mathbf{C}_3\}$ does not span $\mathbf{V}_{xy}$ but only its subspace $\mathbf{V}_1$ with basis $(1/\sqrt{2}\,)(x+iy)$. Hence $\mathbf{V}_{xy}$ is not irreducible under $\mathbf{C}_3$.

However, $\mathbf{V}_{xy}$ is an orthogonally irreducible representation space of $\mathbf{C}_3$. Just as the equation $x^2=-1$ has no real roots, there is no real (orthogonal) transformation of basis which reduces the $\mathbf{V}_{xy}$ part of Example 2.1. The assumption throughout this book is that complex numbers are being used. As with quadratic equations, complex numbers provide the means to prove many completeness theorems. In particular, Schur's second lemma explicitly uses the fact that all matrices have at least one eigenvalue. This is only true if complex numbers are used.

It is clear that irreducibility may not be apparent from a cursory inspection. Character theory does provide us with simple, rigorous tests of reducibility. However, it is usually unnecessary to test for reducibility except in the simplest cases. Coupling and transformation coefficients enable one to produce the irreps directly and explicitly, with prior knowledge of what is reducible and what is irreducible.

The irreducibility concept leads to Schur's lemmas, which in turn lead to the factorization theorems and uniqueness theorems which are the cornerstones of the application of group theory to physical problems. We state and prove Schur's lemmas in terms of operators P_R, Q_R on irrep spaces $\mathbf{V}_\lambda, \mathbf{V}_\mu$ (P_R: $\mathbf{V}_\lambda \to \mathbf{V}_\lambda$; Q_R: $\mathbf{V}_\mu \to \mathbf{V}_\mu$) and an operator A either mapping between them (A: $\mathbf{V}_\mu \to \mathbf{V}_\lambda$) or on $\mathbf{V}_\lambda$ only (A: $\mathbf{V}_\lambda \to \mathbf{V}_\lambda$).

Schur Lemma I. If $\mathbf{V}_\lambda, \mathbf{V}_\mu$ are two irrep spaces of a group $\mathbf{G}$ and if an operator A: $\mathbf{V}_\mu \to \mathbf{V}_\lambda$ satisfies

$$P_R A = A Q_R \qquad \forall R \in \mathbf{G}$$

then either A is zero or $\mathbf{V}_\lambda$ is equivalent to $\mathbf{V}_\mu$.

Schur Lemma II. If $\mathbf{V}_\lambda$ is an irrep space and if an operator A: $\mathbf{V}_\lambda \to \mathbf{V}_\lambda$ satisfies

$$P_R A = A P_R \qquad \forall R \in \mathbf{G}$$

then A is a multiple of the unit operator.

Proof I. Take sets of partners $|\lambda l\rangle$ of $\mathbf{V}_\lambda$, and $|\mu m\rangle$ of $\mathbf{V}_\mu$. The assumption of Lemma I is that for any m,

$$P_R A |\mu m\rangle = A Q_R |\mu m\rangle \tag{2.1.14}$$

This may be written as

$$P_R(A|\mu m\rangle) = \sum_{m'} (A|\mu m'\rangle)\mu(R)_{m'm} \tag{2.1.15}$$

Two possibilities occur. Either $A|\mu m\rangle$ is zero for all m (i.e., A is the zero operator) or each ket $A|\mu m\rangle$ behaves like a partner of a representation $\boldsymbol{\mu}$, for they give rise to representation matrix elements $\mu(R)_{m'm}$. In the second case we argue that $A|\mu m\rangle$ is a ket of $\mathbf{V}_\lambda$ since by assumption A maps $\mathbf{V}_\mu$ to $\mathbf{V}_\lambda$; thus $\mathbf{V}_\lambda$ contains a subspace which is equivalent to $\mathbf{V}_\mu$. Since $\mathbf{V}_\lambda$ is irreducible, the subspace cannot be a proper subspace, but must be the entire space. Finally, if it is the entire space, then the transformation A provides a basis for $\mathbf{V}_\lambda$ which gives rise to the matrix irrep $\boldsymbol{\mu}(\mathbf{G})$, that is, $\mathbf{V}_\lambda$ is equivalent to $\mathbf{V}_\mu$. □

PROOF II. Take a ket $|\lambda a\rangle$ which is an eigenket of A [every operator has at least one (complex) eigenvalue and eigenket, just as does every matrix]

$$A|\lambda a\rangle = |\lambda a\rangle a \tag{2.1.16}$$

Acting with O_R and using the assumption of Lemma II gives

$$A(O_R|\lambda a\rangle) = O_R A|\lambda a\rangle = (O_R|\lambda a\rangle)a \tag{2.1.17}$$

Because $\mathbf{V}_\lambda$ is irreducible, the various $O_R|\lambda a\rangle$ span $\mathbf{V}_\lambda$. Hence all kets in V_λ are eigenkets of A with eigenvalue a. Therefore A is a multiple of the unit operator, $A = aE$. This ends our proof. □

These lemmas say a great deal about basis choices. Lemma II says that given two equivalent irrep spaces $\mathbf{V}_{x\lambda}, \mathbf{V}_{y\lambda}$ and given a set of partners $|x\lambda l\rangle$ of $\mathbf{V}_{x\lambda}$, the corresponding set of partners $|y\lambda l\rangle$ of $\mathbf{V}_{y\lambda}$ are fixed up to a single overall phase, for the normalization of partners is fixed as unity. The argument in full is that if $\mathbf{V}_{x\lambda}, \mathbf{V}_{y\lambda}$ are equivalent, then their respective partners are to be chosen so that if

$$P_R|x\lambda l\rangle = \sum_{l'} |x\lambda l'\rangle \lambda(R)_{l'l} \tag{2.1.18}$$

then

$$Q_R|y\lambda l\rangle = \sum_{l'} |y\lambda l'\rangle \lambda(R)_{l'l} \tag{2.1.19}$$

with the same matrix elements $\lambda(R)_{l'l}$. Take any other set of partners of $\mathbf{V}_{y\lambda}$, say $|y\lambda m\rangle'$, which are related to the first by a unitary matrix $\mathbf{A}$,

$$|y\lambda m\rangle' = \sum_{l} |y\lambda l\rangle A_{lm} \tag{2.1.20}$$

Act on both sides with a group operator Q_R, then use the expansion (2.1.20) again to obtain

$$\sum_{nm'} |y\lambda n\rangle A_{nm'} \lambda(R)_{m'm} = \sum_{ll'} |y\lambda l'\rangle \lambda(R)_{l'l} A_{lm} \tag{2.1.21}$$

We use the orthonormality of the unprimed kets to obtain the commutation property

$$\sum_{m'} A_{nm'}\lambda(R)_{m'm} = \sum_{l} \lambda(R)_{nl} A_{lm} \tag{2.1.22}$$

for every $R \in \mathbf{G}$. Schur's Lemma II may now be used to show that $\mathbf{A}$ is a multiple of the unit matrix

$$A_{lm} = a\,\delta_{lm} \tag{2.1.23}$$

and because $\mathbf{A}$ is unitary, a is a complex number of modulus 1. A similar argument gives that any irrep space $\mathbf{V}_{z\lambda}$ constructed from a linear combination of the partners of two equivalent spaces $\mathbf{V}_{x\lambda}$ and $\mathbf{V}_{y\lambda}$ must have as the combination coefficients numbers $a_{x\lambda}$ and $a_{y\lambda}$ which are independent of the particular partners

$$|z\lambda l\rangle = |x\lambda l\rangle a_{x\lambda} + |y\lambda l\rangle b_{y\lambda} \tag{2.1.24}$$

Schur's lemmas are saying that once one set of partners, transforming as some $\mathbf{V}_{x\lambda}$, has been chosen, then the relationships between the partners belonging to all other $\mathbf{V}_{y\lambda}$ is fixed. Consider the three partners of the $J=1$ irrep of $\mathbf{SO}_3$. Common alternative choices of partners are $\{p_x, p_y, p_z\}$ and $\{p_0, p_+, p_-\}$. Once one set is chosen as standard, the $\mathbf{J=1}$ rotation matrix for a given rotation is fixed. Whenever the $\mathbf{J=1}$ irrep occurs in a calculation, the same relationship between its three partners will have to be chosen or a conflict will arise with the resultant set of rotation matrices. In this way the Schur result informs us how the transformation properties of the system λl are separated from the other properties x.

SCHUR'S LEMMAS IN SUMMARY. The matrix elements of an operator A: $\mathbf{V}_{y\mu} \to \mathbf{V}_{x\lambda}$ are written

$$A_{lm} = \langle x\lambda l | A | y\mu m\rangle \tag{2.1.25}$$

If A commutes with the action of the group, $O_R A = A O_R$ for all $R \in \mathbf{G}$, then Schur's Lemma I says A_{lm} is zero if $\lambda \neq \mu$. Schur's Lemma II says that if $\lambda = \mu$, then A is a multiple a of the unit matrix δ_{lm}. Writing a as the (l, m)-independent number $\langle x\lambda \| A \| y\mu\rangle$ gives Schur's lemmas as

$$O_R A = A O_R \quad \text{if and only if}$$

$$\langle x\lambda l | A | y\mu m\rangle = \delta_{\lambda\mu}\,\delta_{lm}\langle x\lambda \| A \| y\mu\rangle \tag{2.1.26}$$

It is in this form that we shall use Schur's two lemmas to derive our various results. They embody the consequences of the conditions that we have imposed upon the basis kets $|x\lambda l\rangle$. The label λ labels the irreducible representation space to which $|x\lambda l\rangle$ belongs, the same label λ for equivalent irrep spaces. Schur's first lemma says that if A commutes with all O_R, then A does not map between different $\boldsymbol{\lambda}$. The label l labels the specific partner of the irrep space $\boldsymbol{\lambda}$. Schur's second lemma says that if A commutes with all O_R, then A does not map between different l, and further, that it is constant with respect to all l of a given $\boldsymbol{\lambda}$. As a final point, consider a map between reducible representation spaces. Now any linear map between two reducible representation spaces can be written as a sum of maps A between the various irreducible components, and for each such map (2.1.26) holds. We will need Schur's lemmas in this reducible form primarily to deduce that a basis transformation within any representation space (a particular case of a map between spaces) takes on the block diagonal form of (2.1.26).

2.2 Group–Subgroup Bases

Several of the algebraically simple features of usual angular momentum theory stem not so much from the group $\mathbf{SO}_3$ which underlies the theory, but rather from the particular basis chosen. The JM basis of angular momentum theory has been chosen to have well-defined transformation properties for the group of rotations about the z axis, $\mathbf{SO}_2$. In particular, if $R_z(\alpha)$ is a rotation through angle α about the z axis, then

$$R_z(\alpha)|JM\rangle = |JM\rangle e^{-i\alpha M} \tag{2.2.1}$$

Note only do the $2J+1$ kets $|JM\rangle$, $-J \leqslant M \leqslant J$, form the partners of the irrep $\mathbf{J}(\mathbf{SO}_3)$, but also each ket is the single partner of a one-dimensional irrep $\mathbf{M}(\mathbf{SO}_2)$. For example, the five kets

$$\{|p^2 2\pm 2\rangle, |p^2 2\pm 1\rangle, |p^2 20\rangle\}$$

are simultaneously the five partners of one five-dimensional irrep $\mathbf{2}(\mathbf{SO}_3)$ and the single partners of the five one-dimensional irreps $\mathbf{2}(\mathbf{SO}_2), \mathbf{1}(\mathbf{SO}_2)$, $\mathbf{0}(\mathbf{SO}_2)$, $-\mathbf{1}(\mathbf{SO}_2)$, $-\mathbf{2}(\mathbf{SO}_2)$.

The process illustrated above is fundamental. The partners of any irrep of any group can be chosen to be simultaneously partners of the irreps of any specific subgroup. This is a repeat of the process of the previous section. Consider the irrep space $\mathbf{V}_{x\lambda}$ of a group **G** under the action of the operations O_R, where $R \in \mathbf{H}$, a subgroup of **G**. Then $\mathbf{V}_{x\lambda}$ is certainly a representation space of **H** because the elements of **H** are elements of **G** and they map $\mathbf{V}_{x\lambda}$ to itself. On the other hand, $\mathbf{V}_{x\lambda}$ will not, in general, be irreducible under **H**. The $\mathbf{V}_{x\lambda}$ will decompose into a sum of irrep spaces $\mathbf{V}_{x\lambda a\rho}$ of **H**. Two new labels have been introduced here. The label ρ is an irrep label, having the same relationship to the subgroup **H** as λ does to **G**. The label a is known as a branching multiplicity label and is necessary if the irrep $\boldsymbol{\rho}$ occurs more than once in the irrep $\boldsymbol{\lambda}$, in much the same way that x will distinguish kets with the same properties under **G** but are otherwise different.

An example of branching multiplicity is the restriction of the $\mathbf{1}(\mathbf{SO}_3)$ (given by the coordinate vectors $\{x, y, z\}$ of Example 2.1) restricted to $\mathbf{C}_2$, twofold rotations about the z axis. Both x and y—indeed, any linear combination of them—transform identically under $\mathbf{C}_2$. Because they transform the same way under $\mathbf{C}_2$, that is, as equivalent irreps labeled therefore by the same irrep label of $\mathbf{C}_2$, Schur's lemmas for $\mathbf{C}_2$ cannot be used to derive any properties of a map between x and y. Furthermore, because x and y belong to the one irrep of $\mathbf{SO}_3$, Schur's lemmas with $\mathbf{SO}_3$ as the group hold no information either. All is not lost in this example. The group $\mathbf{SO}_3$ contains many subgroups which in turn contain $\mathbf{C}_2$. For some such intermediate groups, e.g., $\mathbf{SO}_2$ and $\mathbf{D}_2$, the kets x and y are distinguished, and useful information can be obtained from Schur's lemmas in the intermediate group. We say we have a resolution of the branching multiplicity.

The functions which are partners simultaneously of the three groups in the chain $\mathbf{SO}_3 \supset \mathbf{SO}_2 \supset \mathbf{C}_2$ are different from those for $\mathbf{SO}_3 \supset \mathbf{D}_2 \supset \mathbf{C}_2$. The tables in this book describe the construction and properties of functions which are simultaneously partners of irreps of all possible point groups. The tables are so constructed that any part of the chain may be used. For example, consider the groups in the chains

$$\mathbf{SO}_3 \supset \mathbf{SO}_2 \supset \mathbf{C}_1 \tag{2.2.2}$$

$$\mathbf{SO}_3 \supset \mathbf{O} \supset \mathbf{D}_3 \supset \mathbf{C}_3 \supset \mathbf{C}_1 \tag{2.2.3}$$

O is the octahedral group, the group consisting of the 24 proper rotations which leave an octahedron (equivalently a cube) invariant. $\mathbf{D}_3$ is the dihedral group, which is the symmetry group of an equilateral triangle, and

$\mathbf{C}_3$ is a subgroup which consists of threefold rotations about one axis. $\mathbf{C}_1$ is the group consisting of only the identity operation. By saying $\mathbf{D}_3 \supset \mathbf{C}_3$ we say also that the threefold axis of $\mathbf{C}_3$ is the threefold axis of $\mathbf{D}_3$.

From the respective tables of Chapter 12 we find that the $J=2$ or D-state five-dimensional irrep of $\mathbf{SO}_3$ decomposes under restriction to $\mathbf{O}$ into two irreps labeled $\mathbf{2(O)}$ and $\tilde{\mathbf{1}}\mathbf{(O)}$ (E and T_2 in Mulliken's notation). We write

$$\mathbf{2(SO_3)} \to \mathbf{2(O)} + \tilde{\mathbf{1}}\mathbf{(O)} \tag{2.2.4}$$

Under the further restriction of $\mathbf{O} \supset \mathbf{D}_3$ the appropriate branching rule table of Chapter 12 shows that $\mathbf{2(O)} \to \mathbf{1(D_3)}$ and $\tilde{\mathbf{1}}\mathbf{(O)} \to \mathbf{0(D_3)} + \mathbf{1(D_3)}$. For $\mathbf{D}_3 \supset \mathbf{C}_3$, $\mathbf{0(D_3)} \to \mathbf{0(C_3)}$ and $\mathbf{1(D_3)} \to \mathbf{1(C_3)} + \mathbf{-1(C_3)}$. All irreps of $\mathbf{C}_3$ are one dimensional and are scalar under (are unchanged by) the one operator of $\mathbf{C}_1$. We shall often omit explicit reference to $\mathbf{C}_1$ because Schur's lemmas applied to $\mathbf{C}_1$ give no information.

Section 5.3 will show how to obtain a set of partners of an irrep $\mathbf{J(SO_3)}$ that are also partners of any specified subgroups of $\mathbf{SO}_3$. Chapter 16 contains many tables of such partners. The explicit form of the partners as functions of x, y, and z depends on the actual relationship of the axes of the finite groups to the standard axes of angular momentum theory. It suffices here to give the two sets of five functions quadratic in x, y, and z which transform as $\mathbf{2(SO_3)}$ and as partners of the irreps of each of the groups in the above chains. We denote partners of $\mathbf{SO}_3 \supset \mathbf{SO}_2 \supset \mathbf{C}_1$ as

$$|J(SO_3)M(SO_2)0(C_1)\rangle = |JM\rangle_a$$

and of $\mathbf{SO}_3 \supset \mathbf{O} \supset \mathbf{D}_3 \supset \mathbf{C}_3 \supset \mathbf{C}_1$ as

$$|J(SO_3)i(O)j(D_3)k(C_3)0(C_1)\rangle = |Jijk\rangle_b$$

Using the normalization of Example 2.1, the $\mathbf{SO}_3 \supset \mathbf{SO}_2 \supset \mathbf{C}_1$ partners of Section 16.2 become

$$|22\rangle_a = (1/2)(x+iy)^2$$

$$|21\rangle_a = -(x+iy)z$$

$$|20\rangle_a = -(1/\sqrt{6})(x^2+y^2-2z^2) \tag{2.2.5}$$

$$|2\,-1\rangle_a = (x-iy)z$$

$$|2\,-2\rangle_a = (1/2)(x-iy)^2$$

The corresponding set of $\mathbf{SO}_3 \supset \mathbf{O} \supset \mathbf{D}_3 \supset \mathbf{C}_3 \supset \mathbf{C}_1$ partners may be read off the appropriate table of Chapter 16 as

$$
\begin{aligned}
|2211\rangle_b &= (2/3)^{1/2}(x+iy)z + \left(1/2\sqrt{3}\,\right)(x-iy)^2 \\
|221-1\rangle_b &= (2/3)^{1/2}(x-iy)z + \left(1/2\sqrt{3}\,\right)(x+iy)^2 \\
|2\tilde{1}11\rangle_b &= \left(1/\sqrt{3}\,\right)(x+iy)z - \left(1/\sqrt{6}\,\right)(x-iy)^2 \qquad (2.2.6) \\
|2\tilde{1}1-1\rangle_b &= \left(1/\sqrt{3}\,\right)(x-iy)z - \left(1/\sqrt{6}\,\right)(x+iy)^2 \\
|2\tilde{1}00\rangle_b &= \left(1/\sqrt{6}\,\right)(x^2+y^2-2z^2)
\end{aligned}
$$

The above two sets, $|JM\rangle_a$ for various M, and $|Jijk\rangle_b$ for various ijk, are alternative bases for the five-dimensional $\mathbf{2}(\mathbf{SO}_3)$ space. The two sets are therefore related by a unitary transformation. The matrix elements $\langle JM|Jijk\rangle$ of this transformation are a special case of what are known as *transformation brackets* or *transformation coefficients*

$$|x\lambda i\rangle_b = \sum_l |x\lambda l\rangle_a \langle \lambda l|\lambda i\rangle \qquad (2.2.7)$$

Chapter 16 contains the various alternative sets of partners of various irreps in this form.

Our approach to the problem of computing the various coefficients is the reverse of that adopted by most authors. The usual method is to first obtain the irrep matrices of the groups, then obtain functions in form of (2.2.6) or (2.2.7), and finally obtain coupling coefficients from them. Both for ease of computation and because we must ensure that properly symmetrized coupling coefficients are tabulated we have chosen to do our calculations in reverse order.

We show in the next chapter that the $6j$ symbols and then the $3jm$ factors may be easily machine-computed. If we then specify the relationship of our group axes to the x, y, z axis system (here we have chosen the $\mathbf{C}_3$ axis to be the z axis), the relationship of the basic spinor partners (or basic x, y, z functions) is fixed. A recoupling calculation then fixes the higher order (e.g., quadratic) functions. Some sample calculations are done in full detail in Section 5.3.

Very few studies of the properties of arbitrary transformation coefficients have been made; for example, one could ask, do they have any factorization properties? On the other hand, the question of which irreps of a subgroup occur within an irrep of a group is an old and well-studied problem. The decomposition rules, e.g.,

$$\mathbf{J}(\mathbf{SO}_3)\rightarrow \sum_{M=-J}^{J} \mathbf{M}(\mathbf{SO}_2) \tag{2.2.8}$$

$$\mathbf{2}(\mathbf{SO}_3)\rightarrow \mathbf{2}(\mathbf{O})+\tilde{\mathbf{1}}(\mathbf{O}) \tag{2.2.9}$$

are usually known as branching rules and are easily computed from purely character-theoretic information, both for finite groups and for continuous groups. An outline of character theory for finite groups is given in Section 2.4 and complete tables for the point groups are included in Chapter 11. Complete tables of branching rules are given in Chapter 12.

The branching $\mathbf{SO}_3\rightarrow\mathbf{SO}_2$ does not exhibit the most general possibility, namely that one irrep of a group may contain an irrep of one of its subgroups more than once. In several cases of physical interest there is no intermediate group to resolve the branching multiplicity. For example, the 11 partners of $\mathbf{5}(\mathbf{SO}_3)$ form partners of four irreps of $\mathbf{O}$,

$$\mathbf{5}(\mathbf{SO}_3)\rightarrow \mathbf{1}(\mathbf{O})+\mathbf{1}(\mathbf{O})+\mathbf{2}(\mathbf{O})+\tilde{\mathbf{1}}(\mathbf{O}) \tag{2.2.10}$$

This branching rule shows that it is possible to choose a set of partners such that two partners transform as $\mathbf{2}(\mathbf{O})$, three partners as $\tilde{\mathbf{1}}(\mathbf{O})$, and two independent sets of three partners transform as $\mathbf{1}(\mathbf{O})$. The existence of two independent sets which transform the same way under a group caused various delays in the development of the theory of coupling coefficients. As in (2.1.24), any linear combination of the partners of one set with the corresponding partners of the other set will transform as the irrep $\mathbf{1}(\mathbf{O})$. Various methods have been suggested in the literature for separating such a multiplicity, that is, for finding suitable orthogonal sets of partners. Most methods depend on the production of an operator which commutes with all operations of $\mathbf{O}$, but whose eigenvalues are different on each of the various irreps forming the multiplicity. The perturbation Hamiltonian of a physical problem is often suitable as such an operator, but as this varies from problem to problem we have used a simpler approach (see Section 3.5). (See also Bickerstaff and Wybourne, 1976.)

In cases of multiplicity we must include an additional label a in the labels of the basis kets, e.g., multiplicities occur for $\mathbf{SO}_3\supset\mathbf{O}$ beginning with $\mathbf{J}=\frac{9}{2}$, so we label the general ket in scheme b above as

$$|x\mathrm{J}a i(O) j(D_3) k(C_3) 0(C_1)\rangle_b = |xJaijk\rangle_b \tag{2.2.11}$$

2.3 Coupling Coefficients

The *tensor product* $\mathbf{V}_1 \otimes \mathbf{V}_2$ of two spaces $\mathbf{V}_1$ and $\mathbf{V}_2$ is defined to be a vector space whose basis is obtained as the product of every basis vector $|1i\rangle$ of the first with every basis vector $|2j\rangle$ of second, written $|1i\rangle|2j\rangle$ or $|1i,2j\rangle$. For example, the tensor product describes the construction of $(n+m)$-particle states from n-particle states and m-particle states, where the n-particle and m-particle states transform as irreps $\boldsymbol{\lambda}$ of, say, the octahedral group

$$|n\lambda_1 l_1, m\lambda_2 l_2\rangle = |n\lambda_1 l_1\rangle |m\lambda_2 l_2\rangle \tag{2.3.1}$$

or, in terms of wave function notation for f-shell states which have known $\mathbf{SO}_3$ properties also,

$$\Psi(f^n J_1 a_1 \lambda_1 l_1, f^m J_2 a_2 \lambda_2 l_2) = \psi(f^n J_1 a_1 \lambda_1 l_1)\phi(f^m J_2 a_2 \lambda_2 l_2)$$

If $\mathbf{V}_1$ is of dimension $|V_1|$ and $\mathbf{V}_2$ of dimension $|V_2|$, the tensor product space is of dimension $|V_1| \times |V_2|$.

Let the spaces $\mathbf{V}_1$ and $\mathbf{V}_2$ be representation spaces of some group; so, continuing the second example, $|f^n J_1 a_1 \lambda_1 l_1\rangle$ is the $a_1 \lambda_1 l_1$ partner of the irrep $\mathbf{J}_1$ of $\mathbf{SO}_3$ and the l_1 partner of the (smaller) irrep $\boldsymbol{\lambda}_1$ of some subgroup $\mathbf{G}$, and similarly for $|f^m J_2 a_2 \lambda_2 l_2\rangle$. The product space is also a representation space of both $\mathbf{SO}_3$ and $\mathbf{G}$, as can be seen by the action of any operation of $\mathbf{SO}_3$

$$\begin{aligned}
O_R|f^n J_1 a_1 \lambda_1 l_1, f^m J_2 a_2 \lambda_2 l_2\rangle \\
&= O_R|f^n J_1 a_1 \lambda_1 l_1\rangle O_R|f^m J_2 a_2 \lambda_2 l_2\rangle \\
&= \sum_{a_1'\lambda_1' l_1'} |f^n J_1 a_1' \lambda_1' l_1'\rangle J_1(R)_{a_1'\lambda_1' l_1', a_1 \lambda_1 l_1} \\
&\quad \times \sum_{a_2'\lambda_2' l_2'} |f^m J_2 a_2' \lambda_2' l_2'\rangle J_2(R)_{a_2'\lambda_2' l_2', a_2 \lambda_2 l_2} \qquad (2.3.2) \\
&= \sum_{a_1'\lambda_1' l_1' a_2'\lambda_2' l_2'} |f^n J_1 a_1' \lambda_1' l_1' f^m J_2 a_2' \lambda_2' l_2'\rangle \\
&\quad \times J_1 \otimes J_2(R)_{a_1'\lambda_1' l_1' a_2'\lambda_2' l_2', a_1 \lambda_1 l_1 a_2 \lambda_2 l_2}
\end{aligned}$$

where the matrix $\mathbf{J}_1\otimes\mathbf{J}_2(R)$ is known as the *matrix direct product* of the matrices $\mathbf{J}_1(R)$ and $\mathbf{J}_2(R)$. The matrix $\mathbf{J}_1\otimes\mathbf{J}_2(R)$ varies as R ranges over all the elements of $\mathbf{SO}_3$, and is easily shown to obey the matrix representation condition (2.1.9). If $R\in\mathbf{G}$, then (2.3.2) still holds because $\mathbf{G}\subset\mathbf{SO}_3$, but we have the restriction that

$$J_1(R)_{a_1'\lambda_1'l_1'a_1\lambda_1l_1}=\delta_{a_1'a_1}\delta_{\lambda_1'\lambda_1}\lambda_1(R)_{l_1'l_1} \tag{2.3.3}$$

since R will only have nonzero matrix elements between partners of single irreps of $\mathbf{G}$. The set of kets for fixed $n\,J_1a_1\lambda_1m\,J_2a_2\lambda_2$ but all possible l_1l_2 in (2.3.2) will form a representation $\boldsymbol{\lambda}_1\otimes\boldsymbol{\lambda}_2$ of $\mathbf{G}$.

Representations formed this way, by taking tensor products of representation spaces or by taking direct products of representation matrices, are known as *Kronecker product representations*. The same construction is sometimes viewed as taking operators O_R and P_R (mapping $\mathbf{V}_1\to\mathbf{V}_1$ and $\mathbf{V}_2\to\mathbf{V}_2$, respectively) and forming an operator $Q_R=O_RP_R$ (mapping $\mathbf{V}_1\otimes\mathbf{V}_2\to\mathbf{V}_1\otimes\mathbf{V}_2$). Some authors then go further and write all three sets of operators as $\mathcal{G}$, with the operator product as $\mathcal{G}\times\mathcal{G}\to\mathcal{G}$, and refer to it as the *inner product* of the group.

Kronecker product representations do not transform irreducibly in general. We may need to take linear combinations to obtain a set of basis vectors that not only transform as irreps but also transform in the standard manner for each particular irrep. The linear combinations need to include all partners of a pair of irreps of the group $\mathbf{J}_1\otimes\mathbf{J}_2$ for $\mathbf{SO}_3$ or $\boldsymbol{\lambda}_1\otimes\boldsymbol{\lambda}_2$ for $\mathbf{G}$, as can be seen by inspection of (2.3.2) and (2.3.3). The tensor product of the partners of a pair of irreps $\boldsymbol{\lambda}_1$, $\boldsymbol{\lambda}_2$ forms a reducible representation of $\mathbf{G}$ which we want to decompose into irreps of $\mathbf{G}$. For this purpose we sum over the appropriate partner labels, here l_1 and l_2. The unitary transformation with coefficients $\langle\lambda_1l_1,\lambda_2l_2|r\lambda l\rangle$ means that the linear combinations

$$|(f^nJ_1a_1\lambda_1,f^mJ_2a_2\lambda_2)r\lambda l\rangle=\sum_{l_1l_2}|f^nJ_1a_1\lambda_1l_1,f^mJ_2a_2\lambda_2l_2\rangle\langle\lambda_1l_1,\lambda_2l_2|r\lambda l\rangle \tag{2.3.4}$$

will be partners of an irrep $\boldsymbol{\lambda}$ of $\mathbf{G}$ belonging to the product $\boldsymbol{\lambda}_1\otimes\boldsymbol{\lambda}_2$.

Likewise the kets

$$|(f^nJ_1, f^mJ_2)Ja\lambda l\rangle = \sum_{a_1\lambda_1 l_1 a_2\lambda_2 l_2} |f^nJ_1a_1\lambda_1 l_1, f^mJ_2a_2\lambda_2 l_2\rangle$$

$$\times\langle J_1a_1\lambda_1 l_1, J_2a_2\lambda_2 l_2|Ja\lambda l\rangle \tag{2.3.5}$$

will be partners of an irrep **J** of $\mathbf{SO_3}$ belonging to the product $\mathbf{J_1 \otimes J_2}$.

We will refer to the coefficients $\langle\lambda_1 l_1, \lambda_2\lambda_2|r\lambda l\rangle$ and $\langle J_1a_1\lambda_1 l_1, J_2a_2\lambda_2 l_2|Ja\lambda l\rangle$ as the *coupling coefficients* for **G** in the G–C_1 basis and for $\mathbf{SO_3}$ in the SO_3–G–C_1 basis, respectively. The fact that the coefficients can be made to depend only on the labels specified may be deduced from the fact that in (2.3.4) the effect of operations of **G** depends only on λl on the left-hand side and only on $\lambda_1 l_1\lambda_2 l_2$ on the right-hand side. It is to be understood that the irrep $\boldsymbol{\lambda}$ may occur several times (up to three for the icosahedral group) in the Kronecker product of $\boldsymbol{\lambda}_1$ and $\boldsymbol{\lambda}_2$. We thus must have an additional index r to specify which of the subspaces with transformation property $\boldsymbol{\lambda}(\mathbf{G})$ we have. As with the branching multiplicity, this *Kronecker product multiplicity* may be resolved in any (essentially arbitrary) fashion. We shall say a little more about our choice later in this section and much more in Sections 3.3–3.5. Part and parcel of this freedom in separating the multiplicity is one overall free phase for the entire set of partners of each occurrence r of each irrep $\boldsymbol{\lambda}$.

Observe that if every partner l of the rth occurrence of an irrep $\boldsymbol{\lambda}$ is changed in phase by the same phase, say $U(\lambda_1\lambda_2\lambda)_{rr}$, then the phase cancels in the equation that defines the representation properties of the partners. If $R \in \mathbf{G}$, then the action of R is given by

$$O_R|f^nJ_1a_1\lambda_1, f^mJ_2a_2\lambda_2)r\lambda l\rangle U(\lambda_1\lambda_2\lambda)_{rr}$$

$$= \sum_{l'} |f^nJ_1a_1\lambda_1, f^mJ_2a_2\lambda_2)r\lambda l'\rangle U(\lambda_1\lambda_2\lambda)_{rr}\lambda(R)_{l'l} \tag{2.3.6}$$

Recall, however, that normalizations are fixed by orthonormality. Thus, taking into account the freedom in separation r, we have a freedom in coupling coefficients up to a unitary matrix in r, elements $U(\lambda_1\lambda_2\lambda)_{r'r}$.

Namely,

$$\langle\lambda_1 l_1, \lambda_2 l_2 | r\lambda l\rangle_{\text{one choice}} = \sum_{r'} \langle\lambda_1 l_1, \lambda_2 l_2 | r'\lambda l\rangle_{\text{another choice}} U(\lambda_1\lambda_2\lambda)_{r'r} \tag{2.3.7}$$

The presence of this freedom means that the partial tables of coupling coefficients previously published have several different choices of multiplicity separation as well as many phase differences. The next chapter ties up many but not all of the phases by requiring that coupling coefficients have the simplest possible symmetry properties.

The properties of coupling coefficients have been thoroughly studied, the original work on symmetries for $\mathbf{SO}_3 \supset \mathbf{SO}_2 \supset \mathbf{C}_1$ by Wigner (1940) and the early work on factorization by Racah (1949), who studied $\mathbf{U}_{14} \supset \mathbf{Sp}_{14} \supset \mathbf{SU}_2 \times (\mathbf{SO}_7 \supset \mathbf{G}_2 \supset \mathbf{SO}_3)$. We postpone a study of the symmetries until the next chapter, dealing only with the factorization property here.

First note that the coupling coefficients are elements of a unitary transformation between alternative bases $|x_1\lambda_1 l_1, x_2\lambda_2 l_2\rangle$ and $|(x_1\lambda_1, x_2\lambda_2)r\lambda l\rangle$ of the product space $\mathbf{V}_{x_1\lambda_1} \otimes \mathbf{V}_{x_2\lambda_2}$. The row and column indices of this matrix are labeled by the sets $l_1 l_2$ or $r\lambda l$, and the matrix elements therefore satisfy the orthonormality conditions:

$$\sum_{r\lambda l} \langle\lambda_1 l_1, \lambda_2 l_2 | r\lambda l\rangle \langle r\lambda l | \lambda_1 l_1', \lambda_2 l_2'\rangle = \delta_{l_1 l_1'} \delta_{l_2 l_2'} \tag{2.3.8}$$

$$\sum_{l_1 l_2} \langle r\lambda l | \lambda_1 l_1, \lambda_2 l_2\rangle \langle\lambda_1 l_1, \lambda_2 l_2 | r'\lambda' l'\rangle = \delta_{rr'} \delta_{\lambda\lambda'} \delta_{ll'} \tag{2.3.9}$$

where, of course, we are using the abbreviated notation $\langle\lambda_1 l_1, \lambda_2 l_2 | r\lambda l\rangle$ for the matrix element

$$\langle x_1\lambda_1 l_1, x_2\lambda_2 l_2 | (x_1\lambda_1, x_2\lambda_2) r\lambda l\rangle$$

because all coefficients are independent of x_i, and where

$$\langle r\lambda l | \lambda_1 l_1, \lambda_2 l_2\rangle = \langle\lambda_1 l_1, \lambda_2 l_2 | r\lambda l\rangle^* \tag{2.3.10}$$

We derive a generalization of *Racah's factorization lemma* (1949) from a study of (2.3.4) and (2.3.5) and the use of Schur's lemmas.

In (2.3.4) the irrep partner $|(J_1a_1\lambda_1, J_2a_2\lambda_2)r\lambda l\rangle$ is constructed out of the partners $|J_1a_1\lambda_1l_1\rangle$ and $|J_2a_2\lambda_2l_2\rangle$ for fixed $J_1a_1\lambda_1J_2a_2\lambda_2$, whereas the irrep partner $|(J_1J_2)Ja\lambda l\rangle$ of (2.3.5) is constructed out of partners $|J_1a_1\lambda_1l_1\rangle$ and $|J_2a_2\lambda_2l_2\rangle$ for only J_1J_2 fixed. If we increase the first set of partners by varying $a_1\lambda_1a_2\lambda_2$ over their range, then the two sets of partners must be unitary linear combinations of each other. Applying Schur's lemmas, (2.1.26) shows that the transform between the two bases of (2.3.4) and (2.3.5) is diagonal in the label λ and independent of l. Because the couplings in $\mathbf{SO}_3$ and $\mathbf{G}$ were chosen to be independent of the non-group-theoretic labels f^n, f^m, we omit unnecessary labels and obtain

$$\langle (f^nJ_1, f^mJ_2)Ja\lambda l|(f^nJ_1a_1\lambda_1, f^mJ_2a_2\lambda_2)r\lambda' l'\rangle$$

$$=\delta_{\lambda'\lambda}\delta_{l'l}\langle Ja\lambda|J_1a_1\lambda_1, J_2a_2\lambda_2\rangle_r \qquad (2.3.11)$$

These coefficients complete the coupling of partners $|J_1a_1\lambda_1l_1\rangle|J_2a_2\lambda_2l_2\rangle$ to a resultant $|Ja\lambda l\rangle$ by a two-stage process. First use (2.3.11),

$$|(f^nJ_1, f^mJ_2)Ja\lambda l\rangle = \sum_{a_1\lambda_1a_2\lambda_2r} |(f^nJ_1a_1\lambda_1, f^mJ_2a_2\lambda_2)r\lambda l\rangle$$

$$\times\langle J_1a_1\lambda_1, J_2a_2\lambda_2|Ja\lambda\rangle_r$$

then (2.3.4),

$$= \sum_{a_1\lambda_1l_1a_2\lambda_2l_2r} |f^nJ_1a_1\lambda_1l_1, f^mJ_2a_2\lambda_2l_2\rangle$$

$$\times\langle\lambda_1l_1, \lambda_2l_2|r\lambda l\rangle\langle J_1a_1\lambda_1, J_2a_2\lambda_2|Ja\lambda\rangle_r \qquad (2.3.12)$$

and then expand the left using (2.3.5) to obtain

$$\sum_r \langle J_1a_1\lambda_1, J_2a_2\lambda_2|Ja\lambda\rangle_r\langle\lambda_1l_1, \lambda_2l_2|r\lambda l\rangle=\langle J_1a_1\lambda_1l_1, J_2a_2\lambda_2l_2|Ja\lambda l\rangle \qquad (2.3.13)$$

Note that in these equations, the multiplicity index r belongs to the coupling $\boldsymbol{\lambda}_1 \times \boldsymbol{\lambda}_2$ giving $\boldsymbol{\lambda}$ and if we had not eliminated all duplicated labels, the coupling factor of (2.3.11) would be written $\langle Ja\lambda|(J_1a_1\lambda_1, J_2a_2\lambda_2)r\lambda\rangle$. We have discontinued our previous practice (Butler 1975) of transferring the r to the $Ja\lambda$ set, as being misleading.

The factorization property expressed by (2.3.13), and which is called the Racah factorization lemma, holds for any group–subgroup pair. One repeats the above argument, the only generalization required being the presence of a multiplicity index in the group as well as the subgroup. Coefficients of the form $\langle J_1a_1\lambda_1, J_2a_2\lambda_2|Ja\lambda\rangle_r$ were called *isoscalar factors* by Racah. We call them *coupling factors*, for (2.3.12) is the statement that the coupling coefficients for a group–subgroup chain, here $\mathbf{SO}_3 \supset \mathbf{G} \supset \mathbf{C}_1$, factorize into largely independent factors, one for each of the subchains $\mathbf{SO}_3 \supset \mathbf{G}$ and $\mathbf{G} \supset \mathbf{C}_1$. As we see later, the interdependence is in the phases and product multiplicity separations of the irreps of $\mathbf{G}$ taken as a whole. The coupling factors take the r irreps $\boldsymbol{\lambda}$ contained in a product $\boldsymbol{\lambda}_1 \otimes \boldsymbol{\lambda}_2$ having one set of phases and separations and forms the Ja irreps $\boldsymbol{\lambda}$ with another set of phases and combinations.

The coupling factors for irreps $\boldsymbol{\lambda}$ of $\mathbf{G}$ and $\boldsymbol{\rho}$ of a subgroup $\mathbf{H}$ satisfy the two orthogonality relations

$$\sum_{r\lambda a} \langle\lambda_1 a_1 \rho_1, \lambda_2 a_2 \rho_2 | r\lambda a\rho\rangle_s \langle r\lambda a\rho|\lambda_1 a_1'\rho_1', \lambda_2 a_2'\rho_2'\rangle_{s'}$$

$$= \delta_{a_1a_1'}\delta_{\rho_1\rho_1'}\delta_{a_2a_2'}\delta_{\rho_2\rho_2'}\delta_{ss'} \tag{2.3.14}$$

$$\sum_{a_1\rho_1a_2\rho_2s} \langle r\lambda a\rho|\lambda_1a_1\rho_1, \lambda_2a_2\rho_2\rangle_s \langle\lambda_1a_1\rho_1, \lambda_2a_2\rho_2|r'\lambda'a'\rho\rangle_s$$

$$= \delta_{rr'}\delta_{\lambda\lambda'}\delta_{aa'} \tag{2.3.15}$$

2.4 Character Theory

This section is somewhat of an aside. The reader who is not interested in deriving the representation structure of groups but wishes only to use the representation structure of groups to simplify or solve for the angular structure of physical or chemical problems will find no results of direct use. This section outlines the tools of character theory so that one could rederive the structural properties of the representations of the point groups. Fuller descriptions of character theory may be found elsewhere (e.g., Hamermesh, 1962, Chapter 3). We provide the tools to answer questions of the type: Why does the tetrahedral group have precisely seven inequivalent irreps (including spin irreps)?

We could stray further from the point by asking a more basic question yet, namely, why are there precisely 32 crystallographic point groups? Hamermesh (1962, Chapter 2) is one of many physics texts which answers such questions. Modern mathematics has devoted much effort to finding general answers to this question of what abstract group structures can exist. Given an abstract group structure, one major method of obtaining the properties of the irreps is character theory. Burnside (1911) contains most presently known results for finite groups, except for the *symmetric groups* $\mathbf{S}_n$, where $\mathbf{S}_n$ is the group of all permutations on n objects. The character theory of symmetric groups is intimately connected to combinatorics, and is also closely related to the theory of compact continuous groups. (See, for example, Wybourne, 1970.)

The term character derives from the old term *characteristic* of a matrix, also called the trace or spur of the matrix, being the sum of the diagonal terms. The characteristic tr($\mathbf{A}$) of a matrix $\mathbf{A}$ is unchanged by a similarity transformation $\mathbf{X}$.

PROOF. If $\mathbf{X}^{-1}=\mathbf{Y}$, then we have

$$\begin{aligned}\operatorname{tr}(\mathbf{XAY}) &= \sum_i \sum_{jk} X_{ij}A_{jk}Y_{ki} \\ &= \sum_{ijk} A_{jk}Y_{ki}X_{ij} \\ &= \sum_{jk} A_{jk}\,\delta_{jk} = \sum_j A_{jj} \\ &= \operatorname{tr}(\mathbf{A}) \qquad (2.4.1)\end{aligned}$$

so in particular the characteristic of a matrix is the sum of its eigenvalues. Changes of basis are similarity transformations, so the characteristic of an operator is defined as the trace of the corresponding matrix in any basis.

We define the characteristic of group operator O_R: $\mathbf{V}_\lambda \to \mathbf{V}_\lambda$ acting on representation space $\mathbf{V}_\lambda$ as

$$\chi^\lambda(R) = \mathrm{tr}(O_R) = \mathrm{tr}(\boldsymbol{\lambda}(R)) \tag{2.4.2}$$

The set of such complex numbers as R varies over the various group elements is known as the *character* of representation $\boldsymbol{\lambda}$ of group $\mathbf{G}$,

$$\chi^\lambda(\mathbf{G}) = \{\chi^\lambda(R): \quad R \in \mathbf{G}\} \tag{2.4.3}$$

We have dealt at length with similar representations $\mathbf{V}_{x\lambda}$ and $\mathbf{V}_{y\lambda}$, which give rise to the same matrix representation $\boldsymbol{\lambda}(\mathbf{G})$ if the partners are chosen appropriately. Because the character $\chi^\lambda(\mathbf{G})$ is independent of the particular choice of partners, equivalent representations have the same character. The converse is also true. Representations with the same character are equivalent. We postpone the proof until (2.4.12).

Consider the group elements $R, S, T, \ldots$. The product of each of the elements by a fixed element S gives rise to the group elements in a different order. If two products SR and ST were to be the same, we could premultiply by S^{-1} to obtain that

$$SR = ST$$

implies

$$S^{-1}SR = S^{-1}ST = R = T$$

If $f(R)$ is any function of the group elements, then

$$\sum_{R \in \mathbf{G}} f(R) = \sum_{R \in \mathbf{G}} f(SR) \tag{2.4.4}$$

because the sum on the right-hand side contains the same terms as the left, only in a different order. Equation (2.4.4) is often called the *group rearrangement theorem*.

We use this and Schur's lemmas to derive a result known as the great orthogonality theorem.

Suppose $\boldsymbol{\lambda}(\mathbf{G})$, $\boldsymbol{\mu}(\mathbf{G})$ are irreps and X maps between them, so that $\mathbf{X}$ is a $|\lambda| \times |\mu|$ matrix. Let a particular function be the matrix

$$\mathbf{f}(R) = \boldsymbol{\lambda}(R)\mathbf{X}\boldsymbol{\mu}(R^{-1}) \tag{2.4.5}$$

and construct the map A with matrix

$$\mathbf{A} = \sum_{R \in \mathbf{G}} \mathbf{f}(R) \tag{2.4.6}$$

which commutes with O_S, for all $S \in \mathbf{G}$. To prove this use the definition

$$\boldsymbol{\lambda}(S)\mathbf{A}\boldsymbol{\mu}(S^{-1}) = \sum_{R \in \mathbf{G}} \boldsymbol{\lambda}(S)\boldsymbol{\lambda}(R)\mathbf{X}\boldsymbol{\mu}(R^{-1})\boldsymbol{\mu}(S^{-1})$$

Now use the fact that $\boldsymbol{\lambda}$ and $\boldsymbol{\mu}$ are representations to combine $\boldsymbol{\lambda}(S)\boldsymbol{\lambda}(R)$ as $\boldsymbol{\lambda}(SR)$; this is

$$= \sum_{R \in \mathbf{G}} \boldsymbol{\lambda}(SR)\mathbf{X}\boldsymbol{\mu}\big((SR)^{-1}\big)$$

Finally, use (2.4.4) because S does not vary inside the sum,

$$= \sum_{R \in \mathbf{G}} \boldsymbol{\lambda}(R)\mathbf{X}\boldsymbol{\mu}(R) = \mathbf{A} \tag{2.4.7}$$

which completes the proof that A commutes with O_S. Application of Schur's lemmas to this shows that $\mathbf{A}$ is a multiple a of the unit matrix if $\lambda = \mu$, or zero otherwise. Choose $\mathbf{X}$ to have all elements zero except for the pq element, $X_{pq} = 1$. Expression (2.4.5) for $\mathbf{f}(R)$ simplifies to a product of one element from $\boldsymbol{\lambda}(R)$ and one from $\boldsymbol{\mu}(R^{-1})$. The number a takes on some value a_{pq}. Thus the ij element of $\mathbf{A}$ can be written

$$A_{ij} = a_{pq}\delta_{\lambda\mu}\delta_{ij}$$

$$= \sum_R \sum_{kl} \lambda(R)_{ik} X_{kl} \mu(R^{-1})_{lj}$$

$$= \sum_R \lambda(R)_{ip} \mu(R^{-1})_{qj} \tag{2.4.8}$$

To evaluate a_{pq}, meaningful only when $\lambda=\mu$ and $i=j$, take such a term, then sum over i,

$$\begin{aligned}\sum_i a_{pq} &= \sum_{Ri} \lambda(R)_{ip}\lambda(R^{-1})_{qi} \\ &= \sum_{Ri} \lambda(R^{-1})_{qi}\lambda(R)_{ip} \\ &= \sum_{R} \lambda(R^{-1}R)_{qp} \\ &= \sum_{R} \lambda(E)_{qp} \end{aligned} \tag{2.4.9}$$

The representation matrix for the identity operation is the unit matrix. The number of terms on the left is the dimension $|\lambda|$ of the matrix, the number on the right the order $|G|$ of the group, so

$$|\lambda| a_{pq} = |G|\,\delta_{pq} \tag{2.4.10}$$

Substituting this value of a_{pq} into (2.4.8) and using the unitarity of the irrep matrices, $\mu(R^{-1})_{qj} = \mu(R)^{*}_{jq}$, one has the *great orthogonality theorem*:

$$\sum_R \lambda(R)_{ij}\lambda'(R)^{*}_{i'j'} = \frac{|G|}{|\lambda|}\delta_{ii'}\delta_{jj'}\delta_{\lambda\lambda'} \tag{2.4.11}$$

This result is very powerful. A matrix irrep is a set of matrices which cannot be simultaneously block-diagonalized. Schur's lemmas said that only diagonal matrices can commute with all the matrices of an irrep. The great orthogonality theorem says that the irrep matrices are such that the sets (for various R) of numbers occurring in a particular position of a particular irrep (the $ij\lambda$ set is $\{\lambda(R)_{ij}\colon R\in G\}$) are orthogonal to all other such sets. The irreducibility of irreps seems a relatively innocuous condition when defined in terms of decomposition of representation space, but the great orthogonality theorem shows that it forces a very strong condition on the set of irrep matrices.

Much of the strength of the great orthogonality theorem is retained in the orthogonality of characters. The trace of a matrix is independent of the basis, that is, the coordinate system chosen, and thus a result on the orthogonality of traces of irrep matrices has certain advantages over the above result. Recall that the great orthogonality theorem assumed that there is a unique set of matrices $\boldsymbol{\lambda}(R)$ for the irrep $\boldsymbol{\lambda}(\mathbf{G})$ [see the requirement in (2.1.10)–(2.1.11)].

The orthogonality of characters is obtained from (2.4.11) by setting $i=j$ and $i'=j'$ and summing over both i and i',

$$\sum_R \sum_{ii'} \lambda(R)_{ii} \lambda'(R)^*_{i'i'} = \frac{|G|}{|\lambda|} \delta_{\lambda\lambda'} \sum_{ii'} \delta^2_{ii'}$$

or

$$\sum_R \chi^\lambda(R) \chi^{\lambda'}(R)^* = |G| \, \delta_{\lambda\lambda'} \tag{2.4.12}$$

The character χ^λ is a complex-valued vector or function. For each $R \in \mathbf{G}$, it takes on the value $\chi^\lambda(R)$, a complex number which is the trace of the matrix $\boldsymbol{\lambda}(R)$. Equation (2.4.12) says that these functions are orthogonal with normalization $|G|^{1/2}$.

The characters of irreps are sometimes called primitive characters or simple characters. Because they are mutually orthogonal, they are limited in number by the dimension of the space to which they belong. The number of independent functions is not the number of group elements $|G|$, but the number of *conjugacy classes* r. The *conjugacy class* $\mathbf{C}_R$ containing R is defined as the set of group elements such that if $S \in \mathbf{C}_R$, then there exists some group element X such that $S = XRX^{-1}$,

$$\mathbf{C}_R = \{S: \quad S = XRX^{-1}, \quad \text{with } S, X \in \mathbf{G}\} \tag{2.4.13}$$

The characteristic of all elements of a class is the same

$$\chi^\lambda(R) = \chi^\lambda_{C_R} \tag{2.4.14}$$

because the irrep matrices are related by a similarity transformation

$$\boldsymbol{\lambda}(S) = \boldsymbol{\lambda}(XRX^{-1}) = \boldsymbol{\lambda}(X)\boldsymbol{\lambda}(R)\boldsymbol{\lambda}(X)^{-1} \tag{2.4.15}$$

In the group $\mathbf{SO}_3$, all elements $R(\theta, \mathbf{n})$ consisting of a rotation through certain angle θ $(0 \leqslant \theta < 2\pi)$ about any axis $\mathbf{n}$ belong to the same class. The elements $R(\theta, \mathbf{n})$ and $R(\theta, \mathbf{n}')$ may be transformed one into the other by the rotation which rotates axis $\mathbf{n}$ into axis $\mathbf{n}'$. The classes and characters of $\mathbf{SO}_3$

may be parametrized by the rotation angle θ of their elements. Much the same is true for $\mathbf{D}_3$, the group consisting of the identity, the threefold rotations by $2\pi/3$ and $4\pi/3$ about the z axis, and the twofold rotations about three axes separated by $2\pi/3$ in the xy plane. The twofold rotations transform the threefold rotations into themselves and vice versa. For $\mathbf{C}_3$, the elements $R(2\pi/3, z)$ and $R(4\pi/3, z) = R(2\pi/3, -z)$ belong to separate classes; there is no group element which transforms one into the other.

For a finite group **G** with r conjugacy classes the character χ^λ takes on at most r different values, (2.4.14). The characters of inequivalent irreps λ and μ are orthogonal functions [see (2.4.12)], so there can be at most r classes. Likewise the great orthogonality theorem (2.4.11) says that the irrep matrix elements $\lambda(R)_{ij}$, regarded as functions of R, are mutually orthogonal. Each irrep λ contributes $|\lambda|^2$ matrix elements and there are $|G|$ different R, so

$$\sum_\lambda |\lambda|^2 \leqslant |G| \tag{2.4.16}$$

It is not easy to prove, but for finite groups the number of inequivalent irreps equals the number of conjugacy classes and, further, the equality holds in (2.4.16).

For continuous groups, such as the group of rotations in [illegible]nsions and the group of translations in n dimensions, and also infinite discrete groups, such as translations of a crystal lattice, the number of classes is not equal to the number of irreps, nor does (2.4.16) hold. For compact continuous groups the sum over group elements becomes an invariant Haar integral, and these groups have a bounded continuity of classes and a countable series of finite-dimensional unitary irreps. Noncompact groups have representations for which the concept of irreducibility needs modification.

The above character orthogonality (2.4.12) may be rewritten in terms of the notation χ_C^λ, where C is a conjugacy class label. If $|C|$ is the number of elements in class **C**, then

$$\sum_C |C| \chi_C^\lambda \chi_C^{\lambda'*} = |G| \, \delta_{\lambda\lambda'} \tag{2.4.17}$$

and for finite groups a second orthogonality relation can be derived

$$\sum_\lambda \chi_C^\lambda \chi_{C'}^{\lambda*} = |G| \, \delta_{CC'} \tag{2.4.18}$$

Chapter 11 gives character tables of all the subgroups of $\mathbf{SO}_3$. For the finite groups one has square tables of χ_C^λ, the rows being labeled by irreps $\boldsymbol{\lambda}$ and the columns by the classes.

Orthogonality (2.4.17) is all that is required to deduce the irreps contained in a given representation $\mathbf{L}(\mathbf{G})$. The representation $\mathbf{L}$ may be given as a vector space or as a set of matrices. In some cases it is necessary to compute the characteristics for each class, but it may be possible to deduce the characteristics by an indirect route. $\mathbf{L}$, with characteristics χ_C^L, must be a sum of irreps

$$\mathbf{L}(\mathbf{G}) = \sum_\lambda a_\lambda \boldsymbol{\lambda}(\mathbf{G}) \tag{2.4.19}$$

and application of (2.4.17) shows

$$\sum_C |C| \chi_C^L \chi_C^{\lambda *} = a_\lambda |G| \tag{2.4.20}$$

Indeed, even if all χ_C^L are not known, with use of the fact that the a_λ are nonnegative integers it is often possible to deduce the a_λ from the summability of traces

$$\chi_C^L = \sum_\lambda a_\lambda \chi_C^\lambda \tag{2.4.21}$$

using very few values of χ_C^L.

The Kronecker product of two irreps $\boldsymbol{\lambda}_1 \times \boldsymbol{\lambda}_2$ is equivalent to the tensor product of irrep spaces $\mathbf{V}_{\lambda_1} \otimes \mathbf{V}_{\lambda_2}$ or to the matrix direct product of the representation matrices $(\boldsymbol{\lambda}_1 \times \boldsymbol{\lambda}_2)(R) = \boldsymbol{\lambda}_1(R) \otimes \boldsymbol{\lambda}_2(R)$. The trace of a direct product is the product of the traces, and hence the irreps contained in the Kronecker product may be obtained by the above orthogonality. If

$$\boldsymbol{\lambda}_1 \times \boldsymbol{\lambda}_2 = \sum_\lambda m_{\lambda_1 \lambda_2}^\lambda \boldsymbol{\lambda} \tag{2.4.22}$$

then

$$m_{\lambda_1 \lambda_2}^\lambda = |G|^{-1} \sum_C |C| \chi_C^{\lambda_1} \chi_C^{\lambda_2} \chi_C^{\lambda *} \tag{2.4.23}$$

The complex conjugate of a characteristic $\chi^\lambda(R)$ is always the characteristic of an irrep. If χ^λ is not real for all R, then the function $(\chi^\lambda)^*$ is different from χ^λ and thus must be the character of some other irrep called the *complex conjugate irrep* $\boldsymbol{\lambda}^*$. Because of the obvious symmetries of the number

$$n_{\lambda_1\lambda_2\lambda_3}=|G|^{-1}\sum_C |C|\chi_C^{\lambda_1}\chi_C^{\lambda_2}\chi_C^{\lambda_3} \tag{2.4.24}$$

it is clear that $m^{\lambda}_{\lambda_1\lambda_2}$ has the symmetries

$$m^{\lambda}_{\lambda_1\lambda_2}=m^{\lambda}_{\lambda_2\lambda_1}=m^{\lambda_2^*}_{\lambda_1\lambda^*}=\text{etc.} \tag{2.4.25}$$

2.5 Complex Conjugation

The kets $|r1\rangle=-(1/\sqrt{2})(x+iy)$ and $|r-1\rangle=(1/\sqrt{2})(x-iy)$ were shown in Example 2.1 to give rise to the one-dimensional irrep matrices $\exp(-i\theta)$ and $\exp(i\theta)$ for the group $\mathbf{SO}_2$. The two kets are related by complex conjugation, as are the two matrices. We make a few remarks in this section on the algebraic structure related to complex conjugation, as this structure is fundamental to the topic of this book. Chapter 8 connects the algebraic operation of complex conjugation with the physical operations of time reversal and reversal of the directions of motion.

The functions x, y, z and $-(1/\sqrt{2})(x+iy), (1/\sqrt{2})(x-iy), z$ are two possible sets of partners for the three-dimensional irrep space $\mathbf{V}_r$ of $\mathbf{SO}_3$. The sets of irrep matrices are written as either $\mathbf{r}(\mathbf{SO}_3)$ or $\mathbf{1}(\mathbf{SO}_3)$. Depending on the basis (set of partners) used, the irrep matrices will be either real or strictly complex (strictly complex means that at least one matrix element has both real and imaginary parts). Complex-conjugating the partners will not affect the space $\mathbf{V}_r$ as a whole, but may reorder the partners and change their phases; for example,

$$|r1\rangle^*=-|r-1\rangle \tag{2.5.1}$$

The various irrep matrices will go into their complex conjugates, but these will be related by a basis transformation $\mathbf{T}$ to themselves (this is nontrivial!)

$$\mathbf{r}(R)=\mathbf{T}\ \mathbf{r}(R)^*\ \mathbf{T}^{-1} \tag{2.5.2}$$

The above paragraphs illustrate two relationships between an irrep $\boldsymbol{\lambda}(\mathbf{G})$ and its complex conjugate $\boldsymbol{\lambda}^*(\mathbf{G})$. The ket $|r1\rangle$ transforms as the single partner of an irrep $\mathbf{1}(\mathbf{SO}_2)$ and the ket $|r-1\rangle$ obtained by complex conjugation transforms as a single partner of a different irrep $-\mathbf{1}(\mathbf{SO}_2)$.

We write

$$\mathbf{1}^*(\mathbf{SO}_2) = -\mathbf{1}(\mathbf{SO}_2) \tag{2.5.3}$$

The three kets $|r1\rangle$, $|r0\rangle$, and $|r-1\rangle$ transform as the irrep $\mathbf{1}(\mathbf{SO}_3)$ or $\mathbf{r}(\mathbf{SO}_3)$. Under complex conjugation (2.5.2) shows that

$$\mathbf{1}^*(\mathbf{SO}_3) = \mathbf{1}(\mathbf{SO}_3) \tag{2.5.4}$$

although with this, the JM basis, the irrep matrices are not real and for many $R \in \mathbf{SO}_3$ the complex conjugate $\mathbf{1}(R)^*$ of the irrep matrix $\mathbf{1}(R)$ is not equal to itself. We write

$$\mathbf{1}(\mathbf{SO}_3)^* \neq \mathbf{1}(\mathbf{SO}_3) \tag{2.5.5}$$

In (2.5.4) we complex-conjugate the label, which means both complex-conjugate the irrep matrices and also perform the change of basis (2.5.2) to give the standard basis (here the JM basis).

A matrix irrep $\boldsymbol{\lambda}(\mathbf{G})$ is a set of matrices $\boldsymbol{\lambda}(R)$ which satisfy (2.1.9). Complex-conjugating gives

$$\boldsymbol{\lambda}(R)^*\boldsymbol{\lambda}(S)^* = \boldsymbol{\lambda}(RS)^* \tag{2.5.6}$$

so the complex conjugate matrices always form a matrix irrep. The matrix irrep $\boldsymbol{\lambda}(\mathbf{G})^*$ will be similar to some matrix irrep $\boldsymbol{\lambda}^*(\mathbf{G})$ in the basis scheme being used,

$$\boldsymbol{\lambda}^*(R) = \mathbf{T}\boldsymbol{\lambda}(R)^*\mathbf{T}^{-1} \tag{2.5.7}$$

Irreps such that $\lambda^* \neq \lambda$, such as both $\mathbf{1}(\mathbf{SO}_2)$ and $-\mathbf{1}(\mathbf{SO}_2)$, are said to be *complex irreps*. Irreps such that $\lambda^* = \lambda$ are said to be *real irreps*. The trace of a matrix is unchanged by a similarity transformation, so (2.5.7) shows real irreps have real characters. All irreps of $\mathbf{SO}_3$ are real. Real irreps subdivide into two classes. *Orthogonal irreps*, such as the true irreps of $\mathbf{SO}_3$, have a basis [such as the x, y, z basis of $\mathbf{1}(\mathbf{SO}_3)$] where the irrep matrices are orthogonal. An *orthogonal matrix* $\mathbf{X}$ satisfies

$$\mathbf{X}\mathbf{X}^{\mathrm{T}} = \mathbf{I} \tag{2.5.8}$$

Where ^{T} denotes transpose and $\mathbf{I}$ is the unit matrix. *Symplectic irreps*, such as the spin irreps of $\mathbf{SO}_3$, are those where a basis exists in which all irrep matrices are symplectic. A *symplectic matrix* $\mathbf{X}$ satisfies

$$\mathbf{X}\mathbf{J}\mathbf{X}^{\mathrm{T}}=\mathbf{J} \tag{2.5.9}$$

where $\mathbf{J}$ is the $2k\times 2k$ matrix of the form

$$\begin{pmatrix} \mathbf{O}_k & \mathbf{I}_k \\ -\mathbf{I}_k & \mathbf{O}_k \end{pmatrix}$$

$\mathbf{O}_k$ and $\mathbf{I}_k$ being the $k\times k$ zero and unit matrices. Symplectic matrices are of even dimension, and while they have some complex elements, their trace is real (Lax 1974 uses the term pseudoreal).

The Frobenius–Schur invariant [Hamermesh 1962, Eq. (5-84)]

$$c_\lambda=\frac{1}{|G|}\sum_{R\in G}\chi^\lambda(R^2) \tag{2.5.10}$$

gives a means of computing this property under complex conjugation. The Frobenius–Schur invariant is 1 for orthogonal irreps, -1 for symplectic irreps, and 0 for complex irreps.

Abelian groups have one-dimensional irreps, so all true and spin irreps are either orthogonal or complex. Non-Abelian point groups have most of their true irreps orthogonal, and most of their spin irreps symplectic, but the group $\mathbf{T}$ is one group that has an example of a pair of complex true irreps and a pair of complex spin irreps.

2.6 Spin Irreps and Labeling Irreps

The discovery of electron spin (especially the Stern–Gerlach experiment of 1922) showed that representations as defined in Section 2.1 are not sufficiently general. Around the turn of the century Schur gave a method for finding representations of a finite group in terms of *fractional linear transformations* (projective transformations). These retain the group multiplication law (2.1.3), namely $O_RO_S=O_{RS}$, but the space on which they act is a projective space (as in projective geometry), not a linear vector space. Such representation matrices do not obey

$$\lambda(R)\lambda(S)=\lambda(RS) \tag{2.6.1}$$

Hamermesh (1962, Chapter 12) shows that fractional linear representations are equivalent to *projective representations* as usually defined, e.g., in space group theory. One asks that the product of the two matrices representing elements R, S is equal to a scalar multiple $\omega_{R,S}$ of the matrix representing the element RS,

$$\boldsymbol{\lambda}(R)\boldsymbol{\lambda}(S)=\omega_{R,S}\boldsymbol{\lambda}(RS) \tag{2.6.2}$$

This in turn is equivalent to having a set of p matrices, scalar multiples of each other, representing each group element. This is a p-valued representation, whereas (2.6.1) is single-valued, a unique matrix for each group element. This projective representation condition (2.6.2) is weaker than the true representation condition (2.6.1), so a group has more projective representations than true representations. Cartan in 1913 developed a general method of constructing projective matrix irreps for the continuous groups. The continuity condition may be used to show that only the group of orthogonal transformations in n dimensions has nontrivial projective representations, and then $\omega_{R,S}=\pm 1$ only, so the representations are at most double-valued (Littlewood 1950, p. 248).

Both kinds of representations have many equivalent names. Representations satisfying (2.6.1) have been called ordinary, true, integer (because $\mathbf{SO}_3$ irreps are so labeled), or vector (because of the vector space linearity). Representations satisfying (2.6.2) have several old names, but it seems worth retaining only projective for the general p-valued case and spin for the double-valued case which is of particular interest to us. A p-valued representation of a group $\mathbf{G}$ of order $|G|$ can be regarded as a true representation of a group of order $p|G|$. The group of rotations and reflections in real 3-space, and hence its subgroups, too, have both true and spin representations. It was not until quantum mechanics that physical systems were discovered which transformed as spin representations of the various point groups. For example, the spin of an electron transforms as a spin irrep of the octahedral group $\mathbf{O}$. Both Schur and Cartan found it useful to work in terms of a group and its projective representations and Altmann (1979) argues the case for continuing this tradition. Many modern authors use the fact that a spin irrep of a group equals a true irrep of a double covering group, and work with the double group. This leads to attempts to find "physical" rotation operators with spinor properties, as well as confusion in terminology: the electron spin transforms as a true (vector) irrep of the double group.

The expression "the irreps of the octahedral group" will encompass the true and the spin irreps, or equivalently, all true irreps of the double octahedral group. Excepting some character theory results which require a little modification (see Altmann 1979), all our results are independent of which viewpoint is taken.

The various equations of Section 3.3 which we solve to obtain the values of the *jm* and *j* symbols, and hence the values of the coupling and transformation coefficients, are quadratic and cubic or worse. However, there is an old result of character theory which allows us to solve those equations directly and exactly for $\mathbf{SO}_3$ and most of the point groups. A *faithful representation* of a group is one in which the action of distinct group elements is distinct. Burnside (1911, p. 299) notes that a representation of a finite group is faithful if and only if all irreps of the group can be obtained as Kronecker powers of the representation, that is, by successive couplings of the representation with itself. For continuous compact groups, for example, the groups $\mathbf{U}_{14}$, $\mathbf{Sp}_{14}$, $\mathbf{O}_7$, and $\mathbf{G}_2$ used by Racah in his *f*-shell structure calculations, a similar result holds if in addition we take Kronecker powers of the complex conjugate of the faithful representation.

We are in fact dealing with both the true and spin irreps of the groups of transformations in 3-space, so only the spin irreps of the groups will contain all irreps in their Kronecker powers. The two basic spin functions,

$$\alpha=|J=\tfrac{1}{2},M_J=\tfrac{1}{2}\rangle,\qquad \beta=|J=\tfrac{1}{2},M_J=-\tfrac{1}{2}\rangle \tag{2.6.3}$$

form a suitable representation of all the point groups. The pair are partners of an irreducible spin representation $\frac{1}{2}(\mathbf{G})$ of all the noncyclic groups. This particular irrep we call the *primitive irrep* of **G**. For the cyclic groups $\mathbf{C}_n$, the two spinor functions α and β are each the single partner of a complex conjugate pair of irreps, $\frac{1}{2}(\mathbf{C}_n)$ and $-\frac{1}{2}(\mathbf{C}_n)$, respectively, where $\frac{1}{2}^*(\mathbf{C}_n)=-\frac{1}{2}(\mathbf{C}_n)$. For these groups we say that both $\frac{1}{2}$ and $-\frac{1}{2}$ are primitive irreps.

The fact that $\frac{1}{2}(\mathbf{G})$ has as its basis one or both of the spin functions of (2.6.2) is of little direct use to us in preparing our tables. However, the fact that it is of small dimension is related to the fact that few simultaneous equations need be solved when calculating the *jm* and *j* symbols (Butler and Wybourne 1976a,b and Butler, Haase, and Wybourne 1978, 1979).

The literature contains a number of labeling schemes for the various irreps of the point groups. We make no apologies for introducing, and using exclusively, a system of labels, to which we refer as the *natural labels*. In the character theory of $\mathbf{SO}_3$ and $\mathbf{SO}_2$ positive and negative, integer and half-integer, labels arise naturally. The various relations between irreps are expressed as algebraic functions of these labels. Similar labels for the other point groups may be obtained either from the imbedding of the group in $\mathbf{SO_3}$ or equivalently by using the concept of the power of an irrep. The *power* p_λ of irrep $\boldsymbol{\lambda}(\mathbf{G})$ is defined to be the lowest number p_λ for which $\boldsymbol{\lambda}$ occurs in the p_λ-fold Kronecker product of the primitive irrep $\frac{1}{2}(\mathbf{G})$, or irreps $(\frac{1}{2}+-\frac{1}{2})$, with itself (Butler and Wybourne, 1976a). For $\mathbf{SO_3}$ there is precisely one irrep $\frac{1}{2}\mathbf{m}$ of each power m. For $\mathbf{SO_2}$ there are two irreps of power m, $\frac{1}{2}\mathbf{m}$ and $-\frac{1}{2}\mathbf{m}$. The dihedral and cyclic groups are similar and we label the two-dimensional irreps of $\mathbf{D}_n$ by integers or half-integers, and most one-dimensional irreps of $\mathbf{D}_n$ and $\mathbf{C}_n$ form complex conjugate pairs $\frac{1}{2}\mathbf{m}$ and $-\frac{1}{2}\mathbf{m}$. For $\mathbf{C}_n$ the irrep of power n is real and is thus labeled $\frac{1}{2}\mathbf{n}$, which is integer or half-integer as n is even or odd. Certain one-dimensional irreps of $\mathbf{D}_n$ must be labeled differently. For $n\geqslant 2$ there are two irreps of power two, the two-dimensional (xy) irrep and the one-dimensional (z) irrep. The former is naturally written as $\mathbf{1}(\mathbf{D}_n)$ and we write the latter as $\tilde{\mathbf{0}}(\mathbf{D}_n)$. For n even, the two $\mathbf{D}_n$ irreps of power n form a real pair we label $\frac{1}{2}\mathbf{n}$ and $\frac{1}{2}\tilde{\mathbf{n}}$. We have that $\frac{1}{2}\tilde{\mathbf{n}}=\tilde{\mathbf{0}}\times\frac{1}{2}\mathbf{n}$. A similar reasoning gives rise to the labels used for $\mathbf{T}$, $\mathbf{O}$, and $\mathbf{K}$. For example, for the tetrahedral group $\mathbf{T}$ we have the correspondence among the natural label, the label of Koster, Dimmock, Wheeler, and Statz (1963), and the Mulliken label as follows:

natural	$\mathbf{0}$	$\frac{1}{2}$	$\mathbf{1}$	$\frac{3}{2}$	$-\frac{3}{2}$	$\mathbf{2}$	$-\mathbf{2}$
Koster *et al.*	Γ_1	Γ_5	Γ_4	Γ_6	Γ_7	Γ_2	Γ_3
Mulliken	A_1	E'	T	V'		E	

Correspondences for all groups are given with the character tables. Inspection of the tables of Chapter 12 will show that the various branching rules follow a regular pattern using these labels. However, it is the nature of the finite groups that no system leads to completely algebraic formulas. Note also that the dimension $|\lambda|$ is not to be confused with the absolute value of the label. The group label can be attached as a subscript to avoid confusion, e.g.,

$$|2|_{SO_3}=5=|2|_K$$

$$|2|_O=2=|2|_{D_5}$$

$$|2|_T=1=|2|_{D_4}=|2|_{SO_2}$$

3

The *jm* Factors and *j* Symbols

Coupling coefficients were defined in the previous chapter. They give the construction of many-particle states of given transformation properties from products of fewer particle states. They also give the matrix elements of operators of known transformation properties. The latter result is the Wigner–Eckart theorem and is proved in the following chapter. This chapter is concerned with the algebraic properties of coupling coefficients.

Much of the notation and terminology is due to Wigner (1940). Derome and Sharp (1965) and Derome (1966) generalized many of Wigner's results, which were for simply reducible groups only. A *simply reducible* group is one with no product multiplicity and where all irreps have real characters. Butler (1975) has given a review of the development. The results are given there in general form, that is, without the assumption used throughout this book that all irreps are simple phase. *A simple-phase* irrep $\boldsymbol{\lambda}$ is one in which the $3j$ phase describing interchange symmetries of the $3jm$ symbol involving λ three times is independent of which interchange is discussed, $1\leftrightarrow 2$, $2\leftrightarrow 3$, or $3\leftrightarrow 1$. The permutation groups $\mathbf{S}_n$ for $n \geqslant 6$, and the matrix groups $\mathbf{SU}_n$ and $\mathbf{Sp}_n$ for $n \geqslant 4$ and $\mathbf{SO}_n$ for $n \geqslant 5$, all have non-simple-phase irreps (Butler and King 1974).

For all groups, it is convenient to define a renormalized set of coupling coefficients, especially the $2jm$ and $3jm$ symbols below (the $1j$ and $3j$ symbols of Wigner). Wigner's notation was chosen to give symbols which display the symmetries in a reasonably obvious manner. We call the generalized coefficients *jm* symbols because the development owes much to the coupling coefficients for the JM basis, the SO_3–SO_2–C_1 basis of angular momentum theory. The next section translates the results proved in Chapter 2 into this notation.

The j symbols for the group **G** together with the j symbols of a subgroup **H** give the symmetries of the *jm* symbols for the chain **G**⊃**H**. After defining the $2j$, $3j$, and $6j$ symbols in Section 3.2, we continue in Section 3.3 by giving the various relations between various j symbols and between *jm* and j symbols.

Section 3.4 shows that these symmetry, orthogonality, and recoupling equations may be used to compute the values of the *jm* and j symbols. In other words, subject to certain phase freedoms intrinsic to coupling coefficients, their values follow from the selection rules of the groups, which in turn may be deduced from character theory.

We end the chapter with a careful statement of how we chose the phases and the multiplicity separations for the tables of this book.

3.1 The *jm* Factors

The relationship between the irrep $\boldsymbol{\lambda}(\mathbf{G})$ and its complex conjugate $\boldsymbol{\lambda}^*(\mathbf{G})$ is central to the analysis of the Racah–Wigner algebra. The coupling of an irrep with its complex conjugate is the only way the scalar irrep $\mathbf{0}(\mathbf{G})$ can be constructed. The character of the scalar irrep is unity, $\chi^0(R)=1$. This, together with the orthogonality of characters (2.4.17) and the decomposition rule for Kronecker products (2.4.23), shows that only $\boldsymbol{\lambda}\times\boldsymbol{\lambda}^*$ gives $\mathbf{0}$. The importance of invariants means that much use is made of coupling coefficients which couple to the scalar.

The 2*jm* Symbol

The 2*jm symbol* is a renormalized coefficient describing the coupling to the scalar,

$$\begin{pmatrix} \lambda_1 & \lambda_2 \\ l_1 & l_2 \end{pmatrix} = |\lambda_1|^{1/2}\langle 0|\lambda_1 l_1, \lambda_2 l_2\rangle \tag{3.1.1}$$

The only product spaces $\mathbf{V}_{x_1\lambda_1,\, x_2\lambda_2}$ in which the scalar irrep $\mathbf{0}(\mathbf{G})$ occurs are where $\boldsymbol{\lambda}_2(\mathbf{G})=\boldsymbol{\lambda}_1^*(\mathbf{G})$. The scalar occurs just once and has the one partner, so the multiplicity and partner labels are omitted. Applying the Racah factorization lemma (2.3.13) for **G**⊃**H** gives a 2*jm* factor which is diagonal in $\lambda_1^*\lambda_2$ and $\rho_1^*\rho_2$,

$$\begin{bmatrix} \lambda_1 & \lambda_2 \\ a_1 & a_2 \\ \rho_1 & \rho_2 \end{bmatrix} = \frac{|\lambda_1|^{1/2}}{|\rho_1|^{1/2}} \langle 0|\lambda_1 a_1 \rho_1, \lambda_2 a_2 \rho_2 \rangle$$

$$= \frac{|\lambda_1|^{1/2}}{|\rho_1|^{1/2}} \langle 0|\lambda_1 a_1 \rho_1, \lambda_1^* a_2 \rho_1^* \rangle \delta_{\lambda_1^*\lambda_2} \delta_{\rho_1^*\rho_2} \qquad (3.1.2)$$

Often there will be no branching multiplicity, but even if there is, Butler (1975) and Butler and Wybourne (1976a) prove that it is always possible to choose the $2jm$ factors to be diagonal in $a_1^* a_2$ and real. Indeed for the point groups where a_1 and a_2 are simple numerical indices we have also $a_1^* = a_1$, so we introduce a simplified single-column notation. The dimension factors of (3.1.2) ensure that the nonzero $2jm$ factors have magnitude unity

$$\begin{pmatrix} \lambda \\ a \\ \rho \end{pmatrix} = \begin{bmatrix} \lambda & \lambda^* \\ a & a^* \\ \rho & \rho^* \end{bmatrix} = \pm 1 \qquad (3.1.3)$$

No formula exists for the sign. It is tabulated in Chapter 13. However, for the special case of the $2jm$ factor for $\mathbf{SO}_3 \supset \mathbf{SO}_2$, and with the Condon and Shortley phases used here, we have

$$\begin{pmatrix} J \\ M \end{pmatrix} = (-1)^{J-M} \qquad (3.1.4)$$

We do not work through the proof of the above results, because they involve using the *jm* factor symmetries, which are described by the *j* symbols of the next section, and which are enumerated in Section 3.3.

The 3*jm* Symbol

The 3*jm* *symbol* is defined by

$$\begin{pmatrix} \lambda_1 & \lambda_2 & \lambda \\ l_1 & l_2 & l \end{pmatrix}^r = |\lambda|^{-1/2} \begin{pmatrix} \lambda^* \\ l^* \end{pmatrix} \langle r\lambda^* l^* | \lambda_1 l_1, \lambda_2 l_2 \rangle H(\lambda_1 \lambda_2 \lambda^*) \qquad (3.1.5)$$

where the term $H(\lambda_1\lambda_2\lambda)$ must be inserted for historical reasons. For the usual angular momentum case $\mathbf{SO}_3 \supset \mathbf{SO}_2 \supset \mathbf{C}_1$, Wigner's choice of phase for the $3jm$ symbols is related to Condon and Shortley's choice of phase for coupling coefficients by the factor

$$H(J_1J_2J)=(-1)^{J_1-J_2+J} \tag{3.1.6}$$

This discrepancy in phase choices is now thoroughly part of physics. It is to be hoped that the similar discrepancies in phases between Griffith (1961, 1962) and between these tables and the various other tables will soon be lost in history. We would strongly recommend that calculations use the phase of these tables for all point calculations, and that except for $\mathbf{SO}_3$, (3.1.5) be used with the phase factor $H(\lambda_1\lambda_2\lambda^*)$ equal to one.

It is one of the tiresome facts of life in quantum mechanical calculations that while overall phases are not observable and thus must cancel, it is usually of critical importance that relative phases be absolutely correct. This tiresome fact is, however, reason for confidence in the accuracy of the computer program which produced these tables, for a simple sign error in any part of the programs tends to propagate rapidly, not as phase errors, but as magnitude errors. Magnitude errors are much more likely to be found.

The $2jm$ in the definition (3.1.5) serves to put the three irreps $\boldsymbol{\lambda}_1$, $\boldsymbol{\lambda}_2$, $\boldsymbol{\lambda}$ on a more symmetric footing than they appear in a coupling coefficient. Consider the double coupling of the triple product $\mathbf{V}_{x_1\lambda_1}\otimes\mathbf{V}_{x_2\lambda_2}\otimes\mathbf{V}_{x_3\lambda_3}$ to the scalar irrep. The first product $\mathbf{V}_{x_1\lambda_1}\otimes\mathbf{V}_{x_2\lambda_2}$ may be coupled to give irrep spaces $\mathbf{V}_{x_1\lambda_1x_2\lambda_2r\lambda}$, which may then be coupled with $\mathbf{V}_{x_3\lambda_3}$ to give the scalar irrep. Only terms for which $\lambda=\lambda_3^*$ contribute, and the transformations are x independent, as usual:

$$\begin{aligned}\langle((\lambda_1\lambda_2)r\lambda,\lambda_3)0|\lambda_1l_1,\lambda_2l_2,\lambda_3l_3\rangle &= \sum_{l}\langle 0|\lambda l,\lambda_3l_3\rangle\langle r\lambda l|\lambda_1l_1,\lambda_2l_2\rangle \\ &= \begin{pmatrix}\lambda_1 & \lambda_2 & \lambda_3\\ l_1 & l_2 & l_3\end{pmatrix}^{r}\delta_{\lambda\lambda_3^*}H(\lambda_1\lambda_2\lambda_3^*)\end{aligned} \tag{3.1.7}$$

The first equality follows from the definition of the twice coupled ket $|((x_1\lambda_1,x_2\lambda_2)r\lambda,x_3\lambda_3)0\rangle$, and the second follows directly from the definition of the $3jm$.

The three irreps $\boldsymbol{\lambda}_1$, $\boldsymbol{\lambda}_2$, $\boldsymbol{\lambda}_3$ and the multiplicity index r are called the *triad* $(\lambda_1\lambda_2\lambda_3 r)$. The irreps $\boldsymbol{\lambda}_1$, $\boldsymbol{\lambda}_2$, $\boldsymbol{\lambda}_3$ are said to form a triad if the scalar occurs at least once in the triple Kronecker product $\boldsymbol{\lambda}_1\times\boldsymbol{\lambda}_2\times\boldsymbol{\lambda}_3$. This is equivalent to $\boldsymbol{\lambda}_3^*$ occurring in the product $\boldsymbol{\lambda}_1\times\boldsymbol{\lambda}_2$ or the number $m_{\lambda_1\lambda_2}^{\lambda_3^*}$ of (2.4.22) being nonzero. For example, the three irreps $\frac{3}{2}$, $\frac{3}{2}$, 1 of the octahedral group **O** contain the scalar twice, and so form the two triads $(\frac{3}{2}\frac{3}{2}10)$ and $(\frac{3}{2}\frac{3}{2}11)$.

In a group–subgroup basis $\mathbf{G}\supset\mathbf{H}$ the kets are labeled $|\lambda(G)a\rho(H)m\rangle$. The label m labels various partners of $\boldsymbol{\rho}(\mathbf{H})$ and the set $a\rho m$ label various partners of $\boldsymbol{\lambda}(\mathbf{G})$. (See Section 2.2.) The coupling coefficients may be written as a product of coupling factors, one for each step in the chain. It follows that $2jm$ and $3jm$ symbols may also be written as a product of jm factors for each step in the chain. For example, for the chain $\mathbf{SO}_3\supset\mathbf{G}\supset\mathbf{C}_1$ with irreps $\mathbf{J}(\mathbf{SO}_3)$, $\boldsymbol{\lambda}(\mathbf{G})$, $\mathbf{0}(\mathbf{C}_1)$ the $2jm$ symbol factorizes into a product of a $2jm$ *factor* for $\mathbf{SO}_3\supset\mathbf{G}$ and a $2jm$ factor for $\mathbf{G}\supset\mathbf{C}_1$:

$$\begin{bmatrix} J \\ a \\ \lambda \\ b \\ 0 \end{bmatrix}\begin{matrix} SO_3 \\ \\ G \\ \\ C_1 \end{matrix} = \begin{pmatrix} J \\ a \\ \lambda \end{pmatrix}\begin{matrix} SO_3 \\ \\ G \end{matrix}\begin{pmatrix} \lambda \\ b \\ 0 \end{pmatrix}\begin{matrix} G \\ \\ C_1 \end{matrix} \tag{3.1.8}$$

where a and b are branching multiplicities.

The $3jm$ symbol factorizes into a sum of products of $3jm$ factors,

$$\begin{bmatrix} J_1 & J_2 & J_3 \\ a_1 & a_2 & a_3 \\ \lambda_1 & \lambda_2 & \lambda_3 \\ b_1 & b_2 & b_3 \\ 0 & 0 & 0 \end{bmatrix}\begin{matrix} SO_3 \\ \\ G \\ \\ C_1 \end{matrix} = \sum_r \begin{pmatrix} J_1 & J_2 & J_3 \\ a_1 & a_2 & a_3 \\ \lambda_1 & \lambda_2 & \lambda_3 \end{pmatrix}\begin{matrix} SO_3 \\ \\ rG \end{matrix}\begin{pmatrix} \lambda_1 & \lambda_2 & \lambda_3 \\ b_1 & b_2 & b_3 \\ 0 & 0 & 0 \end{pmatrix}\begin{matrix} rG \\ \\ C_1 \end{matrix} \tag{3.1.9}$$

where we display the ket labels as columns of the $3jm$ symbol and $3jm$ factors and we label the rows by the group to which they apply. Note the existence of the sum over product multiplicities for the triad $(\lambda_1\lambda_2\lambda_3 r)$ of the intermediate group.

This equation can be simplified by dropping superfluous labels; in particular, the irrep label $0(C_1)$ contains no information because everything transforms as $\mathbf{0}(\mathbf{C}_1)$. The $3jm$ factor for $\mathbf{G}\supset\mathbf{C}_1$ can be regarded as identical to the $3jm$ symbol for $\mathbf{G}$, namely

$$\begin{pmatrix} \lambda_1 & \lambda_2 & \lambda \\ b_1 & b_2 & b \end{pmatrix}^{rG}$$

The b labels are equivalent therefore to the l labels previously used. Irreps of Abelian groups $\mathbf{C}_n$ are one dimensional, so the partner label is superfluous, and the $3jm$ factor for $\mathbf{C}_n\supset\mathbf{C}_m$ is (with the phases of this book) equal to $+1$. For example, the $3jm$ symbol for $\mathbf{SO}_3\supset\mathbf{SO}_2\equiv\mathbf{C}_\infty\supset\mathbf{C}_n\supset\mathbf{C}_1$ is simply the $3jm$ factor for $\mathbf{SO}_3\supset\mathbf{SO}_2$, the $3jm$ symbols for all Abelian groups being trivial. The $3jm$ symbol, given a subgroup basis (and $\mathbf{C}_1$ is always a subgroup) is a $3jm$ factor. For this reason we shall usually use only the term *jm* factor and not *jm* symbol.

In certain cases, e.g., $\mathbf{O}\supset\mathbf{D}_3$, $\mathbf{D}_3\supset\mathbf{C}_3$, and $\mathbf{C}_3\supset\mathbf{C}_1$, there is no branching multiplicity and we may omit the branching multiplicity index a or b. Also, the C_3–C_1 factor can be omitted entirely. In several imbeddings, e.g., $\mathbf{O}\supset\mathbf{T}$, the $3jm$ factor requires two product multiplicity indices, one for the group triad and one for the subgroup triad. The completely general $3jm$ factor thus has three group irrep labels, three subgroup labels, three branching multiplicity labels, one group product index, and one subgroup product index.

3.2 The j Symbols

The j symbols of a group give the symmetry properties of the *jm* symbols of the group. The symmetry properties are independent of the actual choice of partners for the irreps. The symmetries of the *jm* factors of a group–subgroup chain are given by the j symbols of the group and subgroup.

The $2j$ Phase

The group transformation properties of $|x_1\lambda_1 l_1, x_2\lambda_2 l_2\rangle$ and $|x_2\lambda_2 l_2, x_1\lambda_1 l_1\rangle$ are identical. The matrix elements of group operators O_R

are the numbers $\lambda_1(R)_{l_1'l_1}\lambda_1(R)_{l_2'l_2}$ for both. Different $2jm$'s, $\langle 0|\lambda_1 l_1, \lambda_2 l_2\rangle$ and $\langle 0|\lambda_2 l_2, \lambda_1 l_1\rangle$, are used to couple each to give the scalar irrep. The map between the two results will be the simple map interchanging the 1 and 2 indices and will commute with the group operators. By Schur's lemma the matrix elements of the map will be just a single number $\{\lambda_1\lambda_2\}$ independent of l_1 and l_2, the $2j$ *phase*,

$$\langle 0|\lambda_1 l_1, \lambda_2 l_2\rangle = \{\lambda_1\lambda_2\}\langle 0|\lambda_2 l_2, \lambda_1 l_1\rangle \tag{3.2.1}$$

Using the fact that the scalar only occurs when $\lambda_2 = \lambda_1^*$, we have

$$\begin{pmatrix} \lambda & \lambda^* \\ l & l^* \end{pmatrix} = \{\lambda\lambda^*\}\begin{pmatrix} \lambda^* & \lambda \\ l^* & l \end{pmatrix} \tag{3.2.2}$$

we shall write the $2j$ phase $\{\lambda\lambda^*\}$ as $\{\lambda\}$ rather than Derome and Sharp's (1965) ϕ_λ. With the abbreviated notation, (3.2.2) becomes

$$\begin{pmatrix} \lambda \\ l \end{pmatrix} = \{\lambda\}\begin{pmatrix} \lambda^* \\ l^* \end{pmatrix} \tag{3.2.3}$$

The $2j$ phase is related to the Frobenius–Schur invariant c_λ of (2.5.10). The number c_λ describes the reality, symmetry, or antisymmetry of the representation $\boldsymbol{\lambda}(\mathbf{G})$,

$$c_\lambda = \{\lambda\}\,\delta_{\lambda\lambda^*} \tag{3.2.4}$$

When $\boldsymbol{\lambda}(\mathbf{G}) \neq \boldsymbol{\lambda}^*(\mathbf{G})$, that is, for complex irreps, we choose $\{\lambda\}$ as follows. If $\mathbf{G}$ is non-Abelian, that is, noncyclic, choose $\{\lambda\} = 1$ for true irreps and $\{\lambda\} = -1$ for spin irreps. If $\mathbf{G}$ is cyclic, choose $\{\lambda\} = 1$. When $\boldsymbol{\lambda} = \boldsymbol{\lambda}^*$, that is, when there is no choice, (3.2.4) can be used to show that $\{\lambda\}$ takes on values which follow this prescription.

The $3j$ Phase

The above discussion of the reordering symmetry of a $2jm$ symbol generalizes directly to a $3jm$ symbol. The $3jm$ symbol was related to the coupling of the three irreps to a scalar [see (3.1.7)]. There are six different orderings to perform the two couplings. The transformations between the different partners of scalar irreps in the product spaces, such as

$$\{(23), \lambda_1\lambda_2\lambda_3\}_{r_{13}r_{12}}$$
$$=\langle((x_1\lambda_1, x_3\lambda_3)r_{13}\lambda_{13}, x_2\lambda_2)0|((x_1\lambda_1, x_2\lambda_2)r_{12}\lambda_{12}, x_3\lambda_3)0\rangle \tag{3.2.5}$$

will be elements of a unitary transformation of dimension $m_{\lambda_1\lambda_2}^{\lambda_3^*}$.

The $3j$ *phase* $\{\pi, \lambda_1\lambda_2\lambda_3\}_{r'r}$ associated with permutation π of 1, 2, 3 is thus a transformation between the appropriate $3jm$ symbols. If $\pi\,(1,2,3)=(a,b,c)$, then

$$\begin{pmatrix} \lambda_a & \lambda_b & \lambda_c \\ l_a & l_b & l_c \end{pmatrix}^{r'} = \sum_r \{\pi, \lambda_1\lambda_2\lambda_3\}_{r'r} \begin{pmatrix} \lambda_1 & \lambda_2 & \lambda_3 \\ l_1 & l_2 & l_3 \end{pmatrix}^{r} \tag{3.2.6}$$

The $3j$ "phases" are elements of unitary matrices, rows, and columns labeled by the multiplicity labels r, r', etc. For example, if we have a multiplicity 2 product, e.g., $\frac{3}{2}(\mathbf{K})\times\mathbf{2}(\mathbf{K})\times\frac{5}{2}(\mathbf{K})$, there are two independent linear combinations of the various partners of each irrep that transform as $\mathbf{0}(\mathbf{K})$. There is no a priori reason that a table of $3jm$ symbols would lead to the same linear combination of partners if the coupling were now done in the order $\frac{3}{2}(\mathbf{K})\times\frac{5}{2}(\mathbf{K})\times\mathbf{2}(\mathbf{K})$, or any one of the other four orders. If $\boldsymbol{\lambda}_1$, $\boldsymbol{\lambda}_2$, and $\boldsymbol{\lambda}_3$ are three different irreps of a group, then it makes sense to choose the six couplings so that the six $3jm$ symbols are simply related. It is clear that as the $3jm$ symbols are calculated, the first linear combination of kets obtained can be used for the other five linear combinations, leading to the $3j$ phases taking the form

$$\{\pi, \lambda_1\lambda_2\lambda_3\}_{r'r} = \theta\,\delta_{r'r} \tag{3.2.7}$$

where θ is any phase factor. If there is no multiplicity, then there is only one linear combination, so this result is forced upon us.

The situation is different if two or three of the irreps $\boldsymbol{\lambda}_1$, $\boldsymbol{\lambda}_2$, and $\boldsymbol{\lambda}_3$ are equivalent; assume $\boldsymbol{\lambda}_1 = \boldsymbol{\lambda}_2$. The coefficients

$$\begin{pmatrix} \lambda_1 & \lambda_1 & \lambda_3 \\ l_1 & l_2 & l_3 \end{pmatrix}^{r} \quad \text{and} \quad \begin{pmatrix} \lambda_1 & \lambda_1 & \lambda_3 \\ l_2 & l_1 & l_3 \end{pmatrix}^{r'}$$

are not only related by the $3j$ phase of (3.2.6), but also are coefficients of different partners in the one scalar product. Permuting the irrep spaces $\mathbf{V}_{\lambda_1}$

and $\mathbf{V}_{\lambda_2}$ in the product is a symmetry of the equation, the group being $\mathbf{S}_2$, the two-element interchange group of permutations on two objects. We may apply the apparatus of Chapter 2 to the problem and conclude that the scalar functions may be separated into two groupings, those that are scalar (symmetric) under the action of $\mathbf{S}_2$ and those that are antisymmetric. Demanding that the linear combinations have this permutation symmetry will impose restrictions on the linear combinations possible. For example, the product $\tilde{\mathbf{1}}(\mathbf{O})\times\tilde{\mathbf{1}}(\mathbf{O})\times\mathbf{1}(\mathbf{O})$ contains the scalar twice, once as a sum symmetric in $\tilde{\mathbf{1}}(\mathbf{O})\times\tilde{\mathbf{1}}(\mathbf{O})$ and once antisymmetrically. The product $\mathbf{2}(\mathbf{K})\times\mathbf{2}(\mathbf{K})\times\mathbf{2}(\mathbf{K})$ contains the scalar twice symmetrically. Any linear combination of the two symmetric scalars is symmetric. It so happens for the point groups that in cases where all three irreps are the same, the scalars have the same symmetries with respect to interchanging the second and third sets of partners. If this were not so, then the $3j$ phases for the interchange of the first pair of partners $\{(12),\lambda\lambda\lambda\}_{r'r}$ and the second pair of partners $\{(23),\lambda\lambda\lambda\}_{r'r}$ would not be the same and indeed there would be no linear combinations in r that would lead to the $3j$ phase being a phase of the form (3.2.7). Many groups of interest in physics and chemistry, for example, the matrix groups used by Racah (1949), $\mathbf{U}_{14}$, $\mathbf{Sp}_{14}$, $\mathbf{SO}_7$, and $\mathbf{G}_2$, all have triple products which have one symmetry for a (12) interchange and a different symmetry for a (23) interchange. Such groups are known as *non-simple-phase* groups.

The resolution of the Kronecker square of irrep $\boldsymbol{\lambda}$ is written in terms of the irreps of $\mathbf{S}_2$, the group of permutations of two objects. The symmetric part is written $\boldsymbol{\lambda}\otimes\{2\}$ and its character is

$$\chi^{\lambda\otimes\{2\}}(R)=\tfrac{1}{2}\{\chi^{\lambda}(R)^2+\chi^{\lambda}(R^2)\} \tag{3.2.8}$$

while the antisymmetric part is

$$\chi^{\lambda\otimes\{1^2\}}(R)=\tfrac{1}{2}\{\chi^{\lambda}(R)^2-\chi^{\lambda}(R^2)\} \tag{3.2.9}$$

When asking for the symmetry of the triad with three irreps equal, one must ask not only what terms $\boldsymbol{\lambda}^*$ appear in $\boldsymbol{\lambda}\otimes\{2\}$ and $\boldsymbol{\lambda}\otimes\{1^2\}$, but also where the scalar appears in the cube. The number of $\boldsymbol{\lambda}^*$ in $\boldsymbol{\lambda}\otimes\{2\}$ equals the number of $\mathbf{0}$ in the fully symmetric part $\boldsymbol{\lambda}\otimes\{3\}$, plus the number in the mixed symmetry part $\boldsymbol{\lambda}\otimes\{21\}$. The character formula for the number of mixed symmetry scalars is (Derome 1966, Butler and King 1974)

$$n_{\{21\}}=\frac{1}{3|G|}\sum_R\{\chi^{\lambda}(R)^3-\chi^{\lambda}(R^3)\} \tag{3.2.10}$$

It is straightforward to use the various character tables of Chapter 11 to verify that all point groups have no mixed symmetry terms and are thus simple phase.

For all point groups we may and do choose the multiplicity separation and the phases of the set of $3jm$ symbols in such a way that all $3j$ symbols $\{\pi, \lambda_1\lambda_2\lambda_2\}_{r'r}$ are real, are diagonal in r, and, further $\{\pi, \lambda_1\lambda_2\lambda_3\}_{rr'} = \{\pi, \lambda_2\lambda_1\lambda_3\}_{rr'}$, etc. If, for example, $\lambda_1 = \lambda_2$ and $\lambda_2 \neq \lambda_3$, then we force the $3j$ phases for cyclic permutations to be $+\delta_{r'r}$ by taking the appropriate interchange $3j$ phases to be $\pm\delta_{r'r}$. We abbreviate the interchange $3j$ phases as

$$\{i, \lambda_1\lambda_2\lambda_3\}_{r'r} = \{\lambda_1\lambda_2\lambda_3 r\}\,\delta_{r'r} \tag{3.2.11}$$

We use this simple phase assumption in all our equations. The full equations are available elsewhere (Butler 1975, Butler and Wybourne 1976a).

The $2j$ and $3j$ phases give the permutation symmetries of $2jm$ and $3jm$ symbols. The $2j$ takes the values ± 1 only and the $3j$ takes $\pm\delta_{r'r}$ for simple-phase groups (but nondiagonal values for non-simple-phase groups). The value of both sets of phases may be easily deduced from character theory if so fixed, or chosen freely otherwise. For $\mathbf{SO_3}$ we have

$$\{J\} = \{JJ\} = (-1)^{2J} \quad \text{and} \quad \{J_1J_2J_3\} = (-1)^{J_1+J_2+J_3} \tag{3.2.12}$$

The $6j$ Symbol

A $6j$ symbol may be defined by considering the coupling of three irreps to a fourth, nonscalar irrep or, equivalently, coupling four irreps to a scalar. The actual coupling would be described by a $4jm$, a generalization of the $2jm$ and $3jm$ symbols. The coupling may be done by a two-stage process; the various resultant scalars may be labeled by intermediate irreps with their multiplicity labels r. Using the coupling coefficient formalism, we may write

$$\begin{aligned}|\lambda_1 l_1\rangle|\lambda_2 l_2\rangle|\lambda_3 l_3\rangle &= \sum_{r_{12}\lambda_{12}l_{12}} |(\lambda_1\lambda_2)r_{12}\lambda_{12}l_{12}\rangle|\lambda_3 l_3\rangle\langle r_{12}\lambda_{12}l_{12}|\lambda_1 l_1, \lambda_2 l_2\rangle \\ &= \sum_{r_{12}\lambda_{12}l_{12}r\lambda l} |(\lambda_1\lambda_2)r_{12}\lambda_{12}, \lambda_3, r\lambda l\rangle\langle r_{12}\lambda_{12}l_{12}|\lambda_1 l_1, \lambda_2 l_2\rangle \\ &\quad\times\langle r\lambda l|\lambda_{12}l_{12}, \lambda_3 l_3\rangle\end{aligned} \tag{3.2.13}$$

so that the labels $r_{12}\lambda_{12}r$ distinguish the various different resultant kets that transform as the partner $|\lambda l\rangle$. The coupling coefficients ensure that the set of resultant kets

$$|(\lambda_1\lambda_2)r_{12}\lambda_{12}, \lambda_3, r\lambda l\rangle \tag{3.2.14}$$

for varying r_{12}, λ_{12}, r, λ, and l will be orthonormal.

On the other hand, the resultant kets

$$|\lambda_1(\lambda_2\lambda_3)r_{23}\lambda_{23}, r'\lambda' l'\rangle \tag{3.2.15}$$

obtained by using coupling coefficient to couple $|\lambda_2 l_2\rangle$ to $|\lambda_3 l_3\rangle$ first, will form an alternative orthonormal basis for the coupled product space. Once again the Schur result tells us that the transformation between the two sets of partners is diagonal in λl and independent of l. The overlap matrix element

$$\begin{aligned}&\langle(\lambda_1\lambda_2)r_{12}\lambda_{12}, \lambda_3, r\lambda l|\lambda_1(\lambda_2\lambda_3)r_{23}\lambda_{23}, r'\lambda' l'\rangle\\ &\qquad=\langle(\lambda_1\lambda_2)r_{12}\lambda_{12}, \lambda_3; r\lambda|\lambda_1(\lambda_2\lambda_3)r_{23}\lambda_{23}; r'\lambda\rangle\,\delta_{\lambda\lambda'}\delta_{ll'}\end{aligned} \tag{3.2.16}$$

is called a *recoupling coefficient*. Using the linear independence of the uncoupled and variously coupled partners, the above equations give that

$$\begin{aligned}&\sum_{r_{12}\lambda_{12}l_{12}r}\langle r_{12}\lambda_{12}l_{12}|\lambda_1 l_1, \lambda_2 l_2\rangle\langle r\lambda l|\lambda_{12}l_{12}, \lambda_3\lambda_3\rangle\\ &\qquad\times\langle\lambda_1(\lambda_2\lambda_3)r_{23}\lambda_{23}; r'\lambda|(\lambda_1\lambda_2)r_{12}\lambda_{12}, \lambda_3; r\lambda\rangle\\ &\qquad=\sum_{\lambda_{23}l_{23}}\langle r'\lambda l|\lambda_1 l_1, \lambda_{23}l_{23}\rangle\langle r_{23}\lambda_{23}l_{23}|\lambda_2 l_2, \lambda_3 l_3\rangle\end{aligned} \tag{3.2.17}$$

Having shown the relationship of the recoupling coefficient to the alternately coupled partners of (3.2.14) and (3.2.15), we shall now convert to the renormalized recoupling coefficient, the $6j$ *symbol*

$$\begin{aligned}&\begin{Bmatrix}\lambda_1 & \lambda_{23} & \lambda^*\\ \lambda_3^* & \lambda_{12} & \lambda_2\end{Bmatrix}_{r_{12}r_{23}rs}\\ &\qquad=|\lambda_{12}|^{1/2}|\lambda_{23}|^{1/2}\{\lambda_2\}\{\lambda_{12}\lambda_3\lambda^* r\}\{\lambda_1\lambda_2\lambda_{12}^* r_{12}\}\\ &\qquad\times\langle(\lambda_1\lambda_2)r_{12}\lambda_{12}, \lambda_3; r\lambda|\lambda_1(\lambda_2\lambda_3)r_{23}\lambda_{23}; s\lambda\rangle\end{aligned} \tag{3.2.18}$$

The four historical phases of angular momentum theory, one phase for each coupling, must be used for the SO_3 $6j$. They introduce a net additional factor of $(-1)^{2\lambda_1}$.

The $9j$ Symbol

In many recoupling applications one needs a direct recoupling of four kets coupled in one scheme to an alternative scheme. An example of much importance is the comparison of an SL coupled wave function with total angular momentum $\mathbf{J}$, $|(s_1 s_2)S(l_1 l_2)L, J\rangle$, to the j–j coupled wave function $|(s_1 l_1)j_1(s_2 l_2)j_2 J\rangle$ coupled to the same total $\mathbf{J}$.

The recoupling between the schemes may be carried out by a product of three $6j$ symbols, but it is useful to define and tabulate the $9j$ *symbol*

$$\begin{matrix} \left\{\begin{matrix} \lambda_1 & \lambda_2 & \lambda_3 \\ \mu_1 & \mu_2 & \mu_3 \\ \nu_1 & \nu_2 & \nu_3 \end{matrix}\right\} \begin{matrix} r_1 \\ r_2 \\ r_3 \end{matrix} \\ \begin{matrix} s_1 & \quad s_2 & \quad s_3 \end{matrix} \qquad \end{matrix}$$

$$=|\lambda_3|^{-1/2}|\mu_3|^{-1/2}|\nu_1|^{-1/2}|\nu_2|^{-1/2}$$

$$\times\langle(\lambda_1\lambda_2)r_1\lambda_3^*(\mu_1\mu_2)r_2\mu_3^*, s_3\nu_3|(\lambda_1\mu_1)s_1\nu_1^*(\lambda_2\mu_1)s_2\nu_2^*, r_3\nu_3\rangle \tag{3.2.19}$$

The $9j$ symbol arises also in the evaluation of many-body matrix elements in perturbation theories. The historical $\mathbf{SO}_3$ phases introduce no factor here.

Some of these definitions of *jm* and *j* symbols have used the simple phase condition (3.2.11). There is another property of the point and parentage groups which leads to useful simplifications of the Racah–Wigner algebra. This is the property of quasi-ambivalence. Wigner (1940) called a group ambivalent if each element was in the same conjugacy class as its inverse. It is straightforward to use the character orthogonality (2.4.12) to show that ambivalence implies an absence of complex irreps. Butler and King (1974) introduced the concept of quasi-ambivalence for the case when complex irreps occur, but where the complex irreps behave as if they are orthogonal or symplectic. For such groups we classify the complex irreps as quasi-orthogonal or quasisymplectic. For the non-Abelian point groups all true irreps are orthogonal or quasi-orthogonal and all spin irreps are symplectic or quasisymplectic. All irreps of Abelian groups are either orthogonal or quasi-orthogonal. The useful property is that a product of a true irrep with a true irrep gives a set of true irreps, as does the product of a spin irrep with a spin irrep, while the product of true irreps with spin irreps gives spin irreps. This property of true and spin irreps holds for all groups which have both, but it happens for the continuous compact groups that the (stronger) quasi-ambivalent classification holds also. For those large, finite groups (not point groups) that fail this quasi-ambivalence separation, the Derome–Sharp lemma of (3.3.12) requires an additional phase matrix, as do many equations derived from it.

A group is said to be quasi-ambivalent if, whenever three irreps $\lambda_1\lambda_2\lambda_3$ form a triad, there is a choice of $2j$ phases such that the product of the three $2j$ phases is positive

$$\{\lambda_1\}\{\lambda_2\}\{\lambda_3\}=1 \tag{3.2.20}$$

A finite group of order 24 fails this condition (for complex irreps), while another of much larger order is known to fail for symplectic irreps because the square of a symplectic irrep contains itself twice.

3.3 Properties of *jm* Factors and *j* Symbols

We collect together in this section the various algebraic properties of the *jm* and *j* symbols as they apply to the point groups. This involves two simplifications over the results for arbitrary groups. First, no mixed symmetry products occur [the groups are simple phase, see (3.2.11)]; second, quasi-ambivalence holds [see (3.2.20)]. We do not distinguish, in this listing of properties, between those properties which are true in general and those true for the point groups or certain other special cases. Nor do we always distinguish between properties which must hold and those which only hold because of our particular phase and multiplicity conventions. The reader who wishes to use these distinctions can see the full equations in Butler (1975) and Butler and Wybourne (1976a, b).

The labels λ_i, μ_i, and ν_i will denote irreps of a group **G**, whereas ρ_i and σ_i denote irreps of a subgroup **H**. The $2j$ phase has the values

$$\{\lambda\}=\{\lambda^*\}=\begin{cases} +1 & \text{if } \boldsymbol{\lambda} \text{ is a true irrep or } \mathbf{G} \text{ is cyclic} \\ -1 & \text{otherwise} \end{cases} \tag{3.3.1}$$

The $3j$ phase is also real,

$$\{\lambda_1\lambda_2\lambda_3 r\} = \pm 1 \tag{3.3.2}$$

but requires tabulation. It is invariant under permutations and conjugations of the irrep labels, namely

$$\{\lambda_1\lambda_2\lambda_3 r\} = \{\lambda_1^*\lambda_2^*\lambda_3^* r\} = \{\lambda_1\lambda_2\lambda_3 r\} = \text{etc.} \tag{3.3.3}$$

The $2j$ phase is the special case of the $3j$

$$\{\lambda\} = \{\lambda\lambda^*\} = \{\lambda\lambda^* 00\} \tag{3.3.4}$$

For $\mathbf{SO}_3$ we have the algebraic formulas

$$\{J_1 J_2 J_3 0\} = (-1)^{J_1+J_2+J_3} \tag{3.3.5a}$$

$$\{J\} = (-1)^{2J} \tag{3.3.5b}$$

The G–H $3jm$ factor is a renormalized coupling factor,

$$\begin{bmatrix} \lambda_1 & \lambda_2 & \lambda \\ a_1 & a_2 & a \\ \sigma_1 & \sigma_2 & \sigma \end{bmatrix}^{r}_{s} = \frac{|\sigma|^{1/2}}{|\lambda|^{1/2}} \begin{pmatrix} \lambda^* \\ a \\ \sigma^* \end{pmatrix} \langle r\lambda^* a\sigma^* | \lambda_1 a_1 \sigma_1, \lambda_2 a_2 \sigma_2 \rangle_s \tag{3.3.6}$$

but $\mathbf{SO}_3 - \mathbf{SO}_2$ has a different phase, [see (3.1.5)]. The $2jm$ factor is

$$\begin{pmatrix} \lambda \\ a \\ \sigma \end{pmatrix} = \frac{|\lambda|^{1/2}}{|\sigma|^{1/2}} \langle 0 | \lambda a\sigma, \lambda^* a\sigma^* \rangle \tag{3.3.7}$$

$$= \begin{pmatrix} \lambda & \lambda^* \\ a & a \\ \sigma & \sigma^* \end{pmatrix} = \frac{|\lambda|^{1/2}}{|\sigma|^{1/2}} \begin{bmatrix} \lambda & \lambda^* & 0 \\ a & a & 0 \\ \sigma & \sigma^* & 0 \end{bmatrix}^{0}_{0} \tag{3.3.8}$$

where both branching and product multiplicities are labeled by a Latin index a or r, both of which start from zero. We have chosen the complex conjugate of the ath occurrence of symmetry type $\lambda\sigma$ to be the ath occurrence of type $\lambda^*\sigma^*$. The $2jm$ factors are real, and have the symmetry

$$\begin{pmatrix} \lambda^* \\ a \\ \sigma^* \end{pmatrix} = \{\lambda\}\{\sigma\} \begin{pmatrix} \lambda \\ a \\ \sigma \end{pmatrix} \tag{3.3.9}$$

For $\mathbf{SO}_3 \supset \mathbf{SO}_2$

$$\begin{pmatrix} J \\ M \end{pmatrix} = (-1)^{J-M} \tag{3.3.10}$$

The $2jm$ factors for each branching are tabulated at the head of the corresponding $3jm$ table.

Under odd-column permutations the value of the $3jm$ factor changes by a product of the $3j$ phases of the group and subgroup, e.g.,

$$\begin{bmatrix} \lambda_2 & \lambda_1 & \lambda_3 \\ a_2 & a_1 & a_3 \\ \sigma_2 & \sigma_1 & \sigma_3 \end{bmatrix}^{r}_{s} = \{\lambda_1\lambda_2\lambda_3 r\}\{\sigma_1\sigma_2\sigma_3 s\} \begin{bmatrix} \lambda_1 & \lambda_2 & \lambda_3 \\ a_1 & a_2 & a_3 \\ \sigma_1 & \sigma_2 & \sigma_3 \end{bmatrix}^{r}_{s} \tag{3.3.11}$$

Under even permutations (cyclic permutations) the $3jm$ is invariant. The $3jm$ factors are not real; rather, we have the *Derome–Sharp lemma* (1965)

$$\begin{bmatrix} \lambda_1 & \lambda_2 & \lambda_3 \\ a_1 & a_2 & a_3 \\ \sigma_1 & \sigma_2 & \sigma_3 \end{bmatrix}^{*r}_{s} = \begin{bmatrix} \lambda_1 \\ a_1 \\ \sigma_1 \end{bmatrix} \begin{bmatrix} \lambda_2 \\ a_2 \\ \sigma_2 \end{bmatrix} \begin{bmatrix} \lambda_3 \\ a_3 \\ \sigma_3 \end{bmatrix} \begin{bmatrix} \lambda_1^* & \lambda_2^* & \lambda_3^* \\ a_1 & a_2 & a_3 \\ \sigma_1^* & \sigma_2^* & \sigma_3^* \end{bmatrix}^{r}_{s} \tag{3.3.12}$$

which for the $\mathbf{SO}_3 \supset \mathbf{SO}_2$ case reduces to the algebraic form

$$\begin{pmatrix} J_1 & J_2 & J_3 \\ M_1 & M_2 & M_3 \end{pmatrix} = (-1)^{J_1+J_2+J_3} \begin{pmatrix} J_1 & J_2 & J_3 \\ -M_1 & -M_2 & -M_3 \end{pmatrix} \tag{3.3.13}$$

A proof of the Derome–Sharp lemma that avoids group integration can be constructed by applying an antilinear transform (see Section 8.1) to the uncoupled $|x_1\lambda_1 l_1, x_2\lambda_2 l_2\rangle$ and coupled $|(x_1\lambda_1 l_1, x_2\lambda_2)r\lambda l\rangle$ bases of a tensor product and then using Schur's lemma. This gives rise to the $2jm$'s of (3.3.12) and a matrix **A** in the multiplicity index. For quasi-ambivalent groups one may choose **A** the unit matrix.

The $3jm$ factors are not orthonormal, owing to the dimensional factor introduced in (3.1.5) to place the three irreps on an equal footing. Instead, as may be deduced from the coupling factor orthonormalities, they satisfy

$$\sum_{r\lambda a} \frac{|\lambda|}{|\sigma|} \begin{bmatrix} \lambda_1 & \lambda_2 & \lambda \\ a_1 & a_2 & a \\ \sigma_1 & \sigma_2 & \sigma \end{bmatrix}^{r}_{s} \begin{bmatrix} \lambda_1 & \lambda_2 & \lambda \\ a_1' & a_2' & a \\ \sigma_1' & \sigma_2' & \sigma \end{bmatrix}^{*r}_{s'} = \delta_{a_1a_1'}\delta_{a_2a_2'}\delta_{\sigma_1\sigma_1'}\delta_{\sigma_2\sigma_2'}\delta_{ss'} \tag{3.3.14}$$

and

$$\sum_{a_1\sigma_1 a_2\sigma_2 s} \frac{|\lambda|}{|\sigma|} \begin{bmatrix} \lambda_1 & \lambda_2 & \lambda \\ a_1 & a_2 & a \\ \sigma_1 & \sigma_2 & \sigma \end{bmatrix}^{r}_{s} \begin{bmatrix} \lambda_1 & \lambda_2 & \lambda' \\ a_1 & a_2 & a' \\ \sigma_1 & \sigma_2 & \sigma \end{bmatrix}^{*r'}_{s} = \delta_{\lambda\lambda'}\delta_{aa'}\delta_{rr'} \tag{3.3.15}$$

The second relation is subtly different from the corresponding $3jm$ symbol relation because of the omission of the $\mathbf{C}_1$ labels. Standard texts on angular momentum theory have a $\delta_{MM'}$ in the corresponding relation (equivalent to a $\delta_{\sigma\sigma'}$ in our notation). The $\delta_{MM'}$ occurs because $\mathbf{SO}_2$ is Abelian, so the Kronecker product $\mathbf{M}_1 \times \mathbf{M}_2$ gives a single irrep $-\mathbf{M}$. While such a delta occurs in the $3jm$ symbol orthonormality for $\mathbf{SO}_3 \supset \mathbf{SO}_2 \supset \mathbf{C}_1$, it does not occur in the SO_3–SO_2 $3jm$ factor orthonormality, but instead in the omitted SO_2–C_1 factor.

We defined the $6j$ symbol from the recoupling coefficient (3.2.18), which was in turn related to coupling coefficients. That expression could be converted to $3jm$ symbols and manipulated, using the above relations, to give an expression for a $6j$ symbol in terms of four $3jm$ symbols,

$$\begin{Bmatrix} \lambda_1 & \lambda_2 & \lambda_3 \\ \mu_1 & \mu_2 & \mu_3 \end{Bmatrix}_{r_1 r_2 r_3 r_4} = \sum_{m_1 m_2 m_3 l_1 l_2 l_3} \begin{pmatrix} \mu_1 \\ m_1 \end{pmatrix} \begin{pmatrix} \mu_2 \\ m_2 \end{pmatrix} \begin{pmatrix} \mu_3 \\ m_3 \end{pmatrix}$$
$$\times \begin{pmatrix} \lambda_1 & \mu_2^* & \mu_3 \\ l_1 & m_2^* & m_3 \end{pmatrix}^{r_1} \begin{pmatrix} \mu_1 & \lambda_2 & \mu_3^* \\ m_1 & l_2 & m_3^* \end{pmatrix}^{r_2}$$
$$\times \begin{pmatrix} \mu_1^* & \mu_2 & \lambda_3 \\ m_2^* & m_2 & l_3 \end{pmatrix}^{r_3} \begin{pmatrix} \lambda_1 & \lambda_2 & \lambda_3 \\ l_1 & l_2 & l_3 \end{pmatrix}^{*r_4} \tag{3.3.16}$$

The sum is over all partners l_i, m_i of six irreps $\boldsymbol{\lambda}_i$, $\boldsymbol{\mu}_i$. Observe that the four triads, corresponding to the four coupled products in a coupling coefficient, appear in a $6j$ symbol as shown in the diagrams

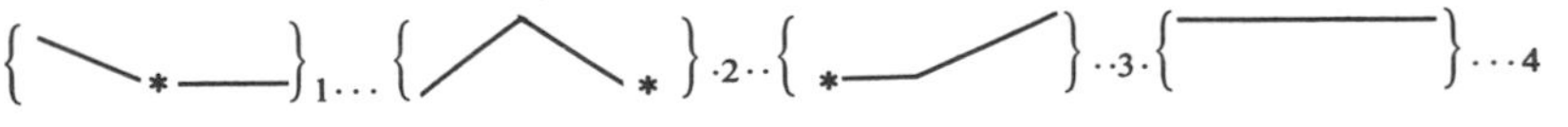

$$\{\diagdown{*}\!\!-\!\!-\}\cdots 1 \cdots \{\diagup\diagdown{*}\}\cdots 2 \cdots \{{*}\!\!-\!\!-\diagup\}\cdots 3 \cdots \{\overline{\qquad\quad}\}\cdots 4 \tag{3.3.17}$$

The triads generalize the $\mathbf{SO}_3$ special case in two ways. There is a multiplicity index with the triad and the complex conjugate of a μ irrep appears in the first three cases. This star appears in the bottom line in the position following cyclicly on from the top line entry.

The symmetries of the $6j$ follow easily from (3.3.16) and the $3jm$ symmetries. The $6j$ is invariant under cyclic permutations of the columns, but changes by a phase factor under column interchanges. The same phase factor occurs for (12), (23), or (31) interchanges, we give the (12) equation,

$$\begin{Bmatrix} \lambda_2 & \lambda_1 & \lambda_3 \\ \mu_2 & \mu_1 & \mu_3 \end{Bmatrix}_{r_2r_1r_3r_4} = \{\mu_1\}\{\mu_2\}\{\mu_3\}\{\lambda_1\mu_2\mu_3^*r_1\}\{\mu_1^*\lambda_2\mu_3r_2\}$$

$$\times\{\mu_1\mu_2^*\lambda_3r_3\}\{\lambda_1\lambda_2\lambda_2r_4\}\begin{Bmatrix} \lambda_1 & \lambda_2 & \lambda_3 \\ \mu_1^* & \mu_2^* & \mu_3^* \end{Bmatrix}_{r_1r_2r_3r_4} \tag{3.3.18}$$

A quick check verifies that the $\mathbf{SO}_3$ result of no net phase change is a special case.

Any of the lower three triads may be interchanged with the upper (fourth) triad. No phase is introduced, but complex conjugation of the entries occurs,

$$\begin{Bmatrix} \lambda_1 & \lambda_2 & \lambda_3 \\ \mu_1 & \mu_2 & \mu_3 \end{Bmatrix}_{r_1r_2r_3r_4} = \begin{Bmatrix} \lambda_1^* & \mu_2 & \mu_3^* \\ \mu_1^* & \lambda_2 & \lambda_3^* \end{Bmatrix}_{r_4r_3r_2r_1} = \begin{Bmatrix} \mu_1^* & \lambda_2^* & \mu_3 \\ \lambda_1^* & \mu_2^* & \lambda_3 \end{Bmatrix}_{r_3r_4r_1r_2}$$

$$= \begin{Bmatrix} \mu_1 & \mu_2^* & \lambda_3^* \\ \lambda_1 & \lambda_2^* & \mu_3^* \end{Bmatrix}_{r_2r_1r_4r_3} \tag{3.3.19}$$

The column, $i+1$, not starred follows cyclicly the column not flipped, i, and the multiplicity indices are swapped in pairs, one pair being r_ir_4. The position of both the stars and the multiplicity labels is quite readily deduced by comparing the triads with the diagrams (3.3.17). Complex conjugation is rather more easily remembered,

$$\begin{Bmatrix} \lambda_1 & \lambda_2 & \lambda_3 \\ \mu_1 & \mu_2 & \mu_3 \end{Bmatrix}^*_{r_1r_2r_3r_4} = \begin{Bmatrix} \lambda_1^* & \lambda_2^* & \lambda_3^* \\ \mu_1^* & \mu_2^* & \mu_3^* \end{Bmatrix}_{r_1r_2r_3r_4} \tag{3.3.20}$$

The above symmetries are used to reduce substantially the size of the tables of $6j$ symbols. They allow one to place the largest label in the λ_1

position, μ_1 thus being fixed, then the largest remaining label in the λ_2 position, fixing the positions of the remaining. The symmetries are a generalization of those of $\mathbf{SO}_3$, but the presence of complex conjugation, especially for the tetrahedral group, can make it not only tedious to apply the symmetries correctly, but also difficult to make the choice of which symmetry to apply. The larger groups (the octahedral and icosahedral groups) have real irreps and thus this complication is removed. Consider, for example, the following $6j$'s of the tetrahedral group:

$$\begin{Bmatrix} \frac{3}{2} & \frac{3}{2} & 1 \\ -\frac{3}{2} & \frac{3}{2} & 1 \end{Bmatrix} \quad \text{and} \quad \begin{Bmatrix} \frac{3}{2} & \frac{3}{2} & 1 \\ \frac{3}{2} & -\frac{3}{2} & 1 \end{Bmatrix}$$

These are not related by symmetries. Interchanging columns 1 and 2 would seem to equate them, but the complex conjugation interchanges irreps $\frac{3}{2}$ and $-\frac{3}{2}$ because $\frac{3}{2}^* = -\frac{3}{2}$. On the other hand, flipping columns 2 and 3 shows the equality of

$$\begin{Bmatrix} -\frac{3}{2} & \frac{3}{2} & 1 \\ -\frac{3}{2} & \frac{3}{2} & 1 \end{Bmatrix} \quad \text{and} \quad \begin{Bmatrix} \frac{3}{2} & \frac{3}{2} & 1 \\ \frac{3}{2} & \frac{3}{2} & 1 \end{Bmatrix}$$

The $6j$ symbol, a renormalized basis transformation (3.2.18), obeys the orthogonality condition

$$\sum_{\mu_3 r_1 r_2} |\lambda_3||\mu_3| \begin{Bmatrix} \lambda_1 & \lambda_2 & \lambda_3 \\ \mu_1 & \mu_2 & \mu_3 \end{Bmatrix}_{r_1 r_2 r_3 r_4} \begin{Bmatrix} \lambda_1 & \lambda_2 & \lambda_3' \\ \mu_1 & \mu_2 & \mu_3 \end{Bmatrix}^*_{r_1 r_2 r_3' r_4'} = \delta_{\lambda_3 \lambda_3'} \delta_{r_3 r_3'} \delta_{r_4 r_4'} \quad (3.3.21)$$

as well as the many others obtained from this by the use of the symmetries.

The trivial $6j$ is related to a $3j$ phase

$$\begin{Bmatrix} \lambda_1 & \lambda_2 & \lambda_3 \\ \lambda_2^* & \lambda_1 & 0 \end{Bmatrix}_{00rs} = |\lambda_1|^{-1/2} |\lambda_2|^{-1/2} \{\lambda_1 \lambda_2 \lambda_3 r\} \delta_{rs} \quad (3.3.22)$$

Additional relations between $6j$ symbols may be obtained by considering multiple recouplings. The generalized Racah backcoupling rule is

$$\begin{Bmatrix} \lambda_1 & \lambda_2 & \lambda_3 \\ \mu_1 & \mu_2 & \mu_3 \end{Bmatrix}_{r_1 r_2 r_3 r_4} = \sum_{\nu rs} |\nu| \{\lambda_3\} \{\lambda_1 \mu_2^* \mu_3 r_1\} \{\mu_1 \lambda_2 \mu_3^* r_2\} \times \{\lambda_1^* \mu_1 \nu r\} \begin{Bmatrix} \mu_3 & \nu & \lambda_3 \\ \mu_1 & \mu_2 & \lambda_1 \end{Bmatrix}_{r_1 r r_3 s} \times \begin{Bmatrix} \lambda_1 & \lambda_2 & \lambda_3 \\ \mu_3^* & \nu & \mu_1^* \end{Bmatrix}_{r r_2 s r_4} \quad (3.3.23)$$

This is easily obtained by using the expression for a $6j$ in terms of recoupling coefficients appearing in the transformation between various bases for the irreps $\boldsymbol{\lambda}_3^*$ in the space $\boldsymbol{\lambda}_1 \times \boldsymbol{\mu}_3 \times \boldsymbol{\mu}_1^*$. Consider a transformation between two schemes as the sum over a possible third scheme. This gives the identity

$$\langle(\lambda_1\mu_3)r_1\mu_2, \mu_1^*; r_3\lambda_3^*|\lambda_1(\mu_3\mu_1^*)r_2\lambda_2; r_4\lambda_3^*\rangle$$

$$= \sum_{\nu rs} \langle(\lambda_1\mu_3)r_1\mu_2, \mu_1^*; r_3\lambda_3|\mu_3(\lambda_1\mu_1^*)r\nu; s\lambda_3^*\rangle$$

$$\times \langle\mu_3(\lambda_1\mu_1^*)r\nu; s\lambda_3^*|\lambda_1(\mu_3\mu_1^*)r_2\lambda_2; r_4\lambda_3^*\rangle \tag{3.3.24}$$

The left side is in the appropriate form for transcription into a $6j$ symbol; the bases on the right have to be modified by changing the ordering of couplings, using the $3j$ phases $\{\lambda_1\mu_3\mu_2^*r\}$ for the first bra, $\{\mu_3\nu\lambda_3r\}$ for the second bra, and $\{\mu_3\mu_1^*\lambda_2^*r_2\}$ for the second ket. The generalization of the Biedenharn (1953)–Elliott (1953) sum rule follows from the similar identity for $\boldsymbol{\lambda}$ in $\boldsymbol{\lambda}_1 \times \boldsymbol{\lambda}_2 \times \boldsymbol{\lambda}_3 \times \boldsymbol{\lambda}_4$. The overlap

$$\langle(((\lambda_1\lambda_2)r_{12}\lambda_{12}, \lambda_3)r_{123}\lambda_{123}, \lambda_4; r\lambda|(\lambda_2\lambda_3)r_{23}\lambda_{23}, (\lambda_1\lambda_4)r_{14}\lambda_{14}; s\lambda\rangle$$

may be written simply as the sum of two successive overlaps:

$$= \sum_t \langle(((\lambda_1\lambda_2)r_{12}\lambda_{12}, \lambda_3)r_{123}\lambda_{123}, \lambda_4; r\lambda|(\lambda_1(\lambda_2\lambda_3)r_{23}\lambda_{23})t\lambda_{123}, \lambda_4; r\lambda\rangle$$

$$\times \langle(\lambda_1(\lambda_2\lambda_3)r_{23}\lambda_{23})t\lambda_{123}, \lambda_4; r\lambda|(\lambda_2\lambda_3)r_{23}\lambda_{23}, (\lambda_1\lambda_4)r_{14}\lambda_{14}; s\lambda\rangle$$

Now the $(\lambda_{123}\lambda_4\lambda^*r)$ triad in the first term appears twice and thus may be omitted, for the overlap of this part of the bases will be unity. Similarly, one omits the $(\lambda_2\lambda_3\lambda_{23}^*r_{23})$ triad in the second

$$= \sum_t \langle(\lambda_1\lambda_2)r_{12}\lambda_{12}, \lambda_3; r_{123}\lambda_{123}|\lambda_1(\lambda_2\lambda_3)r_{23}\lambda_{23}; t\lambda_{123}\rangle$$

$$\times \langle(\lambda_1\lambda_{23})t\lambda_{123}, \lambda_4; r\lambda|\lambda_{23}(\lambda_1\lambda_4)r_{14}\lambda_{14}; s\lambda\rangle \tag{3.3.25}$$

A $\{\lambda_1\lambda_{23}\lambda_{123}^*t\}$ $3j$ phase applied to the second overlap now puts these in standard form for translation into $6j$ symbols. The original overlap may be expanded into three standard recoupling coefficients by working on the right (or ket) side. We have

$$= \sum_{r_{124}\lambda_{124}r'r'_{124}} \langle((\lambda_1\lambda_2)r_{12}\lambda_{12},\lambda_3)r_{123}\lambda_{123},\lambda_4;r\lambda|\lambda_3((\lambda_1\lambda_2)r_{12}\lambda_{12},\lambda_4)r_{124}\lambda_{124};r'\lambda\rangle$$

$$\times\langle\lambda_3((\lambda_1\lambda_2)r_{12}\lambda_{12},\lambda_4)r_{124}\lambda_{124};r'\lambda|\lambda_3(\lambda_2(\lambda_1\lambda_4)r_{14}\lambda_{14})r'_{124}\lambda_{124};r'\lambda\rangle$$

$$\times\langle\lambda_3(\lambda_2(\lambda_1\lambda_4)r_{14}\lambda_{14})r'_{124}\lambda_{124};r'\lambda|(\lambda_2\lambda_3)r_{12}\lambda_{23},(\lambda_1\lambda_4)r_{14}\lambda_{14};s\lambda\rangle \tag{3.3.26}$$

The first overlap is independent of the $(\lambda_1\lambda_2\lambda_{12}^*r_{12})$ coupling, the second of $(\lambda_3\lambda_{124}\lambda^*r')$, and the third of $(\lambda_1\lambda_4\lambda_{14}^*r_{14})$. After writing these in terms of $6j$ symbols and assorted $3j$ phases, one can change variable names and use symmetries to obtain the symmetric result:

$$\sum_r \begin{Bmatrix} \lambda_1 & \lambda_2 & \lambda_3 \\ \mu_1 & \mu_2 & \mu_3 \end{Bmatrix}_{r_1r_2r_3r} \begin{Bmatrix} \lambda_1 & \lambda_2 & \lambda_3 \\ \nu_1 & \nu_2 & \nu_3 \end{Bmatrix}^*_{s_1s_2s_3r}$$

$$= \sum_{\lambda t_1t_2t_3} |\lambda|\{\lambda_1\}\{\nu_1\}\{\lambda_1\mu_2^*\mu_3r_1\}\{\mu_1\lambda_2\mu_3^*r_2\}$$

$$\times\{\mu_1^*\mu_2\lambda_3r_3\}\{\lambda\mu_1^*\nu_1t_1\}\{\lambda\mu_2^*\nu_2t_2\}\{\lambda\mu_3^*\nu_3t\}$$

$$\times\begin{Bmatrix} \nu_2 & \mu_2^* & \lambda \\ \mu_3 & \nu_3 & \lambda_1^* \end{Bmatrix}_{s_1r_1t_3t_2} \begin{Bmatrix} \nu_3 & \mu_3^* & \lambda \\ \mu_1 & \nu_1 & \lambda_2^* \end{Bmatrix}_{s_2r_2t_1t_3}$$

$$\times\begin{Bmatrix} \nu_1 & \mu_1^* & \lambda \\ \mu_2 & \nu_2 & \lambda_3^* \end{Bmatrix}_{s_3r_3t_2t_1} \tag{3.3.27}$$

For the purposes of using the Biedenharn–Elliott sum rule as a recursion relation in the case of multiplicity, one can use the $6j$ orthogonality (3.3.21) to give

$$\begin{Bmatrix} \lambda_1 & \lambda_2 & \lambda_3 \\ \mu_1 & \mu_2 & \mu_3 \end{Bmatrix}_{r_1r_2r_3r_4}$$

$$= \sum_{\substack{\lambda\nu_3 \\ t_1t_2t_3s_1s_2}} |\lambda_3||\nu_3||\lambda|\{\lambda_1\}\{\nu_1\}\{\lambda_1\mu_2^*\mu_3r_1\}$$

$$\times\{\mu_1\lambda_2\mu_3^*t_2\}\{\mu_1^*\mu_2\lambda_3r_3\}\{\lambda\mu_1^*\nu_1t_1\}\{\lambda\mu_2^*\nu_2t_2\}\{\lambda\mu_3^*\nu_3t_3\}$$

$$\times\begin{Bmatrix} \nu_2 & \mu_2^* & \lambda \\ \mu_3 & \nu_3 & \lambda_1^* \end{Bmatrix}_{s_1r_1t_3t_2}\begin{Bmatrix} \nu_3 & \mu_3^* & \lambda \\ \mu_1 & \nu_1 & \lambda_2^* \end{Bmatrix}_{s_2r_2t_1t_3}$$

$$\times\begin{Bmatrix} \nu_1 & \mu_1^* & \lambda \\ \mu_2 & \nu_2 & \lambda_3^* \end{Bmatrix}_{s_3r_3t_2t_1}\begin{Bmatrix} \lambda_1 & \lambda_2 & \lambda_3 \\ \nu_1 & \nu_2 & \nu_3 \end{Bmatrix}_{s_1s_2s_3r_4} \tag{3.3.28}$$

This equation holds providing $\nu_1\nu_2s_3$ are chosen in such a way that the right-hand side does not vanish identically. With the appropriate choice, namely ν_1 chosen the primitive irrep and the power of ν_2 chosen less than the power of λ_3, this equation is a recursive relation for a $6j$ in terms of primitive $6j$ symbols and a "smaller" $6j$ (see p. 70).

The various relationships between the $6j$ symbols of a group find certain uses in applications. More often in applications one wishes to replace sums of $3jm$ symbols by $6j$ symbols. Three important variants exist. The first allows the appropriate product of three $3jm$'s summed over all partners of the three irreps μ_1, μ_2, and μ_3 to be replaced by a single $3jm$ and $6j$:

$$\sum_r \begin{bmatrix} \lambda_1 & \lambda_2 & \lambda_3 \\ a_1 & a_2 & a_3 \\ \sigma_1 & \sigma_2 & \sigma_3 \end{bmatrix}^{rG}_{sH} \begin{Bmatrix} \lambda_1 & \lambda_2 & \lambda_3 \\ \mu_1 & \mu_2 & \mu_3 \end{Bmatrix}^{G}_{r_1r_2r_3r_4}$$

$$= \sum_{b_1\rho_1b_2\rho_2b_3\rho_3s_1s_2s_3} \begin{bmatrix} \mu_1 \\ b_1 \\ \rho_1 \end{bmatrix}\begin{bmatrix} \mu_2 \\ b_2 \\ \rho_2 \end{bmatrix}\begin{bmatrix} \mu_3 \\ b_3 \\ \rho_3 \end{bmatrix}$$

$$\times\begin{bmatrix} \lambda_1 & \mu_2^* & \mu_3 \\ a_1 & b_2 & b_3 \\ \sigma_1 & \rho_2^* & \rho_3 \end{bmatrix}^{r_1}_{s_1}\begin{bmatrix} \mu_1 & \lambda_2 & \mu_3^* \\ b_1 & a_2 & b_3 \\ \rho_1 & \sigma_2 & \rho_3^* \end{bmatrix}^{r_2}_{s_2}\begin{bmatrix} \mu_1^* & \mu_2 & \lambda_3 \\ b_1 & b_2 & a_3 \\ \rho_1^* & \rho_2 & \sigma_3 \end{bmatrix}^{r_3}_{s_3}$$

$$\times\begin{Bmatrix} \sigma_1 & \sigma_2 & \sigma_3 \\ \rho_1 & \rho_2 & \rho_3 \end{Bmatrix}^{H}_{s_1s_2s_3s} \tag{3.3.29}$$

We shall refer to the above as the Wigner relation. A second variant arises from viewing this as a recursion relation for a $3jm$ involving $\lambda_1\lambda_2\lambda_3$ in terms of $3jm$'s for $\lambda_1\mu_2^*\mu_3$, $\mu_1\lambda_2\mu_3^*$, and $\mu_1^*\mu_2\lambda_3$. The irreps $\mu_1\mu_2\mu_3$ can be selected so the $3jm$'s on the right appear earlier in a recursive calculation. The sum over the multiplicity label on the left can be eliminated using the orthogonality (3.3.21) of $6j$'s,

$$\begin{bmatrix} \lambda_1 & \lambda_2 & \lambda_3 \\ a_1 & a_2 & a_3 \\ \sigma_1 & \sigma_2 & \sigma_3 \end{bmatrix}^{rG}_{sH}$$

$$= \sum_{b_1\rho_1 b_2\rho_2 b_3\rho_3 s_1 s_2 s_3 \mu_2 r_1 r_3} |\lambda_1||\mu_1| \begin{Bmatrix} \lambda_1 & \lambda_2 & \lambda_3 \\ \mu_1 & \mu_2 & \mu_3 \end{Bmatrix}^{*G}_{r_1 r_2 r_3 r}$$

$$\times \begin{pmatrix} \mu_1 \\ b_1 \\ \rho_1 \end{pmatrix} \begin{pmatrix} \mu_2 \\ b_2 \\ \rho_2 \end{pmatrix} \begin{pmatrix} \mu_3 \\ b_3 \\ \rho_3 \end{pmatrix} \begin{bmatrix} \lambda_1 & \mu_2^* & \mu_3 \\ a_1 & b_2 & b_3 \\ \sigma_1 & \rho_2^* & \rho_3 \end{bmatrix}^{r_1}_{s_1} \begin{bmatrix} \mu_1 & \lambda_2 & \mu_3^* \\ b_1 & a_2 & b_3 \\ \rho_1 & \sigma_2 & \rho_3^* \end{bmatrix}^{r_2}_{s_2}$$

$$\times \begin{bmatrix} \mu_1^* & \mu_2 & \lambda_3 \\ b_1 & b_2 & a_3 \\ \rho_1^* & \rho_2 & \sigma_3 \end{bmatrix}^{r_3}_{s_3} \begin{Bmatrix} \sigma_1 & \sigma_2 & \sigma_3 \\ \rho_1 & \rho_2 & \rho_3 \end{Bmatrix}^{H}_{s_1 s_2 s_2 s} \tag{3.3.30}$$

Yet another variant of importance is

$$\sum_{r_1 r_4 \lambda_1} \frac{|\lambda_1|}{|\sigma_1|} \begin{bmatrix} \lambda_1 & \mu_2^* & \mu_3 \\ a_1 & b_2 & b_3 \\ \sigma_1 & \rho_2^* & \rho_3 \end{bmatrix}^{r_1^*}_{s_1} \begin{bmatrix} \lambda_1 & \lambda_2 & \lambda_3 \\ a_1 & a_2 & a_3 \\ \sigma_1 & \sigma_2 & \sigma_3 \end{bmatrix}^{r_4}_{s_4} \begin{Bmatrix} \lambda_1 & \lambda_2 & \lambda_3 \\ \mu_1 & \mu_2 & \mu_3 \end{Bmatrix}^{G}_{r_1 r_2 r_3 r_4}$$

$$= \sum_{b_1\rho_1 s_2 s_3} \begin{pmatrix} \mu_1 \\ b_1 \\ \rho_1 \end{pmatrix} \begin{pmatrix} \mu_2 \\ b_2 \\ \rho_2 \end{pmatrix} \begin{pmatrix} \mu_3 \\ b_3 \\ \rho_3 \end{pmatrix} \begin{bmatrix} \mu_1 & \lambda_2 & \mu_3^* \\ b_1 & a_2 & b_3 \\ \rho_1 & \sigma_2 & \rho_3^* \end{bmatrix}^{r_2}_{s_2} \begin{bmatrix} \mu_1^* & \mu_2 & \lambda_3 \\ b_1 & b_2 & a_3 \\ \rho_1^* & \rho_2 & \sigma_3 \end{bmatrix}^{r_3}_{s_3}$$

$$\times \begin{Bmatrix} \sigma_1 & \sigma_2 & \sigma_3 \\ \rho_1 & \rho_2 & \rho_3 \end{Bmatrix}^{H}_{s_1 s_2 s_3 s_4} \tag{3.3.31}$$

These three relations between $6j$ symbols for a group **G**, $6j$ symbols for a subgroup **H**, and G–H $3jm$ factors can be rewritten immediately for $3jm$ symbols. One chooses the subgroup **H** to be $\mathbf{C}_1$. All irrep labels of $\mathbf{C}_1$

can be omitted, as can the $6j$ symbol for $\mathbf{C}_1$. A fourth variant was written this way as (3.3.16). In an application with a sum over four $3jm$'s, one could either use (3.3.16) or obtain the same result by using the Wigner relation (3.3.29) and then the orthogonality (3.3.15) of $3jm$'s.

The $9j$ symbol has a column interchange symmetry given by the appropriate $3j$ phases of the affected triads, the row triads. We have for the (23) interchange

$$\begin{array}{c}\left\{\begin{array}{ccc}\lambda_1 & \lambda_2 & \lambda_3\\ \mu_1 & \mu_2 & \mu_3\\ \nu_1 & \nu_2 & \nu_3\end{array}\right\}\begin{array}{c}r_1\\ r_2\\ r_3\end{array}\\ \begin{array}{ccc}s_1 & s_2 & s_3\end{array}\end{array} = \{\lambda_1\lambda_2\lambda_3 r_1\}\{\mu_1\mu_2\mu_3 r_2\}\{\nu_1\nu_2\nu_3 r_3\}$$

$$\times\begin{array}{c}\left\{\begin{array}{ccc}\lambda_1 & \lambda_3 & \lambda_2\\ \mu_1 & \mu_3 & \mu_2\\ \nu_1 & \nu_3 & \nu_2\end{array}\right\}\begin{array}{c}r_1\\ r_2\\ r_3\end{array}\\ \begin{array}{ccc}s_1 & s_3 & s_2\end{array}\end{array} \tag{3.3.32}$$

The similar property for the rows is

$$\begin{array}{c}\left\{\begin{array}{ccc}\lambda_1 & \lambda_2 & \lambda_3\\ \mu_1 & \mu_2 & \mu_3\\ \nu_1 & \nu_2 & \nu_3\end{array}\right\}\begin{array}{c}r_1\\ r_3\\ r_2\end{array}\\ \begin{array}{ccc}s_1 & s_2 & s_3\end{array}\end{array} = \{\lambda_1\mu_1\nu_1 s_1\}\{\lambda_2\mu_2\nu_2 s_2\}\{\lambda_3\mu_3\nu_3 s_3\}$$

$$\times\begin{array}{c}\left\{\begin{array}{ccc}\lambda_1 & \lambda_2 & \lambda_3\\ \nu_1 & \nu_2 & \nu_3\\ \mu_1 & \mu_2 & \mu_3\end{array}\right\}\begin{array}{c}r_1\\ r_2\\ r_3\end{array}\\ \begin{array}{ccc}s_1 & s_2 & s_3\end{array}\end{array} \tag{3.3.33}$$

The symmetries under complex conjugation of the irreps and transposition about the leading diagonal are equally straightforward

$$\begin{array}{c}\left\{\begin{array}{ccc}\lambda_1 & \lambda_2 & \lambda_3\\ \mu_1 & \mu_2 & \mu_3\\ \nu_1 & \nu_2 & \nu_3\end{array}\right\}\begin{array}{c}r_1\\ r_2\\ r_3\end{array}\\ \begin{array}{ccc}s_1 & s_2 & s_3\end{array}\end{array} = \begin{array}{c}\left\{\begin{array}{ccc}\lambda_1^* & \lambda_2^* & \lambda_3^*\\ \mu_1^* & \mu_2^* & \mu_3^*\\ \nu_1^* & \nu_2^* & \nu_3^*\end{array}\right\}\begin{array}{c}{}^{*}r_1\\ r_2\\ r_3\end{array}\\ \begin{array}{ccc}s_1 & s_2 & s_3\end{array}\end{array} \tag{3.3.34}$$

$$= \begin{array}{c}\left\{\begin{array}{ccc}\lambda_1 & \mu_1 & \nu_1\\ \lambda_2 & \mu_2 & \nu_2\\ \lambda_3 & \mu_3 & \nu_3\end{array}\right\}\begin{array}{c}{}^{*}s_1\\ s_2\\ s_3\end{array}\\ \begin{array}{ccc}r_1 & r_2 & r_3\end{array}\end{array} \tag{3.3.35}$$

The $9j$ with one entry the identity irrep reduces to a $6j$ together with several $2j$ and $3j$ phases,

$$\begin{Bmatrix} \lambda_1 & \lambda_2 & \mu \\ \lambda_3 & \lambda_4 & \mu^* \\ \nu & \nu^* & 0 \end{Bmatrix}\begin{matrix} r_1 \\ r_2 \\ 0 \end{matrix}_{\;r_3\; r_4\; 0} = |\mu|^{-1/2}|\nu|^{-1/2}\{\lambda_4\}\{\lambda_2\lambda_4\nu^* r_4\}\{\lambda_3\lambda_4\mu^* r_2\}$$

$$\times \begin{Bmatrix} \lambda_1 & \lambda_3 & \nu \\ \lambda_4 & \lambda_2^* & \mu \end{Bmatrix}_{r_1 r_2 r_4 r_3} \tag{3.3.36}$$

The $9j$ contains no essentially new information, for it is a sum of products of $6j$ symbols,

$$\begin{Bmatrix} \lambda_1 & \lambda_2 & \lambda_3 \\ \mu_1 & \mu_2 & \mu_3 \\ \nu_1 & \nu_2 & \nu_3 \end{Bmatrix}\begin{matrix} r_1 \\ r_2 \\ r_3 \end{matrix}_{\;s_1\; s_2\; s_3} = \sum_{\kappa t_1 t_2 t_3} |\kappa|\{\kappa\} \begin{Bmatrix} \lambda_1 & \mu_1 & \nu_1 \\ \nu_2^* & \nu_3 & \kappa \end{Bmatrix}_{t_2 t_1 r_3 s_1}$$

$$\times \begin{Bmatrix} \lambda_2 & \mu_2 & \nu_2 \\ \mu_1 & \kappa & \mu_3^* \end{Bmatrix}_{t_3 r_2 t_1 s_2} \begin{Bmatrix} \lambda_3 & \mu_3 & \nu_3 \\ \kappa & \lambda_1^* & \lambda_2 \end{Bmatrix}_{r_1 t_3 t_2 s_3} \tag{3.3.37}$$

This sum occurs less often in applications than the expression of a $9j$ in terms of six $3jm$'s,

$$\begin{Bmatrix} \lambda_1 & \lambda_2 & \lambda_3 \\ \mu_1 & \mu_2 & \mu_3 \\ \nu_1 & \nu_2 & \nu_3 \end{Bmatrix}\begin{matrix} r_1 \\ r_2 \\ r_3 \end{matrix}_{\;s_1\; s_2\; s_3}$$

$$= \sum_{\substack{l_1 l_2 l_3 \\ m_1 m_2 m_3 \\ n_1 n_2 n_3}} \begin{pmatrix} \lambda_1 & \lambda_2 & \lambda_3 \\ l_1 & l_2 & l_3 \end{pmatrix}^{r_1} \begin{pmatrix} \mu_1 & \mu_2 & \mu_3 \\ m_1 & m_2 & m_3 \end{pmatrix}^{r_2} \begin{pmatrix} \nu_1 & \nu_2 & \nu_3 \\ n_1 & n_2 & n_3 \end{pmatrix}^{r_3}$$

$$\times \begin{pmatrix} \lambda_1 & \mu_1 & \nu_1 \\ l_1 & m_1 & n_1 \end{pmatrix}^{*s_1} \begin{pmatrix} \lambda_2 & \mu_2 & \nu_2 \\ l_2 & m_2 & n_2 \end{pmatrix}^{*s_2} \begin{pmatrix} \lambda_3 & \mu_3 & \nu_3 \\ l_3 & m_3 & n_3 \end{pmatrix}^{*s_3} \tag{3.3.38}$$

This relation, like those relating $6j$'s and $3jm$'s, can be written in terms of $3jm$ factors and a $9j$ for a subgroup. Variants also exist with one or more $3jm$'s moved to the left. Complicated expressions of this nature are often better handled by diagram techniques. Stedman (1975, 1976) has generalized the diagram methods used with the JM basis (Jucys, Levinson, and Vanagas, 1960) to include the multiplicities and symmetries of this book.

Finally, we compare the $3jm$ symbols of different schemes. Comparing the result of basis transformation $\langle\lambda i|\lambda l\rangle$ of the form (2.2.7) before and after coupling shows that

$$\begin{pmatrix} \lambda_1 & \lambda_2 & \lambda_3 \\ i_1 & i_2 & i_3 \end{pmatrix}^r = \sum_{l_1 l_2 l_3} \begin{pmatrix} \lambda_1 & \lambda_2 & \lambda_3 \\ l_1 & l_2 & l_3 \end{pmatrix}^r \times \langle\lambda_1 l_1|\lambda_1 i_1\rangle\langle\lambda_2 l_2|\lambda_2 l_2\rangle\langle\lambda_3 l_3|\lambda_3 i_3\rangle \tag{3.3.39}$$

The transformation coefficients $\langle\lambda l|\lambda i\rangle$ are elements of a unitary matrix, dimension $|\lambda|$, row and column labels l and i. We have used an independence of $\langle\lambda l|\lambda i\rangle$ of any additional labels; this is clearly possible because the group structure is the same for both sets of symbols, only the subgroup structure differs. Using the unitarity to move one transformation coefficient to the left, and (3.3.15) with $\mathbf{H}$ the trivial subgroup $\mathbf{C}_1$ to move the $3jm$ to the right, shows

$$\begin{aligned} \langle\lambda_3 i_3|\lambda_3 l_3\rangle &= \langle\lambda_3 l_3|\lambda_3 i_3\rangle^* \\ &= \sum_{l_1 l_2 i_1 i_2} |\lambda_3| \begin{pmatrix} \lambda_1 & \lambda_2 & \lambda_3 \\ l_1 & l_2 & l_3 \end{pmatrix}^r \\ &\quad \times \langle\lambda_1 l_1|\lambda_1 i_1\rangle\langle\lambda_2 l_2|\lambda_2 i_2\rangle \begin{pmatrix} \lambda_1 & \lambda_2 & \lambda_3 \\ i_1 & i_2 & i_3 \end{pmatrix}^{*r} \end{aligned} \tag{3.3.40}$$

These results factorize if the two basis schemes contain a common subgroup, e.g., $\mathbf{O}\supset\mathbf{T}\supset\mathbf{D}_2\supset\mathbf{C}_2$ and $\mathbf{O}\supset\mathbf{D}_4\supset\mathbf{D}_2\supset\mathbf{C}_2$. The general case is exemplified by the diagram

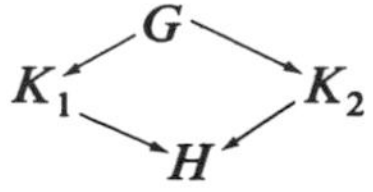

The transformation between the two schemes will be diagonal in the labels of **H**, but only if the group **H** appearing in each scheme is actually the same, not merely isomorphic. For example, there are several $\mathbf{D}_2$ groups in **O**; one must choose precisely the same one for each scheme. Assume this is the case. Collect the labels associated with the $\mathbf{K}_1$ scheme into a label a, those with the $\mathbf{K}_2$ scheme into a label b, and (3.3.39) becomes

$$\begin{Bmatrix} \lambda_1 & \lambda_2 & \lambda_3 \\ a_1 & a_2 & a_3 \\ \rho_1 & \rho_2 & \rho_3 \end{Bmatrix} \begin{matrix} rG \\ K_1 \\ sH \end{matrix} = \sum_{b_1 b_2 b_3} \begin{Bmatrix} \lambda_1 & \lambda_2 & \lambda_3 \\ b_1 & b_2 & b_3 \\ \rho_1 & \rho_2 & \rho_3 \end{Bmatrix} \begin{matrix} rG \\ K_2 \\ sH \end{matrix}$$

$$\times \langle \lambda_1 b_1 \rho_1 | \lambda_1 a_1 \rho_1 \rangle \langle \lambda_2 b_2 \rho_2 | \lambda_2 a_2 \rho_2 \rangle \langle \lambda_3 b_3 \rho_3 | \lambda_3 a_3 \rho_3 \rangle \tag{3.3.41}$$

The case of the two subgroups **H** being merely isomorphic has been the object of some study. The transformation between the two schemes must include the matrix elements of an operator which performs the isomorphism. Most work has involved the symmetric groups (see, for example, Sullivan 1978 and references therein).

3.4 Computational Methods

Many of the equations of the previous section are quadratic and several are cubic, quartic, or quintic. However, they may be used recursively to solve for the j symbols and jm factors from a knowledge of dimensions, product rules, and branching rules. In this section we discuss how the separation of the j symbols and jm factors into three categories—trivial, primitive, and nonprimitive types—reduces the problem to manageable proportions. The trivial $6j$ symbols are the $3j$ and $2j$ phases to within dimension factors and contain reordering information as well as certain complex conjugation information. Either the values of these symbols may be deduced from character theory or a free choice exists. We leave discussion of these choices and of similar choices for $2jm$ symbols to the following section. The primitive $6j$ symbols and $3jm$ factors contain all the other phase and basis information which is contained in the Racah algebra of a specific group. In this section we prove this by showing how to use the trivial and primitive coefficients to calculate all nonprimitive $6j$ symbols, all $9j$ symbols, and all nonprimitive $3jm$ factors.

This method of computing all coefficients from a knowledge of the selection rules, which in turn follows from character theory, was first used for the chain $\mathbf{SO_3} \supset \mathbf{O} \supset \mathbf{T} \supset \mathbf{C_3}$ (Butler and Wybourne 1976a, b) and to obtain Racah's algebraic formulas for the SO_3 $6j$ symbols and the SO_3–SO_2 $3jm$ factors (Butler 1976). A number of different methods have been used for calculating *jm* factors and *j* symbols. The best known is the ladder operator techniques customarily used for angular momentum. Generalizations of ladder operators to other continuous groups abound in the literature, but the explicit form of the ladder operator implies a specific basis, which may not be the subgroup basis required. Another common method of calculating point group coefficients involves first obtaining basis transformation coefficients from the *JM* basis to the point group basis under study, and then using (3.3.39) to obtain the point group *jm* symbols. (See, for example, Harnung and Schäffer 1972, Harnung 1973, Kibler and Grenet 1977, Kibler and Guichon 1976, Konig and Kremer 1973.) Many other methods exist; some, for example, follow Racah's (1949) calculation for the coupling factors appropriate to parentage groups. He used the known matrix elements of certain operators, that is, he used the Wigner–Eckart theorem in reverse (see also Judd 1963, Chapter 7), as well as using symmetries and recursion relations. Fano and Racah (1959, Appendix I) suggested more extensive use of recursion relations. Our method requires only character theory, which gives basis-independent results. This has the advantages of not requiring the representation matrices of the group to be known, not requiring any Gramm–Schmidt orthogonalization of basis functions, nor requiring prior knowledge of *jm* symbols in another basis.

For the purposes of our method of calculation it is useful to divide both the $6j$ symbols and the $3jm$ factors into three categories: (i) the trivial symbols, which have at least one group irrep, the identity irrep $\mathbf{0}$; (ii) the primitive symbols, which have at least one group irrep, the basic spinor or primitive irrep $\frac{1}{2}$; and (iii) the nonprimitive general symbols.

Neither the *jm* and *j* symbols, nor indeed the Racah algebra as a whole, contains all the information about partners. We shall see in Chapter 5 that some aspects of orientation of physical axes and certain overall phases do not enter into the Racah algebra and therefore may be chosen quite independently. The Racah algebra contains structural information about irreps and their partners. For $\mathbf{D_3} \supset \mathbf{C_3}$ the $3jm$ factor table may be used not only with the functions $(1/\sqrt{2})(x+iy)$, z and $-(1/\sqrt{2})(x-iy)$, but also any set obtained from these by rotation of the x, y, z axis system (see Section 5.3).

The next section discusses the specification of those free phases and choices of multiplicity separation that are within the Racah algebra. For pedagodical reasons we are arguing in the reverse order to which the calculations are performed. We begin by showing here that the nonprimitive $6j$ symbols and $3jm$ factors follow from the primitive symbols by recoupling recursively. The following section discusses how the primitive symbols may be obtained by solving the orthogonality and backcoupling equations.

The key concept needed to show that the primitive coefficients enable all the others to be computed without further phase choices is the power of an irrep, defined in Section 2.6. The definition there of the power $p(\lambda)$ of an irrep $\boldsymbol{\lambda}$ was that if one took powers $\boldsymbol{\varepsilon}\times\boldsymbol{\varepsilon}\times\cdots\times\boldsymbol{\varepsilon}=\boldsymbol{\varepsilon}^{\times p}$ of $\boldsymbol{\varepsilon}$, where $\boldsymbol{\varepsilon}$ is either the primitive irrep $\frac{1}{2}$ if $\frac{1}{2}=\frac{1}{2}^*$ or $\boldsymbol{\varepsilon}=\frac{1}{2}+(-\frac{1}{2})$ if $\frac{1}{2}$ is complex, then the lowest power p, in which $\boldsymbol{\lambda}$ occurred is $p(\lambda)$. It follows from this that $p(\lambda^*)=p(\lambda)$. If $\boldsymbol{\mu}$ is contained in a triad $(\varepsilon\lambda\mu r)$, then we need to show that

$$p(\lambda)-1\leqslant p(\mu)\leqslant p(\lambda)+1 \tag{3.4.1}$$

It is sufficient to show that $p(\mu)\leqslant p(\lambda)+1$, because the other inequality then follows from the symmetry of the triad $(\varepsilon\lambda\mu r)$. Now by the definition of $p(\lambda)$, $\boldsymbol{\lambda}$ is contained in the product $\boldsymbol{\varepsilon}^{\times p(\lambda)}$. However, because $\boldsymbol{\mu}^*$ is contained in $\boldsymbol{\lambda}\times\boldsymbol{\varepsilon}$ it must be contained in $\boldsymbol{\varepsilon}^{\times(p(\lambda)+1)}$. Thus the power of $\boldsymbol{\mu}$ is at most $p(\lambda)+1$. This proves the result (3.4.1). We now use it to prove the completeness of the following recursion relation for nonprimitive $6j$ symbols and $3jm$ factors.

Nonprimitive $6j$ symbols may be calculated recursively from primitive $6j$ symbols. The Biedenharn–Elliott sum rule (3.3.28) gives a $6j$ as a sum of ($3j$ phases and) $6j$ symbols, where two irreps $\boldsymbol{\nu}_1$ and $\boldsymbol{\nu}_2$ may be chosen within a range. This restriction in range appears in the derivation as (omitted) delta functions in the multiplicity indices on the left, namely if no terms occur in the sum on the right, then the delta functions on the left would have been zero. Our recursive relation appears if we take $\boldsymbol{\nu}_1=\frac{1}{2}$. With our assumption that all $3j$ phases and primitive $6j$ symbols are known, (3.3.28) simplifies into the equation

$$\begin{Bmatrix} \lambda_1 & \lambda_2 & \lambda_3 \\ \mu_1 & \mu_2 & \mu_3 \end{Bmatrix}_{r_1r_2r_3r_4} = \sum_{\substack{\lambda\nu_3 \\ t_2t_3s_1}} c \begin{Bmatrix} \lambda_1 & \nu_3 & \nu_2^* \\ \lambda & \mu_2 & \mu_3 \end{Bmatrix}_{r_1t_3t_2s_1} \tag{3.4.2}$$

where the number c contains a sum over primitive $6j$ symbols and other known information (we have used the symmetries of the $6j$). Irrep $\boldsymbol{\nu}_2$ is to be chosen subject only to the right-hand side not being identically zero. Use the symmetries of the $6j$ so $\boldsymbol{\lambda}_3$ is the smallest irrep of the unknown $6j$. Choose $\boldsymbol{\nu}_2$ to be an irrep of power $p(\lambda_3)-1$ that occurs in the product $\frac{1}{2}\times\boldsymbol{\lambda}_3^*$. It can be seen that (3.4.2) is a recursion relation, reducing the smallest power of an irrep in a $6j$ by unity, so that after $p(\lambda_3)-1$ applications, one has only primitive $6j$'s.

The nonprimitive $3jm$ symbols are calculated recursively from (3.3.30) by setting $\boldsymbol{\mu}_1=\frac{1}{2}$. Given the $6j$ symbols of group and subgroup and the primitive $3jm$ symbols, the Wigner relation (3.3.27) takes on the form

$$\begin{bmatrix} \lambda_1 & \lambda_2 & \lambda_3 \\ a_1 & a_2 & a_3 \\ \sigma_1 & \sigma_2 & \sigma_3 \end{bmatrix}_s^r = \sum_{\mu_2} d \begin{bmatrix} \lambda_1 & \mu_2^* & \mu_3 \\ a_1 & b_2 & b_3 \\ \sigma_1 & \rho_1^* & \rho_2 \end{bmatrix}_{s_1}^{r_1} \tag{3.4.3}$$

with $\boldsymbol{\mu}_3$ to be chosen. The sufficiency of this relation as a recursion relation is easier to see than in the $6j$ case. One simply chooses a $\boldsymbol{\mu}_3$ contained in $\boldsymbol{\lambda}_2\times\frac{1}{2}$ such that $p(\mu_3)=p(\lambda_2)-1$.

3.5 Phase and Multiplicity Choices

We saw in Chapter 2 that there are at least two kinds of unitary freedoms in the Racah algebra. In this section we continue the discussion on the freedoms and state our choices explicitly as part of a description of how the primitive $6j$ symbols and $3jm$ factors are computed. The two kinds of phase choices previously mentioned do much to fix the irrep matrices and the various partners, but for several imbeddings, an additional freedom arises in the choice of primitive $3jm$'s. This additional freedom is associated with a choice of orientations of the groups.

The first freedom apparent is that of the choice of the irrep matrices. The form of these is partly fixed by the particular subgroup choice, for the subgroup forces one to take those kets that are partners of the subgroup irreps, and the irrep subspaces are unique up to equivalence. Let there be a branching multiplicity α_σ^λ in the labeling of a set of partners of irrep $\boldsymbol{\lambda}(\mathbf{G})$,

$$\{|\lambda a\sigma i\rangle:\ \ \lambda \text{ fixed};\ \ i=0,1,\ldots,(|\sigma|-1);\ \ a=0,1,\ldots,(\alpha_\sigma^\lambda-1)\} \tag{3.5.1}$$

symmetrized in the chain $\mathbf{G} \supset \mathbf{H}$. Assume that the irrep matrices for $\mathbf{H}$, $\boldsymbol{\sigma}(\mathbf{H})$, are fixed so that any freedom in the partners $|\lambda a\sigma i\rangle$ is restricted, by Schur's lemmas applied to $\mathbf{H}$, to being diagonal in $\boldsymbol{\sigma}$ and i and independent of i. Diagonality in $\boldsymbol{\lambda}$ is ensured by Schur's lemmas applied to $\mathbf{G}$. Within these restrictions alternative sets of partners may be formed as the linear combinations

$$|\lambda a'\sigma i\rangle_{\text{alternative}} = \sum_{a} |\lambda a\sigma i\rangle A_{aa'} \tag{3.5.2}$$

where $A_{aa'}$ are elements of a unitary matrix. Such changes of basis will change the irrep matrices $\boldsymbol{\lambda}(\mathbf{G})$ without affecting the $\boldsymbol{\sigma}(\mathbf{H})$. If $\alpha_{\sigma}^{\lambda} = 1$, that is, if a and a' take on the value 0 only, then A_{00} is a phase—a complex number of modulus 1. If $\alpha_{\sigma}^{\lambda} > 1$, then we have the freedom of allowing linear combinations—a multiplicity freedom.

From the point of view of the groups' irrep structures the $\mathbf{A}$ matrices of (3.5.2), for various $\boldsymbol{\sigma}$ or $\boldsymbol{\lambda}$, are unrelated. However, the complex conjugation symmetries imposed in the Racah algebra in the previous sections restrict us to certain A. In other words, some of the free phases and multiplicities have already been fixed by requests of symmetry.

The Racah algebra deals with coupling and recoupling, not with single kets. One ket never appears alone in an equation, so only relative phases will be contained in the Racah algebra. We show below that "relative" means relative to the "first" partners of an irrep. In addition to the ket freedoms, every coupling equation contains a similar unitary freedom $U(\lambda_1\lambda_2\lambda_3)_{r'r}$ in the phase and separation of the product multiplicity, (2.3.7). Again, as we see below, it is relative information that is important.

The situation with regard to the choices entering the j symbol calculation is simpler than that for the jm symbols, so we deal with it first. The j phases and symbols contain only product information and the matrices $U(\lambda_1\lambda_2\lambda_3)$ describe all the phase and multiplicity freedoms. The $2j$ and $3j$ phases give reordering symmetries of the $2jm$ and $3jm$ symbols and correspond to relating $U(\lambda_1\lambda_2 0)$ to $U(\lambda_2\lambda_1 0)$ and to relating the various $U(\lambda_a\lambda_b\lambda_c)$ to each other, for abc the six permutations of 123. If the primitive irrep $\frac{1}{2}$ is symplectic, as it is for the non-Abelian point groups, it may be shown that the formulas

$$\{\lambda\} = \{\lambda\lambda^*\} = (-1)^{p(\lambda)} \tag{3.5.3}$$

$$\{\lambda_1\lambda_2\lambda_3 r\} = (-1)^{[p(\lambda_1)+p(\lambda_2)+p(\lambda_3)]/2+r} \tag{3.5.4}$$

where $p(\lambda)$ is the power of irrep λ, usually satisfy the symmetries of the $3jm$'s as deduced from character theory, (3.2.8)–(3.2.10). Recall that the $3j$ describes whether or not a $3jm$ changes sign upon interchange of its columns. Recall also that if two or more irreps are the same, say $\boldsymbol{\lambda}_1=\boldsymbol{\lambda}_2$, then the interchange of columns 1 and 2 has the effect of interchanging the partner labels of $\boldsymbol{\lambda}_1$ and $\boldsymbol{\lambda}_2$. The coupling operation fixes relationships between partners, and thus the $3j$ is fixed by the nature of the coupling rather than by any choice of our own.

All irreps of the cyclic groups are orthogonal or quasi-orthogonal and a simpler set of phases is possible

$$\{\lambda\}=\{\lambda\lambda^*\}=1=\{\lambda_1\lambda_2\lambda_3\} \qquad \text{for } \mathbf{C}_n \tag{3.5.5}$$

These formulas, (3.5.4)–(3.5.5), for the $3j$ are our choices if the irreps $\boldsymbol{\lambda}_1$, $\boldsymbol{\lambda}_2$, and $\boldsymbol{\lambda}_3$ are all different, and is the actual symmetry derived from character theory otherwise, except for one or two products for the octahedral, icosahedral, and odd-dimensional dihedral groups. Because the formulas fail for special cases, one must use the values tabulated at the head of the $6j$ symbol tables of Chapter 14. The advantage of using these formulas is that, when they do hold, the $6j$ symbol is invariant under column interchanges.

Making these, or indeed any other, choices of the $2j$ and $3j$ phases reduces the freedom in the recoupling (or j symbol) algebra to a free $U(\lambda_1\lambda_2\lambda_3)$ matrix for each set of six permutations of $\lambda_1\lambda_2\lambda_3$. Our choice earlier of asking the $3j$ phases to be diagonal and our choice of taking the Derome–Sharp complex conjugation matrix as the unit matrix in (3.3.12) further restrict the freedom. The matrices $U(\lambda\lambda\lambda')$ must be block diagonal with respect to symmetry type and the matrices $U(\lambda_1\lambda_2\lambda_3)$ and $U(\lambda_1^*\lambda_2^*\lambda_3^*)$ are related. The primitive $6j$'s contain the remaining choices relating the various $U(\lambda_1\lambda_2\lambda_3)$ matrices. They relate the matrices for all triads except for one triad for each irrep. This last statement is proved as follows.

The primitive $6j$'s of the form

$$\begin{Bmatrix} \lambda_1 & \lambda_2 & \lambda_3 \\ \frac{1}{2} & \mu_3 & \mu_2 \end{Bmatrix}_{st_1t_2r}$$

contain two primitive triads $(\frac{1}{2}\lambda_2\mu_2^*t_1)$ and $(\frac{1}{2}{}^*\mu_3\lambda_3t_2)$ and two general triads $(\lambda_1\mu_3^*\mu_2 s)$ and $(\lambda_1\lambda_2\lambda_3 r)$. Consider changing the coupling phase and multiplicity separations of the nonprimitive triads according to the $U(\lambda_1\lambda_2\lambda_3)_{r'r}$ of (2.3.7). It follows from the expression for a $6j$ as a sum of products of four $3jm$'s that the primitive $6j$ above will change as

$$\begin{Bmatrix} \lambda_1 & \lambda_2 & \lambda_3 \\ \frac{1}{2} & \mu_3 & \mu_2 \end{Bmatrix}^{\text{one choice}}_{st_1t_2r} = \sum_{r's'} \begin{Bmatrix} \lambda_1 & \lambda_2 & \lambda_3 \\ \frac{1}{2} & \mu_3 & \mu_2 \end{Bmatrix}^{\text{another choice}}_{s't_1t_2r'} \times U(\lambda_1\mu_3^*\mu_2^*)_{s's} U(\lambda_1\lambda_2\lambda_3^*)^*_{r'r} \qquad (3.5.6)$$

The two choices of primitive $6j$ which are related by (3.5.6) are equally satisfactory.

When a triad first appears once only in a $6j$ we may choose the phase of that $6j$. If the **U** matrix is restricted by the Derome–Sharp complex conjugation symmetry

$$\mathbf{U}(\lambda_1\lambda_2\lambda_3)^* = \mathbf{U}(\lambda_1^*\lambda_2^*\lambda_3^*) \qquad (3.5.7)$$

to be real, then the phase will just be a sign freedom. If there is multiplicity and the different triads have the same permutation symmetry, then the freedom includes a choice of linear combination.

The orthogonality (3.3.21) and Racah backcoupling (3.3.23) equations provide a series of equations between primitive $6j$'s. One may use these to obtain the norm (or square) of a $6j$ for which the phase (and perhaps the multiplicity freedom) of a triad $(\lambda_1\lambda_2\lambda_3 r)$ has not been chosen. Choosing the phase (and multiplicity separation) will fix the relationship of $\mathbf{U}(\lambda_1\lambda_2\lambda_3^*)$ to the matrices $\mathbf{U}(\lambda_1\mu_2\mu_3)$, $\mathbf{U}(\lambda_3\mu_3\frac{1}{2})$, and $\mathbf{U}(\lambda_2\frac{1}{2}\mu_2)$. Consideration of the powers of the irreps in a triad shows for every ordered triad $(\lambda_1\lambda_2\lambda_3 r_4)$ with $\frac{1}{2} < \lambda_1 \leqslant \lambda_2 \leqslant \lambda_3$ in which a particular $\boldsymbol{\lambda}_3$ occurs that there must be a $6j$

$$\begin{Bmatrix} \lambda_1 & \lambda_2 & \lambda_3 \\ \frac{1}{2} & \mu_3 & \mu_2 \end{Bmatrix}_{r_1r_2r_3r_4}$$

in which $p(\mu_2)$ and $p(\mu_3)$ are less than $p(\lambda_3)$. The phase of such a $6j$ can be chosen, fixing the coupling freedom of $(\lambda_1\lambda_2\lambda_3 r_4)$ in terms of the choices for triads involving irreps of lower power. An equivalent result holds for $\boldsymbol{\lambda}_1 = \frac{1}{2}$ if there are two distinct irreps $\boldsymbol{\lambda}_2$ and $\boldsymbol{\mu}_3$ less (or equal) in power to $\boldsymbol{\lambda}_3$, where we have the primitive triads $(\frac{1}{2}\lambda_2\lambda_3 r_4)$ and $(\frac{1}{2}^*\mu_3\lambda_3 r_3)$. Such a multiplicity of primitive triads for a single irrep $\boldsymbol{\lambda}_3$ does not occur for the point groups. For the point groups the number of primitive triads equals the number of irreps (up to complex conjugation). For other groups one must select one's subset of primitive triads (the basis triads) with respect to which one chooses coupling phases and multiplicity separation.

As an example, consider the icosahedral group. There are 59 triads in all, of which 16 are primitive (and basis), leaving 43 phase choices. All representations are real, so complex conjugation symmetry gives that these 43 phase choices are simply choices of sign. There are seven product multiplicities of two and one of three. Reordering symmetries, namely the requirement that we separate symmetric and antisymmetric occurrences of an irrep, reduces our choice of multiplicity separations to three pairs of triads of the same symmetry and two pairs of no symmetry.

The unitarity conditions (3.3.21) and the Racah backcoupling equation (3.3.23) give sufficient equations for the point groups to enable one to deduce all primitive $6j$ symbols up to the above-mentioned phase and multiplicity freedoms. Our computer program cycles through the list of $6j$ symbols repeatedly. For each uncalculated primitive $6j$ symbol it looks at the four triads involved. If exactly one nonbasis triad has its phase not chosen, its normalization conditions are used to attempt to find a norm. After a check that either the $6j$ is real (a $\pm$ choice) or the free triad is complex (a $e^{i\theta}$ phase choice) the positive square root of the norm is stored as the value. If all non-basis triads have chosen phases, then the orthogonality and backcoupling equations are tried in an attempt to find this $6j$'s value. Our computer program is successful only if the $6j$ it is solving for is the only unknown $6j$ in the equation being tried.

Several computational points are worth noting. First, there are 12 different orthogonality and 24 (12 with the $6j$ on the left, 12 on the right) different backcoupling equations involving each primitive $6j$, although symmetries will reduce this number, as will the fact that not all equations will involve only primitive or trivial $6j$ symbols. Second, the program was successful for all the point groups, but the simple-minded approach to solving these simultaneous quadratic equations by searching the list for equations with only one unknown has failed for the $6j$'s of some of the groups suggested for elementary particle theory, namely $\mathbf{SU}_3$, $\mathbf{E}_7$, and $\mathbf{G}_2$ ($\mathbf{G}_2$ was first used by Racah as an algebraic symmetry in *f*-shell rare-earth spectra), although simultaneous solution by hand of the computer-generated equations is usually straightforward. It is clear that the Biedenharn–Elliott sum rule (3.3.27) gives many useful equations in the larger problems. A key point seems to be that for the point groups the product of the primitive irrep with another irrep usually gives two terms only, because the dimension of the primitive is two.

Third, there is the matter of multiplicities. The normalization equations for the $6j$'s having the various triads associated with different multiplicities of a product are such that one obtains the sum of the norms of the $6j$. If one $6j$ is zero for reasons of symmetry the norm of the other $6j$ may be obtained and the phase associated with its triad fixed. A later nonzero $6j$ enables the phase of the first triad to be fixed. The program requires the user to specify which $6j$'s are to be set zero to resolve the various multiplicities. The five nonsymmetry multiplicities of the icosahedral group need five such resolutions. Several possibilities arise as to which particular primitive $6j$ is chosen zero.

For example, for the $\frac{5}{2}\frac{5}{2}\frac{1}{2}$ product of multiplicity two in the icosahedral group one has a range of pairs of $6j$'s to use to define the multiplicity resolution, some of the pairs being ($r=0$ or 1)

$$\begin{Bmatrix} \frac{5}{2} & \frac{5}{2} & 1 \\ \frac{1}{2} & \frac{1}{2} & 2 \end{Bmatrix}^{K}_{000r}, \quad \begin{Bmatrix} \frac{5}{2} & \frac{5}{2} & 1 \\ \frac{3}{2} & \frac{1}{2} & 2 \end{Bmatrix}^{K}_{000r},$$

$$\begin{Bmatrix} \frac{5}{2} & \frac{5}{2} & 1 \\ \frac{3}{2} & \frac{1}{2} & 2 \end{Bmatrix}^{K}_{010r}, \quad \begin{Bmatrix} \frac{5}{2} & \frac{5}{2} & 1 \\ 2 & 2 & \frac{1}{2} \end{Bmatrix}^{K}_{000r}$$

The different pairs of $6j$'s have different sums of squares. The second and third of the pairs will vary again as the $(\frac{5}{2}2\frac{3}{2}\,s)$ separation is varied. One $6j$ of any pair could be set zero and an adequate resolution of the $(\frac{5}{2}\frac{5}{2}1r)$ multiplicity obtained. For the icosahedral group the only reasons known for choosing one separation over another is either that the zero used to fix the separation gives rise to the simplest value for its pair (that is, no large prime numbers) or that as many other zeros appear in the complete table. The tilde symmetry described below gave some guidance for some products, but many separations were tried for the icosahedral group.

Hamermesh (1962, p. 266) gives an additional criterion for separating some multiplicities for symmetric groups $\mathbf{S}_n$. His results generalize (Butler and Ford 1979). The octahedral group is abstractly $\mathbf{S}_4$ and has a one-dimensional irrep $\mathbf{A}_2(\mathbf{O})=[\mathbf{1}^4](\mathbf{S}_4)=\tilde{\mathbf{0}}$ which is not the identity irrep $\mathbf{A}_1(\mathbf{O})=[\mathbf{4}](\mathbf{S}_4)=\mathbf{0}$. For any group with such an irrep, irreps come in pairs $\boldsymbol{\lambda}$ and $\tilde{\boldsymbol{\lambda}}=\tilde{\mathbf{0}}\times\boldsymbol{\lambda}$, where $\tilde{\boldsymbol{\lambda}}$ may or may not equal $\boldsymbol{\lambda}$. We call $\tilde{\boldsymbol{\lambda}}$ the tilde of $\boldsymbol{\lambda}$. For every triad $(\lambda\mu\nu r)$ one has three other triads of the form $(\lambda\tilde{\mu}\tilde{\nu}s)$ and it is easy to show they have the same multiplicity. One would say they have the same multiplicity separation and thus a new symmetry of the Racah–Wigner algebra if the $6j$ relating these two to the tildes of μ, ν is diagonal

$$\begin{Bmatrix} \lambda & \mu & \nu \\ \tilde{0} & \tilde{\nu} & \tilde{\mu} \end{Bmatrix}_{s00r} = \pm|\mu|^{-1/2}|\nu|^{-1/2}\delta_{rs} \tag{3.5.8}$$

For the octahedral group this requirement does not conflict with permutational symmetries, and what is more, gives a separation of the $(\frac{3}{2}\frac{3}{2}1r)$ couplings in terms of the $(\frac{3}{2}\frac{3}{2}\tilde{1}r)$ couplings. The $(\frac{3}{2}\frac{3}{2}\tilde{1}r)$ couplings are separated by permutational symmetry (one even, one odd), but the $(\frac{3}{2}\frac{3}{2}1r)$ couplings are not (both even). Using the tilde gives $\frac{3}{2}=\tilde{\mathbf{0}}\times\frac{3}{2}=\frac{3}{2}$ and $\mathbf{1}=\tilde{\mathbf{0}}\times\mathbf{1}=\tilde{\mathbf{1}}$. Interestingly, the separation in $(\frac{3}{2}\frac{3}{2}1r)$ given through (3.5.8) coincides with the simplicity criteria above and as a result our octahedral $6j$ table has no factors of $\sqrt{5}$ and more zeros than previous tabulations (Dobosh 1972, Harnung 1973).

Consider now the question of the values of the $2jm$'s and the primitive $3jm$'s. The phase and multiplicity freedoms in the kets described by (3.5.2) demand that choices be made also in the $2jm$ factors and the primitive $3jm$ factors. The $2jm$ factor choices are restricted by (3.5.3) and (3.3.9). The primitive $3jm$ factors by definition contain only primitive triads of the group. We have seen that the primitive $6j$ symbols contain all coupling information of nonprimitive triads with respect to the primitive triads. For some non-point groups the $6j$ symbols contain in addition some relations between primitive triads. The primitive $3jm$ symbols, and indeed the $3jm$ factors, for they are what we work with, contain not only a little coupling information of some primitive triads with respect to other primitive triads, and ket information with respect to the primitive kets, but also some "orientation" information.

Equation (3.5.2) specifies the freedom in partners $|\lambda a\sigma i\rangle$ of irrep $\boldsymbol{\lambda}(\mathbf{G})$. Primitive $3jm$ symbols perform couplings of the form

$$|\mu b\rho j\rangle|\tfrac{1}{2}\eta\rangle\rightarrow|\left(\mu\tfrac{1}{2}\right)\lambda a\sigma i\rangle \tag{3.5.9}$$

where $|\frac{1}{2}\eta\rangle$ is a partner of the primitive irrep. There is one phase choice (and corresponding multiplicity choice) for each $a\sigma$ contained in irrep $\boldsymbol{\lambda}(\mathbf{G})$. If $\boldsymbol{\mu}<\boldsymbol{\lambda}$ in our ordering, then this will be relative to the partners $|\mu b\rho j\rangle$ and $|\frac{1}{2}\eta\rangle$. We see that the one free phase in the algebra for each set $\lambda a\sigma$ could be chosen the first time partner $|\lambda a\sigma\rangle$ occurs. This will be in a $3jm$ of the form

$$\begin{bmatrix} \lambda^* & \mu & \frac{1}{2} \\ a & b & 0 \\ \sigma^* & \rho & \eta \end{bmatrix}$$

We must ensure that this phase is chosen to satisfy the complex conjugation symmetry. For $3jm$'s in the chain $\mathbf{O}\supset\mathbf{T}$ every primitive $3jm$ has such a phase freedom, in four cases the complex conjugation symmetry requires sign choices only, and in two cases the complex conjugate pairs of partners are independent, allowing real $3jm$ factors to be chosen. The reality or otherwise of the cases with a sign choice depends on $2jm$ choices; the $2jm$'s are usually chosen at this point so as many as possible primitive $3jm$'s are real. This completes the argument for $\mathbf{SO}_3\supset\mathbf{SO}_2$, $\mathbf{SO}_3\supset\mathbf{D}_\infty$, $\mathbf{SO}_3\supset\mathbf{K}$, $\mathbf{SO}_3\supset\mathbf{O}$, $\mathbf{K}\supset\mathbf{D}_3$, $\mathbf{O}\supset\mathbf{T}$, $\mathbf{O}\supset\mathbf{D}_4$, and all $\mathbf{D}_m\supset\mathbf{D}_n$.

The primitive $3jm$ factors for the remaining point group chains all involve the interplay of the ket freedom described by (3.5.2) and the coupling freedom described by (2.3.7). To this end we write the coupling (3.5.9) in full. For simplicity we consider $3jm$ symbols for a group $\mathbf{G}$ rather than the corresponding G–H $3jm$ factors, and we note that product multiplicity never occurs in primitive couplings. Thus (3.5.9) becomes

$$|(\mu\tfrac{1}{2})\lambda l\rangle = \sum_{mn} |\mu m\rangle|\tfrac{1}{2}n\rangle\langle\lambda l|\mu m,\tfrac{1}{2}n\rangle \tag{3.5.10}$$

The partners $|(\mu\frac{1}{2})\lambda l\rangle$ contain all the transformation properties of the irrep $\boldsymbol{\lambda}$ because they contain the relationship of the different partners of $\boldsymbol{\lambda}$ to each other, as seen in the fundamental transformation equation (2.1.5),

$$O_R|(\mu\tfrac{1}{2})\lambda l\rangle = \sum_{l'} |(\mu\tfrac{1}{2})\lambda l'\rangle\lambda(R)_{l'l} \tag{3.5.11}$$

The choices of phase (and branching multiplicity) discussed in the previous paragraph therefore fix the matrices $\boldsymbol{\lambda}(R)$. It is apparent from (3.5.11) that only the $|\lambda|-1$ relative phases are needed to fix the irrep matrices, but one further choice is required in (3.5.10) to fix the coupling phase. There are therefore precisely $|\lambda|$ choices for every nonprimitive irrep $\boldsymbol{\lambda}(\mathbf{G})$.

In addition to the above choices one additional phase must be chosen for each of the noncyclic$\supset$cyclic group chains. Near the end of the primitive G–C_n $3jm$ calculation the particular nature of the irreps and triads of the two groups interact in such a way that one cannot calculate the phase of one $3jm$. The example of the D_3–C_3 $3jm$ is used in Section 5.3 to illustrate that all possible values for this additional phase are abstractly equivalent. They correspond to different orientations of the D_3–C_3 system. We call this freedom in phase an "orientation freedom." For those D_3–C_3 and D_5–C_5 $3jm$'s which exhibit this orientation phase freedom we choose a phase of i. This, taken with the JM choices, and our many other phase choices, has the effect of making the y axis a twofold axis of $\mathbf{D}_n$.

The above orientation freedom appears in a modified form in the branchings $\mathbf{K}\supset\mathbf{D}_5$, $\mathbf{K}\supset\mathbf{T}$, $\mathbf{O}\supset\mathbf{D}_3$, and $\mathbf{T}\supset\mathbf{D}_2$. It occurs when, but not always when, there is a set of irreps among which the abstract group–subgroup structure does not distinguish, but which the $3jm$'s do. A typical example is that of the T–D_2 $3jm$'s. Inspection shows that the three kets $|1(T)\tilde{0}(D_2)\rangle$, $|11\rangle$, and $|1\tilde{1}\rangle$ are not distinguished by the structure of $\mathbf{T}$ or $\mathbf{D}_2$ or the imbedding. This can be seen indirectly in that the branching rules and product rules remain unchanged by permuting the labels. It can be seen directly from the character table of $\mathbf{D}_2$. In the $3jm$ primitive calculation one finds (choosing positive signs) that

$$\begin{pmatrix} 1 & \frac{1}{2} & \frac{1}{2} \\ \tilde{0} & \frac{1}{2} & \frac{1}{2} \end{pmatrix}^{T}_{D_2} = \begin{pmatrix} 1 & \frac{1}{2} & \frac{1}{2} \\ 1 & \frac{1}{2} & \frac{1}{2} \end{pmatrix} = \begin{pmatrix} 1 & \frac{1}{2} & \frac{1}{2} \\ \tilde{1} & \frac{1}{2} & \frac{1}{2} \end{pmatrix} = \frac{1}{\sqrt{3}} \tag{3.5.12}$$

However, the $3jm$ algebra demands a distinction among the three kets when one calculates the three $3jm$'s

$$\begin{pmatrix} \frac{3}{2} & 1 & \frac{1}{2} \\ \frac{1}{2} & \tilde{0} & \frac{1}{2} \end{pmatrix}, \quad \begin{pmatrix} \frac{3}{2} & 1 & \frac{1}{2} \\ \frac{1}{2} & 1 & \frac{1}{2} \end{pmatrix}, \quad \text{and} \quad \begin{pmatrix} \frac{3}{2} & 1 & \frac{1}{2} \\ \frac{1}{2} & \tilde{1} & \frac{1}{2} \end{pmatrix}$$

All three have norm $1/\sqrt{3}$, but a distinction has to be made with respect to phases. Calling their phases σ, ρ, and τ, respectively, one obtains from the orthogonality

$$\begin{pmatrix} \frac{3}{2} & 1 & \frac{1}{2} \\ \frac{1}{2} & \tilde{0} & \frac{1}{2} \end{pmatrix}\begin{pmatrix} \frac{1}{2} & 1 & \frac{1}{2} \\ \frac{1}{2} & \tilde{0} & \frac{1}{2} \end{pmatrix}^{*} + \begin{pmatrix} \frac{3}{2} & 1 & \frac{1}{2} \\ \frac{1}{2} & 1 & \frac{1}{2} \end{pmatrix}\begin{pmatrix} \frac{1}{2} & 1 & \frac{1}{2} \\ \frac{1}{2} & 1 & \frac{1}{2} \end{pmatrix}^{*}$$
$$+ \begin{pmatrix} \frac{3}{2} & 1 & \frac{1}{2} \\ \frac{1}{2} & \tilde{1} & \frac{1}{2} \end{pmatrix}\begin{pmatrix} \frac{1}{2} & 1 & \frac{1}{2} \\ \frac{1}{2} & \tilde{1} & \frac{1}{2} \end{pmatrix}^{*} = 0 \tag{3.5.13}$$

that

$$\sigma+\rho+\tau=0 \tag{3.5.14}$$

an equation which has no real solution. We choose $\sigma=1$ [using the $|\frac{3}{2}\frac{1}{2}\rangle$ ket freedom of (3.5.2)]. One deduces then that ρ, $\tau=-\frac{1}{2}\pm i\sqrt{3}/2$, where either solution may be chosen for ρ, the other solution for τ. The methods discussed in Section 5.3 show that these alternative solutions are related by a $\pi/2$ rotation about the z axis $(x, y, z)\rightarrow(y, -x, z)$, as one would expect from the character table. Interchanging $|1\tilde{0}\rangle$ and $|11\rangle$ is performed by a $\pi/2$ rotation about the y axis.

Branching multiplicities occur for the $3jm$ factors for $\mathbf{SO_3}$ containing either the octahedral or icosahedral groups, and for $\mathbf{K}\supset\mathbf{D_3}$. Branching multiplicities are separated by the same techniques as used for product multiplicities. We discuss the $\mathbf{SO_3}\supset\mathbf{K}$ and $\mathbf{SO_3}\supset\mathbf{O}$ cases here.

Only two cases of branching multiplicity occur in the SO_3–K $3jm$ tables, the pairs $(r=0, 1)$

$$|\tfrac{15}{2}r\tfrac{5}{2}i\rangle \qquad \text{and} \qquad |8r2i\rangle$$

near the end of the table. The branching multiplicity is separated here and elsewhere in a similar fashion to the product multiplicity. We ask that in the couplings [omitting labels below the icosahedral group, that is, writing the kets as $|J(SO_3)a\lambda(K)\rangle$]

$$|702\rangle|\tfrac{1}{2}0\tfrac{1}{2}\rangle\rightarrow a_0|\tfrac{15}{2}0\tfrac{5}{2}\rangle+a_1|\tfrac{15}{2}1\tfrac{5}{2}\rangle$$

$$|\tfrac{15}{2}0\tfrac{3}{2}\rangle|\tfrac{1}{2}0\tfrac{1}{2}\rangle\rightarrow b_0|802\rangle+b_1|812\rangle$$

the coefficients a_0 and b_0 equal zero. That is, we set

$$\begin{pmatrix} \frac{15}{2} & 7 & \frac{1}{2} \\ 0 & 0 & 0 \\ \frac{5}{2} & 2 & \frac{1}{2} \end{pmatrix}\begin{matrix} SO_3 \\ \\ K \end{matrix} = \begin{pmatrix} 8 & \frac{15}{2} & \frac{1}{2} \\ 0 & 0 & 0 \\ 2 & \frac{3}{2} & \frac{1}{2} \end{pmatrix}\begin{matrix} SO_3 \\ \\ K \end{matrix} =0$$

It was less easy to decide upon the multiplicity equations for $\mathbf{SO_3}\supset\mathbf{O}$. A large number of possibilities were tried; some combinations of choices of vanishing couplings led to many other vanishings and simple values of nonzero $3jm$ symbols, other combinations led to few other zeros and awkward-looking nonzero values. Once again "simple" means that no large primes occur in the values of the primitive $3jm$ factors. For $\mathbf{SO}_3 \supset \mathbf{O}$ the choice first used led to only one primitive $3jm$ factor containing a prime greater than $2J+1$, and that right near the end of our table. This prime was later removed after the full table was calculated. With the full table it was apparent that by choosing separations so that a particular nonprimitive $3jm$ was zero the prime disappeared from the entire table. This was achieved by explicitly computing the matrix $\mathbf{A}$ of (3.5.2) and using it to transform the $3jm$ primitive pair in such a way that when the table was recomputed our chosen nonprimitive $3jm$ was zero. All other available choices led to the much larger primes occurring, and also to fewer zeros.

4

The Wigner–Eckart Theorem

Thus far we have focused on the transformation properties of ket vectors. We have used group theory, often just Schur's lemmas, to derive the properties of the various coupling coefficients and then to calculate their numerical values. However, quantum mechanics is not the theory of wave functions, it is the theory of measurement, the theory of interactions. Interactions are expressed mathematically as operators on a Hilbert space of wave functions, but it is the eigenvalues and matrix elements of the operators that are compared with experiment, not the functional form of the wave functions.

In this chapter we show that an operator of which we know the transformation properties has matrix elements which are given up to normalization by the $3jm$ symbol of the transforming group. The normalization is known as the *reduced matrix element*.

The partners of the irrep spaces of Chapter 2 were chosen to be orthonormal basis functions for the space. Section 4.1 discusses possible basis choices for operators and other linear algebra ideas. Section 4.2 picks out basis operators with particular transformation properties and shows how any tensor operator may be written in this basis, leading immediately to the Wigner–Eckart theorem for any group.

Many of the operators describing the physical interactions within atoms, molecules, and crystals are composed of combinations of operators with various transformation properties. Section 4.3 gives various formulas for such combinations.

4.1 Basis Operators

We begin with the basic linear algebra. A vector space **V** of dimension n with basis $\{|1\rangle, |2\rangle,, \ldots, |n\rangle\}$ may be mapped to itself by linear operators L: $\mathbf{V} \to \mathbf{V}$

$$L|k\rangle = \sum_i |i\rangle\langle i|L|k\rangle \tag{4.1.1}$$

The number of linearly independent operators L is n^2. A suitable basis for these operators is $L_{ij} = |i\rangle\langle j|$. For example, the operator A with matrix elements $A_{ij} = \langle i|A|j\rangle$ may be written in any one of the forms

$$A = \sum_{ij} L_{ij} A_{ij} = \sum_{ij} |i\rangle\langle j| A_{ij}$$

$$= \sum_{ij} |i\rangle A_{ij} \langle j| = \sum_{ij} |i\rangle\langle i|A|j\rangle\langle j| \tag{4.1.2a}$$

so that

$$A|k\rangle = \sum_{ij} |i\rangle\langle j|A_{ij}|k\rangle = \sum_{ij} |i\rangle A_{ij}\langle j|k\rangle$$

$$= \sum_{ij} |i\rangle A_{ij}\delta_{jk} = \sum_i |i\rangle A_{ik} \tag{4.1.2b}$$

In this notation (Dirac's) with a two-dimensional example, our basis kets are $|1\rangle$ and $|2\rangle$ and our 2^2 operators may be written as $|1\rangle\langle 1|$, $|1\rangle\langle 2|$, $|2\rangle\langle 1|$, and $|2\rangle\langle 2|$.

Presume that **V** is the space of our problem, and under a group **G** it may be written as the sum of irrep spaces, with partners $|x\lambda i\rangle$. The equation

$$O_R L|x\lambda i\rangle = O_R L(O_{R^{-1}} O_R)|x\lambda i\rangle = (O_R L O_{R^{-1}}) O_R |x\lambda i\rangle \tag{4.1.3}$$

says that since the transform of $|x\lambda i\rangle$ by group operation R is $O_R|x\lambda i\rangle$, then the transform of L by the group operation is $O_R L O_{R^{-1}}$. This should be familiar.

It follows that the transform of the basis operators, now written as $|x_1\lambda_1 l_1\rangle\langle x_2\lambda_2 l_2|$, is

$$O_R|x_1\lambda_1 l_1\rangle\langle x_2\lambda_2 l_2|O_{R^{-1}}=(O_R|x_1\lambda_1 l_1\rangle)(\langle x_2\lambda_2 l_2|O_{R^{-1}})$$

which is, using the unitarity of O_R, namely $O_{R^{-1}}=O_R^{\dagger}$,

$$=(O_R|x_1\lambda_1 l_1\rangle)(O_R|x_2\lambda_2 l_2\rangle)^{\dagger}$$

$$=\sum_{l_1'l_2'}|x_1\lambda_1 l_1'\rangle\lambda_1(R)_{l_1'l_1}(|x_2\lambda_2 l_2'\rangle\lambda_2(R)_{l_2'l_2})^{\dagger}$$

$$=\sum_{l_1'l_2'}|x_1\lambda_1 l_1'\rangle\langle x_2\lambda_2 l_2'|\lambda_1(R)_{l_1'l_1}\lambda_2(R)^{*}_{l_2'l_2} \tag{4.1.4}$$

The above choice of basis operators is unsatisfactory for our purposes, as the operators can be seen not to transform irreducibly, but rather as the product $\boldsymbol{\lambda}_1\times\boldsymbol{\lambda}_2^*$. This is no obstacle, however, as the orthogonal linear combinations

$$U_l^{r\lambda}(x_1\lambda_1x_2\lambda_2)=\sum_{l_1l_2}|x_1\lambda_1 l_1\rangle\langle x_2\lambda_2 l_2|\begin{pmatrix}\lambda_1\\ l_1\end{pmatrix}\begin{pmatrix}\lambda_1^* & \lambda & \lambda_2\\ l_1^* & l & l_2\end{pmatrix}^r \tag{4.1.5}$$

transform irreducibly as the irrep $\boldsymbol{\lambda}$, indeed exactly as the ket $|\lambda l\rangle$. The properties of the $3jm$ symbols may be used to show that

$$O_RU_l^{r\lambda}(x_1\lambda_1x_2\lambda_2)O_{R^{-1}}=\sum_{l'}U_{l'}^{r\lambda}(x_1\lambda_1x_2\lambda_2)\lambda(R)_{l'l} \tag{4.1.6}$$

Schur's lemmas may be invoked to show that the only sets of operators which transform like the irrep $\mathbf{V}_\lambda$ with basis $\{|\lambda l\rangle\}$ are linear combinations of these operators, where the combination coefficients do not depend on l and are diagonal in $\boldsymbol{\lambda}$. If T_l^λ is such that

$$O_RT_l^\lambda O_{R^{-1}}=\sum_{l'}T_{l'}^\lambda\lambda(R)_{l'l} \tag{4.1.7}$$

then

$$T_l^{\lambda} = \sum_{x_1\lambda_1 x_2\lambda_2 r} U_l^{r\lambda}(x_1\lambda_1, x_2\lambda_2) a_{x_1\lambda_1 x_2\lambda_2 r\lambda} \tag{4.1.8}$$

where the coefficients a may depend on all labels shown but are independent of l.

4.2 The Wigner–Eckart Theorem

A *tensor operator* T_l^{λ} is a member of an irreducible tensorial set $\mathbf{T}^{\lambda} = \{T_1^{\lambda}, T_2^{\lambda}, \ldots, T_{|\lambda|}^{\lambda}\}$ transforming together as the partners of an operator representation $\boldsymbol{\lambda}$,

$$O_R T_l^{\lambda} O_{R^{-1}} = \sum_{l'} T_{l'}^{\lambda} \lambda(R)_{l'l} \tag{4.2.1}$$

Equation (4.1.8) shows that T_l^{λ} may be written in terms of the basis operators $U_l^{r\lambda}(x_1\lambda_1 x_2\lambda_2)$ with coefficients a which we now write as $\langle x_1\lambda_1 \| T^{\lambda} \| x_2\lambda_2 \rangle_r$ and which are independent of l,

$$T_l^{\lambda} = \sum_{x_1\lambda_1 x_2\lambda_2 r} U_l^{r\lambda}(x_1\lambda_1 x_2\lambda_2) \langle x_1\lambda_1 \| T^{\lambda} \| x_2\lambda_2 \rangle_r \tag{4.2.2}$$

The basis operators have simple matrix elements, namely a product of a $2jm$ and a $3jm$ and many δ's. The matrix elements of T_l^{λ} follow by substitution of (4.2.2) and (4.1.5) for T_l^{λ},

$$\langle x_1\lambda_1 l_1 | T_l^{\lambda} | x_2\lambda_2 l_2 \rangle = \sum_r \begin{pmatrix} \lambda_1 \\ l_1 \end{pmatrix} \begin{pmatrix} \lambda_1^* & \lambda & \lambda_2 \\ l_1^* & l & l_2 \end{pmatrix}^{r} \langle x_1\lambda_1 \| T^{\lambda} \| x_2\lambda_2 \rangle_r \tag{4.2.3}$$

This is the *Wigner–Eckart theorem.*

The numbers $\langle x_1\lambda_1 \| T^{\lambda} \| x_2\lambda_2 \rangle_r$ are known as *reduced matrix elements*, for they contain the properties of the operators T_l^{λ} reduced by the extraction of their transformation properties. Our definition of the basis operators and reduced matrix elements has been chosen to agree with Judd (1963) and Nielson and Koster (1963) for the $\mathbf{SO}_3$ case, and with the general definition of Messiah (1965, p. 1094). Several authors have other normalizations or phases, introduced in the definition by using a Condon and Shortley coupling coefficient or by reordering the columns of the $3jm$. Observe that the unit operator 1, defined as $1|x\lambda l\rangle = |x\lambda l\rangle$, transforms as U_0^0 and has a reduced matrix element $\langle x\lambda \| 1 \| x\lambda\rangle = |\lambda|^{1/2}$ because the product of the $2jm$ and the trivial $3jm$ gives this dimension factor.

If the states and operators are partners of various groups in a chain, for example, $\mathbf{G}\supset\mathbf{H}$, then the Wigner–Eckart theorem may be applied to either group. The factorization of the $3jm$ symbol obtained from application of the theorem to $\mathbf{G}$ shows that the H reduced matrix elements are related to the G reduced matrix elements by

$$\langle x_1\lambda_1 a_1\sigma_1 \| T^{\lambda a\sigma} \| x_2\lambda_2 a_2\sigma_2\rangle_s^H$$

$$= \sum_r \begin{Bmatrix} \lambda_1 \\ a_1 \\ \sigma_1 \end{Bmatrix}_H^G \begin{pmatrix} \lambda_1^* & \lambda & \lambda_2 \\ a_1 & a & a_2 \\ \sigma_1^* & \sigma & \sigma_2 \end{pmatrix}_{sH}^{rG} \langle x_1\lambda_1 \| T^\lambda \| x_2\lambda_2\rangle_r^G \tag{4.2.4}$$

This is a very important result, for it relates, for example, octahedral reduced matrix elements to SO_3 reduced matrix elements. Racah's (1949) interest in the result that we call the Racah factorization lemma was in the parentage groups of Chapter 7. He related the various $SU_2^S\times SO_3^L$ reduced matrix elements in the SL scheme for various numbers of f electrons to a few U_{14} reduced matrix elements with an easily computed electron number dependence. Judd (1967) showed that this was not the end of the useful group chain; one could go to $\mathbf{U}_{2^{14}}$, which contains all partially filled f shells in a single irrep. He thus related all matrix elements of all many-electron f-shell operators to a single $U_{2^{14}}$ reduced matrix element.

Although the major part of the discussion of applications is left until Chapter 9, it is worthwhile digressing to give a few examples of the transformation properties of operators. Consider the Hamiltonian of an ion in a cubic crystal field, which has a small $\mathbf{D}_3$ lattice distortion, and which has a small applied magnetic field not aligned with a crystal axis. There is no exact symmetry (except the trivial $\mathbf{C}_1$), but to lesser orders of approximation it is likely that the eigenfunctions transforming as partners of irreps of $\mathbf{C}_1$, $\mathbf{C}_3$, $\mathbf{D}_3$, $\mathbf{O}$, and $\mathbf{SO}_3$ groups, labeled $|Ja\lambda_1(O)\lambda_2(D_3)$ $\lambda_3(C_3)\lambda_4(C_1)\rangle$, will be degenerate. The terms contributing to the Hamiltonian in this basis may be deduced from the following symmetry considerations.

The cubic crystal field is invariant under octahedral transformations but may transform as various values of **J** under $\mathbf{SO_3}$. Inspection of the branching rules for $\mathbf{SO_3} \supset \mathbf{O}$ says that the first three irreps of $\mathbf{SO_3}$ containing octahedral scalars are **J** = **0**, **4**, and **6**, so the principal crystal field Hamiltonian is

$$H_{\mathrm{CF}} = a_0 U^{0(SO_3)} + a_4 U^{4(SO_3)0(O)} + a_6 U^{6(SO_3)0(O)} + \cdots \tag{4.2.5}$$

where the coefficients cannot be determined from symmetry considerations.

The dihedral distortion is by definition invariant under the dihedral operations. Only the irreps **0(O)** and $\tilde{\mathbf{1}}$**(O)** contain $\mathbf{D_3}$ scalars. The **0(O)** part of the $\mathbf{D_3}$ distortion will be indistinguishable from the main crystal field. The $\tilde{\mathbf{1}}$**(O)** must transform as **J** = **4**, **6**, **8**, or higher, giving the distortion Hamiltonian as

$$H_{\mathrm{dist}} = b_4 U^{4(SO_3)\tilde{1}(O)0(T)} + b_6 U^{6(SO_3)\tilde{1}(O)0(C_3)} + \cdots \tag{4.2.6}$$

The external magnetic field Zeeman term must be approached from the other end. It is proportional to

$$J_z = \sum_j \langle j \| J \| j \rangle U_z(j, j) \tag{4.2.7}$$

where

$$U_z(j, j) = U^{1(SO_3)0(SO_2)} \tag{4.2.8}$$

but we need not assume any relationship of the magnetic z axis to the crystal axis. Nevertheless the Zeeman term transforms as $\mathbf{1(SO_3)}$. Branching rule information implies that it transforms as **1(O)** and as a linear combination of $\mathbf{1(D_3)0(C_3)}$, $\mathbf{1(D_3)1(C_3)}$, and $\mathbf{1(D_3)-1(C_3)}$. The combination depends on the relationship of the z axis to the crystal axes,

$$\begin{aligned} U_z = U^{1(SO_3)0(SO_2)} = {} & c_0 U^{1(SO_3)1(O)\tilde{0}(D_3)0(C_3)} \\ & + c_1 U^{1(SO_3)1(O)1(D_3)1(C_3)} \\ & + c_{-1} U^{1(SO_3)1(O)1(D_3)-1(C_3)} \end{aligned} \tag{4.2.9}$$

For certain orientations of the z axis it is trivial to derive the relationship of the U_z operator to the three operators, call them U_0, U_1, and U_{-1}. The important angle is the angle θ between the z axis (the $\mathbf{SO}_2$ axis) and the $\mathbf{C}_3$ axis. If $\theta=0$, then clearly $c_1=c_{-1}=0$, $c_0=1$, for U_z is scalar under $\mathbf{C}_3$ operations. If $\theta=90°$, then U_z will be a linear combination of U_1 and U_{-1} only. We postpone deriving the actual linear combination until we have considered basis choices in detail in the following chapter.

The Wigner–Eckart theorem makes the evaluation of matrix elements absolutely trivial. For the last operator in the above equation, we have

$$\langle x_1 J_1 a_1 \lambda_1(0)\rho_1(D_3)\sigma_1(C_3)|U^{1(SO_3)1(O)1(D_3)-1(C_3)}|x_2 J_2 a_2 \lambda_2(O)\rho_2(D_3)\sigma_2(C_3)\rangle$$

$$=\langle x_1 J_1 \| U^1 \| x_2 J_2\rangle$$

$$\times \sum_r \begin{bmatrix} J_1 & 1 & J_2 \\ a & 0 & a_2 \\ \lambda_1 & 1 & \lambda_2 \end{bmatrix}^{SO_3}_{rO} \begin{pmatrix} \lambda_1 & 1 & \lambda_2 \\ \rho_1^* & 1 & \rho_2 \end{pmatrix}^{rO}_{D_3} \begin{pmatrix} \rho_1 \\ \sigma_1 \end{pmatrix}^{D_3}_{C_3} \begin{pmatrix} \rho_1^* & 1 & \rho_2 \\ \sigma_1^* & -1 & \sigma_2 \end{pmatrix}^{D_3}_{C_3} \tag{4.2.10}$$

Use has been made of the factorization properties of the $3jm$ symbol: it is a product of $3jm$ factors for the groups in the chain. The $2jm$'s for $\mathbf{SO}_3 \supset \mathbf{O}$ and for $\mathbf{O} \supset \mathbf{D}_3$ have been omitted because they are all unity.

Symmetry considerations do much to fix the form of perturbation Hamiltonians; such considerations cannot fix absolute magnitudes, but we have seen that transformation properties may reduce the free parameters to a very small number. Once the Hamiltonian is in tensor operator form, and the above considerations have fixed this form for our example, the Wigner–Eckart theorem and the tables of this book give the matrix elements directly.

4.3 Coupled Tensors

Many interactions of interest are most simply described as a coupled product of operators; for example, the spin–orbit interaction $\mathbf{s}\cdot\mathbf{l}$ is the scalar coupled product of the two $J=1$ operators $\mathbf{s}$ and $\mathbf{l}$.

The matrix elements of a coupled product

$$\{P^{\kappa_1}Q^{\kappa_2}\}^{r\kappa}_k = \sum_{k_1 k_2} P^{\kappa_1}_{k_1} Q^{\kappa_2}_{k_2} \langle \kappa_1 k_1, \kappa_2 k_2 | r\kappa k\rangle \tag{4.3.1}$$

may be evaluated using the Wigner–Eckart theorem, as the coupled product has well-defined transformation properties,

$$\langle x_1\lambda_1 l_1|\{P^{\kappa_1}Q^{\kappa_2}\}^{r\kappa}_k|x_2\lambda_2 l_2\rangle=\sum_s\begin{pmatrix}\lambda_1\\ l_1\end{pmatrix}\begin{pmatrix}\lambda_1^* & \kappa & \lambda_2\\ l_1^* & k & l_2\end{pmatrix}^s$$

$$\times\langle x_1\lambda_1\|\{P^{\kappa_1}Q^{\kappa_2}\}^{r\kappa}\|x_2\lambda_2\rangle_s \quad (4.3.2)$$

Note the presence of multiplicity indices (r and s) in the reduced matrix element of a coupled tensor, r from the coupling of the tensors and s from the Wigner–Eckart theorem.

A word of caution with respect to phases is needed here. We follow Judd and many, but not all, other authors in defining coupled tensors with coupling coefficients and reduced matrix elements with $3jm$ symbols. For $\mathbf{SO}_3$ there is the historical phase difference $H(j_1j_2j_3)=(-1)^{j_1-j_2+j_3}$ relating the two and it will appear in the following results.

The reduced matrix element of a coupled product of tensors may be related to the matrix elements of each tensor by insertion of a sum over the unit operator $\Sigma_{x_3\lambda_3 l_3}|x_3\lambda_3 l_3\rangle\langle x_3\lambda_3 l_3|$. We have from (4.3.1) that

$$\langle x_1\lambda_1 l_1|\{P^{\kappa_1}_{k_1}Q^{\kappa_2}_{k_2}\}^{r\kappa}_k|x_2\lambda_2 l_2\rangle$$

$$=\sum_{k_1k_2}\langle x_1\lambda_1 l_1|P^{\kappa_1}_{k_1}Q^{\kappa_2}_{k_2}|x_2\lambda_2 l_2\rangle\langle\lambda_1 k_1,\kappa_2 k_2|r\kappa k\rangle$$

$$=\sum_{k_1k_2x_3\lambda_3l_3}\langle x_1\lambda_1 l_1|P^{\kappa_1}_{k_1}|x_3\lambda_3 l_3\rangle\langle x_3\lambda_3 l_3|Q^{\kappa_2}_{k_2}|x_2\lambda_2 l_2\rangle$$

$$\times\langle\kappa_1 k_1,\kappa_2 k_2|r\kappa k\rangle \quad (4.3.3)$$

The Wigner–Eckart theorem may now be applied to both sides, the coupling coefficient written as a $3jm$ and other factors, and the orthogonality of the $3jm$ symbol used to move the $3jm$ on the left to the right. After some tedious collection of phases the result becomes completely basis independent:

$$\langle x_1\lambda_1\|\{P^{\kappa_1}Q^{\kappa_2}\}^{r\kappa}\|x_2\lambda_2\rangle_s$$

$$=\sum_{x_3\lambda_3s_1s_2}H(\kappa_1\kappa_2\kappa)|\kappa|^{1/2}\{\lambda_1\}\{\lambda_1^*\kappa_1\lambda_3s_1\}$$

$$\times\{\lambda_3^*\kappa_2\lambda_2s_2\}\begin{Bmatrix}\kappa_2 & \kappa^* & \kappa_1\\ \lambda_1 & \lambda_3 & \lambda_2\end{Bmatrix}_{s_2ss_1r}\langle x_1\lambda_1\|P^{\kappa_1}\|x_3\lambda_3\rangle_{s_1}$$

$$\times\langle x_3\lambda_3\|Q^{\kappa_2}\|x_2\lambda_2\rangle_{s_2} \quad (4.3.4a)$$

In the special case of the coupled tensor being scalar the $6j$ becomes a $3j$ and one has

$$\langle x_1\lambda_1\|\{P^{\kappa}Q^{\kappa^*}\}^0\|x_2\lambda_2\rangle$$
$$=\sum_{x_3\lambda_3 s}\delta_{\lambda_1\lambda_2}|\lambda_1|^{-1/2}|\kappa|^{-1/2}\{\lambda_1\}\{\lambda_1^*\kappa\lambda_3 s\}$$
$$\times\langle x_1\lambda_1\|P^{\kappa}\|x_3\lambda_3\rangle_s\langle x_3\lambda_3\|Q^{\kappa^*}\|x_2\lambda_2\rangle_s \tag{4.3.4b}$$

Another special case of this equation is of great importance. If the operators P^{κ_1} and Q^{κ_2} act on parts 1 and 2 of a system, then the labels x_1x_2 will usually contain more group-theoretic information and the reduced matrix element may be further simplified. For example, let P^{κ_1} act only the orbital part and Q^{κ_2} act on the spin part of the ket, or alternatively let P^{κ_1} act on the first particle and Q^{κ_2} act on the second, where the particles could be two electrons in one atom, or two atoms in a molecule, etc. The matrix element can now be written

$$\langle(\lambda_1\lambda_2)r_1\lambda l|\{P^{\kappa_1}Q^{\kappa_2}\}^{r\kappa}_k|(\mu_1\mu_2)r_2\mu m\rangle \tag{4.3.5}$$

The uncoupled kets $|\lambda_1 l_1\rangle|\lambda_2 l_2\rangle$ will undoubtably require additional labels x for their complete specification. For the sake of brevity we omit them. The assumption that P and Q act on separate parts of the system is expressed mathematically by the equation

$$P^{\kappa_1}_{k_1}Q^{\kappa_2}_{k_2}|\mu_1 m_1,\mu_2 m_2\rangle=(P^{\kappa_1}_{k_1}|\mu_1 m_1\rangle)(Q^{\kappa_2}_{k_2}|\mu_2 m_2\rangle) \tag{4.3.6}$$

On comparing the result of using the Wigner–Eckart theorem on (4.3.5) as it stands with the result of uncoupling both kets and operators and applying (4.3.6) first, we obtain the basis-independent result

$$\langle(\lambda_1\lambda_2)r_1\lambda\|\{P^{\kappa_1}Q^{\kappa_2}\}^{t\kappa}\|(\mu_1\mu_2)r_2\mu\rangle_s$$
$$=\sum_{s_1s_2}H(\lambda_1\lambda_2\lambda)H(\kappa_1\kappa_2\kappa)H(\mu_1\mu_2\mu)$$
$$\times|\lambda|^{1/2}|\mu|^{1/2}|\kappa|^{1/2}\left\{\begin{matrix}\lambda_1 & \lambda_2 & \lambda^* \\ \kappa_1^* & \kappa_2^* & \kappa \\ \mu_1^* & \mu_2^* & \mu \\ s_1 & s_2 & s\end{matrix}\right\}\begin{matrix}r_1\\ t\\ r_2\\ \end{matrix}$$
$$\times\langle\lambda_1\|P^{\kappa_1}\|\mu_1\rangle_{s_1}\langle\lambda_2\|Q^{\kappa_2}\|\mu_2\rangle_{s_2} \tag{4.3.7}$$

Some particular cases are to be noted. In an SL coupled ion a perturbation term may involve an operator $P_{k_1}^{\kappa_1}$ acting only on the spin part of the kets, in which case $Q_{k_1}^{\kappa_2}$ is the unit operator ($\kappa_2=0$) in orbital space, leading to

$$\langle(\lambda_1\lambda_2)r_1\lambda\|P^{\kappa_1}\|(\mu_1\mu_2)r_2\mu\rangle_s$$

$$=\sum_{s_1}\delta_{\lambda_2\mu_2}H(\lambda_1\lambda_2\lambda)H(\mu_1\mu_2\mu)|\lambda|^{1/2}|\mu|^{1/2}$$

$$\times\{\lambda_1\}\{\lambda_1\lambda_2\lambda^*r_1\}\{\lambda_1^*\kappa_1\mu_1s_1\}\begin{Bmatrix}\lambda^* & \kappa_1 & \mu\\ \mu_1 & \lambda_2^* & \lambda_1\end{Bmatrix}_{r_1s_1r_2s}\langle\lambda_1\|p^{\kappa_1}\|\mu_1\rangle_{s_1} \tag{4.3.8}$$

or, on setting $\kappa_1=0$, to

$$\langle(\lambda_1\lambda_2)r_1\lambda\|Q^{\kappa_2}\|(\mu_1\mu_2)r_2\mu\rangle_s$$

$$=\sum_{s_2}\delta_{\lambda_1\mu_1}H(\lambda_1\lambda_2\lambda)H(\mu_1\mu_2\mu)|\lambda|^{1/2}|\mu|^{1/2}$$

$$\times\{\mu_2\}\{\lambda_2^*\kappa_2\mu_2s_2\}\{\mu_1\mu_2\mu^*r_2\}\begin{Bmatrix}\lambda^* & \kappa_2 & \mu\\ \mu_2 & \lambda_1^* & \lambda_2\end{Bmatrix}_{r_1s_2r_2s}\langle\lambda_2\|Q^{\kappa_2}\|\mu_2\rangle_{s_2} \tag{4.3.9}$$

(Recall that the reduced matrix element of the unit operator is a dimension factor.)

The interelectron Coulomb interaction is discussed in Section 9.1. It is usually written as the scalar product of two tensorial sets. The scalar product is defined in the SO_3–SO_2 basis as

$$(P^J\cdot Q^J)=\sum_M(-1)^M P_M^J Q_{-M}^J \tag{4.3.10}$$

which may be written as

$$(P^J\cdot Q^J)=(-1)^J|J|^{1/2}\{P^JQ^J\}_0^0 \tag{4.3.11}$$

This second form is basis independent and so is suitable for all point group basis schemes. The scalar product is an SO_3 scalar product. It involves a natural normalization $|J|^{1/2}$ and a natural phase $(-1)^J$. While these are customary for $\mathbf{SO}_3$, there is no point in endeavoring to choose alternative factors for other G scalar products. We use the coupled scalar $\{P^\kappa Q^{\kappa*}\}_0^0$.

The coupled G scalar product is evaluated for any group using (4.3.7). The full matrix element is basis independent because the $3jm$ symbol and the $2jm$ symbol in the Wigner–Eckart theorem cancel,

$$\langle(\lambda_1\lambda_2)r_1\lambda i|\{P^\kappa Q^{\kappa*}\}_0^0|(\mu_1\mu_2)r_2\mu j\rangle$$

$$=|\lambda|^{-1/2}\langle(\lambda_1\lambda_2)r_1\lambda\|\{P^\kappa Q^{\kappa*}\}^0\|(\mu_1\mu_2)r_2\mu\rangle\delta_{ij}$$

$$=\sum_{s_1s_2}\delta_{ij}\delta_{\lambda\mu}H(\lambda_1\lambda_2\lambda)H(\mu_1\mu_2\mu)|\kappa|^{-1/2}$$

$$\times\{\lambda_2\}\{\lambda_2^*\kappa^*\mu_2s_2\}\{\lambda_1\lambda_2\lambda^*r_1\}\begin{Bmatrix}\mu_1 & \mu_2 & \lambda^*\\ \lambda_2^* & \lambda_1 & \kappa\end{Bmatrix}_{s_1s_2r_1r_2}$$

$$\times\langle\lambda_1\|P^\kappa\|\mu_1\rangle_{s_1}\langle\lambda_2\|Q^{\kappa*}\|\mu_2\rangle_{s_2} \tag{4.3.12}$$

Restricting this result to the $\mathbf{SO}_3$ case and returning to the dot product, one finds that

$$\langle(j_1j_2)JM|(P^k\cdot Q^k)|(j_1'j_2')J'M'\rangle$$

$$=\delta_{JJ'}\delta_{MM'}(-1)^{j_1'+j_2+J}\begin{Bmatrix}j_1' & j_2' & J\\ j_2 & j_1 & k\end{Bmatrix}_{SO_3}\langle j_1\|P^k\|j_1'\rangle\langle j_2\|Q^k\|j_2'\rangle \tag{4.3.13}$$

The $9j$ symbol and the reduced matrix elements appearing in (4.3.7) have an interesting reinterpretation, an interpretation relating to the Kronecker product as an inner product (p. 24). The ket $|(\mu_1\mu_2)r\mu m\rangle$ is a partner m of an irrep $\boldsymbol{\mu}(\mathbf{G})$, where $\mathbf{G}$ is a group which acts on (rotates) the two parts of the system as a single unit. In the SLJ case, $\mathbf{G}$ would be $\mathbf{SO}_3^J$, which rotates the spin and orbital functions together. The ket is also a partner $r\mu m$ of an irrep $\boldsymbol{\mu}_1\boldsymbol{\mu}_2(\mathbf{G}_1\cdot\mathbf{G}_2)=\boldsymbol{\mu}_1(\mathbf{G}_1)\times\boldsymbol{\mu}_2(\mathbf{G}_2)$ of the direct product group $\mathbf{G}_1\cdot\mathbf{G}_2$, which acts on the two parts separately. In the SLJ case, $\mathbf{G}_1\cdot\mathbf{G}_2$ is $\mathbf{SU}_2^S\times\mathbf{SO}_3^L$, which rotates the spin functions and the orbital functions as independent functions. $\mathbf{SO}_3^J$ is a subgroup of $\mathbf{SU}_2^S \times \mathbf{SO}_3^L$. It does nothing but rotate the spin functions and orbital functions as a whole, by the same angles. The ket $|(\mu_1\mu_2)r\mu m\rangle$ may be viewed as the $r\mu m$ partner of irrep $\boldsymbol{\mu}_1\boldsymbol{\mu}_2$ of the covering group $\mathbf{G}_1 \cdot \mathbf{G}_2$, rather than the coupled product in $\boldsymbol{\mu}_1(\mathbf{G}) \times \boldsymbol{\mu}_2(\mathbf{G})$. The Wigner–Eckart theorem says the G reduced matrix element on the left of (4.3.7) equals a $G_1 \cdot G_2$ reduced matrix element and a $G_1 \cdot G_2$–G $3jm$:

$$\langle(\lambda_1\lambda_2)r_1\lambda \| \{P^{\kappa_1}Q^{\kappa_2}\}^{t\kappa} \| (\mu_1\mu_2)r_2\mu\rangle_s^G$$

$$= \sum_{s_1 s_2} \begin{Bmatrix} \lambda_1\lambda_2 \\ r_1 \\ \lambda \end{Bmatrix}_G^{G_1G_2} \begin{pmatrix} \lambda_1^*\lambda_2^* & \kappa_1\kappa_2 & \mu_1\mu_2 \\ r_1^* & t & r_2 \\ \lambda^* & \kappa & \mu \end{pmatrix}_{sG}^{s_1s_2G_1\cdot G_2}$$

$$\times \langle\lambda_1\lambda_2 \| P^{\kappa_1}Q^{\kappa_2} \| \mu_1\mu_2\rangle_{s_1s_2}^{G_1\cdot G_2} \tag{4.3.14}$$

The reduced matrix element for a direct product group is the product of reduced matrix elements for each group, which is why the double multiplicity label s_1s_2 is used. Comparison of (4.3.7) and (4.3.14) shows that a $G_1 \cdot G_2$–G $3jm$ factor equals (to within dimensions, a $2jm$, and historical phases) a G $9j$ symbol. This equivalence is one of the several steps used by Kramer and Seligman (1969) to derive, from a group theory viewpoint, the Regge symmetry of the SO_3–SO_2 $3jm$. Another case of some interest is that of $\mathbf{SO}_4$, which is important as a symmetry of the hydrogenic Hamiltonian. $\mathbf{SO}_4$ is a freakish group, being locally $\mathbf{SO}_3 \times \mathbf{SO}_3$. Biedenharn (1961) showed that the SO_4–SO_3 $3jm$ factor was an SO_3 $9j$.

4.4 The Standard Racah Tensors of $\mathbf{SO}_3$

Racah (1942) defined tensor operator sets $\mathbf{u}^{(k)}$ to have members $u_q^{(k)}$ which transform together under $\mathbf{SO}_3$ rotations in the same way as the partners $|JM\rangle$ of a vector irrep $\mathbf{J}(\mathbf{SO}_3)$, and which have reduced matrix elements of unity between single-electron states of the same fixed orbital:

$$\mathbf{u}^k = \{u_q^k: \quad -k \leqslant q \leqslant k\} \tag{4.4.1}$$

where

$$\langle nl \| u^k \| n'l'\rangle = \delta_{nn'}\delta_{ll'} \tag{4.4.2}$$

It is usual to distinguish between operators written in lower case, which act on a specific electron i, and those which act on any electron

$$U_q^k = \sum_{i=1}^{N} u_q^k(i) \tag{4.4.3}$$

Judd (1963) used operators v_q^k renormalized by multiplication by a factor $(2k+1)^{1/2}$. His normalization has the advantage that it leads to a set of ligand field parameters whose magnitudes give a reasonably direct measure of the effect they cause in a perturbation calculation. We prefer to include the dimension factor explicitly; see Section 9.3. The above *Racah unit tensors* give the angular integrals appropriate for one-electron interactions within a specified configuration. These purely orbital operations generalize to the *double tensor operators* $u_{\pi q}^{\kappa k}$ acting on both the spin and orbital spaces. These operators transform as $\boldsymbol{\kappa}(\mathbf{SU}_2^S)$ in the spin space and $\mathbf{k}(\mathbf{SO}_3^L)$ in the orbital space. The spin–orbit operator $(\mathbf{s}\cdot\mathbf{l})$ is scalar under $\mathbf{SO}_3^J$ and transforms as $\mathbf{1}(\mathbf{SU}_2^S)\times\mathbf{1}(\mathbf{SO}_3^L)$.

Elliott (1958), Feneuille (1967a, b), and Butler and Wybourne (1970a, b) generalized these tensors to operators which act between electrons belonging to different configurations. Their operators $w_{\pi q}^{\kappa k}(nl, n'l')$ have reduced matrix elements which satisfy

$$\langle n_a l_a \| w^{\kappa k}(nl, n'l') \| n_b l_b \rangle = |\kappa|^{1/2} |k|^{1/2} \delta_{n_a n} \delta_{l_a l} \delta_{n_b n'} \delta_{l l'} \tag{4.4.4}$$

Such operators allow the orbital and spin parts of any mixed configuration interaction to be written in terms of operators whose algebraic properties may be exploited by the methods of this book. The matrix elements of such tensors are evaluated for the case of many electrons in a multiconfiguration system of m spatial orbitals by applying the Wigner–Eckart theorem to the continuous groups $\mathbf{SO}_{4m+1} \supset \mathbf{U}_{2m} \supset \mathbf{SP}_{2m} \supset \mathbf{SU}_2^S \times (\mathbf{SO}_m^L \supset \mathbf{SU}_2^S \times \mathbf{SO}_1^L)$ (see also Judd 1967).

These tensors are all various special cases of the general tensor operators defined by (4.1.5). They form tensor operator irreps of $\mathbf{SO}_3$ or of the product $\mathbf{SU}_2^S \times \mathbf{SO}_3^L$. They are expressed in the $\mathbf{SO}_2$ or JM basis, and they are normalized by their action on single-electron wave functions $|nslm_s m_l\rangle$. Their action on N-electron states will be decreed by the way an N-electron state is constructed from one-electron states. Nielson and Koster (1963) tabulate values of N-electron, single-configuration, $SU_2^S \times SO_3^L$ reduced matrix elements $\langle l^N \alpha SL \| U^{(k)} \| l^N \alpha' S' L' \rangle$ and $\langle l^N \alpha SL \| V^{(11)} \|$

$l^N\alpha'S'L'\rangle$ for the s, p, d, and f configurations. They follow Racah's normalizations. Their SU_2–SO_3 tensor operator $V^{(11)}$ is larger than ours, namely

$$U_{\pi q}^{11} = \sqrt{\tfrac{2}{3}}\; V_{\pi q}^{(11)} \qquad \text{(Nielson and Koster)} \tag{4.4.5}$$

The discussion of fractional parentage in Chapter 7 shows that the tables of Nielson and Koster are often useful even when there is no physical $\mathbf{SU}_2^S \times \mathbf{SO}_3^L$ structure in the problem.

Generalization of the Racah tensors from the JM basis to other point group bases is trivial, because such changes may be made by the transformation coefficients (2.2.7). Such a change of basis will not affect the SO_3 reduced matrix elements. Therefore if we can write a particular interaction in terms of the Racah operators $\mathbf{u}^{\kappa k}$ within a single l^N configuration in any point group symmetry scheme, the tables of Nielson and Koster may be used with the present tables to obtain all single configuration matrix elements without further ado.

One particular operator is used often in writing interactions in terms of tensor operators, the spherical tensor C_q^k. It is defined in relation to the spherical harmonics as

$$C_q^k = \left(\frac{4\pi}{2k+1}\right)^{1/2} Y_{kq}(\theta,\phi) \tag{4.4.6}$$

where we use the Condon and Shortley choice of spherical harmonics $(0 \leqslant \theta < \pi, 0 \leqslant \phi < 2\pi)$

$$Y_{lm}(\theta,\phi) = (-1)^m \left[\frac{2l+1}{4\pi}\,\frac{(l-m)!}{(l+m)!}\right]^{1/2} P_l^m(\cos\theta)e^{im\phi} \tag{4.4.7}$$

with the Legendre polynomials as (Arfken 1970, p. 570–1)

$$P_l^m(x) = \frac{(1-x^2)^{m/2}}{2^l l!}\,\frac{d^{l+m}}{dx^{l+m}}(x^2-1)^l \tag{4.4.8}$$

With some considerable algebraic manipulation it is possible to show that the SO_3 reduced matrix elements of the spherical tensors with respect to the single-electron orbital functions $|nlm\rangle$ is an $\mathbf{SO_3}$–$\mathbf{SO_2}$ $3jm$

$$\langle nl\|C^k\|n'l'\rangle=(-1)^l|l|^{1/2}|l'|^{1/2}\begin{pmatrix} l & k & l' \\ 0 & 0 & 0 \end{pmatrix}^{SO_3}_{SO_2} \tag{4.4.9}$$

The symmetry of the SO_3–SO_2 $3jm$ factor shows that this vanishes "because of parity" if $l+k+l'$ is odd. This result is for an SO_3 reduced matrix element, and thus it may be used in any point group basis.

The angular momentum operators J_x, J_y, and J_z together form a tensor operator transforming as $\mathbf{1}(\mathbf{SO_3})$. In the SO_3–SO_2 (JM) basis we have

$$J^1_0=J_z \tag{4.4.10a}$$

$$J^1_{\pm1}=\mp(1/\sqrt{2}\,)(J_x\pm iJ_y) \tag{4.4.10b}$$

Their SO_3 reduced matrix elements are

$$\langle j\|J^1\|j\rangle^{SO_3}=\{j(j+1)(2j+1)\}^{1/2} \tag{4.4.11}$$

so that, for instance,

$$\langle s\|S^1\|s\rangle^{SO_3}=\sqrt{\tfrac{3}{2}} \tag{4.4.12}$$

This, and Racah's definition for $V^{(11)}$ as $\Sigma_i s^1_i u^1_i$, explains the presence of the normalization in (4.4.5).

5

O_3 and Its Subgroups

To a very large extent the previous chapters have been independent of the group or groups chosen. The only modification required to generalize the various equations to the case of an arbitrary finite or compact continuous group is the insertion of more complicated symmetries for non-simple-phase and non-quasi-ambivalent groups. Many of the results of this chapter are equally general, but we do focus our attention on a number of points which may be dismissed lightly in a general treatment but which assume great importance in physical calculations.

The group $\mathbf{O}_3$ is the group of real, orthogonal transformations in 3-space, having determinant ± 1. It is the direct product of two groups, the group of rotations in 3-space $\mathbf{SO}_3$ and the two-element group $\mathbf{C}_i$ consisting of the inversion operation $(x, y, z) \to (-x, -y, -z)$ and the identity. The group $\mathbf{O}_3$ is sometimes called the full rotation group, although it contains operations that are not rotations.

The group $\mathbf{O}_3$, as indeed do all its subgroups, has two kinds of representations, true representations and spin representations. True representations are genuine representations in the mathematical sense; for every group operation R there is a unique transformation O_R which has a well-defined physical action. Spin representations are not representations in this sense, in that there is not a unique transformation for each group element, but instead there are two transformations for each group element (see Section 2.6).

We choose to be consistent with the usual treatment of $\mathbf{SO}_3$ in our notation and terminology. This means that the true and spin representations of the point groups appear together in the tables; when we talk about the representations of group $\mathbf{G}$ we include its spin representations. A little caution is needed sometimes concerning the action of group operators, for we are really applying the tools of group theory to the double groups. In the character tables we denote the true operator by R, and denote the second operator, the one belonging only to the double group, by $\bar{R}$. We follow Koster, Dimmock, Wheeler, and Statz (1963) for the action of the double group operators. Hamermesh (1962) uses another convention for the product of two operators of the double group. This makes no physical difference, nor any difference to the structural properties of the double group, but it does lead to different labeling schemes for the classes of the double groups.

The point groups which contain the inversion operator are all a direct product of the inversion group and a pure rotation group. Section 5.1 discusses the relation of *jm* and *j* symbols of the direct product of two groups to the *jm* and *j* symbols of each of the groups taken individually. The point groups which contain reflections but do not contain the inversion operator are all isomorphic to some point group of pure rotations. Section 5.2 discusses the values of the *j* symbols and *jm* factors of these groups.

Section 5.3 discusses the orientation of axes. Simply requiring that the physical partners transform as irreps of the chain $\mathbf{SO}_3 \supset \mathbf{O} \supset \mathbf{D}_3 \supset \mathbf{C}_3 \supset \mathbf{C}_1$, rather than, say, as $\mathbf{SO}_3 \supset \mathbf{K} \supset \mathbf{T} \supset \mathbf{D}_2 \supset \mathbf{C}_2 \supset \mathbf{C}_1$, specifies much of the basis information. Section 5.3 shows how the remaining information may be fixed.

5.1 Direct Product Groups

There are 32 crystallographic point groups, there being 32 symmetries which may be repeated though a crystal lattice (see Hamermesh 1962, Chapter 2; or Jansen and Boon 1967). There are more possible symmetry groups of an isolated entity, namely the icosahedral symmetries $\mathbf{K}$, the planar symmetries $\mathbf{C}_n$ and $\mathbf{D}_n$, $n \geqslant 5$, and these augmented by a reflection or inversion. Of the 32 crystallographic point groups, 11 are composed of pure rotations only. We include tables of these 11—$\mathbf{D}_6$, $\mathbf{O}$, $\mathbf{T}$, $\mathbf{D}_4$, $\mathbf{D}_3$, $\mathbf{D}_2$, $\mathbf{C}_6$, $\mathbf{C}_4$, $\mathbf{C}_3$, $\mathbf{C}_2$, and $\mathbf{C}_1$—to which we add tables for $\mathbf{C}_5$, $\mathbf{D}_5$, $\mathbf{K}$, and $\mathbf{SO}_3$, giving tables for 15 pure rotation point groups.

The properties of the other point groups follow directly from the properties of these 15 groups. Most of the other point groups are a direct product of one of the rotation groups with the two-element inversion group. A group $\mathbf{G}$ is said to be a direct product of two groups $\mathbf{G}_1$ and $\mathbf{G}_2$, $\mathbf{G}=\mathbf{G}_1\times\mathbf{G}_2$, if the operations of $\mathbf{G}_1$ commute with operations of $\mathbf{G}_2$ and every operation of $\mathbf{G}$ can be written uniquely as a product of an operation of $\mathbf{G}_1$ with an operation of $\mathbf{G}_2$. For example, every orthogonal transformation in 3-space may be written as a pure rotation in 3-space preceded by either the inversion or the identity:

$$\mathbf{O}_3=\mathbf{SO}_3\times\mathbf{C}_i \tag{5.1.1}$$

As written, we act first with the inversion group $\mathbf{C}_i$, but this is irrelevant because both elements of the group commute with pure rotations. Note we avoid the use of the letter I. We have previously written the icosahedral group as $\mathbf{K}$ (some use $\mathbf{Y}$), and written the identity group as $\mathbf{C}_1$. The direct product of $\mathbf{G}$ and $\mathbf{C}_i$ is not usually written $\mathbf{G}_i$: $\mathbf{SO}_{3i}$ is written $\mathbf{O}_3$ because the S indicates no inversion. Other point groups derived from a pure rotation group by addition or mirror planes are labeled by a description of the plane (horizontal, vertical, or diagonal). We use $\mathbf{G}_i$ for an arbitrary point group containing inversions, and $\tilde{\mathbf{G}}$ for any group isomorphic to $\mathbf{G}$ and containing reflections but not inversions. Examples: if $\mathbf{G}=\mathbf{O}$, then $\mathbf{G}_i=\mathbf{O}_h$ and $\tilde{\mathbf{G}}=\mathbf{T}_d$; if $\mathbf{G}=\mathbf{D}_3$, then $\mathbf{G}_i=\mathbf{D}_{3d}$ and $\tilde{\mathbf{G}}=\mathbf{C}_{3v}$; if $\mathbf{G}=\mathbf{D}_4$, then $\mathbf{G}_i=\mathbf{D}_{4h}$ and $\tilde{\mathbf{G}}=\mathbf{C}_{4v}$ or $\mathbf{D}_{2d}$.

Because the operations of $\mathbf{G}_1$ and $\mathbf{G}_2$ commute in the direct product $\mathbf{G}_1\times\mathbf{G}_2$, the partners of the representations may be labeled by a pair of independent labels $|\lambda_1(G_1)\lambda_2(G_2)l_1l_2\rangle$, where $\lambda_1 l_1$ labels partners of $\mathbf{G}_1$ and $\lambda_2 l_2$ those of $\mathbf{G}_2$. The coupling coefficients for each group may be used to decompose a coupled product into irreps,

$$\begin{aligned}|\lambda_1\lambda_2 l_1 l_2\rangle|\mu_1\mu_2 m_1 m_2\rangle &= \sum_{r_1r_2\nu_1\nu_2n_1n_2}|(\lambda_1\lambda_2,\mu_1\mu_2)r_1r_2\nu_1\nu_2n_1n_2\rangle\\ &\quad\times\langle r_1\nu_1n_1|\lambda_1l_1,\mu_1m_1\rangle\langle r_2\nu_2n_2|\lambda_2l_2,\mu_2m_2\rangle\end{aligned} \tag{5.1.2}$$

It is usual to label the two irreps of the inversion group as $-$ and $+$ or u and g (from the German ungerade or gerade, meaning uneven or even) according to whether or not they change sign under the inversion operation. The only triads for $\mathbf{C}_i$ are $(+++0)$ and $(+--0)$. The corresponding $3j$ phases are $+1$, so all nonzero j symbols for the group are $+1$.

Hence for each of the 15 *rotation–inversion groups* $\mathbf{G}_i = \mathbf{G}\times\mathbf{C}_i$, all the j symbols may be obtained from the corresponding table for the pure rotation group.

For a rotation–inversion group $\mathbf{G}_i$ there are three kinds of subgroups: (i) the pure rotation group $\mathbf{G}$; (ii) a rotation–inversion subgroup $\mathbf{H}_i$ corresponding to every subgroup $\mathbf{H}$ of $\mathbf{G}$; and (iii) sometimes, but not always, rotation–reflection groups $\tilde{\mathbf{G}}$ abstractly isomorphic to the pure rotation group $\mathbf{G}$. For $\mathbf{G}_i \supset \mathbf{G}$ the branching is $\boldsymbol{\lambda}^{\pm}(\mathbf{G}_i)\rightarrow\boldsymbol{\lambda}(\mathbf{G})$ and the jm factors are trivial, being unity if the triads and branchings exist. The G_i–G $3jm$ may be written as a product of a G–G $3jm$ and a C_i–C_1 $3jm$:

$$\begin{pmatrix} \lambda^{\pm} & \mu^{\pm} & \nu^{\pm} \\ \lambda & \mu & \nu \end{pmatrix}^{rG_i}_{sG} = \begin{pmatrix} \lambda & \mu & \nu \\ \lambda & \mu & \nu \end{pmatrix}^{rG}_{sG} \begin{pmatrix} \pm & \pm & \pm \\ 0 & 0 & 0 \end{pmatrix}^{C_i}_{C_1} = \delta_{rs}\delta_{\pm\pm\pm} \tag{5.1.3}$$

$\delta_{\pm\pm\pm}$ is unity if one or three signs are +, and zero otherwise. Similarly, for the G_i–H_i $3jm$ we have

$$\begin{pmatrix} \lambda^{\pm} & \mu^{\pm} & \nu^{\pm} \\ \rho^{\pm} & \sigma^{\pm} & \tau^{\pm} \end{pmatrix}^{rG_i}_{sH_i} = \begin{pmatrix} \lambda & \mu & \nu \\ \rho & \sigma & \tau \end{pmatrix}^{rG}_{sH} \begin{pmatrix} \pm & \pm & \pm \\ \pm & \pm & \pm \end{pmatrix}^{C_i}_{C_i} \tag{5.1.4}$$

The C_i–C_i $3jm$ is unity if the various signs satisfy the triad and branching conditions, and zero otherwise. Thus if the signs match up, the G_i–H_i $3jm$ is equal to the G–H $3jm$. The rotation–reflection imbedding, giving rise to a G_i–$\tilde{G}$ $3jm$, requires more study. We begin by considering the physical aspects of the group $\tilde{\mathbf{G}}$.

5.2 Isomorphic Subgroups

Of the 32 crystallographic point groups we have seen that 11 are composed of pure rotations and 11 are composed of the direct product of pure rotations and inversion, leaving ten not containing a pure inversion, but rather inversion composed with various finite rotations. Each of these ten groups is isomorphic to a group of pure rotations, namely

$$\begin{gathered} \mathbf{T}_d\sim\mathbf{O}, \quad \mathbf{D}_{3h}\sim\mathbf{D}_6, \quad \mathbf{C}_{6v}\sim\mathbf{D}_6, \quad \mathbf{C}_{4v}\sim\mathbf{D}_4, \quad \mathbf{D}_{2d}\sim\mathbf{D}_4 \\ \mathbf{C}_{3h}\sim\mathbf{C}_6, \quad \mathbf{S}_4\sim\mathbf{C}_4, \quad \mathbf{C}_{3v}\sim\mathbf{D}_3, \quad \mathbf{C}_{2v}\sim\mathbf{D}_2, \quad \mathbf{C}_s\sim\mathbf{C}_2 \end{gathered} \tag{5.2.1}$$

The j symbols for each of these groups are precisely those of the corresponding pure rotation group, because j symbols are functions of the abstract properties of groups. The jm symbols depend on the abstract properties of the groups together with the abstract properties of the imbedding in the form of the branching rules. The 23 imbeddings given by Koster, Dimmock, Wheeler, and Statz (1963) involving the above ten rotation–reflection groups, one in another, may be obtained from the corresponding 18 imbeddings of pure rotation groups simply by the appropriate interpretation of the abstract operations.

There is one final class of imbeddings which is not trivial; the $3jm$ factors cannot be read off the pure rotation group tables simply by interpreting the abstract group operations in terms of the appropriate physical operations. This is the case of a rotation–inversion group $\mathbf{G}_i$ containing a rotation–reflection subgroup $\tilde{\mathbf{G}}$, for example, $\mathbf{O}_h \supset \mathbf{T}_d$. The difference in the $3jm$'s arises from a difference in the branching rules between $\mathbf{G}_i \supset \mathbf{G}$ and $\mathbf{G}_i \supset \tilde{\mathbf{G}}$. Consider the functions x, y, z as partners of the $\mathbf{1}^-$ or $\mathbf{T}_{1u}$ irrep of $\mathbf{O}_h$. The pure rotation group $\mathbf{O}$ simply ignores the odd (ungerade) nature of the x, y, z functions and they transform as $\mathbf{1}(\mathbf{O})$. The rotation–reflection group $\mathbf{T}_d$ distinguishes *axial* (even) and *polar* (odd) vectors, x, y, z transforming as $\tilde{\mathbf{1}}(\mathbf{T}_d)$.

The distinction between even and odd irreps of $\mathbf{G}_i$ under the rotation–reflection subgroup $\tilde{\mathbf{G}}$ is related to the existence of the one-dimensional real irrep $\tilde{\mathbf{0}}(\tilde{\mathbf{G}})$, other than the scalar irrep $\mathbf{0}(\tilde{\mathbf{G}})$. The odd scalar $\mathbf{0}^-(\mathbf{G}_i)$ branches to $\tilde{\mathbf{0}}(\tilde{\mathbf{G}})$. The even irrep $\boldsymbol{\lambda}^+(\mathbf{G}_i)$ transforms as $\boldsymbol{\lambda}(\tilde{\mathbf{G}})$, but the odd irrep $\boldsymbol{\lambda}^-(\mathbf{G}_i)$ transforms as the irrep $\tilde{\boldsymbol{\lambda}}(\tilde{\mathbf{G}})$ obtained as the product $\tilde{\mathbf{0}} \times \boldsymbol{\lambda} = \tilde{\boldsymbol{\lambda}}$. The pure rotation groups $\mathbf{T}$ and $\mathbf{K}$ have only the one real irrep of dimension one, whereas the groups $\mathbf{D}_2$, $\mathbf{D}_4$, and $\mathbf{D}_6$ have four. As a result, several distinct (but isomorphic) rotation–reflection groups exist. Distinct here means they are not simply related by different choices of the x, y, z axis systems. In all cases, having specified which irrep is $\tilde{\mathbf{0}}$, the other irreps $\tilde{\boldsymbol{\lambda}}$ follow uniquely, as do the branching rules:

$$\tilde{\boldsymbol{\lambda}} = \boldsymbol{\lambda} \times \tilde{\mathbf{0}} \tag{5.2.2}$$

$$\boldsymbol{\lambda}^+(\mathbf{G}_i) \to \boldsymbol{\lambda}(\tilde{\mathbf{G}}) \tag{5.2.3a}$$

$$\boldsymbol{\lambda}^-(\mathbf{G}_i) \to \tilde{\boldsymbol{\lambda}}(\tilde{\mathbf{G}}) \tag{5.2.3b}$$

We emphasize that various irreps can play the role of $\tilde{\mathbf{0}}$, e.g., 2 in C_{4h}–S_4 (p. 210) and 3 in D_{6h}–D_{3h} (p. 213).

Further details on the interpretation of the physical operations may be found in several texts, e.g., Hamermesh (1962, Chapter 2).

The above character theory results are used below to calculate the $3jm$ factors for $\mathbf{G}_i \supset \tilde{\mathbf{G}}$. The free phases occurring are one phase for each nonprimitive ket $|\lambda^+\lambda\rangle$ and $|\lambda^-\tilde{\lambda}\rangle$. It is easier to see what is happening by viewing the primitive irrep $\frac{1}{2}^-$ as the product $\frac{1}{2}^+ \times \mathbf{0}^-$. We choose

$$\begin{pmatrix} \lambda^- & \lambda^{*+} & 0^- \\ \tilde{\lambda} & \lambda^* & \tilde{0} \end{pmatrix}^{G_i}_{\tilde{G}} = 1 \tag{5.2.4}$$

for all $\lambda > 0$, and for every $\lambda > \frac{1}{2}$ choose one $\mu < \lambda$ and set

$$\begin{bmatrix} \lambda^+ & \mu^+ & \frac{1}{2}^+ \\ \lambda & \mu & \frac{1}{2} \end{bmatrix}^{G_i}_{\tilde{G}} = 1 \tag{5.2.5}$$

The general case can now be obtained by recursion: assume that one has shown that (5.2.5) is satisfied for any choice of $\boldsymbol{\mu}$ and that all irreps are real. The recursive recoupling (3.3.30) shows that

$$\begin{pmatrix} \lambda^+ & \mu^+ & \nu^+ \\ \lambda & \mu & \nu \end{pmatrix}^{rG_i}_{s\tilde{G}} = \sum_{\varepsilon\lambda' r' s'} |\lambda||\mu'| \begin{Bmatrix} \lambda & \mu & \nu \\ \mu' & \lambda' & \frac{1}{2} \end{Bmatrix}^{G}_{00r'r} \begin{Bmatrix} + & + & + \\ + & \varepsilon & + \end{Bmatrix}^{C_i}_{0000}$$

$$\times \begin{bmatrix} \lambda^+ & \lambda'^{\varepsilon} & \frac{1}{2}^+ \\ \lambda & \lambda' & \frac{1}{2} \end{bmatrix}^{0}_{0} \begin{bmatrix} \mu'^+ & \mu^+ & \frac{1}{2}^+ \\ \mu' & \mu & \frac{1}{2} \end{bmatrix}^{0}_{0}$$

$$\times \begin{pmatrix} \mu'^+ & \lambda'^{\varepsilon} & \nu^+ \\ \mu' & \lambda' & \nu \end{pmatrix}^{r'}_{s'} \begin{Bmatrix} \lambda & \mu & \nu \\ \mu' & \lambda' & \frac{1}{2} \end{Bmatrix}^{G}_{00s's} \tag{5.2.6}$$

where the branching rules have been used, and the $6j$ of $\mathbf{G}_i$ has been written as the product of a G $6j$ and a C_i $6j$. The equation is valid for any $\boldsymbol{\mu}'$ which satisfies the triad conditions. The sign ε is $+$ only by the triad requirement so that the C_i $6j$ is unity. Arguing recursively from (5.2.5), it follows that if the $3jm$ for any $\mu' < \mu$ is $\delta_{r's'}$, then the sum on the right reduces to a sum over $6j$'s of G. By the orthogonality of $6j$'s (3.3.21), we have

$$\begin{pmatrix} \lambda^+ & \mu^+ & \nu^+ \\ \lambda & \mu & \nu \end{pmatrix}^{rG_i}_{s\tilde{G}} = \delta_{rs} \tag{5.2.7}$$

On the other hand, (3.3.29) gives

$$\sum_r \begin{pmatrix} \lambda^- & \mu^- & \nu^+ \\ \tilde{\lambda} & \tilde{\mu} & \nu \end{pmatrix}^r_s \begin{Bmatrix} \lambda & \mu & \nu \\ \mu & \lambda & 0 \end{Bmatrix}_{00r'r}$$

$$= \sum_{s'} \begin{pmatrix} \lambda^- & \lambda^+ & 0^- \\ \tilde{\lambda} & \lambda & \tilde{0} \end{pmatrix}^0_0 \begin{pmatrix} \mu^+ & \mu^- & 0^- \\ \mu & \tilde{\mu} & \tilde{0} \end{pmatrix}^0_0$$

$$\times \begin{pmatrix} \mu^+ & \lambda^+ & \nu^+ \\ \mu & \lambda & \nu \end{pmatrix}^{r'}_{s'} \begin{Bmatrix} \tilde{\lambda} & \tilde{\mu} & \nu \\ \mu & \lambda & \tilde{0} \end{Bmatrix}_{00s's} \tag{5.2.8}$$

The $6j$ on the left is related to a $3j$, the second $3jm$ is related by $3j$'s to a $3jm$ of the form (5.2.4), and the other $3jm$'s are positive; hence (relaxing the reality assumption)

$$\begin{pmatrix} \lambda^- & \mu^- & \nu^+ \\ \tilde{\lambda} & \tilde{\mu} & \nu \end{pmatrix}^r_s = |\lambda|^{1/2}|\mu|^{1/2}\{\mu\}\{\lambda\mu\nu r\}\{\tilde{\mu}\mu^*\tilde{0}\} \begin{Bmatrix} \tilde{\lambda} & \tilde{\mu} & \nu \\ \mu^* & \lambda & \tilde{0} \end{Bmatrix}_{00rs} \tag{5.2.9}$$

This formula is used to produce several tables of Chapter 13, where we list all the $3jm$ factors involving odd irreps for the seven non-Abelian imbeddings of (5.2.1).

The values of the G_i–$\tilde{G}$ $3jm$ factors are not all positive, but observe that for the only case of product multiplicity, that of $\mathbf{O}_h \supset \mathbf{T}_d$, the $3jm$ is diagonal in the multiplicity labels δ_{rs}. They are diagonal because all $6j$ symbols of $\mathbf{O}$ of the form $\begin{Bmatrix} \lambda\,\mu\,\nu \\ \mu\,\lambda\,\tilde{0} \end{Bmatrix}_{00rs}$ are diagonal. That the $6j$'s of $\mathbf{O}$ can be chosen in this form shown by Butler and Ford (1979) and was used as the criterion for the separation of the $\left(\frac{3}{2}\frac{3}{2}1r\right) = (U'U'T_1r)$ octahedral product multiplicity in Section 3.5.

For some purposes it is necessary to know the action of the reflection operators on the basis functions of the problem. We need the action of the reflection operators on the spherical harmonics, and on the basic angular momentum spinors α and β. Writing σ_h and σ_v for reflections in the xy and xz planes, respectively, we find that (4.4.7) shows that

$$\sigma_h Y_{lm} = (-1)^{l-m} Y_{lm} \tag{5.2.10a}$$

$$\sigma_v Y_{lm} = (-1)^m Y_{l-m} \tag{5.2.10b}$$

while the usual choice of spin functions gives

$$\sigma_h |\tfrac{1}{2}m\rangle = i(-1)^{1/2-m} |\tfrac{1}{2}m\rangle = (-1)^m |\tfrac{1}{2}m\rangle \tag{5.2.11a}$$

$$\sigma_v |\tfrac{1}{2}m\rangle = (-1)^{1/2-m} |\tfrac{1}{2}-m\rangle \tag{5.2.11b}$$

Consider a system with n electrons with various orbital quantum numbers l_i each of parity $(-1)^{l_i}$ contributing to a total state $|\alpha JM\rangle$ of parity

$$p = \sum_{i=1}^{n} l_i \tag{5.2.12}$$

To obtain a general result we argue recursively. We assume that

$$\sigma_h |\alpha JM\rangle = (-1)^{p+M} |\alpha JM\rangle \tag{5.2.13}$$

for all $n_1, n_2 < n$, for this is the result for one electron. The n-particle state is some coupled product of states of n_1 and n_2 particles, with $n_1 + n_2 = n$, so

$$\sigma_h |\alpha JM\rangle = \sum_{\alpha_1 J_1 M_1 \alpha_2 J_2 M_2} \sigma_h |\alpha_1 J_1 M_1, \alpha_2 J_2 M_2\rangle \langle \alpha_1 J_1 M_1, \alpha_2 J_2 M_2 | \alpha JM\rangle$$

(we need to include the labels α_i in the coupling coefficient, for we are dealing with parentage coefficients—see Chapter 7)

$$= \sum (-1)^{p_1+M_1+p_2+M_2} |\alpha_1 J_1 M_1, \alpha_2 J_2 M_2\rangle \langle \alpha_1 J_1 M_1, \alpha_2 J_2 M_2 | \alpha JM\rangle \tag{5.2.14}$$

which is (5.2.13) because $M_1+M_2=M$ and $p=p_1+p_2$. These factors are constant and may be moved outside the sum. This ends the proof.

A similar argument for σ_v requires the complex conjugation symmetry of the JM $3jm$ symbol. Again assume the single electron result holds for $n_1, n_2<n$, that is,

$$\sigma_v|\alpha JM\rangle=(-1)^{p+J-M}|J-M\rangle \tag{5.2.15}$$

Expanding as above gives rise to a sum of terms

$$(-1)^{p_1+J_1-M_1+p_2+J_2-M_2}$$
$$\times|\alpha_1 J_1-M_1, \alpha_2 J_2-M_2\rangle\langle\alpha_1 J_1, \alpha_2 J_2|\alpha J\rangle\langle J_1 M_1, J_2 M_2|JM\rangle$$

Use has been made of the fact that the fractional parentage coefficient factorizes into an SO_3–SO_2 coupling coefficient and an SO_2 independent factor. The complex conjugation symmetry for SO_3–SO_2 $3jm$'s shows that this equals

$$=(-1)^{p+J-M}|\alpha J_1-M_1, \alpha_2 J_2-M_2\rangle$$
$$\times\langle\alpha_1 J_1, \alpha_2 J_2|\alpha J\rangle\langle J_1-M_1, J_2-M_2|J-M\rangle \tag{5.2.16}$$

The sum may now be performed to obtain (5.2.15), ending the proof.

It is to be observed that both derivations rely on the fact that $M=\Sigma m$ for $\mathbf{SO_2}$. The factor $(-1)^{J-M}$ in (5.2.15) did not arise as an SO_3–SO_2 $2jm$. As a result, the matrix elements of the σ_h and σ_v operators do not take parallel forms for other point group chains. This fact is in contrast to the time reversal result (8.1.17), where a $2jm$ arises from the conjugate linear properties of the time reversal operator.

5.3 Specification of x, y, z Axes and Bases

There is often no need to obtain explicit functions which transform according to a specific symmetry. How often does one need the functional form of the $|J=2,\ M_J=1\rangle$ partner of a D-state irrep? Explicit forms do have their uses, however. It is very instructive to obtain several explicit forms for various symmetry chains, and in addition explicit forms are needed to enable one to write physical operators in terms of the standard operators of a given symmetry chain. This section is devoted to a few examples.

In this section we use four examples of symmetry chains to illustrate how the tables may be used with different axis choices and thus different bases. One example is chosen to illustrate the relationship of the rotation–reflection groups to the pure rotation groups. We consider in detail the transformation properties of the spin functions α and β; the rotation operators (axial vectors) J_x, J_y, J_z; and the coordinate (polar) vectors x, y, and z, under the various alternative chains of symmetries:

$$\begin{array}{ll} a & \mathbf{O}_3 \supset \mathbf{SO}_3 \supset \mathbf{SO}_2 \supset \mathbf{C}_1 \\ b & \mathbf{O}_3 \supset \mathbf{SO}_3 \supset \mathbf{O} \supset \mathbf{D}_3 \supset \mathbf{C}_3 \supset \mathbf{C}_1 \\ c & \mathbf{O}_3 \supset \mathbf{SO}_3 \supset \mathbf{O} \supset \mathbf{D}_4 \supset \mathbf{D}_2 \supset \mathbf{C}_2 \supset \mathbf{C}_1 \\ d & \mathbf{O}_3 \supset \mathbf{O}_h \supset \mathbf{T}_d \supset \mathbf{C}_{3v} \supset \mathbf{C}_3 \supset \mathbf{C}_1 \end{array} \tag{5.3.1}$$

Inspection of the various branching rules (Chapter 12) for these four chains gives that one of the two partners α and β of the spinor irrep $[\frac{1}{2}]^+$ of $\mathbf{O}_3$ may be labeled as follows:

$$\begin{aligned} \alpha_a &= |[\tfrac{1}{2}]^+ \tfrac{1}{2}(SO_2)\rangle = |\tfrac{1}{2}^+ \tfrac{1}{2}\rangle_a \\ \alpha_b &= |[\tfrac{1}{2}]^+ \tfrac{1}{2}(O)\tfrac{1}{2}(D_3)\tfrac{1}{2}(C_3)\rangle = |\tfrac{1}{2}^+ \tfrac{1}{2}\tfrac{1}{2}\tfrac{1}{2}\rangle_b \\ \alpha_c &= |[\tfrac{1}{2}]^+ \tfrac{1}{2}(O)\tfrac{1}{2}(D_4)\tfrac{1}{2}(D_2)\tfrac{1}{2}(C_2)\rangle = |\tfrac{1}{2}^+ \tfrac{1}{2}\tfrac{1}{2}\tfrac{1}{2}\tfrac{1}{2}\rangle_c \\ \alpha_d &= |[\tfrac{1}{2}]^+ \tfrac{1}{2}^+(O_h)\tfrac{1}{2}(T_d)\tfrac{1}{2}(C_{3v})\tfrac{1}{2}(C_3)\rangle = |\tfrac{1}{2}^+ \tfrac{1}{2}\tfrac{1}{2}\tfrac{1}{2}\tfrac{1}{2}\rangle_d \end{aligned} \tag{5.3.2}$$

where the groups have been omitted from the labels in the third form and the brief α_i for chain i will be used for brevity in some of what follows. The various second partners for the $[\frac{1}{2}]^+$ irrep are

$$\begin{aligned} \beta_a &= |\tfrac{1}{2}^+ -\tfrac{1}{2}\rangle_a, \qquad \beta_b = |\tfrac{1}{2}^+ \tfrac{1}{2}\tfrac{1}{2} -\tfrac{1}{2}\rangle_b \\ \beta_c &= |\tfrac{1}{2}^+ \tfrac{1}{2}\tfrac{1}{2}\tfrac{1}{2} -\tfrac{1}{2}\rangle_c, \qquad \beta_d = |\tfrac{1}{2}^+ \tfrac{1}{2}\tfrac{1}{2}\tfrac{1}{2} -\tfrac{1}{2}\rangle_d \end{aligned} \tag{5.3.3}$$

Similar use of branching tables gives the transformation properties of the three partners of the axial and polar vectors in the four symmetry schemes. Namely, the branching rules show that the four sets of partners of $1^+(O_3)$, being sets of components of an axial vector, can be labeled as

$$
\begin{array}{lll}
|1^+1\rangle_a & |1^+0\rangle_a & |1^+-1\rangle_a \\
|1^+111\rangle_b & |1^+1\tilde{0}0\rangle_b & |1^+11-1\rangle_b \\
|1^+1111\rangle_c & |1^+1\tilde{0}\tilde{0}0\rangle_c & |1^+11\tilde{1}1\rangle_c \\
|1^+1111\rangle_b & |1^+11\tilde{0}0\rangle_d & |1^+111-1\rangle_d
\end{array}
$$

and as

$$
\begin{array}{lll}
|1^-1\rangle_a & |1^-0\rangle_a & |1^--1\rangle_a \\
|1^-111\rangle_b & |1^-1\tilde{0}0\rangle_b & |1^-11-1\rangle_b \\
|1^-1111\rangle_c & |1^-1\tilde{0}\tilde{0}0\rangle_c & |1^-11\tilde{1}1\rangle_c \\
|1^-1\tilde{1}11\rangle_d & |1^-1\tilde{1}00\rangle_d & |1^-1\tilde{1}1-1\rangle_d
\end{array}
$$

for the components of a polar vector.

The branching rules lead to the above labels for the various sets of partners in the four symmetry chains, but the relationship between the partners of the one chain and those of another depends on the relationship of the axes chosen in the problem and as well as the abstract transformation properties as described by the *jm* symbols. Consider the basic spinor functions first.

Now, J_z is the generator of rotations about the z axis, so that the effect of the rotation operator $R(\theta, z)$, which rotates the system through an angle θ about the $\mathbf{SO}_2$ symmetry axis, labeled the z axis, is given in the standard α_a, β_a basis as [see (2.1.12)]

$$R(\theta, z)\alpha_a = e^{-i\theta J_z}\alpha_a = \alpha_a e^{-i\theta/2} \tag{5.3.4}$$

and

$$R(\theta, z)\beta_a = e^{-i\theta J_z}\beta_a = \beta_a e^{i\theta/2} \tag{5.3.5}$$

The relationship between the partners (α_b, β_b) and (α_a, β_a) depends on the orientation of the axes in the b scheme relative to those of the a scheme. α_b

and β_b are distinguished by their behavior under a rotation about the $\mathbf{C}_3$ axis $R(2\pi/3, C_3)$,

$$R(2\pi/3, C_3)\alpha_b = \alpha_b e^{-2\pi i/3} \tag{5.3.6}$$

$$R(2\pi/3, C_3)\beta_b = \beta_b e^{2\pi i/3} \tag{5.3.7}$$

If the $\mathbf{C}_3$ axis is the z axis, then to within a phase $\alpha_a = \alpha_b$ and $\beta_a = \beta_b$. The phase gives the relationship between the xy axes of the two schemes. In general we have

$$\alpha_b = A\alpha_a + B\beta_a \tag{5.3.8a}$$

$$\beta_b = C\alpha_a + D\beta_a \tag{5.3.8b}$$

where A, B, C, D are elements of the unitary matrix $R(a \to b)$ which rotates the a scheme into the b scheme. Thus $R(a \to b)$ will be the matrix $D^{1/2}(\alpha, \beta, \gamma)$ of angular momentum theory, which rotates the x, y, z axis system into an x', y', z' system such that the z' axis is the C_3 axis and y' is the C_2 axis of $\mathbf{D}_3$.

The three coordinate vectors x, y, and z transform according to the irrep $\mathbf{1}^-(\mathbf{O}_3)$. The standard partners for the symmetry scheme a are

$$|1^- 0\rangle_a = z, \quad |1^- 1\rangle_a = -(1/\sqrt{2})(x + iy), \quad |1^- -1\rangle_a = (1/\sqrt{2})(x - iy) \tag{5.3.9}$$

The various branching rules for schemes b, c, and d label the transformation properties of the functions x, y, and z in these subgroup schemes. The explicit linear combinations of x, y, and z appropriate to these alternative schemes may be derived from the $3jm$ factor tables by making a specific choice of the axis system. The $3jm$'s give the transformation from the product of the partners of two irreps of a group to the partners of the coupled irreps. In particular the $3jm$ tables give the coupled partners of an axial vector $\mathbf{1}^+(\mathbf{O}_3)$ as linear combinations of the quadratic product of spin functions α and β, which are partners of the irrep $\frac{1}{2}^+(\mathbf{O}_3)$. In the following pages [up to Eq. (5.3.14)] we use the $3jm$ tables to produce the linear combinations for each of the above four schemes.

Recall that the general coupling is

$$|(\lambda_1\lambda_2)r\lambda l\rangle = \sum_{l_1 l_2} |\lambda_1 l_1\rangle |\lambda_2 l_2\rangle \langle \lambda_1 l_1, \lambda_2 l_2 | r\lambda l\rangle \tag{5.3.10}$$

where the $\mathbf{SO}_3$ coupling coefficient has no multiplicity but includes the historical Condon and Shortley phase factor $H(J_1J_2J)=(-1)^{J_1-J_2+J}$. In general we have [see (3.1.5)]

$$\langle r\lambda l|\lambda_1 l_1,\lambda_2 l_2\rangle = H(\lambda_1\lambda_2\lambda)|\lambda|^{1/2}\begin{pmatrix}\lambda\\ l\end{pmatrix}\begin{pmatrix}\lambda_1 & \lambda_2 & \lambda^*\\ l_1 & l_2 & l^*\end{pmatrix} \quad (5.3.11)$$

The coupling coefficient factorizes for the various schemes into *jm* factors for each step of the chain.

Hence for scheme a, $\mathbf{SO}_3\supset\mathbf{SO}_2\supset\mathbf{C}_1$,

$$|(\tfrac{1}{2}\tfrac{1}{2})10\rangle_a = |\tfrac{1}{2}\tfrac{1}{2}\rangle_a|\tfrac{1}{2}-\tfrac{1}{2}\rangle_a\langle 10|\tfrac{1}{2}\tfrac{1}{2},\tfrac{1}{2}-\tfrac{1}{2}\rangle^*$$

$$+|\tfrac{1}{2}-\tfrac{1}{2}\rangle|\tfrac{1}{2}\tfrac{1}{2}\rangle\langle 10|\tfrac{1}{2}-\tfrac{1}{2},\tfrac{1}{2}\tfrac{1}{2}\rangle^*$$

which can be written, using (5.3.2), (5.3.3), and (5.3.11), as

$$-\alpha_a\beta_a H(\tfrac{1}{2}\tfrac{1}{2}1)|1|^{1/2}\begin{pmatrix}1\\0\end{pmatrix}\begin{matrix}SO_3\\SO_2\end{matrix}\begin{bmatrix}\tfrac{1}{2} & \tfrac{1}{2} & 1^*\\ \tfrac{1}{2} & -\tfrac{1}{2} & 0^*\end{bmatrix}\begin{matrix}{}^*SO_3\\SO_2\end{matrix}$$

$$+\beta_a\alpha_a H(\tfrac{1}{2}\tfrac{1}{2}1)|1|^{1/2}\begin{pmatrix}1\\0\end{pmatrix}\begin{matrix}SO_3\\SO_2\end{matrix}\begin{bmatrix}\tfrac{1}{2} & \tfrac{1}{2} & 1^*\\ -\tfrac{1}{2} & \tfrac{1}{2} & 0^*\end{bmatrix}\begin{matrix}{}^*SO_3\\SO_2\end{matrix}$$

The value of the historical phase is given above, the dimension of $\mathbf{1}(\mathbf{SO}_3)$ is 3, $\mathbf{1}^*(\mathbf{SO}_3)=\mathbf{1}(\mathbf{SO}_3)$, and $\mathbf{0}^*(\mathbf{SO}_2)=\mathbf{0}(\mathbf{SO}_2)$. Finally the numerical values of the SO_3–SO_2 3*jm* can be found in Rotenberg, Bivins, Metropolis, and Wooten (1959) under the name of 3*j* symbol to give

$$=\alpha_a\beta_a(-1)^{1/2-1/2+1}\sqrt{3}\,(-1)^{1-0}\sqrt{\tfrac{1}{6}}$$

$$+\beta_a\alpha_a(-1)^{1/2-1/2+1}\sqrt{3}\,(-1)^{1-0}\sqrt{\tfrac{1}{6}}$$

$$=(1/\sqrt{2}\,)(\alpha_a\beta_a+\beta_a\alpha_a) \quad (5.3.12a)$$

We leave the result in this form because $\alpha_a\beta_a$ can mean, for distinguishable particles, particle 1 spin up times particle 2 spin down. In this form the result is normalized. If $\alpha_a\beta_a$ is regarded as being the same as $\beta_a\alpha_a$ and not orthogonal to it, then we must renormalize, $\alpha_a\beta_a+\beta_a\alpha_a \to \sqrt{2}\ \alpha_a\beta_a$. In the quadratic functions of x, y, z in Sections 2.2 and 16.2 the normalization can be understood either by making such transformations or by performing the spherical harmonic integration for a single particle.

For scheme b, where $\mathbf{SO}_3 \supset \mathbf{O} \supset \mathbf{D}_3 \supset \mathbf{C}_3 \supset \mathbf{C}_1$, let us calculate $|(\frac{1}{2}\frac{1}{2})11\tilde{0}0\rangle$ first. The $3j$ tables at the head of the $6j$ table of the appropriate groups show that the only nonzero couplings which occur are the two

$$|(\tfrac{1}{2}\tfrac{1}{2})11\tilde{0}0\rangle_b = |\tfrac{1}{2}\tfrac{1}{2}\tfrac{1}{2}\tfrac{1}{2}\rangle_b |\tfrac{1}{2}\tfrac{1}{2}\tfrac{1}{2}-\tfrac{1}{2}\rangle_b \langle 11\tilde{0}0|\tfrac{1}{2}\tfrac{1}{2}\tfrac{1}{2}\tfrac{1}{2}, \tfrac{1}{2}\tfrac{1}{2}\tfrac{1}{2}-\tfrac{1}{2}\rangle^*$$

$$+|\tfrac{1}{2}\tfrac{1}{2}\tfrac{1}{2}-\tfrac{1}{2}\rangle_b |\tfrac{1}{2}\tfrac{1}{2}\tfrac{1}{2}\tfrac{1}{2}\rangle_b \langle 11\tilde{0}0|\tfrac{1}{2}\tfrac{1}{2}\tfrac{1}{2}-1, \tfrac{1}{2}\tfrac{1}{2}\tfrac{1}{2}\tfrac{1}{2}\rangle^*$$

The coupling coefficients are written in terms of SO_3–O–D_3–C_3 jm symbols using (5.3.11) as above.

Changing to a column notation for the ket labels, this is

$$= \left|\begin{matrix}(\frac{1}{2}\frac{1}{2})1\\ 1\\ \tilde{0}\\ 0\end{matrix}\right\rangle = H(\tfrac{1}{2}\tfrac{1}{2}1)|1|^{1/2}_{SO_3}\left\{\left|\begin{matrix}\frac{1}{2}\\ \frac{1}{2}\\ \frac{1}{2}\\ \frac{1}{2}\end{matrix}\right\rangle\left|\begin{matrix}\frac{1}{2}\\ \frac{1}{2}\\ \frac{1}{2}\\ -\frac{1}{2}\end{matrix}\right\rangle\begin{pmatrix}1\\ 1\\ \tilde{0}\\ 0\end{pmatrix}\begin{pmatrix}\frac{1}{2} & \frac{1}{2} & 1^*\\ \frac{1}{2} & \frac{1}{2} & 1^*\\ \frac{1}{2} & \frac{1}{2} & \tilde{0}^*\\ \frac{1}{2} & -\frac{1}{2} & 0^*\end{pmatrix}^*\right.$$

$$\left. + \left|\begin{matrix}\frac{1}{2}\\ \frac{1}{2}\\ \frac{1}{2}\\ -\frac{1}{2}\end{matrix}\right\rangle\left|\begin{matrix}\frac{1}{2}\\ \frac{1}{2}\\ \frac{1}{2}\\ \frac{1}{2}\end{matrix}\right\rangle\begin{pmatrix}1\\ 1\\ 0\\ 0\end{pmatrix}\begin{pmatrix}\frac{1}{2} & \frac{1}{2} & 1^*\\ \frac{1}{2} & \frac{1}{2} & 1^*\\ \frac{1}{2} & \frac{1}{2} & \tilde{0}^*\\ -\frac{1}{2} & \frac{1}{2} & 0^*\end{pmatrix}\begin{matrix}{}^*SO_3\\ O\\ D_3\\ C_3\end{matrix}\right\}$$

The *jm* symbols factorize by the Racah factorization lemma as exemplified in (3.1.9) into SO_3–O *jm*'s, O–D_3 *jm*'s, and D_3–C_3 *jm*'s, but without product multiplicities in the present case. We have

$$=\alpha_b\beta_b(-1)^{1/2-1/2+1}\sqrt{3}\begin{pmatrix}\frac{1}{2} & \frac{1}{2} & 1^* \\ \frac{1}{2} & \frac{1}{2} & 1^*\end{pmatrix}{}^{*SO_3}_{O}\begin{pmatrix}\frac{1}{2} & \frac{1}{2} & 1^* \\ \frac{1}{2} & \frac{1}{2} & \tilde{0}^*\end{pmatrix}{}^{*O}_{D_3}$$

$$\times\begin{pmatrix}1\\0\end{pmatrix}{}^{D_3}_{C_3}\begin{pmatrix}\frac{1}{2} & \frac{1}{2} & \tilde{0}^* \\ \frac{1}{2} & -\frac{1}{2} & 0^*\end{pmatrix}{}^{*D_3}_{C_3}$$

$$+\beta_b\alpha_b(-1)^{1/2-1/2+1}\sqrt{3}\begin{pmatrix}\frac{1}{2} & \frac{1}{2} & 1^* \\ \frac{1}{2} & \frac{1}{2} & 1^*\end{pmatrix}{}^{*SO_3}_{O}\begin{pmatrix}\frac{1}{2} & \frac{1}{2} & 1^* \\ \frac{1}{2} & \frac{1}{2} & \tilde{0}^*\end{pmatrix}{}^{*O}_{D_3}$$

$$\times\begin{pmatrix}1\\0\end{pmatrix}{}^{D_3}_{C_3}\begin{pmatrix}\frac{1}{2} & \frac{1}{2} & \tilde{0}^* \\ -\frac{1}{2} & \frac{1}{2} & 0^*\end{pmatrix}{}^{*D_3}_{C_3}$$

The $2jm$ and $3jm$ factors are found in the appropriate tables in Chapter 13. We have

$$=(\alpha_b\beta_b+\beta_b\alpha_b)(-1)\sqrt{3}\cdot 1\cdot\frac{1}{\sqrt{3}}(-1)\cdot\frac{1}{\sqrt{2}}=\frac{1}{\sqrt{2}}(\alpha_b\beta_b+\beta_b\alpha_b)$$

In both above examples the resultant irreps are real and the * in the third column of each $3jm$ has no effect. Consider the construction of $|(\frac{1}{2}\frac{1}{2})1111\rangle_b$ to illustrate the case of complex irreps. To obtain the irrep $\mathbf{1(C_3)}$ we must look in the $\mathbf{C}_3$ triad table in Chapter 14 for triads involving either **1** or $\mathbf{1^*=-1}$. This is because the triad $(\lambda_1\lambda_2\lambda_3 r)$ describes the Kronecker product $\boldsymbol{\lambda}_1\times\boldsymbol{\lambda}_2\to\boldsymbol{\lambda}_3^*$. In our case the spin-$\frac{1}{2}$ functions $\boldsymbol{\lambda}_1$ and $\boldsymbol{\lambda}_2$ can be $\mathbf{\frac{1}{2}(C_3)}$ or $\mathbf{-\frac{1}{2}(C_3)}$. The only triad is $(1-\frac{1}{2}-\frac{1}{2}0)$, corresponding to the product $-\frac{1}{2}\times-\frac{1}{2}\rightarrow\mathbf{1}^*$ or $-\frac{1}{2}^*\times-\frac{1}{2}^*\rightarrow\mathbf{1}$. Hence

$$|(\tfrac{1}{2}\tfrac{1}{2})1111\rangle_b=|\tfrac{1}{2}\tfrac{1}{2}\tfrac{1}{2}\tfrac{1}{2}\rangle_b|\tfrac{1}{2}\tfrac{1}{2}\tfrac{1}{2}\tfrac{1}{2}\rangle_b\langle\tfrac{1}{2}\tfrac{1}{2}\tfrac{1}{2}\tfrac{1}{2},\tfrac{1}{2}\tfrac{1}{2}\tfrac{1}{2}\tfrac{1}{2}|1111\rangle_b$$

$$=\alpha_b\alpha_b(-1)^{1/2-1/2+1}\sqrt{3}\begin{bmatrix}\frac{1}{2} & \frac{1}{2} & 1^*\\ \frac{1}{2} & \frac{1}{2} & 1^*\end{bmatrix}^{*SO_3}_{O}$$

$$\times\begin{bmatrix}\frac{1}{2} & \frac{1}{2} & 1^*\\ \frac{1}{2} & \frac{1}{2} & 1^*\end{bmatrix}^{*O}_{D_3}\begin{pmatrix}1\\1\end{pmatrix}^{D_3}_{C_3}\begin{bmatrix}\frac{1}{2} & \frac{1}{2} & 1^*\\ \frac{1}{2} & \frac{1}{2} & 1^*\end{bmatrix}^{*D_3}_{C_3}$$

$$=\alpha_b\alpha_b(-1)\sqrt{3}\begin{bmatrix}\frac{1}{2} & \frac{1}{2} & 1\\ \frac{1}{2} & \frac{1}{2} & 1\end{bmatrix}^{*SO_3}_{O}$$

$$\times\begin{bmatrix}\frac{1}{2} & \frac{1}{2} & 1\\ \frac{1}{2} & \frac{1}{2} & 1\end{bmatrix}^{*O}_{D_3}\begin{pmatrix}1\\1\end{pmatrix}^{D_3}_{C_3}\begin{bmatrix}\frac{1}{2} & \frac{1}{2} & 1\\ \frac{1}{2} & \frac{1}{2} & -1\end{bmatrix}^{*D_3}_{C_3}$$

$$=\alpha_b\alpha_b(-1)\sqrt{3}\cdot 1\cdot\frac{\sqrt{2}}{\sqrt{3}}\cdot(+1)\frac{1}{\sqrt{2}}$$

$$=-\alpha_b\alpha_b \tag{5.3.12b}$$

Scheme c, $\mathbf{SO}_3\supset\mathbf{O}\supset\mathbf{D}_4\supset\mathbf{D}_2\supset\mathbf{C}_2\supset\mathbf{C}_1$, results in

$$|(\tfrac{1}{2}\tfrac{1}{2})11\tilde{0}\tilde{0}0\rangle_c=\alpha_c\beta_c(-1)^1\sqrt{3}\begin{bmatrix}\frac{1}{2} & \frac{1}{2} & 1\\ \frac{1}{2} & \frac{1}{2} & 1\end{bmatrix}^{*SO_3}_{O}\begin{bmatrix}\frac{1}{2} & \frac{1}{2} & 1\\ \frac{1}{2} & \frac{1}{2} & \tilde{0}\end{bmatrix}^{*O}_{D_4}$$

$$\times\begin{bmatrix}\frac{1}{2} & \frac{1}{2} & \tilde{0}\\ \frac{1}{2} & \frac{1}{2} & \tilde{0}\end{bmatrix}^{*D_4}_{D_2}\begin{pmatrix}\tilde{0}\\0\end{pmatrix}^{D_2}_{C_2}\begin{bmatrix}\frac{1}{2} & \frac{1}{2} & \tilde{0}\\ \frac{1}{2} & -\frac{1}{2} & 0\end{bmatrix}^{*D_2}_{C_2}$$

$$+\beta_c\alpha_c(-1)^1\sqrt{3}\begin{bmatrix}\frac{1}{2} & \frac{1}{2} & 1\\ \frac{1}{2} & \frac{1}{2} & 1\end{bmatrix}^{*SO_3}\begin{bmatrix}\frac{1}{2} & \frac{1}{2} & 1\\ \frac{1}{2} & \frac{1}{2} & \tilde{0}\end{bmatrix}^{*O}_{D_4}$$

$$\times\begin{bmatrix}\frac{1}{2} & \frac{1}{2} & \tilde{0}\\ \frac{1}{2} & \frac{1}{2} & \tilde{0}\end{bmatrix}^{*D_4}_{D_2}\begin{pmatrix}\tilde{0}\\ \tilde{0}\end{pmatrix}^{D_2}_{C_2}\begin{bmatrix}\frac{1}{2} & \frac{1}{2} & \tilde{0}\\ -\frac{1}{2} & \frac{1}{2} & 0\end{bmatrix}^{*D_2}_{C_2}$$

$$=(\alpha_c\beta_c+\beta_c\alpha_c)(-1)\sqrt{3}\cdot 1\cdot\frac{1}{\sqrt{3}}\cdot 1(-1)\frac{1}{\sqrt{2}}$$

$$=\frac{1}{\sqrt{2}}(\alpha_c\beta_c+\beta_c\alpha_c) \tag{5.3.12c}$$

Finally note that the product of spin functions in scheme d, $\mathbf{O}_3\supset\mathbf{O}_h\supset\mathbf{T}_d\supset\mathbf{C}_{3v}\supset\mathbf{C}_3\supset\mathbf{C}_1$, gives an even-parity $J=1$ ket. The O_3–O_h $3jm$ factor equals the SO_3–O $3jm$ factor without the parity ($\pm$ signs of the irreps), and the O_h–T_d $3jm$ factor is unity here, so the result is the same as the pure rotation scheme b:

$$|(\tfrac{1}{2}^+\tfrac{1}{2}^+)1^+1^+1\tilde{0}0\rangle_d=(1/\sqrt{2})(\alpha_d\beta_d+\beta_d\alpha_d) \tag{5.3.12d}$$

Repeating the calculation for the various other partners gives

$$|(\tfrac{1}{2}\tfrac{1}{2})11\rangle_a=\alpha_a\alpha_a$$

$$|(\tfrac{1}{2}\tfrac{1}{2})1111\rangle_b=-\alpha_b\alpha_b$$

$$|(\tfrac{1}{2}\tfrac{1}{2})11111\rangle_c=-(1/\sqrt{2})(\alpha_c\alpha_c+\beta_c\beta_c)$$

$$|(\tfrac{1}{2}^+\tfrac{1}{2}^+)1^+1111\rangle_d=-\alpha_d\alpha_d \tag{5.3.13}$$

and

$$|(\tfrac{1}{2}\tfrac{1}{2})1-1\rangle_a=\beta_a\beta_a$$

$$|(\tfrac{1}{2}\tfrac{1}{2})111-1\rangle_b=-\beta_b\beta_b$$

$$|(\tfrac{1}{2}\tfrac{1}{2})111\tilde{1}1\rangle_c=(1/\sqrt{2})(\alpha_c\alpha_c-\beta_c\beta_c) \tag{5.3.14}$$

$$|(\tfrac{1}{2}^+\tfrac{1}{2}^+)^+111-1\rangle_d=-\beta_d\beta_d$$

Polar vectors have the same transformation properties as axial vectors for all group elements excepting those containing an inversion or reflection. A polar vector may be obtained from an axial vector by coupling on a ket, $\gamma = |0^-\rangle$ being the solitary partner of an irrep of dimension one. The ket $\gamma = |0^-\rangle$ is absolutely invariant under rotations, changes sign under inversion, and is chosen to be the same in all schemes. For symmetry schemes a, b, and c the coupling coefficient is unity, e.g., for scheme b, $\mathbf{O}_3 \supset \mathbf{SO}_3 \supset \mathbf{O} \supset \mathbf{D}_3 \supset \mathbf{C}_3 \supset \mathbf{C}_1$,

$$|(1^+0^-)1^-1\tilde{0}0\rangle_b = |1^+1\tilde{0}0\rangle_b|0^-000\rangle_b \tag{5.3.15}$$

Under the subgroups of scheme d, $|0^-\rangle$ is not invariant; thus, for $\mathbf{O}_3 \supset \mathbf{O}_h \supset \mathbf{T}_d \supset \mathbf{C}_{3v} \supset \mathbf{C}_3 \supset \mathbf{C}_1$ the resultant may be obtained as

$$\begin{aligned} &|(1^+0^-)1^-1\tilde{1}11\rangle_d \\ &= |1^+1111\rangle_d|0^-0^-\tilde{0}\tilde{0}0\rangle_d\langle 1^+1111,0^-0^-\tilde{0}\tilde{0}0|1^-1\tilde{1}11\rangle \end{aligned} \tag{5.3.16}$$

The coupling coefficient may be factorized as before. It is customary to include the historical $\mathbf{SO}_3$ phase in $\mathbf{O}_3$ couplings; indeed, this is included in the statement that the $\mathbf{O}_3$ coefficients are the $\mathbf{SO}_3$ coefficients times a Kronecker delta, (5.1.2). The coupling coefficient may be evaluated

$$\begin{aligned} &\langle 1^-1\tilde{1}11|1^+1111,0^-0\tilde{0}\tilde{0}0\rangle \\ &= (-1)^{1-0+1}\sqrt{3}\begin{pmatrix} 1^+ & 0^- & 1^- \\ 1^+ & 0^- & 1^- \end{pmatrix}^{O_3}_{O_h}\begin{pmatrix} 1^+ & 0^- & 1^- \\ 1 & \tilde{0} & \tilde{1} \end{pmatrix}^{O_h}_{T_d} \\ &\times \begin{pmatrix} 1 & \tilde{0} & \tilde{1} \\ 1 & \tilde{0} & 1 \end{pmatrix}^{T_d}_{C_{3v}}\begin{pmatrix} 1 \\ 1 \end{pmatrix}^{C_{3v}}_{C_3}\begin{pmatrix} 1 & \tilde{0} & 1 \\ 1 & 0 & 1^* \end{pmatrix}^{C_{3v}}_{C_3} \end{aligned} \tag{5.3.17}$$

The O_3–O_h $3jm$ equals an SO_3–O $3jm$, the O_h–T_d $3jm$ is given by (5.2.9) and tabulated on p. 246, the T_d–C_{3v} $3jm$ equals the O–D_3 $3jm$ and is found on p. 242, and finally the C_{3v}–C_3 $2jm$ and $3jm$ are the D_3–C_3 jm's on p. 222. We have

$$\begin{aligned} &= (+1)\sqrt{3}\begin{pmatrix} 1 & 0 & 1 \\ 1 & 0 & 1 \end{pmatrix}^{SO_3}_{O}\begin{pmatrix} 1^+ & 0^- & 1^- \\ 1 & \tilde{0} & \tilde{1} \end{pmatrix}^{O_h}_{T_d} \\ &\times \begin{pmatrix} 1 & \tilde{0} & \tilde{1} \\ 1 & \tilde{0} & 1 \end{pmatrix}^{O}_{D_3}\begin{pmatrix} 1 \\ 1 \end{pmatrix}^{D_3}_{C_3}\begin{pmatrix} 1 & \tilde{0} & 1 \\ 1 & 0 & 1^* \end{pmatrix}^{D_3}_{C_3} \\ &= (+1)\sqrt{3}\cdot 1\cdot 1\cdot \frac{\sqrt{2}}{\sqrt{3}}\cdot(+1)\cdot\left(-\frac{1}{\sqrt{2}}\right) = -1 \end{aligned} \tag{5.3.18}$$

The coupling coefficients for the other two partners may be evaluated also, showing

$$|(1^+0^-)1^-1\tilde{1}11\rangle_d = -|1^+1111\rangle_d|0^-0^-\tilde{0}\tilde{0}0\rangle_d$$

$$|(1^+0^-)1^-1\tilde{1}1-1\rangle_d = +|1^+111-1\rangle_d|0^-0^-\tilde{0}\tilde{0}0\rangle_d \qquad (5.3.19)$$

$$|(1^+0^-)1^-1\tilde{1}00\rangle_d = +|1^+1100\rangle_d|0^-0^-\tilde{0}\tilde{0}0\rangle_d$$

Equations (5.3.12)–(5.3.14) give the partners of the $\mathbf{1}^+(\mathbf{O}_3)$ axial irrep. The relationship of the coordinate vectors to these coupled polar vectors is standard in the $|JM\rangle$ basis of scheme a, Eq. (5.3.9). Hence we may identify

$$|1^-1\rangle_a = -(1/\sqrt{2})(x+iy) \qquad (5.3.20)$$

with

$$|((\tfrac{1}{2}\tfrac{1}{2})1^+,0^-)1^-1\rangle_a = \alpha_a\alpha_a\gamma \qquad (5.3.21)$$

etc., obtaining (γ the pseudoscalar $|0^-\rangle$)

$$-(1/\sqrt{2})(x+iy) = \alpha_a\alpha_a\gamma = |1^-1\rangle_a$$

$$z = (1/\sqrt{2})(\alpha_a\beta_a+\beta_a\alpha_a)\gamma = |1^-0\rangle_a$$

$$(1/\sqrt{2})(x-iy) = \beta_a\beta_a\gamma = |1^- -1\rangle_a \qquad (5.3.22)$$

The above arguments have all been in terms of the abstract group structure and the imbeddings of one group in another. Now the octahedron associated with the octahedral group used in schemes b, c, and d may be orientated however one likes in 3-space. The abstract properties of the imbedding and thus the relationship of the three $\mathbf{1}^-(\mathbf{O}_3)$ functions to the spinor functions α and β are not affected by our choice of orientation. By the same argument, the preceding calculation in scheme b, $\mathbf{O}_3 \supset \mathbf{SO}_3 \supset \mathbf{O} \supset \mathbf{D}_3 \supset \mathbf{C}_3 \supset \mathbf{C}_1$, is unaffected by our choice of which threefold axis of $\mathbf{O}$ is used for the groups $\mathbf{D}_3$ and $\mathbf{C}_3$. In (5.3.4)–(5.3.7) it was shown that the $M(SO_2)$ label gave the behavior of any partners of the basis of scheme a under rotations about the z axis, and the $\lambda(C_3)$ label of scheme b gave the behavior under rotations about the $\mathbf{C}_3$ axis. The JM or z axis must be fixed if we are to fix the relationship of the partners of scheme b to those of scheme a. For the present choose axis systems in the four schemes so that the z axis is the $\mathbf{SO}_2$ axis of scheme a, the $\mathbf{C}_3$ axis of schemes b and d, and the $\mathbf{C}_2$ axis of scheme c. This choice orients our octahedron differently in the different schemes. For example, in scheme b one of the three-fold rotations of $\mathbf{O}$ is about the z axis, while in scheme c no three fold rotation is about the z axis, for z is a four-fold rotation axis.

With the recognition that the axis systems of the four schemes are distinct, that is, we cannot always simply equate the axes of one scheme with those of another, the above calculations show that we may write the partners of $\mathbf{1}^-(\mathbf{O}_3)$ in terms of the respective x, y, z functions as

$$
\begin{aligned}
|1^-1\rangle_a &= \frac{-1}{\sqrt{2}}(x+iy) & |1^-0\rangle_a &= z \\
|1^-111\rangle_b &= \frac{1}{\sqrt{2}}(x+iy) & |1^-1\tilde{0}0\rangle_b &= z \\
|1^-1111\rangle_c &= iy & |1^-1\tilde{0}\tilde{0}0\rangle_c &= z \\
|1^-1^-\tilde{1}11\rangle_d &= \frac{-1}{\sqrt{2}}(x+iy) & |1^-1^-\tilde{1}00\rangle_d &= z
\end{aligned}
$$

$$
\begin{aligned}
|1^- -1\rangle_a &= \frac{1}{\sqrt{2}}(x-iy) \\
|1^-11-1\rangle_b &= \frac{-1}{\sqrt{2}}(x-iy) \\
|1^-11\tilde{1}1\rangle_c &= -x \\
|1^-1^-\tilde{1}1-1\rangle_d &= \frac{-1}{\sqrt{2}}(x-iy)
\end{aligned}
\tag{5.3.23}
$$

These explicit relationships between the irrep partners and the coordinate functions may be used to further the Zeeman perturbation calculation begun in Section 4.2. That problem was posed in scheme b. The Zeeman term H_{Zeeman} was written as an SO_3 reduced matrix element and the SO_3 unit tensors U^{1^+110}, $U^{1^+1\tilde{0}1}$, and U^{1^+11-1}. The linear combination of the unit tensors depends on the relationship of the laboratory (Zeeman) z axis and the crystal x', y', z' axes. For the sake of definiteness assume that the magnetic field is along the x' crystal axis. Inspection of (5.3.23) shows that

$$x' = z = (1/\sqrt{2})(|1^-111\rangle_b - |1^-11-1\rangle_b) \tag{5.3.24}$$

Axial and polar vectors transform alike in scheme b; hence in (4.2.7), $c_1 = 1/\sqrt{2}$ and $c_{-1} = -1/\sqrt{2}$.

The coefficients describing transformations between two basis schemes were defined in (2.2.7). If it is our purpose to expand the partners of an irrep in one scheme in terms of the partners of the same irrep in another scheme, then the properties of transformation coefficients may be used instead. Indeed, it is much easier to do so. As an example of the calculation, we use (3.3.40) to obtain the relationship between the partners of the $\mathbf{1}^+(\mathbf{O}_3)$ irrep in the a and b schemes by this alternative method.

We showed in (5.3.4)–(5.3.7) that the assumption that the spinor functions $\alpha_a\beta_a$ are the same as the functions $\alpha_b\beta_b$ implies the SO_3 axis is the same as the C_3 axis,

$$|\tfrac{1}{2}\tfrac{1}{2}\rangle_a = |\tfrac{1}{2}\tfrac{1}{2}\tfrac{1}{2}\tfrac{1}{2}\rangle_b \tag{5.3.25a}$$

$$|\tfrac{1}{2}-\tfrac{1}{2}\rangle_a = |\tfrac{1}{2}\tfrac{1}{2}\tfrac{1}{2}-\tfrac{1}{2}\rangle_b \tag{5.3.25b}$$

The second equation follows from the first. Equation (3.3.40) implies that the scalar partners are identical in all point group schemes

$$_a\langle 00|0000\rangle_b = 1 \tag{5.3.26a}$$

Inserting (5.3.25a) into (3.3.40) shows immediately that the relationship between the second partners is fixed as (3.3.25b).

This point arose in our discussion in Section 3.5 concerning phase choices for primitive $3jm$'s. The primitive $3jm$'s fix the relationship between the partners of each irrep; the only freedom remaining is the choice of orientation of the b system with respect to the a system. This freedom is the freedom given by the Euler angles of the $D^{1/2}(\alpha,\beta,\gamma)$ matrix with elements A, B, C, D. [See (5.3.8) and Section 16.1.]

For our present situation we have

$$A = D = {}_a\langle\tfrac{1}{2}\tfrac{1}{2}|\tfrac{1}{2}\tfrac{1}{2}\tfrac{1}{2}\tfrac{1}{2}\rangle_b = {}_a\langle\tfrac{1}{2}-\tfrac{1}{2}|\tfrac{1}{2}\tfrac{1}{2}\tfrac{1}{2}-\tfrac{1}{2}\rangle_b = 1 \tag{5.3.26b}$$

and

$$B = C = {}_a\langle\tfrac{1}{2}\tfrac{1}{2}|\tfrac{1}{2}\tfrac{1}{2}\tfrac{1}{2}-\tfrac{1}{2}\rangle_b = {}_a\langle\tfrac{1}{2}-\tfrac{1}{2}|\tfrac{1}{2}\tfrac{1}{2}\tfrac{1}{2}\tfrac{1}{2}\rangle_b = 0 \tag{5.3.26c}$$

These transformation coefficients may be used in (3.3.40) to give the SO_3–SO_2–C_1 to SO_3–O–D_3–C_3–C_1 transformation as

$$
{}_a\langle(\tfrac{1}{2}\tfrac{1}{2})11|(\tfrac{1}{2}\tfrac{1}{2})1111\rangle_b^* = {}_a\langle 11|1111\rangle_b^*
$$

$$
= \sum_{m_1 m_2 \lambda_1 \lambda_2} |1|_{SO_3} \begin{pmatrix} \frac{1}{2} & \frac{1}{2} & 1 \\ m_1 & m_2 & 1 \end{pmatrix}^{SO_3}_{SO_2} \langle \tfrac{1}{2} m_1 | \tfrac{1}{2}\tfrac{1}{2}\tfrac{1}{2}\lambda_1 \rangle
$$

$$
\times \langle \tfrac{1}{2} m_2 | \tfrac{1}{2}\tfrac{1}{2}\tfrac{1}{2}\lambda_2 \rangle \begin{bmatrix} \frac{1}{2} & \frac{1}{2} & 1 \\ \frac{1}{2} & \frac{1}{2} & 1 \end{bmatrix}^{*SO_3}_{O} \begin{bmatrix} \frac{1}{2} & \frac{1}{2} & 1 \\ \frac{1}{2} & \frac{1}{2} & 1 \end{bmatrix}^{*O}_{D_3} \begin{pmatrix} \frac{1}{2} & \frac{1}{2} & 1 \\ \lambda_1 & \lambda_2 & 1 \end{pmatrix}^{*D_3}_{C_3}
\tag{5.3.27}
$$

because the $3jm$ symbol for the b scheme is a product of the $3jm$ factors and only one term survives in the branching until $\mathbf{SO_2}$ or $\mathbf{C_3}$. The summation is reduced to one term by (5.3.26), hence

$$
{}_b\langle 1111|11\rangle_a = 3 \begin{bmatrix} \frac{1}{2} & \frac{1}{2} & 1 \\ -\frac{1}{2} & -\frac{1}{2} & 1 \end{bmatrix}^{SO_3}_{SO_2} 1\cdot 1 \begin{bmatrix} \frac{1}{2} & \frac{1}{2} & 1 \\ \frac{1}{2} & \frac{1}{2} & 1 \end{bmatrix}^{*SO_3}_{O}
$$

$$
\times \begin{bmatrix} \frac{1}{2} & \frac{1}{2} & 1 \\ \frac{1}{2} & \frac{1}{2} & 1 \end{bmatrix}^{*O}_{D_3} \begin{bmatrix} \frac{1}{2} & \frac{1}{2} & 1 \\ -\frac{1}{2} & -\frac{1}{2} & 1 \end{bmatrix}^{*D_3}_{C_3}
$$

$$
= 3\cdot \frac{-1}{\sqrt{3}} \cdot 1\cdot 1\cdot 1 \frac{\sqrt{2}}{\sqrt{3}} \frac{1}{\sqrt{2}}
$$

$$
= -1 \tag{5.3.28}
$$

which checks the calculation which ended at (5.3.13). Any other transformation coefficients may be obtained by similar calculations, as indeed were the tables of Chapter 16.

The next point to consider is the case when the two axis systems are not coincident but where α, β, γ are the Euler angles relating the axes of the two schemes. The spinor functions α_a and β_a may be rotated by the standard rotation matrices $D^{1/2}(\alpha, \beta, \gamma)$; see Section 16.1. The matrix elements of $D^{1/2}(\alpha, \beta, \gamma)$ appear as the transformation coefficients in

$$|\tfrac{1}{2}\tfrac{1}{2}\tfrac{1}{2}\lambda\rangle_b = |\tfrac{1}{2}\tfrac{1}{2}\rangle_a\langle\tfrac{1}{2}\tfrac{1}{2}|\tfrac{1}{2}\tfrac{1}{2}\tfrac{1}{2}\lambda\rangle + |\tfrac{1}{2}-\tfrac{1}{2}\rangle_a\langle\tfrac{1}{2}-\tfrac{1}{2}|\tfrac{1}{2}\tfrac{1}{2}\tfrac{1}{2}\lambda\rangle \qquad (5.3.29)$$

and these general coefficients $\langle\tfrac{1}{2}\, m|\tfrac{1}{2}\tfrac{1}{2}\tfrac{1}{2}\lambda\rangle$ may be used in (5.3.27).

As an example of noncoincident axis systems let us find a possible relationship between the basis functions of the D_4 and D_3 axes of schemes c and b, namely the schemes $\mathbf{O}_3 \supset \mathbf{O} \supset \mathbf{D}_4 \supset \mathbf{D}_2 \supset \mathbf{C}_2 \supset \mathbf{C}_1$ and $\mathbf{O}_3 \supset \mathbf{O} \supset \mathbf{D}_3 \supset \mathbf{C}_3 \supset \mathbf{C}_1$. Call the axes xyz and XYZ, respectively. We regard xyz as the laboratory and JM axes. It is possible to consider any arbitrary relationship between the xyz and XYZ axis systems, but let us restrict our consideration to the special case of having the octahedron (or cube) associated with scheme b coincident with the octahedron (or cube) of scheme c. The two groups $\mathbf{O}$, one for each scheme, are therefore identical and hence the transformation coefficients between our two schemes will be diagonal in the $\mathbf{O}$ labels. Figures 11.5 and 11.6 are shown superimposed in Figure 5.1, from which it can be seen that the XYZ axes (that is, the b-scheme axes) can be rotated into the xyz axis system (the c scheme) by the following sequence of rotations. First, rotate about the laboratory z axis by angle $\gamma' = \pi/4$. Second, rotate about the laboratory y axis by angle $\beta' = \cos^{-1}(1/\sqrt{3})$. Third, rotate about the laboratory z axis by angle $\alpha' = 0$. In Figure 5.1 we mark the angles and the intermediate position X' of the X axis in the equivalent body axis description (see p. 515), namely

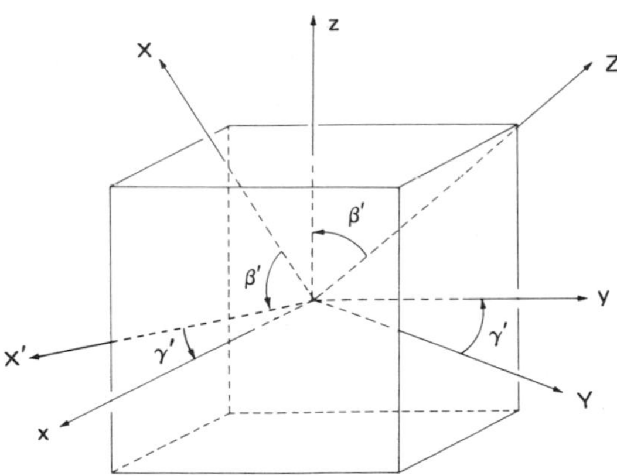

Figure 5.1. The O–D_4–D_2–C_2 axes xyz and the O–D_3–C_3 axes XYZ related by Euler angles $\alpha' = 0$, $\beta' = \cos^{-1}(1/\sqrt{3})$, $\gamma' = \pi/4$. Note that this is the passive viewpoint.

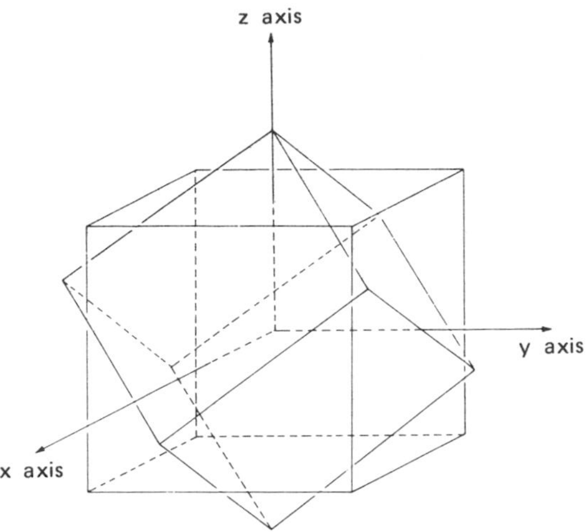

Figure 5.2. A cube in the O–D_4–D_2–C_2 orientation and a cube in the O–D_3–C_3 orientation related by Euler angles $\alpha=-\pi/4$, $\beta=-\cos^{-1}(1/\sqrt{3})$, $\gamma=0$.

$R_{Z''}(\gamma')R_{Y'}(\beta')R_Z(\alpha')$, because this description shows the angle β' most clearly.

In the following discussion we follow our active convention for rotations (Section 16.1) and discuss the transformation between the partners of the one cube expressed in terms of the alternative axis systems. Therefore we must rotate the partners in the opposite direction; $\alpha=-\pi/4$, $\beta=-\cos^{-1}(1/\sqrt{3})$, $\gamma=0$. Figure 5.2 displays this active viewpoint. In this figure the two axis systems coincide, but the cubes do not. We must express the b scheme O–D_3–C_3 partners in the xyz system, and doing this requires rotating the O–D_3–C_3 cube: first, by an angle $\gamma=0$ about z; second, by an angle $\beta=-\cos^{-1}(1/\sqrt{3})$ about y; and third, by an angle $\alpha=-\pi/4$ about z.

For the spinor functions α_a and β_a of the JM basis, one has

$$\begin{aligned}\alpha_a(\mathrm{rot})&=\alpha_a A+\beta_a B\\ \beta_a(\mathrm{rot})&=\alpha_a(-B^*)+\beta_a A^*\end{aligned} \qquad (5.3.30)$$

where (Section 16.1)

$$\begin{aligned}A&=e^{-i\alpha/2}\cos\tfrac{1}{2}\beta e^{-i\gamma/2}\\ B&=e^{i\alpha/2}\sin\tfrac{1}{2}\beta\, e^{-i\gamma/2}\end{aligned} \qquad (5.3.31)$$

Now our c-scheme axes are JM axes; so, as before, $\alpha_a = \alpha_c$, $\beta_a = \beta_c$, but the rotated spinors are the α_b and β_b functions of the XYZ scheme, $\alpha_a(\text{rot}) = \alpha_b$, $\beta_a(\text{rot}) = \beta_b$. Therefore

$$\begin{aligned} \alpha_b &= \alpha_c A + \beta_c B \\ \beta_b &= \alpha_c(-B^*) + \beta_c A^* \end{aligned} \tag{5.3.32}$$

or using the group-theoretic labels of the $\mathbf{O} \supset \mathbf{D}_4 \supset \mathbf{D}_2 \supset \mathbf{C}_2$ and $\mathbf{O} \supset \mathbf{D}_3 \supset \mathbf{C}_3$ schemes, we have

$$\begin{aligned} |\tfrac{1}{2}\tfrac{1}{2}\tfrac{1}{2}\rangle_b &= |\tfrac{1}{2}\tfrac{1}{2}\tfrac{1}{2}\tfrac{1}{2}\rangle_c \langle \tfrac{1}{2}\tfrac{1}{2}\tfrac{1}{2}\tfrac{1}{2}|\tfrac{1}{2}\tfrac{1}{2}\tfrac{1}{2}\rangle + |\tfrac{1}{2}\tfrac{1}{2}\tfrac{1}{2}-\tfrac{1}{2}\rangle_c \langle \tfrac{1}{2}\tfrac{1}{2}\tfrac{1}{2}-\tfrac{1}{2}|\tfrac{1}{2}\tfrac{1}{2}\tfrac{1}{2}\rangle \\ |\tfrac{1}{2}\tfrac{1}{2}-\tfrac{1}{2}\rangle_b &= |\tfrac{1}{2}\tfrac{1}{2}\tfrac{1}{2}\tfrac{1}{2}\rangle_c \langle \tfrac{1}{2}\tfrac{1}{2}\tfrac{1}{2}\tfrac{1}{2}|\tfrac{1}{2}\tfrac{1}{2}-\tfrac{1}{2}\rangle + |\tfrac{1}{2}\tfrac{1}{2}\tfrac{1}{2}-\tfrac{1}{2}\rangle_c \langle \tfrac{1}{2}\tfrac{1}{2}\tfrac{1}{2}-\tfrac{1}{2}|\tfrac{1}{2}\tfrac{1}{2}-\tfrac{1}{2}\rangle \end{aligned} \tag{5.3.33}$$

where

$$\begin{aligned} \langle \tfrac{1}{2}\tfrac{1}{2}\tfrac{1}{2}\tfrac{1}{2}|\tfrac{1}{2}\tfrac{1}{2}\tfrac{1}{2}\rangle &= A = e^{-i\alpha/2}\cos\tfrac{1}{2}\beta\, e^{-i\gamma/2} = \langle \tfrac{1}{2}\tfrac{1}{2}\tfrac{1}{2}-\tfrac{1}{2}|\tfrac{1}{2}\tfrac{1}{2}-\tfrac{1}{2}\rangle^* \\ \langle \tfrac{1}{2}\tfrac{1}{2}\tfrac{1}{2}-\tfrac{1}{2}|\tfrac{1}{2}\tfrac{1}{2}\tfrac{1}{2}\rangle &= B = e^{i\alpha/2}\sin\tfrac{1}{2}\beta\, e^{-i\gamma/2} = -\langle \tfrac{1}{2}\tfrac{1}{2}\tfrac{1}{2}\tfrac{1}{2}|\tfrac{1}{2}\tfrac{1}{2}-\tfrac{1}{2}\rangle^* \end{aligned} \tag{5.3.34}$$

Equation (3.3.40) may now be used to obtain all transformation coefficients. For example, we have in detail

$${}_b\langle 111|1111\rangle_c$$

$$= |1| \left[\begin{matrix} \tfrac{1}{2} & \tfrac{1}{2} & 1 \\ \tfrac{1}{2} & \tfrac{1}{2} & 1 \\ \tfrac{1}{2} & \tfrac{1}{2} & 1 \\ \tfrac{1}{2} & \tfrac{1}{2} & 1 \end{matrix}\right]\begin{matrix} O \\ D_4 \\ D_2 \\ C_2 \end{matrix} \langle \tfrac{1}{2}\tfrac{1}{2}\tfrac{1}{2}\tfrac{1}{2}|\tfrac{1}{2}\tfrac{1}{2}-\tfrac{1}{2}\rangle \langle \tfrac{1}{2}\tfrac{1}{2}\tfrac{1}{2}\tfrac{1}{2}|\tfrac{1}{2}\tfrac{1}{2}-\tfrac{1}{2}\rangle \left[\begin{matrix} \tfrac{1}{2} & \tfrac{1}{2} & 1 \\ \tfrac{1}{2} & \tfrac{1}{2} & 1 \\ -\tfrac{1}{2} & -\tfrac{1}{2} & 1 \end{matrix}\right]\begin{matrix} {}^*O \\ D_3 \\ C_3 \end{matrix}$$

$$+ |1| \left[\begin{matrix} \tfrac{1}{2} & \tfrac{1}{2} & 1 \\ \tfrac{1}{2} & \tfrac{1}{2} & 1 \\ \tfrac{1}{2} & \tfrac{1}{2} & 1 \\ -\tfrac{1}{2} & -\tfrac{1}{2} & 1 \end{matrix}\right]\begin{matrix} O \\ D_4 \\ D_2 \\ C_2 \end{matrix} \langle \tfrac{1}{2}\tfrac{1}{2}\tfrac{1}{2}-\tfrac{1}{2}|\tfrac{1}{2}\tfrac{1}{2}-\tfrac{1}{2}\rangle$$

$$\times \langle \tfrac{1}{2}\tfrac{1}{2}\tfrac{1}{2}-\tfrac{1}{2}|\tfrac{1}{2}\tfrac{1}{2}-\tfrac{1}{2}\rangle \left[\begin{matrix} \tfrac{1}{2} & \tfrac{1}{2} & 1 \\ \tfrac{1}{2} & \tfrac{1}{2} & 1 \\ -\tfrac{1}{2} & -\tfrac{1}{2} & 1 \end{matrix}\right]\begin{matrix} {}^*O \\ D_3 \\ C_3 \end{matrix}$$

$$=3\frac{\sqrt{2}}{\sqrt{3}}\frac{1}{\sqrt{2}}\frac{1}{\sqrt{2}}(-B^*)^2\frac{\sqrt{2}}{\sqrt{3}}\frac{1}{\sqrt{2}}+3\frac{\sqrt{2}}{\sqrt{3}}\frac{1}{\sqrt{2}}\frac{1}{\sqrt{2}}(A^*)^2\frac{\sqrt{2}}{\sqrt{3}}\frac{1}{\sqrt{2}}$$

$$=\frac{1}{\sqrt{2}}(B^*B^*+A^*A^*)$$

$$=\frac{1}{\sqrt{2}}\left(e^{-i\alpha+i\gamma}\sin^2\frac{\beta}{2}+e^{i\alpha+i\gamma}\cos^2\frac{\beta}{2}\right)$$

$$_b\langle 111|11\tilde{1}1\rangle_c=\frac{1}{\sqrt{2}}(B^*B^*-A^*A^*)$$

$$=\frac{1}{\sqrt{2}}\left(e^{-i\alpha+i\gamma}\sin^2\frac{\beta}{2}-e^{i\alpha+i\gamma}\cos^2\frac{\beta}{2}\right)$$

$$_b\langle 111|1\tilde{0}\tilde{0}0\rangle_c=-\sqrt{2}\,e^{i\gamma}\sin\frac{\beta}{2}\cos\frac{\beta}{2} \tag{5.3.35}$$

For the transformation of Figure 5.2 these expressions evaluate as

$$_b\langle 111|1111\rangle_c=\frac{1}{2}\left[(1+i)\sin^2\frac{\beta}{2}+(1-i)\cos^2\frac{\beta}{2}\right]$$

$$=\frac{1}{2}(1-i\cos\beta)=\frac{1}{2}-\frac{i}{2\sqrt{3}}$$

$$_b\langle 111|11\tilde{1}1\rangle_c=-\frac{1}{2\sqrt{3}}+\frac{i}{2}$$

$$_b\langle 111|1\tilde{0}\tilde{0}0\rangle_c=\frac{1}{\sqrt{3}} \tag{5.3.36}$$

These transformation coefficients, and the others for $|11-1\rangle_b$ and $|1\tilde{0}0\rangle_b$, may be checked by explicit comparison of the axis systems themselves. Inspection of Figure 5.1 shows that the XYZ axes are given in terms of xyz axes by

$$X=\frac{1}{\sqrt{6}}(x-y+2z),\qquad Y=\frac{1}{\sqrt{2}}(x+y),\qquad Z=\frac{1}{\sqrt{3}}(-x+y+z) \tag{5.3.37}$$

The basis functions were related to the axes by a previous calculation, in which we showed [Eq. (5.3.23)] that $|111\rangle_b = (1/\sqrt{2})(X+iY)$, etc., and $|1111\rangle_c = iy$, etc. Combining these with (5.3.37) shows

$$
\begin{aligned}
{}_b\langle 111|1111\rangle_c &= \left\langle \frac{1}{\sqrt{2}}(X+iY)|iy \right\rangle \\
&= \left\langle \frac{1}{2\sqrt{3}}(x-y+2z)+\frac{i}{2}(x+y)|iy \right\rangle \\
&= \left\langle \left(-\frac{1}{2\sqrt{3}}+\frac{i}{2}\right)y|iy \right\rangle
\end{aligned}
$$

because x and z are orthogonal to y,

$$
= \left(-\frac{1}{2\sqrt{3}}-\frac{i}{2}\right)i = \frac{1}{2}-\frac{i}{2\sqrt{3}} \tag{5.3.38}
$$

because the scalar product complex-conjugates the coefficients in the bra. This result is in agreement with (5.3.36). The reader may wish to obtain ${}_b\langle 11-1|11\tilde{1}1\rangle_c$ by both methods.

The tables of Chapter 16 were obtained by the first method. Observe that the Euler angle of $\cos^{-1}(1/\sqrt{3})$ does not lead to simple numbers for spin functions. An alternative method of obtaining transformation coefficients between chains rotated with respect to each other is to use the expressions for the partners in terms of the JM basis (Chapter 16) and then to use the JM rotation matrices of Section 16.1. The distinction between these methods is only the question of whether one rotates before building up or vice versa.

The relationship just found between the O–D_4–D_2–C_2 and O–D_3–C_3 chains is not unique. We required the respective cubes to be the same, a requirement which lead to the expressions (5.3.37) relating the two axis systems. However, an operation of $\mathbf{O}$ will return the cube to an equivalent position, and any such operation may be included in the Euler angles α, β, γ. Indeed, the above example differs from the table on p. 549 by $\alpha=\pi/2$. We illustrate this property of the group operators, and at the same time demonstrate the effect of the orientation phase choice of Section 3.5, by considering the chain $\mathbf{D}_3 \supset \mathbf{C}_3$.

Consider two D_3–C_3 systems rotated with respect to one another by an angle α about the z axis. Labeling the systems e and f, (5.3.34) becomes

$$
{}_e\langle \tfrac{1}{2}\tfrac{1}{2}|\tfrac{1}{2}\tfrac{1}{2}\rangle_f = e^{-i\alpha/2} = {}_e\langle \tfrac{1}{2}-\tfrac{1}{2}|\tfrac{1}{2}-\tfrac{1}{2}\rangle_f^* \tag{5.3.39}
$$

Building up with the aid of (3.3.40) shows that

$$ {}_e\langle 11|11\rangle_f = e^{-i\alpha} = {}_e\langle 1-1|1-1\rangle_f^* \tag{5.3.40}$$

while

$$ {}_e\langle \tfrac{3}{2}\tfrac{3}{2}|\tfrac{3}{2}\tfrac{3}{2}\rangle_f = \tfrac{1}{2}e^{-3i\alpha/2}(e^{3i\alpha}+1) \tag{5.3.41}$$

If any SO_3–G–D_3–C_3 chain is used, the off-diagonal terms are nonzero and the two $\mathbf{D}_3$ irreps $\frac{3}{2}$ and $-\frac{3}{2}$ are mixed. This mixing is zero only if the coefficient of (5.3.41) is norm 1, that is, if $e^{3i\alpha}=1$. This is the case for α a $\mathbf{D}_3$ angle, namely a multiple of $2\pi/3$. Recall that we only considered in (5.3.39) rotations about the z axis; equivalent results for two fold rotations about the y axis may be obtained.

The D_3–C_3 system exhibits the orientation phase choice. In the calculation of primitive D_3–C_3 $3jm$'s all the free phases associated with phase freedoms for the partners have been used by the time one computes

$$\begin{pmatrix} \frac{3}{2} & 1 & \frac{1}{2} \\ \frac{3}{2} & -1 & -\frac{1}{2} \end{pmatrix}$$

Taking this $3jm$ to be $i/\sqrt{2}$ as in the tables of Chapter 13 for scheme e, and as $ie^{i\theta}/\sqrt{2}$ for scheme f, one may rederive (5.3.41) from (5.3.40) as

$$\begin{aligned}
{}_e\langle \tfrac{3}{2}\tfrac{3}{2}|\tfrac{3}{2}\tfrac{3}{2}\rangle_f &= \begin{pmatrix} \frac{1}{2} & 1 & \frac{3}{2} \\ -\frac{1}{2} & -1 & \frac{3}{2} \end{pmatrix}_e^* {}_e\langle \tfrac{1}{2}-\tfrac{1}{2}|\tfrac{1}{2}-\tfrac{1}{2}\rangle_f^* \\
&\quad\times {}_e\langle 1-1|1-1\rangle_f^* \begin{pmatrix} \frac{1}{2} & 1 & \frac{3}{2} \\ -\frac{1}{2} & -1 & \frac{3}{2} \end{pmatrix}_f \\
&\quad+ \begin{pmatrix} \frac{1}{2} & 1 & \frac{3}{2} \\ \frac{1}{2} & 1 & \frac{3}{2} \end{pmatrix}_e^* \langle \tfrac{1}{2}\tfrac{1}{2}|\tfrac{1}{2}\tfrac{1}{2}\rangle_f^* \, {}_e\langle 11|11\rangle_f^* \begin{pmatrix} \frac{1}{2} & 1 & \frac{3}{2} \\ \frac{1}{2} & 1 & \frac{3}{2} \end{pmatrix}_f \\
&= \frac{-i}{\sqrt{2}} e^{i\alpha/2} e^{i\alpha} \frac{ie^{i\theta}}{\sqrt{2}} + \frac{1}{\sqrt{2}} e^{-i\alpha/2} e^{-i\alpha} \frac{1}{\sqrt{2}} \\
&= \tfrac{1}{2} e^{-3i\alpha/2}(e^{3i\alpha+i\theta}+1)
\end{aligned} \tag{5.3.42}$$

This final equation shows that changing the D_3–C_3 orientation phase by an angle θ mixes the $\frac{3}{2}$ and $-\frac{3}{2}$ irreps of $\mathbf{D}_3$ in the same way as a rotation about the C_3 axis by an angle 3α. The effect of the particular orientation phase choices made in this book is to imply a default orientation of the D_n and C_n systems as described in the character tables of Chapter 11.

6

Properties of the Dihedral Groups

The group $\mathbf{D}_\infty$ is of considerable interest to us. It is one of the three maximal subgroups of $\mathbf{SO}_3$, the others being **O** and **K**. The group $\mathbf{D}_\infty$ contains all $\mathbf{D}_n$ and $\mathbf{C}_n$ for all n, including $\mathbf{C}_\infty = \mathbf{SO}_2$, the group of all rotations about a single axis. As such, it not only appears in many instances as a covering group for a finite point group, but also appears in the analysis of the JM basis. $\mathbf{D}_\infty$ is a simply reducible group in the sense of Wigner (1940) and this together with the simple algebraic nature of its representations leads to simple algebraic formulas for D_∞ $6j$ symbols.

As we shall see in Section 6.2, the phase choices for JM kets are such that JM kets are not symmetry-adapted to the $\mathbf{SO}_3 \supset \mathbf{D}_\infty \supset \mathbf{SO}_2$ symmetry scheme. However, the symmetry adaptation involves a simple multiplicative phase, so the Racah factorization lemma leads immediately to SO_3–D_∞ $3jm$ factors. The tables of Rotenberg, Bivins, Metropolis, and Wooten (1959) for SO_3–SO_2 $3jm$ factors are easily adapted for our purposes. Our tables therefore include only the different possible subgroup schemes of $\mathbf{D}_\infty$, not $\mathbf{SO}_3 \supset \mathbf{D}_\infty$. Physical symmetries for $\mathbf{D}_n$, $n > 6$, are uncommon, but Section 6.3 shows how the phases may be chosen systematically to give D_n–D_m or D_n–C_n $3jm$'s.

The contents of this chapter together with some algebraic results were published by Butler and Reid (1979). Alternative calculations have been carried out by Harnung and Schäffer (1972), Golding and Newmarch (1977), and Kibler and Grenet (1977).

6.1 The Structure of the Dihedral Groups

In this section we give a brief summary of the class structure and the irrep structure of the dihedral groups. The character tables of Section 11.2 will be referred to several times. The dihedral groups form straightforward examples of a number of the character theory concepts of Section 2.4 and can be used to illustrate the calculation of $6j$ symbols for cases without multiplicity.

The abstract dihedral group $\mathbf{D}_n$ may be realized as the pure rotation group composed of n rotations $\exp(-iJ_z\theta)$ about the z axis together with n twofold rotations R_ϕ about certain axes in the xy plane. The angles θ and ϕ take the values $\theta=2\pi m/n$ and $\phi=\pi m/n$ for $0\leqslant m<n$. The axis of the rotation R_ϕ is in the xy plane at an angle of ϕ to the y axis. A little manipulation of a solid object should convince the reader that

$$R_\phi=R_\phi^{-1}=R_{\phi+\pi} \tag{6.1.1}$$

$$R_\phi \exp(-iJ_z\theta)\,R_\phi=\exp(iJ_z\theta)=\exp[\,iJ_z(\theta-2\pi)\,] \tag{6.1.2}$$

and

$$\exp(-iJ_z\theta)\,R_\phi \exp(iJ_z\theta)=R_{\phi+2\theta} \tag{6.1.3}$$

The first of these equations indicates that the parametrization of the R_ϕ is sufficient as given. The second shows that the elements $\exp(-iJ_z\theta)$ and $\exp(iJ_z\theta)$ belong to the same conjugacy class. (If $\theta=0$ or π they are the one element.) The third shows that the series of elements R_ϕ, $R_{\phi+4\pi/n}, R_{\phi+8\pi/n},\ldots$ belong to the one class. If n is odd ($n=2k+1$), then this series contains all elements R_ϕ, because $R_\phi=R_{\phi+\pi}$. For example, for $\mathbf{D}_5$ we have the $\phi=0$ series as $R_0, R_{4\pi/5}, R_{8\pi/5}=R_{3\pi/5}, R_{12\pi/5}=R_{2\pi/5}, R_{16\pi/5}=R_{\pi/5}$. If n is even, the series generated by $\phi=0$ and $\phi=\pi/n$ are distinct. When the double-valued rotations are included to form the double group, one obtains the class structures as shown in Tables 11.7–11.11.

The character table and hence the irrep structure of the group $\mathbf{D}_n$ may be readily built up from the nature of the irrep $\frac{1}{2}(\mathbf{SO}_3)$ and tools of character theory, especially using the orthogonality (2.4.17) of characters of irreps to prove irreducibility. The pair of kets $|\frac{1}{2}(SO_3)\pm\frac{1}{2}(SO_2)\rangle$ are the two partners of a spin irrep $\frac{1}{2}(\mathbf{D}_n)$. (We assume $n>1$.) The characteristics of this irrep [the traces of the operators $\exp(-iJ_z\theta)$ and R_ϕ] may be obtained as traces of special cases of the $\mathbf{SO}_3$ formula (Section 16.1) or by direct calculation. The character table for $\mathbf{D}_n$ may be readily constructed by multiplying this irrep by itself and reducing the product using the methods on p. 35. If, at the same, one keeps a record of the J and M values of the partners of products, one produces the branching rules for $\mathbf{SO}_3 \supset \mathbf{D}_\infty$:

$$\begin{aligned} \mathbf{J} &\to \mathbf{0}+\mathbf{1}+\mathbf{2}+\cdots+\mathbf{J} && \text{for } J \text{ even} \\ &\to \tilde{\mathbf{0}}+\mathbf{1}+\mathbf{2}+\cdots+\mathbf{J} && \text{for } J \text{ odd} \\ &\to \tfrac{\mathbf{1}}{\mathbf{2}}+\tfrac{\mathbf{3}}{\mathbf{2}}+\cdots+\mathbf{J} && \text{for } J \text{ half-integer} \end{aligned} \tag{6.1.4}$$

and for $\mathbf{D}_\infty \supset \mathbf{SO}_2$,

$$\begin{aligned} &\mathbf{N} \to \mathbf{N}+(-\mathbf{N}) && \text{for } N \neq 0, \tilde{0} \\ &\mathbf{0} \to \mathbf{0} \\ &\tilde{\mathbf{0}} \to \mathbf{0} \end{aligned} \tag{6.1.5}$$

The finite dihedral group $\mathbf{D}_n$ has $n-1$ two-dimensional irreps and four one-dimensional irreps. For n even all irreps are real, while for n odd the two one-dimensional irreps $\frac{1}{2}\mathbf{n}$ and $-\frac{1}{2}\mathbf{n}$ are complex conjugates of each other. The branching rules for $\mathbf{D}_\infty \supset \mathbf{D}_6$, $\mathbf{D}_\infty \supset \mathbf{D}_5$, and $\mathbf{D}_\infty \supset \mathbf{D}_4$ are given in Chapter 12.

Either the character formulas (3.2.8) and (3.2.9) or the knowledge of $\mathbf{SO}_3$ symmetrized products give the values of the $3j$ phases for the various $\mathbf{D}_n$ as tabulated in Chapter 14. Where a choice occurs we have chosen the simple algebraic formula $\{a, b, c\} = (-1)^{a+b+c}$ because it is usually satisfied where there is no choice. The only exceptions occur for n odd for the particular $3j$ phases of the form $\{a, a, n-2a\}$. In such cases the $3j$ is positive but the algebraic formula gives a negative sign. These exceptions show that some D_{odd} $6j$'s are imaginary. Symmetry (3.3.20) shows that for $\mathbf{D}_3$

$$\begin{Bmatrix} 1 & 1 & 1 \\ \frac{3}{2} & \frac{1}{2} & \frac{1}{2} \end{Bmatrix}^* = \begin{Bmatrix} 1^* & 1^* & 1^* \\ \frac{3}{2}^* & \frac{1}{2}^* & \frac{1}{2}^* \end{Bmatrix} = \begin{Bmatrix} 1 & 1 & 1 \\ -\frac{3}{2} & \frac{1}{2} & \frac{1}{2} \end{Bmatrix}$$

while (3.3.18) for a (23) interchange is

$$= \left\{\tfrac{3}{2}\right\}\left\{\tfrac{1}{2}\right\}\left\{\tfrac{1}{2}\right\}\{111\}\left\{1\tfrac{1}{2}\tfrac{1}{2}\right\}\left\{\tfrac{3}{2}1\tfrac{1}{2}\right\}\left\{\tfrac{3}{2}1\tfrac{1}{2}\right\}\begin{Bmatrix} 1 & 1 & 1 \\ -\frac{3}{2}^* & \frac{1}{2}^* & \frac{1}{2}^* \end{Bmatrix}$$

The $3j$ phase $\{111\}=1$, the $3j$ $\{1\,\tfrac{1}{2}\,\tfrac{1}{2}\}=1$ by symmetry constraints, $\{\tfrac{1}{2}\}$ and $\{\tfrac{3}{2}\,1\,\tfrac{1}{2}\}$ occur twice, giving [for the choice $\{\tfrac{3}{2}\}=-1$]

$$= -\begin{Bmatrix} 1 & 1 & 1 \\ \tfrac{3}{2} & \tfrac{1}{2} & \tfrac{1}{2} \end{Bmatrix} \tag{6.1.6}$$

The $3j$ $\{111\}$ is the nonalgebraic factor. Changing $\{\tfrac{3}{2}\}$ merely moves the i elsewhere in the table. The values of the $6j$ symbols for all $\mathbf{D}_n$ are readily computed following the methods of Chapter 3.

6.2 The Chain $\mathbf{SO}_3 \supset \mathbf{D}_\infty \supset \mathbf{SO}_2$

The branching rules for the chain $\mathbf{SO}_3 \supset \mathbf{D}_\infty \supset \mathbf{SO}_2$ were given above, (6.1.4) and (6.1.5). The existence of the group $\mathbf{D}_\infty$ inbetween $\mathbf{SO}_3$ and $\mathbf{SO}_2$ means that if phases are chosen appropriately, the SO_3–SO_2 jm factors can be made to factorize into SO_3–D_∞ jm factors and D_∞–SO_2 jm factors. This factorization is of little intrinsic interest, but it is most useful for our purposes because $\mathbf{D}_\infty$ has many other subgroups, namely $\mathbf{D}_n$ and $\mathbf{C}_n$ for all n. In a study of a system such as benzene (C_6H_6) with $\mathbf{D}_6$ symmetry the coupling coefficients for $\mathbf{SO}_3 \supset \mathbf{D}_6$ will factorize into SO_3–D_∞ and D_∞–D_6 $3jm$ factors. The former factors are very closely related to the SO_3–SO_2 $3jm$ factors as tabulated by Rotenberg, Bivins, Metropolis, and Wooten (1959), so we need only tabulate D_∞–D_6 $3jm$'s for such a problem. The details of the finite subgroups of $\mathbf{D}_\infty$ are to be found in the following section. The present section derives the phase change so that the usual JM $3jm$ symbols factorize.

First consider the proof that the JM $3jm$ symbols do not factorize. Under complex conjugation symmetry (3.3.12) we have for the standard JM basis

$$\begin{pmatrix} 2 & 1 & 1 \\ -1 & 0 & 1 \end{pmatrix}^{*J}_{M} = \begin{pmatrix} 2 \\ -1 \end{pmatrix}\begin{pmatrix} 1 \\ 0 \end{pmatrix}\begin{pmatrix} 1 \\ 1 \end{pmatrix}\begin{pmatrix} 2^* & 1^* & 1^* \\ -1^* & 0^* & 1^* \end{pmatrix}^{J}_{M}$$

$$= (-1)^{2+1+1-0+1-1}\begin{pmatrix} 2 & 1 & 1 \\ 1 & 0 & -1 \end{pmatrix}^{J}_{M} \tag{6.2.1}$$

$$= \begin{pmatrix} 2 & 1 & 1 \\ 1 & 0 & -1 \end{pmatrix}^{J}_{M}$$

In the $\mathbf{SO_3} \supset \mathbf{D_\infty} \supset \mathbf{SO_2}$ basis we have the factorizations

$$\begin{pmatrix} 2 & 1 & 1 \\ -1 & 0 & 1 \end{pmatrix}^{SO_3}_{SO_2} = \begin{pmatrix} 2 & 1 & 1 \\ 1 & \tilde{0} & 1 \end{pmatrix}^{SO_3}_{D_\infty} \begin{pmatrix} 1 & \tilde{0} & 1 \\ -1 & 0 & 1 \end{pmatrix}^{D_\infty}_{SO_2} \tag{6.2.2}$$

$$\begin{pmatrix} 2 & 1 & 1 \\ 1 & 0 & -1 \end{pmatrix}^{SO_3}_{SO_2} = \begin{pmatrix} 2 & 1 & 1 \\ 1 & \tilde{0} & 1 \end{pmatrix}^{SO_3}_{D_\infty} \begin{pmatrix} 1 & \tilde{0} & 1 \\ 1 & 0 & -1 \end{pmatrix}^{D_\infty}_{SO_2} \tag{6.2.3}$$

The SO_3–D_∞ $3jm$ factors in (6.2.2) and (6.2.3) are identical. The D_∞–SO_2 $3jm$ factors in the same equations are related by a (13) column interchange, which introduces the D_∞ $3j$ phase $\{1\tilde{0}1\}$. By the character theory of $\mathbf{D_\infty}$ this $3j$ is -1 because $\tilde{\mathbf{0}}$ occurs in the antisymmetric part of $\mathbf{1(D_\infty)\times 1(D_\infty)}$. This is different from the JM result (6.2.1) and shows that the standard basis for $\mathbf{SO_3} \supset \mathbf{SO_2}$ (the JM basis) is not an $\mathbf{SO_3} \supset \mathbf{D_\infty} \supset \mathbf{SO_2}$ basis.

The difference in the two bases can be traced to the number of free phases in the Racah–Wigner algebra. The JM basis regards the kets $|21\rangle$ and $|2-1\rangle$ as unrelated; the relation is fixed by phase choices during the calculation, for example, by choosing ladder operators $J_\pm = \mp(J_x \pm iJ_y)$. The SO_3–D_∞–SO_2 basis has less freedom and the relationship between $|2(SO_3)1(D_\infty)1(SO_2)\rangle$ and $|21-1\rangle$ is forced to be the "same" as the relationship between $|111\rangle$ and $|11-1\rangle$.

For $\mathbf{D_\infty} \supset \mathbf{SO_2}$ the $2jm$ factors can be chosen as

$$\begin{pmatrix} a \\ M \end{pmatrix}^{D_\infty}_{SO_2} = (-1)^{a-M} \qquad \text{and} \qquad \begin{pmatrix} \tilde{0} \\ 0 \end{pmatrix}^{D_\infty}_{SO_2} = -1 \tag{6.2.4}$$

The latter is essential if the $3jm$ factors are to be real. An appropriate choice for the phases of the primitive $3jm$'s and a recoupling shows that all nontrivial D_∞–SO_2 $3jm$ factors can be cast into one of the simple forms

$$\begin{pmatrix} a+b & a & b \\ a+b & -a & -b \end{pmatrix}^{D_\infty}_{SO_2} = \frac{1}{\sqrt{2}}(-1)^{2b} \tag{6.2.5}$$

$$\begin{pmatrix} a & a & \tilde{0} \\ a & -a & 0 \end{pmatrix}^{D_\infty}_{SO_2} = \frac{1}{\sqrt{2}} \tag{6.2.6}$$

The SO_3–D_∞ jm factors also follow simply from the usual arguments (compare Butler 1976 for the JM basis). For every ket $|J(SO_3)a(D_\infty)\rangle$ there is a free phase. We choose all $2jm$ symbols positive. We use the orthogonalities of $3jm$ factors to obtain the norms of the primitive $3jm$ factors. For example, (3.3.14) gives

$$\frac{2J+2}{2}\begin{Bmatrix} J+\frac{1}{2} & J & \frac{1}{2} \\ a+\frac{1}{2} & a & \frac{1}{2} \end{Bmatrix}^2 + \frac{2J}{2}\begin{Bmatrix} J-\frac{1}{2} & J & \frac{1}{2} \\ a+\frac{1}{2} & a & \frac{1}{2} \end{Bmatrix}^2 = 1 \tag{6.2.7a}$$

and

$$\frac{2J+1}{2}\begin{Bmatrix} J & J-\frac{1}{2} & \frac{1}{2} \\ a & a+\frac{1}{2} & \frac{1}{2} \end{Bmatrix}^2 + \frac{2J-1}{2}\begin{Bmatrix} J-1 & J-\frac{1}{2} & \frac{1}{2} \\ a & a+\frac{1}{2} & \frac{1}{2} \end{Bmatrix}^2 = 1 \tag{6.2.7b}$$

The second term of the first equation is related to the first term of the second by a column permutation. Combining the equations to eliminate these terms gives

$$\frac{(2J+2)\,(2J+1)}{2}\begin{Bmatrix} J+\frac{1}{2} & J & \frac{1}{2} \\ a+\frac{1}{2} & a & \frac{1}{2} \end{Bmatrix}^2$$

$$=\frac{(2J+2)\,(2J)}{2}\begin{Bmatrix} (J-1)+\frac{1}{2} & J-1 & \frac{1}{2} \\ a+\frac{1}{2} & a & \frac{1}{2} \end{Bmatrix}^2 + 1 \tag{6.2.8}$$

This is a recursion relation. The term on the right is zero for $J=a$. It is therefore easily solved. After choosing the positive square root, the orthogonality gives the value and sign of the $3jm$ with $a-\frac{1}{2}$. The results are

$$\begin{Bmatrix} J+\frac{1}{2} & J & \frac{1}{2} \\ a\pm\frac{1}{2} & a & \frac{1}{2} \end{Bmatrix} = \left[\frac{2J\pm 2a+2}{(2J+1)(2J+2)}\right]^{1/2} \tag{6.2.9}$$

Similarly one obtains the norm and chooses the phase for

$$\begin{Bmatrix} J+\frac{1}{2} & J & \frac{1}{2} \\ \eta & \frac{1}{2} & \frac{1}{2} \end{Bmatrix} = \frac{1}{(2J+1)^{1/2}} \tag{6.2.10}$$

and derives the special case as

$$\begin{bmatrix} J+\frac{1}{2} & J & \frac{1}{2} \\ \frac{1}{2} & \eta & \frac{1}{2} \end{bmatrix} = (-1)^{J} \frac{1}{(2J+1)^{1/2}} \tag{6.2.11}$$

where η is 0 or $\tilde{0}$ as $J+\frac{1}{2}$ (or J) is even or odd.

If it is assumed that the scalar and α spin functions are the same in the JM and SO_3–D_∞–SO_2 bases, that is,

$$\langle 000|00\rangle = \langle \tfrac{1}{2}\tfrac{1}{2}\tfrac{1}{2}|\tfrac{1}{2}\tfrac{1}{2}\rangle = 1 \tag{6.2.12}$$

then use of the recursion relation (3.3.40) gives

$$\langle Jaa|Ja\rangle = (-1)^{J-a} \tag{6.2.13}$$

$$\langle Ja-a|J-a\rangle = 1 \tag{6.2.14}$$

for $a>0$. Further, the $\mathbf{SO_2}$ scalars are equal,

$$\langle J\tilde{0}0|00\rangle = \langle J00|00\rangle = 1 \tag{6.2.15}$$

Now that we have the relationship of the standard JM basis kets to the SO_3–D_∞–SO_2 basis kets it is a trivial matter to use (3.3.39) to relate the JM $3jm$ to the SO_3–D_∞–SO_2 $3jm$. The SO_3–D_∞–SO_2 result factorizes with the D_∞–SO_2 $3jm$ being a simple factor, (6.2.4)–(6.2.6). Thus we have the relationships

$$\begin{pmatrix} J_1 & J_2 & J_3 \\ a+b & a & b \end{pmatrix}^{SO_3}_{D_\infty} = \sqrt{2}\,(-1)^{J_1-a+b} \begin{pmatrix} J_1 & J_2 & J_3 \\ a+b & -a & -b \end{pmatrix}^{J}_{M} \tag{6.2.16}$$

a, b nonzero, and for $\eta=\tilde{0}$ or 0

$$\begin{pmatrix} J_1 & J_2 & J_3 \\ a & a & \eta \end{pmatrix}^{SO_3}_{D_\infty} = \sqrt{2}\,(-1)^{J_1-a} \begin{pmatrix} J_1 & J_2 & J_3 \\ a & -a & 0 \end{pmatrix}^{J}_{M} \tag{6.2.17}$$

$$\begin{pmatrix} J_1 & J_2 & J_3 \\ \eta_1 & \eta_2 & \eta_3 \end{pmatrix}^{SO_3}_{D_\infty} = \begin{pmatrix} J_1 & J_2 & J_3 \\ 0 & 0 & 0 \end{pmatrix}^{J}_{M} \tag{6.2.18}$$

All SO_3–D_∞ $3jm$'s are related to these three forms by cyclic permutations of columns (cyclic symmetries introduce no phase change). The reader may wish to check that the phase change introduced by interchanging two columns in (6.2.16), namely $(-1)^{J_1+J_2+J_3}$ for $\mathbf{SO_3}$, $(-1)^{a+b+a+b}$ for $\mathbf{D_\infty}$, and $(-1)^{a+b-a-b}$ for $\mathbf{SO_2}$, is in fact the same on both sides.

6.3 Finite Subgroups of $\mathbf{D_\infty}$

In this section we discuss $3jm$ factors for the various possible imbeddings involving finite subgroups of $\mathbf{D_\infty}$. The group $\mathbf{D_\infty}$ has as subgroups $\mathbf{C_\infty} = \mathbf{SO_2}$, $\mathbf{D}_n$ for all n, and $\mathbf{C}_n$ for all n. The D_∞–SO_2 $3jm$'s were considered in the previous section. The C_∞–C_n $3jm$'s and the C_m–C_n $3jm$'s are either zero or, if triad and branching rules are satisfied, unity. As we shall see, this leaves four categories of $3jm$ factors to consider: D_∞–D_n $3jm$'s for all $n>1$, D_{pn}–D_n $3jm$'s for p prime and $n>1$, D_n–C_n $3jm$'s for all $n>1$, and D_p–C_2 $3jm$'s for $p>2$ and prime. All except the first category cannot be factored by insertion of an intermediate group. It is necessary to say a few words about the properties of dihedral chains under factorization.

Throughout this book we have considered only maximal group chains. For example, in considering $\mathbf{O} \supset \mathbf{C_3}$ we have always chosen a specific one of the two intermediate groups $\mathbf{D_3}$ and $\mathbf{T}$. The different intermediate groups lead to different basis functions and different O–C_3 $3jm$'s. Consideration of the factorization of the $3jm$ resulting from the intermediate group introduces restraints on phases and multiplicities, reducing the number of phase choices and, in the $\mathbf{O} \supset \mathbf{C_3}$ case, eliminating multiplicity choices. The reduction in the number of phases was relevant in the previous section when we proved that the JM phases were such that the JM $3jm$'s did not factorize into a product of SO_3–D_∞ and D_∞–SO_2 $3jm$'s. For the imbeddings $\mathbf{D_\infty} \supset \mathbf{D}_n$ above, there is no set of maximal intermediate groups. For example, given any $\mathbf{D}_n$, not only do we have infinitely many groups $\mathbf{D}_{pn}$, p a prime, of which $\mathbf{D}_n$ is a maximal subgroup, but also for each of these groups there are infinitely many larger groups $\mathbf{D}_{qpn}$ to choose from. The process neither stops (there is no maximal group chain), nor can phases be chosen so that the D_∞–D_n $3jm$'s factorize for each intermediate group (unlike $\mathbf{C}_n$ in $\mathbf{C_\infty}$). For example, one can demonstrate (somewhat tediously) that the freedoms in the phases for D_{36}–D_{18} $3jm$'s, D_{36}–D_{12} $3jm$'s, D_{18}–D_9 $3jm$'s, D_{18}–D_6 $3jm$'s, D_{12}–D_6 $3jm$'s, D_9–D_3 $3jm$'s, and D_6–D_3 $3jm$'s are such that there is no way in which one can have the

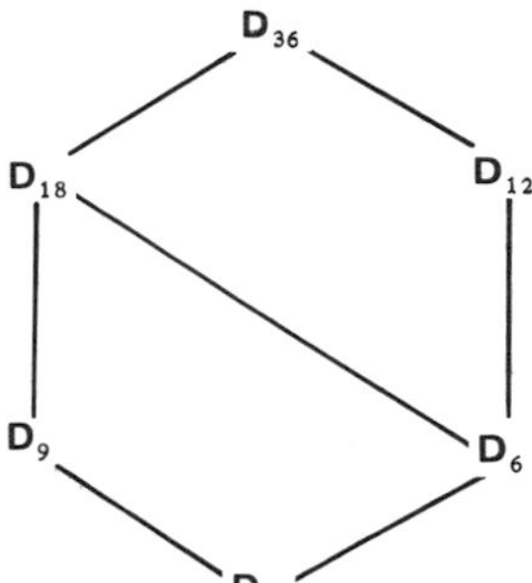

Figure 6.1. The three maximal chains for $\mathbf{D}_{36} \supset \mathbf{D}_3$.

D_{36}–D_3 $3jm$'s simultaneously factorize in each of the three routes available in the diagram shown in Figure 6.1. For this reason we choose phases so that the $3jm$'s in each of the above four categories are as simple as possible. Consider D_∞–D_n $3jm$'s first.

The branching rules for $\mathbf{D}_\infty \supset \mathbf{D}_n$ are given for $n=4,5,6$ in Chapter 12. They follow the same pattern for all n. It is advantageous to choose all $2jm$'s of the form $\begin{pmatrix} A \\ a \end{pmatrix}$ negative for A in all of the ranges $(4k+1)n/2<A<(4k+3)n/2$, k integer, because then all primitive $3jm$'s are either real or may be chosen so. Butler and Reid (1979, Section 5) give the algebraic formulas for primitive D_∞–D_n $3jm$'s, with the phase choices used in the tables of Chapter 13. The same general results for D_∞–D_n $3jm$'s may be used for the D_{pn}–D_n $3jm$'s together with some additional choices for the irreps $\frac{1}{2}\mathbf{pn}(\mathbf{D}_{pn})$ and $-\frac{1}{2}\mathbf{pn}(\mathbf{D}_{pn})$ or $\widetilde{\frac{1}{2}\mathbf{pn}}(\mathbf{D}_{pn})$.

The previous section gave the D_∞–C_∞ $3jm$'s. As with the D_{pn}–D_n case above, the D_n–C_n $3jm$'s can be chosen to have the same values as the D_∞–C_∞ $3jm$'s, except for the irreps $\pm\frac{1}{2}\mathbf{n}$. The $\mathbf{D}_p \supset \mathbf{C}_2$ case, for p an odd prime, gives rise to a larger set of $3jm$'s than for $\mathbf{D}_p \supset \mathbf{C}_p$. The two-dimensional true irreps of $\mathbf{D}_p$ all branch to $\mathbf{0}+\mathbf{1}$ of $\mathbf{C}_2$, whereas they branch to a complex conjugate pair in $\mathbf{C}_p$. This difference leads to different norms (quite a few D_p–C_2 $3jm$'s are of norm $\frac{1}{2}$) and a different availability of phase choices (mere sign choices for the irreps in $\mathbf{D}_p \supset \mathbf{C}_2$ but a full choice including reality for $\mathbf{D}_p \supset \mathbf{C}_p$). The reader who wishes to have tables or formulas beyond those of Chapter 12 will be able to continue the pattern of phase choices used there. For each set of D_n–C_n $3jm$'s and each set of D_{odd}–C_2 $3jm$'s there is an orientation phase. We have chosen all these phases except D_n–C_n for n odd, as $+1$. The exceptions (D_3–C_3 and D_5–C_5) have been chosen as $+i$, so that a twofold rotation about the y axis is in the group $\mathbf{D}_n$ in this C_n orientation (see also Sections 3.5 and 5.3).

7

Fractional Parentage Coefficients

The algebraic techniques of this book owe a very large part of their development to Racah. In 1943 he followed up an idea of Bacher and Goudsmit (1934) in deriving an explicit form for an N-electron state of well-defined S and L values, by taking combinations of $(N-1)$- and one-electron states. In that paper he computed sets of *coefficients of fractional parentage* (cfp) which gave these explicit linear combinations for the p^N and d^N configurations. In 1949 Racah recognized that the nice properties of a *seniority* operator Q, which he had introduced in the 1943 paper, followed from a group-theoretic argument. He then, to use our terminology, showed that the cfp's were coupling coefficients for a certain group chain and proved the Racah factorization lemma for the coefficients. (See Racah's series of papers reprinted in Biedenharn and van Dam 1965).

In Section 7.1 we define the cfp's and show that they are coupling coefficients of certain unitary groups. Section 7.2 contains a brief discussion of the irrep structure of these and other groups which enter into the cfp chain. This knowledge is used in Section 7.3 to discuss the full cfp chain. Nielson and Koster (1963) give tables of cfp's for all p^N, d^N, and f^N configurations. (They also give tables of the $SU_2^S \times SO_3^L$ reduced matrix elements of the Racah unit tensors.) For any point group **G**, the G reduced matrix elements in a weak-field scheme are given by an $SU_2^S \times SO_3^L$ reduced matrix element and an $SU_2^S \times SO_3^L$–G $3jm$ factor (or factors, if intermediate groups occur). Section 7.4 shows that the p^N cfp tables are also relevant for any strong-field states constructed from any spatial triplet, such as strong-field states formed as $\mathbf{t}_1^N(\mathbf{O})$ or $\mathbf{t}_2^N(\mathbf{O})$. This follows because the abstract structures of the groups for t^n and p^n are the same. To illustrate generalizations to other systems, we discuss the structure of cfp's for the mixed configuration $(e_g + t_{2g})^N(\mathbf{O}_h)$.

7.1 The Parentage Concept

In the early years of many-electron quantum mechanics various proposals were put forward for obtaining orthonormal, antisymmetrized, many-electron functions. Bacher and Goudsmit (1934) suggested building up the N-electron state from $(N-1)$-electron functions, the *parents* of the state. Racah showed that this could be done without loss of phase information and that it had the great advantage over simple determinantal methods because it actually used the rotational properties of the system, rather than destroying the $\mathbf{SO}_3$ structure.

The N-electron states $|\Omega\rangle$ may be expanded as a linear combination of products of the $N-1$ electron parents $|\bar{\Omega}\rangle$ with the single-electron states $|\omega\rangle$,

$$|\Omega\rangle = \sum_{\bar{\Omega}\omega} |\bar{\Omega}\rangle|\omega\rangle\langle\bar{\Omega}, \omega|\Omega\rangle \tag{7.1.1}$$

This transformation from the simple unsymmetrized product states $|\bar{\Omega}\rangle|\omega\rangle$ to the symmetrized states $|\Omega\rangle$ has the same form as the coupling equation (2.3.4), but we should not call the cfp $\langle\bar{\Omega}, \omega|\Omega\rangle$ a coupling coefficient until we have shown that the labels $\bar{\Omega}$, ω, and Ω are labels of partners of irreps of some group. In this section we give such a group-theoretic significance to the labels $\bar{\Omega}$, ω, and Ω for every physical situation. It does not matter whether we consider the original single-configuration electron problem, the generalization in the Elliott model of the nucleus to mixed configuration schemes, or problems where the concept of an orbital configuration loses its meaning, for example, in molecules or in strong ligand fields.

Consider first the particular case which was so important historically and which is still a major area for the application of cfp's, the f^N configuration (Judd 1963, Chapter 7). In the Russell–Saunders (SL) coupling scheme the single-electron states are labeled $|sm_s lm_l\rangle$, where $s=\frac{1}{2}$, $l=3$, and the N-electron states are labeled $|f^N\alpha SM_S LM_L\rangle$, where α is some label used to distinguish the various N-electron states with the same S, M_s, L, and M_l quantum numbers. With this labeling for the states Ω, $\bar{\Omega}$, and ω, (7.1.1) becomes

$$|f^N\alpha SM_SLM_L\rangle = \sum_{\alpha'S'M_S'L'M_l'm_sm_l} |f^{N-1}\alpha'S'M_S'L'M_L'\rangle|sm_slm_l\rangle$$

$$\times\langle f^{N-1}\alpha'S'L', sl|f^N\alpha SL\rangle\langle S'M_S', sm_s|SM_S\rangle\langle L'M_L', lM_l|LM_L\rangle \tag{7.1.2}$$

We have used the fact that because we are performing a coupling in $\mathbf{SU}_2^S \times \mathbf{SO}_3^L$, the cfp must be the product of a coefficient-independent M_S and M_L,

$$\langle f^{N-1}\alpha'S'L', sl|f^N\alpha SL\rangle \tag{7.1.3}$$

a coupling coefficient for the $\mathbf{SU}_2^S$ group, $\langle S'M_S', sm_S|SM_S\rangle$, and a coupling coefficient for the $\mathbf{SO}_3^L$ group, $\langle L'M_L', lm_l|LM_L\rangle$.

We now show that the label f^N is in fact an irrep label of $\mathbf{U}_{14}$. In the f shell there are 14 single-electron kets $|sm_slm_l\rangle$. The quantum numbers s and l are fixed at $\frac{1}{2}$ and 3, respectively, while m_s and m_l vary independently over the values $\frac{1}{2}, -\frac{1}{2}$ and $-3, -2, -1, 0, 1, 2, 3$. Consider the set of all operators which mix these 14 kets. It is apparent that the operators are equivalent to the set of 14×14 unitary matrices, and that the matrices form a group. (The product of two unitary matrices is unitary, inverses exist, etc.) The group is the *unitary group* in 14 dimensions, $\mathbf{U}_{14}$. It is also apparent that the 14 kets are partners of an irrep of $\mathbf{U}_{14}$, for they certainiy satisfy the representation property (an operator of $\mathbf{U}_{14}$ acting on any ket gives a linear combination of the 14), and are irreducible (if any ket is removed, one no longer has a representation).

The N-electron f-shell states are antisymmetrized N-fold products of the partners of the 14-dimensional irrep. The usual group-theoretic label for the antisymmetric N-fold irrep of $\mathbf{U}_{14}$ is $\{\mathbf{1}^N\}$. Replacing the spectroscopic label f^N by this group-theoretic label, the cfp factor of (7.1.3) becomes $\langle\{1^{N-1}\}\alpha'S'L', \{1\}sl|\{1^N\}\alpha sL\rangle$. (The fact that the $14!/[N!(14-N)!]$ antisymmetrized functions of the f^N configuration are the partners of the irrep $\{\mathbf{1}^N\}$ of $\mathbf{U}_{14}$ needs some character theory of $\mathbf{U}_{14}$ for its verification. Such results for $\mathbf{U}_{14}$ are discussed and references given in the next section.) Using the $\mathbf{U}_{14}$ irrep property of the set of states of the configuration gives that the cfp factor $\langle\{1^{N-1}\}\alpha'S'L', \{1\}sl|\{1^N\}\alpha SL\rangle$ is a coupling factor for the chain $\mathbf{U}_{14} \supset \mathbf{SU}_2^S \times \mathbf{SO}_3^L$,

$$\langle f^{N-1}\alpha' S' L', sl \,|\, f^N \alpha SL \rangle$$

$$= \langle \{1^{N-1}\}\alpha' S' L', \{1\} sl \,|\, \{1^N\} \alpha SL \rangle^{U_{14}}_{SU_2^S \times SO_3^L} \tag{7.1.4}$$

The labels α and α' are branching multiplicity labels. Racah not only recognized this, but also recognized that the branching multiplicity labels could be given group-theoretic significance and the factors further factorized, because the chain $\mathbf{U}_{14} \supset \mathbf{SU}_2^S \times \mathbf{SO}_3^L$ is not maximal. There exist several groups inside $\mathbf{U}_{14}$ which contain $\mathbf{SU}_2^S \times \mathbf{SO}_3^L$. The extra factorizations naturally provide computational simplifications, but interestingly, the additional labels turn out to be reasonable quantum numbers under the interelectron Coulomb interaction.

We turn now to a discussion of these matrix groups.

7.2 Continuous Matrix Groups

Elie Cartan in 1894 gave a complete classification of the continuous matrix groups (the classical Lie groups). He classified them according to their algebraic structures. We give here a very brief summary of the classical groups and their representations.

Consider all $n \times n$ matrices with complex numbers as entries. The subset of matrices with nonzero determinant form a group $\mathbf{GL}(n, \mathbf{C})$, as may be readily verified—distributive and associative laws automatically hold for matrices, the product of matrices with nonzero determinant gives a matrix with nonzero determinant, and the nonzero determinant condition is precisely the condition that the matrix has an inverse. The matrices correspond to general linear transformations, hence the name. We are interested only in unitary transformations, so we restrict ourselves to the subgroup of unitary matrices, the $\mathbf{GL}(n, \mathbf{C})$ matrices which have their inverse equal to themselves complex conjugated and transposed, called $\mathbf{U}_n$ [or $\mathbf{U}(n, \mathbf{C})$ by some]. The further restriction to the special unitary group $\mathbf{SU}_n$, of unitary matrices of determinant $+1$, is often used. With some simple matrix theory it is simple to show that $\mathbf{U}_n$ is a direct product group

$$\mathbf{U}_n = \mathbf{U}_1 \times \mathbf{SU}_n \tag{7.2.1}$$

where $\mathbf{U}_1$ is the group of complex numbers of modulus 1, the determinants of the various $\mathbf{U}_n$ matrices.

It is perhaps surprising how few subgroups of $\mathbf{U}_n$ exist. Usually, foı n odd, $\mathbf{SU}_n$ has only one maximal *simple* subgroup, namely $\mathbf{SO}_n$, while for n even $\mathbf{SU}_n$ has two, $\mathbf{SO}_n$ and $\mathbf{Sp}_n$. Simple in this mathematical sense means roughly that the subgroup cannot be written as a direct product.

The set of unitary matrices with real elements form the orthogonal group $\mathbf{U}_n \supset \mathbf{O}_n$ or $\mathbf{SU}_n \supset \mathbf{SO}_n$. This condition, that if $\mathbf{X} \in \mathbf{O}_n$, then $\mathbf{X}^{-1} = \mathbf{X}^{\mathrm{T}}$, may be alternatively written

$$\mathbf{X}\mathbf{I}_n\mathbf{X}^{\mathrm{T}} = \mathbf{I}_n \tag{7.2.2}$$

where $\mathbf{I}_n$ is the $n \times n$ unit (identity) matrix consisting of 1's on the diagonal and 0's elsewhere. As a third alternative the condition is equivalent to leaving invariant a symmetric bilinear (quadratic) form on the variables $x_1, \ldots, x_n$. Namely $x_1^2 + \cdots + x_n^2$ is invariant. $\mathbf{SO}_n$ is the group of pure rotations in n dimensions, and is often written $\mathbf{R}_n$ for this reason. $\mathbf{O}_n$ contains reflections.

For even n ($n=2k$) the symplectic group $\mathbf{Sp}_n$ exists. It may be defined as the set of unitary matrices Y which satisfy

$$\mathbf{Y}\mathbf{J}\mathbf{Y}^{\mathrm{T}} = \mathbf{J} \tag{7.2.3}$$

where $\mathbf{J}$ is the $2k \times 2k$ antisymmetric matrix

$$\begin{pmatrix} \mathbf{O}_k & \mathbf{I}_k \\ -\mathbf{I}_k & \mathbf{O}_k \end{pmatrix}$$

The symplectic group leaves invariant an antisymmetric bilinear form on two sets of n variables $x_1, \ldots, x_n$ and $y_1, \ldots, y_n$. Namely $(x_1 y_{k+1} - x_{k+1} y_1) + \cdots + (x_k y_n - x_n y_k)$ is invariant.

Elie Cartan labeled these groups on the grounds of certain algebraic properties. We have the correspondence between his and the present notation as $\mathbf{A}_n = \mathbf{SU}_{n+1}$, $\mathbf{B}_n = \mathbf{SO}_{2n+1}$, $\mathbf{C}_n = \mathbf{Sp}_{2n}$, and $\mathbf{D}_n = \mathbf{SO}_{2n}$. These exist for all n, although for small n we have the isomorphisms $\mathbf{SU}_2 = \mathbf{SO}_3 = \mathbf{Sp}_2$, $\mathbf{SU}_4 = \mathbf{SO}_6$, $\mathbf{SO}_4 = \mathbf{SO}_3 \times \mathbf{SO}_3$, and $\mathbf{Sp}_4 = \mathbf{SO}_5$. [In these isomorphisms, $\mathbf{SO}_n$ is to be understood to mean the spin covering groups of $\mathbf{SO}_n$, not the matrix group defined by (7.2.2).] He further showed that only five other groups exist, the *exceptional groups*, $\mathbf{E}_6$, $\mathbf{E}_7$, $\mathbf{E}_8$, $\mathbf{F}_4$, and $\mathbf{G}_2$. The group $\mathbf{G}_2$ is a subgroup of $\mathbf{SO}_7$ and contains $\mathbf{SO}_3$, so is used for f-shell calculations, but none of the other exceptional groups have found use in physics except in a few postulated fundamental particle models.

Direct product subgroups can be found by inspection using dimensional arguments. For example, $\mathbf{SU}_{14} \supset \mathbf{SU}_2 \times \mathbf{SU}_7$ because the direct product of a set of 2×2 unitary matrices and a set of 7×7 unitary matrices is a subset of all 14×14 unitary matrices. Sometimes it is necessary to use arguments as to whether the product is orthogonal or symplectic, e.g., $\mathbf{Sp}_{14} \supset \mathbf{Sp}_2 \times \mathbf{SO}_7$ for $\mathbf{Sp}_2$ matrices direct-multiplied by $\mathbf{SO}_7$ matrices give 14×14 matrices and these matrices will be symplectic. $\mathbf{Sp}_2$ is isomorphic to $\mathbf{SU}_2$, so one usually writes $\mathbf{Sp}_{14} \supset \mathbf{SU}_2 \times \mathbf{SO}_7$.

With this knowledge two distinct *f*-shell schemes are seen to be possible (Judd 1963): either

$$\mathbf{SU}_{14} \supset \mathbf{SU}_2^S \times (\mathbf{SU}_7 \supset \mathbf{SO}_7 \supset \mathbf{G}_2 \supset \mathbf{SO}_3^L) \tag{7.2.4}$$

or

$$\mathbf{SU}_{14} \supset \mathbf{Sp}_{14} \supset \mathbf{SU}_2^S \times (\mathbf{SO}_7 \supset \mathbf{G}_2 \supset \mathbf{SO}_3^L) \tag{7.2.5}$$

The latter is Racah's parentage scheme. As further subgroups of the lowest subgroup shown, $SU_2^S \times SO_3^L$, we may consider either one of the two schemes, a total angular momentum scheme J,

$$\mathbf{SU}_2^S \times \mathbf{SO}_3^L \supset \mathbf{SO}_3^J \supset (\text{point groups})^J \tag{7.2.6}$$

or a scheme in which the spins and orbitals are independent,

$$\mathbf{SU}_2^S \times \mathbf{SO}_3^L \supset (\text{point groups})^S \times (\text{point groups})^J \tag{7.2.7}$$

where the "point groups" may be the standard $\mathbf{SO}_2$ or any other point group chain discussed elsewhere in this book.

A number of books give descriptions of the classical Lie groups and their representations. We prefer the combinatoric approach of Wybourne (1970), in which he describes Littlewood's (1950) development of the theory of Schur functions.

A *partition* of l, $(\lambda)=(\lambda_1\lambda_2\lambda_3\cdots\lambda_n)$, is an ordered set of positive integers which sum to l,

$$\lambda_1 \geqslant \lambda_2 \geqslant \lambda_3 \cdots \geqslant \lambda_n \geqslant 0 \tag{7.2.8}$$

$$\lambda_1 + \lambda_2 + \cdots + \lambda_n = l \tag{7.2.9}$$

A *Schur function* $\{\lambda; x_1, x_2, \ldots, x_n\}$ for a partition (λ) is a certain polynomial in the variables $x_1, x_2, \ldots, x_n$ that has a leading term $x_1^{\lambda_1} x_2^{\lambda_2} \cdots x_n^{\lambda_n}$:

$$\{\lambda; x\} = \det\left(x_j^{\lambda_i + n - i}\right) / \det\left(x_j^{n-i}\right) \tag{7.2.10}$$

where $\det(x_j^{\lambda_i + n - i})$ indicates an $n \times n$ determinant whose *ij* element is the variable x_j raised to the $\lambda_i + n - i$ power. Such functions were studied throughout the 19th century, but it was not until a period around the turn of the 20th century that it was realized that many of the properties of Schur functions followed from group theory and conversely. In particular, if the variables are the eigenvalues of an element of $\mathbf{U}_n$, then the Schur function is the characteristic (trace) of that element in the $\{\boldsymbol{\lambda}\}$ irrep of $\mathbf{U}_n$. The coefficients in the polynomial $\{\lambda; x_1, x_2, \ldots, x_n\}$ are also closely related to the character of an irrep $[\boldsymbol{\lambda}]$ of $\mathbf{S}_l$, the symmetric (permutation) group on l objects.

Schur functions bring together the properties of polynomials and generating functions, the properties of continuous groups, especially $\mathbf{U}_n$, the properties of finite groups, especially $\mathbf{S}_l$, and the general subject of combinatorics. The machinery of Schur functions reduces the character theory of the continuous matrix groups and of the symmetric groups to simple combinatoric rules.

The Schur-function approach to the character theory of the continuous groups leads to the defining irrep being labeled by the partition (1). The set of $n \times n$ unitary matrices is the defining irrep of $\mathbf{U}_n$, labeled $\{\mathbf{1}\}(\mathbf{U}_n)$; the symplectic matrices belong to $\langle\mathbf{1}\rangle(\mathbf{Sp}_n)$; and the $n \times n$ orthogonal matrices define $[\mathbf{1}](\mathbf{O}_n)$. Occasionally we are interested in spin irreps of the classical groups. Only $\mathbf{O}_n$ (and $\mathbf{SO}_n$) have such irreps (Hamermesh 1962, pp. 319–321; Littlewood 1950, p. 248). For the spin covering group of the orthogonal groups the primitive irrep is an irrep $\boldsymbol{\Delta} = [\frac{1}{2}\frac{1}{2} \cdots \frac{1}{2}]$ or irreps $\boldsymbol{\Delta}_{\pm}(\mathbf{SO}_n, n \text{ even}) = [\frac{1}{2}\frac{1}{2} \cdots \pm\frac{1}{2}]$.

The combinatoric structure of Schur functions means that the l-fold antisymmetric irrep of $\mathbf{U}_n$ is labeled $\{\mathbf{1}, \mathbf{1}, \mathbf{1}, \ldots, \mathbf{1}\}$ with l 1's, abbreviated $\{\mathbf{1}^l\}$. The fully symmetric term in an l-fold product is labeled $\{\mathbf{l}\}$, and there exist mixed symmetry irreps for all other partitions (λ) of l with at most n parts. The reader who is interested performing calculations in the character theory of the continuous groups will find a full description together with many tables in Wybourne (1970), several combinatoric improvements in King (1975), and recent developments in the theory for the exceptional Lie groups ($\mathbf{E}_6$, $\mathbf{E}_7$, $\mathbf{E}_8$, $\mathbf{F}_4$, and $\mathbf{G}_2$) in the references of Butler, Haase, and Wybourne (1978).

7.3 Complete Parentage Schemes

We return to the discussion of fractional parentage groups which we began in Section 7.1. (Judd 1968, Judd in Judd and Elliott 1970, and Wybourne 1973 use the language of second quantization to develop several of the ideas we discuss here.) We ask the questions: What groups other than the symmetry group are relevant? Under which irreps of these groups do states and operators transform? Which *jm* factors are required in constructing many-particle states? Which *jm* factors are required in using the Wigner–Eckart theorem with these new groups? And where can one find or how can one compute such *jm* factors?

We have two aims in finding groups larger than the symmetry group of the problem (say $\mathbf{O}_h$). First, larger groups may lead to physical insight by giving approximate symmetries—this is certainly true in a weak-field problem. Second, larger groups always lead to mathematical advances. They enable one to use directly the mathematical structure implicit in physical statements such as "consider the t^N configuration" or "consider only two-electron operators."

As an illustrative example consider again the f^N configuration. Rotations in spin space $\mathbf{SU}_2^S$ combine in an arbitrary manner the two spinor functions α and β associated with the single-particle states. Rotations in orbital space $\mathbf{SO}_3^L$ mix the spatial (that is, orbital) part, but not arbitrarily. There are seven spatial functions, so $\mathbf{O}_7$ would be the group to give arbitrary (normalized) real combinations; $\mathbf{U}_7$ is needed for all combinations. Thus the group chain $\mathbf{SU}_2^S \times \mathbf{U}_7 \supset \mathbf{SU}_2^S \times \mathbf{O}_7 \supset \mathbf{SU}_2^S \times \mathbf{SO}_3^L$ is relevant. The set of all operators which map all 14 spin orbitals to each other is $\mathbf{U}_{14}$. The symplectic group $\mathbf{Sp}_{14}$, which gives rise to Racah's (1943) seniority quantum number, is a subgroup of $\mathbf{U}_{14}$ and contains $\mathbf{SU}_2 \times \mathbf{SO}_7$. Finally the exceptional group $\mathbf{G}_2$ is a subgroup of $\mathbf{SO}_7$ and contains $\mathbf{SO}_3$. One might guess on purely algebraic grounds that the chain (7.2.5) gives some of the physical symmetries of the rare-earth ion Hamiltonian. This is indeed the case (Judd 1963, p. 209), but the real importance of (7.2.5) is as a computational aid. The group $\mathbf{U}_{14}$ which takes arbitrary combinations of single-particle states is not the largest useful group. The U_{14} reduced matrix elements are of the form of N-particle reduced matrix elements $\langle\{1^N\}\|T^\lambda\|\{1^N\}\rangle$. Consider now the manifold of all f-shell states, for all N. Each of the 14 states $|fm_l m_s\rangle$ can either be occupied or unoccupied, giving 2^{14} possibilities. The group of all unitary transformations between all such states is $\mathbf{U}_{2^{14}}$. This is rather a large group, but this is of little concern, because the character theory of the unitary groups $\mathbf{U}_n$ is largely independent of n (see the previous section and references therein). The appropriate character theory (Wybourne 1970, p. 133–139) shows that

$$\mathbf{U}_{2^{14}} \supset \mathbf{SO}_{29} \supset \mathbf{SO}_{28} \supset \mathbf{U}_{14} \tag{7.3.1}$$

and so the particle number dependence of the U_{14} reduced matrix elements can be given as jm factors for these large groups. Judd (1968) uses his less physical quasiparticle scheme,

$$\mathbf{U}_{2^{14}} \supset \mathbf{SO}_{29} \supset \mathbf{SO}_{28} \supset \mathbf{SU}_2^Q \times \left[\mathbf{Sp}_{14} \supset \mathbf{SU}_2^S \times (\mathbf{SO}_7 \supset \mathbf{G}_2 \supset \mathbf{SO}_3^L)\right] \tag{7.3.2}$$

This has the advantage that the particle number dependence is contained in the quasiparticle group SU_2^Q, whose jm factors are of course well known. It has the disadvantage that its irreps contain parts of different f^N configurations; as a result a transformation is required to a more physical basis where occupation number is a good quantum number.

The above construction of covering groups of the symmetry group carries over immediately to the general case. Let the spatial symmetry group of the problem be a point group $\mathbf{G}$, and let the ground state be λ^N, where $\boldsymbol{\lambda}$ is a representation of $\mathbf{G}$ (not necessarily an irrep). Let $|\lambda|=k$, so there are $2k$ spin orbitals. In the absence of spin–orbit effects the symmetry group is $\mathbf{G}^S \times \mathbf{G}^L$. The spin orbitals are labeled $|\frac{1}{2}(G^S)\lambda(G^L)m_S m_L\rangle$, where m_S and m_L are partner labels in the basis scheme chosen. The matrix group $\mathbf{SO}_k$ of all real, orthogonal transformations between the k spatial functions contains (for $k>3$) the subgroup $\mathbf{O}_3^p$, which in turn contains the symmetry point group $\mathbf{G}$. For $k=3$, such as for t_{1g}^N or t_{2g}^N of $\mathbf{O}_h$, $\mathbf{SO}_3^p$ is actually $\mathbf{SO}_k$. Note that $\mathbf{O}_k$ and $\mathbf{O}_3^p$ do not perform rotations in physical 3-space. We refer to $\mathbf{SO}_k$ and $\mathbf{SO}_3^p$ as parentage groups and instead of attaching a superscript L, which would indicate a physical rotation of the spatial functions, we have attached a superscript p for parentage.

To illustrate the distinction between the parentage group $\mathbf{O}_3^p$ above and the physical rotation–reflection group $\mathbf{O}_3^L$, consider the t_{2g}^N configuration, where the t_{2g} spatial states were derived from d orbitals. $\mathbf{O}_3^p$ mixes the three t_{2g} functions [as the $\mathbf{1}^-(\mathbf{O}_3)$ irrep]. $\mathbf{SO}_3^L$ mixes the five d functions [as the $\mathbf{2}^+(\mathbf{O}_3)$ irrep], that is, mixes the t_{2g} states with the e_g states according to the usual prescription for $J=2$ rotation matrices.

The remainder of the $\lambda(\mathbf{G})$ parentage scheme is not liable to this confusion. Because we consider rotation–reflection or inversion point groups, we must add the extra inversion operator to $\mathbf{SO}_k$ to give $\mathbf{O}_k$, and add to $\mathbf{Sp}_{2k}$ all $\mathbf{U}_1$ phase operators. The group scheme

$$\mathbf{U}_{2^{2k}} \supset \mathbf{O}_{4k+1} \supset \mathbf{SO}_{4k} \supset \mathbf{U}_{2k} \supset \mathbf{U}_1 \times \mathbf{Sp}_{2k} \supset (\mathbf{SU}_2 \supset \mathbf{G}^S) \times (\mathbf{O}_k \supset \cdots \supset \mathbf{G}^L) \tag{7.3.3}$$

means that the single-particle state is labeled

$$|\{1\}(U_{2^{2k}})\Delta(O_{4k+1})\Delta_-(SO_{4k})\{1\}(U_{2k})\langle 1\rangle(Sp_{2k})\tfrac{1}{2}(SU_2)\tfrac{1}{2}(G^S)[1](O_k)\cdots\lambda(G)\cdots\rangle \tag{7.3.4}$$

while the N-particle state is labeled, for N even/odd,

$$|\{1\}\Delta\Delta_{\pm}\{1^N\}\langle 1^v\rangle S\lambda_S W\cdots\lambda(G)\cdots\rangle.$$

The labels v, S, $\lambda_S W\lambda$, etc., take on values derivable from the branching rules (see Wybourne 1970, 1973). The transformation properties of the operators within our λ^N configuration follow from the character theory. For any $\mathbf{U}_n$ the generators transform as $\{21^{n-2}0\}$. (The generators of a continuous group are the generalizations of the J_x, J_y, J_z of $\mathbf{SO}_3$.) Now the generators of $\mathbf{U}_{2^{2k}}$ transform all states of all occupation numbers in the λ shell to all others. The generators of $\mathbf{U}_{2k}$ merely transform any N-particle state to any other N-particle state, so Racah's single-particle unit tensor $u_l^\lambda(\lambda,\lambda)$ transform as $\{21^{2k-2}0\}$ of U_{2k}. Familiarity with the representation structure of the continuous groups is all that is required to write down the parentage irrep labels of Racah unit tensors. (For examples see Judd 1963, pp. 196–223; Butler and Wybourne 1970a, b.)

This section has shown that by starting from the $2k$ states of the one-electron configuration $\lambda(\mathbf{G})$, all states λ^N may be obtained by coupling with cfp coefficients in $\mathbf{U}_{2k}$. Cfp coefficients are coupling factors for $\mathbf{U}_{2k} \supset \mathbf{Sp}_{2k} \supset \cdots \supset \mathbf{G}^S \times \mathbf{G}^L$. The resultant 2^{2k} states of λ^N configurations for all N together form partners of a single irrep $\{\mathbf{1}\}(\mathbf{U}_{2^{2k}})$. This large group would allow all $\mathbf{U}_{2k}$ reduced matrix elements to be related to a single $\mathbf{U}_{2^{2k}}$ reduced matrix element, if the relevant *jm* factors were available. In the absence of such elegant general treatments one has recourse in most instances to tabulations, especially those of Nielson and Koster (1963).

7.4 Strong-Field Parentage Schemes

Fractional parentage coefficients are coupling factors for the appropriate groups and as such they depend only on the abstract properties of these groups. In this section we show that Nielson and Koster's (1963) tables may be used whenever there is a one-, three-, five-, or sevenfold set of degenerate single-particle spatial orbitals. There is no requirement in the cfp method that the orbitals have full $\mathbf{SU}_2^S \times \mathbf{SO}_3^L$ symmetry. In particular the p-electron tables may be used whenever the single-particle states come as in a configuration of six states, e.g., ${}^2\mathbf{t}_{1g}(\mathbf{O}_h)$ or ${}^2\mathbf{t}_{2g}(\mathbf{O}_h)$ or ${}^2\mathbf{t}_1(\mathbf{T}_d)$.

Consider that the ten ${}^2\mathbf{d}(\mathbf{SU}_2^S \times \mathbf{SO}_3^L)$ spin-spatial single electron states of the free ion are perturbed by an octahedral field. The ten will be split into a group of six ${}^2\mathbf{t}_{2g}(\mathbf{O}_h^S \times \mathbf{O}_h^L)$ and a group of four ${}^2\mathbf{e}_g(\mathbf{O}_h^S \times O_h^L)$ spin-spatial single-electron states. If these two groups are sufficiently well separated (or if their respective radial functions are sufficiently different), then the two groups are usually treated as the separate configurations t_{2g}^N and e_g^N. We shall use this example to prove that the p^N cfp's are equal to the t_{2g}^N cfp's (within an O_3–O_h–T_d $3jm$). We end this chapter on groups other than point groups by noting that Elliott's (1958) method of mixed nuclear configurations has been applied to mixed electronic or vibrational problems, in particular to the Jahn–Teller effect.

The determinant of some $\mathbf{O}_h$ operators acting on t_{2g} states is -1. This can be seen from the character table. Some twofold rotations have trace $+1$, but must have their three eigenvalues chosen from $+1, -1$. The determinant, being the product of the three eigenvalues $(1, 1, -1)$ is therefore negative. If one adds to the set of $\mathbf{O}_h$ operators all other operators which take real, linear combinations of the t_{2g} states, one produces $\mathbf{O}_3$. The t_{2g} states transform as $\mathbf{1}^-(\mathbf{O}_3)$. Remember that under physical rotations, a quite different group action, they transform with the e_g states as $\mathbf{2}^+(\mathbf{O}_3^L)$. Abstractly, the present imbedding is the same as $\mathbf{1}^-(\mathbf{O}_3) \rightarrow \mathbf{t}_2(\mathbf{T}_d)$ because the gerade (even) nature of t_{2g} is irrelevant. This abstract isomorphism may be used to obtain the t_{2g}^N terms from the usual p^N terms and from the branchings for $\mathbf{O}_3 \supset \mathbf{O}_h \supset \mathbf{T}_d$ (see Chapter 12). The results are given in Table 7.1. Observe from the table the particle–hole correspondence between states of N and $6-N$ particles. The only difference in the group-theoretic labeling between two such states is that the $\mathbf{U}_6$ irrep label $\{\mathbf{1}^N\}$ is complex-conjugated, $\{\mathbf{1}^N\}^*(\mathbf{U}_6) = \{\mathbf{1}^{6-N}\}$.

Table 7.2 gives a similar table for e_g^N. In this table the fact that e_g is not a faithful irrep of $\mathbf{O}_h$ is very evident. Both t_{2g} and e_g are not affected

Table 7.1. Terms for t_{2g}^N in the Parentage Scheme

U_{64}	SO_{13}	SO_{12}	U_6 occupation	Sp_6 seniority	$SU_2 \times O_3$ ^{2S+1}L	$SU_2 \times O_h$ symmetry
{1}	$[\frac{1}{2}\frac{1}{2}\frac{1}{2}\frac{1}{2}\frac{1}{2}\frac{1}{2}]$	$[\frac{1}{2}\frac{1}{2}\frac{1}{2}\frac{1}{2}\frac{1}{2}\frac{1}{2}]$	{0}	⟨0⟩	$^10^+$	$^1A_{1g}$
			{1²}	⟨0⟩	$^10^+$	$^1A_{1g}$
				⟨1²⟩	$^12^+$	^1E_g
						$^1T_{2g}$
					$^31^+$	$^3T_{1g}$
			{1⁴}	⟨0⟩	$^10^+$	$^1A_{1g}$
				⟨1²⟩	$^12^+$	^1E_g
						$^1T_{2g}$
					$^31^+$	$^3T_{1g}$
			{1⁶}	⟨0⟩	$^10^+$	$^1A_{1g}$
		$[\frac{1}{2}\frac{1}{2}\frac{1}{2}\frac{1}{2}\frac{1}{2}-\frac{1}{2}]$	{1}	⟨1⟩	$^21^-$	$^2T_{2g}$
			{1³}	⟨1⟩	$^21^-$	$^2T_{2g}$
				⟨1³⟩	$^22^-$	^2E_g
						$^2T_{1g}$
					$^40^-$	$^4A_{2g}$
			{1⁵}	⟨1⟩	$^21^-$	$^2T_{2g}$

by inversions of 3-space, and e_g is not affected by several other $\mathbf{O}_h$ operations also. For this reason their parentage groups $\mathbf{O}_3$ and $\mathbf{O}_2$ do not, in these imbeddings, contain $\mathbf{O}_h$. For e_g this is obvious because $\mathbf{O}_2$ (which is isomorphic to $\mathbf{D}_\infty$) does not contain $\mathbf{O}$. For the t_{2g} case $\mathbf{O}_3$ does contain $\mathbf{O}_h$, but not with the imbedding of Table 7.1; rather, $\mathbf{O}_3$ only contains $\mathbf{T}_d \simeq \mathbf{O}$. It is convenient in the following to have some notation for the imbedding of nonfaithful irreps of $\mathbf{O}_h$ in other groups. We write

$$\mathbf{O}_3 \underset{\sim}{\supset} \mathbf{O}_h \quad \text{when} \quad \mathbf{1}^-(\mathbf{O}_3) \rightarrow \mathbf{t}_{2g}(\mathbf{O}_h) \tag{7.4.1}$$

$$\mathbf{O}_2 \underset{\sim}{\supset} \mathbf{O}_h \quad \text{when} \quad \mathbf{1}(\mathbf{O}_2) \rightarrow \mathbf{e}_g(\mathbf{O}_h) \tag{7.4.2}$$

These two imbeddings exist because $\mathbf{O}_h$ has the invariant (or normal) subgroups $\mathbf{O}$ (or $\mathbf{T}_d$) and $\mathbf{D}_2$ (or $\mathbf{C}_{2v}$). This is another illustration of the difference between $\mathbf{O}_3$ (spatial) and $\mathbf{O}_3$ (parentage). In a strong-field case they must not be confused. For example, the dipole operator transforms as $\mathbf{J}_{\text{spatial}} = \mathbf{1}$, but of unknown parentage, whereas a t_{2g} state transforms as $\mathbf{J}_{\text{parentage}} = \mathbf{1}$ but of unknown spatial properties.

The mixed configuration methods used by Elliott (1958) for $(s+d)^N$ nuclear configurations have been used by a number of authors in various atomic (Feneuille 1967a, b, Butler and Wybourne 1970a, b) and molecular situations. Consider the Jahn–Teller effect in circumstances like the F^+

Table 7.2. Terms for e_g^N in the Parentage Scheme

U_{16}	SO_9	SO_8	U_4	Sp_4	$SU_2 \times O_2$	$SU_2 \times O_h$
$\{1\}$	$[\frac{1}{2}\frac{1}{2}\frac{1}{2}\frac{1}{2}]$	$[\frac{1}{2}\frac{1}{2}\frac{1}{2}\frac{1}{2}]$	$\{0\}$	$\langle 0\rangle$	^{10}	${}^1A_{1g}$
			$\{1^2\}$	$\langle 0\rangle$	^{10}	${}^1A_{1g}$
				$\langle 1^2\rangle$	^{12}	${}^1E_g$
					${}^3\tilde{0}$	${}^3A_{2g}$
			$\{1^4\}$	$\langle 0\rangle$	^{10}	${}^1A_{1g}$
		$[\frac{1}{2}\frac{1}{2}\frac{1}{2}-\frac{1}{2}]$	$\{1\}$	$\langle 1\rangle$	^{21}	${}^2E_g$
			$\{1^3\}$	$\langle 1\rangle$	^{21}	${}^2E_g$

center in CaO, which is an octahedral site. The t_{2g} and e_g modes are presumed to be nearly degenerate, so that perturbation treatments consider the modes together. Judd (1974), O'Brien (1976), and Pooler and O'Brien (1977) considered $(e_g+t_{2g})^N$ in the scheme

$$\mathbf{U}_5 \supset \mathbf{SO}_5 \supset \mathbf{SO}_3 \supset \mathbf{O}_h \tag{7.4.3}$$

Other schemes are possible. Kustov (1977) has tabulated a number of cfp's, including those for the chain

$$\mathbf{U}_{10} \supset \mathbf{U}_4 \times \mathbf{U}_6 \supset (\mathbf{SU}_2^S \times \mathbf{SU}_2^e) \times (\mathbf{SU}_2^S \times \mathbf{SU}_3^{t_2}) \tag{7.4.4}$$

where

$$\mathbf{SU}_2^e \underset{\sim}{\supset} \mathbf{O}_h \tag{7.4.5}$$

$$\mathbf{SU}_3^{t_2} \underset{\sim}{\supset} \mathbf{O}_h \tag{7.4.6}$$

The imbeddings (7.4.5) and (7.4.6) are not maximal, e.g., $\mathbf{SU}_2^e \supset \mathbf{O}_2 \underset{\sim}{\supset} \mathbf{O}_h$ is isomorphic to $\mathbf{SO}_3 \supset \mathbf{D}_\infty \supset \mathbf{D}_2$ and it is clear that any $\mathbf{D}_{2n}$ group may be included between $\mathbf{D}_\infty$ and $\mathbf{D}_2$. In the scheme (7.4.3) the partners will not be of the form $e_g^{N_1} t_{2g}^{N_2}$ but will be a linear combination of such terms. In the second scheme, the irrep labels of $\mathbf{U}_4$ and $\mathbf{U}_6$ will, however, carry these N_1, N_2 occupation numbers. Which scheme more closely approximates the eigenfunctions of the Hamiltonian depends on the relative strengths of the various terms in the Hamiltonian.

8

Time Reversal

Time reversal symmetry is an important symmetry in many situations. The difference between the time reversal operator and spatial transformations is discussed in Section 8.1. The relationship between time reversal and complex conjugation is used to derive the time reverse of a state in an atomic orbital in any point group basis. The result is given generally and also in terms of the Condon and Shortley spherical harmonics. We relate these results to Wigner's general results on the doubling of the point group degeneracy.

A particular time reversal signature of an operator can be used to derive relationships between pairs of reduced matrix elements. These relationships, together with Hermiticity requirements, lead in Section 8.2 to various selection rules on reduced matrix elements, selection rules which are in addition to consideration of a point group symmetry. The Wigner–Eckart theorem is thus extended from a group which acts linearly to one which has both linear and conjugate linear operators. Our results are an extension of those of Abragam and Bleaney (1970, Chapter 15). (See also Jansen and Boon 1967, Section II.7; Lax 1974, Chapter 10; Stedman and Butler 1980).

8.1 Time Reversal of States

The time reversal operator is unlike most other operators commonly used in quantum mechanics. Most operators are linear. Consider a ket formed as the linear combination $|1\rangle a+|2\rangle b$ of two kets $|1\rangle$ and $|2\rangle$. An operator L is said to be a *linear operator* if one has the relationship

$$L(|1\rangle a+|2\rangle b)=(L|1\rangle)a+(L|2\rangle)b \tag{8.1.1}$$

An operator A is said to be an *antilinear* or *conjugate linear operator* if instead the coefficients a and b are complex-conjugated by the operator

$$A(|1\rangle a+|2\rangle b)=(A|1\rangle)a^*+(A|2\rangle)b^* \tag{8.1.2}$$

Three conjugate linear operators are relevant to our purposes: the time reversal operator, the complex conjugation operator, and the operator mapping between kets and their corresponding bras. Each of these operators preserves the norm of the vector, a *unitary* property, and thus they are said to be *conjugate unitary*, *antilinear unitary*, or simply *antiunitary*. Although the term antiunitary is often used, it has the disadvantage of suggesting a similarity to the term anti-Hermitian. An *anti-Hermitian operator* B is one whose Hermitian conjugate $B^\dagger$ satisfies

$$B^\dagger=-B \tag{8.1.3}$$

Anti-Hermitian operators are the imaginary number i times a Hermitian operator. Hermitian and anti-Hermitian operators in quantum mechanics are usually linear rather than conjugate linear.

Full discussions of conjugate linear operators can be found in several places. We recommend that the reader who desires a fuller discussion than the present one read the relevant portions of Messiah (1965, pp. 633–642). These sections of Messiah lead on to a full discussion of time reversal (to p. 681). Most of the work on time reversal symmetries is due to Wigner in 1932. Wigner (1959, Chapter 26) is the standard reference for this work. In this section we discuss some of the more important properties of time reversal, leading to Kramers degeneracy.

The operator mapping between bras and kets, the dualizing operator, is written as a right superscript dagger (†). It is conjugate linear

$$(|1\rangle a+|2\rangle b)^\dagger=|1\rangle^\dagger a^*+|2\rangle^\dagger b^*=\langle 1|a^*+\langle 2|b^* \tag{8.1.4}$$

and preserves norms; indeed $|i\rangle^{\dagger\dagger}=|i\rangle$ in finite-dimensional spaces. (Some problems with the operator † arise in general Hilbert spaces; in particular the Dirac δ-function requires careful treatment.) We used property (8.1.4) explicitly in deriving the operator transformation law of (4.1.4), but the conjugate linear relationship between bras and kets is implicit in the entire development of the Racah–Wigner algebra, for the scalar product is conjugate linear in the first argument.

Messiah (1965, p. 641) defines the *complex conjugation operator* K_Q as the operator that transforms the wave functions of the "representation" $\{Q\}$ into their complex conjugates. The term "representation" here means "mode of description," in the sense of Schrödinger representation, Heisenberg representation, etc. (In using quotes for this meaning of representation we follow Messiah, 1965, p. 314.) If the kets $|i\rangle$ are basis vectors in the $\{Q\}$ mode of description, then K_Q is the conjugate linear operator which leaves these vectors invariant,

$$K_Q(|i\rangle a)=(K_Q|i\rangle)a^*=|i\rangle a^* \tag{8.1.5}$$

It is to be emphasized that the action of K_Q on the physical states depends on the mode of description in question and notably on the phases of the basis vectors.

The complex conjugation operator K_Q is important in that every conjugate linear operator A can be written in terms of K_Q and a linear operator (K_QA) or (AK_Q),

$$A=K_Q(K_QA)=(AK_Q)K_Q \tag{8.1.6}$$

Every linear operator considered here can be given a matrix representation. By means of the above result, every conjugate linear operator is equivalent to a product of a matrix and K_Q. Both the matrix and K_Q are basis dependent.

The *time reversal operator* K_t is defined as the antilinear transformation taking the time parameter t to $-t$ but leaving $\mathbf{r}$ unchanged. It transforms a state ψ into a state $K_t\psi$ in which all velocities including electron spin have opposite directions to those in ψ (Messiah 1965, p. 669; Wigner 1959, p. 325 uses θ for our K_t). K_t transforms a position operator $\mathbf{r}$ to itself, and momentum and angular momentum operators to minus themselves:

$$K_t\mathbf{r}K_t^\dagger=\mathbf{r} \tag{8.1.7a}$$

$$K_t\mathbf{p}K_t^\dagger=-\mathbf{p} \tag{8.1.7b}$$

$$K_t\mathbf{J}K_t^\dagger=-\mathbf{J} \tag{8.1.7c}$$

Denote by K_{CS} the complex conjugation operator in the Condon and Shortley basis. A bound state of a single particle with orbital angular momentum l is described in terms of the spherical harmonics Y_{lm} of (4.4.7). In this "representation" the time-reversed Hamiltonian is also the complex-conjugated Hamiltonian and thus the "time reversal" of a state involving Y_{lm} involves $(Y_{lm})^*$. The explicit expression (4.4.7) shows that

$$K_t(Y_{lm})=(Y_{lm})^*=(-1)^m Y_{l-m} \tag{8.1.8}$$

Judd and Runciman (1976) combine this with the usual time reversal properties of spin functions,

$$K_t\alpha=\beta, \qquad K_t\beta=-\alpha \tag{8.1.9}$$

and the reality of the JM coupling coefficients, to obtain

$$K_t|\alpha JM\rangle=|\alpha J-M\rangle(-1)^{p+J-M} \tag{8.1.10}$$

The derivation is similar to our derivation of (5.2.13) and (5.2.15), which give σ_h and σ_v in the JM basis. Once again we have that p is the sum of the parities of the orbitals making up the state $|\alpha JM\rangle$, formed as a coupled product of one-electron states $|l_i m_{s_i}\rangle$, namely

$$p=\sum_{i=1}^{N} l_i \tag{8.1.11}$$

Biedenharn, Blatt, and Rose (1952) used an extra factor of i^l in the definition of spherical harmonics, which has the effect of removing the p term in (8.1.10) (Fano and Racah 1959). Several authors quote this alternative convention and its results in an aside but do not give the result in their (Condon and Shortley) convention (see Brink and Satchler 1962, p. 62; Edmonds 1957, p. 21).

Equation (8.1.10) shows that the conjugate linear operator K_t of time reversal may be written in the Condon and Shortley basis as a product of three operators: the linear unitary parity operator P with the eigenvalue $(-1)^p$ of (8.1.11); a linear unitary $2jm$ operator T, which is skew diagonal and has matrix elements $(-1)^{J-M}$; and the conjugate unitary complex conjugation operator K_{CS},

$$\begin{aligned} K_t|JM\rangle &= PTK_{CS}|JM\rangle \\ &= PT|JM\rangle \\ &= P|J-M\rangle(-1)^{J-M} \\ &= |J-M\rangle(-1)^{p+J-M} \end{aligned} \tag{8.1.12}$$

Consider another point group basis, say $|Ja\lambda l\rangle$, obtained from the above $|JM\rangle$ basis by using the basis transformations of Section 2.2,

$$|Ja\lambda l\rangle = \sum_M |JM\rangle\langle JM|Ja\lambda l\rangle \tag{8.1.13}$$

Combining (8.1.12) with this gives

$$\begin{aligned} K_t|Ja\lambda l\rangle &= PTK_{CS}|Ja\lambda l\rangle \\ &= PTK_{CS}\left(\sum_M |JM\rangle\langle JM|Ja\lambda l\rangle\right) \\ &= PT\left(\sum_M (K_{CS}|JM\rangle)\langle JM|Ja\lambda l\rangle^* \right. \end{aligned}$$

because K_{CS} is conjugate linear, but it does not change the JM basis kets, so

$$\begin{aligned} &= PT\sum_M |JM\rangle\langle Ja\lambda l|JM\rangle \\ &= \sum_M |J-M\rangle(-1)^p\begin{pmatrix} J \\ M \end{pmatrix}\langle Ja\lambda l|JM\rangle \end{aligned} \tag{8.1.14}$$

This may be simplified by using a special case of a relationship intermediate between (3.3.39) and (3.3.40). For the transformation coefficients and $3jm$'s of different subgroup bases of SO_3 one has

$$\sum_M \begin{pmatrix} J & J & 0 \\ M & M' & 0 \end{pmatrix}\langle JM|Ja\lambda l\rangle = \sum_{a'\lambda' l'} \begin{bmatrix} J & J & 0 \\ a & a' & \\ \lambda & \lambda' & 0 \\ l & l' & 0 \end{bmatrix} \times\langle Ja'\lambda' l'|JM'\rangle\langle 00|000\rangle \tag{8.1.15}$$

The nonzero terms in the sums include only the terms M^* and $a^*\lambda^*l^*$, respectively. These trivial $3jm$'s are $|J|^{-1/2}$ times $2jm$'s, which are real. Hence, taking $M'=-M$,

$$\begin{pmatrix} J \\ M \end{pmatrix}\langle JM|Ja\lambda l\rangle = \begin{Bmatrix} J \\ a \\ \lambda \\ l \end{Bmatrix}\langle Ja^*\lambda^*l^*|J-M\rangle \tag{8.1.16}$$

and thus

$$K_t|Ja\lambda l\rangle = \sum_M |J-M\rangle\langle J-M|Ja^*\lambda^*l^*\rangle \begin{Bmatrix} J \\ a \\ \lambda \\ l \end{Bmatrix}(-1)^p$$

$$= |Ja^*\lambda^*l^*\rangle(-1)^p \begin{Bmatrix} J \\ a \\ \lambda \\ l \end{Bmatrix} \tag{8.1.17}$$

The explicit functional form (4.4.7) for spherical harmonics has been used in obtaining this result. If the angular part of the wave functions is constructed by symmetry-adapting the $|JM\rangle$ wave functions to the point group scheme of the problem, one obtains what is usually known as atomic orbitals. Both (8.1.17) and its special case (8.1.10) are valid for such atomic orbitals.

The above action of the time reversal operator contains the atomic orbital statement of *Kramers degeneracy*. If the Hamiltonian is time-reversal invariant, there is a degeneracy in addition to the point group degeneracy associated with irrep $\boldsymbol{\lambda}(\mathbf{G})$, a degeneracy between $|Ja\lambda l\rangle$ and $|Ja^*\lambda^*l^*\rangle$. Wigner was first (see Wigner 1959, Chapter 29) to give a rigorous development of the general case. We outline Wigner's proof below.

Let $\mathbf{G}$ be the spatial symmetry group of the Hamiltonian $\mathcal{H}$ of the problem. If $\mathcal{H}$ is time-reversal invariant, the symmetry group is enlarged to a group $\mathbf{G}^t$ by the addition of the time reversal operator K_t of (8.1.7). $\mathbf{G}$ acts linearly and K_t acts conjugate linearly. The product of two linear or two conjugate linear operators is linear and hence in $\mathbf{G}$ itself. The product of one of each is conjugate linear. This implies the space–time symmetry group is exactly twice the size of the spatial group $\mathbf{G}^t = \mathbf{G} + \mathbf{K}_t\mathbf{G}$. The conjugate linearity of operators involving time requires a modification of the representation condition (2.1.9), for which Wigner introduces the term *co-representation*. All irreducible co-representations of the space–time group either contain one, $\boldsymbol{\lambda}$, or two, $\boldsymbol{\lambda}+\boldsymbol{\lambda}^*$, irreps of the underlying spatial group. Wigner's proof shows that (8.1.17) describes the degeneracy introduced into a system of N fermions by time reversal considerations. This time reversal degeneracy is in addition to the spatial degeneracy of order $|\lambda|$ described by the irrep $\boldsymbol{\lambda}(\mathbf{G})$ of the point symmetry group $\mathbf{G}$.

(a) If $\boldsymbol{\lambda}(\mathbf{G})$ is orthogonal and N is even or $\boldsymbol{\lambda}(\mathbf{G})$ is symplectic and N is odd, no additional degeneracy is introduced.

(b) In all other cases the degeneracy is doubled.

Case (a) is the situation for the real irreps of the non-Abelian point groups. In such cases the point group contains operations which transform all kets and their corresponding time-reversed conjugates as part of one irrep. For example, the time reversal conjugate pair of kets $|JM\rangle$, $|J-M\rangle$ are two of the partners of the irrep $\mathbf{J}(\mathbf{SO}_3)$. Systems with even (odd) numbers of fermions transform as the true (spin) irreps of the non-Abelian point groups and such irreps, if real, are usually orthogonal (symplectic). All irreps of $\mathbf{SO}_3$, $\mathbf{O}$, $\mathbf{K}$, and $\mathbf{D}_n$ for n even come into these categories.

The second case is of more interest. Let $\mathbf{H}$ be a pure rotation or rotation–reflection group that is the symmetry group of the problem. Let $\boldsymbol{\lambda}(\mathbf{G})$ be an irrep which does not satisfy the conditions for case (a). Either $\boldsymbol{\lambda}$ will be complex, $\boldsymbol{\lambda}^*\neq\boldsymbol{\lambda}$, or $\boldsymbol{\lambda}$ will be one of the one-dimensional orthogonal spin irreps of an odd cyclic group, e.g., $\frac{3}{2}(\mathbf{C}_3)$. Wigner shows that the state (or states) time-reversal degenerate with the state (or states) of irrep $\boldsymbol{\lambda}(\mathbf{G})$ transform as $\boldsymbol{\lambda}^*(\mathbf{G})$. In fact the time reversal operator can always be written in much the form used in (8.1.12), a product of $K_{\{Q\}}$, the conjugate linear "complex conjugation" operator, a linear skew $2jm$ operator $T_{\{Q\}}$, and a phase operator $P_{\{Q\}}$, each of which depends on the "representation" chosen for the wave functions of the problem.

The relationship of the degenerate pairs of representations can be understood by comparing the present general case with the atomic orbital case above. In (8.1.17) the time reversal operator connects the kets $|Ja\lambda l\rangle$ and $|Ja^*\lambda^*l^*\rangle$. The space–time group $\mathbf{G}^t$ connects kets labeled by $\boldsymbol{\lambda}(\mathbf{G})$ with others with complex-conjugated parentage labels.

Consider an odd-electron system whose Hamiltonian is time-reversal invariant, where $\mathbf{D}_3$ is an approximate spatial symmetry broken by a weak $\mathbf{C}_3$-invariant term. A pair of kets labeled $|\frac{1}{2}(D_3)\pm\frac{1}{2}(C_3)\rangle$ are degenerate under the approximate $\mathbf{D}_3$ symmetry. The above time reversal argument shows that the $\mathbf{C}_3$ perturbation does not lift the degeneracy—the pair is

Kramers' degenerate. Likewise, a pair of kets labeled $|\pm\frac{3}{2}(D_3)\frac{3}{2}(C_3)\rangle$ would be expected, on the spatial argument, to be nongenerate, but the presence of time reversal symmetry indicates degeneracy. In each case the full space–time group is required to prove degeneracy, but the approximate spatial-symmetry group $\mathbf{D}_3$ provides sufficient labels for the partners of Wigner's "co-representation" even when the exact spatial-symmetry group, $\mathbf{C}_3$, does not.

Observe that our group-theoretic multiplicity separations are such that all multiplicity labels appearing in our tables are real, $a^*=a$. To find the time reversal conjugate of an atomic orbital state in any (complete) point group scheme one needs only to complex-conjugate the ket's irrep labels up to the $\mathbf{SO}_3$ label and use (8.1.17) (note that complex conjugation of parentage irrep labels is quite different, being a particle–hole interchange; see Chapter 7). If one does not know the $\mathbf{SO}_3$ label, that is, if the kets are not known in terms of the spherical harmonics Y_{lm} (for example, with some molecular orbital models), one cannot use the explicit relation (8.1.17). One has the qualitative statements of Wigner, saying how the time-reversed ket transforms, to which we can add time reversal properties of the operators. The consequences of the action of time reversal on operators are discussed in the following section.

8.2 Time Reversal and Selection Rules

The previous section discussed the properties of the time reversal operator K_t and its action on the partners of point groups. It is the purpose of this section to extend this analysis to an analysis of the reduced matrix elements of the point groups. The Wigner–Eckart theorem simplifies matrix elements and gives selection rules when kets and operators have known group properties. The extension of the Wigner–Eckart theorem to the space–time group $\mathbf{G}^t$ is nontrivial because $\mathbf{G}^t$ does not act linearly. The extension follows readily, however, because time reversal is represented in Schrödinger's wave mechanics as the complex conjugation operator and the complex conjugation properties of the Racah–Wigner algebra are known.

The time reversal operator takes the wave function $\psi(\mathbf{r},t)$ into the function $\psi^*(\mathbf{r},-t)$. In terms of the ket notation we shall write the time reverse of $|x\lambda l\rangle$ as $|\overline{x\lambda l}\rangle$:

$$K_t|x\lambda l\rangle=|\overline{x\lambda l}\rangle \tag{8.2.1}$$

In a basis where the $\mathbf{SO}_3$ structure is also known (and where Condon and Shortley spherical harmonics are used), we have [see (8.1.17)]

$$|\overline{\alpha Ja\lambda l}\rangle = |\alpha Ja^*\lambda^* l^*\rangle(-1)^p \begin{Bmatrix} J \\ a \\ \lambda \\ l \end{Bmatrix} \tag{8.2.2}$$

Applying time reversal twice and using the properties of the $2jm$ gives

$$\begin{aligned} K_t^2|\alpha Ja\lambda l\rangle &= K_t(K_t|\alpha Ja\lambda l\rangle) \\ &= |\overline{\overline{\alpha Ja\lambda l}}\rangle = (-1)^{2J}|\alpha Ja\lambda l\rangle \end{aligned} \tag{8.2.3}$$

from the properties of $2jm$ symbols. We shall need that

$$\begin{aligned} K_t^\dagger|\alpha Ja\lambda l\rangle &= K_t^\dagger K_t^2(-1)^{2J}|\alpha Ja\lambda l\rangle \\ &= K_t|\alpha Ja\lambda l\rangle(-1)^{2J} \\ &= |\alpha Ja^*\lambda^* l^*\rangle(-1)^p \begin{Bmatrix} J \\ a^* \\ \lambda^* \\ l^* \end{Bmatrix} \end{aligned} \tag{8.2.4}$$

The *time reversal parity* of a state is defined as the change of sign of the state under two consecutive time reverses; it is $(-1)^{2J}$, which is -1 for spin functions (an odd number of fermions) and $+1$ for true functions (an even number of fermions.)

Let us now consider corresponding properties of operators. First recall that two operators A and B are said to be equal if all their matrix elements are equal:

$$A = B \Leftrightarrow \langle i|A|j\rangle = \langle i|B|j\rangle \tag{8.2.5}$$

B is said to be the Hermitian conjugate of A if all their matrix elements are Hermitian conjugate:

$$A^\dagger = B \Leftrightarrow \langle i|A|j\rangle = \langle j|B|i\rangle^* \tag{8.2.6}$$

We say A and B are *time reversal conjugate* and write $\bar{A}=B$ if one is the time reversal transform of the other [see (8.1.7)]

$$\bar{A}=B \Leftrightarrow B=K_t A K_t^{\dagger} \tag{8.2.7}$$

In writing this in terms of matrix elements, care must be taken with the conjugate linear operator K_t: brackets must be used to show the sequence of operators. In particular for K antilinear one has [Messiah 1965, Eq. (XV.17)]

$$(\langle i|K)|j\rangle=(\langle i|(K|j\rangle))^* \tag{8.2.8}$$

Combining the definitions of time-reversed states and time conjugate operators (8.2.7) with (8.2.8) gives

$$\begin{aligned}\langle\bar{i}|\bar{A}|\bar{j}\rangle&=(\langle i|K_t^{\dagger})\left[K_t A K_t^{\dagger}(K_t|j\rangle)\right]\\&=\left[\langle i|(K_t^{\dagger}K_t A K_t^{\dagger}K_t|j\rangle)\right]^*\\&=\langle i|A|j\rangle^*\end{aligned} \tag{8.2.9}$$

An operator A is said to have a definite *time reversal signature* τ_A (even or odd) if $\bar{A}=\tau_A A$ with $\tau_A=\pm 1$. For example, observe in (8.1.7) that as an operator, $\mathbf{r}$ is time reversal even, while $\mathbf{p}$ and $\mathbf{J}$ are odd. However, tensor operators need not have definite time reversal signature. Time reversal is similar to Hermitian conjugation in this respect, for they are both antilinear. The time reverse of a ket transforming as a partner of $\boldsymbol{\lambda}(\mathbf{G})$ transforms as $\boldsymbol{\lambda}^*(\mathbf{G})$, just as its Hermitian conjugate (a bra) transforms as $\boldsymbol{\lambda}^*(\mathbf{G})$ [see (8.2.2) and an intermediate step in (4.1.4)]. In both cases a $2jm$ is required to give the new entity standard transformation properties.

The Wigner–Eckart theorem may be applied to any tensor operator. Consider the three tensor operators T_l^{λ},

$$S_{l^*}^{\lambda^*}=\begin{pmatrix}\lambda^*\\ l^*\end{pmatrix}(T_l^{\lambda})^{\dagger} \tag{8.2.10}$$

and

$$R_{l*}^{\lambda*} = \begin{pmatrix} \lambda \\ l \end{pmatrix} \overline{T_l^\lambda} \tag{8.2.11}$$

Application of the Wigner–Eckart theorem to these equations gives relationships between the reduced matrix elements of R, S, and T. The relationships reduce to selection rules (which are in addition to the group's selection rules) in cases where the Hermitian or time reversal nature of the interaction is known. The Hermitian conjugate operator has matrix elements

$$\langle x_1\lambda_1 l_1 | S_{l*}^{\lambda*} | x_2\lambda_2 l_2 \rangle$$

$$= \sum_r \langle x_1\lambda_1 \| S^{\lambda*} \| x_2\lambda_2 \rangle_r \begin{pmatrix} \lambda_1 \\ l_1 \end{pmatrix} \begin{pmatrix} \lambda_1^* & \lambda^* & \lambda_2 \\ l_1^* & l^* & l_2 \end{pmatrix}^r$$

which is, using (8.2.10) and (8.2.6),

$$= \begin{pmatrix} \lambda^* \\ l^* \end{pmatrix} \langle x_2\lambda_2 l_2 | T_l^\lambda | x_1\lambda_1 l_1 \rangle^*$$

$$= \sum_s \langle x_2\lambda_2 \| T^\lambda \| x_1\lambda_1 \rangle_s^* \begin{pmatrix} \lambda^* \\ l^* \end{pmatrix} \begin{pmatrix} \lambda_2 \\ l_2 \end{pmatrix} \begin{pmatrix} \lambda_2^* & \lambda & \lambda_1 \\ l_2^* & l & l_1 \end{pmatrix}^{*s}$$

The symmetry properties of the $3jm$, (3.3.11) and (3.3.12), give

$$= \sum_s \langle x_2\lambda_2 \| T^\lambda \| x_1\lambda_1 \rangle_s^* \{\lambda_1\} \{\lambda_1 \lambda \lambda_2^* s\} \begin{pmatrix} \lambda_1 \\ l_1 \end{pmatrix} \begin{pmatrix} \lambda_1^* & \lambda^* & \lambda_2 \\ l_1^* & l^* & l_2 \end{pmatrix}^s \tag{8.2.12}$$

Hence

$$\langle x_1\lambda_1 \| S^{\lambda*} \| x_2\lambda_2 \rangle_r = \langle x_2\lambda_2 \| T^\lambda \| x_1\lambda_1 \rangle_r^* \{\lambda_1\} \{\lambda_1 \lambda \lambda_2^* r\} \tag{8.2.13}$$

A similar calculation for the time reversal conjugate can be carried out, but only if the explicit time reversal properties of the states are known.

Equations (8.2.7), (8.2.8), (8.2.4), and (8.2.2) show that

$$\langle \alpha_1 J_1 a_1 \lambda_1 l_1 | \begin{pmatrix} \lambda \\ l \end{pmatrix} \overline{T_l^\lambda} | \alpha_2 J_2 a_2 \lambda_2 l_2 \rangle$$

$$= \begin{pmatrix} \lambda \\ l \end{pmatrix} \langle \alpha_1 J_1 a_1 \lambda_1 l_1 | (K_t T_l^\lambda K_t^\dagger | \alpha_2 J_2 a_2 \lambda_2 l_2 \rangle)$$

$$= \begin{pmatrix} \lambda \\ l \end{pmatrix} [(\langle \alpha_1 J_1 a_1 \lambda_1 l_1 | K_t) T_l^\lambda (K_t^\dagger | \alpha_2 J_2 a_2 \lambda_2 l_2 \rangle)]^*$$

$$= (-1)^{p_1+p_2} \begin{bmatrix} J_1 \\ a_1^* \\ \lambda_1^* \end{bmatrix} \begin{bmatrix} J_2 \\ a_2^* \\ \lambda_2^* \end{bmatrix} \begin{pmatrix} \lambda_1^* \\ l_1^* \end{pmatrix} \begin{pmatrix} \lambda \\ l \end{pmatrix} \begin{pmatrix} \lambda_2^* \\ l_2^* \end{pmatrix}$$

$$\times \langle \alpha_1 J_1 a_1^* \lambda_1^* l_1^* | T_l^\lambda | \alpha_2 J_2 a_2^* \lambda_2^* l_2^* \rangle^* \quad (8.2.14)$$

We may now use the Wigner–Eckart theorem and the linear independence of the jm's to obtain

$$\langle \alpha_1 J_1 a_1 \lambda_1 \| R^{\lambda^*} \| \alpha_2 J_2 a_2 \lambda_2 \rangle_r$$

$$= (-1)^{p_1+p_2} \begin{bmatrix} J_1 \\ a_1^* \\ \lambda_1^* \end{bmatrix} \begin{bmatrix} J_2 \\ a_2^* \\ \lambda_2^* \end{bmatrix} \langle \alpha_1 J_1 a_1^* \lambda_1^* \| T^\lambda \| \alpha_2 J_2 a_2^* \lambda_2^* \rangle_r^* \quad (8.2.15)$$

Tensors of particular importance are the unit tensor operators of (4.1.5). Their reduced matrix elements are products of Kronecker deltas and thus one has

$$U_l^{r\lambda}(x_1 \lambda_1 x_2 \lambda_2)^\dagger = U_{l^*}^{r\lambda^*}(x_2 \lambda_2 x_1 \lambda_1) \begin{pmatrix} \lambda \\ l \end{pmatrix} \{\lambda_1\} \{\lambda_1^* \lambda \lambda_2 r\} \quad (8.2.16)$$

and

$$\overline{U_l^{r\lambda}(J_1 a_1 \lambda_1 J_2 a_2 \lambda_2)} = U_{l^*}^{r\lambda^*}(J_1 a_1^* \lambda_1^* J_2 a_2^* \lambda_2^*) \begin{pmatrix} \lambda \\ l \end{pmatrix} (-1)^{p_1+p_2} \begin{bmatrix} J_1 \\ a_1^* \\ \lambda_1^* \end{bmatrix} \begin{bmatrix} J_2 \\ a_2^* \\ \lambda_2^* \end{bmatrix} \quad (8.2.17)$$

Any operator V can be written as a linear combination of the unit tensors. Unless the operator V transforms as a particular irrep operator V_l^λ, the coefficients required will depend on all labels. They are not quite the reduced matrix elements corresponding to a tensor operator, as in (4.2.2).

$$V = \sum_{x_1\lambda_1 x_2\lambda_2 r\lambda l} U_l^{r\lambda}(x_1\lambda_1 x_2\lambda_2)\langle x_1\lambda_1 \| V_l^\lambda \| x_2\lambda_2\rangle_r \tag{8.2.18}$$

If V is Hermitian, $V = V^\dagger$, (8.2.18) may be Hermitian-conjugated, (8.2.16) applied, and finally the coefficients compared (for the unit tensors are linearly independent). Thus if $V = V^\dagger$,

$$\langle x_1\lambda_1 \| V_l^\lambda \| x_2\lambda_2\rangle_r^* = \begin{pmatrix}\lambda\\ l\end{pmatrix}\{\lambda_1\}\{\lambda_1^*\lambda\lambda_2 r\}\langle x_2\lambda_2 \| V_{l^*}^{\lambda^*} \| x_1\lambda_1\rangle \tag{8.2.19}$$

Similarly, if V has time reversal signature τ_V, $\overline{V} = \tau_V V$, then

$$\langle J_1 a_1\lambda_1 \| V_l^\lambda \| J_2 a_2\lambda_2\rangle_r^* = \tau_V(-1)^{p_1+p_2}\begin{pmatrix}\lambda\\ l\end{pmatrix}\begin{bmatrix}J_1\\ a_1^*\\ \lambda_1^*\end{bmatrix}\begin{bmatrix}J_2\\ a_2^*\\ \lambda_2^*\end{bmatrix}$$

$$\times\langle J_1 a_1^*\lambda_1^* \| V_{l^*}^{\lambda^*} \| J_2 a_2^*\lambda_2^*\rangle_r \tag{8.2.20}$$

Both the above results relate the complex conjugate of one reduced matrix element of V [being a parameter in the tensor basis expansion (8.2.18)] with another. While these relations therefore reduce the number of free parameters in the expansion of an operator V in terms of basis tensor operators, the relations used separately do not imply that any parameters are zero. However, in cases where they give conflicting restrictions on a parameter, we can deduce that the parameter vanishes. Consider the case of a real, time-reversal-invariant irrep space, that is, where $x_1\lambda_1 = x_2\lambda_2 = x_1^*\lambda_1^*$. The time reversal phases arising in (8.2.20) from the $|J_1 a_1\lambda_1\rangle$ spaces cancel and (8.2.19) and (8.2.20) may be combined to give

$$\langle x_1\lambda_1 \| V_{l^*}^{\lambda^*} \| x_1\lambda_1\rangle_r = \tau_V\{\lambda_1\}\{\lambda_1\lambda\lambda_1 r\}\langle x_1\lambda_1 \| V_{l^*}^{\lambda^*} \| x_1\lambda_1\rangle_r \tag{8.2.21}$$

This diagonal reduced matrix element must vanish whenever

$$\tau_V\{\lambda_1\}\{\lambda_1\lambda\lambda_1 r\} = -1 \tag{8.2.22}$$

Where the group in question is $\mathbf{SO}_3$ the reality and the time-reversal invariance of the irrep $\mathbf{J}_1$ are guaranteed and (8.2.22) reduces to an operator V_M^J having zero diagonal reduced matrix elements whenever

$$\tau_V(-1)^J=-1 \tag{8.2.23}$$

Hence a time-reversal-even operator transforming as $\mathbf{J}$ odd (such as the Stark effect operator $\mathcal{E}_z$) has zero diagonal reduced matrix elements, while any time-reversal-odd operator (such as many magnetic effects) must be expanded in terms of unit tensor operators of odd $\mathbf{J}$ only. Similar results may be derived by the alternative assumption of a well-defined parity for the $\mathbf{J}_1$ manifold. The explicit time reversal (even or odd) nature of the manifold does not enter our result, (8.2.22), but the time-reversal invariance of $\mathbf{J}_1$ is required in its derivation. In the corresponding parity argument an equivalent assumption concerning parity is required. The two assumptions are different, although they lead to the same conclusions about diagonal matrix elements [see (4.4.9)].

Consider now a second example where the irrep of the group, $\boldsymbol{\lambda}_1(\mathbf{G})$, is again invariant under the time reversal group $\mathbf{G}'$, but where product multiplicity occurs. A typical group is the octahedral group $\mathbf{O}$; the square of the spinor irrep $\frac{3}{2}$ contains the two irreps $\mathbf{1}$ and $\tilde{\mathbf{1}}$ twice. Consider a Hermitian, time-even operator V which transforms under $\mathbf{O}$ as a partner l of irrep $\boldsymbol{\lambda}$. For $\boldsymbol{\lambda}=\mathbf{1}$ ($r=0$ or 1), $\boldsymbol{\lambda}=\tilde{\mathbf{0}}$, or $\boldsymbol{\lambda}=\tilde{\mathbf{1}}$ ($r=1$) the product $\tau_V\{\frac{3}{2}\}\{\frac{3}{2}\frac{3}{2}\lambda r\}$ is negative, implying that the reduced matrix element $\langle x_1\frac{3}{2}\|V_{l^*}^{\lambda^*}\|x_1\frac{3}{2}\rangle_r$ vanishes, while for $\boldsymbol{\lambda}=\mathbf{0}$, $\boldsymbol{\lambda}=\mathbf{2}$, or $\boldsymbol{\lambda}=\tilde{\mathbf{1}}$ ($r=0$) the reduced matrix element exists, and is either real or imaginary as given by (8.2.19). The selection rules for time-odd operators are the reverse of the above, with the reality of the reduced matrix elements again given by (8.2.19). Observe, though, that if a time-odd V transforms as $\mathbf{1(O)}$, both reduced matrix elements $\langle x_1\frac{3}{2}\|V_l^1\|x_1\frac{3}{2}\rangle_r$ (for $r=0$ and 1) are independent parameters about which the space–time group only gives conditions on their reality or otherwise.

Similar conclusions to these octahedral results follow for most irreps of most point groups. In the cases where the operator transforms as a sum of complex irreps [e.g., $\mathbf{2(T)}$ and $-\mathbf{2(T)}$] the complex conjugation condition (8.2.13) relates a pair of reduced matrix elements. The case of states transforming as a complex irrep [e.g., $\frac{1}{2}(\mathbf{C}_3)$] or transforming as an orthogonal but spinor irrep [e.g., $\frac{3}{2}(\mathbf{C}_3)$] requires separate treatment.

Time reversal symmetry $\mathbf{G}'$ doubles the spatial degeneracy resulting from a symmetry $\mathbf{G}$ if the irreducible co-representation $\Lambda(\mathbf{G}')$ contains a complex conjugate pair of irreps of $\mathbf{G}$ (see our statement of Wigner's result on p. 159). As discussed in the previous section, the degenerate states will be labeled either by the irreps $\boldsymbol{\lambda}$ and $\boldsymbol{\lambda}^*$ of a complex pair or by the orthogonal spin irrep $\frac{1}{2}\mathbf{n}$ of some $\mathbf{C}_n$ together with a multiplicity label. We may label the states as $|x_1\Lambda_1a_1\lambda_1l_1\rangle$, where $x_1^*=x_1$, $\Lambda_1^*=\Lambda_1$. Unless $\boldsymbol{\lambda}_1=\frac{1}{2}\mathbf{n}(\mathbf{C}_n)$, n odd, we have also $a_1^*=a_1=0$, $\boldsymbol{\lambda}_1^*\neq\boldsymbol{\lambda}_1$. As before, any Hermiticity and time reversal parity of an operator may be used in (8.2.19) and (8.2.20) to obtain restrictions on the diagonal reduced matrix elements. Use of the co-representation condition $x_1^*\Lambda_1^*=x_1\Lambda_1$ and restriction to diagonal co-representations $x_1\Lambda_1=x_2\Lambda_2$ gives

$$\begin{aligned}\langle x_1\Lambda_1a_1\lambda_1\|V_l^\lambda\|x_1\Lambda_1a_2\lambda_2\rangle_r^*\\&=\begin{pmatrix}\lambda\\l\end{pmatrix}\{\lambda_1\}\{\lambda_1^*\lambda\lambda_2r\}\langle x_1\Lambda_1a_2\lambda_2\|V_{l^*}^{\lambda^*}\|x_1\Lambda_1a_1\lambda_1\rangle_r\\&=\tau_V\begin{pmatrix}\lambda\\l\end{pmatrix}\begin{bmatrix}\Lambda_1\\a_1\\\lambda_1\end{bmatrix}\begin{bmatrix}\Lambda_1\\a_2\\\lambda_2\end{bmatrix}\langle x_1\Lambda_1a_1^*\lambda_1^*\|V_{l^*}^{\lambda^*}\|x_1\Lambda_1a_2^*\lambda_2^*\rangle_r\end{aligned}\tag{8.2.24}$$

The structure of the point groups may be used explicitly to further simplify this result because the time reversal doubling of spatial degeneracy only occurs for certain one-dimensional irreps of $\mathbf{T}$, $\mathbf{D}_{\text{odd}}$, and $\mathbf{C}_n$, and for the pair of two-dimensional irreps $\pm\frac{3}{2}(\mathbf{T})$. No product multiplicity occurs for any of these irreps. For the diagonal elements (8.2.24) implies

$$\langle\Lambda_1a_1\lambda_1\|V_l^\lambda\|\Lambda_1a_1\lambda_1\rangle=\tau_V\{\lambda_1\}\{\lambda_1^*\lambda\lambda_1\}\langle\Lambda_1a_1^*\lambda_1^*\|V_l^\lambda\|\Lambda_1a_1^*\lambda_1^*\rangle\tag{8.2.25}$$

That is, the two diagonal reduced matrix elements are either equal or opposite. For almost all cases only the irrep $\mathbf{0}$ occurs and thus the $3j$ cancels the $2j$, giving the simplified result

$$\langle\Lambda_1a_1\lambda_1\|V^0\|\Lambda_1a_1\lambda_1\rangle=\tau_V\langle\Lambda_1a_1^*\lambda_1^*\|V^0\|\Lambda_1a_1^*\lambda_1^*\rangle\tag{8.2.25a}$$

The only exceptions are for the case when $\boldsymbol{\lambda}_1(\mathbf{G})$ is $\pm\frac{3}{2}(\mathbf{T})$. These reduced matrix elements are real, but unlike the case of real irreps, (8.2.22), no information on vanishings is available. On the other hand, (8.2.24) indicates vanishings of the off-diagonal reduced matrix elements. With our $2jm$ phases, the G^t–G $2jm$'s are related by the product $\{\Lambda_1\}\{\lambda_1\}$, where $\{\Lambda_1\}=\pm 1$ as Λ_1 is true or spin. Thus

$$\langle \Lambda_1 a_1^* \lambda_1^* \| V_l^\lambda \| \Lambda_1 a_1 \lambda_1 \rangle = 0 \qquad \text{when } \tau_V\{\Lambda_1\}\{\lambda_1\lambda_1\lambda\} = -1 \quad (8.2.26)$$

In this case the first line of (8.2.24) gives the relationship between off-diagonal reduced matrix elements.

As an example of Kramers degeneracy, consider a ligand field V which has symmetry $\mathbf{C}_n$. The ligand field, being electrostatic in origin (in a typical model) is time-even, $\tau_V=1$. Take an odd number of electrons. Their wave functions will transform as spinor irreps $\boldsymbol{\lambda}_1(\mathbf{C}_n)$. All C_n jm symbols are unity, so the matrix elements of V are the C_n reduced matrix elements. V has symmetry $\mathbf{C}_n$, that is, V transforms as $\mathbf{0}(\mathbf{C}_n)$. The diagonal matrix elements are equal by (8.2.25a) and the off-diagonal, when allowed by the Wigner–Eckart theorem, are disallowed by the time reversal result, (8.2.26). In this case the Wigner–Eckart theorem gives a useful selection rule for V^0 except when $\boldsymbol{\lambda}_1$ is $\frac{1}{2}\mathbf{n}(\mathbf{C}_n)$, which is the exceptional case of an orthogonal spinor irrep. In the absence of any point group symmetry in the ligand field, that is, when no information on the spatial transformation properties of V is obtainable, (8.2.25) and (8.2.26) still retain their force: The Kramers degeneracy of spinor states remains.

The two-dimensional complex irreps $\pm\frac{3}{2}(\mathbf{T})$ bring to the fore the difference between reduced matrix elements and the matrix elements themselves. Consider a time-even ligand field V transforming as $\mathbf{1}(\mathbf{T})\mathbf{0}(\mathbf{C}_3)$. It occurs in the symmetric part of $\frac{3}{2}\times\frac{3}{2}$ of $\mathbf{T}$; thus (8.2.25) gives

$$\langle \Lambda_1 \tfrac{3}{2} \| V^1 \| \Lambda_1 \tfrac{3}{2} \rangle^T = -\langle \Lambda_1 -\tfrac{3}{2} \| V^1 \| \Lambda_1 -\tfrac{3}{2} \rangle^T \qquad (8.2.27)$$

Although the reduced matrix elements for the time reversal states are of opposite sign, the actual matrix elements $\langle \Lambda_1 \frac{3}{2}\frac{3}{2} | V_0^1 | \Lambda_1 \frac{3}{2}\frac{3}{2} \rangle$ and $\langle \Lambda_1 -\frac{3}{2}\frac{3}{2} | V_0^1 | \Lambda_1 -\frac{3}{2}\frac{3}{2} \rangle$ are equal, because

$$\begin{pmatrix} \frac{3}{2} \\ \frac{3}{2} \end{pmatrix} \begin{pmatrix} -\frac{3}{2} & 1 & \frac{3}{2} \\ \frac{3}{2} & 0 & \frac{3}{2} \end{pmatrix}_{C_3}^{T} = -\begin{pmatrix} -\frac{3}{2} \\ \frac{3}{2} \end{pmatrix} \begin{pmatrix} \frac{3}{2} & 1 & -\frac{3}{2} \\ \frac{3}{2} & 0 & \frac{3}{2} \end{pmatrix}_{C_3}^{T} \qquad (8.2.28)$$

and therefore the two states $|\Lambda_1 \frac{3}{2} \frac{3}{2}\rangle$ and $|\Lambda_1 - \frac{3}{2} \frac{3}{2}\rangle$ are indeed Kramers degenerate. This is an example of the second case of Wigner, Section 8.1, for the kets transform as the same C_3 irrep, the orthogonal spinor irrep $\frac{3}{2}(\mathbf{C}_3)$. However, they are distinguished by their T labels as $\pm\frac{3}{2}(\mathbf{T})$, and the apparatus of the Racah algebra may be applied. Of course, with our assumption that the ligand field transforms as **1(T)**, the irreps of T will be split. The states $|\Lambda_1 \frac{3}{2} \frac{1}{2}\rangle$ and $|\Lambda_1 - \frac{3}{2} - \frac{1}{2}\rangle$ will form a second Kramers doublet, split from the above doublet by the ligand field.

9

Applications

This chapter discusses the application of our methods to physical problems. In the first section we quote the results of writing the various interaction terms of the free-ion Hamiltonian in terms of standard Racah tensor operators. We show how the standard results may be transferred into any point group basis. Section 9.2 gives the standard results for the Zeeman interaction in the SL coupled scheme and then shows how these results may be used unaltered in any point group scheme.

Perturbation fields external to the ion do not usually have $\mathbf{SO}_3$ symmetry. It is here that tables of this book provide fundamental simplifications. Section 9.3 is devoted to a discussion of how the effect of the environment of the ion may be written in terms of tensor operators which have point group symmetry. Particularly powerful results are obtained if an approximation is used in which the effect of the environment is written in terms of tensors which transform not only as irreps of the point group but also as irreps of $\mathbf{SO}_3$.

9.1 The Free-Ion Hamiltonian

The Hamiltonian of an N-electron free ion with nuclear charge Ze cannot be written down exactly. If the nuclear mass is assumed infinite, the Coulomb part of the Hamiltonian may be written

$$H_{\text{nonrel}} = -\frac{\hbar^2}{2m}\sum_{i=1}^{N}\nabla_i^2 - \sum_{i=1}^{N}\frac{Ze^2}{r_i} + \sum_{i<j}^{N}\frac{e^2}{r_{ij}} \tag{9.1.1}$$

Exact solutions for this simplified equation are not possible for systems of more than one electron. The most common further approximation used is the *central field approximation* (Condon and Shortley 1935), in which each electron is assumed to move independently in a potential field $-U(r_i)/e$ arising as an average of the other electrons. The resulting central field Hamiltonian

$$H_{\text{central}} = \sum_{i=1}^{N} \left[\frac{-\hbar^2}{2m} \nabla_i^2 + U(r_i) \right] \tag{9.1.2}$$

is readily separable into radial and angular equations for each electron. The normalized solutions for the bound states of (9.1.2) take the form of antisymmetrized products of single-electron functions,

$$\phi = r^{-1} R_{nl}(r) Y_{lm_l}(\theta, \phi) \sigma_s \tag{9.1.3}$$

where σ_s is a spin function.

The difference between the original Hamiltonian and the central field Hamiltonian is composed of two parts: an angularly independent term H_{radial}, which is a simple sum of one-electron potentials,

$$H_{\text{radial}} = \sum_{i=1}^{N} \left[-\frac{Ze^2}{r_i} - U(r_i) \right] \tag{9.1.4}$$

and the interelectron Coulomb potential,

$$H_{\text{Coul}} = \sum_{i<j}^{N} \frac{e^2}{r_{ij}} \tag{9.1.5}$$

Calculations of $U(r_i)$ are outside the scope of this book, but the principle used is to obtain a set of $U(r_i)$ such that the sum $H_{\text{radial}} + H_{\text{Coul}}$ is of a size that may be treated by perturbation theory. The terms H_{central} and H_{radial} are both scalar under rotations of any electron. These terms commute with all operations of the group $\mathbf{SU}_2^i \times \mathbf{SO}_3^i$ for each electron i. Recall that $\mathbf{SU}_2 \times \mathbf{SO}_3$ is a direct product group, $\mathbf{SU}_2$ acts on the spin coordinates and $\mathbf{SO}_3$ on the angular coordinates of the system, and that $\mathbf{SU}_2$ is the same abstract group as the spin covering group of $\mathbf{SO}_3$. Thus, by the Wigner–Eckart theorem the single-electron quantum numbers s, m_s, l, and m_l are all good quantum numbers with respect to $H_{\text{central}} + H_{\text{radial}}$. If the quantum number n is given the hydrogenic interpretation, then the properties of $U(r_i)$ will mean that n is not a good quantum number for many-electron atoms. It is usual, however, to define modified radial functions, indexed by some new n, such that the new n is a reasonable quantum number.

The properties of Legendre polynomials and spherical harmonics may be used to write the Coulomb interaction between electrons i and j in terms of radial integrals and of scalar products of tensors for electrons i and j. (For details of the method see, for example, Judd 1963.) The interelectron separation may be written in terms of Legendre polynomials for the angle ω_{ij} between r_i and r_j as

$$\frac{1}{r_{ij}} = \sum_k \frac{r_<^k}{r_>^{k+1}} P_k(\cos\omega_{ij}) \tag{9.1.6}$$

This in turn can be written as the scalar product (4.3.10) of the spherical tensors of (4.4.6),

$$H_{\text{Coul}} = e^2 \sum_k \frac{r_<^k}{r_>^{k+1}} \left[C^k(i) \cdot C^k(j) \right] \tag{9.1.7}$$

The reduced matrix elements of the spherical tensors with respect to the single-electron functions $\psi_{nlm}(i)$ and $\psi_{n'l'n'}(i)$ are the $\mathbf{SO}_3 \supset \mathbf{SO}_2$ $3jm$ factors of (4.4.9). The radial integrals which arise for electrons i and j are

$$R^k(n_1 l_1 n_2 l_2; n_3 l_3 n_4 l_4)$$
$$= e \int_0^\infty \int_0^\infty R^*_{n_1 l_1}(r_i) R^*_{n_2 l_2}(r_j) \frac{r_<^k}{r_>^{k+1}} R_{n_3 l_3}(r_i) R_{n_4 l_4}(r_j)\, dr_i\, dr_j \tag{9.1.8}$$

where $r_<$ ($r_>$) is the lesser (greater) of r_i and r_j. Extensive tables of these integrals for hydrogenic radial functions are available (Butler, Minchin, and Wybourne 1971), as are many Hartree–Fock calculations.

The scalar product occurring in (9.1.7) is a scalar product in $\mathbf{SO}_3^L$, the group which rotates as an entity the orbital functions of the various electrons. The product is not scalar under rotations by each of the separate groups $\mathbf{SO}_3^i$ which rotate a single-electron function. Thus, to obtain eigenfunctions of the Coulomb interaction a change of basis is required. The change is discussed in Chapter 7 under the title of coefficients of fractional parentage (cfp). The cfp transform the product functions $\Pi_i|nm_s lm_l\rangle^i$ into functions with given $\mathbf{SU}_2^S \times \mathbf{SO}_3^L$ irrep structure, together with certain covering group structure, subsumed under the label α. The resulting functions are labeled $|\alpha SM_S LM_L\rangle$.

The Coulomb interaction is scalar under $\mathbf{SU}_2^S \times \mathbf{SO}_3^L$ and thus the Wigner–Eckart theorem tells us that the $(2S+1)\times(2L+1)$ partners of each $V_{\alpha SL}$ are degenerate and the only Coulomb mixing between such multiplets is where the S and L quantum numbers are equal.

The spin–orbit interaction is the largest of the relativistic correction terms to be added. It may be written as a perturbation

$$H_{S\text{-}O} = \sum_{i=1} \xi(r_i)(\mathbf{s}_i \cdot \mathbf{l}_i) \tag{9.1.9}$$

to the above central field Hamiltonian, where the radial function is given by

$$\xi(r) = \frac{\hbar^2}{2m^2c^2}\frac{1}{r}\frac{dU}{dr} \tag{9.1.10}$$

The spin–orbit term does not commute with $\mathbf{SU}_2^S \times \mathbf{SO}_3^L$, but only with the subgroup $\mathbf{SO}_3^J$, which rotates the system as a whole. The change of basis from $|\alpha SM_S LM_L\rangle$ to $|\alpha SL\, JM_J\rangle$ is performed by SO_3–SO_2 coupling coefficients. The new states will be diagonal in $\mathbf{J}(\mathbf{SO}_3^J)$ and degenerate in $\mathbf{M}_J(\mathbf{SO}_2^J)$.

The above changes of basis have been expressed in terms of the SO_3–SO_2 basis scheme. It should be obvious that because these interactions are all $\mathbf{SO}_3$ scalars, the point group basis chosen is immaterial. The SO_3 reduced matrix elements will be the same in all point group schemes. Usual fractional parentage coefficients give SL coupled kets, and the $9j$ relation (4.3.7) between the reduced matrix elements of coupled and uncoupled tensors may be used to write the angular matrix elements of the spin–orbit interaction in terms of $\mathbf{SU}_2^S \times \mathbf{SO}_3^L$ reduced matrix elements. A particular term $(\mathbf{s}_i \cdot \mathbf{l}_i)$ in (9.1.9) is independent of i because the wave functions are antisymmetric functions of the electrons. The matrix elements of the $(\mathbf{s} \cdot \mathbf{l})$ operator may be simplified by (4.3.13). Although s and l act on different parts of the system, namely the spin and orbital parts, the total spin S and total orbital angular momentum L are not simply related to the quantum numbers of the single electron, but depend on the parentage of the state:

$$\langle \alpha SLJM|(\mathbf{s}\cdot\mathbf{l})|\alpha' S'L'J'M'\rangle$$

$$=(-1)^{S'+L+J}\delta_{JJ'}\delta_{MM'}\begin{Bmatrix} S & S' & 1 \\ L' & L & J \end{Bmatrix}\langle \alpha SL\|sl\|\alpha' S'L'\rangle \tag{9.1.11}$$

The operator sl is an example of a double tensor operator defined in Section 4.4. The single-electron operator sl has reduced matrix elements between single-electron states as in (4.4.11),

$$\langle n_a\tfrac{1}{2}l_a\|sl\|n_b\tfrac{1}{2}l_b\rangle=\langle\tfrac{1}{2}\|s\|\tfrac{1}{2}\rangle\langle l_a\|l\|l_b\rangle$$

$$=\delta_{n_an_b}\delta_{l_al_b}\left[\tfrac{3}{2}l_a(l_a+1)(2l_a+1)\right]^{1/2} \tag{9.1.12}$$

Nielson and Koster (1963) tabulate the reduced matrix element of Racah's operator $V^{(11)}$ in the p^N, d^N, and f^N configurations. Their $V^{(11)}$ is the sum of N operators $s(i)u^{(1)}(i)$ for each electron i, where u has single-electron reduced matrix elements of unity. Nielson and Koster's $V^{(11)}$ is thus related to the operator $sl=\Sigma_i s(i)l(i)$ by the numerical factor $[l_a(l_a+1)(2l_a+1)]^{1/2}$, and their operator is related to a unit tensor operator by the factor $(\frac{2}{3})^{1/2}$.

The above methods give the entire angular dependence of the free-ion Hamiltonian,

$$H_{\text{free}}=H_{\text{central}}+H_{\text{radial}}+H_{\text{Coul}}+H_{S\text{-}O} \tag{9.1.13}$$

The method does not require that all electrons belong to a single nl configuration, although the tables of cfp's given by Nielson and Koster (1963) do make this assumption.

For certain cases of physical interest further relativistic correction terms to this H_{free} give rise to measurable contributions to the energy levels. Wybourne (1965, pp. 78–82) gives the orbit–orbit, the spin–spin, and spin–other-orbit interactions in terms of the above $\mathbf{SU}_2\times\mathbf{SO}_3$ tensor operators. The magnetic field at the nucleus produced by an orbital electron gives rise to the magnetic hyperfine interaction which Judd (1963, pp. 85–87) shows how to treat by the tensor operator algebra.

9.2 The Zeeman Interaction

Perturbation fields external to an ion rarely have $\mathbf{SO_3}$ symmetry. However, the angular momentum techniques familiar to physicists and chemists may be used with very few modifications in the presence of external perturbation fields. The tables of this book enable a suitably symmetrized basis to be chosen for the atomic or molecular wave functions. The symmetrized basis means that both the zeroth-order and the major perturbation terms are diagonal at the outset of a calculation.

Once again we use as a model perturbation the external magnetic field of the Zeeman interaction. In Sections 4.2 and 5.3 we used the fact that the Zeeman interaction was proportional (within a **J** multiplet) to the operator J_z if the z axis was the magnetic axis. Condon and Shortley (1935) give the first-order result for the interaction between the N electrons and an external magnetic field $\mathcal{H}$ along the z axis as

$$H_{\text{Zeeman}} = \beta\mathcal{H}\sum_{i=1}^{N}\left[l_z(i) + g_s s_z(i)\right] \tag{9.2.1}$$

where β is the Bohr magneton

$$\beta = e\hbar/2mc \tag{9.2.2}$$

and g_s is the gyromagnetic ratio as modified by Schwinger to include quantum electrodynamic effects,

$$g_s \approx 2.002320 \tag{9.2.3}$$

The Zeeman Hamiltonian is manifestly invariant under $\mathbf{SO}_2^J$, the group of rotations of the system about the z axis. Also, both operators $l(i)$ and $s(i)$ appearing in (9.2.1) transform as the irrep $\mathbf{1}(\mathbf{SO}_3^J)$. The Wigner–Eckart theorem applied to $\mathbf{SO}_3^J$ therefore gives

$$\langle \alpha_1 J_1 M_1 | H_{\text{Zeeman}} | \alpha_2 J_2 M_2 \rangle = \langle \alpha_1 J_1 \| H_{\text{Zeeman}} \| \alpha_2 J_2 \rangle_{SO_3^J} \times \begin{pmatrix} J_1 \\ M_1 \end{pmatrix} \begin{pmatrix} J_1 & 1 & J_2 \\ -M_1 & 0 & M_2 \end{pmatrix}^{SO_3}_{SO_2} \tag{9.2.4}$$

The M_J dependence of this expression is precisely that of the operator J_z within a diagonal manifold of fixed $J_1 = J_2$. This is the result used in Sections 4.2 and 5.3.

However, this result does not exhaust the $\mathbf{SO}_3$ structure of (9.2.1). Consider the SL scheme in which the states are labeled $|\alpha SLJM_J\rangle$. (Wybourne 1965, pp. 99–103, also considers Jl and Jj coupling.) The SL states have well-defined transformation properties under $\mathbf{SU}_2^S\times\mathbf{SO}_3^L$. The $\mathbf{SO}_3^J$ reduced matrix element may be evaluated by noting that $L=\Sigma_i l(i)$ transforms as $\mathbf{0}(\mathbf{SU}_2^S)$ and $\mathbf{1}(\mathbf{SO}_3^L)$ and S as $\mathbf{1}(\mathbf{SU}_2^S)\times\mathbf{0}(\mathbf{SO}_3^L)$, so that, using (4.3.8) and (4.3.9),

$$\begin{aligned}
&\langle\alpha_1S_1L_1J_1\|(L+g_sS)\|\alpha_2S_2L_2J_2\rangle\\
&\quad=\langle\alpha_1S_1L_1J_1\|L\|\alpha_2S_2L_2J_2\rangle+g_s\langle\alpha_1S_1L_1J_1\|S\|\alpha_2S_2L_2J_2\rangle\\
&\quad=(-1)^{S_2+L_2+J_1+1}|J_1|^{1/2}|J_2|^{1/2}\begin{Bmatrix} J_1 & 1 & J_2\\ L_2 & S_1 & L_1\end{Bmatrix}\langle\alpha_1S_1L_1\|L\|\alpha_2S_2L_2\rangle\\
&\qquad+g_s(-1)^{S_1+L_1+J_2+1}|J_1|^{1/2}|J_2|^{1/2}\begin{Bmatrix} J_1 & 1 & J_2\\ S_2 & L_1 & S_1\end{Bmatrix}\\
&\qquad\times\langle\alpha_1S_1L_1\|S\|\alpha_2S_2L_2\rangle
\end{aligned}\tag{9.2.5}$$

An angular momentum operator, e.g., L_z, has SO_3 reduced matrix elements given by

$$\langle L_1\|L\|L_2\rangle=\delta_{L_1L_2}\{L_1(L_1+1)(2L_1+1)\}^{1/2}\tag{9.2.6}$$

and a simple algebraic formula exists for SO_3 $6j$ symbols with one entry unity, so an algebraic formula for the matrix elements of the Zeeman operator may be obtained. This formula shows that in the SL scheme used above, the matrix elements are independent of α_1 and α_2, and diagonal in αL and S. They are not diagonal in J, but have nonzero off-diagonal elements when $J_2=J_1\pm1$. Evaluation of the SO_3 $6j$ symbols shows that

$$\langle\alpha SLJ\|L+g_sS\|\alpha SLJ\rangle=g_{SLJ}\{J(J+1)(2J+1)\}^{1/2}\tag{9.2.7}$$

where the Landé g-factor is

$$g_{SLJ}=1+(g_s-1)\frac{J(J+1)-L(L+1)+S(S+1)}{2J(J+1)}\tag{9.2.8}$$

The off-diagonal reduced matrix elements

$$\langle \alpha SLJ \| L + g_s S \| \alpha SLJ-1 \rangle$$

$$= (1-g_s)\left[\frac{(J+L+S+1)(J+S-L)(J+L-S)(L+S+1-J)}{4J}\right]^{1/2} \tag{9.2.9}$$

are less important, for energy levels which differ in J are often well separated in energy.

In summary, the Zeeman interaction is proportional to an SO_3^J unit tensor $U^{1(SO_3)0(SO_2)}$, the proportionality constant being an SO_3^J reduced matrix element

$$H_{\text{Zeeman}} = \sum_{\alpha_1 J_1 \alpha_2 J_2} U^{1(SO_3)0(SO_2)}(\alpha_1 J_1, \alpha_2 J_2)\langle \alpha_1 J_1 \| H_{\text{Zeeman}} \| \alpha_2 J_2 \rangle \tag{9.2.10}$$

The Zeeman operator is in fact given in terms of the operator $(L_z + g_s S_z)$, so if the $|\alpha J\rangle$ states have SL parentage, we have

$$\langle \alpha SLJ \| H_{\text{Zeeman}} \| \alpha' S' L' J' \rangle^{SO_3^J}$$

$$= \beta \mathcal{H} \delta_{\alpha\alpha'} \delta_{SS'} \delta_{LL'} \langle \alpha SLJ \| L + g_s S \| \alpha SLJ' \rangle^{SO_3^J} \tag{9.2.11}$$

where the reduced matrix elements of $(L + g_s S)$ are given by (9.2.7)–(9.2.9).

Using the approximation that the terms off-diagonal in J do not contribute, these results may be combined together as

$$\langle \alpha SLJM_J | H_{\text{Zeeman}} | \alpha SLJM_J \rangle$$

$$= \beta \mathcal{H}(-1)^{J-M}\begin{pmatrix} J & L & J \\ -M_J & 0 & M_J \end{pmatrix}^{SO_3}_{SO_2} \langle \alpha SLJ \| L + g_s S \| \alpha SLJ \rangle$$

$$= \beta \mathcal{H} M_J \{J(J+1)(2J+1)\}^{-1/2} g_{SLJ} \{J(J+1)(2J+1)\}^{1/2}$$

$$= \beta \mathcal{H} g_{SLJ} M_J \tag{9.2.12}$$

showing that the resultant energy levels are equally spaced, the separation being $\beta \mathcal{H} g_{SLJ}$.

The advantage of giving the SO_3^J reduced matrix elements as above is that the Zeeman matrix elements are now available in any orientation and in any point group basis. Consider an SL-coupled ion in a weak octahedral crystal field environment such that the energy level of interest is an octahedral quartet, that is, the irrep $\frac{3}{2}(\mathbf{O})$. Let the magnetic field of intensity $\mathcal{H}$ be directed along the crystal z axis. Choose the SO_3–O–D_3–C_3 basis of scheme b. Inspection of the tables at the end of Chapter 16 shows that

$$U^{1(SO_3)0(SO_2)}(\alpha_1 J_1 \alpha_2 J_2) = U^{1(SO_3)1(O)\tilde{0}(D_3)0(C_3)}(\alpha_1 J_1, \alpha_2 J_2) \quad (9.2.13)$$

which, together with (9.2.10), gives a ready formula for the Zeeman interaction in this basis. A typical calculation is given below.

The branching rules for the appropriate groups in the tables of Chapter 12 show that the four quartet ($J=3/2$) states are labeled $|\alpha SL\,\frac{3}{2}\frac{3}{2}\frac{3}{2}\frac{3}{2}\rangle_b$, $|\alpha SL\,\frac{3}{2}\frac{3}{2}\frac{1}{2}\frac{1}{2}\rangle_b$, $|\alpha SL\,\frac{3}{2}\frac{3}{2}\frac{1}{2}-\frac{1}{2}\rangle_b$, and $|\alpha SL\,\frac{3}{2}\frac{3}{2}-\frac{3}{2}\frac{3}{2}\rangle_b$. The matrix elements may be calculated directly using the Wigner–Eckart theorem (4.2.3) and the jm tables of Chapter 13. Consider the general $J=\frac{3}{2}$ matrix element. Equations (9.2.10) and (9.2.13) give

$$\langle \alpha SL\tfrac{3}{2}\tfrac{3}{2}\lambda\mu | H_{\text{Zeeman}} | \alpha SL\tfrac{3}{2}\tfrac{3}{2}\lambda'\mu' \rangle$$

$$= \sum_{\alpha_1 J_1 \alpha_2 J_2} \langle \alpha SL\tfrac{3}{2}\tfrac{3}{2}\lambda\mu | U^{11\tilde{0}0}(\alpha_1 J_1 \alpha_2 J_2) | \alpha_{SL}\tfrac{3}{2}\tfrac{3}{2}\lambda'\mu' \rangle$$

$$\times \langle \alpha_1 J_1 \| H_{\text{Zeeman}} \| \alpha_2 J_2 \rangle$$

The matrix elements of the unit tensor operator $U^{11\tilde{0}0}(\alpha_1 J_1, \alpha_2 J_2)$ are zero unless all operator parentage labels match the state parentage labels. Using this, the summation vanishes:

$$= \langle \alpha SL\tfrac{3}{2}\tfrac{3}{2}\lambda\mu | U^{11\tilde{0}0}\big(\alpha SL\tfrac{3}{2}, \alpha SL\tfrac{3}{2}\big) | \alpha SL\tfrac{3}{2}\tfrac{3}{2}\lambda'\mu' \rangle$$

$$\times \langle \alpha SL\tfrac{3}{2} \| H_{\text{Zeeman}} \| \alpha SL\tfrac{3}{2} \rangle$$

which with the Wigner–Eckart theorem (4.2.3) factorizes as

$$=\langle \alpha SL\tfrac{3}{2}\|U^1(\alpha SL\tfrac{3}{2},\alpha SL\tfrac{3}{2})\|\alpha SL\tfrac{3}{2}\rangle\langle \alpha SL\tfrac{3}{2}\|H_{\text{Zeeman}}\|\alpha SL\tfrac{3}{2}\rangle$$

$$\times\begin{pmatrix}\frac{3}{2}\\ \frac{3}{2}\\ \lambda\\ \mu\end{pmatrix}\begin{pmatrix}\frac{3}{2}* & 1 & \frac{3}{2}\\ \frac{3}{2}* & 1 & \frac{3}{2}\\ \lambda* & \tilde{0} & \lambda'\\ \mu* & 0 & \mu'\end{pmatrix}\begin{matrix}SO_3\\ O\\ D_3\\ C_3\end{matrix}$$

Now the reduced matrix element of the unit tensor operator is unity, and thus may be omitted. The SO_3–O–D_3–C_3 jm's factorize (introducing the sum over a multiplicity for the octahedral coupling):

$$=\sum_r\langle \alpha SL\tfrac{3}{2}\|H_{\text{Zeeman}}\|\alpha SL\tfrac{3}{2}\rangle\begin{pmatrix}\frac{3}{2}\\ \frac{3}{2}\end{pmatrix}\begin{pmatrix}\frac{3}{2}* & 1 & \frac{3}{2}\\ \frac{3}{2}* & 1 & \frac{3}{2}\end{pmatrix}\begin{matrix}SO_3\\ rO\end{matrix}$$

$$\times\begin{pmatrix}\frac{3}{2}\\ \lambda\end{pmatrix}\begin{pmatrix}\frac{3}{2}* & 1 & \frac{3}{2}\\ \lambda* & \tilde{0} & \lambda'\end{pmatrix}\begin{matrix}rO\\ D_3\end{matrix}\begin{pmatrix}\lambda\\ \mu\end{pmatrix}\begin{pmatrix}\lambda* & \tilde{0} & \lambda'\\ \mu* & 0 & \mu'\end{pmatrix}\begin{matrix}D_3\\ C_3\end{matrix} \qquad (9.2.14)$$

In practice the intermediate steps in the above calculation would not need to be written out. Observe that the Zeeman interaction is a $\mathbf{C}_3$ scalar, so a $\delta_{\mu\mu'}$ arises in the above. Further observe that in $\mathbf{D}_3$, $\tilde{\mathbf{0}}\times\boldsymbol{\lambda}=\boldsymbol{\lambda}$ for $\boldsymbol{\lambda}$ a spin irrep. As a result the Zeeman interaction within the $\mathbf{J}=\tfrac{3}{2}$ quartet is diagonal. Let us consider a typical diagonal term and use the tables of Chapter 13. We have

$$\langle \alpha SL\tfrac{3}{2}\tfrac{3}{2}\tfrac{1}{2}\tfrac{1}{2}|H_{\text{Zeeman}}|\alpha SL\tfrac{3}{2}\tfrac{3}{2}\tfrac{1}{2}\tfrac{1}{2}\rangle$$

$$=\langle \alpha SL\tfrac{3}{2}\|H_{\text{Zeeman}}\|\alpha SL\tfrac{3}{2}\rangle\sum_r\left\{\begin{pmatrix}\frac{3}{2}\\ \frac{3}{2}\end{pmatrix}\begin{pmatrix}\frac{3}{2} & 1 & \frac{3}{2}\\ \frac{3}{2} & 1 & \frac{3}{2}\end{pmatrix}\begin{matrix}SO_3\\ rO\end{matrix}\right.$$

$$\left.\times\begin{pmatrix}\frac{3}{2}\\ \frac{1}{2}\end{pmatrix}\begin{pmatrix}\frac{3}{2} & 1 & \frac{3}{2}\\ \frac{1}{2} & \tilde{0} & \frac{1}{2}\end{pmatrix}\begin{matrix}rO\\ D_3\end{matrix}\right\}\begin{pmatrix}\frac{1}{2}\\ \frac{1}{2}\end{pmatrix}\begin{pmatrix}\frac{1}{2} & \tilde{0} & \frac{1}{2}\\ -\frac{1}{2} & 0 & \frac{1}{2}\end{pmatrix}\begin{matrix}D_3\\ C_3\end{matrix}$$

$$=\beta\mathcal{H}g_{SL3/2}\left\{\frac{3}{2}\times\frac{5}{2}\times 4\right\}^{1/2}\left\{(+1)\frac{1}{\sqrt{5}}(+1)\frac{1}{\sqrt{6}}+(+1)\frac{2}{\sqrt{5}}(+1)0\right\}$$

$$\times(+1)\frac{1}{\sqrt{2}}=\frac{1}{2}\beta\mathcal{H}g_{SL3/2}$$

Matrix elements between different octahedral irreps may be computed directly, as may be the matrix elements between different **J**. For the latter one uses the reduced matrix element (9.2.9). As an example of such a calculation, consider the case of interaction of the above $|\alpha SL\tfrac{5}{2}\tfrac{3}{2}\tfrac{3}{2}\tfrac{1}{2}\rangle$ crystal field state with an $|\alpha SL\tfrac{3}{2}\tfrac{3}{2}\tfrac{1}{2}-\tfrac{1}{2}\rangle$ state, induced by having the magnetic field along the crystal x' axis. With this change of crystal axes with respect to the Zeeman axis, (9.2.13) is changed to (see Section 5.3 and the tables at the end of Chapter 16).

$$U^{1(SO_3)0(SO_3)} = \frac{1}{\sqrt{2}}\{U^{1(SO_3)1(O)1(D_3)1(C_3)} - U^{1(SO_3)1(O)1(D_3)-1(C_3)}\} \tag{9.2.15}$$

so this Zeeman element is expanded as

$$\begin{aligned}
&\langle \alpha SL\tfrac{5}{2}\tfrac{3}{2}\tfrac{3}{2}\tfrac{3}{2}|H_{\text{Zeeman}}|\alpha SL\tfrac{3}{2}\tfrac{3}{2}\tfrac{1}{2}-\tfrac{1}{2}\rangle \\
&\quad = \langle \alpha SL\tfrac{5}{2}\|H_{\text{Zeeman}}\|\alpha SL\tfrac{3}{2}\rangle \\
&\quad \times\left\{\sum_r \begin{pmatrix}\tfrac{5}{2}\\ \tfrac{3}{2}\end{pmatrix}\begin{pmatrix}\tfrac{5}{2} & 1 & \tfrac{3}{2}\\ \tfrac{3}{2} & 1 & \tfrac{3}{2}\end{pmatrix}^{SO_3}_{rO}\begin{pmatrix}\tfrac{3}{2}\\ \tfrac{3}{2}\end{pmatrix}\begin{pmatrix}\tfrac{3}{2} & 1 & \tfrac{3}{2}\\ \tfrac{3}{2} & 1 & \tfrac{1}{2}\end{pmatrix}^{rO}_{D_3}\right\} \\
&\quad \times\frac{1}{\sqrt{2}}\left\{\begin{pmatrix}\tfrac{3}{2}\\ \tfrac{3}{2}\end{pmatrix}\begin{pmatrix}\tfrac{3}{2} & 1 & \tfrac{1}{2}\\ \tfrac{3}{2} & 1 & -\tfrac{1}{2}\end{pmatrix}^{D_3}_{C_3} - \begin{pmatrix}\tfrac{3}{2}\\ \tfrac{3}{2}\end{pmatrix}\begin{pmatrix}\tfrac{3}{2} & 1 & \tfrac{1}{2}\\ \tfrac{3}{2} & -1 & -\tfrac{1}{2}\end{pmatrix}^{D_3}_{C_3}\right\} \\
&\quad = \langle \alpha SL\tfrac{5}{2}\|H_{\text{Zeeman}}\|\alpha SL\tfrac{3}{2}\rangle \\
&\quad \times\left\{(+1)\left(-\frac{2\sqrt{2}}{\sqrt{15}}\right)(+1)\left(-\frac{1}{3\sqrt{2}}+\frac{i}{3}\right)\right. \\
&\quad \left.+(+1)\frac{\sqrt{2}}{\sqrt{15}}(+1)\left(-\frac{1}{3\sqrt{2}}-\frac{i}{6}\right)\right\}\frac{1}{\sqrt{2}}\left\{(+1)0-(+1)\frac{i}{\sqrt{2}}\right\}
\end{aligned}$$

for the first D_3–C_3 $3jm$ does not satisfy the triad conditions,

$$= \left(\frac{\sqrt{5}}{6\sqrt{6}} - \frac{i}{6\sqrt{15}}\right)\beta\mathcal{H}\langle \alpha SL\tfrac{5}{2}\|L+g_sS\|\alpha SL\tfrac{3}{2}\rangle \tag{9.2.16}$$

The off-diagonal SO_3^J reduced matrix element of $L+g_sS$ is given by (9.2.9).

9.3 Ligand Fields

Bethe (1929) developed crystal field theory by thinking in terms of a purely electrostatic field perturbing the wave function of the central ion. He replaced the surrounding ions by point charges at the crystal sites. More generally, one may represent the ions by any charge distribution $\rho(\mathbf{R})$, $\mathbf{R}$ being the position within the crystal. The potential energy for an electron of the central ion is thus

$$H_{\text{env}} = -\int \frac{e\rho(\mathbf{R})}{|\mathbf{R}-\mathbf{r}|} d\mathbf{R} \tag{9.3.1}$$

This potential may be treated in the same manner as the interelectron Coulomb term H_{Coulomb} of Section 9.1 to produce an expansion in terms of the ith electron spherical tensors $C_q^k(i)$,

$$H_{\text{env}} = \sum_{kq} B_q^k \sum_{i}^{N} C_q^k(i) \tag{9.3.2}$$

where B_q^k contains all the radial information in the form of a radial expectation value

$$B_q^k = A_q^k \langle r^k \rangle \tag{9.3.3}$$

and the spatial integral of the charge distribution,

$$A_q^k = -e\int \frac{\rho(\mathbf{R})}{R^{k+1}} (-1)^q C_q^k(\Theta, \Phi)\, d\mathbf{R} \tag{9.3.4}$$

where R, Θ and Φ are the spherical polar coordinates of $\mathbf{R}$ (see Wybourne 1965, p. 214).

Such an ab initio calculation is unsatisfactory except in strongly ionic crystals where a fixed external charge distribution is a reasonable construct. However, as a phenomenological starting point (9.3.2) is very useful. The radial parameters may be fitted to experimental data and excellent fits to the data obtained.

The number of free parameters B_q^k may be deduced from symmetry considerations. Consider first the $\mathbf{SO}_3$ transformation properties of the basis functions. The spherical tensors C_q^k have zero matrix elements between single-electron wave functions $|nlm_l\rangle$ and $|n'l'm'_l\rangle$ if lkl' does not form a triad or if $l+k+l'$ is odd [see (4.4.9)]. For example, for f electrons one is restricted to k values of 0, 2, 4, and 6, while for an interaction between p and f electrons (a configuration interaction) one has $k=2$ or 4 only.

The environment will have a certain point group symmetry. In a crystal the environment can have only a crystallographic point group as an exact symmetry group, but molecules may have higher symmetry axes than these. For example, the polyboranes form icosahedral molecules. If the environment has **G** as an exact symmetry group, then H_{env} is invariant under the action of the elements of **G**, namely H_{env} transforms as **0(G)**. Thus H_{env} must be composed of those particular linear combinations of the C_q^k that transform as **0(G)**.

Transformation coefficients allow one to transform between the $\mathbf{SO}_3 \supset \mathbf{SO}_2$ basis in which the C_q^k are expressed and the $\mathbf{SO}_3 \supset \mathbf{G}$ chain. One has

$$C^{k(SO_3)a0(G)} = \sum_q C_q^k \langle kq(SO_2)|ka0(G)\rangle \tag{9.3.5}$$

where we note there may be a multiplicity of scalars of **G** for given **k**.

To satisfy the requirement that H_{env} is scalar, it is necessary that H_{env} takes the form

$$H_{\text{env}} = \sum_{ka} B^{ka0} C^{ka0} \tag{9.3.6}$$

where B^{ka0} are arbitrary. The B^{ka0} may be related to the B_q^k by the inverse to (9.3.5),

$$B^{ka0} = \sum_q B_q^k \langle ka0(G)|kq(SO_2)\rangle \tag{9.3.7}$$

if one desires to compare the two sets of parameters. The B^{ka0} have the advantage for phenomenological calculations in that they are precisely the set of independent parameters. It is useful to use renormalized parameters. We suggest rewriting (9.3.6) in terms of the unit tensor operators of Section 4.1, but including an explicit dimension factor,

$$H_{\text{env}} = \sum_{ka} |k|^{1/2} X^{ka0} U^{ka0} \tag{9.3.8}$$

The parameters X^{ka0} incorporate the reduced matrix elements of the spherical tensors (4.4.9),

$$X^{ka0} = B^{ka0}|l||k|^{1/2}\begin{pmatrix} l & k & l \\ 0 & 0 & 0 \end{pmatrix} \tag{9.3.9}$$

The dimension factor in (9.3.8) is equivalent to using Judd's (1963) standard tensors and has the very desirable effect of making the splittings caused by each of the terms in the expansion approximately proportional to the magnitude of the parameters.

Consider briefly a transition metal ion in a cubic crystal lattice distorted around the site such that the exact symmetry of the site is reduced to $\mathbf{C}_3$. A natural basis to take is $\mathbf{SO}_3 \supset \mathbf{O} \supset \mathbf{D}_3 \supset \mathbf{C}_3$, where we include the group $\mathbf{D}_3$ so as to have a maximal chain, but if there is any inversion symmetry one could use $\mathbf{O}_3 \supset \mathbf{O}_h \supset \mathbf{T}_d \supset \mathbf{C}_{3v} \supset \mathbf{C}_3$. The single-electron orbitals are d orbitals, so k is restricted to 0, 2, and 4. The branching rules show that the $\mathbf{C}_3$ parameters in the $\mathbf{SO}_3 \supset \mathbf{O} \supset \mathbf{D}_3 \supset \mathbf{C}_3$ basis are

$$X^{0000}, \quad X^{2\bar{1}00}, \quad X^{4000}, \quad X^{41\tilde{0}0}, \quad X^{4\bar{1}00} \tag{9.3.10}$$

whereas in the $\mathbf{O}_3 \supset \mathbf{O}_h \supset \mathbf{T}_d \supset \mathbf{C}_{3v} \supset \mathbf{C}_3$ basis the parameters are

$$X^{0^+0^+000}, \quad X^{2^+\bar{1}^+\bar{1}00}, \quad X^{4^+0^+000}, \quad X^{4^+1^+1\tilde{0}0}, \quad X^{4^+\bar{1}^+\bar{1}00} \tag{9.3.11}$$

Note, however, that the $X^{41\tilde{0}0}$ and $X^{4\bar{1}00}$ are not independently free parameters, for they are related by the choice of the C_2 axis. Either parameter may be chosen zero. This is because a rotation through an angle θ about the C_3 axis mixes the $\mathbf{1(O)\tilde{0}(D_3)}$ and $\mathbf{1(O)0(D_3)}$ partners; see the orientation phase discussion near the end of Section 5.3.

As an example of how to evaluate the phenomenological H_{env}, consider the only part that survives under the assumption of pure octahedral symmetry, namely the $\mathbf{SO}_3$ invariant and a single octapole term,

$$H_{\text{env}}^{\text{approx}} = X^{0000}U^{0000} + 3X^{4000}U^{4000} \tag{9.3.12}$$

The states are labeled $|\alpha SL\, J\lambda(O)\mu(D_3)\nu(C_3)\rangle$. The reduced matrix elements of the Racah unit tensors $U^k = \Sigma_i u^k(i)$ in the SL scheme are tabulated by Nielson and Koster (1963). U^0 is the identity operator, merely moving the entire configuration. The general matrix element thus depends only on U^{4000} and is obtained by use of the Wigner–Eckart theorem (4.2.3) applied in the point group scheme of interest,

$$\langle \alpha SLJ\lambda\mu\beta | U^{4000} | \alpha' S' L' J' \lambda' \mu' \nu' \rangle$$

$$= \langle \alpha SLJ \| U^4 \| \alpha' S' L' J' \rangle$$

$$\times \begin{pmatrix} J \\ \lambda \end{pmatrix} \begin{pmatrix} J & 4 & J' \\ \lambda & 0 & \lambda' \end{pmatrix}^{SO_3}_{O} \begin{pmatrix} \lambda \\ \mu \end{pmatrix} \begin{pmatrix} \lambda & 0 & \lambda' \\ \mu^* & 0 & \mu' \end{pmatrix}^{O}_{D_3} \begin{pmatrix} \mu \\ \nu \end{pmatrix} \begin{pmatrix} \mu^* & 0 & \mu' \\ \nu^* & 0 & \nu' \end{pmatrix}^{D_3}_{C_3} \tag{9.3.13}$$

Two of the $3jm$ factors are always $2jm$'s. The U^4 operator is a purely orbital tensor transforming as $\mathbf{0}(\mathbf{SU}_2^S)\times\mathbf{4}(\mathbf{SO}_3^L)$, so its J reduced matrix element is related to its SL reduced matrix element by (4.3.9). Inserting this gives

$$= \delta_{SS'}(-1)^{S'+L'+J}|J|^{1/2}|J'|^{1/2}\begin{Bmatrix} J & 4 & J' \\ L' & S & L \end{Bmatrix}\langle \alpha SL \| U^4 \| \alpha' S' L' \rangle$$

$$\times \begin{pmatrix} J & 4 & J' \\ \lambda & 0 & \lambda \end{pmatrix}^{SO_3}_{O} |\lambda|^{-1/2}\delta_{\lambda\lambda'}\delta_{\mu\mu'}\delta_{\nu\nu'} \tag{9.3.14}$$

The SO_3 $6j$ may be looked up in the tables of Rotenberg, Bivins, Metropolis, and Wooten (1959), the reduced matrix element in Nielson and Koster (1963), and the SO_3–O $3jm$ in the tables in Chapter 13.

Our interpretation of the operator U^{ka0} in the phenomenological equations, (9.3.6) or (9.3.7), was strong. We assumed that it was composed of a sum of single-electron operators which acted only on the orbital part of the given electronic configuration. Without these assumptions, the algebraic part of the calculation would stop at either the $|\alpha SLJ\rangle$ or $|\alpha SL\rangle$ reduced matrix elements, and the reduced matrix elements would be incorporated into the set of free parameters X^{ka0}. These parameters would thus become dependent on either the αSLJ or the αSL quantum numbers.

The above description completes our description of how the tables of this book may be used to make weak crystal field perturbation calculations. The final result, (9.3.14), gives the single octahedral perturbation term for d electrons; a similar calculation would give the $\mathbf{D}_3$ and $\mathbf{C}_3$ terms, namely the three other parameters in (9.3.10) or (9.3.11). Generally one would expect the parameters relating to the larger group symmetries, here X^{4000}, to be of more importance physically, that is, to have more effect on the energy levels, than the parameters corresponding to the lesser symmetries and that, after this, terms with a smaller rank (the irrep label k) be of more importance than those with higher rank. Our choice of normalization of the parameters gives a direct relationship between parameter size and resultant energy level splitting; therefore in this example of a $\mathbf{D}_3$ split octahedral symmetry we would expect that $X^{4000} > X^{2\bar{1}00} > X^{4\bar{1}00}$.

10

Programming Notes

Most of the tables of this book have been generated by machine (a Burroughs B6700 computer) from a small amount of input data. The resultant tables have then been arranged into lines, columns, and pages within the computer in preparation for typesetting. The typesetter (for the tables) was a minicomputer-controlled phototypesetter. The data transfer between machines was by paper tape. The user of the tables has two interests in our computer implementation of the methods herein: that of the accuracy of the printed tables of Chapters 12–16; and that of the availability of the tables in machine-readable form. We discuss these matters separately.

Modern electronic computers are very reliable because they have various internal checks, such as parity, by which they check every operation. Unfortunately, the transfer of data to the phototypesetter is not similarly checked. Some indication of the likelihood of errors occurring in the transfer is provided by the fact that the data transferred included the various typesetting commands; indeed, these are approximately five times more numerous than the print characters themselves. Wrong control characters give rise to incorrect fonts, alignment, etc., which are easily checked. Few such errors occurred. Input data to the program is checked for self-consistency, and most data are simply the $3j$ phases and $2jm$ factors which appear at the head of the respective $6j$ and $3jm$ tables. The programs were checked against the equations that they implement, they were checked against many hand calculations, and the results were also checked for self-consistency.

The entire program is written in "Burroughs 6700 Extended Algol," which is an extension of standard ALGOL 60 primarily in its character handling and file handling constructs. These features of Burrough's Algol facilitated the writing of control structures within the program to allow communication to be by mnemonics and to be interactive. Certain features of our problem make our program rather unusual in relation to other scientific programs; for example, there are no floating point numbers in the program. Our arithmetic routines had to be specially written to handle the various operations of numbers of the general form $\Sigma_i p_i r_i^{1/2}/q_i$, where p_i, q_i, and r_i are positive or negative integers.

The calculation of a typical $6j$ table would involve the following steps:

1. A disk file containing the information about the group is read. The dimensions, labels for the irreps, the $2j$ symbols, and the triads are required.
2. The triads are checked for self-consistency by dimensional means.
3. The table of $6j$ symbols is generated.
4. Primitive $6j$ symbols not involving multiplicity separations are calculated by using the orthogonality and Racah backcoupling equations.
5. Zeros or occasionally other values, are stored in the appropriate places to resolve multiplicities.
6. The remaining primitive $6j$ symbols are calculated.
7. All other $6j$ symbols are evaluated using the Biedenharn–Elliott sum rule.
8. The tables are saved on disk.
9. Selected results are checked using the orthogonality relations.
10. The results are printed on the lineprinter or typeset on the phototypesetter.

The interactive nature of the program allows these steps to be attempted in any order, and indeed a calculation to be stopped in the middle, saved on disk, and continued at a later time.

The program has been used for various continuous groups (Butler, Haase, and Wybourne 1978, 1979). Reid (1979) has added further routines to the package. His routines use the tables to construct the transformation coefficients which give rise to the tables of Chapter 16. His routines will also construct and then diagonalize the matrix of any Hamiltonian in any point group basis. The package of programs will be made generally available when various cfp tables are included and when some of the machine dependence of the programs is removed. Requests should be directed to the author.

11

Group Information Tables

This chapter contains the character tables of the various point groups, together with an assortment of information about each particular group. Figure 11.1 is a guide to the presentation of this material. For each pure

Table number. Group name(s) **Names of the corresponding rotation–inversion group**		
↑ ⋮ Group symbol	↑ ⋮ Class labels for rotation–reflection group	↑ ⋮ Subgroup basis
Group symbol	Class labels for pure rotation group (including the number of elements)	Subgroup basis (specifies orientation)
{ Schönflies notation { Koster *et al*. notation { Our notation irreps	Character values	

Notes on orientations and character values.

Figure 11.1. Guide to the presentation of material in Tables 11.1–11.14.

rotation point group we give its character table, including its spin irreps. On the left of the table we give the irrep labels. Along the top we give the class labels, usually for several orientations. The corresponding partners (basis functions) may be found in Chapter 16. The assignment of the class labels may be checked by applying the appropriate rotation (in terms of Euler angles) to those partners. Various labeling systems are in use for the group itself, for its irreps, and for its operators. Physicists tend to follow Bethe's (1929) numbering of the irreps, while chemists tend to use the letters introduced by Mulliken (1933). Variations have occurred in both systems. Lax (1974, p. 416) gives a useful description of the variations in the literature. We have adapted the integer–half-integer labeling universally used for $\mathbf{SO}_3$ and $\mathbf{SO}_2$ and used by some for the cyclic groups. The three sets of labels are given for each point group.

Koster, Dimmock, Wheeler, and Statz (1963), use a passive convention for rotations. As with most recent texts, our convention is active, resulting in our characters being the complex conjugates of those using the passive convention. For example, the C_{3z} operator is the rotation operator $\exp(-\frac{2}{3}\pi i J_z)$; see below and Section 16.1 for further details. Lax avoids the issue of convention with regard to the double group operators (see Section 2.6). We have no need to use the double group operators, but for the sake of completeness we include them in our character tables, following the convention of Koster *et al.* rather than that of Hamermesh (1962). The conventions differ simply in the naming of "group operators R," versus "double group operators $\overline{R}$." In our $\mathbf{D}_3$ character table the notation $2C_{3z}$ labeling a class indicates that C_{3z} and C_{3z}^{-1} belong to that class. Note that for true functions, $C_{3z}^2 = C_{3z}^{-1}$, but for spin functions, J_z has half-integral eigenvalues, so we may write

$$
\begin{aligned}
C_{3z}^2 &= e^{-4\pi i J_z/3} \\
&= -e^{\pi i}e^{-4\pi i J_z/3} \\
&= -e^{(2\pi - 4\pi/3)iJ_z} \\
&= -e^{2\pi i J_z/3} \\
&= \overline{C_{3z}^{-1}}
\end{aligned}
$$

The relative orientation of the point group's axes and the laboratory axes depends both on the abstract group chain and the choice made for the special roots in Section 3.5. Consider the group $\mathbf{D}_3$. The character table of Table 11.8 when labeled in the $\mathbf{C}_3$ orientation has as the threefold axis the crystal z axis. Because of a choice concerning reality in the D_3–C_3 $3jm$ calculation, our three twofold rotations are aligned along the y axis and at $2\pi/3$ relative to it, rather than along the x axis. This is the usual choice. For all D_n–C_n branchings we choose C_{2y}, being a twofold rotation about the y axis, to be a group element. Further references to orientation are to be found in the index.

The character tables are given for all groups other than the rotation–inversion groups $\mathbf{G}_i = \mathbf{G} \times \mathbf{C}_i$. These character tables are a product of the $\mathbf{G}$ and $\mathbf{C}_i$ tables; the $\mathbf{C}_i$ table is

	$\mathbf{C}_i$	1	C_i
even	g +	1	1
odd	u −	1	−1

Each irrep $\lambda(\mathbf{G})$ gives rise to the two irreps λ^+ and λ^- of $\mathbf{G}_i$, and each operator R of $\mathbf{G}$ gives rise to the two operators R and RC_i of $\mathbf{G}_i$.

11.1 The Cyclic Groups (Tables 11.1–11.6)

The cyclic group $\mathbf{C}_n$ contains n operators C_{nz}^k for $k=0,1,\ldots,n-1$, being rotations about some axis (usually taken to be the z axis). The groups $\mathbf{C}_1$, $\mathbf{C}_2$, $\mathbf{C}_3$, $\mathbf{C}_4$, and $\mathbf{C}_6$ are crystallographic point groups. Every even cyclic group is isomorphic to a rotation–reflection group whose mirror plane is the xy (horizontal) plane (labeled σ_z or σ_h):

$$\mathbf{C}_2 \cong \mathbf{C}_s$$

$$\mathbf{C}_4 \cong \mathbf{S}_4$$

$$\mathbf{C}_6 \cong \mathbf{C}_{3h}$$

Tables 11.1–11.6 begin on the next page.

Table 11.1 The Cyclic Group $C_1 = (1)$
$C_1 \times C_i = C_i = S_2 = (\bar{1})$

	C_1	E	$\bar{E}$	C_1 basis
Γ_1	0	1	1	
Γ_2	$\frac{1}{2}$	1	-1	

Table 11.2 The Cyclic Group $C_2 = (2)$, and $C_s = C_{1h} = (m)$
$C_2 \times C_i = C_{2h} = (2/m) = \left(\frac{2}{m}\right)$

	C_s	E	σ_z	$\bar{E}$	$\overline{\sigma_z}$	C_1 basis
	C_2	E	C_{2z}	$\bar{E}$	$\overline{C_{2z}}$	C_1 basis
Γ_1	0	1	1	1	1	
Γ_3	$\frac{1}{2}$	1	$-i$	-1	i	
Γ_4	$-\frac{1}{2}$	1	i	-1	$-i$	
Γ_2	1	1	-1	1	-1	

Table 11.3 The Cyclic Group $C_3 = (3)$
$C_3 \times C_i = C_{3i} = S_6 = (\bar{3})$

	C_3	E	C_{3z}	C_{3z}^{-1}	$\bar{E}$	$\overline{C_{3z}}$	$\overline{C_{3z}^{-1}}$	C_1 basis
Γ_1	0	1	1	1	1	1	1	
Γ_4	$\frac{1}{2}$	1	ω	ω^5	-1	ω^4	ω^2	
Γ_5	$-\frac{1}{2}$	1	ω^5	ω	-1	ω^2	ω^4	
Γ_2	1	1	ω^2	ω^4	1	ω^2	ω^4	
Γ_3	-1	1	ω^4	ω^2	1	ω^4	ω^2	
Γ_6	$\frac{3}{2}$	1	-1	-1	-1	1	1	

$\omega = e^{-i\pi/3}$, $\omega^3 = -1$.

Table 11.4 The Cyclic Group $C_4=(4)$, and $S_4=(\bar{4})$

$$C_4\times C_i=C_{4h}=(4/m)=\left(\frac{4}{m}\right)$$

	S_4	E	S_4^{-1}	C_{2z}	S_4	$\bar{E}$	$\overline{S_4^3}$	$\overline{C_{2z}}$	$\overline{S_4}$	C_2 basis
	C_4	E	C_{4z}	C_{2z}	C_{4z}^{-1}	$\bar{E}$	$\overline{C_{4z}}$	$\overline{C_{2z}}$	$\overline{C_{4z}^{-1}}$	C_2 basis
Γ_1	0	1	1	1	1	1	1	1	1	
Γ_5	$\frac{1}{2}$	1	ω	$-i$	ω^7	-1	ω^5	i	ω^3	
Γ_6	$-\frac{1}{2}$	1	ω^7	i	ω	-1	ω^3	$-i$	ω^5	
Γ_3	1	1	$-i$	-1	i	1	$-i$	-1	i	
Γ_4	-1	1	i	-1	$-i$	1	i	-1	$-i$	
Γ_8	$\frac{3}{2}$	1	ω^3	i	ω^5	-1	ω^7	$-i$	ω	
Γ_7	$-\frac{3}{2}$	1	ω^5	$-i$	ω^3	-1	ω	i	ω^7	
Γ_2	2	1	-1	1	-1	1	-1	1	-1	

$\omega=e^{-i\pi/4}$, $\omega^2=-i$.

Table 11.5 The Cyclic Group C_5

$$C_5\times C_i=C_{5i}=S_{10}$$

C_5	E	C_{5z}	C_{5z}^2	C_{5z}^{-2}	C_{5z}^{-1}	$\bar{E}$	$\overline{C_{5z}}$	$\overline{C_{5z}^2}$	$\overline{C_{5z}^{-2}}$	$\overline{C_{5z}^{-1}}$	C_1 basis
0	1	1	1	1	1	1	1	1	1	1	
$\frac{1}{2}$	1	ω	ω^2	ω^8	ω^9	-1	ω^6	ω^7	ω^3	ω^4	
$-\frac{1}{2}$	1	ω^9	ω^8	ω^2	ω	-1	ω^4	ω^3	ω^7	ω^6	
1	1	ω^2	ω^4	ω^6	ω^8	1	ω^2	ω^4	ω^6	ω^8	
-1	1	ω^8	ω^6	ω^4	ω^2	1	ω^8	ω^6	ω^4	ω^2	
$\frac{3}{2}$	1	ω^3	ω^6	ω^4	ω^7	-1	ω^8	ω	ω^9	ω^2	
$-\frac{3}{2}$	1	ω^7	ω^4	ω^6	ω^3	-1	ω^2	ω^9	ω	ω^8	
2	1	ω^4	ω^8	ω^2	ω^6	1	ω^4	ω^8	ω^2	ω^6	
-2	1	ω^6	ω^2	ω^8	ω^4	1	ω^6	ω^2	ω^8	ω^4	
$\frac{5}{2}$	1	-1	1	1	-1	-1	1	-1	-1	1	

C_5 is not a crystallographic group. $\omega=e^{-i\pi/5}$, $\omega^5=-1$.

Table 11.6 The Cyclic Group $C_6 = (6)$, and $C_{3h} = (\bar{6})$

$$C_6 \times C_i = C_{6h} = (6/m) = \left(\frac{6}{m}\right)$$

		C_{3h}	E	S_{3z}^{-1}	C_{3z}	σ_z	C_{3z}^{-1}	S_{3z}	$\bar{E}$	$\overline{S_{3z}^{-1}}$	$\overline{C_{3z}}$	$\overline{\sigma_z}$	$\overline{C_{3z}^{-1}}$	$\overline{S_{3z}}$	C_3 and C_2 bases
		C_6	E	C_{6z}	C_{3z}	C_{2z}	C_{3z}^{-1}	C_{6z}^{-1}	$\bar{E}$	$\overline{C_{6z}}$	$\overline{C_{3z}}$	$\overline{C_{2z}}$	$\overline{C_{3z}^{-1}}$	$\overline{C_{6z}^{-1}}$	C_3 and C_2 bases
A_1	Γ_1	0	1	1	1	1	1	1	1	1	1	1	1	1	
	Γ_7	$\frac{1}{2}$	1	ω	ω^2	$-i$	ω^{10}	ω^{11}	-1	ω^7	ω^8	i	ω^4	ω^5	
	Γ_8	$-\frac{1}{2}$	1	ω^{11}	ω^{10}	i	ω^2	ω	-1	ω^5	ω^4	$-i$	ω^8	ω^7	
E_1	Γ_5	1	1	ω^2	ω^4	-1	ω^8	ω^{10}	1	ω^2	ω^4	-1	ω^8	ω^{10}	
	Γ_6	-1	1	ω^{10}	ω^8	-1	ω^4	ω^2	1	ω^{10}	ω^8	-1	ω^4	ω^2	
	Γ_{12}	$\frac{3}{2}$	1	$-i$	-1	i	-1	i	-1	i	1	$-i$	1	$-i$	
	Γ_{11}	$-\frac{3}{2}$	1	i	-1	$-i$	-1	$-i$	-1	$-i$	1	i	1	i	
E_2	Γ_3	2	1	ω^4	ω^8	1	ω^4	ω^8	1	ω^4	ω^8	1	ω^4	ω^8	
	Γ_2	-2	1	ω^8	ω^4	1	ω^8	ω^4	1	ω^8	ω^4	1	ω^8	ω^4	
	Γ_{10}	$\frac{5}{2}$	1	ω^5	ω^{10}	$-i$	ω^2	ω^7	-1	ω^{11}	ω^4	i	ω^8	ω	
	Γ_9	$-\frac{5}{2}$	1	ω^7	ω^2	i	ω^{10}	ω^5	-1	ω	ω^8	$-i$	ω^4	ω^{11}	
A_2	Γ_4	3	1	-1	1	-1	1	-1	1	-1	1	-1	1	-1	

$\omega = e^{-i\pi/6}$, $\omega^3 = -i$, $C_{3z} = C_{6z}^2$, $C_{2z} = C_{6z}^3$.

11.2 The Dihedral Groups (Tables 11.7–11.11)

The dihedral group $\mathbf{D}_n$ contains n operators C_{nz}^k, $k=0,1,\ldots,n-1$, being rotations about some axis (usually taken to be the z axis), and n operators C_2 being rotations by π about axes lying in the xy plane. Because of a particular orientation phase chosen in the D_3–C_3 and D_5–C_5 $3jm$ calculations, we have one of the C_2 rotations about the y axis for all $\mathbf{D}_n$ in this $\mathbf{C}_n$ orientation. The D_{odd}–C_2 has as the C_2 axis the z axis, and the orientation phase is chosen so that the n-fold axis is along the y axis.

Two C_2 bases of $\mathbf{D}_2$ are given, one following the D_{even}–C_{even} branching rule prescription and one following the D_{odd}–C_2 prescription. An equivalent way of obtaining the same result is to rotate about the x axis. The two imbeddings are clearly distinguished in the two D_4–D_2–C_2 and D_4–D_2–C_{2y} chains, where the fourfold axis is the z axis and the y axis, respectively. The C_{2v}–C_s imbedding is equivalent to the D_2–C_{2y} imbedding. The pure rotation group $\mathbf{D}_n$ is isomorphic to the rotation–reflection group $\mathbf{C}_{nv}$, being the group generated by an n-fold rotation about the z axis and a mirror plane in the xz (vertical) plane.

The orientation in the D_6–D_3, C_{6v}–C_{3v}, D_{3h}–D_3, and D_{3h}–C_{3v} bases depends on the further subgroups chosen. The labeling of the classes is unchanged in the C_3 basis, but the role of y and z is interchanged in the C_2 or C_s bases. Similar remarks apply to the D_6–D_2, C_{6v}–C_{2v}, and D_{3h}–C_{2v} bases; see the $\mathbf{D}_2$ table for the appropriate labels. In addition, for rotation–reflection groups the orientation (and thus the class labels) can depend on the choice of covering groups. As a result, the labeling of classes shown in the character tables is to be taken as representative only. The labels apppropriate to a particular group chain may be found by constructing the transformation of a small number of odd partners of the chain, to those of a known chain such as the JM basis (see Section 5.3 and Chapter 16, also Reid and Butler, 1980).

Tables 11.7–11.11 begin on the next page.

Table 11.7 The Dihedral Group $D_2 = V = (222)$, and $C_{2v} = (mm2)$

$$D_2 \times C_i = D_{2i} = D_{2h} = V_h(mmm) = \left(\frac{2}{m}\frac{2}{m}\frac{2}{m}\right)$$

			E	$C_{2y}, \bar{C}_{2y}$	$\sigma_z, \bar{\sigma}_z$	$\sigma_x, \bar{\sigma}_x$	$\bar{E}$	C_s basis
		C_{2v}	E	$C_{2z}, \bar{C}_{2z}$	$\sigma_y, \bar{\sigma}_y$	$\sigma_x, \bar{\sigma}_x$	$\bar{E}$	C_2 basis
			E	$C_{2y}, \bar{C}_{2y}$	$C_{2z}, \bar{C}_{2z}$	$C_{2x}, \bar{C}_{2x}$	$\bar{E}$	C_{2y} basis
		D_2	E	$C_{2z}, \bar{C}_{2z}$	$C_{2y}, \bar{C}_{2y}$	$C_{2x}, \bar{C}_{2x}$	$\bar{E}$	C_2 basis
A_1	Γ_1	0	1	1	1	1	1	
E'	Γ_5	$\frac{1}{2}$	2	0	0	0	−2	
B_1	Γ_3	$\tilde{0}$	1	1	−1	−1	1	
B_2	Γ_2	1	1	−1	1	−1	1	
B_3	Γ_4	$\tilde{1}$	1	−1	−1	1	1	

Table 11.8 The Dihedral Group $D_3 = (32)$, and $C_{3v} = (3m)$

$$D_3 \times C_i = D_{3i} = D_{3d} = (\bar{3}\,m) = \left(\bar{3}\frac{2}{m}\right)$$

		C_{3v}	E	$2C_{3z}$	$3\sigma_y$	$\bar{E}$	$2\bar{C}_{3z}$	$3\bar{\sigma}_y$	C_3 basis
			E	$2C_{3y}$	$3\sigma_z$	$\bar{E}$	$2\bar{C}_{3y}$	$3\bar{\sigma}_z$	C_s basis
		D_3	E	$2C_{3z}$	$3C_{2y}$	$\bar{E}$	$2\bar{C}_{3z}$	$3\bar{C}_{2y}$	C_3 basis
			E	$2C_{3y}$	$3C_{2z}$	$\bar{E}$	$2\bar{C}_{3y}$	$3\bar{C}_{2z}$	C_2 basis
A_1	Γ_1	0	1	1	1	1	1	1	
E'	Γ_4	$\frac{1}{2}$	2	1	0	−2	−1	0	
A_2	Γ_2	$\tilde{0}$	1	1	−1	1	1	−1	
E	Γ_3	1	2	−1	0	2	−1	0	
E''	Γ_5	$\frac{3}{2}$	1	−1	$-i$	−1	1	i	
	Γ_6	$-\frac{3}{2}$	1	−1	i	−1	1	$-i$	

Table 11.9 The Dihedral Group $\mathbf{D}_4 = (422)$, $\mathbf{C}_{4v} = (4mm)$, and $\mathbf{D}_{2d} = \mathbf{V}_d = (\bar{4}2m)$

$$\mathbf{D}_4 \times \mathbf{C}_i = \mathbf{D}_{4i} = \mathbf{D}_{4h} = (4/mmm) = \left(\frac{4}{m}\frac{2}{m}\frac{2}{m}\right)$$

		$\mathbf{D}_{2d}$	E	$2S_{4z}$	$C_{2z}, \bar{C}_{2z}$	$2C_{2y}, 2\bar{C}_{2y}$	$2\sigma_{xy}, 2\bar{\sigma}_{xy}$	$\bar{E}$	$2\bar{S}_{4z}$	$\mathbf{D}_2$ and $\mathbf{S}_4$ bases
		$\mathbf{C}_{4v}$	E	$2C_{4z}$	$C_{2z}, \bar{C}_{2z}$	$2\sigma_y, 2\bar{\sigma}_y$	$2\sigma_{xy}, 2\bar{\sigma}_{xy}$	$\bar{E}$	$2\bar{C}_{4z}$	$\mathbf{C}_4$ and $\mathbf{C}_{2v}$ bases
		$\mathbf{D}_4$	E	$2C_{4z}$	$C_{2z}, \bar{C}_{2z}$	$2C_{2y}, 2\bar{C}_{2y}$	$2C_{2xy}, 2\bar{C}_{2xy}$	$\bar{E}$	$2\bar{C}_{4z}$	$\mathbf{D}_2$ and $\mathbf{C}_4$ bases
A_1	Γ_1	0	1	1	1	1	1	1	1	
E'	Γ_6	$\frac{1}{2}$	2	$\sqrt{2}$	0	0	0	-2	$-\sqrt{2}$	
A_2	Γ_2	$\tilde{0}$	1	1	1	-1	-1	1	1	
E	Γ_5	1	2	0	-2	0	0	2	0	
E''	Γ_7	$\frac{3}{2}$	2	$-\sqrt{2}$	0	0	0	-2	$\sqrt{2}$	
B_1	Γ_3	2	1	-1	1	1	-1	1	-1	
B_2	Γ_4	$\tilde{2}$	1	-1	1	-1	1	1	-1	

For the $\mathbf{C}_{2v}$ basis of $\mathbf{D}_{2d}$ one swaps the irrep labels 2 and $\tilde{2}$ or, equivalently, swaps the class labels C_{2y} and σ_{xy}.

Table 11.10 The Dihedral Group D_5 and C_{5v}
$D_5 \times C_i = D_{5i} = D_{5d}$

	C_{5v}	E	$2C_{5z}$	$2C_{5z}^2$	$5\sigma_y$	$\bar{E}$	$2\bar{C}_{5z}$	$2\bar{C}_{5z}^2$	$5\bar{\sigma}_y$	C_5 basis
	D_5	E	$2C_{5z}$	$2C_{5z}^2$	$5C_{2y}$	$\bar{E}$	$2\bar{C}_{5z}$	$2\bar{C}_{5z}^2$	$5\bar{C}_{2y}$	C_5 basis
A_1	0	1	1	1	1	1	1	1	1	
E'	$\frac{1}{2}$	2	$\frac{1}{2}(\sqrt{5}+1)$	$\frac{1}{2}(\sqrt{5}-1)$	0	-2	$-\frac{1}{2}(\sqrt{5}+1)$	$-\frac{1}{2}(\sqrt{5}-1)$	0	
A_2	$\tilde{0}$	1	1	1	-1	1	1	1	-1	
E_1	1	2	$\frac{1}{2}(\sqrt{5}-1)$	$-\frac{1}{2}(\sqrt{5}+1)$	0	2	$\frac{1}{2}(\sqrt{5}-1)$	$-\frac{1}{2}(\sqrt{5}+1)$	0	
E''	$\frac{3}{2}$	2	$-\frac{1}{2}(\sqrt{5}-1)$	$-\frac{1}{2}(\sqrt{5}+1)$	0	-2	$\frac{1}{2}(\sqrt{5}-1)$	$\frac{1}{2}(\sqrt{5}+1)$	0	
E_2	2	2	$-\frac{1}{2}(\sqrt{5}+1)$	$\frac{1}{2}(\sqrt{5}-1)$	0	2	$-\frac{1}{2}(\sqrt{5}+1)$	$\frac{1}{2}(\sqrt{5}-1)$	0	
E'''	$\frac{5}{2}$	1	-1	1	$-i$	-1	1	-1	i	
	$-\frac{5}{2}$	1	-1	1	i	-1	1	-1	$-i$	

The C_s basis of C_{5v} and the C_2 basis of D_5 are obtained by interchanging y and z in the class labels. Neither group is a crystallographic group.

Table 11.11 The Dihedral Group $D_6 = (622)$, $C_{6v} = (6mm)$, and $D_{3h} = (\bar{6}m2)$

$$D_6 \times C_i = D_{6i} = D_{6h} = (6/mmm) = \left(\frac{6}{m}\frac{2}{m}\frac{2}{m}\right)$$

		D_{3h}	E	$2S_{3z}$	$2C_{3z}$	$\sigma_z, \bar{\sigma}_z$	$3C_{2y}, 3\bar{C}_{2y}$	$3\sigma_x, 3\bar{\sigma}_x$	E	$2\bar{S}_{2z}$	$2\bar{C}_{3z}$	C_{3i} basis
		C_{6v}	E	$2C_{6z}$	$2C_{3z}$	$C_{2z}, \bar{C}_{2z}$	$3\sigma_y, 3\bar{\sigma}_y$	$3\sigma_x, 3\bar{\sigma}_x$	$\bar{E}$	$2\bar{C}_{6z}$	$2\bar{C}_{3z}$	C_6 basis
		D_6	E	$2C_{6z}$	$2C_{3z}$	$C_{2z}, \bar{C}_{2z}$	$3C_{2y}, 3\bar{C}_{2y}$	$3C_{2x}, 3\bar{C}_{2x}$	$\bar{E}$	$2\bar{C}_{6z}$	$2\bar{C}_{3z}$	C_6 basis
A_1	Γ_1	0	1	1	1	1	1	1	1	1	1	
E'	Γ_7	$\frac{1}{2}$	2	$\sqrt{3}$	1	0	0	0	−2	$-\sqrt{3}$	−1	
A_2	Γ_2	$\tilde{0}$	1	1	1	1	−1	−1	1	1	1	
E_1	Γ_5	1	2	1	−1	−2	0	0	2	1	−1	
E'''	Γ_9	$\frac{3}{2}$	2	0	−2	0	0	0	−2	0	2	
E_2	Γ_6	2	2	−1	−1	2	0	0	2	−1	−1	
E''	Γ_8	$\frac{5}{2}$	2	$-\sqrt{3}$	1	0	0	0	−2	$\sqrt{3}$	−1	
B_1	Γ_3	3	1	−1	1	−1	1	−1	1	−1	1	
B_2	Γ_4	$\tilde{3}$	1	−1	1	−1	−1	1	1	−1	1	

The three twofold axes in the class labeled C_{2y} (or C_{2x}) are along the y (or x) axis and at $\pi/3$, $2\pi/3$ with respect to it. For the C_{2v} and C_{3v} bases of D_{3h} one swaps the irrep labels 3 and $\tilde{3}$.

11.3 The Tetrahedral Groups (Table 11.12; Figures 11.2, 11.3)

The tetrahedral group **T** is the pure rotation symmetry group of a tetrahedron. It is isomorphic to the group of even permutations on four objects, namely the alternating group on four objects $\mathbf{A}_4$. It is interesting from a mathematical viewpoint in that it is the smallest group to display both Kronecker product multiplicity and have complex irreps. As such it forms a useful example of various calculational complexities which arise in the various point groups but do not arise in $\mathbf{SO}_3$ itself. From the physical viewpoint it is less interesting because a simple tetrahedron has a larger symmetry, that of the rotation–reflection group $\mathbf{T}_d$ (isomorphic to **O**).

Table 11.12 The Tetrahedral Group T = (23)

$$\mathbf{T}\times\mathbf{C}_i=\mathbf{T}_i=\mathbf{T}_h=(m3)=\left(\frac{2}{m}\bar{3}\right)$$

		T	E	$4C_{3xyz}$	$4C_{3xyz}^{-1}$	$3C_{2z}, 3\bar{C}_{2z}$	$\bar{E}$	$4\overline{C_{3xyz}}$	$4\overline{C_{3xyz}^{-1}}$	$\mathbf{D}_2$ basis
			E	$4C_{3z}$	$4C_{3z}^{-1}$	$3C_{2a}, 3\bar{C}_{2a}$	$\bar{E}$	$4\overline{C_{3z}}$	$4\overline{C_{3z}^{-1}}$	$\mathbf{C}_3$ basis
A_1	Γ_1	0	1	1	1	1	1	1	1	
E'	Γ_5	$\frac{1}{2}$	2	1	1	0	-1	-1	-1	
T	Γ_4	1	3	0	0	-1	3	0	0	
E''	Γ_6	$\frac{3}{2}$	2	ω^2	ω^4	0	-2	ω^5	ω	
E'''	Γ_7	$-\frac{3}{2}$	2	ω^4	ω^2	0	-2	ω	ω^5	
E	Γ_2	2	1	ω^2	ω^4	1	1	ω^5	ω	
	Γ_3	-2	1	ω^4	ω^2	1	1	ω	ω^5	

The $\mathbf{D}_2$ basis is that of Koster *et al.* (1962) except that their diagram has the C_{3xyz} labeled as C_{3xyz}^2. The a axis of the $\mathbf{C}_3$ basis is in the xz plane at an angle of $\cos^{-1}\sqrt{\frac{1}{3}}$ to the z axis, over the $-x$ axis. $\omega=e^{-i\pi/3}$, $\omega^3=-1$.

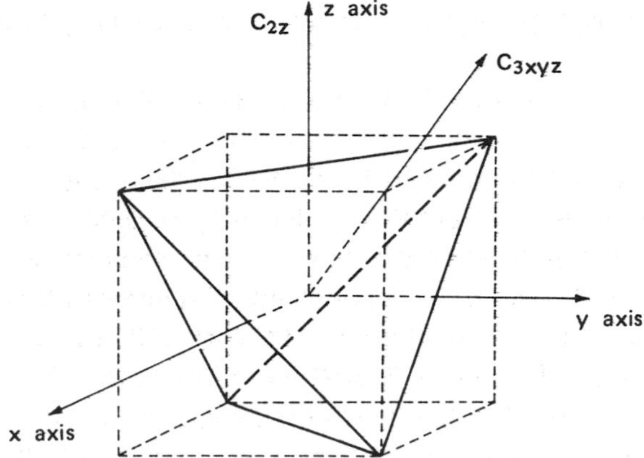

Figure 11.2. The tetrahedron in the D_2 orientation.

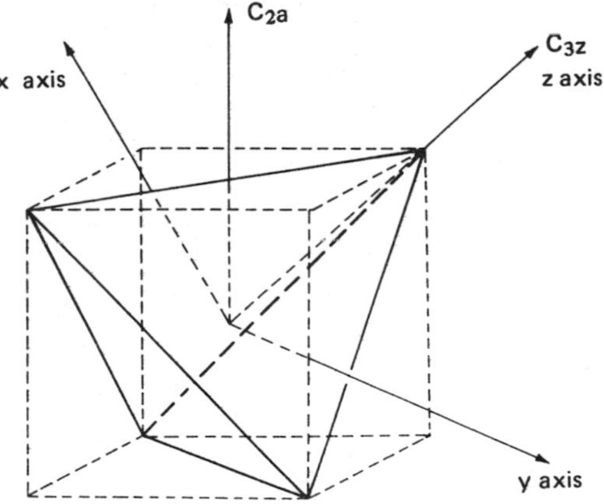

Figure 11.3. The tetrahedron in the C_3 orientation.

11.4 The Octahedral Groups (Table 11.13; Figures 11.4–11.7)

The octahedral group **O** is the pure rotation symmetry group of both the cube and its dual figure, the octahedron. The six faces (eight vertices) of the cube are dual to the six vertices (eight faces) of the octahedron. The site symmetry of many crystals is octahedral, as is the symmetry of the environment of a (transition) metal ion in many complexes. The rotation–reflection group $\mathbf{T}_d$ is a common symmetry of systems with a coordination number of four, for these often form tetrahedra. The full symmetry group of a regular tetrahedron includes reflections but no inversion, so $\mathbf{T}_d$ tends to find more use than the "tetrahedral groups" **T** and $\mathbf{T}_h$.

The groups **O** and $\mathbf{T}_d$ are isomorphic to the group of all permutations on four objects, the symmetric group $\mathbf{S}_4$ (sometimes denoted $\mathbf{\Sigma}_4$ to avoid confusion with the rotation–reflection group isomorphic to $\mathbf{C}_4$).

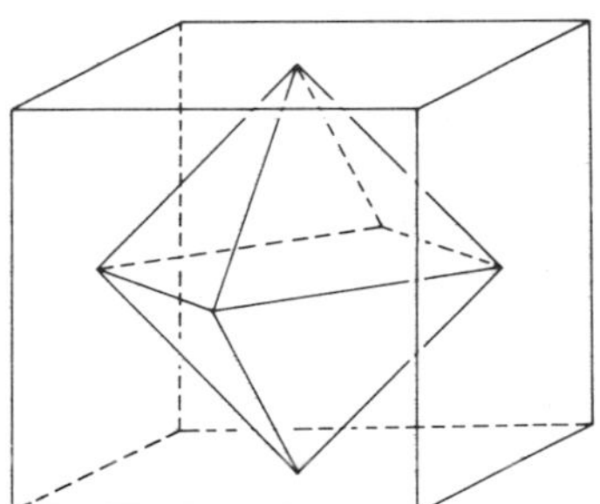

Figure 11.4. The octahedron/cube duality.

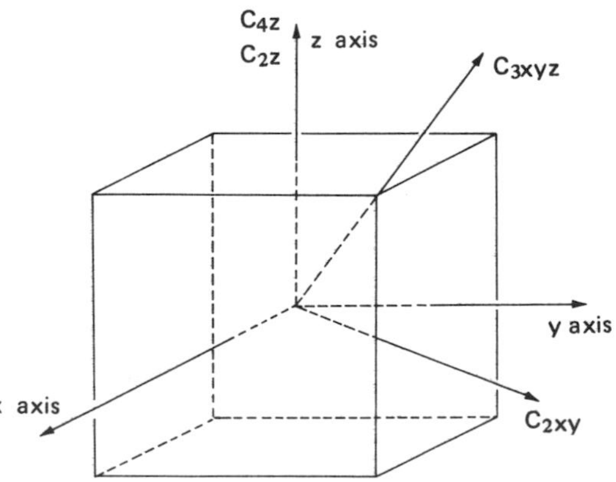

Figure 11.5. The cube in the O–D_4–D_2–C_2, O–D_4–C_4, and O–T–D_2–C_2 orientations.

Table 11.13 The Octahedral Group $\mathbf{O}=(432)$, and $\mathbf{T}_d=(\bar{4}3m)$
$\mathbf{O}\times\mathbf{C}_i=\mathbf{O}_i=\mathbf{O}_h=(m3m)$

		$\mathbf{T}_d$	E	$8C_{3xyz}$	$6\sigma_{4z}$	$3C_{2z},3\bar{C}_{2z}$	$6\sigma_{xy},6\bar{\sigma}_{xy}$	$\bar{E}$	$8\bar{C}_{3xyz}$	$6\bar{\sigma}_{4z}$	D_{2d}–D_2 and D_{2d}–S_4
			E	$8C_{3z}$	$6\sigma_{4a}$	$3C_{2a},3\bar{C}_{2a}$	$6\sigma_z,6\bar{\sigma}_z$	$\bar{E}$	$8\bar{C}_{3z}$	$6\bar{\sigma}_{4a}$	T–C_3 and C_{3v}–C_3
		$\mathbf{O}$	E	$8C_{3xyz}$	$6C_{4z}$	$3C_{2z},3\bar{C}_{2z}$	$6C_{2xy},6\bar{C}_{2xy}$	$\bar{E}$	$8\bar{C}_{xyz}$	$6\bar{C}_{4z}$	D_4 and T–D_2
			E	$8C_{3z}$	$6C_{4a}$	$3C_{2a},3\bar{C}_{2a}$	$6C_{2z},6C_{2z}$	$\bar{E}$	$8\bar{C}_{3z}$	$6\bar{C}_{4a}$	D_3–C_3 and T–C_3
			E	$8C_{3y}$	$6C_{4b}$	$3C_{2b},3\bar{C}_{2b}$	$6C_{2z},6\bar{C}_{2z}$	$\bar{E}$	$8\bar{C}_{3y}$	$6\bar{C}_{4b}$	D_3–C_2
A_1	Γ_1	0	1	1	1	1	1	1	1	1	
E'	Γ_6	$\frac{1}{2}$	2	1	$\sqrt{2}$	0	0	-2	-1	$-\sqrt{2}$	
T_1	Γ_4	1	3	0	1	-1	-1	3	0	1	
U'	Γ_8	$\frac{3}{2}$	4	-1	0	0	0	-4	1	0	
E	Γ_3	2	2	-1	0	2	0	2	-1	0	
T_2	Γ_5	$\tilde{1}$	3	0	-1	-1	1	3	0	-1	
E''	Γ_7	$\tilde{\frac{1}{2}}$	2	1	$-\sqrt{2}$	0	0	-2	-1	$\sqrt{2}$	
A_2	Γ_2	$\tilde{0}$	1	1	-1	1	-1	1	1	-1	

The a axis of the $\mathbf{T}_d\supset\mathbf{C}_3$ and $\mathbf{O}\supset\mathbf{C}_3$ bases is the same as for the $\mathbf{T}\supset\mathbf{C}_3$ basis; it has (x,y,z) coordinates $\left(\sqrt{\frac{2}{3}}\,,0,1/\sqrt{3}\,\right)$. The b axis has coordinates $\left(\sqrt{\frac{2}{3}}\,,-1/\sqrt{3}\,,0\right)$. The labeling of the classes is valid for all chains which contain the subgroups indicated. Note, though, that different subgroup chains (e.g., $\mathbf{O}\supset\mathbf{D}_3\supset\mathbf{C}_3$ and $\mathbf{O}\supset\mathbf{T}\supset\mathbf{C}_3$) have different linear combinations of the JM kets as their partners.

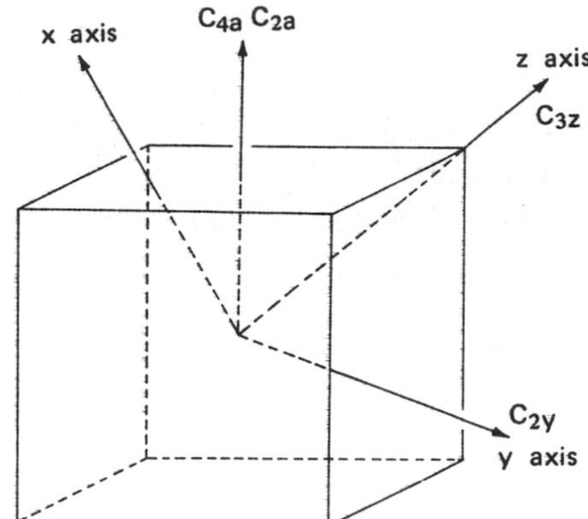

Figure 11.6. The cube in the O–D_3–C_3 and O–T–C_3 orientations. $C_{2a} = R(0, 2\cos^{-1} 1/\sqrt{3}, \pi)$.

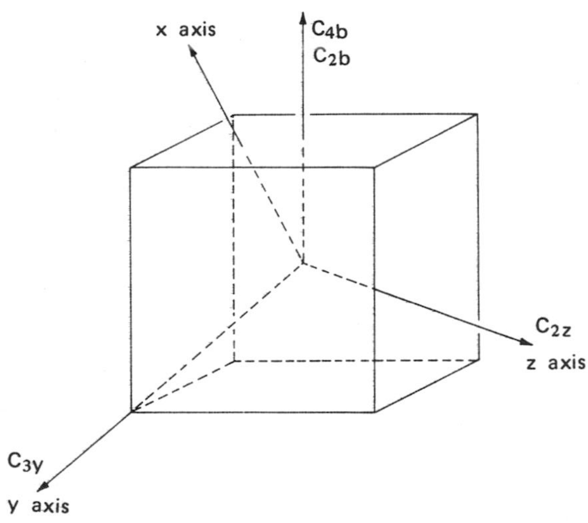

Figure 11.7. The cube in the O–D_3–C_2 orientation. $C_{2a} = R(2\cos^{-1} 1/\sqrt{3}, \pi, 0)$.

11.5 The Icosahedral Groups (Table 11.14; Figures 11.8, 11.9)

The icosahedral group **K** is the pure rotation symmetry group of the regular icosahedron and of its face–vertex dual figure, the dodecahedron. It is isomorphic to $\mathbf{A}_5$, the group of even permutations of five objects. It is not a crystallographic point group and molecular systems displaying exact icosahedral symmetry are rare. It is occasionally suggested that certain impurity sites in crystals may show an approximate fivefold axis and thus also approximate icosahedral symmetry.

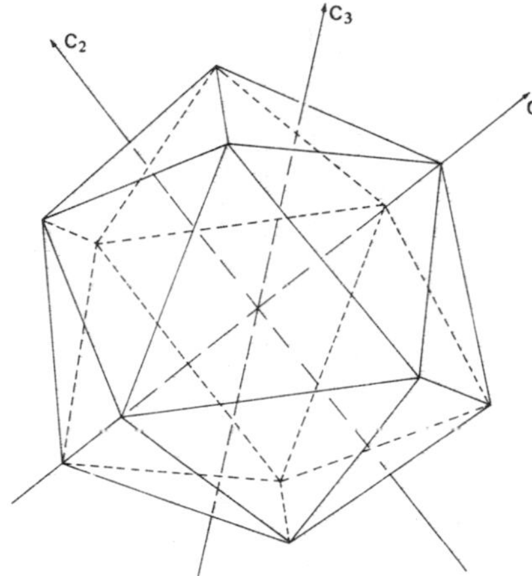

Figure 11.8. The icosahedron, showing various axes.

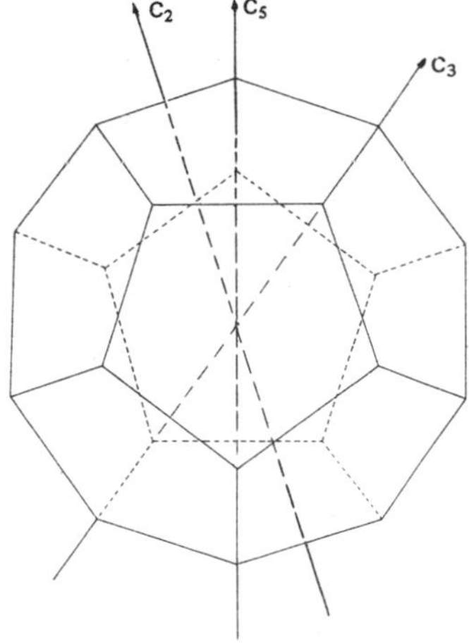

Figure 11.9. The dodecahedron, showing various axes.

Table 11.14 The Icosahedral Group K = I = Y
K × C$_i$ = K$_i$ = K$_h$

	K	E	$12C_5$	$12C_5^2$	$20C_3$	$15C_2, 15\bar{C}_2$	$\bar{E}$	$12\bar{C}_5$	$12\bar{C}_5^2$	$20\bar{C}_3$
A_1	0	1	1	1	1	1	1	1	1	1
E'	$\frac{1}{2}$	2	$\frac{1}{2}(\sqrt{5}+1)$	$\frac{1}{2}(\sqrt{5}-1)$	0	0	-2	$-\frac{1}{2}(\sqrt{5}+1)$	$-\frac{1}{2}(\sqrt{5}-1)$	-1
T_1	1	3	$\frac{1}{2}(\sqrt{5}+1)$	$-\frac{1}{2}(\sqrt{5}-1)$	-1	-1	3	$\frac{1}{2}(\sqrt{5}+1)$	$-\frac{1}{2}(\sqrt{5}-1)$	0
U'	$\frac{3}{2}$	4	1	-1	0	0	-4	-1	1	1
V	2	5	0	0	1	1	5	0	0	-1
W'	$\frac{5}{2}$	6	-1	1	0	0	-6	1	-1	0
U	3	4	-1	-1	0	0	4	-1	-1	1
T_2	$\tilde{1}$	3	$-\frac{1}{2}(\sqrt{5}-1)$	$\frac{1}{2}(\sqrt{5}+1)$	-1	-1	3	$-\frac{1}{2}(\sqrt{5}-1)$	$\frac{1}{2}(\sqrt{5}+1)$	0
E''	$\frac{\tilde{1}}{2}$	2	$-\frac{1}{2}(\sqrt{5}-1)$	$-\frac{1}{2}(\sqrt{5}+1)$	0	0	-2	$\frac{1}{2}(\sqrt{5}+1)$	$\frac{1}{2}(\sqrt{5}+1)$	-1

Many orientations occur. The reader can deduce several from the tables of Chapter 16. Those axes which belong to the various subgroups may be read off the table of the appropriate subgroup.

12

Branching Rule Tables

Introduction

The various possibilities of imbedding point groups in each other are given in this chapter in the form of two diagrams. Figure 12.1 shows the relationship among the various pure rotation point groups. The finite groups $\mathbf{D}_n$ and $\mathbf{C}_n$ for $n>6$ are not shown. The large number of imbeddings of crystallographic point groups one in another is shown in Figure 12.2. Each step of a chain of imbeddings falls into one of six categories:

1. Pure rotation **H** in pure rotation **G** (18 such; denoted by the heavy solid lines in the diagram).
2. Rotation-inversion $\mathbf{H}_i$ in rotation–inversion $\mathbf{G}_i$ (18; denoted by thin solid lines).
3. Pure rotation **G** in rotation–inversion $\mathbf{G}_i$ (11; denoted by the heavy dashed lines).
4. Rotation–reflection $\tilde{\mathbf{G}}$ in rotation–inversion $\mathbf{G}_i$ (10; denoted by dotted lines).
5. Rotation–reflection $\tilde{\mathbf{H}}$ in rotation–reflection $\tilde{\mathbf{G}}$ (13; denoted by the thin dashed lines).
6. Pure rotation **H** in rotation–reflection $\tilde{\mathbf{G}}$ (10; denoted by lines of circles).

The properties of imbeddings are related to pure rotation group imbeddings. (See the heading of the table of interest and Chapter 5.)

The tables which follow give the branching rules for the various imbeddings. As a heading to each set of branching rules we give the group–subgroup names. In most cases the pure-rotation pair is given first,

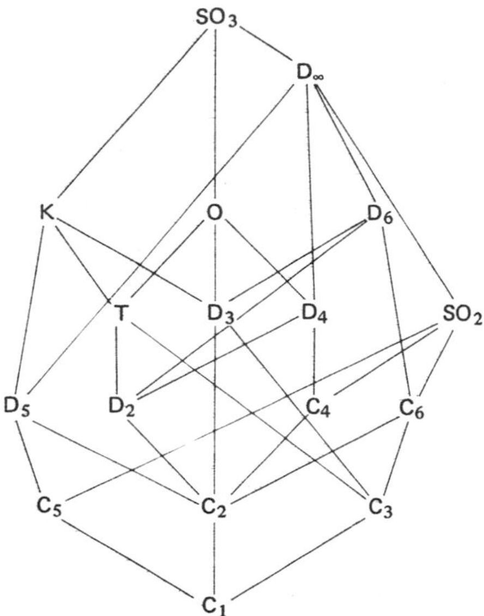

Figure 12.1. Imbeddings of rotation point groups.

then all isomorphic rotation–reflection pairs, and finally (in parentheses) the corresponding rotation–inversion names. In such cases, the rotation–inversion irrep labels are obtained by adding a ± superscript to the irrep labels given. The branchings from a rotation–inversion group to a rotation–reflection subgroup are tabulated in full. The rotation–inversion group to pure-rotation subgroup branching rules are not given because they consist of simply omitting the superscript ± sign labeling the parity of the irrep. In reading off a branching rule the + or − sign before a subgroup irrep is to be ignored because it is the sign of the corresponding $2jm$. In cases of branching multiplicity, the multiplicity is indicated by a digit in small type. For example, on page 216 under $\mathbf{SO_3}\rightarrow\mathbf{O}$, the entry

$$5\rightarrow+2\,1+2+\tilde{1}$$

means that the irrep $\mathbf{5(SO_3)}$ branches to two copies of irrep $\mathbf{1(O)}$ and one copy of each of $\mathbf{2(O)}$ and $\mathbf{\tilde{1}(O)}$.

It is to be noted that the branching rules specify some orientation information, especially for $\mathbf{D_2}$, $\mathbf{D_4}$, $\mathbf{D_6}$, and the corresponding rotation–reflection groups. Two tables of D_2–C_2-branchings are given to illustrate this. As can be seen in the D_2-character table, the irreps $\mathbf{\tilde{0}}$, $\mathbf{1}$, and $\mathbf{\tilde{1}}$ are distinguished only by orientation information. The choice that $\mathbf{\tilde{0}(D_2)}$ shall be invariant under the operations of $\mathbf{C_2}$ gives an orientation we call the z-orientation and regard as standard, the $\mathbf{D_2}\rightarrow\mathbf{C_2}$ table. The choice of $\mathbf{1(D_2)}$ as the $\mathbf{C_2}$-invariant gives the $\mathbf{D_2}\rightarrow\mathbf{C_{2y}}$ table. The latter table gives

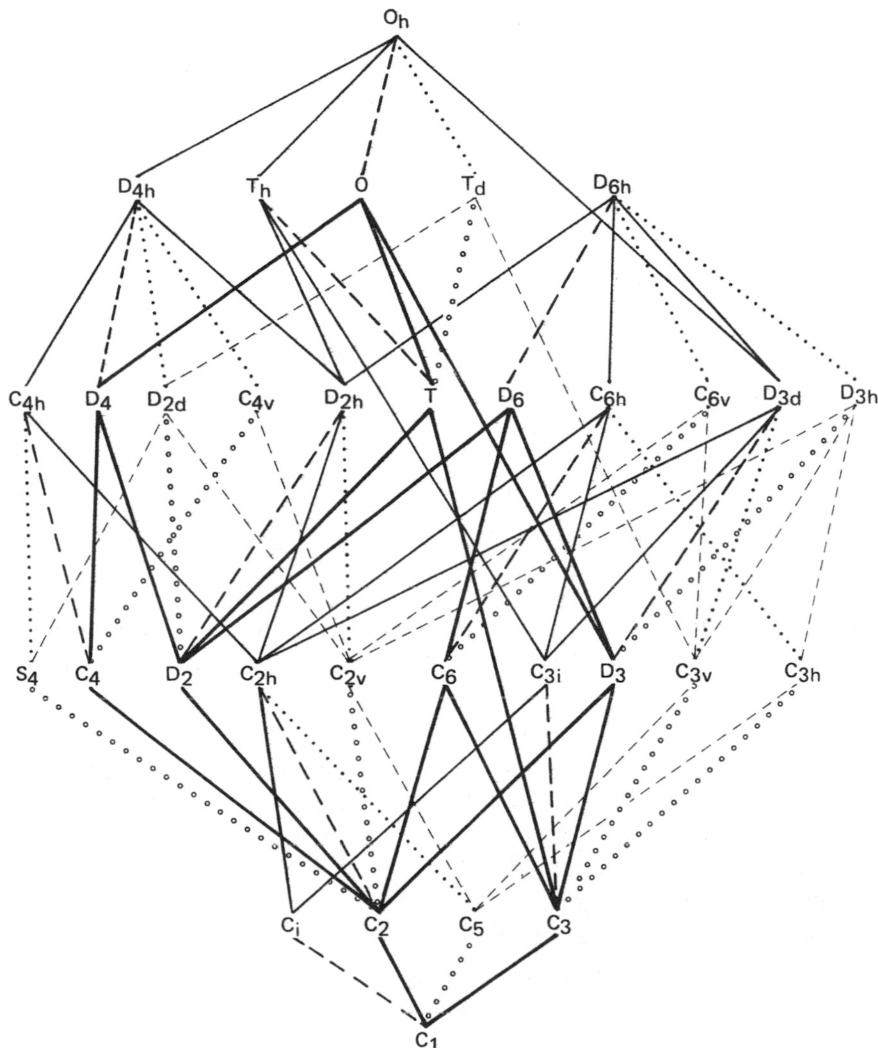

Figure 12.2. Imbeddings of the crystallographic point groups.

the usual branching rules for the $\mathbf{C}_{2v} \rightarrow \mathbf{C}_s$ imbedding. The two tables are interchanged on switching the $\tilde{0}$ and 1 labels.

The irreps are ordered in the same fashion as in the character tables, namely a negative label follows directly after its complex conjugate (positive) label, $\tilde{\mathbf{0}}(\mathbf{D}_n)$ comes between $\frac{1}{2}(\mathbf{D}_n)$ and $\mathbf{1}(\mathbf{D}_n)$, and the tilde labels of **O** and **K** come after the non-tilde (and in reverse order, because that is power order).

$C_2 \rightarrow C_1$ $C_s \rightarrow C_1$ $(C_{2h} \rightarrow C_i)$
$0 \rightarrow +\ 0$
$\frac{1}{2} \rightarrow +\ \frac{1}{2}$
$-\frac{1}{2} \rightarrow +\ \frac{1}{2}$
$1 \rightarrow +\ 0$

$C_{2h} \rightarrow C_s$
$0^{+} \rightarrow +\ 0$
$0^{-} \rightarrow +\ 1$
$\frac{1}{2}^{+} \rightarrow +\ \frac{1}{2}$
$\frac{1}{2}^{-} \rightarrow +\ -\frac{1}{2}$
$-\frac{1}{2}^{+} \rightarrow +\ -\frac{1}{2}$
$-\frac{1}{2}^{-} \rightarrow +\ \frac{1}{2}$
$1^{+} \rightarrow +\ 1$
$1^{-} \rightarrow +\ 0$

$C_3 \rightarrow C_1$ $(C_{3i} \rightarrow C_i)$
$0 \rightarrow +\ 0$
$\frac{1}{2} \rightarrow +\ \frac{1}{2}$
$-\frac{1}{2} \rightarrow +\ \frac{1}{2}$
$1 \rightarrow +\ 0$
$-1 \rightarrow +\ 0$
$\frac{3}{2} \rightarrow +\ \frac{1}{2}$

$C_4 \rightarrow C_2$ $S_4 \rightarrow C_2$ $(C_{4h} \rightarrow C_{2h})$
$0 \rightarrow +\ 0$
$\frac{1}{2} \rightarrow +\ \frac{1}{2}$
$-\frac{1}{2} \rightarrow +\ -\frac{1}{2}$
$1 \rightarrow +\ 1$
$-1 \rightarrow +\ 1$
$\frac{3}{2} \rightarrow +\ -\frac{1}{2}$
$-\frac{3}{2} \rightarrow +\ \frac{1}{2}$
$2 \rightarrow +\ 0$

$C_{4h} \rightarrow S_4$
$0^{+} \rightarrow +\ 0$
$0^{-} \rightarrow +\ 2$
$\frac{1}{2}^{+} \rightarrow +\ \frac{1}{2}$
$\frac{1}{2}^{-} \rightarrow +\ -\frac{3}{2}$
$-\frac{1}{2}^{+} \rightarrow +\ -\frac{1}{2}$
$-\frac{1}{2}^{-} \rightarrow +\ \frac{3}{2}$
$1^{+} \rightarrow +\ 1$
$1^{-} \rightarrow +\ -1$
$-1^{+} \rightarrow +\ -1$
$-1^{-} \rightarrow +\ 1$
$\frac{3}{2}^{+} \rightarrow +\ \frac{3}{2}$
$\frac{3}{2}^{-} \rightarrow +\ -\frac{1}{2}$
$-\frac{3}{2}^{+} \rightarrow +\ -\frac{3}{2}$
$-\frac{3}{2}^{-} \rightarrow +\ \frac{1}{2}$
$2^{+} \rightarrow +\ 2$
$2^{-} \rightarrow +\ 0$

$C_5 \rightarrow C_1$ $(C_{5i} \rightarrow C_i)$
$0 \rightarrow +\ 0$
$\frac{1}{2} \rightarrow +\ \frac{1}{2}$
$-\frac{1}{2} \rightarrow +\ \frac{1}{2}$
$1 \rightarrow +\ 0$
$-1 \rightarrow +\ 0$
$\frac{3}{2} \rightarrow +\ \frac{1}{2}$
$-\frac{3}{2} \rightarrow +\ \frac{1}{2}$
$2 \rightarrow +\ 0$
$-2 \rightarrow +\ 0$
$\frac{5}{2} \rightarrow +\ \frac{1}{2}$

$C_6 \rightarrow C_2$ $C_{3h} \rightarrow C_s$ $(C_{6h} \rightarrow C_{2h})$
$0 \rightarrow +\ 0$
$\frac{1}{2} \rightarrow +\ \frac{1}{2}$
$-\frac{1}{2} \rightarrow +\ -\frac{1}{2}$
$1 \rightarrow +\ 1$
$-1 \rightarrow +\ 1$
$\frac{3}{2} \rightarrow +\ -\frac{1}{2}$
$-\frac{3}{2} \rightarrow +\ \frac{1}{2}$
$2 \rightarrow +\ 0$
$-2 \rightarrow +\ 0$
$\frac{5}{2} \rightarrow +\ \frac{1}{2}$
$-\frac{5}{2} \rightarrow +\ -\frac{1}{2}$
$3 \rightarrow +\ 1$

$C_6 \rightarrow C_3$ $C_{3h} \rightarrow C_3$ $(C_{6h} \rightarrow C_{3i})$
$0 \rightarrow +\ 0$
$\frac{1}{2} \rightarrow +\ \frac{1}{2}$
$-\frac{1}{2} \rightarrow +\ -\frac{1}{2}$
$1 \rightarrow +\ 1$
$-1 \rightarrow +\ -1$
$\frac{3}{2} \rightarrow +\ \frac{3}{2}$
$-\frac{3}{2} \rightarrow +\ \frac{3}{2}$
$2 \rightarrow +\ -1$
$-2 \rightarrow +\ 1$
$\frac{5}{2} \rightarrow +\ -\frac{1}{2}$
$-\frac{5}{2} \rightarrow +\ \frac{1}{2}$
$3 \rightarrow +\ 0$

$C_{6h} \rightarrow C_{3h}$
$0^{+} \Rightarrow +\ 0$
$0^{-} \Rightarrow +\ 3$
$\frac{1}{2}^{+} \Rightarrow +\ \frac{1}{2}$
$\frac{1}{2}^{-} \Rightarrow +\ \text{-}\frac{5}{2}$
$\text{-}\frac{1}{2}^{+} \Rightarrow +\ \text{-}\frac{1}{2}$
$\text{-}\frac{1}{2}^{-} \Rightarrow +\ \frac{5}{2}$
$1^{+} \Rightarrow +\ 1$
$1^{-} \Rightarrow +\ \text{-}2$
$\text{-}1^{+} \Rightarrow +\ \text{-}1$
$\text{-}1^{-} \Rightarrow +\ 2$
$\frac{3}{2}^{+} \Rightarrow +\ \frac{3}{2}$
$\frac{3}{2}^{-} \Rightarrow +\ \text{-}\frac{3}{2}$
$\text{-}\frac{3}{2}^{+} \Rightarrow +\ \text{-}\frac{3}{2}$
$\text{-}\frac{3}{2}^{-} \Rightarrow +\ \frac{3}{2}$
$2^{+} \Rightarrow +\ 2$
$2^{-} \Rightarrow +\ \text{-}1$
$\text{-}2^{+} \Rightarrow +\ \text{-}2$
$\text{-}2^{-} \Rightarrow +\ 1$
$\frac{5}{2}^{+} \Rightarrow +\ \frac{5}{2}$
$\frac{5}{2}^{-} \Rightarrow +\ \text{-}\frac{1}{2}$
$\text{-}\frac{5}{2}^{+} \Rightarrow +\ \text{-}\frac{5}{2}$
$\text{-}\frac{5}{2}^{-} \Rightarrow +\ \frac{1}{2}$
$3^{+} \Rightarrow +\ 3$
$3^{-} \Rightarrow +\ 0$

$D_2 \rightarrow C_2$ $C_{2v} \rightarrow C_2$ $(D_{2h} \rightarrow C_{2h})$
$0 \Rightarrow +\ 0$
$\frac{1}{2} \Rightarrow +\ \frac{1}{2}\ -\ \text{-}\frac{1}{2}$
$\tilde{0} \Rightarrow -\ 0$
$1 \Rightarrow +\ 1$
$\tilde{1} \Rightarrow -\ 1$

$D_2 \rightarrow C_{2y}$ $C_{2v} \rightarrow C_s$
$0 \Rightarrow +\ 0$
$\frac{1}{2} \Rightarrow +\ \frac{1}{2}\ -\ \text{-}\frac{1}{2}$
$\tilde{0} \Rightarrow +\ 1$
$1 \Rightarrow -\ 0$
$\tilde{1} \Rightarrow -\ 1$

$D_{2h} \rightarrow C_{2v}$
$0^{+} \Rightarrow +\ 0$
$0^{-} \Rightarrow +\ \tilde{0}$
$\frac{1}{2}^{+} \Rightarrow +\ \frac{1}{2}$
$\frac{1}{2}^{-} \Rightarrow +\ \frac{1}{2}$
$\tilde{0}^{+} \Rightarrow +\ \tilde{0}$
$\tilde{0}^{-} \Rightarrow +\ 0$
$1^{+} \Rightarrow +\ 1$
$1^{-} \Rightarrow +\ \tilde{1}$
$\tilde{1}^{+} \Rightarrow +\ \tilde{1}$
$\tilde{1}^{-} \Rightarrow +\ 1$

$D_3 \rightarrow C_2$ $C_{3v} \rightarrow C_s$ $(D_{3d} \rightarrow C_{2h})$
$0 \Rightarrow +\ 0$
$\frac{1}{2} \Rightarrow +\ \frac{1}{2}\ -\ \text{-}\frac{1}{2}$
$\tilde{0} \Rightarrow +\ 1$
$1 \Rightarrow -\ 0\ -\ 1$
$\frac{3}{2} \Rightarrow +\ \frac{1}{2}$
$\text{-}\frac{3}{2} \Rightarrow -\ \text{-}\frac{1}{2}$

$D_3 \rightarrow C_3$ $C_{3v} \rightarrow C_3$ $(D_{3d} \rightarrow C_{3i})$
$0 \Rightarrow +\ 0$
$\frac{1}{2} \Rightarrow +\ \frac{1}{2}\ -\ \text{-}\frac{1}{2}$
$\tilde{0} \Rightarrow -\ 0$
$1 \Rightarrow +\ 1\ +\ \text{-}1$
$\frac{3}{2} \Rightarrow +\ \frac{3}{2}$
$\text{-}\frac{3}{2} \Rightarrow -\ \frac{3}{2}$

$D_{3d} \rightarrow C_{3v}$
$0^{+} \Rightarrow +\ 0$
$0^{-} \Rightarrow +\ \tilde{0}$
$\frac{1}{2}^{+} \Rightarrow +\ \frac{1}{2}$
$\frac{1}{2}^{-} \Rightarrow +\ \frac{1}{2}$
$\tilde{0}^{+} \Rightarrow +\ \tilde{0}$
$\tilde{0}^{-} \Rightarrow +\ 0$
$1^{+} \Rightarrow +\ 1$
$1^{-} \Rightarrow +\ 1$
$\frac{3}{2}^{+} \Rightarrow +\ \frac{3}{2}$
$\frac{3}{2}^{-} \Rightarrow +\ \text{-}\frac{3}{2}$
$\text{-}\frac{3}{2}^{+} \Rightarrow +\ \text{-}\frac{3}{2}$
$\text{-}\frac{3}{2}^{-} \Rightarrow +\ \frac{3}{2}$

$D_4 \rightarrow C_4$ $C_{4v} \rightarrow C_4$ $D_{2d} \rightarrow S_4$ $(D_{4h} \rightarrow C_{4h})$
$0 \Rightarrow +\ 0$
$\frac{1}{2} \Rightarrow +\ \frac{1}{2}\ -\ \text{-}\frac{1}{2}$
$\tilde{0} \Rightarrow -\ 0$
$1 \Rightarrow +\ 1\ +\ \text{-}1$
$\frac{3}{2} \Rightarrow +\ \frac{3}{2}\ -\ \text{-}\frac{3}{2}$
$2 \Rightarrow +\ 2$
$\tilde{2} \Rightarrow -\ 2$

$D_4 \rightarrow D_2$ $C_{4v} \rightarrow C_{2v}$ $D_{2d} \rightarrow D_2$ $D_{2d} \rightarrow C_{2v}^{\dagger}$ $(D_{4h} \rightarrow D_{2h})$
$0 \Rightarrow +\ 0$
$\frac{1}{2} \Rightarrow +\ \frac{1}{2}$
$\tilde{0} \Rightarrow +\ \tilde{0}$
$1 \Rightarrow +\ 1\ +\ \tilde{1}$
$\frac{3}{2} \Rightarrow +\ \frac{1}{2}$
$2 \Rightarrow +\ 0$
$\tilde{2} \Rightarrow +\ \tilde{0}$

†For D_{2d}–C_{2v} the irreps 2 and $\tilde{2}$ are interchanged.

$D_{4h} \rightarrow C_{4v}$
$0^+ \rightarrow +\ 0$
$0^- \rightarrow +\ \tilde{0}$
$\frac{1}{2}^+ \rightarrow +\ \frac{1}{2}$
$\frac{1}{2}^- \rightarrow +\ \frac{1}{2}$
$\tilde{0}^+ \rightarrow +\ \tilde{0}$
$\tilde{0}^- \rightarrow +\ 0$
$1^+ \rightarrow +\ 1$
$1^- \rightarrow +\ 1$
$\frac{3}{2}^+ \rightarrow +\ \frac{3}{2}$
$\frac{3}{2}^- \rightarrow +\ \frac{3}{2}$
$2^+ \rightarrow +\ 2$
$2^- \rightarrow +\ \tilde{2}$
$\tilde{2}^+ \rightarrow +\ \tilde{2}$
$\tilde{2}^- \rightarrow +\ 2$

$D_{4h} \rightarrow D_{2d}$
$0^+ \rightarrow +\ 0$
$0^- \rightarrow +\ 2$
$\frac{1}{2}^+ \rightarrow +\ \frac{1}{2}$
$\frac{1}{2}^- \rightarrow +\ \frac{3}{2}$
$\tilde{0}^+ \rightarrow +\ \tilde{0}$
$\tilde{0}^- \rightarrow +\ \tilde{2}$
$1^+ \rightarrow +\ 1$
$1^- \rightarrow +\ 1$
$\frac{3}{2}^+ \rightarrow +\ \frac{3}{2}$
$\frac{3}{2}^- \rightarrow +\ \frac{1}{2}$
$2^+ \rightarrow +\ 2$
$2^- \rightarrow +\ 0$
$\tilde{2}^+ \rightarrow +\ \tilde{2}$
$\tilde{2}^- \rightarrow +\ \tilde{0}$

$D_5 \rightarrow C_2$ $C_{5v} \rightarrow C_s$ $(D_{5d} \rightarrow C_{2h})$
$0 \rightarrow +\ 0$
$\frac{1}{2} \rightarrow +\ \frac{1}{2}\ -\ \text{-}\frac{1}{2}$
$\tilde{0} \rightarrow +\ 1$
$1 \rightarrow -\ 0\ -\ 1$
$\frac{3}{2} \rightarrow +\ \frac{1}{2}\ -\ \text{-}\frac{1}{2}$
$2 \rightarrow +\ 0\ +\ 1$
$\frac{5}{2} \rightarrow +\ \frac{1}{2}$
$\text{-}\frac{5}{2} \rightarrow -\ \text{-}\frac{1}{2}$

$D_5 \rightarrow C_5$ $C_{5v} \rightarrow C_5$ $(D_{5d} \rightarrow C_{5i})$
$0 \rightarrow +\ 0$
$\frac{1}{2} \rightarrow +\ \frac{1}{2}\ -\ \text{-}\frac{1}{2}$
$\tilde{0} \rightarrow -\ 0$
$1 \rightarrow +\ 1\ +\ \text{-}1$
$\frac{3}{2} \rightarrow +\ \frac{3}{2}\ -\ \text{-}\frac{3}{2}$
$2 \rightarrow +\ 2\ +\ \text{-}2$
$\frac{5}{2} \rightarrow +\ \frac{5}{2}$
$\text{-}\frac{5}{2} \rightarrow -\ \frac{5}{2}$

$D_{5d} \rightarrow C_{5v}$
$0^+ \rightarrow +\ 0$
$0^- \rightarrow +\ \tilde{0}$
$\frac{1}{2}^+ \rightarrow +\ \frac{1}{2}$
$\frac{1}{2}^- \rightarrow +\ \frac{1}{2}$
$\tilde{0}^+ \rightarrow +\ \tilde{0}$
$\tilde{0}^- \rightarrow +\ 0$
$1^+ \rightarrow +\ 1$
$1^- \rightarrow +\ 1$
$\frac{3}{2}^+ \rightarrow +\ \frac{3}{2}$
$\frac{3}{2}^- \rightarrow +\ \frac{3}{2}$
$2^+ \rightarrow +\ 2$
$2^- \rightarrow +\ 2$
$\frac{5}{2}^+ \rightarrow +\ \frac{5}{2}$
$\frac{5}{2}^- \rightarrow +\ \text{-}\frac{5}{2}$
$\text{-}\frac{5}{2}^+ \rightarrow +\ \text{-}\frac{5}{2}$
$\text{-}\frac{5}{2}^- \rightarrow +\ \frac{5}{2}$

$D_6 \rightarrow C_6$ $C_{6v} \rightarrow C_6$ $D_{3h} \rightarrow C_{3h}$ $(D_{6h} \rightarrow C_{6h})$
$0 \rightarrow +\ 0$
$\frac{1}{2} \rightarrow +\ \frac{1}{2}\ -\ \text{-}\frac{1}{2}$
$\tilde{0} \rightarrow -\ 0$
$1 \rightarrow +\ 1\ +\ \text{-}1$
$\frac{3}{2} \rightarrow +\ \frac{3}{2}\ -\ \text{-}\frac{3}{2}$
$2 \rightarrow +\ 2\ +\ \text{-}2$
$\frac{5}{2} \rightarrow +\ \frac{5}{2}\ -\ \text{-}\frac{5}{2}$
$3 \rightarrow +\ 3$
$\tilde{3} \rightarrow -\ 3$

$D_6 \rightarrow D_2$ $C_{6v} \rightarrow C_{2v}$ $D_{3h} \rightarrow C_{2v}^{\dagger}$ $(D_{6h} \rightarrow D_{2h})$
$0 \rightarrow +\ 0$
$\frac{1}{2} \rightarrow +\ \frac{1}{2}$
$\tilde{0} \rightarrow +\ \tilde{0}$
$1 \rightarrow +\ 1\ +\ \tilde{1}$
$\frac{3}{2} \rightarrow +\ \frac{1}{2}$
$2 \rightarrow +\ 0\ +\ \tilde{0}$
$\frac{5}{2} \rightarrow +\ \frac{1}{2}$
$3 \rightarrow +\ 1$
$\tilde{3} \rightarrow +\ \tilde{1}$

$D_6 \rightarrow D_3$ $C_{6v} \rightarrow C_{3v}$ $D_{3h} \rightarrow D_3$ $D_{3h} \rightarrow C_{3v}^{\dagger}$ $(D_{6h} \rightarrow D_{3d})$
$0 \rightarrow +\ 0$
$\frac{1}{2} \rightarrow +\ \frac{1}{2}$
$\tilde{0} \rightarrow +\ \tilde{0}$
$1 \rightarrow +\ 1$
$\frac{3}{2} \rightarrow +\ \frac{3}{2}\ +\ \text{-}\frac{3}{2}$
$2 \rightarrow -\ 1$
$\frac{5}{2} \rightarrow -\ \frac{1}{2}$
$3 \rightarrow -\ 0$
$\tilde{3} \rightarrow -\ \tilde{0}$

†For D_{3h}–C_{2v} and D_{3h}–C_{3v} the irreps 3 and $\tilde{3}$ are interchanged.

$D_{6h} \to C_{6v}$	$D_{6h} \to D_{3h}$	$D_\infty \to D_4$ $C_{\infty v} \to C_{4v}$ $(D_{\infty h} \to D_{4h})$	$D_\infty \to D_5$ $C_{\infty v} \to C_{5v}$ $(D_{\infty h} \to D_{5d})$
$0^+ \to +\ 0$	$0^+ \to +\ 0$	$0 \to +\ 0$	$0 \to +\ 0$
$0^- \to +\ \tilde{0}$	$0^- \to +\ 3$	$\frac{1}{2} \to +\ \frac{1}{2}$	$\frac{1}{2} \to +\ \frac{1}{2}$
$\frac{1}{2}^+ \to +\ \frac{1}{2}$	$\frac{1}{2}^+ \to +\ \frac{1}{2}$	$\tilde{0} \to +\ \tilde{0}$	$\tilde{0} \to +\ \tilde{0}$
$\frac{1}{2}^- \to +\ \frac{1}{2}$	$\frac{1}{2}^- \to +\ \frac{5}{2}$	$1 \to +\ 1$	$1 \to +\ 1$
$\tilde{0}^+ \to +\ \tilde{0}$	$\tilde{0}^+ \to +\ \tilde{0}$	$\frac{3}{2} \to +\ \frac{3}{2}$	$\frac{3}{2} \to +\ \frac{3}{2}$
$\tilde{0}^- \to +\ 0$	$\tilde{0}^- \to +\ \tilde{3}$	$2 \to +\ 2 +\ \tilde{2}$	$2 \to +\ 2$
$1^+ \to +\ 1$	$1^+ \to +\ 1$	$\frac{5}{2} \to +\ \frac{3}{2}$	$\frac{5}{2} \to +\ \frac{5}{2} + -\frac{5}{2}$
$1^- \to +\ 1$	$1^- \to +\ 2$	$3 \to +\ 1$	$3 \to -\ 2$
$\frac{3}{2}^+ \to +\ \frac{3}{2}$	$\frac{3}{2}^+ \to +\ \frac{3}{2}$	$\frac{7}{2} \to +\ \frac{1}{2}$	$\frac{7}{2} \to -\ \frac{3}{2}$
$\frac{3}{2}^- \to +\ \frac{3}{2}$	$\frac{3}{2}^- \to +\ \frac{3}{2}$	$4 \to +\ 0 +\ \tilde{0}$	$4 \to -\ 1$
$2^+ \to +\ 2$	$2^+ \to +\ 2$	$\frac{9}{2} \to +\ \frac{1}{2}$	$\frac{9}{2} \to -\ \frac{1}{2}$
$2^- \to +\ 2$	$2^- \to +\ 1$	$5 \to +\ 1$	$5 \to -\ 0 -\ \tilde{0}$
$\frac{5}{2}^+ \to +\ \frac{5}{2}$	$\frac{5}{2}^+ \to +\ \frac{5}{2}$	$\frac{11}{2} \to +\ \frac{3}{2}$	$\frac{11}{2} \to -\ \frac{1}{2}$
$\frac{5}{2}^- \to +\ \frac{5}{2}$	$\frac{5}{2}^- \to +\ \frac{1}{2}$	$6 \to +\ 2 +\ \tilde{2}$	$6 \to -\ 1$
$3^+ \to +\ 3$	$3^+ \to +\ 3$	$\frac{13}{2} \to +\ \frac{3}{2}$	$\frac{13}{2} \to -\ \frac{3}{2}$
$3^- \to +\ \tilde{3}$	$3^- \to +\ 0$	$7 \to +\ 1$	$7 \to -\ 2$
$\tilde{3}^+ \to +\ \tilde{3}$	$\tilde{3}^+ \to +\ \tilde{3}$	$\frac{15}{2} \to +\ \frac{1}{2}$	$\frac{15}{2} \to +\ \frac{5}{2} + -\frac{5}{2}$
$\tilde{3}^- \to +\ 3$	$\tilde{3}^- \to +\ \tilde{0}$	$8 \to +\ 0 +\ \tilde{0}$	$8 \to +\ 2$

$D_\infty \to D_6$
$D_{\infty v} \to C_{6v}$
$(D_{\infty h} \to D_{6h})$

$0 \to +\ 0$	$\frac{5}{2} \to +\ \frac{5}{2}$	$\frac{11}{2} \to +\ \frac{1}{2}$
$\frac{1}{2} \to +\ \frac{1}{2}$	$3 \to +\ 3 +\ \tilde{3}$	$6 \to +\ 0 +\ \tilde{0}$
$\tilde{0} \to +\ \tilde{0}$	$\frac{7}{2} \to +\ \frac{5}{2}$	$\frac{13}{2} \to +\ \frac{1}{2}$
$1 \to +\ 1$	$4 \to +\ 2$	$7 \to +\ 1$
$\frac{3}{2} \to +\ \frac{3}{2}$	$\frac{9}{2} \to +\ \frac{3}{2}$	$\frac{15}{2} \to +\ \frac{3}{2}$
$2 \to +\ 2$	$5 \to +\ 1$	$8 \to +\ 2$

$T \to C_3$ ($T_h \to C_{3i}$)	$T \to D_2$ ($T_h \to D_{2h}$)	$O \to D_3$, $T_d \to C_{3v}$ ($O_h \to D_{3d}$)
$0 \to +\,0$	$0 \to +\,0$	$0 \to +\,0$
$\frac{1}{2} \to +\,\frac{1}{2} \,-\, \text{-}\frac{1}{2}$	$\frac{1}{2} \to +\,\frac{1}{2}$	$\frac{1}{2} \to +\,\frac{1}{2}$
$1 \to -\,0 + 1 + \text{-}1$	$1 \to +\,\tilde{0} + 1 + \tilde{1}$	$1 \to +\,\tilde{0} + 1$
$\frac{3}{2} \to +\,\frac{1}{2} + \frac{3}{2}$	$\frac{3}{2} \to +\,\frac{1}{2}$	$\frac{3}{2} \to +\,\frac{1}{2} + \frac{3}{2} + \text{-}\frac{3}{2}$
$\text{-}\frac{3}{2} \to -\,\text{-}\frac{1}{2} - \frac{3}{2}$	$\text{-}\frac{3}{2} \to +\,\frac{1}{2}$	$2 \to +\,1$
$2 \to +\,1$	$2 \to +\,0$	$\tilde{1} \to +\,0 + 1$
$\text{-}2 \to +\,\text{-}1$	$\text{-}2 \to +\,0$	$\tilde{\frac{1}{2}} \to +\,\frac{1}{2}$
		$\tilde{0} \to +\,\tilde{0}$

$O_h \to T_d$		
$0^+ \to +\,0$	$1^- \to +\,\tilde{1}$	$\tilde{1}^+ \to +\,\tilde{1}$
$0^- \to +\,\tilde{0}$	$\frac{3}{2}^+ \to +\,\frac{3}{2}$	$\tilde{1}^- \to +\,1$
$\frac{1}{2}^+ \to +\,\frac{1}{2}$	$\frac{3}{2}^- \to +\,\frac{3}{2}$	$\tilde{\frac{1}{2}}^+ \to +\,\tilde{\frac{1}{2}}$
$\frac{1}{2}^- \to +\,\tilde{\frac{1}{2}}$	$2^+ \to +\,2$	$\tilde{\frac{1}{2}}^- \to +\,\frac{1}{2}$
$1^+ \to +\,1$	$2^- \to +\,2$	$\tilde{0}^+ \to +\,\tilde{0}$
		$\tilde{0}^- \to +\,0$

$O \to D_4$, $T_d \to D_{2d}$ ($O_h \to D_{4h}$)	$O \to T$, $T_d \to T$ ($O_h \to T_h$)
$0 \to +\,0$	$0 \to +\,0$
$\frac{1}{2} \to +\,\frac{1}{2}$	$\frac{1}{2} \to +\,\frac{1}{2}$
$1 \to +\,\tilde{0} + 1$	$1 \to +\,1$
$\frac{3}{2} \to +\,\frac{1}{2} + \frac{3}{2}$	$\frac{3}{2} \to +\,\frac{3}{2} + \text{-}\frac{3}{2}$
$2 \to +\,0 + 2$	$2 \to +\,2 + \text{-}2$
$\tilde{1} \to +\,1 + \tilde{2}$	$\tilde{1} \to -\,1$
$\tilde{\frac{1}{2}} \to +\,\frac{3}{2}$	$\tilde{\frac{1}{2}} \to -\,\frac{1}{2}$
$\tilde{0} \to +\,2$	$\tilde{0} \to -\,0$

$K \rightarrow D_3$ ($K_h \rightarrow D_{3d}$)
$0 \rightarrow +\ 0$
$\frac{1}{2} \rightarrow +\ \frac{1}{2}$
$1 \rightarrow +\ \tilde{0} +\ 1$
$\frac{3}{2} \rightarrow +\ \frac{1}{2} +\ \frac{3}{2} +\ {-\frac{3}{2}}$
$2 \rightarrow +\ 0 +\ 2\ 1$
$\frac{5}{2} \rightarrow +\ 2\ \frac{1}{2} +\ \frac{3}{2} +\ {-\frac{3}{2}}$
$3 \rightarrow +\ 0 +\ \tilde{0} +\ 1$
$\tilde{1} \rightarrow +\ \tilde{0} +\ 1$
$\tilde{\frac{1}{2}} \rightarrow +\ \frac{1}{2}$

$K \rightarrow D_5$ ($K_h \rightarrow D_{5d}$)
$0 \rightarrow +\ 0$
$\frac{1}{2} \rightarrow +\ \frac{1}{2}$
$1 \rightarrow +\ \tilde{0} +\ 1$
$\frac{3}{2} \rightarrow +\ \frac{1}{2} +\ \frac{3}{2}$
$2 \rightarrow +\ 0 +\ 1 +\ 2$
$\frac{5}{2} \rightarrow +\ \frac{1}{2} +\ \frac{3}{2} +\ \frac{5}{2} +\ {-\frac{5}{2}}$
$3 \rightarrow +\ 1 +\ 2$
$\tilde{1} \rightarrow +\ \tilde{0} +\ 2$
$\tilde{\frac{1}{2}} \rightarrow +\ \frac{3}{2}$

$K \rightarrow T$ ($K_h \rightarrow T_h$)
$0 \rightarrow +\ 0$
$\frac{1}{2} \rightarrow +\ \frac{1}{2}$
$1 \rightarrow +\ 1$
$\frac{3}{2} \rightarrow +\ \frac{3}{2} +\ {-\frac{3}{2}}$
$2 \rightarrow +\ 1 +\ 2 +\ {-2}$
$\frac{5}{2} \rightarrow +\ \frac{1}{2} +\ \frac{3}{2} +\ {-\frac{3}{2}}$
$3 \rightarrow +\ 0 +\ 1$
$\tilde{1} \rightarrow +\ 1$
$\tilde{\frac{1}{2}} \rightarrow +\ \frac{1}{2}$

$SO_3 \rightarrow D_\infty$ ($O_3 \rightarrow D_{\infty h}$)
$0 \rightarrow +\ 0$
$\frac{1}{2} \rightarrow +\ \frac{1}{2}$
$1 \rightarrow +\ \tilde{0} +\ 1$
$\frac{3}{2} \rightarrow +\ \frac{1}{2} +\ \frac{3}{2}$
$2 \rightarrow +\ 0 +\ 1 +\ 2$
$\frac{5}{2} \rightarrow +\ \frac{1}{2} +\ \frac{3}{2} +\ \frac{5}{2}$
$3 \rightarrow +\ \tilde{0} +\ 1 +\ 2 +\ 3$
$\frac{7}{2} \rightarrow +\ \frac{1}{2} +\ \frac{3}{2} +\ \frac{5}{2} +\ \frac{7}{2}$
$4 \rightarrow +\ 0 +\ 1 +\ 2 +\ 3 +\ 4$
$\frac{9}{2} \rightarrow +\ \frac{1}{2} +\ \frac{3}{2} +\ \frac{5}{2} +\ \frac{7}{2} +\ \frac{9}{2}$
$5 \rightarrow +\ \tilde{0} +\ 1 +\ 2 +\ 3 +\ 4 +\ 5$
$\frac{11}{2} \rightarrow +\ \frac{1}{2} +\ \frac{3}{2} +\ \frac{5}{2} +\ \frac{7}{2} +\ \frac{9}{2} +\ \frac{11}{2}$
$6 \rightarrow +\ 0 +\ 1 +\ 2 +\ 3 +\ 4 +\ 5 +\ 6$
$\frac{13}{2} \rightarrow +\ \frac{1}{2} +\ \frac{3}{2} +\ \frac{5}{2} +\ \frac{7}{2} +\ \frac{9}{2} +\ \frac{11}{2} +\ \frac{13}{2}$
$7 \rightarrow +\ \tilde{0} +\ 1 +\ 2 +\ 3 +\ 4 +\ 5 +\ 6 +\ 7$
$\frac{15}{2} \rightarrow +\ \frac{1}{2} +\ \frac{3}{2} +\ \frac{5}{2} +\ \frac{7}{2} +\ \frac{9}{2} +\ \frac{11}{2} +\ \frac{13}{2} +\ \frac{15}{2}$
$8 \rightarrow +\ 0 +\ 1 +\ 2 +\ 3 +\ 4 +\ 5 +\ 6 +\ 7 +\ 8$

$SO_3 \to O$ $(O_3 \to O_h)$	$SO_3 \to K$ $(O_3 \to K_h)$
$0 \to +0$	$0 \to +0$
$\frac{1}{2} \to +\frac{1}{2}$	$\frac{1}{2} \to +\frac{1}{2}$
$1 \to +1$	$1 \to +1$
$\frac{3}{2} \to +\frac{3}{2}$	$\frac{3}{2} \to +\frac{3}{2}$
$2 \to +2+\tilde{1}$	$2 \to +2$
$\frac{5}{2} \to +\frac{3}{2}+\tilde{\frac{1}{2}}$	$\frac{5}{2} \to +\frac{5}{2}$
$3 \to +1+\tilde{1}+\tilde{0}$	$3 \to +3+\tilde{1}$
$\frac{7}{2} \to +\frac{1}{2}+\frac{3}{2}+\tilde{\frac{1}{2}}$	$\frac{7}{2} \to +\frac{5}{2}+\tilde{\frac{1}{2}}$
$4 \to +0+1+2+\tilde{1}$	$4 \to +2+3$
$\frac{9}{2} \to +\frac{1}{2}+{}_{2}\frac{3}{2}$	$\frac{9}{2} \to +\frac{3}{2}+\frac{5}{2}$
$5 \to +{}_{2}1+2+\tilde{1}$	$5 \to +1+2+\tilde{1}$
$\frac{11}{2} \to +\frac{1}{2}+{}_{2}\frac{3}{2}+\tilde{\frac{1}{2}}$	$\frac{11}{2} \to +\frac{1}{2}+\frac{3}{2}+\frac{5}{2}$
$6 \to +0+1+2+{}_{2}\tilde{1}+\tilde{0}$	$6 \to +0+1+2+3$
$\frac{13}{2} \to +\frac{1}{2}+{}_{2}\frac{3}{2}+{}_{2}\tilde{\frac{1}{2}}$	$\frac{13}{2} \to +\frac{1}{2}+\frac{3}{2}+\frac{5}{2}+\tilde{\frac{1}{2}}$
$7 \to +{}_{2}1+2+{}_{2}\tilde{1}+\tilde{0}$	$7 \to +1+2+3+\tilde{1}$
$\frac{15}{2} \to +\frac{1}{2}+{}_{3}\frac{3}{2}+\tilde{\frac{1}{2}}$	$\frac{15}{2} \to +\frac{3}{2}+{}_{2}\frac{5}{2}$
$8 \to +0+{}_{2}1+{}_{2}2+{}_{2}\tilde{1}$	$8 \to +{}_{2}2+3+\tilde{1}$

13

jm Factor Tables

Introduction

This chapter contains tables of the $2jm$ and $3jm$ factors of most point group chains. The $3jm$ factors are the generalization of Wigner's "$3j$ symbols" of angular momentum. The "$3j$ symbol" is an SO_3–SO_2 $3jm$ factor, depending on three "j" labels (being irrep labels of $\mathbf{SO_3}$) and three "m" labels (labels of $\mathbf{SO_2}$). The SO_3–SO_2 $3jm$ factors are tabulated by Rotenberg *et al.* (1959) for $J \leqslant 8$. The SO_3–D_∞ $3jm$ factors are easily obtained from these [see equations (6.2.16)–(6.2.18)]. Those for D_∞–SO_2 are given by a simple formula [(6.2.5)–(6.2.6)]. We restrict our D_∞–D_n, SO_3–O, and SO_3–K $3jm$ tables to $J \leqslant 8$, and for the dihedral groups $\mathbf{D}_n$, to $n \leqslant 6$. The C_m–C_n $3jm$ factors are unity whenever the triad and branching selection rules are satisfied, and are not included.

Denoting the group and subgroup irrep labels by λ_i and μ_i respectively, one writes the general G–H $3jm$ factor as

$$\begin{Bmatrix} \lambda_1 & \lambda_2 & \lambda_3 \\ a_1 & a_2 & a_3 \\ \mu_1 & \mu_2 & \mu_3 \end{Bmatrix}^{rG}_{sH}$$

where the a_i denotes branching multiplicity labels, r denotes the coupling multiplicity in $\mathbf{G}$, and s denotes it in $\mathbf{H}$. No point group pair requires all the multiplicity labels and many (including SO_3–SO_2) require none. The entries at the top of the left column of page 263 may be used to illustrate our tabulation of $3jm$ values. The bold heading $\mathbf{3\,\frac{5}{2}\,\frac{3}{2}}$ on the first line gives the group labels $\lambda_1 \lambda_2 \lambda_3$. Under this heading we have the entry

$$1\tfrac{\bar{1}}{2}\tfrac{3}{2}0 \quad 000+ \quad +\sqrt{3}/\sqrt{2.7}$$

where the symbols are

subgroup labels: $\mu_1 = 1,\ \mu_2 = \frac{\bar{1}}{2},\ \mu_3 = \frac{3}{2}$
subgroup multiplicity: $s = 0$
branching multiplicity: $a_1 = 0,\ a_2 = 0,\ a_3 = 0$
column interchange sign: $+$
value: $+\sqrt{3}\,/\sqrt{2.7}\ (=\sqrt{3}\,/\sqrt{14}\,)$

Note that the dots used in the $3jm$ values are multiplication dots, not decimal points.

The next two entries show that

$$\begin{pmatrix} 3 & \frac{5}{2} & \frac{3}{2} \\ 0 & 0 & 0 \\ \tilde{1} & \frac{3}{2} & \frac{3}{2} \end{pmatrix}^{SO_3}_{0\,0} = \frac{5}{2\sqrt{21}} \tag{13.0.1}$$

and

$$\begin{pmatrix} 3 & \frac{5}{2} & \frac{3}{2} \\ 0 & 0 & 0 \\ \tilde{1} & \frac{3}{2} & \frac{3}{2} \end{pmatrix}^{SO_3}_{1\,0} = \frac{-1}{2\sqrt{21}} \tag{13.0.2}$$

For those cases where there is a group multiplicity, namely **K**, **O**, and **T**, the index r is included as part of the bold header. In those cases where there is no subgroup multiplicity or no branching multiplicity, the respective labels (zero) are omitted. The other aspects of the present tabulation are most simply illustrated by reference to the tables on page 220, which have no multiplicity indices.

$3jm$ factors have two kinds of symmetry, a column interchange symmetry and a complex conjugation symmetry, each of which may introduce a negative sign [see (3.3.11) and (3.3.12)]. The sign on column interchange (a product of two $3j$ phases) is given in the tables after the branching indices. For example the D_2–C_2 $3jm$

$$\begin{pmatrix} \frac{1}{2} & \frac{1}{2} & 0 \\ -\frac{1}{2} & \frac{1}{2} & 0 \end{pmatrix}\begin{matrix} D_2 \\ C_2 \end{matrix} = -\frac{1}{\sqrt{2}} \tag{13.0.3}$$

appears in the tables. Inspection of the $3j$ phases of Chapter 14 shows that $\{\frac{1}{2}\frac{1}{2}0\}D_2 = -1$ and $\{-\frac{1}{2}\frac{1}{2}0\}C_2 = +1$. The sign of the product, $-$, occurs in the entry for the above $3jm$; one has

$$\begin{pmatrix} \frac{1}{2} & 0 & \frac{1}{2} \\ -\frac{1}{2} & 0 & \frac{1}{2} \end{pmatrix} = \begin{pmatrix} 0 & \frac{1}{2} & \frac{1}{2} \\ 0 & \frac{1}{2} & -\frac{1}{2} \end{pmatrix} = \begin{pmatrix} \frac{1}{2} & \frac{1}{2} & 0 \\ \frac{1}{2} & -\frac{1}{2} & 0 \end{pmatrix} = \frac{1}{\sqrt{2}} \tag{13.0.4}$$

Note that those $3jm$'s obtained by cyclic permutations of columns (two successive interchanges) never require a sign change. Under complex conjugation symmetry a product of three $2jm$ factors enters. If this product is negative an asterisk is included after the interchange sign. The symmetry states that the complex conjugate of a $3jm$ factor is this product of $2jm$'s times the $3jm$ factor of the complex conjugate irreps. Reading again from the D_2–C_2 $3jm$ table one finds that

$$\begin{pmatrix} \frac{1}{2} & \frac{1}{2} & 0 \\ -\frac{1}{2} & \frac{1}{2} & 0 \end{pmatrix}^* = \left(\frac{-1}{\sqrt{2}}\right)^* = -\begin{pmatrix} \frac{1}{2} & \frac{1}{2} & 0 \\ \frac{1}{2} & -\frac{1}{2} & 0 \end{pmatrix} \tag{13.0.5}$$

because $\frac{1}{2}^*(D_2) = \frac{1}{2}(D_2)$ and $\frac{1}{2}^*(C_2) = -\frac{1}{2}(C_2)$. The result agrees with the result obtained by column interchange. Less trivially one has the following D_3–C_3 $3jm$ factors (p. 222):

$$\begin{pmatrix} -\frac{3}{2} & 1 & \frac{1}{2} \\ \frac{3}{2} & -1 & -\frac{1}{2} \end{pmatrix} = +\begin{pmatrix} \frac{3}{2} & 1 & \frac{1}{2} \\ \frac{3}{2} & 1 & \frac{1}{2} \end{pmatrix}^* = \left(\frac{1}{\sqrt{2}}\right)^* = \frac{1}{\sqrt{2}} \tag{13.0.6}$$

and

$$\begin{pmatrix} -\frac{3}{2} & 1 & \frac{1}{2} \\ \frac{3}{2} & 1 & \frac{1}{2} \end{pmatrix} = -\begin{pmatrix} \frac{3}{2} & 1 & \frac{1}{2} \\ \frac{3}{2} & -1 & -\frac{1}{2} \end{pmatrix}^* = -\left(\frac{i}{\sqrt{2}}\right)^* = \frac{i}{\sqrt{2}} \tag{13.0.7}$$

These two symmetries have been used to reduce the size of the tabulations. To find a particular 3*jm* one should order the columns with the larger labels on the left, and use the complex conjugation symmetry to make the labels positive where possible. *The table is in dictionary order* and all 3*jm* factors which satisfy the triad and branching selection rules are included.

Preceding each 3*jm* table is a corresponding 2*jm* table. The 2*jm* is a special case of the 3*jm*—see (3.3.8). Wigner called the 2*jm* a 1*j* and showed that the $\binom{J}{M}$ of the SO_3–SO_2 2*jm* has the value $(-1)^{J-M}$. Our 2*jm* tables are a repeat of our branching rule tables (see Chapter 12). The tables include the sign of the 2*jm* before the subgroup irrep. The sign is independent of the branching multiplicity label.

D_2–C_2 and C_{2v}–C_2 2jm Factors (D_{2i}–C_{2i} = D_{2h}–C_{2h})

$0 \to +0$	$\tilde{0} \to -0$	$\tilde{1} \to -1$
$\frac{1}{2} \to +\frac{1}{2}\ -{-\frac{1}{2}}$	$1 \to +1$	

D_2–C_2 and C_{2v}–C_2 3jm Factors

0 0 0	**$\tilde{0}$ $\tilde{0}$ 0**	**$\tilde{1}$ ½ ½**
0 0 0 + +1	0 0 0 + −1	1 ½ ½ +* $+1/\sqrt{2}$
½ ½ 0	**1 ½ ½**	**$\tilde{1}$ 1 $\tilde{0}$**
−½ ½ 0 −* $-1/\sqrt{2}$	1 ½ ½ + $+1/\sqrt{2}$	1 1 0 − −1
$\tilde{0}$ ½ ½	**1 1 0**	**$\tilde{1}$ $\tilde{1}$ 0**
0 −½ ½ + $+1/\sqrt{2}$	1 1 0 + +1	1 1 0 + −1

D_2–C_{2y} and C_{2v}–C_s 2jm Factors

$0 \to +0$	$\tilde{0} \to +1$	$\tilde{1} \to -1$
$\frac{1}{2} \to +\frac{1}{2}\ -{-\frac{1}{2}}$	$1 \to -0$	

D_2–C_{2y} and C_{2v}–C_s 3jm Factors

0 0 0	**$\tilde{0}$ $\tilde{0}$ 0**	**$\tilde{1}$ $\frac{1}{2}$ $\frac{1}{2}$**
0 0 0 + +1	1 1 0 + +1	1 $\frac{1}{2}$ $\frac{1}{2}$ +* $+1/\sqrt{2}$
$\frac{1}{2}$ $\frac{1}{2}$ 0	**1 $\frac{1}{2}$ $\frac{1}{2}$**	**$\tilde{1}$ 1 $\tilde{0}$**
-$\frac{1}{2}$ $\frac{1}{2}$ 0 −* $-1/\sqrt{2}$	0-$\frac{1}{2}$ $\frac{1}{2}$ + $+1/\sqrt{2}$	1 0 1 − +1
$\tilde{0}$ $\frac{1}{2}$ $\frac{1}{2}$	**1 1 0**	**$\tilde{1}$ $\tilde{1}$ 0**
1 $\frac{1}{2}$ $\frac{1}{2}$ + $+1/\sqrt{2}$	0 0 0 + −1	1 1 0 + −1

D_{2h}–C_{2v} 2jm Factors

$0^+ \Rightarrow +\ 0$	$\tilde{0}^+ \Rightarrow +\ \tilde{0}$	$\tilde{1}^+ \Rightarrow +\ \tilde{1}$
$0^- \Rightarrow +\ \tilde{0}$	$\tilde{0}^- \Rightarrow +\ 0$	$\tilde{1}^- \Rightarrow +\ 1$
$\frac{1}{2}^+ \Rightarrow +\ \frac{1}{2}$	$1^+ \Rightarrow +\ 1$	
$\frac{1}{2}^- \Rightarrow +\ \frac{1}{2}$	$1^- \Rightarrow +\ \tilde{1}$	

D_{2h}–C_{2v} 3jm Factors

$0^+\ 0^+\ 0^+$	**$\tilde{0}^+\ \frac{1}{2}^-\ \frac{1}{2}^-$**	**$1^+\ 1^+\ 0^+$**	**$\tilde{1}^-\ 1^-\ \tilde{0}^+$**
0 0 0 + +1	$\tilde{0}$ $\frac{1}{2}$ $\frac{1}{2}$ + +1	1 1 0 + +1	1 $\tilde{1}$ $\tilde{0}$ + +1
$0^-\ 0^-\ 0^+$	**$\tilde{0}^+\ \tilde{0}^+\ 0^+$**	**$1^-\ 1^-\ 0^+$**	**$\tilde{1}^-\ 1^+\ \tilde{0}^-$**
$\tilde{0}$ $\tilde{0}$ 0 + +1	$\tilde{0}$ $\tilde{0}$ 0 + +1	$\tilde{1}$ $\tilde{1}$ 0 + +1	1 1 0 − −1
$\frac{1}{2}^+\ \frac{1}{2}^+\ 0^+$	**$\tilde{0}^-\ \tilde{0}^-\ 0^+$**	**$1^-\ 1^+\ 0^-$**	**$\tilde{1}^+\ 1^-\ \tilde{0}^-$**
$\frac{1}{2}$ $\frac{1}{2}$ 0 + +1	0 0 0 + +1	$\tilde{1}$ 1 $\tilde{0}$ − +1	$\tilde{1}$ $\tilde{1}$ 0 − +1
$\frac{1}{2}^-\ \frac{1}{2}^-\ 0^+$	**$\tilde{0}^-\ \tilde{0}^+\ 0^-$**	**$\tilde{1}^+\ \frac{1}{2}^+\ \frac{1}{2}^+$**	**$\tilde{1}^+\ \tilde{1}^+\ 0^+$**
$\frac{1}{2}$ $\frac{1}{2}$ 0 + +1	0 $\tilde{0}$ $\tilde{0}$ + +1	$\tilde{1}$ $\frac{1}{2}$ $\frac{1}{2}$ + +1	$\tilde{1}$ $\tilde{1}$ 0 + +1
$\frac{1}{2}^-\ \frac{1}{2}^+\ 0^-$	**$1^+\ \frac{1}{2}^+\ \frac{1}{2}^+$**	**$\tilde{1}^-\ \frac{1}{2}^-\ \frac{1}{2}^+$**	**$\tilde{1}^-\ \tilde{1}^-\ 0^+$**
$\frac{1}{2}$ $\frac{1}{2}$ $\tilde{0}$ − +1	1 $\frac{1}{2}$ $\frac{1}{2}$ + +1	1 $\frac{1}{2}$ $\frac{1}{2}$ + +1	1 1 0 + +1
$\tilde{0}^+\ \frac{1}{2}^+\ \frac{1}{2}^+$	**$1^-\ \frac{1}{2}^-\ \frac{1}{2}^+$**	**$\tilde{1}^+\ \frac{1}{2}^-\ \frac{1}{2}^-$**	**$\tilde{1}^-\ \tilde{1}^+\ 0^-$**
$\tilde{0}$ $\frac{1}{2}$ $\frac{1}{2}$ + +1	$\tilde{1}$ $\frac{1}{2}$ $\frac{1}{2}$ + +1	$\tilde{1}$ $\frac{1}{2}$ $\frac{1}{2}$ + −1	1 $\tilde{1}$ $\tilde{0}$ − +1
$\tilde{0}^-\ \frac{1}{2}^-\ \frac{1}{2}^+$	**$1^+\ \frac{1}{2}^-\ \frac{1}{2}^-$**	**$\tilde{1}^+\ 1^+\ \tilde{0}^+$**	
0 $\frac{1}{2}$ $\frac{1}{2}$ − −1	1 $\frac{1}{2}$ $\frac{1}{2}$ + −1	$\tilde{1}$ 1 $\tilde{0}$ + +1	

D_3–C_2 and C_{3v}–C_s 2jm Factors (D_{3i}–C_{2i} = D_{3d}–C_{2h})

$0 \Rightarrow +\ 0$	$\tilde{0} \Rightarrow +\ 1$	$\frac{3}{2} \Rightarrow +\ \frac{1}{2}$
$\frac{1}{2} \Rightarrow +\ \frac{1}{2}\ -\ \text{-}\frac{1}{2}$	$1 \Rightarrow -\ 0\ -\ 1$	$\text{-}\frac{3}{2} \Rightarrow -\ \text{-}\frac{1}{2}$

D_3–C_2 and C_{3v}–C_3 3jm Factors

$\mathbf{0\ 0\ 0}$	$\mathbf{1\ \frac{1}{2}\ \frac{1}{2}}$	$\mathbf{\frac{3}{2}\ 1\ \frac{1}{2}}$
$0\ 0\ 0\ +\quad +1$	$1\ \frac{1}{2}\ \frac{1}{2}\ +^*\ +1/2$	$\frac{1}{2}\ 0\ -\frac{1}{2}\ -\quad +1/\sqrt{2}$
$\mathbf{\frac{1}{2}\ \frac{1}{2}\ 0}$	$\mathbf{1\ 1\ 0}$	$\frac{1}{2}\ 1\ \frac{1}{2}\ -^*\ -1/\sqrt{2}$
$-\frac{1}{2}\ \frac{1}{2}\ 0\ -^*\ -1/\sqrt{2}$	$0\ 0\ 0\ +\quad -1/\sqrt{2}$	$\mathbf{\frac{3}{2}\ \frac{3}{2}\ \tilde{0}}$
$\mathbf{\tilde{0}\ \frac{1}{2}\ \frac{1}{2}}$	$1\ 1\ 0\ +\quad -1/\sqrt{2}$	$\frac{1}{2}\ \frac{1}{2}\ 1\ +\quad -1$
$1\ \frac{1}{2}\ \frac{1}{2}\ +\quad +1/\sqrt{2}$	$\mathbf{1\ 1\ \tilde{0}}$	$\mathbf{-\frac{3}{2}\ \frac{3}{2}\ 0}$
$\mathbf{\tilde{0}\ \tilde{0}\ 0}$	$1\ 0\ 1\ -\quad +1/\sqrt{2}$	$-\frac{1}{2}\ \frac{1}{2}\ 0\ -^*\ -1$
$1\ 1\ 0\ +\quad +1$	$\mathbf{1\ 1\ 1}$	
$\mathbf{1\ \frac{1}{2}\ \frac{1}{2}}$	$0\ 0\ 0\ +^*\ +i/2$	
$0\ -\frac{1}{2}\ \frac{1}{2}\ +\quad +1/2$	$1\ 1\ 0\ +^*\ -i/2$	

D_3–C_3 and C_{3v}–C_3 2jm Factors (D_{3i}–C_{3i} = D_{3d}–C_{3i})

$0 \to +\ 0$	$\tilde{0} \to -\ 0$	$\frac{3}{2} \to +\ \frac{3}{2}$
$\frac{1}{2} \to +\ \frac{1}{2}\ -\ -\frac{1}{2}$	$1 \to +\ 1\ +\ -1$	$-\frac{3}{2} \to -\ \frac{3}{2}$

D_3–C_3 and C_{3v}–C_3 3jm Factors

$\mathbf{0\ 0\ 0}$	$\mathbf{1\ \frac{1}{2}\ \frac{1}{2}}$	$\mathbf{\frac{3}{2}\ 1\ \frac{1}{2}}$
$0\ 0\ 0\ +\quad +1$	$1\ -\frac{1}{2}\ -\frac{1}{2}\ +\quad +1/\sqrt{2}$	$\frac{3}{2}\ 1\ \frac{1}{2}\ -\quad +1/\sqrt{2}$
$\mathbf{\frac{1}{2}\ \frac{1}{2}\ 0}$	$\mathbf{1\ 1\ 0}$	$\frac{3}{2}\ -1\ -\frac{1}{2}\ -^*\ +i/\sqrt{2}$
$-\frac{1}{2}\ \frac{1}{2}\ 0\ -^*\ -1/\sqrt{2}$	$-1\ 1\ 0\ +\quad +1/\sqrt{2}$	$\mathbf{\frac{3}{2}\ \frac{3}{2}\ \tilde{0}}$
$\mathbf{\tilde{0}\ \frac{1}{2}\ \frac{1}{2}}$	$\mathbf{1\ 1\ \tilde{0}}$	$\frac{3}{2}\ \frac{3}{2}\ 0\ +^*\ -i$
$0\ -\frac{1}{2}\ \frac{1}{2}\ +\quad +1/\sqrt{2}$	$-1\ 1\ 0\ -^*\ -1/\sqrt{2}$	$\mathbf{-\frac{3}{2}\ \frac{3}{2}\ 0}$
$\mathbf{\tilde{0}\ \tilde{0}\ 0}$	$\mathbf{1\ 1\ 1}$	$\frac{3}{2}\ \frac{3}{2}\ 0\ -^*\ -1$
$0\ 0\ 0\ +\quad -1$	$1\ 1\ 1\ +\quad +1/\sqrt{2}$	

D_{3d}–C_{3v} 2jm Factors

$0^+ \to +\ 0$	$\frac{1}{2}^- \to +\ \frac{1}{2}$	$1^+ \to +\ 1$	$\frac{3}{2}^- \to +\ -\frac{3}{2}$
$0^- \to +\ \tilde{0}$	$\tilde{0}^+ \to +\ \tilde{0}$	$1^- \to +\ 1$	$-\frac{3}{2}^+ \to +\ -\frac{3}{2}$
$\frac{1}{2}^+ \to +\ \frac{1}{2}$	$\tilde{0}^- \to +\ 0$	$\frac{3}{2}^+ \to +\ \frac{3}{2}$	$-\frac{3}{2}^- \to +\ \frac{3}{2}$

D_{3d}–C_{3v} 3jm Factors

$\mathbf{0^+\ 0^+\ 0^+}$	$\mathbf{\frac{1}{2}^-\ \frac{1}{2}^-\ 0^+}$	$\mathbf{\tilde{0}^-\ \frac{1}{2}^-\ \frac{1}{2}^+}$	$\mathbf{\tilde{0}^-\ \tilde{0}^-\ 0^+}$
$0\ 0\ 0\ +\ +1$	$\frac{1}{2}\ \frac{1}{2}\ 0\ +\ +1$	$0\ \frac{1}{2}\ \frac{1}{2}\ -\ -1$	$0\ 0\ 0\ +\ +1$
$\mathbf{0^-\ 0^-\ 0^+}$	$\mathbf{\frac{1}{2}^-\ \frac{1}{2}^+\ 0^-}$	$\mathbf{\tilde{0}^+\ \frac{1}{2}^-\ \frac{1}{2}^-}$	$\mathbf{\tilde{0}^-\ \tilde{0}^+\ 0^-}$
$\tilde{0}\ \tilde{0}\ 0\ +\ +1$	$\frac{1}{2}\ \frac{1}{2}\ \tilde{0}\ -\ +1$	$\tilde{0}\ \frac{1}{2}\ \frac{1}{2}\ +\ +1$	$0\ \tilde{0}\ \tilde{0}\ +\ +1$
$\mathbf{\frac{1}{2}^+\ \frac{1}{2}^+\ 0^+}$	$\mathbf{\tilde{0}^+\ \frac{1}{2}^+\ \frac{1}{2}^+}$	$\mathbf{\tilde{0}^+\ \tilde{0}^+\ 0^+}$	$\mathbf{1^+\ \frac{1}{2}^+\ \frac{1}{2}^+}$
$\frac{1}{2}\ \frac{1}{2}\ 0\ +\ +1$	$\tilde{0}\ \frac{1}{2}\ \frac{1}{2}\ +\ +1$	$\tilde{0}\ \tilde{0}\ 0\ +\ +1$	$1\ \frac{1}{2}\ \frac{1}{2}\ +\ +1$

D_{3d}–C_{3v} 3jm Factors (cont.)

$\mathbf{1^- \tfrac{1}{2}^- \tfrac{1}{2}^+}$	$\mathbf{1^- 1^- \tilde{0}^+}$	$\mathbf{\tfrac{3}{2}^- 1^+ \tfrac{1}{2}^-}$	$\mathbf{-\tfrac{3}{2}^- \tfrac{3}{2}^- 0^+}$
$1\ \tfrac{1}{2}\ \tfrac{1}{2}\ +\ +1$	$1\ 1\ \tilde{0}\ +\ +1$	$-\tfrac{3}{2}\ 1\ \tfrac{1}{2}\ +\ +1$	$\tfrac{3}{2}\ -\tfrac{3}{2}\ 0\ +\ +1$
$\mathbf{1^+ \tfrac{1}{2}^- \tfrac{1}{2}^-}$	$\mathbf{1^- 1^+ \tilde{0}^-}$	$\mathbf{\tfrac{3}{2}^+ 1^- \tfrac{1}{2}^-}$	$\mathbf{-\tfrac{3}{2}^- \tfrac{3}{2}^+ 0^-}$
$1\ \tfrac{1}{2}\ \tfrac{1}{2}\ +\ -1$	$1\ 1\ 0\ -\ -1$	$\tfrac{3}{2}\ 1\ \tfrac{1}{2}\ +\ -1$	$\tfrac{3}{2}\ \tfrac{3}{2}\ \tilde{0}\ -\ +1$
$\mathbf{1^+ 1^+ 0^+}$	$\mathbf{1^+ 1^+ 1^+}$	$\mathbf{\tfrac{3}{2}^+ \tfrac{3}{2}^+ \tilde{0}^+}$	$\mathbf{-\tfrac{3}{2}^+ \tfrac{3}{2}^- 0^-}$
$1\ 1\ 0\ +\ +1$	$1\ 1\ 1\ +\ +1$	$\tfrac{3}{2}\ \tfrac{3}{2}\ \tilde{0}\ +\ +1$	$-\tfrac{3}{2}\ -\tfrac{3}{2}\ \tilde{0}\ -\ -1$
$\mathbf{1^- 1^- 0^+}$	$\mathbf{1^- 1^- 1^+}$	$\mathbf{\tfrac{3}{2}^- \tfrac{3}{2}^- \tilde{0}^+}$	
$1\ 1\ 0\ +\ +1$	$1\ 1\ 1\ +\ -1$	$-\tfrac{3}{2}\ -\tfrac{3}{2}\ \tilde{0}\ +\ +1$	
$\mathbf{1^- 1^+ 0^-}$	$\mathbf{\tfrac{3}{2}^+ 1^+ \tfrac{1}{2}^+}$	$\mathbf{\tfrac{3}{2}^- \tfrac{3}{2}^+ \tilde{0}^-}$	
$1\ 1\ \tilde{0}\ -\ +1$	$\tfrac{3}{2}\ 1\ \tfrac{1}{2}\ +\ +1$	$-\tfrac{3}{2}\ \tfrac{3}{2}\ 0\ -\ -1$	
$\mathbf{1^+ 1^+ \tilde{0}^+}$	$\mathbf{\tfrac{3}{2}^- 1^- \tfrac{1}{2}^+}$	$\mathbf{-\tfrac{3}{2}^+ \tfrac{3}{2}^+ 0^+}$	
$1\ 1\ \tilde{0}\ +\ +1$	$-\tfrac{3}{2}\ 1\ \tfrac{1}{2}\ +\ +1$	$-\tfrac{3}{2}\ \tfrac{3}{2}\ 0\ +\ +1$	

D_4–C_4, C_{4v}–C_4, and D_{2d}–S_4 2jm Factors (D_{4i}–C_{4i} = D_{4h}–C_{4h})

$0 \Rightarrow +\ 0$	$1 \Rightarrow +\ 1\ +\text{-}1$	$2 \Rightarrow +\ 2$
$\tfrac{1}{2} \Rightarrow +\ \tfrac{1}{2}\ -\ \text{-}\tfrac{1}{2}$	$\tfrac{3}{2} \Rightarrow +\ \tfrac{3}{2}\ -\ \text{-}\tfrac{3}{2}$	$\tilde{2} \Rightarrow -\ 2$
$\tilde{0} \Rightarrow -\ 0$		

D_4–C_4, C_{4v}–C_4, and D_{2d}–S_4 3jm Factors

$\mathbf{0\ 0\ 0}$	$\mathbf{1\ 1\ \tilde{0}}$	$\mathbf{2\ \tfrac{3}{2}\ \tfrac{1}{2}}$
$0\ 0\ 0\ +\quad +1$	$\text{-}1\ 1\ 0\ -{*}\quad -1/\sqrt{2}$	$2\ \tfrac{3}{2}\ \tfrac{1}{2}\ +\quad +1/\sqrt{2}$
$\mathbf{\tfrac{1}{2}\ \tfrac{1}{2}\ 0}$	$\mathbf{\tfrac{3}{2}\ 1\ \tfrac{1}{2}}$	$\mathbf{2\ 2\ 0}$
$\text{-}\tfrac{1}{2}\ \tfrac{1}{2}\ 0\ -{*}\quad -1/\sqrt{2}$	$\tfrac{3}{2}\ \text{-}1\ \text{-}\tfrac{1}{2}\ -{*}\quad +1/\sqrt{2}$	$2\ 2\ 0\ +\quad +1$
$\mathbf{\tilde{0}\ \tfrac{1}{2}\ \tfrac{1}{2}}$	$\mathbf{\tfrac{3}{2}\ \tfrac{3}{2}\ 0}$	$\mathbf{\tilde{2}\ 1\ 1}$
$0\ \text{-}\tfrac{1}{2}\ \tfrac{1}{2}\ +\quad +1/\sqrt{2}$	$\text{-}\tfrac{3}{2}\ \tfrac{3}{2}\ 0\ -{*}\quad -1/\sqrt{2}$	$2\ 1\ 1\ +{*}\quad -1/\sqrt{2}$
$\mathbf{\tilde{0}\ \tilde{0}\ 0}$	$\mathbf{\tfrac{3}{2}\ \tfrac{3}{2}\ \tilde{0}}$	$\mathbf{\tilde{2}\ \tfrac{3}{2}\ \tfrac{1}{2}}$
$0\ 0\ 0\ +\quad -1$	$\text{-}\tfrac{3}{2}\ \tfrac{3}{2}\ 0\ +\quad +1/\sqrt{2}$	$2\ \tfrac{3}{2}\ \tfrac{1}{2}\ +{*}\quad +1/\sqrt{2}$
$\mathbf{1\ \tfrac{1}{2}\ \tfrac{1}{2}}$	$\mathbf{\tfrac{3}{2}\ \tfrac{3}{2}\ 1}$	$\mathbf{\tilde{2}\ 2\ \tilde{0}}$
$1\ \text{-}\tfrac{1}{2}\ \text{-}\tfrac{1}{2}\ +\quad +1/\sqrt{2}$	$\tfrac{3}{2}\ \tfrac{3}{2}\ 1\ +\quad +1/\sqrt{2}$	$2\ 2\ 0\ -\quad -1$
$\mathbf{1\ 1\ 0}$	$\mathbf{2\ 1\ 1}$	$\mathbf{\tilde{2}\ \tilde{2}\ 0}$
$\text{-}1\ 1\ 0\ +\quad +1/\sqrt{2}$	$2\ 1\ 1\ +\quad -1/\sqrt{2}$	$2\ 2\ 0\ +\quad -1$

D_4–D_2, C_{4v}–C_{2v}, D_{2d}–D_2, and D_{2d}–$C_{2v}^{\dagger}$ 2jm Factors ($D_{41}-D_{21}=D_{4h}-D_{2h}$)

$0 \Rightarrow +\ 0$	$\tilde{0} \Rightarrow +\ \tilde{0}$	$\tfrac{3}{2} \Rightarrow +\ \tfrac{1}{2}$	$\tilde{2} \Rightarrow +\ \tilde{0}$
$\tfrac{1}{2} \Rightarrow +\ \tfrac{1}{2}$	$1 \Rightarrow +\ 1\ +\ \tilde{1}$	$2 \Rightarrow +\ 0$	

† For D_{2d}–C_{2v} the irreps 2 and $\tilde{2}$ are interchanged.

D_4–D_2, C_{4v}–C_{2v}, D_{2d}–D_2, and D_{2d}–$C_{2v}^{\dagger}$ 3jm Factors

$\mathbf{0\ 0\ 0}$		$\mathbf{1\ 1\ \tilde{0}}$		$\mathbf{2\ 1\ 1}$	
$0\ 0\ 0$ +	$+1$	$\tilde{1}\ 1\ \tilde{0}$ +	$+1/\sqrt{2}$	$0\ \tilde{1}\ \tilde{1}$ +	$-1/\sqrt{2}$
$\mathbf{\frac{1}{2}\ \frac{1}{2}\ 0}$		$\mathbf{\frac{3}{2}\ 1\ \frac{1}{2}}$		$\mathbf{2\ \frac{3}{2}\ \frac{1}{2}}$	
$\frac{1}{2}\ \frac{1}{2}\ 0$ +	$+1$	$\frac{1}{2}\ 1\ \frac{1}{2}$ −	$+1/\sqrt{2}$	$0\ \frac{1}{2}\ \frac{1}{2}$ −	$+1$
$\mathbf{\tilde{0}\ \frac{1}{2}\ \frac{1}{2}}$		$\frac{1}{2}\ \tilde{1}\ \frac{1}{2}$ −	$-1/\sqrt{2}$	$\mathbf{2\ 2\ 0}$	
$\tilde{0}\ \frac{1}{2}\ \frac{1}{2}$ +	$+1$	$\mathbf{\frac{3}{2}\ \frac{3}{2}\ 0}$		$0\ 0\ 0$ +	$+1$
$\mathbf{\tilde{0}\ \tilde{0}\ 0}$		$\frac{1}{2}\ \frac{1}{2}\ 0$ +	$+1$	$\mathbf{\tilde{2}\ 1\ 1}$	
$\tilde{0}\ \tilde{0}\ 0$ +	$+1$	$\mathbf{\frac{3}{2}\ \frac{3}{2}\ \tilde{0}}$		$\tilde{0}\ \tilde{1}\ 1$ −	$-1/\sqrt{2}$
$\mathbf{1\ \frac{1}{2}\ \frac{1}{2}}$		$\frac{1}{2}\ \frac{1}{2}\ \tilde{0}$ +	-1	$\mathbf{\tilde{2}\ \frac{3}{2}\ \frac{1}{2}}$	
$1\ \frac{1}{2}\ \frac{1}{2}$ +	$+1/\sqrt{2}$	$\mathbf{\frac{3}{2}\ \frac{3}{2}\ 1}$		$\tilde{0}\ \frac{1}{2}\ \frac{1}{2}$ +	$+1$
$\tilde{1}\ \frac{1}{2}\ \frac{1}{2}$ +	$+1/\sqrt{2}$	$\frac{1}{2}\ \frac{1}{2}\ 1$ +	$+1/\sqrt{2}^{\dagger}$	$\mathbf{\tilde{2}\ 2\ \tilde{0}}$	
$\mathbf{1\ 1\ 0}$		$\frac{1}{2}\ \frac{1}{2}\ \tilde{1}$ +	$+1/\sqrt{2}^{\dagger}$	$\tilde{0}\ 0\ \tilde{0}$ −	-1
$1\ 1\ 0$ +	$+1/\sqrt{2}$	$\mathbf{2\ 1\ 1}$		$\mathbf{\tilde{2}\ \tilde{2}\ 0}$	
$\tilde{1}\ \tilde{1}\ 0$ +	$+1/\sqrt{2}$	$0\ 1\ 1$ +	$+1/\sqrt{2}$	$\tilde{0}\ \tilde{0}\ 0$ +	$+1$

D_{4h}–C_{4v} 2jm Factors

$0^+ \rightarrow +\ 0$	$\tilde{0}^+ \rightarrow +\ \tilde{0}$	$\frac{3}{2}^+ \rightarrow +\ \frac{3}{2}$	$\tilde{2}^+ \rightarrow +\ \tilde{2}$
$0^- \rightarrow +\ \tilde{0}$	$\tilde{0}^- \rightarrow +\ 0$	$\frac{3}{2}^- \rightarrow +\ \frac{3}{2}$	$\tilde{2}^- \rightarrow +\ 2$
$\frac{1}{2}^+ \rightarrow +\ \frac{1}{2}$	$1^+ \rightarrow +\ 1$	$2^+ \rightarrow +\ 2$	
$\frac{1}{2}^- \rightarrow +\ \frac{1}{2}$	$1^- \rightarrow +\ 1$	$2^- \rightarrow +\ \tilde{2}$	

D_{4h}–C_{4v} 3jm Factors

$\mathbf{0^+\ 0^+\ 0^+}$		$\mathbf{\tilde{0}^+\ \tilde{0}^+\ 0^+}$		$\mathbf{1^-\ 1^+\ 0^-}$	
$0\ 0\ 0$ +	$+1$	$\tilde{0}\ \tilde{0}\ 0$ +	$+1$	$1\ 1\ \tilde{0}$ −	$+1$
$\mathbf{0^-\ 0^-\ 0^+}$		$\mathbf{\tilde{0}^-\ \tilde{0}^-\ 0^+}$		$\mathbf{1^+\ 1^+\ \tilde{0}^+}$	
$\tilde{0}\ \tilde{0}\ 0$ +	$+1$	$0\ 0\ 0$ +	$+1$	$1\ 1\ \tilde{0}$ +	$+1$
$\mathbf{\frac{1}{2}^+\ \frac{1}{2}^+\ 0^+}$		$\mathbf{\tilde{0}^-\ \tilde{0}^+\ 0^-}$		$\mathbf{1^-\ 1^-\ \tilde{0}^+}$	
$\frac{1}{2}\ \frac{1}{2}\ 0$ +	$+1$	$0\ \tilde{0}\ \tilde{0}$ +	$+1$	$1\ 1\ \tilde{0}$ +	$+1$
$\mathbf{\frac{1}{2}^-\ \frac{1}{2}^-\ 0^+}$		$\mathbf{1^+\ \frac{1}{2}^+\ \frac{1}{2}^+}$		$\mathbf{1^-\ 1^+\ \tilde{0}^-}$	
$\frac{1}{2}\ \frac{1}{2}\ 0$ +	$+1$	$1\ \frac{1}{2}\ \frac{1}{2}$ +	$+1$	$1\ 1\ 0$ −	-1
$\mathbf{\frac{1}{2}^-\ \frac{1}{2}^+\ 0^-}$		$\mathbf{1^-\ \frac{1}{2}^-\ \frac{1}{2}^+}$		$\mathbf{\frac{3}{2}^+\ 1^+\ \frac{1}{2}^+}$	
$\frac{1}{2}\ \frac{1}{2}\ \tilde{0}$ −	$+1$	$1\ \frac{1}{2}\ \frac{1}{2}$ +	$+1$	$\frac{3}{2}\ 1\ \frac{1}{2}$ +	$+1$
$\mathbf{\tilde{0}^+\ \frac{1}{2}^+\ \frac{1}{2}^+}$		$\mathbf{1^+\ \frac{1}{2}^-\ \frac{1}{2}^-}$		$\mathbf{\frac{3}{2}^-\ 1^-\ \frac{1}{2}^+}$	
$\tilde{0}\ \frac{1}{2}\ \frac{1}{2}$ +	$+1$	$1\ \frac{1}{2}\ \frac{1}{2}$ +	-1	$\frac{3}{2}\ 1\ \frac{1}{2}$ +	$+1$
$\mathbf{\tilde{0}^-\ \frac{1}{2}^-\ \frac{1}{2}^+}$		$\mathbf{1^+\ 1^+\ 0^+}$		$\mathbf{\frac{3}{2}^-\ 1^+\ \frac{1}{2}^-}$	
$0\ \frac{1}{2}\ \frac{1}{2}$ −	-1	$1\ 1\ 0$ +	$+1$	$\frac{3}{2}\ 1\ \frac{1}{2}$ +	$+1$
$\mathbf{\tilde{0}^+\ \frac{1}{2}^-\ \frac{1}{2}^-}$		$\mathbf{1^-\ 1^-\ 0^+}$		$\mathbf{\frac{3}{2}^+\ 1^-\ \frac{1}{2}^-}$	
$\tilde{0}\ \frac{1}{2}\ \frac{1}{2}$ +	$+1$	$1\ 1\ 0$ +	$+1$	$\frac{3}{2}\ 1\ \frac{1}{2}$ +	-1

† For D_{2d}–C_{2v} the irreps 2 and $\tilde{2}$ are interchanged, and the $3jm$'s marked by a dagger (†) change sign.

D_{4h}–C_{4v} 3jm Factors (cont.)

$\mathbf{\frac{3}{2}^{+}\ \frac{3}{2}^{+}\ 0^{+}}$	$\mathbf{2^{+}\ 1^{-}\ 1^{-}}$	$\mathbf{\tilde{2}^{+}\ \frac{3}{2}^{+}\ \frac{1}{2}^{+}}$
$\frac{3}{2}\ \frac{3}{2}\ 0\ +\ +1$	$2\ 1\ 1\ +\ -1$	$\tilde{2}\ \frac{3}{2}\ \frac{1}{2}\ +\ +1$
$\mathbf{\frac{3}{2}^{-}\ \frac{3}{2}^{-}\ 0^{+}}$	$\mathbf{2^{+}\ \frac{3}{2}^{+}\ \frac{1}{2}^{+}}$	$\mathbf{\tilde{2}^{-}\ \frac{3}{2}^{-}\ \frac{1}{2}^{+}}$
$\frac{3}{2}\ \frac{3}{2}\ 0\ +\ +1$	$2\ \frac{3}{2}\ \frac{1}{2}\ +\ +1$	$2\ \frac{3}{2}\ \frac{1}{2}\ +\ +1$
$\mathbf{\frac{3}{2}^{-}\ \frac{3}{2}^{+}\ 0^{-}}$	$\mathbf{2^{-}\ \frac{3}{2}^{-}\ \frac{1}{2}^{+}}$	$\mathbf{\tilde{2}^{-}\ \frac{3}{2}^{+}\ \frac{1}{2}^{-}}$
$\frac{3}{2}\ \frac{3}{2}\ \tilde{0}\ -\ +1$	$\tilde{2}\ \frac{3}{2}\ \frac{1}{2}\ +\ +1$	$2\ \frac{3}{2}\ \frac{1}{2}\ +\ +1$
$\mathbf{\frac{3}{2}^{+}\ \frac{3}{2}^{+}\ \tilde{0}^{+}}$	$\mathbf{2^{-}\ \frac{3}{2}^{+}\ \frac{1}{2}^{-}}$	$\mathbf{\tilde{2}^{+}\ \frac{3}{2}^{-}\ \frac{1}{2}^{-}}$
$\frac{3}{2}\ \frac{3}{2}\ \tilde{0}\ +\ +1$	$\tilde{2}\ \frac{3}{2}\ \frac{1}{2}\ +\ +1$	$\tilde{2}\ \frac{3}{2}\ \frac{1}{2}\ +\ -1$
$\mathbf{\frac{3}{2}^{-}\ \frac{3}{2}^{-}\ \tilde{0}^{+}}$	$\mathbf{2^{+}\ \frac{3}{2}^{-}\ \frac{1}{2}^{-}}$	$\mathbf{\tilde{2}^{+}\ 2^{+}\ \tilde{0}^{+}}$
$\frac{3}{2}\ \frac{3}{2}\ \tilde{0}\ +\ +1$	$2\ \frac{3}{2}\ \frac{1}{2}\ +\ -1$	$\tilde{2}\ 2\ \tilde{0}\ +\ +1$
$\mathbf{\frac{3}{2}^{-}\ \frac{3}{2}^{+}\ \tilde{0}^{-}}$	$\mathbf{2^{+}\ 2^{+}\ 0^{+}}$	$\mathbf{\tilde{2}^{-}\ 2^{-}\ \tilde{0}^{+}}$
$\frac{3}{2}\ \frac{3}{2}\ 0\ -\ -1$	$2\ 2\ 0\ +\ +1$	$2\ \tilde{2}\ \tilde{0}\ +\ +1$
$\mathbf{\frac{3}{2}^{+}\ \frac{3}{2}^{+}\ 1^{+}}$	$\mathbf{2^{-}\ 2^{-}\ 0^{+}}$	$\mathbf{\tilde{2}^{-}\ 2^{+}\ \tilde{0}^{-}}$
$\frac{3}{2}\ \frac{3}{2}\ 1\ +\ +1$	$\tilde{2}\ \tilde{2}\ 0\ +\ +1$	$2\ 2\ 0\ -\ -1$
$\mathbf{\frac{3}{2}^{-}\ \frac{3}{2}^{-}\ 1^{+}}$	$\mathbf{2^{-}\ 2^{+}\ 0^{-}}$	$\mathbf{\tilde{2}^{+}\ 2^{-}\ \tilde{0}^{-}}$
$\frac{3}{2}\ \frac{3}{2}\ 1\ +\ -1$	$\tilde{2}\ 2\ \tilde{0}\ -\ +1$	$\tilde{2}\ \tilde{2}\ 0\ -\ +1$
$\mathbf{\frac{3}{2}^{-}\ \frac{3}{2}^{+}\ 1^{-}}$	$\mathbf{\tilde{2}^{+}\ 1^{+}\ 1^{+}}$	$\mathbf{\tilde{2}^{+}\ \tilde{2}^{+}\ 0^{+}}$
$\frac{3}{2}\ \frac{3}{2}\ 1\ +\ -1$	$\tilde{2}\ 1\ 1\ +\ +1$	$\tilde{2}\ \tilde{2}\ 0\ +\ +1$
$\mathbf{2^{+}\ 1^{+}\ 1^{+}}$	$\mathbf{\tilde{2}^{-}\ 1^{-}\ 1^{+}}$	$\mathbf{\tilde{2}^{-}\ \tilde{2}^{-}\ 0^{+}}$
$2\ 1\ 1\ +\ +1$	$2\ 1\ 1\ +\ +1$	$2\ 2\ 0\ +\ +1$
$\mathbf{2^{-}\ 1^{-}\ 1^{+}}$	$\mathbf{\tilde{2}^{+}\ 1^{-}\ 1^{-}}$	$\mathbf{\tilde{2}^{-}\ \tilde{2}^{+}\ 0^{-}}$
$\tilde{2}\ 1\ 1\ +\ +1$	$\tilde{2}\ 1\ 1\ +\ -1$	$2\ \tilde{2}\ \tilde{0}\ -\ +1$

D_{4h}–D_{2d} 2jm Factors

$0^{+} \rightarrow +\ 0$	$\tilde{0}^{+} \rightarrow +\ \tilde{0}$	$\frac{3}{2}^{+} \rightarrow +\ \frac{3}{2}$	$\tilde{2}^{+} \rightarrow +\ \tilde{2}$
$0^{-} \rightarrow +\ 2$	$\tilde{0}^{-} \rightarrow +\ \tilde{2}$	$\frac{3}{2}^{-} \rightarrow +\ \frac{1}{2}$	$\tilde{2}^{-} \rightarrow +\ \tilde{0}$
$\frac{1}{2}^{+} \rightarrow +\ \frac{1}{2}$	$1^{+} \rightarrow +\ 1$	$2^{+} \rightarrow +\ 2$	
$\frac{1}{2}^{-} \rightarrow +\ \frac{3}{2}$	$1^{-} \rightarrow +\ 1$	$2^{-} \rightarrow +\ 0$	

D_{4h}–D_{2d} 3jm Factors

$\mathbf{0^{+}\ 0^{+}\ 0^{+}}$	$\mathbf{\frac{1}{2}^{-}\ \frac{1}{2}^{+}\ 0^{-}}$	$\mathbf{\tilde{0}^{+}\ \tilde{0}^{+}\ 0^{+}}$	$\mathbf{1^{-}\ \frac{1}{2}^{-}\ \frac{1}{2}^{+}}$
$0\ 0\ 0\ +\ +1$	$\frac{3}{2}\ \frac{1}{2}\ 2\ -\ +1$	$\tilde{0}\ \tilde{0}\ 0\ +\ +1$	$1\ \frac{3}{2}\ \frac{1}{2}\ -\ -1$
$\mathbf{0^{-}\ 0^{-}\ 0^{+}}$	$\mathbf{\tilde{0}^{+}\ \frac{1}{2}^{+}\ \frac{1}{2}^{+}}$	$\mathbf{\tilde{0}^{-}\ \tilde{0}^{-}\ 0^{+}}$	$\mathbf{1^{+}\ \frac{1}{2}^{-}\ \frac{1}{2}^{-}}$
$2\ 2\ 0\ +\ +1$	$\tilde{0}\ \frac{1}{2}\ \frac{1}{2}\ +\ +1$	$\tilde{2}\ \tilde{2}\ 0\ +\ +1$	$1\ \frac{3}{2}\ \frac{3}{2}\ +\ +1$
$\mathbf{\frac{1}{2}^{+}\ \frac{1}{2}^{+}\ 0^{+}}$	$\mathbf{\tilde{0}^{-}\ \frac{1}{2}^{-}\ \frac{1}{2}^{+}}$	$\mathbf{\tilde{0}^{-}\ \tilde{0}^{+}\ 0^{-}}$	$\mathbf{1^{+}\ 1^{+}\ 0^{+}}$
$\frac{1}{2}\ \frac{1}{2}\ 0\ +\ +1$	$\tilde{2}\ \frac{3}{2}\ \frac{1}{2}\ +\ +1$	$\tilde{2}\ \tilde{0}\ 2\ -\ +1$	$1\ 1\ 0\ +\ +1$
$\mathbf{\frac{1}{2}^{-}\ \frac{1}{2}^{-}\ 0^{+}}$	$\mathbf{\tilde{0}^{+}\ \frac{1}{2}^{-}\ \frac{1}{2}^{-}}$	$\mathbf{1^{+}\ \frac{1}{2}^{+}\ \frac{1}{2}^{+}}$	$\mathbf{1^{-}\ 1^{-}\ 0^{+}}$
$\frac{3}{2}\ \frac{3}{2}\ 0\ +\ +1$	$\tilde{0}\ \frac{3}{2}\ \frac{3}{2}\ +\ -1$	$1\ \frac{1}{2}\ \frac{1}{2}\ +\ +1$	$1\ 1\ 0\ +\ +1$

$D_{4h}-D_{2d}$ 3jm Factors (cont.)

$1^-\ 1^+\ 0^-$	**$\frac{3}{2}^+\ \frac{3}{2}^+\ \tilde{0}^+$**	**$2^-\ \frac{3}{2}^-\ \frac{1}{2}^+$**	**$\tilde{2}^-\ \frac{3}{2}^-\ \frac{1}{2}^+$**
$1\ 1\ 2\ +\ +1$	$\frac{3}{2}\ \frac{3}{2}\ \tilde{0}\ +\ +1$	$0\ \frac{1}{2}\ \frac{1}{2}\ -\ -1$	$\tilde{0}\ \frac{1}{2}\ \frac{1}{2}\ +\ +1$
$1^+\ 1^+\ \tilde{0}^+$	**$\frac{3}{2}^-\ \frac{3}{2}^-\ \tilde{0}^+$**	**$2^-\ \frac{3}{2}^+\ \frac{1}{2}^-$**	**$\tilde{2}^-\ \frac{3}{2}^+\ \frac{1}{2}^-$**
$1\ 1\ \tilde{0}\ +\ +1$	$\frac{1}{2}\ \frac{1}{2}\ \tilde{0}\ +\ -1$	$0\ \frac{3}{2}\ \frac{3}{2}\ -\ +1$	$\tilde{0}\ \frac{3}{2}\ \frac{3}{2}\ +\ +1$
$1^-\ 1^-\ \tilde{0}^+$	**$\frac{3}{2}^-\ \frac{3}{2}^+\ \tilde{0}^-$**	**$2^+\ \frac{3}{2}^-\ \frac{1}{2}^-$**	**$\tilde{2}^+\ \frac{3}{2}^-\ \frac{1}{2}^-$**
$1\ 1\ \tilde{0}\ +\ -1$	$\frac{1}{2}\ \frac{3}{2}\ \tilde{2}\ +\ +1$	$2\ \frac{1}{2}\ \frac{3}{2}\ +\ +1$	$\tilde{2}\ \frac{1}{2}\ \frac{3}{2}\ +\ -1$
$1^-\ 1^+\ \tilde{0}^-$	**$\frac{3}{2}^+\ \frac{3}{2}^+\ 1^+$**	**$2^+\ 2^+\ 0^+$**	**$\tilde{2}^+\ 2^+\ \tilde{0}^+$**
$1\ 1\ \tilde{2}\ -\ +1$	$\frac{3}{2}\ \frac{3}{2}\ 1\ +\ +1$	$2\ 2\ 0\ +\ +1$	$\tilde{2}\ 2\ \tilde{0}\ +\ +1$
$\frac{3}{2}^+\ 1^+\ \frac{1}{2}^+$	**$\frac{3}{2}^-\ \frac{3}{2}^-\ 1^+$**	**$2^-\ 2^-\ 0^+$**	**$\tilde{2}^-\ 2^-\ \tilde{0}^+$**
$\frac{3}{2}\ 1\ \frac{1}{2}\ +\ +1$	$\frac{1}{2}\ \frac{1}{2}\ 1\ +\ +1$	$0\ 0\ 0\ +\ +1$	$\tilde{0}\ 0\ \tilde{0}\ -\ +1$
$\frac{3}{2}^-\ 1^-\ \frac{1}{2}^+$	**$\frac{3}{2}^-\ \frac{3}{2}^+\ 1^-$**	**$2^-\ 2^+\ 0^-$**	**$\tilde{2}^-\ 2^+\ \tilde{0}^-$**
$\frac{1}{2}\ 1\ \frac{1}{2}\ -\ -1$	$\frac{1}{2}\ \frac{3}{2}\ 1\ -\ -1$	$0\ 2\ 2\ +\ +1$	$\tilde{0}\ 2\ \tilde{2}\ +\ +1$
$\frac{3}{2}^-\ 1^+\ \frac{1}{2}^-$	**$2^+\ 1^+\ 1^+$**	**$\tilde{2}^+\ 1^+\ 1^+$**	**$\tilde{2}^+\ 2^-\ \tilde{0}^-$**
$\frac{1}{2}\ 1\ \frac{3}{2}\ +\ +1$	$2\ 1\ 1\ +\ +1$	$\tilde{2}\ 1\ 1\ +\ +1$	$\tilde{2}\ 0\ \tilde{2}\ -\ -1$
$\frac{3}{2}^+\ 1^-\ \frac{1}{2}^-$	**$2^-\ 1^-\ 1^+$**	**$\tilde{2}^-\ 1^-\ 1^+$**	**$\tilde{2}^+\ \tilde{2}^+\ 0^+$**
$\frac{3}{2}\ 1\ \frac{3}{2}\ -\ +1$	$0\ 1\ 1\ +\ +1$	$\tilde{0}\ 1\ 1\ -\ -1$	$\tilde{2}\ \tilde{2}\ 0\ +\ +1$
$\frac{3}{2}^+\ \frac{3}{2}^+\ 0^+$	**$2^+\ 1^-\ 1^-$**	**$\tilde{2}^+\ 1^-\ 1^-$**	**$\tilde{2}^-\ \tilde{2}^-\ 0^+$**
$\frac{3}{2}\ \frac{3}{2}\ 0\ +\ +1$	$2\ 1\ 1\ +\ +1$	$\tilde{2}\ 1\ 1\ +\ -1$	$\tilde{0}\ \tilde{0}\ 0\ +\ +1$
$\frac{3}{2}^-\ \frac{3}{2}^-\ 0^+$	**$2^+\ \frac{3}{2}^+\ \frac{1}{2}^+$**	**$\tilde{2}^+\ \frac{3}{2}^+\ \frac{1}{2}^+$**	**$\tilde{2}^-\ \tilde{2}^+\ 0^-$**
$\frac{1}{2}\ \frac{1}{2}\ 0\ +\ +1$	$2\ \frac{3}{2}\ \frac{1}{2}\ +\ +1$	$\tilde{2}\ \frac{3}{2}\ \frac{1}{2}\ +\ +1$	$\tilde{0}\ \tilde{2}\ 2\ -\ +1$
$\frac{3}{2}^-\ \frac{3}{2}^+\ 0^-$			
$\frac{1}{2}\ \frac{3}{2}\ 2\ -\ +1$			

D_5-C_2 and $C_{5v}-C_s$ 2jm Factors ($D_{5i}-C_{2i}=D_{5d}-C_{2h}$)

$0 \to +\ 0$	$1 \to -\ 0\ -\ 1$	$\frac{5}{2} \to +\ \frac{1}{2}$
$\frac{1}{2} \to +\ \frac{1}{2}\ -\ \text{-}\frac{1}{2}$	$\frac{3}{2} \to +\ \frac{1}{2}\ -\ \text{-}\frac{1}{2}$	$\text{-}\frac{5}{2} \to -\ \text{-}\frac{1}{2}$
$\tilde{0} \to +\ 1$	$2 \to +\ 0\ +\ 1$	

D_5-C_2 and $C_{5v}-C_s$ 3jm Factors

$0\ 0\ 0$	**$1\ 1\ 0$**	**$\frac{3}{2}\ \frac{3}{2}\ \tilde{0}$**
$0\ 0\ 0\ +\quad +1$	$0\ 0\ 0\ +\quad -1/\sqrt{2}$	$\frac{1}{2}\ \frac{1}{2}\ 1\ +\quad -1/\sqrt{2}$
$\frac{1}{2}\ \frac{1}{2}\ 0$	$1\ 1\ 0\ +\quad -1/\sqrt{2}$	**$2\ 1\ 1$**
$\text{-}\frac{1}{2}\ \frac{1}{2}\ 0\ -^*\quad -1/\sqrt{2}$	**$1\ 1\ \tilde{0}$**	$0\ 0\ 0\ +\quad -1/2$
$\tilde{0}\ \frac{1}{2}\ \frac{1}{2}$	$1\ 0\ 1\ -\quad +1/\sqrt{2}$	$0\ 1\ 1\ +\quad +1/2$
$1\ \frac{1}{2}\ \frac{1}{2}\ +\quad +1/\sqrt{2}$	**$\frac{3}{2}\ 1\ \frac{1}{2}$**	$1\ 1\ 0\ +\quad -1/2$
$\tilde{0}\ \tilde{0}\ 0$	$\frac{1}{2}\ 0\ \text{-}\frac{1}{2}\ -\quad +1/2$	**$2\ \frac{3}{2}\ \frac{1}{2}$**
$1\ 1\ 0\ +\quad +1$	$\frac{1}{2}\ 1\ \frac{1}{2}\ -^*\quad -1/2$	$0\ \frac{1}{2}\ \text{-}\frac{1}{2}\ +^*\quad +1/2$
$1\ \frac{1}{2}\ \frac{1}{2}$	**$\frac{3}{2}\ \frac{3}{2}\ 0$**	$1\ \frac{1}{2}\ \frac{1}{2}\ +\quad +1/2$
$0\ \text{-}\frac{1}{2}\ \frac{1}{2}\ +\quad +1/2$	$\text{-}\frac{1}{2}\ \frac{1}{2}\ 0\ -^*\quad -1/\sqrt{2}$	**$2\ \frac{3}{2}\ \frac{3}{2}$**
$1\ \frac{1}{2}\ \frac{1}{2}\ +^*\quad +1/2$		$0\ \text{-}\frac{1}{2}\ \frac{1}{2}\ +^*\quad +i/2$
		$1\ \frac{1}{2}\ \frac{1}{2}\ +\quad +i/2$

D_5–C_2 and C_{5v}–C_s 3jm Factors (cont.)

2 2 0	**2 2 1**	**$\frac{5}{2}\ 2\ \frac{1}{2}$**
$0\ 0\ 0\ +\quad +1/\sqrt{2}$	$0\ 0\ 0\ +^{*}\ -i/2$	$\frac{1}{2}\ 0\ {-\frac{1}{2}}\ -^{*}\ +1/\sqrt{2}$
$1\ 1\ 0\ +\quad +1/\sqrt{2}$	$1\ 0\ 1\ +^{*}\ +i/2$	$\frac{1}{2}\ 1\ \frac{1}{2}\ -\quad -1/\sqrt{2}$
$2\ 2\ \tilde{0}$	$1\ 1\ 0\ +^{*}\ +i/2$	**$\frac{5}{2}\ \frac{5}{2}\ \tilde{0}$**
$1\ 0\ 1\ -\quad -1/\sqrt{2}$	**$\frac{5}{2}\ \frac{3}{2}\ 1$**	$\frac{1}{2}\ \frac{1}{2}\ 1\ +\quad +1$
	$\frac{1}{2}\ \frac{1}{2}\ 1\ -^{*}\ -1/\sqrt{2}$	**${-\frac{5}{2}}\ \frac{5}{2}\ 0$**
	$\frac{1}{2}\ {-\frac{1}{2}}\ 0\ -\quad -1/\sqrt{2}$	${-\frac{1}{2}}\ \frac{1}{2}\ 0\ -^{*}\ -1$

D_5–C_5 and C_{5v}–C_5 2jm Factors (D_{5i}–C_{5i} = D_{5d}–C_{5i})

$0 \Rightarrow +\ 0$	$1 \Rightarrow +\ 1\ +\text{-}1$	$\frac{5}{2} \Rightarrow +\ \frac{5}{2}$
$\frac{1}{2} \Rightarrow +\ \frac{1}{2}\ -\text{-}\frac{1}{2}$	$\frac{3}{2} \Rightarrow +\ \frac{3}{2}\ -\text{-}\frac{3}{2}$	${-\frac{5}{2}} \Rightarrow -\ \frac{5}{2}$
$\tilde{0} \Rightarrow -\ 0$	$2 \Rightarrow +\ 2\ +\text{-}2$	

D_5–C_5 and C_{5v}–C_5 3jm Factors

0 0 0	**$\frac{3}{2}\ 1\ \frac{1}{2}$**	**$2\ 2\ \tilde{0}$**
$0\ 0\ 0\ +\quad +1$	$\frac{3}{2}\ {-1}\ {-\frac{1}{2}}\ -^{*}\ +1/\sqrt{2}$	${-2}\ 2\ 0\ -^{*}\ -1/\sqrt{2}$
$\frac{1}{2}\ \frac{1}{2}\ 0$	**$\frac{3}{2}\ \frac{3}{2}\ 0$**	**2 2 1**
${-\frac{1}{2}}\ \frac{1}{2}\ 0\ -^{*}\ -1/\sqrt{2}$	${-\frac{3}{2}}\ \frac{3}{2}\ 0\ -^{*}\ -1/\sqrt{2}$	$2\ 2\ 1\ +\quad +1/\sqrt{2}$
$\tilde{0}\ \frac{1}{2}\ \frac{1}{2}$	**$\frac{3}{2}\ \frac{3}{2}\ \tilde{0}$**	**$\frac{5}{2}\ \frac{3}{2}\ 1$**
$0\ {-\frac{1}{2}}\ \frac{1}{2}\ +\quad +1/\sqrt{2}$	${-\frac{3}{2}}\ \frac{3}{2}\ 0\ +\quad +1/\sqrt{2}$	$\frac{5}{2}\ \frac{3}{2}\ 1\ -\quad -1/\sqrt{2}$
$\tilde{0}\ \tilde{0}\ 0$	**2 1 1**	$\frac{5}{2}\ {-\frac{3}{2}}\ {-1}\ -^{*}\ -i/\sqrt{2}$
$0\ 0\ 0\ +\quad -1$	$2\ {-1}\ {-1}\ +\quad -1/\sqrt{2}$	**$\frac{5}{2}\ 2\ \frac{1}{2}$**
$1\ \frac{1}{2}\ \frac{1}{2}$	**$2\ \frac{3}{2}\ \frac{1}{2}$**	$\frac{5}{2}\ 2\ \frac{1}{2}\ -\quad +1/\sqrt{2}$
$1\ {-\frac{1}{2}}\ {-\frac{1}{2}}\ +\quad +1/\sqrt{2}$	$2\ {-\frac{3}{2}}\ {-\frac{1}{2}}\ +\quad +1/\sqrt{2}$	$\frac{5}{2}\ {-2}\ {-\frac{1}{2}}\ -^{*}\ +i/\sqrt{2}$
1 1 0	**$2\ \frac{3}{2}\ \frac{3}{2}$**	**$\frac{5}{2}\ \frac{5}{2}\ \tilde{0}$**
${-1}\ 1\ 0\ +\quad +1/\sqrt{2}$	$2\ \frac{3}{2}\ \frac{3}{2}\ +\quad -1/\sqrt{2}$	$\frac{5}{2}\ \frac{5}{2}\ 0\ +^{*}\ -i$
$1\ 1\ \tilde{0}$	**2 2 0**	**${-\frac{5}{2}}\ \frac{5}{2}\ 0$**
${-1}\ 1\ 0\ -^{*}\ -1/\sqrt{2}$	${-2}\ 2\ 0\ +\quad +1/\sqrt{2}$	$\frac{5}{2}\ \frac{5}{2}\ 0\ -^{*}\ -1$

D_{5d}–C_{5v} 2jm Factors

$0^{+} \Rightarrow +\ 0$	$\tilde{0}^{+} \Rightarrow +\ \tilde{0}$	$\frac{3}{2}^{+} \Rightarrow +\ \frac{3}{2}$	$\frac{5}{2}^{+} \Rightarrow +\ \frac{5}{2}$
$0^{-} \Rightarrow +\ \tilde{0}$	$\tilde{0}^{-} \Rightarrow +\ 0$	$\frac{3}{2}^{-} \Rightarrow +\ \frac{3}{2}$	$\frac{5}{2}^{-} \Rightarrow +\ {-\frac{5}{2}}$
$\frac{1}{2}^{+} \Rightarrow +\ \frac{1}{2}$	$1^{+} \Rightarrow +\ 1$	$2^{+} \Rightarrow +\ 2$	${-\frac{5}{2}}^{+} \Rightarrow +\ {-\frac{5}{2}}$
$\frac{1}{2}^{-} \Rightarrow +\ \frac{1}{2}$	$1^{-} \Rightarrow +\ 1$	$2^{-} \Rightarrow +\ 2$	${-\frac{5}{2}}^{-} \Rightarrow +\ \frac{5}{2}$

D_{5d}–C_{5v} 3jm Factors

$\mathbf{0^+\ 0^+\ 0^+}$	$\mathbf{1^-\ 1^+\ 0^-}$	$\mathbf{2^+\ 1^-\ 1^-}$	$\mathbf{2^-\ 2^+\ 1^-}$
$0\ \ 0\ \ 0\ \ +\ \ +1$	$1\ \ 1\ \ \tilde{0}\ \ -\ \ +1$	$2\ \ 1\ \ 1\ \ +\ \ -1$	$2\ \ 2\ \ 1\ \ +\ \ -1$
$\mathbf{0^-\ 0^-\ 0^+}$	$\mathbf{1^+\ 1^+\ \tilde{0}^+}$	$\mathbf{2^+\ \tfrac{3}{2}^+\ \tfrac{1}{2}^+}$	$\mathbf{\tfrac{5}{2}^+\ \tfrac{3}{2}^+\ 1^+}$
$\tilde{0}\ \ \tilde{0}\ \ 0\ \ +\ \ +1$	$1\ \ 1\ \ \tilde{0}\ \ +\ \ +1$	$2\ \ \tfrac{3}{2}\ \ \tfrac{1}{2}\ \ +\ \ +1$	$\tfrac{5}{2}\ \ \tfrac{3}{2}\ \ 1\ \ +\ \ +1$
$\mathbf{\tfrac{1}{2}^+\ \tfrac{1}{2}^+\ 0^+}$	$\mathbf{1^-\ 1^-\ \tilde{0}^+}$	$\mathbf{2^-\ \tfrac{3}{2}^-\ \tfrac{1}{2}^+}$	$\mathbf{\tfrac{5}{2}^-\ \tfrac{3}{2}^-\ 1^+}$
$\tfrac{1}{2}\ \ \tfrac{1}{2}\ \ 0\ \ +\ \ +1$	$1\ \ 1\ \ \tilde{0}\ \ +\ \ +1$	$2\ \ \tfrac{3}{2}\ \ \tfrac{1}{2}\ \ +\ \ +1$	$-\tfrac{5}{2}\ \ \tfrac{3}{2}\ \ 1\ \ +\ \ +1$
$\mathbf{\tfrac{1}{2}^-\ \tfrac{1}{2}^-\ 0^+}$	$\mathbf{1^-\ 1^+\ \tilde{0}^-}$	$\mathbf{2^-\ \tfrac{3}{2}^+\ \tfrac{1}{2}^-}$	$\mathbf{\tfrac{5}{2}^-\ \tfrac{3}{2}^+\ 1^-}$
$\tfrac{1}{2}\ \ \tfrac{1}{2}\ \ 0\ \ +\ \ +1$	$1\ \ 1\ \ 0\ \ -\ \ -1$	$2\ \ \tfrac{3}{2}\ \ \tfrac{1}{2}\ \ +\ \ +1$	$-\tfrac{5}{2}\ \ \tfrac{3}{2}\ \ 1\ \ +\ \ +1$
$\mathbf{\tfrac{1}{2}^-\ \tfrac{1}{2}^+\ 0^-}$	$\mathbf{\tfrac{3}{2}^+\ 1^+\ \tfrac{1}{2}^+}$	$\mathbf{2^+\ \tfrac{3}{2}^-\ \tfrac{1}{2}^-}$	$\mathbf{\tfrac{5}{2}^+\ \tfrac{3}{2}^-\ 1^-}$
$\tfrac{1}{2}\ \ \tfrac{1}{2}\ \ \tilde{0}\ \ -\ \ +1$	$\tfrac{3}{2}\ \ 1\ \ \tfrac{1}{2}\ \ +\ \ +1$	$2\ \ \tfrac{3}{2}\ \ \tfrac{1}{2}\ \ +\ \ -1$	$\tfrac{5}{2}\ \ \tfrac{3}{2}\ \ 1\ \ +\ \ -1$
$\mathbf{\tilde{0}^+\ \tfrac{1}{2}^+\ \tfrac{1}{2}^+}$	$\mathbf{\tfrac{3}{2}^-\ 1^-\ \tfrac{1}{2}^+}$	$\mathbf{2^+\ \tfrac{3}{2}^+\ \tfrac{3}{2}^+}$	$\mathbf{\tfrac{5}{2}^+\ 2^+\ \tfrac{1}{2}^+}$
$\tilde{0}\ \ \tfrac{1}{2}\ \ \tfrac{1}{2}\ \ +\ \ +1$	$\tfrac{3}{2}\ \ 1\ \ \tfrac{1}{2}\ \ +\ \ +1$	$2\ \ \tfrac{3}{2}\ \ \tfrac{3}{2}\ \ +\ \ +1$	$\tfrac{5}{2}\ \ 2\ \ \tfrac{1}{2}\ \ +\ \ +1$
$\mathbf{\tilde{0}^-\ \tfrac{1}{2}^-\ \tfrac{1}{2}^+}$	$\mathbf{\tfrac{3}{2}^-\ 1^+\ \tfrac{1}{2}^-}$	$\mathbf{2^-\ \tfrac{3}{2}^-\ \tfrac{3}{2}^+}$	$\mathbf{\tfrac{5}{2}^-\ 2^-\ \tfrac{1}{2}^+}$
$0\ \ \tfrac{1}{2}\ \ \tfrac{1}{2}\ \ -\ \ -1$	$\tfrac{3}{2}\ \ 1\ \ \tfrac{1}{2}\ \ +\ \ +1$	$2\ \ \tfrac{3}{2}\ \ \tfrac{3}{2}\ \ +\ \ -1$	$-\tfrac{5}{2}\ \ 2\ \ \tfrac{1}{2}\ \ +\ \ +1$
$\mathbf{\tilde{0}^+\ \tfrac{1}{2}^-\ \tfrac{1}{2}^-}$	$\mathbf{\tfrac{3}{2}^+\ 1^-\ \tfrac{1}{2}^-}$	$\mathbf{2^+\ \tfrac{3}{2}^-\ \tfrac{3}{2}^-}$	$\mathbf{\tfrac{5}{2}^-\ 2^+\ \tfrac{1}{2}^-}$
$\tilde{0}\ \ \tfrac{1}{2}\ \ \tfrac{1}{2}\ \ +\ \ +1$	$\tfrac{3}{2}\ \ 1\ \ \tfrac{1}{2}\ \ +\ \ -1$	$2\ \ \tfrac{3}{2}\ \ \tfrac{3}{2}\ \ +\ \ -1$	$-\tfrac{5}{2}\ \ 2\ \ \tfrac{1}{2}\ \ +\ \ +1$
$\mathbf{\tilde{0}^+\ \tilde{0}^+\ 0^+}$	$\mathbf{\tfrac{3}{2}^+\ \tfrac{3}{2}^+\ 0^+}$	$\mathbf{2^+\ 2^+\ 0^+}$	$\mathbf{\tfrac{5}{2}^+\ 2^-\ \tfrac{1}{2}^-}$
$\tilde{0}\ \ \tilde{0}\ \ 0\ \ +\ \ +1$	$\tfrac{3}{2}\ \ \tfrac{3}{2}\ \ 0\ \ +\ \ +1$	$2\ \ 2\ \ 0\ \ +\ \ +1$	$\tfrac{5}{2}\ \ 2\ \ \tfrac{1}{2}\ \ +\ \ -1$
$\mathbf{\tilde{0}^-\ \tilde{0}^-\ 0^+}$	$\mathbf{\tfrac{3}{2}^-\ \tfrac{3}{2}^-\ 0^+}$	$\mathbf{2^-\ 2^-\ 0^+}$	$\mathbf{\tfrac{5}{2}^+\ \tfrac{5}{2}^+\ \tilde{0}^+}$
$0\ \ 0\ \ 0\ \ +\ \ +1$	$\tfrac{3}{2}\ \ \tfrac{3}{2}\ \ 0\ \ +\ \ +1$	$2\ \ 2\ \ 0\ \ +\ \ +1$	$\tfrac{5}{2}\ \ \tfrac{5}{2}\ \ \tilde{0}\ \ +\ \ +1$
$\mathbf{\tilde{0}^-\ \tilde{0}^+\ 0^-}$	$\mathbf{\tfrac{3}{2}^-\ \tfrac{3}{2}^+\ 0^-}$	$\mathbf{2^-\ 2^+\ 0^-}$	$\mathbf{\tfrac{5}{2}^-\ \tfrac{5}{2}^-\ \tilde{0}^+}$
$0\ \ \tilde{0}\ \ \tilde{0}\ \ +\ \ +1$	$\tfrac{3}{2}\ \ \tfrac{3}{2}\ \ \tilde{0}\ \ -\ \ +1$	$2\ \ 2\ \ \tilde{0}\ \ -\ \ +1$	$-\tfrac{5}{2}\ \ -\tfrac{5}{2}\ \ \tilde{0}\ \ +\ \ +1$
$\mathbf{1^+\ \tfrac{1}{2}^+\ \tfrac{1}{2}^+}$	$\mathbf{\tfrac{3}{2}^+\ \tfrac{3}{2}^+\ \tilde{0}^+}$	$\mathbf{2^+\ 2^+\ \tilde{0}^+}$	$\mathbf{\tfrac{5}{2}^-\ \tfrac{5}{2}^+\ \tilde{0}^-}$
$1\ \ \tfrac{1}{2}\ \ \tfrac{1}{2}\ \ +\ \ +1$	$\tfrac{3}{2}\ \ \tfrac{3}{2}\ \ \tilde{0}\ \ +\ \ +1$	$2\ \ 2\ \ \tilde{0}\ \ +\ \ +1$	$-\tfrac{5}{2}\ \ \tfrac{5}{2}\ \ 0\ \ -\ \ -1$
$\mathbf{1^-\ \tfrac{1}{2}^-\ \tfrac{1}{2}^+}$	$\mathbf{\tfrac{3}{2}^-\ \tfrac{3}{2}^-\ \tilde{0}^+}$	$\mathbf{2^-\ 2^-\ \tilde{0}^+}$	$\mathbf{-\tfrac{5}{2}^+\ \tfrac{5}{2}^+\ 0^+}$
$1\ \ \tfrac{1}{2}\ \ \tfrac{1}{2}\ \ +\ \ +1$	$\tfrac{3}{2}\ \ \tfrac{3}{2}\ \ \tilde{0}\ \ +\ \ +1$	$2\ \ 2\ \ \tilde{0}\ \ +\ \ +1$	$-\tfrac{5}{2}\ \ \tfrac{5}{2}\ \ 0\ \ +\ \ +1$
$\mathbf{1^+\ \tfrac{1}{2}^-\ \tfrac{1}{2}^-}$	$\mathbf{\tfrac{3}{2}^-\ \tfrac{3}{2}^+\ \tilde{0}^-}$	$\mathbf{2^-\ 2^+\ \tilde{0}^-}$	$\mathbf{-\tfrac{5}{2}^-\ \tfrac{5}{2}^-\ 0^+}$
$1\ \ \tfrac{1}{2}\ \ \tfrac{1}{2}\ \ +\ \ -1$	$\tfrac{3}{2}\ \ \tfrac{3}{2}\ \ 0\ \ -\ \ -1$	$2\ \ 2\ \ 0\ \ -\ \ -1$	$\tfrac{5}{2}\ \ -\tfrac{5}{2}\ \ 0\ \ +\ \ +1$
$\mathbf{1^+\ 1^+\ 0^+}$	$\mathbf{2^+\ 1^+\ 1^+}$	$\mathbf{2^+\ 2^+\ 1^+}$	$\mathbf{-\tfrac{5}{2}^-\ \tfrac{5}{2}^+\ 0^-}$
$1\ \ 1\ \ 0\ \ +\ \ +1$	$2\ \ 1\ \ 1\ \ +\ \ +1$	$2\ \ 2\ \ 1\ \ +\ \ +1$	$\tfrac{5}{2}\ \ \tfrac{5}{2}\ \ \tilde{0}\ \ -\ \ +1$
$\mathbf{1^-\ 1^-\ 0^+}$	$\mathbf{2^-\ 1^-\ 1^+}$	$\mathbf{2^-\ 2^-\ 1^+}$	$\mathbf{-\tfrac{5}{2}^+\ \tfrac{5}{2}^-\ 0^-}$
$1\ \ 1\ \ 0\ \ +\ \ +1$	$2\ \ 1\ \ 1\ \ +\ \ +1$	$2\ \ 2\ \ 1\ \ +\ \ -1$	$-\tfrac{5}{2}\ \ -\tfrac{5}{2}\ \ \tilde{0}\ \ -\ \ -1$

D_6–C_6, C_{6v}–C_6, and D_{3h}–C_{3h} 2jm Factors (D_{6i}–C_{6i} = D_{6h}–C_{6h})

$0 \Rightarrow +\ 0$	$1 \Rightarrow +\ 1\ +\ \text{-}1$	$\tfrac{5}{2} \Rightarrow +\ \tfrac{5}{2}\ -\ \text{-}\tfrac{5}{2}$
$\tfrac{1}{2} \Rightarrow +\ \tfrac{1}{2}\ -\ \text{-}\tfrac{1}{2}$	$\tfrac{3}{2} \Rightarrow +\ \tfrac{3}{2}\ -\ \text{-}\tfrac{3}{2}$	$3 \Rightarrow +\ 3$
$\tilde{0} \Rightarrow -\ 0$	$2 \Rightarrow +\ 2\ +\ \text{-}2$	$\tilde{3} \Rightarrow -\ 3$

D_6–C_6, C_{6v}–C_6, and D_{3h}–C_{3h} 3jm Factors

0 0 0	**2 1 1**	**$\frac{5}{2}$ $\frac{5}{2}$ 1**
0 0 0 + +1	2 -1 -1 + $-1/\sqrt{2}$	$\frac{5}{2}$ $\frac{5}{2}$ 1 + $+1/\sqrt{2}$
$\frac{1}{2}$ $\frac{1}{2}$ 0	**2 $\frac{3}{2}$ $\frac{1}{2}$**	**3 $\frac{3}{2}$ $\frac{3}{2}$**
$-\frac{1}{2}$ $\frac{1}{2}$ 0 −* $-1/\sqrt{2}$	2 $-\frac{3}{2}$ $-\frac{1}{2}$ + $+1/\sqrt{2}$	3 $\frac{3}{2}$ $\frac{3}{2}$ + $+1/\sqrt{2}$
$\tilde{0}$ $\frac{1}{2}$ $\frac{1}{2}$	**2 2 0**	**3 2 1**
0 $-\frac{1}{2}$ $\frac{1}{2}$ + $+1/\sqrt{2}$	-2 2 0 + $+1/\sqrt{2}$	3 2 1 + $-1/\sqrt{2}$
$\tilde{0}$ $\tilde{0}$ 0	**2 2 $\tilde{0}$**	**3 $\frac{5}{2}$ $\frac{1}{2}$**
0 0 0 + -1	-2 2 0 −* $-1/\sqrt{2}$	3 $\frac{5}{2}$ $\frac{1}{2}$ + $+1/\sqrt{2}$
1 $\frac{1}{2}$ $\frac{1}{2}$	**2 2 2**	**3 3 0**
1 $-\frac{1}{2}$ $-\frac{1}{2}$ + $+1/\sqrt{2}$	2 2 2 + $+1/\sqrt{2}$	3 3 0 + +1
1 1 0	**$\frac{5}{2}$ $\frac{3}{2}$ 1**	**$\tilde{3}$ $\frac{3}{2}$ $\frac{3}{2}$**
-1 1 0 + $+1/\sqrt{2}$	$\frac{5}{2}$ $-\frac{3}{2}$ -1 −* $-1/\sqrt{2}$	3 $\frac{3}{2}$ $\frac{3}{2}$ +* $+1/\sqrt{2}$
1 1 $\tilde{0}$	**$\frac{5}{2}$ 2 $\frac{1}{2}$**	**$\tilde{3}$ 2 1**
-1 1 0 −* $-1/\sqrt{2}$	$\frac{5}{2}$ -2 $-\frac{1}{2}$ −* $+1/\sqrt{2}$	3 2 1 +* $-1/\sqrt{2}$
$\frac{3}{2}$ 1 $\frac{1}{2}$	**$\frac{5}{2}$ 2 $\frac{3}{2}$**	**$\tilde{3}$ $\frac{5}{2}$ $\frac{1}{2}$**
$\frac{3}{2}$ -1 $-\frac{1}{2}$ −* $+1/\sqrt{2}$	$\frac{5}{2}$ 2 $\frac{3}{2}$ + $-1/\sqrt{2}$	3 $\frac{5}{2}$ $\frac{1}{2}$ +* $+1/\sqrt{2}$
$\frac{3}{2}$ $\frac{3}{2}$ 0	**$\frac{5}{2}$ $\frac{5}{2}$ 0**	**$\tilde{3}$ 3 $\tilde{0}$**
$-\frac{3}{2}$ $\frac{3}{2}$ 0 −* $-1/\sqrt{2}$	$-\frac{5}{2}$ $\frac{5}{2}$ 0 −* $-1/\sqrt{2}$	3 3 0 − -1
$\frac{3}{2}$ $\frac{3}{2}$ $\tilde{0}$	**$\frac{5}{2}$ $\frac{5}{2}$ $\tilde{0}$**	**$\tilde{3}$ $\tilde{3}$ 0**
$-\frac{3}{2}$ $\frac{3}{2}$ 0 + $+1/\sqrt{2}$	$-\frac{5}{2}$ $\frac{5}{2}$ 0 + $+1/\sqrt{2}$	3 3 0 + -1

D_6–D_2, C_{6v}–C_{2v}, and D_{3h}–$C_{2v}^{\dagger}$ 2jm Factors (D_{6i}–D_{2i} = D_{6h}–D_{2h})

0 → + 0	1 → + 1 + $\tilde{1}$	$\frac{5}{2}$ → + $\frac{1}{2}$
$\frac{1}{2}$ → + $\frac{1}{2}$	$\frac{3}{2}$ → + $\frac{1}{2}$	3 → + 1
$\tilde{0}$ → + $\tilde{0}$	2 → + 0 + $\tilde{0}$	$\tilde{3}$ → + $\tilde{1}$

D_6–D_2, C_{6v}–C_{2v}, and D_{3h}–$C_{2v}^{\dagger}$ 3jm Factors

0 0 0	**1 1 0**	**2 1 1**
0 0 0 + +1	1 1 0 + $+1/\sqrt{2}$	0 1 1 + $+1/2$
$\frac{1}{2}$ $\frac{1}{2}$ 0	$\tilde{1}$ $\tilde{1}$ 0 + $+1/\sqrt{2}$	0 $\tilde{1}$ $\tilde{1}$ + $-1/2$
$\frac{1}{2}$ $\frac{1}{2}$ 0 + +1	**1 1 $\tilde{0}$**	$\tilde{0}$ $\tilde{1}$ 1 − $-1/2$
$\tilde{0}$ $\frac{1}{2}$ $\frac{1}{2}$	$\tilde{1}$ 1 $\tilde{0}$ + $+1/\sqrt{2}$	**2 $\frac{3}{2}$ $\frac{1}{2}$**
$\tilde{0}$ $\frac{1}{2}$ $\frac{1}{2}$ + +1	**$\frac{3}{2}$ 1 $\frac{1}{2}$**	0 $\frac{1}{2}$ $\frac{1}{2}$ − $+1/\sqrt{2}$
$\tilde{0}$ $\tilde{0}$ 0	$\frac{1}{2}$ 1 $\frac{1}{2}$ − $+1/\sqrt{2}$	$\tilde{0}$ $\frac{1}{2}$ $\frac{1}{2}$ + $+1/\sqrt{2}$
$\tilde{0}$ $\tilde{0}$ 0 + +1	$\frac{1}{2}$ $\tilde{1}$ $\frac{1}{2}$ − $-1/\sqrt{2}$	**2 2 0**
1 $\frac{1}{2}$ $\frac{1}{2}$	**$\frac{3}{2}$ $\frac{3}{2}$ 0**	0 0 0 + $+1/\sqrt{2}$
1 $\frac{1}{2}$ $\frac{1}{2}$ + $+1/\sqrt{2}$	$\frac{1}{2}$ $\frac{1}{2}$ 0 + +1	$\tilde{0}$ $\tilde{0}$ 0 + $+1/\sqrt{2}$
$\tilde{1}$ $\frac{1}{2}$ $\frac{1}{2}$ + $+1/\sqrt{2}$	**$\frac{3}{2}$ $\frac{3}{2}$ $\tilde{0}$**	**2 2 $\tilde{0}$**
	$\frac{1}{2}$ $\frac{1}{2}$ $\tilde{0}$ + -1	$\tilde{0}$ 0 $\tilde{0}$ − $-1/\sqrt{2}$

† For D_{3h}–C_{2v} the irreps 3 and $\tilde{3}$ are interchanged and the 3*jm*'s marked by a dagger (†) change sign.

D_6–D_2, C_{6v}–C_{2v}, and D_{3h}–$C_{2v}^{\dagger}$ 3jm Factors (cont.)

$2\ 2\ 2$		**$\frac{5}{2}\ \frac{5}{2}\ \tilde{0}$**		**$\tilde{3}\ \frac{3}{2}\ \frac{3}{2}$**	
$0\ 0\ 0$ +	$-1/2^{\dagger}$	$\frac{1}{2}\ \frac{1}{2}\ \tilde{0}$ +	$+1$	$\tilde{1}\ \frac{1}{2}\ \frac{1}{2}$ +	-1
$\tilde{0}\ \tilde{0}\ 0$ +	$+1/2^{\dagger}$	**$\frac{5}{2}\ \frac{5}{2}\ 1$**		**$\tilde{3}\ 2\ 1$**	
$\frac{5}{2}\ \frac{3}{2}\ 1$		$\frac{1}{2}\ \frac{1}{2}\ 1$ +	$+1/\sqrt{2}^{\dagger}$	$\tilde{1}\ 0\ \tilde{1}$ +	$+1/\sqrt{2}$
$\frac{1}{2}\ \frac{1}{2}\ 1$ −	$+1/\sqrt{2}$	$\frac{1}{2}\ \frac{1}{2}\ \tilde{1}$ +	$-1/\sqrt{2}^{\dagger}$	$\tilde{1}\ \tilde{0}\ 1$ −	$+1/\sqrt{2}$
$\frac{1}{2}\ \frac{1}{2}\ \tilde{1}$ −	$+1/\sqrt{2}$	**$3\ \frac{3}{2}\ \frac{3}{2}$**		**$\tilde{3}\ \frac{5}{2}\ \frac{1}{2}$**	
$\frac{5}{2}\ 2\ \frac{1}{2}$		$1\ \frac{1}{2}\ \frac{1}{2}$ +	$+1$	$\tilde{1}\ \frac{1}{2}\ \frac{1}{2}$ +	$+1$
$\frac{1}{2}\ 0\ \frac{1}{2}$ +	$+1/\sqrt{2}$	**$3\ 2\ 1$**		**$\tilde{3}\ 3\ \tilde{0}$**	
$\frac{1}{2}\ \tilde{0}\ \frac{1}{2}$ −	$+1/\sqrt{2}$	$1\ 0\ 1$ +	$+1/\sqrt{2}$	$\tilde{1}\ 1\ \tilde{0}$ +	$+1$
$\frac{5}{2}\ 2\ \frac{3}{2}$		$1\ \tilde{0}\ \tilde{1}$ −	$+1/\sqrt{2}$	**$\tilde{3}\ \tilde{3}\ 0$**	
$\frac{1}{2}\ 0\ \frac{1}{2}$ −	$-1/\sqrt{2}^{\dagger}$	**$3\ \frac{5}{2}\ \frac{1}{2}$**		$\tilde{1}\ \tilde{1}\ 0$ +	$+1$
$\frac{1}{2}\ \tilde{0}\ \frac{1}{2}$ +	$+1/\sqrt{2}^{\dagger}$	$1\ \frac{1}{2}\ \frac{1}{2}$ +	$+1$		
$\frac{5}{2}\ \frac{5}{2}\ 0$		**$3\ 3\ 0$**			
$\frac{1}{2}\ \frac{1}{2}\ 0$ +	$+1$	$1\ 1\ 0$ +	$+1$		

D_6–D_3, C_{6v}–C_{3v}, D_{3h}–D_3, and D_{3h}–$C_{3v}^{\dagger}$ 2jm Factors (D_{6i}–D_{3i} = D_{6h}–D_{3d})

$0 \rightarrow +\ 0$	$1 \rightarrow +\ 1$	$\frac{5}{2} \rightarrow -\ \frac{1}{2}$
$\frac{1}{2} \rightarrow +\ \frac{1}{2}$	$\frac{3}{2} \rightarrow +\ \frac{3}{2}\ + -\frac{3}{2}$	$3 \rightarrow -\ 0$
$\tilde{0} \rightarrow +\ \tilde{0}$	$2 \rightarrow -\ 1$	$\tilde{3} \rightarrow -\ \tilde{0}$

D_6–D_3, C_{6v}–C_{3v}, D_{3h}–D_3, and D_{3h}–$C_{3v}^{\dagger}$ 3jm Factors

$0\ 0\ 0$		**$2\ 1\ 1$**		**$\frac{5}{2}\ \frac{5}{2}\ 1$**	
$0\ 0\ 0$ +	$+1$	$1\ 1\ 1$ +*	$-i$	$\frac{1}{2}\ \frac{1}{2}\ 1$ +	$-1^{\dagger}$
$\frac{1}{2}\ \frac{1}{2}\ 0$		**$2\ \frac{3}{2}\ \frac{1}{2}$**		**$3\ \frac{3}{2}\ \frac{3}{2}$**	
$\frac{1}{2}\ \frac{1}{2}\ 0$ +	$+1$	$1\ \frac{3}{2}\ \frac{1}{2}$ −*	$+1/\sqrt{2}$	$0\ -\frac{3}{2}\ \frac{3}{2}$ −*	$-1/\sqrt{2}$
$\tilde{0}\ \frac{1}{2}\ \frac{1}{2}$		**$2\ 2\ 0$**		**$3\ 2\ 1$**	
$\tilde{0}\ \frac{1}{2}\ \frac{1}{2}$ +	$+1$	$1\ 1\ 0$ +	-1	$0\ 1\ 1$ +	-1
$\tilde{0}\ \tilde{0}\ 0$		**$2\ 2\ \tilde{0}$**		**$3\ \frac{5}{2}\ \frac{1}{2}$**	
$\tilde{0}\ \tilde{0}\ 0$ +	$+1$	$1\ 1\ \tilde{0}$ +	$+1$	$0\ \frac{1}{2}\ \frac{1}{2}$ −	$+1$
$1\ \frac{1}{2}\ \frac{1}{2}$		**$2\ 2\ 2$**		**$3\ 3\ 0$**	
$1\ \frac{1}{2}\ \frac{1}{2}$ +	$+1$	$1\ 1\ 1$ +*	$-i^{\dagger}$	$0\ 0\ 0$ +	-1
$1\ 1\ 0$		**$\frac{5}{2}\ \frac{3}{2}\ 1$**		**$\tilde{3}\ \frac{3}{2}\ \frac{3}{2}$**	
$1\ 1\ 0$ +	$+1$	$\frac{1}{2}\ \frac{3}{2}\ 1$ +*	$-1/\sqrt{2}$	$\tilde{0}\ \frac{3}{2}\ \frac{3}{2}$ +*	$+1/\sqrt{2}$
$1\ 1\ \tilde{0}$		**$\frac{5}{2}\ 2\ \frac{1}{2}$**		**$\tilde{3}\ 2\ 1$**	
$1\ 1\ \tilde{0}$ +	$+1$	$\frac{1}{2}\ 1\ \frac{1}{2}$ −	$+1$	$\tilde{0}\ 1\ 1$ −	-1
$\frac{3}{2}\ 1\ \frac{1}{2}$		**$\frac{5}{2}\ 2\ \frac{3}{2}$**		**$\tilde{3}\ \frac{5}{2}\ \frac{1}{2}$**	
$\frac{3}{2}\ 1\ \frac{1}{2}$ +	$+1/\sqrt{2}$	$\frac{1}{2}\ 1\ \frac{3}{2}$ −	$+1/\sqrt{2}^{\dagger}$	$\tilde{0}\ \frac{1}{2}\ \frac{1}{2}$ +	$+1$
$\frac{3}{2}\ \frac{3}{2}\ 0$		**$\frac{5}{2}\ \frac{5}{2}\ 0$**		**$\tilde{3}\ 3\ \tilde{0}$**	
$-\frac{3}{2}\ \frac{3}{2}\ 0$ +	$+1/\sqrt{2}$	$\frac{1}{2}\ \frac{1}{2}\ 0$ +	-1	$\tilde{0}\ 0\ \tilde{0}$ −	$+1$
$\frac{3}{2}\ \frac{3}{2}\ \tilde{0}$		**$\frac{5}{2}\ \frac{5}{2}\ \tilde{0}$**		**$\tilde{3}\ \tilde{3}\ 0$**	
$\frac{3}{2}\ \frac{3}{2}\ \tilde{0}$ +	$+1/\sqrt{2}$	$\frac{1}{2}\ \frac{1}{2}\ \tilde{0}$ +	$+1$	$\tilde{0}\ \tilde{0}\ 0$ +	-1

† For D_{3h}–C_{2v} and D_{3h}–C_{3v} the irreps 3 and $\tilde{3}$ are interchanged and the $3jm$'s marked by a dagger (†) change sign.

D_{6h}–C_{6v} 2jm Factors

$0^{+} \Rightarrow + \ 0$	$\tilde{0}^{-} \Rightarrow + \ 0$	$2^{+} \Rightarrow + \ 2$	$3^{-} \Rightarrow + \ \tilde{3}$
$0^{-} \Rightarrow + \ \tilde{0}$	$1^{+} \Rightarrow + \ 1$	$2^{-} \Rightarrow + \ 2$	$\tilde{3}^{+} \Rightarrow + \ \tilde{3}$
$\tfrac{1}{2}^{+} \Rightarrow + \ \tfrac{1}{2}$	$1^{-} \Rightarrow + \ 1$	$\tfrac{5}{2}^{+} \Rightarrow + \ \tfrac{5}{2}$	$\tilde{3}^{-} \Rightarrow + \ 3$
$\tfrac{1}{2}^{-} \Rightarrow + \ \tfrac{1}{2}$	$\tfrac{3}{2}^{+} \Rightarrow + \ \tfrac{3}{2}$	$\tfrac{5}{2}^{-} \Rightarrow + \ \tfrac{5}{2}$	
$\tilde{0}^{+} \Rightarrow + \ \tilde{0}$	$\tfrac{3}{2}^{-} \Rightarrow + \ \tfrac{3}{2}$	$3^{+} \Rightarrow + \ 3$	

D_{6h}–C_{6v} 3jm Factors

$0^{+}\ 0^{+}\ 0^{+}$	$1^{-}\ 1^{-}\ 0^{+}$	$2^{+}\ 1^{+}\ 1^{+}$	$\tfrac{5}{2}^{+}\ \tfrac{3}{2}^{+}\ 1^{+}$
$0\ 0\ 0\ \ +\ \ +1$	$1\ 1\ 0\ \ +\ \ +1$	$2\ 1\ 1\ \ +\ \ +1$	$\tfrac{5}{2}\ \tfrac{3}{2}\ 1\ \ +\ \ +1$
$0^{-}\ 0^{-}\ 0^{+}$	$1^{-}\ 1^{+}\ 0^{-}$	$2^{-}\ 1^{-}\ 1^{+}$	$\tfrac{5}{2}^{-}\ \tfrac{3}{2}^{-}\ 1^{+}$
$\tilde{0}\ \tilde{0}\ 0\ \ +\ \ +1$	$1\ 1\ \tilde{0}\ \ -\ \ +1$	$2\ 1\ 1\ \ +\ \ +1$	$\tfrac{5}{2}\ \tfrac{3}{2}\ 1\ \ +\ \ +1$
$\tfrac{1}{2}^{+}\ \tfrac{1}{2}^{+}\ 0^{+}$	$1^{+}\ 1^{+}\ \tilde{0}^{+}$	$2^{+}\ 1^{-}\ 1^{-}$	$\tfrac{5}{2}^{-}\ \tfrac{3}{2}^{+}\ 1^{-}$
$\tfrac{1}{2}\ \tfrac{1}{2}\ 0\ \ +\ \ +1$	$1\ 1\ \tilde{0}\ \ +\ \ +1$	$2\ 1\ 1\ \ +\ \ -1$	$\tfrac{5}{2}\ \tfrac{3}{2}\ 1\ \ +\ \ +1$
$\tfrac{1}{2}^{-}\ \tfrac{1}{2}^{-}\ 0^{+}$	$1^{-}\ 1^{-}\ \tilde{0}^{+}$	$2^{+}\ \tfrac{3}{2}^{+}\ \tfrac{1}{2}^{+}$	$\tfrac{5}{2}^{+}\ \tfrac{3}{2}^{-}\ 1^{-}$
$\tfrac{1}{2}\ \tfrac{1}{2}\ 0\ \ +\ \ +1$	$1\ 1\ \tilde{0}\ \ +\ \ +1$	$2\ \tfrac{3}{2}\ \tfrac{1}{2}\ \ +\ \ +1$	$\tfrac{5}{2}\ \tfrac{3}{2}\ 1\ \ +\ \ -1$
$\tfrac{1}{2}^{-}\ \tfrac{1}{2}^{+}\ 0^{-}$	$1^{-}\ 1^{+}\ \tilde{0}^{-}$	$2^{-}\ \tfrac{3}{2}^{-}\ \tfrac{1}{2}^{+}$	$\tfrac{5}{2}^{+}\ 2^{+}\ \tfrac{1}{2}^{+}$
$\tfrac{1}{2}\ \tfrac{1}{2}\ \tilde{0}\ \ -\ \ +1$	$1\ 1\ 0\ \ -\ \ -1$	$2\ \tfrac{3}{2}\ \tfrac{1}{2}\ \ +\ \ +1$	$\tfrac{5}{2}\ 2\ \tfrac{1}{2}\ \ +\ \ +1$
$\tilde{0}^{+}\ \tfrac{1}{2}^{+}\ \tfrac{1}{2}^{+}$	$\tfrac{3}{2}^{+}\ 1^{+}\ \tfrac{1}{2}^{+}$	$2^{-}\ \tfrac{3}{2}^{+}\ \tfrac{1}{2}^{-}$	$\tfrac{5}{2}^{-}\ 2^{-}\ \tfrac{1}{2}^{+}$
$\tilde{0}\ \tfrac{1}{2}\ \tfrac{1}{2}\ \ +\ \ +1$	$\tfrac{3}{2}\ 1\ \tfrac{1}{2}\ \ +\ \ +1$	$2\ \tfrac{3}{2}\ \tfrac{1}{2}\ \ +\ \ +1$	$\tfrac{5}{2}\ 2\ \tfrac{1}{2}\ \ +\ \ +1$
$\tilde{0}^{-}\ \tfrac{1}{2}^{-}\ \tfrac{1}{2}^{+}$	$\tfrac{3}{2}^{-}\ 1^{-}\ \tfrac{1}{2}^{+}$	$2^{+}\ \tfrac{3}{2}^{-}\ \tfrac{1}{2}^{-}$	$\tfrac{5}{2}^{-}\ 2^{+}\ \tfrac{1}{2}^{-}$
$0\ \tfrac{1}{2}\ \tfrac{1}{2}\ \ -\ \ -1$	$\tfrac{3}{2}\ 1\ \tfrac{1}{2}\ \ +\ \ +1$	$2\ \tfrac{3}{2}\ \tfrac{1}{2}\ \ +\ \ -1$	$\tfrac{5}{2}\ 2\ \tfrac{1}{2}\ \ +\ \ +1$
$\tilde{0}^{+}\ \tfrac{1}{2}^{-}\ \tfrac{1}{2}^{-}$	$\tfrac{3}{2}^{-}\ 1^{+}\ \tfrac{1}{2}^{-}$	$2^{+}\ 2^{+}\ 0^{+}$	$\tfrac{5}{2}^{+}\ 2^{-}\ \tfrac{1}{2}^{-}$
$\tilde{0}\ \tfrac{1}{2}\ \tfrac{1}{2}\ \ +\ \ +1$	$\tfrac{3}{2}\ 1\ \tfrac{1}{2}\ \ +\ \ +1$	$2\ 2\ 0\ \ +\ \ +1$	$\tfrac{5}{2}\ 2\ \tfrac{1}{2}\ \ +\ \ -1$
$\tilde{0}^{+}\ \tilde{0}^{+}\ 0^{+}$	$\tfrac{3}{2}^{+}\ 1^{-}\ \tfrac{1}{2}^{-}$	$2^{-}\ 2^{-}\ 0^{+}$	$\tfrac{5}{2}^{+}\ 2^{+}\ \tfrac{3}{2}^{+}$
$\tilde{0}\ \tilde{0}\ 0\ \ +\ \ +1$	$\tfrac{3}{2}\ 1\ \tfrac{1}{2}\ \ +\ \ -1$	$2\ 2\ 0\ \ +\ \ +1$	$\tfrac{5}{2}\ 2\ \tfrac{3}{2}\ \ +\ \ +1$
$\tilde{0}^{-}\ \tilde{0}^{-}\ 0^{+}$	$\tfrac{3}{2}^{+}\ \tfrac{3}{2}^{+}\ 0^{+}$	$2^{-}\ 2^{+}\ 0^{-}$	$\tfrac{5}{2}^{-}\ 2^{-}\ \tfrac{3}{2}^{+}$
$0\ 0\ 0\ \ +\ \ +1$	$\tfrac{3}{2}\ \tfrac{3}{2}\ 0\ \ +\ \ +1$	$2\ 2\ \tilde{0}\ \ -\ \ +1$	$\tfrac{5}{2}\ 2\ \tfrac{3}{2}\ \ +\ \ -1$
$\tilde{0}^{-}\ \tilde{0}^{+}\ 0^{-}$	$\tfrac{3}{2}^{-}\ \tfrac{3}{2}^{-}\ 0^{+}$	$2^{+}\ 2^{+}\ \tilde{0}^{+}$	$\tfrac{5}{2}^{-}\ 2^{+}\ \tfrac{3}{2}^{-}$
$0\ \tilde{0}\ \tilde{0}\ \ +\ \ +1$	$\tfrac{3}{2}\ \tfrac{3}{2}\ 0\ \ +\ \ +1$	$2\ 2\ \tilde{0}\ \ +\ \ +1$	$\tfrac{5}{2}\ 2\ \tfrac{3}{2}\ \ +\ \ -1$
$1^{+}\ \tfrac{1}{2}^{+}\ \tfrac{1}{2}^{+}$	$\tfrac{3}{2}^{-}\ \tfrac{3}{2}^{+}\ 0^{-}$	$2^{-}\ 2^{-}\ \tilde{0}^{+}$	$\tfrac{5}{2}^{+}\ 2^{-}\ \tfrac{3}{2}^{-}$
$1\ \tfrac{1}{2}\ \tfrac{1}{2}\ \ +\ \ +1$	$\tfrac{3}{2}\ \tfrac{3}{2}\ \tilde{0}\ \ -\ \ +1$	$2\ 2\ \tilde{0}\ \ +\ \ +1$	$\tfrac{5}{2}\ 2\ \tfrac{3}{2}\ \ +\ \ -1$
$1^{-}\ \tfrac{1}{2}^{-}\ \tfrac{1}{2}^{+}$	$\tfrac{3}{2}^{+}\ \tfrac{3}{2}^{+}\ \tilde{0}^{+}$	$2^{-}\ 2^{+}\ \tilde{0}^{-}$	$\tfrac{5}{2}^{+}\ \tfrac{5}{2}^{+}\ 0^{+}$
$1\ \tfrac{1}{2}\ \tfrac{1}{2}\ \ +\ \ +1$	$\tfrac{3}{2}\ \tfrac{3}{2}\ \tilde{0}\ \ +\ \ +1$	$2\ 2\ 0\ \ -\ \ -1$	$\tfrac{5}{2}\ \tfrac{5}{2}\ 0\ \ +\ \ +1$
$1^{+}\ \tfrac{1}{2}^{-}\ \tfrac{1}{2}^{-}$	$\tfrac{3}{2}^{-}\ \tfrac{3}{2}^{-}\ \tilde{0}^{+}$	$2^{+}\ 2^{+}\ 2^{+}$	$\tfrac{5}{2}^{-}\ \tfrac{5}{2}^{-}\ 0^{+}$
$1\ \tfrac{1}{2}\ \tfrac{1}{2}\ \ +\ \ -1$	$\tfrac{3}{2}\ \tfrac{3}{2}\ \tilde{0}\ \ +\ \ +1$	$2\ 2\ 2\ \ +\ \ +1$	$\tfrac{5}{2}\ \tfrac{5}{2}\ 0\ \ +\ \ +1$
$1^{+}\ 1^{+}\ 0^{+}$	$\tfrac{3}{2}^{-}\ \tfrac{3}{2}^{+}\ \tilde{0}^{-}$	$2^{-}\ 2^{-}\ 2^{+}$	$\tfrac{5}{2}^{-}\ \tfrac{5}{2}^{+}\ 0^{-}$
$1\ 1\ 0\ \ +\ \ +1$	$\tfrac{3}{2}\ \tfrac{3}{2}\ 0\ \ -\ \ -1$	$2\ 2\ 2\ \ +\ \ -1$	$\tfrac{5}{2}\ \tfrac{5}{2}\ \tilde{0}\ \ -\ \ +1$

D_{6h}–C_{6v} 3jm Factors (cont.)

$\frac{5}{2}^{+}\ \frac{5}{2}^{+}\ \tilde{0}^{+}$ $\frac{5}{2}\ \frac{5}{2}\ \tilde{0}\ \ +\ \ +1$	$3^{-}\ 2^{-}\ 1^{+}$ $\tilde{3}\ 2\ 1\ \ +\ \ +1$	$\tilde{3}^{+}\ \frac{3}{2}^{+}\ \frac{3}{2}^{+}$ $\tilde{3}\ \frac{3}{2}\ \frac{3}{2}\ \ +\ \ +1$	$\tilde{3}^{+}\ \frac{5}{2}^{-}\ \frac{1}{2}^{-}$ $\tilde{3}\ \frac{5}{2}\ \frac{1}{2}\ \ +\ \ -1$
$\frac{5}{2}^{-}\ \frac{5}{2}^{-}\ \tilde{0}^{+}$ $\frac{5}{2}\ \frac{5}{2}\ \tilde{0}\ \ +\ \ +1$	$3^{-}\ 2^{+}\ 1^{-}$ $\tilde{3}\ 2\ 1\ \ +\ \ +1$	$\tilde{3}^{-}\ \frac{3}{2}^{-}\ \frac{3}{2}^{+}$ $3\ \frac{3}{2}\ \frac{3}{2}\ \ +\ \ +1$	$\tilde{3}^{+}\ 3^{+}\ \tilde{0}^{+}$ $\tilde{3}\ 3\ \tilde{0}\ \ +\ \ +1$
$\frac{5}{2}^{-}\ \frac{5}{2}^{+}\ \tilde{0}^{-}$ $\frac{5}{2}\ \frac{5}{2}\ 0\ \ -\ \ -1$	$3^{+}\ 2^{-}\ 1^{-}$ $3\ 2\ 1\ \ +\ \ -1$	$\tilde{3}^{+}\ \frac{3}{2}^{-}\ \frac{3}{2}^{-}$ $\tilde{3}\ \frac{3}{2}\ \frac{3}{2}\ \ +\ \ -1$	$\tilde{3}^{-}\ 3^{-}\ \tilde{0}^{+}$ $3\ \tilde{3}\ \tilde{0}\ \ +\ \ +1$
$\frac{5}{2}^{+}\ \frac{5}{2}^{+}\ 1^{+}$ $\frac{5}{2}\ \frac{5}{2}\ 1\ \ +\ \ +1$	$3^{+}\ \frac{5}{2}^{+}\ \frac{1}{2}^{+}$ $3\ \frac{5}{2}\ \frac{1}{2}\ \ +\ \ +1$	$\tilde{3}^{+}\ 2^{+}\ 1^{+}$ $\tilde{3}\ 2\ 1\ \ +\ \ +1$	$\tilde{3}^{-}\ 3^{+}\ \tilde{0}^{-}$ $3\ 3\ 0\ \ -\ \ -1$
$\frac{5}{2}^{-}\ \frac{5}{2}^{-}\ 1^{+}$ $\frac{5}{2}\ \frac{5}{2}\ 1\ \ +\ \ -1$	$3^{-}\ \frac{5}{2}^{-}\ \frac{1}{2}^{+}$ $\tilde{3}\ \frac{5}{2}\ \frac{1}{2}\ \ +\ \ +1$	$\tilde{3}^{-}\ 2^{-}\ 1^{+}$ $3\ 2\ 1\ \ +\ \ +1$	$\tilde{3}^{+}\ 3^{-}\ \tilde{0}^{-}$ $\tilde{3}\ \tilde{3}\ 0\ \ -\ \ +1$
$\frac{5}{2}^{-}\ \frac{5}{2}^{+}\ 1^{-}$ $\frac{5}{2}\ \frac{5}{2}\ 1\ \ +\ \ -1$	$3^{-}\ \frac{5}{2}^{+}\ \frac{1}{2}^{-}$ $\tilde{3}\ \frac{5}{2}\ \frac{1}{2}\ \ +\ \ +1$	$\tilde{3}^{-}\ 2^{+}\ 1^{-}$ $3\ 2\ 1\ \ +\ \ +1$	$\tilde{3}^{+}\ \tilde{3}^{+}\ 0^{+}$ $\tilde{3}\ \tilde{3}\ 0\ \ +\ \ +1$
$3^{+}\ \frac{3}{2}^{+}\ \frac{3}{2}^{+}$ $3\ \frac{3}{2}\ \frac{3}{2}\ \ +\ \ +1$	$3^{+}\ \frac{5}{2}^{-}\ \frac{1}{2}^{-}$ $3\ \frac{5}{2}\ \frac{1}{2}\ \ +\ \ -1$	$\tilde{3}^{+}\ 2^{-}\ 1^{-}$ $\tilde{3}\ 2\ 1\ \ +\ \ -1$	$\tilde{3}^{-}\ \tilde{3}^{-}\ 0^{+}$ $3\ 3\ 0\ \ +\ \ +1$
$3^{-}\ \frac{3}{2}^{-}\ \frac{3}{2}^{+}$ $\tilde{3}\ \frac{3}{2}\ \frac{3}{2}\ \ +\ \ +1$	$3^{+}\ 3^{+}\ 0^{+}$ $3\ 3\ 0\ \ +\ \ +1$	$\tilde{3}^{+}\ \frac{5}{2}^{+}\ \frac{1}{2}^{+}$ $\tilde{3}\ \frac{5}{2}\ \frac{1}{2}\ \ +\ \ +1$	$\tilde{3}^{-}\ \tilde{3}^{+}\ 0^{-}$ $3\ \tilde{3}\ \tilde{0}\ \ -\ \ +1$
$3^{+}\ \frac{3}{2}^{-}\ \frac{3}{2}^{-}$ $3\ \frac{3}{2}\ \frac{3}{2}\ \ +\ \ -1$	$3^{-}\ 3^{-}\ 0^{+}$ $\tilde{3}\ \tilde{3}\ 0\ \ +\ \ +1$	$\tilde{3}^{-}\ \frac{5}{2}^{-}\ \frac{1}{2}^{+}$ $3\ \frac{5}{2}\ \frac{1}{2}\ \ +\ \ +1$	
$3^{+}\ 2^{+}\ 1^{+}$ $3\ 2\ 1\ \ +\ \ +1$	$3^{-}\ 3^{+}\ 0^{-}$ $\tilde{3}\ 3\ \tilde{0}\ \ -\ \ +1$	$\tilde{3}^{-}\ \frac{5}{2}^{+}\ \frac{1}{2}^{-}$ $3\ \frac{5}{2}\ \frac{1}{2}\ \ +\ \ +1$	

D_{6h}–D_{3h} 2jm Factors

$0^{+} \Rightarrow +\ 0$	$\tilde{0}^{-} \Rightarrow +\ \tilde{3}$	$2^{+} \Rightarrow +\ 2$	$3^{-} \Rightarrow +\ 0$
$0^{-} \Rightarrow +\ 3$	$1^{+} \Rightarrow +\ 1$	$2^{-} \Rightarrow +\ 1$	$\tilde{3}^{+} \Rightarrow +\ \tilde{3}$
$\frac{1}{2}^{+} \Rightarrow +\ \frac{1}{2}$	$1^{-} \Rightarrow +\ 2$	$\frac{5}{2}^{+} \Rightarrow +\ \frac{5}{2}$	$\tilde{3}^{-} \Rightarrow +\ \tilde{0}$
$\frac{1}{2}^{-} \Rightarrow +\ \frac{5}{2}$	$\frac{3}{2}^{+} \Rightarrow +\ \frac{3}{2}$	$\frac{5}{2}^{-} \Rightarrow +\ \frac{1}{2}$	
$\tilde{0}^{+} \Rightarrow +\ \tilde{0}$	$\frac{3}{2}^{-} \Rightarrow +\ \frac{3}{2}$	$3^{+} \Rightarrow +\ 3$	

D_{6h}–D_{3h} 3jm Factors

$0^{+}\ 0^{+}\ 0^{+}$ $0\ 0\ 0\ \ +\ \ +1$	$\tilde{0}^{+}\ \frac{1}{2}^{+}\ \frac{1}{2}^{+}$ $\tilde{0}\ \frac{1}{2}\ \frac{1}{2}\ \ +\ \ +1$	$\tilde{0}^{-}\ \tilde{0}^{+}\ 0^{-}$ $\tilde{3}\ \tilde{0}\ 3\ \ -\ \ +1$	$1^{-}\ 1^{-}\ 0^{+}$ $2\ 2\ 0\ \ +\ \ +1$
$0^{-}\ 0^{-}\ 0^{+}$ $3\ 3\ 0\ \ +\ \ +1$	$\tilde{0}^{-}\ \frac{1}{2}^{-}\ \frac{1}{2}^{+}$ $\tilde{3}\ \frac{5}{2}\ \frac{1}{2}\ \ +\ \ +1$	$1^{+}\ \frac{1}{2}^{+}\ \frac{1}{2}^{+}$ $1\ \frac{1}{2}\ \frac{1}{2}\ \ +\ \ +1$	$1^{-}\ 1^{+}\ 0^{-}$ $2\ 1\ 3\ \ +\ \ +1$
$\frac{1}{2}^{+}\ \frac{1}{2}^{+}\ 0^{+}$ $\frac{1}{2}\ \frac{1}{2}\ 0\ \ +\ \ +1$	$\tilde{0}^{+}\ \frac{1}{2}^{-}\ \frac{1}{2}^{-}$ $\tilde{0}\ \frac{5}{2}\ \frac{5}{2}\ \ +\ \ -1$	$1^{-}\ \frac{1}{2}^{-}\ \frac{1}{2}^{+}$ $2\ \frac{5}{2}\ \frac{1}{2}\ \ -\ \ -1$	$1^{+}\ 1^{+}\ \tilde{0}^{+}$ $1\ 1\ \tilde{0}\ \ +\ \ +1$
$\frac{1}{2}^{-}\ \frac{1}{2}^{-}\ 0^{+}$ $\frac{5}{2}\ \frac{5}{2}\ 0\ \ +\ \ +1$	$\tilde{0}^{+}\ \tilde{0}^{+}\ 0^{+}$ $\tilde{0}\ \tilde{0}\ 0\ \ +\ \ +1$	$1^{+}\ \frac{1}{2}^{-}\ \frac{1}{2}^{-}$ $1\ \frac{5}{2}\ \frac{5}{2}\ \ +\ \ +1$	$1^{-}\ 1^{-}\ \tilde{0}^{+}$ $2\ 2\ \tilde{0}\ \ +\ \ -1$
$\frac{1}{2}^{-}\ \frac{1}{2}^{+}\ 0^{-}$ $\frac{5}{2}\ \frac{1}{2}\ 3\ \ -\ \ +1$	$\tilde{0}^{-}\ \tilde{0}^{-}\ 0^{+}$ $\tilde{3}\ \tilde{3}\ 0\ \ +\ \ +1$	$1^{+}\ 1^{+}\ 0^{+}$ $1\ 1\ 0\ \ +\ \ +1$	$1^{-}\ 1^{+}\ \tilde{0}^{-}$ $2\ 1\ \tilde{3}\ \ -\ \ +1$

D_{6h}–D_{3h} 3jm Factors (cont.)

$\frac{3}{2}^{+}\ 1^{+}\ \frac{1}{2}^{+}$	**$2^{+}\ 2^{+}\ \tilde{0}^{+}$**	**$\frac{5}{2}^{+}\ \frac{5}{2}^{+}\ \tilde{0}^{+}$**	**$\tilde{3}^{+}\ \frac{3}{2}^{+}\ \frac{3}{2}^{+}$**
$\frac{3}{2}$ 1 $\frac{1}{2}$ + +1	2 2 $\tilde{0}$ + +1	$\frac{5}{2}$ $\frac{5}{2}$ $\tilde{0}$ + +1	$\tilde{3}$ $\frac{3}{2}$ $\frac{3}{2}$ + +1
$\frac{3}{2}^{-}\ 1^{-}\ \frac{1}{2}^{+}$	**$2^{-}\ 2^{-}\ \tilde{0}^{+}$**	**$\frac{5}{2}^{-}\ \frac{5}{2}^{-}\ \tilde{0}^{+}$**	**$\tilde{3}^{-}\ \frac{3}{2}^{-}\ \frac{3}{2}^{+}$**
$\frac{3}{2}$ 2 $\frac{1}{2}$ − −1	1 1 $\tilde{0}$ + −1	$\frac{1}{2}$ $\frac{1}{2}$ $\tilde{0}$ + −1	$\tilde{0}$ $\frac{3}{2}$ $\frac{3}{2}$ + +1
$\frac{3}{2}^{-}\ 1^{+}\ \frac{1}{2}^{-}$	**$2^{-}\ 2^{+}\ \tilde{0}^{-}$**	**$\frac{5}{2}^{-}\ \frac{5}{2}^{+}\ \tilde{0}^{-}$**	**$\tilde{3}^{+}\ \frac{3}{2}^{-}\ \frac{3}{2}^{-}$**
$\frac{3}{2}$ 1 $\frac{5}{2}$ + +1	1 2 $\tilde{3}$ − +1	$\frac{1}{2}$ $\frac{5}{2}$ $\tilde{3}$ + +1	$\tilde{3}$ $\frac{3}{2}$ $\frac{3}{2}$ + −1
$\frac{3}{2}^{+}\ 1^{-}\ \frac{1}{2}^{-}$	**$2^{+}\ 2^{+}\ 2^{+}$**	**$\frac{5}{2}^{+}\ \frac{5}{2}^{+}\ 1^{+}$**	**$\tilde{3}^{+}\ 2^{+}\ 1^{+}$**
$\frac{3}{2}$ 2 $\frac{5}{2}$ − −1	2 2 2 + +1	$\frac{5}{2}$ $\frac{5}{2}$ 1 + +1	$\tilde{3}$ 2 1 + +1
$\frac{3}{2}^{+}\ \frac{3}{2}^{+}\ 0^{+}$	**$2^{-}\ 2^{-}\ 2^{+}$**	**$\frac{5}{2}^{-}\ \frac{5}{2}^{-}\ 1^{+}$**	**$\tilde{3}^{-}\ 2^{-}\ 1^{+}$**
$\frac{3}{2}$ $\frac{3}{2}$ 0 + +1	1 1 2 + −1	$\frac{1}{2}$ $\frac{1}{2}$ 1 + +1	$\tilde{0}$ 1 1 − −1
$\frac{3}{2}^{-}\ \frac{3}{2}^{-}\ 0^{+}$	**$\frac{5}{2}^{+}\ \frac{3}{2}^{+}\ 1^{+}$**	**$\frac{5}{2}^{-}\ \frac{5}{2}^{+}\ 1^{-}$**	**$\tilde{3}^{-}\ 2^{+}\ 1^{-}$**
$\frac{3}{2}$ $\frac{3}{2}$ 0 + +1	$\frac{5}{2}$ $\frac{3}{2}$ 1 + +1	$\frac{1}{2}$ $\frac{5}{2}$ 2 − −1	$\tilde{0}$ 2 2 − +1
$\frac{3}{2}^{-}\ \frac{3}{2}^{+}\ 0^{-}$	**$\frac{5}{2}^{-}\ \frac{3}{2}^{-}\ 1^{+}$**	**$3^{+}\ \frac{3}{2}^{+}\ \frac{3}{2}^{+}$**	**$\tilde{3}^{+}\ 2^{-}\ 1^{-}$**
$\frac{3}{2}$ $\frac{3}{2}$ 3 − +1	$\frac{1}{2}$ $\frac{3}{2}$ 1 + +1	3 $\frac{3}{2}$ $\frac{3}{2}$ + +1	$\tilde{3}$ 1 2 + −1
$\frac{3}{2}^{+}\ \frac{3}{2}^{+}\ \tilde{0}^{+}$	**$\frac{5}{2}^{-}\ \frac{3}{2}^{+}\ 1^{-}$**	**$3^{-}\ \frac{3}{2}^{-}\ \frac{3}{2}^{+}$**	**$\tilde{3}^{+}\ \frac{5}{2}^{+}\ \frac{1}{2}^{+}$**
$\frac{3}{2}$ $\frac{3}{2}$ $\tilde{0}$ + +1	$\frac{1}{2}$ $\frac{3}{2}$ 2 − +1	0 $\frac{3}{2}$ $\frac{3}{2}$ − −1	$\tilde{3}$ $\frac{5}{2}$ $\frac{1}{2}$ + +1
$\frac{3}{2}^{-}\ \frac{3}{2}^{-}\ \tilde{0}^{+}$	**$\frac{5}{2}^{+}\ \frac{3}{2}^{-}\ 1^{-}$**	**$3^{+}\ \frac{3}{2}^{-}\ \frac{3}{2}^{-}$**	**$\tilde{3}^{-}\ \frac{5}{2}^{-}\ \frac{1}{2}^{+}$**
$\frac{3}{2}$ $\frac{3}{2}$ $\tilde{0}$ + −1	$\frac{5}{2}$ $\frac{3}{2}$ 2 − +1	3 $\frac{3}{2}$ $\frac{3}{2}$ + +1	$\tilde{0}$ $\frac{1}{2}$ $\frac{1}{2}$ + +1
$\frac{3}{2}^{-}\ \frac{3}{2}^{+}\ \tilde{0}^{-}$	**$\frac{5}{2}^{+}\ 2^{+}\ \frac{1}{2}^{+}$**	**$3^{+}\ 2^{+}\ 1^{+}$**	**$\tilde{3}^{-}\ \frac{5}{2}^{+}\ \frac{1}{2}^{-}$**
$\frac{3}{2}$ $\frac{3}{2}$ $\tilde{3}$ + +1	$\frac{5}{2}$ 2 $\frac{1}{2}$ + +1	3 2 1 + +1	$\tilde{0}$ $\frac{5}{2}$ $\frac{5}{2}$ + +1
$2^{+}\ 1^{+}\ 1^{+}$	**$\frac{5}{2}^{-}\ 2^{-}\ \frac{1}{2}^{+}$**	**$3^{-}\ 2^{-}\ 1^{+}$**	**$\tilde{3}^{+}\ \frac{5}{2}^{-}\ \frac{1}{2}^{-}$**
2 1 1 + +1	$\frac{1}{2}$ 1 $\frac{1}{2}$ − −1	0 1 1 + +1	$\tilde{3}$ $\frac{1}{2}$ $\frac{5}{2}$ + −1
$2^{-}\ 1^{-}\ 1^{+}$	**$\frac{5}{2}^{-}\ 2^{+}\ \frac{1}{2}^{-}$**	**$3^{-}\ 2^{+}\ 1^{-}$**	**$\tilde{3}^{+}\ 3^{+}\ \tilde{0}^{+}$**
1 2 1 + +1	$\frac{1}{2}$ 2 $\frac{5}{2}$ + +1	0 2 2 + +1	$\tilde{3}$ 3 $\tilde{0}$ + +1
$2^{+}\ 1^{-}\ 1^{-}$	**$\frac{5}{2}^{+}\ 2^{-}\ \frac{1}{2}^{-}$**	**$3^{+}\ 2^{-}\ 1^{-}$**	**$\tilde{3}^{-}\ 3^{-}\ \tilde{0}^{+}$**
2 2 2 + −1	$\frac{5}{2}$ 1 $\frac{5}{2}$ − +1	3 1 2 + +1	$\tilde{0}$ 0 $\tilde{0}$ − +1
$2^{+}\ \frac{3}{2}^{+}\ \frac{1}{2}^{+}$	**$\frac{5}{2}^{+}\ 2^{+}\ \frac{3}{2}^{+}$**	**$3^{+}\ \frac{5}{2}^{+}\ \frac{1}{2}^{+}$**	**$\tilde{3}^{-}\ 3^{+}\ \tilde{0}^{-}$**
2 $\frac{3}{2}$ $\frac{1}{2}$ + +1	$\frac{5}{2}$ 2 $\frac{3}{2}$ + +1	3 $\frac{5}{2}$ $\frac{1}{2}$ + +1	$\tilde{0}$ 3 $\tilde{3}$ + +1
$2^{-}\ \frac{3}{2}^{-}\ \frac{1}{2}^{+}$	**$\frac{5}{2}^{-}\ 2^{-}\ \frac{3}{2}^{+}$**	**$3^{-}\ \frac{5}{2}^{-}\ \frac{1}{2}^{+}$**	**$\tilde{3}^{+}\ 3^{-}\ \tilde{0}^{-}$**
1 $\frac{3}{2}$ $\frac{1}{2}$ − −1	$\frac{1}{2}$ 1 $\frac{3}{2}$ − −1	0 $\frac{1}{2}$ $\frac{1}{2}$ − −1	$\tilde{3}$ 0 $\tilde{3}$ − −1
$2^{-}\ \frac{3}{2}^{+}\ \frac{1}{2}^{-}$	**$\frac{5}{2}^{-}\ 2^{+}\ \frac{3}{2}^{-}$**	**$3^{-}\ \frac{5}{2}^{+}\ \frac{1}{2}^{-}$**	**$\tilde{3}^{+}\ \tilde{3}^{+}\ 0^{+}$**
1 $\frac{3}{2}$ $\frac{5}{2}$ − +1	$\frac{1}{2}$ 2 $\frac{3}{2}$ + −1	0 $\frac{5}{2}$ $\frac{5}{2}$ − +1	$\tilde{3}$ $\tilde{3}$ 0 + +1
$2^{+}\ \frac{3}{2}^{-}\ \frac{1}{2}^{-}$	**$\frac{5}{2}^{+}\ 2^{-}\ \frac{3}{2}^{-}$**	**$3^{+}\ \frac{5}{2}^{-}\ \frac{1}{2}^{-}$**	**$\tilde{3}^{-}\ \tilde{3}^{-}\ 0^{+}$**
2 $\frac{3}{2}$ $\frac{5}{2}$ + −1	$\frac{5}{2}$ 1 $\frac{3}{2}$ − +1	3 $\frac{1}{2}$ $\frac{5}{2}$ + +1	$\tilde{0}$ $\tilde{0}$ 0 + +1
$2^{+}\ 2^{+}\ 0^{+}$	**$\frac{5}{2}^{+}\ \frac{5}{2}^{+}\ 0^{+}$**	**$3^{+}\ 3^{+}\ 0^{+}$**	**$\tilde{3}^{-}\ \tilde{3}^{+}\ 0^{-}$**
2 2 0 + +1	$\frac{5}{2}$ $\frac{5}{2}$ 0 + +1	3 3 0 + +1	$\tilde{0}$ $\tilde{3}$ 3 − +1
$2^{-}\ 2^{-}\ 0^{+}$	**$\frac{5}{2}^{-}\ \frac{5}{2}^{-}\ 0^{+}$**	**$3^{-}\ 3^{-}\ 0^{+}$**	
1 1 0 + +1	$\frac{1}{2}$ $\frac{1}{2}$ 0 + +1	0 0 0 + +1	
$2^{-}\ 2^{+}\ 0^{-}$	**$\frac{5}{2}^{-}\ \frac{5}{2}^{+}\ 0^{-}$**	**$3^{-}\ 3^{+}\ 0^{-}$**	
1 2 3 + +1	$\frac{1}{2}$ $\frac{5}{2}$ 3 − +1	0 3 3 + +1	

D_∞–D_4 and $C_{\infty v}$–C_{4v} 2jm Factors ($D_{\infty i}$–D_{4i} = $D_{\infty h}$–D_{4h})

$0 \Rightarrow +\ 0$	$\frac{5}{2} \Rightarrow +\ \frac{3}{2}$	$\frac{11}{2} \Rightarrow +\ \frac{3}{2}$
$\frac{1}{2} \Rightarrow +\ \frac{1}{2}$	$3 \Rightarrow +\ 1$	$6 \Rightarrow +\ 2 +\ \tilde{2}$
$\tilde{0} \Rightarrow +\ \tilde{0}$	$\frac{7}{2} \Rightarrow +\ \frac{1}{2}$	$\frac{13}{2} \Rightarrow +\ \frac{3}{2}$
$1 \Rightarrow +\ 1$	$4 \Rightarrow +\ 0 +\ \tilde{0}$	$7 \Rightarrow +\ 1$
$\frac{3}{2} \Rightarrow +\ \frac{3}{2}$	$\frac{9}{2} \Rightarrow +\ \frac{1}{2}$	$\frac{15}{2} \Rightarrow +\ \frac{1}{2}$
$2 \Rightarrow +\ 2 +\ \tilde{2}$	$5 \Rightarrow +\ 1$	$8 \Rightarrow +\ 0 +\ \tilde{0}$

D_∞–D_4 and $C_{\infty v}$–C_{4v} 3jm Factors

$0\ 0\ 0$
$0\ 0\ 0\ +\quad +1$
$\frac{1}{2}\ \frac{1}{2}\ 0$
$\frac{1}{2}\ \frac{1}{2}\ 0\ +\quad +1$
$\tilde{0}\ \frac{1}{2}\ \frac{1}{2}$
$\tilde{0}\ \frac{1}{2}\ \frac{1}{2}\ +\quad +1$
$\tilde{0}\ \tilde{0}\ 0$
$\tilde{0}\ \tilde{0}\ 0\ +\quad +1$
$1\ \frac{1}{2}\ \frac{1}{2}$
$1\ \frac{1}{2}\ \frac{1}{2}\ +\quad +1$
$1\ 1\ 0$
$1\ 1\ 0\ +\quad +1$
$1\ 1\ \tilde{0}$
$1\ 1\ \tilde{0}\ +\quad +1$
$\frac{3}{2}\ 1\ \frac{1}{2}$
$\frac{3}{2}\ 1\ \frac{1}{2}\ +\quad +1$
$\frac{3}{2}\ \frac{3}{2}\ 0$
$\frac{3}{2}\ \frac{3}{2}\ 0\ +\quad +1$
$\frac{3}{2}\ \frac{3}{2}\ \tilde{0}$
$\frac{3}{2}\ \frac{3}{2}\ \tilde{0}\ +\quad +1$
$2\ 1\ 1$
$2\ 1\ 1\ +\quad +1/\sqrt{2}$
$\tilde{2}\ 1\ 1\ +\quad +1/\sqrt{2}$
$2\ \frac{3}{2}\ \frac{1}{2}$
$2\ \frac{3}{2}\ \frac{1}{2}\ +\quad +1/\sqrt{2}$
$\tilde{2}\ \frac{3}{2}\ \frac{1}{2}\ +\quad +1/\sqrt{2}$
$2\ 2\ 0$
$2\ 2\ 0\ +\quad +1/\sqrt{2}$
$\tilde{2}\ \tilde{2}\ 0\ +\quad +1/\sqrt{2}$
$2\ 2\ \tilde{0}$
$\tilde{2}\ 2\ \tilde{0}\ +\quad +1/\sqrt{2}$

$\frac{5}{2}\ \frac{3}{2}\ 1$
$\frac{3}{2}\ \frac{3}{2}\ 1\ -\quad -1$
$\frac{5}{2}\ 2\ \frac{1}{2}$
$\frac{3}{2}\ 2\ \frac{1}{2}\ -\quad +1/\sqrt{2}$
$\frac{3}{2}\ \tilde{2}\ \frac{1}{2}\ -\quad -1/\sqrt{2}$
$\frac{5}{2}\ \frac{5}{2}\ 0$
$\frac{3}{2}\ \frac{3}{2}\ 0\ +\quad +1$
$\frac{5}{2}\ \frac{5}{2}\ \tilde{0}$
$\frac{3}{2}\ \frac{3}{2}\ \tilde{0}\ +\quad -1$
$3\ \frac{3}{2}\ \frac{3}{2}$
$1\ \frac{3}{2}\ \frac{3}{2}\ +\quad +1$
$3\ 2\ 1$
$1\ 2\ 1\ +\quad +1/\sqrt{2}$
$1\ \tilde{2}\ 1\ +\quad -1/\sqrt{2}$
$3\ \frac{5}{2}\ \frac{1}{2}$
$1\ \frac{3}{2}\ \frac{1}{2}\ -\quad +1$
$3\ 3\ 0$
$1\ 1\ 0\ +\quad +1$
$3\ 3\ \tilde{0}$
$1\ 1\ \tilde{0}\ +\quad -1$
$\frac{7}{2}\ 2\ \frac{3}{2}$
$\frac{1}{2}\ 2\ \frac{3}{2}\ -\quad +1/\sqrt{2}$
$\frac{1}{2}\ \tilde{2}\ \frac{3}{2}\ -\quad -1/\sqrt{2}$
$\frac{7}{2}\ \frac{5}{2}\ 1$
$\frac{1}{2}\ \frac{3}{2}\ 1\ +\quad +1$
$\frac{7}{2}\ 3\ \frac{1}{2}$
$\frac{1}{2}\ 1\ \frac{1}{2}\ -\quad +1$
$\frac{7}{2}\ \frac{7}{2}\ 0$
$\frac{1}{2}\ \frac{1}{2}\ 0\ +\quad +1$
$\frac{7}{2}\ \frac{7}{2}\ \tilde{0}$
$\frac{1}{2}\ \frac{1}{2}\ \tilde{0}\ +\quad -1$

$4\ 2\ 2$
$0\ 2\ 2\ +\quad +1/2$
$0\ \tilde{2}\ \tilde{2}\ +\quad -1/2$
$\tilde{0}\ \tilde{2}\ 2\ -\quad -1/2$
$4\ \frac{5}{2}\ \frac{3}{2}$
$0\ \frac{3}{2}\ \frac{3}{2}\ -\quad +1/\sqrt{2}$
$\tilde{0}\ \frac{3}{2}\ \frac{3}{2}\ +\quad +1/\sqrt{2}$
$4\ 3\ 1$
$0\ 1\ 1\ +\quad +1/\sqrt{2}$
$\tilde{0}\ 1\ 1\ -\quad +1/\sqrt{2}$
$4\ \frac{7}{2}\ \frac{1}{2}$
$0\ \frac{1}{2}\ \frac{1}{2}\ -\quad +1/\sqrt{2}$
$\tilde{0}\ \frac{1}{2}\ \frac{1}{2}\ +\quad +1/\sqrt{2}$
$4\ 4\ 0$
$0\ 0\ 0\ +\quad +1/\sqrt{2}$
$\tilde{0}\ \tilde{0}\ 0\ +\quad +1/\sqrt{2}$
$4\ 4\ \tilde{0}$
$\tilde{0}\ 0\ \tilde{0}\ -\quad -1/\sqrt{2}$
$\frac{9}{2}\ \frac{5}{2}\ 2$
$\frac{1}{2}\ \frac{3}{2}\ 2\ -\quad +1/\sqrt{2}$
$\frac{1}{2}\ \frac{3}{2}\ \tilde{2}\ -\quad +1/\sqrt{2}$
$\frac{9}{2}\ 3\ \frac{3}{2}$
$\frac{1}{2}\ 1\ \frac{3}{2}\ +\quad +1$
$\frac{9}{2}\ \frac{7}{2}\ 1$
$\frac{1}{2}\ \frac{1}{2}\ 1\ -\quad +1$
$\frac{9}{2}\ 4\ \frac{1}{2}$
$\frac{1}{2}\ 0\ \frac{1}{2}\ +\quad +1/\sqrt{2}$
$\frac{1}{2}\ \tilde{0}\ \frac{1}{2}\ -\quad +1/\sqrt{2}$
$\frac{9}{2}\ \frac{9}{2}\ 0$
$\frac{1}{2}\ \frac{1}{2}\ 0\ +\quad +1$

$D_\infty - D_4$ and $C_{\infty v} - C_{4v}$ 3jm Factors (cont.)

$\frac{9}{2}\ \frac{9}{2}\ \tilde{0}$
$\frac{1}{2}\ \frac{1}{2}\ \tilde{0}\ +\quad +1$
$5\ \frac{5}{2}\ \frac{5}{2}$
$1\ \frac{3}{2}\ \frac{3}{2}\ +\quad +1$
$5\ 3\ 2$
$1\ 1\ 2\ +\quad +1/\sqrt{2}$
$1\ 1\ \tilde{2}\ +\quad +1/\sqrt{2}$
$5\ \frac{7}{2}\ \frac{3}{2}$
$1\ \frac{1}{2}\ \frac{3}{2}\ -\quad +1$
$5\ 4\ 1$
$1\ 0\ 1\ +\quad +1/\sqrt{2}$
$1\ \tilde{0}\ 1\ -\quad +1/\sqrt{2}$
$5\ \frac{9}{2}\ \frac{1}{2}$
$1\ \frac{1}{2}\ \frac{1}{2}\ +\quad +1$
$5\ 5\ 0$
$1\ 1\ 0\ +\quad +1$
$5\ 5\ \tilde{0}$
$1\ 1\ \tilde{0}\ +\quad +1$
$\frac{11}{2}\ 3\ \frac{5}{2}$
$\frac{3}{2}\ 1\ \frac{3}{2}\ -\quad -1$
$\frac{11}{2}\ \frac{7}{2}\ 2$
$\frac{3}{2}\ \frac{1}{2}\ 2\ -\quad +1/\sqrt{2}$
$\frac{3}{2}\ \frac{1}{2}\ \tilde{2}\ -\quad +1/\sqrt{2}$
$\frac{11}{2}\ 4\ \frac{3}{2}$
$\frac{3}{2}\ 0\ \frac{3}{2}\ +\quad +1/\sqrt{2}$
$\frac{3}{2}\ \tilde{0}\ \frac{3}{2}\ -\quad +1/\sqrt{2}$
$\frac{11}{2}\ \frac{9}{2}\ 1$
$\frac{3}{2}\ \frac{1}{2}\ 1\ +\quad +1$
$\frac{11}{2}\ 5\ \frac{1}{2}$
$\frac{3}{2}\ 1\ \frac{1}{2}\ +\quad +1$
$\frac{11}{2}\ \frac{11}{2}\ 0$
$\frac{3}{2}\ \frac{3}{2}\ 0\ +\quad +1$
$\frac{11}{2}\ \frac{11}{2}\ \tilde{0}$
$\frac{3}{2}\ \frac{3}{2}\ \tilde{0}\ +\quad +1$
$6\ 3\ 3$
$2\ 1\ 1\ +\quad +1/\sqrt{2}$
$\tilde{2}\ 1\ 1\ +\quad -1/\sqrt{2}$
$6\ \frac{7}{2}\ \frac{5}{2}$
$2\ \frac{1}{2}\ \frac{3}{2}\ +\quad +1/\sqrt{2}$
$\tilde{2}\ \frac{1}{2}\ \frac{3}{2}\ +\quad -1/\sqrt{2}$

$6\ 4\ 2$
$2\ 0\ 2\ +\quad +1/2$
$2\ \tilde{0}\ \tilde{2}\ -\quad +1/2$
$\tilde{2}\ 0\ \tilde{2}\ +\quad +1/2$
$\tilde{2}\ \tilde{0}\ 2\ -\quad +1/2$
$6\ \frac{9}{2}\ \frac{3}{2}$
$2\ \frac{1}{2}\ \frac{3}{2}\ +\quad +1/\sqrt{2}$
$\tilde{2}\ \frac{1}{2}\ \frac{3}{2}\ +\quad +1/\sqrt{2}$
$6\ 5\ 1$
$2\ 1\ 1\ +\quad +1/\sqrt{2}$
$\tilde{2}\ 1\ 1\ +\quad +1/\sqrt{2}$
$6\ \frac{11}{2}\ \frac{1}{2}$
$2\ \frac{3}{2}\ \frac{1}{2}\ +\quad +1/\sqrt{2}$
$\tilde{2}\ \frac{3}{2}\ \frac{1}{2}\ +\quad +1/\sqrt{2}$
$6\ 6\ 0$
$2\ 2\ 0\ +\quad +1/\sqrt{2}$
$\tilde{2}\ \tilde{2}\ 0\ +\quad +1/\sqrt{2}$
$6\ 6\ \tilde{0}$
$\tilde{2}\ 2\ \tilde{0}\ +\quad +1/\sqrt{2}$
$\frac{13}{2}\ \frac{7}{2}\ 3$
$\frac{3}{2}\ \frac{1}{2}\ 1\ +\quad +1$
$\frac{13}{2}\ 4\ \frac{5}{2}$
$\frac{3}{2}\ 0\ \frac{3}{2}\ +\quad +1/\sqrt{2}$
$\frac{3}{2}\ \tilde{0}\ \frac{3}{2}\ -\quad -1/\sqrt{2}$
$\frac{13}{2}\ \frac{9}{2}\ 2$
$\frac{3}{2}\ \frac{1}{2}\ 2\ -\quad -1/\sqrt{2}$
$\frac{3}{2}\ \frac{1}{2}\ \tilde{2}\ -\quad +1/\sqrt{2}$
$\frac{13}{2}\ 5\ \frac{3}{2}$
$\frac{3}{2}\ 1\ \frac{3}{2}\ -\quad +1$
$\frac{13}{2}\ \frac{11}{2}\ 1$
$\frac{3}{2}\ \frac{3}{2}\ 1\ -\quad -1$
$\frac{13}{2}\ 6\ \frac{1}{2}$
$\frac{3}{2}\ 2\ \frac{1}{2}\ -\quad +1/\sqrt{2}$
$\frac{3}{2}\ \tilde{2}\ \frac{1}{2}\ -\quad -1/\sqrt{2}$
$\frac{13}{2}\ \frac{13}{2}\ 0$
$\frac{3}{2}\ \frac{3}{2}\ 0\ +\quad +1$
$\frac{13}{2}\ \frac{13}{2}\ \tilde{0}$
$\frac{3}{2}\ \frac{3}{2}\ \tilde{0}\ +\quad -1$
$7\ \frac{7}{2}\ \frac{7}{2}$
$1\ \frac{1}{2}\ \frac{1}{2}\ +\quad +1$

$7\ 4\ 3$
$1\ 0\ 1\ +\quad +1/\sqrt{2}$
$1\ \tilde{0}\ 1\ -\quad -1/\sqrt{2}$
$7\ \frac{9}{2}\ \frac{5}{2}$
$1\ \frac{1}{2}\ \frac{3}{2}\ -\quad -1$
$7\ 5\ 2$
$1\ 1\ 2\ +\quad +1/\sqrt{2}$
$1\ 1\ \tilde{2}\ +\quad -1/\sqrt{2}$
$7\ \frac{11}{2}\ \frac{3}{2}$
$1\ \frac{3}{2}\ \frac{3}{2}\ +\quad +1$
$7\ 6\ 1$
$1\ 2\ 1\ +\quad +1/\sqrt{2}$
$1\ \tilde{2}\ 1\ +\quad -1/\sqrt{2}$
$7\ \frac{13}{2}\ \frac{1}{2}$
$1\ \frac{3}{2}\ \frac{1}{2}\ -\quad +1$
$7\ 7\ 0$
$1\ 1\ 0\ +\quad +1$
$7\ 7\ \tilde{0}$
$1\ 1\ \tilde{0}\ +\quad -1$
$\frac{15}{2}\ 4\ \frac{7}{2}$
$\frac{1}{2}\ 0\ \frac{1}{2}\ +\quad +1/\sqrt{2}$
$\frac{1}{2}\ \tilde{0}\ \frac{1}{2}\ -\quad -1/\sqrt{2}$
$\frac{15}{2}\ \frac{9}{2}\ 3$
$\frac{1}{2}\ \frac{1}{2}\ 1\ -\quad -1$
$\frac{15}{2}\ 5\ \frac{5}{2}$
$\frac{1}{2}\ 1\ \frac{3}{2}\ +\quad +1$
$\frac{15}{2}\ \frac{11}{2}\ 2$
$\frac{1}{2}\ \frac{3}{2}\ 2\ -\quad -1/\sqrt{2}$
$\frac{1}{2}\ \frac{3}{2}\ \tilde{2}\ -\quad +1/\sqrt{2}$
$\frac{15}{2}\ 6\ \frac{3}{2}$
$\frac{1}{2}\ 2\ \frac{3}{2}\ -\quad +1/\sqrt{2}$
$\frac{1}{2}\ \tilde{2}\ \frac{3}{2}\ -\quad -1/\sqrt{2}$
$\frac{15}{2}\ \frac{13}{2}\ 1$
$\frac{1}{2}\ \frac{3}{2}\ 1\ +\quad +1$
$\frac{15}{2}\ 7\ \frac{1}{2}$
$\frac{1}{2}\ 1\ \frac{1}{2}\ -\quad +1$
$\frac{15}{2}\ \frac{15}{2}\ 0$
$\frac{1}{2}\ \frac{1}{2}\ 0\ +\quad +1$
$\frac{15}{2}\ \frac{15}{2}\ \tilde{0}$
$\frac{1}{2}\ \frac{1}{2}\ \tilde{0}\ +\quad -1$

D_∞–D_4 and $C_{\infty v}$–C_{4v} 3jm Factors (cont.)

$8\ 4\ 4$
$0\ 0\ 0\ +\quad +1/2$
$0\ \tilde{0}\ \tilde{0}\ +\quad -1/2$
$\tilde{0}\ \tilde{0}\ 0\ +\quad +1/2$
$8\ \frac{9}{2}\ \frac{7}{2}$
$0\ \frac{1}{2}\ \frac{1}{2}\ -\quad -1/\sqrt{2}$
$\tilde{0}\ \frac{1}{2}\ \frac{1}{2}\ +\quad +1/\sqrt{2}$
$8\ 5\ 3$
$0\ 1\ 1\ +\quad +1/\sqrt{2}$
$\tilde{0}\ 1\ 1\ -\quad -1/\sqrt{2}$

$8\ \frac{11}{2}\ \frac{5}{2}$
$0\ \frac{3}{2}\ \frac{3}{2}\ -\quad -1/\sqrt{2}$
$\tilde{0}\ \frac{3}{2}\ \frac{3}{2}\ +\quad +1/\sqrt{2}$
$8\ 6\ 2$
$0\ 2\ 2\ +\quad +1/2$
$0\ \tilde{2}\ \tilde{2}\ +\quad -1/2$
$\tilde{0}\ 2\ \tilde{2}\ -\quad +1/2$
$\tilde{0}\ \tilde{2}\ 2\ -\quad -1/2$
$8\ \frac{13}{2}\ \frac{3}{2}$
$0\ \frac{3}{2}\ \frac{3}{2}\ -\quad +1/\sqrt{2}$
$\tilde{0}\ \frac{3}{2}\ \frac{3}{2}\ +\quad +1/\sqrt{2}$

$8\ 7\ 1$
$0\ 1\ 1\ +\quad +1/\sqrt{2}$
$\tilde{0}\ 1\ 1\ -\quad +1/\sqrt{2}$
$8\ \frac{15}{2}\ \frac{1}{2}$
$0\ \frac{1}{2}\ \frac{1}{2}\ -\quad +1/\sqrt{2}$
$\tilde{0}\ \frac{1}{2}\ \frac{1}{2}\ +\quad +1/\sqrt{2}$
$8\ 8\ 0$
$0\ 0\ 0\ +\quad +1/\sqrt{2}$
$\tilde{0}\ \tilde{0}\ 0\ +\quad +1/\sqrt{2}$
$8\ 8\ \tilde{0}$
$\tilde{0}\ 0\ \tilde{0}\ -\quad -1/\sqrt{2}$

D_∞–D_5 and $C_{\infty v}$–C_{5v} 2jm Factors ($D_{\infty i}-D_{5i} = D_{\infty h}-D_{5d}$)

$0 \to +\ 0$	$\frac{5}{2} \to +\ \frac{5}{2}\ +\text{-}\frac{5}{2}$	$\frac{11}{2} \to -\ \frac{1}{2}$
$\frac{1}{2} \to +\ \frac{1}{2}$	$3 \to -\ 2$	$6 \to -\ 1$
$\tilde{0} \to +\ \tilde{0}$	$\frac{7}{2} \to -\ \frac{3}{2}$	$\frac{13}{2} \to -\ \frac{3}{2}$
$1 \to +\ 1$	$4 \to -\ 1$	$7 \to -\ 2$
$\frac{3}{2} \to +\ \frac{3}{2}$	$\frac{9}{2} \to -\ \frac{1}{2}$	$\frac{15}{2} \to +\ \frac{5}{2}\ +\text{-}\frac{5}{2}$
$2 \to +\ 2$	$5 \to -\ 0\ -\ \tilde{0}$	$8 \to +\ 2$

D_∞–D_5 and $C_{\infty v}$–C_{5v} 3jm Factors

$0\ 0\ 0$
$0\ 0\ 0\ +\quad +1$
$\frac{1}{2}\ \frac{1}{2}\ 0$
$\frac{1}{2}\ \frac{1}{2}\ 0\ +\quad +1$
$\tilde{0}\ \frac{1}{2}\ \frac{1}{2}$
$\tilde{0}\ \frac{1}{2}\ \frac{1}{2}\ +\quad +1$
$\tilde{0}\ \tilde{0}\ 0$
$\tilde{0}\ \tilde{0}\ 0\ +\quad +1$
$1\ \frac{1}{2}\ \frac{1}{2}$
$1\ \frac{1}{2}\ \frac{1}{2}\ +\quad +1$
$1\ 1\ 0$
$1\ 1\ 0\ +\quad +1$
$1\ 1\ \tilde{0}$
$1\ 1\ \tilde{0}\ +\quad +1$
$\frac{3}{2}\ 1\ \frac{1}{2}$
$\frac{3}{2}\ 1\ \frac{1}{2}\ +\quad +1$
$\frac{3}{2}\ \frac{3}{2}\ 0$
$\frac{3}{2}\ \frac{3}{2}\ 0\ +\quad +1$

$\frac{3}{2}\ \frac{3}{2}\ \tilde{0}$
$\frac{3}{2}\ \frac{3}{2}\ \tilde{0}\ +\quad +1$
$2\ 1\ 1$
$2\ 1\ 1\ +\quad +1$
$2\ \frac{3}{2}\ \frac{1}{2}$
$2\ \frac{3}{2}\ \frac{1}{2}\ +\quad +1$
$2\ 2\ 0$
$2\ 2\ 0\ +\quad +1$
$2\ 2\ \tilde{0}$
$2\ 2\ \tilde{0}\ +\quad +1$
$\frac{5}{2}\ \frac{3}{2}\ 1$
$\frac{5}{2}\ \frac{3}{2}\ 1\ +\quad +1/\sqrt{2}$
$\frac{5}{2}\ 2\ \frac{1}{2}$
$\frac{5}{2}\ 2\ \frac{1}{2}\ +\quad +1/\sqrt{2}$
$\frac{5}{2}\ \frac{5}{2}\ 0$
$\text{-}\frac{5}{2}\ \frac{5}{2}\ 0\ +\quad +1/\sqrt{2}$
$\frac{5}{2}\ \frac{5}{2}\ \tilde{0}$
$\frac{5}{2}\ \frac{5}{2}\ \tilde{0}\ +\quad +1/\sqrt{2}$

$3\ \frac{3}{2}\ \frac{3}{2}$
$2\ \frac{3}{2}\ \frac{3}{2}\ +^*\quad -i$
$3\ 2\ 1$
$2\ 2\ 1\ +^*\quad -i$
$3\ \frac{5}{2}\ \frac{1}{2}$
$2\ \frac{5}{2}\ \frac{1}{2}\ -^*\quad +1/\sqrt{2}$
$3\ 3\ 0$
$2\ 2\ 0\ +\quad -1$
$3\ 3\ \tilde{0}$
$2\ 2\ \tilde{0}\ +\quad +1$
$\frac{7}{2}\ 2\ \frac{3}{2}$
$\frac{3}{2}\ 2\ \frac{3}{2}\ -^*\quad +i$
$\frac{7}{2}\ \frac{5}{2}\ 1$
$\frac{3}{2}\ \frac{5}{2}\ 1\ +^*\quad -1/\sqrt{2}$
$\frac{7}{2}\ 3\ \frac{1}{2}$
$\frac{3}{2}\ 2\ \frac{1}{2}\ -\quad +1$
$\frac{7}{2}\ \frac{7}{2}\ 0$
$\frac{3}{2}\ \frac{3}{2}\ 0\ +\quad -1$

D_∞–D_5 and $C_{\infty v}$–C_{5v} 3jm Factors (cont.)

$\frac{7}{2}$	**$\frac{7}{2}$**	**$\tilde{0}$**		
$\frac{3}{2}$	$\frac{3}{2}$	$\tilde{0}$	+	$+1$
4	**2**	**2**		
1	2	2	+*	$-i$
4	**$\frac{5}{2}$**	**$\frac{3}{2}$**		
1	$\frac{5}{2}$	$\frac{3}{2}$	−*	$+1/\sqrt{2}$
4	**3**	**1**		
1	2	1	+	-1
4	**$\frac{7}{2}$**	**$\frac{1}{2}$**		
1	$\frac{3}{2}$	$\frac{1}{2}$	−	$+1$
4	**4**	**0**		
1	1	0	+	-1
4	**4**	**$\tilde{0}$**		
1	1	$\tilde{0}$	+	$+1$
$\frac{9}{2}$	**$\frac{5}{2}$**	**2**		
$\frac{1}{2}$	$\frac{5}{2}$	2	+*	$-1/\sqrt{2}$
$\frac{9}{2}$	**3**	**$\frac{3}{2}$**		
$\frac{1}{2}$	2	$\frac{3}{2}$	−	$+1$
$\frac{9}{2}$	**$\frac{7}{2}$**	**1**		
$\frac{1}{2}$	$\frac{3}{2}$	1	+	-1
$\frac{9}{2}$	**4**	**$\frac{1}{2}$**		
$\frac{1}{2}$	1	$\frac{1}{2}$	−	$+1$
$\frac{9}{2}$	**$\frac{9}{2}$**	**0**		
$\frac{1}{2}$	$\frac{1}{2}$	0	+	-1
$\frac{9}{2}$	**$\frac{9}{2}$**	**$\tilde{0}$**		
$\frac{1}{2}$	$\frac{1}{2}$	$\tilde{0}$	+	$+1$
5	**$\frac{5}{2}$**	**$\frac{5}{2}$**		
0	$-\frac{5}{2}$	$\frac{5}{2}$	−*	$-1/2$
$\tilde{0}$	$\frac{5}{2}$	$\frac{5}{2}$	+*	$+1/2$
5	**3**	**2**		
0	2	2	+	$-1/\sqrt{2}$
$\tilde{0}$	2	2	−	$-1/\sqrt{2}$
5	**$\frac{7}{2}$**	**$\frac{3}{2}$**		
0	$\frac{3}{2}$	$\frac{3}{2}$	−	$+1/\sqrt{2}$
$\tilde{0}$	$\frac{3}{2}$	$\frac{3}{2}$	+	$+1/\sqrt{2}$
5	**4**	**1**		
0	1	1	+	$-1/\sqrt{2}$
$\tilde{0}$	1	1	−	$-1/\sqrt{2}$
5	**$\frac{9}{2}$**	**$\frac{1}{2}$**		
0	$\frac{1}{2}$	$\frac{1}{2}$	−	$+1/\sqrt{2}$

5	**$\frac{9}{2}$**	**$\frac{1}{2}$**		
$\tilde{0}$	$\frac{1}{2}$	$\frac{1}{2}$	+	$+1/\sqrt{2}$
5	**5**	**0**		
0	0	0	+	$-1/\sqrt{2}$
$\tilde{0}$	$\tilde{0}$	0	+	$-1/\sqrt{2}$
5	**5**	**$\tilde{0}$**		
$\tilde{0}$	0	$\tilde{0}$	−	$+1/\sqrt{2}$
$\frac{11}{2}$	**3**	**$\frac{5}{2}$**		
$\frac{1}{2}$	2	$\frac{5}{2}$	+	$+1/\sqrt{2}$
$\frac{11}{2}$	**$\frac{7}{2}$**	**2**		
$\frac{1}{2}$	$\frac{3}{2}$	2	−	-1
$\frac{11}{2}$	**4**	**$\frac{3}{2}$**		
$\frac{1}{2}$	1	$\frac{3}{2}$	+	$+1$
$\frac{11}{2}$	**$\frac{9}{2}$**	**1**		
$\frac{1}{2}$	$\frac{1}{2}$	1	−	-1
$\frac{11}{2}$	**5**	**$\frac{1}{2}$**		
$\frac{1}{2}$	0	$\frac{1}{2}$	+	$+1/\sqrt{2}$
$\frac{1}{2}$	$\tilde{0}$	$\frac{1}{2}$	−	$+1/\sqrt{2}$
$\frac{11}{2}$	**$\frac{11}{2}$**	**0**		
$\frac{1}{2}$	$\frac{1}{2}$	0	+	-1
$\frac{11}{2}$	**$\frac{11}{2}$**	**$\tilde{0}$**		
$\frac{1}{2}$	$\frac{1}{2}$	$\tilde{0}$	+	-1
6	**3**	**3**		
1	2	2	+*	$+i$
6	**$\frac{7}{2}$**	**$\frac{5}{2}$**		
1	$\frac{3}{2}$	$\frac{5}{2}$	−	$+1/\sqrt{2}$
6	**4**	**2**		
1	1	2	+	-1
6	**$\frac{9}{2}$**	**$\frac{3}{2}$**		
1	$\frac{1}{2}$	$\frac{3}{2}$	−	$+1$
6	**5**	**1**		
1	0	1	+	$-1/\sqrt{2}$
1	$\tilde{0}$	1	−	$-1/\sqrt{2}$
6	**$\frac{11}{2}$**	**$\frac{1}{2}$**		
1	$\frac{1}{2}$	$\frac{1}{2}$	+	$+1$
6	**6**	**0**		
1	1	0	+	-1
6	**6**	**$\tilde{0}$**		
1	1	$\tilde{0}$	+	-1

$\frac{13}{2}$	**$\frac{7}{2}$**	**3**		
$\frac{3}{2}$	$\frac{3}{2}$	2	−*	$+i$
$\frac{13}{2}$	**4**	**$\frac{5}{2}$**		
$\frac{3}{2}$	1	$\frac{5}{2}$	+	$+1/\sqrt{2}$
$\frac{13}{2}$	**$\frac{9}{2}$**	**2**		
$\frac{3}{2}$	$\frac{1}{2}$	2	−	-1
$\frac{13}{2}$	**5**	**$\frac{3}{2}$**		
$\frac{3}{2}$	0	$\frac{3}{2}$	+	$+1/\sqrt{2}$
$\frac{3}{2}$	$\tilde{0}$	$\frac{3}{2}$	−	$+1/\sqrt{2}$
$\frac{13}{2}$	**$\frac{11}{2}$**	**1**		
$\frac{3}{2}$	$\frac{1}{2}$	1	+	-1
$\frac{13}{2}$	**6**	**$\frac{1}{2}$**		
$\frac{3}{2}$	1	$\frac{1}{2}$	+	$+1$
$\frac{13}{2}$	**$\frac{13}{2}$**	**0**		
$\frac{3}{2}$	$\frac{3}{2}$	0	+	-1
$\frac{13}{2}$	**$\frac{13}{2}$**	**$\tilde{0}$**		
$\frac{3}{2}$	$\frac{3}{2}$	$\tilde{0}$	+	-1
7	**$\frac{7}{2}$**	**$\frac{7}{2}$**		
2	$\frac{3}{2}$	$\frac{3}{2}$	+*	$+i$
7	**4**	**3**		
2	1	2	+*	$+i$
7	**$\frac{9}{2}$**	**$\frac{5}{2}$**		
2	$\frac{1}{2}$	$\frac{5}{2}$	−	$+1/\sqrt{2}$
7	**5**	**2**		
2	0	2	+	$-1/\sqrt{2}$
2	$\tilde{0}$	2	−	$-1/\sqrt{2}$
7	**$\frac{11}{2}$**	**$\frac{3}{2}$**		
2	$\frac{1}{2}$	$\frac{3}{2}$	+	$+1$
7	**6**	**1**		
2	1	1	+	-1
7	**$\frac{13}{2}$**	**$\frac{1}{2}$**		
2	$\frac{3}{2}$	$\frac{1}{2}$	+	$+1$
7	**7**	**0**		
2	2	0	+	-1
7	**7**	**$\tilde{0}$**		
2	2	$\tilde{0}$	+	-1
$\frac{15}{2}$	**4**	**$\frac{7}{2}$**		
$\frac{5}{2}$	1	$\frac{3}{2}$	+	$-1/\sqrt{2}$
$\frac{15}{2}$	**$\frac{9}{2}$**	**3**		
$\frac{5}{2}$	$\frac{1}{2}$	2	+	$-1/\sqrt{2}$

D_∞–D_5 and $C_{\infty v}$–C_{5v} 3jm Factors (cont.)

$\frac{15}{2}$ 5 $\frac{5}{2}$
$\frac{5}{2}$ 0 $-\frac{5}{2}$ +* $+1/2$
$\frac{5}{2}$ $\tilde{0}$ $\frac{5}{2}$ −* $+1/2$
$\frac{15}{2}$ $\frac{11}{2}$ 2
$\frac{5}{2}$ $\frac{1}{2}$ 2 +* $-1/\sqrt{2}$
$\frac{15}{2}$ 6 $\frac{3}{2}$
$\frac{5}{2}$ 1 $\frac{3}{2}$ +* $+1/\sqrt{2}$
$\frac{15}{2}$ $\frac{13}{2}$ 1
$\frac{5}{2}$ $\frac{3}{2}$ 1 +* $-1/\sqrt{2}$
$\frac{15}{2}$ 7 $\frac{1}{2}$
$\frac{5}{2}$ 2 $\frac{1}{2}$ +* $+1/\sqrt{2}$
$\frac{15}{2}$ $\frac{15}{2}$ 0
$-\frac{5}{2}$ $\frac{5}{2}$ 0 + $+1/\sqrt{2}$
$\frac{15}{2}$ $\frac{15}{2}$ $\tilde{0}$
$\frac{5}{2}$ $\frac{5}{2}$ $\tilde{0}$ + $-1/\sqrt{2}$
8 4 4
2 1 1 + $+1$
8 $\frac{9}{2}$ $\frac{7}{2}$
2 $\frac{1}{2}$ $\frac{3}{2}$ + $+1$
8 5 3
2 0 2 + $+1/\sqrt{2}$
2 $\tilde{0}$ 2 − $-1/\sqrt{2}$
8 $\frac{11}{2}$ $\frac{5}{2}$
2 $\frac{1}{2}$ $\frac{5}{2}$ −* $-1/\sqrt{2}$
8 6 2
2 1 2 +* $+i$
8 $\frac{13}{2}$ $\frac{3}{2}$
2 $\frac{3}{2}$ $\frac{3}{2}$ +* $-i$
8 7 1
2 2 1 +* $+i$
8 $\frac{15}{2}$ $\frac{1}{2}$
2 $\frac{5}{2}$ $\frac{1}{2}$ − $+1/\sqrt{2}$
8 8 0
2 2 0 + $+1$
8 8 $\tilde{0}$
2 2 $\tilde{0}$ + -1

D_∞–D_6 and $C_{\infty v}$–C_{6v} 2jm Factors ($D_{\infty i}$–D_{6i} = $D_{\infty h}$–D_{6h})

0 → + 0	$\frac{5}{2}$ → + $\frac{5}{2}$	$\frac{11}{2}$ → + $\frac{1}{2}$
$\frac{1}{2}$ → + $\frac{1}{2}$	3 → + 3 + $\tilde{3}$	6 → + 0 + $\tilde{0}$
$\tilde{0}$ → + $\tilde{0}$	$\frac{7}{2}$ → + $\frac{5}{2}$	$\frac{13}{2}$ → + $\frac{1}{2}$
1 → + 1	4 → + 2	7 → + 1
$\frac{3}{2}$ → + $\frac{3}{2}$	$\frac{9}{2}$ → + $\frac{3}{2}$	$\frac{15}{2}$ → + $\frac{3}{2}$
2 → + 2	5 → + 1	8 → + 2

D_∞–D_6 and $C_{\infty v}$–C_{6v} 3jm Factors

0 0 0
0 0 0 + $+1$
$\frac{1}{2}$ $\frac{1}{2}$ 0
$\frac{1}{2}$ $\frac{1}{2}$ 0 + $+1$
$\tilde{0}$ $\frac{1}{2}$ $\frac{1}{2}$
$\tilde{0}$ $\frac{1}{2}$ $\frac{1}{2}$ + $+1$
$\tilde{0}$ $\tilde{0}$ 0
$\tilde{0}$ $\tilde{0}$ 0 + $+1$
1 $\frac{1}{2}$ $\frac{1}{2}$
1 $\frac{1}{2}$ $\frac{1}{2}$ + $+1$
1 1 0
1 1 0 + $+1$
1 1 $\tilde{0}$
1 1 $\tilde{0}$ + $+1$
$\frac{3}{2}$ 1 $\frac{1}{2}$
$\frac{3}{2}$ 1 $\frac{1}{2}$ + $+1$
$\frac{3}{2}$ $\frac{3}{2}$ 0
$\frac{3}{2}$ $\frac{3}{2}$ 0 + $+1$
$\frac{3}{2}$ $\frac{3}{2}$ $\tilde{0}$
$\frac{3}{2}$ $\frac{3}{2}$ $\tilde{0}$ + $+1$
2 1 1
2 1 1 + $+1$
2 $\frac{3}{2}$ $\frac{1}{2}$
2 $\frac{3}{2}$ $\frac{1}{2}$ + $+1$
2 2 0
2 2 0 + $+1$
2 2 $\tilde{0}$
2 2 $\tilde{0}$ + $+1$
$\frac{5}{2}$ $\frac{3}{2}$ 1
$\frac{5}{2}$ $\frac{3}{2}$ 1 + $+1$
$\frac{5}{2}$ 2 $\frac{1}{2}$
$\frac{5}{2}$ 2 $\frac{1}{2}$ + $+1$
$\frac{5}{2}$ $\frac{5}{2}$ 0
$\frac{5}{2}$ $\frac{5}{2}$ 0 + $+1$
$\frac{5}{2}$ $\frac{5}{2}$ $\tilde{0}$
$\frac{5}{2}$ $\frac{5}{2}$ $\tilde{0}$ + $+1$
3 $\frac{3}{2}$ $\frac{3}{2}$
3 $\frac{3}{2}$ $\frac{3}{2}$ + $+1/\sqrt{2}$
$\tilde{3}$ $\frac{3}{2}$ $\frac{3}{2}$ + $+1/\sqrt{2}$
3 2 1
3 2 1 + $+1/\sqrt{2}$
$\tilde{3}$ 2 1 + $+1/\sqrt{2}$

D_∞–D_6 and $C_{\infty v}$–C_{6v} 3jm Factors (cont.)

$\mathbf{3\ \frac{5}{2}\ \frac{1}{2}}$

$3\ \frac{5}{2}\ \frac{1}{2}\ +\quad +1/\sqrt{2}$

$\tilde{3}\ \frac{5}{2}\ \frac{1}{2}\ +\quad +1/\sqrt{2}$

$\mathbf{3\ 3\ 0}$

$3\ 3\ 0\ +\quad +1/\sqrt{2}$

$\tilde{3}\ \tilde{3}\ 0\ +\quad +1/\sqrt{2}$

$\mathbf{3\ 3\ \tilde{0}}$

$\tilde{3}\ 3\ \tilde{0}\ +\quad +1/\sqrt{2}$

$\mathbf{\frac{7}{2}\ 2\ \frac{3}{2}}$

$\frac{5}{2}\ 2\ \frac{3}{2}\ -\quad -1$

$\mathbf{\frac{7}{2}\ \frac{5}{2}\ 1}$

$\frac{5}{2}\ \frac{5}{2}\ 1\ -\quad -1$

$\mathbf{\frac{7}{2}\ 3\ \frac{1}{2}}$

$\frac{5}{2}\ 3\ \frac{1}{2}\ -\quad +1/\sqrt{2}$

$\frac{5}{2}\ \tilde{3}\ \frac{1}{2}\ -\quad -1/\sqrt{2}$

$\mathbf{\frac{7}{2}\ \frac{7}{2}\ 0}$

$\frac{5}{2}\ \frac{5}{2}\ 0\ +\quad +1$

$\mathbf{\frac{7}{2}\ \frac{7}{2}\ \tilde{0}}$

$\frac{5}{2}\ \frac{5}{2}\ \tilde{0}\ +\quad -1$

$\mathbf{4\ 2\ 2}$

$2\ 2\ 2\ +\quad -1$

$\mathbf{4\ \frac{5}{2}\ \frac{3}{2}}$

$2\ \frac{5}{2}\ \frac{3}{2}\ +\quad -1$

$\mathbf{4\ 3\ 1}$

$2\ 3\ 1\ +\quad +1/\sqrt{2}$

$2\ \tilde{3}\ 1\ +\quad -1/\sqrt{2}$

$\mathbf{4\ \frac{7}{2}\ \frac{1}{2}}$

$2\ \frac{5}{2}\ \frac{1}{2}\ -\quad +1$

$\mathbf{4\ 4\ 0}$

$2\ 2\ 0\ +\quad +1$

$\mathbf{4\ 4\ \tilde{0}}$

$2\ 2\ \tilde{0}\ +\quad -1$

$\mathbf{\frac{9}{2}\ \frac{5}{2}\ 2}$

$\frac{3}{2}\ \frac{5}{2}\ 2\ -\quad +1$

$\mathbf{\frac{9}{2}\ 3\ \frac{3}{2}}$

$\frac{3}{2}\ 3\ \frac{3}{2}\ -\quad +1/\sqrt{2}$

$\frac{3}{2}\ \tilde{3}\ \frac{3}{2}\ -\quad -1/\sqrt{2}$

$\mathbf{\frac{9}{2}\ \frac{7}{2}\ 1}$

$\frac{3}{2}\ \frac{5}{2}\ 1\ +\quad +1$

$\mathbf{\frac{9}{2}\ 4\ \frac{1}{2}}$

$\frac{3}{2}\ 2\ \frac{1}{2}\ -\quad +1$

$\mathbf{\frac{9}{2}\ \frac{9}{2}\ 0}$

$\frac{3}{2}\ \frac{3}{2}\ 0\ +\quad +1$

$\mathbf{\frac{9}{2}\ \frac{9}{2}\ \tilde{0}}$

$\frac{3}{2}\ \frac{3}{2}\ \tilde{0}\ +\quad -1$

$\mathbf{5\ \frac{5}{2}\ \frac{5}{2}}$

$1\ \frac{5}{2}\ \frac{5}{2}\ +\quad +1$

$\mathbf{5\ 3\ 2}$

$1\ 3\ 2\ +\quad +1/\sqrt{2}$

$1\ \tilde{3}\ 2\ +\quad -1/\sqrt{2}$

$\mathbf{5\ \frac{7}{2}\ \frac{3}{2}}$

$1\ \frac{5}{2}\ \frac{3}{2}\ -\quad +1$

$\mathbf{5\ 4\ 1}$

$1\ 2\ 1\ +\quad +1$

$\mathbf{5\ \frac{9}{2}\ \frac{1}{2}}$

$1\ \frac{3}{2}\ \frac{1}{2}\ -\quad +1$

$\mathbf{5\ 5\ 0}$

$1\ 1\ 0\ +\quad +1$

$\mathbf{5\ 5\ \tilde{0}}$

$1\ 1\ \tilde{0}\ +\quad -1$

$\mathbf{\frac{11}{2}\ 3\ \frac{5}{2}}$

$\frac{1}{2}\ 3\ \frac{5}{2}\ -\quad +1/\sqrt{2}$

$\frac{1}{2}\ \tilde{3}\ \frac{5}{2}\ -\quad -1/\sqrt{2}$

$\mathbf{\frac{11}{2}\ \frac{7}{2}\ 2}$

$\frac{1}{2}\ \frac{5}{2}\ 2\ +\quad +1$

$\mathbf{\frac{11}{2}\ 4\ \frac{3}{2}}$

$\frac{1}{2}\ 2\ \frac{3}{2}\ -\quad +1$

$\mathbf{\frac{11}{2}\ \frac{9}{2}\ 1}$

$\frac{1}{2}\ \frac{3}{2}\ 1\ +\quad +1$

$\mathbf{\frac{11}{2}\ 5\ \frac{1}{2}}$

$\frac{1}{2}\ 1\ \frac{1}{2}\ -\quad +1$

$\mathbf{\frac{11}{2}\ \frac{11}{2}\ 0}$

$\frac{1}{2}\ \frac{1}{2}\ 0\ +\quad +1$

$\mathbf{\frac{11}{2}\ \frac{11}{2}\ \tilde{0}}$

$\frac{1}{2}\ \frac{1}{2}\ \tilde{0}\ +\quad -1$

$\mathbf{6\ 3\ 3}$

$0\ 3\ 3\ +\quad +1/2$

$0\ \tilde{3}\ \tilde{3}\ +\quad -1/2$

$\tilde{0}\ \tilde{3}\ 3\ -\quad -1/2$

$\mathbf{6\ \frac{7}{2}\ \frac{5}{2}}$

$0\ \frac{5}{2}\ \frac{5}{2}\ -\quad +1/\sqrt{2}$

$\tilde{0}\ \frac{5}{2}\ \frac{5}{2}\ +\quad +1/\sqrt{2}$

$\mathbf{6\ 4\ 2}$

$0\ 2\ 2\ +\quad +1/\sqrt{2}$

$\tilde{0}\ 2\ 2\ -\quad +1/\sqrt{2}$

$\mathbf{6\ \frac{9}{2}\ \frac{3}{2}}$

$0\ \frac{3}{2}\ \frac{3}{2}\ -\quad +1/\sqrt{2}$

$\tilde{0}\ \frac{3}{2}\ \frac{3}{2}\ +\quad +1/\sqrt{2}$

$\mathbf{6\ 5\ 1}$

$0\ 1\ 1\ +\quad +1/\sqrt{2}$

$\tilde{0}\ 1\ 1\ -\quad +1/\sqrt{2}$

$\mathbf{6\ \frac{11}{2}\ \frac{1}{2}}$

$0\ \frac{1}{2}\ \frac{1}{2}\ -\quad +1/\sqrt{2}$

$\tilde{0}\ \frac{1}{2}\ \frac{1}{2}\ +\quad +1/\sqrt{2}$

$\mathbf{6\ 6\ 0}$

$0\ 0\ 0\ +\quad +1/\sqrt{2}$

$\tilde{0}\ \tilde{0}\ 0\ +\quad +1/\sqrt{2}$

$\mathbf{6\ 6\ \tilde{0}}$

$\tilde{0}\ 0\ \tilde{0}\ -\quad -1/\sqrt{2}$

$\mathbf{\frac{13}{2}\ \frac{7}{2}\ 3}$

$\frac{1}{2}\ \frac{5}{2}\ 3\ -\quad +1/\sqrt{2}$

$\frac{1}{2}\ \frac{5}{2}\ \tilde{3}\ -\quad +1/\sqrt{2}$

$\mathbf{\frac{13}{2}\ 4\ \frac{5}{2}}$

$\frac{1}{2}\ 2\ \frac{5}{2}\ +\quad +1$

$\mathbf{\frac{13}{2}\ \frac{9}{2}\ 2}$

$\frac{1}{2}\ \frac{3}{2}\ 2\ -\quad +1$

$\mathbf{\frac{13}{2}\ 5\ \frac{3}{2}}$

$\frac{1}{2}\ 1\ \frac{3}{2}\ +\quad +1$

$\mathbf{\frac{13}{2}\ \frac{11}{2}\ 1}$

$\frac{1}{2}\ \frac{1}{2}\ 1\ -\quad +1$

$\mathbf{\frac{13}{2}\ 6\ \frac{1}{2}}$

$\frac{1}{2}\ 0\ \frac{1}{2}\ +\quad +1/\sqrt{2}$

$\frac{1}{2}\ \tilde{0}\ \frac{1}{2}\ -\quad +1/\sqrt{2}$

$\mathbf{\frac{13}{2}\ \frac{13}{2}\ 0}$

$\frac{1}{2}\ \frac{1}{2}\ 0\ +\quad +1$

$\mathbf{\frac{13}{2}\ \frac{13}{2}\ \tilde{0}}$

$\frac{1}{2}\ \frac{1}{2}\ \tilde{0}\ +\quad +1$

$\mathbf{7\ \frac{7}{2}\ \frac{7}{2}}$

$1\ \frac{5}{2}\ \frac{5}{2}\ +\quad +1$

D_∞–D_6 and $C_{\infty v}$–C_{6v} 3jm Factors (cont.)

7 4 3	
$1\ 2\ 3\ +$	$+1/\sqrt{2}$
$1\ 2\ \tilde{3}\ +$	$+1/\sqrt{2}$
7 $\frac{9}{2}$ $\frac{5}{2}$	
$1\ \frac{3}{2}\ \frac{5}{2}\ -$	$+1$
7 5 2	
$1\ 1\ 2\ +$	$+1$
7 $\frac{11}{2}$ $\frac{3}{2}$	
$1\ \frac{1}{2}\ \frac{3}{2}\ -$	$+1$
7 6 1	
$1\ 0\ 1\ +$	$+1/\sqrt{2}$
$1\ \tilde{0}\ 1\ -$	$+1/\sqrt{2}$
7 $\frac{13}{2}$ $\frac{1}{2}$	
$1\ \frac{1}{2}\ \frac{1}{2}\ +$	$+1$
7 7 0	
$1\ 1\ 0\ +$	$+1$
7 7 $\tilde{0}$	
$1\ 1\ \tilde{0}\ +$	$+1$
$\frac{15}{2}$ 4 $\frac{7}{2}$	
$\frac{3}{2}\ 2\ \frac{5}{2}\ -$	$+1$
$\frac{15}{2}$ $\frac{9}{2}$ 3	
$\frac{3}{2}\ \frac{3}{2}\ 3\ -$	$+1/\sqrt{2}$
$\frac{3}{2}\ \frac{3}{2}\ \tilde{3}\ -$	$+1/\sqrt{2}$
$\frac{15}{2}$ 5 $\frac{5}{2}$	
$\frac{3}{2}\ 1\ \frac{5}{2}\ +$	$+1$
$\frac{15}{2}$ $\frac{11}{2}$ 2	
$\frac{3}{2}\ \frac{1}{2}\ 2\ -$	$+1$
$\frac{15}{2}$ 6 $\frac{3}{2}$	
$\frac{3}{2}\ 0\ \frac{3}{2}\ +$	$+1/\sqrt{2}$
$\frac{3}{2}\ \tilde{0}\ \frac{3}{2}\ -$	$+1/\sqrt{2}$
$\frac{15}{2}$ $\frac{13}{2}$ 1	
$\frac{3}{2}\ \frac{1}{2}\ 1\ +$	$+1$
$\frac{15}{2}$ 7 $\frac{1}{2}$	
$\frac{3}{2}\ 1\ \frac{1}{2}\ +$	$+1$
$\frac{15}{2}$ $\frac{15}{2}$ 0	
$\frac{3}{2}\ \frac{3}{2}\ 0\ +$	$+1$
$\frac{15}{2}$ $\frac{15}{2}$ $\tilde{0}$	
$\frac{3}{2}\ \frac{3}{2}\ \tilde{0}\ +$	$+1$
8 4 4	
$2\ 2\ 2\ +$	-1
8 $\frac{9}{2}$ $\frac{7}{2}$	
$2\ \frac{3}{2}\ \frac{5}{2}\ +$	-1
8 5 3	
$2\ 1\ 3\ +$	$+1/\sqrt{2}$
$2\ 1\ \tilde{3}\ +$	$+1/\sqrt{2}$
8 $\frac{11}{2}$ $\frac{5}{2}$	
$2\ \frac{1}{2}\ \frac{5}{2}\ -$	$+1$
8 6 2	
$2\ 0\ 2\ +$	$+1/\sqrt{2}$
$2\ \tilde{0}\ 2\ -$	$+1/\sqrt{2}$
8 $\frac{13}{2}$ $\frac{3}{2}$	
$2\ \frac{1}{2}\ \frac{3}{2}\ +$	$+1$
8 7 1	
$2\ 1\ 1\ +$	$+1$
8 $\frac{15}{2}$ $\frac{1}{2}$	
$2\ \frac{3}{2}\ \frac{1}{2}\ +$	$+1$
8 8 0	
$2\ 2\ 0\ +$	$+1$
8 8 $\tilde{0}$	
$2\ 2\ \tilde{0}\ +$	$+1$

T–C_3 2jm Factors (T_i–C_{3i} = T_h–C_{3i})

$0 \rightarrow +\ 0$

$\frac{1}{2} \rightarrow +\ \frac{1}{2}\ -\ \text{-}\frac{1}{2}$

$1 \rightarrow -\ 0\ +\ 1\ +\ \text{-}1$

$\frac{3}{2} \rightarrow +\ \frac{1}{2}\ +\ \frac{3}{2}$

$\text{-}\frac{3}{2} \rightarrow -\ \text{-}\frac{1}{2}\ -\ \frac{3}{2}$

$2 \rightarrow +\ 1$

$\text{-}2 \rightarrow +\ \text{-}1$

T–C_3 3jm Factors

0 0 0 0
0 0 0 + $+1$
$\frac{1}{2}$ $\frac{1}{2}$ 0 0
-$\frac{1}{2}$ $\frac{1}{2}$ 0 −* $-1/\sqrt{2}$
1 $\frac{1}{2}$ $\frac{1}{2}$ 0
0 -$\frac{1}{2}$ $\frac{1}{2}$ + $+1/\sqrt{2.3}$
1 -$\frac{1}{2}$ -$\frac{1}{2}$ + $+1/\sqrt{3}$
1 1 0 0
0 0 0 + $-1/\sqrt{3}$
-1 1 0 + $+1/\sqrt{3}$
1 1 1 0
0 0 0 −* 0
1 1 1 − 0
-1 1 0 −* $+1/\sqrt{2.3}$
1 1 1 1
0 0 0 +* $+i\sqrt{2/3}$

1 1 1 1
1 1 1 + $+i\sqrt{2/3}$
-1 1 0 +* $+i/3\sqrt{2}$
$\frac{3}{2}$ 1 $\frac{1}{2}$ 0
$\frac{1}{2}$ 0 -$\frac{1}{2}$ + $+1/\sqrt{3}$
$\frac{1}{2}$ -1 $\frac{1}{2}$ + $-1/\sqrt{2.3}$
$\frac{3}{2}$ 1 $\frac{1}{2}$ + $+1/\sqrt{3}$
$\frac{3}{2}$ -1 -$\frac{1}{2}$ +* $+1/\sqrt{2.3}$
$\frac{3}{2}$ $\frac{3}{2}$ 1 0
$\frac{1}{2}$ $\frac{1}{2}$ -1 + $+1/\sqrt{3}$
$\frac{3}{2}$ $\frac{1}{2}$ 1 + $+1/\sqrt{2.3}$
$\frac{3}{2}$ $\frac{3}{2}$ 0 +* $+1/\sqrt{3}$
-$\frac{3}{2}$ $\frac{3}{2}$ 0 0
-$\frac{1}{2}$ $\frac{1}{2}$ 0 −* $-1/\sqrt{2}$
$\frac{3}{2}$ $\frac{3}{2}$ 0 −* $-1/\sqrt{2}$

-$\frac{3}{2}$ $\frac{3}{2}$ 1 0
-$\frac{1}{2}$ $\frac{1}{2}$ 0 + $-1/\sqrt{2.3}$
-$\frac{1}{2}$ $\frac{3}{2}$ -1 +* $+1/\sqrt{3}$
$\frac{3}{2}$ $\frac{3}{2}$ 0 + $+1/\sqrt{2.3}$
2 1 1 0
1 1 1 + $+1/\sqrt{3}$
1 -1 0 +* $-1/\sqrt{3}$
2 $\frac{3}{2}$ $\frac{3}{2}$ 0
1 $\frac{3}{2}$ $\frac{1}{2}$ − $-1/\sqrt{2}$
2 -$\frac{3}{2}$ $\frac{1}{2}$ 0
1 -$\frac{1}{2}$ -$\frac{1}{2}$ − $+1/\sqrt{2}$
1 $\frac{3}{2}$ $\frac{1}{2}$ −* $-1/\sqrt{2}$
2 2 2 0
1 1 1 + -1
-2 2 0 0
-1 1 0 + $+1$

T–D_2 2jm Factors (T_i–D_{2i} = T_h–D_{2h})

0 > + 0
$\frac{1}{2}$ > + $\frac{1}{2}$
1 > + $\tilde{0}$ + 1 + $\tilde{1}$
$\frac{3}{2}$ > + $\frac{1}{2}$
-$\frac{3}{2}$ > + $\frac{1}{2}$
2 > + 0
-2 > + 0

T–D_2 3jm Factors

0 0 0 0
0 0 0 + $+1$
$\frac{1}{2}$ $\frac{1}{2}$ 0 0
$\frac{1}{2}$ $\frac{1}{2}$ 0 + $+1$
1 $\frac{1}{2}$ $\frac{1}{2}$ 0
$\tilde{0}$ $\frac{1}{2}$ $\frac{1}{2}$ + $+1/\sqrt{3}$
1 $\frac{1}{2}$ $\frac{1}{2}$ + $+1/\sqrt{3}$
$\tilde{1}$ $\frac{1}{2}$ $\frac{1}{2}$ + $+1/\sqrt{3}$
1 1 0 0
$\tilde{0}$ $\tilde{0}$ 0 + $+1/\sqrt{3}$
1 1 0 + $+1/\sqrt{3}$
$\tilde{1}$ $\tilde{1}$ 0 + $+1/\sqrt{3}$
1 1 1 0
$\tilde{1}$ 1 $\tilde{0}$ + $-1/\sqrt{2.3}$

1 1 1 1
$\tilde{1}$ 1 $\tilde{0}$ − $+1/\sqrt{2.3}$
$\frac{3}{2}$ 1 $\frac{1}{2}$ 0
$\frac{1}{2}$ $\tilde{0}$ $\frac{1}{2}$ + $+1/\sqrt{3}$
$\frac{1}{2}$ 1 $\frac{1}{2}$ + $-1/2\sqrt{3}+i/2$
$\frac{1}{2}$ $\tilde{1}$ $\frac{1}{2}$ + $-1/2\sqrt{3}-i/2$
$\frac{3}{2}$ $\frac{3}{2}$ 1 0
$\frac{1}{2}$ $\frac{1}{2}$ $\tilde{0}$ + $-1/\sqrt{3}$
$\frac{1}{2}$ $\frac{1}{2}$ 1 + $+1/2\sqrt{3}+i/2$
$\frac{1}{2}$ $\frac{1}{2}$ $\tilde{1}$ + $+1/2\sqrt{3}-i/2$
-$\frac{3}{2}$ $\frac{3}{2}$ 0 0
$\frac{1}{2}$ $\frac{1}{2}$ 0 + $+1$
-$\frac{3}{2}$ $\frac{3}{2}$ 1 0
$\frac{1}{2}$ $\frac{1}{2}$ $\tilde{0}$ + $+1/\sqrt{3}$
$\frac{1}{2}$ $\frac{1}{2}$ 1 + $+1/\sqrt{3}$

-$\frac{3}{2}$ $\frac{3}{2}$ 1 0
$\frac{1}{2}$ $\frac{1}{2}$ $\tilde{1}$ + $+1/\sqrt{3}$
2 1 1 0
0 $\tilde{0}$ $\tilde{0}$ + $+1/\sqrt{3}$
0 1 1 + $-1/2\sqrt{3}+i/2$
0 $\tilde{1}$ $\tilde{1}$ + $-1/2\sqrt{3}-i/2$
2 $\frac{3}{2}$ $\frac{3}{2}$ 0
0 $\frac{1}{2}$ $\frac{1}{2}$ + -1
2 -$\frac{3}{2}$ $\frac{1}{2}$ 0
0 $\frac{1}{2}$ $\frac{1}{2}$ + $+1$
2 2 2 0
0 0 0 + $+1$
-2 2 0 0
0 0 0 + $+1$

O–D_3 and T_d–C_{3v} 2jm Factors (O_i–D_{3i} = O_h–D_{3d})

$0 \rightarrow +\,0$

$\tfrac{1}{2} \rightarrow +\,\tfrac{1}{2}$

$1 \rightarrow +\,\tilde{0} + 1$

$\tfrac{3}{2} \rightarrow +\,\tfrac{1}{2} + \tfrac{3}{2} + \text{-}\tfrac{3}{2}$

$2 \rightarrow +\,1$

$\tilde{1} \rightarrow +\,0 + 1$

$\tilde{\tfrac{1}{2}} \rightarrow +\,\tfrac{1}{2}$

$\tilde{0} \rightarrow +\,\tilde{0}$

O–D_3 and T_d–C_{3v} 3jm Factors

0 0 0 0
$0\ 0\ 0\ +\quad +1$
$\tfrac{1}{2}$ $\tfrac{1}{2}$ 0 0
$\tfrac{1}{2}\ \tfrac{1}{2}\ 0\ +\quad +1$
1 $\tfrac{1}{2}$ $\tfrac{1}{2}$ 0
$\tilde{0}\ \tfrac{1}{2}\ \tfrac{1}{2}\ +\quad +1/\sqrt{3}$
$1\ \tfrac{1}{2}\ \tfrac{1}{2}\ +\quad +\sqrt{2}/\sqrt{3}$
1 1 0 0
$\tilde{0}\ \tilde{0}\ 0\ +\quad +1/\sqrt{3}$
$1\ 1\ 0\ +\quad +\sqrt{2}/\sqrt{3}$
1 1 1 0
$1\ 1\ \tilde{0}\ +\quad -1/\sqrt{3}$
$1\ 1\ 1\ -\quad 0$
$\tfrac{3}{2}$ 1 $\tfrac{1}{2}$ 0
$\tfrac{1}{2}\ \tilde{0}\ \tfrac{1}{2}\ -\quad +1/\sqrt{3}$
$\tfrac{1}{2}\ 1\ \tfrac{1}{2}\ -\quad -1/\sqrt{2.3}$
$\tfrac{3}{2}\ 1\ \tfrac{1}{2}\ +\quad +1/2$
$\tfrac{3}{2}$ $\tfrac{3}{2}$ 0 0
$\tfrac{1}{2}\ \tfrac{1}{2}\ 0\ +\quad +1/\sqrt{2}$
$\text{-}\tfrac{3}{2}\ \tfrac{3}{2}\ 0\ +\quad +1/2$
$\tfrac{3}{2}$ $\tfrac{3}{2}$ 1 0
$\tfrac{1}{2}\ \tfrac{1}{2}\ \tilde{0}\ +\quad +1/\sqrt{2.3}$
$\tfrac{1}{2}\ \tfrac{1}{2}\ 1\ +\quad 0$
$\tfrac{3}{2}\ \tfrac{1}{2}\ 1\ -\quad +1/3\sqrt{2}-i/3$
$\tfrac{3}{2}\ \tfrac{3}{2}\ \tilde{0}\ +\quad +1/2.3\sqrt{3}+i\sqrt{2}/3\sqrt{3}$
$\tfrac{3}{2}$ $\tfrac{3}{2}$ 1 1
$\tfrac{1}{2}\ \tfrac{1}{2}\ \tilde{0}\ +\quad 0$
$\tfrac{1}{2}\ \tfrac{1}{2}\ 1\ +\quad -1/\sqrt{3}$
$\tfrac{3}{2}$ $\tfrac{3}{2}$ 1 1
$\tfrac{3}{2}\ \tfrac{1}{2}\ 1\ -\quad +1/3\sqrt{2}+i/2.3$
$\tfrac{3}{2}\ \tfrac{3}{2}\ \tilde{0}\ +\quad +2/3\sqrt{3}-i/3\sqrt{2.3}$
2 1 1 0
$1\ 1\ \tilde{0}\ -\quad -1/\sqrt{3}$
$1\ 1\ 1\ +\quad +1/\sqrt{3}$
2 $\tfrac{3}{2}$ $\tfrac{1}{2}$ 0
$1\ \tfrac{1}{2}\ \tfrac{1}{2}\ +\quad +1/\sqrt{2}$
$1\ \tfrac{3}{2}\ \tfrac{1}{2}\ -\quad -1/2\sqrt{3}+i/\sqrt{2.3}$
2 $\tfrac{3}{2}$ $\tfrac{3}{2}$ 0
$1\ \tfrac{1}{2}\ \tfrac{1}{2}\ -\quad 0$
$1\ \tfrac{3}{2}\ \tfrac{1}{2}\ +\quad +1/\sqrt{2.3}+i/2\sqrt{3}$
2 2 0 0
$1\ 1\ 0\ +\quad +1$
2 2 2 0
$1\ 1\ 1\ +\quad +1$
$\tilde{1}$ 1 1 0
$0\ \tilde{0}\ \tilde{0}\ +\quad +\sqrt{2}/3$
$0\ 1\ 1\ +\quad -1/3$
$1\ 1\ \tilde{0}\ -\quad -1/3$
$1\ 1\ 1\ +\quad -2/3$
$\tilde{1}$ $\tfrac{3}{2}$ $\tfrac{1}{2}$ 0
$0\ \tfrac{1}{2}\ \tfrac{1}{2}\ -\quad +1/\sqrt{3}$
$1\ \tfrac{1}{2}\ \tfrac{1}{2}\ +\quad +1/\sqrt{2.3}$
$1\ \tfrac{3}{2}\ \tfrac{1}{2}\ -\quad -1/2.3-i\sqrt{2}/3$
$\tilde{1}$ $\tfrac{3}{2}$ $\tfrac{3}{2}$ 0
$0\ \tfrac{1}{2}\ \tfrac{1}{2}\ +\quad -1/\sqrt{2.3}$
$0\ \text{-}\tfrac{3}{2}\ \tfrac{3}{2}\ +\quad +1/2\sqrt{3}$

O–D_3 and T_d–C_{3v} 3jm Factors (cont.)

$\tilde{1}\ \frac{3}{2}\ \frac{3}{2}$ 0

$1\ \frac{1}{2}\ \frac{1}{2}\ -\quad 0$

$1\ \frac{3}{2}\ \frac{1}{2}\ +\quad +1/3\sqrt{2}-i/3$

$\tilde{1}\ \frac{3}{2}\ \frac{3}{2}$ 1

$0\ \frac{1}{2}\ \frac{1}{2}\ -\quad 0$

$0\ -\frac{3}{2}\ \frac{3}{2}\ -\quad +i/\sqrt{2.3}$

$1\ \frac{1}{2}\ \frac{1}{2}\ +\quad +1/\sqrt{3}$

$1\ \frac{3}{2}\ \frac{1}{2}\ -\quad +1/3\sqrt{2}+i/2.3$

$\tilde{1}\ 2\ 1$ 0

$0\ 1\ 1\ -\quad -1/\sqrt{3}$

$1\ 1\ \tilde{0}\ +\quad -1/\sqrt{3}$

$1\ 1\ 1\ -\quad +1/\sqrt{3}$

$\tilde{1}\ \tilde{1}\ 0$ 0

$0\ 0\ 0\ +\quad +1/\sqrt{3}$

$1\ 1\ 0\ +\quad +\sqrt{2}/\sqrt{3}$

$\tilde{1}\ \tilde{1}\ 1$ 0

$1\ 0\ 1\ -\quad +1/\sqrt{3}$

$1\ 1\ \tilde{0}\ +\quad +1/\sqrt{3}$

$1\ 1\ 1\ -\quad 0$

$\tilde{1}\ \tilde{1}\ 2$ 0

$1\ 0\ 1\ +\quad -1/\sqrt{3}$

$1\ 1\ 1\ +\quad -1/\sqrt{3}$

$\tilde{1}\ \tilde{1}\ \tilde{1}$ 0

$0\ 0\ 0\ +\quad -\sqrt{2}/3$

$1\ 1\ 0\ +\quad +1/3$

$1\ 1\ 1\ +\quad -2/3$

$\tilde{\frac{1}{2}}\ \frac{3}{2}\ 1$ 0

$\frac{1}{2}\ \frac{1}{2}\ \tilde{0}\ -\quad -1/\sqrt{3}$

$\frac{1}{2}\ \frac{1}{2}\ 1\ -\quad -1/\sqrt{2.3}$

$\tilde{\frac{1}{2}}\ \frac{3}{2}\ 1$ 0

$\frac{1}{2}\ \frac{3}{2}\ 1\ +\quad +1/2.3+i\sqrt{2/3}$

$\tilde{\frac{1}{2}}\ 2\ \frac{3}{2}$ 0

$\frac{1}{2}\ 1\ \frac{1}{2}\ +\quad -1/\sqrt{2}$

$\frac{1}{2}\ 1\ \frac{3}{2}\ -\quad -1/2\sqrt{3}+i/\sqrt{2.3}$

$\tilde{\frac{1}{2}}\ \tilde{1}\ \frac{1}{2}$ 0

$\frac{1}{2}\ 0\ \frac{1}{2}\ +\quad +1/\sqrt{3}$

$\frac{1}{2}\ 1\ \frac{1}{2}\ -\quad +\sqrt{2}/\sqrt{3}$

$\tilde{\frac{1}{2}}\ \tilde{1}\ \frac{3}{2}$ 0

$\frac{1}{2}\ 0\ \frac{1}{2}\ -\quad -1/\sqrt{3}$

$\frac{1}{2}\ 1\ \frac{1}{2}\ +\quad +1/\sqrt{2.3}$

$\frac{1}{2}\ 1\ \frac{3}{2}\ -\quad -1/2$

$\tilde{\frac{1}{2}}\ \tilde{\frac{1}{2}}\ 0$ 0

$\frac{1}{2}\ \frac{1}{2}\ 0\ +\quad +1$

$\tilde{\frac{1}{2}}\ \tilde{\frac{1}{2}}\ 1$ 0

$\frac{1}{2}\ \frac{1}{2}\ \tilde{0}\ +\quad -1/\sqrt{3}$

$\frac{1}{2}\ \frac{1}{2}\ 1\ +\quad +\sqrt{2}/\sqrt{3}$

$\tilde{0}\ \frac{3}{2}\ \frac{3}{2}$ 0

$\tilde{0}\ \frac{1}{2}\ \frac{1}{2}\ +\quad -1/\sqrt{2}$

$\tilde{0}\ \frac{3}{2}\ \frac{3}{2}\ +\quad +1/2.3+i\sqrt{2/3}$

$\tilde{0}\ 2\ 2$ 0

$\tilde{0}\ 1\ 1\ +\quad +1$

$\tilde{0}\ \tilde{1}\ 1$ 0

$\tilde{0}\ 0\ \tilde{0}\ +\quad +1/\sqrt{3}$

$\tilde{0}\ 1\ 1\ -\quad +\sqrt{2}/\sqrt{3}$

$\tilde{0}\ \tilde{\frac{1}{2}}\ \frac{1}{2}$ 0

$\tilde{0}\ \frac{1}{2}\ \frac{1}{2}\ +\quad +1$

$\tilde{0}\ \tilde{0}\ 0$ 0

$\tilde{0}\ \tilde{0}\ 0\ +\quad +1$

O–D_4 and T_d–D_{2d} 2jm Factors (O_i–D_{4i} = O_h–D_{4h})

$0 \Rightarrow +\ 0$	$2 \Rightarrow +\ 0\ +\ 2$
$\frac{1}{2} \Rightarrow +\ \frac{1}{2}$	$\tilde{1} \Rightarrow +\ 1\ +\ \tilde{2}$
$1 \Rightarrow +\ \tilde{0}\ +\ 1$	$\tilde{\frac{1}{2}} \Rightarrow +\ \frac{3}{2}$
$\frac{3}{2} \Rightarrow +\ \frac{1}{2}\ +\ \frac{3}{2}$	$\tilde{0} \Rightarrow +\ 2$

O–D_4 and T_d–D_{2d} 3jm Factors

$0\ 0\ 0\ 0$

$0\ 0\ 0\ +\quad +1$

$\tfrac{1}{2}\ \tfrac{1}{2}\ 0\ 0$

$\tfrac{1}{2}\ \tfrac{1}{2}\ 0\ +\quad +1$

$1\ \tfrac{1}{2}\ \tfrac{1}{2}\ 0$

$\tilde{0}\ \tfrac{1}{2}\ \tfrac{1}{2}\ +\quad +1/\sqrt{3}$

$1\ \tfrac{1}{2}\ \tfrac{1}{2}\ +\quad +\sqrt{2}/\sqrt{3}$

$1\ 1\ 0\ 0$

$\tilde{0}\ \tilde{0}\ 0\ +\quad +1/\sqrt{3}$

$1\ 1\ 0\ +\quad +\sqrt{2}/\sqrt{3}$

$1\ 1\ 1\ 0$

$1\ 1\ \tilde{0}\ +\quad -1/\sqrt{3}$

$\tfrac{3}{2}\ 1\ \tfrac{1}{2}\ 0$

$\tfrac{1}{2}\ \tilde{0}\ \tfrac{1}{2}\ -\quad +1/\sqrt{3}$

$\tfrac{1}{2}\ 1\ \tfrac{1}{2}\ -\quad -1/\sqrt{2.3}$

$\tfrac{3}{2}\ 1\ \tfrac{1}{2}\ +\quad +1/\sqrt{2}$

$\tfrac{3}{2}\ \tfrac{3}{2}\ 0\ 0$

$\tfrac{1}{2}\ \tfrac{1}{2}\ 0\ +\quad +1/\sqrt{2}$

$\tfrac{3}{2}\ \tfrac{3}{2}\ 0\ +\quad +1/\sqrt{2}$

$\tfrac{3}{2}\ \tfrac{3}{2}\ 1\ 0$

$\tfrac{1}{2}\ \tfrac{1}{2}\ \tilde{0}\ +\quad -1/\sqrt{2.3}$

$\tfrac{1}{2}\ \tfrac{1}{2}\ 1\ +\quad -1/\sqrt{3}$

$\tfrac{3}{2}\ \tfrac{1}{2}\ 1\ -\quad 0$

$\tfrac{3}{2}\ \tfrac{3}{2}\ \tilde{0}\ +\quad +1/\sqrt{2.3}$

$\tfrac{3}{2}\ \tfrac{3}{2}\ 1\ +\quad -1/\sqrt{3}$

$\tfrac{3}{2}\ \tfrac{3}{2}\ 1\ 1$

$\tfrac{1}{2}\ \tfrac{1}{2}\ \tilde{0}\ +\quad +1/\sqrt{2.3}$

$\tfrac{1}{2}\ \tfrac{1}{2}\ 1\ +\quad -1/2\sqrt{3}$

$\tfrac{3}{2}\ \tfrac{1}{2}\ 1\ -\quad +1/2$

$\tfrac{3}{2}\ \tfrac{3}{2}\ \tilde{0}\ +\quad +1/\sqrt{2.3}$

$\tfrac{3}{2}\ \tfrac{3}{2}\ 1\ +\quad +1/2\sqrt{3}$

$2\ 1\ 1\ 0$

$0\ \tilde{0}\ \tilde{0}\ +\quad +1/\sqrt{3}$

$0\ 1\ 1\ +\quad -1/\sqrt{2.3}$

$2\ 1\ 1\ +\quad +1/\sqrt{2}$

$2\ \tfrac{3}{2}\ \tfrac{1}{2}\ 0$

$0\ \tfrac{1}{2}\ \tfrac{1}{2}\ -\quad +1/\sqrt{2}$

$2\ \tfrac{3}{2}\ \tfrac{1}{2}\ +\quad +1/\sqrt{2}$

$2\ \tfrac{3}{2}\ \tfrac{3}{2}\ 0$

$0\ \tfrac{1}{2}\ \tfrac{1}{2}\ +\quad -1/2$

$0\ \tfrac{3}{2}\ \tfrac{3}{2}\ +\quad +1/2$

$2\ \tfrac{3}{2}\ \tfrac{1}{2}\ -\quad +1/2$

$2\ 2\ 0\ 0$

$0\ 0\ 0\ +\quad +1/\sqrt{2}$

$2\ 2\ 0\ +\quad +1/\sqrt{2}$

$2\ 2\ 2\ 0$

$0\ 0\ 0\ +\quad -1/2$

$2\ 2\ 0\ +\quad +1/2$

$\tilde{1}\ 1\ 1\ 0$

$1\ 1\ \tilde{0}\ -\quad -1/\sqrt{3}$

$\tilde{2}\ 1\ 1\ +\quad +1/\sqrt{3}$

$\tilde{1}\ \tfrac{3}{2}\ \tfrac{1}{2}\ 0$

$1\ \tfrac{1}{2}\ \tfrac{1}{2}\ +\quad +1/\sqrt{2}$

$1\ \tfrac{3}{2}\ \tfrac{1}{2}\ -\quad -1/\sqrt{2.3}$

$\tilde{2}\ \tfrac{3}{2}\ \tfrac{1}{2}\ +\quad +1/\sqrt{3}$

$\tilde{1}\ \tfrac{3}{2}\ \tfrac{3}{2}\ 0$

$1\ \tfrac{1}{2}\ \tfrac{1}{2}\ -\quad 0$

$1\ \tfrac{3}{2}\ \tfrac{1}{2}\ +\quad +1/\sqrt{3}$

$1\ \tfrac{3}{2}\ \tfrac{3}{2}\ -\quad 0$

$\tilde{2}\ \tfrac{3}{2}\ \tfrac{1}{2}\ -\quad +1/\sqrt{2.3}$

$\tilde{1}\ \tfrac{3}{2}\ \tfrac{3}{2}\ 1$

$1\ \tfrac{1}{2}\ \tfrac{1}{2}\ +\quad -1/2$

$1\ \tfrac{3}{2}\ \tfrac{1}{2}\ -\quad -1/2\sqrt{3}$

$1\ \tfrac{3}{2}\ \tfrac{3}{2}\ +\quad +1/2$

$\tilde{2}\ \tfrac{3}{2}\ \tfrac{1}{2}\ +\quad +1/\sqrt{2.3}$

$\tilde{1}\ 2\ 1\ 0$

$1\ 0\ 1\ -\quad +1/\sqrt{2}$

$1\ 2\ 1\ -\quad +1/\sqrt{2.3}$

$\tilde{2}\ 2\ \tilde{0}\ +\quad -1/\sqrt{3}$

$\tilde{1}\ \tilde{1}\ 0\ 0$

$1\ 1\ 0\ +\quad +\sqrt{2}/\sqrt{3}$

$\tilde{2}\ \tilde{2}\ 0\ +\quad +1/\sqrt{3}$

$\tilde{1}\ \tilde{1}\ 1\ 0$

$1\ 1\ \tilde{0}\ +\quad -1/\sqrt{3}$

$\tilde{2}\ 1\ 1\ -\quad -1/\sqrt{3}$

$\tilde{1}\ \tilde{1}\ 2\ 0$

$1\ 1\ 0\ +\quad -1/\sqrt{2.3}$

$1\ 1\ 2\ +\quad +1/\sqrt{2}$

$\tilde{2}\ \tilde{2}\ 0\ +\quad +1/\sqrt{3}$

$\tilde{1}\ \tilde{1}\ \tilde{1}\ 0$

$\tilde{2}\ 1\ 1\ +\quad +1/\sqrt{3}$

$\widetilde{\tfrac{3}{2}}\ \tfrac{3}{2}\ 1\ 0$

$\tfrac{3}{2}\ \tfrac{1}{2}\ 1\ +\quad -1/\sqrt{2}$

$\tfrac{3}{2}\ \tfrac{3}{2}\ \tilde{0}\ -\quad +1/\sqrt{3}$

$\tfrac{3}{2}\ \tfrac{3}{2}\ 1\ -\quad +1/\sqrt{2.3}$

$\widetilde{\tfrac{3}{2}}\ 2\ \tfrac{3}{2}\ 0$

$\tfrac{3}{2}\ 0\ \tfrac{3}{2}\ -\quad +1/\sqrt{2}$

$\tfrac{3}{2}\ 2\ \tfrac{1}{2}\ +\quad +1/\sqrt{2}$

$\widetilde{\tfrac{3}{2}}\ \tilde{1}\ \tfrac{1}{2}\ 0$

$\tfrac{3}{2}\ 1\ \tfrac{1}{2}\ +\quad +\sqrt{2}/\sqrt{3}$

$\tfrac{3}{2}\ \tilde{2}\ \tfrac{1}{2}\ -\quad -1/\sqrt{3}$

$\widetilde{\tfrac{3}{2}}\ \tilde{1}\ \tfrac{3}{2}\ 0$

$\tfrac{3}{2}\ 1\ \tfrac{1}{2}\ -\quad +1/\sqrt{2.3}$

$\tfrac{3}{2}\ 1\ \tfrac{3}{2}\ +\quad -1/\sqrt{2}$

$\tfrac{3}{2}\ \tilde{2}\ \tfrac{1}{2}\ +\quad +1/\sqrt{3}$

$\widetilde{\tfrac{3}{2}}\ \widetilde{\tfrac{3}{2}}\ 0\ 0$

$\tfrac{3}{2}\ \tfrac{3}{2}\ 0\ +\quad +1$

$\widetilde{\tfrac{3}{2}}\ \widetilde{\tfrac{3}{2}}\ 1\ 0$

$\tfrac{3}{2}\ \tfrac{3}{2}\ \tilde{0}\ +\quad +1/\sqrt{3}$

$\tfrac{3}{2}\ \tfrac{3}{2}\ 1\ +\quad -\sqrt{2}/\sqrt{3}$

$\tilde{0}\ \tfrac{3}{2}\ \tfrac{3}{2}\ 0$

$2\ \tfrac{3}{2}\ \tfrac{1}{2}\ +\quad -1/\sqrt{2}$

$\tilde{0}\ 2\ 2\ 0$

$2\ 2\ 0\ -\quad +1/\sqrt{2}$

$\tilde{0}\ \tilde{1}\ 1\ 0$

$2\ 1\ 1\ +\quad +\sqrt{2}/\sqrt{3}$

$2\ \tilde{2}\ \tilde{0}\ -\quad -1/\sqrt{3}$

$\tilde{0}\ \widetilde{\tfrac{3}{2}}\ \tfrac{1}{2}\ 0$

$2\ \tfrac{3}{2}\ \tfrac{1}{2}\ +\quad +1$

$\tilde{0}\ \tilde{0}\ 0\ 0$

$2\ 2\ 0\ +\quad +1$

O–T and T_d–T 2jm Factors (O_i–T_i = O_h–T_h)

$0 \Rightarrow +\ 0$	$\frac{3}{2} \Rightarrow +\ \frac{3}{2}\ +\text{-}\frac{3}{2}$	$\frac{\tilde{1}}{2} \Rightarrow -\ \frac{1}{2}$
$\frac{1}{2} \Rightarrow +\ \frac{1}{2}$	$2 \Rightarrow +\ 2\ +\text{-}2$	$\tilde{0} \Rightarrow -\ 0$
$1 \Rightarrow +\ 1$	$1 \Rightarrow -\ 1$	

O–T and T_d–T 3jm Factors

0 0 0 0	**2 $\frac{3}{2}$ $\frac{3}{2}$ 0**	**$\tilde{1}$ $\tilde{1}$ $\tilde{1}$ 0**
0 0 0 0 + $+1$	2 $\frac{3}{2}$ $\frac{3}{2}$ 0 + $+1/\sqrt{2}$	1 1 1 0 −* 0
$\frac{1}{2}$ $\frac{1}{2}$ 0 0	**2 2 0 0**	1 1 1 1 +* $+i$
$\frac{1}{2}$ $\frac{1}{2}$ 0 0 + $+1$	-2 2 0 0 + $+1/\sqrt{2}$	**$\frac{\tilde{1}}{2}$ $\frac{3}{2}$ 1 0**
1 $\frac{1}{2}$ $\frac{1}{2}$ 0	**2 2 2 0**	$\frac{1}{2}$ $\frac{3}{2}$ 1 0 −* $-1/\sqrt{2}$
1 $\frac{1}{2}$ $\frac{1}{2}$ 0 + $+1$	2 2 2 0 + $-1/\sqrt{2}$	**$\frac{\tilde{1}}{2}$ 2 $\frac{3}{2}$ 0**
1 1 0 0	**$\tilde{1}$ 1 1 0**	$\frac{1}{2}$ 2 -$\frac{3}{2}$ 0 −* $-1/\sqrt{2}$
1 1 0 0 + $+1$	1 1 1 0 −* 0	**$\frac{\tilde{1}}{2}$ $\tilde{1}$ $\frac{1}{2}$ 0**
1 1 1 0	1 1 1 1 +* $+i$	$\frac{1}{2}$ 1 $\frac{1}{2}$ 0 − $+1$
1 1 1 0 + $+1$	**$\tilde{1}$ $\frac{3}{2}$ $\frac{1}{2}$ 0**	**$\frac{\tilde{1}}{2}$ $\tilde{1}$ $\frac{3}{2}$ 0**
1 1 1 1 − 0	1 $\frac{3}{2}$ $\frac{1}{2}$ 0 +* $+1/\sqrt{2}$	$\frac{1}{2}$ 1 $\frac{3}{2}$ 0 + $-1/\sqrt{2}$
$\frac{3}{2}$ 1 $\frac{1}{2}$ 0	**$\tilde{1}$ $\frac{3}{2}$ $\frac{3}{2}$ 0**	**$\frac{\tilde{1}}{2}$ $\frac{\tilde{1}}{2}$ 0 0**
$\frac{3}{2}$ 1 $\frac{1}{2}$ 0 − $+1/\sqrt{2}$	1 $\frac{3}{2}$ $\frac{3}{2}$ 0 −* 0	$\frac{1}{2}$ $\frac{1}{2}$ 0 0 + -1
$\frac{3}{2}$ $\frac{3}{2}$ 0 0	1 -$\frac{3}{2}$ $\frac{3}{2}$ 0 −* $-1/\sqrt{2}$	**$\frac{\tilde{1}}{2}$ $\frac{\tilde{1}}{2}$ 1 0**
-$\frac{3}{2}$ $\frac{3}{2}$ 0 0 + $+1/\sqrt{2}$	**$\tilde{1}$ $\frac{3}{2}$ $\frac{3}{2}$ 1**	$\frac{1}{2}$ $\frac{1}{2}$ 1 0 + $+1$
$\frac{3}{2}$ $\frac{3}{2}$ 1 0	1 $\frac{3}{2}$ $\frac{3}{2}$ 0 +* $-1/\sqrt{2}$	**$\tilde{0}$ $\frac{3}{2}$ $\frac{3}{2}$ 0**
$\frac{3}{2}$ $\frac{3}{2}$ 1 0 + 0	1 -$\frac{3}{2}$ $\frac{3}{2}$ 0 +* 0	0 -$\frac{3}{2}$ $\frac{3}{2}$ 0 −* $-1/\sqrt{2}$
-$\frac{3}{2}$ $\frac{3}{2}$ 1 0 + $-1/\sqrt{2}$	**$\tilde{1}$ 2 1 0**	**$\tilde{0}$ 2 2 0**
$\frac{3}{2}$ $\frac{3}{2}$ 1 1	1 2 1 0 −* $+1/\sqrt{2}$	0 -2 2 0 −* $+1/\sqrt{2}$
$\frac{3}{2}$ $\frac{3}{2}$ 1 0 + $-1/\sqrt{2}$	**$\tilde{1}$ $\tilde{1}$ 0 0**	**$\tilde{0}$ $\tilde{1}$ 1 0**
-$\frac{3}{2}$ $\frac{3}{2}$ 1 0 + 0	1 1 0 0 + -1	0 1 1 0 + -1
2 1 1 0	**$\tilde{1}$ $\tilde{1}$ 1 0**	**$\tilde{0}$ $\frac{\tilde{1}}{2}$ $\frac{1}{2}$ 0**
2 1 1 0 + $+1/\sqrt{2}$	1 1 1 0 + $+1$	0 $\frac{1}{2}$ $\frac{1}{2}$ 0 − $+1$
2 $\frac{3}{2}$ $\frac{1}{2}$ 0	1 1 1 1 − 0	**$\tilde{0}$ $\tilde{0}$ 0 0**
2 -$\frac{3}{2}$ $\frac{1}{2}$ 0 − $+1/\sqrt{2}$	**$\tilde{1}$ $\tilde{1}$ 2 0**	0 0 0 0 + -1
	1 1 2 0 + $-1/\sqrt{2}$	

O_h–T_d 2jm Factors

$0^+ \Rightarrow +\ 0$	$1^+ \Rightarrow +\ 1$	$2^+ \Rightarrow +\ 2$	$\frac{\tilde{1}}{2}^+ \Rightarrow +\ \frac{\tilde{1}}{2}$
$0^- \Rightarrow +\ \tilde{0}$	$1^- \Rightarrow +\ \tilde{1}$	$2^- \Rightarrow +\ 2$	$\frac{\tilde{1}}{2}^- \Rightarrow +\ \frac{1}{2}$
$\frac{1}{2}^+ \Rightarrow +\ \frac{1}{2}$	$\frac{3}{2}^+ \Rightarrow +\ \frac{3}{2}$	$\tilde{1}^+ \Rightarrow +\ \tilde{1}$	$\tilde{0}^+ \Rightarrow +\ \tilde{0}$
$\frac{1}{2}^- \Rightarrow +\ \frac{\tilde{1}}{2}$	$\frac{3}{2}^- \Rightarrow +\ \frac{3}{2}$	$\tilde{1}^- \Rightarrow +\ 1$	$\tilde{0}^- \Rightarrow +\ 0$

O_h–T_d 3jm Factors

0^+	**0^+**	**0^+**	**0**	
0	0	0	$0+$	$+1$
0^-	**0^-**	**0^+**	**0**	
$\tilde{0}$	$\tilde{0}$	0	$0+$	$+1$
$\frac{1}{2}^+$	**$\frac{1}{2}^+$**	**0^+**	**0**	
$\frac{1}{2}$	$\frac{1}{2}$	0	$0+$	$+1$
$\frac{1}{2}^-$	**$\frac{1}{2}^-$**	**0^+**	**0**	
$\tilde{\frac{1}{2}}$	$\tilde{\frac{1}{2}}$	0	$0+$	$+1$
$\frac{1}{2}^-$	**$\frac{1}{2}^+$**	**0^-**	**0**	
$\tilde{\frac{1}{2}}$	$\frac{1}{2}$	$\tilde{0}$	$0-$	$+1$
1^+	**$\frac{1}{2}^+$**	**$\frac{1}{2}^+$**	**0**	
1	$\frac{1}{2}$	$\frac{1}{2}$	$0+$	$+1$
1^-	**$\frac{1}{2}^-$**	**$\frac{1}{2}^+$**	**0**	
$\tilde{1}$	$\tilde{\frac{1}{2}}$	$\frac{1}{2}$	$0-$	-1
1^+	**$\frac{1}{2}^-$**	**$\frac{1}{2}^-$**	**0**	
1	$\tilde{\frac{1}{2}}$	$\tilde{\frac{1}{2}}$	$0+$	-1
1^+	**1^+**	**0^+**	**0**	
1	1	0	$0+$	$+1$
1^-	**1^-**	**0^+**	**0**	
$\tilde{1}$	$\tilde{1}$	0	$0+$	$+1$
1^-	**1^+**	**0^-**	**0**	
$\tilde{1}$	1	$\tilde{0}$	$0+$	$+1$
1^+	**1^+**	**1^+**	**0**	
1	1	1	$0+$	$+1$
1^-	**1^-**	**1^+**	**0**	
$\tilde{1}$	$\tilde{1}$	1	$0+$	-1
$\frac{3}{2}^+$	**1^+**	**$\frac{1}{2}^+$**	**0**	
$\frac{3}{2}$	1	$\frac{1}{2}$	$0+$	$+1$
$\frac{3}{2}^-$	**1^-**	**$\frac{1}{2}^+$**	**0**	
$\frac{3}{2}$	$\tilde{1}$	$\frac{1}{2}$	$0-$	$+1$
$\frac{3}{2}^-$	**1^+**	**$\frac{1}{2}^-$**	**0**	
$\frac{3}{2}$	1	$\tilde{\frac{1}{2}}$	$0+$	$+1$
$\frac{3}{2}^+$	**1^-**	**$\frac{1}{2}^-$**	**0**	
$\frac{3}{2}$	$\tilde{1}$	$\tilde{\frac{1}{2}}$	$0-$	-1
$\frac{3}{2}^+$	**$\frac{3}{2}^+$**	**0^+**	**0**	
$\frac{3}{2}$	$\frac{3}{2}$	0	$0+$	$+1$
$\frac{3}{2}^-$	**$\frac{3}{2}^-$**	**0^+**	**0**	
$\frac{3}{2}$	$\frac{3}{2}$	0	$0+$	$+1$
$\frac{3}{2}^-$	**$\frac{3}{2}^+$**	**0^-**	**0**	
$\frac{3}{2}$	$\frac{3}{2}$	$\tilde{0}$	$0-$	$+1$
$\frac{3}{2}^+$	**$\frac{3}{2}^+$**	**1^+**	**0**	
$\frac{3}{2}$	$\frac{3}{2}$	1	$0+$	$+1$
$\frac{3}{2}$	$\frac{3}{2}$	1	$1+$	0
$\frac{3}{2}^-$	**$\frac{3}{2}^-$**	**1^+**	**0**	
$\frac{3}{2}$	$\frac{3}{2}$	1	$0+$	$+1$
$\frac{3}{2}$	$\frac{3}{2}$	1	$1+$	0
$\frac{3}{2}^-$	**$\frac{3}{2}^+$**	**1^-**	**0**	
$\frac{3}{2}$	$\frac{3}{2}$	$\tilde{1}$	$0-$	-1
$\frac{3}{2}$	$\frac{3}{2}$	$\tilde{1}$	$1+$	0
$\frac{3}{2}^+$	**$\frac{3}{2}^+$**	**1^+**	**1**	
$\frac{3}{2}$	$\frac{3}{2}$	1	$0+$	0
$\frac{3}{2}$	$\frac{3}{2}$	1	$1+$	$+1$
$\frac{3}{2}^-$	**$\frac{3}{2}^-$**	**1^+**	**1**	
$\frac{3}{2}$	$\frac{3}{2}$	1	$0+$	0
$\frac{3}{2}$	$\frac{3}{2}$	1	$1+$	-1
$\frac{3}{2}^-$	**$\frac{3}{2}^+$**	**1^-**	**1**	
$\frac{3}{2}$	$\frac{3}{2}$	$\tilde{1}$	$0-$	0
$\frac{3}{2}$	$\frac{3}{2}$	$\tilde{1}$	$1+$	$+1$
2^+	**1^+**	**1^+**	**0**	
2	1	1	$0+$	$+1$
2^-	**1^-**	**1^+**	**0**	
2	$\tilde{1}$	1	$0-$	$+1$
2^+	**1^-**	**1^-**	**0**	
2	$\tilde{1}$	$\tilde{1}$	$0+$	$+1$
2^+	**$\frac{3}{2}^+$**	**$\frac{1}{2}^+$**	**0**	
2	$\frac{3}{2}$	$\frac{1}{2}$	$0+$	$+1$
2^-	**$\frac{3}{2}^-$**	**$\frac{1}{2}^+$**	**0**	
2	$\frac{3}{2}$	$\frac{1}{2}$	$0+$	-1
2^-	**$\frac{3}{2}^+$**	**$\frac{1}{2}^-$**	**0**	
2	$\frac{3}{2}$	$\tilde{\frac{1}{2}}$	$0+$	-1
2^+	**$\frac{3}{2}^-$**	**$\frac{1}{2}^-$**	**0**	
2	$\frac{3}{2}$	$\tilde{\frac{1}{2}}$	$0+$	-1
2^+	**$\frac{3}{2}^+$**	**$\frac{3}{2}^+$**	**0**	
2	$\frac{3}{2}$	$\frac{3}{2}$	$0+$	$+1$
2^-	**$\frac{3}{2}^-$**	**$\frac{3}{2}^+$**	**0**	
2	$\frac{3}{2}$	$\frac{3}{2}$	$0+$	$+1$
2^+	**$\frac{3}{2}^-$**	**$\frac{3}{2}^-$**	**0**	
2	$\frac{3}{2}$	$\frac{3}{2}$	$0+$	-1
2^+	**2^+**	**0^+**	**0**	
2	2	0	$0+$	$+1$
2^-	**2^-**	**0^+**	**0**	
2	2	0	$0+$	$+1$
2^-	**2^+**	**0^-**	**0**	
2	2	$\tilde{0}$	$0-$	$+1$
2^+	**2^+**	**2^+**	**0**	
2	2	2	$0+$	$+1$
2^-	**2^-**	**2^+**	**0**	
2	2	2	$0+$	-1
$\tilde{1}^+$	**1^+**	**1^+**	**0**	
$\tilde{1}$	1	1	$0+$	$+1$
$\tilde{1}^-$	**1^-**	**1^+**	**0**	
1	$\tilde{1}$	1	$0+$	$+1$
$\tilde{1}^+$	**1^-**	**1^-**	**0**	
$\tilde{1}$	$\tilde{1}$	$\tilde{1}$	$0+$	-1
$\tilde{1}^+$	**$\frac{3}{2}^+$**	**$\frac{1}{2}^+$**	**0**	
$\tilde{1}$	$\frac{3}{2}$	$\frac{1}{2}$	$0+$	$+1$
$\tilde{1}^-$	**$\frac{3}{2}^-$**	**$\frac{1}{2}^+$**	**0**	
1	$\frac{3}{2}$	$\frac{1}{2}$	$0-$	$+1$
$\tilde{1}^-$	**$\frac{3}{2}^+$**	**$\frac{1}{2}^-$**	**0**	
1	$\frac{3}{2}$	$\tilde{\frac{1}{2}}$	$0-$	-1
$\tilde{1}^+$	**$\frac{3}{2}^-$**	**$\frac{1}{2}^-$**	**0**	
$\tilde{1}$	$\frac{3}{2}$	$\tilde{\frac{1}{2}}$	$0+$	$+1$
$\tilde{1}^+$	**$\frac{3}{2}^+$**	**$\frac{3}{2}^+$**	**0**	
$\tilde{1}$	$\frac{3}{2}$	$\frac{3}{2}$	$0+$	$+1$
$\tilde{1}$	$\frac{3}{2}$	$\frac{3}{2}$	$1+$	0
$\tilde{1}^-$	**$\frac{3}{2}^-$**	**$\frac{3}{2}^+$**	**0**	
1	$\frac{3}{2}$	$\frac{3}{2}$	$0-$	$+1$
1	$\frac{3}{2}$	$\frac{3}{2}$	$1-$	0
$\tilde{1}^+$	**$\frac{3}{2}^-$**	**$\frac{3}{2}^-$**	**0**	
$\tilde{1}$	$\frac{3}{2}$	$\frac{3}{2}$	$0+$	$+1$
$\tilde{1}$	$\frac{3}{2}$	$\frac{3}{2}$	$1-$	0
$\tilde{1}^+$	**$\frac{3}{2}^+$**	**$\frac{3}{2}^+$**	**1**	
$\tilde{1}$	$\frac{3}{2}$	$\frac{3}{2}$	$0+$	0
$\tilde{1}$	$\frac{3}{2}$	$\frac{3}{2}$	$1+$	$+1$
$\tilde{1}^-$	**$\frac{3}{2}^-$**	**$\frac{3}{2}^+$**	**1**	
1	$\frac{3}{2}$	$\frac{3}{2}$	$0+$	0
1	$\frac{3}{2}$	$\frac{3}{2}$	$1+$	-1
$\tilde{1}^+$	**$\frac{3}{2}^-$**	**$\frac{3}{2}^-$**	**1**	
$\tilde{1}$	$\frac{3}{2}$	$\frac{3}{2}$	$0-$	0
$\tilde{1}$	$\frac{3}{2}$	$\frac{3}{2}$	$1+$	-1

O_h–T_d 3jm Factors (cont.)

$\tilde{1}^{+}\ 2^{+}\ 1^{+}\ 0$

$\tilde{1}\ 2\ 1\ 0+\ +1$

$\tilde{1}^{-}\ 2^{-}\ 1^{+}\ 0$

$1\ 2\ 1\ 0-\ +1$

$\tilde{1}^{-}\ 2^{+}\ 1^{-}\ 0$

$1\ 2\ \tilde{1}\ 0+\ -1$

$\tilde{1}^{+}\ 2^{-}\ 1^{-}\ 0$

$\tilde{1}\ 2\ \tilde{1}\ 0-\ +1$

$\tilde{1}^{+}\ \tilde{1}^{+}\ 0^{+}\ 0$

$\tilde{1}\ \tilde{1}\ 0\ 0+\ +1$

$\tilde{1}^{-}\ \tilde{1}^{-}\ 0^{+}\ 0$

$1\ 1\ 0\ 0+\ +1$

$\tilde{1}^{-}\ \tilde{1}^{+}\ 0^{-}\ 0$

$1\ \tilde{1}\ \tilde{0}\ 0+\ +1$

$\tilde{1}^{+}\ \tilde{1}^{+}\ 1^{+}\ 0$

$\tilde{1}\ \tilde{1}\ 1\ 0+\ +1$

$\tilde{1}^{-}\ \tilde{1}^{-}\ 1^{+}\ 0$

$1\ 1\ 1\ 0+\ -1$

$\tilde{1}^{-}\ \tilde{1}^{+}\ 1^{-}\ 0$

$1\ \tilde{1}\ \tilde{1}\ 0+\ +1$

$\tilde{1}^{+}\ \tilde{1}^{+}\ 2^{+}\ 0$

$\tilde{1}\ \tilde{1}\ 2\ 0+\ +1$

$\tilde{1}^{-}\ \tilde{1}^{-}\ 2^{+}\ 0$

$1\ 1\ 2\ 0+\ +1$

$\tilde{1}^{-}\ \tilde{1}^{+}\ 2^{-}\ 0$

$1\ \tilde{1}\ 2\ 0-\ -1$

$\tilde{1}^{+}\ \tilde{1}^{+}\ \tilde{1}^{+}\ 0$

$\tilde{1}\ \tilde{1}\ \tilde{1}\ 0+\ +1$

$\tilde{1}^{-}\ \tilde{1}^{-}\ \tilde{1}^{+}\ 0$

$1\ 1\ \tilde{1}\ 0+\ -1$

$\tilde{\frac{1}{2}}^{+}\ \frac{3}{2}^{+}\ 1^{+}\ 0$

$\tilde{\frac{1}{2}}\ \frac{3}{2}\ 1\ 0+\ +1$

$\tilde{\frac{1}{2}}^{-}\ \frac{3}{2}^{-}\ 1^{+}\ 0$

$\frac{1}{2}\ \frac{3}{2}\ 1\ 0+\ +1$

$\tilde{\frac{1}{2}}^{-}\ \frac{3}{2}^{+}\ 1^{-}\ 0$

$\frac{1}{2}\ \frac{3}{2}\ \tilde{1}\ 0-\ -1$

$\tilde{\frac{1}{2}}^{+}\ \frac{3}{2}^{-}\ 1^{-}\ 0$

$\tilde{\frac{1}{2}}\ \frac{3}{2}\ \tilde{1}\ 0-\ +1$

$\tilde{\frac{1}{2}}^{+}\ 2^{+}\ \frac{3}{2}^{+}\ 0$

$\tilde{\frac{1}{2}}\ 2\ \frac{3}{2}\ 0+\ +1$

$\tilde{\frac{1}{2}}^{-}\ 2^{-}\ \frac{3}{2}^{+}\ 0$

$\frac{1}{2}\ 2\ \frac{3}{2}\ 0+\ -1$

$\tilde{\frac{1}{2}}^{-}\ 2^{+}\ \frac{3}{2}^{-}\ 0$

$\frac{1}{2}\ 2\ \frac{3}{2}\ 0+\ -1$

$\tilde{\frac{1}{2}}^{+}\ 2^{-}\ \frac{3}{2}^{-}\ 0$

$\tilde{\frac{1}{2}}\ 2\ \frac{3}{2}\ 0+\ -1$

$\tilde{\frac{1}{2}}^{+}\ \tilde{1}^{+}\ \frac{1}{2}^{+}\ 0$

$\tilde{\frac{1}{2}}\ \tilde{1}\ \frac{1}{2}\ 0+\ +1$

$\tilde{\frac{1}{2}}^{-}\ \tilde{1}^{-}\ \frac{1}{2}^{+}\ 0$

$\frac{1}{2}\ 1\ \frac{1}{2}\ 0-\ -1$

$\tilde{\frac{1}{2}}^{-}\ \tilde{1}^{+}\ \frac{1}{2}^{-}\ 0$

$\frac{1}{2}\ \tilde{1}\ \tilde{\frac{1}{2}}\ 0+\ +1$

$\tilde{\frac{1}{2}}^{+}\ \tilde{1}^{-}\ \frac{1}{2}^{-}\ 0$

$\tilde{\frac{1}{2}}\ 1\ \tilde{\frac{1}{2}}\ 0-\ -1$

$\tilde{\frac{1}{2}}^{+}\ \tilde{1}^{+}\ \frac{3}{2}^{+}\ 0$

$\tilde{\frac{1}{2}}\ \tilde{1}\ \frac{3}{2}\ 0+\ +1$

$\tilde{\frac{1}{2}}^{-}\ \tilde{1}^{-}\ \frac{3}{2}^{+}\ 0$

$\frac{1}{2}\ 1\ \frac{3}{2}\ 0-\ -1$

$\tilde{\frac{1}{2}}^{-}\ \tilde{1}^{+}\ \frac{3}{2}^{-}\ 0$

$\frac{1}{2}\ \tilde{1}\ \frac{3}{2}\ 0+\ +1$

$\tilde{\frac{1}{2}}^{+}\ \tilde{1}^{-}\ \frac{3}{2}^{-}\ 0$

$\tilde{\frac{1}{2}}\ 1\ \frac{3}{2}\ 0-\ +1$

$\tilde{\frac{1}{2}}^{+}\ \tilde{\frac{1}{2}}^{+}\ 0^{+}\ 0$

$\tilde{\frac{1}{2}}\ \tilde{\frac{1}{2}}\ 0\ 0+\ +1$

$\tilde{\frac{1}{2}}^{-}\ \tilde{\frac{1}{2}}^{-}\ 0^{+}\ 0$

$\frac{1}{2}\ \frac{1}{2}\ 0\ 0+\ +1$

$\tilde{\frac{1}{2}}^{-}\ \tilde{\frac{1}{2}}^{+}\ 0^{-}\ 0$

$\frac{1}{2}\ \tilde{\frac{1}{2}}\ \tilde{0}\ 0-\ +1$

$\tilde{\frac{1}{2}}^{+}\ \tilde{\frac{1}{2}}^{+}\ 1^{+}\ 0$

$\tilde{\frac{1}{2}}\ \tilde{\frac{1}{2}}\ 1\ 0+\ +1$

$\tilde{\frac{1}{2}}^{-}\ \tilde{\frac{1}{2}}^{-}\ 1^{+}\ 0$

$\frac{1}{2}\ \frac{1}{2}\ 1\ 0+\ -1$

$\tilde{\frac{1}{2}}^{-}\ \tilde{\frac{1}{2}}^{+}\ 1^{-}\ 0$

$\frac{1}{2}\ \tilde{\frac{1}{2}}\ \tilde{1}\ 0-\ +1$

$\tilde{0}^{+}\ \frac{3}{2}^{+}\ \frac{3}{2}^{+}\ 0$

$\tilde{0}\ \frac{3}{2}\ \frac{3}{2}\ 0+\ +1$

$\tilde{0}^{-}\ \frac{3}{2}^{-}\ \frac{3}{2}^{+}\ 0$

$0\ \frac{3}{2}\ \frac{3}{2}\ 0-\ -1$

$\tilde{0}^{+}\ \frac{3}{2}^{-}\ \frac{3}{2}^{-}\ 0$

$\tilde{0}\ \frac{3}{2}\ \frac{3}{2}\ 0+\ +1$

$\tilde{0}^{+}\ 2^{+}\ 2^{+}\ 0$

$\tilde{0}\ 2\ 2\ 0+\ +1$

$\tilde{0}^{-}\ 2^{-}\ 2^{+}\ 0$

$0\ 2\ 2\ 0-\ -1$

$\tilde{0}^{+}\ 2^{-}\ 2^{-}\ 0$

$\tilde{0}\ 2\ 2\ 0+\ +1$

$\tilde{0}^{+}\ \tilde{1}^{+}\ 1^{+}\ 0$

$\tilde{0}\ \tilde{1}\ 1\ 0+\ +1$

$\tilde{0}^{-}\ \tilde{1}^{-}\ 1^{+}\ 0$

$0\ 1\ 1\ 0+\ +1$

$\tilde{0}^{-}\ \tilde{1}^{+}\ 1^{-}\ 0$

$0\ \tilde{1}\ \tilde{1}\ 0+\ +1$

$\tilde{0}^{+}\ \tilde{1}^{-}\ 1^{-}\ 0$

$\tilde{0}\ 1\ \tilde{1}\ 0+\ +1$

$\tilde{0}^{+}\ \tilde{\frac{1}{2}}^{+}\ \frac{1}{2}^{+}\ 0$

$\tilde{0}\ \tilde{\frac{1}{2}}\ \frac{1}{2}\ 0+\ +1$

$\tilde{0}^{-}\ \tilde{\frac{1}{2}}^{-}\ \frac{1}{2}^{+}\ 0$

$0\ \frac{1}{2}\ \frac{1}{2}\ 0-\ -1$

$\tilde{0}^{-}\ \tilde{\frac{1}{2}}^{+}\ \frac{1}{2}^{-}\ 0$

$0\ \tilde{\frac{1}{2}}\ \tilde{\frac{1}{2}}\ 0-\ +1$

$\tilde{0}^{+}\ \tilde{\frac{1}{2}}^{-}\ \frac{1}{2}^{-}\ 0$

$\tilde{0}\ \frac{1}{2}\ \tilde{\frac{1}{2}}\ 0+\ +1$

$\tilde{0}^{+}\ \tilde{0}^{+}\ 0^{+}\ 0$

$\tilde{0}\ \tilde{0}\ 0\ 0+\ +1$

$\tilde{0}^{-}\ \tilde{0}^{-}\ 0^{+}\ 0$

$0\ 0\ 0\ 0+\ +1$

$\tilde{0}^{-}\ \tilde{0}^{+}\ 0^{-}\ 0$

$0\ \tilde{0}\ \tilde{0}\ 0+\ +1$

K–D_3 2jm Factors (K_i–D_{3i} = K_h–D_{3d})

$0 \Rightarrow +\,0$

$\frac{1}{2} \Rightarrow +\,\frac{1}{2}$

$1 \Rightarrow +\,\tilde{0} + 1$

$\frac{3}{2} \Rightarrow +\,\frac{1}{2} + \frac{3}{2} + \text{-}\frac{3}{2}$

$2 \Rightarrow +\,0 + {}_2 1$

$\frac{5}{2} \Rightarrow +\,{}_2\frac{1}{2} + \frac{3}{2} + \text{-}\frac{3}{2}$

$3 \Rightarrow +\,0 + \tilde{0} + 1$

$\tilde{1} \Rightarrow +\,\tilde{0} + 1$

$\tilde{\frac{1}{2}} \Rightarrow +\,\frac{1}{2}$

K–D_3 3jm Factors

0 0 0 0

$0\ 0\ 0\ 000+ \quad +1$

$\frac{1}{2}\ \frac{1}{2}$ 0 0

$\frac{1}{2}\ \frac{1}{2}\ 0\ 000+ \quad +1$

1 $\frac{1}{2}\ \frac{1}{2}$ 0

$\tilde{0}\ \frac{1}{2}\ \frac{1}{2}\ 000+ \quad +1/\sqrt{3}$

$1\ \frac{1}{2}\ \frac{1}{2}\ 000+ \quad +\sqrt{2}/\sqrt{3}$

1 1 0 0

$\tilde{0}\ \tilde{0}\ 0\ 000+ \quad +1/\sqrt{3}$

$1\ 1\ 0\ 000+ \quad +\sqrt{2}/\sqrt{3}$

1 1 1 0

$1\ 1\ \tilde{0}\ 000+ \quad -1/\sqrt{3}$

$1\ 1\ 1\ 000- \quad 0$

$\frac{3}{2}$ 1 $\frac{1}{2}$ 0

$\frac{1}{2}\ \tilde{0}\ \frac{1}{2}\ 000- \quad +1/\sqrt{3}$

$\frac{1}{2}\ 1\ \frac{1}{2}\ 000- \quad -1/\sqrt{2.3}$

$\frac{3}{2}\ 1\ \frac{1}{2}\ 000+ \quad +1/2$

$\frac{3}{2}\ \frac{3}{2}$ 0 0

$\frac{1}{2}\ \frac{1}{2}\ 0\ 000+ \quad +1/\sqrt{2}$

$\text{-}\frac{3}{2}\ \frac{3}{2}\ 0\ 000+ \quad +1/2$

$\frac{3}{2}\ \frac{3}{2}$ 1 0

$\frac{1}{2}\ \frac{1}{2}\ \tilde{0}\ 000+ \quad +1/\sqrt{2.3.5}$

$\frac{1}{2}\ \frac{1}{2}\ 1\ 000+ \quad -2/\sqrt{3.5}$

$\frac{3}{2}\ \frac{1}{2}\ 1\ 000- \quad +1/\sqrt{2.5}$

$\frac{3}{2}\ \frac{3}{2}\ \tilde{0}\ 000+ \quad +\sqrt{3}/2\sqrt{5}$

2 1 1 0

$0\ \tilde{0}\ \tilde{0}\ 000+ \quad +\sqrt{2}/\sqrt{3.5}$

$0\ 1\ 1\ 000+ \quad -1/\sqrt{3.5}$

$1\ 1\ \tilde{0}\ 000- \quad -1/\sqrt{5}$

$1\ 1\ \tilde{0}\ 100- \quad 0$

2 1 1 0

$1\ 1\ 1\ 000+ \quad 0$

$1\ 1\ 1\ 100+ \quad +\sqrt{2}/\sqrt{5}$

2 $\frac{3}{2}\ \frac{1}{2}$ 0

$0\ \frac{1}{2}\ \frac{1}{2}\ 000- \quad +1/\sqrt{5}$

$1\ \frac{1}{2}\ \frac{1}{2}\ 000+ \quad +\sqrt{3}/\sqrt{2.5}$

$1\ \frac{1}{2}\ \frac{1}{2}\ 100+ \quad 0$

$1\ \frac{3}{2}\ \frac{1}{2}\ 000- \quad -1/2\sqrt{5}$

$1\ \frac{3}{2}\ \frac{1}{2}\ 100- \quad +i/\sqrt{5}$

2 $\frac{3}{2}\ \frac{3}{2}$ 0

$0\ \frac{1}{2}\ \frac{1}{2}\ 000+ \quad -1/\sqrt{2.5}$

$0\ \text{-}\frac{3}{2}\ \frac{3}{2}\ 000+ \quad +1/2\sqrt{5}$

$1\ \frac{1}{2}\ \frac{1}{2}\ 000- \quad 0$

$1\ \frac{1}{2}\ \frac{1}{2}\ 100- \quad 0$

$1\ \frac{3}{2}\ \frac{1}{2}\ 000+ \quad +1/\sqrt{2.5}$

$1\ \frac{3}{2}\ \frac{1}{2}\ 100+ \quad +i/\sqrt{2.5}$

2 2 0 0

$0\ 0\ 0\ 000+ \quad +1/\sqrt{5}$

$1\ 1\ 0\ 000+ \quad +\sqrt{2}/\sqrt{5}$

$1\ 1\ 0\ 100+ \quad 0$

$1\ 1\ 0\ 110+ \quad +\sqrt{2}/\sqrt{5}$

2 2 1 0

$1\ 0\ 1\ 000- \quad +1/\sqrt{5}$

$1\ 0\ 1\ 100- \quad 0$

$1\ 1\ \tilde{0}\ 000+ \quad -1/\sqrt{3.5}$

$1\ 1\ \tilde{0}\ 100+ \quad 0$

$1\ 1\ \tilde{0}\ 110+ \quad +2/\sqrt{3.5}$

$1\ 1\ 1\ 000- \quad 0$

$1\ 1\ 1\ 100- \quad -\sqrt{2}/\sqrt{3.5}$

$1\ 1\ 1\ 110- \quad 0$

K–D_3 3jm Factors (cont.)

2 2 2 0

0 0 0 000+ 0
1 1 0 000+ $-1/3\sqrt{2}$
1 1 0 100+ $+\sqrt{2}/3\sqrt{5}$
1 1 0 110+ $+1/3\sqrt{2}$
1 1 1 000+ $-4/3\sqrt{3.5}$
1 1 1 100+ $+1/3\sqrt{3}$
1 1 1 110+ $+1/3\sqrt{3.5}$
1 1 1 111+ $+2/3\sqrt{3}$

2 2 2 1

0 0 0 000+ $+2/3\sqrt{5}$
1 1 0 000+ $+1/9\sqrt{2.5}$
1 1 0 100+ $+\sqrt{2/9}$
1 1 0 110+ $-7/9\sqrt{2.5}$
1 1 1 000+ $-4/9\sqrt{3}$
1 1 1 100+ $-13/9\sqrt{3.5}$
1 1 1 110+ $+1/9\sqrt{3}$
1 1 1 111+ $+2\sqrt{5}/9\sqrt{3}$

$\frac{5}{2}$ $\frac{3}{2}$ 1 0

$\frac{1}{2}$ $\frac{1}{2}$ $\tilde{0}$ 000− $-1/\sqrt{5}$
$\frac{1}{2}$ $\frac{1}{2}$ $\tilde{0}$ 100− 0
$\frac{1}{2}$ $\frac{1}{2}$ 1 000− $-1/\sqrt{2.5}$
$\frac{1}{2}$ $\frac{1}{2}$ 1 100− 0
$\frac{1}{2}$ $\frac{3}{2}$ 1 000+ $+1/2\sqrt{3.5}$
$\frac{1}{2}$ $\frac{3}{2}$ 1 100+ $-i/\sqrt{2.3}$
$\frac{3}{2}$ $\frac{1}{2}$ 1 000+ $-1/\sqrt{2.5}$
$\frac{3}{2}$ $\frac{3}{2}$ $\tilde{0}$ 000− $+1/\sqrt{3.5}$

$\frac{5}{2}$ 2 $\frac{1}{2}$ 0

$\frac{1}{2}$ 0 $\frac{1}{2}$ 000+ $+1/\sqrt{5}$
$\frac{1}{2}$ 0 $\frac{1}{2}$ 100+ 0
$\frac{1}{2}$ 1 $\frac{1}{2}$ 000− $+\sqrt{2}/\sqrt{3.5}$
$\frac{1}{2}$ 1 $\frac{1}{2}$ 010− 0
$\frac{1}{2}$ 1 $\frac{1}{2}$ 100− 0
$\frac{1}{2}$ 1 $\frac{1}{2}$ 110− $+1/\sqrt{3}$
$\frac{3}{2}$ 1 $\frac{1}{2}$ 000+ $+\sqrt{2}/\sqrt{3.5}$
$\frac{3}{2}$ 1 $\frac{1}{2}$ 010+ $+i/\sqrt{2.3.5}$

$\frac{5}{2}$ 2 $\frac{3}{2}$ 0

$\frac{1}{2}$ 0 $\frac{1}{2}$ 000− $+1/3\sqrt{5}$
$\frac{1}{2}$ 0 $\frac{1}{2}$ 100− $-\sqrt{2}/3\sqrt{5}$
$\frac{1}{2}$ 1 $\frac{1}{2}$ 000+ $-\sqrt{5}/3\sqrt{2.3}$
$\frac{1}{2}$ 1 $\frac{1}{2}$ 010+ $+\sqrt{2}/3\sqrt{3}$

$\frac{5}{2}$ 2 $\frac{3}{2}$ 0

$\frac{1}{2}$ 1 $\frac{1}{2}$ 100+ $+2/3\sqrt{3.5}$
$\frac{1}{2}$ 1 $\frac{1}{2}$ 110+ $+1/3\sqrt{3}$
$\frac{1}{2}$ 1 $\frac{3}{2}$ 000− $-1/2.9\sqrt{5}+2i/9$
$\frac{1}{2}$ 1 $\frac{3}{2}$ 010− $-1/9+2i/9\sqrt{5}$
$\frac{1}{2}$ 1 $\frac{3}{2}$ 100− $-\sqrt{2}/9\sqrt{5}-i/9\sqrt{2}$
$\frac{1}{2}$ 1 $\frac{3}{2}$ 110− $-2\sqrt{2}/9-i/9\sqrt{2.5}$
$\frac{3}{2}$ 0 -$\frac{3}{2}$ 000− $+2/3\sqrt{3.5}-i/3\sqrt{3}$
$\frac{3}{2}$ 1 $\frac{1}{2}$ 000− $-1/3\sqrt{2.3.5}+i\sqrt{2}/3\sqrt{3}$
$\frac{3}{2}$ 1 $\frac{1}{2}$ 010− $+1/3\sqrt{2.3}-i/3\sqrt{2.3.5}$

$\frac{5}{2}$ 2 $\frac{3}{2}$ 1

$\frac{1}{2}$ 0 $\frac{1}{2}$ 000+ $-2\sqrt{2}/3\sqrt{3.5}$
$\frac{1}{2}$ 0 $\frac{1}{2}$ 100+ $+1/3\sqrt{3.5}$
$\frac{1}{2}$ 1 $\frac{1}{2}$ 000− $-\sqrt{5}/9$
$\frac{1}{2}$ 1 $\frac{1}{2}$ 010− $-1/9$
$\frac{1}{2}$ 1 $\frac{1}{2}$ 100− $-\sqrt{2}/9\sqrt{5}$
$\frac{1}{2}$ 1 $\frac{1}{2}$ 110− $+5/9\sqrt{2}$
$\frac{1}{2}$ 1 $\frac{3}{2}$ 000+ $-13/9\sqrt{2.3.5}-i\sqrt{2}/9\sqrt{3}$
$\frac{1}{2}$ 1 $\frac{3}{2}$ 010+ $+1/9\sqrt{2.3}+7i/9\sqrt{2.3.5}$
$\frac{1}{2}$ 1 $\frac{3}{2}$ 100+ $+1/9\sqrt{3.5}-4i/9\sqrt{3}$
$\frac{1}{2}$ 1 $\frac{3}{2}$ 110+ $+2/9\sqrt{3}+i/2.9\sqrt{3.5}$
$\frac{3}{2}$ 0 -$\frac{3}{2}$ 000+ $+7/9\sqrt{2.5}+i/9\sqrt{2}$
$\frac{3}{2}$ 1 $\frac{1}{2}$ 000+ $+2/9\sqrt{5}-i/9$
$\frac{3}{2}$ 1 $\frac{1}{2}$ 010+ $-1/2.9-11i/2.9\sqrt{5}$

$\frac{5}{2}$ $\frac{5}{2}$ 0 0

$\frac{1}{2}$ $\frac{1}{2}$ 0 000+ $+1/\sqrt{3}$
$\frac{1}{2}$ $\frac{1}{2}$ 0 100+ 0
$\frac{1}{2}$ $\frac{1}{2}$ 0 110+ $+1/\sqrt{3}$
-$\frac{3}{2}$ $\frac{3}{2}$ 0 000+ $+1/\sqrt{2.3}$

$\frac{5}{2}$ $\frac{5}{2}$ 1 0

$\frac{1}{2}$ $\frac{1}{2}$ $\tilde{0}$ 000+ $-1/3\sqrt{2.3}$
$\frac{1}{2}$ $\frac{1}{2}$ $\tilde{0}$ 100+ $+1/3\sqrt{3}$
$\frac{1}{2}$ $\frac{1}{2}$ $\tilde{0}$ 110+ $-\sqrt{2}/3\sqrt{3}$
$\frac{1}{2}$ $\frac{1}{2}$ 1 000+ $+2/3\sqrt{3}$
$\frac{1}{2}$ $\frac{1}{2}$ 1 100+ $-1/3\sqrt{2.3}$
$\frac{1}{2}$ $\frac{1}{2}$ 1 110+ $-2/3\sqrt{3}$
$\frac{3}{2}$ $\frac{1}{2}$ 1 000− $-1/2.3\sqrt{3}-i\sqrt{5}/2.3\sqrt{3}$
$\frac{3}{2}$ $\frac{1}{2}$ 1 010− $-1/2.3\sqrt{2.3}-i\sqrt{5}/2.3\sqrt{2.3}$
$\frac{3}{2}$ $\frac{3}{2}$ $\tilde{0}$ 000+ $+1/3\sqrt{3}-i\sqrt{5}/2.3\sqrt{3}$

$\frac{5}{2}$ $\frac{5}{2}$ 1 1

$\frac{1}{2}$ $\frac{1}{2}$ $\tilde{0}$ 000+ $+11/9\sqrt{2.3.5}$

K–D_3 3jm Factors (cont.)

$\frac{5}{2}\ \frac{5}{2}$ 1 1

$\frac12\ \frac12\ \tilde{0}$ 100+ $-\sqrt{5}/9\sqrt{3}$
$\frac12\ \frac12\ \tilde{0}$ 110+ $-2\sqrt{2.5}/9\sqrt{3}$
$\frac12\ \frac12\ 1$ 000+ $+8/9\sqrt{3.5}$
$\frac12\ \frac12\ 1$ 100+ $+\sqrt{5}/9\sqrt{2.3}$
$\frac12\ \frac12\ 1$ 110+ $+2\sqrt{5}/9\sqrt{3}$
$\frac32\ \frac12\ 1$ 000− $-19/2.9\sqrt{3.5}+5\,i/2.9\sqrt{3}$
$\frac32\ \frac12\ 1$ 010− $+\sqrt{5}/2.9\sqrt{2.3}-7\,i/2.9\sqrt{2.3}$
$\frac32\ \frac32\ \tilde{0}$ 000+ $+4/9\sqrt{3.5}+5\,i/2.9\sqrt{3}$

$\frac{5}{2}\ \frac{5}{2}$ 2 0

$\frac12\ \frac12\ 0$ 000+ $-\sqrt{2}/3\sqrt{3.5}$
$\frac12\ \frac12\ 0$ 100+ $+2/3\sqrt{3.5}$
$\frac12\ \frac12\ 0$ 110+ $+\sqrt{5}/3\sqrt{2.3}$
$\frac12\ \frac12\ 1$ 000− 0
$\frac12\ \frac12\ 1$ 001− 0
$\frac12\ \frac12\ 1$ 100− $-\sqrt{2}/3\sqrt{5}$
$\frac12\ \frac12\ 1$ 101− $-1/3\sqrt{2}$
$\frac12\ \frac12\ 1$ 110− 0
$\frac12\ \frac12\ 1$ 111− 0
$\frac32\ \frac12\ 1$ 000+ $-4/9\sqrt{5}-i/9$
$\frac32\ \frac12\ 1$ 001+ $+1/9+7\,i/2.9\sqrt{5}$
$\frac32\ \frac12\ 1$ 010+ $-1/9\sqrt{2.5}+i\sqrt{2}/9$
$\frac32\ \frac12\ 1$ 011+ $-\sqrt{2}/9+i\sqrt{2}/9\sqrt{5}$
$-\frac32\ \frac32\ 0$ 000+ $-1/2\sqrt{3.5}$

$\frac{5}{2}\ \frac{5}{2}$ 2 1

$\frac12\ \frac12\ 0$ 000− 0
$\frac12\ \frac12\ 0$ 100− $-\sqrt{2}/3\sqrt{5}$
$\frac12\ \frac12\ 0$ 110− 0
$\frac12\ \frac12\ 1$ 000+ $+\sqrt{2.5}/9\sqrt{3}$
$\frac12\ \frac12\ 1$ 001+ $+4\sqrt{2}/9\sqrt{3}$
$\frac12\ \frac12\ 1$ 100+ $-1/9\sqrt{3.5}$
$\frac12\ \frac12\ 1$ 101+ $+1/9\sqrt{3}$
$\frac12\ \frac12\ 1$ 110+ $+2\sqrt{2.5}/9\sqrt{3}$
$\frac12\ \frac12\ 1$ 111+ $-\sqrt{2}/9\sqrt{3}$
$\frac32\ \frac12\ 1$ 000− $+\sqrt{5}/9\sqrt{2.3}-i/9\sqrt{2.3}$
$\frac32\ \frac12\ 1$ 001− $+2\sqrt{2}/9\sqrt{3}+i\sqrt{5}/9\sqrt{2.3}$
$\frac32\ \frac12\ 1$ 010− $-13/2.9\sqrt{3.5}-i/2.9\sqrt{3}$
$\frac32\ \frac12\ 1$ 011− $+2/9\sqrt{3}-2\,i/9\sqrt{3.5}$
$-\frac32\ \frac32\ 0$ 000− $-i/3\sqrt{2}$

$\frac{5}{2}\ \frac{5}{2}$ 2 2

$\frac12\ \frac12\ 0$ 000+ $-4\sqrt{2}/9\sqrt{5}$
$\frac12\ \frac12\ 0$ 100+ $-4/9\sqrt{5}$
$\frac12\ \frac12\ 0$ 110+ $+\sqrt{5}/9\sqrt{2}$
$\frac12\ \frac12\ 1$ 000− 0
$\frac12\ \frac12\ 1$ 001− 0
$\frac12\ \frac12\ 1$ 100− $+2\sqrt{2}/3\sqrt{3.5}$
$\frac12\ \frac12\ 1$ 101− $-1/3\sqrt{2.3}$
$\frac12\ \frac12\ 1$ 110− 0
$\frac12\ \frac12\ 1$ 111− 0
$\frac32\ \frac12\ 1$ 000+ $-1/9\sqrt{3.5}+2\,i/9\sqrt{3}$
$\frac32\ \frac12\ 1$ 001+ $-2/9\sqrt{3}+13\,i/2.9\sqrt{3.5}$
$\frac32\ \frac12\ 1$ 010+ $+\sqrt{2}/9\sqrt{3.5}+5\,i/9\sqrt{2.3}$
$\frac32\ \frac12\ 1$ 011+ $+2\sqrt{2}/9\sqrt{3}-2\,i\sqrt{2}/9\sqrt{3.5}$
$-\frac32\ \frac32\ 0$ 000+ $+1/2.3\sqrt{5}$

3 $\frac{3}{2}\ \frac{3}{2}$ 0

$0\ \frac12\ \frac12$ 000− 0
$0\ {-\frac32}\ \frac32$ 000− $+i/2\sqrt{2}$
$\tilde{0}\ \frac12\ \frac12$ 000+ $+1/\sqrt{5}$
$\tilde{0}\ \frac32\ \frac32$ 000+ $-1/3\sqrt{2.5}-i/2.3\sqrt{2}$
$1\ \frac12\ \frac12$ 000+ $+1/\sqrt{2.5}$
$1\ \frac32\ \frac12$ 000− $+1/2\sqrt{3.5}-i/2\sqrt{3}$

3 2 1 0

$0\ 1\ 1$ 000+ 0
$0\ 1\ 1$ 010+ $+1/2$
$\tilde{0}\ 0\ \tilde{0}$ 000+ $-\sqrt{2}/\sqrt{3.5}$
$\tilde{0}\ 1\ 1$ 000− $-2/3\sqrt{5}$
$\tilde{0}\ 1\ 1$ 010− $+1/2.3$
$1\ 0\ 1$ 000+ $+1/\sqrt{3.5}$
$1\ 1\ \tilde{0}$ 000− $+2/3\sqrt{5}$
$1\ 1\ \tilde{0}$ 010− $+1/3$
$1\ 1\ 1$ 000+ $+\sqrt{2}/3$
$1\ 1\ 1$ 010+ $-1/3\sqrt{2.5}$

3 2 2 0

$0\ 0\ 0$ 000− 0
$0\ 1\ 1$ 000− 0
$0\ 1\ 1$ 010− $-1/2\sqrt{2}$
$0\ 1\ 1$ 011− 0
$\tilde{0}\ 1\ 1$ 000+ $-2\sqrt{2}/3\sqrt{5}$
$\tilde{0}\ 1\ 1$ 010+ $-1/2.3\sqrt{2}$
$\tilde{0}\ 1\ 1$ 011+ $-\sqrt{2}/3\sqrt{5}$
$1\ 1\ 0$ 000− $-1/\sqrt{2.3.5}$

K–D_3 3jm Factors (cont.)

3 2 2 0	
1 1 0 010−	$+1/\sqrt{2.3}$
1 1 1 000−	0
1 1 1 010−	$-1/2\sqrt{5}$
1 1 1 011−	0
3 2 2 1	
0 0 0 000+	$-2/3\sqrt{5}$
0 1 1 000+	$+4\sqrt{2}/9\sqrt{5}$
0 1 1 010+	$+5/2.9\sqrt{2}$
0 1 1 011+	$-\sqrt{2}/9\sqrt{5}$
$\tilde{0}$ 1 1 000−	0
$\tilde{0}$ 1 1 010−	$-1/2\sqrt{2}$
$\tilde{0}$ 1 1 011−	0
1 1 0 000+	$+\sqrt{5}/3\sqrt{2.3}$
1 1 0 010+	$+1/3\sqrt{2.3}$
1 1 1 000+	$-2/9$
1 1 1 010+	$+\sqrt{5}/2.9$
1 1 1 011+	$-4/9$
3 $\frac{5}{2}$ $\frac{1}{2}$ 0	
0 $\frac{1}{2}$ $\frac{1}{2}$ 000−	0
0 $\frac{1}{2}$ $\frac{1}{2}$ 010−	$+1/2$
$\tilde{0}$ $\frac{1}{2}$ $\frac{1}{2}$ 000+	$-\sqrt{2}/3$
$\tilde{0}$ $\frac{1}{2}$ $\frac{1}{2}$ 010+	$+1/2.3$
1 $\frac{1}{2}$ $\frac{1}{2}$ 000+	$+1/3$
1 $\frac{1}{2}$ $\frac{1}{2}$ 010+	$+1/3\sqrt{2}$
1 $\frac{3}{2}$ $\frac{1}{2}$ 000−	$+1/2.3+i\sqrt{5}/2.3$
3 $\frac{5}{2}$ $\frac{3}{2}$ 0	
0 $\frac{1}{2}$ $\frac{1}{2}$ 000+	$+1/2\sqrt{2}$
0 $\frac{1}{2}$ $\frac{1}{2}$ 010+	0
0 $\frac{3}{2}$ -$\frac{3}{2}$ 000+	$+1/4\sqrt{2.3}+i\sqrt{5}/4\sqrt{2.3}$
$\tilde{0}$ $\frac{1}{2}$ $\frac{1}{2}$ 000−	$-1/2.3\sqrt{2}$
$\tilde{0}$ $\frac{1}{2}$ $\frac{1}{2}$ 010−	$-1/3$
$\tilde{0}$ $\frac{3}{2}$ $\frac{3}{2}$ 000−	$-1/4\sqrt{2.3}-i\sqrt{5}/4\sqrt{2.3}$
1 $\frac{1}{2}$ $\frac{1}{2}$ 000−	$+1/3$
1 $\frac{1}{2}$ $\frac{1}{2}$ 010−	$+1/3\sqrt{2}$
1 $\frac{1}{2}$ $\frac{3}{2}$ 000+	$+1/2\sqrt{2.3}$
1 $\frac{1}{2}$ $\frac{3}{2}$ 010+	$-1/2\sqrt{3}$
1 $\frac{3}{2}$ $\frac{1}{2}$ 000+	$-1/4.3-i\sqrt{5}/4.3$
3 $\frac{5}{2}$ $\frac{3}{2}$ 1	
0 $\frac{1}{2}$ $\frac{1}{2}$ 000−	$+\sqrt{5}/2.3\sqrt{2.3}$
0 $\frac{1}{2}$ $\frac{1}{2}$ 010−	$+\sqrt{5}/3\sqrt{3}$
0 $\frac{3}{2}$ -$\frac{3}{2}$ 000−	$+\sqrt{5}/4.9\sqrt{2}-7i/4.9\sqrt{2}$
$\tilde{0}$ $\frac{1}{2}$ $\frac{1}{2}$ 000+	$+\sqrt{3}/2\sqrt{2.5}$
$\tilde{0}$ $\frac{1}{2}$ $\frac{1}{2}$ 010+	0
$\tilde{0}$ $\frac{3}{2}$ $\frac{3}{2}$ 000+	$+3/4\sqrt{2.5}-i/4\sqrt{2}$
1 $\frac{1}{2}$ $\frac{1}{2}$ 000+	$+1/3\sqrt{3.5}$
1 $\frac{1}{2}$ $\frac{1}{2}$ 010+	$-\sqrt{5}/3\sqrt{2.3}$
1 $\frac{1}{2}$ $\frac{3}{2}$ 000−	$-7/2.9\sqrt{2.5}-2i\sqrt{2}/9$
1 $\frac{1}{2}$ $\frac{3}{2}$ 010−	$-\sqrt{5}/2.9+i/9$
1 $\frac{3}{2}$ $\frac{1}{2}$ 000−	$-11/4.3\sqrt{3.5}+i/4.3\sqrt{3}$
3 $\frac{5}{2}$ $\frac{5}{2}$ 0	
0 $\frac{1}{2}$ $\frac{1}{2}$ 000−	0
0 $\frac{1}{2}$ $\frac{1}{2}$ 010−	$-\sqrt{5}/2.3\sqrt{2}$
0 $\frac{1}{2}$ $\frac{1}{2}$ 011−	0
0 -$\frac{3}{2}$ $\frac{3}{2}$ 000−	$+i/3\sqrt{2}$
$\tilde{0}$ $\frac{1}{2}$ $\frac{1}{2}$ 000+	$-4/9\sqrt{5}$
$\tilde{0}$ $\frac{1}{2}$ $\frac{1}{2}$ 010+	$-\sqrt{5}/2.9\sqrt{2}$
$\tilde{0}$ $\frac{1}{2}$ $\frac{1}{2}$ 011+	$-\sqrt{5}/9$
$\tilde{0}$ $\frac{3}{2}$ $\frac{3}{2}$ 000+	$-7/9\sqrt{2.5}-i/9\sqrt{2}$
1 $\frac{1}{2}$ $\frac{1}{2}$ 000+	$+\sqrt{2}/3\sqrt{5}$
1 $\frac{1}{2}$ $\frac{1}{2}$ 010+	$-\sqrt{5}/2.3$
1 $\frac{1}{2}$ $\frac{1}{2}$ 011+	0
1 $\frac{3}{2}$ $\frac{1}{2}$ 000−	$-1/2.3\sqrt{2.5}+i/2.3\sqrt{2}$
1 $\frac{3}{2}$ $\frac{1}{2}$ 001−	$+i/2.3$
3 $\frac{5}{2}$ $\frac{5}{2}$ 1	
0 $\frac{1}{2}$ $\frac{1}{2}$ 000+	$+2/9$
0 $\frac{1}{2}$ $\frac{1}{2}$ 010+	$-5/2.9\sqrt{2}$
0 $\frac{1}{2}$ $\frac{1}{2}$ 011+	$+1/9$
0 -$\frac{3}{2}$ $\frac{3}{2}$ 000+	$-1/3\sqrt{2}$
$\tilde{0}$ $\frac{1}{2}$ $\frac{1}{2}$ 000−	0
$\tilde{0}$ $\frac{1}{2}$ $\frac{1}{2}$ 010−	$+1/2\sqrt{2}$
$\tilde{0}$ $\frac{1}{2}$ $\frac{1}{2}$ 011−	0
$\tilde{0}$ $\frac{3}{2}$ $\frac{3}{2}$ 000−	0
1 $\frac{1}{2}$ $\frac{1}{2}$ 000−	0
1 $\frac{1}{2}$ $\frac{1}{2}$ 010−	$+1/2.3$
1 $\frac{1}{2}$ $\frac{1}{2}$ 011−	0
1 $\frac{3}{2}$ $\frac{1}{2}$ 000+	$-5/2.9\sqrt{2}+i\sqrt{5}/2.9\sqrt{2}$
1 $\frac{3}{2}$ $\frac{1}{2}$ 001+	$-2/9-i\sqrt{5}/2.9$
3 3 0 0	
0 0 0 000+	$+1/2$
$\tilde{0}$ $\tilde{0}$ 0 000+	$+1/2$

K–D_3 3jm Factors (cont.)

3 3 0 0
1 1 0 000+ $+1/\sqrt{2}$
3 3 1 0
$\tilde{0}$ 0 $\tilde{0}$ 000− $+1/2\sqrt{3}$
1 0 1 000− $+1/\sqrt{2.3}$
1 $\tilde{0}$ 1 000+ $+1/\sqrt{2.3}$
1 1 $\tilde{0}$ 000+ $+1/\sqrt{2.3}$
1 1 1 000− 0
3 3 2 0
0 0 0 000+ $+\sqrt{5}/2.3$
$\tilde{0}$ $\tilde{0}$ 0 000+ $-1/2\sqrt{5}$
1 0 1 000+ $+\sqrt{5}/3\sqrt{2.3}$
1 0 1 001+ $+1/3\sqrt{2.3}$
1 $\tilde{0}$ 1 000− $-1/\sqrt{2.3.5}$
1 $\tilde{0}$ 1 001− $+1/\sqrt{2.3}$
1 1 0 000+ $-1/3\sqrt{2.5}$
1 1 1 000+ $-2/3\sqrt{3}$
1 1 1 001+ $-2/3\sqrt{3.5}$
3 3 3 0
0 0 0 000+ $-1/2.3\sqrt{2}$
$\tilde{0}$ $\tilde{0}$ 0 000+ $-1/2\sqrt{2}$
1 1 0 000+ $+1/3$
1 1 $\tilde{0}$ 000− 0
1 1 1 000+ $+\sqrt{5}/3\sqrt{2}$
$\tilde{\mathbf{1}}$ $\frac{3}{2}$ $\frac{3}{2}$ 0
$\tilde{0}$ $\frac{1}{2}$ $\frac{1}{2}$ 000+ $-1/\sqrt{2.3.5}$
$\tilde{0}$ $\frac{3}{2}$ $\frac{3}{2}$ 000+ $+1/2.3\sqrt{3.5}-2i/3\sqrt{3}$
1 $\frac{1}{2}$ $\frac{1}{2}$ 000+ $-2/\sqrt{3.5}$
1 $\frac{3}{2}$ $\frac{1}{2}$ 000− $-\sqrt{2}/3\sqrt{5}-i/3\sqrt{2}$
$\tilde{\mathbf{1}}$ **2 1** 0
$\tilde{0}$ 0 $\tilde{0}$ 000+ $+1/3\sqrt{5}$
$\tilde{0}$ 1 1 000− $+\sqrt{2}/3\sqrt{3.5}$
$\tilde{0}$ 1 1 010− $+2\sqrt{2}/3\sqrt{3}$
1 0 1 000+ $-2\sqrt{2}/3\sqrt{5}$
1 1 $\tilde{0}$ 000− $-4\sqrt{2}/3\sqrt{3.5}$
1 1 $\tilde{0}$ 010− $+\sqrt{2}/3\sqrt{3}$
1 1 1 000+ $+2/3\sqrt{3}$
1 1 1 010+ $+2/3\sqrt{3.5}$
$\tilde{\mathbf{1}}$ **2 2** 0
$\tilde{0}$ 1 1 000+ $+2/3\sqrt{3.5}$
$\tilde{0}$ 1 1 010+ $-2/3\sqrt{3}$

$\tilde{\mathbf{1}}$ **2 2** 0
$\tilde{0}$ 1 1 011+ $+1/3\sqrt{3.5}$
1 1 0 000− $+2/3\sqrt{5}$
1 1 0 010− $+1/3$
1 1 1 000− 0
1 1 1 010− $+\sqrt{2}/\sqrt{3.5}$
1 1 1 011− 0
$\tilde{\mathbf{1}}$ $\frac{5}{2}$ $\frac{1}{2}$ 0
$\tilde{0}$ $\frac{1}{2}$ $\frac{1}{2}$ 000+ $+1/3\sqrt{3}$
$\tilde{0}$ $\frac{1}{2}$ $\frac{1}{2}$ 010+ $+2\sqrt{2}/3\sqrt{3}$
1 $\frac{1}{2}$ $\frac{1}{2}$ 000+ $-2\sqrt{2}/3\sqrt{3}$
1 $\frac{1}{2}$ $\frac{1}{2}$ 010+ $+1/3\sqrt{3}$
1 $\frac{3}{2}$ $\frac{1}{2}$ 000− $-\sqrt{2}/3\sqrt{3}+i\sqrt{5}/3\sqrt{2.3}$
$\tilde{\mathbf{1}}$ $\frac{5}{2}$ $\frac{3}{2}$ 0
$\tilde{0}$ $\frac{1}{2}$ $\frac{1}{2}$ 000− $-\sqrt{2}/3\sqrt{3.5}$
$\tilde{0}$ $\frac{1}{2}$ $\frac{1}{2}$ 010− $+\sqrt{5}/3\sqrt{3}$
$\tilde{0}$ $\frac{3}{2}$ $\frac{3}{2}$ 000− $-1/3\sqrt{2.5}-i/3\sqrt{2}$
1 $\frac{1}{2}$ $\frac{1}{2}$ 000− $+1/3\sqrt{3.5}$
1 $\frac{1}{2}$ $\frac{1}{2}$ 010− $-\sqrt{5}/3\sqrt{2.3}$
1 $\frac{1}{2}$ $\frac{3}{2}$ 000+ $+1/\sqrt{2.5}-i/3\sqrt{2}$
1 $\frac{1}{2}$ $\frac{3}{2}$ 010+ $-i/2.3$
1 $\frac{3}{2}$ $\frac{1}{2}$ 000+ $+7/2.3\sqrt{3.5}+i/2.3\sqrt{3}$
$\tilde{\mathbf{1}}$ $\frac{5}{2}$ $\frac{5}{2}$ 0
$\tilde{0}$ $\frac{1}{2}$ $\frac{1}{2}$ 000+ $+7\sqrt{2}/9\sqrt{3.5}$
$\tilde{0}$ $\frac{1}{2}$ $\frac{1}{2}$ 010+ $+2\sqrt{5}/9\sqrt{3}$
$\tilde{0}$ $\frac{1}{2}$ $\frac{1}{2}$ 011+ $-\sqrt{5}/9\sqrt{2.3}$
$\tilde{0}$ $\frac{3}{2}$ $\frac{3}{2}$ 000+ $-13/2.9\sqrt{3.5}+i/9\sqrt{3}$
1 $\frac{1}{2}$ $\frac{1}{2}$ 000+ $+8/9\sqrt{3.5}$
1 $\frac{1}{2}$ $\frac{1}{2}$ 010+ $+\sqrt{5}/9\sqrt{2.3}$
1 $\frac{1}{2}$ $\frac{1}{2}$ 011+ $+2\sqrt{5}/9\sqrt{3}$
1 $\frac{3}{2}$ $\frac{1}{2}$ 000− $-2/9\sqrt{3.5}-7i/2.9\sqrt{3}$
1 $\frac{3}{2}$ $\frac{1}{2}$ 001− $+\sqrt{2.5}/9\sqrt{3}+2i\sqrt{2}/9\sqrt{3}$
$\tilde{\mathbf{1}}$ $\frac{5}{2}$ $\frac{5}{2}$ 1
$\tilde{0}$ $\frac{1}{2}$ $\frac{1}{2}$ 000+ $+4\sqrt{2}/9\sqrt{3}$
$\tilde{0}$ $\frac{1}{2}$ $\frac{1}{2}$ 010+ $-2/9\sqrt{3}$
$\tilde{0}$ $\frac{1}{2}$ $\frac{1}{2}$ 011+ $+1/9\sqrt{2.3}$
$\tilde{0}$ $\frac{3}{2}$ $\frac{3}{2}$ 000+ $-1/2.9\sqrt{3}-2i\sqrt{5}/9\sqrt{3}$
1 $\frac{1}{2}$ $\frac{1}{2}$ 000+ $-4/9\sqrt{3}$
1 $\frac{1}{2}$ $\frac{1}{2}$ 010+ $-7/9\sqrt{2.3}$
1 $\frac{1}{2}$ $\frac{1}{2}$ 011+ $+4/9\sqrt{3}$
1 $\frac{3}{2}$ $\frac{1}{2}$ 000− $+1/9\sqrt{3}-i\sqrt{5}/2.9\sqrt{3}$

K–D_3 3jm Factors (cont.)

$\tilde{1}\ \frac{5}{2}\ \frac{5}{2}\ 1$		
$1\ \frac{3}{2}\ \frac{1}{2}$	001−	$+2\sqrt{2}/9\sqrt{3}-i\sqrt{2.5}/9\sqrt{3}$
$\tilde{1}\ 3\ 1\ 0$		
$\tilde{0}\ 0\ \tilde{0}$	000−	$+\sqrt{2/3}$
$\tilde{0}\ 1\ 1$	000+	$-1/3$
$1\ 0\ 1$	000−	$+1/2.3$
$1\ \tilde{0}\ 1$	000+	$-1/2$
$1\ 1\ \tilde{0}$	000+	$+1/3$
$1\ 1\ 1$	000−	$+\sqrt{5}/3\sqrt{2}$
$\tilde{1}\ 3\ 2\ 0$		
$\tilde{0}\ \tilde{0}\ 0$	000+	$+\sqrt{2}/\sqrt{3.5}$
$\tilde{0}\ 1\ 1$	000−	$+1/\sqrt{5}$
$\tilde{0}\ 1\ 1$	001−	0
$1\ 0\ 1$	000+	$+\sqrt{5}/2.3$
$1\ 0\ 1$	001+	$-1/3$
$1\ \tilde{0}\ 1$	000−	$+1/2.3\sqrt{5}$
$1\ \tilde{0}\ 1$	001−	$+1/3$
$1\ 1\ 0$	000+	$+1/\sqrt{3.5}$
$1\ 1\ 1$	000+	$+1/3\sqrt{2}$
$1\ 1\ 1$	001+	$+2\sqrt{2}/3\sqrt{5}$
$\tilde{1}\ 3\ 3\ 0$		
$\tilde{0}\ \tilde{0}\ 0$	000−	$-1/2\sqrt{3}$
$\tilde{0}\ 1\ 1$	000+	$+1/\sqrt{2.3}$
$1\ 1\ 0$	000−	$+1/\sqrt{2.3}$
$1\ 1\ \tilde{0}$	000+	$-1/\sqrt{2.3}$
$1\ 1\ 1$	000−	0
$\tilde{1}\ \tilde{1}\ 0\ 0$		
$\tilde{0}\ \tilde{0}\ 0$	000+	$+1/\sqrt{3}$
$1\ 1\ 0$	000+	$+\sqrt{2}/\sqrt{3}$
$\tilde{1}\ \tilde{1}\ 2\ 0$		
$\tilde{0}\ \tilde{0}\ 0$	000+	$+\sqrt{2}/\sqrt{3.5}$
$1\ \tilde{0}\ 1$	000−	$-2/3\sqrt{5}$
$1\ \tilde{0}\ 1$	001−	$-1/3$
$1\ 1\ 0$	000+	$-1/\sqrt{3.5}$
$1\ 1\ 1$	000+	$+\sqrt{2}/3$
$1\ 1\ 1$	001+	$-2\sqrt{2}/3\sqrt{5}$
$\tilde{1}\ \tilde{1}\ \tilde{1}\ 0$		
$1\ 1\ \tilde{0}$	000+	$-1/\sqrt{3}$
$1\ 1\ 1$	000−	0
$\widetilde{\tfrac{1}{2}}\ 2\ \frac{3}{2}\ 0$		
$\frac{1}{2}\ 0\ \frac{1}{2}$	000+	$+1/\sqrt{5}$
$\widetilde{\tfrac{1}{2}}\ 2\ \frac{3}{2}\ 0$		
$\frac{1}{2}\ 1\ \frac{1}{2}$	000−	$-\sqrt{2}/\sqrt{3.5}$
$\frac{1}{2}\ 1\ \frac{1}{2}$	010−	$-1/\sqrt{2.3}$
$\frac{1}{2}\ 1\ \frac{3}{2}$	000+	$+1/3\sqrt{5}-i/3$
$\frac{1}{2}\ 1\ \frac{3}{2}$	010+	$-1/3+i/2.3\sqrt{5}$
$\widetilde{\tfrac{1}{2}}\ \frac{5}{2}\ 1\ 0$		
$\frac{1}{2}\ \frac{1}{2}\ \tilde{0}$	000−	$-\sqrt{2}/3$
$\frac{1}{2}\ \frac{1}{2}\ \tilde{0}$	010−	$-1/3$
$\frac{1}{2}\ \frac{1}{2}\ 1$	000−	$+1/3$
$\frac{1}{2}\ \frac{1}{2}\ 1$	010−	$-\sqrt{2}/3$
$\frac{1}{2}\ \frac{3}{2}\ 1$	000+	$-1/2.3-i\sqrt{5}/2.3$
$\widetilde{\tfrac{1}{2}}\ \frac{5}{2}\ 2\ 0$		
$\frac{1}{2}\ \frac{1}{2}\ 0$	000−	$-\sqrt{2}/3\sqrt{3.5}$
$\frac{1}{2}\ \frac{1}{2}\ 0$	010−	$+\sqrt{5}/3\sqrt{3}$
$\frac{1}{2}\ \frac{1}{2}\ 1$	000+	$-7/9\sqrt{5}$
$\frac{1}{2}\ \frac{1}{2}\ 1$	001+	$+4/9$
$\frac{1}{2}\ \frac{1}{2}\ 1$	010+	$-\sqrt{2.5}/9$
$\frac{1}{2}\ \frac{1}{2}\ 1$	011+	$-\sqrt{2}/9$
$\frac{1}{2}\ \frac{3}{2}\ 1$	000−	$+11/2.9\sqrt{5}-i/2.9$
$\frac{1}{2}\ \frac{3}{2}\ 1$	001−	$+2/9+4i/9\sqrt{5}$
$\widetilde{\tfrac{1}{2}}\ 3\ \frac{1}{2}\ 0$		
$\frac{1}{2}\ 0\ \frac{1}{2}$	000+	$+1/2$
$\frac{1}{2}\ \tilde{0}\ \frac{1}{2}$	000−	$+1/2$
$\frac{1}{2}\ 1\ \frac{1}{2}$	000−	$+1/\sqrt{2}$
$\widetilde{\tfrac{1}{2}}\ 3\ \frac{5}{2}\ 0$		
$\frac{1}{2}\ 0\ \frac{1}{2}$	000+	$+5/2.3\sqrt{3}$
$\frac{1}{2}\ 0\ \frac{1}{2}$	001+	$+1/3\sqrt{2.3}$
$\frac{1}{2}\ \tilde{0}\ \frac{1}{2}$	000−	$-1/2\sqrt{3}$
$\frac{1}{2}\ \tilde{0}\ \frac{1}{2}$	001−	$+1/\sqrt{2.3}$
$\frac{1}{2}\ 1\ \frac{1}{2}$	000−	$-1/3\sqrt{2.3}$
$\frac{1}{2}\ 1\ \frac{1}{2}$	001−	$-2/3\sqrt{3}$
$\frac{1}{2}\ 1\ \frac{3}{2}$	000+	$+\sqrt{2}/3\sqrt{3}-i\sqrt{5}/3\sqrt{2.3}$
$\widetilde{\tfrac{1}{2}}\ \tilde{1}\ \frac{3}{2}\ 0$		
$\frac{1}{2}\ \tilde{0}\ \frac{1}{2}$	000+	$+1/\sqrt{3}$
$\frac{1}{2}\ 1\ \frac{1}{2}$	000+	$+1/\sqrt{2.3}$
$\frac{1}{2}\ 1\ \frac{3}{2}$	000−	$-1/3-i\sqrt{5}/2.3$
$\widetilde{\tfrac{1}{2}}\ \widetilde{\tfrac{1}{2}}\ 0\ 0$		
$\frac{1}{2}\ \frac{1}{2}\ 0$	000+	$+1$
$\widetilde{\tfrac{1}{2}}\ \widetilde{\tfrac{1}{2}}\ \tilde{1}\ 0$		
$\frac{1}{2}\ \frac{1}{2}\ \tilde{0}$	000+	$+1/\sqrt{3}$
$\frac{1}{2}\ \frac{1}{2}\ 1$	000+	$-\sqrt{2}/\sqrt{3}$

K–D_5 2jm Factors (K_i–D_{5i} = K_h–D_{5d})

$0 \to +0$

$\frac{1}{2} \to +\frac{1}{2}$

$1 \to +\tilde{0} + 1$

$\frac{3}{2} \to +\frac{1}{2} + \frac{3}{2}$

$2 \to +0 + 1 + 2$

$\frac{5}{2} \to +\frac{1}{2} + \frac{3}{2} + \frac{5}{2} + \text{-}\frac{5}{2}$

$3 \to +1 + 2$

$\tilde{1} \to +\tilde{0} + 2$

$\widetilde{\frac{1}{2}} \to +\frac{3}{2}$

K–D_5 3jm Factors

$0\ 0\ 0$ 0
$0\ 0\ 0\ +\quad +1$

$\frac{1}{2}\ \frac{1}{2}\ 0$ 0
$\frac{1}{2}\ \frac{1}{2}\ 0\ +\quad +1$

$1\ \frac{1}{2}\ \frac{1}{2}$ 0
$\tilde{0}\ \frac{1}{2}\ \frac{1}{2}\ +\quad +1/\sqrt{3}$
$1\ \frac{1}{2}\ \frac{1}{2}\ +\quad +\sqrt{2}/\sqrt{3}$

$1\ 1\ 0$ 0
$\tilde{0}\ \tilde{0}\ 0\ +\quad +1/\sqrt{3}$
$1\ 1\ 0\ +\quad +\sqrt{2}/\sqrt{3}$

$1\ 1\ 1$ 0
$1\ 1\ \tilde{0}\ +\quad -1/\sqrt{3}$

$\frac{3}{2}\ 1\ \frac{1}{2}$ 0
$\frac{1}{2}\ \tilde{0}\ \frac{1}{2}\ -\quad +1/\sqrt{3}$
$\frac{1}{2}\ 1\ \frac{1}{2}\ -\quad -1/\sqrt{2.3}$
$\frac{3}{2}\ 1\ \frac{1}{2}\ +\quad +1/\sqrt{2}$

$\frac{3}{2}\ \frac{3}{2}\ 0$ 0
$\frac{1}{2}\ \frac{1}{2}\ 0\ +\quad +1/\sqrt{2}$
$\frac{3}{2}\ \frac{3}{2}\ 0\ +\quad +1/\sqrt{2}$

$\frac{3}{2}\ \frac{3}{2}\ 1$ 0
$\frac{1}{2}\ \frac{1}{2}\ \tilde{0}\ +\quad +1/\sqrt{2.3.5}$
$\frac{1}{2}\ \frac{1}{2}\ 1\ +\quad -2/\sqrt{3.5}$
$\frac{3}{2}\ \frac{1}{2}\ 1\ -\quad +1/\sqrt{5}$
$\frac{3}{2}\ \frac{3}{2}\ \tilde{0}\ +\quad +\sqrt{3}/\sqrt{2.5}$

$2\ 1\ 1$ 0
$0\ \tilde{0}\ \tilde{0}\ +\quad +\sqrt{2}/\sqrt{3.5}$
$0\ 1\ 1\ +\quad -1/\sqrt{3.5}$
$1\ 1\ \tilde{0}\ -\quad -1/\sqrt{5}$
$2\ 1\ 1\ +\quad +\sqrt{2}/\sqrt{5}$

$2\ \frac{3}{2}\ \frac{1}{2}$ 0
$0\ \frac{1}{2}\ \frac{1}{2}\ -\quad +1/\sqrt{5}$
$1\ \frac{1}{2}\ \frac{1}{2}\ +\quad +\sqrt{3}/\sqrt{2.5}$
$1\ \frac{3}{2}\ \frac{1}{2}\ -\quad -1/\sqrt{2.5}$
$2\ \frac{3}{2}\ \frac{1}{2}\ +\quad +\sqrt{2}/\sqrt{5}$

$2\ \frac{3}{2}\ \frac{3}{2}$ 0
$0\ \frac{1}{2}\ \frac{1}{2}\ +\quad -1/\sqrt{2.5}$
$0\ \frac{3}{2}\ \frac{3}{2}\ +\quad +1/\sqrt{2.5}$
$1\ \frac{1}{2}\ \frac{1}{2}\ -\quad 0$
$1\ \frac{3}{2}\ \frac{1}{2}\ +\quad +1/\sqrt{5}$
$2\ \frac{3}{2}\ \frac{1}{2}\ -\quad +1/\sqrt{5}$
$2\ \frac{3}{2}\ \frac{3}{2}\ -\quad 0$

K–D_5 3jm Factors (cont.)

2 2 0 0

0 0 0 + $+1/\sqrt{5}$

1 1 0 + $+\sqrt{2}/\sqrt{5}$

2 2 0 + $+\sqrt{2}/\sqrt{5}$

2 2 1 0

1 0 1 − $+1/\sqrt{5}$

1 1 $\tilde{0}$ + $-1/\sqrt{3.5}$

2 1 1 − $-\sqrt{2}/\sqrt{3.5}$

2 2 $\tilde{0}$ + $-2/\sqrt{3.5}$

2 2 1 − 0

2 2 2 0

0 0 0 + $-2/5$

1 1 0 + $+1/5\sqrt{2}$

2 1 1 + $+\sqrt{3/5}$

2 2 0 + $+1/5\sqrt{2}$

2 2 1 + $+\sqrt{3/5}$

2 2 2 1

0 0 0 + 0

1 1 0 + $+1/\sqrt{2.5}$

2 1 1 + $-1/\sqrt{3.5}$

2 2 0 + $-1/\sqrt{2.5}$

2 2 1 + $+1/\sqrt{3.5}$

$\frac{5}{2}$ $\frac{3}{2}$ 1 0

$\frac{1}{2}$ $\frac{1}{2}$ $\tilde{0}$ − $-1/\sqrt{5}$

$\frac{1}{2}$ $\frac{1}{2}$ 1 − $-1/\sqrt{2.5}$

$\frac{1}{2}$ $\frac{3}{2}$ 1 + $+1/\sqrt{2.3.5}$

$\frac{3}{2}$ $\frac{1}{2}$ 1 + $-1/\sqrt{5}$

$\frac{3}{2}$ $\frac{3}{2}$ $\tilde{0}$ − $+\sqrt{2}/\sqrt{3.5}$

$\frac{5}{2}$ $\frac{3}{2}$ 1 + $-1/\sqrt{2.3}$

$\frac{5}{2}$ 2 $\frac{1}{2}$ 0

$\frac{1}{2}$ 0 $\frac{1}{2}$ + $+1/\sqrt{5}$

$\frac{1}{2}$ 1 $\frac{1}{2}$ − $+\sqrt{2}/\sqrt{3.5}$

$\frac{3}{2}$ 1 $\frac{1}{2}$ + $+2/\sqrt{3.5}$

$\frac{3}{2}$ 2 $\frac{1}{2}$ − $-1/\sqrt{3.5}$

$\frac{5}{2}$ 2 $\frac{1}{2}$ + $+1/\sqrt{2.3}$

$\frac{5}{2}$ 2 $\frac{3}{2}$ 0

$\frac{1}{2}$ 0 $\frac{1}{2}$ − $-1/\sqrt{5}$

$\frac{1}{2}$ 1 $\frac{1}{2}$ + $+1/\sqrt{2.3.5}$

$\frac{1}{2}$ 1 $\frac{3}{2}$ − $-1/\sqrt{2.5}$

$\frac{1}{2}$ 2 $\frac{3}{2}$ + 0

$\frac{3}{2}$ 0 $\frac{3}{2}$ − 0

$\frac{5}{2}$ 2 $\frac{3}{2}$ 0

$\frac{3}{2}$ 1 $\frac{1}{2}$ − $+1/\sqrt{3.5}$

$\frac{3}{2}$ 2 $\frac{1}{2}$ + $+1/\sqrt{3.5}$

$\frac{3}{2}$ 2 $\frac{3}{2}$ + $+1/\sqrt{5}$

$\frac{5}{2}$ 1 $\frac{3}{2}$ − $-1/5\sqrt{2}-i\sqrt{2/5}$

$\frac{5}{2}$ 2 $\frac{1}{2}$ − $+1/5\sqrt{2.3}-i\sqrt{3}/5\sqrt{2}$

$\frac{5}{2}$ 2 $\frac{3}{2}$ 1

$\frac{1}{2}$ 0 $\frac{1}{2}$ + 0

$\frac{1}{2}$ 1 $\frac{1}{2}$ − $-1/\sqrt{5}$

$\frac{1}{2}$ 1 $\frac{3}{2}$ + $-1/\sqrt{3.5}$

$\frac{1}{2}$ 2 $\frac{3}{2}$ − $+1/\sqrt{3.5}$

$\frac{3}{2}$ 0 $\frac{3}{2}$ + $+1/\sqrt{5}$

$\frac{3}{2}$ 1 $\frac{1}{2}$ + 0

$\frac{3}{2}$ 2 $\frac{1}{2}$ − $+1/\sqrt{2.5}$

$\frac{3}{2}$ 2 $\frac{3}{2}$ − $-1/\sqrt{2.3.5}$

$\frac{5}{2}$ 1 $\frac{3}{2}$ + $-2/5\sqrt{3}+i/5\sqrt{3}$

$\frac{5}{2}$ 2 $\frac{1}{2}$ + $+3/2.5+i/2.5$

$\frac{5}{2}$ $\frac{5}{2}$ 0 0

$\frac{1}{2}$ $\frac{1}{2}$ 0 + $+1/\sqrt{3}$

$\frac{3}{2}$ $\frac{3}{2}$ 0 + $+1/\sqrt{3}$

-$\frac{5}{2}$ $\frac{5}{2}$ 0 + $+1/\sqrt{2.3}$

$\frac{5}{2}$ $\frac{5}{2}$ 1 0

$\frac{1}{2}$ $\frac{1}{2}$ $\tilde{0}$ + $+1/\sqrt{2.3}$

$\frac{1}{2}$ $\frac{1}{2}$ 1 + 0

$\frac{3}{2}$ $\frac{1}{2}$ 1 − $-1/\sqrt{2.3}$

$\frac{3}{2}$ $\frac{3}{2}$ $\tilde{0}$ + 0

$\frac{5}{2}$ $\frac{3}{2}$ 1 − $+1/2\sqrt{2.3.5}-i\sqrt{3}/2\sqrt{2.5}$

$\frac{5}{2}$ $\frac{5}{2}$ $\tilde{0}$ + $+2/5\sqrt{3}+i\sqrt{3/2.5}$

$\frac{5}{2}$ $\frac{5}{2}$ 1 1

$\frac{1}{2}$ $\frac{1}{2}$ $\tilde{0}$ + $-1/\sqrt{2.3.5}$

$\frac{1}{2}$ $\frac{1}{2}$ 1 + $+2/\sqrt{3.5}$

$\frac{3}{2}$ $\frac{1}{2}$ 1 − $-1/\sqrt{2.3.5}$

$\frac{3}{2}$ $\frac{3}{2}$ $\tilde{0}$ + $+\sqrt{2}/\sqrt{3.5}$

$\frac{5}{2}$ $\frac{3}{2}$ 1 − $+1/2\sqrt{2.3}+i/2\sqrt{2.3}$

$\frac{5}{2}$ $\frac{5}{2}$ $\tilde{0}$ + $+1/\sqrt{3.5}-i/2\sqrt{3.5}$

$\frac{5}{2}$ $\frac{5}{2}$ 2 0

$\frac{1}{2}$ $\frac{1}{2}$ 0 + $-\sqrt{2}/\sqrt{3.5}$

$\frac{1}{2}$ $\frac{1}{2}$ 1 − 0

$\frac{3}{2}$ $\frac{1}{2}$ 1 + 0

$\frac{3}{2}$ $\frac{1}{2}$ 2 − $+1/\sqrt{2.5}$

$\frac{3}{2}$ $\frac{3}{2}$ 0 + $+1/\sqrt{2.3.5}$

K–D_5 3jm Factors (cont.)

$\frac{5}{2}\ \frac{5}{2}\ 2\ 0$

$\frac{3}{2}\ \frac{3}{2}\ 2\ -\quad 0$

$\frac{5}{2}\ \frac{1}{2}\ 2\ +\quad +1/2.5+i/5$

$\frac{5}{2}\ \frac{3}{2}\ 1\ +\quad +\sqrt{2/5}-i/5\sqrt{2}$

$-\frac{5}{2}\ \frac{5}{2}\ 0\ +\quad +1/2\sqrt{3.5}$

$\frac{5}{2}\ \frac{5}{2}\ 2\ 1$

$\frac{1}{2}\ \frac{1}{2}\ 0\ -\quad 0$

$\frac{1}{2}\ \frac{1}{2}\ 1\ +\quad -\sqrt{2}/\sqrt{3.5}$

$\frac{3}{2}\ \frac{1}{2}\ 1\ -\quad -1/\sqrt{3.5}$

$\frac{3}{2}\ \frac{1}{2}\ 2\ +\quad -1/\sqrt{3.5}$

$\frac{3}{2}\ \frac{3}{2}\ 0\ -\quad 0$

$\frac{3}{2}\ \frac{3}{2}\ 2\ +\quad -\sqrt{2}/\sqrt{3.5}$

$\frac{5}{2}\ \frac{1}{2}\ 2\ -\quad +1/5\sqrt{2.3}+i\sqrt{2}/5\sqrt{3}$

$\frac{5}{2}\ \frac{3}{2}\ 1\ -\quad -1/2.5\sqrt{3}+i\sqrt{3/2.5}$

$-\frac{5}{2}\ \frac{5}{2}\ 0\ -\quad +i/\sqrt{2.5}$

$\frac{5}{2}\ \frac{5}{2}\ 2\ 2$

$\frac{1}{2}\ \frac{1}{2}\ 0\ +\quad 0$

$\frac{1}{2}\ \frac{1}{2}\ 1\ -\quad 0$

$\frac{3}{2}\ \frac{1}{2}\ 1\ +\quad -\sqrt{2}/\sqrt{3.5}$

$\frac{3}{2}\ \frac{1}{2}\ 2\ -\quad +1/\sqrt{2.3.5}$

$\frac{3}{2}\ \frac{3}{2}\ 0\ +\quad -1/\sqrt{2.5}$

$\frac{3}{2}\ \frac{3}{2}\ 2\ -\quad 0$

$\frac{5}{2}\ \frac{1}{2}\ 2\ +\quad +\sqrt{3/2.5}-2i/5\sqrt{3}$

$\frac{5}{2}\ \frac{3}{2}\ 1\ +\quad +1/5\sqrt{2.3}+i\sqrt{2}/5\sqrt{3}$

$-\frac{5}{2}\ \frac{5}{2}\ 0\ +\quad +1/2\sqrt{5}$

$3\ \frac{3}{2}\ \frac{3}{2}\ 0$

$1\ \frac{1}{2}\ \frac{1}{2}\ +\quad +\sqrt{3}/\sqrt{2.5}$

$1\ \frac{3}{2}\ \frac{1}{2}\ -\quad +1/\sqrt{2.5}$

$2\ \frac{3}{2}\ \frac{1}{2}\ +\quad -1/\sqrt{2.5}$

$2\ \frac{3}{2}\ \frac{3}{2}\ +\quad -\sqrt{3}/\sqrt{2.5}$

$3\ 2\ 1\ 0$

$1\ 0\ 1\ +\quad +1/\sqrt{5}$

$1\ 1\ \tilde{0}\ -\quad +2/\sqrt{3.5}$

$1\ 2\ 1\ +\quad -1/\sqrt{2.3.5}$

$2\ 1\ 1\ +\quad +\sqrt{2}/\sqrt{3.5}$

$2\ 2\ \tilde{0}\ -\quad -1/\sqrt{3.5}$

$2\ 2\ 1\ +\quad +\sqrt{3}/\sqrt{2.5}$

$3\ 2\ 2\ 0$

$1\ 1\ 0\ -\quad -1/\sqrt{2.5}$

$1\ 2\ 1\ -\quad -\sqrt{3}/2\sqrt{5}$

$1\ 2\ 2\ -\quad 0$

$3\ 2\ 2\ 0$

$2\ 1\ 1\ -\quad 0$

$2\ 2\ 0\ -\quad +1/\sqrt{2.5}$

$2\ 2\ 1\ -\quad -\sqrt{3}/2\sqrt{5}$

$3\ 2\ 2\ 1$

$1\ 1\ 0\ +\quad -1/\sqrt{2.5}$

$1\ 2\ 1\ +\quad -1/2\sqrt{3.5}$

$1\ 2\ 2\ +\quad +2/\sqrt{3.5}$

$2\ 1\ 1\ +\quad +2/\sqrt{3.5}$

$2\ 2\ 0\ +\quad -1/\sqrt{2.5}$

$2\ 2\ 1\ +\quad -1/2\sqrt{3.5}$

$3\ \frac{5}{2}\ \frac{1}{2}\ 0$

$1\ \frac{1}{2}\ \frac{1}{2}\ +\quad +1/\sqrt{3}$

$1\ \frac{3}{2}\ \frac{1}{2}\ -\quad +1/\sqrt{2.3}$

$2\ \frac{3}{2}\ \frac{1}{2}\ +\quad +1/\sqrt{2.3}$

$2\ \frac{5}{2}\ \frac{1}{2}\ -\quad -1/2\sqrt{3.5}+i\sqrt{3}/2\sqrt{5}$

$3\ \frac{5}{2}\ \frac{3}{2}\ 0$

$1\ \frac{1}{2}\ \frac{1}{2}\ -\quad -1/2\sqrt{3}$

$1\ \frac{1}{2}\ \frac{3}{2}\ +\quad 0$

$1\ \frac{3}{2}\ \frac{1}{2}\ +\quad +1/\sqrt{2.3}$

$1\ \frac{5}{2}\ \frac{3}{2}\ +\quad -1/4\sqrt{5}+3i/4\sqrt{5}$

$2\ \frac{1}{2}\ \frac{3}{2}\ -\quad -1/2$

$2\ \frac{3}{2}\ \frac{1}{2}\ -\quad +1/\sqrt{2.3}$

$2\ \frac{3}{2}\ \frac{3}{2}\ -\quad 0$

$2\ \frac{5}{2}\ \frac{1}{2}\ +\quad +1/4\sqrt{3.5}-i\sqrt{3}/4\sqrt{5}$

$3\ \frac{5}{2}\ \frac{3}{2}\ 1$

$1\ \frac{1}{2}\ \frac{1}{2}\ +\quad -1/2\sqrt{5}$

$1\ \frac{1}{2}\ \frac{3}{2}\ -\quad -2/\sqrt{3.5}$

$1\ \frac{3}{2}\ \frac{1}{2}\ -\quad -1/\sqrt{2.5}$

$1\ \frac{5}{2}\ \frac{3}{2}\ -\quad +1/4\sqrt{3}+i/4\sqrt{3}$

$2\ \frac{1}{2}\ \frac{3}{2}\ +\quad +1/2\sqrt{3.5}$

$2\ \frac{3}{2}\ \frac{1}{2}\ +\quad +1/\sqrt{2.5}$

$2\ \frac{3}{2}\ \frac{3}{2}\ +\quad -\sqrt{2}/\sqrt{3.5}$

$2\ \frac{5}{2}\ \frac{1}{2}\ -\quad +1/4+i/4$

$3\ \frac{5}{2}\ \frac{5}{2}\ 0$

$1\ \frac{1}{2}\ \frac{1}{2}\ +\quad +\sqrt{2}/\sqrt{3.5}$

$1\ \frac{3}{2}\ \frac{1}{2}\ -\quad -1/2\sqrt{3.5}$

$1\ \frac{5}{2}\ \frac{3}{2}\ -\quad -1/2\sqrt{3}$

$2\ \frac{3}{2}\ \frac{1}{2}\ +\quad +1/2\sqrt{3.5}$

$2\ \frac{3}{2}\ \frac{3}{2}\ +\quad -\sqrt{2}/\sqrt{3.5}$

$2\ \frac{5}{2}\ \frac{1}{2}\ -\quad +1/2\sqrt{2.3}-i/2\sqrt{2.3}$

K–D_5 3jm Factors (cont.)

$\mathbf{3\ \frac{5}{2}\ \frac{5}{2}\ 1}$
$1\ \frac{1}{2}\ \frac{1}{2}\ -\quad 0$
$1\ \frac{3}{2}\ \frac{1}{2}\ +\quad +1/2\sqrt{3}$
$1\ \frac{5}{2}\ \frac{3}{2}\ +\quad +1/2\sqrt{3.5}+i/\sqrt{3.5}$
$2\ \frac{3}{2}\ \frac{1}{2}\ -\quad -1/2\sqrt{3}$
$2\ \frac{3}{2}\ \frac{3}{2}\ -\quad 0$
$2\ \frac{5}{2}\ \frac{1}{2}\ +\quad +\sqrt{3}/2\sqrt{2.5}+i/2\sqrt{2.3.5}$
$\mathbf{3\ 3\ 0\ 0}$
$1\ 1\ 0\ +\quad +1/\sqrt{2}$
$2\ 2\ 0\ +\quad +1/\sqrt{2}$
$\mathbf{3\ 3\ 1\ 0}$
$1\ 1\ \tilde{0}\ +\quad -1/\sqrt{2.3}$
$2\ 1\ 1\ -\quad +1/\sqrt{3}$
$2\ 2\ \tilde{0}\ +\quad +1/\sqrt{2.3}$
$2\ 2\ 1\ -\quad 0$
$\mathbf{3\ 3\ 2\ 0}$
$1\ 1\ 0\ +\quad -1/\sqrt{2.5}$
$1\ 1\ 2\ +\quad -2/\sqrt{3.5}$
$2\ 1\ 1\ +\quad -1/\sqrt{3.5}$
$2\ 1\ 2\ +\quad +1/\sqrt{3.5}$
$2\ 2\ 0\ +\quad +1/\sqrt{2.5}$
$2\ 2\ 1\ +\quad +2/\sqrt{3.5}$
$\mathbf{3\ 3\ 3\ 0}$
$2\ 1\ 1\ +\quad -1/\sqrt{2.3}$
$2\ 2\ 1\ +\quad -1/\sqrt{2.3}$
$\mathbf{\tilde{1}\ \frac{3}{2}\ \frac{3}{2}\ 0}$
$\tilde{0}\ \frac{1}{2}\ \frac{1}{2}\ +\quad -\sqrt{3}/\sqrt{2.5}$
$\tilde{0}\ \frac{3}{2}\ \frac{3}{2}\ +\quad +1/\sqrt{2.3.5}$
$2\ \frac{3}{2}\ \frac{1}{2}\ +\quad -1/\sqrt{5}$
$2\ \frac{3}{2}\ \frac{3}{2}\ +\quad +2/\sqrt{3.5}$
$\mathbf{\tilde{1}\ 2\ 1\ 0}$
$\tilde{0}\ 0\ \tilde{0}\ +\quad +1/\sqrt{5}$
$\tilde{0}\ 1\ 1\ -\quad +\sqrt{2}/\sqrt{3.5}$
$2\ 1\ 1\ +\quad +2/\sqrt{3.5}$
$2\ 2\ \tilde{0}\ -\quad -\sqrt{2}/\sqrt{3.5}$
$2\ 2\ 1\ +\quad -2/\sqrt{3.5}$
$\mathbf{\tilde{1}\ 2\ 2\ 0}$
$\tilde{0}\ 1\ 1\ +\quad +2/\sqrt{3.5}$
$\tilde{0}\ 2\ 2\ +\quad -1/\sqrt{3.5}$
$2\ 1\ 1\ -\quad 0$
$2\ 2\ 0\ -\quad +1/\sqrt{5}$

$\mathbf{\tilde{1}\ 2\ 2\ 0}$
$2\ 2\ 1\ -\quad +\sqrt{2}/\sqrt{3.5}$
$\mathbf{\tilde{1}\ \frac{5}{2}\ \frac{1}{2}\ 0}$
$\tilde{0}\ \frac{1}{2}\ \frac{1}{2}\ +\quad +1/\sqrt{3}$
$2\ \frac{3}{2}\ \frac{1}{2}\ +\quad +1/\sqrt{3}$
$2\ \frac{5}{2}\ \frac{1}{2}\ -\quad -1/\sqrt{2.3.5}-i\sqrt{2}/\sqrt{3.5}$
$\mathbf{\tilde{1}\ \frac{5}{2}\ \frac{3}{2}\ 0}$
$\tilde{0}\ \frac{1}{2}\ \frac{1}{2}\ -\quad -\sqrt{2}/\sqrt{3.5}$
$\tilde{0}\ \frac{3}{2}\ \frac{3}{2}\ -\quad -1/\sqrt{5}$
$2\ \frac{1}{2}\ \frac{3}{2}\ -\quad +1/\sqrt{5}$
$2\ \frac{3}{2}\ \frac{1}{2}\ -\quad +1/\sqrt{2.3.5}$
$2\ \frac{3}{2}\ \frac{3}{2}\ -\quad +1/\sqrt{2.5}$
$2\ \frac{5}{2}\ \frac{1}{2}\ +\quad +1/2\sqrt{3}-i/2\sqrt{3}$
$\mathbf{\tilde{1}\ \frac{5}{2}\ \frac{5}{2}\ 0}$
$\tilde{0}\ \frac{1}{2}\ \frac{1}{2}\ +\quad -\sqrt{2}/\sqrt{3.5}$
$\tilde{0}\ \frac{3}{2}\ \frac{3}{2}\ +\quad -1/\sqrt{2.3.5}$
$\tilde{0}\ \frac{5}{2}\ \frac{5}{2}\ +\quad +1/2\sqrt{3.5}+i/\sqrt{3.5}$
$2\ \frac{3}{2}\ \frac{1}{2}\ +\quad +1/\sqrt{2.3.5}$
$2\ \frac{3}{2}\ \frac{3}{2}\ +\quad -2/\sqrt{3.5}$
$2\ \frac{5}{2}\ \frac{1}{2}\ -\quad -1/2\sqrt{3}$
$\mathbf{\tilde{1}\ \frac{5}{2}\ \frac{5}{2}\ 1}$
$\tilde{0}\ \frac{1}{2}\ \frac{1}{2}\ +\quad 0$
$\tilde{0}\ \frac{3}{2}\ \frac{3}{2}\ +\quad +1/\sqrt{2.3}$
$\tilde{0}\ \frac{5}{2}\ \frac{5}{2}\ +\quad -\sqrt{3}/2.5+2i/5\sqrt{3}$
$2\ \frac{3}{2}\ \frac{1}{2}\ +\quad +1/\sqrt{2.3}$
$2\ \frac{3}{2}\ \frac{3}{2}\ +\quad 0$
$2\ \frac{5}{2}\ \frac{1}{2}\ -\quad +1/2\sqrt{3.5}+i/\sqrt{3.5}$
$\mathbf{\tilde{1}\ 3\ 1\ 0}$
$\tilde{0}\ 1\ 1\ +\quad -1/\sqrt{3}$
$2\ 1\ 1\ -\quad +1/\sqrt{2.3}$
$2\ 2\ \tilde{0}\ +\quad -1/\sqrt{3}$
$2\ 2\ 1\ -\quad +1/\sqrt{2.3}$
$\mathbf{\tilde{1}\ 3\ 2\ 0}$
$\tilde{0}\ 1\ 1\ -\quad -1/\sqrt{3.5}$
$\tilde{0}\ 2\ 2\ -\quad -2/\sqrt{3.5}$
$2\ 1\ 1\ +\quad -\sqrt{3}/\sqrt{2.5}$
$2\ 1\ 2\ +\quad -\sqrt{2}/\sqrt{3.5}$
$2\ 2\ 0\ +\quad -1/\sqrt{5}$
$2\ 2\ 1\ +\quad +1/\sqrt{2.3.5}$
$\mathbf{\tilde{1}\ 3\ 3\ 0}$
$\tilde{0}\ 1\ 1\ +\quad -1/\sqrt{2.3}$

K–D_5 3jm Factors (cont.)

$\tilde{1}$ 3 3 0

$\tilde{0}$ 2 2 + $-1/\sqrt{2.3}$

2 1 1 − 0

2 2 1 − $-1/\sqrt{3}$

$\tilde{1}$ $\tilde{1}$ 0 0

$\tilde{0}$ $\tilde{0}$ 0 + $+1/\sqrt{3}$

2 2 0 + $+\sqrt{2}/\sqrt{3}$

$\tilde{1}$ $\tilde{1}$ 2 0

$\tilde{0}$ $\tilde{0}$ 0 + $-\sqrt{2}/\sqrt{3.5}$

2 $\tilde{0}$ 2 − $+1/\sqrt{5}$

2 2 0 + $+1/\sqrt{3.5}$

2 2 1 + $-\sqrt{2}/\sqrt{5}$

$\tilde{1}$ $\tilde{1}$ $\tilde{1}$ 0

2 2 $\tilde{0}$ + $-1/\sqrt{3}$

$\frac{\tilde{1}}{2}$ 2 $\frac{3}{2}$ 0

$\frac{3}{2}$ 0 $\frac{3}{2}$ + $-1/\sqrt{5}$

$\frac{3}{2}$ 1 $\frac{1}{2}$ + $+\sqrt{2}/\sqrt{5}$

$\frac{3}{2}$ 2 $\frac{1}{2}$ − $+1/\sqrt{2.5}$

$\frac{3}{2}$ 2 $\frac{3}{2}$ − $-\sqrt{3}/\sqrt{2.5}$

$\frac{\tilde{1}}{2}$ $\frac{5}{2}$ 1 0

$\frac{3}{2}$ $\frac{1}{2}$ 1 + $-1/\sqrt{3}$

$\frac{3}{2}$ $\frac{3}{2}$ $\tilde{0}$ − $-1/\sqrt{3}$

$\frac{3}{2}$ $\frac{5}{2}$ 1 + $+1/2\sqrt{3.5}-i\sqrt{3}/2\sqrt{5}$

$\frac{\tilde{1}}{2}$ $\frac{5}{2}$ 2 0

$\frac{3}{2}$ $\frac{1}{2}$ 1 − $+1/\sqrt{3.5}$

$\frac{3}{2}$ $\frac{1}{2}$ 2 + $+2/\sqrt{3.5}$

$\frac{3}{2}$ $\frac{3}{2}$ 0 − $-1/\sqrt{5}$

$\frac{3}{2}$ $\frac{3}{2}$ 2 + $-\sqrt{2}/\sqrt{3.5}$

$\frac{3}{2}$ $\frac{5}{2}$ 1 − $-1/2\sqrt{3}-i/2\sqrt{3}$

$\frac{\tilde{1}}{2}$ 3 $\frac{1}{2}$ 0

$\frac{3}{2}$ 1 $\frac{1}{2}$ + $+1/\sqrt{2}$

$\frac{3}{2}$ 2 $\frac{1}{2}$ − $+1/\sqrt{2}$

$\frac{\tilde{1}}{2}$ 3 $\frac{5}{2}$ 0

$\frac{3}{2}$ 1 $\frac{1}{2}$ + $-1/\sqrt{2.3}$

$\frac{3}{2}$ 1 $\frac{5}{2}$ + $-\sqrt{2}/\sqrt{3.5}+i/\sqrt{2.3.5}$

$\frac{3}{2}$ 2 $\frac{1}{2}$ − $+1/\sqrt{2.3}$

$\frac{3}{2}$ 2 $\frac{3}{2}$ − $+1/\sqrt{3}$

$\frac{\tilde{1}}{2}$ $\tilde{1}$ $\frac{3}{2}$ 0

$\frac{3}{2}$ $\tilde{0}$ $\frac{3}{2}$ + $-1/\sqrt{3}$

$\frac{3}{2}$ 2 $\frac{1}{2}$ + $-1/\sqrt{2}$

$\frac{3}{2}$ 2 $\frac{3}{2}$ + $-1/\sqrt{2.3}$

$\frac{\tilde{1}}{2}$ $\frac{\tilde{1}}{2}$ 0 0

$\frac{3}{2}$ $\frac{3}{2}$ 0 + $+1$

$\frac{\tilde{1}}{2}$ $\frac{\tilde{1}}{2}$ $\tilde{1}$ 0

$\frac{3}{2}$ $\frac{3}{2}$ $\tilde{0}$ + $-1/\sqrt{3}$

$\frac{3}{2}$ $\frac{3}{2}$ 2 + $+\sqrt{2}/\sqrt{3}$

K–T 2jm Factors (K_i–T_i = K_h–T_h)

0 → + 0

$\frac{1}{2}$ → + $\frac{1}{2}$

1 → + 1

$\frac{3}{2}$ → + $\frac{3}{2}$ + -$\frac{3}{2}$

2 → + 1 + 2 + -2

$\frac{5}{2}$ → + $\frac{1}{2}$ + $\frac{3}{2}$ + -$\frac{3}{2}$

3 → + 0 + 1

$\tilde{1}$ → + 1

$\frac{\tilde{1}}{2}$ → + $\frac{1}{2}$

K–T 3jm Factors

0 0 0 0

0 0 0 0 + $+1$

$\frac{1}{2}$ $\frac{1}{2}$ 0 0

$\frac{1}{2}$ $\frac{1}{2}$ 0 0 + $+1$

1 $\frac{1}{2}$ $\frac{1}{2}$ 0

1 $\frac{1}{2}$ $\frac{1}{2}$ 0 + $+1$

1 1 0 0

1 1 0 0 + $+1$

1 1 1 0

1 1 1 0 + $+1$

1 1 1 1 − 0

$\frac{3}{2}$ 1 $\frac{1}{2}$ 0

$\frac{3}{2}$ 1 $\frac{1}{2}$ 0 − $+1/\sqrt{2}$

$\frac{3}{2}$ $\frac{3}{2}$ 0 0

-$\frac{3}{2}$ $\frac{3}{2}$ 0 0 + $+1/\sqrt{2}$

$\frac{3}{2}$ $\frac{3}{2}$ 1 0

$\frac{3}{2}$ $\frac{3}{2}$ 1 0 + $-\sqrt{2}/\sqrt{5}$

-$\frac{3}{2}$ $\frac{3}{2}$ 1 0 + $-1/\sqrt{2.5}$

2 1 1 0

1 1 1 0 − 0

1 1 1 1 + $-\sqrt{3}/\sqrt{5}$

2 1 1 0 + $+1/\sqrt{5}$

2 $\frac{3}{2}$ $\frac{1}{2}$ 0

1 $\frac{3}{2}$ $\frac{1}{2}$ 0 + $+i\sqrt{3}/\sqrt{2.5}$

2 -$\frac{3}{2}$ $\frac{1}{2}$ 0 − $+1/\sqrt{5}$

2 $\frac{3}{2}$ $\frac{3}{2}$ 0

1 $\frac{3}{2}$ $\frac{3}{2}$ 0 − 0

1 -$\frac{3}{2}$ $\frac{3}{2}$ 0 − $-i\sqrt{3}/\sqrt{2.5}$

2 $\frac{3}{2}$ $\frac{3}{2}$ 0 + $+1/\sqrt{5}$

2 2 0 0

1 1 0 0 + $+\sqrt{3}/\sqrt{5}$

-2 2 0 0 + $+1/\sqrt{5}$

2 2 1 0

1 1 1 0 + $-1/\sqrt{5}$

1 1 1 1 − 0

2 1 1 0 − $-i/\sqrt{5}$

2 2 2 0

1 1 1 0 − 0

1 1 1 1 + 0

2 1 1 0 + $+\sqrt{3}/4\sqrt{2}+3i/4\sqrt{2.5}$

2 2 2 0 + $-1/8\sqrt{2}+3i\sqrt{3}/8\sqrt{2.5}$

2 2 2 1

1 1 1 0 − 0

1 1 1 1 + $-\sqrt{2}/\sqrt{5}$

2 1 1 0 + $-\sqrt{3}/4\sqrt{2.5}+i/4\sqrt{2}$

2 2 2 0 + $+9/8\sqrt{2.5}+i\sqrt{3}/8\sqrt{2}$

$\frac{5}{2}$ $\frac{3}{2}$ 1 0

$\frac{1}{2}$ $\frac{3}{2}$ 1 0 − $+i/\sqrt{2.3}$

$\frac{3}{2}$ $\frac{3}{2}$ 1 0 − $-i/\sqrt{3.5}$

$\frac{3}{2}$ -$\frac{3}{2}$ 1 0 − $+2i/\sqrt{3.5}$

$\frac{5}{2}$ 2 $\frac{1}{2}$ 0

$\frac{1}{2}$ 1 $\frac{1}{2}$ 0 − $+1/\sqrt{3}$

$\frac{3}{2}$ 1 $\frac{1}{2}$ 0 − $+\sqrt{2}/\sqrt{3.5}$

$\frac{3}{2}$ -2 $\frac{1}{2}$ 0 + $-i/\sqrt{5}$

$\frac{5}{2}$ 2 $\frac{3}{2}$ 0

$\frac{1}{2}$ 1 $\frac{3}{2}$ 0 + $+1/2\sqrt{2.3}+i/2\sqrt{2.5}$

$\frac{1}{2}$ 2 -$\frac{3}{2}$ 0 − $-\sqrt{3}/4\sqrt{5}+i/4$

$\frac{3}{2}$ 1 $\frac{3}{2}$ 0 + $-\sqrt{5}/4\sqrt{3}-i/4$

$\frac{3}{2}$ 1 -$\frac{3}{2}$ 0 + $-1/4\sqrt{3.5}-i/4$

$\frac{3}{2}$ 2 $\frac{3}{2}$ 0 − $-\sqrt{3}/4\sqrt{2}+i/4\sqrt{2.5}$

$\frac{5}{2}$ 2 $\frac{3}{2}$ 1

$\frac{1}{2}$ 1 $\frac{3}{2}$ 0 − $-1/4-i/4\sqrt{3.5}$

$\frac{1}{2}$ 2 -$\frac{3}{2}$ 0 + $+1/4\sqrt{2.5}+i\sqrt{3}/4\sqrt{2}$

$\frac{3}{2}$ 1 $\frac{3}{2}$ 0 − $-\sqrt{5}/4\sqrt{2}+i/4\sqrt{2.3}$

$\frac{3}{2}$ 1 -$\frac{3}{2}$ 0 − $+3/4\sqrt{2.5}+i/4\sqrt{2.3}$

$\frac{3}{2}$ 2 $\frac{3}{2}$ 0 + $+1/8-3i\sqrt{3}/8\sqrt{5}$

$\frac{5}{2}$ $\frac{5}{2}$ 0 0

$\frac{1}{2}$ $\frac{1}{2}$ 0 0 + $+1/\sqrt{3}$

-$\frac{3}{2}$ $\frac{3}{2}$ 0 0 + $+1/\sqrt{3}$

$\frac{5}{2}$ $\frac{5}{2}$ 1 0

$\frac{1}{2}$ $\frac{1}{2}$ 1 0 + $-\sqrt{2/3}$

$\frac{3}{2}$ $\frac{1}{2}$ 1 0 + $+\sqrt{5/8.3}-i\sqrt{3/8}$

$\frac{3}{2}$ $\frac{3}{2}$ 1 0 + $-11/8.3\sqrt{2}-i\sqrt{3.5}/8\sqrt{2}$

-$\frac{3}{2}$ $\frac{3}{2}$ 1 0 + $+1/3\sqrt{2}$

$\frac{5}{2}$ $\frac{5}{2}$ 1 1

$\frac{1}{2}$ $\frac{1}{2}$ 1 0 + 0

$\frac{3}{2}$ $\frac{1}{2}$ 1 0 + $+3/8+i\sqrt{5}/8\sqrt{3}$

$\frac{3}{2}$ $\frac{3}{2}$ 1 0 + $-1/8\sqrt{2.5}+5i/8\sqrt{2.3}$

-$\frac{3}{2}$ $\frac{3}{2}$ 1 0 + $+1/\sqrt{2.5}$

$\frac{5}{2}$ $\frac{5}{2}$ 2 0

$\frac{1}{2}$ $\frac{1}{2}$ 1 0 − 0

$\frac{3}{2}$ $\frac{1}{2}$ 1 0 − $+\sqrt{3}/4\sqrt{5}-i/4$

K–T 3jm Factors (cont.)

$\mathbf{\frac{5}{2}\ \frac{5}{2}\ 2\ 0}$	
$\frac{3}{2}\ \frac{1}{2}\ {-2}\ 0\ +$	$+1/2\sqrt{2.3}-i/2\sqrt{2.5}$
$\frac{3}{2}\ \frac{3}{2}\ 1\ 0\ -$	0
$\frac{3}{2}\ \frac{3}{2}\ 2\ 0\ +$	$-7/8\sqrt{3.5}+i/8$
$-\frac{3}{2}\ \frac{3}{2}\ 1\ 0\ -$	$+i/\sqrt{2.5}$
$\mathbf{\frac{5}{2}\ \frac{5}{2}\ 2\ 1}$	
$\frac{1}{2}\ \frac{1}{2}\ 1\ 0\ +$	$+1/3$
$\frac{3}{2}\ \frac{1}{2}\ 1\ 0\ +$	$-1/4.3\sqrt{2.5}-i/4\sqrt{2.3}$
$\frac{3}{2}\ \frac{1}{2}\ {-2}\ 0\ -$	$-1/4-i\sqrt{3}/4\sqrt{5}$
$\frac{3}{2}\ \frac{3}{2}\ 1\ 0\ +$	$+7/8.3-i\sqrt{5}/8\sqrt{3}$
$\frac{3}{2}\ \frac{3}{2}\ 2\ 0\ -$	0
$-\frac{3}{2}\ \frac{3}{2}\ 1\ 0\ +$	$+1/3$
$\mathbf{\frac{5}{2}\ \frac{5}{2}\ 2\ 2}$	
$\frac{1}{2}\ \frac{1}{2}\ 1\ 0\ -$	0
$\frac{3}{2}\ \frac{1}{2}\ 1\ 0\ -$	$-1/2\sqrt{5}-i/2\sqrt{3}$
$\frac{3}{2}\ \frac{1}{2}\ {-2}\ 0\ +$	$+i/\sqrt{2.3.5}$
$\frac{3}{2}\ \frac{3}{2}\ 1\ 0\ -$	0
$\frac{3}{2}\ \frac{3}{2}\ 2\ 0\ +$	$-3/4\sqrt{5}-i/4\sqrt{3}$
$-\frac{3}{2}\ \frac{3}{2}\ 1\ 0\ -$	$-i/\sqrt{2.3.5}$
$\mathbf{3\ \frac{3}{2}\ \frac{3}{2}\ 0}$	
$0\ {-\frac{3}{2}}\ \frac{3}{2}\ 0\ -$	$-i/2\sqrt{2}$
$1\ \frac{3}{2}\ \frac{3}{2}\ 0\ +$	$-3/4\sqrt{2.5}-i\sqrt{3}/4\sqrt{2}$
$1\ {-\frac{3}{2}}\ \frac{3}{2}\ 0\ +$	$+3/2\sqrt{2.5}$
$\mathbf{3\ 2\ 1\ 0}$	
$0\ 1\ 1\ 0\ +$	$+1/2$
$1\ 1\ 1\ 0\ -$	$-1/2\sqrt{2}$
$1\ 1\ 1\ 1\ +$	$-3/2\sqrt{2.5}$
$1\ 2\ 1\ 0\ +$	$-3\sqrt{3}/4\sqrt{2.5}-i/4\sqrt{2}$
$\mathbf{3\ 2\ 2\ 0}$	
$0\ 1\ 1\ 0\ -$	0
$0\ {-2}\ 2\ 0\ -$	$+i/2\sqrt{2}$
$1\ 1\ 1\ 0\ +$	$-3/2\sqrt{5}$
$1\ 1\ 1\ 1\ -$	0
$1\ 2\ 1\ 0\ -$	$+\sqrt{3}/8+3i/8\sqrt{5}$
$\mathbf{3\ 2\ 2\ 1}$	
$0\ 1\ 1\ 0\ +$	$+1/\sqrt{2.5}$
$0\ {-2}\ 2\ 0\ +$	$-\sqrt{3}/2\sqrt{2.5}$
$1\ 1\ 1\ 0\ -$	0
$1\ 1\ 1\ 1\ +$	$+1/2$
$1\ 2\ 1\ 0\ +$	$-\sqrt{3}/8+i\sqrt{5}/8$
$\mathbf{3\ \frac{5}{2}\ \frac{1}{2}\ 0}$	
$0\ \frac{1}{2}\ \frac{1}{2}\ 0\ -$	$+1/2$
$1\ \frac{1}{2}\ \frac{1}{2}\ 0\ +$	$+1/2\sqrt{3}$
$1\ \frac{3}{2}\ \frac{1}{2}\ 0\ +$	$-\sqrt{5}/4\sqrt{2.3}+3i/4\sqrt{2}$
$\mathbf{3\ \frac{5}{2}\ \frac{3}{2}\ 0}$	
$0\ \frac{3}{2}\ {-\frac{3}{2}}\ 0\ +$	$+\sqrt{5}/8.2-3i\sqrt{3}/8.2$
$1\ \frac{1}{2}\ \frac{3}{2}\ 0\ -$	$-1/\sqrt{2.3}$
$1\ \frac{3}{2}\ \frac{3}{2}\ 0\ -$	$+\sqrt{5}/8\sqrt{3}-3i/8$
$1\ \frac{3}{2}\ {-\frac{3}{2}}\ 0\ -$	$+\sqrt{5}/8.2\sqrt{3}-3i/8.2$
$\mathbf{3\ \frac{5}{2}\ \frac{3}{2}\ 1}$	
$0\ \frac{3}{2}\ {-\frac{3}{2}}\ 0\ -$	$-3\sqrt{3}/8.2-i\sqrt{5}/8.2$
$1\ \frac{1}{2}\ \frac{3}{2}\ 0\ +$	$+i/\sqrt{2.3}$
$1\ \frac{3}{2}\ \frac{3}{2}\ 0\ +$	$+1/8+7i/8\sqrt{3.5}$
$1\ \frac{3}{2}\ {-\frac{3}{2}}\ 0\ +$	$+5/8.2-13i/8.2\sqrt{3.5}$
$\mathbf{3\ \frac{5}{2}\ \frac{5}{2}\ 0}$	
$0\ \frac{1}{2}\ \frac{1}{2}\ 0\ -$	0
$0\ {-\frac{3}{2}}\ \frac{3}{2}\ 0\ -$	$+i/2\sqrt{2}$
$1\ \frac{1}{2}\ \frac{1}{2}\ 0\ +$	$+\sqrt{5}/3\sqrt{2}$
$1\ \frac{3}{2}\ \frac{1}{2}\ 0\ +$	$-1/8.3-i\sqrt{5}/8\sqrt{3}$
$1\ \frac{3}{2}\ \frac{3}{2}\ 0\ +$	$-7/2.3\sqrt{2.5}+i/2\sqrt{2.3}$
$1\ {-\frac{3}{2}}\ \frac{3}{2}\ 0\ +$	$+1/2.3\sqrt{2.5}$
$\mathbf{3\ \frac{5}{2}\ \frac{5}{2}\ 1}$	
$0\ \frac{1}{2}\ \frac{1}{2}\ 0\ +$	$-1/\sqrt{2.3}$
$0\ {-\frac{3}{2}}\ \frac{3}{2}\ 0\ +$	$+1/2\sqrt{2.3}$
$1\ \frac{1}{2}\ \frac{1}{2}\ 0\ -$	0
$1\ \frac{3}{2}\ \frac{1}{2}\ 0\ -$	$-\sqrt{5}/8+i/8\sqrt{3}$
$1\ \frac{3}{2}\ \frac{3}{2}\ 0\ -$	0
$1\ {-\frac{3}{2}}\ \frac{3}{2}\ 0\ -$	$+i\sqrt{5}/2\sqrt{2.3}$
$\mathbf{3\ 3\ 0\ 0}$	
$0\ 0\ 0\ 0\ +$	$+1/2$
$1\ 1\ 0\ 0\ +$	$+\sqrt{3}/2$
$\mathbf{3\ 3\ 1\ 0}$	
$1\ 0\ 1\ 0\ -$	$+1/2$
$1\ 1\ 1\ 0\ +$	$-1/\sqrt{2}$
$1\ 1\ 1\ 1\ -$	0
$\mathbf{3\ 3\ 2\ 0}$	
$1\ 0\ 1\ 0\ +$	$-1/2$
$1\ 1\ 1\ 0\ -$	0
$1\ 1\ 1\ 1\ +$	$-1/\sqrt{2.5}$
$1\ 1\ 2\ 0\ +$	$-\sqrt{3}/2\sqrt{2.5}+i/2\sqrt{2}$

K–T 3jm Factors (cont.)

3 3 3 0		
0 0 0 0	+	$+\sqrt{3}/4$
1 1 0 0	+	$-1/4$
1 1 1 0	−	0
1 1 1 1	+	$+\sqrt{5}/2\sqrt{2}$
$\tilde{1}$ $\frac{3}{2}$ $\frac{3}{2}$ 0		
1 $\frac{3}{2}$ $\frac{3}{2}$ 0	+	$+1/2\sqrt{2.5}-i\sqrt{3}/2\sqrt{2}$
1 $-\frac{3}{2}$ $\frac{3}{2}$ 0	+	$-1/\sqrt{2.5}$
$\tilde{1}$ 2 1 0		
1 1 1 0	−	$-1/\sqrt{2}$
1 1 1 1	+	$+1/\sqrt{2.5}$
1 2 1 0	+	$+\sqrt{3}/2\sqrt{2.5}-i/2\sqrt{2}$
$\tilde{1}$ 2 2 0		
1 1 1 0	+	$+1/\sqrt{5}$
1 1 1 1	−	0
1 2 1 0	−	$+\sqrt{3}/4-i/4\sqrt{5}$
$\tilde{1}$ $\frac{5}{2}$ $\frac{1}{2}$ 0		
1 $\frac{1}{2}$ $\frac{1}{2}$ 0	+	$+1/\sqrt{3}$
1 $\frac{3}{2}$ $\frac{1}{2}$ 0	+	$-\sqrt{5}/2\sqrt{2.3}-i/2\sqrt{2}$
$\tilde{1}$ $\frac{5}{2}$ $\frac{3}{2}$ 0		
1 $\frac{1}{2}$ $\frac{3}{2}$ 0	−	$+\sqrt{5}/4\sqrt{3}-i/4$
1 $\frac{3}{2}$ $\frac{3}{2}$ 0	−	$+1/4\sqrt{2.3}-3i/4\sqrt{2.5}$
1 $\frac{3}{2}$ $-\frac{3}{2}$ 0	−	$+5/4\sqrt{2.3}+i/4\sqrt{2.5}$
$\tilde{1}$ $\frac{5}{2}$ $\frac{5}{2}$ 0		
1 $\frac{1}{2}$ $\frac{1}{2}$ 0	+	0
1 $\frac{3}{2}$ $\frac{1}{2}$ 0	+	$-1/4+i\sqrt{5}/4\sqrt{3}$
1 $\frac{3}{2}$ $\frac{3}{2}$ 0	+	$-3/4\sqrt{2.5}-i/4\sqrt{2.3}$
1 $-\frac{3}{2}$ $\frac{3}{2}$ 0	+	$+1/\sqrt{2.5}$
$\tilde{1}$ $\frac{5}{2}$ $\frac{5}{2}$ 1		
1 $\frac{1}{2}$ $\frac{1}{2}$ 0	+	$-\sqrt{2/3}$
1 $\frac{3}{2}$ $\frac{1}{2}$ 0	+	$-\sqrt{5}/4.3-i/4\sqrt{3}$
1 $\frac{3}{2}$ $\frac{3}{2}$ 0	+	$+1/2.3\sqrt{2}+i\sqrt{5}/2\sqrt{2.3}$
1 $-\frac{3}{2}$ $\frac{3}{2}$ 0	+	$+1/3\sqrt{2}$
$\tilde{1}$ 3 1 0		
1 0 1 0	−	$+1/2$
1 1 1 0	+	$+1/2\sqrt{2}$
1 1 1 1	−	$+\sqrt{5}/2\sqrt{2}$
$\tilde{1}$ 3 2 0		
1 0 1 0	+	$+1/2$
1 1 1 0	−	$+1/2\sqrt{2}$
1 1 1 1	+	$-3/2\sqrt{2.5}$
$\tilde{1}$ 3 2 0		
1 1 2 0	+	$+\sqrt{3}/2\sqrt{2.5}+i/2\sqrt{2}$
$\tilde{1}$ 3 3 0		
1 1 0 0	−	$+1/2$
1 1 1 0	+	$+1/\sqrt{2}$
1 1 1 1	−	0
$\tilde{1}$ $\tilde{1}$ 0 0		
1 1 0 0	+	$+1$
$\tilde{1}$ $\tilde{1}$ 2 0		
1 1 1 0	−	0
1 1 1 1	+	$-\sqrt{3}/\sqrt{5}$
1 1 2 0	+	$-1/4\sqrt{5}-i\sqrt{3}/4$
$\tilde{1}$ $\tilde{1}$ $\tilde{1}$ 0		
1 1 1 0	+	-1
1 1 1 1	−	0
$\tilde{\frac{1}{2}}$ 2 $\frac{3}{2}$ 0		
$\frac{1}{2}$ 1 $\frac{3}{2}$ 0	−	$+\sqrt{3}/4-3i/4\sqrt{5}$
$\frac{1}{2}$ 2 $-\frac{3}{2}$ 0	+	$+3\sqrt{3}/4\sqrt{2.5}+i/4\sqrt{2}$
$\tilde{\frac{1}{2}}$ $\frac{5}{2}$ 1 0		
$\frac{1}{2}$ $\frac{1}{2}$ 1 0	−	$-1/\sqrt{3}$
$\frac{1}{2}$ $\frac{3}{2}$ 1 0	−	$-\sqrt{5}/4\sqrt{2.3}+3i/4\sqrt{2}$
$\tilde{\frac{1}{2}}$ $\frac{5}{2}$ 2 0		
$\frac{1}{2}$ $\frac{1}{2}$ 1 0	+	$+1/\sqrt{3}$
$\frac{1}{2}$ $\frac{3}{2}$ 1 0	+	$-7/4\sqrt{2.3.5}+i/4\sqrt{2}$
$\frac{1}{2}$ $\frac{3}{2}$ -2 0	−	$-\sqrt{3}/4+i/4\sqrt{5}$
$\tilde{\frac{1}{2}}$ 3 $\frac{1}{2}$ 0		
$\frac{1}{2}$ 0 $\frac{1}{2}$ 0	+	$+1/2$
$\frac{1}{2}$ 1 $\frac{1}{2}$ 0	−	$+\sqrt{3}/2$
$\tilde{\frac{1}{2}}$ 3 $\frac{5}{2}$ 0		
$\frac{1}{2}$ 0 $\frac{1}{2}$ 0	+	$-1/2$
$\frac{1}{2}$ 1 $\frac{1}{2}$ 0	−	$+1/2\sqrt{3}$
$\frac{1}{2}$ 1 $\frac{3}{2}$ 0	−	$+\sqrt{5}/2\sqrt{2.3}+i/2\sqrt{2}$
$\tilde{\frac{1}{2}}$ $\tilde{1}$ $\frac{3}{2}$ 0		
$\frac{1}{2}$ 1 $\frac{3}{2}$ 0	+	$+\sqrt{3}/4+i\sqrt{5}/4$
$\tilde{\frac{1}{2}}$ $\tilde{\frac{1}{2}}$ 0 0		
$\frac{1}{2}$ $\frac{1}{2}$ 0 0	+	$+1$
$\tilde{\frac{1}{2}}$ $\tilde{\frac{1}{2}}$ $\tilde{1}$ 0		
$\frac{1}{2}$ $\frac{1}{2}$ 1 0	+	-1

SO_3–O 2jm Factors ($SO_{3i}-O_i = O_3-O_h$)

$0 \rightarrow +0$

$\frac{1}{2} \rightarrow +\frac{1}{2}$

$1 \rightarrow +1$

$\frac{3}{2} \rightarrow +\frac{3}{2}$

$2 \rightarrow +2+\tilde{1}$

$\frac{5}{2} \rightarrow +\frac{3}{2}+\tilde{\frac{1}{2}}$

$3 \rightarrow +1+\tilde{1}+\tilde{0}$

$\frac{7}{2} \rightarrow +\frac{1}{2}+\frac{3}{2}+\tilde{\frac{1}{2}}$

$4 \rightarrow +0+1+2+\tilde{1}$

$\frac{9}{2} \rightarrow +\frac{1}{2}+2\,\frac{3}{2}$

$5 \rightarrow +2\,1+2+\tilde{1}$

$\frac{11}{2} \rightarrow +\frac{1}{2}+2\,\frac{3}{2}+\tilde{\frac{1}{2}}$

$6 \rightarrow +0+1+2+2\,\tilde{1}+\tilde{0}$

$\frac{13}{2} \rightarrow +\frac{1}{2}+2\,\frac{3}{2}+2\,\tilde{\frac{1}{2}}$

$7 \rightarrow +2\,1+2+2\,\tilde{1}+\tilde{0}$

$\frac{15}{2} \rightarrow +\frac{1}{2}+3\,\frac{3}{2}+\tilde{\frac{1}{2}}$

$8 \rightarrow +0+2\,1+2\,2+2\,\tilde{1}$

SO_3–O 3jm Factors

$0\ 0\ 0$
$0\ 0\ 0$ 0 000+ $+1$
$\frac{1}{2}\ \frac{1}{2}\ 0$
$\frac{1}{2}\ \frac{1}{2}\ 0$ 0 000+ $+1$
$1\ \frac{1}{2}\ \frac{1}{2}$
$1\ \frac{1}{2}\ \frac{1}{2}$ 0 000+ $+1$
$1\ 1\ 0$
$1\ 1\ 0$ 0 000+ $+1$
$1\ 1\ 1$
$1\ 1\ 1$ 0 000+ -1
$\frac{3}{2}\ 1\ \frac{1}{2}$
$\frac{3}{2}\ 1\ \frac{1}{2}$ 0 000+ $+1$
$\frac{3}{2}\ \frac{3}{2}\ 0$
$\frac{3}{2}\ \frac{3}{2}\ 0$ 0 000+ $+1$
$\frac{3}{2}\ \frac{3}{2}\ 1$
$\frac{3}{2}\ \frac{3}{2}\ 1$ 0 000+ $+1/\sqrt{5}$
$\frac{3}{2}\ \frac{3}{2}\ 1$ 1 000+ $+2/\sqrt{5}$
$2\ 1\ 1$
$2\ 1\ 1$ 0 000+ $+\sqrt{2}/\sqrt{5}$
$\tilde{1}\ 1\ 1$ 0 000+ $+\sqrt{3}/\sqrt{5}$
$2\ \frac{3}{2}\ \frac{1}{2}$
$2\ \frac{3}{2}\ \frac{1}{2}$ 0 000+ $+\sqrt{2}/\sqrt{5}$
$\tilde{1}\ \frac{3}{2}\ \frac{1}{2}$ 0 000+ $+\sqrt{3}/\sqrt{5}$
$2\ \frac{3}{2}\ \frac{3}{2}$
$2\ \frac{3}{2}\ \frac{3}{2}$ 0 000+ $-\sqrt{2}/\sqrt{5}$
$\tilde{1}\ \frac{3}{2}\ \frac{3}{2}$ 0 000+ $-\sqrt{3}/\sqrt{5}$
$\tilde{1}\ \frac{3}{2}\ \frac{3}{2}$ 1 000− 0
$2\ 2\ 0$
$2\ 2\ 0$ 0 000+ $+\sqrt{2}/\sqrt{5}$
$\tilde{1}\ \tilde{1}\ 0$ 0 000+ $+\sqrt{3}/\sqrt{5}$
$2\ 2\ 1$
$\tilde{1}\ 2\ 1$ 0 000+ $-\sqrt{2}/\sqrt{5}$
$\tilde{1}\ \tilde{1}\ 1$ 0 000+ $-1/\sqrt{5}$
$2\ 2\ 2$
$2\ 2\ 2$ 0 000+ $-2\sqrt{2}/\sqrt{5.7}$
$\tilde{1}\ \tilde{1}\ 2$ 0 000+ $-\sqrt{2.3}/\sqrt{5.7}$
$\tilde{1}\ \tilde{1}\ \tilde{1}$ 0 000+ $-3/\sqrt{5.7}$
$\frac{5}{2}\ \frac{3}{2}\ 1$
$\frac{3}{2}\ \frac{3}{2}\ 1$ 0 000− $+2\sqrt{2}/\sqrt{3.5}$
$\frac{3}{2}\ \frac{3}{2}\ 1$ 1 000− $-\sqrt{2}/\sqrt{3.5}$
$\tilde{\frac{1}{2}}\ \frac{3}{2}\ 1$ 0 000+ $-1/\sqrt{3}$
$\frac{5}{2}\ 2\ \frac{1}{2}$
$\frac{3}{2}\ 2\ \frac{1}{2}$ 0 000− $+\sqrt{2}/\sqrt{5}$
$\frac{3}{2}\ \tilde{1}\ \frac{1}{2}$ 0 000− $-2/\sqrt{3.5}$
$\tilde{\frac{1}{2}}\ \tilde{1}\ \frac{1}{2}$ 0 000+ $+1/\sqrt{3}$
$\frac{5}{2}\ 2\ \frac{3}{2}$
$\frac{3}{2}\ 2\ \frac{3}{2}$ 0 000− $+2/\sqrt{5.7}$
$\frac{3}{2}\ \tilde{1}\ \frac{3}{2}$ 0 000− $-2\sqrt{2}/\sqrt{3.5.7}$
$\frac{3}{2}\ \tilde{1}\ \frac{3}{2}$ 1 000+ $+\sqrt{2.5}/\sqrt{3.7}$
$\tilde{\frac{1}{2}}\ 2\ \frac{3}{2}$ 0 000+ $-\sqrt{2}/\sqrt{7}$
$\tilde{\frac{1}{2}}\ \tilde{1}\ \frac{3}{2}$ 0 000+ $-1/\sqrt{3.7}$
$\frac{5}{2}\ \frac{5}{2}\ 0$
$\frac{3}{2}\ \frac{3}{2}\ 0$ 0 000+ $+\sqrt{2}/\sqrt{3}$
$\tilde{\frac{1}{2}}\ \tilde{\frac{1}{2}}\ 0$ 0 000+ $+1/\sqrt{3}$
$\frac{5}{2}\ \frac{5}{2}\ 1$
$\frac{3}{2}\ \frac{3}{2}\ 1$ 0 000+ $-\sqrt{2.7}/3\sqrt{5}$
$\frac{3}{2}\ \frac{3}{2}\ 1$ 1 000+ $-4\sqrt{2}/3\sqrt{5.7}$
$\tilde{\frac{1}{2}}\ \frac{3}{2}\ 1$ 0 000− $-4/3\sqrt{7}$
$\tilde{\frac{1}{2}}\ \tilde{\frac{1}{2}}\ 1$ 0 000+ $+\sqrt{5}/3\sqrt{7}$
$\frac{5}{2}\ \frac{5}{2}\ 2$
$\frac{3}{2}\ \frac{3}{2}\ 2$ 0 000+ $-4\sqrt{2}/\sqrt{3.5.7}$
$\frac{3}{2}\ \frac{3}{2}\ \tilde{1}$ 0 000+ $+1/\sqrt{5.7}$
$\frac{3}{2}\ \frac{3}{2}\ \tilde{1}$ 1 000− 0
$\tilde{\frac{1}{2}}\ \frac{3}{2}\ 2$ 0 000− $+1/\sqrt{3.7}$
$\tilde{\frac{1}{2}}\ \frac{3}{2}\ \tilde{1}$ 0 000− $+\sqrt{2}/\sqrt{7}$
$3\ \frac{3}{2}\ \frac{3}{2}$
$1\ \frac{3}{2}\ \frac{3}{2}$ 0 000+ $-2\sqrt{3}/\sqrt{5.7}$
$1\ \frac{3}{2}\ \frac{3}{2}$ 1 000+ $+\sqrt{3}/\sqrt{5.7}$
$\tilde{1}\ \frac{3}{2}\ \frac{3}{2}$ 0 000− 0
$\tilde{1}\ \frac{3}{2}\ \frac{3}{2}$ 1 000+ $-\sqrt{3}/\sqrt{7}$
$\tilde{0}\ \frac{3}{2}\ \frac{3}{2}$ 0 000+ $-1/\sqrt{7}$
$3\ 2\ 1$
$1\ 2\ 1$ 0 000+ $+3/\sqrt{5.7}$
$1\ \tilde{1}\ 1$ 0 000+ $-\sqrt{2.3}/\sqrt{5.7}$
$\tilde{1}\ 2\ 1$ 0 000− $-1/\sqrt{7}$
$\tilde{1}\ \tilde{1}\ 1$ 0 000− $+\sqrt{2}/\sqrt{7}$
$\tilde{0}\ \tilde{1}\ 1$ 0 000+ $+1/\sqrt{7}$
$3\ 2\ 2$
$1\ \tilde{1}\ 2$ 0 000+ $+\sqrt{3}/\sqrt{2.5.7}$
$1\ \tilde{1}\ \tilde{1}$ 0 000+ $-2\sqrt{3}/\sqrt{5.7}$
$\tilde{1}\ \tilde{1}\ 2$ 0 000− $-\sqrt{3}/\sqrt{2.7}$
$\tilde{1}\ \tilde{1}\ \tilde{1}$ 0 000− 0
$\tilde{0}\ 2\ 2$ 0 000+ $+1/\sqrt{7}$
$3\ \frac{5}{2}\ \frac{1}{2}$
$1\ \frac{3}{2}\ \frac{1}{2}$ 0 000− $+\sqrt{3}/\sqrt{7}$
$\tilde{1}\ \frac{3}{2}\ \frac{1}{2}$ 0 000+ $+\sqrt{5}/\sqrt{3.7}$
$\tilde{1}\ \tilde{\frac{1}{2}}\ \frac{1}{2}$ 0 000− $-2/\sqrt{3.7}$
$\tilde{0}\ \tilde{\frac{1}{2}}\ \frac{1}{2}$ 0 000+ $+1/\sqrt{7}$
$3\ \frac{5}{2}\ \frac{3}{2}$
$1\ \frac{3}{2}\ \frac{3}{2}$ 0 000− $+\sqrt{3}/2\sqrt{5.7}$
$1\ \frac{3}{2}\ \frac{3}{2}$ 1 000− $-3\sqrt{3}/2\sqrt{5.7}$

SO_3–O 3jm Factors (cont.)

3 $\frac{5}{2}$ $\frac{3}{2}$

$1\ \tilde{\frac{1}{2}}\ \frac{3}{2}\ 0\ 000+\quad +\sqrt{3}/\sqrt{2.7}$

$\tilde{1}\ \frac{3}{2}\ \frac{3}{2}\ 0\ 000+\quad +5/2\sqrt{3.7}$

$\tilde{1}\ \frac{3}{2}\ \frac{3}{2}\ 1\ 000-\quad -1/2\sqrt{3.7}$

$\tilde{1}\ \tilde{\frac{1}{2}}\ \frac{3}{2}\ 0\ 000-\quad +\sqrt{5}/\sqrt{2.3.7}$

$\tilde{0}\ \frac{3}{2}\ \frac{3}{2}\ 0\ 000-\quad -1/\sqrt{7}$

3 $\frac{5}{2}$ $\frac{5}{2}$

$1\ \frac{3}{2}\ \frac{3}{2}\ 0\ 000+\quad +2/3\sqrt{3.5.7}$

$1\ \frac{3}{2}\ \frac{3}{2}\ 1\ 000+\quad +2\sqrt{7}/3\sqrt{3.5}$

$1\ \tilde{\frac{1}{2}}\ \frac{3}{2}\ 0\ 000-\quad -1/3\sqrt{2.3.7}$

$1\ \tilde{\frac{1}{2}}\ \tilde{\frac{1}{2}}\ 0\ 000+\quad +2\sqrt{2.5}/3\sqrt{3.7}$

$\tilde{1}\ \frac{3}{2}\ \frac{3}{2}\ 0\ 000-\quad 0$

$\tilde{1}\ \frac{3}{2}\ \frac{3}{2}\ 1\ 000+\quad +2/\sqrt{3.7}$

$\tilde{1}\ \tilde{\frac{1}{2}}\ \frac{3}{2}\ 0\ 000+\quad +\sqrt{5}/\sqrt{2.3.7}$

$\tilde{0}\ \frac{3}{2}\ \frac{3}{2}\ 0\ 000+\quad -1/\sqrt{7}$

3 3 0

$1\ 1\ 0\ 0\ 000+\quad +\sqrt{3}/\sqrt{7}$

$\tilde{1}\ \tilde{1}\ 0\ 0\ 000+\quad +\sqrt{3}/\sqrt{7}$

$\tilde{0}\ \tilde{0}\ 0\ 0\ 000+\quad +1/\sqrt{7}$

3 3 1

$1\ 1\ 1\ 0\ 000+\quad +3/2\sqrt{2.7}$

$\tilde{1}\ 1\ 1\ 0\ 000-\quad -\sqrt{3.5}/2\sqrt{2.7}$

$\tilde{1}\ \tilde{1}\ 1\ 0\ 000+\quad +1/2\sqrt{2.7}$

$\tilde{0}\ \tilde{1}\ 1\ 0\ 000-\quad +1/\sqrt{7}$

3 3 2

$1\ 1\ 2\ 0\ 000+\quad +2/\sqrt{5.7}$

$1\ 1\ \tilde{1}\ 0\ 000+\quad -\sqrt{3}/2\sqrt{2.5.7}$

$\tilde{1}\ 1\ 2\ 0\ 000-\quad -1/\sqrt{7}$

$\tilde{1}\ 1\ \tilde{1}\ 0\ 000-\quad -1/2\sqrt{2.7}$

$\tilde{1}\ \tilde{1}\ 2\ 0\ 000+\quad 0$

$\tilde{1}\ \tilde{1}\ \tilde{1}\ 0\ 000+\quad +\sqrt{3.5}/2\sqrt{2.7}$

$\tilde{0}\ 1\ \tilde{1}\ 0\ 000+\quad -1/\sqrt{7}$

3 3 3

$1\ 1\ 1\ 0\ 000+\quad -1/\sqrt{7}$

$\tilde{1}\ 1\ 1\ 0\ 000-\quad 0$

$\tilde{1}\ \tilde{1}\ 1\ 0\ 000+\quad -1/\sqrt{7}$

$\tilde{1}\ \tilde{1}\ \tilde{1}\ 0\ 000-\quad 0$

$\tilde{0}\ \tilde{1}\ 1\ 0\ 000-\quad +1/\sqrt{2.7}$

$\frac{7}{2}$ 2 $\frac{3}{2}$

$\frac{1}{2}\ 2\ \frac{3}{2}\ 0\ 000-\quad +\sqrt{3}/2\sqrt{5}$

$\frac{1}{2}\ \tilde{1}\ \frac{3}{2}\ 0\ 000-\quad -1/\sqrt{2.5}$

$\frac{3}{2}\ 2\ \frac{3}{2}\ 0\ 000+\quad -\sqrt{3}/\sqrt{2.7}$

$\frac{3}{2}\ \tilde{1}\ \frac{3}{2}\ 0\ 000+\quad +1/\sqrt{7}$

$\frac{3}{2}\ \tilde{1}\ \frac{3}{2}\ 1\ 000-\quad +1/\sqrt{7}$

$\tilde{\frac{1}{2}}\ 2\ \frac{3}{2}\ 0\ 000-\quad -1/2\sqrt{7}$

$\tilde{\frac{1}{2}}\ \tilde{1}\ \frac{3}{2}\ 0\ 000-\quad +\sqrt{3}/\sqrt{2.7}$

$\frac{7}{2}$ $\frac{5}{2}$ 1

$\frac{1}{2}\ \frac{3}{2}\ 1\ 0\ 000+\quad +1/2$

$\frac{3}{2}\ \frac{3}{2}\ 1\ 0\ 000-\quad 0$

$\frac{3}{2}\ \frac{3}{2}\ 1\ 1\ 000-\quad +\sqrt{5}/\sqrt{2.7}$

$\frac{3}{2}\ \tilde{\frac{1}{2}}\ 1\ 0\ 000+\quad +1/\sqrt{7}$

$\tilde{\frac{1}{2}}\ \frac{3}{2}\ 1\ 0\ 000+\quad -\sqrt{5}/2\sqrt{3.7}$

$\tilde{\frac{1}{2}}\ \tilde{\frac{1}{2}}\ 1\ 0\ 000-\quad -2/\sqrt{3.7}$

$\frac{7}{2}$ $\frac{5}{2}$ 2

$\frac{1}{2}\ \frac{3}{2}\ 2\ 0\ 000+\quad +1/3\sqrt{2.5}$

$\frac{1}{2}\ \frac{3}{2}\ \tilde{1}\ 0\ 000+\quad -7/2.3\sqrt{3.5}$

$\frac{1}{2}\ \tilde{\frac{1}{2}}\ \tilde{1}\ 0\ 000-\quad -2/3\sqrt{3}$

$\frac{3}{2}\ \frac{3}{2}\ 2\ 0\ 000-\quad +1/3\sqrt{7}$

$\frac{3}{2}\ \frac{3}{2}\ \tilde{1}\ 0\ 000-\quad +4\sqrt{2}/3\sqrt{3.7}$

$\frac{3}{2}\ \frac{3}{2}\ \tilde{1}\ 1\ 000+\quad +\sqrt{7}/3\sqrt{2.3}$

$\frac{3}{2}\ \tilde{\frac{1}{2}}\ 2\ 0\ 000+\quad +\sqrt{2.5}/3\sqrt{7}$

$\frac{3}{2}\ \tilde{\frac{1}{2}}\ \tilde{1}\ 0\ 000+\quad -\sqrt{5}/3\sqrt{3.7}$

$\tilde{\frac{1}{2}}\ \frac{3}{2}\ 2\ 0\ 000+\quad +\sqrt{3}/\sqrt{2.7}$

$\tilde{\frac{1}{2}}\ \frac{3}{2}\ \tilde{1}\ 0\ 000+\quad -1/2\sqrt{7}$

$\frac{7}{2}$ 3 $\frac{1}{2}$

$\frac{1}{2}\ 1\ \frac{1}{2}\ 0\ 000-\quad +1/2$

$\frac{3}{2}\ 1\ \frac{1}{2}\ 0\ 000+\quad +\sqrt{5}/2\sqrt{7}$

$\frac{3}{2}\ \tilde{1}\ \frac{1}{2}\ 0\ 000-\quad +3/2\sqrt{7}$

$\tilde{\frac{1}{2}}\ \tilde{1}\ \frac{1}{2}\ 0\ 000+\quad +\sqrt{3}/2\sqrt{7}$

$\tilde{\frac{1}{2}}\ \tilde{0}\ \frac{1}{2}\ 0\ 000-\quad -1/\sqrt{7}$

$\frac{7}{2}$ 3 $\frac{3}{2}$

$\frac{1}{2}\ 1\ \frac{3}{2}\ 0\ 000-\quad -1/2\sqrt{2.3}$

$\frac{1}{2}\ \tilde{1}\ \frac{3}{2}\ 0\ 000+\quad +\sqrt{5}/2\sqrt{2.3}$

$\frac{3}{2}\ 1\ \frac{3}{2}\ 0\ 000+\quad -\sqrt{5}/\sqrt{3.7}$

$\frac{3}{2}\ 1\ \frac{3}{2}\ 1\ 000+\quad -\sqrt{5}/2\sqrt{3.7}$

$\frac{3}{2}\ \tilde{1}\ \frac{3}{2}\ 0\ 000-\quad -1/\sqrt{3.7}$

$\frac{3}{2}\ \tilde{1}\ \frac{3}{2}\ 1\ 000+\quad +1/2\sqrt{3.7}$

$\frac{3}{2}\ \tilde{0}\ \frac{3}{2}\ 0\ 000+\quad +1/\sqrt{7}$

$\tilde{\frac{1}{2}}\ 1\ \frac{3}{2}\ 0\ 000-\quad +\sqrt{5}/2\sqrt{2.7}$

$\tilde{\frac{1}{2}}\ \tilde{1}\ \frac{3}{2}\ 0\ 000+\quad +3/2\sqrt{2.7}$

$\frac{7}{2}$ 3 $\frac{5}{2}$

$\frac{1}{2}\ 1\ \frac{3}{2}\ 0\ 000+\quad +1/2\sqrt{3}$

$\frac{1}{2}\ \tilde{1}\ \frac{3}{2}\ 0\ 000-\quad -\sqrt{5}/2.3\sqrt{3}$

$\frac{1}{2}\ \tilde{1}\ \tilde{\frac{1}{2}}\ 0\ 000+\quad -1/2.3\sqrt{3}$

$\frac{1}{2}\ \tilde{0}\ \tilde{\frac{1}{2}}\ 0\ 000-\quad -1/3$

$\frac{3}{2}\ 1\ \frac{3}{2}\ 0\ 000-\quad +\sqrt{5}/\sqrt{2.3.7}$

$\frac{3}{2}\ 1\ \frac{3}{2}\ 1\ 000-\quad 0$

$\frac{3}{2}\ 1\ \tilde{\frac{1}{2}}\ 0\ 000+\quad +1/2\sqrt{3.7}$

$\frac{3}{2}\ \tilde{1}\ \frac{3}{2}\ 0\ 000+\quad -\sqrt{7}/3\sqrt{2.3}$

$\frac{3}{2}\ \tilde{1}\ \frac{3}{2}\ 1\ 000-\quad -2\sqrt{2}/3\sqrt{3.7}$

$\frac{3}{2}\ \tilde{1}\ \tilde{\frac{1}{2}}\ 0\ 000-\quad +5\sqrt{5}/2.3\sqrt{3.7}$

$\frac{3}{2}\ \tilde{0}\ \frac{3}{2}\ 0\ 000-\quad +\sqrt{2}/3\sqrt{7}$

$\tilde{\frac{1}{2}}\ 1\ \frac{3}{2}\ 0\ 000+\quad +\sqrt{5}/2\sqrt{7}$

$\tilde{\frac{1}{2}}\ 1\ \tilde{\frac{1}{2}}\ 0\ 000-\quad -1/2\sqrt{7}$

$\tilde{\frac{1}{2}}\ \tilde{1}\ \frac{3}{2}\ 0\ 000-\quad +1/2\sqrt{7}$

$\frac{7}{2}$ $\frac{7}{2}$ 0

$\frac{1}{2}\ \frac{1}{2}\ 0\ 0\ 000+\quad +1/2$

$\frac{3}{2}\ \frac{3}{2}\ 0\ 0\ 000+\quad +1/\sqrt{2}$

$\tilde{\frac{1}{2}}\ \tilde{\frac{1}{2}}\ 0\ 0\ 000+\quad +1/2$

$\frac{7}{2}$ $\frac{7}{2}$ 1

$\frac{1}{2}\ \frac{1}{2}\ 1\ 0\ 000+\quad -\sqrt{7}/2.3\sqrt{3}$

$\frac{3}{2}\ \frac{1}{2}\ 1\ 0\ 000-\quad -\sqrt{5}/3\sqrt{3}$

$\frac{3}{2}\ \frac{3}{2}\ 1\ 0\ 000+\quad +\sqrt{7}/3\sqrt{2.3}$

$\frac{3}{2}\ \frac{3}{2}\ 1\ 1\ 000+\quad -2\sqrt{2}/3\sqrt{3.7}$

$\tilde{\frac{1}{2}}\ \frac{3}{2}\ 1\ 0\ 000-\quad +1/\sqrt{7}$

$\tilde{\frac{1}{2}}\ \tilde{\frac{1}{2}}\ 1\ 0\ 000+\quad -\sqrt{3}/2\sqrt{7}$

$\frac{7}{2}$ $\frac{7}{2}$ 2

$\frac{3}{2}\ \frac{1}{2}\ 2\ 0\ 000-\quad +1/\sqrt{2.3}$

$\frac{3}{2}\ \frac{1}{2}\ \tilde{1}\ 0\ 000-\quad 0$

$\frac{3}{2}\ \frac{3}{2}\ 2\ 0\ 000+\quad +2/\sqrt{3.5.7}$

$\frac{3}{2}\ \frac{3}{2}\ \tilde{1}\ 0\ 000+\quad +3/\sqrt{2.5.7}$

$\frac{3}{2}\ \frac{3}{2}\ \tilde{1}\ 1\ 000-\quad 0$

$\tilde{\frac{1}{2}}\ \frac{1}{2}\ \tilde{1}\ 0\ 000+\quad +1/2\sqrt{3}$

$\tilde{\frac{1}{2}}\ \frac{3}{2}\ 2\ 0\ 000-\quad -1/\sqrt{2.5.7}$

$\tilde{\frac{1}{2}}\ \frac{3}{2}\ \tilde{1}\ 0\ 000-\quad +4/\sqrt{3.5.7}$

$\frac{7}{2}$ $\frac{7}{2}$ 3

$\frac{1}{2}\ \frac{1}{2}\ 1\ 0\ 000+\quad +\sqrt{7}/3\sqrt{11}$

$\frac{3}{2}\ \frac{1}{2}\ 1\ 0\ 000-\quad +\sqrt{5}/2.3\sqrt{11}$

$\frac{3}{2}\ \frac{1}{2}\ \tilde{1}\ 0\ 000+\quad +1/2\sqrt{11}$

$\frac{3}{2}\ \frac{3}{2}\ 1\ 0\ 000+\quad -\sqrt{2}/3\sqrt{7.11}$

$\frac{3}{2}\ \frac{3}{2}\ 1\ 1\ 000+\quad +\sqrt{2.7}/3\sqrt{11}$

$\frac{3}{2}\ \frac{3}{2}\ \tilde{1}\ 0\ 000-\quad 0$

$\frac{3}{2}\ \frac{3}{2}\ \tilde{1}\ 1\ 000+\quad -\sqrt{2.5}/\sqrt{7.11}$

$\frac{3}{2}\ \frac{3}{2}\ \tilde{0}\ 0\ 000+\quad +\sqrt{3.5}/\sqrt{2.7.11}$

$\tilde{\frac{1}{2}}\ \frac{1}{2}\ \tilde{1}\ 0\ 000-\quad +2/\sqrt{3.11}$

$\tilde{\frac{1}{2}}\ \frac{1}{2}\ \tilde{0}\ 0\ 000+\quad +1/2\sqrt{11}$

$\tilde{\frac{1}{2}}\ \frac{3}{2}\ 1\ 0\ 000-\quad +3\sqrt{3}/2\sqrt{7.11}$

$\tilde{\frac{1}{2}}\ \frac{3}{2}\ \tilde{1}\ 0\ 000+\quad +\sqrt{5}/2\sqrt{3.7.11}$

$\tilde{\frac{1}{2}}\ \tilde{\frac{1}{2}}\ 1\ 0\ 000+\quad -1/\sqrt{7.11}$

4 2 2

$0\ 2\ 2\ 0\ 000+\quad +1/\sqrt{3.5}$

$0\ \tilde{1}\ \tilde{1}\ 0\ 000+\quad -\sqrt{2}/3\sqrt{5}$

$1\ \tilde{1}\ 2\ 0\ 000-\quad +1/\sqrt{2.3}$

$1\ \tilde{1}\ \tilde{1}\ 0\ 000-\quad 0$

$2\ 2\ 2\ 0\ 000+\quad -\sqrt{2}/\sqrt{3.7}$

$2\ \tilde{1}\ \tilde{1}\ 0\ 000+\quad +2\sqrt{2}/3\sqrt{7}$

$\tilde{1}\ \tilde{1}\ 2\ 0\ 000+\quad -1/\sqrt{2.7}$

$\tilde{1}\ \tilde{1}\ \tilde{1}\ 0\ 000+\quad +2/\sqrt{3.7}$

4 $\frac{5}{2}$ $\frac{3}{2}$

$0\ \frac{3}{2}\ \frac{3}{2}\ 0\ 000-\quad +1/3$

$1\ \frac{3}{2}\ \frac{3}{2}\ 0\ 000+\quad +\sqrt{5}/2.3$

$1\ \frac{3}{2}\ \frac{3}{2}\ 1\ 000+\quad +\sqrt{5}/2.3$

$1\ \tilde{\frac{1}{2}}\ \frac{3}{2}\ 0\ 000-\quad +1/3\sqrt{2}$

$2\ \frac{3}{2}\ \frac{3}{2}\ 0\ 000-\quad -\sqrt{2.5}/3\sqrt{7}$

$2\ \tilde{\frac{1}{2}}\ \frac{3}{2}\ 0\ 000+\quad +2/3\sqrt{7}$

$\tilde{1}\ \frac{3}{2}\ \frac{3}{2}\ 0\ 000-\quad -\sqrt{5}/2\sqrt{3.7}$

$\tilde{1}\ \frac{3}{2}\ \frac{3}{2}\ 1\ 000+\quad +\sqrt{5}/2\sqrt{3.7}$

$\tilde{1}\ \tilde{\frac{1}{2}}\ \frac{3}{2}\ 0\ 000+\quad +\sqrt{3}/\sqrt{2.7}$

SO_3–O 3jm Factors (cont.)

4 $\frac{5}{2}$ $\frac{5}{2}$

$0\ \tfrac{3}{2}\ \tfrac{3}{2}$ 0 000+ $+1/3\sqrt{3}$

$0\ \tilde{\tfrac{1}{2}}\ \tilde{\tfrac{1}{2}}$ 0 000+ $-\sqrt{2}/3\sqrt{3}$

$1\ \tfrac{3}{2}\ \tfrac{3}{2}$ 0 000− 0

$1\ \tfrac{3}{2}\ \tfrac{3}{2}$ 1 000− 0

$1\ \tilde{\tfrac{1}{2}}\ \tfrac{3}{2}$ 0 000+ $+1/\sqrt{2.3}$

$1\ \tilde{\tfrac{1}{2}}\ \tilde{\tfrac{1}{2}}$ 0 000− 0

$2\ \tfrac{3}{2}\ \tfrac{3}{2}$ 0 000+ $-\sqrt{2.5}/3\sqrt{3.7}$

$2\ \tilde{\tfrac{1}{2}}\ \tfrac{3}{2}$ 0 000− $-4/3\sqrt{3.7}$

$\tilde{1}\ \tfrac{3}{2}\ \tfrac{3}{2}$ 0 000+ $+2\sqrt{5}/3\sqrt{7}$

$\tilde{1}\ \tfrac{3}{2}\ \tfrac{3}{2}$ 1 000− 0

$\tilde{1}\ \tilde{\tfrac{1}{2}}\ \tfrac{3}{2}$ 0 000− $-1/3\sqrt{2.7}$

4 3 1

$0\ 1\ 1$ 0 000+ $+1/3$

$1\ 1\ 1$ 0 000− $-\sqrt{5}/2\sqrt{2.3}$

$1\ \tilde{1}\ 1$ 0 000+ $+1/2\sqrt{2}$

$2\ 1\ 1$ 0 000+ $+\sqrt{5}/3\sqrt{7}$

$2\ \tilde{1}\ 1$ 0 000− $-1/\sqrt{7}$

$\tilde{1}\ 1\ 1$ 0 000+ $+\sqrt{5}/2\sqrt{2.3.7}$

$\tilde{1}\ \tilde{1}\ 1$ 0 000− $-3/2\sqrt{2.7}$

$\tilde{1}\ \tilde{0}\ 1$ 0 000+ $-1/\sqrt{7}$

4 3 2

$0\ \tilde{1}\ \tilde{1}$ 0 000− $+1/3$

$1\ 1\ 2$ 0 000− 0

$1\ 1\ \tilde{1}$ 0 000− $-1/2\sqrt{2}$

$1\ \tilde{1}\ 2$ 0 000+ $-1/\sqrt{3.5}$

$1\ \tilde{1}\ \tilde{1}$ 0 000+ $-\sqrt{3}/2\sqrt{2.5}$

$1\ \tilde{0}\ \tilde{1}$ 0 000− $-1/\sqrt{3.5}$

$2\ 1\ \tilde{1}$ 0 000+ $+1/\sqrt{7}$

$2\ \tilde{1}\ \tilde{1}$ 0 000− $-1/3\sqrt{5.7}$

$2\ \tilde{0}\ 2$ 0 000+ $-2\sqrt{2}/\sqrt{3.5.7}$

$\tilde{1}\ 1\ 2$ 0 000+ $+1/\sqrt{7}$

$\tilde{1}\ 1\ \tilde{1}$ 0 000+ $-1/2\sqrt{2.7}$

$\tilde{1}\ \tilde{1}\ 2$ 0 000− $-2/\sqrt{5.7}$

$\tilde{1}\ \tilde{1}\ \tilde{1}$ 0 000− $+\sqrt{7}/2\sqrt{2.3.5}$

4 3 3

$0\ 1\ 1$ 0 000+ $+1/\sqrt{2.11}$

$0\ \tilde{1}\ \tilde{1}$ 0 000+ $-1/3\sqrt{2.11}$

$0\ \tilde{0}\ \tilde{0}$ 0 000+ $-\sqrt{2}/\sqrt{3.11}$

$1\ 1\ 1$ 0 000− 0

$1\ \tilde{1}\ 1$ 0 000+ $+1/\sqrt{11}$

$1\ \tilde{1}\ \tilde{1}$ 0 000− 0

$1\ \tilde{0}\ \tilde{1}$ 0 000+ $-\sqrt{5}/\sqrt{2.3.11}$

$2\ 1\ 1$ 0 000+ $+\sqrt{5}/\sqrt{2.7.11}$

$2\ \tilde{1}\ 1$ 0 000− $+1/\sqrt{2.7.11}$

$2\ \tilde{1}\ \tilde{1}$ 0 000+ $-\sqrt{5.7}/3\sqrt{2.11}$

$\tilde{1}\ 1\ 1$ 0 000+ $-\sqrt{3.5}/\sqrt{7.11}$

$\tilde{1}\ \tilde{1}\ 1$ 0 000− $+2/\sqrt{7.11}$

$\tilde{1}\ \tilde{1}\ \tilde{1}$ 0 000+ $+\sqrt{5}/\sqrt{3.7.11}$

$\tilde{1}\ \tilde{0}\ 1$ 0 000+ $+1/\sqrt{2.7.11}$

4 $\frac{7}{2}$ $\frac{1}{2}$

$0\ \tfrac{1}{2}\ \tfrac{1}{2}$ 0 000− $+1/3$

$1\ \tfrac{1}{2}\ \tfrac{1}{2}$ 0 000+ $+\sqrt{5}/2.3$

$1\ \tfrac{3}{2}\ \tfrac{1}{2}$ 0 000− $+\sqrt{7}/2.3$

$2\ \tfrac{3}{2}\ \tfrac{1}{2}$ 0 000+ $+\sqrt{2}/3$

$\tilde{1}\ \tfrac{3}{2}\ \tfrac{1}{2}$ 0 000+ $+1/2\sqrt{3}$

$\tilde{1}\ \tilde{\tfrac{1}{2}}\ \tfrac{1}{2}$ 0 000− $+1/2$

4 $\frac{7}{2}$ $\frac{3}{2}$

$0\ \tfrac{3}{2}\ \tfrac{3}{2}$ 0 000+ $+1/3$

$1\ \tfrac{1}{2}\ \tfrac{3}{2}$ 0 000+ $-\sqrt{7}/2.3\sqrt{2}$

$1\ \tfrac{3}{2}\ \tfrac{3}{2}$ 0 000− $-1/3\sqrt{5}$

$1\ \tfrac{3}{2}\ \tfrac{3}{2}$ 1 000− $+\sqrt{5}/2.3$

$1\ \tilde{\tfrac{1}{2}}\ \tfrac{3}{2}$ 0 000+ $-\sqrt{3}/2\sqrt{2.5}$

$2\ \tfrac{1}{2}\ \tfrac{3}{2}$ 0 000− $-1/3$

$2\ \tfrac{3}{2}\ \tfrac{3}{2}$ 0 000+ $+2\sqrt{2}/3\sqrt{5.7}$

$2\ \tilde{\tfrac{1}{2}}\ \tfrac{3}{2}$ 0 000− $-\sqrt{3}/\sqrt{5.7}$

$\tilde{1}\ \tfrac{1}{2}\ \tfrac{3}{2}$ 0 000− $-1/2\sqrt{2.3}$

$\tilde{1}\ \tfrac{3}{2}\ \tfrac{3}{2}$ 0 000+ $+1/\sqrt{3.5.7}$

$\tilde{1}\ \tfrac{3}{2}\ \tfrac{3}{2}$ 1 000− $-3\sqrt{3}/2\sqrt{5.7}$

$\tilde{1}\ \tilde{\tfrac{1}{2}}\ \tfrac{3}{2}$ 0 000− $+\sqrt{5}/2\sqrt{2.7}$

4 $\frac{7}{2}$ $\frac{5}{2}$

$0\ \tfrac{3}{2}\ \tfrac{3}{2}$ 0 000− $-\sqrt{2}/\sqrt{3.11}$

$0\ \tilde{\tfrac{1}{2}}\ \tilde{\tfrac{1}{2}}$ 0 000− $+\sqrt{5}/3\sqrt{11}$

$1\ \tfrac{1}{2}\ \tfrac{3}{2}$ 0 000− $+\sqrt{7}/2\sqrt{3.11}$

$1\ \tfrac{3}{2}\ \tfrac{3}{2}$ 0 000+ $+7/\sqrt{2.3.5.11}$

$1\ \tfrac{3}{2}\ \tfrac{3}{2}$ 1 000+ 0

$1\ \tfrac{3}{2}\ \tilde{\tfrac{1}{2}}$ 0 000− $+1/2\sqrt{3.11}$

$1\ \tilde{\tfrac{1}{2}}\ \tfrac{3}{2}$ 0 000− $-1/2.3\sqrt{5.11}$

$1\ \tilde{\tfrac{1}{2}}\ \tilde{\tfrac{1}{2}}$ 0 000+ $+7/2.3\sqrt{11}$

$2\ \tfrac{1}{2}\ \tfrac{3}{2}$ 0 000+ $+\sqrt{2}/\sqrt{3.11}$

$2\ \tfrac{3}{2}\ \tfrac{3}{2}$ 0 000− $-4/\sqrt{3.5.7.11}$

$2\ \tfrac{3}{2}\ \tilde{\tfrac{1}{2}}$ 0 000+ $+\sqrt{2.3}/\sqrt{7.11}$

$2\ \tilde{\tfrac{1}{2}}\ \tfrac{3}{2}$ 0 000+ $-\sqrt{2.11}/3\sqrt{5.7}$

$\tilde{1}\ \tfrac{1}{2}\ \tfrac{3}{2}$ 0 000+ $-7/2.3\sqrt{11}$

$\tilde{1}\ \tfrac{1}{2}\ \tilde{\tfrac{1}{2}}$ 0 000− $-\sqrt{5}/2.3\sqrt{11}$

$\tilde{1}\ \tfrac{3}{2}\ \tfrac{3}{2}$ 0 000− $-\sqrt{11}/3\sqrt{2.5.7}$

$\tilde{1}\ \tfrac{3}{2}\ \tfrac{3}{2}$ 1 000+ $-2\sqrt{2.7}/3\sqrt{5.11}$

$\tilde{1}\ \tfrac{3}{2}\ \tilde{\tfrac{1}{2}}$ 0 000+ $-13/2.3\sqrt{7.11}$

$\tilde{1}\ \tilde{\tfrac{1}{2}}\ \tfrac{3}{2}$ 0 000+ $+\sqrt{5}/2\sqrt{3.7.11}$

4 $\frac{7}{2}$ $\frac{7}{2}$

$0\ \tfrac{1}{2}\ \tfrac{1}{2}$ 0 000+ $+7/2.3\sqrt{3.11}$

$0\ \tfrac{3}{2}\ \tfrac{3}{2}$ 0 000+ $+1/3\sqrt{2.3.11}$

$0\ \tilde{\tfrac{1}{2}}\ \tilde{\tfrac{1}{2}}$ 0 000+ $-\sqrt{3}/2\sqrt{11}$

$1\ \tfrac{1}{2}\ \tfrac{1}{2}$ 0 000− 0

$1\ \tfrac{3}{2}\ \tfrac{1}{2}$ 0 000+ $+\sqrt{7}/2\sqrt{3.11}$

$1\ \tfrac{3}{2}\ \tfrac{3}{2}$ 0 000− 0

$1\ \tfrac{3}{2}\ \tfrac{3}{2}$ 1 000− 0

$1\ \tilde{\tfrac{1}{2}}\ \tfrac{3}{2}$ 0 000+ $-\sqrt{5}/2\sqrt{11}$

$1\ \tilde{\tfrac{1}{2}}\ \tilde{\tfrac{1}{2}}$ 0 000− 0

$2\ \tfrac{3}{2}\ \tfrac{1}{2}$ 0 000− $+1/3\sqrt{2.3.11}$

$2\ \tfrac{3}{2}\ \tfrac{3}{2}$ 0 000+ $-8\sqrt{5}/3\sqrt{3.7.11}$

$2\ \tilde{\tfrac{1}{2}}\ \tfrac{3}{2}$ 0 000− $-\sqrt{5}/\sqrt{2.7.11}$

$\tilde{1}\ \tfrac{3}{2}\ \tfrac{1}{2}$ 0 000− $-7/2.3\sqrt{11}$

$\tilde{1}\ \tfrac{3}{2}\ \tfrac{3}{2}$ 0 000+ $+\sqrt{2.5}/3\sqrt{7.11}$

$\tilde{1}\ \tfrac{3}{2}\ \tfrac{3}{2}$ 1 000− 0

$\tilde{1}\ \tilde{\tfrac{1}{2}}\ \tfrac{1}{2}$ 0 000+ $-1/\sqrt{3.11}$

$\tilde{1}\ \tilde{\tfrac{1}{2}}\ \tfrac{3}{2}$ 0 000− $+\sqrt{5}/2\sqrt{3.7.11}$

4 4 0

$0\ 0\ 0$ 0 000+ $+1/3$

$1\ 1\ 0$ 0 000+ $+1/\sqrt{3}$

$2\ 2\ 0$ 0 000+ $+\sqrt{2}/3$

$\tilde{1}\ \tilde{1}\ 0$ 0 000+ $+1/\sqrt{3}$

4 4 1

$1\ 0\ 1$ 0 000− $-1/3$

$1\ 1\ 1$ 0 000+ $-1/2\sqrt{2.3.5}$

$2\ 1\ 1$ 0 000− $-\sqrt{7}/3\sqrt{5}$

$\tilde{1}\ 1\ 1$ 0 000− $-\sqrt{7}/2\sqrt{2.3.5}$

$\tilde{1}\ 2\ 1$ 0 000+ $-1/\sqrt{3.5}$

$\tilde{1}\ \tilde{1}\ 1$ 0 000+ $+\sqrt{5}/2\sqrt{2.3}$

4 4 2

$1\ 1\ 2$ 0 000+ $-2\sqrt{7}/\sqrt{3.5.11}$

$1\ 1\ \tilde{1}$ 0 000+ $-\sqrt{7}/2\sqrt{2.5.11}$

$2\ 0\ 2$ 0 000+ $+2\sqrt{2}/3\sqrt{11}$

$2\ 1\ \tilde{1}$ 0 000− $+1/\sqrt{5.11}$

$2\ 2\ 2$ 0 000+ $+8/3\sqrt{5.7.11}$

$\tilde{1}\ 0\ \tilde{1}$ 0 000+ $+1/\sqrt{3.11}$

$\tilde{1}\ 1\ 2$ 0 000− $-1/\sqrt{5.11}$

$\tilde{1}\ 1\ \tilde{1}$ 0 000− $+7/2\sqrt{2.5.11}$

$\tilde{1}\ 2\ \tilde{1}$ 0 000+ $+\sqrt{11}/\sqrt{3.5.7}$

$\tilde{1}\ \tilde{1}\ 2$ 0 000+ $+4/\sqrt{3.5.7.11}$

$\tilde{1}\ \tilde{1}\ \tilde{1}$ 0 000+ $-13/2\sqrt{2.5.7.11}$

4 4 3

$1\ 0\ 1$ 0 000− $+\sqrt{7}/3\sqrt{2.11}$

$1\ 1\ 1$ 0 000+ $-\sqrt{7}/\sqrt{3.5.11}$

$1\ 1\ \tilde{1}$ 0 000− 0

$2\ 1\ 1$ 0 000− $+7/3\sqrt{2.5.11}$

$2\ 1\ \tilde{1}$ 0 000+ $-1/\sqrt{2.11}$

$2\ 2\ \tilde{0}$ 0 000+ $-2/\sqrt{7.11}$

$\tilde{1}\ 0\ \tilde{1}$ 0 000− $+\sqrt{5}/\sqrt{2.3.11}$

$\tilde{1}\ 1\ 1$ 0 000− $-2/\sqrt{3.5.11}$

$\tilde{1}\ 1\ \tilde{1}$ 0 000+ $+1/\sqrt{11}$

$\tilde{1}\ 1\ \tilde{0}$ 0 000− $-1/\sqrt{2.11}$

$\tilde{1}\ 2\ 1$ 0 000+ $-13/\sqrt{2.3.5.7.11}$

$\tilde{1}\ 2\ \tilde{1}$ 0 000− $-1/\sqrt{2.3.7.11}$

$\tilde{1}\ \tilde{1}\ 1$ 0 000+ $+\sqrt{5}/\sqrt{3.7.11}$

$\tilde{1}\ \tilde{1}\ \tilde{1}$ 0 000− 0

4 4 4

$0\ 0\ 0$ 0 000+ $+7\sqrt{2}/3\sqrt{3.11.13}$

$1\ 1\ 0$ 0 000+ $+7/3\sqrt{2.11.13}$

SO_3–O 3jm Factors (cont.)

4 4 4

$1\ 1\ 1\ 0\ 000-\quad 0$

$2\ 1\ 1\ 0\ 000+\quad +\sqrt{5.7}/3\sqrt{2.11.13}$

$2\ 2\ 0\ 0\ 000+\quad +2/3\sqrt{3.11.13}$

$2\ 2\ 2\ 0\ 000+\quad -8.2\sqrt{2.5}/3\sqrt{3.7.11.13}$

$\tilde{1}\ 1\ 1\ 0\ 000+\quad +\sqrt{5.7}/\sqrt{3.11.13}$

$\tilde{1}\ 2\ 1\ 0\ 000-\quad +\sqrt{3.5}/\sqrt{2.11.13}$

$\tilde{1}\ \tilde{1}\ 0\ 0\ 000+\quad -\sqrt{13}/3\sqrt{2.11}$

$\tilde{1}\ \tilde{1}\ 1\ 0\ 000-\quad 0$

$\tilde{1}\ \tilde{1}\ 2\ 0\ 000+\quad +5\sqrt{5}/3\sqrt{2.7.11.13}$

$\tilde{1}\ \tilde{1}\ \tilde{1}\ 0\ 000+\quad -\sqrt{5}/\sqrt{3.7.11.13}$

$\tfrac{9}{2}\ \tfrac{5}{2}\ 2$

$\tfrac{1}{2}\ \tfrac{3}{2}\ 2\ 0\ 000-\quad +1/3$

$\tfrac{1}{2}\ \tfrac{3}{2}\ \tilde{1}\ 0\ 000-\quad +\sqrt{2}/3\sqrt{3}$

$\tfrac{1}{2}\ \widetilde{\tfrac{1}{2}}\ \tilde{1}\ 0\ 000+\quad -\sqrt{2}/3\sqrt{3.5}$

$\tfrac{3}{2}\ \tfrac{3}{2}\ 2\ 0\ 000+\quad 0$

$\tfrac{3}{2}\ \tfrac{3}{2}\ 2\ 0\ 100+\quad -\sqrt{2.5}/3\sqrt{7}$

$\tfrac{3}{2}\ \tfrac{3}{2}\ \tilde{1}\ 0\ 000+\quad -1/\sqrt{5}$

$\tfrac{3}{2}\ \tfrac{3}{2}\ \tilde{1}\ 0\ 100+\quad -4/3\sqrt{3.5.7}$

$\tfrac{3}{2}\ \tfrac{3}{2}\ \tilde{1}\ 1\ 000-\quad 0$

$\tfrac{3}{2}\ \tfrac{3}{2}\ \tilde{1}\ 1\ 100-\quad +2\sqrt{5}/3\sqrt{3.7}$

$\tfrac{3}{2}\ \widetilde{\tfrac{1}{2}}\ 2\ 0\ 000-\quad +\sqrt{3/5}$

$\tfrac{3}{2}\ \widetilde{\tfrac{1}{2}}\ 2\ 0\ 100-\quad +4/3.5\sqrt{7}$

$\tfrac{3}{2}\ \widetilde{\tfrac{1}{2}}\ \tilde{1}\ 0\ 000-\quad -\sqrt{2/5}$

$\tfrac{3}{2}\ \widetilde{\tfrac{1}{2}}\ \tilde{1}\ 0\ 100-\quad +8.2\sqrt{2}/3.5\sqrt{3.7}$

$\tfrac{9}{2}\ 3\ \tfrac{3}{2}$

$\tfrac{1}{2}\ 1\ \tfrac{3}{2}\ 0\ 000+\quad +1/\sqrt{2.3}$

$\tfrac{1}{2}\ \tilde{1}\ \tfrac{3}{2}\ 0\ 000-\quad +1/\sqrt{2.3.5}$

$\tfrac{3}{2}\ 1\ \tfrac{3}{2}\ 0\ 000-\quad +1/2\sqrt{5}$

$\tfrac{3}{2}\ 1\ \tfrac{3}{2}\ 0\ 100-\quad -1/\sqrt{3.5.7}$

$\tfrac{3}{2}\ 1\ \tfrac{3}{2}\ 1\ 000-\quad +1/2\sqrt{5}$

$\tfrac{3}{2}\ 1\ \tfrac{3}{2}\ 1\ 100-\quad +4/\sqrt{3.5.7}$

$\tfrac{3}{2}\ \tilde{1}\ \tfrac{3}{2}\ 0\ 000+\quad -3/2.5$

$\tfrac{3}{2}\ \tilde{1}\ \tfrac{3}{2}\ 0\ 100+\quad -\sqrt{7}/5\sqrt{3}$

$\tfrac{3}{2}\ \tilde{1}\ \tfrac{3}{2}\ 1\ 000-\quad -3/2.5$

$\tfrac{3}{2}\ \tilde{1}\ \tfrac{3}{2}\ 1\ 100-\quad +8/5\sqrt{3.7}$

$\tfrac{3}{2}\ \tilde{0}\ \tfrac{3}{2}\ 0\ 000-\quad +\sqrt{3/5}$

$\tfrac{3}{2}\ \tilde{0}\ \tfrac{3}{2}\ 0\ 100-\quad -2/5\sqrt{7}$

$\tfrac{9}{2}\ 3\ \tfrac{5}{2}$

$\tfrac{1}{2}\ 1\ \tfrac{3}{2}\ 0\ 000-\quad +1/\sqrt{2.3.11}$

$\tfrac{1}{2}\ \tilde{1}\ \tfrac{3}{2}\ 0\ 000+\quad +13/3\sqrt{2.3.5.11}$

$\tfrac{1}{2}\ \tilde{1}\ \widetilde{\tfrac{1}{2}}\ 0\ 000-\quad +4\sqrt{2}/3\sqrt{3.11}$

$\tfrac{1}{2}\ \tilde{0}\ \widetilde{\tfrac{1}{2}}\ 0\ 000+\quad -\sqrt{2}/3\sqrt{11}$

$\tfrac{3}{2}\ 1\ \tfrac{3}{2}\ 0\ 000+\quad -2/\sqrt{5.11}$

$\tfrac{3}{2}\ 1\ \tfrac{3}{2}\ 0\ 100+\quad +4/\sqrt{3.5.7.11}$

$\tfrac{3}{2}\ 1\ \tfrac{3}{2}\ 1\ 000+\quad -2/\sqrt{5.11}$

$\tfrac{3}{2}\ 1\ \tfrac{3}{2}\ 1\ 100+\quad +3\sqrt{3}/\sqrt{5.7.11}$

$\tfrac{3}{2}\ 1\ \widetilde{\tfrac{1}{2}}\ 0\ 000-\quad +1/\sqrt{2.11}$

$\tfrac{3}{2}\ 1\ \widetilde{\tfrac{1}{2}}\ 0\ 100-\quad +4\sqrt{2}/\sqrt{3.7.11}$

$\tfrac{3}{2}\ \tilde{1}\ \tfrac{3}{2}\ 0\ 000-\quad -4/5\sqrt{11}$

$\tfrac{9}{2}\ 3\ \tfrac{5}{2}$

$\tfrac{3}{2}\ \tilde{1}\ \tfrac{3}{2}\ 0\ 100-\quad +4\sqrt{11}/3.5\sqrt{3.7}$

$\tfrac{3}{2}\ \tilde{1}\ \tfrac{3}{2}\ 1\ 000+\quad +2.3/5\sqrt{11}$

$\tfrac{3}{2}\ \tilde{1}\ \tfrac{3}{2}\ 1\ 100+\quad +29/3.5\sqrt{3.7.11}$

$\tfrac{3}{2}\ \tilde{1}\ \widetilde{\tfrac{1}{2}}\ 0\ 000+\quad +1/\sqrt{2.5.11}$

$\tfrac{3}{2}\ \tilde{1}\ \widetilde{\tfrac{1}{2}}\ 0\ 100+\quad -8\sqrt{2}/3\sqrt{3.5.7.11}$

$\tfrac{3}{2}\ \tilde{0}\ \tfrac{3}{2}\ 0\ 000+\quad +\sqrt{3}/5\sqrt{11}$

$\tfrac{3}{2}\ \tilde{0}\ \tfrac{3}{2}\ 0\ 100+\quad +4\sqrt{11}/3.5\sqrt{7}$

$\tfrac{9}{2}\ \tfrac{7}{2}\ 1$

$\tfrac{1}{2}\ \tfrac{1}{2}\ 1\ 0\ 000-\quad +2/3\sqrt{3}$

$\tfrac{1}{2}\ \tfrac{3}{2}\ 1\ 0\ 000+\quad +\sqrt{7}/3\sqrt{3.5}$

$\tfrac{3}{2}\ \tfrac{1}{2}\ 1\ 0\ 000+\quad +1/2\sqrt{5}$

$\tfrac{3}{2}\ \tfrac{1}{2}\ 1\ 0\ 100+\quad +\sqrt{7}/3\sqrt{3.5}$

$\tfrac{3}{2}\ \tfrac{3}{2}\ 1\ 0\ 000-\quad 0$

$\tfrac{3}{2}\ \tfrac{3}{2}\ 1\ 0\ 100-\quad +2\sqrt{2}/3\sqrt{3}$

$\tfrac{3}{2}\ \tfrac{3}{2}\ 1\ 1\ 000-\quad +\sqrt{7}/5\sqrt{2}$

$\tfrac{3}{2}\ \tfrac{3}{2}\ 1\ 1\ 100-\quad +2\sqrt{2}/3.5\sqrt{3}$

$\tfrac{3}{2}\ \widetilde{\tfrac{1}{2}}\ 1\ 0\ 000+\quad -\sqrt{3.7}/2.5$

$\tfrac{3}{2}\ \widetilde{\tfrac{1}{2}}\ 1\ 0\ 100+\quad +1/5$

$\tfrac{9}{2}\ \tfrac{7}{2}\ 2$

$\tfrac{1}{2}\ \tfrac{3}{2}\ 2\ 0\ 000+\quad +\sqrt{2}/\sqrt{3.5.11}$

$\tfrac{1}{2}\ \tfrac{3}{2}\ \tilde{1}\ 0\ 000+\quad +3/\sqrt{5.11}$

$\tfrac{1}{2}\ \widetilde{\tfrac{1}{2}}\ \tilde{1}\ 0\ 000-\quad +2/\sqrt{3.5.11}$

$\tfrac{3}{2}\ \tfrac{1}{2}\ 2\ 0\ 000+\quad -\sqrt{7}/\sqrt{2.5.11}$

$\tfrac{3}{2}\ \tfrac{1}{2}\ 2\ 0\ 100+\quad +\sqrt{2}/\sqrt{3.5.11}$

$\tfrac{3}{2}\ \tfrac{1}{2}\ \tilde{1}\ 0\ 000+\quad -\sqrt{7}/2\sqrt{3.5.11}$

$\tfrac{3}{2}\ \tfrac{1}{2}\ \tilde{1}\ 0\ 100+\quad -3/\sqrt{5.11}$

$\tfrac{3}{2}\ \tfrac{3}{2}\ 2\ 0\ 000-\quad -7/5\sqrt{11}$

$\tfrac{3}{2}\ \tfrac{3}{2}\ 2\ 0\ 100-\quad +4/5\sqrt{3.7.11}$

$\tfrac{3}{2}\ \tfrac{3}{2}\ \tilde{1}\ 0\ 000-\quad +4\sqrt{2}/5\sqrt{3.11}$

$\tfrac{3}{2}\ \tfrac{3}{2}\ \tilde{1}\ 0\ 100-\quad -2.3\sqrt{2}/5\sqrt{7.11}$

$\tfrac{3}{2}\ \tfrac{3}{2}\ \tilde{1}\ 1\ 000+\quad -7/5\sqrt{2.3.11}$

$\tfrac{3}{2}\ \tfrac{3}{2}\ \tilde{1}\ 1\ 100+\quad -2.3\sqrt{2}/5\sqrt{7.11}$

$\tfrac{3}{2}\ \widetilde{\tfrac{1}{2}}\ 2\ 0\ 000+\quad +\sqrt{3}/5\sqrt{2.11}$

$\tfrac{3}{2}\ \widetilde{\tfrac{1}{2}}\ 2\ 0\ 100+\quad -\sqrt{2.11}/5\sqrt{7}$

$\tfrac{3}{2}\ \widetilde{\tfrac{1}{2}}\ \tilde{1}\ 0\ 000+\quad -9/2.5\sqrt{11}$

$\tfrac{3}{2}\ \widetilde{\tfrac{1}{2}}\ \tilde{1}\ 0\ 100+\quad -\sqrt{11}/5\sqrt{3.7}$

$\tfrac{9}{2}\ \tfrac{7}{2}\ 3$

$\tfrac{1}{2}\ \tfrac{1}{2}\ 1\ 0\ 000-\quad +\sqrt{7}/2.3\sqrt{11}$

$\tfrac{1}{2}\ \tfrac{3}{2}\ 1\ 0\ 000+\quad +\sqrt{11}/2.3\sqrt{5}$

$\tfrac{1}{2}\ \tfrac{3}{2}\ \tilde{1}\ 0\ 000-\quad -1/2\sqrt{11}$

$\tfrac{1}{2}\ \widetilde{\tfrac{1}{2}}\ \tilde{1}\ 0\ 000+\quad +1/2\sqrt{3.11}$

$\tfrac{1}{2}\ \widetilde{\tfrac{1}{2}}\ \tilde{0}\ 0\ 000-\quad +1/\sqrt{11}$

$\tfrac{3}{2}\ \tfrac{1}{2}\ 1\ 0\ 000+\quad -\sqrt{3.7}/2\sqrt{5.11}$

$\tfrac{3}{2}\ \tfrac{1}{2}\ 1\ 0\ 100+\quad +\sqrt{11}/2.3\sqrt{5}$

$\tfrac{3}{2}\ \tfrac{1}{2}\ \tilde{1}\ 0\ 000-\quad +\sqrt{7}/2\sqrt{3.11}$

$\tfrac{3}{2}\ \tfrac{1}{2}\ \tilde{1}\ 0\ 100-\quad +1/2\sqrt{11}$

$\tfrac{3}{2}\ \tfrac{3}{2}\ 1\ 0\ 000-\quad +\sqrt{3}/\sqrt{2.11}$

$\tfrac{3}{2}\ \tfrac{3}{2}\ 1\ 0\ 100-\quad +1/3\sqrt{2.7.11}$

$\tfrac{3}{2}\ \tfrac{3}{2}\ 1\ 1\ 000-\quad -\sqrt{2.3}/5\sqrt{11}$

SO_3–O 3jm Factors (cont.)

$\frac{9}{2}$ $\frac{7}{2}$ 3
$\frac{3}{2}$ $\frac{3}{2}$ 1 1 100− $-4\sqrt{2}/3.5\sqrt{7.11}$
$\frac{3}{2}$ $\frac{3}{2}$ $\tilde{1}$ 0 000+ $+1/\sqrt{2.3.5.11}$
$\frac{3}{2}$ $\frac{3}{2}$ $\tilde{1}$ 0 100+ $-9/\sqrt{2.5.7.11}$
$\frac{3}{2}$ $\frac{3}{2}$ $\tilde{1}$ 1 000− $-\sqrt{2}/\sqrt{3.5.11}$
$\frac{3}{2}$ $\frac{3}{2}$ $\tilde{1}$ 1 100− $+4\sqrt{2}/\sqrt{5.7.11}$
$\frac{3}{2}$ $\frac{3}{2}$ $\tilde{0}$ 0 000− $+\sqrt{2}/\sqrt{5.11}$
$\frac{3}{2}$ $\frac{3}{2}$ $\tilde{0}$ 0 100− $+\sqrt{2.3}/\sqrt{5.7.11}$
$\frac{3}{2}$ $\widetilde{\frac{1}{2}}$ 1 0 000+ $+1/2.5\sqrt{11}$
$\frac{3}{2}$ $\widetilde{\frac{1}{2}}$ 1 0 100+ $-9\sqrt{3}/2.5\sqrt{7.11}$
$\frac{3}{2}$ $\widetilde{\frac{1}{2}}$ $\tilde{1}$ 0 000− $-3/2\sqrt{5.11}$
$\frac{3}{2}$ $\widetilde{\frac{1}{2}}$ $\tilde{1}$ 0 100− $-19/2\sqrt{3.5.7.11}$

$\frac{9}{2}$ 4 $\frac{1}{2}$
$\frac{1}{2}$ 0 $\frac{1}{2}$ 0 000+ $+1/3$
$\frac{1}{2}$ 1 $\frac{1}{2}$ 0 000− $+2/3\sqrt{5}$
$\frac{3}{2}$ 1 $\frac{1}{2}$ 0 000+ $+\sqrt{3/5}$
$\frac{3}{2}$ 1 $\frac{1}{2}$ 0 100+ $+2\sqrt{7/3.5}$
$\frac{3}{2}$ 2 $\frac{1}{2}$ 0 000− 0
$\frac{3}{2}$ 2 $\frac{1}{2}$ 0 100− $+\sqrt{2/3}$
$\frac{3}{2}$ $\tilde{1}$ $\frac{1}{2}$ 0 000− $+\sqrt{7/5}$
$\frac{3}{2}$ $\tilde{1}$ $\frac{1}{2}$ 0 100− $-2/5\sqrt{3}$

$\frac{9}{2}$ 4 $\frac{3}{2}$
$\frac{1}{2}$ 1 $\frac{3}{2}$ 0 000− $+\sqrt{11}/3\sqrt{2.5}$
$\frac{1}{2}$ 2 $\frac{3}{2}$ 0 000+ $+2\sqrt{7}/3\sqrt{5.11}$
$\frac{1}{2}$ $\tilde{1}$ $\frac{3}{2}$ 0 000+ $+\sqrt{7}/\sqrt{2.3.5.11}$
$\frac{3}{2}$ 0 $\frac{3}{2}$ 0 000− $-\sqrt{3}/\sqrt{5.11}$
$\frac{3}{2}$ 0 $\frac{3}{2}$ 0 100− $-2\sqrt{7}/3\sqrt{5.11}$
$\frac{3}{2}$ 1 $\frac{3}{2}$ 0 000+ $-3\sqrt{3}/2.5\sqrt{11}$
$\frac{3}{2}$ 1 $\frac{3}{2}$ 0 100+ $+7\sqrt{7}/3.5\sqrt{11}$
$\frac{3}{2}$ 1 $\frac{3}{2}$ 1 000+ $+\sqrt{3}/2.5\sqrt{11}$
$\frac{3}{2}$ 1 $\frac{3}{2}$ 1 100+ $-4\sqrt{7}/3.5\sqrt{11}$
$\frac{3}{2}$ 2 $\frac{3}{2}$ 0 000− $-\sqrt{2.3.7}/5\sqrt{11}$
$\frac{3}{2}$ 2 $\frac{3}{2}$ 0 100− $-4\sqrt{2}/3.5\sqrt{11}$
$\frac{3}{2}$ $\tilde{1}$ $\frac{3}{2}$ 0 000− $+\sqrt{7}/2.5\sqrt{11}$
$\frac{3}{2}$ $\tilde{1}$ $\frac{3}{2}$ 0 100− $-\sqrt{11}/5\sqrt{3}$
$\frac{3}{2}$ $\tilde{1}$ $\frac{3}{2}$ 1 000+ $+\sqrt{7}/2\sqrt{11}$
$\frac{3}{2}$ $\tilde{1}$ $\frac{3}{2}$ 1 100+ 0

$\frac{9}{2}$ 4 $\frac{5}{2}$
$\frac{1}{2}$ 1 $\frac{3}{2}$ 0 000+ $-\sqrt{7}/\sqrt{2.3.5.11}$
$\frac{1}{2}$ 2 $\frac{3}{2}$ 0 000− $-4/\sqrt{3.5.11}$
$\frac{1}{2}$ $\tilde{1}$ $\frac{3}{2}$ 0 000− $-1/3\sqrt{2.5.11}$
$\frac{1}{2}$ $\tilde{1}$ $\widetilde{\frac{1}{2}}$ 0 000+ $+2\sqrt{2}/3\sqrt{11}$
$\frac{3}{2}$ 0 $\frac{3}{2}$ 0 000+ $-\sqrt{7}/3\sqrt{5.11}$
$\frac{3}{2}$ 0 $\frac{3}{2}$ 0 100+ $+4/\sqrt{3.5.11}$
$\frac{3}{2}$ 1 $\frac{3}{2}$ 0 000− $-4\sqrt{7}/3.5\sqrt{11}$
$\frac{3}{2}$ 1 $\frac{3}{2}$ 0 100− $-4/5\sqrt{3.11}$
$\frac{3}{2}$ 1 $\frac{3}{2}$ 1 000− $+8\sqrt{7}/3.5\sqrt{11}$
$\frac{3}{2}$ 1 $\frac{3}{2}$ 1 100− $+\sqrt{3}/5\sqrt{11}$
$\frac{3}{2}$ 1 $\widetilde{\frac{1}{2}}$ 0 000+ $-\sqrt{7}/3\sqrt{2.5.11}$
$\frac{3}{2}$ 1 $\widetilde{\frac{1}{2}}$ 0 100+ $+2\sqrt{2}/\sqrt{3.5.11}$
$\frac{3}{2}$ 2 $\frac{3}{2}$ 0 000+ $-7\sqrt{2}/3.5\sqrt{11}$
$\frac{3}{2}$ 2 $\frac{3}{2}$ 0 100+ $+8\sqrt{2}/5\sqrt{3.7.11}$
$\frac{3}{2}$ 2 $\widetilde{\frac{1}{2}}$ 0 000− $+4/3\sqrt{5.11}$
$\frac{3}{2}$ 2 $\widetilde{\frac{1}{2}}$ 0 100− $-2\sqrt{3}/\sqrt{5.7.11}$
$\frac{3}{2}$ $\tilde{1}$ $\frac{3}{2}$ 0 000+ $+2/5\sqrt{3.11}$
$\frac{3}{2}$ $\tilde{1}$ $\frac{3}{2}$ 0 100+ $-4\sqrt{11}/3.5\sqrt{7}$
$\frac{3}{2}$ $\tilde{1}$ $\frac{3}{2}$ 1 000− 0
$\frac{3}{2}$ $\tilde{1}$ $\frac{3}{2}$ 1 100− $-1/3\sqrt{7.11}$
$\frac{3}{2}$ $\tilde{1}$ $\widetilde{\frac{1}{2}}$ 0 000− $-\sqrt{5}/\sqrt{2.3.11}$
$\frac{3}{2}$ $\tilde{1}$ $\widetilde{\frac{1}{2}}$ 0 100− $-2\sqrt{2.5}/3\sqrt{7.11}$

$\frac{9}{2}$ 4 $\frac{7}{2}$
$\frac{1}{2}$ 0 $\frac{1}{2}$ 0 000− $+7/3\sqrt{3.11.13}$
$\frac{1}{2}$ 1 $\frac{1}{2}$ 0 000+ $+7\sqrt{3}/2\sqrt{5.11.13}$
$\frac{1}{2}$ 1 $\frac{3}{2}$ 0 000− $-\sqrt{7}/2\sqrt{3.11.13}$
$\frac{1}{2}$ 2 $\frac{3}{2}$ 0 000+ $+\sqrt{2}/3\sqrt{3.11.13}$
$\frac{1}{2}$ $\tilde{1}$ $\frac{3}{2}$ 0 000+ $+\sqrt{13}/2.3\sqrt{11}$
$\frac{1}{2}$ $\tilde{1}$ $\widetilde{\frac{1}{2}}$ 0 000− $-\sqrt{13}/2\sqrt{3.11}$
$\frac{3}{2}$ 0 $\frac{3}{2}$ 0 000− $+\sqrt{2.7}/\sqrt{11.13}$
$\frac{3}{2}$ 0 $\frac{3}{2}$ 0 100− $-\sqrt{2}/3\sqrt{3.11.13}$
$\frac{3}{2}$ 1 $\frac{1}{2}$ 0 000− $-3.7/2.5\sqrt{11.13}$
$\frac{3}{2}$ 1 $\frac{1}{2}$ 0 100− $+\sqrt{7.11}/2.5\sqrt{3.13}$
$\frac{3}{2}$ 1 $\frac{3}{2}$ 0 000+ $+\sqrt{7}/\sqrt{2.5.11.13}$
$\frac{3}{2}$ 1 $\frac{3}{2}$ 0 100+ $+\sqrt{3}/\sqrt{2.5.11.13}$
$\frac{3}{2}$ 1 $\frac{3}{2}$ 1 000+ $-\sqrt{2.7}/\sqrt{5.11.13}$
$\frac{3}{2}$ 1 $\frac{3}{2}$ 1 100+ $+4\sqrt{2.3}/\sqrt{5.11.13}$
$\frac{3}{2}$ 1 $\widetilde{\frac{1}{2}}$ 0 000− $-3\sqrt{3.7}/2\sqrt{5.11.13}$
$\frac{3}{2}$ 1 $\widetilde{\frac{1}{2}}$ 0 100− $+1/2\sqrt{5.11.13}$
$\frac{3}{2}$ 2 $\frac{1}{2}$ 0 000+ $+\sqrt{2.7}/\sqrt{11.13}$
$\frac{3}{2}$ 2 $\frac{1}{2}$ 0 100+ $+\sqrt{2}/3\sqrt{3.11.13}$
$\frac{3}{2}$ 2 $\frac{3}{2}$ 0 000− 0
$\frac{3}{2}$ 2 $\frac{3}{2}$ 0 100− $+8.2\sqrt{5}/3\sqrt{3.7.11.13}$
$\frac{3}{2}$ 2 $\widetilde{\frac{1}{2}}$ 0 000+ $-\sqrt{2.3}/\sqrt{5.11.13}$
$\frac{3}{2}$ 2 $\widetilde{\frac{1}{2}}$ 0 100+ $-\sqrt{2.13}/\sqrt{5.7.11}$
$\frac{3}{2}$ $\tilde{1}$ $\frac{1}{2}$ 0 000+ $-\sqrt{7.11}/2.5\sqrt{3.13}$
$\frac{3}{2}$ $\tilde{1}$ $\frac{1}{2}$ 0 100+ $+47/2.3.5\sqrt{11.13}$
$\frac{3}{2}$ $\tilde{1}$ $\frac{3}{2}$ 0 000− $-\sqrt{5}/\sqrt{2.3.11.13}$
$\frac{3}{2}$ $\tilde{1}$ $\frac{3}{2}$ 0 100− $-\sqrt{5.13}/3\sqrt{2.7.11}$
$\frac{3}{2}$ $\tilde{1}$ $\frac{3}{2}$ 1 000+ $+7\sqrt{2}/\sqrt{3.5.11.13}$
$\frac{3}{2}$ $\tilde{1}$ $\frac{3}{2}$ 1 100+ $+4.3\sqrt{2}/\sqrt{5.7.11.13}$
$\frac{3}{2}$ $\tilde{1}$ $\widetilde{\frac{1}{2}}$ 0 000+ $-3/2\sqrt{5.11.13}$
$\frac{3}{2}$ $\tilde{1}$ $\widetilde{\frac{1}{2}}$ 0 100+ $-19/2\sqrt{3.5.7.11.13}$

$\frac{9}{2}$ $\frac{9}{2}$ 0
$\frac{1}{2}$ $\frac{1}{2}$ 0 0 000+ $+1/\sqrt{5}$
$\frac{3}{2}$ $\frac{3}{2}$ 0 0 000+ $+\sqrt{2}/\sqrt{5}$
$\frac{3}{2}$ $\frac{3}{2}$ 0 0 100+ 0
$\frac{3}{2}$ $\frac{3}{2}$ 0 0 110+ $+\sqrt{2}/\sqrt{5}$

$\frac{9}{2}$ $\frac{9}{2}$ 1
$\frac{1}{2}$ $\frac{1}{2}$ 1 0 000+ $+\sqrt{11}/3\sqrt{3.5}$
$\frac{3}{2}$ $\frac{1}{2}$ 1 0 000− $-4/5\sqrt{11}$

SO_3–O 3jm Factors (cont.)

$\frac{9}{2}$ $\frac{9}{2}$ 1

$\frac{3}{2}$ $\frac{1}{2}$ 1 0 100− $-8\sqrt{7}/3.5\sqrt{3.11}$

$\frac{3}{2}$ $\frac{3}{2}$ 1 0 000+ $+\sqrt{2.3}/\sqrt{5.11}$

$\frac{3}{2}$ $\frac{3}{2}$ 1 0 100+ 0

$\frac{3}{2}$ $\frac{3}{2}$ 1 0 110+ $-\sqrt{2.11}/3\sqrt{3.5}$

$\frac{3}{2}$ $\frac{3}{2}$ 1 1 000+ $-4\sqrt{2.3}/5\sqrt{5.11}$

$\frac{3}{2}$ $\frac{3}{2}$ 1 1 100+ $+4\sqrt{2.7}/5\sqrt{5.11}$

$\frac{3}{2}$ $\frac{3}{2}$ 1 1 110+ $+8.2\sqrt{2}/3.5\sqrt{3.5.11}$

$\frac{9}{2}$ $\frac{9}{2}$ 2

$\frac{3}{2}$ $\frac{1}{2}$ 2 0 000− $-1/5\sqrt{11}$

$\frac{3}{2}$ $\frac{1}{2}$ 2 0 100− $-4\sqrt{7}/5\sqrt{3.11}$

$\frac{3}{2}$ $\frac{1}{2}$ $\tilde{1}$ 0 000− $-\sqrt{2}/\sqrt{3.11}$

$\frac{3}{2}$ $\frac{1}{2}$ $\tilde{1}$ 0 100− 0

$\frac{3}{2}$ $\frac{3}{2}$ 2 0 000+ $+4\sqrt{2.3}/5\sqrt{5.11}$

$\frac{3}{2}$ $\frac{3}{2}$ 2 0 100+ $+\sqrt{2.7}/5\sqrt{5.11}$

$\frac{3}{2}$ $\frac{3}{2}$ 2 0 110+ $+8\sqrt{2}/5\sqrt{3.5.11}$

$\frac{3}{2}$ $\frac{3}{2}$ $\tilde{1}$ 0 000+ $-9/5\sqrt{5.11}$

$\frac{3}{2}$ $\frac{3}{2}$ $\tilde{1}$ 0 100+ $+4\sqrt{7}/5\sqrt{3.5.11}$

$\frac{3}{2}$ $\frac{3}{2}$ $\tilde{1}$ 0 110+ $-2.3/5\sqrt{5.11}$

$\frac{3}{2}$ $\frac{3}{2}$ $\tilde{1}$ 1 000− 0

$\frac{3}{2}$ $\frac{3}{2}$ $\tilde{1}$ 1 100− $-2\sqrt{7}/\sqrt{3.5.11}$

$\frac{3}{2}$ $\frac{3}{2}$ $\tilde{1}$ 1 110− 0

$\frac{9}{2}$ $\frac{9}{2}$ 3

$\frac{1}{2}$ $\frac{1}{2}$ 1 0 000+ $-2\sqrt{2.7}/3\sqrt{5.11.13}$

$\frac{3}{2}$ $\frac{1}{2}$ 1 0 000− $+\sqrt{3.7}/5\sqrt{2.11.13}$

$\frac{3}{2}$ $\frac{1}{2}$ 1 0 100− $+8.4\sqrt{2}/3.5\sqrt{11.13}$

$\frac{3}{2}$ $\frac{1}{2}$ $\tilde{1}$ 0 000+ $-7\sqrt{7}/\sqrt{2.3.5.11.13}$

$\frac{3}{2}$ $\frac{1}{2}$ $\tilde{1}$ 0 100+ $+4\sqrt{2}/\sqrt{5.11.13}$

$\frac{3}{2}$ $\frac{3}{2}$ 1 0 000+ $+2\sqrt{7}/\sqrt{5.11.13}$

$\frac{3}{2}$ $\frac{3}{2}$ 1 0 100+ $+4\sqrt{3}/\sqrt{5.11.13}$

$\frac{3}{2}$ $\frac{3}{2}$ 1 0 110+ $+4/3\sqrt{5.7.11.13}$

$\frac{3}{2}$ $\frac{3}{2}$ 1 1 000+ $-2\sqrt{7}/5\sqrt{5.11.13}$

$\frac{3}{2}$ $\frac{3}{2}$ 1 1 100+ $+\sqrt{3.13}/5\sqrt{5.11}$

$\frac{3}{2}$ $\frac{3}{2}$ 1 1 110+ $-8.31/3.5\sqrt{5.7.11.13}$

$\frac{3}{2}$ $\frac{3}{2}$ $\tilde{1}$ 0 000− 0

$\frac{3}{2}$ $\frac{3}{2}$ $\tilde{1}$ 0 100− $+4/\sqrt{3.11.13}$

$\frac{3}{2}$ $\frac{3}{2}$ $\tilde{1}$ 0 110− 0

$\frac{3}{2}$ $\frac{3}{2}$ $\tilde{1}$ 1 000+ $+2.3\sqrt{7}/5\sqrt{11.13}$

$\frac{3}{2}$ $\frac{3}{2}$ $\tilde{1}$ 1 100+ $+\sqrt{13}/5\sqrt{3.11}$

$\frac{3}{2}$ $\frac{3}{2}$ $\tilde{1}$ 1 110+ $+8/5\sqrt{7.11.13}$

$\frac{3}{2}$ $\frac{3}{2}$ $\tilde{0}$ 0 000+ $-3\sqrt{3.7}/5\sqrt{11.13}$

$\frac{3}{2}$ $\frac{3}{2}$ $\tilde{0}$ 0 100+ $-4/5\sqrt{11.13}$

$\frac{3}{2}$ $\frac{3}{2}$ $\tilde{0}$ 0 110+ $+2\sqrt{3.13}/5\sqrt{7.11}$

$\frac{9}{2}$ $\frac{9}{2}$ 4

$\frac{1}{2}$ $\frac{1}{2}$ 0 0 000+ $+7\sqrt{2.7}/3\sqrt{3.5.11.13}$

$\frac{1}{2}$ $\frac{1}{2}$ 1 0 000− 0

$\frac{3}{2}$ $\frac{1}{2}$ 1 0 000+ $+3\sqrt{7}/\sqrt{2.5.11.13}$

$\frac{3}{2}$ $\frac{1}{2}$ 1 0 100+ $+2\sqrt{2}/\sqrt{3.5.11.13}$

$\frac{3}{2}$ $\frac{1}{2}$ 2 0 000− $+4/\sqrt{5.11.13}$

$\frac{3}{2}$ $\frac{1}{2}$ 2 0 100− $-2\sqrt{7}/3\sqrt{3.5.11.13}$

$\frac{9}{2}$ $\frac{9}{2}$ 4

$\frac{3}{2}$ $\frac{1}{2}$ $\tilde{1}$ 0 000− $+19/\sqrt{2.3.5.11.13}$

$\frac{3}{2}$ $\frac{1}{2}$ $\tilde{1}$ 0 100− $-2\sqrt{2.7}/3\sqrt{5.11.13}$

$\frac{3}{2}$ $\frac{3}{2}$ 0 0 000+ $-\sqrt{3.7}/\sqrt{5.11.13}$

$\frac{3}{2}$ $\frac{3}{2}$ 0 0 100+ $+4/\sqrt{5.11.13}$

$\frac{3}{2}$ $\frac{3}{2}$ 0 0 110+ $+2\sqrt{7}/3\sqrt{3.5.11.13}$

$\frac{3}{2}$ $\frac{3}{2}$ 1 0 000− 0

$\frac{3}{2}$ $\frac{3}{2}$ 1 0 100− $+4/\sqrt{11.13}$

$\frac{3}{2}$ $\frac{3}{2}$ 1 0 110− 0

$\frac{3}{2}$ $\frac{3}{2}$ 1 1 000− 0

$\frac{3}{2}$ $\frac{3}{2}$ 1 1 100− $+1/\sqrt{11.13}$

$\frac{3}{2}$ $\frac{3}{2}$ 1 1 110− 0

$\frac{3}{2}$ $\frac{3}{2}$ 2 0 000+ $+\sqrt{2.3}/\sqrt{11.13}$

$\frac{3}{2}$ $\frac{3}{2}$ 2 0 100+ 0

$\frac{3}{2}$ $\frac{3}{2}$ 2 0 110+ $-8.2\sqrt{2}/3\sqrt{3.11.13}$

$\frac{3}{2}$ $\frac{3}{2}$ $\tilde{1}$ 0 000+ $-2.3/5\sqrt{11.13}$

$\frac{3}{2}$ $\frac{3}{2}$ $\tilde{1}$ 0 100+ $-4\sqrt{7}/5\sqrt{3.11.13}$

$\frac{3}{2}$ $\frac{3}{2}$ $\tilde{1}$ 0 110+ $-4\sqrt{13}/3.5\sqrt{11}$

$\frac{3}{2}$ $\frac{3}{2}$ $\tilde{1}$ 1 000− 0

$\frac{3}{2}$ $\frac{3}{2}$ $\tilde{1}$ 1 100− $-\sqrt{7}/\sqrt{3.11.13}$

$\frac{3}{2}$ $\frac{3}{2}$ $\tilde{1}$ 1 110− 0

5 $\frac{5}{2}$ $\frac{5}{2}$

1 $\frac{3}{2}$ $\frac{3}{2}$ 0 000+ 0

1 $\frac{3}{2}$ $\frac{3}{2}$ 0 100+ $-\sqrt{5}/3\sqrt{3}$

1 $\frac{3}{2}$ $\frac{3}{2}$ 1 000+ $+5\sqrt{3}/11\sqrt{7}$

1 $\frac{3}{2}$ $\frac{3}{2}$ 1 100+ $+4\sqrt{5}/3.11\sqrt{3}$

1 $\tilde{\frac{1}{2}}$ $\frac{3}{2}$ 0 000− $-\sqrt{2.3.5}/11\sqrt{7}$

1 $\tilde{\frac{1}{2}}$ $\frac{3}{2}$ 0 100− $+7\sqrt{2}/3.11\sqrt{3}$

1 $\tilde{\frac{1}{2}}$ $\tilde{\frac{1}{2}}$ 0 000+ $-4\sqrt{2.3}/11\sqrt{7}$

1 $\tilde{\frac{1}{2}}$ $\tilde{\frac{1}{2}}$ 0 100+ $-\sqrt{2.5}/3.11\sqrt{3}$

2 $\frac{3}{2}$ $\frac{3}{2}$ 0 000− 0

2 $\tilde{\frac{1}{2}}$ $\frac{3}{2}$ 0 000+ $-1/\sqrt{11}$

$\tilde{1}$ $\frac{3}{2}$ $\frac{3}{2}$ 0 000− 0

$\tilde{1}$ $\frac{3}{2}$ $\frac{3}{2}$ 1 000+ $+\sqrt{5}/\sqrt{3.11}$

$\tilde{1}$ $\tilde{\frac{1}{2}}$ $\frac{3}{2}$ 0 000+ $-\sqrt{2}/\sqrt{3.11}$

5 3 2

1 1 2 0 000+ $+3\sqrt{5}/11\sqrt{2.7}$

1 1 2 0 100+ $+5/11\sqrt{2}$

1 1 $\tilde{1}$ 0 000+ $-\sqrt{3.5}/11\sqrt{7}$

1 1 $\tilde{1}$ 0 100+ $+2\sqrt{3}/11$

1 $\tilde{1}$ 2 0 000− $+9/11\sqrt{2.7}$

1 $\tilde{1}$ 2 0 100− $-7/11\sqrt{2.5}$

1 $\tilde{1}$ $\tilde{1}$ 0 000− $-9/11\sqrt{7}$

1 $\tilde{1}$ $\tilde{1}$ 0 100− $-4/11\sqrt{5}$

1 $\tilde{0}$ $\tilde{1}$ 0 000+ $+2.3\sqrt{2}/11\sqrt{7}$

1 $\tilde{0}$ $\tilde{1}$ 0 100+ $-\sqrt{2}/11\sqrt{5}$

2 1 $\tilde{1}$ 0 000− $-1/\sqrt{2.11}$

2 $\tilde{1}$ $\tilde{1}$ 0 000+ $-3/\sqrt{2.5.11}$

2 $\tilde{0}$ 2 0 000− $-\sqrt{3}/\sqrt{5.11}$

$\tilde{1}$ 1 2 0 000− $-1/\sqrt{2.11}$

SO_3–O 3jm Factors (cont.)

5 3 2

$\tilde{1}$ 1 $\tilde{1}$ 0 000− $-1/\sqrt{11}$

$\tilde{1}$ $\tilde{1}$ 2 0 000+ $-3/\sqrt{2.5.11}$

$\tilde{1}$ $\tilde{1}$ $\tilde{1}$ 0 000+ $-\sqrt{3}/\sqrt{5.11}$

5 3 3

1 1 1 0 000+ $-3.5/11\sqrt{2.7}$

1 1 1 0 100+ $+\sqrt{5}/2.11\sqrt{2}$

1 $\tilde{1}$ 1 0 000− $-\sqrt{2.3.5}/11\sqrt{7}$

1 $\tilde{1}$ 1 0 100− $-3\sqrt{3}/2.11\sqrt{2}$

1 $\tilde{1}$ $\tilde{1}$ 0 000+ $+9/11\sqrt{2.7}$

1 $\tilde{1}$ $\tilde{1}$ 0 100+ $-5\sqrt{5}/2.11\sqrt{2}$

1 $\tilde{0}$ $\tilde{1}$ 0 000− $-3/11\sqrt{7}$

1 $\tilde{0}$ $\tilde{1}$ 0 100− $-\sqrt{5}/11$

2 1 1 0 000− 0

2 $\tilde{1}$ 1 0 000+ $+1/\sqrt{11}$

2 $\tilde{1}$ $\tilde{1}$ 0 000− 0

$\tilde{1}$ 1 1 0 000− 0

$\tilde{1}$ $\tilde{1}$ 1 0 000+ $-1/\sqrt{2.11}$

$\tilde{1}$ $\tilde{1}$ $\tilde{1}$ 0 000− 0

$\tilde{1}$ $\tilde{0}$ 1 0 000− $+1/\sqrt{11}$

5 $\frac{7}{2}$ $\frac{3}{2}$

1 $\frac{1}{2}$ $\frac{3}{2}$ 0 000− 0

1 $\frac{1}{2}$ $\frac{3}{2}$ 0 100− $+1/\sqrt{2.3}$

1 $\frac{3}{2}$ $\frac{3}{2}$ 0 000+ $-2\sqrt{3}/11$

1 $\frac{3}{2}$ $\frac{3}{2}$ 0 100+ $+\sqrt{7}/11\sqrt{3.5}$

1 $\frac{3}{2}$ $\frac{3}{2}$ 1 000+ $+\sqrt{3}/11$

1 $\frac{3}{2}$ $\frac{3}{2}$ 1 100+ $+\sqrt{5.7}/11\sqrt{3}$

1 $\tilde{\frac{1}{2}}$ $\frac{3}{2}$ 0 000− $+3\sqrt{2}/11$

1 $\tilde{\frac{1}{2}}$ $\frac{3}{2}$ 0 100− $-\sqrt{7}/11\sqrt{2.5}$

2 $\frac{1}{2}$ $\frac{3}{2}$ 0 000+ $+1/2\sqrt{11}$

2 $\frac{3}{2}$ $\frac{3}{2}$ 0 000− $-\sqrt{7}/\sqrt{2.5.11}$

2 $\tilde{\frac{1}{2}}$ $\frac{3}{2}$ 0 000+ $-\sqrt{3.7}/2\sqrt{5.11}$

$\tilde{1}$ $\frac{1}{2}$ $\frac{3}{2}$ 0 000+ $+\sqrt{2}/\sqrt{3.11}$

$\tilde{1}$ $\frac{3}{2}$ $\frac{3}{2}$ 0 000− $-2\sqrt{7}/\sqrt{3.5.11}$

$\tilde{1}$ $\frac{3}{2}$ $\frac{3}{2}$ 1 000+ $-\sqrt{7}/\sqrt{3.5.11}$

$\tilde{1}$ $\tilde{\frac{1}{2}}$ $\frac{3}{2}$ 0 000+ 0

5 $\frac{7}{2}$ $\frac{5}{2}$

1 $\frac{1}{2}$ $\frac{3}{2}$ 0 000+ $+5\sqrt{5}/2.11\sqrt{3}$

1 $\frac{1}{2}$ $\frac{3}{2}$ 0 100+ $+\sqrt{7}/2.11\sqrt{3}$

1 $\frac{3}{2}$ $\frac{3}{2}$ 0 000− $-\sqrt{7}/11\sqrt{2.3}$

1 $\frac{3}{2}$ $\frac{3}{2}$ 0 100− $-4\sqrt{2}/11\sqrt{3.5}$

1 $\frac{3}{2}$ $\frac{3}{2}$ 1 000− $-4\sqrt{2.3}/11\sqrt{7}$

1 $\frac{3}{2}$ $\frac{3}{2}$ 1 100− $+\sqrt{3.5}/11\sqrt{2}$

1 $\frac{3}{2}$ $\tilde{\frac{1}{2}}$ 0 000+ $-2\sqrt{5}/11\sqrt{3.7}$

1 $\frac{3}{2}$ $\tilde{\frac{1}{2}}$ 0 100+ $-7/11\sqrt{3}$

1 $\tilde{\frac{1}{2}}$ $\frac{3}{2}$ 0 000+ $-13/2.11\sqrt{7}$

1 $\tilde{\frac{1}{2}}$ $\frac{3}{2}$ 0 100+ $-7/2.11\sqrt{5}$

1 $\tilde{\frac{1}{2}}$ $\tilde{\frac{1}{2}}$ 0 000− $-\sqrt{5}/11\sqrt{7}$

1 $\tilde{\frac{1}{2}}$ $\tilde{\frac{1}{2}}$ 0 100− $+2/11$

2 $\frac{1}{2}$ $\frac{3}{2}$ 0 000− $-\sqrt{7}/3\sqrt{2.11}$

5 $\frac{7}{2}$ $\frac{5}{2}$

2 $\frac{3}{2}$ $\frac{3}{2}$ 0 000+ $+7/3\sqrt{5.11}$

2 $\frac{3}{2}$ $\tilde{\frac{1}{2}}$ 0 000− $+\sqrt{2}/3\sqrt{11}$

2 $\tilde{\frac{1}{2}}$ $\frac{3}{2}$ 0 000− $-\sqrt{3}/\sqrt{2.5.11}$

$\tilde{1}$ $\frac{1}{2}$ $\frac{3}{2}$ 0 000− $+\sqrt{7}/2.3\sqrt{3.11}$

$\tilde{1}$ $\frac{1}{2}$ $\tilde{\frac{1}{2}}$ 0 000+ $-\sqrt{5.7}/3\sqrt{3.11}$

$\tilde{1}$ $\frac{3}{2}$ $\frac{3}{2}$ 0 000+ $-7/3\sqrt{2.3.5.11}$

$\tilde{1}$ $\frac{3}{2}$ $\frac{3}{2}$ 1 000− $+2\sqrt{2}/3\sqrt{3.5.11}$

$\tilde{1}$ $\frac{3}{2}$ $\tilde{\frac{1}{2}}$ 0 000− $+2/3\sqrt{3.11}$

$\tilde{1}$ $\tilde{\frac{1}{2}}$ $\frac{3}{2}$ 0 000− $+\sqrt{5}/2\sqrt{11}$

5 $\frac{7}{2}$ $\frac{7}{2}$

1 $\frac{1}{2}$ $\frac{1}{2}$ 0 000+ $+5\sqrt{7}/11\sqrt{13}$

1 $\frac{1}{2}$ $\frac{1}{2}$ 0 100+ $-7\sqrt{5}/2.3.11\sqrt{13}$

1 $\frac{3}{2}$ $\frac{1}{2}$ 0 000− $-\sqrt{5}/2.11\sqrt{13}$

1 $\frac{3}{2}$ $\frac{1}{2}$ 0 100− $-8\sqrt{7}/3.11\sqrt{13}$

1 $\frac{3}{2}$ $\frac{3}{2}$ 0 000+ $+4\sqrt{2.7}/11\sqrt{13}$

1 $\frac{3}{2}$ $\frac{3}{2}$ 0 100+ $+\sqrt{5}/3.11\sqrt{2.13}$

1 $\frac{3}{2}$ $\frac{3}{2}$ 1 000+ $-\sqrt{2}/11\sqrt{7.13}$

1 $\frac{3}{2}$ $\frac{3}{2}$ 1 100+ $+2.5\sqrt{2.5}/3.11\sqrt{13}$

1 $\tilde{\frac{1}{2}}$ $\frac{3}{2}$ 0 000− $-\sqrt{3}/2\sqrt{7.13}$

1 $\tilde{\frac{1}{2}}$ $\frac{3}{2}$ 0 100− 0

1 $\tilde{\frac{1}{2}}$ $\tilde{\frac{1}{2}}$ 0 000+ $+3/11\sqrt{7.13}$

1 $\tilde{\frac{1}{2}}$ $\tilde{\frac{1}{2}}$ 0 100+ $+\sqrt{5.13}/2.11$

2 $\frac{3}{2}$ $\frac{1}{2}$ 0 000+ $-\sqrt{3.7}/\sqrt{2.11.13}$

2 $\frac{3}{2}$ $\frac{3}{2}$ 0 000− 0

2 $\tilde{\frac{1}{2}}$ $\frac{3}{2}$ 0 000+ $+\sqrt{5}/\sqrt{2.11.13}$

$\tilde{1}$ $\frac{3}{2}$ $\frac{1}{2}$ 0 000+ $+\sqrt{7}/2\sqrt{11.13}$

$\tilde{1}$ $\frac{3}{2}$ $\frac{3}{2}$ 0 000− 0

$\tilde{1}$ $\frac{3}{2}$ $\frac{3}{2}$ 1 000+ $-\sqrt{2.5}/\sqrt{11.13}$

$\tilde{1}$ $\tilde{\frac{1}{2}}$ $\frac{1}{2}$ 0 000− $-\sqrt{7}/\sqrt{3.11.13}$

$\tilde{1}$ $\tilde{\frac{1}{2}}$ $\frac{3}{2}$ 0 000+ $-5\sqrt{5}/2\sqrt{3.11.13}$

5 4 1

1 0 1 0 000+ 0

1 0 1 0 100+ $+1/3$

1 1 1 0 000− 0

1 1 1 0 100− $-\sqrt{2}/\sqrt{3.5}$

1 2 1 0 000+ $+3/11$

1 2 1 0 100+ $+4\sqrt{7}/3.11\sqrt{5}$

1 $\tilde{1}$ 1 0 000+ $-2\sqrt{2.3}/11$

1 $\tilde{1}$ 1 0 100+ $+\sqrt{2.7}/11\sqrt{3.5}$

2 1 1 0 000+ $+\sqrt{3}/\sqrt{5.11}$

2 $\tilde{1}$ 1 0 000− $-\sqrt{7}/\sqrt{5.11}$

$\tilde{1}$ 1 1 0 000+ $+2\sqrt{2}/\sqrt{5.11}$

$\tilde{1}$ 2 1 0 000− $-\sqrt{7}/\sqrt{5.11}$

$\tilde{1}$ $\tilde{1}$ 1 0 000− 0

5 4 2

1 1 2 0 000− $-\sqrt{3.7}/11\sqrt{2}$

1 1 2 0 100− $+3\sqrt{3}/11\sqrt{2.5}$

1 1 $\tilde{1}$ 0 000− $+\sqrt{7}/11$

1 1 $\tilde{1}$ 0 100− $+8/11\sqrt{5}$

SO_3–O 3jm Factors (cont.)

5 4 2	
$1\ 2\ \tilde{1}\ 0\ 000+$	$+\sqrt{2}/11$
$1\ 2\ \tilde{1}\ 0\ 100+$	$-2\sqrt{2.7}/11\sqrt{5}$
$1\ \tilde{1}\ 2\ 0\ 000+$	$-5/11\sqrt{2}$
$1\ \tilde{1}\ 2\ 0\ 100+$	$-\sqrt{7}/11\sqrt{2.5}$
$1\ \tilde{1}\ \tilde{1}\ 0\ 000+$	$+1/11$
$1\ \tilde{1}\ \tilde{1}\ 0\ 100+$	$-2\sqrt{7}/11\sqrt{5}$
$2\ 0\ 2\ 0\ 000-$	$-1/\sqrt{3.11}$
$2\ 1\ \tilde{1}\ 0\ 000+$	$+1/\sqrt{2.3.5.11}$
$2\ 2\ 2\ 0\ 000-$	$-\sqrt{2.7}/\sqrt{3.5.11}$
$2\ \tilde{1}\ \tilde{1}\ 0\ 000-$	$+\sqrt{7}/\sqrt{2.5.11}$
$\tilde{1}\ 0\ \tilde{1}\ 0\ 000-$	$-2\sqrt{2}/3\sqrt{11}$
$\tilde{1}\ 1\ 2\ 0\ 000+$	$-1/\sqrt{2.3.5.11}$
$\tilde{1}\ 1\ \tilde{1}\ 0\ 000+$	$+\sqrt{3}/\sqrt{5.11}$
$\tilde{1}\ 2\ \tilde{1}\ 0\ 000-$	$-\sqrt{2.7}/3\sqrt{5.11}$
$\tilde{1}\ \tilde{1}\ 2\ 0\ 000-$	$-\sqrt{7}/\sqrt{2.5.11}$
$\tilde{1}\ \tilde{1}\ \tilde{1}\ 0\ 000-$	$-\sqrt{7}/\sqrt{3.5.11}$
5 4 3	
$1\ 0\ 1\ 0\ 000+$	$+5\sqrt{5}/11\sqrt{13}$
$1\ 0\ 1\ 0\ 100+$	$+\sqrt{7}/11\sqrt{13}$
$1\ 1\ 1\ 0\ 000-$	$+5\sqrt{3}/11\sqrt{2.13}$
$1\ 1\ 1\ 0\ 100-$	$-9\sqrt{3.7}/2.11\sqrt{2.5.13}$
$1\ 1\ \tilde{1}\ 0\ 000+$	$+2\sqrt{2.5}/11\sqrt{13}$
$1\ 1\ \tilde{1}\ 0\ 100+$	$-5\sqrt{7}/2.11\sqrt{2.13}$
$1\ 2\ 1\ 0\ 000+$	$-17/11\sqrt{7.13}$
$1\ 2\ 1\ 0\ 100+$	$+23/11\sqrt{5.13}$
$1\ 2\ \tilde{1}\ 0\ 000-$	$+\sqrt{5.13}/11\sqrt{7}$
$1\ 2\ \tilde{1}\ 0\ 100-$	$+7/11\sqrt{13}$
$1\ \tilde{1}\ 1\ 0\ 000+$	$+8\sqrt{2.3}/11\sqrt{7.13}$
$1\ \tilde{1}\ 1\ 0\ 100+$	$+\sqrt{3.13}/2.11\sqrt{2.5}$
$1\ \tilde{1}\ \tilde{1}\ 0\ 000-$	$-5\sqrt{5}/11\sqrt{2.7.13}$
$1\ \tilde{1}\ \tilde{1}\ 0\ 100-$	$-\sqrt{13}/2.11\sqrt{2}$
$1\ \tilde{1}\ \tilde{0}\ 0\ 000+$	$-\sqrt{5}/11\sqrt{7.13}$
$1\ \tilde{1}\ \tilde{0}\ 0\ 100+$	$+\sqrt{13}/11$
$2\ 1\ 1\ 0\ 000+$	$-2\sqrt{3.7}/\sqrt{5.11.13}$
$2\ 1\ \tilde{1}\ 0\ 000-$	$-\sqrt{7}/\sqrt{3.11.13}$
$2\ 2\ \tilde{0}\ 0\ 000-$	$-2\sqrt{2}/\sqrt{3.11.13}$
$2\ \tilde{1}\ 1\ 0\ 000-$	$-1/\sqrt{5.11.13}$
$2\ \tilde{1}\ \tilde{1}\ 0\ 000+$	$+2/\sqrt{11.13}$
$\tilde{1}\ 0\ \tilde{1}\ 0\ 000+$	$+\sqrt{5.7}/3\sqrt{11.13}$
$\tilde{1}\ 1\ 1\ 0\ 000+$	$+\sqrt{2.7}/\sqrt{5.11.13}$
$\tilde{1}\ 1\ \tilde{1}\ 0\ 000-$	$-\sqrt{3.7}/\sqrt{2.11.13}$
$\tilde{1}\ 1\ \tilde{0}\ 0\ 000+$	$-\sqrt{7}/\sqrt{3.11.13}$
$\tilde{1}\ 2\ 1\ 0\ 000-$	$-1/\sqrt{5.11.13}$
$\tilde{1}\ 2\ \tilde{1}\ 0\ 000+$	$-1/3\sqrt{11.13}$
$\tilde{1}\ \tilde{1}\ 1\ 0\ 000-$	$+\sqrt{5}/\sqrt{2.11.13}$
$\tilde{1}\ \tilde{1}\ \tilde{1}\ 0\ 000+$	$-5\sqrt{2}/\sqrt{3.11.13}$
5 4 4	
$1\ 1\ 0\ 0\ 000-$	$+3\sqrt{5}/11\sqrt{13}$
$1\ 1\ 0\ 0\ 100-$	$-7\sqrt{7}/3.11\sqrt{13}$
5 4 4	
$1\ 1\ 1\ 0\ 000+$	$-9\sqrt{3}/11\sqrt{2.13}$
$1\ 1\ 1\ 0\ 100+$	$-7\sqrt{5.7}/2.11\sqrt{2.3.13}$
$1\ 2\ 1\ 0\ 000-$	$-3\sqrt{7}/11\sqrt{13}$
$1\ 2\ 1\ 0\ 100-$	$+\sqrt{5}/3.11\sqrt{13}$
$1\ \tilde{1}\ 1\ 0\ 000-$	$+\sqrt{2.3.7}/11\sqrt{13}$
$1\ \tilde{1}\ 1\ 0\ 100-$	$-5\sqrt{5}/2.11\sqrt{2.3.13}$
$1\ \tilde{1}\ 2\ 0\ 000+$	$-\sqrt{3}/11\sqrt{13}$
$1\ \tilde{1}\ 2\ 0\ 100+$	$-\sqrt{5.7}/11\sqrt{3.13}$
$1\ \tilde{1}\ \tilde{1}\ 0\ 000+$	$-\sqrt{3}/11\sqrt{2.13}$
$1\ \tilde{1}\ \tilde{1}\ 0\ 100+$	$-\sqrt{5.7.13}/2.11\sqrt{2.3}$
$2\ 1\ 1\ 0\ 000-$	0
$2\ 2\ 0\ 0\ 000-$	$-2\sqrt{2}/\sqrt{11.13}$
$2\ 2\ 2\ 0\ 000-$	0
$2\ \tilde{1}\ 1\ 0\ 000+$	$+\sqrt{5}/\sqrt{11.13}$
$2\ \tilde{1}\ \tilde{1}\ 0\ 000-$	0
$\tilde{1}\ 1\ 1\ 0\ 000-$	0
$\tilde{1}\ 2\ 1\ 0\ 000+$	$-\sqrt{5}/\sqrt{11.13}$
$\tilde{1}\ \tilde{1}\ 0\ 0\ 000-$	$-1/\sqrt{3.11.13}$
$\tilde{1}\ \tilde{1}\ 1\ 0\ 000+$	$-\sqrt{5}/\sqrt{2.11.13}$
$\tilde{1}\ \tilde{1}\ 2\ 0\ 000-$	$-\sqrt{5.7}/\sqrt{3.11.13}$
$\tilde{1}\ \tilde{1}\ \tilde{1}\ 0\ 000-$	0
5 $\frac{9}{2}$ $\frac{1}{2}$	
$1\ \frac{1}{2}\ \frac{1}{2}\ 0\ 000+$	0
$1\ \frac{1}{2}\ \frac{1}{2}\ 0\ 100+$	$+1/\sqrt{5}$
$1\ \frac{3}{2}\ \frac{1}{2}\ 0\ 000-$	$-2\sqrt{3.7}/11\sqrt{5}$
$1\ \frac{3}{2}\ \frac{1}{2}\ 0\ 010-$	$+9/11\sqrt{5}$
$1\ \frac{3}{2}\ \frac{1}{2}\ 0\ 100-$	$+2.3\sqrt{3}/5.11$
$1\ \frac{3}{2}\ \frac{1}{2}\ 0\ 110-$	$+4\sqrt{7}/5.11$
$2\ \frac{3}{2}\ \frac{1}{2}\ 0\ 000+$	$+\sqrt{2}/\sqrt{11}$
$2\ \frac{3}{2}\ \frac{1}{2}\ 0\ 010+$	0
$\tilde{1}\ \frac{3}{2}\ \frac{1}{2}\ 0\ 000+$	$+2\sqrt{3}/5\sqrt{11}$
$\tilde{1}\ \frac{3}{2}\ \frac{1}{2}\ 0\ 010+$	$+3\sqrt{7}/5\sqrt{11}$
5 $\frac{9}{2}$ $\frac{3}{2}$	
$1\ \frac{1}{2}\ \frac{3}{2}\ 0\ 000+$	0
$1\ \frac{1}{2}\ \frac{3}{2}\ 0\ 100+$	$+1/\sqrt{3.5}$
$1\ \frac{3}{2}\ \frac{3}{2}\ 0\ 000-$	$+3\sqrt{2.7}/11\sqrt{5}$
$1\ \frac{3}{2}\ \frac{3}{2}\ 0\ 010-$	$+\sqrt{3}/11\sqrt{2.5}$
$1\ \frac{3}{2}\ \frac{3}{2}\ 0\ 100-$	$+2\sqrt{2}/5.11$
$1\ \frac{3}{2}\ \frac{3}{2}\ 0\ 110-$	$+8\sqrt{2.7}/5.11\sqrt{3}$
$1\ \frac{3}{2}\ \frac{3}{2}\ 1\ 000-$	0
$1\ \frac{3}{2}\ \frac{3}{2}\ 1\ 010-$	$-\sqrt{3.5}/11\sqrt{2}$
$1\ \frac{3}{2}\ \frac{3}{2}\ 1\ 100-$	$+\sqrt{2}/5$
$1\ \frac{3}{2}\ \frac{3}{2}\ 1\ 110-$	$+4\sqrt{2.7}/5.11\sqrt{3}$
$2\ \frac{1}{2}\ \frac{3}{2}\ 0\ 000-$	$-\sqrt{2}/\sqrt{5.11}$
$2\ \frac{3}{2}\ \frac{3}{2}\ 0\ 000+$	$+2\sqrt{3}/5\sqrt{11}$
$2\ \frac{3}{2}\ \frac{3}{2}\ 0\ 010+$	$-2\sqrt{7}/5\sqrt{11}$
$\tilde{1}\ \frac{1}{2}\ \frac{3}{2}\ 0\ 000-$	$-4/\sqrt{3.5.11}$
$\tilde{1}\ \frac{3}{2}\ \frac{3}{2}\ 0\ 000+$	$-3\sqrt{2}/5\sqrt{11}$
$\tilde{1}\ \frac{3}{2}\ \frac{3}{2}\ 0\ 010+$	$+\sqrt{7}/5\sqrt{2.3.11}$

SO_3–O 3jm Factors (cont.)

5	$\frac{9}{2}$	$\frac{3}{2}$			
$\tilde{1}$	$\frac{3}{2}$	$\frac{3}{2}$	1	000−	0
$\tilde{1}$	$\frac{3}{2}$	$\frac{3}{2}$	1	010−	$+\sqrt{7}/\sqrt{2.3.11}$

5	$\frac{9}{2}$	$\frac{5}{2}$			
1	$\frac{1}{2}$	$\frac{3}{2}$	0	000−	$-5\sqrt{2.7}/11\sqrt{3.13}$
1	$\frac{1}{2}$	$\frac{3}{2}$	0	100−	$-7\sqrt{2}/11\sqrt{3.5.13}$
1	$\frac{3}{2}$	$\frac{3}{2}$	0	000+	$+8\sqrt{7}/11\sqrt{5.13}$
1	$\frac{3}{2}$	$\frac{3}{2}$	0	010+	$+4/11\sqrt{3.5.13}$
1	$\frac{3}{2}$	$\frac{3}{2}$	0	100+	$-\sqrt{13/5.11}$
1	$\frac{3}{2}$	$\frac{3}{2}$	0	110+	$-4\sqrt{7.13}/5.11\sqrt{3}$
1	$\frac{3}{2}$	$\frac{3}{2}$	1	000+	$+\sqrt{5.7}/11\sqrt{13}$
1	$\frac{3}{2}$	$\frac{3}{2}$	1	010+	$-2\sqrt{3.5}/11\sqrt{13}$
1	$\frac{3}{2}$	$\frac{3}{2}$	1	100+	$-4/5.11\sqrt{13}$
1	$\frac{3}{2}$	$\frac{3}{2}$	1	110+	$-2\sqrt{3.7}/5.11\sqrt{13}$
1	$\frac{3}{2}$	$\tilde{\frac{1}{2}}$	0	000−	$-\sqrt{2.7}/11\sqrt{13}$
1	$\frac{3}{2}$	$\tilde{\frac{1}{2}}$	0	010−	$-\sqrt{2.13}/11\sqrt{3}$
1	$\frac{3}{2}$	$\tilde{\frac{1}{2}}$	0	100−	$-19\sqrt{2}/11\sqrt{5.13}$
1	$\frac{3}{2}$	$\tilde{\frac{1}{2}}$	0	110−	$+4\sqrt{2.7}/11\sqrt{3.5.13}$
2	$\frac{1}{2}$	$\frac{3}{2}$	0	000+	$-\sqrt{13}/3\sqrt{5.11}$
2	$\frac{3}{2}$	$\frac{3}{2}$	0	000−	$-4\sqrt{2.3}/5\sqrt{11.13}$
2	$\frac{3}{2}$	$\frac{3}{2}$	0	010−	$-\sqrt{2.7.13}/3.5\sqrt{11}$
2	$\frac{3}{2}$	$\tilde{\frac{1}{2}}$	0	000+	$-3\sqrt{3}/\sqrt{5.11.13}$
2	$\frac{3}{2}$	$\tilde{\frac{1}{2}}$	0	010+	$+4\sqrt{7}/3\sqrt{5.11.13}$
$\tilde{1}$	$\frac{1}{2}$	$\frac{3}{2}$	0	000+	$+29\sqrt{2}/3\sqrt{3.5.11.13}$
$\tilde{1}$	$\frac{1}{2}$	$\tilde{\frac{1}{2}}$	0	000−	$-4\sqrt{2}/3\sqrt{3.11.13}$
$\tilde{1}$	$\frac{3}{2}$	$\frac{3}{2}$	0	000−	$-8/5\sqrt{11.13}$
$\tilde{1}$	$\frac{3}{2}$	$\frac{3}{2}$	0	010−	$+4\sqrt{7}/3.5\sqrt{3.11.13}$
$\tilde{1}$	$\frac{3}{2}$	$\frac{3}{2}$	1	000+	$-3/\sqrt{11.13}$
$\tilde{1}$	$\frac{3}{2}$	$\frac{3}{2}$	1	010+	$-2\sqrt{7}/3\sqrt{3.11.13}$
$\tilde{1}$	$\frac{3}{2}$	$\tilde{\frac{1}{2}}$	0	000+	$+\sqrt{2.5}/\sqrt{11.13}$
$\tilde{1}$	$\frac{3}{2}$	$\tilde{\frac{1}{2}}$	0	010+	$-\sqrt{2.5.7}/3\sqrt{3.11.13}$

5	$\frac{9}{2}$	$\frac{7}{2}$			
1	$\frac{1}{2}$	$\frac{1}{2}$	0	000−	$+5/11\sqrt{13}$
1	$\frac{1}{2}$	$\frac{1}{2}$	0	100−	$+2.7\sqrt{7}/3.11\sqrt{5.13}$
1	$\frac{1}{2}$	$\frac{3}{2}$	0	000+	$+2\sqrt{5.7}/11\sqrt{13}$
1	$\frac{1}{2}$	$\frac{3}{2}$	0	100+	$-7/3.11\sqrt{13}$
1	$\frac{3}{2}$	$\frac{1}{2}$	0	000+	$+\sqrt{3}/2\sqrt{5.13}$
1	$\frac{3}{2}$	$\frac{1}{2}$	0	010+	$-8\sqrt{7}/11\sqrt{5.13}$
1	$\frac{3}{2}$	$\frac{1}{2}$	0	100+	$+\sqrt{3.7}/2.5\sqrt{13}$
1	$\frac{3}{2}$	$\frac{1}{2}$	0	110+	$+127/3.5.11\sqrt{13}$
1	$\frac{3}{2}$	$\frac{3}{2}$	0	000−	$-\sqrt{3.7}/11\sqrt{2.13}$
1	$\frac{3}{2}$	$\frac{3}{2}$	0	010−	$-4\sqrt{2}/11\sqrt{13}$
1	$\frac{3}{2}$	$\frac{3}{2}$	0	100−	$-4\sqrt{2.3}/11\sqrt{5.13}$
1	$\frac{3}{2}$	$\frac{3}{2}$	0	110−	$+2\sqrt{2.7}/3.11\sqrt{5.13}$
1	$\frac{3}{2}$	$\frac{3}{2}$	1	000−	0
1	$\frac{3}{2}$	$\frac{3}{2}$	1	010−	$-5\sqrt{2}/11\sqrt{13}$
1	$\frac{3}{2}$	$\frac{3}{2}$	1	100−	$+\sqrt{3}/\sqrt{2.5.13}$
1	$\frac{3}{2}$	$\frac{3}{2}$	1	110−	$-2.7\sqrt{2.7}/3.11\sqrt{5.13}$
1	$\frac{3}{2}$	$\tilde{\frac{1}{2}}$	0	000+	$-3\sqrt{7}/2.11\sqrt{13}$
1	$\frac{3}{2}$	$\tilde{\frac{1}{2}}$	0	010+	$-4\sqrt{3}/11\sqrt{13}$

5	$\frac{9}{2}$	$\frac{7}{2}$			
1	$\frac{3}{2}$	$\tilde{\frac{1}{2}}$	0	100+	$+53/2.11\sqrt{5.13}$
1	$\frac{3}{2}$	$\tilde{\frac{1}{2}}$	0	110+	$-3\sqrt{3.7}/11\sqrt{5.13}$
2	$\frac{1}{2}$	$\frac{3}{2}$	0	000−	$+\sqrt{2.3}/\sqrt{11.13}$
2	$\frac{3}{2}$	$\frac{1}{2}$	0	000−	$-\sqrt{7}/\sqrt{2.11.13}$
2	$\frac{3}{2}$	$\frac{1}{2}$	0	010−	$-\sqrt{2.3}/\sqrt{11.13}$
2	$\frac{3}{2}$	$\frac{3}{2}$	0	000+	$+\sqrt{5}/\sqrt{11.13}$
2	$\frac{3}{2}$	$\frac{3}{2}$	0	010+	0
2	$\frac{3}{2}$	$\tilde{\frac{1}{2}}$	0	000−	$-3\sqrt{3}/\sqrt{2.5.11.13}$
2	$\frac{3}{2}$	$\tilde{\frac{1}{2}}$	0	010−	$-\sqrt{2.7}/\sqrt{5.11.13}$
$\tilde{1}$	$\frac{1}{2}$	$\frac{3}{2}$	0	000−	$+2/\sqrt{11.13}$
$\tilde{1}$	$\frac{1}{2}$	$\tilde{\frac{1}{2}}$	0	000+	$+1/\sqrt{3.11.13}$
$\tilde{1}$	$\frac{3}{2}$	$\frac{1}{2}$	0	000−	$+\sqrt{7}/2.5\sqrt{3.11.13}$
$\tilde{1}$	$\frac{3}{2}$	$\frac{1}{2}$	0	010−	$+8/5\sqrt{11.13}$
$\tilde{1}$	$\frac{3}{2}$	$\frac{3}{2}$	0	000+	$-5\sqrt{5}/\sqrt{2.3.11.13}$
$\tilde{1}$	$\frac{3}{2}$	$\frac{3}{2}$	0	010+	0
$\tilde{1}$	$\frac{3}{2}$	$\frac{3}{2}$	1	000−	$-2\sqrt{2}/\sqrt{3.5.11.13}$
$\tilde{1}$	$\frac{3}{2}$	$\frac{3}{2}$	1	010−	$-\sqrt{2.7}/\sqrt{5.11.13}$
$\tilde{1}$	$\frac{3}{2}$	$\tilde{\frac{1}{2}}$	0	000−	$+3/2\sqrt{5.11.13}$
$\tilde{1}$	$\frac{3}{2}$	$\tilde{\frac{1}{2}}$	0	010−	$-4\sqrt{7}/\sqrt{3.5.11.13}$

5	$\frac{9}{2}$	$\frac{9}{2}$			
1	$\frac{1}{2}$	$\frac{1}{2}$	0	000+	$-4\sqrt{2}/11\sqrt{5.13}$
1	$\frac{1}{2}$	$\frac{1}{2}$	0	100+	$+7\sqrt{2.7}/3.11\sqrt{13}$
1	$\frac{3}{2}$	$\frac{1}{2}$	0	000−	$+19\sqrt{2.3}/5.11\sqrt{13}$
1	$\frac{3}{2}$	$\frac{1}{2}$	0	010−	$+\sqrt{2.7}/5.11\sqrt{13}$
1	$\frac{3}{2}$	$\frac{1}{2}$	0	100−	$-\sqrt{2.3.7}/11\sqrt{5.13}$
1	$\frac{3}{2}$	$\frac{1}{2}$	0	110−	$-4\sqrt{2}/3.11\sqrt{5.13}$
1	$\frac{3}{2}$	$\frac{3}{2}$	0	000+	$-8.2.3/5.11\sqrt{5.13}$
1	$\frac{3}{2}$	$\frac{3}{2}$	0	010+	$-4\sqrt{3.7}/5.11\sqrt{5.13}$
1	$\frac{3}{2}$	$\frac{3}{2}$	0	011+	$-8.4/5.11\sqrt{5.13}$
1	$\frac{3}{2}$	$\frac{3}{2}$	0	100+	$-\sqrt{7}/11\sqrt{13}$
1	$\frac{3}{2}$	$\frac{3}{2}$	0	110+	$+4\sqrt{3}/11\sqrt{13}$
1	$\frac{3}{2}$	$\frac{3}{2}$	0	111+	$-2\sqrt{7}/3.11\sqrt{13}$
1	$\frac{3}{2}$	$\frac{3}{2}$	1	000+	$+3.17/5.11\sqrt{5.13}$
1	$\frac{3}{2}$	$\frac{3}{2}$	1	010+	$+2.9\sqrt{3.7}/5.11\sqrt{5.13}$
1	$\frac{3}{2}$	$\frac{3}{2}$	1	011+	$+8\sqrt{13}/5.11\sqrt{5}$
1	$\frac{3}{2}$	$\frac{3}{2}$	1	100+	$+4.7\sqrt{7}/5.11\sqrt{13}$
1	$\frac{3}{2}$	$\frac{3}{2}$	1	110+	$-2\sqrt{3}/5.11\sqrt{13}$
1	$\frac{3}{2}$	$\frac{3}{2}$	1	111+	$+8.2\sqrt{7}/3.5.11\sqrt{13}$
2	$\frac{3}{2}$	$\frac{1}{2}$	0	000+	$-\sqrt{7}/\sqrt{5.11.13}$
2	$\frac{3}{2}$	$\frac{1}{2}$	0	010+	$+4\sqrt{3}/\sqrt{5.11.13}$
2	$\frac{3}{2}$	$\frac{3}{2}$	0	000−	0
2	$\frac{3}{2}$	$\frac{3}{2}$	0	010−	$+\sqrt{2}/\sqrt{11.13}$
2	$\frac{3}{2}$	$\frac{3}{2}$	0	011−	0
$\tilde{1}$	$\frac{3}{2}$	$\frac{1}{2}$	0	000+	$+\sqrt{2.7}/\sqrt{3.5.11.13}$
$\tilde{1}$	$\frac{3}{2}$	$\frac{1}{2}$	0	010+	$+\sqrt{2}/\sqrt{5.11.13}$
$\tilde{1}$	$\frac{3}{2}$	$\frac{3}{2}$	0	000−	0
$\tilde{1}$	$\frac{3}{2}$	$\frac{3}{2}$	0	010−	$+4/\sqrt{3.11.13}$
$\tilde{1}$	$\frac{3}{2}$	$\frac{3}{2}$	0	011−	0
$\tilde{1}$	$\frac{3}{2}$	$\frac{3}{2}$	1	000+	$-3\sqrt{7}/5\sqrt{11.13}$

SO_3–O 3jm Factors (cont.)

5 $\frac{9}{2}$ $\frac{9}{2}$	
$\tilde{1}$ $\frac{3}{2}$ $\frac{3}{2}$ 1 010+	$-2.7/5\sqrt{3.11.13}$
$\tilde{1}$ $\frac{3}{2}$ $\frac{3}{2}$ 1 011+	$+8\sqrt{7}/5\sqrt{11.13}$
5 5 0	
1 1 0 0 000+	$+\sqrt{3}/\sqrt{11}$
1 1 0 0 100+	0
1 1 0 0 110+	$+\sqrt{3}/\sqrt{11}$
2 2 0 0 000+	$+\sqrt{2}/\sqrt{11}$
$\tilde{1}$ $\tilde{1}$ 0 0 000+	$+\sqrt{3}/\sqrt{11}$
5 5 1	
1 1 1 0 000+	$+\sqrt{5}/2\sqrt{11}$
1 1 1 0 100+	0
1 1 1 0 110+	$-3/\sqrt{5.11}$
2 1 1 0 000−	$+\sqrt{2.7/11}$
2 1 1 0 010−	$-3\sqrt{2/11}\sqrt{5}$
$\tilde{1}$ 1 1 0 000−	$-\sqrt{3.7/2.11}$
$\tilde{1}$ 1 1 0 010−	$-4\sqrt{3/11}\sqrt{5}$
$\tilde{1}$ 2 1 0 000+	$-\sqrt{2}/\sqrt{5.11}$
$\tilde{1}$ $\tilde{1}$ 1 0 000+	$-\sqrt{5}/2\sqrt{11}$
5 5 2	
1 1 2 0 000+	$-3\sqrt{2.5}/11\sqrt{11.13}$
1 1 2 0 100+	$+2.3\sqrt{2.7}/11\sqrt{11.13}$
1 1 2 0 110+	$+37\sqrt{2}/11\sqrt{5.11.13}$
1 1 $\tilde{1}$ 0 000+	$+3.5\sqrt{3.5/2.11}\sqrt{11.13}$
1 1 $\tilde{1}$ 0 100+	$-4\sqrt{3.7}/11\sqrt{11.13}$
1 1 $\tilde{1}$ 0 110+	$+23\sqrt{3}/11\sqrt{5.11.13}$
2 1 $\tilde{1}$ 0 000−	$+3\sqrt{2.7}/11\sqrt{13}$
2 1 $\tilde{1}$ 0 010−	$+\sqrt{2.13}/11\sqrt{5}$
2 2 2 0 000+	$+2\sqrt{2.3}/\sqrt{5.11.13}$
$\tilde{1}$ 1 2 0 000−	0
$\tilde{1}$ 1 2 0 010−	$+2\sqrt{2}/\sqrt{5.13}$
$\tilde{1}$ 1 $\tilde{1}$ 0 000−	$-9\sqrt{7/2.11}\sqrt{13}$
$\tilde{1}$ 1 $\tilde{1}$ 0 010−	$+8/11\sqrt{5.13}$
$\tilde{1}$ 2 $\tilde{1}$ 0 000+	$-3\sqrt{2}/\sqrt{5.11.13}$
$\tilde{1}$ $\tilde{1}$ 2 0 000+	$+3\sqrt{2}/\sqrt{5.11.13}$
$\tilde{1}$ $\tilde{1}$ $\tilde{1}$ 0 000+	$-\sqrt{3}/2\sqrt{5.11.13}$
5 5 3	
1 1 1 0 000+	$+2.3\sqrt{5.7}/11\sqrt{11.13}$
1 1 1 0 100+	$+3.5/11\sqrt{11.13}$
1 1 1 0 110+	$+2.7\sqrt{7}/11\sqrt{5.11.13}$
1 1 $\tilde{1}$ 0 000−	0
1 1 $\tilde{1}$ 0 100−	$+\sqrt{3.5}/\sqrt{11.13}$
1 1 $\tilde{1}$ 0 110−	0
2 1 1 0 000−	$+3/11\sqrt{2.13}$
2 1 1 0 010−	$-9/11\sqrt{2.5.7.13}$
2 1 $\tilde{1}$ 0 000+	$-3\sqrt{5}/11\sqrt{2.13}$
2 1 $\tilde{1}$ 0 010+	$+31/11\sqrt{2.7.13}$
2 2 $\tilde{0}$ 0 000+	$+3\sqrt{3}/\sqrt{7.11.13}$
$\tilde{1}$ 1 1 0 000−	$-2\sqrt{3}/11\sqrt{13}$
$\tilde{1}$ 1 1 0 010−	$+3\sqrt{3.13}/11\sqrt{5.7}$
5 5 3	
$\tilde{1}$ 1 $\tilde{1}$ 0 000+	0
$\tilde{1}$ 1 $\tilde{1}$ 0 010+	$-1/\sqrt{7.13}$
$\tilde{1}$ 1 $\tilde{0}$ 0 000−	$-3\sqrt{2.5}/11\sqrt{13}$
$\tilde{1}$ 1 $\tilde{0}$ 0 010−	$-2\sqrt{2}/11\sqrt{7.13}$
$\tilde{1}$ 2 1 0 000+	$-23/\sqrt{2.5.7.11.13}$
$\tilde{1}$ 2 $\tilde{1}$ 0 000−	$+9/\sqrt{2.7.11.13}$
$\tilde{1}$ $\tilde{1}$ 1 0 000+	$+2\sqrt{5}/\sqrt{7.11.13}$
$\tilde{1}$ $\tilde{1}$ $\tilde{1}$ 0 000−	0
5 5 4	
1 1 0 0 000+	$-7\sqrt{2.7}/11\sqrt{11.13}$
1 1 0 0 100+	$+2\sqrt{2.5}/11\sqrt{11.13}$
1 1 0 0 110+	$+7\sqrt{2.7}/11\sqrt{11.13}$
1 1 1 0 000−	0
1 1 1 0 100−	$+\sqrt{3}/\sqrt{11.13}$
1 1 1 0 110−	0
1 1 2 0 000+	$+2\sqrt{2.5}/11\sqrt{11.13}$
1 1 2 0 100+	$+7\sqrt{2.7}/11\sqrt{11.13}$
1 1 2 0 110+	$-2\sqrt{2.5}/11\sqrt{11.13}$
1 1 $\tilde{1}$ 0 000+	$+2\sqrt{3.5}/11\sqrt{11.13}$
1 1 $\tilde{1}$ 0 100+	$+7\sqrt{3.7}/11\sqrt{11.13}$
1 1 $\tilde{1}$ 0 110+	$-2\sqrt{3.5}/11\sqrt{11.13}$
2 1 1 0 000+	$+3\sqrt{3}/11\sqrt{2.13}$
2 1 1 0 010+	$+\sqrt{3.5.7}/11\sqrt{2.13}$
2 1 $\tilde{1}$ 0 000−	$+\sqrt{7}/11\sqrt{2.13}$
2 1 $\tilde{1}$ 0 010−	$-5\sqrt{5}/11\sqrt{2.13}$
2 2 0 0 000+	$-\sqrt{7}/\sqrt{3.11.13}$
2 2 2 0 000+	$+\sqrt{2.5}/\sqrt{3.11.13}$
$\tilde{1}$ 1 1 0 000+	$-8/11\sqrt{13}$
$\tilde{1}$ 1 1 0 010+	$+\sqrt{5.7}/11\sqrt{13}$
$\tilde{1}$ 1 2 0 000−	$-2\sqrt{2.7}/11\sqrt{13}$
$\tilde{1}$ 1 2 0 010−	$-\sqrt{2.5}/11\sqrt{13}$
$\tilde{1}$ 1 $\tilde{1}$ 0 000−	$-2\sqrt{7}/11\sqrt{13}$
$\tilde{1}$ 1 $\tilde{1}$ 0 010−	$-\sqrt{5}/11\sqrt{13}$
$\tilde{1}$ 2 1 0 000−	$-\sqrt{5.7}/\sqrt{2.3.11.13}$
$\tilde{1}$ 2 $\tilde{1}$ 0 000+	$-\sqrt{5}/\sqrt{2.11.13}$
$\tilde{1}$ $\tilde{1}$ 0 0 000+	$+\sqrt{2.7}/3\sqrt{11.13}$
$\tilde{1}$ $\tilde{1}$ 1 0 000−	0
$\tilde{1}$ $\tilde{1}$ 2 0 000+	$-2\sqrt{2.5}/3\sqrt{11.13}$
$\tilde{1}$ $\tilde{1}$ $\tilde{1}$ 0 000+	$-2\sqrt{5}/\sqrt{3.11.13}$
5 5 5	
1 1 1 0 000+	$+9\sqrt{2}/11.11\sqrt{13}$
1 1 1 0 100+	$-3.7\sqrt{5.7/2}.11.11\sqrt{2.13}$
1 1 1 0 110+	$+8.3\sqrt{2}/11.11\sqrt{13}$
1 1 1 0 111+	$-2.7\sqrt{2.5.7}/11.11\sqrt{13}$
2 1 1 0 000−	0
2 1 1 0 010−	$+3/\sqrt{11.13}$
2 1 1 0 011−	0
2 2 2 0 000−	0
$\tilde{1}$ 1 1 0 000−	0

SO_3–O 3jm Factors (cont.)

5	**5**	**5**			
$\tilde{1}$	1	1	0	010−	$+\sqrt{3}/2\sqrt{2.11.13}$
$\tilde{1}$	1	1	0	011−	0
$\tilde{1}$	2	1	0	000+	$-3/11\sqrt{13}$
$\tilde{1}$	2	1	0	001+	$-\sqrt{5.7}/11\sqrt{13}$
$\tilde{1}$	$\tilde{1}$	1	0	000+	$-9\sqrt{2}/11\sqrt{13}$
$\tilde{1}$	$\tilde{1}$	1	0	001+	$-\sqrt{5.7}/2.11\sqrt{2.13}$
$\tilde{1}$	$\tilde{1}$	2	0	000−	0
$\tilde{1}$	$\tilde{1}$	$\tilde{1}$	0	000−	0
$\frac{11}{2}$	**3**	$\frac{5}{2}$			
$\frac{1}{2}$	1	$\frac{3}{2}$	0	000+	$+5/2\sqrt{3.7.11}$
$\frac{1}{2}$	$\tilde{1}$	$\frac{3}{2}$	0	000−	$-\sqrt{3.5}/2\sqrt{7.11}$
$\frac{1}{2}$	$\tilde{1}$	$\tilde{\frac{1}{2}}$	0	000+	$+\sqrt{3}/\sqrt{7.11}$
$\frac{1}{2}$	$\tilde{0}$	$\tilde{\frac{1}{2}}$	0	000−	$+2/\sqrt{7.11}$
$\frac{3}{2}$	1	$\frac{3}{2}$	0	000−	$-5/11\sqrt{2.3}$
$\frac{3}{2}$	1	$\frac{3}{2}$	0	100−	$+19\sqrt{5}/2.3.11\sqrt{3}$
$\frac{3}{2}$	1	$\frac{3}{2}$	1	000−	0
$\frac{3}{2}$	1	$\frac{3}{2}$	1	100−	$-\sqrt{5}/2.3\sqrt{3}$
$\frac{3}{2}$	1	$\tilde{\frac{1}{2}}$	0	000+	$-2\sqrt{5}/11\sqrt{3}$
$\frac{3}{2}$	1	$\tilde{\frac{1}{2}}$	0	100+	$-17/3.11\sqrt{2.3}$
$\frac{3}{2}$	$\tilde{1}$	$\frac{3}{2}$	0	000+	$+\sqrt{3.5}/11\sqrt{2}$
$\frac{3}{2}$	$\tilde{1}$	$\frac{3}{2}$	0	100+	$+\sqrt{3}/2.11$
$\frac{3}{2}$	$\tilde{1}$	$\frac{3}{2}$	1	000−	0
$\frac{3}{2}$	$\tilde{1}$	$\frac{3}{2}$	1	100−	$+1/2\sqrt{3}$
$\frac{3}{2}$	$\tilde{1}$	$\tilde{\frac{1}{2}}$	0	000−	$+2\sqrt{3}/11$
$\frac{3}{2}$	$\tilde{1}$	$\tilde{\frac{1}{2}}$	0	100−	$-\sqrt{5}/11\sqrt{2.3}$
$\frac{3}{2}$	$\tilde{0}$	$\frac{3}{2}$	0	000−	$+\sqrt{2.5}/11$
$\frac{3}{2}$	$\tilde{0}$	$\frac{3}{2}$	0	100−	$+1/11$
$\tilde{\frac{1}{2}}$	1	$\frac{3}{2}$	0	000+	$+\sqrt{5}/2.3\sqrt{3}$
$\tilde{\frac{1}{2}}$	1	$\tilde{\frac{1}{2}}$	0	000−	$+1/3\sqrt{3}$
$\tilde{\frac{1}{2}}$	$\tilde{1}$	$\frac{3}{2}$	0	000−	$-1/2\sqrt{3}$
$\frac{11}{2}$	$\frac{7}{2}$	**2**			
$\frac{1}{2}$	$\frac{3}{2}$	2	0	000−	$-1/\sqrt{2.11}$
$\frac{1}{2}$	$\frac{3}{2}$	$\tilde{1}$	0	000−	$+1/\sqrt{3.11}$
$\frac{1}{2}$	$\tilde{\frac{1}{2}}$	$\tilde{1}$	0	000+	$-1/\sqrt{11}$
$\frac{3}{2}$	$\frac{1}{2}$	2	0	000−	$-\sqrt{5}/11\sqrt{2}$
$\frac{3}{2}$	$\frac{1}{2}$	2	0	100−	$+19/2.3.11$
$\frac{3}{2}$	$\frac{1}{2}$	$\tilde{1}$	0	000−	$+\sqrt{5}/11\sqrt{3}$
$\frac{3}{2}$	$\frac{1}{2}$	$\tilde{1}$	0	100−	$+5.5/3.11\sqrt{2.3}$
$\frac{3}{2}$	$\frac{3}{2}$	2	0	000+	0
$\frac{3}{2}$	$\frac{3}{2}$	2	0	100+	$-\sqrt{7}/3\sqrt{2.5}$
$\frac{3}{2}$	$\frac{3}{2}$	$\tilde{1}$	0	000+	0
$\frac{3}{2}$	$\frac{3}{2}$	$\tilde{1}$	0	100+	$-\sqrt{7}/3\sqrt{3.5}$
$\frac{3}{2}$	$\frac{3}{2}$	$\tilde{1}$	1	000−	$+2\sqrt{2.7}/11\sqrt{3}$
$\frac{3}{2}$	$\frac{3}{2}$	$\tilde{1}$	1	100−	$-\sqrt{5.7}/3.11\sqrt{3}$
$\frac{3}{2}$	$\tilde{\frac{1}{2}}$	2	0	000−	$-\sqrt{3.7}/11\sqrt{2}$
$\frac{3}{2}$	$\tilde{\frac{1}{2}}$	2	0	100−	$-\sqrt{3.7}/2.11\sqrt{5}$
$\frac{3}{2}$	$\tilde{\frac{1}{2}}$	$\tilde{1}$	0	000−	$-\sqrt{7}/11$
$\frac{3}{2}$	$\tilde{\frac{1}{2}}$	$\tilde{1}$	0	100−	$-\sqrt{7}/11\sqrt{2.5}$
$\tilde{\frac{1}{2}}$	$\frac{1}{2}$	$\tilde{1}$	0	000+	$+1/3\sqrt{3}$

$\frac{11}{2}$	$\frac{7}{2}$	**2**			
$\tilde{\frac{1}{2}}$	$\frac{3}{2}$	2	0	000−	$-\sqrt{7}/3\sqrt{2.5}$
$\tilde{\frac{1}{2}}$	$\frac{3}{2}$	$\tilde{1}$	0	000−	$-\sqrt{7}/3\sqrt{3.5}$
$\frac{11}{2}$	$\frac{7}{2}$	**3**			
$\frac{1}{2}$	$\frac{1}{2}$	1	0	000+	$+5/\sqrt{3.11.13}$
$\frac{1}{2}$	$\frac{3}{2}$	1	0	000−	$-\sqrt{5.11}/2\sqrt{3.7.13}$
$\frac{1}{2}$	$\frac{3}{2}$	$\tilde{1}$	0	000+	$-19/2\sqrt{3.7.11.13}$
$\frac{1}{2}$	$\tilde{\frac{1}{2}}$	$\tilde{1}$	0	000−	$+5/\sqrt{7.11.13}$
$\frac{1}{2}$	$\tilde{\frac{1}{2}}$	$\tilde{0}$	0	000+	$-\sqrt{3}/\sqrt{7.11.13}$
$\frac{3}{2}$	$\frac{1}{2}$	1	0	000−	$-5\sqrt{7}/2.11\sqrt{3.13}$
$\frac{3}{2}$	$\frac{1}{2}$	1	0	100−	$-\sqrt{5.7}/2.11\sqrt{2.3.13}$
$\frac{3}{2}$	$\frac{1}{2}$	$\tilde{1}$	0	000+	$-7\sqrt{5.7}/2.11\sqrt{3.13}$
$\frac{3}{2}$	$\frac{1}{2}$	$\tilde{1}$	0	100+	$+23\sqrt{7}/2.3.11\sqrt{2.3.13}$
$\frac{3}{2}$	$\frac{3}{2}$	1	0	000+	$-2\sqrt{2.5}/11\sqrt{3.13}$
$\frac{3}{2}$	$\frac{3}{2}$	1	0	100+	$-\sqrt{13}/11\sqrt{3}$
$\frac{3}{2}$	$\frac{3}{2}$	1	1	000+	$-7\sqrt{2.5}/11\sqrt{3.13}$
$\frac{3}{2}$	$\frac{3}{2}$	1	1	100+	$-\sqrt{3}/2.11\sqrt{13}$
$\frac{3}{2}$	$\frac{3}{2}$	$\tilde{1}$	0	000−	$-2\sqrt{2.3}/11\sqrt{13}$
$\frac{3}{2}$	$\frac{3}{2}$	$\tilde{1}$	0	100−	$-19\sqrt{5}/3.11\sqrt{3.13}$
$\frac{3}{2}$	$\frac{3}{2}$	$\tilde{1}$	1	000+	$-7\sqrt{2}/11\sqrt{3.13}$
$\frac{3}{2}$	$\frac{3}{2}$	$\tilde{1}$	1	100+	$-59\sqrt{5}/2.3.11\sqrt{3.13}$
$\frac{3}{2}$	$\frac{3}{2}$	$\tilde{0}$	0	000+	$+4\sqrt{2}/11\sqrt{13}$
$\frac{3}{2}$	$\frac{3}{2}$	$\tilde{0}$	0	100+	$-\sqrt{5.13}/3.11$
$\frac{3}{2}$	$\tilde{\frac{1}{2}}$	1	0	000−	$-9\sqrt{5}/2.11\sqrt{13}$
$\frac{3}{2}$	$\tilde{\frac{1}{2}}$	1	0	100−	$+\sqrt{13}/2.11\sqrt{2}$
$\frac{3}{2}$	$\tilde{\frac{1}{2}}$	$\tilde{1}$	0	000+	$+1/2.11\sqrt{13}$
$\frac{3}{2}$	$\tilde{\frac{1}{2}}$	$\tilde{1}$	0	100+	$-\sqrt{5.13}/2.11\sqrt{2}$
$\tilde{\frac{1}{2}}$	$\frac{1}{2}$	$\tilde{1}$	0	000−	$+\sqrt{7}/3\sqrt{3.13}$
$\tilde{\frac{1}{2}}$	$\frac{1}{2}$	$\tilde{0}$	0	000+	$-\sqrt{7}/3\sqrt{13}$
$\tilde{\frac{1}{2}}$	$\frac{3}{2}$	1	0	000−	$+1/2\sqrt{3.13}$
$\tilde{\frac{1}{2}}$	$\frac{3}{2}$	$\tilde{1}$	0	000+	$-\sqrt{5}/2.3\sqrt{3.13}$
$\tilde{\frac{1}{2}}$	$\tilde{\frac{1}{2}}$	1	0	000+	$+1/\sqrt{13}$
$\frac{11}{2}$	**4**	$\frac{3}{2}$			
$\frac{1}{2}$	1	$\frac{3}{2}$	0	000+	0
$\frac{1}{2}$	2	$\frac{3}{2}$	0	000−	$+1/\sqrt{2.11}$
$\frac{1}{2}$	$\tilde{1}$	$\frac{3}{2}$	0	000−	$-2/\sqrt{3.11}$
$\frac{3}{2}$	0	$\frac{3}{2}$	0	000+	0
$\frac{3}{2}$	0	$\frac{3}{2}$	0	100+	$+1/3$
$\frac{3}{2}$	1	$\frac{3}{2}$	0	000−	$-2\sqrt{2}/11$
$\frac{3}{2}$	1	$\frac{3}{2}$	0	100−	$+8.2/3.11\sqrt{5}$
$\frac{3}{2}$	1	$\frac{3}{2}$	1	000−	$+\sqrt{2}/11$
$\frac{3}{2}$	1	$\frac{3}{2}$	1	100−	$+5\sqrt{5}/3.11$
$\frac{3}{2}$	2	$\frac{3}{2}$	0	000+	$+\sqrt{7}/11$
$\frac{3}{2}$	2	$\frac{3}{2}$	0	100+	$-4\sqrt{2.7}/3.11\sqrt{5}$
$\frac{3}{2}$	$\tilde{1}$	$\frac{3}{2}$	0	000+	$-2\sqrt{2.7}/11\sqrt{3}$
$\frac{3}{2}$	$\tilde{1}$	$\frac{3}{2}$	0	100+	$-2\sqrt{7}/11\sqrt{3.5}$
$\frac{3}{2}$	$\tilde{1}$	$\frac{3}{2}$	1	000−	$+\sqrt{2.7}/11\sqrt{3}$
$\frac{3}{2}$	$\tilde{1}$	$\frac{3}{2}$	1	100−	$+\sqrt{7}/11\sqrt{3.5}$
$\tilde{\frac{1}{2}}$	1	$\frac{3}{2}$	0	000+	$-2/3\sqrt{5}$
$\tilde{\frac{1}{2}}$	2	$\frac{3}{2}$	0	000−	$-\sqrt{7}/3\sqrt{2.5}$

SO_3–O 3jm Factors (cont.)

$\frac{11}{2}\ 4\ \frac{3}{2}$

$\tilde{\frac{1}{2}}\ \tilde{1}\ \frac{3}{2}$ 0 000− 0

$\frac{11}{2}\ 4\ \frac{5}{2}$

$\frac{1}{2}\ 1\ \frac{3}{2}$ 0 000− $-5\sqrt{7}/2\sqrt{3.11.13}$

$\frac{1}{2}\ 2\ \frac{3}{2}$ 0 000+ $+\sqrt{2}/\sqrt{3.11.13}$

$\frac{1}{2}\ \tilde{1}\ \frac{3}{2}$ 0 000+ $+17/2.3\sqrt{11.13}$

$\frac{1}{2}\ \tilde{1}\ \tilde{\frac{1}{2}}$ 0 000− $-\sqrt{5}/3\sqrt{11.13}$

$\frac{3}{2}\ 0\ \frac{3}{2}$ 0 000− $-5\sqrt{2.5}/11\sqrt{3.13}$

$\frac{3}{2}\ 0\ \frac{3}{2}$ 0 100− $+7/3.11\sqrt{3.13}$

$\frac{3}{2}\ 1\ \frac{3}{2}$ 0 000+ $+7/11\sqrt{2.3.13}$

$\frac{3}{2}\ 1\ \frac{3}{2}$ 0 100+ $+73/2.11\sqrt{3.5.13}$

$\frac{3}{2}\ 1\ \frac{3}{2}$ 1 000+ $-8\sqrt{2}/11\sqrt{3.13}$

$\frac{3}{2}\ 1\ \frac{3}{2}$ 1 100+ $-\sqrt{5}/2.11\sqrt{3.13}$

$\frac{3}{2}\ 1\ \tilde{\frac{1}{2}}$ 0 000− $+2\sqrt{3.5}/11\sqrt{13}$

$\frac{3}{2}\ 1\ \tilde{\frac{1}{2}}$ 0 100− $-9\sqrt{3}/11\sqrt{2.13}$

$\frac{3}{2}\ 2\ \frac{3}{2}$ 0 000− $-8\sqrt{7}/11\sqrt{3.13}$

$\frac{3}{2}\ 2\ \frac{3}{2}$ 0 100− $-23\sqrt{2.7}/3.11\sqrt{3.5.13}$

$\frac{3}{2}\ 2\ \tilde{\frac{1}{2}}$ 0 000+ 0

$\frac{3}{2}\ 2\ \tilde{\frac{1}{2}}$ 0 100+ $-2\sqrt{7}/3\sqrt{3.13}$

$\frac{3}{2}\ \tilde{1}\ \frac{3}{2}$ 0 000− $-\sqrt{7.13}/3.11\sqrt{2}$

$\frac{3}{2}\ \tilde{1}\ \frac{3}{2}$ 0 100− $-\sqrt{7.13}/2.3.11\sqrt{5}$

$\frac{3}{2}\ \tilde{1}\ \frac{3}{2}$ 1 000+ $-8\sqrt{2.7}/3.11\sqrt{13}$

$\frac{3}{2}\ \tilde{1}\ \frac{3}{2}$ 1 100+ $+\sqrt{7.13}/2.11\sqrt{5}$

$\frac{3}{2}\ \tilde{1}\ \tilde{\frac{1}{2}}$ 0 000+ $-2\sqrt{5.7}/3.11\sqrt{13}$

$\frac{3}{2}\ \tilde{1}\ \tilde{\frac{1}{2}}$ 0 100+ $-\sqrt{7.13}/3.11\sqrt{2}$

$\tilde{\frac{1}{2}}\ 0\ \tilde{\frac{1}{2}}$ 0 000− $-2\sqrt{5}/3\sqrt{3.13}$

$\tilde{\frac{1}{2}}\ 1\ \frac{3}{2}$ 0 000− $-1/2\sqrt{3.5.13}$

$\tilde{\frac{1}{2}}\ 1\ \tilde{\frac{1}{2}}$ 0 000+ $-1/\sqrt{3.13}$

$\tilde{\frac{1}{2}}\ 2\ \frac{3}{2}$ 0 000+ $-\sqrt{2.7}/3\sqrt{3.5.13}$

$\tilde{\frac{1}{2}}\ \tilde{1}\ \frac{3}{2}$ 0 000+ $-\sqrt{5.7}/2.3\sqrt{13}$

$\frac{11}{2}\ 4\ \frac{7}{2}$

$\frac{1}{2}\ 0\ \frac{1}{2}$ 0 000+ $+5/\sqrt{3.11.13}$

$\frac{1}{2}\ 1\ \frac{1}{2}$ 0 000− $-\sqrt{5}/\sqrt{3.11.13}$

$\frac{1}{2}\ 1\ \frac{3}{2}$ 0 000+ $+\sqrt{7}/2\sqrt{3.11.13}$

$\frac{1}{2}\ 2\ \frac{3}{2}$ 0 000− $-4\sqrt{2}/\sqrt{3.11.13}$

$\frac{1}{2}\ \tilde{1}\ \frac{3}{2}$ 0 000− $+3/2\sqrt{11.13}$

$\frac{1}{2}\ \tilde{1}\ \tilde{\frac{1}{2}}$ 0 000+ $+1/\sqrt{3.11.13}$

$\frac{3}{2}\ 0\ \frac{3}{2}$ 0 000+ $-4\sqrt{2.5}/11\sqrt{3.13}$

$\frac{3}{2}\ 0\ \frac{3}{2}$ 0 100+ $+7/11\sqrt{3.13}$

$\frac{3}{2}\ 1\ \frac{1}{2}$ 0 000+ $-7\sqrt{5.7}/2.11\sqrt{3.13}$

$\frac{3}{2}\ 1\ \frac{1}{2}$ 0 100+ $-7\sqrt{7}/2.11\sqrt{2.3.13}$

$\frac{3}{2}\ 1\ \frac{3}{2}$ 0 000− $+2.7\sqrt{2}/11\sqrt{3.13}$

$\frac{3}{2}\ 1\ \frac{3}{2}$ 0 100− $+5\sqrt{5}/11\sqrt{3.13}$

$\frac{3}{2}\ 1\ \frac{3}{2}$ 1 000− $-5\sqrt{2}/11\sqrt{3.13}$

$\frac{3}{2}\ 1\ \frac{3}{2}$ 1 100− $+3\sqrt{3.5}/2.11\sqrt{13}$

$\frac{3}{2}\ 1\ \tilde{\frac{1}{2}}$ 0 000+ $+1/2.11\sqrt{13}$

$\frac{3}{2}\ 1\ \tilde{\frac{1}{2}}$ 0 100+ $+5\sqrt{5}/2.3.11\sqrt{2.13}$

$\frac{3}{2}\ 2\ \frac{1}{2}$ 0 000− $-2\sqrt{2.5}/11\sqrt{3.13}$

$\frac{3}{2}\ 2\ \frac{1}{2}$ 0 100− $-\sqrt{13}/11\sqrt{3}$

$\frac{3}{2}\ 2\ \frac{3}{2}$ 0 000+ $+2\sqrt{7}/11\sqrt{3.13}$

$\frac{11}{2}\ 4\ \frac{7}{2}$

$\frac{3}{2}\ 2\ \frac{3}{2}$ 0 100+ $-2\sqrt{2.5.7}/11\sqrt{3.13}$

$\frac{3}{2}\ 2\ \tilde{\frac{1}{2}}$ 0 000− $-2\sqrt{2.7}/11\sqrt{13}$

$\frac{3}{2}\ 2\ \tilde{\frac{1}{2}}$ 0 100− $+\sqrt{5.7}/3.11\sqrt{13}$

$\frac{3}{2}\ \tilde{1}\ \frac{1}{2}$ 0 000− $+5\sqrt{5}/2.11\sqrt{13}$

$\frac{3}{2}\ \tilde{1}\ \frac{1}{2}$ 0 100− $-29/2.3.11\sqrt{2.13}$

$\frac{3}{2}\ \tilde{1}\ \frac{3}{2}$ 0 000+ $+2\sqrt{2.7}/11\sqrt{13}$

$\frac{3}{2}\ \tilde{1}\ \frac{3}{2}$ 0 100+ $-\sqrt{5.7}/3.11\sqrt{13}$

$\frac{3}{2}\ \tilde{1}\ \frac{3}{2}$ 1 000− $+\sqrt{2.7}/11\sqrt{13}$

$\frac{3}{2}\ \tilde{1}\ \frac{3}{2}$ 1 100− $-\sqrt{5.7}/2.3.11\sqrt{13}$

$\frac{3}{2}\ \tilde{1}\ \tilde{\frac{1}{2}}$ 0 000− $+\sqrt{7}/2.11\sqrt{3.13}$

$\frac{3}{2}\ \tilde{1}\ \tilde{\frac{1}{2}}$ 0 100− $-\sqrt{5.7.13}/2.11\sqrt{2.3}$

$\tilde{\frac{1}{2}}\ 0\ \tilde{\frac{1}{2}}$ 0 000+ $+1/3\sqrt{13}$

$\tilde{\frac{1}{2}}\ 1\ \frac{3}{2}$ 0 000+ $-\sqrt{5}/2\sqrt{3.13}$

$\tilde{\frac{1}{2}}\ 1\ \tilde{\frac{1}{2}}$ 0 000− $+\sqrt{5}/3\sqrt{13}$

$\tilde{\frac{1}{2}}\ 2\ \frac{3}{2}$ 0 000− 0

$\tilde{\frac{1}{2}}\ \tilde{1}\ \frac{1}{2}$ 0 000+ $+1/3\sqrt{13}$

$\tilde{\frac{1}{2}}\ \tilde{1}\ \frac{3}{2}$ 0 000− $+\sqrt{5.7}/2.3\sqrt{13}$

$\frac{11}{2}\ \frac{9}{2}\ 1$

$\frac{1}{2}\ \frac{1}{2}\ 1$ 0 000− 0

$\frac{1}{2}\ \frac{3}{2}\ 1$ 0 000+ $-\sqrt{2.7}/\sqrt{3.5.11}$

$\frac{1}{2}\ \frac{3}{2}\ 1$ 0 010+ $+3/\sqrt{2.5.11}$

$\frac{3}{2}\ \frac{1}{2}\ 1$ 0 000+ 0

$\frac{3}{2}\ \frac{1}{2}\ 1$ 0 100+ $+1/\sqrt{5}$

$\frac{3}{2}\ \frac{3}{2}\ 1$ 0 000− $+4\sqrt{5}/11\sqrt{3}$

$\frac{3}{2}\ \frac{3}{2}\ 1$ 0 010− 0

$\frac{3}{2}\ \frac{3}{2}\ 1$ 0 100− $+2\sqrt{2}/11\sqrt{3}$

$\frac{3}{2}\ \frac{3}{2}\ 1$ 0 110− 0

$\frac{3}{2}\ \frac{3}{2}\ 1$ 1 000− $+4/11\sqrt{3.5}$

$\frac{3}{2}\ \frac{3}{2}\ 1$ 1 010− $-3\sqrt{7}/11\sqrt{5}$

$\frac{3}{2}\ \frac{3}{2}\ 1$ 1 100− $+13\sqrt{2}/5.11\sqrt{3}$

$\frac{3}{2}\ \frac{3}{2}\ 1$ 1 110− $+4\sqrt{2.7}/5.11$

$\tilde{\frac{1}{2}}\ \frac{3}{2}\ 1$ 0 000+ $-\sqrt{2}/5\sqrt{3}$

$\tilde{\frac{1}{2}}\ \frac{3}{2}\ 1$ 0 010+ $-\sqrt{7}/5\sqrt{2}$

$\frac{11}{2}\ \frac{9}{2}\ 2$

$\frac{1}{2}\ \frac{3}{2}\ 2$ 0 000+ $+2\sqrt{3.7}/\sqrt{5.11.13}$

$\frac{1}{2}\ \frac{3}{2}\ 2$ 0 010+ $+1/\sqrt{5.11.13}$

$\frac{1}{2}\ \frac{3}{2}\ \tilde{1}$ 0 000+ $-\sqrt{2.7}/\sqrt{5.11.13}$

$\frac{1}{2}\ \frac{3}{2}\ \tilde{1}$ 0 010+ $-\sqrt{11}/\sqrt{2.3.5.13}$

$\frac{3}{2}\ \frac{1}{2}\ 2$ 0 000+ $+8/11\sqrt{13}$

$\frac{3}{2}\ \frac{1}{2}\ 2$ 0 100+ $+23\sqrt{2}/3.11\sqrt{5.13}$

$\frac{3}{2}\ \frac{1}{2}\ \tilde{1}$ 0 000+ $-8\sqrt{2}/11\sqrt{3.13}$

$\frac{3}{2}\ \frac{1}{2}\ \tilde{1}$ 0 100+ $+97/3.11\sqrt{3.5.13}$

$\frac{3}{2}\ \frac{3}{2}\ 2$ 0 000− $+4\sqrt{2.3}/11\sqrt{5.13}$

$\frac{3}{2}\ \frac{3}{2}\ 2$ 0 010− $-9\sqrt{2.7}/11\sqrt{5.13}$

$\frac{3}{2}\ \frac{3}{2}\ 2$ 0 100− $-2.9\sqrt{3}/5.11\sqrt{13}$

$\frac{3}{2}\ \frac{3}{2}\ 2$ 0 110− $-8.2\sqrt{7}/3.5.11\sqrt{13}$

$\frac{3}{2}\ \frac{3}{2}\ \tilde{1}$ 0 000− $+4\sqrt{5}/11\sqrt{13}$

$\frac{3}{2}\ \frac{3}{2}\ \tilde{1}$ 0 010− 0

$\frac{3}{2}\ \frac{3}{2}\ \tilde{1}$ 0 100− $-2.17\sqrt{2}/5.11\sqrt{13}$

SO_3–O 3jm Factors (cont.)

j_1	j_2	j_3	r		
$\frac{11}{2}$	$\frac{9}{2}$	**2**			
$\frac{3}{2}$	$\frac{3}{2}$	$\tilde{1}$	0	110−	$-8\sqrt{2.7}/3.5\sqrt{3.13}$
$\frac{3}{2}$	$\frac{3}{2}$	$\tilde{1}$	1	000+	$-4\sqrt{5}/11\sqrt{13}$
$\frac{3}{2}$	$\frac{3}{2}$	$\tilde{1}$	1	010+	$+\sqrt{5.7}/11\sqrt{3.13}$
$\frac{3}{2}$	$\frac{3}{2}$	$\tilde{1}$	1	100+	$-3.7\sqrt{2}/5.11\sqrt{13}$
$\frac{3}{2}$	$\frac{3}{2}$	$\tilde{1}$	1	110+	$+4.17\sqrt{2.7}/3.5.11\sqrt{3.13}$
$\widetilde{\frac{1}{2}}$	$\frac{1}{2}$	$\tilde{1}$	0	000−	$-8\sqrt{2}/3\sqrt{3.5.13}$
$\widetilde{\frac{1}{2}}$	$\frac{3}{2}$	2	0	000+	$-2\sqrt{3}/5\sqrt{13}$
$\widetilde{\frac{1}{2}}$	$\frac{3}{2}$	2	0	010+	$+\sqrt{7}/3.5\sqrt{13}$
$\widetilde{\frac{1}{2}}$	$\frac{3}{2}$	$\tilde{1}$	0	000+	$-\sqrt{2}/5\sqrt{13}$
$\widetilde{\frac{1}{2}}$	$\frac{3}{2}$	$\tilde{1}$	0	010+	$+11\sqrt{7}/3.5\sqrt{2.3.13}$

j_1	j_2	j_3	r		
$\frac{11}{2}$	$\frac{9}{2}$	**3**			
$\frac{1}{2}$	$\frac{1}{2}$	1	0	000−	$-5/\sqrt{3.11.13}$
$\frac{1}{2}$	$\frac{3}{2}$	1	0	000+	$+9/2\sqrt{5.11.13}$
$\frac{1}{2}$	$\frac{3}{2}$	1	0	010+	$-2\sqrt{7}/\sqrt{3.5.11.13}$
$\frac{1}{2}$	$\frac{3}{2}$	$\tilde{1}$	0	000−	$-1/2\sqrt{11.13}$
$\frac{1}{2}$	$\frac{3}{2}$	$\tilde{1}$	0	010−	$-2\sqrt{7}/\sqrt{3.11.13}$
$\frac{3}{2}$	$\frac{1}{2}$	1	0	000+	$-2.5.5/11\sqrt{3.7.13}$
$\frac{3}{2}$	$\frac{1}{2}$	1	0	100+	$+7\sqrt{7}/11\sqrt{2.3.5.13}$
$\frac{3}{2}$	$\frac{1}{2}$	$\tilde{1}$	0	000−	$+2\sqrt{5}/11\sqrt{3.7.13}$
$\frac{3}{2}$	$\frac{1}{2}$	$\tilde{1}$	0	100−	$-7\sqrt{7}/3.11\sqrt{2.3.13}$
$\frac{3}{2}$	$\frac{3}{2}$	1	0	000−	$-9\sqrt{5}/11\sqrt{2.7.13}$
$\frac{3}{2}$	$\frac{3}{2}$	1	0	010−	$-2\sqrt{2.5}/11\sqrt{3.13}$
$\frac{3}{2}$	$\frac{3}{2}$	1	0	100−	$+\sqrt{13}/2.11\sqrt{7}$
$\frac{3}{2}$	$\frac{3}{2}$	1	0	110−	$-\sqrt{13}/11\sqrt{3}$
$\frac{3}{2}$	$\frac{3}{2}$	1	1	000−	$+4.9\sqrt{2}/11\sqrt{5.7.13}$
$\frac{3}{2}$	$\frac{3}{2}$	1	1	010−	$+19\sqrt{2}/11\sqrt{3.5.13}$
$\frac{3}{2}$	$\frac{3}{2}$	1	1	100−	$+83/2.5.11\sqrt{7.13}$
$\frac{3}{2}$	$\frac{3}{2}$	1	1	110−	$+8.4\sqrt{3}/5.11\sqrt{13}$
$\frac{3}{2}$	$\frac{3}{2}$	$\tilde{1}$	0	000+	$+37/11\sqrt{2.7.13}$
$\frac{3}{2}$	$\frac{3}{2}$	$\tilde{1}$	0	010+	$-2\sqrt{2.3}/11\sqrt{13}$
$\frac{3}{2}$	$\frac{3}{2}$	$\tilde{1}$	0	100+	$-\sqrt{7}/2.11\sqrt{5.13}$
$\frac{3}{2}$	$\frac{3}{2}$	$\tilde{1}$	0	110+	$+103/3.11\sqrt{3.5.13}$
$\frac{3}{2}$	$\frac{3}{2}$	$\tilde{1}$	1	000−	$+4\sqrt{2}/11\sqrt{7.13}$
$\frac{3}{2}$	$\frac{3}{2}$	$\tilde{1}$	1	010−	$+\sqrt{2}/\sqrt{3.13}$
$\frac{3}{2}$	$\frac{3}{2}$	$\tilde{1}$	1	100−	$-3.23/2.11\sqrt{5.7.13}$
$\frac{3}{2}$	$\frac{3}{2}$	$\tilde{1}$	1	110−	$-4/3\sqrt{3.5.13}$
$\frac{3}{2}$	$\frac{3}{2}$	$\tilde{0}$	0	000−	$+\sqrt{2.3}/11\sqrt{7.13}$
$\frac{3}{2}$	$\frac{3}{2}$	$\tilde{0}$	0	010−	$+4\sqrt{2}/11\sqrt{13}$
$\frac{3}{2}$	$\frac{3}{2}$	$\tilde{0}$	0	100−	$-43\sqrt{3}/11\sqrt{5.7.13}$
$\frac{3}{2}$	$\frac{3}{2}$	$\tilde{0}$	0	110−	$+2.17/3.11\sqrt{5.13}$
$\widetilde{\frac{1}{2}}$	$\frac{1}{2}$	$\tilde{1}$	0	000+	$+11/3\sqrt{3.7.13}$
$\widetilde{\frac{1}{2}}$	$\frac{1}{2}$	$\tilde{0}$	0	000−	$-2/3\sqrt{7.13}$
$\widetilde{\frac{1}{2}}$	$\frac{3}{2}$	1	0	000+	$-11/2.5\sqrt{7.13}$
$\widetilde{\frac{1}{2}}$	$\frac{3}{2}$	1	0	010+	$-2/5\sqrt{3.13}$
$\widetilde{\frac{1}{2}}$	$\frac{3}{2}$	$\tilde{1}$	0	000−	$-\sqrt{13}/2\sqrt{5.7}$
$\widetilde{\frac{1}{2}}$	$\frac{3}{2}$	$\tilde{1}$	0	010−	$+2/3\sqrt{3.5.13}$

j_1	j_2	j_3	r		
$\frac{11}{2}$	$\frac{9}{2}$	**4**			
$\frac{1}{2}$	$\frac{1}{2}$	0	0	000−	$+2/\sqrt{3.11.13}$
$\frac{1}{2}$	$\frac{1}{2}$	1	0	000+	$+\sqrt{11}/\sqrt{3.5.13}$
$\frac{1}{2}$	$\frac{3}{2}$	1	0	000−	$+1/2\sqrt{11.13}$
$\frac{1}{2}$	$\frac{3}{2}$	1	0	010−	$-2\sqrt{7}/\sqrt{3.11.13}$
$\frac{1}{2}$	$\frac{3}{2}$	2	0	000+	$-\sqrt{2.7}/5\sqrt{11.13}$
$\frac{1}{2}$	$\frac{3}{2}$	2	0	010+	$-8\sqrt{2}/5\sqrt{3.11.13}$
$\frac{1}{2}$	$\frac{3}{2}$	$\tilde{1}$	0	000+	$+7\sqrt{7}/2.5\sqrt{3.11.13}$
$\frac{1}{2}$	$\frac{3}{2}$	$\tilde{1}$	0	010+	$+2.3/5\sqrt{11.13}$
$\frac{3}{2}$	$\frac{1}{2}$	1	0	000−	$+2\sqrt{7}/11\sqrt{3.5.13}$
$\frac{3}{2}$	$\frac{1}{2}$	1	0	100−	$+7\sqrt{7}/11\sqrt{2.3.13}$
$\frac{3}{2}$	$\frac{1}{2}$	2	0	000+	$-8.4\sqrt{2}/11\sqrt{3.5.13}$
$\frac{3}{2}$	$\frac{1}{2}$	2	0	100+	$-2/11\sqrt{3.13}$
$\frac{3}{2}$	$\frac{1}{2}$	$\tilde{1}$	0	000+	$-2/11\sqrt{5.13}$
$\frac{3}{2}$	$\frac{1}{2}$	$\tilde{1}$	0	100+	$+1/3.11\sqrt{2.13}$
$\frac{3}{2}$	$\frac{3}{2}$	0	0	000−	$-\sqrt{2.7}/11\sqrt{5.13}$
$\frac{3}{2}$	$\frac{3}{2}$	0	0	010−	$-4.7\sqrt{2}/11\sqrt{3.5.13}$
$\frac{3}{2}$	$\frac{3}{2}$	0	0	100−	$-5\sqrt{7}/3.11\sqrt{13}$
$\frac{3}{2}$	$\frac{3}{2}$	0	0	110−	$-2.5/11\sqrt{3.13}$
$\frac{3}{2}$	$\frac{3}{2}$	1	0	000+	$-19\sqrt{7}/5.11\sqrt{2.13}$
$\frac{3}{2}$	$\frac{3}{2}$	1	0	010+	$-2\sqrt{2.13}/5.11\sqrt{3}$
$\frac{3}{2}$	$\frac{3}{2}$	1	0	100+	$-7\sqrt{7}/2.3.11\sqrt{5.13}$
$\frac{3}{2}$	$\frac{3}{2}$	1	0	110+	$+19/11\sqrt{3.5.13}$
$\frac{3}{2}$	$\frac{3}{2}$	1	1	000+	$-4.3\sqrt{2.7}/5.11\sqrt{13}$
$\frac{3}{2}$	$\frac{3}{2}$	1	1	010+	$-\sqrt{2}/5.11\sqrt{3.13}$
$\frac{3}{2}$	$\frac{3}{2}$	1	1	100+	$+23\sqrt{7}/2.3.11\sqrt{5.13}$
$\frac{3}{2}$	$\frac{3}{2}$	1	1	110+	$+8\sqrt{3}/11\sqrt{5.13}$
$\frac{3}{2}$	$\frac{3}{2}$	2	0	000−	$+8.3/5.11\sqrt{13}$
$\frac{3}{2}$	$\frac{3}{2}$	2	0	010−	$-4\sqrt{7}/5.11\sqrt{3.13}$
$\frac{3}{2}$	$\frac{3}{2}$	2	0	100−	$-17\sqrt{2}/3.11\sqrt{5.13}$
$\frac{3}{2}$	$\frac{3}{2}$	2	0	110−	$+4\sqrt{2.7}/11\sqrt{3.5.13}$
$\frac{3}{2}$	$\frac{3}{2}$	$\tilde{1}$	0	000−	$-1/\sqrt{2.3.13}$
$\frac{3}{2}$	$\frac{3}{2}$	$\tilde{1}$	0	010−	$-2\sqrt{2.7}/11\sqrt{13}$
$\frac{3}{2}$	$\frac{3}{2}$	$\tilde{1}$	0	100−	$-\sqrt{5}/2\sqrt{3.13}$
$\frac{3}{2}$	$\frac{3}{2}$	$\tilde{1}$	0	110−	$+\sqrt{5.7}/3.11\sqrt{13}$
$\frac{3}{2}$	$\frac{3}{2}$	$\tilde{1}$	1	000+	$-4\sqrt{2}/5.11\sqrt{3.13}$
$\frac{3}{2}$	$\frac{3}{2}$	$\tilde{1}$	1	010+	$-7\sqrt{2.7}/5.11\sqrt{13}$
$\frac{3}{2}$	$\frac{3}{2}$	$\tilde{1}$	1	100+	$-7\sqrt{3.5}/2.11\sqrt{13}$
$\frac{3}{2}$	$\frac{3}{2}$	$\tilde{1}$	1	110+	$+4\sqrt{5.7}/3.11\sqrt{13}$
$\widetilde{\frac{1}{2}}$	$\frac{1}{2}$	$\tilde{1}$	0	000−	$+1/3\sqrt{13}$
$\widetilde{\frac{1}{2}}$	$\frac{3}{2}$	1	0	000−	$+\sqrt{7}/2.3\sqrt{5.13}$
$\widetilde{\frac{1}{2}}$	$\frac{3}{2}$	1	0	010−	$-2/\sqrt{3.5.13}$
$\widetilde{\frac{1}{2}}$	$\frac{3}{2}$	2	0	000+	$-\sqrt{2.5}/3\sqrt{13}$
$\widetilde{\frac{1}{2}}$	$\frac{3}{2}$	2	0	010+	0
$\widetilde{\frac{1}{2}}$	$\frac{3}{2}$	$\tilde{1}$	0	000+	$-1/2\sqrt{3.5.13}$
$\widetilde{\frac{1}{2}}$	$\frac{3}{2}$	$\tilde{1}$	0	010+	$-2\sqrt{7}/3\sqrt{5.13}$

j_1	j_2	j_3	r		
$\frac{11}{2}$	**5**	$\frac{1}{2}$			
$\frac{1}{2}$	1	$\frac{1}{2}$	0	000−	$+1/\sqrt{2.3}$
$\frac{1}{2}$	1	$\frac{1}{2}$	0	010−	0
$\frac{3}{2}$	1	$\frac{1}{2}$	0	000+	$-\sqrt{7}/\sqrt{2.3.11}$
$\frac{3}{2}$	1	$\frac{1}{2}$	0	010+	0
$\frac{3}{2}$	1	$\frac{1}{2}$	0	100+	0

SO_3–O 3jm Factors (cont.)

$\frac{11}{2}$ 5 $\frac{1}{2}$
$\frac{3}{2}$ 1 $\frac{1}{2}$ 0 110+ $+\sqrt{3}/\sqrt{11}$
$\frac{3}{2}$ 2 $\frac{1}{2}$ 0 000− $+2\sqrt{5/11}$
$\frac{3}{2}$ 2 $\frac{1}{2}$ 0 100− $+\sqrt{2/11}$
$\frac{3}{2}$ $\tilde{1}$ $\frac{1}{2}$ 0 000− $-\sqrt{3.5}/11\sqrt{2}$
$\frac{3}{2}$ $\tilde{1}$ $\frac{1}{2}$ 0 100− $+4/11\sqrt{3}$
$\tilde{\frac{1}{2}}$ $\tilde{1}$ $\frac{1}{2}$ 0 000+ $+1/\sqrt{2.3}$

$\frac{11}{2}$ 5 $\frac{3}{2}$
$\frac{1}{2}$ 1 $\frac{3}{2}$ 0 000− $-\sqrt{5}/2\sqrt{3.13}$
$\frac{1}{2}$ 1 $\frac{3}{2}$ 0 010− 0
$\frac{1}{2}$ 2 $\frac{3}{2}$ 0 000+ $-\sqrt{2.7}/\sqrt{11.13}$
$\frac{1}{2}$ $\tilde{1}$ $\frac{3}{2}$ 0 000+ $+\sqrt{3.7}/2\sqrt{11.13}$
$\frac{3}{2}$ 1 $\frac{3}{2}$ 0 000+ $-\sqrt{2.5.7}/11\sqrt{3.11.13}$
$\frac{3}{2}$ 1 $\frac{3}{2}$ 0 010+ $+4.3\sqrt{2.3}/11\sqrt{11.13}$
$\frac{3}{2}$ 1 $\frac{3}{2}$ 0 100+ $-4\sqrt{3.7}/11\sqrt{11.13}$
$\frac{3}{2}$ 1 $\frac{3}{2}$ 0 110+ $+23\sqrt{3}/11\sqrt{5.11.13}$
$\frac{3}{2}$ 1 $\frac{3}{2}$ 1 000+ $+23\sqrt{5.7}/11\sqrt{2.3.11.13}$
$\frac{3}{2}$ 1 $\frac{3}{2}$ 1 010+ $-2.3\sqrt{2.3}/11\sqrt{11.13}$
$\frac{3}{2}$ 1 $\frac{3}{2}$ 1 100+ $+2\sqrt{3.7}/11\sqrt{11.13}$
$\frac{3}{2}$ 1 $\frac{3}{2}$ 1 110+ $+4.3\sqrt{3.5}/11\sqrt{11.13}$
$\frac{3}{2}$ 2 $\frac{3}{2}$ 0 000− $-4/11\sqrt{13}$
$\frac{3}{2}$ 2 $\frac{3}{2}$ 0 100− $-\sqrt{2.13}/11\sqrt{5}$
$\frac{3}{2}$ $\tilde{1}$ $\frac{3}{2}$ 0 000− $+\sqrt{2.3}/11\sqrt{13}$
$\frac{3}{2}$ $\tilde{1}$ $\frac{3}{2}$ 0 100− $-4\sqrt{13}/11\sqrt{3.5}$
$\frac{3}{2}$ $\tilde{1}$ $\frac{3}{2}$ 1 000+ $+9\sqrt{3}/11\sqrt{2.13}$
$\frac{3}{2}$ $\tilde{1}$ $\frac{3}{2}$ 1 100+ $-2.7/11\sqrt{3.5.13}$
$\tilde{\frac{1}{2}}$ 1 $\frac{3}{2}$ 0 000− $+\sqrt{3.7}/2\sqrt{11.13}$
$\tilde{\frac{1}{2}}$ 1 $\frac{3}{2}$ 0 010− $+4\sqrt{3}/\sqrt{5.11.13}$
$\tilde{\frac{1}{2}}$ 2 $\frac{3}{2}$ 0 000+ $-\sqrt{2}/\sqrt{5.13}$
$\tilde{\frac{1}{2}}$ $\tilde{1}$ $\frac{3}{2}$ 0 000+ $-\sqrt{5}/2\sqrt{3.13}$

$\frac{11}{2}$ 5 $\frac{5}{2}$
$\frac{1}{2}$ 1 $\frac{3}{2}$ 0 000+ $-\sqrt{3.5.7}/11\sqrt{2.13}$
$\frac{1}{2}$ 1 $\frac{3}{2}$ 0 010+ $+2.5\sqrt{2}/11\sqrt{3.13}$
$\frac{1}{2}$ 2 $\frac{3}{2}$ 0 000− $-2/\sqrt{11.13}$
$\frac{1}{2}$ $\tilde{1}$ $\frac{3}{2}$ 0 000− $+\sqrt{3}/\sqrt{2.11.13}$
$\frac{1}{2}$ $\tilde{1}$ $\tilde{\frac{1}{2}}$ 0 000+ $-\sqrt{3.5}/\sqrt{2.11.13}$
$\frac{3}{2}$ 1 $\frac{3}{2}$ 0 000− $+\sqrt{3.5}/11\sqrt{11.13}$
$\frac{3}{2}$ 1 $\frac{3}{2}$ 0 010− $+8.2\sqrt{7}/11\sqrt{3.11.13}$
$\frac{3}{2}$ 1 $\frac{3}{2}$ 0 100− $+2.3\sqrt{2.3}/11\sqrt{11.13}$
$\frac{3}{2}$ 1 $\frac{3}{2}$ 0 110− $+2.23\sqrt{2.7/3.11}\sqrt{3.5.11.13}$
$\frac{3}{2}$ 1 $\frac{3}{2}$ 1 000− $+2.3\sqrt{3.5}/11\sqrt{11.13}$
$\frac{3}{2}$ 1 $\frac{3}{2}$ 1 010− $+8.2.3\sqrt{3}/11\sqrt{7.11.13}$
$\frac{3}{2}$ 1 $\frac{3}{2}$ 1 100− $-8\sqrt{2.3}/11\sqrt{11.13}$
$\frac{3}{2}$ 1 $\frac{3}{2}$ 1 110− $-\sqrt{2.5.7.13/3.11}\sqrt{3.11}$
$\frac{3}{2}$ 1 $\tilde{\frac{1}{2}}$ 0 000+ $+5.5\sqrt{3}/11\sqrt{2.11.13}$
$\frac{3}{2}$ 1 $\tilde{\frac{1}{2}}$ 0 010+ $+8.2\sqrt{2.5}/11\sqrt{3.7.11.13}$
$\frac{3}{2}$ 1 $\tilde{\frac{1}{2}}$ 0 100+ $+8\sqrt{3.5}/11\sqrt{11.13}$
$\frac{3}{2}$ 1 $\tilde{\frac{1}{2}}$ 0 110+ $-23\sqrt{7/3.11}\sqrt{3.11.13}$
$\frac{3}{2}$ 2 $\frac{3}{2}$ 0 000+ $+8.2\sqrt{2}/11\sqrt{7.13}$
$\frac{3}{2}$ 2 $\frac{3}{2}$ 0 100+ $-2.3/11\sqrt{5.7.13}$

$\frac{11}{2}$ 5 $\frac{5}{2}$
$\frac{3}{2}$ 2 $\tilde{\frac{1}{2}}$ 0 000− $+2.5\sqrt{5}/11\sqrt{7.13}$
$\frac{3}{2}$ 2 $\tilde{\frac{1}{2}}$ 0 100− $-17\sqrt{2}/11\sqrt{7.13}$
$\frac{3}{2}$ $\tilde{1}$ $\frac{3}{2}$ 0 000+ $-3\sqrt{3}/11\sqrt{7.13}$
$\frac{3}{2}$ $\tilde{1}$ $\frac{3}{2}$ 0 100+ $+2\sqrt{2.3.13}/11\sqrt{5.7}$
$\frac{3}{2}$ $\tilde{1}$ $\frac{3}{2}$ 1 000− $+2\sqrt{3.7}/11\sqrt{13}$
$\frac{3}{2}$ $\tilde{1}$ $\frac{3}{2}$ 1 100− $+4.19\sqrt{2}/11\sqrt{3.5.7.13}$
$\frac{3}{2}$ $\tilde{1}$ $\tilde{\frac{1}{2}}$ 0 000− $+\sqrt{3.5}/11\sqrt{2.7.13}$
$\frac{3}{2}$ $\tilde{1}$ $\tilde{\frac{1}{2}}$ 0 100− $-4/11\sqrt{3.7.13}$
$\tilde{\frac{1}{2}}$ 1 $\frac{3}{2}$ 0 000+ $+\sqrt{3}/\sqrt{2.11.13}$
$\tilde{\frac{1}{2}}$ 1 $\frac{3}{2}$ 0 010+ $-2.37\sqrt{2/3}\sqrt{3.5.7.11.13}$
$\tilde{\frac{1}{2}}$ 1 $\tilde{\frac{1}{2}}$ 0 000− $+\sqrt{3.5}/\sqrt{2.11.13}$
$\tilde{\frac{1}{2}}$ 1 $\tilde{\frac{1}{2}}$ 0 010− $-8\sqrt{2/3}\sqrt{3.7.11.13}$
$\tilde{\frac{1}{2}}$ 2 $\frac{3}{2}$ 0 000− $-2/\sqrt{5.7.13}$
$\tilde{\frac{1}{2}}$ $\tilde{1}$ $\frac{3}{2}$ 0 000− $+\sqrt{5}/\sqrt{2.3.7.13}$

$\frac{11}{2}$ 5 $\frac{7}{2}$
$\frac{1}{2}$ 1 $\frac{1}{2}$ 0 000+ $-2.7/11\sqrt{3.13}$
$\frac{1}{2}$ 1 $\frac{1}{2}$ 0 010+ $-\sqrt{5.7}/11\sqrt{3.13}$
$\frac{1}{2}$ 1 $\frac{3}{2}$ 0 000− $-\sqrt{5.7}/11\sqrt{3.13}$
$\frac{1}{2}$ 1 $\frac{3}{2}$ 0 010− $-19/11\sqrt{3.13}$
$\frac{1}{2}$ 2 $\frac{3}{2}$ 0 000+ $+1/\sqrt{2.11.13}$
$\frac{1}{2}$ $\tilde{1}$ $\frac{3}{2}$ 0 000+ $+1/\sqrt{3.11.13}$
$\frac{1}{2}$ $\tilde{1}$ $\tilde{\frac{1}{2}}$ 0 000− $-2/\sqrt{11.13}$
$\frac{3}{2}$ 1 $\frac{1}{2}$ 0 000− $-\sqrt{7}/11\sqrt{3.11.13}$
$\frac{3}{2}$ 1 $\frac{1}{2}$ 0 010− $-17\sqrt{5}/11\sqrt{3.11.13}$
$\frac{3}{2}$ 1 $\frac{1}{2}$ 0 100− $+\sqrt{2.5.7}/11\sqrt{3.11.13}$
$\frac{3}{2}$ 1 $\frac{1}{2}$ 0 110− $+7.7/11\sqrt{2.3.11.13}$
$\frac{3}{2}$ 1 $\frac{3}{2}$ 0 000+ $+8\sqrt{2.5}/11\sqrt{3.11.13}$
$\frac{3}{2}$ 1 $\frac{3}{2}$ 0 010+ $-8.2\sqrt{2.7}/11\sqrt{3.11.13}$
$\frac{3}{2}$ 1 $\frac{3}{2}$ 0 100+ $-2.29/11\sqrt{3.11.13}$
$\frac{3}{2}$ 1 $\frac{3}{2}$ 0 110+ $-\sqrt{5.7}/11\sqrt{3.11.13}$
$\frac{3}{2}$ 1 $\frac{3}{2}$ 1 000+ $+\sqrt{2.5.13}/11\sqrt{3.11}$
$\frac{3}{2}$ 1 $\frac{3}{2}$ 1 010+ $-2.5.5\sqrt{2}/11\sqrt{3.7.11.13}$
$\frac{3}{2}$ 1 $\frac{3}{2}$ 1 100+ $-3\sqrt{3}/11\sqrt{11.13}$
$\frac{3}{2}$ 1 $\frac{3}{2}$ 1 110+ $-\sqrt{3.5.7}/11\sqrt{11.13}$
$\frac{3}{2}$ 1 $\tilde{\frac{1}{2}}$ 0 000− $+3\sqrt{5}/11\sqrt{11.13}$
$\frac{3}{2}$ 1 $\tilde{\frac{1}{2}}$ 0 010− $-9/11\sqrt{7.11.13}$
$\frac{3}{2}$ 1 $\tilde{\frac{1}{2}}$ 0 100− $-2\sqrt{2.13}/11\sqrt{11}$
$\frac{3}{2}$ 1 $\tilde{\frac{1}{2}}$ 0 110− $+\sqrt{5.7}/11\sqrt{2.11.13}$
$\frac{3}{2}$ 2 $\frac{1}{2}$ 0 000+ $-3\sqrt{5}/11\sqrt{2.13}$
$\frac{3}{2}$ 2 $\frac{1}{2}$ 0 100+ $+\sqrt{13/2.3.11}$
$\frac{3}{2}$ 2 $\frac{3}{2}$ 0 000− $-8/11\sqrt{7.13}$
$\frac{3}{2}$ 2 $\frac{3}{2}$ 0 100− $-29\sqrt{5/3.11}\sqrt{2.7.13}$
$\frac{3}{2}$ 2 $\tilde{\frac{1}{2}}$ 0 000+ $+\sqrt{3}/\sqrt{2.7.13}$
$\frac{3}{2}$ 2 $\tilde{\frac{1}{2}}$ 0 100+ $+\sqrt{3.5}/2\sqrt{7.13}$
$\frac{3}{2}$ $\tilde{1}$ $\frac{1}{2}$ 0 000+ $+7\sqrt{5}/11\sqrt{3.13}$
$\frac{3}{2}$ $\tilde{1}$ $\frac{1}{2}$ 0 100+ $+2.19\sqrt{2/3.11}\sqrt{3.13}$
$\frac{3}{2}$ $\tilde{1}$ $\frac{3}{2}$ 0 000− $-8\sqrt{2.3}/11\sqrt{7.13}$
$\frac{3}{2}$ $\tilde{1}$ $\frac{3}{2}$ 0 100− $+2.17\sqrt{5/3.11}\sqrt{3.7.13}$
$\frac{3}{2}$ $\tilde{1}$ $\frac{3}{2}$ 1 000+ $-\sqrt{2.7}/11\sqrt{3.13}$

SO_3–O 3jm Factors (cont.)

$\frac{11}{2}\ 5\ \frac{7}{2}$			
$\frac{3}{2}\ \tilde{1}\ \frac{3}{2}$	1	100+	$-\sqrt{5.13}/3.11\sqrt{3.7}$
$\frac{3}{2}\ \tilde{1}\ \tilde{\frac{1}{2}}$	0	000+	$+23/11\sqrt{7.13}$
$\frac{3}{2}\ \tilde{1}\ \tilde{\frac{1}{2}}$	0	100+	$-\sqrt{2.5}/11\sqrt{7.13}$
$\tilde{\frac{1}{2}}\ 1\ \frac{3}{2}$	0	000−	$+1/\sqrt{3.11.13}$
$\tilde{\frac{1}{2}}\ 1\ \frac{3}{2}$	0	010−	$-5\sqrt{5}/\sqrt{3.7.11.13}$
$\tilde{\frac{1}{2}}\ 1\ \tilde{\frac{1}{2}}$	0	000+	$+2/\sqrt{11.13}$
$\tilde{\frac{1}{2}}\ 1\ \tilde{\frac{1}{2}}$	0	010+	$+\sqrt{5}/\sqrt{7.11.13}$
$\tilde{\frac{1}{2}}\ 2\ \frac{3}{2}$	0	000+	$-5\sqrt{5}/3\sqrt{2.7.13}$
$\tilde{\frac{1}{2}}\ \tilde{1}\ \frac{1}{2}$	0	000−	$+2/3\sqrt{3.13}$
$\tilde{\frac{1}{2}}\ \tilde{1}\ \frac{3}{2}$	0	000+	$+\sqrt{5}/3\sqrt{3.7.13}$

$\frac{11}{2}\ 5\ \frac{9}{2}$			
$\frac{1}{2}\ 1\ \frac{1}{2}$	0	000−	$-7\sqrt{7}/2.11\sqrt{3.13}$
$\frac{1}{2}\ 1\ \frac{1}{2}$	0	010−	$+8.2/11\sqrt{3.5.13}$
$\frac{1}{2}\ 1\ \frac{3}{2}$	0	000+	$+3\sqrt{7}/2.11\sqrt{5.13}$
$\frac{1}{2}\ 1\ \frac{3}{2}$	0	001+	$+2/11\sqrt{3.5.13}$
$\frac{1}{2}\ 1\ \frac{3}{2}$	0	010+	$+9/11\sqrt{13}$
$\frac{1}{2}\ 1\ \frac{3}{2}$	0	011+	$+5\sqrt{7}/2.11\sqrt{3.13}$
$\frac{1}{2}\ 2\ \frac{3}{2}$	0	000−	$+2\sqrt{2.3}/5\sqrt{11.13}$
$\frac{1}{2}\ 2\ \frac{3}{2}$	0	001−	$+\sqrt{7}/5\sqrt{2.11.13}$
$\frac{1}{2}\ \tilde{1}\ \frac{3}{2}$	0	000−	$-19/2.5\sqrt{11.13}$
$\frac{1}{2}\ \tilde{1}\ \frac{3}{2}$	0	001−	$-8\sqrt{7}/5\sqrt{3.11.13}$
$\frac{3}{2}\ 1\ \frac{1}{2}$	0	000+	$+7\sqrt{13}/2.11\sqrt{3.11}$
$\frac{3}{2}\ 1\ \frac{1}{2}$	0	010+	$+8\sqrt{7}/11\sqrt{3.5.11.13}$
$\frac{3}{2}\ 1\ \frac{1}{2}$	0	100+	$+4\sqrt{2.5}/11\sqrt{3.11.13}$
$\frac{3}{2}\ 1\ \frac{1}{2}$	0	110+	$+2.7\sqrt{2.7}/11\sqrt{3.11.13}$
$\frac{3}{2}\ 1\ \frac{3}{2}$	0	000−	$+3\sqrt{5}/11\sqrt{2.11.13}$
$\frac{3}{2}\ 1\ \frac{3}{2}$	0	001−	$-2\sqrt{2.5.7}/11\sqrt{3.11.13}$
$\frac{3}{2}\ 1\ \frac{3}{2}$	0	010−	$+2.9\sqrt{2.7}/5.11\sqrt{11.13}$
$\frac{3}{2}\ 1\ \frac{3}{2}$	0	011−	$-8.8.4\sqrt{2}/5.11\sqrt{3.11.13}$
$\frac{3}{2}\ 1\ \frac{3}{2}$	0	100−	$-37/11\sqrt{11.13}$
$\frac{3}{2}\ 1\ \frac{3}{2}$	0	101−	$+29\sqrt{7}/2.11\sqrt{3.11.13}$
$\frac{3}{2}\ 1\ \frac{3}{2}$	0	110−	$+8\sqrt{7}/11\sqrt{5.11.13}$
$\frac{3}{2}\ 1\ \frac{3}{2}$	0	111−	$-8.2/11\sqrt{3.5.11.13}$
$\frac{3}{2}\ 1\ \frac{3}{2}$	1	000−	$-9\sqrt{2}/11\sqrt{5.11.13}$
$\frac{3}{2}\ 1\ \frac{3}{2}$	1	001−	$+4\sqrt{2.7}/11\sqrt{3.5.11.13}$
$\frac{3}{2}\ 1\ \frac{3}{2}$	1	010−	$+2.9\sqrt{2.7}/5.11\sqrt{11.13}$
$\frac{3}{2}\ 1\ \frac{3}{2}$	1	011−	$+53/5.11\sqrt{2.3.11.13}$
$\frac{3}{2}\ 1\ \frac{3}{2}$	1	100−	$-4/11\sqrt{11.13}$
$\frac{3}{2}\ 1\ \frac{3}{2}$	1	101−	$-9\sqrt{3.7}/2.11\sqrt{11.13}$
$\frac{3}{2}\ 1\ \frac{3}{2}$	1	110−	$+8\sqrt{7}/11\sqrt{5.11.13}$
$\frac{3}{2}\ 1\ \frac{3}{2}$	1	111−	$+8\sqrt{3}/11\sqrt{5.11.13}$
$\frac{3}{2}\ 2\ \frac{1}{2}$	0	000−	$-3\sqrt{2.7}/11\sqrt{5.13}$
$\frac{3}{2}\ 2\ \frac{1}{2}$	0	100−	$-4\sqrt{7}/3.11\sqrt{13}$
$\frac{3}{2}\ 2\ \frac{3}{2}$	0	000+	$+4\sqrt{3.7}/5.11\sqrt{13}$
$\frac{3}{2}\ 2\ \frac{3}{2}$	0	001+	$+37/5.11\sqrt{13}$
$\frac{3}{2}\ 2\ \frac{3}{2}$	0	100+	$-4\sqrt{2.3.7}/11\sqrt{5.13}$
$\frac{3}{2}\ 2\ \frac{3}{2}$	0	101+	$-\sqrt{2}/3.11\sqrt{5.13}$
$\frac{3}{2}\ \tilde{1}\ \frac{1}{2}$	0	000−	$-23\sqrt{7}/2.11\sqrt{3.5.13}$
$\frac{3}{2}\ \tilde{1}\ \frac{1}{2}$	0	100−	$-4\sqrt{2.7}/3.11\sqrt{3.13}$

$\frac{11}{2}\ 5\ \frac{9}{2}$			
$\frac{3}{2}\ \tilde{1}\ \frac{3}{2}$	0	000+	$+\sqrt{7}/11\sqrt{2.13}$
$\frac{3}{2}\ \tilde{1}\ \frac{3}{2}$	0	001+	$+2\sqrt{2.3}/11\sqrt{13}$
$\frac{3}{2}\ \tilde{1}\ \frac{3}{2}$	0	100+	$-\sqrt{5.7}/11\sqrt{13}$
$\frac{3}{2}\ \tilde{1}\ \frac{3}{2}$	0	101+	$-17\sqrt{5}/2.3.11\sqrt{3.13}$
$\frac{3}{2}\ \tilde{1}\ \frac{3}{2}$	1	000−	$-\sqrt{2.7}/5\sqrt{13}$
$\frac{3}{2}\ \tilde{1}\ \frac{3}{2}$	1	001−	$+2\sqrt{2}/5.11\sqrt{3.13}$
$\frac{3}{2}\ \tilde{1}\ \frac{3}{2}$	1	100−	0
$\frac{3}{2}\ \tilde{1}\ \frac{3}{2}$	1	101−	$-7\sqrt{5}/2.3.11\sqrt{3.13}$
$\tilde{\frac{1}{2}}\ 1\ \frac{3}{2}$	0	000+	$+1/2\sqrt{11.13}$
$\tilde{\frac{1}{2}}\ 1\ \frac{3}{2}$	0	001+	$+2\sqrt{7}/\sqrt{3.11.13}$
$\tilde{\frac{1}{2}}\ 1\ \frac{3}{2}$	0	010+	$-\sqrt{7}/\sqrt{5.11.13}$
$\tilde{\frac{1}{2}}\ 1\ \frac{3}{2}$	0	011+	$-1/2\sqrt{3.5.11.13}$
$\tilde{\frac{1}{2}}\ 2\ \frac{3}{2}$	0	000−	0
$\tilde{\frac{1}{2}}\ 2\ \frac{3}{2}$	0	001−	$-\sqrt{5}/3\sqrt{2.13}$
$\tilde{\frac{1}{2}}\ \tilde{1}\ \frac{1}{2}$	0	000+	$+\sqrt{7}/2.3\sqrt{3.13}$
$\tilde{\frac{1}{2}}\ \tilde{1}\ \frac{3}{2}$	0	000−	$+\sqrt{7}/2\sqrt{5.13}$
$\tilde{\frac{1}{2}}\ \tilde{1}\ \frac{3}{2}$	0	001−	$-8/3\sqrt{3.5.13}$

$\frac{11}{2}\ \frac{11}{2}\ 0$			
$\frac{1}{2}\ \frac{1}{2}\ 0$	0	000+	$+1/\sqrt{2.3}$
$\frac{3}{2}\ \frac{3}{2}\ 0$	0	000+	$+1/\sqrt{3}$
$\frac{3}{2}\ \frac{3}{2}\ 0$	0	100+	0
$\frac{3}{2}\ \frac{3}{2}\ 0$	0	110+	$+1/\sqrt{3}$
$\tilde{\frac{1}{2}}\ \tilde{\frac{1}{2}}\ 0$	0	000+	$+1/\sqrt{2.3}$

$\frac{11}{2}\ \frac{11}{2}\ 1$			
$\frac{1}{2}\ \frac{1}{2}\ 1$	0	000+	$-\sqrt{11}/3\sqrt{2.13}$
$\frac{3}{2}\ \frac{1}{2}\ 1$	0	000−	$+\sqrt{2.7}/3\sqrt{13}$
$\frac{3}{2}\ \frac{1}{2}\ 1$	0	100−	0
$\frac{3}{2}\ \frac{3}{2}\ 1$	0	000+	$-7.17/3.11\sqrt{11.13}$
$\frac{3}{2}\ \frac{3}{2}\ 1$	0	100+	$-8\sqrt{2.5}/11\sqrt{11.13}$
$\frac{3}{2}\ \frac{3}{2}\ 1$	0	110+	$+97/3.11\sqrt{11.13}$
$\frac{3}{2}\ \frac{3}{2}\ 1$	1	000+	$+8.4/3.11\sqrt{11.13}$
$\frac{3}{2}\ \frac{3}{2}\ 1$	1	100+	$+4\sqrt{2.5}/11\sqrt{11.13}$
$\frac{3}{2}\ \frac{3}{2}\ 1$	1	110+	$+2.83/3.11\sqrt{11.13}$
$\tilde{\frac{1}{2}}\ \frac{3}{2}\ 1$	0	000−	$-\sqrt{2.5}/\sqrt{11.13}$
$\tilde{\frac{1}{2}}\ \frac{3}{2}\ 1$	0	010−	$+8/3\sqrt{11.13}$
$\tilde{\frac{1}{2}}\ \tilde{\frac{1}{2}}\ 1$	0	000+	$+\sqrt{11}/3\sqrt{2.13}$

$\frac{11}{2}\ \frac{11}{2}\ 2$			
$\frac{3}{2}\ \frac{1}{2}\ 2$	0	000−	$+\sqrt{5}/11\sqrt{3.13}$
$\frac{3}{2}\ \frac{1}{2}\ 2$	0	100−	$+2\sqrt{2.3}/11\sqrt{13}$
$\frac{3}{2}\ \frac{1}{2}\ \tilde{1}$	0	000−	$-4\sqrt{2.5}/11\sqrt{13}$
$\frac{3}{2}\ \frac{1}{2}\ \tilde{1}$	0	100−	$-4/11\sqrt{13}$
$\frac{3}{2}\ \frac{3}{2}\ 2$	0	000+	$+8.4\sqrt{2.5}/11\sqrt{3.7.11.13}$
$\frac{3}{2}\ \frac{3}{2}\ 2$	0	100+	$+8.5\sqrt{3}/11\sqrt{7.11.13}$
$\frac{3}{2}\ \frac{3}{2}\ 2$	0	110+	$-37\sqrt{2.7}/11\sqrt{3.5.11.13}$
$\frac{3}{2}\ \frac{3}{2}\ \tilde{1}$	0	000+	$+41\sqrt{5}/11\sqrt{7.11.13}$
$\frac{3}{2}\ \frac{3}{2}\ \tilde{1}$	0	100+	$-8.5\sqrt{2}/11\sqrt{7.11.13}$
$\frac{3}{2}\ \frac{3}{2}\ \tilde{1}$	0	110+	$-23\sqrt{7}/11\sqrt{5.11.13}$
$\frac{3}{2}\ \frac{3}{2}\ \tilde{1}$	1	000−	0
$\frac{3}{2}\ \frac{3}{2}\ \tilde{1}$	1	100−	$+4\sqrt{2}/\sqrt{7.11.13}$

SO_3–O 3jm Factors (cont.)

$\frac{11}{2}\ \frac{11}{2}\ 2$

$\frac{3}{2}\ \frac{3}{2}\ \tilde{1}$ 1 110− 0
$\widetilde{\tfrac{1}{2}}\ \frac{1}{2}\ \tilde{1}$ 0 000+ $+1/\sqrt{2.13}$
$\widetilde{\tfrac{1}{2}}\ \frac{3}{2}\ 2$ 0 000− $-\sqrt{3}/\sqrt{7.11.13}$
$\widetilde{\tfrac{1}{2}}\ \frac{3}{2}\ 2$ 0 010− $+2\sqrt{2.13}/\sqrt{3.5.7.11}$
$\widetilde{\tfrac{1}{2}}\ \frac{3}{2}\ \tilde{1}$ 0 000− $+4\sqrt{2}/\sqrt{7.11.13}$
$\widetilde{\tfrac{1}{2}}\ \frac{3}{2}\ \tilde{1}$ 0 010− $+4/\sqrt{5.7.11.13}$

$\frac{11}{2}\ \frac{11}{2}\ 3$

$\frac{1}{2}\ \frac{1}{2}\ 1$ 0 000+ $-\sqrt{2.7}/\sqrt{3.11.13}$
$\frac{3}{2}\ \frac{1}{2}\ 1$ 0 000− $+\sqrt{13}/11\sqrt{2.3}$
$\frac{3}{2}\ \frac{1}{2}\ 1$ 0 100− $-2\sqrt{5}/11\sqrt{3.13}$
$\frac{3}{2}\ \frac{1}{2}\ \tilde{1}$ 0 000+ $+\sqrt{3.5}/11\sqrt{2.13}$
$\frac{3}{2}\ \frac{1}{2}\ \tilde{1}$ 0 100+ $+2.3\sqrt{3}/11\sqrt{13}$
$\frac{3}{2}\ \frac{3}{2}\ 1$ 0 000+ $-2.43/11\sqrt{3.7.11.13}$
$\frac{3}{2}\ \frac{3}{2}\ 1$ 0 100+ $+4.5\sqrt{2.5}/11\sqrt{3.7.11.13}$
$\frac{3}{2}\ \frac{3}{2}\ 1$ 0 110+ $-2.157/3.11\sqrt{3.7.11.13}$
$\frac{3}{2}\ \frac{3}{2}\ 1$ 1 000+ $-8.2\sqrt{7}/11\sqrt{3.11.13}$
$\frac{3}{2}\ \frac{3}{2}\ 1$ 1 100+ $+8\sqrt{2.3.5}/11\sqrt{7.11.13}$
$\frac{3}{2}\ \frac{3}{2}\ 1$ 1 110+ $-137/3.11\sqrt{3.7.11.13}$
$\frac{3}{2}\ \frac{3}{2}\ \tilde{1}$ 0 000− 0
$\frac{3}{2}\ \frac{3}{2}\ \tilde{1}$ 0 100− $+4\sqrt{2.3}/\sqrt{7.11.13}$
$\frac{3}{2}\ \frac{3}{2}\ \tilde{1}$ 0 110− 0
$\frac{3}{2}\ \frac{3}{2}\ \tilde{1}$ 1 000+ $-8.2\sqrt{3.5}/11\sqrt{7.11.13}$
$\frac{3}{2}\ \frac{3}{2}\ \tilde{1}$ 1 100+ $-8\sqrt{2.3}/11\sqrt{7.11.13}$
$\frac{3}{2}\ \frac{3}{2}\ \tilde{1}$ 1 110+ $-29\sqrt{5}/11\sqrt{3.7.11.13}$
$\frac{3}{2}\ \frac{3}{2}\ \tilde{0}$ 0 000+ $-3\sqrt{5.13}/11\sqrt{7.11}$
$\frac{3}{2}\ \frac{3}{2}\ \tilde{0}$ 0 100+ $+8\sqrt{2}/11\sqrt{7.11.13}$
$\frac{3}{2}\ \frac{3}{2}\ \tilde{0}$ 0 110+ $-5\sqrt{5}/11\sqrt{7.11.13}$
$\widetilde{\tfrac{1}{2}}\ \frac{1}{2}\ \tilde{1}$ 0 000− 0
$\widetilde{\tfrac{1}{2}}\ \frac{1}{2}\ \tilde{0}$ 0 000+ $-1/\sqrt{2.13}$
$\widetilde{\tfrac{1}{2}}\ \frac{3}{2}\ 1$ 0 000− $-5\sqrt{5}/\sqrt{2.3.7.11.13}$
$\widetilde{\tfrac{1}{2}}\ \frac{3}{2}\ 1$ 0 010− $-2.23/3\sqrt{3.7.11.13}$
$\widetilde{\tfrac{1}{2}}\ \frac{3}{2}\ \tilde{1}$ 0 000+ $-3\sqrt{3}/\sqrt{2.7.11.13}$
$\widetilde{\tfrac{1}{2}}\ \frac{3}{2}\ \tilde{1}$ 0 010+ $-2\sqrt{5}/\sqrt{3.7.11.13}$
$\widetilde{\tfrac{1}{2}}\ \widetilde{\tfrac{1}{2}}\ 1$ 0 000+ $-\sqrt{2.11}/3\sqrt{3.7.13}$

$\frac{11}{2}\ \frac{11}{2}\ 4$

$\frac{1}{2}\ \frac{1}{2}\ 0$ 0 000+ $-7/2\sqrt{3.11.13}$
$\frac{1}{2}\ \frac{1}{2}\ 1$ 0 000− 0
$\frac{3}{2}\ \frac{1}{2}\ 1$ 0 000+ $+\sqrt{3.5.7}/2.11\sqrt{13}$
$\frac{3}{2}\ \frac{1}{2}\ 1$ 0 100+ $-2\sqrt{2.7}/11\sqrt{3.13}$
$\frac{3}{2}\ \frac{1}{2}\ 2$ 0 000− $-2\sqrt{2.5}/11\sqrt{3.13}$
$\frac{3}{2}\ \frac{1}{2}\ 2$ 0 100− $+29/2.11\sqrt{3.13}$
$\frac{3}{2}\ \frac{1}{2}\ \tilde{1}$ 0 000− $-3\sqrt{5}/2.11\sqrt{13}$
$\frac{3}{2}\ \frac{1}{2}\ \tilde{1}$ 0 100− $+17\sqrt{2}/3.11\sqrt{13}$
$\frac{3}{2}\ \frac{3}{2}\ 0$ 0 000+ $-47/11\sqrt{2.3.11.13}$
$\frac{3}{2}\ \frac{3}{2}\ 0$ 0 100+ $-8\sqrt{5}/11\sqrt{3.11.13}$
$\frac{3}{2}\ \frac{3}{2}\ 0$ 0 110+ $+2.7.7\sqrt{2}/3.11\sqrt{3.11.13}$
$\frac{3}{2}\ \frac{3}{2}\ 1$ 0 000− 0
$\frac{3}{2}\ \frac{3}{2}\ 1$ 0 100− $-2/\sqrt{3.11.13}$
$\frac{3}{2}\ \frac{3}{2}\ 1$ 0 110− 0

$\frac{11}{2}\ \frac{11}{2}\ 4$

$\frac{3}{2}\ \frac{3}{2}\ 1$ 1 000− 0
$\frac{3}{2}\ \frac{3}{2}\ 1$ 1 100− $+5/\sqrt{3.11.13}$
$\frac{3}{2}\ \frac{3}{2}\ 1$ 1 110− 0
$\frac{3}{2}\ \frac{3}{2}\ 2$ 0 000+ $+8.5\sqrt{5}/11\sqrt{3.7.11.13}$
$\frac{3}{2}\ \frac{3}{2}\ 2$ 0 100+ $+73/11\sqrt{2.3.7.11.13}$
$\frac{3}{2}\ \frac{3}{2}\ 2$ 0 110+ $+8\sqrt{5.7}/3.11\sqrt{3.11.13}$
$\frac{3}{2}\ \frac{3}{2}\ \tilde{1}$ 0 000+ $+9\sqrt{2.5}/11\sqrt{7.11.13}$
$\frac{3}{2}\ \frac{3}{2}\ \tilde{1}$ 0 100+ $+2.107/3.11\sqrt{7.11.13}$
$\frac{3}{2}\ \frac{3}{2}\ \tilde{1}$ 0 110+ $+4\sqrt{2.5.7}/3.11\sqrt{11.13}$
$\frac{3}{2}\ \frac{3}{2}\ \tilde{1}$ 1 000− 0
$\frac{3}{2}\ \frac{3}{2}\ \tilde{1}$ 1 100− $-\sqrt{13}/3\sqrt{7.11}$
$\frac{3}{2}\ \frac{3}{2}\ \tilde{1}$ 1 110− 0
$\widetilde{\tfrac{1}{2}}\ \frac{1}{2}\ \tilde{1}$ 0 000+ $+1/3\sqrt{13}$
$\widetilde{\tfrac{1}{2}}\ \frac{3}{2}\ 1$ 0 000+ $-3\sqrt{3}/2\sqrt{11.13}$
$\widetilde{\tfrac{1}{2}}\ \frac{3}{2}\ 1$ 0 010+ $-\sqrt{2.5}/\sqrt{3.11.13}$
$\widetilde{\tfrac{1}{2}}\ \frac{3}{2}\ 2$ 0 000− $+2\sqrt{2.3}/\sqrt{7.11.13}$
$\widetilde{\tfrac{1}{2}}\ \frac{3}{2}\ 2$ 0 010− $-17\sqrt{5}/2.3\sqrt{3.7.11.13}$
$\widetilde{\tfrac{1}{2}}\ \frac{3}{2}\ \tilde{1}$ 0 000− $+37/2.3\sqrt{7.11.13}$
$\widetilde{\tfrac{1}{2}}\ \frac{3}{2}\ \tilde{1}$ 0 010− $-2\sqrt{2.5}/3\sqrt{7.11.13}$
$\widetilde{\tfrac{1}{2}}\ \widetilde{\tfrac{1}{2}}\ 0$ 0 000+ $+\sqrt{11}/2.3\sqrt{3.13}$
$\widetilde{\tfrac{1}{2}}\ \widetilde{\tfrac{1}{2}}\ 1$ 0 000− 0

$\frac{11}{2}\ \frac{11}{2}\ 5$

$\frac{1}{2}\ \frac{1}{2}\ 1$ 0 000+ $-2\sqrt{3}/11\sqrt{13.17}$
$\frac{1}{2}\ \frac{1}{2}\ 1$ 0 001+ $+7\sqrt{5.7}/2.11\sqrt{3.13.17}$
$\frac{3}{2}\ \frac{1}{2}\ 1$ 0 000− $+\sqrt{3.7}/11\sqrt{11.13.17}$
$\frac{3}{2}\ \frac{1}{2}\ 1$ 0 001− $-2.5.7\sqrt{5}/11\sqrt{3.11.13.17}$
$\frac{3}{2}\ \frac{1}{2}\ 1$ 0 100− $+7\sqrt{3.5.7}/2.11\sqrt{2.11.13.17}$
$\frac{3}{2}\ \frac{1}{2}\ 1$ 0 101− $-4\sqrt{2.17}/11\sqrt{3.11.13}$
$\frac{3}{2}\ \frac{1}{2}\ 2$ 0 000+ $-3\sqrt{5}/11\sqrt{2.13.17}$
$\frac{3}{2}\ \frac{1}{2}\ 2$ 0 100+ $+37/11\sqrt{13.17}$
$\frac{3}{2}\ \frac{1}{2}\ \tilde{1}$ 0 000+ $-\sqrt{3.5}/11\sqrt{13.17}$
$\frac{3}{2}\ \frac{1}{2}\ \tilde{1}$ 0 100+ $+31\sqrt{3}/2.11\sqrt{2.13.17}$
$\frac{3}{2}\ \frac{3}{2}\ 1$ 0 000+ $-8.7\sqrt{2.3}/11.11\sqrt{13.17}$
$\frac{3}{2}\ \frac{3}{2}\ 1$ 0 001+ $+97\sqrt{5.7}/11.11\sqrt{2.3.13.17}$
$\frac{3}{2}\ \frac{3}{2}\ 1$ 0 100+ $+79\sqrt{3.5}/11.11\sqrt{13.17}$
$\frac{3}{2}\ \frac{3}{2}\ 1$ 0 101+ $+4.19\sqrt{7}/11.11\sqrt{3.13.17}$
$\frac{3}{2}\ \frac{3}{2}\ 1$ 0 110+ $+4.3\sqrt{2.3}/11.11\sqrt{13.17}$
$\frac{3}{2}\ \frac{3}{2}\ 1$ 0 111+ $+2.73\sqrt{2.5.7}/3.11.11\sqrt{3.13.17}$
$\frac{3}{2}\ \frac{3}{2}\ 1$ 1 000+ $+8\sqrt{2.3}/11.11\sqrt{13.17}$
$\frac{3}{2}\ \frac{3}{2}\ 1$ 1 001+ $-8.2.5\sqrt{2.5.7}/11.11\sqrt{3.13.17}$
$\frac{3}{2}\ \frac{3}{2}\ 1$ 1 100+ $-3.73\sqrt{3.5}/2.11.11\sqrt{13.17}$
$\frac{3}{2}\ \frac{3}{2}\ 1$ 1 101+ $+2.5\sqrt{3.7}/11.11\sqrt{13.17}$
$\frac{3}{2}\ \frac{3}{2}\ 1$ 1 110+ $-2.37\sqrt{2.3}/11.11\sqrt{13.17}$
$\frac{3}{2}\ \frac{3}{2}\ 1$ 1 111+ $+4.71\sqrt{2.5.7}/3.11.11\sqrt{3.13.17}$
$\frac{3}{2}\ \frac{3}{2}\ 2$ 0 000− 0
$\frac{3}{2}\ \frac{3}{2}\ 2$ 0 100− $+2\sqrt{2.7}/\sqrt{11.13.17}$
$\frac{3}{2}\ \frac{3}{2}\ 2$ 0 110− 0
$\frac{3}{2}\ \frac{3}{2}\ \tilde{1}$ 0 000− 0
$\frac{3}{2}\ \frac{3}{2}\ \tilde{1}$ 0 100− $-\sqrt{3.7}/\sqrt{11.13.17}$

SO_3–O 3jm Factors (cont.)

$\mathbf{\frac{11}{2}}$	$\mathbf{\frac{11}{2}}$	**5**			
$\frac{3}{2}$	$\frac{3}{2}$	$\tilde{1}$	0	110−	0
$\frac{3}{2}$	$\frac{3}{2}$	$\tilde{1}$	1	000+	$-8\sqrt{2.3.5.7}/11\sqrt{11.13.17}$
$\frac{3}{2}$	$\frac{3}{2}$	$\tilde{1}$	1	100+	$+3\sqrt{3.7.13}/2.11\sqrt{11.17}$
$\frac{3}{2}$	$\frac{3}{2}$	$\tilde{1}$	1	110+	$+2\sqrt{2.5.7}/11\sqrt{3.11.13.17}$
$\tilde{\frac{1}{2}}$	$\frac{1}{2}$	$\tilde{1}$	0	000−	$+2\sqrt{3}/\sqrt{13.17}$
$\tilde{\frac{1}{2}}$	$\frac{3}{2}$	1	0	000−	$-5\sqrt{3.5}/11\sqrt{13.17}$
$\tilde{\frac{1}{2}}$	$\frac{3}{2}$	1	0	001−	$-2.7\sqrt{7}/11\sqrt{3.13.17}$
$\tilde{\frac{1}{2}}$	$\frac{3}{2}$	1	0	010−	$+\sqrt{3}/2.11\sqrt{2.13.17}$
$\tilde{\frac{1}{2}}$	$\frac{3}{2}$	1	0	011−	$-2\sqrt{2.5.7}/3.11\sqrt{3.13.17}$
$\tilde{\frac{1}{2}}$	$\frac{3}{2}$	2	0	000+	$+\sqrt{7}/\sqrt{2.11.13.17}$
$\tilde{\frac{1}{2}}$	$\frac{3}{2}$	2	0	010+	$-\sqrt{5.7}/\sqrt{11.13.17}$
$\tilde{\frac{1}{2}}$	$\frac{3}{2}$	$\tilde{1}$	0	000+	$+\sqrt{3.7}/\sqrt{11.13.17}$
$\tilde{\frac{1}{2}}$	$\frac{3}{2}$	$\tilde{1}$	0	010+	$-\sqrt{5.7}/2\sqrt{2.3.11.13.17}$
$\tilde{\frac{1}{2}}$	$\tilde{\frac{1}{2}}$	1	0	000+	$+2\sqrt{3}/\sqrt{13.17}$
$\tilde{\frac{1}{2}}$	$\tilde{\frac{1}{2}}$	1	0	001+	$+\sqrt{5.7}/2.3\sqrt{3.13.17}$
6	**3**	**3**			
0	1	1	0	000+	$+5/\sqrt{2.7.11.13}$
0	$\tilde{1}$	$\tilde{1}$	0	000+	$-9/\sqrt{2.7.11.13}$
0	$\tilde{0}$	$\tilde{0}$	0	000+	$+2\sqrt{2.3}/\sqrt{7.11.13}$
1	1	1	0	000−	0
1	$\tilde{1}$	1	0	000+	$+\sqrt{3.5}/\sqrt{2.11.13}$
1	$\tilde{1}$	$\tilde{1}$	0	000−	0
1	$\tilde{0}$	$\tilde{1}$	0	000+	$+3/\sqrt{11.13}$
2	1	1	0	000+	$-5/\sqrt{2.11.13}$
2	$\tilde{1}$	1	0	000−	$-\sqrt{5}/\sqrt{2.11.13}$
2	$\tilde{1}$	$\tilde{1}$	0	000+	$-3/\sqrt{2.11.13}$
$\tilde{1}$	1	1	0	000+	$-\sqrt{3.5}/\sqrt{2.7.13}$
$\tilde{1}$	1	1	0	100+	$-5\sqrt{3}/2\sqrt{2.7.11.13}$
$\tilde{1}$	$\tilde{1}$	1	0	000−	$-\sqrt{2}/\sqrt{7.13}$
$\tilde{1}$	$\tilde{1}$	1	0	100−	$-9\sqrt{5}/2\sqrt{2.7.11.13}$
$\tilde{1}$	$\tilde{1}$	$\tilde{1}$	0	000+	$-\sqrt{3.5}/\sqrt{2.7.13}$
$\tilde{1}$	$\tilde{1}$	$\tilde{1}$	0	100+	$+9\sqrt{3}/2\sqrt{2.7.11.13}$
$\tilde{1}$	$\tilde{0}$	1	0	000+	$-1/\sqrt{7.13}$
$\tilde{1}$	$\tilde{0}$	1	0	100+	$+3\sqrt{5}/\sqrt{7.11.13}$
$\tilde{0}$	$\tilde{1}$	1	0	000+	$-1/\sqrt{2.13}$
6	$\mathbf{\frac{7}{2}}$	$\mathbf{\frac{5}{2}}$			
0	$\frac{3}{2}$	$\frac{3}{2}$	0	000−	$-\sqrt{5}/\sqrt{11.13}$
0	$\tilde{\frac{1}{2}}$	$\tilde{\frac{1}{2}}$	0	000−	$-\sqrt{2.3}/\sqrt{11.13}$
1	$\frac{1}{2}$	$\frac{3}{2}$	0	000−	$+5/2\sqrt{3.11.13}$
1	$\frac{3}{2}$	$\frac{3}{2}$	0	000+	$+\sqrt{5.7}/\sqrt{2.3.11.13}$
1	$\frac{3}{2}$	$\frac{3}{2}$	1	000+	0
1	$\frac{3}{2}$	$\tilde{\frac{1}{2}}$	0	000−	$-2\sqrt{7}/\sqrt{3.11.13}$
1	$\tilde{\frac{1}{2}}$	$\frac{3}{2}$	0	000−	$+\sqrt{5.7}/2\sqrt{11.13}$
1	$\tilde{\frac{1}{2}}$	$\tilde{\frac{1}{2}}$	0	000+	$-\sqrt{7}/\sqrt{11.13}$
2	$\frac{1}{2}$	$\frac{3}{2}$	0	000+	$-2.5/3\sqrt{11.13}$
2	$\frac{3}{2}$	$\frac{3}{2}$	0	000−	$-\sqrt{2.5.7}/3\sqrt{11.13}$
2	$\frac{3}{2}$	$\tilde{\frac{1}{2}}$	0	000+	$+2\sqrt{7}/3\sqrt{11.13}$
2	$\tilde{\frac{1}{2}}$	$\frac{3}{2}$	0	000+	0
$\tilde{1}$	$\frac{1}{2}$	$\frac{3}{2}$	0	000+	$-11\sqrt{5}/2.3\sqrt{3.7.13}$
$\tilde{1}$	$\frac{1}{2}$	$\frac{3}{2}$	0	100+	$-5\sqrt{3}/2\sqrt{7.11.13}$
$\tilde{1}$	$\frac{1}{2}$	$\tilde{\frac{1}{2}}$	0	000−	$+1/3\sqrt{3.7.13}$
$\tilde{1}$	$\frac{1}{2}$	$\tilde{\frac{1}{2}}$	0	100−	$+2\sqrt{3.5}/\sqrt{7.11.13}$
$\tilde{1}$	$\frac{3}{2}$	$\frac{3}{2}$	0	000−	$-7/3\sqrt{2.3.13}$
$\tilde{1}$	$\frac{3}{2}$	$\frac{3}{2}$	0	100−	0
$\tilde{1}$	$\frac{3}{2}$	$\frac{3}{2}$	1	000+	$+2\sqrt{2}/3\sqrt{3.13}$
$\tilde{1}$	$\frac{3}{2}$	$\frac{3}{2}$	1	100+	$+\sqrt{3.5}/\sqrt{2.11.13}$
$\tilde{1}$	$\frac{3}{2}$	$\tilde{\frac{1}{2}}$	0	000+	$+2\sqrt{5}/3\sqrt{3.13}$
$\tilde{1}$	$\frac{3}{2}$	$\tilde{\frac{1}{2}}$	0	100+	$-\sqrt{3}/\sqrt{11.13}$
$\tilde{1}$	$\tilde{\frac{1}{2}}$	$\frac{3}{2}$	0	000+	$-1/2\sqrt{13}$
$\tilde{1}$	$\tilde{\frac{1}{2}}$	$\frac{3}{2}$	0	100+	$+3\sqrt{5}/2\sqrt{11.13}$
$\tilde{0}$	$\frac{1}{2}$	$\tilde{\frac{1}{2}}$	0	000+	$+\sqrt{2}/3\sqrt{13}$
$\tilde{0}$	$\frac{3}{2}$	$\frac{3}{2}$	0	000+	$+\sqrt{7}/3\sqrt{13}$
6	$\mathbf{\frac{7}{2}}$	$\mathbf{\frac{7}{2}}$			
0	$\frac{1}{2}$	$\frac{1}{2}$	0	000+	$+5/\sqrt{2.3.11.13}$
0	$\frac{3}{2}$	$\frac{3}{2}$	0	000+	$-4/\sqrt{3.11.13}$
0	$\tilde{\frac{1}{2}}$	$\tilde{\frac{1}{2}}$	0	000+	$+\sqrt{3}/\sqrt{2.11.13}$
1	$\frac{1}{2}$	$\frac{1}{2}$	0	000−	0
1	$\frac{3}{2}$	$\frac{1}{2}$	0	000+	$+3\sqrt{5}/2\sqrt{11.13}$
1	$\frac{3}{2}$	$\frac{3}{2}$	0	000−	0
1	$\frac{3}{2}$	$\frac{3}{2}$	1	000−	0
1	$\tilde{\frac{1}{2}}$	$\frac{3}{2}$	0	000+	$+\sqrt{3.7}/2\sqrt{11.13}$
1	$\tilde{\frac{1}{2}}$	$\tilde{\frac{1}{2}}$	0	000−	0
2	$\frac{3}{2}$	$\frac{1}{2}$	0	000−	$-\sqrt{5}/\sqrt{3.11.13}$
2	$\frac{3}{2}$	$\frac{3}{2}$	0	000+	$+\sqrt{2.7}/\sqrt{3.11.13}$
2	$\tilde{\frac{1}{2}}$	$\frac{3}{2}$	0	000−	$-\sqrt{7}/\sqrt{11.13}$
$\tilde{1}$	$\frac{3}{2}$	$\frac{1}{2}$	0	000−	$-3/2\sqrt{7.13}$
$\tilde{1}$	$\frac{3}{2}$	$\frac{1}{2}$	0	100−	$+\sqrt{5}/\sqrt{7.11.13}$
$\tilde{1}$	$\frac{3}{2}$	$\frac{3}{2}$	0	000+	0
$\tilde{1}$	$\frac{3}{2}$	$\frac{3}{2}$	0	100+	$+7/\sqrt{2.11.13}$
$\tilde{1}$	$\frac{3}{2}$	$\frac{3}{2}$	1	000−	0
$\tilde{1}$	$\frac{3}{2}$	$\frac{3}{2}$	1	100−	0
$\tilde{1}$	$\tilde{\frac{1}{2}}$	$\frac{1}{2}$	0	000+	$+4/\sqrt{3.7.13}$
$\tilde{1}$	$\tilde{\frac{1}{2}}$	$\frac{1}{2}$	0	100+	$-\sqrt{3.5}/2\sqrt{7.11.13}$
$\tilde{1}$	$\tilde{\frac{1}{2}}$	$\frac{3}{2}$	0	000−	$-\sqrt{5}/2\sqrt{3.13}$
$\tilde{1}$	$\tilde{\frac{1}{2}}$	$\frac{3}{2}$	0	100−	$-\sqrt{3}/\sqrt{11.13}$
$\tilde{0}$	$\frac{3}{2}$	$\frac{3}{2}$	0	000−	0
$\tilde{0}$	$\tilde{\frac{1}{2}}$	$\frac{1}{2}$	0	000−	$-1/\sqrt{2.13}$
6	**4**	**2**			
0	2	2	0	000+	$+\sqrt{3}/\sqrt{11.13}$
0	$\tilde{1}$	$\tilde{1}$	0	000+	$-2\sqrt{2}/\sqrt{11.13}$
1	1	2	0	000+	$-\sqrt{3}/\sqrt{11.13}$
1	1	$\tilde{1}$	0	000+	$+\sqrt{2}/\sqrt{11.13}$
1	2	$\tilde{1}$	0	000−	$-\sqrt{7}/\sqrt{11.13}$
1	$\tilde{1}$	2	0	000−	$+\sqrt{7}/\sqrt{11.13}$
1	$\tilde{1}$	$\tilde{1}$	0	000−	$+\sqrt{2.7}/\sqrt{11.13}$
2	0	2	0	000+	$-2\sqrt{5}/\sqrt{3.11.13}$
2	1	$\tilde{1}$	0	000−	$+4\sqrt{2}/\sqrt{3.11.13}$
2	2	2	0	000+	$+\sqrt{2.7}/\sqrt{3.11.13}$

SO_3–O 3jm Factors (cont.)

6 4 2

2	$\tilde{1}$	$\tilde{1}$	0	000+	0
$\tilde{1}$	0	$\tilde{1}$	0	000+	$+4/3\sqrt{7.13}$
$\tilde{1}$	0	$\tilde{1}$	0	100+	$+3\sqrt{5}/\sqrt{7.11.13}$
$\tilde{1}$	1	2	0	000−	$-11/\sqrt{3.5.7.13}$
$\tilde{1}$	1	2	0	100−	$-3\sqrt{3}/\sqrt{7.11.13}$
$\tilde{1}$	1	$\tilde{1}$	0	000−	$+\sqrt{2.3}/\sqrt{5.7.13}$
$\tilde{1}$	1	$\tilde{1}$	0	100−	$-3\sqrt{2.3}/\sqrt{7.11.13}$
$\tilde{1}$	2	$\tilde{1}$	0	000+	$-7/3\sqrt{5.13}$
$\tilde{1}$	2	$\tilde{1}$	0	100+	0
$\tilde{1}$	$\tilde{1}$	2	0	000+	$+1/\sqrt{5.13}$
$\tilde{1}$	$\tilde{1}$	2	0	100+	$-3/\sqrt{11.13}$
$\tilde{1}$	$\tilde{1}$	$\tilde{1}$	0	000+	$+\sqrt{2}/\sqrt{3.5.13}$
$\tilde{1}$	$\tilde{1}$	$\tilde{1}$	0	100+	$-\sqrt{2.3}/\sqrt{11.13}$
$\tilde{0}$	1	$\tilde{1}$	0	000+	$+2\sqrt{2}/\sqrt{3.5.13}$
$\tilde{0}$	2	2	0	000−	$+\sqrt{7}/\sqrt{3.5.13}$

6 4 3

0	1	1	0	000−	$-\sqrt{3.5}/\sqrt{2.11.13}$
0	$\tilde{1}$	$\tilde{1}$	0	000−	$+\sqrt{7}/\sqrt{2.11.13}$
1	0	1	0	000−	$+5/\sqrt{7.11.13}$
1	1	1	0	000+	$-\sqrt{2.3.5}/\sqrt{7.11.13}$
1	1	$\tilde{1}$	0	000−	$-\sqrt{11}/\sqrt{2.7.13}$
1	2	1	0	000−	$-\sqrt{5}/\sqrt{11.13}$
1	2	$\tilde{1}$	0	000+	$+1/\sqrt{11.13}$
1	$\tilde{1}$	1	0	000−	$-\sqrt{3.5}/\sqrt{2.11.13}$
1	$\tilde{1}$	$\tilde{1}$	0	000+	$-\sqrt{2}/\sqrt{11.13}$
1	$\tilde{1}$	$\tilde{0}$	0	000−	$-1/\sqrt{11.13}$
2	1	1	0	000−	$-\sqrt{3.5}/\sqrt{2.7.11.13}$
2	1	$\tilde{1}$	0	000+	$+19/\sqrt{2.3.7.11.13}$
2	2	$\tilde{0}$	0	000+	$-4/\sqrt{3.11.13}$
2	$\tilde{1}$	1	0	000+	$-\sqrt{5}/\sqrt{2.11.13}$
2	$\tilde{1}$	$\tilde{1}$	0	000−	$-3/\sqrt{2.11.13}$
$\tilde{1}$	0	$\tilde{1}$	0	000−	$+17/3.7\sqrt{13}$
$\tilde{1}$	0	$\tilde{1}$	0	100−	$-3\sqrt{5}/7\sqrt{11.13}$
$\tilde{1}$	1	1	0	000−	$-1/7\sqrt{2.13}$
$\tilde{1}$	1	1	0	100−	$-3.5\sqrt{5}/2.7\sqrt{2.11.13}$
$\tilde{1}$	1	$\tilde{1}$	0	000+	0
$\tilde{1}$	1	$\tilde{1}$	0	100+	$+3\sqrt{3}/2\sqrt{2.11.13}$
$\tilde{1}$	1	$\tilde{0}$	0	000−	$+5\sqrt{5}/7\sqrt{3.13}$
$\tilde{1}$	1	$\tilde{0}$	0	100−	$+3\sqrt{3}/7\sqrt{11.13}$
$\tilde{1}$	2	1	0	000+	$-1/\sqrt{7.13}$
$\tilde{1}$	2	1	0	100+	$+3\sqrt{5}/\sqrt{7.11.13}$
$\tilde{1}$	2	$\tilde{1}$	0	000−	$-\sqrt{5}/3\sqrt{7.13}$
$\tilde{1}$	2	$\tilde{1}$	0	100−	$-9/\sqrt{7.11.13}$
$\tilde{1}$	$\tilde{1}$	1	0	000+	$-2\sqrt{2}/\sqrt{7.13}$
$\tilde{1}$	$\tilde{1}$	1	0	100+	$+3\sqrt{5}/2\sqrt{2.7.11.13}$
$\tilde{1}$	$\tilde{1}$	$\tilde{1}$	0	000−	$-\sqrt{5}/\sqrt{2.3.7.13}$
$\tilde{1}$	$\tilde{1}$	$\tilde{1}$	0	100−	$-\sqrt{3.11}/2\sqrt{2.7.13}$
$\tilde{0}$	0	$\tilde{0}$	0	000−	$-2\sqrt{2}/\sqrt{3.7.13}$
$\tilde{0}$	1	$\tilde{1}$	0	000−	$+\sqrt{5}/\sqrt{2.3.7.13}$
$\tilde{0}$	$\tilde{1}$	1	0	000−	$-1/\sqrt{2.13}$

6 4 4

0	0	0	0	000+	$+2\sqrt{2.5}/3\sqrt{11.13}$
0	1	1	0	000+	$-1/\sqrt{2.3.5.11.13}$
0	2	2	0	000+	$-8.2/3\sqrt{5.11.13}$
0	$\tilde{1}$	$\tilde{1}$	0	000+	$+\sqrt{5}/\sqrt{2.3.11.13}$
1	1	0	0	000+	$+\sqrt{7}/\sqrt{11.13}$
1	1	1	0	000−	0
1	2	1	0	000+	$+\sqrt{5}/\sqrt{11.13}$
1	$\tilde{1}$	1	0	000+	$-\sqrt{3}/\sqrt{2.5.11.13}$
1	$\tilde{1}$	2	0	000−	$-\sqrt{3.7}/\sqrt{5.11.13}$
1	$\tilde{1}$	$\tilde{1}$	0	000−	0
2	1	1	0	000+	$+\sqrt{7}/\sqrt{2.3.5.11.13}$
2	2	0	0	000+	$-4/3\sqrt{11.13}$
2	2	2	0	000+	$+4\sqrt{2.7}/3\sqrt{5.11.13}$
2	$\tilde{1}$	1	0	000−	$-3/\sqrt{2.5.11.13}$
2	$\tilde{1}$	$\tilde{1}$	0	000+	$+7\sqrt{7}/\sqrt{2.3.5.11.13}$
$\tilde{1}$	1	1	0	000+	$+1/\sqrt{2.13}$
$\tilde{1}$	1	1	0	100+	$-23/2\sqrt{2.5.11.13}$
$\tilde{1}$	2	1	0	000−	$+1/\sqrt{7.13}$
$\tilde{1}$	2	1	0	100−	$+\sqrt{13}/\sqrt{5.7.11}$
$\tilde{1}$	$\tilde{1}$	0	0	000+	$+\sqrt{5}/\sqrt{3.7.13}$
$\tilde{1}$	$\tilde{1}$	0	0	100+	$-1/\sqrt{3.7.11.13}$
$\tilde{1}$	$\tilde{1}$	1	0	000−	$+\sqrt{2}/\sqrt{7.13}$
$\tilde{1}$	$\tilde{1}$	1	0	100−	$-\sqrt{11}/2\sqrt{2.5.7.13}$
$\tilde{1}$	$\tilde{1}$	2	0	000+	$-1/\sqrt{3.13}$
$\tilde{1}$	$\tilde{1}$	2	0	100+	$+1/\sqrt{3.5.11.13}$
$\tilde{1}$	$\tilde{1}$	$\tilde{1}$	0	000+	$+1/\sqrt{2.13}$
$\tilde{1}$	$\tilde{1}$	$\tilde{1}$	0	100+	$+19/2\sqrt{2.5.11.13}$
$\tilde{0}$	2	2	0	000−	0
$\tilde{0}$	$\tilde{1}$	1	0	000+	$+1/\sqrt{2.13}$

6 $\frac{9}{2}$ $\frac{3}{2}$

0	$\frac{3}{2}$	$\frac{3}{2}$	0	000−	$-2\sqrt{7}/\sqrt{5.11.13}$
0	$\frac{3}{2}$	$\frac{3}{2}$	0	010−	$+3\sqrt{3}/\sqrt{5.11.13}$
1	$\frac{1}{2}$	$\frac{3}{2}$	0	000−	0
1	$\frac{3}{2}$	$\frac{3}{2}$	0	000+	$+\sqrt{2.3}/\sqrt{5.11.13}$
1	$\frac{3}{2}$	$\frac{3}{2}$	0	010+	$+3\sqrt{7}/\sqrt{2.5.11.13}$
1	$\frac{3}{2}$	$\frac{3}{2}$	1	000+	$-4\sqrt{2.3}/\sqrt{5.11.13}$
1	$\frac{3}{2}$	$\frac{3}{2}$	1	010+	$+3\sqrt{7}/\sqrt{2.5.11.13}$
2	$\frac{1}{2}$	$\frac{3}{2}$	0	000+	$-2\sqrt{3}/\sqrt{11.13}$
2	$\frac{3}{2}$	$\frac{3}{2}$	0	000−	$+2\sqrt{2}/\sqrt{5.11.13}$
2	$\frac{3}{2}$	$\frac{3}{2}$	0	010−	$+\sqrt{2.3.7}/\sqrt{5.11.13}$
$\tilde{1}$	$\frac{1}{2}$	$\frac{3}{2}$	0	000+	$+4/\sqrt{5.7.13}$
$\tilde{1}$	$\frac{1}{2}$	$\frac{3}{2}$	0	100+	$+9/\sqrt{7.11.13}$
$\tilde{1}$	$\frac{3}{2}$	$\frac{3}{2}$	0	000−	$+\sqrt{2.3}/5\sqrt{7.13}$
$\tilde{1}$	$\frac{3}{2}$	$\frac{3}{2}$	0	010−	$-7/5\sqrt{2.13}$
$\tilde{1}$	$\frac{3}{2}$	$\frac{3}{2}$	0	100−	$-2\sqrt{2.3.5}/\sqrt{7.11.13}$
$\tilde{1}$	$\frac{3}{2}$	$\frac{3}{2}$	0	110−	0
$\tilde{1}$	$\frac{3}{2}$	$\frac{3}{2}$	1	000+	$+4\sqrt{2.3}/5\sqrt{7.13}$

SO_3–O 3jm Factors (cont.)

6 $\frac{9}{2}$ $\frac{3}{2}$			
$\tilde{1}$ $\frac{3}{2}$ $\frac{3}{2}$	1	010+	$+7/5\sqrt{2.13}$
$\tilde{1}$ $\frac{3}{2}$ $\frac{3}{2}$	1	100+	$-\sqrt{2.3.5}/\sqrt{7.11.13}$
$\tilde{1}$ $\frac{3}{2}$ $\frac{3}{2}$	1	110+	0
$\tilde{0}$ $\frac{3}{2}$ $\frac{3}{2}$	0	000+	$-2/5\sqrt{13}$
$\tilde{0}$ $\frac{3}{2}$ $\frac{3}{2}$	0	010+	$-\sqrt{3.7}/5\sqrt{13}$

6 $\frac{9}{2}$ $\frac{5}{2}$			
0 $\frac{3}{2}$ $\frac{3}{2}$	0	000+	$+4\sqrt{3}/\sqrt{5.11.13}$
0 $\frac{3}{2}$ $\frac{3}{2}$	0	010+	$+\sqrt{7}/\sqrt{5.11.13}$
1 $\frac{1}{2}$ $\frac{3}{2}$	0	000+	$-2.5/\sqrt{3.7.11.13}$
1 $\frac{3}{2}$ $\frac{3}{2}$	0	000−	$-8.2\sqrt{2}/\sqrt{5.7.11.13}$
1 $\frac{3}{2}$ $\frac{3}{2}$	0	010−	$+\sqrt{2}/\sqrt{3.5.11.13}$
1 $\frac{3}{2}$ $\frac{3}{2}$	1	000−	$-\sqrt{2}/\sqrt{5.7.11.13}$
1 $\frac{3}{2}$ $\frac{3}{2}$	1	010−	$-3\sqrt{2.3}/\sqrt{5.11.13}$
1 $\frac{3}{2}$ $\tilde{\frac{1}{2}}$	0	000+	$+4/\sqrt{7.11.13}$
1 $\frac{3}{2}$ $\tilde{\frac{1}{2}}$	0	010+	$+1/\sqrt{3.11.13}$
2 $\frac{1}{2}$ $\frac{3}{2}$	0	000−	$-2\sqrt{7}/3\sqrt{11.13}$
2 $\frac{3}{2}$ $\frac{3}{2}$	0	000+	$-4\sqrt{2.3}/\sqrt{5.7.11.13}$
2 $\frac{3}{2}$ $\frac{3}{2}$	0	010+	$+7\sqrt{2}/3\sqrt{5.11.13}$
2 $\frac{3}{2}$ $\tilde{\frac{1}{2}}$	0	000−	$+4\sqrt{3}/\sqrt{7.11.13}$
2 $\frac{3}{2}$ $\tilde{\frac{1}{2}}$	0	010−	$+8/3\sqrt{11.13}$
$\tilde{1}$ $\frac{1}{2}$ $\frac{3}{2}$	0	000−	$+2.19/3.7\sqrt{3.5.13}$
$\tilde{1}$ $\frac{1}{2}$ $\frac{3}{2}$	0	100−	$-8\sqrt{3}/7\sqrt{11.13}$
$\tilde{1}$ $\frac{1}{2}$ $\tilde{\frac{1}{2}}$	0	000+	$+8.4/3.7\sqrt{3.13}$
$\tilde{1}$ $\frac{1}{2}$ $\tilde{\frac{1}{2}}$	0	100+	$+\sqrt{3.5}/7\sqrt{11.13}$
$\tilde{1}$ $\frac{3}{2}$ $\frac{3}{2}$	0	000+	$-8.2\sqrt{2}/5.7\sqrt{13}$
$\tilde{1}$ $\frac{3}{2}$ $\frac{3}{2}$	0	010+	$-\sqrt{2.7}/3.5\sqrt{3.13}$
$\tilde{1}$ $\frac{3}{2}$ $\frac{3}{2}$	0	100+	$-3\sqrt{2.5}/7\sqrt{11.13}$
$\tilde{1}$ $\frac{3}{2}$ $\frac{3}{2}$	0	110+	0
$\tilde{1}$ $\frac{3}{2}$ $\frac{3}{2}$	1	000−	$+3\sqrt{2}/5\sqrt{13}$
$\tilde{1}$ $\frac{3}{2}$ $\frac{3}{2}$	1	010−	$-23\sqrt{2}/3.5\sqrt{3.7.13}$
$\tilde{1}$ $\frac{3}{2}$ $\frac{3}{2}$	1	100−	0
$\tilde{1}$ $\frac{3}{2}$ $\frac{3}{2}$	1	110−	$+2\sqrt{2.3.5}/\sqrt{7.11.13}$
$\tilde{1}$ $\frac{3}{2}$ $\tilde{\frac{1}{2}}$	0	000−	$-4/7\sqrt{5.13}$
$\tilde{1}$ $\frac{3}{2}$ $\tilde{\frac{1}{2}}$	0	010−	$-23/3\sqrt{3.5.7.13}$
$\tilde{1}$ $\frac{3}{2}$ $\tilde{\frac{1}{2}}$	0	100−	$+4.3/7\sqrt{11.13}$
$\tilde{1}$ $\frac{3}{2}$ $\tilde{\frac{1}{2}}$	0	110−	$-4\sqrt{3}/\sqrt{7.11.13}$
$\tilde{0}$ $\frac{1}{2}$ $\tilde{\frac{1}{2}}$	0	000−	$-4\sqrt{2}/3\sqrt{7.13}$
$\tilde{0}$ $\frac{3}{2}$ $\frac{3}{2}$	0	000−	$+4\sqrt{3}/5\sqrt{7.13}$
$\tilde{0}$ $\frac{3}{2}$ $\frac{3}{2}$	0	010−	$-7/3.5\sqrt{13}$

6 $\frac{9}{2}$ $\frac{7}{2}$			
0 $\frac{1}{2}$ $\frac{1}{2}$	0	000−	$-\sqrt{2.7}/\sqrt{3.11.13}$
0 $\frac{3}{2}$ $\frac{3}{2}$	0	000−	$+3/5\sqrt{11.13}$
0 $\frac{3}{2}$ $\frac{3}{2}$	0	010−	$-8\sqrt{7}/5\sqrt{3.11.13}$
1 $\frac{1}{2}$ $\frac{1}{2}$	0	000+	$+3/\sqrt{11.13}$
1 $\frac{1}{2}$ $\frac{3}{2}$	0	000−	$-2.3/\sqrt{5.7.11.13}$
1 $\frac{3}{2}$ $\frac{1}{2}$	0	000−	$-\sqrt{3.5}/2\sqrt{11.13}$
1 $\frac{3}{2}$ $\frac{1}{2}$	0	010−	0
1 $\frac{3}{2}$ $\frac{3}{2}$	0	000+	$-\sqrt{3}/5\sqrt{2.7.11.13}$
1 $\frac{3}{2}$ $\frac{3}{2}$	0	010+	$-4.3\sqrt{2}/5\sqrt{11.13}$

6 $\frac{9}{2}$ $\frac{7}{2}$			
1 $\frac{3}{2}$ $\frac{3}{2}$	1	000+	$+4.3\sqrt{2.3}/5\sqrt{7.11.13}$
1 $\frac{3}{2}$ $\frac{3}{2}$	1	010+	$+3\sqrt{2}/5\sqrt{11.13}$
1 $\frac{3}{2}$ $\tilde{\frac{1}{2}}$	0	000−	$-9/2.5\sqrt{7.11.13}$
1 $\frac{3}{2}$ $\tilde{\frac{1}{2}}$	0	010−	$+4\sqrt{3}/5\sqrt{11.13}$
2 $\frac{1}{2}$ $\frac{3}{2}$	0	000+	$-2\sqrt{7}/\sqrt{3.5.11.13}$
2 $\frac{3}{2}$ $\frac{1}{2}$	0	000+	$+2/\sqrt{5.11.13}$
2 $\frac{3}{2}$ $\frac{1}{2}$	0	010+	$-2\sqrt{7}/\sqrt{3.5.11.13}$
2 $\frac{3}{2}$ $\frac{3}{2}$	0	000−	$+\sqrt{2.11}/5\sqrt{7.13}$
2 $\frac{3}{2}$ $\frac{3}{2}$	0	010−	$+2.7\sqrt{2}/5\sqrt{3.11.13}$
2 $\frac{3}{2}$ $\tilde{\frac{1}{2}}$	0	000+	$+4\sqrt{3}/\sqrt{7.11.13}$
2 $\frac{3}{2}$ $\tilde{\frac{1}{2}}$	0	010+	$-2/\sqrt{11.13}$
$\tilde{1}$ $\frac{1}{2}$ $\frac{3}{2}$	0	000+	$+2.3/7\sqrt{13}$
$\tilde{1}$ $\frac{1}{2}$ $\frac{3}{2}$	0	100+	$-\sqrt{13}/7\sqrt{5.11}$
$\tilde{1}$ $\frac{1}{2}$ $\tilde{\frac{1}{2}}$	0	000−	$-1/7\sqrt{3.13}$
$\tilde{1}$ $\frac{1}{2}$ $\tilde{\frac{1}{2}}$	0	100−	$-2\sqrt{3}/7\sqrt{5.11.13}$
$\tilde{1}$ $\frac{3}{2}$ $\frac{1}{2}$	0	000+	$+5/2\sqrt{3.7.13}$
$\tilde{1}$ $\frac{3}{2}$ $\frac{1}{2}$	0	010+	0
$\tilde{1}$ $\frac{3}{2}$ $\frac{1}{2}$	0	100+	$-\sqrt{3.11}/2\sqrt{5.7.13}$
$\tilde{1}$ $\frac{3}{2}$ $\frac{1}{2}$	0	110+	$-7/\sqrt{5.11.13}$
$\tilde{1}$ $\frac{3}{2}$ $\frac{3}{2}$	0	000−	$+\sqrt{5}/7\sqrt{2.3.13}$
$\tilde{1}$ $\frac{3}{2}$ $\frac{3}{2}$	0	010−	0
$\tilde{1}$ $\frac{3}{2}$ $\frac{3}{2}$	0	100−	$-8.2\sqrt{2.3}/5.7\sqrt{11.13}$
$\tilde{1}$ $\frac{3}{2}$ $\frac{3}{2}$	0	110−	$-2\sqrt{2.7}/5\sqrt{11.13}$
$\tilde{1}$ $\frac{3}{2}$ $\frac{3}{2}$	1	000+	$-2\sqrt{2}/\sqrt{3.5.13}$
$\tilde{1}$ $\frac{3}{2}$ $\frac{3}{2}$	1	010+	$+3\sqrt{2}/\sqrt{5.7.13}$
$\tilde{1}$ $\frac{3}{2}$ $\frac{3}{2}$	1	100+	$+7\sqrt{3}/5\sqrt{2.11.13}$
$\tilde{1}$ $\frac{3}{2}$ $\frac{3}{2}$	1	110+	$+2.9\sqrt{2}/5\sqrt{7.11.13}$
$\tilde{1}$ $\frac{3}{2}$ $\tilde{\frac{1}{2}}$	0	000+	$+9.3/2.7\sqrt{5.13}$
$\tilde{1}$ $\frac{3}{2}$ $\tilde{\frac{1}{2}}$	0	010+	$+4/\sqrt{3.5.7.13}$
$\tilde{1}$ $\frac{3}{2}$ $\tilde{\frac{1}{2}}$	0	100+	$+3.61/2.5.7\sqrt{11.13}$
$\tilde{1}$ $\frac{3}{2}$ $\tilde{\frac{1}{2}}$	0	110+	$-\sqrt{3.13}/5\sqrt{7.11}$
$\tilde{0}$ $\frac{1}{2}$ $\tilde{\frac{1}{2}}$	0	000+	$+\sqrt{2}/\sqrt{7.13}$
$\tilde{0}$ $\frac{3}{2}$ $\frac{3}{2}$	0	000+	$-\sqrt{5}/\sqrt{7.13}$
$\tilde{0}$ $\frac{3}{2}$ $\frac{3}{2}$	0	010+	0

6 $\frac{9}{2}$ $\frac{9}{2}$			
0 $\frac{1}{2}$ $\frac{1}{2}$	0	000+	$+8/\sqrt{3.5.11.13}$
0 $\frac{3}{2}$ $\frac{3}{2}$	0	000+	$+4\sqrt{2.3}/5\sqrt{5.11.13}$
0 $\frac{3}{2}$ $\frac{3}{2}$	0	010+	$-3\sqrt{7}/5\sqrt{2.5.11.13}$
0 $\frac{3}{2}$ $\frac{3}{2}$	0	011+	$-8.4\sqrt{2}/5\sqrt{3.5.11.13}$
1 $\frac{1}{2}$ $\frac{1}{2}$	0	000−	0
1 $\frac{3}{2}$ $\frac{1}{2}$	0	000+	$+\sqrt{2.3.7}/5\sqrt{11.13}$
1 $\frac{3}{2}$ $\frac{1}{2}$	0	010+	$+3.7/5\sqrt{2.11.13}$
1 $\frac{3}{2}$ $\frac{3}{2}$	0	000−	0
1 $\frac{3}{2}$ $\frac{3}{2}$	0	010−	$-\sqrt{3}/\sqrt{5.11.13}$
1 $\frac{3}{2}$ $\frac{3}{2}$	0	011−	0
1 $\frac{3}{2}$ $\frac{3}{2}$	1	000−	0
1 $\frac{3}{2}$ $\frac{3}{2}$	1	010−	$-3\sqrt{3}/\sqrt{5.11.13}$
1 $\frac{3}{2}$ $\frac{3}{2}$	1	011−	0
2 $\frac{3}{2}$ $\frac{1}{2}$	0	000−	$+\sqrt{2.7}/5\sqrt{11.13}$

SO_3–O 3jm Factors (cont.)

$6\ \frac{9}{2}\ \frac{9}{2}$	
$2\ \frac{3}{2}\ \frac{1}{2}\ 0\ 010-$	$+8\sqrt{2}/5\sqrt{3.11.13}$
$2\ \frac{3}{2}\ \frac{3}{2}\ 0\ 000+$	$+8\sqrt{3.7}/5\sqrt{5.11.13}$
$2\ \frac{3}{2}\ \frac{3}{2}\ 0\ 010+$	$-\sqrt{11}/5\sqrt{5.13}$
$2\ \frac{3}{2}\ \frac{3}{2}\ 0\ 011+$	$+8.2\sqrt{7}/5\sqrt{3.5.11.13}$
$\tilde{1}\ \frac{3}{2}\ \frac{1}{2}\ 0\ 000-$	$-\sqrt{2}/\sqrt{3.5.13}$
$\tilde{1}\ \frac{3}{2}\ \frac{1}{2}\ 0\ 010-$	$+3/\sqrt{2.5.7.13}$
$\tilde{1}\ \frac{3}{2}\ \frac{1}{2}\ 0\ 100-$	$-2\sqrt{2.3}/5\sqrt{11.13}$
$\tilde{1}\ \frac{3}{2}\ \frac{1}{2}\ 0\ 110-$	$-2\sqrt{2.13}/5\sqrt{7.11}$
$\tilde{1}\ \frac{3}{2}\ \frac{3}{2}\ 0\ 000+$	0
$\tilde{1}\ \frac{3}{2}\ \frac{3}{2}\ 0\ 010+$	$-5/\sqrt{3.7.13}$
$\tilde{1}\ \frac{3}{2}\ \frac{3}{2}\ 0\ 011+$	0
$\tilde{1}\ \frac{3}{2}\ \frac{3}{2}\ 0\ 100+$	$+2.3.7/5\sqrt{5.11.13}$
$\tilde{1}\ \frac{3}{2}\ \frac{3}{2}\ 0\ 110+$	$-8\sqrt{3}/5\sqrt{5.7.11.13}$
$\tilde{1}\ \frac{3}{2}\ \frac{3}{2}\ 0\ 111+$	$-7/5\sqrt{5.11.13}$
$\tilde{1}\ \frac{3}{2}\ \frac{3}{2}\ 1\ 000-$	0
$\tilde{1}\ \frac{3}{2}\ \frac{3}{2}\ 1\ 010-$	$-1/\sqrt{3.7.13}$
$\tilde{1}\ \frac{3}{2}\ \frac{3}{2}\ 1\ 011-$	0
$\tilde{1}\ \frac{3}{2}\ \frac{3}{2}\ 1\ 100-$	0
$\tilde{1}\ \frac{3}{2}\ \frac{3}{2}\ 1\ 110-$	$-2\sqrt{3}/\sqrt{5.7.11.13}$
$\tilde{1}\ \frac{3}{2}\ \frac{3}{2}\ 1\ 111-$	0
$\tilde{0}\ \frac{3}{2}\ \frac{3}{2}\ 0\ 000-$	0
$\tilde{0}\ \frac{3}{2}\ \frac{3}{2}\ 0\ 010-$	$-1/\sqrt{2.13}$
$\tilde{0}\ \frac{3}{2}\ \frac{3}{2}\ 0\ 011-$	0
6 5 1	
$0\ 1\ 1\ 0\ 000+$	$+1/\sqrt{13}$
$0\ 1\ 1\ 0\ 010+$	0
$1\ 1\ 1\ 0\ 000-$	$-\sqrt{7}/2\sqrt{13}$
$1\ 1\ 1\ 0\ 010-$	0
$1\ 2\ 1\ 0\ 000+$	$-\sqrt{2.5}/\sqrt{11.13}$
$1\ \tilde{1}\ 1\ 0\ 000+$	$+\sqrt{3.5}/2\sqrt{11.13}$
$2\ 1\ 1\ 0\ 000+$	$-\sqrt{7}/11\sqrt{13}$
$2\ 1\ 1\ 0\ 010+$	$-2.3\sqrt{5}/11\sqrt{13}$
$2\ \tilde{1}\ 1\ 0\ 000-$	$+\sqrt{5}/\sqrt{11.13}$
$\tilde{1}\ 1\ 1\ 0\ 000+$	$+\sqrt{3.5}/2\sqrt{11.13}$
$\tilde{1}\ 1\ 1\ 0\ 010+$	$+4\sqrt{3}/\sqrt{7.11.13}$
$\tilde{1}\ 1\ 1\ 0\ 100+$	$-4\sqrt{3}/11\sqrt{13}$
$\tilde{1}\ 1\ 1\ 0\ 110+$	$+9\sqrt{3.5}/11\sqrt{7.13}$
$\tilde{1}\ 2\ 1\ 0\ 000-$	$+\sqrt{2}/\sqrt{7.13}$
$\tilde{1}\ 2\ 1\ 0\ 100-$	$-3\sqrt{2.5}/\sqrt{7.11.13}$
$\tilde{1}\ \tilde{1}\ 1\ 0\ 000-$	$-\sqrt{7}/2\sqrt{13}$
$\tilde{1}\ \tilde{1}\ 1\ 0\ 100-$	0
$\tilde{0}\ \tilde{1}\ 1\ 0\ 000+$	$+1/\sqrt{13}$
6 5 2	
$0\ 2\ 2\ 0\ 000-$	$-2\sqrt{2}/\sqrt{11.13}$
$0\ \tilde{1}\ \tilde{1}\ 0\ 000-$	$+\sqrt{3}/\sqrt{11.13}$
$1\ 1\ 2\ 0\ 000-$	$-2\sqrt{2.3.5}/11\sqrt{13}$
$1\ 1\ 2\ 0\ 010-$	$+2.3\sqrt{2.3}/11\sqrt{7.13}$
$1\ 1\ \tilde{1}\ 0\ 000-$	$-3\sqrt{5}/2.11\sqrt{13}$
$1\ 1\ \tilde{1}\ 0\ 010-$	$-4.3/11\sqrt{7.13}$
6 5 2	
$1\ 2\ \tilde{1}\ 0\ 000+$	$+3\sqrt{2.3}/\sqrt{7.11.13}$
$1\ \tilde{1}\ 2\ 0\ 000+$	$-3\sqrt{2.3}/\sqrt{7.11.13}$
$1\ \tilde{1}\ \tilde{1}\ 0\ 000+$	$+3\sqrt{3}/2\sqrt{7.11.13}$
$2\ 1\ \tilde{1}\ 0\ 000+$	$-\sqrt{3.5}/11\sqrt{13}$
$2\ 1\ \tilde{1}\ 0\ 010+$	$+2\sqrt{3.7}/11\sqrt{13}$
$2\ 2\ 2\ 0\ 000-$	$+4/\sqrt{7.11.13}$
$2\ \tilde{1}\ \tilde{1}\ 0\ 000-$	$+5\sqrt{3}/\sqrt{7.11.13}$
$\tilde{1}\ 1\ 2\ 0\ 000+$	$+3\sqrt{2.3}/\sqrt{7.11.13}$
$\tilde{1}\ 1\ 2\ 0\ 010+$	$+2\sqrt{2.3}/7\sqrt{5.11.13}$
$\tilde{1}\ 1\ 2\ 0\ 100+$	$-2\sqrt{2.3.5}/11\sqrt{7.13}$
$\tilde{1}\ 1\ 2\ 0\ 110+$	$-9.3\sqrt{2.3}/7.11\sqrt{13}$
$\tilde{1}\ 1\ \tilde{1}\ 0\ 000+$	$+3\sqrt{3}/2\sqrt{7.11.13}$
$\tilde{1}\ 1\ \tilde{1}\ 0\ 010+$	$-8.4\sqrt{3}/7\sqrt{5.11.13}$
$\tilde{1}\ 1\ \tilde{1}\ 0\ 100+$	$-8\sqrt{3.5}/11\sqrt{7.13}$
$\tilde{1}\ 1\ \tilde{1}\ 0\ 110+$	$-9\sqrt{3}/7.11\sqrt{13}$
$\tilde{1}\ 2\ \tilde{1}\ 0\ 000-$	$-3\sqrt{2.3}/7\sqrt{5.13}$
$\tilde{1}\ 2\ \tilde{1}\ 0\ 100-$	$-5\sqrt{2.3}/7\sqrt{11.13}$
$\tilde{1}\ \tilde{1}\ 2\ 0\ 000-$	$+2\sqrt{2.3}/7\sqrt{5.13}$
$\tilde{1}\ \tilde{1}\ 2\ 0\ 100-$	$-2.3\sqrt{2.3}/7\sqrt{11.13}$
$\tilde{1}\ \tilde{1}\ \tilde{1}\ 0\ 000-$	$-9.3/2.7\sqrt{5.13}$
$\tilde{1}\ \tilde{1}\ \tilde{1}\ 0\ 100-$	$-4.3/7\sqrt{11.13}$
$\tilde{0}\ 1\ \tilde{1}\ 0\ 000-$	$-\sqrt{3}/\sqrt{11.13}$
$\tilde{0}\ 1\ \tilde{1}\ 0\ 010-$	$-8\sqrt{3}/\sqrt{5.7.11.13}$
$\tilde{0}\ 2\ 2\ 0\ 000+$	$+2\sqrt{2}/\sqrt{5.7.13}$
6 5 3	
$0\ 1\ 1\ 0\ 000+$	$-3\sqrt{7}/11\sqrt{2.13}$
$0\ 1\ 1\ 0\ 010+$	$+2\sqrt{2.5}/11\sqrt{13}$
$0\ \tilde{1}\ \tilde{1}\ 0\ 000+$	$+3/\sqrt{2.11.13}$
$1\ 1\ 1\ 0\ 000-$	$+3/11\sqrt{2.13}$
$1\ 1\ 1\ 0\ 010-$	$+9\sqrt{2.5}/11\sqrt{7.13}$
$1\ 1\ \tilde{1}\ 0\ 000+$	$-3\sqrt{3.5}/11\sqrt{2.13}$
$1\ 1\ \tilde{1}\ 0\ 010+$	$-\sqrt{2.3}/11\sqrt{7.13}$
$1\ 2\ 1\ 0\ 000+$	$+3\sqrt{5}/\sqrt{7.11.13}$
$1\ 2\ \tilde{1}\ 0\ 000-$	$+3/\sqrt{7.11.13}$
$1\ \tilde{1}\ 1\ 0\ 000+$	$-\sqrt{3.5}/\sqrt{2.7.11.13}$
$1\ \tilde{1}\ \tilde{1}\ 0\ 000-$	$+9/\sqrt{2.7.11.13}$
$1\ \tilde{1}\ \tilde{0}\ 0\ 000+$	$-3/\sqrt{7.11.13}$
$2\ 1\ 1\ 0\ 000+$	$-9/11\sqrt{2.13}$
$2\ 1\ 1\ 0\ 010+$	$-5\sqrt{2.5}/11\sqrt{7.13}$
$2\ 1\ \tilde{1}\ 0\ 000-$	$-3\sqrt{5}/11\sqrt{2.13}$
$2\ 1\ \tilde{1}\ 0\ 010-$	$-\sqrt{2}/11\sqrt{7.13}$
$2\ 2\ \tilde{0}\ 0\ 000-$	$-4\sqrt{3}/\sqrt{7.11.13}$
$2\ \tilde{1}\ 1\ 0\ 000-$	$+\sqrt{5}/\sqrt{2.7.11.13}$
$2\ \tilde{1}\ \tilde{1}\ 0\ 000+$	$-9/\sqrt{2.7.11.13}$
$\tilde{1}\ 1\ 1\ 0\ 000+$	$-\sqrt{3.5}/\sqrt{2.7.11.13}$
$\tilde{1}\ 1\ 1\ 0\ 010+$	$+\sqrt{2.3}/\sqrt{11.13}$
$\tilde{1}\ 1\ 1\ 0\ 100+$	$-3\sqrt{2.3}/11\sqrt{7.13}$
$\tilde{1}\ 1\ 1\ 0\ 110+$	$-\sqrt{2.3.5}/11\sqrt{13}$
$\tilde{1}\ 1\ \tilde{1}\ 0\ 000-$	$+9/\sqrt{2.7.11.13}$

SO_3–O 3jm Factors (cont.)

6 5 3

1̃ 1 1̃ 0 010−	+5√2.5/7√11.13
1̃ 1 1̃ 0 100−	−3√2.5/11√7.13
1̃ 1 1̃ 0 110−	+9√2/7.11√13
1̃ 1 0̃ 0 000+	−3/√7.11.13
1̃ 1 0̃ 0 010+	+4√5/7√11.13
1̃ 1 0̃ 0 100+	−4.3√5/11√7.13
1̃ 1 0̃ 0 110+	+3/7.11√13
1̃ 2 1 0 000−	+1/7√13
1̃ 2 1 0 100−	−3√5/7√11.13
1̃ 2 1̃ 0 000+	−3√5/7√13
1̃ 2 1̃ 0 100+	+3/7√11.13
1̃ 1̃ 1 0 000−	−1/7√2.13
1̃ 1̃ 1 0 100−	−9√2.5/7√11.13
1̃ 1̃ 1̃ 0 000+	+√3.5/7√2.13
1̃ 1̃ 1̃ 0 100+	+3√2.3/7√11.13
0̃ 1 1̃ 0 000+	−3/√2.11.13
0̃ 1 1̃ 0 010+	+2√2.5/√7.11.13
0̃ 1̃ 1 0 000+	+1/√2.7.13

6 5 4

0 1 1 0 000−	−7√3/2.11√13
0 1 1 0 010−	−4√3.7/11√5.13
0 2 2 0 000−	+√3/√2.5.11.13
0 1̃ 1̃ 0 000−	−√5/2√11.13
1 1 0 0 000−	+√5.7/11√2.13
1 1 0 0 010−	+4√2/11√13
1 1 1 0 000+	+3√3.7/2.11√13
1 1 1 0 010+	−2√3.5/11√13
1 1 2 0 000−	−2√2/11√13
1 1 2 0 010−	−√5.13/11√2.7
1 1 1̃ 0 000−	−√3/2.11√13
1 1 1̃ 0 010−	−4√3/11√5.7.13
1 2 1 0 000−	0
1 2 1̃ 0 000+	−4√2/√5.7.11.13
1 1̃ 1 0 000−	−√11/2√5.13
1 1̃ 2 0 000+	+4√2/√5.7.11.13
1 1̃ 1̃ 0 000+	−5√5/2√7.11.13
2 1 1 0 000−	+√3.7/2.11√13
2 1 1 0 010−	−7√3/11√5.13
2 1 1̃ 0 000+	−23/2.11√13
2 1 1̃ 0 010+	+√7/11√5.13
2 2 0 0 000−	+√2/√3.11.13
2 2 2 0 000−	+√11/√3.5.7.13
2 1̃ 1 0 000+	+17/2√3.5.11.13
2 1̃ 1̃ 0 000−	+3/2√5.7.11.13
1̃ 1 1 0 000−	+√5/2√11.13
1̃ 1 1 0 010−	+4/√7.11.13
1̃ 1 1 0 100−	+3/11√13
1̃ 1 1 0 110−	+8.3/11√5.7.13
1̃ 1 2 0 000+	+2√2.5/√7.11.13

6 5 4

1̃ 1 2 0 010+	−23/7√2.11.13
1̃ 1 2 0 100+	+3/11√2.7.13
1̃ 1 2 0 110+	+2.3√2.13/7.11√5
1̃ 1 1̃ 0 000+	−5√5/2√7.11.13
1̃ 1 1̃ 0 010+	+2/7√11.13
1̃ 1 1̃ 0 100+	−3.5/11√7.13
1̃ 1 1̃ 0 110+	+4.3/7.11√5.13
1̃ 2 1 0 000+	−2√2/√3.7.13
1̃ 2 1 0 100+	+3√2.3/√5.7.11.13
1̃ 2 1̃ 0 000−	−2√2/7√13
1̃ 2 1̃ 0 100−	−3√2.11/7√5.13
1̃ 1̃ 0 0 000−	+√5/3√2.7.13
1̃ 1̃ 0 0 100−	−2.3√2/√7.11.13
1̃ 1̃ 1 0 000+	−√3/2√7.13
1̃ 1̃ 1 0 100+	−3√3/√5.7.11.13
1̃ 1̃ 2 0 000−	+4√2/3.7√13
1̃ 1̃ 2 0 100−	+9/7√2.5.11.13
1̃ 1̃ 1̃ 0 000−	+17/2.7√3.13
1̃ 1̃ 1̃ 0 100−	−√3.11/7√5.13
0̃ 1 1̃ 0 000−	+√5/2√11.13
0̃ 1 1̃ 0 010−	+4/√7.11.13
0̃ 2 2 0 000+	+5/√2.3.7.13
0̃ 1̃ 1 0 000−	+1/2√3.13

6 5 5

0 1 1 0 000+	−3√2.5/11√11.13.17
0 1 1 0 010+	−3.7√7/11√2.11.13.17
0 1 1 0 011+	+8√2.17/11√5.11.13
0 2 2 0 000+	+4√3/√5.11.13.17
0 1̃ 1̃ 0 000+	−3√2.5/√11.13.17
1 1 1 0 000−	0
1 1 1 0 010−	−3.7/2√2.11.13.17
1 1 1 0 011−	0
1 2 1 0 000+	+9/11√13.17
1 2 1 0 001+	+3√5.7/11√13.17
1 1̃ 1 0 000+	+3√2.3/11√13.17
1 1̃ 1 0 001+	+53√3.7/2.11√2.5.13.17
1 1̃ 2 0 000−	+2.3√7/√5.11.13.17
1 1̃ 1̃ 0 000−	0
2 1 1 0 000+	−3.5√2.5.7/11√11.13.17
2 1 1 0 010+	−9/11√2.11.13.17
2 1 1 0 011+	−2.23√2.7/11√5.11.13.17
2 2 2 0 000+	+4√2.3.7/√5.11.13.17
2 1̃ 1 0 000−	+2.3√2/11√13.17
2 1̃ 1 0 001−	−√7.13/11√2.5.17
2 1̃ 1̃ 0 000+	+3√2.7/√5.11.13.17
1̃ 1 1 0 000+	−3.5√2.3/11√13.17
1̃ 1 1 0 010+	−41√3.5/2.11√2.7.13.17
1̃ 1 1 0 011+	+4√2.3/11√13.17
1̃ 1 1 0 100+	−3√3.5.17/2.11√2.11.13

SO_3–O 3jm Factors (cont.)

6 5 5

$\tilde{1}\ 1\ 1$ 0 110+ $+8.4.3\sqrt{2.3}/11\sqrt{7.11.13.17}$

$\tilde{1}\ 1\ 1$ 0 111+ $+2.47\sqrt{2.3}/11\sqrt{5.11.13.17}$

$\tilde{1}\ 2\ 1$ 0 000− $-2.3\sqrt{5}/\sqrt{7.11.13.17}$

$\tilde{1}\ 2\ 1$ 0 001− $+1/\sqrt{11.13.17}$

$\tilde{1}\ 2\ 1$ 0 100− $-3.19/11\sqrt{7.13.17}$

$\tilde{1}\ 2\ 1$ 0 101− $+8.2.3/11\sqrt{5.13.17}$

$\tilde{1}\ \tilde{1}\ 1$ 0 000− 0

$\tilde{1}\ \tilde{1}\ 1$ 0 001− $-\sqrt{11}/2\sqrt{2.13.17}$

$\tilde{1}\ \tilde{1}\ 1$ 0 100− $-3.5\sqrt{7}/2.11\sqrt{2.13.17}$

$\tilde{1}\ \tilde{1}\ 1$ 0 101− $+4.9\sqrt{2}/11\sqrt{5.13.17}$

$\tilde{1}\ \tilde{1}\ 2$ 0 000+ $-3/\sqrt{13.17}$

$\tilde{1}\ \tilde{1}\ 2$ 0 100+ $+3/\sqrt{5.11.13.17}$

$\tilde{1}\ \tilde{1}\ \tilde{1}$ 0 000+ $-\sqrt{2.3}/\sqrt{13.17}$

$\tilde{1}\ \tilde{1}\ \tilde{1}$ 0 100+ $-3\sqrt{3}/2\sqrt{2.5.11.13.17}$

$\tilde{0}\ 2\ 2$ 0 000− 0

$\tilde{0}\ \tilde{1}\ 1$ 0 000+ $-3\sqrt{2.5}/\sqrt{11.13.17}$

$\tilde{0}\ \tilde{1}\ 1$ 0 001+ $+\sqrt{7}/\sqrt{2.11.13.17}$

$6\ \frac{11}{2}\ \frac{1}{2}$

$0\ \frac{1}{2}\ \frac{1}{2}$ 0 000− $+1/\sqrt{13}$

$1\ \frac{1}{2}\ \frac{1}{2}$ 0 000+ $+\sqrt{7}/\sqrt{2.3.13}$

$1\ \frac{3}{2}\ \frac{1}{2}$ 0 000− $-\sqrt{11}/\sqrt{2.3.13}$

$1\ \frac{3}{2}\ \frac{1}{2}$ 0 010− 0

$2\ \frac{3}{2}\ \frac{1}{2}$ 0 000+ $+\sqrt{2}/\sqrt{11.13}$

$2\ \frac{3}{2}\ \frac{1}{2}$ 0 010+ $-2\sqrt{5}/\sqrt{11.13}$

$\tilde{1}\ \frac{3}{2}\ \frac{1}{2}$ 0 000+ $-\sqrt{3.5}/\sqrt{2.7.13}$

$\tilde{1}\ \frac{3}{2}\ \frac{1}{2}$ 0 010+ $+4/\sqrt{3.7.13}$

$\tilde{1}\ \frac{3}{2}\ \frac{1}{2}$ 0 100+ $+4\sqrt{2.3}/\sqrt{7.11.13}$

$\tilde{1}\ \frac{3}{2}\ \frac{1}{2}$ 0 110+ $+3\sqrt{3.5}/\sqrt{7.11.13}$

$\tilde{1}\ \tilde{\frac{1}{2}}\ \frac{1}{2}$ 0 000− $+\sqrt{7}/\sqrt{2.3.13}$

$\tilde{1}\ \tilde{\frac{1}{2}}\ \frac{1}{2}$ 0 100− 0

$\tilde{0}\ \tilde{\frac{1}{2}}\ \frac{1}{2}$ 0 000+ $+1/\sqrt{13}$

$6\ \frac{11}{2}\ \frac{3}{2}$

$0\ \frac{3}{2}\ \frac{3}{2}$ 0 000+ $-1/\sqrt{13}$

$0\ \frac{3}{2}\ \frac{3}{2}$ 0 010+ 0

$1\ \frac{1}{2}\ \frac{3}{2}$ 0 000+ $-\sqrt{11}/2\sqrt{3.13}$

$1\ \frac{3}{2}\ \frac{3}{2}$ 0 000− $-2.19\sqrt{2}/11\sqrt{3.7.13}$

$1\ \frac{3}{2}\ \frac{3}{2}$ 0 010− $-4\sqrt{3.5}/11\sqrt{7.13}$

$1\ \frac{3}{2}\ \frac{3}{2}$ 1 000− $-17/11\sqrt{2.3.7.13}$

$1\ \frac{3}{2}\ \frac{3}{2}$ 1 010− $+2\sqrt{3.5}/11\sqrt{7.13}$

$1\ \tilde{\frac{1}{2}}\ \frac{3}{2}$ 0 000+ $-\sqrt{3.5}/2\sqrt{7.13}$

$2\ \frac{1}{2}\ \frac{3}{2}$ 0 000− $+1/\sqrt{11.13}$

$2\ \frac{3}{2}\ \frac{3}{2}$ 0 000+ $+4\sqrt{2}/11\sqrt{7.13}$

$2\ \frac{3}{2}\ \frac{3}{2}$ 0 010+ $+2\sqrt{5.7}/11\sqrt{13}$

$2\ \tilde{\frac{1}{2}}\ \frac{3}{2}$ 0 000− $+\sqrt{5}/\sqrt{7.13}$

$\tilde{1}\ \frac{1}{2}\ \frac{3}{2}$ 0 000− $-\sqrt{3.5}/2\sqrt{7.13}$

$\tilde{1}\ \frac{1}{2}\ \frac{3}{2}$ 0 100− $+4\sqrt{3}/\sqrt{7.11.13}$

$\tilde{1}\ \frac{3}{2}\ \frac{3}{2}$ 0 000+ $-2\sqrt{2.3.5}/7\sqrt{11.13}$

$\tilde{1}\ \frac{3}{2}\ \frac{3}{2}$ 0 010+ $-4/\sqrt{3.11.13}$

$\tilde{1}\ \frac{3}{2}\ \frac{3}{2}$ 0 100+ $+8.2\sqrt{2.3}/7.11\sqrt{13}$

$6\ \frac{11}{2}\ \frac{3}{2}$

$\tilde{1}\ \frac{3}{2}\ \frac{3}{2}$ 0 110+ $-3\sqrt{3.5}/11\sqrt{13}$

$\tilde{1}\ \frac{3}{2}\ \frac{3}{2}$ 1 000− $+3\sqrt{3.5}/7\sqrt{2.11.13}$

$\tilde{1}\ \frac{3}{2}\ \frac{3}{2}$ 1 010− $-2.17/7\sqrt{3.11.13}$

$\tilde{1}\ \frac{3}{2}\ \frac{3}{2}$ 1 100− $+2.3.5\sqrt{2.3}/7.11\sqrt{13}$

$\tilde{1}\ \frac{3}{2}\ \frac{3}{2}$ 1 110− $+2.3\sqrt{3.5}/7.11\sqrt{13}$

$\tilde{1}\ \tilde{\frac{1}{2}}\ \frac{3}{2}$ 0 000− $-\sqrt{11}/2\sqrt{3.13}$

$\tilde{1}\ \tilde{\frac{1}{2}}\ \frac{3}{2}$ 0 100− 0

$\tilde{0}\ \frac{3}{2}\ \frac{3}{2}$ 0 000− $-3\sqrt{5}/\sqrt{7.11.13}$

$\tilde{0}\ \frac{3}{2}\ \frac{3}{2}$ 0 010− $+4\sqrt{2}/\sqrt{7.11.13}$

$6\ \frac{11}{2}\ \frac{5}{2}$

$0\ \frac{3}{2}\ \frac{3}{2}$ 0 000− $-3\sqrt{2.3}/11\sqrt{13}$

$0\ \frac{3}{2}\ \frac{3}{2}$ 0 010− $-4\sqrt{5}/11\sqrt{3.13}$

$0\ \tilde{\frac{1}{2}}\ \tilde{\frac{1}{2}}$ 0 000− $+1/\sqrt{3.13}$

$1\ \frac{1}{2}\ \frac{3}{2}$ 0 000− $-3/\sqrt{2.11.13}$

$1\ \frac{3}{2}\ \frac{3}{2}$ 0 000+ $-\sqrt{13}/11\sqrt{7}$

$1\ \frac{3}{2}\ \frac{3}{2}$ 0 010+ $+2\sqrt{2.5}/11\sqrt{7.13}$

$1\ \frac{3}{2}\ \frac{3}{2}$ 1 000+ $+2\sqrt{7}/11\sqrt{13}$

$1\ \frac{3}{2}\ \frac{3}{2}$ 1 010+ $+8\sqrt{2.5}/11\sqrt{7.13}$

$1\ \frac{3}{2}\ \tilde{\frac{1}{2}}$ 0 000− $-17\sqrt{5}/11\sqrt{2.7.13}$

$1\ \frac{3}{2}\ \tilde{\frac{1}{2}}$ 0 010− $+8/11\sqrt{7.13}$

$1\ \tilde{\frac{1}{2}}\ \frac{3}{2}$ 0 000− $-\sqrt{5}/\sqrt{2.7.13}$

$1\ \tilde{\frac{1}{2}}\ \tilde{\frac{1}{2}}$ 0 000+ $-1/\sqrt{2.7.13}$

$2\ \frac{1}{2}\ \frac{3}{2}$ 0 000+ $-\sqrt{2}/\sqrt{3.11.13}$

$2\ \frac{3}{2}\ \frac{3}{2}$ 0 000− $+8.4/11\sqrt{3.7.13}$

$2\ \frac{3}{2}\ \frac{3}{2}$ 0 010− $-2.5\sqrt{2.5}/11\sqrt{3.7.13}$

$2\ \frac{3}{2}\ \tilde{\frac{1}{2}}$ 0 000+ $+\sqrt{2.5}/\sqrt{3.7.13}$

$2\ \frac{3}{2}\ \tilde{\frac{1}{2}}$ 0 010+ $+2/\sqrt{3.7.13}$

$2\ \tilde{\frac{1}{2}}\ \frac{3}{2}$ 0 000+ $-\sqrt{2.5}/\sqrt{3.7.13}$

$\tilde{1}\ \frac{1}{2}\ \frac{3}{2}$ 0 000+ $-\sqrt{5}/\sqrt{2.7.13}$

$\tilde{1}\ \frac{1}{2}\ \frac{3}{2}$ 0 100+ $-2\sqrt{2}/3\sqrt{7.11.13}$

$\tilde{1}\ \frac{1}{2}\ \tilde{\frac{1}{2}}$ 0 000− $+1/\sqrt{2.7.13}$

$\tilde{1}\ \frac{1}{2}\ \tilde{\frac{1}{2}}$ 0 100− $-8\sqrt{2.5}/3\sqrt{7.11.13}$

$\tilde{1}\ \frac{3}{2}\ \frac{3}{2}$ 0 000− $-5\sqrt{5}/7\sqrt{11.13}$

$\tilde{1}\ \frac{3}{2}\ \frac{3}{2}$ 0 010− $+2\sqrt{2}/\sqrt{11.13}$

$\tilde{1}\ \frac{3}{2}\ \frac{3}{2}$ 0 100− $+8.8/3.7.11\sqrt{13}$

$\tilde{1}\ \frac{3}{2}\ \frac{3}{2}$ 0 110− $-2\sqrt{2.5}/11\sqrt{13}$

$\tilde{1}\ \frac{3}{2}\ \frac{3}{2}$ 1 000+ $+2\sqrt{5}/7\sqrt{11.13}$

$\tilde{1}\ \frac{3}{2}\ \frac{3}{2}$ 1 010+ $+4.3\sqrt{2}/7\sqrt{11.13}$

$\tilde{1}\ \frac{3}{2}\ \frac{3}{2}$ 1 100+ $-8.4.5/3.7.11\sqrt{13}$

$\tilde{1}\ \frac{3}{2}\ \frac{3}{2}$ 1 110+ $+\sqrt{2.5.13}/7.11$

$\tilde{1}\ \frac{3}{2}\ \tilde{\frac{1}{2}}$ 0 000+ $-29/7\sqrt{2.11.13}$

$\tilde{1}\ \frac{3}{2}\ \tilde{\frac{1}{2}}$ 0 010+ $-4\sqrt{5}/7\sqrt{11.13}$

$\tilde{1}\ \frac{3}{2}\ \tilde{\frac{1}{2}}$ 0 100+ $-4.5\sqrt{2.5}/3.7.11\sqrt{13}$

$\tilde{1}\ \frac{3}{2}\ \tilde{\frac{1}{2}}$ 0 110+ $-3/7.11\sqrt{13}$

$\tilde{1}\ \tilde{\frac{1}{2}}\ \frac{3}{2}$ 0 000+ $+\sqrt{11}/7\sqrt{2.13}$

$\tilde{1}\ \tilde{\frac{1}{2}}\ \frac{3}{2}$ 0 100+ $+2\sqrt{2.5}/7\sqrt{13}$

$\tilde{0}\ \frac{1}{2}\ \tilde{\frac{1}{2}}$ 0 000+ $+1/\sqrt{3.13}$

$\tilde{0}\ \frac{3}{2}\ \frac{3}{2}$ 0 000+ $-\sqrt{2.5}/\sqrt{3.7.11.13}$

$\tilde{0}\ \frac{3}{2}\ \frac{3}{2}$ 0 010+ $-4\sqrt{3}/\sqrt{7.11.13}$

SO_3–O 3jm Factors (cont.)

$6\ \frac{11}{2}\ \frac{7}{2}$	
$0\ \frac{1}{2}\ \frac{1}{2}\ 0\ 000+$	$-7/4\sqrt{11.13}$
$0\ \frac{3}{2}\ \frac{3}{2}\ 0\ 000+$	$-\sqrt{5}/2.11\sqrt{2.13}$
$0\ \frac{3}{2}\ \frac{3}{2}\ 0\ 010+$	$+8/11\sqrt{13}$
$0\ \tilde{\frac{1}{2}}\ \tilde{\frac{1}{2}}\ 0\ 000+$	$+\sqrt{3}/4\sqrt{13}$
$1\ \frac{1}{2}\ \frac{1}{2}\ 0\ 000-$	$-\sqrt{7}/2\sqrt{2.3.11.13}$
$1\ \frac{1}{2}\ \frac{3}{2}\ 0\ 000+$	$-\sqrt{2.5}/\sqrt{3.11.13}$
$1\ \frac{3}{2}\ \frac{1}{2}\ 0\ 000+$	$+41/2.11\sqrt{2.3.13}$
$1\ \frac{3}{2}\ \frac{1}{2}\ 0\ 010+$	$-4\sqrt{5}/11\sqrt{3.13}$
$1\ \frac{3}{2}\ \frac{3}{2}\ 0\ 000-$	$+\sqrt{5.13}/2.11\sqrt{3.7}$
$1\ \frac{3}{2}\ \frac{3}{2}\ 0\ 010-$	$-5\sqrt{2}/11\sqrt{3.7.13}$
$1\ \frac{3}{2}\ \frac{3}{2}\ 1\ 000-$	$-\sqrt{5.7}/2.11\sqrt{3.13}$
$1\ \frac{3}{2}\ \frac{3}{2}\ 1\ 010-$	$-5.5\sqrt{3}/11\sqrt{2.7.13}$
$1\ \frac{3}{2}\ \tilde{\frac{1}{2}}\ 0\ 000+$	$-3.5\sqrt{5}/2.11\sqrt{2.7.13}$
$1\ \frac{3}{2}\ \tilde{\frac{1}{2}}\ 0\ 010+$	$-1/11\sqrt{7.13}$
$1\ \tilde{\frac{1}{2}}\ \frac{3}{2}\ 0\ 000+$	$-1/\sqrt{2.3.7.13}$
$1\ \tilde{\frac{1}{2}}\ \tilde{\frac{1}{2}}\ 0\ 000-$	$-5/2\sqrt{2.7.13}$
$2\ \frac{1}{2}\ \frac{3}{2}\ 0\ 000-$	$+3\sqrt{5}/2\sqrt{2.11.13}$
$2\ \frac{3}{2}\ \frac{1}{2}\ 0\ 000-$	$+\sqrt{13}/2.11\sqrt{2}$
$2\ \frac{3}{2}\ \frac{1}{2}\ 0\ 010-$	$-17\sqrt{5}/2.3.11\sqrt{13}$
$2\ \frac{3}{2}\ \frac{3}{2}\ 0\ 000+$	$+4\sqrt{5}/11\sqrt{7.13}$
$2\ \frac{3}{2}\ \frac{3}{2}\ 0\ 010+$	$+1/3.11\sqrt{2.7.13}$
$2\ \frac{3}{2}\ \tilde{\frac{1}{2}}\ 0\ 000-$	$-17\sqrt{3.5}/2.11\sqrt{2.7.13}$
$2\ \frac{3}{2}\ \tilde{\frac{1}{2}}\ 0\ 010-$	$-3\sqrt{3}/2.11\sqrt{7.13}$
$2\ \tilde{\frac{1}{2}}\ \frac{3}{2}\ 0\ 000-$	$-11/2.3\sqrt{2.7.13}$
$\tilde{1}\ \frac{1}{2}\ \frac{3}{2}\ 0\ 000-$	$-1/\sqrt{2.3.7.13}$
$\tilde{1}\ \frac{1}{2}\ \frac{3}{2}\ 0\ 100-$	$+\sqrt{3.5}/\sqrt{2.7.11.13}$
$\tilde{1}\ \frac{1}{2}\ \tilde{\frac{1}{2}}\ 0\ 000+$	$+5/2\sqrt{2.7.13}$
$\tilde{1}\ \frac{1}{2}\ \tilde{\frac{1}{2}}\ 0\ 100+$	$+3\sqrt{5}/\sqrt{2.7.11.13}$
$\tilde{1}\ \frac{3}{2}\ \frac{1}{2}\ 0\ 000-$	$+\sqrt{5}/2\sqrt{2.3.7.11.13}$
$\tilde{1}\ \frac{3}{2}\ \frac{1}{2}\ 0\ 010-$	$+31/3\sqrt{3.7.11.13}$
$\tilde{1}\ \frac{3}{2}\ \frac{1}{2}\ 0\ 100-$	$+\sqrt{3}/11\sqrt{2.7.13}$
$\tilde{1}\ \frac{3}{2}\ \frac{1}{2}\ 0\ 110-$	$-\sqrt{3.5}/11\sqrt{7.13}$
$\tilde{1}\ \frac{3}{2}\ \frac{3}{2}\ 0\ 000+$	$-29\sqrt{3}/2.7\sqrt{11.13}$
$\tilde{1}\ \frac{3}{2}\ \frac{3}{2}\ 0\ 010+$	$-\sqrt{2.5}/3\sqrt{3.11.13}$
$\tilde{1}\ \frac{3}{2}\ \frac{3}{2}\ 0\ 100+$	$-2\sqrt{3.5}/7.11\sqrt{13}$
$\tilde{1}\ \frac{3}{2}\ \frac{3}{2}\ 0\ 110+$	$+3\sqrt{2.3}/11\sqrt{13}$
$\tilde{1}\ \frac{3}{2}\ \frac{3}{2}\ 1\ 000-$	$+\sqrt{11}/2.7\sqrt{3.13}$
$\tilde{1}\ \frac{3}{2}\ \frac{3}{2}\ 1\ 010-$	$+5\sqrt{5.11}/3.7\sqrt{2.3.13}$
$\tilde{1}\ \frac{3}{2}\ \frac{3}{2}\ 1\ 100-$	$+4.3\sqrt{3.5}/7.11\sqrt{13}$
$\tilde{1}\ \frac{3}{2}\ \frac{3}{2}\ 1\ 110-$	$-5\sqrt{2.3}/7.11\sqrt{13}$
$\tilde{1}\ \frac{3}{2}\ \tilde{\frac{1}{2}}\ 0\ 000-$	$-19/2.7\sqrt{2.11.13}$
$\tilde{1}\ \frac{3}{2}\ \tilde{\frac{1}{2}}\ 0\ 010-$	$+4\sqrt{5}/7\sqrt{11.13}$
$\tilde{1}\ \frac{3}{2}\ \tilde{\frac{1}{2}}\ 0\ 100-$	$+3\sqrt{5.13}/7.11\sqrt{2}$
$\tilde{1}\ \frac{3}{2}\ \tilde{\frac{1}{2}}\ 0\ 110-$	$+3/7.11\sqrt{13}$
$\tilde{1}\ \tilde{\frac{1}{2}}\ \frac{1}{2}\ 0\ 000+$	$+\sqrt{11}/2.3\sqrt{2.3.7.13}$
$\tilde{1}\ \tilde{\frac{1}{2}}\ \frac{1}{2}\ 0\ 100+$	$+\sqrt{3.5}/\sqrt{2.7.13}$
$\tilde{1}\ \tilde{\frac{1}{2}}\ \frac{3}{2}\ 0\ 000-$	$+\sqrt{2.5.11}/3.7\sqrt{3.13}$
$\tilde{1}\ \tilde{\frac{1}{2}}\ \frac{3}{2}\ 0\ 100-$	$-\sqrt{3}/7\sqrt{2.13}$
$\tilde{0}\ \frac{1}{2}\ \tilde{\frac{1}{2}}\ 0\ 000-$	$+\sqrt{3}/4\sqrt{13}$
$\tilde{0}\ \frac{3}{2}\ \frac{3}{2}\ 0\ 000-$	$-\sqrt{13}/2\sqrt{2.7.11}$
$\tilde{0}\ \frac{3}{2}\ \frac{3}{2}\ 0\ 010-$	$-8\sqrt{5}/3\sqrt{7.11.13}$
$\tilde{0}\ \tilde{\frac{1}{2}}\ \frac{1}{2}\ 0\ 000-$	$+\sqrt{11}/4.3\sqrt{13}$

$6\ \frac{11}{2}\ \frac{9}{2}$	
$0\ \frac{1}{2}\ \frac{1}{2}\ 0\ 000-$	$-7/\sqrt{2.11.13.17}$
$0\ \frac{3}{2}\ \frac{3}{2}\ 0\ 000-$	$+9\sqrt{3.7}/11\sqrt{5.13.17}$
$0\ \frac{3}{2}\ \frac{3}{2}\ 0\ 001-$	$+8.4/11\sqrt{5.13.17}$
$0\ \frac{3}{2}\ \frac{3}{2}\ 0\ 010-$	$-23\sqrt{2.3.7}/5.11\sqrt{13.17}$
$0\ \frac{3}{2}\ \frac{3}{2}\ 0\ 011-$	$-7.19/5.11\sqrt{2.13.17}$
$1\ \frac{1}{2}\ \frac{1}{2}\ 0\ 000+$	$-\sqrt{7.11}/2\sqrt{3.13.17}$
$1\ \frac{1}{2}\ \frac{3}{2}\ 0\ 000-$	$+3\sqrt{7}/2\sqrt{5.11.13.17}$
$1\ \frac{1}{2}\ \frac{3}{2}\ 0\ 001-$	$+2/\sqrt{3.5.11.13.17}$
$1\ \frac{3}{2}\ \frac{1}{2}\ 0\ 000-$	$+19/2.11\sqrt{3.13.17}$
$1\ \frac{3}{2}\ \frac{1}{2}\ 0\ 010-$	$+4\sqrt{2.17}/11\sqrt{3.5.13}$
$1\ \frac{3}{2}\ \frac{3}{2}\ 0\ 000+$	$+3\sqrt{13}/11\sqrt{2.5.17}$
$1\ \frac{3}{2}\ \frac{3}{2}\ 0\ 001+$	$-2.37\sqrt{2}/11\sqrt{3.5.7.13.17}$
$1\ \frac{3}{2}\ \frac{3}{2}\ 0\ 010+$	$+47/5.11\sqrt{13.17}$
$1\ \frac{3}{2}\ \frac{3}{2}\ 0\ 011+$	$+1667/2.5.11\sqrt{3.7.13.17}$
$1\ \frac{3}{2}\ \frac{3}{2}\ 1\ 000+$	$+9\sqrt{2}/11\sqrt{5.13.17}$
$1\ \frac{3}{2}\ \frac{3}{2}\ 1\ 001+$	$+8.19\sqrt{2}/11\sqrt{3.5.7.13.17}$
$1\ \frac{3}{2}\ \frac{3}{2}\ 1\ 010+$	$+8.19/5.11\sqrt{13.17}$
$1\ \frac{3}{2}\ \frac{3}{2}\ 1\ 011+$	$-181\sqrt{3}/2.5.11\sqrt{7.13.17}$
$1\ \tilde{\frac{1}{2}}\ \frac{3}{2}\ 0\ 000-$	$+29/2.5\sqrt{13.17}$
$1\ \tilde{\frac{1}{2}}\ \frac{3}{2}\ 0\ 001-$	$+2.23/5\sqrt{3.7.13.17}$
$2\ \frac{1}{2}\ \frac{3}{2}\ 0\ 000+$	$+\sqrt{3.5.7}/\sqrt{11.13.17}$
$2\ \frac{1}{2}\ \frac{3}{2}\ 0\ 001+$	$-3\sqrt{5}/\sqrt{11.13.17}$
$2\ \frac{3}{2}\ \frac{1}{2}\ 0\ 000+$	$-5/11\sqrt{13.17}$
$2\ \frac{3}{2}\ \frac{1}{2}\ 0\ 010+$	$-8.2.7\sqrt{2}/3.11\sqrt{5.13.17}$
$2\ \frac{3}{2}\ \frac{3}{2}\ 0\ 000-$	$+4.7\sqrt{2.3}/11\sqrt{5.13.17}$
$2\ \frac{3}{2}\ \frac{3}{2}\ 0\ 001-$	$+59\sqrt{2}/11\sqrt{5.7.13.17}$
$2\ \frac{3}{2}\ \frac{3}{2}\ 0\ 010-$	$+4.7\sqrt{3}/5.11\sqrt{13.17}$
$2\ \frac{3}{2}\ \frac{3}{2}\ 0\ 011-$	$+41\sqrt{7}/3.5.11\sqrt{13.17}$
$2\ \tilde{\frac{1}{2}}\ \frac{3}{2}\ 0\ 000+$	$+\sqrt{3}/\sqrt{13.17}$
$2\ \tilde{\frac{1}{2}}\ \frac{3}{2}\ 0\ 001+$	$+11/3\sqrt{7.13.17}$
$\tilde{1}\ \frac{1}{2}\ \frac{3}{2}\ 0\ 000+$	$+1/2\sqrt{13.17}$
$\tilde{1}\ \frac{1}{2}\ \frac{3}{2}\ 0\ 001+$	$-8.2/\sqrt{3.7.13.17}$
$\tilde{1}\ \frac{1}{2}\ \frac{3}{2}\ 0\ 100+$	$+3/\sqrt{5.11.13.17}$
$\tilde{1}\ \frac{1}{2}\ \frac{3}{2}\ 0\ 101+$	$-29\sqrt{3}/2\sqrt{5.7.11.13.17}$
$\tilde{1}\ \frac{3}{2}\ \frac{1}{2}\ 0\ 000+$	$+\sqrt{5.11}/2\sqrt{3.7.13.17}$
$\tilde{1}\ \frac{3}{2}\ \frac{1}{2}\ 0\ 010+$	$+4\sqrt{2.11}/3\sqrt{3.7.13.17}$
$\tilde{1}\ \frac{3}{2}\ \frac{1}{2}\ 0\ 100+$	$-4\sqrt{3.13}/11\sqrt{7.17}$
$\tilde{1}\ \frac{3}{2}\ \frac{1}{2}\ 0\ 110+$	$+2.31\sqrt{2.3}/11\sqrt{5.7.13.17}$
$\tilde{1}\ \frac{3}{2}\ \frac{3}{2}\ 0\ 000-$	$-1/\sqrt{2.7.11.13.17}$
$\tilde{1}\ \frac{3}{2}\ \frac{3}{2}\ 0\ 001-$	$+2\sqrt{2.3}/7\sqrt{11.13.17}$
$\tilde{1}\ \frac{3}{2}\ \frac{3}{2}\ 0\ 010-$	$+\sqrt{5}/\sqrt{7.11.13.17}$
$\tilde{1}\ \frac{3}{2}\ \frac{3}{2}\ 0\ 011-$	$-37\sqrt{5}/2.3\sqrt{3.11.13.17}$
$\tilde{1}\ \frac{3}{2}\ \frac{3}{2}\ 0\ 100-$	$-2.3\sqrt{2}/\sqrt{5.7.13.17}$
$\tilde{1}\ \frac{3}{2}\ \frac{3}{2}\ 0\ 101-$	$-8.2\sqrt{2.3}/7.11\sqrt{5.13.17}$
$\tilde{1}\ \frac{3}{2}\ \frac{3}{2}\ 0\ 110-$	$-8.3/5\sqrt{7.13.17}$

SO_3–O 3jm Factors (cont.)

$\mathbf{6\ \frac{11}{2}\ \frac{9}{2}}$	
$\tilde{1}\ \frac{3}{2}\ \frac{3}{2}$ 0 111−	$+8.2.3\sqrt{3}/5.11\sqrt{13.17}$
$\tilde{1}\ \frac{3}{2}\ \frac{3}{2}$ 1 000+	$+23\sqrt{2}/\sqrt{7.11.13.17}$
$\tilde{1}\ \frac{3}{2}\ \frac{3}{2}$ 1 001+	$+2\sqrt{2.17}/7\sqrt{3.11.13}$
$\tilde{1}\ \frac{3}{2}\ \frac{3}{2}$ 1 010+	$+4.3/\sqrt{5.7.11.13.17}$
$\tilde{1}\ \frac{3}{2}\ \frac{3}{2}$ 1 011+	$+1051/2.3.7\sqrt{3.5.11.13.17}$
$\tilde{1}\ \frac{3}{2}\ \frac{3}{2}$ 1 100+	$+2.3.23\sqrt{2}/11\sqrt{5.7.13.17}$
$\tilde{1}\ \frac{3}{2}\ \frac{3}{2}$ 1 101+	$-3.83\sqrt{3}/7.11\sqrt{2.5.13.17}$
$\tilde{1}\ \frac{3}{2}\ \frac{3}{2}$ 1 110+	$+8.9/5.11\sqrt{7.13.17}$
$\tilde{1}\ \frac{3}{2}\ \frac{3}{2}$ 1 111+	$-4.137\sqrt{3}/5.7.11\sqrt{13.17}$
$\tilde{1}\ \tilde{\frac{1}{2}}\ \frac{1}{2}$ 0 000−	$+11\sqrt{11}/2.3\sqrt{3.7.13.17}$
$\tilde{1}\ \tilde{\frac{1}{2}}\ \frac{1}{2}$ 0 100−	$-8\sqrt{3}/\sqrt{5.7.13.17}$
$\tilde{1}\ \tilde{\frac{1}{2}}\ \frac{3}{2}$ 0 000+	$+\sqrt{7.11}/2\sqrt{5.13.17}$
$\tilde{1}\ \tilde{\frac{1}{2}}\ \frac{3}{2}$ 0 001+	$-8\sqrt{11}/3\sqrt{3.5.13.17}$
$\tilde{1}\ \tilde{\frac{1}{2}}\ \frac{3}{2}$ 0 100+	$-3\sqrt{7}/5\sqrt{13.17}$
$\tilde{1}\ \tilde{\frac{1}{2}}\ \frac{3}{2}$ 0 101+	$-\sqrt{3}/2.5\sqrt{13.17}$
$\tilde{0}\ \frac{3}{2}\ \frac{3}{2}$ 0 000+	$+\sqrt{3}/\sqrt{11.13.17}$
$\tilde{0}\ \frac{3}{2}\ \frac{3}{2}$ 0 001+	$-8.4/\sqrt{7.11.13.17}$
$\tilde{0}\ \frac{3}{2}\ \frac{3}{2}$ 0 010+	$-\sqrt{2.3.5}/\sqrt{11.13.17}$
$\tilde{0}\ \frac{3}{2}\ \frac{3}{2}$ 0 011+	$+5\sqrt{5}/3\sqrt{2.7.11.13.17}$
$\tilde{0}\ \tilde{\frac{1}{2}}\ \frac{1}{2}$ 0 000+	$+\sqrt{11}/3\sqrt{2.13.17}$

$\mathbf{6\ \frac{11}{2}\ \frac{11}{2}}$	
$0\ \frac{1}{2}\ \frac{1}{2}$ 0 000+	$-\sqrt{2}/\sqrt{3.11.13.17}$
$0\ \frac{3}{2}\ \frac{3}{2}$ 0 000+	$-4/11\sqrt{3.11.13.17}$
$0\ \frac{3}{2}\ \frac{3}{2}$ 0 010+	$+3.7\sqrt{3.5}/11\sqrt{2.11.13.17}$
$0\ \frac{3}{2}\ \frac{3}{2}$ 0 011+	$+8\sqrt{17}/11\sqrt{3.11.13}$
$0\ \tilde{\frac{1}{2}}\ \tilde{\frac{1}{2}}$ 0 000+	$-\sqrt{2.11}/\sqrt{3.13.17}$
$1\ \frac{1}{2}\ \frac{1}{2}$ 0 000−	0
$1\ \frac{3}{2}\ \frac{1}{2}$ 0 000+	$+3/11\sqrt{13.17}$
$1\ \frac{3}{2}\ \frac{1}{2}$ 0 010+	$+3.7\sqrt{5}/2.11\sqrt{2.13.17}$
$1\ \frac{3}{2}\ \frac{3}{2}$ 0 000−	0
$1\ \frac{3}{2}\ \frac{3}{2}$ 0 010−	$-2\sqrt{5.7}/\sqrt{11.13.17}$
$1\ \frac{3}{2}\ \frac{3}{2}$ 0 011−	0
$1\ \frac{3}{2}\ \frac{3}{2}$ 1 000−	0
$1\ \frac{3}{2}\ \frac{3}{2}$ 1 010−	$-\sqrt{5.7}/2\sqrt{11.13.17}$
$1\ \frac{3}{2}\ \frac{3}{2}$ 1 011−	0
$1\ \tilde{\frac{1}{2}}\ \frac{3}{2}$ 0 000+	$+\sqrt{5.7}/\sqrt{11.13.17}$
$1\ \tilde{\frac{1}{2}}\ \frac{3}{2}$ 0 001+	$-9\sqrt{7}/2\sqrt{2.11.13.17}$
$1\ \tilde{\frac{1}{2}}\ \tilde{\frac{1}{2}}$ 0 000−	0
$2\ \frac{3}{2}\ \frac{1}{2}$ 0 000−	$+41/11\sqrt{3.13.17}$
$2\ \frac{3}{2}\ \frac{1}{2}$ 0 010−	$-5\sqrt{5}/11\sqrt{2.3.13.17}$
$2\ \frac{3}{2}\ \frac{3}{2}$ 0 000+	$+8.8\sqrt{2.7}/11\sqrt{3.11.13.17}$
$2\ \frac{3}{2}\ \frac{3}{2}$ 0 010+	$+4\sqrt{5.7}/11\sqrt{3.11.13.17}$
$2\ \frac{3}{2}\ \frac{3}{2}$ 0 011+	$+2.23\sqrt{2.7}/11\sqrt{3.11.13.17}$
$2\ \tilde{\frac{1}{2}}\ \frac{3}{2}$ 0 000−	$-\sqrt{5.7}/\sqrt{3.11.13.17}$
$2\ \tilde{\frac{1}{2}}\ \frac{3}{2}$ 0 001−	$-\sqrt{7}/\sqrt{2.3.11.13.17}$
$\tilde{1}\ \frac{3}{2}\ \frac{1}{2}$ 0 000−	$+\sqrt{5.11}/\sqrt{7.13.17}$
$\tilde{1}\ \frac{3}{2}\ \frac{1}{2}$ 0 010−	$-3\sqrt{11}/2\sqrt{2.7.13.17}$
$\tilde{1}\ \frac{3}{2}\ \frac{1}{2}$ 0 100−	$+59/11\sqrt{7.13.17}$
$\tilde{1}\ \frac{3}{2}\ \frac{1}{2}$ 0 110−	$+4.19\sqrt{2.5}/3.11\sqrt{7.13.17}$
$\tilde{1}\ \frac{3}{2}\ \frac{3}{2}$ 0 000+	$+4\sqrt{2.5}/11\sqrt{13.17}$
$\tilde{1}\ \frac{3}{2}\ \frac{3}{2}$ 0 010+	$-2.9/11\sqrt{13.17}$
$\tilde{1}\ \frac{3}{2}\ \frac{3}{2}$ 0 011+	$-4\sqrt{2.5}/11\sqrt{13.17}$
$\tilde{1}\ \frac{3}{2}\ \frac{3}{2}$ 0 100+	$-43/11\sqrt{2.11.13.17}$
$\tilde{1}\ \frac{3}{2}\ \frac{3}{2}$ 0 110+	$+8\sqrt{5}/3.11\sqrt{11.13.17}$
$\tilde{1}\ \frac{3}{2}\ \frac{3}{2}$ 0 111+	$-2.47\sqrt{2}/11\sqrt{11.13.17}$
$\tilde{1}\ \frac{3}{2}\ \frac{3}{2}$ 1 000−	0
$\tilde{1}\ \frac{3}{2}\ \frac{3}{2}$ 1 010−	$-3/2\sqrt{13.17}$
$\tilde{1}\ \frac{3}{2}\ \frac{3}{2}$ 1 011−	0
$\tilde{1}\ \frac{3}{2}\ \frac{3}{2}$ 1 100−	0
$\tilde{1}\ \frac{3}{2}\ \frac{3}{2}$ 1 110−	$+2.5\sqrt{5}/3\sqrt{11.13.17}$
$\tilde{1}\ \frac{3}{2}\ \frac{3}{2}$ 1 111−	0
$\tilde{1}\ \tilde{\frac{1}{2}}\ \frac{1}{2}$ 0 000+	0
$\tilde{1}\ \tilde{\frac{1}{2}}\ \frac{1}{2}$ 0 100+	$+\sqrt{5.7}/2.3\sqrt{13.17}$
$\tilde{1}\ \tilde{\frac{1}{2}}\ \frac{3}{2}$ 0 000−	$+3/\sqrt{13.17}$
$\tilde{1}\ \tilde{\frac{1}{2}}\ \frac{3}{2}$ 0 001−	$-\sqrt{5}/2\sqrt{2.13.17}$
$\tilde{1}\ \tilde{\frac{1}{2}}\ \frac{3}{2}$ 0 100−	$+\sqrt{5}/3\sqrt{11.13.17}$
$\tilde{1}\ \tilde{\frac{1}{2}}\ \frac{3}{2}$ 0 101−	$+2\sqrt{2}/\sqrt{11.13.17}$
$\tilde{0}\ \frac{3}{2}\ \frac{3}{2}$ 0 000−	0
$\tilde{0}\ \frac{3}{2}\ \frac{3}{2}$ 0 010−	$+\sqrt{7}/\sqrt{2.3.13.17}$
$\tilde{0}\ \frac{3}{2}\ \frac{3}{2}$ 0 011−	0
$\tilde{0}\ \tilde{\frac{1}{2}}\ \frac{1}{2}$ 0 000−	$-\sqrt{2.11}/\sqrt{3.13.17}$

$\mathbf{6\ 6\ 0}$	
$0\ 0\ 0$ 0 000+	$+1/\sqrt{13}$
$1\ 1\ 0$ 0 000+	$+\sqrt{3}/\sqrt{13}$
$2\ 2\ 0$ 0 000+	$+\sqrt{2}/\sqrt{13}$
$\tilde{1}\ \tilde{1}\ 0$ 0 000+	$+\sqrt{3}/\sqrt{13}$
$\tilde{1}\ \tilde{1}\ 0$ 0 100+	0
$\tilde{1}\ \tilde{1}\ 0$ 0 110+	$+\sqrt{3}/\sqrt{13}$
$\tilde{0}\ \tilde{0}\ 0$ 0 000+	$+1/\sqrt{13}$

$\mathbf{6\ 6\ 1}$	
$1\ 0\ 1$ 0 000−	$-1/\sqrt{13}$
$1\ 1\ 1$ 0 000+	$-1/2\sqrt{7.13}$
$2\ 1\ 1$ 0 000−	$+1/\sqrt{7.13}$
$\tilde{1}\ 1\ 1$ 0 000−	$-\sqrt{3.5.11}/2.7\sqrt{13}$
$\tilde{1}\ 1\ 1$ 0 100−	$+4\sqrt{3}/7\sqrt{13}$
$\tilde{1}\ 2\ 1$ 0 000+	$+\sqrt{5.11}/7\sqrt{13}$
$\tilde{1}\ 2\ 1$ 0 100+	$+2.3/7\sqrt{13}$
$\tilde{1}\ \tilde{1}\ 1$ 0 000+	$+1/2\sqrt{7.13}$
$\tilde{1}\ \tilde{1}\ 1$ 0 100+	0
$\tilde{1}\ \tilde{1}\ 1$ 0 110+	$-3/\sqrt{7.13}$
$\tilde{0}\ \tilde{1}\ 1$ 0 000−	$-1/\sqrt{13}$
$\tilde{0}\ \tilde{1}\ 1$ 0 010−	0

$\mathbf{6\ 6\ 2}$	
$1\ 1\ 2$ 0 000+	$-\sqrt{2.3}/\sqrt{5.7.11.13}$
$1\ 1\ \tilde{1}$ 0 000+	$-3.17/2\sqrt{5.7.11.13}$
$2\ 0\ 2$ 0 000+	$-2/\sqrt{5.11.13}$
$2\ 1\ \tilde{1}$ 0 000−	$-\sqrt{3}/\sqrt{5.7.11.13}$
$2\ 2\ 2$ 0 000+	$-4.3\sqrt{2}/\sqrt{5.7.11.13}$

SO_3–O 3jm Factors (cont.)

6 6 2		
$\tilde{1}\ 0\ \tilde{1}\ 0$	000+	$+\sqrt{3}/\sqrt{7.13}$
$\tilde{1}\ 0\ \tilde{1}\ 0$	100+	$-8\sqrt{3}/\sqrt{5.7.11.13}$
$\tilde{1}\ 1\ 2\ 0$	000−	0
$\tilde{1}\ 1\ 2\ 0$	100−	$-2\sqrt{2.3}/\sqrt{5.11.13}$
$\tilde{1}\ 1\ \tilde{1}\ 0$	000−	$+3\sqrt{3}/2.7\sqrt{13}$
$\tilde{1}\ 1\ \tilde{1}\ 0$	100−	$+8.2\sqrt{3}/7\sqrt{5.11.13}$
$\tilde{1}\ 2\ \tilde{1}\ 0$	000+	$-\sqrt{3}/7\sqrt{13}$
$\tilde{1}\ 2\ \tilde{1}\ 0$	100+	$+2\sqrt{3.11}/7\sqrt{5.13}$
$\tilde{1}\ \tilde{1}\ 2\ 0$	000+	$-\sqrt{2.3.5.11}/7\sqrt{7.13}$
$\tilde{1}\ \tilde{1}\ 2\ 0$	100+	$-2.3\sqrt{2.3}/7\sqrt{7.13}$
$\tilde{1}\ \tilde{1}\ 2\ 0$	110+	$+\sqrt{2.3.13}/7\sqrt{5.7.11}$
$\tilde{1}\ \tilde{1}\ \tilde{1}\ 0$	000+	$+3\sqrt{5.11}/2.7\sqrt{7.13}$
$\tilde{1}\ \tilde{1}\ \tilde{1}\ 0$	100+	$-4.3/7\sqrt{7.13}$
$\tilde{1}\ \tilde{1}\ \tilde{1}\ 0$	110+	$-3.17/7\sqrt{5.7.11.13}$
$\tilde{0}\ 1\ \tilde{1}\ 0$	000+	$+\sqrt{3}/\sqrt{7.13}$
$\tilde{0}\ 2\ 2\ 0$	000−	$+2/\sqrt{7.13}$

6 6 3		
$1\ 0\ 1\ 0$	000−	$-3\sqrt{3}/2\sqrt{2.11.13}$
$1\ 1\ 1\ 0$	000+	$+3\sqrt{3}/\sqrt{2.7.11.13}$
$1\ 1\ \tilde{1}\ 0$	000−	0
$2\ 1\ 1\ 0$	000−	$+3\sqrt{3}/2\sqrt{2.7.11.13}$
$2\ 1\ \tilde{1}\ 0$	000+	$+5\sqrt{3.5}/2\sqrt{2.7.11.13}$
$2\ 2\ \tilde{0}\ 0$	000+	$+\sqrt{5}/\sqrt{2.7.11.13}$
$\tilde{1}\ 0\ \tilde{1}\ 0$	000−	$+\sqrt{3}/2\sqrt{2.7.13}$
$\tilde{1}\ 0\ \tilde{1}\ 0$	100−	$+\sqrt{2.3.5}/\sqrt{7.11.13}$
$\tilde{1}\ 1\ 1\ 0$	000−	$-3\sqrt{5}/7\sqrt{2.13}$
$\tilde{1}\ 1\ 1\ 0$	100−	$+3/7\sqrt{2.11.13}$
$\tilde{1}\ 1\ \tilde{1}\ 0$	000+	0
$\tilde{1}\ 1\ \tilde{1}\ 0$	100+	$-\sqrt{3.5}/\sqrt{2.11.13}$
$\tilde{1}\ 1\ \tilde{0}\ 0$	000−	$+3\sqrt{3}/7\sqrt{13}$
$\tilde{1}\ 1\ \tilde{0}\ 0$	100−	$-2\sqrt{3.5}/7\sqrt{11.13}$
$\tilde{1}\ 2\ 1\ 0$	000+	$-5\sqrt{3.5}/2.7\sqrt{2.13}$
$\tilde{1}\ 2\ 1\ 0$	100+	$-\sqrt{3}/7\sqrt{2.11.13}$
$\tilde{1}\ 2\ \tilde{1}\ 0$	000−	$-\sqrt{3}/2.7\sqrt{2.13}$
$\tilde{1}\ 2\ \tilde{1}\ 0$	100−	$+5\sqrt{3.5}/7\sqrt{2.11.13}$
$\tilde{1}\ \tilde{1}\ 1\ 0$	000+	$+\sqrt{3.11}/7\sqrt{2.7.13}$
$\tilde{1}\ \tilde{1}\ 1\ 0$	100+	$-3\sqrt{3.5}/7\sqrt{2.7.13}$
$\tilde{1}\ \tilde{1}\ 1\ 0$	110+	$-2\sqrt{2.3}/7\sqrt{7.11.13}$
$\tilde{1}\ \tilde{1}\ \tilde{1}\ 0$	000−	0
$\tilde{1}\ \tilde{1}\ \tilde{1}\ 0$	100−	$-3/\sqrt{2.7.13}$
$\tilde{1}\ \tilde{1}\ \tilde{1}\ 0$	110−	0
$\tilde{0}\ 0\ \tilde{0}\ 0$	000−	$-1/2\sqrt{13}$
$\tilde{0}\ 1\ \tilde{1}\ 0$	000−	$-\sqrt{3}/2\sqrt{2.7.13}$
$\tilde{0}\ \tilde{1}\ 1\ 0$	000−	$+\sqrt{3.11}/2.7\sqrt{2.13}$
$\tilde{0}\ \tilde{1}\ 1\ 0$	010−	$+\sqrt{2.3.5}/7\sqrt{13}$

6 6 4		
$0\ 0\ 0\ 0$	000+	$-7\sqrt{3}/2\sqrt{11.13.17}$
$1\ 0\ 1\ 0$	000+	$-\sqrt{3.5.7}/2\sqrt{2.11.13.17}$
$1\ 1\ 0\ 0$	000+	$-8/\sqrt{11.13.17}$

6 6 4		
$1\ 1\ 1\ 0$	000−	0
$1\ 1\ 2\ 0$	000+	$-8\sqrt{5}/\sqrt{7.11.13.17}$
$1\ 1\ \tilde{1}\ 0$	000+	$-5\sqrt{3.5}/\sqrt{2.7.11.13.17}$
$2\ 0\ 2\ 0$	000+	$-3\sqrt{3.5}/\sqrt{2.11.13.17}$
$2\ 1\ 1\ 0$	000+	$+7\sqrt{3.5}/2\sqrt{2.11.13.17}$
$2\ 1\ \tilde{1}\ 0$	000−	$+19\sqrt{5}/2\sqrt{2.7.11.13.17}$
$2\ 2\ 0\ 0$	000+	$+19/\sqrt{2.3.11.13.17}$
$2\ 2\ 2\ 0$	000+	$+2\sqrt{5}/\sqrt{3.7.11.13.17}$
$\tilde{1}\ 0\ \tilde{1}\ 0$	000+	$+\sqrt{17}/2\sqrt{2.7.13}$
$\tilde{1}\ 0\ \tilde{1}\ 0$	100+	$+3\sqrt{2.5}/\sqrt{7.11.13.17}$
$\tilde{1}\ 1\ 1\ 0$	000+	$-4\sqrt{2}/\sqrt{7.13.17}$
$\tilde{1}\ 1\ 1\ 0$	100+	$+3\sqrt{5}/\sqrt{2.7.11.13.17}$
$\tilde{1}\ 1\ 2\ 0$	000−	$-2/\sqrt{13.17}$
$\tilde{1}\ 1\ 2\ 0$	100−	$-3\sqrt{5}/\sqrt{11.13.17}$
$\tilde{1}\ 1\ \tilde{1}\ 0$	000−	$-\sqrt{17}/7\sqrt{2.13}$
$\tilde{1}\ 1\ \tilde{1}\ 0$	100−	$-3\sqrt{5.11}/7\sqrt{2.13.17}$
$\tilde{1}\ 2\ 1\ 0$	000−	$-1/2\sqrt{2.3.7.13.17}$
$\tilde{1}\ 2\ 1\ 0$	100−	$-3\sqrt{3.5}/\sqrt{2.7.11.13.17}$
$\tilde{1}\ 2\ \tilde{1}\ 0$	000+	$+3\sqrt{13}/2.7\sqrt{2.17}$
$\tilde{1}\ 2\ \tilde{1}\ 0$	100+	$-9.3\sqrt{5}/7\sqrt{2.11.13.17}$
$\tilde{1}\ \tilde{1}\ 0\ 0$	000+	$+8\sqrt{11}/3.7\sqrt{13.17}$
$\tilde{1}\ \tilde{1}\ 0\ 0$	100+	$+2.3\sqrt{5}/7\sqrt{13.17}$
$\tilde{1}\ \tilde{1}\ 0\ 0$	110+	$-2.3/7\sqrt{11.13.17}$
$\tilde{1}\ \tilde{1}\ 1\ 0$	000−	0
$\tilde{1}\ \tilde{1}\ 1\ 0$	100−	$+3\sqrt{3}/\sqrt{2.13.17}$
$\tilde{1}\ \tilde{1}\ 1\ 0$	110−	0
$\tilde{1}\ \tilde{1}\ 2\ 0$	000+	$+4.5\sqrt{5.11}/3.7\sqrt{7.13.17}$
$\tilde{1}\ \tilde{1}\ 2\ 0$	100+	$-9/7\sqrt{7.13.17}$
$\tilde{1}\ \tilde{1}\ 2\ 0$	110+	$+8.2.3\sqrt{5}/7\sqrt{7.11.13.17}$
$\tilde{1}\ \tilde{1}\ \tilde{1}\ 0$	000+	$+11\sqrt{5.11}/7\sqrt{2.3.7.13.17}$
$\tilde{1}\ \tilde{1}\ \tilde{1}\ 0$	100+	$-\sqrt{3.13}/7\sqrt{2.7.17}$
$\tilde{1}\ \tilde{1}\ \tilde{1}\ 0$	110+	$+2.3.5\sqrt{2.3.5}/7\sqrt{7.11.13.17}$
$\tilde{0}\ 1\ \tilde{1}\ 0$	000+	$+\sqrt{17}/2\sqrt{2.7.13}$
$\tilde{0}\ 2\ 2\ 0$	000−	$-11/\sqrt{2.3.7.13.17}$
$\tilde{0}\ \tilde{1}\ 1\ 0$	000+	$-\sqrt{5.11}/2\sqrt{2.3.7.13.17}$
$\tilde{0}\ \tilde{1}\ 1\ 0$	010+	$+3\sqrt{2.3}/\sqrt{7.13.17}$
$\tilde{0}\ \tilde{0}\ 0\ 0$	000+	$+\sqrt{11}/2\sqrt{3.13.17}$

6 6 5		
$1\ 0\ 1\ 0$	000−	$-\sqrt{2.3}/11\sqrt{13.17}$
$1\ 0\ 1\ 0$	001−	$+3\sqrt{3.5.7}/11\sqrt{2.13.17}$
$1\ 1\ 1\ 0$	000+	$+3\sqrt{2.3}/11\sqrt{7.13.17}$
$1\ 1\ 1\ 0$	001+	$+3.5\sqrt{3.5}/2.11\sqrt{2.13.17}$
$1\ 1\ 2\ 0$	000−	0
$1\ 1\ \tilde{1}\ 0$	000−	0
$2\ 0\ 2\ 0$	000−	$-\sqrt{2.5.7}/\sqrt{11.13.17}$
$2\ 1\ 1\ 0$	000−	$-2\sqrt{2.3.17}/11\sqrt{7.13}$
$2\ 1\ 1\ 0$	001−	$+3\sqrt{3.5}/11\sqrt{2.13.17}$
$2\ 1\ \tilde{1}\ 0$	000+	$+\sqrt{2.3.5}/\sqrt{11.13.17}$
$2\ 2\ 2\ 0$	000−	0

SO_3–O 3jm Factors (cont.)

6 6 5	
$\tilde{1}\ 0\ \tilde{1}\ 0\ 000-$	$+\sqrt{2.3}/\sqrt{13.17}$
$\tilde{1}\ 0\ \tilde{1}\ 0\ 100-$	$+\sqrt{3.5}/\sqrt{2.11.13.17}$
$\tilde{1}\ 1\ 1\ 0\ 000-$	$-4.3\sqrt{2.5}/7\sqrt{11.13.17}$
$\tilde{1}\ 1\ 1\ 0\ 001-$	$-3/2\sqrt{2.7.11.13.17}$
$\tilde{1}\ 1\ 1\ 0\ 100-$	$-3.19/2.7.11\sqrt{2.13.17}$
$\tilde{1}\ 1\ 1\ 0\ 101-$	$-8.3\sqrt{2.5}/11\sqrt{7.13.17}$
$\tilde{1}\ 1\ 2\ 0\ 000+$	$-3\sqrt{3}/\sqrt{7.13.17}$
$\tilde{1}\ 1\ 2\ 0\ 100+$	$-5\sqrt{3.5}/\sqrt{7.11.13.17}$
$\tilde{1}\ 1\ \tilde{1}\ 0\ 000+$	$+3\sqrt{2.3}/\sqrt{7.13.17}$
$\tilde{1}\ 1\ \tilde{1}\ 0\ 100+$	$-\sqrt{3.5}/2\sqrt{2.7.11.13.17}$
$\tilde{1}\ 2\ 1\ 0\ 000+$	$+\sqrt{2.3.5}/7\sqrt{11.13.17}$
$\tilde{1}\ 2\ 1\ 0\ 001+$	$+\sqrt{3.7}/\sqrt{2.11.13.17}$
$\tilde{1}\ 2\ 1\ 0\ 100+$	$-37\sqrt{3}/7.11\sqrt{2.13.17}$
$\tilde{1}\ 2\ 1\ 0\ 101+$	$+2\sqrt{2.3.5.7}/11\sqrt{13.17}$
$\tilde{1}\ 2\ \tilde{1}\ 0\ 000-$	$+2\sqrt{2.3}/\sqrt{7.13.17}$
$\tilde{1}\ 2\ \tilde{1}\ 0\ 100-$	$-5\sqrt{3.5}/\sqrt{2.7.11.13.17}$
$\tilde{1}\ \tilde{1}\ 1\ 0\ 000+$	$-3\sqrt{2.3}/\sqrt{7.13.17}$
$\tilde{1}\ \tilde{1}\ 1\ 0\ 001+$	$+5\sqrt{3.5}/2.7\sqrt{2.13.17}$
$\tilde{1}\ \tilde{1}\ 1\ 0\ 100+$	$-\sqrt{3.5}/2\sqrt{2.7.11.13.17}$
$\tilde{1}\ \tilde{1}\ 1\ 0\ 101+$	$-8.3\sqrt{2.3}/7\sqrt{11.13.17}$
$\tilde{1}\ \tilde{1}\ 1\ 0\ 110+$	$+8.2.3\sqrt{2.3}/11\sqrt{7.13.17}$
$\tilde{1}\ \tilde{1}\ 1\ 0\ 111+$	$+2\sqrt{2.3.5}/7.11\sqrt{13.17}$
$\tilde{1}\ \tilde{1}\ 2\ 0\ 000-$	0
$\tilde{1}\ \tilde{1}\ 2\ 0\ 100-$	$-\sqrt{3}/\sqrt{13.17}$
$\tilde{1}\ \tilde{1}\ 2\ 0\ 110-$	0
$\tilde{1}\ \tilde{1}\ \tilde{1}\ 0\ 000-$	0
$\tilde{1}\ \tilde{1}\ \tilde{1}\ 0\ 100-$	$+3/2\sqrt{2.13.17}$
$\tilde{1}\ \tilde{1}\ \tilde{1}\ 0\ 110-$	0
$\tilde{0}\ 1\ \tilde{1}\ 0\ 000-$	$+\sqrt{2.3}/\sqrt{13.17}$
$\tilde{0}\ 2\ 2\ 0\ 000+$	$-\sqrt{2}/\sqrt{13.17}$
$\tilde{0}\ \tilde{1}\ 1\ 0\ 000-$	$-\sqrt{2.3}/\sqrt{13.17}$
$\tilde{0}\ \tilde{1}\ 1\ 0\ 001-$	$-\sqrt{3.5}/\sqrt{2.7.13.17}$
$\tilde{0}\ \tilde{1}\ 1\ 0\ 010-$	$+\sqrt{3.5}/\sqrt{2.11.13.17}$
$\tilde{0}\ \tilde{1}\ 1\ 0\ 011-$	$+4\sqrt{2.3}/\sqrt{7.11.13.17}$
6 6 6	
$0\ 0\ 0\ 0\ 000+$	$-2\sqrt{2}/\sqrt{11.13.17.19}$
$1\ 1\ 0\ 0\ 000+$	$-\sqrt{2.3}/\sqrt{11.13.17.19}$
$1\ 1\ 1\ 0\ 000-$	0
$2\ 1\ 1\ 0\ 000+$	$+\sqrt{2.3.7}/\sqrt{11.13.17.19}$
$2\ 2\ 0\ 0\ 000+$	$+4.3/\sqrt{11.13.17.19}$
$2\ 2\ 2\ 0\ 000+$	$+8\sqrt{2.7}/\sqrt{11.13.17.19}$
$\tilde{1}\ 1\ 1\ 0\ 000+$	$+3\sqrt{2.5}/\sqrt{13.17.19}$
$\tilde{1}\ 1\ 1\ 0\ 100+$	$+3.31/2\sqrt{2.11.13.17.19}$
$\tilde{1}\ 2\ 1\ 0\ 000-$	$-2\sqrt{2.3.5}/\sqrt{13.17.19}$
$\tilde{1}\ 2\ 1\ 0\ 100-$	$+5.5\sqrt{3}/\sqrt{2.11.13.17.19}$
$\tilde{1}\ \tilde{1}\ 0\ 0\ 000+$	$-\sqrt{2.3.11}/\sqrt{13.17.19}$
$\tilde{1}\ \tilde{1}\ 0\ 0\ 100+$	$+\sqrt{3.5}/\sqrt{2.13.17.19}$
$\tilde{1}\ \tilde{1}\ 0\ 0\ 110+$	$+8.2\sqrt{2.3}/\sqrt{11.13.17.19}$
$\tilde{1}\ \tilde{1}\ 1\ 0\ 000-$	0
6 6 6	
$\tilde{1}\ \tilde{1}\ 1\ 0\ 100-$	$+\sqrt{3.5.7}/2\sqrt{2.13.17.19}$
$\tilde{1}\ \tilde{1}\ 1\ 0\ 110-$	0
$\tilde{1}\ \tilde{1}\ 2\ 0\ 000+$	$+\sqrt{2.3.11}/\sqrt{7.13.17.19}$
$\tilde{1}\ \tilde{1}\ 2\ 0\ 100+$	$+\sqrt{3.5}/\sqrt{2.7.13.17.19}$
$\tilde{1}\ \tilde{1}\ 2\ 0\ 110+$	$+2\sqrt{2.3.19}/\sqrt{7.11.13.17}$
$\tilde{1}\ \tilde{1}\ \tilde{1}\ 0\ 000+$	$+3.11\sqrt{2.5}/7\sqrt{13.17.19}$
$\tilde{1}\ \tilde{1}\ \tilde{1}\ 0\ 100+$	$-3.11\sqrt{11}/2.7\sqrt{2.13.17.19}$
$\tilde{1}\ \tilde{1}\ \tilde{1}\ 0\ 110+$	$-4.3\sqrt{2.5}/7\sqrt{13.17.19}$
$\tilde{1}\ \tilde{1}\ \tilde{1}\ 0\ 111+$	$-2.3\sqrt{2.19}/7\sqrt{11.13.17}$
$\tilde{0}\ 2\ 2\ 0\ 000-$	0
$\tilde{0}\ \tilde{1}\ 1\ 0\ 000+$	$+\sqrt{2.3.11}/\sqrt{13.17.19}$
$\tilde{0}\ \tilde{1}\ 1\ 0\ 010+$	$+\sqrt{3.5}/\sqrt{2.13.17.19}$
$\tilde{0}\ \tilde{0}\ 0\ 0\ 000+$	$-2\sqrt{2.11}/\sqrt{13.17.19}$
$\frac{13}{2}\ \frac{7}{2}\ 3$	
$\frac{1}{2}\ \frac{1}{2}\ 1\ 0\ 000-$	$+5/\sqrt{2.3.7.11.13}$
$\frac{1}{2}\ \frac{3}{2}\ 1\ 0\ 000+$	$+\sqrt{2.5}/\sqrt{3.11.13}$
$\frac{1}{2}\ \frac{3}{2}\ \tilde{1}\ 0\ 000-$	$-\sqrt{2.3}/\sqrt{11.13}$
$\frac{1}{2}\ \tilde{\frac{1}{2}}\ \tilde{1}\ 0\ 000+$	$-3/\sqrt{2.11.13}$
$\frac{1}{2}\ \tilde{\frac{1}{2}}\ \tilde{0}\ 0\ 000-$	$-\sqrt{2.3}/\sqrt{11.13}$
$\frac{3}{2}\ \frac{1}{2}\ 1\ 0\ 000+$	$-5.5/2.7\sqrt{3.13}$
$\frac{3}{2}\ \frac{1}{2}\ 1\ 0\ 100+$	$-2.5\sqrt{2}/7\sqrt{11.13}$
$\frac{3}{2}\ \frac{1}{2}\ \tilde{1}\ 0\ 000-$	$-\sqrt{5}/2.7\sqrt{3.13}$
$\frac{3}{2}\ \frac{1}{2}\ \tilde{1}\ 0\ 100-$	$-2.3\sqrt{2.5}/7\sqrt{11.13}$
$\frac{3}{2}\ \frac{3}{2}\ 1\ 0\ 000-$	$+\sqrt{5}/\sqrt{2.3.7.13}$
$\frac{3}{2}\ \frac{3}{2}\ 1\ 0\ 100-$	$-2\sqrt{5}/\sqrt{7.11.13}$
$\frac{3}{2}\ \frac{3}{2}\ 1\ 1\ 000-$	$+\sqrt{2.5}/\sqrt{3.7.13}$
$\frac{3}{2}\ \frac{3}{2}\ 1\ 1\ 100-$	$+3\sqrt{5}/\sqrt{7.11.13}$
$\frac{3}{2}\ \frac{3}{2}\ \tilde{1}\ 0\ 000+$	$-\sqrt{7}/\sqrt{2.3.13}$
$\frac{3}{2}\ \frac{3}{2}\ \tilde{1}\ 0\ 100+$	0
$\frac{3}{2}\ \frac{3}{2}\ \tilde{1}\ 1\ 000-$	$+\sqrt{2}/\sqrt{3.7.13}$
$\frac{3}{2}\ \frac{3}{2}\ \tilde{1}\ 1\ 100-$	$+3/\sqrt{7.11.13}$
$\frac{3}{2}\ \frac{3}{2}\ \tilde{0}\ 0\ 000-$	$-\sqrt{2}/\sqrt{7.13}$
$\frac{3}{2}\ \frac{3}{2}\ \tilde{0}\ 0\ 100-$	$+4\sqrt{3}/\sqrt{7.11.13}$
$\frac{3}{2}\ \tilde{\frac{1}{2}}\ 1\ 0\ 000+$	$-\sqrt{5}/2\sqrt{7.13}$
$\frac{3}{2}\ \tilde{\frac{1}{2}}\ 1\ 0\ 100+$	$+\sqrt{2.3.5}/\sqrt{7.11.13}$
$\frac{3}{2}\ \tilde{\frac{1}{2}}\ \tilde{1}\ 0\ 000-$	$+3/2\sqrt{7.13}$
$\frac{3}{2}\ \tilde{\frac{1}{2}}\ \tilde{1}\ 0\ 100-$	$-3\sqrt{2.3}/\sqrt{7.11.13}$
$\tilde{\frac{1}{2}}\ \frac{1}{2}\ \tilde{1}\ 0\ 000+$	$-11/7\sqrt{2.3.13}$
$\tilde{\frac{1}{2}}\ \frac{1}{2}\ \tilde{1}\ 0\ 100+$	$-3\sqrt{5}/2.7\sqrt{11}$
$\tilde{\frac{1}{2}}\ \frac{1}{2}\ \tilde{0}\ 0\ 000-$	$-\sqrt{2}/7\sqrt{13}$
$\tilde{\frac{1}{2}}\ \frac{1}{2}\ \tilde{0}\ 0\ 100-$	$+\sqrt{3.5}/7\sqrt{11}$
$\tilde{\frac{1}{2}}\ \frac{3}{2}\ 1\ 0\ 000+$	$+\sqrt{2.3}/\sqrt{7.13}$
$\tilde{\frac{1}{2}}\ \frac{3}{2}\ 1\ 0\ 100+$	$+\sqrt{5}/2\sqrt{7.11}$
$\tilde{\frac{1}{2}}\ \frac{3}{2}\ \tilde{1}\ 0\ 000-$	$-\sqrt{2.5}/\sqrt{3.7.13}$
$\tilde{\frac{1}{2}}\ \frac{3}{2}\ \tilde{1}\ 0\ 100-$	$+3/2\sqrt{7.11}$
$\tilde{\frac{1}{2}}\ \tilde{\frac{1}{2}}\ 1\ 0\ 000-$	$-1/\sqrt{2.7.13}$
$\tilde{\frac{1}{2}}\ \tilde{\frac{1}{2}}\ 1\ 0\ 100-$	$+\sqrt{3.5}/2\sqrt{7.11}$
$\frac{13}{2}\ 4\ \frac{5}{2}$	
$\frac{1}{2}\ 1\ \frac{3}{2}\ 0\ 000+$	$-\sqrt{2.5}/\sqrt{7.11.13}$

SO_3–O 3jm Factors (cont.)

$\frac{13}{2}$	4	$\frac{5}{2}$			
$\frac12$	2	$\frac32$	0	000−	$-\sqrt{5}/\sqrt{11.13}$
$\frac12$	$\tilde{1}$	$\frac32$	0	000−	$-\sqrt{2.5}/\sqrt{3.11.13}$
$\frac12$	$\tilde{1}$	$\tilde{\frac12}$	0	000+	$-4\sqrt{2}/\sqrt{3.11.13}$
$\frac32$	0	$\frac32$	0	000+	$-2.5\sqrt{2}/3.7\sqrt{13}$
$\frac32$	0	$\frac32$	0	100+	$-2.5\sqrt{3}/7\sqrt{11.13}$
$\frac32$	1	$\frac32$	0	000−	$+4\sqrt{2.5}/3.7\sqrt{13}$
$\frac32$	1	$\frac32$	0	100−	$+4\sqrt{3.5}/7\sqrt{11.13}$
$\frac32$	1	$\frac32$	1	000−	$+\sqrt{2.5}/3\sqrt{13}$
$\frac32$	1	$\frac32$	1	100−	0
$\frac32$	1	$\tilde{\frac12}$	0	000+	$+2/3.7\sqrt{13}$
$\frac32$	1	$\tilde{\frac12}$	0	100+	$+8\sqrt{2.3}/7\sqrt{11.13}$
$\frac32$	2	$\frac32$	0	000+	$+2\sqrt{5}/3\sqrt{7.13}$
$\frac32$	2	$\frac32$	0	100+	$+\sqrt{2.3.5}/\sqrt{7.11.13}$
$\frac32$	2	$\tilde{\frac12}$	0	000−	$-5\sqrt{2}/3\sqrt{7.13}$
$\frac32$	2	$\tilde{\frac12}$	0	100−	$+2\sqrt{3}/\sqrt{7.11.13}$
$\frac32$	$\tilde{1}$	$\frac32$	0	000+	0
$\frac32$	$\tilde{1}$	$\frac32$	0	100+	0
$\frac32$	$\tilde{1}$	$\frac32$	1	000−	$-\sqrt{2.5}/\sqrt{3.7.13}$
$\frac32$	$\tilde{1}$	$\frac32$	1	100−	$+4\sqrt{5}/\sqrt{7.11.13}$
$\frac32$	$\tilde{1}$	$\tilde{\frac12}$	0	000−	$+2/\sqrt{3.7.13}$
$\frac32$	$\tilde{1}$	$\tilde{\frac12}$	0	100−	$-4\sqrt{2}/\sqrt{7.11.13}$
$\tilde{\frac12}$	0	$\tilde{\frac12}$	0	000+	$+4\sqrt{2}/3.7\sqrt{13}$
$\tilde{\frac12}$	0	$\tilde{\frac12}$	0	100+	$+\sqrt{3.5}/7\sqrt{11}$
$\tilde{\frac12}$	1	$\frac32$	0	000+	$+11\sqrt{2}/3.7\sqrt{13}$
$\tilde{\frac12}$	1	$\frac32$	0	100+	$+\sqrt{3.5}/7\sqrt{11}$
$\tilde{\frac12}$	1	$\tilde{\frac12}$	0	000−	$-4\sqrt{2.5}/3.7\sqrt{13}$
$\tilde{\frac12}$	1	$\tilde{\frac12}$	0	100−	$+2\sqrt{3}/7\sqrt{11}$
$\tilde{\frac12}$	2	$\frac32$	0	000−	$-\sqrt{7}/3\sqrt{13}$
$\tilde{\frac12}$	2	$\frac32$	0	100−	0
$\tilde{\frac12}$	$\tilde{1}$	$\frac32$	0	000−	$+\sqrt{2}/\sqrt{3.7.13}$
$\tilde{\frac12}$	$\tilde{1}$	$\frac32$	0	100−	$-\sqrt{5}/\sqrt{7.11}$

$\frac{13}{2}$	4	$\frac{7}{2}$			
$\frac12$	0	$\frac12$	0	000+	$+\sqrt{2.5}/3\sqrt{11.13}$
$\frac12$	1	$\frac12$	0	000+	$+\sqrt{11}/3\sqrt{2.13}$
$\frac12$	1	$\frac32$	0	000−	$-29\sqrt{2}/3\sqrt{5.7.11.13}$
$\frac12$	2	$\frac32$	0	000+	$-8/3\sqrt{5.11.13}$
$\frac12$	$\tilde{1}$	$\frac32$	0	000+	$-\sqrt{2.5}/\sqrt{3.11.13}$
$\frac12$	$\tilde{1}$	$\tilde{\frac12}$	0	000−	$+\sqrt{5}/\sqrt{2.11.13}$
$\frac32$	0	$\frac32$	0	000−	$+11\sqrt{2}/3.7\sqrt{13}$
$\frac32$	0	$\frac32$	0	100−	$-8.2/7\sqrt{3.11.13}$
$\frac32$	1	$\frac12$	0	000−	$-5/2.3\sqrt{7.13}$
$\frac32$	1	$\frac12$	0	100−	$-2\sqrt{2}/\sqrt{3.7.11.13}$
$\frac32$	1	$\frac32$	0	000+	$+\sqrt{5}/3.7\sqrt{2.13}$
$\frac32$	1	$\frac32$	0	100+	$-2\sqrt{3.11}/7\sqrt{5.13}$
$\frac32$	1	$\frac32$	1	000+	$-\sqrt{2}/3\sqrt{5.13}$
$\frac32$	1	$\frac32$	1	100+	$-\sqrt{11}/\sqrt{3.5.13}$
$\frac32$	1	$\tilde{\frac12}$	0	000−	$+19\sqrt{3}/2.7\sqrt{5.13}$
$\frac32$	1	$\tilde{\frac12}$	0	100−	$-\sqrt{2}/7\sqrt{5.11.13}$
$\frac32$	2	$\frac12$	0	000+	$+\sqrt{2}/3\sqrt{13}$
$\frac32$	2	$\frac12$	0	100+	$-4/\sqrt{3.11.13}$
$\frac32$	2	$\frac32$	0	000−	$-4/3\sqrt{5.7.13}$
$\frac32$	2	$\frac32$	0	100−	$-2\sqrt{2.3}/\sqrt{5.7.11.13}$
$\frac32$	2	$\tilde{\frac12}$	0	000+	$-\sqrt{2.3}/\sqrt{5.7.13}$
$\frac32$	2	$\tilde{\frac12}$	0	100+	$-8.2/\sqrt{5.7.11.13}$
$\frac32$	$\tilde{1}$	$\frac12$	0	000+	$+\sqrt{3}/2\sqrt{13}$
$\frac32$	$\tilde{1}$	$\frac12$	0	100+	$-2\sqrt{2}/3\sqrt{11.13}$
$\frac32$	$\tilde{1}$	$\frac32$	0	000−	$-\sqrt{7}/\sqrt{2.3.5.13}$
$\frac32$	$\tilde{1}$	$\frac32$	0	100−	$-8\sqrt{7}/3\sqrt{5.11.13}$
$\frac32$	$\tilde{1}$	$\frac32$	1	000+	$-\sqrt{2.5}/\sqrt{3.7.13}$
$\frac32$	$\tilde{1}$	$\frac32$	1	100+	$+\sqrt{11}/3\sqrt{5.7.13}$
$\frac32$	$\tilde{1}$	$\tilde{\frac12}$	0	000+	$+1/2\sqrt{5.7.13}$
$\frac32$	$\tilde{1}$	$\tilde{\frac12}$	0	100+	$-\sqrt{2.3}/\sqrt{5.7.11.13}$
$\tilde{\frac12}$	0	$\tilde{\frac12}$	0	000−	$+\sqrt{2.3.5}/7\sqrt{13}$
$\tilde{\frac12}$	0	$\tilde{\frac12}$	0	100−	$-1/7\sqrt{11}$
$\tilde{\frac12}$	1	$\frac32$	0	000−	$-\sqrt{2}/7\sqrt{13}$
$\tilde{\frac12}$	1	$\frac32$	0	100−	$+23/2.7\sqrt{3.5.11}$
$\tilde{\frac12}$	1	$\tilde{\frac12}$	0	000+	$-3\sqrt{3}/7\sqrt{2.13}$
$\tilde{\frac12}$	1	$\tilde{\frac12}$	0	100+	$-\sqrt{11}/2.7\sqrt{5}$
$\tilde{\frac12}$	2	$\frac32$	0	000+	0
$\tilde{\frac12}$	2	$\frac32$	0	100+	$+\sqrt{2.7}/\sqrt{3.5.11}$
$\tilde{\frac12}$	$\tilde{1}$	$\frac12$	0	000−	$-\sqrt{5}/\sqrt{2.3.13}$
$\tilde{\frac12}$	$\tilde{1}$	$\frac12$	0	100−	$+1/2.3\sqrt{11}$
$\tilde{\frac12}$	$\tilde{1}$	$\frac32$	0	000+	$+\sqrt{2}/\sqrt{3.7.13}$
$\tilde{\frac12}$	$\tilde{1}$	$\frac32$	0	100+	$+19/2.3\sqrt{5.7.11}$

$\frac{13}{2}$	$\frac{9}{2}$	2			
$\frac12$	$\frac32$	2	0	000−	$-4/\sqrt{5.7.11.13}$
$\frac12$	$\frac32$	2	0	010−	$+3\sqrt{3}/\sqrt{5.11.13}$
$\frac12$	$\frac32$	$\tilde{1}$	0	000−	$-8\sqrt{2.3}/\sqrt{5.7.11.13}$
$\frac12$	$\frac32$	$\tilde{1}$	0	010−	$+3\sqrt{2}/\sqrt{5.11.13}$
$\frac32$	$\frac12$	2	0	000−	$+2\sqrt{2.3}/7\sqrt{13}$
$\frac32$	$\frac12$	2	0	100−	$+2.9/7\sqrt{11.13}$
$\frac32$	$\frac12$	$\tilde{1}$	0	000−	$-4/7\sqrt{13}$
$\frac32$	$\frac12$	$\tilde{1}$	0	100−	$-2.3\sqrt{2.3}/7\sqrt{11.13}$
$\frac32$	$\frac32$	2	0	000+	$+2/\sqrt{5.13}$
$\frac32$	$\frac32$	2	0	010+	$+2\sqrt{3}/\sqrt{5.7.13}$
$\frac32$	$\frac32$	2	0	100+	$-2\sqrt{2.3}/\sqrt{5.11.13}$
$\frac32$	$\frac32$	2	0	110+	$+9\sqrt{2}/\sqrt{5.7.11.13}$
$\frac32$	$\frac32$	$\tilde{1}$	0	000+	0
$\frac32$	$\frac32$	$\tilde{1}$	0	010+	$-\sqrt{2.5}/\sqrt{7.13}$
$\frac32$	$\frac32$	$\tilde{1}$	0	100+	$-4/\sqrt{5.11.13}$
$\frac32$	$\frac32$	$\tilde{1}$	0	110+	$+2.3\sqrt{3}/\sqrt{5.7.11.13}$
$\frac32$	$\frac32$	$\tilde{1}$	1	000−	$+2\sqrt{2.3}/7\sqrt{5.13}$
$\frac32$	$\frac32$	$\tilde{1}$	1	010−	$-2\sqrt{2}/\sqrt{5.7.13}$
$\frac32$	$\frac32$	$\tilde{1}$	1	100−	$-4\sqrt{13}/7\sqrt{5.11}$
$\frac32$	$\frac32$	$\tilde{1}$	1	110−	$-2.3\sqrt{3}/\sqrt{5.7.11.13}$
$\tilde{\frac12}$	$\frac12$	$\tilde{1}$	0	000+	$+4\sqrt{2}/7\sqrt{5.13}$
$\tilde{\frac12}$	$\frac12$	$\tilde{1}$	0	100+	$+3\sqrt{3}/7\sqrt{11}$
$\tilde{\frac12}$	$\frac32$	2	0	000−	$-4/5.7\sqrt{13}$

SO_3–O 3jm Factors (cont.)

$\frac{13}{2}$	$\frac{9}{2}$	**2**			
$\tilde{\frac{1}{2}}$	$\frac{3}{2}$	2	0	010−	$-\sqrt{3.7}/5\sqrt{13}$
$\tilde{\frac{1}{2}}$	$\frac{3}{2}$	2	0	100−	$-\sqrt{2.3.5}/7\sqrt{11}$
$\tilde{\frac{1}{2}}$	$\frac{3}{2}$	2	0	110−	0
$\tilde{\frac{1}{2}}$	$\frac{3}{2}$	$\tilde{1}$	0	000−	$+8\sqrt{2.3}/5.7\sqrt{13}$
$\tilde{\frac{1}{2}}$	$\frac{3}{2}$	$\tilde{1}$	0	010−	$+\sqrt{2.7}/5\sqrt{13}$
$\tilde{\frac{1}{2}}$	$\frac{3}{2}$	$\tilde{1}$	0	100−	$-2\sqrt{5}/7\sqrt{11}$
$\tilde{\frac{1}{2}}$	$\frac{3}{2}$	$\tilde{1}$	0	110−	0
$\frac{13}{2}$	$\frac{9}{2}$	**3**			
$\frac{1}{2}$	$\frac{1}{2}$	1	0	000+	$-4\sqrt{2}/\sqrt{3.7.11.13}$
$\frac{1}{2}$	$\frac{3}{2}$	1	0	000−	$-4.3\sqrt{2}/\sqrt{5.7.11.13}$
$\frac{1}{2}$	$\frac{3}{2}$	1	0	010−	$-\sqrt{11}/\sqrt{2.3.5.13}$
$\frac{1}{2}$	$\frac{3}{2}$	$\tilde{1}$	0	000+	$+8.3\sqrt{2}/5\sqrt{7.11.13}$
$\frac{1}{2}$	$\frac{3}{2}$	$\tilde{1}$	0	010+	$-\sqrt{3}/5\sqrt{2.11.13}$
$\frac{3}{2}$	$\frac{1}{2}$	1	0	000−	$+2/7\sqrt{3.13}$
$\frac{3}{2}$	$\frac{1}{2}$	1	0	100−	$-\sqrt{2.11}/7\sqrt{13}$
$\frac{3}{2}$	$\frac{1}{2}$	$\tilde{1}$	0	000+	$-2\sqrt{13}/7\sqrt{3.5}$
$\frac{3}{2}$	$\frac{1}{2}$	$\tilde{1}$	0	100+	$+3\sqrt{2}/7\sqrt{5.11.13}$
$\frac{3}{2}$	$\frac{3}{2}$	1	0	000+	$-4\sqrt{2}/7\sqrt{5.13}$
$\frac{3}{2}$	$\frac{3}{2}$	1	0	010+	$+4\sqrt{2}/\sqrt{3.5.7.13}$
$\frac{3}{2}$	$\frac{3}{2}$	1	0	100+	$+8.2\sqrt{3}/7\sqrt{5.11.13}$
$\frac{3}{2}$	$\frac{3}{2}$	1	0	110+	$-8.2/\sqrt{5.7.11.13}$
$\frac{3}{2}$	$\frac{3}{2}$	1	1	000+	$+11\sqrt{2}/7\sqrt{5.13}$
$\frac{3}{2}$	$\frac{3}{2}$	1	1	010+	$-\sqrt{2}/\sqrt{3.5.7.13}$
$\frac{3}{2}$	$\frac{3}{2}$	1	1	100+	$-2\sqrt{3}/7\sqrt{5.11.13}$
$\frac{3}{2}$	$\frac{3}{2}$	1	1	110+	$-3/\sqrt{5.7.11.13}$
$\frac{3}{2}$	$\frac{3}{2}$	$\tilde{1}$	0	000−	$-4.3\sqrt{2}/5.7\sqrt{13}$
$\frac{3}{2}$	$\frac{3}{2}$	$\tilde{1}$	0	010−	$+2\sqrt{2}/5\sqrt{3.7.13}$
$\frac{3}{2}$	$\frac{3}{2}$	$\tilde{1}$	0	100−	$-8.8\sqrt{3}/5.7\sqrt{11.13}$
$\frac{3}{2}$	$\frac{3}{2}$	$\tilde{1}$	0	110−	$-4.9/5\sqrt{7.11.13}$
$\frac{3}{2}$	$\frac{3}{2}$	$\tilde{1}$	1	000+	$-3\sqrt{2}/5.7\sqrt{13}$
$\frac{3}{2}$	$\frac{3}{2}$	$\tilde{1}$	1	010+	$-\sqrt{2.7}/5\sqrt{3.13}$
$\frac{3}{2}$	$\frac{3}{2}$	$\tilde{1}$	1	100+	$+2\sqrt{3.13}/5.7\sqrt{11}$
$\frac{3}{2}$	$\frac{3}{2}$	$\tilde{1}$	1	110+	$-3\sqrt{7}/5\sqrt{11.13}$
$\frac{3}{2}$	$\frac{3}{2}$	$\tilde{0}$	0	000+	$+8\sqrt{2.3}/5.7\sqrt{13}$
$\frac{3}{2}$	$\frac{3}{2}$	$\tilde{0}$	0	010+	$+\sqrt{2.7}/5\sqrt{13}$
$\frac{3}{2}$	$\frac{3}{2}$	$\tilde{0}$	0	100+	$-4.3/5.7\sqrt{11.13}$
$\frac{3}{2}$	$\frac{3}{2}$	$\tilde{0}$	0	110+	$+2\sqrt{3.7}/5\sqrt{11.13}$
$\tilde{\frac{1}{2}}$	$\frac{1}{2}$	$\tilde{1}$	0	000−	$+4\sqrt{2}/7\sqrt{3.13}$
$\tilde{\frac{1}{2}}$	$\frac{1}{2}$	$\tilde{1}$	0	100−	$-2.3/7\sqrt{5.11}$
$\tilde{\frac{1}{2}}$	$\frac{1}{2}$	$\tilde{0}$	0	000+	$+4\sqrt{2}/7\sqrt{13}$
$\tilde{\frac{1}{2}}$	$\frac{1}{2}$	$\tilde{0}$	0	100+	$+\sqrt{3}/7\sqrt{5.11}$
$\tilde{\frac{1}{2}}$	$\frac{3}{2}$	1	0	000−	$+4\sqrt{2}/7\sqrt{13}$
$\tilde{\frac{1}{2}}$	$\frac{3}{2}$	1	0	010−	$-\sqrt{3}/\sqrt{2.7.13}$
$\tilde{\frac{1}{2}}$	$\frac{3}{2}$	1	0	100−	$+\sqrt{3}/7\sqrt{5.11}$
$\tilde{\frac{1}{2}}$	$\frac{3}{2}$	1	0	110−	$+4/\sqrt{5.7.11}$
$\tilde{\frac{1}{2}}$	$\frac{3}{2}$	$\tilde{1}$	0	000+	0
$\tilde{\frac{1}{2}}$	$\frac{3}{2}$	$\tilde{1}$	0	010+	$+\sqrt{5}/\sqrt{2.3.7.13}$
$\tilde{\frac{1}{2}}$	$\frac{3}{2}$	$\tilde{1}$	0	100+	$-\sqrt{3}/5\sqrt{11}$
$\tilde{\frac{1}{2}}$	$\frac{3}{2}$	$\tilde{1}$	0	110+	$+4.3/5\sqrt{7.11}$
$\frac{13}{2}$	$\frac{9}{2}$	**4**			
$\frac{1}{2}$	$\frac{1}{2}$	0	0	000+	$+8/3\sqrt{11.13}$
$\frac{1}{2}$	$\frac{1}{2}$	1	0	000−	$-8/3\sqrt{5.11.13}$
$\frac{1}{2}$	$\frac{3}{2}$	1	0	000+	$-\sqrt{3}/5\sqrt{11.13}$
$\frac{1}{2}$	$\frac{3}{2}$	1	0	010+	$+\sqrt{7.11}/2.3.5\sqrt{13}$
$\frac{1}{2}$	$\frac{3}{2}$	2	0	000−	$+3\sqrt{3}/5\sqrt{2.7.11.13}$
$\frac{1}{2}$	$\frac{3}{2}$	2	0	010−	$-8.4\sqrt{2}/3.5\sqrt{11.13}$
$\frac{1}{2}$	$\frac{3}{2}$	$\tilde{1}$	0	000−	$+\sqrt{11}/5\sqrt{7.13}$
$\frac{1}{2}$	$\frac{3}{2}$	$\tilde{1}$	0	010−	$-17/2.5\sqrt{3.11.13}$
$\frac{3}{2}$	$\frac{1}{2}$	1	0	000+	$+8\sqrt{2}/3\sqrt{5.7.13}$
$\frac{3}{2}$	$\frac{1}{2}$	1	0	100+	$-5\sqrt{5}/\sqrt{3.7.11.13}$
$\frac{3}{2}$	$\frac{1}{2}$	2	0	000−	$+5\sqrt{5}/3.7\sqrt{13}$
$\frac{3}{2}$	$\frac{1}{2}$	2	0	100−	$+\sqrt{2.13}/7\sqrt{3.5.11}$
$\frac{3}{2}$	$\frac{1}{2}$	$\tilde{1}$	0	000−	$-2\sqrt{2.3}/7\sqrt{5.13}$
$\frac{3}{2}$	$\frac{1}{2}$	$\tilde{1}$	0	100−	$-\sqrt{5}/3.7\sqrt{11.13}$
$\frac{3}{2}$	$\frac{3}{2}$	0	0	000+	$+2\sqrt{3}/\sqrt{5.7.13}$
$\frac{3}{2}$	$\frac{3}{2}$	0	0	010+	$-1/3\sqrt{5.13}$
$\frac{3}{2}$	$\frac{3}{2}$	0	0	100+	$-\sqrt{2.5}/\sqrt{7.11.13}$
$\frac{3}{2}$	$\frac{3}{2}$	0	0	110+	$-\sqrt{2.5}/\sqrt{3.11.13}$
$\frac{3}{2}$	$\frac{3}{2}$	1	0	000−	0
$\frac{3}{2}$	$\frac{3}{2}$	1	0	010−	$-2/3\sqrt{13}$
$\frac{3}{2}$	$\frac{3}{2}$	1	0	100−	$+2\sqrt{2.7}/5\sqrt{11.13}$
$\frac{3}{2}$	$\frac{3}{2}$	1	0	110−	$+2\sqrt{2.3}/5\sqrt{11.13}$
$\frac{3}{2}$	$\frac{3}{2}$	1	1	000−	$-2\sqrt{3}/5\sqrt{7.13}$
$\frac{3}{2}$	$\frac{3}{2}$	1	1	010−	$+1/3.5\sqrt{13}$
$\frac{3}{2}$	$\frac{3}{2}$	1	1	100−	$+\sqrt{2}/\sqrt{7.11.13}$
$\frac{3}{2}$	$\frac{3}{2}$	1	1	110−	$-5/\sqrt{2.3.11.13}$
$\frac{3}{2}$	$\frac{3}{2}$	2	0	000+	$-\sqrt{2.3}/5\sqrt{13}$
$\frac{3}{2}$	$\frac{3}{2}$	2	0	010+	$+8.2\sqrt{2}/3.5\sqrt{7.13}$
$\frac{3}{2}$	$\frac{3}{2}$	2	0	100+	$+\sqrt{11}/5\sqrt{13}$
$\frac{3}{2}$	$\frac{3}{2}$	2	0	110+	$+8.2\sqrt{3}/5\sqrt{7.11.13}$
$\frac{3}{2}$	$\frac{3}{2}$	$\tilde{1}$	0	000+	$+4/5\sqrt{13}$
$\frac{3}{2}$	$\frac{3}{2}$	$\tilde{1}$	0	010+	$-8/5\sqrt{3.7.13}$
$\frac{3}{2}$	$\frac{3}{2}$	$\tilde{1}$	0	100+	$+4\sqrt{2.3}/5\sqrt{11.13}$
$\frac{3}{2}$	$\frac{3}{2}$	$\tilde{1}$	0	110+	$-8.8\sqrt{2}/3.5\sqrt{7.11.13}$
$\frac{3}{2}$	$\frac{3}{2}$	$\tilde{1}$	1	000−	$-2.11/5.7\sqrt{13}$
$\frac{3}{2}$	$\frac{3}{2}$	$\tilde{1}$	1	010−	$-\sqrt{13}/5\sqrt{3.7}$
$\frac{3}{2}$	$\frac{3}{2}$	$\tilde{1}$	1	100−	$-19\sqrt{2.3}/5.7\sqrt{11.13}$
$\frac{3}{2}$	$\frac{3}{2}$	$\tilde{1}$	1	110−	$+79/3.5\sqrt{2.7.11.13}$
$\tilde{\frac{1}{2}}$	$\frac{1}{2}$	$\tilde{1}$	0	000+	$+8/7\sqrt{3.13}$
$\tilde{\frac{1}{2}}$	$\frac{1}{2}$	$\tilde{1}$	0	100+	$-4\sqrt{2}/3.7\sqrt{5.11}$
$\tilde{\frac{1}{2}}$	$\frac{3}{2}$	1	0	000+	$-3\sqrt{3}/\sqrt{5.7.13}$
$\tilde{\frac{1}{2}}$	$\frac{3}{2}$	1	0	010+	$+1/2\sqrt{5.13}$
$\tilde{\frac{1}{2}}$	$\frac{3}{2}$	1	0	100+	$+\sqrt{2}/\sqrt{7.11}$
$\tilde{\frac{1}{2}}$	$\frac{3}{2}$	1	0	110+	$+\sqrt{2}/\sqrt{3.11}$
$\tilde{\frac{1}{2}}$	$\frac{3}{2}$	2	0	000−	$-\sqrt{3.5}/7\sqrt{2.13}$
$\tilde{\frac{1}{2}}$	$\frac{3}{2}$	2	0	010−	0
$\tilde{\frac{1}{2}}$	$\frac{3}{2}$	2	0	100−	$-8/5.7\sqrt{11}$
$\tilde{\frac{1}{2}}$	$\frac{3}{2}$	2	0	110−	$-\sqrt{7}/5\sqrt{3.11}$
$\tilde{\frac{1}{2}}$	$\frac{3}{2}$	$\tilde{1}$	0	000−	$+9/7\sqrt{5.13}$

SO_3–O 3jm Factors (cont.)

$\frac{13}{2}$ $\frac{9}{2}$ 4

$\tilde{\frac{1}{2}}$	$\frac{3}{2}$	$\tilde{1}$	0	010−	$+\sqrt{7}/2\sqrt{3.5.13}$
$\tilde{\frac{1}{2}}$	$\frac{3}{2}$	$\tilde{1}$	0	100−	$+9\sqrt{2.3}/5.7\sqrt{11}$
$\tilde{\frac{1}{2}}$	$\frac{3}{2}$	$\tilde{1}$	0	110−	$-\sqrt{2.7}/3.5\sqrt{11}$

$\frac{13}{2}$ 5 $\frac{3}{2}$

$\frac{1}{2}$	1	$\frac{3}{2}$	0	000+	$+\sqrt{3}/\sqrt{2.13}$
$\frac{1}{2}$	1	$\frac{3}{2}$	0	010+	0
$\frac{1}{2}$	2	$\frac{3}{2}$	0	000−	$-2\sqrt{5}/\sqrt{7.11.13}$
$\frac{1}{2}$	$\tilde{1}$	$\frac{3}{2}$	0	000−	$+\sqrt{3.5}/\sqrt{2.7.11.13}$
$\frac{3}{2}$	1	$\frac{3}{2}$	0	000−	$+5\sqrt{3}/\sqrt{2.7.11.13}$
$\frac{3}{2}$	1	$\frac{3}{2}$	0	010−	$+4\sqrt{2.3.5}/7\sqrt{11.13}$
$\frac{3}{2}$	1	$\frac{3}{2}$	0	100−	$-8.2/11\sqrt{7.13}$
$\frac{3}{2}$	1	$\frac{3}{2}$	0	110−	$+4.9\sqrt{5}/7.11\sqrt{13}$
$\frac{3}{2}$	1	$\frac{3}{2}$	1	000−	$+3\sqrt{3}/\sqrt{2.7.11.13}$
$\frac{3}{2}$	1	$\frac{3}{2}$	1	010−	$-2\sqrt{2.3.5}/7\sqrt{11.13}$
$\frac{3}{2}$	1	$\frac{3}{2}$	1	100−	$-5.5/11\sqrt{7.13}$
$\frac{3}{2}$	1	$\frac{3}{2}$	1	110−	$-2.9\sqrt{5}/7.11\sqrt{13}$
$\frac{3}{2}$	2	$\frac{3}{2}$	0	000+	$+2\sqrt{5}/7\sqrt{13}$
$\frac{3}{2}$	2	$\frac{3}{2}$	0	100+	$-4\sqrt{2.3.5}/7\sqrt{11.13}$
$\frac{3}{2}$	$\tilde{1}$	$\frac{3}{2}$	0	000+	$-\sqrt{3.5}/7\sqrt{2.13}$
$\frac{3}{2}$	$\tilde{1}$	$\frac{3}{2}$	0	100+	$+2.3\sqrt{5}/7\sqrt{11.13}$
$\frac{3}{2}$	$\tilde{1}$	$\frac{3}{2}$	1	000−	$-3\sqrt{3.5}/7\sqrt{2.13}$
$\frac{3}{2}$	$\tilde{1}$	$\frac{3}{2}$	1	100−	$-3\sqrt{5}/7\sqrt{11.13}$
$\tilde{\frac{1}{2}}$	1	$\frac{3}{2}$	0	000+	$-\sqrt{3.5}/\sqrt{2.7.11.13}$
$\tilde{\frac{1}{2}}$	1	$\frac{3}{2}$	0	010+	$-4\sqrt{2.3}/7\sqrt{11.13}$
$\tilde{\frac{1}{2}}$	1	$\frac{3}{2}$	0	100+	$+4/11\sqrt{7}$
$\tilde{\frac{1}{2}}$	1	$\frac{3}{2}$	0	110+	$-9\sqrt{5}/7.11$
$\tilde{\frac{1}{2}}$	2	$\frac{3}{2}$	0	000−	$+2/7\sqrt{13}$
$\tilde{\frac{1}{2}}$	2	$\frac{3}{2}$	0	100−	$-\sqrt{2.3.5}/7\sqrt{11}$
$\tilde{\frac{1}{2}}$	$\tilde{1}$	$\frac{3}{2}$	0	000−	$-\sqrt{3}/\sqrt{2.13}$
$\tilde{\frac{1}{2}}$	$\tilde{1}$	$\frac{3}{2}$	0	100−	0

$\frac{13}{2}$ 5 $\frac{5}{2}$

$\frac{1}{2}$	1	$\frac{3}{2}$	0	000−	$-3\sqrt{3}/11\sqrt{2.13}$
$\frac{1}{2}$	1	$\frac{3}{2}$	0	010−	$+2\sqrt{2.3.5}/11\sqrt{7.13}$
$\frac{1}{2}$	2	$\frac{3}{2}$	0	000+	$+4\sqrt{5}/\sqrt{7.11.13}$
$\frac{1}{2}$	$\tilde{1}$	$\frac{3}{2}$	0	000+	$+\sqrt{5.7}/\sqrt{2.3.11.13}$
$\frac{1}{2}$	$\tilde{1}$	$\tilde{\frac{1}{2}}$	0	000−	$+2\sqrt{2}/\sqrt{3.7.11.13}$
$\frac{3}{2}$	1	$\frac{3}{2}$	0	000+	$-2.5\sqrt{2}/\sqrt{3.7.11.13}$
$\frac{3}{2}$	1	$\frac{3}{2}$	0	010+	$+2\sqrt{2.3.5}/7\sqrt{11.13}$
$\frac{3}{2}$	1	$\frac{3}{2}$	0	100+	$+8.8/3.11\sqrt{7.13}$
$\frac{3}{2}$	1	$\frac{3}{2}$	0	110+	$+2.9\sqrt{5}/7.11\sqrt{13}$
$\frac{3}{2}$	1	$\frac{3}{2}$	1	000+	$-\sqrt{2}/\sqrt{3.7.11.13}$
$\frac{3}{2}$	1	$\frac{3}{2}$	1	010+	$-2\sqrt{2.3.5}/7\sqrt{11.13}$
$\frac{3}{2}$	1	$\frac{3}{2}$	1	100+	$-5\sqrt{13}/3.11\sqrt{7}$
$\frac{3}{2}$	1	$\frac{3}{2}$	1	110+	$+8.3\sqrt{5}/7.11\sqrt{13}$
$\frac{3}{2}$	1	$\tilde{\frac{1}{2}}$	0	000−	$-\sqrt{5}/\sqrt{3.7.11.13}$
$\frac{3}{2}$	1	$\tilde{\frac{1}{2}}$	0	010−	$+4.3\sqrt{3}/7\sqrt{11.13}$
$\frac{3}{2}$	1	$\tilde{\frac{1}{2}}$	0	100−	$-\sqrt{2.5}/3.11\sqrt{7.13}$
$\frac{3}{2}$	1	$\tilde{\frac{1}{2}}$	0	110−	$+4.3\sqrt{2}/7.11\sqrt{13}$
$\frac{3}{2}$	2	$\frac{3}{2}$	0	000−	$+2\sqrt{5}/7\sqrt{13}$

$\frac{13}{2}$ 5 $\frac{5}{2}$

$\frac{3}{2}$	2	$\frac{3}{2}$	0	100−	$+2\sqrt{2.5}/7\sqrt{3.11.13}$
$\frac{3}{2}$	2	$\tilde{\frac{1}{2}}$	0	000+	0
$\frac{3}{2}$	2	$\tilde{\frac{1}{2}}$	0	100+	$-4/\sqrt{3.11.13}$
$\frac{3}{2}$	$\tilde{1}$	$\frac{3}{2}$	0	000−	$+2\sqrt{2.5}/7\sqrt{3.13}$
$\frac{3}{2}$	$\tilde{1}$	$\frac{3}{2}$	0	100−	$-8\sqrt{5}/7\sqrt{11.13}$
$\frac{3}{2}$	$\tilde{1}$	$\frac{3}{2}$	1	000+	$-\sqrt{2.5}/7\sqrt{3.13}$
$\frac{3}{2}$	$\tilde{1}$	$\frac{3}{2}$	1	100+	$-3\sqrt{5}/7\sqrt{11.13}$
$\frac{3}{2}$	$\tilde{1}$	$\tilde{\frac{1}{2}}$	0	000+	$-11/7\sqrt{3.13}$
$\frac{3}{2}$	$\tilde{1}$	$\tilde{\frac{1}{2}}$	0	100+	$-\sqrt{2.13}/7\sqrt{11}$
$\tilde{\frac{1}{2}}$	1	$\frac{3}{2}$	0	000−	$-\sqrt{5.7}/\sqrt{2.3.11.13}$
$\tilde{\frac{1}{2}}$	1	$\frac{3}{2}$	0	010−	$-2\sqrt{2.3}/7\sqrt{11.13}$
$\tilde{\frac{1}{2}}$	1	$\frac{3}{2}$	0	100−	$+2\sqrt{7}/3.11$
$\tilde{\frac{1}{2}}$	1	$\frac{3}{2}$	0	110−	$+2.3\sqrt{5}/7.11$
$\tilde{\frac{1}{2}}$	1	$\tilde{\frac{1}{2}}$	0	000+	$+2\sqrt{2}/\sqrt{3.7.11.13}$
$\tilde{\frac{1}{2}}$	1	$\tilde{\frac{1}{2}}$	0	010+	$+4\sqrt{2.3.5}/7\sqrt{11.13}$
$\tilde{\frac{1}{2}}$	1	$\tilde{\frac{1}{2}}$	0	100+	$+8\sqrt{5}/3.11\sqrt{7}$
$\tilde{\frac{1}{2}}$	1	$\tilde{\frac{1}{2}}$	0	110+	$+3/7.11$
$\tilde{\frac{1}{2}}$	2	$\frac{3}{2}$	0	000+	$-4/7\sqrt{13}$
$\tilde{\frac{1}{2}}$	2	$\frac{3}{2}$	0	100+	$-\sqrt{2.5}/7\sqrt{3.11}$
$\tilde{\frac{1}{2}}$	$\tilde{1}$	$\frac{3}{2}$	0	000+	$-\sqrt{3}/7\sqrt{2.13}$
$\tilde{\frac{1}{2}}$	$\tilde{1}$	$\frac{3}{2}$	0	100+	$-2\sqrt{5}/7\sqrt{11}$

$\frac{13}{2}$ 5 $\frac{7}{2}$

$\frac{1}{2}$	1	$\frac{1}{2}$	0	000−	$+\sqrt{3.5.7}/4.11\sqrt{13}$
$\frac{1}{2}$	1	$\frac{1}{2}$	0	010−	$-8.2/11\sqrt{3.13}$
$\frac{1}{2}$	1	$\frac{3}{2}$	0	000+	$-17\sqrt{3}/4.11\sqrt{13}$
$\frac{1}{2}$	1	$\frac{3}{2}$	0	010+	$+8/11\sqrt{3.5.7.13}$
$\frac{1}{2}$	2	$\frac{3}{2}$	0	000−	$+3/\sqrt{2.5.7.11.13}$
$\frac{1}{2}$	$\tilde{1}$	$\frac{3}{2}$	0	000−	$-\sqrt{3.5.7}/4\sqrt{11.13}$
$\frac{1}{2}$	$\tilde{1}$	$\tilde{\frac{1}{2}}$	0	000+	$+3\sqrt{5}/4\sqrt{7.11.13}$
$\frac{3}{2}$	1	$\frac{1}{2}$	0	000+	$-\sqrt{3.5}/2\sqrt{2.11.13}$
$\frac{3}{2}$	1	$\frac{1}{2}$	0	010+	$+5\sqrt{2}/\sqrt{3.7.11.13}$
$\frac{3}{2}$	1	$\frac{1}{2}$	0	100+	$-9\sqrt{5}/4.11\sqrt{13}$
$\frac{3}{2}$	1	$\frac{1}{2}$	0	110+	$-47/2.11\sqrt{7.13}$
$\frac{3}{2}$	1	$\frac{3}{2}$	0	000−	$+3\sqrt{3}/\sqrt{7.11.13}$
$\frac{3}{2}$	1	$\frac{3}{2}$	0	010−	$+2.5\sqrt{5}/7\sqrt{3.11.13}$
$\frac{3}{2}$	1	$\frac{3}{2}$	0	100−	$-9/2.11\sqrt{2.7.13}$
$\frac{3}{2}$	1	$\frac{3}{2}$	0	110−	$+2\sqrt{2.13}/7.11\sqrt{5}$
$\frac{3}{2}$	1	$\frac{3}{2}$	1	000−	$-5\sqrt{3}/2\sqrt{7.11.13}$
$\frac{3}{2}$	1	$\frac{3}{2}$	1	010−	$-71/7\sqrt{3.5.11.13}$
$\frac{3}{2}$	1	$\frac{3}{2}$	1	100−	$+9/11\sqrt{2.7.13}$
$\frac{3}{2}$	1	$\frac{3}{2}$	1	110−	$+3\sqrt{13}/7.11\sqrt{2.5}$
$\frac{3}{2}$	1	$\tilde{\frac{1}{2}}$	0	000+	$-9/2\sqrt{2.7.11.13}$
$\frac{3}{2}$	1	$\tilde{\frac{1}{2}}$	0	010+	$+4\sqrt{2}/7\sqrt{5.11.13}$
$\frac{3}{2}$	1	$\tilde{\frac{1}{2}}$	0	100+	$-5\sqrt{3}/4\sqrt{7.13}$
$\frac{3}{2}$	1	$\tilde{\frac{1}{2}}$	0	110+	$-\sqrt{3}/2.7\sqrt{5.13}$
$\frac{3}{2}$	2	$\frac{1}{2}$	0	000−	$-1/\sqrt{7.13}$
$\frac{3}{2}$	2	$\frac{1}{2}$	0	100−	$+2\sqrt{2.3}/\sqrt{7.11.13}$
$\frac{3}{2}$	2	$\frac{3}{2}$	0	000+	$-8\sqrt{2}/7\sqrt{5.13}$
$\frac{3}{2}$	2	$\frac{3}{2}$	0	100+	$+\sqrt{3.11}/7\sqrt{5.13}$

SO_3–O 3jm Factors (cont.)

$\frac{13}{2}$	5	$\frac{7}{2}$			
$\frac{3}{2}$	2	$\tilde{\frac{1}{2}}$	0	000−	$-\sqrt{3}/\sqrt{5.13}$
$\frac{3}{2}$	2	$\tilde{\frac{1}{2}}$	0	100−	$-3\sqrt{2}/\sqrt{5.11.13}$
$\frac{3}{2}$	$\tilde{1}$	$\frac{1}{2}$	0	000−	$+1/2\sqrt{2.3.7.13}$
$\frac{3}{2}$	$\tilde{1}$	$\frac{1}{2}$	0	100−	$+3/4\sqrt{7.11.13}$
$\frac{3}{2}$	$\tilde{1}$	$\frac{3}{2}$	0	000+	$-2/7\sqrt{3.5.13}$
$\frac{3}{2}$	$\tilde{1}$	$\frac{3}{2}$	0	100+	$-3\sqrt{11}/2.7\sqrt{2.5.13}$
$\frac{3}{2}$	$\tilde{1}$	$\frac{3}{2}$	1	000−	$-\sqrt{5}/2.7\sqrt{3.13}$
$\frac{3}{2}$	$\tilde{1}$	$\frac{3}{2}$	1	100−	$-9.9/7\sqrt{2.5.11.13}$
$\frac{3}{2}$	$\tilde{1}$	$\tilde{\frac{1}{2}}$	0	000−	$+9.3/2.7\sqrt{2.5.13}$
$\frac{3}{2}$	$\tilde{1}$	$\tilde{\frac{1}{2}}$	0	100−	$-3\sqrt{3}/4.7\sqrt{5.11.13}$
$\tilde{\frac{1}{2}}$	1	$\frac{3}{2}$	0	000+	$+\sqrt{3.5.7}/4\sqrt{11.13}$
$\tilde{\frac{1}{2}}$	1	$\frac{3}{2}$	0	010+	$-8\sqrt{3}/7\sqrt{11.13}$
$\tilde{\frac{1}{2}}$	1	$\frac{3}{2}$	0	100+	0
$\tilde{\frac{1}{2}}$	1	$\frac{3}{2}$	0	110+	$+\sqrt{2}/7\sqrt{5}$
$\tilde{\frac{1}{2}}$	1	$\tilde{\frac{1}{2}}$	0	000−	$+3\sqrt{5}/4\sqrt{7.11.13}$
$\tilde{\frac{1}{2}}$	1	$\tilde{\frac{1}{2}}$	0	010−	$-8.2/7\sqrt{11.13}$
$\tilde{\frac{1}{2}}$	1	$\tilde{\frac{1}{2}}$	0	100−	$+5\sqrt{3}/11\sqrt{2.7}$
$\tilde{\frac{1}{2}}$	1	$\tilde{\frac{1}{2}}$	0	110−	$-2\sqrt{2.3}/7.11\sqrt{5}$
$\tilde{\frac{1}{2}}$	2	$\frac{3}{2}$	0	000−	$+5/7\sqrt{2.13}$
$\tilde{\frac{1}{2}}$	2	$\frac{3}{2}$	0	100−	$-2\sqrt{3}/7\sqrt{5.11}$
$\tilde{\frac{1}{2}}$	$\tilde{1}$	$\frac{1}{2}$	0	000+	$+\sqrt{5}/4\sqrt{3.7.13}$
$\tilde{\frac{1}{2}}$	$\tilde{1}$	$\frac{1}{2}$	0	100+	$-3/\sqrt{2.7.11}$
$\tilde{\frac{1}{2}}$	$\tilde{1}$	$\frac{3}{2}$	0	000−	$-17/4.7\sqrt{3.13}$
$\tilde{\frac{1}{2}}$	$\tilde{1}$	$\frac{3}{2}$	0	100−	$-2.3\sqrt{2}/7\sqrt{5.11}$
$\frac{13}{2}$	**5**	**$\frac{9}{2}$**			
$\frac{1}{2}$	1	$\frac{1}{2}$	0	000+	$+2\sqrt{3.7}/11\sqrt{13.17}$
$\frac{1}{2}$	1	$\frac{1}{2}$	0	010+	$+8\sqrt{17}/11\sqrt{3.5.13}$
$\frac{1}{2}$	1	$\frac{3}{2}$	0	000−	$-3\sqrt{7}/2\sqrt{5.13.17}$
$\frac{1}{2}$	1	$\frac{3}{2}$	0	001−	$-29\sqrt{3}/11\sqrt{5.13.17}$
$\frac{1}{2}$	1	$\frac{3}{2}$	0	010−	$-2.3/5\sqrt{13.17}$
$\frac{1}{2}$	1	$\frac{3}{2}$	0	011−	$-79\sqrt{7}/5.11\sqrt{3.13.17}$
$\frac{1}{2}$	2	$\frac{3}{2}$	0	000+	$+4\sqrt{2.3}/5\sqrt{11.13.17}$
$\frac{1}{2}$	2	$\frac{3}{2}$	0	001+	$-3\sqrt{17}/5\sqrt{2.7.11.13}$
$\frac{1}{2}$	$\tilde{1}$	$\frac{3}{2}$	0	000+	$+9.3/2.5\sqrt{11.13.17}$
$\frac{1}{2}$	$\tilde{1}$	$\frac{3}{2}$	0	001+	$-59\sqrt{3}/5\sqrt{7.11.13.17}$
$\frac{3}{2}$	1	$\frac{1}{2}$	0	000−	$+5\sqrt{3}/\sqrt{2.11.13.17}$
$\frac{3}{2}$	1	$\frac{1}{2}$	0	010−	$+8.2\sqrt{2}/\sqrt{3.5.7.11.13.17}$
$\frac{3}{2}$	1	$\frac{1}{2}$	0	100−	$-9/11\sqrt{13.17}$
$\frac{3}{2}$	1	$\frac{1}{2}$	0	110−	$-2.5\sqrt{5}/11\sqrt{7.13.17}$
$\frac{3}{2}$	1	$\frac{3}{2}$	0	000+	$+2.3/\sqrt{5.11.13.17}$
$\frac{3}{2}$	1	$\frac{3}{2}$	0	001+	$-8.3\sqrt{3}/\sqrt{5.7.11.13.17}$
$\frac{3}{2}$	1	$\frac{3}{2}$	0	010+	$+2\sqrt{7}/\sqrt{11.13.17}$
$\frac{3}{2}$	1	$\frac{3}{2}$	0	011+	$+5\sqrt{17}/7\sqrt{3.11.13}$
$\frac{3}{2}$	1	$\frac{3}{2}$	0	100+	$+8.2\sqrt{2.3}/11\sqrt{5.13.17}$
$\frac{3}{2}$	1	$\frac{3}{2}$	0	101+	$+2.9\sqrt{2}/11\sqrt{5.7.13.17}$
$\frac{3}{2}$	1	$\frac{3}{2}$	0	110+	$+\sqrt{2.3.7}/5.11\sqrt{13.17}$
$\frac{3}{2}$	1	$\frac{3}{2}$	0	111+	$+\sqrt{2.13.17}/5.7.11$
$\frac{3}{2}$	1	$\frac{3}{2}$	1	000+	$+3.7/\sqrt{5.11.13.17}$
$\frac{3}{2}$	1	$\frac{3}{2}$	1	001+	$-2\sqrt{3.17}/\sqrt{5.7.11.13}$
$\frac{13}{2}$	**5**	**$\frac{9}{2}$**			
$\frac{3}{2}$	1	$\frac{3}{2}$	1	010+	$-8/5\sqrt{7.11.13.17}$
$\frac{3}{2}$	1	$\frac{3}{2}$	1	011+	$+2.263/5.7\sqrt{3.11.13.17}$
$\frac{3}{2}$	1	$\frac{3}{2}$	1	100+	$+7\sqrt{3}/\sqrt{2.5.13.17}$
$\frac{3}{2}$	1	$\frac{3}{2}$	1	101+	$-9.3\sqrt{2}/11\sqrt{5.7.13.17}$
$\frac{3}{2}$	1	$\frac{3}{2}$	1	110+	$-2\sqrt{2.3}/\sqrt{7.13.17}$
$\frac{3}{2}$	1	$\frac{3}{2}$	1	111+	$-3.5\sqrt{2.13}/7.11\sqrt{17}$
$\frac{3}{2}$	2	$\frac{1}{2}$	0	000+	$+2\sqrt{5}/\sqrt{7.13.17}$
$\frac{3}{2}$	2	$\frac{1}{2}$	0	100+	$+\sqrt{2.3}/\sqrt{5.7.11.13.17}$
$\frac{3}{2}$	2	$\frac{3}{2}$	0	000−	$-8\sqrt{2.3}/5\sqrt{7.13.17}$
$\frac{3}{2}$	2	$\frac{3}{2}$	0	001−	$+\sqrt{2}/5.7\sqrt{13.17}$
$\frac{3}{2}$	2	$\frac{3}{2}$	0	100−	$-8.9/5\sqrt{7.11.13.17}$
$\frac{3}{2}$	2	$\frac{3}{2}$	0	101−	$+\sqrt{3.11.13}/5.7\sqrt{17}$
$\frac{3}{2}$	$\tilde{1}$	$\frac{1}{2}$	0	000+	$+\sqrt{13}/\sqrt{2.3.5.7.17}$
$\frac{3}{2}$	$\tilde{1}$	$\frac{1}{2}$	0	100+	$-3.5\sqrt{5}/\sqrt{7.11.13.17}$
$\frac{3}{2}$	$\tilde{1}$	$\frac{3}{2}$	0	000−	$+2.3\sqrt{7}/5\sqrt{13.17}$
$\frac{3}{2}$	$\tilde{1}$	$\frac{3}{2}$	0	001−	$+8.2/5.7\sqrt{3.13.17}$
$\frac{3}{2}$	$\tilde{1}$	$\frac{3}{2}$	0	100−	$-4\sqrt{2.3.7}/5\sqrt{11.13.17}$
$\frac{3}{2}$	$\tilde{1}$	$\frac{3}{2}$	0	101−	$+2.3\sqrt{2.11}/5.7\sqrt{13.17}$
$\frac{3}{2}$	$\tilde{1}$	$\frac{3}{2}$	1	000+	$-9.3/5\sqrt{7.13.17}$
$\frac{3}{2}$	$\tilde{1}$	$\frac{3}{2}$	1	001+	$-2.53/5.7\sqrt{3.13.17}$
$\frac{3}{2}$	$\tilde{1}$	$\frac{3}{2}$	1	100+	$+\sqrt{3.17}/5\sqrt{2.7.11.13}$
$\frac{3}{2}$	$\tilde{1}$	$\frac{3}{2}$	1	101+	$-9\sqrt{2.13}/5.7\sqrt{11.17}$
$\tilde{\frac{1}{2}}$	1	$\frac{3}{2}$	0	000−	$-3.5/2\sqrt{11.13.17}$
$\tilde{\frac{1}{2}}$	1	$\frac{3}{2}$	0	001−	$-5\sqrt{3}/\sqrt{7.11.13.17}$
$\tilde{\frac{1}{2}}$	1	$\frac{3}{2}$	0	010−	$-2.19/\sqrt{5.7.11.13.17}$
$\tilde{\frac{1}{2}}$	1	$\frac{3}{2}$	0	011−	$+37\sqrt{3}/7\sqrt{5.11.13.17}$
$\tilde{\frac{1}{2}}$	1	$\frac{3}{2}$	0	100−	$-4\sqrt{2.3}/11\sqrt{5.17}$
$\tilde{\frac{1}{2}}$	1	$\frac{3}{2}$	0	101−	$-9/11\sqrt{2.5.7.17}$
$\tilde{\frac{1}{2}}$	1	$\frac{3}{2}$	0	110−	$-2\sqrt{2.3}/11\sqrt{7.17}$
$\tilde{\frac{1}{2}}$	1	$\frac{3}{2}$	0	111−	$-2.19\sqrt{2}/7.11\sqrt{17}$
$\tilde{\frac{1}{2}}$	2	$\frac{3}{2}$	0	000+	0
$\tilde{\frac{1}{2}}$	2	$\frac{3}{2}$	0	001+	$-\sqrt{5.17}/7\sqrt{2.13}$
$\tilde{\frac{1}{2}}$	2	$\frac{3}{2}$	0	100+	$+2.3\sqrt{7}/5\sqrt{11.17}$
$\tilde{\frac{1}{2}}$	2	$\frac{3}{2}$	0	101+	$-8\sqrt{3}/5.7\sqrt{11.17}$
$\tilde{\frac{1}{2}}$	$\tilde{1}$	$\frac{1}{2}$	0	000−	$+2/\sqrt{3.7.13.17}$
$\tilde{\frac{1}{2}}$	$\tilde{1}$	$\frac{1}{2}$	0	100−	$-4.3\sqrt{2}/\sqrt{5.7.11.17}$
$\tilde{\frac{1}{2}}$	$\tilde{1}$	$\frac{3}{2}$	0	000+	$-9.3/2\sqrt{5.7.13.17}$
$\tilde{\frac{1}{2}}$	$\tilde{1}$	$\frac{3}{2}$	0	001+	$+1/\sqrt{3.5.13.17}$
$\tilde{\frac{1}{2}}$	$\tilde{1}$	$\frac{3}{2}$	0	100+	$+4\sqrt{2.3}/5\sqrt{7.11.17}$
$\tilde{\frac{1}{2}}$	$\tilde{1}$	$\frac{3}{2}$	0	101+	$-3/5\sqrt{2.11.17}$
$\frac{13}{2}$	**$\frac{11}{2}$**	**1**			
$\frac{1}{2}$	$\frac{1}{2}$	1	0	000−	$+2/\sqrt{3.13}$
$\frac{1}{2}$	$\frac{3}{2}$	1	0	000+	$-\sqrt{11}/\sqrt{3.7.13}$
$\frac{1}{2}$	$\frac{3}{2}$	1	0	010+	0
$\frac{3}{2}$	$\frac{1}{2}$	1	0	000+	$+\sqrt{11}/\sqrt{2.3.7.13}$
$\frac{3}{2}$	$\frac{1}{2}$	1	0	100+	$-2/\sqrt{7.13}$
$\frac{3}{2}$	$\frac{3}{2}$	1	0	000−	$+4/7\sqrt{3.13}$
$\frac{3}{2}$	$\frac{3}{2}$	1	0	010−	$-4\sqrt{2.5}/7\sqrt{3.13}$
$\frac{3}{2}$	$\frac{3}{2}$	1	0	100−	$+2.3\sqrt{2}/7\sqrt{11.13}$

SO_3–O 3jm Factors (cont.)

$\frac{13}{2}$	$\frac{11}{2}$	1			
$\frac{3}{2}$	$\frac{3}{2}$	1	0	110−	$-4.3\sqrt{5}/7\sqrt{11.13}$
$\frac{3}{2}$	$\frac{3}{2}$	1	1	000−	$-\sqrt{13}/7\sqrt{3}$
$\frac{3}{2}$	$\frac{3}{2}$	1	1	010−	$+2\sqrt{2.5}/7\sqrt{3.13}$
$\frac{3}{2}$	$\frac{3}{2}$	1	1	100−	$+19\sqrt{2}/7\sqrt{11.13}$
$\frac{3}{2}$	$\frac{3}{2}$	1	1	110−	$+2.3\sqrt{5}/7\sqrt{11.13}$
$\frac{3}{2}$	$\tilde{\frac{1}{2}}$	1	0	000+	$-\sqrt{5}/\sqrt{2.3.13}$
$\frac{3}{2}$	$\tilde{\frac{1}{2}}$	1	0	100+	0
$\tilde{\frac{1}{2}}$	$\frac{3}{2}$	1	0	000+	$+\sqrt{3.5}/7\sqrt{13}$
$\tilde{\frac{1}{2}}$	$\frac{3}{2}$	1	0	010+	$-4\sqrt{2}/7\sqrt{3.13}$
$\tilde{\frac{1}{2}}$	$\frac{3}{2}$	1	0	100+	$-4\sqrt{2}/7\sqrt{11}$
$\tilde{\frac{1}{2}}$	$\frac{3}{2}$	1	0	110+	$-3\sqrt{5}/7\sqrt{11}$
$\tilde{\frac{1}{2}}$	$\tilde{\frac{1}{2}}$	1	0	000−	$+2/\sqrt{3.13}$
$\tilde{\frac{1}{2}}$	$\tilde{\frac{1}{2}}$	1	0	100−	0

$\frac{13}{2}$	$\frac{11}{2}$	2			
$\frac{1}{2}$	$\frac{3}{2}$	2	0	000+	$-3.17\sqrt{2}/11\sqrt{5.7.13}$
$\frac{1}{2}$	$\frac{3}{2}$	2	0	010+	$-8/11\sqrt{7.13}$
$\frac{1}{2}$	$\frac{3}{2}$	$\tilde{1}$	0	000+	$-3\sqrt{3.7}/11\sqrt{5.13}$
$\frac{1}{2}$	$\frac{3}{2}$	$\tilde{1}$	0	010+	$+8\sqrt{2}/11\sqrt{3.7.13}$
$\frac{1}{2}$	$\tilde{\frac{1}{2}}$	$\tilde{1}$	0	000−	$+2/\sqrt{3.7.13}$
$\frac{3}{2}$	$\frac{1}{2}$	2	0	000+	$-\sqrt{5}/\sqrt{7.13}$
$\frac{3}{2}$	$\frac{1}{2}$	2	0	100+	$+8.2\sqrt{2}/\sqrt{3.5.7.11.13}$
$\frac{3}{2}$	$\frac{1}{2}$	$\tilde{1}$	0	000+	$-1/\sqrt{2.3.5.7.13}$
$\frac{3}{2}$	$\frac{1}{2}$	$\tilde{1}$	0	100+	$+2.17/3\sqrt{5.7.11.13}$
$\frac{3}{2}$	$\frac{3}{2}$	2	0	000−	$+\sqrt{2.13}/7\sqrt{5.11}$
$\frac{3}{2}$	$\frac{3}{2}$	2	0	010−	$-4/7\sqrt{11.13}$
$\frac{3}{2}$	$\frac{3}{2}$	2	0	100−	$+2\sqrt{5}/7\sqrt{3.13}$
$\frac{3}{2}$	$\frac{3}{2}$	2	0	110−	$+2\sqrt{2.3}/7\sqrt{13}$
$\frac{3}{2}$	$\frac{3}{2}$	$\tilde{1}$	0	000−	$-4/7\sqrt{3.5.11.13}$
$\frac{3}{2}$	$\frac{3}{2}$	$\tilde{1}$	0	010−	$+4\sqrt{2}/7\sqrt{3.11.13}$
$\frac{3}{2}$	$\frac{3}{2}$	$\tilde{1}$	0	100−	$-2.11\sqrt{2}/3.7\sqrt{5.13}$
$\frac{3}{2}$	$\frac{3}{2}$	$\tilde{1}$	0	110−	$-4/7\sqrt{13}$
$\frac{3}{2}$	$\frac{3}{2}$	$\tilde{1}$	1	000+	$-43/7\sqrt{3.5.11.13}$
$\frac{3}{2}$	$\frac{3}{2}$	$\tilde{1}$	1	010+	$-2.17\sqrt{2}/7\sqrt{3.11.13}$
$\frac{3}{2}$	$\frac{3}{2}$	$\tilde{1}$	1	100+	$+223\sqrt{2}/3.7.11\sqrt{5.13}$
$\frac{3}{2}$	$\frac{3}{2}$	$\tilde{1}$	1	110+	$-2.23/7.11\sqrt{13}$
$\frac{3}{2}$	$\tilde{\frac{1}{2}}$	2	0	000+	$-\sqrt{11}/7\sqrt{13}$
$\frac{3}{2}$	$\tilde{\frac{1}{2}}$	2	0	100+	$+2\sqrt{2.3}/7\sqrt{13}$
$\frac{3}{2}$	$\tilde{\frac{1}{2}}$	$\tilde{1}$	0	000+	$+5\sqrt{11}/7\sqrt{2.3.13}$
$\frac{3}{2}$	$\tilde{\frac{1}{2}}$	$\tilde{1}$	0	100+	$+4/7\sqrt{13}$
$\tilde{\frac{1}{2}}$	$\frac{1}{2}$	$\tilde{1}$	0	000−	$-2/\sqrt{3.7.13}$
$\tilde{\frac{1}{2}}$	$\frac{1}{2}$	$\tilde{1}$	0	100−	$+8\sqrt{2}/3\sqrt{5.7.11}$
$\tilde{\frac{1}{2}}$	$\frac{3}{2}$	2	0	000+	$-\sqrt{2.11}/7\sqrt{13}$
$\tilde{\frac{1}{2}}$	$\frac{3}{2}$	2	0	010+	0
$\tilde{\frac{1}{2}}$	$\frac{3}{2}$	2	0	100+	$+8.4/7.11\sqrt{3.5}$
$\tilde{\frac{1}{2}}$	$\frac{3}{2}$	2	0	110+	$-\sqrt{2.3}/11$
$\tilde{\frac{1}{2}}$	$\frac{3}{2}$	$\tilde{1}$	0	000+	$+23/7\sqrt{3.11.13}$
$\tilde{\frac{1}{2}}$	$\frac{3}{2}$	$\tilde{1}$	0	010+	$-4\sqrt{2.3.5}/7\sqrt{11.13}$
$\tilde{\frac{1}{2}}$	$\frac{3}{2}$	$\tilde{1}$	0	100+	$+4.19\sqrt{2}/3.7.11\sqrt{5}$
$\tilde{\frac{1}{2}}$	$\frac{3}{2}$	$\tilde{1}$	0	110+	$-3/7.11$

$\frac{13}{2}$	$\frac{11}{2}$	3			
$\frac{1}{2}$	$\frac{1}{2}$	1	0	000−	$-3\sqrt{3}/4\sqrt{11.13}$
$\frac{1}{2}$	$\frac{3}{2}$	1	0	000+	$+3\sqrt{3.13}/4.11\sqrt{7}$
$\frac{1}{2}$	$\frac{3}{2}$	1	0	010+	$+4\sqrt{2.3.5}/11\sqrt{7.13}$
$\frac{1}{2}$	$\frac{3}{2}$	$\tilde{1}$	0	000−	$-3.5\sqrt{3.5}/4.11\sqrt{7.13}$
$\frac{1}{2}$	$\frac{3}{2}$	$\tilde{1}$	0	010−	$+4\sqrt{2}/11\sqrt{3.7.13}$
$\frac{1}{2}$	$\tilde{\frac{1}{2}}$	$\tilde{1}$	0	000+	$+\sqrt{13}/4\sqrt{3.7}$
$\frac{1}{2}$	$\tilde{\frac{1}{2}}$	$\tilde{0}$	0	000−	$+1/2\sqrt{7.13}$
$\frac{3}{2}$	$\frac{1}{2}$	1	0	000+	$-\sqrt{3}/\sqrt{2.7.13}$
$\frac{3}{2}$	$\frac{1}{2}$	1	0	100+	$-\sqrt{11}/4\sqrt{7.13}$
$\frac{3}{2}$	$\frac{1}{2}$	$\tilde{1}$	0	000−	$-\sqrt{5}/\sqrt{2.3.7.13}$
$\frac{3}{2}$	$\frac{1}{2}$	$\tilde{1}$	0	100−	$+31\sqrt{5}/4.3\sqrt{7.11.13}$
$\frac{3}{2}$	$\frac{3}{2}$	1	0	000−	$-3\sqrt{3}/2.7\sqrt{11.13}$
$\frac{3}{2}$	$\frac{3}{2}$	1	0	010−	$-4\sqrt{2.3.5}/7\sqrt{11.13}$
$\frac{3}{2}$	$\frac{3}{2}$	1	0	100−	$+113/2.7.11\sqrt{2.13}$
$\frac{3}{2}$	$\frac{3}{2}$	1	0	110−	$+19\sqrt{5}/2.7.11\sqrt{13}$
$\frac{3}{2}$	$\frac{3}{2}$	1	1	000−	$+\sqrt{3}/2.7\sqrt{11.13}$
$\frac{3}{2}$	$\frac{3}{2}$	1	1	010−	$-\sqrt{3.5}/7\sqrt{2.11.13}$
$\frac{3}{2}$	$\frac{3}{2}$	1	1	100−	$-3.23/7.11\sqrt{2.13}$
$\frac{3}{2}$	$\frac{3}{2}$	1	1	110−	$+61\sqrt{5}/2.7.11\sqrt{13}$
$\frac{3}{2}$	$\frac{3}{2}$	$\tilde{1}$	0	000+	$-17\sqrt{5}/2.7\sqrt{3.11.13}$
$\frac{3}{2}$	$\frac{3}{2}$	$\tilde{1}$	0	010+	$-\sqrt{2}/\sqrt{3.11.13}$
$\frac{3}{2}$	$\frac{3}{2}$	$\tilde{1}$	0	100+	$+43\sqrt{5}/2.3.7.11\sqrt{2.13}$
$\frac{3}{2}$	$\frac{3}{2}$	$\tilde{1}$	0	110+	$-5/2.11\sqrt{13}$
$\frac{3}{2}$	$\frac{3}{2}$	$\tilde{1}$	1	000−	$-41\sqrt{5}/2.7\sqrt{3.11.13}$
$\frac{3}{2}$	$\frac{3}{2}$	$\tilde{1}$	1	010−	$+29/7\sqrt{2.3.11.13}$
$\frac{3}{2}$	$\frac{3}{2}$	$\tilde{1}$	1	100−	$+29\sqrt{5}/3.7.11\sqrt{2.13}$
$\frac{3}{2}$	$\frac{3}{2}$	$\tilde{1}$	1	110−	$+17/2.7.11\sqrt{13}$
$\frac{3}{2}$	$\frac{3}{2}$	$\tilde{0}$	0	000−	$+5\sqrt{5}/7\sqrt{11.13}$
$\frac{3}{2}$	$\frac{3}{2}$	$\tilde{0}$	0	010−	$+8\sqrt{2}/7\sqrt{11.13}$
$\frac{3}{2}$	$\frac{3}{2}$	$\tilde{0}$	0	100−	$+73\sqrt{5}/7.11\sqrt{2.3.13}$
$\frac{3}{2}$	$\frac{3}{2}$	$\tilde{0}$	0	110−	$+3\sqrt{3}/7.11\sqrt{13}$
$\frac{3}{2}$	$\tilde{\frac{1}{2}}$	1	0	000+	0
$\frac{3}{2}$	$\tilde{\frac{1}{2}}$	1	0	100+	$-\sqrt{5}/4\sqrt{13}$
$\frac{3}{2}$	$\tilde{\frac{1}{2}}$	$\tilde{1}$	0	000−	$-\sqrt{2.11}/7\sqrt{3.13}$
$\frac{3}{2}$	$\tilde{\frac{1}{2}}$	$\tilde{1}$	0	100−	$-19/4.7\sqrt{13}$
$\tilde{\frac{1}{2}}$	$\frac{1}{2}$	$\tilde{1}$	0	000+	$-\sqrt{13}/4\sqrt{3.7}$
$\tilde{\frac{1}{2}}$	$\frac{1}{2}$	$\tilde{1}$	0	100+	$+\sqrt{5}/3\sqrt{2.7.11}$
$\tilde{\frac{1}{2}}$	$\frac{1}{2}$	$\tilde{0}$	0	000−	$-1/2\sqrt{7.13}$
$\tilde{\frac{1}{2}}$	$\frac{1}{2}$	$\tilde{0}$	0	100−	$-\sqrt{2.5}/\sqrt{3.7.11}$
$\tilde{\frac{1}{2}}$	$\frac{3}{2}$	1	0	000+	$+\sqrt{3.5}/4.7\sqrt{11.13}$
$\tilde{\frac{1}{2}}$	$\frac{3}{2}$	1	0	010+	$-4\sqrt{2.3}/7\sqrt{11.13}$
$\tilde{\frac{1}{2}}$	$\frac{3}{2}$	1	0	100+	$-8\sqrt{2}/7.11$
$\tilde{\frac{1}{2}}$	$\frac{3}{2}$	1	0	110+	$+5\sqrt{5}/7.11$
$\tilde{\frac{1}{2}}$	$\frac{3}{2}$	$\tilde{1}$	0	000−	$+73/4.7\sqrt{3.11.13}$
$\tilde{\frac{1}{2}}$	$\frac{3}{2}$	$\tilde{1}$	0	010−	$+4\sqrt{2.3.5}/7\sqrt{11.13}$
$\tilde{\frac{1}{2}}$	$\frac{3}{2}$	$\tilde{1}$	0	100−	$-2\sqrt{2.5}/3.7.11$
$\tilde{\frac{1}{2}}$	$\frac{3}{2}$	$\tilde{1}$	0	110−	$+3/7.11$
$\tilde{\frac{1}{2}}$	$\tilde{\frac{1}{2}}$	1	0	000−	$+\sqrt{3.11}/4.7\sqrt{13}$
$\tilde{\frac{1}{2}}$	$\tilde{\frac{1}{2}}$	1	0	100−	$+\sqrt{5}/7\sqrt{2}$

SO_3–O 3jm Factors (cont.)

$\frac{13}{2}\ \frac{11}{2}\ 4$

$\frac{1}{2}\ \frac{1}{2}$ 0 0 000− $+\sqrt{5.7}/2\sqrt{11.13.17}$
$\frac{1}{2}\ \frac{1}{2}$ 1 0 000+ $-\sqrt{7.13}/4\sqrt{11.17}$
$\frac{1}{2}\ \frac{3}{2}$ 1 0 000− $+61/4.11\sqrt{13.17}$
$\frac{1}{2}\ \frac{3}{2}$ 1 0 010− $-4\sqrt{2.17}/11\sqrt{5.13}$
$\frac{1}{2}\ \frac{3}{2}$ 2 0 000+ $+19/11\sqrt{2.7.13.17}$
$\frac{1}{2}\ \frac{3}{2}$ 2 0 010+ $+4.23/11\sqrt{5.7.13.17}$
$\frac{1}{2}\ \frac{3}{2}$ $\tilde{1}$ 0 000+ $+5\sqrt{3.7}/4.11\sqrt{13.17}$
$\frac{1}{2}\ \frac{3}{2}$ $\tilde{1}$ 0 010+ $+4\sqrt{2.5}/11\sqrt{3.7.13.17}$
$\frac{1}{2}\ \tilde{\frac{1}{2}}$ $\tilde{1}$ 0 000− $-\sqrt{5.17}/4\sqrt{3.7.13}$
$\frac{3}{2}\ \frac{1}{2}$ 1 0 000− $+1/\sqrt{2.13.17}$
$\frac{3}{2}\ \frac{1}{2}$ 1 0 100− $+\sqrt{3.13}/4\sqrt{11.17}$
$\frac{3}{2}\ \frac{1}{2}$ 2 0 000+ $-2.3/\sqrt{7.13.17}$
$\frac{3}{2}\ \frac{1}{2}$ 2 0 100+ $+3\sqrt{3}/\sqrt{2.7.11.13.17}$
$\frac{3}{2}\ \frac{1}{2}$ $\tilde{1}$ 0 000+ $+\sqrt{17}/\sqrt{2.3.7.13}$
$\frac{3}{2}\ \frac{1}{2}$ $\tilde{1}$ 0 100+ $+5.19/4\sqrt{7.11.13.17}$
$\frac{3}{2}\ \frac{3}{2}$ 0 0 000− $-\sqrt{5}/\sqrt{7.11.13.17}$
$\frac{3}{2}\ \frac{3}{2}$ 0 0 010− $+2\sqrt{2.13}/3\sqrt{7.11.17}$
$\frac{3}{2}\ \frac{3}{2}$ 0 0 100− $+5.5\sqrt{3.5}/11\sqrt{2.7.13.17}$
$\frac{3}{2}\ \frac{3}{2}$ 0 0 110− $-37\sqrt{3}/11\sqrt{7.13.17}$
$\frac{3}{2}\ \frac{3}{2}$ 1 0 000+ $+3\sqrt{17}/2\sqrt{7.11.13}$
$\frac{3}{2}\ \frac{3}{2}$ 1 0 010+ $-8.2\sqrt{2.5}/3\sqrt{7.11.13.17}$
$\frac{3}{2}\ \frac{3}{2}$ 1 0 100+ $-43\sqrt{3}/2.11\sqrt{2.7.13.17}$
$\frac{3}{2}\ \frac{3}{2}$ 1 0 110+ $-181\sqrt{3}/2.11\sqrt{5.7.13.17}$
$\frac{3}{2}\ \frac{3}{2}$ 1 1 000+ $+5\sqrt{7}/2\sqrt{11.13.17}$
$\frac{3}{2}\ \frac{3}{2}$ 1 1 010+ $+179/3\sqrt{2.5.7.11.13.17}$
$\frac{3}{2}\ \frac{3}{2}$ 1 1 100+ $-3\sqrt{3.7}/11\sqrt{2.13.17}$
$\frac{3}{2}\ \frac{3}{2}$ 1 1 110+ $-3.29\sqrt{3}/2.11\sqrt{5.7.13.17}$
$\frac{3}{2}\ \frac{3}{2}$ 2 0 000− $+\sqrt{2.17}/7\sqrt{11.13}$
$\frac{3}{2}\ \frac{3}{2}$ 2 0 010− $+2.251/3.7\sqrt{5.11.13.17}$
$\frac{3}{2}\ \frac{3}{2}$ 2 0 100− $+2.29\sqrt{3}/7.11\sqrt{13.17}$
$\frac{3}{2}\ \frac{3}{2}$ 2 0 110− $-13\sqrt{2.3.13}/7.11\sqrt{5.17}$
$\frac{3}{2}\ \frac{3}{2}$ $\tilde{1}$ 0 000− $-263/2.7\sqrt{3.11.13.17}$
$\frac{3}{2}\ \frac{3}{2}$ $\tilde{1}$ 0 010− $+\sqrt{2.3}/7\sqrt{5.11.13.17}$
$\frac{3}{2}\ \frac{3}{2}$ $\tilde{1}$ 0 100− $-37\sqrt{13}/2.7.11\sqrt{2.17}$
$\frac{3}{2}\ \frac{3}{2}$ $\tilde{1}$ 0 110− $-3.71/2.7.11\sqrt{5.13.17}$
$\frac{3}{2}\ \frac{3}{2}$ $\tilde{1}$ 1 000+ $+\sqrt{13}/2.7\sqrt{3.11.17}$
$\frac{3}{2}\ \frac{3}{2}$ $\tilde{1}$ 1 010+ $-\sqrt{5.13}/7\sqrt{2.3.11.17}$
$\frac{3}{2}\ \frac{3}{2}$ $\tilde{1}$ 1 100+ $-23\sqrt{13}/7.11\sqrt{2.17}$
$\frac{3}{2}\ \frac{3}{2}$ $\tilde{1}$ 1 110+ $-\sqrt{13}/2.7.11\sqrt{5.17}$
$\frac{3}{2}\ \tilde{\frac{1}{2}}$ 1 0 000− $-2\sqrt{2.11}/3\sqrt{5.7.13.17}$
$\frac{3}{2}\ \tilde{\frac{1}{2}}$ 1 0 100− $-43\sqrt{3}/4\sqrt{5.7.13.17}$
$\frac{3}{2}\ \tilde{\frac{1}{2}}$ 2 0 000+ $-4\sqrt{11}/3.7\sqrt{5.13.17}$
$\frac{3}{2}\ \tilde{\frac{1}{2}}$ 2 0 100+ $-11\sqrt{3}/7\sqrt{2.5.13.17}$
$\frac{3}{2}\ \tilde{\frac{1}{2}}$ $\tilde{1}$ 0 000+ $-\sqrt{2.11.13}/7\sqrt{3.5.17}$
$\frac{3}{2}\ \tilde{\frac{1}{2}}$ $\tilde{1}$ 0 100+ $+103/4.7\sqrt{5.13.17}$
$\tilde{\frac{1}{2}}\ \frac{1}{2}$ $\tilde{1}$ 0 000− $+\sqrt{5.17}/4\sqrt{3.7.13}$
$\tilde{\frac{1}{2}}\ \frac{1}{2}$ $\tilde{1}$ 0 100− $+5/\sqrt{2.7.11.17}$
$\tilde{\frac{1}{2}}\ \frac{3}{2}$ 1 0 000− $-5\sqrt{5}/4\sqrt{7.11.13.17}$
$\tilde{\frac{1}{2}}\ \frac{3}{2}$ 1 0 010− $-4\sqrt{2.7}/3\sqrt{11.13.17}$
$\tilde{\frac{1}{2}}\ \frac{3}{2}$ 1 0 100− $+4\sqrt{2.3}/11\sqrt{7.17}$
$\tilde{\frac{1}{2}}\ \frac{3}{2}$ 1 0 110− $-\sqrt{3.7}/11\sqrt{5.17}$
$\tilde{\frac{1}{2}}\ \frac{3}{2}$ 2 0 000+ $-53\sqrt{5}/7\sqrt{2.11.13.17}$
$\tilde{\frac{1}{2}}\ \frac{3}{2}$ 2 0 010+ $-4.5/3\sqrt{11.13.17}$
$\tilde{\frac{1}{2}}\ \frac{3}{2}$ 2 0 100+ $-4\sqrt{3}/7.11\sqrt{17}$
$\tilde{\frac{1}{2}}\ \frac{3}{2}$ 2 0 110+ $+2.3\sqrt{2.3}/11\sqrt{5.17}$
$\tilde{\frac{1}{2}}\ \frac{3}{2}$ $\tilde{1}$ 0 000+ $-\sqrt{5.17}/4.7\sqrt{3.11.13}$
$\tilde{\frac{1}{2}}\ \frac{3}{2}$ $\tilde{1}$ 0 010+ $+4\sqrt{2.17}/7\sqrt{3.11.13}$
$\tilde{\frac{1}{2}}\ \frac{3}{2}$ $\tilde{1}$ 0 100+ $+2.5.5\sqrt{2}/7.11\sqrt{17}$
$\tilde{\frac{1}{2}}\ \frac{3}{2}$ $\tilde{1}$ 0 110+ $+\sqrt{17}/7.11\sqrt{5}$
$\tilde{\frac{1}{2}}\ \tilde{\frac{1}{2}}$ 0 0 000− $+\sqrt{5.11}/2.3\sqrt{7.13.17}$
$\tilde{\frac{1}{2}}\ \tilde{\frac{1}{2}}$ 0 0 100− $-\sqrt{2.3}/\sqrt{7.17}$
$\tilde{\frac{1}{2}}\ \tilde{\frac{1}{2}}$ 1 0 000+ $+\sqrt{11.13}/4.3\sqrt{7.17}$
$\tilde{\frac{1}{2}}\ \tilde{\frac{1}{2}}$ 1 0 100+ $+\sqrt{3}/\sqrt{2.5.7.17}$

$\frac{13}{2}\ \frac{11}{2}\ 5$

$\frac{1}{2}\ \frac{1}{2}$ 1 0 000− $-\sqrt{2.5}/11\sqrt{3.13.17}$
$\frac{1}{2}\ \frac{1}{2}$ 1 0 001− $-3\sqrt{2.3.7}/11\sqrt{13.17}$
$\frac{1}{2}\ \frac{3}{2}$ 1 0 000+ $+8.2\sqrt{2.5}/11\sqrt{3.7.11.13.17}$
$\frac{1}{2}\ \frac{3}{2}$ 1 0 001+ $-3\sqrt{3}/11\sqrt{2.11.13.17}$
$\frac{1}{2}\ \frac{3}{2}$ 1 0 010+ $-3.7\sqrt{3.7}/2.11\sqrt{11.13.17}$
$\frac{1}{2}\ \frac{3}{2}$ 1 0 011+ $+8\sqrt{3.17}/11\sqrt{5.11.13}$
$\frac{1}{2}\ \frac{3}{2}$ 2 0 000− $+\sqrt{17}/11\sqrt{13}$
$\frac{1}{2}\ \frac{3}{2}$ 2 0 010− $-19\sqrt{2}/11\sqrt{5.13.17}$
$\frac{1}{2}\ \frac{3}{2}$ $\tilde{1}$ 0 000− $+2.5\sqrt{2.3}/11\sqrt{13.17}$
$\frac{1}{2}\ \frac{3}{2}$ $\tilde{1}$ 0 010− $-43\sqrt{5}/2.11\sqrt{3.13.17}$
$\frac{1}{2}\ \tilde{\frac{1}{2}}$ $\tilde{1}$ 0 000+ $-\sqrt{2.5}/\sqrt{3.13.17}$
$\frac{3}{2}\ \frac{1}{2}$ 1 0 000+ $-5.5\sqrt{5}/\sqrt{3.7.11.13.17}$
$\frac{3}{2}\ \frac{1}{2}$ 1 0 001+ $-7\sqrt{3}/2\sqrt{11.13.17}$
$\frac{3}{2}\ \frac{1}{2}$ 1 0 100+ $-47\sqrt{5}/11\sqrt{2.7.13.17}$
$\frac{3}{2}\ \frac{1}{2}$ 1 0 101+ $+2.7\sqrt{2}/11\sqrt{13.17}$
$\frac{3}{2}\ \frac{1}{2}$ 2 0 000− $-1/\sqrt{2.13.17}$
$\frac{3}{2}\ \frac{1}{2}$ 2 0 100− $-8/\sqrt{3.11.13.17}$
$\frac{3}{2}\ \frac{1}{2}$ $\tilde{1}$ 0 000− $+1/\sqrt{3.13.17}$
$\frac{3}{2}\ \frac{1}{2}$ $\tilde{1}$ 0 100− $-5/3\sqrt{2.11.13.17}$
$\frac{3}{2}\ \frac{3}{2}$ 1 0 000− $+5\sqrt{2.5}/7.11\sqrt{3.13.17}$
$\frac{3}{2}\ \frac{3}{2}$ 1 0 001− $-2.9\sqrt{2.3}/11\sqrt{7.13.17}$
$\frac{3}{2}\ \frac{3}{2}$ 1 0 010− $+439/2.7.11\sqrt{3.13.17}$
$\frac{3}{2}\ \frac{3}{2}$ 1 0 011− $-8\sqrt{3.5}/11\sqrt{7.13.17}$
$\frac{3}{2}\ \frac{3}{2}$ 1 0 100− $+211\sqrt{5}/7.11\sqrt{11.13.17}$
$\frac{3}{2}\ \frac{3}{2}$ 1 0 101− $+2.31/11\sqrt{7.11.13.17}$
$\frac{3}{2}\ \frac{3}{2}$ 1 0 110− $-1009\sqrt{2}/3.7.11\sqrt{11.13.17}$
$\frac{3}{2}\ \frac{3}{2}$ 1 0 111− $-8.47\sqrt{2}/11\sqrt{5.7.11.13.17}$
$\frac{3}{2}\ \frac{3}{2}$ 1 1 000− $+2.5\sqrt{2.5}/7.11\sqrt{3.13.17}$
$\frac{3}{2}\ \frac{3}{2}$ 1 1 001− $+5\sqrt{3}/11\sqrt{2.7.13.17}$
$\frac{3}{2}\ \frac{3}{2}$ 1 1 010− $+5.59/2.7.11\sqrt{3.13.17}$
$\frac{3}{2}\ \frac{3}{2}$ 1 1 011− $+4\sqrt{3.13}/11\sqrt{5.7.17}$
$\frac{3}{2}\ \frac{3}{2}$ 1 1 100− $-2.41\sqrt{5}/7.11\sqrt{11.13.17}$
$\frac{3}{2}\ \frac{3}{2}$ 1 1 101− $-3.59/11\sqrt{7.11.13.17}$
$\frac{3}{2}\ \frac{3}{2}$ 1 1 110− $-1049/3.7.11\sqrt{2.11.13.17}$

SO_3–O 3jm Factors (cont.)

$\frac{13}{2}$	$\frac{11}{2}$	5			
$\frac{3}{2}$	$\frac{3}{2}$	1	1	111−	$+2.19\sqrt{2}/11\sqrt{5.7.11.13.17}$
$\frac{3}{2}$	$\frac{3}{2}$	2	0	000+	$+9.3/\sqrt{7.11.13.17}$
$\frac{3}{2}$	$\frac{3}{2}$	2	0	010+	$-3\sqrt{2}/\sqrt{5.7.11.13.17}$
$\frac{3}{2}$	$\frac{3}{2}$	2	0	100+	$+97\sqrt{2}/11\sqrt{3.7.13.17}$
$\frac{3}{2}$	$\frac{3}{2}$	2	0	110+	$+2.109/11\sqrt{3.5.7.13.17}$
$\frac{3}{2}$	$\frac{3}{2}$	$\tilde{1}$	0	000+	$-\sqrt{2.17}/\sqrt{3.7.11.13}$
$\frac{3}{2}$	$\frac{3}{2}$	$\tilde{1}$	0	010+	$-199/2\sqrt{3.5.7.11.13.17}$
$\frac{3}{2}$	$\frac{3}{2}$	$\tilde{1}$	0	100+	$+43/3.11\sqrt{7.13.17}$
$\frac{3}{2}$	$\frac{3}{2}$	$\tilde{1}$	0	110+	$+4.9.3\sqrt{2}/11\sqrt{5.7.13.17}$
$\frac{3}{2}$	$\frac{3}{2}$	$\tilde{1}$	1	000−	$-8\sqrt{2}/\sqrt{3.7.11.13.17}$
$\frac{3}{2}$	$\frac{3}{2}$	$\tilde{1}$	1	010−	$-\sqrt{5}/2\sqrt{3.7.11.13.17}$
$\frac{3}{2}$	$\frac{3}{2}$	$\tilde{1}$	1	100−	$-2.43/3.11\sqrt{7.13.17}$
$\frac{3}{2}$	$\frac{3}{2}$	$\tilde{1}$	1	110−	$-179/11\sqrt{2.5.7.13.17}$
$\frac{3}{2}$	$\tilde{\frac{1}{2}}$	1	0	000+	$-1/\sqrt{3.13.17}$
$\frac{3}{2}$	$\tilde{\frac{1}{2}}$	1	0	001+	$-9\sqrt{3}/2\sqrt{5.7.13.17}$
$\frac{3}{2}$	$\tilde{\frac{1}{2}}$	1	0	100+	$-5/3\sqrt{2.11.13.17}$
$\frac{3}{2}$	$\tilde{\frac{1}{2}}$	1	0	101+	$+41\sqrt{2}/\sqrt{5.7.11.13.17}$
$\frac{3}{2}$	$\tilde{\frac{1}{2}}$	2	0	000−	$+\sqrt{7.11}/\sqrt{2.5.13.17}$
$\frac{3}{2}$	$\tilde{\frac{1}{2}}$	2	0	100−	$-2\sqrt{7}/\sqrt{3.5.13.17}$
$\frac{3}{2}$	$\tilde{\frac{1}{2}}$	$\tilde{1}$	0	000−	$+\sqrt{7.11}/\sqrt{3.5.13.17}$
$\frac{3}{2}$	$\tilde{\frac{1}{2}}$	$\tilde{1}$	0	100−	$+\sqrt{7}/\sqrt{2.5.13.17}$
$\tilde{\frac{1}{2}}$	$\frac{1}{2}$	$\tilde{1}$	0	000+	$-\sqrt{2.5}/\sqrt{3.13.17}$
$\tilde{\frac{1}{2}}$	$\frac{1}{2}$	$\tilde{1}$	0	100+	$-5/2.3\sqrt{11.17}$
$\tilde{\frac{1}{2}}$	$\frac{3}{2}$	1	0	000+	$-2.9.5\sqrt{2.3}/7.11\sqrt{13.17}$
$\tilde{\frac{1}{2}}$	$\frac{3}{2}$	1	0	001+	$-5\sqrt{3.5}/11\sqrt{2.7.13.17}$
$\tilde{\frac{1}{2}}$	$\frac{3}{2}$	1	0	010+	$+211\sqrt{5}/2.7.11\sqrt{3.13.17}$
$\tilde{\frac{1}{2}}$	$\frac{3}{2}$	1	0	011+	$-8\sqrt{3}/11\sqrt{7.13.17}$
$\tilde{\frac{1}{2}}$	$\frac{3}{2}$	1	0	100+	$-9\sqrt{5.17}/2.7.11\sqrt{11}$
$\tilde{\frac{1}{2}}$	$\frac{3}{2}$	1	0	101+	$+4.19/11\sqrt{7.11.17}$
$\tilde{\frac{1}{2}}$	$\frac{3}{2}$	1	0	110+	$-2.5.31\sqrt{2}/3.7.11\sqrt{11.17}$
$\tilde{\frac{1}{2}}$	$\frac{3}{2}$	1	0	111+	$-2.47\sqrt{2}/11\sqrt{5.7.11.17}$
$\tilde{\frac{1}{2}}$	$\frac{3}{2}$	2	0	000−	$+\sqrt{5}/\sqrt{7.11.13.17}$
$\tilde{\frac{1}{2}}$	$\frac{3}{2}$	2	0	010−	$-5\sqrt{2}/\sqrt{7.11.13.17}$
$\tilde{\frac{1}{2}}$	$\frac{3}{2}$	2	0	100−	$-\sqrt{2}/\sqrt{3.7.17}$
$\tilde{\frac{1}{2}}$	$\frac{3}{2}$	2	0	110−	$-4/\sqrt{3.5.7.17}$
$\tilde{\frac{1}{2}}$	$\frac{3}{2}$	$\tilde{1}$	0	000−	$+8.2\sqrt{2.5}/\sqrt{3.7.11.13.17}$
$\tilde{\frac{1}{2}}$	$\frac{3}{2}$	$\tilde{1}$	0	010−	$+\sqrt{3.7}/2\sqrt{11.13.17}$
$\tilde{\frac{1}{2}}$	$\frac{3}{2}$	$\tilde{1}$	0	100−	$-5.5/2.3.11\sqrt{7.17}$
$\tilde{\frac{1}{2}}$	$\frac{3}{2}$	$\tilde{1}$	0	110−	$-2\sqrt{2.7}/11\sqrt{5.17}$
$\tilde{\frac{1}{2}}$	$\tilde{\frac{1}{2}}$	1	0	000−	$-\sqrt{2.5}/\sqrt{3.13.17}$
$\tilde{\frac{1}{2}}$	$\tilde{\frac{1}{2}}$	1	0	001−	$+\sqrt{2.3}/\sqrt{7.13.17}$
$\tilde{\frac{1}{2}}$	$\tilde{\frac{1}{2}}$	1	0	100−	$+5/2.3\sqrt{11.17}$
$\tilde{\frac{1}{2}}$	$\tilde{\frac{1}{2}}$	1	0	101−	$-8/\sqrt{5.7.11.17}$

$\frac{13}{2}$	6	$\frac{1}{2}$			
$\frac{1}{2}$	0	$\frac{1}{2}$	0	000+	$+1/\sqrt{13}$
$\frac{1}{2}$	1	$\frac{1}{2}$	0	000−	$+\sqrt{2.3}/\sqrt{7.13}$
$\frac{3}{2}$	1	$\frac{1}{2}$	0	000+	$+\sqrt{3.11}/7\sqrt{13}$
$\frac{3}{2}$	1	$\frac{1}{2}$	0	100+	$-2.3\sqrt{2}/7\sqrt{13}$
$\frac{3}{2}$	2	$\frac{1}{2}$	0	000−	$+2\sqrt{11}/7\sqrt{13}$

$\frac{13}{2}$	6	$\frac{1}{2}$			
$\frac{3}{2}$	2	$\frac{1}{2}$	0	100−	$+3\sqrt{2.3}/7\sqrt{13}$
$\frac{3}{2}$	$\tilde{1}$	$\frac{1}{2}$	0	000−	$+\sqrt{3.5}/\sqrt{7.13}$
$\frac{3}{2}$	$\tilde{1}$	$\frac{1}{2}$	0	010−	0
$\frac{3}{2}$	$\tilde{1}$	$\frac{1}{2}$	0	100−	0
$\frac{3}{2}$	$\tilde{1}$	$\frac{1}{2}$	0	110−	$+2\sqrt{2}/\sqrt{7.13}$
$\tilde{\frac{1}{2}}$	$\tilde{1}$	$\frac{1}{2}$	0	000+	$+\sqrt{2.3}/\sqrt{7.13}$
$\tilde{\frac{1}{2}}$	$\tilde{1}$	$\frac{1}{2}$	0	010+	0
$\tilde{\frac{1}{2}}$	$\tilde{1}$	$\frac{1}{2}$	0	100+	0
$\tilde{\frac{1}{2}}$	$\tilde{1}$	$\frac{1}{2}$	0	110+	$+1/\sqrt{7}$
$\tilde{\frac{1}{2}}$	$\tilde{0}$	$\frac{1}{2}$	0	000−	$+1/\sqrt{13}$
$\tilde{\frac{1}{2}}$	$\tilde{0}$	$\frac{1}{2}$	0	100−	0

$\frac{13}{2}$	6	$\frac{3}{2}$			
$\frac{1}{2}$	1	$\frac{3}{2}$	0	000−	$+\sqrt{3.5}/\sqrt{2.7.13}$
$\frac{1}{2}$	2	$\frac{3}{2}$	0	000+	$-\sqrt{2}/\sqrt{5.7.13}$
$\frac{1}{2}$	$\tilde{1}$	$\frac{3}{2}$	0	000+	$+\sqrt{3.11}/7\sqrt{2.13}$
$\frac{1}{2}$	$\tilde{1}$	$\frac{3}{2}$	0	010+	$-4\sqrt{2.3}/7\sqrt{5.13}$
$\frac{3}{2}$	0	$\frac{3}{2}$	0	000−	$-\sqrt{11}/\sqrt{5.7.13}$
$\frac{3}{2}$	0	$\frac{3}{2}$	0	100−	$+2\sqrt{2.3}/\sqrt{5.7.13}$
$\frac{3}{2}$	1	$\frac{3}{2}$	0	000+	$-3\sqrt{3.11}/7\sqrt{2.5.13}$
$\frac{3}{2}$	1	$\frac{3}{2}$	0	100+	$+4/7\sqrt{5.13}$
$\frac{3}{2}$	1	$\frac{3}{2}$	1	000+	$+\sqrt{3.11}/7\sqrt{2.5.13}$
$\frac{3}{2}$	1	$\frac{3}{2}$	1	100+	$+1/7\sqrt{5.13}$
$\frac{3}{2}$	2	$\frac{3}{2}$	0	000−	$+3\sqrt{2.11}/7\sqrt{5.13}$
$\frac{3}{2}$	2	$\frac{3}{2}$	0	100−	$+2\sqrt{3}/7\sqrt{5.13}$
$\frac{3}{2}$	$\tilde{1}$	$\frac{3}{2}$	0	000−	$+3\sqrt{3}/7\sqrt{2.7.13}$
$\frac{3}{2}$	$\tilde{1}$	$\frac{3}{2}$	0	010−	$+4\sqrt{2.3.11}/7\sqrt{5.7.13}$
$\frac{3}{2}$	$\tilde{1}$	$\frac{3}{2}$	0	100−	$+2.3\sqrt{11}/7\sqrt{7.13}$
$\frac{3}{2}$	$\tilde{1}$	$\frac{3}{2}$	0	110−	$-4\sqrt{5}/7\sqrt{7.13}$
$\frac{3}{2}$	$\tilde{1}$	$\frac{3}{2}$	1	000+	$-5\sqrt{3}/7\sqrt{2.7.13}$
$\frac{3}{2}$	$\tilde{1}$	$\frac{3}{2}$	1	010+	$-2\sqrt{2.3.11}/7\sqrt{5.7.13}$
$\frac{3}{2}$	$\tilde{1}$	$\frac{3}{2}$	1	100+	$-3\sqrt{11}/7\sqrt{7.13}$
$\frac{3}{2}$	$\tilde{1}$	$\frac{3}{2}$	1	110+	$-2.23/7\sqrt{5.7.13}$
$\frac{3}{2}$	$\tilde{0}$	$\frac{3}{2}$	0	000+	$-1/\sqrt{13}$
$\frac{3}{2}$	$\tilde{0}$	$\frac{3}{2}$	0	100+	0
$\tilde{\frac{1}{2}}$	1	$\frac{3}{2}$	0	000−	$+\sqrt{3.11}/7\sqrt{2.13}$
$\tilde{\frac{1}{2}}$	1	$\frac{3}{2}$	0	100−	$-4/7\sqrt{5}$
$\tilde{\frac{1}{2}}$	2	$\frac{3}{2}$	0	000+	$+\sqrt{2.11}/7\sqrt{13}$
$\tilde{\frac{1}{2}}$	2	$\frac{3}{2}$	0	100+	$+2\sqrt{3}/7\sqrt{5}$
$\tilde{\frac{1}{2}}$	$\tilde{1}$	$\frac{3}{2}$	0	000+	$+\sqrt{3.5}/\sqrt{2.7.13}$
$\tilde{\frac{1}{2}}$	$\tilde{1}$	$\frac{3}{2}$	0	010+	0
$\tilde{\frac{1}{2}}$	$\tilde{1}$	$\frac{3}{2}$	0	100+	0
$\tilde{\frac{1}{2}}$	$\tilde{1}$	$\frac{3}{2}$	0	110+	$-1/\sqrt{5.7}$

$\frac{13}{2}$	6	$\frac{5}{2}$			
$\frac{1}{2}$	1	$\frac{3}{2}$	0	000+	$+3\sqrt{3.5}/2\sqrt{2.7.11.13}$
$\frac{1}{2}$	2	$\frac{3}{2}$	0	000−	$+4\sqrt{2}/\sqrt{5.7.11.13}$
$\frac{1}{2}$	$\tilde{1}$	$\frac{3}{2}$	0	000−	$+11/2.7\sqrt{2.3.13}$
$\frac{1}{2}$	$\tilde{1}$	$\frac{3}{2}$	0	010−	$-19\sqrt{2.3}/7\sqrt{5.11.13}$
$\frac{1}{2}$	$\tilde{1}$	$\tilde{\frac{1}{2}}$	0	000+	$+5\sqrt{5}/7\sqrt{2.3.13}$
$\frac{1}{2}$	$\tilde{1}$	$\tilde{\frac{1}{2}}$	0	010+	$-2\sqrt{2.3}/7\sqrt{11.13}$

SO_3–O 3jm Factors (cont.)

$\frac{13}{2}\ 6\ \frac{5}{2}$		
$\frac{1}{2}\ \tilde{0}\ \tilde{\frac{1}{2}}$	0 000−	$+\sqrt{5}/2\sqrt{7.13}$
$\frac{3}{2}\ 0\ \frac{3}{2}$	0 000+	$-\sqrt{7}/2\sqrt{5.13}$
$\frac{3}{2}\ 0\ \frac{3}{2}$	0 100+	$+\sqrt{7}/\sqrt{2.3.5.11.13}$
$\frac{3}{2}\ 1\ \frac{3}{2}$	0 000−	$-19/7\sqrt{2.3.5.13}$
$\frac{3}{2}\ 1\ \frac{3}{2}$	0 100−	$+107/3.7\sqrt{5.11.13}$
$\frac{3}{2}\ 1\ \frac{3}{2}$	1 000−	$+4\sqrt{2}/7\sqrt{3.5.13}$
$\frac{3}{2}\ 1\ \frac{3}{2}$	1 100−	$+11\sqrt{11}/2.3.7\sqrt{5.13}$
$\frac{3}{2}\ 1\ \tilde{\frac{1}{2}}$	0 000+	$+4/7\sqrt{3.13}$
$\frac{3}{2}\ 1\ \tilde{\frac{1}{2}}$	0 100+	$-97/3.7\sqrt{2.11.13}$
$\frac{3}{2}\ 2\ \frac{3}{2}$	0 000+	$-9/7\sqrt{2.5.13}$
$\frac{3}{2}\ 2\ \frac{3}{2}$	0 100+	$-79/7\sqrt{3.5.11.13}$
$\frac{3}{2}\ 2\ \tilde{\frac{1}{2}}$	0 000−	$-1/7\sqrt{13}$
$\frac{3}{2}\ 2\ \tilde{\frac{1}{2}}$	0 100−	$+19/7\sqrt{2.3.11.13}$
$\frac{3}{2}\ \tilde{1}\ \frac{3}{2}$	0 000+	$-11\sqrt{11}/7\sqrt{2.3.7.13}$
$\frac{3}{2}\ \tilde{1}\ \frac{3}{2}$	0 010+	$-11\sqrt{2.3}/7\sqrt{5.7.13}$
$\frac{3}{2}\ \tilde{1}\ \frac{3}{2}$	0 100+	$+1/7\sqrt{7.13}$
$\frac{3}{2}\ \tilde{1}\ \frac{3}{2}$	0 110+	$+\sqrt{5.11}/7\sqrt{7.13}$
$\frac{3}{2}\ \tilde{1}\ \frac{3}{2}$	1 000−	$+5\sqrt{2.11}/7\sqrt{3.7.13}$
$\frac{3}{2}\ \tilde{1}\ \frac{3}{2}$	1 010−	$+3\sqrt{2.3}/7\sqrt{5.7.13}$
$\frac{3}{2}\ \tilde{1}\ \frac{3}{2}$	1 100−	$+9/2.7\sqrt{7.13}$
$\frac{3}{2}\ \tilde{1}\ \frac{3}{2}$	1 110−	$-4.9/7\sqrt{5.7.11.13}$
$\frac{3}{2}\ \tilde{1}\ \tilde{\frac{1}{2}}$	0 000−	$+2\sqrt{5.11}/7\sqrt{3.7.13}$
$\frac{3}{2}\ \tilde{1}\ \tilde{\frac{1}{2}}$	0 010−	$-2.5\sqrt{3}/7\sqrt{7.13}$
$\frac{3}{2}\ \tilde{1}\ \tilde{\frac{1}{2}}$	0 100−	$-\sqrt{5}/7\sqrt{2.7.13}$
$\frac{3}{2}\ \tilde{1}\ \tilde{\frac{1}{2}}$	0 110−	$-17\sqrt{2}/7\sqrt{7.11.13}$
$\frac{3}{2}\ \tilde{0}\ \frac{3}{2}$	0 000−	$+\sqrt{11}/2.7\sqrt{13}$
$\frac{3}{2}\ \tilde{0}\ \frac{3}{2}$	0 100−	$+5\sqrt{3}/7\sqrt{2.13}$
$\tilde{\frac{1}{2}}\ 0\ \tilde{\frac{1}{2}}$	0 000+	$+\sqrt{5}/2\sqrt{7.13}$
$\tilde{\frac{1}{2}}\ 0\ \tilde{\frac{1}{2}}$	0 100+	$-2\sqrt{2}/\sqrt{3.7.11}$
$\tilde{\frac{1}{2}}\ 1\ \frac{3}{2}$	0 000+	$+11/2.7\sqrt{2.3.13}$
$\tilde{\frac{1}{2}}\ 1\ \frac{3}{2}$	0 100+	$+13/3.7\sqrt{5.11}$
$\tilde{\frac{1}{2}}\ 1\ \tilde{\frac{1}{2}}$	0 000−	$-5\sqrt{5}/7\sqrt{2.3.13}$
$\tilde{\frac{1}{2}}\ 1\ \tilde{\frac{1}{2}}$	0 100−	$-8/3.7\sqrt{11}$
$\tilde{\frac{1}{2}}\ 2\ \frac{3}{2}$	0 000−	$-4\sqrt{2}/7\sqrt{13}$
$\tilde{\frac{1}{2}}\ 2\ \frac{3}{2}$	0 100−	$+\sqrt{11}/7\sqrt{3.5}$
$\tilde{\frac{1}{2}}\ \tilde{1}\ \frac{3}{2}$	0 000−	$-\sqrt{3.5.11}/2.7\sqrt{2.7.13}$
$\tilde{\frac{1}{2}}\ \tilde{1}\ \frac{3}{2}$	0 010−	$-5\sqrt{2.3}/7\sqrt{7.13}$
$\tilde{\frac{1}{2}}\ \tilde{1}\ \frac{3}{2}$	0 100−	$-5/7\sqrt{7}$
$\tilde{\frac{1}{2}}\ \tilde{1}\ \frac{3}{2}$	0 110−	$-2/7\sqrt{5.7.11}$

$\frac{13}{2}\ 6\ \frac{7}{2}$		
$\frac{1}{2}\ 0\ \frac{1}{2}$	0 000−	$-3\sqrt{3.7}/2\sqrt{2.11.13.17}$
$\frac{1}{2}\ 1\ \frac{1}{2}$	0 000+	$-3\sqrt{11}/4\sqrt{13.17}$
$\frac{1}{2}\ 1\ \frac{3}{2}$	0 000−	$+3.5\sqrt{5}/4\sqrt{7.11.13.17}$
$\frac{1}{2}\ 2\ \frac{3}{2}$	0 000+	$-9\sqrt{3.5}/2\sqrt{7.11.13.17}$
$\frac{1}{2}\ \tilde{1}\ \frac{3}{2}$	0 000+	$+3\sqrt{17}/4.7\sqrt{13}$
$\frac{1}{2}\ \tilde{1}\ \frac{3}{2}$	0 010+	$+2.3.5\sqrt{5}/7\sqrt{11.13.17}$
$\frac{1}{2}\ \tilde{1}\ \tilde{\frac{1}{2}}$	0 000−	$+\sqrt{3.17}/4.7\sqrt{13}$
$\frac{1}{2}\ \tilde{1}\ \tilde{\frac{1}{2}}$	0 010−	$-4\sqrt{3.5}/7\sqrt{11.13.17}$
$\frac{1}{2}\ \tilde{0}\ \tilde{\frac{1}{2}}$	0 000+	$+\sqrt{17}/2\sqrt{2.7.13}$

$\frac{13}{2}\ 6\ \frac{7}{2}$		
$\frac{3}{2}\ 0\ \frac{3}{2}$	0 000−	0
$\frac{3}{2}\ 0\ \frac{3}{2}$	0 100−	$+3\sqrt{5.7}/2\sqrt{11.13.17}$
$\frac{3}{2}\ 1\ \frac{1}{2}$	0 000−	$-9/2\sqrt{2.7.13.17}$
$\frac{3}{2}\ 1\ \frac{1}{2}$	0 100−	$+\sqrt{3}/4\sqrt{7.11.13.17}$
$\frac{3}{2}\ 1\ \frac{3}{2}$	0 000+	$-9\sqrt{5}/7\sqrt{13.17}$
$\frac{3}{2}\ 1\ \frac{3}{2}$	0 100+	$-3\sqrt{3.5.11}/2.7\sqrt{2.13.17}$
$\frac{3}{2}\ 1\ \frac{3}{2}$	1 000+	$-9\sqrt{5}/2.7\sqrt{13.17}$
$\frac{3}{2}\ 1\ \frac{3}{2}$	1 100+	$+\sqrt{3.5.11}/7\sqrt{2.13.17}$
$\frac{3}{2}\ 1\ \tilde{\frac{1}{2}}$	0 000−	$+\sqrt{3.5.17}/2.7\sqrt{2.13}$
$\frac{3}{2}\ 1\ \tilde{\frac{1}{2}}$	0 100−	$+41\sqrt{5}/4.7\sqrt{11.13.17}$
$\frac{3}{2}\ 2\ \frac{1}{2}$	0 000+	$+\sqrt{3.7}/\sqrt{2.13.17}$
$\frac{3}{2}\ 2\ \frac{1}{2}$	0 100+	$-\sqrt{7}/2\sqrt{11.13.17}$
$\frac{3}{2}\ 2\ \frac{3}{2}$	0 000−	$-5\sqrt{3.5}/7\sqrt{13.17}$
$\frac{3}{2}\ 2\ \frac{3}{2}$	0 100−	$+8.2\sqrt{2.5}/7\sqrt{11.13.17}$
$\frac{3}{2}\ 2\ \tilde{\frac{1}{2}}$	0 000+	$+5\sqrt{5}/7\sqrt{2.13.17}$
$\frac{3}{2}\ 2\ \tilde{\frac{1}{2}}$	0 100+	$+\sqrt{3.5}/2.7\sqrt{11.13.17}$
$\frac{3}{2}\ \tilde{1}\ \frac{1}{2}$	0 000+	$+\sqrt{5.11}/2.7\sqrt{2.13.17}$
$\frac{3}{2}\ \tilde{1}\ \frac{1}{2}$	0 010+	$+5\sqrt{2}/7\sqrt{13.17}$
$\frac{3}{2}\ \tilde{1}\ \frac{1}{2}$	0 100+	$+31\sqrt{3.5}/4.7\sqrt{13.17}$
$\frac{3}{2}\ \tilde{1}\ \frac{1}{2}$	0 110+	$-1/2.7\sqrt{3.11.13.17}$
$\frac{3}{2}\ \tilde{1}\ \frac{3}{2}$	0 000−	$-8\sqrt{11}/7\sqrt{7.13.17}$
$\frac{3}{2}\ \tilde{1}\ \frac{3}{2}$	0 010−	$+2.5\sqrt{5}/7\sqrt{7.13.17}$
$\frac{3}{2}\ \tilde{1}\ \frac{3}{2}$	0 100−	$-5.11\sqrt{3}/2.7\sqrt{2.7.13.17}$
$\frac{3}{2}\ \tilde{1}\ \frac{3}{2}$	0 110−	$-2.37\sqrt{2.5}/7\sqrt{3.7.11.13.17}$
$\frac{3}{2}\ \tilde{1}\ \frac{3}{2}$	1 000+	$+19\sqrt{11}/2.7\sqrt{7.13.17}$
$\frac{3}{2}\ \tilde{1}\ \frac{3}{2}$	1 010+	$-5.5\sqrt{5}/7\sqrt{7.13.17}$
$\frac{3}{2}\ \tilde{1}\ \frac{3}{2}$	1 100+	$+11\sqrt{3}/7\sqrt{2.7.13.17}$
$\frac{3}{2}\ \tilde{1}\ \frac{3}{2}$	1 110+	$+5\sqrt{5.11}/7\sqrt{2.3.7.13.17}$
$\frac{3}{2}\ \tilde{1}\ \tilde{\frac{1}{2}}$	0 000+	$+11\sqrt{3.11}/2.7\sqrt{2.7.13.17}$
$\frac{3}{2}\ \tilde{1}\ \tilde{\frac{1}{2}}$	0 010+	$+4\sqrt{2.3.5}/7\sqrt{7.13.17}$
$\frac{3}{2}\ \tilde{1}\ \tilde{\frac{1}{2}}$	0 100+	$-3.79/4.7\sqrt{7.13.17}$
$\frac{3}{2}\ \tilde{1}\ \tilde{\frac{1}{2}}$	0 110+	$+9.3.5\sqrt{5}/2.7\sqrt{7.11.13.17}$
$\frac{3}{2}\ \tilde{0}\ \frac{3}{2}$	0 000+	$+\sqrt{2.3.11}/7\sqrt{13.17}$
$\frac{3}{2}\ \tilde{0}\ \frac{3}{2}$	0 100+	$+11/2.7\sqrt{13.17}$
$\tilde{\frac{1}{2}}\ 0\ \tilde{\frac{1}{2}}$	0 000−	$+\sqrt{17}/2\sqrt{2.7.13}$
$\tilde{\frac{1}{2}}\ 0\ \tilde{\frac{1}{2}}$	0 100−	$+\sqrt{3.5}/\sqrt{7.11.17}$
$\tilde{\frac{1}{2}}\ 1\ \frac{3}{2}$	0 000−	$+3\sqrt{17}/4.7\sqrt{13}$
$\tilde{\frac{1}{2}}\ 1\ \frac{3}{2}$	0 100−	$-2\sqrt{2.3.5}/7\sqrt{11.17}$
$\tilde{\frac{1}{2}}\ 1\ \tilde{\frac{1}{2}}$	0 000+	$-\sqrt{3.17}/4.7\sqrt{13}$
$\tilde{\frac{1}{2}}\ 1\ \tilde{\frac{1}{2}}$	0 100+	$+\sqrt{5.11}/7\sqrt{2.17}$
$\tilde{\frac{1}{2}}\ 2\ \frac{3}{2}$	0 000+	$-11\sqrt{3}/2.7\sqrt{13.17}$
$\tilde{\frac{1}{2}}\ 2\ \frac{3}{2}$	0 100+	$-4\sqrt{2.5}/7\sqrt{11.17}$
$\tilde{\frac{1}{2}}\ \tilde{1}\ \frac{1}{2}$	0 000−	$+11\sqrt{11}/4.7\sqrt{13.17}$
$\tilde{\frac{1}{2}}\ \tilde{1}\ \frac{1}{2}$	0 010−	$+4\sqrt{5}/7\sqrt{13.17}$
$\tilde{\frac{1}{2}}\ \tilde{1}\ \frac{1}{2}$	0 100−	$-\sqrt{3.5}/7\sqrt{2.17}$
$\tilde{\frac{1}{2}}\ \tilde{1}\ \frac{1}{2}$	0 110−	$-2\sqrt{2}/7\sqrt{3.11.17}$
$\tilde{\frac{1}{2}}\ \tilde{1}\ \frac{3}{2}$	0 000+	$-5\sqrt{5.11}/4.7\sqrt{7.13.17}$
$\tilde{\frac{1}{2}}\ \tilde{1}\ \frac{3}{2}$	0 010+	$-2.37/7\sqrt{7.13.17}$
$\tilde{\frac{1}{2}}\ \tilde{1}\ \frac{3}{2}$	0 100+	$+4\sqrt{2.3}/7\sqrt{7.17}$

SO_3–O 3jm Factors (cont.)

$\frac{13}{2}\ 6\ \frac{7}{2}$

$\tilde{\frac{1}{2}}\ \tilde{1}\ \frac{3}{2}$ 0 110+ $-\sqrt{2.5}/7\sqrt{3.7.11.17}$
$\tilde{\frac{1}{2}}\ \tilde{0}\ \frac{1}{2}$ 0 000+ $+\sqrt{3.11}/2\sqrt{2.7.13.17}$
$\tilde{\frac{1}{2}}\ \tilde{0}\ \frac{1}{2}$ 0 100+ $+\sqrt{5}/\sqrt{7.17}$

$\frac{13}{2}\ 6\ \frac{9}{2}$

$\frac{1}{2}\ 0\ \frac{1}{2}$ 0 000+ $-3\sqrt{3.7}/2\sqrt{11.13.17}$
$\frac{1}{2}\ 1\ \frac{1}{2}$ 0 000− $-3/\sqrt{2.11.13.17}$
$\frac{1}{2}\ 1\ \frac{3}{2}$ 0 000+ $-\sqrt{3.17}/2\sqrt{2.5.11.13}$
$\frac{1}{2}\ 1\ \frac{3}{2}$ 0 001+ $-3\sqrt{11}/\sqrt{2.5.7.13.17}$
$\frac{1}{2}\ 2\ \frac{3}{2}$ 0 000− $+2\sqrt{2.5}/\sqrt{11.13.17}$
$\frac{1}{2}\ 2\ \frac{3}{2}$ 0 001− $+9\sqrt{3.5}/\sqrt{2.7.11.13.17}$
$\frac{1}{2}\ \tilde{1}\ \frac{3}{2}$ 0 000− $-67\sqrt{3}/2.5\sqrt{2.7.13.17}$
$\frac{1}{2}\ \tilde{1}\ \frac{3}{2}$ 0 001− $-3.11/5.7\sqrt{2.13.17}$
$\frac{1}{2}\ \tilde{1}\ \frac{3}{2}$ 0 010− $-\sqrt{2.3.11}/\sqrt{5.7.13.17}$
$\frac{1}{2}\ \tilde{1}\ \frac{3}{2}$ 0 011− $-3\sqrt{2}/7\sqrt{5.11.13.17}$
$\frac{3}{2}\ 0\ \frac{3}{2}$ 0 000+ $+7/2\sqrt{5.13.17}$
$\frac{3}{2}\ 0\ \frac{3}{2}$ 0 001+ $-9\sqrt{3}/\sqrt{5.7.13.17}$
$\frac{3}{2}\ 0\ \frac{3}{2}$ 0 100+ $+7\sqrt{3}/\sqrt{2.5.11.13.17}$
$\frac{3}{2}\ 0\ \frac{3}{2}$ 0 101+ $-3\sqrt{13}/\sqrt{2.5.7.11.17}$
$\frac{3}{2}\ 1\ \frac{1}{2}$ 0 000+ $-3/\sqrt{7.13.17}$
$\frac{3}{2}\ 1\ \frac{1}{2}$ 0 100+ $+19\sqrt{3}/\sqrt{2.7.11.13.17}$
$\frac{3}{2}\ 1\ \frac{3}{2}$ 0 000− $-11\sqrt{3}/\sqrt{2.5.7.13.17}$
$\frac{3}{2}\ 1\ \frac{3}{2}$ 0 001− $-9\sqrt{2}/7\sqrt{5.13.17}$
$\frac{3}{2}\ 1\ \frac{3}{2}$ 0 100− $-53/\sqrt{5.7.11.13.17}$
$\frac{3}{2}\ 1\ \frac{3}{2}$ 0 101− $+3.19\sqrt{3}/7\sqrt{5.11.13.17}$
$\frac{3}{2}\ 1\ \frac{3}{2}$ 1 000− $+3\sqrt{2.3}/\sqrt{5.7.13.17}$
$\frac{3}{2}\ 1\ \frac{3}{2}$ 1 001− $+8.3\sqrt{2}/7\sqrt{5.13.17}$
$\frac{3}{2}\ 1\ \frac{3}{2}$ 1 100− $+47/2\sqrt{5.7.11.13.17}$
$\frac{3}{2}\ 1\ \frac{3}{2}$ 1 101− $+8.2\sqrt{3}/7\sqrt{5.11.13.17}$
$\frac{3}{2}\ 2\ \frac{1}{2}$ 0 000− $+\sqrt{3}/\sqrt{7.13.17}$
$\frac{3}{2}\ 2\ \frac{1}{2}$ 0 100− $+37/\sqrt{2.7.11.13.17}$
$\frac{3}{2}\ 2\ \frac{3}{2}$ 0 000+ $+9/\sqrt{2.5.7.13.17}$
$\frac{3}{2}\ 2\ \frac{3}{2}$ 0 001+ $-8\sqrt{2.3}/7\sqrt{5.13.17}$
$\frac{3}{2}\ 2\ \frac{3}{2}$ 0 100+ $-\sqrt{3.11}/\sqrt{5.7.13.17}$
$\frac{3}{2}\ 2\ \frac{3}{2}$ 0 101+ $+2.41/7\sqrt{5.11.13.17}$
$\frac{3}{2}\ \tilde{1}\ \frac{1}{2}$ 0 000− $+\sqrt{11}/\sqrt{5.13.17}$
$\frac{3}{2}\ \tilde{1}\ \frac{1}{2}$ 0 010− $+2/\sqrt{13.17}$
$\frac{3}{2}\ \tilde{1}\ \frac{1}{2}$ 0 100− $-\sqrt{3}/\sqrt{2.5.13.17}$
$\frac{3}{2}\ \tilde{1}\ \frac{1}{2}$ 0 110− $-\sqrt{2}/\sqrt{3.11.13.17}$
$\frac{3}{2}\ \tilde{1}\ \frac{3}{2}$ 0 000+ $+\sqrt{3.11.17}/5.7\sqrt{2.13}$
$\frac{3}{2}\ \tilde{1}\ \frac{3}{2}$ 0 001+ $+\sqrt{2.11.13}/5.7\sqrt{7.17}$
$\frac{3}{2}\ \tilde{1}\ \frac{3}{2}$ 0 010+ $-3\sqrt{2.3.5}/7\sqrt{13.17}$
$\frac{3}{2}\ \tilde{1}\ \frac{3}{2}$ 0 011+ $-5\sqrt{2.5}/7\sqrt{7.13.17}$
$\frac{3}{2}\ \tilde{1}\ \frac{3}{2}$ 0 100+ $+31/5.7\sqrt{13.17}$
$\frac{3}{2}\ \tilde{1}\ \frac{3}{2}$ 0 101+ $+11\sqrt{3}/5.7\sqrt{7.13.17}$
$\frac{3}{2}\ \tilde{1}\ \frac{3}{2}$ 0 110+ $+3.5\sqrt{5}/7\sqrt{11.13.17}$
$\frac{3}{2}\ \tilde{1}\ \frac{3}{2}$ 0 111+ $+2.37\sqrt{5}/7\sqrt{3.7.11.13.17}$
$\frac{3}{2}\ \tilde{1}\ \frac{3}{2}$ 1 000− $+2.3\sqrt{2.3.11}/5.7\sqrt{13.17}$
$\frac{3}{2}\ \tilde{1}\ \frac{3}{2}$ 1 001− $-4\sqrt{2.11.13}/5.7\sqrt{7.17}$
$\frac{3}{2}\ \tilde{1}\ \frac{3}{2}$ 1 010− $-11\sqrt{2.3}/7\sqrt{5.13.17}$
$\frac{3}{2}\ \tilde{1}\ \frac{3}{2}$ 1 011− $-4\sqrt{2.17}/7\sqrt{5.7.13}$
$\frac{3}{2}\ \tilde{1}\ \frac{3}{2}$ 1 100− $+73/2.5.7\sqrt{13.17}$
$\frac{3}{2}\ \tilde{1}\ \frac{3}{2}$ 1 101− $+4.31\sqrt{3}/5.7\sqrt{7.13.17}$
$\frac{3}{2}\ \tilde{1}\ \frac{3}{2}$ 1 110− $-8.3\sqrt{5}/7\sqrt{11.13.17}$
$\frac{3}{2}\ \tilde{1}\ \frac{3}{2}$ 1 111− $+8.5\sqrt{5}/7\sqrt{3.7.11.13.17}$
$\frac{3}{2}\ \tilde{0}\ \frac{3}{2}$ 0 000− $+11\sqrt{11}/2.5\sqrt{7.13.17}$
$\frac{3}{2}\ \tilde{0}\ \frac{3}{2}$ 0 001− $-\sqrt{3.11}/5\sqrt{13.17}$
$\frac{3}{2}\ \tilde{0}\ \frac{3}{2}$ 0 100− $-29\sqrt{3}/5\sqrt{2.7.13.17}$
$\frac{3}{2}\ \tilde{0}\ \frac{3}{2}$ 0 101− $-11/5\sqrt{2.13.17}$
$\tilde{\frac{1}{2}}\ 1\ \frac{3}{2}$ 0 000+ $-\sqrt{3.7}/2\sqrt{2.13.17}$
$\tilde{\frac{1}{2}}\ 1\ \frac{3}{2}$ 0 001+ $+9.3/7\sqrt{2.13.17}$
$\tilde{\frac{1}{2}}\ 1\ \frac{3}{2}$ 0 100+ $-\sqrt{7}/\sqrt{5.11.17}$
$\tilde{\frac{1}{2}}\ 1\ \frac{3}{2}$ 0 101+ $+13\sqrt{3}/7\sqrt{5.11.17}$
$\tilde{\frac{1}{2}}\ 2\ \frac{3}{2}$ 0 000− $-2\sqrt{2}/\sqrt{7.13.17}$
$\tilde{\frac{1}{2}}\ 2\ \frac{3}{2}$ 0 001− $+11\sqrt{3}/7\sqrt{2.13.17}$
$\tilde{\frac{1}{2}}\ 2\ \frac{3}{2}$ 0 100− $+3\sqrt{3}/\sqrt{5.7.11.17}$
$\tilde{\frac{1}{2}}\ 2\ \frac{3}{2}$ 0 101− $-8.2/7\sqrt{5.11.17}$
$\tilde{\frac{1}{2}}\ \tilde{1}\ \frac{1}{2}$ 0 000+ $+\sqrt{11}/7\sqrt{2.13.17}$
$\tilde{\frac{1}{2}}\ \tilde{1}\ \frac{1}{2}$ 0 010+ $+2\sqrt{2.5}/7\sqrt{13.17}$
$\tilde{\frac{1}{2}}\ \tilde{1}\ \frac{1}{2}$ 0 100+ $+8\sqrt{3}/7\sqrt{5.17}$
$\tilde{\frac{1}{2}}\ \tilde{1}\ \frac{1}{2}$ 0 110+ $-2/7\sqrt{3.11.17}$
$\tilde{\frac{1}{2}}\ \tilde{1}\ \frac{3}{2}$ 0 000− $-\sqrt{3.5.11}/2.7\sqrt{2.13.17}$
$\tilde{\frac{1}{2}}\ \tilde{1}\ \frac{3}{2}$ 0 001− $+\sqrt{5.11}/\sqrt{2.7.13.17}$
$\tilde{\frac{1}{2}}\ \tilde{1}\ \frac{3}{2}$ 0 010− $-5\sqrt{2.3}/7\sqrt{13.17}$
$\tilde{\frac{1}{2}}\ \tilde{1}\ \frac{3}{2}$ 0 011− $-5\sqrt{2}/\sqrt{7.13.17}$
$\tilde{\frac{1}{2}}\ \tilde{1}\ \frac{3}{2}$ 0 100− $+\sqrt{17}/5.7$
$\tilde{\frac{1}{2}}\ \tilde{1}\ \frac{3}{2}$ 0 101− $+\sqrt{3}/5\sqrt{7.17}$
$\tilde{\frac{1}{2}}\ \tilde{1}\ \frac{3}{2}$ 0 110− $-2.3\sqrt{5}/7\sqrt{11.17}$
$\tilde{\frac{1}{2}}\ \tilde{1}\ \frac{3}{2}$ 0 111− $+4\sqrt{5}/\sqrt{3.7.11.17}$
$\tilde{\frac{1}{2}}\ \tilde{0}\ \frac{1}{2}$ 0 000− $+\sqrt{3.11}/2\sqrt{7.13.17}$
$\tilde{\frac{1}{2}}\ \tilde{0}\ \frac{1}{2}$ 0 100− $-2\sqrt{2}/\sqrt{5.7.17}$

$\frac{13}{2}\ 6\ \frac{11}{2}$

$\frac{1}{2}\ 0\ \frac{1}{2}$ 0 000− $-2/\sqrt{11.13.17.19}$
$\frac{1}{2}\ 1\ \frac{1}{2}$ 0 000+ $-\sqrt{2.13}/\sqrt{3.7.11.17.19}$
$\frac{1}{2}\ 1\ \frac{3}{2}$ 0 000− $-4\sqrt{2}/11\sqrt{3.13.17.19}$
$\frac{1}{2}\ 1\ \frac{3}{2}$ 0 001− $-3\sqrt{3.5.19}/2.11\sqrt{13.17}$
$\frac{1}{2}\ 2\ \frac{3}{2}$ 0 000+ $-89\sqrt{2}/11\sqrt{13.17.19}$
$\frac{1}{2}\ 2\ \frac{3}{2}$ 0 001+ $-31\sqrt{5}/11\sqrt{13.17.19}$
$\frac{1}{2}\ \tilde{1}\ \frac{3}{2}$ 0 000+ $-2\sqrt{2.3.5}/\sqrt{7.11.13.17.19}$
$\frac{1}{2}\ \tilde{1}\ \frac{3}{2}$ 0 001+ $+109/2\sqrt{3.7.11.13.17.19}$
$\frac{1}{2}\ \tilde{1}\ \frac{3}{2}$ 0 010+ $+53\sqrt{3}/11\sqrt{2.7.13.17.19}$
$\frac{1}{2}\ \tilde{1}\ \frac{3}{2}$ 0 011+ $+4\sqrt{3.5.17}/11\sqrt{7.13.19}$
$\frac{1}{2}\ \tilde{1}\ \tilde{\frac{1}{2}}$ 0 000− $-\sqrt{2.11.13}/\sqrt{3.7.17.19}$
$\frac{1}{2}\ \tilde{1}\ \tilde{\frac{1}{2}}$ 0 010− $-\sqrt{2.3.5}/\sqrt{7.13.17.19}$
$\frac{1}{2}\ \tilde{0}\ \tilde{\frac{1}{2}}$ 0 000+ $-2\sqrt{11}/\sqrt{13.17.19}$
$\frac{3}{2}\ 0\ \frac{3}{2}$ 0 000− $-2\sqrt{17}/\sqrt{11.13.19}$
$\frac{3}{2}\ 0\ \frac{3}{2}$ 0 001− $+\sqrt{5.13}/\sqrt{2.11.17.19}$
$\frac{3}{2}\ 0\ \frac{3}{2}$ 0 100− $-37\sqrt{2.3}/11\sqrt{13.17.19}$
$\frac{3}{2}\ 0\ \frac{3}{2}$ 0 101− $-8.8\sqrt{5}/11\sqrt{3.13.17.19}$

SO_3–O 3jm Factors (cont.)

$\frac{13}{2}$	6	$\frac{11}{2}$			
$\frac{3}{2}$	1	$\frac{1}{2}$	0	000−	$-5.5.5/7\sqrt{3.13.17.19}$
$\frac{3}{2}$	1	$\frac{1}{2}$	0	100−	$-5.47/7\sqrt{2.11.13.17.19}$
$\frac{3}{2}$	1	$\frac{3}{2}$	0	000+	$+5\sqrt{2.7}/\sqrt{3.11.13.17.19}$
$\frac{3}{2}$	1	$\frac{3}{2}$	0	001+	$-\sqrt{5.19}/2\sqrt{3.7.11.13.17}$
$\frac{3}{2}$	1	$\frac{3}{2}$	0	100+	$+7\sqrt{7}/11\sqrt{13.17.19}$
$\frac{3}{2}$	1	$\frac{3}{2}$	0	101+	$-5.5.5\sqrt{2.5}/3.11\sqrt{7.13.17.19}$
$\frac{3}{2}$	1	$\frac{3}{2}$	1	000+	$+2.71\sqrt{2}/\sqrt{3.7.11.13.17.19}$
$\frac{3}{2}$	1	$\frac{3}{2}$	1	001+	$-73\sqrt{5}/2\sqrt{3.7.11.13.17.19}$
$\frac{3}{2}$	1	$\frac{3}{2}$	1	100+	$+2.5.23/11\sqrt{7.13.17.19}$
$\frac{3}{2}$	1	$\frac{3}{2}$	1	101+	$+347\sqrt{5}/3.11\sqrt{2.7.13.17.19}$
$\frac{3}{2}$	1	$\tilde{\frac{1}{2}}$	0	000−	$-\sqrt{5.11}/\sqrt{3.7.13.17.19}$
$\frac{3}{2}$	1	$\tilde{\frac{1}{2}}$	0	100−	$+\sqrt{5.19}/3\sqrt{2.7.13.17}$
$\frac{3}{2}$	2	$\frac{1}{2}$	0	000+	$-3\sqrt{13}/7\sqrt{17.19}$
$\frac{3}{2}$	2	$\frac{1}{2}$	0	100+	$+241\sqrt{2}/7\sqrt{3.11.13.17.19}$
$\frac{3}{2}$	2	$\frac{3}{2}$	0	000−	$+\sqrt{2.7}/\sqrt{11.13.17.19}$
$\frac{3}{2}$	2	$\frac{3}{2}$	0	001−	$+\sqrt{5.19}/\sqrt{7.11.13.17}$
$\frac{3}{2}$	2	$\frac{3}{2}$	0	100−	$-2\sqrt{7.13}/11\sqrt{3.17.19}$
$\frac{3}{2}$	2	$\frac{3}{2}$	0	101−	$+61\sqrt{2.5}/11\sqrt{3.7.13.17.19}$
$\frac{3}{2}$	2	$\tilde{\frac{1}{2}}$	0	000+	$+3\sqrt{5.11}/\sqrt{7.13.17.19}$
$\frac{3}{2}$	2	$\tilde{\frac{1}{2}}$	0	100+	$-11\sqrt{2.5}/\sqrt{3.7.13.17.19}$
$\frac{3}{2}$	$\tilde{1}$	$\frac{1}{2}$	0	000+	$+\sqrt{5.11}/\sqrt{3.7.13.17.19}$
$\frac{3}{2}$	$\tilde{1}$	$\frac{1}{2}$	0	010+	$+3\sqrt{3}/2\sqrt{7.13.17.19}$
$\frac{3}{2}$	$\tilde{1}$	$\frac{1}{2}$	0	100+	$+\sqrt{5.19}/3\sqrt{2.7.13.17}$
$\frac{3}{2}$	$\tilde{1}$	$\frac{1}{2}$	0	110+	$-4.5\sqrt{2}/\sqrt{7.11.13.17.19}$
$\frac{3}{2}$	$\tilde{1}$	$\frac{3}{2}$	0	000−	$+5\sqrt{2.5}/\sqrt{3.13.17.19}$
$\frac{3}{2}$	$\tilde{1}$	$\frac{3}{2}$	0	001−	$+37/2.7\sqrt{3.13.17.19}$
$\frac{3}{2}$	$\tilde{1}$	$\frac{3}{2}$	0	010−	$+2.5\sqrt{2.3}/\sqrt{11.13.17.19}$
$\frac{3}{2}$	$\tilde{1}$	$\frac{3}{2}$	0	011−	$-8\sqrt{3.5}/7\sqrt{11.13.17.19}$
$\frac{3}{2}$	$\tilde{1}$	$\frac{3}{2}$	0	100−	$-43\sqrt{5}/3\sqrt{11.13.17.19}$
$\frac{3}{2}$	$\tilde{1}$	$\frac{3}{2}$	0	101−	$-2.3.5\sqrt{2}/7\sqrt{11.13.17.19}$
$\frac{3}{2}$	$\tilde{1}$	$\frac{3}{2}$	0	110−	$-2.43/11\sqrt{13.17.19}$
$\frac{3}{2}$	$\tilde{1}$	$\frac{3}{2}$	0	111−	$+8.2\sqrt{2.5.17}/7.11\sqrt{13.19}$
$\frac{3}{2}$	$\tilde{1}$	$\frac{3}{2}$	1	000+	$+4\sqrt{2.5}/\sqrt{3.13.17.19}$
$\frac{3}{2}$	$\tilde{1}$	$\frac{3}{2}$	1	001+	$+67/2.7\sqrt{3.13.17.19}$
$\frac{3}{2}$	$\tilde{1}$	$\frac{3}{2}$	1	010+	$-7\sqrt{3}/\sqrt{2.11.13.17.19}$
$\frac{3}{2}$	$\tilde{1}$	$\frac{3}{2}$	1	011+	$-8.9\sqrt{3.5}/7\sqrt{11.13.17.19}$
$\frac{3}{2}$	$\tilde{1}$	$\frac{3}{2}$	1	100+	$+2\sqrt{5.11}/3\sqrt{13.17.19}$
$\frac{3}{2}$	$\tilde{1}$	$\frac{3}{2}$	1	101+	$-37\sqrt{11}/7\sqrt{2.13.17.19}$
$\frac{3}{2}$	$\tilde{1}$	$\frac{3}{2}$	1	110+	$-7\sqrt{19}/11\sqrt{13.17}$
$\frac{3}{2}$	$\tilde{1}$	$\frac{3}{2}$	1	111+	$-2.3\sqrt{2.5.19}/7.11\sqrt{13.17}$
$\frac{3}{2}$	$\tilde{1}$	$\tilde{\frac{1}{2}}$	0	000+	$+11/\sqrt{3.13.17.19}$
$\frac{3}{2}$	$\tilde{1}$	$\tilde{\frac{1}{2}}$	0	010+	$-\sqrt{3.5.11}/2\sqrt{13.17.19}$
$\frac{3}{2}$	$\tilde{1}$	$\tilde{\frac{1}{2}}$	0	100+	$+\sqrt{11}/\sqrt{2.13.17.19}$
$\frac{3}{2}$	$\tilde{1}$	$\tilde{\frac{1}{2}}$	0	110+	$-\sqrt{2.5}/\sqrt{13.17.19}$
$\frac{3}{2}$	$\tilde{0}$	$\frac{3}{2}$	0	000+	$+2\sqrt{5.7}/\sqrt{13.17.19}$
$\frac{3}{2}$	$\tilde{0}$	$\frac{3}{2}$	0	001+	$-\sqrt{19}/\sqrt{2.7.13.17}$
$\frac{3}{2}$	$\tilde{0}$	$\frac{3}{2}$	0	100+	$-\sqrt{2.5.7}/\sqrt{3.11.13.17.19}$
$\frac{3}{2}$	$\tilde{0}$	$\frac{3}{2}$	0	101+	$+2\sqrt{3.19}/\sqrt{7.11.13.17}$
$\tilde{\frac{1}{2}}$	0	$\tilde{\frac{1}{2}}$	0	000−	$+2\sqrt{11}/\sqrt{13.17.19}$
$\tilde{\frac{1}{2}}$	0	$\tilde{\frac{1}{2}}$	0	100−	$+\sqrt{5}/\sqrt{2.3.17.19}$
$\tilde{\frac{1}{2}}$	1	$\frac{3}{2}$	0	000−	$+2.3\sqrt{2.3.5}/\sqrt{7.11.13.17.19}$
$\tilde{\frac{1}{2}}$	1	$\frac{3}{2}$	0	001−	$-\sqrt{19}/2\sqrt{3.7.11.13.17}$
$\tilde{\frac{1}{2}}$	1	$\frac{3}{2}$	0	100−	$+3.31/2.11\sqrt{7.17.19}$
$\tilde{\frac{1}{2}}$	1	$\frac{3}{2}$	0	101−	$+2.43\sqrt{2.5}/3.11\sqrt{7.17.19}$
$\tilde{\frac{1}{2}}$	1	$\tilde{\frac{1}{2}}$	0	000+	$-\sqrt{2.11.13}/\sqrt{3.7.17.19}$
$\tilde{\frac{1}{2}}$	1	$\tilde{\frac{1}{2}}$	0	100+	$+\sqrt{5}/2.3\sqrt{7.17.19}$
$\tilde{\frac{1}{2}}$	2	$\frac{3}{2}$	0	000+	$+\sqrt{2.5.19}/\sqrt{7.11.13.17}$
$\tilde{\frac{1}{2}}$	2	$\frac{3}{2}$	0	001+	$+\sqrt{19}/\sqrt{7.11.13.17}$
$\tilde{\frac{1}{2}}$	2	$\frac{3}{2}$	0	100+	$-\sqrt{19}/11\sqrt{3.7.17}$
$\tilde{\frac{1}{2}}$	2	$\frac{3}{2}$	0	101+	$+4.13\sqrt{2.5}/11\sqrt{3.7.17.19}$
$\tilde{\frac{1}{2}}$	$\tilde{1}$	$\frac{1}{2}$	0	000−	$-\sqrt{2.11.13}/\sqrt{3.7.17.19}$
$\tilde{\frac{1}{2}}$	$\tilde{1}$	$\frac{1}{2}$	0	010−	$+\sqrt{2.3.5}/\sqrt{7.13.17.19}$
$\tilde{\frac{1}{2}}$	$\tilde{1}$	$\frac{1}{2}$	0	100−	$-\sqrt{5}/2.3\sqrt{7.17.19}$
$\tilde{\frac{1}{2}}$	$\tilde{1}$	$\frac{1}{2}$	0	110−	$+8.4/\sqrt{7.11.17.19}$
$\tilde{\frac{1}{2}}$	$\tilde{1}$	$\frac{3}{2}$	0	000+	$-4\sqrt{2}/\sqrt{3.13.17.19}$
$\tilde{\frac{1}{2}}$	$\tilde{1}$	$\frac{3}{2}$	0	001+	$+\sqrt{3.5.19}/2.7\sqrt{13.17}$
$\tilde{\frac{1}{2}}$	$\tilde{1}$	$\frac{3}{2}$	0	010+	$+5\sqrt{3.5}/\sqrt{2.11.13.17.19}$
$\tilde{\frac{1}{2}}$	$\tilde{1}$	$\frac{3}{2}$	0	011+	$-4\sqrt{3.19}/7\sqrt{11.13.17}$
$\tilde{\frac{1}{2}}$	$\tilde{1}$	$\frac{3}{2}$	0	100+	$-7\sqrt{5}/2.3\sqrt{11.17.19}$
$\tilde{\frac{1}{2}}$	$\tilde{1}$	$\frac{3}{2}$	0	101+	$-2\sqrt{2.19}/7\sqrt{11.17}$
$\tilde{\frac{1}{2}}$	$\tilde{1}$	$\frac{3}{2}$	0	110+	$-8.2/11\sqrt{17.19}$
$\tilde{\frac{1}{2}}$	$\tilde{1}$	$\frac{3}{2}$	0	111+	$+2\sqrt{2.5}/7.11\sqrt{17.19}$
$\tilde{\frac{1}{2}}$	$\tilde{0}$	$\frac{1}{2}$	0	000+	$-2\sqrt{11}/\sqrt{13.17.19}$
$\tilde{\frac{1}{2}}$	$\tilde{0}$	$\frac{1}{2}$	0	100+	$+\sqrt{5}/\sqrt{2.3.17.19}$

$\frac{13}{2}$	$\frac{13}{2}$	0			
$\frac{1}{2}$	$\frac{1}{2}$	0	0	000+	$+1/\sqrt{7}$
$\frac{3}{2}$	$\frac{3}{2}$	0	0	000+	$+\sqrt{2}/\sqrt{7}$
$\frac{3}{2}$	$\frac{3}{2}$	0	0	100+	0
$\frac{3}{2}$	$\frac{3}{2}$	0	0	110+	$+\sqrt{2}/\sqrt{7}$
$\tilde{\frac{1}{2}}$	$\tilde{\frac{1}{2}}$	0	0	000+	$+1/\sqrt{7}$
$\tilde{\frac{1}{2}}$	$\tilde{\frac{1}{2}}$	0	0	100+	0
$\tilde{\frac{1}{2}}$	$\tilde{\frac{1}{2}}$	0	0	110+	$+1/\sqrt{7}$

$\frac{13}{2}$	$\frac{13}{2}$	1			
$\frac{1}{2}$	$\frac{1}{2}$	1	0	000+	$+\sqrt{5}/\sqrt{7.13}$
$\frac{3}{2}$	$\frac{1}{2}$	1	0	000−	$-2\sqrt{2.11}/7\sqrt{5.13}$
$\frac{3}{2}$	$\frac{1}{2}$	1	0	100−	$+8\sqrt{3}/7\sqrt{5.13}$
$\frac{3}{2}$	$\frac{3}{2}$	1	0	000+	$+\sqrt{2}/7\sqrt{5.7.13}$
$\frac{3}{2}$	$\frac{3}{2}$	1	0	100+	$-8\sqrt{3.11}/7\sqrt{5.7.13}$
$\frac{3}{2}$	$\frac{3}{2}$	1	0	110+	$-17\sqrt{2}/3.7\sqrt{5.7.13}$
$\frac{3}{2}$	$\frac{3}{2}$	1	1	000+	$-8.3\sqrt{2}/7\sqrt{5.7.13}$
$\frac{3}{2}$	$\frac{3}{2}$	1	1	100+	$-4\sqrt{3.11}/7\sqrt{5.7.13}$
$\frac{3}{2}$	$\frac{3}{2}$	1	1	110+	$+8.2\sqrt{2}/3.7\sqrt{5.7.13}$
$\tilde{\frac{1}{2}}$	$\frac{3}{2}$	1	0	000−	$+2\sqrt{2}/\sqrt{7.13}$
$\tilde{\frac{1}{2}}$	$\frac{3}{2}$	1	0	010−	0
$\tilde{\frac{1}{2}}$	$\frac{3}{2}$	1	0	100−	0
$\tilde{\frac{1}{2}}$	$\frac{3}{2}$	1	0	110−	$+4\sqrt{2}/3\sqrt{5.7}$
$\tilde{\frac{1}{2}}$	$\tilde{\frac{1}{2}}$	1	0	000+	$-\sqrt{5}/\sqrt{7.13}$
$\tilde{\frac{1}{2}}$	$\tilde{\frac{1}{2}}$	1	0	100+	0

SO_3–O 3jm Factors (cont.)

$\frac{13}{2}\ \frac{13}{2}$ **1**

$\tilde{\frac{1}{2}}\ \tilde{\frac{1}{2}}$ 1 0 110+ $+\sqrt{13}/3\sqrt{5.7}$

$\frac{13}{2}\ \frac{13}{2}$ **2**

$\frac{3}{2}\ \frac{1}{2}$ 2 0 000− $+\sqrt{11}/2.7\sqrt{5.13}$

$\frac{3}{2}\ \frac{1}{2}$ 2 0 100− $+\sqrt{3.5}/7\sqrt{2.13}$

$\frac{3}{2}\ \frac{1}{2}$ $\tilde{1}$ 0 000− $-3\sqrt{3.11}/7\sqrt{2.5.13}$

$\frac{3}{2}\ \frac{1}{2}$ $\tilde{1}$ 0 100− $+11/7\sqrt{5.13}$

$\frac{3}{2}\ \frac{3}{2}$ 2 0 000+ $-2.3\sqrt{5}/7\sqrt{7.13}$

$\frac{3}{2}\ \frac{3}{2}$ 2 0 100+ $-\sqrt{3.5.11}/7\sqrt{2.7.13}$

$\frac{3}{2}\ \frac{3}{2}$ 2 0 110+ $-2\sqrt{13}/7\sqrt{5.7}$

$\frac{3}{2}\ \frac{3}{2}$ $\tilde{1}$ 0 000+ $+4\sqrt{2.3}/7\sqrt{5.7.13}$

$\frac{3}{2}\ \frac{3}{2}$ $\tilde{1}$ 0 100+ $-\sqrt{5.11}/7\sqrt{7.13}$

$\frac{3}{2}\ \frac{3}{2}$ $\tilde{1}$ 0 110+ $+9.3\sqrt{3}/7\sqrt{2.5.7.13}$

$\frac{3}{2}\ \frac{3}{2}$ $\tilde{1}$ 1 000− 0

$\frac{3}{2}\ \frac{3}{2}$ $\tilde{1}$ 1 100− $+\sqrt{11}/\sqrt{5.7.13}$

$\frac{3}{2}\ \frac{3}{2}$ $\tilde{1}$ 1 110− 0

$\tilde{\frac{1}{2}}\ \frac{1}{2}$ $\tilde{1}$ 0 000+ $+\sqrt{3.11}/2.7\sqrt{13}$

$\tilde{\frac{1}{2}}\ \frac{1}{2}$ $\tilde{1}$ 0 100+ $-2\sqrt{2}/7\sqrt{5}$

$\tilde{\frac{1}{2}}\ \frac{3}{2}$ 2 0 000− $-29/2.7\sqrt{7.13}$

$\tilde{\frac{1}{2}}\ \frac{3}{2}$ 2 0 010− $-3\sqrt{3.11}/7\sqrt{2.7.13}$

$\tilde{\frac{1}{2}}\ \frac{3}{2}$ 2 0 100− $-\sqrt{2.3.11}/7\sqrt{5.7}$

$\tilde{\frac{1}{2}}\ \frac{3}{2}$ 2 0 110− $+\sqrt{5}/7\sqrt{7}$

$\tilde{\frac{1}{2}}\ \frac{3}{2}$ $\tilde{1}$ 0 000− $+9\sqrt{3}/7\sqrt{2.7.13}$

$\tilde{\frac{1}{2}}\ \frac{3}{2}$ $\tilde{1}$ 0 010− $-3\sqrt{11}/7\sqrt{7.13}$

$\tilde{\frac{1}{2}}\ \frac{3}{2}$ $\tilde{1}$ 0 100− $-2\sqrt{11}/7\sqrt{5.7}$

$\tilde{\frac{1}{2}}\ \frac{3}{2}$ $\tilde{1}$ 0 110− $-3\sqrt{2.3}/7\sqrt{5.7}$

$\frac{13}{2}\ \frac{13}{2}$ **3**

$\frac{1}{2}\ \frac{1}{2}$ 1 0 000+ $+3\sqrt{3.5}/\sqrt{7.11.13.17}$

$\frac{3}{2}\ \frac{1}{2}$ 1 0 000− $-3\sqrt{3.17}/2.7\sqrt{2.5.13}$

$\frac{3}{2}\ \frac{1}{2}$ 1 0 100− $+3.67/2.7\sqrt{5.11.13.17}$

$\frac{3}{2}\ \frac{1}{2}$ $\tilde{1}$ 0 000+ $-\sqrt{3.17}/2.7\sqrt{2.13}$

$\frac{3}{2}\ \frac{1}{2}$ $\tilde{1}$ 0 100+ $-73/2.7\sqrt{11.13.17}$

$\frac{3}{2}\ \frac{3}{2}$ 1 0 000+ $+2.11\sqrt{2.3.11}/7\sqrt{5.7.13.17}$

$\frac{3}{2}\ \frac{3}{2}$ 1 0 100+ $+2/7\sqrt{5.7.13.17}$

$\frac{3}{2}\ \frac{3}{2}$ 1 0 110+ $-89\sqrt{2}/7\sqrt{3.5.7.11.13.17}$

$\frac{3}{2}\ \frac{3}{2}$ 1 1 000+ $+\sqrt{2.3.11}/\sqrt{5.7.13.17}$

$\frac{3}{2}\ \frac{3}{2}$ 1 1 100+ $+31/2\sqrt{5.7.13.17}$

$\frac{3}{2}\ \frac{3}{2}$ 1 1 110+ $+\sqrt{2}/\sqrt{3.5.7.11.13.17}$

$\frac{3}{2}\ \frac{3}{2}$ $\tilde{1}$ 0 000− 0

$\frac{3}{2}\ \frac{3}{2}$ $\tilde{1}$ 0 100− $-4/\sqrt{7.13.17}$

$\frac{3}{2}\ \frac{3}{2}$ $\tilde{1}$ 0 110− 0

$\frac{3}{2}\ \frac{3}{2}$ $\tilde{1}$ 1 000+ $-3\sqrt{2.3.11}/7\sqrt{7.13.17}$

$\frac{3}{2}\ \frac{3}{2}$ $\tilde{1}$ 1 100+ $-5/2.7\sqrt{7.13.17}$

$\frac{3}{2}\ \frac{3}{2}$ $\tilde{1}$ 1 110+ $+131\sqrt{2.3}/7\sqrt{7.11.13.17}$

$\frac{3}{2}\ \frac{3}{2}$ $\tilde{0}$ 0 000+ $+8\sqrt{2.11}/7\sqrt{7.13.17}$

$\frac{3}{2}\ \frac{3}{2}$ $\tilde{0}$ 0 100+ $-5.5\sqrt{3}/7\sqrt{7.13.17}$

$\frac{3}{2}\ \frac{3}{2}$ $\tilde{0}$ 0 110+ $+3.23/7\sqrt{2.7.11.13.17}$

$\tilde{\frac{1}{2}}\ \frac{1}{2}$ $\tilde{1}$ 0 000− 0

$\tilde{\frac{1}{2}}\ \frac{1}{2}$ $\tilde{1}$ 0 100− $+2\sqrt{2}/\sqrt{11.17}$

$\tilde{\frac{1}{2}}\ \frac{1}{2}$ $\tilde{0}$ 0 000+ $+\sqrt{5.17}/2.7\sqrt{13}$

$\frac{13}{2}\ \frac{13}{2}$ **3**

$\tilde{\frac{1}{2}}\ \frac{1}{2}$ $\tilde{0}$ 0 100+ $-2\sqrt{2.3}/7\sqrt{11.17}$

$\tilde{\frac{1}{2}}\ \frac{3}{2}$ 1 0 000− $-\sqrt{3.11.13}/2.7\sqrt{2.7.17}$

$\tilde{\frac{1}{2}}\ \frac{3}{2}$ 1 0 010− $-5.5.5/2.7\sqrt{7.13.17}$

$\tilde{\frac{1}{2}}\ \frac{3}{2}$ 1 0 100− $+\sqrt{5}/7\sqrt{7.17}$

$\tilde{\frac{1}{2}}\ \frac{3}{2}$ 1 0 110− $+\sqrt{17}/7\sqrt{2.3.5.7.11}$

$\tilde{\frac{1}{2}}\ \frac{3}{2}$ $\tilde{1}$ 0 000+ $-\sqrt{3.5.11}/2.7\sqrt{2.7.13.17}$

$\tilde{\frac{1}{2}}\ \frac{3}{2}$ $\tilde{1}$ 0 010+ $+9.3\sqrt{5}/2.7\sqrt{7.13.17}$

$\tilde{\frac{1}{2}}\ \frac{3}{2}$ $\tilde{1}$ 0 100+ $-19/7\sqrt{7.17}$

$\tilde{\frac{1}{2}}\ \frac{3}{2}$ $\tilde{1}$ 0 110+ $-\sqrt{3}/7\sqrt{2.7.11.17}$

$\tilde{\frac{1}{2}}\ \tilde{\frac{1}{2}}$ 1 0 000+ $+\sqrt{3.5.11}/7\sqrt{7.13.17}$

$\tilde{\frac{1}{2}}\ \tilde{\frac{1}{2}}$ 1 0 100+ $+2.5\sqrt{2}/7\sqrt{7.17}$

$\tilde{\frac{1}{2}}\ \tilde{\frac{1}{2}}$ 1 0 110+ $+4\sqrt{13}/7\sqrt{3.5.7.11.17}$

$\frac{13}{2}\ \frac{13}{2}$ **4**

$\frac{1}{2}\ \frac{1}{2}$ 0 0 000+ $-9\sqrt{3}/2\sqrt{11.13.17}$

$\frac{1}{2}\ \frac{1}{2}$ 1 0 000− 0

$\frac{3}{2}\ \frac{1}{2}$ 1 0 000+ $-\sqrt{3.5}/2\sqrt{2.7.13.17}$

$\frac{3}{2}\ \frac{1}{2}$ 1 0 100+ $+\sqrt{5.13}/2\sqrt{7.11.17}$

$\frac{3}{2}\ \frac{1}{2}$ 2 0 000− $+3\sqrt{3.5}/7\sqrt{13.17}$

$\frac{3}{2}\ \frac{1}{2}$ 2 0 100− $+41\sqrt{5}/7\sqrt{2.11.13.17}$

$\frac{3}{2}\ \frac{1}{2}$ $\tilde{1}$ 0 000− $-\sqrt{5.17}/2.7\sqrt{2.13}$

$\frac{3}{2}\ \frac{1}{2}$ $\tilde{1}$ 0 100− $-\sqrt{3.5}/2.7\sqrt{11.13.17}$

$\frac{3}{2}\ \frac{3}{2}$ 0 0 000+ $+\sqrt{2.3.11}/7\sqrt{13.17}$

$\frac{3}{2}\ \frac{3}{2}$ 0 0 100+ $+5.11/3.7\sqrt{13.17}$

$\frac{3}{2}\ \frac{3}{2}$ 0 0 110+ $-37/3.7\sqrt{2.3.11.13.17}$

$\frac{3}{2}\ \frac{3}{2}$ 1 0 000− 0

$\frac{3}{2}\ \frac{3}{2}$ 1 0 100− $+4\sqrt{5}/3\sqrt{13.17}$

$\frac{3}{2}\ \frac{3}{2}$ 1 0 110− 0

$\frac{3}{2}\ \frac{3}{2}$ 1 1 000− 0

$\frac{3}{2}\ \frac{3}{2}$ 1 1 100− $-7\sqrt{5}/2.3\sqrt{13.17}$

$\frac{3}{2}\ \frac{3}{2}$ 1 1 110− 0

$\frac{3}{2}\ \frac{3}{2}$ 2 0 000+ $+4\sqrt{3.5.11}/7\sqrt{7.13.17}$

$\frac{3}{2}\ \frac{3}{2}$ 2 0 100+ $-8.2\sqrt{2.5}/3.7\sqrt{7.13.17}$

$\frac{3}{2}\ \frac{3}{2}$ 2 0 110+ $+2.5\sqrt{5}/3.7\sqrt{3.7.11.13.17}$

$\frac{3}{2}\ \frac{3}{2}$ $\tilde{1}$ 0 000+ $+4\sqrt{2.5.11}/7\sqrt{7.13.17}$

$\frac{3}{2}\ \frac{3}{2}$ $\tilde{1}$ 0 100+ $-2\sqrt{3.5}/7\sqrt{7.13.17}$

$\frac{3}{2}\ \frac{3}{2}$ $\tilde{1}$ 0 110+ $+29\sqrt{2.5}/3.7\sqrt{7.11.13.17}$

$\frac{3}{2}\ \frac{3}{2}$ $\tilde{1}$ 1 000− 0

$\frac{3}{2}\ \frac{3}{2}$ $\tilde{1}$ 1 100− $-\sqrt{5.13}/2\sqrt{3.7.17}$

$\frac{3}{2}\ \frac{3}{2}$ $\tilde{1}$ 1 110− 0

$\tilde{\frac{1}{2}}\ \frac{1}{2}$ $\tilde{1}$ 0 000+ $+\sqrt{17}/7\sqrt{13}$

$\tilde{\frac{1}{2}}\ \frac{1}{2}$ $\tilde{1}$ 0 100+ $+2\sqrt{2.3.5}/7\sqrt{11.17}$

$\tilde{\frac{1}{2}}\ \frac{3}{2}$ 1 0 000+ $+3\sqrt{3.11}/2.7\sqrt{2.13.17}$

$\tilde{\frac{1}{2}}\ \frac{3}{2}$ 1 0 010+ $-61/2.3.7\sqrt{13.17}$

$\tilde{\frac{1}{2}}\ \frac{3}{2}$ 1 0 100+ $-5\sqrt{5}/3.7\sqrt{17}$

$\tilde{\frac{1}{2}}\ \frac{3}{2}$ 1 0 110+ $-\sqrt{5}/7\sqrt{2.3.11.17}$

$\tilde{\frac{1}{2}}\ \frac{3}{2}$ 2 0 000− $+5\sqrt{3.11}/7\sqrt{7.13.17}$

$\tilde{\frac{1}{2}}\ \frac{3}{2}$ 2 0 010− $+11.11/3.7\sqrt{2.7.13.17}$

$\tilde{\frac{1}{2}}\ \frac{3}{2}$ 2 0 100− $-5\sqrt{2.5}/3.7\sqrt{7.17}$

$\tilde{\frac{1}{2}}\ \frac{3}{2}$ 2 0 110− $+2.37\sqrt{5}/3.7\sqrt{3.7.11.17}$

SO_3–O 3jm Factors (cont.)

$\frac{13}{2}$	$\frac{13}{2}$	**4**			
$\tilde{\frac{1}{2}}$	$\frac{3}{2}$	$\tilde{1}$	0	000−	$+\sqrt{11.13}/2.7\sqrt{2.7.17}$
$\tilde{\frac{1}{2}}$	$\frac{3}{2}$	$\tilde{1}$	0	010−	$-113/2.7\sqrt{3.7.13.17}$
$\tilde{\frac{1}{2}}$	$\frac{3}{2}$	$\tilde{1}$	0	100−	$-\sqrt{5}/7\sqrt{3.7.17}$
$\tilde{\frac{1}{2}}$	$\frac{3}{2}$	$\tilde{1}$	0	110−	$+89\sqrt{5}/3.7\sqrt{2.7.11.17}$
$\tilde{\frac{1}{2}}$	$\tilde{\frac{1}{2}}$	0	0	000+	$+3\sqrt{3.11}/2.7\sqrt{13.17}$
$\tilde{\frac{1}{2}}$	$\tilde{\frac{1}{2}}$	0	0	100+	$+2\sqrt{2.5}/3.7\sqrt{17}$
$\tilde{\frac{1}{2}}$	$\tilde{\frac{1}{2}}$	0	0	110+	$-2\sqrt{13}/3.7\sqrt{3.11.17}$
$\tilde{\frac{1}{2}}$	$\tilde{\frac{1}{2}}$	1	0	000−	0
$\tilde{\frac{1}{2}}$	$\tilde{\frac{1}{2}}$	1	0	100−	$-2\sqrt{2}/3\sqrt{17}$
$\tilde{\frac{1}{2}}$	$\tilde{\frac{1}{2}}$	1	0	110−	0

$\frac{13}{2}$	$\frac{13}{2}$	**5**			
$\frac{1}{2}$	$\frac{1}{2}$	1	0	000+	$+8\sqrt{3}/11\sqrt{7.13.17.19}$
$\frac{1}{2}$	$\frac{1}{2}$	1	0	001+	$-3\sqrt{3.5.19}/2.11\sqrt{13.17}$
$\frac{3}{2}$	$\frac{1}{2}$	1	0	000−	$-3.23\sqrt{2.3}/7\sqrt{11.13.17.19}$
$\frac{3}{2}$	$\frac{1}{2}$	1	0	001−	$+9\sqrt{3.5}/\sqrt{2.7.11.13.17.19}$
$\frac{3}{2}$	$\frac{1}{2}$	1	0	100−	$-9.61/2.7.11\sqrt{13.17.19}$
$\frac{3}{2}$	$\frac{1}{2}$	1	0	101−	$-3.5.5\sqrt{5}/11\sqrt{7.13.17.19}$
$\frac{3}{2}$	$\frac{1}{2}$	2	0	000+	$+23\sqrt{5}/2\sqrt{7.13.17.19}$
$\frac{3}{2}$	$\frac{1}{2}$	2	0	100+	$+31\sqrt{3.5}/\sqrt{2.7.11.13.17.19}$
$\frac{3}{2}$	$\frac{1}{2}$	$\tilde{1}$	0	000+	$-2.3\sqrt{2.3.5}/\sqrt{7.13.17.19}$
$\frac{3}{2}$	$\frac{1}{2}$	$\tilde{1}$	0	100+	$-31\sqrt{5}/2\sqrt{7.11.13.17.19}$
$\frac{3}{2}$	$\frac{3}{2}$	1	0	000+	$-4.9\sqrt{2.3}/7\sqrt{7.13.17.19}$
$\frac{3}{2}$	$\frac{3}{2}$	1	0	001+	$-\sqrt{2.3.5}/7\sqrt{13.17.19}$
$\frac{3}{2}$	$\frac{3}{2}$	1	0	100+	$-4.5.5/7\sqrt{7.11.13.17.19}$
$\frac{3}{2}$	$\frac{3}{2}$	1	0	101+	$-37\sqrt{5}/7\sqrt{11.13.17.19}$
$\frac{3}{2}$	$\frac{3}{2}$	1	0	110+	$+8.109\sqrt{2.3}/7.11\sqrt{7.13.17.19}$
$\frac{3}{2}$	$\frac{3}{2}$	1	0	111+	$-37.37\sqrt{5}/7.11\sqrt{2.3.13.17.19}$
$\frac{3}{2}$	$\frac{3}{2}$	1	1	000+	$+8.2.5\sqrt{2.3}/7\sqrt{7.13.17.19}$
$\frac{3}{2}$	$\frac{3}{2}$	1	1	001+	$-2.5\sqrt{2.3.5}/7\sqrt{13.17.19}$
$\frac{3}{2}$	$\frac{3}{2}$	1	1	100+	$-1/7\sqrt{7.11.13.17.19}$
$\frac{3}{2}$	$\frac{3}{2}$	1	1	101+	$+29\sqrt{5}/7\sqrt{11.13.17.19}$
$\frac{3}{2}$	$\frac{3}{2}$	1	1	110+	$-5.197\sqrt{2.3}/7.11\sqrt{7.13.17.19}$
$\frac{3}{2}$	$\frac{3}{2}$	1	1	111+	$-2.157\sqrt{2.5}/7.11\sqrt{3.13.17.19}$
$\frac{3}{2}$	$\frac{3}{2}$	2	0	000−	0
$\frac{3}{2}$	$\frac{3}{2}$	2	0	100−	$+\sqrt{3.5}/\sqrt{2.13.17.19}$
$\frac{3}{2}$	$\frac{3}{2}$	2	0	110−	0
$\frac{3}{2}$	$\frac{3}{2}$	$\tilde{1}$	0	000−	0
$\frac{3}{2}$	$\frac{3}{2}$	$\tilde{1}$	0	100−	$-4\sqrt{5}/\sqrt{13.17.19}$
$\frac{3}{2}$	$\frac{3}{2}$	$\tilde{1}$	0	110−	0
$\frac{3}{2}$	$\frac{3}{2}$	$\tilde{1}$	1	000+	$+4\sqrt{2.3.5.11}/7\sqrt{13.17.19}$
$\frac{3}{2}$	$\frac{3}{2}$	$\tilde{1}$	1	100+	$-\sqrt{5.13}/7\sqrt{17.19}$
$\frac{3}{2}$	$\frac{3}{2}$	$\tilde{1}$	1	110+	$+5\sqrt{2.3.5}/7\sqrt{11.13.17.19}$
$\tilde{\frac{1}{2}}$	$\frac{1}{2}$	$\tilde{1}$	0	000−	$+8\sqrt{3}/\sqrt{7.13.17.19}$
$\tilde{\frac{1}{2}}$	$\frac{1}{2}$	$\tilde{1}$	0	100−	$+2\sqrt{2.5}/\sqrt{7.11.17.19}$
$\tilde{\frac{1}{2}}$	$\frac{3}{2}$	1	0	000−	$+2.3\sqrt{2.3.5}/\sqrt{7.13.17.19}$
$\tilde{\frac{1}{2}}$	$\frac{3}{2}$	1	0	001−	$+\sqrt{3.19}/7\sqrt{2.13.17}$
$\tilde{\frac{1}{2}}$	$\frac{3}{2}$	1	0	010−	$-31\sqrt{5}/2\sqrt{7.11.13.17.19}$
$\tilde{\frac{1}{2}}$	$\frac{3}{2}$	1	0	011−	$+\sqrt{19}/7\sqrt{11.13.17}$
$\tilde{\frac{1}{2}}$	$\frac{3}{2}$	1	0	100−	$-1/\sqrt{7.11.17.19}$
$\tilde{\frac{1}{2}}$	$\frac{3}{2}$	1	0	101−	$+2\sqrt{5.19}/7\sqrt{11.17}$
$\tilde{\frac{1}{2}}$	$\frac{3}{2}$	1	0	110−	$-2\sqrt{2.3}/\sqrt{7.17.19}$
$\tilde{\frac{1}{2}}$	$\frac{3}{2}$	1	0	111−	$-\sqrt{2.5}/7\sqrt{3.17.19}$
$\tilde{\frac{1}{2}}$	$\frac{3}{2}$	2	0	000+	$+5\sqrt{11}/2.7\sqrt{13.17.19}$
$\tilde{\frac{1}{2}}$	$\frac{3}{2}$	2	0	010+	$+53\sqrt{3}/7\sqrt{2.13.17.19}$
$\tilde{\frac{1}{2}}$	$\frac{3}{2}$	2	0	100+	$-3\sqrt{2.3.5}/7\sqrt{17.19}$
$\tilde{\frac{1}{2}}$	$\frac{3}{2}$	2	0	110+	$+3.5\sqrt{5}/7\sqrt{11.17.19}$
$\tilde{\frac{1}{2}}$	$\frac{3}{2}$	$\tilde{1}$	0	000+	$+3\sqrt{2.3.11}/7\sqrt{13.17.19}$
$\tilde{\frac{1}{2}}$	$\frac{3}{2}$	$\tilde{1}$	0	010+	$-3.59/2.7\sqrt{13.17.19}$
$\tilde{\frac{1}{2}}$	$\frac{3}{2}$	$\tilde{1}$	0	100+	$+5\sqrt{5}/7\sqrt{17.19}$
$\tilde{\frac{1}{2}}$	$\frac{3}{2}$	$\tilde{1}$	0	110+	$+4\sqrt{2.3.5}/7\sqrt{11.17.19}$
$\tilde{\frac{1}{2}}$	$\tilde{\frac{1}{2}}$	1	0	000+	$-8\sqrt{3}/\sqrt{7.13.17.19}$
$\tilde{\frac{1}{2}}$	$\tilde{\frac{1}{2}}$	1	0	001+	$-\sqrt{3.5.19}/2.7\sqrt{13.17}$
$\tilde{\frac{1}{2}}$	$\tilde{\frac{1}{2}}$	1	0	100+	$+2\sqrt{2.5}/\sqrt{7.11.17.19}$
$\tilde{\frac{1}{2}}$	$\tilde{\frac{1}{2}}$	1	0	101+	$+2\sqrt{2.19}/7\sqrt{11.17}$
$\tilde{\frac{1}{2}}$	$\tilde{\frac{1}{2}}$	1	0	110+	$-8.2\sqrt{3.13}/11\sqrt{7.17.19}$
$\tilde{\frac{1}{2}}$	$\tilde{\frac{1}{2}}$	1	0	111+	$-2\sqrt{5.13}/7.11\sqrt{3.17.19}$

$\frac{13}{2}$	$\frac{13}{2}$	**6**			
$\frac{1}{2}$	$\frac{1}{2}$	0	0	000+	$-4\sqrt{5}/\sqrt{7.11.13.17.19}$
$\frac{1}{2}$	$\frac{1}{2}$	1	0	000−	0
$\frac{3}{2}$	$\frac{1}{2}$	1	0	000+	$+8.2\sqrt{3}/\sqrt{5.7.13.17.19}$
$\frac{3}{2}$	$\frac{1}{2}$	1	0	100+	$+3.41/\sqrt{2.5.7.11.13.17.19}$
$\frac{3}{2}$	$\frac{1}{2}$	2	0	000−	$-3\sqrt{13}/\sqrt{5.7.17.19}$
$\frac{3}{2}$	$\frac{1}{2}$	2	0	100−	$+\sqrt{2.3}/\sqrt{5.7.11.13.17.19}$
$\frac{3}{2}$	$\frac{1}{2}$	$\tilde{1}$	0	000−	$+2\sqrt{3.11}/\sqrt{13.17.19}$
$\frac{3}{2}$	$\frac{1}{2}$	$\tilde{1}$	0	001−	$-4\sqrt{3}/\sqrt{5.13.17.19}$
$\frac{3}{2}$	$\frac{1}{2}$	$\tilde{1}$	0	100−	$-1/\sqrt{2.13.17.19}$
$\frac{3}{2}$	$\frac{1}{2}$	$\tilde{1}$	0	101−	$-4\sqrt{2.17}/\sqrt{5.11.13.19}$
$\frac{3}{2}$	$\frac{3}{2}$	0	0	000+	$-2.3\sqrt{2.11}/\sqrt{5.7.13.17.19}$
$\frac{3}{2}$	$\frac{3}{2}$	0	0	100+	$+11\sqrt{3}/\sqrt{5.7.13.17.19}$
$\frac{3}{2}$	$\frac{3}{2}$	0	0	110+	$+2.41\sqrt{2}/\sqrt{5.7.11.13.17.19}$
$\frac{3}{2}$	$\frac{3}{2}$	1	0	000−	0
$\frac{3}{2}$	$\frac{3}{2}$	1	0	100−	$-\sqrt{2.17}/\sqrt{5.13.19}$
$\frac{3}{2}$	$\frac{3}{2}$	1	0	110−	0
$\frac{3}{2}$	$\frac{3}{2}$	1	1	000−	0
$\frac{3}{2}$	$\frac{3}{2}$	1	1	100−	$-2.7\sqrt{2}/\sqrt{5.13.17.19}$
$\frac{3}{2}$	$\frac{3}{2}$	1	1	110−	0
$\frac{3}{2}$	$\frac{3}{2}$	2	0	000+	$-8\sqrt{11}/7\sqrt{5.13.17.19}$
$\frac{3}{2}$	$\frac{3}{2}$	2	0	100+	$+8.9\sqrt{2.3}/7\sqrt{5.13.17.19}$
$\frac{3}{2}$	$\frac{3}{2}$	2	0	110+	$+8.4/7\sqrt{5.11.13.17.19}$
$\frac{3}{2}$	$\frac{3}{2}$	$\tilde{1}$	0	000+	$+4.3.11\sqrt{3}/7\sqrt{7.13.17.19}$
$\frac{3}{2}$	$\frac{3}{2}$	$\tilde{1}$	0	001+	$-61\sqrt{3.11}/7\sqrt{5.7.13.17.19}$
$\frac{3}{2}$	$\frac{3}{2}$	$\tilde{1}$	0	100+	$+5\sqrt{2.11}/7\sqrt{7.13.17.19}$
$\frac{3}{2}$	$\frac{3}{2}$	$\tilde{1}$	0	101+	$+2.71\sqrt{2}/7\sqrt{5.7.13.17.19}$
$\frac{3}{2}$	$\frac{3}{2}$	$\tilde{1}$	0	110+	$-4.3.11\sqrt{3}/7\sqrt{7.13.17.19}$
$\frac{3}{2}$	$\frac{3}{2}$	$\tilde{1}$	0	111+	$+5\sqrt{3.5.19}/7\sqrt{7.11.13.17}$
$\frac{3}{2}$	$\frac{3}{2}$	$\tilde{1}$	1	000−	0
$\frac{3}{2}$	$\frac{3}{2}$	$\tilde{1}$	1	001−	0
$\frac{3}{2}$	$\frac{3}{2}$	$\tilde{1}$	1	100−	$+2\sqrt{2.11}/\sqrt{7.13.17.19}$

SO_3–O 3jm Factors (cont.)

$\frac{13}{2}\ \frac{13}{2}\ 6$		
$\frac{3}{2}\ \frac{3}{2}\ \tilde{1}$	1 101−	$-2\sqrt{2}/\sqrt{5.7.13.17.19}$
$\frac{3}{2}\ \frac{3}{2}\ \tilde{1}$	1 110−	0
$\frac{3}{2}\ \frac{3}{2}\ \tilde{1}$	1 111−	0
$\frac{3}{2}\ \frac{3}{2}\ \tilde{0}$	0 000−	0
$\frac{3}{2}\ \frac{3}{2}\ \tilde{0}$	0 100−	$+\sqrt{3.11}/\sqrt{13.17.19}$
$\frac{3}{2}\ \frac{3}{2}\ \tilde{0}$	0 110−	0
$\tilde{\frac{1}{2}}\ \frac{1}{2}\ \tilde{1}$	0 000+	0
$\tilde{\frac{1}{2}}\ \frac{1}{2}\ \tilde{1}$	0 001+	$+\sqrt{3}/\sqrt{2.13.17.19}$
$\tilde{\frac{1}{2}}\ \frac{1}{2}\ \tilde{1}$	0 100+	$+1/\sqrt{17.19}$
$\tilde{\frac{1}{2}}\ \frac{1}{2}\ \tilde{1}$	0 101+	$+8.2/\sqrt{5.11.17.19}$
$\tilde{\frac{1}{2}}\ \frac{1}{2}\ \tilde{0}$	0 000−	$+4\sqrt{5.11}/\sqrt{7.13.17.19}$
$\tilde{\frac{1}{2}}\ \frac{1}{2}\ \tilde{0}$	0 100−	$+\sqrt{3}/\sqrt{2.7.17.19}$
$\tilde{\frac{1}{2}}\ \frac{3}{2}\ 1$	0 000+	$-2\sqrt{3.11}/\sqrt{13.17.19}$
$\tilde{\frac{1}{2}}\ \frac{3}{2}\ 1$	0 010+	$-1/\sqrt{2.13.17.19}$
$\tilde{\frac{1}{2}}\ \frac{3}{2}\ 1$	0 100+	$-1/\sqrt{2.5.17.19}$
$\tilde{\frac{1}{2}}\ \frac{3}{2}\ 1$	0 110+	$+4.3\sqrt{3}/\sqrt{5.11.17.19}$
$\tilde{\frac{1}{2}}\ \frac{3}{2}\ 2$	0 000−	$+9\sqrt{11}/7\sqrt{13.17.19}$
$\tilde{\frac{1}{2}}\ \frac{3}{2}\ 2$	0 010−	$-11\sqrt{2.3}/7\sqrt{13.17.19}$
$\tilde{\frac{1}{2}}\ \frac{3}{2}\ 2$	0 100−	$-\sqrt{2.3.5}/7\sqrt{17.19}$
$\tilde{\frac{1}{2}}\ \frac{3}{2}\ 2$	0 110−	$+151/7\sqrt{5.11.17.19}$
$\tilde{\frac{1}{2}}\ \frac{3}{2}\ \tilde{1}$	0 000−	$+4.11\sqrt{3.5}/7\sqrt{7.13.17.19}$
$\tilde{\frac{1}{2}}\ \frac{3}{2}\ \tilde{1}$	0 001−	$-4\sqrt{3.11}/7\sqrt{7.13.17.19}$
$\tilde{\frac{1}{2}}\ \frac{3}{2}\ \tilde{1}$	0 010−	$-9.3\sqrt{5.11}/7\sqrt{2.7.13.17.19}$
$\tilde{\frac{1}{2}}\ \frac{3}{2}\ \tilde{1}$	0 011−	$-4.3.5\sqrt{2}/7\sqrt{7.13.17.19}$
$\tilde{\frac{1}{2}}\ \frac{3}{2}\ \tilde{1}$	0 100−	$-11\sqrt{11}/7\sqrt{2.7.17.19}$
$\tilde{\frac{1}{2}}\ \frac{3}{2}\ \tilde{1}$	0 101−	$-4\sqrt{2.17}/7\sqrt{5.7.19}$
$\tilde{\frac{1}{2}}\ \frac{3}{2}\ \tilde{1}$	0 110−	$-2.3\sqrt{3}/7\sqrt{7.17.19}$
$\tilde{\frac{1}{2}}\ \frac{3}{2}\ \tilde{1}$	0 111−	$-2\sqrt{3.5.19}/7\sqrt{7.11.17}$
$\tilde{\frac{1}{2}}\ \tilde{\frac{1}{2}}\ 0$	0 000+	$-4\sqrt{5.11}/\sqrt{7.13.17.19}$
$\tilde{\frac{1}{2}}\ \tilde{\frac{1}{2}}\ 0$	0 100+	$+\sqrt{3}/\sqrt{2.7.17.19}$
$\tilde{\frac{1}{2}}\ \tilde{\frac{1}{2}}\ 0$	0 110+	$+8.2\sqrt{13}/\sqrt{5.7.11.17.19}$
$\tilde{\frac{1}{2}}\ \tilde{\frac{1}{2}}\ 1$	0 000−	0
$\tilde{\frac{1}{2}}\ \tilde{\frac{1}{2}}\ 1$	0 100−	$-1/\sqrt{17.19}$
$\tilde{\frac{1}{2}}\ \tilde{\frac{1}{2}}\ 1$	0 110−	0

$7\ \frac{7}{2}\ \frac{7}{2}$		
$1\ \frac{1}{2}\ \frac{1}{2}$	0 000+	$-\sqrt{7}/3\sqrt{2.3.11.13}$
$1\ \frac{1}{2}\ \frac{1}{2}$	0 100+	$-2\sqrt{2}/3\sqrt{13}$
$1\ \frac{3}{2}\ \frac{1}{2}$	0 000−	$-19/3\sqrt{2.3.5.11.13}$
$1\ \frac{3}{2}\ \frac{1}{2}$	0 100−	$+\sqrt{2.7}/3\sqrt{5.13}$
$1\ \frac{3}{2}\ \frac{3}{2}$	0 000+	$-4\sqrt{7}/3.5\sqrt{3.11.13}$
$1\ \frac{3}{2}\ \frac{3}{2}$	0 100+	$+2.7/3.5\sqrt{13}$
$1\ \frac{3}{2}\ \frac{3}{2}$	1 000+	$+4.7\sqrt{7}/3.5\sqrt{3.11.13}$
$1\ \frac{3}{2}\ \frac{3}{2}$	1 100+	$+7/3.5\sqrt{13}$
$1\ \tilde{\frac{1}{2}}\ \frac{3}{2}$	0 000−	$-\sqrt{7}/\sqrt{2.11.13}$
$1\ \tilde{\frac{1}{2}}\ \frac{3}{2}$	0 100−	0
$1\ \tilde{\frac{1}{2}}\ \tilde{\frac{1}{2}}$	0 000+	$-\sqrt{3.7}/\sqrt{2.11.13}$
$1\ \tilde{\frac{1}{2}}\ \tilde{\frac{1}{2}}$	0 100+	0
$2\ \frac{3}{2}\ \frac{1}{2}$	0 000+	$-1/\sqrt{2.3.13}$
$2\ \frac{3}{2}\ \frac{3}{2}$	0 000−	0
$2\ \tilde{\frac{1}{2}}\ \frac{3}{2}$	0 000+	$-\sqrt{7}/\sqrt{2.5.13}$
$\tilde{1}\ \frac{3}{2}\ \frac{1}{2}$	0 000+	$-5/\sqrt{2.3.7.13}$
$\tilde{1}\ \frac{3}{2}\ \frac{1}{2}$	0 100+	$-2\sqrt{2}/\sqrt{3.7.11}$
$\tilde{1}\ \frac{3}{2}\ \frac{3}{2}$	0 000−	0
$\tilde{1}\ \frac{3}{2}\ \frac{3}{2}$	0 100−	0
$\tilde{1}\ \frac{3}{2}\ \frac{3}{2}$	1 000+	$-4/\sqrt{3.5.13}$
$\tilde{1}\ \frac{3}{2}\ \frac{3}{2}$	1 100+	$+1/\sqrt{3.5.11}$
$\tilde{1}\ \tilde{\frac{1}{2}}\ \frac{1}{2}$	0 000−	$-1/\sqrt{2.7.13}$
$\tilde{1}\ \tilde{\frac{1}{2}}\ \frac{1}{2}$	0 100−	$+\sqrt{2}/\sqrt{7.11}$
$\tilde{1}\ \tilde{\frac{1}{2}}\ \frac{3}{2}$	0 000+	$+1/\sqrt{2.5.13}$
$\tilde{1}\ \tilde{\frac{1}{2}}\ \frac{3}{2}$	0 100+	$-\sqrt{2}/\sqrt{5.11}$
$\tilde{0}\ \frac{3}{2}\ \frac{3}{2}$	0 000+	$+\sqrt{7}/\sqrt{2.3.5.11}$
$\tilde{0}\ \tilde{\frac{1}{2}}\ \frac{1}{2}$	0 000+	$-1/2\sqrt{11}$

$7\ 4\ 3$		
$1\ 0\ 1$	0 000+	$+2\sqrt{2}/3\sqrt{7.11.13}$
$1\ 0\ 1$	0 100+	$-\sqrt{2}/\sqrt{3.13}$
$1\ 1\ 1$	0 000−	$-\sqrt{11}/\sqrt{3.5.7.13}$
$1\ 1\ 1$	0 100−	$-2/\sqrt{5.13}$
$1\ 1\ \tilde{1}$	0 000+	$-19/5\sqrt{7.11.13}$
$1\ 1\ \tilde{1}$	0 100+	$+2\sqrt{3}/5\sqrt{13}$
$1\ 2\ 1$	0 000+	$+17/3\sqrt{2.5.11.13}$
$1\ 2\ 1$	0 100+	$-\sqrt{7}/\sqrt{2.3.5.13}$
$1\ 2\ \tilde{1}$	0 000−	$+\sqrt{13}/5\sqrt{2.11}$
$1\ 2\ \tilde{1}$	0 100−	$+\sqrt{3.7}/5\sqrt{2.13}$
$1\ \tilde{1}\ 1$	0 000+	$+\sqrt{5}/\sqrt{3.11.13}$
$1\ \tilde{1}\ 1$	0 100+	0
$1\ \tilde{1}\ \tilde{1}$	0 000−	$+3/\sqrt{11.13}$
$1\ \tilde{1}\ \tilde{1}$	0 100−	0
$1\ \tilde{1}\ \tilde{0}$	0 000+	$-2\sqrt{2}/\sqrt{11.13}$
$1\ \tilde{1}\ \tilde{0}$	0 100+	0
$2\ 1\ 1$	0 000+	$-1/\sqrt{7.13}$
$2\ 1\ \tilde{1}$	0 000−	$-3/\sqrt{5.7.13}$
$2\ 2\ \tilde{0}$	0 000−	$+\sqrt{2}/\sqrt{5.13}$
$2\ \tilde{1}\ 1$	0 000−	$-1/\sqrt{3.13}$
$2\ \tilde{1}\ \tilde{1}$	0 000+	$-\sqrt{3}/\sqrt{5.13}$
$\tilde{1}\ 0\ \tilde{1}$	0 000+	$-2\sqrt{2}/7\sqrt{13}$
$\tilde{1}\ 0\ \tilde{1}$	0 100+	$-3\sqrt{2}/7\sqrt{11}$
$\tilde{1}\ 1\ 1$	0 000+	$-5/7\sqrt{13}$
$\tilde{1}\ 1\ 1$	0 100+	$-4/7\sqrt{11}$
$\tilde{1}\ 1\ \tilde{1}$	0 000−	$-\sqrt{3}/\sqrt{5.13}$
$\tilde{1}\ 1\ \tilde{1}$	0 100−	0
$\tilde{1}\ 1\ \tilde{0}$	0 000+	$+2\sqrt{2.3}/7\sqrt{5.13}$
$\tilde{1}\ 1\ \tilde{0}$	0 100+	$-4\sqrt{2.3}/7\sqrt{5.11}$
$\tilde{1}\ 2\ 1$	0 000−	$-3/\sqrt{2.7.13}$
$\tilde{1}\ 2\ 1$	0 100−	$-1/\sqrt{2.7.11}$
$\tilde{1}\ 2\ \tilde{1}$	0 000+	$-\sqrt{5}/\sqrt{2.7.13}$
$\tilde{1}\ 2\ \tilde{1}$	0 100+	$+3/\sqrt{2.5.7.11}$
$\tilde{1}\ \tilde{1}\ 1$	0 000−	$+1/\sqrt{7.13}$
$\tilde{1}\ \tilde{1}\ 1$	0 100−	$-2/\sqrt{7.11}$

SO_3–O 3jm Factors (cont.)

7 4 3	
$\tilde{1}\ \tilde{1}\ \tilde{1}$ 0 000+	$+\sqrt{3}/\sqrt{5.7.13}$
$\tilde{1}\ \tilde{1}\ \tilde{1}$ 0 100+	$-2\sqrt{3}/\sqrt{5.7.11}$
$\tilde{0}\ 0\ \tilde{0}$ 0 000+	$+1/\sqrt{7.11}$
$\tilde{0}\ 1\ \tilde{1}$ 0 000+	$-3/\sqrt{5.7.11}$
$\tilde{0}\ \tilde{1}\ 1$ 0 000+	$-1/\sqrt{3.11}$

7 4 4	
$1\ 1\ 0$ 0 000−	$-4\sqrt{7}/3\sqrt{5.11.13}$
$1\ 1\ 0$ 0 100−	$-1/\sqrt{3.5.13}$
$1\ 1\ 1$ 0 000+	$-7\sqrt{2.7}/5\sqrt{3.11.13}$
$1\ 1\ 1$ 0 100+	$+2\sqrt{2}/5\sqrt{13}$
$1\ 2\ 1$ 0 000−	$+43/2.3.5\sqrt{11.13}$
$1\ 2\ 1$ 0 100−	$+\sqrt{7}/2.5\sqrt{3.13}$
$1\ \tilde{1}\ 1$ 0 000−	$+8\sqrt{2}/5\sqrt{3.11.13}$
$1\ \tilde{1}\ 1$ 0 100−	$+\sqrt{2.7}/5\sqrt{13}$
$1\ \tilde{1}\ 2$ 0 000+	$+7\sqrt{7}/2.5\sqrt{3.11.13}$
$1\ \tilde{1}\ 2$ 0 100+	$-7/2.5\sqrt{13}$
$1\ \tilde{1}\ \tilde{1}$ 0 000+	$+\sqrt{2.7}/\sqrt{3.11.13}$
$1\ \tilde{1}\ \tilde{1}$ 0 100+	0
$2\ 1\ 1$ 0 000−	0
$2\ 2\ 0$ 0 000−	$-1/\sqrt{3.13}$
$2\ 2\ 2$ 0 000−	0
$2\ \tilde{1}\ 1$ 0 000+	$-2\sqrt{2}/\sqrt{3.5.13}$
$2\ \tilde{1}\ \tilde{1}$ 0 000−	0
$\tilde{1}\ 1\ 1$ 0 000−	0
$\tilde{1}\ 1\ 1$ 0 100−	0
$\tilde{1}\ 2\ 1$ 0 000+	$-1/2\sqrt{5.7.13}$
$\tilde{1}\ 2\ 1$ 0 100+	$+9/2\sqrt{5.7.11}$
$\tilde{1}\ \tilde{1}\ 0$ 0 000−	$-4/\sqrt{3.7.13}$
$\tilde{1}\ \tilde{1}\ 0$ 0 100−	$+1/\sqrt{3.7.11}$
$\tilde{1}\ \tilde{1}\ 1$ 0 000+	$+3\sqrt{2}/\sqrt{5.7.13}$
$\tilde{1}\ \tilde{1}\ 1$ 0 100+	$+\sqrt{2}/\sqrt{5.7.11}$
$\tilde{1}\ \tilde{1}\ 2$ 0 000−	$-1/2\sqrt{3.5.13}$
$\tilde{1}\ \tilde{1}\ 2$ 0 100−	$-\sqrt{5}/2\sqrt{3.11}$
$\tilde{1}\ \tilde{1}\ \tilde{1}$ 0 000−	0
$\tilde{1}\ \tilde{1}\ \tilde{1}$ 0 100−	0
$\tilde{0}\ 2\ 2$ 0 000+	$-\sqrt{7}/\sqrt{3.5.11}$
$\tilde{0}\ \tilde{1}\ 1$ 0 000−	$+\sqrt{2}/\sqrt{3.5.11}$

7 $\frac{9}{2}$ $\frac{5}{2}$	
$1\ \frac{1}{2}\ \frac{3}{2}$ 0 000−	$-2\sqrt{2}/\sqrt{5.7.11.13}$
$1\ \frac{1}{2}\ \frac{3}{2}$ 0 100−	$+\sqrt{2.3}/\sqrt{5.13}$
$1\ \frac{3}{2}\ \frac{3}{2}$ 0 000+	$+4/\sqrt{3.7.11.13}$
$1\ \frac{3}{2}\ \frac{3}{2}$ 0 010+	$-3/\sqrt{11.13}$
$1\ \frac{3}{2}\ \frac{3}{2}$ 0 100+	0
$1\ \frac{3}{2}\ \frac{3}{2}$ 0 110+	0
$1\ \frac{3}{2}\ \frac{3}{2}$ 1 000+	$-2.23/5\sqrt{3.7.11.13}$
$1\ \frac{3}{2}\ \frac{3}{2}$ 1 010+	$+2/5\sqrt{11.13}$
$1\ \frac{3}{2}\ \frac{3}{2}$ 1 100+	$-2/5\sqrt{13}$
$1\ \frac{3}{2}\ \frac{3}{2}$ 1 110+	$-\sqrt{3.7}/5\sqrt{13}$
$1\ \frac{3}{2}\ \tilde{\frac{1}{2}}$ 0 000−	$+4.31\sqrt{2}/5\sqrt{3.5.7.11.13}$
$1\ \frac{3}{2}\ \tilde{\frac{1}{2}}$ 0 010−	$-\sqrt{2.13}/5\sqrt{5.11}$
$1\ \frac{3}{2}\ \tilde{\frac{1}{2}}$ 0 100−	$-2\sqrt{2}/5\sqrt{5.13}$
$1\ \frac{3}{2}\ \tilde{\frac{1}{2}}$ 0 110−	$-\sqrt{2.3.7}/5\sqrt{5.13}$
$2\ \frac{1}{2}\ \frac{3}{2}$ 0 000+	0
$2\ \frac{3}{2}\ \frac{3}{2}$ 0 000−	$-2\sqrt{5}/\sqrt{3.7.13}$
$2\ \frac{3}{2}\ \frac{3}{2}$ 0 010−	0
$2\ \frac{3}{2}\ \tilde{\frac{1}{2}}$ 0 000+	$+4\sqrt{2}/5\sqrt{3.7.13}$
$2\ \frac{3}{2}\ \tilde{\frac{1}{2}}$ 0 010+	$-3\sqrt{2}/5\sqrt{13}$
$\tilde{1}\ \frac{1}{2}\ \frac{3}{2}$ 0 000+	$-2\sqrt{2}/7\sqrt{13}$
$\tilde{1}\ \frac{1}{2}\ \frac{3}{2}$ 0 100+	$-3\sqrt{2}/7\sqrt{11}$
$\tilde{1}\ \frac{1}{2}\ \tilde{\frac{1}{2}}$ 0 000−	$-4\sqrt{2}/7\sqrt{5.13}$
$\tilde{1}\ \frac{1}{2}\ \tilde{\frac{1}{2}}$ 0 100−	$-2.3\sqrt{2}/7\sqrt{5.11}$
$\tilde{1}\ \frac{3}{2}\ \frac{3}{2}$ 0 000−	$-4/7\sqrt{3.5.13}$
$\tilde{1}\ \frac{3}{2}\ \frac{3}{2}$ 0 010−	$-\sqrt{7}/\sqrt{5.13}$
$\tilde{1}\ \frac{3}{2}\ \frac{3}{2}$ 0 100−	$-4\sqrt{5}/7\sqrt{3.11}$
$\tilde{1}\ \frac{3}{2}\ \frac{3}{2}$ 0 110−	0
$\tilde{1}\ \frac{3}{2}\ \frac{3}{2}$ 1 000+	$-2/\sqrt{3.5.13}$
$\tilde{1}\ \frac{3}{2}\ \frac{3}{2}$ 1 010+	$-2/\sqrt{5.7.13}$
$\tilde{1}\ \frac{3}{2}\ \frac{3}{2}$ 1 100+	$+2/\sqrt{3.5.11}$
$\tilde{1}\ \frac{3}{2}\ \frac{3}{2}$ 1 110+	$-3/\sqrt{5.7.11}$
$\tilde{1}\ \frac{3}{2}\ \tilde{\frac{1}{2}}$ 0 000+	$-4\sqrt{2}/7\sqrt{3.13}$
$\tilde{1}\ \frac{3}{2}\ \tilde{\frac{1}{2}}$ 0 010+	$-\sqrt{2}/\sqrt{7.13}$
$\tilde{1}\ \frac{3}{2}\ \tilde{\frac{1}{2}}$ 0 100+	$+2.13\sqrt{2}/5.7\sqrt{3.11}$
$\tilde{1}\ \frac{3}{2}\ \tilde{\frac{1}{2}}$ 0 110+	$+3\sqrt{2}/5\sqrt{7.11}$
$\tilde{0}\ \frac{1}{2}\ \tilde{\frac{1}{2}}$ 0 000+	$+3/\sqrt{5.7.11}$
$\tilde{0}\ \frac{3}{2}\ \frac{3}{2}$ 0 000+	$+\sqrt{2.5}/\sqrt{3.7.11}$
$\tilde{0}\ \frac{3}{2}\ \frac{3}{2}$ 0 010+	0

7 $\frac{9}{2}$ $\frac{7}{2}$	
$1\ \frac{1}{2}\ \frac{1}{2}$ 0 000−	$+8.2/3\sqrt{3.5.11.13}$
$1\ \frac{1}{2}\ \frac{1}{2}$ 0 100−	$-\sqrt{7}/2.3\sqrt{5.13}$
$1\ \frac{1}{2}\ \frac{3}{2}$ 0 000+	$-8\sqrt{11}/3.5\sqrt{3.7.13}$
$1\ \frac{1}{2}\ \frac{3}{2}$ 0 100+	$-17/2.3.5\sqrt{13}$
$1\ \frac{3}{2}\ \frac{1}{2}$ 0 000+	$+23/2.5\sqrt{11.13}$
$1\ \frac{3}{2}\ \frac{1}{2}$ 0 010+	$+7\sqrt{7.11}/4.3.5\sqrt{3.13}$
$1\ \frac{3}{2}\ \frac{1}{2}$ 0 100+	$+\sqrt{3.7}/2.5\sqrt{13}$
$1\ \frac{3}{2}\ \frac{1}{2}$ 0 110+	$-7/4.3.5\sqrt{13}$
$1\ \frac{3}{2}\ \frac{3}{2}$ 0 000−	$+3\sqrt{2}/5\sqrt{5.7.11.13}$
$1\ \frac{3}{2}\ \frac{3}{2}$ 0 010−	$-8.8\sqrt{2}/3.5\sqrt{3.5.11.13}$
$1\ \frac{3}{2}\ \frac{3}{2}$ 0 100−	$+7\sqrt{3}/5\sqrt{2.5.13}$
$1\ \frac{3}{2}\ \frac{3}{2}$ 0 110−	$-7\sqrt{7}/2.3.5\sqrt{2.5.13}$
$1\ \frac{3}{2}\ \frac{3}{2}$ 1 000−	$-\sqrt{2.13}/\sqrt{5.7.11}$
$1\ \frac{3}{2}\ \frac{3}{2}$ 1 010−	$-5\sqrt{5}/2.3\sqrt{2.3.11.13}$
$1\ \frac{3}{2}\ \frac{3}{2}$ 1 100−	$+\sqrt{3}/\sqrt{2.5.13}$
$1\ \frac{3}{2}\ \frac{3}{2}$ 1 110−	$+\sqrt{7}/3\sqrt{2.5.13}$
$1\ \frac{3}{2}\ \tilde{\frac{1}{2}}$ 0 000+	$-59\sqrt{3}/2.5\sqrt{5.7.11.13}$
$1\ \frac{3}{2}\ \tilde{\frac{1}{2}}$ 0 010+	$+23/4.5\sqrt{5.11.13}$
$1\ \frac{3}{2}\ \tilde{\frac{1}{2}}$ 0 100+	$-9/2.5\sqrt{5.13}$
$1\ \frac{3}{2}\ \tilde{\frac{1}{2}}$ 0 110+	$-9\sqrt{3.7}/4.5\sqrt{5.13}$
$2\ \frac{1}{2}\ \frac{3}{2}$ 0 000−	$+\sqrt{7}/\sqrt{3.5.13}$

SO_3–O 3jm Factors (cont.)

7	$\frac{9}{2}$	$\frac{7}{2}$			
2	$\frac{3}{2}$	$\frac{1}{2}$	0	000−	$+1/\sqrt{5.13}$
2	$\frac{3}{2}$	$\frac{1}{2}$	0	010−	$-\sqrt{7}/\sqrt{3.5.13}$
2	$\frac{3}{2}$	$\frac{3}{2}$	0	000+	$-\sqrt{2}/\sqrt{7.13}$
2	$\frac{3}{2}$	$\frac{3}{2}$	0	010+	0
2	$\frac{3}{2}$	$\widetilde{\frac{3}{2}}$	0	000−	$+3\sqrt{3}/5\sqrt{7.13}$
2	$\frac{3}{2}$	$\widetilde{\frac{3}{2}}$	0	010−	$+2/5\sqrt{13}$
$\tilde{1}$	$\frac{1}{2}$	$\frac{3}{2}$	0	000−	$+8/7\sqrt{3.5.13}$
$\tilde{1}$	$\frac{1}{2}$	$\frac{3}{2}$	0	100−	$-5\sqrt{5}/2.7\sqrt{3.11}$
$\tilde{1}$	$\frac{1}{2}$	$\widetilde{\frac{3}{2}}$	0	000+	$-8.2/7\sqrt{5.13}$
$\tilde{1}$	$\frac{1}{2}$	$\widetilde{\frac{3}{2}}$	0	100+	$+1/2.7\sqrt{5.11}$
$\tilde{1}$	$\frac{3}{2}$	$\frac{1}{2}$	0	000−	$-11/2\sqrt{5.7.13}$
$\tilde{1}$	$\frac{3}{2}$	$\frac{1}{2}$	0	010−	$+7/4\sqrt{3.5.13}$
$\tilde{1}$	$\frac{3}{2}$	$\frac{1}{2}$	0	100−	$+1/2\sqrt{5.7.11}$
$\tilde{1}$	$\frac{3}{2}$	$\frac{1}{2}$	0	110−	$-7/4\sqrt{3.5.11}$
$\tilde{1}$	$\frac{3}{2}$	$\frac{3}{2}$	0	000+	$+\sqrt{2}/7\sqrt{13}$
$\tilde{1}$	$\frac{3}{2}$	$\frac{3}{2}$	0	010+	0
$\tilde{1}$	$\frac{3}{2}$	$\frac{3}{2}$	0	100+	$+1/5.7\sqrt{2.11}$
$\tilde{1}$	$\frac{3}{2}$	$\frac{3}{2}$	0	110+	$+3\sqrt{3.7}/2.5\sqrt{2.11}$
$\tilde{1}$	$\frac{3}{2}$	$\frac{3}{2}$	1	000−	$-\sqrt{2}/5\sqrt{13}$
$\tilde{1}$	$\frac{3}{2}$	$\frac{3}{2}$	1	010−	$+23/2.5\sqrt{2.3.7.13}$
$\tilde{1}$	$\frac{3}{2}$	$\frac{3}{2}$	1	100−	$-1/\sqrt{2.11}$
$\tilde{1}$	$\frac{3}{2}$	$\frac{3}{2}$	1	110−	$+1/\sqrt{2.3.7.11}$
$\tilde{1}$	$\frac{3}{2}$	$\widetilde{\frac{3}{2}}$	0	000−	$-9\sqrt{3}/2.5.7\sqrt{13}$
$\tilde{1}$	$\frac{3}{2}$	$\widetilde{\frac{3}{2}}$	0	010−	$-17/4.5\sqrt{7.13}$
$\tilde{1}$	$\frac{3}{2}$	$\widetilde{\frac{3}{2}}$	0	100−	$-3\sqrt{3}/2.5.7\sqrt{11}$
$\tilde{1}$	$\frac{3}{2}$	$\widetilde{\frac{3}{2}}$	0	110−	$-29/4.5\sqrt{7.11}$
$\tilde{0}$	$\frac{1}{2}$	$\widetilde{\frac{3}{2}}$	0	000−	$-\sqrt{2}/\sqrt{5.7.11}$
$\tilde{0}$	$\frac{3}{2}$	$\frac{3}{2}$	0	000−	$-2/5\sqrt{7.11}$
$\tilde{0}$	$\frac{3}{2}$	$\frac{3}{2}$	0	010−	$+7/5\sqrt{3.11}$

7	$\frac{9}{2}$	$\frac{9}{2}$			
1	$\frac{1}{2}$	$\frac{1}{2}$	0	000+	$+8\sqrt{7.17}/3.5\sqrt{3.11.13}$
1	$\frac{1}{2}$	$\frac{1}{2}$	0	100+	$+4/3.5\sqrt{13.17}$
1	$\frac{3}{2}$	$\frac{1}{2}$	0	000−	$+2\sqrt{7}/\sqrt{5.11.13.17}$
1	$\frac{3}{2}$	$\frac{1}{2}$	0	010−	$+7/3\sqrt{3.5.11.13.17}$
1	$\frac{3}{2}$	$\frac{1}{2}$	0	100−	$+\sqrt{3}/\sqrt{5.13.17}$
1	$\frac{3}{2}$	$\frac{1}{2}$	0	110−	$-\sqrt{5.7}/3\sqrt{13.17}$
1	$\frac{3}{2}$	$\frac{3}{2}$	0	000+	$+4\sqrt{2.3.7}/5.5\sqrt{11.13.17}$
1	$\frac{3}{2}$	$\frac{3}{2}$	0	010+	$-3\sqrt{17}/5.5\sqrt{2.11.13}$
1	$\frac{3}{2}$	$\frac{3}{2}$	0	011+	$+8.4\sqrt{2.7.17}/3.5.5\sqrt{3.11.13}$
1	$\frac{3}{2}$	$\frac{3}{2}$	0	100+	$-8.9\sqrt{2}/5.5\sqrt{13.17}$
1	$\frac{3}{2}$	$\frac{3}{2}$	0	110+	$+4\sqrt{2.3.7}/5.5\sqrt{13.17}$
1	$\frac{3}{2}$	$\frac{3}{2}$	0	111+	$-2.7\sqrt{2}/3.5.5\sqrt{13.17}$
1	$\frac{3}{2}$	$\frac{3}{2}$	1	000+	$-2.7\sqrt{2.3.7}/5.5\sqrt{11.13.17}$
1	$\frac{3}{2}$	$\frac{3}{2}$	1	010+	$+53\sqrt{2}/5.5\sqrt{11.13.17}$
1	$\frac{3}{2}$	$\frac{3}{2}$	1	011+	$+2\sqrt{2.7.13}/3.5.5\sqrt{3.11.17}$
1	$\frac{3}{2}$	$\frac{3}{2}$	1	100+	$-2.9\sqrt{2}/5.5\sqrt{13.17}$
1	$\frac{3}{2}$	$\frac{3}{2}$	1	110+	$+7\sqrt{3.7}/5.5\sqrt{2.13.17}$
1	$\frac{3}{2}$	$\frac{3}{2}$	1	111+	$-4.7.7\sqrt{2}/3.5.5\sqrt{13.17}$
2	$\frac{3}{2}$	$\frac{1}{2}$	0	000+	$+4\sqrt{7}/5\sqrt{13.17}$
2	$\frac{3}{2}$	$\frac{1}{2}$	0	010+	$+\sqrt{17}/5\sqrt{3.13}$
2	$\frac{3}{2}$	$\frac{3}{2}$	0	000−	0
2	$\frac{3}{2}$	$\frac{3}{2}$	0	010−	$-4\sqrt{2}/\sqrt{5.13.17}$
2	$\frac{3}{2}$	$\frac{3}{2}$	0	011−	0
$\tilde{1}$	$\frac{3}{2}$	$\frac{1}{2}$	0	000+	$+2.7/5\sqrt{13.17}$
$\tilde{1}$	$\frac{3}{2}$	$\frac{1}{2}$	0	010+	$-23/5\sqrt{3.7.13.17}$
$\tilde{1}$	$\frac{3}{2}$	$\frac{1}{2}$	0	100+	$-7/5\sqrt{11.17}$
$\tilde{1}$	$\frac{3}{2}$	$\frac{1}{2}$	0	110+	$-31/5\sqrt{3.7.11.17}$
$\tilde{1}$	$\frac{3}{2}$	$\frac{3}{2}$	0	000−	0
$\tilde{1}$	$\frac{3}{2}$	$\frac{3}{2}$	0	010−	$+\sqrt{17}/\sqrt{2.5.7.13}$
$\tilde{1}$	$\frac{3}{2}$	$\frac{3}{2}$	0	011−	0
$\tilde{1}$	$\frac{3}{2}$	$\frac{3}{2}$	0	100−	0
$\tilde{1}$	$\frac{3}{2}$	$\frac{3}{2}$	0	110−	$+4\sqrt{2}/\sqrt{5.7.11.17}$
$\tilde{1}$	$\frac{3}{2}$	$\frac{3}{2}$	0	111−	0
$\tilde{1}$	$\frac{3}{2}$	$\frac{3}{2}$	1	000+	$+2.9\sqrt{2.3}/5\sqrt{5.13.17}$
$\tilde{1}$	$\frac{3}{2}$	$\frac{3}{2}$	1	010+	$+19\sqrt{2}/5\sqrt{5.7.13.17}$
$\tilde{1}$	$\frac{3}{2}$	$\frac{3}{2}$	1	011+	$-2.7\sqrt{2}/5\sqrt{3.5.13.17}$
$\tilde{1}$	$\frac{3}{2}$	$\frac{3}{2}$	1	100+	$+2.3\sqrt{2.3}/5\sqrt{5.11.17}$
$\tilde{1}$	$\frac{3}{2}$	$\frac{3}{2}$	1	110+	$-139/5\sqrt{2.5.7.11.17}$
$\tilde{1}$	$\frac{3}{2}$	$\frac{3}{2}$	1	111+	$-4.7\sqrt{2}/5\sqrt{3.5.11.17}$
$\tilde{0}$	$\frac{3}{2}$	$\frac{3}{2}$	0	000+	$+2.3\sqrt{3.7}/5\sqrt{5.11.17}$
$\tilde{0}$	$\frac{3}{2}$	$\frac{3}{2}$	0	010+	$+8/5\sqrt{5.11.17}$
$\tilde{0}$	$\frac{3}{2}$	$\frac{3}{2}$	0	011+	$+\sqrt{7.17}/5\sqrt{3.5.11}$

7	**5**	**2**			
1	1	2	0	000+	$+3.19/5.11\sqrt{2.13}$
1	1	2	0	010+	$+2.3\sqrt{2}/11\sqrt{5.7.13}$
1	1	2	0	100+	$-\sqrt{3.7}/5\sqrt{2.11.13}$
1	1	2	0	110+	$-3\sqrt{2.3}/\sqrt{5.11.13}$
1	1	$\tilde{1}$	0	000+	$+4.9\sqrt{3}/5.11\sqrt{13}$
1	1	$\tilde{1}$	0	010+	$-4\sqrt{3}/11\sqrt{5.7.13}$
1	1	$\tilde{1}$	0	100+	$+\sqrt{7}/5\sqrt{11.13}$
1	1	$\tilde{1}$	0	110+	$+2.3/\sqrt{5.11.13}$
1	2	$\tilde{1}$	0	000−	$+2\sqrt{2.5}/\sqrt{7.11.13}$
1	2	$\tilde{1}$	0	100−	0
1	$\tilde{1}$	2	0	000−	$-\sqrt{13}/\sqrt{2.5.7.11}$
1	$\tilde{1}$	2	0	100−	$-\sqrt{3}/\sqrt{2.5.13}$
1	$\tilde{1}$	$\tilde{1}$	0	000−	$-2/\sqrt{5.7.11.13}$
1	$\tilde{1}$	$\tilde{1}$	0	100−	$+\sqrt{3}/\sqrt{5.13}$
2	1	$\tilde{1}$	0	000−	$-\sqrt{11}/\sqrt{3.5.13}$
2	1	$\tilde{1}$	0	010−	0
2	2	2	0	000+	$+2/\sqrt{7.13}$
2	$\tilde{1}$	$\tilde{1}$	0	000+	$-\sqrt{3}/\sqrt{7.13}$
$\tilde{1}$	1	2	0	000−	$-\sqrt{13}/\sqrt{2.5.7.11}$
$\tilde{1}$	1	2	0	010−	$-2.3\sqrt{2}/7\sqrt{11.13}$
$\tilde{1}$	1	2	0	100−	$+19/11\sqrt{2.5.7}$
$\tilde{1}$	1	2	0	110−	$-9\sqrt{2}/7.11$
$\tilde{1}$	1	$\tilde{1}$	0	000−	$+2/\sqrt{5.7.11.13}$
$\tilde{1}$	1	$\tilde{1}$	0	010−	$+4.3/7\sqrt{11.13}$
$\tilde{1}$	1	$\tilde{1}$	0	100−	$+3/11\sqrt{5.7}$

SO_3–O 3jm Factors (cont.)

7 5 2	
1̃ 1 1̃ 0 110−	+2.9/7.11
1̃ 2 1̃ 0 000+	+2√2/7√13
1̃ 2 1̃ 0 100+	−4√2/7√11
1̃ 1̃ 2 0 000+	+9/7√2.13
1̃ 1̃ 2 0 100+	+3/7√2.11
1̃ 1̃ 1̃ 0 000+	−4√3/7√13
1̃ 1̃ 1̃ 0 100+	+√3/7√11
0̃ 1 1̃ 0 000+	−4/11√3.5
0̃ 1 1̃ 0 010+	+3√3/11√7
0̃ 2 2 0 000−	+√2/√7.11

7 5 3	
1 1 1 0 000+	−9/4.11√2.13
1 1 1 0 010+	+8.3√2/11√5.7.13
1 1 1 0 100+	+√3.7/2√2.11.13
1 1 1 0 110+	−3√3/2√2.5.11.13
1 1 1̃ 0 000−	−3√3.13/4.11√2.5
1 1 1̃ 0 010−	+8.2√2.3/5.11√7.13
1 1 1̃ 0 100−	−√7/√2.5.11.13
1 1 1̃ 0 110−	+3√13/2.5√2.11
1 2 1 0 000−	−3√11/2√5.7.13
1 2 1 0 100−	−√3/2√5.13
1 2 1̃ 0 000+	−61/2.5√7.11.13
1 2 1̃ 0 100+	+3√3/2.5√13
1 1̃ 1 0 000−	−√3.5.11/4√2.7.13
1 1̃ 1 0 100−	0
1 1̃ 1̃ 0 000+	−√13/4√2.7.11
1 1̃ 1̃ 0 100+	+√3/2√2.13
1 1̃ 0̃ 0 000−	+1/2√7.11.13
1 1̃ 0̃ 0 100−	+√3/2√13
2 1 1 0 000−	−√3.5/√2.11.13
2 1 1 0 010−	+3√3/√2.7.11.13
2 1 1̃ 0 000+	+√2/√3.11.13
2 1 1̃ 0 010+	+9√3/√2.5.7.11.13
2 2 0̃ 0 000+	−2/√5.7.13
2 1̃ 1 0 000+	−√3/√2.7.13
2 1̃ 1̃ 0 000−	+√2.3/√5.7.13
1̃ 1 1 0 000−	+√3.5.11/4√2.7.13
1̃ 1 1 0 010−	0
1̃ 1 1 0 100−	−√2.3.5/11√7
1̃ 1 1 0 110−	−3√3/2.11√2
1̃ 1 1̃ 0 000+	+√13/4√2.7.11
1̃ 1 1̃ 0 010+	+8.3√2/7√5.11.13
1̃ 1 1̃ 0 100+	+3.5/2.11√2.7
1̃ 1 1̃ 0 110+	−9√5/2.7.11√2
1̃ 1 0̃ 0 000−	−1/2√7.11.13
1̃ 1 0̃ 0 010−	−8.2.3/7√5.11.13
1̃ 1 0̃ 0 100−	−5/2.11√7
1̃ 1 0̃ 0 110−	−9/7.11√5
1̃ 2 1 0 000+	−9/2.7√13

7 5 3	
1̃ 2 1 0 100+	−3/2.7√11
1̃ 2 1̃ 0 000−	−√5/2.7√13
1̃ 2 1̃ 0 100−	−√11/2.7√5
1̃ 1̃ 1 0 000+	−3/4.7√2.13
1̃ 1̃ 1 0 100+	+3/2.7√2.11
1̃ 1̃ 1̃ 0 000−	−√3.13/4.7√2.5
1̃ 1̃ 1̃ 0 100−	−√3.11/7√2.5
0̃ 1 1̃ 0 000−	−5/11√2.3
0̃ 1 1̃ 0 010−	−3√2.3/11√5.7
0̃ 1̃ 1 0 000−	−√3/√2.7.11

7 5 4	
1 1 0 0 000+	−3√7/2.11√13.17
1 1 0 0 010+	+8.2√17/3.11√5.13
1 1 0 0 100+	−7√3/2√11.13.17
1 1 0 0 110+	−√7/√3.5.11.13.17
1 1 1 0 000−	−3√3.7.13/4.11√2.5.17
1 1 1 0 010−	+8√2.17/5.11√3.13
1 1 1 0 100−	+3.7/2√2.5.11.13.17
1 1 1 0 110−	+√7.13/2.5√2.11.17
1 1 2 0 000+	−9.3√5/2.11√13.17
1 1 2 0 010+	−2.139/3.5.11√7.13.17
1 1 2 0 100+	−√3.5.7/2√11.13.17
1 1 2 0 110+	−71/5√3.11.13.17
1 1 1̃ 0 000+	−127√3/4.11√2.5.13.17
1 1 1̃ 0 010+	+4√2/5.11√3.7.13.17
1 1 1̃ 0 100+	+2.3√2.7/√5.11.13.17
1 1 1̃ 0 110+	+9.3/2.5√2.11.13.17
1 2 1 0 000+	−7√3/2.5√11.13.17
1 2 1 0 100+	+3√7/2.5√13.17
1 2 1̃ 0 000−	−53/2.5√7.11.13.17
1 2 1̃ 0 100−	−3.7√3/2.5√13.17
1 1̃ 1 0 000+	+157/4.5√2.11.13.17
1 1̃ 1 0 100+	+√3.7/5√2.13.17
1 1̃ 2 0 000−	+277/2.5√7.11.13.17
1 1̃ 2 0 100−	−11√3/2.5√13.17
1 1̃ 1̃ 0 000−	+3√17/4√2.7.11.13
1 1̃ 1̃ 0 100−	+3√3/2√2.13.17
2 1 1 0 000+	+3√7/√2.11.13.17
2 1 1 0 010+	+7√5/√2.11.13.17
2 1 1̃ 0 000−	−√2.3/√11.13.17
2 1 1̃ 0 010−	+√7/√2.3.5.11.13.17
2 2 0 0 000+	+2/√13.17
2 2 2 0 000+	−2√2.5/√7.13.17
2 1̃ 1 0 000−	+3/√2.5.13.17
2 1̃ 1̃ 0 000+	+√2.3.5/√7.13.17
1̃ 1 1 0 000+	−3.19/4√2.11.13.17
1̃ 1 1 0 010+	+4√2.5/√7.11.13.17
1̃ 1 1 0 100+	−3/11√2.17
1̃ 1 1 0 110+	−5.5√5/2.11√2.7.17

SO_3–O 3jm Factors (cont.)

7 5 4	
$\tilde{1}$ 1 2 0 000−	$+3\sqrt{11}/2\sqrt{7.13.17}$
$\tilde{1}$ 1 2 0 010−	$-2.3\sqrt{11}/7\sqrt{5.13.17}$
$\tilde{1}$ 1 2 0 100−	$-3/2.11\sqrt{7.17}$
$\tilde{1}$ 1 2 0 110−	$-1/7.11\sqrt{5.17}$
$\tilde{1}$ 1 $\tilde{1}$ 0 000−	$-3\sqrt{17}/4\sqrt{2.7.11.13}$
$\tilde{1}$ 1 $\tilde{1}$ 0 010−	$+8\sqrt{2.17}/7\sqrt{5.11.13}$
$\tilde{1}$ 1 $\tilde{1}$ 0 100−	$-3.5.5/2.11\sqrt{2.7.17}$
$\tilde{1}$ 1 $\tilde{1}$ 0 110−	$-\sqrt{17}/2.7.11\sqrt{2.5}$
$\tilde{1}$ 2 1 0 000−	$+\sqrt{3.17}/2\sqrt{5.7.13}$
$\tilde{1}$ 2 1 0 100−	$-13\sqrt{3}/2\sqrt{5.7.11.17}$
$\tilde{1}$ 2 $\tilde{1}$ 0 000+	$+3\sqrt{17}/2.7\sqrt{5.13}$
$\tilde{1}$ 2 $\tilde{1}$ 0 100+	$+9\sqrt{5}/2.7\sqrt{11.17}$
$\tilde{1}$ $\tilde{1}$ 0 0 000+	$+1/2\sqrt{7.13.17}$
$\tilde{1}$ $\tilde{1}$ 0 0 100+	$-9/2\sqrt{7.11.17}$
$\tilde{1}$ $\tilde{1}$ 1 0 000−	$-\sqrt{3.13}/4\sqrt{2.5.7.17}$
$\tilde{1}$ $\tilde{1}$ 1 0 100−	$+9.3\sqrt{3}/2\sqrt{2.5.7.11.17}$
$\tilde{1}$ $\tilde{1}$ 2 0 000+	$+29/2.7\sqrt{5.13.17}$
$\tilde{1}$ $\tilde{1}$ 2 0 100+	$+3.5\sqrt{5}/2.7\sqrt{11.17}$
$\tilde{1}$ $\tilde{1}$ $\tilde{1}$ 0 000+	$-\sqrt{3.5.17}/4.7\sqrt{2.13}$
$\tilde{1}$ $\tilde{1}$ $\tilde{1}$ 0 100+	$-\sqrt{2.3.5}/7\sqrt{11.17}$
$\tilde{0}$ 1 $\tilde{1}$ 0 000+	$+5\sqrt{3}/11\sqrt{2.17}$
$\tilde{0}$ 1 $\tilde{1}$ 0 010+	$-\sqrt{2.17}/11\sqrt{3.5.7}$
$\tilde{0}$ 2 2 0 000−	$+4/\sqrt{5.7.11.17}$
$\tilde{0}$ $\tilde{1}$ 1 0 000+	$-9/\sqrt{2.5.11.17}$

7 5 5	
1 1 1 0 000+	$-\sqrt{2.7}/11\sqrt{11.13.17}$
1 1 1 0 010+	$-9.7\sqrt{2}/11\sqrt{5.11.13.17}$
1 1 1 0 011+	$-4.3\sqrt{2.7.17}/5.11\sqrt{11.13}$
1 1 1 0 100+	$+3.7\sqrt{2.3}/11\sqrt{13.17}$
1 1 1 0 110+	$-7\sqrt{3.7}/11\sqrt{2.5.13.17}$
1 1 1 0 111+	$-2.3\sqrt{2.3}/5.11\sqrt{13.17}$
1 2 1 0 000−	$+\sqrt{5.13}/2.11\sqrt{17}$
1 2 1 0 001−	$+4.3\sqrt{7}/5.11\sqrt{13.17}$
1 2 1 0 100−	$-\sqrt{3.5.7}/2\sqrt{11.13.17}$
1 2 1 0 101−	$+\sqrt{3.13}/5\sqrt{11.17}$
1 $\tilde{1}$ 1 0 000−	$+2\sqrt{2.3.13}/11\sqrt{5.17}$
1 $\tilde{1}$ 1 0 001−	$+2.7\sqrt{2.3.7}/5.11\sqrt{13.17}$
1 $\tilde{1}$ 1 0 100−	$+\sqrt{2.7}/\sqrt{5.11.13.17}$
1 $\tilde{1}$ 1 0 101−	$-3\sqrt{13}/5\sqrt{2.11.17}$
1 $\tilde{1}$ 2 0 000+	$-\sqrt{7}/2.5\sqrt{11.13.17}$
1 $\tilde{1}$ 2 0 100+	$+\sqrt{3}/2.5\sqrt{13.17}$
1 $\tilde{1}$ $\tilde{1}$ 0 000+	$+\sqrt{2.7}/\sqrt{11.13.17}$
1 $\tilde{1}$ $\tilde{1}$ 0 100+	$+\sqrt{2.3}/\sqrt{13.17}$
2 1 1 0 000−	0
2 1 1 0 010−	$-\sqrt{3}/\sqrt{2.13.17}$
2 1 1 0 011−	0
2 2 2 0 000−	0
2 $\tilde{1}$ 1 0 000+	$+8\sqrt{2}/\sqrt{3.11.13.17}$
2 $\tilde{1}$ 1 0 001+	$-7\sqrt{3.7}/\sqrt{2.5.11.13.17}$

7 5 5	
2 $\tilde{1}$ $\tilde{1}$ 0 000−	0
$\tilde{1}$ 1 1 0 000−	0
$\tilde{1}$ 1 1 0 010−	$-2\sqrt{2.3}/\sqrt{7.13.17}$
$\tilde{1}$ 1 1 0 011−	0
$\tilde{1}$ 1 1 0 100−	0
$\tilde{1}$ 1 1 0 110−	$+\sqrt{3}/\sqrt{2.7.11.17}$
$\tilde{1}$ 1 1 0 111−	0
$\tilde{1}$ 2 1 0 000+	$-43/2\sqrt{7.11.13.17}$
$\tilde{1}$ 2 1 0 001+	$-4.3/\sqrt{5.11.13.17}$
$\tilde{1}$ 2 1 0 100+	$-47/2.11\sqrt{7.17}$
$\tilde{1}$ 2 1 0 101+	$+3/11\sqrt{5.17}$
$\tilde{1}$ $\tilde{1}$ 1 0 000+	$+\sqrt{2.7}/\sqrt{11.13.17}$
$\tilde{1}$ $\tilde{1}$ 1 0 001+	$-3\sqrt{2}/\sqrt{5.11.13.17}$
$\tilde{1}$ $\tilde{1}$ 1 0 100+	0
$\tilde{1}$ $\tilde{1}$ 1 0 101+	$+3/\sqrt{2.5.17}$
$\tilde{1}$ $\tilde{1}$ 2 0 000−	$+\sqrt{13}/2\sqrt{5.17}$
$\tilde{1}$ $\tilde{1}$ 2 0 100−	$-\sqrt{5}/2\sqrt{11.17}$
$\tilde{1}$ $\tilde{1}$ $\tilde{1}$ 0 000−	0
$\tilde{1}$ $\tilde{1}$ $\tilde{1}$ 0 100−	0
$\tilde{0}$ 2 2 0 000+	$-2\sqrt{7}/\sqrt{5.11.17}$
$\tilde{0}$ $\tilde{1}$ 1 0 000−	$-5/11\sqrt{2.3.17}$
$\tilde{0}$ $\tilde{1}$ 1 0 001−	$-2\sqrt{2.3.7}/11\sqrt{5.17}$

7 $\frac{11}{2}$ $\frac{3}{2}$	
1 $\frac{1}{2}$ $\frac{3}{2}$ 0 000−	$+\sqrt{3}/\sqrt{2.13}$
1 $\frac{1}{2}$ $\frac{3}{2}$ 0 100−	0
1 $\frac{3}{2}$ $\frac{3}{2}$ 0 000+	$-17\sqrt{3}/5\sqrt{7.11.13}$
1 $\frac{3}{2}$ $\frac{3}{2}$ 0 010+	$-4\sqrt{2}/\sqrt{3.5.7.11.13}$
1 $\frac{3}{2}$ $\frac{3}{2}$ 0 100+	$-2/5\sqrt{13}$
1 $\frac{3}{2}$ $\frac{3}{2}$ 0 110+	$+2\sqrt{2}/\sqrt{5.13}$
1 $\frac{3}{2}$ $\frac{3}{2}$ 1 000+	$-19\sqrt{3}/5\sqrt{7.11.13}$
1 $\frac{3}{2}$ $\frac{3}{2}$ 1 010+	$+2\sqrt{2}/\sqrt{3.5.7.11.13}$
1 $\frac{3}{2}$ $\frac{3}{2}$ 1 100+	$+1/5\sqrt{13}$
1 $\frac{3}{2}$ $\frac{3}{2}$ 1 110+	$-\sqrt{2}/\sqrt{5.13}$
1 $\tilde{\frac{1}{2}}$ $\frac{3}{2}$ 0 000−	$-\sqrt{11}/\sqrt{2.3.5.7.13}$
1 $\tilde{\frac{1}{2}}$ $\frac{3}{2}$ 0 100−	$-\sqrt{2}/\sqrt{5.13}$
2 $\frac{1}{2}$ $\frac{3}{2}$ 0 000+	$+\sqrt{11}/\sqrt{2.3.5.13}$
2 $\frac{3}{2}$ $\frac{3}{2}$ 0 000−	$+11/\sqrt{3.5.7.13}$
2 $\frac{3}{2}$ $\frac{3}{2}$ 0 010−	0
2 $\tilde{\frac{1}{2}}$ $\frac{3}{2}$ 0 000+	$-\sqrt{3}/\sqrt{2.7.13}$
$\tilde{1}$ $\frac{1}{2}$ $\frac{3}{2}$ 0 000+	$+\sqrt{11}/\sqrt{2.3.5.7.13}$
$\tilde{1}$ $\frac{1}{2}$ $\frac{3}{2}$ 0 100+	$-\sqrt{2}/\sqrt{3.5.7}$
$\tilde{1}$ $\frac{3}{2}$ $\frac{3}{2}$ 0 000−	$+11/7\sqrt{3.5.13}$
$\tilde{1}$ $\frac{3}{2}$ $\frac{3}{2}$ 0 010−	0
$\tilde{1}$ $\frac{3}{2}$ $\frac{3}{2}$ 0 100−	$-2\sqrt{11}/7\sqrt{3.5}$
$\tilde{1}$ $\frac{3}{2}$ $\frac{3}{2}$ 0 110−	0
$\tilde{1}$ $\frac{3}{2}$ $\frac{3}{2}$ 1 000+	$-17/7\sqrt{3.5.13}$
$\tilde{1}$ $\frac{3}{2}$ $\frac{3}{2}$ 1 010+	$+2\sqrt{2.3}/7\sqrt{13}$
$\tilde{1}$ $\frac{3}{2}$ $\frac{3}{2}$ 1 100+	$+13/7\sqrt{3.5.11}$
$\tilde{1}$ $\frac{3}{2}$ $\frac{3}{2}$ 1 110+	$+3\sqrt{2.3}/7\sqrt{11}$

SO_3–O 3jm Factors (cont.)

7	$\frac{11}{2}$	$\frac{3}{2}$			
$\tilde{1}$	$\tilde{\frac{1}{2}}$	$\frac{3}{2}$	0	000+	$-\sqrt{3}/\sqrt{2.13}$
$\tilde{1}$	$\tilde{\frac{1}{2}}$	$\frac{3}{2}$	0	100+	0
$\tilde{0}$	$\frac{3}{2}$	$\frac{3}{2}$	0	000+	$-4\sqrt{2}/\sqrt{3.5.7.11}$
$\tilde{0}$	$\frac{3}{2}$	$\frac{3}{2}$	0	010+	$-\sqrt{3}/\sqrt{7.11}$
7	$\frac{11}{2}$	$\frac{5}{2}$			
1	$\frac{1}{2}$	$\frac{3}{2}$	0	000+	$-3/4\sqrt{11.13}$
1	$\frac{1}{2}$	$\frac{3}{2}$	0	100+	$-\sqrt{7}/4\sqrt{3.13}$
1	$\frac{3}{2}$	$\frac{3}{2}$	0	000−	$+8.17\sqrt{2}/5.11\sqrt{7.13}$
1	$\frac{3}{2}$	$\frac{3}{2}$	0	010−	$+8.8/3.11\sqrt{5.7.13}$
1	$\frac{3}{2}$	$\frac{3}{2}$	0	100−	$-\sqrt{11}/2.5\sqrt{2.3.13}$
1	$\frac{3}{2}$	$\frac{3}{2}$	0	110−	$+\sqrt{11}/2\sqrt{3.5.13}$
1	$\frac{3}{2}$	$\frac{3}{2}$	1	000−	$-\sqrt{7}/2.5\sqrt{2.13}$
1	$\frac{3}{2}$	$\frac{3}{2}$	1	010−	$+8/3\sqrt{5.7.13}$
1	$\frac{3}{2}$	$\frac{3}{2}$	1	100−	$-7\sqrt{3}/5\sqrt{2.11.13}$
1	$\frac{3}{2}$	$\frac{3}{2}$	1	110−	$-\sqrt{13}/2\sqrt{3.5.11}$
1	$\frac{3}{2}$	$\tilde{\frac{1}{2}}$	0	000+	$+31/2.11\sqrt{5.7.13}$
1	$\frac{3}{2}$	$\tilde{\frac{1}{2}}$	0	010+	$-8\sqrt{2}/3.11\sqrt{7.13}$
1	$\frac{3}{2}$	$\tilde{\frac{1}{2}}$	0	100+	$-2/\sqrt{3.5.11.13}$
1	$\frac{3}{2}$	$\tilde{\frac{1}{2}}$	0	110+	$-7/\sqrt{2.3.11.13}$
1	$\tilde{\frac{1}{2}}$	$\frac{3}{2}$	0	000+	$-43/4.3\sqrt{5.7.13}$
1	$\tilde{\frac{1}{2}}$	$\frac{3}{2}$	0	100+	$-\sqrt{11}/4\sqrt{3.5.13}$
1	$\tilde{\frac{1}{2}}$	$\tilde{\frac{1}{2}}$	0	000−	$+1/3\sqrt{7.13}$
1	$\tilde{\frac{1}{2}}$	$\tilde{\frac{1}{2}}$	0	100−	$-\sqrt{11}/2\sqrt{3.13}$
2	$\frac{1}{2}$	$\frac{3}{2}$	0	000−	$-11/2.3\sqrt{5.13}$
2	$\frac{3}{2}$	$\frac{3}{2}$	0	000+	$+29/3\sqrt{2.5.7.11.13}$
2	$\frac{3}{2}$	$\frac{3}{2}$	0	010+	$+5/\sqrt{7.11.13}$
2	$\frac{3}{2}$	$\tilde{\frac{1}{2}}$	0	000−	$+8/3\sqrt{7.11.13}$
2	$\frac{3}{2}$	$\tilde{\frac{1}{2}}$	0	010−	$+\sqrt{2.5}/\sqrt{7.11.13}$
2	$\tilde{\frac{1}{2}}$	$\frac{3}{2}$	0	000−	$+\sqrt{11}/2\sqrt{7.13}$
$\tilde{1}$	$\frac{1}{2}$	$\frac{3}{2}$	0	000−	$+43/4.3\sqrt{5.7.13}$
$\tilde{1}$	$\frac{1}{2}$	$\frac{3}{2}$	0	100−	$-17/4\sqrt{5.7.11}$
$\tilde{1}$	$\frac{1}{2}$	$\tilde{\frac{1}{2}}$	0	000+	$+1/3\sqrt{7.13}$
$\tilde{1}$	$\frac{1}{2}$	$\tilde{\frac{1}{2}}$	0	100+	$+1/2\sqrt{7.11}$
$\tilde{1}$	$\frac{3}{2}$	$\frac{3}{2}$	0	000+	$-8\sqrt{2.11}/3.7\sqrt{5.13}$
$\tilde{1}$	$\frac{3}{2}$	$\frac{3}{2}$	0	010+	0
$\tilde{1}$	$\frac{3}{2}$	$\frac{3}{2}$	0	100+	$+13/2.7.11\sqrt{2.5}$
$\tilde{1}$	$\frac{3}{2}$	$\frac{3}{2}$	0	110+	$-5/2.11$
$\tilde{1}$	$\frac{3}{2}$	$\frac{3}{2}$	1	000−	$-191/2.3.7\sqrt{2.5.11.13}$
$\tilde{1}$	$\frac{3}{2}$	$\frac{3}{2}$	1	010−	$+8/7\sqrt{11.13}$
$\tilde{1}$	$\frac{3}{2}$	$\frac{3}{2}$	1	100−	$-1/7\sqrt{2.5}$
$\tilde{1}$	$\frac{3}{2}$	$\frac{3}{2}$	1	110−	$-1/2.7$
$\tilde{1}$	$\frac{3}{2}$	$\tilde{\frac{1}{2}}$	0	000−	$-37/2.3.7\sqrt{11.13}$
$\tilde{1}$	$\frac{3}{2}$	$\tilde{\frac{1}{2}}$	0	010−	$+8\sqrt{2.5}/7\sqrt{11.13}$
$\tilde{1}$	$\frac{3}{2}$	$\tilde{\frac{1}{2}}$	0	100−	$-2.9/7.11$
$\tilde{1}$	$\frac{3}{2}$	$\tilde{\frac{1}{2}}$	0	110−	$+3\sqrt{5}/7.11\sqrt{2}$
$\tilde{1}$	$\tilde{\frac{1}{2}}$	$\frac{3}{2}$	0	000−	$-\sqrt{11}/4.7\sqrt{13}$
$\tilde{1}$	$\tilde{\frac{1}{2}}$	$\frac{3}{2}$	0	100−	$-5/4.7$
$\tilde{0}$	$\frac{1}{2}$	$\tilde{\frac{1}{2}}$	0	000−	$+\sqrt{2}/3\sqrt{11}$
$\tilde{0}$	$\frac{3}{2}$	$\frac{3}{2}$	0	000−	$-31/3.11\sqrt{5.7}$
7	$\frac{11}{2}$	$\frac{5}{2}$			
$\tilde{0}$	$\frac{3}{2}$	$\frac{3}{2}$	0	010−	$+3\sqrt{2}/11\sqrt{7}$
7	$\frac{11}{2}$	$\frac{7}{2}$			
1	$\frac{1}{2}$	$\frac{1}{2}$	0	000+	$+3\sqrt{7}/4\sqrt{11.13.17}$
1	$\frac{1}{2}$	$\frac{1}{2}$	0	100+	$-7/2\sqrt{3.13.17}$
1	$\frac{1}{2}$	$\frac{3}{2}$	0	000−	$-3\sqrt{11}/2\sqrt{5.13.17}$
1	$\frac{1}{2}$	$\frac{3}{2}$	0	100−	$-\sqrt{7}/4\sqrt{3.5.13.17}$
1	$\frac{3}{2}$	$\frac{1}{2}$	0	000−	$-3.5/11\sqrt{13.17}$
1	$\frac{3}{2}$	$\frac{1}{2}$	0	010−	$-4\sqrt{2.17}/11\sqrt{5.13}$
1	$\frac{3}{2}$	$\frac{1}{2}$	0	100−	$-5\sqrt{7}/4\sqrt{3.11.13.17}$
1	$\frac{3}{2}$	$\frac{1}{2}$	0	110−	$+\sqrt{3.7}/2\sqrt{2.5.11.13.17}$
1	$\frac{3}{2}$	$\frac{3}{2}$	0	000+	$+3.19/2.11\sqrt{2.5.7.13.17}$
1	$\frac{3}{2}$	$\frac{3}{2}$	0	010+	$+2.3.23/5.11\sqrt{7.13.17}$
1	$\frac{3}{2}$	$\frac{3}{2}$	0	100+	$-\sqrt{2.17}/\sqrt{3.5.11.13}$
1	$\frac{3}{2}$	$\frac{3}{2}$	0	110+	$+31\sqrt{3}/5\sqrt{11.13.17}$
1	$\frac{3}{2}$	$\frac{3}{2}$	1	000+	$+3\sqrt{7.13}/11\sqrt{2.5.17}$
1	$\frac{3}{2}$	$\frac{3}{2}$	1	010+	$-1/5.11\sqrt{7.13.17}$
1	$\frac{3}{2}$	$\frac{3}{2}$	1	100+	$+2.7\sqrt{2}/\sqrt{3.5.11.13.17}$
1	$\frac{3}{2}$	$\frac{3}{2}$	1	110+	$-9\sqrt{3}/2.5\sqrt{11.13.17}$
1	$\frac{3}{2}$	$\tilde{\frac{1}{2}}$	0	000−	$+2.3\sqrt{3.5}/11\sqrt{7.13.17}$
1	$\frac{3}{2}$	$\tilde{\frac{1}{2}}$	0	010−	$-\sqrt{2.13}/11\sqrt{3.7.17}$
1	$\frac{3}{2}$	$\tilde{\frac{1}{2}}$	0	100−	$-5.5\sqrt{5}/4\sqrt{11.13.17}$
1	$\frac{3}{2}$	$\tilde{\frac{1}{2}}$	0	110−	$-7/2\sqrt{2.11.13.17}$
1	$\tilde{\frac{1}{2}}$	$\frac{3}{2}$	0	000−	$+\sqrt{17}/2\sqrt{7.13}$
1	$\tilde{\frac{1}{2}}$	$\frac{3}{2}$	0	100−	$-\sqrt{3.11}/4\sqrt{13.17}$
1	$\tilde{\frac{1}{2}}$	$\tilde{\frac{1}{2}}$	0	000+	$+\sqrt{17}/4\sqrt{3.7.13}$
1	$\tilde{\frac{1}{2}}$	$\tilde{\frac{1}{2}}$	0	100+	$+\sqrt{11}/2\sqrt{13.17}$
2	$\frac{1}{2}$	$\frac{3}{2}$	0	000+	$+3/2\sqrt{13.17}$
2	$\frac{3}{2}$	$\frac{1}{2}$	0	000+	$+2\sqrt{5}/\sqrt{11.13.17}$
2	$\frac{3}{2}$	$\frac{1}{2}$	0	010+	$+\sqrt{13}/\sqrt{2.11.17}$
2	$\frac{3}{2}$	$\frac{3}{2}$	0	000−	$-41/\sqrt{2.7.11.13.17}$
2	$\frac{3}{2}$	$\frac{3}{2}$	0	010−	$+8\sqrt{5}/\sqrt{7.11.13.17}$
2	$\frac{3}{2}$	$\tilde{\frac{1}{2}}$	0	000+	$-2\sqrt{11}/\sqrt{3.7.13.17}$
2	$\frac{3}{2}$	$\tilde{\frac{1}{2}}$	0	010+	$+\sqrt{3.11}/\sqrt{2.5.7.13.17}$
2	$\tilde{\frac{1}{2}}$	$\frac{3}{2}$	0	000+	$+\sqrt{11}/2\sqrt{5.7.13.17}$
$\tilde{1}$	$\frac{1}{2}$	$\frac{3}{2}$	0	000+	$-\sqrt{17}/2\sqrt{7.13}$
$\tilde{1}$	$\frac{1}{2}$	$\frac{3}{2}$	0	100+	$+5/4\sqrt{7.11.17}$
$\tilde{1}$	$\frac{1}{2}$	$\tilde{\frac{1}{2}}$	0	000−	$+\sqrt{17}/4\sqrt{3.7.13}$
$\tilde{1}$	$\frac{1}{2}$	$\tilde{\frac{1}{2}}$	0	100−	$+5.5/2\sqrt{3.7.11.17}$
$\tilde{1}$	$\frac{3}{2}$	$\frac{1}{2}$	0	000+	$-\sqrt{5}/\sqrt{7.11.13.17}$
$\tilde{1}$	$\frac{3}{2}$	$\frac{1}{2}$	0	010+	$+5\sqrt{2}/\sqrt{7.11.13.17}$
$\tilde{1}$	$\frac{3}{2}$	$\frac{1}{2}$	0	100+	$+43\sqrt{5}/4.11\sqrt{7.17}$
$\tilde{1}$	$\frac{3}{2}$	$\frac{1}{2}$	0	110+	$-61/2.11\sqrt{2.7.17}$
$\tilde{1}$	$\frac{3}{2}$	$\frac{3}{2}$	0	000−	$+3.53/2.7\sqrt{2.11.13.17}$
$\tilde{1}$	$\frac{3}{2}$	$\frac{3}{2}$	0	010−	$+2\sqrt{5}/\sqrt{11.13.17}$
$\tilde{1}$	$\frac{3}{2}$	$\frac{3}{2}$	0	100−	$-13\sqrt{2}/7.11\sqrt{17}$
$\tilde{1}$	$\frac{3}{2}$	$\frac{3}{2}$	0	110−	$-\sqrt{5}/11\sqrt{17}$
$\tilde{1}$	$\frac{3}{2}$	$\frac{3}{2}$	1	000+	$+41/7\sqrt{2.11.13.17}$
$\tilde{1}$	$\frac{3}{2}$	$\frac{3}{2}$	1	010+	$+13\sqrt{13}/7\sqrt{5.11.17}$
$\tilde{1}$	$\frac{3}{2}$	$\frac{3}{2}$	1	100+	$-2.3\sqrt{2}/7.11\sqrt{17}$

SO_3–O 3jm Factors (cont.)

7	$\frac{11}{2}$	$\frac{7}{2}$			
$\tilde{1}$	$\frac{3}{2}$	$\frac{3}{2}$	1	110+	$-67/2.7.11\sqrt{5.17}$
$\tilde{1}$	$\frac{3}{2}$	$\tilde{\frac{1}{2}}$	0	000+	$-2\sqrt{17}/7\sqrt{3.11.13}$
$\tilde{1}$	$\frac{3}{2}$	$\tilde{\frac{1}{2}}$	0	010+	$-4\sqrt{2.3.17}/7\sqrt{5.11.13}$
$\tilde{1}$	$\frac{3}{2}$	$\tilde{\frac{1}{2}}$	0	100+	$-5.59/4.7.11\sqrt{3.17}$
$\tilde{1}$	$\frac{3}{2}$	$\tilde{\frac{1}{2}}$	0	110+	$-3\sqrt{3.17}/2.7.11\sqrt{2.5}$
$\tilde{1}$	$\tilde{\frac{1}{2}}$	$\frac{1}{2}$	0	000−	$+\sqrt{11}/4\sqrt{7.13.17}$
$\tilde{1}$	$\tilde{\frac{1}{2}}$	$\frac{1}{2}$	0	100−	$-1/2\sqrt{7.17}$
$\tilde{1}$	$\tilde{\frac{1}{2}}$	$\frac{3}{2}$	0	000+	$-11\sqrt{11}/2.7\sqrt{5.13.17}$
$\tilde{1}$	$\tilde{\frac{1}{2}}$	$\frac{3}{2}$	0	100+	$-61/4.7\sqrt{5.17}$
$\tilde{0}$	$\frac{1}{2}$	$\tilde{\frac{1}{2}}$	0	000+	$+5/\sqrt{2.3.11.17}$
$\tilde{0}$	$\frac{3}{2}$	$\frac{3}{2}$	0	000+	$-2.5/11\sqrt{7.17}$
$\tilde{0}$	$\frac{3}{2}$	$\frac{3}{2}$	0	010+	$+\sqrt{2.17}/11\sqrt{5.7}$
$\tilde{0}$	$\tilde{\frac{1}{2}}$	$\frac{1}{2}$	0	000+	$-1/\sqrt{2.17}$

7	$\frac{11}{2}$	$\frac{9}{2}$			
1	$\frac{1}{2}$	$\frac{1}{2}$	0	000−	$+3\sqrt{7}/\sqrt{5.11.13.17}$
1	$\frac{1}{2}$	$\frac{1}{2}$	0	100−	$+7/2\sqrt{3.5.13.17}$
1	$\frac{1}{2}$	$\frac{3}{2}$	0	000+	$-83\sqrt{7}/4.5\sqrt{3.11.13.17}$
1	$\frac{1}{2}$	$\frac{3}{2}$	0	001+	$-3.23/2.5\sqrt{11.13.17}$
1	$\frac{1}{2}$	$\frac{3}{2}$	0	100+	$+9.7/4.5\sqrt{13.17}$
1	$\frac{1}{2}$	$\frac{3}{2}$	0	101+	$-7\sqrt{7}/5\sqrt{3.13.17}$
1	$\frac{3}{2}$	$\frac{1}{2}$	0	000+	$-3\sqrt{13}/2.11\sqrt{5.17}$
1	$\frac{3}{2}$	$\frac{1}{2}$	0	010+	$+8\sqrt{2.17}/5.11\sqrt{13}$
1	$\frac{3}{2}$	$\frac{1}{2}$	0	100+	$+8\sqrt{7}/\sqrt{3.5.11.13.17}$
1	$\frac{3}{2}$	$\frac{1}{2}$	0	110+	$-\sqrt{3.7}/5\sqrt{2.11.13.17}$
1	$\frac{3}{2}$	$\frac{3}{2}$	0	000−	$+4\sqrt{2.17}/5.11\sqrt{3.13}$
1	$\frac{3}{2}$	$\frac{3}{2}$	0	001−	$-3.19\sqrt{2}/5.11\sqrt{7.13.17}$
1	$\frac{3}{2}$	$\frac{3}{2}$	0	010−	$-8.19/5.11\sqrt{3.5.13.17}$
1	$\frac{3}{2}$	$\frac{3}{2}$	0	011−	$-8.3.23/5.11\sqrt{5.7.13.17}$
1	$\frac{3}{2}$	$\frac{3}{2}$	0	100−	$+19\sqrt{7}/2.5\sqrt{2.11.13.17}$
1	$\frac{3}{2}$	$\frac{3}{2}$	0	101−	$-\sqrt{2.17}/5\sqrt{3.11.13}$
1	$\frac{3}{2}$	$\frac{3}{2}$	0	110−	$+37\sqrt{7}/2.5\sqrt{5.11.13.17}$
1	$\frac{3}{2}$	$\frac{3}{2}$	0	111−	$+31\sqrt{3}/5\sqrt{5.11.13.17}$
1	$\frac{3}{2}$	$\frac{3}{2}$	1	000−	$+139/2.5.11\sqrt{2.3.13.17}$
1	$\frac{3}{2}$	$\frac{3}{2}$	1	001−	$+9.19\sqrt{2}/5.11\sqrt{7.13.17}$
1	$\frac{3}{2}$	$\frac{3}{2}$	1	010−	$-8.8/11\sqrt{3.5.13.17}$
1	$\frac{3}{2}$	$\frac{3}{2}$	1	011−	$-2.83/11\sqrt{5.7.13.17}$
1	$\frac{3}{2}$	$\frac{3}{2}$	1	100−	$+29\sqrt{7}/5\sqrt{2.11.13.17}$
1	$\frac{3}{2}$	$\frac{3}{2}$	1	101−	$-53/5\sqrt{2.3.11.13.17}$
1	$\frac{3}{2}$	$\frac{3}{2}$	1	110−	$-3\sqrt{7}/2\sqrt{5.11.13.17}$
1	$\frac{3}{2}$	$\frac{3}{2}$	1	111−	$-8\sqrt{3}/\sqrt{5.11.13.17}$
1	$\tilde{\frac{1}{2}}$	$\frac{3}{2}$	0	000+	$-11.11/4.5\sqrt{3.5.13.17}$
1	$\tilde{\frac{1}{2}}$	$\frac{3}{2}$	0	001+	$+107/2.5\sqrt{5.7.13.17}$
1	$\tilde{\frac{1}{2}}$	$\frac{3}{2}$	0	100+	$-7\sqrt{7.11}/4.5\sqrt{5.13.17}$
1	$\tilde{\frac{1}{2}}$	$\frac{3}{2}$	0	101+	$-3\sqrt{3.11}/5\sqrt{5.13.17}$
2	$\frac{1}{2}$	$\frac{3}{2}$	0	000−	$-\sqrt{7}/2\sqrt{3.5.13.17}$
2	$\frac{1}{2}$	$\frac{3}{2}$	0	001−	$-3/\sqrt{5.13.17}$
2	$\frac{3}{2}$	$\frac{1}{2}$	0	000−	$-8/\sqrt{11.13.17}$
2	$\frac{3}{2}$	$\frac{1}{2}$	0	010−	$+7\sqrt{2}/\sqrt{5.11.13.17}$
2	$\frac{3}{2}$	$\frac{3}{2}$	0	000+	$+7/\sqrt{2.3.5.11.13.17}$

7	$\frac{11}{2}$	$\frac{9}{2}$			
2	$\frac{3}{2}$	$\frac{3}{2}$	0	001+	$-19\sqrt{2}/\sqrt{5.7.11.13.17}$
2	$\frac{3}{2}$	$\frac{3}{2}$	0	010+	$-7/\sqrt{3.11.13.17}$
2	$\frac{3}{2}$	$\frac{3}{2}$	0	011+	$-4\sqrt{7}/\sqrt{11.13.17}$
2	$\tilde{\frac{1}{2}}$	$\frac{3}{2}$	0	000−	$+\sqrt{11.13}/2.5\sqrt{3.17}$
2	$\tilde{\frac{1}{2}}$	$\frac{3}{2}$	0	001−	$-\sqrt{11}/5\sqrt{7.13.17}$
$\tilde{1}$	$\frac{1}{2}$	$\frac{3}{2}$	0	000−	$+37/4\sqrt{3.5.13.17}$
$\tilde{1}$	$\frac{1}{2}$	$\frac{3}{2}$	0	001−	$+1/2\sqrt{5.7.13.17}$
$\tilde{1}$	$\frac{1}{2}$	$\frac{3}{2}$	0	100−	$+31/4\sqrt{3.5.11.17}$
$\tilde{1}$	$\frac{1}{2}$	$\frac{3}{2}$	0	101−	$-1/\sqrt{5.7.11.17}$
$\tilde{1}$	$\frac{3}{2}$	$\frac{1}{2}$	0	000−	$-1/2\sqrt{7.11.13.17}$
$\tilde{1}$	$\frac{3}{2}$	$\frac{1}{2}$	0	010−	$+8\sqrt{2}/\sqrt{5.7.11.13.17}$
$\tilde{1}$	$\frac{3}{2}$	$\frac{1}{2}$	0	100−	$+8/11\sqrt{7.17}$
$\tilde{1}$	$\frac{3}{2}$	$\frac{1}{2}$	0	110−	$-5\sqrt{5}/11\sqrt{2.7.17}$
$\tilde{1}$	$\frac{3}{2}$	$\frac{3}{2}$	0	000+	$-4\sqrt{2}/\sqrt{3.5.7.11.13.17}$
$\tilde{1}$	$\frac{3}{2}$	$\frac{3}{2}$	0	001+	$-3.53\sqrt{2}/7\sqrt{5.11.13.17}$
$\tilde{1}$	$\frac{3}{2}$	$\frac{3}{2}$	0	010+	$+8/\sqrt{3.7.11.13.17}$
$\tilde{1}$	$\frac{3}{2}$	$\frac{3}{2}$	0	011+	$-8/\sqrt{11.13.17}$
$\tilde{1}$	$\frac{3}{2}$	$\frac{3}{2}$	0	100+	$+61\sqrt{5}/2.11\sqrt{2.3.7.17}$
$\tilde{1}$	$\frac{3}{2}$	$\frac{3}{2}$	0	101+	$-5\sqrt{2.5}/7.11\sqrt{17}$
$\tilde{1}$	$\frac{3}{2}$	$\frac{3}{2}$	0	110+	$+13.13/2.5.11\sqrt{3.7.17}$
$\tilde{1}$	$\frac{3}{2}$	$\frac{3}{2}$	0	111+	$+31/5.11\sqrt{17}$
$\tilde{1}$	$\frac{3}{2}$	$\frac{3}{2}$	1	000−	$-5.5\sqrt{5}/2\sqrt{2.3.7.11.13.17}$
$\tilde{1}$	$\frac{3}{2}$	$\frac{3}{2}$	1	001−	$-\sqrt{2.5}/7\sqrt{11.13.17}$
$\tilde{1}$	$\frac{3}{2}$	$\frac{3}{2}$	1	010−	$+8.2\sqrt{13}/5\sqrt{3.7.11.17}$
$\tilde{1}$	$\frac{3}{2}$	$\frac{3}{2}$	1	011−	$-2.41/5.7\sqrt{11.13.17}$
$\tilde{1}$	$\frac{3}{2}$	$\frac{3}{2}$	1	100−	$+41/11\sqrt{2.3.5.7.17}$
$\tilde{1}$	$\frac{3}{2}$	$\frac{3}{2}$	1	101−	$-9/7\sqrt{2.5.17}$
$\tilde{1}$	$\frac{3}{2}$	$\frac{3}{2}$	1	110−	$+\sqrt{17}/2.11\sqrt{3.7}$
$\tilde{1}$	$\frac{3}{2}$	$\frac{3}{2}$	1	111−	$+4/7\sqrt{17}$
$\tilde{1}$	$\tilde{\frac{1}{2}}$	$\frac{1}{2}$	0	000+	$+\sqrt{11}/\sqrt{5.7.13.17}$
$\tilde{1}$	$\tilde{\frac{1}{2}}$	$\frac{1}{2}$	0	100+	$-11/2\sqrt{5.7.17}$
$\tilde{1}$	$\tilde{\frac{1}{2}}$	$\frac{3}{2}$	0	000−	$-7\sqrt{7.11}/4.5\sqrt{3.13.17}$
$\tilde{1}$	$\tilde{\frac{1}{2}}$	$\frac{3}{2}$	0	001−	$-\sqrt{11}/2.5\sqrt{13.17}$
$\tilde{1}$	$\tilde{\frac{1}{2}}$	$\frac{3}{2}$	0	100−	$+\sqrt{7}/4.5\sqrt{3.17}$
$\tilde{1}$	$\tilde{\frac{1}{2}}$	$\frac{3}{2}$	0	101−	$-1/5\sqrt{17}$
$\tilde{0}$	$\frac{3}{2}$	$\frac{3}{2}$	0	000−	$+31/11\sqrt{3.5.17}$
$\tilde{0}$	$\frac{3}{2}$	$\frac{3}{2}$	0	001−	$-4/11\sqrt{5.7.17}$
$\tilde{0}$	$\frac{3}{2}$	$\frac{3}{2}$	0	010−	$-\sqrt{2}/5.11\sqrt{3.17}$
$\tilde{0}$	$\frac{3}{2}$	$\frac{3}{2}$	0	011−	$-2.23\sqrt{2}/5.11\sqrt{7.17}$
$\tilde{0}$	$\tilde{\frac{1}{2}}$	$\frac{1}{2}$	0	000−	$-\sqrt{2}/\sqrt{5.17}$

7	$\frac{11}{2}$	$\frac{11}{2}$			
1	$\frac{1}{2}$	$\frac{1}{2}$	0	000+	$+\sqrt{2.7}/3\sqrt{11.13.17.19}$
1	$\frac{1}{2}$	$\frac{1}{2}$	0	100+	$-7\sqrt{2.3}/\sqrt{13.17.19}$
1	$\frac{3}{2}$	$\frac{1}{2}$	0	000−	$-23\sqrt{2}/3.11\sqrt{13.17.19}$
1	$\frac{3}{2}$	$\frac{1}{2}$	0	010−	$+3.7\sqrt{19}/2.11\sqrt{5.13.17}$
1	$\frac{3}{2}$	$\frac{1}{2}$	0	100−	$+\sqrt{3.7}/\sqrt{2.11.13.17.19}$
1	$\frac{3}{2}$	$\frac{1}{2}$	0	110−	$+7\sqrt{7}/\sqrt{3.5.11.13.17.19}$
1	$\frac{3}{2}$	$\frac{3}{2}$	0	000+	$-4.103\sqrt{7}/3.11\sqrt{11.13.17.19}$
1	$\frac{3}{2}$	$\frac{3}{2}$	0	010+	$+9.3.7\sqrt{7}/11\sqrt{2.5.11.13.17.19}$

SO_3–O 3jm Factors (cont.)

7 $\frac{11}{2}$ $\frac{11}{2}$	
1 $\frac{3}{2}$ $\frac{3}{2}$ 0 011+	$+8.2.71\sqrt{7}/3.5.11\sqrt{11.13.17.19}$
1 $\frac{3}{2}$ $\frac{3}{2}$ 0 100+	$+2.7\sqrt{3}/11\sqrt{13.17.19}$
1 $\frac{3}{2}$ $\frac{3}{2}$ 0 110+	$-\sqrt{2}/11\sqrt{3.5.13.17.19}$
1 $\frac{3}{2}$ $\frac{3}{2}$ 0 111+	$+8.103/5.11\sqrt{3.13.17.19}$
1 $\frac{3}{2}$ $\frac{3}{2}$ 1 000+	$+4.37\sqrt{7}/3.11\sqrt{11.13.17.19}$
1 $\frac{3}{2}$ $\frac{3}{2}$ 1 010+	$-3.41\sqrt{7}/11\sqrt{2.5.11.13.17.19}$
1 $\frac{3}{2}$ $\frac{3}{2}$ 1 011+	$+4.827\sqrt{7}/3.5.11\sqrt{11.13.17.19}$
1 $\frac{3}{2}$ $\frac{3}{2}$ 1 100+	$-8.4.3\sqrt{3}/11\sqrt{13.17.19}$
1 $\frac{3}{2}$ $\frac{3}{2}$ 1 110+	$-107\sqrt{3}/11\sqrt{2.5.13.17.19}$
1 $\frac{3}{2}$ $\frac{3}{2}$ 1 111+	$-2.149/5.11\sqrt{3.13.17.19}$
1 $\tilde{\frac{1}{2}}$ $\frac{3}{2}$ 0 000−	$+3\sqrt{2.5.7}/\sqrt{11.13.17.19}$
1 $\tilde{\frac{1}{2}}$ $\frac{3}{2}$ 0 001−	$-59\sqrt{7}/2.3\sqrt{11.13.17.19}$
1 $\tilde{\frac{1}{2}}$ $\frac{3}{2}$ 0 100−	$+5\sqrt{5}/\sqrt{2.3.13.17.19}$
1 $\tilde{\frac{1}{2}}$ $\frac{3}{2}$ 0 101−	$+8/\sqrt{3.13.17.19}$
1 $\tilde{\frac{1}{2}}$ $\tilde{\frac{1}{2}}$ 0 000+	$-\sqrt{2.7.11}/3\sqrt{13.17.19}$
1 $\tilde{\frac{1}{2}}$ $\tilde{\frac{1}{2}}$ 0 100+	$-11\sqrt{2}/\sqrt{3.13.17.19}$
2 $\frac{3}{2}$ $\frac{1}{2}$ 0 000+	$+4\sqrt{2.5}/\sqrt{11.13.17.19}$
2 $\frac{3}{2}$ $\frac{1}{2}$ 0 010+	$-163/2.3\sqrt{11.13.17.19}$
2 $\frac{3}{2}$ $\frac{3}{2}$ 0 000−	0
2 $\frac{3}{2}$ $\frac{3}{2}$ 0 010−	$+7\sqrt{7}/3\sqrt{2.13.17.19}$
2 $\frac{3}{2}$ $\frac{3}{2}$ 0 011−	0
2 $\tilde{\frac{1}{2}}$ $\frac{3}{2}$ 0 000+	$-4\sqrt{2.7}/3\sqrt{13.17.19}$
2 $\tilde{\frac{1}{2}}$ $\frac{3}{2}$ 0 001+	$3.7\sqrt{7}/2\sqrt{5.13.17.19}$
$\tilde{1}$ $\frac{3}{2}$ $\frac{1}{2}$ 0 000+	$+5.5\sqrt{2.5}/\sqrt{7.11.13.17.19}$
$\tilde{1}$ $\frac{3}{2}$ $\frac{1}{2}$ 0 010+	$-\sqrt{17.19}/2.3\sqrt{7.11.13}$
$\tilde{1}$ $\frac{3}{2}$ $\frac{1}{2}$ 0 100+	$+47\sqrt{5}/11\sqrt{2.7.17.19}$
$\tilde{1}$ $\frac{3}{2}$ $\frac{1}{2}$ 0 110+	$+8.5/11\sqrt{7.17.19}$
$\tilde{1}$ $\frac{3}{2}$ $\frac{3}{2}$ 0 000−	0
$\tilde{1}$ $\frac{3}{2}$ $\frac{3}{2}$ 0 010−	$-\sqrt{19}/3\sqrt{2.13.17}$
$\tilde{1}$ $\frac{3}{2}$ $\frac{3}{2}$ 0 011−	0
$\tilde{1}$ $\frac{3}{2}$ $\frac{3}{2}$ 0 100−	0
$\tilde{1}$ $\frac{3}{2}$ $\frac{3}{2}$ 0 110−	$-3\sqrt{2}/\sqrt{11.17.19}$
$\tilde{1}$ $\frac{3}{2}$ $\frac{3}{2}$ 0 111−	0
$\tilde{1}$ $\frac{3}{2}$ $\frac{3}{2}$ 1 000+	$-4\sqrt{5.17}/11\sqrt{13.19}$
$\tilde{1}$ $\frac{3}{2}$ $\frac{3}{2}$ 1 010+	$+\sqrt{13.19}/3.11\sqrt{2.17}$
$\tilde{1}$ $\frac{3}{2}$ $\frac{3}{2}$ 1 011+	$+4\sqrt{19}/11\sqrt{5.13.17}$
$\tilde{1}$ $\frac{3}{2}$ $\frac{3}{2}$ 1 100+	$-8.4\sqrt{5}/11\sqrt{11.17.19}$
$\tilde{1}$ $\frac{3}{2}$ $\frac{3}{2}$ 1 110+	$-9\sqrt{17}/11\sqrt{2.11.19}$
$\tilde{1}$ $\frac{3}{2}$ $\frac{3}{2}$ 1 111+	$+2.113/11\sqrt{5.11.17.19}$
$\tilde{1}$ $\tilde{\frac{1}{2}}$ $\frac{1}{2}$ 0 000−	$-\sqrt{2.7.11}/3\sqrt{13.17.19}$
$\tilde{1}$ $\tilde{\frac{1}{2}}$ $\frac{1}{2}$ 0 100−	0
$\tilde{1}$ $\tilde{\frac{1}{2}}$ $\frac{3}{2}$ 0 000+	$-23\sqrt{2}/3\sqrt{13.17.19}$
$\tilde{1}$ $\tilde{\frac{1}{2}}$ $\frac{3}{2}$ 0 001+	$-\sqrt{19}/2\sqrt{5.13.17}$
$\tilde{1}$ $\tilde{\frac{1}{2}}$ $\frac{3}{2}$ 0 100+	$+5/\sqrt{2.11.17.19}$
$\tilde{1}$ $\tilde{\frac{1}{2}}$ $\frac{3}{2}$ 0 101+	$+\sqrt{19}/\sqrt{5.11.17}$
$\tilde{0}$ $\frac{3}{2}$ $\frac{3}{2}$ 0 000+	$+\sqrt{5.7.19}/11\sqrt{2.11.17}$
$\tilde{0}$ $\frac{3}{2}$ $\frac{3}{2}$ 0 010+	$+4.5.5\sqrt{7}/3.11\sqrt{11.17.19}$
$\tilde{0}$ $\frac{3}{2}$ $\frac{3}{2}$ 0 011+	$-2.3.7\sqrt{2.7}/11\sqrt{5.11.17.19}$
$\tilde{0}$ $\tilde{\frac{1}{2}}$ $\frac{1}{2}$ 0 000+	$-5/2.3\sqrt{17.19}$

7 6 1	
1 0 1 0 000+	$+1/\sqrt{13}$
1 0 1 0 100+	0
1 1 1 0 000−	$-3/\sqrt{7.13}$
1 1 1 0 100−	0
1 2 1 0 000+	$-2/5\sqrt{7.13}$
1 2 1 0 100+	$+\sqrt{3.11}/5\sqrt{13}$
1 $\tilde{1}$ 1 0 000+	$+\sqrt{3.11}/7\sqrt{5.13}$
1 $\tilde{1}$ 1 0 010+	$-8\sqrt{3}/5.7\sqrt{13}$
1 $\tilde{1}$ 1 0 100+	$+2.3/\sqrt{5.7.13}$
1 $\tilde{1}$ 1 0 110+	$+2\sqrt{11}/5\sqrt{7.13}$
2 1 1 0 000+	$+\sqrt{3.11}/\sqrt{5.7.13}$
2 $\tilde{1}$ 1 0 000−	$-3\sqrt{3}/7\sqrt{13}$
2 $\tilde{1}$ 1 0 010−	$+4\sqrt{11}/7\sqrt{3.5.13}$
$\tilde{1}$ 1 1 0 000+	$+\sqrt{3.11}/7\sqrt{5.13}$
$\tilde{1}$ 1 1 0 100+	$-2\sqrt{3}/7\sqrt{5}$
$\tilde{1}$ 2 1 0 000−	$-2\sqrt{11}/7\sqrt{5.13}$
$\tilde{1}$ 2 1 0 100−	$-3/7\sqrt{5}$
$\tilde{1}$ $\tilde{1}$ 1 0 000−	$-3/\sqrt{7.13}$
$\tilde{1}$ $\tilde{1}$ 1 0 010−	0
$\tilde{1}$ $\tilde{1}$ 1 0 100−	0
$\tilde{1}$ $\tilde{1}$ 1 0 110−	$-2/\sqrt{5.7}$
$\tilde{1}$ $\tilde{0}$ 1 0 000+	$+1/\sqrt{13}$
$\tilde{1}$ $\tilde{0}$ 1 0 100+	0
$\tilde{0}$ $\tilde{1}$ 1 0 000+	0
$\tilde{0}$ $\tilde{1}$ 1 0 010+	$+1/\sqrt{3.5}$

7 6 2	
1 1 2 0 000−	$+3.11/2.5\sqrt{2.7.13}$
1 1 2 0 100−	$+\sqrt{3.11}/2.5\sqrt{2.13}$
1 1 $\tilde{1}$ 0 000−	$+9\sqrt{3}/2.5\sqrt{7.13}$
1 1 $\tilde{1}$ 0 100−	$-\sqrt{11}/2.5\sqrt{13}$
1 2 $\tilde{1}$ 0 000+	$+4/5\sqrt{7.13}$
1 2 $\tilde{1}$ 0 100+	$+\sqrt{3.11}/2.5\sqrt{13}$
1 $\tilde{1}$ 2 0 000+	$-\sqrt{11}/2\sqrt{2.5.13}$
1 $\tilde{1}$ 2 0 010+	$+3\sqrt{2}/5\sqrt{13}$
1 $\tilde{1}$ 2 0 100+	$+\sqrt{3}/2\sqrt{2.5.7.13}$
1 $\tilde{1}$ 2 0 110+	$-\sqrt{3.11}/5\sqrt{2.7.13}$
1 $\tilde{1}$ $\tilde{1}$ 0 000+	$-\sqrt{5.11}/2.7\sqrt{13}$
1 $\tilde{1}$ $\tilde{1}$ 0 010+	$+2.3/5.7\sqrt{13}$
1 $\tilde{1}$ $\tilde{1}$ 0 100+	$+\sqrt{3.5}/2\sqrt{7.13}$
1 $\tilde{1}$ $\tilde{1}$ 0 110+	$+2\sqrt{3.11}/5\sqrt{7.13}$
1 $\tilde{0}$ $\tilde{1}$ 0 000−	$+\sqrt{11}/2\sqrt{5.7.13}$
1 $\tilde{0}$ $\tilde{1}$ 0 100−	$+\sqrt{3}/\sqrt{5.13}$
2 0 2 0 000−	$-\sqrt{11}/2\sqrt{5.13}$
2 1 $\tilde{1}$ 0 000+	$+\sqrt{11}/\sqrt{3.5.7.13}$
2 2 2 0 000−	$+\sqrt{11}/\sqrt{2.5.7.13}$
2 $\tilde{1}$ $\tilde{1}$ 0 000−	$-\sqrt{3}/7\sqrt{13}$
2 $\tilde{1}$ $\tilde{1}$ 0 010−	$+2\sqrt{3.11}/7\sqrt{5.13}$
2 $\tilde{0}$ 2 0 000+	$+3/2\sqrt{7.13}$
$\tilde{1}$ 0 $\tilde{1}$ 0 000−	$-\sqrt{11}/2\sqrt{5.7.13}$

SO_3–O 3jm Factors (cont.)

7 6 2

1̃ 0 1̃ 0 100− $+1/\sqrt{5.7}$
1̃ 1 2 0 000+ $+\sqrt{11}/2\sqrt{2.5.13}$
1̃ 1 2 0 100+ $-1/2\sqrt{2.5}$
1̃ 1 1̃ 0 000+ $-\sqrt{5.11}/2.7\sqrt{13}$
1̃ 1 1̃ 0 100+ $+3/2.7\sqrt{5}$
1̃ 2 1̃ 0 000− $+4\sqrt{11}/7\sqrt{5.13}$
1̃ 2 1̃ 0 100− $+\sqrt{5}/2.7$
1̃ 1̃ 2 0 000− $-9.3/2.7\sqrt{2.7.13}$
1̃ 1̃ 2 0 010− $+3\sqrt{2.11}/7\sqrt{5.7.13}$
1̃ 1̃ 2 0 100− $+3\sqrt{11}/2.7\sqrt{2.7}$
1̃ 1̃ 2 0 110− $+9/7\sqrt{2.5.7}$
1̃ 1̃ 1̃ 0 000− $+19\sqrt{3}/2.7\sqrt{7.13}$
1̃ 1̃ 1̃ 0 010− $+2\sqrt{3.11}/7\sqrt{5.7.13}$
1̃ 1̃ 1̃ 0 100− $+\sqrt{3.11}/2.7\sqrt{7}$
1̃ 1̃ 1̃ 0 110− $-4\sqrt{3}/7\sqrt{5.7}$
0̃ 1 1̃ 0 000− $+2/\sqrt{3.5.7}$
0̃ 2 2 0 000+ $-1/\sqrt{5.7}$

7 6 3

1 0 1 0 000+ $-9/2\sqrt{2.11.13.17}$
1 0 1 0 100+ $-\sqrt{3.7}/2\sqrt{2.13.17}$
1 1 1 0 000− $+9\sqrt{11}/4\sqrt{2.7.13.17}$
1 1 1 0 100− $-\sqrt{3}/\sqrt{2.13.17}$
1 1 1̃ 0 000+ $+3\sqrt{3.5.7}/4\sqrt{2.11.13.17}$
1 1 1̃ 0 100+ $-\sqrt{5}/2\sqrt{2.13.17}$
1 2 1 0 000+ $-3.71/2.5\sqrt{2.7.11.13.17}$
1 2 1 0 100+ $+19\sqrt{3}/2.5\sqrt{2.13.17}$
1 2 1̃ 0 000− $+19/2\sqrt{2.5.7.11.13.17}$
1 2 1̃ 0 100− $-7\sqrt{3}/2\sqrt{2.5.13.17}$
1 1̃ 1 0 000+ $-\sqrt{3.17}/4.7\sqrt{2.5.13}$
1 1̃ 1 0 010+ $+8.8.3\sqrt{2.3}/5.7\sqrt{11.13.17}$
1 1̃ 1 0 100+ $+9\sqrt{11}/2\sqrt{2.5.7.13.17}$
1 1̃ 1 0 110+ $+3.11/2.5\sqrt{2.7.13.17}$
1 1̃ 1̃ 0 000− $-\sqrt{17}/4\sqrt{2.13}$
1 1̃ 1̃ 0 010− $+2.3\sqrt{2}/\sqrt{5.11.13.17}$
1 1̃ 1̃ 0 100− $-\sqrt{2.3.11}/\sqrt{7.13.17}$
1 1̃ 1̃ 0 110− $+11\sqrt{3}/2\sqrt{2.5.7.13.17}$
1 1̃ 0̃ 0 000+ $+\sqrt{17}/2.7\sqrt{13}$
1 1̃ 0̃ 0 010+ $-4.3/7\sqrt{5.11.13.17}$
1 1̃ 0̃ 0 100+ $-\sqrt{3.11}/2\sqrt{7.13.17}$
1 1̃ 0̃ 0 110+ $+11\sqrt{3}/\sqrt{5.7.13.17}$
1 0̃ 1̃ 0 000+ $+\sqrt{17}/2\sqrt{2.7.13}$
1 0̃ 1̃ 0 100+ $-\sqrt{3.11}/2\sqrt{2.13.17}$
2 1 1 0 000+ $-4\sqrt{2.3}/\sqrt{5.7.13.17}$
2 1 1̃ 0 000− $+23/\sqrt{2.3.7.13.17}$
2 2 0̃ 0 000− $+2\sqrt{2}/\sqrt{7.13.17}$
2 1̃ 1 0 000− $+2\sqrt{2.3.11}/7\sqrt{13.17}$
2 1̃ 1 0 010− $+31\sqrt{3}/7\sqrt{2.5.13.17}$
2 1̃ 1̃ 0 000+ $-\sqrt{3.5.11}/7\sqrt{2.13.17}$
2 1̃ 1̃ 0 010+ $+\sqrt{3}/7\sqrt{2.13.17}$

7 6 3

1̃ 0 1̃ 0 000+ $-\sqrt{17}/2\sqrt{2.7.13}$
1̃ 0 1̃ 0 100+ $+13/2\sqrt{2.7.11.17}$
1̃ 1 1 0 000+ $-\sqrt{3.17}/4.7\sqrt{2.5.13}$
1̃ 1 1 0 100+ $-53\sqrt{3}/2.7\sqrt{2.5.11.17}$
1̃ 1 1̃ 0 000− $-\sqrt{17}/4\sqrt{2.13}$
1̃ 1 1̃ 0 100− 0
1̃ 1 0̃ 0 000+ $+\sqrt{17}/2.7\sqrt{13}$
1̃ 1 0̃ 0 100+ $+43/2.7\sqrt{11.17}$
1̃ 2 1 0 000− $-3.31/2.7\sqrt{2.5.13.17}$
1̃ 2 1 0 100− $+3.41/2.7\sqrt{2.5.11.17}$
1̃ 2 1̃ 0 000+ $-53/2.7\sqrt{2.13.17}$
1̃ 2 1̃ 0 100+ $+1/2.7\sqrt{2.11.17}$
1̃ 1̃ 1 0 000− $-3.11\sqrt{11}/4.7\sqrt{2.7.13.17}$
1̃ 1̃ 1 0 010− $-2.3\sqrt{2.5}/7\sqrt{7.13.17}$
1̃ 1̃ 1 0 100− $-3.5/7\sqrt{2.7.17}$
1̃ 1̃ 1 0 110− $+9/2.7\sqrt{2.5.7.11.17}$
1̃ 1̃ 1̃ 0 000+ $-\sqrt{3.5.11}/4\sqrt{2.7.13.17}$
1̃ 1̃ 1̃ 0 010+ $-4\sqrt{2.3}/\sqrt{7.13.17}$
1̃ 1̃ 1̃ 0 100+ $-\sqrt{3.5}/2\sqrt{2.7.17}$
1̃ 1̃ 1̃ 0 110+ $-\sqrt{3}/2\sqrt{2.7.11.17}$
1̃ 0̃ 1 0 000+ $+3\sqrt{11}/2.7\sqrt{2.13.17}$
1̃ 0̃ 1 0 100+ $+3.5/2.7\sqrt{2.17}$
0̃ 0 0̃ 0 000+ $-2/\sqrt{11.17}$
0̃ 1 1̃ 0 000+ $+1/\sqrt{2.3.7.11.17}$
0̃ 1̃ 1 0 000+ $-5\sqrt{3}/7\sqrt{2.17}$
0̃ 1̃ 1 0 010+ $-\sqrt{2.3}/7\sqrt{5.11.17}$

7 6 4

1 0 1 0 000− $-9\sqrt{3.7}/2\sqrt{2.5.11.13.17}$
1 0 1 0 100− $+7/2\sqrt{2.5.13.17}$
1 1 0 0 000− $+9/2\sqrt{11.13.17}$
1 1 0 0 100− $-\sqrt{7}/2\sqrt{3.13.17}$
1 1 1 0 000+ $+9.7\sqrt{3}/4\sqrt{2.5.11.13.17}$
1 1 1 0 100+ $+\sqrt{2.7}/\sqrt{5.13.17}$
1 1 2 0 000− $+3/2\sqrt{5.7.11.13.17}$
1 1 2 0 100− $+\sqrt{13}/2\sqrt{3.5.17}$
1 1 1̃ 0 000− $-\sqrt{3.5}/4\sqrt{2.7.11.13.17}$
1 1 1̃ 0 100− $-5\sqrt{5}/2\sqrt{2.13.17}$
1 2 1 0 000− $+7\sqrt{3.5}/2\sqrt{2.11.13.17}$
1 2 1 0 100− $+\sqrt{5.7}/2\sqrt{2.13.17}$
1 2 1̃ 0 000+ $+5\sqrt{5}/2\sqrt{2.7.11.13.17}$
1 2 1̃ 0 100+ $+\sqrt{5}/2\sqrt{2.3.13.17}$
1 1̃ 1 0 000− $-3.61/4.5\sqrt{2.7.13.17}$
1 1̃ 1 0 010− $-8.4\sqrt{2}/\sqrt{5.7.11.13.17}$
1 1̃ 1 0 100− $+7\sqrt{3.11}/2.5\sqrt{2.13.17}$
1 1̃ 1 0 110− $+7/2\sqrt{2.3.5.13.17}$
1 1̃ 2 0 000+ $-9/2.5\sqrt{13.17}$
1 1̃ 2 0 010+ $-2.7/\sqrt{5.11.13.17}$
1 1̃ 2 0 100+ $-\sqrt{3.11}/2.5\sqrt{7.13.17}$
1 1̃ 2 0 110+ $-\sqrt{13}/\sqrt{3.5.7.17}$

SO_3–O 3jm Factors (cont.)

7	**6**	**4**			
1	$\tilde{1}$	$\tilde{1}$	0	000+	$+3\sqrt{17}/4.7\sqrt{2.13}$
1	$\tilde{1}$	$\tilde{1}$	0	010+	$-2.3\sqrt{2.5}/7\sqrt{11.13.17}$
1	$\tilde{1}$	$\tilde{1}$	0	100+	$-\sqrt{3.11}/\sqrt{2.7.13.17}$
1	$\tilde{1}$	$\tilde{1}$	0	110+	$+11\sqrt{5}/2\sqrt{2.3.7.13.17}$
1	$\tilde{0}$	$\tilde{1}$	0	000−	$-\sqrt{17}/2\sqrt{2.7.13}$
1	$\tilde{0}$	$\tilde{1}$	0	100−	$-\sqrt{11}/2\sqrt{2.3.13.17}$
2	0	2	0	000−	$-\sqrt{2}/\sqrt{13.17}$
2	1	1	0	000−	0
2	1	$\tilde{1}$	0	000+	$-\sqrt{3}/\sqrt{2.7.13.17}$
2	2	0	0	000−	$-2\sqrt{2.5}/3\sqrt{13.17}$
2	2	2	0	000−	$-2.11/3\sqrt{7.13.17}$
2	$\tilde{1}$	1	0	000+	$+2\sqrt{2.11}/\sqrt{5.7.13.17}$
2	$\tilde{1}$	1	0	010+	$-29/3\sqrt{2.7.13.17}$
2	$\tilde{1}$	$\tilde{1}$	0	000−	$+9\sqrt{3.11}/7\sqrt{2.5.13.17}$
2	$\tilde{1}$	$\tilde{1}$	0	010−	$-11/7\sqrt{2.3.13.17}$
2	$\tilde{0}$	2	0	000+	$-\sqrt{2.11}/3\sqrt{5.7.13.17}$
$\tilde{1}$	0	$\tilde{1}$	0	000−	$+\sqrt{17}/2\sqrt{2.7.13}$
$\tilde{1}$	0	$\tilde{1}$	0	100−	$+3.5/2\sqrt{2.7.11.17}$
$\tilde{1}$	1	1	0	000−	$-11/4\sqrt{2.7.13.17}$
$\tilde{1}$	1	1	0	100−	$+\sqrt{11}/2\sqrt{2.7.17}$
$\tilde{1}$	1	2	0	000+	$+5/2\sqrt{13.17}$
$\tilde{1}$	1	2	0	100+	$-1/2\sqrt{11.17}$
$\tilde{1}$	1	$\tilde{1}$	0	000+	$+3\sqrt{17}/4.7\sqrt{2.13}$
$\tilde{1}$	1	$\tilde{1}$	0	100+	$-3.5/7\sqrt{2.11.17}$
$\tilde{1}$	2	1	0	000+	$-5\sqrt{3}/2\sqrt{2.7.13.17}$
$\tilde{1}$	2	1	0	100+	$-19/2\sqrt{2.3.7.11.17}$
$\tilde{1}$	2	$\tilde{1}$	0	000−	$-31/2.7\sqrt{2.13.17}$
$\tilde{1}$	2	$\tilde{1}$	0	100−	$-1/2.7\sqrt{2.11.17}$
$\tilde{1}$	$\tilde{1}$	0	0	000	$+3\sqrt{11}/2.7\sqrt{13.17}$
$\tilde{1}$	$\tilde{1}$	0	0	010−	$+4\sqrt{5}/3.7\sqrt{13.17}$
$\tilde{1}$	$\tilde{1}$	0	0	100−	$-13/2.7\sqrt{17}$
$\tilde{1}$	$\tilde{1}$	0	0	110−	$-\sqrt{5}/3.7\sqrt{11.17}$
$\tilde{1}$	$\tilde{1}$	1	0	000+	$-3\sqrt{3.11}/4\sqrt{2.5.13.17}$
$\tilde{1}$	$\tilde{1}$	1	0	010+	$-2\sqrt{2}/\sqrt{3.13.17}$
$\tilde{1}$	$\tilde{1}$	1	0	100+	0
$\tilde{1}$	$\tilde{1}$	1	0	110+	$+1/2\sqrt{2.3.11.17}$
$\tilde{1}$	$\tilde{1}$	2	0	000−	$-3\sqrt{5.11}/2.7\sqrt{7.13.17}$
$\tilde{1}$	$\tilde{1}$	2	0	010−	$+2.109/3.7\sqrt{7.13.17}$
$\tilde{1}$	$\tilde{1}$	2	0	100−	$-47/2.7\sqrt{5.7.17}$
$\tilde{1}$	$\tilde{1}$	2	0	110−	$-5.13/3.7\sqrt{7.11.17}$
$\tilde{1}$	$\tilde{1}$	$\tilde{1}$	0	000−	$-41\sqrt{3.11}/4.7\sqrt{2.5.7.13.17}$
$\tilde{1}$	$\tilde{1}$	$\tilde{1}$	0	010−	$-4.11\sqrt{2}/7\sqrt{3.7.13.17}$
$\tilde{1}$	$\tilde{1}$	$\tilde{1}$	0	100−	$+41\sqrt{3}/2.7\sqrt{2.5.7.17}$
$\tilde{1}$	$\tilde{1}$	$\tilde{1}$	0	110−	$-\sqrt{11.17}/2.7\sqrt{2.3.7}$
$\tilde{1}$	$\tilde{0}$	1	0	000−	$+3\sqrt{3.11}/2\sqrt{2.5.7.13.17}$
$\tilde{1}$	$\tilde{0}$	1	0	100−	$+\sqrt{17}/2\sqrt{2.3.5.7}$
$\tilde{0}$	1	$\tilde{1}$	0	000−	$+5\sqrt{3}/\sqrt{2.7.11.17}$
$\tilde{0}$	2	2	0	000+	$+8\sqrt{2}/3\sqrt{7.11.17}$
$\tilde{0}$	$\tilde{1}$	1	0	000−	$+1/\sqrt{2.5.7.17}$
7	**6**	**4**			
$\tilde{0}$	$\tilde{1}$	1	0	010−	$-\sqrt{2}/3\sqrt{7.11.17}$
$\tilde{0}$	$\tilde{0}$	0	0	000−	$-2/3\sqrt{17}$
7	**6**	**5**			
1	0	1	0	000+	$-4/11\sqrt{13.17.19}$
1	0	1	0	001+	$-9\sqrt{7.19}/2.11\sqrt{5.13.17}$
1	0	1	0	100+	$-3.7\sqrt{3.7}/2\sqrt{11.13.17.19}$
1	0	1	0	101+	$-7\sqrt{3}/\sqrt{5.11.13.17.19}$
1	1	1	0	000−	$+2\sqrt{13}/11\sqrt{7.17.19}$
1	1	1	0	001−	$+9\sqrt{19}/2.11\sqrt{5.13.17}$
1	1	1	0	100−	$-3.7\sqrt{3}/2\sqrt{11.13.17.19}$
1	1	1	0	101−	$-\sqrt{3.7.13}/2\sqrt{5.11.17.19}$
1	1	2	0	000+	$-7\sqrt{17}/2\sqrt{2.5.11.13.19}$
1	1	2	0	100+	$+\sqrt{3.7.17}/2\sqrt{2.5.13.19}$
1	1	$\tilde{1}$	0	000+	$-7\sqrt{3.5}/\sqrt{11.13.17.19}$
1	1	$\tilde{1}$	0	100+	$-\sqrt{5.7}/2\sqrt{13.17.19}$
1	2	1	0	000+	$+97/2.11\sqrt{7.13.17.19}$
1	2	1	0	001+	$-8.2.3\sqrt{5}/11\sqrt{13.17.19}$
1	2	1	0	100+	$+2\sqrt{3}/\sqrt{11.13.17.19}$
1	2	1	0	101+	$+5\sqrt{3.5.7}/2\sqrt{11.13.17.19}$
1	2	$\tilde{1}$	0	000−	$-41\sqrt{5}/2\sqrt{11.13.17.19}$
1	2	$\tilde{1}$	0	100−	0
1	$\tilde{1}$	1	0	000+	$-179\sqrt{3}/7\sqrt{5.11.13.17.19}$
1	$\tilde{1}$	1	0	001+	$+353\sqrt{3}/2.5\sqrt{7.11.13.17.19}$
1	$\tilde{1}$	1	0	010+	$-8.23\sqrt{3}/7.11\sqrt{13.17.19}$
1	$\tilde{1}$	1	0	011+	$+2.3.43\sqrt{3}/11\sqrt{5.7.13.17.19}$
1	$\tilde{1}$	1	0	100+	$+37/2\sqrt{5.7.13.17.19}$
1	$\tilde{1}$	1	0	101+	$+3/2.5\sqrt{13.17.19}$
1	$\tilde{1}$	1	0	110+	$-8.3/\sqrt{7.11.13.17.19}$
1	$\tilde{1}$	1	0	111+	$+8.2.3/\sqrt{5.11.13.17.19}$
1	$\tilde{1}$	2	0	000−	$+349/2.5\sqrt{2.7.13.17.19}$
1	$\tilde{1}$	2	0	010−	$+9.3\sqrt{2}/\sqrt{5.7.11.13.17.19}$
1	$\tilde{1}$	2	0	100−	$+23\sqrt{3.11}/2.5\sqrt{2.13.17.19}$
1	$\tilde{1}$	2	0	110−	$-\sqrt{3.17}/\sqrt{2.5.13.19}$
1	$\tilde{1}$	$\tilde{1}$	0	000−	$+2\sqrt{13}/\sqrt{7.17.19}$
1	$\tilde{1}$	$\tilde{1}$	0	010−	$+2.3\sqrt{5}/\sqrt{7.11.13.17.19}$
1	$\tilde{1}$	$\tilde{1}$	0	100−	$+\sqrt{3.11}/2\sqrt{13.17.19}$
1	$\tilde{1}$	$\tilde{1}$	0	110−	$+2\sqrt{3.5}/\sqrt{13.17.19}$
1	$\tilde{0}$	$\tilde{1}$	0	000+	$-4/\sqrt{13.17.19}$
1	$\tilde{0}$	$\tilde{1}$	0	100+	$+\sqrt{3.7.11}/2\sqrt{13.17.19}$
2	0	2	0	000+	$+\sqrt{7}/2\sqrt{13.17.19}$
2	1	1	0	000+	$-4.3\sqrt{3.5}/\sqrt{7.11.13.17.19}$
2	1	1	0	001+	$-4.5\sqrt{3}/\sqrt{11.13.17.19}$
2	1	$\tilde{1}$	0	000−	$+2/\sqrt{3.13.17.19}$
2	2	2	0	000+	$+7/\sqrt{2.13.17.19}$
2	$\tilde{1}$	1	0	000−	$-4\sqrt{17}/7\sqrt{3.13.19}$
2	$\tilde{1}$	1	0	001−	$-4\sqrt{3.7}/\sqrt{5.13.17.19}$
2	$\tilde{1}$	1	0	010−	$-37\sqrt{3.5}/7\sqrt{11.13.17.19}$
2	$\tilde{1}$	1	0	011−	$-2\sqrt{3.7}/\sqrt{11.13.17.19}$
2	$\tilde{1}$	$\tilde{1}$	0	000+	$-2.3\sqrt{3.11}/\sqrt{5.7.13.17.19}$

SO_3–O 3jm Factors (cont.)

7 6 5

2 1̃ 1̃ 0 010+	$-\sqrt{3.17}/\sqrt{7.13.19}$
2 0̃ 2 0 000−	$-\sqrt{11.13}/2\sqrt{5.17.19}$
1̃ 0 1̃ 0 000+	$-4/\sqrt{13.17.19}$
1̃ 0 1̃ 0 100+	$-5/2\sqrt{11.17.19}$
1̃ 1 1 0 000+	$-3.5.5\sqrt{3.5}/7\sqrt{11.13.17.19}$
1̃ 1 1 0 001+	$-\sqrt{3.19}/2\sqrt{7.11.13.17}$
1̃ 1 1 0 100+	$-3.47\sqrt{3.5}/2.7.11\sqrt{17.19}$
1̃ 1 1 0 101+	$+73\sqrt{3}/2.11\sqrt{7.17.19}$
1̃ 1 2 0 000−	$-5.5/2\sqrt{2.7.13.17.19}$
1̃ 1 2 0 100−	$+1/2\sqrt{2.7.11.17.19}$
1̃ 1 1̃ 0 000−	$-2\sqrt{13}/\sqrt{7.17.19}$
1̃ 1 1̃ 0 100−	$-3.5/2\sqrt{7.11.17.19}$
1̃ 2 1 0 000−	$+\sqrt{5.11.19}/2.7\sqrt{13.17}$
1̃ 2 1 0 001−	0
1̃ 2 1 0 100−	$-4\sqrt{5.19}/7.11\sqrt{17}$
1̃ 2 1 0 101−	$-3.7\sqrt{7}/2.11\sqrt{17.19}$
1̃ 2 1̃ 0 000+	$+\sqrt{19}/2\sqrt{7.13.17}$
1̃ 2 1̃ 0 100+	$+2/\sqrt{7.11.17.19}$
1̃ 1̃ 1 0 000−	$+2\sqrt{13}/\sqrt{7.17.19}$
1̃ 1̃ 1 0 001−	$-3\sqrt{19}/2.7\sqrt{5.13.17}$
1̃ 1̃ 1 0 010−	$-2.3\sqrt{5}/\sqrt{7.11.13.17.19}$
1̃ 1̃ 1 0 011−	$-2.3\sqrt{19}/7\sqrt{11.13.17}$
1̃ 1̃ 1 0 100−	$-3.5/2\sqrt{7.11.17.19}$
1̃ 1̃ 1 0 101−	$-3\sqrt{5.19}/2.7\sqrt{11.17}$
1̃ 1̃ 1 0 110−	$-2.3\sqrt{5}/11\sqrt{7.17.19}$
1̃ 1̃ 1 0 111−	$+2.9/7.11\sqrt{17.19}$
1̃ 1̃ 2 0 000+	$+\sqrt{5.11}/2\sqrt{2.13.17.19}$
1̃ 1̃ 2 0 010+	$+3\sqrt{2}/\sqrt{13.17.19}$
1̃ 1̃ 2 0 100+	$-7/2\sqrt{2.5.17.19}$
1̃ 1̃ 2 0 110+	$+3.5/\sqrt{2.11.17.19}$
1̃ 1̃ 1̃ 0 000+	$+3\sqrt{3.11}/\sqrt{5.13.17.19}$
1̃ 1̃ 1̃ 0 010+	$-8\sqrt{3}/\sqrt{13.17.19}$
1̃ 1̃ 1̃ 0 100+	$+3\sqrt{3}/2\sqrt{5.17.19}$
1̃ 1̃ 1̃ 0 110+	$+4\sqrt{3}/\sqrt{11.17.19}$
1̃ 0̃ 1 0 000+	$-4/\sqrt{13.17.19}$
1̃ 0̃ 1 0 001+	$+3\sqrt{19}/2\sqrt{5.7.13.17}$
1̃ 0̃ 1 0 100+	$+5/2\sqrt{11.17.19}$
1̃ 0̃ 1 0 101+	$-3\sqrt{19}/\sqrt{5.7.11.17}$
0̃ 1 1̃ 0 000+	$-5/\sqrt{3.11.17.19}$
0̃ 2 2 0 000−	$-3/\sqrt{11.17.19}$
0̃ 1̃ 1 0 000+	$-5/\sqrt{3.11.17.19}$
0̃ 1̃ 1 0 001+	$+2\sqrt{3.19}/\sqrt{5.7.11.17}$
0̃ 1̃ 1 0 010+	$+8\sqrt{3.5}/11\sqrt{17.19}$
0̃ 1̃ 1 0 011+	$-2\sqrt{3}/11\sqrt{7.17.19}$

7 6 6

1 1 0 0 000−	$+2\sqrt{2}/\sqrt{11.13.17.19}$
1 1 0 0 100−	$-3\sqrt{3.7}/\sqrt{2.13.17.19}$
1 1 1 0 000+	$+\sqrt{2.7}/\sqrt{11.13.17.19}$
1 1 1 0 100+	$+2.3\sqrt{2.3}/\sqrt{13.17.19}$

7 6 6

1 2 1 0 000−	$-83\sqrt{7}/5\sqrt{2.11.13.17.19}$
1 2 1 0 100−	$-\sqrt{2.3}/5\sqrt{13.17.19}$
1 1̃ 1 0 000−	$+2\sqrt{2.3}/\sqrt{5.13.17.19}$
1 1̃ 1 0 010−	$+73\sqrt{3}/5\sqrt{2.11.13.17.19}$
1 1̃ 1 0 100−	$+4\sqrt{2.11}/\sqrt{5.7.13.17.19}$
1 1̃ 1 0 110−	$+3\sqrt{2}/5\sqrt{7.13.17.19}$
1 1̃ 2 0 000+	$-1/\sqrt{2.5.13.17.19}$
1 1̃ 2 0 010+	$+3.41\sqrt{2}/5\sqrt{11.13.17.19}$
1 1̃ 2 0 100+	$+3\sqrt{2.3.11}/\sqrt{5.7.13.17.19}$
1 1̃ 2 0 110+	$+29\sqrt{3}/5\sqrt{2.7.13.17.19}$
1 1̃ 1̃ 0 000+	$-\sqrt{2.7.11}/\sqrt{13.17.19}$
1 1̃ 1̃ 0 010+	$-3\sqrt{7}/\sqrt{2.5.13.17.19}$
1 1̃ 1̃ 0 011+	$-8.3\sqrt{2.7}/5\sqrt{11.13.17.19}$
1 1̃ 1̃ 0 100+	$+2.11\sqrt{2.3}/7\sqrt{13.17.19}$
1 1̃ 1̃ 0 110+	$-2\sqrt{2.3.11}/7\sqrt{5.13.17.19}$
1 1̃ 1̃ 0 111+	$+2.9.11\sqrt{2.3}/5.7\sqrt{13.17.19}$
1 0̃ 1̃ 0 000−	$+2\sqrt{2.11}/\sqrt{13.17.19}$
1 0̃ 1̃ 0 001−	$+3/\sqrt{2.5.13.17.19}$
1 0̃ 1̃ 0 100−	$+11\sqrt{3}/\sqrt{2.7.13.17.19}$
1 0̃ 1̃ 0 101−	$-4\sqrt{2.3.11}/\sqrt{5.7.13.17.19}$
2 1 1 0 000−	0
2 2 0 0 000−	$+\sqrt{17}/\sqrt{5.13.19}$
2 2 2 0 000−	0
2 1̃ 1 0 000+	$+2\sqrt{2.11}/\sqrt{3.13.17.19}$
2 1̃ 1 0 010+	$+\sqrt{3}/\sqrt{2.5.13.17.19}$
2 1̃ 1̃ 0 000−	0
2 1̃ 1̃ 0 010−	$-\sqrt{3.7.11}/\sqrt{2.13.17.19}$
2 1̃ 1̃ 0 011−	0
2 0̃ 2 0 000+	$-\sqrt{7.11}/\sqrt{13.17.19}$
1̃ 1 1 0 000−	0
1̃ 1 1 0 100−	0
1̃ 2 1 0 000+	$-7/\sqrt{2.5.13.17.19}$
1̃ 2 1 0 100+	$-7\sqrt{2}/\sqrt{5.11.17.19}$
1̃ 1̃ 0 0 000−	$-2\sqrt{2.11}/\sqrt{13.17.19}$
1̃ 1̃ 0 0 010−	$+3/\sqrt{2.5.13.17.19}$
1̃ 1̃ 0 0 100−	$-1/\sqrt{2.17.19}$
1̃ 1̃ 0 0 110−	$+4.3\sqrt{2}/\sqrt{5.11.17.19}$
1̃ 1̃ 1 0 000+	$+\sqrt{2.7.11}/\sqrt{13.17.19}$
1̃ 1̃ 1 0 010+	$-3\sqrt{7}/\sqrt{2.5.13.17.19}$
1̃ 1̃ 1 0 100+	0
1̃ 1̃ 1 0 110+	$-2.3\sqrt{2.7}/\sqrt{5.11.17.19}$
1̃ 1̃ 2 0 000−	$+5\sqrt{11}/\sqrt{2.7.13.17.19}$
1̃ 1̃ 2 0 010−	$-9.3\sqrt{2}/\sqrt{5.7.13.17.19}$
1̃ 1̃ 2 0 100−	$-5\sqrt{2}/\sqrt{7.17.19}$
1̃ 1̃ 2 0 110−	$-3.13/\sqrt{2.5.7.11.17.19}$
1̃ 1̃ 1̃ 0 000−	0
1̃ 1̃ 1̃ 0 010−	$-\sqrt{3.11}/\sqrt{2.13.17.19}$
1̃ 1̃ 1̃ 0 011−	0
1̃ 1̃ 1̃ 0 100−	0

SO_3–O 3jm Factors (cont.)

$\mathbf{7\ 6\ 6}$

$\tilde{1}\ \tilde{1}\ \tilde{1}\ 0\ 110-\quad +\sqrt{2.3}/\sqrt{17.19}$

$\tilde{1}\ \tilde{1}\ \tilde{1}\ 0\ 111-\quad 0$

$\tilde{1}\ \tilde{0}\ 1\ 0\ 000-\quad -2\sqrt{2.11}/\sqrt{13.17.19}$

$\tilde{1}\ \tilde{0}\ 1\ 0\ 100-\quad +1/\sqrt{2.17.19}$

$\tilde{0}\ 2\ 2\ 0\ 000+\quad +7\sqrt{7}/\sqrt{5.11.17.19}$

$\tilde{0}\ \tilde{1}\ 1\ 0\ 000-\quad +1/\sqrt{2.3.17.19}$

$\tilde{0}\ \tilde{1}\ 1\ 0\ 010-\quad +8\sqrt{2.3}/\sqrt{5.11.17.19}$

$\tilde{0}\ \tilde{0}\ 0\ 0\ 000-\quad -1/\sqrt{2.17.19}$

$\mathbf{7\ \frac{13}{2}\ \frac{1}{2}}$

$1\ \frac{1}{2}\ \frac{1}{2}\ 0\ 000+\quad +1/\sqrt{7}$

$1\ \frac{1}{2}\ \frac{1}{2}\ 0\ 100+\quad 0$

$1\ \frac{3}{2}\ \frac{1}{2}\ 0\ 000-\quad +\sqrt{2.11}/5.7$

$1\ \frac{3}{2}\ \frac{1}{2}\ 0\ 010-\quad -4\sqrt{3}/5.7$

$1\ \frac{3}{2}\ \frac{1}{2}\ 0\ 100-\quad +2\sqrt{2.3}/5\sqrt{7}$

$1\ \frac{3}{2}\ \frac{1}{2}\ 0\ 110-\quad +\sqrt{11}/5\sqrt{7}$

$2\ \frac{3}{2}\ \frac{1}{2}\ 0\ 000+\quad +3\sqrt{2}/7\sqrt{5}$

$2\ \frac{3}{2}\ \frac{1}{2}\ 0\ 010+\quad -2\sqrt{11}/7\sqrt{3.5}$

$\tilde{1}\ \frac{3}{2}\ \frac{1}{2}\ 0\ 000+\quad +\sqrt{2}/\sqrt{5.7}$

$\tilde{1}\ \frac{3}{2}\ \frac{1}{2}\ 0\ 010+\quad 0$

$\tilde{1}\ \frac{3}{2}\ \frac{1}{2}\ 0\ 100+\quad 0$

$\tilde{1}\ \frac{3}{2}\ \frac{1}{2}\ 0\ 110+\quad +\sqrt{13}/\sqrt{3.5.7}$

$\tilde{1}\ \tilde{\frac{1}{2}}\ \frac{1}{2}\ 0\ 000-\quad +1/\sqrt{7}$

$\tilde{1}\ \tilde{\frac{1}{2}}\ \frac{1}{2}\ 0\ 010-\quad 0$

$\tilde{1}\ \tilde{\frac{1}{2}}\ \frac{1}{2}\ 0\ 100-\quad 0$

$\tilde{1}\ \tilde{\frac{1}{2}}\ \frac{1}{2}\ 0\ 110-\quad +2\sqrt{2}/\sqrt{3.5.7}$

$\tilde{0}\ \tilde{\frac{1}{2}}\ \frac{1}{2}\ 0\ 000+\quad 0$

$\tilde{0}\ \tilde{\frac{1}{2}}\ \frac{1}{2}\ 0\ 010+\quad +1/\sqrt{3.5}$

$\mathbf{7\ \frac{13}{2}\ \frac{3}{2}}$

$1\ \frac{1}{2}\ \frac{3}{2}\ 0\ 000+\quad +2/\sqrt{7.13}$

$1\ \frac{1}{2}\ \frac{3}{2}\ 0\ 100+\quad 0$

$1\ \frac{3}{2}\ \frac{3}{2}\ 0\ 000-\quad -\sqrt{11}/5.7\sqrt{13}$

$1\ \frac{3}{2}\ \frac{3}{2}\ 0\ 010-\quad -\sqrt{2.3}/7\sqrt{13}$

$1\ \frac{3}{2}\ \frac{3}{2}\ 0\ 100-\quad -2\sqrt{3}/5\sqrt{7.13}$

$1\ \frac{3}{2}\ \frac{3}{2}\ 0\ 110-\quad +\sqrt{11}/2\sqrt{2.7.13}$

$1\ \frac{3}{2}\ \frac{3}{2}\ 1\ 000-\quad +17\sqrt{11}/2.5.7\sqrt{13}$

$1\ \frac{3}{2}\ \frac{3}{2}\ 1\ 010-\quad -9.3\sqrt{3}/5.7\sqrt{2.13}$

$1\ \frac{3}{2}\ \frac{3}{2}\ 1\ 100-\quad -11\sqrt{3}/2.5\sqrt{7.13}$

$1\ \frac{3}{2}\ \frac{3}{2}\ 1\ 110-\quad -9\sqrt{11}/2.5\sqrt{2.7.13}$

$1\ \tilde{\frac{1}{2}}\ \frac{3}{2}\ 0\ 000+\quad -3\sqrt{11}/2.7\sqrt{5.13}$

$1\ \tilde{\frac{1}{2}}\ \frac{3}{2}\ 0\ 010+\quad +2\sqrt{2.3}/5.7$

$1\ \tilde{\frac{1}{2}}\ \frac{3}{2}\ 0\ 100+\quad -3\sqrt{3}/\sqrt{5.7.13}$

$1\ \tilde{\frac{1}{2}}\ \frac{3}{2}\ 0\ 110+\quad -\sqrt{11}/5\sqrt{2.7}$

$2\ \frac{1}{2}\ \frac{3}{2}\ 0\ 000-\quad -3\sqrt{11}/2\sqrt{5.7.13}$

$2\ \frac{3}{2}\ \frac{3}{2}\ 0\ 000+\quad +3/7\sqrt{5.13}$

$2\ \frac{3}{2}\ \frac{3}{2}\ 0\ 010+\quad +\sqrt{5.11}/7\sqrt{2.3.13}$

$2\ \tilde{\frac{1}{2}}\ \frac{3}{2}\ 0\ 000-\quad -9/2.7\sqrt{13}$

$2\ \tilde{\frac{1}{2}}\ \frac{3}{2}\ 0\ 010-\quad +\sqrt{2.11}/7\sqrt{3.5}$

$\tilde{1}\ \frac{1}{2}\ \frac{3}{2}\ 0\ 000-\quad -3\sqrt{11}/2.7\sqrt{5.13}$

$\tilde{1}\ \frac{1}{2}\ \frac{3}{2}\ 0\ 100-\quad +3/7\sqrt{5}$

$\mathbf{7\ \frac{13}{2}\ \frac{3}{2}}$

$\tilde{1}\ \frac{3}{2}\ \frac{3}{2}\ 0\ 000+\quad +29/7\sqrt{5.7.13}$

$\tilde{1}\ \frac{3}{2}\ \frac{3}{2}\ 0\ 010+\quad +3\sqrt{2.3.11}/7\sqrt{5.7.13}$

$\tilde{1}\ \frac{3}{2}\ \frac{3}{2}\ 0\ 100+\quad +3\sqrt{11}/7\sqrt{5.7}$

$\tilde{1}\ \frac{3}{2}\ \frac{3}{2}\ 0\ 110+\quad +19/2.7\sqrt{2.3.5.7}$

$\tilde{1}\ \frac{3}{2}\ \frac{3}{2}\ 1\ 000-\quad -83/2.7\sqrt{5.7.13}$

$\tilde{1}\ \frac{3}{2}\ \frac{3}{2}\ 1\ 010-\quad +3\sqrt{3.11}/7\sqrt{2.5.7.13}$

$\tilde{1}\ \frac{3}{2}\ \frac{3}{2}\ 1\ 100-\quad +3\sqrt{11}/2.7\sqrt{5.7}$

$\tilde{1}\ \frac{3}{2}\ \frac{3}{2}\ 1\ 110-\quad -29/2.7\sqrt{2.3.5.7}$

$\tilde{1}\ \tilde{\frac{1}{2}}\ \frac{3}{2}\ 0\ 000-\quad +2/\sqrt{7.13}$

$\tilde{1}\ \tilde{\frac{1}{2}}\ \frac{3}{2}\ 0\ 010-\quad 0$

$\tilde{1}\ \tilde{\frac{1}{2}}\ \frac{3}{2}\ 0\ 100-\quad 0$

$\tilde{1}\ \tilde{\frac{1}{2}}\ \frac{3}{2}\ 0\ 110-\quad -\sqrt{13}/\sqrt{2.3.5.7}$

$\tilde{0}\ \frac{3}{2}\ \frac{3}{2}\ 0\ 000-\quad 0$

$\tilde{0}\ \frac{3}{2}\ \frac{3}{2}\ 0\ 010-\quad +1/\sqrt{3.5}$

$\mathbf{7\ \frac{13}{2}\ \frac{5}{2}}$

$1\ \frac{1}{2}\ \frac{3}{2}\ 0\ 000-\quad +3\sqrt{3}/2\sqrt{7.13.17}$

$1\ \frac{1}{2}\ \frac{3}{2}\ 0\ 100-\quad +\sqrt{11}/2\sqrt{13.17}$

$1\ \frac{3}{2}\ \frac{3}{2}\ 0\ 000+\quad +\sqrt{11.17}/2.5.7\sqrt{3.13}$

$1\ \frac{3}{2}\ \frac{3}{2}\ 0\ 010+\quad +\sqrt{17}/7\sqrt{2.13}$

$1\ \frac{3}{2}\ \frac{3}{2}\ 0\ 100+\quad -8/5\sqrt{7.13.17}$

$1\ \frac{3}{2}\ \frac{3}{2}\ 0\ 110+\quad +\sqrt{2.11}/\sqrt{3.7.13.17}$

$1\ \frac{3}{2}\ \frac{3}{2}\ 1\ 000+\quad +2\sqrt{11.17}/5.7\sqrt{3.13}$

$1\ \frac{3}{2}\ \frac{3}{2}\ 1\ 010+\quad -11\sqrt{13}/5.7\sqrt{2.17}$

$1\ \frac{3}{2}\ \frac{3}{2}\ 1\ 100+\quad +9.9/2.5\sqrt{7.13.17}$

$1\ \frac{3}{2}\ \frac{3}{2}\ 1\ 110+\quad +11\sqrt{2.11}/5\sqrt{3.7.13.17}$

$1\ \frac{3}{2}\ \tilde{\frac{1}{2}}\ 0\ 000-\quad -\sqrt{2.11.17}/7\sqrt{3.5.13}$

$1\ \frac{3}{2}\ \tilde{\frac{1}{2}}\ 0\ 010-\quad +19/7\sqrt{5.13.17}$

$1\ \frac{3}{2}\ \tilde{\frac{1}{2}}\ 0\ 100-\quad +\sqrt{7}/\sqrt{2.5.13.17}$

$1\ \frac{3}{2}\ \tilde{\frac{1}{2}}\ 0\ 110-\quad -\sqrt{7.11}/2\sqrt{3.5.13.17}$

$1\ \tilde{\frac{1}{2}}\ \frac{3}{2}\ 0\ 000-\quad -\sqrt{11.17}/2.7\sqrt{3.5.13}$

$1\ \tilde{\frac{1}{2}}\ \frac{3}{2}\ 0\ 010-\quad +23\sqrt{2}/5.7\sqrt{17}$

$1\ \tilde{\frac{1}{2}}\ \frac{3}{2}\ 0\ 100-\quad +41/2\sqrt{5.7.13.17}$

$1\ \tilde{\frac{1}{2}}\ \frac{3}{2}\ 0\ 110-\quad +\sqrt{11}/5\sqrt{2.3.7.17}$

$1\ \tilde{\frac{1}{2}}\ \tilde{\frac{1}{2}}\ 0\ 000+\quad +\sqrt{11.17}/2.7\sqrt{3.13}$

$1\ \tilde{\frac{1}{2}}\ \tilde{\frac{1}{2}}\ 0\ 010+\quad -2\sqrt{2}/7\sqrt{5.17}$

$1\ \tilde{\frac{1}{2}}\ \tilde{\frac{1}{2}}\ 0\ 100+\quad +2/\sqrt{7.13.17}$

$1\ \tilde{\frac{1}{2}}\ \tilde{\frac{1}{2}}\ 0\ 110+\quad -2\sqrt{2.11}/\sqrt{3.5.7.17}$

$2\ \frac{1}{2}\ \frac{3}{2}\ 0\ 000+\quad -\sqrt{11.17}/2\sqrt{3.5.7.13}$

$2\ \frac{3}{2}\ \frac{3}{2}\ 0\ 000-\quad +37/7\sqrt{3.5.13.17}$

$2\ \frac{3}{2}\ \frac{3}{2}\ 0\ 010-\quad +\sqrt{5.11}/7\sqrt{2.13.17}$

$2\ \frac{3}{2}\ \tilde{\frac{1}{2}}\ 0\ 000+\quad +\sqrt{2.13}/7\sqrt{3.17}$

$2\ \frac{3}{2}\ \tilde{\frac{1}{2}}\ 0\ 010+\quad -2.3\sqrt{11}/7\sqrt{13.17}$

$2\ \tilde{\frac{1}{2}}\ \frac{3}{2}\ 0\ 000+\quad -\sqrt{3.17}/2.7\sqrt{13}$

$2\ \tilde{\frac{1}{2}}\ \frac{3}{2}\ 0\ 010+\quad -2\sqrt{2.11}/7\sqrt{5.17}$

$\tilde{1}\ \frac{1}{2}\ \frac{3}{2}\ 0\ 000+\quad -\sqrt{11.17}/2.7\sqrt{3.5.13}$

$\tilde{1}\ \frac{1}{2}\ \frac{3}{2}\ 0\ 100+\quad -1/2.7\sqrt{3.5.17}$

$\tilde{1}\ \frac{1}{2}\ \tilde{\frac{1}{2}}\ 0\ 000-\quad -\sqrt{11.17}/2.7\sqrt{3.13}$

$\tilde{1}\ \frac{1}{2}\ \tilde{\frac{1}{2}}\ 0\ 100-\quad +2.5/7\sqrt{3.17}$

$\tilde{1}\ \frac{3}{2}\ \frac{3}{2}\ 0\ 000-\quad -29\sqrt{17}/2.7\sqrt{3.5.7.13}$

SO_3–O 3jm Factors (cont.)

$7\ \frac{13}{2}\ \frac{5}{2}$	
$\tilde{1}\ \frac{3}{2}\ \frac{3}{2}\ 0\ 010-$	$-3\sqrt{11.17}/7\sqrt{2.5.7.13}$
$\tilde{1}\ \frac{3}{2}\ \frac{3}{2}\ 0\ 100-$	$-8\sqrt{11}/7\sqrt{3.5.7.17}$
$\tilde{1}\ \frac{3}{2}\ \frac{3}{2}\ 0\ 110-$	$+23\sqrt{2}/7\sqrt{5.7.17}$
$\tilde{1}\ \frac{3}{2}\ \frac{3}{2}\ 1\ 000+$	$-8.2/7\sqrt{3.5.7.13.17}$
$\tilde{1}\ \frac{3}{2}\ \frac{3}{2}\ 1\ 010+$	$+37\sqrt{11}/7\sqrt{2.5.7.13.17}$
$\tilde{1}\ \frac{3}{2}\ \frac{3}{2}\ 1\ 100+$	$+37\sqrt{11}/2.7\sqrt{3.5.7.17}$
$\tilde{1}\ \frac{3}{2}\ \frac{3}{2}\ 1\ 110+$	$+4.3\sqrt{2}/7\sqrt{5.7.17}$
$\tilde{1}\ \frac{3}{2}\ \tilde{\frac{1}{2}}\ 0\ 000+$	$-41\sqrt{2}/7\sqrt{3.7.13.17}$
$\tilde{1}\ \frac{3}{2}\ \tilde{\frac{1}{2}}\ 0\ 010+$	$+\sqrt{11.13}/7\sqrt{7.17}$
$\tilde{1}\ \frac{3}{2}\ \tilde{\frac{1}{2}}\ 0\ 100+$	$-\sqrt{11}/7\sqrt{2.3.7.17}$
$\tilde{1}\ \frac{3}{2}\ \tilde{\frac{1}{2}}\ 0\ 110+$	$+5.5/2.7\sqrt{7.17}$
$\tilde{1}\ \tilde{\frac{1}{2}}\ \frac{3}{2}\ 0\ 000+$	$-11\sqrt{3}/2.7\sqrt{7.13.17}$
$\tilde{1}\ \tilde{\frac{1}{2}}\ \frac{3}{2}\ 0\ 010+$	$-\sqrt{2.5.11}/7\sqrt{7.17}$
$\tilde{1}\ \tilde{\frac{1}{2}}\ \frac{3}{2}\ 0\ 100+$	$-5\sqrt{3.11}/2.7\sqrt{7.17}$
$\tilde{1}\ \tilde{\frac{1}{2}}\ \frac{3}{2}\ 0\ 110+$	$-\sqrt{13}/7\sqrt{2.5.7.17}$
$\tilde{0}\ \frac{1}{2}\ \tilde{\frac{1}{2}}\ 0\ 000+$	$-2\sqrt{2}/\sqrt{3.7.17}$
$\tilde{0}\ \frac{3}{2}\ \frac{3}{2}\ 0\ 000+$	$+\sqrt{2.5.11}/7\sqrt{3.17}$
$\tilde{0}\ \frac{3}{2}\ \frac{3}{2}\ 0\ 010+$	$+1/7\sqrt{5.17}$

$7\ \frac{13}{2}\ \frac{7}{2}$	
$1\ \frac{1}{2}\ \frac{1}{2}\ 0\ 000-$	$-9/\sqrt{2.11.13.17}$
$1\ \frac{1}{2}\ \frac{1}{2}\ 0\ 100-$	$-\sqrt{7}/2\sqrt{2.3.13.17}$
$1\ \frac{1}{2}\ \frac{3}{2}\ 0\ 000+$	$+9\sqrt{5}/2\sqrt{2.7.11.13.17}$
$1\ \frac{1}{2}\ \frac{3}{2}\ 0\ 100+$	$-\sqrt{2.5}/\sqrt{3.13.17}$
$1\ \frac{3}{2}\ \frac{1}{2}\ 0\ 000+$	$-89/4.5\sqrt{7.13.17}$
$1\ \frac{3}{2}\ \frac{1}{2}\ 0\ 010+$	$+4.29\sqrt{2}/5\sqrt{3.7.11.13.17}$
$1\ \frac{3}{2}\ \frac{1}{2}\ 0\ 100+$	$+31\sqrt{11}/4.5\sqrt{3.13.17}$
$1\ \frac{3}{2}\ \frac{1}{2}\ 0\ 110+$	$+59/2.3.5\sqrt{2.13.17}$
$1\ \frac{3}{2}\ \frac{3}{2}\ 0\ 000-$	$-3\sqrt{2.5}/7\sqrt{13.17}$
$1\ \frac{3}{2}\ \frac{3}{2}\ 0\ 010-$	$-41\sqrt{5}/7\sqrt{3.11.13.17}$
$1\ \frac{3}{2}\ \frac{3}{2}\ 0\ 100-$	$-\sqrt{5.11}/2\sqrt{2.3.7.13.17}$
$1\ \frac{3}{2}\ \frac{3}{2}\ 0\ 110-$	$+\sqrt{5.13}/2.3\sqrt{7.17}$
$1\ \frac{3}{2}\ \frac{3}{2}\ 1\ 000-$	$+89/2.7\sqrt{2.5.13.17}$
$1\ \frac{3}{2}\ \frac{3}{2}\ 1\ 010-$	$+131/2.7\sqrt{3.5.11.13.17}$
$1\ \frac{3}{2}\ \frac{3}{2}\ 1\ 100-$	$+2\sqrt{2.3.11}/\sqrt{5.7.13.17}$
$1\ \frac{3}{2}\ \frac{3}{2}\ 1\ 110-$	$-\sqrt{13}/2\sqrt{5.7.17}$
$1\ \frac{3}{2}\ \tilde{\frac{1}{2}}\ 0\ 000+$	$-\sqrt{3.17}/4.7\sqrt{5.13}$
$1\ \frac{3}{2}\ \tilde{\frac{1}{2}}\ 0\ 010+$	$+3\sqrt{17}/7\sqrt{2.5.11.13}$
$1\ \frac{3}{2}\ \tilde{\frac{1}{2}}\ 0\ 100+$	$-\sqrt{7.11}/4\sqrt{5.13.17}$
$1\ \frac{3}{2}\ \tilde{\frac{1}{2}}\ 0\ 110+$	$+11\sqrt{7}/2\sqrt{2.3.5.13.17}$
$1\ \tilde{\frac{1}{2}}\ \frac{3}{2}\ 0\ 000+$	$-\sqrt{17}/2.7\sqrt{2.13}$
$1\ \tilde{\frac{1}{2}}\ \frac{3}{2}\ 0\ 010+$	$-2.13\sqrt{3}/7\sqrt{5.11.17}$
$1\ \tilde{\frac{1}{2}}\ \frac{3}{2}\ 0\ 100+$	$+2\sqrt{2.11}/\sqrt{3.7.13.17}$
$1\ \tilde{\frac{1}{2}}\ \frac{3}{2}\ 0\ 110+$	$-11/2.3\sqrt{5.7.17}$
$1\ \tilde{\frac{1}{2}}\ \tilde{\frac{1}{2}}\ 0\ 000-$	$+\sqrt{3.17}/7\sqrt{2.13}$
$1\ \tilde{\frac{1}{2}}\ \tilde{\frac{1}{2}}\ 0\ 010-$	$-4.3/7\sqrt{5.11.17}$
$1\ \tilde{\frac{1}{2}}\ \tilde{\frac{1}{2}}\ 0\ 100-$	$-\sqrt{11}/2\sqrt{2.7.13.17}$
$1\ \tilde{\frac{1}{2}}\ \tilde{\frac{1}{2}}\ 0\ 110-$	$+11/2\sqrt{3.5.7.17}$
$2\ \frac{1}{2}\ \frac{3}{2}\ 0\ 000-$	$-5/\sqrt{2.7.13.17}$
$2\ \frac{3}{2}\ \frac{1}{2}\ 0\ 000-$	$-2\sqrt{7.11}/3\sqrt{5.13.17}$
$2\ \frac{3}{2}\ \frac{1}{2}\ 0\ 010-$	$-\sqrt{2.7}/3\sqrt{3.5.13.17}$
$2\ \frac{3}{2}\ \frac{3}{2}\ 0\ 000+$	$-\sqrt{2.11}/3.7\sqrt{13.17}$
$2\ \frac{3}{2}\ \frac{3}{2}\ 0\ 010+$	$-11.11/3.7\sqrt{3.13.17}$
$2\ \frac{3}{2}\ \tilde{\frac{1}{2}}\ 0\ 000-$	$+3\sqrt{3.11}/7\sqrt{13.17}$
$2\ \frac{3}{2}\ \tilde{\frac{1}{2}}\ 0\ 010-$	$-4\sqrt{2}/7\sqrt{13.17}$
$2\ \tilde{\frac{1}{2}}\ \frac{3}{2}\ 0\ 000-$	$+\sqrt{5.11}/3.7\sqrt{2.13.17}$
$2\ \tilde{\frac{1}{2}}\ \frac{3}{2}\ 0\ 010-$	$+\sqrt{17}/3.7\sqrt{3}$
$\tilde{1}\ \frac{1}{2}\ \frac{3}{2}\ 0\ 000-$	$-\sqrt{17}/2.7\sqrt{2.13}$
$\tilde{1}\ \frac{1}{2}\ \frac{3}{2}\ 0\ 100-$	$+4.3\sqrt{2}/7\sqrt{11.17}$
$\tilde{1}\ \frac{1}{2}\ \tilde{\frac{1}{2}}\ 0\ 000+$	$-\sqrt{3.17}/7\sqrt{2.13}$
$\tilde{1}\ \frac{1}{2}\ \tilde{\frac{1}{2}}\ 0\ 100+$	$-9\sqrt{3}/2.7\sqrt{2.11.17}$
$\tilde{1}\ \frac{3}{2}\ \frac{1}{2}\ 0\ 000-$	$+9.3\sqrt{11}/4.7\sqrt{5.13.17}$
$\tilde{1}\ \frac{3}{2}\ \frac{1}{2}\ 0\ 010-$	$+41\sqrt{5}/3.7\sqrt{2.3.13.17}$
$\tilde{1}\ \frac{3}{2}\ \frac{1}{2}\ 0\ 100-$	$-\sqrt{5.17}/4.3.7$
$\tilde{1}\ \frac{3}{2}\ \frac{1}{2}\ 0\ 110-$	$-19/2.7\sqrt{2.3.5.11.17}$
$\tilde{1}\ \frac{3}{2}\ \frac{3}{2}\ 0\ 000+$	$-5\sqrt{2.11}/7\sqrt{7.13.17}$
$\tilde{1}\ \frac{3}{2}\ \frac{3}{2}\ 0\ 010+$	$-11.11/3.7\sqrt{3.7.13.17}$
$\tilde{1}\ \frac{3}{2}\ \frac{3}{2}\ 0\ 100+$	$-41/2.3.7\sqrt{2.7.17}$
$\tilde{1}\ \frac{3}{2}\ \frac{3}{2}\ 0\ 110+$	$-11\sqrt{11}/2.7\sqrt{3.7.17}$
$\tilde{1}\ \frac{3}{2}\ \frac{3}{2}\ 1\ 000-$	$-\sqrt{11}/2.7\sqrt{2.7.13.17}$
$\tilde{1}\ \frac{3}{2}\ \frac{3}{2}\ 1\ 010-$	$+5.131/2.3.7\sqrt{3.7.13.17}$
$\tilde{1}\ \frac{3}{2}\ \frac{3}{2}\ 1\ 100-$	$-23\sqrt{2}/3.7\sqrt{7.17}$
$\tilde{1}\ \frac{3}{2}\ \frac{3}{2}\ 1\ 110-$	$-61/2.7\sqrt{3.7.11.17}$
$\tilde{1}\ \frac{3}{2}\ \tilde{\frac{1}{2}}\ 0\ 000-$	$-5\sqrt{3.11}/4.7\sqrt{7.13.17}$
$\tilde{1}\ \frac{3}{2}\ \tilde{\frac{1}{2}}\ 0\ 010-$	$+4\sqrt{2}/7\sqrt{7.13.17}$
$\tilde{1}\ \frac{3}{2}\ \tilde{\frac{1}{2}}\ 0\ 100-$	$+31\sqrt{3}/4.7\sqrt{7.17}$
$\tilde{1}\ \frac{3}{2}\ \tilde{\frac{1}{2}}\ 0\ 110-$	$-109/2.7\sqrt{2.7.11.17}$
$\tilde{1}\ \tilde{\frac{1}{2}}\ \frac{1}{2}\ 0\ 000+$	$+3\sqrt{11}/7\sqrt{2.13.17}$
$\tilde{1}\ \tilde{\frac{1}{2}}\ \frac{1}{2}\ 0\ 010+$	$+4\sqrt{5}/3.7\sqrt{3.17}$
$\tilde{1}\ \tilde{\frac{1}{2}}\ \frac{1}{2}\ 0\ 100+$	$+5.11/2.3.7\sqrt{2.17}$
$\tilde{1}\ \tilde{\frac{1}{2}}\ \frac{1}{2}\ 0\ 110+$	$-\sqrt{13}/2.7\sqrt{3.5.11.17}$
$\tilde{1}\ \tilde{\frac{1}{2}}\ \frac{3}{2}\ 0\ 000-$	$-3\sqrt{5.11}/2.7\sqrt{2.7.13.17}$
$\tilde{1}\ \tilde{\frac{1}{2}}\ \frac{3}{2}\ 0\ 010-$	$-2.47/3.7\sqrt{3.7.17}$
$\tilde{1}\ \tilde{\frac{1}{2}}\ \frac{3}{2}\ 0\ 100-$	$+\sqrt{2.5}/3.7\sqrt{7.17}$
$\tilde{1}\ \tilde{\frac{1}{2}}\ \frac{3}{2}\ 0\ 110-$	$-\sqrt{3.13}/2.7\sqrt{7.11.17}$
$\tilde{0}\ \frac{1}{2}\ \tilde{\frac{1}{2}}\ 0\ 000-$	$+3\sqrt{3}/\sqrt{7.11.17}$
$\tilde{0}\ \frac{3}{2}\ \frac{3}{2}\ 0\ 000-$	$-13/3.7\sqrt{17}$
$\tilde{0}\ \frac{3}{2}\ \frac{3}{2}\ 0\ 010-$	$+\sqrt{2}/3.7\sqrt{3.11.17}$
$\tilde{0}\ \tilde{\frac{1}{2}}\ \frac{1}{2}\ 0\ 000-$	$-5/3\sqrt{7.17}$
$\tilde{0}\ \tilde{\frac{1}{2}}\ \frac{1}{2}\ 0\ 010-$	$-\sqrt{2.13}/3\sqrt{3.5.7.11.17}$

$7\ \frac{13}{2}\ \frac{9}{2}$	
$1\ \frac{1}{2}\ \frac{1}{2}\ 0\ 000+$	$-9\sqrt{19}/2\sqrt{5.11.13.17}$
$1\ \frac{1}{2}\ \frac{1}{2}\ 0\ 100+$	$+2\sqrt{7}/\sqrt{3.5.13.17.19}$
$1\ \frac{1}{2}\ \frac{3}{2}\ 0\ 000-$	$-3\sqrt{3.17}/2.5\sqrt{11.13.19}$
$1\ \frac{1}{2}\ \frac{3}{2}\ 0\ 001-$	$-9\sqrt{11}/5\sqrt{7.13.17.19}$
$1\ \frac{1}{2}\ \frac{3}{2}\ 0\ 100-$	$-3.7\sqrt{7}/2.5\sqrt{13.17.19}$
$1\ \frac{1}{2}\ \frac{3}{2}\ 0\ 101-$	$+7.31/2.5\sqrt{3.13.17.19}$
$1\ \frac{3}{2}\ \frac{1}{2}\ 0\ 000-$	$-2\sqrt{2}/\sqrt{5.7.13.17.19}$
$1\ \frac{3}{2}\ \frac{1}{2}\ 0\ 010-$	$-\sqrt{11}/\sqrt{3.5.7.13.17.19}$

SO_3–O 3jm Factors (cont.)

7 $\frac{13}{2}$ $\frac{9}{2}$

1 $\frac{3}{2}$ $\frac{1}{2}$ 0 100− $+11\sqrt{11}/\sqrt{2.3.5.13.17.19}$

1 $\frac{3}{2}$ $\frac{1}{2}$ 0 110− $+113/2.3\sqrt{5.13.17.19}$

1 $\frac{3}{2}$ $\frac{3}{2}$ 0 000+ $+23\sqrt{3}/2\sqrt{7.13.17.19}$

1 $\frac{3}{2}$ $\frac{3}{2}$ 0 001+ $+3\sqrt{19}/7\sqrt{13.17}$

1 $\frac{3}{2}$ $\frac{3}{2}$ 0 010+ $+3.31/\sqrt{2.7.11.13.17.19}$

1 $\frac{3}{2}$ $\frac{3}{2}$ 0 011+ $+41\sqrt{19}/7\sqrt{2.3.11.13.17}$

1 $\frac{3}{2}$ $\frac{3}{2}$ 0 100+ $-4\sqrt{11}/\sqrt{13.17.19}$

1 $\frac{3}{2}$ $\frac{3}{2}$ 0 101+ $-2\sqrt{11}/\sqrt{3.7.13.17.19}$

1 $\frac{3}{2}$ $\frac{3}{2}$ 0 110+ $+\sqrt{2.17}/\sqrt{3.13.19}$

1 $\frac{3}{2}$ $\frac{3}{2}$ 0 111+ $+2\sqrt{2.13}/3\sqrt{7.17.19}$

1 $\frac{3}{2}$ $\frac{3}{2}$ 1 000+ $-8.11\sqrt{3}/5\sqrt{7.13.17.19}$

1 $\frac{3}{2}$ $\frac{3}{2}$ 1 001+ $-2.37/5.7\sqrt{13.17.19}$

1 $\frac{3}{2}$ $\frac{3}{2}$ 1 010+ $+3.37/5\sqrt{2.7.11.13.17.19}$

1 $\frac{3}{2}$ $\frac{3}{2}$ 1 011+ $+2.257\sqrt{2}/5.7\sqrt{3.11.13.17.19}$

1 $\frac{3}{2}$ $\frac{3}{2}$ 1 100+ $-\sqrt{11}/2.5\sqrt{13.17.19}$

1 $\frac{3}{2}$ $\frac{3}{2}$ 1 101+ $-8\sqrt{3.11}/5\sqrt{7.13.17.19}$

1 $\frac{3}{2}$ $\frac{3}{2}$ 1 110+ $-8.4\sqrt{2}/5\sqrt{3.13.17.19}$

1 $\frac{3}{2}$ $\frac{3}{2}$ 1 111+ $+\sqrt{2.13}/5\sqrt{7.17.19}$

1 $\tilde{\frac{1}{2}}$ $\frac{3}{2}$ 0 000− $+53\sqrt{3.7}/2.5\sqrt{5.13.17.19}$

1 $\tilde{\frac{1}{2}}$ $\frac{3}{2}$ 0 001− $-71/5.7\sqrt{5.13.17.19}$

1 $\tilde{\frac{1}{2}}$ $\frac{3}{2}$ 0 010− $+9\sqrt{2.7}/5\sqrt{11.17.19}$

1 $\tilde{\frac{1}{2}}$ $\frac{3}{2}$ 0 011− $-\sqrt{2.3.17}/5.7\sqrt{11.19}$

1 $\tilde{\frac{1}{2}}$ $\frac{3}{2}$ 0 100− $+\sqrt{11}/2.5\sqrt{5.13.17.19}$

1 $\tilde{\frac{1}{2}}$ $\frac{3}{2}$ 0 101− $-227\sqrt{11}/2.5\sqrt{3.5.7.13.17.19}$

1 $\tilde{\frac{1}{2}}$ $\frac{3}{2}$ 0 110− $-\sqrt{19}/5\sqrt{2.3.17}$

1 $\tilde{\frac{1}{2}}$ $\frac{3}{2}$ 0 111− $-31\sqrt{2}/3.5\sqrt{7.17.19}$

2 $\frac{1}{2}$ $\frac{3}{2}$ 0 000+ $+3\sqrt{3}/2\sqrt{5.13.17.19}$

2 $\frac{1}{2}$ $\frac{3}{2}$ 0 001+ $-2\sqrt{19}/\sqrt{5.7.13.17}$

2 $\frac{3}{2}$ $\frac{1}{2}$ 0 000+ $+4\sqrt{2.11}/3\sqrt{7.13.17.19}$

2 $\frac{3}{2}$ $\frac{1}{2}$ 0 010+ $+4.37/3\sqrt{3.7.13.17.19}$

2 $\frac{3}{2}$ $\frac{3}{2}$ 0 000− $+3\sqrt{3.11}/\sqrt{5.7.13.17.19}$

2 $\frac{3}{2}$ $\frac{3}{2}$ 0 001− $-4.47\sqrt{11}/3.7\sqrt{5.13.17.19}$

2 $\frac{3}{2}$ $\frac{3}{2}$ 0 010− $+9.3/\sqrt{2.5.7.13.17.19}$

2 $\frac{3}{2}$ $\frac{3}{2}$ 0 011− $-2.11\sqrt{2.13}/3.7\sqrt{3.5.17.19}$

2 $\tilde{\frac{1}{2}}$ $\frac{3}{2}$ 0 000+ $+3\sqrt{3.11.13}/2.5\sqrt{7.17.19}$

2 $\tilde{\frac{1}{2}}$ $\frac{3}{2}$ 0 001+ $+2\sqrt{11.19}/3.5.7\sqrt{13.17}$

2 $\tilde{\frac{1}{2}}$ $\frac{3}{2}$ 0 010+ $-2\sqrt{2.5}/\sqrt{7.17.19}$

2 $\tilde{\frac{1}{2}}$ $\frac{3}{2}$ 0 011+ $-\sqrt{2.5.17}/3.7\sqrt{3.19}$

$\tilde{1}$ $\frac{1}{2}$ $\frac{3}{2}$ 0 000+ $+11\sqrt{3.5}/2\sqrt{7.13.17.19}$

$\tilde{1}$ $\frac{1}{2}$ $\frac{3}{2}$ 0 001+ $-23\sqrt{5}/7\sqrt{13.17.19}$

$\tilde{1}$ $\frac{1}{2}$ $\frac{3}{2}$ 0 100+ $+3\sqrt{3.17}/2\sqrt{5.7.11.19}$

$\tilde{1}$ $\frac{1}{2}$ $\frac{3}{2}$ 0 101+ $-3.59/2.7\sqrt{5.11.17.19}$

$\tilde{1}$ $\frac{3}{2}$ $\frac{1}{2}$ 0 000+ 0

$\tilde{1}$ $\frac{3}{2}$ $\frac{1}{2}$ 0 010+ $+\sqrt{19}/3\sqrt{3.13.17}$

$\tilde{1}$ $\frac{3}{2}$ $\frac{1}{2}$ 0 100+ $+\sqrt{19}/3\sqrt{2.17}$

$\tilde{1}$ $\frac{3}{2}$ $\frac{1}{2}$ 0 110+ $-1/2\sqrt{3.11.17.19}$

$\tilde{1}$ $\frac{3}{2}$ $\frac{3}{2}$ 0 000− $-\sqrt{3.5.11}/2.7\sqrt{13.17.19}$

$\tilde{1}$ $\frac{3}{2}$ $\frac{3}{2}$ 0 001− $+\sqrt{5.11.19}/7\sqrt{7.13.17}$

$\tilde{1}$ $\frac{3}{2}$ $\frac{3}{2}$ 0 010− $-3.53/7\sqrt{2.5.13.17.19}$

7 $\frac{13}{2}$ $\frac{9}{2}$

$\tilde{1}$ $\frac{3}{2}$ $\frac{3}{2}$ 0 011− $+11.11\sqrt{19}/3.7\sqrt{2.3.5.7.13.17}$

$\tilde{1}$ $\frac{3}{2}$ $\frac{3}{2}$ 0 100− $+4.3\sqrt{3}/7\sqrt{5.17.19}$

$\tilde{1}$ $\frac{3}{2}$ $\frac{3}{2}$ 0 101− $-2.131/3.7\sqrt{5.7.17.19}$

$\tilde{1}$ $\frac{3}{2}$ $\frac{3}{2}$ 0 110− $+\sqrt{2.19}/7\sqrt{5.11.17}$

$\tilde{1}$ $\frac{3}{2}$ $\frac{3}{2}$ 0 111− $+2.101\sqrt{2}/7\sqrt{3.5.7.11.17.19}$

$\tilde{1}$ $\frac{3}{2}$ $\frac{3}{2}$ 1 000+ $+2\sqrt{3.5.11}/7\sqrt{13.17.19}$

$\tilde{1}$ $\frac{3}{2}$ $\frac{3}{2}$ 1 001+ $-2.11\sqrt{5.11}/7\sqrt{7.13.17.19}$

$\tilde{1}$ $\frac{3}{2}$ $\frac{3}{2}$ 1 010+ $-29/7\sqrt{2.5.13.17.19}$

$\tilde{1}$ $\frac{3}{2}$ $\frac{3}{2}$ 1 011+ $+2.67\sqrt{2.13}/3.7\sqrt{3.5.7.17.19}$

$\tilde{1}$ $\frac{3}{2}$ $\frac{3}{2}$ 1 100+ $-\sqrt{3.5}/2.7\sqrt{17.19}$

$\tilde{1}$ $\frac{3}{2}$ $\frac{3}{2}$ 1 101+ $-8\sqrt{5}/3.7\sqrt{7.17.19}$

$\tilde{1}$ $\frac{3}{2}$ $\frac{3}{2}$ 1 110+ $+71\sqrt{2}/7\sqrt{5.11.17.19}$

$\tilde{1}$ $\frac{3}{2}$ $\frac{3}{2}$ 1 111+ $-137\sqrt{2}/7\sqrt{3.5.7.11.17.19}$

$\tilde{1}$ $\tilde{\frac{1}{2}}$ $\frac{1}{2}$ 0 000− $+3\sqrt{11.19}/2.7\sqrt{5.13.17}$

$\tilde{1}$ $\tilde{\frac{1}{2}}$ $\frac{1}{2}$ 0 010− $+2\sqrt{2.19}/3.7\sqrt{3.17}$

$\tilde{1}$ $\tilde{\frac{1}{2}}$ $\frac{1}{2}$ 0 100− $-2\sqrt{19}/3.7\sqrt{5.17}$

$\tilde{1}$ $\tilde{\frac{1}{2}}$ $\frac{1}{2}$ 0 110− $-2\sqrt{2.13}/7\sqrt{3.11.17.19}$

$\tilde{1}$ $\tilde{\frac{1}{2}}$ $\frac{3}{2}$ 0 000+ $-3\sqrt{3.11}/2.7\sqrt{13.17.19}$

$\tilde{1}$ $\tilde{\frac{1}{2}}$ $\frac{3}{2}$ 0 001+ $+3\sqrt{11}/\sqrt{7.13.17.19}$

$\tilde{1}$ $\tilde{\frac{1}{2}}$ $\frac{3}{2}$ 0 010+ $-29\sqrt{2}/7\sqrt{5.17.19}$

$\tilde{1}$ $\tilde{\frac{1}{2}}$ $\frac{3}{2}$ 0 011+ $-73\sqrt{2}/3\sqrt{3.5.7.17.19}$

$\tilde{1}$ $\tilde{\frac{1}{2}}$ $\frac{3}{2}$ 0 100+ $+11.11\sqrt{3}/2.5.7\sqrt{17.19}$

$\tilde{1}$ $\tilde{\frac{1}{2}}$ $\frac{3}{2}$ 0 101+ $+167/2.3.5\sqrt{7.17.19}$

$\tilde{1}$ $\tilde{\frac{1}{2}}$ $\frac{3}{2}$ 0 110+ $-9\sqrt{5.13}/7\sqrt{2.11.17.19}$

$\tilde{1}$ $\tilde{\frac{1}{2}}$ $\frac{3}{2}$ 0 111+ $+\sqrt{2.3.5.13}/\sqrt{7.11.17.19}$

$\tilde{0}$ $\frac{3}{2}$ $\frac{3}{2}$ 0 000+ $-3\sqrt{2.3}/\sqrt{5.7.17.19}$

$\tilde{0}$ $\frac{3}{2}$ $\frac{3}{2}$ 0 001+ $-\sqrt{2}/3\sqrt{5.17.19}$

$\tilde{0}$ $\frac{3}{2}$ $\frac{3}{2}$ 0 010+ $+43/\sqrt{5.7.11.17.19}$

$\tilde{0}$ $\frac{3}{2}$ $\frac{3}{2}$ 0 011+ $-2.47/3\sqrt{3.5.11.17.19}$

$\tilde{0}$ $\tilde{\frac{1}{2}}$ $\frac{1}{2}$ 0 000+ $+2\sqrt{2.19}/3\sqrt{5.7.17}$

$\tilde{0}$ $\tilde{\frac{1}{2}}$ $\frac{1}{2}$ 0 010+ $-2\sqrt{13}/3\sqrt{3.7.11.17.19}$

7 $\frac{13}{2}$ $\frac{11}{2}$

1 $\frac{1}{2}$ $\frac{1}{2}$ 0 000− $-8\sqrt{2}/\sqrt{3.7.11.13.17.19}$

1 $\frac{1}{2}$ $\frac{1}{2}$ 0 100− $-3.7/2\sqrt{2.13.17.19}$

1 $\frac{1}{2}$ $\frac{3}{2}$ 0 000+ $-4\sqrt{2}/11\sqrt{3.13.17.19}$

1 $\frac{1}{2}$ $\frac{3}{2}$ 0 001+ $-3\sqrt{3.5.19}/2.11\sqrt{13.17}$

1 $\frac{1}{2}$ $\frac{3}{2}$ 0 100+ $+3\sqrt{7.19}/2\sqrt{2.11.13.17}$

1 $\frac{1}{2}$ $\frac{3}{2}$ 0 101+ $-\sqrt{5.7}/\sqrt{11.13.17.19}$

1 $\frac{3}{2}$ $\frac{1}{2}$ 0 000+ $-233/5.7\sqrt{3.13.17.19}$

1 $\frac{3}{2}$ $\frac{1}{2}$ 0 010+ $-31\sqrt{11}/2.5.7\sqrt{2.13.17.19}$

1 $\frac{3}{2}$ $\frac{1}{2}$ 0 100+ $+3.11\sqrt{11}/2.5\sqrt{7.13.17.19}$

1 $\frac{3}{2}$ $\frac{1}{2}$ 0 110+ $-31\sqrt{3}/5\sqrt{2.7.13.17.19}$

1 $\frac{3}{2}$ $\frac{3}{2}$ 0 000− $-71\sqrt{2.7}/5\sqrt{3.11.13.17.19}$

1 $\frac{3}{2}$ $\frac{3}{2}$ 0 001− $-271/\sqrt{3.5.7.11.13.17.19}$

1 $\frac{3}{2}$ $\frac{3}{2}$ 0 010− $-2.3.37\sqrt{7}/5.11\sqrt{13.17.19}$

1 $\frac{3}{2}$ $\frac{3}{2}$ 0 011− $-151\sqrt{2}/11\sqrt{5.7.13.17.19}$

1 $\frac{3}{2}$ $\frac{3}{2}$ 0 100− $-29\sqrt{2}/5\sqrt{13.17.19}$

1 $\frac{3}{2}$ $\frac{3}{2}$ 0 101− $+3/\sqrt{5.13.17.19}$

1 $\frac{3}{2}$ $\frac{3}{2}$ 0 110− $+67\sqrt{3}/5\sqrt{11.13.17.19}$

SO_3–O 3jm Factors (cont.)

7	$\frac{13}{2}$	$\frac{11}{2}$			
1	$\frac{3}{2}$	$\frac{3}{2}$	0	111−	$+131/2\sqrt{2.3.5.11.13.17.19}$
1	$\frac{3}{2}$	$\frac{3}{2}$	1	000−	$-109/\sqrt{2.3.7.11.13.17.19}$
1	$\frac{3}{2}$	$\frac{3}{2}$	1	001−	$+1/2\sqrt{3.5.7.11.13.17.19}$
1	$\frac{3}{2}$	$\frac{3}{2}$	1	010−	$-263/2.5.11\sqrt{7.13.17.19}$
1	$\frac{3}{2}$	$\frac{3}{2}$	1	011−	$+53\sqrt{17}/11\sqrt{2.5.7.13.19}$
1	$\frac{3}{2}$	$\frac{3}{2}$	1	100−	$-\sqrt{2}/\sqrt{13.17.19}$
1	$\frac{3}{2}$	$\frac{3}{2}$	1	101−	$-\sqrt{13}/2\sqrt{5.17.19}$
1	$\frac{3}{2}$	$\frac{3}{2}$	1	110−	$+9.3\sqrt{3}/5\sqrt{11.13.17.19}$
1	$\frac{3}{2}$	$\frac{3}{2}$	1	111−	$-313/2\sqrt{2.3.5.11.13.17.19}$
1	$\frac{3}{2}$	$\tilde{\frac{1}{2}}$	0	000+	$+4.11\sqrt{11}/\sqrt{3.5.7.13.17.19}$
1	$\frac{3}{2}$	$\tilde{\frac{1}{2}}$	0	010+	$+47/2\sqrt{2.5.7.13.17.19}$
1	$\frac{3}{2}$	$\tilde{\frac{1}{2}}$	0	100+	$+11/2\sqrt{5.13.17.19}$
1	$\frac{3}{2}$	$\tilde{\frac{1}{2}}$	0	110+	$+\sqrt{11}/\sqrt{2.3.5.13.17.19}$
1	$\tilde{\frac{1}{2}}$	$\frac{3}{2}$	0	000+	$+8.2\sqrt{2.3}/\sqrt{5.7.11.13.17.19}$
1	$\tilde{\frac{1}{2}}$	$\frac{3}{2}$	0	001+	$-41/\sqrt{3.7.11.13.17.19}$
1	$\tilde{\frac{1}{2}}$	$\frac{3}{2}$	0	010+	$-23/5.11\sqrt{7.17.19}$
1	$\tilde{\frac{1}{2}}$	$\frac{3}{2}$	0	011+	$-4\sqrt{2.19}/11\sqrt{5.7.17}$
1	$\tilde{\frac{1}{2}}$	$\frac{3}{2}$	0	100+	$+59/2\sqrt{2.5.13.17.19}$
1	$\tilde{\frac{1}{2}}$	$\frac{3}{2}$	0	101+	$-2/\sqrt{13.17.19}$
1	$\tilde{\frac{1}{2}}$	$\frac{3}{2}$	0	110+	$-8.2\sqrt{3}/5\sqrt{11.17.19}$
1	$\tilde{\frac{1}{2}}$	$\frac{3}{2}$	0	111+	$-47/\sqrt{2.3.5.11.17.19}$
1	$\tilde{\frac{1}{2}}$	$\tilde{\frac{1}{2}}$	0	000−	$-8\sqrt{2.11}/\sqrt{3.7.13.17.19}$
1	$\tilde{\frac{1}{2}}$	$\tilde{\frac{1}{2}}$	0	010−	$-2/\sqrt{5.7.17.19}$
1	$\tilde{\frac{1}{2}}$	$\tilde{\frac{1}{2}}$	0	100−	$+11/2\sqrt{2.13.17.19}$
1	$\tilde{\frac{1}{2}}$	$\tilde{\frac{1}{2}}$	0	110−	$-2\sqrt{11}/\sqrt{3.5.17.19}$
2	$\frac{1}{2}$	$\frac{3}{2}$	0	000−	$-3\sqrt{3}/\sqrt{2.5.11.13.17.19}$
2	$\frac{1}{2}$	$\frac{3}{2}$	0	001−	$+73/2\sqrt{3.11.13.17.19}$
2	$\frac{3}{2}$	$\frac{1}{2}$	0	000−	$+8.2\sqrt{11}/7\sqrt{3.5.13.17.19}$
2	$\frac{3}{2}$	$\frac{1}{2}$	0	010−	$+73\sqrt{2}/7\sqrt{5.13.17.19}$
2	$\frac{3}{2}$	$\frac{3}{2}$	0	000+	$+\sqrt{2.7}/\sqrt{3.5.13.17.19}$
2	$\frac{3}{2}$	$\frac{3}{2}$	0	001+	$+41/\sqrt{3.7.13.17.19}$
2	$\frac{3}{2}$	$\frac{3}{2}$	0	010+	$+\sqrt{7.17}/\sqrt{5.11.13.19}$
2	$\frac{3}{2}$	$\frac{3}{2}$	0	011+	$-73/\sqrt{2.7.11.13.17.19}$
2	$\frac{3}{2}$	$\tilde{\frac{1}{2}}$	0	000−	$+2.11/\sqrt{3.7.13.17.19}$
2	$\frac{3}{2}$	$\tilde{\frac{1}{2}}$	0	010−	$+3\sqrt{2.11}/\sqrt{7.13.17.19}$
2	$\tilde{\frac{1}{2}}$	$\frac{3}{2}$	0	000−	$+79/\sqrt{2.3.7.13.17.19}$
2	$\tilde{\frac{1}{2}}$	$\frac{3}{2}$	0	001−	$-5\sqrt{3.5}/2\sqrt{7.13.17.19}$
2	$\tilde{\frac{1}{2}}$	$\frac{3}{2}$	0	010−	$+4\sqrt{11}/\sqrt{5.7.17.19}$
2	$\tilde{\frac{1}{2}}$	$\frac{3}{2}$	0	011−	$-\sqrt{2.11}/\sqrt{7.17.19}$
$\tilde{1}$	$\frac{1}{2}$	$\frac{3}{2}$	0	000−	$-8.3\sqrt{2.3}/\sqrt{5.7.11.13.17.19}$
$\tilde{1}$	$\frac{1}{2}$	$\frac{3}{2}$	0	001−	$+23/\sqrt{3.7.11.13.17.19}$
$\tilde{1}$	$\frac{1}{2}$	$\frac{3}{2}$	0	100−	$-3.41\sqrt{3}/2.11\sqrt{2.5.7.17.19}$
$\tilde{1}$	$\frac{1}{2}$	$\frac{3}{2}$	0	101−	$-2.37/11\sqrt{3.7.17.19}$
$\tilde{1}$	$\frac{1}{2}$	$\tilde{\frac{1}{2}}$	0	000+	$-8\sqrt{2.11}/\sqrt{3.7.13.17.19}$
$\tilde{1}$	$\frac{1}{2}$	$\tilde{\frac{1}{2}}$	0	100+	$-13/2\sqrt{2.3.7.17.19}$
$\tilde{1}$	$\frac{3}{2}$	$\frac{1}{2}$	0	000−	$-4.11\sqrt{11}/\sqrt{3.5.7.13.17.19}$
$\tilde{1}$	$\frac{3}{2}$	$\frac{1}{2}$	0	010−	$+47/2\sqrt{2.5.7.13.17.19}$
$\tilde{1}$	$\frac{3}{2}$	$\frac{1}{2}$	0	100−	$-1/2\sqrt{3.5.7.17.19}$
$\tilde{1}$	$\frac{3}{2}$	$\frac{1}{2}$	0	110−	$+97/\sqrt{2.5.7.11.17.19}$

7	$\frac{13}{2}$	$\frac{11}{2}$			
$\tilde{1}$	$\frac{3}{2}$	$\frac{3}{2}$	0	000+	$+\sqrt{2}/\sqrt{3.5.13.17.19}$
$\tilde{1}$	$\frac{3}{2}$	$\frac{3}{2}$	0	001+	$+41/7\sqrt{3.13.17.19}$
$\tilde{1}$	$\frac{3}{2}$	$\frac{3}{2}$	0	010+	$+2.41/\sqrt{5.11.13.17.19}$
$\tilde{1}$	$\frac{3}{2}$	$\frac{3}{2}$	0	011+	$+9\sqrt{2}/7\sqrt{11.13.17.19}$
$\tilde{1}$	$\frac{3}{2}$	$\frac{3}{2}$	0	100+	$-23\sqrt{2}/\sqrt{3.5.11.17.19}$
$\tilde{1}$	$\frac{3}{2}$	$\frac{3}{2}$	0	101+	$+2.29/7\sqrt{3.11.17.19}$
$\tilde{1}$	$\frac{3}{2}$	$\frac{3}{2}$	0	110+	$+5\sqrt{5}/11\sqrt{17.19}$
$\tilde{1}$	$\frac{3}{2}$	$\frac{3}{2}$	0	111+	$+719/2.7.11\sqrt{2.17.19}$
$\tilde{1}$	$\frac{3}{2}$	$\frac{3}{2}$	1	000−	$+\sqrt{17}/\sqrt{2.3.5.13.19}$
$\tilde{1}$	$\frac{3}{2}$	$\frac{3}{2}$	1	001−	$-73/2.7\sqrt{3.13.17.19}$
$\tilde{1}$	$\frac{3}{2}$	$\frac{3}{2}$	1	010−	$+23/2\sqrt{5.11.13.17.19}$
$\tilde{1}$	$\frac{3}{2}$	$\frac{3}{2}$	1	011−	$-29/7\sqrt{2.11.13.17.19}$
$\tilde{1}$	$\frac{3}{2}$	$\frac{3}{2}$	1	100−	$-\sqrt{2.5}/\sqrt{3.11.17.19}$
$\tilde{1}$	$\frac{3}{2}$	$\frac{3}{2}$	1	101−	$-113/2.7\sqrt{3.11.17.19}$
$\tilde{1}$	$\frac{3}{2}$	$\frac{3}{2}$	1	110−	$+13/11\sqrt{5.17.19}$
$\tilde{1}$	$\frac{3}{2}$	$\frac{3}{2}$	1	111−	$+3.139/2.7.11\sqrt{2.17.19}$
$\tilde{1}$	$\frac{3}{2}$	$\tilde{\frac{1}{2}}$	0	000−	$-11/\sqrt{3.13.17.19}$
$\tilde{1}$	$\frac{3}{2}$	$\tilde{\frac{1}{2}}$	0	010−	$+11\sqrt{11}/2\sqrt{2.13.17.19}$
$\tilde{1}$	$\frac{3}{2}$	$\tilde{\frac{1}{2}}$	0	100−	$-\sqrt{11}/2\sqrt{3.17.19}$
$\tilde{1}$	$\frac{3}{2}$	$\tilde{\frac{1}{2}}$	0	110−	$-1/\sqrt{2.17.19}$
$\tilde{1}$	$\tilde{\frac{1}{2}}$	$\frac{1}{2}$	0	000+	$-8\sqrt{2.11}/\sqrt{3.7.13.17.19}$
$\tilde{1}$	$\tilde{\frac{1}{2}}$	$\frac{1}{2}$	0	010+	$+2/\sqrt{5.7.17.19}$
$\tilde{1}$	$\tilde{\frac{1}{2}}$	$\frac{1}{2}$	0	100+	$+13/2\sqrt{2.3.7.17.19}$
$\tilde{1}$	$\tilde{\frac{1}{2}}$	$\frac{1}{2}$	0	110+	$+2\sqrt{13}/\sqrt{5.7.11.17.19}$
$\tilde{1}$	$\tilde{\frac{1}{2}}$	$\frac{3}{2}$	0	000−	$-4\sqrt{2}/\sqrt{3.13.17.19}$
$\tilde{1}$	$\tilde{\frac{1}{2}}$	$\frac{3}{2}$	0	001−	$+\sqrt{3.5.19}/2.7\sqrt{13.17}$
$\tilde{1}$	$\tilde{\frac{1}{2}}$	$\frac{3}{2}$	0	010−	$+7/\sqrt{5.11.17.19}$
$\tilde{1}$	$\tilde{\frac{1}{2}}$	$\frac{3}{2}$	0	011−	$-2\sqrt{2.19}/7\sqrt{11.17}$
$\tilde{1}$	$\tilde{\frac{1}{2}}$	$\frac{3}{2}$	0	100−	$-1/2\sqrt{2.3.11.17.19}$
$\tilde{1}$	$\tilde{\frac{1}{2}}$	$\frac{3}{2}$	0	101−	$-\sqrt{3.5.19}/7\sqrt{11.17}$
$\tilde{1}$	$\tilde{\frac{1}{2}}$	$\frac{3}{2}$	0	110−	$-4\sqrt{5.13}/11\sqrt{17.19}$
$\tilde{1}$	$\tilde{\frac{1}{2}}$	$\frac{3}{2}$	0	111−	$+5\sqrt{13}/7.11\sqrt{2.17.19}$
$\tilde{0}$	$\frac{1}{2}$	$\tilde{\frac{1}{2}}$	0	000−	$+1/\sqrt{3.17.19}$
$\tilde{0}$	$\frac{3}{2}$	$\frac{3}{2}$	0	000−	$+4\sqrt{7}/\sqrt{3.5.11.17.19}$
$\tilde{0}$	$\frac{3}{2}$	$\frac{3}{2}$	0	001−	$+2\sqrt{2.19}/\sqrt{3.7.11.17}$
$\tilde{0}$	$\frac{3}{2}$	$\frac{3}{2}$	0	010−	$-2\sqrt{2.7}/11\sqrt{5.17.19}$
$\tilde{0}$	$\frac{3}{2}$	$\frac{3}{2}$	0	011−	$-5/11\sqrt{7.17.19}$
$\tilde{0}$	$\tilde{\frac{1}{2}}$	$\frac{1}{2}$	0	000−	$-1/\sqrt{3.17.19}$
$\tilde{0}$	$\tilde{\frac{1}{2}}$	$\frac{1}{2}$	0	010−	$+4\sqrt{2.13}/\sqrt{5.11.17.19}$

7	$\frac{13}{2}$	$\frac{13}{2}$			
1	$\frac{1}{2}$	$\frac{1}{2}$	0	000+	$-4/\sqrt{11.13.17.19}$
1	$\frac{1}{2}$	$\frac{1}{2}$	0	100+	$+2.3\sqrt{3}/\sqrt{7.13.17.19}$
1	$\frac{3}{2}$	$\frac{1}{2}$	0	000−	$-8.2\sqrt{2}/5\sqrt{7.13.17.19}$
1	$\frac{3}{2}$	$\frac{1}{2}$	0	010−	$-41\sqrt{3}/5\sqrt{7.11.13.17.19}$
1	$\frac{3}{2}$	$\frac{1}{2}$	0	100−	$-43\sqrt{3.11}/5.7\sqrt{2.13.17.19}$
1	$\frac{3}{2}$	$\frac{1}{2}$	0	110−	$+8.41/5.7\sqrt{13.17.19}$
1	$\frac{3}{2}$	$\frac{3}{2}$	0	000+	$-2.3.11\sqrt{2.11}/5.7\sqrt{13.17.19}$
1	$\frac{3}{2}$	$\frac{3}{2}$	0	010+	$-3\sqrt{3.17}/5.7\sqrt{13.19}$
1	$\frac{3}{2}$	$\frac{3}{2}$	0	011+	$+2.61\sqrt{2}/3.7\sqrt{11.13.17.19}$

SO_3–O 3jm Factors (cont.)

7	$\mathbf{\frac{13}{2}}$	$\mathbf{\frac{13}{2}}$			
1	$\frac{3}{2}$	$\frac{3}{2}$	0	100+	$-2.11\sqrt{2.3}/5\sqrt{7.13.17.19}$
1	$\frac{3}{2}$	$\frac{3}{2}$	0	110+	$+4\sqrt{11}/5\sqrt{7.13.17.19}$
1	$\frac{3}{2}$	$\frac{3}{2}$	0	111+	$-2\sqrt{2.3}/\sqrt{7.13.17.19}$
1	$\frac{3}{2}$	$\frac{3}{2}$	1	000+	$+2.43\sqrt{2.11}/5.7\sqrt{13.17.19}$
1	$\frac{3}{2}$	$\frac{3}{2}$	1	010+	$-4.9\sqrt{3}/5.7\sqrt{13.17.19}$
1	$\frac{3}{2}$	$\frac{3}{2}$	1	011+	$-2.1321\sqrt{2}/3.5.7\sqrt{11.13.17.19}$
1	$\frac{3}{2}$	$\frac{3}{2}$	1	100+	$-4.9.11\sqrt{2.3}/5.7\sqrt{7.13.17.19}$
1	$\frac{3}{2}$	$\frac{3}{2}$	1	110+	$+4\sqrt{11.17}/5.7\sqrt{7.13.19}$
1	$\frac{3}{2}$	$\frac{3}{2}$	1	111+	$-2.9.9.3\sqrt{2.3}/5.7\sqrt{7.13.17.19}$
1	$\tilde{\frac{1}{2}}$	$\frac{3}{2}$	0	000−	$-2\sqrt{2.11}/\sqrt{5.13.17.19}$
1	$\tilde{\frac{1}{2}}$	$\frac{3}{2}$	0	001−	$+\sqrt{3}/\sqrt{5.13.17.19}$
1	$\tilde{\frac{1}{2}}$	$\frac{3}{2}$	0	010−	$+3\sqrt{3}/5\sqrt{17.19}$
1	$\tilde{\frac{1}{2}}$	$\frac{3}{2}$	0	011−	$+4.5\sqrt{2}/3\sqrt{11.17.19}$
1	$\tilde{\frac{1}{2}}$	$\frac{3}{2}$	0	100−	$-11\sqrt{3.17}/7\sqrt{2.5.7.13.19}$
1	$\tilde{\frac{1}{2}}$	$\frac{3}{2}$	0	101−	$+2.37\sqrt{11}/7\sqrt{5.7.13.17.19}$
1	$\tilde{\frac{1}{2}}$	$\frac{3}{2}$	0	110−	$+4.3\sqrt{11}/5.7\sqrt{7.17.19}$
1	$\tilde{\frac{1}{2}}$	$\frac{3}{2}$	0	111−	$-11\sqrt{3}/7\sqrt{2.7.17.19}$
1	$\tilde{\frac{1}{2}}$	$\tilde{\frac{1}{2}}$	0	000+	$+4\sqrt{11}/\sqrt{13.17.19}$
1	$\tilde{\frac{1}{2}}$	$\tilde{\frac{1}{2}}$	0	010+	$+\sqrt{3}/\sqrt{2.5.17.19}$
1	$\tilde{\frac{1}{2}}$	$\tilde{\frac{1}{2}}$	0	011+	$+8.2\sqrt{13}/3.5\sqrt{11.17.19}$
1	$\tilde{\frac{1}{2}}$	$\tilde{\frac{1}{2}}$	0	100+	$+2.11\sqrt{3}/7\sqrt{7.13.17.19}$
1	$\tilde{\frac{1}{2}}$	$\tilde{\frac{1}{2}}$	0	110+	$-8\sqrt{2.11}/7\sqrt{5.7.17.19}$
1	$\tilde{\frac{1}{2}}$	$\tilde{\frac{1}{2}}$	0	111+	$-4.11\sqrt{3.13}/5.7\sqrt{7.17.19}$
2	$\frac{3}{2}$	$\frac{1}{2}$	0	000+	$-9\sqrt{2.11}/\sqrt{5.7.13.17.19}$
2	$\frac{3}{2}$	$\frac{1}{2}$	0	010+	$-53/\sqrt{3.5.7.13.17.19}$
2	$\frac{3}{2}$	$\frac{3}{2}$	0	000−	0
2	$\frac{3}{2}$	$\frac{3}{2}$	0	010−	$-8\sqrt{11}/\sqrt{3.5.13.17.19}$
2	$\frac{3}{2}$	$\frac{3}{2}$	0	011−	0
2	$\tilde{\frac{1}{2}}$	$\frac{3}{2}$	0	000+	$+3.11\sqrt{2}/7\sqrt{13.17.19}$
2	$\tilde{\frac{1}{2}}$	$\frac{3}{2}$	0	001+	$+9\sqrt{3.11}/7\sqrt{13.17.19}$
2	$\tilde{\frac{1}{2}}$	$\frac{3}{2}$	0	010+	$-\sqrt{11.17}/7\sqrt{3.5.19}$
2	$\tilde{\frac{1}{2}}$	$\frac{3}{2}$	0	011+	$-\sqrt{2}/7\sqrt{5.17.19}$
$\tilde{1}$	$\frac{3}{2}$	$\frac{1}{2}$	0	000+	$+2\sqrt{2.11}/\sqrt{5.13.17.19}$
$\tilde{1}$	$\frac{3}{2}$	$\frac{1}{2}$	0	010+	$+\sqrt{3}/\sqrt{5.13.17.19}$
$\tilde{1}$	$\frac{3}{2}$	$\frac{1}{2}$	0	100+	$+3/\sqrt{2.5.17.19}$
$\tilde{1}$	$\frac{3}{2}$	$\frac{1}{2}$	0	110+	$-2/\sqrt{3.5.11.17.19}$
$\tilde{1}$	$\frac{3}{2}$	$\frac{3}{2}$	0	000−	0
$\tilde{1}$	$\frac{3}{2}$	$\frac{3}{2}$	0	010−	$-\sqrt{3.7.11}/\sqrt{5.13.17.19}$
$\tilde{1}$	$\frac{3}{2}$	$\frac{3}{2}$	0	011−	0
$\tilde{1}$	$\frac{3}{2}$	$\frac{3}{2}$	0	100−	0
$\tilde{1}$	$\frac{3}{2}$	$\frac{3}{2}$	0	110−	$+4\sqrt{7}/\sqrt{3.5.17.19}$
$\tilde{1}$	$\frac{3}{2}$	$\frac{3}{2}$	0	111−	0
$\tilde{1}$	$\frac{3}{2}$	$\frac{3}{2}$	1	000+	$+2.11\sqrt{2.17}/7\sqrt{5.7.13.19}$
$\tilde{1}$	$\frac{3}{2}$	$\frac{3}{2}$	1	010+	$-4\sqrt{3.11.13}/7\sqrt{5.7.17.19}$
$\tilde{1}$	$\frac{3}{2}$	$\frac{3}{2}$	1	011+	$-2.47\sqrt{2}/7\sqrt{5.7.13.17.19}$
$\tilde{1}$	$\frac{3}{2}$	$\frac{3}{2}$	1	100+	$-4.3\sqrt{2.11}/7\sqrt{5.7.17.19}$
$\tilde{1}$	$\frac{3}{2}$	$\frac{3}{2}$	1	110+	$-4.41/7\sqrt{3.5.7.17.19}$
$\tilde{1}$	$\frac{3}{2}$	$\frac{3}{2}$	1	111+	$+2.31\sqrt{2}/7\sqrt{5.7.11.17.19}$
$\tilde{1}$	$\tilde{\frac{1}{2}}$	$\frac{1}{2}$	0	000−	$-4\sqrt{11}/\sqrt{13.17.19}$

7	$\mathbf{\frac{13}{2}}$	$\mathbf{\frac{13}{2}}$			
$\tilde{1}$	$\tilde{\frac{1}{2}}$	$\frac{1}{2}$	0	010−	$+\sqrt{3}/\sqrt{2.5.17.19}$
$\tilde{1}$	$\tilde{\frac{1}{2}}$	$\frac{1}{2}$	0	100−	0
$\tilde{1}$	$\tilde{\frac{1}{2}}$	$\frac{1}{2}$	0	110−	$+8\sqrt{2.13}/\sqrt{3.5.11.17.19}$
$\tilde{1}$	$\tilde{\frac{1}{2}}$	$\frac{3}{2}$	0	000+	$-4.11\sqrt{2}/7\sqrt{7.13.17.19}$
$\tilde{1}$	$\tilde{\frac{1}{2}}$	$\frac{3}{2}$	0	001+	$+9\sqrt{3.11}/7\sqrt{7.13.17.19}$
$\tilde{1}$	$\tilde{\frac{1}{2}}$	$\frac{3}{2}$	0	010+	$-\sqrt{3.11}/7\sqrt{5.7.17.19}$
$\tilde{1}$	$\tilde{\frac{1}{2}}$	$\frac{3}{2}$	0	011+	$+2\sqrt{2.17}/7\sqrt{5.7.19}$
$\tilde{1}$	$\tilde{\frac{1}{2}}$	$\frac{3}{2}$	0	100+	$+9\sqrt{11}/7\sqrt{2.7.17.19}$
$\tilde{1}$	$\tilde{\frac{1}{2}}$	$\frac{3}{2}$	0	101+	$-4\sqrt{3}/7\sqrt{7.17.19}$
$\tilde{1}$	$\tilde{\frac{1}{2}}$	$\frac{3}{2}$	0	110+	$+2\sqrt{13.17}/7\sqrt{3.5.7.19}$
$\tilde{1}$	$\tilde{\frac{1}{2}}$	$\frac{3}{2}$	0	111+	$+\sqrt{5.13.19}/7\sqrt{2.7.11.17}$
$\tilde{0}$	$\frac{3}{2}$	$\frac{3}{2}$	0	000+	$-3\sqrt{11}/7\sqrt{5.17.19}$
$\tilde{0}$	$\frac{3}{2}$	$\frac{3}{2}$	0	010+	$-2\sqrt{2.5}/7\sqrt{3.17.19}$
$\tilde{0}$	$\frac{3}{2}$	$\frac{3}{2}$	0	011+	$-191/7\sqrt{5.11.17.19}$
$\tilde{0}$	$\tilde{\frac{1}{2}}$	$\frac{1}{2}$	0	000+	$+3/\sqrt{2.7.17.19}$
$\tilde{0}$	$\tilde{\frac{1}{2}}$	$\frac{1}{2}$	0	010+	$+8.2\sqrt{13}/\sqrt{3.5.7.11.17.19}$
7	**7**	**0**			
1	1	0	0	000+	$+1/\sqrt{5}$
1	1	0	0	100+	0
1	1	0	0	110+	$+1/\sqrt{5}$
2	2	0	0	000+	$+\sqrt{2}/\sqrt{3.5}$
$\tilde{1}$	$\tilde{1}$	0	0	000+	$+1/\sqrt{5}$
$\tilde{1}$	$\tilde{1}$	0	0	100+	0
$\tilde{1}$	$\tilde{1}$	0	0	110+	$+1/\sqrt{5}$
$\tilde{0}$	$\tilde{0}$	0	0	000+	$+1/\sqrt{3.5}$
7	**7**	**1**			
1	1	1	0	000+	$-2/\sqrt{5.7}$
1	1	1	0	100+	0
1	1	1	0	110+	$+\sqrt{7}/4\sqrt{5}$
2	1	1	0	000−	$-\sqrt{3.11}/2.5\sqrt{7}$
2	1	1	0	010−	$-1/2.5$
$\tilde{1}$	1	1	0	000−	$-\sqrt{3.11}/2.5.7$
$\tilde{1}$	1	1	0	010−	$-3/5\sqrt{7}$
$\tilde{1}$	1	1	0	100−	$+\sqrt{3.13}/5.7$
$\tilde{1}$	1	1	0	110−	$-\sqrt{11.13}/4.5\sqrt{7}$
$\tilde{1}$	2	1	0	000+	$-3\sqrt{3}/2.7\sqrt{5}$
$\tilde{1}$	2	1	0	100+	$+\sqrt{11.13}/2.7\sqrt{3.5}$
$\tilde{1}$	$\tilde{1}$	1	0	000+	$+2/\sqrt{5.7}$
$\tilde{1}$	$\tilde{1}$	1	0	100+	0
$\tilde{1}$	$\tilde{1}$	1	0	110+	$+1/4\sqrt{5.7}$
$\tilde{0}$	$\tilde{1}$	1	0	000−	0
$\tilde{0}$	$\tilde{1}$	1	0	010−	$-1/\sqrt{3.5}$
7	**7**	**2**			
1	1	2	0	000+	$+41/5\sqrt{2.5.7.13.17}$
1	1	2	0	100+	$-3\sqrt{3.11}/5\sqrt{2.5.13.17}$
1	1	2	0	110+	$+\sqrt{7.17}/5\sqrt{2.5.13}$
1	1	$\tilde{1}$	0	000+	$+43\sqrt{3}/5\sqrt{5.7.13.17}$
1	1	$\tilde{1}$	0	100+	$+3\sqrt{11}/5\sqrt{5.13.17}$
1	1	$\tilde{1}$	0	110+	$-\sqrt{3.7}/4.5\sqrt{5.13.17}$

SO_3–O 3jm Factors (cont.)

7 7 2

2 1 $\tilde{1}$ 0 000−	$+\sqrt{3.11.17}/2.5\sqrt{7.13}$
2 1 $\tilde{1}$ 0 010−	$-\sqrt{13}/2.5\sqrt{17}$
2 2 2 0 000+	$-4\sqrt{2}/\sqrt{3.5.7.13.17}$
$\tilde{1}$ 1 2 0 000−	0
$\tilde{1}$ 1 2 0 010−	$-9\sqrt{3}/\sqrt{2.7.13.17}$
$\tilde{1}$ 1 2 0 100−	$-3/5\sqrt{2.17}$
$\tilde{1}$ 1 2 0 110−	$-\sqrt{2.3.11}/5\sqrt{7.17}$
$\tilde{1}$ 1 $\tilde{1}$ 0 000−	$+3\sqrt{11.17}/2.5.7\sqrt{13}$
$\tilde{1}$ 1 $\tilde{1}$ 0 010−	$+2.3\sqrt{3}/5\sqrt{7.13.17}$
$\tilde{1}$ 1 $\tilde{1}$ 0 100−	$-2.3/7\sqrt{17}$
$\tilde{1}$ 1 $\tilde{1}$ 0 110−	$-\sqrt{3.11}/4\sqrt{7.17}$
$\tilde{1}$ 2 $\tilde{1}$ 0 000+	$+3\sqrt{3.17}/2.7\sqrt{5.13}$
$\tilde{1}$ 2 $\tilde{1}$ 0 100+	$-\sqrt{3.11}/2.7\sqrt{5.17}$
$\tilde{1}$ $\tilde{1}$ 2 0 000+	$+23\sqrt{5}/7\sqrt{2.7.13.17}$
$\tilde{1}$ $\tilde{1}$ 2 0 100+	$+9\sqrt{11}/7\sqrt{2.5.7.17}$
$\tilde{1}$ $\tilde{1}$ 2 0 110+	$-11\sqrt{13}/7\sqrt{2.5.7.17}$
$\tilde{1}$ $\tilde{1}$ $\tilde{1}$ 0 000+	$-41\sqrt{3}/7\sqrt{5.7.13.17}$
$\tilde{1}$ $\tilde{1}$ $\tilde{1}$ 0 100+	$+3\sqrt{3.11}/7\sqrt{5.7.17}$
$\tilde{1}$ $\tilde{1}$ $\tilde{1}$ 0 110+	$+5\sqrt{3.5.13}/4.7\sqrt{7.17}$
$\tilde{0}$ 1 $\tilde{1}$ 0 000+	$-4\sqrt{3}/5\sqrt{7.17}$
$\tilde{0}$ 1 $\tilde{1}$ 0 010+	$+\sqrt{11}/5\sqrt{17}$
$\tilde{0}$ 2 2 0 000−	$+2\sqrt{11}/\sqrt{3.5.7.17}$

7 7 3

1 1 1 0 000+	$-9\sqrt{3}/\sqrt{2.5.7.13.17}$
1 1 1 0 100+	$-\sqrt{11}/2\sqrt{2.5.13.17}$
1 1 1 0 110+	$-7\sqrt{7}/\sqrt{2.3.5.13.17}$
1 1 $\tilde{1}$ 0 000−	0
1 1 $\tilde{1}$ 0 100−	$+\sqrt{3.11}/2\sqrt{2.13.17}$
1 1 $\tilde{1}$ 0 110−	0
2 1 1 0 000−	$-29\sqrt{11}/2.5\sqrt{2.7.13.17}$
2 1 1 0 010−	$-29/2.5\sqrt{2.3.13.17}$
2 1 $\tilde{1}$ 0 000+	$+3\sqrt{11}/2\sqrt{2.5.7.13.17}$
2 1 $\tilde{1}$ 0 010+	$-7/2\sqrt{2.3.5.13.17}$
2 2 $\tilde{0}$ 0 000+	$-5\sqrt{2}/\sqrt{7.13.17}$
$\tilde{1}$ 1 1 0 000−	$-\sqrt{11.17}/5.7\sqrt{2.13}$
$\tilde{1}$ 1 1 0 010−	$+\sqrt{13.17}/2.5\sqrt{2.3.7}$
$\tilde{1}$ 1 1 0 100−	$+3.11/2.5.7\sqrt{2.17}$
$\tilde{1}$ 1 1 0 110−	$-\sqrt{11}/5\sqrt{2.3.7.17}$
$\tilde{1}$ 1 $\tilde{1}$ 0 000+	0
$\tilde{1}$ 1 $\tilde{1}$ 0 010+	$+3\sqrt{5}/2\sqrt{2.7.13.17}$
$\tilde{1}$ 1 $\tilde{1}$ 0 100+	$+3\sqrt{3}/2\sqrt{2.5.17}$
$\tilde{1}$ 1 $\tilde{1}$ 0 110+	$-\sqrt{2.11}/\sqrt{5.7.17}$
$\tilde{1}$ 1 $\tilde{0}$ 0 000−	$-\sqrt{3.11.17}/2.7\sqrt{5.13}$
$\tilde{1}$ 1 $\tilde{0}$ 0 010−	$-1/\sqrt{5.7.13.17}$
$\tilde{1}$ 1 $\tilde{0}$ 0 100−	$+3\sqrt{3}/7\sqrt{5.17}$
$\tilde{1}$ 1 $\tilde{0}$ 0 110−	$+\sqrt{11}/2\sqrt{5.7.17}$
$\tilde{1}$ 2 1 0 000+	$-181/2.3.7\sqrt{2.5.13.17}$
$\tilde{1}$ 2 1 0 100+	$-11\sqrt{11}/2.3.7\sqrt{2.5.17}$
$\tilde{1}$ 2 $\tilde{1}$ 0 000−	$-5.5/2.7\sqrt{2.13.17}$

7 7 3

$\tilde{1}$ 2 $\tilde{1}$ 0 100−	$-5\sqrt{11}/2.7\sqrt{2.17}$
$\tilde{1}$ $\tilde{1}$ 1 0 000+	$-3.11\sqrt{3}/7\sqrt{2.5.7.13.17}$
$\tilde{1}$ $\tilde{1}$ 1 0 100+	$-11\sqrt{5.11}/2.7\sqrt{2.3.7.17}$
$\tilde{1}$ $\tilde{1}$ 1 0 110+	$-\sqrt{13}/7\sqrt{2.3.5.7.17}$
$\tilde{1}$ $\tilde{1}$ $\tilde{1}$ 0 000−	0
$\tilde{1}$ $\tilde{1}$ $\tilde{1}$ 0 100−	$+\sqrt{11}/2\sqrt{2.7.17}$
$\tilde{1}$ $\tilde{1}$ $\tilde{1}$ 0 110−	0
$\tilde{0}$ 1 $\tilde{1}$ 0 000−	$+3\sqrt{2}/\sqrt{5.7.17}$
$\tilde{0}$ 1 $\tilde{1}$ 0 010−	$+\sqrt{11}/\sqrt{2.3.5.17}$
$\tilde{0}$ $\tilde{1}$ 1 0 000−	$-\sqrt{2.5.11}/3.7\sqrt{17}$
$\tilde{0}$ $\tilde{1}$ 1 0 010−	$-\sqrt{13}/3.7\sqrt{2.5.17}$

7 7 4

1 1 0 0 000+	$-9\sqrt{19}/2\sqrt{5.11.13.17}$
1 1 0 0 100+	$-\sqrt{7}/\sqrt{3.5.13.17.19}$
1 1 0 0 110+	$+7.7\sqrt{11}/2.3\sqrt{5.13.17.19}$
1 1 1 0 000−	0
1 1 1 0 100−	$+3\sqrt{7}/2\sqrt{2.13.17.19}$
1 1 1 0 110−	0
1 1 2 0 000+	$+2.3.5/\sqrt{7.11.13.17.19}$
1 1 2 0 100+	$-41/2\sqrt{3.13.17.19}$
1 1 2 0 110+	$-\sqrt{7.11}/3\sqrt{13.17.19}$
1 1 $\tilde{1}$ 0 000+	$+\sqrt{3.17}/\sqrt{2.7.11.13.19}$
1 1 $\tilde{1}$ 0 100+	$+11/2\sqrt{2.13.17.19}$
1 1 $\tilde{1}$ 0 110+	$-\sqrt{7.11}/\sqrt{2.3.13.17.19}$
2 1 1 0 000+	$-7\sqrt{5}/2\sqrt{2.13.17.19}$
2 1 1 0 010+	$+\sqrt{3.5.7.11}/2\sqrt{2.13.17.19}$
2 1 $\tilde{1}$ 0 000−	$+9\sqrt{3.5}/2\sqrt{2.7.13.17.19}$
2 1 $\tilde{1}$ 0 010−	$+11\sqrt{5.11}/2.3\sqrt{2.13.17.19}$
2 2 0 0 000+	$-8.2\sqrt{2.11}/3\sqrt{3.5.13.17.19}$
2 2 2 0 000+	$+8.4\sqrt{11}/3\sqrt{3.7.13.17.19}$
$\tilde{1}$ 1 1 0 000+	$-8\sqrt{2}/\sqrt{5.7.13.17.19}$
$\tilde{1}$ 1 1 0 010+	$-7\sqrt{11}/2\sqrt{2.3.5.13.17.19}$
$\tilde{1}$ 1 1 0 100+	$+113/2\sqrt{2.5.7.11.17.19}$
$\tilde{1}$ 1 1 0 110+	$+7\sqrt{2}/\sqrt{3.5.17.19}$
$\tilde{1}$ 1 2 0 000−	$-4/\sqrt{5.13.17.19}$
$\tilde{1}$ 1 2 0 010−	$+59\sqrt{11}/2\sqrt{3.5.7.13.17.19}$
$\tilde{1}$ 1 2 0 100−	$-43/2\sqrt{5.11.17.19}$
$\tilde{1}$ 1 2 0 110−	$-\sqrt{3}/\sqrt{5.7.17.19}$
$\tilde{1}$ 1 $\tilde{1}$ 0 000−	$+\sqrt{17.19}/7\sqrt{2.5.13}$
$\tilde{1}$ 1 $\tilde{1}$ 0 010−	$-\sqrt{11.19}/2\sqrt{2.3.5.7.13.17}$
$\tilde{1}$ 1 $\tilde{1}$ 0 100−	$+9\sqrt{19}/2.7\sqrt{2.5.11.17}$
$\tilde{1}$ 1 $\tilde{1}$ 0 110−	$+11\sqrt{3}/\sqrt{2.5.7.17.19}$
$\tilde{1}$ 2 1 0 000−	$-31\sqrt{11}/2.3\sqrt{2.7.13.17.19}$
$\tilde{1}$ 2 1 0 100−	$+13/2.3\sqrt{2.7.17.19}$
$\tilde{1}$ 2 $\tilde{1}$ 0 000+	$+47\sqrt{11}/2.7\sqrt{2.3.13.17.19}$
$\tilde{1}$ 2 $\tilde{1}$ 0 100+	$-29\sqrt{3}/2.7\sqrt{2.17.19}$
$\tilde{1}$ $\tilde{1}$ 0 0 000+	$+3\sqrt{11.19}/2.7\sqrt{5.13.17}$
$\tilde{1}$ $\tilde{1}$ 0 0 100+	$+\sqrt{5.19}/3.7\sqrt{17}$
$\tilde{1}$ $\tilde{1}$ 0 0 110+	$-23\sqrt{13}/2.3.7\sqrt{5.11.17.19}$

SO_3–O 3jm Factors (cont.)

7 7 4	
$\tilde{1}\ \tilde{1}\ 1\ 0\ 000-$	0
$\tilde{1}\ \tilde{1}\ 1\ 0\ 100-$	$+\sqrt{19}/2\sqrt{2.3.17}$
$\tilde{1}\ \tilde{1}\ 1\ 0\ 110-$	0
$\tilde{1}\ \tilde{1}\ 2\ 0\ 000+$	$-2.9\sqrt{11}/7\sqrt{7.13.17.19}$
$\tilde{1}\ \tilde{1}\ 2\ 0\ 100+$	$-113/2.3.7\sqrt{7.17.19}$
$\tilde{1}\ \tilde{1}\ 2\ 0\ 110+$	$-97\sqrt{13}/3.7\sqrt{7.11.17.19}$
$\tilde{1}\ \tilde{1}\ \tilde{1}\ 0\ 000+$	$-5\sqrt{3.11}/7\sqrt{2.7.13.17.19}$
$\tilde{1}\ \tilde{1}\ \tilde{1}\ 0\ 100+$	$+109/2.7\sqrt{2.3.7.17.19}$
$\tilde{1}\ \tilde{1}\ \tilde{1}\ 0\ 110+$	$-109\sqrt{13}/7\sqrt{2.3.7.11.17.19}$
$\tilde{0}\ 1\ \tilde{1}\ 0\ 000+$	$+\sqrt{2.3.19}/\sqrt{5.7.11.17}$
$\tilde{0}\ 1\ \tilde{1}\ 0\ 010+$	$-11/3\sqrt{2.5.17.19}$
$\tilde{0}\ 2\ 2\ 0\ 000-$	$+\sqrt{2.17}/3\sqrt{3.7.19}$
$\tilde{0}\ \tilde{1}\ 1\ 0\ 000+$	$+\sqrt{2.19}/3\sqrt{7.17}$
$\tilde{0}\ \tilde{1}\ 1\ 0\ 010+$	$+\sqrt{13}/3\sqrt{2.7.11.17.19}$
$\tilde{0}\ \tilde{0}\ 0\ 0\ 000+$	$-2\sqrt{13}/3\sqrt{3.5.11.17.19}$
7 7 5	
$1\ 1\ 1\ 0\ 000+$	$-8.3\sqrt{3}/11\sqrt{5.7.13.17.19}$
$1\ 1\ 1\ 0\ 001+$	$+9\sqrt{3.19}/2.11\sqrt{13.17}$
$1\ 1\ 1\ 0\ 100+$	$-9.3.7/4\sqrt{5.11.13.17.19}$
$1\ 1\ 1\ 0\ 101+$	$-2\sqrt{7}/\sqrt{11.13.17.19}$
$1\ 1\ 1\ 0\ 110+$	$-\sqrt{3.7}/\sqrt{5.13.17.19}$
$1\ 1\ 1\ 0\ 111+$	$+7.7/4\sqrt{3.13.17.19}$
$1\ 1\ 2\ 0\ 000-$	0
$1\ 1\ 2\ 0\ 100-$	$-3\sqrt{2.7}/\sqrt{13.17.19}$
$1\ 1\ 2\ 0\ 110-$	0
$1\ 1\ \tilde{1}\ 0\ 000-$	0
$1\ 1\ \tilde{1}\ 0\ 100-$	$+11\sqrt{3.7}/4\sqrt{13.17.19}$
$1\ 1\ \tilde{1}\ 0\ 110-$	0
$2\ 1\ 1\ 0\ 000-$	$-9.3\sqrt{13}/2.5\sqrt{7.11.17.19}$
$2\ 1\ 1\ 0\ 001-$	$+37/2\sqrt{5.11.13.17.19}$
$2\ 1\ 1\ 0\ 010-$	$+53\sqrt{3}/2.5\sqrt{13.17.19}$
$2\ 1\ 1\ 0\ 011-$	$-\sqrt{7.13}/2\sqrt{3.5.17.19}$
$2\ 1\ \tilde{1}\ 0\ 000+$	$+9.3/2\sqrt{5.13.17.19}$
$2\ 1\ \tilde{1}\ 0\ 010+$	$+7\sqrt{7.11}/2\sqrt{3.5.13.17.19}$
$2\ 2\ 2\ 0\ 000-$	0
$\tilde{1}\ 1\ 1\ 0\ 000-$	$-2.3\sqrt{11.13}/5.7\sqrt{17.19}$
$\tilde{1}\ 1\ 1\ 0\ 001-$	$+2\sqrt{11}/\sqrt{5.7.13.17.19}$
$\tilde{1}\ 1\ 1\ 0\ 010-$	$-239\sqrt{3}/4.5\sqrt{7.13.17.19}$
$\tilde{1}\ 1\ 1\ 0\ 011-$	$-\sqrt{13}/\sqrt{3.5.17.19}$
$\tilde{1}\ 1\ 1\ 0\ 100-$	$-9.13.23/4.5.7.11\sqrt{17.19}$
$\tilde{1}\ 1\ 1\ 0\ 101-$	$-3\sqrt{17}/11\sqrt{5.7.19}$
$\tilde{1}\ 1\ 1\ 0\ 110-$	$+8.4\sqrt{3}/5\sqrt{7.11.17.19}$
$\tilde{1}\ 1\ 1\ 0\ 111-$	$-13.13/4\sqrt{3.5.11.17.19}$
$\tilde{1}\ 1\ 2\ 0\ 000+$	$-31\sqrt{3}/\sqrt{2.5.7.13.17.19}$
$\tilde{1}\ 1\ 2\ 0\ 010+$	$+2\sqrt{2.11}/\sqrt{5.13.17.19}$
$\tilde{1}\ 1\ 2\ 0\ 100+$	$-2.9\sqrt{2.3}/\sqrt{5.7.11.17.19}$
$\tilde{1}\ 1\ 2\ 0\ 110+$	$-1/\sqrt{2.5.17.19}$
$\tilde{1}\ 1\ \tilde{1}\ 0\ 000+$	$+8.3\sqrt{3}/\sqrt{5.7.13.17.19}$
$\tilde{1}\ 1\ \tilde{1}\ 0\ 010+$	$-9\sqrt{11}/4\sqrt{5.13.17.19}$

7 7 5	
$\tilde{1}\ 1\ \tilde{1}\ 0\ 100+$	$+3.13\sqrt{3}/4\sqrt{5.7.11.17.19}$
$\tilde{1}\ 1\ \tilde{1}\ 0\ 110+$	$+1/\sqrt{5.17.19}$
$\tilde{1}\ 2\ 1\ 0\ 000+$	$-9.31/2.7\sqrt{5.13.17.19}$
$\tilde{1}\ 2\ 1\ 0\ 001+$	$+\sqrt{7}/2.3\sqrt{13.17.19}$
$\tilde{1}\ 2\ 1\ 0\ 100+$	$-3\sqrt{17}/2.7\sqrt{5.11.19}$
$\tilde{1}\ 2\ 1\ 0\ 101+$	$-23\sqrt{7}/2.3\sqrt{11.17.19}$
$\tilde{1}\ 2\ \tilde{1}\ 0\ 000-$	$+\sqrt{11.13}/2\sqrt{7.17.19}$
$\tilde{1}\ 2\ \tilde{1}\ 0\ 100-$	$-5/2\sqrt{7.17.19}$
$\tilde{1}\ \tilde{1}\ 1\ 0\ 000+$	$+8.3\sqrt{3}/\sqrt{5.7.13.17.19}$
$\tilde{1}\ \tilde{1}\ 1\ 0\ 001+$	$+3\sqrt{3.19}/2.7\sqrt{13.17}$
$\tilde{1}\ \tilde{1}\ 1\ 0\ 100+$	$-3.13\sqrt{3}/4\sqrt{5.7.11.17.19}$
$\tilde{1}\ \tilde{1}\ 1\ 0\ 101+$	$-2\sqrt{19}/7\sqrt{3.11.17}$
$\tilde{1}\ \tilde{1}\ 1\ 0\ 110+$	$+9.3\sqrt{3.13}/11\sqrt{5.7.17.19}$
$\tilde{1}\ \tilde{1}\ 1\ 0\ 111+$	$-5.5\sqrt{13}/4.7.11\sqrt{3.17.19}$
$\tilde{1}\ \tilde{1}\ 2\ 0\ 000-$	0
$\tilde{1}\ \tilde{1}\ 2\ 0\ 100-$	$+\sqrt{2.3}/\sqrt{17.19}$
$\tilde{1}\ \tilde{1}\ 2\ 0\ 110-$	0
$\tilde{1}\ \tilde{1}\ \tilde{1}\ 0\ 000-$	0
$\tilde{1}\ \tilde{1}\ \tilde{1}\ 0\ 100-$	$-13/4\sqrt{17.19}$
$\tilde{1}\ \tilde{1}\ \tilde{1}\ 0\ 110-$	0
$\tilde{0}\ 1\ \tilde{1}\ 0\ 000-$	$+3/\sqrt{5.11.17.19}$
$\tilde{0}\ 1\ \tilde{1}\ 0\ 010-$	$-2\sqrt{7}/\sqrt{3.5.17.19}$
$\tilde{0}\ 2\ 2\ 0\ 000+$	$+2/\sqrt{17.19}$
$\tilde{0}\ \tilde{1}\ 1\ 0\ 000-$	$-3/\sqrt{5.11.17.19}$
$\tilde{0}\ \tilde{1}\ 1\ 0\ 001-$	$-4\sqrt{19}/3\sqrt{7.11.17}$
$\tilde{0}\ \tilde{1}\ 1\ 0\ 010-$	$+2.9\sqrt{13}/11\sqrt{5.17.19}$
$\tilde{0}\ \tilde{1}\ 1\ 0\ 011-$	$+5\sqrt{13}/3.11\sqrt{7.17.19}$
7 7 6	
$1\ 1\ 0\ 0\ 000+$	$-8/\sqrt{5.11.13.17.19}$
$1\ 1\ 0\ 0\ 100+$	$-3\sqrt{3.7}/2\sqrt{5.13.17.19}$
$1\ 1\ 0\ 0\ 110+$	$+\sqrt{11}/\sqrt{5.13.17.19}$
$1\ 1\ 1\ 0\ 000-$	0
$1\ 1\ 1\ 0\ 100-$	$+9\sqrt{3.5}/4\sqrt{13.17.19}$
$1\ 1\ 1\ 0\ 110-$	0
$1\ 1\ 2\ 0\ 000+$	$+73\sqrt{7}/5\sqrt{5.11.13.17.19}$
$1\ 1\ 2\ 0\ 100+$	$+79\sqrt{3}/2.5\sqrt{5.13.17.19}$
$1\ 1\ 2\ 0\ 110+$	$-8\sqrt{7.11}/5\sqrt{5.13.17.19}$
$1\ 1\ \tilde{1}\ 0\ 000+$	$-2.7\sqrt{3}/5\sqrt{13.17.19}$
$1\ 1\ \tilde{1}\ 0\ 001+$	$-7.43\sqrt{3}/2.5\sqrt{5.11.13.17.19}$
$1\ 1\ \tilde{1}\ 0\ 100+$	$-3.59\sqrt{11}/4.5\sqrt{7.13.17.19}$
$1\ 1\ \tilde{1}\ 0\ 101+$	$+11.29/5\sqrt{5.7.13.17.19}$
$1\ 1\ \tilde{1}\ 0\ 110+$	$-11\sqrt{3}/5\sqrt{13.17.19}$
$1\ 1\ \tilde{1}\ 0\ 111+$	$-3\sqrt{3.11}/4.5\sqrt{5.13.17.19}$
$2\ 1\ 1\ 0\ 000+$	$+\sqrt{3.7}/2.5\sqrt{13.17.19}$
$2\ 1\ 1\ 0\ 010+$	$-3\sqrt{11}/2.5\sqrt{13.17.19}$
$2\ 1\ \tilde{1}\ 0\ 000-$	$-\sqrt{3.5.11}/2\sqrt{13.17.19}$
$2\ 1\ \tilde{1}\ 0\ 001-$	$-97/2.5\sqrt{3.13.17.19}$
$2\ 1\ \tilde{1}\ 0\ 010-$	$+11\sqrt{5}/2\sqrt{7.13.17.19}$
$2\ 1\ \tilde{1}\ 0\ 011-$	$-3.11\sqrt{11}/2.5\sqrt{7.13.17.19}$

SO_3–O 3jm Factors (cont.)

7 7 6		
2 2 0	0 000+	$-2\sqrt{2.11}/\sqrt{3.5.13.17.19}$
2 2 2	0 000+	$-4\sqrt{7.11}/\sqrt{3.5.13.17.19}$
2 2 $\tilde{0}$	0 000−	0
$\tilde{1}$ 1 1	0 000+	$+8\sqrt{3}/5\sqrt{13.17.19}$
$\tilde{1}$ 1 1	0 010+	$+3.53\sqrt{11}/4.5\sqrt{7.13.17.19}$
$\tilde{1}$ 1 1	0 100+	$+41\sqrt{3}/4.5\sqrt{11.17.19}$
$\tilde{1}$ 1 1	0 110+	$-9.3/5\sqrt{7.17.19}$
$\tilde{1}$ 1 2	0 000−	$+8.2/5\sqrt{13.17.19}$
$\tilde{1}$ 1 2	0 010−	$+\sqrt{3.11.13}/2.5\sqrt{7.17.19}$
$\tilde{1}$ 1 2	0 100−	$-3/2\sqrt{11.17.19}$
$\tilde{1}$ 1 2	0 110−	$+3\sqrt{3}/\sqrt{7.17.19}$
$\tilde{1}$ 1 $\tilde{1}$	0 000−	0
$\tilde{1}$ 1 $\tilde{1}$	0 001−	$+3\sqrt{7}/5\sqrt{13.17.19}$
$\tilde{1}$ 1 $\tilde{1}$	0 010−	$+3.11\sqrt{3.5}/4.7\sqrt{13.17.19}$
$\tilde{1}$ 1 $\tilde{1}$	0 011−	$-8.2\sqrt{3.11}/5.7\sqrt{13.17.19}$
$\tilde{1}$ 1 $\tilde{1}$	0 100−	$+3\sqrt{7}/4\sqrt{5.17.19}$
$\tilde{1}$ 1 $\tilde{1}$	0 101−	$+8.2\sqrt{7}/5\sqrt{11.17.19}$
$\tilde{1}$ 1 $\tilde{1}$	0 110−	$-4\sqrt{3.11}/7\sqrt{5.17.19}$
$\tilde{1}$ 1 $\tilde{1}$	0 111−	$-3.11\sqrt{3}/4.5.7\sqrt{17.19}$
$\tilde{1}$ 1 $\tilde{0}$	0 000+	$-8\sqrt{11}/\sqrt{5.13.17.19}$
$\tilde{1}$ 1 $\tilde{0}$	0 010+	$+11\sqrt{3}/2\sqrt{5.7.13.17.19}$
$\tilde{1}$ 1 $\tilde{0}$	0 100+	$-3/2\sqrt{5.17.19}$
$\tilde{1}$ 1 $\tilde{0}$	0 110+	$-\sqrt{3.11}/\sqrt{5.7.17.19}$
$\tilde{1}$ 2 1	0 000−	$+7\sqrt{3.11}/2\sqrt{5.13.17.19}$
$\tilde{1}$ 2 1	0 100−	$+7\sqrt{3}/2\sqrt{5.17.19}$
$\tilde{1}$ 2 $\tilde{1}$	0 000+	$+3.11\sqrt{3}/2\sqrt{7.13.17.19}$
$\tilde{1}$ 2 $\tilde{1}$	0 001+	$-3\sqrt{3.11}/2\sqrt{5.7.13.17.19}$
$\tilde{1}$ 2 $\tilde{1}$	0 100+	$+\sqrt{3.11}/2\sqrt{7.17.19}$
$\tilde{1}$ 2 $\tilde{1}$	0 101+	$-\sqrt{3}/2\sqrt{5.7.17.19}$
$\tilde{1}$ $\tilde{1}$ 0	0 000+	$-8\sqrt{11}/\sqrt{5.13.17.19}$
$\tilde{1}$ $\tilde{1}$ 0	0 100+	$+3/2\sqrt{5.17.19}$
$\tilde{1}$ $\tilde{1}$ 0	0 110+	$+\sqrt{5.13}/\sqrt{11.17.19}$
$\tilde{1}$ $\tilde{1}$ 1	0 000−	0
$\tilde{1}$ $\tilde{1}$ 1	0 100−	$+3\sqrt{7}/4\sqrt{5.17.19}$
$\tilde{1}$ $\tilde{1}$ 1	0 110−	0
$\tilde{1}$ $\tilde{1}$ 2	0 000+	$-\sqrt{7.11}/\sqrt{5.13.17.19}$
$\tilde{1}$ $\tilde{1}$ 2	0 100+	$+3\sqrt{7}/2\sqrt{5.17.19}$
$\tilde{1}$ $\tilde{1}$ 2	0 110+	$-2\sqrt{7.13}/\sqrt{5.11.17.19}$
$\tilde{1}$ $\tilde{1}$ $\tilde{1}$	0 000+	$-2.11\sqrt{3}/7\sqrt{13.17.19}$
$\tilde{1}$ $\tilde{1}$ $\tilde{1}$	0 001+	$-3\sqrt{3.11}/2.7\sqrt{5.13.17.19}$
$\tilde{1}$ $\tilde{1}$ $\tilde{1}$	0 100+	$+9\sqrt{3.11}/4.7\sqrt{17.19}$
$\tilde{1}$ $\tilde{1}$ $\tilde{1}$	0 101+	$+\sqrt{3.17}/7\sqrt{5.19}$
$\tilde{1}$ $\tilde{1}$ $\tilde{1}$	0 110+	$-\sqrt{3.13}/7\sqrt{17.19}$
$\tilde{1}$ $\tilde{1}$ $\tilde{1}$	0 111+	$+\sqrt{3.5.13.19}/4.7\sqrt{11.17}$
$\tilde{0}$ 1 $\tilde{1}$	0 000+	$+\sqrt{3}/\sqrt{5.17.19}$
$\tilde{0}$ 1 $\tilde{1}$	0 001+	$+8.4/5\sqrt{3.11.17.19}$
$\tilde{0}$ 1 $\tilde{1}$	0 010+	$+2\sqrt{11}/\sqrt{5.7.17.19}$
$\tilde{0}$ 1 $\tilde{1}$	0 011+	$+3.11/5\sqrt{7.17.19}$
$\tilde{0}$ 2 2	0 000−	$+2\sqrt{2.7}/\sqrt{3.5.17.19}$
7 7 6		
$\tilde{0}$ $\tilde{1}$ 1	0 000+	$+\sqrt{3}/\sqrt{5.17.19}$
$\tilde{0}$ $\tilde{1}$ 1	0 010+	$-2\sqrt{3.13}/\sqrt{5.11.17.19}$
$\tilde{0}$ $\tilde{0}$ 0	0 000+	$+8\sqrt{13}/\sqrt{3.5.11.17.19}$
7 7 7		
1 1 1	0 000+	$+2\sqrt{2.7}/5\sqrt{13.17.19}$
1 1 1	0 100+	$-3\sqrt{2.3.11}/5\sqrt{13.17.19}$
1 1 1	0 110+	$+\sqrt{2.7}/5\sqrt{13.17.19}$
1 1 1	0 111+	$+\sqrt{2.3.11}/\sqrt{13.17.19}$
2 1 1	0 000−	0
2 1 1	0 010−	$+31/\sqrt{2.5.13.17.19}$
2 1 1	0 011−	0
2 2 2	0 000−	0
$\tilde{1}$ 1 1	0 000−	0
$\tilde{1}$ 1 1	0 010−	$-3\sqrt{2}/\sqrt{5.7.13.17.19}$
$\tilde{1}$ 1 1	0 011−	0
$\tilde{1}$ 1 1	0 100−	0
$\tilde{1}$ 1 1	0 110−	$-2\sqrt{2.11}/\sqrt{5.7.17.19}$
$\tilde{1}$ 1 1	0 111−	0
$\tilde{1}$ 2 1	0 000+	$+2\sqrt{2.3}/5\sqrt{13.17.19}$
$\tilde{1}$ 2 1	0 001+	$+53\sqrt{11}/5\sqrt{2.7.13.17.19}$
$\tilde{1}$ 2 1	0 100+	$+\sqrt{11}/5\sqrt{2.3.17.19}$
$\tilde{1}$ 2 1	0 101+	$-2.9\sqrt{2}/5\sqrt{7.17.19}$
$\tilde{1}$ $\tilde{1}$ 1	0 000+	$-2.11\sqrt{2.7}/5\sqrt{13.17.19}$
$\tilde{1}$ $\tilde{1}$ 1	0 001+	$-11\sqrt{2.3.11}/5.7\sqrt{13.17.19}$
$\tilde{1}$ $\tilde{1}$ 1	0 100+	0
$\tilde{1}$ $\tilde{1}$ 1	0 101+	$+3\sqrt{2.3}/7\sqrt{17.19}$
$\tilde{1}$ $\tilde{1}$ 1	0 110+	$+\sqrt{2.7.13}/5\sqrt{17.19}$
$\tilde{1}$ $\tilde{1}$ 1	0 111+	$+3\sqrt{2.3.11.13}/5.7\sqrt{17.19}$
$\tilde{1}$ $\tilde{1}$ 2	0 000−	0
$\tilde{1}$ $\tilde{1}$ 2	0 100−	$+\sqrt{3.7}/\sqrt{2.5.17.19}$
$\tilde{1}$ $\tilde{1}$ 2	0 110−	0
$\tilde{1}$ $\tilde{1}$ $\tilde{1}$	0 000−	0
$\tilde{1}$ $\tilde{1}$ $\tilde{1}$	0 100−	0
$\tilde{1}$ $\tilde{1}$ $\tilde{1}$	0 110−	0
$\tilde{1}$ $\tilde{1}$ $\tilde{1}$	0 111−	0
$\tilde{0}$ 2 2	0 000+	$+\sqrt{7}/\sqrt{5.17.19}$
$\tilde{0}$ $\tilde{1}$ 1	0 000−	$-\sqrt{3.11}/5\sqrt{2.17.19}$
$\tilde{0}$ $\tilde{1}$ 1	0 001−	$+4\sqrt{2}/5\sqrt{7.17.19}$
$\tilde{0}$ $\tilde{1}$ 1	0 010−	$-4\sqrt{2.13}/5\sqrt{3.17.19}$
$\tilde{0}$ $\tilde{1}$ 1	0 011−	$+3\sqrt{11.13}/5\sqrt{2.7.17.19}$
$\frac{15}{2}$ 4 $\frac{7}{2}$		
$\frac{1}{2}$ 0 $\frac{1}{2}$	0 000+	$-\sqrt{5}/2\sqrt{3.13}$
$\frac{1}{2}$ 1 $\frac{1}{2}$	0 000−	$+1/\sqrt{3.13}$
$\frac{1}{2}$ 1 $\frac{3}{2}$	0 000+	$+\sqrt{7}/\sqrt{3.5.13}$
$\frac{1}{2}$ 2 $\frac{3}{2}$	0 000−	$-7/2\sqrt{2.3.5.13}$
$\frac{1}{2}$ $\tilde{1}$ $\frac{3}{2}$	0 000−	0
$\frac{1}{2}$ $\tilde{1}$ $\tilde{\frac{3}{2}}$	0 000+	0
$\frac{3}{2}$ 0 $\frac{3}{2}$	0 000+	$+4/3\sqrt{5.11.13}$
$\frac{3}{2}$ 0 $\frac{3}{2}$	0 100+	$-\sqrt{7}/\sqrt{2.5.13}$

SO_3–O 3jm Factors (cont.)

$\frac{15}{2}$ 4 $\frac{7}{2}$

$\frac{3}{2}$	0	$\frac{3}{2}$	0	200+	0
$\frac{3}{2}$	1	$\frac{1}{2}$	0	000+	$-\sqrt{2.7}/3\sqrt{5.11.13}$
$\frac{3}{2}$	1	$\frac{1}{2}$	0	100+	$-2/\sqrt{5.13}$
$\frac{3}{2}$	1	$\frac{1}{2}$	0	200+	0
$\frac{3}{2}$	1	$\frac{3}{2}$	0	000−	$+7.7/2.3.5\sqrt{11.13}$
$\frac{3}{2}$	1	$\frac{3}{2}$	0	100−	$-2\sqrt{2.7}/3.5\sqrt{13}$
$\frac{3}{2}$	1	$\frac{3}{2}$	0	200−	$+1/2.3$
$\frac{3}{2}$	1	$\frac{3}{2}$	1	000−	$+\sqrt{11}/4.3\sqrt{13}$
$\frac{3}{2}$	1	$\frac{3}{2}$	1	100−	$+\sqrt{7}/3\sqrt{2.13}$
$\frac{3}{2}$	1	$\frac{3}{2}$	1	200−	$-1/4.3$
$\frac{3}{2}$	1	$\tilde{\frac{1}{2}}$	0	000+	$+9\sqrt{3}/2.5\sqrt{2.11.13}$
$\frac{3}{2}$	1	$\tilde{\frac{1}{2}}$	0	100+	$-\sqrt{7}/5\sqrt{3.13}$
$\frac{3}{2}$	1	$\tilde{\frac{1}{2}}$	0	200+	$-1/2\sqrt{2.3}$
$\frac{3}{2}$	2	$\frac{1}{2}$	0	000−	$-\sqrt{5.13}/8.3\sqrt{11}$
$\frac{3}{2}$	2	$\frac{1}{2}$	0	100−	$+\sqrt{5.7}/2.3\sqrt{2.13}$
$\frac{3}{2}$	2	$\frac{1}{2}$	0	200−	$-\sqrt{5}/8.3$
$\frac{3}{2}$	2	$\frac{3}{2}$	0	000+	$-71\sqrt{7}/4.3.5\sqrt{2.11.13}$
$\frac{3}{2}$	2	$\frac{3}{2}$	0	100+	$-7/3.5\sqrt{13}$
$\frac{3}{2}$	2	$\frac{3}{2}$	0	200+	$+\sqrt{7}/4.3\sqrt{2}$
$\frac{3}{2}$	2	$\tilde{\frac{1}{2}}$	0	000−	$+\sqrt{3.7.11}/8.5\sqrt{13}$
$\frac{3}{2}$	2	$\tilde{\frac{1}{2}}$	0	100−	$+7/2.5\sqrt{2.3.13}$
$\frac{3}{2}$	2	$\tilde{\frac{1}{2}}$	0	200−	$+\sqrt{7}/8\sqrt{3}$
$\frac{3}{2}$	$\tilde{1}$	$\frac{1}{2}$	0	000−	$-3\sqrt{3}/2\sqrt{2.5.11.13}$
$\frac{3}{2}$	$\tilde{1}$	$\frac{1}{2}$	0	100−	$+\sqrt{7}/3\sqrt{3.5.13}$
$\frac{3}{2}$	$\tilde{1}$	$\frac{1}{2}$	0	200−	$+\sqrt{5}/2.3\sqrt{2.3}$
$\frac{3}{2}$	$\tilde{1}$	$\frac{3}{2}$	0	000+	$-\sqrt{7.11}/2.5\sqrt{3.13}$
$\frac{3}{2}$	$\tilde{1}$	$\frac{3}{2}$	0	100+	$-7\sqrt{2}/3.5\sqrt{3.13}$
$\frac{3}{2}$	$\tilde{1}$	$\frac{3}{2}$	0	200+	$-\sqrt{7}/2.3\sqrt{3}$
$\frac{3}{2}$	$\tilde{1}$	$\frac{3}{2}$	1	000−	$+29\sqrt{7}/4.5\sqrt{3.11.13}$
$\frac{3}{2}$	$\tilde{1}$	$\frac{3}{2}$	1	100−	$-7/3.5\sqrt{2.3.13}$
$\frac{3}{2}$	$\tilde{1}$	$\frac{3}{2}$	1	200−	$-\sqrt{7}/4.3\sqrt{3}$
$\frac{3}{2}$	$\tilde{1}$	$\tilde{\frac{1}{2}}$	0	000−	$+\sqrt{2.7}/\sqrt{11.13}$
$\frac{3}{2}$	$\tilde{1}$	$\tilde{\frac{1}{2}}$	0	100−	0
$\frac{3}{2}$	$\tilde{1}$	$\tilde{\frac{1}{2}}$	0	200−	0
$\tilde{\frac{1}{2}}$	0	$\tilde{\frac{1}{2}}$	0	000+	$-1/2\sqrt{11}$
$\tilde{\frac{1}{2}}$	1	$\frac{3}{2}$	0	000+	$+2/\sqrt{3.5.11}$
$\tilde{\frac{1}{2}}$	1	$\tilde{\frac{1}{2}}$	0	000−	$+1/\sqrt{5.11}$
$\tilde{\frac{1}{2}}$	2	$\frac{3}{2}$	0	000−	$-\sqrt{7}/2\sqrt{2.3.5.11}$
$\tilde{\frac{1}{2}}$	$\tilde{1}$	$\frac{1}{2}$	0	000+	$-2/3\sqrt{11}$
$\tilde{\frac{1}{2}}$	$\tilde{1}$	$\frac{3}{2}$	0	000−	$+\sqrt{7}/3\sqrt{5.11}$

$\frac{15}{2}$ $\frac{9}{2}$ 3

$\frac{1}{2}$	$\frac{1}{2}$	1	0	000−	$+\sqrt{3}/2\sqrt{13}$
$\frac{1}{2}$	$\frac{3}{2}$	1	0	000+	$-1/2\sqrt{5.13}$
$\frac{1}{2}$	$\frac{3}{2}$	1	0	010+	$-\sqrt{3.7}/4\sqrt{5.13}$
$\frac{1}{2}$	$\frac{3}{2}$	$\tilde{1}$	0	000−	$-3/2.5\sqrt{13}$
$\frac{1}{2}$	$\frac{3}{2}$	$\tilde{1}$	0	010−	$-3\sqrt{3.7}/4.5\sqrt{13}$
$\frac{3}{2}$	$\frac{1}{2}$	1	0	000+	$+2\sqrt{2}/\sqrt{5.7.11.13}$
$\frac{3}{2}$	$\frac{1}{2}$	1	0	100+	$-3/2\sqrt{5.13}$
$\frac{3}{2}$	$\frac{1}{2}$	1	0	200+	0

$\frac{15}{2}$ $\frac{9}{2}$ 3

$\frac{3}{2}$	$\frac{1}{2}$	$\tilde{1}$	0	000−	$-2.3\sqrt{2}/5\sqrt{7.11.13}$
$\frac{3}{2}$	$\frac{1}{2}$	$\tilde{1}$	0	100−	$+9/2.5\sqrt{13}$
$\frac{3}{2}$	$\frac{1}{2}$	$\tilde{1}$	0	200−	0
$\frac{3}{2}$	$\frac{3}{2}$	1	0	000−	$+9\sqrt{3}/4\sqrt{7.11.13}$
$\frac{3}{2}$	$\frac{3}{2}$	1	0	010−	$-\sqrt{13}/8\sqrt{11}$
$\frac{3}{2}$	$\frac{3}{2}$	1	0	100−	$+1/\sqrt{2.3.13}$
$\frac{3}{2}$	$\frac{3}{2}$	1	0	110−	$+\sqrt{7}/2\sqrt{2.13}$
$\frac{3}{2}$	$\frac{3}{2}$	1	0	200−	$-1/4\sqrt{3.7}$
$\frac{3}{2}$	$\frac{3}{2}$	1	0	210−	$-1/8$
$\frac{3}{2}$	$\frac{3}{2}$	1	1	000−	$+3\sqrt{3}/5\sqrt{7.11.13}$
$\frac{3}{2}$	$\frac{3}{2}$	1	1	010−	$+71/8.5\sqrt{11.13}$
$\frac{3}{2}$	$\frac{3}{2}$	1	1	100−	$+7/5\sqrt{2.3.13}$
$\frac{3}{2}$	$\frac{3}{2}$	1	1	110−	$+\sqrt{7}/5\sqrt{2.13}$
$\frac{3}{2}$	$\frac{3}{2}$	1	1	200−	$+1/\sqrt{3.7}$
$\frac{3}{2}$	$\frac{3}{2}$	1	1	210−	$-1/8$
$\frac{3}{2}$	$\frac{3}{2}$	$\tilde{1}$	0	000+	$+3.41\sqrt{3}/4.5\sqrt{5.7.11.13}$
$\frac{3}{2}$	$\frac{3}{2}$	$\tilde{1}$	0	010+	$+9\sqrt{11}/8.5\sqrt{5.13}$
$\frac{3}{2}$	$\frac{3}{2}$	$\tilde{1}$	0	100+	$-7/5\sqrt{2.3.5.13}$
$\frac{3}{2}$	$\frac{3}{2}$	$\tilde{1}$	0	110+	$+3\sqrt{7}/2.5\sqrt{2.5.13}$
$\frac{3}{2}$	$\frac{3}{2}$	$\tilde{1}$	0	200+	$-\sqrt{7}/4\sqrt{3.5}$
$\frac{3}{2}$	$\frac{3}{2}$	$\tilde{1}$	0	210+	$+3/8\sqrt{5}$
$\frac{3}{2}$	$\frac{3}{2}$	$\tilde{1}$	1	000−	$+2.23\sqrt{3}/5\sqrt{5.7.11.13}$
$\frac{3}{2}$	$\frac{3}{2}$	$\tilde{1}$	1	010−	$-3.71/8.5\sqrt{5.11.13}$
$\frac{3}{2}$	$\frac{3}{2}$	$\tilde{1}$	1	100−	$-1/5\sqrt{2.3.5.13}$
$\frac{3}{2}$	$\frac{3}{2}$	$\tilde{1}$	1	110−	$-3\sqrt{7}/5\sqrt{2.5.13}$
$\frac{3}{2}$	$\frac{3}{2}$	$\tilde{1}$	1	200−	$+2/\sqrt{3.5.7}$
$\frac{3}{2}$	$\frac{3}{2}$	$\tilde{1}$	1	210−	$+3/8\sqrt{5}$
$\frac{3}{2}$	$\frac{3}{2}$	$\tilde{0}$	0	000−	$+167/2.5\sqrt{5.7.11.13}$
$\frac{3}{2}$	$\frac{3}{2}$	$\tilde{0}$	0	010−	$-3\sqrt{3.11}/4.5\sqrt{5.13}$
$\frac{3}{2}$	$\frac{3}{2}$	$\tilde{0}$	0	100−	$-\sqrt{2}/5\sqrt{5.13}$
$\frac{3}{2}$	$\frac{3}{2}$	$\tilde{0}$	0	110−	$-\sqrt{3.7}/5\sqrt{2.5.13}$
$\frac{3}{2}$	$\frac{3}{2}$	$\tilde{0}$	0	200−	$-1/2\sqrt{5.7}$
$\frac{3}{2}$	$\frac{3}{2}$	$\tilde{0}$	0	210−	$-\sqrt{3}/4\sqrt{5}$
$\tilde{\frac{1}{2}}$	$\frac{1}{2}$	$\tilde{1}$	0	000+	$-3\sqrt{3}/2\sqrt{5.7.11}$
$\tilde{\frac{1}{2}}$	$\frac{1}{2}$	$\tilde{0}$	0	000−	$-3/\sqrt{5.7.11}$
$\tilde{\frac{1}{2}}$	$\frac{3}{2}$	1	0	000+	$+9/2\sqrt{5.7.11}$
$\tilde{\frac{1}{2}}$	$\frac{3}{2}$	1	0	010+	$-\sqrt{3}/4\sqrt{5.11}$
$\tilde{\frac{1}{2}}$	$\frac{3}{2}$	$\tilde{1}$	0	000−	$-13/2.5\sqrt{7.11}$
$\tilde{\frac{1}{2}}$	$\frac{3}{2}$	$\tilde{1}$	0	010−	$-3\sqrt{3}/4.5\sqrt{11}$

$\frac{15}{2}$ $\frac{9}{2}$ 4

$\frac{1}{2}$	$\frac{1}{2}$	0	0	000−	$-1/\sqrt{3.13.17}$
$\frac{1}{2}$	$\frac{1}{2}$	1	0	000+	$-\sqrt{13}/2\sqrt{3.5.17}$
$\frac{1}{2}$	$\frac{3}{2}$	1	0	000−	$+9.3/2.5\sqrt{13.17}$
$\frac{1}{2}$	$\frac{3}{2}$	1	0	010−	$+\sqrt{7}/4.5\sqrt{3.13.17}$
$\frac{1}{2}$	$\frac{3}{2}$	2	0	000+	$+3\sqrt{2.7}/5\sqrt{13.17}$
$\frac{1}{2}$	$\frac{3}{2}$	2	0	010+	$-7/5\sqrt{2.3.13.17}$
$\frac{1}{2}$	$\frac{3}{2}$	$\tilde{1}$	0	000+	$+3\sqrt{3.7}/2.5\sqrt{13.17}$
$\frac{1}{2}$	$\frac{3}{2}$	$\tilde{1}$	0	010+	$+9.7/4.5\sqrt{13.17}$
$\frac{3}{2}$	$\frac{1}{2}$	1	0	000−	$+2\sqrt{2.7.17}/3.5\sqrt{11.13}$

SO_3–O 3jm Factors (cont.)

$\frac{15}{2}$	**$\frac{9}{2}$**	**4**			
$\frac{3}{2}$	$\frac{1}{2}$	1	0	100−	$+1/2.5\sqrt{13.17}$
$\frac{3}{2}$	$\frac{1}{2}$	1	0	200−	0
$\frac{3}{2}$	$\frac{1}{2}$	2	0	000+	$-2.29/3.5\sqrt{11.13.17}$
$\frac{3}{2}$	$\frac{1}{2}$	2	0	100+	$-7\sqrt{2.7}/3.5\sqrt{13.17}$
$\frac{3}{2}$	$\frac{1}{2}$	2	0	200+	$-2/3\sqrt{17}$
$\frac{3}{2}$	$\frac{1}{2}$	$\tilde{1}$	0	000+	$-2.3\sqrt{2.3}/5\sqrt{11.13.17}$
$\frac{3}{2}$	$\frac{1}{2}$	$\tilde{1}$	0	100+	$-7.7\sqrt{7}/2.3.5\sqrt{3.13.17}$
$\frac{3}{2}$	$\frac{1}{2}$	$\tilde{1}$	0	200+	$+4\sqrt{2}/3\sqrt{3.17}$
$\frac{3}{2}$	$\frac{3}{2}$	0	0	000−	$-\sqrt{3.7.13}/2.5\sqrt{11.17}$
$\frac{3}{2}$	$\frac{3}{2}$	0	0	010−	$-7.47/4.3.5\sqrt{11.13.17}$
$\frac{3}{2}$	$\frac{3}{2}$	0	0	100−	$-11\sqrt{2}/5\sqrt{3.13.17}$
$\frac{3}{2}$	$\frac{3}{2}$	0	0	110−	$-\sqrt{7}/5\sqrt{2.13.17}$
$\frac{3}{2}$	$\frac{3}{2}$	0	0	200−	$+\sqrt{7}/2\sqrt{3.17}$
$\frac{3}{2}$	$\frac{3}{2}$	0	0	210−	$-3/4\sqrt{17}$
$\frac{3}{2}$	$\frac{3}{2}$	1	0	000+	$+83\sqrt{3.7}/4.5\sqrt{5.11.13.17}$
$\frac{3}{2}$	$\frac{3}{2}$	1	0	010+	$-211/8.3.5\sqrt{5.11.13.17}$
$\frac{3}{2}$	$\frac{3}{2}$	1	0	100+	$+61/5\sqrt{2.3.5.13.17}$
$\frac{3}{2}$	$\frac{3}{2}$	1	0	110+	$-97\sqrt{7}/2.3.5\sqrt{2.5.13.17}$
$\frac{3}{2}$	$\frac{3}{2}$	1	0	200+	$+\sqrt{7}/4\sqrt{3.5.17}$
$\frac{3}{2}$	$\frac{3}{2}$	1	0	210+	$-19/8.3\sqrt{5.17}$
$\frac{3}{2}$	$\frac{3}{2}$	1	1	000+	$+19\sqrt{3.7}/5\sqrt{5.11.13.17}$
$\frac{3}{2}$	$\frac{3}{2}$	1	1	010+	$+2713/8.3.5\sqrt{5.11.13.17}$
$\frac{3}{2}$	$\frac{3}{2}$	1	1	100+	$-53/5\sqrt{2.3.5.13.17}$
$\frac{3}{2}$	$\frac{3}{2}$	1	1	110+	$-\sqrt{7.17}/3.5\sqrt{2.5.13}$
$\frac{3}{2}$	$\frac{3}{2}$	1	1	200+	$+\sqrt{7}/\sqrt{3.5.17}$
$\frac{3}{2}$	$\frac{3}{2}$	1	1	210+	$-31/8.3\sqrt{5.17}$
$\frac{3}{2}$	$\frac{3}{2}$	2	0	000−	$+37\sqrt{2.3}/5\sqrt{5.11.13.17}$
$\frac{3}{2}$	$\frac{3}{2}$	2	0	010−	$+71\sqrt{7}/2.3.5\sqrt{2.5.11.13.17}$
$\frac{3}{2}$	$\frac{3}{2}$	2	0	100−	$-2.11\sqrt{7}/5\sqrt{3.5.13.17}$
$\frac{3}{2}$	$\frac{3}{2}$	2	0	110−	$+2.7/3.5\sqrt{5.13.17}$
$\frac{3}{2}$	$\frac{3}{2}$	2	0	200−	$-\sqrt{2}/\sqrt{3.5.17}$
$\frac{3}{2}$	$\frac{3}{2}$	2	0	210−	$-\sqrt{7}/2.3\sqrt{2.5.17}$
$\frac{3}{2}$	$\frac{3}{2}$	$\tilde{1}$	0	000−	$+23\sqrt{13}/4.5\sqrt{5.11.17}$
$\frac{3}{2}$	$\frac{3}{2}$	$\tilde{1}$	0	010−	$+\sqrt{7.11.17}/8.5\sqrt{3.5.13}$
$\frac{3}{2}$	$\frac{3}{2}$	$\tilde{1}$	0	100−	$+3\sqrt{7}/5\sqrt{2.5.13.17}$
$\frac{3}{2}$	$\frac{3}{2}$	$\tilde{1}$	0	110−	$+7\sqrt{17}/2.3.5\sqrt{2.3.5.13}$
$\frac{3}{2}$	$\frac{3}{2}$	$\tilde{1}$	0	200−	$+3/4\sqrt{5.17}$
$\frac{3}{2}$	$\frac{3}{2}$	$\tilde{1}$	0	210−	$+\sqrt{7.17}/8.3\sqrt{3.5}$
$\frac{3}{2}$	$\frac{3}{2}$	$\tilde{1}$	1	000+	$+8\sqrt{11}/5\sqrt{5.13.17}$
$\frac{3}{2}$	$\frac{3}{2}$	$\tilde{1}$	1	010+	$-17\sqrt{7.17}/8.5\sqrt{3.5.11.13}$
$\frac{3}{2}$	$\frac{3}{2}$	$\tilde{1}$	1	100+	$+9\sqrt{7}/5\sqrt{2.5.13.17}$
$\frac{3}{2}$	$\frac{3}{2}$	$\tilde{1}$	1	110+	$+7.43/3.5\sqrt{2.3.5.13.17}$
$\frac{3}{2}$	$\frac{3}{2}$	$\tilde{1}$	1	200+	0
$\frac{3}{2}$	$\frac{3}{2}$	$\tilde{1}$	1	210+	$+\sqrt{5.7}/8.3\sqrt{3.17}$
$\tilde{\frac{1}{2}}$	$\frac{1}{2}$	$\tilde{1}$	0	000−	$-\sqrt{17}/2.3\sqrt{5.11}$
$\tilde{\frac{1}{2}}$	$\frac{3}{2}$	1	0	000−	$-\sqrt{7}/2\sqrt{11.17}$
$\tilde{\frac{1}{2}}$	$\frac{3}{2}$	1	0	010−	$+13/4\sqrt{3.11.17}$
$\tilde{\frac{1}{2}}$	$\frac{3}{2}$	2	0	000+	$-\sqrt{2}/5\sqrt{11.17}$
$\tilde{\frac{1}{2}}$	$\frac{3}{2}$	2	0	010+	$+\sqrt{7.17}/5\sqrt{2.3.11}$
$\frac{15}{2}$	**$\frac{9}{2}$**	**4**			
$\tilde{\frac{1}{2}}$	$\frac{3}{2}$	$\tilde{1}$	0	000+	$-3\sqrt{3}/2.5\sqrt{11.17}$
$\tilde{\frac{1}{2}}$	$\frac{3}{2}$	$\tilde{1}$	0	010+	$-37\sqrt{7}/4.3.5\sqrt{11.17}$
$\frac{15}{2}$	**5**	**$\frac{5}{2}$**			
$\frac{1}{2}$	1	$\frac{3}{2}$	0	000+	$-\sqrt{7}/4\sqrt{11.13}$
$\frac{1}{2}$	1	$\frac{3}{2}$	0	010+	$-3\sqrt{5}/2\sqrt{11.13}$
$\frac{1}{2}$	2	$\frac{3}{2}$	0	000−	0
$\frac{1}{2}$	$\tilde{1}$	$\frac{3}{2}$	0	000−	$+\sqrt{5}/4\sqrt{13}$
$\frac{1}{2}$	$\tilde{1}$	$\tilde{\frac{1}{2}}$	0	000+	$-1/2\sqrt{13}$
$\frac{3}{2}$	1	$\frac{3}{2}$	0	000−	$+29/8\sqrt{3.5.13}$
$\frac{3}{2}$	1	$\frac{3}{2}$	0	010−	0
$\frac{3}{2}$	1	$\frac{3}{2}$	0	100−	$-\sqrt{7.11}/2.3\sqrt{2.3.5.13}$
$\frac{3}{2}$	1	$\frac{3}{2}$	0	110−	0
$\frac{3}{2}$	1	$\frac{3}{2}$	0	200−	$-\sqrt{5.11}/8.3\sqrt{3}$
$\frac{3}{2}$	1	$\frac{3}{2}$	0	210−	0
$\frac{3}{2}$	1	$\frac{3}{2}$	1	000−	$-137/8.11\sqrt{3.5.13}$
$\frac{3}{2}$	1	$\frac{3}{2}$	1	010−	$-4\sqrt{3}/11\sqrt{7.13}$
$\frac{3}{2}$	1	$\frac{3}{2}$	1	100−	$-\sqrt{7}/3\sqrt{2.3.5.11.13}$
$\frac{3}{2}$	1	$\frac{3}{2}$	1	110−	$+3\sqrt{3}/\sqrt{2.11.13}$
$\frac{3}{2}$	1	$\frac{3}{2}$	1	200−	$-\sqrt{5.11}/8.3\sqrt{3}$
$\frac{3}{2}$	1	$\frac{3}{2}$	1	210−	0
$\frac{3}{2}$	1	$\tilde{\frac{1}{2}}$	0	000+	$-109/4.11\sqrt{2.3.13}$
$\frac{3}{2}$	1	$\tilde{\frac{1}{2}}$	0	010+	$+4\sqrt{2.3}/11\sqrt{5.7.13}$
$\frac{3}{2}$	1	$\tilde{\frac{1}{2}}$	0	100+	$-2\sqrt{7}/3\sqrt{3.11.13}$
$\frac{3}{2}$	1	$\tilde{\frac{1}{2}}$	0	110+	$-3\sqrt{3}/\sqrt{5.11.13}$
$\frac{3}{2}$	1	$\tilde{\frac{1}{2}}$	0	200+	$-\sqrt{11}/4.3\sqrt{2.3}$
$\frac{3}{2}$	1	$\tilde{\frac{1}{2}}$	0	210+	0
$\frac{3}{2}$	2	$\frac{3}{2}$	0	000+	$+\sqrt{11}/2\sqrt{2.7.13}$
$\frac{3}{2}$	2	$\frac{3}{2}$	0	100+	$+1/3\sqrt{13}$
$\frac{3}{2}$	2	$\frac{3}{2}$	0	200+	$+5/2.3\sqrt{2.7}$
$\frac{3}{2}$	2	$\tilde{\frac{1}{2}}$	0	000−	$+29/2\sqrt{5.7.11.13}$
$\frac{3}{2}$	2	$\tilde{\frac{1}{2}}$	0	100−	$-\sqrt{2}/3\sqrt{5.13}$
$\frac{3}{2}$	2	$\tilde{\frac{1}{2}}$	0	200−	$-\sqrt{5}/2.3\sqrt{7}$
$\frac{3}{2}$	$\tilde{1}$	$\frac{3}{2}$	0	000+	$+5\sqrt{11}/8\sqrt{3.7.13}$
$\frac{3}{2}$	$\tilde{1}$	$\frac{3}{2}$	0	100+	$+5/2\sqrt{2.3.13}$
$\frac{3}{2}$	$\tilde{1}$	$\frac{3}{2}$	0	200+	$-5/8\sqrt{3.7}$
$\frac{3}{2}$	$\tilde{1}$	$\frac{3}{2}$	1	000−	$+7\sqrt{7}/8\sqrt{3.11.13}$
$\frac{3}{2}$	$\tilde{1}$	$\frac{3}{2}$	1	100−	$-1/\sqrt{2.3.13}$
$\frac{3}{2}$	$\tilde{1}$	$\frac{3}{2}$	1	200−	$+5/8\sqrt{3.7}$
$\frac{3}{2}$	$\tilde{1}$	$\tilde{\frac{1}{2}}$	0	000−	$-\sqrt{5}/4\sqrt{2.3.7.11.13}$
$\frac{3}{2}$	$\tilde{1}$	$\tilde{\frac{1}{2}}$	0	100−	0
$\frac{3}{2}$	$\tilde{1}$	$\tilde{\frac{1}{2}}$	0	200−	$+\sqrt{3.5}/4\sqrt{2.7}$
$\tilde{\frac{1}{2}}$	1	$\frac{3}{2}$	0	000+	$-19/4.3.11$
$\tilde{\frac{1}{2}}$	1	$\frac{3}{2}$	0	010+	$+3\sqrt{5}/2.11\sqrt{7}$
$\tilde{\frac{1}{2}}$	1	$\tilde{\frac{1}{2}}$	0	000−	$+\sqrt{5}/2.3.11$
$\tilde{\frac{1}{2}}$	1	$\tilde{\frac{1}{2}}$	0	010−	$-2.3/11\sqrt{7}$
$\tilde{\frac{1}{2}}$	2	$\frac{3}{2}$	0	000−	$-\sqrt{2.5}/\sqrt{3.7.11}$
$\tilde{\frac{1}{2}}$	$\tilde{1}$	$\frac{3}{2}$	0	000−	$+\sqrt{5}/4\sqrt{7.11}$
$\frac{15}{2}$	**5**	**$\frac{7}{2}$**			
$\frac{1}{2}$	1	$\frac{1}{2}$	0	000+	$-7\sqrt{5}/2\sqrt{3.11.13.17}$

SO_3–O 3jm Factors (cont.)

$\frac{15}{2}$	5	$\frac{7}{2}$			
$\frac{1}{2}$	1	$\frac{1}{2}$	0	010+	$+\sqrt{3.7}/2\sqrt{11.13.17}$
$\frac{1}{2}$	1	$\frac{3}{2}$	0	000−	$-\sqrt{7.11}/4\sqrt{3.13.17}$
$\frac{1}{2}$	1	$\frac{3}{2}$	0	010−	$+\sqrt{3.11}/\sqrt{5.13.17}$
$\frac{1}{2}$	2	$\frac{3}{2}$	0	000+	$-7/\sqrt{2.5.13.17}$
$\frac{1}{2}$	$\tilde{1}$	$\frac{3}{2}$	0	000+	$-\sqrt{3.5}/4\sqrt{13.17}$
$\frac{1}{2}$	$\tilde{1}$	$\tilde{\frac{1}{2}}$	0	000−	$-3\sqrt{5}/2\sqrt{13.17}$
$\frac{3}{2}$	1	$\frac{1}{2}$	0	000−	$-9\sqrt{7}/8.11\sqrt{2.13.17}$
$\frac{3}{2}$	1	$\frac{1}{2}$	0	010−	$+2\sqrt{2.17}/11\sqrt{5.13}$
$\frac{3}{2}$	1	$\frac{1}{2}$	0	100−	$+7/4.3\sqrt{11.13.17}$
$\frac{3}{2}$	1	$\frac{1}{2}$	0	110−	$-3\sqrt{7}/\sqrt{5.11.13.17}$
$\frac{3}{2}$	1	$\frac{1}{2}$	0	200−	$-5\sqrt{7.11}/8.3\sqrt{2.17}$
$\frac{3}{2}$	1	$\frac{1}{2}$	0	210−	0
$\frac{3}{2}$	1	$\frac{3}{2}$	0	000+	$+3.67/4.11\sqrt{5.13.17}$
$\frac{3}{2}$	1	$\frac{3}{2}$	0	010+	$-29\sqrt{7}/5.11\sqrt{13.17}$
$\frac{3}{2}$	1	$\frac{3}{2}$	0	100+	$+8\sqrt{2.7}/3\sqrt{5.11.13.17}$
$\frac{3}{2}$	1	$\frac{3}{2}$	0	110+	$-7.7/5\sqrt{2.11.13.17}$
$\frac{3}{2}$	1	$\frac{3}{2}$	0	200+	$-5\sqrt{5}/4.3\sqrt{11.17}$
$\frac{3}{2}$	1	$\frac{3}{2}$	0	210+	$-\sqrt{7}/\sqrt{11.17}$
$\frac{3}{2}$	1	$\frac{3}{2}$	1	000+	$-3/8.11\sqrt{5.13.17}$
$\frac{3}{2}$	1	$\frac{3}{2}$	1	010+	$-109/2.11\sqrt{7.13.17}$
$\frac{3}{2}$	1	$\frac{3}{2}$	1	100+	$-\sqrt{2.7.17}/3\sqrt{5.11.13}$
$\frac{3}{2}$	1	$\frac{3}{2}$	1	110+	$-\sqrt{2}/\sqrt{11.13.17}$
$\frac{3}{2}$	1	$\frac{3}{2}$	1	200+	$-\sqrt{5.17}/8.3\sqrt{11}$
$\frac{3}{2}$	1	$\frac{3}{2}$	1	210+	$+\sqrt{7}/2\sqrt{11.17}$
$\frac{3}{2}$	1	$\tilde{\frac{1}{2}}$	0	000−	$+9.29\sqrt{3}/8.11\sqrt{2.5.13.17}$
$\frac{3}{2}$	1	$\tilde{\frac{1}{2}}$	0	010−	$-3.23\sqrt{3}/5.11\sqrt{2.7.13.17}$
$\frac{3}{2}$	1	$\tilde{\frac{1}{2}}$	0	100−	$-19\sqrt{7}/4\sqrt{3.5.11.13.17}$
$\frac{3}{2}$	1	$\tilde{\frac{1}{2}}$	0	110−	$-31\sqrt{3}/5\sqrt{11.13.17}$
$\frac{3}{2}$	1	$\tilde{\frac{1}{2}}$	0	200−	$-\sqrt{5}/8\sqrt{2.3.11.17}$
$\frac{3}{2}$	1	$\tilde{\frac{1}{2}}$	0	210−	$+\sqrt{3.7}/\sqrt{2.11.17}$
$\frac{3}{2}$	2	$\frac{1}{2}$	0	000+	$+9\sqrt{3.5}/4\sqrt{11.13.17}$
$\frac{3}{2}$	2	$\frac{1}{2}$	0	100+	$+\sqrt{5.7}/\sqrt{2.3.13.17}$
$\frac{3}{2}$	2	$\frac{1}{2}$	0	200+	$-\sqrt{5}/4\sqrt{3.17}$
$\frac{3}{2}$	2	$\frac{3}{2}$	0	000−	$+3.41\sqrt{3}/5\sqrt{2.7.11.13.17}$
$\frac{3}{2}$	2	$\frac{3}{2}$	0	100−	$-19/5\sqrt{3.13.17}$
$\frac{3}{2}$	2	$\frac{3}{2}$	0	200−	$+1/\sqrt{2.3.7.17}$
$\frac{3}{2}$	2	$\tilde{\frac{1}{2}}$	0	000+	$-23\sqrt{13}/4.5\sqrt{7.11.17}$
$\frac{3}{2}$	2	$\tilde{\frac{1}{2}}$	0	100+	$-3/5\sqrt{2.13.17}$
$\frac{3}{2}$	2	$\tilde{\frac{1}{2}}$	0	200+	$-3/4\sqrt{7.17}$
$\frac{3}{2}$	$\tilde{1}$	$\frac{1}{2}$	0	000+	$+3.7\sqrt{13}/8\sqrt{2.5.11.17}$
$\frac{3}{2}$	$\tilde{1}$	$\frac{1}{2}$	0	100+	$+\sqrt{7}/4\sqrt{5.13.17}$
$\frac{3}{2}$	$\tilde{1}$	$\frac{1}{2}$	0	200+	$+\sqrt{5}/8\sqrt{2.17}$
$\frac{3}{2}$	$\tilde{1}$	$\frac{3}{2}$	0	000−	$+3.149/4.5\sqrt{7.11.13.17}$
$\frac{3}{2}$	$\tilde{1}$	$\frac{3}{2}$	0	100−	$-8\sqrt{2}/5\sqrt{13.17}$
$\frac{3}{2}$	$\tilde{1}$	$\frac{3}{2}$	0	200−	$-1/4\sqrt{7.17}$
$\frac{3}{2}$	$\tilde{1}$	$\frac{3}{2}$	1	000+	$-9\sqrt{7}/8.5\sqrt{11.13.17}$
$\frac{3}{2}$	$\tilde{1}$	$\frac{3}{2}$	1	100+	$+2.3\sqrt{2}/5\sqrt{13.17}$
$\frac{3}{2}$	$\tilde{1}$	$\frac{3}{2}$	1	200+	$+9/8\sqrt{7.17}$
$\frac{3}{2}$	$\tilde{1}$	$\tilde{\frac{1}{2}}$	0	000+	$-\sqrt{3.17}/8\sqrt{2.7.11.13}$
$\frac{3}{2}$	$\tilde{1}$	$\tilde{\frac{1}{2}}$	0	100+	$-7\sqrt{3}/4\sqrt{13.17}$
$\frac{3}{2}$	$\tilde{1}$	$\tilde{\frac{1}{2}}$	0	200+	$-5\sqrt{3}/8\sqrt{2.7.17}$
$\tilde{\frac{1}{2}}$	1	$\frac{3}{2}$	0	000−	$+5.5\sqrt{3}/4.11\sqrt{17}$
$\tilde{\frac{1}{2}}$	1	$\frac{3}{2}$	0	010−	$+\sqrt{3.17}/11\sqrt{5.7}$
$\tilde{\frac{1}{2}}$	1	$\tilde{\frac{1}{2}}$	0	000+	$+5/2.11\sqrt{17}$
$\tilde{\frac{1}{2}}$	1	$\tilde{\frac{1}{2}}$	0	010+	$+3\sqrt{17}/2.11\sqrt{5.7}$
$\tilde{\frac{1}{2}}$	2	$\frac{3}{2}$	0	000+	$+3/\sqrt{2.5.7.11.17}$
$\tilde{\frac{1}{2}}$	$\tilde{1}$	$\frac{1}{2}$	0	000−	$-\sqrt{3}/2\sqrt{11.17}$
$\tilde{\frac{1}{2}}$	$\tilde{1}$	$\frac{3}{2}$	0	000+	$+37\sqrt{3}/4\sqrt{5.7.11.17}$
$\frac{15}{2}$	5	$\frac{9}{2}$			
$\frac{1}{2}$	1	$\frac{1}{2}$	0	000−	$-7\sqrt{7}/2\sqrt{3.11.13.17}$
$\frac{1}{2}$	1	$\frac{1}{2}$	0	010−	$-2\sqrt{3}/\sqrt{5.11.13.17}$
$\frac{1}{2}$	1	$\frac{3}{2}$	0	000+	$+3.7\sqrt{7}/4\sqrt{5.11.13.17}$
$\frac{1}{2}$	1	$\frac{3}{2}$	0	001+	$-4.7/\sqrt{3.5.11.13.17}$
$\frac{1}{2}$	1	$\frac{3}{2}$	0	010+	$+7/2.5\sqrt{11.13.17}$
$\frac{1}{2}$	1	$\frac{3}{2}$	0	011+	$-7\sqrt{3.7}/5\sqrt{11.13.17}$
$\frac{1}{2}$	2	$\frac{3}{2}$	0	000−	$+2.3\sqrt{2.3}/5\sqrt{13.17}$
$\frac{1}{2}$	2	$\frac{3}{2}$	0	001−	$-\sqrt{2.7}/5\sqrt{13.17}$
$\frac{1}{2}$	$\tilde{1}$	$\frac{3}{2}$	0	000−	$-9/4.5\sqrt{13.17}$
$\frac{1}{2}$	$\tilde{1}$	$\frac{3}{2}$	0	001−	$+2\sqrt{3.7}/5\sqrt{13.17}$
$\frac{3}{2}$	1	$\frac{1}{2}$	0	000+	$-9.7/4.11\sqrt{2.5.13.17}$
$\frac{3}{2}$	1	$\frac{1}{2}$	0	010+	$+4\sqrt{2.7.17}/5.11\sqrt{13}$
$\frac{3}{2}$	1	$\frac{1}{2}$	0	100+	$-2.7\sqrt{7}/3\sqrt{5.11.13.17}$
$\frac{3}{2}$	1	$\frac{1}{2}$	0	110+	$+3/5\sqrt{11.13.17}$
$\frac{3}{2}$	1	$\frac{1}{2}$	0	200+	$+\sqrt{5.11}/4.3\sqrt{2.17}$
$\frac{3}{2}$	1	$\frac{1}{2}$	0	210+	0
$\frac{3}{2}$	1	$\frac{3}{2}$	0	000−	$+367\sqrt{3}/8.5.11\sqrt{13.17}$
$\frac{3}{2}$	1	$\frac{3}{2}$	0	001−	$-3.67\sqrt{7}/4.5.11\sqrt{13.17}$
$\frac{3}{2}$	1	$\frac{3}{2}$	0	010−	$-4.3\sqrt{3.7}/5.11\sqrt{5.13.17}$
$\frac{3}{2}$	1	$\frac{3}{2}$	0	011−	$-4.29/5.11\sqrt{5.13.17}$
$\frac{3}{2}$	1	$\frac{3}{2}$	0	100−	$-103\sqrt{7}/2.5\sqrt{2.3.11.13.17}$
$\frac{3}{2}$	1	$\frac{3}{2}$	0	101−	$-8.7\sqrt{2}/3.5\sqrt{11.13.17}$
$\frac{3}{2}$	1	$\frac{3}{2}$	0	110−	$-8\sqrt{2}/5\sqrt{3.5.11.13.17}$
$\frac{3}{2}$	1	$\frac{3}{2}$	0	111−	$-2.7\sqrt{2.7}/5\sqrt{5.11.13.17}$
$\frac{3}{2}$	1	$\frac{3}{2}$	0	200−	$+5/8\sqrt{3.11.17}$
$\frac{3}{2}$	1	$\frac{3}{2}$	0	201−	$+5\sqrt{7}/4.3\sqrt{11.17}$
$\frac{3}{2}$	1	$\frac{3}{2}$	0	210−	$-4\sqrt{7}/\sqrt{3.5.11.17}$
$\frac{3}{2}$	1	$\frac{3}{2}$	0	211−	$-4/\sqrt{5.11.17}$
$\frac{3}{2}$	1	$\frac{3}{2}$	1	000−	$+107\sqrt{3}/8.5.11\sqrt{13.17}$
$\frac{3}{2}$	1	$\frac{3}{2}$	1	001−	$+3\sqrt{7.17}/5.11\sqrt{13}$
$\frac{3}{2}$	1	$\frac{3}{2}$	1	010−	$-8.3\sqrt{3.7}/5.11\sqrt{5.13.17}$
$\frac{3}{2}$	1	$\frac{3}{2}$	1	011−	$+23/5.11\sqrt{5.13.17}$
$\frac{3}{2}$	1	$\frac{3}{2}$	1	100−	$+31\sqrt{7}/5\sqrt{2.3.11.13.17}$
$\frac{3}{2}$	1	$\frac{3}{2}$	1	101−	$-7\sqrt{11}/3.5\sqrt{2.13.17}$
$\frac{3}{2}$	1	$\frac{3}{2}$	1	110−	$-137/5\sqrt{2.3.5.11.13.17}$
$\frac{3}{2}$	1	$\frac{3}{2}$	1	111−	$+2\sqrt{2.7.11}/5\sqrt{5.13.17}$
$\frac{3}{2}$	1	$\frac{3}{2}$	1	200−	$+\sqrt{17}/8\sqrt{3.11}$
$\frac{3}{2}$	1	$\frac{3}{2}$	1	201−	$+\sqrt{7}/3\sqrt{11.17}$
$\frac{3}{2}$	1	$\frac{3}{2}$	1	210−	$+4\sqrt{7}/\sqrt{3.5.11.17}$

SO_3–O 3jm Factors (cont.)

$\frac{15}{2}$	5	$\frac{9}{2}$			
$\frac{3}{2}$	1	$\frac{3}{2}$	1	211−	$-1/\sqrt{5.11.17}$
$\frac{3}{2}$	2	$\frac{1}{2}$	0	000−	$+3\sqrt{3.7}/2.5\sqrt{11.13.17}$
$\frac{3}{2}$	2	$\frac{1}{2}$	0	100−	$+\sqrt{2}/5\sqrt{3.13.17}$
$\frac{3}{2}$	2	$\frac{1}{2}$	0	200−	$+\sqrt{7}/2\sqrt{3.17}$
$\frac{3}{2}$	2	$\frac{3}{2}$	0	000+	$+31\sqrt{7}/2.5\sqrt{2.5.11.13.17}$
$\frac{3}{2}$	2	$\frac{3}{2}$	0	001+	$-2.3\sqrt{2.3}/5\sqrt{5.11.13.17}$
$\frac{3}{2}$	2	$\frac{3}{2}$	0	100+	$+29/5\sqrt{5.13.17}$
$\frac{3}{2}$	2	$\frac{3}{2}$	0	101+	$-2.7\sqrt{7}/5\sqrt{3.5.13.17}$
$\frac{3}{2}$	2	$\frac{3}{2}$	0	200+	$-\sqrt{7}/2\sqrt{2.5.17}$
$\frac{3}{2}$	2	$\frac{3}{2}$	0	201+	$-2\sqrt{2}/\sqrt{3.5.17}$
$\frac{3}{2}$	$\tilde{1}$	$\frac{1}{2}$	0	000−	$-9\sqrt{7}/4.5\sqrt{2.11.13.17}$
$\frac{3}{2}$	$\tilde{1}$	$\frac{1}{2}$	0	100−	$-8/5\sqrt{13.17}$
$\frac{3}{2}$	$\tilde{1}$	$\frac{1}{2}$	0	200−	$-\sqrt{7}/4\sqrt{2.17}$
$\frac{3}{2}$	$\tilde{1}$	$\frac{3}{2}$	0	000+	$-43\sqrt{3.7}/8.5\sqrt{5.11.13.17}$
$\frac{3}{2}$	$\tilde{1}$	$\frac{3}{2}$	0	001+	$-3.149/4.5\sqrt{5.11.13.17}$
$\frac{3}{2}$	$\tilde{1}$	$\frac{3}{2}$	0	100+	$-41/2.5\sqrt{2.3.5.13.17}$
$\frac{3}{2}$	$\tilde{1}$	$\frac{3}{2}$	0	101+	$+8\sqrt{2.7}/5\sqrt{5.13.17}$
$\frac{3}{2}$	$\tilde{1}$	$\frac{3}{2}$	0	200+	$+7\sqrt{7}/8\sqrt{3.5.17}$
$\frac{3}{2}$	$\tilde{1}$	$\frac{3}{2}$	0	201+	$+1/4\sqrt{5.17}$
$\frac{3}{2}$	$\tilde{1}$	$\frac{3}{2}$	1	000−	$+3\sqrt{3.7.13}/8.5\sqrt{5.11.17}$
$\frac{3}{2}$	$\tilde{1}$	$\frac{3}{2}$	1	001−	$-8.4.3/5\sqrt{5.11.13.17}$
$\frac{3}{2}$	$\tilde{1}$	$\frac{3}{2}$	1	100−	$+7.7/5\sqrt{2.3.5.13.17}$
$\frac{3}{2}$	$\tilde{1}$	$\frac{3}{2}$	1	101−	$-3\sqrt{7}/5\sqrt{2.5.13.17}$
$\frac{3}{2}$	$\tilde{1}$	$\frac{3}{2}$	1	200−	$+\sqrt{5.7}/8\sqrt{3.17}$
$\frac{3}{2}$	$\tilde{1}$	$\frac{3}{2}$	1	201−	0
$\tilde{\frac{1}{2}}$	1	$\frac{3}{2}$	0	000+	$-59/4.11\sqrt{5.17}$
$\tilde{\frac{1}{2}}$	1	$\frac{3}{2}$	0	001+	$+2\sqrt{3.7}/11\sqrt{5.17}$
$\tilde{\frac{1}{2}}$	1	$\frac{3}{2}$	0	010+	$+3\sqrt{7}/2.11\sqrt{17}$
$\tilde{\frac{1}{2}}$	1	$\frac{3}{2}$	0	011+	$+\sqrt{3}/11\sqrt{17}$
$\tilde{\frac{1}{2}}$	2	$\frac{3}{2}$	0	000−	$-\sqrt{2.3.7}/5\sqrt{11.17}$
$\tilde{\frac{1}{2}}$	2	$\frac{3}{2}$	0	001−	$-3\sqrt{2}/5\sqrt{11.17}$
$\tilde{\frac{1}{2}}$	$\tilde{1}$	$\frac{1}{2}$	0	000+	$-\sqrt{3.7}/2\sqrt{5.11.17}$
$\tilde{\frac{1}{2}}$	$\tilde{1}$	$\frac{3}{2}$	0	000−	$+\sqrt{7.17}/4.5\sqrt{11}$
$\tilde{\frac{1}{2}}$	$\tilde{1}$	$\frac{3}{2}$	0	001−	$+8\sqrt{3}/5\sqrt{11.17}$

$\frac{15}{2}$	$\frac{11}{2}$	2			
$\frac{1}{2}$	$\frac{3}{2}$	2	0	000−	$-\sqrt{3}/2\sqrt{2.5.13}$
$\frac{1}{2}$	$\frac{3}{2}$	2	0	010−	$+\sqrt{3}/2\sqrt{13}$
$\frac{1}{2}$	$\frac{3}{2}$	$\tilde{1}$	0	000−	$+1/2\sqrt{5.13}$
$\frac{1}{2}$	$\frac{3}{2}$	$\tilde{1}$	0	010−	$-1/\sqrt{2.13}$
$\frac{1}{2}$	$\tilde{\frac{1}{2}}$	$\tilde{1}$	0	000+	$+1/2\sqrt{13}$
$\frac{3}{2}$	$\frac{1}{2}$	2	0	000−	$+29/8.5\sqrt{13}$
$\frac{3}{2}$	$\frac{1}{2}$	2	0	100−	$-\sqrt{7.11}/2.3.5\sqrt{2.13}$
$\frac{3}{2}$	$\frac{1}{2}$	2	0	200−	$-\sqrt{11}/8.3$
$\frac{3}{2}$	$\frac{1}{2}$	$\tilde{1}$	0	000−	$+17\sqrt{3}/4.5\sqrt{2.13}$
$\frac{3}{2}$	$\frac{1}{2}$	$\tilde{1}$	0	100−	$+\sqrt{7.11}/2.3.5\sqrt{3.13}$
$\frac{3}{2}$	$\frac{1}{2}$	$\tilde{1}$	0	200−	$+\sqrt{11}/4.3\sqrt{2.3}$
$\frac{3}{2}$	$\frac{3}{2}$	2	0	000+	$+137/4.5\sqrt{2.7.11.13}$
$\frac{3}{2}$	$\frac{3}{2}$	2	0	010+	$-4/\sqrt{5.7.11.13}$
$\frac{3}{2}$	$\frac{3}{2}$	2	0	100+	$+1/3.5\sqrt{13}$

$\frac{15}{2}$	$\frac{11}{2}$	2			
$\frac{3}{2}$	$\frac{3}{2}$	2	0	110+	$+3/\sqrt{2.5.13}$
$\frac{3}{2}$	$\frac{3}{2}$	2	0	200+	$+11/4.3\sqrt{2.7}$
$\frac{3}{2}$	$\frac{3}{2}$	2	0	210+	0
$\frac{3}{2}$	$\frac{3}{2}$	$\tilde{1}$	0	000+	$+101\sqrt{3}/4.5\sqrt{7.11.13}$
$\frac{3}{2}$	$\frac{3}{2}$	$\tilde{1}$	0	010+	$+4\sqrt{2}/\sqrt{3.5.7.11.13}$
$\frac{3}{2}$	$\frac{3}{2}$	$\tilde{1}$	0	100+	$-\sqrt{2}/3.5\sqrt{3.13}$
$\frac{3}{2}$	$\frac{3}{2}$	$\tilde{1}$	0	110+	$-\sqrt{3}/\sqrt{5.13}$
$\frac{3}{2}$	$\frac{3}{2}$	$\tilde{1}$	0	200+	$-11/4.3\sqrt{3.7}$
$\frac{3}{2}$	$\frac{3}{2}$	$\tilde{1}$	0	210+	0
$\frac{3}{2}$	$\frac{3}{2}$	$\tilde{1}$	1	000−	$-\sqrt{3.7}/4\sqrt{11.13}$
$\frac{3}{2}$	$\frac{3}{2}$	$\tilde{1}$	1	010−	$-4\sqrt{2}/\sqrt{3.5.7.11.13}$
$\frac{3}{2}$	$\frac{3}{2}$	$\tilde{1}$	1	100−	$-2\sqrt{2}/3\sqrt{3.13}$
$\frac{3}{2}$	$\frac{3}{2}$	$\tilde{1}$	1	110−	$+\sqrt{3}/\sqrt{5.13}$
$\frac{3}{2}$	$\frac{3}{2}$	$\tilde{1}$	1	200−	$-11/4.3\sqrt{3.7}$
$\frac{3}{2}$	$\frac{3}{2}$	$\tilde{1}$	1	210−	0
$\frac{3}{2}$	$\tilde{\frac{1}{2}}$	2	0	000−	$-\sqrt{5.11}/8\sqrt{7.13}$
$\frac{3}{2}$	$\tilde{\frac{1}{2}}$	2	0	100−	$-\sqrt{5}/2\sqrt{2.13}$
$\frac{3}{2}$	$\tilde{\frac{1}{2}}$	2	0	200−	$+\sqrt{5}/8\sqrt{7}$
$\frac{3}{2}$	$\tilde{\frac{1}{2}}$	$\tilde{1}$	0	000−	$-\sqrt{11}/4\sqrt{2.3.5.7.13}$
$\frac{3}{2}$	$\tilde{\frac{1}{2}}$	$\tilde{1}$	0	100−	$+\sqrt{3}/2\sqrt{5.13}$
$\frac{3}{2}$	$\tilde{\frac{1}{2}}$	$\tilde{1}$	0	200−	$-\sqrt{3.5}/4\sqrt{2.7}$
$\tilde{\frac{1}{2}}$	$\frac{1}{2}$	$\tilde{1}$	0	000+	$-1/2.3\sqrt{5}$
$\tilde{\frac{1}{2}}$	$\frac{3}{2}$	2	0	000−	$-19/2\sqrt{2.3.5.7.11}$
$\tilde{\frac{1}{2}}$	$\frac{3}{2}$	2	0	010−	$-\sqrt{3}/2\sqrt{7.11}$
$\tilde{\frac{1}{2}}$	$\frac{3}{2}$	$\tilde{1}$	0	000−	$+13/2.3\sqrt{5.7.11}$
$\tilde{\frac{1}{2}}$	$\frac{3}{2}$	$\tilde{1}$	0	010−	$+3/\sqrt{2.7.11}$

$\frac{15}{2}$	$\frac{11}{2}$	3			
$\frac{1}{2}$	$\frac{1}{2}$	1	0	000+	$-\sqrt{7}/2\sqrt{13.17}$
$\frac{1}{2}$	$\frac{3}{2}$	1	0	000−	$-19/4\sqrt{11.13.17}$
$\frac{1}{2}$	$\frac{3}{2}$	1	0	010−	$-3\sqrt{5}/2\sqrt{2.11.13.17}$
$\frac{1}{2}$	$\frac{3}{2}$	$\tilde{1}$	0	000+	$-7\sqrt{5}/4\sqrt{11.13.17}$
$\frac{1}{2}$	$\frac{3}{2}$	$\tilde{1}$	0	010+	$-31/2\sqrt{2.11.13.17}$
$\frac{1}{2}$	$\tilde{\frac{1}{2}}$	$\tilde{1}$	0	000−	$+\sqrt{11}/2\sqrt{13.17}$
$\frac{1}{2}$	$\tilde{\frac{1}{2}}$	$\tilde{0}$	0	000+	$-\sqrt{3.11}/2\sqrt{13.17}$
$\frac{3}{2}$	$\frac{1}{2}$	1	0	000−	$+3\sqrt{3.5}/8\sqrt{2.11.13.17}$
$\frac{3}{2}$	$\frac{1}{2}$	1	0	100−	$-\sqrt{5.7}/4\sqrt{3.13.17}$
$\frac{3}{2}$	$\frac{1}{2}$	1	0	200−	$-11\sqrt{5}/8\sqrt{2.3.17}$
$\frac{3}{2}$	$\frac{1}{2}$	$\tilde{1}$	0	000+	$-31\sqrt{3}/8\sqrt{2.11.13.17}$
$\frac{3}{2}$	$\frac{1}{2}$	$\tilde{1}$	0	100+	$-\sqrt{7.13}/4.3\sqrt{3.17}$
$\frac{3}{2}$	$\frac{1}{2}$	$\tilde{1}$	0	200+	$-11/8.3\sqrt{2.3.17}$
$\frac{3}{2}$	$\frac{3}{2}$	1	0	000+	$+547\sqrt{3}/4.11\sqrt{5.7.13.17}$
$\frac{3}{2}$	$\frac{3}{2}$	1	0	010+	$-\sqrt{2.3.13}/11\sqrt{7.17}$
$\frac{3}{2}$	$\frac{3}{2}$	1	0	100+	$+7\sqrt{2.3}/\sqrt{5.11.13.17}$
$\frac{3}{2}$	$\frac{3}{2}$	1	0	110+	$+\sqrt{13}/\sqrt{3.11.17}$
$\frac{3}{2}$	$\frac{3}{2}$	1	0	200+	$-5\sqrt{3.5}/4\sqrt{7.11.17}$
$\frac{3}{2}$	$\frac{3}{2}$	1	0	210+	$+5\sqrt{2}/\sqrt{3.7.11.17}$
$\frac{3}{2}$	$\frac{3}{2}$	1	1	000+	$-11\sqrt{3.7}/8\sqrt{5.13.17}$
$\frac{3}{2}$	$\frac{3}{2}$	1	1	010+	$-5\sqrt{3}/\sqrt{2.7.13.17}$
$\frac{3}{2}$	$\frac{3}{2}$	1	1	100+	$+7\sqrt{2}/\sqrt{3.5.11.13.17}$

SO_3–O 3jm Factors (cont.)

$\frac{15}{2}\ \frac{11}{2}\ 3$

$\frac{3}{2}\ \frac{3}{2}\ 1\ 1\ 110+\quad +5/2\sqrt{3.11.13.17}$

$\frac{3}{2}\ \frac{3}{2}\ 1\ 1\ 200+\quad -29\sqrt{5}/8\sqrt{3.7.11.17}$

$\frac{3}{2}\ \frac{3}{2}\ 1\ 1\ 210+\quad -5/\sqrt{2.3.7.11.17}$

$\frac{3}{2}\ \frac{3}{2}\ \tilde{1}\ 0\ 000-\quad -23\sqrt{3.7}/4.11\sqrt{13.17}$

$\frac{3}{2}\ \frac{3}{2}\ \tilde{1}\ 0\ 010-\quad -8\sqrt{2.17}/11\sqrt{3.5.7.13}$

$\frac{3}{2}\ \frac{3}{2}\ \tilde{1}\ 0\ 100-\quad +\sqrt{2.11}/3\sqrt{3.13.17}$

$\frac{3}{2}\ \frac{3}{2}\ \tilde{1}\ 0\ 110-\quad -\sqrt{3.11}/\sqrt{5.13.17}$

$\frac{3}{2}\ \frac{3}{2}\ \tilde{1}\ 0\ 200-\quad +\sqrt{11}/4.3\sqrt{3.7.17}$

$\frac{3}{2}\ \frac{3}{2}\ \tilde{1}\ 0\ 210-\quad 0$

$\frac{3}{2}\ \frac{3}{2}\ \tilde{1}\ 1\ 000+\quad +3.11\sqrt{3}/8\sqrt{7.13.17}$

$\frac{3}{2}\ \frac{3}{2}\ \tilde{1}\ 1\ 010+\quad +1/\sqrt{2.3.5.7.13.17}$

$\frac{3}{2}\ \frac{3}{2}\ \tilde{1}\ 1\ 100+\quad -2\sqrt{2.17}/3\sqrt{3.11.13}$

$\frac{3}{2}\ \frac{3}{2}\ \tilde{1}\ 1\ 110+\quad +\sqrt{3}/2\sqrt{5.11.13.17}$

$\frac{3}{2}\ \frac{3}{2}\ \tilde{1}\ 1\ 200+\quad +31/8.3\sqrt{3.7.11.17}$

$\frac{3}{2}\ \frac{3}{2}\ \tilde{1}\ 1\ 210+\quad -3\sqrt{3.5}/\sqrt{2.7.11.17}$

$\frac{3}{2}\ \frac{3}{2}\ \tilde{0}\ 0\ 000+\quad +41/4.11\sqrt{7.13.17}$

$\frac{3}{2}\ \frac{3}{2}\ \tilde{0}\ 0\ 010+\quad -2\sqrt{2.5}/11\sqrt{7.13.17}$

$\frac{3}{2}\ \frac{3}{2}\ \tilde{0}\ 0\ 100+\quad +\sqrt{2.13}/3\sqrt{11.17}$

$\frac{3}{2}\ \frac{3}{2}\ \tilde{0}\ 0\ 110+\quad -5\sqrt{5}/\sqrt{11.13.17}$

$\frac{3}{2}\ \frac{3}{2}\ \tilde{0}\ 0\ 200+\quad +31/4.3\sqrt{7.11.17}$

$\frac{3}{2}\ \frac{3}{2}\ \tilde{0}\ 0\ 210+\quad +2\sqrt{2.5}/\sqrt{7.11.17}$

$\frac{3}{2}\ \tilde{\frac{1}{2}}\ 1\ 0\ 000-\quad +3\sqrt{3.17}/8\sqrt{2.7.13}$

$\frac{3}{2}\ \tilde{\frac{1}{2}}\ 1\ 0\ 100-\quad +\sqrt{11}/4\sqrt{3.13.17}$

$\frac{3}{2}\ \tilde{\frac{1}{2}}\ 1\ 0\ 200-\quad +5\sqrt{11}/8\sqrt{2.3.7.17}$

$\frac{3}{2}\ \tilde{\frac{1}{2}}\ \tilde{1}\ 0\ 000+\quad -\sqrt{5.17}/8\sqrt{2.3.7.13}$

$\frac{3}{2}\ \tilde{\frac{1}{2}}\ \tilde{1}\ 0\ 100+\quad +\sqrt{3.5.11}/4\sqrt{13.17}$

$\frac{3}{2}\ \tilde{\frac{1}{2}}\ \tilde{1}\ 0\ 200+\quad +\sqrt{3.5.11}/8\sqrt{2.7.17}$

$\tilde{\frac{1}{2}}\ \frac{1}{2}\ \tilde{1}\ 0\ 000-\quad -7\sqrt{5}/2.3\sqrt{11.17}$

$\tilde{\frac{1}{2}}\ \frac{1}{2}\ \tilde{0}\ 0\ 000+\quad -\sqrt{5}/2\sqrt{3.11.17}$

$\tilde{\frac{1}{2}}\ \frac{3}{2}\ 1\ 0\ 000-\quad -61/4.11\sqrt{7.17}$

$\tilde{\frac{1}{2}}\ \frac{3}{2}\ 1\ 0\ 010-\quad +\sqrt{5.17}/2.11\sqrt{2.7}$

$\tilde{\frac{1}{2}}\ \frac{3}{2}\ \tilde{1}\ 0\ 000+\quad +101\sqrt{5}/4.3.11\sqrt{7.17}$

$\tilde{\frac{1}{2}}\ \frac{3}{2}\ \tilde{1}\ 0\ 010+\quad -3\sqrt{17}/2.11\sqrt{2.7}$

$\tilde{\frac{1}{2}}\ \tilde{\frac{1}{2}}\ 1\ 0\ 000+\quad -\sqrt{5}/2\sqrt{7.17}$

$\frac{15}{2}\ \frac{11}{2}\ 4$

$\frac{1}{2}\ \frac{1}{2}\ 0\ 0\ 000+\quad -7\sqrt{5}/2.3\sqrt{3.13.17}$

$\frac{1}{2}\ \frac{1}{2}\ 1\ 0\ 000-\quad -7/2.3\sqrt{3.13.17}$

$\frac{1}{2}\ \frac{3}{2}\ 1\ 0\ 000+\quad -\sqrt{7}/4.3\sqrt{3.11.13.17}$

$\frac{1}{2}\ \frac{3}{2}\ 1\ 0\ 010+\quad +3\sqrt{3.7}/2\sqrt{2.5.11.13.17}$

$\frac{1}{2}\ \frac{3}{2}\ 2\ 0\ 000-\quad -\sqrt{2.17}/3\sqrt{3.11.13}$

$\frac{1}{2}\ \frac{3}{2}\ 2\ 0\ 010-\quad +31/\sqrt{3.5.11.13.17}$

$\frac{1}{2}\ \frac{3}{2}\ \tilde{1}\ 0\ 000-\quad +5.31/4.3\sqrt{11.13.17}$

$\frac{1}{2}\ \frac{3}{2}\ \tilde{1}\ 0\ 010-\quad -\sqrt{5}/2.3\sqrt{2.11.13.17}$

$\frac{1}{2}\ \tilde{\frac{1}{2}}\ \tilde{1}\ 0\ 000+\quad +\sqrt{5.11}/2.3\sqrt{13.17}$

$\frac{3}{2}\ \frac{1}{2}\ 1\ 0\ 000+\quad +9.3\sqrt{7}/8\sqrt{2.5.11.13.17}$

$\frac{3}{2}\ \frac{1}{2}\ 1\ 0\ 100+\quad -7.19/4.9\sqrt{5.13.17}$

$\frac{3}{2}\ \frac{1}{2}\ 1\ 0\ 200+\quad +11\sqrt{5.7}/8.9\sqrt{2.17}$

$\frac{3}{2}\ \frac{1}{2}\ 2\ 0\ 000-\quad -67/4\sqrt{5.11.13.17}$

$\frac{3}{2}\ \frac{1}{2}\ 2\ 0\ 100-\quad -8\sqrt{2.7}/9\sqrt{5.13.17}$

$\frac{15}{2}\ \frac{11}{2}\ 4$

$\frac{3}{2}\ \frac{1}{2}\ 2\ 0\ 200-\quad +5\sqrt{5}/4.9\sqrt{17}$

$\frac{3}{2}\ \frac{1}{2}\ \tilde{1}\ 0\ 000-\quad -193/8\sqrt{2.3.5.11.13.17}$

$\frac{3}{2}\ \frac{1}{2}\ \tilde{1}\ 0\ 100-\quad +47\sqrt{7}/4.3\sqrt{3.5.13.17}$

$\frac{3}{2}\ \frac{1}{2}\ \tilde{1}\ 0\ 200-\quad +5\sqrt{5}/8.3\sqrt{2.3.17}$

$\frac{3}{2}\ \frac{3}{2}\ 0\ 0\ 000+\quad +9.3/4.11\sqrt{13.17}$

$\frac{3}{2}\ \frac{3}{2}\ 0\ 0\ 010+\quad +4\sqrt{2.17}/11\sqrt{5.13}$

$\frac{3}{2}\ \frac{3}{2}\ 0\ 0\ 100+\quad +\sqrt{2.7.13}/9\sqrt{11.17}$

$\frac{3}{2}\ \frac{3}{2}\ 0\ 0\ 110+\quad -\sqrt{7}/\sqrt{5.11.13.17}$

$\frac{3}{2}\ \frac{3}{2}\ 0\ 0\ 200+\quad +5\sqrt{11}/4.9\sqrt{17}$

$\frac{3}{2}\ \frac{3}{2}\ 0\ 0\ 210+\quad 0$

$\frac{3}{2}\ \frac{3}{2}\ 1\ 0\ 000-\quad -7\sqrt{17}/4.11\sqrt{5.13}$

$\frac{3}{2}\ \frac{3}{2}\ 1\ 0\ 010-\quad -19\sqrt{2}/5.11\sqrt{13.17}$

$\frac{3}{2}\ \frac{3}{2}\ 1\ 0\ 100-\quad -31\sqrt{2.7}/9\sqrt{5.11.13.17}$

$\frac{3}{2}\ \frac{3}{2}\ 1\ 0\ 110-\quad +19\sqrt{7}/3.5\sqrt{11.13.17}$

$\frac{3}{2}\ \frac{3}{2}\ 1\ 0\ 200-\quad +5.5\sqrt{5}/4.9\sqrt{11.17}$

$\frac{3}{2}\ \frac{3}{2}\ 1\ 0\ 210-\quad +7\sqrt{2}/3\sqrt{11.17}$

$\frac{3}{2}\ \frac{3}{2}\ 1\ 1\ 000-\quad -7.37/8.11\sqrt{5.13.17}$

$\frac{3}{2}\ \frac{3}{2}\ 1\ 1\ 010-\quad -37/11\sqrt{2.13.17}$

$\frac{3}{2}\ \frac{3}{2}\ 1\ 1\ 100-\quad +2\sqrt{2.7.11}/9\sqrt{5.13.17}$

$\frac{3}{2}\ \frac{3}{2}\ 1\ 1\ 110-\quad -\sqrt{7.11}/2.3\sqrt{13.17}$

$\frac{3}{2}\ \frac{3}{2}\ 1\ 1\ 200-\quad +5\sqrt{5.17}/8.9\sqrt{11}$

$\frac{3}{2}\ \frac{3}{2}\ 1\ 1\ 210-\quad -7/3\sqrt{2.11.17}$

$\frac{3}{2}\ \frac{3}{2}\ 2\ 0\ 000+\quad -239/11\sqrt{2.5.7.13.17}$

$\frac{3}{2}\ \frac{3}{2}\ 2\ 0\ 010+\quad +139/5.11\sqrt{7.13.17}$

$\frac{3}{2}\ \frac{3}{2}\ 2\ 0\ 100+\quad -59/9\sqrt{5.11.13.17}$

$\frac{3}{2}\ \frac{3}{2}\ 2\ 0\ 110+\quad +2.29\sqrt{2}/3.5\sqrt{11.13.17}$

$\frac{3}{2}\ \frac{3}{2}\ 2\ 0\ 200+\quad -31\sqrt{5}/9\sqrt{2.7.11.17}$

$\frac{3}{2}\ \frac{3}{2}\ 2\ 0\ 210+\quad -\sqrt{7}/3\sqrt{11.17}$

$\frac{3}{2}\ \frac{3}{2}\ \tilde{1}\ 0\ 000+\quad -653/4.11\sqrt{3.5.7.13.17}$

$\frac{3}{2}\ \frac{3}{2}\ \tilde{1}\ 0\ 010+\quad -2\sqrt{2.3}/5.11\sqrt{7.13.17}$

$\frac{3}{2}\ \frac{3}{2}\ \tilde{1}\ 0\ 100+\quad +\sqrt{2}/3\sqrt{3.5.11.13.17}$

$\frac{3}{2}\ \frac{3}{2}\ \tilde{1}\ 0\ 110+\quad -109/3.5\sqrt{3.11.13.17}$

$\frac{3}{2}\ \frac{3}{2}\ \tilde{1}\ 0\ 200+\quad -31\sqrt{5}/4.3\sqrt{3.7.11.17}$

$\frac{3}{2}\ \frac{3}{2}\ \tilde{1}\ 0\ 210+\quad +2\sqrt{2.7}/3\sqrt{3.11.17}$

$\frac{3}{2}\ \frac{3}{2}\ \tilde{1}\ 1\ 000-\quad -61\sqrt{7}/8.11\sqrt{3.5.13.17}$

$\frac{3}{2}\ \frac{3}{2}\ \tilde{1}\ 1\ 010-\quad +67\sqrt{3}/5.11\sqrt{2.7.13.17}$

$\frac{3}{2}\ \frac{3}{2}\ \tilde{1}\ 1\ 100-\quad +29\sqrt{2}/\sqrt{3.5.11.13.17}$

$\frac{3}{2}\ \frac{3}{2}\ \tilde{1}\ 1\ 110-\quad +449/2.3.5\sqrt{3.11.13.17}$

$\frac{3}{2}\ \frac{3}{2}\ \tilde{1}\ 1\ 200-\quad +5\sqrt{5}/8\sqrt{3.7.11.17}$

$\frac{3}{2}\ \frac{3}{2}\ \tilde{1}\ 1\ 210-\quad -7\sqrt{7}/3\sqrt{2.3.11.17}$

$\frac{3}{2}\ \tilde{\frac{1}{2}}\ 1\ 0\ 000+\quad -109/8.5\sqrt{2.13.17}$

$\frac{3}{2}\ \tilde{\frac{1}{2}}\ 1\ 0\ 100+\quad +\sqrt{7.11}/4.3.5\sqrt{13.17}$

$\frac{3}{2}\ \tilde{\frac{1}{2}}\ 1\ 0\ 200+\quad +\sqrt{11}/8.3\sqrt{2.17}$

$\frac{3}{2}\ \tilde{\frac{1}{2}}\ 2\ 0\ 000-\quad +149/4.5\sqrt{7.13.17}$

$\frac{3}{2}\ \tilde{\frac{1}{2}}\ 2\ 0\ 100-\quad -8\sqrt{2.11}/3.5\sqrt{13.17}$

$\frac{3}{2}\ \tilde{\frac{1}{2}}\ 2\ 0\ 200-\quad -\sqrt{11}/4.3\sqrt{7.17}$

$\frac{3}{2}\ \tilde{\frac{1}{2}}\ \tilde{1}\ 0\ 000-\quad +\sqrt{3.17}/8\sqrt{2.7.13}$

$\frac{3}{2}\ \tilde{\frac{1}{2}}\ \tilde{1}\ 0\ 100-\quad +\sqrt{11.13}/4.3\sqrt{3.17}$

$\frac{3}{2}\ \tilde{\frac{1}{2}}\ \tilde{1}\ 0\ 200-\quad +5\sqrt{11}/8.3\sqrt{2.3.7.17}$

SO_3–O 3jm Factors (cont.)

$\frac{15}{2}$	$\frac{11}{2}$	4		
$\tilde{\frac{1}{2}}$	$\frac{1}{2}$	$\tilde{1}$	0 000+	$+5/2\sqrt{11.17}$
$\tilde{\frac{1}{2}}$	$\frac{3}{2}$	1	0 000+	$-37/4.11\sqrt{3.17}$
$\tilde{\frac{1}{2}}$	$\frac{3}{2}$	1	0 010+	$+53/2.11\sqrt{2.3.5.17}$
$\tilde{\frac{1}{2}}$	$\frac{3}{2}$	2	0 000−	$-\sqrt{2}/\sqrt{3.7.17}$
$\tilde{\frac{1}{2}}$	$\frac{3}{2}$	2	0 010−	$+1/\sqrt{3.5.7.17}$
$\tilde{\frac{1}{2}}$	$\frac{3}{2}$	$\tilde{1}$	0 000−	$-5.5/4.11\sqrt{7.17}$
$\tilde{\frac{1}{2}}$	$\frac{3}{2}$	$\tilde{1}$	0 010−	$-3\sqrt{17}/2.11\sqrt{2.5.7}$
$\tilde{\frac{1}{2}}$	$\tilde{\frac{1}{2}}$	0	0 000+	$-1/2\sqrt{3.17}$
$\tilde{\frac{1}{2}}$	$\tilde{\frac{1}{2}}$	1	0 000−	$-7/2\sqrt{3.5.17}$

$\frac{15}{2}$	$\frac{11}{2}$	5		
$\frac{1}{2}$	$\frac{1}{2}$	1	0 000+	$-3.7\sqrt{5}/\sqrt{11.13.17.19}$
$\frac{1}{2}$	$\frac{1}{2}$	1	0 001+	$+7\sqrt{7}/2\sqrt{11.13.17.19}$
$\frac{1}{2}$	$\frac{3}{2}$	1	0 000−	$-3\sqrt{5.7.13}/2.11\sqrt{17.19}$
$\frac{1}{2}$	$\frac{3}{2}$	1	0 001−	$+7.23/2.11\sqrt{13.17.19}$
$\frac{1}{2}$	$\frac{3}{2}$	1	0 010−	$-7\sqrt{2.7}/11\sqrt{13.17.19}$
$\frac{1}{2}$	$\frac{3}{2}$	1	0 011−	$-3\sqrt{19}/11\sqrt{2.5.13.17}$
$\frac{1}{2}$	$\frac{3}{2}$	2	0 000+	$+\sqrt{3.19}/2\sqrt{2.11.13.17}$
$\frac{1}{2}$	$\frac{3}{2}$	2	0 010+	$+37\sqrt{3}/2\sqrt{5.11.13.17.19}$
$\frac{1}{2}$	$\frac{3}{2}$	$\tilde{1}$	0 000+	$+5/2\sqrt{11.13.17.19}$
$\frac{1}{2}$	$\frac{3}{2}$	$\tilde{1}$	0 010+	$-8\sqrt{2.5}/\sqrt{11.13.17.19}$
$\frac{1}{2}$	$\tilde{\frac{1}{2}}$	$\tilde{1}$	0 000−	$+\sqrt{5.11}/\sqrt{13.17.19}$
$\frac{3}{2}$	$\frac{1}{2}$	1	0 000−	$+\sqrt{2.7}/11\sqrt{3.13.17.19}$
$\frac{3}{2}$	$\frac{1}{2}$	1	0 001−	$+3.7\sqrt{3.19}/4.11\sqrt{2.5.13.17}$
$\frac{3}{2}$	$\frac{1}{2}$	1	0 100−	$-3.7\sqrt{3}/2\sqrt{11.13.17.19}$
$\frac{3}{2}$	$\frac{1}{2}$	1	0 101−	$+7.7\sqrt{7}/2\sqrt{3.5.11.13.17.19}$
$\frac{3}{2}$	$\frac{1}{2}$	1	0 200−	0
$\frac{3}{2}$	$\frac{1}{2}$	1	0 201−	$-\sqrt{5.11}/4\sqrt{2.3.17.19}$
$\frac{3}{2}$	$\frac{1}{2}$	2	0 000+	$-367/8\sqrt{5.11.13.17.19}$
$\frac{3}{2}$	$\frac{1}{2}$	2	0 100+	$+103\sqrt{7}/2.3\sqrt{2.5.13.17.19}$
$\frac{3}{2}$	$\frac{1}{2}$	2	0 200+	$-5\sqrt{5}/8.3\sqrt{17.19}$
$\frac{3}{2}$	$\frac{1}{2}$	$\tilde{1}$	0 000+	$-137\sqrt{3}/2\sqrt{2.5.11.13.17.19}$
$\frac{3}{2}$	$\frac{1}{2}$	$\tilde{1}$	0 100+	$-37\sqrt{7}/2.3\sqrt{3.5.13.17.19}$
$\frac{3}{2}$	$\frac{1}{2}$	$\tilde{1}$	0 200+	$-5\sqrt{5}/2.3\sqrt{2.3.17.19}$
$\frac{3}{2}$	$\frac{3}{2}$	1	0 000+	$-7.131/2.11\sqrt{3.11.13.17.19}$
$\frac{3}{2}$	$\frac{3}{2}$	1	0 001+	$-7.7.7\sqrt{3.7}/4.11\sqrt{5.11.13.17.19}$
$\frac{3}{2}$	$\frac{3}{2}$	1	0 010+	$+2.137\sqrt{2}/11\sqrt{3.5.11.13.17.19}$
$\frac{3}{2}$	$\frac{3}{2}$	1	0 011+	$+8.2\sqrt{2.3.7.17}/5.11\sqrt{11.13.19}$
$\frac{3}{2}$	$\frac{3}{2}$	1	0 100+	$+2\sqrt{2.3.7}/\sqrt{13.17.19}$
$\frac{3}{2}$	$\frac{3}{2}$	1	0 101+	$-\sqrt{2.3}/\sqrt{5.13.17.19}$
$\frac{3}{2}$	$\frac{3}{2}$	1	0 110+	$+2\sqrt{7.13}/3\sqrt{3.5.17.19}$
$\frac{3}{2}$	$\frac{3}{2}$	1	0 111+	$+11/5\sqrt{3.13.17.19}$
$\frac{3}{2}$	$\frac{3}{2}$	1	0 200+	$+5\sqrt{3}/2.11\sqrt{17.19}$
$\frac{3}{2}$	$\frac{3}{2}$	1	0 201+	$-3.5\sqrt{3.5.7}/4.11\sqrt{17.19}$
$\frac{3}{2}$	$\frac{3}{2}$	1	0 210+	$-2.5.5\sqrt{2.5}/3.11\sqrt{3.17.19}$
$\frac{3}{2}$	$\frac{3}{2}$	1	0 211+	$+4\sqrt{2.7}/11\sqrt{3.17.19}$
$\frac{3}{2}$	$\frac{3}{2}$	1	1 000+	$+733/4.11\sqrt{3.11.13.17.19}$
$\frac{3}{2}$	$\frac{3}{2}$	1	1 001+	$-199\sqrt{3.7}/4.11\sqrt{5.11.13.17.19}$
$\frac{3}{2}$	$\frac{3}{2}$	1	1 010+	$+71\sqrt{13}/11\sqrt{2.3.5.11.17.19}$
$\frac{3}{2}$	$\frac{3}{2}$	1	1 011+	$+2.3\sqrt{2.3.7.17}/11\sqrt{11.13.19}$
$\frac{3}{2}$	$\frac{3}{2}$	1	1 100+	$+47\sqrt{3.7}/11\sqrt{2.13.17.19}$
$\frac{3}{2}$	$\frac{3}{2}$	1	1 101+	$+2.73\sqrt{2}/11\sqrt{3.5.13.17.19}$
$\frac{3}{2}$	$\frac{3}{2}$	1	1 110+	$+151\sqrt{7}/3.11\sqrt{3.5.13.17.19}$
$\frac{3}{2}$	$\frac{3}{2}$	1	1 111+	$+5/11\sqrt{3.13.17.19}$
$\frac{3}{2}$	$\frac{3}{2}$	1	1 200+	$-5\sqrt{3}/4.11\sqrt{17.19}$
$\frac{3}{2}$	$\frac{3}{2}$	1	1 201+	$-5\sqrt{5.7}/4.11\sqrt{3.17.19}$
$\frac{3}{2}$	$\frac{3}{2}$	1	1 210+	$+\sqrt{5.17}/3.11\sqrt{2.3.19}$
$\frac{3}{2}$	$\frac{3}{2}$	1	1 211+	$-2\sqrt{2.7}/11\sqrt{3.17.19}$
$\frac{3}{2}$	$\frac{3}{2}$	2	0 000−	$-3.43\sqrt{7}/4.11\sqrt{2.5.13.17.19}$
$\frac{3}{2}$	$\frac{3}{2}$	2	0 010−	$+4\sqrt{7.19}/5.11\sqrt{13.17}$
$\frac{3}{2}$	$\frac{3}{2}$	2	0 100−	$-139/3\sqrt{5.11.13.17.19}$
$\frac{3}{2}$	$\frac{3}{2}$	2	0 110−	$+433/3.5\sqrt{2.11.13.17.19}$
$\frac{3}{2}$	$\frac{3}{2}$	2	0 200−	$+13\sqrt{5.7}/4.3\sqrt{2.11.17.19}$
$\frac{3}{2}$	$\frac{3}{2}$	2	0 210−	$-8\sqrt{7}/3\sqrt{11.17.19}$
$\frac{3}{2}$	$\frac{3}{2}$	$\tilde{1}$	0 000−	$-67\sqrt{3.7}/2.11\sqrt{5.13.17.19}$
$\frac{3}{2}$	$\frac{3}{2}$	$\tilde{1}$	0 010−	$+2.7\sqrt{2.7.19}/5.11\sqrt{3.13.17}$
$\frac{3}{2}$	$\frac{3}{2}$	$\tilde{1}$	0 100−	$-2.31\sqrt{2}/3\sqrt{3.5.11.13.17.19}$
$\frac{3}{2}$	$\frac{3}{2}$	$\tilde{1}$	0 110−	$-2.47/5\sqrt{3.11.13.17.19}$
$\frac{3}{2}$	$\frac{3}{2}$	$\tilde{1}$	0 200−	$+13\sqrt{5.7}/2.3\sqrt{3.11.17.19}$
$\frac{3}{2}$	$\frac{3}{2}$	$\tilde{1}$	0 210−	$+2\sqrt{2.7}/\sqrt{3.11.17.19}$
$\frac{3}{2}$	$\frac{3}{2}$	$\tilde{1}$	1 000+	$-53\sqrt{3.7}/4.11\sqrt{5.13.17.19}$
$\frac{3}{2}$	$\frac{3}{2}$	$\tilde{1}$	1 010+	$+29\sqrt{7}/5.11\sqrt{2.3.13.17.19}$
$\frac{3}{2}$	$\frac{3}{2}$	$\tilde{1}$	1 100+	$-313/3\sqrt{2.3.5.11.13.17.19}$
$\frac{3}{2}$	$\frac{3}{2}$	$\tilde{1}$	1 110+	$+7.41/5\sqrt{3.11.13.17.19}$
$\frac{3}{2}$	$\frac{3}{2}$	$\tilde{1}$	1 200+	$+5.7\sqrt{5.7}/4.3\sqrt{3.11.17.19}$
$\frac{3}{2}$	$\frac{3}{2}$	$\tilde{1}$	1 210+	$+13\sqrt{7}/\sqrt{2.3.11.17.19}$
$\frac{3}{2}$	$\tilde{\frac{1}{2}}$	1	0 000−	$-389/2\sqrt{2.3.5.11.13.17.19}$
$\frac{3}{2}$	$\tilde{\frac{1}{2}}$	1	0 001−	$-79\sqrt{3.7}/4.5\sqrt{2.11.13.17.19}$
$\frac{3}{2}$	$\tilde{\frac{1}{2}}$	1	0 100−	$-\sqrt{7.13}/2.3\sqrt{3.5.17.19}$
$\frac{3}{2}$	$\tilde{\frac{1}{2}}$	1	0 101−	$+167/2.5\sqrt{3.13.17.19}$
$\frac{3}{2}$	$\tilde{\frac{1}{2}}$	1	0 200−	$-11\sqrt{5}/2.3\sqrt{2.3.17.19}$
$\frac{3}{2}$	$\tilde{\frac{1}{2}}$	1	0 201−	$+11\sqrt{7}/4\sqrt{2.3.17.19}$
$\frac{3}{2}$	$\tilde{\frac{1}{2}}$	2	0 000+	$-43\sqrt{7}/8.5\sqrt{13.17.19}$
$\frac{3}{2}$	$\tilde{\frac{1}{2}}$	2	0 100+	$-41\sqrt{11}/2.3.5\sqrt{2.13.17.19}$
$\frac{3}{2}$	$\tilde{\frac{1}{2}}$	2	0 200+	$+7\sqrt{7.11}/8.3\sqrt{17.19}$
$\frac{3}{2}$	$\tilde{\frac{1}{2}}$	$\tilde{1}$	0 000+	$+\sqrt{2.7}/\sqrt{3.13.17.19}$
$\frac{3}{2}$	$\tilde{\frac{1}{2}}$	$\tilde{1}$	0 100+	$+\sqrt{3.11}/2\sqrt{13.17.19}$
$\frac{3}{2}$	$\tilde{\frac{1}{2}}$	$\tilde{1}$	0 200+	0
$\tilde{\frac{1}{2}}$	$\frac{1}{2}$	$\tilde{1}$	0 000−	$+5/3\sqrt{11.17.19}$
$\tilde{\frac{1}{2}}$	$\frac{3}{2}$	1	0 000−	$-3\sqrt{5.19}/2.11\sqrt{11.17}$
$\tilde{\frac{1}{2}}$	$\frac{3}{2}$	1	0 001−	$-7\sqrt{7}/2.11\sqrt{11.17.19}$
$\tilde{\frac{1}{2}}$	$\frac{3}{2}$	1	0 010−	$-8.5\sqrt{2}/3.11\sqrt{11.17.19}$
$\tilde{\frac{1}{2}}$	$\frac{3}{2}$	1	0 011−	$+13\sqrt{7}/11\sqrt{2.5.11.17.19}$
$\tilde{\frac{1}{2}}$	$\frac{3}{2}$	2	0 000+	$+79\sqrt{7}/2.11\sqrt{2.3.17.19}$
$\tilde{\frac{1}{2}}$	$\frac{3}{2}$	2	0 010+	$+23\sqrt{7}/2.11\sqrt{3.5.17.19}$
$\tilde{\frac{1}{2}}$	$\frac{3}{2}$	$\tilde{1}$	0 000+	$-5.5\sqrt{7}/2.3.11\sqrt{17.19}$
$\tilde{\frac{1}{2}}$	$\frac{3}{2}$	$\tilde{1}$	0 010+	$+\sqrt{2.7.19}/11\sqrt{5.17}$
$\tilde{\frac{1}{2}}$	$\tilde{\frac{1}{2}}$	1	0 000+	$-5/3\sqrt{11.17.19}$
$\tilde{\frac{1}{2}}$	$\tilde{\frac{1}{2}}$	1	0 001+	$-\sqrt{7.19}/2\sqrt{5.11.17}$

SO_3–O 3jm Factors (cont.)

$\frac{15}{2}$	6	$\frac{3}{2}$		
$\frac{1}{2}$	1	$\frac{3}{2}$	0 000+	0
$\frac{1}{2}$	2	$\frac{3}{2}$	0 000−	$+\sqrt{3.11}/2\sqrt{2.5.13}$
$\frac{1}{2}$	$\tilde{1}$	$\frac{3}{2}$	0 000−	$+3/\sqrt{2.7.13}$
$\frac{1}{2}$	$\tilde{1}$	$\frac{3}{2}$	0 010−	$+\sqrt{11}/\sqrt{2.5.7.13}$
$\frac{3}{2}$	0	$\frac{3}{2}$	0 000+	$+1/\sqrt{13}$
$\frac{3}{2}$	0	$\frac{3}{2}$	0 100+	0
$\frac{3}{2}$	0	$\frac{3}{2}$	0 200+	0
$\frac{3}{2}$	1	$\frac{3}{2}$	0 000−	$+9\sqrt{3}/2.5\sqrt{2.7.13}$
$\frac{3}{2}$	1	$\frac{3}{2}$	0 100−	$-\sqrt{11}/5\sqrt{3.13}$
$\frac{3}{2}$	1	$\frac{3}{2}$	0 200−	$-\sqrt{11}/2\sqrt{2.3.7}$
$\frac{3}{2}$	1	$\frac{3}{2}$	1 000−	$+3.17\sqrt{3}/4.5\sqrt{2.7.13}$
$\frac{3}{2}$	1	$\frac{3}{2}$	1 100−	$+\sqrt{11}/2.5\sqrt{3.13}$
$\frac{3}{2}$	1	$\frac{3}{2}$	1 200−	$+\sqrt{11}/4\sqrt{2.3.7}$
$\frac{3}{2}$	2	$\frac{3}{2}$	0 000+	$+2\sqrt{2}/5\sqrt{7.13}$
$\frac{3}{2}$	2	$\frac{3}{2}$	0 100+	$-3\sqrt{11}/2.5\sqrt{13}$
$\frac{3}{2}$	2	$\frac{3}{2}$	0 200+	0
$\frac{3}{2}$	$\tilde{1}$	$\frac{3}{2}$	0 000+	$-\sqrt{2.3.11}/7\sqrt{5.13}$
$\frac{3}{2}$	$\tilde{1}$	$\frac{3}{2}$	0 010+	$+8\sqrt{2.3}/5.7\sqrt{13}$
$\frac{3}{2}$	$\tilde{1}$	$\frac{3}{2}$	0 100+	$-3\sqrt{3}/\sqrt{5.7.13}$
$\frac{3}{2}$	$\tilde{1}$	$\frac{3}{2}$	0 110+	$-\sqrt{3.11}/5\sqrt{7.13}$
$\frac{3}{2}$	$\tilde{1}$	$\frac{3}{2}$	0 200+	0
$\frac{3}{2}$	$\tilde{1}$	$\frac{3}{2}$	0 210+	0
$\frac{3}{2}$	$\tilde{1}$	$\frac{3}{2}$	1 000−	$+\sqrt{3.11}/4.7\sqrt{2.5.13}$
$\frac{3}{2}$	$\tilde{1}$	$\frac{3}{2}$	1 010−	$+\sqrt{3}/7\sqrt{2.13}$
$\frac{3}{2}$	$\tilde{1}$	$\frac{3}{2}$	1 100−	$-3\sqrt{3}/2\sqrt{5.7.13}$
$\frac{3}{2}$	$\tilde{1}$	$\frac{3}{2}$	1 110−	$-\sqrt{11}/\sqrt{3.7.13}$
$\frac{3}{2}$	$\tilde{1}$	$\frac{3}{2}$	1 200−	$+3\sqrt{3.5}/4.7\sqrt{2}$
$\frac{3}{2}$	$\tilde{1}$	$\frac{3}{2}$	1 210−	$-\sqrt{11}/7\sqrt{2.3}$
$\frac{3}{2}$	$\tilde{0}$	$\frac{3}{2}$	0 000−	$-\sqrt{11}/4\sqrt{5.7.13}$
$\frac{3}{2}$	$\tilde{0}$	$\frac{3}{2}$	0 100−	$-\sqrt{2}/\sqrt{5.13}$
$\frac{3}{2}$	$\tilde{0}$	$\frac{3}{2}$	0 200−	$+\sqrt{5}/4\sqrt{7}$
$\tilde{\frac{1}{2}}$	1	$\frac{3}{2}$	0 000+	$+1/\sqrt{2.5.7}$
$\tilde{\frac{1}{2}}$	2	$\frac{3}{2}$	0 000−	$-\sqrt{3}/2\sqrt{2.5.7}$
$\tilde{\frac{1}{2}}$	$\tilde{1}$	$\frac{3}{2}$	0 000−	0
$\tilde{\frac{1}{2}}$	$\tilde{1}$	$\frac{3}{2}$	0 010−	$-1/\sqrt{2.5}$

$\frac{15}{2}$	6	$\frac{5}{2}$		
$\frac{1}{2}$	1	$\frac{3}{2}$	0 000−	$+\sqrt{5.11}/4\sqrt{13.17}$
$\frac{1}{2}$	2	$\frac{3}{2}$	0 000+	$+\sqrt{3.11}/2\sqrt{5.13.17}$
$\frac{1}{2}$	$\tilde{1}$	$\frac{3}{2}$	0 000+	$+\sqrt{7}/4\sqrt{13.17}$
$\frac{1}{2}$	$\tilde{1}$	$\frac{3}{2}$	0 010+	$+\sqrt{7.11}/2\sqrt{5.13.17}$
$\frac{1}{2}$	$\tilde{1}$	$\tilde{\frac{1}{2}}$	0 000−	$-5\sqrt{5}/2\sqrt{7.13.17}$
$\frac{1}{2}$	$\tilde{1}$	$\tilde{\frac{1}{2}}$	0 010−	$-2\sqrt{11}/\sqrt{7.13.17}$
$\frac{1}{2}$	$\tilde{0}$	$\tilde{\frac{1}{2}}$	0 000+	$+\sqrt{3.5}/\sqrt{2.13.17}$
$\frac{3}{2}$	0	$\frac{3}{2}$	0 000−	$-3/4\sqrt{2.13.17}$
$\frac{3}{2}$	0	$\frac{3}{2}$	0 100−	$-\sqrt{7.11}/2.3\sqrt{13.17}$
$\frac{3}{2}$	0	$\frac{3}{2}$	0 200−	$-5\sqrt{11}/4.3\sqrt{2.17}$
$\frac{3}{2}$	1	$\frac{3}{2}$	0 000+	$+11.53/8.5\sqrt{3.7.13.17}$
$\frac{3}{2}$	1	$\frac{3}{2}$	0 100+	$+\sqrt{11.13}/2.3.5\sqrt{2.3.17}$
$\frac{3}{2}$	1	$\frac{3}{2}$	0 200+	$+13\sqrt{11}/8.3\sqrt{3.7.17}$
$\frac{3}{2}$	1	$\frac{3}{2}$	1 000+	$-7.11\sqrt{7}/8.5\sqrt{3.13.17}$
$\frac{3}{2}$	1	$\frac{3}{2}$	1 100+	$-7\sqrt{11}/3.5\sqrt{2.3.13.17}$
$\frac{3}{2}$	1	$\frac{3}{2}$	1 200+	$+31\sqrt{11}/8.3\sqrt{3.7.17}$
$\frac{3}{2}$	1	$\tilde{\frac{1}{2}}$	0 000−	$-67/4\sqrt{2.3.5.7.13.17}$
$\frac{3}{2}$	1	$\tilde{\frac{1}{2}}$	0 100−	$+4\sqrt{11}/3\sqrt{3.5.13.17}$
$\frac{3}{2}$	1	$\tilde{\frac{1}{2}}$	0 200−	$-\sqrt{5.11}/4.3\sqrt{2.3.7.17}$
$\frac{3}{2}$	2	$\frac{3}{2}$	0 000−	$+71/4.5\sqrt{7.13.17}$
$\frac{3}{2}$	2	$\frac{3}{2}$	0 100−	$-2\sqrt{2.11}/3.5\sqrt{13.17}$
$\frac{3}{2}$	2	$\frac{3}{2}$	0 200−	$+5\sqrt{11}/4.3\sqrt{7.17}$
$\frac{3}{2}$	2	$\tilde{\frac{1}{2}}$	0 000+	$+\sqrt{13}/2\sqrt{2.5.7.17}$
$\frac{3}{2}$	2	$\tilde{\frac{1}{2}}$	0 100+	$+8\sqrt{11}/3\sqrt{5.13.17}$
$\frac{3}{2}$	2	$\tilde{\frac{1}{2}}$	0 200+	$-\sqrt{5.11}/2.3\sqrt{2.7.17}$
$\frac{3}{2}$	$\tilde{1}$	$\frac{3}{2}$	0 000−	$+\sqrt{11.17}/8.7\sqrt{3.5.13}$
$\frac{3}{2}$	$\tilde{1}$	$\frac{3}{2}$	0 010−	$-8.8\sqrt{3}/5.7\sqrt{13.17}$
$\frac{3}{2}$	$\tilde{1}$	$\frac{3}{2}$	0 100−	$-31/2\sqrt{2.3.5.7.13.17}$
$\frac{3}{2}$	$\tilde{1}$	$\frac{3}{2}$	0 110−	$+4\sqrt{2.3.11}/5\sqrt{7.13.17}$
$\frac{3}{2}$	$\tilde{1}$	$\frac{3}{2}$	0 200−	$+5\sqrt{5}/8\sqrt{3.17}$
$\frac{3}{2}$	$\tilde{1}$	$\frac{3}{2}$	0 210−	0
$\frac{3}{2}$	$\tilde{1}$	$\frac{3}{2}$	1 000+	$+\sqrt{11.13.17}/8.7\sqrt{3.5}$
$\frac{3}{2}$	$\tilde{1}$	$\frac{3}{2}$	1 010+	$-11\sqrt{3}/7\sqrt{13.17}$
$\frac{3}{2}$	$\tilde{1}$	$\frac{3}{2}$	1 100+	$-19/\sqrt{2.3.5.7.13.17}$
$\frac{3}{2}$	$\tilde{1}$	$\frac{3}{2}$	1 110+	$-5\sqrt{11}/\sqrt{2.3.7.13.17}$
$\frac{3}{2}$	$\tilde{1}$	$\frac{3}{2}$	1 200+	$-\sqrt{5.17}/8.7\sqrt{3}$
$\frac{3}{2}$	$\tilde{1}$	$\frac{3}{2}$	1 210+	$-\sqrt{11}/7\sqrt{3.17}$
$\frac{3}{2}$	$\tilde{1}$	$\tilde{\frac{1}{2}}$	0 000+	$-\sqrt{11.17}/4.7\sqrt{2.3.13}$
$\frac{3}{2}$	$\tilde{1}$	$\tilde{\frac{1}{2}}$	0 010+	$+\sqrt{2.3}/7\sqrt{5.13.17}$
$\frac{3}{2}$	$\tilde{1}$	$\tilde{\frac{1}{2}}$	0 100+	$+2.3\sqrt{3}/\sqrt{7.13.17}$
$\frac{3}{2}$	$\tilde{1}$	$\tilde{\frac{1}{2}}$	0 110+	$-\sqrt{11}/\sqrt{3.5.7.13.17}$
$\frac{3}{2}$	$\tilde{1}$	$\tilde{\frac{1}{2}}$	0 200+	$-5\sqrt{3}/4.7\sqrt{2.17}$
$\frac{3}{2}$	$\tilde{1}$	$\tilde{\frac{1}{2}}$	0 210+	$-\sqrt{2.5.11}/7\sqrt{3.17}$
$\frac{3}{2}$	$\tilde{0}$	$\frac{3}{2}$	0 000+	$-\sqrt{11.17}/4\sqrt{2.5.7.13}$
$\frac{3}{2}$	$\tilde{0}$	$\frac{3}{2}$	0 100+	$-9/2\sqrt{5.13.17}$
$\frac{3}{2}$	$\tilde{0}$	$\frac{3}{2}$	0 200+	$-3\sqrt{5}/4\sqrt{2.7.17}$
$\tilde{\frac{1}{2}}$	0	$\tilde{\frac{1}{2}}$	0 000−	$+1/\sqrt{2.3.17}$
$\tilde{\frac{1}{2}}$	1	$\frac{3}{2}$	0 000−	$+47/4.3\sqrt{5.7.17}$
$\tilde{\frac{1}{2}}$	1	$\tilde{\frac{1}{2}}$	0 000+	$-11/2.3\sqrt{7.17}$
$\tilde{\frac{1}{2}}$	2	$\frac{3}{2}$	0 000+	$+\sqrt{17}/2\sqrt{3.5.7}$
$\tilde{\frac{1}{2}}$	$\tilde{1}$	$\frac{3}{2}$	0 000+	$+5\sqrt{11}/4.7\sqrt{17}$
$\tilde{\frac{1}{2}}$	$\tilde{1}$	$\frac{3}{2}$	0 010+	$+1/2.7\sqrt{5.17}$

$\frac{15}{2}$	6	$\frac{7}{2}$		
$\frac{1}{2}$	0	$\frac{1}{2}$	0 000+	$-7/2\sqrt{2.3.13.17}$
$\frac{1}{2}$	1	$\frac{1}{2}$	0 000−	$+\sqrt{7}/3\sqrt{13.17}$
$\frac{1}{2}$	1	$\frac{3}{2}$	0 000+	$-\sqrt{5}/4.3\sqrt{13.17}$
$\frac{1}{2}$	2	$\frac{3}{2}$	0 000−	$+2\sqrt{5}/\sqrt{3.13.17}$
$\frac{1}{2}$	$\tilde{1}$	$\frac{3}{2}$	0 000−	$+\sqrt{7.11}/4\sqrt{13.17}$
$\frac{1}{2}$	$\tilde{1}$	$\frac{3}{2}$	0 010−	0
$\frac{1}{2}$	$\tilde{1}$	$\tilde{\frac{1}{2}}$	0 000+	$+\sqrt{3.11}/\sqrt{7.13.17}$
$\frac{1}{2}$	$\tilde{1}$	$\tilde{\frac{1}{2}}$	0 010+	$-11\sqrt{5}/2\sqrt{3.7.13.17}$
$\frac{1}{2}$	$\tilde{0}$	$\tilde{\frac{1}{2}}$	0 000−	$-\sqrt{11}/2\sqrt{2.13.17}$

SO_3–O 3jm Factors (cont.)

$\mathbf{\tfrac{15}{2}\ 6\ \tfrac{7}{2}}$

$\tfrac{3}{2}\ 0\ \tfrac{3}{2}\ 0\ 000+\quad -9.3/4\sqrt{2.11.13.17}$

$\tfrac{3}{2}\ 0\ \tfrac{3}{2}\ 0\ 100+\quad -\sqrt{7}/3\sqrt{13.17}$

$\tfrac{3}{2}\ 0\ \tfrac{3}{2}\ 0\ 200+\quad +11/4.3\sqrt{2.17}$

$\tfrac{3}{2}\ 1\ \tfrac{1}{2}\ 0\ 000+\quad +9\sqrt{3.5}/8\sqrt{2.11.13.17}$

$\tfrac{3}{2}\ 1\ \tfrac{1}{2}\ 0\ 100+\quad -7\sqrt{5.7}/4.3\sqrt{3.13.17}$

$\tfrac{3}{2}\ 1\ \tfrac{1}{2}\ 0\ 200+\quad -11\sqrt{5}/8.3\sqrt{2.3.17}$

$\tfrac{3}{2}\ 1\ \tfrac{3}{2}\ 0\ 000-\quad -2.5\sqrt{3}/\sqrt{7.11.13.17}$

$\tfrac{3}{2}\ 1\ \tfrac{3}{2}\ 0\ 100-\quad -5\sqrt{2}/3\sqrt{3.13.17}$

$\tfrac{3}{2}\ 1\ \tfrac{3}{2}\ 0\ 200-\quad -2/3\sqrt{3.7.17}$

$\tfrac{3}{2}\ 1\ \tfrac{3}{2}\ 1\ 000-\quad -\sqrt{3.7.11}/8\sqrt{13.17}$

$\tfrac{3}{2}\ 1\ \tfrac{3}{2}\ 1\ 100-\quad +7/3\sqrt{2.3.13.17}$

$\tfrac{3}{2}\ 1\ \tfrac{3}{2}\ 1\ 200-\quad -113/8.3\sqrt{3.7.17}$

$\tfrac{3}{2}\ 1\ \tilde{\tfrac{1}{2}}\ 0\ 000+\quad -3.5.5/8\sqrt{2.7.11.13.17}$

$\tfrac{3}{2}\ 1\ \tilde{\tfrac{1}{2}}\ 0\ 100+\quad +3.5/4\sqrt{13.17}$

$\tfrac{3}{2}\ 1\ \tilde{\tfrac{1}{2}}\ 0\ 200+\quad +9/8\sqrt{2.7.17}$

$\tfrac{3}{2}\ 2\ \tfrac{1}{2}\ 0\ 000-\quad +97/4\sqrt{2.5.11.13.17}$

$\tfrac{3}{2}\ 2\ \tfrac{1}{2}\ 0\ 100-\quad -4\sqrt{7}/3\sqrt{5.13.17}$

$\tfrac{3}{2}\ 2\ \tfrac{1}{2}\ 0\ 200-\quad -5\sqrt{5}/4.3\sqrt{2.17}$

$\tfrac{3}{2}\ 2\ \tfrac{3}{2}\ 0\ 000+\quad -2/\sqrt{7.11.13.17}$

$\tfrac{3}{2}\ 2\ \tfrac{3}{2}\ 0\ 100+\quad -1/\sqrt{2.13.17}$

$\tfrac{3}{2}\ 2\ \tfrac{3}{2}\ 0\ 200+\quad +2/\sqrt{7.17}$

$\tfrac{3}{2}\ 2\ \tilde{\tfrac{1}{2}}\ 0\ 000-\quad +\sqrt{3.11}/4\sqrt{2.7.13.17}$

$\tfrac{3}{2}\ 2\ \tilde{\tfrac{1}{2}}\ 0\ 100-\quad -2/\sqrt{3.13.17}$

$\tfrac{3}{2}\ 2\ \tilde{\tfrac{1}{2}}\ 0\ 200-\quad +13/4\sqrt{2.3.7.17}$

$\tfrac{3}{2}\ \tilde{1}\ \tfrac{1}{2}\ 0\ 000-\quad -11\sqrt{3}/8\sqrt{2.7.13.17}$

$\tfrac{3}{2}\ \tilde{1}\ \tfrac{1}{2}\ 0\ 010-\quad -4.23\sqrt{2}/\sqrt{3.5.7.11.13.17}$

$\tfrac{3}{2}\ \tilde{1}\ \tfrac{1}{2}\ 0\ 100-\quad -\sqrt{11}/4\sqrt{3.13.17}$

$\tfrac{3}{2}\ \tilde{1}\ \tfrac{1}{2}\ 0\ 110-\quad -8/3\sqrt{3.5.13.17}$

$\tfrac{3}{2}\ \tilde{1}\ \tfrac{1}{2}\ 0\ 200-\quad -5\sqrt{11}/8\sqrt{2.3.7.17}$

$\tfrac{3}{2}\ \tilde{1}\ \tfrac{1}{2}\ 0\ 210-\quad -2\sqrt{2.5}/3\sqrt{3.7.17}$

$\tfrac{3}{2}\ \tilde{1}\ \tfrac{3}{2}\ 0\ 000+\quad +8.2\sqrt{3}/7\sqrt{5.13.17}$

$\tfrac{3}{2}\ \tilde{1}\ \tfrac{3}{2}\ 0\ 010+\quad -8\sqrt{11}/7\sqrt{3.13.17}$

$\tfrac{3}{2}\ \tilde{1}\ \tfrac{3}{2}\ 0\ 100+\quad -\sqrt{2.3.11}/\sqrt{5.7.13.17}$

$\tfrac{3}{2}\ \tilde{1}\ \tfrac{3}{2}\ 0\ 110+\quad -37/3\sqrt{2.3.7.13.17}$

$\tfrac{3}{2}\ \tilde{1}\ \tfrac{3}{2}\ 0\ 200+\quad 0$

$\tfrac{3}{2}\ \tilde{1}\ \tfrac{3}{2}\ 0\ 210+\quad +4/3\sqrt{3.17}$

$\tfrac{3}{2}\ \tilde{1}\ \tfrac{3}{2}\ 1\ 000-\quad -191\sqrt{3}/8.7\sqrt{5.13.17}$

$\tfrac{3}{2}\ \tilde{1}\ \tfrac{3}{2}\ 1\ 010-\quad +29/2.7\sqrt{3.11.13.17}$

$\tfrac{3}{2}\ \tilde{1}\ \tfrac{3}{2}\ 1\ 100-\quad -\sqrt{11.13}/\sqrt{2.3.5.7.17}$

$\tfrac{3}{2}\ \tilde{1}\ \tfrac{3}{2}\ 1\ 110-\quad -\sqrt{2}/3\sqrt{3.7.13.17}$

$\tfrac{3}{2}\ \tilde{1}\ \tfrac{3}{2}\ 1\ 200-\quad -5\sqrt{5.11}/8.7\sqrt{3.17}$

$\tfrac{3}{2}\ \tilde{1}\ \tfrac{3}{2}\ 1\ 210-\quad -5/2.3.7\sqrt{3.17}$

$\tfrac{3}{2}\ \tilde{1}\ \tilde{\tfrac{1}{2}}\ 0\ 000-\quad -3\sqrt{5.17}/8.7\sqrt{2.13}$

$\tfrac{3}{2}\ \tilde{1}\ \tilde{\tfrac{1}{2}}\ 0\ 010-\quad +9/7\sqrt{2.11.13.17}$

$\tfrac{3}{2}\ \tilde{1}\ \tilde{\tfrac{1}{2}}\ 0\ 100-\quad -\sqrt{5.11}/4\sqrt{7.13.17}$

$\tfrac{3}{2}\ \tilde{1}\ \tilde{\tfrac{1}{2}}\ 0\ 110-\quad -2.11/3\sqrt{7.13.17}$

$\tfrac{3}{2}\ \tilde{1}\ \tilde{\tfrac{1}{2}}\ 0\ 200-\quad +\sqrt{5.11}/8.7\sqrt{2.17}$

$\tfrac{3}{2}\ \tilde{1}\ \tilde{\tfrac{1}{2}}\ 0\ 210-\quad +11/3.7\sqrt{2.17}$

$\tfrac{3}{2}\ \tilde{0}\ \tfrac{3}{2}\ 0\ 000-\quad +\sqrt{5.17}/4\sqrt{2.7.13}$

$\tfrac{3}{2}\ \tilde{0}\ \tfrac{3}{2}\ 0\ 100-\quad -\sqrt{5.11}/3\sqrt{13.17}$

$\tfrac{3}{2}\ \tilde{0}\ \tfrac{3}{2}\ 0\ 200-\quad -\sqrt{5.11}/4.3\sqrt{2.7.17}$

$\tilde{\tfrac{1}{2}}\ 0\ \tilde{\tfrac{1}{2}}\ 0\ 000+\quad +3\sqrt{5}/2\sqrt{2.11.17}$

$\tilde{\tfrac{1}{2}}\ 1\ \tfrac{3}{2}\ 0\ 000+\quad +3\sqrt{5}/4\sqrt{7.11.17}$

$\tilde{\tfrac{1}{2}}\ 1\ \tilde{\tfrac{1}{2}}\ 0\ 000-\quad -\sqrt{3.5}/\sqrt{7.11.17}$

$\tilde{\tfrac{1}{2}}\ 2\ \tfrac{3}{2}\ 0\ 000-\quad +2\sqrt{5}/\sqrt{3.7.11.17}$

$\tilde{\tfrac{1}{2}}\ \tilde{1}\ \tfrac{1}{2}\ 0\ 000+\quad -\sqrt{5}/\sqrt{7.17}$

$\tilde{\tfrac{1}{2}}\ \tilde{1}\ \tfrac{1}{2}\ 0\ 010+\quad -1/2.3\sqrt{7.11.17}$

$\tilde{\tfrac{1}{2}}\ \tilde{1}\ \tfrac{3}{2}\ 0\ 000-\quad +\sqrt{17}/4.7$

$\tilde{\tfrac{1}{2}}\ \tilde{1}\ \tfrac{3}{2}\ 0\ 010-\quad +2\sqrt{5}/3.7\sqrt{11.17}$

$\tilde{\tfrac{1}{2}}\ \tilde{0}\ \tfrac{1}{2}\ 0\ 000-\quad -\sqrt{5}/2\sqrt{2.3.17}$

$\mathbf{\tfrac{15}{2}\ 6\ \tfrac{9}{2}}$

$\tfrac{1}{2}\ 0\ \tfrac{1}{2}\ 0\ 000-\quad +7/\sqrt{2.3.13.17.19}$

$\tfrac{1}{2}\ 1\ \tfrac{1}{2}\ 0\ 000+\quad -\sqrt{7.13}/2.3\sqrt{17.19}$

$\tfrac{1}{2}\ 1\ \tfrac{3}{2}\ 0\ 000-\quad -29\sqrt{3.7}/4\sqrt{5.13.17.19}$

$\tfrac{1}{2}\ 1\ \tfrac{3}{2}\ 0\ 001-\quad +8.7/3\sqrt{5.13.17.19}$

$\tfrac{1}{2}\ 2\ \tfrac{3}{2}\ 0\ 000+\quad +\sqrt{5.7}/2\sqrt{13.17.19}$

$\tfrac{1}{2}\ 2\ \tfrac{3}{2}\ 0\ 001+\quad +4\sqrt{5}/\sqrt{3.13.17.19}$

$\tfrac{1}{2}\ \tilde{1}\ \tfrac{3}{2}\ 0\ 000+\quad +31\sqrt{3.11}/4.5\sqrt{13.17.19}$

$\tfrac{1}{2}\ \tilde{1}\ \tfrac{3}{2}\ 0\ 001+\quad +2.11\sqrt{11}/5\sqrt{7.13.17.19}$

$\tfrac{1}{2}\ \tilde{1}\ \tfrac{3}{2}\ 0\ 010+\quad -\sqrt{17}/2\sqrt{3.5.13.19}$

$\tfrac{1}{2}\ \tilde{1}\ \tfrac{3}{2}\ 0\ 011+\quad +3\sqrt{13}/\sqrt{5.7.17.19}$

$\tfrac{3}{2}\ 0\ \tfrac{3}{2}\ 0\ 000-\quad -137\sqrt{3.7}/4.5\sqrt{2.11.13.17.19}$

$\tfrac{3}{2}\ 0\ \tfrac{3}{2}\ 0\ 001-\quad -9.37/2.5\sqrt{2.11.13.17.19}$

$\tfrac{3}{2}\ 0\ \tfrac{3}{2}\ 0\ 100-\quad +7.11\sqrt{3}/2.5\sqrt{13.17.19}$

$\tfrac{3}{2}\ 0\ \tfrac{3}{2}\ 0\ 101-\quad -4.7\sqrt{7}/3.5\sqrt{13.17.19}$

$\tfrac{3}{2}\ 0\ \tfrac{3}{2}\ 0\ 200-\quad -\sqrt{3.7}/4\sqrt{2.17.19}$

$\tfrac{3}{2}\ 0\ \tfrac{3}{2}\ 0\ 201-\quad -7/2.3\sqrt{2.17.19}$

$\tfrac{3}{2}\ 1\ \tfrac{1}{2}\ 0\ 000-\quad +9\sqrt{3.19}/4\sqrt{2.5.11.13.17}$

$\tfrac{3}{2}\ 1\ \tfrac{1}{2}\ 0\ 100-\quad +4\sqrt{7}/3\sqrt{3.5.13.17.19}$

$\tfrac{3}{2}\ 1\ \tfrac{1}{2}\ 0\ 200-\quad +11\sqrt{5}/4.3\sqrt{2.3.17.19}$

$\tfrac{3}{2}\ 1\ \tfrac{3}{2}\ 0\ 000+\quad -9.31/8.5\sqrt{11.13.17.19}$

$\tfrac{3}{2}\ 1\ \tfrac{3}{2}\ 0\ 001+\quad -607\sqrt{3}/4.5\sqrt{7.11.13.17.19}$

$\tfrac{3}{2}\ 1\ \tfrac{3}{2}\ 0\ 100+\quad +37\sqrt{7}/2.5\sqrt{2.13.17.19}$

$\tfrac{3}{2}\ 1\ \tfrac{3}{2}\ 0\ 101+\quad -4\sqrt{2.17}/3.5\sqrt{3.13.19}$

$\tfrac{3}{2}\ 1\ \tfrac{3}{2}\ 0\ 200+\quad +1/8\sqrt{17.19}$

$\tfrac{3}{2}\ 1\ \tfrac{3}{2}\ 0\ 201+\quad +137/4.3\sqrt{3.7.17.19}$

$\tfrac{3}{2}\ 1\ \tfrac{3}{2}\ 1\ 000+\quad -3.331/8.5\sqrt{11.13.17.19}$

$\tfrac{3}{2}\ 1\ \tfrac{3}{2}\ 1\ 001+\quad -2.83\sqrt{3}/5\sqrt{7.11.13.17.19}$

$\tfrac{3}{2}\ 1\ \tfrac{3}{2}\ 1\ 100+\quad -3\sqrt{7}/5\sqrt{2.13.17.19}$

$\tfrac{3}{2}\ 1\ \tfrac{3}{2}\ 1\ 101+\quad -317/3.5\sqrt{2.3.13.17.19}$

$\tfrac{3}{2}\ 1\ \tfrac{3}{2}\ 1\ 200+\quad +3.5/8\sqrt{17.19}$

$\tfrac{3}{2}\ 1\ \tfrac{3}{2}\ 1\ 201+\quad +2.5/3\sqrt{3.7.17.19}$

$\tfrac{3}{2}\ 2\ \tfrac{1}{2}\ 0\ 000+\quad -229/2\sqrt{2.5.11.13.17.19}$

$\tfrac{3}{2}\ 2\ \tfrac{1}{2}\ 0\ 100+\quad -4\sqrt{7}/3\sqrt{5.13.17.19}$

$\tfrac{3}{2}\ 2\ \tfrac{1}{2}\ 0\ 200+\quad -7\sqrt{5}/2.3\sqrt{2.17.19}$

$\tfrac{3}{2}\ 2\ \tfrac{3}{2}\ 0\ 000-\quad -7.37\sqrt{3}/4.5\sqrt{11.13.17.19}$

$\tfrac{3}{2}\ 2\ \tfrac{3}{2}\ 0\ 001-\quad -383/5\sqrt{7.11.13.17.19}$

$\tfrac{3}{2}\ 2\ \tfrac{3}{2}\ 0\ 100-\quad -2.11\sqrt{2.7}/5\sqrt{3.13.17.19}$

SO_3–O 3jm Factors (cont.)

$\frac{15}{2}$	6	$\frac{9}{2}$			
$\frac{3}{2}$	2	$\frac{3}{2}$	0	101−	$-9.3\sqrt{2}/5\sqrt{13.17.19}$
$\frac{3}{2}$	2	$\frac{3}{2}$	0	200−	$+7/4\sqrt{3.17.19}$
$\frac{3}{2}$	2	$\frac{3}{2}$	0	201−	$-\sqrt{7}/\sqrt{17.19}$
$\frac{3}{2}$	$\tilde{1}$	$\frac{1}{2}$	0	000+	$+373\sqrt{3}/4.5\sqrt{2.7.13.17.19}$
$\frac{3}{2}$	$\tilde{1}$	$\frac{1}{2}$	0	010+	$+199\sqrt{2}/\sqrt{3.5.7.11.13.17.19}$
$\frac{3}{2}$	$\tilde{1}$	$\frac{1}{2}$	0	100+	$-2\sqrt{11.13}/5\sqrt{3.17.19}$
$\frac{3}{2}$	$\tilde{1}$	$\frac{1}{2}$	0	110+	$+31/3\sqrt{3.5.13.17.19}$
$\frac{3}{2}$	$\tilde{1}$	$\frac{1}{2}$	0	200+	$-5.5\sqrt{11}/4\sqrt{2.3.7.17.19}$
$\frac{3}{2}$	$\tilde{1}$	$\frac{1}{2}$	0	210+	$+\sqrt{2.5}/3\sqrt{3.7.17.19}$
$\frac{3}{2}$	$\tilde{1}$	$\frac{3}{2}$	0	000−	$+3.281/8.5\sqrt{5.7.13.17.19}$
$\frac{3}{2}$	$\tilde{1}$	$\frac{3}{2}$	0	001−	$+43.43\sqrt{3}/4.5.7\sqrt{5.13.17.19}$
$\frac{3}{2}$	$\tilde{1}$	$\frac{3}{2}$	0	010−	$+4.9/\sqrt{7.11.13.17.19}$
$\frac{3}{2}$	$\tilde{1}$	$\frac{3}{2}$	0	011−	$+8.4\sqrt{11}/7\sqrt{3.13.17.19}$
$\frac{3}{2}$	$\tilde{1}$	$\frac{3}{2}$	0	100−	$+3.7\sqrt{11}/2.5\sqrt{2.5.13.17.19}$
$\frac{3}{2}$	$\tilde{1}$	$\frac{3}{2}$	0	101−	$-4.9\sqrt{2.3.11}/5\sqrt{5.7.13.17.19}$
$\frac{3}{2}$	$\tilde{1}$	$\frac{3}{2}$	0	110−	$-4.7\sqrt{2}/3\sqrt{13.17.19}$
$\frac{3}{2}$	$\tilde{1}$	$\frac{3}{2}$	0	111−	$+2.37\sqrt{2}/3\sqrt{3.7.13.17.19}$
$\frac{3}{2}$	$\tilde{1}$	$\frac{3}{2}$	0	200−	$+3\sqrt{11.17}/8\sqrt{5.7.19}$
$\frac{3}{2}$	$\tilde{1}$	$\frac{3}{2}$	0	201−	$-\sqrt{3.11}/4\sqrt{5.17.19}$
$\frac{3}{2}$	$\tilde{1}$	$\frac{3}{2}$	0	210−	$-4/3\sqrt{7.17.19}$
$\frac{3}{2}$	$\tilde{1}$	$\frac{3}{2}$	0	211−	$-8.2/3\sqrt{3.17.19}$
$\frac{3}{2}$	$\tilde{1}$	$\frac{3}{2}$	1	000+	$+9.139/8.5\sqrt{5.7.13.17.19}$
$\frac{3}{2}$	$\tilde{1}$	$\frac{3}{2}$	1	001+	$-163\sqrt{3}/5.7\sqrt{5.13.17.19}$
$\frac{3}{2}$	$\tilde{1}$	$\frac{3}{2}$	1	010+	$-9\sqrt{11}/5\sqrt{7.13.17.19}$
$\frac{3}{2}$	$\tilde{1}$	$\frac{3}{2}$	1	011+	$+1033/5.7\sqrt{3.11.13.17.19}$
$\frac{3}{2}$	$\tilde{1}$	$\frac{3}{2}$	1	100+	$-\sqrt{11.19}/5\sqrt{2.5.13.17}$
$\frac{3}{2}$	$\tilde{1}$	$\frac{3}{2}$	1	101+	$+53\sqrt{11}/5\sqrt{2.3.5.7.13.17.19}$
$\frac{3}{2}$	$\tilde{1}$	$\frac{3}{2}$	1	110+	$+59/3.5\sqrt{2.13.17.19}$
$\frac{3}{2}$	$\tilde{1}$	$\frac{3}{2}$	1	111+	$2.107\sqrt{2}/3.5\sqrt{3.7.13.17.19}$
$\frac{3}{2}$	$\tilde{1}$	$\frac{3}{2}$	1	200+	$+\sqrt{11.19}/8\sqrt{5.7.17}$
$\frac{3}{2}$	$\tilde{1}$	$\frac{3}{2}$	1	201+	$-\sqrt{11}/7\sqrt{3.5.17.19}$
$\frac{3}{2}$	$\tilde{1}$	$\frac{3}{2}$	1	210+	$+11/3\sqrt{7.17.19}$
$\frac{3}{2}$	$\tilde{1}$	$\frac{3}{2}$	1	211+	$+163/3.7\sqrt{3.17.19}$
$\frac{3}{2}$	$\tilde{0}$	$\frac{3}{2}$	0	000+	$-3.97\sqrt{3}/4.5\sqrt{2.5.13.17.19}$
$\frac{3}{2}$	$\tilde{0}$	$\frac{3}{2}$	0	001+	$+491/2.5\sqrt{2.5.7.13.17.19}$
$\frac{3}{2}$	$\tilde{0}$	$\frac{3}{2}$	0	100+	$-31\sqrt{7.11}/2.5\sqrt{3.5.13.17.19}$
$\frac{3}{2}$	$\tilde{0}$	$\frac{3}{2}$	0	101+	$-4.11\sqrt{11}/3.5\sqrt{5.13.17.19}$
$\frac{3}{2}$	$\tilde{0}$	$\frac{3}{2}$	0	200+	$-\sqrt{11}/4\sqrt{2.3.5.17.19}$
$\frac{3}{2}$	$\tilde{0}$	$\frac{3}{2}$	0	201+	$+\sqrt{11}/2.3\sqrt{2.5.7.17.19}$
$\tilde{\frac{1}{2}}$	1	$\frac{3}{2}$	0	000−	$-61\sqrt{3}/4\sqrt{5.11.17.19}$
$\tilde{\frac{1}{2}}$	1	$\frac{3}{2}$	0	001−	$+2.9/\sqrt{5.7.11.17.19}$
$\tilde{\frac{1}{2}}$	2	$\frac{3}{2}$	0	000+	$-9/2\sqrt{5.11.17.19}$
$\tilde{\frac{1}{2}}$	2	$\frac{3}{2}$	0	001+	$-4\sqrt{19}/\sqrt{3.5.7.11.17}$
$\tilde{\frac{1}{2}}$	$\tilde{1}$	$\frac{1}{2}$	0	000−	$-\sqrt{19}/2\sqrt{5.7.17}$
$\tilde{\frac{1}{2}}$	$\tilde{1}$	$\frac{1}{2}$	0	010−	$-2/3\sqrt{7.11.17.19}$
$\tilde{\frac{1}{2}}$	$\tilde{1}$	$\frac{3}{2}$	0	000+	$-7\sqrt{3.7}/4.5\sqrt{17.19}$
$\tilde{\frac{1}{2}}$	$\tilde{1}$	$\frac{3}{2}$	0	001+	$-4/5\sqrt{17.19}$
$\tilde{\frac{1}{2}}$	$\tilde{1}$	$\frac{3}{2}$	0	010+	$+\sqrt{3.5.7}/2\sqrt{11.17.19}$
$\tilde{\frac{1}{2}}$	$\tilde{1}$	$\frac{3}{2}$	0	011+	$-5\sqrt{5}/3\sqrt{11.17.19}$
$\frac{15}{2}$	6	$\frac{9}{2}$			
$\tilde{\frac{1}{2}}$	$\tilde{0}$	$\frac{1}{2}$	0	000+	$-\sqrt{19}/\sqrt{2.3.5.17}$
$\frac{15}{2}$	6	$\frac{11}{2}$			
$\frac{1}{2}$	0	$\frac{1}{2}$	0	000+	$-3.7\sqrt{3}/2\sqrt{2.13.17.19}$
$\frac{1}{2}$	1	$\frac{1}{2}$	0	000−	$+3\sqrt{7}/2\sqrt{13.17.19}$
$\frac{1}{2}$	1	$\frac{3}{2}$	0	000+	$+8.3/\sqrt{11.13.17.19}$
$\frac{1}{2}$	1	$\frac{3}{2}$	0	001+	$+7\sqrt{5}/\sqrt{2.11.13.17.19}$
$\frac{1}{2}$	2	$\frac{3}{2}$	0	000−	$-\sqrt{3}/\sqrt{11.13.17.19}$
$\frac{1}{2}$	2	$\frac{3}{2}$	0	001−	$+3.5\sqrt{3.5}/2\sqrt{2.11.13.17.19}$
$\frac{1}{2}$	$\tilde{1}$	$\frac{3}{2}$	0	000−	$-8\sqrt{5}/\sqrt{7.13.17.19}$
$\frac{1}{2}$	$\tilde{1}$	$\frac{3}{2}$	0	001−	$-4\sqrt{2}/\sqrt{7.13.17.19}$
$\frac{1}{2}$	$\tilde{1}$	$\frac{3}{2}$	0	010−	$+2.3\sqrt{13}/\sqrt{7.11.17.19}$
$\frac{1}{2}$	$\tilde{1}$	$\frac{3}{2}$	0	011−	$+31\sqrt{5}/\sqrt{2.7.11.13.17.19}$
$\frac{1}{2}$	$\tilde{1}$	$\tilde{\frac{1}{2}}$	0	000+	$-11/2\sqrt{7.13.17.19}$
$\frac{1}{2}$	$\tilde{1}$	$\tilde{\frac{1}{2}}$	0	010+	$-2\sqrt{5.11}/\sqrt{7.13.17.19}$
$\frac{1}{2}$	$\tilde{0}$	$\tilde{\frac{1}{2}}$	0	000−	$+11\sqrt{3}/2\sqrt{2.13.17.19}$
$\frac{3}{2}$	0	$\frac{3}{2}$	0	000+	$+2\sqrt{2.5}/11\sqrt{13.17.19}$
$\frac{3}{2}$	0	$\frac{3}{2}$	0	001+	$-3.7\sqrt{19}/4.11\sqrt{13.17}$
$\frac{3}{2}$	0	$\frac{3}{2}$	0	100+	$+9\sqrt{5.7}/2\sqrt{11.13.17.19}$
$\frac{3}{2}$	0	$\frac{3}{2}$	0	101+	$-7\sqrt{2.7}/3\sqrt{11.13.17.19}$
$\frac{3}{2}$	0	$\frac{3}{2}$	0	200+	0
$\frac{3}{2}$	0	$\frac{3}{2}$	0	201+	$+\sqrt{11}/4.3\sqrt{17.19}$
$\frac{3}{2}$	1	$\frac{1}{2}$	0	000+	$+\sqrt{2.5}/\sqrt{3.11.13.17.19}$
$\frac{3}{2}$	1	$\frac{1}{2}$	0	100+	$-3\sqrt{3.5.7}/2\sqrt{13.17.19}$
$\frac{3}{2}$	1	$\frac{1}{2}$	0	200+	0
$\frac{3}{2}$	1	$\frac{3}{2}$	0	000−	$-227\sqrt{7}/2.11\sqrt{3.5.13.17.19}$
$\frac{3}{2}$	1	$\frac{3}{2}$	0	001−	$+7\sqrt{7}/11\sqrt{2.3.13.17.19}$
$\frac{3}{2}$	1	$\frac{3}{2}$	0	100−	$+73\sqrt{3}/\sqrt{2.5.11.13.17.19}$
$\frac{3}{2}$	1	$\frac{3}{2}$	0	101−	$+113/3\sqrt{3.11.13.17.19}$
$\frac{3}{2}$	1	$\frac{3}{2}$	0	200−	$-\sqrt{3.5.7}/2\sqrt{11.17.19}$
$\frac{3}{2}$	1	$\frac{3}{2}$	0	201−	$+23\sqrt{7}/3\sqrt{2.3.11.17.19}$
$\frac{3}{2}$	1	$\frac{3}{2}$	1	000−	$+307\sqrt{7}/4.11\sqrt{3.5.13.17.19}$
$\frac{3}{2}$	1	$\frac{3}{2}$	1	001−	$-5.37\sqrt{7}/4.11\sqrt{2.3.13.17.19}$
$\frac{3}{2}$	1	$\frac{3}{2}$	1	100−	$-2\sqrt{2.3.11}/\sqrt{5.13.17.19}$
$\frac{3}{2}$	1	$\frac{3}{2}$	1	101−	$+5\sqrt{11}/2.3\sqrt{3.13.17.19}$
$\frac{3}{2}$	1	$\frac{3}{2}$	1	200−	$+\sqrt{3.5.7}/4\sqrt{11.17.19}$
$\frac{3}{2}$	1	$\frac{3}{2}$	1	201−	$-79\sqrt{7}/4.3\sqrt{2.3.11.17.19}$
$\frac{3}{2}$	1	$\tilde{\frac{1}{2}}$	0	000+	$+\sqrt{7.13}/2\sqrt{2.3.17.19}$
$\frac{3}{2}$	1	$\tilde{\frac{1}{2}}$	0	100+	$+\sqrt{11.17}/2.3\sqrt{3.13.19}$
$\frac{3}{2}$	1	$\tilde{\frac{1}{2}}$	0	200+	$-\sqrt{7.11}/2.3\sqrt{2.3.17.19}$
$\frac{3}{2}$	2	$\frac{1}{2}$	0	000−	$-3.61/4\sqrt{2.5.11.13.17.19}$
$\frac{3}{2}$	2	$\frac{1}{2}$	0	100−	$-4\sqrt{7}/3\sqrt{5.13.17.19}$
$\frac{3}{2}$	2	$\frac{1}{2}$	0	200−	$+\sqrt{5.19}/4.3\sqrt{2.17}$
$\frac{3}{2}$	2	$\frac{3}{2}$	0	000+	$-149\sqrt{7}/4.11\sqrt{5.13.17.19}$
$\frac{3}{2}$	2	$\frac{3}{2}$	0	001+	$+8.5\sqrt{2.7}/11\sqrt{13.17.19}$
$\frac{3}{2}$	2	$\frac{3}{2}$	0	100+	$+\sqrt{2}/3\sqrt{5.11.13.17.19}$
$\frac{3}{2}$	2	$\frac{3}{2}$	0	101+	$-5.5.5/2.3\sqrt{11.13.17.19}$
$\frac{3}{2}$	2	$\frac{3}{2}$	0	200+	$+\sqrt{5.7.17}/4.3\sqrt{11.19}$
$\frac{3}{2}$	2	$\frac{3}{2}$	0	201+	$-2\sqrt{2.7}/3\sqrt{11.17.19}$
$\frac{3}{2}$	2	$\tilde{\frac{1}{2}}$	0	000−	$-23\sqrt{7}/4\sqrt{2.13.17.19}$

SO_3–O 3jm Factors (cont.)

$\frac{15}{2}$ 6 $\frac{11}{2}$

$\frac{3}{2}$	2	$\tilde{\frac{1}{2}}$	0	100−	$+2\sqrt{11}/3\sqrt{13.17.19}$
$\frac{3}{2}$	2	$\tilde{\frac{1}{2}}$	0	200−	$-\sqrt{7.11}/4.3\sqrt{2.17.19}$
$\frac{3}{2}$	$\tilde{1}$	$\frac{1}{2}$	0	000−	$+23\sqrt{3}/2\sqrt{2.7.13.17.19}$
$\frac{3}{2}$	$\tilde{1}$	$\frac{1}{2}$	0	010−	$+3.53\sqrt{3}/2\sqrt{2.5.7.11.13.17.19}$
$\frac{3}{2}$	$\tilde{1}$	$\frac{1}{2}$	0	100−	$-7\sqrt{11}/2.3\sqrt{3.13.17.19}$
$\frac{3}{2}$	$\tilde{1}$	$\frac{1}{2}$	0	110−	$+2.7/\sqrt{3.5.13.17.19}$
$\frac{3}{2}$	$\tilde{1}$	$\frac{1}{2}$	0	200−	$+23\sqrt{11}/2.3\sqrt{2.3.7.17.19}$
$\frac{3}{2}$	$\tilde{1}$	$\frac{1}{2}$	0	210−	$+5\sqrt{5}/2\sqrt{2.3.7.17.19}$
$\frac{3}{2}$	$\tilde{1}$	$\frac{3}{2}$	0	000+	$-3.5\sqrt{3}/2\sqrt{11.13.17.19}$
$\frac{3}{2}$	$\tilde{1}$	$\frac{3}{2}$	0	001+	$-353/2\sqrt{2.3.5.11.13.17.19}$
$\frac{3}{2}$	$\tilde{1}$	$\frac{3}{2}$	0	010+	$-\sqrt{3.13}/2.11\sqrt{5.17.19}$
$\frac{3}{2}$	$\tilde{1}$	$\frac{3}{2}$	0	011+	$-43\sqrt{2.3}/11\sqrt{13.17.19}$
$\frac{3}{2}$	$\tilde{1}$	$\frac{3}{2}$	0	100+	$-5\sqrt{19}/3\sqrt{2.3.7.13.17}$
$\frac{3}{2}$	$\tilde{1}$	$\frac{3}{2}$	0	101+	$-47/\sqrt{3.5.7.13.17.19}$
$\frac{3}{2}$	$\tilde{1}$	$\frac{3}{2}$	0	110+	$+8.2\sqrt{2.19}/\sqrt{3.5.7.11.13.17}$
$\frac{3}{2}$	$\tilde{1}$	$\frac{3}{2}$	0	111+	$-23\sqrt{3}/\sqrt{7.11.13.17.19}$
$\frac{3}{2}$	$\tilde{1}$	$\frac{3}{2}$	0	200+	$+13/2.3\sqrt{3.17.19}$
$\frac{3}{2}$	$\tilde{1}$	$\frac{3}{2}$	0	201+	$-5\sqrt{5}/2\sqrt{2.3.17.19}$
$\frac{3}{2}$	$\tilde{1}$	$\frac{3}{2}$	0	210+	$+\sqrt{5}/2\sqrt{3.11.17.19}$
$\frac{3}{2}$	$\tilde{1}$	$\frac{3}{2}$	0	211+	$+\sqrt{2.3}/\sqrt{11.17.19}$
$\frac{3}{2}$	$\tilde{1}$	$\frac{3}{2}$	1	000−	$-3.29\sqrt{3}/4\sqrt{11.13.17.19}$
$\frac{3}{2}$	$\tilde{1}$	$\frac{3}{2}$	1	001−	$+29\sqrt{13}/4\sqrt{2.3.5.11.17.19}$
$\frac{3}{2}$	$\tilde{1}$	$\frac{3}{2}$	1	010−	$+\sqrt{3.5}/2.11\sqrt{13.17.19}$
$\frac{3}{2}$	$\tilde{1}$	$\frac{3}{2}$	1	011−	$-5\sqrt{3}/11\sqrt{2.13.17.19}$
$\frac{3}{2}$	$\tilde{1}$	$\frac{3}{2}$	1	100−	$-8\sqrt{2.7}/3\sqrt{3.13.17.19}$
$\frac{3}{2}$	$\tilde{1}$	$\frac{3}{2}$	1	101−	$-107/2\sqrt{3.5.7.13.17.19}$
$\frac{3}{2}$	$\tilde{1}$	$\frac{3}{2}$	1	110−	$-2\sqrt{2.5.7}/\sqrt{3.11.13.17.19}$
$\frac{3}{2}$	$\tilde{1}$	$\frac{3}{2}$	1	111−	$-5.5/\sqrt{3.7.11.13.17.19}$
$\frac{3}{2}$	$\tilde{1}$	$\frac{3}{2}$	1	200−	$+37/4.3\sqrt{3.17.19}$
$\frac{3}{2}$	$\tilde{1}$	$\frac{3}{2}$	1	201−	$+\sqrt{5}/4\sqrt{2.3.17.19}$
$\frac{3}{2}$	$\tilde{1}$	$\frac{3}{2}$	1	210−	$-\sqrt{5}/2\sqrt{3.11.17.19}$
$\frac{3}{2}$	$\tilde{1}$	$\frac{3}{2}$	1	211−	$+5/\sqrt{2.3.11.17.19}$
$\frac{3}{2}$	$\tilde{1}$	$\tilde{\frac{1}{2}}$	0	000−	$+\sqrt{2.5.11}/\sqrt{3.13.17.19}$
$\frac{3}{2}$	$\tilde{1}$	$\tilde{\frac{1}{2}}$	0	010−	$+\sqrt{3}/2\sqrt{2.13.17.19}$
$\frac{3}{2}$	$\tilde{1}$	$\tilde{\frac{1}{2}}$	0	100−	$+11\sqrt{3.5}/2\sqrt{7.13.17.19}$
$\frac{3}{2}$	$\tilde{1}$	$\tilde{\frac{1}{2}}$	0	110−	$+2\sqrt{11}/\sqrt{3.7.13.17.19}$
$\frac{3}{2}$	$\tilde{1}$	$\tilde{\frac{1}{2}}$	0	200−	0
$\frac{3}{2}$	$\tilde{1}$	$\tilde{\frac{1}{2}}$	0	210−	$+7\sqrt{11}/2\sqrt{2.3.17.19}$
$\frac{3}{2}$	$\tilde{0}$	$\frac{3}{2}$	0	000−	$+4\sqrt{2.7}/\sqrt{11.13.17.19}$
$\frac{3}{2}$	$\tilde{0}$	$\frac{3}{2}$	0	001−	$-5\sqrt{5.7}/2\sqrt{11.13.17.19}$
$\frac{3}{2}$	$\tilde{0}$	$\frac{3}{2}$	0	100−	$-1/2.3\sqrt{13.17.19}$
$\frac{3}{2}$	$\tilde{0}$	$\frac{3}{2}$	0	101−	$+2\sqrt{2.5}/\sqrt{13.17.19}$
$\frac{3}{2}$	$\tilde{0}$	$\frac{3}{2}$	0	200−	$+2\sqrt{2.7}/3\sqrt{17.19}$
$\frac{3}{2}$	$\tilde{0}$	$\frac{3}{2}$	0	201−	$+\sqrt{5.7}/2\sqrt{17.19}$
$\tilde{\frac{1}{2}}$	0	$\tilde{\frac{1}{2}}$	0	000+	$-\sqrt{5}/2\sqrt{2.3.17.19}$
$\tilde{\frac{1}{2}}$	1	$\frac{3}{2}$	0	000+	$-2.3\sqrt{7}/11\sqrt{17.19}$
$\tilde{\frac{1}{2}}$	1	$\frac{3}{2}$	0	001+	$-4\sqrt{2.5.7}/3.11\sqrt{17.19}$
$\tilde{\frac{1}{2}}$	1	$\tilde{\frac{1}{2}}$	0	000−	$+\sqrt{5.7}/2.3\sqrt{17.19}$
$\tilde{\frac{1}{2}}$	2	$\frac{3}{2}$	0	000−	$+\sqrt{7}/\sqrt{3.17.19}$
$\tilde{\frac{1}{2}}$	2	$\frac{3}{2}$	0	001−	$-\sqrt{5.7}/2\sqrt{2.3.17.19}$
$\tilde{\frac{1}{2}}$	$\tilde{1}$	$\frac{1}{2}$	0	000+	$-\sqrt{5.7}/2.3\sqrt{17.19}$
$\tilde{\frac{1}{2}}$	$\tilde{1}$	$\frac{1}{2}$	0	010+	$+2\sqrt{7}/\sqrt{11.17.19}$
$\tilde{\frac{1}{2}}$	$\tilde{1}$	$\frac{3}{2}$	0	000−	$-2\sqrt{5}/3\sqrt{11.17.19}$
$\tilde{\frac{1}{2}}$	$\tilde{1}$	$\frac{3}{2}$	0	001−	$-\sqrt{19}/\sqrt{2.11.17}$
$\tilde{\frac{1}{2}}$	$\tilde{1}$	$\frac{3}{2}$	0	010−	$+2\sqrt{19}/11\sqrt{17}$
$\tilde{\frac{1}{2}}$	$\tilde{1}$	$\frac{3}{2}$	0	011−	$+\sqrt{5}/11\sqrt{2.17.19}$
$\tilde{\frac{1}{2}}$	$\tilde{0}$	$\frac{1}{2}$	0	000−	$-\sqrt{5}/2\sqrt{2.3.17.19}$

$\frac{15}{2}$ $\frac{13}{2}$ 1

$\frac{1}{2}$	$\frac{1}{2}$	1	0	000−	0
$\frac{1}{2}$	$\frac{3}{2}$	1	0	000+	$+\sqrt{3}/\sqrt{5.7}$
$\frac{1}{2}$	$\frac{3}{2}$	1	0	010+	$+\sqrt{11}/2\sqrt{2.5.7}$
$\frac{3}{2}$	$\frac{1}{2}$	1	0	000+	$+1/\sqrt{7}$
$\frac{3}{2}$	$\frac{1}{2}$	1	0	100+	0
$\frac{3}{2}$	$\frac{1}{2}$	1	0	200+	0
$\frac{3}{2}$	$\frac{3}{2}$	1	0	000−	$+3\sqrt{11}/2.5.7$
$\frac{3}{2}$	$\frac{3}{2}$	1	0	010−	$-11/5.7\sqrt{2.3}$
$\frac{3}{2}$	$\frac{3}{2}$	1	0	100−	$+\sqrt{2}/5\sqrt{7}$
$\frac{3}{2}$	$\frac{3}{2}$	1	0	110−	$-2\sqrt{11}/3.5\sqrt{3.7}$
$\frac{3}{2}$	$\frac{3}{2}$	1	0	200−	$+\sqrt{13}/2.7$
$\frac{3}{2}$	$\frac{3}{2}$	1	0	210−	$-\sqrt{11.13}/3.7\sqrt{2.3}$
$\frac{3}{2}$	$\frac{3}{2}$	1	1	000−	$+\sqrt{11}/4.7$
$\frac{3}{2}$	$\frac{3}{2}$	1	1	010−	$-37/2.5.7\sqrt{2.3}$
$\frac{3}{2}$	$\frac{3}{2}$	1	1	100−	$+1/\sqrt{2.7}$
$\frac{3}{2}$	$\frac{3}{2}$	1	1	110−	$+11\sqrt{11}/2.3.5\sqrt{3.7}$
$\frac{3}{2}$	$\frac{3}{2}$	1	1	200−	$-\sqrt{13}/4.7$
$\frac{3}{2}$	$\frac{3}{2}$	1	1	210−	$+\sqrt{11.13}/2.3.7\sqrt{2.3}$
$\frac{3}{2}$	$\tilde{\frac{1}{2}}$	1	0	000+	$-\sqrt{11}/4.7\sqrt{5}$
$\frac{3}{2}$	$\tilde{\frac{1}{2}}$	1	0	010+	$+\sqrt{2.13}/5.7\sqrt{3}$
$\frac{3}{2}$	$\tilde{\frac{1}{2}}$	1	0	100+	$-\sqrt{2}/\sqrt{5.7}$
$\frac{3}{2}$	$\tilde{\frac{1}{2}}$	1	0	110+	$-\sqrt{11.13}/3.5\sqrt{3.7}$
$\frac{3}{2}$	$\tilde{\frac{1}{2}}$	1	0	200+	$+\sqrt{5.13}/4.7$
$\frac{3}{2}$	$\tilde{\frac{1}{2}}$	1	0	210+	$-\sqrt{2.11}/3.7\sqrt{3}$
$\tilde{\frac{1}{2}}$	$\frac{3}{2}$	1	0	000+	0
$\tilde{\frac{1}{2}}$	$\frac{3}{2}$	1	0	010+	$-\sqrt{13}/2.3\sqrt{2.5}$
$\tilde{\frac{1}{2}}$	$\tilde{\frac{1}{2}}$	1	0	000−	0
$\tilde{\frac{1}{2}}$	$\tilde{\frac{1}{2}}$	1	0	010−	$+2/3\sqrt{5}$

$\frac{15}{2}$ $\frac{13}{2}$ 2

$\frac{1}{2}$	$\frac{3}{2}$	2	0	000+	$-2\sqrt{2.3}/\sqrt{5.7.13.17}$
$\frac{1}{2}$	$\frac{3}{2}$	2	0	010+	$+\sqrt{5.11}/2\sqrt{7.13.17}$
$\frac{1}{2}$	$\frac{3}{2}$	$\tilde{1}$	0	000+	$-9/\sqrt{5.7.13.17}$
$\frac{1}{2}$	$\frac{3}{2}$	$\tilde{1}$	0	010+	$-23\sqrt{11}/2\sqrt{2.3.5.7.13.17}$
$\frac{1}{2}$	$\tilde{\frac{1}{2}}$	$\tilde{1}$	0	000−	$+2.3\sqrt{2}/\sqrt{7.13.17}$
$\frac{1}{2}$	$\tilde{\frac{1}{2}}$	$\tilde{1}$	0	010−	$+2\sqrt{11}/\sqrt{3.5.7.17}$
$\frac{3}{2}$	$\frac{1}{2}$	2	0	000+	$+43/5\sqrt{2.7.13.17}$
$\frac{3}{2}$	$\frac{1}{2}$	2	0	100+	$+2\sqrt{11}/5\sqrt{13.17}$
$\frac{3}{2}$	$\frac{1}{2}$	2	0	200+	$+\sqrt{11}/\sqrt{2.7.17}$
$\frac{3}{2}$	$\frac{1}{2}$	$\tilde{1}$	0	000+	$+2\sqrt{3.7}/5\sqrt{13.17}$
$\frac{3}{2}$	$\frac{1}{2}$	$\tilde{1}$	0	100+	$-2\sqrt{2.11}/5\sqrt{3.13.17}$

SO_3–O 3jm Factors (cont.)

$\frac{15}{2}$	$\frac{13}{2}$	2			
$\frac{3}{2}$	$\frac{1}{2}$	$\tilde{1}$	0	200+	$-\sqrt{11}/\sqrt{3.7.17}$
$\frac{3}{2}$	$\frac{3}{2}$	2	0	000−	$-\sqrt{11.17}/2.7\sqrt{2.13}$
$\frac{3}{2}$	$\frac{3}{2}$	2	0	010−	$+149\sqrt{3}/2.5.7\sqrt{13.17}$
$\frac{3}{2}$	$\frac{3}{2}$	2	0	100−	$+1/\sqrt{7.13.17}$
$\frac{3}{2}$	$\frac{3}{2}$	2	0	110−	$+\sqrt{11.13}/5\sqrt{2.3.7.17}$
$\frac{3}{2}$	$\frac{3}{2}$	2	0	200−	$+\sqrt{17}/2.7\sqrt{2}$
$\frac{3}{2}$	$\frac{3}{2}$	2	0	210−	$-\sqrt{11}/2.7\sqrt{3.17}$
$\frac{3}{2}$	$\frac{3}{2}$	$\tilde{1}$	0	000−	$-\sqrt{3.11.17}/2.5.7\sqrt{13}$
$\frac{3}{2}$	$\frac{3}{2}$	$\tilde{1}$	0	010−	$+3.41/5.7\sqrt{2.13.17}$
$\frac{3}{2}$	$\frac{3}{2}$	$\tilde{1}$	0	100−	$+73\sqrt{2}/5\sqrt{3.7.13.17}$
$\frac{3}{2}$	$\frac{3}{2}$	$\tilde{1}$	0	110−	$+2\sqrt{11.13}/3.5\sqrt{7.17}$
$\frac{3}{2}$	$\frac{3}{2}$	$\tilde{1}$	0	200−	$-\sqrt{17}/2.7\sqrt{3}$
$\frac{3}{2}$	$\frac{3}{2}$	$\tilde{1}$	0	210−	$+\sqrt{11}/3.7\sqrt{2.17}$
$\frac{3}{2}$	$\frac{3}{2}$	$\tilde{1}$	1	000+	$-3\sqrt{3.11.17}/4.5.7\sqrt{13}$
$\frac{3}{2}$	$\frac{3}{2}$	$\tilde{1}$	1	010+	$-9/2.5.7\sqrt{2.13.17}$
$\frac{3}{2}$	$\frac{3}{2}$	$\tilde{1}$	1	100+	$+59/5\sqrt{2.3.7.13.17}$
$\frac{3}{2}$	$\frac{3}{2}$	$\tilde{1}$	1	110+	$+47\sqrt{11}/2.3.5\sqrt{7.13.17}$
$\frac{3}{2}$	$\frac{3}{2}$	$\tilde{1}$	1	200+	$+5/4.7\sqrt{3.17}$
$\frac{3}{2}$	$\frac{3}{2}$	$\tilde{1}$	1	210+	$-11\sqrt{11}/2.3.7\sqrt{2.17}$
$\frac{3}{2}$	$\widetilde{\frac{1}{2}}$	2	0	000+	$-\sqrt{11.17}/2.7\sqrt{2.5.13}$
$\frac{3}{2}$	$\widetilde{\frac{1}{2}}$	2	0	010+	$+8.2\sqrt{3}/5.7\sqrt{17}$
$\frac{3}{2}$	$\widetilde{\frac{1}{2}}$	2	0	100+	$-4/\sqrt{5.7.13.17}$
$\frac{3}{2}$	$\widetilde{\frac{1}{2}}$	2	0	110+	$-\sqrt{2.3.11}/5\sqrt{7.17}$
$\frac{3}{2}$	$\widetilde{\frac{1}{2}}$	2	0	200+	$-\sqrt{5}/2\sqrt{2.17}$
$\frac{3}{2}$	$\widetilde{\frac{1}{2}}$	2	0	210+	0
$\frac{3}{2}$	$\widetilde{\frac{1}{2}}$	$\tilde{1}$	0	000+	$-\sqrt{3.11.17}/4.7\sqrt{5.13}$
$\frac{3}{2}$	$\widetilde{\frac{1}{2}}$	$\tilde{1}$	0	010+	$+3\sqrt{2}/5.7\sqrt{17}$
$\frac{3}{2}$	$\widetilde{\frac{1}{2}}$	$\tilde{1}$	0	100+	$-\sqrt{2.3.7}/\sqrt{5.13.17}$
$\frac{3}{2}$	$\widetilde{\frac{1}{2}}$	$\tilde{1}$	0	110+	$+\sqrt{7.11}/3.5\sqrt{17}$
$\frac{3}{2}$	$\widetilde{\frac{1}{2}}$	$\tilde{1}$	0	200+	$+\sqrt{3.5}/4.7\sqrt{17}$
$\frac{3}{2}$	$\widetilde{\frac{1}{2}}$	$\tilde{1}$	0	210+	$+\sqrt{2.11.13}/3.7\sqrt{17}$
$\widetilde{\frac{1}{2}}$	$\frac{1}{2}$	$\tilde{1}$	0	000−	$+2\sqrt{2}/\sqrt{5.7.17}$
$\widetilde{\frac{1}{2}}$	$\frac{3}{2}$	2	0	000+	$+\sqrt{2.3.11}/7\sqrt{5.17}$
$\widetilde{\frac{1}{2}}$	$\frac{3}{2}$	2	0	010+	$+5\sqrt{5}/2.7\sqrt{17}$
$\widetilde{\frac{1}{2}}$	$\frac{3}{2}$	$\tilde{1}$	0	000+	$+2\sqrt{11}/7\sqrt{5.17}$
$\widetilde{\frac{1}{2}}$	$\frac{3}{2}$	$\tilde{1}$	0	010+	$-23\sqrt{3}/2.7\sqrt{2.5.17}$

$\frac{15}{2}$	$\frac{13}{2}$	3			
$\frac{1}{2}$	$\frac{1}{2}$	1	0	000−	$+\sqrt{5.11}/2\sqrt{2.3.13.17}$
$\frac{1}{2}$	$\frac{3}{2}$	1	0	000+	$+73/4\sqrt{3.5.7.13.17}$
$\frac{1}{2}$	$\frac{3}{2}$	1	0	010+	$+9\sqrt{11}/2\sqrt{2.5.7.13.17}$
$\frac{1}{2}$	$\frac{3}{2}$	$\tilde{1}$	0	000−	$+\sqrt{3.13}/4\sqrt{7.17}$
$\frac{1}{2}$	$\frac{3}{2}$	$\tilde{1}$	0	010−	$+\sqrt{11}/2.3\sqrt{2.7.13.17}$
$\frac{1}{2}$	$\widetilde{\frac{1}{2}}$	$\tilde{1}$	0	000+	$-5\sqrt{3.5}/2\sqrt{2.7.13.17}$
$\frac{1}{2}$	$\widetilde{\frac{1}{2}}$	$\tilde{1}$	0	010+	$+\sqrt{11}/2.3\sqrt{7.17}$
$\frac{1}{2}$	$\widetilde{\frac{1}{2}}$	$\tilde{0}$	0	000−	$+\sqrt{5}/\sqrt{2.7.13.17}$
$\frac{1}{2}$	$\widetilde{\frac{1}{2}}$	$\tilde{0}$	0	010−	$-\sqrt{11}/\sqrt{3.7.17}$
$\frac{3}{2}$	$\frac{1}{2}$	1	0	000+	$-9.3/8\sqrt{7.13.17}$
$\frac{3}{2}$	$\frac{1}{2}$	1	0	100+	$-\sqrt{2.11}/3\sqrt{13.17}$
$\frac{3}{2}$	$\frac{1}{2}$	1	0	200+	$-5\sqrt{11}/8.3\sqrt{7.17}$

$\frac{15}{2}$	$\frac{13}{2}$	3			
$\frac{3}{2}$	$\frac{1}{2}$	$\tilde{1}$	0	000−	$+9\sqrt{5}/8\sqrt{7.13.17}$
$\frac{3}{2}$	$\frac{1}{2}$	$\tilde{1}$	0	100−	0
$\frac{3}{2}$	$\frac{1}{2}$	$\tilde{1}$	0	200−	$-3\sqrt{5.11}/8\sqrt{7.17}$
$\frac{3}{2}$	$\frac{3}{2}$	1	0	000−	$-\sqrt{11.17}/8.5.7\sqrt{13}$
$\frac{3}{2}$	$\frac{3}{2}$	1	0	010−	$+199\sqrt{3}/2.5.7\sqrt{2.13.17}$
$\frac{3}{2}$	$\frac{3}{2}$	1	0	100−	$-67/2.3.5\sqrt{2.7.13.17}$
$\frac{3}{2}$	$\frac{3}{2}$	1	0	110−	$-71\sqrt{11}/2.3.5\sqrt{3.7.13.17}$
$\frac{3}{2}$	$\frac{3}{2}$	1	0	200−	$-97/8.3.7\sqrt{17}$
$\frac{3}{2}$	$\frac{3}{2}$	1	0	210−	$-13\sqrt{11}/2.3.7\sqrt{2.3.17}$
$\frac{3}{2}$	$\frac{3}{2}$	1	1	000−	$-5.5\sqrt{11}/8.7\sqrt{13.17}$
$\frac{3}{2}$	$\frac{3}{2}$	1	1	010−	$+3.11\sqrt{3}/4.5.7\sqrt{2.13.17}$
$\frac{3}{2}$	$\frac{3}{2}$	1	1	100−	$+3\sqrt{2}/\sqrt{7.13.17}$
$\frac{3}{2}$	$\frac{3}{2}$	1	1	110−	$-11\sqrt{11}/2.3.5\sqrt{3.7.13.17}$
$\frac{3}{2}$	$\frac{3}{2}$	1	1	200−	$-23/8.7\sqrt{17}$
$\frac{3}{2}$	$\frac{3}{2}$	1	1	210−	$-47\sqrt{11}/4.3.7\sqrt{2.3.17}$
$\frac{3}{2}$	$\frac{3}{2}$	$\tilde{1}$	0	000+	$-3.23\sqrt{11}/8.7\sqrt{5.13.17}$
$\frac{3}{2}$	$\frac{3}{2}$	$\tilde{1}$	0	010+	$+\sqrt{3.17}/7\sqrt{2.5.13}$
$\frac{3}{2}$	$\frac{3}{2}$	$\tilde{1}$	0	100+	$-43/2\sqrt{2.5.7.13.17}$
$\frac{3}{2}$	$\frac{3}{2}$	$\tilde{1}$	0	110+	$-41\sqrt{11}/2.3\sqrt{3.5.7.13.17}$
$\frac{3}{2}$	$\frac{3}{2}$	$\tilde{1}$	0	200+	$-\sqrt{5}/8\sqrt{17}$
$\frac{3}{2}$	$\frac{3}{2}$	$\tilde{1}$	0	210+	$+\sqrt{5.11}/3\sqrt{2.3.17}$
$\frac{3}{2}$	$\frac{3}{2}$	$\tilde{1}$	1	000−	$-3.31\sqrt{11}/8.7\sqrt{5.13.17}$
$\frac{3}{2}$	$\frac{3}{2}$	$\tilde{1}$	1	010−	$+53\sqrt{3}/4.7\sqrt{2.5.13.17}$
$\frac{3}{2}$	$\frac{3}{2}$	$\tilde{1}$	1	100−	$-9\sqrt{2}/\sqrt{5.7.13.17}$
$\frac{3}{2}$	$\frac{3}{2}$	$\tilde{1}$	1	110−	$+29\sqrt{11}/2.3\sqrt{3.5.7.13.17}$
$\frac{3}{2}$	$\frac{3}{2}$	$\tilde{1}$	1	200−	$+9\sqrt{5}/8.7\sqrt{17}$
$\frac{3}{2}$	$\frac{3}{2}$	$\tilde{1}$	1	210−	$+5\sqrt{5.11}/4.3.7\sqrt{2.3.17}$
$\frac{3}{2}$	$\frac{3}{2}$	$\tilde{0}$	0	000−	$-\sqrt{3.11.17}/4.7\sqrt{5.13}$
$\frac{3}{2}$	$\frac{3}{2}$	$\tilde{0}$	0	010−	$+9/7\sqrt{2.5.13.17}$
$\frac{3}{2}$	$\frac{3}{2}$	$\tilde{0}$	0	100−	$+23/\sqrt{2.3.5.7.13.17}$
$\frac{3}{2}$	$\frac{3}{2}$	$\tilde{0}$	0	110−	$-\sqrt{11.17}/3\sqrt{5.7.13}$
$\frac{3}{2}$	$\frac{3}{2}$	$\tilde{0}$	0	200−	$-\sqrt{5}/4.7\sqrt{3.17}$
$\frac{3}{2}$	$\frac{3}{2}$	$\tilde{0}$	0	210−	$-\sqrt{5.11}/3.7\sqrt{2.17}$
$\frac{3}{2}$	$\widetilde{\frac{1}{2}}$	1	0	000+	$+\sqrt{11.17}/8.7\sqrt{5.13}$
$\frac{3}{2}$	$\widetilde{\frac{1}{2}}$	1	0	010+	$-29\sqrt{3}/5.7\sqrt{2.17}$
$\frac{3}{2}$	$\widetilde{\frac{1}{2}}$	1	0	100+	$-37\sqrt{2}/3\sqrt{5.7.13.17}$
$\frac{3}{2}$	$\widetilde{\frac{1}{2}}$	1	0	110+	$-13\sqrt{11}/2.3.5\sqrt{3.7.17}$
$\frac{3}{2}$	$\widetilde{\frac{1}{2}}$	1	0	200+	$-31\sqrt{5}/8.3.7\sqrt{17}$
$\frac{3}{2}$	$\widetilde{\frac{1}{2}}$	1	0	210+	$-\sqrt{11.13}/3.7\sqrt{2.3.17}$
$\frac{3}{2}$	$\widetilde{\frac{1}{2}}$	$\tilde{1}$	0	000−	$-3\sqrt{11.17}/8.7\sqrt{13}$
$\frac{3}{2}$	$\widetilde{\frac{1}{2}}$	$\tilde{1}$	0	010−	$+3\sqrt{3}/7\sqrt{2.5.17}$
$\frac{3}{2}$	$\widetilde{\frac{1}{2}}$	$\tilde{1}$	0	100−	$+2\sqrt{2}/\sqrt{7.13.17}$
$\frac{3}{2}$	$\widetilde{\frac{1}{2}}$	$\tilde{1}$	0	110−	$+11\sqrt{11}/2.3\sqrt{3.5.7.17}$
$\frac{3}{2}$	$\widetilde{\frac{1}{2}}$	$\tilde{1}$	0	200−	$-5/8.7\sqrt{17}$
$\frac{3}{2}$	$\widetilde{\frac{1}{2}}$	$\tilde{1}$	0	210−	$-\sqrt{5.11.13}/3.7\sqrt{2.3.17}$
$\widetilde{\frac{1}{2}}$	$\frac{1}{2}$	$\tilde{1}$	0	000+	$+\sqrt{3}/2\sqrt{2.7.17}$
$\widetilde{\frac{1}{2}}$	$\frac{1}{2}$	$\tilde{0}$	0	000−	$+3/\sqrt{2.7.17}$
$\widetilde{\frac{1}{2}}$	$\frac{3}{2}$	1	0	000+	$+\sqrt{5.11}/4\sqrt{3.17}$
$\widetilde{\frac{1}{2}}$	$\frac{3}{2}$	1	0	010+	$+1/2.3\sqrt{2.5.17}$

SO_3–O 3jm Factors (cont.)

$\frac{15}{2}\ \frac{13}{2}\ 3$

$\tilde{\frac{1}{2}}\ \frac{3}{2}\ \tilde{1}\ 0\ 000-\quad -\sqrt{3.11}/4.7\sqrt{17}$
$\tilde{\frac{1}{2}}\ \frac{3}{2}\ \tilde{1}\ 0\ 010-\quad -1/2.7\sqrt{2.17}$
$\tilde{\frac{1}{2}}\ \tilde{\frac{1}{2}}\ 1\ 0\ 000-\quad -5\sqrt{11}/2.7\sqrt{2.3.17}$
$\tilde{\frac{1}{2}}\ \tilde{\frac{1}{2}}\ 1\ 0\ 010-\quad -\sqrt{13}/2.3.7\sqrt{5.17}$

$\frac{15}{2}\ \frac{13}{2}\ 4$

$\frac{1}{2}\ \frac{1}{2}\ 0\ 0\ 000-\quad -\sqrt{7}/\sqrt{2.13.17.19}$
$\frac{1}{2}\ \frac{1}{2}\ 1\ 0\ 000+\quad -3\sqrt{5.7}/2\sqrt{2.13.17.19}$
$\frac{1}{2}\ \frac{3}{2}\ 1\ 0\ 000-\quad +7\sqrt{5.11}/4\sqrt{13.17.19}$
$\frac{1}{2}\ \frac{3}{2}\ 1\ 0\ 010-\quad +7\sqrt{5}/2\sqrt{2.3.13.17.19}$
$\frac{1}{2}\ \frac{3}{2}\ 2\ 0\ 000+\quad -\sqrt{5.11}/\sqrt{2.7.13.17.19}$
$\frac{1}{2}\ \frac{3}{2}\ 2\ 0\ 010+\quad +\sqrt{5.13}/\sqrt{3.7.17.19}$
$\frac{1}{2}\ \frac{3}{2}\ \tilde{1}\ 0\ 000+\quad +\sqrt{5.11.19}/4\sqrt{3.7.13.17}$
$\frac{1}{2}\ \frac{3}{2}\ \tilde{1}\ 0\ 010+\quad -11.11\sqrt{5}/2.3\sqrt{2.7.13.17.19}$
$\frac{1}{2}\ \tilde{\frac{1}{2}}\ \tilde{1}\ 0\ 000-\quad -\sqrt{11.19}/2\sqrt{2.3.7.13.17}$
$\frac{1}{2}\ \tilde{\frac{1}{2}}\ \tilde{1}\ 0\ 010-\quad +11\sqrt{5}/2.3\sqrt{7.17.19}$
$\frac{3}{2}\ \frac{1}{2}\ 1\ 0\ 000-\quad -9\sqrt{3.19}/8\sqrt{11.13.17}$
$\frac{3}{2}\ \frac{1}{2}\ 1\ 0\ 100-\quad +\sqrt{2.7}/\sqrt{3.13.17.19}$
$\frac{3}{2}\ \frac{1}{2}\ 1\ 0\ 200-\quad +11/8\sqrt{3.17.19}$
$\frac{3}{2}\ \frac{1}{2}\ 2\ 0\ 000+\quad +\sqrt{3.17}/2\sqrt{2.7.11.13.19}$
$\frac{3}{2}\ \frac{1}{2}\ 2\ 0\ 100+\quad +2/\sqrt{3.13.17.19}$
$\frac{3}{2}\ \frac{1}{2}\ 2\ 0\ 200+\quad +31/2\sqrt{2.3.7.17.19}$
$\frac{3}{2}\ \frac{1}{2}\ \tilde{1}\ 0\ 000+\quad +5\sqrt{7}/8\sqrt{11.13.17.19}$
$\frac{3}{2}\ \frac{1}{2}\ \tilde{1}\ 0\ 100+\quad -2.5\sqrt{2}/\sqrt{13.17.19}$
$\frac{3}{2}\ \frac{1}{2}\ \tilde{1}\ 0\ 200+\quad +31/8\sqrt{7.17.19}$
$\frac{3}{2}\ \frac{3}{2}\ 0\ 0\ 000-\quad +179/4\sqrt{3.5.7.13.17.19}$
$\frac{3}{2}\ \frac{3}{2}\ 0\ 0\ 010-\quad -251\sqrt{2}/3\sqrt{5.7.11.13.17.19}$
$\frac{3}{2}\ \frac{3}{2}\ 0\ 0\ 100-\quad -41\sqrt{11}/3\sqrt{2.3.5.13.17.19}$
$\frac{3}{2}\ \frac{3}{2}\ 0\ 0\ 110-\quad +1/9\sqrt{5.13.17.19}$
$\frac{3}{2}\ \frac{3}{2}\ 0\ 0\ 200-\quad -23\sqrt{5.11}/4.3\sqrt{3.7.17.19}$
$\frac{3}{2}\ \frac{3}{2}\ 0\ 0\ 210-\quad -11\sqrt{2.5}/9\sqrt{7.17.19}$
$\frac{3}{2}\ \frac{3}{2}\ 1\ 0\ 000+\quad -5.103/8\sqrt{3.7.13.17.19}$
$\frac{3}{2}\ \frac{3}{2}\ 1\ 0\ 010+\quad -5.97/2.3\sqrt{2.7.11.13.17.19}$
$\frac{3}{2}\ \frac{3}{2}\ 1\ 0\ 100+\quad +5\sqrt{11}/2.3\sqrt{2.3.13.17.19}$
$\frac{3}{2}\ \frac{3}{2}\ 1\ 0\ 110+\quad +5.5/2.3\sqrt{13.17.19}$
$\frac{3}{2}\ \frac{3}{2}\ 1\ 0\ 200+\quad -29\sqrt{11}/8.3\sqrt{3.7.17.19}$
$\frac{3}{2}\ \frac{3}{2}\ 1\ 0\ 210+\quad +43/2.3\sqrt{2.7.17.19}$
$\frac{3}{2}\ \frac{3}{2}\ 1\ 1\ 000+\quad -5\sqrt{7}/8\sqrt{3.13.17.19}$
$\frac{3}{2}\ \frac{3}{2}\ 1\ 1\ 010+\quad +5.5\sqrt{7.11}/4.3\sqrt{2.13.17.19}$
$\frac{3}{2}\ \frac{3}{2}\ 1\ 1\ 100+\quad +5\sqrt{2.11}/3\sqrt{3.13.17.19}$
$\frac{3}{2}\ \frac{3}{2}\ 1\ 1\ 110+\quad +5\sqrt{13}/2.3\sqrt{17.19}$
$\frac{3}{2}\ \frac{3}{2}\ 1\ 1\ 200+\quad -53\sqrt{11}/8.3\sqrt{3.7.17.19}$
$\frac{3}{2}\ \frac{3}{2}\ 1\ 1\ 210+\quad +1/4.3\sqrt{2.7.17.19}$
$\frac{3}{2}\ \frac{3}{2}\ 2\ 0\ 000-\quad -5.31/2.7\sqrt{2.3.13.17.19}$
$\frac{3}{2}\ \frac{3}{2}\ 2\ 0\ 010-\quad -5.101/2.3.7\sqrt{11.13.17.19}$
$\frac{3}{2}\ \frac{3}{2}\ 2\ 0\ 100-\quad -8.5\sqrt{11}/3\sqrt{3.7.13.17.19}$
$\frac{3}{2}\ \frac{3}{2}\ 2\ 0\ 110-\quad +5\sqrt{2.13}/9\sqrt{7.17.19}$
$\frac{3}{2}\ \frac{3}{2}\ 2\ 0\ 200-\quad -29\sqrt{11}/2.3.7\sqrt{2.3.17.19}$
$\frac{3}{2}\ \frac{3}{2}\ 2\ 0\ 210-\quad -5.5.23/2.9.7\sqrt{17.19}$
$\frac{3}{2}\ \frac{3}{2}\ \tilde{1}\ 0\ 000-\quad +11.11/8.7\sqrt{13.17.19}$

$\frac{15}{2}\ \frac{13}{2}\ 4$

$\frac{3}{2}\ \frac{3}{2}\ \tilde{1}\ 0\ 010-\quad +\sqrt{11}/7\sqrt{2.3.13.17.19}$
$\frac{3}{2}\ \frac{3}{2}\ \tilde{1}\ 0\ 100-\quad +31\sqrt{11}/2.3\sqrt{2.7.13.17.19}$
$\frac{3}{2}\ \frac{3}{2}\ \tilde{1}\ 0\ 110-\quad -97/2\sqrt{3.7.13.17.19}$
$\frac{3}{2}\ \frac{3}{2}\ \tilde{1}\ 0\ 200-\quad +167\sqrt{11}/8.3.7\sqrt{17.19}$
$\frac{3}{2}\ \frac{3}{2}\ \tilde{1}\ 0\ 210-\quad -\sqrt{3}/7\sqrt{2.17.19}$
$\frac{3}{2}\ \frac{3}{2}\ \tilde{1}\ 1\ 000+\quad +101/8.7\sqrt{13.17.19}$
$\frac{3}{2}\ \frac{3}{2}\ \tilde{1}\ 1\ 010+\quad -991/4.7\sqrt{2.3.11.13.17.19}$
$\frac{3}{2}\ \frac{3}{2}\ \tilde{1}\ 1\ 100+\quad -11\sqrt{2.11}/3\sqrt{7.13.17.19}$
$\frac{3}{2}\ \frac{3}{2}\ \tilde{1}\ 1\ 110+\quad +\sqrt{19}/2.3\sqrt{3.7.13.17}$
$\frac{3}{2}\ \frac{3}{2}\ \tilde{1}\ 1\ 200+\quad -61\sqrt{11}/8.3.7\sqrt{17.19}$
$\frac{3}{2}\ \frac{3}{2}\ \tilde{1}\ 1\ 210+\quad +359/4.3.7\sqrt{2.3.17.19}$
$\frac{3}{2}\ \tilde{\frac{1}{2}}\ 1\ 0\ 000-\quad +11.41/8\sqrt{3.5.7.13.17.19}$
$\frac{3}{2}\ \tilde{\frac{1}{2}}\ 1\ 0\ 010-\quad +103/3\sqrt{2.7.11.17.19}$
$\frac{3}{2}\ \tilde{\frac{1}{2}}\ 1\ 0\ 100-\quad -31\sqrt{2.11}/3\sqrt{3.5.13.17.19}$
$\frac{3}{2}\ \tilde{\frac{1}{2}}\ 1\ 0\ 110-\quad -1/2.3\sqrt{17.19}$
$\frac{3}{2}\ \tilde{\frac{1}{2}}\ 1\ 0\ 200-\quad -\sqrt{5.7.11}/8.3\sqrt{3.17.19}$
$\frac{3}{2}\ \tilde{\frac{1}{2}}\ 1\ 0\ 210-\quad +\sqrt{7.13}/3\sqrt{2.17.19}$
$\frac{3}{2}\ \tilde{\frac{1}{2}}\ 2\ 0\ 000+\quad -211/2.7\sqrt{2.3.5.13.17.19}$
$\frac{3}{2}\ \tilde{\frac{1}{2}}\ 2\ 0\ 010+\quad -8.2\sqrt{11}/3.7\sqrt{17.19}$
$\frac{3}{2}\ \tilde{\frac{1}{2}}\ 2\ 0\ 100+\quad +2\sqrt{11.17}/3\sqrt{3.5.7.13.19}$
$\frac{3}{2}\ \tilde{\frac{1}{2}}\ 2\ 0\ 110+\quad -37\sqrt{2}/9\sqrt{7.17.19}$
$\frac{3}{2}\ \tilde{\frac{1}{2}}\ 2\ 0\ 200+\quad +\sqrt{5.11}/2.3\sqrt{2.3.17.19}$
$\frac{3}{2}\ \tilde{\frac{1}{2}}\ 2\ 0\ 210+\quad +8\sqrt{13}/9\sqrt{17.19}$
$\frac{3}{2}\ \tilde{\frac{1}{2}}\ \tilde{1}\ 0\ 000+\quad +\sqrt{5.17.19}/8.7\sqrt{13}$
$\frac{3}{2}\ \tilde{\frac{1}{2}}\ \tilde{1}\ 0\ 010+\quad -\sqrt{3.19}/7\sqrt{2.11.17}$
$\frac{3}{2}\ \tilde{\frac{1}{2}}\ \tilde{1}\ 0\ 100+\quad 0$
$\frac{3}{2}\ \tilde{\frac{1}{2}}\ \tilde{1}\ 0\ 110+\quad +11\sqrt{7}/2.3\sqrt{3.17.19}$
$\frac{3}{2}\ \tilde{\frac{1}{2}}\ \tilde{1}\ 0\ 200+\quad -\sqrt{5.11}/8.7\sqrt{17.19}$
$\frac{3}{2}\ \tilde{\frac{1}{2}}\ \tilde{1}\ 0\ 210+\quad -11\sqrt{13}/3.7\sqrt{2.3.17.19}$
$\tilde{\frac{1}{2}}\ \frac{1}{2}\ \tilde{1}\ 0\ 000-\quad +\sqrt{3.5.19}/2\sqrt{2.7.11.17}$
$\tilde{\frac{1}{2}}\ \frac{3}{2}\ 1\ 0\ 000-\quad -\sqrt{5.19}/4.3\sqrt{7.17}$
$\tilde{\frac{1}{2}}\ \frac{3}{2}\ 1\ 0\ 010-\quad +\sqrt{3.5}/2\sqrt{2.7.11.17.19}$
$\tilde{\frac{1}{2}}\ \frac{3}{2}\ 2\ 0\ 000+\quad -29\sqrt{5}/3.7\sqrt{2.17.19}$
$\tilde{\frac{1}{2}}\ \frac{3}{2}\ 2\ 0\ 010+\quad -71\sqrt{5}/3.7\sqrt{3.11.17.19}$
$\tilde{\frac{1}{2}}\ \frac{3}{2}\ \tilde{1}\ 0\ 000+\quad +43\sqrt{5}/4.7\sqrt{3.17.19}$
$\tilde{\frac{1}{2}}\ \frac{3}{2}\ \tilde{1}\ 0\ 010+\quad -181\sqrt{5}/2.3.7\sqrt{2.11.17.19}$
$\tilde{\frac{1}{2}}\ \tilde{\frac{1}{2}}\ 0\ 0\ 000-\quad -\sqrt{5.19}/3\sqrt{2.7.17}$
$\tilde{\frac{1}{2}}\ \tilde{\frac{1}{2}}\ 0\ 0\ 010-\quad -\sqrt{13}/3\sqrt{3.7.11.17.19}$
$\tilde{\frac{1}{2}}\ \tilde{\frac{1}{2}}\ 1\ 0\ 000+\quad -\sqrt{19}/2.3\sqrt{2.7.17}$
$\tilde{\frac{1}{2}}\ \tilde{\frac{1}{2}}\ 1\ 0\ 010+\quad -\sqrt{5.13}/2\sqrt{3.7.11.17.19}$

$\frac{15}{2}\ \frac{13}{2}\ 5$

$\frac{1}{2}\ \frac{1}{2}\ 1\ 0\ 000-\quad +9.7\sqrt{3}/2\sqrt{2.11.13.17.19}$
$\frac{1}{2}\ \frac{1}{2}\ 1\ 0\ 001-\quad +2\sqrt{2.5.7}/\sqrt{3.11.13.17.19}$
$\frac{1}{2}\ \frac{3}{2}\ 1\ 0\ 000+\quad +2\sqrt{3}/\sqrt{7.13.17.19}$
$\frac{1}{2}\ \frac{3}{2}\ 1\ 0\ 001+\quad +2\sqrt{5}/\sqrt{3.13.17.19}$
$\frac{1}{2}\ \frac{3}{2}\ 1\ 0\ 010+\quad +\sqrt{11}/\sqrt{2.7.13.17.19}$
$\frac{1}{2}\ \frac{3}{2}\ 1\ 0\ 011+\quad +3\sqrt{5.11}/2\sqrt{2.13.17.19}$
$\frac{1}{2}\ \frac{3}{2}\ 2\ 0\ 000-\quad -\sqrt{2.5.11}/\sqrt{13.17.19}$
$\frac{1}{2}\ \frac{3}{2}\ 2\ 0\ 010-\quad +\sqrt{5.17}/2\sqrt{3.13.19}$

SO_3–O 3jm Factors (cont.)

$\frac{15}{2}$	$\frac{13}{2}$	5		
$\frac{1}{2}$	$\frac{3}{2}$	$\tilde{1}$	0 000−	0
$\frac{1}{2}$	$\frac{3}{2}$	$\tilde{1}$	0 010−	$-7\sqrt{5}/3\sqrt{2.13.17.19}$
$\frac{1}{2}$	$\tilde{\frac{1}{2}}$	$\tilde{1}$	0 000+	$+3\sqrt{3.11}/2\sqrt{2.13.17.19}$
$\frac{1}{2}$	$\tilde{\frac{1}{2}}$	$\tilde{1}$	0 010+	$+2\sqrt{5}/3\sqrt{17.19}$
$\frac{3}{2}$	$\frac{1}{2}$	1	0 000+	$-2.3\sqrt{5}/11\sqrt{7.13.17.19}$
$\frac{3}{2}$	$\frac{1}{2}$	1	0 001+	$-9.3\sqrt{19}/4.11\sqrt{13.17}$
$\frac{3}{2}$	$\frac{1}{2}$	1	0 100+	$-9.7\sqrt{5}/2\sqrt{2.11.13.17.19}$
$\frac{3}{2}$	$\frac{1}{2}$	1	0 101+	$+\sqrt{2.7}/3\sqrt{11.13.17.19}$
$\frac{3}{2}$	$\frac{1}{2}$	1	0 200+	0
$\frac{3}{2}$	$\frac{1}{2}$	1	0 201+	$-\sqrt{11}/4.3\sqrt{17.19}$
$\frac{3}{2}$	$\frac{1}{2}$	2	0 000−	$-\sqrt{3.17}/2\sqrt{2.11.13.19}$
$\frac{3}{2}$	$\frac{1}{2}$	2	0 100−	$+\sqrt{3.7}/\sqrt{13.17.19}$
$\frac{3}{2}$	$\frac{1}{2}$	2	0 200−	$-\sqrt{3}/2\sqrt{2.17.19}$
$\frac{3}{2}$	$\frac{1}{2}$	$\tilde{1}$	0 000−	$-3.5/\sqrt{11.13.17.19}$
$\frac{3}{2}$	$\frac{1}{2}$	$\tilde{1}$	0 100−	$+5\sqrt{7}/2\sqrt{2.13.17.19}$
$\frac{3}{2}$	$\frac{1}{2}$	$\tilde{1}$	0 200−	$-1/\sqrt{17.19}$
$\frac{3}{2}$	$\frac{3}{2}$	1	0 000−	$-9.29/7\sqrt{5.11.13.17.19}$
$\frac{3}{2}$	$\frac{3}{2}$	1	0 001−	$+5.29/2\sqrt{7.11.13.17.19}$
$\frac{3}{2}$	$\frac{3}{2}$	1	0 010−	$-23\sqrt{3.19}/4.7.11\sqrt{2.5.13.17}$
$\frac{3}{2}$	$\frac{3}{2}$	1	0 011−	$+179\sqrt{3}/11\sqrt{2.7.13.17.19}$
$\frac{3}{2}$	$\frac{3}{2}$	1	0 100−	$+67/\sqrt{2.5.7.13.17.19}$
$\frac{3}{2}$	$\frac{3}{2}$	1	0 101−	$-5\sqrt{2}/3\sqrt{13.17.19}$
$\frac{3}{2}$	$\frac{3}{2}$	1	0 110−	$-311/\sqrt{3.5.7.11.13.17.19}$
$\frac{3}{2}$	$\frac{3}{2}$	1	0 111−	$+2.37/3\sqrt{3.11.13.17.19}$
$\frac{3}{2}$	$\frac{3}{2}$	1	0 200−	$+\sqrt{5}/7\sqrt{17.19}$
$\frac{3}{2}$	$\frac{3}{2}$	1	0 201−	$-1/2.3\sqrt{7.17.19}$
$\frac{3}{2}$	$\frac{3}{2}$	1	0 210−	$+281\sqrt{5}/4.7\sqrt{2.3.11.17.19}$
$\frac{3}{2}$	$\frac{3}{2}$	1	0 211−	$+167/3\sqrt{2.3.7.11.17.19}$
$\frac{3}{2}$	$\frac{3}{2}$	1	1 000−	$-3.59/4.7\sqrt{5.11.13.17.19}$
$\frac{3}{2}$	$\frac{3}{2}$	1	1 001−	$-13\sqrt{13}/4\sqrt{7.11.17.19}$
$\frac{3}{2}$	$\frac{3}{2}$	1	1 010−	$-293\sqrt{3}/4.7.11\sqrt{2.5.13.17.19}$
$\frac{3}{2}$	$\frac{3}{2}$	1	1 011−	$-3.67\sqrt{3}/2.11\sqrt{2.7.13.17.19}$
$\frac{3}{2}$	$\frac{3}{2}$	1	1 100−	$-2\sqrt{2}/\sqrt{5.7.13.17.19}$
$\frac{3}{2}$	$\frac{3}{2}$	1	1 101−	$+\sqrt{17}/\sqrt{2.13.19}$
$\frac{3}{2}$	$\frac{3}{2}$	1	1 110−	$-\sqrt{3.13}/\sqrt{5.7.11.17.19}$
$\frac{3}{2}$	$\frac{3}{2}$	1	1 111−	$+109/2.3\sqrt{3.11.13.17.19}$
$\frac{3}{2}$	$\frac{3}{2}$	1	1 200−	$+47\sqrt{5}/4.7\sqrt{17.19}$
$\frac{3}{2}$	$\frac{3}{2}$	1	1 201−	$+9.3/4\sqrt{7.17.19}$
$\frac{3}{2}$	$\frac{3}{2}$	1	1 210−	$+43\sqrt{3.5}/4.7\sqrt{2.11.17.19}$
$\frac{3}{2}$	$\frac{3}{2}$	1	1 211−	$-101/2.3\sqrt{2.3.7.11.17.19}$
$\frac{3}{2}$	$\frac{3}{2}$	2	0 000+	$+41\sqrt{3}/2\sqrt{2.7.13.17.19}$
$\frac{3}{2}$	$\frac{3}{2}$	2	0 010+	$-2.9/\sqrt{7.11.13.17.19}$
$\frac{3}{2}$	$\frac{3}{2}$	2	0 100+	$+\sqrt{11}/\sqrt{3.13.17.19}$
$\frac{3}{2}$	$\frac{3}{2}$	2	0 110+	$+23/3\sqrt{2.13.17.19}$
$\frac{3}{2}$	$\frac{3}{2}$	2	0 200+	$-\sqrt{11}/2\sqrt{2.3.7.17.19}$
$\frac{3}{2}$	$\frac{3}{2}$	2	0 210+	$+2.5/3\sqrt{7.17.19}$
$\frac{3}{2}$	$\frac{3}{2}$	$\tilde{1}$	0 000+	$-3/2\sqrt{7.13.17.19}$
$\frac{3}{2}$	$\frac{3}{2}$	$\tilde{1}$	0 010+	$-19\sqrt{3.19}/4\sqrt{2.7.11.13.17}$
$\frac{3}{2}$	$\frac{3}{2}$	$\tilde{1}$	0 100+	$+3\sqrt{11}/\sqrt{2.13.17.19}$

$\frac{15}{2}$	$\frac{13}{2}$	5		
$\frac{3}{2}$	$\frac{3}{2}$	$\tilde{1}$	0 110+	$-\sqrt{19}/3\sqrt{3.13.17}$
$\frac{3}{2}$	$\frac{3}{2}$	$\tilde{1}$	0 200+	$-3\sqrt{11}/2\sqrt{7.17.19}$
$\frac{3}{2}$	$\frac{3}{2}$	$\tilde{1}$	0 210+	$+83/4.3\sqrt{2.3.7.17.19}$
$\frac{3}{2}$	$\frac{3}{2}$	$\tilde{1}$	1 000−	$+9.3.5/4\sqrt{7.13.17.19}$
$\frac{3}{2}$	$\frac{3}{2}$	$\tilde{1}$	1 010−	$+251\sqrt{3}/4\sqrt{2.7.11.13.17.19}$
$\frac{3}{2}$	$\frac{3}{2}$	$\tilde{1}$	1 100−	0
$\frac{3}{2}$	$\frac{3}{2}$	$\tilde{1}$	1 110−	$+7.7/3\sqrt{3.13.17.19}$
$\frac{3}{2}$	$\frac{3}{2}$	$\tilde{1}$	1 200−	$+3\sqrt{11}/4\sqrt{7.17.19}$
$\frac{3}{2}$	$\frac{3}{2}$	$\tilde{1}$	1 210−	$-\sqrt{17}/4.3\sqrt{2.3.7.19}$
$\frac{3}{2}$	$\tilde{\frac{1}{2}}$	1	0 000+	$+2.3.23/7\sqrt{11.13.17.19}$
$\frac{3}{2}$	$\tilde{\frac{1}{2}}$	1	0 001+	$-167/2\sqrt{5.7.11.13.17.19}$
$\frac{3}{2}$	$\tilde{\frac{1}{2}}$	1	0 010+	$+23\sqrt{3.17}/2.7.11\sqrt{2.5.19}$
$\frac{3}{2}$	$\tilde{\frac{1}{2}}$	1	0 011+	$-8.2\sqrt{2.3}/11\sqrt{7.17.19}$
$\frac{3}{2}$	$\tilde{\frac{1}{2}}$	1	0 100+	$+\sqrt{13}/2\sqrt{2.7.17.19}$
$\frac{3}{2}$	$\tilde{\frac{1}{2}}$	1	0 101+	$+8\sqrt{2}/3\sqrt{5.13.17.19}$
$\frac{3}{2}$	$\tilde{\frac{1}{2}}$	1	0 110+	$+8.2/\sqrt{3.5.7.11.17.19}$
$\frac{3}{2}$	$\tilde{\frac{1}{2}}$	1	0 111+	$-\sqrt{17}/3\sqrt{3.11.19}$
$\frac{3}{2}$	$\tilde{\frac{1}{2}}$	1	0 200+	$-8/7\sqrt{17.19}$
$\frac{3}{2}$	$\tilde{\frac{1}{2}}$	1	0 201+	$+\sqrt{5.17}/2.3\sqrt{7.19}$
$\frac{3}{2}$	$\tilde{\frac{1}{2}}$	1	0 210+	$+13\sqrt{5.13}/2.7\sqrt{2.3.11.17.19}$
$\frac{3}{2}$	$\tilde{\frac{1}{2}}$	1	0 211+	$+8\sqrt{2.13}/3\sqrt{3.7.11.17.19}$
$\frac{3}{2}$	$\tilde{\frac{1}{2}}$	2	0 000−	$+2.9\sqrt{2.3}/\sqrt{5.7.13.17.19}$
$\frac{3}{2}$	$\tilde{\frac{1}{2}}$	2	0 010−	$+9/\sqrt{7.11.17.19}$
$\frac{3}{2}$	$\tilde{\frac{1}{2}}$	2	0 100−	$+7\sqrt{11}/\sqrt{3.5.13.17.19}$
$\frac{3}{2}$	$\tilde{\frac{1}{2}}$	2	0 110−	$-7\sqrt{2}/3\sqrt{17.19}$
$\frac{3}{2}$	$\tilde{\frac{1}{2}}$	2	0 200−	$+\sqrt{2.5.11}/\sqrt{3.7.17.19}$
$\frac{3}{2}$	$\tilde{\frac{1}{2}}$	2	0 210−	$-\sqrt{13}/3\sqrt{7.17.19}$
$\frac{3}{2}$	$\tilde{\frac{1}{2}}$	$\tilde{1}$	0 000−	$+2.3\sqrt{5}/\sqrt{7.13.17.19}$
$\frac{3}{2}$	$\tilde{\frac{1}{2}}$	$\tilde{1}$	0 010−	$+3\sqrt{3}/2\sqrt{2.7.11.17.19}$
$\frac{3}{2}$	$\tilde{\frac{1}{2}}$	$\tilde{1}$	0 100−	$-3\sqrt{5.11}/2\sqrt{2.13.17.19}$
$\frac{3}{2}$	$\tilde{\frac{1}{2}}$	$\tilde{1}$	0 110−	$+4/3\sqrt{3.17.19}$
$\frac{3}{2}$	$\tilde{\frac{1}{2}}$	$\tilde{1}$	0 200−	0
$\frac{3}{2}$	$\tilde{\frac{1}{2}}$	$\tilde{1}$	0 210−	$-7\sqrt{7.13}/2.3\sqrt{2.3.17.19}$
$\tilde{\frac{1}{2}}$	$\frac{1}{2}$	$\tilde{1}$	0 000+	$-\sqrt{3.5}/2\sqrt{2.11.17.19}$
$\tilde{\frac{1}{2}}$	$\frac{3}{2}$	1	0 000+	$+2\sqrt{3}/\sqrt{11.17.19}$
$\tilde{\frac{1}{2}}$	$\frac{3}{2}$	1	0 001+	$-\sqrt{5.19}/\sqrt{3.7.11.17}$
$\tilde{\frac{1}{2}}$	$\frac{3}{2}$	1	0 010+	$-3\sqrt{17}/11\sqrt{2.19}$
$\tilde{\frac{1}{2}}$	$\frac{3}{2}$	1	0 011+	$+\sqrt{5}/2.3.11\sqrt{2.7.17.19}$
$\tilde{\frac{1}{2}}$	$\frac{3}{2}$	2	0 000−	0
$\tilde{\frac{1}{2}}$	$\frac{3}{2}$	2	0 010−	$+\sqrt{3.5.7}/2\sqrt{11.17.19}$
$\tilde{\frac{1}{2}}$	$\frac{3}{2}$	$\tilde{1}$	0 000−	0
$\tilde{\frac{1}{2}}$	$\frac{3}{2}$	$\tilde{1}$	0 010−	$+\sqrt{5.7}/\sqrt{2.11.17.19}$
$\tilde{\frac{1}{2}}$	$\tilde{\frac{1}{2}}$	1	0 000−	$-\sqrt{3.5}/2\sqrt{2.11.17.19}$
$\tilde{\frac{1}{2}}$	$\tilde{\frac{1}{2}}$	1	0 001−	$+2\sqrt{2.19}/\sqrt{3.7.11.17}$
$\tilde{\frac{1}{2}}$	$\tilde{\frac{1}{2}}$	1	0 010−	$+2.3\sqrt{13}/11\sqrt{17.19}$
$\tilde{\frac{1}{2}}$	$\tilde{\frac{1}{2}}$	1	0 011−	$-2\sqrt{5.13}/3.11\sqrt{7.17.19}$
$\frac{15}{2}$	**$\frac{13}{2}$**	**6**		
$\frac{1}{2}$	$\frac{1}{2}$	0	0 000−	$-3\sqrt{3.5}/2\sqrt{13.17.19}$
$\frac{1}{2}$	$\frac{1}{2}$	1	0 000+	$-9.5\sqrt{5}/2\sqrt{2.7.13.17.19}$

SO_3–O 3jm Factors (cont.)

$\frac{15}{2}$	$\frac{13}{2}$	6			
$\frac{1}{2}$	$\frac{3}{2}$	1	0	000−	+8.3√11/7√5.13.17.19
$\frac{1}{2}$	$\frac{3}{2}$	1	0	010−	−61√3/7√2.5.13.17.19
$\frac{1}{2}$	$\frac{3}{2}$	2	0	000+	−8√3.11/7√5.13.17.19
$\frac{1}{2}$	$\frac{3}{2}$	2	0	010+	−23/7√2.5.13.17.19
$\frac{1}{2}$	$\frac{3}{2}$	$\tilde{1}$	0	000+	−2.3.11/7√7.13.17.19
$\frac{1}{2}$	$\frac{3}{2}$	$\tilde{1}$	0	001+	−8√11/7√5.7.13.17.19
$\frac{1}{2}$	$\frac{3}{2}$	$\tilde{1}$	0	010+	−47√11/7√2.3.7.13.17.19
$\frac{1}{2}$	$\frac{3}{2}$	$\tilde{1}$	0	011+	−3.11.11√3/2.7√2.5.7.13.17.19
$\frac{1}{2}$	$\tilde{\frac{1}{2}}$	$\tilde{1}$	0	000−	+3.5.11√5/2.7√2.7.13.17.19
$\frac{1}{2}$	$\tilde{\frac{1}{2}}$	$\tilde{1}$	0	001−	−8.2√2.11/7√7.13.17.19
$\frac{1}{2}$	$\tilde{\frac{1}{2}}$	$\tilde{1}$	0	010−	−2√11/7√3.7.17.19
$\frac{1}{2}$	$\tilde{\frac{1}{2}}$	$\tilde{1}$	0	011−	+2.3.11√3/7√5.7.17.19
$\frac{1}{2}$	$\tilde{\frac{1}{2}}$	$\tilde{0}$	0	000+	+11√3.5/2.7√13.17.19
$\frac{1}{2}$	$\tilde{\frac{1}{2}}$	$\tilde{0}$	0	010+	−2√2.11/7√17.19
$\frac{3}{2}$	$\frac{1}{2}$	1	0	000−	−2√3/√11.13.17.19
$\frac{3}{2}$	$\frac{1}{2}$	1	0	100−	+9√3/2√2.7.13.17.19
$\frac{3}{2}$	$\frac{1}{2}$	1	0	200−	0
$\frac{3}{2}$	$\frac{1}{2}$	2	0	000+	+239/2.5√11.13.17.19
$\frac{3}{2}$	$\frac{1}{2}$	2	0	100+	+3.29/5√2.7.13.17.19
$\frac{3}{2}$	$\frac{1}{2}$	2	0	200+	+3/2√17.19
$\frac{3}{2}$	$\frac{1}{2}$	$\tilde{1}$	0	000+	−11√3/√5.7.13.17.19
$\frac{3}{2}$	$\frac{1}{2}$	$\tilde{1}$	0	001+	−593√3/4.5√7.11.13.17.19
$\frac{3}{2}$	$\frac{1}{2}$	$\tilde{1}$	0	100+	−241√11/2.7√2.3.5.13.17.19
$\frac{3}{2}$	$\frac{1}{2}$	$\tilde{1}$	0	101+	+307√2/5.7√3.13.17.19
$\frac{3}{2}$	$\frac{1}{2}$	$\tilde{1}$	0	200+	+√5.11/√3.7.17.19
$\frac{3}{2}$	$\frac{1}{2}$	$\tilde{1}$	0	201+	+√17/4√3.7.19
$\frac{3}{2}$	$\frac{3}{2}$	0	0	000−	+√13/5√2.17.19
$\frac{3}{2}$	$\frac{3}{2}$	0	0	010−	+√3/4.5√11.13.17.19
$\frac{3}{2}$	$\frac{3}{2}$	0	0	100−	−31√11/5√7.13.17.19
$\frac{3}{2}$	$\frac{3}{2}$	0	0	110−	+151√2/5√3.7.13.17.19
$\frac{3}{2}$	$\frac{3}{2}$	0	0	200−	+√11/√2.17.19
$\frac{3}{2}$	$\frac{3}{2}$	0	0	210−	+11/4√3.17.19
$\frac{3}{2}$	$\frac{3}{2}$	1	0	000+	+4√3.7/5√13.17.19
$\frac{3}{2}$	$\frac{3}{2}$	1	0	010+	+9.9.3√7/4.5√2.11.13.17.19
$\frac{3}{2}$	$\frac{3}{2}$	1	0	100+	+79√11/5.7√2.3.13.17.19
$\frac{3}{2}$	$\frac{3}{2}$	1	0	110+	+3.29/5.7√13.17.19
$\frac{3}{2}$	$\frac{3}{2}$	1	0	200+	−2√11/√3.7.17.19
$\frac{3}{2}$	$\frac{3}{2}$	1	0	210+	−3/4√2.7.17.19
$\frac{3}{2}$	$\frac{3}{2}$	1	1	000+	−211√3/4.5√7.13.17.19
$\frac{3}{2}$	$\frac{3}{2}$	1	1	010+	−3.41√11/4.5√2.7.13.17.19
$\frac{3}{2}$	$\frac{3}{2}$	1	1	100+	−2.53√2.11/5.7√3.13.17.19
$\frac{3}{2}$	$\frac{3}{2}$	1	1	110+	+389/5.7√13.17.19
$\frac{3}{2}$	$\frac{3}{2}$	1	1	200+	+√7.11/4√3.17.19
$\frac{3}{2}$	$\frac{3}{2}$	1	1	210+	+√7/4√2.17.19
$\frac{3}{2}$	$\frac{3}{2}$	2	0	000−	−3√7/5√13.17.19
$\frac{3}{2}$	$\frac{3}{2}$	2	0	010−	+103√3.7/2.5√2.11.13.17.19
$\frac{3}{2}$	$\frac{3}{2}$	2	0	100−	+√2.11/5√13.17.19
$\frac{3}{2}$	$\frac{3}{2}$	2	0	110−	−√3.13/5√17.19
$\frac{3}{2}$	$\frac{3}{2}$	2	0	200−	−√11/√7.17.19

$\frac{15}{2}$	$\frac{13}{2}$	6			
$\frac{3}{2}$	$\frac{3}{2}$	2	0	210−	+√3/2√2.7.17.19
$\frac{3}{2}$	$\frac{3}{2}$	$\tilde{1}$	0	000−	−3.11√3.11/2.7√5.13.17.19
$\frac{3}{2}$	$\frac{3}{2}$	$\tilde{1}$	0	001−	+11.31√3/2.5.7√13.17.19
$\frac{3}{2}$	$\frac{3}{2}$	$\tilde{1}$	0	010−	−3.149/4.7√2.5.13.17.19
$\frac{3}{2}$	$\frac{3}{2}$	$\tilde{1}$	0	011−	+√11.17/5.7√2.13.19
$\frac{3}{2}$	$\frac{3}{2}$	$\tilde{1}$	0	100−	+11.23/7√2.3.5.7.13.17.19
$\frac{3}{2}$	$\frac{3}{2}$	$\tilde{1}$	0	101−	+167√2.11/5.7√3.7.13.17.19
$\frac{3}{2}$	$\frac{3}{2}$	$\tilde{1}$	0	110−	+163√11/3.7√5.7.13.17.19
$\frac{3}{2}$	$\frac{3}{2}$	$\tilde{1}$	0	111−	+2.163/5.7√7.13.17.19
$\frac{3}{2}$	$\frac{3}{2}$	$\tilde{1}$	0	200−	+11√5/2.7√3.17.19
$\frac{3}{2}$	$\frac{3}{2}$	$\tilde{1}$	0	201−	−√11/2.7√3.17.19
$\frac{3}{2}$	$\frac{3}{2}$	$\tilde{1}$	0	210−	+11√5.11/4.3.7√2.17.19
$\frac{3}{2}$	$\frac{3}{2}$	$\tilde{1}$	0	211−	−5/7√2.17.19
$\frac{3}{2}$	$\frac{3}{2}$	$\tilde{1}$	1	000+	−107√3.11/4.7√5.13.17.19
$\frac{3}{2}$	$\frac{3}{2}$	$\tilde{1}$	1	001+	−593√3/4.5.7√13.17.19
$\frac{3}{2}$	$\frac{3}{2}$	$\tilde{1}$	1	010+	−9.9/4.7√2.5.13.17.19
$\frac{3}{2}$	$\frac{3}{2}$	$\tilde{1}$	1	011+	−29.103/2.5.7√2.11.13.17.19
$\frac{3}{2}$	$\frac{3}{2}$	$\tilde{1}$	1	100+	+4.11√2/√3.5.7.13.17.19
$\frac{3}{2}$	$\frac{3}{2}$	$\tilde{1}$	1	101+	−11√3.11/5√2.7.13.17.19
$\frac{3}{2}$	$\frac{3}{2}$	$\tilde{1}$	1	110+	−√11.13/3√5.7.17.19
$\frac{3}{2}$	$\frac{3}{2}$	$\tilde{1}$	1	111+	+307/2.5√7.13.17.19
$\frac{3}{2}$	$\frac{3}{2}$	$\tilde{1}$	1	200+	−11√5/4.7√3.17.19
$\frac{3}{2}$	$\frac{3}{2}$	$\tilde{1}$	1	201+	−9√3.11/4.7√17.19
$\frac{3}{2}$	$\frac{3}{2}$	$\tilde{1}$	1	210+	+61√5.11/4.3.7√2.17.19
$\frac{3}{2}$	$\frac{3}{2}$	$\tilde{1}$	1	211+	+5/2.7√2.17.19
$\frac{3}{2}$	$\frac{3}{2}$	$\tilde{0}$	0	000+	+2√2.7.11/√5.13.17.19
$\frac{3}{2}$	$\frac{3}{2}$	$\tilde{0}$	0	010+	+3√3.7/4√5.13.17.19
$\frac{3}{2}$	$\frac{3}{2}$	$\tilde{0}$	0	100+	+9.11/7√5.13.17.19
$\frac{3}{2}$	$\frac{3}{2}$	$\tilde{0}$	0	110+	−√2.11.19/7√3.5.13.17
$\frac{3}{2}$	$\frac{3}{2}$	$\tilde{0}$	0	200+	0
$\frac{3}{2}$	$\frac{3}{2}$	$\tilde{0}$	0	210+	+√5.7.11/4√3.17.19
$\frac{3}{2}$	$\tilde{\frac{1}{2}}$	1	0	000−	0
$\frac{3}{2}$	$\tilde{\frac{1}{2}}$	1	0	010−	−3√7/2√2.11.17.19
$\frac{3}{2}$	$\tilde{\frac{1}{2}}$	1	0	100−	+√5.11.19/2.7√2.3.13.17
$\frac{3}{2}$	$\tilde{\frac{1}{2}}$	1	0	110−	−8/7√17.19
$\frac{3}{2}$	$\tilde{\frac{1}{2}}$	1	0	200−	−2√5.11/√3.7.17.19
$\frac{3}{2}$	$\tilde{\frac{1}{2}}$	1	0	210−	−√13/2√2.7.17.19
$\frac{3}{2}$	$\tilde{\frac{1}{2}}$	2	0	000+	+√7/√5.13.17.19
$\frac{3}{2}$	$\tilde{\frac{1}{2}}$	2	0	010+	+√3.7.11/5√2.17.19
$\frac{3}{2}$	$\tilde{\frac{1}{2}}$	2	0	100+	+37√11/7√2.5.13.17.19
$\frac{3}{2}$	$\tilde{\frac{1}{2}}$	2	0	110+	+2.59/5.7√3.17.19
$\frac{3}{2}$	$\tilde{\frac{1}{2}}$	2	0	200+	+√5.11/√7.17.19
$\frac{3}{2}$	$\tilde{\frac{1}{2}}$	2	0	210+	−√13/√2.3.7.17.19
$\frac{3}{2}$	$\tilde{\frac{1}{2}}$	$\tilde{1}$	0	000+	−2√3.11/√13.17.19
$\frac{3}{2}$	$\tilde{\frac{1}{2}}$	$\tilde{1}$	0	001+	−3√3/2√5.13.17.19
$\frac{3}{2}$	$\tilde{\frac{1}{2}}$	$\tilde{1}$	0	010+	−3/2√2.5.17.19
$\frac{3}{2}$	$\tilde{\frac{1}{2}}$	$\tilde{1}$	0	011+	−8√2/5√11.17.19
$\frac{3}{2}$	$\tilde{\frac{1}{2}}$	$\tilde{1}$	0	100+	−3.11√3/2.7√2.7.13.17.19
$\frac{3}{2}$	$\tilde{\frac{1}{2}}$	$\tilde{1}$	0	101+	+8.11√2.11/7√3.5.7.13.17.19

SO_3–O 3jm Factors (cont.)

$\frac{15}{2}\ \frac{13}{2}\ 6$

$\frac{3}{2}\ \tilde{\tfrac{1}{2}}\ \tilde{1}$ 0 110+ $-4.13\sqrt{11}/3.7\sqrt{5.7.17.19}$

$\frac{3}{2}\ \tilde{\tfrac{1}{2}}\ \tilde{1}$ 0 111+ $+9.11/5.7\sqrt{7.17.19}$

$\frac{3}{2}\ \tilde{\tfrac{1}{2}}\ \tilde{1}$ 0 200+ 0

$\frac{3}{2}\ \tilde{\tfrac{1}{2}}\ \tilde{1}$ 0 201+ $+\sqrt{5.11}/2\sqrt{3.17.19}$

$\frac{3}{2}\ \tilde{\tfrac{1}{2}}\ \tilde{1}$ 0 210+ $-\sqrt{5.11.13}/2.3\sqrt{2.17.19}$

$\frac{3}{2}\ \tilde{\tfrac{1}{2}}\ \tilde{1}$ 0 211+ 0

$\tilde{\tfrac{1}{2}}\ \frac{1}{2}\ \tilde{1}$ 0 000− $+1/2\sqrt{2.7.17.19}$

$\tilde{\tfrac{1}{2}}\ \frac{1}{2}\ \tilde{1}$ 0 001− $-8.2\sqrt{2}/\sqrt{5.7.11.17.19}$

$\tilde{\tfrac{1}{2}}\ \frac{1}{2}\ \tilde{0}$ 0 000+ $-\sqrt{3}/2\sqrt{17.19}$

$\tilde{\tfrac{1}{2}}\ \frac{3}{2}\ 1$ 0 000− $-4/\sqrt{5.7.17.19}$

$\tilde{\tfrac{1}{2}}\ \frac{3}{2}\ 1$ 0 010− $-\sqrt{3.19}/\sqrt{2.5.7.11.17}$

$\tilde{\tfrac{1}{2}}\ \frac{3}{2}\ 2$ 0 000+ $-2\sqrt{3.5}/\sqrt{7.17.19}$

$\tilde{\tfrac{1}{2}}\ \frac{3}{2}\ 2$ 0 010+ $-41/\sqrt{2.5.7.11.17.19}$

$\tilde{\tfrac{1}{2}}\ \frac{3}{2}\ \tilde{1}$ 0 000+ $-2\sqrt{11}/7\sqrt{17.19}$

$\tilde{\tfrac{1}{2}}\ \frac{3}{2}\ \tilde{1}$ 0 001+ $-47/7\sqrt{5.17.19}$

$\tilde{\tfrac{1}{2}}\ \frac{3}{2}\ \tilde{1}$ 0 010+ $+\sqrt{3}/7\sqrt{2.17.19}$

$\tilde{\tfrac{1}{2}}\ \frac{3}{2}\ \tilde{1}$ 0 011+ $+29\sqrt{3.5}/2.7\sqrt{2.11.17.19}$

$\tilde{\tfrac{1}{2}}\ \tilde{\tfrac{1}{2}}\ 0$ 0 000− $-\sqrt{3}/2\sqrt{17.19}$

$\tilde{\tfrac{1}{2}}\ \tilde{\tfrac{1}{2}}\ 0$ 0 010− $+2\sqrt{2.13}/\sqrt{5.11.17.19}$

$\tilde{\tfrac{1}{2}}\ \tilde{\tfrac{1}{2}}\ 1$ 0 000+ $-1/2\sqrt{2.7.17.19}$

$\tilde{\tfrac{1}{2}}\ \tilde{\tfrac{1}{2}}\ 1$ 0 010+ $+2\sqrt{3.5.13}/\sqrt{7.11.17.19}$

$\frac{15}{2}\ 7\ \frac{1}{2}$

$\frac{1}{2}\ 1\ \frac{1}{2}$ 0 000− 0

$\frac{1}{2}\ 1\ \frac{1}{2}$ 0 010− $+1/2\sqrt{2}$

$\frac{3}{2}\ 1\ \frac{1}{2}$ 0 000+ $+1/\sqrt{5}$

$\frac{3}{2}\ 1\ \frac{1}{2}$ 0 010+ 0

$\frac{3}{2}\ 1\ \frac{1}{2}$ 0 100+ 0

$\frac{3}{2}\ 1\ \frac{1}{2}$ 0 110+ $+\sqrt{3}/2\sqrt{2.5}$

$\frac{3}{2}\ 1\ \frac{1}{2}$ 0 200+ 0

$\frac{3}{2}\ 1\ \frac{1}{2}$ 0 210+ 0

$\frac{3}{2}\ 2\ \frac{1}{2}$ 0 000− $+\sqrt{11}/4.5$

$\frac{3}{2}\ 2\ \frac{1}{2}$ 0 100− $+\sqrt{7}/3.5\sqrt{2}$

$\frac{3}{2}\ 2\ \frac{1}{2}$ 0 200− $+\sqrt{13}/4.3$

$\frac{3}{2}\ \tilde{1}\ \frac{1}{2}$ 0 000− $+\sqrt{11}/4.5\sqrt{7}$

$\frac{3}{2}\ \tilde{1}\ \frac{1}{2}$ 0 010− $-\sqrt{13}/2.5\sqrt{7}$

$\frac{3}{2}\ \tilde{1}\ \frac{1}{2}$ 0 100− $+\sqrt{2}/5$

$\frac{3}{2}\ \tilde{1}\ \frac{1}{2}$ 0 110− $+\sqrt{11.13}/2.3.5\sqrt{2}$

$\frac{3}{2}\ \tilde{1}\ \frac{1}{2}$ 0 200− $-\sqrt{13}/4\sqrt{7}$

$\frac{3}{2}\ \tilde{1}\ \frac{1}{2}$ 0 210− $+\sqrt{11}/2.3\sqrt{7}$

$\tilde{\tfrac{1}{2}}\ \tilde{1}\ \frac{1}{2}$ 0 000+ 0

$\tilde{\tfrac{1}{2}}\ \tilde{1}\ \frac{1}{2}$ 0 010+ $+\sqrt{7}/2\sqrt{2.3.5}$

$\tilde{\tfrac{1}{2}}\ \tilde{0}\ \frac{1}{2}$ 0 000− $+1/\sqrt{3.5}$

$\frac{15}{2}\ 7\ \frac{3}{2}$

$\frac{1}{2}\ 1\ \frac{3}{2}$ 0 000− 0

$\frac{1}{2}\ 1\ \frac{3}{2}$ 0 010− $-\sqrt{7}/4\sqrt{17}$

$\frac{1}{2}\ 2\ \frac{3}{2}$ 0 000+ $+\sqrt{3}/2\sqrt{5.17}$

$\frac{1}{2}\ \tilde{1}\ \frac{3}{2}$ 0 000+ $+3\sqrt{3}/\sqrt{5.7.17}$

$\frac{1}{2}\ \tilde{1}\ \frac{3}{2}$ 0 010+ $+\sqrt{3.11.13}/4\sqrt{5.7.17}$

$\frac{3}{2}\ 1\ \frac{3}{2}$ 0 000+ $+43/2.5\sqrt{5.7.17}$

$\frac{15}{2}\ 7\ \frac{3}{2}$

$\frac{3}{2}\ 1\ \frac{3}{2}$ 0 010+ $+\sqrt{3.11}/2.5\sqrt{5.17}$

$\frac{3}{2}\ 1\ \frac{3}{2}$ 0 100+ $+\sqrt{2.11}/5\sqrt{5.17}$

$\frac{3}{2}\ 1\ \frac{3}{2}$ 0 110+ $-13\sqrt{7}/5\sqrt{2.3.5.17}$

$\frac{3}{2}\ 1\ \frac{3}{2}$ 0 200+ $+\sqrt{11.13}/2\sqrt{5.7.17}$

$\frac{3}{2}\ 1\ \frac{3}{2}$ 0 210+ $+\sqrt{13}/2\sqrt{3.5.17}$

$\frac{3}{2}\ 1\ \frac{3}{2}$ 1 000+ $+127/4.5\sqrt{5.7.17}$

$\frac{3}{2}\ 1\ \frac{3}{2}$ 1 010+ $-\sqrt{3.11}/4.5\sqrt{5.17}$

$\frac{3}{2}\ 1\ \frac{3}{2}$ 1 100+ $-\sqrt{11}/5\sqrt{2.5.17}$

$\frac{3}{2}\ 1\ \frac{3}{2}$ 1 110+ $-\sqrt{7.17}/2.5\sqrt{2.3.5}$

$\frac{3}{2}\ 1\ \frac{3}{2}$ 1 200+ $-\sqrt{11.13}/4\sqrt{5.7.17}$

$\frac{3}{2}\ 1\ \frac{3}{2}$ 1 210+ $-\sqrt{13}/4\sqrt{3.5.17}$

$\frac{3}{2}\ 2\ \frac{3}{2}$ 0 000− $-\sqrt{11.17}/4.5\sqrt{7}$

$\frac{3}{2}\ 2\ \frac{3}{2}$ 0 100− $-\sqrt{2}/3.5\sqrt{17}$

$\frac{3}{2}\ 2\ \frac{3}{2}$ 0 200− $+\sqrt{7.13}/4.3\sqrt{17}$

$\frac{3}{2}\ \tilde{1}\ \frac{3}{2}$ 0 000− $-\sqrt{11.17}/4.5.7$

$\frac{3}{2}\ \tilde{1}\ \frac{3}{2}$ 0 010− $+\sqrt{13.17}/2.5.7$

$\frac{3}{2}\ \tilde{1}\ \frac{3}{2}$ 0 100− $-2\sqrt{2}/5\sqrt{7.17}$

$\frac{3}{2}\ \tilde{1}\ \frac{3}{2}$ 0 110− $-\sqrt{11.13}/3.5\sqrt{2.7.17}$

$\frac{3}{2}\ \tilde{1}\ \frac{3}{2}$ 0 200− $-\sqrt{13}/4\sqrt{17}$

$\frac{3}{2}\ \tilde{1}\ \frac{3}{2}$ 0 210− $+\sqrt{11}/2.3\sqrt{17}$

$\frac{3}{2}\ \tilde{1}\ \frac{3}{2}$ 1 000+ $+\sqrt{11.17}/4.5.7$

$\frac{3}{2}\ \tilde{1}\ \frac{3}{2}$ 1 010+ $-13\sqrt{13}/4.5.7\sqrt{17}$

$\frac{3}{2}\ \tilde{1}\ \frac{3}{2}$ 1 100+ $+19/5\sqrt{2.7.17}$

$\frac{3}{2}\ \tilde{1}\ \frac{3}{2}$ 1 110+ $-\sqrt{11.13}/2.3.5\sqrt{2.7.17}$

$\frac{3}{2}\ \tilde{1}\ \frac{3}{2}$ 1 200+ $-5\sqrt{13}/4.7\sqrt{17}$

$\frac{3}{2}\ \tilde{1}\ \frac{3}{2}$ 1 210+ $-11\sqrt{11}/4.3.7\sqrt{17}$

$\frac{3}{2}\ \tilde{0}\ \frac{3}{2}$ 0 000+ $+\sqrt{2.13}/5\sqrt{7.17}$

$\frac{3}{2}\ \tilde{0}\ \frac{3}{2}$ 0 100+ $-\sqrt{11.13}/3.5\sqrt{17}$

$\frac{3}{2}\ \tilde{0}\ \frac{3}{2}$ 0 200+ $-\sqrt{2.11}/3\sqrt{7.17}$

$\tilde{\tfrac{1}{2}}\ 1\ \frac{3}{2}$ 0 000− $-\sqrt{3.13}/5\sqrt{7.17}$

$\tilde{\tfrac{1}{2}}\ 1\ \frac{3}{2}$ 0 010− $+\sqrt{11.13}/4.5\sqrt{17}$

$\tilde{\tfrac{1}{2}}\ 2\ \frac{3}{2}$ 0 000+ $+\sqrt{11.13}/2\sqrt{3.5.7.17}$

$\tilde{\tfrac{1}{2}}\ \tilde{1}\ \frac{3}{2}$ 0 000+ 0

$\tilde{\tfrac{1}{2}}\ \tilde{1}\ \frac{3}{2}$ 0 010+ $+\sqrt{17}/4\sqrt{3.5}$

$\frac{15}{2}\ 7\ \frac{5}{2}$

$\frac{1}{2}\ 1\ \frac{3}{2}$ 0 000+ $-\sqrt{11}/\sqrt{2.3.13.17}$

$\frac{1}{2}\ 1\ \frac{3}{2}$ 0 010+ $+7\sqrt{7}/2.3\sqrt{2.13.17}$

$\frac{1}{2}\ 2\ \frac{3}{2}$ 0 000− $+\sqrt{13}/3\sqrt{2.3.5.17}$

$\frac{1}{2}\ \tilde{1}\ \frac{3}{2}$ 0 000− $-7\sqrt{7}/\sqrt{2.3.5.13.17}$

$\frac{1}{2}\ \tilde{1}\ \frac{3}{2}$ 0 010− $-\sqrt{7.11}/2.3\sqrt{2.3.5.17}$

$\frac{1}{2}\ \tilde{1}\ \tilde{\tfrac{1}{2}}$ 0 000+ $+2\sqrt{2}/\sqrt{3.7.13.17}$

$\frac{1}{2}\ \tilde{1}\ \tilde{\tfrac{1}{2}}$ 0 010+ $-5\sqrt{11}/2.3\sqrt{2.3.7.17}$

$\frac{1}{2}\ \tilde{0}\ \tilde{\tfrac{1}{2}}$ 0 000− $+\sqrt{11}/3\sqrt{3.17}$

$\frac{3}{2}\ 1\ \frac{3}{2}$ 0 000− $+43\sqrt{2}/5\sqrt{5.7.13.17}$

$\frac{3}{2}\ 1\ \frac{3}{2}$ 0 010− $-\sqrt{11.13}/2.5\sqrt{2.3.5.17}$

$\frac{3}{2}\ 1\ \frac{3}{2}$ 0 100− $+4\sqrt{11}/5\sqrt{5.13.17}$

$\frac{3}{2}\ 1\ \frac{3}{2}$ 0 110− $+13\sqrt{7.13}/2.3.5\sqrt{3.5.17}$

$\frac{3}{2}\ 1\ \frac{3}{2}$ 0 200− $+\sqrt{2.11}/\sqrt{5.7.17}$

$\frac{3}{2}\ 1\ \frac{3}{2}$ 0 210− $-13/2.3\sqrt{2.3.5.17}$

SO_3–O 3jm Factors (cont.)

$\frac{15}{2}$ 7 $\frac{5}{2}$

$\frac{3}{2}$ 1 $\frac{3}{2}$ 1 000− $+7\sqrt{7}/2.5\sqrt{2.5.13.17}$
$\frac{3}{2}$ 1 $\frac{3}{2}$ 1 010− $+7\sqrt{11}/5\sqrt{2.3.5.13.17}$
$\frac{3}{2}$ 1 $\frac{3}{2}$ 1 100− $+2.7\sqrt{11}/3.5\sqrt{5.13.17}$
$\frac{3}{2}$ 1 $\frac{3}{2}$ 1 110− $+19\sqrt{7}/3.5\sqrt{3.5.13.17}$
$\frac{3}{2}$ 1 $\frac{3}{2}$ 1 200− $-\sqrt{11}/2.3\sqrt{2.5.7.17}$
$\frac{3}{2}$ 1 $\frac{3}{2}$ 1 210− $-13/3\sqrt{2.3.5.17}$
$\frac{3}{2}$ 1 $\tilde{\frac{1}{2}}$ 0 000+ $-43/2.5\sqrt{7.13.17}$
$\frac{3}{2}$ 1 $\tilde{\frac{1}{2}}$ 0 010+ $-4\sqrt{11}/5\sqrt{3.13.17}$
$\frac{3}{2}$ 1 $\tilde{\frac{1}{2}}$ 0 100+ $+2\sqrt{2.11}/3.5\sqrt{13.17}$
$\frac{3}{2}$ 1 $\tilde{\frac{1}{2}}$ 0 110+ $+\sqrt{7}/2.5\sqrt{2.3.13.17}$
$\frac{3}{2}$ 1 $\tilde{\frac{1}{2}}$ 0 200+ $+5\sqrt{11}/2.3\sqrt{7.17}$
$\frac{3}{2}$ 1 $\tilde{\frac{1}{2}}$ 0 210+ 0
$\frac{3}{2}$ 2 $\frac{3}{2}$ 0 000+ $-29\sqrt{11}/2.5\sqrt{2.7.13.17}$
$\frac{3}{2}$ 2 $\frac{3}{2}$ 0 100+ $-2.11/9.5\sqrt{13.17}$
$\frac{3}{2}$ 2 $\frac{3}{2}$ 0 200+ $+\sqrt{17}/2.9\sqrt{2.7}$
$\frac{3}{2}$ 2 $\tilde{\frac{1}{2}}$ 0 000− $-19\sqrt{11}/4\sqrt{5.7.13.17}$
$\frac{3}{2}$ 2 $\tilde{\frac{1}{2}}$ 0 100− $+53/9\sqrt{2.5.13.17}$
$\frac{3}{2}$ 2 $\tilde{\frac{1}{2}}$ 0 200− $-5.5\sqrt{5}/4.9\sqrt{7.17}$
$\frac{3}{2}$ $\tilde{1}$ $\frac{3}{2}$ 0 000+ $-\sqrt{11.17}/5.7\sqrt{2.13}$
$\frac{3}{2}$ $\tilde{1}$ $\frac{3}{2}$ 0 010+ $+3.11/2.5.7\sqrt{2.17}$
$\frac{3}{2}$ $\tilde{1}$ $\frac{3}{2}$ 0 100+ $-8/5\sqrt{7.13.17}$
$\frac{3}{2}$ $\tilde{1}$ $\frac{3}{2}$ 0 110+ $-47\sqrt{11}/2.9.5\sqrt{7.17}$
$\frac{3}{2}$ $\tilde{1}$ $\frac{3}{2}$ 0 200+ $-1/\sqrt{2.17}$
$\frac{3}{2}$ $\tilde{1}$ $\frac{3}{2}$ 0 210+ $-\sqrt{11.13}/2.9\sqrt{2.17}$
$\frac{3}{2}$ $\tilde{1}$ $\frac{3}{2}$ 1 000− $+\sqrt{11.17}/5.7\sqrt{2.13}$
$\frac{3}{2}$ $\tilde{1}$ $\frac{3}{2}$ 1 010− $-8.3\sqrt{2}/5.7\sqrt{17}$
$\frac{3}{2}$ $\tilde{1}$ $\frac{3}{2}$ 1 100− $-8\sqrt{17}/3.5\sqrt{7.13}$
$\frac{3}{2}$ $\tilde{1}$ $\frac{3}{2}$ 1 110− $-13\sqrt{11}/9.5\sqrt{7.17}$
$\frac{3}{2}$ $\tilde{1}$ $\frac{3}{2}$ 1 200− $-5/3.7\sqrt{2.17}$
$\frac{3}{2}$ $\tilde{1}$ $\frac{3}{2}$ 1 210− $+2\sqrt{2.11.13}/9.7\sqrt{17}$
$\frac{3}{2}$ $\tilde{1}$ $\tilde{\frac{1}{2}}$ 0 000− $-3\sqrt{11.17}/4.7\sqrt{5.13}$
$\frac{3}{2}$ $\tilde{1}$ $\tilde{\frac{1}{2}}$ 0 010− $+9/2.7\sqrt{5.17}$
$\frac{3}{2}$ $\tilde{1}$ $\tilde{\frac{1}{2}}$ 0 100− $-\sqrt{2.13}/3\sqrt{5.7.17}$
$\frac{3}{2}$ $\tilde{1}$ $\tilde{\frac{1}{2}}$ 0 110− $+43\sqrt{11}/2.9\sqrt{2.5.7.17}$
$\frac{3}{2}$ $\tilde{1}$ $\tilde{\frac{1}{2}}$ 0 200− $+\sqrt{5}/4.3.7\sqrt{17}$
$\frac{3}{2}$ $\tilde{1}$ $\tilde{\frac{1}{2}}$ 0 210− $+\sqrt{5.11.13}/2.9.7\sqrt{17}$
$\frac{3}{2}$ $\tilde{0}$ $\frac{3}{2}$ 0 000− $+9/5\sqrt{7.17}$
$\frac{3}{2}$ $\tilde{0}$ $\frac{3}{2}$ 0 100− $-\sqrt{2.11}/9.5\sqrt{17}$
$\frac{3}{2}$ $\tilde{0}$ $\frac{3}{2}$ 0 200− $+\sqrt{11.13}/9\sqrt{7.17}$
$\tilde{\frac{1}{2}}$ 1 $\frac{3}{2}$ 0 000+ $+\sqrt{3}/5\sqrt{2.7.17}$
$\tilde{\frac{1}{2}}$ 1 $\frac{3}{2}$ 0 010+ $+11\sqrt{11}/2.3.5\sqrt{2.17}$
$\tilde{\frac{1}{2}}$ 1 $\tilde{\frac{1}{2}}$ 0 000− $+2\sqrt{2.3}/\sqrt{5.7.17}$
$\tilde{\frac{1}{2}}$ 1 $\tilde{\frac{1}{2}}$ 0 010− $-\sqrt{11}/2.3\sqrt{2.5.17}$
$\tilde{\frac{1}{2}}$ 2 $\frac{3}{2}$ 0 000− $-\sqrt{3.11}/\sqrt{2.5.7.17}$
$\tilde{\frac{1}{2}}$ $\tilde{1}$ $\frac{3}{2}$ 0 000− $+\sqrt{5.11}/7\sqrt{2.3.17}$
$\tilde{\frac{1}{2}}$ $\tilde{1}$ $\frac{3}{2}$ 0 010− $+\sqrt{13}/2.7\sqrt{2.3.5.17}$

$\frac{15}{2}$ 7 $\frac{7}{2}$

$\frac{1}{2}$ 1 $\frac{1}{2}$ 0 000+ $+\sqrt{7.11}/2\sqrt{2.3.13.17.19}$
$\frac{1}{2}$ 1 $\frac{1}{2}$ 0 010+ $+7.7/3\sqrt{2.13.17.19}$

$\frac{15}{2}$ 7 $\frac{7}{2}$

$\frac{1}{2}$ 1 $\frac{3}{2}$ 0 000− $+5\sqrt{5.11}/2\sqrt{2.3.13.17.19}$
$\frac{1}{2}$ 1 $\frac{3}{2}$ 0 010− $+7\sqrt{5.7}/2.3\sqrt{2.13.17.19}$
$\frac{1}{2}$ 2 $\frac{3}{2}$ 0 000+ $+5.5/2\sqrt{2.3.13.17.19}$
$\frac{1}{2}$ $\tilde{1}$ $\frac{3}{2}$ 0 000+ $-\sqrt{7.19}/2\sqrt{2.3.13.17}$
$\frac{1}{2}$ $\tilde{1}$ $\frac{3}{2}$ 0 010+ $+\sqrt{7.11}/2\sqrt{2.3.17.19}$
$\frac{1}{2}$ $\tilde{1}$ $\tilde{\frac{1}{2}}$ 0 000− $-\sqrt{19}/2\sqrt{2.7.13.17}$
$\frac{1}{2}$ $\tilde{1}$ $\tilde{\frac{1}{2}}$ 0 010− $+3\sqrt{11}/\sqrt{2.7.17.19}$
$\frac{1}{2}$ $\tilde{0}$ $\tilde{\frac{1}{2}}$ 0 000+ $-\sqrt{11}/2\sqrt{17.19}$
$\frac{3}{2}$ 1 $\frac{1}{2}$ 0 000− $-9\sqrt{19}/8\sqrt{5.13.17}$
$\frac{3}{2}$ 1 $\frac{1}{2}$ 0 010− $-\sqrt{7.11}/4\sqrt{3.5.13.17.19}$
$\frac{3}{2}$ 1 $\frac{1}{2}$ 0 100− $-7\sqrt{7.11}/2.3\sqrt{2.5.13.17.19}$
$\frac{3}{2}$ 1 $\frac{1}{2}$ 0 110− $+7.7/2.3\sqrt{2.3.5.13.17.19}$
$\frac{3}{2}$ 1 $\frac{1}{2}$ 0 200− $-\sqrt{5.11}/8.3\sqrt{17.19}$
$\frac{3}{2}$ 1 $\frac{1}{2}$ 0 210− $-13\sqrt{5.7}/4.3\sqrt{3.17.19}$
$\frac{3}{2}$ 1 $\frac{3}{2}$ 0 000+ $+\sqrt{17}/4\sqrt{2.7.13.19}$
$\frac{3}{2}$ 1 $\frac{3}{2}$ 0 010+ $+\sqrt{11.17}/2\sqrt{2.3.13.19}$
$\frac{3}{2}$ 1 $\frac{3}{2}$ 0 100+ $+\sqrt{11}/3\sqrt{13.17.19}$
$\frac{3}{2}$ 1 $\frac{3}{2}$ 0 110+ $+4\sqrt{7}/\sqrt{3.13.17.19}$
$\frac{3}{2}$ 1 $\frac{3}{2}$ 0 200+ $+31\sqrt{11}/4.3\sqrt{2.7.17.19}$
$\frac{3}{2}$ 1 $\frac{3}{2}$ 0 210+ $-3\sqrt{3}/2\sqrt{2.17.19}$
$\frac{3}{2}$ 1 $\frac{3}{2}$ 1 000+ $+11\sqrt{7}/4\sqrt{2.13.17.19}$
$\frac{3}{2}$ 1 $\frac{3}{2}$ 1 010+ $-7\sqrt{11}/4\sqrt{2.3.13.17.19}$
$\frac{3}{2}$ 1 $\frac{3}{2}$ 1 100+ $-7\sqrt{11}/3\sqrt{13.17.19}$
$\frac{3}{2}$ 1 $\frac{3}{2}$ 1 110+ $+37\sqrt{7}/2.3\sqrt{3.13.17.19}$
$\frac{3}{2}$ 1 $\frac{3}{2}$ 1 200+ $-37\sqrt{11}/4.3\sqrt{2.7.17.19}$
$\frac{3}{2}$ 1 $\frac{3}{2}$ 1 210+ $+53/4.3\sqrt{2.3.17.19}$
$\frac{3}{2}$ 1 $\tilde{\frac{1}{2}}$ 0 000− $-\sqrt{3.17}/8\sqrt{7.13.19}$
$\frac{3}{2}$ 1 $\tilde{\frac{1}{2}}$ 0 010− $+\sqrt{11}/2\sqrt{13.17.19}$
$\frac{3}{2}$ 1 $\tilde{\frac{1}{2}}$ 0 100− $-5\sqrt{3.11}/2\sqrt{2.13.17.19}$
$\frac{3}{2}$ 1 $\tilde{\frac{1}{2}}$ 0 110− $+11\sqrt{7}/2.3\sqrt{2.13.17.19}$
$\frac{3}{2}$ 1 $\tilde{\frac{1}{2}}$ 0 200− $+11\sqrt{3.11}/8\sqrt{7.17.19}$
$\frac{3}{2}$ 1 $\tilde{\frac{1}{2}}$ 0 210− $+7/2.3\sqrt{17.19}$
$\frac{3}{2}$ 2 $\frac{1}{2}$ 0 000+ $-11.11\sqrt{11}/8.5\sqrt{13.17.19}$
$\frac{3}{2}$ 2 $\frac{1}{2}$ 0 100+ $+137\sqrt{7}/2.9.5\sqrt{2.13.17.19}$
$\frac{3}{2}$ 2 $\frac{1}{2}$ 0 200+ $-43/8.9\sqrt{17.19}$
$\frac{3}{2}$ 2 $\frac{3}{2}$ 0 000− $-3\sqrt{11.17}/4\sqrt{2.5.7.13.19}$
$\frac{3}{2}$ 2 $\frac{3}{2}$ 0 100− $-239/9\sqrt{5.13.17.19}$
$\frac{3}{2}$ 2 $\frac{3}{2}$ 0 200− $-\sqrt{5}/4.9\sqrt{2.7.17.19}$
$\frac{3}{2}$ 2 $\tilde{\frac{1}{2}}$ 0 000+ $-\sqrt{3.11.13}/8\sqrt{5.7.17.19}$
$\frac{3}{2}$ 2 $\tilde{\frac{1}{2}}$ 0 100+ $-41\sqrt{3}/2\sqrt{2.5.13.17.19}$
$\frac{3}{2}$ 2 $\tilde{\frac{1}{2}}$ 0 200+ $-3.5\sqrt{3.5}/8\sqrt{7.17.19}$
$\frac{3}{2}$ $\tilde{1}$ $\frac{1}{2}$ 0 000+ $-31\sqrt{11.13}/8.3.5\sqrt{7.17.19}$
$\frac{3}{2}$ $\tilde{1}$ $\frac{1}{2}$ 0 010+ $+149/2.3.5\sqrt{7.17.19}$
$\frac{3}{2}$ $\tilde{1}$ $\frac{1}{2}$ 0 100+ $+1907/2.9.5\sqrt{2.13.17.19}$
$\frac{3}{2}$ $\tilde{1}$ $\frac{1}{2}$ 0 110+ $+\sqrt{11.17}/2.3.5\sqrt{2.19}$
$\frac{3}{2}$ $\tilde{1}$ $\frac{1}{2}$ 0 200+ $+269/8.9\sqrt{7.17.19}$
$\frac{3}{2}$ $\tilde{1}$ $\frac{1}{2}$ 0 210+ $-\sqrt{11.13}/2.3\sqrt{7.17.19}$
$\frac{3}{2}$ $\tilde{1}$ $\frac{3}{2}$ 0 000− $-211\sqrt{11}/4.3.7\sqrt{2.5.13.17.19}$
$\frac{3}{2}$ $\tilde{1}$ $\frac{3}{2}$ 0 010− $-41/2.3.7\sqrt{2.5.17.19}$

SO_3–O 3jm Factors (cont.)

$\frac{15}{2}$	7	$\frac{7}{2}$		
$\frac{3}{2}$	$\tilde{1}$	$\frac{3}{2}$	0 100−	$+11\sqrt{17}/9\sqrt{5.7.13.19}$
$\frac{3}{2}$	$\tilde{1}$	$\frac{3}{2}$	0 110−	$-2\sqrt{11}/3\sqrt{5.7.17.19}$
$\frac{3}{2}$	$\tilde{1}$	$\frac{3}{2}$	0 200−	$+11\sqrt{5}/4.9\sqrt{2.17.19}$
$\frac{3}{2}$	$\tilde{1}$	$\frac{3}{2}$	0 210−	$+\sqrt{5.11.13}/2.3\sqrt{2.17.19}$
$\frac{3}{2}$	$\tilde{1}$	$\frac{3}{2}$	1 000+	$+379\sqrt{11}/4.3.7\sqrt{2.5.13.17.19}$
$\frac{3}{2}$	$\tilde{1}$	$\frac{3}{2}$	1 010+	$+197\sqrt{5}/4.3.7\sqrt{2.17.19}$
$\frac{3}{2}$	$\tilde{1}$	$\frac{3}{2}$	1 100+	$-163/9\sqrt{5.7.13.17.19}$
$\frac{3}{2}$	$\tilde{1}$	$\frac{3}{2}$	1 110+	$-5\sqrt{5.11}/2.3\sqrt{7.17.19}$
$\frac{3}{2}$	$\tilde{1}$	$\frac{3}{2}$	1 200+	$+5.23\sqrt{5}/4.9.7\sqrt{2.17.19}$
$\frac{3}{2}$	$\tilde{1}$	$\frac{3}{2}$	1 210+	$+\sqrt{5.11.13.17}/4.3.7\sqrt{2.19}$
$\frac{3}{2}$	$\tilde{1}$	$\tilde{\frac{1}{2}}$	0 000+	$-\sqrt{3.11.17.19}/8.7\sqrt{5.13}$
$\frac{3}{2}$	$\tilde{1}$	$\tilde{\frac{1}{2}}$	0 010+	$+3\sqrt{3.19}/4.7\sqrt{5.17}$
$\frac{3}{2}$	$\tilde{1}$	$\tilde{\frac{1}{2}}$	0 100+	$+\sqrt{3.19}/2\sqrt{2.5.7.13.17}$
$\frac{3}{2}$	$\tilde{1}$	$\tilde{\frac{1}{2}}$	0 110+	$-11\sqrt{11}/2\sqrt{2.3.5.7.17.19}$
$\frac{3}{2}$	$\tilde{1}$	$\tilde{\frac{1}{2}}$	0 200+	$-\sqrt{3.5}/8.7\sqrt{17.19}$
$\frac{3}{2}$	$\tilde{1}$	$\tilde{\frac{1}{2}}$	0 210+	$-\sqrt{5.11.13}/4.7\sqrt{3.17.19}$
$\frac{3}{2}$	$\tilde{0}$	$\frac{3}{2}$	0 000+	$-\sqrt{19}/\sqrt{5.7.17}$
$\frac{3}{2}$	$\tilde{0}$	$\frac{3}{2}$	0 100+	$-23\sqrt{11}/9\sqrt{2.5.17.19}$
$\frac{3}{2}$	$\tilde{0}$	$\frac{3}{2}$	0 200+	$+\sqrt{5.11.13}/9\sqrt{7.17.19}$
$\tilde{\frac{1}{2}}$	1	$\frac{3}{2}$	0 000−	$-\sqrt{3.19}/2\sqrt{2.5.7.17}$
$\tilde{\frac{1}{2}}$	1	$\frac{3}{2}$	0 010−	$-\sqrt{11.17}/2.3\sqrt{2.5.19}$
$\tilde{\frac{1}{2}}$	1	$\tilde{\frac{1}{2}}$	0 000+	$-3\sqrt{19}/2\sqrt{2.5.7.17}$
$\tilde{\frac{1}{2}}$	1	$\tilde{\frac{1}{2}}$	0 010+	$-\sqrt{11}/\sqrt{2.3.5.17.19}$
$\tilde{\frac{1}{2}}$	2	$\frac{3}{2}$	0 000+	$+\sqrt{11.17}/2.3\sqrt{2.3.7.19}$
$\tilde{\frac{1}{2}}$	$\tilde{1}$	$\frac{1}{2}$	0 000−	$-\sqrt{5.11.19}/2.3\sqrt{2.3.7.17}$
$\tilde{\frac{1}{2}}$	$\tilde{1}$	$\frac{1}{2}$	0 010−	$-\sqrt{13}/\sqrt{2.3.5.7.17.19}$
$\tilde{\frac{1}{2}}$	$\tilde{1}$	$\frac{3}{2}$	0 000+	$-\sqrt{11.19}/2.3.7\sqrt{2.3.17}$
$\tilde{\frac{1}{2}}$	$\tilde{1}$	$\frac{3}{2}$	0 010+	$+\sqrt{13}/2.7\sqrt{2.3.17.19}$
$\tilde{\frac{1}{2}}$	$\tilde{0}$	$\frac{1}{2}$	0 000+	$-\sqrt{13}/2.3\sqrt{3.5.17.19}$
$\frac{15}{2}$	7	$\frac{9}{2}$		
$\frac{1}{2}$	1	$\frac{1}{2}$	0 000−	$-2\sqrt{2.7}/\sqrt{3.5.13.17.19}$
$\frac{1}{2}$	1	$\frac{1}{2}$	0 010−	$+7.7\sqrt{11}/2.3\sqrt{2.5.13.17.19}$
$\frac{1}{2}$	1	$\frac{3}{2}$	0 000+	$+3\sqrt{2.7}/5\sqrt{13.17.19}$
$\frac{1}{2}$	1	$\frac{3}{2}$	0 001+	$-7.37/2.5\sqrt{2.3.13.17.19}$
$\frac{1}{2}$	1	$\frac{3}{2}$	0 010+	$-7\sqrt{3.11}/2.5\sqrt{2.13.17.19}$
$\frac{1}{2}$	1	$\frac{3}{2}$	0 011+	$-\sqrt{2.7.11}/3.5\sqrt{13.17.19}$
$\frac{1}{2}$	2	$\frac{3}{2}$	0 000−	$+\sqrt{5.7.11}/\sqrt{2.13.17.19}$
$\frac{1}{2}$	2	$\frac{3}{2}$	0 001−	$+\sqrt{5.11}/\sqrt{2.3.13.17.19}$
$\frac{1}{2}$	$\tilde{1}$	$\frac{3}{2}$	0 000−	$-\sqrt{2.11}/\sqrt{5.13.17.19}$
$\frac{1}{2}$	$\tilde{1}$	$\frac{3}{2}$	0 001−	$+53\sqrt{11}/2\sqrt{2.3.5.7.13.17.19}$
$\frac{1}{2}$	$\tilde{1}$	$\frac{3}{2}$	0 010−	$+3\sqrt{5}/2\sqrt{2.17.19}$
$\frac{1}{2}$	$\tilde{1}$	$\frac{3}{2}$	0 011−	$+\sqrt{2.5}/\sqrt{3.7.17.19}$
$\frac{3}{2}$	1	$\frac{1}{2}$	0 000+	$-9\sqrt{19}/4\sqrt{11.13.17}$
$\frac{3}{2}$	1	$\frac{1}{2}$	0 010+	$-\sqrt{7}/2\sqrt{3.13.17.19}$
$\frac{3}{2}$	1	$\frac{1}{2}$	0 100+	$+\sqrt{2.7}/3\sqrt{13.17.19}$
$\frac{3}{2}$	1	$\frac{1}{2}$	0 110+	$+7.7\sqrt{11}/2.3\sqrt{2.3.13.17.19}$
$\frac{3}{2}$	1	$\frac{1}{2}$	0 200+	$+11/4.3.5\sqrt{17.19}$
$\frac{3}{2}$	1	$\frac{1}{2}$	0 210+	$+13\sqrt{7.11}/2.3.5\sqrt{3.17.19}$
$\frac{3}{2}$	1	$\frac{3}{2}$	0 000−	$-\sqrt{3.17}/\sqrt{2.5.11.13.19}$
$\frac{3}{2}$	1	$\frac{3}{2}$	0 001−	$+\sqrt{2.17}/\sqrt{5.7.11.13.19}$
$\frac{3}{2}$	1	$\frac{3}{2}$	0 010−	$-11\sqrt{7}/\sqrt{2.5.13.17.19}$
$\frac{3}{2}$	1	$\frac{3}{2}$	0 011−	$-11\sqrt{17}/4\sqrt{2.3.5.13.19}$
$\frac{3}{2}$	1	$\frac{3}{2}$	0 100−	$+2\sqrt{3.7}/\sqrt{5.13.17.19}$
$\frac{3}{2}$	1	$\frac{3}{2}$	0 101−	$+8/3\sqrt{5.13.17.19}$
$\frac{3}{2}$	1	$\frac{3}{2}$	0 110−	$+7\sqrt{11}/2.3\sqrt{5.13.17.19}$
$\frac{3}{2}$	1	$\frac{3}{2}$	0 111−	$-2\sqrt{7.11}/\sqrt{3.5.13.17.19}$
$\frac{3}{2}$	1	$\frac{3}{2}$	0 200−	$-\sqrt{3}/\sqrt{2.5.17.19}$
$\frac{3}{2}$	1	$\frac{3}{2}$	0 201−	$+31\sqrt{2}/3\sqrt{5.7.17.19}$
$\frac{3}{2}$	1	$\frac{3}{2}$	0 210−	$-\sqrt{7.11}/3\sqrt{2.5.17.19}$
$\frac{3}{2}$	1	$\frac{3}{2}$	0 211−	$+3\sqrt{3.11}/4\sqrt{2.5.17.19}$
$\frac{3}{2}$	1	$\frac{3}{2}$	1 000−	$-\sqrt{3.17}/\sqrt{2.5.11.13.19}$
$\frac{3}{2}$	1	$\frac{3}{2}$	1 001−	$-2.29\sqrt{2}/\sqrt{5.7.11.13.17.19}$
$\frac{3}{2}$	1	$\frac{3}{2}$	1 010−	$+2\sqrt{2.7}/\sqrt{5.13.17.19}$
$\frac{3}{2}$	1	$\frac{3}{2}$	1 011−	$+223/4\sqrt{2.3.5.13.17.19}$
$\frac{3}{2}$	1	$\frac{3}{2}$	1 100−	$-2.3\sqrt{3.7}/\sqrt{5.13.17.19}$
$\frac{3}{2}$	1	$\frac{3}{2}$	1 101−	$+167/2.3\sqrt{5.13.17.19}$
$\frac{3}{2}$	1	$\frac{3}{2}$	1 110−	$-7\sqrt{11}/3\sqrt{5.13.17.19}$
$\frac{3}{2}$	1	$\frac{3}{2}$	1 111−	$+2\sqrt{7.11}/3\sqrt{3.5.13.17.19}$
$\frac{3}{2}$	1	$\frac{3}{2}$	1 200−	$+9\sqrt{3}/5\sqrt{2.5.17.19}$
$\frac{3}{2}$	1	$\frac{3}{2}$	1 201−	$-8.4\sqrt{2}/3.5\sqrt{5.7.17.19}$
$\frac{3}{2}$	1	$\frac{3}{2}$	1 210−	$-4\sqrt{2.7.11}/3.5\sqrt{5.17.19}$
$\frac{3}{2}$	1	$\frac{3}{2}$	1 211−	$+349\sqrt{11}/4.3.5\sqrt{2.3.5.17.19}$
$\frac{3}{2}$	2	$\frac{1}{2}$	0 000−	$+11/4\sqrt{5.13.17.19}$
$\frac{3}{2}$	2	$\frac{1}{2}$	0 100−	$-\sqrt{7.11.17}/9\sqrt{2.5.13.19}$
$\frac{3}{2}$	2	$\frac{1}{2}$	0 200−	$-7.7\sqrt{11}/4.9\sqrt{5.17.19}$
$\frac{3}{2}$	2	$\frac{3}{2}$	0 000+	$+9.7\sqrt{3}/2.5\sqrt{2.13.17.19}$
$\frac{3}{2}$	2	$\frac{3}{2}$	0 001+	$+9.11/2.5\sqrt{2.7.13.17.19}$
$\frac{3}{2}$	2	$\frac{3}{2}$	0 100+	$-2\sqrt{7.11}/5\sqrt{3.13.17.19}$
$\frac{3}{2}$	2	$\frac{3}{2}$	0 101+	$-2\sqrt{11.19}/9.5\sqrt{13.17}$
$\frac{3}{2}$	2	$\frac{3}{2}$	0 200+	$-7\sqrt{11}/2.5\sqrt{2.3.17.19}$
$\frac{3}{2}$	2	$\frac{3}{2}$	0 201+	$+41\sqrt{7.11}/2.9.5\sqrt{2.17.19}$
$\frac{3}{2}$	$\tilde{1}$	$\frac{1}{2}$	0 000−	$+\sqrt{5}/3\sqrt{7.13.17.19}$
$\frac{3}{2}$	$\tilde{1}$	$\frac{1}{2}$	0 010−	$-31/3\sqrt{5.7.11.17.19}$
$\frac{3}{2}$	$\tilde{1}$	$\frac{1}{2}$	0 100−	$+4\sqrt{2.5.11}/9\sqrt{13.17.19}$
$\frac{3}{2}$	$\tilde{1}$	$\frac{1}{2}$	0 110−	$+\sqrt{19}/2.3\sqrt{2.5.17}$
$\frac{3}{2}$	$\tilde{1}$	$\frac{1}{2}$	0 200−	$-2.11\sqrt{11}/9\sqrt{5.7.17.19}$
$\frac{3}{2}$	$\tilde{1}$	$\frac{1}{2}$	0 210−	$-11\sqrt{13}/3\sqrt{5.7.17.19}$
$\frac{3}{2}$	$\tilde{1}$	$\frac{3}{2}$	0 000+	$+4.9\sqrt{2.3}/5\sqrt{7.13.17.19}$
$\frac{3}{2}$	$\tilde{1}$	$\frac{3}{2}$	0 001+	$-211\sqrt{2}/3.5.7\sqrt{13.17.19}$
$\frac{3}{2}$	$\tilde{1}$	$\frac{3}{2}$	0 010+	$-5\sqrt{3}/\sqrt{2.7.11.17.19}$
$\frac{3}{2}$	$\tilde{1}$	$\frac{3}{2}$	0 011+	$-107\sqrt{11}/4.3.7\sqrt{2.17.19}$
$\frac{3}{2}$	$\tilde{1}$	$\frac{3}{2}$	0 100+	$+2.7\sqrt{11}/5\sqrt{3.13.17.19}$
$\frac{3}{2}$	$\tilde{1}$	$\frac{3}{2}$	0 101+	$+8\sqrt{11.17}/9.5\sqrt{7.13.19}$
$\frac{3}{2}$	$\tilde{1}$	$\frac{3}{2}$	0 110+	$-7/2\sqrt{3.17.19}$
$\frac{3}{2}$	$\tilde{1}$	$\frac{3}{2}$	0 111+	$-4.5/3\sqrt{7.17.19}$
$\frac{3}{2}$	$\tilde{1}$	$\frac{3}{2}$	0 200+	$+2\sqrt{2.11}/\sqrt{3.7.17.19}$
$\frac{3}{2}$	$\tilde{1}$	$\frac{3}{2}$	0 201+	$+\sqrt{2.11}/9\sqrt{17.19}$
$\frac{3}{2}$	$\tilde{1}$	$\frac{3}{2}$	0 210+	$-\sqrt{13.19}/5\sqrt{2.3.7.17}$

SO_3–O 3jm Factors (cont.)

$\frac{15}{2}$	7	$\frac{9}{2}$			
$\frac{3}{2}$	$\tilde{1}$	$\frac{3}{2}$	0	211+	$+41\sqrt{13}/4.3.5\sqrt{2.17.19}$
$\frac{3}{2}$	$\tilde{1}$	$\frac{3}{2}$	1	000−	$-23\sqrt{3.13}/2.5\sqrt{2.7.17.19}$
$\frac{3}{2}$	$\tilde{1}$	$\frac{3}{2}$	1	001−	$+\sqrt{2}/3.5.7\sqrt{13.17.19}$
$\frac{3}{2}$	$\tilde{1}$	$\frac{3}{2}$	1	010−	$-13\sqrt{3}/\sqrt{2.7.11.17.19}$
$\frac{3}{2}$	$\tilde{1}$	$\frac{3}{2}$	1	011−	$-37\sqrt{17}/4.3.7\sqrt{2.11.19}$
$\frac{3}{2}$	$\tilde{1}$	$\frac{3}{2}$	1	100−	$+8.2\sqrt{11}/5\sqrt{3.13.17.19}$
$\frac{3}{2}$	$\tilde{1}$	$\frac{3}{2}$	1	101−	$+37\sqrt{11.19}/2.9.5\sqrt{7.13.17}$
$\frac{3}{2}$	$\tilde{1}$	$\frac{3}{2}$	1	110−	$-5/\sqrt{3.17.19}$
$\frac{3}{2}$	$\tilde{1}$	$\frac{3}{2}$	1	111−	$+4/3\sqrt{7.17.19}$
$\frac{3}{2}$	$\tilde{1}$	$\frac{3}{2}$	1	200−	$+23\sqrt{11}/2.5\sqrt{2.3.7.17.19}$
$\frac{3}{2}$	$\tilde{1}$	$\frac{3}{2}$	1	201−	$+13\sqrt{2.11}/9.5.7\sqrt{17.19}$
$\frac{3}{2}$	$\tilde{1}$	$\frac{3}{2}$	1	210−	$+\sqrt{13}/\sqrt{2.3.7.17.19}$
$\frac{3}{2}$	$\tilde{1}$	$\frac{3}{2}$	1	211−	$+79\sqrt{13}/4.3.7\sqrt{2.17.19}$
$\frac{3}{2}$	$\tilde{0}$	$\frac{3}{2}$	0	000−	$+4\sqrt{3}/\sqrt{11.17.19}$
$\frac{3}{2}$	$\tilde{0}$	$\frac{3}{2}$	0	001−	$-\sqrt{11}/2\sqrt{7.17.19}$
$\frac{3}{2}$	$\tilde{0}$	$\frac{3}{2}$	0	100−	$-\sqrt{2.7}/\sqrt{3.17.19}$
$\frac{3}{2}$	$\tilde{0}$	$\frac{3}{2}$	0	101−	$-4\sqrt{2}/9\sqrt{17.19}$
$\frac{3}{2}$	$\tilde{0}$	$\frac{3}{2}$	0	200−	$+4\sqrt{13}/5\sqrt{3.17.19}$
$\frac{3}{2}$	$\tilde{0}$	$\frac{3}{2}$	0	201−	$+197\sqrt{13}/2.9.5\sqrt{7.17.19}$
$\tilde{\frac{1}{2}}$	1	$\frac{3}{2}$	0	000+	$-9.3\sqrt{2}/5\sqrt{11.17.19}$
$\tilde{\frac{1}{2}}$	1	$\frac{3}{2}$	0	001+	$+97\sqrt{3}/2.5\sqrt{2.7.11.17.19}$
$\tilde{\frac{1}{2}}$	1	$\frac{3}{2}$	0	010+	$+7\sqrt{7}/2.5\sqrt{2.3.17.19}$
$\tilde{\frac{1}{2}}$	1	$\frac{3}{2}$	0	011+	$-8.2\sqrt{2}/3.5\sqrt{17.19}$
$\tilde{\frac{1}{2}}$	2	$\frac{3}{2}$	0	000−	$+\sqrt{5}/\sqrt{2.17.19}$
$\tilde{\frac{1}{2}}$	2	$\frac{3}{2}$	0	001−	$-\sqrt{5.17}/3\sqrt{2.3.7.19}$
$\tilde{\frac{1}{2}}$	$\tilde{1}$	$\frac{1}{2}$	0	000+	$+2\sqrt{2.19}/3\sqrt{3.7.17}$
$\tilde{\frac{1}{2}}$	$\tilde{1}$	$\frac{1}{2}$	0	010+	$-\sqrt{13}/2\sqrt{2.3.7.11.17.19}$
$\tilde{\frac{1}{2}}$	$\tilde{1}$	$\frac{3}{2}$	0	000−	$-\sqrt{2.7}/\sqrt{5.17.19}$
$\tilde{\frac{1}{2}}$	$\tilde{1}$	$\frac{3}{2}$	0	001−	$-41/2.3\sqrt{2.3.5.17.19}$
$\tilde{\frac{1}{2}}$	$\tilde{1}$	$\frac{3}{2}$	0	010−	$+3\sqrt{7\ 13}/2\sqrt{2.5.11.17.19}$
$\tilde{\frac{1}{2}}$	$\tilde{1}$	$\frac{3}{2}$	0	011−	$-4\sqrt{2.13}/\sqrt{3.5.11.17.19}$
$\tilde{\frac{1}{2}}$	$\tilde{0}$	$\frac{1}{2}$	0	000−	$-\sqrt{13}/3\sqrt{3.11.17.19}$
$\frac{15}{2}$	7	$\frac{11}{2}$			
$\frac{1}{2}$	1	$\frac{1}{2}$	0	000+	$+3\sqrt{3.7}/2\sqrt{2.13.17.19}$
$\frac{1}{2}$	1	$\frac{1}{2}$	0	010+	$+\sqrt{2.11}/3\sqrt{13.17.19}$
$\frac{1}{2}$	1	$\frac{3}{2}$	0	000−	$-3\sqrt{3.13}/\sqrt{2.11.17.19}$
$\frac{1}{2}$	1	$\frac{3}{2}$	0	001−	$+7\sqrt{5}/\sqrt{3.11.13.17.19}$
$\frac{1}{2}$	1	$\frac{3}{2}$	0	010−	$-\sqrt{7}/3\sqrt{2.13.17.19}$
$\frac{1}{2}$	1	$\frac{3}{2}$	0	011−	$-7\sqrt{5.7}/4.3\sqrt{13.17.19}$
$\frac{1}{2}$	2	$\frac{3}{2}$	0	000+	$+2.11\sqrt{2}/\sqrt{3.5.13.17.19}$
$\frac{1}{2}$	2	$\frac{3}{2}$	0	001+	$+11/2.3\sqrt{3.13.17.19}$
$\frac{1}{2}$	$\tilde{1}$	$\frac{3}{2}$	0	000+	$-37\sqrt{3}/\sqrt{2.5.7.13.17.19}$
$\frac{1}{2}$	$\tilde{1}$	$\frac{3}{2}$	0	001+	$+4\sqrt{3}/\sqrt{7.13.17.19}$
$\frac{1}{2}$	$\tilde{1}$	$\frac{3}{2}$	0	010+	$+47/\sqrt{2.3.5.7.11.17.19}$
$\frac{1}{2}$	$\tilde{1}$	$\frac{3}{2}$	0	011+	$+503/4.3\sqrt{3.7.11.17.19}$
$\frac{1}{2}$	$\tilde{1}$	$\tilde{\frac{1}{2}}$	0	000−	$+11\sqrt{3}/2\sqrt{2.7.13.17.19}$
$\frac{1}{2}$	$\tilde{1}$	$\tilde{\frac{1}{2}}$	0	010−	$-\sqrt{2.11}/3\sqrt{3.7.17.19}$
$\frac{1}{2}$	$\tilde{0}$	$\tilde{\frac{1}{2}}$	0	000+	$+2\sqrt{11}/3\sqrt{3.17.19}$
$\frac{3}{2}$	1	$\frac{1}{2}$	0	000−	$-2.3/\sqrt{5.11.13.17.19}$
$\frac{15}{2}$	7	$\frac{11}{2}$			
$\frac{3}{2}$	1	$\frac{1}{2}$	0	010−	$-9\sqrt{3.7}/8\sqrt{5.13.17.19}$
$\frac{3}{2}$	1	$\frac{1}{2}$	0	100−	$-2.3\sqrt{2.7}/\sqrt{5.13.17.19}$
$\frac{3}{2}$	1	$\frac{1}{2}$	0	110−	$-\sqrt{11}/3\sqrt{2.3.5.13.17.19}$
$\frac{3}{2}$	1	$\frac{1}{2}$	0	200−	0
$\frac{3}{2}$	1	$\frac{1}{2}$	0	210−	$+7\sqrt{5.7.11}/8.3\sqrt{3.17.19}$
$\frac{3}{2}$	1	$\frac{3}{2}$	0	000+	$-3.31\sqrt{7}/5.11\sqrt{2.5.13.17.19}$
$\frac{3}{2}$	1	$\frac{3}{2}$	0	001+	$-563\sqrt{7}/4.5.11\sqrt{13.17.19}$
$\frac{3}{2}$	1	$\frac{3}{2}$	0	010+	$+3.7\sqrt{3.11}/2.5\sqrt{2.5.13.17.19}$
$\frac{3}{2}$	1	$\frac{3}{2}$	0	011+	$+47\sqrt{11}/2.5\sqrt{3.13.17.19}$
$\frac{3}{2}$	1	$\frac{3}{2}$	0	100+	$+23\sqrt{19}/2.5\sqrt{5.11.13.17}$
$\frac{3}{2}$	1	$\frac{3}{2}$	0	101+	$-2\sqrt{2.13}/5\sqrt{11.17.19}$
$\frac{3}{2}$	1	$\frac{3}{2}$	0	110+	$+2\sqrt{7.19}/3.5\sqrt{3.5.13.17}$
$\frac{3}{2}$	1	$\frac{3}{2}$	0	111+	$-131\sqrt{7}/3.5\sqrt{2.3.13.17.19}$
$\frac{3}{2}$	1	$\frac{3}{2}$	0	200+	$-\sqrt{7}/\sqrt{2.5.11.17.19}$
$\frac{3}{2}$	1	$\frac{3}{2}$	0	201+	$-7\sqrt{7}/4\sqrt{11.17.19}$
$\frac{3}{2}$	1	$\frac{3}{2}$	0	210+	$+23/2.3\sqrt{2.3.5.17.19}$
$\frac{3}{2}$	1	$\frac{3}{2}$	0	211+	$-1/2.3\sqrt{3.17.19}$
$\frac{3}{2}$	1	$\frac{3}{2}$	1	000+	$+3\sqrt{7}/2.5\sqrt{2.5.13.17.19}$
$\frac{3}{2}$	1	$\frac{3}{2}$	1	001+	$-7\sqrt{7.13}/4.5\sqrt{17.19}$
$\frac{3}{2}$	1	$\frac{3}{2}$	1	010+	$-3.467\sqrt{3}/4.5\sqrt{2.5.11.13.17.19}$
$\frac{3}{2}$	1	$\frac{3}{2}$	1	011+	$-307/4.5\sqrt{3.11.13.17.19}$
$\frac{3}{2}$	1	$\frac{3}{2}$	1	100+	$+4.103/5\sqrt{5.11.13.17.19}$
$\frac{3}{2}$	1	$\frac{3}{2}$	1	101+	$+113/3.5\sqrt{2.11.13.17.19}$
$\frac{3}{2}$	1	$\frac{3}{2}$	1	110+	$-59\sqrt{7}/3.5\sqrt{3.5.13.17.19}$
$\frac{3}{2}$	1	$\frac{3}{2}$	1	111+	$-149\sqrt{7}/2.3.5\sqrt{2.3.13.17.19}$
$\frac{3}{2}$	1	$\frac{3}{2}$	1	200+	$+\sqrt{7}/2\sqrt{2.5.11.17.19}$
$\frac{3}{2}$	1	$\frac{3}{2}$	1	201+	$+5\sqrt{7}/4.3\sqrt{11.17.19}$
$\frac{3}{2}$	1	$\frac{3}{2}$	1	210+	$+47/4.3\sqrt{2.3.5.17.19}$
$\frac{3}{2}$	1	$\frac{3}{2}$	1	211+	$-5.5/4.3\sqrt{3.17.19}$
$\frac{3}{2}$	1	$\tilde{\frac{1}{2}}$	0	000−	$+7\sqrt{7}/2.5\sqrt{13.17.19}$
$\frac{3}{2}$	1	$\tilde{\frac{1}{2}}$	0	010−	$+281\sqrt{11}/8.5\sqrt{3.13.17.19}$
$\frac{3}{2}$	1	$\tilde{\frac{1}{2}}$	0	100−	$-2\sqrt{2.11.13}/3.5\sqrt{17.19}$
$\frac{3}{2}$	1	$\tilde{\frac{1}{2}}$	0	110−	$+11\sqrt{7}/5\sqrt{2.3.13.17.19}$
$\frac{3}{2}$	1	$\tilde{\frac{1}{2}}$	0	200−	$+\sqrt{7.11}/2.3\sqrt{17.19}$
$\frac{3}{2}$	1	$\tilde{\frac{1}{2}}$	0	210−	$+11/8\sqrt{3.17.19}$
$\frac{3}{2}$	2	$\frac{1}{2}$	0	000+	$-11/4\sqrt{13.17.19}$
$\frac{3}{2}$	2	$\frac{1}{2}$	0	100+	$+5\sqrt{7.11}/3\sqrt{2.13.17.19}$
$\frac{3}{2}$	2	$\frac{1}{2}$	0	200+	$-5\sqrt{11}/4.3\sqrt{17.19}$
$\frac{3}{2}$	2	$\frac{3}{2}$	0	000−	$+8.2\sqrt{2.7}/5\sqrt{11.13.17.19}$
$\frac{3}{2}$	2	$\frac{3}{2}$	0	001−	$+37\sqrt{7}/4\sqrt{5.11.13.17.19}$
$\frac{3}{2}$	2	$\frac{3}{2}$	0	100−	$+2.3/5\sqrt{13.17.19}$
$\frac{3}{2}$	2	$\frac{3}{2}$	0	101−	$-109\sqrt{2}/9\sqrt{5.13.17.19}$
$\frac{3}{2}$	2	$\frac{3}{2}$	0	200−	0
$\frac{3}{2}$	2	$\frac{3}{2}$	0	201−	$-11\sqrt{5.7}/4.9\sqrt{17.19}$
$\frac{3}{2}$	2	$\tilde{\frac{1}{2}}$	0	000+	$+7\sqrt{7.11}/4\sqrt{5.13.17.19}$
$\frac{3}{2}$	2	$\tilde{\frac{1}{2}}$	0	100+	$+11\sqrt{17}/9\sqrt{2.5.13.19}$
$\frac{3}{2}$	2	$\tilde{\frac{1}{2}}$	0	200+	$-11\sqrt{5.7}/4.9\sqrt{17.19}$
$\frac{3}{2}$	$\tilde{1}$	$\frac{1}{2}$	0	000+	$-23/2\sqrt{7.13.17.19}$
$\frac{3}{2}$	$\tilde{1}$	$\frac{1}{2}$	0	010+	$-61/8\sqrt{7.11.17.19}$

SO_3–O 3jm Factors (cont.)

$\frac{15}{2}$	7	$\frac{11}{2}$		
$\frac{3}{2}$	$\tilde{1}$	$\frac{1}{2}$	0 100+	$-2\sqrt{2.11}/3\sqrt{13.17.19}$
$\frac{3}{2}$	$\tilde{1}$	$\frac{1}{2}$	0 110+	$+1/3\sqrt{2.17.19}$
$\frac{3}{2}$	$\tilde{1}$	$\frac{1}{2}$	0 200+	$+11\sqrt{11}/2.3\sqrt{7.17.19}$
$\frac{3}{2}$	$\tilde{1}$	$\frac{1}{2}$	0 210+	$-11\sqrt{13}/8.3\sqrt{7.17.19}$
$\frac{3}{2}$	$\tilde{1}$	$\frac{3}{2}$	0 000−	$+\sqrt{11.17}/5\sqrt{2.13.19}$
$\frac{3}{2}$	$\tilde{1}$	$\frac{3}{2}$	0 001−	$+\sqrt{11}/\sqrt{5.13.17.19}$
$\frac{3}{2}$	$\tilde{1}$	$\frac{3}{2}$	0 010−	$+337/2.5.11\sqrt{2.17.19}$
$\frac{3}{2}$	$\tilde{1}$	$\frac{3}{2}$	0 011−	$-3\sqrt{17}/2.11\sqrt{5.19}$
$\frac{3}{2}$	$\tilde{1}$	$\frac{3}{2}$	0 100−	$+601/2.3.5\sqrt{7.13.17.19}$
$\frac{3}{2}$	$\tilde{1}$	$\frac{3}{2}$	0 101−	$-31\sqrt{2}/3\sqrt{5.7.13.17.19}$
$\frac{3}{2}$	$\tilde{1}$	$\frac{3}{2}$	0 110−	$-2.13\sqrt{11}/3.5\sqrt{7.17.19}$
$\frac{3}{2}$	$\tilde{1}$	$\frac{3}{2}$	0 111−	$-47\sqrt{11}/9\sqrt{2.5.7.17.19}$
$\frac{3}{2}$	$\tilde{1}$	$\frac{3}{2}$	0 200−	$+13/3\sqrt{2.17.19}$
$\frac{3}{2}$	$\tilde{1}$	$\frac{3}{2}$	0 201−	$-2\sqrt{5}/3\sqrt{17.19}$
$\frac{3}{2}$	$\tilde{1}$	$\frac{3}{2}$	0 210−	$+\sqrt{11.13}/2.3\sqrt{2.17.19}$
$\frac{3}{2}$	$\tilde{1}$	$\frac{3}{2}$	0 211−	$-\sqrt{5.11.13}/2.9\sqrt{17.19}$
$\frac{3}{2}$	$\tilde{1}$	$\frac{3}{2}$	1 000+	$+43/2.5\sqrt{2.11.13.17.19}$
$\frac{3}{2}$	$\tilde{1}$	$\frac{3}{2}$	1 001+	$+29\sqrt{5}/4\sqrt{11.13.17.19}$
$\frac{3}{2}$	$\tilde{1}$	$\frac{3}{2}$	1 010+	$+41/4.5\sqrt{2.17.19}$
$\frac{3}{2}$	$\tilde{1}$	$\frac{3}{2}$	1 011+	$+3\sqrt{5}/4\sqrt{17.19}$
$\frac{3}{2}$	$\tilde{1}$	$\frac{3}{2}$	1 100+	$-8.4\sqrt{7}/3.5\sqrt{13.17.19}$
$\frac{3}{2}$	$\tilde{1}$	$\frac{3}{2}$	1 101+	$-5\sqrt{5}/3\sqrt{2.7.13.17.19}$
$\frac{3}{2}$	$\tilde{1}$	$\frac{3}{2}$	1 110+	$+\sqrt{7}/5\sqrt{11.17.19}$
$\frac{3}{2}$	$\tilde{1}$	$\frac{3}{2}$	1 111+	$+53\sqrt{5}/2.9\sqrt{2.7.11.17.19}$
$\frac{3}{2}$	$\tilde{1}$	$\frac{3}{2}$	1 200+	$+1/2.3\sqrt{2.17.19}$
$\frac{3}{2}$	$\tilde{1}$	$\frac{3}{2}$	1 201+	$-13\sqrt{5}/4.3\sqrt{17.19}$
$\frac{3}{2}$	$\tilde{1}$	$\frac{3}{2}$	1 210+	$-13\sqrt{13}/4\sqrt{2.11.17.19}$
$\frac{3}{2}$	$\tilde{1}$	$\frac{3}{2}$	1 211+	$+29\sqrt{5.13}/4.9\sqrt{11.17.19}$
$\frac{3}{2}$	$\tilde{1}$	$\tilde{\frac{1}{2}}$	0 000+	$+2.3\sqrt{11}/\sqrt{5.13.17.19}$
$\frac{3}{2}$	$\tilde{1}$	$\tilde{\frac{1}{2}}$	0 010+	$+9/8\sqrt{5.17.19}$
$\frac{3}{2}$	$\tilde{1}$	$\tilde{\frac{1}{2}}$	0 100+	$-2.11\sqrt{2}/\sqrt{5.7.13.17.19}$
$\frac{3}{2}$	$\tilde{1}$	$\tilde{\frac{1}{2}}$	0 110+	$+37\sqrt{11}/9\sqrt{2.5.7.17.19}$
$\frac{3}{2}$	$\tilde{1}$	$\tilde{\frac{1}{2}}$	0 200+	0
$\frac{3}{2}$	$\tilde{1}$	$\tilde{\frac{1}{2}}$	0 210+	$-7\sqrt{5.11.13}/8.9\sqrt{17.19}$
$\frac{3}{2}$	$\tilde{0}$	$\frac{3}{2}$	0 000+	$+\sqrt{7}/2.5.11\sqrt{17.19}$
$\frac{3}{2}$	$\tilde{0}$	$\frac{3}{2}$	0 001+	$-2.3\sqrt{2.7}/11\sqrt{5.17.19}$
$\frac{3}{2}$	$\tilde{0}$	$\frac{3}{2}$	0 100+	$-4\sqrt{2.17}/3.5\sqrt{11.19}$
$\frac{3}{2}$	$\tilde{0}$	$\frac{3}{2}$	0 101+	$-109/9\sqrt{5.11.17.19}$
$\frac{3}{2}$	$\tilde{0}$	$\frac{3}{2}$	0 200+	$-5\sqrt{7.13}/2.3\sqrt{11.17.19}$
$\frac{3}{2}$	$\tilde{0}$	$\frac{3}{2}$	0 201+	$+2\sqrt{2.5.7.13}/9\sqrt{11.17.19}$
$\tilde{\frac{1}{2}}$	1	$\frac{3}{2}$	0 000−	$+7\sqrt{3.7}/5.11\sqrt{2.17.19}$
$\tilde{\frac{1}{2}}$	1	$\frac{3}{2}$	0 001−	$+4\sqrt{3.7}/11\sqrt{5.17.19}$
$\tilde{\frac{1}{2}}$	1	$\frac{3}{2}$	0 010−	$-1/5\sqrt{2.11.17.19}$
$\tilde{\frac{1}{2}}$	1	$\frac{3}{2}$	0 011−	$+13\sqrt{17}/4.3\sqrt{5.11.19}$
$\tilde{\frac{1}{2}}$	1	$\tilde{\frac{1}{2}}$	0 000+	$-\sqrt{3.7}/2\sqrt{2.5.17.19}$
$\tilde{\frac{1}{2}}$	1	$\tilde{\frac{1}{2}}$	0 010+	$+\sqrt{2.11}/3\sqrt{5.17.19}$
$\tilde{\frac{1}{2}}$	2	$\frac{3}{2}$	0 000+	$-2\sqrt{2.3.7}/\sqrt{5.11.17.19}$
$\tilde{\frac{1}{2}}$	2	$\frac{3}{2}$	0 001+	$-3\sqrt{3.7}/2\sqrt{11.17.19}$
$\tilde{\frac{1}{2}}$	$\tilde{1}$	$\frac{1}{2}$	0 000−	$-\sqrt{3.7}/2\sqrt{2.5.17.19}$
$\tilde{\frac{1}{2}}$	$\tilde{1}$	$\frac{1}{2}$	0 010−	$+\sqrt{2.3.7.13}/\sqrt{5.11.17.19}$
$\tilde{\frac{1}{2}}$	$\tilde{1}$	$\frac{3}{2}$	0 000+	$+\sqrt{3}/\sqrt{2.5.11.17.19}$
$\tilde{\frac{1}{2}}$	$\tilde{1}$	$\frac{3}{2}$	0 001+	$+\sqrt{19}/\sqrt{3.11.17}$
$\tilde{\frac{1}{2}}$	$\tilde{1}$	$\frac{3}{2}$	0 010+	$+\sqrt{3.5.13}/11\sqrt{2.17.19}$
$\tilde{\frac{1}{2}}$	$\tilde{1}$	$\frac{3}{2}$	0 011+	$-5\sqrt{13}/4.11\sqrt{3.17.19}$
$\tilde{\frac{1}{2}}$	$\tilde{0}$	$\frac{1}{2}$	0 000+	$+2\sqrt{3.13}/\sqrt{5.11.17.19}$
$\frac{15}{2}$	7	$\frac{13}{2}$		
$\frac{1}{2}$	1	$\frac{1}{2}$	0 000−	$-3\sqrt{3.11}/\sqrt{7.13.17.19}$
$\frac{1}{2}$	1	$\frac{1}{2}$	0 010−	$+1/\sqrt{13.17.19}$
$\frac{1}{2}$	1	$\frac{3}{2}$	0 000+	$-397\sqrt{3}/2.5.7\sqrt{2.13.17.19}$
$\frac{1}{2}$	1	$\frac{3}{2}$	0 001+	$+11.11\sqrt{11}/5.7\sqrt{13.17.19}$
$\frac{1}{2}$	1	$\frac{3}{2}$	0 010+	$-31\sqrt{11}/5\sqrt{2.7.13.17.19}$
$\frac{1}{2}$	1	$\frac{3}{2}$	0 011+	$-3\sqrt{3.17}/2.5\sqrt{7.13.19}$
$\frac{1}{2}$	2	$\frac{3}{2}$	0 000−	$+11\sqrt{3.11}/7\sqrt{2.5.13.17.19}$
$\frac{1}{2}$	2	$\frac{3}{2}$	0 001−	$-8.3/7\sqrt{5.13.17.19}$
$\frac{1}{2}$	$\tilde{1}$	$\frac{3}{2}$	0 000−	$+11\sqrt{3.11.13}/2.7\sqrt{2.5.7.17.19}$
$\frac{1}{2}$	$\tilde{1}$	$\frac{3}{2}$	0 001−	$-4.53/7\sqrt{5.7.13.17.19}$
$\frac{1}{2}$	$\tilde{1}$	$\frac{3}{2}$	0 010−	$-\sqrt{3.17}/7\sqrt{2.5.7.19}$
$\frac{1}{2}$	$\tilde{1}$	$\frac{3}{2}$	0 011−	$+3\sqrt{11.17}/2.7\sqrt{5.7.19}$
$\frac{1}{2}$	$\tilde{1}$	$\tilde{\frac{1}{2}}$	0 000+	$+11\sqrt{3.11}/7\sqrt{7.13.17.19}$
$\frac{1}{2}$	$\tilde{1}$	$\tilde{\frac{1}{2}}$	0 001+	$-8.2\sqrt{2}/7\sqrt{5.7.17.19}$
$\frac{1}{2}$	$\tilde{1}$	$\tilde{\frac{1}{2}}$	0 010+	$-3.5\sqrt{3}/7\sqrt{7.17.19}$
$\frac{1}{2}$	$\tilde{1}$	$\tilde{\frac{1}{2}}$	0 011+	$-3\sqrt{11.13}/2.7\sqrt{2.5.7.17.19}$
$\frac{1}{2}$	$\tilde{0}$	$\tilde{\frac{1}{2}}$	0 000−	$+2\sqrt{2.3}/7\sqrt{17.19}$
$\frac{1}{2}$	$\tilde{0}$	$\tilde{\frac{1}{2}}$	0 001−	$+3\sqrt{11.13}/7\sqrt{5.17.19}$
$\frac{3}{2}$	1	$\frac{1}{2}$	0 000+	$-2\sqrt{2}/\sqrt{5.13.17.19}$
$\frac{3}{2}$	1	$\frac{1}{2}$	0 010+	$-3\sqrt{3.7.11}/4\sqrt{2.5.13.17.19}$
$\frac{3}{2}$	1	$\frac{1}{2}$	0 100+	$+9\sqrt{11}/2\sqrt{5.7.13.17.19}$
$\frac{3}{2}$	1	$\frac{1}{2}$	0 110+	$+2\sqrt{3}/\sqrt{5.13.17.19}$
$\frac{3}{2}$	1	$\frac{1}{2}$	0 200+	0
$\frac{3}{2}$	1	$\frac{1}{2}$	0 210+	$-\sqrt{3.5.7}/4\sqrt{2.17.19}$
$\frac{3}{2}$	1	$\frac{3}{2}$	0 000−	$+\sqrt{2.7.11}/\sqrt{5.13.17.19}$
$\frac{3}{2}$	1	$\frac{3}{2}$	0 001−	$+43\sqrt{7}/5\sqrt{3.5.13.17.19}$
$\frac{3}{2}$	1	$\frac{3}{2}$	0 010−	$+\sqrt{3.5.19}/4\sqrt{2.13.17}$
$\frac{3}{2}$	1	$\frac{3}{2}$	0 011−	$-2.9\sqrt{11}/5\sqrt{5.13.17.19}$
$\frac{3}{2}$	1	$\frac{3}{2}$	0 100−	$+2.59/7\sqrt{5.13.17.19}$
$\frac{3}{2}$	1	$\frac{3}{2}$	0 101−	$+\sqrt{2.11.19}/3.5.7\sqrt{3.5.13.17}$
$\frac{3}{2}$	1	$\frac{3}{2}$	0 110−	$-3\sqrt{3.11}/\sqrt{5.7.13.17.19}$
$\frac{3}{2}$	1	$\frac{3}{2}$	0 111−	$-233/5\sqrt{2.5.7.13.17.19}$
$\frac{3}{2}$	1	$\frac{3}{2}$	0 200−	$+\sqrt{2.5}/\sqrt{7.17.19}$
$\frac{3}{2}$	1	$\frac{3}{2}$	0 201−	$+11\sqrt{11}/3\sqrt{3.5.7.17.19}$
$\frac{3}{2}$	1	$\frac{3}{2}$	0 210−	$-\sqrt{3.5.11}/4\sqrt{2.17.19}$
$\frac{3}{2}$	1	$\frac{3}{2}$	0 211−	$+4/\sqrt{5.17.19}$
$\frac{3}{2}$	1	$\frac{3}{2}$	1 000−	$-3\sqrt{11}/5\sqrt{2.5.7.13.17.19}$
$\frac{3}{2}$	1	$\frac{3}{2}$	1 001−	$-11/2\sqrt{3.5.7.13.17.19}$
$\frac{3}{2}$	1	$\frac{3}{2}$	1 010−	$-3\sqrt{3.13}/4.5\sqrt{2.5.17.19}$
$\frac{3}{2}$	1	$\frac{3}{2}$	1 011−	$-\sqrt{11.17}/4\sqrt{5.13.19}$
$\frac{3}{2}$	1	$\frac{3}{2}$	1 100−	$-683/2.5.7\sqrt{5.13.17.19}$
$\frac{3}{2}$	1	$\frac{3}{2}$	1 101−	$-\sqrt{5.11.13}/3.7\sqrt{2.3.17.19}$

SO_3–O 3jm Factors (cont.)

$\frac{15}{2}$	7	$\frac{13}{2}$			
$\frac{3}{2}$	1	$\frac{3}{2}$	1	110−	$-8\sqrt{3.11}/5\sqrt{5.7.13.17.19}$
$\frac{3}{2}$	1	$\frac{3}{2}$	1	111−	$+4\sqrt{2}/\sqrt{5.7.13.17.19}$
$\frac{3}{2}$	1	$\frac{3}{2}$	1	200−	$+\sqrt{7}/\sqrt{2.5.17.19}$
$\frac{3}{2}$	1	$\frac{3}{2}$	1	201−	$+\sqrt{7.11}/2.3\sqrt{3.5.17.19}$
$\frac{3}{2}$	1	$\frac{3}{2}$	1	210−	$+\sqrt{3.11}/4\sqrt{2.5.17.19}$
$\frac{3}{2}$	1	$\frac{3}{2}$	1	211−	$+1/4\sqrt{5.17.19}$
$\frac{3}{2}$	1	$\tilde{\frac{1}{2}}$	0	000+	$+\sqrt{2.7.11}/5\sqrt{13.17.19}$
$\frac{3}{2}$	1	$\tilde{\frac{1}{2}}$	0	001+	$+43\sqrt{7}/4.5\sqrt{3.5.17.19}$
$\frac{3}{2}$	1	$\tilde{\frac{1}{2}}$	0	010+	$+3.29\sqrt{3}/4.5\sqrt{2.13.17.19}$
$\frac{3}{2}$	1	$\tilde{\frac{1}{2}}$	0	011+	$-\sqrt{11.19}/2.5\sqrt{5.17}$
$\frac{3}{2}$	1	$\tilde{\frac{1}{2}}$	0	100+	$+53\sqrt{13}/2.5.7\sqrt{17.19}$
$\frac{3}{2}$	1	$\tilde{\frac{1}{2}}$	0	101+	$-239\sqrt{2.11}/3.5.7\sqrt{3.5.17.19}$
$\frac{3}{2}$	1	$\tilde{\frac{1}{2}}$	0	110+	$+2.11\sqrt{11}/5\sqrt{3.7.13.17.19}$
$\frac{3}{2}$	1	$\tilde{\frac{1}{2}}$	0	111+	$+11/2.5\sqrt{2.5.7.17.19}$
$\frac{3}{2}$	1	$\tilde{\frac{1}{2}}$	0	200+	$-5\sqrt{2}/\sqrt{7.17.19}$
$\frac{3}{2}$	1	$\tilde{\frac{1}{2}}$	0	201+	$-13\sqrt{11.13}/4.3\sqrt{3.5.7.17.19}$
$\frac{3}{2}$	1	$\tilde{\frac{1}{2}}$	0	210+	$+5\sqrt{11}/4\sqrt{2.3.17.19}$
$\frac{3}{2}$	1	$\tilde{\frac{1}{2}}$	0	211+	$-\sqrt{13}/2\sqrt{5.17.19}$
$\frac{3}{2}$	2	$\frac{1}{2}$	0	000−	$-\sqrt{11}/2.5\sqrt{2.13.17.19}$
$\frac{3}{2}$	2	$\frac{1}{2}$	0	100−	$-277/3.5\sqrt{7.13.17.19}$
$\frac{3}{2}$	2	$\frac{1}{2}$	0	200−	$-1/2.3\sqrt{2.17.19}$
$\frac{3}{2}$	2	$\frac{3}{2}$	0	000+	$-37\sqrt{7}/2.5\sqrt{2.13.17.19}$
$\frac{3}{2}$	2	$\frac{3}{2}$	0	001+	$-4\sqrt{7.11}/5\sqrt{3.13.17.19}$
$\frac{3}{2}$	2	$\frac{3}{2}$	0	100+	$+2.23\sqrt{11}/3.5\sqrt{13.17.19}$
$\frac{3}{2}$	2	$\frac{3}{2}$	0	101+	$-2.31\sqrt{2}/5\sqrt{3.13.17.19}$
$\frac{3}{2}$	2	$\frac{3}{2}$	0	200+	$-\sqrt{11.17}/2.3\sqrt{2.7.19}$
$\frac{3}{2}$	2	$\frac{3}{2}$	0	201+	$-4/\sqrt{3.7.17.19}$
$\frac{3}{2}$	2	$\tilde{\frac{1}{2}}$	0	000−	$+9\sqrt{7}/2\sqrt{2.5.13.17.19}$
$\frac{3}{2}$	2	$\tilde{\frac{1}{2}}$	0	001−	$+\sqrt{7.11}/4\sqrt{3.17.19}$
$\frac{3}{2}$	2	$\tilde{\frac{1}{2}}$	0	100−	$+\sqrt{11.17}/7\sqrt{5.13.19}$
$\frac{3}{2}$	2	$\tilde{\frac{1}{2}}$	0	101−	$+5/7\sqrt{2.3.17.19}$
$\frac{3}{2}$	2	$\tilde{\frac{1}{2}}$	0	200−	$-\sqrt{5.11}/2\sqrt{2.7.17.19}$
$\frac{3}{2}$	2	$\tilde{\frac{1}{2}}$	0	201−	$+\sqrt{13}/4\sqrt{3.7.17.19}$
$\frac{3}{2}$	$\tilde{1}$	$\frac{1}{2}$	0	000−	$-4\sqrt{2.11}/5\sqrt{7.13.17.19}$
$\frac{3}{2}$	$\tilde{1}$	$\frac{1}{2}$	0	010−	$-41/4.5\sqrt{2.7.17.19}$
$\frac{3}{2}$	$\tilde{1}$	$\frac{1}{2}$	0	100−	$-3.181/2.5.7\sqrt{13.17.19}$
$\frac{3}{2}$	$\tilde{1}$	$\frac{1}{2}$	0	110−	$+2.47\sqrt{11}/3.5.7\sqrt{17.19}$
$\frac{3}{2}$	$\tilde{1}$	$\frac{1}{2}$	0	200−	$+2.3\sqrt{2}/\sqrt{7.17.19}$
$\frac{3}{2}$	$\tilde{1}$	$\frac{1}{2}$	0	210−	$+5\sqrt{11.13}/4.3\sqrt{2.7.17.19}$
$\frac{3}{2}$	$\tilde{1}$	$\frac{3}{2}$	0	000+	$-8.4\sqrt{2}/5.7\sqrt{13.17.19}$
$\frac{3}{2}$	$\tilde{1}$	$\frac{3}{2}$	0	001+	$+2.9\sqrt{3.11}/5.7\sqrt{13.17.19}$
$\frac{3}{2}$	$\tilde{1}$	$\frac{3}{2}$	0	010+	$+23\sqrt{11}/4.5.7\sqrt{2.17.19}$
$\frac{3}{2}$	$\tilde{1}$	$\frac{3}{2}$	0	011+	$+211/2.5.7\sqrt{3.17.19}$
$\frac{3}{2}$	$\tilde{1}$	$\frac{3}{2}$	0	100+	$+2.29\sqrt{11}/5.7\sqrt{7.13.17.19}$
$\frac{3}{2}$	$\tilde{1}$	$\frac{3}{2}$	0	101+	$+11.67\sqrt{2}/5.7\sqrt{3.7.13.17.19}$
$\frac{3}{2}$	$\tilde{1}$	$\frac{3}{2}$	0	110+	$+163/3.5.7\sqrt{7.17.19}$
$\frac{3}{2}$	$\tilde{1}$	$\frac{3}{2}$	0	111+	$+103\sqrt{11}/5.7\sqrt{2.3.7.17.19}$
$\frac{3}{2}$	$\tilde{1}$	$\frac{3}{2}$	0	200+	$+4\sqrt{2.11}/7\sqrt{17.19}$
$\frac{3}{2}$	$\tilde{1}$	$\frac{3}{2}$	0	201+	$+2.11/7\sqrt{3.17.19}$
$\frac{3}{2}$	$\tilde{1}$	$\frac{3}{2}$	0	210+	$-73\sqrt{13}/4.3.7\sqrt{2.17.19}$
$\frac{3}{2}$	$\tilde{1}$	$\frac{3}{2}$	0	211+	$-\sqrt{11.13}/2.7\sqrt{3.17.19}$
$\frac{3}{2}$	$\tilde{1}$	$\frac{3}{2}$	1	000−	$+2.3.23\sqrt{2}/5.7\sqrt{13.17.19}$
$\frac{3}{2}$	$\tilde{1}$	$\frac{3}{2}$	1	001−	$+2\sqrt{3.11}/5.7\sqrt{13.17.19}$
$\frac{3}{2}$	$\tilde{1}$	$\frac{3}{2}$	1	010−	$-41\sqrt{11}/4.5.7\sqrt{2.17.19}$
$\frac{3}{2}$	$\tilde{1}$	$\frac{3}{2}$	1	011−	$-601/4.5.7\sqrt{3.17.19}$
$\frac{3}{2}$	$\tilde{1}$	$\frac{3}{2}$	1	100−	$+\sqrt{11.13}/2.5\sqrt{7.17.19}$
$\frac{3}{2}$	$\tilde{1}$	$\frac{3}{2}$	1	101−	$-\sqrt{3.19}/5\sqrt{2.7.13.17}$
$\frac{3}{2}$	$\tilde{1}$	$\frac{3}{2}$	1	110−	$+2.31/3.5\sqrt{7.17.19}$
$\frac{3}{2}$	$\tilde{1}$	$\frac{3}{2}$	1	111−	$+4\sqrt{2.11}/5\sqrt{3.7.17.19}$
$\frac{3}{2}$	$\tilde{1}$	$\frac{3}{2}$	1	200−	$+2\sqrt{2.11}/7\sqrt{17.19}$
$\frac{3}{2}$	$\tilde{1}$	$\frac{3}{2}$	1	201−	$-8\sqrt{3}/7\sqrt{17.19}$
$\frac{3}{2}$	$\tilde{1}$	$\frac{3}{2}$	1	210−	$+79\sqrt{13}/4.3.7\sqrt{2.17.19}$
$\frac{3}{2}$	$\tilde{1}$	$\frac{3}{2}$	1	211−	$-\sqrt{11.13}/4.7\sqrt{3.17.19}$
$\frac{3}{2}$	$\tilde{1}$	$\tilde{\frac{1}{2}}$	0	000−	$-2.11\sqrt{2}/\sqrt{5.13.17.19}$
$\frac{3}{2}$	$\tilde{1}$	$\tilde{\frac{1}{2}}$	0	001−	$-\sqrt{3.11}/2.5\sqrt{17.19}$
$\frac{3}{2}$	$\tilde{1}$	$\tilde{\frac{1}{2}}$	0	010−	$-3\sqrt{11}/4\sqrt{2.5.17.19}$
$\frac{3}{2}$	$\tilde{1}$	$\tilde{\frac{1}{2}}$	0	011−	$-4\sqrt{13}/5\sqrt{3.17.19}$
$\frac{3}{2}$	$\tilde{1}$	$\tilde{\frac{1}{2}}$	0	100−	$-3.11\sqrt{11}/2.7\sqrt{5.7.13.17.19}$
$\frac{3}{2}$	$\tilde{1}$	$\tilde{\frac{1}{2}}$	0	101−	$+8\sqrt{2.17}/5.7\sqrt{3.7.19}$
$\frac{3}{2}$	$\tilde{1}$	$\tilde{\frac{1}{2}}$	0	110−	$-2/7\sqrt{5.7.17.19}$
$\frac{3}{2}$	$\tilde{1}$	$\tilde{\frac{1}{2}}$	0	111−	$+3.11\sqrt{3.11.13}/2.5.7\sqrt{2.7.17.19}$
$\frac{3}{2}$	$\tilde{1}$	$\tilde{\frac{1}{2}}$	0	200−	0
$\frac{3}{2}$	$\tilde{1}$	$\tilde{\frac{1}{2}}$	0	201−	$+\sqrt{13}/2\sqrt{3.17.19}$
$\frac{3}{2}$	$\tilde{1}$	$\tilde{\frac{1}{2}}$	0	210−	$-\sqrt{5.13}/4\sqrt{2.17.19}$
$\frac{3}{2}$	$\tilde{1}$	$\tilde{\frac{1}{2}}$	0	211−	0
$\frac{3}{2}$	$\tilde{0}$	$\frac{3}{2}$	0	000−	$+\sqrt{7.11}/2.5\sqrt{17.19}$
$\frac{3}{2}$	$\tilde{0}$	$\frac{3}{2}$	0	001−	$+4\sqrt{2.7}/5\sqrt{3.17.19}$
$\frac{3}{2}$	$\tilde{0}$	$\frac{3}{2}$	0	100−	$+2.47\sqrt{2}/3.5.7\sqrt{17.19}$
$\frac{3}{2}$	$\tilde{0}$	$\frac{3}{2}$	0	101−	$+2.3\sqrt{3.11}/5.7\sqrt{17.19}$
$\frac{3}{2}$	$\tilde{0}$	$\frac{3}{2}$	0	200−	$+\sqrt{7.13}/2.3\sqrt{17.19}$
$\frac{3}{2}$	$\tilde{0}$	$\frac{3}{2}$	0	201−	0
$\tilde{\frac{1}{2}}$	1	$\frac{3}{2}$	0	000+	$-3\sqrt{3.11}/2.5\sqrt{2.7.17.19}$
$\tilde{\frac{1}{2}}$	1	$\frac{3}{2}$	0	001+	$+4/3\sqrt{7.17.19}$
$\tilde{\frac{1}{2}}$	1	$\frac{3}{2}$	0	010+	$-9/5\sqrt{2.17.19}$
$\tilde{\frac{1}{2}}$	1	$\frac{3}{2}$	0	011+	$-\sqrt{3.11}/2\sqrt{17.19}$
$\tilde{\frac{1}{2}}$	1	$\tilde{\frac{1}{2}}$	0	000−	$-\sqrt{3.11}/\sqrt{5.7.17.19}$
$\tilde{\frac{1}{2}}$	1	$\tilde{\frac{1}{2}}$	0	001−	$-8.2\sqrt{2.13}/3.5\sqrt{7.17.19}$
$\tilde{\frac{1}{2}}$	1	$\tilde{\frac{1}{2}}$	0	010−	$-1/\sqrt{5.17.19}$
$\tilde{\frac{1}{2}}$	1	$\tilde{\frac{1}{2}}$	0	011−	$-\sqrt{3.11.13}/2.5\sqrt{2.17.19}$
$\tilde{\frac{1}{2}}$	2	$\frac{3}{2}$	0	000−	$-13/\sqrt{2.3.5.7.17.19}$
$\tilde{\frac{1}{2}}$	2	$\frac{3}{2}$	0	001−	$-4\sqrt{11}/\sqrt{5.7.17.19}$
$\tilde{\frac{1}{2}}$	$\tilde{1}$	$\frac{1}{2}$	0	000+	$+\sqrt{3.11}/\sqrt{5.7.17.19}$
$\tilde{\frac{1}{2}}$	$\tilde{1}$	$\frac{1}{2}$	0	010+	$+\sqrt{13}/\sqrt{3.5.7.17.19}$
$\tilde{\frac{1}{2}}$	$\tilde{1}$	$\frac{3}{2}$	0	000−	$-13\sqrt{3}/2.7\sqrt{2.5.17.19}$
$\tilde{\frac{1}{2}}$	$\tilde{1}$	$\frac{3}{2}$	0	001−	$+\sqrt{11}/7\sqrt{5.17.19}$
$\tilde{\frac{1}{2}}$	$\tilde{1}$	$\frac{3}{2}$	0	010−	$+\sqrt{11.13}/7\sqrt{2.3.5.17.19}$
$\tilde{\frac{1}{2}}$	$\tilde{1}$	$\frac{3}{2}$	0	011−	$-5\sqrt{5.13}/2.7\sqrt{17.19}$
$\tilde{\frac{1}{2}}$	$\tilde{0}$	$\frac{1}{2}$	0	000−	$+2\sqrt{2.13}/\sqrt{3.5.17.19}$

SO_3–O 3jm Factors (cont.)

$\frac{15}{2}$	**$\frac{15}{2}$**	**0**			
$\frac{1}{2}$	$\frac{1}{2}$	0	0	000+	$+1/2\sqrt{2}$
$\frac{3}{2}$	$\frac{3}{2}$	0	0	000+	$+1/2$
$\frac{3}{2}$	$\frac{3}{2}$	0	0	100+	0
$\frac{3}{2}$	$\frac{3}{2}$	0	0	110+	$+1/2$
$\frac{3}{2}$	$\frac{3}{2}$	0	0	200+	0
$\frac{3}{2}$	$\frac{3}{2}$	0	0	210+	0
$\frac{3}{2}$	$\frac{3}{2}$	0	0	220+	$+1/2$
$\tilde{\frac{1}{2}}$	$\tilde{\frac{1}{2}}$	0	0	000+	$+1/2\sqrt{2}$
$\frac{15}{2}$	**$\frac{15}{2}$**	**1**			
$\frac{1}{2}$	$\frac{1}{2}$	1	0	000+	$-\sqrt{5}/2\sqrt{2.17}$
$\frac{3}{2}$	$\frac{1}{2}$	1	0	000−	0
$\frac{3}{2}$	$\frac{1}{2}$	1	0	100−	$-\sqrt{3}/\sqrt{2.17}$
$\frac{3}{2}$	$\frac{1}{2}$	1	0	200−	0
$\frac{3}{2}$	$\frac{3}{2}$	1	0	000+	$+7/5\sqrt{5.17}$
$\frac{3}{2}$	$\frac{3}{2}$	1	0	100+	$-\sqrt{2.7.11}/3.5\sqrt{5.17}$
$\frac{3}{2}$	$\frac{3}{2}$	1	0	110+	$+13.13/2.9.5\sqrt{5.17}$
$\frac{3}{2}$	$\frac{3}{2}$	1	0	200+	$-\sqrt{11.13}/2.3\sqrt{5.17}$
$\frac{3}{2}$	$\frac{3}{2}$	1	0	210+	$-\sqrt{2.7.13}/9\sqrt{5.17}$
$\frac{3}{2}$	$\frac{3}{2}$	1	0	220+	$-2\sqrt{5}/9\sqrt{17}$
$\frac{3}{2}$	$\frac{3}{2}$	1	1	000+	$+71/4.5\sqrt{5.17}$
$\frac{3}{2}$	$\frac{3}{2}$	1	1	100+	$+\sqrt{7.11}/3.5\sqrt{2.5.17}$
$\frac{3}{2}$	$\frac{3}{2}$	1	1	110+	$-4.19/9.5\sqrt{5.17}$
$\frac{3}{2}$	$\frac{3}{2}$	1	1	200+	$+\sqrt{11.13}/4.3\sqrt{5.17}$
$\frac{3}{2}$	$\frac{3}{2}$	1	1	210+	$+\sqrt{7.13}/9\sqrt{2.5.17}$
$\frac{3}{2}$	$\frac{3}{2}$	1	1	220+	$-23\sqrt{5}/4.9\sqrt{17}$
$\tilde{\frac{1}{2}}$	$\frac{3}{2}$	1	0	000−	$-\sqrt{13}/5\sqrt{3.17}$
$\tilde{\frac{1}{2}}$	$\frac{3}{2}$	1	0	010−	$+\sqrt{7.11.13}/3.5\sqrt{2.3.17}$
$\tilde{\frac{1}{2}}$	$\frac{3}{2}$	1	0	020−	$+\sqrt{11}/3\sqrt{3.17}$
$\tilde{\frac{1}{2}}$	$\tilde{\frac{1}{2}}$	1	0	000+	$-\sqrt{17}/2.3\sqrt{2.5}$
$\frac{15}{2}$	**$\frac{15}{2}$**	**2**			
$\frac{3}{2}$	$\frac{1}{2}$	2	0	000−	$-\sqrt{11}/2.5\sqrt{2.17}$
$\frac{3}{2}$	$\frac{1}{2}$	2	0	100−	$+13\sqrt{7}/2.3.5\sqrt{17}$
$\frac{3}{2}$	$\frac{1}{2}$	2	0	200−	$-\sqrt{13}/2.3\sqrt{2.17}$
$\frac{3}{2}$	$\frac{1}{2}$	$\tilde{1}$	0	000−	$+\sqrt{11}/2.5\sqrt{3.17}$
$\frac{3}{2}$	$\frac{1}{2}$	$\tilde{1}$	0	100−	$+\sqrt{2.7}/3.5\sqrt{3.17}$
$\frac{3}{2}$	$\frac{1}{2}$	$\tilde{1}$	0	200−	$+\sqrt{13}/2.3\sqrt{3.17}$
$\frac{3}{2}$	$\frac{3}{2}$	2	0	000+	$-41\sqrt{3}/4.5\sqrt{2.5.7.17}$
$\frac{3}{2}$	$\frac{3}{2}$	2	0	100+	$+2\sqrt{11}/5\sqrt{3.5.17}$
$\frac{3}{2}$	$\frac{3}{2}$	2	0	110+	$+2.7\sqrt{2.7}/3.5\sqrt{3.5.17}$
$\frac{3}{2}$	$\frac{3}{2}$	2	0	200+	$+\sqrt{11.13}/4\sqrt{2.3.5.7.17}$
$\frac{3}{2}$	$\frac{3}{2}$	2	0	210+	$-2\sqrt{13}/3\sqrt{3.5.17}$
$\frac{3}{2}$	$\frac{3}{2}$	2	0	220+	$-\sqrt{5.7}/4.3\sqrt{2.3.17}$
$\frac{3}{2}$	$\frac{3}{2}$	$\tilde{1}$	0	000+	$-3.43/4.5\sqrt{5.7.17}$
$\frac{3}{2}$	$\frac{3}{2}$	$\tilde{1}$	0	100+	$-2\sqrt{2.11}/3.5\sqrt{5.17}$
$\frac{3}{2}$	$\frac{3}{2}$	$\tilde{1}$	0	110+	$+7.7\sqrt{7}/2.9.5\sqrt{5.17}$
$\frac{3}{2}$	$\frac{3}{2}$	$\tilde{1}$	0	200+	$-\sqrt{11.13}/4.3\sqrt{5.7.17}$
$\frac{3}{2}$	$\frac{3}{2}$	$\tilde{1}$	0	210+	$+2\sqrt{2.13}/9\sqrt{5.17}$
$\frac{3}{2}$	$\frac{3}{2}$	$\tilde{1}$	0	220+	$-5\sqrt{5.7}/4.9\sqrt{17}$
$\frac{3}{2}$	$\frac{3}{2}$	$\tilde{1}$	1	000−	0
$\frac{15}{2}$	**$\frac{15}{2}$**	**2**			
$\frac{3}{2}$	$\frac{3}{2}$	$\tilde{1}$	1	100−	$-\sqrt{11}/3\sqrt{2.5.17}$
$\frac{3}{2}$	$\frac{3}{2}$	$\tilde{1}$	1	110−	0
$\frac{3}{2}$	$\frac{3}{2}$	$\tilde{1}$	1	200−	$-2\sqrt{11.13}/3\sqrt{5.7.17}$
$\frac{3}{2}$	$\frac{3}{2}$	$\tilde{1}$	1	210−	$-\sqrt{13}/3\sqrt{2.5.17}$
$\frac{3}{2}$	$\frac{3}{2}$	$\tilde{1}$	1	220−	0
$\tilde{\frac{1}{2}}$	$\frac{1}{2}$	$\tilde{1}$	0	000+	$+\sqrt{11.13}/2.3\sqrt{2.5.17}$
$\tilde{\frac{1}{2}}$	$\frac{3}{2}$	2	0	000−	$-3\sqrt{13}/2.5\sqrt{2.7.17}$
$\tilde{\frac{1}{2}}$	$\frac{3}{2}$	2	0	010−	$-\sqrt{11.13}/2.3.5\sqrt{17}$
$\tilde{\frac{1}{2}}$	$\frac{3}{2}$	2	0	020−	$-5\sqrt{11}/2.3\sqrt{2.7.17}$
$\tilde{\frac{1}{2}}$	$\frac{3}{2}$	$\tilde{1}$	0	000−	$-3\sqrt{3.13}/2.5\sqrt{7.17}$
$\tilde{\frac{1}{2}}$	$\frac{3}{2}$	$\tilde{1}$	0	010−	$+\sqrt{2.11.13}/3.5\sqrt{3.17}$
$\tilde{\frac{1}{2}}$	$\frac{3}{2}$	$\tilde{1}$	0	020−	$+\sqrt{11}/2.3\sqrt{3.7.17}$
$\frac{15}{2}$	**$\frac{15}{2}$**	**3**			
$\frac{1}{2}$	$\frac{1}{2}$	1	0	000+	$+7\sqrt{5.7}/3\sqrt{2.13.17.19}$
$\frac{3}{2}$	$\frac{1}{2}$	1	0	000−	$+\sqrt{3.11}/4\sqrt{13.17.19}$
$\frac{3}{2}$	$\frac{1}{2}$	1	0	100−	$+7.7\sqrt{7}/2.3\sqrt{2.3.13.17.19}$
$\frac{3}{2}$	$\frac{1}{2}$	1	0	200−	$-5.13/4.3\sqrt{3.17.19}$
$\frac{3}{2}$	$\frac{1}{2}$	$\tilde{1}$	0	000+	$-7\sqrt{5.11}/4\sqrt{3.13.17.19}$
$\frac{3}{2}$	$\frac{1}{2}$	$\tilde{1}$	0	100+	$+7\sqrt{5.7}/2.3\sqrt{2.3.13.17.19}$
$\frac{3}{2}$	$\frac{1}{2}$	$\tilde{1}$	0	200+	$+13\sqrt{5}/4.3\sqrt{3.17.19}$
$\frac{3}{2}$	$\frac{3}{2}$	1	0	000+	$+47/2.5\sqrt{5.7.13.17.19}$
$\frac{3}{2}$	$\frac{3}{2}$	1	0	100+	$-7\sqrt{11}/5\sqrt{2.5.13.17.19}$
$\frac{3}{2}$	$\frac{3}{2}$	1	0	110+	$-317\sqrt{7}/9.5\sqrt{5.13.17.19}$
$\frac{3}{2}$	$\frac{3}{2}$	1	0	200+	$-11\sqrt{11}/2\sqrt{5.7.17.19}$
$\frac{3}{2}$	$\frac{3}{2}$	1	0	210+	$+59/9\sqrt{2.5.17.19}$
$\frac{3}{2}$	$\frac{3}{2}$	1	0	220+	$+\sqrt{5.13}/2.9\sqrt{7.17.19}$
$\frac{3}{2}$	$\frac{3}{2}$	1	1	000+	$+353\sqrt{7}/8.5\sqrt{5.13.17.19}$
$\frac{3}{2}$	$\frac{3}{2}$	1	1	100+	$+3.7\sqrt{11}/5\sqrt{2.5.13.17.19}$
$\frac{3}{2}$	$\frac{3}{2}$	1	1	110+	$+2.37\sqrt{7}/5\sqrt{5.13.17.19}$
$\frac{3}{2}$	$\frac{3}{2}$	1	1	200+	$+9.3\sqrt{11}/8\sqrt{5.7.17.19}$
$\frac{3}{2}$	$\frac{3}{2}$	1	1	210+	$+1/3\sqrt{2.5.17.19}$
$\frac{3}{2}$	$\frac{3}{2}$	1	1	220+	$-\sqrt{5.13.17}/8.3\sqrt{7.19}$
$\frac{3}{2}$	$\frac{3}{2}$	$\tilde{1}$	0	000−	0
$\frac{3}{2}$	$\frac{3}{2}$	$\tilde{1}$	0	100−	$-11\sqrt{11}/3\sqrt{2.13.17.19}$
$\frac{3}{2}$	$\frac{3}{2}$	$\tilde{1}$	0	110−	0
$\frac{3}{2}$	$\frac{3}{2}$	$\tilde{1}$	0	200−	$-4\sqrt{11}/3\sqrt{7.17.19}$
$\frac{3}{2}$	$\frac{3}{2}$	$\tilde{1}$	0	210−	$+13/3\sqrt{2.17.19}$
$\frac{3}{2}$	$\frac{3}{2}$	$\tilde{1}$	0	220−	0
$\frac{3}{2}$	$\frac{3}{2}$	$\tilde{1}$	1	000+	$-3.197/8.5\sqrt{7.13.17.19}$
$\frac{3}{2}$	$\frac{3}{2}$	$\tilde{1}$	1	100+	$+37\sqrt{11}/3.5\sqrt{2.13.17.19}$
$\frac{3}{2}$	$\frac{3}{2}$	$\tilde{1}$	1	110+	$-2.73\sqrt{7}/9.5\sqrt{13.17.19}$
$\frac{3}{2}$	$\frac{3}{2}$	$\tilde{1}$	1	200+	$+23\sqrt{11}/8.3\sqrt{7.17.19}$
$\frac{3}{2}$	$\frac{3}{2}$	$\tilde{1}$	1	210+	$-31/9\sqrt{2.17.19}$
$\frac{3}{2}$	$\frac{3}{2}$	$\tilde{1}$	1	220+	$-5.5\sqrt{13}/8.9\sqrt{7.17.19}$
$\frac{3}{2}$	$\frac{3}{2}$	$\tilde{0}$	0	000+	$-3.47\sqrt{3}/4.5\sqrt{7.13.17.19}$
$\frac{3}{2}$	$\frac{3}{2}$	$\tilde{0}$	0	100+	$-8\sqrt{2.11}/5\sqrt{3.13.17.19}$
$\frac{3}{2}$	$\frac{3}{2}$	$\tilde{0}$	0	110+	$+71\sqrt{7}/2.3.5\sqrt{3.13.17.19}$
$\frac{3}{2}$	$\frac{3}{2}$	$\tilde{0}$	0	200+	$+23\sqrt{11}/4\sqrt{3.7.17.19}$
$\frac{3}{2}$	$\frac{3}{2}$	$\tilde{0}$	0	210+	$+4\sqrt{2}/3\sqrt{3.17.19}$

SO_3–O 3jm Factors (cont.)

$\frac{15}{2}$ $\frac{15}{2}$ 3
$\frac{3}{2}$ $\frac{3}{2}$ $\tilde{0}$ 0 220+ $+5.5\sqrt{13}/4.3\sqrt{3.7.17.19}$
$\tilde{\frac{1}{2}}$ $\frac{1}{2}$ $\tilde{1}$ 0 000− $+2\sqrt{2.11}/3\sqrt{17.19}$
$\tilde{\frac{1}{2}}$ $\frac{1}{2}$ $\tilde{0}$ 0 000+ $+\sqrt{11}/2\sqrt{2.3.17.19}$
$\tilde{\frac{1}{2}}$ $\frac{3}{2}$ 1 0 000− $-\sqrt{3.19}/4.5\sqrt{7.17}$
$\tilde{\frac{1}{2}}$ $\frac{3}{2}$ 1 0 010− $+41\sqrt{11}/2.3.5\sqrt{2.3.17.19}$
$\tilde{\frac{1}{2}}$ $\frac{3}{2}$ 1 0 020− $-11\sqrt{11.13}/4.3\sqrt{3.7.17.19}$
$\tilde{\frac{1}{2}}$ $\frac{3}{2}$ $\tilde{1}$ 0 000+ $+3\sqrt{3.19}/4\sqrt{5.7.17}$
$\tilde{\frac{1}{2}}$ $\frac{3}{2}$ $\tilde{1}$ 0 010+ $+\sqrt{11.17}/2.3\sqrt{2.3.5.19}$
$\tilde{\frac{1}{2}}$ $\frac{3}{2}$ $\tilde{1}$ 0 020+ $+\sqrt{5.11.13}/4.3\sqrt{3.7.17.19}$
$\tilde{\frac{1}{2}}$ $\tilde{\frac{1}{2}}$ 1 0 000+ $-\sqrt{13}/3\sqrt{2.5.7.17.19}$

$\frac{15}{2}$ $\frac{15}{2}$ 4
$\frac{1}{2}$ $\frac{1}{2}$ 0 0 000+ $+7.7/2\sqrt{2.3.13.17.19}$
$\frac{1}{2}$ $\frac{1}{2}$ 1 0 000− 0
$\frac{3}{2}$ $\frac{1}{2}$ 1 0 000+ $+\sqrt{7.11}/4\sqrt{13.17.19}$
$\frac{3}{2}$ $\frac{1}{2}$ 1 0 100+ $+7.7/2.3\sqrt{2.13.17.19}$
$\frac{3}{2}$ $\frac{1}{2}$ 1 0 200+ $+13\sqrt{7}/4.3\sqrt{17.19}$
$\frac{3}{2}$ $\frac{1}{2}$ 2 0 000− $-\sqrt{11.17}/4\sqrt{2.13.19}$
$\frac{3}{2}$ $\frac{1}{2}$ 2 0 100− $-2\sqrt{7}/\sqrt{13.17.19}$
$\frac{3}{2}$ $\frac{1}{2}$ 2 0 200− $+9/4\sqrt{2.17.19}$
$\frac{3}{2}$ $\frac{1}{2}$ $\tilde{1}$ 0 000− $+\sqrt{3.11}/4\sqrt{13.17.19}$
$\frac{3}{2}$ $\frac{1}{2}$ $\tilde{1}$ 0 100− $-29\sqrt{7}/2.3\sqrt{2.3.13.17.19}$
$\frac{3}{2}$ $\frac{1}{2}$ $\tilde{1}$ 0 200− $-41/4.3\sqrt{3.17.19}$
$\frac{3}{2}$ $\frac{3}{2}$ 0 0 000+ $-3\sqrt{3.19}/8\sqrt{13.17}$
$\frac{3}{2}$ $\frac{3}{2}$ 0 0 100+ $-\sqrt{7.11}/3\sqrt{2.3.13.17.19}$
$\frac{3}{2}$ $\frac{3}{2}$ 0 0 110+ $+7.7.7/2.9\sqrt{3.13.17.19}$
$\frac{3}{2}$ $\frac{3}{2}$ 0 0 200+ $-\sqrt{11}/8.3\sqrt{3.17.19}$
$\frac{3}{2}$ $\frac{3}{2}$ 0 0 210+ $-13\sqrt{7}/9\sqrt{2.3.17.19}$
$\frac{3}{2}$ $\frac{3}{2}$ 0 0 220+ $-7.7\sqrt{13}/8.9\sqrt{3.17.19}$
$\frac{3}{2}$ $\frac{3}{2}$ 1 0 000− 0
$\frac{3}{2}$ $\frac{3}{2}$ 1 0 100− $-\sqrt{2.5.7.11}/3\sqrt{3.13.17.19}$
$\frac{3}{2}$ $\frac{3}{2}$ 1 0 110− 0
$\frac{3}{2}$ $\frac{3}{2}$ 1 0 200− $-8\sqrt{11}/3\sqrt{3.5.17.19}$
$\frac{3}{2}$ $\frac{3}{2}$ 1 0 210− $+\sqrt{2.7}/\sqrt{3.5.17.19}$
$\frac{3}{2}$ $\frac{3}{2}$ 1 0 220− 0
$\frac{3}{2}$ $\frac{3}{2}$ 1 1 000− 0
$\frac{3}{2}$ $\frac{3}{2}$ 1 1 100− $-\sqrt{5.7.11}/2.3\sqrt{2.3.13.17.19}$
$\frac{3}{2}$ $\frac{3}{2}$ 1 1 110− 0
$\frac{3}{2}$ $\frac{3}{2}$ 1 1 200− $+\sqrt{5.11}/2.3\sqrt{3.17.19}$
$\frac{3}{2}$ $\frac{3}{2}$ 1 1 210− $-\sqrt{3.5.7}/2\sqrt{2.17.19}$
$\frac{3}{2}$ $\frac{3}{2}$ 1 1 220− 0
$\frac{3}{2}$ $\frac{3}{2}$ 2 0 000+ $-5\sqrt{3.5}/\sqrt{2.7.13.17.19}$
$\frac{3}{2}$ $\frac{3}{2}$ 2 0 100+ $+31\sqrt{5.11}/4.3\sqrt{3.13.17.19}$
$\frac{3}{2}$ $\frac{3}{2}$ 2 0 110+ $-2.7\sqrt{2.5.7}/9\sqrt{3.13.17.19}$
$\frac{3}{2}$ $\frac{3}{2}$ 2 0 200+ $+\sqrt{11.17}/3\sqrt{2.3.5.7.19}$
$\frac{3}{2}$ $\frac{3}{2}$ 2 0 210+ $+191/4.9\sqrt{3.5.17.19}$
$\frac{3}{2}$ $\frac{3}{2}$ 2 0 220+ $+\sqrt{5.7.13}/9\sqrt{2.3.17.19}$
$\frac{3}{2}$ $\frac{3}{2}$ $\tilde{1}$ 0 000+ $-\sqrt{5.17}/4\sqrt{7.13.19}$
$\frac{3}{2}$ $\frac{3}{2}$ $\tilde{1}$ 0 100+ $-2\sqrt{2.5.11}/3\sqrt{13.17.19}$
$\frac{3}{2}$ $\frac{3}{2}$ $\tilde{1}$ 0 110+ $-7\sqrt{5.7}/9\sqrt{13.17.19}$

$\frac{15}{2}$ $\frac{15}{2}$ 4
$\frac{3}{2}$ $\frac{3}{2}$ $\tilde{1}$ 0 200+ $+\sqrt{11.17}/4.3\sqrt{5.7.19}$
$\frac{3}{2}$ $\frac{3}{2}$ $\tilde{1}$ 0 210+ $-2\sqrt{2.17}/9\sqrt{5.19}$
$\frac{3}{2}$ $\frac{3}{2}$ $\tilde{1}$ 0 220+ $-5\sqrt{5.7.13}/4.9\sqrt{17.19}$
$\frac{3}{2}$ $\frac{3}{2}$ $\tilde{1}$ 1 000− 0
$\frac{3}{2}$ $\frac{3}{2}$ $\tilde{1}$ 1 100− $+11\sqrt{5.11}/2.3\sqrt{2.13.17.19}$
$\frac{3}{2}$ $\frac{3}{2}$ $\tilde{1}$ 1 110− 0
$\frac{3}{2}$ $\frac{3}{2}$ $\tilde{1}$ 1 200− $-37\sqrt{11}/2.3\sqrt{5.7.17.19}$
$\frac{3}{2}$ $\frac{3}{2}$ $\tilde{1}$ 1 210− $-1/2.3\sqrt{2.5.17.19}$
$\frac{3}{2}$ $\frac{3}{2}$ $\tilde{1}$ 1 220− 0
$\tilde{\frac{1}{2}}$ $\frac{1}{2}$ $\tilde{1}$ 0 000+ $-\sqrt{5.11}/3\sqrt{2.17.19}$
$\tilde{\frac{1}{2}}$ $\frac{3}{2}$ 1 0 000+ $-\sqrt{17}/4.3\sqrt{19}$
$\tilde{\frac{1}{2}}$ $\frac{3}{2}$ 1 0 010+ $-\sqrt{7.11}/2\sqrt{2.17.19}$
$\tilde{\frac{1}{2}}$ $\frac{3}{2}$ 1 0 020+ $+\sqrt{11.13}/4\sqrt{17.19}$
$\tilde{\frac{1}{2}}$ $\frac{3}{2}$ 2 0 000− $-83/4.3\sqrt{2.7.17.19}$
$\tilde{\frac{1}{2}}$ $\frac{3}{2}$ 2 0 010− $-4\sqrt{11}/9\sqrt{17.19}$
$\tilde{\frac{1}{2}}$ $\frac{3}{2}$ 2 0 020− $-13\sqrt{11.13}/4.9\sqrt{2.7.17.19}$
$\tilde{\frac{1}{2}}$ $\frac{3}{2}$ $\tilde{1}$ 0 000− $+\sqrt{3.19}/4\sqrt{7.17}$
$\tilde{\frac{1}{2}}$ $\frac{3}{2}$ $\tilde{1}$ 0 010− $-\sqrt{11}/2.3\sqrt{2.3.17.19}$
$\tilde{\frac{1}{2}}$ $\frac{3}{2}$ $\tilde{1}$ 0 020− $-\sqrt{11.13}/4.3\sqrt{3.7.17.19}$
$\tilde{\frac{1}{2}}$ $\tilde{\frac{1}{2}}$ 0 0 000+ $-\sqrt{13}/2.3\sqrt{2.3.17.19}$
$\tilde{\frac{1}{2}}$ $\tilde{\frac{1}{2}}$ 1 0 000− 0

$\frac{15}{2}$ $\frac{15}{2}$ 5
$\frac{1}{2}$ $\frac{1}{2}$ 1 0 000+ $+\sqrt{2}/\sqrt{13.17.19}$
$\frac{1}{2}$ $\frac{1}{2}$ 1 0 001+ $-7\sqrt{5.7}/2.3\sqrt{2.13.17.19}$
$\frac{3}{2}$ $\frac{1}{2}$ 1 0 000− $+9\sqrt{3.5.7}/8\sqrt{11.13.17.19}$
$\frac{3}{2}$ $\frac{1}{2}$ 1 0 001− $+7\sqrt{3}/2\sqrt{11.13.17.19}$
$\frac{3}{2}$ $\frac{1}{2}$ 1 0 100− $+\sqrt{5}/\sqrt{2.3.13.17.19}$
$\frac{3}{2}$ $\frac{1}{2}$ 1 0 101− $-2.7\sqrt{2.7}/3\sqrt{3.13.17.19}$
$\frac{3}{2}$ $\frac{1}{2}$ 1 0 200− $+7\sqrt{5.7}/8\sqrt{3.17.19}$
$\frac{3}{2}$ $\frac{1}{2}$ 1 0 201− $-13/2.3\sqrt{3.17.19}$
$\frac{3}{2}$ $\frac{1}{2}$ 2 0 000+ $+11/\sqrt{2.13.17.19}$
$\frac{3}{2}$ $\frac{1}{2}$ 2 0 100+ $-\sqrt{7.11}/2.3\sqrt{13.17.19}$
$\frac{3}{2}$ $\frac{1}{2}$ 2 0 200+ $+\sqrt{11}/3\sqrt{2.17.19}$
$\frac{3}{2}$ $\frac{1}{2}$ $\tilde{1}$ 0 000+ $-157/8\sqrt{3.13.17.19}$
$\frac{3}{2}$ $\frac{1}{2}$ $\tilde{1}$ 0 100+ $-\sqrt{7.11}/3\sqrt{2.3.13.17.19}$
$\frac{3}{2}$ $\frac{1}{2}$ $\tilde{1}$ 0 200+ $-\sqrt{11.19}/8.3\sqrt{3.17}$
$\frac{3}{2}$ $\frac{3}{2}$ 1 0 000+ $-3.7/2.5\sqrt{13.17.19}$
$\frac{3}{2}$ $\frac{3}{2}$ 1 0 001+ $-47\sqrt{7}/8\sqrt{5.13.17.19}$
$\frac{3}{2}$ $\frac{3}{2}$ 1 0 100+ $+\sqrt{7.11.17}/2.5\sqrt{2.13.19}$
$\frac{3}{2}$ $\frac{3}{2}$ 1 0 101+ $+\sqrt{11}/\sqrt{2.5.13.17.19}$
$\frac{3}{2}$ $\frac{3}{2}$ 1 0 110+ $+4.31/3.5\sqrt{13.17.19}$
$\frac{3}{2}$ $\frac{3}{2}$ 1 0 111+ $+127\sqrt{7}/2.9\sqrt{5.13.17.19}$
$\frac{3}{2}$ $\frac{3}{2}$ 1 0 200+ $-\sqrt{11}/2\sqrt{17.19}$
$\frac{3}{2}$ $\frac{3}{2}$ 1 0 201+ $+\sqrt{7.11}/8\sqrt{5.17.19}$
$\frac{3}{2}$ $\frac{3}{2}$ 1 0 210+ $+\sqrt{7}/2.3\sqrt{2.17.19}$
$\frac{3}{2}$ $\frac{3}{2}$ 1 0 211+ $+23/9\sqrt{2.5.17.19}$
$\frac{3}{2}$ $\frac{3}{2}$ 1 0 220+ $-\sqrt{13}/2.3\sqrt{17.19}$
$\frac{3}{2}$ $\frac{3}{2}$ 1 0 221+ $+7\sqrt{5.7.13}/8.9\sqrt{17.19}$
$\frac{3}{2}$ $\frac{3}{2}$ 1 1 000+ $+3.137/4.5.11\sqrt{13.17.19}$

SO_3–O 3jm Factors (cont.)

$\frac{15}{2}\ \frac{15}{2}\ 5$

$\frac{3}{2}\ \frac{3}{2}\ 1\ 1\ 001+\quad -8.8.2\sqrt{7}/11\sqrt{5.13.17.19}$

$\frac{3}{2}\ \frac{3}{2}\ 1\ 1\ 100+\quad +173\sqrt{7}/4.5\sqrt{2.11.13.17.19}$

$\frac{3}{2}\ \frac{3}{2}\ 1\ 1\ 101+\quad +2.3\sqrt{2}/\sqrt{5.11.13.17.19}$

$\frac{3}{2}\ \frac{3}{2}\ 1\ 1\ 110+\quad -8.4/3.5\sqrt{13.17.19}$

$\frac{3}{2}\ \frac{3}{2}\ 1\ 1\ 111+\quad +4\sqrt{7}/3\sqrt{5.13.17.19}$

$\frac{3}{2}\ \frac{3}{2}\ 1\ 1\ 200+\quad +\sqrt{11}/4\sqrt{17.19}$

$\frac{3}{2}\ \frac{3}{2}\ 1\ 1\ 201+\quad 0$

$\frac{3}{2}\ \frac{3}{2}\ 1\ 1\ 210+\quad -\sqrt{7}/4.3\sqrt{2.17.19}$

$\frac{3}{2}\ \frac{3}{2}\ 1\ 1\ 211+\quad +2\sqrt{2.5}/3\sqrt{17.19}$

$\frac{3}{2}\ \frac{3}{2}\ 1\ 1\ 220+\quad -11\sqrt{13}/4.3\sqrt{17.19}$

$\frac{3}{2}\ \frac{3}{2}\ 1\ 1\ 221+\quad 0$

$\frac{3}{2}\ \frac{3}{2}\ 2\ 0\ 000-\quad 0$

$\frac{3}{2}\ \frac{3}{2}\ 2\ 0\ 100-\quad +2.5\sqrt{5}/\sqrt{3.13.17.19}$

$\frac{3}{2}\ \frac{3}{2}\ 2\ 0\ 110-\quad 0$

$\frac{3}{2}\ \frac{3}{2}\ 2\ 0\ 200-\quad -\sqrt{7}/\sqrt{2.3.5.17.19}$

$\frac{3}{2}\ \frac{3}{2}\ 2\ 0\ 210-\quad +2\sqrt{11}/\sqrt{3.5.17.19}$

$\frac{3}{2}\ \frac{3}{2}\ 2\ 0\ 220-\quad 0$

$\frac{3}{2}\ \frac{3}{2}\ \tilde{1}\ 0\ 000-\quad 0$

$\frac{3}{2}\ \frac{3}{2}\ \tilde{1}\ 0\ 100-\quad -37\sqrt{5}/2.3\sqrt{2.13.17.19}$

$\frac{3}{2}\ \frac{3}{2}\ \tilde{1}\ 0\ 110-\quad 0$

$\frac{3}{2}\ \frac{3}{2}\ \tilde{1}\ 0\ 200-\quad -2\sqrt{7}/3\sqrt{5.17.19}$

$\frac{3}{2}\ \frac{3}{2}\ \tilde{1}\ 0\ 210-\quad -\sqrt{11.19}/2.3\sqrt{2.5.17}$

$\frac{3}{2}\ \frac{3}{2}\ \tilde{1}\ 0\ 220-\quad 0$

$\frac{3}{2}\ \frac{3}{2}\ \tilde{1}\ 1\ 000+\quad -3.7\sqrt{7}/4\sqrt{5.11.13.17.19}$

$\frac{3}{2}\ \frac{3}{2}\ \tilde{1}\ 1\ 100+\quad +127/4.3\sqrt{2.5.13.17.19}$

$\frac{3}{2}\ \frac{3}{2}\ \tilde{1}\ 1\ 110+\quad -4\sqrt{7.11}/9\sqrt{5.13.17.19}$

$\frac{3}{2}\ \frac{3}{2}\ \tilde{1}\ 1\ 200+\quad -\sqrt{7.19}/4.3\sqrt{5.17}$

$\frac{3}{2}\ \frac{3}{2}\ \tilde{1}\ 1\ 210+\quad -7\sqrt{11}/4.9\sqrt{2.5.17.19}$

$\frac{3}{2}\ \frac{3}{2}\ \tilde{1}\ 1\ 220+\quad -\sqrt{5.7.11.13}/4.9\sqrt{17.19}$

$\tilde{\frac{1}{2}}\ \frac{1}{2}\ \tilde{1}\ 0\ 000-\quad -\sqrt{2.5}/3\sqrt{17.19}$

$\tilde{\frac{1}{2}}\ \frac{3}{2}\ 1\ 0\ 000-\quad +199\sqrt{3}/8.11\sqrt{5.17.19}$

$\tilde{\frac{1}{2}}\ \frac{3}{2}\ 1\ 0\ 001-\quad +\sqrt{3.7}/2.11\sqrt{17.19}$

$\tilde{\frac{1}{2}}\ \frac{3}{2}\ 1\ 0\ 010-\quad -13\sqrt{7}/\sqrt{2.3.5.11.17.19}$

$\tilde{\frac{1}{2}}\ \frac{3}{2}\ 1\ 0\ 011-\quad +2.13\sqrt{2}/3\sqrt{3.11.17.19}$

$\tilde{\frac{1}{2}}\ \frac{3}{2}\ 1\ 0\ 020-\quad -\sqrt{5.11.13}/8\sqrt{3.17.19}$

$\tilde{\frac{1}{2}}\ \frac{3}{2}\ 1\ 0\ 021-\quad -\sqrt{7.11.13}/2.3\sqrt{3.17.19}$

$\tilde{\frac{1}{2}}\ \frac{3}{2}\ 2\ 0\ 000+\quad -3\sqrt{7}/\sqrt{2.11.17.19}$

$\tilde{\frac{1}{2}}\ \frac{3}{2}\ 2\ 0\ 010+\quad +1/2.3\sqrt{17.19}$

$\tilde{\frac{1}{2}}\ \frac{3}{2}\ 2\ 0\ 020+\quad -\sqrt{7.13}/3\sqrt{2.17.19}$

$\tilde{\frac{1}{2}}\ \frac{3}{2}\ \tilde{1}\ 0\ 000+\quad +3\sqrt{3.7}/8\sqrt{11.17.19}$

$\tilde{\frac{1}{2}}\ \frac{3}{2}\ \tilde{1}\ 0\ 010+\quad -11/3\sqrt{2.3.17.19}$

$\tilde{\frac{1}{2}}\ \frac{3}{2}\ \tilde{1}\ 0\ 020+\quad +7\sqrt{7.13}/8.3\sqrt{3.17.19}$

$\tilde{\frac{1}{2}}\ \tilde{\frac{1}{2}}\ 1\ 0\ 000+\quad +3\sqrt{2.13}/11\sqrt{17.19}$

$\tilde{\frac{1}{2}}\ \tilde{\frac{1}{2}}\ 1\ 0\ 001+\quad +\sqrt{5.7.13}/2.3.11\sqrt{2.17.19}$

$\frac{15}{2}\ \frac{15}{2}\ 6$

$\frac{1}{2}\ \frac{1}{2}\ 0\ 0\ 000+\quad +\sqrt{5}/\sqrt{3.13.17.19}$

$\frac{1}{2}\ \frac{1}{2}\ 1\ 0\ 000-\quad 0$

$\frac{3}{2}\ \frac{1}{2}\ 1\ 0\ 000+\quad +9\sqrt{3.11}/8\sqrt{13.17.19}$

$\frac{3}{2}\ \frac{1}{2}\ 1\ 0\ 100+\quad +\sqrt{7}/\sqrt{2.3.13.17.19}$

$\frac{3}{2}\ \frac{1}{2}\ 1\ 0\ 200+\quad -5.7/8\sqrt{3.17.19}$

$\frac{3}{2}\ \frac{1}{2}\ 2\ 0\ 000-\quad +37\sqrt{11}/4.5\sqrt{13.17.19}$

$\frac{3}{2}\ \frac{1}{2}\ 2\ 0\ 100-\quad -3.7\sqrt{7}/5\sqrt{2.13.17.19}$

$\frac{3}{2}\ \frac{1}{2}\ 2\ 0\ 200-\quad -1/4\sqrt{17.19}$

$\frac{3}{2}\ \frac{1}{2}\ \tilde{1}\ 0\ 000-\quad -53\sqrt{17}/8\sqrt{3.5.7.13.19}$

$\frac{3}{2}\ \frac{1}{2}\ \tilde{1}\ 0\ 001-\quad +2.11\sqrt{3.11}/5\sqrt{7.13.17.19}$

$\frac{3}{2}\ \frac{1}{2}\ \tilde{1}\ 0\ 100-\quad -\sqrt{11.13}/3\sqrt{2.3.5.17.19}$

$\frac{3}{2}\ \frac{1}{2}\ \tilde{1}\ 0\ 101-\quad +23/5\sqrt{2.3.13.17.19}$

$\frac{3}{2}\ \frac{1}{2}\ \tilde{1}\ 0\ 200-\quad +\sqrt{5.11}/8.3\sqrt{3.7.17.19}$

$\frac{3}{2}\ \frac{1}{2}\ \tilde{1}\ 0\ 201-\quad +2/\sqrt{3.7.17.19}$

$\frac{3}{2}\ \frac{3}{2}\ 0\ 0\ 000+\quad -\sqrt{2.3}/\sqrt{5.13.17.19}$

$\frac{3}{2}\ \frac{3}{2}\ 0\ 0\ 100+\quad -3\sqrt{3.7.11}/4\sqrt{5.13.17.19}$

$\frac{3}{2}\ \frac{3}{2}\ 0\ 0\ 110+\quad +8\sqrt{2}/3\sqrt{3.5.13.17.19}$

$\frac{3}{2}\ \frac{3}{2}\ 0\ 0\ 200+\quad 0$

$\frac{3}{2}\ \frac{3}{2}\ 0\ 0\ 210+\quad +7\sqrt{5.7}/4.3\sqrt{3.17.19}$

$\frac{3}{2}\ \frac{3}{2}\ 0\ 0\ 220+\quad -\sqrt{2.5.13}/3\sqrt{3.17.19}$

$\frac{3}{2}\ \frac{3}{2}\ 1\ 0\ 000-\quad 0$

$\frac{3}{2}\ \frac{3}{2}\ 1\ 0\ 100-\quad -3\sqrt{2.11}/\sqrt{5.13.17.19}$

$\frac{3}{2}\ \frac{3}{2}\ 1\ 0\ 110-\quad 0$

$\frac{3}{2}\ \frac{3}{2}\ 1\ 0\ 200-\quad 0$

$\frac{3}{2}\ \frac{3}{2}\ 1\ 0\ 210-\quad +\sqrt{2.5}/\sqrt{17.19}$

$\frac{3}{2}\ \frac{3}{2}\ 1\ 0\ 220-\quad 0$

$\frac{3}{2}\ \frac{3}{2}\ 1\ 1\ 000-\quad 0$

$\frac{3}{2}\ \frac{3}{2}\ 1\ 1\ 100-\quad -3.11\sqrt{11}/4\sqrt{2.5.13.17.19}$

$\frac{3}{2}\ \frac{3}{2}\ 1\ 1\ 110-\quad 0$

$\frac{3}{2}\ \frac{3}{2}\ 1\ 1\ 200-\quad 0$

$\frac{3}{2}\ \frac{3}{2}\ 1\ 1\ 210-\quad +3\sqrt{5}/4\sqrt{2.17.19}$

$\frac{3}{2}\ \frac{3}{2}\ 1\ 1\ 220-\quad 0$

$\frac{3}{2}\ \frac{3}{2}\ 2\ 0\ 000+\quad -73\sqrt{3.7}/4.5\sqrt{5.13.17.19}$

$\frac{3}{2}\ \frac{3}{2}\ 2\ 0\ 100+\quad -\sqrt{2.11.13}/5\sqrt{3.5.17.19}$

$\frac{3}{2}\ \frac{3}{2}\ 2\ 0\ 110+\quad -2\sqrt{7.17}/5\sqrt{3.5.13.19}$

$\frac{3}{2}\ \frac{3}{2}\ 2\ 0\ 200+\quad +\sqrt{7.11}/4\sqrt{3.5.17.19}$

$\frac{3}{2}\ \frac{3}{2}\ 2\ 0\ 210+\quad -\sqrt{2.3}/\sqrt{5.17.19}$

$\frac{3}{2}\ \frac{3}{2}\ 2\ 0\ 220+\quad -\sqrt{5.7.13}/4\sqrt{3.17.19}$

$\frac{3}{2}\ \frac{3}{2}\ \tilde{1}\ 0\ 000+\quad +3.7\sqrt{11}/2.5\sqrt{13.17.19}$

$\frac{3}{2}\ \frac{3}{2}\ \tilde{1}\ 0\ 001+\quad +3.7.43/8.5\sqrt{5.13.17.19}$

$\frac{3}{2}\ \frac{3}{2}\ \tilde{1}\ 0\ 100+\quad +167\sqrt{2}/3.5\sqrt{7.13.17.19}$

$\frac{3}{2}\ \frac{3}{2}\ \tilde{1}\ 0\ 101+\quad -9.11\sqrt{11}/5\sqrt{2.5.7.13.17.19}$

$\frac{3}{2}\ \frac{3}{2}\ \tilde{1}\ 0\ 110+\quad +4\sqrt{11.17}/9.5\sqrt{13.19}$

$\frac{3}{2}\ \frac{3}{2}\ \tilde{1}\ 0\ 111+\quad +43\sqrt{13}/2.3.5\sqrt{5.17.19}$

$\frac{3}{2}\ \frac{3}{2}\ \tilde{1}\ 0\ 200+\quad +5/2.3\sqrt{17.19}$

$\frac{3}{2}\ \frac{3}{2}\ \tilde{1}\ 0\ 201+\quad +3\sqrt{11}/8\sqrt{5.17.19}$

$\frac{3}{2}\ \frac{3}{2}\ \tilde{1}\ 0\ 210+\quad +5\sqrt{2.11}/9\sqrt{7.17.19}$

$\frac{3}{2}\ \frac{3}{2}\ \tilde{1}\ 0\ 211+\quad +13/3\sqrt{2.5.7.17.19}$

$\frac{3}{2}\ \frac{3}{2}\ \tilde{1}\ 0\ 220+\quad -5\sqrt{11.13}/2.9\sqrt{17.19}$

$\frac{3}{2}\ \frac{3}{2}\ \tilde{1}\ 0\ 221+\quad -\sqrt{5.13}/8.3\sqrt{17.19}$

$\frac{3}{2}\ \frac{3}{2}\ \tilde{1}\ 1\ 000-\quad 0$

$\frac{3}{2}\ \frac{3}{2}\ \tilde{1}\ 1\ 001-\quad 0$

$\frac{3}{2}\ \frac{3}{2}\ \tilde{1}\ 1\ 100-\quad +31\sqrt{7}/4.3\sqrt{2.13.17.19}$

SO_3–O 3jm Factors (cont.)

$\frac{15}{2}\ \frac{15}{2}\ 6$

$\frac{3}{2}\ \frac{3}{2}\ \tilde{1}$ 1 101− $+\sqrt{2.7.11}/\sqrt{5.13.17.19}$

$\frac{3}{2}\ \frac{3}{2}\ \tilde{1}$ 1 110− 0

$\frac{3}{2}\ \frac{3}{2}\ \tilde{1}$ 1 111− 0

$\frac{3}{2}\ \frac{3}{2}\ \tilde{1}$ 1 200− $+2/3\sqrt{17.19}$

$\frac{3}{2}\ \frac{3}{2}\ \tilde{1}$ 1 201− $+\sqrt{11}/\sqrt{5.17.19}$

$\frac{3}{2}\ \frac{3}{2}\ \tilde{1}$ 1 210− $-37\sqrt{11}/4.3\sqrt{2.7.17.19}$

$\frac{3}{2}\ \frac{3}{2}\ \tilde{1}$ 1 211− $+3\sqrt{2}/\sqrt{5.7.17.19}$

$\frac{3}{2}\ \frac{3}{2}\ \tilde{1}$ 1 220− 0

$\frac{3}{2}\ \frac{3}{2}\ \tilde{1}$ 1 221− 0

$\frac{3}{2}\ \frac{3}{2}\ \tilde{0}$ 0 000− 0

$\frac{3}{2}\ \frac{3}{2}\ \tilde{0}$ 0 100− $-67/4\sqrt{3.13.17.19}$

$\frac{3}{2}\ \frac{3}{2}\ \tilde{0}$ 0 110− 0

$\frac{3}{2}\ \frac{3}{2}\ \tilde{0}$ 0 200− $+\sqrt{2.7}/\sqrt{3.17.19}$

$\frac{3}{2}\ \frac{3}{2}\ \tilde{0}$ 0 210− $-\sqrt{11}/4\sqrt{3.17.19}$

$\frac{3}{2}\ \frac{3}{2}\ \tilde{0}$ 0 220− 0

$\tilde{\frac{1}{2}}\ \frac{1}{2}\ \tilde{1}$ 0 000+ $+8\sqrt{2}/3\sqrt{7.17.19}$

$\tilde{\frac{1}{2}}\ \frac{1}{2}\ \tilde{1}$ 0 001+ $+9\sqrt{11}/2\sqrt{2.5.7.17.19}$

$\tilde{\frac{1}{2}}\ \frac{1}{2}\ \tilde{0}$ 0 000− $-1/\sqrt{3.17.19}$

$\tilde{\frac{1}{2}}\ \frac{3}{2}\ 1$ 0 000+ $-\sqrt{3.7}/8\sqrt{17.19}$

$\tilde{\frac{1}{2}}\ \frac{3}{2}\ 1$ 0 010+ $+\sqrt{11}/\sqrt{2.3.17.19}$

$\tilde{\frac{1}{2}}\ \frac{3}{2}\ 1$ 0 020+ $+\sqrt{7.11.13}/8\sqrt{3.17.19}$

$\tilde{\frac{1}{2}}\ \frac{3}{2}\ 2$ 0 000− $-3\sqrt{7}/4.5\sqrt{17.19}$

$\tilde{\frac{1}{2}}\ \frac{3}{2}\ 2$ 0 010− $-\sqrt{11}/3.5\sqrt{2.17.19}$

$\tilde{\frac{1}{2}}\ \frac{3}{2}\ 2$ 0 020− $-\sqrt{7.11.13}/4.3\sqrt{17.19}$

$\tilde{\frac{1}{2}}\ \frac{3}{2}\ \tilde{1}$ 0 000− $+3\sqrt{3.11}/8\sqrt{5.17.19}$

$\tilde{\frac{1}{2}}\ \frac{3}{2}\ \tilde{1}$ 0 001− $+4\sqrt{3}/5\sqrt{17.19}$

$\tilde{\frac{1}{2}}\ \frac{3}{2}\ \tilde{1}$ 0 010− $+\sqrt{19}/3\sqrt{2.3.5.7.17}$

$\tilde{\frac{1}{2}}\ \frac{3}{2}\ \tilde{1}$ 0 011− $+9\sqrt{3.11}/5\sqrt{2.7.17.19}$

$\tilde{\frac{1}{2}}\ \frac{3}{2}\ \tilde{1}$ 0 020− $-7\sqrt{5.13}/8.3\sqrt{3.17.19}$

$\tilde{\frac{1}{2}}\ \frac{3}{2}\ \tilde{1}$ 0 021− 0

$\tilde{\frac{1}{2}}\ \tilde{\frac{1}{2}}\ 0$ 0 000+ $+\sqrt{3.13}/\sqrt{5.17.19}$

$\tilde{\frac{1}{2}}\ \tilde{\frac{1}{2}}\ 1$ 0 000− 0

$\frac{15}{2}\ \frac{15}{2}\ 7$

$\frac{1}{2}\ \frac{1}{2}\ 1$ 0 000+ $-\sqrt{7}/\sqrt{13.17.19.23}$

$\frac{1}{2}\ \frac{1}{2}\ 1$ 0 001+ $-5\sqrt{3.11}/\sqrt{13.17.19.23}$

$\frac{3}{2}\ \frac{1}{2}\ 1$ 0 000− $+3\sqrt{3.11.23}/4\sqrt{2.5.13.17.19}$

$\frac{3}{2}\ \frac{1}{2}\ 1$ 0 001− $-\sqrt{2.7}/\sqrt{5.13.17.19.23}$

$\frac{3}{2}\ \frac{1}{2}\ 1$ 0 100− $-3\sqrt{3.7}/\sqrt{5.13.17.19.23}$

$\frac{3}{2}\ \frac{1}{2}\ 1$ 0 101− $-3\sqrt{5.11}/2\sqrt{13.17.19.23}$

$\frac{3}{2}\ \frac{1}{2}\ 1$ 0 200− $+7\sqrt{3.5}/4\sqrt{2.17.19.23}$

$\frac{3}{2}\ \frac{1}{2}\ 1$ 0 201− 0

$\frac{3}{2}\ \frac{1}{2}\ 2$ 0 000+ $-9.3.31\sqrt{3}/4.5\sqrt{2.13.17.19.23}$

$\frac{3}{2}\ \frac{1}{2}\ 2$ 0 100+ $-2\sqrt{3.7.11}/5\sqrt{13.17.19.23}$

$\frac{3}{2}\ \frac{1}{2}\ 2$ 0 200+ $+11\sqrt{3.11}/4\sqrt{2.17.19.23}$

$\frac{3}{2}\ \frac{1}{2}\ \tilde{1}$ 0 000+ $-317\sqrt{3}/4.5\sqrt{2.7.13.17.19.23}$

$\frac{3}{2}\ \frac{1}{2}\ \tilde{1}$ 0 001+ $+67\sqrt{3.11}/2.5\sqrt{2.7.17.19.23}$

$\frac{3}{2}\ \frac{1}{2}\ \tilde{1}$ 0 100+ $-31\sqrt{11}/5\sqrt{3.13.17.19.23}$

$\frac{3}{2}\ \frac{1}{2}\ \tilde{1}$ 0 101+ $-\sqrt{3.17}/2.5\sqrt{19.23}$

$\frac{3}{2}\ \frac{1}{2}\ \tilde{1}$ 0 200+ $-71\sqrt{11}/4\sqrt{2.3.7.17.19.23}$

$\frac{15}{2}\ \frac{15}{2}\ 7$

$\frac{3}{2}\ \frac{1}{2}\ \tilde{1}$ 0 201+ $+5\sqrt{3.13}/2\sqrt{2.7.17.19.23}$

$\frac{3}{2}\ \frac{3}{2}\ 1$ 0 000+ $-\sqrt{7.19}/5.5\sqrt{2.13.17.23}$

$\frac{3}{2}\ \frac{3}{2}\ 1$ 0 001+ $-3.7.7\sqrt{3.11}/2.5.5\sqrt{2.13.17.19.23}$

$\frac{3}{2}\ \frac{3}{2}\ 1$ 0 100+ $+191\sqrt{11}/4.3.5.5\sqrt{13.17.19.23}$

$\frac{3}{2}\ \frac{3}{2}\ 1$ 0 101+ $-1129\sqrt{7}/2.5.5\sqrt{3.13.17.19.23}$

$\frac{3}{2}\ \frac{3}{2}\ 1$ 0 110+ $+4.307\sqrt{2.7}/9.5.5\sqrt{13.17.19.23}$

$\frac{3}{2}\ \frac{3}{2}\ 1$ 0 111+ $+\sqrt{2.11.19}/5.5\sqrt{3.13.17.23}$

$\frac{3}{2}\ \frac{3}{2}\ 1$ 0 200+ $+\sqrt{7.11}/3.5\sqrt{2.17.19.23}$

$\frac{3}{2}\ \frac{3}{2}\ 1$ 0 201+ $-167/2.5\sqrt{2.3.17.19.23}$

$\frac{3}{2}\ \frac{3}{2}\ 1$ 0 210+ $+1871/4.9.5\sqrt{17.19.23}$

$\frac{3}{2}\ \frac{3}{2}\ 1$ 0 211+ $+7\sqrt{7.11}/2.5\sqrt{3.17.19.23}$

$\frac{3}{2}\ \frac{3}{2}\ 1$ 0 220+ $-5\sqrt{7.13}/9\sqrt{2.17.19.23}$

$\frac{3}{2}\ \frac{3}{2}\ 1$ 0 221+ $+5\sqrt{11.13}/2\sqrt{2.3.17.19.23}$

$\frac{3}{2}\ \frac{3}{2}\ 1$ 1 000+ $-211\sqrt{7}/2.5.5\sqrt{2.13.17.19.23}$

$\frac{3}{2}\ \frac{3}{2}\ 1$ 1 001+ $+9.3.31\sqrt{3.11}/4.5.5\sqrt{2.13.17.19.23}$

$\frac{3}{2}\ \frac{3}{2}\ 1$ 1 100+ $+31.47\sqrt{11}/4.3.5.5\sqrt{13.17.19.23}$

$\frac{3}{2}\ \frac{3}{2}\ 1$ 1 101+ $+769\sqrt{7}/4.5.5\sqrt{3.13.17.19.23}$

$\frac{3}{2}\ \frac{3}{2}\ 1$ 1 110+ $-2.127\sqrt{2.7}/9.5.5\sqrt{13.17.19.23}$

$\frac{3}{2}\ \frac{3}{2}\ 1$ 1 111+ $-347\sqrt{2.11}/5.5\sqrt{3.13.17.19.23}$

$\frac{3}{2}\ \frac{3}{2}\ 1$ 1 200+ $-\sqrt{7.11}/2.3.5\sqrt{2.17.19.23}$

$\frac{3}{2}\ \frac{3}{2}\ 1$ 1 201+ $+13.29/4.5\sqrt{2.3.17.19.23}$

$\frac{3}{2}\ \frac{3}{2}\ 1$ 1 210+ $-157\sqrt{19}/4.9.5\sqrt{17.23}$

$\frac{3}{2}\ \frac{3}{2}\ 1$ 1 211+ $-7\sqrt{7.11}/4.5\sqrt{3.17.19.23}$

$\frac{3}{2}\ \frac{3}{2}\ 1$ 1 220+ $+\sqrt{7.13.23}/2.9\sqrt{2.17.19}$

$\frac{3}{2}\ \frac{3}{2}\ 1$ 1 221+ $+13\sqrt{11.13}/4\sqrt{2.3.17.19.23}$

$\frac{3}{2}\ \frac{3}{2}\ 2$ 0 000− 0

$\frac{3}{2}\ \frac{3}{2}\ 2$ 0 100− $-11.31/4.3\sqrt{5.13.17.19.23}$

$\frac{3}{2}\ \frac{3}{2}\ 2$ 0 110− 0

$\frac{3}{2}\ \frac{3}{2}\ 2$ 0 200− $-\sqrt{2.7}/3\sqrt{5.17.19.23}$

$\frac{3}{2}\ \frac{3}{2}\ 2$ 0 210− $+9.3\sqrt{11}/4\sqrt{5.17.19.23}$

$\frac{3}{2}\ \frac{3}{2}\ 2$ 0 220− 0

$\frac{3}{2}\ \frac{3}{2}\ \tilde{1}$ 0 000− 0

$\frac{3}{2}\ \frac{3}{2}\ \tilde{1}$ 0 001− 0

$\frac{3}{2}\ \frac{3}{2}\ \tilde{1}$ 0 100− $+3.167/4\sqrt{5.7.13.17.19.23}$

$\frac{3}{2}\ \frac{3}{2}\ \tilde{1}$ 0 101− $-\sqrt{11.17}/2.3\sqrt{5.7.19.23}$

$\frac{3}{2}\ \frac{3}{2}\ \tilde{1}$ 0 110− 0

$\frac{3}{2}\ \frac{3}{2}\ \tilde{1}$ 0 111− 0

$\frac{3}{2}\ \frac{3}{2}\ \tilde{1}$ 0 200− $-3\sqrt{2.5}/\sqrt{17.19.23}$

$\frac{3}{2}\ \frac{3}{2}\ \tilde{1}$ 0 201− $-2\sqrt{2.11.13}/3\sqrt{5.17.19.23}$

$\frac{3}{2}\ \frac{3}{2}\ \tilde{1}$ 0 210− $-5\sqrt{5.11}/4\sqrt{7.17.19.23}$

$\frac{3}{2}\ \frac{3}{2}\ \tilde{1}$ 0 211− $-3\sqrt{13}/2\sqrt{5.7.17.19.23}$

$\frac{3}{2}\ \frac{3}{2}\ \tilde{1}$ 0 220− 0

$\frac{3}{2}\ \frac{3}{2}\ \tilde{1}$ 0 221− 0

$\frac{3}{2}\ \frac{3}{2}\ \tilde{1}$ 1 000+ $-7\sqrt{11.23}/2.5\sqrt{2.5.13.17.19}$

$\frac{3}{2}\ \frac{3}{2}\ \tilde{1}$ 1 001+ $-7.61/4.5\sqrt{2.5.17.19.23}$

$\frac{3}{2}\ \frac{3}{2}\ \tilde{1}$ 1 100+ $-953\sqrt{7}/4.5\sqrt{5.13.17.19.23}$

$\frac{3}{2}\ \frac{3}{2}\ \tilde{1}$ 1 101+ $+11.11\sqrt{7.11}/4.3.5\sqrt{5.17.19.23}$

$\frac{3}{2}\ \frac{3}{2}\ \tilde{1}$ 1 110+ $-2.7.7\sqrt{2.11}/3.5\sqrt{5.13.17.19.23}$

$\frac{3}{2}\ \frac{3}{2}\ \tilde{1}$ 1 111+ $+7\sqrt{2.17}/5\sqrt{5.19.23}$

SO_3–O 3jm Factors (cont.)

$\frac{15}{2}$ $\frac{15}{2}$ 7	
$\frac{3}{2}$ $\frac{3}{2}$ $\tilde{1}$ 1 200+	$+\sqrt{5.17}/2\sqrt{2.19.23}$
$\frac{3}{2}$ $\frac{3}{2}$ $\tilde{1}$ 1 201+	$+11\sqrt{11.13}/4.3\sqrt{2.5.17.19.23}$
$\frac{3}{2}$ $\frac{3}{2}$ $\tilde{1}$ 1 210+	$-\sqrt{5.11.17}/4.3\sqrt{7.19.23}$
$\frac{3}{2}$ $\frac{3}{2}$ $\tilde{1}$ 1 211+	$+31\sqrt{13}/4\sqrt{5.7.17.19.23}$
$\frac{3}{2}$ $\frac{3}{2}$ $\tilde{1}$ 1 220+	$-\sqrt{5.11.13}/2.3\sqrt{2.17.19.23}$
$\frac{3}{2}$ $\frac{3}{2}$ $\tilde{1}$ 1 221+	$-13\sqrt{5}/4\sqrt{2.17.19.23}$
$\frac{3}{2}$ $\frac{3}{2}$ $\tilde{0}$ 0 000+	$+167\sqrt{7}/8.5\sqrt{5.17.19.23}$
$\frac{3}{2}$ $\frac{3}{2}$ $\tilde{0}$ 0 100+	$-173\sqrt{11}/3.5\sqrt{2.5.17.19.23}$
$\frac{3}{2}$ $\frac{3}{2}$ $\tilde{0}$ 0 110+	$+211\sqrt{7}/2.3.5\sqrt{5.17.19.23}$
$\frac{3}{2}$ $\frac{3}{2}$ $\tilde{0}$ 0 200+	$-7\sqrt{7.11.13}/8.3\sqrt{5.17.19.23}$
$\frac{3}{2}$ $\frac{3}{2}$ $\tilde{0}$ 0 210+	$+\sqrt{13}/3\sqrt{2.5.17.19.23}$
$\frac{3}{2}$ $\frac{3}{2}$ $\tilde{0}$ 0 220+	$-13\sqrt{5.7}/8.3\sqrt{17.19.23}$
$\tilde{\frac{1}{2}}$ $\frac{1}{2}$ $\tilde{1}$ 0 000−	$-\sqrt{23}/\sqrt{5.7.17.19}$
$\tilde{\frac{1}{2}}$ $\frac{1}{2}$ $\tilde{1}$ 0 001−	$-2.3\sqrt{11.13}/\sqrt{5.7.17.19.23}$
$\tilde{\frac{1}{2}}$ $\frac{1}{2}$ $\tilde{0}$ 0 000+	$+3\sqrt{11.13}/2\sqrt{2.5.17.19.23}$
$\tilde{\frac{1}{2}}$ $\frac{3}{2}$ 1 0 000−	$-37\sqrt{7}/4.5\sqrt{2.3.5.17.19.23}$
$\tilde{\frac{1}{2}}$ $\frac{3}{2}$ 1 0 001−	$-29\sqrt{11}/2.5\sqrt{2.5.17.19.23}$
$\tilde{\frac{1}{2}}$ $\frac{3}{2}$ 1 0 010−	$+11.13\sqrt{11}/3.5\sqrt{3.5.17.19.23}$
$\tilde{\frac{1}{2}}$ $\frac{3}{2}$ 1 0 011−	$-7.13\sqrt{7}/2.5\sqrt{5.17.19.23}$
$\tilde{\frac{1}{2}}$ $\frac{3}{2}$ 1 0 020−	$+\sqrt{7.11.13}/4.3\sqrt{2.3.5.17.19.23}$
$\tilde{\frac{1}{2}}$ $\frac{3}{2}$ 1 0 021−	$-11\sqrt{13}/2\sqrt{2.5.17.19.23}$
$\tilde{\frac{1}{2}}$ $\frac{3}{2}$ 2 0 000+	$-\sqrt{7.11}/4\sqrt{2.3.17.19.23}$
$\tilde{\frac{1}{2}}$ $\frac{3}{2}$ 2 0 010+	$-4/\sqrt{3.17.19.23}$
$\tilde{\frac{1}{2}}$ $\frac{3}{2}$ 2 0 020+	$-7\sqrt{7.13}/4\sqrt{2.3.17.19.23}$
$\tilde{\frac{1}{2}}$ $\frac{3}{2}$ $\tilde{1}$ 0 000+	$+\sqrt{3.11.23}/4.5\sqrt{2.17.19}$
$\tilde{\frac{1}{2}}$ $\frac{3}{2}$ $\tilde{1}$ 0 001+	$+\sqrt{2.13.23}/5\sqrt{3.17.19}$
$\tilde{\frac{1}{2}}$ $\frac{3}{2}$ $\tilde{1}$ 0 010+	$+\sqrt{23}/5\sqrt{3.7.17.19}$
$\tilde{\frac{1}{2}}$ $\frac{3}{2}$ $\tilde{1}$ 0 011+	$-3\sqrt{3.11.13}/2.5\sqrt{7.17.19.23}$
$\tilde{\frac{1}{2}}$ $\frac{3}{2}$ $\tilde{1}$ 0 020+	$+7\sqrt{13}/4\sqrt{2.3.17.19.23}$
$\tilde{\frac{1}{2}}$ $\frac{3}{2}$ $\tilde{1}$ 0 021+	0
$\tilde{\frac{1}{2}}$ $\tilde{\frac{1}{2}}$ 1 0 000+	$-\sqrt{7.13.23}/3.5\sqrt{17.19}$
$\tilde{\frac{1}{2}}$ $\tilde{\frac{1}{2}}$ 1 0 001+	$+\sqrt{3.11.13}/5\sqrt{17.19.23}$
8 4 4	
0 0 0 0 000+	$-\sqrt{2.5}/\sqrt{3.13.17}$
0 1 1 0 000+	$+4\sqrt{2}/\sqrt{5.13.17}$
0 2 2 0 000+	$-7/\sqrt{3.5.13.17}$
0 $\tilde{1}$ $\tilde{1}$ 0 000+	0
1 1 0 0 000+	$-3/\sqrt{13.17}$
1 1 0 0 100+	0
1 1 1 0 000−	0
1 1 1 0 100−	0
1 2 1 0 000+	$+\sqrt{5.7}/2\sqrt{13.17}$
1 2 1 0 100+	$-1/2\sqrt{17}$
1 $\tilde{1}$ 1 0 000+	$+\sqrt{2.7}/\sqrt{3.5.13.17}$
1 $\tilde{1}$ 1 0 100+	$+\sqrt{2}/\sqrt{3.17}$
1 $\tilde{1}$ 2 0 000−	$-7/2\sqrt{3.5.13.17}$
1 $\tilde{1}$ 2 0 100−	$-\sqrt{7}/2\sqrt{3.17}$
1 $\tilde{1}$ $\tilde{1}$ 0 000−	0
1 $\tilde{1}$ $\tilde{1}$ 0 100−	0
8 4 4	
2 1 1 0 000+	$-9\sqrt{7}/2\sqrt{11.13.17}$
2 1 1 0 100+	$+1/2\sqrt{3.5.17}$
2 2 0 0 000+	$-5\sqrt{3.5}/2\sqrt{2.11.13.17}$
2 2 0 0 100+	$+\sqrt{7}/2.3\sqrt{2.17}$
2 2 2 0 000+	$-3\sqrt{3.7}/2\sqrt{11.13.17}$
2 2 2 0 100+	$-7/2.3\sqrt{5.17}$
2 $\tilde{1}$ 1 0 000−	$+\sqrt{3}/2\sqrt{11.13.17}$
2 $\tilde{1}$ 1 0 100−	$+\sqrt{7}/2\sqrt{5.17}$
2 $\tilde{1}$ $\tilde{1}$ 0 000+	$+\sqrt{7}/2\sqrt{11.13.17}$
2 $\tilde{1}$ $\tilde{1}$ 0 100+	$+7/2\sqrt{3.5.17}$
$\tilde{1}$ 1 1 0 000+	$-4\sqrt{2.7}/17\sqrt{5.11}$
$\tilde{1}$ 1 1 0 100+	$-3\sqrt{2.3}/17\sqrt{13}$
$\tilde{1}$ 2 1 0 000−	$-23/2.17\sqrt{5.11}$
$\tilde{1}$ 2 1 0 100−	$+3\sqrt{3.7}/2.17\sqrt{13}$
$\tilde{1}$ $\tilde{1}$ 0 0 000+	$-1/\sqrt{3.11}$
$\tilde{1}$ $\tilde{1}$ 0 0 100+	0
$\tilde{1}$ $\tilde{1}$ 1 0 000−	$+\sqrt{2}/\sqrt{5.11}$
$\tilde{1}$ $\tilde{1}$ 1 0 100−	0
$\tilde{1}$ $\tilde{1}$ 2 0 000+	$-\sqrt{7.11}/2.17\sqrt{3.5}$
$\tilde{1}$ $\tilde{1}$ 2 0 100+	$-3.7/2.17\sqrt{13}$
$\tilde{1}$ $\tilde{1}$ $\tilde{1}$ 0 000+	$+2\sqrt{2.7}/17\sqrt{5.11}$
$\tilde{1}$ $\tilde{1}$ $\tilde{1}$ 0 100+	$-7\sqrt{2.3}/17\sqrt{13}$
8 $\frac{9}{2}$ $\frac{7}{2}$	
0 $\frac{1}{2}$ $\frac{1}{2}$ 0 000−	$+\sqrt{2.3}/\sqrt{13.17}$
0 $\frac{3}{2}$ $\frac{3}{2}$ 0 000−	$-2\sqrt{7}/5\sqrt{13.17}$
0 $\frac{3}{2}$ $\frac{3}{2}$ 0 010−	$-7\sqrt{3}/5\sqrt{13.17}$
1 $\frac{1}{2}$ $\frac{1}{2}$ 0 000+	$-3\sqrt{3}/2\sqrt{13.17}$
1 $\frac{1}{2}$ $\frac{1}{2}$ 0 100+	0
1 $\frac{1}{2}$ $\frac{3}{2}$ 0 000−	$+3\sqrt{3.7}/2\sqrt{5.13.17}$
1 $\frac{1}{2}$ $\frac{3}{2}$ 0 100−	0
1 $\frac{3}{2}$ $\frac{1}{2}$ 0 000−	$+5\sqrt{5}/2.3\sqrt{13.17}$
1 $\frac{3}{2}$ $\frac{1}{2}$ 0 010−	$+\sqrt{3.5.7}/4\sqrt{13.17}$
1 $\frac{3}{2}$ $\frac{1}{2}$ 0 100−	$+\sqrt{7}/2.3\sqrt{17}$
1 $\frac{3}{2}$ $\frac{1}{2}$ 0 110−	$-\sqrt{3}/4\sqrt{17}$
1 $\frac{3}{2}$ $\frac{3}{2}$ 0 000+	$+\sqrt{7}/3.5\sqrt{2.13.17}$
1 $\frac{3}{2}$ $\frac{3}{2}$ 0 010+	$-3.7\sqrt{3}/2.5\sqrt{2.13.17}$
1 $\frac{3}{2}$ $\frac{3}{2}$ 0 100+	$+\sqrt{2.5}/3\sqrt{17}$
1 $\frac{3}{2}$ $\frac{3}{2}$ 0 110+	0
1 $\frac{3}{2}$ $\frac{3}{2}$ 1 000+	$+11\sqrt{7}/3.5\sqrt{2.13.17}$
1 $\frac{3}{2}$ $\frac{3}{2}$ 1 010+	$+7\sqrt{3}/5\sqrt{2.13.17}$
1 $\frac{3}{2}$ $\frac{3}{2}$ 1 100+	$+\sqrt{2}/3\sqrt{5.17}$
1 $\frac{3}{2}$ $\frac{3}{2}$ 1 110+	$-\sqrt{3.7}/2\sqrt{2.5.17}$
1 $\frac{3}{2}$ $\tilde{\frac{1}{2}}$ 0 000−	$+\sqrt{3.7}/2.5\sqrt{13.17}$
1 $\frac{3}{2}$ $\tilde{\frac{1}{2}}$ 0 010−	$+3.7/4.5\sqrt{13.17}$
1 $\frac{3}{2}$ $\tilde{\frac{1}{2}}$ 0 100−	$+\sqrt{3}/2\sqrt{5.17}$
1 $\frac{3}{2}$ $\tilde{\frac{1}{2}}$ 0 110−	$+3\sqrt{7}/4\sqrt{5.17}$
2 $\frac{1}{2}$ $\frac{3}{2}$ 0 000+	$-3.5\sqrt{3}/2\sqrt{2.11.13.17}$
2 $\frac{1}{2}$ $\frac{3}{2}$ 0 100+	$+\sqrt{7}/2\sqrt{2.5.17}$
2 $\frac{3}{2}$ $\frac{1}{2}$ 0 000+	$-5\sqrt{7}/2\sqrt{2.11.13.17}$

SO_3–O 3jm Factors (cont.)

$8\ \frac{9}{2}\ \frac{7}{2}$

$2\ \frac{3}{2}\ \frac{1}{2}\ 0\ 010+$	$-3.5\sqrt{3}/2\sqrt{2.11.13.17}$
$2\ \frac{3}{2}\ \frac{1}{2}\ 0\ 100+$	$-\sqrt{3}/2\sqrt{2.5.17}$
$2\ \frac{3}{2}\ \frac{1}{2}\ 0\ 110+$	$+\sqrt{7}/2\sqrt{2.5.17}$
$2\ \frac{3}{2}\ \frac{3}{2}\ 0\ 000-$	$-\sqrt{13}/2\sqrt{5.11.17}$
$2\ \frac{3}{2}\ \frac{3}{2}\ 0\ 010-$	$-9\sqrt{3.7}/2\sqrt{5.11.13.17}$
$2\ \frac{3}{2}\ \frac{3}{2}\ 0\ 100-$	$+\sqrt{3.7}/2.5\sqrt{17}$
$2\ \frac{3}{2}\ \frac{3}{2}\ 0\ 110-$	$-7/2.5\sqrt{17}$
$2\ \frac{3}{2}\ \tilde{\frac{1}{2}}\ 0\ 000+$	$+\sqrt{3.5}/2\sqrt{2.11.13.17}$
$2\ \frac{3}{2}\ \tilde{\frac{1}{2}}\ 0\ 010+$	0
$2\ \frac{3}{2}\ \tilde{\frac{1}{2}}\ 0\ 100+$	$+\sqrt{7}/2\sqrt{2.17}$
$2\ \frac{3}{2}\ \tilde{\frac{1}{2}}\ 0\ 110+$	0
$\tilde{1}\ \frac{1}{2}\ \frac{3}{2}\ 0\ 000+$	$-\sqrt{3}/2\sqrt{5.11}$
$\tilde{1}\ \frac{1}{2}\ \frac{3}{2}\ 0\ 100+$	0
$\tilde{1}\ \frac{1}{2}\ \tilde{\frac{1}{2}}\ 0\ 000-$	$+3/2\sqrt{5.11}$
$\tilde{1}\ \frac{1}{2}\ \tilde{\frac{1}{2}}\ 0\ 100-$	0
$\tilde{1}\ \frac{3}{2}\ \frac{1}{2}\ 0\ 000+$	$-3.7\sqrt{7}/2.17\sqrt{5.11}$
$\tilde{1}\ \frac{3}{2}\ \frac{1}{2}\ 0\ 010+$	$+23\sqrt{3}/4.17\sqrt{5.11}$
$\tilde{1}\ \frac{3}{2}\ \frac{1}{2}\ 0\ 100+$	$-3\sqrt{3}/2.17\sqrt{13}$
$\tilde{1}\ \frac{3}{2}\ \frac{1}{2}\ 0\ 110+$	$-9\sqrt{7}/4.17\sqrt{13}$
$\tilde{1}\ \frac{3}{2}\ \frac{3}{2}\ 0\ 000-$	$+3/5.17\sqrt{2.11}$
$\tilde{1}\ \frac{3}{2}\ \frac{3}{2}\ 0\ 010-$	$-\sqrt{3.7}/2.5\sqrt{2.11}$
$\tilde{1}\ \frac{3}{2}\ \frac{3}{2}\ 0\ 100-$	$-\sqrt{2.3.5.7}/17\sqrt{13}$
$\tilde{1}\ \frac{3}{2}\ \frac{3}{2}\ 0\ 110-$	0
$\tilde{1}\ \frac{3}{2}\ \frac{3}{2}\ 1\ 000+$	$-9.9/5.17\sqrt{2.11}$
$\tilde{1}\ \frac{3}{2}\ \frac{3}{2}\ 1\ 010+$	$-3\sqrt{3.7}/5.17\sqrt{2.11}$
$\tilde{1}\ \frac{3}{2}\ \frac{3}{2}\ 1\ 100+$	$-\sqrt{2.3.7}/17\sqrt{5.13}$
$\tilde{1}\ \frac{3}{2}\ \frac{3}{2}\ 1\ 110+$	$+9.7/2.17\sqrt{2.5.13}$
$\tilde{1}\ \frac{3}{2}\ \tilde{\frac{1}{2}}\ 0\ 000+$	$+31\sqrt{3}/2.5.17\sqrt{11}$
$\tilde{1}\ \frac{3}{2}\ \tilde{\frac{1}{2}}\ 0\ 010+$	$+3\sqrt{7.11}/4.5.17$
$\tilde{1}\ \frac{3}{2}\ \tilde{\frac{1}{2}}\ 0\ 100+$	$-3.7\sqrt{7}/2.17\sqrt{5.13}$
$\tilde{1}\ \frac{3}{2}\ \tilde{\frac{1}{2}}\ 0\ 110+$	$+9.7\sqrt{3}/4.17\sqrt{5.13}$

$8\ \frac{9}{2}\ \frac{9}{2}$

$0\ \frac{1}{2}\ \frac{1}{2}\ 0\ 000+$	$-\sqrt{2.3}/\sqrt{5.13.17}$
$0\ \frac{3}{2}\ \frac{3}{2}\ 0\ 000+$	$+4.3\sqrt{3}/5\sqrt{5.13.17}$
$0\ \frac{3}{2}\ \frac{3}{2}\ 0\ 010+$	$+8\sqrt{7}/5\sqrt{5.13.17}$
$0\ \frac{3}{2}\ \frac{3}{2}\ 0\ 011+$	$-7\sqrt{3}/5\sqrt{5.13.17}$
$1\ \frac{1}{2}\ \frac{1}{2}\ 0\ 000-$	0
$1\ \frac{1}{2}\ \frac{1}{2}\ 0\ 100-$	0
$1\ \frac{3}{2}\ \frac{1}{2}\ 0\ 000+$	$-31/3.5\sqrt{13.17}$
$1\ \frac{3}{2}\ \frac{1}{2}\ 0\ 010+$	$-\sqrt{3.7}/5\sqrt{13.17}$
$1\ \frac{3}{2}\ \frac{1}{2}\ 0\ 100+$	$+2\sqrt{7}/3\sqrt{5.17}$
$1\ \frac{3}{2}\ \frac{1}{2}\ 0\ 110+$	$-\sqrt{3}/\sqrt{5.17}$
$1\ \frac{3}{2}\ \frac{3}{2}\ 0\ 000-$	0
$1\ \frac{3}{2}\ \frac{3}{2}\ 0\ 010-$	$+4\sqrt{2.7}/3\sqrt{5.13.17}$
$1\ \frac{3}{2}\ \frac{3}{2}\ 0\ 011-$	0
$1\ \frac{3}{2}\ \frac{3}{2}\ 0\ 100-$	0
$1\ \frac{3}{2}\ \frac{3}{2}\ 0\ 110-$	$-1/3\sqrt{2.17}$
$1\ \frac{3}{2}\ \frac{3}{2}\ 0\ 111-$	0
$1\ \frac{3}{2}\ \frac{3}{2}\ 1\ 000-$	0

$8\ \frac{9}{2}\ \frac{9}{2}$

$1\ \frac{3}{2}\ \frac{3}{2}\ 1\ 010-$	$-11\sqrt{7}/3\sqrt{2.5.13.17}$
$1\ \frac{3}{2}\ \frac{3}{2}\ 1\ 011-$	0
$1\ \frac{3}{2}\ \frac{3}{2}\ 1\ 100-$	0
$1\ \frac{3}{2}\ \frac{3}{2}\ 1\ 110-$	$-\sqrt{2}/3\sqrt{17}$
$1\ \frac{3}{2}\ \frac{3}{2}\ 1\ 111-$	0
$2\ \frac{3}{2}\ \frac{1}{2}\ 0\ 000-$	$-\sqrt{2.5.7}/\sqrt{11.13.17}$
$2\ \frac{3}{2}\ \frac{1}{2}\ 0\ 010-$	$+3\sqrt{3.5}/2\sqrt{2.11.13.17}$
$2\ \frac{3}{2}\ \frac{1}{2}\ 0\ 100-$	$-\sqrt{2.3}/5\sqrt{17}$
$2\ \frac{3}{2}\ \frac{1}{2}\ 0\ 110-$	$-\sqrt{7}/2.5\sqrt{2.17}$
$2\ \frac{3}{2}\ \frac{3}{2}\ 0\ 000+$	$+7\sqrt{3.7}/5\sqrt{11.13.17}$
$2\ \frac{3}{2}\ \frac{3}{2}\ 0\ 010+$	$+2\sqrt{13}/5\sqrt{11.17}$
$2\ \frac{3}{2}\ \frac{3}{2}\ 0\ 011+$	$-9\sqrt{3.7}/2.5\sqrt{11.13.17}$
$2\ \frac{3}{2}\ \frac{3}{2}\ 0\ 100+$	$+1/5\sqrt{5.17}$
$2\ \frac{3}{2}\ \frac{3}{2}\ 0\ 110+$	$-2\sqrt{3.7}/5\sqrt{5.17}$
$2\ \frac{3}{2}\ \frac{3}{2}\ 0\ 111+$	$-7/2.5\sqrt{5.17}$
$\tilde{1}\ \frac{3}{2}\ \frac{1}{2}\ 0\ 000-$	$+9\sqrt{7}/5.17\sqrt{11}$
$\tilde{1}\ \frac{3}{2}\ \frac{1}{2}\ 0\ 010-$	$-\sqrt{3.11}/5.17$
$\tilde{1}\ \frac{3}{2}\ \frac{1}{2}\ 0\ 100-$	$-2.3\sqrt{3}/17\sqrt{5.13}$
$\tilde{1}\ \frac{3}{2}\ \frac{1}{2}\ 0\ 110-$	$-9\sqrt{7}/17\sqrt{5.13}$
$\tilde{1}\ \frac{3}{2}\ \frac{3}{2}\ 0\ 000+$	$-4\sqrt{2.3.7}/5.17\sqrt{5.11}$
$\tilde{1}\ \frac{3}{2}\ \frac{3}{2}\ 0\ 010+$	$-4.3\sqrt{2}/5.17\sqrt{5.11}$
$\tilde{1}\ \frac{3}{2}\ \frac{3}{2}\ 0\ 011+$	$+2\sqrt{2.3.7}/5\sqrt{5.11}$
$\tilde{1}\ \frac{3}{2}\ \frac{3}{2}\ 0\ 100+$	$-4.3\sqrt{2}/17\sqrt{13}$
$\tilde{1}\ \frac{3}{2}\ \frac{3}{2}\ 0\ 110+$	$-\sqrt{3.7}/17\sqrt{2.13}$
$\tilde{1}\ \frac{3}{2}\ \frac{3}{2}\ 0\ 111+$	0
$\tilde{1}\ \frac{3}{2}\ \frac{3}{2}\ 1\ 000-$	0
$\tilde{1}\ \frac{3}{2}\ \frac{3}{2}\ 1\ 010-$	$-3.7/17\sqrt{2.5.11}$
$\tilde{1}\ \frac{3}{2}\ \frac{3}{2}\ 1\ 011-$	0
$\tilde{1}\ \frac{3}{2}\ \frac{3}{2}\ 1\ 100-$	0
$\tilde{1}\ \frac{3}{2}\ \frac{3}{2}\ 1\ 110-$	$+\sqrt{2.3.7}/17\sqrt{13}$
$\tilde{1}\ \frac{3}{2}\ \frac{3}{2}\ 1\ 111-$	0

$8\ 5\ 3$

$0\ 1\ 1\ 0\ 000+$	$-\sqrt{7}/\sqrt{2.11.13.17}$
$0\ 1\ 1\ 0\ 010+$	$-3\sqrt{2.5}/\sqrt{11.13.17}$
$0\ \tilde{1}\ \tilde{1}\ 0\ 000+$	$+3/\sqrt{2.13.17}$
$1\ 1\ 1\ 0\ 000-$	$+\sqrt{7}/2\sqrt{2.3.11.13.17}$
$1\ 1\ 1\ 0\ 010-$	$-9\sqrt{3.5}/2\sqrt{2.11.13.17}$
$1\ 1\ 1\ 0\ 100-$	$+\sqrt{5.11}/4\sqrt{2.3.17}$
$1\ 1\ 1\ 0\ 110-$	0
$1\ 1\ \tilde{1}\ 0\ 000+$	$-\sqrt{5.7}/\sqrt{2.11.13.17}$
$1\ 1\ \tilde{1}\ 0\ 010+$	$-9.3/2\sqrt{2.11.13.17}$
$1\ 1\ \tilde{1}\ 0\ 100+$	$-\sqrt{11}/4\sqrt{2.17}$
$1\ 1\ \tilde{1}\ 0\ 110+$	0
$1\ 2\ 1\ 0\ 000+$	$-\sqrt{5}/2\sqrt{3.13.17}$
$1\ 2\ 1\ 0\ 100+$	$-5/2\sqrt{3.7.17}$
$1\ 2\ \tilde{1}\ 0\ 000-$	$+\sqrt{3}/2\sqrt{13.17}$
$1\ 2\ \tilde{1}\ 0\ 100-$	$+\sqrt{3.5}/2\sqrt{7.17}$
$1\ \tilde{1}\ 1\ 0\ 000+$	$-\sqrt{2.5}/\sqrt{13.17}$
$1\ \tilde{1}\ 1\ 0\ 100+$	$+5/4\sqrt{2.7.17}$

SO_3–O 3jm Factors (cont.)

8 5 3	
1 1̃ 1̃ 0 000−	$+3\sqrt{3}/2\sqrt{2.13.17}$
1 1̃ 1̃ 0 100−	$-3\sqrt{3.5}/4\sqrt{2.7.17}$
1 1̃ 0̃ 0 000+	$-\sqrt{3}/2\sqrt{13.17}$
1 1̃ 0̃ 0 100+	$-\sqrt{3.5}/2\sqrt{7.17}$
2 1 1 0 000+	$-\sqrt{5}/2.11\sqrt{13.17}$
2 1 1 0 010+	$-9.5.5/4.11\sqrt{7.13.17}$
2 1 1 0 100+	$+\sqrt{3.7}/2\sqrt{11.17}$
2 1 1 0 110+	$+\sqrt{3.5}/4\sqrt{11.17}$
2 1 1̃ 0 000−	$-3\sqrt{13}/4.11\sqrt{17}$
2 1 1̃ 0 010−	$-9.3.5\sqrt{5}/4.11\sqrt{7.13.17}$
2 1 1̃ 0 100−	$-\sqrt{3.5.7}/4\sqrt{11.17}$
2 1 1̃ 0 110−	$+3\sqrt{3}/4\sqrt{11.17}$
2 2 0̃ 0 000−	$-\sqrt{3.5}/\sqrt{2.7.11.13.17}$
2 2 0̃ 0 100−	$-1/\sqrt{2.17}$
2 1̃ 1 0 000−	$-2.3.5/\sqrt{7.11.13.17}$
2 1̃ 1 0 100−	0
2 1̃ 1̃ 0 000+	$+9.3\sqrt{5}/4\sqrt{7.11.13.17}$
2 1̃ 1̃ 0 100+	$+\sqrt{3}/4\sqrt{17}$
1̃ 1 1 0 000+	$+8.3\sqrt{2}/11.17$
1̃ 1 1 0 010+	$-3\sqrt{5}/2.11\sqrt{2.7}$
1̃ 1 1 0 100+	$-\sqrt{3.5.7.11}/4.17\sqrt{2.13}$
1̃ 1 1 0 110+	0
1̃ 1 1̃ 0 000−	$+7\sqrt{3.5}/2.11.17\sqrt{2}$
1̃ 1 1̃ 0 010−	$-9\sqrt{3}/2.11\sqrt{2.7}$
1̃ 1 1̃ 0 100−	$+9\sqrt{7.11}/4.17\sqrt{2.13}$
1̃ 1 1̃ 0 110−	0
1̃ 1 0̃ 0 000+	$-\sqrt{3.5}/2.11.17$
1̃ 1 0̃ 0 010+	$+3\sqrt{3}/11\sqrt{7}$
1̃ 1 0̃ 0 100+	$+3\sqrt{7.11}/2.17\sqrt{13}$
1̃ 1 0̃ 0 110+	0
1̃ 2 1 0 000−	$-3.5\sqrt{3.5}/2.17\sqrt{7.11}$
1̃ 2 1 0 100−	$-3.5/2.17\sqrt{13}$
1̃ 2 1̃ 0 000+	$-23\sqrt{3}/2.17\sqrt{7.11}$
1̃ 2 1̃ 0 100+	$+9\sqrt{5}/2.17\sqrt{13}$
1̃ 1̃ 1 0 000−	$+3\sqrt{3.5}/2.17\sqrt{2.7.11}$
1̃ 1̃ 1 0 100−	$-9.5/4.17\sqrt{2.13}$
1̃ 1̃ 1̃ 0 000+	$+3\sqrt{7}/17\sqrt{2.11}$
1̃ 1̃ 1̃ 0 100+	$+9\sqrt{3.5}/4.17\sqrt{2.13}$

8 5 4	
0 1 1 0 000−	$-7\sqrt{7}/3\sqrt{2.11.13.17}$
0 1 1 0 010−	$+9\sqrt{2}/\sqrt{5.11.13.17}$
0 2 2 0 000−	$-4\sqrt{7}/3\sqrt{5.13.17}$
0 1̃ 1̃ 0 000−	$-\sqrt{5.7}/\sqrt{2.3.13.17}$
1 1 0 0 000−	$+7\sqrt{5.7}/2.9\sqrt{11.13.17}$
1 1 0 0 010−	$-3/\sqrt{11.13.17}$
1 1 0 0 100−	$-5\sqrt{11}/2.9\sqrt{17}$
1 1 0 0 110−	0
1 1 1 0 000+	$+7\sqrt{7}/2\sqrt{2.3.11.13.17}$
1 1 1 0 010+	$+3\sqrt{3.5}/2\sqrt{2.11.13.17}$

8 5 4	
1 1 1 0 100+	$+\sqrt{5.11}/4\sqrt{2.3.17}$
1 1 1 0 110+	0
1 1 2 0 000−	$-7.19/2.9\sqrt{11.13.17}$
1 1 2 0 010−	$+\sqrt{5.7}/\sqrt{11.13.17}$
1 1 2 0 100−	$-\sqrt{5.7}/2.9\sqrt{11.17}$
1 1 2 0 110−	$+2/\sqrt{11.17}$
1 1 1̃ 0 000−	$+7\sqrt{2}/3\sqrt{3.11.13.17}$
1 1 1̃ 0 010−	$+29\sqrt{7}/2\sqrt{2.3.5.11.13.17}$
1 1 1̃ 0 100−	$-\sqrt{5.7}/4.3\sqrt{2.3.11.17}$
1 1 1̃ 0 110−	$-4\sqrt{2}/\sqrt{3.11.17}$
1 2 1 0 000−	$+\sqrt{5.13}/2.3\sqrt{3.17}$
1 2 1 0 100−	$-\sqrt{7}/2.3\sqrt{3.17}$
1 2 1̃ 0 000+	$-\sqrt{7}/2.3\sqrt{5.13.17}$
1 2 1̃ 0 100+	$-1/2.3\sqrt{17}$
1 1̃ 1 0 000−	$+\sqrt{13}/3\sqrt{2.5.17}$
1 1̃ 1 0 100−	$+\sqrt{7}/4.3\sqrt{2.17}$
1 1̃ 2 0 000+	$+\sqrt{7}/2.3\sqrt{5.13.17}$
1 1̃ 2 0 100+	$+1/2.3\sqrt{17}$
1 1̃ 1̃ 0 000+	$-5\sqrt{5.7}/2.3\sqrt{2.13.17}$
1 1̃ 1̃ 0 100+	$-5/4.3\sqrt{2.17}$
2 1 1 0 000−	$+71\sqrt{5}/2.3.11\sqrt{13.17}$
2 1 1 0 010−	$-3\sqrt{7}/4.11\sqrt{13.17}$
2 1 1 0 100−	$+\sqrt{7}/2\sqrt{3.11.17}$
2 1 1 0 110−	$+9\sqrt{3}/4\sqrt{5.11.17}$
2 1 1̃ 0 000+	$+5\sqrt{5.7}/4.11\sqrt{3.13.17}$
2 1 1̃ 0 010+	$+83\sqrt{3}/4.11\sqrt{13.17}$
2 1 1̃ 0 100+	$-7/4\sqrt{11.17}$
2 1 1̃ 0 110+	$+3\sqrt{7}/4\sqrt{5.11.17}$
2 2 0 0 000−	$-5\sqrt{5.7}/3\sqrt{2.11.13.17}$
2 2 0 0 100−	$-1/\sqrt{2.3.17}$
2 2 2 0 000−	$-\sqrt{13}/3\sqrt{11.17}$
2 2 2 0 100−	$+\sqrt{7}/\sqrt{3.5.17}$
2 1̃ 1 0 000+	$+\sqrt{7}/\sqrt{11.13.17}$
2 1̃ 1 0 100+	$-\sqrt{3}/\sqrt{5.17}$
2 1̃ 1̃ 0 000−	$-37/4\sqrt{3.11.13.17}$
2 1̃ 1̃ 0 100−	$+\sqrt{7}/4\sqrt{5.17}$
1̃ 1 1 0 000−	$-47/11.17\sqrt{2}$
1̃ 1 1 0 010−	$-3.7\sqrt{7}/2.11.17\sqrt{2.5}$
1̃ 1 1 0 100−	$-41\sqrt{5.7}/4.17\sqrt{2.3.11.13}$
1̃ 1 1 0 110−	$+4.3\sqrt{2.3}/17\sqrt{11.13}$
1̃ 1 2 0 000+	$+29\sqrt{7}/2.11.17$
1̃ 1 2 0 010+	$+3.13/11.17\sqrt{5}$
1̃ 1 2 0 100+	$-7\sqrt{5}/2.17\sqrt{3.11.13}$
1̃ 1 2 0 110+	$+2.3\sqrt{3.7}/17\sqrt{11.13}$
1̃ 1 1̃ 0 000+	$-3.5\sqrt{7}/2.11.17\sqrt{2}$
1̃ 1 1̃ 0 010+	$-9/2.11\sqrt{2.5}$
1̃ 1 1̃ 0 100+	$+7\sqrt{5.11}/4.17\sqrt{2.3.13}$
1̃ 1 1̃ 0 110+	0
1̃ 2 1 0 000+	$+3\sqrt{3.7}/2.17\sqrt{5.11}$

SO_3–O 3jm Factors (cont.)

8 5 4	
$\tilde{1}$ 2 1 0 100+	$-23/2.17\sqrt{13}$
$\tilde{1}$ 2 $\tilde{1}$ 0 000−	$-3/2.17\sqrt{5.11}$
$\tilde{1}$ 2 $\tilde{1}$ 0 100−	$+\sqrt{7.13}/2.17\sqrt{3}$
$\tilde{1}$ $\tilde{1}$ 0 0 000−	$-\sqrt{7}/2.17\sqrt{11}$
$\tilde{1}$ $\tilde{1}$ 0 0 100−	$-5\sqrt{3.5}/2.17\sqrt{13}$
$\tilde{1}$ $\tilde{1}$ 1 0 000+	$+7\sqrt{3.7}/2.17\sqrt{2.5.11}$
$\tilde{1}$ $\tilde{1}$ 1 0 100+	$-9.5/4.17\sqrt{2.13}$
$\tilde{1}$ $\tilde{1}$ 2 0 000−	$-53/2.17\sqrt{5.11}$
$\tilde{1}$ $\tilde{1}$ 2 0 100−	$+\sqrt{3.7}/2.17\sqrt{13}$
$\tilde{1}$ $\tilde{1}$ $\tilde{1}$ 0 000−	$+2.3\sqrt{2.3}/17\sqrt{5.11}$
$\tilde{1}$ $\tilde{1}$ $\tilde{1}$ 0 100−	$+\sqrt{7.13}/4.17\sqrt{2}$
8 5 5	
0 1 1 0 000+	$-9.7\sqrt{5}/11\sqrt{2.13.17.19}$
0 1 1 0 010+	$-2.7\sqrt{2.7}/11\sqrt{13.17.19}$
0 1 1 0 011+	$-3\sqrt{2.19}/11\sqrt{5.13.17}$
0 2 2 0 000+	$+4.3\sqrt{3}/\sqrt{5.13.17.19}$
0 $\tilde{1}$ $\tilde{1}$ 0 000+	$+3\sqrt{5}/\sqrt{2.13.17.19}$
1 1 1 0 000−	0
1 1 1 0 010−	$-7\sqrt{7}/\sqrt{2.3.13.17.19}$
1 1 1 0 011−	0
1 1 1 0 100−	0
1 1 1 0 110−	$+\sqrt{2.5}/\sqrt{3.17.19}$
1 1 1 0 111−	0
1 2 1 0 000+	$+\sqrt{3.7.19}/2\sqrt{11.13.17}$
1 2 1 0 001+	$-5\sqrt{5}/\sqrt{3.11.13.17.19}$
1 2 1 0 100+	$+\sqrt{3.5}/2\sqrt{11.17.19}$
1 2 1 0 101+	$+8\sqrt{7}/\sqrt{3.11.17.19}$
1 $\tilde{1}$ 1 0 000+	$-8\sqrt{2.7}/\sqrt{11.13.17.19}$
1 $\tilde{1}$ 1 0 001+	$-73/\sqrt{2.5.11.13.17.19}$
1 $\tilde{1}$ 1 0 100+	$+\sqrt{2.5}/\sqrt{11.17.19}$
1 $\tilde{1}$ 1 0 101+	$-2\sqrt{2.7}/\sqrt{11.17.19}$
1 $\tilde{1}$ 2 0 000−	$+7\sqrt{3}/2\sqrt{5.13.17.19}$
1 $\tilde{1}$ 2 0 100−	$+\sqrt{3.7}/2\sqrt{17.19}$
1 $\tilde{1}$ $\tilde{1}$ 0 000−	0
1 $\tilde{1}$ $\tilde{1}$ 0 100−	0
2 1 1 0 000+	$+3.5\sqrt{7.19}/4.11\sqrt{11.13.17}$
2 1 1 0 010+	$-677\sqrt{5}/4.11\sqrt{11.13.17.19}$
2 1 1 0 011+	$+4.9\sqrt{7}/11\sqrt{11.13.17.19}$
2 1 1 0 100+	$-5.7\sqrt{3.5}/4.11\sqrt{17.19}$
2 1 1 0 110+	$-7\sqrt{3.7}/4.11\sqrt{17.19}$
2 1 1 0 111+	$+4.13\sqrt{3}/11\sqrt{5.17.19}$
2 2 2 0 000+	$+7\sqrt{3.7}/\sqrt{11.13.17.19}$
2 2 2 0 100+	$+1/\sqrt{5.17.19}$
2 $\tilde{1}$ 1 0 000−	$-3.5\sqrt{5}/4\sqrt{13.17.19}$
2 $\tilde{1}$ 1 0 001−	$+3\sqrt{7}/4\sqrt{13.17.19}$
2 $\tilde{1}$ 1 0 100−	$+\sqrt{3.7}/4\sqrt{11.17.19}$
2 $\tilde{1}$ 1 0 101−	$-3\sqrt{3}/4\sqrt{5.11.17.19}$
2 $\tilde{1}$ $\tilde{1}$ 0 000+	$+9\sqrt{7}/4\sqrt{11.13.17.19}$
2 $\tilde{1}$ $\tilde{1}$ 0 100+	$-\sqrt{3.19}/4\sqrt{5.17}$

8 5 5	
$\tilde{1}$ 1 1 0 000+	$-3\sqrt{2.5.7.19}/11.17\sqrt{11}$
$\tilde{1}$ 1 1 0 010+	$+3.37/11.17\sqrt{2.11.19}$
$\tilde{1}$ 1 1 0 011+	$-2.3.13\sqrt{2.7}/11.17\sqrt{5.11.19}$
$\tilde{1}$ 1 1 0 100+	$+3.5.7\sqrt{2.3}/11.17\sqrt{13.19}$
$\tilde{1}$ 1 1 0 110+	$+2\sqrt{2.3.5.7}/11.17\sqrt{13.19}$
$\tilde{1}$ 1 1 0 111+	$+4.9.9\sqrt{2.3}/11.17\sqrt{13.19}$
$\tilde{1}$ 2 1 0 000−	$+71\sqrt{3}/2.11.17\sqrt{19}$
$\tilde{1}$ 2 1 0 001−	$-9.3\sqrt{3.7}/11.17\sqrt{5.19}$
$\tilde{1}$ 2 1 0 100−	$+3.7\sqrt{5.7}/2.17\sqrt{11.13.19}$
$\tilde{1}$ 2 1 0 101−	$-8.2.3/17\sqrt{11.13.19}$
$\tilde{1}$ $\tilde{1}$ 1 0 000−	$+5\sqrt{2.3}/11.17\sqrt{19}$
$\tilde{1}$ $\tilde{1}$ 1 0 001−	$+3\sqrt{3.7.19}/11.17\sqrt{2.5}$
$\tilde{1}$ $\tilde{1}$ 1 0 100−	$+2.9\sqrt{2.5.7}/17\sqrt{11.13.19}$
$\tilde{1}$ $\tilde{1}$ 1 0 101−	$+9.5\sqrt{2}/17\sqrt{11.13.19}$
$\tilde{1}$ $\tilde{1}$ 2 0 000+	$-29\sqrt{3.7}/2.17\sqrt{5.11.19}$
$\tilde{1}$ $\tilde{1}$ 2 0 100+	$-9.3/2.17\sqrt{13.19}$
$\tilde{1}$ $\tilde{1}$ $\tilde{1}$ 0 000+	$+3.13\sqrt{2.7}/17\sqrt{5.11.19}$
$\tilde{1}$ $\tilde{1}$ $\tilde{1}$ 0 100+	$-9\sqrt{2.3}/17\sqrt{13.19}$
8 $\frac{11}{2}$ $\frac{5}{2}$	
0 $\frac{3}{2}$ $\frac{3}{2}$ 0 000−	$-1/\sqrt{13.17}$
0 $\frac{3}{2}$ $\frac{3}{2}$ 0 010−	$+\sqrt{2.5}/\sqrt{13.17}$
0 $\tilde{\frac{1}{2}}$ $\tilde{\frac{1}{2}}$ 0 000−	$+\sqrt{2}/\sqrt{13.17}$
1 $\frac{1}{2}$ $\frac{3}{2}$ 0 000−	$-\sqrt{7.11}/4.3\sqrt{13.17}$
1 $\frac{1}{2}$ $\frac{3}{2}$ 0 100−	$-\sqrt{5.11}/4.3\sqrt{17}$
1 $\frac{3}{2}$ $\frac{3}{2}$ 0 000+	$+3/2\sqrt{2.13.17}$
1 $\frac{3}{2}$ $\frac{3}{2}$ 0 010+	$-3\sqrt{5}/2\sqrt{13.17}$
1 $\frac{3}{2}$ $\frac{3}{2}$ 0 100+	0
1 $\frac{3}{2}$ $\frac{3}{2}$ 0 110+	0
1 $\frac{3}{2}$ $\frac{3}{2}$ 1 000+	$-1/3\sqrt{2.13.17}$
1 $\frac{3}{2}$ $\frac{3}{2}$ 1 010+	$-3\sqrt{5}/2\sqrt{13.17}$
1 $\frac{3}{2}$ $\frac{3}{2}$ 1 100+	$-11\sqrt{5}/2.3\sqrt{2.7.17}$
1 $\frac{3}{2}$ $\frac{3}{2}$ 1 110+	0
1 $\frac{3}{2}$ $\tilde{\frac{1}{2}}$ 0 000−	$-2\sqrt{5}/3\sqrt{13.17}$
1 $\frac{3}{2}$ $\tilde{\frac{1}{2}}$ 0 010−	$+3/\sqrt{2.13.17}$
1 $\frac{3}{2}$ $\tilde{\frac{1}{2}}$ 0 100−	$-11/2.3\sqrt{7.17}$
1 $\frac{3}{2}$ $\tilde{\frac{1}{2}}$ 0 110−	0
1 $\tilde{\frac{1}{2}}$ $\frac{3}{2}$ 0 000−	$-5\sqrt{5}/4\sqrt{13.17}$
1 $\tilde{\frac{1}{2}}$ $\frac{3}{2}$ 0 100−	$+5/4\sqrt{7.17}$
1 $\tilde{\frac{1}{2}}$ $\tilde{\frac{1}{2}}$ 0 000+	$-1/2\sqrt{13.17}$
1 $\tilde{\frac{1}{2}}$ $\tilde{\frac{1}{2}}$ 0 100+	$+\sqrt{5}/\sqrt{7.17}$
2 $\frac{1}{2}$ $\frac{3}{2}$ 0 000+	$+\sqrt{5}/4\sqrt{2.13.17}$
2 $\frac{1}{2}$ $\frac{3}{2}$ 0 100+	$+\sqrt{7.11}/4\sqrt{2.3.17}$
2 $\frac{3}{2}$ $\frac{3}{2}$ 0 000−	$+3.5\sqrt{5}/4\sqrt{7.11.13.17}$
2 $\frac{3}{2}$ $\frac{3}{2}$ 0 010−	$-3.5.5/2\sqrt{2.7.11.13.17}$
2 $\frac{3}{2}$ $\frac{3}{2}$ 0 100−	$-1/4\sqrt{3.17}$
2 $\frac{3}{2}$ $\frac{3}{2}$ 0 110−	$+\sqrt{5}/2\sqrt{2.3.17}$
2 $\frac{3}{2}$ $\tilde{\frac{1}{2}}$ 0 000+	$+\sqrt{13}/\sqrt{2.7.11.17}$
2 $\frac{3}{2}$ $\tilde{\frac{1}{2}}$ 0 010+	$-3.5\sqrt{5}/2\sqrt{7.11.13.17}$
2 $\frac{3}{2}$ $\tilde{\frac{1}{2}}$ 0 100+	$+\sqrt{5}/\sqrt{2.3.17}$

SO_3–O 3jm Factors (cont.)

8	**$\frac{11}{2}$**	**$\frac{5}{2}$**			
2	$\frac{3}{2}$	$\frac{\tilde{1}}{2}$	0	110+	$+1/2\sqrt{3.17}$
2	$\frac{\tilde{1}}{2}$	$\frac{3}{2}$	0	000+	$+3.5\sqrt{11}/4\sqrt{2.7.13.17}$
2	$\frac{\tilde{1}}{2}$	$\frac{3}{2}$	0	100+	$+\sqrt{5}/4\sqrt{2.3.17}$
$\tilde{1}$	$\frac{1}{2}$	$\frac{3}{2}$	0	000+	$+23/4.3.17$
$\tilde{1}$	$\frac{1}{2}$	$\frac{3}{2}$	0	100+	$-\sqrt{3.5.7.11}/4.17\sqrt{13}$
$\tilde{1}$	$\frac{1}{2}$	$\frac{\tilde{1}}{2}$	0	000−	$+\sqrt{5}/2.3.17$
$\tilde{1}$	$\frac{1}{2}$	$\frac{\tilde{1}}{2}$	0	100−	$+\sqrt{3.7.11}/17\sqrt{13}$
$\tilde{1}$	$\frac{3}{2}$	$\frac{3}{2}$	0	000−	$-19/2.3\sqrt{2.7.11}$
$\tilde{1}$	$\frac{3}{2}$	$\frac{3}{2}$	0	010−	$-\sqrt{5}/2\sqrt{7.11}$
$\tilde{1}$	$\frac{3}{2}$	$\frac{3}{2}$	0	100−	0
$\tilde{1}$	$\frac{3}{2}$	$\frac{3}{2}$	0	110−	0
$\tilde{1}$	$\frac{3}{2}$	$\frac{3}{2}$	1	000+	$-5\sqrt{7}/3.17\sqrt{2.11}$
$\tilde{1}$	$\frac{3}{2}$	$\frac{3}{2}$	1	010+	$-\sqrt{5}/2\sqrt{7.11}$
$\tilde{1}$	$\frac{3}{2}$	$\frac{3}{2}$	1	100+	$-11\sqrt{3.5}/2.17\sqrt{2.13}$
$\tilde{1}$	$\frac{3}{2}$	$\frac{3}{2}$	1	110+	0
$\tilde{1}$	$\frac{3}{2}$	$\frac{\tilde{1}}{2}$	0	000+	$-2.5\sqrt{5}/3.17\sqrt{7.11}$
$\tilde{1}$	$\frac{3}{2}$	$\frac{\tilde{1}}{2}$	0	010+	$-3/\sqrt{2.7.11}$
$\tilde{1}$	$\frac{3}{2}$	$\frac{\tilde{1}}{2}$	0	100+	$+11\sqrt{3}/2.17\sqrt{13}$
$\tilde{1}$	$\frac{3}{2}$	$\frac{\tilde{1}}{2}$	0	110+	0
$\tilde{1}$	$\frac{\tilde{1}}{2}$	$\frac{3}{2}$	0	000+	$+\sqrt{5.11}/4.17\sqrt{7}$
$\tilde{1}$	$\frac{\tilde{1}}{2}$	$\frac{3}{2}$	0	100+	$+3.5\sqrt{3}/4.17\sqrt{13}$
8	**$\frac{11}{2}$**	**$\frac{7}{2}$**			
0	$\frac{1}{2}$	$\frac{1}{2}$	0	000+	$-7/3\sqrt{2.3.13.17}$
0	$\frac{3}{2}$	$\frac{3}{2}$	0	000+	$-2.5\sqrt{5}/3\sqrt{3.11.13.17}$
0	$\frac{3}{2}$	$\frac{3}{2}$	0	010+	$-3\sqrt{2.3}/\sqrt{11.13.17}$
0	$\frac{\tilde{1}}{2}$	$\frac{\tilde{1}}{2}$	0	000+	$+\sqrt{11}/\sqrt{2.13.17}$
1	$\frac{1}{2}$	$\frac{1}{2}$	0	000−	$-7/2.9\sqrt{3.13.17}$
1	$\frac{1}{2}$	$\frac{1}{2}$	0	100−	$-11\sqrt{5.7}/4.9\sqrt{3.17}$
1	$\frac{1}{2}$	$\frac{3}{2}$	0	000+	$-\sqrt{5.7.17}/4.9\sqrt{3.13}$
1	$\frac{1}{2}$	$\frac{3}{2}$	0	100+	$-11/2.9\sqrt{3.17}$
1	$\frac{3}{2}$	$\frac{1}{2}$	0	000+	$+127\sqrt{7}/4.9\sqrt{3.11.13.17}$
1	$\frac{3}{2}$	$\frac{1}{2}$	0	010+	$+\sqrt{3.5.7}/2\sqrt{2.11.13.17}$
1	$\frac{3}{2}$	$\frac{1}{2}$	0	100+	$-\sqrt{5.11}/9\sqrt{3.17}$
1	$\frac{3}{2}$	$\frac{1}{2}$	0	110+	0
1	$\frac{3}{2}$	$\frac{3}{2}$	0	000−	$-37\sqrt{2.5}/9\sqrt{3.11.13.17}$
1	$\frac{3}{2}$	$\frac{3}{2}$	0	010−	$-5/3\sqrt{3.11.13.17}$
1	$\frac{3}{2}$	$\frac{3}{2}$	0	100−	$+\sqrt{7}/2.9\sqrt{2.3.11.17}$
1	$\frac{3}{2}$	$\frac{3}{2}$	0	110−	$-2\sqrt{5.7}/3\sqrt{3.11.17}$
1	$\frac{3}{2}$	$\frac{3}{2}$	1	000−	$+8.2\sqrt{2.5}/9\sqrt{3.11.13.17}$
1	$\frac{3}{2}$	$\frac{3}{2}$	1	010−	$+5.19/2.3\sqrt{3.11.13.17}$
1	$\frac{3}{2}$	$\frac{3}{2}$	1	100−	$-5.13/9\sqrt{2.3.7.11.17}$
1	$\frac{3}{2}$	$\frac{3}{2}$	1	110−	$+\sqrt{5.7}/3\sqrt{3.11.17}$
1	$\frac{3}{2}$	$\frac{\tilde{1}}{2}$	0	000+	$-23\sqrt{5}/4.3\sqrt{11.13.17}$
1	$\frac{3}{2}$	$\frac{\tilde{1}}{2}$	0	010+	$+71/2.3\sqrt{2.11.13.17}$
1	$\frac{3}{2}$	$\frac{\tilde{1}}{2}$	0	100+	$-4/3\sqrt{7.11.17}$
1	$\frac{3}{2}$	$\frac{\tilde{1}}{2}$	0	110+	$-\sqrt{2.5.7}/3\sqrt{11.17}$
1	$\frac{\tilde{1}}{2}$	$\frac{3}{2}$	0	000+	$+11\sqrt{11}/4.3\sqrt{3.13.17}$
1	$\frac{\tilde{1}}{2}$	$\frac{3}{2}$	0	100+	$+\sqrt{5.11}/2.3\sqrt{3.7.17}$
1	$\frac{\tilde{1}}{2}$	$\frac{\tilde{1}}{2}$	0	000−	$-5\sqrt{11}/2.3\sqrt{13.17}$
1	$\frac{\tilde{1}}{2}$	$\frac{\tilde{1}}{2}$	0	100−	$-\sqrt{5.11}/4.3\sqrt{7.17}$
2	$\frac{1}{2}$	$\frac{3}{2}$	0	000−	$+127/4.3\sqrt{2.3.11.13.17}$
2	$\frac{1}{2}$	$\frac{3}{2}$	0	100−	$+\sqrt{5.7}/4.3\sqrt{2.17}$
2	$\frac{3}{2}$	$\frac{1}{2}$	0	000−	$-173\sqrt{5}/2.3.11\sqrt{2.3.13.17}$
2	$\frac{3}{2}$	$\frac{1}{2}$	0	010−	$-5\sqrt{13}/4.11\sqrt{3.17}$
2	$\frac{3}{2}$	$\frac{1}{2}$	0	100−	$-7\sqrt{7}/2.3\sqrt{2.11.17}$
2	$\frac{3}{2}$	$\frac{1}{2}$	0	110−	$+\sqrt{5.7}/4\sqrt{11.17}$
2	$\frac{3}{2}$	$\frac{3}{2}$	0	000+	$-103/4.3\sqrt{3.7.13.17}$
2	$\frac{3}{2}$	$\frac{3}{2}$	0	010+	$+\sqrt{2.5}/\sqrt{3.7.13.17}$
2	$\frac{3}{2}$	$\frac{3}{2}$	0	100+	$+13\sqrt{5}/4.3\sqrt{11.17}$
2	$\frac{3}{2}$	$\frac{3}{2}$	0	110+	$+\sqrt{2}/\sqrt{11.17}$
2	$\frac{3}{2}$	$\frac{\tilde{1}}{2}$	0	000−	$+73/2.11\sqrt{2.7.13.17}$
2	$\frac{3}{2}$	$\frac{\tilde{1}}{2}$	0	010−	$-179\sqrt{5}/4.11\sqrt{7.13.17}$
2	$\frac{3}{2}$	$\frac{\tilde{1}}{2}$	0	100−	$-\sqrt{3.5}/2\sqrt{2.11.17}$
2	$\frac{3}{2}$	$\frac{\tilde{1}}{2}$	0	110−	$-\sqrt{3}/4\sqrt{11.17}$
2	$\frac{\tilde{1}}{2}$	$\frac{3}{2}$	0	000−	$-\sqrt{5.13}/4\sqrt{2.3.7.17}$
2	$\frac{\tilde{1}}{2}$	$\frac{3}{2}$	0	100−	$+\sqrt{11}/4\sqrt{2.17}$
$\tilde{1}$	$\frac{1}{2}$	$\frac{3}{2}$	0	000−	$-13\sqrt{3.5}/4.17\sqrt{11}$
$\tilde{1}$	$\frac{1}{2}$	$\frac{3}{2}$	0	100−	$-11\sqrt{7}/2.3.17\sqrt{13}$
$\tilde{1}$	$\frac{1}{2}$	$\frac{\tilde{1}}{2}$	0	000+	$+3\sqrt{5}/2.17\sqrt{11}$
$\tilde{1}$	$\frac{1}{2}$	$\frac{\tilde{1}}{2}$	0	100+	$-11\sqrt{7}/4.17\sqrt{3.13}$
$\tilde{1}$	$\frac{3}{2}$	$\frac{1}{2}$	0	000−	$+7.13/4.11.17\sqrt{3}$
$\tilde{1}$	$\frac{3}{2}$	$\frac{1}{2}$	0	010−	$-5.5\sqrt{5}/2.11.17\sqrt{2.3}$
$\tilde{1}$	$\frac{3}{2}$	$\frac{1}{2}$	0	100−	$+19\sqrt{5.7}/3.17\sqrt{11.13}$
$\tilde{1}$	$\frac{3}{2}$	$\frac{1}{2}$	0	110−	$+5\sqrt{2.7}/17\sqrt{11.13}$
$\tilde{1}$	$\frac{3}{2}$	$\frac{3}{2}$	0	000+	$+\sqrt{2.5}/17\sqrt{3.7}$
$\tilde{1}$	$\frac{3}{2}$	$\frac{3}{2}$	0	010+	$-1/17\sqrt{3.7}$
$\tilde{1}$	$\frac{3}{2}$	$\frac{3}{2}$	0	100+	$-7/2.3.17\sqrt{2.11.13}$
$\tilde{1}$	$\frac{3}{2}$	$\frac{3}{2}$	0	110+	$+2.7\sqrt{5}/17\sqrt{11.13}$
$\tilde{1}$	$\frac{3}{2}$	$\frac{3}{2}$	1	000−	$+8\sqrt{2.5.7}/11.17\sqrt{3}$
$\tilde{1}$	$\frac{3}{2}$	$\frac{3}{2}$	1	010−	$-5.59/2.11.17\sqrt{3.7}$
$\tilde{1}$	$\frac{3}{2}$	$\frac{3}{2}$	1	100−	$-59/17\sqrt{2.11.13}$
$\tilde{1}$	$\frac{3}{2}$	$\frac{3}{2}$	1	110−	$-7\sqrt{5}/17\sqrt{11.13}$
$\tilde{1}$	$\frac{3}{2}$	$\frac{\tilde{1}}{2}$	0	000−	$-3.19\sqrt{5}/4.11.17\sqrt{7}$
$\tilde{1}$	$\frac{3}{2}$	$\frac{\tilde{1}}{2}$	0	010−	$+9/2.11\sqrt{2.7}$
$\tilde{1}$	$\frac{3}{2}$	$\frac{\tilde{1}}{2}$	0	100−	$-4\sqrt{11}/17\sqrt{3.13}$
$\tilde{1}$	$\frac{3}{2}$	$\frac{\tilde{1}}{2}$	0	110−	0
$\tilde{1}$	$\frac{\tilde{1}}{2}$	$\frac{1}{2}$	0	000+	$-\sqrt{5}/2.17\sqrt{3}$
$\tilde{1}$	$\frac{\tilde{1}}{2}$	$\frac{1}{2}$	0	100+	$+5\sqrt{7.11}/4.17\sqrt{13}$
$\tilde{1}$	$\frac{\tilde{1}}{2}$	$\frac{3}{2}$	0	000−	$+47/4.17\sqrt{3.7}$
$\tilde{1}$	$\frac{\tilde{1}}{2}$	$\frac{3}{2}$	0	100−	$+\sqrt{5.11}/2.17\sqrt{13}$
8	**$\frac{11}{2}$**	**$\frac{9}{2}$**			
0	$\frac{1}{2}$	$\frac{1}{2}$	0	000−	$-7\sqrt{2.7}/3\sqrt{3.13.17.19}$
0	$\frac{3}{2}$	$\frac{3}{2}$	0	000−	$+7\sqrt{11}/\sqrt{5.13.17.19}$
0	$\frac{3}{2}$	$\frac{3}{2}$	0	001−	$-4.29\sqrt{7}/3\sqrt{3.5.11.13.17.19}$
0	$\frac{3}{2}$	$\frac{3}{2}$	0	010−	$+\sqrt{2.11}/5\sqrt{13.17.19}$
0	$\frac{3}{2}$	$\frac{3}{2}$	0	011−	$+2.9\sqrt{2.3.7}/5\sqrt{11.13.17.19}$
1	$\frac{1}{2}$	$\frac{1}{2}$	0	000+	$-7.11\sqrt{7}/2.9\sqrt{3.13.17.19}$
1	$\frac{1}{2}$	$\frac{1}{2}$	0	100+	$+11\sqrt{5}/9\sqrt{3.17.19}$

SO_3–O 3jm Factors (cont.)

$8\ \frac{11}{2}\ \frac{9}{2}$	
$1\ \frac{1}{2}\ \frac{3}{2}\ 0\ 000-$	$+7.7\sqrt{7}/4.3\sqrt{5.13.17.19}$
$1\ \frac{1}{2}\ \frac{3}{2}\ 0\ 001-$	$-7.43/9\sqrt{3.5.13.17.19}$
$1\ \frac{1}{2}\ \frac{3}{2}\ 0\ 100-$	$+13/4.3\sqrt{17.19}$
$1\ \frac{1}{2}\ \frac{3}{2}\ 0\ 101-$	$+13\sqrt{7}/2.9\sqrt{3.17.19}$
$1\ \frac{3}{2}\ \frac{1}{2}\ 0\ 000-$	$+8.4.7/9\sqrt{3.11.13.17.19}$
$1\ \frac{3}{2}\ \frac{1}{2}\ 0\ 010-$	$-\sqrt{3.19}/\sqrt{2.5.11.13.17}$
$1\ \frac{3}{2}\ \frac{1}{2}\ 0\ 100-$	$+5\sqrt{5.7.11}/2.9\sqrt{3.17.19}$
$1\ \frac{3}{2}\ \frac{1}{2}\ 0\ 110-$	0
$1\ \frac{3}{2}\ \frac{3}{2}\ 0\ 000+$	$+787/2.3\sqrt{2.5.11.13.17.19}$
$1\ \frac{3}{2}\ \frac{3}{2}\ 0\ 001+$	$+7\sqrt{2.7.13}/9\sqrt{3.5.11.17.19}$
$1\ \frac{3}{2}\ \frac{3}{2}\ 0\ 010+$	$+421/2.3.5\sqrt{11.13.17.19}$
$1\ \frac{3}{2}\ \frac{3}{2}\ 0\ 011+$	$-373\sqrt{7}/3.5\sqrt{3.11.13.17.19}$
$1\ \frac{3}{2}\ \frac{3}{2}\ 0\ 100+$	$+4\sqrt{2.7}/3\sqrt{11.17.19}$
$1\ \frac{3}{2}\ \frac{3}{2}\ 0\ 101+$	$+\sqrt{2}/9\sqrt{3.11.17.19}$
$1\ \frac{3}{2}\ \frac{3}{2}\ 0\ 110+$	$-8\sqrt{5.7}/3\sqrt{11.17.19}$
$1\ \frac{3}{2}\ \frac{3}{2}\ 0\ 111+$	$-8\sqrt{5}/3\sqrt{3.11.17.19}$
$1\ \frac{3}{2}\ \frac{3}{2}\ 1\ 000+$	$-11\sqrt{11}/\sqrt{2.5.13.17.19}$
$1\ \frac{3}{2}\ \frac{3}{2}\ 1\ 001+$	$+29\sqrt{7.11}/9\sqrt{2.3.5.13.17.19}$
$1\ \frac{3}{2}\ \frac{3}{2}\ 1\ 010+$	$+11\sqrt{11}/2.3.5\sqrt{13.17.19}$
$1\ \frac{3}{2}\ \frac{3}{2}\ 1\ 011+$	$-8\sqrt{7.11}/3.5\sqrt{3.13.17.19}$
$1\ \frac{3}{2}\ \frac{3}{2}\ 1\ 100+$	$+7\sqrt{7}/2\sqrt{2.11.17.19}$
$1\ \frac{3}{2}\ \frac{3}{2}\ 1\ 101+$	$+101\sqrt{2}/9\sqrt{3.11.17.19}$
$1\ \frac{3}{2}\ \frac{3}{2}\ 1\ 110+$	$+8\sqrt{7}/3\sqrt{5.11.17.19}$
$1\ \frac{3}{2}\ \frac{3}{2}\ 1\ 111+$	$+2.37/3\sqrt{3.5.11.17.19}$
$1\ \tilde{\frac{1}{2}}\ \frac{3}{2}\ 0\ 000-$	$+97\sqrt{11}/4.3.5\sqrt{13.17.19}$
$1\ \tilde{\frac{1}{2}}\ \frac{3}{2}\ 0\ 001-$	$-23\sqrt{7.11}/3.5\sqrt{3.13.17.19}$
$1\ \tilde{\frac{1}{2}}\ \frac{3}{2}\ 0\ 100-$	$+\sqrt{7.11}/4.3\sqrt{5.17.19}$
$1\ \tilde{\frac{1}{2}}\ \frac{3}{2}\ 0\ 101-$	$-\sqrt{11}/2.3\sqrt{3.5.17.19}$
$2\ \frac{1}{2}\ \frac{3}{2}\ 0\ 000+$	$-3.59/4\sqrt{2.11.13.17.19}$
$2\ \frac{1}{2}\ \frac{3}{2}\ 0\ 001+$	$-127\sqrt{7}/2.3\sqrt{2.3.11.13.17.19}$
$2\ \frac{1}{2}\ \frac{3}{2}\ 0\ 100+$	$+\sqrt{3.5.7}/4\sqrt{2.17.19}$
$2\ \frac{1}{2}\ \frac{3}{2}\ 0\ 101+$	$-7\sqrt{5}/2.3\sqrt{2.17.19}$
$2\ \frac{3}{2}\ \frac{1}{2}\ 0\ 000+$	$+97\sqrt{5.7}/3.11\sqrt{2.3.13.17.19}$
$2\ \frac{3}{2}\ \frac{1}{2}\ 0\ 010+$	$+5.5\sqrt{7}/2.11\sqrt{3.13.17.19}$
$2\ \frac{3}{2}\ \frac{1}{2}\ 0\ 100+$	$+5/3\sqrt{2.11.17.19}$
$2\ \frac{3}{2}\ \frac{1}{2}\ 0\ 110+$	$+53/2\sqrt{5.11.17.19}$
$2\ \frac{3}{2}\ \frac{3}{2}\ 0\ 000-$	$+127\sqrt{7}/4.11\sqrt{13.17.19}$
$2\ \frac{3}{2}\ \frac{3}{2}\ 0\ 001-$	$-271/2.3.11\sqrt{3.13.17.19}$
$2\ \frac{3}{2}\ \frac{3}{2}\ 0\ 010-$	$+17\sqrt{7.17}/2.11\sqrt{2.5.13.19}$
$2\ \frac{3}{2}\ \frac{3}{2}\ 0\ 011-$	$-227\sqrt{2}/11\sqrt{3.5.13.17.19}$
$2\ \frac{3}{2}\ \frac{3}{2}\ 0\ 100-$	$+3\sqrt{3}/4\sqrt{5.11.17.19}$
$2\ \frac{3}{2}\ \frac{3}{2}\ 0\ 101-$	$+43\sqrt{7}/2.3\sqrt{5.11.17.19}$
$2\ \frac{3}{2}\ \frac{3}{2}\ 0\ 110-$	$+29\sqrt{3}/2.5\sqrt{2.11.17.19}$
$2\ \frac{3}{2}\ \frac{3}{2}\ 0\ 111-$	$-\sqrt{2.7}/5\sqrt{11.17.19}$
$2\ \tilde{\frac{1}{2}}\ \frac{3}{2}\ 0\ 000+$	$-5\sqrt{5.7}/4\sqrt{2.13.17.19}$
$2\ \tilde{\frac{1}{2}}\ \frac{3}{2}\ 0\ 001+$	$+\sqrt{5.13}/2\sqrt{2.3.17.19}$
$2\ \tilde{\frac{1}{2}}\ \frac{3}{2}\ 0\ 100+$	$-\sqrt{3.11}/4\sqrt{2.17.19}$
$2\ \tilde{\frac{1}{2}}\ \frac{3}{2}\ 0\ 101+$	$-\sqrt{7.11}/2\sqrt{2.17.19}$
$\tilde{1}\ \frac{1}{2}\ \frac{3}{2}\ 0\ 000+$	$+9.43/4.17\sqrt{5.11.19}$
$\tilde{1}\ \frac{1}{2}\ \frac{3}{2}\ 0\ 001+$	$-\sqrt{3.7.11}/17\sqrt{5.19}$
$\tilde{1}\ \frac{1}{2}\ \frac{3}{2}\ 0\ 100+$	$+23\sqrt{7}/4.17\sqrt{3.13.19}$
$\tilde{1}\ \frac{1}{2}\ \frac{3}{2}\ 0\ 101+$	$+7.31/2.3.17\sqrt{13.19}$
$\tilde{1}\ \frac{3}{2}\ \frac{1}{2}\ 0\ 000+$	$+8.4\sqrt{7}/11.17\sqrt{3.19}$
$\tilde{1}\ \frac{3}{2}\ \frac{1}{2}\ 0\ 010+$	$-89\sqrt{7}/11.17\sqrt{2.3.5.19}$
$\tilde{1}\ \frac{3}{2}\ \frac{1}{2}\ 0\ 100+$	$+127\sqrt{5}/2.3.17\sqrt{11.13.19}$
$\tilde{1}\ \frac{3}{2}\ \frac{1}{2}\ 0\ 110+$	$+8.4.5\sqrt{2}/17\sqrt{11.13.19}$
$\tilde{1}\ \frac{3}{2}\ \frac{3}{2}\ 0\ 000-$	$-3.79\sqrt{7}/2.11.17\sqrt{2.5.19}$
$\tilde{1}\ \frac{3}{2}\ \frac{3}{2}\ 0\ 001-$	$-\sqrt{2.19}/17\sqrt{3.5}$
$\tilde{1}\ \frac{3}{2}\ \frac{3}{2}\ 0\ 010-$	$-3.23\sqrt{7}/2.5.11.17\sqrt{19}$
$\tilde{1}\ \frac{3}{2}\ \frac{3}{2}\ 0\ 011-$	$+\sqrt{19}/5.17\sqrt{3}$
$\tilde{1}\ \frac{3}{2}\ \frac{3}{2}\ 0\ 100-$	$+4\sqrt{2.19}/17\sqrt{3.11.13}$
$\tilde{1}\ \frac{3}{2}\ \frac{3}{2}\ 0\ 101-$	$-\sqrt{2.7}/3.17\sqrt{11.13.19}$
$\tilde{1}\ \frac{3}{2}\ \frac{3}{2}\ 0\ 110-$	$+8\sqrt{3.5}/17\sqrt{11.13.19}$
$\tilde{1}\ \frac{3}{2}\ \frac{3}{2}\ 0\ 111-$	$+8\sqrt{5.7}/17\sqrt{11.13.19}$
$\tilde{1}\ \frac{3}{2}\ \frac{3}{2}\ 1\ 000+$	$+3\sqrt{7}/11.17\sqrt{2.5.19}$
$\tilde{1}\ \frac{3}{2}\ \frac{3}{2}\ 1\ 001+$	$+509/11.17\sqrt{2.3.5.19}$
$\tilde{1}\ \frac{3}{2}\ \frac{3}{2}\ 1\ 010+$	$-3.131\sqrt{7}/2.5.11.17\sqrt{19}$
$\tilde{1}\ \frac{3}{2}\ \frac{3}{2}\ 1\ 011+$	$-4.7\sqrt{19}/5.11.17\sqrt{3}$
$\tilde{1}\ \frac{3}{2}\ \frac{3}{2}\ 1\ 100+$	$-199/2.17\sqrt{2.3.11.13.19}$
$\tilde{1}\ \frac{3}{2}\ \frac{3}{2}\ 1\ 101+$	$+47\sqrt{2.7}/17\sqrt{11.13.19}$
$\tilde{1}\ \frac{3}{2}\ \frac{3}{2}\ 1\ 110+$	$-8.7\sqrt{3}/17\sqrt{5.11.13.19}$
$\tilde{1}\ \frac{3}{2}\ \frac{3}{2}\ 1\ 111+$	$-2.37\sqrt{7}/17\sqrt{5.11.13.19}$
$\tilde{1}\ \tilde{\frac{1}{2}}\ \frac{1}{2}\ 0\ 000-$	$-\sqrt{7.19}/2.17\sqrt{3.5}$
$\tilde{1}\ \tilde{\frac{1}{2}}\ \frac{1}{2}\ 0\ 100-$	$-5\sqrt{11}/17\sqrt{13.19}$
$\tilde{1}\ \tilde{\frac{1}{2}}\ \frac{3}{2}\ 0\ 000+$	$+3.31\sqrt{7}/4.5.17\sqrt{19}$
$\tilde{1}\ \tilde{\frac{1}{2}}\ \frac{3}{2}\ 0\ 001+$	$+11.11/5.17\sqrt{3.19}$
$\tilde{1}\ \tilde{\frac{1}{2}}\ \frac{3}{2}\ 0\ 100+$	$+23\sqrt{3.11}/4.17\sqrt{5.13.19}$
$\tilde{1}\ \tilde{\frac{1}{2}}\ \frac{3}{2}\ 0\ 101+$	$-\sqrt{7.11.19}/2.17\sqrt{5.13}$
$8\ \frac{11}{2}\ \frac{11}{2}$	
$0\ \frac{1}{2}\ \frac{1}{2}\ 0\ 000+$	$-3.7/2\sqrt{13.17.19}$
$0\ \frac{3}{2}\ \frac{3}{2}\ 0\ 000+$	$+3.31/11\sqrt{2.13.17.19}$
$0\ \frac{3}{2}\ \frac{3}{2}\ 0\ 010+$	$+4.7\sqrt{5}/11\sqrt{13.17.19}$
$0\ \frac{3}{2}\ \frac{3}{2}\ 0\ 011+$	$-\sqrt{2.19}/11\sqrt{13.17}$
$0\ \tilde{\frac{1}{2}}\ \tilde{\frac{1}{2}}\ 0\ 000+$	$+11/2\sqrt{13.17.19}$
$1\ \frac{1}{2}\ \frac{1}{2}\ 0\ 000-$	0
$1\ \frac{1}{2}\ \frac{1}{2}\ 0\ 100-$	0
$1\ \frac{3}{2}\ \frac{1}{2}\ 0\ 000+$	$+9.3\sqrt{7}/\sqrt{2.11.13.17.19}$
$1\ \frac{3}{2}\ \frac{1}{2}\ 0\ 010+$	$+7\sqrt{5.7}/3\sqrt{11.13.17.19}$
$1\ \frac{3}{2}\ \frac{1}{2}\ 0\ 100+$	0
$1\ \frac{3}{2}\ \frac{1}{2}\ 0\ 110+$	$-\sqrt{11}/2.3\sqrt{17.19}$
$1\ \frac{3}{2}\ \frac{3}{2}\ 0\ 000-$	0
$1\ \frac{3}{2}\ \frac{3}{2}\ 0\ 010-$	$-\sqrt{2.5}/\sqrt{13.17.19}$
$1\ \frac{3}{2}\ \frac{3}{2}\ 0\ 011-$	0
$1\ \frac{3}{2}\ \frac{3}{2}\ 0\ 100-$	0
$1\ \frac{3}{2}\ \frac{3}{2}\ 0\ 110-$	$-\sqrt{7}/\sqrt{2.17.19}$
$1\ \frac{3}{2}\ \frac{3}{2}\ 0\ 111-$	0
$1\ \frac{3}{2}\ \frac{3}{2}\ 1\ 000-$	0
$1\ \frac{3}{2}\ \frac{3}{2}\ 1\ 010-$	$-11\sqrt{5}/3\sqrt{2.13.17.19}$

SO_3–O 3jm Factors (cont.)

8 $\frac{11}{2}$ $\frac{11}{2}$

$1\ \frac{3}{2}\ \frac{3}{2}$ 1 011− 0
$1\ \frac{3}{2}\ \frac{3}{2}$ 1 100− 0
$1\ \frac{3}{2}\ \frac{3}{2}$ 1 110− $+5\sqrt{7}/3\sqrt{2.17.19}$
$1\ \frac{3}{2}\ \frac{3}{2}$ 1 111− 0
$1\ \tilde{\frac{1}{2}}\ \frac{3}{2}$ 0 000+ $-7\sqrt{5}/3\sqrt{2.13.17.19}$
$1\ \tilde{\frac{1}{2}}\ \frac{3}{2}$ 0 001+ $+8/\sqrt{13.17.19}$
$1\ \tilde{\frac{1}{2}}\ \frac{3}{2}$ 0 100+ $+2\sqrt{2.7}/3\sqrt{17.19}$
$1\ \tilde{\frac{1}{2}}\ \frac{3}{2}$ 0 101+ $+\sqrt{5.7}/2\sqrt{17.19}$
$1\ \tilde{\frac{1}{2}}\ \tilde{\frac{1}{2}}$ 0 000− 0
$1\ \tilde{\frac{1}{2}}\ \tilde{\frac{1}{2}}$ 0 100− 0
$2\ \frac{3}{2}\ \frac{1}{2}$ 0 000− $-149\sqrt{5}/4.11\sqrt{13.17.19}$
$2\ \frac{3}{2}\ \frac{1}{2}$ 0 010− $-457/4.11\sqrt{2.13.17.19}$
$2\ \frac{3}{2}\ \frac{1}{2}$ 0 100− $+\sqrt{7.19}/4\sqrt{3.11.17}$
$2\ \frac{3}{2}\ \frac{1}{2}$ 0 110− $-5\sqrt{5.7}/4\sqrt{2.3.11.17.19}$
$2\ \frac{3}{2}\ \frac{3}{2}$ 0 000+ $+8.7\sqrt{2.5.7}/11\sqrt{11.13.17.19}$
$2\ \frac{3}{2}\ \frac{3}{2}$ 0 010+ $-3.83\sqrt{7}/4.11\sqrt{11.13.17.19}$
$2\ \frac{3}{2}\ \frac{3}{2}$ 0 011+ $-4.3\sqrt{2.5.7}/11\sqrt{11.13.17.19}$
$2\ \frac{3}{2}\ \frac{3}{2}$ 0 100+ $+8\sqrt{2}/11\sqrt{3.17.19}$
$2\ \frac{3}{2}\ \frac{3}{2}$ 0 110+ $+13\sqrt{5}/4.11\sqrt{3.17.19}$
$2\ \frac{3}{2}\ \frac{3}{2}$ 0 111+ $-4.13\sqrt{2}/11\sqrt{3.17.19}$
$2\ \tilde{\frac{1}{2}}\ \frac{3}{2}$ 0 000− $+23\sqrt{7}/4\sqrt{11.13.17.19}$
$2\ \tilde{\frac{1}{2}}\ \frac{3}{2}$ 0 001− $+9\sqrt{5.7}/4\sqrt{2.11.13.17.19}$
$2\ \tilde{\frac{1}{2}}\ \frac{3}{2}$ 0 100− $+5\sqrt{5}/4\sqrt{3.17.19}$
$2\ \tilde{\frac{1}{2}}\ \frac{3}{2}$ 0 101− $-\sqrt{17}/4\sqrt{2.3.19}$
$\tilde{1}\ \frac{3}{2}\ \frac{1}{2}$ 0 000− $+67/11.17\sqrt{2.19}$
$\tilde{1}\ \frac{3}{2}\ \frac{1}{2}$ 0 010− $+8\sqrt{5}/3.11.17\sqrt{19}$
$\tilde{1}\ \frac{3}{2}\ \frac{1}{2}$ 0 100− $-4\sqrt{2.3.5.7}/17\sqrt{11.13.19}$
$\tilde{1}\ \frac{3}{2}\ \frac{1}{2}$ 0 110− $-\sqrt{3.7.19}/2.17\sqrt{11.13}$
$\tilde{1}\ \frac{3}{2}\ \frac{3}{2}$ 0 000+ $+2.139\sqrt{7}/11.17\sqrt{11.19}$
$\tilde{1}\ \frac{3}{2}\ \frac{3}{2}$ 0 010+ $+5.7\sqrt{2.5.7}/3.11.17\sqrt{11.19}$
$\tilde{1}\ \frac{3}{2}\ \frac{3}{2}$ 0 011+ $+4.13\sqrt{7}/11.17\sqrt{11.19}$
$\tilde{1}\ \frac{3}{2}\ \frac{3}{2}$ 0 100+ $-4\sqrt{3.5}/11.17\sqrt{13.19}$
$\tilde{1}\ \frac{3}{2}\ \frac{3}{2}$ 0 110+ $+3.83\sqrt{3}/11.17\sqrt{2.13.19}$
$\tilde{1}\ \frac{3}{2}\ \frac{3}{2}$ 0 111+ $-8.9.3\sqrt{3.5}/11.17\sqrt{13.19}$
$\tilde{1}\ \frac{3}{2}\ \frac{3}{2}$ 1 000− 0
$\tilde{1}\ \frac{3}{2}\ \frac{3}{2}$ 1 010− $-7\sqrt{5.7}/3.17\sqrt{2.11.19}$
$\tilde{1}\ \frac{3}{2}\ \frac{3}{2}$ 1 011− 0
$\tilde{1}\ \frac{3}{2}\ \frac{3}{2}$ 1 100− 0
$\tilde{1}\ \frac{3}{2}\ \frac{3}{2}$ 1 110− $-\sqrt{3.13}/17\sqrt{2.19}$
$\tilde{1}\ \frac{3}{2}\ \frac{3}{2}$ 1 111− 0
$\tilde{1}\ \tilde{\frac{1}{2}}\ \frac{1}{2}$ 0 000+ $-\sqrt{2.5}/3.17\sqrt{19}$
$\tilde{1}\ \tilde{\frac{1}{2}}\ \frac{1}{2}$ 0 100+ $-2\sqrt{2.3.7.11}/17\sqrt{13.19}$
$\tilde{1}\ \tilde{\frac{1}{2}}\ \frac{3}{2}$ 0 000− $-7\sqrt{5.7}/3.17\sqrt{2.11.19}$
$\tilde{1}\ \tilde{\frac{1}{2}}\ \frac{3}{2}$ 0 001− $+\sqrt{7.19}/17\sqrt{11}$
$\tilde{1}\ \tilde{\frac{1}{2}}\ \frac{3}{2}$ 0 100− $+2\sqrt{2.3}/17\sqrt{13.19}$
$\tilde{1}\ \tilde{\frac{1}{2}}\ \frac{3}{2}$ 0 101 $+3\sqrt{3.5}/2.17\sqrt{13.19}$

8 6 2

$0\ 2\ 2$ 0 000+ $+\sqrt{3.11}/\sqrt{5.13.17}$
$0\ \tilde{1}\ \tilde{1}$ 0 000+ $+2.3/\sqrt{7.13.17}$
$0\ \tilde{1}\ \tilde{1}$ 0 010+ $+2\sqrt{11}/\sqrt{5.7.13.17}$
$1\ 1\ 2$ 0 000+ $+\sqrt{3.11}/2\sqrt{2.5.13.17}$
$1\ 1\ 2$ 0 100+ $+\sqrt{3.11}/2\sqrt{2.7.17}$
$1\ 1\ \tilde{1}$ 0 000+ $-\sqrt{11}/2\sqrt{5.13.17}$
$1\ 1\ \tilde{1}$ 0 100+ $-\sqrt{11}/2\sqrt{7.17}$
$1\ 2\ \tilde{1}$ 0 000− $-3\sqrt{3.11}/2\sqrt{5.13.17}$
$1\ 2\ \tilde{1}$ 0 100− 0
$1\ \tilde{1}\ 2$ 0 000− $-3.5\sqrt{3}/2\sqrt{2.7.13.17}$
$1\ \tilde{1}\ 2$ 0 010− $-11\sqrt{11}/\sqrt{2.3.5.7.13.17}$
$1\ \tilde{1}\ 2$ 0 100− $+3\sqrt{3.5}/2.7\sqrt{2.17}$
$1\ \tilde{1}\ 2$ 0 110− $-\sqrt{2.11}/7\sqrt{3.17}$
$1\ \tilde{1}\ \tilde{1}$ 0 000− $-3\sqrt{3}/2\sqrt{7.13.17}$
$1\ \tilde{1}\ \tilde{1}$ 0 010− $+2\sqrt{11}/\sqrt{3.5.7.13.17}$
$1\ \tilde{1}\ \tilde{1}$ 0 100− $-3\sqrt{3.5}/2.7\sqrt{17}$
$1\ \tilde{1}\ \tilde{1}$ 0 110− $+2\sqrt{11}/7\sqrt{3.17}$
$1\ \tilde{0}\ \tilde{1}$ 0 000+ $+\sqrt{3}/\sqrt{13.17}$
$1\ \tilde{0}\ \tilde{1}$ 0 100+ $-\sqrt{3.5}/2\sqrt{7.17}$
$2\ 0\ 2$ 0 000+ $+\sqrt{3}/4\sqrt{2.13.17}$
$2\ 0\ 2$ 0 100+ $+\sqrt{7.11}/4\sqrt{2.5.17}$
$2\ 1\ \tilde{1}$ 0 000− $-3/2\sqrt{2.7.13.17}$
$2\ 1\ \tilde{1}$ 0 100− $-\sqrt{3.11}/2\sqrt{2.5.17}$
$2\ 2\ 2$ 0 000+ $-3.5\sqrt{3}/4\sqrt{7.13.17}$
$2\ 2\ 2$ 0 100+ $+\sqrt{11}/4\sqrt{5.17}$
$2\ \tilde{1}\ \tilde{1}$ 0 000+ $-9\sqrt{5.11}/2.7\sqrt{2.13.17}$
$2\ \tilde{1}\ \tilde{1}$ 0 010+ $-\sqrt{13}/7\sqrt{2.17}$
$2\ \tilde{1}\ \tilde{1}$ 0 100+ $-\sqrt{3}/2\sqrt{2.7.17}$
$2\ \tilde{1}\ \tilde{1}$ 0 110+ $+\sqrt{3.11}/\sqrt{2.5.7.17}$
$2\ \tilde{0}\ 2$ 0 000− $+3\sqrt{3.5.11}/4\sqrt{2.7.13.17}$
$2\ \tilde{0}\ 2$ 0 100− $+1/4\sqrt{2.17}$
$\tilde{1}\ 0\ \tilde{1}$ 0 000+ $-\sqrt{3}/17\sqrt{5}$
$\tilde{1}\ 0\ \tilde{1}$ 0 100+ $+3\sqrt{7.11}/2.17\sqrt{13}$
$\tilde{1}\ 1\ 2$ 0 000− $+23\sqrt{3}/2.17\sqrt{2.5.7}$
$\tilde{1}\ 1\ 2$ 0 100− $-9\sqrt{11}/2.17\sqrt{2.13}$
$\tilde{1}\ 1\ \tilde{1}$ 0 000− $-11\sqrt{3}/2.17\sqrt{5.7}$
$\tilde{1}\ 1\ \tilde{1}$ 0 100− $-9\sqrt{11}/2.17\sqrt{13}$
$\tilde{1}\ 2\ \tilde{1}$ 0 000+ $-\sqrt{3}/2\sqrt{5.7}$
$\tilde{1}\ 2\ \tilde{1}$ 0 100+ 0
$\tilde{1}\ \tilde{1}\ 2$ 0 000+ $-3\sqrt{3.11}/2.7.17\sqrt{2}$
$\tilde{1}\ \tilde{1}\ 2$ 0 010+ $+47\sqrt{3}/7.17\sqrt{2.5}$
$\tilde{1}\ \tilde{1}\ 2$ 0 100+ $-9.3\sqrt{5}/2.17\sqrt{2.7.13}$
$\tilde{1}\ \tilde{1}\ 2$ 0 110+ $+3\sqrt{2.11}/17\sqrt{7.13}$
$\tilde{1}\ \tilde{1}\ \tilde{1}$ 0 000+ $-3\sqrt{11}/2.7.17$
$\tilde{1}\ \tilde{1}\ \tilde{1}$ 0 010+ $-8.9/7.17\sqrt{5}$
$\tilde{1}\ \tilde{1}\ \tilde{1}$ 0 100+ $-9\sqrt{3.5}/2.17\sqrt{7.13}$
$\tilde{1}\ \tilde{1}\ \tilde{1}$ 0 110+ $+2\sqrt{3.11}/17\sqrt{7.13}$

8 6 3

$0\ 1\ 1$ 0 000− $+\sqrt{11}/\sqrt{2.3.13.17}$
$0\ \tilde{1}\ \tilde{1}$ 0 000− $+3\sqrt{3}/\sqrt{2.7.13.17}$
$0\ \tilde{1}\ \tilde{1}$ 0 010− $+\sqrt{2.5.11}/\sqrt{3.7.13.17}$

SO_3–O 3jm Factors (cont.)

8 6 3

0 $\tilde{0}$ $\tilde{0}$ 0 000−	$+2/\sqrt{13.17}$
1 0 1 0 000−	$-\sqrt{7.11}/2.3\sqrt{2.13.17}$
1 0 1 0 100−	$-\sqrt{5.11}/2.3\sqrt{2.17}$
1 1 1 0 000+	$+\sqrt{11}/3\sqrt{2.13.17}$
1 1 1 0 100+	$-5\sqrt{5.11}/4.3\sqrt{2.7.17}$
1 1 $\tilde{1}$ 0 000−	$+\sqrt{5.11}/2\sqrt{2.3.13.17}$
1 1 $\tilde{1}$ 0 100−	$+\sqrt{11}/4\sqrt{2.3.7.17}$
1 2 1 0 000−	$-\sqrt{11}/2.3\sqrt{2.13.17}$
1 2 1 0 100−	$-\sqrt{5.11}/2.3\sqrt{2.7.17}$
1 2 $\tilde{1}$ 0 000+	$-\sqrt{5.11}/2\sqrt{2.13.17}$
1 2 $\tilde{1}$ 0 100+	$+\sqrt{11}/2\sqrt{2.7.17}$
1 $\tilde{1}$ 1 0 000−	$+\sqrt{5}/2\sqrt{2.3.7.13.17}$
1 $\tilde{1}$ 1 0 010−	$-3\sqrt{3.11}/2\sqrt{2.7.13.17}$
1 $\tilde{1}$ 1 0 100−	$-5/4\sqrt{2.3.17}$
1 $\tilde{1}$ 1 0 110−	0
1 $\tilde{1}$ $\tilde{1}$ 0 000+	0
1 $\tilde{1}$ $\tilde{1}$ 0 010+	$+\sqrt{5.7.11}/2.3\sqrt{2.13.17}$
1 $\tilde{1}$ $\tilde{1}$ 0 100+	$+9\sqrt{5}/4.7\sqrt{2.17}$
1 $\tilde{1}$ $\tilde{1}$ 0 110+	$+2\sqrt{2.11}/3.7\sqrt{17}$
1 $\tilde{1}$ $\tilde{0}$ 0 000−	$-\sqrt{17}/2\sqrt{7.13}$
1 $\tilde{1}$ $\tilde{0}$ 0 010−	$-\sqrt{5.11}/3\sqrt{7.13.17}$
1 $\tilde{1}$ $\tilde{0}$ 0 100−	$+\sqrt{5}/2.7\sqrt{17}$
1 $\tilde{1}$ $\tilde{0}$ 0 110−	$+4\sqrt{11}/3.7\sqrt{17}$
1 $\tilde{0}$ $\tilde{1}$ 0 000−	$+7/2\sqrt{2.13.17}$
1 $\tilde{0}$ $\tilde{1}$ 0 100−	$+\sqrt{5}/2\sqrt{2.7.17}$
2 1 1 0 000−	$+11\sqrt{5}/4\sqrt{3.7.13.17}$
2 1 1 0 100−	$+\sqrt{11}/4\sqrt{17}$
2 1 $\tilde{1}$ 0 000+	$+2\sqrt{3}/\sqrt{7.13.17}$
2 1 $\tilde{1}$ 0 100+	0
2 2 $\tilde{0}$ 0 000+	$-8/\sqrt{7.13.17}$
2 2 $\tilde{0}$ 0 100+	0
2 $\tilde{1}$ 1 0 000+	$+5\sqrt{3.11}/4.7\sqrt{13.17}$
2 $\tilde{1}$ 1 0 010+	$-3.5\sqrt{3.5}/4.7\sqrt{13.17}$
2 $\tilde{1}$ 1 0 100+	$+3\sqrt{5}/4\sqrt{7.17}$
2 $\tilde{1}$ 1 0 110+	$+\sqrt{11}/4\sqrt{7.17}$
2 $\tilde{1}$ $\tilde{1}$ 0 000−	$-3\sqrt{3.5.11}/2.7\sqrt{13.17}$
2 $\tilde{1}$ $\tilde{1}$ 0 010−	$-31/4.7\sqrt{3.13.17}$
2 $\tilde{1}$ $\tilde{1}$ 0 100−	$+3/2\sqrt{7.17}$
2 $\tilde{1}$ $\tilde{1}$ 0 110−	$-\sqrt{5.11}/4\sqrt{7.17}$
$\tilde{1}$ 0 $\tilde{1}$ 0 000−	$-3\sqrt{5}/2.17\sqrt{2}$
$\tilde{1}$ 0 $\tilde{1}$ 0 100−	$-\sqrt{3.7.11}/2.17\sqrt{2.13}$
$\tilde{1}$ 1 1 0 000−	$-9\sqrt{3}/2.17\sqrt{2.7}$
$\tilde{1}$ 1 1 0 100−	$-7\sqrt{5.11}/4.17\sqrt{2.13}$
$\tilde{1}$ 1 $\tilde{1}$ 0 000+	$-3\sqrt{2.5}/17\sqrt{7}$
$\tilde{1}$ 1 $\tilde{1}$ 0 100+	$+9\sqrt{3.11}/4.17\sqrt{2.13}$
$\tilde{1}$ 1 $\tilde{0}$ 0 000−	$+3\sqrt{5}/2.17\sqrt{7}$
$\tilde{1}$ 1 $\tilde{0}$ 0 100−	$+\sqrt{3.11}/2.17\sqrt{13}$
$\tilde{1}$ 2 1 0 000+	$+19/2.17\sqrt{2.7}$
$\tilde{1}$ 2 1 0 100+	$-\sqrt{3.5.11}/2.17\sqrt{2.13}$

8 6 3

$\tilde{1}$ 2 $\tilde{1}$ 0 000−	$+9\sqrt{5}/2.17\sqrt{2.7}$
$\tilde{1}$ 2 $\tilde{1}$ 0 100−	$+3\sqrt{3.11}/2.17\sqrt{2.13}$
$\tilde{1}$ $\tilde{1}$ 1 0 000+	$+3\sqrt{5.11}/7.17\sqrt{2}$
$\tilde{1}$ $\tilde{1}$ 1 0 010+	$+1/2.7.17\sqrt{2}$
$\tilde{1}$ $\tilde{1}$ 1 0 100+	$-3.5.5\sqrt{3}/4.17\sqrt{2.7.13}$
$\tilde{1}$ $\tilde{1}$ 1 0 110+	$+2\sqrt{2.3.5.11}/17\sqrt{7.13}$
$\tilde{1}$ $\tilde{1}$ $\tilde{1}$ 0 000−	$+9\sqrt{3.11}/2.7.17\sqrt{2}$
$\tilde{1}$ $\tilde{1}$ $\tilde{1}$ 0 010−	$+\sqrt{3.5}/2.7.17\sqrt{2}$
$\tilde{1}$ $\tilde{1}$ $\tilde{1}$ 0 100−	$+9\sqrt{5}/4.17\sqrt{2.7.13}$
$\tilde{1}$ $\tilde{1}$ $\tilde{1}$ 0 110−	$-4\sqrt{2.11}/17\sqrt{7.13}$
$\tilde{1}$ $\tilde{0}$ 1 0 000−	$-\sqrt{5.11}/2.17\sqrt{2.7}$
$\tilde{1}$ $\tilde{0}$ 1 0 100−	$-3.5\sqrt{3}/2.17\sqrt{2.13}$

8 6 4

0 0 0 0 000+	$-2.7/3\sqrt{13.17.19}$
0 1 1 0 000+	$+\sqrt{5.7}/3\sqrt{2.13.17.19}$
0 2 2 0 000+	$+8\sqrt{2.5}/3\sqrt{13.17.19}$
0 $\tilde{1}$ $\tilde{1}$ 0 000+	$+\sqrt{11.19}/\sqrt{2.3.7.13.17}$
0 $\tilde{1}$ $\tilde{1}$ 0 010+	$-11\sqrt{2.5}/\sqrt{3.7.13.17.19}$
1 0 1 0 000+	$-5.7\sqrt{5}/2.3\sqrt{2.3.13.17.19}$
1 0 1 0 100+	$+11\sqrt{7}/2.3\sqrt{2.3.17.19}$
1 1 0 0 000+	$-47\sqrt{7}/2.9\sqrt{13.17.19}$
1 1 0 0 100+	$-11\sqrt{5}/2.9\sqrt{17.19}$
1 1 1 0 000−	0
1 1 1 0 100−	$+11\sqrt{3}/4\sqrt{2.17.19}$
1 1 2 0 000+	$+31\sqrt{5}/2.9\sqrt{13.17.19}$
1 1 2 0 100+	$-97/2.9\sqrt{7.17.19}$
1 1 $\tilde{1}$ 0 000+	$-79\sqrt{5}/2.3\sqrt{2.3.13.17.19}$
1 1 $\tilde{1}$ 0 100+	$-97/4.3\sqrt{2.3.7.17.19}$
1 2 1 0 000+	$-\sqrt{5.7.17}/2.3\sqrt{2.3.13.19}$
1 2 1 0 100+	$-7.7/2.3\sqrt{2.3.17.19}$
1 2 $\tilde{1}$ 0 000−	$-9\sqrt{5}/2\sqrt{2.13.17.19}$
1 2 $\tilde{1}$ 0 100−	$+3\sqrt{7}/2\sqrt{2.17.19}$
1 $\tilde{1}$ 1 0 000+	$-7\sqrt{11}/2.3\sqrt{2.13.17.19}$
1 $\tilde{1}$ 1 0 010+	$+7\sqrt{5}/2.3\sqrt{2.13.17.19}$
1 $\tilde{1}$ 1 0 100+	$-5\sqrt{5.11}/4.3\sqrt{2.7.17.19}$
1 $\tilde{1}$ 1 0 110+	$-8.2\sqrt{2}/3\sqrt{7.17.19}$
1 $\tilde{1}$ 2 0 000−	$-59\sqrt{11}/2.3\sqrt{7.13.17.19}$
1 $\tilde{1}$ 2 0 010−	$-\sqrt{5.13}/3\sqrt{7.17.19}$
1 $\tilde{1}$ 2 0 100−	$-5\sqrt{5.11}/2.3.7\sqrt{17.19}$
1 $\tilde{1}$ 2 0 110−	$+2\sqrt{17}/3.7\sqrt{19}$
1 $\tilde{1}$ $\tilde{1}$ 0 000−	$+\sqrt{11.19}/3\sqrt{2.7.13.17}$
1 $\tilde{1}$ $\tilde{1}$ 0 010−	$-11.11\sqrt{5}/2.3\sqrt{2.7.13.17.19}$
1 $\tilde{1}$ $\tilde{1}$ 0 100−	$+13\sqrt{5.11}/4.3.7\sqrt{2.17.19}$
1 $\tilde{1}$ $\tilde{1}$ 0 110−	$+2.11\sqrt{2}/3.7\sqrt{17.19}$
1 $\tilde{0}$ $\tilde{1}$ 0 000+	$+\sqrt{11.19}/2.3\sqrt{2.13.17}$
1 $\tilde{0}$ $\tilde{1}$ 0 100+	$+\sqrt{5.11}/2.3\sqrt{2.7.17.19}$
2 0 2 0 000+	$-127/2.3\sqrt{11.13.17.19}$
2 0 2 0 100+	$-\sqrt{5.7}/2\sqrt{3.17.19}$
2 1 1 0 000+	$-7.47/4.3\sqrt{11.13.17.19}$

SO_3–O 3jm Factors (cont.)

8 6 4	
2 1 1 0 100+	$+\sqrt{5.7}/4\sqrt{3.17.19}$
2 1 1̃ 0 000−	$-139/\sqrt{3.7.11.13.17.19}$
2 1 1̃ 0 100−	$+\sqrt{5}/\sqrt{17.19}$
2 2 0 0 000+	$+2\sqrt{5}/3\sqrt{11.13.17.19}$
2 2 0 0 100+	$-2\sqrt{7}/\sqrt{3.17.19}$
2 2 2 0 000+	$-223/3\sqrt{2.7.11.13.17.19}$
2 2 2 0 100+	$-\sqrt{5}/\sqrt{2.3.17.19}$
2 1̃ 1 0 000−	$-11\sqrt{5}/4\sqrt{7.13.17.19}$
2 1̃ 1 0 010−	$+\sqrt{13.17}/4\sqrt{7.11.19}$
2 1̃ 1 0 100−	$-\sqrt{3.11}/4\sqrt{17.19}$
2 1̃ 1 0 110−	$+3\sqrt{3.5}/4\sqrt{17.19}$
2 1̃ 1̃ 0 000+	$-71\sqrt{5}/2.7\sqrt{3.13.17.19}$
2 1̃ 1̃ 0 010+	$-1951/4.7\sqrt{3.11.13.17.19}$
2 1̃ 1̃ 0 100+	$-\sqrt{11}/2\sqrt{7.17.19}$
2 1̃ 1̃ 0 110+	$-\sqrt{5}/4\sqrt{7.17.19}$
2 0̃ 2 0 000−	$+\sqrt{5.13}/2\sqrt{7.17.19}$
2 0̃ 2 0 100−	$-\sqrt{3.11}/2\sqrt{17.19}$
1̃ 0 1̃ 0 000+	$+3\sqrt{5.19}/2.17\sqrt{2.11}$
1̃ 0 1̃ 0 100+	$-11\sqrt{7}/2.17\sqrt{2.3.13.19}$
1̃ 1 1 0 000+	$-23\sqrt{5}/2.17\sqrt{2.11.19}$
1̃ 1 1 0 100+	$+73\sqrt{7}/4.17\sqrt{2.3.13.19}$
1̃ 1 2 0 000−	$+31\sqrt{5}/2.17\sqrt{7.11.19}$
1̃ 1 2 0 100−	$+167/2.17\sqrt{3.13.19}$
1̃ 1 1̃ 0 000−	$+3\sqrt{5.19}/17\sqrt{2.7.11}$
1̃ 1 1̃ 0 100−	$+11\sqrt{13}/4.17\sqrt{2.3.19}$
1̃ 2 1 0 000−	$+\sqrt{3.5}/2\sqrt{2.11.19}$
1̃ 2 1 0 100−	$-\sqrt{7}/2\sqrt{2.13.19}$
1̃ 2 1̃ 0 000+	$+3\sqrt{5.11}/2.17\sqrt{2.7.19}$
1̃ 2 1̃ 0 100+	$-5.7/2.17\sqrt{2.3.13.19}$
1̃ 1̃ 0 0 000+	$-\sqrt{5.19}/2.17\sqrt{7}$
1̃ 1̃ 0 0 010+	$-1/17\sqrt{7.11.19}$
1̃ 1̃ 0 0 100+	$+5\sqrt{3.11}/2.17\sqrt{13.19}$
1̃ 1̃ 0 0 110+	$+4.11\sqrt{5}/17\sqrt{3.13.19}$
1̃ 1̃ 1 0 000−	$+\sqrt{2.3.19}/17\sqrt{7}$
1̃ 1̃ 1 0 010−	$+3\sqrt{3.5}/2.17\sqrt{2.7.11.19}$
1̃ 1̃ 1 0 100−	$+9.3\sqrt{5.11}/4.17\sqrt{2.13.19}$
1̃ 1̃ 1 0 110−	$+2.11\sqrt{2}/17\sqrt{13.19}$
1̃ 1̃ 2 0 000+	$-89/2.7.17\sqrt{19}$
1̃ 1̃ 2 0 010+	$+\sqrt{5.11}/7.17\sqrt{19}$
1̃ 1̃ 2 0 100+	$+\sqrt{3.5.11.19}/2.17\sqrt{7.13}$
1̃ 1̃ 2 0 110+	$-2.37/17\sqrt{3.7.13.19}$
1̃ 1̃ 1̃ 0 000+	$+9.3\sqrt{3}/2.7.17\sqrt{2.19}$
1̃ 1̃ 1̃ 0 010+	$+73\sqrt{3.5}/2.7.17\sqrt{2.11.19}$
1̃ 1̃ 1̃ 0 100+	$+5.11\sqrt{5.11}/4.17\sqrt{2.7.13.19}$
1̃ 1̃ 1̃ 0 110+	$-4.5\sqrt{2}/17\sqrt{7.13.19}$
1̃ 0̃ 1 0 000+	$-\sqrt{3.19}/2.17\sqrt{2}$
1̃ 0̃ 1 0 100+	$+3\sqrt{5.7.11}/2.17\sqrt{2.13.19}$

8 6 5	
0 1 1 0 000−	$+9\sqrt{3.7}/\sqrt{11.13.17.19}$
0 1 1 0 001−	$-2.7\sqrt{5}/\sqrt{3.11.13.17.19}$
0 2 2 0 000−	$+\sqrt{5}/\sqrt{13.17.19}$
0 1̃ 1̃ 0 000−	$+3\sqrt{3.11}/\sqrt{7.13.17.19}$
0 1̃ 1̃ 0 010−	$+8\sqrt{5}/\sqrt{3.7.13.17.19}$
1 0 1 0 000−	$-9.7/2\sqrt{11.13.17.19}$
1 0 1 0 001−	$+7\sqrt{5.7}/3\sqrt{11.13.17.19}$
1 0 1 0 100−	0
1 0 1 0 101−	$-\sqrt{11}/2.3\sqrt{17.19}$
1 1 1 0 000+	$+9.3\sqrt{7}/2\sqrt{11.13.17.19}$
1 1 1 0 001+	$+5.7\sqrt{5}/2.3\sqrt{11.13.17.19}$
1 1 1 0 100+	0
1 1 1 0 101+	$-\sqrt{7.11}/2.3\sqrt{17.19}$
1 1 2 0 000−	$-\sqrt{5.13}/2\sqrt{2.17.19}$
1 1 2 0 100−	$+\sqrt{7}/2\sqrt{2.17.19}$
1 1 1̃ 0 000−	$+\sqrt{5}/2\sqrt{3.13.17.19}$
1 1 1̃ 0 100−	$+\sqrt{7}/\sqrt{3.17.19}$
1 2 1 0 000−	$-2\sqrt{7}/\sqrt{11.13.17.19}$
1 2 1 0 001−	$+5\sqrt{5}/2.3\sqrt{11.13.17.19}$
1 2 1 0 100−	$+5\sqrt{5}/2\sqrt{11.17.19}$
1 2 1 0 101−	$-4\sqrt{7}/3\sqrt{11.17.19}$
1 2 1̃ 0 000+	$-4\sqrt{5}/\sqrt{13.17.19}$
1 2 1̃ 0 100+	$-\sqrt{7}/2\sqrt{17.19}$
1 1̃ 1 0 000−	$+3\sqrt{3.5}/2\sqrt{13.17.19}$
1 1̃ 1 0 001−	$-71/2\sqrt{3.7.13.17.19}$
1 1̃ 1 0 010−	$-8.8/\sqrt{3.11.13.17.19}$
1 1̃ 1 0 011−	$+4\sqrt{3.5}/\sqrt{7.11.13.17.19}$
1 1̃ 1 0 100−	$+3\sqrt{3}/\sqrt{7.17.19}$
1 1̃ 1 0 101−	$-5\sqrt{5}/2\sqrt{3.17.19}$
1 1̃ 1 0 110−	$+8\sqrt{5}/\sqrt{3.7.11.17.19}$
1 1̃ 1 0 111−	$+2\sqrt{3}/\sqrt{11.17.19}$
1 1̃ 2 0 000+	$-11\sqrt{11}/2\sqrt{2.7.13.17.19}$
1 1̃ 2 0 010+	$-\sqrt{5.17}/3\sqrt{2.7.13.19}$
1 1̃ 2 0 100+	$+\sqrt{5.11}/2\sqrt{2.17.19}$
1 1̃ 2 0 110+	$+\sqrt{2}/3\sqrt{17.19}$
1 1̃ 1̃ 0 000+	$-9\sqrt{11}/2\sqrt{7.13.17.19}$
1 1̃ 1̃ 0 010+	$+2\sqrt{5}/3\sqrt{7.13.17.19}$
1 1̃ 1̃ 0 100+	0
1 1̃ 1̃ 0 110+	$-2.7/3\sqrt{17.19}$
1 0̃ 1̃ 0 000−	$+3\sqrt{11}/2\sqrt{13.17.19}$
1 0̃ 1̃ 0 100−	0
2 0 2 0 000−	$+3.59/4\sqrt{2.11.13.17.19}$
2 0 2 0 100−	$-\sqrt{3.5.7}/4\sqrt{2.17.19}$
2 1 1 0 000−	$-9.3\sqrt{3.5}/2.11\sqrt{2.13.17.19}$
2 1 1 0 001−	$+\sqrt{2.7.13}/11\sqrt{3.17.19}$
2 1 1 0 100−	$-3\sqrt{7}/2\sqrt{2.11.17.19}$
2 1 1 0 101−	$+\sqrt{2.5}/\sqrt{11.17.19}$
2 1 1̃ 0 000+	$+\sqrt{3.7.17}/2\sqrt{2.11.13.19}$
2 1 1̃ 0 100+	$+3\sqrt{5}/2\sqrt{2.17.19}$
2 2 2 0 000−	$-41\sqrt{7}/4\sqrt{11.13.17.19}$

SO_3–O 3jm Factors (cont.)

8 6 5

$2\ 2\ 2\ 0$ 100−	$-\sqrt{3.5}/4\sqrt{17.19}$
$2\ \tilde{1}\ 1\ 0$ 000+	$+3\sqrt{3.19}/2\sqrt{2.7.11.13.17}$
$2\ \tilde{1}\ 1\ 0$ 001+	$+2\sqrt{2.3.5}/\sqrt{11.13.17.19}$
$2\ \tilde{1}\ 1\ 0$ 010+	$-269\sqrt{3.5}/2.11\sqrt{2.7.13.17.19}$
$2\ \tilde{1}\ 1\ 0$ 011+	$-3\sqrt{3.13}/11\sqrt{2.17.19}$
$2\ \tilde{1}\ 1\ 0$ 100+	$-3\sqrt{5}/2\sqrt{2.17.19}$
$2\ \tilde{1}\ 1\ 0$ 101+	$-2.3\sqrt{2}/\sqrt{7.17.19}$
$2\ \tilde{1}\ 1\ 0$ 110+	$+3/2\sqrt{2.11.17.19}$
$2\ \tilde{1}\ 1\ 0$ 111+	$-13\sqrt{5}/\sqrt{2.7.11.17.19}$
$2\ \tilde{1}\ \tilde{1}\ 0$ 000−	$-3\sqrt{3.5}/2\sqrt{2.13.17.19}$
$2\ \tilde{1}\ \tilde{1}\ 0$ 010−	$-\sqrt{19}/2\sqrt{2.3.11.13.17}$
$2\ \tilde{1}\ \tilde{1}\ 0$ 100−	$+3\sqrt{11}/2\sqrt{2.7.17.19}$
$2\ \tilde{1}\ \tilde{1}\ 0$ 110−	$+5\sqrt{5}/2\sqrt{2.7.17.19}$
$2\ \tilde{0}\ 2\ 0$ 000+	$+5\sqrt{5.7}/4\sqrt{2.13.17.19}$
$2\ \tilde{0}\ 2\ 0$ 100+	$+\sqrt{3.11}/4\sqrt{2.17.19}$
$\tilde{1}\ 0\ \tilde{1}\ 0$ 000−	$+3\sqrt{5}/2.17\sqrt{11.19}$
$\tilde{1}\ 0\ \tilde{1}\ 0$ 100−	$-8\sqrt{3.7}/17\sqrt{13.19}$
$\tilde{1}\ 1\ 1\ 0$ 000−	$-3.41\sqrt{3}/2.11.17\sqrt{19}$
$\tilde{1}\ 1\ 1\ 0$ 001−	$+3\sqrt{3.5.7}/2.11.17\sqrt{19}$
$\tilde{1}\ 1\ 1\ 0$ 100−	$+3\sqrt{5.7}/17\sqrt{11.13.19}$
$\tilde{1}\ 1\ 1\ 0$ 101−	$+191/2.17\sqrt{11.13.19}$
$\tilde{1}\ 1\ 2\ 0$ 000+	$-3.7\sqrt{5.7}/2.17\sqrt{2.11.19}$
$\tilde{1}\ 1\ 2\ 0$ 100+	$+\sqrt{3.19}/2.17\sqrt{2.13}$
$\tilde{1}\ 1\ \tilde{1}\ 0$ 000+	$+3\sqrt{5.7}/2.17\sqrt{11.19}$
$\tilde{1}\ 1\ \tilde{1}\ 0$ 100+	$+4.3\sqrt{3}/17\sqrt{13.19}$
$\tilde{1}\ 2\ 1\ 0$ 000+	$-8.3/11.17\sqrt{19}$
$\tilde{1}\ 2\ 1\ 0$ 001+	$-5\sqrt{5.7}/2.11.17\sqrt{19}$
$\tilde{1}\ 2\ 1\ 0$ 100+	$+7\sqrt{3.5.7}/2.17\sqrt{11.13.19}$
$\tilde{1}\ 2\ 1\ 0$ 101+	$+4.29\sqrt{3}/17\sqrt{11.13.19}$
$\tilde{1}\ 2\ \tilde{1}\ 0$ 000−	$-2.3\sqrt{5.7}/17\sqrt{11.19}$
$\tilde{1}\ 2\ \tilde{1}\ 0$ 100−	$-9.5\sqrt{3}/2.17\sqrt{13.19}$
$\tilde{1}\ \tilde{1}\ 1\ 0$ 000+	$+3\sqrt{5.7}/2.17\sqrt{11.19}$
$\tilde{1}\ \tilde{1}\ 1\ 0$ 001+	$+3\sqrt{19}/2.17\sqrt{11}$
$\tilde{1}\ \tilde{1}\ 1\ 0$ 010+	$-2.9\sqrt{7}/11.17\sqrt{19}$
$\tilde{1}\ \tilde{1}\ 1\ 0$ 011+	$+2\sqrt{5}/11.17\sqrt{19}$
$\tilde{1}\ \tilde{1}\ 1\ 0$ 100+	$+4.3\sqrt{3}/17\sqrt{13.19}$
$\tilde{1}\ \tilde{1}\ 1\ 0$ 101+	$-3\sqrt{3.5.7}/2.17\sqrt{13.19}$
$\tilde{1}\ \tilde{1}\ 1\ 0$ 110+	$-2\sqrt{3.5.11}/17\sqrt{13.19}$
$\tilde{1}\ \tilde{1}\ 1\ 0$ 111+	$-2\sqrt{3.7.11}/17\sqrt{13.19}$
$\tilde{1}\ \tilde{1}\ 2\ 0$ 000−	$+3.7/2.17\sqrt{2.19}$
$\tilde{1}\ \tilde{1}\ 2\ 0$ 010−	$-3.7\sqrt{5}/17\sqrt{2.11.19}$
$\tilde{1}\ \tilde{1}\ 2\ 0$ 100−	$+3\sqrt{3.5.11}/2.17\sqrt{2.7.13.19}$
$\tilde{1}\ \tilde{1}\ 2\ 0$ 110−	$+\sqrt{2.3}/17\sqrt{7.13.19}$
$\tilde{1}\ \tilde{1}\ \tilde{1}\ 0$ 000−	$-9\sqrt{3}/2.17\sqrt{19}$
$\tilde{1}\ \tilde{1}\ \tilde{1}\ 0$ 010−	$-4\sqrt{3.5}/17\sqrt{11.19}$
$\tilde{1}\ \tilde{1}\ \tilde{1}\ 0$ 100−	$+9\sqrt{5.11}/17\sqrt{7.13.19}$
$\tilde{1}\ \tilde{1}\ \tilde{1}\ 0$ 110−	$+8.5/17\sqrt{7.13.19}$
$\tilde{1}\ \tilde{0}\ 1\ 0$ 000−	$-3\sqrt{5}/2.17\sqrt{11.19}$
$\tilde{1}\ \tilde{0}\ 1\ 0$ 001−	$-\sqrt{7.19}/17\sqrt{11}$

8 6 5

$\tilde{1}\ \tilde{0}\ 1\ 0$ 100−	$+8\sqrt{3.7}/17\sqrt{13.19}$
$\tilde{1}\ \tilde{0}\ 1\ 0$ 101−	$-3\sqrt{3.5}/2.17\sqrt{13.19}$

8 6 6

$0\ 0\ 0\ 0$ 000+	$-3\sqrt{3.7}/\sqrt{2.13.17.19}$
$0\ 1\ 1\ 0$ 000+	$-9/\sqrt{2.7.13.17.19}$
$0\ 2\ 2\ 0$ 000+	$-\sqrt{3.7}/\sqrt{13.17.19}$
$0\ \tilde{1}\ \tilde{1}\ 0$ 000+	$+3.11/7\sqrt{2.7.13.17.19}$
$0\ \tilde{1}\ \tilde{1}\ 0$ 010+	$-8\sqrt{2.5.11}/7\sqrt{7.13.17.19}$
$0\ \tilde{1}\ \tilde{1}\ 0$ 011+	$+9.11\sqrt{2}/7\sqrt{7.13.17.19}$
$0\ \tilde{0}\ \tilde{0}\ 0$ 000+	$+11\sqrt{3}/\sqrt{2.7.13.17.19}$
$1\ 1\ 0\ 0$ 000+	$-9\sqrt{3}/\sqrt{2.13.17.19}$
$1\ 1\ 0\ 0$ 100+	0
$1\ 1\ 1\ 0$ 000−	0
$1\ 1\ 1\ 0$ 100−	0
$1\ 2\ 1\ 0$ 000+	$-\sqrt{2.3}/\sqrt{7.13.17.19}$
$1\ 2\ 1\ 0$ 100+	$+\sqrt{3.5}/\sqrt{2.17.19}$
$1\ \tilde{1}\ 1\ 0$ 000+	$-4\sqrt{2.5.11}/7\sqrt{13.17.19}$
$1\ \tilde{1}\ 1\ 0$ 010+	$+53\sqrt{2}/7\sqrt{13.17.19}$
$1\ \tilde{1}\ 1\ 0$ 100+	$+\sqrt{2.11}/\sqrt{7.17.19}$
$1\ \tilde{1}\ 1\ 0$ 110+	$+\sqrt{5}/\sqrt{2.7.17.19}$
$1\ \tilde{1}\ 2\ 0$ 000−	$+\sqrt{2.3.5.11}/7\sqrt{13.17.19}$
$1\ \tilde{1}\ 2\ 0$ 010−	$+5\sqrt{17}/7\sqrt{2.3.13.19}$
$1\ \tilde{1}\ 2\ 0$ 100−	$+\sqrt{3.11}/\sqrt{2.7.17.19}$
$1\ \tilde{1}\ 2\ 0$ 110−	$-\sqrt{2.5}/\sqrt{3.7.17.19}$
$1\ \tilde{1}\ \tilde{1}\ 0$ 000−	0
$1\ \tilde{1}\ \tilde{1}\ 0$ 010−	$-2\sqrt{2.5.11}/\sqrt{3.7.13.17.19}$
$1\ \tilde{1}\ \tilde{1}\ 0$ 011−	0
$1\ \tilde{1}\ \tilde{1}\ 0$ 100−	0
$1\ \tilde{1}\ \tilde{1}\ 0$ 110−	$+\sqrt{11}/\sqrt{2.3.17.19}$
$1\ \tilde{1}\ \tilde{1}\ 0$ 111−	0
$1\ \tilde{0}\ \tilde{1}\ 0$ 000+	$-3.11\sqrt{3}/7\sqrt{2.13.17.19}$
$1\ \tilde{0}\ \tilde{1}\ 0$ 001+	$-4\sqrt{2.5.11}/7\sqrt{3.13.17.19}$
$1\ \tilde{0}\ \tilde{1}\ 0$ 100+	0
$1\ \tilde{0}\ \tilde{1}\ 0$ 101+	$-\sqrt{7.11}/\sqrt{2.3.17.19}$
$2\ 1\ 1\ 0$ 000+	$-3.29\sqrt{5}/4\sqrt{11.13.17.19}$
$2\ 1\ 1\ 0$ 100+	$+13\sqrt{3}/4\sqrt{7.17.19}$
$2\ 2\ 0\ 0$ 000+	$+\sqrt{2.3.5.7}/\sqrt{11.13.17.19}$
$2\ 2\ 0\ 0$ 100+	$+\sqrt{2}/\sqrt{17.19}$
$2\ 2\ 2\ 0$ 000+	$+\sqrt{3.5}/\sqrt{11.13.17.19}$
$2\ 2\ 2\ 0$ 100+	$-9/\sqrt{7.17.19}$
$2\ \tilde{1}\ 1\ 0$ 000−	$+3.11/4\sqrt{7.13.17.19}$
$2\ \tilde{1}\ 1\ 0$ 010−	$+3.23\sqrt{5}/4\sqrt{7.11.13.17.19}$
$2\ \tilde{1}\ 1\ 0$ 100−	$+\sqrt{3.5.11}/4.7\sqrt{17.19}$
$2\ \tilde{1}\ 1\ 0$ 110−	$-3.5\sqrt{3}/4.7\sqrt{17.19}$
$2\ \tilde{1}\ \tilde{1}\ 0$ 000+	$+9.3\sqrt{5.11}/4.7\sqrt{13.17.19}$
$2\ \tilde{1}\ \tilde{1}\ 0$ 010+	$-223/4.7\sqrt{13.17.19}$
$2\ \tilde{1}\ \tilde{1}\ 0$ 011+	$-2.3\sqrt{5}/7\sqrt{11.13.17.19}$
$2\ \tilde{1}\ \tilde{1}\ 0$ 100+	$+11\sqrt{3}/4\sqrt{7.17.19}$
$2\ \tilde{1}\ \tilde{1}\ 0$ 110+	$-\sqrt{3.5.11}/4\sqrt{7.17.19}$

SO_3–O 3jm Factors (cont.)

8 6 6

$2\ \tilde{1}\ \tilde{1}$	0	111+	$-2\sqrt{3}/\sqrt{7.17.19}$
$2\ \tilde{0}\ 2$	0	000−	$+\sqrt{3}/\sqrt{2.13.17.19}$
$2\ \tilde{0}\ 2$	0	100−	$-\sqrt{5.11}/\sqrt{2.7.17.19}$
$\tilde{1}\ 1\ 1$	0	000+	$-4.3\sqrt{2}/17\sqrt{11.19}$
$\tilde{1}\ 1\ 1$	0	100+	$-9\sqrt{2.3.5}/17\sqrt{7.13.19}$
$\tilde{1}\ 2\ 1$	0	000−	$-5\sqrt{2.3}/17\sqrt{11.19}$
$\tilde{1}\ 2\ 1$	0	100−	$+3\sqrt{5}/17\sqrt{2.7.13.19}$
$\tilde{1}\ \tilde{1}\ 0$	0	000+	$-\sqrt{3.5}/17\sqrt{2.19}$
$\tilde{1}\ \tilde{1}\ 0$	0	010+	$+4\sqrt{2.3}/17\sqrt{11.19}$
$\tilde{1}\ \tilde{1}\ 0$	0	100+	$-4.3\sqrt{2.11}/17\sqrt{7.13.19}$
$\tilde{1}\ \tilde{1}\ 0$	0	110+	$-3.11\sqrt{5}/17\sqrt{2.7.13.19}$
$\tilde{1}\ \tilde{1}\ 1$	0	000−	$+2\sqrt{2.3.5}/17\sqrt{7.19}$
$\tilde{1}\ \tilde{1}\ 1$	0	010−	$-2.3.5\sqrt{2.3}/17\sqrt{7.11.19}$
$\tilde{1}\ \tilde{1}\ 1$	0	100−	$-2.9.3\sqrt{2.11}/7.17\sqrt{13.19}$
$\tilde{1}\ \tilde{1}\ 1$	0	110−	$-3.11\sqrt{5}/7.17\sqrt{2.13.19}$
$\tilde{1}\ \tilde{1}\ 2$	0	000+	$-5\sqrt{2.3.5}/17\sqrt{7.19}$
$\tilde{1}\ \tilde{1}\ 2$	0	010+	$-\sqrt{3.11}/17\sqrt{2.7.19}$
$\tilde{1}\ \tilde{1}\ 2$	0	100+	$+9\sqrt{11.13}/7.17\sqrt{2.19}$
$\tilde{1}\ \tilde{1}\ 2$	0	110+	$+9.5\sqrt{2.5}/7.17\sqrt{13.19}$
$\tilde{1}\ \tilde{1}\ \tilde{1}$	0	000+	$-4.3\sqrt{2.11}/7.17\sqrt{19}$
$\tilde{1}\ \tilde{1}\ \tilde{1}$	0	010+	$-9.3\sqrt{2.5}/7.17\sqrt{19}$
$\tilde{1}\ \tilde{1}\ \tilde{1}$	0	011+	$+2.3.29\sqrt{2}/7.17\sqrt{11.19}$
$\tilde{1}\ \tilde{1}\ \tilde{1}$	0	100+	$-9.11\sqrt{2.3.5}/7.17\sqrt{7.13.19}$
$\tilde{1}\ \tilde{1}\ \tilde{1}$	0	110+	$-11\sqrt{3.11.13}/7.17\sqrt{2.7.19}$
$\tilde{1}\ \tilde{1}\ \tilde{1}$	0	111+	$-8.2.3\sqrt{2.3.5}/7.17\sqrt{7.13.19}$
$\tilde{1}\ \tilde{0}\ 1$	0	000+	$-\sqrt{3.5}/17\sqrt{2.19}$
$\tilde{1}\ \tilde{0}\ 1$	0	100+	$-4.3\sqrt{2.11}/17\sqrt{7.13.19}$

8 $\frac{13}{2}$ $\frac{3}{2}$

$0\ \frac{3}{2}\ \frac{3}{2}$	0	000−	$+2\sqrt{2.3}/\sqrt{5.7.17}$
$0\ \frac{3}{2}\ \frac{3}{2}$	0	010−	$+\sqrt{11}/\sqrt{5.7.17}$
$1\ \frac{1}{2}\ \frac{3}{2}$	0	000−	0
$1\ \frac{1}{2}\ \frac{3}{2}$	0	100−	0
$1\ \frac{3}{2}\ \frac{3}{2}$	0	000+	$+2\sqrt{3}/\sqrt{5.7.17}$
$1\ \frac{3}{2}\ \frac{3}{2}$	0	010+	$+13\sqrt{11}/2.3\sqrt{2.5.7.17}$
$1\ \frac{3}{2}\ \frac{3}{2}$	0	100+	$-\sqrt{3.13}/7\sqrt{17}$
$1\ \frac{3}{2}\ \frac{3}{2}$	0	110+	$+\sqrt{2.11.13}/3.7\sqrt{17}$
$1\ \frac{3}{2}\ \frac{3}{2}$	1	000+	$+\sqrt{3.7}/2\sqrt{5.17}$
$1\ \frac{3}{2}\ \frac{3}{2}$	1	010+	$+\sqrt{7.11}/2.3\sqrt{2.5.17}$
$1\ \frac{3}{2}\ \frac{3}{2}$	1	100+	$+\sqrt{3.13}/2.7\sqrt{17}$
$1\ \frac{3}{2}\ \frac{3}{2}$	1	110+	$-\sqrt{11.13}/3.7\sqrt{2.17}$
$1\ \tilde{\frac{1}{2}}\ \frac{3}{2}$	0	000−	$-\sqrt{3}/\sqrt{7.17}$
$1\ \tilde{\frac{1}{2}}\ \frac{3}{2}$	0	010−	$-\sqrt{11.13}/3\sqrt{2.5.7.17}$
$1\ \tilde{\frac{1}{2}}\ \frac{3}{2}$	0	100−	$+\sqrt{3.5.13}/2.7\sqrt{17}$
$1\ \tilde{\frac{1}{2}}\ \frac{3}{2}$	0	110−	$-2\sqrt{2.11}/3.7\sqrt{17}$
$2\ \frac{1}{2}\ \frac{3}{2}$	0	000+	$+\sqrt{3}/4\sqrt{2.7.17}$
$2\ \frac{1}{2}\ \frac{3}{2}$	0	100+	$+\sqrt{11.13}/4\sqrt{2.5.17}$
$2\ \frac{3}{2}\ \frac{3}{2}$	0	000−	$-3\sqrt{3.11}/2.7\sqrt{2.17}$
$2\ \frac{3}{2}\ \frac{3}{2}$	0	010−	$-13/4.7\sqrt{17}$
$2\ \frac{3}{2}\ \frac{3}{2}$	0	100−	$-\sqrt{13}/2\sqrt{2.5.7.17}$
$2\ \frac{3}{2}\ \frac{3}{2}$	0	110−	$+\sqrt{3.11.13}/4\sqrt{5.7.17}$
$2\ \tilde{\frac{1}{2}}\ \frac{3}{2}$	0	000+	$-3\sqrt{3.5.11}/4.7\sqrt{2.17}$
$2\ \tilde{\frac{1}{2}}\ \frac{3}{2}$	0	010+	$-\sqrt{13}/2.7\sqrt{17}$
$2\ \tilde{\frac{1}{2}}\ \frac{3}{2}$	0	100+	$-\sqrt{13}/4\sqrt{2.7.17}$
$2\ \tilde{\frac{1}{2}}\ \frac{3}{2}$	0	110+	$+\sqrt{3.11}/2\sqrt{5.7.17}$
$\tilde{1}\ \frac{1}{2}\ \frac{3}{2}$	0	000+	$-\sqrt{3.13}/17\sqrt{5.7}$
$\tilde{1}\ \frac{1}{2}\ \frac{3}{2}$	0	100+	$+3\sqrt{11}/2.17$
$\tilde{1}\ \frac{3}{2}\ \frac{3}{2}$	0	000−	$-\sqrt{3.11.13}/7.17\sqrt{5}$
$\tilde{1}\ \frac{3}{2}\ \frac{3}{2}$	0	010−	$-5\sqrt{5.13}/2.7.17\sqrt{2}$
$\tilde{1}\ \frac{3}{2}\ \frac{3}{2}$	0	100−	$-9/17\sqrt{7}$
$\tilde{1}\ \frac{3}{2}\ \frac{3}{2}$	0	110−	$+\sqrt{2.3.11}/17\sqrt{7}$
$\tilde{1}\ \frac{3}{2}\ \frac{3}{2}$	1	000+	$-\sqrt{3.11.13}/2.7.17\sqrt{5}$
$\tilde{1}\ \frac{3}{2}\ \frac{3}{2}$	1	010+	$+47\sqrt{13}/2.7.17\sqrt{2.5}$
$\tilde{1}\ \frac{3}{2}\ \frac{3}{2}$	1	100+	$-9/2.17\sqrt{7}$
$\tilde{1}\ \frac{3}{2}\ \frac{3}{2}$	1	110+	$+\sqrt{3.11}/17\sqrt{2.7}$
$\tilde{1}\ \tilde{\frac{1}{2}}\ \frac{3}{2}$	0	000+	0
$\tilde{1}\ \tilde{\frac{1}{2}}\ \frac{3}{2}$	0	010+	$-1/\sqrt{2.5}$
$\tilde{1}\ \tilde{\frac{1}{2}}\ \frac{3}{2}$	0	100+	0
$\tilde{1}\ \tilde{\frac{1}{2}}\ \frac{3}{2}$	0	110+	0

8 $\frac{13}{2}$ $\frac{5}{2}$

$0\ \frac{3}{2}\ \frac{3}{2}$	0	000+	$+\sqrt{2}/\sqrt{5.7.13.17}$
$0\ \frac{3}{2}\ \frac{3}{2}$	0	010+	$+19\sqrt{11}/3\sqrt{3.5.7.13.17}$
$0\ \tilde{\frac{1}{2}}\ \tilde{\frac{1}{2}}$	0	000+	$+2\sqrt{2.5}/\sqrt{7.13.17}$
$0\ \tilde{\frac{1}{2}}\ \tilde{\frac{1}{2}}$	0	010+	$+2\sqrt{11}/3\sqrt{3.7.17}$
$1\ \frac{1}{2}\ \frac{3}{2}$	0	000+	$+\sqrt{5.11}/2.3\sqrt{13.17}$
$1\ \frac{1}{2}\ \frac{3}{2}$	0	100+	$+5\sqrt{11}/2.3\sqrt{7.17}$
$1\ \frac{3}{2}\ \frac{3}{2}$	0	000−	$+4\sqrt{7}/3\sqrt{5.13.17}$
$1\ \frac{3}{2}\ \frac{3}{2}$	0	010−	$-\sqrt{2.7.11}/9\sqrt{3.5.13.17}$
$1\ \frac{3}{2}\ \frac{3}{2}$	0	100−	$-13/2.3.7\sqrt{17}$
$1\ \frac{3}{2}\ \frac{3}{2}$	0	110−	$+13\sqrt{11}/9.7\sqrt{2.3.17}$
$1\ \frac{3}{2}\ \frac{3}{2}$	1	000−	$+11/2\sqrt{5.7.13.17}$
$1\ \frac{3}{2}\ \frac{3}{2}$	1	010−	$-\sqrt{2.11.13}/9\sqrt{3.5.7.17}$
$1\ \frac{3}{2}\ \frac{3}{2}$	1	100−	$-2.3/7\sqrt{17}$
$1\ \frac{3}{2}\ \frac{3}{2}$	1	110−	$+\sqrt{11}/9.7\sqrt{2.3.17}$
$1\ \frac{3}{2}\ \tilde{\frac{1}{2}}$	0	000+	$+31/3\sqrt{2.7.13.17}$
$1\ \frac{3}{2}\ \tilde{\frac{1}{2}}$	0	010+	$+71\sqrt{11}/2.9\sqrt{3.7.13.17}$
$1\ \frac{3}{2}\ \tilde{\frac{1}{2}}$	0	100+	$-\sqrt{2.5}/3.7\sqrt{17}$
$1\ \frac{3}{2}\ \tilde{\frac{1}{2}}$	0	110+	$-5\sqrt{5.11}/9.7\sqrt{3.17}$
$1\ \tilde{\frac{1}{2}}\ \frac{3}{2}$	0	000+	$-23/2.3\sqrt{7.13.17}$
$1\ \tilde{\frac{1}{2}}\ \frac{3}{2}$	0	010+	$-29\sqrt{11}/9\sqrt{2.3.5.7.17}$
$1\ \tilde{\frac{1}{2}}\ \frac{3}{2}$	0	100+	$-11\sqrt{5}/2.3.7\sqrt{17}$
$1\ \tilde{\frac{1}{2}}\ \frac{3}{2}$	0	110+	$-\sqrt{2.11.13}/9.7\sqrt{3.17}$
$1\ \tilde{\frac{1}{2}}\ \tilde{\frac{1}{2}}$	0	000−	$+2.5\sqrt{5}/3\sqrt{7.13.17}$
$1\ \tilde{\frac{1}{2}}\ \tilde{\frac{1}{2}}$	0	010−	$-2\sqrt{2.11}/9\sqrt{3.7.17}$
$1\ \tilde{\frac{1}{2}}\ \tilde{\frac{1}{2}}$	0	100−	$-5/2.3.7\sqrt{17}$
$1\ \tilde{\frac{1}{2}}\ \tilde{\frac{1}{2}}$	0	110−	$-2\sqrt{2.5.11.13}/9.7\sqrt{3.17}$
$2\ \frac{1}{2}\ \frac{3}{2}$	0	000−	$+3/4\sqrt{2.7.13.17}$
$2\ \frac{1}{2}\ \frac{3}{2}$	0	100−	$+\sqrt{3.11}/4\sqrt{2.5.17}$
$2\ \frac{3}{2}\ \frac{3}{2}$	0	000+	$+\sqrt{11}/2.7\sqrt{2.13.17}$

SO_3–O 3jm Factors (cont.)

$8\ \frac{13}{2}\ \frac{5}{2}$

$2\ \frac{3}{2}\ \frac{3}{2}\ 0\ 010+\quad -337/4.3.7\sqrt{3.13.17}$
$2\ \frac{3}{2}\ \frac{3}{2}\ 0\ 100+\quad +9\sqrt{3}/2\sqrt{2.5.7.17}$
$2\ \frac{3}{2}\ \frac{3}{2}\ 0\ 110+\quad -11\sqrt{11}/4.3\sqrt{5.7.17}$
$2\ \frac{3}{2}\ \tilde{\frac{1}{2}}\ 0\ 000-\quad +\sqrt{5.11}/2\sqrt{13.17}$
$2\ \frac{3}{2}\ \tilde{\frac{1}{2}}\ 0\ 010-\quad +5\sqrt{5}/3\sqrt{2.3.13.17}$
$2\ \frac{3}{2}\ \tilde{\frac{1}{2}}\ 0\ 100-\quad -\sqrt{3}/2\sqrt{7.17}$
$2\ \frac{3}{2}\ \tilde{\frac{1}{2}}\ 0\ 110-\quad +\sqrt{11}/3\sqrt{2.7.17}$
$2\ \tilde{\frac{1}{2}}\ \frac{3}{2}\ 0\ 000-\quad -9\sqrt{5.11}/4.7\sqrt{2.13.17}$
$2\ \tilde{\frac{1}{2}}\ \frac{3}{2}\ 0\ 010-\quad +13/3.7\sqrt{3.17}$
$2\ \tilde{\frac{1}{2}}\ \frac{3}{2}\ 0\ 100-\quad -\sqrt{3}/4\sqrt{2.7.17}$
$2\ \tilde{\frac{1}{2}}\ \frac{3}{2}\ 0\ 110-\quad -\sqrt{11.13}/3\sqrt{5.7.17}$
$\tilde{1}\ \frac{1}{2}\ \frac{3}{2}\ 0\ 000-\quad -3\sqrt{7}/2.17\sqrt{5}$
$\tilde{1}\ \frac{1}{2}\ \frac{3}{2}\ 0\ 100-\quad -11\sqrt{11}/2.17\sqrt{3.13}$
$\tilde{1}\ \frac{1}{2}\ \tilde{\frac{1}{2}}\ 0\ 000+\quad -2.3/17\sqrt{7}$
$\tilde{1}\ \frac{1}{2}\ \tilde{\frac{1}{2}}\ 0\ 100+\quad +\sqrt{5.11}/2.17\sqrt{3.13}$
$\tilde{1}\ \frac{3}{2}\ \frac{3}{2}\ 0\ 000+\quad +4.3\sqrt{11}/7.17\sqrt{5}$
$\tilde{1}\ \frac{3}{2}\ \frac{3}{2}\ 0\ 010+\quad +5\sqrt{2.3.5}/7.17$
$\tilde{1}\ \frac{3}{2}\ \frac{3}{2}\ 0\ 100+\quad -\sqrt{3.13}/2.17\sqrt{7}$
$\tilde{1}\ \frac{3}{2}\ \frac{3}{2}\ 0\ 110+\quad +\sqrt{11.13}/3.17\sqrt{2.7}$
$\tilde{1}\ \frac{3}{2}\ \frac{3}{2}\ 1\ 000-\quad -23\sqrt{11}/2.7.17\sqrt{5}$
$\tilde{1}\ \frac{3}{2}\ \frac{3}{2}\ 1\ 010-\quad +2\sqrt{2}/7\sqrt{3.5}$
$\tilde{1}\ \frac{3}{2}\ \frac{3}{2}\ 1\ 100-\quad +8\sqrt{3}/17\sqrt{7.13}$
$\tilde{1}\ \frac{3}{2}\ \frac{3}{2}\ 1\ 110-\quad -\sqrt{11}/\sqrt{2.7.13}$
$\tilde{1}\ \frac{3}{2}\ \tilde{\frac{1}{2}}\ 0\ 000-\quad -5\sqrt{11}/7.17\sqrt{2}$
$\tilde{1}\ \frac{3}{2}\ \tilde{\frac{1}{2}}\ 0\ 010-\quad +67/2.7.17\sqrt{3}$
$\tilde{1}\ \frac{3}{2}\ \tilde{\frac{1}{2}}\ 0\ 100-\quad +\sqrt{2.3.5}/17\sqrt{7.13}$
$\tilde{1}\ \frac{3}{2}\ \tilde{\frac{1}{2}}\ 0\ 110-\quad +5\sqrt{5.11}/3.17\sqrt{7.13}$
$\tilde{1}\ \tilde{\frac{1}{2}}\ \frac{3}{2}\ 0\ 000-\quad +5\sqrt{11}/2.7.17$
$\tilde{1}\ \tilde{\frac{1}{2}}\ \frac{3}{2}\ 0\ 010-\quad +\sqrt{13}/7.17\sqrt{2.3.5}$
$\tilde{1}\ \tilde{\frac{1}{2}}\ \frac{3}{2}\ 0\ 100-\quad +3.5\sqrt{3.5}/2.17\sqrt{7.13}$
$\tilde{1}\ \tilde{\frac{1}{2}}\ \frac{3}{2}\ 0\ 110-\quad -5\sqrt{2.11}/3.17\sqrt{7}$

$8\ \frac{13}{2}\ \frac{7}{2}$

$0\ \frac{1}{2}\ \frac{1}{2}\ 0\ 000-\quad +\sqrt{7.11}/\sqrt{3.13.17.19}$
$0\ \frac{3}{2}\ \frac{3}{2}\ 0\ 000-\quad +\sqrt{5.19}/\sqrt{3.7.13.17}$
$0\ \frac{3}{2}\ \frac{3}{2}\ 0\ 010-\quad +\sqrt{2.5.11}/\sqrt{7.13.17.19}$
$0\ \tilde{\frac{1}{2}}\ \tilde{\frac{1}{2}}\ 0\ 000-\quad -\sqrt{19}/\sqrt{7.13.17}$
$0\ \tilde{\frac{1}{2}}\ \tilde{\frac{1}{2}}\ 0\ 010-\quad +\sqrt{2.5.11}/\sqrt{3.7.17.19}$
$1\ \frac{1}{2}\ \frac{1}{2}\ 0\ 000+\quad -11\sqrt{7.11}/2.3\sqrt{2.3.13.17.19}$
$1\ \frac{1}{2}\ \frac{1}{2}\ 0\ 100+\quad -\sqrt{5.11}/3\sqrt{2.3.17.19}$
$1\ \frac{1}{2}\ \frac{3}{2}\ 0\ 000-\quad +\sqrt{2.5.11}/3\sqrt{3.13.17.19}$
$1\ \frac{1}{2}\ \frac{3}{2}\ 0\ 100-\quad -43\sqrt{11}/2.3\sqrt{2.3.7.17.19}$
$1\ \frac{3}{2}\ \frac{1}{2}\ 0\ 000-\quad -29/4.3\sqrt{3.13.17.19}$
$1\ \frac{3}{2}\ \frac{1}{2}\ 0\ 010-\quad -\sqrt{11.17}/2.3\sqrt{2.13.19}$
$1\ \frac{3}{2}\ \frac{1}{2}\ 0\ 100-\quad -47\sqrt{5}/4.3\sqrt{3.7.17.19}$
$1\ \frac{3}{2}\ \frac{1}{2}\ 0\ 110-\quad -2\sqrt{2.5.11}/3\sqrt{7.17.19}$
$1\ \frac{3}{2}\ \frac{3}{2}\ 0\ 000+\quad +\sqrt{5.7.17}/2.3\sqrt{2.3.13.19}$
$1\ \frac{3}{2}\ \frac{3}{2}\ 0\ 010+\quad +\sqrt{5.7.11}/2\sqrt{13.17.19}$
$1\ \frac{3}{2}\ \frac{3}{2}\ 0\ 100+\quad +43\sqrt{2}/3.7\sqrt{3.17.19}$
$1\ \frac{3}{2}\ \frac{3}{2}\ 0\ 110+\quad -5\sqrt{11}/7\sqrt{17.19}$

$8\ \frac{13}{2}\ \frac{7}{2}$

$1\ \frac{3}{2}\ \frac{3}{2}\ 1\ 000+\quad +2.11\sqrt{2.5}/3\sqrt{3.7.13.17.19}$
$1\ \frac{3}{2}\ \frac{3}{2}\ 1\ 010+\quad +5\sqrt{5.11}/2.3\sqrt{7.13.17.19}$
$1\ \frac{3}{2}\ \frac{3}{2}\ 1\ 100+\quad -109/2.3.7\sqrt{2.3.17.19}$
$1\ \frac{3}{2}\ \frac{3}{2}\ 1\ 110+\quad -47\sqrt{11}/2.3.7\sqrt{17.19}$
$1\ \frac{3}{2}\ \tilde{\frac{1}{2}}\ 0\ 000-\quad -3\sqrt{5.19}/4\sqrt{7.13.17}$
$1\ \frac{3}{2}\ \tilde{\frac{1}{2}}\ 0\ 010-\quad +43\sqrt{5.11}/2.3\sqrt{2.3.7.13.17.19}$
$1\ \frac{3}{2}\ \tilde{\frac{1}{2}}\ 0\ 100-\quad +3.5/4.7\sqrt{17.19}$
$1\ \frac{3}{2}\ \tilde{\frac{1}{2}}\ 0\ 110-\quad +31\sqrt{11}/3.7\sqrt{2.3.17.19}$
$1\ \tilde{\frac{1}{2}}\ \frac{3}{2}\ 0\ 000-\quad -2\sqrt{2.19}/\sqrt{3.7.13.17}$
$1\ \tilde{\frac{1}{2}}\ \frac{3}{2}\ 0\ 010-\quad +\sqrt{5.11}/2.3\sqrt{7.17.19}$
$1\ \tilde{\frac{1}{2}}\ \frac{3}{2}\ 0\ 100-\quad -\sqrt{5.17}/2.7\sqrt{2.3.19}$
$1\ \tilde{\frac{1}{2}}\ \frac{3}{2}\ 0\ 110-\quad -2\sqrt{11.13}/3.7\sqrt{17.19}$
$1\ \tilde{\frac{1}{2}}\ \tilde{\frac{1}{2}}\ 0\ 000+\quad -\sqrt{19}/2\sqrt{2.7.13.17}$
$1\ \tilde{\frac{1}{2}}\ \tilde{\frac{1}{2}}\ 0\ 010+\quad -11\sqrt{5.11}/2.3\sqrt{3.7.17.19}$
$1\ \tilde{\frac{1}{2}}\ \tilde{\frac{1}{2}}\ 0\ 100+\quad +\sqrt{5}/7\sqrt{2.17.19}$
$1\ \tilde{\frac{1}{2}}\ \tilde{\frac{1}{2}}\ 0\ 110+\quad +4\sqrt{11.13}/3.7\sqrt{3.17.19}$
$2\ \frac{1}{2}\ \frac{3}{2}\ 0\ 000+\quad -127/4\sqrt{3.7.13.17.19}$
$2\ \frac{1}{2}\ \frac{3}{2}\ 0\ 100+\quad -\sqrt{5.11}/4\sqrt{17.19}$
$2\ \frac{3}{2}\ \frac{1}{2}\ 0\ 000+\quad -\sqrt{5.11}/\sqrt{2.3.7.13.17.19}$
$2\ \frac{3}{2}\ \frac{1}{2}\ 0\ 010+\quad +2\sqrt{5}/\sqrt{7.13.17.19}$
$2\ \frac{3}{2}\ \frac{1}{2}\ 0\ 100+\quad -1/\sqrt{2.17.19}$
$2\ \frac{3}{2}\ \frac{1}{2}\ 0\ 110+\quad -2\sqrt{11}/\sqrt{3.17.19}$
$2\ \frac{3}{2}\ \frac{3}{2}\ 0\ 000-\quad -41\sqrt{11}/7\sqrt{3.13.17.19}$
$2\ \frac{3}{2}\ \frac{3}{2}\ 0\ 010-\quad +9.3/2.7\sqrt{2.13.17.19}$
$2\ \frac{3}{2}\ \frac{3}{2}\ 0\ 100-\quad -\sqrt{5}/\sqrt{7.17.19}$
$2\ \frac{3}{2}\ \frac{3}{2}\ 0\ 110-\quad -\sqrt{5.11}/2\sqrt{2.3.7.17.19}$
$2\ \frac{3}{2}\ \tilde{\frac{1}{2}}\ 0\ 000+\quad -\sqrt{11}/2\sqrt{2.13.17.19}$
$2\ \frac{3}{2}\ \tilde{\frac{1}{2}}\ 0\ 010+\quad -47/2\sqrt{3.13.17.19}$
$2\ \frac{3}{2}\ \tilde{\frac{1}{2}}\ 0\ 100+\quad -3\sqrt{3.5}/2\sqrt{2.7.17.19}$
$2\ \frac{3}{2}\ \tilde{\frac{1}{2}}\ 0\ 110+\quad +\sqrt{5.11}/2\sqrt{7.17.19}$
$2\ \tilde{\frac{1}{2}}\ \frac{3}{2}\ 0\ 000+\quad -\sqrt{3.5.11.13}/4.7\sqrt{17.19}$
$2\ \tilde{\frac{1}{2}}\ \frac{3}{2}\ 0\ 010+\quad -53/2.7\sqrt{2.17.19}$
$2\ \tilde{\frac{1}{2}}\ \frac{3}{2}\ 0\ 100+\quad +3.11/4\sqrt{7.17.19}$
$2\ \tilde{\frac{1}{2}}\ \frac{3}{2}\ 0\ 110+\quad -\sqrt{5.11.13}/2\sqrt{2.3.7.17.19}$
$\tilde{1}\ \frac{1}{2}\ \frac{3}{2}\ 0\ 000+\quad 0$
$\tilde{1}\ \frac{1}{2}\ \frac{3}{2}\ 0\ 100+\quad -\sqrt{11}/2\sqrt{2.13.19}$
$\tilde{1}\ \frac{1}{2}\ \tilde{\frac{1}{2}}\ 0\ 000-\quad -3\sqrt{5.19}/2.17\sqrt{2.7}$
$\tilde{1}\ \frac{1}{2}\ \tilde{\frac{1}{2}}\ 0\ 100-\quad -\sqrt{3.11}/17\sqrt{2.13.19}$
$\tilde{1}\ \frac{3}{2}\ \frac{1}{2}\ 0\ 000+\quad -5\sqrt{11.19}/4.17\sqrt{3.7}$
$\tilde{1}\ \frac{3}{2}\ \frac{1}{2}\ 0\ 010+\quad -11/2.3.17\sqrt{2.7.19}$
$\tilde{1}\ \frac{3}{2}\ \frac{1}{2}\ 0\ 100+\quad +3.7\sqrt{5}/4.17\sqrt{13.19}$
$\tilde{1}\ \frac{3}{2}\ \frac{1}{2}\ 0\ 110+\quad +7\sqrt{5.11}/17\sqrt{2.3.13.19}$
$\tilde{1}\ \frac{3}{2}\ \frac{3}{2}\ 0\ 000-\quad +29\sqrt{5.11}/2.7.17\sqrt{2.3.19}$
$\tilde{1}\ \frac{3}{2}\ \frac{3}{2}\ 0\ 010-\quad +71\sqrt{5}/2.3.7.17\sqrt{19}$
$\tilde{1}\ \frac{3}{2}\ \frac{3}{2}\ 0\ 100-\quad +113\sqrt{2}/17\sqrt{7.13.19}$
$\tilde{1}\ \frac{3}{2}\ \frac{3}{2}\ 0\ 110-\quad -59\sqrt{11}/17\sqrt{3.7.13.19}$
$\tilde{1}\ \frac{3}{2}\ \frac{3}{2}\ 1\ 000+\quad -13\sqrt{2.5.11}/7.17\sqrt{3.19}$
$\tilde{1}\ \frac{3}{2}\ \frac{3}{2}\ 1\ 010+\quad +11\sqrt{5}/2.3.7.17\sqrt{19}$
$\tilde{1}\ \frac{3}{2}\ \frac{3}{2}\ 1\ 100+\quad -83/2.17\sqrt{2.7.13.19}$

SO_3–O 3jm Factors (cont.)

$8\ \frac{13}{2}\ \frac{7}{2}$					
$\tilde{1}$	$\frac{3}{2}$	$\frac{3}{2}$	1	110+	$-\sqrt{11.19}/2.17\sqrt{3.7.13}$
$\tilde{1}$	$\frac{3}{2}$	$\widetilde{\frac{1}{2}}$	0	000+	$-9\sqrt{5.11}/4.7.17\sqrt{19}$
$\tilde{1}$	$\frac{3}{2}$	$\widetilde{\frac{1}{2}}$	0	010+	$+5.5\sqrt{3.5}/2.7.17\sqrt{2.19}$
$\tilde{1}$	$\frac{3}{2}$	$\widetilde{\frac{1}{2}}$	0	100+	$+137\sqrt{3}/4.17\sqrt{7.13.19}$
$\tilde{1}$	$\frac{3}{2}$	$\widetilde{\frac{1}{2}}$	0	110+	$+2\sqrt{2.11}/17\sqrt{7.13.19}$
$\tilde{1}$	$\widetilde{\frac{1}{2}}$	$\frac{1}{2}$	0	000−	$-\sqrt{5.11.19}/2.17\sqrt{2.3.7}$
$\tilde{1}$	$\widetilde{\frac{1}{2}}$	$\frac{1}{2}$	0	010−	$-\sqrt{13}/2.3.17\sqrt{7.19}$
$\tilde{1}$	$\widetilde{\frac{1}{2}}$	$\frac{1}{2}$	0	100−	$-3.5/17\sqrt{2.13.19}$
$\tilde{1}$	$\widetilde{\frac{1}{2}}$	$\frac{1}{2}$	0	110−	$-4\sqrt{5.11}/17\sqrt{3.19}$
$\tilde{1}$	$\widetilde{\frac{1}{2}}$	$\frac{3}{2}$	0	000+	$+\sqrt{2.11.19}/7.17\sqrt{3}$
$\tilde{1}$	$\widetilde{\frac{1}{2}}$	$\frac{3}{2}$	0	010+	$+5\sqrt{5.13}/2.3.7.17\sqrt{19}$
$\tilde{1}$	$\widetilde{\frac{1}{2}}$	$\frac{3}{2}$	0	100+	$-3.43\sqrt{5}/2.17\sqrt{2.7.13.19}$
$\tilde{1}$	$\widetilde{\frac{1}{2}}$	$\frac{3}{2}$	0	110+	$-2\sqrt{11}/17\sqrt{3.7.19}$

$8\ \frac{13}{2}\ \frac{9}{2}$					
0	$\frac{1}{2}$	$\frac{1}{2}$	0	000+	$-2\sqrt{2.7}/\sqrt{3.13.17.19}$
0	$\frac{3}{2}$	$\frac{3}{2}$	0	000+	$+\sqrt{2.5.11}/\sqrt{13.17.19}$
0	$\frac{3}{2}$	$\frac{3}{2}$	0	001+	$+\sqrt{2.5.11}/\sqrt{3.7.13.17.19}$
0	$\frac{3}{2}$	$\frac{3}{2}$	0	010+	$-5\sqrt{5}/\sqrt{3.13.17.19}$
0	$\frac{3}{2}$	$\frac{3}{2}$	0	011+	$+2.5\sqrt{5}/\sqrt{7.13.17.19}$
1	$\frac{1}{2}$	$\frac{1}{2}$	0	000−	$+2\sqrt{7}/3\sqrt{3.13.17.19}$
1	$\frac{1}{2}$	$\frac{1}{2}$	0	100−	$+11/2.3\sqrt{3.5.17.19}$
1	$\frac{1}{2}$	$\frac{3}{2}$	0	000+	$-\sqrt{7.19}/2\sqrt{5.13.17}$
1	$\frac{1}{2}$	$\frac{3}{2}$	0	001+	$+7.7/2.3\sqrt{3.5.13.17.19}$
1	$\frac{1}{2}$	$\frac{3}{2}$	0	100+	$+13/2.5\sqrt{17.19}$
1	$\frac{1}{2}$	$\frac{3}{2}$	0	101+	$+13\sqrt{7}/3.5\sqrt{3.17.19}$
1	$\frac{3}{2}$	$\frac{1}{2}$	0	000+	$-\sqrt{11.17}/3\sqrt{2.3.13.19}$
1	$\frac{3}{2}$	$\frac{1}{2}$	0	010+	$+\sqrt{19}/2.3\sqrt{13.17}$
1	$\frac{3}{2}$	$\frac{1}{2}$	0	100+	$-4\sqrt{2.5.11}/3\sqrt{3.7.17.19}$
1	$\frac{3}{2}$	$\frac{1}{2}$	0	110+	$-11/3\sqrt{5.7.17.19}$
1	$\frac{3}{2}$	$\frac{3}{2}$	0	000−	0
1	$\frac{3}{2}$	$\frac{3}{2}$	0	001−	$+2.5\sqrt{5.11}/3\sqrt{3.7.13.17.19}$
1	$\frac{3}{2}$	$\frac{3}{2}$	0	010−	$+7\sqrt{2.5}/3\sqrt{3.13.17.19}$
1	$\frac{3}{2}$	$\frac{3}{2}$	0	011−	$-2\sqrt{2.5}/\sqrt{7.13.17.19}$
1	$\frac{3}{2}$	$\frac{3}{2}$	0	100−	$-3\sqrt{7.11}/2.5\sqrt{17.19}$
1	$\frac{3}{2}$	$\frac{3}{2}$	0	101	$+43\sqrt{11}/3.5.7\sqrt{3.17.19}$
1	$\frac{3}{2}$	$\frac{3}{2}$	0	110−	$+\sqrt{7}/3\sqrt{2.3.17.19}$
1	$\frac{3}{2}$	$\frac{3}{2}$	0	111−	$-11/7\sqrt{2.17.19}$
1	$\frac{3}{2}$	$\frac{3}{2}$	1	000−	$+\sqrt{11}/2\sqrt{5.13.17.19}$
1	$\frac{3}{2}$	$\frac{3}{2}$	1	001−	$+8.2\sqrt{7.11}/3\sqrt{3.5.13.17.19}$
1	$\frac{3}{2}$	$\frac{3}{2}$	1	010−	$+2.11\sqrt{2.5}/3\sqrt{3.13.17.19}$
1	$\frac{3}{2}$	$\frac{3}{2}$	1	011−	$-\sqrt{2.5.7}/3\sqrt{13.17.19}$
1	$\frac{3}{2}$	$\frac{3}{2}$	1	100−	$-8\sqrt{11}/5\sqrt{7.17.19}$
1	$\frac{3}{2}$	$\frac{3}{2}$	1	101−	$+2\sqrt{11.19}/3.5.7\sqrt{3.17}$
1	$\frac{3}{2}$	$\frac{3}{2}$	1	110−	$+29/3.5\sqrt{2.3.7.17.19}$
1	$\frac{3}{2}$	$\frac{3}{2}$	1	111−	$+2.151\sqrt{2}/3.5.7\sqrt{17.19}$
1	$\widetilde{\frac{1}{2}}$	$\frac{3}{2}$	0	000+	$-5\sqrt{11}/2\sqrt{13.17.19}$
1	$\widetilde{\frac{1}{2}}$	$\frac{3}{2}$	0	001+	$-5\sqrt{11}/2\sqrt{3.7.13.17.19}$
1	$\widetilde{\frac{1}{2}}$	$\frac{3}{2}$	0	010+	$+5\sqrt{5}/3\sqrt{2.3.17.19}$
1	$\widetilde{\frac{1}{2}}$	$\frac{3}{2}$	0	011+	$-5\sqrt{2.5}/3\sqrt{7.17.19}$
1	$\widetilde{\frac{1}{2}}$	$\frac{3}{2}$	0	100+	$-\sqrt{11}/2\sqrt{5.7.17.19}$
1	$\widetilde{\frac{1}{2}}$	$\frac{3}{2}$	0	101+	$+\sqrt{11}/7\sqrt{3.5.17.19}$
1	$\widetilde{\frac{1}{2}}$	$\frac{3}{2}$	0	110+	$+13\sqrt{2.13}/3.5\sqrt{3.7.17.19}$
1	$\widetilde{\frac{1}{2}}$	$\frac{3}{2}$	0	111+	$+73\sqrt{2.13}/3.5.7\sqrt{17.19}$
2	$\frac{1}{2}$	$\frac{3}{2}$	0	000−	$+3.59/4\sqrt{2.11.13.17.19}$
2	$\frac{1}{2}$	$\frac{3}{2}$	0	001−	$-127/\sqrt{2.3.7.11.13.17.19}$
2	$\frac{1}{2}$	$\frac{3}{2}$	0	100−	$-\sqrt{3.5.7}/4\sqrt{2.17.19}$
2	$\frac{1}{2}$	$\frac{3}{2}$	0	101−	$-\sqrt{5}/\sqrt{2.17.19}$
2	$\frac{3}{2}$	$\frac{1}{2}$	0	000−	$-\sqrt{17}/\sqrt{3.5.7.13.19}$
2	$\frac{3}{2}$	$\frac{1}{2}$	0	010−	$+\sqrt{2.13}/\sqrt{5.7.11.17.19}$
2	$\frac{3}{2}$	$\frac{1}{2}$	0	100−	$+\sqrt{11}/\sqrt{17.19}$
2	$\frac{3}{2}$	$\frac{1}{2}$	0	110−	$+\sqrt{2}/\sqrt{3.17.19}$
2	$\frac{3}{2}$	$\frac{3}{2}$	0	000+	$-193/2.5\sqrt{2.7.13.17.19}$
2	$\frac{3}{2}$	$\frac{3}{2}$	0	001+	$+233/5.7\sqrt{2.3.13.17.19}$
2	$\frac{3}{2}$	$\frac{3}{2}$	0	010+	$-1933/4.5\sqrt{3.7.11.13.17.19}$
2	$\frac{3}{2}$	$\frac{3}{2}$	0	011+	$-8.8.2.9/5.7\sqrt{11.13.17.19}$
2	$\frac{3}{2}$	$\frac{3}{2}$	0	100+	$+\sqrt{3.11}/2\sqrt{2.5.17.19}$
2	$\frac{3}{2}$	$\frac{3}{2}$	0	101+	$-\sqrt{11}/\sqrt{2.5.7.17.19}$
2	$\frac{3}{2}$	$\frac{3}{2}$	0	110+	$-\sqrt{19}/4\sqrt{5.17}$
2	$\frac{3}{2}$	$\frac{3}{2}$	0	111+	$-8/\sqrt{3.5.7.17.19}$
2	$\widetilde{\frac{1}{2}}$	$\frac{3}{2}$	0	000−	$-5\sqrt{5.7}/4\sqrt{2.13.17.19}$
2	$\widetilde{\frac{1}{2}}$	$\frac{3}{2}$	0	001−	$-\sqrt{3.5.13}/7\sqrt{2.17.19}$
2	$\widetilde{\frac{1}{2}}$	$\frac{3}{2}$	0	010−	$-23\sqrt{7}/5\sqrt{3.11.17.19}$
2	$\widetilde{\frac{1}{2}}$	$\frac{3}{2}$	0	011−	$+53\sqrt{11}/2.5.7\sqrt{17.19}$
2	$\widetilde{\frac{1}{2}}$	$\frac{3}{2}$	0	100−	$-\sqrt{3.11}/4\sqrt{2.17.19}$
2	$\widetilde{\frac{1}{2}}$	$\frac{3}{2}$	0	101−	$+3\sqrt{11}/\sqrt{2.7.17.19}$
2	$\widetilde{\frac{1}{2}}$	$\frac{3}{2}$	0	110−	$+\sqrt{13}/\sqrt{5.17.19}$
2	$\widetilde{\frac{1}{2}}$	$\frac{3}{2}$	0	111−	$+11\sqrt{13}/2\sqrt{3.5.7.17.19}$
$\tilde{1}$	$\frac{1}{2}$	$\frac{3}{2}$	0	000−	$+3.59/2.17\sqrt{5.11.19}$
$\tilde{1}$	$\frac{1}{2}$	$\frac{3}{2}$	0	001−	$-3.31\sqrt{3}/2.17\sqrt{5.7.11.19}$
$\tilde{1}$	$\frac{1}{2}$	$\frac{3}{2}$	0	100−	$-43\sqrt{3.7}/2.5.17\sqrt{13.19}$
$\tilde{1}$	$\frac{1}{2}$	$\frac{3}{2}$	0	101−	$-157/5.17\sqrt{13.19}$
$\tilde{1}$	$\frac{3}{2}$	$\frac{1}{2}$	0	000−	$-\sqrt{19}/17\sqrt{2.3.7}$
$\tilde{1}$	$\frac{3}{2}$	$\frac{1}{2}$	0	010−	$-1/2.3\sqrt{7.11.19}$
$\tilde{1}$	$\frac{3}{2}$	$\frac{1}{2}$	0	100−	$-2.3\sqrt{2.11}/17\sqrt{5.13.19}$
$\tilde{1}$	$\frac{3}{2}$	$\frac{1}{2}$	0	110−	$-11/\sqrt{3.5.13.19}$
$\tilde{1}$	$\frac{3}{2}$	$\frac{3}{2}$	0	000+	0
$\tilde{1}$	$\frac{3}{2}$	$\frac{3}{2}$	0	001+	$-2.29\sqrt{5}/7.17\sqrt{3.19}$
$\tilde{1}$	$\frac{3}{2}$	$\frac{3}{2}$	0	010+	$-\sqrt{2.3.5.7}/17\sqrt{11.19}$
$\tilde{1}$	$\frac{3}{2}$	$\frac{3}{2}$	0	011+	$-2.71\sqrt{2.5}/3.7.17\sqrt{11.19}$
$\tilde{1}$	$\frac{3}{2}$	$\frac{3}{2}$	0	100+	$+\sqrt{3.11}/2.5\sqrt{13.19}$
$\tilde{1}$	$\frac{3}{2}$	$\frac{3}{2}$	0	101+	$+113\sqrt{11}/5.17\sqrt{7.13.19}$
$\tilde{1}$	$\frac{3}{2}$	$\frac{3}{2}$	0	110+	$-53/5.17\sqrt{2.13.19}$
$\tilde{1}$	$\frac{3}{2}$	$\frac{3}{2}$	0	111+	$-11.59/5.17\sqrt{2.3.7.13.19}$
$\tilde{1}$	$\frac{3}{2}$	$\frac{3}{2}$	1	000−	$+3.43/2.17\sqrt{5.7.19}$
$\tilde{1}$	$\frac{3}{2}$	$\frac{3}{2}$	1	001−	$+8.2/7.17\sqrt{3.5.19}$
$\tilde{1}$	$\frac{3}{2}$	$\frac{3}{2}$	1	010−	$-3.29\sqrt{2.3}/17\sqrt{5.7.11.19}$
$\tilde{1}$	$\frac{3}{2}$	$\frac{3}{2}$	1	011−	$+457\sqrt{2}/3.7.17\sqrt{5.11.19}$
$\tilde{1}$	$\frac{3}{2}$	$\frac{3}{2}$	1	100−	$+2.3\sqrt{3.11}/17\sqrt{13.19}$

SO_3–O 3jm Factors (cont.)

$8\ \frac{13}{2}\ \frac{9}{2}$

$\tilde{1}\ \frac{3}{2}\ \frac{3}{2}\ 1\ 101-\quad +2\sqrt{11.13}/17\sqrt{7.19}$
$\tilde{1}\ \frac{3}{2}\ \frac{3}{2}\ 1\ 110-\quad +9.9/5.17\sqrt{2.13.19}$
$\tilde{1}\ \frac{3}{2}\ \frac{3}{2}\ 1\ 111-\quad +2.317\sqrt{2}/5.17\sqrt{3.7.13.19}$
$\tilde{1}\ \tilde{\frac{1}{2}}\ \frac{1}{2}\ 0\ 000+\quad -2\sqrt{5.19}/17\sqrt{3.7}$
$\tilde{1}\ \tilde{\frac{1}{2}}\ \frac{1}{2}\ 0\ 010+\quad -2\sqrt{2.13}/3.17\sqrt{7.11.19}$
$\tilde{1}\ \tilde{\frac{1}{2}}\ \frac{1}{2}\ 0\ 100+\quad +3\sqrt{11}/2.17\sqrt{13.19}$
$\tilde{1}\ \tilde{\frac{1}{2}}\ \frac{1}{2}\ 0\ 110+\quad +2.11\sqrt{2}/17\sqrt{3.5.19}$
$\tilde{1}\ \tilde{\frac{1}{2}}\ \frac{3}{2}\ 0\ 000-\quad -9/2.17\sqrt{7.19}$
$\tilde{1}\ \tilde{\frac{1}{2}}\ \frac{3}{2}\ 0\ 001-\quad -1/2.17\sqrt{3.19}$
$\tilde{1}\ \tilde{\frac{1}{2}}\ \frac{3}{2}\ 0\ 010-\quad +\sqrt{3.13.19}/17\sqrt{2.5.7.11}$
$\tilde{1}\ \tilde{\frac{1}{2}}\ \frac{3}{2}\ 0\ 011-\quad -\sqrt{2.11.13}/3.17\sqrt{5.19}$
$\tilde{1}\ \tilde{\frac{1}{2}}\ \frac{3}{2}\ 0\ 100-\quad +31\sqrt{3.11}/2.17\sqrt{5.13.19}$
$\tilde{1}\ \tilde{\frac{1}{2}}\ \frac{3}{2}\ 0\ 101-\quad -3\sqrt{7.11}/17\sqrt{5.13.19}$
$\tilde{1}\ \tilde{\frac{1}{2}}\ \frac{3}{2}\ 0\ 110-\quad +\sqrt{2}/5.17\sqrt{19}$
$\tilde{1}\ \tilde{\frac{1}{2}}\ \frac{3}{2}\ 0\ 111-\quad +\sqrt{2.7.19}/5.17\sqrt{3}$

$8\ \frac{13}{2}\ \frac{11}{2}$

$0\ \frac{1}{2}\ \frac{1}{2}\ 0\ 000-\quad +9/\sqrt{13.17.19}$
$0\ \frac{3}{2}\ \frac{3}{2}\ 0\ 000-\quad -4.11/7\sqrt{13.17.19}$
$0\ \frac{3}{2}\ \frac{3}{2}\ 0\ 001-\quad 0$
$0\ \frac{3}{2}\ \frac{3}{2}\ 0\ 010-\quad +2.97\sqrt{2}/7\sqrt{3.11.13.17.19}$
$0\ \frac{3}{2}\ \frac{3}{2}\ 0\ 011-\quad +7.7\sqrt{5}/3\sqrt{3.11.13.17.19}$
$0\ \tilde{\frac{1}{2}}\ \tilde{\frac{1}{2}}\ 0\ 000-\quad +3.11/7\sqrt{13.17.19}$
$0\ \tilde{\frac{1}{2}}\ \tilde{\frac{1}{2}}\ 0\ 010-\quad +4\sqrt{2.5.11}/3.7\sqrt{3.17.19}$
$1\ \frac{1}{2}\ \frac{1}{2}\ 0\ 000+\quad -3\sqrt{13}/2\sqrt{2.17.19}$
$1\ \frac{1}{2}\ \frac{1}{2}\ 0\ 100+\quad 0$
$1\ \frac{1}{2}\ \frac{3}{2}\ 0\ 000-\quad -3\sqrt{17}/2\sqrt{2.7.11.13.19}$
$1\ \frac{1}{2}\ \frac{3}{2}\ 0\ 001-\quad +7\sqrt{5.7}/3\sqrt{11.13.17.19}$
$1\ \frac{1}{2}\ \frac{3}{2}\ 0\ 100-\quad 0$
$1\ \frac{1}{2}\ \frac{3}{2}\ 0\ 101-\quad -\sqrt{11}/2.3\sqrt{17.19}$
$1\ \frac{3}{2}\ \frac{1}{2}\ 0\ 000-\quad +5\sqrt{11}/2.3\sqrt{7.13.17.19}$
$1\ \frac{3}{2}\ \frac{1}{2}\ 0\ 010-\quad -5\sqrt{13}/3\sqrt{2.3.7.17.19}$
$1\ \frac{3}{2}\ \frac{1}{2}\ 0\ 100-\quad +\sqrt{5.11}/3\sqrt{17.19}$
$1\ \frac{3}{2}\ \frac{1}{2}\ 0\ 110-\quad +11\sqrt{5}/2.3\sqrt{2.3.17.19}$
$1\ \frac{3}{2}\ \frac{3}{2}\ 0\ 000+\quad +5.29\sqrt{2}/3.7\sqrt{13.17.19}$
$1\ \frac{3}{2}\ \frac{3}{2}\ 0\ 001+\quad +73\sqrt{5}/3.7\sqrt{13.17.19}$
$1\ \frac{3}{2}\ \frac{3}{2}\ 0\ 010+\quad -149\sqrt{3}/7\sqrt{11.13.17.19}$
$1\ \frac{3}{2}\ \frac{3}{2}\ 0\ 011+\quad +659\sqrt{5}/2.9.7\sqrt{2.3.11.13.17.19}$
$1\ \frac{3}{2}\ \frac{3}{2}\ 0\ 100+\quad -5\sqrt{2.5}/3\sqrt{7.17.19}$
$1\ \frac{3}{2}\ \frac{3}{2}\ 0\ 101+\quad +\sqrt{17}/3\sqrt{7.19}$
$1\ \frac{3}{2}\ \frac{3}{2}\ 0\ 110+\quad +2\sqrt{3.5}/\sqrt{7.11.17.19}$
$1\ \frac{3}{2}\ \frac{3}{2}\ 0\ 111+\quad -5.43\sqrt{2}/9\sqrt{3.7.11.17.19}$
$1\ \frac{3}{2}\ \frac{3}{2}\ 1\ 000+\quad -31\sqrt{2}/3.7\sqrt{13.17.19}$
$1\ \frac{3}{2}\ \frac{3}{2}\ 1\ 001+\quad +5.5\sqrt{5}/2.3.7\sqrt{13.17.19}$
$1\ \frac{3}{2}\ \frac{3}{2}\ 1\ 010+\quad -5.5.5/3.7\sqrt{3.11.13.17.19}$
$1\ \frac{3}{2}\ \frac{3}{2}\ 1\ 011+\quad +149\sqrt{5.19}/2.9.7\sqrt{2.3.11.13.17}$
$1\ \frac{3}{2}\ \frac{3}{2}\ 1\ 100+\quad -5\sqrt{5}/3\sqrt{2.7.17.19}$
$1\ \frac{3}{2}\ \frac{3}{2}\ 1\ 101+\quad +\sqrt{17}/2.3\sqrt{7.19}$
$1\ \frac{3}{2}\ \frac{3}{2}\ 1\ 110+\quad -73\sqrt{5}/2.3\sqrt{3.7.11.17.19}$
$1\ \frac{3}{2}\ \frac{3}{2}\ 1\ 111+\quad +149/9\sqrt{2.3.7.11.17.19}$

$8\ \frac{13}{2}\ \frac{11}{2}$

$1\ \frac{3}{2}\ \tilde{\frac{1}{2}}\ 0\ 000-\quad +11\sqrt{5}/2.7\sqrt{13.17.19}$
$1\ \frac{3}{2}\ \tilde{\frac{1}{2}}\ 0\ 010-\quad +11.11\sqrt{5.11}/9.7\sqrt{2.3.13.17.19}$
$1\ \frac{3}{2}\ \tilde{\frac{1}{2}}\ 0\ 100-\quad 0$
$1\ \frac{3}{2}\ \tilde{\frac{1}{2}}\ 0\ 110-\quad +\sqrt{7.11}/2.9\sqrt{2.3.17.19}$
$1\ \tilde{\frac{1}{2}}\ \frac{3}{2}\ 0\ 000-\quad +11\sqrt{5}/2.7\sqrt{2.13.17.19}$
$1\ \tilde{\frac{1}{2}}\ \frac{3}{2}\ 0\ 001-\quad +2.11/7\sqrt{13.17.19}$
$1\ \tilde{\frac{1}{2}}\ \frac{3}{2}\ 0\ 010-\quad -8.2\sqrt{19}/3.7\sqrt{3.11.17}$
$1\ \tilde{\frac{1}{2}}\ \frac{3}{2}\ 0\ 011-\quad -167\sqrt{5}/9.7\sqrt{2.3.11.17.19}$
$1\ \tilde{\frac{1}{2}}\ \frac{3}{2}\ 0\ 100-\quad +4\sqrt{2}/\sqrt{7.17.19}$
$1\ \tilde{\frac{1}{2}}\ \frac{3}{2}\ 0\ 101-\quad +3\sqrt{5}/\sqrt{7.17.19}$
$1\ \tilde{\frac{1}{2}}\ \frac{3}{2}\ 0\ 110-\quad -\sqrt{5.13}/3\sqrt{3.7.11.17.19}$
$1\ \tilde{\frac{1}{2}}\ \frac{3}{2}\ 0\ 111-\quad +4\sqrt{2.13}/9\sqrt{3.7.11.17.19}$
$1\ \tilde{\frac{1}{2}}\ \tilde{\frac{1}{2}}\ 0\ 000+\quad +11\sqrt{13}/2.7\sqrt{2.17.19}$
$1\ \tilde{\frac{1}{2}}\ \tilde{\frac{1}{2}}\ 0\ 010+\quad -2\sqrt{5.11}/9.7\sqrt{3.17.19}$
$1\ \tilde{\frac{1}{2}}\ \tilde{\frac{1}{2}}\ 0\ 100+\quad 0$
$1\ \tilde{\frac{1}{2}}\ \tilde{\frac{1}{2}}\ 0\ 110+\quad +2\sqrt{7.11.13}/9\sqrt{3.17.19}$
$2\ \frac{1}{2}\ \frac{3}{2}\ 0\ 000+\quad +3.67\sqrt{5}/4.11\sqrt{13.17.19}$
$2\ \frac{1}{2}\ \frac{3}{2}\ 0\ 001+\quad -9.7/4.11\sqrt{2.13.17.19}$
$2\ \frac{1}{2}\ \frac{3}{2}\ 0\ 100+\quad -9.3\sqrt{3}/4\sqrt{7.11.17.19}$
$2\ \frac{1}{2}\ \frac{3}{2}\ 0\ 101+\quad -23\sqrt{3.5}/4\sqrt{2.7.11.17.19}$
$2\ \frac{3}{2}\ \frac{1}{2}\ 0\ 000+\quad -3\sqrt{5}/\sqrt{2.13.17.19}$
$2\ \frac{3}{2}\ \frac{1}{2}\ 0\ 010+\quad +5.5\sqrt{5}/\sqrt{3.11.13.17.19}$
$2\ \frac{3}{2}\ \frac{1}{2}\ 0\ 100+\quad +\sqrt{3.11}/\sqrt{2.7.17.19}$
$2\ \frac{3}{2}\ \frac{1}{2}\ 0\ 110+\quad +1/\sqrt{7.17.19}$
$2\ \frac{3}{2}\ \frac{3}{2}\ 0\ 000-\quad -5\sqrt{5.7}/2\sqrt{11.13.17.19}$
$2\ \frac{3}{2}\ \frac{3}{2}\ 0\ 001-\quad -3.41/2\sqrt{2.7.11.13.17.19}$
$2\ \frac{3}{2}\ \frac{3}{2}\ 0\ 010-\quad -5\sqrt{5.7}/2\sqrt{2.3.13.17.19}$
$2\ \frac{3}{2}\ \frac{3}{2}\ 0\ 011-\quad -\sqrt{13}/4.3\sqrt{3.7.17.19}$
$2\ \frac{3}{2}\ \frac{3}{2}\ 0\ 100-\quad -23\sqrt{3}/2.7\sqrt{17.19}$
$2\ \frac{3}{2}\ \frac{3}{2}\ 0\ 101-\quad +13\sqrt{3.5}/2.7\sqrt{2.17.19}$
$2\ \frac{3}{2}\ \frac{3}{2}\ 0\ 110-\quad +3/2.7\sqrt{2.11.17.19}$
$2\ \frac{3}{2}\ \frac{3}{2}\ 0\ 111-\quad +13.13\sqrt{5}/4.3.7\sqrt{11.17.19}$
$2\ \frac{3}{2}\ \tilde{\frac{1}{2}}\ 0\ 000+\quad +4\sqrt{2.11}/\sqrt{7.13.17.19}$
$2\ \frac{3}{2}\ \tilde{\frac{1}{2}}\ 0\ 010+\quad -67/3\sqrt{3.7.13.17.19}$
$2\ \frac{3}{2}\ \tilde{\frac{1}{2}}\ 0\ 100+\quad 0$
$2\ \frac{3}{2}\ \tilde{\frac{1}{2}}\ 0\ 110+\quad -\sqrt{5.11}/3\sqrt{17.19}$
$2\ \tilde{\frac{1}{2}}\ \frac{3}{2}\ 0\ 000+\quad -\sqrt{11.17}/4\sqrt{7.13.19}$
$2\ \tilde{\frac{1}{2}}\ \frac{3}{2}\ 0\ 001+\quad -\sqrt{5.11}/4\sqrt{2.7.13.17.19}$
$2\ \tilde{\frac{1}{2}}\ \frac{3}{2}\ 0\ 010+\quad +4.3\sqrt{2.3.5}/11\sqrt{7.17.19}$
$2\ \tilde{\frac{1}{2}}\ \frac{3}{2}\ 0\ 011+\quad -389/2.3.11\sqrt{3.7.17.19}$
$2\ \tilde{\frac{1}{2}}\ \frac{3}{2}\ 0\ 100+\quad -3\sqrt{3.5}/4.7\sqrt{17.19}$
$2\ \tilde{\frac{1}{2}}\ \frac{3}{2}\ 0\ 101+\quad -47\sqrt{3}/4.7\sqrt{2.17.19}$
$2\ \tilde{\frac{1}{2}}\ \frac{3}{2}\ 0\ 110+\quad -4\sqrt{2.13}/7\sqrt{11.17.19}$
$2\ \tilde{\frac{1}{2}}\ \frac{3}{2}\ 0\ 111+\quad -29\sqrt{5.13}/2.3.7\sqrt{11.17.19}$
$\tilde{1}\ \frac{1}{2}\ \frac{3}{2}\ 0\ 000+\quad +3.29/2.11.17\sqrt{2.19}$
$\tilde{1}\ \frac{1}{2}\ \frac{3}{2}\ 0\ 001+\quad +2.3\sqrt{5}/11.17\sqrt{19}$
$\tilde{1}\ \frac{1}{2}\ \frac{3}{2}\ 0\ 100+\quad +4\sqrt{2.3.5.7}/17\sqrt{11.13.19}$
$\tilde{1}\ \frac{1}{2}\ \frac{3}{2}\ 0\ 101+\quad -587/17\sqrt{3.7.11.13.19}$
$\tilde{1}\ \frac{1}{2}\ \tilde{\frac{1}{2}}\ 0\ 000-\quad +3\sqrt{5}/2.17\sqrt{2.19}$

SO_3–O 3jm Factors (cont.)

8 7 6	
$2\ \tilde{1}\ \tilde{1}\ 0\ 111-$	$+3\sqrt{3.11.13}/4\sqrt{2.5.7.17.19}$
$2\ \tilde{0}\ 2\ 0\ 000+$	$-5/3\sqrt{17.19}$
$2\ \tilde{0}\ 2\ 0\ 100+$	$-\sqrt{11.13}/\sqrt{3.5.7.17.19}$
$\tilde{1}\ 1\ 1\ 0\ 000-$	$-3/4.5.17\sqrt{19}$
$\tilde{1}\ 1\ 1\ 0\ 010-$	$+\sqrt{3.7.11}/5.17\sqrt{19}$
$\tilde{1}\ 1\ 1\ 0\ 100-$	$-8\sqrt{3.11}/17\sqrt{5.7.13.19}$
$\tilde{1}\ 1\ 1\ 0\ 110-$	$+7\sqrt{19}/4.17\sqrt{5.13}$
$\tilde{1}\ 1\ 2\ 0\ 000+$	$-9\sqrt{3}/2.5.17\sqrt{19}$
$\tilde{1}\ 1\ 2\ 0\ 010+$	$-7\sqrt{7.11}/5.17\sqrt{19}$
$\tilde{1}\ 1\ 2\ 0\ 100+$	$-4\sqrt{11.19}/17\sqrt{5.7.13}$
$\tilde{1}\ 1\ 2\ 0\ 110+$	$+53/2.17\sqrt{3.5.13.19}$
$\tilde{1}\ 1\ \tilde{1}\ 0\ 000+$	$+3\sqrt{3.11}/4.17\sqrt{5.7.19}$
$\tilde{1}\ 1\ \tilde{1}\ 0\ 001+$	$-8.2.3\sqrt{3}/5.17\sqrt{7.19}$
$\tilde{1}\ 1\ \tilde{1}\ 0\ 010+$	$-8/17\sqrt{5.19}$
$\tilde{1}\ 1\ \tilde{1}\ 0\ 011+$	$-9.3\sqrt{11}/4.5.17\sqrt{19}$
$\tilde{1}\ 1\ \tilde{1}\ 0\ 100+$	$+8.8.2/7.17\sqrt{13.19}$
$\tilde{1}\ 1\ \tilde{1}\ 0\ 101+$	$+9\sqrt{5.11}/7.17\sqrt{13.19}$
$\tilde{1}\ 1\ \tilde{1}\ 0\ 110+$	$+47\sqrt{11}/4.17\sqrt{3.7.13.19}$
$\tilde{1}\ 1\ \tilde{1}\ 0\ 111+$	$+8.4\sqrt{3.5}/17\sqrt{7.13.19}$
$\tilde{1}\ 1\ \tilde{0}\ 0\ 000-$	$+3\sqrt{3.11}/2.17\sqrt{5.19}$
$\tilde{1}\ 1\ \tilde{0}\ 0\ 010-$	$-\sqrt{7}/17\sqrt{5.19}$
$\tilde{1}\ 1\ \tilde{0}\ 0\ 100-$	$-8.2/17\sqrt{7.13.19}$
$\tilde{1}\ 1\ \tilde{0}\ 0\ 110-$	$-47\sqrt{11}/2.17\sqrt{3.13.19}$
$\tilde{1}\ 2\ 1\ 0\ 000+$	$+3\sqrt{11}/2.17\sqrt{5.19}$
$\tilde{1}\ 2\ 1\ 0\ 100+$	$+11.11/2.17\sqrt{3.7.13.19}$
$\tilde{1}\ 2\ \tilde{1}\ 0\ 000-$	$-9/2.17\sqrt{7.19}$
$\tilde{1}\ 2\ \tilde{1}\ 0\ 001-$	$-9\sqrt{5.11}/2.17\sqrt{7.19}$
$\tilde{1}\ 2\ \tilde{1}\ 0\ 100-$	$-67\sqrt{5.11}/2.7.17\sqrt{3.13.19}$
$\tilde{1}\ 2\ \tilde{1}\ 0\ 101-$	$+41\sqrt{3}/2.7.17\sqrt{13.19}$
$\tilde{1}\ \tilde{1}\ 0\ 0\ 000-$	$3\sqrt{3.11}/2.17\sqrt{5.19}$
$\tilde{1}\ \tilde{1}\ 0\ 0\ 010-$	$+3\sqrt{3.13}/17\sqrt{5.19}$
$\tilde{1}\ \tilde{1}\ 0\ 0\ 100-$	$+8.2/17\sqrt{7.13.19}$
$\tilde{1}\ \tilde{1}\ 0\ 0\ 110-$	$+5\sqrt{11}/2.17\sqrt{7.19}$
$\tilde{1}\ \tilde{1}\ 1\ 0\ 000+$	$+3\sqrt{3.11}/4.17\sqrt{5.7.19}$
$\tilde{1}\ \tilde{1}\ 1\ 0\ 010+$	$-4.3\sqrt{3.13}/17\sqrt{5.7.19}$
$\tilde{1}\ \tilde{1}\ 1\ 0\ 100+$	$+8.8.2/7.17\sqrt{13.19}$
$\tilde{1}\ \tilde{1}\ 1\ 0\ 110+$	$+9.5\sqrt{11}/4.7.17\sqrt{19}$
$\tilde{1}\ \tilde{1}\ 2\ 0\ 000-$	$+\sqrt{3.11.19}/2.17\sqrt{5.7}$
$\tilde{1}\ \tilde{1}\ 2\ 0\ 010-$	$+2\sqrt{3.13}/17\sqrt{5.7.19}$
$\tilde{1}\ \tilde{1}\ 2\ 0\ 100-$	$+97/7.17\sqrt{13.19}$
$\tilde{1}\ \tilde{1}\ 2\ 0\ 110-$	$+43\sqrt{11}/2.7.17\sqrt{19}$
$\tilde{1}\ \tilde{1}\ \tilde{1}\ 0\ 000-$	$+127/4.7.17\sqrt{19}$
$\tilde{1}\ \tilde{1}\ \tilde{1}\ 0\ 001-$	$+9\sqrt{5.11}/7.17\sqrt{19}$
$\tilde{1}\ \tilde{1}\ \tilde{1}\ 0\ 010-$	$-\sqrt{11.13}/7.17\sqrt{19}$
$\tilde{1}\ \tilde{1}\ \tilde{1}\ 0\ 011-$	$-9\sqrt{5.13}/4.7.17\sqrt{19}$
$\tilde{1}\ \tilde{1}\ \tilde{1}\ 0\ 100-$	$-2.29\sqrt{3.5.11}/7.17\sqrt{7.13.19}$
$\tilde{1}\ \tilde{1}\ \tilde{1}\ 0\ 101-$	$+5.89\sqrt{3}/2.7.17\sqrt{7.13.19}$
$\tilde{1}\ \tilde{1}\ \tilde{1}\ 0\ 110-$	$-337\sqrt{5}/4.7.17\sqrt{3.7.19}$
$\tilde{1}\ \tilde{1}\ \tilde{1}\ 0\ 111-$	$-\sqrt{3.11}/7.17\sqrt{7.19}$

8 7 6	
$\tilde{1}\ \tilde{0}\ 1\ 0\ 000-$	$+2.3\sqrt{13}/17\sqrt{5.19}$
$\tilde{1}\ \tilde{0}\ 1\ 0\ 100-$	$+5\sqrt{11}/17\sqrt{3.7.19}$

8 7 7	
$0\ 1\ 1\ 0\ 000+$	$-3\sqrt{3.11.23}/\sqrt{2.5.7.13.17.19}$
$0\ 1\ 1\ 0\ 010+$	$+4\sqrt{2}/\sqrt{5.13.17.19.23}$
$0\ 1\ 1\ 0\ 011+$	$-\sqrt{3.5.7.11}/\sqrt{2.13.17.19.23}$
$0\ 2\ 2\ 0\ 000+$	$+3\sqrt{7.11}/\sqrt{5.13.17.19.23}$
$0\ \tilde{1}\ \tilde{1}\ 0\ 000+$	$+11\sqrt{3.11.23}/7\sqrt{2.5.7.13.17.19}$
$0\ \tilde{1}\ \tilde{1}\ 0\ 010+$	$-4\sqrt{2.3.23}/7\sqrt{5.7.17.19}$
$0\ \tilde{1}\ \tilde{1}\ 0\ 011+$	$+3\sqrt{3.5.11.13}/7\sqrt{2.7.17.19.23}$
$0\ \tilde{0}\ \tilde{0}\ 0\ 000+$	$+3\sqrt{2.11.13}/\sqrt{5.7.17.19.23}$
$1\ 1\ 1\ 0\ 000-$	0
$1\ 1\ 1\ 0\ 010-$	$-\sqrt{2.3.5}/\sqrt{13.17.19.23}$
$1\ 1\ 1\ 0\ 011-$	0
$1\ 1\ 1\ 0\ 100-$	0
$1\ 1\ 1\ 0\ 110-$	$+\sqrt{2.3.7}/\sqrt{17.19.23}$
$1\ 1\ 1\ 0\ 111-$	0
$1\ 2\ 1\ 0\ 000+$	$-31.47/5\sqrt{2.3.7.13.17.19.23}$
$1\ 2\ 1\ 0\ 001+$	$-4.3\sqrt{2.11}/5\sqrt{13.17.19.23}$
$1\ 2\ 1\ 0\ 100+$	$+2\sqrt{2}/\sqrt{3.5.17.19.23}$
$1\ 2\ 1\ 0\ 101+$	$+3\sqrt{7.11}/\sqrt{2.5.17.19.23}$
$1\ \tilde{1}\ 1\ 0\ 000+$	$-8.2.9.3\sqrt{2.3}/5.7\sqrt{13.17.19.23}$
$1\ \tilde{1}\ 1\ 0\ 001+$	$-2.31\sqrt{2.11}/5\sqrt{7.13.17.19.23}$
$1\ \tilde{1}\ 1\ 0\ 010+$	$+4.43\sqrt{2.11}/5.7\sqrt{3.17.19.23}$
$1\ \tilde{1}\ 1\ 0\ 011+$	$-3\sqrt{2.17}/5\sqrt{7.19.23}$
$1\ \tilde{1}\ 1\ 0\ 100+$	$+2.3\sqrt{2.3.5}/\sqrt{7.17.19.23}$
$1\ \tilde{1}\ 1\ 0\ 101+$	$-\sqrt{2.5.11}/\sqrt{17.19.23}$
$1\ \tilde{1}\ 1\ 0\ 110+$	$+4\sqrt{2.11.13}/\sqrt{3.5.7.17.19.23}$
$1\ \tilde{1}\ 1\ 0\ 111+$	$+3\sqrt{2.13}/\sqrt{5.17.19.23}$
$1\ \tilde{1}\ 2\ 0\ 000$	$-239\sqrt{11}/7\sqrt{2.3.5.13.17.19.23}$
$1\ \tilde{1}\ 2\ 0\ 010-$	$+2.13\sqrt{2.3}/7\sqrt{5.17.19.23}$
$1\ \tilde{1}\ 2\ 0\ 100-$	$+8\sqrt{2.11}/\sqrt{3.7.17.19.23}$
$1\ \tilde{1}\ 2\ 0\ 110-$	$+\sqrt{3.13}/\sqrt{2.7.17.19.23}$
$1\ \tilde{1}\ \tilde{1}\ 0\ 000-$	0
$1\ \tilde{1}\ \tilde{1}\ 0\ 010-$	$-\sqrt{2.23}/\sqrt{5.7.17.19}$
$1\ \tilde{1}\ \tilde{1}\ 0\ 011-$	0
$1\ \tilde{1}\ \tilde{1}\ 0\ 100-$	0
$1\ \tilde{1}\ \tilde{1}\ 0\ 110-$	$-2\sqrt{2.13}/\sqrt{17.19.23}$
$1\ \tilde{1}\ \tilde{1}\ 0\ 111-$	0
$1\ \tilde{0}\ \tilde{1}\ 0\ 000+$	$-4\sqrt{2.23}/7\sqrt{3.5.17.19}$
$1\ \tilde{0}\ \tilde{1}\ 0\ 001+$	$-9\sqrt{3.11.13}/7\sqrt{2.5.17.19.23}$
$1\ \tilde{0}\ \tilde{1}\ 0\ 100+$	$-\sqrt{7.13}/\sqrt{2.3.17.19.23}$
$1\ \tilde{0}\ \tilde{1}\ 0\ 101+$	0
$2\ 1\ 1\ 0\ 000+$	$+79\sqrt{3}/2.5\sqrt{13.17.19.23}$
$2\ 1\ 1\ 0\ 010+$	$+29\sqrt{7.11}/2.5\sqrt{13.17.19.23}$
$2\ 1\ 1\ 0\ 011+$	$-7\sqrt{3}/5\sqrt{13.17.19.23}$
$2\ 1\ 1\ 0\ 100+$	$-47\sqrt{11}/2.5\sqrt{5.7.17.19.23}$
$2\ 1\ 1\ 0\ 110+$	$+31\sqrt{3}/2.5\sqrt{5.17.19.23}$
$2\ 1\ 1\ 0\ 111+$	$+13\sqrt{7.11}/5\sqrt{5.17.19.23}$

SO_3–O 3jm Factors (cont.)

8 7 5

$\tilde{1}\ \tilde{1}\ 2\ 0\ 100+$	$+2.29\sqrt{2.11}/17\sqrt{5.7.13.19}$
$\tilde{1}\ \tilde{1}\ 2\ 0\ 110+$	$+3\sqrt{2}/17\sqrt{5.7.19}$
$\tilde{1}\ \tilde{1}\ \tilde{1}\ 0\ 000+$	$-13/4.17\sqrt{19}$
$\tilde{1}\ \tilde{1}\ \tilde{1}\ 0\ 010+$	$-4\sqrt{13}/17\sqrt{11.19}$
$\tilde{1}\ \tilde{1}\ \tilde{1}\ 0\ 100+$	$+8\sqrt{3.11}/17\sqrt{5.7.13.19}$
$\tilde{1}\ \tilde{1}\ \tilde{1}\ 0\ 110+$	$-367/4.17\sqrt{3.5.7.19}$
$\tilde{1}\ \tilde{0}\ 1\ 0\ 000+$	$+2.3\sqrt{5.13}/11.17\sqrt{19}$
$\tilde{1}\ \tilde{0}\ 1\ 0\ 001+$	$-\sqrt{7.13}/11.17\sqrt{19}$
$\tilde{1}\ \tilde{0}\ 1\ 0\ 100+$	$-\sqrt{7.11}/17\sqrt{3.19}$
$\tilde{1}\ \tilde{0}\ 1\ 0\ 101+$	$+4\sqrt{11}/17\sqrt{3.5.19}$

8 7 6

$0\ 1\ 1\ 0\ 000-$	$+9\sqrt{11}/\sqrt{7.13.17.19}$
$0\ 1\ 1\ 0\ 010-$	$+2/\sqrt{3.13.17.19}$
$0\ 2\ 2\ 0\ 000-$	$+2\sqrt{2.7}/3\sqrt{5.13.17.19}$
$0\ \tilde{1}\ \tilde{1}\ 0\ 000-$	$+3.11\sqrt{11}/7\sqrt{7.13.17.19}$
$0\ \tilde{1}\ \tilde{1}\ 0\ 001-$	$-8.4.3/7\sqrt{5.7.13.17.19}$
$0\ \tilde{1}\ \tilde{1}\ 0\ 010-$	$+2.13/3.7\sqrt{7.17.19}$
$0\ \tilde{1}\ \tilde{1}\ 0\ 011-$	$+9.3\sqrt{11}/7\sqrt{5.7.17.19}$
$0\ \tilde{0}\ \tilde{0}\ 0\ 000-$	$+8/3\sqrt{7.17.19}$
$1\ 1\ 0\ 0\ 000-$	$-3\sqrt{3.11}/2\sqrt{13.17.19}$
$1\ 1\ 0\ 0\ 010-$	$-\sqrt{7}/3\sqrt{13.17.19}$
$1\ 1\ 0\ 0\ 100-$	0
$1\ 1\ 0\ 0\ 110-$	$+7\sqrt{5}/2.3\sqrt{17.19}$
$1\ 1\ 1\ 0\ 000+$	$+3\sqrt{3.7.11}/4\sqrt{13.17.19}$
$1\ 1\ 1\ 0\ 010+$	$-4/3\sqrt{13.17.19}$
$1\ 1\ 1\ 0\ 100+$	0
$1\ 1\ 1\ 0\ 110+$	$-\sqrt{5.7}/4.3\sqrt{17.19}$
$1\ 1\ 2\ 0\ 000-$	$-59\sqrt{11}/2.5\sqrt{3.7.13.17.19}$
$1\ 1\ 2\ 0\ 010-$	$-4/5\sqrt{13.17.19}$
$1\ 1\ 2\ 0\ 100-$	$-\sqrt{11}/\sqrt{3.5.17.19}$
$1\ 1\ 2\ 0\ 110-$	$+\sqrt{7}/2\sqrt{5.17.19}$
$1\ 1\ \tilde{1}\ 0\ 000-$	$+1367/4.3.7\sqrt{5.13.17.19}$
$1\ 1\ \tilde{1}\ 0\ 001-$	$-3.11\sqrt{11}/5.7\sqrt{13.17.19}$
$1\ 1\ \tilde{1}\ 0\ 010-$	$-47\sqrt{11}/3\sqrt{3.5.7.13.17.19}$
$1\ 1\ \tilde{1}\ 0\ 011-$	$-467/4.5\sqrt{3.7.13.17.19}$
$1\ 1\ \tilde{1}\ 0\ 100-$	$-2.5/3\sqrt{7.17.19}$
$1\ 1\ \tilde{1}\ 0\ 101-$	$-3\sqrt{11}/2\sqrt{5.7.17.19}$
$1\ 1\ \tilde{1}\ 0\ 110-$	$-5\sqrt{11}/4.3\sqrt{3.17.19}$
$1\ 1\ \tilde{1}\ 0\ 111-$	$+1/\sqrt{3.5.17.19}$
$1\ 2\ 1\ 0\ 000-$	$+5\sqrt{5.13}/2.3\sqrt{7.17.19}$
$1\ 2\ 1\ 0\ 100-$	$-5/2.3\sqrt{17.19}$
$1\ 2\ \tilde{1}\ 0\ 000+$	$+11.11\sqrt{11}/2.9.7\sqrt{13.17.19}$
$1\ 2\ \tilde{1}\ 0\ 001+$	$-197/2.7\sqrt{5.13.17.19}$
$1\ 2\ \tilde{1}\ 0\ 100+$	$-11\sqrt{5.11}/2.9\sqrt{7.17.19}$
$1\ 2\ \tilde{1}\ 0\ 101+$	$-5/2\sqrt{7.17.19}$
$1\ \tilde{1}\ 1\ 0\ 000-$	$-23\sqrt{5}/4.7\sqrt{13.17.19}$
$1\ \tilde{1}\ 1\ 0\ 010-$	$-\sqrt{5.11}/3.7\sqrt{17.19}$
$1\ \tilde{1}\ 1\ 0\ 100-$	$+8/\sqrt{7.17.19}$
$1\ \tilde{1}\ 1\ 0\ 110-$	$+\sqrt{11.13}/4.3\sqrt{7.17.19}$
$1\ \tilde{1}\ 2\ 0\ 000+$	$+73\sqrt{3}/2.7\sqrt{5.13.17.19}$
$1\ \tilde{1}\ 2\ 0\ 010+$	$+37\sqrt{11}/3.7\sqrt{3.5.17.19}$
$1\ \tilde{1}\ 2\ 0\ 100+$	$+4\sqrt{3}/\sqrt{7.17.19}$
$1\ \tilde{1}\ 2\ 0\ 110+$	$-\sqrt{11.13}/2.3\sqrt{3.7.17.19}$
$1\ \tilde{1}\ \tilde{1}\ 0\ 000+$	$-11\sqrt{3.11}/4\sqrt{7.13.17.19}$
$1\ \tilde{1}\ \tilde{1}\ 0\ 001+$	$+8.2\sqrt{3}/\sqrt{5.7.13.17.19}$
$1\ \tilde{1}\ \tilde{1}\ 0\ 010+$	0
$1\ \tilde{1}\ \tilde{1}\ 0\ 011+$	$-9\sqrt{3.11}/4\sqrt{5.7.17.19}$
$1\ \tilde{1}\ \tilde{1}\ 0\ 100+$	0
$1\ \tilde{1}\ \tilde{1}\ 0\ 101+$	$+\sqrt{3}/\sqrt{17.19}$
$1\ \tilde{1}\ \tilde{1}\ 0\ 110+$	$-\sqrt{3.5.13}/4\sqrt{17.19}$
$1\ \tilde{1}\ \tilde{1}\ 0\ 111+$	0
$1\ \tilde{1}\ \tilde{0}\ 0\ 000-$	$+11\sqrt{3.11}/2.7\sqrt{13.17.19}$
$1\ \tilde{1}\ \tilde{0}\ 0\ 010-$	$-43/3.7\sqrt{3.17.19}$
$1\ \tilde{1}\ \tilde{0}\ 0\ 100-$	0
$1\ \tilde{1}\ \tilde{0}\ 0\ 110-$	$+\sqrt{5.7.13}/2.3\sqrt{3.17.19}$
$1\ \tilde{0}\ \tilde{1}\ 0\ 000-$	$-2.29/9.7\sqrt{17.19}$
$1\ \tilde{0}\ \tilde{1}\ 0\ 001-$	$+9\sqrt{11}/7\sqrt{5.17.19}$
$1\ \tilde{0}\ \tilde{1}\ 0\ 100-$	$-\sqrt{5.7.13}/9\sqrt{17.19}$
$1\ \tilde{0}\ \tilde{1}\ 0\ 101-$	0
$2\ 1\ 1\ 0\ 000-$	$-3.11/4\sqrt{2.5.13.17.19}$
$2\ 1\ 1\ 0\ 010-$	$-\sqrt{7.11.13}/4\sqrt{2.3.5.17.19}$
$2\ 1\ 1\ 0\ 100-$	$+3\sqrt{3.11}/4.5\sqrt{2.7.17.19}$
$2\ 1\ 1\ 0\ 110-$	$-67/4.5\sqrt{2.17.19}$
$2\ 1\ \tilde{1}\ 0\ 000+$	$+9.5\sqrt{11}/4\sqrt{2.7.13.17.19}$
$2\ 1\ \tilde{1}\ 0\ 001+$	$-9\sqrt{19}/4\sqrt{2.5.7.13.17}$
$2\ 1\ \tilde{1}\ 0\ 010+$	$+5/4\sqrt{2.3.13.17.19}$
$2\ 1\ \tilde{1}\ 0\ 011+$	$+\sqrt{3.11.13}/4\sqrt{2.5.17.19}$
$2\ 1\ \tilde{1}\ 0\ 100+$	$+\sqrt{3.19}/4.7\sqrt{2.5.17}$
$2\ 1\ \tilde{1}\ 0\ 101+$	$+9.9\sqrt{3.11}/4.5.7\sqrt{2.17.19}$
$2\ 1\ \tilde{1}\ 0\ 110+$	$-31\sqrt{11}/4\sqrt{2.5.7.17.19}$
$2\ 1\ \tilde{1}\ 0\ 111+$	$-3.13/4.5\sqrt{2.7.17.19}$
$2\ 2\ 0\ 0\ 000-$	$-\sqrt{7.11}/\sqrt{13.17.19}$
$2\ 2\ 0\ 0\ 100-$	$+\sqrt{3}/\sqrt{5.17.19}$
$2\ 2\ 2\ 0\ 000-$	$+2\sqrt{2.11}/3\sqrt{13.17.19}$
$2\ 2\ 2\ 0\ 100-$	$-2\sqrt{2.5}/\sqrt{3.7.17.19}$
$2\ 2\ \tilde{0}\ 0\ 000+$	$-2.5\sqrt{5}/3\sqrt{13.17.19}$
$2\ 2\ \tilde{0}\ 0\ 100+$	$-2\sqrt{11}/\sqrt{3.7.17.19}$
$2\ \tilde{1}\ 1\ 0\ 000+$	$+31\sqrt{11}/4\sqrt{2.7.13.17.19}$
$2\ \tilde{1}\ 1\ 0\ 010+$	$-9.3/4\sqrt{2.7.17.19}$
$2\ \tilde{1}\ 1\ 0\ 100+$	$+3.11\sqrt{3}/4.7\sqrt{2.5.17.19}$
$2\ \tilde{1}\ 1\ 0\ 110+$	$+\sqrt{3.11.13}/4.7\sqrt{2.5.17.19}$
$2\ \tilde{1}\ \tilde{1}\ 0\ 000-$	$-\sqrt{5}/4.7\sqrt{2.13.17.19}$
$2\ \tilde{1}\ \tilde{1}\ 0\ 001-$	$+9\sqrt{11}/4.7\sqrt{2.13.17.19}$
$2\ \tilde{1}\ \tilde{1}\ 0\ 010-$	$+\sqrt{5.11.19}/4.3.7\sqrt{2.17}$
$2\ \tilde{1}\ \tilde{1}\ 0\ 011-$	$-3.13/4.7\sqrt{2.17.19}$
$2\ \tilde{1}\ \tilde{1}\ 0\ 100-$	$+5\sqrt{3.11}/4\sqrt{2.7.17.19}$
$2\ \tilde{1}\ \tilde{1}\ 0\ 101-$	$+3.11\sqrt{3}/4\sqrt{2.5.7.17.19}$
$2\ \tilde{1}\ \tilde{1}\ 0\ 110-$	$-5\sqrt{13}/4\sqrt{2.3.7.17.19}$

SO_3–O 3jm Factors (cont.)

8 7 5	
1 1 1 0 010−	$-1/3\sqrt{13.17.19}$
1 1 1 0 011−	$-7\sqrt{7.13}/4.3\sqrt{5.17.19}$
1 1 1 0 100−	0
1 1 1 0 101−	$-\sqrt{7.11}/2.5\sqrt{3.17.19}$
1 1 1 0 110−	$-7\sqrt{5.7}/4.3\sqrt{17.19}$
1 1 1 0 111−	$-2.13/3.5\sqrt{17.19}$
1 1 2 0 000+	$-\sqrt{2.17}/\sqrt{3.5.13.19}$
1 1 2 0 010+	$-2\sqrt{2.7.11}/3\sqrt{5.13.17.19}$
1 1 2 0 100+	$+2\sqrt{2.7}/5\sqrt{3.17.19}$
1 1 2 0 110+	$-\sqrt{2.11}/3.5\sqrt{17.19}$
1 1 $\tilde{1}$ 0 000+	$-101/4.3\sqrt{5.13.17.19}$
1 1 $\tilde{1}$ 0 010+	$-4\sqrt{7.11}/3\sqrt{3.5.13.17.19}$
1 1 $\tilde{1}$ 0 100+	$+8\sqrt{7}/3.5\sqrt{17.19}$
1 1 $\tilde{1}$ 0 110+	$+\sqrt{11.19}/4.3.5\sqrt{3.17}$
1 2 1 0 000+	$+5\sqrt{5.7}/2.3\sqrt{13.17.19}$
1 2 1 0 001+	$-37/2.3\sqrt{13.17.19}$
1 2 1 0 100+	$-11/2.3\sqrt{17.19}$
1 2 1 0 101+	$-11\sqrt{7}/2.3\sqrt{5.17.19}$
1 2 $\tilde{1}$ 0 000−	$-37\sqrt{11}/2.9\sqrt{13.17.19}$
1 2 $\tilde{1}$ 0 100−	$+7\sqrt{7.11}/2.9\sqrt{5.17.19}$
1 $\tilde{1}$ 1 0 000+	$-5\sqrt{5}/4\sqrt{13.17.19}$
1 $\tilde{1}$ 1 0 001+	$+1/\sqrt{7.13.17.19}$
1 $\tilde{1}$ 1 0 010+	$+4\sqrt{5}/3\sqrt{11.17.19}$
1 $\tilde{1}$ 1 0 011+	$+5\sqrt{19}/4.3\sqrt{7.11.17}$
1 $\tilde{1}$ 1 0 100+	$+2/\sqrt{7.17.19}$
1 $\tilde{1}$ 1 0 101+	$-4/\sqrt{5.17.19}$
1 $\tilde{1}$ 1 0 110+	$-5\sqrt{11.13}/4.3\sqrt{7.17.19}$
1 $\tilde{1}$ 1 0 111+	$-\sqrt{11.13}/3\sqrt{5.17.19}$
1 $\tilde{1}$ 2 0 000−	$+2\sqrt{2.3.11}/\sqrt{7.13.17.19}$
1 $\tilde{1}$ 2 0 010−	$-5.13/3\sqrt{2.3.7.17.19}$
1 $\tilde{1}$ 2 0 100−	$+\sqrt{3.11}/\sqrt{2.5.17.19}$
1 $\tilde{1}$ 2 0 110−	$-2\sqrt{2.13}/3\sqrt{3.5.17.19}$
1 $\tilde{1}$ $\tilde{1}$ 0 000−	$-\sqrt{3.11.13}/4\sqrt{7.17.19}$
1 $\tilde{1}$ $\tilde{1}$ 0 010−	$-5/\sqrt{3.7.17.19}$
1 $\tilde{1}$ $\tilde{1}$ 0 100−	0
1 $\tilde{1}$ $\tilde{1}$ 0 110−	$-7\sqrt{13}/4\sqrt{3.5.17.19}$
1 $\tilde{0}$ $\tilde{1}$ 0 000+	$+2.5/9\sqrt{17.19}$
1 $\tilde{0}$ $\tilde{1}$ 0 100+	$-7\sqrt{7.13}/9\sqrt{5.17.19}$
2 1 1 0 000+	$-3.577/4.11\sqrt{2.5.13.17.19}$
2 1 1 0 001+	$+\sqrt{7.19}/4.5.11\sqrt{2.13.17}$
2 1 1 0 010+	$-71\sqrt{7}/4\sqrt{2.3.5.11.13.17.19}$
2 1 1 0 011+	$+653/4.5\sqrt{2.3.11.13.17.19}$
2 1 1 0 100+	$+3\sqrt{3.7}/4\sqrt{2.11.17.19}$
2 1 1 0 101+	$-53\sqrt{3}/4\sqrt{2.5.11.17.19}$
2 1 1 0 110+	$+1/4\sqrt{2.17.19}$
2 1 1 0 111+	$-13\sqrt{7}/4\sqrt{2.5.17.19}$
2 1 $\tilde{1}$ 0 000−	$-3.31\sqrt{7}/4.5\sqrt{2.11.13.17.19}$
2 1 $\tilde{1}$ 0 010−	$+239/4.5\sqrt{2.3.13.17.19}$
2 1 $\tilde{1}$ 0 100−	$+11\sqrt{3}/4\sqrt{2.5.17.19}$

8 7 5	
2 1 $\tilde{1}$ 0 110−	$+\sqrt{7.11}/4\sqrt{2.5.17.19}$
2 2 2 0 000+	$+\sqrt{7}/3\sqrt{5.13.17.19}$
2 2 2 0 100+	$+\sqrt{11}/\sqrt{3.17.19}$
2 $\tilde{1}$ 1 0 000−	$+5.5.5/4\sqrt{2.7.11.13.17.19}$
2 $\tilde{1}$ 1 0 001−	$-3.43/4\sqrt{2.5.11.13.17.19}$
2 $\tilde{1}$ 1 0 010−	$+3.101/4.11\sqrt{2.7.17.19}$
2 $\tilde{1}$ 1 0 011−	$-31\sqrt{5}/4.11\sqrt{2.17.19}$
2 $\tilde{1}$ 1 0 100−	$+3\sqrt{3.5}/4\sqrt{2.17.19}$
2 $\tilde{1}$ 1 0 101−	$-9\sqrt{3}/4\sqrt{2.7.17.19}$
2 $\tilde{1}$ 1 0 110−	$+\sqrt{3.5.13}/4\sqrt{2.11.17.19}$
2 $\tilde{1}$ 1 0 111−	$-3\sqrt{3.13}/4\sqrt{2.7.11.17.19}$
2 $\tilde{1}$ $\tilde{1}$ 0 000+	$+83/4\sqrt{2.5.13.17.19}$
2 $\tilde{1}$ $\tilde{1}$ 0 010+	$-379/4.3\sqrt{2.5.11.17.19}$
2 $\tilde{1}$ $\tilde{1}$ 0 100+	$-5\sqrt{3.11}/4\sqrt{2.7.17.19}$
2 $\tilde{1}$ $\tilde{1}$ 0 110+	$-\sqrt{13.19}/4\sqrt{2.3.7.17}$
2 $\tilde{0}$ 2 0 000−	$+23\sqrt{7}/3\sqrt{2.5.11.17.19}$
2 $\tilde{0}$ 2 0 100−	$-\sqrt{13}/\sqrt{2.3.17.19}$
$\tilde{1}$ 1 1 0 000+	$-3.139/4.11.17\sqrt{19}$
$\tilde{1}$ 1 1 0 001+	$+9.3\sqrt{7}/11.17\sqrt{5.19}$
$\tilde{1}$ 1 1 0 010+	$+4\sqrt{3.7}/17\sqrt{11.19}$
$\tilde{1}$ 1 1 0 011+	$+9.13\sqrt{3}/4.17\sqrt{5.11.19}$
$\tilde{1}$ 1 1 0 100+	$-2\sqrt{3.7.11}/17\sqrt{5.13.19}$
$\tilde{1}$ 1 1 0 101+	$-4.9\sqrt{3.11}/5.17\sqrt{13.19}$
$\tilde{1}$ 1 1 0 110+	$-41/4.17\sqrt{5.13.19}$
$\tilde{1}$ 1 1 0 111+	$+59\sqrt{7}/5.17\sqrt{13.19}$
$\tilde{1}$ 1 2 0 000−	$-2.3\sqrt{2.3.7}/17\sqrt{5.11.19}$
$\tilde{1}$ 1 2 0 010−	$+11/17\sqrt{2.5.19}$
$\tilde{1}$ 1 2 0 100−	$+103/5.17\sqrt{2.13.19}$
$\tilde{1}$ 1 2 0 110−	$+2.7\sqrt{2.7.11}/5.17\sqrt{3.13.19}$
$\tilde{1}$ 1 $\tilde{1}$ 0 000−	$-3\sqrt{3.5.7}/4.17\sqrt{11.19}$
$\tilde{1}$ 1 $\tilde{1}$ 0 010−	$+\sqrt{5}/17\sqrt{19}$
$\tilde{1}$ 1 $\tilde{1}$ 0 100−	$+8.2/17\sqrt{13.19}$
$\tilde{1}$ 1 $\tilde{1}$ 0 110−	$-47\sqrt{7.11}/4.17\sqrt{3.13.19}$
$\tilde{1}$ 2 1 0 000−	$-3\sqrt{5.11}/2.17\sqrt{19}$
$\tilde{1}$ 2 1 0 001−	$+\sqrt{7.11}/2.17\sqrt{19}$
$\tilde{1}$ 2 1 0 100−	$+\sqrt{7}/2.17\sqrt{3.13.19}$
$\tilde{1}$ 2 1 0 101−	$-5\sqrt{5}/2.17\sqrt{3.13.19}$
$\tilde{1}$ 2 $\tilde{1}$ 0 000+	$-3\sqrt{7}/2.17\sqrt{19}$
$\tilde{1}$ 2 $\tilde{1}$ 0 100+	$+47\sqrt{11}/2.17\sqrt{3.5.13.19}$
$\tilde{1}$ $\tilde{1}$ 1 0 000−	$+3\sqrt{3.5.7}/4.17\sqrt{11.19}$
$\tilde{1}$ $\tilde{1}$ 1 0 001−	$-2\sqrt{3.19}/17\sqrt{11}$
$\tilde{1}$ $\tilde{1}$ 1 0 010−	$-3\sqrt{3.5.7.13}/11.17\sqrt{19}$
$\tilde{1}$ $\tilde{1}$ 1 0 011−	$+3\sqrt{3.13}/4.11.17\sqrt{19}$
$\tilde{1}$ $\tilde{1}$ 1 0 100−	$-8.2/17\sqrt{13.19}$
$\tilde{1}$ $\tilde{1}$ 1 0 101−	$+9\sqrt{7}/2.17\sqrt{5.13.19}$
$\tilde{1}$ $\tilde{1}$ 1 0 110−	$-3\sqrt{11}/4.17\sqrt{19}$
$\tilde{1}$ $\tilde{1}$ 1 0 111−	$+2\sqrt{7.11}/17\sqrt{5.19}$
$\tilde{1}$ $\tilde{1}$ 2 0 000+	$+\sqrt{2.3}/17\sqrt{19}$
$\tilde{1}$ $\tilde{1}$ 2 0 010+	$-2\sqrt{2.3.13}/17\sqrt{11.19}$

SO_3–O 3jm Factors (cont.)

8 7 4

1 1 0 0 100−	$-\sqrt{11}/2.3\sqrt{5.17.19}$
1 1 0 0 110−	$-13\sqrt{7}/3\sqrt{3.5.17.19}$
1 1 1 0 000+	$+7\sqrt{7.11}/2\sqrt{2.3.5.13.17.19}$
1 1 1 0 010+	$-7.7\sqrt{2}/3\sqrt{5.13.17.19}$
1 1 1 0 100+	$+2\sqrt{2.11}/5\sqrt{3.17.19}$
1 1 1 0 110+	$+13\sqrt{7}/2.3.5\sqrt{2.17.19}$
1 1 2 0 000−	$-\sqrt{11}/2.3\sqrt{5.13.17.19}$
1 1 2 0 010−	$+\sqrt{7.17}/3\sqrt{3.5.13.19}$
1 1 2 0 100−	$-2\sqrt{11.17}/3.5\sqrt{7.19}$
1 1 2 0 110−	$+107/2.3.5\sqrt{3.17.19}$
1 1 $\tilde{1}$ 0 000−	$+31\sqrt{11}/2\sqrt{2.3.5.13.17.19}$
1 1 $\tilde{1}$ 0 010−	$+\sqrt{7}/\sqrt{2.5.13.17.19}$
1 1 $\tilde{1}$ 0 100−	$-\sqrt{11.17}/5\sqrt{2.3.7.19}$
1 1 $\tilde{1}$ 0 110−	$-23/2.5\sqrt{2.17.19}$
1 2 1 0 000−	$+\sqrt{7.13}/2.3\sqrt{2.17.19}$
1 2 1 0 100−	$-7/2.3\sqrt{2.5.17.19}$
1 2 $\tilde{1}$ 0 000+	$-181/2.3\sqrt{2.3.13.17.19}$
1 2 $\tilde{1}$ 0 100+	$-29\sqrt{7}/2.3\sqrt{2.3.5.17.19}$
1 $\tilde{1}$ 1 0 000−	$+7\sqrt{13}/2.3\sqrt{2.17.19}$
1 $\tilde{1}$ 1 0 010−	0
1 $\tilde{1}$ 1 0 100−	$+8\sqrt{2}/3\sqrt{5.7.17.19}$
1 $\tilde{1}$ 1 0 110−	$-3\sqrt{11.13}/2\sqrt{2.5.7.17.19}$
1 $\tilde{1}$ 2 0 000+	$+83/2.3\sqrt{7.13.17.19}$
1 $\tilde{1}$ 2 0 010+	$-5\sqrt{11}/3\sqrt{7.17.19}$
1 $\tilde{1}$ 2 0 100+	$+8.4/3.7\sqrt{5.17.19}$
1 $\tilde{1}$ 2 0 110+	$+31\sqrt{11.13}/2.3.7\sqrt{5.17.19}$
1 $\tilde{1}$ $\tilde{1}$ 0 000+	$+\sqrt{19}/2\sqrt{2.7.13.17}$
1 $\tilde{1}$ $\tilde{1}$ 0 010+	$+5\sqrt{11}/3\sqrt{2.7.17.19}$
1 $\tilde{1}$ $\tilde{1}$ 0 100+	$-\sqrt{5}/7\sqrt{2.17.19}$
1 $\tilde{1}$ $\tilde{1}$ 0 110+	$-13\sqrt{11.13}/2.3.7\sqrt{2.5.17.19}$
1 $\tilde{0}$ $\tilde{1}$ 0 000−	$+5\sqrt{11}/3\sqrt{2.3.17.19}$
1 $\tilde{0}$ $\tilde{1}$ 0 100−	$-\sqrt{2.11.13}/3\sqrt{3.5.7.17.19}$
2 1 1 0 000−	$-7.11/8.5\sqrt{13.17.19}$
2 1 1 0 010−	$+43\sqrt{7.11}/8.5\sqrt{3.13.17.19}$
2 1 1 0 100−	$-3\sqrt{3.7.11}/8\sqrt{5.17.19}$
2 1 1 0 110−	$+7.7/8\sqrt{5.17.19}$
2 1 $\tilde{1}$ 0 000+	$+23.29\sqrt{3}/8.5\sqrt{7.13.17.19}$
2 1 $\tilde{1}$ 0 010+	$-9.7\sqrt{11}/8.5\sqrt{13.17.19}$
2 1 $\tilde{1}$ 0 100+	$-\sqrt{11}/8\sqrt{5.17.19}$
2 1 $\tilde{1}$ 0 110+	$-\sqrt{7}/8\sqrt{3.5.17.19}$
2 2 0 0 000−	$-\sqrt{11}/\sqrt{3.13.17.19}$
2 2 0 0 100−	$+\sqrt{5.7}/3\sqrt{17.19}$
2 2 2 0 000−	$-\sqrt{3.11.13}/\sqrt{2.5.7.17.19}$
2 2 2 0 100−	$-1/3\sqrt{2.17.19}$
2 $\tilde{1}$ 1 0 000+	$-113\sqrt{11}/8\sqrt{5.7.13.17.19}$
2 $\tilde{1}$ 1 0 010+	$-3.5\sqrt{5}/8\sqrt{7.17.19}$
2 $\tilde{1}$ 1 0 100+	$+11\sqrt{3}/8\sqrt{17.19}$
2 $\tilde{1}$ 1 0 110+	$+\sqrt{11.13}/8\sqrt{3.17.19}$
2 $\tilde{1}$ $\tilde{1}$ 0 000−	$-11\sqrt{11}/8.7\sqrt{3.5.13.17.19}$

8 7 4

2 $\tilde{1}$ $\tilde{1}$ 0 010−	$-419/8.7\sqrt{3.5.17.19}$
2 $\tilde{1}$ $\tilde{1}$ 0 100−	$-5.11/8\sqrt{7.17.19}$
2 $\tilde{1}$ $\tilde{1}$ 0 110−	$+3\sqrt{11.13}/8\sqrt{7.17.19}$
2 $\tilde{0}$ 2 0 000+	$+53/2\sqrt{3.5.7.17.19}$
2 $\tilde{0}$ 2 0 100+	$+\sqrt{11.13}/2.3\sqrt{17.19}$
$\tilde{1}$ 1 1 0 000−	$+23/2.17\sqrt{2.5.19}$
$\tilde{1}$ 1 1 0 010−	$+2\sqrt{2.7.11}/17\sqrt{3.5.19}$
$\tilde{1}$ 1 1 0 100−	$-8\sqrt{2.3.7.11}/5.17\sqrt{13.19}$
$\tilde{1}$ 1 1 0 110−	$-7.67/2.5.17\sqrt{2.13.19}$
$\tilde{1}$ 1 2 0 000+	$-31/2.17\sqrt{5.7.19}$
$\tilde{1}$ 1 2 0 010+	$-13\sqrt{11}/17\sqrt{3.5.19}$
$\tilde{1}$ 1 2 0 100+	$-8\sqrt{3.11}/5.17\sqrt{13.19}$
$\tilde{1}$ 1 2 0 110+	$-101\sqrt{7}/2.5.17\sqrt{13.19}$
$\tilde{1}$ 1 $\tilde{1}$ 0 000+	$+3\sqrt{5.19}/2.17\sqrt{2.7}$
$\tilde{1}$ 1 $\tilde{1}$ 0 010+	$+\sqrt{5.11}/17\sqrt{2.3.19}$
$\tilde{1}$ 1 $\tilde{1}$ 0 100+	$+\sqrt{3.11}/17\sqrt{2.13.19}$
$\tilde{1}$ 1 $\tilde{1}$ 0 110+	$+\sqrt{7.13}/2.17\sqrt{2.19}$
$\tilde{1}$ 2 1 0 000+	$+11\sqrt{11}/2.3.17\sqrt{2.19}$
$\tilde{1}$ 2 1 0 100+	$+11\sqrt{3.5.7}/2.17\sqrt{2.13.19}$
$\tilde{1}$ 2 $\tilde{1}$ 0 000−	$+31\sqrt{11}/2.17\sqrt{2.3.7.19}$
$\tilde{1}$ 2 $\tilde{1}$ 0 100−	$+7\sqrt{13}/2.17\sqrt{2.5.19}$
$\tilde{1}$ $\tilde{1}$ 0 0 000−	$-\sqrt{5.11.19}/3.17\sqrt{7}$
$\tilde{1}$ $\tilde{1}$ 0 0 010−	$-\sqrt{5.13}/2.3.17\sqrt{7.19}$
$\tilde{1}$ $\tilde{1}$ 0 0 100−	$-3\sqrt{3}/2.17\sqrt{13.19}$
$\tilde{1}$ $\tilde{1}$ 0 0 110−	$-\sqrt{3.11}/17\sqrt{19}$
$\tilde{1}$ $\tilde{1}$ 1 0 000+	$+\sqrt{11.19}/2.17\sqrt{2.3.7}$
$\tilde{1}$ $\tilde{1}$ 1 0 010+	$+\sqrt{2.13}/17\sqrt{3.7.19}$
$\tilde{1}$ $\tilde{1}$ 1 0 100+	$-2.9\sqrt{2}/17\sqrt{5.13.19}$
$\tilde{1}$ $\tilde{1}$ 1 0 110+	$-9.3\sqrt{11}/2.17\sqrt{2.5.19}$
$\tilde{1}$ $\tilde{1}$ 2 0 000−	$+53\sqrt{11}/2.3.7.17\sqrt{19}$
$\tilde{1}$ $\tilde{1}$ 2 0 010−	$+\sqrt{13}/3.7\sqrt{19}$
$\tilde{1}$ $\tilde{1}$ 2 0 100−	$-2.23\sqrt{3}/17\sqrt{5.7.13.19}$
$\tilde{1}$ $\tilde{1}$ 2 0 110−	$-\sqrt{3.11}/2\sqrt{5.7.19}$
$\tilde{1}$ $\tilde{1}$ $\tilde{1}$ 0 000−	$-37\sqrt{11}/2.7.17\sqrt{2.3.19}$
$\tilde{1}$ $\tilde{1}$ $\tilde{1}$ 0 010−	$+23\sqrt{13}/7.17\sqrt{2.3.19}$
$\tilde{1}$ $\tilde{1}$ $\tilde{1}$ 0 100−	$+29\sqrt{13}/17\sqrt{2.5.7.19}$
$\tilde{1}$ $\tilde{1}$ $\tilde{1}$ 0 110−	$-13\sqrt{11}/2.17\sqrt{2.5.7.19}$
$\tilde{1}$ $\tilde{0}$ 1 0 000−	$-\sqrt{13}/3.17\sqrt{2.19}$
$\tilde{1}$ $\tilde{0}$ 1 0 100−	$-\sqrt{2.3.7.11}/17\sqrt{5.19}$

8 7 5

0 1 1 0 000+	$-9\sqrt{7}/\sqrt{11.13.17.19}$
0 1 1 0 001+	$-4.7/\sqrt{5.11.13.17.19}$
0 1 1 0 010+	$+2/\sqrt{3.13.17.19}$
0 1 1 0 011+	$+7\sqrt{7}/\sqrt{3.5.13.17.19}$
0 2 2 0 000+	$+2.5\sqrt{11}/3\sqrt{13.17.19}$
0 $\tilde{1}$ $\tilde{1}$ 0 000+	$+3\sqrt{11}/\sqrt{7.13.17.19}$
0 $\tilde{1}$ $\tilde{1}$ 0 010+	$+2.5/3\sqrt{7.17.19}$
1 1 1 0 000−	$-3\sqrt{3.7.13}/4\sqrt{11.17.19}$
1 1 1 0 001−	$-2.7/\sqrt{3.5.11.13.17.19}$

SO_3–O 3jm Factors (cont.)

8 7 2
$\tilde{1}\ 1\ \tilde{1}\ 0\ 000+\quad +2.3\sqrt{3.13}/5.17\sqrt{7}$
$\tilde{1}\ 1\ \tilde{1}\ 0\ 010+\quad -\sqrt{11.13}/4.5.17$
$\tilde{1}\ 1\ \tilde{1}\ 0\ 100+\quad -\sqrt{11}/2.17\sqrt{5}$
$\tilde{1}\ 1\ \tilde{1}\ 0\ 110+\quad +2\sqrt{7}/17\sqrt{3.5}$
$\tilde{1}\ 2\ \tilde{1}\ 0\ 000-\quad +3\sqrt{11.13}/2.17\sqrt{5.7}$
$\tilde{1}\ 2\ \tilde{1}\ 0\ 100-\quad -7/2.17\sqrt{3}$
$\tilde{1}\ \tilde{1}\ 2\ 0\ 000-\quad -\sqrt{3.11.13}/7.17\sqrt{2.5}$
$\tilde{1}\ \tilde{1}\ 2\ 0\ 010-\quad -37\sqrt{3}/7.17\sqrt{2.5}$
$\tilde{1}\ \tilde{1}\ 2\ 0\ 100-\quad -9/17\sqrt{2.7}$
$\tilde{1}\ \tilde{1}\ 2\ 0\ 110-\quad +\sqrt{11.13}/17\sqrt{2.7}$
$\tilde{1}\ \tilde{1}\ \tilde{1}\ 0\ 000-\quad -\sqrt{11.13}/7.17\sqrt{5}$
$\tilde{1}\ \tilde{1}\ \tilde{1}\ 0\ 010-\quad +11.19/4.7.17\sqrt{5}$
$\tilde{1}\ \tilde{1}\ \tilde{1}\ 0\ 100-\quad -3\sqrt{3}/17\sqrt{7}$
$\tilde{1}\ \tilde{1}\ \tilde{1}\ 0\ 110-\quad +\sqrt{11.13}/17\sqrt{3.7}$

8 7 3
$0\ 1\ 1\ 0\ 000+\quad -\sqrt{2.11}/\sqrt{13.17.19}$
$0\ 1\ 1\ 0\ 010+\quad +7\sqrt{7}/\sqrt{2.3.13.17.19}$
$0\ \tilde{1}\ \tilde{1}\ 0\ 000+\quad -\sqrt{2.19}/\sqrt{7.13.17}$
$0\ \tilde{1}\ \tilde{1}\ 0\ 010+\quad +\sqrt{11}/3\sqrt{2.7.17.19}$
$0\ \tilde{0}\ \tilde{0}\ 0\ 000+\quad +2\sqrt{11}/3\sqrt{17.19}$
$1\ 1\ 1\ 0\ 000-\quad -11\sqrt{11}/2\sqrt{2.3.13.17.19}$
$1\ 1\ 1\ 0\ 010-\quad -7\sqrt{7}/3\sqrt{2.13.17.19}$
$1\ 1\ 1\ 0\ 100-\quad -\sqrt{5.11}/\sqrt{2.3.7.17.19}$
$1\ 1\ 1\ 0\ 110-\quad +13\sqrt{5}/2.3\sqrt{2.17.19}$
$1\ 1\ \tilde{1}\ 0\ 000+\quad +\sqrt{5.11}/2.3\sqrt{2.13.17.19}$
$1\ 1\ \tilde{1}\ 0\ 010+\quad +7\sqrt{2.5.7}/3\sqrt{3.13.17.19}$
$1\ 1\ \tilde{1}\ 0\ 100+\quad +8\sqrt{2.11}/3\sqrt{7.17.19}$
$1\ 1\ \tilde{1}\ 0\ 110+\quad -13/2.3\sqrt{2.3.17.19}$
$1\ 2\ 1\ 0\ 000+\quad -\sqrt{17}/2.3\sqrt{2.5.13.19}$
$1\ 2\ 1\ 0\ 100+\quad -\sqrt{17}/2.3\sqrt{2.7.19}$
$1\ 2\ \tilde{1}\ 0\ 000-\quad -167/2.9\sqrt{2.13.17.19}$
$1\ 2\ \tilde{1}\ 0\ 100-\quad +67\sqrt{5}/2.9\sqrt{2.7.17.19}$
$1\ \tilde{1}\ 1\ 0\ 000+\quad +9\sqrt{19}/2\sqrt{2.5.7.13.17}$
$1\ \tilde{1}\ 1\ 0\ 010+\quad +\sqrt{11.17}/3\sqrt{2.5.7.19}$
$1\ \tilde{1}\ 1\ 0\ 100+\quad +3/\sqrt{2.17.19}$
$1\ \tilde{1}\ 1\ 0\ 110+\quad +\sqrt{11.13}/2.3\sqrt{2.17.19}$
$1\ \tilde{1}\ \tilde{1}\ 0\ 000-\quad +\sqrt{7.19}/2\sqrt{2.3.13.17}$
$1\ \tilde{1}\ \tilde{1}\ 0\ 010-\quad 0$
$1\ \tilde{1}\ \tilde{1}\ 0\ 100-\quad -2\sqrt{2.5}/7\sqrt{3.17.19}$
$1\ \tilde{1}\ \tilde{1}\ 0\ 110-\quad -\sqrt{3.5.11.13}/2.7\sqrt{2.17.19}$
$1\ \tilde{1}\ \tilde{0}\ 0\ 000+\quad +\sqrt{19}/\sqrt{3.7.13.17}$
$1\ \tilde{1}\ \tilde{0}\ 0\ 010+\quad -29\sqrt{11}/2.3\sqrt{3.7.17.19}$
$1\ \tilde{1}\ \tilde{0}\ 0\ 100+\quad -\sqrt{5}/2.7\sqrt{3.17.19}$
$1\ \tilde{1}\ \tilde{0}\ 0\ 110+\quad -\sqrt{5.11.13}/3.7\sqrt{3.17.19}$
$1\ \tilde{0}\ \tilde{1}\ 0\ 000+\quad +7\sqrt{11}/9\sqrt{2.17.19}$
$1\ \tilde{0}\ \tilde{1}\ 0\ 100+\quad -\sqrt{2.5.11.13}/9\sqrt{7.17.19}$
$2\ 1\ 1\ 0\ 000+\quad -157/8\sqrt{5.7.13.17.19}$
$2\ 1\ 1\ 0\ 010+\quad +43\sqrt{11}/8\sqrt{3.5.13.17.19}$
$2\ 1\ 1\ 0\ 100+\quad +\sqrt{3.11}/8.5\sqrt{17.19}$

8 7 3
$2\ 1\ 1\ 0\ 110+\quad -29\sqrt{7}/8.5\sqrt{17.19}$
$2\ 1\ \tilde{1}\ 0\ 000-\quad -9\sqrt{13}/8\sqrt{7.17.19}$
$2\ 1\ \tilde{1}\ 0\ 010-\quad -47\sqrt{11}/8\sqrt{3.13.17.19}$
$2\ 1\ \tilde{1}\ 0\ 100-\quad -7\sqrt{3.11}/8\sqrt{5.17.19}$
$2\ 1\ \tilde{1}\ 0\ 110-\quad -7\sqrt{7}/8\sqrt{5.17.19}$
$2\ 2\ \tilde{0}\ 0\ 000-\quad -\sqrt{5.11.17}/2.3\sqrt{7.13.19}$
$2\ 2\ \tilde{0}\ 0\ 100-\quad +7/2\sqrt{3.17.19}$
$2\ \tilde{1}\ 1\ 0\ 000-\quad -3.43\sqrt{11}/8.7\sqrt{13.17.19}$
$2\ \tilde{1}\ 1\ 0\ 010-\quad -43/8.7\sqrt{17.19}$
$2\ \tilde{1}\ 1\ 0\ 100-\quad +3.13\sqrt{3}/8\sqrt{5.7.17.19}$
$2\ \tilde{1}\ 1\ 0\ 110-\quad -9\sqrt{3.11.13}/8\sqrt{5.7.17.19}$
$2\ \tilde{1}\ \tilde{1}\ 0\ 000+\quad -9\sqrt{5.11}/8.7\sqrt{13.17.19}$
$2\ \tilde{1}\ \tilde{1}\ 0\ 010+\quad -5.5.11\sqrt{5}/8.3.7\sqrt{17.19}$
$2\ \tilde{1}\ \tilde{1}\ 0\ 100+\quad +31\sqrt{3}/8\sqrt{7.17.19}$
$2\ \tilde{1}\ \tilde{1}\ 0\ 110+\quad -\sqrt{11.13}/8\sqrt{3.7.17.19}$
$\tilde{1}\ 1\ 1\ 0\ 000+\quad -9\sqrt{19}/2.5.17\sqrt{2.7}$
$\tilde{1}\ 1\ 1\ 0\ 010+\quad -3.7\sqrt{3.11}/5.17\sqrt{2.19}$
$\tilde{1}\ 1\ 1\ 0\ 100+\quad -3.7\sqrt{3.11}/17\sqrt{2.5.13.19}$
$\tilde{1}\ 1\ 1\ 0\ 110+\quad +\sqrt{7.13}/2.17\sqrt{2.5.19}$
$\tilde{1}\ 1\ \tilde{1}\ 0\ 000-\quad +3\sqrt{3.19}/2.17\sqrt{2.5.7}$
$\tilde{1}\ 1\ \tilde{1}\ 0\ 010-\quad -2\sqrt{2.11}/17\sqrt{5.19}$
$\tilde{1}\ 1\ \tilde{1}\ 0\ 100-\quad +2\sqrt{2.11}/17\sqrt{13.19}$
$\tilde{1}\ 1\ \tilde{1}\ 0\ 110-\quad +\sqrt{7.13}/2.17\sqrt{2.3.19}$
$\tilde{1}\ 1\ \tilde{0}\ 0\ 000+\quad -3\sqrt{3.19}/17\sqrt{5.7}$
$\tilde{1}\ 1\ \tilde{0}\ 0\ 010+\quad +\sqrt{11}/2.17\sqrt{5.19}$
$\tilde{1}\ 1\ \tilde{0}\ 0\ 100+\quad +\sqrt{11}/2.17\sqrt{13.19}$
$\tilde{1}\ 1\ \tilde{0}\ 0\ 110+\quad +\sqrt{7.13}/17\sqrt{3.19}$
$\tilde{1}\ 2\ 1\ 0\ 000-\quad -61\sqrt{11}/2.17\sqrt{2.5.7.19}$
$\tilde{1}\ 2\ 1\ 0\ 100-\quad +131/2.17\sqrt{2.3.13.19}$
$\tilde{1}\ 2\ \tilde{1}\ 0\ 000+\quad -9\sqrt{11}/2.17\sqrt{2.7.19}$
$\tilde{1}\ 2\ \tilde{1}\ 0\ 100+\quad -5.23\sqrt{5}/2.17\sqrt{2.3.13.19}$
$\tilde{1}\ \tilde{1}\ 1\ 0\ 000-\quad +\sqrt{3.5.11.19}/2.7.17\sqrt{2}$
$\tilde{1}\ \tilde{1}\ 1\ 0\ 010-\quad +3\sqrt{3.13}/7.17\sqrt{2.5.19}$
$\tilde{1}\ \tilde{1}\ 1\ 0\ 100-\quad +9.5/17\sqrt{2.7.13.19}$
$\tilde{1}\ \tilde{1}\ 1\ 0\ 110-\quad +5.5\sqrt{11}/2.17\sqrt{2.7.19}$
$\tilde{1}\ \tilde{1}\ \tilde{1}\ 0\ 000+\quad +\sqrt{11.19}/2.7.17\sqrt{2}$
$\tilde{1}\ \tilde{1}\ \tilde{1}\ 0\ 010+\quad +\sqrt{2.13}/7.17\sqrt{19}$
$\tilde{1}\ \tilde{1}\ \tilde{1}\ 0\ 100+\quad -8.3\sqrt{2.3.5}/17\sqrt{7.13.19}$
$\tilde{1}\ \tilde{1}\ \tilde{1}\ 0\ 110+\quad +11\sqrt{5.11}/2.17\sqrt{2.3.7.19}$
$\tilde{1}\ \tilde{0}\ 1\ 0\ 000+\quad -\sqrt{13}/17\sqrt{2.5.7.19}$
$\tilde{1}\ \tilde{0}\ 1\ 0\ 100+\quad +5\sqrt{2.11}/17\sqrt{3.19}$

8 7 4
$0\ 1\ 1\ 0\ 000-\quad +\sqrt{2.7.11}/\sqrt{5.13.17.19}$
$0\ 1\ 1\ 0\ 010-\quad +7.7/\sqrt{2.3.5.13.17.19}$
$0\ 2\ 2\ 0\ 000-\quad +5\sqrt{2}/\sqrt{3.13.17.19}$
$0\ \tilde{1}\ \tilde{1}\ 0\ 000-\quad -\sqrt{2.19}/\sqrt{3.7.13.17}$
$0\ \tilde{1}\ \tilde{1}\ 0\ 010-\quad +5\sqrt{11}/\sqrt{2.3.7.17.19}$
$1\ 1\ 0\ 0\ 000-\quad -\sqrt{7.11}/3\sqrt{13.17.19}$
$1\ 1\ 0\ 0\ 010-\quad -7.7/2.3\sqrt{3.13.17.19}$

SO_3–O 3jm Factors (cont.)

8	$\mathbf{\frac{13}{2}}$	$\mathbf{\frac{13}{2}}$			
$\tilde{1}$	$\frac{3}{2}$	$\frac{1}{2}$	0	110−	$+5.5\sqrt{3.11}/7.17\sqrt{13.19}$
$\tilde{1}$	$\frac{3}{2}$	$\frac{3}{2}$	0	000+	$-2.37\sqrt{2.3}/7.17\sqrt{5.19}$
$\tilde{1}$	$\frac{3}{2}$	$\frac{3}{2}$	0	010+	$-4.11\sqrt{11}/7.17\sqrt{5.19}$
$\tilde{1}$	$\frac{3}{2}$	$\frac{3}{2}$	0	011+	$-2\sqrt{2.3.5}/7.17\sqrt{19}$
$\tilde{1}$	$\frac{3}{2}$	$\frac{3}{2}$	0	100+	$+2.3\sqrt{2.11.19}/7.17\sqrt{7.13}$
$\tilde{1}$	$\frac{3}{2}$	$\frac{3}{2}$	0	110+	$-3\sqrt{3.19}/7.17\sqrt{7.13}$
$\tilde{1}$	$\frac{3}{2}$	$\frac{3}{2}$	0	111+	$+2.9.3\sqrt{2.11}/7.17\sqrt{7.13.19}$
$\tilde{1}$	$\frac{3}{2}$	$\frac{3}{2}$	1	000−	0
$\tilde{1}$	$\frac{3}{2}$	$\frac{3}{2}$	1	010−	$+4\sqrt{11}/17\sqrt{5.19}$
$\tilde{1}$	$\frac{3}{2}$	$\frac{3}{2}$	1	011−	0
$\tilde{1}$	$\frac{3}{2}$	$\frac{3}{2}$	1	100−	0
$\tilde{1}$	$\frac{3}{2}$	$\frac{3}{2}$	1	110−	$-4\sqrt{3.13}/17\sqrt{7.19}$
$\tilde{1}$	$\frac{3}{2}$	$\frac{3}{2}$	1	111−	0
$\tilde{1}$	$\widetilde{\frac{1}{2}}$	$\frac{1}{2}$	0	000+	$-2\sqrt{3.11}/17\sqrt{7.19}$
$\tilde{1}$	$\widetilde{\frac{1}{2}}$	$\frac{1}{2}$	0	010+	$+8\sqrt{2.13}/17\sqrt{5.7.19}$
$\tilde{1}$	$\widetilde{\frac{1}{2}}$	$\frac{1}{2}$	0	100+	$-8.3\sqrt{5}/7.17\sqrt{13.19}$
$\tilde{1}$	$\widetilde{\frac{1}{2}}$	$\frac{1}{2}$	0	110+	$-5\sqrt{3.11}/7.17\sqrt{2.19}$
$\tilde{1}$	$\widetilde{\frac{1}{2}}$	$\frac{3}{2}$	0	000−	$-\sqrt{3}/17\sqrt{2.19}$
$\tilde{1}$	$\widetilde{\frac{1}{2}}$	$\frac{3}{2}$	0	001−	0
$\tilde{1}$	$\widetilde{\frac{1}{2}}$	$\frac{3}{2}$	0	010−	$-2\sqrt{11.13}/17\sqrt{5.19}$
$\tilde{1}$	$\widetilde{\frac{1}{2}}$	$\frac{3}{2}$	0	011−	$+\sqrt{3.5.13}/17\sqrt{2.19}$
$\tilde{1}$	$\widetilde{\frac{1}{2}}$	$\frac{3}{2}$	0	100−	$-2.3\sqrt{2.5.11.13}/7.17\sqrt{7.19}$
$\tilde{1}$	$\widetilde{\frac{1}{2}}$	$\frac{3}{2}$	0	101−	$-\sqrt{3.5}/7\sqrt{7.13.19}$
$\tilde{1}$	$\widetilde{\frac{1}{2}}$	$\frac{3}{2}$	0	110−	$-107\sqrt{3}/7.17\sqrt{7.19}$
$\tilde{1}$	$\widetilde{\frac{1}{2}}$	$\frac{3}{2}$	0	111−	$-2.9\sqrt{2.11}/7.17\sqrt{7.19}$

8	**7**	**1**			
0	1	1	0	000+	0
0	1	1	0	010+	$+1/\sqrt{17}$
1	1	1	0	000−	0
1	1	1	0	010−	$-3\sqrt{3}/4\sqrt{17}$
1	1	1	0	100−	0
1	1	1	0	110−	0
1	2	1	0	000+	$+\sqrt{7}/2\sqrt{3.5.17}$
1	2	1	0	100+	$+\sqrt{13}/2\sqrt{3.17}$
1	$\tilde{1}$	1	0	000+	$+\sqrt{3}/\sqrt{5.17}$
1	$\tilde{1}$	1	0	010+	$+\sqrt{11.13}/4\sqrt{3.5.17}$
1	$\tilde{1}$	1	0	100+	$-\sqrt{3.13}/2\sqrt{7.17}$
1	$\tilde{1}$	1	0	110+	$+\sqrt{11}/\sqrt{3.7.17}$
2	1	1	0	000+	$+\sqrt{3}/4\sqrt{2.5.17}$
2	1	1	0	010+	$+\sqrt{7.11}/4\sqrt{2.5.17}$
2	1	1	0	100+	$+\sqrt{7.11.13}/4.5\sqrt{2.17}$
2	1	1	0	110+	$-\sqrt{3.13}/4.5\sqrt{2.17}$
2	$\tilde{1}$	1	0	000−	$-3\sqrt{3.11}/4\sqrt{2.7.17}$
2	$\tilde{1}$	1	0	010−	$-\sqrt{3.13}/4\sqrt{2.7.17}$
2	$\tilde{1}$	1	0	100−	$-\sqrt{13}/4\sqrt{2.5.17}$
2	$\tilde{1}$	1	0	110−	$+3\sqrt{11}/4\sqrt{2.5.17}$
$\tilde{1}$	1	1	0	000+	$-\sqrt{3.13/5.17}$
$\tilde{1}$	1	1	0	010+	$+\sqrt{7.11.13/4.5.17}$
$\tilde{1}$	1	1	0	100+	$+3\sqrt{7.11}/2.17\sqrt{5}$
$\tilde{1}$	1	1	0	110+	$+\sqrt{3}/17\sqrt{5}$
$\tilde{1}$	2	1	0	000−	$-\sqrt{11.13}/2.17\sqrt{3.5}$
$\tilde{1}$	2	1	0	100−	$-3\sqrt{7}/2.17$
$\tilde{1}$	$\tilde{1}$	1	0	000−	0
$\tilde{1}$	$\tilde{1}$	1	0	010−	$-\sqrt{7}/4\sqrt{5}$
$\tilde{1}$	$\tilde{1}$	1	0	100−	0
$\tilde{1}$	$\tilde{1}$	1	0	110−	0
$\tilde{1}$	$\tilde{0}$	1	0	000+	$+1/\sqrt{3.5}$
$\tilde{1}$	$\tilde{0}$	1	0	100+	0

8	**7**	**2**			
0	2	2	0	000−	$+2/3\sqrt{5.17}$
0	$\tilde{1}$	$\tilde{1}$	0	000−	$+4/\sqrt{5.7.17}$
0	$\tilde{1}$	$\tilde{1}$	0	010−	$+\sqrt{11.13}/3\sqrt{5.7.17}$
1	1	2	0	000−	$-\sqrt{11}/5\sqrt{2.3.17}$
1	1	2	0	010−	$-\sqrt{7}/3.5\sqrt{2.17}$
1	1	2	0	100−	$-\sqrt{11.13}/\sqrt{2.3.5.7.17}$
1	1	2	0	110−	$-\sqrt{13}/3\sqrt{2.5.17}$
1	1	$\tilde{1}$	0	000−	$+\sqrt{11}/3.5\sqrt{17}$
1	1	$\tilde{1}$	0	010−	$-41\sqrt{7}/4.3.5\sqrt{3.17}$
1	1	$\tilde{1}$	0	100−	$+\sqrt{11.13}/3\sqrt{5.7.17}$
1	1	$\tilde{1}$	0	110−	$+\sqrt{13}/3\sqrt{3.5.17}$
1	2	$\tilde{1}$	0	000+	$-11/2.9\sqrt{5.17}$
1	2	$\tilde{1}$	0	100+	$+\sqrt{7.13}/2.9\sqrt{17}$
1	$\tilde{1}$	2	0	000+	$-1/\sqrt{2.3.5.7.17}$
1	$\tilde{1}$	2	0	010+	$-\sqrt{2.11.13}/3\sqrt{3.5.7.17}$
1	$\tilde{1}$	2	0	100+	$-2\sqrt{2.13}/7\sqrt{3.17}$
1	$\tilde{1}$	2	0	110+	$+\sqrt{11}/3.7\sqrt{2.3.17}$
1	$\tilde{1}$	$\tilde{1}$	0	000+	$-2\sqrt{5}/\sqrt{3.7.17}$
1	$\tilde{1}$	$\tilde{1}$	0	010+	$-\sqrt{11.13}/4\sqrt{3.5.7.17}$
1	$\tilde{1}$	$\tilde{1}$	0	100+	$+\sqrt{13}/2.7\sqrt{3.17}$
1	$\tilde{1}$	$\tilde{1}$	0	110+	$+2\sqrt{11}/7\sqrt{3.17}$
1	$\tilde{0}$	$\tilde{1}$	0	000−	$+\sqrt{11.13}/9\sqrt{5.17}$
1	$\tilde{0}$	$\tilde{1}$	0	100−	$+4\sqrt{11}/9\sqrt{7.17}$
2	1	$\tilde{1}$	0	000+	$-3/4\sqrt{2.5.7.17}$
2	1	$\tilde{1}$	0	010+	$+7\sqrt{11}/4\sqrt{2.3.5.17}$
2	1	$\tilde{1}$	0	100+	$-\sqrt{3.11.13}/4.5\sqrt{2.17}$
2	1	$\tilde{1}$	0	110+	$-\sqrt{7.13}/4.5\sqrt{2.17}$
2	2	2	0	000−	$-\sqrt{11}/3\sqrt{7.17}$
2	2	2	0	100−	$-\sqrt{13}/\sqrt{3.5.17}$
2	$\tilde{1}$	$\tilde{1}$	0	000−	$+9\sqrt{11}/4.7\sqrt{2.17}$
2	$\tilde{1}$	$\tilde{1}$	0	010−	$-5\sqrt{13}/4.3.7\sqrt{2.17}$
2	$\tilde{1}$	$\tilde{1}$	0	100−	$+\sqrt{3.13}/4\sqrt{2.5.7.17}$
2	$\tilde{1}$	$\tilde{1}$	0	110−	$+\sqrt{5.11}/4\sqrt{2.3.7.17}$
2	$\tilde{0}$	2	0	000+	$+\sqrt{13}/3\sqrt{2.7.17}$
2	$\tilde{0}$	2	0	100+	$-\sqrt{11}/\sqrt{2.3.5.17}$
$\tilde{1}$	1	2	0	000+	$-3\sqrt{3.13}/5.17\sqrt{2.7}$
$\tilde{1}$	1	2	0	010+	$+\sqrt{2.11.13}/5.17$
$\tilde{1}$	1	2	0	100+	$-2\sqrt{2.11}/17\sqrt{5}$
$\tilde{1}$	1	2	0	110+	$+\sqrt{7}/17\sqrt{2.3.5}$

SO_3–O 3jm Factors (cont.)

8 $\frac{13}{2}$ $\frac{11}{2}$	
$\tilde{1}$ $\frac{1}{2}$ $\tilde{\frac{1}{2}}$ 0 100−	$-8.2\sqrt{2.11}/17\sqrt{3.7.13.19}$
$\tilde{1}$ $\frac{3}{2}$ $\frac{1}{2}$ 0 000+	$-9/2.17\sqrt{19}$
$\tilde{1}$ $\frac{3}{2}$ $\frac{1}{2}$ 0 010+	$+5.5\sqrt{3}/17\sqrt{2.11.19}$
$\tilde{1}$ $\frac{3}{2}$ $\frac{1}{2}$ 0 100+	$-8\sqrt{5.11}/17\sqrt{3.7.13.19}$
$\tilde{1}$ $\frac{3}{2}$ $\frac{1}{2}$ 0 110+	$-11\sqrt{5}/2.3.17\sqrt{2.7.13.19}$
$\tilde{1}$ $\frac{3}{2}$ $\frac{3}{2}$ 0 000−	$+3.5\sqrt{2}/17\sqrt{7.11.19}$
$\tilde{1}$ $\frac{3}{2}$ $\frac{3}{2}$ 0 001−	$-2.3.5\sqrt{5}/17\sqrt{7.11.19}$
$\tilde{1}$ $\frac{3}{2}$ $\frac{3}{2}$ 0 010−	$+13\sqrt{3}/17\sqrt{7.19}$
$\tilde{1}$ $\frac{3}{2}$ $\frac{3}{2}$ 0 011−	$-3\sqrt{3.5}/2.17\sqrt{2.7.19}$
$\tilde{1}$ $\frac{3}{2}$ $\frac{3}{2}$ 0 100−	$+5.11\sqrt{2.5}/7.17\sqrt{3.13.19}$
$\tilde{1}$ $\frac{3}{2}$ $\frac{3}{2}$ 0 101−	$-11.11\sqrt{3}/7.17\sqrt{13.19}$
$\tilde{1}$ $\frac{3}{2}$ $\frac{3}{2}$ 0 110−	$+2\sqrt{5.13}/3.7.17\sqrt{11.19}$
$\tilde{1}$ $\frac{3}{2}$ $\frac{3}{2}$ 0 111−	$-23.31\sqrt{2}/3.7.17\sqrt{11.13.19}$
$\tilde{1}$ $\frac{3}{2}$ $\frac{3}{2}$ 1 000+	$+3\sqrt{2.7}/17\sqrt{11.19}$
$\tilde{1}$ $\frac{3}{2}$ $\frac{3}{2}$ 1 001+	$+103\sqrt{5}/2.17\sqrt{7.11.19}$
$\tilde{1}$ $\frac{3}{2}$ $\frac{3}{2}$ 1 010+	$-9.3\sqrt{3.7}/11.17\sqrt{19}$
$\tilde{1}$ $\frac{3}{2}$ $\frac{3}{2}$ 1 011+	$-53\sqrt{5}/2.11.17\sqrt{2.3.7.19}$
$\tilde{1}$ $\frac{3}{2}$ $\frac{3}{2}$ 1 100+	$-13\sqrt{5.13}/7.17\sqrt{2.3.19}$
$\tilde{1}$ $\frac{3}{2}$ $\frac{3}{2}$ 1 101+	$-\sqrt{3.13.19}/2.7.17$
$\tilde{1}$ $\frac{3}{2}$ $\frac{3}{2}$ 1 110+	$+79\sqrt{5.13}/2.3.7.17\sqrt{11.19}$
$\tilde{1}$ $\frac{3}{2}$ $\frac{3}{2}$ 1 111+	$-9.13\sqrt{13}/7.17\sqrt{2.11.19}$
$\tilde{1}$ $\frac{3}{2}$ $\tilde{\frac{1}{2}}$ 0 000+	$+\sqrt{5.11}/2.17\sqrt{7.19}$
$\tilde{1}$ $\frac{3}{2}$ $\tilde{\frac{1}{2}}$ 0 010+	$-13\sqrt{5}/17\sqrt{2.3.7.19}$
$\tilde{1}$ $\frac{3}{2}$ $\tilde{\frac{1}{2}}$ 0 100+	$-11\sqrt{3.13}/7.17\sqrt{19}$
$\tilde{1}$ $\frac{3}{2}$ $\tilde{\frac{1}{2}}$ 0 110+	$-709\sqrt{11}/2.3.7.17\sqrt{2.13.19}$
$\tilde{1}$ $\tilde{\frac{1}{2}}$ $\frac{1}{2}$ 0 000−	$-3\sqrt{5}/2.17\sqrt{2.19}$
$\tilde{1}$ $\tilde{\frac{1}{2}}$ $\frac{1}{2}$ 0 010−	$+2\sqrt{3.13}/17\sqrt{11.19}$
$\tilde{1}$ $\tilde{\frac{1}{2}}$ $\frac{1}{2}$ 0 100−	$+8.2\sqrt{2.11}/17\sqrt{3.7.13.19}$
$\tilde{1}$ $\tilde{\frac{1}{2}}$ $\frac{1}{2}$ 0 110−	$+2.11\sqrt{5}/3.17\sqrt{7.19}$
$\tilde{1}$ $\tilde{\frac{1}{2}}$ $\frac{3}{2}$ 0 000+	$-3.5\sqrt{5}/2.17\sqrt{2.7.11.19}$
$\tilde{1}$ $\tilde{\frac{1}{2}}$ $\frac{3}{2}$ 0 001+	$-\sqrt{7.19}/17\sqrt{11}$
$\tilde{1}$ $\tilde{\frac{1}{2}}$ $\frac{3}{2}$ 0 010+	$+4.13\sqrt{3.13}/11.17\sqrt{7.19}$
$\tilde{1}$ $\tilde{\frac{1}{2}}$ $\frac{3}{2}$ 0 011+	$+\sqrt{5.7.13}/11.17\sqrt{2.3.19}$
$\tilde{1}$ $\tilde{\frac{1}{2}}$ $\frac{3}{2}$ 0 100+	$-8.41\sqrt{2}/7.17\sqrt{3.13.19}$
$\tilde{1}$ $\tilde{\frac{1}{2}}$ $\frac{3}{2}$ 0 101+	$-3\sqrt{3.5}/2.17\sqrt{13.19}$
$\tilde{1}$ $\tilde{\frac{1}{2}}$ $\frac{3}{2}$ 0 110+	$+\sqrt{5.11}/3.7.17\sqrt{19}$
$\tilde{1}$ $\tilde{\frac{1}{2}}$ $\frac{3}{2}$ 0 111+	$-2\sqrt{2.11}/3.17\sqrt{19}$

8 $\frac{13}{2}$ $\frac{13}{2}$	
0 $\frac{1}{2}$ $\frac{1}{2}$ 0 000+	$-3\sqrt{3.5.11}/\sqrt{2.7.13.17.19}$
0 $\frac{3}{2}$ $\frac{3}{2}$ 0 000+	$-11\sqrt{3.11}/7\sqrt{5.7.13.17.19}$
0 $\frac{3}{2}$ $\frac{3}{2}$ 0 010+	$-4.29\sqrt{2}/7\sqrt{5.7.13.17.19}$
0 $\frac{3}{2}$ $\frac{3}{2}$ 0 011+	$-3\sqrt{3.11}/7\sqrt{5.7.13.17.19}$
0 $\tilde{\frac{1}{2}}$ $\tilde{\frac{1}{2}}$ 0 000+	$+11\sqrt{3.5.11}/7\sqrt{2.7.13.17.19}$
0 $\tilde{\frac{1}{2}}$ $\tilde{\frac{1}{2}}$ 0 010+	$-8.2/7\sqrt{7.17.19}$
0 $\tilde{\frac{1}{2}}$ $\tilde{\frac{1}{2}}$ 0 011+	$+3\sqrt{2.3.11.13}/7\sqrt{5.7.17.19}$
1 $\frac{1}{2}$ $\frac{1}{2}$ 0 000−	0
1 $\frac{1}{2}$ $\frac{1}{2}$ 0 100−	0
1 $\frac{3}{2}$ $\frac{1}{2}$ 0 000+	$-107\sqrt{3}/7\sqrt{2.5.13.17.19}$
1 $\frac{3}{2}$ $\frac{1}{2}$ 0 010+	$+4\sqrt{11.13}/7\sqrt{5.17.19}$

8 $\frac{13}{2}$ $\frac{13}{2}$	
1 $\frac{3}{2}$ $\frac{1}{2}$ 0 100+	$+2\sqrt{2.3}/\sqrt{7.17.19}$
1 $\frac{3}{2}$ $\frac{1}{2}$ 0 110+	$+\sqrt{11}/\sqrt{7.17.19}$
1 $\frac{3}{2}$ $\frac{3}{2}$ 0 000−	0
1 $\frac{3}{2}$ $\frac{3}{2}$ 0 010−	$-4\sqrt{7}/3\sqrt{5.13.17.19}$
1 $\frac{3}{2}$ $\frac{3}{2}$ 0 011−	0
1 $\frac{3}{2}$ $\frac{3}{2}$ 0 100−	0
1 $\frac{3}{2}$ $\frac{3}{2}$ 0 110−	$+5/3\sqrt{17.19}$
1 $\frac{3}{2}$ $\frac{3}{2}$ 0 111−	0
1 $\frac{3}{2}$ $\frac{3}{2}$ 1 000−	0
1 $\frac{3}{2}$ $\frac{3}{2}$ 1 010−	$+4.11/3\sqrt{5.7.13.17.19}$
1 $\frac{3}{2}$ $\frac{3}{2}$ 1 011−	0
1 $\frac{3}{2}$ $\frac{3}{2}$ 1 100−	0
1 $\frac{3}{2}$ $\frac{3}{2}$ 1 110−	$-4/3\sqrt{17.19}$
1 $\frac{3}{2}$ $\frac{3}{2}$ 1 111−	0
1 $\tilde{\frac{1}{2}}$ $\frac{3}{2}$ 0 000+	$+3.11\sqrt{3.11}/7\sqrt{2.7.13.17.19}$
1 $\tilde{\frac{1}{2}}$ $\frac{3}{2}$ 0 001+	$-2\sqrt{17}/3.7\sqrt{7.13.19}$
1 $\tilde{\frac{1}{2}}$ $\frac{3}{2}$ 0 010+	$-4/3.7\sqrt{5.7.17.19}$
1 $\tilde{\frac{1}{2}}$ $\frac{3}{2}$ 0 011+	$+9.3\sqrt{3.11}/7\sqrt{2.5.7.17.19}$
1 $\tilde{\frac{1}{2}}$ $\frac{3}{2}$ 0 100+	0
1 $\tilde{\frac{1}{2}}$ $\frac{3}{2}$ 0 101+	$+5\sqrt{5}/3\sqrt{17.19}$
1 $\tilde{\frac{1}{2}}$ $\frac{3}{2}$ 0 110+	$-\sqrt{13}/3\sqrt{17.19}$
1 $\tilde{\frac{1}{2}}$ $\frac{3}{2}$ 0 111+	0
1 $\tilde{\frac{1}{2}}$ $\tilde{\frac{1}{2}}$ 0 000−	0
1 $\tilde{\frac{1}{2}}$ $\tilde{\frac{1}{2}}$ 0 010−	$+8\sqrt{2}/3\sqrt{7.17.19}$
1 $\tilde{\frac{1}{2}}$ $\tilde{\frac{1}{2}}$ 0 011−	0
1 $\tilde{\frac{1}{2}}$ $\tilde{\frac{1}{2}}$ 0 100−	0
1 $\tilde{\frac{1}{2}}$ $\tilde{\frac{1}{2}}$ 0 110−	$+\sqrt{5.13}/3\sqrt{2.17.19}$
1 $\tilde{\frac{1}{2}}$ $\tilde{\frac{1}{2}}$ 0 111−	0
2 $\frac{3}{2}$ $\frac{1}{2}$ 0 000−	$-4\sqrt{3.11}/\sqrt{7.13.17.19}$
2 $\frac{3}{2}$ $\frac{1}{2}$ 0 010−	$+3\sqrt{2}/\sqrt{7.13.17.19}$
2 $\frac{3}{2}$ $\frac{1}{2}$ 0 100−	$-8/7\sqrt{5.17.19}$
2 $\frac{3}{2}$ $\frac{1}{2}$ 0 110−	$-3\sqrt{2.3.11}/7\sqrt{5.17.19}$
2 $\frac{3}{2}$ $\frac{3}{2}$ 0 000+	$+43\sqrt{3}/2.7\sqrt{13.17.19}$
2 $\frac{3}{2}$ $\frac{3}{2}$ 0 010+	$-\sqrt{11.19}/7\sqrt{2.13.17}$
2 $\frac{3}{2}$ $\frac{3}{2}$ 0 011+	$+2\sqrt{3.19}/7\sqrt{13.17}$
2 $\frac{3}{2}$ $\frac{3}{2}$ 0 100+	$-9.3\sqrt{11}/2.7\sqrt{5.7.17.19}$
2 $\frac{3}{2}$ $\frac{3}{2}$ 0 110+	$-37\sqrt{3}/7\sqrt{2.5.7.17.19}$
2 $\frac{3}{2}$ $\frac{3}{2}$ 0 111+	$-2.23\sqrt{11}/7\sqrt{5.7.17.19}$
2 $\tilde{\frac{1}{2}}$ $\frac{3}{2}$ 0 000−	$+\sqrt{3.5}/\sqrt{13.17.19}$
2 $\tilde{\frac{1}{2}}$ $\frac{3}{2}$ 0 001−	$-\sqrt{5.11}/\sqrt{2.13.17.19}$
2 $\tilde{\frac{1}{2}}$ $\frac{3}{2}$ 0 010−	$-\sqrt{11}/2\sqrt{2.17.19}$
2 $\tilde{\frac{1}{2}}$ $\frac{3}{2}$ 0 011−	$-\sqrt{3}/2\sqrt{17.19}$
2 $\tilde{\frac{1}{2}}$ $\frac{3}{2}$ 0 100−	$+13\sqrt{11}/7\sqrt{7.17.19}$
2 $\tilde{\frac{1}{2}}$ $\frac{3}{2}$ 0 101−	$+11\sqrt{3}/7\sqrt{2.7.17.19}$
2 $\tilde{\frac{1}{2}}$ $\frac{3}{2}$ 0 110−	$-5\sqrt{3.5.13}/2.7\sqrt{2.7.17.19}$
2 $\tilde{\frac{1}{2}}$ $\frac{3}{2}$ 0 111−	$-\sqrt{11.13}/2.7\sqrt{5.7.17.19}$
$\tilde{1}$ $\frac{3}{2}$ $\frac{1}{2}$ 0 000−	$+\sqrt{3.11}/17\sqrt{2.5.7.19}$
$\tilde{1}$ $\frac{3}{2}$ $\frac{1}{2}$ 0 010−	$+2.11/17\sqrt{5.7.19}$
$\tilde{1}$ $\frac{3}{2}$ $\frac{1}{2}$ 0 100−	$+4.9.3\sqrt{2}/7.17\sqrt{13.19}$

SO_3–O 3jm Factors (cont.)

8 7 7

2 2 2 0 000+ $+109/2\sqrt{13.17.19.23}$
2 2 2 0 100+ $-29\sqrt{11}/2\sqrt{3.5.7.17.19.23}$
2 $\tilde{1}$ 1 0 000− $+3.29\sqrt{3.11}/2\sqrt{5.7.13.17.19.23}$
2 $\tilde{1}$ 1 0 001− $+2.3/\sqrt{5.13.17.19.23}$
2 $\tilde{1}$ 1 0 010− $-41\sqrt{3}/2\sqrt{5.7.17.19.23}$
2 $\tilde{1}$ 1 0 011− $-3\sqrt{11}/\sqrt{5.17.19.23}$
2 $\tilde{1}$ 1 0 100− $+211/2.5.7\sqrt{17.19.23}$
2 $\tilde{1}$ 1 0 101− $+8.2\sqrt{3.11}/5\sqrt{7.17.19.23}$
2 $\tilde{1}$ 1 0 110− $+9\sqrt{11.13}/2.7\sqrt{17.19.23}$
2 $\tilde{1}$ 1 0 111− $-5\sqrt{3.13}/\sqrt{7.17.19.23}$
2 $\tilde{1}$ $\tilde{1}$ 0 000+ $-97\sqrt{3}/2.7\sqrt{13.17.19.23}$
2 $\tilde{1}$ $\tilde{1}$ 0 010+ $+5\sqrt{3.11}/7\sqrt{17.19.23}$
2 $\tilde{1}$ $\tilde{1}$ 0 011+ $+11\sqrt{3.13}/7\sqrt{17.19.23}$
2 $\tilde{1}$ $\tilde{1}$ 0 100+ $-41\sqrt{11}/2\sqrt{5.7.17.19.23}$
2 $\tilde{1}$ $\tilde{1}$ 0 110+ $-3\sqrt{13}/\sqrt{5.7.17.19.23}$
2 $\tilde{1}$ $\tilde{1}$ 0 111+ $-13\sqrt{11}/\sqrt{5.7.17.19.23}$
2 $\tilde{0}$ 2 0 000− $+\sqrt{11}/2\sqrt{2.17.19.23}$
2 $\tilde{0}$ 2 0 100− $+29\sqrt{13}/2\sqrt{2.3.5.7.17.19.23}$
$\tilde{1}$ 1 1 0 000+ $+37\sqrt{2.3}/5.17\sqrt{5.19.23}$
$\tilde{1}$ 1 1 0 010+ $-8\sqrt{2.7.11}/5.17\sqrt{5.19.23}$
$\tilde{1}$ 1 1 0 011+ $+3.7.13\sqrt{2.3}/5.17\sqrt{5.19.23}$
$\tilde{1}$ 1 1 0 100+ $-2.3\sqrt{2.11}/5.17\sqrt{7.13.19.23}$
$\tilde{1}$ 1 1 0 110+ $-7.41\sqrt{2.3}/5.17\sqrt{13.19.23}$
$\tilde{1}$ 1 1 0 111+ $-9.3\sqrt{2.7.11}/5.17\sqrt{13.19.23}$
$\tilde{1}$ 2 1 0 000− $+53\sqrt{11}/5.17\sqrt{2.3.19.23}$
$\tilde{1}$ 2 1 0 001− $-2.3.7\sqrt{2.7}/5.17\sqrt{19.23}$
$\tilde{1}$ 2 1 0 100− $+8.9\sqrt{2/17}\sqrt{5.7.13.19.23}$
$\tilde{1}$ 2 1 0 101− $+73\sqrt{3.11}/17\sqrt{2.5.13.19.23}$
$\tilde{1}$ $\tilde{1}$ 1 0 000− $-3\sqrt{2.11.23}/5.17\sqrt{7.19}$
$\tilde{1}$ $\tilde{1}$ 1 0 001− $-\sqrt{2.3.23}/5.17\sqrt{19}$
$\tilde{1}$ $\tilde{1}$ 1 0 010− $-\sqrt{2.13.23}/5.17\sqrt{7.19}$
$\tilde{1}$ $\tilde{1}$ 1 0 011− $-2.3\sqrt{2.3.11.13}/5.17\sqrt{19.23}$
$\tilde{1}$ $\tilde{1}$ 1 0 100− $-4.3\sqrt{2.3.23}/7.17\sqrt{5.13.19}$
$\tilde{1}$ $\tilde{1}$ 1 0 101− $-3.47\sqrt{2.11}/17\sqrt{5.7.13.19.23}$
$\tilde{1}$ $\tilde{1}$ 1 0 110− $-2.3\sqrt{2.3.5.11}/7.17\sqrt{19.23}$
$\tilde{1}$ $\tilde{1}$ 1 0 111− $+2.3\sqrt{2.5}/17\sqrt{7.19.23}$
$\tilde{1}$ $\tilde{1}$ 2 0 000+ $-3\sqrt{3}/17\sqrt{2.5.7.19.23}$
$\tilde{1}$ $\tilde{1}$ 2 0 010+ $-8\sqrt{2.3.11.13}/17\sqrt{5.7.19.23}$
$\tilde{1}$ $\tilde{1}$ 2 0 100+ $-2.3.5.5\sqrt{2.11}/7.17\sqrt{13.19.23}$
$\tilde{1}$ $\tilde{1}$ 2 0 110+ $-3.5\sqrt{23}/7.17\sqrt{2.19}$
$\tilde{1}$ $\tilde{1}$ $\tilde{1}$ 0 000+ $-3.13\sqrt{2.3}/7.17\sqrt{5.19.23}$
$\tilde{1}$ $\tilde{1}$ $\tilde{1}$ 0 010+ $+2\sqrt{2.3.11.13}/7.17\sqrt{5.19.23}$
$\tilde{1}$ $\tilde{1}$ $\tilde{1}$ 0 011+ $-5.13\sqrt{2.3.5}/7.17\sqrt{19.23}$
$\tilde{1}$ $\tilde{1}$ $\tilde{1}$ 0 100+ $+2.3.149\sqrt{2.11}/7.17\sqrt{7.13.19.23}$
$\tilde{1}$ $\tilde{1}$ $\tilde{1}$ 0 110+ $+4.3\sqrt{2.19}/7.17\sqrt{7.23}$
$\tilde{1}$ $\tilde{1}$ $\tilde{1}$ 0 111+ $+3.11\sqrt{2.11.13}/7.17\sqrt{7.19.23}$
$\tilde{1}$ $\tilde{0}$ 1 0 000+ $+4\sqrt{2.13.23}/5.17\sqrt{3.19}$
$\tilde{1}$ $\tilde{0}$ 1 0 001+ $+3\sqrt{7.11.13}/5.17\sqrt{2.19.23}$
$\tilde{1}$ $\tilde{0}$ 1 0 100+ $-3\sqrt{5.11}/17\sqrt{2.7.19.23}$
$\tilde{1}$ $\tilde{0}$ 1 0 101+ $+8\sqrt{2.3.5}/17\sqrt{19.23}$

8 $\frac{15}{2}$ $\frac{1}{2}$

0 $\frac{1}{2}$ $\frac{1}{2}$ 0 000− $+1/\sqrt{17}$
1 $\frac{1}{2}$ $\frac{1}{2}$ 0 000+ $+3/2\sqrt{2.17}$
1 $\frac{1}{2}$ $\frac{1}{2}$ 0 100+ 0
1 $\frac{3}{2}$ $\frac{1}{2}$ 0 000− 0
1 $\frac{3}{2}$ $\frac{1}{2}$ 0 010− $+\sqrt{3.5}/2\sqrt{2.17}$
1 $\frac{3}{2}$ $\frac{1}{2}$ 0 020− 0
1 $\frac{3}{2}$ $\frac{1}{2}$ 0 100− 0
1 $\frac{3}{2}$ $\frac{1}{2}$ 0 110− 0
1 $\frac{3}{2}$ $\frac{1}{2}$ 0 120− $+\sqrt{3}/\sqrt{17}$
2 $\frac{3}{2}$ $\frac{1}{2}$ 0 000+ $+\sqrt{3}/8\sqrt{2.17}$
2 $\frac{3}{2}$ $\frac{1}{2}$ 0 010+ $+\sqrt{7.11}/4\sqrt{3.17}$
2 $\frac{3}{2}$ $\frac{1}{2}$ 0 020+ $-\sqrt{11.13}/8\sqrt{2.3.17}$
2 $\frac{3}{2}$ $\frac{1}{2}$ 0 100+ $+\sqrt{7.11.13}/8\sqrt{2.5.17}$
2 $\frac{3}{2}$ $\frac{1}{2}$ 0 110+ $-\sqrt{13}/4\sqrt{5.17}$
2 $\frac{3}{2}$ $\frac{1}{2}$ 0 120+ $-\sqrt{5.7}/8\sqrt{2.17}$
$\tilde{1}$ $\frac{3}{2}$ $\frac{1}{2}$ 0 000+ $-\sqrt{3.13}/2.17\sqrt{5}$
$\tilde{1}$ $\frac{3}{2}$ $\frac{1}{2}$ 0 010+ $+\sqrt{7.11.13}/2.17\sqrt{2.3.5}$
$\tilde{1}$ $\frac{3}{2}$ $\frac{1}{2}$ 0 020+ $+\sqrt{5.11}/2.17\sqrt{3}$
$\tilde{1}$ $\frac{3}{2}$ $\frac{1}{2}$ 0 100+ $+3\sqrt{7.11}/4.17$
$\tilde{1}$ $\frac{3}{2}$ $\frac{1}{2}$ 0 110+ $+\sqrt{2/17}$
$\tilde{1}$ $\frac{3}{2}$ $\frac{1}{2}$ 0 120+ $+\sqrt{7.13}/4.17$
$\tilde{1}$ $\tilde{\frac{1}{2}}$ $\frac{1}{2}$ 0 000− $+1/2\sqrt{2}$
$\tilde{1}$ $\tilde{\frac{1}{2}}$ $\frac{1}{2}$ 0 100− 0

8 $\frac{15}{2}$ $\frac{3}{2}$

0 $\frac{3}{2}$ $\frac{3}{2}$ 0 000+ 0
0 $\frac{3}{2}$ $\frac{3}{2}$ 0 010+ $+1/\sqrt{17}$
0 $\frac{3}{2}$ $\frac{3}{2}$ 0 020+ 0
1 $\frac{1}{2}$ $\frac{3}{2}$ 0 000+ $-\sqrt{3.5}/4\sqrt{17}$
1 $\frac{1}{2}$ $\frac{3}{2}$ 0 100+ 0
1 $\frac{3}{2}$ $\frac{3}{2}$ 0 000− $+\sqrt{7.11}/2.3.5\sqrt{17}$
1 $\frac{3}{2}$ $\frac{3}{2}$ 0 010− $-31/9.5\sqrt{2.17}$
1 $\frac{3}{2}$ $\frac{3}{2}$ 0 020− $+\sqrt{7.13}/2.9\sqrt{17}$
1 $\frac{3}{2}$ $\frac{3}{2}$ 0 100− $+\sqrt{11.13}/2.3\sqrt{5.17}$
1 $\frac{3}{2}$ $\frac{3}{2}$ 0 110− $+\sqrt{2.7.13}/9\sqrt{5.17}$
1 $\frac{3}{2}$ $\frac{3}{2}$ 0 120− $-5\sqrt{5}/2.9\sqrt{17}$
1 $\frac{3}{2}$ $\frac{3}{2}$ 1 000− $-\sqrt{7.11}/4.3.5\sqrt{17}$
1 $\frac{3}{2}$ $\frac{3}{2}$ 1 010− $+11.11/2.9.5\sqrt{2.17}$
1 $\frac{3}{2}$ $\frac{3}{2}$ 1 020− $-\sqrt{7.13}/4.9\sqrt{17}$
1 $\frac{3}{2}$ $\frac{3}{2}$ 1 100− $-\sqrt{11.13}/4.3\sqrt{5.17}$
1 $\frac{3}{2}$ $\frac{3}{2}$ 1 110− $-\sqrt{7.13}/9\sqrt{2.5.17}$
1 $\frac{3}{2}$ $\frac{3}{2}$ 1 120− $-13\sqrt{5}/4.9\sqrt{17}$
1 $\tilde{\frac{1}{2}}$ $\frac{3}{2}$ 0 000+ $-\sqrt{7.11.13}/4.3\sqrt{3.5.17}$
1 $\tilde{\frac{1}{2}}$ $\frac{3}{2}$ 0 100+ $-\sqrt{11}/3\sqrt{3.17}$
2 $\frac{1}{2}$ $\frac{3}{2}$ 0 000− $-\sqrt{7.11}/4\sqrt{2.3.17}$
2 $\frac{1}{2}$ $\frac{3}{2}$ 0 100− $+\sqrt{13}/4\sqrt{2.5.17}$
2 $\frac{3}{2}$ $\frac{3}{2}$ 0 000+ $-3/8\sqrt{2.5.17}$
2 $\frac{3}{2}$ $\frac{3}{2}$ 0 010+ $+\sqrt{7.11}/2.3\sqrt{5.17}$

SO_3–O 3jm Factors (cont.)

$\mathbf{8\ \tfrac{15}{2}\ \tfrac{3}{2}}$	
$2\ \tfrac{3}{2}\ \tfrac{3}{2}\ 0\ 020+$	$-\sqrt{5.11.13}/8.3\sqrt{2.17}$
$2\ \tfrac{3}{2}\ \tfrac{3}{2}\ 0\ 100+$	$-\sqrt{3.7.11.13}/8.5\sqrt{2.17}$
$2\ \tfrac{3}{2}\ \tfrac{3}{2}\ 0\ 110+$	$-\sqrt{13}/2.5\sqrt{3.17}$
$2\ \tfrac{3}{2}\ \tfrac{3}{2}\ 0\ 120+$	$-5\sqrt{7}/8\sqrt{2.3.17}$
$2\ \tilde{\tfrac{1}{2}}\ \tfrac{3}{2}\ 0\ 000-$	$-\sqrt{13}/4\sqrt{2.3.17}$
$2\ \tilde{\tfrac{1}{2}}\ \tfrac{3}{2}\ 0\ 100-$	$+\sqrt{7.11}/4\sqrt{2.5.17}$
$\tilde{1}\ \tfrac{1}{2}\ \tfrac{3}{2}\ 0\ 000-$	$-\sqrt{7.11.13}/4.17\sqrt{3.5}$
$\tilde{1}\ \tfrac{1}{2}\ \tfrac{3}{2}\ 0\ 100-$	$-1/17$
$\tilde{1}\ \tfrac{3}{2}\ \tfrac{3}{2}\ 0\ 000+$	$+3\sqrt{13}/2.5.17$
$\tilde{1}\ \tfrac{3}{2}\ \tfrac{3}{2}\ 0\ 010+$	$+\sqrt{7.11.13}/3.5.17\sqrt{2}$
$\tilde{1}\ \tfrac{3}{2}\ \tfrac{3}{2}\ 0\ 020+$	$+5\sqrt{11}/2.3.17$
$\tilde{1}\ \tfrac{3}{2}\ \tfrac{3}{2}\ 0\ 100+$	$-3\sqrt{3.7.11}/4.17\sqrt{5}$
$\tilde{1}\ \tfrac{3}{2}\ \tfrac{3}{2}\ 0\ 110+$	$+2\sqrt{2}/17\sqrt{3.5}$
$\tilde{1}\ \tfrac{3}{2}\ \tfrac{3}{2}\ 0\ 120+$	$+\sqrt{5.7.13}/4.17\sqrt{3}$
$\tilde{1}\ \tfrac{3}{2}\ \tfrac{3}{2}\ 1\ 000-$	$+3\sqrt{13}/4.17$
$\tilde{1}\ \tfrac{3}{2}\ \tfrac{3}{2}\ 1\ 010-$	$-\sqrt{7.11.13}/2.3.17\sqrt{2}$
$\tilde{1}\ \tfrac{3}{2}\ \tfrac{3}{2}\ 1\ 020-$	$-7\sqrt{11}/4.3.17$
$\tilde{1}\ \tfrac{3}{2}\ \tfrac{3}{2}\ 1\ 100-$	$+\sqrt{3.7.11}/4.17\sqrt{5}$
$\tilde{1}\ \tfrac{3}{2}\ \tfrac{3}{2}\ 1\ 110-$	$+7/17\sqrt{2.3.5}$
$\tilde{1}\ \tfrac{3}{2}\ \tfrac{3}{2}\ 1\ 120-$	$+\sqrt{5.7.13}/4.17\sqrt{3}$
$\tilde{1}\ \tilde{\tfrac{1}{2}}\ \tfrac{3}{2}\ 0\ 000-$	$+\sqrt{3}/4\sqrt{5}$
$\tilde{1}\ \tilde{\tfrac{1}{2}}\ \tfrac{3}{2}\ 0\ 100-$	0
$\mathbf{8\ \tfrac{15}{2}\ \tfrac{5}{2}}$	
$0\ \tfrac{3}{2}\ \tfrac{3}{2}\ 0\ 000-$	$+\sqrt{11}/3\sqrt{17.19}$
$0\ \tfrac{3}{2}\ \tfrac{3}{2}\ 0\ 010-$	$-7\sqrt{2.7}/9\sqrt{17.19}$
$0\ \tfrac{3}{2}\ \tfrac{3}{2}\ 0\ 020-$	$+5\sqrt{13}/9\sqrt{17.19}$
$0\ \tilde{\tfrac{1}{2}}\ \tilde{\tfrac{1}{2}}\ 0\ 000-$	$+\sqrt{11.13}/3\sqrt{3.17.19}$
$1\ \tfrac{1}{2}\ \tfrac{3}{2}\ 0\ 000-$	$+7\sqrt{5.7}/2.3\sqrt{2.3.17.19}$
$1\ \tfrac{1}{2}\ \tfrac{3}{2}\ 0\ 100-$	$-5\sqrt{13}/3\sqrt{2.3.17.19}$
$1\ \tfrac{3}{2}\ \tfrac{3}{2}\ 0\ 000+$	$-\sqrt{11.17}/2.9.5\sqrt{2.19}$
$1\ \tfrac{3}{2}\ \tfrac{3}{2}\ 0\ 010+$	$+461\sqrt{7}/2.9.3.5\sqrt{17.19}$
$1\ \tfrac{3}{2}\ \tfrac{3}{2}\ 0\ 020+$	$-\sqrt{13.17}/2.9.3\sqrt{2.19}$
$1\ \tfrac{3}{2}\ \tfrac{3}{2}\ 0\ 100+$	$+\sqrt{2.7.11.13}/9\sqrt{5.17.19}$
$1\ \tfrac{3}{2}\ \tfrac{3}{2}\ 0\ 110+$	$+4.7\sqrt{13}/9.3\sqrt{5.17.19}$
$1\ \tfrac{3}{2}\ \tfrac{3}{2}\ 0\ 120+$	$-5\sqrt{2.5.7}/9.3\sqrt{17.19}$
$1\ \tfrac{3}{2}\ \tfrac{3}{2}\ 1\ 000+$	$-37\sqrt{11}/9.5\sqrt{2.17.19}$
$1\ \tfrac{3}{2}\ \tfrac{3}{2}\ 1\ 010+$	$-89\sqrt{7}/9.3.5\sqrt{17.19}$
$1\ \tfrac{3}{2}\ \tfrac{3}{2}\ 1\ 020+$	$+23\sqrt{13}/9.3\sqrt{2.17.19}$
$1\ \tfrac{3}{2}\ \tfrac{3}{2}\ 1\ 100+$	$-29\sqrt{11.13}/2.9\sqrt{2.5.7.17.19}$
$1\ \tfrac{3}{2}\ \tfrac{3}{2}\ 1\ 110+$	$+2.23\sqrt{13}/9.3\sqrt{5.17.19}$
$1\ \tfrac{3}{2}\ \tfrac{3}{2}\ 1\ 120+$	$-11\sqrt{5.7}/2.9.3\sqrt{2.17.19}$
$1\ \tfrac{3}{2}\ \tilde{\tfrac{1}{2}}\ 0\ 000-$	$-4\sqrt{11}/9\sqrt{5.17.19}$
$1\ \tfrac{3}{2}\ \tilde{\tfrac{1}{2}}\ 0\ 010-$	$+73\sqrt{7}/2.9.3\sqrt{2.5.17.19}$
$1\ \tfrac{3}{2}\ \tilde{\tfrac{1}{2}}\ 0\ 020-$	$+8\sqrt{5.13}/9.3\sqrt{17.19}$
$1\ \tfrac{3}{2}\ \tilde{\tfrac{1}{2}}\ 0\ 100-$	$-\sqrt{11.13.17}/2.9\sqrt{7.19}$
$1\ \tfrac{3}{2}\ \tilde{\tfrac{1}{2}}\ 0\ 110-$	$-2\sqrt{2.13}/9.3\sqrt{17.19}$
$1\ \tfrac{3}{2}\ \tilde{\tfrac{1}{2}}\ 0\ 120-$	$-5.5\sqrt{7}/2.9.3\sqrt{17.19}$
$1\ \tilde{\tfrac{1}{2}}\ \tfrac{3}{2}\ 0\ 000-$	$+\sqrt{11.13}/2.9\sqrt{2.3.5.17.19}$
$1\ \tilde{\tfrac{1}{2}}\ \tfrac{3}{2}\ 0\ 100-$	$+47\sqrt{11}/9\sqrt{2.3.7.17.19}$
$\mathbf{8\ \tfrac{15}{2}\ \tfrac{5}{2}}$	
$1\ \tilde{\tfrac{1}{2}}\ \tilde{\tfrac{1}{2}}\ 0\ 000+$	$+11\sqrt{11.13}/2.9\sqrt{2.3.17.19}$
$1\ \tilde{\tfrac{1}{2}}\ \tilde{\tfrac{1}{2}}\ 0\ 100+$	$+2\sqrt{2.5.11}/9\sqrt{3.7.17.19}$
$2\ \tfrac{1}{2}\ \tfrac{3}{2}\ 0\ 000+$	$+7\sqrt{11}/4\sqrt{3.17.19}$
$2\ \tfrac{1}{2}\ \tfrac{3}{2}\ 0\ 100+$	$-\sqrt{7.13}/4\sqrt{5.17.19}$
$2\ \tfrac{3}{2}\ \tfrac{3}{2}\ 0\ 000-$	$-157/8.3\sqrt{5.7.17.19}$
$2\ \tfrac{3}{2}\ \tfrac{3}{2}\ 0\ 010-$	$-31\sqrt{11}/9\sqrt{2.5.17.19}$
$2\ \tfrac{3}{2}\ \tfrac{3}{2}\ 0\ 020-$	$+\sqrt{5.11.13}/8.9\sqrt{7.17.19}$
$2\ \tfrac{3}{2}\ \tfrac{3}{2}\ 0\ 100-$	$+\sqrt{11.13}/8.5\sqrt{3.17.19}$
$2\ \tfrac{3}{2}\ \tfrac{3}{2}\ 0\ 110-$	$-7\sqrt{7.13}/3.5\sqrt{2.3.17.19}$
$2\ \tfrac{3}{2}\ \tfrac{3}{2}\ 0\ 120-$	$-5.31/8.3\sqrt{3.17.19}$
$2\ \tfrac{3}{2}\ \tilde{\tfrac{1}{2}}\ 0\ 000+$	$-97/8.3\sqrt{2.7.17.19}$
$2\ \tfrac{3}{2}\ \tilde{\tfrac{1}{2}}\ 0\ 010+$	$+13\sqrt{11}/4.9\sqrt{17.19}$
$2\ \tfrac{3}{2}\ \tilde{\tfrac{1}{2}}\ 0\ 020+$	$+5.13\sqrt{11.13}/8.9\sqrt{2.7.17.19}$
$2\ \tfrac{3}{2}\ \tilde{\tfrac{1}{2}}\ 0\ 100+$	$-11\sqrt{11.13}/8\sqrt{2.3.5.17.19}$
$2\ \tfrac{3}{2}\ \tilde{\tfrac{1}{2}}\ 0\ 110+$	$-11\sqrt{7.13}/4.3\sqrt{3.5.17.19}$
$2\ \tfrac{3}{2}\ \tilde{\tfrac{1}{2}}\ 0\ 120+$	$+\sqrt{5}/8.3\sqrt{2.3.17.19}$
$2\ \tilde{\tfrac{1}{2}}\ \tfrac{3}{2}\ 0\ 000+$	$-\sqrt{13.19}/4.3\sqrt{3.7.17}$
$2\ \tilde{\tfrac{1}{2}}\ \tfrac{3}{2}\ 0\ 100+$	$+\sqrt{11.19}/4.3\sqrt{5.17}$
$\tilde{1}\ \tfrac{1}{2}\ \tfrac{3}{2}\ 0\ 000+$	$-13\sqrt{11.13}/2.17\sqrt{2.3.5.19}$
$\tilde{1}\ \tfrac{1}{2}\ \tfrac{3}{2}\ 0\ 100+$	$+\sqrt{7}/3.17\sqrt{2.19}$
$\tilde{1}\ \tfrac{1}{2}\ \tilde{\tfrac{1}{2}}\ 0\ 000-$	$-\sqrt{11.13}/2.17\sqrt{2.3.19}$
$\tilde{1}\ \tfrac{1}{2}\ \tilde{\tfrac{1}{2}}\ 0\ 100-$	$+2\sqrt{2.5.7}/3.17\sqrt{19}$
$\tilde{1}\ \tfrac{3}{2}\ \tfrac{3}{2}\ 0\ 000-$	$-3\sqrt{13.19}/2.5.17\sqrt{2.7}$
$\tilde{1}\ \tfrac{3}{2}\ \tfrac{3}{2}\ 0\ 010-$	$-\sqrt{11.13.19}/2.3.5.17$
$\tilde{1}\ \tfrac{3}{2}\ \tfrac{3}{2}\ 0\ 020-$	$-5\sqrt{11.19}/2.3.17\sqrt{2.7}$
$\tilde{1}\ \tfrac{3}{2}\ \tfrac{3}{2}\ 0\ 100-$	$-7\sqrt{3.11}/17\sqrt{2.5.19}$
$\tilde{1}\ \tfrac{3}{2}\ \tfrac{3}{2}\ 0\ 110-$	$+8\sqrt{7}/3.17\sqrt{3.5.19}$
$\tilde{1}\ \tfrac{3}{2}\ \tfrac{3}{2}\ 0\ 120-$	$+7\sqrt{5.13}/3.17\sqrt{2.3.19}$
$\tilde{1}\ \tfrac{3}{2}\ \tfrac{3}{2}\ 1\ 000+$	0
$\tilde{1}\ \tfrac{3}{2}\ \tfrac{3}{2}\ 1\ 010+$	$-\sqrt{11.13}/17\sqrt{19}$
$\tilde{1}\ \tfrac{3}{2}\ \tfrac{3}{2}\ 1\ 020+$	$+2\sqrt{2.11}/17\sqrt{7.19}$
$\tilde{1}\ \tfrac{3}{2}\ \tfrac{3}{2}\ 1\ 100+$	$+\sqrt{11}/\sqrt{2.3.5.19}$
$\tilde{1}\ \tfrac{3}{2}\ \tfrac{3}{2}\ 1\ 110+$	$-8.2\sqrt{7}/3.17\sqrt{3.5.19}$
$\tilde{1}\ \tfrac{3}{2}\ \tfrac{3}{2}\ 1\ 120+$	$+7\sqrt{5.13}/3.17\sqrt{2.3.19}$
$\tilde{1}\ \tfrac{3}{2}\ \tilde{\tfrac{1}{2}}\ 0\ 000+$	$+3\sqrt{13.19}/2.17\sqrt{5.7}$
$\tilde{1}\ \tfrac{3}{2}\ \tilde{\tfrac{1}{2}}\ 0\ 010+$	$-7\sqrt{11.13}/2.3.17\sqrt{2.5.19}$
$\tilde{1}\ \tfrac{3}{2}\ \tilde{\tfrac{1}{2}}\ 0\ 020+$	$-\sqrt{5.11}/2.3.17\sqrt{7.19}$
$\tilde{1}\ \tfrac{3}{2}\ \tilde{\tfrac{1}{2}}\ 0\ 100+$	$-\sqrt{11}/4.17\sqrt{3.19}$
$\tilde{1}\ \tfrac{3}{2}\ \tilde{\tfrac{1}{2}}\ 0\ 110+$	$-\sqrt{2.7}/17\sqrt{3.19}$
$\tilde{1}\ \tfrac{3}{2}\ \tilde{\tfrac{1}{2}}\ 0\ 120+$	$-5.7\sqrt{13}/4.17\sqrt{3.19}$
$\tilde{1}\ \tilde{\tfrac{1}{2}}\ \tfrac{3}{2}\ 0\ 000+$	$+13/2.17\sqrt{2.3.5.7.19}$
$\tilde{1}\ \tilde{\tfrac{1}{2}}\ \tfrac{3}{2}\ 0\ 100+$	$-5\sqrt{11.13}/3.17\sqrt{2.19}$
$\mathbf{8\ \tfrac{15}{2}\ \tfrac{7}{2}}$	
$0\ \tfrac{1}{2}\ \tfrac{1}{2}\ 0\ 000+$	$+7.7/2.3\sqrt{13.17.19}$
$0\ \tfrac{3}{2}\ \tfrac{3}{2}\ 0\ 000+$	$-\sqrt{3.11}/\sqrt{13.17.19}$
$0\ \tfrac{3}{2}\ \tfrac{3}{2}\ 0\ 010+$	$-7\sqrt{7}/3\sqrt{2.3.13.17.19}$
$0\ \tfrac{3}{2}\ \tfrac{3}{2}\ 0\ 020+$	$+13/3\sqrt{3.17.19}$
$0\ \tilde{\tfrac{1}{2}}\ \tilde{\tfrac{1}{2}}\ 0\ 000+$	$-\sqrt{5.11}/2\sqrt{3.17.19}$
$1\ \tfrac{1}{2}\ \tfrac{1}{2}\ 0\ 000-$	$+7.7/9\sqrt{2.13.17.19}$

SO_3–O 3jm Factors (cont.)

$\mathbf{8\ \frac{15}{2}\ \frac{7}{2}}$

$1\ \frac{1}{2}\ \frac{1}{2}\ 0\ 100-\quad -13\sqrt{5.7}/2.9\sqrt{2.17.19}$

$1\ \frac{1}{2}\ \frac{3}{2}\ 0\ 000+\quad +5.7\sqrt{5.7}/2.9\sqrt{2.13.17.19}$

$1\ \frac{1}{2}\ \frac{3}{2}\ 0\ 100+\quad +13/2.9\sqrt{2.17.19}$

$1\ \frac{3}{2}\ \frac{1}{2}\ 0\ 000+\quad +\sqrt{5.7.11}/4\sqrt{3.13.17.19}$

$1\ \frac{3}{2}\ \frac{1}{2}\ 0\ 010+\quad +5.7.7\sqrt{5}/2.9\sqrt{2.3.13.17.19}$

$1\ \frac{3}{2}\ \frac{1}{2}\ 0\ 020+\quad -13\sqrt{5.7}/4.9\sqrt{3.17.19}$

$1\ \frac{3}{2}\ \frac{1}{2}\ 0\ 100+\quad +\sqrt{11}/8\sqrt{3.17.19}$

$1\ \frac{3}{2}\ \frac{1}{2}\ 0\ 110+\quad +13\sqrt{7}/2.9\sqrt{2.3.17.19}$

$1\ \frac{3}{2}\ \frac{1}{2}\ 0\ 120+\quad -7.7\sqrt{13}/8.9\sqrt{3.17.19}$

$1\ \frac{3}{2}\ \frac{3}{2}\ 0\ 000-\quad +3\sqrt{3.11}/2\sqrt{2.13.17.19}$

$1\ \frac{3}{2}\ \frac{3}{2}\ 0\ 010-\quad +2.11\sqrt{7}/9\sqrt{3.13.17.19}$

$1\ \frac{3}{2}\ \frac{3}{2}\ 0\ 020-\quad +23/2.9\sqrt{2.3.17.19}$

$1\ \frac{3}{2}\ \frac{3}{2}\ 0\ 100-\quad +3\sqrt{3.7.11}/4\sqrt{2.5.17.19}$

$1\ \frac{3}{2}\ \frac{3}{2}\ 0\ 110-\quad -7\sqrt{19}/9\sqrt{3.5.17}$

$1\ \frac{3}{2}\ \frac{3}{2}\ 0\ 120-\quad -5\sqrt{5.7.13}/4.9\sqrt{2.3.17.19}$

$1\ \frac{3}{2}\ \frac{3}{2}\ 1\ 000-\quad +\sqrt{11.17}/4\sqrt{2.3.13.19}$

$1\ \frac{3}{2}\ \frac{3}{2}\ 1\ 010-\quad -29\sqrt{7}/2.9\sqrt{3.13.17.19}$

$1\ \frac{3}{2}\ \frac{3}{2}\ 1\ 020-\quad +263/4.9\sqrt{2.3.17.19}$

$1\ \frac{3}{2}\ \frac{3}{2}\ 1\ 100-\quad -\sqrt{5.11}/4\sqrt{2.3.7.17.19}$

$1\ \frac{3}{2}\ \frac{3}{2}\ 1\ 110-\quad -\sqrt{5}/9\sqrt{3.17.19}$

$1\ \frac{3}{2}\ \frac{3}{2}\ 1\ 120-\quad -\sqrt{5.7.13}/4.9\sqrt{2.3.17.19}$

$1\ \frac{3}{2}\ \tilde{\frac{1}{2}}\ 0\ 000+\quad +11\sqrt{11}/2.3\sqrt{13.17.19}$

$1\ \frac{3}{2}\ \tilde{\frac{1}{2}}\ 0\ 010+\quad -61\sqrt{7}/2.9\sqrt{2.13.17.19}$

$1\ \frac{3}{2}\ \tilde{\frac{1}{2}}\ 0\ 020+\quad -47/2.9\sqrt{17.19}$

$1\ \frac{3}{2}\ \tilde{\frac{1}{2}}\ 0\ 100+\quad -131\sqrt{11}/8.3\sqrt{5.7.17.19}$

$1\ \frac{3}{2}\ \tilde{\frac{1}{2}}\ 0\ 110+\quad -71/2.9\sqrt{2.5.17.19}$

$1\ \frac{3}{2}\ \tilde{\frac{1}{2}}\ 0\ 120+\quad -5\sqrt{5.7.13}/8.9\sqrt{17.19}$

$1\ \tilde{\frac{1}{2}}\ \frac{3}{2}\ 0\ 000+\quad -\sqrt{5.11}/2\sqrt{2.17.19}$

$1\ \tilde{\frac{1}{2}}\ \frac{3}{2}\ 0\ 100+\quad +\sqrt{11.13}/2\sqrt{2.7.17.19}$

$1\ \tilde{\frac{1}{2}}\ \tilde{\frac{1}{2}}\ 0\ 000-\quad -\sqrt{5.11}/3\sqrt{2.3.17.19}$

$1\ \tilde{\frac{1}{2}}\ \tilde{\frac{1}{2}}\ 0\ 100-\quad -\sqrt{11.13}/2.3\sqrt{2.3.7.17.19}$

$2\ \frac{1}{2}\ \frac{3}{2}\ 0\ 000-\quad +7.7\sqrt{11}/8.3\sqrt{13.17.19}$

$2\ \frac{1}{2}\ \frac{3}{2}\ 0\ 100-\quad +5\sqrt{5.7}/8\sqrt{3.17.19}$

$2\ \frac{3}{2}\ \frac{1}{2}\ 0\ 000-\quad -3\sqrt{3.13}/8.2\sqrt{2.17.19}$

$2\ \frac{3}{2}\ \frac{1}{2}\ 0\ 010-\quad +43\sqrt{7.11}/8.3\sqrt{3.13.17.19}$

$2\ \frac{3}{2}\ \frac{1}{2}\ 0\ 020-\quad -43\sqrt{11}/8.2.3\sqrt{2.3.17.19}$

$2\ \frac{3}{2}\ \frac{1}{2}\ 0\ 100-\quad -7\sqrt{7.11}/8.2\sqrt{2.5.17.19}$

$2\ \frac{3}{2}\ \frac{1}{2}\ 0\ 110-\quad +7/8\sqrt{5.17.19}$

$2\ \frac{3}{2}\ \frac{1}{2}\ 0\ 120-\quad +3\sqrt{5.7.13}/8.2\sqrt{2.17.19}$

$2\ \frac{3}{2}\ \frac{3}{2}\ 0\ 000+\quad +41\sqrt{3}/8.2\sqrt{5.7.13.17.19}$

$2\ \frac{3}{2}\ \frac{3}{2}\ 0\ 010+\quad -101\sqrt{11}/2.3\sqrt{2.3.5.13.17.19}$

$2\ \frac{3}{2}\ \frac{3}{2}\ 0\ 020+\quad -5.11\sqrt{5.11}/8.2.3\sqrt{3.7.17.19}$

$2\ \frac{3}{2}\ \frac{3}{2}\ 0\ 100+\quad -11\sqrt{11}/8.2\sqrt{17.19}$

$2\ \frac{3}{2}\ \frac{3}{2}\ 0\ 110+\quad +7\sqrt{7}/2.3\sqrt{2.17.19}$

$2\ \frac{3}{2}\ \frac{3}{2}\ 0\ 120+\quad +\sqrt{13}/8.2.3\sqrt{17.19}$

$2\ \frac{3}{2}\ \tilde{\frac{1}{2}}\ 0\ 000-\quad -3.431/8.2\sqrt{2.5.7.13.17.19}$

$2\ \frac{3}{2}\ \tilde{\frac{1}{2}}\ 0\ 010-\quad -23\sqrt{11}/8.3\sqrt{5.13.17.19}$

$2\ \frac{3}{2}\ \tilde{\frac{1}{2}}\ 0\ 020-\quad -23\sqrt{5.11}/8.2.3\sqrt{2.7.17.19}$

$2\ \frac{3}{2}\ \tilde{\frac{1}{2}}\ 0\ 100-\quad -3\sqrt{3.11}/8.2\sqrt{2.17.19}$

$\mathbf{8\ \frac{15}{2}\ \frac{7}{2}}$

$2\ \frac{3}{2}\ \tilde{\frac{1}{2}}\ 0\ 110-\quad -\sqrt{3.7}/8\sqrt{17.19}$

$2\ \frac{3}{2}\ \tilde{\frac{1}{2}}\ 0\ 120-\quad +11\sqrt{3.13}/8.2\sqrt{2.17.19}$

$2\ \tilde{\frac{1}{2}}\ \frac{3}{2}\ 0\ 000-\quad -53/8\sqrt{7.17.19}$

$2\ \tilde{\frac{1}{2}}\ \frac{3}{2}\ 0\ 100-\quad -\sqrt{5.11.13}/8\sqrt{3.17.19}$

$\tilde{1}\ \frac{1}{2}\ \frac{3}{2}\ 0\ 000-\quad +5\sqrt{5.11}/2.17\sqrt{2.19}$

$\tilde{1}\ \frac{1}{2}\ \frac{3}{2}\ 0\ 100-\quad -\sqrt{7.13}/2.17\sqrt{2.3.19}$

$\tilde{1}\ \frac{1}{2}\ \tilde{\frac{1}{2}}\ 0\ 000+\quad -\sqrt{5.11}/17\sqrt{2.3.19}$

$\tilde{1}\ \frac{1}{2}\ \tilde{\frac{1}{2}}\ 0\ 100+\quad -\sqrt{7.13}/2.17\sqrt{2.19}$

$\tilde{1}\ \frac{3}{2}\ \frac{1}{2}\ 0\ 000-\quad -7/2.17\sqrt{3.5.19}$

$\tilde{1}\ \frac{3}{2}\ \frac{1}{2}\ 0\ 010-\quad -43\sqrt{7.11}/2.3.17\sqrt{2.3.5.19}$

$\tilde{1}\ \frac{3}{2}\ \frac{1}{2}\ 0\ 020-\quad +5\sqrt{5.11.13}/2.3.17\sqrt{3.19}$

$\tilde{1}\ \frac{3}{2}\ \frac{1}{2}\ 0\ 100-\quad -37\sqrt{7.11}/8.17\sqrt{13.19}$

$\tilde{1}\ \frac{3}{2}\ \frac{1}{2}\ 0\ 110-\quad -9.7/2.17\sqrt{2.13.19}$

$\tilde{1}\ \frac{3}{2}\ \frac{1}{2}\ 0\ 120-\quad -\sqrt{7.19}/8.17$

$\tilde{1}\ \frac{3}{2}\ \frac{3}{2}\ 0\ 000+\quad +83/2.17\sqrt{2.3.7.19}$

$\tilde{1}\ \frac{3}{2}\ \frac{3}{2}\ 0\ 010+\quad +8\sqrt{11}/3.17\sqrt{3.19}$

$\tilde{1}\ \frac{3}{2}\ \frac{3}{2}\ 0\ 020+\quad +13\sqrt{11.13}/2.3.17\sqrt{2.3.7.19}$

$\tilde{1}\ \frac{3}{2}\ \frac{3}{2}\ 0\ 100+\quad -7\sqrt{11}/4.17\sqrt{2.5.13.19}$

$\tilde{1}\ \frac{3}{2}\ \frac{3}{2}\ 0\ 110+\quad -101\sqrt{7}/3.17\sqrt{5.13.19}$

$\tilde{1}\ \frac{3}{2}\ \frac{3}{2}\ 0\ 120+\quad -7\sqrt{5}/4.3.17\sqrt{2.19}$

$\tilde{1}\ \frac{3}{2}\ \frac{3}{2}\ 1\ 000-\quad +37\sqrt{7}/4.17\sqrt{2.3.19}$

$\tilde{1}\ \frac{3}{2}\ \frac{3}{2}\ 1\ 010-\quad +\sqrt{11.19}/2.3.17\sqrt{3}$

$\tilde{1}\ \frac{3}{2}\ \frac{3}{2}\ 1\ 020-\quad +5\sqrt{11.13}/4.3.17\sqrt{2.3.7.19}$

$\tilde{1}\ \frac{3}{2}\ \frac{3}{2}\ 1\ 100-\quad +41\sqrt{11}/4.17\sqrt{2.5.13.19}$

$\tilde{1}\ \frac{3}{2}\ \frac{3}{2}\ 1\ 110-\quad +83\sqrt{7}/3.17\sqrt{5.13.19}$

$\tilde{1}\ \frac{3}{2}\ \frac{3}{2}\ 1\ 120-\quad +7.7\sqrt{5}/4.3.17\sqrt{2.19}$

$\tilde{1}\ \frac{3}{2}\ \tilde{\frac{1}{2}}\ 0\ 000-\quad -9\sqrt{19}/4.17\sqrt{7}$

$\tilde{1}\ \frac{3}{2}\ \tilde{\frac{1}{2}}\ 0\ 010-\quad -7\sqrt{11}/2.3.17\sqrt{2.19}$

$\tilde{1}\ \frac{3}{2}\ \tilde{\frac{1}{2}}\ 0\ 020-\quad -\sqrt{11.13}/4.3.17\sqrt{7.19}$

$\tilde{1}\ \frac{3}{2}\ \tilde{\frac{1}{2}}\ 0\ 100-\quad +\sqrt{3.5.11}/8.17\sqrt{13.19}$

$\tilde{1}\ \frac{3}{2}\ \tilde{\frac{1}{2}}\ 0\ 110-\quad +\sqrt{5.7.13}/2.17\sqrt{2.3.19}$

$\tilde{1}\ \frac{3}{2}\ \tilde{\frac{1}{2}}\ 0\ 120-\quad -7.13\sqrt{5}/8.17\sqrt{3.19}$

$\tilde{1}\ \tilde{\frac{1}{2}}\ \frac{1}{2}\ 0\ 000+\quad -\sqrt{13}/3.17\sqrt{2.19}$

$\tilde{1}\ \tilde{\frac{1}{2}}\ \frac{1}{2}\ 0\ 100+\quad +\sqrt{5.7.11}/2.17\sqrt{2.3.19}$

$\tilde{1}\ \tilde{\frac{1}{2}}\ \frac{3}{2}\ 0\ 000-\quad +\sqrt{5.13}/2.3.17\sqrt{2.7.19}$

$\tilde{1}\ \tilde{\frac{1}{2}}\ \frac{3}{2}\ 0\ 100-\quad +23\sqrt{11}/2.17\sqrt{2.3.19}$

$\mathbf{8\ \frac{15}{2}\ \frac{9}{2}}$

$0\ \frac{1}{2}\ \frac{1}{2}\ 0\ 000-\quad +7\sqrt{7}/3\sqrt{13.17.19}$

$0\ \frac{3}{2}\ \frac{3}{2}\ 0\ 000-\quad 0$

$0\ \frac{3}{2}\ \frac{3}{2}\ 0\ 001-\quad +\sqrt{3.7.11}/2\sqrt{13.17.19}$

$0\ \frac{3}{2}\ \frac{3}{2}\ 0\ 010-\quad +5\sqrt{2.7}/3\sqrt{13.17.19}$

$0\ \frac{3}{2}\ \frac{3}{2}\ 0\ 011-\quad +8.2\sqrt{2}/3\sqrt{3.13.17.19}$

$0\ \frac{3}{2}\ \frac{3}{2}\ 0\ 020-\quad +8/3\sqrt{17.19}$

$0\ \frac{3}{2}\ \frac{3}{2}\ 0\ 021-\quad -\sqrt{7}/2.3\sqrt{3.17.19}$

$1\ \frac{1}{2}\ \frac{1}{2}\ 0\ 000+\quad +7\sqrt{7.13}/2.9\sqrt{2.17.19}$

$1\ \frac{1}{2}\ \frac{1}{2}\ 0\ 100+\quad +2.13\sqrt{2}/9\sqrt{5.17.19}$

$1\ \frac{1}{2}\ \frac{3}{2}\ 0\ 000-\quad -11\sqrt{7}/2\sqrt{2.3.5.13.17.19}$

$1\ \frac{1}{2}\ \frac{3}{2}\ 0\ 001-\quad -11\sqrt{2}/9\sqrt{5.13.17.19}$

$1\ \frac{1}{2}\ \frac{3}{2}\ 0\ 100-\quad -7\sqrt{2}/5\sqrt{3.17.19}$

SO_3–O 3jm Factors (cont.)

$8\ \frac{15}{2}\ \frac{9}{2}$	
$1\ \frac{1}{2}\ \frac{3}{2}\ 0\ 101-$	$+7\sqrt{7.17}/2.9.5\sqrt{2.19}$
$1\ \frac{3}{2}\ \frac{1}{2}\ 0\ 000-$	$-7\sqrt{11}/2\sqrt{3.5.13.17.19}$
$1\ \frac{3}{2}\ \frac{1}{2}\ 0\ 010-$	$+7.23\sqrt{7}/2.9\sqrt{2.3.5.13.17.19}$
$1\ \frac{3}{2}\ \frac{1}{2}\ 0\ 020-$	$-13\sqrt{5}/2.9\sqrt{3.17.19}$
$1\ \frac{3}{2}\ \frac{1}{2}\ 0\ 100-$	$-\sqrt{7.11}/4.5\sqrt{3.17.19}$
$1\ \frac{3}{2}\ \frac{1}{2}\ 0\ 110-$	$-11.13\sqrt{2}/9.5\sqrt{3.17.19}$
$1\ \frac{3}{2}\ \frac{1}{2}\ 0\ 120-$	$-7\sqrt{7.13}/4.9\sqrt{3.17.19}$
$1\ \frac{3}{2}\ \frac{3}{2}\ 0\ 000+$	$+11\sqrt{11}/3\sqrt{2.13.17.19}$
$1\ \frac{3}{2}\ \frac{3}{2}\ 0\ 001+$	$-3\sqrt{3.7.11}/4\sqrt{2.13.17.19}$
$1\ \frac{3}{2}\ \frac{3}{2}\ 0\ 010+$	$+29\sqrt{7}/2.9\sqrt{13.17.19}$
$1\ \frac{3}{2}\ \frac{3}{2}\ 0\ 011+$	$-2.31/9\sqrt{3.13.17.19}$
$1\ \frac{3}{2}\ \frac{3}{2}\ 0\ 020+$	$+43/9\sqrt{2.17.19}$
$1\ \frac{3}{2}\ \frac{3}{2}\ 0\ 021+$	$+73\sqrt{7}/4.9\sqrt{2.3.17.19}$
$1\ \frac{3}{2}\ \frac{3}{2}\ 0\ 100+$	$-7\sqrt{7.11}/3.5\sqrt{2.5.17.19}$
$1\ \frac{3}{2}\ \frac{3}{2}\ 0\ 101+$	$+3\sqrt{2.3.11}/5\sqrt{5.17.19}$
$1\ \frac{3}{2}\ \frac{3}{2}\ 0\ 110+$	$-2\sqrt{19}/9.5\sqrt{5.17}$
$1\ \frac{3}{2}\ \frac{3}{2}\ 0\ 111+$	$-8\sqrt{7.19}/9.5\sqrt{3.5.17}$
$1\ \frac{3}{2}\ \frac{3}{2}\ 0\ 120+$	$+\sqrt{5.7.13}/9\sqrt{2.17.19}$
$1\ \frac{3}{2}\ \frac{3}{2}\ 0\ 121+$	$-\sqrt{2.5.13}/9\sqrt{3.17.19}$
$1\ \frac{3}{2}\ \frac{3}{2}\ 1\ 000+$	$+2\sqrt{2.11}/3.5\sqrt{13.17.19}$
$1\ \frac{3}{2}\ \frac{3}{2}\ 1\ 001+$	$-47\sqrt{7.11}/4.5\sqrt{2.3.13.17.19}$
$1\ \frac{3}{2}\ \frac{3}{2}\ 1\ 010+$	$+23\sqrt{7}/9.5\sqrt{13.17.19}$
$1\ \frac{3}{2}\ \frac{3}{2}\ 1\ 011+$	$+2.257/9.5\sqrt{3.13.17.19}$
$1\ \frac{3}{2}\ \frac{3}{2}\ 1\ 020+$	$+8\sqrt{2}/9\sqrt{17.19}$
$1\ \frac{3}{2}\ \frac{3}{2}\ 1\ 021+$	$+11\sqrt{7}/4.9\sqrt{2.3.17.19}$
$1\ \frac{3}{2}\ \frac{3}{2}\ 1\ 100+$	$-\sqrt{7.11}/3.5\sqrt{2.5.17.19}$
$1\ \frac{3}{2}\ \frac{3}{2}\ 1\ 101+$	$-8\sqrt{2.11}/5\sqrt{3.5.17.19}$
$1\ \frac{3}{2}\ \frac{3}{2}\ 1\ 110+$	$+2.53/9.5\sqrt{5.17.19}$
$1\ \frac{3}{2}\ \frac{3}{2}\ 1\ 111+$	$-257\sqrt{7}/2.9.5\sqrt{3.5.17.19}$
$1\ \frac{3}{2}\ \frac{3}{2}\ 1\ 120+$	$+\sqrt{5.7.13}/9\sqrt{2.17.19}$
$1\ \frac{3}{2}\ \frac{3}{2}\ 1\ 121+$	$-2\sqrt{2.5.13}/9\sqrt{3.17.19}$
$1\ \tilde{\frac{1}{2}}\ \frac{3}{2}\ 0\ 000-$	$-5\sqrt{5.11}/2.3\sqrt{2.3.17.19}$
$1\ \tilde{\frac{1}{2}}\ \frac{3}{2}\ 0\ 001-$	0
$1\ \tilde{\frac{1}{2}}\ \frac{3}{2}\ 0\ 100-$	$+\sqrt{2.7.11.13}/3.5\sqrt{3.17.19}$
$1\ \tilde{\frac{1}{2}}\ \frac{3}{2}\ 0\ 101-$	$+3\sqrt{11.13}/2.5\sqrt{2.17.19}$
$2\ \frac{1}{2}\ \frac{3}{2}\ 0\ 000+$	$+\sqrt{3.11}/4\sqrt{13.17.19}$
$2\ \frac{1}{2}\ \frac{3}{2}\ 0\ 001+$	$-7\sqrt{7.11}/4.3\sqrt{13.17.19}$
$2\ \frac{1}{2}\ \frac{3}{2}\ 0\ 100+$	$-\sqrt{5.7}/4\sqrt{17.19}$
$2\ \frac{1}{2}\ \frac{3}{2}\ 0\ 101+$	$-5\sqrt{5}/4\sqrt{3.17.19}$
$2\ \frac{3}{2}\ \frac{1}{2}\ 0\ 000+$	$-9\sqrt{3.7}/8.5\sqrt{2.13.17.19}$
$2\ \frac{3}{2}\ \frac{1}{2}\ 0\ 010+$	$+251\sqrt{11}/4.3.5\sqrt{3.13.17.19}$
$2\ \frac{3}{2}\ \frac{1}{2}\ 0\ 020+$	$-\sqrt{7.11}/8.3\sqrt{2.3.17.19}$
$2\ \frac{3}{2}\ \frac{1}{2}\ 0\ 100+$	$-\sqrt{11.19}/8\sqrt{2.5.17}$
$2\ \frac{3}{2}\ \frac{1}{2}\ 0\ 110+$	$-3\sqrt{7}/4\sqrt{5.17.19}$
$2\ \frac{3}{2}\ \frac{1}{2}\ 0\ 120+$	$-\sqrt{5.13}/8\sqrt{2.17.19}$
$2\ \frac{3}{2}\ \frac{3}{2}\ 0\ 000-$	$+3.11.11\sqrt{7}/8.5\sqrt{5.13.17.19}$
$2\ \frac{3}{2}\ \frac{3}{2}\ 0\ 001-$	$-397\sqrt{3}/8.5\sqrt{5.13.17.19}$
$2\ \frac{3}{2}\ \frac{3}{2}\ 0\ 010-$	$-47\sqrt{11}/3.5\sqrt{2.5.13.17.19}$
$2\ \frac{3}{2}\ \frac{3}{2}\ 0\ 011-$	$+11\sqrt{2.7.11}/3.5\sqrt{3.5.13.17.19}$

$8\ \frac{15}{2}\ \frac{9}{2}$	
$2\ \frac{3}{2}\ \frac{3}{2}\ 0\ 020-$	$-7\sqrt{7.11}/8.3\sqrt{5.17.19}$
$2\ \frac{3}{2}\ \frac{3}{2}\ 0\ 021-$	$-61\sqrt{11}/8.3\sqrt{3.5.17.19}$
$2\ \frac{3}{2}\ \frac{3}{2}\ 0\ 100-$	$+3\sqrt{3.11}/8.5\sqrt{17.19}$
$2\ \frac{3}{2}\ \frac{3}{2}\ 0\ 101-$	$+7\sqrt{7.11}/8.5\sqrt{17.19}$
$2\ \frac{3}{2}\ \frac{3}{2}\ 0\ 110-$	$-11\sqrt{7}/5\sqrt{2.3.17.19}$
$2\ \frac{3}{2}\ \frac{3}{2}\ 0\ 111-$	$-\sqrt{2.19}/3.5\sqrt{17}$
$2\ \frac{3}{2}\ \frac{3}{2}\ 0\ 120-$	$+11\sqrt{13}/8\sqrt{3.17.19}$
$2\ \frac{3}{2}\ \frac{3}{2}\ 0\ 121-$	$-\sqrt{7.13}/8.3\sqrt{17.19}$
$2\ \tilde{\frac{1}{2}}\ \frac{3}{2}\ 0\ 000+$	$-23\sqrt{7}/4.5\sqrt{3.17.19}$
$2\ \tilde{\frac{1}{2}}\ \frac{3}{2}\ 0\ 001+$	$-53/4.5\sqrt{17.19}$
$2\ \tilde{\frac{1}{2}}\ \frac{3}{2}\ 0\ 100+$	$+\sqrt{11.13}/4\sqrt{5.17.19}$
$2\ \tilde{\frac{1}{2}}\ \frac{3}{2}\ 0\ 101+$	$-\sqrt{7.11.13}/4\sqrt{3.5.17.19}$
$\tilde{1}\ \frac{1}{2}\ \frac{3}{2}\ 0\ 000+$	$-\sqrt{11}/2\sqrt{2.3.5.19}$
$\tilde{1}\ \frac{1}{2}\ \frac{3}{2}\ 0\ 001+$	$+\sqrt{2.7.11}/17\sqrt{5.19}$
$\tilde{1}\ \frac{1}{2}\ \frac{3}{2}\ 0\ 100+$	$+\sqrt{2.7}/5\sqrt{13.19}$
$\tilde{1}\ \frac{1}{2}\ \frac{3}{2}\ 0\ 101+$	$+1039/2.5.17\sqrt{2.3.13.19}$
$\tilde{1}\ \frac{3}{2}\ \frac{1}{2}\ 0\ 000+$	$+\sqrt{5.7}/17\sqrt{3.19}$
$\tilde{1}\ \frac{3}{2}\ \frac{1}{2}\ 0\ 010+$	$+23\sqrt{5.11}/2.3.17\sqrt{2.3.19}$
$\tilde{1}\ \frac{3}{2}\ \frac{1}{2}\ 0\ 020+$	$-\sqrt{5.7.11.13}/3.17\sqrt{3.19}$
$\tilde{1}\ \frac{3}{2}\ \frac{1}{2}\ 0\ 100+$	$-2.31\sqrt{11}/5.17\sqrt{13.19}$
$\tilde{1}\ \frac{3}{2}\ \frac{1}{2}\ 0\ 110+$	$+4.3\sqrt{2.7}/5.17\sqrt{13.19}$
$\tilde{1}\ \frac{3}{2}\ \frac{1}{2}\ 0\ 120+$	$-1/17\sqrt{19}$
$\tilde{1}\ \frac{3}{2}\ \frac{3}{2}\ 0\ 000-$	$+3\sqrt{7}/17\sqrt{2.19}$
$\tilde{1}\ \frac{3}{2}\ \frac{3}{2}\ 0\ 001-$	$-83/4.17\sqrt{2.3.19}$
$\tilde{1}\ \frac{3}{2}\ \frac{3}{2}\ 0\ 010-$	$-\sqrt{11}/2.3.17\sqrt{19}$
$\tilde{1}\ \frac{3}{2}\ \frac{3}{2}\ 0\ 011-$	$-4\sqrt{7.11}/3.17\sqrt{3.19}$
$\tilde{1}\ \frac{3}{2}\ \frac{3}{2}\ 0\ 020-$	$+\sqrt{7.11.13}/3.17\sqrt{2.19}$
$\tilde{1}\ \frac{3}{2}\ \frac{3}{2}\ 0\ 021-$	$-13\sqrt{11.13}/4.3.17\sqrt{2.3.19}$
$\tilde{1}\ \frac{3}{2}\ \frac{3}{2}\ 0\ 100-$	$+2\sqrt{2.3.11.13}/5.17\sqrt{5.19}$
$\tilde{1}\ \frac{3}{2}\ \frac{3}{2}\ 0\ 101-$	$-\sqrt{2.7.11}/5.17\sqrt{5.13.19}$
$\tilde{1}\ \frac{3}{2}\ \frac{3}{2}\ 0\ 110-$	$+2.7.7\sqrt{3.7}/5.17\sqrt{5.13.19}$
$\tilde{1}\ \frac{3}{2}\ \frac{3}{2}\ 0\ 111-$	$-8.101/3.5.17\sqrt{5.13.19}$
$\tilde{1}\ \frac{3}{2}\ \frac{3}{2}\ 0\ 120-$	$-8\sqrt{2.3}/17\sqrt{5.19}$
$\tilde{1}\ \frac{3}{2}\ \frac{3}{2}\ 0\ 121-$	$-\sqrt{2.7}/3.17\sqrt{5.19}$
$\tilde{1}\ \frac{3}{2}\ \frac{3}{2}\ 1\ 000+$	$+3.11\sqrt{7}/5.17\sqrt{2.19}$
$\tilde{1}\ \frac{3}{2}\ \frac{3}{2}\ 1\ 001+$	$-229/4.5.17\sqrt{2.3.19}$
$\tilde{1}\ \frac{3}{2}\ \frac{3}{2}\ 1\ 010+$	$+7\sqrt{11}/3.5.17\sqrt{19}$
$\tilde{1}\ \frac{3}{2}\ \frac{3}{2}\ 1\ 011+$	$+4.7\sqrt{7.11}/3.5.17\sqrt{3.19}$
$\tilde{1}\ \frac{3}{2}\ \frac{3}{2}\ 1\ 020+$	$-\sqrt{7.11.13}/3.17\sqrt{2.19}$
$\tilde{1}\ \frac{3}{2}\ \frac{3}{2}\ 1\ 021+$	$-7\sqrt{11.13}/4.3.17\sqrt{2.3.19}$
$\tilde{1}\ \frac{3}{2}\ \frac{3}{2}\ 1\ 100+$	$+\sqrt{3.11}/2.5.17\sqrt{2.5.13.19}$
$\tilde{1}\ \frac{3}{2}\ \frac{3}{2}\ 1\ 101+$	$+11\sqrt{2.7.11}/5.17\sqrt{5.13.19}$
$\tilde{1}\ \frac{3}{2}\ \frac{3}{2}\ 1\ 110+$	$-8.23\sqrt{7}/5.17\sqrt{3.5.13.19}$
$\tilde{1}\ \frac{3}{2}\ \frac{3}{2}\ 1\ 111+$	$-7.367/2.3.5.17\sqrt{5.13.19}$
$\tilde{1}\ \frac{3}{2}\ \frac{3}{2}\ 1\ 120+$	$-7.13/2.17\sqrt{2.3.5.19}$
$\tilde{1}\ \frac{3}{2}\ \frac{3}{2}\ 1\ 121+$	$-11\sqrt{2.7}/3.17\sqrt{5.19}$
$\tilde{1}\ \tilde{\frac{1}{2}}\ \frac{1}{2}\ 0\ 000-$	$-\sqrt{7.13}/2.3.17\sqrt{2.19}$
$\tilde{1}\ \tilde{\frac{1}{2}}\ \frac{1}{2}\ 0\ 100-$	$-2\sqrt{2.11}/17\sqrt{3.5.19}$
$\tilde{1}\ \tilde{\frac{1}{2}}\ \frac{3}{2}\ 0\ 000+$	$+\sqrt{3.7.13}/2.17\sqrt{2.5.19}$

SO_3–O 3jm Factors (cont.)

8	**$\frac{15}{2}$**	**$\frac{9}{2}$**			
$\tilde{1}$	$\tilde{\frac{1}{2}}$	$\frac{3}{2}$	0	001+	$-4\sqrt{2.13}/3.17\sqrt{5.19}$
$\tilde{1}$	$\tilde{\frac{1}{2}}$	$\frac{3}{2}$	0	100+	$-11\sqrt{2.11}/5.17\sqrt{19}$
$\tilde{1}$	$\tilde{\frac{1}{2}}$	$\frac{3}{2}$	0	101+	$-11\sqrt{7.11}/2.5.17\sqrt{2.3.19}$
8	**$\frac{15}{2}$**	**$\frac{11}{2}$**			
0	$\frac{1}{2}$	$\frac{1}{2}$	0	000+	$+2/\sqrt{3.13.17.19}$
0	$\frac{3}{2}$	$\frac{3}{2}$	0	000+	$+9\sqrt{5}/2\sqrt{13.17.19}$
0	$\frac{3}{2}$	$\frac{3}{2}$	0	001+	$-7\sqrt{2}/3\sqrt{13.17.19}$
0	$\frac{3}{2}$	$\frac{3}{2}$	0	010+	0
0	$\frac{3}{2}$	$\frac{3}{2}$	0	011+	$+7\sqrt{7.11}/9\sqrt{13.17.19}$
0	$\frac{3}{2}$	$\frac{3}{2}$	0	020+	$+7\sqrt{5}/2\sqrt{11.17.19}$
0	$\frac{3}{2}$	$\frac{3}{2}$	0	021+	$+13\sqrt{2}/9\sqrt{11.17.19}$
0	$\tilde{\frac{1}{2}}$	$\tilde{\frac{1}{2}}$	0	000+	$+2\sqrt{5}/3\sqrt{3.17.19}$
1	$\frac{1}{2}$	$\frac{1}{2}$	0	000−	$+\sqrt{2}/3\sqrt{3.13.17.19}$
1	$\frac{1}{2}$	$\frac{1}{2}$	0	100−	$+7\sqrt{5.7}/2.3\sqrt{2.3.17.19}$
1	$\frac{1}{2}$	$\frac{3}{2}$	0	000+	$-\sqrt{7.11}/3\sqrt{2.3.13.17.19}$
1	$\frac{1}{2}$	$\frac{3}{2}$	0	001+	$-7\sqrt{5.7.11}/4.3\sqrt{3.13.17.19}$
1	$\frac{1}{2}$	$\frac{3}{2}$	0	100+	$+7\sqrt{5}/3\sqrt{2.3.11.17.19}$
1	$\frac{1}{2}$	$\frac{3}{2}$	0	101+	$-13/3\sqrt{3.11.17.19}$
1	$\frac{3}{2}$	$\frac{1}{2}$	0	000+	$+3\sqrt{5.7.11}/8\sqrt{13.17.19}$
1	$\frac{3}{2}$	$\frac{1}{2}$	0	010+	$+\sqrt{5}/\sqrt{2.13.17.19}$
1	$\frac{3}{2}$	$\frac{1}{2}$	0	020+	$+7\sqrt{5.7}/8.3\sqrt{17.19}$
1	$\frac{3}{2}$	$\frac{1}{2}$	0	100+	0
1	$\frac{3}{2}$	$\frac{1}{2}$	0	110+	0
1	$\frac{3}{2}$	$\frac{1}{2}$	0	120+	$-2\sqrt{13}/3\sqrt{17.19}$
1	$\frac{3}{2}$	$\frac{3}{2}$	0	000−	$-7.11/2.3\sqrt{2.5.13.17.19}$
1	$\frac{3}{2}$	$\frac{3}{2}$	0	001−	$-11/2.9\sqrt{13.17.19}$
1	$\frac{3}{2}$	$\frac{3}{2}$	0	010−	$+2.59\sqrt{7}/9\sqrt{5.11.13.17.19}$
1	$\frac{3}{2}$	$\frac{3}{2}$	0	011−	$-37\sqrt{7}/9.3\sqrt{2.11.13.17.19}$
1	$\frac{3}{2}$	$\frac{3}{2}$	0	020−	$+43\sqrt{5}/2.9\sqrt{2.11.17.19}$
1	$\frac{3}{2}$	$\frac{3}{2}$	0	021−	$-7/2.9.3\sqrt{11.17.19}$
1	$\frac{3}{2}$	$\frac{3}{2}$	0	100−	$-\sqrt{7}/3\sqrt{2.17.19}$
1	$\frac{3}{2}$	$\frac{3}{2}$	0	101−	$-\sqrt{7.17}/4.9\sqrt{5.19}$
1	$\frac{3}{2}$	$\frac{3}{2}$	0	110−	$-43/2.9\sqrt{11.17.19}$
1	$\frac{3}{2}$	$\frac{3}{2}$	0	111−	$+2.109\sqrt{2}/9.3\sqrt{5.11.17.19}$
1	$\frac{3}{2}$	$\frac{3}{2}$	0	120−	$+7\sqrt{7.13}/9\sqrt{2.11.17.19}$
1	$\frac{3}{2}$	$\frac{3}{2}$	0	121−	$-29\sqrt{5.7.13}/4.9.3\sqrt{11.17.19}$
1	$\frac{3}{2}$	$\frac{3}{2}$	1	000−	$-7.79/4.3\sqrt{2.5.13.17.19}$
1	$\frac{3}{2}$	$\frac{3}{2}$	1	001−	$+53/4.9\sqrt{13.17.19}$
1	$\frac{3}{2}$	$\frac{3}{2}$	1	010−	$-59\sqrt{7}/9\sqrt{5.11.13.17.19}$
1	$\frac{3}{2}$	$\frac{3}{2}$	1	011−	$+31.31\sqrt{7}/2.9.3\sqrt{2.11.13.17.19}$
1	$\frac{3}{2}$	$\frac{3}{2}$	1	020−	$+167\sqrt{5}/4.9\sqrt{2.11.17.19}$
1	$\frac{3}{2}$	$\frac{3}{2}$	1	021−	$+7.79/4.9.3\sqrt{11.17.19}$
1	$\frac{3}{2}$	$\frac{3}{2}$	1	100−	$+\sqrt{7}/2.3\sqrt{2.17.19}$
1	$\frac{3}{2}$	$\frac{3}{2}$	1	101−	$+5\sqrt{5.7}/4.9\sqrt{17.19}$
1	$\frac{3}{2}$	$\frac{3}{2}$	1	110−	$-4\sqrt{17}/9\sqrt{11.19}$
1	$\frac{3}{2}$	$\frac{3}{2}$	1	111−	$+89\sqrt{5}/9.3\sqrt{2.11.17.19}$
1	$\frac{3}{2}$	$\frac{3}{2}$	1	120−	$+5\sqrt{7.13}/2.9\sqrt{2.11.17.19}$
1	$\frac{3}{2}$	$\frac{3}{2}$	1	121−	$+5.5\sqrt{5.7.13}/4.9.3\sqrt{11.17.19}$
1	$\frac{3}{2}$	$\tilde{\frac{1}{2}}$	0	000+	$+7.59/8.9\sqrt{13.17.19}$
8	**$\frac{15}{2}$**	**$\frac{11}{2}$**			
1	$\frac{3}{2}$	$\tilde{\frac{1}{2}}$	0	010+	$-\sqrt{7.11}/9.3\sqrt{2.13.17.19}$
1	$\frac{3}{2}$	$\tilde{\frac{1}{2}}$	0	020+	$+107\sqrt{11}/8.9.3\sqrt{17.19}$
1	$\frac{3}{2}$	$\tilde{\frac{1}{2}}$	0	100+	$-23\sqrt{7}/2.9\sqrt{5.17.19}$
1	$\frac{3}{2}$	$\tilde{\frac{1}{2}}$	0	110+	$-8.2\sqrt{2.11}/9.3\sqrt{5.17.19}$
1	$\frac{3}{2}$	$\tilde{\frac{1}{2}}$	0	120+	$-\sqrt{5.7.11.13}/2.9.3\sqrt{17.19}$
1	$\tilde{\frac{1}{2}}$	$\frac{3}{2}$	0	000+	$-7/3\sqrt{2.3.17.19}$
1	$\tilde{\frac{1}{2}}$	$\frac{3}{2}$	0	001+	$-37\sqrt{5}/4.9\sqrt{3.17.19}$
1	$\tilde{\frac{1}{2}}$	$\frac{3}{2}$	0	100+	$-\sqrt{5.7.13}/3\sqrt{2.3.17.19}$
1	$\tilde{\frac{1}{2}}$	$\frac{3}{2}$	0	101+	$+4\sqrt{7.13}/9\sqrt{3.17.19}$
1	$\tilde{\frac{1}{2}}$	$\tilde{\frac{1}{2}}$	0	000−	$+7\sqrt{2.5}/9\sqrt{3.17.19}$
1	$\tilde{\frac{1}{2}}$	$\tilde{\frac{1}{2}}$	0	100−	$-7\sqrt{7.13}/2.9\sqrt{2.3.17.19}$
2	$\frac{1}{2}$	$\frac{3}{2}$	0	000−	$+\sqrt{5}/2\sqrt{3.13.17.19}$
2	$\frac{1}{2}$	$\frac{3}{2}$	0	001−	$-53/4\sqrt{2.3.13.17.19}$
2	$\frac{1}{2}$	$\frac{3}{2}$	0	100−	$+\sqrt{7}/2\sqrt{11.17.19}$
2	$\frac{1}{2}$	$\frac{3}{2}$	0	101−	$+7\sqrt{5.7}/4\sqrt{2.11.17.19}$
2	$\frac{3}{2}$	$\frac{1}{2}$	0	000−	$-109/8\sqrt{2.13.17.19}$
2	$\frac{3}{2}$	$\frac{1}{2}$	0	010−	$-3\sqrt{7.11}/4\sqrt{13.17.19}$
2	$\frac{3}{2}$	$\frac{1}{2}$	0	020−	$-\sqrt{11}/8\sqrt{2.17.19}$
2	$\frac{3}{2}$	$\frac{1}{2}$	0	100−	$+\sqrt{3.7.11}/8\sqrt{2.5.17.19}$
2	$\frac{3}{2}$	$\frac{1}{2}$	0	110−	$-23/4\sqrt{3.5.17.19}$
2	$\frac{3}{2}$	$\frac{1}{2}$	0	120−	$+\sqrt{5.7.13}/8\sqrt{2.3.17.19}$
2	$\frac{3}{2}$	$\frac{3}{2}$	0	000+	$-\sqrt{7.19}/2\sqrt{11.13.17}$
2	$\frac{3}{2}$	$\frac{3}{2}$	0	001+	$-\sqrt{7.19}/8.3\sqrt{2.5.11.13.17}$
2	$\frac{3}{2}$	$\frac{3}{2}$	0	010+	$+3/\sqrt{2.13.17.19}$
2	$\frac{3}{2}$	$\frac{3}{2}$	0	011+	$+71/2.9\sqrt{5.13.17.19}$
2	$\frac{3}{2}$	$\frac{3}{2}$	0	020+	$+\sqrt{7}/2\sqrt{17.19}$
2	$\frac{3}{2}$	$\frac{3}{2}$	0	021+	$-\sqrt{5.7.17}/8.9\sqrt{2.19}$
2	$\frac{3}{2}$	$\frac{3}{2}$	0	100+	$+\sqrt{3}/2\sqrt{5.17.19}$
2	$\frac{3}{2}$	$\frac{3}{2}$	0	101+	$+53/8\sqrt{2.3.17.19}$
2	$\frac{3}{2}$	$\frac{3}{2}$	0	110+	$+\sqrt{7.17}/\sqrt{2.3.5.11.19}$
2	$\frac{3}{2}$	$\frac{3}{2}$	0	111+	$+\sqrt{7.19}/2.3\sqrt{3.11.17}$
2	$\frac{3}{2}$	$\frac{3}{2}$	0	120+	$+\sqrt{5.13}/2\sqrt{3.11.17.19}$
2	$\frac{3}{2}$	$\frac{3}{2}$	0	121+	$+\sqrt{13}/8.3\sqrt{2.3.11.17.19}$
2	$\frac{3}{2}$	$\tilde{\frac{1}{2}}$	0	000−	$-53\sqrt{7.11}/8.3\sqrt{2.5.13.17.19}$
2	$\frac{3}{2}$	$\tilde{\frac{1}{2}}$	0	010−	$+409/4.9\sqrt{5.13.17.19}$
2	$\frac{3}{2}$	$\tilde{\frac{1}{2}}$	0	020−	$+43\sqrt{5.7}/8.9\sqrt{2.17.19}$
2	$\frac{3}{2}$	$\tilde{\frac{1}{2}}$	0	100−	$+\sqrt{17}/8\sqrt{2.3.19}$
2	$\frac{3}{2}$	$\tilde{\frac{1}{2}}$	0	110−	$+7\sqrt{7.11}/4.3\sqrt{3.17.19}$
2	$\frac{3}{2}$	$\tilde{\frac{1}{2}}$	0	120−	$-\sqrt{11.13}/8.3\sqrt{2.3.17.19}$
2	$\tilde{\frac{1}{2}}$	$\frac{3}{2}$	0	000−	$-\sqrt{3.5.7}/2\sqrt{11.17.19}$
2	$\tilde{\frac{1}{2}}$	$\frac{3}{2}$	0	001−	$-73\sqrt{7}/4.3\sqrt{2.3.11.17.19}$
2	$\tilde{\frac{1}{2}}$	$\frac{3}{2}$	0	100−	$+\sqrt{13}/2\sqrt{17.19}$
2	$\tilde{\frac{1}{2}}$	$\frac{3}{2}$	0	101−	$-\sqrt{5.13}/4.3\sqrt{2.17.19}$
$\tilde{1}$	$\frac{1}{2}$	$\frac{3}{2}$	0	000−	$+\sqrt{3}/17\sqrt{2.19}$
$\tilde{1}$	$\frac{1}{2}$	$\frac{3}{2}$	0	001−	$-23\sqrt{5}/4.17\sqrt{3.19}$
$\tilde{1}$	$\frac{1}{2}$	$\frac{3}{2}$	0	100−	$-5.5\sqrt{5.7}/17\sqrt{2.11.13.19}$
$\tilde{1}$	$\frac{1}{2}$	$\frac{3}{2}$	0	101−	$-4\sqrt{7.19}/3.17\sqrt{11.13}$
$\tilde{1}$	$\frac{1}{2}$	$\tilde{\frac{1}{2}}$	0	000+	$+\sqrt{2.5}/17\sqrt{3.19}$
$\tilde{1}$	$\frac{1}{2}$	$\tilde{\frac{1}{2}}$	0	100+	$-47\sqrt{7.11}/2.3.17\sqrt{2.13.19}$

SO_3–O 3jm Factors (cont.)

$\mathbf{8\ \frac{15}{2}\ \frac{11}{2}}$	
$\tilde{1}\ \frac{3}{2}\ \frac{1}{2}\ 0\ 000-$	$-9.13/8.17\sqrt{5.19}$
$\tilde{1}\ \frac{3}{2}\ \frac{1}{2}\ 0\ 010-$	$+3\sqrt{7.11}/17\sqrt{2.5.19}$
$\tilde{1}\ \frac{3}{2}\ \frac{1}{2}\ 0\ 020-$	$+3\sqrt{5.11.13}/8.17\sqrt{19}$
$\tilde{1}\ \frac{3}{2}\ \frac{1}{2}\ 0\ 100-$	$-5\sqrt{7.11}/2.17\sqrt{3.13.19}$
$\tilde{1}\ \frac{3}{2}\ \frac{1}{2}\ 0\ 110-$	$+4.5\sqrt{2}/3.17\sqrt{3.13.19}$
$\tilde{1}\ \frac{3}{2}\ \frac{1}{2}\ 0\ 120-$	$-5\sqrt{7}/2.3.17\sqrt{3.19}$
$\tilde{1}\ \frac{3}{2}\ \frac{3}{2}\ 0\ 000+$	$-3\sqrt{7.19}/2.17\sqrt{2.5.11}$
$\tilde{1}\ \frac{3}{2}\ \frac{3}{2}\ 0\ 001+$	$-9\sqrt{7}/2.17\sqrt{11.19}$
$\tilde{1}\ \frac{3}{2}\ \frac{3}{2}\ 0\ 010+$	$+2/17\sqrt{5.19}$
$\tilde{1}\ \frac{3}{2}\ \frac{3}{2}\ 0\ 011+$	$-1/3.17\sqrt{2.19}$
$\tilde{1}\ \frac{3}{2}\ \frac{3}{2}\ 0\ 020+$	$-\sqrt{5.7.13}/2.17\sqrt{2.19}$
$\tilde{1}\ \frac{3}{2}\ \frac{3}{2}\ 0\ 021+$	$-7\sqrt{7.13}/2.3.17\sqrt{19}$
$\tilde{1}\ \frac{3}{2}\ \frac{3}{2}\ 0\ 100+$	$-11/17\sqrt{2.3.13.19}$
$\tilde{1}\ \frac{3}{2}\ \frac{3}{2}\ 0\ 101+$	$+2.3.11\sqrt{3}/17\sqrt{5.13.19}$
$\tilde{1}\ \frac{3}{2}\ \frac{3}{2}\ 0\ 110+$	$+431\sqrt{7}/2.3.17\sqrt{3.11.13.19}$
$\tilde{1}\ \frac{3}{2}\ \frac{3}{2}\ 0\ 111+$	$-269\sqrt{2.7}/3.17\sqrt{3.5.11.13.19}$
$\tilde{1}\ \frac{3}{2}\ \frac{3}{2}\ 0\ 120+$	$-107/3.17\sqrt{2.3.11.19}$
$\tilde{1}\ \frac{3}{2}\ \frac{3}{2}\ 0\ 121+$	$-5.5\sqrt{5}/3.17\sqrt{3.11.19}$
$\tilde{1}\ \frac{3}{2}\ \frac{3}{2}\ 1\ 000-$	$+3\sqrt{5.7}/4.17\sqrt{2.11.19}$
$\tilde{1}\ \frac{3}{2}\ \frac{3}{2}\ 1\ 001-$	$-3.5\sqrt{7}/4.17\sqrt{11.19}$
$\tilde{1}\ \frac{3}{2}\ \frac{3}{2}\ 1\ 010-$	$-\sqrt{5}/17\sqrt{19}$
$\tilde{1}\ \frac{3}{2}\ \frac{3}{2}\ 1\ 011-$	$-5.7/2.17\sqrt{2.19}$
$\tilde{1}\ \frac{3}{2}\ \frac{3}{2}\ 1\ 020-$	$-\sqrt{5.7.13}/4.17\sqrt{2.19}$
$\tilde{1}\ \frac{3}{2}\ \frac{3}{2}\ 1\ 021-$	$+5\sqrt{7.13}/4.17\sqrt{19}$
$\tilde{1}\ \frac{3}{2}\ \frac{3}{2}\ 1\ 100-$	$-7.7/2.17\sqrt{2.3.13.19}$
$\tilde{1}\ \frac{3}{2}\ \frac{3}{2}\ 1\ 101-$	$-71/4.17\sqrt{3.5.13.19}$
$\tilde{1}\ \frac{3}{2}\ \frac{3}{2}\ 1\ 110-$	$+8.8.4\sqrt{7}/3.17\sqrt{3.11.13.19}$
$\tilde{1}\ \frac{3}{2}\ \frac{3}{2}\ 1\ 111-$	$+179\sqrt{7}/3.17\sqrt{2.3.5.11.13.19}$
$\tilde{1}\ \frac{3}{2}\ \frac{3}{2}\ 1\ 120-$	$-277/2.3.17\sqrt{2.3.11.19}$
$\tilde{1}\ \frac{3}{2}\ \frac{3}{2}\ 1\ 121-$	$-7\sqrt{5}/4.3.17\sqrt{3.11.19}$
$\tilde{1}\ \frac{3}{2}\ \tilde{\frac{1}{2}}\ 0\ 000-$	$-3\sqrt{7.11}/8.17\sqrt{19}$
$\tilde{1}\ \frac{3}{2}\ \tilde{\frac{1}{2}}\ 0\ 010-$	$+1/3\sqrt{2.19}$
$\tilde{1}\ \frac{3}{2}\ \tilde{\frac{1}{2}}\ 0\ 020-$	$-7\sqrt{7.13}/8.3.17\sqrt{19}$
$\tilde{1}\ \frac{3}{2}\ \tilde{\frac{1}{2}}\ 0\ 100-$	$+4\sqrt{5}/17\sqrt{3.13.19}$
$\tilde{1}\ \frac{3}{2}\ \tilde{\frac{1}{2}}\ 0\ 110-$	0
$\tilde{1}\ \frac{3}{2}\ \tilde{\frac{1}{2}}\ 0\ 120-$	$+2\sqrt{5.11}/17\sqrt{3.19}$
$\tilde{1}\ \tilde{\frac{1}{2}}\ \frac{1}{2}\ 0\ 000+$	$+\sqrt{2.3.13}/17\sqrt{19}$
$\tilde{1}\ \tilde{\frac{1}{2}}\ \frac{1}{2}\ 0\ 100+$	$-\sqrt{5.7.11}/2.3.17\sqrt{2.19}$
$\tilde{1}\ \tilde{\frac{1}{2}}\ \frac{3}{2}\ 0\ 000-$	$+\sqrt{3.7.13}/17\sqrt{2.11.19}$
$\tilde{1}\ \tilde{\frac{1}{2}}\ \frac{3}{2}\ 0\ 001-$	$-\sqrt{5.7.13}/4.17\sqrt{3.11.19}$
$\tilde{1}\ \tilde{\frac{1}{2}}\ \frac{3}{2}\ 0\ 100-$	$+11\sqrt{5}/3.17\sqrt{2.19}$
$\tilde{1}\ \tilde{\frac{1}{2}}\ \frac{3}{2}\ 0\ 101-$	$+11/3.17\sqrt{19}$

$\mathbf{8\ \frac{15}{2}\ \frac{13}{2}}$	
$0\ \frac{1}{2}\ \frac{1}{2}\ 0\ 000-$	$+2\sqrt{2.5}/\sqrt{3.13.17.19.23}$
$0\ \frac{3}{2}\ \frac{3}{2}\ 0\ 000-$	$+3.727/2.5.7\sqrt{13.17.19.23}$
$0\ \frac{3}{2}\ \frac{3}{2}\ 0\ 001-$	$-211\sqrt{2.3.11}/5.7\sqrt{13.17.19.23}$
$0\ \frac{3}{2}\ \frac{3}{2}\ 0\ 010-$	$+2.103\sqrt{2.11}/3.5\sqrt{7.13.17.19.23}$
$0\ \frac{3}{2}\ \frac{3}{2}\ 0\ 011-$	$+2.113/5\sqrt{3.7.13.17.19.23}$
$0\ \frac{3}{2}\ \frac{3}{2}\ 0\ 020-$	$-7\sqrt{11}/2.3\sqrt{17.19.23}$
$0\ \frac{3}{2}\ \frac{3}{2}\ 0\ 021-$	$+7\sqrt{2}/\sqrt{3.17.19.23}$
$0\ \tilde{\frac{1}{2}}\ \tilde{\frac{1}{2}}\ 0\ 000-$	$+2\sqrt{2.23}/7\sqrt{3.17.19}$
$0\ \tilde{\frac{1}{2}}\ \tilde{\frac{1}{2}}\ 0\ 001-$	$+9\sqrt{11.13}/7\sqrt{5.17.19.23}$
$1\ \frac{1}{2}\ \frac{1}{2}\ 0\ 000+$	$+5\sqrt{5}/\sqrt{3.13.17.19.23}$
$1\ \frac{1}{2}\ \frac{1}{2}\ 0\ 100+$	$-5\sqrt{7}/\sqrt{3.17.19.23}$
$1\ \frac{1}{2}\ \frac{3}{2}\ 0\ 000-$	$-31\sqrt{3.11}/\sqrt{2.5.7.13.17.19.23}$
$1\ \frac{1}{2}\ \frac{3}{2}\ 0\ 001-$	$-9\sqrt{17}/2\sqrt{5.7.13.19.23}$
$1\ \frac{1}{2}\ \frac{3}{2}\ 0\ 100-$	$-\sqrt{3.11}/2\sqrt{2.17.19.23}$
$1\ \frac{1}{2}\ \frac{3}{2}\ 0\ 101-$	$+3/\sqrt{17.19.23}$
$1\ \frac{3}{2}\ \frac{1}{2}\ 0\ 000-$	$-9\sqrt{11.23}/4\sqrt{2.7.13.17.19}$
$1\ \frac{3}{2}\ \frac{1}{2}\ 0\ 010-$	$+2/3\sqrt{13.17.19.23}$
$1\ \frac{3}{2}\ \frac{1}{2}\ 0\ 020-$	$+5.7\sqrt{7}/4.3\sqrt{2.17.19.23}$
$1\ \frac{3}{2}\ \frac{1}{2}\ 0\ 100-$	0
$1\ \frac{3}{2}\ \frac{1}{2}\ 0\ 110-$	$+13\sqrt{5.7}/2.3\sqrt{17.19.23}$
$1\ \frac{3}{2}\ \frac{1}{2}\ 0\ 120-$	$-2\sqrt{2.5.13}/3\sqrt{17.19.23}$
$1\ \frac{3}{2}\ \frac{3}{2}\ 0\ 000+$	$+41.331/4.3.5.7\sqrt{2.13.17.19.23}$
$1\ \frac{3}{2}\ \frac{3}{2}\ 0\ 001+$	$+2.211\sqrt{11}/5.7\sqrt{3.13.17.19.23}$
$1\ \frac{3}{2}\ \frac{3}{2}\ 0\ 010+$	$-787\sqrt{11}/9.5\sqrt{7.13.17.19.23}$
$1\ \frac{3}{2}\ \frac{3}{2}\ 0\ 011+$	$-11.277/3.5\sqrt{2.3.7.13.17.19.23}$
$1\ \frac{3}{2}\ \frac{3}{2}\ 0\ 020+$	$+19\sqrt{11.19}/4.9.7\sqrt{2.17.23}$
$1\ \frac{3}{2}\ \frac{3}{2}\ 0\ 021+$	$-8.8.2/3.7\sqrt{3.17.19.23}$
$1\ \frac{3}{2}\ \frac{3}{2}\ 0\ 100+$	$-\sqrt{2.5}/3\sqrt{7.17.19.23}$
$1\ \frac{3}{2}\ \frac{3}{2}\ 0\ 101+$	$+\sqrt{5.11}/\sqrt{3.7.17.19.23}$
$1\ \frac{3}{2}\ \frac{3}{2}\ 0\ 110+$	$-2.7\sqrt{5.11}/9\sqrt{17.19.23}$
$1\ \frac{3}{2}\ \frac{3}{2}\ 0\ 111+$	$+7\sqrt{2.5}/3\sqrt{3.17.19.23}$
$1\ \frac{3}{2}\ \frac{3}{2}\ 0\ 120+$	$+5\sqrt{2.5.11.13}/9\sqrt{7.17.19.23}$
$1\ \frac{3}{2}\ \frac{3}{2}\ 0\ 121+$	$+\sqrt{5.13}/3\sqrt{3.7.17.19.23}$
$1\ \frac{3}{2}\ \frac{3}{2}\ 1\ 000+$	$-11.1973/4.3.5.7\sqrt{2.13.17.19.23}$
$1\ \frac{3}{2}\ \frac{3}{2}\ 1\ 001+$	$+2441\sqrt{11}/4.5.7\sqrt{3.13.17.19.23}$
$1\ \frac{3}{2}\ \frac{3}{2}\ 1\ 010+$	$-8\sqrt{11.23}/9.5\sqrt{7.13.17.19}$
$1\ \frac{3}{2}\ \frac{3}{2}\ 1\ 011+$	$-4\sqrt{2.13}/3.5\sqrt{3.7.17.19.23}$
$1\ \frac{3}{2}\ \frac{3}{2}\ 1\ 020+$	$-43\sqrt{11.23}/4.9.7\sqrt{2.17.19}$
$1\ \frac{3}{2}\ \frac{3}{2}\ 1\ 021+$	$+1873/4.3.7\sqrt{3.17.19.23}$
$1\ \frac{3}{2}\ \frac{3}{2}\ 1\ 100+$	$-103/3\sqrt{2.5.7.17.19.23}$
$1\ \frac{3}{2}\ \frac{3}{2}\ 1\ 101+$	$-\sqrt{11.23}/2\sqrt{3.5.7.17.19}$
$1\ \frac{3}{2}\ \frac{3}{2}\ 1\ 110+$	$+103\sqrt{11}/2.9\sqrt{5.17.19.23}$
$1\ \frac{3}{2}\ \frac{3}{2}\ 1\ 111+$	$-101/3\sqrt{2.3.5.17.19.23}$
$1\ \frac{3}{2}\ \frac{3}{2}\ 1\ 120+$	$+\sqrt{5.11.13.19}/9\sqrt{2.7.17.23}$
$1\ \frac{3}{2}\ \frac{3}{2}\ 1\ 121+$	$+29\sqrt{5.13}/2.3\sqrt{3.7.17.19.23}$
$1\ \frac{3}{2}\ \tilde{\frac{1}{2}}\ 0\ 000-$	$-1787\sqrt{5}/4.3.7\sqrt{2.13.17.19.23}$
$1\ \frac{3}{2}\ \tilde{\frac{1}{2}}\ 0\ 001-$	$+61\sqrt{11}/2.7\sqrt{3.17.19.23}$
$1\ \frac{3}{2}\ \tilde{\frac{1}{2}}\ 0\ 010-$	$-2\sqrt{5.11}/9\sqrt{7.13.17.19.23}$
$1\ \frac{3}{2}\ \tilde{\frac{1}{2}}\ 0\ 011-$	$+167/2.3\sqrt{2.3.7.17.19.23}$
$1\ \frac{3}{2}\ \tilde{\frac{1}{2}}\ 0\ 020-$	$+331\sqrt{5.11}/4.9.7\sqrt{2.17.19.23}$
$1\ \frac{3}{2}\ \tilde{\frac{1}{2}}\ 0\ 021-$	$+\sqrt{13}/2.3.7\sqrt{3.17.19.23}$
$1\ \frac{3}{2}\ \tilde{\frac{1}{2}}\ 0\ 100-$	$+5.11\sqrt{2}/3\sqrt{7.17.19.23}$
$1\ \frac{3}{2}\ \tilde{\frac{1}{2}}\ 0\ 101-$	$+\sqrt{11.13.19}/4\sqrt{3.5.7.17.23}$
$1\ \frac{3}{2}\ \tilde{\frac{1}{2}}\ 0\ 110-$	$-5.7\sqrt{11}/2.9\sqrt{17.19.23}$
$1\ \frac{3}{2}\ \tilde{\frac{1}{2}}\ 0\ 111-$	$+7\sqrt{2.13}/3\sqrt{3.5.17.19.23}$

SO_3–O 3jm Factors (cont.)

8	$\frac{15}{2}$	$\frac{13}{2}$			
1	$\frac{3}{2}$	$\tilde{\frac{1}{2}}$	0	120−	$-5\sqrt{2.11.13}/9\sqrt{7.17.19.23}$
1	$\frac{3}{2}$	$\tilde{\frac{1}{2}}$	0	121−	$-13\sqrt{5}/4.3\sqrt{3.7.17.19.23}$
1	$\tilde{\frac{1}{2}}$	$\frac{3}{2}$	0	000−	$+31\sqrt{23}/3.7\sqrt{2.3.5.17.19}$
1	$\tilde{\frac{1}{2}}$	$\frac{3}{2}$	0	001−	$-9\sqrt{11}/2.7\sqrt{5.17.19.23}$
1	$\tilde{\frac{1}{2}}$	$\frac{3}{2}$	0	100−	$+\sqrt{7.13.19}/2.3\sqrt{2.3.17.23}$
1	$\tilde{\frac{1}{2}}$	$\frac{3}{2}$	0	101−	0
1	$\tilde{\frac{1}{2}}$	$\tilde{\frac{1}{2}}$	0	000+	$+\sqrt{23}/3.7\sqrt{3.17.19}$
1	$\tilde{\frac{1}{2}}$	$\tilde{\frac{1}{2}}$	0	001+	$+9\sqrt{5.11.13}/2.7\sqrt{2.17.19.23}$
1	$\tilde{\frac{1}{2}}$	$\tilde{\frac{1}{2}}$	0	100+	$-\sqrt{5.7.13}/3\sqrt{3.17.19.23}$
1	$\tilde{\frac{1}{2}}$	$\tilde{\frac{1}{2}}$	0	101+	0
2	$\frac{1}{2}$	$\frac{3}{2}$	0	000+	$+59/4\sqrt{3.13.17.19.23}$
2	$\frac{1}{2}$	$\frac{3}{2}$	0	001+	$-\sqrt{11.13}/\sqrt{2.17.19.23}$
2	$\frac{1}{2}$	$\frac{3}{2}$	0	100+	$+3.37\sqrt{11}/4\sqrt{5.7.17.19.23}$
2	$\frac{1}{2}$	$\frac{3}{2}$	0	101+	$-9.3\sqrt{3}/\sqrt{2.5.7.17.19.23}$
2	$\frac{3}{2}$	$\frac{1}{2}$	0	000+	$+9.3/8\sqrt{5.13.17.19.23}$
2	$\frac{3}{2}$	$\frac{1}{2}$	0	010+	$+7\sqrt{7.11}/2.3\sqrt{2.5.13.17.19.23}$
2	$\frac{3}{2}$	$\frac{1}{2}$	0	020+	$+7.7\sqrt{5.11}/8.3\sqrt{17.19.23}$
2	$\frac{3}{2}$	$\frac{1}{2}$	0	100+	$-\sqrt{3.11.17}/8.5\sqrt{7.19.23}$
2	$\frac{3}{2}$	$\frac{1}{2}$	0	110+	$-167/2.5\sqrt{2.3.17.19.23}$
2	$\frac{3}{2}$	$\frac{1}{2}$	0	120+	$-\sqrt{13.17}/8\sqrt{3.7.19.23}$
2	$\frac{3}{2}$	$\frac{3}{2}$	0	000−	$+9.7\sqrt{7.11}/8\sqrt{5.13.17.19.23}$
2	$\frac{3}{2}$	$\frac{3}{2}$	0	001−	$-\sqrt{3.5.7}/2\sqrt{2.13.17.19.23}$
2	$\frac{3}{2}$	$\frac{3}{2}$	0	010−	$+4.7\sqrt{2}/\sqrt{5.13.17.19.23}$
2	$\frac{3}{2}$	$\frac{3}{2}$	0	011−	0
2	$\frac{3}{2}$	$\frac{3}{2}$	0	020−	$+3.5.5\sqrt{5}/8\sqrt{7.17.19.23}$
2	$\frac{3}{2}$	$\frac{3}{2}$	0	021−	$-3\sqrt{3.5.11}/2\sqrt{2.7.17.19.23}$
2	$\frac{3}{2}$	$\frac{3}{2}$	0	100−	$+71\sqrt{3}/8.5.7\sqrt{17.19.23}$
2	$\frac{3}{2}$	$\frac{3}{2}$	0	101−	$+9.11\sqrt{11}/2.5.7\sqrt{2.17.19.23}$
2	$\frac{3}{2}$	$\frac{3}{2}$	0	110−	$-8.2\sqrt{2.11}/5\sqrt{3.7.17.19.23}$
2	$\frac{3}{2}$	$\frac{3}{2}$	0	111−	$+8/5\sqrt{7.17.19.23}$
2	$\frac{3}{2}$	$\frac{3}{2}$	0	120−	$-67\sqrt{11.13}/8.7\sqrt{3.17.19.23}$
2	$\frac{3}{2}$	$\frac{3}{2}$	0	121−	$-31\sqrt{13}/2.7\sqrt{2.17.19.23}$
2	$\frac{3}{2}$	$\tilde{\frac{1}{2}}$	0	000+	$+9.3.11\sqrt{11}/8\sqrt{7.13.17.19.23}$
2	$\frac{3}{2}$	$\tilde{\frac{1}{2}}$	0	001+	$-281\sqrt{3}/8\sqrt{2.5.7.17.19.23}$
2	$\frac{3}{2}$	$\tilde{\frac{1}{2}}$	0	010+	$+37/2.3\sqrt{2.13.17.19.23}$
2	$\frac{3}{2}$	$\tilde{\frac{1}{2}}$	0	011+	$+29\sqrt{11}/4\sqrt{3.5.17.19.23}$
2	$\frac{3}{2}$	$\tilde{\frac{1}{2}}$	0	020+	$+5.73/8.3\sqrt{7.17.19.23}$
2	$\frac{3}{2}$	$\tilde{\frac{1}{2}}$	0	021+	$+5\sqrt{5.11.13}/8\sqrt{2.3.7.17.19.23}$
2	$\frac{3}{2}$	$\tilde{\frac{1}{2}}$	0	100+	$+3.61\sqrt{3}/8.7\sqrt{5.17.19.23}$
2	$\frac{3}{2}$	$\tilde{\frac{1}{2}}$	0	101+	$+9.41\sqrt{11.13}/8.5.7\sqrt{2.17.19.23}$
2	$\frac{3}{2}$	$\tilde{\frac{1}{2}}$	0	110+	$-47\sqrt{11}/2\sqrt{2.3.5.7.17.19.23}$
2	$\frac{3}{2}$	$\tilde{\frac{1}{2}}$	0	111+	$-157\sqrt{13}/4.5\sqrt{7.17.19.23}$
2	$\frac{3}{2}$	$\tilde{\frac{1}{2}}$	0	120+	$+37\sqrt{5.11.13}/8.7\sqrt{3.17.19.23}$
2	$\frac{3}{2}$	$\tilde{\frac{1}{2}}$	0	121+	$+11.13/8.7\sqrt{2.17.19.23}$
2	$\tilde{\frac{1}{2}}$	$\frac{3}{2}$	0	000+	$+29\sqrt{11}/4\sqrt{3.7.17.19.23}$
2	$\tilde{\frac{1}{2}}$	$\frac{3}{2}$	0	001+	$+9.3/\sqrt{2.7.17.19.23}$
2	$\tilde{\frac{1}{2}}$	$\frac{3}{2}$	0	100+	$+\sqrt{5.13}/4\sqrt{17.19.23}$
2	$\tilde{\frac{1}{2}}$	$\frac{3}{2}$	0	101+	$+\sqrt{3.11.13}/\sqrt{2.5.17.19.23}$
$\tilde{1}$	$\frac{1}{2}$	$\frac{3}{2}$	0	000+	$+7\sqrt{23}/17\sqrt{2.3.5.19}$

8	$\frac{15}{2}$	$\frac{13}{2}$			
$\tilde{1}$	$\frac{1}{2}$	$\frac{3}{2}$	0	001+	$+9.7\sqrt{11}/2.17\sqrt{5.19.23}$
$\tilde{1}$	$\frac{1}{2}$	$\frac{3}{2}$	0	100+	$-3.47\sqrt{11}/2.17\sqrt{2.7.13.19.23}$
$\tilde{1}$	$\frac{1}{2}$	$\frac{3}{2}$	0	101+	$-8.5.5/17\sqrt{3.7.13.19.23}$
$\tilde{1}$	$\frac{1}{2}$	$\tilde{\frac{1}{2}}$	0	000−	$-\sqrt{23}/17\sqrt{3.19}$
$\tilde{1}$	$\frac{1}{2}$	$\tilde{\frac{1}{2}}$	0	001−	$-9\sqrt{11.13}/2.17\sqrt{2.5.19.23}$
$\tilde{1}$	$\frac{1}{2}$	$\tilde{\frac{1}{2}}$	0	100−	$-47\sqrt{5.11}/17\sqrt{7.13.19.23}$
$\tilde{1}$	$\frac{1}{2}$	$\tilde{\frac{1}{2}}$	0	101−	$+8.4.5\sqrt{2}/17\sqrt{3.7.19.23}$
$\tilde{1}$	$\frac{3}{2}$	$\frac{1}{2}$	0	000+	$+9.13/4.5.17\sqrt{2.19.23}$
$\tilde{1}$	$\frac{3}{2}$	$\frac{1}{2}$	0	010+	$+2\sqrt{7.11}/5.17\sqrt{19.23}$
$\tilde{1}$	$\frac{3}{2}$	$\frac{1}{2}$	0	020+	$-11\sqrt{11.13}/4.17\sqrt{2.19.23}$
$\tilde{1}$	$\frac{3}{2}$	$\frac{1}{2}$	0	100+	$-2\sqrt{2.3.7.11}/17\sqrt{5.13.19.23}$
$\tilde{1}$	$\frac{3}{2}$	$\frac{1}{2}$	0	110+	$-547/2.17\sqrt{3.5.13.19.23}$
$\tilde{1}$	$\frac{3}{2}$	$\frac{1}{2}$	0	120+	$-8\sqrt{2.5}/17\sqrt{3.7.19.23}$
$\tilde{1}$	$\frac{3}{2}$	$\frac{3}{2}$	0	000−	$-3.31\sqrt{11}/4.5.17\sqrt{2.7.19.23}$
$\tilde{1}$	$\frac{3}{2}$	$\frac{3}{2}$	0	001−	$-131\sqrt{3}/2.5.17\sqrt{7.19.23}$
$\tilde{1}$	$\frac{3}{2}$	$\frac{3}{2}$	0	010−	$+41/3.5.17\sqrt{19.23}$
$\tilde{1}$	$\frac{3}{2}$	$\frac{3}{2}$	0	011−	$-9\sqrt{3.11}/5.17\sqrt{2.19.23}$
$\tilde{1}$	$\frac{3}{2}$	$\frac{3}{2}$	0	020−	$-317\sqrt{13}/4.3.17\sqrt{2.7.19.23}$
$\tilde{1}$	$\frac{3}{2}$	$\frac{3}{2}$	0	021−	$-3.5\sqrt{3.11.13}/2.17\sqrt{7.19.23}$
$\tilde{1}$	$\frac{3}{2}$	$\frac{3}{2}$	0	100−	$-4.137\sqrt{2.3}/7.17\sqrt{5.13.19.23}$
$\tilde{1}$	$\frac{3}{2}$	$\frac{3}{2}$	0	101−	$-2.3.83\sqrt{11}/7.17\sqrt{5.13.19.23}$
$\tilde{1}$	$\frac{3}{2}$	$\frac{3}{2}$	0	110−	$-2\sqrt{7.11.23}/17\sqrt{3.5.13.19}$
$\tilde{1}$	$\frac{3}{2}$	$\frac{3}{2}$	0	111−	$+397\sqrt{2.7}/3.17\sqrt{5.13.19.23}$
$\tilde{1}$	$\frac{3}{2}$	$\frac{3}{2}$	0	120−	$+8\sqrt{2.5.11}/7.17\sqrt{3.19.23}$
$\tilde{1}$	$\frac{3}{2}$	$\frac{3}{2}$	0	121−	$+2\sqrt{5}/3.7.17\sqrt{19.23}$
$\tilde{1}$	$\frac{3}{2}$	$\frac{3}{2}$	1	000+	$+3\sqrt{7.11}/4.5\sqrt{2.19.23}$
$\tilde{1}$	$\frac{3}{2}$	$\frac{3}{2}$	1	001+	$-47\sqrt{3.7}/4.5.17\sqrt{19.23}$
$\tilde{1}$	$\frac{3}{2}$	$\frac{3}{2}$	1	010+	$+2.53/3.5.17\sqrt{19.23}$
$\tilde{1}$	$\frac{3}{2}$	$\frac{3}{2}$	1	011+	$+4.9\sqrt{2.3.11}/5.17\sqrt{19.23}$
$\tilde{1}$	$\frac{3}{2}$	$\frac{3}{2}$	1	020+	$-307\sqrt{13}/4.3.17\sqrt{2.7.19.23}$
$\tilde{1}$	$\frac{3}{2}$	$\frac{3}{2}$	1	021+	$+3.5\sqrt{3.11.13}/4.17\sqrt{7.19.23}$
$\tilde{1}$	$\frac{3}{2}$	$\frac{3}{2}$	1	100+	$+2.9.3\sqrt{2.3}/7\sqrt{5.13.19.23}$
$\tilde{1}$	$\frac{3}{2}$	$\frac{3}{2}$	1	101+	$+4.3.47\sqrt{11}/7.17\sqrt{5.13.19.23}$
$\tilde{1}$	$\frac{3}{2}$	$\frac{3}{2}$	1	110+	$+839\sqrt{11}/2.17\sqrt{3.5.7.13.19.23}$
$\tilde{1}$	$\frac{3}{2}$	$\frac{3}{2}$	1	111+	$+1861/3.17\sqrt{2.5.7.13.19.23}$
$\tilde{1}$	$\frac{3}{2}$	$\frac{3}{2}$	1	120+	$+2.5\sqrt{2.5.11}/7.17\sqrt{3.19.23}$
$\tilde{1}$	$\frac{3}{2}$	$\frac{3}{2}$	1	121+	$+2.59\sqrt{5}/3.7.17\sqrt{19.23}$
$\tilde{1}$	$\frac{3}{2}$	$\tilde{\frac{1}{2}}$	0	000+	$-9\sqrt{11.23}/4.17\sqrt{2.5.7.19}$
$\tilde{1}$	$\frac{3}{2}$	$\tilde{\frac{1}{2}}$	0	001+	$-4\sqrt{3.13.23}/5.17\sqrt{7.19}$
$\tilde{1}$	$\frac{3}{2}$	$\tilde{\frac{1}{2}}$	0	010+	$-2\sqrt{23}/3.17\sqrt{5.19}$
$\tilde{1}$	$\frac{3}{2}$	$\tilde{\frac{1}{2}}$	0	011+	$-9.3\sqrt{3.11.13}/2.5.17\sqrt{2.19.23}$
$\tilde{1}$	$\frac{3}{2}$	$\tilde{\frac{1}{2}}$	0	020+	$+7\sqrt{5.7.13}/4.3.17\sqrt{2.19.23}$
$\tilde{1}$	$\frac{3}{2}$	$\tilde{\frac{1}{2}}$	0	021+	0
$\tilde{1}$	$\frac{3}{2}$	$\tilde{\frac{1}{2}}$	0	100+	$+4\sqrt{2.3.23}/7.17\sqrt{13.19}$
$\tilde{1}$	$\frac{3}{2}$	$\tilde{\frac{1}{2}}$	0	101+	$+9\sqrt{5.11}/2.7.17\sqrt{19.23}$
$\tilde{1}$	$\frac{3}{2}$	$\tilde{\frac{1}{2}}$	0	110+	$+47\sqrt{11.13}/2.17\sqrt{3.7.19.23}$
$\tilde{1}$	$\frac{3}{2}$	$\tilde{\frac{1}{2}}$	0	111+	$+8.2\sqrt{2.5}/3.17\sqrt{7.19.23}$
$\tilde{1}$	$\frac{3}{2}$	$\tilde{\frac{1}{2}}$	0	120+	$-2.5\sqrt{2.11}/17\sqrt{3.19.23}$
$\tilde{1}$	$\frac{3}{2}$	$\tilde{\frac{1}{2}}$	0	121+	$+\sqrt{5.13}/2.3\sqrt{19.23}$

SO_3–O 3jm Factors (cont.)

$8\ \tfrac{15}{2}\ \tfrac{13}{2}$	
$\tilde{1}\ \tilde{\tfrac{1}{2}}\ \tfrac{1}{2}\ 0\ 000-$	$+\sqrt{3.13.23}/17\sqrt{5.19}$
$\tilde{1}\ \tilde{\tfrac{1}{2}}\ \tfrac{1}{2}\ 0\ 100-$	$+5\sqrt{11}/17\sqrt{7.19.23}$
$\tilde{1}\ \tilde{\tfrac{1}{2}}\ \tfrac{3}{2}\ 0\ 000+$	$+\sqrt{3.11.13}/17\sqrt{2.5.7.19.23}$
$\tilde{1}\ \tilde{\tfrac{1}{2}}\ \tfrac{3}{2}\ 0\ 001+$	$-3.5\sqrt{5.13}/2.17\sqrt{7.19.23}$
$\tilde{1}\ \tilde{\tfrac{1}{2}}\ \tfrac{3}{2}\ 0\ 100+$	$-367/2.7.17\sqrt{2.19.23}$
$\tilde{1}\ \tilde{\tfrac{1}{2}}\ \tfrac{3}{2}\ 0\ 101+$	$-43\sqrt{3.11}/7.17\sqrt{19.23}$
$8\ \tfrac{15}{2}\ \tfrac{15}{2}$	
$0\ \tfrac{1}{2}\ \tfrac{1}{2}\ 0\ 000+$	$-5\sqrt{3.5.11}/2\sqrt{2.13.17.19.23}$
$0\ \tfrac{3}{2}\ \tfrac{3}{2}\ 0\ 000+$	$-3\sqrt{3.11.23}/4\sqrt{5.13.17.19}$
$0\ \tfrac{3}{2}\ \tfrac{3}{2}\ 0\ 010+$	$+4\sqrt{2.7}/\sqrt{3.5.13.17.19.23}$
$0\ \tfrac{3}{2}\ \tfrac{3}{2}\ 0\ 011+$	$-\sqrt{3.5.11}/2\sqrt{13.17.19.23}$
$0\ \tfrac{3}{2}\ \tfrac{3}{2}\ 0\ 020+$	$-7\sqrt{5}/4\sqrt{3.17.19.23}$
$0\ \tfrac{3}{2}\ \tfrac{3}{2}\ 0\ 021+$	0
$0\ \tfrac{3}{2}\ \tfrac{3}{2}\ 0\ 022+$	$+\sqrt{3.5.11.13}/4\sqrt{17.19.23}$
$0\ \tilde{\tfrac{1}{2}}\ \tilde{\tfrac{1}{2}}\ 0\ 000+$	$+3\sqrt{3.11.13}/2\sqrt{2.5.17.19.23}$
$1\ \tfrac{1}{2}\ \tfrac{1}{2}\ 0\ 000-$	0
$1\ \tfrac{1}{2}\ \tfrac{1}{2}\ 0\ 100-$	0
$1\ \tfrac{3}{2}\ \tfrac{1}{2}\ 0\ 000+$	$-\sqrt{2.7}/\sqrt{13.17.19.23}$
$1\ \tfrac{3}{2}\ \tfrac{1}{2}\ 0\ 010+$	$-3.5\sqrt{11}/2\sqrt{13.17.19.23}$
$1\ \tfrac{3}{2}\ \tfrac{1}{2}\ 0\ 020+$	0
$1\ \tfrac{3}{2}\ \tfrac{1}{2}\ 0\ 100+$	$+7\sqrt{5}/4\sqrt{2.17.19.23}$
$1\ \tfrac{3}{2}\ \tfrac{1}{2}\ 0\ 110+$	0
$1\ \tfrac{3}{2}\ \tfrac{1}{2}\ 0\ 120+$	$+3\sqrt{5.11.13}/4\sqrt{2.17.19.23}$
$1\ \tfrac{3}{2}\ \tfrac{3}{2}\ 0\ 000-$	0
$1\ \tfrac{3}{2}\ \tfrac{3}{2}\ 0\ 010-$	$-47\sqrt{7}/2\sqrt{3.5.13.17.19.23}$
$1\ \tfrac{3}{2}\ \tfrac{3}{2}\ 0\ 011-$	0
$1\ \tfrac{3}{2}\ \tfrac{3}{2}\ 0\ 020-$	$-2.5\sqrt{2.5}/\sqrt{3.17.19.23}$
$1\ \tfrac{3}{2}\ \tfrac{3}{2}\ 0\ 021-$	$-\sqrt{5.7.11}/2\sqrt{3.17.19.23}$
$1\ \tfrac{3}{2}\ \tfrac{3}{2}\ 0\ 022-$	0
$1\ \tfrac{3}{2}\ \tfrac{3}{2}\ 0\ 100-$	0
$1\ \tfrac{3}{2}\ \tfrac{3}{2}\ 0\ 110-$	$-5\sqrt{23}/4\sqrt{3.17.19}$
$1\ \tfrac{3}{2}\ \tfrac{3}{2}\ 0\ 111-$	0
$1\ \tfrac{3}{2}\ \tfrac{3}{2}\ 0\ 120-$	$+\sqrt{2.7.13}/\sqrt{3.17.19.23}$
$1\ \tfrac{3}{2}\ \tfrac{3}{2}\ 0\ 121-$	$-\sqrt{11.13}/4\sqrt{3.17.19.23}$
$1\ \tfrac{3}{2}\ \tfrac{3}{2}\ 0\ 122-$	0
$1\ \tfrac{3}{2}\ \tfrac{3}{2}\ 1\ 000-$	0
$1\ \tfrac{3}{2}\ \tfrac{3}{2}\ 1\ 010-$	$+29\sqrt{3.7}/4\sqrt{5.13.17.19.23}$
$1\ \tfrac{3}{2}\ \tfrac{3}{2}\ 1\ 011-$	0
$1\ \tfrac{3}{2}\ \tfrac{3}{2}\ 1\ 020-$	$+\sqrt{3.5}/\sqrt{2.17.19.23}$
$1\ \tfrac{3}{2}\ \tfrac{3}{2}\ 1\ 021-$	$+\sqrt{5.7.11}/4\sqrt{3.17.19.23}$
$1\ \tfrac{3}{2}\ \tfrac{3}{2}\ 1\ 022-$	0
$1\ \tfrac{3}{2}\ \tfrac{3}{2}\ 1\ 100-$	0
$1\ \tfrac{3}{2}\ \tfrac{3}{2}\ 1\ 110-$	$-3\sqrt{3}/4\sqrt{17.19.23}$
$1\ \tfrac{3}{2}\ \tfrac{3}{2}\ 1\ 111-$	0
$1\ \tfrac{3}{2}\ \tfrac{3}{2}\ 1\ 120-$	0
$1\ \tfrac{3}{2}\ \tfrac{3}{2}\ 1\ 121-$	$+5\sqrt{11.13}/4\sqrt{3.17.19.23}$
$1\ \tfrac{3}{2}\ \tfrac{3}{2}\ 1\ 122-$	0
$1\ \tilde{\tfrac{1}{2}}\ \tfrac{3}{2}\ 0\ 000+$	$-7\sqrt{11}/2\sqrt{2.17.19.23}$
$1\ \tilde{\tfrac{1}{2}}\ \tfrac{3}{2}\ 0\ 001+$	$+5.5\sqrt{7}/2.3\sqrt{17.19.23}$
$8\ \tfrac{15}{2}\ \tfrac{15}{2}$	
$1\ \tilde{\tfrac{1}{2}}\ \tfrac{3}{2}\ 0\ 002+$	$+\sqrt{13}/2.3\sqrt{2.17.19.23}$
$1\ \tilde{\tfrac{1}{2}}\ \tfrac{3}{2}\ 0\ 100+$	$-\sqrt{7.11.13}/4\sqrt{2.5.17.19.23}$
$1\ \tilde{\tfrac{1}{2}}\ \tfrac{3}{2}\ 0\ 101+$	$+4\sqrt{13}/3\sqrt{5.17.19.23}$
$1\ \tilde{\tfrac{1}{2}}\ \tfrac{3}{2}\ 0\ 102+$	$-13\sqrt{5.7}/4.3\sqrt{2.17.19.23}$
$1\ \tilde{\tfrac{1}{2}}\ \tilde{\tfrac{1}{2}}\ 0\ 000-$	0
$1\ \tilde{\tfrac{1}{2}}\ \tilde{\tfrac{1}{2}}\ 0\ 100-$	0
$2\ \tfrac{3}{2}\ \tfrac{1}{2}\ 0\ 000-$	$+\sqrt{11.23}/2\sqrt{5.13.17.19}$
$2\ \tfrac{3}{2}\ \tfrac{1}{2}\ 0\ 010-$	$+7\sqrt{7}/\sqrt{2.5.13.17.19.23}$
$2\ \tfrac{3}{2}\ \tfrac{1}{2}\ 0\ 020-$	$+\sqrt{5}/\sqrt{17.19.23}$
$2\ \tfrac{3}{2}\ \tfrac{1}{2}\ 0\ 100-$	$-2\sqrt{3.7}/5\sqrt{17.19.23}$
$2\ \tfrac{3}{2}\ \tfrac{1}{2}\ 0\ 110-$	$+29\sqrt{11}/5\sqrt{2.3.17.19.23}$
$2\ \tfrac{3}{2}\ \tfrac{1}{2}\ 0\ 120-$	$+\sqrt{7.11.13}/2\sqrt{3.17.19.23}$
$2\ \tfrac{3}{2}\ \tfrac{3}{2}\ 0\ 000+$	$-79\sqrt{3.7}/4.5\sqrt{13.17.19.23}$
$2\ \tfrac{3}{2}\ \tfrac{3}{2}\ 0\ 010+$	$-4.11\sqrt{2.11}/5\sqrt{3.13.17.19.23}$
$2\ \tfrac{3}{2}\ \tfrac{3}{2}\ 0\ 011+$	$+2.7.7\sqrt{7}/5\sqrt{3.13.17.19.23}$
$2\ \tfrac{3}{2}\ \tfrac{3}{2}\ 0\ 020+$	$+11\sqrt{7.11}/4\sqrt{3.17.19.23}$
$2\ \tfrac{3}{2}\ \tfrac{3}{2}\ 0\ 021+$	$+\sqrt{19}/\sqrt{2.3.17.23}$
$2\ \tfrac{3}{2}\ \tfrac{3}{2}\ 0\ 022+$	$+5\sqrt{7.13}/4\sqrt{3.17.19.23}$
$2\ \tfrac{3}{2}\ \tfrac{3}{2}\ 0\ 100+$	$+47\sqrt{11}/4.5\sqrt{5.17.19.23}$
$2\ \tfrac{3}{2}\ \tfrac{3}{2}\ 0\ 110+$	$-3\sqrt{7.17}/5\sqrt{2.5.19.23}$
$2\ \tfrac{3}{2}\ \tfrac{3}{2}\ 0\ 111+$	$-2.37\sqrt{11}/3.5\sqrt{5.17.19.23}$
$2\ \tfrac{3}{2}\ \tfrac{3}{2}\ 0\ 120+$	$-3\sqrt{13}/4\sqrt{5.17.19.23}$
$2\ \tfrac{3}{2}\ \tfrac{3}{2}\ 0\ 121+$	$+\sqrt{2.7.11.13}/3\sqrt{5.17.19.23}$
$2\ \tfrac{3}{2}\ \tfrac{3}{2}\ 0\ 122+$	$-13\sqrt{5.11}/4.3\sqrt{17.19.23}$
$2\ \tilde{\tfrac{1}{2}}\ \tfrac{3}{2}\ 0\ 000-$	$-9\sqrt{7}/2\sqrt{5.17.19.23}$
$2\ \tilde{\tfrac{1}{2}}\ \tfrac{3}{2}\ 0\ 001-$	$+3\sqrt{11}/\sqrt{2.5.17.19.23}$
$2\ \tilde{\tfrac{1}{2}}\ \tfrac{3}{2}\ 0\ 002-$	0
$2\ \tilde{\tfrac{1}{2}}\ \tfrac{3}{2}\ 0\ 100-$	$+\sqrt{3.11.13}/5\sqrt{17.19.23}$
$2\ \tilde{\tfrac{1}{2}}\ \tfrac{3}{2}\ 0\ 101-$	$-11\sqrt{7.13}/5\sqrt{2.3.17.19.23}$
$2\ \tilde{\tfrac{1}{2}}\ \tfrac{3}{2}\ 0\ 102-$	$-13/2\sqrt{3.17.19.23}$
$\tilde{1}\ \tfrac{3}{2}\ \tfrac{1}{2}\ 0\ 000-$	$-9\sqrt{11}/2.5.17\sqrt{2.19.23}$
$\tilde{1}\ \tfrac{3}{2}\ \tfrac{1}{2}\ 0\ 010-$	$+199\sqrt{7}/2.5.17\sqrt{19.23}$
$\tilde{1}\ \tfrac{3}{2}\ \tfrac{1}{2}\ 0\ 020-$	$+\sqrt{13}/2\sqrt{2.19.23}$
$\tilde{1}\ \tfrac{3}{2}\ \tfrac{1}{2}\ 0\ 100-$	$-3.137\sqrt{3.7}/4.17\sqrt{2.5.13.19.23}$
$\tilde{1}\ \tfrac{3}{2}\ \tfrac{1}{2}\ 0\ 110-$	$-2\sqrt{11.23}/17\sqrt{3.5.13.19}$
$\tilde{1}\ \tfrac{3}{2}\ \tfrac{1}{2}\ 0\ 120-$	$-\sqrt{5.7.11}/4\sqrt{2.3.19.23}$
$\tilde{1}\ \tfrac{3}{2}\ \tfrac{3}{2}\ 0\ 000+$	$-37\sqrt{3.7}/5.17\sqrt{2.5.19.23}$
$\tilde{1}\ \tfrac{3}{2}\ \tfrac{3}{2}\ 0\ 010+$	$+179\sqrt{11}/2.5.17\sqrt{3.5.19.23}$
$\tilde{1}\ \tfrac{3}{2}\ \tfrac{3}{2}\ 0\ 011+$	$-97\sqrt{2.7}/5.17\sqrt{3.5.19.23}$
$\tilde{1}\ \tfrac{3}{2}\ \tfrac{3}{2}\ 0\ 020+$	$-\sqrt{7.11.13}/17\sqrt{2.3.5.19.23}$
$\tilde{1}\ \tfrac{3}{2}\ \tfrac{3}{2}\ 0\ 021+$	$-31\sqrt{13}/2.17\sqrt{3.5.19.23}$
$\tilde{1}\ \tfrac{3}{2}\ \tfrac{3}{2}\ 0\ 022+$	$-13\sqrt{5.7}/17\sqrt{2.3.19.23}$
$\tilde{1}\ \tfrac{3}{2}\ \tfrac{3}{2}\ 0\ 100+$	$+3\sqrt{2.11}/5.17\sqrt{13.19.23}$
$\tilde{1}\ \tfrac{3}{2}\ \tfrac{3}{2}\ 0\ 110+$	$+241\sqrt{7}/4.5.17\sqrt{13.19.23}$
$\tilde{1}\ \tfrac{3}{2}\ \tfrac{3}{2}\ 0\ 111+$	$+8.37\sqrt{2.11}/3.5.17\sqrt{13.19.23}$
$\tilde{1}\ \tfrac{3}{2}\ \tfrac{3}{2}\ 0\ 120+$	$-2\sqrt{2}/17\sqrt{19.23}$
$\tilde{1}\ \tfrac{3}{2}\ \tfrac{3}{2}\ 0\ 121+$	$-7.11\sqrt{7.11}/4.3.17\sqrt{19.23}$
$\tilde{1}\ \tfrac{3}{2}\ \tfrac{3}{2}\ 0\ 122+$	$+5\sqrt{2.11.13}/3.17\sqrt{19.23}$
$\tilde{1}\ \tfrac{3}{2}\ \tfrac{3}{2}\ 1\ 000-$	0

SO_3–O 3jm Factors (cont.)

8	**$\frac{15}{2}$**	**$\frac{15}{2}$**			
$\tilde{1}$	$\frac{3}{2}$	$\frac{3}{2}$	1	010−	$-193\sqrt{11}/4.17\sqrt{3.5.19.23}$
$\tilde{1}$	$\frac{3}{2}$	$\frac{3}{2}$	1	011−	0
$\tilde{1}$	$\frac{3}{2}$	$\frac{3}{2}$	1	020−	$-\sqrt{7.11.13}/17\sqrt{2.3.5.19.23}$
$\tilde{1}$	$\frac{3}{2}$	$\frac{3}{2}$	1	021−	$+3.7\sqrt{3.13}/4.17\sqrt{5.19.23}$
$\tilde{1}$	$\frac{3}{2}$	$\frac{3}{2}$	1	022−	0
$\tilde{1}$	$\frac{3}{2}$	$\frac{3}{2}$	1	100−	0
$\tilde{1}$	$\frac{3}{2}$	$\frac{3}{2}$	1	110−	$-103\sqrt{7}/4.17\sqrt{13.19.23}$
$\tilde{1}$	$\frac{3}{2}$	$\frac{3}{2}$	1	111−	0
$\tilde{1}$	$\frac{3}{2}$	$\frac{3}{2}$	1	120−	$-2\sqrt{2}/17\sqrt{19.23}$
$\tilde{1}$	$\frac{3}{2}$	$\frac{3}{2}$	1	121−	$+11\sqrt{7.11}/4.17\sqrt{19.23}$
$\tilde{1}$	$\frac{3}{2}$	$\frac{3}{2}$	1	122−	0
$\tilde{1}$	$\tilde{\frac{1}{2}}$	$\frac{1}{2}$	0	000+	$+3\sqrt{3.11.13}/17\sqrt{5.19.23}$
$\tilde{1}$	$\tilde{\frac{1}{2}}$	$\frac{1}{2}$	0	100+	$+2.5\sqrt{7}/17\sqrt{19.23}$
$\tilde{1}$	$\tilde{\frac{1}{2}}$	$\frac{3}{2}$	0	000−	$+\sqrt{2.7.13.23}/5.17\sqrt{19}$
$\tilde{1}$	$\tilde{\frac{1}{2}}$	$\frac{3}{2}$	0	001−	$-9\sqrt{11.13}/2.5.17\sqrt{19.23}$
$\tilde{1}$	$\tilde{\frac{1}{2}}$	$\frac{3}{2}$	0	002−	0
$\tilde{1}$	$\tilde{\frac{1}{2}}$	$\frac{3}{2}$	0	100−	$-3\sqrt{3.5.11}/4.17\sqrt{2.19.23}$
$\tilde{1}$	$\tilde{\frac{1}{2}}$	$\frac{3}{2}$	0	101−	$+2.7\sqrt{5.7}/17\sqrt{3.19.23}$
$\tilde{1}$	$\tilde{\frac{1}{2}}$	$\frac{3}{2}$	0	102−	$+\sqrt{5.13}/4\sqrt{2.3.19.23}$

8	**8**	**0**			
0	0	0	0	000+	$+1/\sqrt{17}$
1	1	0	0	000+	$+\sqrt{3}/\sqrt{17}$
1	1	0	0	100+	0
1	1	0	0	110+	$+\sqrt{3}/\sqrt{17}$
2	2	0	0	000+	$+\sqrt{2}/\sqrt{17}$
2	2	0	0	100+	0
2	2	0	0	110+	$+\sqrt{2}/\sqrt{17}$
$\tilde{1}$	$\tilde{1}$	0	0	000+	$+\sqrt{3}/\sqrt{17}$
$\tilde{1}$	$\tilde{1}$	0	0	100+	0
$\tilde{1}$	$\tilde{1}$	0	0	110+	$+\sqrt{3}/\sqrt{17}$

8	**8**	**1**			
1	0	1	0	000−	$-1/\sqrt{17}$
1	0	1	0	100−	0
1	1	1	0	000+	$-1/4\sqrt{3.17}$
1	1	1	0	100+	0
1	1	1	0	110+	$+2/\sqrt{3.17}$
2	1	1	0	000−	$-\sqrt{5.7.11}/4.3\sqrt{2.17}$
2	1	1	0	010−	$+\sqrt{11.13}/4.3\sqrt{2.17}$
2	1	1	0	100−	$+\sqrt{13}/4\sqrt{2.3.17}$
2	1	1	0	110−	$+\sqrt{5.7}/4\sqrt{2.3.17}$
$\tilde{1}$	1	1	0	000−	$-\sqrt{7.11.13}/4.3.17$
$\tilde{1}$	1	1	0	010−	$-\sqrt{5.11}/3.17$
$\tilde{1}$	1	1	0	100−	$-\sqrt{5}/17\sqrt{3}$
$\tilde{1}$	1	1	0	110−	$-\sqrt{7.13}/2.17\sqrt{3}$
$\tilde{1}$	2	1	0	000+	$-\sqrt{5.13}/4.17\sqrt{2}$
$\tilde{1}$	2	1	0	010+	$+\sqrt{3.7.11}/4.17\sqrt{2}$
$\tilde{1}$	2	1	0	100+	$-\sqrt{7.11}/4.17\sqrt{2.3}$
$\tilde{1}$	2	1	0	110+	$-3\sqrt{5.13}/4.17\sqrt{2}$
$\tilde{1}$	$\tilde{1}$	1	0	000+	$+3\sqrt{3}/4\sqrt{17}$
$\tilde{1}$	$\tilde{1}$	1	0	100+	0
$\tilde{1}$	$\tilde{1}$	1	0	110+	$-2/\sqrt{3.17}$

8	**8**	**2**			
1	1	2	0	000+	$-53/3\sqrt{2.5.17.19}$
1	1	2	0	100+	$+\sqrt{7.13}/3\sqrt{2.17.19}$
1	1	2	0	110+	$+\sqrt{5}/3\sqrt{2.17.19}$
1	1	$\tilde{1}$	0	000+	$-73/4.3\sqrt{3.5.17.19}$
1	1	$\tilde{1}$	0	100+	$-\sqrt{7.13}/3\sqrt{3.17.19}$
1	1	$\tilde{1}$	0	110+	$+5\sqrt{5}/3\sqrt{3.17.19}$
2	0	2	0	000+	$+\sqrt{7.11}/\sqrt{2.3.17.19}$
2	0	2	0	100+	$-\sqrt{13}/\sqrt{2.5.17.19}$
2	1	$\tilde{1}$	0	000−	$+\sqrt{7.11}/4\sqrt{2.3.17.19}$
2	1	$\tilde{1}$	0	010−	$+\sqrt{5.11.13}/4\sqrt{2.3.17.19}$
2	1	$\tilde{1}$	0	100−	$-\sqrt{13}/4\sqrt{2.5.17.19}$
2	1	$\tilde{1}$	0	110−	$+5\sqrt{7}/4\sqrt{2.17.19}$
2	2	2	0	000+	$+2\sqrt{2.5}/\sqrt{3.17.19}$
2	2	2	0	100+	0
2	2	2	0	110+	$-2\sqrt{2.3}/\sqrt{5.17.19}$
$\tilde{1}$	0	$\tilde{1}$	0	000+	$+\sqrt{7.11.13}/17\sqrt{3.5.19}$
$\tilde{1}$	0	$\tilde{1}$	0	100+	$+4/17\sqrt{19}$
$\tilde{1}$	1	2	0	000−	$-\sqrt{2.7.11.13}/3.17\sqrt{5.19}$
$\tilde{1}$	1	2	0	010−	$-\sqrt{11}/3\sqrt{2.19}$
$\tilde{1}$	1	2	0	100−	$+3\sqrt{3}/17\sqrt{2.19}$
$\tilde{1}$	1	2	0	110−	0
$\tilde{1}$	1	$\tilde{1}$	0	000−	$+11\sqrt{7.11.13}/4.3.17\sqrt{5.19}$
$\tilde{1}$	1	$\tilde{1}$	0	010−	$+2\sqrt{11}/3.17\sqrt{19}$
$\tilde{1}$	1	$\tilde{1}$	0	100−	$-2\sqrt{3}/17\sqrt{19}$
$\tilde{1}$	1	$\tilde{1}$	0	110−	$-\sqrt{3.5.7.13}/2.17\sqrt{19}$
$\tilde{1}$	2	$\tilde{1}$	0	000+	$+\sqrt{13.19}/4.17\sqrt{2.3}$
$\tilde{1}$	2	$\tilde{1}$	0	010+	$-\sqrt{7.11.19}/4.17\sqrt{2.5}$
$\tilde{1}$	2	$\tilde{1}$	0	100+	$-\sqrt{5.7.11}/4.17\sqrt{2.19}$
$\tilde{1}$	2	$\tilde{1}$	0	110+	$-3.5\sqrt{3.13}/4.17\sqrt{2.19}$
$\tilde{1}$	$\tilde{1}$	2	0	000+	$+3.31/17\sqrt{2.5.17.19}$
$\tilde{1}$	$\tilde{1}$	2	0	100+	$+\sqrt{3.7.11.13}/17\sqrt{2.17.19}$
$\tilde{1}$	$\tilde{1}$	2	0	110+	$-5\sqrt{5}/17\sqrt{2.17.19}$
$\tilde{1}$	$\tilde{1}$	$\tilde{1}$	0	000+	$-199\sqrt{3}/4.17\sqrt{5.17.19}$
$\tilde{1}$	$\tilde{1}$	$\tilde{1}$	0	100+	$+\sqrt{7.11.13}/17\sqrt{17.19}$
$\tilde{1}$	$\tilde{1}$	$\tilde{1}$	0	110+	$-13\sqrt{3.5}/17\sqrt{17.19}$

8	**8**	**3**			
1	0	1	0	000−	$+7\sqrt{7}/3\sqrt{2.3.17.19}$
1	0	1	0	100−	$-\sqrt{2.5.13}/3\sqrt{3.17.19}$
1	1	1	0	000+	$-7\sqrt{7}/9\sqrt{2.17.19}$
1	1	1	0	100+	$-5\sqrt{5.13}/2.9\sqrt{2.17.19}$
1	1	1	0	110+	$+7\sqrt{7}/9\sqrt{2.17.19}$
1	1	$\tilde{1}$	0	000−	0
1	1	$\tilde{1}$	0	100−	$-\sqrt{3.13}/2\sqrt{2.17.19}$
1	1	$\tilde{1}$	0	110−	0
2	1	1	0	000−	$+5.5\sqrt{5.11}/8.3\sqrt{3.17.19}$
2	1	1	0	010−	$-\sqrt{11.13}/8.3\sqrt{3.7.17.19}$

SO_3–O 3jm Factors (cont.)

8 8 3	
2 1 1 0 100−	$+\sqrt{7.13}/8.3\sqrt{17.19}$
2 1 1 0 110−	$+31\sqrt{5}/8.3\sqrt{17.19}$
2 1 $\tilde{1}$ 0 000+	$-\sqrt{3.11}/8\sqrt{17.19}$
2 1 $\tilde{1}$ 0 010+	$-\sqrt{3.5.11.13}/8\sqrt{7.17.19}$
2 1 $\tilde{1}$ 0 100+	$+\sqrt{5.7.13}/8\sqrt{17.19}$
2 1 $\tilde{1}$ 0 110+	$+\sqrt{17}/8\sqrt{19}$
2 2 $\tilde{0}$ 0 000+	$+\sqrt{3.5}/4\sqrt{2.7.17.19}$
2 2 $\tilde{0}$ 0 100+	$+\sqrt{11.13}/4\sqrt{2.17.19}$
2 2 $\tilde{0}$ 0 110+	$+\sqrt{3.5.7}/4\sqrt{2.17.19}$
$\tilde{1}$ 0 $\tilde{1}$ 0 000−	$+\sqrt{5.11.13}/17\sqrt{2.3.19}$
$\tilde{1}$ 0 $\tilde{1}$ 0 100−	$-\sqrt{2.7}/17\sqrt{19}$
$\tilde{1}$ 1 1 0 000−	$-\sqrt{11.13}/17\sqrt{2.3.19}$
$\tilde{1}$ 1 1 0 010−	$+\sqrt{5.11.19}/2.17\sqrt{2.3.7}$
$\tilde{1}$ 1 1 0 100−	$-\sqrt{5.7}/2.3.17\sqrt{2.19}$
$\tilde{1}$ 1 1 0 110−	$-7\sqrt{13}/3.17\sqrt{2.19}$
$\tilde{1}$ 1 $\tilde{1}$ 0 000+	$+\sqrt{2.5.11.13}/3.17\sqrt{19}$
$\tilde{1}$ 1 $\tilde{1}$ 0 010+	$+\sqrt{11}/2.3\sqrt{2.7.19}$
$\tilde{1}$ 1 $\tilde{1}$ 0 100+	$+3\sqrt{3.7}/2.17\sqrt{2.19}$
$\tilde{1}$ 1 $\tilde{1}$ 0 110+	0
$\tilde{1}$ 1 $\tilde{0}$ 0 000−	$-\sqrt{5.11.13}/2.3.17\sqrt{19}$
$\tilde{1}$ 1 $\tilde{0}$ 0 010−	$-\sqrt{11}/3.17\sqrt{7.19}$
$\tilde{1}$ 1 $\tilde{0}$ 0 100−	$+\sqrt{7}/17\sqrt{3.19}$
$\tilde{1}$ 1 $\tilde{0}$ 0 110−	$-7\sqrt{5.13}/2.17\sqrt{3.19}$
$\tilde{1}$ 2 1 0 000+	$-13\sqrt{3.5.13}/8.17\sqrt{7.19}$
$\tilde{1}$ 2 1 0 010+	$+7\sqrt{11}/8.17\sqrt{19}$
$\tilde{1}$ 2 1 0 100+	$-13\sqrt{11}/8.17\sqrt{19}$
$\tilde{1}$ 2 1 0 110+	$-\sqrt{3.5.7.13}/8.17\sqrt{19}$
$\tilde{1}$ 2 $\tilde{1}$ 0 000−	$+41\sqrt{13}/8.17\sqrt{3.7.19}$
$\tilde{1}$ 2 $\tilde{1}$ 0 010−	$+11\sqrt{5.11}/8.17\sqrt{19}$
$\tilde{1}$ 2 $\tilde{1}$ 0 100−	$+5\sqrt{5.11}/8.17\sqrt{19}$
$\tilde{1}$ 2 $\tilde{1}$ 0 110−	$-3\sqrt{3.7.13}/8.17\sqrt{19}$
$\tilde{1}$ $\tilde{1}$ 1 0 000+	$+13/17\sqrt{2.7.17.19}$
$\tilde{1}$ $\tilde{1}$ 1 0 100+	$-\sqrt{3.5.11.13}/2.17\sqrt{2.17.19}$
$\tilde{1}$ $\tilde{1}$ 1 0 110+	$-31\sqrt{7}/17\sqrt{2.17.19}$
$\tilde{1}$ $\tilde{1}$ $\tilde{1}$ 0 000−	0
$\tilde{1}$ $\tilde{1}$ $\tilde{1}$ 0 100−	$+\sqrt{11.13}/2\sqrt{2.17.19}$
$\tilde{1}$ $\tilde{1}$ $\tilde{1}$ 0 110−	0
8 8 4	
0 0 0 0 000+	$+2.7\sqrt{7}/3\sqrt{3.13.17.19}$
1 0 1 0 000+	$+7\sqrt{5.7}/9\sqrt{2.13.17.19}$
1 0 1 0 100+	$+13\sqrt{2}/9\sqrt{17.19}$
1 1 0 0 000+	$+7.31\sqrt{7}/2.9.3\sqrt{13.17.19}$
1 1 0 0 100+	$-13\sqrt{5}/9.3\sqrt{17.19}$
1 1 0 0 110+	$-7\sqrt{7.13}/2.9.3\sqrt{17.19}$
1 1 1 0 000−	0
1 1 1 0 100−	$+13/2\sqrt{2.3.17.19}$
1 1 1 0 110−	0
1 1 2 0 000+	$+89\sqrt{5}/9.3\sqrt{13.17.19}$
1 1 2 0 100+	$-31\sqrt{7}/2.9.3\sqrt{17.19}$
8 8 4	
1 1 2 0 110+	$-2\sqrt{5.13}/9.3\sqrt{17.19}$
1 1 $\tilde{1}$ 0 000+	$+61\sqrt{5}/9\sqrt{2.3.13.17.19}$
1 1 $\tilde{1}$ 0 100+	$+37\sqrt{7}/2.9\sqrt{2.3.17.19}$
1 1 $\tilde{1}$ 0 110+	$+5\sqrt{5.13}/9\sqrt{2.3.17.19}$
2 0 2 0 000+	$-7\sqrt{7.11}/2.3\sqrt{3.13.17.19}$
2 0 2 0 100+	$-5\sqrt{5}/2.3\sqrt{17.19}$
2 1 1 0 000+	$+67\sqrt{11}/8.9\sqrt{13.17.19}$
2 1 1 0 010+	$+7\sqrt{5.7.11}/8.9\sqrt{17.19}$
2 1 1 0 100+	$-\sqrt{5.7}/8.3\sqrt{3.17.19}$
2 1 1 0 110+	$-23\sqrt{13}/8.3\sqrt{3.17.19}$
2 1 $\tilde{1}$ 0 000−	$-\sqrt{7.11}/8.3\sqrt{3.13.17.19}$
2 1 $\tilde{1}$ 0 010−	$-\sqrt{5.11}/8.3\sqrt{3.17.19}$
2 1 $\tilde{1}$ 0 100−	$-\sqrt{5.19}/8.3\sqrt{17}$
2 1 $\tilde{1}$ 0 110−	$+7\sqrt{7.13}/8.3\sqrt{17.19}$
2 2 0 0 000+	$+47\sqrt{7}/2.3\sqrt{2.3.13.17.19}$
2 2 0 0 100+	$-\sqrt{5.11}/2.3\sqrt{2.17.19}$
2 2 0 0 110+	$-\sqrt{7.13}/2\sqrt{2.3.17.19}$
2 2 2 0 000+	$-5.5\sqrt{5}/2.3\sqrt{3.13.17.19}$
2 2 2 0 100+	$+\sqrt{7.11}/2.3\sqrt{17.19}$
2 2 2 0 110+	$-\sqrt{5.13}/2\sqrt{3.17.19}$
$\tilde{1}$ 0 $\tilde{1}$ 0 000+	$-\sqrt{5.7.11}/17\sqrt{2.3.19}$
$\tilde{1}$ 0 $\tilde{1}$ 0 100+	$-\sqrt{2.13}/3.17\sqrt{19}$
$\tilde{1}$ 1 1 0 000+	$+\sqrt{2.5.11}/17\sqrt{19}$
$\tilde{1}$ 1 1 0 010+	$-\sqrt{7.11.13}/2.17\sqrt{2.19}$
$\tilde{1}$ 1 1 0 100+	$+149\sqrt{7}/2.3.17\sqrt{2.3.13.19}$
$\tilde{1}$ 1 1 0 110+	$+8\sqrt{2.5}/3.17\sqrt{3.19}$
$\tilde{1}$ 1 2 0 000−	$-\sqrt{5.7.11}/3.17\sqrt{19}$
$\tilde{1}$ 1 2 0 010−	$-\sqrt{11.13}/2.3.17\sqrt{19}$
$\tilde{1}$ 1 2 0 100−	$+11.37/2.3.17\sqrt{3.13.19}$
$\tilde{1}$ 1 2 0 110−	$+4\sqrt{5.7}/3.17\sqrt{3.19}$
$\tilde{1}$ 1 $\tilde{1}$ 0 000−	$-\sqrt{5.7.11}/3.17\sqrt{2.19}$
$\tilde{1}$ 1 $\tilde{1}$ 0 010−	$-\sqrt{11.13}/2.3.17\sqrt{2.19}$
$\tilde{1}$ 1 $\tilde{1}$ 0 100−	$+13\sqrt{13}/2.3.17\sqrt{2.3.19}$
$\tilde{1}$ 1 $\tilde{1}$ 0 110−	$-13\sqrt{5.7}/3.17\sqrt{2.3.19}$
$\tilde{1}$ 2 1 0 000−	$+41\sqrt{7}/8.17\sqrt{19}$
$\tilde{1}$ 2 1 0 010−	$+\sqrt{3.5.11.13}/8.17\sqrt{19}$
$\tilde{1}$ 2 1 0 100−	$+\sqrt{3.5.11}/8.17\sqrt{13.19}$
$\tilde{1}$ 2 1 0 110−	$+11\sqrt{7}/8.17\sqrt{19}$
$\tilde{1}$ 2 $\tilde{1}$ 0 000+	$+31/8.17\sqrt{3.19}$
$\tilde{1}$ 2 $\tilde{1}$ 0 010+	$+\sqrt{5.7.11.13}/8.17\sqrt{19}$
$\tilde{1}$ 2 $\tilde{1}$ 0 100+	$-5.11\sqrt{5.7.11}/8.3.17\sqrt{13.19}$
$\tilde{1}$ 2 $\tilde{1}$ 0 110+	$-5.5/8.17\sqrt{3.19}$
$\tilde{1}$ $\tilde{1}$ 0 0 000+	$-\sqrt{7.13}/2.17\sqrt{17.19}$
$\tilde{1}$ $\tilde{1}$ 0 0 100+	$+\sqrt{5.11}/17\sqrt{3.17.19}$
$\tilde{1}$ $\tilde{1}$ 0 0 110+	$-31\sqrt{7.13}/2.3.17\sqrt{17.19}$
$\tilde{1}$ $\tilde{1}$ 1 0 000−	0
$\tilde{1}$ $\tilde{1}$ 1 0 100−	$+\sqrt{11}/2\sqrt{2.17.19}$
$\tilde{1}$ $\tilde{1}$ 1 0 110−	0
$\tilde{1}$ $\tilde{1}$ 2 0 000+	$+\sqrt{5.13}/17\sqrt{17.19}$

SO_3–O 3jm Factors (cont.)

8 8 4	
$\tilde{1}\ \tilde{1}\ 2\ 0\ 100+$	$+\sqrt{7.11.19}/2.17\sqrt{3.17}$
$\tilde{1}\ \tilde{1}\ 2\ 0\ 110+$	$+2.23\sqrt{5}/3.17\sqrt{13.17.19}$
$\tilde{1}\ \tilde{1}\ \tilde{1}\ 0\ 000+$	$+\sqrt{3.5.13}/17\sqrt{2.17.19}$
$\tilde{1}\ \tilde{1}\ \tilde{1}\ 0\ 100+$	$-3.5\sqrt{7.11}/2.17\sqrt{2.17.19}$
$\tilde{1}\ \tilde{1}\ \tilde{1}\ 0\ 110+$	$-73\sqrt{5}/17\sqrt{2.3.13.17.19}$
8 8 5	
$1\ 0\ 1\ 0\ 000-$	$+2/\sqrt{3.13.17.19}$
$1\ 0\ 1\ 0\ 001-$	$-7\sqrt{5.7}/3\sqrt{3.13.17.19}$
$1\ 0\ 1\ 0\ 100-$	$+7\sqrt{5.7}/11\sqrt{3.17.19}$
$1\ 0\ 1\ 0\ 101-$	$-4.13/3.11\sqrt{3.17.19}$
$1\ 1\ 1\ 0\ 000+$	$-1/\sqrt{13.17.19}$
$1\ 1\ 1\ 0\ 001+$	$-5.7\sqrt{5.7}/4.9\sqrt{13.17.19}$
$1\ 1\ 1\ 0\ 100+$	$+7\sqrt{5.7}/4.11\sqrt{17.19}$
$1\ 1\ 1\ 0\ 101+$	$-2.7.13/9.11\sqrt{17.19}$
$1\ 1\ 1\ 0\ 110+$	$+8\sqrt{13}/11\sqrt{17.19}$
$1\ 1\ 1\ 0\ 111+$	$-7\sqrt{5.7.13}/2.9.11\sqrt{17.19}$
$1\ 1\ 2\ 0\ 000-$	0
$1\ 1\ 2\ 0\ 100-$	$-3\sqrt{2}/\sqrt{11.17.19}$
$1\ 1\ 2\ 0\ 110-$	0
$1\ 1\ \tilde{1}\ 0\ 000-$	0
$1\ 1\ \tilde{1}\ 0\ 100-$	$+\sqrt{3.19}/4\sqrt{11.17}$
$1\ 1\ \tilde{1}\ 0\ 110-$	0
$2\ 0\ 2\ 0\ 000-$	$-\sqrt{3}/\sqrt{2.13.17.19}$
$2\ 0\ 2\ 0\ 100-$	$+\sqrt{5.7}/\sqrt{2.11.17.19}$
$2\ 1\ 1\ 0\ 000-$	$-29\sqrt{5.7}/4\sqrt{2.3.11.13.17.19}$
$2\ 1\ 1\ 0\ 001-$	$-619/4.3\sqrt{2.3.11.13.17.19}$
$2\ 1\ 1\ 0\ 010-$	$+\sqrt{17}/4\sqrt{2.3.11.19}$
$2\ 1\ 1\ 0\ 011-$	$+\sqrt{5.7.17}/4.3\sqrt{2.3.11.19}$
$2\ 1\ 1\ 0\ 100-$	$-\sqrt{19}/4.11\sqrt{2.17}$
$2\ 1\ 1\ 0\ 101-$	$+5.5\sqrt{5.7}/4.3.11\sqrt{2.17.19}$
$2\ 1\ 1\ 0\ 110-$	$+5\sqrt{5.7.13}/4.11\sqrt{2.17.19}$
$2\ 1\ 1\ 0\ 111-$	$-\sqrt{13}/4.3.11\sqrt{2.17.19}$
$2\ 1\ \tilde{1}\ 0\ 000+$	$+7\sqrt{3}/4\sqrt{2.13.17.19}$
$2\ 1\ \tilde{1}\ 0\ 010+$	$+\sqrt{3.5.7}/4\sqrt{2.17.19}$
$2\ 1\ \tilde{1}\ 0\ 100+$	$+9\sqrt{5.7}/4\sqrt{2.11.17.19}$
$2\ 1\ \tilde{1}\ 0\ 110+$	$-3\sqrt{13}/4\sqrt{2.11.17.19}$
$2\ 2\ 2\ 0\ 000-$	0
$2\ 2\ 2\ 0\ 100-$	$-2\sqrt{2}/\sqrt{17.19}$
$2\ 2\ 2\ 0\ 110-$	0
$\tilde{1}\ 0\ \tilde{1}\ 0\ 000-$	$-2\sqrt{5}/17\sqrt{3.19}$
$\tilde{1}\ 0\ \tilde{1}\ 0\ 100-$	$-47\sqrt{7}/17\sqrt{11.13.19}$
$\tilde{1}\ 1\ 1\ 0\ 000-$	$+8.2\sqrt{7}/17\sqrt{3.11.19}$
$\tilde{1}\ 1\ 1\ 0\ 001-$	$-67\sqrt{5}/4.17\sqrt{3.11.19}$
$\tilde{1}\ 1\ 1\ 0\ 010-$	$+\sqrt{5.13.19}/4.17\sqrt{3.11}$
$\tilde{1}\ 1\ 1\ 0\ 011-$	$+7\sqrt{7.13}/17\sqrt{3.11.19}$
$\tilde{1}\ 1\ 1\ 0\ 100-$	$+9.3.29\sqrt{5}/4.11.17\sqrt{13.19}$
$\tilde{1}\ 1\ 1\ 0\ 101-$	$+41\sqrt{7}/3.11.17\sqrt{13.19}$
$\tilde{1}\ 1\ 1\ 0\ 110-$	$-2.9\sqrt{7}/11.17\sqrt{19}$
$\tilde{1}\ 1\ 1\ 0\ 111-$	$+2.5.5\sqrt{5}/3.11.17\sqrt{19}$
8 8 5	
$\tilde{1}\ 1\ 2\ 0\ 000+$	$+7\sqrt{5}/3.17\sqrt{2.19}$
$\tilde{1}\ 1\ 2\ 0\ 010+$	$-2\sqrt{2.7.13}/3.17\sqrt{19}$
$\tilde{1}\ 1\ 2\ 0\ 100+$	$-2.29\sqrt{2.7}/17\sqrt{3.11.13.19}$
$\tilde{1}\ 1\ 2\ 0\ 110+$	$+\sqrt{5}/17\sqrt{2.3.11.19}$
$\tilde{1}\ 1\ \tilde{1}\ 0\ 000+$	$-7\sqrt{5}/3.17\sqrt{19}$
$\tilde{1}\ 1\ \tilde{1}\ 0\ 010+$	$+7\sqrt{7.13}/4.3.17\sqrt{19}$
$\tilde{1}\ 1\ \tilde{1}\ 0\ 100+$	$+47\sqrt{3.7}/4.17\sqrt{11.13.19}$
$\tilde{1}\ 1\ \tilde{1}\ 0\ 110+$	$+8\sqrt{3.5}/17\sqrt{11.19}$
$\tilde{1}\ 2\ 1\ 0\ 000+$	$+\sqrt{3.5}/4.17\sqrt{2.19}$
$\tilde{1}\ 2\ 1\ 0\ 001+$	$-13\sqrt{3.7}/4.17\sqrt{2.19}$
$\tilde{1}\ 2\ 1\ 0\ 010+$	$-3.7\sqrt{7.13}/4.17\sqrt{2.11.19}$
$\tilde{1}\ 2\ 1\ 0\ 011+$	$-5\sqrt{5.13}/4.17\sqrt{2.11.19}$
$\tilde{1}\ 2\ 1\ 0\ 100+$	$-9.29\sqrt{7}/4.17\sqrt{2.11.13.19}$
$\tilde{1}\ 2\ 1\ 0\ 101+$	$-37\sqrt{5}/4.17\sqrt{2.11.13.19}$
$\tilde{1}\ 2\ 1\ 0\ 110+$	$+\sqrt{3.5}/4.11.17\sqrt{2.19}$
$\tilde{1}\ 2\ 1\ 0\ 111+$	$+163\sqrt{3.7}/4.11.17\sqrt{2.19}$
$\tilde{1}\ 2\ \tilde{1}\ 0\ 000-$	$-11\sqrt{7.11}/4.17\sqrt{2.3.19}$
$\tilde{1}\ 2\ \tilde{1}\ 0\ 010-$	$-\sqrt{5.13}/4.17\sqrt{2.19}$
$\tilde{1}\ 2\ \tilde{1}\ 0\ 100-$	$+41\sqrt{5}/4.17\sqrt{2.13.19}$
$\tilde{1}\ 2\ \tilde{1}\ 0\ 110-$	$-3.5\sqrt{3.7}/4.17\sqrt{2.11.19}$
$\tilde{1}\ \tilde{1}\ 1\ 0\ 000+$	$-9\sqrt{13}/17\sqrt{17.19}$
$\tilde{1}\ \tilde{1}\ 1\ 0\ 001+$	$-\sqrt{5.7.13}/4.17\sqrt{17.19}$
$\tilde{1}\ \tilde{1}\ 1\ 0\ 100+$	$+\sqrt{3.5.7.11}/4.17\sqrt{17.19}$
$\tilde{1}\ \tilde{1}\ 1\ 0\ 101+$	$-2\sqrt{3.11}/17\sqrt{17.19}$
$\tilde{1}\ \tilde{1}\ 1\ 0\ 110+$	$-8.3.5.5/11.17\sqrt{13.17.19}$
$\tilde{1}\ \tilde{1}\ 1\ 0\ 111+$	$+31\sqrt{5.7.13}/2.11.17\sqrt{17.19}$
$\tilde{1}\ \tilde{1}\ 2\ 0\ 000-$	0
$\tilde{1}\ \tilde{1}\ 2\ 0\ 100-$	$-\sqrt{2.3}/\sqrt{17.19}$
$\tilde{1}\ \tilde{1}\ 2\ 0\ 110-$	0
$\tilde{1}\ \tilde{1}\ \tilde{1}\ 0\ 000-$	0
$\tilde{1}\ \tilde{1}\ \tilde{1}\ 0\ 100-$	$+5/4\sqrt{17.19}$
$\tilde{1}\ \tilde{1}\ \tilde{1}\ 0\ 110-$	0
8 8 6	
$0\ 0\ 0\ 0\ 000+$	$+8/\sqrt{3.13.17.19.23}$
$1\ 0\ 1\ 0\ 000+$	$+2\sqrt{7}/\sqrt{3.13.17.19.23}$
$1\ 0\ 1\ 0\ 100+$	$-7\sqrt{5}/\sqrt{3.17.19.23}$
$1\ 1\ 0\ 0\ 000+$	$+\sqrt{17}/3\sqrt{13.19.23}$
$1\ 1\ 0\ 0\ 100+$	$+7\sqrt{5.7}/2.3\sqrt{17.19.23}$
$1\ 1\ 0\ 0\ 110+$	$-8\sqrt{13}/3\sqrt{17.19.23}$
$1\ 1\ 1\ 0\ 000-$	0
$1\ 1\ 1\ 0\ 100-$	$-3.7\sqrt{5}/4\sqrt{17.19.23}$
$1\ 1\ 1\ 0\ 110-$	0
$1\ 1\ 2\ 0\ 000+$	$+8.2\sqrt{7}/\sqrt{13.17.19.23}$
$1\ 1\ 2\ 0\ 100+$	$+3\sqrt{5}/2\sqrt{17.19.23}$
$1\ 1\ 2\ 0\ 110+$	$+\sqrt{7.13}/\sqrt{17.19.23}$
$1\ 1\ \tilde{1}\ 0\ 000+$	$+\sqrt{5.11}/3\sqrt{3.13.17.19.23}$
$1\ 1\ \tilde{1}\ 0\ 001+$	$-167/4\sqrt{3.13.17.19.23}$
$1\ 1\ \tilde{1}\ 0\ 100+$	$-43\sqrt{11}/4.3\sqrt{3.7.17.19.23}$
$1\ 1\ \tilde{1}\ 0\ 101+$	$-5\sqrt{5}/\sqrt{3.7.17.19.23}$

SO_3–O 3jm Factors (cont.)

8 8 6

1 1 1̃ 0 110+	+2√5.11.13/3√3.17.19.23
1 1 1̃ 0 111+	+√13/2√3.17.19.23
2 0 2 0 000+	−5√5.11/√3.13.17.19.23
2 0 2 0 100+	+3√7/√17.19.23
2 1 1 0 000+	+5.5√5.11/4√2.3.13.17.19.23
2 1 1 0 010+	+5√7.11/4√2.3.17.19.23
2 1 1 0 100+	+√7.17/4√2.19.23
2 1 1 0 110+	−√5.13/4√2.17.19.23
2 1 1̃ 0 000−	+587/4√2.3.7.13.17.19.23
2 1 1̃ 0 001−	+√5.11.13/4√2.3.7.17.19.23
2 1 1̃ 0 010−	+53√5/4√2.3.17.19.23
2 1 1̃ 0 011−	−√11/4√2.3.17.19.23
2 1 1̃ 0 100−	+5√5.11/4√2.17.19.23
2 1 1̃ 0 101−	−3.5.5/4√2.17.19.23
2 1 1̃ 0 110−	+√11.13/4√2.7.17.19.23
2 1 1̃ 0 111−	−3√5.13/4√2.7.17.19.23
2 2 0 0 000+	−127/2√2.3.13.17.19.23
2 2 0 0 100+	−√5.7.11/2√2.17.19.23
2 2 0 0 110+	+√3.13/2√2.17.19.23
2 2 2 0 000+	−√7.17/2√3.13.19.23
2 2 2 0 100+	−√5.11/2√17.19.23
2 2 2 0 110+	−√3.7.13/2√17.19.23
2 2 0̃ 0 000−	0
2 2 0̃ 0 100−	−8√2/√17.19.23
2 2 0̃ 0 110−	0
1̃ 0 1̃ 0 000+	+2√5.23/17√3.7.19
1̃ 0 1̃ 0 001+	+9√3.11/17√7.19.23
1̃ 0 1̃ 0 100+	−47√11/17√13.19.23
1̃ 0 1̃ 0 101+	+8.8√5/17√13.19.23
1̃ 1 1 0 000+	−11√11/17√3.19.23
1̃ 1 1 0 010+	−5√5.7.11.13/4.17√3.19.23
1̃ 1 1 0 100+	+89√5.7/4.17√13.19.23
1̃ 1 1 0 110+	−8/17√19.23
1̃ 1 2 0 000−	+5√11/3.17√19.23
1̃ 1 2 0 010−	−5√5.7.11.13/2.3.17√19.23
1̃ 1 2 0 100−	−5.11√5.7/2.17√3.13.19.23
1̃ 1 2 0 110−	−8.2/17√3.19.23
1̃ 1 1̃ 0 000−	+4√5.23/3.17√7.19
1̃ 1 1̃ 0 001−	+9.3.5√11/4.17√7.19.23
1̃ 1 1̃ 0 010−	−7√13/4.3√19.23
1̃ 1 1̃ 0 011−	0
1̃ 1 1̃ 0 100−	+3.47√3.11/4.17√13.19.23
1̃ 1 1̃ 0 101−	−8.4√5/17√3.13.19.23
1̃ 1 1̃ 0 110−	0
1̃ 1 1̃ 0 111−	+√7/√3.19.23
1̃ 1 0̃ 0 000+	+√5.23/3.17√19
1̃ 1 0̃ 0 010+	−7√7.13/2.3.17√19.23
1̃ 1 0̃ 0 100+	−47√7.11/2.17√3.13.19.23
1̃ 1 0̃ 0 110+	+8√5.11/17√3.19.23

8 8 6

1̃ 2 1 0 000−	−7√3.5.7/4.17√2.19.23
1̃ 2 1 0 010−	+3.5√11.13/4.17√2.19.23
1̃ 2 1 0 100−	−3.107√11/4.17√2.13.19.23
1̃ 2 1 0 110−	+11√3.5.7/4.17√2.19.23
1̃ 2 1̃ 0 000+	+37√11/4.17√2.3.19.23
1̃ 2 1̃ 0 001+	+3.5.5√3.5/4.17√2.19.23
1̃ 2 1̃ 0 010+	+47√5.13/4.17√2.7.19.23
1̃ 2 1̃ 0 011+	−3√11.13/4.17√2.7.19.23
1̃ 2 1̃ 0 100+	+97√5/4.17√2.7.13.19.23
1̃ 2 1̃ 0 101+	−257√11/4.17√2.7.13.19.23
1̃ 2 1̃ 0 110+	+9.3√3.11/4.17√2.19.23
1̃ 2 1̃ 0 111+	+59√3.5/4.17√2.19.23
1̃ 1̃ 0 0 000+	+3√13.23/17√17.19
1̃ 1̃ 0 0 100+	−√3.5.7.11/2.17√17.19.23
1̃ 1̃ 0 0 110+	−8.5.5/17√13.17.19.23
1̃ 1̃ 1 0 000−	0
1̃ 1̃ 1 0 100−	−√3.5.11/4√17.19.23
1̃ 1̃ 1 0 110−	0
1̃ 1̃ 2 0 000+	+2.3√7.13/17√17.19.23
1̃ 1̃ 2 0 100+	+9√3.5.11/2.17√17.19.23
1̃ 1̃ 2 0 110+	−163√7/17√13.17.19.23
1̃ 1̃ 1̃ 0 000+	+√3.5.11.13/17√17.19.23
1̃ 1̃ 1̃ 0 001+	−3.13√3.13/4.17√17.19.23
1̃ 1̃ 1̃ 0 100+	−151/4.17√7.17.19.23
1̃ 1̃ 1̃ 0 101+	−3√5.11.19/17√7.17.23
1̃ 1̃ 1̃ 0 110+	+2√3.5.11.19/17√13.17.23
1̃ 1̃ 1̃ 0 111+	+449√3/2.17√13.17.19.23

8 8 7

1 0 1 0 000−	−4√2.7/3√13.17.19.23
1 0 1 0 001−	−5√3.11/√2.13.17.19.23
1 0 1 0 100−	+7√5/3√2.17.19.23
1 0 1 0 101−	0
1 1 1 0 000+	−7√2.7/3√3.13.17.19.23
1 1 1 0 001+	+5√2.11/√13.17.19.23
1 1 1 0 100+	+2.7√2.5/3√3.17.19.23
1 1 1 0 101+	0
1 1 1 0 110+	−2√2.7.13/3√3.17.19.23
1 1 1 0 111+	−√2.11.13/√17.19.23
1 1 2 0 000−	0
1 1 2 0 100−	−5√11/√2.17.19.23
1 1 2 0 110−	0
1 1 1̃ 0 000−	0
1 1 1̃ 0 001−	0
1 1 1̃ 0 100−	+4√2.11/√7.17.19.23
1 1 1̃ 0 101−	−√2.13/√7.17.19.23
1 1 1̃ 0 110−	0
1 1 1̃ 0 111−	0
2 0 2 0 000−	−107/2√2.3.13.17.19.23
2 0 2 0 100−	−9√7.11/2√2.5.17.19.23

SO_3–O 3jm Factors (cont.)

8 8 7

2 1 1 0 000−	$+\sqrt{11.19}/2.3\sqrt{5.13.17.23}$
2 1 1 0 001−	$-7\sqrt{3.7}/\sqrt{5.13.17.19.23}$
2 1 1 0 010−	$-11\sqrt{7.11}/2.3\sqrt{17.19.23}$
2 1 1 0 011−	$-3\sqrt{3}/2\sqrt{17.19.23}$
2 1 1 0 100−	$+3.7\sqrt{3.7}/2.5\sqrt{17.19.23}$
2 1 1 0 101−	$+11\sqrt{11}/5\sqrt{17.19.23}$
2 1 1 0 110−	$+\sqrt{3.13}/2\sqrt{5.17.19.23}$
2 1 1 0 111−	$+\sqrt{7.11.13}/2\sqrt{5.17.19.23}$
2 1 $\tilde{1}$ 0 000+	$-41/3\sqrt{7.13.17.19.23}$
2 1 $\tilde{1}$ 0 001+	$+\sqrt{11.17}/3\sqrt{7.19.23}$
2 1 $\tilde{1}$ 0 010+	$+5\sqrt{5}/2.3\sqrt{17.19.23}$
2 1 $\tilde{1}$ 0 011+	$+\sqrt{5.11.13}/2.3\sqrt{17.19.23}$
2 1 $\tilde{1}$ 0 100+	$-3\sqrt{3.11}/\sqrt{5.17.19.23}$
2 1 $\tilde{1}$ 0 101+	$-\sqrt{3.13}/\sqrt{5.17.19.23}$
2 1 $\tilde{1}$ 0 110+	$+3\sqrt{3.11.13}/2\sqrt{7.17.19.23}$
2 1 $\tilde{1}$ 0 111+	$+13\sqrt{3}/2\sqrt{7.17.19.23}$
2 2 2 0 000−	0
2 2 2 0 100−	$-\sqrt{2}/\sqrt{17.19.23}$
2 2 2 0 110−	0
2 2 $\tilde{0}$ 0 000+	$-7\sqrt{5.7}/8\sqrt{3.17.19.23}$
2 2 $\tilde{0}$ 0 100+	$+5\sqrt{11.13}/8\sqrt{17.19.23}$
2 2 $\tilde{0}$ 0 110+	$-3.13\sqrt{3.7}/8\sqrt{5.17.19.23}$
$\tilde{1}$ 0 $\tilde{1}$ 0 000−	$-4\sqrt{2.23}/17\sqrt{5.7.19}$
$\tilde{1}$ 0 $\tilde{1}$ 0 001−	$-9.3\sqrt{11.13}/17\sqrt{2.5.7.19.23}$
$\tilde{1}$ 0 $\tilde{1}$ 0 100−	$-47\sqrt{11}/17\sqrt{2.3.13.19.23}$
$\tilde{1}$ 0 $\tilde{1}$ 0 101−	$+8.5\sqrt{2}/17\sqrt{3.19.23}$
$\tilde{1}$ 1 1 0 000−	$-2\sqrt{2.11.19}/3.5.17\sqrt{23}$
$\tilde{1}$ 1 1 0 001−	$+2.37\sqrt{2.7}/5.17\sqrt{3.19.23}$
$\tilde{1}$ 1 1 0 010−	$+\sqrt{2.7.11.13}/3.17\sqrt{5.19.23}$
$\tilde{1}$ 1 1 0 011−	$+29\sqrt{2.13}/17\sqrt{3.5.19.23}$
$\tilde{1}$ 1 1 0 100−	$+4.31\sqrt{2.7}/17\sqrt{3.5.13.19.23}$
$\tilde{1}$ 1 1 0 101−	$-97\sqrt{2.11}/3.17\sqrt{5.13.19.23}$
$\tilde{1}$ 1 1 0 110−	$+4\sqrt{2}/17\sqrt{3.19.23}$
$\tilde{1}$ 1 1 0 111−	$-\sqrt{2.7.11}/3.17\sqrt{19.23}$
$\tilde{1}$ 1 2 0 000+	$-2.11\sqrt{2}/17\sqrt{5.19.23}$
$\tilde{1}$ 1 2 0 010+	$-5\sqrt{7.13}/17\sqrt{2.19.23}$
$\tilde{1}$ 1 2 0 100+	$+113\sqrt{7.11}/3.17\sqrt{2.3.13.19.23}$
$\tilde{1}$ 1 2 0 110+	$-2\sqrt{2.5.11}/3.17\sqrt{3.19.23}$
$\tilde{1}$ 1 $\tilde{1}$ 0 000+	$-\sqrt{2.23}/17\sqrt{3.5.7.19}$
$\tilde{1}$ 1 $\tilde{1}$ 0 001+	$-9\sqrt{2.3.11.13}/17\sqrt{5.7.19.23}$
$\tilde{1}$ 1 $\tilde{1}$ 0 010+	$+7\sqrt{2.13}/17\sqrt{3.19.23}$
$\tilde{1}$ 1 $\tilde{1}$ 0 011+	0
$\tilde{1}$ 1 $\tilde{1}$ 0 100+	$+2.47\sqrt{2.11}/3.17\sqrt{13.19.23}$
$\tilde{1}$ 1 $\tilde{1}$ 0 101+	$-2.3.5\sqrt{2}/17\sqrt{19.23}$
$\tilde{1}$ 1 $\tilde{1}$ 0 110+	$-2\sqrt{2.5.7.11}/3.17\sqrt{19.23}$
$\tilde{1}$ 1 $\tilde{1}$ 0 111+	0
$\tilde{1}$ 1 $\tilde{0}$ 0 000−	$-9\sqrt{11.13}/17\sqrt{2.5.19.23}$
$\tilde{1}$ 1 $\tilde{0}$ 0 010−	0
$\tilde{1}$ 1 $\tilde{0}$ 0 100−	$+8.5\sqrt{2.7}/3.17\sqrt{3.19.23}$

8 8 7

$\tilde{1}$ 1 $\tilde{0}$ 0 110−	$+\sqrt{5.13}/3\sqrt{2.3.19.23}$
$\tilde{1}$ 2 1 0 000+	$-31\sqrt{7}/2.17\sqrt{5.19.23}$
$\tilde{1}$ 2 1 0 001+	$+13\sqrt{3.11}/17\sqrt{5.19.23}$
$\tilde{1}$ 2 1 0 010+	$+\sqrt{3.11.13}/2.5.17\sqrt{19.23}$
$\tilde{1}$ 2 1 0 011+	$+3\sqrt{7.13}/5.17\sqrt{19.23}$
$\tilde{1}$ 2 1 0 100+	$-\sqrt{11.23}/2.17\sqrt{3.13.19}$
$\tilde{1}$ 2 1 0 101+	$+8.4\sqrt{7}/17\sqrt{13.19.23}$
$\tilde{1}$ 2 1 0 110+	$+9\sqrt{7}/2.17\sqrt{5.19.23}$
$\tilde{1}$ 2 1 0 111+	$-2\sqrt{3.11}/17\sqrt{5.19.23}$
$\tilde{1}$ 2 $\tilde{1}$ 0 000−	$+2\sqrt{11}/17\sqrt{19.23}$
$\tilde{1}$ 2 $\tilde{1}$ 0 001−	$+3\sqrt{13}/17\sqrt{19.23}$
$\tilde{1}$ 2 $\tilde{1}$ 0 010−	$+8\sqrt{3.13}/17\sqrt{5.7.19.23}$
$\tilde{1}$ 2 $\tilde{1}$ 0 011−	$+3.13\sqrt{3.11}/17\sqrt{5.7.19.23}$
$\tilde{1}$ 2 $\tilde{1}$ 0 100−	$-43\sqrt{5.23}/2.17\sqrt{3.7.13.19}$
$\tilde{1}$ 2 $\tilde{1}$ 0 101−	$+11\sqrt{5.11}/17\sqrt{3.7.19.23}$
$\tilde{1}$ 2 $\tilde{1}$ 0 110−	$-7\sqrt{11}/2.17\sqrt{19.23}$
$\tilde{1}$ 2 $\tilde{1}$ 0 111−	$-7\sqrt{13}/17\sqrt{19.23}$
$\tilde{1}$ $\tilde{1}$ 1 0 000+	$+3\sqrt{2.3.7.13.23}/5.17\sqrt{17.19}$
$\tilde{1}$ $\tilde{1}$ 1 0 001+	$-9.3\sqrt{2.11.13}/5.17\sqrt{17.19.23}$
$\tilde{1}$ $\tilde{1}$ 1 0 100+	$+3\sqrt{2.5.11}/17\sqrt{17.19.23}$
$\tilde{1}$ $\tilde{1}$ 1 0 101+	$-2.3\sqrt{2.3.5.7}/17\sqrt{17.19.23}$
$\tilde{1}$ $\tilde{1}$ 1 0 110+	$+2.5.5\sqrt{2.7}/17\sqrt{3.13.17.19.23}$
$\tilde{1}$ $\tilde{1}$ 1 0 111+	$-3.5\sqrt{2.11.13}/17\sqrt{17.19.23}$
$\tilde{1}$ $\tilde{1}$ 2 0 000−	0
$\tilde{1}$ $\tilde{1}$ 2 0 100−	$-\sqrt{3}/\sqrt{2.17.19.23}$
$\tilde{1}$ $\tilde{1}$ 2 0 110−	0
$\tilde{1}$ $\tilde{1}$ $\tilde{1}$ 0 000−	0
$\tilde{1}$ $\tilde{1}$ $\tilde{1}$ 0 001−	0
$\tilde{1}$ $\tilde{1}$ $\tilde{1}$ 0 100−	$+5\sqrt{2.3}/\sqrt{7.17.19.23}$
$\tilde{1}$ $\tilde{1}$ $\tilde{1}$ 0 101−	$+\sqrt{2.3.11.13}/\sqrt{7.17.19.23}$
$\tilde{1}$ $\tilde{1}$ $\tilde{1}$ 0 110−	0
$\tilde{1}$ $\tilde{1}$ $\tilde{1}$ 0 111−	0

8 8 8

0 0 0 0 000+	$-\sqrt{2.3.5.11}/\sqrt{13.17.19.23}$
1 1 0 0 000+	$-3\sqrt{5.11}/\sqrt{2.13.17.19.23}$
1 1 0 0 100+	0
1 1 0 0 110+	$+3\sqrt{11.13}/\sqrt{2.5.17.19.23}$
1 1 1 0 000−	0
1 1 1 0 100−	0
1 1 1 0 110−	0
1 1 1 0 111−	0
2 1 1 0 000+	$-7\sqrt{7}/\sqrt{13.17.19.23}$
2 1 1 0 010+	$-5\sqrt{5}/2\sqrt{17.19.23}$
2 1 1 0 011+	$-\sqrt{7.13}/2\sqrt{17.19.23}$
2 1 1 0 100+	$-\sqrt{5.11}/\sqrt{3.17.19.23}$
2 1 1 0 110+	$-\sqrt{7.11.13}/2\sqrt{3.17.19.23}$
2 1 1 0 111+	$+13\sqrt{11}/2\sqrt{3.5.17.19.23}$
2 2 0 0 000+	$+9\sqrt{3.11}/8\sqrt{5.13.17.19.23}$
2 2 0 0 100+	$+7\sqrt{7}/8\sqrt{17.19.23}$

SO_3–O 3jm Factors (cont.)

8 8 8	
2 2 0 0 110+	$-\sqrt{3.5.11.13}/8\sqrt{17.19.23}$
2 2 2 0 000+	$-7\sqrt{3.7}/8.2\sqrt{13.17.19.23}$
2 2 2 0 100+	$-11.11\sqrt{11}/8.2\sqrt{5.17.19.23}$
2 2 2 0 110+	$-5\sqrt{3.7.13}/8.2\sqrt{17.19.23}$
2 2 2 0 111+	$+13\sqrt{5.11}/8.2\sqrt{17.19.23}$
$\tilde{1}$ 1 1 0 000+	$-5\sqrt{2.5.7}/17\sqrt{19.23}$
$\tilde{1}$ 1 1 0 010+	$-\sqrt{2.13}/17\sqrt{19.23}$
$\tilde{1}$ 1 1 0 011+	$+13\sqrt{2.7}/17\sqrt{5.19.23}$
$\tilde{1}$ 1 1 0 100+	$-\sqrt{2.11}/17\sqrt{3.13.19.23}$
$\tilde{1}$ 1 1 0 110+	$+8.2\sqrt{2.7.11}/17\sqrt{3.5.19.23}$
$\tilde{1}$ 1 1 0 111+	$-2\sqrt{2.11.13}/17\sqrt{3.19.23}$
$\tilde{1}$ 2 1 0 000−	$-7\sqrt{11}/17\sqrt{19.23}$
$\tilde{1}$ 2 1 0 001−	$-\sqrt{7.11.13}/2.17\sqrt{5.19.23}$
$\tilde{1}$ 2 1 0 010−	$+\sqrt{3.5.7.13}/17\sqrt{19.23}$
$\tilde{1}$ 2 1 0 011−	$+13\sqrt{3}/2.17\sqrt{19.23}$
$\tilde{1}$ 2 1 0 100−	$+31\sqrt{3.7}/17\sqrt{5.13.19.23}$
$\tilde{1}$ 2 1 0 101−	$-5.11\sqrt{3}/2.17\sqrt{19.23}$
$\tilde{1}$ 2 1 0 110−	$+7\sqrt{11}/17\sqrt{19.23}$
$\tilde{1}$ 2 1 0 111−	$+\sqrt{7.11.13}/2.17\sqrt{5.19.23}$
$\tilde{1}$ $\tilde{1}$ 0 0 000+	$+9\sqrt{5.11.13}/17\sqrt{2.17.19.23}$
$\tilde{1}$ $\tilde{1}$ 0 0 100+	$+8\sqrt{2.3.7}/17\sqrt{17.19.23}$
$\tilde{1}$ $\tilde{1}$ 0 0 110+	$-9\sqrt{5.11.13}/17\sqrt{2.17.19.23}$
$\tilde{1}$ $\tilde{1}$ 1 0 000−	0
$\tilde{1}$ $\tilde{1}$ 1 0 001−	0
$\tilde{1}$ $\tilde{1}$ 1 0 100−	$+2\sqrt{2.7}/\sqrt{17.19.23}$
$\tilde{1}$ $\tilde{1}$ 1 0 101−	$+\sqrt{2.13}/\sqrt{5.17.19.23}$
$\tilde{1}$ $\tilde{1}$ 1 0 110−	0
$\tilde{1}$ $\tilde{1}$ 1 0 111−	0
$\tilde{1}$ $\tilde{1}$ 2 0 000+	$-3\sqrt{7.13}/17\sqrt{17.19.23}$
$\tilde{1}$ $\tilde{1}$ 2 0 001+	$+13\sqrt{3.5.11}/17\sqrt{17.19.23}$
$\tilde{1}$ $\tilde{1}$ 2 0 100+	$+4.11\sqrt{3.11}/17\sqrt{5.17.19.23}$
$\tilde{1}$ $\tilde{1}$ 2 0 101+	$-2.3\sqrt{7.13}/17\sqrt{17.19.23}$
$\tilde{1}$ $\tilde{1}$ 2 0 110+	$+3.43\sqrt{7}/2.17\sqrt{13.17.19.23}$
$\tilde{1}$ $\tilde{1}$ 2 0 111+	$-11\sqrt{3.11}/2.17\sqrt{5.17.19.23}$
$\tilde{1}$ $\tilde{1}$ $\tilde{1}$ 0 000+	$+3.13\sqrt{2.5.7}/17.17\sqrt{19.23}$
$\tilde{1}$ $\tilde{1}$ $\tilde{1}$ 0 100+	$-9.3\sqrt{2.3.11.13}/17.17\sqrt{19.23}$
$\tilde{1}$ $\tilde{1}$ $\tilde{1}$ 0 110+	$-3.31\sqrt{2.7}/17.17\sqrt{5.19.23}$
$\tilde{1}$ $\tilde{1}$ $\tilde{1}$ 0 111+	$-2.3\sqrt{2.3.11}/17.17\sqrt{13.19.23}$

SO_3–K 2jm Factors (SO_{3i}–K_i = O_3–O_h)

$0 \to +\ 0$
$\frac{1}{2} \to +\ \frac{1}{2}$
$1 \to +\ 1$
$\frac{3}{2} \to +\ \frac{3}{2}$
$2 \to +\ 2$
$\frac{5}{2} \to +\ \frac{5}{2}$
$3 \to +\ 3 +\ \tilde{1}$
$\frac{7}{2} \to +\ \frac{5}{2} +\ \tilde{\frac{1}{2}}$
$4 \to +\ 2 +\ 3$
$\frac{9}{2} \to +\ \frac{3}{2} +\ \frac{5}{2}$
$5 \to +\ 1 +\ 2 +\ \tilde{1}$
$\frac{11}{2} \to +\ \frac{1}{2} +\ \frac{3}{2} +\ \frac{5}{2}$
$6 \to +\ 0 +\ 1 +\ 2 +\ 3$
$\frac{13}{2} \to +\ \frac{1}{2} +\ \frac{3}{2} +\ \frac{5}{2} +\ \tilde{\frac{1}{2}}$
$7 \to +\ 1 +\ 2 +\ 3 +\ \tilde{1}$
$\frac{15}{2} \to +\ \frac{3}{2} +\ 2\,\frac{5}{2}$
$8 \to +\ 2\,2 +\ 3 +\ \tilde{1}$
$4 \to +\ 2 +\ 3$

SO_3–K 3jm Factors

0 0 0
0 0 0 0 000+ $+1$
$\frac{1}{2}$ $\frac{1}{2}$ 0
$\frac{1}{2}$ $\frac{1}{2}$ 0 0 000+ $+1$
1 $\frac{1}{2}$ $\frac{1}{2}$
1 $\frac{1}{2}$ $\frac{1}{2}$ 0 000+ $+1$
1 1 0
1 1 0 0 000+ $+1$
1 1 1
1 1 1 0 000+ -1
$\frac{3}{2}$ 1 $\frac{1}{2}$
$\frac{3}{2}$ 1 $\frac{1}{2}$ 0 000+ $+1$
$\frac{3}{2}$ $\frac{3}{2}$ 0
$\frac{3}{2}$ $\frac{3}{2}$ 0 0 000+ $+1$
$\frac{3}{2}$ $\frac{3}{2}$ 1
$\frac{3}{2}$ $\frac{3}{2}$ 1 0 000+ $+1$
2 1 1
2 1 1 0 000+ $+1$
2 $\frac{3}{2}$ $\frac{1}{2}$
2 $\frac{3}{2}$ $\frac{1}{2}$ 0 000+ $+1$
2 $\frac{3}{2}$ $\frac{3}{2}$
2 $\frac{3}{2}$ $\frac{3}{2}$ 0 000+ -1
2 2 0
2 2 0 0 000+ $+1$
2 2 1
2 2 1 0 000+ -1
2 2 2
2 2 2 0 000+ $-\sqrt{5}/\sqrt{2.7}$
2 2 2 1 000+ $+3/\sqrt{2.7}$
$\frac{5}{2}$ $\frac{3}{2}$ 1
$\frac{5}{2}$ $\frac{3}{2}$ 1 0 000+ -1
$\frac{5}{2}$ 2 $\frac{1}{2}$
$\frac{5}{2}$ 2 $\frac{1}{2}$ 0 000+ $+1$
$\frac{5}{2}$ 2 $\frac{3}{2}$
$\frac{5}{2}$ 2 $\frac{3}{2}$ 0 000+ $-1/\sqrt{7}$
$\frac{5}{2}$ 2 $\frac{3}{2}$ 1 000− $-\sqrt{2.3}/\sqrt{7}$
$\frac{5}{2}$ $\frac{5}{2}$ 0
$\frac{5}{2}$ $\frac{5}{2}$ 0 0 000+ $+1$
$\frac{5}{2}$ $\frac{5}{2}$ 1
$\frac{5}{2}$ $\frac{5}{2}$ 1 0 000+ $+\sqrt{5}/\sqrt{2.7}$
$\frac{5}{2}$ $\frac{5}{2}$ 1 1 000+ $+3/\sqrt{2.7}$
$\frac{5}{2}$ $\frac{5}{2}$ 2
$\frac{5}{2}$ $\frac{5}{2}$ 2 0 000+ $-2/\sqrt{7}$
$\frac{5}{2}$ $\frac{5}{2}$ 2 1 000− 0
$\frac{5}{2}$ $\frac{5}{2}$ 2 2 000+ $-\sqrt{3}/\sqrt{7}$
3 $\frac{3}{2}$ $\frac{3}{2}$
3 $\frac{3}{2}$ $\frac{3}{2}$ 0 000+ $-2/\sqrt{7}$
$\tilde{1}$ $\frac{3}{2}$ $\frac{3}{2}$ 0 000+ $-\sqrt{3}/\sqrt{7}$
3 2 1
3 2 1 0 000+ $+2/\sqrt{7}$
$\tilde{1}$ 2 1 0 000+ $+\sqrt{3}/\sqrt{7}$
3 2 2
3 2 2 0 000+ $+2/\sqrt{7}$
3 2 2 1 000− 0
$\tilde{1}$ 2 2 0 000+ $+\sqrt{3}/\sqrt{7}$
3 $\frac{5}{2}$ $\frac{1}{2}$
3 $\frac{5}{2}$ $\frac{1}{2}$ 0 000+ $+2/\sqrt{7}$
$\tilde{1}$ $\frac{5}{2}$ $\frac{1}{2}$ 0 000+ $+\sqrt{3}/\sqrt{7}$
3 $\frac{5}{2}$ $\frac{3}{2}$
3 $\frac{5}{2}$ $\frac{3}{2}$ 0 000+ $+\sqrt{5}/2\sqrt{2.7}$
3 $\frac{5}{2}$ $\frac{3}{2}$ 1 000− $-3\sqrt{3}/2\sqrt{2.7}$
$\tilde{1}$ $\frac{5}{2}$ $\frac{3}{2}$ 0 000+ $-\sqrt{3}/\sqrt{7}$
3 $\frac{5}{2}$ $\frac{5}{2}$
3 $\frac{5}{2}$ $\frac{5}{2}$ 0 000+ $+2/\sqrt{7}$
3 $\frac{5}{2}$ $\frac{5}{2}$ 1 000− 0
$\tilde{1}$ $\frac{5}{2}$ $\frac{5}{2}$ 0 000+ $-2/\sqrt{3.7}$
$\tilde{1}$ $\frac{5}{2}$ $\frac{5}{2}$ 1 000+ $+\sqrt{5}/\sqrt{3.7}$
3 3 0
3 3 0 0 000+ $+2/\sqrt{7}$
$\tilde{1}$ $\tilde{1}$ 0 0 000+ $+\sqrt{3}/\sqrt{7}$
3 3 1
3 3 1 0 000+ $-1/\sqrt{7}$
$\tilde{1}$ 3 1 0 000+ $-\sqrt{3}/\sqrt{7}$
3 3 2
3 3 2 0 000+ $-\sqrt{3}/\sqrt{7}$
$\tilde{1}$ 3 2 0 000+ $-1/\sqrt{7}$
$\tilde{1}$ $\tilde{1}$ 2 0 000+ $-\sqrt{2}/\sqrt{7}$
3 3 3
3 3 3 0 000− 0
$\tilde{1}$ 3 3 0 000+ $-\sqrt{2}/\sqrt{7}$
$\tilde{1}$ $\tilde{1}$ $\tilde{1}$ 0 000+ $-1/\sqrt{7}$
$\frac{7}{2}$ 2 $\frac{3}{2}$
$\frac{5}{2}$ 2 $\frac{3}{2}$ 0 000− $+3/\sqrt{2.7}$
$\frac{5}{2}$ 2 $\frac{3}{2}$ 1 000+ $-\sqrt{3}/2\sqrt{7}$
$\tilde{\frac{1}{2}}$ 2 $\frac{3}{2}$ 0 000+ $-1/2$
$\frac{7}{2}$ $\frac{5}{2}$ 1
$\frac{5}{2}$ $\frac{5}{2}$ 1 0 000− $+3\sqrt{3}/2\sqrt{2.7}$
$\frac{5}{2}$ $\frac{5}{2}$ 1 1 000− $-\sqrt{3.5}/2\sqrt{2.7}$
$\tilde{\frac{1}{2}}$ $\frac{5}{2}$ 1 0 000+ $-1/2$
$\frac{7}{2}$ $\frac{5}{2}$ 2
$\frac{5}{2}$ $\frac{5}{2}$ 2 0 000− $+1/\sqrt{2.7}$
$\frac{5}{2}$ $\frac{5}{2}$ 2 1 000+ $+\sqrt{7}/2\sqrt{3}$
$\frac{5}{2}$ $\frac{5}{2}$ 2 2 000− $-\sqrt{2}/\sqrt{3.7}$

SO_3–K 3jm Factors (cont.)

$\mathbf{\tfrac{7}{2}\ \tfrac{5}{2}\ 2}$

$\bar{\tfrac{7}{2}}\ \tfrac{5}{2}\ 2\ 0\ 000+\quad +1/2$

$\mathbf{\tfrac{7}{2}\ 3\ \tfrac{1}{2}}$

$\tfrac{5}{2}\ 3\ \tfrac{1}{2}\ 0\ 000-\quad +3/2\sqrt{7}$

$\tfrac{5}{2}\ \bar{1}\ \tfrac{1}{2}\ 0\ 000-\quad -\sqrt{3}/\sqrt{7}$

$\bar{\tfrac{7}{2}}\ 3\ \tfrac{1}{2}\ 0\ 000+\quad +1/2$

$\mathbf{\tfrac{7}{2}\ 3\ \tfrac{3}{2}}$

$\tfrac{5}{2}\ 3\ \tfrac{3}{2}\ 0\ 000-\quad +\sqrt{3}/\sqrt{2.7}$

$\tfrac{5}{2}\ 3\ \tfrac{3}{2}\ 1\ 000+\quad -\sqrt{5}/\sqrt{2.7}$

$\tfrac{5}{2}\ \bar{1}\ \tfrac{3}{2}\ 0\ 000-\quad -\sqrt{5}/2\sqrt{7}$

$\bar{\tfrac{7}{2}}\ \bar{1}\ \tfrac{3}{2}\ 0\ 000+\quad -1/2$

$\mathbf{\tfrac{7}{2}\ 3\ \tfrac{5}{2}}$

$\tfrac{5}{2}\ 3\ \tfrac{5}{2}\ 0\ 000-\quad +\sqrt{5}/2\sqrt{2.3.7}$

$\tfrac{5}{2}\ 3\ \tfrac{5}{2}\ 1\ 000+\quad -\sqrt{7}/2\sqrt{2.3}$

$\tfrac{5}{2}\ \bar{1}\ \tfrac{5}{2}\ 0\ 000-\quad +\sqrt{5}/\sqrt{2.7}$

$\tfrac{5}{2}\ \bar{1}\ \tfrac{5}{2}\ 1\ 000-\quad +1/\sqrt{2.7}$

$\bar{\tfrac{7}{2}}\ 3\ \tfrac{5}{2}\ 0\ 000+\quad -1/2$

$\mathbf{\tfrac{7}{2}\ \tfrac{7}{2}\ 0}$

$\tfrac{5}{2}\ \tfrac{5}{2}\ 0\ 0\ 000+\quad +\sqrt{3}/2$

$\bar{\tfrac{7}{2}}\ \bar{\tfrac{7}{2}}\ 0\ 0\ 000+\quad +1/2$

$\mathbf{\tfrac{7}{2}\ \tfrac{7}{2}\ 1}$

$\tfrac{5}{2}\ \tfrac{5}{2}\ 1\ 0\ 000+\quad -1/\sqrt{2.3.7}$

$\tfrac{5}{2}\ \tfrac{5}{2}\ 1\ 1\ 000+\quad -\sqrt{2.5}/\sqrt{3.7}$

$\bar{\tfrac{7}{2}}\ \tfrac{5}{2}\ 1\ 0\ 000-\quad +1/2$

$\mathbf{\tfrac{7}{2}\ \tfrac{7}{2}\ 2}$

$\tfrac{5}{2}\ \tfrac{5}{2}\ 2\ 0\ 000+\quad -5/2\sqrt{2.7}$

$\tfrac{5}{2}\ \tfrac{5}{2}\ 2\ 1\ 000-\quad 0$

$\tfrac{5}{2}\ \tfrac{5}{2}\ 2\ 2\ 000+\quad +\sqrt{3}/2\sqrt{2.7}$

$\bar{\tfrac{7}{2}}\ \tfrac{5}{2}\ 2\ 0\ 000-\quad +1/2$

$\mathbf{\tfrac{7}{2}\ \tfrac{7}{2}\ 3}$

$\tfrac{5}{2}\ \tfrac{5}{2}\ 3\ 0\ 000+\quad -\sqrt{2.3.5}/\sqrt{7.11}$

$\tfrac{5}{2}\ \tfrac{5}{2}\ 3\ 1\ 000-\quad 0$

$\tfrac{5}{2}\ \tfrac{5}{2}\ \bar{1}\ 0\ 000+\quad -3\sqrt{5}/2\sqrt{2.7.11}$

$\tfrac{5}{2}\ \tfrac{5}{2}\ \bar{1}\ 1\ 000+\quad +\sqrt{11}/2\sqrt{2.7}$

$\bar{\tfrac{7}{2}}\ \tfrac{5}{2}\ 3\ 0\ 000-\quad +1/\sqrt{11}$

$\bar{\tfrac{7}{2}}\ \bar{\tfrac{7}{2}}\ \bar{1}\ 0\ 000+\quad -\sqrt{7}/2\sqrt{11}$

$\mathbf{4\ 2\ 2}$

$2\ 2\ 2\ 0\ 000+\quad +\sqrt{5}/\sqrt{2.7}$

$2\ 2\ 2\ 1\ 000+\quad +5/3\sqrt{2.7}$

$3\ 2\ 2\ 0\ 000-\quad 0$

$3\ 2\ 2\ 1\ 000+\quad +2/3$

$\mathbf{4\ \tfrac{5}{2}\ \tfrac{3}{2}}$

$2\ \tfrac{5}{2}\ \tfrac{3}{2}\ 0\ 000+\quad +\sqrt{2.5}/\sqrt{3.7}$

$2\ \tfrac{5}{2}\ \tfrac{3}{2}\ 1\ 000-\quad +\sqrt{5}/3\sqrt{7}$

$3\ \tfrac{5}{2}\ \tfrac{3}{2}\ 0\ 000-\quad +\sqrt{3}/2\sqrt{2}$

$3\ \tfrac{5}{2}\ \tfrac{3}{2}\ 1\ 000+\quad +\sqrt{5}/2.3\sqrt{2}$

$\mathbf{4\ \tfrac{5}{2}\ \tfrac{5}{2}}$

$2\ \tfrac{5}{2}\ \tfrac{5}{2}\ 0\ 000+\quad +\sqrt{5}/\sqrt{3.7}$

$2\ \tfrac{5}{2}\ \tfrac{5}{2}\ 1\ 000-\quad 0$

$2\ \tfrac{5}{2}\ \tfrac{5}{2}\ 2\ 000+\quad -2\sqrt{5}/3\sqrt{7}$

$\mathbf{4\ \tfrac{5}{2}\ \tfrac{5}{2}}$

$3\ \tfrac{5}{2}\ \tfrac{5}{2}\ 0\ 000-\quad 0$

$3\ \tfrac{5}{2}\ \tfrac{5}{2}\ 1\ 000+\quad -2/3$

$\mathbf{4\ 3\ 1}$

$2\ 3\ 1\ 0\ 000+\quad +\sqrt{5}/\sqrt{3.7}$

$2\ \bar{1}\ 1\ 0\ 000+\quad -2\sqrt{5}/3\sqrt{7}$

$3\ 3\ 1\ 0\ 000-\quad -1/\sqrt{3}$

$3\ \bar{1}\ 1\ 0\ 000-\quad +1/3$

$\mathbf{4\ 3\ 2}$

$2\ 3\ 2\ 0\ 000+\quad -1/\sqrt{2.7}$

$2\ 3\ 2\ 1\ 000-\quad -\sqrt{7}/3\sqrt{2}$

$2\ \bar{1}\ 2\ 0\ 000+\quad +\sqrt{2}/\sqrt{3.7}$

$3\ 3\ 2\ 0\ 000-\quad +1/3$

$3\ \bar{1}\ 2\ 0\ 000-\quad -1/\sqrt{3}$

$\mathbf{4\ 3\ 3}$

$2\ 3\ 3\ 0\ 000+\quad +\sqrt{2.5}/3\sqrt{7.11}$

$2\ \bar{1}\ 3\ 0\ 000+\quad -2\sqrt{2.5}/\sqrt{3.7.11}$

$2\ \bar{1}\ \bar{1}\ 0\ 000+\quad +\sqrt{3.5}/\sqrt{7.11}$

$3\ 3\ 3\ 0\ 000+\quad +4\sqrt{2}/3\sqrt{11}$

$3\ \bar{1}\ 3\ 0\ 000-\quad +\sqrt{2}/\sqrt{3.11}$

$\mathbf{4\ \tfrac{7}{2}\ \tfrac{1}{2}}$

$2\ \tfrac{5}{2}\ \tfrac{1}{2}\ 0\ 000-\quad +\sqrt{5}/3$

$3\ \tfrac{5}{2}\ \tfrac{1}{2}\ 0\ 000+\quad +\sqrt{7}/2.3$

$3\ \bar{\tfrac{7}{2}}\ \tfrac{1}{2}\ 0\ 000-\quad +1/2$

$\mathbf{4\ \tfrac{7}{2}\ \tfrac{3}{2}}$

$2\ \tfrac{5}{2}\ \tfrac{3}{2}\ 0\ 000-\quad +5/3\sqrt{2.7}$

$2\ \tfrac{5}{2}\ \tfrac{3}{2}\ 1\ 000+\quad -\sqrt{3}/2\sqrt{7}$

$2\ \bar{\tfrac{7}{2}}\ \tfrac{3}{2}\ 0\ 000+\quad -1/2$

$3\ \tfrac{5}{2}\ \tfrac{3}{2}\ 0\ 000+\quad -\sqrt{5}/3\sqrt{2}$

$3\ \tfrac{5}{2}\ \tfrac{3}{2}\ 1\ 000\quad +1/\sqrt{2.3}$

$\mathbf{4\ \tfrac{7}{2}\ \tfrac{5}{2}}$

$2\ \tfrac{5}{2}\ \tfrac{5}{2}\ 0\ 000-\quad -5/\sqrt{2.7.11}$

$2\ \tfrac{5}{2}\ \tfrac{5}{2}\ 1\ 000+\quad +\sqrt{7}/\sqrt{3.11}$

$2\ \tfrac{5}{2}\ \tfrac{5}{2}\ 2\ 000-\quad +\sqrt{3}/\sqrt{2.7.11}$

$2\ \bar{\tfrac{7}{2}}\ \tfrac{5}{2}\ 0\ 000+\quad -4/3\sqrt{11}$

$3\ \tfrac{5}{2}\ \tfrac{5}{2}\ 0\ 000+\quad +7/2\sqrt{2.3.11}$

$3\ \tfrac{5}{2}\ \tfrac{5}{2}\ 1\ 000-\quad -\sqrt{3.5}/2\sqrt{2.11}$

$3\ \bar{\tfrac{7}{2}}\ \tfrac{5}{2}\ 0\ 000-\quad +\sqrt{5.7}/2.3\sqrt{11}$

$\mathbf{4\ \tfrac{7}{2}\ \tfrac{7}{2}}$

$2\ \tfrac{5}{2}\ \tfrac{5}{2}\ 0\ 000+\quad +3\sqrt{3.5}/2\sqrt{2.7.11}$

$2\ \tfrac{5}{2}\ \tfrac{5}{2}\ 1\ 000-\quad 0$

$2\ \tfrac{5}{2}\ \tfrac{5}{2}\ 2\ 000+\quad +17\sqrt{5}/2.3\sqrt{2.7.11}$

$2\ \bar{\tfrac{7}{2}}\ \tfrac{5}{2}\ 0\ 000-\quad +\sqrt{5}/2\sqrt{3.11}$

$3\ \tfrac{5}{2}\ \tfrac{5}{2}\ 0\ 000-\quad 0$

$3\ \tfrac{5}{2}\ \tfrac{5}{2}\ 1\ 000+\quad -\sqrt{2}/3\sqrt{11}$

$3\ \bar{\tfrac{7}{2}}\ \tfrac{5}{2}\ 0\ 000+\quad +\sqrt{7}/\sqrt{3.11}$

$\mathbf{4\ 4\ 0}$

$2\ 2\ 0\ 0\ 000+\quad +\sqrt{5}/3$

$3\ 3\ 0\ 0\ 000+\quad +2/3$

SO_3–K 3jm Factors (cont.)

4 4 1
2 2 1 0 000+ $+2\sqrt{2}/3\sqrt{3}$
3 2 1 0 000− $-\sqrt{7}/3\sqrt{3}$
3 3 1 0 000+ $+\sqrt{5}/3\sqrt{3}$
4 4 2
2 2 2 0 000+ $-5\sqrt{5}/3\sqrt{7.11}$
2 2 2 1 000+ $+1/3\sqrt{7.11}$
3 2 2 0 000− $+7/3\sqrt{2.11}$
3 2 2 1 000+ $+5/3\sqrt{2.11}$
3 3 2 0 000+ $+\sqrt{7}/3\sqrt{11}$
4 4 3
2 2 3 0 000+ $-4/\sqrt{7.11}$
2 2 3 1 000− 0
2 2 $\tilde{1}$ 0 000+ $+17/3\sqrt{3.7.11}$
3 2 3 0 000− $+\sqrt{2}/\sqrt{11}$
3 2 $\tilde{1}$ 0 000− $-2\sqrt{2}/3\sqrt{3.11}$
3 3 3 0 000− 0
3 3 $\tilde{1}$ 0 000+ $-\sqrt{2.5.7}/3\sqrt{3.11}$
4 4 4
2 2 2 0 000+ $+3.5/\sqrt{2.7.11.13}$
2 2 2 1 000+ $-5.19\sqrt{5}/9\sqrt{2.7.11.13}$
3 2 2 0 000− 0
3 2 2 1 000+ $+4\sqrt{5}/9\sqrt{11.13}$
3 3 2 0 000+ $+5\sqrt{2.5.7}/9\sqrt{11.13}$
3 3 3 0 000+ $-4.7\sqrt{2}/9\sqrt{11.13}$
$\frac{9}{2}$ $\frac{5}{2}$ 2
$\frac{3}{2}$ $\frac{5}{2}$ 2 0 000− $-2\sqrt{3}/\sqrt{5.7}$
$\frac{3}{2}$ $\frac{5}{2}$ 2 1 000+ $+\sqrt{2}/\sqrt{5.7}$
$\frac{5}{2}$ $\frac{5}{2}$ 2 0 000+ $+1/\sqrt{5}$
$\frac{5}{2}$ $\frac{5}{2}$ 2 1 000− $-\sqrt{2}/\sqrt{3.5}$
$\frac{5}{2}$ $\frac{5}{2}$ 2 2 000+ $-2/\sqrt{3.5}$
$\frac{9}{2}$ 3 $\frac{3}{2}$
$\frac{3}{2}$ 3 $\frac{3}{2}$ 0 000− $-\sqrt{2.3}/\sqrt{5.7}$
$\frac{3}{2}$ $\tilde{1}$ $\frac{3}{2}$ 0 000− $+2\sqrt{2}/\sqrt{5.7}$
$\frac{5}{2}$ 3 $\frac{3}{2}$ 0 000+ $+\sqrt{3}/2\sqrt{2}$
$\frac{5}{2}$ 3 $\frac{3}{2}$ 1 000− $+1/2\sqrt{2.5}$
$\frac{5}{2}$ $\tilde{1}$ $\frac{3}{2}$ 0 000+ $+1/\sqrt{5}$
$\frac{9}{2}$ 3 $\frac{5}{2}$
$\frac{3}{2}$ 3 $\frac{5}{2}$ 0 000− $-3\sqrt{3}/\sqrt{7.11}$
$\frac{3}{2}$ 3 $\frac{5}{2}$ 1 000+ $-1/\sqrt{5.7.11}$
$\frac{3}{2}$ $\tilde{1}$ $\frac{5}{2}$ 0 000− $-3\sqrt{2}/\sqrt{5.7.11}$
$\frac{5}{2}$ 3 $\frac{5}{2}$ 0 000+ $-4/\sqrt{3.5.11}$
$\frac{5}{2}$ 3 $\frac{5}{2}$ 1 000− $+2/\sqrt{3.11}$
$\frac{5}{2}$ $\tilde{1}$ $\frac{5}{2}$ 0 000+ $+1/\sqrt{5.11}$
$\frac{5}{2}$ $\tilde{1}$ $\frac{5}{2}$ 1 000+ $+2/\sqrt{11}$
$\frac{9}{2}$ $\frac{7}{2}$ 1
$\frac{3}{2}$ $\frac{5}{2}$ 1 0 000+ $-\sqrt{2}/\sqrt{5}$
$\frac{5}{2}$ $\frac{5}{2}$ 1 0 000− $+\sqrt{7}/2\sqrt{2.3}$
$\frac{5}{2}$ $\frac{5}{2}$ 1 1 000− $+\sqrt{7}/2\sqrt{2.3.5}$
$\frac{5}{2}$ $\tilde{\frac{1}{2}}$ 1 0 000+ $-1/2$
$\frac{9}{2}$ $\frac{7}{2}$ 2
$\frac{3}{2}$ $\frac{5}{2}$ 2 0 000+ $-\sqrt{2.3}/\sqrt{5.7.11}$
$\frac{3}{2}$ $\frac{5}{2}$ 2 1 000− $+8/\sqrt{5.7.11}$
$\frac{3}{2}$ $\tilde{\frac{1}{2}}$ 2 0 000− $-2\sqrt{3}/\sqrt{5.11}$
$\frac{5}{2}$ $\frac{5}{2}$ 2 0 000− $+1/\sqrt{2.5.11}$
$\frac{5}{2}$ $\frac{5}{2}$ 2 1 000+ $-3\sqrt{3}/2\sqrt{5.11}$
$\frac{5}{2}$ $\frac{5}{2}$ 2 2 000− $+2\sqrt{2.3}/\sqrt{5.11}$
$\frac{5}{2}$ $\tilde{\frac{1}{2}}$ 2 0 000+ $+\sqrt{7}/2\sqrt{5.11}$
$\frac{9}{2}$ $\frac{7}{2}$ 3
$\frac{3}{2}$ $\frac{5}{2}$ 3 0 000+ $-\sqrt{2.3}/\sqrt{7.11}$
$\frac{3}{2}$ $\frac{5}{2}$ 3 1 000− $-2\sqrt{2}/\sqrt{5.7.11}$
$\frac{3}{2}$ $\frac{5}{2}$ $\tilde{1}$ 0 000+ $-9/\sqrt{5.7.11}$
$\frac{3}{2}$ $\tilde{\frac{1}{2}}$ $\tilde{1}$ 0 000− $+1/\sqrt{11}$
$\frac{5}{2}$ $\frac{5}{2}$ 3 0 000− $+\sqrt{3}/2\sqrt{2.5.11}$
$\frac{5}{2}$ $\frac{5}{2}$ 3 1 000+ $-3\sqrt{3}/2\sqrt{2.11}$
$\frac{5}{2}$ $\frac{5}{2}$ $\tilde{1}$ 0 000− $-3/\sqrt{2.5.11}$
$\frac{5}{2}$ $\frac{5}{2}$ $\tilde{1}$ 1 000− $+1/\sqrt{2.11}$
$\frac{5}{2}$ $\tilde{\frac{1}{2}}$ 3 0 000+ $+\sqrt{7}/2\sqrt{11}$
$\frac{9}{2}$ 4 $\frac{1}{2}$
$\frac{3}{2}$ 2 $\frac{1}{2}$ 0 000− $+\sqrt{2}/\sqrt{5}$
$\frac{5}{2}$ 2 $\frac{1}{2}$ 0 000+ $+\sqrt{7}/3\sqrt{5}$
$\frac{5}{2}$ 3 $\frac{1}{2}$ 0 000− $+2/3$
$\frac{9}{2}$ 4 $\frac{3}{2}$
$\frac{3}{2}$ 2 $\frac{3}{2}$ 0 000− $+2\sqrt{2}/\sqrt{5.11}$
$\frac{3}{2}$ 3 $\frac{3}{2}$ 0 000+ $-\sqrt{2.7}/\sqrt{5.11}$
$\frac{5}{2}$ 2 $\frac{3}{2}$ 0 000+ $-\sqrt{2.7}/3\sqrt{5.11}$
$\frac{5}{2}$ 2 $\frac{3}{2}$ 1 000− $+\sqrt{3.7}/\sqrt{5.11}$
$\frac{5}{2}$ 3 $\frac{3}{2}$ 0 000− $-\sqrt{11}/2.3\sqrt{2}$
$\frac{5}{2}$ 3 $\frac{3}{2}$ 1 000+ $+7/2\sqrt{2.3.5.11}$
$\frac{9}{2}$ 4 $\frac{5}{2}$
$\frac{3}{2}$ 2 $\frac{5}{2}$ 0 000− $-4\sqrt{3}/\sqrt{5.7.11}$
$\frac{3}{2}$ 2 $\frac{5}{2}$ 1 000+ $+\sqrt{2}/3\sqrt{5.7.11}$
$\frac{3}{2}$ 3 $\frac{5}{2}$ 0 000+ $+1/\sqrt{3.11}$
$\frac{3}{2}$ 3 $\frac{5}{2}$ 1 000− $+\sqrt{11}/3\sqrt{5}$
$\frac{5}{2}$ 2 $\frac{5}{2}$ 0 000+ $-2/\sqrt{5.11}$
$\frac{5}{2}$ 2 $\frac{5}{2}$ 1 000− $+\sqrt{2.5}/\sqrt{3.11}$
$\frac{5}{2}$ 2 $\frac{5}{2}$ 2 000+ $-\sqrt{3}/\sqrt{5.11}$
$\frac{5}{2}$ 3 $\frac{5}{2}$ 0 000− $-2\sqrt{7}/\sqrt{3.5.11}$
$\frac{5}{2}$ 3 $\frac{5}{2}$ 1 000+ 0
$\frac{9}{2}$ 4 $\frac{7}{2}$
$\frac{3}{2}$ 2 $\frac{5}{2}$ 0 000+ $+9\sqrt{2}/\sqrt{5.7.11.13}$
$\frac{3}{2}$ 2 $\frac{5}{2}$ 1 000− $-61/\sqrt{3.5.7.11.13}$
$\frac{3}{2}$ 2 $\tilde{\frac{1}{2}}$ 0 000− $+3/\sqrt{5.11.13}$
$\frac{3}{2}$ 3 $\frac{5}{2}$ 0 000− $+\sqrt{2}/\sqrt{11.13}$
$\frac{3}{2}$ 3 $\frac{5}{2}$ 1 000+ $-2\sqrt{2.5}/\sqrt{3.11.13}$
$\frac{5}{2}$ 2 $\frac{5}{2}$ 0 000− $+3\sqrt{3}/\sqrt{2.5.11.13}$
$\frac{5}{2}$ 2 $\frac{5}{2}$ 1 000+ $+3/\sqrt{5.11.13}$
$\frac{5}{2}$ 2 $\frac{5}{2}$ 2 000− $+17/3\sqrt{2.5.11.13}$
$\frac{5}{2}$ 2 $\tilde{\frac{1}{2}}$ 0 000+ $-8\sqrt{7}/\sqrt{3.5.11.13}$
$\frac{5}{2}$ 3 $\frac{5}{2}$ 0 000+ $-3\sqrt{5.7}/2\sqrt{2.11.13}$

SO_3–K 3jm Factors (cont.)

$\frac{9}{2}$ 4 $\frac{7}{2}$
$\frac{5}{2}$ 3 $\frac{5}{2}$ 1 000− $-7\sqrt{7}/2.3\sqrt{2.11.13}$
$\frac{5}{2}$ 3 $\tilde{\frac{1}{2}}$ 0 000− $-7/2\sqrt{3.11.13}$
$\frac{9}{2}$ $\frac{9}{2}$ 0
$\frac{3}{2}$ $\frac{3}{2}$ 0 0 000+ $+\sqrt{2}/\sqrt{5}$
$\frac{5}{2}$ $\frac{5}{2}$ 0 0 000+ $+\sqrt{3}/\sqrt{5}$
$\frac{9}{2}$ $\frac{9}{2}$ 1
$\frac{3}{2}$ $\frac{3}{2}$ 1 0 000+ $-3\sqrt{2.3}/5\sqrt{11}$
$\frac{5}{2}$ $\frac{3}{2}$ 1 0 000− $+2\sqrt{2.7}/5\sqrt{11}$
$\frac{5}{2}$ $\frac{5}{2}$ 1 0 000+ $-\sqrt{11}/\sqrt{2.3.5}$
$\frac{5}{2}$ $\frac{5}{2}$ 1 1 000+ $+7/5\sqrt{2.3.11}$
$\frac{9}{2}$ $\frac{9}{2}$ 2
$\frac{3}{2}$ $\frac{3}{2}$ 2 0 000+ $-2\sqrt{3}/5\sqrt{11}$
$\frac{5}{2}$ $\frac{3}{2}$ 2 0 000− $-2\sqrt{3.7}/5\sqrt{11}$
$\frac{5}{2}$ $\frac{3}{2}$ 2 1 000+ $+\sqrt{2.7}/5\sqrt{11}$
$\frac{5}{2}$ $\frac{5}{2}$ 2 0 000+ $+8/5\sqrt{11}$
$\frac{5}{2}$ $\frac{5}{2}$ 2 1 000− 0
$\frac{5}{2}$ $\frac{5}{2}$ 2 2 000+ $-\sqrt{3}/5\sqrt{11}$
$\frac{9}{2}$ $\frac{9}{2}$ 3
$\frac{3}{2}$ $\frac{3}{2}$ 3 0 000+ $+8.3\sqrt{3}/5\sqrt{7.11.13}$
$\frac{3}{2}$ $\frac{3}{2}$ $\tilde{1}$ 0 000+ $-2.17/5\sqrt{7.11.13}$
$\frac{5}{2}$ $\frac{3}{2}$ 3 0 000− $-\sqrt{3}/\sqrt{5.11.13}$
$\frac{5}{2}$ $\frac{3}{2}$ 3 1 000+ $+29/5\sqrt{11.13}$
$\frac{5}{2}$ $\frac{3}{2}$ $\tilde{1}$ 0 000− $-9\sqrt{2}/5\sqrt{11.13}$
$\frac{5}{2}$ $\frac{5}{2}$ 3 0 000+ $+2\sqrt{3.7}/5\sqrt{11.13}$
$\frac{5}{2}$ $\frac{5}{2}$ 3 1 000− 0
$\frac{5}{2}$ $\frac{5}{2}$ $\tilde{1}$ 0 000+ $-4.3\sqrt{7}/5\sqrt{11.13}$
$\frac{5}{2}$ $\frac{5}{2}$ $\tilde{1}$ 1 000+ $+\sqrt{7}/\sqrt{5.11.13}$
$\frac{9}{2}$ $\frac{9}{2}$ 4
$\frac{3}{2}$ $\frac{3}{2}$ 2 0 000+ $+2.3/\sqrt{11.13}$
$\frac{3}{2}$ $\frac{3}{2}$ 3 0 000− 0
$\frac{5}{2}$ $\frac{3}{2}$ 2 0 000− 0
$\frac{5}{2}$ $\frac{3}{2}$ 2 1 000+ $-\sqrt{2.7}/\sqrt{3.11.13}$
$\frac{5}{2}$ $\frac{3}{2}$ 3 0 000+ $+1/\sqrt{5.11.13}$
$\frac{5}{2}$ $\frac{3}{2}$ 3 1 000− $+7/\sqrt{3.11.13}$
$\frac{5}{2}$ $\frac{5}{2}$ 2 0 000+ $+3\sqrt{3}/\sqrt{11.13}$
$\frac{5}{2}$ $\frac{5}{2}$ 2 1 000− 0
$\frac{5}{2}$ $\frac{5}{2}$ 2 2 000+ $+8/3\sqrt{11.13}$
$\frac{5}{2}$ $\frac{5}{2}$ 3 0 000− 0
$\frac{5}{2}$ $\frac{5}{2}$ 3 1 000+ $+2.7\sqrt{7}/3\sqrt{5.11.13}$
5 $\frac{5}{2}$ $\frac{5}{2}$
1 $\frac{5}{2}$ $\frac{5}{2}$ 0 000+ $-3\sqrt{3}/\sqrt{2.7.11}$
1 $\frac{5}{2}$ $\frac{5}{2}$ 1 000+ $+\sqrt{3.5}/\sqrt{2.7.11}$
2 $\frac{5}{2}$ $\frac{5}{2}$ 0 000− 0
2 $\frac{5}{2}$ $\frac{5}{2}$ 1 000+ $-\sqrt{5}/\sqrt{11}$
2 $\frac{5}{2}$ $\frac{5}{2}$ 2 000− 0
$\tilde{1}$ $\frac{5}{2}$ $\frac{5}{2}$ 0 000+ $+\sqrt{5}/\sqrt{3.11}$
$\tilde{1}$ $\frac{5}{2}$ $\frac{5}{2}$ 1 000+ $+2/\sqrt{3.11}$
5 3 2
1 3 2 0 000+ $+3/\sqrt{7.11}$
1 $\tilde{1}$ 2 0 000+ $-2\sqrt{3}/\sqrt{7.11}$
2 3 2 0 000− $+\sqrt{3}/\sqrt{2.11}$
2 3 2 1 000+ $-\sqrt{3}/\sqrt{2.11}$
2 $\tilde{1}$ 2 0 000− $-\sqrt{2}/\sqrt{11}$
$\tilde{1}$ 3 2 0 000+ $-\sqrt{2}/\sqrt{11}$
$\tilde{1}$ $\tilde{1}$ 2 0 000+ $+1/\sqrt{11}$
5 3 3
1 3 3 0 000+ $+3\sqrt{2}/\sqrt{7.11}$
1 $\tilde{1}$ 3 0 000+ $-\sqrt{3}/\sqrt{2.7.11}$
2 3 3 0 000− 0
2 $\tilde{1}$ 3 0 000− $-\sqrt{5}/\sqrt{2.11}$
2 $\tilde{1}$ $\tilde{1}$ 0 000− 0
$\tilde{1}$ 3 3 0 000+ $+1/\sqrt{11}$
$\tilde{1}$ $\tilde{1}$ $\tilde{1}$ 0 000+ $-\sqrt{2}/\sqrt{11}$
5 $\frac{7}{2}$ $\frac{3}{2}$
1 $\frac{5}{2}$ $\frac{3}{2}$ 0 000− $-\sqrt{3}/\sqrt{11}$
2 $\frac{5}{2}$ $\frac{3}{2}$ 0 000+ 0
2 $\frac{5}{2}$ $\frac{3}{2}$ 1 000− $+\sqrt{7}/\sqrt{2.11}$
2 $\tilde{\frac{1}{2}}$ $\frac{3}{2}$ 0 000− $-\sqrt{3}/\sqrt{2.11}$
$\tilde{1}$ $\frac{5}{2}$ $\frac{3}{2}$ 0 000− $-\sqrt{7}/2\sqrt{11}$
$\tilde{1}$ $\tilde{\frac{1}{2}}$ $\frac{3}{2}$ 0 000+ $-\sqrt{5}/2\sqrt{11}$
5 $\frac{7}{2}$ $\frac{5}{2}$
1 $\frac{5}{2}$ $\frac{5}{2}$ 0 000− $-5\sqrt{5}/4\sqrt{7.11}$
1 $\frac{5}{2}$ $\frac{5}{2}$ 1 000− $-1/4\sqrt{7.11}$
1 $\tilde{\frac{1}{2}}$ $\frac{5}{2}$ 0 000+ $-\sqrt{3.5}/2\sqrt{2.11}$
2 $\frac{5}{2}$ $\frac{5}{2}$ 0 000+ 0
2 $\frac{5}{2}$ $\frac{5}{2}$ 1 000− $+1/2\sqrt{2.3.11}$
2 $\frac{5}{2}$ $\frac{5}{2}$ 2 000+ $+7/2\sqrt{3.11}$
2 $\tilde{\frac{1}{2}}$ $\frac{5}{2}$ 0 000− $+\sqrt{7}/2\sqrt{2.11}$
$\tilde{1}$ $\frac{5}{2}$ $\frac{5}{2}$ 0 000− $+1/\sqrt{2.11}$
$\tilde{1}$ $\frac{5}{2}$ $\frac{5}{2}$ 1 000− $-\sqrt{5}/\sqrt{2.11}$
5 $\frac{7}{2}$ $\frac{7}{2}$
1 $\frac{5}{2}$ $\frac{5}{2}$ 0 000+ $-4.3/\sqrt{7.11.13}$
1 $\frac{5}{2}$ $\frac{5}{2}$ 1 000+ $-3\sqrt{5}/\sqrt{7.11.13}$
1 $\tilde{\frac{1}{2}}$ $\frac{5}{2}$ 0 000− $-\sqrt{2.3}/\sqrt{11.13}$
2 $\frac{5}{2}$ $\frac{5}{2}$ 0 000− 0
2 $\frac{5}{2}$ $\frac{5}{2}$ 1 000+ $+\sqrt{2.3.5}/\sqrt{11.13}$
2 $\frac{5}{2}$ $\frac{5}{2}$ 2 000− 0
2 $\tilde{\frac{1}{2}}$ $\frac{5}{2}$ 0 000+ $-\sqrt{5.7}/\sqrt{2.11.13}$
$\tilde{1}$ $\frac{5}{2}$ $\frac{5}{2}$ 0 000+ $+3\sqrt{5}/2\sqrt{2.11.13}$
$\tilde{1}$ $\frac{5}{2}$ $\frac{5}{2}$ 1 000+ $+\sqrt{13}/2\sqrt{2.11}$
$\tilde{1}$ $\tilde{\frac{1}{2}}$ $\tilde{\frac{1}{2}}$ 0 000+ $+7/2\sqrt{11.13}$
5 4 1
1 2 1 0 000+ $+\sqrt{3}/\sqrt{11}$
2 2 1 0 000− $-\sqrt{7}/\sqrt{3.11}$
2 3 1 0 000+ $+2\sqrt{2}/\sqrt{3.11}$
$\tilde{1}$ 2 1 0 000+ $+\sqrt{7}/3\sqrt{11}$
$\tilde{1}$ 3 1 0 000− $-2\sqrt{5}/3\sqrt{11}$

SO_3–K 3jm Factors (cont.)

5 4 2

1 2 2 0 000+ $+\sqrt{2}/\sqrt{3.11}$

1 3 2 0 000− $+\sqrt{7}/\sqrt{3.11}$

2 2 2 0 000− 0

2 2 2 1 000− $-2\sqrt{7}/3\sqrt{11}$

2 3 2 0 000+ $+1/\sqrt{2.11}$

2 3 2 1 000− $+5/3\sqrt{2.11}$

$\tilde{1}$ 2 2 0 000+ $-\sqrt{7}/\sqrt{3.11}$

$\tilde{1}$ 3 2 0 000− $+\sqrt{2}/\sqrt{3.11}$

5 4 3

1 2 3 0 000+ $-\sqrt{11}/\sqrt{2.3.7.13}$

1 2 $\tilde{1}$ 0 000+ $+2.3\sqrt{2}/\sqrt{7.11.13}$

1 3 3 0 000− $-\sqrt{2.5}/\sqrt{3.11.13}$

1 3 $\tilde{1}$ 0 000− $+3\sqrt{5}/\sqrt{2.11.13}$

2 2 3 0 000− $-1/2\sqrt{11.13}$

2 2 3 1 000+ $+5.5/2.3\sqrt{11.13}$

2 2 $\tilde{1}$ 0 000− $+8/\sqrt{3.11.13}$

2 3 3 0 000+ $+4\sqrt{2.7}/3\sqrt{11.13}$

2 3 $\tilde{1}$ 0 000+ $+\sqrt{7}/\sqrt{2.3.11.13}$

$\tilde{1}$ 2 3 0 000+ $-8/\sqrt{3.11.13}$

$\tilde{1}$ 2 $\tilde{1}$ 0 000+ $-\sqrt{2.3}/\sqrt{11.13}$

$\tilde{1}$ 3 3 0 000− $+\sqrt{5.7}/\sqrt{3.11.13}$

5 4 4

1 2 2 0 000+ $+4\sqrt{5}/\sqrt{3.11.13}$

1 3 2 0 000− $+\sqrt{5.7}/\sqrt{2.3.11.13}$

1 3 3 0 000+ $-\sqrt{2}/\sqrt{3.11.13}$

2 2 2 0 000− 0

2 2 2 1 000− 0

2 3 2 0 000+ $-5\sqrt{5}/2\sqrt{11.13}$

2 3 2 1 000− $+\sqrt{5}/2\sqrt{11.13}$

2 3 3 0 000− 0

$\tilde{1}$ 2 2 0 000+ $+\sqrt{2.5.7}/3\sqrt{3.11.13}$

$\tilde{1}$ 3 2 0 000− $-8\sqrt{5}/3\sqrt{3.11.13}$

$\tilde{1}$ 3 3 0 000+ $+7\sqrt{7}/3\sqrt{3.11.13}$

5 $\frac{9}{2}$ $\frac{1}{2}$

1 $\frac{3}{2}$ $\frac{1}{2}$ 0 000− $+\sqrt{3}/\sqrt{11}$

2 $\frac{3}{2}$ $\frac{1}{2}$ 0 000+ $+\sqrt{7}/\sqrt{5.11}$

2 $\frac{5}{2}$ $\frac{1}{2}$ 0 000− $+3\sqrt{2}/\sqrt{5.11}$

$\tilde{1}$ $\frac{5}{2}$ $\frac{1}{2}$ 0 000+ $+\sqrt{3}/\sqrt{11}$

5 $\frac{9}{2}$ $\frac{3}{2}$

1 $\frac{3}{2}$ $\frac{3}{2}$ 0 000− $-3/\sqrt{2.5.11}$

1 $\frac{5}{2}$ $\frac{3}{2}$ 0 000+ $-\sqrt{3.7}/\sqrt{2.5.11}$

2 $\frac{3}{2}$ $\frac{3}{2}$ 0 000+ $+\sqrt{3.7}/\sqrt{2.5.11}$

2 $\frac{5}{2}$ $\frac{3}{2}$ 0 000− $-3\sqrt{3}/\sqrt{2.5.11}$

2 $\frac{5}{2}$ $\frac{3}{2}$ 1 000+ $-1/\sqrt{5.11}$

$\tilde{1}$ $\frac{3}{2}$ $\frac{3}{2}$ 0 000− $+\sqrt{7}/\sqrt{5.11}$

$\tilde{1}$ $\frac{5}{2}$ $\frac{3}{2}$ 0 000+ $+2\sqrt{2}/\sqrt{5.11}$

5 $\frac{9}{2}$ $\frac{5}{2}$

1 $\frac{3}{2}$ $\frac{5}{2}$ 0 000− $-\sqrt{2.3}/\sqrt{5.11.13}$

1 $\frac{5}{2}$ $\frac{5}{2}$ 0 000+ $+\sqrt{7}/\sqrt{2.11.13}$

1 $\frac{5}{2}$ $\frac{5}{2}$ 1 000+ $-7\sqrt{7}/\sqrt{2.5.11.13}$

2 $\frac{3}{2}$ $\frac{5}{2}$ 0 000+ $+\sqrt{2.3.7}/\sqrt{5.11.13}$

2 $\frac{3}{2}$ $\frac{5}{2}$ 1 000− $-4\sqrt{7}/\sqrt{5.11.13}$

2 $\frac{5}{2}$ $\frac{5}{2}$ 0 000− $+9\sqrt{2}/\sqrt{5.11.13}$

2 $\frac{5}{2}$ $\frac{5}{2}$ 1 000+ $-\sqrt{5}/\sqrt{3.11.13}$

2 $\frac{5}{2}$ $\frac{5}{2}$ 2 000− $+\sqrt{2}/\sqrt{3.5.11.13}$

$\tilde{1}$ $\frac{3}{2}$ $\frac{5}{2}$ 0 000− $-3\sqrt{2.7}/\sqrt{5.11.13}$

$\tilde{1}$ $\frac{5}{2}$ $\frac{5}{2}$ 0 000+ $+8/\sqrt{5.11.13}$

$\tilde{1}$ $\frac{5}{2}$ $\frac{5}{2}$ 1 000+ $+1/\sqrt{11.13}$

5 $\frac{9}{2}$ $\frac{7}{2}$

1 $\frac{3}{2}$ $\frac{5}{2}$ 0 000+ $+\sqrt{3.5}/\sqrt{11.13}$

1 $\frac{5}{2}$ $\frac{5}{2}$ 0 000− $+3\sqrt{7}/4\sqrt{11.13}$

1 $\frac{5}{2}$ $\frac{5}{2}$ 1 000− $+3\sqrt{5.7}/4\sqrt{11.13}$

1 $\frac{5}{2}$ $\tilde{\frac{1}{2}}$ 0 000+ $-\sqrt{3}/2\sqrt{2.11.13}$

2 $\frac{3}{2}$ $\frac{5}{2}$ 0 000− $+\sqrt{3.7}/\sqrt{5.11.13}$

2 $\frac{3}{2}$ $\frac{5}{2}$ 1 000+ $+2\sqrt{2.7}/\sqrt{5.11.13}$

2 $\frac{3}{2}$ $\tilde{\frac{1}{2}}$ 0 000+ $+3\sqrt{2.3}/\sqrt{5.11.13}$

2 $\frac{5}{2}$ $\frac{5}{2}$ 0 000+ $+9/\sqrt{5.11.13}$

2 $\frac{5}{2}$ $\frac{5}{2}$ 1 000− $+\sqrt{3.13}/2\sqrt{2.5.11}$

2 $\frac{5}{2}$ $\frac{5}{2}$ 2 000+ $+3\sqrt{3}/2\sqrt{5.11.13}$

2 $\frac{5}{2}$ $\tilde{\frac{1}{2}}$ 0 000− $+7\sqrt{7}/2\sqrt{2.5.11.13}$

$\tilde{1}$ $\frac{3}{2}$ $\frac{5}{2}$ 0 000+ 0

$\tilde{1}$ $\frac{3}{2}$ $\tilde{\frac{1}{2}}$ 0 000− $+4/\sqrt{11.13}$

$\tilde{1}$ $\frac{5}{2}$ $\frac{5}{2}$ 0 000− $+3\sqrt{5}/\sqrt{2.11.13}$

$\tilde{1}$ $\frac{5}{2}$ $\frac{5}{2}$ 1 000− $+1/\sqrt{2.11.13}$

5 $\frac{9}{2}$ $\frac{9}{2}$

1 $\frac{3}{2}$ $\frac{3}{2}$ 0 000+ $-2.9\sqrt{2}/5\sqrt{11.13}$

1 $\frac{5}{2}$ $\frac{3}{2}$ 0 000− $-\sqrt{2.3.7}/5\sqrt{11.13}$

1 $\frac{5}{2}$ $\frac{5}{2}$ 0 000+ $+3/\sqrt{2.5.11.13}$

1 $\frac{5}{2}$ $\frac{5}{2}$ 1 000+ $-3.7/5\sqrt{2.11.13}$

2 $\frac{3}{2}$ $\frac{3}{2}$ 0 000− 0

2 $\frac{5}{2}$ $\frac{3}{2}$ 0 000+ $+\sqrt{2.3}/\sqrt{11.13}$

2 $\frac{5}{2}$ $\frac{3}{2}$ 1 000− $-4/\sqrt{11.13}$

2 $\frac{5}{2}$ $\frac{5}{2}$ 0 000− 0

2 $\frac{5}{2}$ $\frac{5}{2}$ 1 000+ $-\sqrt{3.7}/\sqrt{11.13}$

2 $\frac{5}{2}$ $\frac{5}{2}$ 2 000− 0

$\tilde{1}$ $\frac{3}{2}$ $\frac{3}{2}$ 0 000+ $-2\sqrt{7}/5\sqrt{11.13}$

$\tilde{1}$ $\frac{5}{2}$ $\frac{3}{2}$ 0 000− $-9\sqrt{2}/5\sqrt{11.13}$

$\tilde{1}$ $\frac{5}{2}$ $\frac{5}{2}$ 0 000+ $+3\sqrt{7}/5\sqrt{11.13}$

$\tilde{1}$ $\frac{5}{2}$ $\frac{5}{2}$ 1 000+ $-4\sqrt{7}/\sqrt{5.11.13}$

5 5 0

1 1 0 0 000+ $+\sqrt{3}/\sqrt{11}$

2 2 0 0 000+ $+\sqrt{5}/\sqrt{11}$

$\tilde{1}$ $\tilde{1}$ 0 0 000+ $+\sqrt{3}/\sqrt{11}$

5 5 1

1 1 1 0 000+ $+\sqrt{5}/2\sqrt{11}$

2 1 1 0 000− $-\sqrt{7}/2\sqrt{11}$

2 2 1 0 000+ $+1/2\sqrt{11}$

$\tilde{1}$ 2 1 0 000− $-\sqrt{3}/\sqrt{11}$

SO_3–K 3jm Factors (cont.)

5 5 2
1 1 2 0 000+ +3/2√11.13
2 1 2 0 000− −3√7/2√11.13
2 2 2 0 000+ +5√3.5/2√2.11.13
2 2 2 1 000+ +√3/2√2.11.13
1̃ 1 2 0 000+ +√3.7/√11.13
1̃ 2 2 0 000− +√2/√11.13
1̃ 1̃ 2 0 000+ +4/√11.13
5 5 3
2 1 3 0 000− −3√2/√11.13
2 1 1̃ 0 000− +√2.3/√11.13
2 2 3 0 000+ +2.3√3/√7.11.13
2 2 3 1 000− 0
2 2 1̃ 0 000+ +4/√7.11.13
1̃ 1 3 0 000+ +√3.5/√11.13
1̃ 2 3 0 000− +1/√7.11.13
1̃ 2 1̃ 0 000− −9√2/√7.11.13
1̃ 1̃ 1̃ 0 000+ −√5/√7.11.13
5 5 4
1 1 2 0 000+ −√3.5/√11.13
2 1 2 0 000− +√5.7/√3.11.13
2 1 3 0 000+ −√2.5/√3.11.13
2 2 2 0 000+ +5/√2.11.13
2 2 2 1 000+ −5√5/3√2.11.13
2 2 3 0 000− 0
2 2 3 1 000+ −2√5.7/3√11.13
1̃ 1 2 0 000+ 0
1̃ 1 3 0 000− +3/√11.13
1̃ 2 2 0 000− −√2.5/√3.11.13
1̃ 2 3 0 000+ +√5.7/√3.11.13
1̃ 1̃ 2 0 000+ +√3.5/√11.13
5 5 5
1 1 1 0 000+ +9/2√11.13
2 1 1 0 000− 0
2 2 1 0 000+ +3√5/2√11.13
2 2 2 0 000− 0
2 2 2 1 000− 0
1̃ 2 1 0 000− +√3.5/2√11.13
1̃ 2 2 0 000+ −√5.7/√2.11.13
1̃ 1̃ 2 0 000− 0
1̃ 1̃ 1̃ 0 000+ +√2.7/√11.13
11/2 3 5/2
1/2 3 5/2 0 000− +1/√2.7
1/2 1̃ 5/2 0 000− −√2/√3.7
3/2 3 5/2 0 000+ −1/2√2.11
3/2 3 5/2 1 000− −√3.5/2√2.11
3/2 1̃ 5/2 0 000+ +√5/√3.11
5/2 3 5/2 0 000− +√5/2√11
5/2 3 5/2 1 000+ +3/2√11
5/2 1̃ 5/2 0 000− +√5/√3.11
11/2 3 5/2
5/2 1̃ 5/2 1 000− −1/√3.11
11/2 7/2 2
1/2 5/2 2 0 000+ +1/√2.3
3/2 5/2 2 0 000− −√7/√2.3.11
3/2 5/2 2 1 000+ −√7/2√11
3/2 1̃/2 2 0 000+ −√3/2√11
5/2 5/2 2 0 000+ −√7/2√11
5/2 5/2 2 1 000− −√7/√2.3.11
5/2 5/2 2 2 000+ −√7/2√3.11
5/2 1̃/2 2 0 000− +√2/√11
11/2 7/2 3
1/2 5/2 3 0 000+ −5/√2.3.7.13
1/2 5/2 1̃ 0 000+ +1/√2.7.13
1/2 1̃/2 3 0 000− +√3/√2.13
3/2 5/2 3 0 000− −√13/√2.3.11
3/2 5/2 3 1 000+ −√5/√2.11.13
3/2 5/2 1̃ 0 000− +√5/2√11.13
3/2 1̃/2 1̃ 0 000+ −3√7/2√11.13
5/2 5/2 3 0 000+ −5√5/2√3.11.13
5/2 5/2 3 1 000− −√13/2√3.11
5/2 5/2 1̃ 0 000+ −√5/2√11.13
5/2 5/2 1̃ 1 000+ −√13/2√11
5/2 1̃/2 3 0 000− −√7/√2.11.13
11/2 4 3/2
1/2 2 3/2 0 000− +1/√2.3
3/2 2 3/2 0 000+ −√7/√3.11
3/2 3 3/2 0 000− −2/√3.11
5/2 2 3/2 0 000− −√7/√2.3.11
5/2 2 3/2 1 000+ −√7/3√11
5/2 3 3/2 0 000+ 0
5/2 3 3/2 1 000− +4√2/3√11
11/2 4 5/2
1/2 2 5/2 0 000− −√2/3√13
1/2 3 5/2 0 000+ +√5.7/3√2.13
3/2 2 5/2 0 000+ −√2.7/3√11.13
3/2 2 5/2 1 000− −√3.7/√11.13
3/2 3 5/2 0 000− +17√5/2.3√2.11.13
3/2 3 5/2 1 000+ −√11/2√2.3.13
5/2 2 5/2 0 000− +√7/√3.11.13
5/2 2 5/2 1 000+ +√2.7/√11.13
5/2 2 5/2 2 000− +7√7/3√11.13
5/2 3 5/2 0 000+ +1/2√11.13
5/2 3 5/2 1 000− −√5.11/2.3√13
11/2 4 7/2
1/2 2 5/2 0 000+ −√5/√2.3.13
1/2 3 5/2 0 000− +√7/√2.3.13
1/2 3 1̃/2 0 000+ +1/√2.3.13
3/2 2 5/2 0 000− −√5.7/√2.3.11.13
3/2 2 5/2 1 000+ −√5.7/2√11.13

SO_3–K 3jm Factors (cont.)

$\frac{11}{2}$ 4 $\frac{7}{2}$

$\frac{3}{2}$ 2 $\tilde{\frac{1}{2}}$ 0 000+ $+5\sqrt{5}/2\sqrt{3.11.13}$

$\frac{3}{2}$ 3 $\frac{5}{2}$ 0 000+ $+1/\sqrt{2.3.11.13}$

$\frac{3}{2}$ 3 $\frac{5}{2}$ 1 000− $+3\sqrt{5}/\sqrt{2.11.13}$

$\frac{5}{2}$ 2 $\frac{5}{2}$ 0 000+ $+\sqrt{5.7}/2\sqrt{11.13}$

$\frac{5}{2}$ 2 $\frac{5}{2}$ 1 000− $+\sqrt{5.7}/\sqrt{2.3.11.13}$

$\frac{5}{2}$ 2 $\frac{5}{2}$ 2 000+ $-\sqrt{3.5.7}/2\sqrt{11.13}$

$\frac{5}{2}$ 2 $\tilde{\frac{1}{2}}$ 0 000− $-2\sqrt{2.5}/3\sqrt{11.13}$

$\frac{5}{2}$ 3 $\frac{5}{2}$ 0 000− $-\sqrt{5}/2\sqrt{3.11.13}$

$\frac{5}{2}$ 3 $\frac{5}{2}$ 1 000+ $+3\sqrt{3}/2\sqrt{11.13}$

$\frac{5}{2}$ 3 $\tilde{\frac{1}{2}}$ 0 000+ $+7\sqrt{7}/3\sqrt{2.11.13}$

$\frac{11}{2}$ $\frac{9}{2}$ 1

$\frac{1}{2}$ $\frac{3}{2}$ 1 0 000+ $+1/\sqrt{2.3}$

$\frac{3}{2}$ $\frac{3}{2}$ 1 0 000− $+2\sqrt{7}/\sqrt{3.5.11}$

$\frac{3}{2}$ $\frac{5}{2}$ 1 0 000+ $-3/\sqrt{5.11}$

$\frac{5}{2}$ $\frac{3}{2}$ 1 0 000+ $-\sqrt{7}/\sqrt{2.5.11}$

$\frac{5}{2}$ $\frac{5}{2}$ 1 0 000− 0

$\frac{5}{2}$ $\frac{5}{2}$ 1 1 000− $+2\sqrt{2.3}/\sqrt{5.11}$

$\frac{11}{2}$ $\frac{9}{2}$ 2

$\frac{1}{2}$ $\frac{3}{2}$ 2 0 000+ $-\sqrt{3}/\sqrt{2.5.13}$

$\frac{1}{2}$ $\frac{5}{2}$ 2 0 000− $+2\sqrt{7}/\sqrt{3.5.13}$

$\frac{3}{2}$ $\frac{3}{2}$ 2 0 000− $+2\sqrt{3.7}/\sqrt{5.11.13}$

$\frac{3}{2}$ $\frac{5}{2}$ 2 0 000+ $+\sqrt{13}/\sqrt{3.5.11}$

$\frac{3}{2}$ $\frac{5}{2}$ 2 1 000− $-7\sqrt{2}/\sqrt{5.11.13}$

$\frac{5}{2}$ $\frac{3}{2}$ 2 0 000+ $-\sqrt{3.7}/\sqrt{2.5.11.13}$

$\frac{5}{2}$ $\frac{3}{2}$ 2 1 000− $-\sqrt{5.7}/\sqrt{11.13}$

$\frac{5}{2}$ $\frac{5}{2}$ 2 0 000− $+2\sqrt{2}/\sqrt{5.11.13}$

$\frac{5}{2}$ $\frac{5}{2}$ 2 1 000+ $+2\sqrt{11}/\sqrt{3.5.13}$

$\frac{5}{2}$ $\frac{5}{2}$ 2 2 000− $+2\sqrt{2}/\sqrt{3.5.11.13}$

$\frac{11}{2}$ $\frac{9}{2}$ 3

$\frac{1}{2}$ $\frac{5}{2}$ 3 0 000− $+1/\sqrt{2.3.13}$

$\frac{1}{2}$ $\frac{5}{2}$ $\tilde{1}$ 0 000− $-\sqrt{2}/\sqrt{13}$

$\frac{3}{2}$ $\frac{3}{2}$ 3 0 000− $+\sqrt{2.3}/\sqrt{5.11.13}$

$\frac{3}{2}$ $\frac{3}{2}$ $\tilde{1}$ 0 000− $-2.3\sqrt{2}/\sqrt{5.11.13}$

$\frac{3}{2}$ $\frac{5}{2}$ 3 0 000+ $+31/2\sqrt{2.3.7.11.13}$

$\frac{3}{2}$ $\frac{5}{2}$ 3 1 000− $-67/2\sqrt{2.5.7.11.13}$

$\frac{3}{2}$ $\frac{5}{2}$ $\tilde{1}$ 0 000+ $+19/\sqrt{5.7.11.13}$

$\frac{5}{2}$ $\frac{3}{2}$ 3 0 000+ $+3\sqrt{3}/\sqrt{11.13}$

$\frac{5}{2}$ $\frac{3}{2}$ 3 1 000− $+1/\sqrt{5.11.13}$

$\frac{5}{2}$ $\frac{3}{2}$ $\tilde{1}$ 0 000+ $-2.3\sqrt{2}/\sqrt{5.11.13}$

$\frac{5}{2}$ $\frac{5}{2}$ 3 0 000− $-109/2\sqrt{3.5.7.11.13}$

$\frac{5}{2}$ $\frac{5}{2}$ 3 1 000+ $+\sqrt{11}/2\sqrt{3.7.13}$

$\frac{5}{2}$ $\frac{5}{2}$ $\tilde{1}$ 0 000− $-1/\sqrt{5.7.11.13}$

$\frac{5}{2}$ $\frac{5}{2}$ $\tilde{1}$ 1 000− $+1/\sqrt{7.11.13}$

$\frac{11}{2}$ $\frac{9}{2}$ 4

$\frac{1}{2}$ $\frac{3}{2}$ 2 0 000+ $-4/\sqrt{3.5.13}$

$\frac{1}{2}$ $\frac{5}{2}$ 2 0 000− $+\sqrt{2.7}/\sqrt{3.5.13}$

$\frac{1}{2}$ $\frac{5}{2}$ 3 0 000+ $+1/\sqrt{2.3.13}$

$\frac{3}{2}$ $\frac{3}{2}$ 2 0 000− $+2\sqrt{2.7}/\sqrt{3.5.11.13}$

$\frac{3}{2}$ $\frac{3}{2}$ 3 0 000+ $+\sqrt{2.11}/\sqrt{3.5.13}$

$\frac{11}{2}$ $\frac{9}{2}$ 4

$\frac{3}{2}$ $\frac{5}{2}$ 2 0 000+ $+\sqrt{2.11}/\sqrt{3.5.13}$

$\frac{3}{2}$ $\frac{5}{2}$ 2 1 000− $+7/\sqrt{5.11.13}$

$\frac{3}{2}$ $\frac{5}{2}$ 3 0 000− $+\sqrt{7}/2\sqrt{2.3.11.13}$

$\frac{3}{2}$ $\frac{5}{2}$ 3 1 000+ $+3\sqrt{7}/2\sqrt{2.5.11.13}$

$\frac{5}{2}$ $\frac{3}{2}$ 2 0 000+ 0

$\frac{5}{2}$ $\frac{3}{2}$ 2 1 000− $+2\sqrt{2.7}/3\sqrt{5.11.13}$

$\frac{5}{2}$ $\frac{3}{2}$ 3 0 000− $-7/\sqrt{3.11.13}$

$\frac{5}{2}$ $\frac{3}{2}$ 3 1 000+ $+19/3\sqrt{5.11.13}$

$\frac{5}{2}$ $\frac{5}{2}$ 2 0 000− $+\sqrt{5}/\sqrt{11.13}$

$\frac{5}{2}$ $\frac{5}{2}$ 2 1 000+ $-\sqrt{2.11}/\sqrt{3.5.13}$

$\frac{5}{2}$ $\frac{5}{2}$ 2 2 000− $-3\sqrt{3}/\sqrt{5.11.13}$

$\frac{5}{2}$ $\frac{5}{2}$ 3 0 000+ $-\sqrt{7.11}/2\sqrt{3.5.13}$

$\frac{5}{2}$ $\frac{5}{2}$ 3 1 000− $-\sqrt{3.7}/2\sqrt{11.13}$

$\frac{11}{2}$ 5 $\frac{1}{2}$

$\frac{1}{2}$ 1 $\frac{1}{2}$ 0 000− $+1/\sqrt{2.3}$

$\frac{3}{2}$ 1 $\frac{1}{2}$ 0 000+ $+\sqrt{7}/\sqrt{2.3.11}$

$\frac{3}{2}$ 2 $\frac{1}{2}$ 0 000− $+\sqrt{5}/\sqrt{2.11}$

$\frac{5}{2}$ 2 $\frac{1}{2}$ 0 000+ $+\sqrt{5}/\sqrt{2.11}$

$\frac{5}{2}$ $\tilde{1}$ $\frac{1}{2}$ 0 000− $+\sqrt{3}/\sqrt{11}$

$\frac{11}{2}$ 5 $\frac{3}{2}$

$\frac{1}{2}$ 1 $\frac{3}{2}$ 0 000− $-\sqrt{5}/2\sqrt{3.13}$

$\frac{1}{2}$ 2 $\frac{3}{2}$ 0 000+ $+\sqrt{7}/2\sqrt{13}$

$\frac{3}{2}$ 1 $\frac{3}{2}$ 0 000+ $-2\sqrt{2.7}/\sqrt{3.11.13}$

$\frac{3}{2}$ 2 $\frac{3}{2}$ 0 000− $-\sqrt{2}/\sqrt{11.13}$

$\frac{3}{2}$ $\tilde{1}$ $\frac{3}{2}$ 0·000+ $-3\sqrt{3}/\sqrt{11.13}$

$\frac{5}{2}$ 1 $\frac{3}{2}$ 0 000− $+3\sqrt{7}/2\sqrt{11.13}$

$\frac{5}{2}$ 2 $\frac{3}{2}$ 0 000+ $+\sqrt{13}/2\sqrt{11}$

$\frac{5}{2}$ 2 $\frac{3}{2}$ 1 000− $+\sqrt{3}/\sqrt{2.11.13}$

$\frac{5}{2}$ $\tilde{1}$ $\frac{3}{2}$ 0 000− $-2\sqrt{3}/\sqrt{11.13}$

$\frac{11}{2}$ 5 $\frac{5}{2}$

$\frac{1}{2}$ 2 $\frac{5}{2}$ 0 000+ $-1/\sqrt{2.13}$

$\frac{1}{2}$ $\tilde{1}$ $\frac{5}{2}$ 0 000− $+\sqrt{5}/\sqrt{3.13}$

$\frac{3}{2}$ 1 $\frac{5}{2}$ 0 000+ $-3/\sqrt{2.11.13}$

$\frac{3}{2}$ 2 $\frac{5}{2}$ 0 000− $+23/\sqrt{2.7.11.13}$

$\frac{3}{2}$ 2 $\frac{5}{2}$ 1 000+ $-3\sqrt{3}/\sqrt{7.11.13}$

$\frac{3}{2}$ $\tilde{1}$ $\frac{5}{2}$ 0 000+ $-4\sqrt{2}/\sqrt{3.7.11.13}$

$\frac{5}{2}$ 1 $\frac{5}{2}$ 0 000− $+3\sqrt{3.5}/2\sqrt{11.13}$

$\frac{5}{2}$ 1 $\frac{5}{2}$ 1 000− $+\sqrt{3}/2\sqrt{11.13}$

$\frac{5}{2}$ 2 $\frac{5}{2}$ 0 000+ $+2.3\sqrt{3}/\sqrt{7.11.13}$

$\frac{5}{2}$ 2 $\frac{5}{2}$ 1 000− $-2\sqrt{2}/\sqrt{7.11.13}$

$\frac{5}{2}$ 2 $\frac{5}{2}$ 000+ $-3/\sqrt{7.11.13}$

$\frac{5}{2}$ $\tilde{1}$ $\frac{5}{2}$ 0 000− $+2\sqrt{2.7}/\sqrt{3.11.13}$

$\frac{5}{2}$ $\tilde{1}$ $\frac{5}{2}$ 1 000− $-\sqrt{2.5}/\sqrt{3.7.11.13}$

$\frac{11}{2}$ 5 $\frac{7}{2}$

$\frac{1}{2}$ 2 $\frac{5}{2}$ 0 000− $+\sqrt{5}/\sqrt{3.13}$

$\frac{1}{2}$ $\tilde{1}$ $\frac{5}{2}$ 0 000+ $-1/\sqrt{2.13}$

$\frac{3}{2}$ 1 $\frac{5}{2}$ 0 000− $+\sqrt{3.5}/\sqrt{11.13}$

$\frac{3}{2}$ 2 $\frac{5}{2}$ 0 000+ $-2\sqrt{5}/\sqrt{3.7.11.13}$

$\frac{3}{2}$ 2 $\frac{5}{2}$ 1 000− $-\sqrt{5}/\sqrt{2.7.11.13}$

SO_3–K 3jm Factors (cont.)

$\frac{11}{2}\ 5\ \frac{7}{2}$

$\frac{3}{2}\ 2\ \tilde{\frac{1}{2}}\ 0\ 000-\quad -\sqrt{3.5}/\sqrt{2.11.13}$
$\frac{3}{2}\ \tilde{1}\ \frac{5}{2}\ 0\ 000-\quad -\sqrt{5.11}/2\sqrt{7.13}$
$\frac{3}{2}\ \tilde{1}\ \tilde{\frac{1}{2}}\ 0\ 000+\quad +3/2\sqrt{11.13}$
$\frac{5}{2}\ 1\ \frac{5}{2}\ 0\ 000+\quad -1/\sqrt{2.11.13}$
$\frac{5}{2}\ 1\ \frac{5}{2}\ 1\ 000+\quad -\sqrt{5}/\sqrt{2.11.13}$
$\frac{5}{2}\ 1\ \tilde{\frac{1}{2}}\ 0\ 000-\quad +\sqrt{3.7}/\sqrt{11.13}$
$\frac{5}{2}\ 2\ \frac{5}{2}\ 0\ 000-\quad +9\sqrt{5}/\sqrt{2.7.11.13}$
$\frac{5}{2}\ 2\ \frac{5}{2}\ 1\ 000+\quad +4\sqrt{5}/\sqrt{3.7.11.13}$
$\frac{5}{2}\ 2\ \frac{5}{2}\ 2\ 000-\quad -\sqrt{5}/\sqrt{2.3.7.11.13}$
$\frac{5}{2}\ 2\ \tilde{\frac{1}{2}}\ 0\ 000+\quad +\sqrt{5}/\sqrt{11.13}$
$\frac{5}{2}\ \tilde{1}\ \frac{5}{2}\ 0\ 000+\quad +\sqrt{5.7}/2\sqrt{11.13}$
$\frac{5}{2}\ \tilde{1}\ \frac{5}{2}\ 1\ 000+\quad -5/2\sqrt{7.11.13}$

$\frac{11}{2}\ 5\ \frac{9}{2}$

$\frac{1}{2}\ 1\ \frac{3}{2}\ 0\ 000+\quad -3/2\sqrt{2.13}$
$\frac{1}{2}\ 2\ \frac{3}{2}\ 0\ 000-\quad +\sqrt{3.7}/2\sqrt{2.5.13}$
$\frac{1}{2}\ 2\ \frac{5}{2}\ 0\ 000+\quad -1/2\sqrt{3.5.13}$
$\frac{1}{2}\ \tilde{1}\ \frac{5}{2}\ 0\ 000-\quad +1/\sqrt{2.13}$
$\frac{3}{2}\ 1\ \frac{3}{2}\ 0\ 000-\quad -3\sqrt{7}/2\sqrt{5.11.13}$
$\frac{3}{2}\ 1\ \frac{5}{2}\ 0\ 000+\quad +4\sqrt{3}/\sqrt{5.11.13}$
$\frac{3}{2}\ 2\ \frac{3}{2}\ 0\ 000+\quad +\sqrt{3.11}/2\sqrt{5.13}$
$\frac{3}{2}\ 2\ \frac{5}{2}\ 0\ 000-\quad +4\sqrt{7}/\sqrt{3.5.11.13}$
$\frac{3}{2}\ 2\ \frac{5}{2}\ 1\ 000+\quad +\sqrt{2.7}/\sqrt{5.11.13}$
$\frac{3}{2}\ \tilde{1}\ \frac{3}{2}\ 0\ 000-\quad -3/\sqrt{2.5.11.13}$
$\frac{3}{2}\ \tilde{1}\ \frac{5}{2}\ 0\ 000+\quad +2\sqrt{7}/\sqrt{5.11.13}$
$\frac{5}{2}\ 1\ \frac{3}{2}\ 0\ 000+\quad -\sqrt{3.7}/2\sqrt{2.5.11.13}$
$\frac{5}{2}\ 1\ \frac{5}{2}\ 0\ 000-\quad -7/2\sqrt{2.11.13}$
$\frac{5}{2}\ 1\ \frac{5}{2}\ 1\ 000-\quad -17/2\sqrt{2.5.11.13}$
$\frac{5}{2}\ 2\ \frac{3}{2}\ 0\ 000-\quad -\sqrt{3.5}/2\sqrt{2.11.13}$
$\frac{5}{2}\ 2\ \frac{3}{2}\ 1\ 000+\quad +1/2\sqrt{5.11.13}$
$\frac{5}{2}\ 2\ \frac{5}{2}\ 0\ 000+\quad 0$
$\frac{5}{2}\ 2\ \frac{5}{2}\ 1\ 000-\quad +\sqrt{7}/\sqrt{3.5.11.13}$
$\frac{5}{2}\ 2\ \frac{5}{2}\ 2\ 000+\quad +\sqrt{7.11}/\sqrt{2.3.5.13}$
$\frac{5}{2}\ \tilde{1}\ \frac{3}{2}\ 0\ 000+\quad +2.3\sqrt{2}/\sqrt{5.11.13}$
$\frac{5}{2}\ \tilde{1}\ \frac{5}{2}\ 0\ 000-\quad +2\sqrt{7}/\sqrt{5.11.13}$
$\frac{5}{2}\ \tilde{1}\ \frac{5}{2}\ 1\ 000-\quad +\sqrt{7}/\sqrt{11.13}$

$\frac{11}{2}\ \frac{11}{2}\ 0$

$\frac{1}{2}\ \frac{1}{2}\ 0\ 0\ 000+\quad +1/\sqrt{2.3}$
$\frac{3}{2}\ \frac{3}{2}\ 0\ 0\ 000+\quad +1/\sqrt{3}$
$\frac{5}{2}\ \frac{5}{2}\ 0\ 0\ 000+\quad +1/\sqrt{2}$

$\frac{11}{2}\ \frac{11}{2}\ 1$

$\frac{1}{2}\ \frac{1}{2}\ 1\ 0\ 000+\quad -\sqrt{11}/3\sqrt{2.13}$
$\frac{3}{2}\ \frac{1}{2}\ 1\ 0\ 000-\quad -\sqrt{2.7}/3\sqrt{13}$
$\frac{3}{2}\ \frac{3}{2}\ 1\ 0\ 000+\quad -\sqrt{5}/3\sqrt{11.13}$
$\frac{5}{2}\ \frac{3}{2}\ 1\ 0\ 000-\quad +\sqrt{2.3.5}/\sqrt{11.13}$
$\frac{5}{2}\ \frac{5}{2}\ 1\ 0\ 000+\quad +\sqrt{11}/2\sqrt{13}$
$\frac{5}{2}\ \frac{5}{2}\ 1\ 1\ 000+\quad -3\sqrt{5}/2\sqrt{11.13}$

$\frac{11}{2}\ \frac{11}{2}\ 2$

$\frac{3}{2}\ \frac{1}{2}\ 2\ 0\ 000-\quad +\sqrt{2}/\sqrt{3.13}$
$\frac{3}{2}\ \frac{3}{2}\ 2\ 0\ 000+\quad +17/\sqrt{3.7.11.13}$

$\frac{11}{2}\ \frac{11}{2}\ 2$

$\frac{5}{2}\ \frac{1}{2}\ 2\ 0\ 000+\quad +\sqrt{3}/\sqrt{2.13}$
$\frac{5}{2}\ \frac{3}{2}\ 2\ 0\ 000-\quad +5\sqrt{2.3}/\sqrt{7.11.13}$
$\frac{5}{2}\ \frac{3}{2}\ 2\ 1\ 000+\quad +2.3/\sqrt{7.11.13}$
$\frac{5}{2}\ \frac{5}{2}\ 2\ 0\ 000+\quad +2\sqrt{7}/\sqrt{11.13}$
$\frac{5}{2}\ \frac{5}{2}\ 2\ 1\ 000-\quad 0$
$\frac{5}{2}\ \frac{5}{2}\ 2\ 2\ 000+\quad +\sqrt{3}/\sqrt{7.11.13}$

$\frac{11}{2}\ \frac{11}{2}\ 3$

$\frac{3}{2}\ \frac{3}{2}\ 3\ 0\ 000+\quad -2.3\sqrt{5}/\sqrt{7.11.13}$
$\frac{3}{2}\ \frac{3}{2}\ \tilde{1}\ 0\ 000+\quad +\sqrt{3.5}/\sqrt{7.11.13}$
$\frac{5}{2}\ \frac{1}{2}\ 3\ 0\ 000+\quad -\sqrt{2}/\sqrt{13}$
$\frac{5}{2}\ \frac{1}{2}\ \tilde{1}\ 0\ 000+\quad +1/\sqrt{2.3.13}$
$\frac{5}{2}\ \frac{3}{2}\ 3\ 0\ 000-\quad -\sqrt{2}/\sqrt{7.11.13}$
$\frac{5}{2}\ \frac{3}{2}\ 3\ 1\ 000+\quad -\sqrt{2.3.5}/\sqrt{7.11.13}$
$\frac{5}{2}\ \frac{3}{2}\ \tilde{1}\ 0\ 000-\quad -8\sqrt{5}/\sqrt{3.7.11.13}$
$\frac{5}{2}\ \frac{5}{2}\ 3\ 0\ 000+\quad +2\sqrt{5}/\sqrt{7.11.13}$
$\frac{5}{2}\ \frac{5}{2}\ 3\ 1\ 000-\quad 0$
$\frac{5}{2}\ \frac{5}{2}\ \tilde{1}\ 0\ 000+\quad +2.5\sqrt{5}/\sqrt{3.7.11.13}$
$\frac{5}{2}\ \frac{5}{2}\ \tilde{1}\ 1\ 000+\quad +5/\sqrt{3.7.11.13}$

$\frac{11}{2}\ \frac{11}{2}\ 4$

$\frac{3}{2}\ \frac{1}{2}\ 2\ 0\ 000-\quad -\sqrt{5}/2\sqrt{13}$
$\frac{3}{2}\ \frac{3}{2}\ 2\ 0\ 000+\quad +\sqrt{2.5}/\sqrt{7.11.13}$
$\frac{3}{2}\ \frac{3}{2}\ 3\ 0\ 000-\quad 0$
$\frac{5}{2}\ \frac{1}{2}\ 2\ 0\ 000+\quad -\sqrt{5}/2.3\sqrt{13}$
$\frac{5}{2}\ \frac{1}{2}\ 3\ 0\ 000-\quad -\sqrt{7}/3\sqrt{13}$
$\frac{5}{2}\ \frac{3}{2}\ 2\ 0\ 000-\quad +\sqrt{5.11}/2.3\sqrt{7.13}$
$\frac{5}{2}\ \frac{3}{2}\ 2\ 1\ 000+\quad -3\sqrt{3.5}/\sqrt{2.7.11.13}$
$\frac{5}{2}\ \frac{3}{2}\ 3\ 0\ 000+\quad -\sqrt{13}/3\sqrt{11}$
$\frac{5}{2}\ \frac{3}{2}\ 3\ 1\ 000-\quad -\sqrt{5}/\sqrt{3.11.13}$
$\frac{5}{2}\ \frac{5}{2}\ 2\ 0\ 000+\quad +\sqrt{2.5.7}/\sqrt{3.11.13}$
$\frac{5}{2}\ \frac{5}{2}\ 2\ 1\ 000-\quad 0$
$\frac{5}{2}\ \frac{5}{2}\ 2\ 2\ 000+\quad -\sqrt{5}/3\sqrt{2.7.11.13}$
$\frac{5}{2}\ \frac{5}{2}\ 3\ 0\ 000-\quad 0$
$\frac{5}{2}\ \frac{5}{2}\ 3\ 1\ 000+\quad -5\sqrt{2}/3\sqrt{11.13}$

$\frac{11}{2}\ \frac{11}{2}\ 5$

$\frac{1}{2}\ \frac{1}{2}\ 1\ 0\ 000+\quad -\sqrt{3.11}/\sqrt{2.13.17}$
$\frac{3}{2}\ \frac{1}{2}\ 1\ 0\ 000-\quad -\sqrt{3.7}/2\sqrt{2.13.17}$
$\frac{3}{2}\ \frac{1}{2}\ 2\ 0\ 000+\quad -3\sqrt{5}/2\sqrt{2.13.17}$
$\frac{3}{2}\ \frac{3}{2}\ 1\ 0\ 000+\quad -4\sqrt{3.5}/\sqrt{11.13.17}$
$\frac{3}{2}\ \frac{3}{2}\ 2\ 0\ 000-\quad 0$
$\frac{3}{2}\ \frac{3}{2}\ \tilde{1}\ 0\ 000+\quad +\sqrt{2.3.5.7}/\sqrt{11.13.17}$
$\frac{5}{2}\ \frac{1}{2}\ 2\ 0\ 000-\quad -\sqrt{2.5}/\sqrt{13.17}$
$\frac{5}{2}\ \frac{1}{2}\ \tilde{1}\ 0\ 000+\quad +5/2\sqrt{3.13.17}$
$\frac{5}{2}\ \frac{3}{2}\ 1\ 0\ 000-\quad -3\sqrt{5}/2\sqrt{2.11.13.17}$
$\frac{5}{2}\ \frac{3}{2}\ 2\ 0\ 000+\quad -\sqrt{5.7}/2\sqrt{2.11.13.17}$
$\frac{5}{2}\ \frac{3}{2}\ 2\ 1\ 000-\quad -3\sqrt{3.5.7}/2\sqrt{11.13.17}$
$\frac{5}{2}\ \frac{3}{2}\ \tilde{1}\ 0\ 000-\quad +\sqrt{2.5.7}/\sqrt{3.11.13.17}$
$\frac{5}{2}\ \frac{5}{2}\ 1\ 0\ 000+\quad -3.5\sqrt{3}/2\sqrt{11.13.17}$
$\frac{5}{2}\ \frac{5}{2}\ 1\ 1\ 000+\quad -\sqrt{3.5}/2\sqrt{11.13.17}$
$\frac{5}{2}\ \frac{5}{2}\ 2\ 0\ 000-\quad 0$

SO_3–K 3jm Factors (cont.)

$\frac{11}{2}$ $\frac{11}{2}$ 5

$\frac{5}{2}$ $\frac{5}{2}$ 2 1 000+ $+2\sqrt{2.5.7}/\sqrt{11.13.17}$

$\frac{5}{2}$ $\frac{5}{2}$ 2 2 000− 0

$\frac{5}{2}$ $\frac{5}{2}$ $\tilde{1}$ 0 000+ $+\sqrt{2.5.7}/\sqrt{3.11.13.17}$

$\frac{5}{2}$ $\frac{5}{2}$ $\tilde{1}$ 1 000+ $-\sqrt{7.17}/\sqrt{2.3.11.13}$

6 3 3

0 3 3 0 000+ $+\sqrt{3}/\sqrt{7.13}$

0 $\tilde{1}$ $\tilde{1}$ 0 000+ $-2/\sqrt{7.13}$

1 3 3 0 000− 0

1 $\tilde{1}$ 3 0 000− $+\sqrt{3}/\sqrt{2.13}$

2 3 3 0 000+ $-\sqrt{2.3.5}/\sqrt{11.13}$

2 $\tilde{1}$ 3 0 000+ $+\sqrt{5}/\sqrt{2.11.13}$

2 $\tilde{1}$ $\tilde{1}$ 0 000+ $+2\sqrt{5}/\sqrt{11.13}$

3 3 3 0 000+ $-2\sqrt{3}/\sqrt{11.13}$

3 $\tilde{1}$ 3 0 000− $+4/\sqrt{11.13}$

6 $\frac{7}{2}$ $\frac{5}{2}$

0 $\frac{5}{2}$ $\frac{5}{2}$ 0 000− $+1/\sqrt{13}$

1 $\frac{5}{2}$ $\frac{5}{2}$ 0 000+ $+\sqrt{7}/4\sqrt{13}$

1 $\frac{5}{2}$ $\frac{5}{2}$ 1 000+ $+\sqrt{5.7}/4\sqrt{13}$

1 $\frac{\tilde{1}}{2}$ $\frac{5}{2}$ 0 000− $-\sqrt{3}/2\sqrt{2.13}$

2 $\frac{5}{2}$ $\frac{5}{2}$ 0 000− $-\sqrt{5.7}/\sqrt{11.13}$

2 $\frac{5}{2}$ $\frac{5}{2}$ 1 000+ $-\sqrt{5.7}/2\sqrt{2.3.11.13}$

2 $\frac{5}{2}$ $\frac{5}{2}$ 2 000− $-\sqrt{5.7}/2\sqrt{3.11.13}$

2 $\frac{\tilde{1}}{2}$ $\frac{5}{2}$ 0 000+ $+5\sqrt{5}/2\sqrt{2.11.13}$

3 $\frac{5}{2}$ $\frac{5}{2}$ 0 000+ $+\sqrt{2.5.7}/\sqrt{3.11.13}$

3 $\frac{5}{2}$ $\frac{5}{2}$ 1 000− $+\sqrt{2.7}/\sqrt{3.11.13}$

3 $\frac{\tilde{1}}{2}$ $\frac{5}{2}$ 0 000− $-4/\sqrt{11.13}$

6 $\frac{7}{2}$ $\frac{7}{2}$

0 $\frac{5}{2}$ $\frac{5}{2}$ 0 000+ $-1/2\sqrt{13}$

0 $\frac{\tilde{1}}{2}$ $\frac{\tilde{1}}{2}$ 0 000+ $+\sqrt{3}/2\sqrt{13}$

1 $\frac{5}{2}$ $\frac{5}{2}$ 0 000− 0

1 $\frac{5}{2}$ $\frac{5}{2}$ 1 000− 0

1 $\frac{\tilde{1}}{2}$ $\frac{5}{2}$ 0 000+ $+\sqrt{3}/\sqrt{2.13}$

2 $\frac{5}{2}$ $\frac{5}{2}$ 0 000+ $+\sqrt{5.7}/2\sqrt{11.13}$

2 $\frac{5}{2}$ $\frac{5}{2}$ 1 000− 0

2 $\frac{5}{2}$ $\frac{5}{2}$ 2 000+ $-\sqrt{3.5.7}/2\sqrt{11.13}$

2 $\frac{\tilde{1}}{2}$ $\frac{5}{2}$ 0 000− $+\sqrt{2.5}/\sqrt{11.13}$

3 $\frac{5}{2}$ $\frac{5}{2}$ 0 000− 0

3 $\frac{5}{2}$ $\frac{5}{2}$ 1 000+ $+\sqrt{2.3.7}/\sqrt{11.13}$

3 $\frac{\tilde{1}}{2}$ $\frac{5}{2}$ 0 000+ $+1/\sqrt{11.13}$

6 4 2

0 2 2 0 000+ $+1/\sqrt{13}$

1 2 2 0 000− $-\sqrt{7}/\sqrt{3.13}$

1 3 2 0 000+ $+\sqrt{2}/\sqrt{3.13}$

2 2 2 0 000+ $-\sqrt{5.7}/\sqrt{2.11.13}$

2 2 2 1 000+ $+5\sqrt{7}/3\sqrt{2.11.13}$

2 3 2 0 000− $+5/\sqrt{11.13}$

2 3 2 1 000+ $+5/3\sqrt{11.13}$

3 2 2 0 000− $+\sqrt{2.7}/\sqrt{11.13}$

3 2 2 1 000+ $-\sqrt{2.7}/3\sqrt{11.13}$

6 4 2

3 3 2 0 000+ $-8.2/3\sqrt{11.13}$

6 4 3

0 3 3 0 000− $+1/\sqrt{13}$

1 2 3 0 000− $-\sqrt{5}/\sqrt{2.3.13}$

1 2 $\tilde{1}$ 0 000− 0

1 3 3 0 000+ $-4\sqrt{2}/\sqrt{3.7.13}$

1 3 $\tilde{1}$ 0 000+ $-3/\sqrt{2.7.13}$

2 2 3 0 000+ $-3\sqrt{5}/2\sqrt{11.13}$

2 2 3 1 000− $+\sqrt{5.11}/2.3\sqrt{13}$

2 2 $\tilde{1}$ 0 000+ $-2\sqrt{5}/\sqrt{3.11.13}$

2 3 3 0 000− $-\sqrt{2.5}/3\sqrt{7.11.13}$

2 3 $\tilde{1}$ 0 000− $-\sqrt{5.13}/\sqrt{2.3.7.11}$

3 2 3 0 000− $+4\sqrt{5}/3\sqrt{11.13}$

3 2 $\tilde{1}$ 0 000− $+4\sqrt{5}/\sqrt{3.11.13}$

3 3 3 0 000− $+2\sqrt{11}/3\sqrt{7.13}$

3 3 $\tilde{1}$ 0 000+ $+4/\sqrt{3.7.11.13}$

6 4 4

0 2 2 0 000+ $-2/3\sqrt{13}$

0 3 3 0 000+ $+\sqrt{5}/3\sqrt{13}$

1 2 2 0 000− 0

1 3 2 0 000+ $-\sqrt{3}/\sqrt{2.13}$

1 3 3 0 000− 0

2 2 2 0 000+ $+\sqrt{2.5.7}/3\sqrt{11.13}$

2 2 2 1 000+ $+\sqrt{2.7}/\sqrt{11.13}$

2 3 2 0 000− $+\sqrt{11}/2.3\sqrt{13}$

2 3 2 1 000+ $-5/2\sqrt{11.13}$

2 3 3 0 000+ $+\sqrt{2.7}/\sqrt{11.13}$

3 2 2 0 000− 0

3 2 2 1 000+ $-4\sqrt{2.7}/3\sqrt{11.13}$

3 3 2 0 000+ $-4/3\sqrt{11.13}$

3 3 3 0 000+ $-2\sqrt{5.7}/3\sqrt{11.13}$

6 $\frac{9}{2}$ $\frac{3}{2}$

0 $\frac{3}{2}$ $\frac{3}{2}$ 0 000− $+1/\sqrt{13}$

1 $\frac{3}{2}$ $\frac{3}{2}$ 0 000+ $+\sqrt{3.7}/\sqrt{2.5.13}$

1 $\frac{5}{2}$ $\frac{3}{2}$ 0 000− $-3/\sqrt{2.5.13}$

2 $\frac{3}{2}$ $\frac{3}{2}$ 0 000− $-\sqrt{5.7}/\sqrt{2.11.13}$

2 $\frac{5}{2}$ $\frac{3}{2}$ 0 000+ $-3\sqrt{5}/\sqrt{2.11.13}$

2 $\frac{5}{2}$ $\frac{3}{2}$ 1 000− $+\sqrt{3.5}/\sqrt{11.13}$

3 $\frac{3}{2}$ $\frac{3}{2}$ 0 000+ $-2\sqrt{7}/\sqrt{5.11.13}$

3 $\frac{5}{2}$ $\frac{3}{2}$ 0 000− 0

3 $\frac{5}{2}$ $\frac{3}{2}$ 1 000+ $-8\sqrt{3}/\sqrt{5.11.13}$

6 $\frac{9}{2}$ $\frac{5}{2}$

0 $\frac{5}{2}$ $\frac{5}{2}$ 0 000+ $+1/\sqrt{13}$

1 $\frac{3}{2}$ $\frac{5}{2}$ 0 000+ $+\sqrt{3}/\sqrt{5.13}$

1 $\frac{5}{2}$ $\frac{5}{2}$ 0 000− $+4/\sqrt{7.13}$

1 $\frac{5}{2}$ $\frac{5}{2}$ 1 000− $+2/\sqrt{5.7.13}$

2 $\frac{3}{2}$ $\frac{5}{2}$ 0 000− $+\sqrt{3.5}/\sqrt{11.13}$

2 $\frac{3}{2}$ $\frac{5}{2}$ 1 000+ $+\sqrt{2.5}/\sqrt{11.13}$

2 $\frac{5}{2}$ $\frac{5}{2}$ 0 000+ $+2\sqrt{5}/\sqrt{7.11.13}$

SO_3–K 3jm Factors (cont.)

6 $\frac{9}{2}$ $\frac{5}{2}$	
2 $\frac{5}{2}$ $\frac{5}{2}$ 1 000−	$+5\sqrt{2.5}/\sqrt{3.7.11.13}$
2 $\frac{5}{2}$ $\frac{5}{2}$ 2 000+	$-8\sqrt{5}/\sqrt{3.7.11.13}$
3 $\frac{3}{2}$ $\frac{5}{2}$ 0 000+	$-3\sqrt{3}/\sqrt{2.11.13}$
3 $\frac{3}{2}$ $\frac{5}{2}$ 1 000−	$-\sqrt{11}/\sqrt{2.5.13}$
3 $\frac{5}{2}$ $\frac{5}{2}$ 0 000−	$-19\sqrt{2}/\sqrt{3.5.7.11.13}$
3 $\frac{5}{2}$ $\frac{5}{2}$ 1 000+	$-\sqrt{2.11}/\sqrt{3.7.13}$
6 $\frac{9}{2}$ $\frac{7}{2}$	
0 $\frac{5}{2}$ $\frac{5}{2}$ 0 000−	$+1/\sqrt{13}$
1 $\frac{3}{2}$ $\frac{5}{2}$ 0 000−	$+\sqrt{3}/\sqrt{5.13}$
1 $\frac{5}{2}$ $\frac{5}{2}$ 0 000+	$+9/4\sqrt{7.13}$
1 $\frac{5}{2}$ $\frac{5}{2}$ 1 000+	$-9.3/4\sqrt{5.7.13}$
1 $\frac{5}{2}$ $\tilde{\frac{1}{2}}$ 0 000−	$-\sqrt{3}/2\sqrt{2.13}$
2 $\frac{3}{2}$ $\frac{5}{2}$ 0 000+	$+\sqrt{3.5}/\sqrt{11.13}$
2 $\frac{3}{2}$ $\frac{5}{2}$ 1 000−	$-2\sqrt{2}/\sqrt{5.11.13}$
2 $\frac{3}{2}$ $\tilde{\frac{1}{2}}$ 0 000−	$+\sqrt{2.3.7}/\sqrt{5.11.13}$
2 $\frac{5}{2}$ $\frac{5}{2}$ 0 000−	$+2\sqrt{5}/\sqrt{7.11.13}$
2 $\frac{5}{2}$ $\frac{5}{2}$ 1 000+	$-3\sqrt{3.13}/2\sqrt{2.5.7.11}$
2 $\frac{5}{2}$ $\frac{5}{2}$ 2 000−	$+\sqrt{3.13}/2\sqrt{5.7.11}$
2 $\frac{5}{2}$ $\tilde{\frac{1}{2}}$ 0 000+	$-17/2\sqrt{2.5.11.13}$
3 $\frac{3}{2}$ $\frac{5}{2}$ 0 000−	$+2\sqrt{2.3}/\sqrt{11.13}$
3 $\frac{3}{2}$ $\frac{5}{2}$ 1 000+	$-2\sqrt{2}/\sqrt{5.11.13}$
3 $\frac{5}{2}$ $\frac{5}{2}$ 0 000+	$+3\sqrt{2.3}/\sqrt{5.7.11.13}$
3 $\frac{5}{2}$ $\frac{5}{2}$ 1 000−	$+\sqrt{2.3}/\sqrt{7.11.13}$
3 $\frac{5}{2}$ $\tilde{\frac{1}{2}}$ 0 000−	$+4/\sqrt{11.13}$
6 $\frac{9}{2}$ $\frac{9}{2}$	
0 $\frac{3}{2}$ $\frac{3}{2}$ 0 000+	$-\sqrt{3}/\sqrt{5.13}$
0 $\frac{5}{2}$ $\frac{5}{2}$ 0 000+	$+\sqrt{2}/\sqrt{5.13}$
1 $\frac{3}{2}$ $\frac{3}{2}$ 0 000−	0
1 $\frac{5}{2}$ $\frac{3}{2}$ 0 000+	$+\sqrt{3}/\sqrt{2.13}$
1 $\frac{5}{2}$ $\frac{5}{2}$ 0 000−	0
1 $\frac{5}{2}$ $\frac{5}{2}$ 1 000−	0
2 $\frac{3}{2}$ $\frac{3}{2}$ 0 000+	$-\sqrt{2.3.7}/5\sqrt{11.13}$
2 $\frac{5}{2}$ $\frac{3}{2}$ 0 000−	$+\sqrt{3.11}/5\sqrt{2.13}$
2 $\frac{5}{2}$ $\frac{3}{2}$ 1 000+	$+17/5\sqrt{11.13}$
2 $\frac{5}{2}$ $\frac{5}{2}$ 0 000+	$+4\sqrt{2.7}/5\sqrt{11.13}$
2 $\frac{5}{2}$ $\frac{5}{2}$ 1 000−	0
2 $\frac{5}{2}$ $\frac{5}{2}$ 2 000+	$+2\sqrt{2.3.7}/5\sqrt{11.13}$
3 $\frac{3}{2}$ $\frac{3}{2}$ 0 000−	0
3 $\frac{5}{2}$ $\frac{3}{2}$ 0 000+	$-7\sqrt{3}/2\sqrt{5.11.13}$
3 $\frac{5}{2}$ $\frac{3}{2}$ 1 000−	$-5/2\sqrt{11.13}$
3 $\frac{5}{2}$ $\frac{5}{2}$ 0 000−	0
3 $\frac{5}{2}$ $\frac{5}{2}$ 1 000+	$+2\sqrt{3.7}/\sqrt{5.11.13}$
6 5 1	
0 1 1 0 000+	$+1/\sqrt{13}$
1 1 1 0 000−	$-\sqrt{7}/2\sqrt{13}$
1 2 1 0 000+	$+\sqrt{5}/2\sqrt{13}$
2 1 1 0 000+	$+\sqrt{5.7}/2\sqrt{11.13}$
2 2 1 0 000−	$-5\sqrt{5}/2\sqrt{11.13}$
2 $\tilde{1}$ 1 0 000+	$+\sqrt{3.5}/\sqrt{11.13}$
6 5 1	
3 2 1 0 000+	$+2\sqrt{5}/\sqrt{11.13}$
3 $\tilde{1}$ 1 0 000−	$-2\sqrt{2.3}/\sqrt{11.13}$
6 5 2	
0 2 2 0 000−	$+1/\sqrt{13}$
1 1 2 0 000−	$-\sqrt{3}/2\sqrt{13}$
1 2 2 0 000+	$-3\sqrt{3}/2\sqrt{7.13}$
1 $\tilde{1}$ 2 0 000−	$+3/\sqrt{7.13}$
2 1 2 0 000+	$+5\sqrt{3}/2\sqrt{11.13}$
2 2 2 0 000−	$+\sqrt{5.11}/2\sqrt{2.7.13}$
2 2 2 1 000−	$+3.5/2\sqrt{2.7.11.13}$
2 $\tilde{1}$ 2 0 000+	$+5\sqrt{2.3}/\sqrt{7.11.13}$
3 1 2 0 000−	$-2\sqrt{3}/\sqrt{11.13}$
3 2 2 0 000+	$-8\sqrt{2}/\sqrt{7.11.13}$
3 2 2 1 000−	$+2.3\sqrt{2}/\sqrt{7.11.13}$
3 $\tilde{1}$ 2 0 000−	$-2\sqrt{2.3}/\sqrt{7.11.13}$
6 5 3	
0 $\tilde{1}$ $\tilde{1}$ 0 000+	$+1/\sqrt{13}$
1 2 3 0 000+	0
1 2 $\tilde{1}$ 0 000+	$-\sqrt{3.5}/\sqrt{7.13}$
1 $\tilde{1}$ 3 0 000−	$-\sqrt{2.3}/\sqrt{7.13}$
2 1 3 0 000+	0
2 1 $\tilde{1}$ 0 000+	$+\sqrt{3.5}/\sqrt{11.13}$
2 2 3 0 000−	$-2\sqrt{2.3.5}/\sqrt{7.11.13}$
2 2 3 1 000+	$-2\sqrt{2.3.5}/\sqrt{7.11.13}$
2 2 $\tilde{1}$ 0 000−	$-\sqrt{2.5}/\sqrt{7.11.13}$
2 $\tilde{1}$ 3 0 000+	$-\sqrt{2.5}/\sqrt{7.11.13}$
2 $\tilde{1}$ $\tilde{1}$ 0 000+	$+2\sqrt{5}/\sqrt{7.11.13}$
3 1 3 0 000−	$-3\sqrt{2}/\sqrt{11.13}$
3 1 $\tilde{1}$ 0 000−	$-\sqrt{2.3}/\sqrt{11.13}$
3 2 3 0 000+	$+\sqrt{2.3.5}/\sqrt{7.11.13}$
3 2 $\tilde{1}$ 0 000+	$-\sqrt{2.5}/\sqrt{7.11.13}$
3 $\tilde{1}$ 3 0 000−	$+2.5/\sqrt{7.11.13}$
6 5 4	
0 2 2 0 000−	$+1/\sqrt{13}$
1 1 2 0 000−	$-\sqrt{3}/2\sqrt{13}$
1 2 2 0 000+	$+5/2\sqrt{3.7.13}$
1 2 3 0 000−	$+\sqrt{2}/\sqrt{3.13}$
1 $\tilde{1}$ 2 0 000−	$+3/\sqrt{7.13}$
1 $\tilde{1}$ 3 0 000+	0
2 1 2 0 000+	$+1/2\sqrt{3.11.13}$
2 1 3 0 000−	$-2\sqrt{2.7}/\sqrt{3.11.13}$
2 2 2 0 000−	$+\sqrt{5.11}/2\sqrt{2.7.13}$
2 2 2 1 000−	$-\sqrt{11}/2.3\sqrt{2.7.13}$
2 2 3 0 000+	$-1/\sqrt{11.13}$
2 2 3 1 000−	$+5/3\sqrt{11.13}$
2 $\tilde{1}$ 2 0 000+	$-\sqrt{2.13}/\sqrt{3.7.11}$
2 $\tilde{1}$ 3 0 000−	$-4/\sqrt{3.11.13}$
3 1 2 0 000−	$+1/\sqrt{3.11.13}$
3 1 3 0 000+	$+\sqrt{5.7}/\sqrt{3.11.13}$

SO_3–K 3jm Factors (cont.)

6 5 4
3 2 2 0 000+ $+5/\sqrt{2.7.11.13}$
3 2 2 1 000− $-41/3\sqrt{2.7.11.13}$
3 2 3 0 000− $+\sqrt{11}/3\sqrt{13}$
3 $\tilde{1}$ 2 0 000− $+\sqrt{2}/\sqrt{3.7.11.13}$
3 $\tilde{1}$ 3 0 000+ $+\sqrt{2.5}/\sqrt{3.11.13}$
6 5 5
0 1 1 0 000+ $-3\sqrt{5}/2\sqrt{13.17}$
0 2 2 0 000+ $+\sqrt{3}/2\sqrt{13.17}$
0 $\tilde{1}$ $\tilde{1}$ 0 000+ $+\sqrt{5}/\sqrt{13.17}$
1 1 1 0 000− 0
1 2 1 0 000+ $-9/2\sqrt{13.17}$
1 2 2 0 000− 0
1 $\tilde{1}$ 2 0 000+ $-\sqrt{3.7}/2\sqrt{13.17}$
2 1 1 0 000+ $+3\sqrt{7}/2\sqrt{11.13.17}$
2 2 1 0 000− $-9/\sqrt{11.13.17}$
2 2 2 0 000+ $-\sqrt{3.5.7}/2\sqrt{2.11.13.17}$
2 2 2 1 000+ $-\sqrt{3.7.13}/2\sqrt{2.11.17}$
2 $\tilde{1}$ 1 0 000+ $-\sqrt{3.11}/2\sqrt{13.17}$
2 $\tilde{1}$ 2 0 000− $+\sqrt{7}/\sqrt{2.11.13.17}$
2 $\tilde{1}$ $\tilde{1}$ 0 000+ $+4\sqrt{7}/\sqrt{11.13.17}$
3 2 1 0 000+ $+3/\sqrt{11.13.17}$
3 2 2 0 000− 0
3 2 2 1 000+ $+\sqrt{2.3.7}/\sqrt{11.13.17}$
3 $\tilde{1}$ 1 0 000− $-2\sqrt{2.3.5}/\sqrt{11.13.17}$
3 $\tilde{1}$ 2 0 000+ $+4\sqrt{2.7}/\sqrt{11.13.17}$
6 $\frac{11}{2}$ $\frac{1}{2}$
0 $\frac{1}{2}$ $\frac{1}{2}$ 0 000− $+1/\sqrt{13}$
1 $\frac{1}{2}$ $\frac{1}{2}$ 0 000+ $+\sqrt{7}/\sqrt{2.3.13}$
1 $\frac{3}{2}$ $\frac{1}{2}$ 0 000− $+\sqrt{11}/\sqrt{2.3.13}$
2 $\frac{3}{2}$ $\frac{1}{2}$ 0 000+ $+\sqrt{5}/\sqrt{2.13}$
2 $\frac{5}{2}$ $\frac{1}{2}$ 0 000− $+\sqrt{5}/\sqrt{2.13}$
3 $\frac{5}{2}$ $\frac{1}{2}$ 0 000+ $+2/\sqrt{13}$
6 $\frac{11}{2}$ $\frac{3}{2}$
0 $\frac{3}{2}$ $\frac{3}{2}$ 0 000+ $+1/\sqrt{13}$
1 $\frac{1}{2}$ $\frac{3}{2}$ 0 000+ $-\sqrt{11}/2\sqrt{3.13}$
1 $\frac{3}{2}$ $\frac{3}{2}$ 0 000− $+\sqrt{2.5}/\sqrt{3.7.13}$
1 $\frac{5}{2}$ $\frac{3}{2}$ 0 000+ $-3\sqrt{5}/2\sqrt{7.13}$
2 $\frac{1}{2}$ $\frac{3}{2}$ 0 000− $-\sqrt{5}/2\sqrt{13}$
2 $\frac{3}{2}$ $\frac{3}{2}$ 0 000+ $+2\sqrt{2.5}/\sqrt{7.11.13}$
2 $\frac{5}{2}$ $\frac{3}{2}$ 0 000− $+\sqrt{5.7}/2\sqrt{11.13}$
2 $\frac{5}{2}$ $\frac{3}{2}$ 1 000+ $+5\sqrt{3.5}/\sqrt{2.7.11.13}$
3 $\frac{3}{2}$ $\frac{3}{2}$ 0 000− $+2.3\sqrt{5}/\sqrt{7.11.13}$
3 $\frac{5}{2}$ $\frac{3}{2}$ 0 000+ $+\sqrt{11}/\sqrt{2.7.13}$
3 $\frac{5}{2}$ $\frac{3}{2}$ 1 000− $+3\sqrt{3.5}/\sqrt{2.7.11.13}$
6 $\frac{11}{2}$ $\frac{5}{2}$
0 $\frac{5}{2}$ $\frac{5}{2}$ 0 000− $+1/\sqrt{13}$
1 $\frac{3}{2}$ $\frac{5}{2}$ 0 000− $+\sqrt{3.5}/\sqrt{2.7.13}$
1 $\frac{5}{2}$ $\frac{5}{2}$ 0 000+ $-3/2\sqrt{7.13}$
1 $\frac{5}{2}$ $\frac{5}{2}$ 1 000+ $+3\sqrt{5}/2\sqrt{7.13}$
6 $\frac{11}{2}$ $\frac{5}{2}$
2 $\frac{1}{2}$ $\frac{5}{2}$ 0 000− $+\sqrt{5}/\sqrt{2.3.13}$
2 $\frac{3}{2}$ $\frac{5}{2}$ 0 000+ $-\sqrt{5.7}/\sqrt{2.3.11.13}$
2 $\frac{3}{2}$ $\frac{5}{2}$ 1 000− $+5\sqrt{5}/\sqrt{7.11.13}$
2 $\frac{5}{2}$ $\frac{5}{2}$ 0 000− $+2\sqrt{5}/\sqrt{7.11.13}$
2 $\frac{5}{2}$ $\frac{5}{2}$ 1 000+ $-2\sqrt{2.3.5}/\sqrt{7.11.13}$
2 $\frac{5}{2}$ $\frac{5}{2}$ 2 000− $+\sqrt{3.5}/\sqrt{7.11.13}$
3 $\frac{1}{2}$ $\frac{5}{2}$ 0 000+ $+2/\sqrt{3.13}$
3 $\frac{3}{2}$ $\frac{5}{2}$ 0 000− $+8.2/\sqrt{3.7.11.13}$
3 $\frac{3}{2}$ $\frac{5}{2}$ 1 000+ 0
3 $\frac{5}{2}$ $\frac{5}{2}$ 0 000+ $-2\sqrt{2.3.5}/\sqrt{7.11.13}$
3 $\frac{5}{2}$ $\frac{5}{2}$ 1 000− 0
6 $\frac{11}{2}$ $\frac{7}{2}$
0 $\frac{5}{2}$ $\frac{5}{2}$ 0 000+ $-1/\sqrt{13}$
1 $\frac{3}{2}$ $\frac{5}{2}$ 0 000+ $-\sqrt{3.5}/\sqrt{2.7.13}$
1 $\frac{5}{2}$ $\frac{5}{2}$ 0 000− $+\sqrt{13}/4\sqrt{7}$
1 $\frac{5}{2}$ $\frac{5}{2}$ 1 000− $+\sqrt{5}/4\sqrt{7.13}$
1 $\frac{5}{2}$ $\tilde{\frac{7}{2}}$ 0 000+ $+\sqrt{3}/2\sqrt{2.13}$
2 $\frac{1}{2}$ $\frac{5}{2}$ 0 000+ $-\sqrt{5}/\sqrt{2.3.13}$
2 $\frac{3}{2}$ $\frac{5}{2}$ 0 000− $+\sqrt{5.7}/\sqrt{2.3.11.13}$
2 $\frac{3}{2}$ $\frac{5}{2}$ 1 000+ $+2\sqrt{5}/\sqrt{7.11.13}$
2 $\frac{3}{2}$ $\tilde{\frac{7}{2}}$ 0 000+ $-\sqrt{3.5}/\sqrt{11.13}$
2 $\frac{5}{2}$ $\frac{5}{2}$ 0 000+ $-2\sqrt{5}/\sqrt{7.11.13}$
2 $\frac{5}{2}$ $\frac{5}{2}$ 1 000− $-5.5\sqrt{5}/2\sqrt{2.3.7.11.13}$
2 $\frac{5}{2}$ $\frac{5}{2}$ 2 000+ $+\sqrt{5}/2\sqrt{3.7.11.13}$
2 $\frac{5}{2}$ $\tilde{\frac{7}{2}}$ 0 000− $-\sqrt{5}/2\sqrt{2.11.13}$
3 $\frac{1}{2}$ $\frac{5}{2}$ 0 000− $-1/4\sqrt{3.13}$
3 $\frac{1}{2}$ $\tilde{\frac{7}{2}}$ 0 000+ $-\sqrt{3.7}/4\sqrt{13}$
3 $\frac{3}{2}$ $\frac{5}{2}$ 0 000+ $+17/2\sqrt{3.7.11.13}$
3 $\frac{3}{2}$ $\frac{5}{2}$ 1 000− $-\sqrt{5.7}/2\sqrt{11.13}$
3 $\frac{5}{2}$ $\frac{5}{2}$ 0 000− $-43\sqrt{5}/4\sqrt{2.3.7.11.13}$
3 $\frac{5}{2}$ $\frac{5}{2}$ 1 000+ $-5\sqrt{7}/4\sqrt{2.3.11.13}$
3 $\frac{5}{2}$ $\tilde{\frac{7}{2}}$ 0 000+ $-5/4\sqrt{11.13}$
6 $\frac{11}{2}$ $\frac{9}{2}$
0 $\frac{3}{2}$ $\frac{3}{2}$ 0 000− $+3\sqrt{3}/\sqrt{2.13.17}$
0 $\frac{5}{2}$ $\frac{5}{2}$ 0 000− $+\sqrt{7}/\sqrt{2.13.17}$
1 $\frac{1}{2}$ $\frac{3}{2}$ 0 000− $-3\sqrt{11}/2\sqrt{2.13.17}$
1 $\frac{3}{2}$ $\frac{3}{2}$ 0 000+ $-3.11/2\sqrt{5.7.13.17}$
1 $\frac{3}{2}$ $\frac{5}{2}$ 0 000− $-2\sqrt{3}/\sqrt{5.13.17}$
1 $\frac{5}{2}$ $\frac{3}{2}$ 0 000− $-31\sqrt{3}/2\sqrt{2.5.7.13.17}$
1 $\frac{5}{2}$ $\frac{5}{2}$ 0 000+ $-11/2\sqrt{2.13.17}$
1 $\frac{5}{2}$ $\frac{5}{2}$ 1 000+ $+11/2\sqrt{2.5.13.17}$
2 $\frac{1}{2}$ $\frac{3}{2}$ 0 000+ $-\sqrt{3}/2\sqrt{2.5.13.17}$
2 $\frac{1}{2}$ $\frac{5}{2}$ 0 000− $-\sqrt{7.13}/2\sqrt{3.5.17}$
2 $\frac{3}{2}$ $\frac{3}{2}$ 0 000− $+53\sqrt{3}/2\sqrt{5.7.11.13.17}$
2 $\frac{3}{2}$ $\frac{5}{2}$ 0 000+ $-2\sqrt{11}/\sqrt{3.5.13.17}$
2 $\frac{3}{2}$ $\frac{5}{2}$ 1 000− $+\sqrt{2.13}/\sqrt{5.11.17}$
2 $\frac{5}{2}$ $\frac{3}{2}$ 0 000+ $+\sqrt{3.7.11}/2\sqrt{2.5.13.17}$
2 $\frac{5}{2}$ $\frac{3}{2}$ 1 000− $-167/2\sqrt{5.7.11.13.17}$
2 $\frac{5}{2}$ $\frac{5}{2}$ 0 000− $+2.7\sqrt{2}/\sqrt{5.11.13.17}$

SO_3–K 3jm Factors (cont.)

6 $\frac{11}{2}$ $\frac{9}{2}$	
2 $\frac{5}{2}$ $\frac{5}{2}$ 1 000+	$+\sqrt{11}/\sqrt{3.5.13.17}$
2 $\frac{5}{2}$ $\frac{5}{2}$ 2 000−	$-79/\sqrt{2.3.5.11.13.17}$
3 $\frac{1}{2}$ $\frac{5}{2}$ 0 000+	$-\sqrt{2.7}/\sqrt{3.13.17}$
3 $\frac{3}{2}$ $\frac{3}{2}$ 0 000+	$-\sqrt{2.3.11}/\sqrt{5.7.13.17}$
3 $\frac{3}{2}$ $\frac{5}{2}$ 0 000−	$-4.5\sqrt{2}/\sqrt{3.11.13.17}$
3 $\frac{3}{2}$ $\frac{5}{2}$ 1 000+	$-8.2\sqrt{2}/\sqrt{5.11.13.17}$
3 $\frac{5}{2}$ $\frac{3}{2}$ 0 000−	$-3.5\sqrt{3}/\sqrt{7.11.13.17}$
3 $\frac{5}{2}$ $\frac{3}{2}$ 1 000+	$-41/\sqrt{5.7.11.13.17}$
3 $\frac{5}{2}$ $\frac{5}{2}$ 0 000+	$-4\sqrt{11}/\sqrt{3.5.13.17}$
3 $\frac{5}{2}$ $\frac{5}{2}$ 1 000−	$+2.5/\sqrt{3.11.13.17}$
6 $\frac{11}{2}$ $\frac{11}{2}$	
0 $\frac{1}{2}$ $\frac{1}{2}$ 0 000+	$-11/2\sqrt{3.13.17}$
0 $\frac{3}{2}$ $\frac{3}{2}$ 0 000+	$-\sqrt{2}/\sqrt{3.13.17}$
0 $\frac{5}{2}$ $\frac{5}{2}$ 0 000+	$+5/2\sqrt{13.17}$
1 $\frac{1}{2}$ $\frac{1}{2}$ 0 000−	0
1 $\frac{3}{2}$ $\frac{1}{2}$ 0 000+	$-3\sqrt{11}/2\sqrt{2.13.17}$
1 $\frac{3}{2}$ $\frac{3}{2}$ 0 000−	0
1 $\frac{5}{2}$ $\frac{3}{2}$ 0 000+	$+\sqrt{3.5.7}/2\sqrt{2.13.17}$
1 $\frac{5}{2}$ $\frac{5}{2}$ 0 000−	0
1 $\frac{5}{2}$ $\frac{5}{2}$ 1 000−	0
2 $\frac{3}{2}$ $\frac{1}{2}$ 0 000−	$+5\sqrt{5}/2\sqrt{2.3.13.17}$
2 $\frac{3}{2}$ $\frac{3}{2}$ 0 000+	$+4\sqrt{5.7}/\sqrt{3.11.13.17}$
2 $\frac{5}{2}$ $\frac{1}{2}$ 0 000+	$-\sqrt{5}/\sqrt{2.3.13.17}$
2 $\frac{5}{2}$ $\frac{3}{2}$ 0 000−	$-\sqrt{5.7.11}/2\sqrt{2.3.13.17}$
2 $\frac{5}{2}$ $\frac{3}{2}$ 1 000+	$-\sqrt{5.7}/2\sqrt{11.13.17}$
2 $\frac{5}{2}$ $\frac{5}{2}$ 0 000+	$+2\sqrt{5.7}/\sqrt{11.13.17}$
2 $\frac{5}{2}$ $\frac{5}{2}$ 1 000−	0
2 $\frac{5}{2}$ $\frac{5}{2}$ 2 000+	$+\sqrt{3.5.7}/\sqrt{11.13.17}$
3 $\frac{3}{2}$ $\frac{3}{2}$ 0 000−	0
3 $\frac{5}{2}$ $\frac{1}{2}$ 0 000−	$+5/\sqrt{3.13.17}$
3 $\frac{5}{2}$ $\frac{3}{2}$ 0 000+	$-5\sqrt{7}/2\sqrt{3.11.13.17}$
3 $\frac{5}{2}$ $\frac{3}{2}$ 1 000−	$+3\sqrt{5.7}/2\sqrt{11.13.17}$
3 $\frac{5}{2}$ $\frac{5}{2}$ 0 000−	0
3 $\frac{5}{2}$ $\frac{5}{2}$ 1 000+	$-3\sqrt{2.3.7}/\sqrt{11.13.17}$
6 6 0	
0 0 0 0 000+	$+1/\sqrt{13}$
1 1 0 0 000+	$+\sqrt{3}/\sqrt{13}$
2 2 0 0 000+	$+\sqrt{5}/\sqrt{13}$
3 3 0 0 000+	$+2/\sqrt{13}$
6 6 1	
1 0 1 0 000−	$-1/\sqrt{13}$
1 1 1 0 000+	$-1/2\sqrt{7.13}$
2 1 1 0 000−	$-\sqrt{5.11}/2\sqrt{7.13}$
2 2 1 0 000+	$-\sqrt{5}/2\sqrt{7.13}$
3 2 1 0 000−	$-2\sqrt{5}/\sqrt{7.13}$
3 3 1 0 000+	$-2\sqrt{2}/\sqrt{7.13}$
6 6 2	
1 1 2 0 000+	$-\sqrt{3.11}/2\sqrt{7.13}$
2 0 2 0 000+	$+1/\sqrt{13}$
6 6 2	
2 1 2 0 000−	$+\sqrt{3}/2\sqrt{7.13}$
2 2 2 0 000+	$-3\sqrt{5}/2\sqrt{2.7.11.13}$
2 2 2 1 000+	$-3\sqrt{13}/2\sqrt{2.7.11}$
3 1 2 0 000+	$+2\sqrt{3}/\sqrt{7.13}$
3 2 2 0 000−	$-4\sqrt{2}/\sqrt{7.11.13}$
3 2 2 1 000+	$-2.3\sqrt{2}/\sqrt{7.11.13}$
3 3 2 0 000+	$+2.3\sqrt{2}/\sqrt{7.11.13}$
6 6 3	
2 1 3 0 000−	$+\sqrt{3.5}/\sqrt{7.13}$
2 1 $\tilde{1}$ 0 000−	0
2 2 3 0 000+	$+\sqrt{2.5}/\sqrt{7.11.13}$
2 2 3 1 000−	0
2 2 $\tilde{1}$ 0 000+	$-2\sqrt{2.3.5}/\sqrt{7.11.13}$
3 0 3 0 000−	$-1/\sqrt{13}$
3 1 3 0 000+	$+\sqrt{3}/\sqrt{2.7.13}$
3 1 $\tilde{1}$ 0 000+	$-3/\sqrt{2.7.13}$
3 2 3 0 000−	$-3\sqrt{5}/\sqrt{2.7.11.13}$
3 2 $\tilde{1}$ 0 000−	$-3\sqrt{3.5}/\sqrt{2.7.11.13}$
3 3 3 0 000−	0
3 3 $\tilde{1}$ 0 000+	$+5\sqrt{3}/\sqrt{7.11.13}$
6 6 4	
1 1 2 0 000+	$+\sqrt{3.5.11}/\sqrt{2.7.13.17}$
2 0 2 0 000+	$-\sqrt{2.5}/\sqrt{13.17}$
2 1 2 0 000−	$+11\sqrt{5}/\sqrt{2.3.7.13.17}$
2 1 3 0 000+	$+\sqrt{5}/\sqrt{3.13.17}$
2 2 2 0 000+	$+3.5/2\sqrt{7.11.13.17}$
2 2 2 1 000+	$+5\sqrt{5}/2.3\sqrt{7.11.13.17}$
2 2 3 0 000−	0
2 2 3 1 000+	$+19\sqrt{2.5}/3\sqrt{11.13.17}$
3 0 3 0 000+	$-\sqrt{7}/\sqrt{13.17}$
3 1 2 0 000+	$-5\sqrt{5}/\sqrt{2.3.7.13.17}$
3 1 3 0 000−	$+11/\sqrt{2.3.13.17}$
3 2 2 0 000−	$-3\sqrt{5.11}/2\sqrt{7.13.17}$
3 2 2 1 000+	$-47\sqrt{5}/2.3\sqrt{7.11.13.17}$
3 2 3 0 000+	$-\sqrt{5}/3\sqrt{2.11.13.17}$
3 3 2 0 000+	$+41\sqrt{5}/3\sqrt{7.11.13.17}$
3 3 3 0 000+	$+4.5/3\sqrt{11.13.17}$
6 6 5	
1 0 1 0 000−	$-\sqrt{3.11}/2\sqrt{13.17}$
1 1 1 0 000+	$+3\sqrt{3.11}/2\sqrt{7.13.17}$
1 1 2 0 000−	0
2 0 2 0 000−	$-\sqrt{5.7}/2\sqrt{13.17}$
2 1 1 0 000−	$-\sqrt{3.5}/\sqrt{7.13.17}$
2 1 2 0 000+	$+\sqrt{3.5}/2\sqrt{13.17}$
2 1 $\tilde{1}$ 0 000−	$-3\sqrt{5}/2\sqrt{13.17}$
2 2 1 0 000+	$+19\sqrt{3.5}/2\sqrt{7.11.13.17}$
2 2 2 0 000−	0
2 2 2 1 000−	0
2 2 $\tilde{1}$ 0 000+	$-3\sqrt{3.5}/\sqrt{2.11.13.17}$

SO_3–K 3jm Factors (cont.)

6 6 5

3 1 2 0 000− $-\sqrt{3.5}/\sqrt{13.17}$

3 1 $\tilde{1}$ 0 000+ 0

3 2 1 0 000− $+3\sqrt{3.5}/\sqrt{7.11.13.17}$

3 2 2 0 000+ $-4\sqrt{2.5}/\sqrt{11.13.17}$

3 2 2 1 000− $+3\sqrt{2.5}/\sqrt{11.13.17}$

3 2 $\tilde{1}$ 0 000− $+2\sqrt{2.3.5}/\sqrt{11.13.17}$

3 3 1 0 000+ $+2.5\sqrt{2.3}/\sqrt{7.11.13.17}$

3 3 2 0 000− 0

3 3 $\tilde{1}$ 0 000+ $+2.3\sqrt{3}/\sqrt{11.13.17}$

6 6 6

0 0 0 0 000+ $-11/\sqrt{13.17.19}$

1 1 0 0 000+ $-11\sqrt{3}/2\sqrt{13.17.19}$

1 1 1 0 000− 0

2 1 1 0 000+ $-\sqrt{3.5.7.11}/2\sqrt{13.17.19}$

2 2 0 0 000+ $+3\sqrt{5}/2\sqrt{13.17.19}$

2 2 1 0 000− 0

2 2 2 0 000+ $+5\sqrt{7.19}/2\sqrt{2.11.13.17}$

2 2 2 1 000+ $-9\sqrt{5.7}/2\sqrt{2.11.13.17.19}$

3 2 1 0 000+ $-\sqrt{3.5.7}/\sqrt{13.17.19}$

3 2 2 0 000− 0

3 2 2 1 000+ $+3\sqrt{2.5.7}/\sqrt{11.13.17.19}$

3 3 0 0 000+ $+2.5/\sqrt{13.17.19}$

3 3 1 0 000− 0

3 3 2 0 000+ $+2.3\sqrt{2.5.7}/\sqrt{11.13.17.19}$

3 3 3 0 000+ $+8.3\sqrt{7}/\sqrt{11.13.17.19}$

$\frac{13}{2}$ $\frac{7}{2}$ 3

$\frac{1}{2}$ $\frac{5}{2}$ 3 0 000− $+\sqrt{3}/2\sqrt{13}$

$\frac{1}{2}$ $\frac{5}{2}$ $\tilde{1}$ 0 000− $+1/\sqrt{13}$

$\frac{1}{2}$ $\tilde{\frac{1}{2}}$ 3 0 000+ $+\sqrt{3}/2\sqrt{7.13}$

$\frac{3}{2}$ $\frac{5}{2}$ 3 0 000+ 0

$\frac{3}{2}$ $\frac{5}{2}$ 3 1 000− $+\sqrt{2}/\sqrt{13}$

$\frac{3}{2}$ $\frac{5}{2}$ $\tilde{1}$ 0 000+ $-1/\sqrt{13}$

$\frac{3}{2}$ $\tilde{\frac{1}{2}}$ $\tilde{1}$ 0 000− $-\sqrt{5}/\sqrt{7.13}$

$\frac{5}{2}$ $\frac{5}{2}$ 3 0 000− $+3\sqrt{2}/\sqrt{11.13}$

$\frac{5}{2}$ $\frac{5}{2}$ 3 1 000+ 0

$\frac{5}{2}$ $\frac{5}{2}$ $\tilde{1}$ 0 000− $-2\sqrt{2.3}/\sqrt{11.13}$

$\frac{5}{2}$ $\frac{5}{2}$ $\tilde{1}$ 1 000− 0

$\frac{5}{2}$ $\tilde{\frac{1}{2}}$ 3 0 000+ $-3\sqrt{3.5}/\sqrt{7.11.13}$

$\tilde{\frac{1}{2}}$ $\frac{5}{2}$ 3 0 000+ $-1/\sqrt{11}$

$\tilde{\frac{1}{2}}$ $\tilde{\frac{1}{2}}$ $\tilde{1}$ 0 000− $-2/\sqrt{7.11}$

$\frac{13}{2}$ 4 $\frac{5}{2}$

$\frac{1}{2}$ 2 $\frac{5}{2}$ 0 000+ $+\sqrt{5}/\sqrt{3.13}$

$\frac{1}{2}$ 3 $\frac{5}{2}$ 0 000− $+2/\sqrt{3.7.13}$

$\frac{3}{2}$ 2 $\frac{5}{2}$ 0 000− $-\sqrt{2}/\sqrt{3.13}$

$\frac{3}{2}$ 2 $\frac{5}{2}$ 1 000+ $+4/3\sqrt{13}$

$\frac{3}{2}$ 3 $\frac{5}{2}$ 0 000+ $+\sqrt{2.5}/\sqrt{3.7.13}$

$\frac{3}{2}$ 3 $\frac{5}{2}$ 1 000− $-5\sqrt{2}/3\sqrt{7.13}$

$\frac{5}{2}$ 2 $\frac{5}{2}$ 0 000+ $-2\sqrt{2.3}/\sqrt{11.13}$

$\frac{5}{2}$ 2 $\frac{5}{2}$ 1 000− $+1/\sqrt{11.13}$

$\frac{13}{2}$ 4 $\frac{5}{2}$

$\frac{5}{2}$ 2 $\frac{5}{2}$ 2 000+ $-\sqrt{2}/\sqrt{11.13}$

$\frac{5}{2}$ 3 $\frac{5}{2}$ 0 000− $+2.5\sqrt{2}/\sqrt{7.11.13}$

$\frac{5}{2}$ 3 $\frac{5}{2}$ 1 000+ $-2\sqrt{2.5}/\sqrt{7.11.13}$

$\tilde{\frac{1}{2}}$ 2 $\frac{5}{2}$ 0 000− $+\sqrt{5}/3\sqrt{11}$

$\tilde{\frac{1}{2}}$ 3 $\frac{5}{2}$ 0 000+ $-8/3\sqrt{7.11}$

$\frac{13}{2}$ 4 $\frac{7}{2}$

$\frac{1}{2}$ 2 $\frac{5}{2}$ 0 000− $-1/3\sqrt{13}$

$\frac{1}{2}$ 3 $\frac{5}{2}$ 0 000+ $-5\sqrt{5}/2.3\sqrt{7.13}$

$\frac{1}{2}$ 3 $\tilde{\frac{1}{2}}$ 0 000− $+\sqrt{5}/2\sqrt{13}$

$\frac{3}{2}$ 2 $\frac{5}{2}$ 0 000+ $+\sqrt{2}/3\sqrt{5.13}$

$\frac{3}{2}$ 2 $\frac{5}{2}$ 1 000− $+1/\sqrt{3.5.13}$

$\frac{3}{2}$ 2 $\tilde{\frac{1}{2}}$ 0 000− $+\sqrt{7}/\sqrt{5.13}$

$\frac{3}{2}$ 3 $\frac{5}{2}$ 0 000− $-8\sqrt{2}/3\sqrt{7.13}$

$\frac{3}{2}$ 3 $\frac{5}{2}$ 1 000+ $-\sqrt{2.3}/\sqrt{5.7.13}$

$\frac{5}{2}$ 2 $\frac{5}{2}$ 0 000− $+2\sqrt{2}/\sqrt{5.11.13}$

$\frac{5}{2}$ 2 $\frac{5}{2}$ 1 000+ $-2\sqrt{11}/\sqrt{3.5.13}$

$\frac{5}{2}$ 2 $\frac{5}{2}$ 2 000− $-2\sqrt{2.3}/\sqrt{5.11.13}$

$\frac{5}{2}$ 2 $\tilde{\frac{1}{2}}$ 0 000+ $-2\sqrt{7}/\sqrt{5.11.13}$

$\frac{5}{2}$ 3 $\frac{5}{2}$ 0 000+ $+\sqrt{2.11}/\sqrt{3.5.7.13}$

$\frac{5}{2}$ 3 $\frac{5}{2}$ 1 000− $-4\sqrt{2.3}/\sqrt{7.11.13}$

$\frac{5}{2}$ 3 $\tilde{\frac{1}{2}}$ 0 000− $+1/\sqrt{11.13}$

$\tilde{\frac{1}{2}}$ 2 $\frac{5}{2}$ 0 000+ $+2/\sqrt{3.11}$

$\tilde{\frac{1}{2}}$ 3 $\frac{5}{2}$ 0 000− $+\sqrt{5}/\sqrt{3.7.11}$

$\frac{13}{2}$ $\frac{9}{2}$ 2

$\frac{1}{2}$ $\frac{3}{2}$ 2 0 000− $+2\sqrt{2}/\sqrt{5.13}$

$\frac{1}{2}$ $\frac{5}{2}$ 2 0 000+ $+3/\sqrt{5.7.13}$

$\frac{3}{2}$ $\frac{3}{2}$ 2 0 000+ $-\sqrt{2}/\sqrt{13}$

$\frac{3}{2}$ $\frac{5}{2}$ 2 0 000− 0

$\frac{3}{2}$ $\frac{5}{2}$ 2 1 000+ $-2\sqrt{3}/\sqrt{7.13}$

$\frac{5}{2}$ $\frac{3}{2}$ 2 0 000− $-\sqrt{2.3}/\sqrt{11.13}$

$\frac{5}{2}$ $\frac{3}{2}$ 2 1 000+ $+3/\sqrt{11.13}$

$\frac{5}{2}$ $\frac{5}{2}$ 2 0 000+ $-9\sqrt{2}/\sqrt{7.11.13}$

$\frac{5}{2}$ $\frac{5}{2}$ 2 1 000− $+2.3\sqrt{3}/\sqrt{7.11.13}$

$\frac{5}{2}$ $\frac{5}{2}$ 2 2 000+ $-3\sqrt{2.3}/\sqrt{7.11.13}$

$\tilde{\frac{1}{2}}$ $\frac{3}{2}$ 2 0 000+ $-1/\sqrt{5.11}$

$\tilde{\frac{1}{2}}$ $\frac{5}{2}$ 2 0 000− $+4\sqrt{3}/\sqrt{5.7.11}$

$\frac{13}{2}$ $\frac{9}{2}$ 3

$\frac{1}{2}$ $\frac{5}{2}$ 3 0 000+ $+2\sqrt{3}/\sqrt{7.13}$

$\frac{1}{2}$ $\frac{5}{2}$ $\tilde{1}$ 0 000+ $+1/\sqrt{7.13}$

$\frac{3}{2}$ $\frac{3}{2}$ 3 0 000+ $+2\sqrt{2.3}/5\sqrt{13}$

$\frac{3}{2}$ $\frac{3}{2}$ $\tilde{1}$ 0 000+ $+\sqrt{2}/5\sqrt{13}$

$\frac{3}{2}$ $\frac{5}{2}$ 3 0 000− $+3\sqrt{2.3}/\sqrt{5.7.13}$

$\frac{3}{2}$ $\frac{5}{2}$ 3 1 000+ $-\sqrt{2}/5\sqrt{7.13}$

$\frac{3}{2}$ $\frac{5}{2}$ $\tilde{1}$ 0 000− $-2\sqrt{7}/5\sqrt{13}$

$\frac{5}{2}$ $\frac{3}{2}$ 3 0 000− $+3\sqrt{2}/\sqrt{5.11.13}$

$\frac{5}{2}$ $\frac{3}{2}$ 3 1 000+ $+9\sqrt{2.3}/5\sqrt{11.13}$

$\frac{5}{2}$ $\frac{3}{2}$ $\tilde{1}$ 0 000− $+9\sqrt{3}/5\sqrt{11.13}$

$\frac{5}{2}$ $\frac{5}{2}$ 3 0 000+ $-3\sqrt{2}/5\sqrt{7.11.13}$

$\frac{5}{2}$ $\frac{5}{2}$ 3 1 000− $-9\sqrt{2}/\sqrt{5.7.11.13}$

SO_3–K 3jm Factors (cont.)

$\frac{13}{2}\ \frac{9}{2}\ 3$

$\frac{5}{2}\ \frac{5}{2}\ \tilde{1}$ 0 000+ $+17\sqrt{2.3}/5\sqrt{7.11.13}$

$\frac{5}{2}\ \frac{5}{2}\ \tilde{1}$ 1 000+ $+9\sqrt{2.3}/\sqrt{5.7.11.13}$

$\tilde{\frac{1}{2}}\ \frac{3}{2}\ \tilde{1}$ 0 000+ $+1/\sqrt{11}$

$\tilde{\frac{1}{2}}\ \frac{5}{2}\ 3$ 0 000− $+2/\sqrt{7.11}$

$\frac{13}{2}\ \frac{9}{2}\ 4$

$\frac{1}{2}\ \frac{3}{2}\ 2$ 0 000− $-1/\sqrt{5.13}$

$\frac{1}{2}\ \frac{5}{2}\ 2$ 0 000+ $-11\sqrt{2}/3\sqrt{5.7.13}$

$\frac{1}{2}\ \frac{5}{2}\ 3$ 0 000− $+2\sqrt{2}/3\sqrt{13}$

$\frac{3}{2}\ \frac{3}{2}\ 2$ 0 000+ $-1/5\sqrt{13}$

$\frac{3}{2}\ \frac{3}{2}\ 3$ 0 000− $+2\sqrt{7}/5\sqrt{13}$

$\frac{3}{2}\ \frac{5}{2}\ 2$ 0 000− $-7\sqrt{7}/3.5\sqrt{13}$

$\frac{3}{2}\ \frac{5}{2}\ 2$ 1 000+ $+11\sqrt{2}/5\sqrt{3.7.13}$

$\frac{3}{2}\ \frac{5}{2}\ 3$ 0 000+ $-2/3\sqrt{5.13}$

$\frac{3}{2}\ \frac{5}{2}\ 3$ 1 000− $-2\sqrt{3}/5\sqrt{13}$

$\frac{5}{2}\ \frac{3}{2}\ 2$ 0 000− $+\sqrt{3.13}/5\sqrt{11}$

$\frac{5}{2}\ \frac{3}{2}\ 2$ 1 000+ $+17/5\sqrt{2.11.13}$

$\frac{5}{2}\ \frac{3}{2}\ 3$ 0 000+ $-\sqrt{3.7}/\sqrt{5.11.13}$

$\frac{5}{2}\ \frac{3}{2}\ 3$ 1 000− $-\sqrt{7}/5\sqrt{11.13}$

$\frac{5}{2}\ \frac{5}{2}\ 2$ 0 000+ $-37/5\sqrt{7.11.13}$

$\frac{5}{2}\ \frac{5}{2}\ 2$ 1 000− $+5/\sqrt{2.3.7.11.13}$

$\frac{5}{2}\ \frac{5}{2}\ 2$ 2 000+ $+4\sqrt{3}/5\sqrt{7.11.13}$

$\frac{5}{2}\ \frac{5}{2}\ 3$ 0 000− $-23/5\sqrt{3.11.13}$

$\frac{5}{2}\ \frac{5}{2}\ 3$ 1 000+ $+\sqrt{3.5}/\sqrt{11.13}$

$\tilde{\frac{1}{2}}\ \frac{3}{2}\ 2$ 0 000+ $3/\sqrt{2.5.11}$

$\tilde{\frac{1}{2}}\ \frac{5}{2}\ 2$ 0 000− $+1/\sqrt{2.3.5.7.11}$

$\tilde{\frac{1}{2}}\ \frac{5}{2}\ 3$ 0 000+ $-\sqrt{2}/\sqrt{3.11}$

$\frac{13}{2}\ 5\ \frac{3}{2}$

$\frac{1}{2}\ 1\ \frac{3}{2}$ 0 000+ $+\sqrt{3}/\sqrt{2.13}$

$\frac{1}{2}\ 2\ \frac{3}{2}$ 0 000− $+\sqrt{5}/\sqrt{2.7.13}$

$\frac{3}{2}\ 1\ \frac{3}{2}$ 0 000− $+\sqrt{3}/\sqrt{2.13}$

$\frac{3}{2}\ 2\ \frac{3}{2}$ 0 000+ $-5/\sqrt{2.7.13}$

$\frac{3}{2}\ \tilde{1}\ \frac{3}{2}$ 0 000− $-\sqrt{3}/\sqrt{7.13}$

$\frac{5}{2}\ 1\ \frac{3}{2}$ 0 000+ $-\sqrt{2.3}/\sqrt{11.13}$

$\frac{5}{2}\ 2\ \frac{3}{2}$ 0 000− 0

$\frac{5}{2}\ 2\ \frac{3}{2}$ 1 000+ $-3.5/\sqrt{7.11.13}$

$\frac{5}{2}\ \tilde{1}\ \frac{3}{2}$ 0 000+ $-9\sqrt{2}/\sqrt{7.11.13}$

$\tilde{\frac{1}{2}}\ 2\ \frac{3}{2}$ 0 000+ $-\sqrt{5}/\sqrt{7.11}$

$\tilde{\frac{1}{2}}\ \tilde{1}\ \frac{3}{2}$ 0 000− $-\sqrt{2.3}/\sqrt{7.11}$

$\frac{13}{2}\ 5\ \frac{5}{2}$

$\frac{1}{2}\ 2\ \frac{5}{2}$ 0 000− $+\sqrt{2.5}/\sqrt{7.13}$

$\frac{1}{2}\ \tilde{1}\ \frac{5}{2}$ 0 000+ $+\sqrt{3}/\sqrt{7.13}$

$\frac{3}{2}\ 1\ \frac{5}{2}$ 0 000− $+1/\sqrt{13}$

$\frac{3}{2}\ 2\ \frac{5}{2}$ 0 000+ $+1/\sqrt{7.13}$

$\frac{3}{2}\ 2\ \frac{5}{2}$ 1 000− $+\sqrt{2.3}/\sqrt{7.13}$

$\frac{3}{2}\ \tilde{1}\ \frac{5}{2}$ 0 000− $-2\sqrt{3}/\sqrt{7.13}$

$\frac{5}{2}\ 1\ \frac{5}{2}$ 0 000+ $-\sqrt{5}/2\sqrt{11.13}$

$\frac{5}{2}\ 1\ \frac{5}{2}$ 1 000+ $-9/2\sqrt{11.13}$

$\frac{5}{2}\ 2\ \frac{5}{2}$ 0 000− $+\sqrt{11}/\sqrt{7.13}$

$\frac{5}{2}\ 2\ \frac{5}{2}$ 1 000+ $-\sqrt{3}/\sqrt{2.7.11.13}$

$\frac{13}{2}\ 5\ \frac{5}{2}$

$\frac{5}{2}\ 2\ \frac{5}{2}$ 2 000− $-4\sqrt{3}/\sqrt{7.11.13}$

$\frac{5}{2}\ \tilde{1}\ \frac{5}{2}$ 0 000+ $-3\sqrt{2}/\sqrt{7.11.13}$

$\frac{5}{2}\ \tilde{1}\ \frac{5}{2}$ 1 000+ $+3\sqrt{2.5}/\sqrt{7.11.13}$

$\tilde{\frac{1}{2}}\ 1\ \frac{5}{2}$ 0 000− $+1/\sqrt{2.11}$

$\tilde{\frac{1}{2}}\ 2\ \frac{5}{2}$ 0 000+ $-\sqrt{3.5}/\sqrt{2.7.11}$

$\frac{13}{2}\ 5\ \frac{7}{2}$

$\frac{1}{2}\ 2\ \frac{5}{2}$ 0 000+ $+\sqrt{3}/\sqrt{7.13}$

$\frac{1}{2}\ \tilde{1}\ \frac{5}{2}$ 0 000− $+\sqrt{2.5}/\sqrt{7.13}$

$\frac{3}{2}\ 1\ \frac{5}{2}$ 0 000+ $+\sqrt{3}/\sqrt{2.5.13}$

$\frac{3}{2}\ 2\ \frac{5}{2}$ 0 000− $+4\sqrt{2.3}/\sqrt{5.7.13}$

$\frac{3}{2}\ 2\ \frac{5}{2}$ 1 000+ $-1/2\sqrt{5.7.13}$

$\frac{3}{2}\ 2\ \tilde{\frac{1}{2}}$ 0 000+ $-\sqrt{3}/2\sqrt{5.13}$

$\frac{3}{2}\ \tilde{1}\ \frac{5}{2}$ 0 000+ $+1/\sqrt{2.5.7.13}$

$\frac{3}{2}\ \tilde{1}\ \tilde{\frac{1}{2}}$ 0 000− $-1/\sqrt{2.13}$

$\frac{5}{2}\ 1\ \frac{5}{2}$ 0 000− $+3\sqrt{3}/\sqrt{2.11.13}$

$\frac{5}{2}\ 1\ \frac{5}{2}$ 1 000− $-9\sqrt{3}/2\sqrt{2.5.11.13}$

$\frac{5}{2}\ 1\ \tilde{\frac{1}{2}}$ 0 000+ $+3\sqrt{7}/4\sqrt{11.13}$

$\frac{5}{2}\ 2\ \frac{5}{2}$ 0 000+ $-3\sqrt{3.11}/4\sqrt{2.5.7.13}$

$\frac{5}{2}\ 2\ \frac{5}{2}$ 1 000− $-3/2\sqrt{5.7.11.13}$

$\frac{5}{2}\ 2\ \frac{5}{2}$ 2 000+ $+9\sqrt{11}/4\sqrt{2.5.7.13}$

$\frac{5}{2}\ 2\ \tilde{\frac{1}{2}}$ 0 000− $-3.7\sqrt{3}/4\sqrt{5.11.13}$

$\frac{5}{2}\ \tilde{1}\ \frac{5}{2}$ 0 000− $+37\sqrt{3}/4\sqrt{5.7.11.13}$

$\frac{5}{2}\ \tilde{1}\ \frac{5}{2}$ 1 000− $-9\sqrt{3}/4\sqrt{7.11.13}$

$\tilde{\frac{1}{2}}\ 1\ \frac{5}{2}$ 0 000+ $-\sqrt{3.5}/4\sqrt{11}$

$\tilde{\frac{1}{2}}\ 2\ \frac{5}{2}$ 0 000− $+1/4\sqrt{7.11}$

$\tilde{\frac{1}{2}}\ \tilde{1}\ \tilde{\frac{1}{2}}$ 0 000− $+\sqrt{5}/2\sqrt{2.11}$

$\frac{13}{2}\ 5\ \frac{9}{2}$

$\frac{1}{2}\ 1\ \frac{3}{2}$ 0 000− $-3/\sqrt{2.13.17}$

$\frac{1}{2}\ 2\ \frac{3}{2}$ 0 000+ $-19\sqrt{3}/\sqrt{2.5.7.13.17}$

$\frac{1}{2}\ 2\ \frac{5}{2}$ 0 000− $+4\sqrt{3}/\sqrt{5.13.17}$

$\frac{1}{2}\ \tilde{1}\ \frac{5}{2}$ 0 000+ $+\sqrt{2}/\sqrt{13.17}$

$\frac{3}{2}\ 1\ \frac{3}{2}$ 0 000+ $+3\sqrt{2}/5\sqrt{13.17}$

$\frac{3}{2}\ 1\ \frac{5}{2}$ 0 000− $+7\sqrt{3.7}/5\sqrt{2.13.17}$

$\frac{3}{2}\ 2\ \frac{3}{2}$ 0 000− $+2\sqrt{2.3}/5\sqrt{7.13.17}$

$\frac{3}{2}\ 2\ \frac{5}{2}$ 0 000+ $-\sqrt{3}/5\sqrt{2.13.17}$

$\frac{3}{2}\ 2\ \frac{5}{2}$ 1 000− $-\sqrt{17}/5\sqrt{13}$

$\frac{3}{2}\ \tilde{1}\ \frac{3}{2}$ 0 000+ $+71/5\sqrt{7.13.17}$

$\frac{3}{2}\ \tilde{1}\ \frac{5}{2}$ 0 000− $-4\sqrt{2}/5\sqrt{13.17}$

$\frac{5}{2}\ 1\ \frac{3}{2}$ 0 000− $-2.3\sqrt{2}/5\sqrt{11.13.17}$

$\frac{5}{2}\ 1\ \frac{5}{2}$ 0 000+ $-3.7\sqrt{3.7}/2\sqrt{2.5.11.13.17}$

$\frac{5}{2}\ 1\ \frac{5}{2}$ 1 000+ $+9.3\sqrt{3.7}/2.5\sqrt{2.11.13.17}$

$\frac{5}{2}\ 2\ \frac{3}{2}$ 0 000+ $-3.7\sqrt{2.7}/5\sqrt{11.13.17}$

$\frac{5}{2}\ 2\ \frac{3}{2}$ 1 000− $-9\sqrt{3}/5\sqrt{7.11.13.17}$

$\frac{5}{2}\ 2\ \frac{5}{2}$ 0 000− $+3\sqrt{3.13}/5\sqrt{2.11.17}$

$\frac{5}{2}\ 2\ \frac{5}{2}$ 1 000+ $-3.5/2\sqrt{11.13.17}$

$\frac{5}{2}\ 2\ \frac{5}{2}$ 2 000− $+2.9\sqrt{2}/5\sqrt{11.13.17}$

$\frac{5}{2}\ \tilde{1}\ \frac{3}{2}$ 0 000− $-9\sqrt{3.11}/5\sqrt{2.7.13.17}$

$\frac{5}{2}\ \tilde{1}\ \frac{5}{2}$ 0 000+ $+8.2\sqrt{3}/5\sqrt{11.13.17}$

$\frac{5}{2}\ \tilde{1}\ \frac{5}{2}$ 1 000+ $-2.9\sqrt{3}/\sqrt{5.11.13.17}$

SO_3–K 3jm Factors (cont.)

$\frac{13}{2}$ 5 $\frac{9}{2}$
$\frac{\tilde{1}}{2}$ 1 $\frac{5}{2}$ 0 000− $+\sqrt{3.7}/2\sqrt{11.17}$
$\frac{\tilde{1}}{2}$ 2 $\frac{3}{2}$ 0 000− $-9\sqrt{3}/\sqrt{5.7.11.17}$
$\frac{\tilde{1}}{2}$ 2 $\frac{5}{2}$ 0 000+ $+\sqrt{17}/2\sqrt{5.11}$
$\frac{\tilde{1}}{2}$ $\tilde{1}$ $\frac{3}{2}$ 0 000+ $-1/\sqrt{2.7.11.17}$

$\frac{13}{2}$ $\frac{11}{2}$ 1
$\frac{1}{2}$ $\frac{1}{2}$ 1 0 000− $+2/\sqrt{3.13}$
$\frac{1}{2}$ $\frac{3}{2}$ 1 0 000+ $+\sqrt{11}/\sqrt{3.7.13}$
$\frac{3}{2}$ $\frac{1}{2}$ 1 0 000+ $+\sqrt{5}/\sqrt{2.3.13}$
$\frac{3}{2}$ $\frac{3}{2}$ 1 0 000− $+2\sqrt{11}/\sqrt{3.7.13}$
$\frac{3}{2}$ $\frac{5}{2}$ 1 0 000+ $-\sqrt{11}/\sqrt{2.7.13}$
$\frac{5}{2}$ $\frac{3}{2}$ 1 0 000+ $-2\sqrt{3}/\sqrt{7.13}$
$\frac{5}{2}$ $\frac{5}{2}$ 1 0 000− $+3\sqrt{5}/\sqrt{2.7.13}$
$\frac{5}{2}$ $\frac{5}{2}$ 1 1 000− $+3/\sqrt{2.7.13}$
$\frac{\tilde{1}}{2}$ $\frac{5}{2}$ 1 0 000+ $-1/\sqrt{7}$

$\frac{13}{2}$ $\frac{11}{2}$ 2
$\frac{1}{2}$ $\frac{3}{2}$ 2 0 000+ $+3/\sqrt{7.13}$
$\frac{1}{2}$ $\frac{5}{2}$ 2 0 000− $+2/\sqrt{7.13}$
$\frac{3}{2}$ $\frac{1}{2}$ 2 0 000+ $-\sqrt{11}/\sqrt{2.5.13}$
$\frac{3}{2}$ $\frac{3}{2}$ 2 0 000− $-2/\sqrt{5.7.13}$
$\frac{3}{2}$ $\frac{5}{2}$ 2 0 000+ $-11/\sqrt{2.5.7.13}$
$\frac{3}{2}$ $\frac{5}{2}$ 2 1 000− $-3\sqrt{3}/\sqrt{5.7.13}$
$\frac{5}{2}$ $\frac{1}{2}$ 2 0 000− $-4/\sqrt{3.5.13}$
$\frac{5}{2}$ $\frac{3}{2}$ 2 0 000+ $+2.19/\sqrt{3.5.7.11.13}$
$\frac{5}{2}$ $\frac{3}{2}$ 2 1 000− $-8\sqrt{2}/\sqrt{5.7.11.13}$
$\frac{5}{2}$ $\frac{5}{2}$ 2 0 000− $+3\sqrt{2}/\sqrt{5.7.11.13}$
$\frac{5}{2}$ $\frac{5}{2}$ 2 1 000+ $-\sqrt{3.13}/\sqrt{5.7.11}$
$\frac{5}{2}$ $\frac{5}{2}$ 2 2 000− $-2\sqrt{2.3.5}/\sqrt{7.11.13}$
$\frac{\tilde{1}}{2}$ $\frac{3}{2}$ 2 0 000− $+2\sqrt{2}/\sqrt{7.11}$
$\frac{\tilde{1}}{2}$ $\frac{5}{2}$ 2 0 000+ $-\sqrt{3}/\sqrt{7.11}$

$\frac{13}{2}$ $\frac{11}{2}$ 3
$\frac{1}{2}$ $\frac{5}{2}$ 3 0 000− $+1/\sqrt{7.13}$
$\frac{1}{2}$ $\frac{5}{2}$ $\tilde{1}$ 0 000− $+2\sqrt{3}/\sqrt{7.13}$
$\frac{3}{2}$ $\frac{3}{2}$ 3 0 000− $+1/\sqrt{7.13}$
$\frac{3}{2}$ $\frac{3}{2}$ $\tilde{1}$ 0 000− $+2\sqrt{3}/\sqrt{7.13}$
$\frac{3}{2}$ $\frac{5}{2}$ 3 0 000+ $-\sqrt{5}/2\sqrt{2.7.13}$
$\frac{3}{2}$ $\frac{5}{2}$ 3 1 000− $+5\sqrt{3}/2\sqrt{2.7.13}$
$\frac{3}{2}$ $\frac{5}{2}$ $\tilde{1}$ 0 000+ $+\sqrt{3}/\sqrt{7.13}$
$\frac{5}{2}$ $\frac{1}{2}$ 3 0 000− $+\sqrt{5}/4\sqrt{3.13}$
$\frac{5}{2}$ $\frac{1}{2}$ $\tilde{1}$ 0 000− $+\sqrt{5}/2\sqrt{13}$
$\frac{5}{2}$ $\frac{3}{2}$ 3 0 000+ $-19\sqrt{5}/2\sqrt{3.7.11.13}$
$\frac{5}{2}$ $\frac{3}{2}$ 3 1 000− $+9/2\sqrt{7.11.13}$
$\frac{5}{2}$ $\frac{3}{2}$ $\tilde{1}$ 0 000+ $-1/\sqrt{2.7.11.13}$
$\frac{5}{2}$ $\frac{5}{2}$ 3 0 000− $-5.5\sqrt{3}/4\sqrt{2.7.11.13}$
$\frac{5}{2}$ $\frac{5}{2}$ 3 1 000+ $+\sqrt{3.5.13}/4\sqrt{2.7.11}$
$\frac{5}{2}$ $\frac{5}{2}$ $\tilde{1}$ 0 000− $+9/2\sqrt{2.7.11.13}$
$\frac{5}{2}$ $\frac{5}{2}$ $\tilde{1}$ 1 000− $-3\sqrt{5}/2\sqrt{2.7.11.13}$
$\frac{\tilde{1}}{2}$ $\frac{1}{2}$ 3 0 000+ $+1/4$
$\frac{\tilde{1}}{2}$ $\frac{3}{2}$ $\tilde{1}$ 0 000− $-\sqrt{3}/\sqrt{2.7.11}$
$\frac{\tilde{1}}{2}$ $\frac{5}{2}$ 3 0 000+ $+5\sqrt{3}/4\sqrt{7.11}$

$\frac{13}{2}$ $\frac{11}{2}$ 4
$\frac{1}{2}$ $\frac{3}{2}$ 2 0 000+ $+4\sqrt{3}/\sqrt{7.13.17}$
$\frac{1}{2}$ $\frac{5}{2}$ 2 0 000− $+2.11/\sqrt{3.7.13.17}$
$\frac{1}{2}$ $\frac{5}{2}$ 3 0 000+ $-\sqrt{5}/\sqrt{3.13.17}$
$\frac{3}{2}$ $\frac{1}{2}$ 2 0 000+ $-2\sqrt{2.11}/\sqrt{3.5.13.17}$
$\frac{3}{2}$ $\frac{3}{2}$ 2 0 000− $+2\sqrt{17}/\sqrt{3.5.7.13}$
$\frac{3}{2}$ $\frac{3}{2}$ 3 0 000+ $-\sqrt{13}/\sqrt{3.5.17}$
$\frac{3}{2}$ $\frac{5}{2}$ 2 0 000+ $-29\sqrt{2}/\sqrt{3.5.7.13.17}$
$\frac{3}{2}$ $\frac{5}{2}$ 2 1 000− $+11\sqrt{5}/3\sqrt{7.13.17}$
$\frac{3}{2}$ $\frac{5}{2}$ 3 0 000− $+\sqrt{13}/2\sqrt{2.3.17}$
$\frac{3}{2}$ $\frac{5}{2}$ 3 1 000+ $+29/2.3\sqrt{2.5.13.17}$
$\frac{5}{2}$ $\frac{1}{2}$ 2 0 000− $+1/2\sqrt{5.13.17}$
$\frac{5}{2}$ $\frac{1}{2}$ 3 0 000+ $-7\sqrt{7}/4\sqrt{13.17}$
$\frac{5}{2}$ $\frac{3}{2}$ 2 0 000+ $+53/\sqrt{5.7.11.13.17}$
$\frac{5}{2}$ $\frac{3}{2}$ 2 1 000− $+9\sqrt{3}/\sqrt{2.5.7.11.13.17}$
$\frac{5}{2}$ $\frac{3}{2}$ 3 0 000− $+\sqrt{13}/2\sqrt{11.17}$
$\frac{5}{2}$ $\frac{3}{2}$ 3 1 000+ $+41\sqrt{3}/2\sqrt{5.11.13.17}$
$\frac{5}{2}$ $\frac{5}{2}$ 2 0 000− $+37\sqrt{3}/2\sqrt{2.5.7.11.13.17}$
$\frac{5}{2}$ $\frac{5}{2}$ 2 1 000+ $+13\sqrt{13}/2\sqrt{5.7.11.17}$
$\frac{5}{2}$ $\frac{5}{2}$ 2 2 000− $+197/2\sqrt{2.5.7.11.13.17}$
$\frac{5}{2}$ $\frac{5}{2}$ 3 0 000+ $+\sqrt{13}/4\sqrt{2.5.11.17}$
$\frac{5}{2}$ $\frac{5}{2}$ 3 1 000− $+47/4\sqrt{2.11.13.17}$
$\frac{\tilde{1}}{2}$ $\frac{1}{2}$ 3 0 000− $-\sqrt{5.7}/4\sqrt{3.17}$
$\frac{\tilde{1}}{2}$ $\frac{3}{2}$ 2 0 000− $-19/\sqrt{2.3.7.11.17}$
$\frac{\tilde{1}}{2}$ $\frac{5}{2}$ 2 0 000+ $+2/3\sqrt{7.11.17}$
$\frac{\tilde{1}}{2}$ $\frac{5}{2}$ 3 0 000− $+\sqrt{5.17}/4.3\sqrt{11}$

$\frac{13}{2}$ $\frac{11}{2}$ 5
$\frac{1}{2}$ $\frac{1}{2}$ 1 0 000− $-\sqrt{5.11}/2\sqrt{3.13.17}$
$\frac{1}{2}$ $\frac{3}{2}$ 1 0 000+ $-8\sqrt{5}/\sqrt{3.7.13.17}$
$\frac{1}{2}$ $\frac{3}{2}$ 2 0 000− $+2/\sqrt{13.17}$
$\frac{1}{2}$ $\frac{5}{2}$ 2 0 000+ $-1/2\sqrt{13.17}$
$\frac{1}{2}$ $\frac{5}{2}$ $\tilde{1}$ 0 000− $+\sqrt{3.5}/\sqrt{2.13.17}$
$\frac{3}{2}$ $\frac{1}{2}$ 1 0 000+ $+\sqrt{11}/2\sqrt{2.3.13.17}$
$\frac{3}{2}$ $\frac{1}{2}$ 2 0 000− $-3\sqrt{7.11}/2\sqrt{2.5.13.17}$
$\frac{3}{2}$ $\frac{3}{2}$ 1 0 000− $+1/2\sqrt{3.5.7.13.17}$
$\frac{3}{2}$ $\frac{3}{2}$ 2 0 000+ $+\sqrt{13}/2\sqrt{5.17}$
$\frac{3}{2}$ $\frac{3}{2}$ $\tilde{1}$ 0 000− $+\sqrt{3}/\sqrt{2.5.13.17}$
$\frac{3}{2}$ $\frac{5}{2}$ 1 0 000+ $+73/2\sqrt{2.5.7.13.17}$
$\frac{3}{2}$ $\frac{5}{2}$ 2 0 000− $+19/2\sqrt{2.5.13.17}$
$\frac{3}{2}$ $\frac{5}{2}$ 2 1 000+ $-\sqrt{3.5}/2\sqrt{13.17}$
$\frac{3}{2}$ $\frac{5}{2}$ $\tilde{1}$ 0 000+ $+2\sqrt{2.3}/\sqrt{5.13.17}$
$\frac{5}{2}$ $\frac{1}{2}$ 2 0 000+ $-\sqrt{7}/\sqrt{3.5.13.17}$
$\frac{5}{2}$ $\frac{1}{2}$ $\tilde{1}$ 0 000− $-\sqrt{2.7}/\sqrt{13.17}$
$\frac{5}{2}$ $\frac{3}{2}$ 1 0 000+ $-2.9\sqrt{3}/\sqrt{5.7.11.13.17}$
$\frac{5}{2}$ $\frac{3}{2}$ 2 0 000− $-31/\sqrt{3.5.11.13.17}$
$\frac{5}{2}$ $\frac{3}{2}$ 2 1 000+ $-43/\sqrt{2.5.11.13.17}$
$\frac{5}{2}$ $\frac{3}{2}$ $\tilde{1}$ 0 000+ $+\sqrt{11}/\sqrt{5.13.17}$
$\frac{5}{2}$ $\frac{5}{2}$ 1 0 000− $+5/2\sqrt{2.7.11.13.17}$
$\frac{5}{2}$ $\frac{5}{2}$ 1 1 000− $+3.67/2\sqrt{2.5.7.11.13.17}$
$\frac{5}{2}$ $\frac{5}{2}$ 2 0 000+ $+\sqrt{13}/\sqrt{2.5.11.17}$

SO_3–K 3jm Factors (cont.)

$\frac{13}{2}$	$\frac{11}{2}$	5		
$\frac{5}{2}$	$\frac{5}{2}$	2	1 000−	$-3.7\sqrt{3}/2\sqrt{5.11.13.17}$
$\frac{5}{2}$	$\frac{5}{2}$	2	2 000+	$+\sqrt{2.3.13}/\sqrt{5.11.17}$
$\frac{5}{2}$	$\frac{5}{2}$	$\tilde{1}$	0 000−	$+8.3/\sqrt{5.11.13.17}$
$\frac{5}{2}$	$\frac{5}{2}$	$\tilde{1}$	1 000−	$+2.3/\sqrt{11.13.17}$
$\tilde{\frac{1}{2}}$	$\frac{3}{2}$	2	0 000+	$+1/\sqrt{2.11.17}$
$\tilde{\frac{1}{2}}$	$\frac{3}{2}$	$\tilde{1}$	0 000−	$+\sqrt{3.5}/\sqrt{11.17}$
$\tilde{\frac{1}{2}}$	$\frac{5}{2}$	1	0 000+	$-5\sqrt{5}/2\sqrt{7.11.17}$
$\tilde{\frac{1}{2}}$	$\frac{5}{2}$	2	0 000−	$+3\sqrt{3}/2\sqrt{11.17}$

$\frac{13}{2}$	6	$\frac{1}{2}$		
$\frac{1}{2}$	0	$\frac{1}{2}$	0 000+	$+1/\sqrt{13}$
$\frac{1}{2}$	1	$\frac{1}{2}$	0 000−	$+\sqrt{2.3}/\sqrt{7.13}$
$\frac{3}{2}$	1	$\frac{1}{2}$	0 000+	$+\sqrt{3.5}/\sqrt{7.13}$
$\frac{3}{2}$	2	$\frac{1}{2}$	0 000−	$+\sqrt{11}/\sqrt{7.13}$
$\frac{5}{2}$	2	$\frac{1}{2}$	0 000+	$+2\sqrt{2.3}/\sqrt{7.13}$
$\frac{5}{2}$	3	$\frac{1}{2}$	0 000−	$+\sqrt{3.5}/\sqrt{7.13}$
$\tilde{\frac{1}{2}}$	3	$\frac{1}{2}$	0 000+	$+1/\sqrt{7}$

$\frac{13}{2}$	6	$\frac{3}{2}$		
$\frac{1}{2}$	1	$\frac{3}{2}$	0 000−	$+\sqrt{3.5}/\sqrt{2.7.13}$
$\frac{1}{2}$	2	$\frac{3}{2}$	0 000+	$+\sqrt{11}/\sqrt{2.7.13}$
$\frac{3}{2}$	0	$\frac{3}{2}$	0 000−	$-1/\sqrt{13}$
$\frac{3}{2}$	1	$\frac{3}{2}$	0 000+	$-\sqrt{3}/\sqrt{2.5.7.13}$
$\frac{3}{2}$	2	$\frac{3}{2}$	0 000−	$-3\sqrt{11}/\sqrt{2.5.7.13}$
$\frac{3}{2}$	3	$\frac{3}{2}$	0 000+	$-2\sqrt{11}/\sqrt{5.7.13}$
$\frac{5}{2}$	1	$\frac{3}{2}$	0 000−	$+\sqrt{2.3.11}/\sqrt{5.7.13}$
$\frac{5}{2}$	2	$\frac{3}{2}$	0 000+	$-2\sqrt{2.3}/\sqrt{5.7.13}$
$\frac{5}{2}$	2	$\frac{3}{2}$	1 000−	$+3/\sqrt{5.7.13}$
$\frac{5}{2}$	3	$\frac{3}{2}$	0 000−	$+2\sqrt{3}/\sqrt{7.13}$
$\frac{5}{2}$	3	$\frac{3}{2}$	1 000+	$+2.3/\sqrt{5.7.13}$
$\tilde{\frac{1}{2}}$	2	$\frac{3}{2}$	0 000−	$+1/\sqrt{7}$

$\frac{13}{2}$	6	$\frac{5}{2}$		
$\frac{1}{2}$	2	$\frac{5}{2}$	0 000+	$+2\sqrt{2}/\sqrt{7.13}$
$\frac{1}{2}$	3	$\frac{5}{2}$	0 000−	$+\sqrt{5}/\sqrt{7.13}$
$\frac{3}{2}$	1	$\frac{5}{2}$	0 000+	$+2\sqrt{11}/\sqrt{5.7.13}$
$\frac{3}{2}$	2	$\frac{5}{2}$	0 000−	$-4/\sqrt{5.7.13}$
$\frac{3}{2}$	2	$\frac{5}{2}$	1 000+	$-\sqrt{2.3}/\sqrt{5.7.13}$
$\frac{3}{2}$	3	$\frac{5}{2}$	0 000+	$-1/2\sqrt{2.7.13}$
$\frac{3}{2}$	3	$\frac{5}{2}$	1 000−	$-\sqrt{3.13}/2\sqrt{2.5.7}$
$\frac{5}{2}$	0	$\frac{5}{2}$	0 000+	$+1/\sqrt{13}$
$\frac{5}{2}$	1	$\frac{5}{2}$	0 000−	$-\sqrt{7}/4\sqrt{13}$
$\frac{5}{2}$	1	$\frac{5}{2}$	1 000−	$+9/4\sqrt{5.7.13}$
$\frac{5}{2}$	2	$\frac{5}{2}$	0 000+	$+\sqrt{7}/2\sqrt{5.11.13}$
$\frac{5}{2}$	2	$\frac{5}{2}$	1 000−	$-5\sqrt{3.5}/2\sqrt{2.7.11.13}$
$\frac{5}{2}$	2	$\frac{5}{2}$	2 000+	$+2.9\sqrt{3}/\sqrt{5.7.11.13}$
$\frac{5}{2}$	3	$\frac{5}{2}$	0 000−	$+\sqrt{3}/2\sqrt{2.5.7.11.13}$
$\frac{5}{2}$	3	$\frac{5}{2}$	1 000+	$+\sqrt{3.13}/2\sqrt{2.7.11}$
$\tilde{\frac{1}{2}}$	1	$\frac{5}{2}$	0 000+	$-\sqrt{5}/2\sqrt{2.7}$
$\tilde{\frac{1}{2}}$	2	$\frac{5}{2}$	0 000−	$-\sqrt{3}/2\sqrt{2.7.11}$
$\tilde{\frac{1}{2}}$	3	$\frac{5}{2}$	0 000+	$+\sqrt{3.5}/2\sqrt{7.11}$

$\frac{13}{2}$	6	$\frac{7}{2}$		
$\frac{1}{2}$	2	$\frac{5}{2}$	0 000−	$-3\sqrt{5}/\sqrt{7.13.17}$
$\frac{1}{2}$	3	$\frac{5}{2}$	0 000+	$-9\sqrt{2}/\sqrt{7.13.17}$
$\frac{1}{2}$	3	$\tilde{\frac{1}{2}}$	0 000−	$+\sqrt{2}/\sqrt{13.17}$
$\frac{3}{2}$	1	$\frac{5}{2}$	0 000−	$-3\sqrt{11}/\sqrt{2.7.13.17}$
$\frac{3}{2}$	2	$\frac{5}{2}$	0 000+	$+3\sqrt{2}/\sqrt{7.13.17}$
$\frac{3}{2}$	2	$\frac{5}{2}$	1 000−	$+\sqrt{3.17}/2\sqrt{7.13}$
$\frac{3}{2}$	2	$\tilde{\frac{1}{2}}$	0 000−	$+7/2\sqrt{13.17}$
$\frac{3}{2}$	3	$\frac{5}{2}$	0 000−	$+3\sqrt{5}/\sqrt{7.13.17}$
$\frac{3}{2}$	3	$\frac{5}{2}$	1 000+	$-3\sqrt{3}/\sqrt{7.13.17}$
$\frac{5}{2}$	0	$\frac{5}{2}$	0 000−	$-3\sqrt{5}/2\sqrt{2.13.17}$
$\frac{5}{2}$	1	$\frac{5}{2}$	0 000+	0
$\frac{5}{2}$	1	$\frac{5}{2}$	1 000+	$-3.11/2\sqrt{2.7.13.17}$
$\frac{5}{2}$	1	$\tilde{\frac{1}{2}}$	0 000−	$-5\sqrt{3.5}/4\sqrt{13.17}$
$\frac{5}{2}$	2	$\frac{5}{2}$	0 000−	$-3\sqrt{7}/4\sqrt{2.11.13.17}$
$\frac{5}{2}$	2	$\frac{5}{2}$	1 000+	$-\sqrt{3.11}/2\sqrt{7.13.17}$
$\frac{5}{2}$	2	$\frac{5}{2}$	2 000−	$+109\sqrt{3}/4\sqrt{2.7.11.13.17}$
$\frac{5}{2}$	2	$\tilde{\frac{1}{2}}$	0 000+	$+9.3/4\sqrt{11.13.17}$
$\frac{5}{2}$	3	$\frac{5}{2}$	0 000+	$+\sqrt{3.11}/\sqrt{7.13.17}$
$\frac{5}{2}$	3	$\frac{5}{2}$	1 000−	$+2.5\sqrt{3.5}/\sqrt{7.11.13.17}$
$\frac{5}{2}$	3	$\tilde{\frac{1}{2}}$	0 000−	$+3\sqrt{5}/\sqrt{2.11.13.17}$
$\tilde{\frac{1}{2}}$	0	$\tilde{\frac{1}{2}}$	0 000−	$-\sqrt{7}/2\sqrt{2.17}$
$\tilde{\frac{1}{2}}$	1	$\frac{5}{2}$	0 000−	$-3/4\sqrt{7.17}$
$\tilde{\frac{1}{2}}$	2	$\frac{5}{2}$	0 000+	$-9\sqrt{3.5}/4\sqrt{7.11.17}$
$\tilde{\frac{1}{2}}$	3	$\frac{5}{2}$	0 000−	$+5\sqrt{3}/\sqrt{2.7.11.17}$

$\frac{13}{2}$	6	$\frac{9}{2}$		
$\frac{1}{2}$	1	$\frac{3}{2}$	0 000+	$+\sqrt{3.11}/\sqrt{7.13.17}$
$\frac{1}{2}$	2	$\frac{3}{2}$	0 000−	$+5\sqrt{5}/\sqrt{7.13.17}$
$\frac{1}{2}$	2	$\frac{5}{2}$	0 000+	0
$\frac{1}{2}$	3	$\frac{5}{2}$	0 000−	$+3/\sqrt{13.17}$
$\frac{3}{2}$	0	$\frac{3}{2}$	0 000+	$-\sqrt{2.11}/\sqrt{5.13.17}$
$\frac{3}{2}$	1	$\frac{3}{2}$	0 000−	$-8\sqrt{3.11}/5\sqrt{7.13.17}$
$\frac{3}{2}$	1	$\frac{5}{2}$	0 000+	$-3\sqrt{11}/5\sqrt{13.17}$
$\frac{3}{2}$	2	$\frac{3}{2}$	0 000+	$+2.3/\sqrt{7.13.17}$
$\frac{3}{2}$	2	$\frac{5}{2}$	0 000−	$-3/\sqrt{13.17}$
$\frac{3}{2}$	2	$\frac{5}{2}$	1 000+	0
$\frac{3}{2}$	3	$\frac{3}{2}$	0 000−	$-29\sqrt{2}/5\sqrt{7.13.17}$
$\frac{3}{2}$	3	$\frac{5}{2}$	0 000+	$-9.3/2\sqrt{2.5.13.17}$
$\frac{3}{2}$	3	$\frac{5}{2}$	1 000−	$-7\sqrt{3}/2.5\sqrt{2.13.17}$
$\frac{5}{2}$	0	$\frac{5}{2}$	0 000+	$-3\sqrt{7}/\sqrt{5.13.17}$
$\frac{5}{2}$	1	$\frac{3}{2}$	0 000+	$+\sqrt{3}/5\sqrt{7.13.17}$
$\frac{5}{2}$	1	$\frac{5}{2}$	0 000−	$-9.3/4\sqrt{5.13.17}$
$\frac{5}{2}$	1	$\frac{5}{2}$	1 000−	$-3\sqrt{17}/4.5\sqrt{13}$
$\frac{5}{2}$	2	$\frac{3}{2}$	0 000−	$+\sqrt{3}/\sqrt{7.11.13.17}$
$\frac{5}{2}$	2	$\frac{3}{2}$	1 000+	$+3.5\sqrt{2}/\sqrt{7.11.13.17}$
$\frac{5}{2}$	2	$\frac{5}{2}$	0 000+	$+3.7/2\sqrt{11.13.17}$
$\frac{5}{2}$	2	$\frac{5}{2}$	1 000−	$-5.5\sqrt{3}/2\sqrt{2.11.13.17}$
$\frac{5}{2}$	2	$\frac{5}{2}$	2 000+	$-5\sqrt{3}/\sqrt{11.13.17}$
$\frac{5}{2}$	3	$\frac{3}{2}$	0 000+	$-4\sqrt{2.3.5}/\sqrt{7.11.13.17}$
$\frac{5}{2}$	3	$\frac{3}{2}$	1 000−	$-3.41\sqrt{2}/5\sqrt{7.11.13.17}$

SO_3–K 3jm Factors (cont.)

$\frac{13}{2}$ 6 $\frac{9}{2}$
$\frac{5}{2}$ 3 $\frac{5}{2}$ 0 000− $+19\sqrt{3}/2.5\sqrt{2.11.13.17}$
$\frac{5}{2}$ 3 $\frac{5}{2}$ 1 000+ $-\sqrt{3}/2\sqrt{2.5.11.13.17}$
$\frac{\tilde{1}}{2}$ 1 $\frac{5}{2}$ 0 000+ $+3/2\sqrt{2.17}$
$\frac{\tilde{1}}{2}$ 2 $\frac{3}{2}$ 0 000+ $-2\sqrt{2.5}/\sqrt{7.11.17}$
$\frac{\tilde{1}}{2}$ 2 $\frac{5}{2}$ 0 000− $-\sqrt{3.5}/2\sqrt{2.11.17}$
$\frac{\tilde{1}}{2}$ 3 $\frac{5}{2}$ 0 000+ $+3\sqrt{3}/2\sqrt{11.17}$
$\frac{13}{2}$ 6 $\frac{11}{2}$
$\frac{1}{2}$ 0 $\frac{1}{2}$ 0 000− $-11/\sqrt{2.13.17.19}$
$\frac{1}{2}$ 1 $\frac{1}{2}$ 0 000+ $-11\sqrt{13}/2\sqrt{3.7.17.19}$
$\frac{1}{2}$ 1 $\frac{3}{2}$ 0 000− $+2\sqrt{11}/\sqrt{3.13.17.19}$
$\frac{1}{2}$ 2 $\frac{3}{2}$ 0 000+ $-4\sqrt{5}/\sqrt{13.17.19}$
$\frac{1}{2}$ 2 $\frac{5}{2}$ 0 000− $+11\sqrt{5}/2\sqrt{13.17.19}$
$\frac{1}{2}$ 3 $\frac{5}{2}$ 0 000+ $+5\sqrt{2}/\sqrt{13.17.19}$
$\frac{3}{2}$ 0 $\frac{3}{2}$ 0 000− $-\sqrt{5.7.11}/\sqrt{2.13.17.19}$
$\frac{3}{2}$ 1 $\frac{1}{2}$ 0 000− $+11\sqrt{5}/2\sqrt{2.3.7.13.17.19}$
$\frac{3}{2}$ 1 $\frac{3}{2}$ 0 000+ $-\sqrt{11.13}/2\sqrt{3.17.19}$
$\frac{3}{2}$ 1 $\frac{5}{2}$ 0 000− $+7\sqrt{11}/2\sqrt{2.13.17.19}$
$\frac{3}{2}$ 2 $\frac{1}{2}$ 0 000+ $-29\sqrt{11}/2\sqrt{2.7.13.17.19}$
$\frac{3}{2}$ 2 $\frac{3}{2}$ 0 000− $-3/2\sqrt{13.17.19}$
$\frac{3}{2}$ 2 $\frac{5}{2}$ 0 000+ $+5.7/2\sqrt{2.13.17.19}$
$\frac{3}{2}$ 2 $\frac{5}{2}$ 1 000− $-11\sqrt{3}/2\sqrt{13.17.19}$
$\frac{3}{2}$ 3 $\frac{3}{2}$ 0 000+ $+\sqrt{2.13}/\sqrt{17.19}$
$\frac{3}{2}$ 3 $\frac{5}{2}$ 0 000− $+\sqrt{5}/\sqrt{13.17.19}$
$\frac{3}{2}$ 3 $\frac{5}{2}$ 1 000+ $-3\sqrt{3}/\sqrt{13.17.19}$
$\frac{5}{2}$ 0 $\frac{5}{2}$ 0 000− $-\sqrt{2.5.7}/\sqrt{13.17.19}$
$\frac{5}{2}$ 1 $\frac{3}{2}$ 0 000− $+2.3\sqrt{3}/\sqrt{13.17.19}$
$\frac{5}{2}$ 1 $\frac{5}{2}$ 0 000+ $+\sqrt{5.13}/2\sqrt{2.17.19}$
$\frac{5}{2}$ 1 $\frac{5}{2}$ 1 000+ $-3\sqrt{13}/2\sqrt{2.17.19}$
$\frac{5}{2}$ 2 $\frac{1}{2}$ 0 000− $+53/\sqrt{3.7.13.17.19}$
$\frac{5}{2}$ 2 $\frac{3}{2}$ 0 000+ $+79/\sqrt{3.11.13.17.19}$
$\frac{5}{2}$ 2 $\frac{3}{2}$ 1 000− $+53/\sqrt{2.11.13.17.19}$
$\frac{5}{2}$ 2 $\frac{5}{2}$ 0 000− $+5.7/\sqrt{2.11.13.17.19}$
$\frac{5}{2}$ 2 $\frac{5}{2}$ 1 000+ $+3\sqrt{3.13}/2\sqrt{11.17.19}$
$\frac{5}{2}$ 2 $\frac{5}{2}$ 2 000− $-\sqrt{2.3}/\sqrt{11.13.17.19}$
$\frac{5}{2}$ 3 $\frac{1}{2}$ 0 000+ $+\sqrt{2.5}/\sqrt{3.7.13.17.19}$
$\frac{5}{2}$ 3 $\frac{3}{2}$ 0 000− $-\sqrt{2.5.19}/\sqrt{3.11.13.17}$
$\frac{5}{2}$ 3 $\frac{3}{2}$ 1 000+ $+3.7\sqrt{2}/\sqrt{11.13.17.19}$
$\frac{5}{2}$ 3 $\frac{5}{2}$ 0 000+ $-3\sqrt{3.13}/\sqrt{11.17.19}$
$\frac{5}{2}$ 3 $\frac{5}{2}$ 1 000− $+9\sqrt{3.5}/\sqrt{11.13.17.19}$
$\frac{\tilde{1}}{2}$ 1 $\frac{5}{2}$ 0 000− $+5/2\sqrt{17.19}$
$\frac{\tilde{1}}{2}$ 2 $\frac{3}{2}$ 0 000− $+\sqrt{5}/\sqrt{2.11.17.19}$
$\frac{\tilde{1}}{2}$ 2 $\frac{5}{2}$ 0 000+ $-9\sqrt{3.5}/2\sqrt{11.17.19}$
$\frac{\tilde{1}}{2}$ 3 $\frac{1}{2}$ 0 000− $+5\sqrt{2}/\sqrt{7.17.19}$
$\frac{\tilde{1}}{2}$ 3 $\frac{5}{2}$ 0 000− $-3\sqrt{2.3}/\sqrt{11.17.19}$
$\frac{13}{2}$ $\frac{13}{2}$ 0
$\frac{1}{2}$ $\frac{1}{2}$ 0 0 000+ $+1/\sqrt{7}$
$\frac{3}{2}$ $\frac{3}{2}$ 0 0 000+ $+\sqrt{2}/\sqrt{7}$
$\frac{5}{2}$ $\frac{5}{2}$ 0 0 000+ $+\sqrt{3}/\sqrt{7}$
$\frac{\tilde{1}}{2}$ $\frac{\tilde{1}}{2}$ 0 0 000+ $+1/\sqrt{7}$

$\frac{13}{2}$ $\frac{13}{2}$ 1
$\frac{1}{2}$ $\frac{1}{2}$ 1 0 000+ $+\sqrt{5}/\sqrt{7.13}$
$\frac{3}{2}$ $\frac{1}{2}$ 1 0 000− $-2\sqrt{2}/\sqrt{7.13}$
$\frac{3}{2}$ $\frac{3}{2}$ 1 0 000+ $+\sqrt{2.7}/5\sqrt{13}$
$\frac{5}{2}$ $\frac{3}{2}$ 1 0 000− $+4\sqrt{2.11}/5\sqrt{7.13}$
$\frac{5}{2}$ $\frac{5}{2}$ 1 0 000+ $-\sqrt{2}/\sqrt{5.7.13}$
$\frac{5}{2}$ $\frac{5}{2}$ 1 1 000+ $+4.3\sqrt{2}/5\sqrt{7.13}$
$\frac{\tilde{1}}{2}$ $\frac{5}{2}$ 1 0 000− $+1/\sqrt{7}$
$\frac{13}{2}$ $\frac{13}{2}$ 2
$\frac{3}{2}$ $\frac{1}{2}$ 2 0 000− $-4\sqrt{2}/\sqrt{5.7.13}$
$\frac{3}{2}$ $\frac{3}{2}$ 2 0 000+ $+2.3\sqrt{2}/5\sqrt{7.13}$
$\frac{5}{2}$ $\frac{1}{2}$ 2 0 000+ $+\sqrt{3.11}/\sqrt{5.7.13}$
$\frac{5}{2}$ $\frac{3}{2}$ 2 0 000− $-\sqrt{3.11}/5\sqrt{2.7.13}$
$\frac{5}{2}$ $\frac{3}{2}$ 2 1 000+ $-9\sqrt{11}/2.5\sqrt{7.13}$
$\frac{5}{2}$ $\frac{5}{2}$ 2 0 000+ $-\sqrt{3.7}/2.5\sqrt{2.13}$
$\frac{5}{2}$ $\frac{5}{2}$ 2 1 000− 0
$\frac{5}{2}$ $\frac{5}{2}$ 2 2 000+ $+3.19/2.5\sqrt{2.7.13}$
$\frac{\tilde{1}}{2}$ $\frac{3}{2}$ 2 0 000+ $-\sqrt{11}/2\sqrt{5.7}$
$\frac{\tilde{1}}{2}$ $\frac{5}{2}$ 2 0 000− $-3/2\sqrt{5.7}$
$\frac{13}{2}$ $\frac{13}{2}$ 3
$\frac{3}{2}$ $\frac{3}{2}$ 3 0 000+ $+2\sqrt{2.7.11}/5\sqrt{13.17}$
$\frac{3}{2}$ $\frac{3}{2}$ $\tilde{1}$ 0 000+ $+2\sqrt{2.3.11}/5\sqrt{7.13.17}$
$\frac{5}{2}$ $\frac{1}{2}$ 3 0 000+ $+\sqrt{3.7}/\sqrt{13.17}$
$\frac{5}{2}$ $\frac{1}{2}$ $\tilde{1}$ 0 000+ $+3/\sqrt{7.13.17}$
$\frac{5}{2}$ $\frac{3}{2}$ 3 0 000− $+11\sqrt{3}/\sqrt{2.5.7.13.17}$
$\frac{5}{2}$ $\frac{3}{2}$ 3 1 000+ $-3.23/5\sqrt{2.7.13.17}$
$\frac{5}{2}$ $\frac{3}{2}$ $\tilde{1}$ 0 000− $+9\sqrt{7}/2.5\sqrt{13.17}$
$\frac{5}{2}$ $\frac{5}{2}$ 3 0 000+ $-3\sqrt{2.7}/5\sqrt{11.13.17}$
$\frac{5}{2}$ $\frac{5}{2}$ 3 1 000− 0
$\frac{5}{2}$ $\frac{5}{2}$ $\tilde{1}$ 0 000+ $+19\sqrt{3.7}/2.5\sqrt{2.11.13.17}$
$\frac{5}{2}$ $\frac{5}{2}$ $\tilde{1}$ 1 000+ $-19\sqrt{3.11}/2\sqrt{2.5.7.13.17}$
$\frac{\tilde{1}}{2}$ $\frac{1}{2}$ 3 0 000− $-\sqrt{5}/\sqrt{7.17}$
$\frac{\tilde{1}}{2}$ $\frac{3}{2}$ $\tilde{1}$ 0 000+ $-3\sqrt{3}/2\sqrt{7.17}$
$\frac{\tilde{1}}{2}$ $\frac{5}{2}$ 3 0 000− $-3/\sqrt{7.11.17}$
$\frac{\tilde{1}}{2}$ $\frac{\tilde{1}}{2}$ $\tilde{1}$ 0 000+ $+\sqrt{3.5.13}/2\sqrt{7.11.17}$
$\frac{13}{2}$ $\frac{13}{2}$ 4
$\frac{3}{2}$ $\frac{1}{2}$ 2 0 000− $+\sqrt{2.11}/\sqrt{7.13.17}$
$\frac{3}{2}$ $\frac{3}{2}$ 2 0 000+ $-\sqrt{2.5.11}/\sqrt{7.13.17}$
$\frac{3}{2}$ $\frac{3}{2}$ 3 0 000− 0
$\frac{5}{2}$ $\frac{1}{2}$ 2 0 000+ $-\sqrt{17}/\sqrt{3.7.13}$
$\frac{5}{2}$ $\frac{1}{2}$ 3 0 000− $+\sqrt{5}/\sqrt{3.13.17}$
$\frac{5}{2}$ $\frac{3}{2}$ 2 0 000− $+3\sqrt{3.5}/\sqrt{2.7.13.17}$
$\frac{5}{2}$ $\frac{3}{2}$ 2 1 000+ $-\sqrt{5.17}/2.3\sqrt{7.13}$
$\frac{5}{2}$ $\frac{3}{2}$ 3 0 000+ $+\sqrt{13}/\sqrt{2.3.17}$
$\frac{5}{2}$ $\frac{3}{2}$ 3 1 000− $+\sqrt{5}/3\sqrt{2.13.17}$
$\frac{5}{2}$ $\frac{5}{2}$ 2 0 000+ $-5\sqrt{5.7}/2\sqrt{2.3.11.13.17}$
$\frac{5}{2}$ $\frac{5}{2}$ 2 1 000− 0
$\frac{5}{2}$ $\frac{5}{2}$ 2 2 000+ $+61\sqrt{5}/2.3\sqrt{2.7.11.13.17}$
$\frac{5}{2}$ $\frac{5}{2}$ 3 0 000− 0
$\frac{5}{2}$ $\frac{5}{2}$ 3 1 000+ $-23\sqrt{2}/3\sqrt{11.13.17}$

SO_3–K 3jm Factors (cont.)

$\frac{13}{2}$ $\frac{13}{2}$ 4

$\tilde{\frac{1}{2}}$ $\frac{1}{2}$ 3 0 000+ $-1/\sqrt{17}$
$\tilde{\frac{1}{2}}$ $\frac{3}{2}$ 2 0 000+ $+1/2\sqrt{7.17}$
$\tilde{\frac{1}{2}}$ $\frac{5}{2}$ 2 0 000− $-61/2.3\sqrt{7.11.17}$
$\tilde{\frac{1}{2}}$ $\frac{5}{2}$ 3 0 000+ $+\sqrt{5}/3\sqrt{11.17}$

$\frac{13}{2}$ $\frac{13}{2}$ 5

$\frac{1}{2}$ $\frac{1}{2}$ 1 0 000+ $+2\sqrt{2.3.11}/\sqrt{7.13.17.19}$
$\frac{3}{2}$ $\frac{1}{2}$ 1 0 000− $-3\sqrt{3.5.11}/\sqrt{7.13.17.19}$
$\frac{3}{2}$ $\frac{1}{2}$ 2 0 000+ $+\sqrt{11}/\sqrt{13.17.19}$
$\frac{3}{2}$ $\frac{3}{2}$ 1 0 000+ $-2\sqrt{3.7.11}/\sqrt{5.13.17.19}$
$\frac{3}{2}$ $\frac{3}{2}$ 2 0 000− 0
$\frac{3}{2}$ $\frac{3}{2}$ $\tilde{1}$ 0 000+ $-3\sqrt{2.3.11}/\sqrt{5.13.17.19}$
$\frac{5}{2}$ $\frac{1}{2}$ 2 0 000− $-7\sqrt{2.3}/\sqrt{13.17.19}$
$\frac{5}{2}$ $\frac{1}{2}$ $\tilde{1}$ 0 000+ $-3\sqrt{5}/\sqrt{13.17.19}$
$\frac{5}{2}$ $\frac{3}{2}$ 1 0 000− $+\sqrt{3.17}/\sqrt{5.7.13.19}$
$\frac{5}{2}$ $\frac{3}{2}$ 2 0 000+ $+2\sqrt{3.5}/\sqrt{13.17.19}$
$\frac{5}{2}$ $\frac{3}{2}$ 2 1 000− $-3\sqrt{5}/\sqrt{2.13.17.19}$
$\frac{5}{2}$ $\frac{3}{2}$ $\tilde{1}$ 0 000− $+9.7/2\sqrt{5.13.17.19}$
$\frac{5}{2}$ $\frac{5}{2}$ 1 0 000+ $-8.2.3\sqrt{3}/\sqrt{7.11.13.17.19}$
$\frac{5}{2}$ $\frac{5}{2}$ 1 1 000+ $-3.53\sqrt{3}/\sqrt{5.7.11.13.17.19}$
$\frac{5}{2}$ $\frac{5}{2}$ 2 0 000− 0
$\frac{5}{2}$ $\frac{5}{2}$ 2 1 000+ $-3.7\sqrt{2.5}/\sqrt{11.13.17.19}$
$\frac{5}{2}$ $\frac{5}{2}$ 2 2 000− 0
$\frac{5}{2}$ $\frac{5}{2}$ $\tilde{1}$ 0 000+ $+7.29\sqrt{3}/2\sqrt{2.5.11.13.17.19}$
$\frac{5}{2}$ $\frac{5}{2}$ $\tilde{1}$ 1 000+ $+41\sqrt{3}/2\sqrt{2.11.13.17.19}$
$\tilde{\frac{1}{2}}$ $\frac{3}{2}$ 2 0 000− $+7/\sqrt{2.17.19}$
$\tilde{\frac{1}{2}}$ $\frac{3}{2}$ $\tilde{1}$ 0 000+ $+\sqrt{3.5}/2\sqrt{17.19}$
$\tilde{\frac{1}{2}}$ $\frac{5}{2}$ 1 0 000− $-4\sqrt{2.3.5}/\sqrt{7.11.17.19}$
$\tilde{\frac{1}{2}}$ $\frac{5}{2}$ 2 0 000+ $-9/\sqrt{2.11.17.19}$
$\tilde{\frac{1}{2}}$ $\tilde{\frac{1}{2}}$ $\tilde{1}$ 0 000+ $+3\sqrt{3.13}/2\sqrt{11.17.19}$

$\frac{13}{2}$ $\frac{13}{2}$ 6

$\frac{1}{2}$ $\frac{1}{2}$ 0 0 000+ $-11\sqrt{2.5}/\sqrt{7.13.17.19}$
$\frac{1}{2}$ $\frac{1}{2}$ 1 0 000− 0
$\frac{3}{2}$ $\frac{1}{2}$ 1 0 000+ $-11\sqrt{3}/\sqrt{2.13.17.19}$
$\frac{3}{2}$ $\frac{1}{2}$ 2 0 000− $-9\sqrt{11}/\sqrt{2.5.13.17.19}$
$\frac{3}{2}$ $\frac{3}{2}$ 0 0 000+ $-3.11/\sqrt{5.7.13.17.19}$
$\frac{3}{2}$ $\frac{3}{2}$ 1 0 000− 0
$\frac{3}{2}$ $\frac{3}{2}$ 2 0 000+ $+\sqrt{2.11.17}/5\sqrt{13.19}$
$\frac{3}{2}$ $\frac{3}{2}$ 3 0 000− 0
$\frac{5}{2}$ $\frac{1}{2}$ 2 0 000+ $-2\sqrt{3}/\sqrt{5.13.17.19}$
$\frac{5}{2}$ $\frac{1}{2}$ 3 0 000− $-9\sqrt{3}/\sqrt{2.13.17.19}$
$\frac{5}{2}$ $\frac{3}{2}$ 1 0 000+ $+2\sqrt{2.3.11}/\sqrt{13.17.19}$
$\frac{5}{2}$ $\frac{3}{2}$ 2 0 000− $+8.3\sqrt{2.3}/5\sqrt{13.17.19}$
$\frac{5}{2}$ $\frac{3}{2}$ 2 1 000+ $+2.3/5\sqrt{13.17.19}$
$\frac{5}{2}$ $\frac{3}{2}$ 3 0 000+ $+2\sqrt{3}/\sqrt{5.13.17.19}$
$\frac{5}{2}$ $\frac{3}{2}$ 3 1 000− $+4.3/\sqrt{13.17.19}$
$\frac{5}{2}$ $\frac{5}{2}$ 0 0 000+ $+37\sqrt{3}/\sqrt{2.5.7.13.17.19}$
$\frac{5}{2}$ $\frac{5}{2}$ 1 0 000− 0
$\frac{5}{2}$ $\frac{5}{2}$ 1 1 000− 0
$\frac{5}{2}$ $\frac{5}{2}$ 2 0 000+ $+7.7.7\sqrt{3}/5\sqrt{2.11.13.17.19}$

$\frac{13}{2}$ $\frac{13}{2}$ 6

$\frac{5}{2}$ $\frac{5}{2}$ 2 1 000− 0
$\frac{5}{2}$ $\frac{5}{2}$ 2 2 000+ $+9.3/5\sqrt{2.11.13.17.19}$
$\frac{5}{2}$ $\frac{5}{2}$ 3 0 000− 0
$\frac{5}{2}$ $\frac{5}{2}$ 3 1 000+ $-8.3/\sqrt{5.11.13.17.19}$
$\tilde{\frac{1}{2}}$ $\frac{1}{2}$ 3 0 000+ $+\sqrt{5}/\sqrt{2.17.19}$
$\tilde{\frac{1}{2}}$ $\frac{3}{2}$ 2 0 000+ $+2.3/\sqrt{5.17.19}$
$\tilde{\frac{1}{2}}$ $\frac{5}{2}$ 1 0 000+ $+\sqrt{3}/\sqrt{17.19}$
$\tilde{\frac{1}{2}}$ $\frac{5}{2}$ 2 0 000− $-4.3/\sqrt{5.11.17.19}$
$\tilde{\frac{1}{2}}$ $\frac{5}{2}$ 3 0 000+ $+4.3\sqrt{2}/\sqrt{11.17.19}$
$\tilde{\frac{1}{2}}$ $\tilde{\frac{1}{2}}$ 0 0 000+ $+\sqrt{5.13}/\sqrt{2.7.17.19}$

7 $\frac{7}{2}$ $\frac{7}{2}$

1 $\frac{5}{2}$ $\frac{5}{2}$ 0 000+ $-\sqrt{7}/\sqrt{2.3.13}$
1 $\frac{5}{2}$ $\frac{5}{2}$ 1 000+ $+\sqrt{2.7}/\sqrt{3.5.13}$
1 $\tilde{\frac{1}{2}}$ $\frac{5}{2}$ 0 000− $+1/2\sqrt{13}$
2 $\frac{5}{2}$ $\frac{5}{2}$ 0 000− 0
2 $\frac{5}{2}$ $\frac{5}{2}$ 1 000+ $+\sqrt{7}/\sqrt{3.13}$
2 $\frac{5}{2}$ $\frac{5}{2}$ 2 000− 0
2 $\tilde{\frac{1}{2}}$ $\frac{5}{2}$ 0 000+ $+1/\sqrt{13}$
3 $\frac{5}{2}$ $\frac{5}{2}$ 0 000+ $+\sqrt{2.7}/\sqrt{3.5.11}$
3 $\frac{5}{2}$ $\frac{5}{2}$ 1 000− 0
3 $\tilde{\frac{1}{2}}$ $\frac{5}{2}$ 0 000− $+1/\sqrt{11}$
$\tilde{1}$ $\frac{5}{2}$ $\frac{5}{2}$ 0 000+ $+7\sqrt{2}/\sqrt{5.11.13}$
$\tilde{1}$ $\frac{5}{2}$ $\frac{5}{2}$ 1 000+ 0
$\tilde{1}$ $\tilde{\frac{1}{2}}$ $\tilde{\frac{1}{2}}$ 0 000+ $-3/\sqrt{11.13}$

7 4 3

1 2 3 0 000+ $+4/\sqrt{3.5.13}$
1 2 $\tilde{1}$ 0 000+ $+7/3\sqrt{5.13}$
1 3 3 0 000− $-2/\sqrt{3.7.13}$
1 3 $\tilde{1}$ 0 000− $-4/3\sqrt{7.13}$
2 2 3 0 000− $+\sqrt{3}/\sqrt{2.13}$
2 2 3 1 000+ $+1/\sqrt{2.3.13}$
2 2 $\tilde{1}$ 0 000− $+2\sqrt{2}/3\sqrt{13}$
2 3 3 0 000+ $-4/\sqrt{3.7.13}$
2 3 $\tilde{1}$ 0 000+ $-8/3\sqrt{7.13}$
3 2 3 0 000+ $-4/\sqrt{3.5.11}$
3 2 $\tilde{1}$ 0 000+ $+2/3\sqrt{5.11}$
3 3 3 0 000+ $+4/\sqrt{3.7.11}$
3 3 $\tilde{1}$ 0 000− $-8/3\sqrt{7.11}$
$\tilde{1}$ 2 3 0 000+ $-\sqrt{7}/\sqrt{5.11.13}$
$\tilde{1}$ 2 $\tilde{1}$ 0 000+ $-2\sqrt{2.7}/\sqrt{5.11.13}$
$\tilde{1}$ 3 3 0 000− $-4/\sqrt{11.13}$

7 4 4

1 2 2 0 000+ $-\sqrt{7}/3\sqrt{3.5.13}$
1 3 2 0 000− $+4\sqrt{2}/3\sqrt{3.5.13}$
1 3 3 0 000+ $+2\sqrt{2.7}/3\sqrt{3.13}$
2 2 2 0 000− 0
2 2 2 1 000− 0
2 3 2 0 000+ $+1/2\sqrt{3.13}$
2 3 2 1 000− $+5/2\sqrt{3.13}$

SO_3–K 3jm Factors (cont.)

7 4 4

2 3 3 0 000−	0
3 2 2 0 000+	$+2\sqrt{7}/\sqrt{3.5.11}$
3 2 2 1 000−	0
3 3 2 0 000−	$+2\sqrt{2}/\sqrt{3.5.11}$
3 3 3 0 000−	0
$\tilde{1}$ 2 2 0 000+	$-4.7/3\sqrt{5.11.13}$
$\tilde{1}$ 3 2 0 000−	$-7\sqrt{7}/3\sqrt{2.5.11.13}$
$\tilde{1}$ 3 3 0 000+	$-4\sqrt{2}/3\sqrt{11.13}$

7 $\frac{9}{2}$ $\frac{5}{2}$

1 $\frac{3}{2}$ $\frac{5}{2}$ 0 000−	$-\sqrt{2}/\sqrt{13}$
1 $\frac{5}{2}$ $\frac{5}{2}$ 0 000+	$+3\sqrt{3}/\sqrt{2.5.7.13}$
1 $\frac{5}{2}$ $\frac{5}{2}$ 1 000+	$+\sqrt{3}/\sqrt{2.7.13}$
2 $\frac{3}{2}$ $\frac{5}{2}$ 0 000+	$-\sqrt{2}/\sqrt{3.5.13}$
2 $\frac{3}{2}$ $\frac{5}{2}$ 1 000−	$-3/\sqrt{5.13}$
2 $\frac{5}{2}$ $\frac{5}{2}$ 0 000−	$-3\sqrt{2}/\sqrt{5.7.13}$
2 $\frac{5}{2}$ $\frac{5}{2}$ 1 000+	$+2\sqrt{3}/\sqrt{5.7.13}$
2 $\frac{5}{2}$ $\frac{5}{2}$ 2 000−	$-3\sqrt{2.3}/\sqrt{5.7.13}$
3 $\frac{3}{2}$ $\frac{5}{2}$ 0 000−	$-1/\sqrt{2.3.5.11}$
3 $\frac{3}{2}$ $\frac{5}{2}$ 1 000+	$+1/\sqrt{2.11}$
3 $\frac{5}{2}$ $\frac{5}{2}$ 0 000+	$+\sqrt{2.3}/\sqrt{7.11}$
3 $\frac{5}{2}$ $\frac{5}{2}$ 1 000−	$-3\sqrt{2.3}/\sqrt{5.7.11}$
$\tilde{1}$ $\frac{3}{2}$ $\frac{5}{2}$ 0 000−	$-\sqrt{7}/\sqrt{11.13}$
$\tilde{1}$ $\frac{5}{2}$ $\frac{5}{2}$ 0 000+	$-3\sqrt{2}/\sqrt{11.13}$
$\tilde{1}$ $\frac{5}{2}$ $\frac{5}{2}$ 1 000+	$+3\sqrt{2}/\sqrt{5.11.13}$

7 $\frac{9}{2}$ $\frac{7}{2}$

1 $\frac{3}{2}$ $\frac{5}{2}$ 0 000+	$+1/5\sqrt{13}$
1 $\frac{5}{2}$ $\frac{5}{2}$ 0 000−	$-2\sqrt{5}/\sqrt{3.7.13}$
1 $\frac{5}{2}$ $\frac{5}{2}$ 1 000−	$+2/5\sqrt{3.7.13}$
1 $\frac{5}{2}$ $\tilde{\frac{1}{2}}$ 0 000+	$-2\sqrt{2}/\sqrt{5.13}$
2 $\frac{3}{2}$ $\frac{5}{2}$ 0 000−	$-3\sqrt{3}/4\sqrt{5.13}$
2 $\frac{3}{2}$ $\frac{5}{2}$ 1 000+	$+1/4\sqrt{2.5.13}$
2 $\frac{3}{2}$ $\tilde{\frac{1}{2}}$ 0 000+	$+3\sqrt{3.7}/4\sqrt{2.5.13}$
2 $\frac{5}{2}$ $\frac{5}{2}$ 0 000+	$-9.3/4\sqrt{5.7.13}$
2 $\frac{5}{2}$ $\frac{5}{2}$ 1 000−	$+\sqrt{13}/2\sqrt{2.3.5.7}$
2 $\frac{5}{2}$ $\frac{5}{2}$ 2 000+	$+3\sqrt{3.5}/4\sqrt{7.13}$
2 $\frac{5}{2}$ $\tilde{\frac{1}{2}}$ 0 000−	$-1/\sqrt{2.5.13}$
3 $\frac{3}{2}$ $\frac{5}{2}$ 0 000+	$+\sqrt{3}/2\sqrt{5.11}$
3 $\frac{3}{2}$ $\frac{5}{2}$ 1 000−	$+13/2.5\sqrt{11}$
3 $\frac{5}{2}$ $\frac{5}{2}$ 0 000−	$-41/2.5\sqrt{3.7.11}$
3 $\frac{5}{2}$ $\frac{5}{2}$ 1 000+	$-3\sqrt{3}/2\sqrt{5.7.11}$
3 $\frac{5}{2}$ $\tilde{\frac{1}{2}}$ 0 000+	$+1/\sqrt{2.5.11}$
$\tilde{1}$ $\frac{3}{2}$ $\frac{5}{2}$ 0 000+	$+3\sqrt{7.13}/4.5\sqrt{2.11}$
$\tilde{1}$ $\frac{3}{2}$ $\tilde{\frac{1}{2}}$ 0 000−	$+3.7/4\sqrt{2.5.11.13}$
$\tilde{1}$ $\frac{5}{2}$ $\frac{5}{2}$ 0 000−	$-37/4.5\sqrt{11.13}$
$\tilde{1}$ $\frac{5}{2}$ $\frac{5}{2}$ 1 000−	$+9.3/4\sqrt{5.11.13}$

7 $\frac{9}{2}$ $\frac{9}{2}$

1 $\frac{3}{2}$ $\frac{3}{2}$ 0 000+	$+3\sqrt{3.7}/5\sqrt{5.13.17}$
1 $\frac{5}{2}$ $\frac{3}{2}$ 0 000−	$-29/5\sqrt{5.13.17}$
1 $\frac{5}{2}$ $\frac{5}{2}$ 0 000+	$-\sqrt{7.17}/5\sqrt{3.13}$

7 $\frac{9}{2}$ $\frac{9}{2}$

1 $\frac{5}{2}$ $\frac{5}{2}$ 1 000+	$-11\sqrt{7}/5\sqrt{3.5.13.17}$
2 $\frac{3}{2}$ $\frac{3}{2}$ 0 000−	0
2 $\frac{5}{2}$ $\frac{3}{2}$ 0 000+	$-3\sqrt{3}/\sqrt{13.17}$
2 $\frac{5}{2}$ $\frac{3}{2}$ 1 000−	$-1/\sqrt{2.13.17}$
2 $\frac{5}{2}$ $\frac{5}{2}$ 0 000−	0
2 $\frac{5}{2}$ $\frac{5}{2}$ 1 000+	$-2\sqrt{2.7}/\sqrt{3.13.17}$
2 $\frac{5}{2}$ $\frac{5}{2}$ 2 000−	0
3 $\frac{3}{2}$ $\frac{3}{2}$ 0 000+	$-4.3\sqrt{3.7}/5\sqrt{5.11.17}$
3 $\frac{5}{2}$ $\frac{3}{2}$ 0 000−	$+7\sqrt{3}/2.5\sqrt{11.17}$
3 $\frac{5}{2}$ $\frac{3}{2}$ 1 000+	$+\sqrt{17}/2.5\sqrt{5.11}$
3 $\frac{5}{2}$ $\frac{5}{2}$ 0 000+	$+2\sqrt{7.17}/5\sqrt{3.5.11}$
3 $\frac{5}{2}$ $\frac{5}{2}$ 1 000−	0
$\tilde{1}$ $\frac{3}{2}$ $\frac{3}{2}$ 0 000+	$+8.3.7/5\sqrt{5.11.13.17}$
$\tilde{1}$ $\frac{5}{2}$ $\frac{3}{2}$ 0 000−	$+3.7\sqrt{7}/5\sqrt{2.5.11.13.17}$
$\tilde{1}$ $\frac{5}{2}$ $\frac{5}{2}$ 0 000+	$-2.61/5\sqrt{5.11.13.17}$
$\tilde{1}$ $\frac{5}{2}$ $\frac{5}{2}$ 1 000+	$-2.9.3/5\sqrt{11.13.17}$

7 5 2

1 1 2 0 000+	$+3/\sqrt{5.13}$
1 2 2 0 000−	$-\sqrt{5}/\sqrt{7.13}$
1 $\tilde{1}$ 2 0 000+	$+\sqrt{3}/\sqrt{5.7.13}$
2 1 2 0 000−	$-2/\sqrt{3.13}$
2 2 2 0 000+	$-\sqrt{5}/\sqrt{2.7.13}$
2 2 2 1 000+	$+5/\sqrt{2.7.13}$
2 $\tilde{1}$ 2 0 000−	$+\sqrt{2.3}/\sqrt{7.13}$
3 1 2 0 000+	$+\sqrt{2}/\sqrt{3.5.11}$
3 2 2 0 000−	$+\sqrt{5}/\sqrt{7.11}$
3 2 2 1 000+	$+\sqrt{5}/\sqrt{7.11}$
3 $\tilde{1}$ 2 0 000+	$-4\sqrt{3}/\sqrt{5.7.11}$
$\tilde{1}$ 1 2 0 000+	$+\sqrt{2.7}/\sqrt{5.11.13}$
$\tilde{1}$ 2 2 0 000−	$+\sqrt{3.5}/\sqrt{11.13}$
$\tilde{1}$ $\tilde{1}$ 2 0 000+	$-3\sqrt{2.3}/\sqrt{5.11.13}$

7 5 3

1 2 3 0 000−	$+\sqrt{7}/\sqrt{5.13}$
1 2 $\tilde{1}$ 0 000−	$+2\sqrt{3}/\sqrt{5.7.13}$
1 $\tilde{1}$ 3 0 000+	$-\sqrt{2.3}/\sqrt{7.13}$
2 1 3 0 000−	$-7/4\sqrt{3.13}$
2 1 $\tilde{1}$ 0 000−	$-1/2\sqrt{13}$
2 2 3 0 000+	$+1/4\sqrt{2.7.13}$
2 2 3 1 000−	$-\sqrt{13}/4\sqrt{2.7}$
2 2 $\tilde{1}$ 0 000+	$+3\sqrt{3}/2\sqrt{2.7.13}$
2 $\tilde{1}$ 3 0 000−	$-\sqrt{2.3}/\sqrt{7.13}$
2 $\tilde{1}$ $\tilde{1}$ 0 000−	$-3\sqrt{3}/2\sqrt{7.13}$
3 1 3 0 000+	$-1/2\sqrt{3.11}$
3 1 $\tilde{1}$ 0 000+	$+1/\sqrt{11}$
3 2 3 0 000−	$+\sqrt{11}/2\sqrt{5.7}$
3 2 $\tilde{1}$ 0 000−	$-3\sqrt{3}/\sqrt{5.7.11}$
3 $\tilde{1}$ 3 0 000+	$-\sqrt{3}/\sqrt{2.7.11}$
$\tilde{1}$ 1 3 0 000+	$+5\sqrt{7}/4\sqrt{11.13}$
$\tilde{1}$ 2 3 0 000−	$-7\sqrt{3}/4\sqrt{5.11.13}$

SO_3–K 3jm Factors (cont.)

7 5 3

$\tilde{1}$ 2 $\tilde{1}$	0 000−	$+\sqrt{3.11}/2\sqrt{2.5.13}$
$\tilde{1}$ $\tilde{1}$ $\tilde{1}$	0 000+	$-3\sqrt{3}/2\sqrt{11.13}$

7 5 4

1 1 2	0 000+	$-\sqrt{2.3}/\sqrt{5.13.17}$
1 2 2	0 000−	$+19\sqrt{2}/\sqrt{3.5.7.13.17}$
1 2 3	0 000+	$+11/\sqrt{3.5.13.17}$
1 $\tilde{1}$ 2	0 000+	$-4\sqrt{2.13}/3\sqrt{5.7.17}$
1 $\tilde{1}$ 3	0 000−	$-7\sqrt{2}/3\sqrt{13.17}$
2 1 2	0 000−	$-1/\sqrt{2.13.17}$
2 1 3	0 000+	$-5\sqrt{7}/4\sqrt{13.17}$
2 2 2	0 000+	$-3.5\sqrt{3.5}/4\sqrt{7.13.17}$
2 2 2	1 000+	$-5.11/4\sqrt{3.7.13.17}$
2 2 3	0 000−	$-\sqrt{3.13}/4\sqrt{2.17}$
2 2 3	1 000+	$-5/4\sqrt{2.3.13.17}$
2 $\tilde{1}$ 2	0 000−	$-19/2.3\sqrt{7.13.17}$
2 $\tilde{1}$ 3	0 000+	$+5\sqrt{2}/3\sqrt{13.17}$
3 1 2	0 000+	$+\sqrt{11}/\sqrt{5.17}$
3 1 3	0 000−	$+\sqrt{7}/2\sqrt{11.17}$
3 2 2	0 000−	$-\sqrt{3}/\sqrt{2.5.7.11.17}$
3 2 2	1 000+	$+5\sqrt{5}/\sqrt{2.3.7.11.17}$
3 2 3	0 000+	$-\sqrt{17}/2\sqrt{3.5.11}$
3 $\tilde{1}$ 2	0 000+	$-2\sqrt{2}/3\sqrt{5.7.11.17}$
3 $\tilde{1}$ 3	0 000−	$+\sqrt{17}/3\sqrt{2.11}$
$\tilde{1}$ 1 2	0 000+	$-\sqrt{3.7.13}/2\sqrt{5.11.17}$
$\tilde{1}$ 1 3	0 000−	$-7\sqrt{3}/4\sqrt{11.13.17}$
$\tilde{1}$ 2 2	0 000−	$+29/2\sqrt{2.5.11.13.17}$
$\tilde{1}$ 2 3	0 000+	$+7.7\sqrt{7}/4\sqrt{5.11.13.17}$
$\tilde{1}$ $\tilde{1}$ 2	0 000+	$-37/2\sqrt{5.11.13.17}$

7 5 5

1 1 1	0 000+	$-\sqrt{7}/2\sqrt{13.17}$
1 2 1	0 000−	$+\sqrt{13}/2\sqrt{5.17}$
1 2 2	0 000+	$+7\sqrt{7}/2\sqrt{5.13.17}$
1 $\tilde{1}$ 2	0 000−	$-\sqrt{3.7}/\sqrt{5.13.17}$
2 1 1	0 000−	0
2 2 1	0 000+	$-2/\sqrt{3.13.17}$
2 2 2	0 000−	0
2 2 2	1 000−	0
2 $\tilde{1}$ 1	0 000−	$+5/\sqrt{13.17}$
2 $\tilde{1}$ 2	0 000+	$-\sqrt{3.7}/\sqrt{2.13.17}$
2 $\tilde{1}$ $\tilde{1}$	0 000−	0
3 2 1	0 000−	$+8\sqrt{2}/\sqrt{3.5.11.17}$
3 2 2	0 000+	$+4\sqrt{7}/\sqrt{5.11.17}$
3 2 2	1 000−	0
3 $\tilde{1}$ 1	0 000+	$+1/\sqrt{11.17}$
3 $\tilde{1}$ 2	0 000−	$-\sqrt{3.7}/\sqrt{5.11.17}$
$\tilde{1}$ 2 1	0 000−	$-7\sqrt{2.7}/\sqrt{5.11.13.17}$
$\tilde{1}$ 2 2	0 000+	$-2\sqrt{3}/\sqrt{5.11.13.17}$
$\tilde{1}$ $\tilde{1}$ 2	0 000−	$-\sqrt{3.13}/\sqrt{2.5.11.17}$
$\tilde{1}$ $\tilde{1}$ $\tilde{1}$	0 000+	$+2.3\sqrt{3}/\sqrt{11.13.17}$

7 $\frac{11}{2}$ $\frac{3}{2}$

1 $\frac{1}{2}$ $\frac{3}{2}$	0 000−	$+\sqrt{3}/\sqrt{2.13}$
1 $\frac{3}{2}$ $\frac{3}{2}$	0 000+	$+\sqrt{3.11}/\sqrt{5.7.13}$
1 $\frac{5}{2}$ $\frac{3}{2}$	0 000−	$-\sqrt{11}/\sqrt{2.5.7.13}$
2 $\frac{1}{2}$ $\frac{3}{2}$	0 000+	$+\sqrt{2}/\sqrt{3.13}$
2 $\frac{3}{2}$ $\frac{3}{2}$	0 000−	$-2\sqrt{11}/\sqrt{3.7.13}$
2 $\frac{5}{2}$ $\frac{3}{2}$	0 000+	0
2 $\frac{5}{2}$ $\frac{3}{2}$	1 000−	$+\sqrt{11}/\sqrt{7.13}$
3 $\frac{3}{2}$ $\frac{3}{2}$	0 000+	$-2/\sqrt{3.5.7}$
3 $\frac{5}{2}$ $\frac{3}{2}$	0 000−	$+\sqrt{3}/\sqrt{2.7}$
3 $\frac{5}{2}$ $\frac{3}{2}$	1 000+	$-1/\sqrt{2.5.7}$
$\tilde{1}$ $\frac{3}{2}$ $\frac{3}{2}$	0 000+	$-2/\sqrt{5.13}$
$\tilde{1}$ $\frac{5}{2}$ $\frac{3}{2}$	0 000−	$-3/\sqrt{5.13}$

7 $\frac{11}{2}$ $\frac{5}{2}$

1 $\frac{3}{2}$ $\frac{5}{2}$	0 000+	$-4\sqrt{3}/\sqrt{5.7.13}$
1 $\frac{5}{2}$ $\frac{5}{2}$	0 000−	$+1/\sqrt{2.7.13}$
1 $\frac{5}{2}$ $\frac{5}{2}$	1 000−	$+9/\sqrt{2.5.7.13}$
2 $\frac{1}{2}$ $\frac{5}{2}$	0 000+	$-\sqrt{11}/3\sqrt{13}$
2 $\frac{3}{2}$ $\frac{5}{2}$	0 000−	$-\sqrt{7}/2.3\sqrt{13}$
2 $\frac{3}{2}$ $\frac{5}{2}$	1 000+	$+1/2\sqrt{2.3.7.13}$
2 $\frac{5}{2}$ $\frac{5}{2}$	0 000+	$-5\sqrt{3}/2\sqrt{2.7.13}$
2 $\frac{5}{2}$ $\frac{5}{2}$	1 000−	$-3/\sqrt{7.13}$
2 $\frac{5}{2}$ $\frac{5}{2}$	2 000+	$-\sqrt{2}/\sqrt{7.13}$
3 $\frac{1}{2}$ $\frac{5}{2}$	0 000−	$-1/2.3$
3 $\frac{3}{2}$ $\frac{5}{2}$	0 000+	$+13/2.3\sqrt{7.11}$
3 $\frac{3}{2}$ $\frac{5}{2}$	1 000−	$-\sqrt{3.7}/2\sqrt{5.11}$
3 $\frac{5}{2}$ $\frac{5}{2}$	0 000−	$+3/2\sqrt{2.5.7.11}$
3 $\frac{5}{2}$ $\frac{5}{2}$	1 000+	$+\sqrt{7}/2\sqrt{2.11}$
$\tilde{1}$ $\frac{1}{2}$ $\frac{5}{2}$	0 000−	$-\sqrt{7}/2\sqrt{3.13}$
$\tilde{1}$ $\frac{3}{2}$ $\frac{5}{2}$	0 000+	$+31/2\sqrt{2.3.5.11.13}$
$\tilde{1}$ $\frac{5}{2}$ $\frac{5}{2}$	0 000−	$+2\sqrt{2.3}/\sqrt{5.11.13}$
$\tilde{1}$ $\frac{5}{2}$ $\frac{5}{2}$	1 000−	$+5\sqrt{3}/2\sqrt{2.11.13}$

7 $\frac{11}{2}$ $\frac{7}{2}$

1 $\frac{3}{2}$ $\frac{5}{2}$	0 000−	$-9/\sqrt{5.7.13.17}$
1 $\frac{5}{2}$ $\frac{5}{2}$	0 000+	$-9\sqrt{3}/2\sqrt{2.7.13.17}$
1 $\frac{5}{2}$ $\frac{5}{2}$	1 000+	$+59\sqrt{3}/2\sqrt{2.5.7.13.17}$
1 $\frac{5}{2}$ $\tilde{\frac{1}{2}}$	0 000−	$-1/2\sqrt{13.17}$
2 $\frac{1}{2}$ $\frac{5}{2}$	0 000−	$-\sqrt{3.11}/4\sqrt{13.17}$
2 $\frac{3}{2}$ $\frac{5}{2}$	0 000+	$-\sqrt{3.7}/\sqrt{13.17}$
2 $\frac{3}{2}$ $\frac{5}{2}$	1 000−	$+\sqrt{17}/\sqrt{2.7.13}$
2 $\frac{3}{2}$ $\tilde{\frac{1}{2}}$	0 000−	$-1/\sqrt{2.3.13.17}$
2 $\frac{5}{2}$ $\frac{5}{2}$	0 000−	$+3\sqrt{17}/4\sqrt{2.7.13}$
2 $\frac{5}{2}$ $\frac{5}{2}$	1 000+	$-9\sqrt{3}/4\sqrt{7.13.17}$
2 $\frac{5}{2}$ $\frac{5}{2}$	2 000−	$+\sqrt{3}/4\sqrt{2.7.13.17}$
2 $\frac{5}{2}$ $\tilde{\frac{1}{2}}$	0 000+	$+4/\sqrt{13.17}$
3 $\frac{1}{2}$ $\frac{5}{2}$	0 000+	$+3\sqrt{3}/4\sqrt{17}$
3 $\frac{1}{2}$ $\tilde{\frac{1}{2}}$	0 000−	$-\sqrt{7}/4\sqrt{3.17}$
3 $\frac{3}{2}$ $\frac{5}{2}$	0 000−	$-9\sqrt{3}/2\sqrt{7.11.17}$
3 $\frac{3}{2}$ $\frac{5}{2}$	1 000+	$-\sqrt{7}/2\sqrt{5.11.17}$
3 $\frac{5}{2}$ $\frac{5}{2}$	0 000+	$-9.3\sqrt{3}/4\sqrt{2.5.7.11.17}$

SO_3–K 3jm Factors (cont.)

7 $\frac{11}{2}$ $\frac{7}{2}$

3 $\frac{5}{2}$ $\frac{5}{2}$ 1 000− $-\sqrt{3.7}/4\sqrt{2.11.17}$
3 $\frac{5}{2}$ $\tilde{\frac{1}{2}}$ 0 000− $+\sqrt{17}/4\sqrt{11}$
$\tilde{1}$ $\frac{1}{2}$ $\frac{5}{2}$ 0 000+ $-5\sqrt{7}/4\sqrt{13.17}$
$\tilde{1}$ $\frac{3}{2}$ $\frac{5}{2}$ 0 000− $-19/\sqrt{2.5.11.13.17}$
$\tilde{1}$ $\frac{3}{2}$ $\tilde{\frac{1}{2}}$ 0 000+ $+7\sqrt{7}/\sqrt{2.11.13.17}$
$\tilde{1}$ $\frac{5}{2}$ $\frac{5}{2}$ 0 000+ $-3.37/4\sqrt{2.5.11.13.17}$
$\tilde{1}$ $\frac{5}{2}$ $\frac{5}{2}$ 1 000+ $-3\sqrt{17}/4\sqrt{2.11.13}$

7 $\frac{11}{2}$ $\frac{9}{2}$

1 $\frac{1}{2}$ $\frac{3}{2}$ 0 000+ $-\sqrt{2.11}/\sqrt{3.5.13.17}$
1 $\frac{3}{2}$ $\frac{3}{2}$ 0 000− $-8.8/5\sqrt{3.7.13.17}$
1 $\frac{3}{2}$ $\frac{5}{2}$ 0 000+ $-4.3/5\sqrt{13.17}$
1 $\frac{5}{2}$ $\frac{3}{2}$ 0 000+ $+41\sqrt{2}/5\sqrt{7.13.17}$
1 $\frac{5}{2}$ $\frac{5}{2}$ 0 000− $+3\sqrt{3}/\sqrt{2.5.13.17}$
1 $\frac{5}{2}$ $\frac{5}{2}$ 1 000− $+11\sqrt{3}/5\sqrt{2.13.17}$
2 $\frac{1}{2}$ $\frac{3}{2}$ 0 000− $+\sqrt{11}/\sqrt{2.3.5.13.17}$
2 $\frac{1}{2}$ $\frac{5}{2}$ 0 000+ $-\sqrt{3.7.11}/2\sqrt{5.13.17}$
2 $\frac{3}{2}$ $\frac{3}{2}$ 0 000+ $-11/\sqrt{3.5.7.13.17}$
2 $\frac{3}{2}$ $\frac{5}{2}$ 0 000− $+9\sqrt{3}/2\sqrt{5.13.17}$
2 $\frac{3}{2}$ $\frac{5}{2}$ 1 000+ $+19/2\sqrt{2.5.13.17}$
2 $\frac{5}{2}$ $\frac{3}{2}$ 0 000− $-\sqrt{2.5.7}/\sqrt{3.13.17}$
2 $\frac{5}{2}$ $\frac{3}{2}$ 1 000+ $-31/2\sqrt{5.7.13.17}$
2 $\frac{5}{2}$ $\frac{5}{2}$ 0 000+ 0
2 $\frac{5}{2}$ $\frac{5}{2}$ 1 000− $-\sqrt{3}/2\sqrt{5.13.17}$
2 $\frac{5}{2}$ $\frac{5}{2}$ 2 000+ $+11\sqrt{3}/2\sqrt{2.5.13.17}$
3 $\frac{1}{2}$ $\frac{5}{2}$ 0 000− $-\sqrt{3.7}/2\sqrt{5.17}$
3 $\frac{3}{2}$ $\frac{3}{2}$ 0 000− $-2.37/5\sqrt{3.7.11.17}$
3 $\frac{3}{2}$ $\frac{5}{2}$ 0 000+ $+\sqrt{3}/2\sqrt{5.11.17}$
3 $\frac{3}{2}$ $\frac{5}{2}$ 1 000− $+29/2.5\sqrt{11.17}$
3 $\frac{5}{2}$ $\frac{3}{2}$ 0 000+ $+2\sqrt{2.5}/\sqrt{3.7.11.17}$
3 $\frac{5}{2}$ $\frac{3}{2}$ 1 000− $+2.3\sqrt{2}/5\sqrt{7.11.17}$
3 $\frac{5}{2}$ $\frac{5}{2}$ 0 000− $-23\sqrt{3}/2.5\sqrt{2.11.17}$
3 $\frac{5}{2}$ $\frac{5}{2}$ 1 000+ $-\sqrt{3.11}/2\sqrt{2.5.17}$
$\tilde{1}$ $\frac{1}{2}$ $\frac{5}{2}$ 0 000− $+7/\sqrt{5.13.17}$
$\tilde{1}$ $\frac{3}{2}$ $\frac{3}{2}$ 0 000− $+31/5\sqrt{11.13.17}$
$\tilde{1}$ $\frac{3}{2}$ $\frac{5}{2}$ 0 000+ $+61\sqrt{7}/2.5\sqrt{2.11.13.17}$
$\tilde{1}$ $\frac{5}{2}$ $\frac{3}{2}$ 0 000+ $-103/2.5\sqrt{11.13.17}$
$\tilde{1}$ $\frac{5}{2}$ $\frac{5}{2}$ 0 000− $-9\sqrt{7}/2.5\sqrt{2.11.13.17}$
$\tilde{1}$ $\frac{5}{2}$ $\frac{5}{2}$ 1 000− $-2.3\sqrt{2.7}/\sqrt{5.11.13.17}$

7 $\frac{11}{2}$ $\frac{11}{2}$

1 $\frac{1}{2}$ $\frac{1}{2}$ 0 000+ $+11\sqrt{7}/2.3\sqrt{13.17.19}$
1 $\frac{3}{2}$ $\frac{1}{2}$ 0 000− $+23\sqrt{11}/2.3\sqrt{13.17.19}$
1 $\frac{3}{2}$ $\frac{3}{2}$ 0 000+ $-29\sqrt{2.7}/3\sqrt{5.13.17.19}$
1 $\frac{5}{2}$ $\frac{3}{2}$ 0 000− $+3\sqrt{3.7}/2\sqrt{5.13.17.19}$
1 $\frac{5}{2}$ $\frac{5}{2}$ 0 000+ $+5\sqrt{7}/2\sqrt{2.13.17.19}$
1 $\frac{5}{2}$ $\frac{5}{2}$ 1 000+ $-3.11\sqrt{7}/2\sqrt{2.5.13.17.19}$
2 $\frac{3}{2}$ $\frac{1}{2}$ 0 000+ $+\sqrt{11}/\sqrt{13.17.19}$
2 $\frac{3}{2}$ $\frac{3}{2}$ 0 000− 0
2 $\frac{5}{2}$ $\frac{1}{2}$ 0 000− $+\sqrt{11.17}/3\sqrt{13.19}$
2 $\frac{5}{2}$ $\frac{3}{2}$ 0 000+ $+2.7\sqrt{7}/3\sqrt{13.17.19}$
2 $\frac{5}{2}$ $\frac{3}{2}$ 1 000− $-7\sqrt{7}/\sqrt{2.3.13.17.19}$
2 $\frac{5}{2}$ $\frac{5}{2}$ 0 000− 0
2 $\frac{5}{2}$ $\frac{5}{2}$ 1 000+ $+2.3\sqrt{7}/\sqrt{13.17.19}$
2 $\frac{5}{2}$ $\frac{5}{2}$ 2 000− 0
3 $\frac{3}{2}$ $\frac{3}{2}$ 0 000+ $+2\sqrt{2.7}/\sqrt{5.11.17.19}$
3 $\frac{5}{2}$ $\frac{1}{2}$ 0 000+ $+5/3\sqrt{17.19}$
3 $\frac{5}{2}$ $\frac{3}{2}$ 0 000− $+5.5\sqrt{7}/2.3\sqrt{11.17.19}$
3 $\frac{5}{2}$ $\frac{3}{2}$ 1 000+ $-\sqrt{3.7.17}/2\sqrt{5.11.19}$
3 $\frac{5}{2}$ $\frac{5}{2}$ 0 000+ $+3\sqrt{2.7}/\sqrt{5.11.17.19}$
3 $\frac{5}{2}$ $\frac{5}{2}$ 1 000− 0
$\tilde{1}$ $\frac{3}{2}$ $\frac{3}{2}$ 0 000+ $-2.23\sqrt{2.3}/\sqrt{5.11.13.17.19}$
$\tilde{1}$ $\frac{5}{2}$ $\frac{1}{2}$ 0 000+ $-7\sqrt{7}/\sqrt{3.13.17.19}$
$\tilde{1}$ $\frac{5}{2}$ $\frac{3}{2}$ 0 000− $-1/\sqrt{2.3.5.11.13.17.19}$
$\tilde{1}$ $\frac{5}{2}$ $\frac{5}{2}$ 0 000+ $+41\sqrt{2.3}/\sqrt{5.11.13.17.19}$
$\tilde{1}$ $\frac{5}{2}$ $\frac{5}{2}$ 1 000+ $-\sqrt{2.3.19}/\sqrt{11.13.17}$

7 6 1

1 0 1 0 000+ $+1/\sqrt{13}$
1 1 1 0 000− $-3/\sqrt{7.13}$
1 2 1 0 000+ $+\sqrt{11}/\sqrt{5.7.13}$
2 1 1 0 000+ $+2\sqrt{3}/\sqrt{7.13}$
2 2 1 0 000− $-2\sqrt{11}/\sqrt{3.7.13}$
2 3 1 0 000+ $+\sqrt{11}/\sqrt{3.7.13}$
3 2 1 0 000+ $+2\sqrt{2}/\sqrt{3.5.7}$
3 3 1 0 000− $-2/\sqrt{3.7}$
$\tilde{1}$ 2 1 0 000+ $+2\sqrt{2}/\sqrt{5.13}$
$\tilde{1}$ 3 1 0 000− $-1/\sqrt{13}$

7 6 2

1 1 2 0 000− $+2.3/\sqrt{5.7.13}$
1 2 2 0 000+ $-2\sqrt{11}/\sqrt{5.7.13}$
1 3 2 0 000− $+\sqrt{11}/\sqrt{5.7.13}$
2 0 2 0 000− $-1/\sqrt{13}$
2 1 2 0 000+ $+2/\sqrt{3.7.13}$
2 2 2 0 000− $-\sqrt{5.11}/2\sqrt{2.7.13}$
2 2 2 1 000− $+\sqrt{11}/2\sqrt{2.7.13}$
2 3 2 0 000+ $+3\sqrt{11}/2\sqrt{2.7.13}$
2 3 2 1 000− $-\sqrt{11}/2\sqrt{2.7.13}$
3 1 2 0 000− $-\sqrt{11}/\sqrt{2.3.5.7}$
3 2 2 0 000+ $-1/2\sqrt{5.7}$
3 2 2 1 000− $-\sqrt{5}/2\sqrt{7}$
3 3 2 0 000− $-1/\sqrt{5.7}$
$\tilde{1}$ 1 2 0 000− $-\sqrt{11}/\sqrt{2.5.13}$
$\tilde{1}$ 2 2 0 000+ $-\sqrt{3}/2\sqrt{5.13}$
$\tilde{1}$ 3 2 0 000− $-3\sqrt{3}/2\sqrt{5.13}$

7 6 3

1 2 3 0 000+ $+\sqrt{13}/\sqrt{5.7.17}$
1 2 $\tilde{1}$ 0 000+ $+2\sqrt{3.7}/\sqrt{5.13.17}$
1 3 3 0 000− $-2\sqrt{2}/\sqrt{7.13.17}$
1 3 $\tilde{1}$ 0 000− $-5\sqrt{2.3}/\sqrt{7.13.17}$
2 1 3 0 000+ $-\sqrt{11.13}/4\sqrt{3.7.17}$

SO_3–K 3jm Factors (cont.)

7 6 3	
2 1 $\tilde{1}$ 0 000+	$-\sqrt{7.11}/2\sqrt{13.17}$
2 2 3 0 000−	$+\sqrt{17}/4\sqrt{2.7.13}$
2 2 3 1 000+	$+5\sqrt{7}/4\sqrt{2.13.17}$
2 2 $\tilde{1}$ 0 000−	$-\sqrt{3.17}/2\sqrt{2.7.13}$
2 3 3 0 000+	$-19/\sqrt{2.7.13.17}$
2 3 $\tilde{1}$ 0 000+	$+\sqrt{2.3}/\sqrt{7.13.17}$
3 0 3 0 000+	$+1/\sqrt{2.17}$
3 1 3 0 000−	$+11/2\sqrt{3.7.17}$
3 1 $\tilde{1}$ 0 000−	$-2/\sqrt{7.17}$
3 2 3 0 000+	$+29/2\sqrt{5.7.11.17}$
3 2 $\tilde{1}$ 0 000+	$+4\sqrt{3}/\sqrt{5.7.11.17}$
3 3 3 0 000+	$-\sqrt{2.7}/\sqrt{11.17}$
3 3 $\tilde{1}$ 0 000−	$+\sqrt{2.3}/\sqrt{7.11.17}$
$\tilde{1}$ 0 $\tilde{1}$ 0 000+	$+\sqrt{3.7}/\sqrt{2.13.17}$
$\tilde{1}$ 1 3 0 000−	$-1/4\sqrt{13.17}$
$\tilde{1}$ 2 3 0 000+	$+97\sqrt{3}/4\sqrt{5.11.13.17}$
$\tilde{1}$ 2 $\tilde{1}$ 0 000+	$+7\sqrt{3}/2\sqrt{2.5.11.13.17}$
$\tilde{1}$ 3 3 0 000−	$+3\sqrt{3}/\sqrt{2.11.13.17}$

7 6 4	
1 1 2 0 000−	$+\sqrt{2.3.11}/\sqrt{5.7.13.17}$
1 2 2 0 000+	$-5\sqrt{2.5}/\sqrt{3.7.13.17}$
1 2 3 0 000−	$-\sqrt{5}/\sqrt{3.13.17}$
1 3 2 0 000−	$+37\sqrt{2}/\sqrt{3.5.7.13.17}$
1 3 3 0 000+	$+2\sqrt{2}/\sqrt{3.13.17}$
2 0 2 0 000−	$-\sqrt{11}/\sqrt{2.3.13.17}$
2 1 2 0 000+	$+3\sqrt{11}/\sqrt{2.7.13.17}$
2 1 3 0 000−	$+3\sqrt{11}/4\sqrt{13.17}$
2 2 2 0 000−	$-11\sqrt{5}/4\sqrt{3.7.13.17}$
2 2 2 1 000−	$-5.5.11/4.3\sqrt{3.7.13.17}$
2 2 3 0 000+	$+5.7/4\sqrt{2.3.13.17}$
2 2 3 1 000−	$-5\sqrt{17}/4.3\sqrt{2.3.13}$
2 3 2 0 000+	$-4/\sqrt{3.7.13.17}$
2 3 2 1 000−	$-2.19/3\sqrt{3.7.13.17}$
2 3 3 0 000−	$-11/3\sqrt{2.3.13.17}$
3 0 3 0 000−	$-\sqrt{7}/\sqrt{2.3.17}$
3 1 2 0 000−	$+4/\sqrt{5.7.17}$
3 1 3 0 000+	$-1/2\sqrt{17}$
3 2 2 0 000+	$-\sqrt{2.3.5}/\sqrt{7.11.17}$
3 2 2 1 000−	$+\sqrt{2.5.11}/3\sqrt{3.7.17}$
3 2 3 0 000−	$+13\sqrt{5}/2.3\sqrt{3.11.17}$
3 3 2 0 000−	$+29\sqrt{2}/3\sqrt{3.5.7.11.17}$
3 3 3 0 000−	$-\sqrt{2.11}/3\sqrt{3.17}$
$\tilde{1}$ 1 2 0 000−	$-11/2\sqrt{3.5.13.17}$
$\tilde{1}$ 1 3 0 000+	$+\sqrt{7.13}/4\sqrt{3.17}$
$\tilde{1}$ 2 2 0 000+	$+\sqrt{5.13}/2.3\sqrt{2.11.17}$
$\tilde{1}$ 2 3 0 000−	$+5\sqrt{5.7}/4.3\sqrt{11.13.17}$
$\tilde{1}$ 3 2 0 000−	$+61\sqrt{2}/3\sqrt{5.11.13.17}$
$\tilde{1}$ 3 3 0 000+	$-5\sqrt{7}/3\sqrt{2.11.13.17}$

7 6 5	
1 0 1 0 000+	$-\sqrt{2.11}/\sqrt{13.17.19}$
1 1 1 0 000−	$+\sqrt{11.13}/\sqrt{2.7.17.19}$
1 1 2 0 000+	$+7\sqrt{11}/\sqrt{2.5.13.17.19}$
1 2 1 0 000+	$-31\sqrt{5}/\sqrt{2.7.13.17.19}$
1 2 2 0 000−	$-5\sqrt{5}/\sqrt{2.13.17.19}$
1 2 $\tilde{1}$ 0 000+	$+\sqrt{2.3.5}/\sqrt{13.17.19}$
1 3 2 0 000+	$-11\sqrt{2}/\sqrt{5.13.17.19}$
1 3 $\tilde{1}$ 0 000−	$-7\sqrt{3}/\sqrt{13.17.19}$
2 0 2 0 000+	$-\sqrt{2.7.11}/\sqrt{13.17.19}$
2 1 1 0 000+	$+3\sqrt{3.11}/\sqrt{2.7.13.17.19}$
2 1 2 0 000−	$+\sqrt{11.13}/\sqrt{2.3.17.19}$
2 1 $\tilde{1}$ 0 000+	$+\sqrt{11}/\sqrt{2.13.17.19}$
2 2 1 0 000−	$-5/\sqrt{2.3.7.13.17.19}$
2 2 2 0 000+	$+7\sqrt{5}/\sqrt{13.17.19}$
2 2 2 1 000+	0
2 2 $\tilde{1}$ 0 000−	$-5\sqrt{3}/2\sqrt{13.17.19}$
2 3 1 0 000+	$-233/2\sqrt{2.3.7.13.17.19}$
2 3 2 0 000−	$+\sqrt{13}/4\sqrt{17.19}$
2 3 2 1 000+	$+3\sqrt{19}/4\sqrt{13.17}$
2 3 $\tilde{1}$ 0 000+	$+2.3\sqrt{3}/\sqrt{13.17.19}$
3 1 2 0 000+	$-11/\sqrt{3.5.17.19}$
3 1 $\tilde{1}$ 0 000−	$+7/\sqrt{2.17.19}$
3 2 1 0 000+	$+2\sqrt{5.11}/\sqrt{3.7.17.19}$
3 2 2 0 000−	$+3\sqrt{5}/\sqrt{2.11.17.19}$
3 2 2 1 000+	$+3\sqrt{5}/\sqrt{2.11.17.19}$
3 2 $\tilde{1}$ 0 000+	$-5\sqrt{3.5}/\sqrt{2.11.17.19}$
3 3 1 0 000−	$-4.5\sqrt{2}/\sqrt{3.7.11.17.19}$
3 3 2 0 000+	$+3.7\sqrt{2}/\sqrt{5.11.17.19}$
3 3 $\tilde{1}$ 0 000−	$+3\sqrt{3}/\sqrt{11.17.19}$
$\tilde{1}$ 0 $\tilde{1}$ 0 000+	$-7\sqrt{3}/\sqrt{13.17.19}$
$\tilde{1}$ 1 2 0 000+	$+4\sqrt{7}/\sqrt{5.13.17.19}$
$\tilde{1}$ 2 1 0 000+	$+\sqrt{5}/\sqrt{11.13.17.19}$
$\tilde{1}$ 2 2 0 000−	$-9\sqrt{3.5.7}/\sqrt{2.11.13.17.19}$
$\tilde{1}$ 2 $\tilde{1}$ 0 000+	$+7\sqrt{3.5.7}/2\sqrt{11.13.17.19}$
$\tilde{1}$ 3 1 0 000−	$-103/2\sqrt{2.11.13.17.19}$
$\tilde{1}$ 3 2 0 000+	$+31\sqrt{3.7}/2\sqrt{2.5.11.13.17.19}$

7 6 6	
1 1 0 0 000−	$+11/\sqrt{13.17.19}$
1 1 1 0 000+	$+11\sqrt{7}/2\sqrt{13.17.19}$
1 2 1 0 000−	$+\sqrt{7.11}/2\sqrt{5.13.17.19}$
1 2 2 0 000+	$+23\sqrt{7}/2\sqrt{5.13.17.19}$
1 3 2 0 000−	$-8\sqrt{7}/\sqrt{5.13.17.19}$
1 3 3 0 000+	$-\sqrt{2.7}/\sqrt{13.17.19}$
2 1 1 0 000−	0
2 2 0 0 000−	$+4\sqrt{11}/\sqrt{13.17.19}$
2 2 1 0 000+	$+2\sqrt{7.11}/\sqrt{3.13.17.19}$
2 2 2 0 000−	0
2 2 2 1 000−	0
2 3 1 0 000−	$+5\sqrt{7.11}/2\sqrt{3.13.17.19}$

SO_3–K 3jm Factors (cont.)

7 6 6

$2\ 3\ 2\ 0\ 000+\quad -\sqrt{7.17}/2\sqrt{2.13.19}$
$2\ 3\ 2\ 1\ 000-\quad +3\sqrt{7}/2\sqrt{2.13.17.19}$
$2\ 3\ 3\ 0\ 000-\quad 0$
$3\ 2\ 1\ 0\ 000-\quad +\sqrt{2.7}/\sqrt{3.5.17.19}$
$3\ 2\ 2\ 0\ 000+\quad -2.7\sqrt{7}/\sqrt{5.11.17.19}$
$3\ 2\ 2\ 1\ 000-\quad 0$
$3\ 3\ 0\ 0\ 000-\quad +\sqrt{2}/\sqrt{17.19}$
$3\ 3\ 1\ 0\ 000+\quad -2\sqrt{7}/\sqrt{3.17.19}$
$3\ 3\ 2\ 0\ 000-\quad -4.3\sqrt{7}/\sqrt{5.11.17.19}$
$3\ 3\ 3\ 0\ 000-\quad 0$
$\tilde{1}\ 2\ 1\ 0\ 000-\quad +2.11\sqrt{2}/\sqrt{5.13.17.19}$
$\tilde{1}\ 2\ 2\ 0\ 000+\quad -8\sqrt{3}/\sqrt{5.11.13.17.19}$
$\tilde{1}\ 3\ 1\ 0\ 000+\quad +\sqrt{13}/2\sqrt{17.19}$
$\tilde{1}\ 3\ 2\ 0\ 000-\quad +43\sqrt{3}/2\sqrt{5.11.13.17.19}$
$\tilde{1}\ 3\ 3\ 0\ 000+\quad +8.3\sqrt{2.3}/\sqrt{11.13.17.19}$

$7\ \frac{13}{2}\ \frac{1}{2}$

$1\ \tfrac{1}{2}\ \tfrac{1}{2}\ 0\ 000+\quad +1/\sqrt{7}$
$1\ \tfrac{3}{2}\ \tfrac{1}{2}\ 0\ 000-\quad +\sqrt{2}/\sqrt{5.7}$
$2\ \tfrac{3}{2}\ \tfrac{1}{2}\ 0\ 000+\quad +2\sqrt{2}/\sqrt{5.7}$
$2\ \tfrac{5}{2}\ \tfrac{1}{2}\ 0\ 000-\quad +\sqrt{11}/\sqrt{3.5.7}$
$3\ \tfrac{5}{2}\ \tfrac{1}{2}\ 0\ 000+\quad +\sqrt{13}/\sqrt{3.5.7}$
$3\ \tilde{\tfrac{1}{2}}\ \tfrac{1}{2}\ 0\ 000-\quad +1/\sqrt{7}$
$\tilde{1}\ \tfrac{5}{2}\ \tfrac{1}{2}\ 0\ 000+\quad +1/\sqrt{5}$

$7\ \frac{13}{2}\ \frac{3}{2}$

$1\ \tfrac{1}{2}\ \tfrac{3}{2}\ 0\ 000+\quad +2/\sqrt{7.13}$
$1\ \tfrac{3}{2}\ \tfrac{3}{2}\ 0\ 000-\quad +8.2/5\sqrt{7.13}$
$1\ \tfrac{5}{2}\ \tfrac{3}{2}\ 0\ 000+\quad -3\sqrt{11}/5\sqrt{7.13}$
$2\ \tfrac{1}{2}\ \tfrac{3}{2}\ 0\ 000-\quad -3/\sqrt{7.13}$
$2\ \tfrac{3}{2}\ \tfrac{3}{2}\ 0\ 000+\quad -2/\sqrt{5.7.13}$
$2\ \tfrac{5}{2}\ \tfrac{3}{2}\ 0\ 000-\quad -\sqrt{11.13}/4\sqrt{3.5.7}$
$2\ \tfrac{5}{2}\ \tfrac{3}{2}\ 1\ 000+\quad +11\sqrt{11}/4\sqrt{2.5.7.13}$
$2\ \tilde{\tfrac{1}{2}}\ \tfrac{3}{2}\ 0\ 000+\quad -\sqrt{11}/4\sqrt{2.7}$
$3\ \tfrac{3}{2}\ \tfrac{3}{2}\ 0\ 000-\quad +\sqrt{11}/5\sqrt{7}$
$3\ \tfrac{5}{2}\ \tfrac{3}{2}\ 0\ 000+\quad -4/\sqrt{3.5.7}$
$3\ \tfrac{5}{2}\ \tfrac{3}{2}\ 1\ 000-\quad -3/5\sqrt{7}$
$\tilde{1}\ \tfrac{3}{2}\ \tfrac{3}{2}\ 0\ 000-\quad +\sqrt{3.11}/5\sqrt{13}$
$\tilde{1}\ \tfrac{5}{2}\ \tfrac{3}{2}\ 0\ 000+\quad -7/4.5\sqrt{2.13}$
$\tilde{1}\ \tilde{\tfrac{1}{2}}\ \tfrac{3}{2}\ 0\ 000-\quad -\sqrt{3}/4\sqrt{2}$

$7\ \frac{13}{2}\ \frac{5}{2}$

$1\ \tfrac{3}{2}\ \tfrac{5}{2}\ 0\ 000-\quad -2\sqrt{2.17}/5\sqrt{7.13}$
$1\ \tfrac{5}{2}\ \tfrac{5}{2}\ 0\ 000+\quad +3\sqrt{7.11}/2\sqrt{2.5.13.17}$
$1\ \tfrac{5}{2}\ \tfrac{5}{2}\ 1\ 000+\quad +3.11\sqrt{11}/2.5\sqrt{2.7.13.17}$
$1\ \tilde{\tfrac{1}{2}}\ \tfrac{5}{2}\ 0\ 000-\quad -\sqrt{11}/2\sqrt{7.17}$
$2\ \tfrac{1}{2}\ \tfrac{5}{2}\ 0\ 000-\quad -\sqrt{17}/\sqrt{3.7.13}$
$2\ \tfrac{3}{2}\ \tfrac{5}{2}\ 0\ 000+\quad +43/\sqrt{2.3.5.7.13.17}$
$2\ \tfrac{3}{2}\ \tfrac{5}{2}\ 1\ 000-\quad +31/2\sqrt{5.7.13.17}$
$2\ \tfrac{5}{2}\ \tfrac{5}{2}\ 0\ 000-\quad -\sqrt{3.7.11}/2\sqrt{2.5.13.17}$
$2\ \tfrac{5}{2}\ \tfrac{5}{2}\ 1\ 000+\quad +3\sqrt{5.11}/2\sqrt{7.13.17}$
$2\ \tfrac{5}{2}\ \tfrac{5}{2}\ 2\ 000-\quad -3\sqrt{11}/2\sqrt{2.5.7.13.17}$

$7\ \frac{13}{2}\ \frac{5}{2}$

$2\ \tilde{\tfrac{1}{2}}\ \tfrac{5}{2}\ 0\ 000+\quad +\sqrt{11}/\sqrt{7.17}$
$3\ \tfrac{1}{2}\ \tfrac{5}{2}\ 0\ 000+\quad +\sqrt{11}/\sqrt{3.7.17}$
$3\ \tfrac{3}{2}\ \tfrac{5}{2}\ 0\ 000-\quad -13\sqrt{11}/2\sqrt{2.3.5.7.17}$
$3\ \tfrac{3}{2}\ \tfrac{5}{2}\ 1\ 000+\quad +\sqrt{11}/2.5\sqrt{2.7.17}$
$3\ \tfrac{5}{2}\ \tfrac{5}{2}\ 0\ 000+\quad -9.3/2.5\sqrt{2.7.17}$
$3\ \tfrac{5}{2}\ \tfrac{5}{2}\ 1\ 000-\quad +3\sqrt{5}/2\sqrt{2.7.17}$
$3\ \tilde{\tfrac{1}{2}}\ \tfrac{5}{2}\ 0\ 000-\quad -\sqrt{13}/2\sqrt{7.17}$
$\tilde{1}\ \tfrac{1}{2}\ \tfrac{5}{2}\ 0\ 000+\quad +\sqrt{11}/\sqrt{13.17}$
$\tilde{1}\ \tfrac{3}{2}\ \tfrac{5}{2}\ 0\ 000-\quad +7\sqrt{11}/2.5\sqrt{13.17}$
$\tilde{1}\ \tfrac{5}{2}\ \tfrac{5}{2}\ 0\ 000+\quad -7\sqrt{3}/2.5\sqrt{2.13.17}$
$\tilde{1}\ \tfrac{5}{2}\ \tfrac{5}{2}\ 1\ 000+\quad -19\sqrt{3}/2\sqrt{2.5.13.17}$

$7\ \frac{13}{2}\ \frac{7}{2}$

$1\ \tfrac{3}{2}\ \tfrac{5}{2}\ 0\ 000+\quad +\sqrt{11}/\sqrt{7.13.17}$
$1\ \tfrac{5}{2}\ \tfrac{5}{2}\ 0\ 000-\quad -4\sqrt{7}/\sqrt{5.13.17}$
$1\ \tfrac{5}{2}\ \tfrac{5}{2}\ 1\ 000-\quad -2/\sqrt{7.13.17}$
$1\ \tfrac{5}{2}\ \tilde{\tfrac{1}{2}}\ 0\ 000+\quad -2\sqrt{2.3}/\sqrt{5.13.17}$
$1\ \tilde{\tfrac{1}{2}}\ \tfrac{5}{2}\ 0\ 000+\quad +2\sqrt{2}/\sqrt{7.17}$
$2\ \tfrac{1}{2}\ \tfrac{5}{2}\ 0\ 000+\quad +5\sqrt{11}/2\sqrt{2.3.7.13.17}$
$2\ \tfrac{3}{2}\ \tfrac{5}{2}\ 0\ 000-\quad -53\sqrt{11}/4\sqrt{3.5.7.13.17}$
$2\ \tfrac{3}{2}\ \tfrac{5}{2}\ 1\ 000+\quad -19\sqrt{11}/4\sqrt{2.5.7.13.17}$
$2\ \tfrac{3}{2}\ \tilde{\tfrac{1}{2}}\ 0\ 000+\quad +7\sqrt{3.11}/4\sqrt{2.5.13.17}$
$2\ \tfrac{5}{2}\ \tfrac{5}{2}\ 0\ 000+\quad -11\sqrt{7}/4\sqrt{3.5.13.17}$
$2\ \tfrac{5}{2}\ \tfrac{5}{2}\ 1\ 000-\quad -19/3\sqrt{2.5.7.13.17}$
$2\ \tfrac{5}{2}\ \tfrac{5}{2}\ 2\ 000+\quad -11.31/4.3\sqrt{5.7.13.17}$
$2\ \tfrac{5}{2}\ \tilde{\tfrac{1}{2}}\ 0\ 000-\quad -29/2\sqrt{2.3.5.13.17}$
$2\ \tilde{\tfrac{1}{2}}\ \tfrac{5}{2}\ 0\ 000-\quad +5/2\sqrt{2.7.17}$
$3\ \tfrac{1}{2}\ \tfrac{5}{2}\ 0\ 000-\quad -1/\sqrt{2.3.7.17}$
$3\ \tfrac{1}{2}\ \tilde{\tfrac{1}{2}}\ 0\ 000+\quad +\sqrt{3}/\sqrt{2.17}$
$3\ \tfrac{3}{2}\ \tfrac{5}{2}\ 0\ 000+\quad -11/2\sqrt{3.5.7.17}$
$3\ \tfrac{3}{2}\ \tfrac{5}{2}\ 1\ 000-\quad -5/2\sqrt{7.17}$
$3\ \tfrac{5}{2}\ \tfrac{5}{2}\ 0\ 000-\quad -1/2.3\sqrt{7.11.17}$
$3\ \tfrac{5}{2}\ \tfrac{5}{2}\ 1\ 000+\quad +11\sqrt{11}/2.3\sqrt{5.7.17}$
$3\ \tfrac{5}{2}\ \tilde{\tfrac{1}{2}}\ 0\ 000+\quad -\sqrt{2}/\sqrt{3.5.11.17}$
$3\ \tilde{\tfrac{1}{2}}\ \tfrac{5}{2}\ 0\ 000+\quad -3\sqrt{13}/\sqrt{2.7.11.17}$
$\tilde{1}\ \tfrac{1}{2}\ \tfrac{5}{2}\ 0\ 000-\quad -9/2\sqrt{2.13.17}$
$\tilde{1}\ \tfrac{3}{2}\ \tfrac{5}{2}\ 0\ 000+\quad +7/4\sqrt{2.13.17}$
$\tilde{1}\ \tfrac{3}{2}\ \tilde{\tfrac{1}{2}}\ 0\ 000-\quad -\sqrt{7.17}/4\sqrt{2.5.13}$
$\tilde{1}\ \tfrac{5}{2}\ \tfrac{5}{2}\ 0\ 000-\quad -5.7/4\sqrt{3.11.13.17}$
$\tilde{1}\ \tfrac{5}{2}\ \tfrac{5}{2}\ 1\ 000-\quad -19\sqrt{11}/4\sqrt{3.5.13.17}$
$\tilde{1}\ \tilde{\tfrac{1}{2}}\ \tilde{\tfrac{1}{2}}\ 0\ 000-\quad +\sqrt{7}/2\sqrt{2.11.17}$

$7\ \frac{13}{2}\ \frac{9}{2}$

$1\ \tfrac{1}{2}\ \tfrac{3}{2}\ 0\ 000-\quad +3\sqrt{2.3.11}/\sqrt{5.7.13.17.19}$
$1\ \tfrac{3}{2}\ \tfrac{3}{2}\ 0\ 000+\quad +9\sqrt{2.3.11}/\sqrt{5.7.13.17.19}$
$1\ \tfrac{3}{2}\ \tfrac{5}{2}\ 0\ 000-\quad +\sqrt{2.11}/\sqrt{5.13.17.19}$
$1\ \tfrac{5}{2}\ \tfrac{3}{2}\ 0\ 000-\quad -47\sqrt{2.3}/\sqrt{5.7.13.17.19}$
$1\ \tfrac{5}{2}\ \tfrac{5}{2}\ 0\ 000+\quad +\sqrt{19}/2\sqrt{2.13.17}$
$1\ \tfrac{5}{2}\ \tfrac{5}{2}\ 1\ 000+\quad -\sqrt{17}/2\sqrt{2.5.13.19}$
$1\ \tilde{\tfrac{1}{2}}\ \tfrac{5}{2}\ 0\ 000-\quad -\sqrt{19}/2\sqrt{5.17}$
$2\ \tfrac{1}{2}\ \tfrac{3}{2}\ 0\ 000+\quad -8\sqrt{2.3.11}/\sqrt{5.7.13.17.19}$

SO_3–K 3jm Factors (cont.)

7	$\frac{13}{2}$	$\frac{9}{2}$			
2	$\frac{1}{2}$	$\frac{5}{2}$	0	000−	$+7\sqrt{11}/\sqrt{3.5.13.17.19}$
2	$\frac{3}{2}$	$\frac{3}{2}$	0	000−	$+5\sqrt{2.3.11}/\sqrt{7.13.17.19}$
2	$\frac{3}{2}$	$\frac{5}{2}$	0	000+	$+\sqrt{11}/\sqrt{2.3.13.17.19}$
2	$\frac{3}{2}$	$\frac{5}{2}$	1	000−	$-7\sqrt{11}/2\sqrt{13.17.19}$
2	$\frac{5}{2}$	$\frac{3}{2}$	0	000+	$-\sqrt{19}/\sqrt{2.7.13.17}$
2	$\frac{5}{2}$	$\frac{3}{2}$	1	000−	$-109/2\sqrt{3.7.13.17.19}$
2	$\frac{5}{2}$	$\frac{5}{2}$	0	000−	$-7\sqrt{13}/2\sqrt{2.3.17.19}$
2	$\frac{5}{2}$	$\frac{5}{2}$	1	000+	$+23/2.3\sqrt{13.17.19}$
2	$\frac{5}{2}$	$\frac{5}{2}$	2	000−	$+137/2.3\sqrt{2.13.17.19}$
2	$\frac{\tilde{1}}{2}$	$\frac{3}{2}$	0	000−	$-\sqrt{3.17}/2\sqrt{5.7.19}$
2	$\frac{\tilde{1}}{2}$	$\frac{5}{2}$	0	000+	$+1/\sqrt{5.17.19}$
3	$\frac{1}{2}$	$\frac{5}{2}$	0	000+	$-\sqrt{19}/\sqrt{3.5.17}$
3	$\frac{3}{2}$	$\frac{3}{2}$	0	000+	$-3\sqrt{2.3}/\sqrt{5.7.17.19}$
3	$\frac{3}{2}$	$\frac{5}{2}$	0	000−	$+\sqrt{19}/2\sqrt{2.3.17}$
3	$\frac{3}{2}$	$\frac{5}{2}$	1	000+	$-7/2\sqrt{2.5.17.19}$
3	$\frac{5}{2}$	$\frac{3}{2}$	0	000−	$+4.3\sqrt{2}/\sqrt{7.11.17.19}$
3	$\frac{5}{2}$	$\frac{3}{2}$	1	000+	$+97\sqrt{2}/\sqrt{3.5.7.11.17.19}$
3	$\frac{5}{2}$	$\frac{5}{2}$	0	000+	$-173/2.3\sqrt{2.5.11.17.19}$
3	$\frac{5}{2}$	$\frac{5}{2}$	1	000−	$+97/2.3\sqrt{2.11.17.19}$
3	$\frac{\tilde{1}}{2}$	$\frac{5}{2}$	0	000−	$-9\sqrt{13}/2\sqrt{5.11.17.19}$
$\tilde{1}$	$\frac{1}{2}$	$\frac{5}{2}$	0	000+	$-9\sqrt{7}/\sqrt{5.13.17.19}$
$\tilde{1}$	$\frac{3}{2}$	$\frac{3}{2}$	0	000+	$+\sqrt{2.5}/\sqrt{13.17.19}$
$\tilde{1}$	$\frac{3}{2}$	$\frac{5}{2}$	0	000−	$+7\sqrt{5.7}/2\sqrt{13.17.19}$
$\tilde{1}$	$\frac{5}{2}$	$\frac{3}{2}$	0	000−	$+7\sqrt{3.5}/2\sqrt{11.13.17.19}$
$\tilde{1}$	$\frac{5}{2}$	$\frac{5}{2}$	0	000+	$-7\sqrt{5.7}/2\sqrt{2.3.11.13.17.19}$
$\tilde{1}$	$\frac{5}{2}$	$\frac{5}{2}$	1	000+	$+41\sqrt{7}/2\sqrt{2.3.11.13.17.19}$
$\tilde{1}$	$\frac{\tilde{1}}{2}$	$\frac{3}{2}$	0	000+	$+61/2\sqrt{5.11.17.19}$

7	$\frac{13}{2}$	$\frac{11}{2}$			
1	$\frac{1}{2}$	$\frac{1}{2}$	0	000−	$-4.11/\sqrt{3.7.13.17.19}$
1	$\frac{1}{2}$	$\frac{3}{2}$	0	000+	$+2\sqrt{11}/\sqrt{3.13.17.19}$
1	$\frac{3}{2}$	$\frac{1}{2}$	0	000+	$-11.11\sqrt{2}/\sqrt{3.5.7.13.17.19}$
1	$\frac{3}{2}$	$\frac{3}{2}$	0	000−	$+8\sqrt{11}/5\sqrt{3.13.17.19}$
1	$\frac{3}{2}$	$\frac{5}{2}$	0	000+	$-7\sqrt{2.11}/5\sqrt{13.17.19}$
1	$\frac{5}{2}$	$\frac{3}{2}$	0	000+	$+37\sqrt{3}/5\sqrt{13.17.19}$
1	$\frac{5}{2}$	$\frac{5}{2}$	0	000−	$+9.3/\sqrt{2.5.13.17.19}$
1	$\frac{5}{2}$	$\frac{5}{2}$	1	000−	$+3.29/5\sqrt{2.13.17.19}$
1	$\frac{\tilde{1}}{2}$	$\frac{5}{2}$	0	000+	$-1/\sqrt{17.19}$
2	$\frac{1}{2}$	$\frac{3}{2}$	0	000−	$-3\sqrt{3.11}/\sqrt{13.17.19}$
2	$\frac{1}{2}$	$\frac{5}{2}$	0	000+	$-\sqrt{11}/\sqrt{3.13.17.19}$
2	$\frac{3}{2}$	$\frac{1}{2}$	0	000−	$+11\sqrt{5}/\sqrt{2.3.7.13.17.19}$
2	$\frac{3}{2}$	$\frac{3}{2}$	0	000+	$+\sqrt{11.13}/\sqrt{3.5.17.19}$
2	$\frac{3}{2}$	$\frac{5}{2}$	0	000−	$-7\sqrt{11}/\sqrt{2.3.5.13.17.19}$
2	$\frac{3}{2}$	$\frac{5}{2}$	1	000+	$+8\sqrt{11}/\sqrt{5.13.17.19}$
2	$\frac{5}{2}$	$\frac{1}{2}$	0	000+	$-3.5\sqrt{5.11}/4\sqrt{7.13.17.19}$
2	$\frac{5}{2}$	$\frac{3}{2}$	0	000−	$+3.23/4\sqrt{5.13.17.19}$
2	$\frac{5}{2}$	$\frac{3}{2}$	1	000+	$+37\sqrt{3}/4\sqrt{2.5.13.17.19}$
2	$\frac{5}{2}$	$\frac{5}{2}$	0	000+	$+7\sqrt{3.13}/4\sqrt{2.5.17.19}$
2	$\frac{5}{2}$	$\frac{5}{2}$	1	000−	$-9\sqrt{5}/4\sqrt{13.17.19}$
2	$\frac{5}{2}$	$\frac{5}{2}$	2	000+	$+3\sqrt{13}/4\sqrt{2.5.17.19}$
2	$\frac{\tilde{1}}{2}$	$\frac{3}{2}$	0	000+	$+47/4\sqrt{2.3.17.19}$
2	$\frac{\tilde{1}}{2}$	$\frac{5}{2}$	0	000−	$+3/\sqrt{17.19}$
3	$\frac{1}{2}$	$\frac{5}{2}$	0	000−	$-2/\sqrt{3.17.19}$
3	$\frac{3}{2}$	$\frac{3}{2}$	0	000−	$+23/5\sqrt{3.17.19}$
3	$\frac{3}{2}$	$\frac{5}{2}$	0	000+	$+8\sqrt{2}/\sqrt{3.5.17.19}$
3	$\frac{3}{2}$	$\frac{5}{2}$	1	000−	$-2.9\sqrt{2}/5\sqrt{17.19}$
3	$\frac{5}{2}$	$\frac{1}{2}$	0	000−	$+2.9/\sqrt{5.7.17.19}$
3	$\frac{5}{2}$	$\frac{3}{2}$	0	000+	$+2.3/\sqrt{5.11.17.19}$
3	$\frac{5}{2}$	$\frac{3}{2}$	1	000−	$+7\sqrt{3}/5\sqrt{11.17.19}$
3	$\frac{5}{2}$	$\frac{5}{2}$	0	000−	$+3\sqrt{19}/5\sqrt{2.11.17}$
3	$\frac{5}{2}$	$\frac{5}{2}$	1	000+	$+3.13/\sqrt{2.5.11.17.19}$
3	$\frac{\tilde{1}}{2}$	$\frac{1}{2}$	0	000+	$+2\sqrt{13}/\sqrt{3.7.17.19}$
3	$\frac{\tilde{1}}{2}$	$\frac{5}{2}$	0	000+	$+3\sqrt{13}/\sqrt{11.17.19}$
$\tilde{1}$	$\frac{1}{2}$	$\frac{5}{2}$	0	000−	$-5\sqrt{7}/\sqrt{13.17.19}$
$\tilde{1}$	$\frac{3}{2}$	$\frac{3}{2}$	0	000−	$+8\sqrt{7}/5\sqrt{13.17.19}$
$\tilde{1}$	$\frac{3}{2}$	$\frac{5}{2}$	0	000+	$-7\sqrt{7}/5\sqrt{13.17.19}$
$\tilde{1}$	$\frac{5}{2}$	$\frac{1}{2}$	0	000−	$+37\sqrt{3}/4\sqrt{5.13.17.19}$
$\tilde{1}$	$\frac{5}{2}$	$\frac{3}{2}$	0	000+	$+7.7.7\sqrt{3.7}/4.5\sqrt{2.11.13.17.19}$
$\tilde{1}$	$\frac{5}{2}$	$\frac{5}{2}$	0	000−	$+7.7.7\sqrt{3.7}/4.5\sqrt{2.11.13.17.19}$
$\tilde{1}$	$\frac{5}{2}$	$\frac{5}{2}$	1	000−	$-41\sqrt{3.7}/4\sqrt{2.5.11.13.17.19}$
$\tilde{1}$	$\frac{\tilde{1}}{2}$	$\frac{3}{2}$	0	000−	$-\sqrt{7}/4\sqrt{2.11.17.19}$

7	$\frac{13}{2}$	$\frac{13}{2}$			
1	$\frac{1}{2}$	$\frac{1}{2}$	0	000+	$-11\sqrt{2}/\sqrt{13.17.19}$
1	$\frac{3}{2}$	$\frac{1}{2}$	0	000−	$+11/\sqrt{5.13.17.19}$
1	$\frac{3}{2}$	$\frac{3}{2}$	0	000+	$-11\sqrt{17}/5\sqrt{5.13.19}$
1	$\frac{5}{2}$	$\frac{3}{2}$	0	000−	$+8.2\sqrt{11}/5\sqrt{5.13.17.19}$
1	$\frac{5}{2}$	$\frac{5}{2}$	0	000+	$-7.7/5\sqrt{13.17.19}$
1	$\frac{5}{2}$	$\frac{5}{2}$	1	000+	$-4.3/5\sqrt{5.13.17.19}$
1	$\frac{\tilde{1}}{2}$	$\frac{5}{2}$	0	000−	$+7/\sqrt{2.5.17.19}$
2	$\frac{3}{2}$	$\frac{1}{2}$	0	000+	$-2.11/\sqrt{5.13.17.19}$
2	$\frac{3}{2}$	$\frac{3}{2}$	0	000−	0
2	$\frac{5}{2}$	$\frac{1}{2}$	0	000−	$+\sqrt{11.19}/\sqrt{2.3.5.13.17}$
2	$\frac{5}{2}$	$\frac{3}{2}$	0	000+	$+\sqrt{11}/\sqrt{3.13.17.19}$
2	$\frac{5}{2}$	$\frac{3}{2}$	1	000−	$-7\sqrt{11}/\sqrt{2.13.17.19}$
2	$\frac{5}{2}$	$\frac{5}{2}$	0	000−	0
2	$\frac{5}{2}$	$\frac{5}{2}$	1	000+	$-2.5\sqrt{2}/\sqrt{13.17.19}$
2	$\frac{5}{2}$	$\frac{5}{2}$	2	000−	0
2	$\frac{\tilde{1}}{2}$	$\frac{3}{2}$	0	000−	$-\sqrt{11}/\sqrt{2.5.17.19}$
2	$\frac{\tilde{1}}{2}$	$\frac{5}{2}$	0	000+	$-4\sqrt{2}/\sqrt{5.17.19}$
3	$\frac{3}{2}$	$\frac{3}{2}$	0	000+	$+8\sqrt{11}/5\sqrt{5.17.19}$
3	$\frac{5}{2}$	$\frac{1}{2}$	0	000+	$+1/\sqrt{2.3.5.17.19}$
3	$\frac{5}{2}$	$\frac{3}{2}$	0	000−	$-2.11/5\sqrt{3.17.19}$
3	$\frac{5}{2}$	$\frac{3}{2}$	1	000+	$-4/5\sqrt{5.17.19}$
3	$\frac{5}{2}$	$\frac{5}{2}$	0	000+	$-8.29/5\sqrt{5.11.17.19}$
3	$\frac{5}{2}$	$\frac{5}{2}$	1	000−	0
3	$\frac{\tilde{1}}{2}$	$\frac{1}{2}$	0	000−	$+\sqrt{13}/\sqrt{2.17.19}$
3	$\frac{\tilde{1}}{2}$	$\frac{5}{2}$	0	000−	$-4\sqrt{2.13}/\sqrt{5.11.17.19}$
$\tilde{1}$	$\frac{3}{2}$	$\frac{3}{2}$	0	000+	$+4\sqrt{3.11.19}/5\sqrt{5.7.13.17}$
$\tilde{1}$	$\frac{5}{2}$	$\frac{1}{2}$	0	000+	$+37/\sqrt{2.5.7.13.17.19}$

SO_3–K 3jm Factors (cont.)

7 $\frac{13}{2}$ $\frac{13}{2}$

$\tilde{1}$ $\frac{5}{2}$ $\frac{3}{2}$ 0 000−	$-7.7\sqrt{7}/5\sqrt{2.5.13.17.19}$
$\tilde{1}$ $\frac{5}{2}$ $\frac{5}{2}$ 0 000+	$+2.7.7\sqrt{3.7}/5\sqrt{5.11.13.17.19}$
$\tilde{1}$ $\frac{5}{2}$ $\frac{5}{2}$ 1 000+	$+2\sqrt{3.11}/5\sqrt{7.13.17.19}$
$\tilde{1}$ $\tilde{\frac{1}{2}}$ $\frac{3}{2}$ 0 000+	$+9\sqrt{3}/\sqrt{2.5.7.17.19}$
$\tilde{1}$ $\tilde{\frac{1}{2}}$ $\tilde{\frac{1}{2}}$ 0 000+	$+4\sqrt{2.3.13}/\sqrt{7.11.17.19}$

7 7 0

1 1 0 0 000+	$+1/\sqrt{5}$
2 2 0 0 000+	$+1/\sqrt{3}$
3 3 0 0 000+	$+2/\sqrt{3.5}$
$\tilde{1}$ $\tilde{1}$ 0 0 000+	$+1/\sqrt{5}$

7 7 1

1 1 1 0 000+	$-2/\sqrt{5.7}$
2 1 1 0 000−	$-\sqrt{3}/\sqrt{5.7}$
2 2 1 0 000+	$-5/2.3\sqrt{7}$
3 2 1 0 000−	$-\sqrt{11.13}/2.3\sqrt{2.5.7}$
3 3 1 0 000+	$+4\sqrt{2}/3\sqrt{5.7}$
$\tilde{1}$ 2 1 0 000−	$-\sqrt{11}/2\sqrt{2.3.5}$
$\tilde{1}$ 3 1 0 000+	$-\sqrt{13}/2\sqrt{2.3.5}$

7 7 2

1 1 2 0 000+	$+2\sqrt{17}/5\sqrt{7.13}$
2 1 2 0 000−	$+\sqrt{3.17}/\sqrt{5.7.13}$
2 2 2 0 000+	$+19\sqrt{5}/4\sqrt{2.3.7.13.17}$
2 2 2 1 000+	$+3\sqrt{3}/4\sqrt{2.7.13.17}$
3 1 2 0 000+	$+\sqrt{2.3.11}/5\sqrt{7.17}$
3 2 2 0 000−	$-11\sqrt{11}/4\sqrt{3.5.7.17}$
3 2 2 1 000+	$+\sqrt{3.5.11}/4\sqrt{7.17}$
3 3 2 0 000+	$-\sqrt{2.3.13}/5\sqrt{7.17}$
$\tilde{1}$ 1 2 0 000+	$+3\sqrt{2.11}/5\sqrt{13.17}$
$\tilde{1}$ 2 2 0 000−	$-11\sqrt{11}/4\sqrt{5.13.17}$
$\tilde{1}$ 3 2 0 000+	$+\sqrt{17}/2.5\sqrt{2}$
$\tilde{1}$ $\tilde{1}$ 2 0 000+	$-7\sqrt{7}/4.5\sqrt{13.17}$

7 7 3

2 1 3 0 000−	$-37/2\sqrt{5.7.13.17}$
2 1 $\tilde{1}$ 0 000−	$-2\sqrt{7}/\sqrt{3.5.13.17}$
2 2 3 0 000+	$-2.5\sqrt{2}/\sqrt{7.13.17}$
2 2 3 1 000−	0
2 2 $\tilde{1}$ 0 000+	$-5/2\sqrt{2.3.7.13.17}$
3 1 3 0 000+	$-2\sqrt{11}/\sqrt{5.7.17}$
3 1 $\tilde{1}$ 0 000+	$-\sqrt{11}/\sqrt{3.5.7.17}$
3 2 3 0 000−	$-\sqrt{11}/\sqrt{5.7.17}$
3 2 $\tilde{1}$ 0 000−	$+4\sqrt{11}/\sqrt{3.5.7.17}$
3 3 3 0 000−	0
3 3 $\tilde{1}$ 0 000+	$+\sqrt{3.13}/\sqrt{5.7.17}$
$\tilde{1}$ 1 3 0 000+	$-3\sqrt{3.11}/2\sqrt{5.13.17}$
$\tilde{1}$ 2 3 0 000−	$-2\sqrt{11}/\sqrt{3.5.13.17}$
$\tilde{1}$ 2 $\tilde{1}$ 0 000−	$+7\sqrt{11}/2\sqrt{2.3.5.13.17}$
$\tilde{1}$ 3 3 0 000+	$+1/\sqrt{3.5.17}$
$\tilde{1}$ $\tilde{1}$ $\tilde{1}$ 0 000+	$+19\sqrt{7}/2\sqrt{2.3.5.13.17}$

7 7 4

1 1 2 0 000+	$-\sqrt{2.3.11}/\sqrt{7.13.17.19}$
2 1 2 0 000−	$-8\sqrt{2.5.11}/3\sqrt{7.13.17.19}$
2 1 3 0 000+	$-\sqrt{5.11}/2.3\sqrt{13.17.19}$
2 2 2 0 000+	$-79\sqrt{5.11}/4.3\sqrt{7.13.17.19}$
2 2 2 1 000+	$+5.53\sqrt{11}/4.9\sqrt{7.13.17.19}$
2 2 3 0 000−	0
2 2 3 1 000+	$-8.4\sqrt{2.11}/9\sqrt{13.17.19}$
3 1 2 0 000+	$-\sqrt{19}/3\sqrt{7.17}$
3 1 3 0 000−	$-2\sqrt{19}/3\sqrt{5.17}$
3 2 2 0 000−	$-\sqrt{5.17}/3\sqrt{2.7.19}$
3 2 2 1 000+	$-\sqrt{5.17}/9\sqrt{2.7.19}$
3 2 3 0 000+	$+5\sqrt{5}/9\sqrt{17.19}$
3 3 2 0 000+	$-97\sqrt{13}/9\sqrt{7.11.17.19}$
3 3 3 0 000+	$+8\sqrt{13}/9\sqrt{5.11.17.19}$
$\tilde{1}$ 1 2 0 000+	$-2\sqrt{13}/\sqrt{3.17.19}$
$\tilde{1}$ 1 3 0 000−	$-\sqrt{7}/2\sqrt{3.5.13.17.19}$
$\tilde{1}$ 2 2 0 000−	$-53\sqrt{5}/2.3\sqrt{2.3.13.17.19}$
$\tilde{1}$ 2 3 0 000+	$-8.2\sqrt{5.7}/3\sqrt{3.13.17.19}$
$\tilde{1}$ 3 2 0 000+	$+61/3\sqrt{3.11.17.19}$
$\tilde{1}$ 3 3 0 000−	$+\sqrt{7.19}/3\sqrt{3.5.11.17}$
$\tilde{1}$ $\tilde{1}$ 2 0 000+	$-5.7\sqrt{7}/2\sqrt{2.3.11.13.17.19}$

7 7 5

1 1 1 0 000+	$-2.3\sqrt{2.3.11}/\sqrt{5.7.13.17.19}$
1 1 2 0 000−	0
2 1 1 0 000−	$-3\sqrt{11.13}/\sqrt{2.5.7.17.19}$
2 1 2 0 000+	$-\sqrt{5.11}/\sqrt{2.13.17.19}$
2 1 $\tilde{1}$ 0 000−	$+4\sqrt{2.11}/\sqrt{3.5.13.17.19}$
2 2 1 0 000+	$+9\sqrt{3.11}/\sqrt{2.7.13.17.19}$
2 2 2 0 000−	0
2 2 2 1 000−	0
2 2 $\tilde{1}$ 0 000+	$+29\sqrt{11}/4\sqrt{3.13.17.19}$
3 1 2 0 000−	$-4/\sqrt{17.19}$
3 1 $\tilde{1}$ 0 000+	$+7\sqrt{2}/\sqrt{3.5.17.19}$
3 2 1 0 000−	$-\sqrt{3}/4\sqrt{5.7.17.19}$
3 2 2 0 000+	$+\sqrt{5}/4\sqrt{2.17.19}$
3 2 2 1 000−	$-13\sqrt{5}/4\sqrt{2.17.19}$
3 2 $\tilde{1}$ 0 000−	$-4\sqrt{2}/\sqrt{3.5.17.19}$
3 3 1 0 000+	$-4\sqrt{3.13}/\sqrt{5.7.11.17.19}$
3 3 2 0 000−	0
3 3 $\tilde{1}$ 0 000+	$+3\sqrt{2.3.13}/\sqrt{5.11.17.19}$
$\tilde{1}$ 1 2 0 000−	$-3\sqrt{3.7}/\sqrt{13.17.19}$
$\tilde{1}$ 2 1 0 000−	$-\sqrt{19}/2\sqrt{5.13.17}$
$\tilde{1}$ 2 2 0 000+	$-2\sqrt{2.5.7}/\sqrt{3.13.17.19}$
$\tilde{1}$ 2 $\tilde{1}$ 0 000−	$+7\sqrt{7.13}/4\sqrt{3.5.17.19}$
$\tilde{1}$ 3 1 0 000+	$+109/4\sqrt{5.11.17.19}$
$\tilde{1}$ 3 2 0 000−	$-23\sqrt{7}/4\sqrt{3.11.17.19}$
$\tilde{1}$ $\tilde{1}$ 2 0 000−	0
$\tilde{1}$ $\tilde{1}$ $\tilde{1}$ 0 000+	$-7.41/4\sqrt{3.5.11.13.17.19}$

SO_3–K 3jm Factors (cont.)

7 7 6	
1 1 0 0 000+	$-2.11\sqrt{2}/\sqrt{5.13.17.19}$
1 1 1 0 000−	0
1 1 2 0 000+	$+2\sqrt{2.7.11}/5\sqrt{13.17.19}$
2 1 1 0 000+	$-11\sqrt{3.7}/\sqrt{2.5.13.17.19}$
2 1 2 0 000−	$+7\sqrt{7.11}/\sqrt{2.3.5.13.17.19}$
2 1 3 0 000+	$+\sqrt{2.7.11}/\sqrt{3.5.13.17.19}$
2 2 0 0 000+	$-11/2\sqrt{2.3.13.17.19}$
2 2 1 0 000−	0
2 2 2 0 000+	$+\sqrt{5.7.11}/4\sqrt{3.13.17.19}$
2 2 2 1 000+	$-\sqrt{7.11.13}/4\sqrt{3.17.19}$
2 2 3 0 000−	0
2 2 3 1 000+	$-\sqrt{7.11}/\sqrt{3.13.17.19}$
3 1 2 0 000+	$-2\sqrt{7}/5\sqrt{3.17.19}$
3 1 3 0 000−	$+2\sqrt{2.7}/\sqrt{3.5.17.19}$
3 2 1 0 000+	$-3\sqrt{7.11}/4\sqrt{5.17.19}$
3 2 2 0 000−	$-\sqrt{7.17}/4\sqrt{2.3.5.19}$
3 2 2 1 000+	$-7\sqrt{5.7}/4\sqrt{2.3.17.19}$
3 2 3 0 000+	$+4\sqrt{2.7}/\sqrt{3.5.17.19}$
3 3 0 0 000+	$+2\sqrt{2.13}/\sqrt{3.5.17.19}$
3 3 1 0 000−	0
3 3 2 0 000+	$+2\sqrt{7.13}/5\sqrt{3.11.17.19}$
3 3 3 0 000+	$-4\sqrt{2.7.13}/\sqrt{3.5.11.17.19}$
$\tilde{1}$ 1 2 0 000+	$-29/5\sqrt{13.17.19}$
$\tilde{1}$ 1 3 0 000−	$+\sqrt{2.19}/\sqrt{5.13.17}$
$\tilde{1}$ 2 1 0 000+	$-4\sqrt{3.11}/\sqrt{5.13.17.19}$
$\tilde{1}$ 2 2 0 000−	$+3/2\sqrt{2.5.13.17.19}$
$\tilde{1}$ 2 3 0 000+	$-1/\sqrt{2.5.13.17.19}$
$\tilde{1}$ 3 1 0 000−	$+\sqrt{3.5}/4\sqrt{17.19}$
$\tilde{1}$ 3 2 0 000+	$+3.23/4.5\sqrt{11.17.19}$
$\tilde{1}$ 3 3 0 000−	$+8.2\sqrt{2}/\sqrt{5.11.17.19}$
$\tilde{1}$ $\tilde{1}$ 0 0 000+	$+37/2\sqrt{2.5.13.17.19}$
$\tilde{1}$ $\tilde{1}$ 2 0 000+	$+7.7.7\sqrt{7}/2.5\sqrt{2.11.13.17.19}$
7 7 7	
1 1 1 0 000+	$+11\sqrt{7.11}/5\sqrt{13.17.19}$
2 1 1 0 000−	0
2 2 1 0 000+	$+11\sqrt{7.11}/3\sqrt{5.13.17.19}$
2 2 2 0 000−	0
2 2 2 1 000−	0
3 2 1 0 000−	$-\sqrt{2.7}/3\sqrt{17.19}$
3 2 2 0 000+	$-2\sqrt{7}/\sqrt{5.17.19}$
3 2 2 1 000−	0
3 3 1 0 000+	$+\sqrt{2.7.11.13}/3.5\sqrt{17.19}$
3 3 2 0 000−	0
3 3 3 0 000−	0
$\tilde{1}$ 2 1 0 000−	$-\sqrt{17}/\sqrt{2.3.13.19}$
$\tilde{1}$ 2 2 0 000+	$-2.31/\sqrt{3.5.13.17.19}$
$\tilde{1}$ 3 1 0 000+	$-\sqrt{2.11}/5\sqrt{3.17.19}$
$\tilde{1}$ 3 2 0 000−	$-\sqrt{2.11}/\sqrt{3.17.19}$
$\tilde{1}$ 3 3 0 000+	$+8\sqrt{2.13}/5\sqrt{3.17.19}$
7 7 7	
$\tilde{1}$ $\tilde{1}$ 2 0 000−	0
$\tilde{1}$ $\tilde{1}$ $\tilde{1}$ 0 000+	$-2\sqrt{3}/5\sqrt{13.17.19}$
$\frac{15}{2}$ 4 $\tilde{\frac{7}{2}}$	
$\frac{3}{2}$ 2 $\frac{5}{2}$ 0 000−	$-7\sqrt{7}/2.3\sqrt{5.13}$
$\frac{3}{2}$ 2 $\frac{5}{2}$ 1 000+	$-\sqrt{7}/\sqrt{2.3.5.13}$
$\frac{3}{2}$ 2 $\tilde{\frac{1}{2}}$ 0 000+	$-\sqrt{2}/\sqrt{5.13}$
$\frac{3}{2}$ 3 $\frac{5}{2}$ 0 000+	$-2/3\sqrt{13}$
$\frac{3}{2}$ 3 $\frac{5}{2}$ 1 000−	$-2/\sqrt{3.5.13}$
$\frac{5}{2}$ 2 $\frac{5}{2}$ 0 000+	$+7\sqrt{2}/\sqrt{5.11.13}$
$\frac{5}{2}$ 2 $\frac{5}{2}$ 0 100+	0
$\frac{5}{2}$ 2 $\frac{5}{2}$ 1 000−	$+7.19/8.2\sqrt{3.5.11.13}$
$\frac{5}{2}$ 2 $\frac{5}{2}$ 1 100−	$+\sqrt{7}/8.2$
$\frac{5}{2}$ 2 $\frac{5}{2}$ 2 000+	$-7\sqrt{3}/8\sqrt{2.5.11.13}$
$\frac{5}{2}$ 2 $\frac{5}{2}$ 2 100+	$-\sqrt{7}/8\sqrt{2}$
$\frac{5}{2}$ 2 $\tilde{\frac{1}{2}}$ 0 000−	$-7\sqrt{7}/8.2\sqrt{5.11.13}$
$\frac{5}{2}$ 2 $\tilde{\frac{1}{2}}$ 0 100−	$+3\sqrt{3}/8.2$
$\frac{5}{2}$ 3 $\frac{5}{2}$ 0 000−	$+29\sqrt{7}/4\sqrt{2.3.5.11.13}$
$\frac{5}{2}$ 3 $\frac{5}{2}$ 0 100−	$-1/4\sqrt{2}$
$\frac{5}{2}$ 3 $\frac{5}{2}$ 1 000+	$\sqrt{3.7}/4\sqrt{2.11.13}$
$\frac{5}{2}$ 3 $\frac{5}{2}$ 1 100+	$-\sqrt{5}/4\sqrt{2}$
$\frac{5}{2}$ 3 $\tilde{\frac{1}{2}}$ 0 000+	$-4/\sqrt{11.13}$
$\frac{5}{2}$ 3 $\tilde{\frac{1}{2}}$ 0 100+	0
$\frac{15}{2}$ $\frac{9}{2}$ 3	
$\frac{3}{2}$ $\frac{3}{2}$ 3 0 000−	$-11/2.5\sqrt{13}$
$\frac{3}{2}$ $\frac{3}{2}$ $\tilde{1}$ 0 000−	$-3\sqrt{3}/5\sqrt{13}$
$\frac{3}{2}$ $\frac{5}{2}$ 3 0 000+	$+3/\sqrt{5.7.13}$
$\frac{3}{2}$ $\frac{5}{2}$ 3 1 000−	$+3\sqrt{3}/5\sqrt{7.13}$
$\frac{3}{2}$ $\frac{5}{2}$ $\tilde{1}$ 0 000+	$-4\sqrt{2.3}/5\sqrt{7.13}$
$\frac{5}{2}$ $\frac{3}{2}$ 3 0 000+	$+19\sqrt{3.7}/8.2\sqrt{2.5.11.13}$
$\frac{5}{2}$ $\frac{3}{2}$ 3 0 100+	$+3/8.2\sqrt{2}$
$\frac{5}{2}$ $\frac{3}{2}$ 3 1 000−	$+3.7\sqrt{7}/8.2.5\sqrt{2.11.13}$
$\frac{5}{2}$ $\frac{3}{2}$ 3 1 100−	$-9\sqrt{3}/8.2\sqrt{2.5}$
$\frac{5}{2}$ $\frac{3}{2}$ $\tilde{1}$ 0 000+	$-3.17\sqrt{7}/8.5\sqrt{11.13}$
$\frac{5}{2}$ $\frac{3}{2}$ $\tilde{1}$ 0 100+	$-\sqrt{3}/8\sqrt{5}$
$\frac{5}{2}$ $\frac{5}{2}$ 3 0 000	$+9.17\sqrt{3}/8.5\sqrt{2.11.13}$
$\frac{5}{2}$ $\frac{5}{2}$ 3 0 100−	$+9/8\sqrt{2.5.7}$
$\frac{5}{2}$ $\frac{5}{2}$ 3 1 000+	$+3\sqrt{3.13}/8\sqrt{2.5.11}$
$\frac{5}{2}$ $\frac{5}{2}$ 3 1 100+	$-9/8\sqrt{2.7}$
$\frac{5}{2}$ $\frac{5}{2}$ $\tilde{1}$ 0 000−	$+3.29/4.5\sqrt{2.11.13}$
$\frac{5}{2}$ $\frac{5}{2}$ $\tilde{1}$ 0 100−	$-3\sqrt{3}/4\sqrt{2.5.7}$
$\frac{5}{2}$ $\frac{5}{2}$ $\tilde{1}$ 1 000−	$+9/4\sqrt{2.5.11.13}$
$\frac{5}{2}$ $\frac{5}{2}$ $\tilde{1}$ 1 100−	$+3\sqrt{3}/4\sqrt{2.7}$
$\frac{15}{2}$ $\frac{9}{2}$ 4	
$\frac{3}{2}$ $\frac{3}{2}$ 2 0 000−	$-\sqrt{7}/5\sqrt{13.17}$
$\frac{3}{2}$ $\frac{3}{2}$ 3 0 000+	$+\sqrt{13}/2.5\sqrt{17}$
$\frac{3}{2}$ $\frac{5}{2}$ 2 0 000+	$+2\sqrt{13}/3.5\sqrt{17}$
$\frac{3}{2}$ $\frac{5}{2}$ 2 1 000−	$-2.7\sqrt{2}/5\sqrt{3.13.17}$
$\frac{3}{2}$ $\frac{5}{2}$ 3 0 000−	$-\sqrt{7.17}/3\sqrt{5.13}$
$\frac{3}{2}$ $\frac{5}{2}$ 3 1 000+	$-\sqrt{7}/5\sqrt{3.13.17}$

SO_3–K 3jm Factors (cont.)

$\frac{15}{2}\ \frac{9}{2}\ 4$
$\frac{5}{2}\ \frac{3}{2}\ 2\ 0\ 000+\quad +7\sqrt{3.13}/4.5\sqrt{2.11.17}$
$\frac{5}{2}\ \frac{3}{2}\ 2\ 0\ 100+\quad -3\sqrt{7}/4\sqrt{2.5.17}$
$\frac{5}{2}\ \frac{3}{2}\ 2\ 1\ 000-\quad -7.103/8.5\sqrt{11.13.17}$
$\frac{5}{2}\ \frac{3}{2}\ 2\ 1\ 100-\quad +3\sqrt{3.7}/8\sqrt{5.17}$
$\frac{5}{2}\ \frac{3}{2}\ 3\ 0\ 000-\quad -7.19\sqrt{3.7}/8.2\sqrt{2.5.11.13.17}$
$\frac{5}{2}\ \frac{3}{2}\ 3\ 0\ 100-\quad -3.7/8.2\sqrt{2.17}$
$\frac{5}{2}\ \frac{3}{2}\ 3\ 1\ 000+\quad +7\sqrt{7.11}/8.2.5\sqrt{2.13.17}$
$\frac{5}{2}\ \frac{3}{2}\ 3\ 1\ 100+\quad -3.11\sqrt{3}/8.2\sqrt{2.5.17}$
$\frac{5}{2}\ \frac{5}{2}\ 2\ 0\ 000-\quad -37\sqrt{7}/4.5\sqrt{2.11.13.17}$
$\frac{5}{2}\ \frac{5}{2}\ 2\ 0\ 100-\quad +9\sqrt{3}/4\sqrt{2.5.17}$
$\frac{5}{2}\ \frac{5}{2}\ 2\ 1\ 000+\quad +\sqrt{7.13}/4\sqrt{3.11.17}$
$\frac{5}{2}\ \frac{5}{2}\ 2\ 1\ 100+\quad +11/4\sqrt{5.17}$
$\frac{5}{2}\ \frac{5}{2}\ 2\ 2\ 000-\quad -101\sqrt{3.7}/4.5\sqrt{2.11.13.17}$
$\frac{5}{2}\ \frac{5}{2}\ 2\ 2\ 100-\quad -\sqrt{5}/4\sqrt{2.17}$
$\frac{5}{2}\ \frac{5}{2}\ 3\ 0\ 000+\quad +71\sqrt{13}/8.5\sqrt{2.3.11.17}$
$\frac{5}{2}\ \frac{5}{2}\ 3\ 0\ 100+\quad -\sqrt{7}/8\sqrt{2.5.17}$
$\frac{5}{2}\ \frac{5}{2}\ 3\ 1\ 000-\quad +\sqrt{3.5.11}/8\sqrt{2.13.17}$
$\frac{5}{2}\ \frac{5}{2}\ 3\ 1\ 100-\quad +\sqrt{7}/8\sqrt{2.17}$

$\frac{15}{2}\ 5\ \frac{5}{2}$
$\frac{3}{2}\ 1\ \frac{5}{2}\ 0\ 000+\quad -\sqrt{2}/\sqrt{13}$
$\frac{3}{2}\ 2\ \frac{5}{2}\ 0\ 000-\quad -1/\sqrt{2.7.13}$
$\frac{3}{2}\ 2\ \frac{5}{2}\ 1\ 000+\quad +3\sqrt{3}/2\sqrt{7.13}$
$\frac{3}{2}\ \tilde{1}\ \frac{5}{2}\ 0\ 000+\quad -\sqrt{3}/\sqrt{2.7.13}$
$\frac{5}{2}\ 1\ \frac{5}{2}\ 0\ 000-\quad +7\sqrt{5.7}/8.2\sqrt{11.13}$
$\frac{5}{2}\ 1\ \frac{5}{2}\ 0\ 100-\quad +\sqrt{3}/8.2$
$\frac{5}{2}\ 1\ \frac{5}{2}\ 1\ 000-\quad +3\sqrt{7}/8.2\sqrt{11.13}$
$\frac{5}{2}\ 1\ \frac{5}{2}\ 1\ 100-\quad +\sqrt{3.5}/8.2$
$\frac{5}{2}\ 2\ \frac{5}{2}\ 0\ 000+\quad -\sqrt{11}/8\sqrt{13}$
$\frac{5}{2}\ 2\ \frac{5}{2}\ 0\ 100+\quad -\sqrt{3.5}/8\sqrt{7}$
$\frac{5}{2}\ 2\ \frac{5}{2}\ 1\ 000-\quad -29\sqrt{3}/8\sqrt{2.11.13}$
$\frac{5}{2}\ 2\ \frac{5}{2}\ 1\ 100-\quad +3\sqrt{5}/8\sqrt{2.7}$
$\frac{5}{2}\ 2\ \frac{5}{2}\ 2\ 000+\quad -\sqrt{3.11}/8\sqrt{13}$
$\frac{5}{2}\ 2\ \frac{5}{2}\ 2\ 100+\quad -3\sqrt{5}/8\sqrt{7}$
$\frac{5}{2}\ \tilde{1}\ \frac{5}{2}\ 0\ 000-\quad -3\sqrt{2}/\sqrt{11.13}$
$\frac{5}{2}\ \tilde{1}\ \frac{5}{2}\ 0\ 100-\quad 0$
$\frac{5}{2}\ \tilde{1}\ \frac{5}{2}\ 1\ 000-\quad -3\sqrt{5}/4\sqrt{2.11.13}$
$\frac{5}{2}\ \tilde{1}\ \frac{5}{2}\ 1\ 100-\quad +3\sqrt{3}/4\sqrt{2.7}$

$\frac{15}{2}\ 5\ \frac{7}{2}$
$\frac{3}{2}\ 1\ \frac{5}{2}\ 0\ 000-\quad +3/2\sqrt{2.5.13.17}$
$\frac{3}{2}\ 2\ \frac{5}{2}\ 0\ 000+\quad -3\sqrt{17}/2\sqrt{2.5.7.13}$
$\frac{3}{2}\ 2\ \frac{5}{2}\ 1\ 000-\quad -9\sqrt{3}/\sqrt{5.7.13.17}$
$\frac{3}{2}\ 2\ \frac{\tilde{1}}{2}\ 0\ 000-\quad -19/2\sqrt{5.13.17}$
$\frac{3}{2}\ \tilde{1}\ \frac{5}{2}\ 0\ 000-\quad +\sqrt{3.13}/\sqrt{2.5.7.17}$
$\frac{3}{2}\ \tilde{1}\ \frac{\tilde{1}}{2}\ 0\ 000+\quad -3\sqrt{3}/\sqrt{2.13.17}$
$\frac{5}{2}\ 1\ \frac{5}{2}\ 0\ 000+\quad -3.7.7\sqrt{7}/8.4\sqrt{11.13.17}$
$\frac{5}{2}\ 1\ \frac{5}{2}\ 0\ 100+\quad -\sqrt{3.5}/8.4\sqrt{17}$
$\frac{5}{2}\ 1\ \frac{5}{2}\ 1\ 000+\quad -3.89\sqrt{7}/8.4\sqrt{5.11.13.17}$
$\frac{5}{2}\ 1\ \frac{5}{2}\ 1\ 100+\quad +23\sqrt{3}/8.4\sqrt{17}$
$\frac{5}{2}\ 1\ \frac{\tilde{1}}{2}\ 0\ 000-\quad +7.19\sqrt{3}/8.2\sqrt{2.11.13.17}$
$\frac{5}{2}\ 1\ \frac{\tilde{1}}{2}\ 0\ 100-\quad +3\sqrt{5.7}/8.2\sqrt{2.17}$
$\frac{5}{2}\ 2\ \frac{5}{2}\ 0\ 000-\quad -3\sqrt{11.17}/8.2\sqrt{5.13}$
$\frac{5}{2}\ 2\ \frac{5}{2}\ 0\ 100-\quad +5\sqrt{3}/8.2\sqrt{7.17}$
$\frac{5}{2}\ 2\ \frac{5}{2}\ 1\ 000+\quad +9\sqrt{3.13}/8.2\sqrt{2.5.11.17}$
$\frac{5}{2}\ 2\ \frac{5}{2}\ 1\ 100+\quad +3.23/8.2\sqrt{2.7.17}$
$\frac{5}{2}\ 2\ \frac{5}{2}\ 2\ 000-\quad -3\sqrt{3.17}/4\sqrt{5.11.13}$
$\frac{5}{2}\ 2\ \frac{5}{2}\ 2\ 100-\quad +9/4\sqrt{7.17}$
$\frac{5}{2}\ 2\ \frac{\tilde{1}}{2}\ 0\ 000+\quad +3.7\sqrt{7}/8.2\sqrt{2.5.11.13.17}$
$\frac{5}{2}\ 2\ \frac{\tilde{1}}{2}\ 0\ 100+\quad -9\sqrt{3}/8.2\sqrt{2.17}$
$\frac{5}{2}\ \tilde{1}\ \frac{5}{2}\ 0\ 000+\quad -3.37/8\sqrt{2.5.11.13.17}$
$\frac{5}{2}\ \tilde{1}\ \frac{5}{2}\ 0\ 100+\quad -3\sqrt{3.7}/8\sqrt{2.17}$
$\frac{5}{2}\ \tilde{1}\ \frac{5}{2}\ 1\ 000+\quad -9\sqrt{17}/8\sqrt{2.11.13}$
$\frac{5}{2}\ \tilde{1}\ \frac{5}{2}\ 1\ 100+\quad -3\sqrt{3.5}/8\sqrt{2.7.17}$

$\frac{15}{2}\ 5\ \frac{9}{2}$
$\frac{3}{2}\ 1\ \frac{3}{2}\ 0\ 000-\quad +\sqrt{3.7}/5\sqrt{2.13.17}$
$\frac{3}{2}\ 1\ \frac{5}{2}\ 0\ 000+\quad +2.3\sqrt{2}/5\sqrt{13.17}$
$\frac{3}{2}\ 2\ \frac{3}{2}\ 0\ 000+\quad -\sqrt{13}/5\sqrt{2.17}$
$\frac{3}{2}\ 2\ \frac{5}{2}\ 0\ 000-\quad -3\sqrt{7}/5\sqrt{2.13.17}$
$\frac{3}{2}\ 2\ \frac{5}{2}\ 1\ 000+\quad -9\sqrt{3.7}/2.5\sqrt{13.17}$
$\frac{3}{2}\ \tilde{1}\ \frac{3}{2}\ 0\ 000-\quad -9\sqrt{3}/5\sqrt{13.17}$
$\frac{3}{2}\ \tilde{1}\ \frac{5}{2}\ 0\ 000+\quad +7\sqrt{3.7}/5\sqrt{2.13.17}$
$\frac{5}{2}\ 1\ \frac{3}{2}\ 0\ 000+\quad +7.29\sqrt{3}/4.5\sqrt{11.13.17}$
$\frac{5}{2}\ 1\ \frac{3}{2}\ 0\ 100+\quad -3\sqrt{7}/4\sqrt{5.17}$
$\frac{5}{2}\ 1\ \frac{5}{2}\ 0\ 000-\quad -3.7\sqrt{7}/8.2\sqrt{5.11.13.17}$
$\frac{5}{2}\ 1\ \frac{5}{2}\ 0\ 100-\quad -7\sqrt{3}/8.2\sqrt{17}$
$\frac{5}{2}\ 1\ \frac{5}{2}\ 1\ 000-\quad +3.7.7\sqrt{7}/8.2.5\sqrt{11.13.17}$
$\frac{5}{2}\ 1\ \frac{5}{2}\ 1\ 100-\quad +\sqrt{3.17}/8.2\sqrt{5}$
$\frac{5}{2}\ 2\ \frac{3}{2}\ 0\ 000-\quad +7\sqrt{3.7}/5\sqrt{11.13.17}$
$\frac{5}{2}\ 2\ \frac{3}{2}\ 0\ 100-\quad +3/\sqrt{5.17}$
$\frac{5}{2}\ 2\ \frac{3}{2}\ 1\ 000+\quad -3.19\sqrt{7}/4.5\sqrt{2.11.13.17}$
$\frac{5}{2}\ 2\ \frac{3}{2}\ 1\ 100+\quad -3\sqrt{3}/4\sqrt{2.5.17}$
$\frac{5}{2}\ 2\ \frac{5}{2}\ 0\ 000+\quad -3\sqrt{13}/2.5\sqrt{11.17}$
$\frac{5}{2}\ 2\ \frac{5}{2}\ 0\ 100+\quad -\sqrt{3.7}/2\sqrt{5.17}$
$\frac{5}{2}\ 2\ \frac{5}{2}\ 1\ 000-\quad -9.9\sqrt{3}/8\sqrt{2.11.13.17}$
$\frac{5}{2}\ 2\ \frac{5}{2}\ 1\ 100-\quad -3\sqrt{7}/8\sqrt{2.5.17}$
$\frac{5}{2}\ 2\ \frac{5}{2}\ 2\ 000+\quad +3\sqrt{3.13}/5\sqrt{11.17}$
$\frac{5}{2}\ 2\ \frac{5}{2}\ 2\ 100+\quad 0$
$\frac{5}{2}\ \tilde{1}\ \frac{3}{2}\ 0\ 000+\quad -3\sqrt{7.17}/4.5\sqrt{11.13}$
$\frac{5}{2}\ \tilde{1}\ \frac{3}{2}\ 0\ 100+\quad -\sqrt{3}/4\sqrt{5.17}$
$\frac{5}{2}\ \tilde{1}\ \frac{5}{2}\ 0\ 000-\quad -3.83/4.5\sqrt{2.11.13.17}$
$\frac{5}{2}\ \tilde{1}\ \frac{5}{2}\ 0\ 100-\quad +3\sqrt{3.7}/4\sqrt{2.5.17}$
$\frac{5}{2}\ \tilde{1}\ \frac{5}{2}\ 1\ 000-\quad +9\sqrt{2}/\sqrt{5.11.13.17}$
$\frac{5}{2}\ \tilde{1}\ \frac{5}{2}\ 1\ 100-\quad 0$

$\frac{15}{2}\ \frac{11}{2}\ 2$
$\frac{3}{2}\ \frac{1}{2}\ 2\ 0\ 000-\quad +2\sqrt{2}/\sqrt{5.13}$
$\frac{3}{2}\ \frac{3}{2}\ 2\ 0\ 000+\quad -2\sqrt{11}/\sqrt{5.7.13}$
$\frac{3}{2}\ \frac{5}{2}\ 2\ 0\ 000-\quad -\sqrt{11}/\sqrt{2.5.7.13}$
$\frac{3}{2}\ \frac{5}{2}\ 2\ 1\ 000+\quad -\sqrt{3.11}/2\sqrt{5.7.13}$
$\frac{5}{2}\ \frac{1}{2}\ 2\ 0\ 000+\quad +\sqrt{7.11}/4\sqrt{2.3.5.13}$

SO_3–K 3jm Factors (cont.)

$\frac{15}{2}$ $\frac{11}{2}$ 2
$\frac{5}{2}$ $\frac{1}{2}$ 2 0 100+ $+1/4\sqrt{2}$
$\frac{5}{2}$ $\frac{3}{2}$ 2 0 000− $+\sqrt{13}/4\sqrt{2.3.5}$
$\frac{5}{2}$ $\frac{3}{2}$ 2 0 100− $-\sqrt{11}/4\sqrt{2.7}$
$\frac{5}{2}$ $\frac{3}{2}$ 2 1 000+ $+19/8\sqrt{5.13}$
$\frac{5}{2}$ $\frac{3}{2}$ 2 1 100+ $+\sqrt{3.11}/8\sqrt{7}$
$\frac{5}{2}$ $\frac{5}{2}$ 2 0 000+ $-3/\sqrt{5.13}$
$\frac{5}{2}$ $\frac{5}{2}$ 2 0 100+ 0
$\frac{5}{2}$ $\frac{5}{2}$ 2 1 000− $-\sqrt{2.3}/\sqrt{5.13}$
$\frac{5}{2}$ $\frac{5}{2}$ 2 1 100− 0
$\frac{5}{2}$ $\frac{5}{2}$ 2 2 000+ $+\sqrt{3.5}/8\sqrt{13}$
$\frac{5}{2}$ $\frac{5}{2}$ 2 2 100+ $-3\sqrt{11}/8\sqrt{7}$

$\frac{15}{2}$ $\frac{11}{2}$ 3
$\frac{3}{2}$ $\frac{3}{2}$ 3 0 000+ $-11/\sqrt{7.13.17}$
$\frac{3}{2}$ $\frac{3}{2}$ $\tilde{1}$ 0 000+ $-4\sqrt{3}/\sqrt{7.13.17}$
$\frac{3}{2}$ $\frac{5}{2}$ 3 0 000− $+\sqrt{5}/2\sqrt{2.7.13.17}$
$\frac{3}{2}$ $\frac{5}{2}$ 3 1 000+ $+23\sqrt{3}/2\sqrt{2.7.13.17}$
$\frac{3}{2}$ $\frac{5}{2}$ $\tilde{1}$ 0 000− $-5\sqrt{3}/2\sqrt{7.13.17}$
$\frac{5}{2}$ $\frac{1}{2}$ 3 0 000+ $-7\sqrt{5.7}/4\sqrt{2.3.13.17}$
$\frac{5}{2}$ $\frac{1}{2}$ 3 0 100+ $-\sqrt{11}/4\sqrt{2.17}$
$\frac{5}{2}$ $\frac{1}{2}$ $\tilde{1}$ 0 000+ $+\sqrt{5.7}/4\sqrt{2.13.17}$
$\frac{5}{2}$ $\frac{1}{2}$ $\tilde{1}$ 0 100+ $-\sqrt{3.11}/4\sqrt{2.17}$
$\frac{5}{2}$ $\frac{3}{2}$ 3 0 000− $-\sqrt{5.17}/8\sqrt{2.3.11.13}$
$\frac{5}{2}$ $\frac{3}{2}$ 3 0 100− $-23/8\sqrt{2.7.17}$
$\frac{5}{2}$ $\frac{3}{2}$ 3 1 000+ $-3.43/8\sqrt{2.11.13.17}$
$\frac{5}{2}$ $\frac{3}{2}$ 3 1 100+ $+5\sqrt{3.5}/8\sqrt{2.7.17}$
$\frac{5}{2}$ $\frac{3}{2}$ $\tilde{1}$ 0 000− $+97/8\sqrt{11.13.17}$
$\frac{5}{2}$ $\frac{3}{2}$ $\tilde{1}$ 0 100− $-5\sqrt{3.5}/8\sqrt{7.17}$
$\frac{5}{2}$ $\frac{5}{2}$ 3 0 000+ $-23\sqrt{3}/8\sqrt{11.13.17}$
$\frac{5}{2}$ $\frac{5}{2}$ 3 0 100+ $+9\sqrt{5}/8\sqrt{7.17}$
$\frac{5}{2}$ $\frac{5}{2}$ 3 1 000− $+\sqrt{3.5.17}/8\sqrt{11.13}$
$\frac{5}{2}$ $\frac{5}{2}$ 3 1 100− $+3\sqrt{7}/8\sqrt{17}$
$\frac{5}{2}$ $\frac{5}{2}$ $\tilde{1}$ 0 000+ $-9.5/4\sqrt{11.13.17}$
$\frac{5}{2}$ $\frac{5}{2}$ $\tilde{1}$ 0 100+ $-3\sqrt{3.5}/4\sqrt{7.17}$
$\frac{5}{2}$ $\frac{5}{2}$ $\tilde{1}$ 1 000+ $+3\sqrt{5.17}/8\sqrt{11.13}$
$\frac{5}{2}$ $\frac{5}{2}$ $\tilde{1}$ 1 100+ $-3\sqrt{3}/8\sqrt{7.17}$

$\frac{15}{2}$ $\frac{11}{2}$ 4
$\frac{3}{2}$ $\frac{1}{2}$ 2 0 000− $-\sqrt{11}/\sqrt{2.3.5.13.17}$
$\frac{3}{2}$ $\frac{3}{2}$ 2 0 000+ $+8.2/\sqrt{3.5.7.13.17}$
$\frac{3}{2}$ $\frac{3}{2}$ 3 0 000− $-7/\sqrt{3.5.13.17}$
$\frac{3}{2}$ $\frac{5}{2}$ 2 0 000− $+\sqrt{2.3.13}/\sqrt{5.7.17}$
$\frac{3}{2}$ $\frac{5}{2}$ 2 1 000+ $+3\sqrt{5}/2\sqrt{7.13.17}$
$\frac{3}{2}$ $\frac{5}{2}$ 3 0 000+ $-\sqrt{3}/2\sqrt{2.13.17}$
$\frac{3}{2}$ $\frac{5}{2}$ 3 1 000− $+9.3/2\sqrt{2.5.13.17}$
$\frac{5}{2}$ $\frac{1}{2}$ 2 0 000+ $-8\sqrt{2.7}/3\sqrt{5.13.17}$
$\frac{5}{2}$ $\frac{1}{2}$ 2 0 100+ $+\sqrt{2.11}/3\sqrt{3.17}$
$\frac{5}{2}$ $\frac{1}{2}$ 3 0 000− $-7/4.3\sqrt{2.13.17}$
$\frac{5}{2}$ $\frac{1}{2}$ 3 0 100− $-\sqrt{5.7.11}/4.3\sqrt{2.3.17}$
$\frac{5}{2}$ $\frac{3}{2}$ 2 0 000− $-349/4.3\sqrt{2.5.11.13.17}$
$\frac{5}{2}$ $\frac{3}{2}$ 2 0 100− $-109/4.3\sqrt{2.3.7.17}$

$\frac{15}{2}$ $\frac{11}{2}$ 4
$\frac{5}{2}$ $\frac{3}{2}$ 2 1 000+ $+163/8\sqrt{3.5.11.13.17}$
$\frac{5}{2}$ $\frac{3}{2}$ 2 1 100+ $-3.5/8\sqrt{7.17}$
$\frac{5}{2}$ $\frac{3}{2}$ 3 0 000+ $-149\sqrt{7}/8.3\sqrt{2.11.13.17}$
$\frac{5}{2}$ $\frac{3}{2}$ 3 0 100+ $-5\sqrt{5}/8.3\sqrt{2.3.17}$
$\frac{5}{2}$ $\frac{3}{2}$ 3 1 000− $+\sqrt{7.11.17}/8\sqrt{2.3.5.13}$
$\frac{5}{2}$ $\frac{3}{2}$ 3 1 100− $+11/8.3\sqrt{2.17}$
$\frac{5}{2}$ $\frac{5}{2}$ 2 0 000+ $-3.37\sqrt{3}/8\sqrt{5.11.13.17}$
$\frac{5}{2}$ $\frac{5}{2}$ 2 0 100+ $-\sqrt{7}/8\sqrt{17}$
$\frac{5}{2}$ $\frac{5}{2}$ 2 1 000− $+\sqrt{2}/\sqrt{5.11.13.17}$
$\frac{5}{2}$ $\frac{5}{2}$ 2 1 100− 0
$\frac{5}{2}$ $\frac{5}{2}$ 2 2 000+ $+3.19/4\sqrt{5.11.13.17}$
$\frac{5}{2}$ $\frac{5}{2}$ 2 2 100+ $+5/4\sqrt{3.7.17}$
$\frac{5}{2}$ $\frac{5}{2}$ 3 0 000− $+43\sqrt{7}/8\sqrt{5.11.13.17}$
$\frac{5}{2}$ $\frac{5}{2}$ 3 0 100− $-7\sqrt{3}/8\sqrt{17}$
$\frac{5}{2}$ $\frac{5}{2}$ 3 1 000+ $-3\sqrt{7.11}/8\sqrt{13.17}$
$\frac{5}{2}$ $\frac{5}{2}$ 3 1 100+ $-\sqrt{5}/8\sqrt{3.17}$

$\frac{15}{2}$ $\frac{11}{2}$ 5
$\frac{3}{2}$ $\frac{1}{2}$ 1 0 000− $+\sqrt{7.11}/2\sqrt{3.13.17.19}$
$\frac{3}{2}$ $\frac{1}{2}$ 2 0 000+ $-11\sqrt{11}/2\sqrt{5.13.17.19}$
$\frac{3}{2}$ $\frac{3}{2}$ 1 0 000+ $+23\sqrt{2}/\sqrt{3.5.13.17.19}$
$\frac{3}{2}$ $\frac{3}{2}$ 2 0 000− $+9\sqrt{2.7}/\sqrt{5.13.17.19}$
$\frac{3}{2}$ $\frac{3}{2}$ $\tilde{1}$ 0 000+ $-4\sqrt{3.7}/\sqrt{5.13.17.19}$
$\frac{3}{2}$ $\frac{5}{2}$ 1 0 000− $-67/2\sqrt{5.13.17.19}$
$\frac{3}{2}$ $\frac{5}{2}$ 2 0 000+ $+\sqrt{7.19}/2\sqrt{5.13.17}$
$\frac{3}{2}$ $\frac{5}{2}$ 2 1 000− $+\sqrt{2.3.5.7}/\sqrt{13.17.19}$
$\frac{3}{2}$ $\frac{5}{2}$ $\tilde{1}$ 0 000− $-7\sqrt{3.7}/2\sqrt{5.13.17.19}$
$\frac{5}{2}$ $\frac{1}{2}$ 2 0 000− $+7.7\sqrt{7}/2\sqrt{3.5.13.17.19}$
$\frac{5}{2}$ $\frac{1}{2}$ 2 0 100− $-\sqrt{11}/2\sqrt{17.19}$
$\frac{5}{2}$ $\frac{1}{2}$ $\tilde{1}$ 0 000+ $-7\sqrt{7}/2\sqrt{2.13.17.19}$
$\frac{5}{2}$ $\frac{1}{2}$ $\tilde{1}$ 0 100+ $-\sqrt{3.5.11}/2\sqrt{2.17.19}$
$\frac{5}{2}$ $\frac{3}{2}$ 1 0 000− $+3.29\sqrt{3.7}/4\sqrt{5.11.13.17.19}$
$\frac{5}{2}$ $\frac{3}{2}$ 1 0 100− $-9/4\sqrt{17.19}$
$\frac{5}{2}$ $\frac{3}{2}$ 2 0 000+ $+7.37/2\sqrt{3.5.11.13.17.19}$
$\frac{5}{2}$ $\frac{3}{2}$ 2 0 100+ $-\sqrt{7}/2\sqrt{17.19}$
$\frac{5}{2}$ $\frac{3}{2}$ 2 1 000− $-23\sqrt{17}/4\sqrt{2.5.11.13.19}$
$\frac{5}{2}$ $\frac{3}{2}$ 2 1 100− $+5\sqrt{3.7}/4\sqrt{2.17.19}$
$\frac{5}{2}$ $\frac{3}{2}$ $\tilde{1}$ 0 000− $+739/8\sqrt{5.11.13.17.19}$
$\frac{5}{2}$ $\frac{3}{2}$ $\tilde{1}$ 0 100− $+7\sqrt{3.7}/8\sqrt{17.19}$
$\frac{5}{2}$ $\frac{5}{2}$ 1 0 000+ $+5.43\sqrt{7}/8\sqrt{2.11.13.17.19}$
$\frac{5}{2}$ $\frac{5}{2}$ 1 0 100+ $+13\sqrt{3.5}/8\sqrt{2.17.19}$
$\frac{5}{2}$ $\frac{5}{2}$ 1 1 000+ $-9\sqrt{7}/8\sqrt{2.5.11.13.17.19}$
$\frac{5}{2}$ $\frac{5}{2}$ 1 1 100+ $-\sqrt{3.19}/8\sqrt{2.17}$
$\frac{5}{2}$ $\frac{5}{2}$ 2 0 000− $-\sqrt{13.19}/2\sqrt{2.5.11.17}$
$\frac{5}{2}$ $\frac{5}{2}$ 2 0 100− $+\sqrt{3.7}/2\sqrt{2.17.19}$
$\frac{5}{2}$ $\frac{5}{2}$ 2 1 000+ $-3.29\sqrt{3}/8\sqrt{5.11.13.17.19}$
$\frac{5}{2}$ $\frac{5}{2}$ 2 1 100+ $+3.5\sqrt{7}/8\sqrt{17.19}$
$\frac{5}{2}$ $\frac{5}{2}$ 2 2 000− $+\sqrt{2.3.19}/\sqrt{5.11.13.17}$
$\frac{5}{2}$ $\frac{5}{2}$ 2 2 100− 0
$\frac{5}{2}$ $\frac{5}{2}$ $\tilde{1}$ 0 000+ $-3.83/8\sqrt{5.11.13.17.19}$

SO_3–K 3jm Factors (cont.)

$\frac{15}{2}$	**$\frac{11}{2}$**	**5**			
$\frac{5}{2}$	$\frac{5}{2}$	$\tilde{1}$	0	100+	$+3\sqrt{3.7}/8\sqrt{17.19}$
$\frac{5}{2}$	$\frac{5}{2}$	$\tilde{1}$	1	000+	$-3\sqrt{19}/\sqrt{11.13.17}$
$\frac{5}{2}$	$\frac{5}{2}$	$\tilde{1}$	1	100+	0
$\frac{15}{2}$	**6**	**$\frac{3}{2}$**			
$\frac{3}{2}$	0	$\frac{3}{2}$	0	000+	$+1/\sqrt{13}$
$\frac{3}{2}$	1	$\frac{3}{2}$	0	000−	$+3\sqrt{2.3}/\sqrt{5.7.13}$
$\frac{3}{2}$	2	$\frac{3}{2}$	0	000+	$-\sqrt{2.11}/\sqrt{5.7.13}$
$\frac{3}{2}$	3	$\frac{3}{2}$	0	000−	$-\sqrt{11}/2\sqrt{5.7.13}$
$\frac{5}{2}$	1	$\frac{3}{2}$	0	000+	$-\sqrt{3.11}/4\sqrt{5.13}$
$\frac{5}{2}$	1	$\frac{3}{2}$	0	100+	$-3/4\sqrt{7}$
$\frac{5}{2}$	2	$\frac{3}{2}$	0	000−	$-2\sqrt{3}/\sqrt{5.13}$
$\frac{5}{2}$	2	$\frac{3}{2}$	0	100−	0
$\frac{5}{2}$	2	$\frac{3}{2}$	1	000+	$-3/4\sqrt{2.5.13}$
$\frac{5}{2}$	2	$\frac{3}{2}$	1	100+	$-\sqrt{3.11}/4\sqrt{2.7}$
$\frac{5}{2}$	3	$\frac{3}{2}$	0	000+	$-\sqrt{3.13}/8.2\sqrt{2}$
$\frac{5}{2}$	3	$\frac{3}{2}$	0	100+	$+3\sqrt{5.11}/8.2\sqrt{2.7}$
$\frac{5}{2}$	3	$\frac{3}{2}$	1	000−	$+3.17/8.2\sqrt{2.5.13}$
$\frac{5}{2}$	3	$\frac{3}{2}$	1	100−	$+\sqrt{3.11}/8.2\sqrt{2.7}$
$\frac{15}{2}$	**6**	**$\frac{5}{2}$**			
$\frac{3}{2}$	1	$\frac{5}{2}$	0	000−	$-\sqrt{2.17}/\sqrt{5.7.13}$
$\frac{3}{2}$	2	$\frac{5}{2}$	0	000+	$-\sqrt{7.11}/\sqrt{2.5.13.17}$
$\frac{3}{2}$	2	$\frac{5}{2}$	1	000−	$+9\sqrt{3.11}/2\sqrt{5.7.13.17}$
$\frac{3}{2}$	3	$\frac{5}{2}$	0	000−	$+\sqrt{11}/2\sqrt{7.13.17}$
$\frac{3}{2}$	3	$\frac{5}{2}$	1	000+	$-\sqrt{3.7.11}/2\sqrt{5.13.17}$
$\frac{5}{2}$	0	$\frac{5}{2}$	0	000−	$-\sqrt{7.11}/4\sqrt{13.17}$
$\frac{5}{2}$	0	$\frac{5}{2}$	0	100−	$-\sqrt{3.5}/4\sqrt{17}$
$\frac{5}{2}$	1	$\frac{5}{2}$	0	000+	$-23\sqrt{11}/8.2\sqrt{13.17}$
$\frac{5}{2}$	1	$\frac{5}{2}$	0	100+	$+\sqrt{3.5}/8.2\sqrt{7.17}$
$\frac{5}{2}$	1	$\frac{5}{2}$	1	000+	$+3.7\sqrt{11}/8.2\sqrt{5.13.17}$
$\frac{5}{2}$	1	$\frac{5}{2}$	1	100+	$-19\sqrt{3}/8.2\sqrt{7.17}$
$\frac{5}{2}$	2	$\frac{5}{2}$	0	000−	$-7/8\sqrt{5.13.17}$
$\frac{5}{2}$	2	$\frac{5}{2}$	0	100−	$-5\sqrt{3.11}/8\sqrt{7.17}$
$\frac{5}{2}$	2	$\frac{5}{2}$	1	000+	$+11\sqrt{3.5}/8\sqrt{2.13.17}$
$\frac{5}{2}$	2	$\frac{5}{2}$	1	100+	$-9\sqrt{11}/8\sqrt{2.7.17}$
$\frac{5}{2}$	2	$\frac{5}{2}$	2	000−	$-3.7\sqrt{3}/8\sqrt{5.13.17}$
$\frac{5}{2}$	2	$\frac{5}{2}$	2	100−	$+3\sqrt{11}/8\sqrt{7.17}$
$\frac{5}{2}$	3	$\frac{5}{2}$	0	000+	$+11\sqrt{3}/2\sqrt{2.5.13.17}$
$\frac{5}{2}$	3	$\frac{5}{2}$	0	100+	$+3\sqrt{11}/2\sqrt{2.7.17}$
$\frac{5}{2}$	3	$\frac{5}{2}$	1	000−	$+2\sqrt{2.3}/\sqrt{13.17}$
$\frac{5}{2}$	3	$\frac{5}{2}$	1	100−	0
$\frac{15}{2}$	**6**	**$\frac{7}{2}$**			
$\frac{3}{2}$	1	$\frac{5}{2}$	0	000+	$-\sqrt{3.11}/2\sqrt{2.7.13.17}$
$\frac{3}{2}$	2	$\frac{5}{2}$	0	000−	$-3\sqrt{3.7}/2\sqrt{2.13.17}$
$\frac{3}{2}$	2	$\frac{5}{2}$	1	000+	$+1/\sqrt{7.13.17}$
$\frac{3}{2}$	2	$\frac{\tilde{1}}{2}$	0	000+	$-\sqrt{3}/2\sqrt{13.17}$
$\frac{3}{2}$	3	$\frac{5}{2}$	0	000+	$-\sqrt{3.5}/\sqrt{7.13.17}$
$\frac{3}{2}$	3	$\frac{5}{2}$	1	000−	$-2\sqrt{7}/\sqrt{13.17}$
$\frac{5}{2}$	0	$\frac{5}{2}$	0	000+	$-3\sqrt{3.5.7}/8\sqrt{13.17}$
$\frac{5}{2}$	0	$\frac{5}{2}$	0	100+	$+\sqrt{11}/8\sqrt{17}$
$\frac{15}{2}$	**6**	**$\frac{7}{2}$**			
$\frac{5}{2}$	1	$\frac{5}{2}$	0	000−	$+\sqrt{3.5}/8.4\sqrt{13.17}$
$\frac{5}{2}$	1	$\frac{5}{2}$	0	100−	$-43\sqrt{11}/8.4\sqrt{7.17}$
$\frac{5}{2}$	1	$\frac{5}{2}$	1	000−	$+7.7\sqrt{3}/8.4\sqrt{13.17}$
$\frac{5}{2}$	1	$\frac{5}{2}$	1	100−	$+\sqrt{5.11}/8.4\sqrt{7.17}$
$\frac{5}{2}$	1	$\frac{\tilde{1}}{2}$	0	000+	$+9\sqrt{5.7}/8.2\sqrt{2.13.17}$
$\frac{5}{2}$	1	$\frac{\tilde{1}}{2}$	0	100+	$-\sqrt{3.11}/8.2\sqrt{2.17}$
$\frac{5}{2}$	2	$\frac{5}{2}$	0	000+	$-3.7\sqrt{3}/8.2\sqrt{11.13.17}$
$\frac{5}{2}$	2	$\frac{5}{2}$	0	100+	$+11\sqrt{5}/8.2\sqrt{7.17}$
$\frac{5}{2}$	2	$\frac{5}{2}$	1	000−	$-19\sqrt{13}/8.2\sqrt{2.11.17}$
$\frac{5}{2}$	2	$\frac{5}{2}$	1	100−	$-13\sqrt{3.5}/8.2\sqrt{2.7.17}$
$\frac{5}{2}$	2	$\frac{5}{2}$	2	000+	$+7\sqrt{13}/8\sqrt{11.17}$
$\frac{5}{2}$	2	$\frac{5}{2}$	2	100+	$+\sqrt{3.5}/8\sqrt{7.17}$
$\frac{5}{2}$	2	$\frac{\tilde{1}}{2}$	0	000−	$+79\sqrt{3.7}/8.2\sqrt{2.11.13.17}$
$\frac{5}{2}$	2	$\frac{\tilde{1}}{2}$	0	100−	$-\sqrt{5}/8.2\sqrt{2.17}$
$\frac{5}{2}$	3	$\frac{5}{2}$	0	000−	$-89/8\sqrt{2.11.13.17}$
$\frac{5}{2}$	3	$\frac{5}{2}$	0	100−	$-3\sqrt{3.5}/8\sqrt{2.7.17}$
$\frac{5}{2}$	3	$\frac{5}{2}$	1	000+	$-31\sqrt{5}/8\sqrt{2.11.13.17}$
$\frac{5}{2}$	3	$\frac{5}{2}$	1	100+	$+\sqrt{3.7}/8\sqrt{2.17}$
$\frac{5}{2}$	3	$\frac{\tilde{1}}{2}$	0	000+	$+\sqrt{3.5.7}/8\sqrt{11.13.17}$
$\frac{5}{2}$	3	$\frac{\tilde{1}}{2}$	0	100+	$-11/8\sqrt{17}$
$\frac{15}{2}$	**6**	**$\frac{9}{2}$**			
$\frac{3}{2}$	0	$\frac{3}{2}$	0	000−	$-\sqrt{3.11}/\sqrt{5.13.17.19}$
$\frac{3}{2}$	1	$\frac{3}{2}$	0	000+	$-3\sqrt{11.13}/5\sqrt{2.7.17.19}$
$\frac{3}{2}$	1	$\frac{5}{2}$	0	000−	$-4\sqrt{2.3.11}/5\sqrt{13.17.19}$
$\frac{3}{2}$	2	$\frac{3}{2}$	0	000−	$+31\sqrt{3}/\sqrt{2.7.13.17.19}$
$\frac{3}{2}$	2	$\frac{5}{2}$	0	000+	$-3\sqrt{3}/\sqrt{2.13.17.19}$
$\frac{3}{2}$	2	$\frac{5}{2}$	1	000−	$+5.5/2\sqrt{13.17.19}$
$\frac{3}{2}$	3	$\frac{3}{2}$	0	000+	$+4.3\sqrt{3.13}/5\sqrt{7.17.19}$
$\frac{3}{2}$	3	$\frac{5}{2}$	0	000−	$+\sqrt{3.19}/2\sqrt{5.13.17}$
$\frac{3}{2}$	3	$\frac{5}{2}$	1	000+	$-103/2.5\sqrt{13.17.19}$
$\frac{5}{2}$	0	$\frac{5}{2}$	0	000−	$+3.7\sqrt{3}/2\sqrt{5.13.17.19}$
$\frac{5}{2}$	0	$\frac{5}{2}$	0	100−	$-\sqrt{7.11}/2\sqrt{17.19}$
$\frac{5}{2}$	1	$\frac{3}{2}$	0	000−	$+3.53/4.5\sqrt{13.17.19}$
$\frac{5}{2}$	1	$\frac{3}{2}$	0	100−	$-11\sqrt{3.11}/4\sqrt{5.7.17.19}$
$\frac{5}{2}$	1	$\frac{5}{2}$	0	000+	$-7\sqrt{3.7.13}/8.2\sqrt{5.17.19}$
$\frac{5}{2}$	1	$\frac{5}{2}$	0	100+	$+5\sqrt{11}/8.2\sqrt{17.19}$
$\frac{5}{2}$	1	$\frac{5}{2}$	1	000+	$-31\sqrt{3.7.13}/8.2.5\sqrt{17.19}$
$\frac{5}{2}$	1	$\frac{5}{2}$	1	100+	$-\sqrt{11.19}/8.2\sqrt{5.17}$
$\frac{5}{2}$	2	$\frac{3}{2}$	0	000+	$-3/2\sqrt{11.13.17.19}$
$\frac{5}{2}$	2	$\frac{3}{2}$	0	100+	$+\sqrt{3.5.7}/2\sqrt{17.19}$
$\frac{5}{2}$	2	$\frac{3}{2}$	1	000−	$-5.5.5\sqrt{3}/4\sqrt{2.11.13.17.19}$
$\frac{5}{2}$	2	$\frac{3}{2}$	1	100−	$-13\sqrt{5}/4\sqrt{2.7.17.19}$
$\frac{5}{2}$	2	$\frac{5}{2}$	0	000−	$-3.7\sqrt{3.7}/4\sqrt{11.13.17.19}$
$\frac{5}{2}$	2	$\frac{5}{2}$	0	100−	$+7\sqrt{5}/4\sqrt{17.19}$
$\frac{5}{2}$	2	$\frac{5}{2}$	1	000+	$+5\sqrt{7.13}/8\sqrt{2.11.17.19}$
$\frac{5}{2}$	2	$\frac{5}{2}$	1	100+	$+11\sqrt{3.5}/8\sqrt{2.17.19}$
$\frac{5}{2}$	2	$\frac{5}{2}$	2	000−	$+2.5\sqrt{7}/\sqrt{11.13.17.19}$
$\frac{5}{2}$	2	$\frac{5}{2}$	2	100−	$+\sqrt{3.5}/\sqrt{17.19}$
$\frac{5}{2}$	3	$\frac{3}{2}$	0	000−	$-9.3.5\sqrt{5}/4\sqrt{2.11.13.17.19}$

SO_3–K 3jm Factors (cont.)

$\frac{15}{2}\ 6\ \frac{9}{2}$

$\frac{5}{2}\ 3\ \frac{3}{2}\ 0\ 100-\quad -\sqrt{3.17}/4\sqrt{2.7.19}$

$\frac{5}{2}\ 3\ \frac{3}{2}\ 1\ 000+\quad +\sqrt{3.11.17}/4.5\sqrt{2.13.19}$

$\frac{5}{2}\ 3\ \frac{3}{2}\ 1\ 100+\quad +11/4\sqrt{2.5.7.17.19}$

$\frac{5}{2}\ 3\ \frac{5}{2}\ 0\ 000+\quad -\sqrt{7.13}/5\sqrt{2.11.17.19}$

$\frac{5}{2}\ 3\ \frac{5}{2}\ 0\ 100+\quad -3\sqrt{3}/\sqrt{2.5.17.19}$

$\frac{5}{2}\ 3\ \frac{5}{2}\ 1\ 000-\quad -11\sqrt{7.11}/2\sqrt{2.5.13.17.19}$

$\frac{5}{2}\ 3\ \frac{5}{2}\ 1\ 100-\quad +5\sqrt{3}/2\sqrt{2.17.19}$

$\frac{15}{2}\ 6\ \frac{11}{2}$

$\frac{3}{2}\ 0\ \frac{3}{2}\ 0\ 000+\quad -\sqrt{5.11}/\sqrt{13.17.19}$

$\frac{3}{2}\ 1\ \frac{1}{2}\ 0\ 000+\quad +11\sqrt{5}/2\sqrt{3.13.17.19}$

$\frac{3}{2}\ 1\ \frac{3}{2}\ 0\ 000-\quad -2\sqrt{2.7.11}/\sqrt{3.13.17.19}$

$\frac{3}{2}\ 1\ \frac{5}{2}\ 0\ 000+\quad -\sqrt{7.11}/2\sqrt{13.17.19}$

$\frac{3}{2}\ 2\ \frac{1}{2}\ 0\ 000-\quad +9\sqrt{11}/2\sqrt{13.17.19}$

$\frac{3}{2}\ 2\ \frac{3}{2}\ 0\ 000+\quad +4\sqrt{2.7}/\sqrt{13.17.19}$

$\frac{3}{2}\ 2\ \frac{5}{2}\ 0\ 000-\quad +5\sqrt{7}/2\sqrt{13.17.19}$

$\frac{3}{2}\ 2\ \frac{5}{2}\ 1\ 000+\quad +\sqrt{3.7}/\sqrt{2.13.17.19}$

$\frac{3}{2}\ 3\ \frac{3}{2}\ 0\ 000-\quad -\sqrt{7}/2\sqrt{13.17.19}$

$\frac{3}{2}\ 3\ \frac{5}{2}\ 0\ 000+\quad +\sqrt{5.7}/2\sqrt{2.13.17.19}$

$\frac{3}{2}\ 3\ \frac{5}{2}\ 1\ 000-\quad +9\sqrt{3.7}/2\sqrt{2.13.17.19}$

$\frac{5}{2}\ 0\ \frac{5}{2}\ 0\ 000+\quad -7\sqrt{5.7}/2\sqrt{2.13.17.19}$

$\frac{5}{2}\ 0\ \frac{5}{2}\ 0\ 100+\quad -\sqrt{3.11}/2\sqrt{2.17.19}$

$\frac{5}{2}\ 1\ \frac{3}{2}\ 0\ 000+\quad -2.3\sqrt{3}/\sqrt{13.17.19}$

$\frac{5}{2}\ 1\ \frac{3}{2}\ 0\ 100+\quad 0$

$\frac{5}{2}\ 1\ \frac{5}{2}\ 0\ 000-\quad -\sqrt{5}/4\sqrt{2.13.17.19}$

$\frac{5}{2}\ 1\ \frac{5}{2}\ 0\ 100-\quad -\sqrt{3.7.11}/4\sqrt{2.17.19}$

$\frac{5}{2}\ 1\ \frac{5}{2}\ 1\ 000-\quad +9.3/4\sqrt{2.13.17.19}$

$\frac{5}{2}\ 1\ \frac{5}{2}\ 1\ 100-\quad -\sqrt{3.5.7.11}/4\sqrt{2.17.19}$

$\frac{5}{2}\ 2\ \frac{1}{2}\ 0\ 000+\quad -\sqrt{7.13}/4\sqrt{3.17.19}$

$\frac{5}{2}\ 2\ \frac{1}{2}\ 0\ 100+\quad +\sqrt{5.11}/4\sqrt{17.19}$

$\frac{5}{2}\ 2\ \frac{3}{2}\ 0\ 000-\quad -173/4\sqrt{3.11.13.17.19}$

$\frac{5}{2}\ 2\ \frac{3}{2}\ 0\ 100-\quad -\sqrt{5.7}/4\sqrt{17.19}$

$\frac{5}{2}\ 2\ \frac{3}{2}\ 1\ 000+\quad +23\sqrt{13}/4\sqrt{2.11.17.19}$

$\frac{5}{2}\ 2\ \frac{3}{2}\ 1\ 100+\quad -\sqrt{3.5.7}/4\sqrt{2.17.19}$

$\frac{5}{2}\ 2\ \frac{5}{2}\ 0\ 000+\quad +5.7.7/4\sqrt{2.11.13.17.19}$

$\frac{5}{2}\ 2\ \frac{5}{2}\ 0\ 100+\quad +\sqrt{3.5.7}/4\sqrt{2.17.19}$

$\frac{5}{2}\ 2\ \frac{5}{2}\ 1\ 000-\quad +8.3\sqrt{3}/\sqrt{11.13.17.19}$

$\frac{5}{2}\ 2\ \frac{5}{2}\ 1\ 100-\quad 0$

$\frac{5}{2}\ 2\ \frac{5}{2}\ 2\ 000+\quad +139\sqrt{3}/4\sqrt{2.11.13.17.19}$

$\frac{5}{2}\ 2\ \frac{5}{2}\ 2\ 100+\quad -3\sqrt{5.7}/4\sqrt{2.17.19}$

$\frac{5}{2}\ 3\ \frac{1}{2}\ 0\ 000-\quad +43\sqrt{5.7}/8\sqrt{2.3.13.17.19}$

$\frac{5}{2}\ 3\ \frac{1}{2}\ 0\ 100-\quad +13\sqrt{11}/8\sqrt{2.17.19}$

$\frac{5}{2}\ 3\ \frac{3}{2}\ 0\ 000+\quad -269\sqrt{5}/8.2\sqrt{2.3.11.13.17.19}$

$\frac{5}{2}\ 3\ \frac{3}{2}\ 0\ 100+\quad +\sqrt{7.19}/8.2\sqrt{2.17}$

$\frac{5}{2}\ 3\ \frac{3}{2}\ 1\ 000-\quad +9.3\sqrt{11}/8.2\sqrt{2.13.17.19}$

$\frac{5}{2}\ 3\ \frac{3}{2}\ 1\ 100-\quad -11\sqrt{3.5.7}/8.2\sqrt{2.17.19}$

$\frac{5}{2}\ 3\ \frac{5}{2}\ 0\ 000-\quad +3.73\sqrt{3}/8.2\sqrt{11.13.17.19}$

$\frac{5}{2}\ 3\ \frac{5}{2}\ 0\ 100-\quad -3\sqrt{5.7}/8.2\sqrt{17.19}$

$\frac{5}{2}\ 3\ \frac{5}{2}\ 1\ 000+\quad -3.5\sqrt{3.5.11}/8.2\sqrt{13.17.19}$

$\frac{5}{2}\ 3\ \frac{5}{2}\ 1\ 100+\quad -3.5\sqrt{7}/8.2\sqrt{17.19}$

$\frac{15}{2}\ \frac{13}{2}\ 1$

$\frac{3}{2}\ \frac{1}{2}\ 1\ 0\ 000+\quad +1/\sqrt{7}$

$\frac{3}{2}\ \frac{3}{2}\ 1\ 0\ 000-\quad +4/5\sqrt{7}$

$\frac{3}{2}\ \frac{5}{2}\ 1\ 0\ 000+\quad -\sqrt{11}/2.5\sqrt{7}$

$\frac{5}{2}\ \frac{3}{2}\ 1\ 0\ 000+\quad -\sqrt{11}/2.5\sqrt{2}$

$\frac{5}{2}\ \frac{3}{2}\ 1\ 0\ 100+\quad -\sqrt{3.13}/2\sqrt{2.5.7}$

$\frac{5}{2}\ \frac{5}{2}\ 1\ 0\ 000-\quad -13/8.2\sqrt{2.5}$

$\frac{5}{2}\ \frac{5}{2}\ 1\ 0\ 100-\quad +\sqrt{3.11.13}/8.2\sqrt{2.7}$

$\frac{5}{2}\ \frac{5}{2}\ 1\ 1\ 000-\quad +3.17/8.2.5\sqrt{2}$

$\frac{5}{2}\ \frac{5}{2}\ 1\ 1\ 100-\quad +\sqrt{3.11.13}/8.2\sqrt{2.5.7}$

$\frac{5}{2}\ \frac{\tilde{1}}{2}\ 1\ 0\ 000+\quad +\sqrt{13}/8.2$

$\frac{5}{2}\ \frac{\tilde{1}}{2}\ 1\ 0\ 100+\quad -\sqrt{3.5.11}/8.2\sqrt{7}$

$\frac{15}{2}\ \frac{13}{2}\ 2$

$\frac{3}{2}\ \frac{1}{2}\ 2\ 0\ 000+\quad +\sqrt{17}/\sqrt{5.7.13}$

$\frac{3}{2}\ \frac{3}{2}\ 2\ 0\ 000-\quad -4\sqrt{17}/5\sqrt{7.13}$

$\frac{3}{2}\ \frac{5}{2}\ 2\ 0\ 000+\quad -3\sqrt{3.11}/2.5\sqrt{7.13.17}$

$\frac{3}{2}\ \frac{5}{2}\ 2\ 1\ 000-\quad -4.3\sqrt{2.11}/5\sqrt{7.13.17}$

$\frac{3}{2}\ \frac{\tilde{1}}{2}\ 2\ 0\ 000-\quad -\sqrt{11}/\sqrt{2.5.7.17}$

$\frac{5}{2}\ \frac{1}{2}\ 2\ 0\ 000-\quad -\sqrt{3.11}/\sqrt{5.13.17}$

$\frac{5}{2}\ \frac{1}{2}\ 2\ 0\ 100-\quad -3/\sqrt{7.17}$

$\frac{5}{2}\ \frac{3}{2}\ 2\ 0\ 000+\quad -\sqrt{3.11.13}/2.5\sqrt{2.17}$

$\frac{5}{2}\ \frac{3}{2}\ 2\ 0\ 100+\quad +9/2\sqrt{2.5.7.17}$

$\frac{5}{2}\ \frac{3}{2}\ 2\ 1\ 000-\quad -3\sqrt{11}/5\sqrt{13.17}$

$\frac{5}{2}\ \frac{3}{2}\ 2\ 1\ 100-\quad +\sqrt{3}/\sqrt{5.7.17}$

$\frac{5}{2}\ \frac{5}{2}\ 2\ 0\ 000-\quad -7\sqrt{3}/8.5\sqrt{2.13.17}$

$\frac{5}{2}\ \frac{5}{2}\ 2\ 0\ 100-\quad -3.13\sqrt{11}/8\sqrt{2.5.7.17}$

$\frac{5}{2}\ \frac{5}{2}\ 2\ 1\ 000+\quad -3.7/8.2\sqrt{13.17}$

$\frac{5}{2}\ \frac{5}{2}\ 2\ 1\ 100+\quad +13\sqrt{3.11}/8.2\sqrt{5.7.17}$

$\frac{5}{2}\ \frac{5}{2}\ 2\ 2\ 000-\quad -3.19/2.5\sqrt{2.13.17}$

$\frac{5}{2}\ \frac{5}{2}\ 2\ 2\ 100-\quad +\sqrt{3.11}/2\sqrt{2.5.7.17}$

$\frac{5}{2}\ \frac{\tilde{1}}{2}\ 2\ 0\ 000+\quad +3\sqrt{17}/8.2\sqrt{5}$

$\frac{5}{2}\ \frac{\tilde{1}}{2}\ 2\ 0\ 100+\quad +\sqrt{3.11.13}/8.2\sqrt{7.17}$

$\frac{15}{2}\ \frac{13}{2}\ 3$

$\frac{3}{2}\ \frac{3}{2}\ 3\ 0\ 000-\quad -31\sqrt{3}/2.5\sqrt{7.13.17}$

$\frac{3}{2}\ \frac{3}{2}\ \tilde{1}\ 0\ 000-\quad -37/5\sqrt{7.13.17}$

$\frac{3}{2}\ \frac{5}{2}\ 3\ 0\ 000+\quad +3\sqrt{11}/\sqrt{5.7.13.17}$

$\frac{3}{2}\ \frac{5}{2}\ 3\ 1\ 000-\quad +4\sqrt{3.11}/5\sqrt{7.13.17}$

$\frac{3}{2}\ \frac{5}{2}\ \tilde{1}\ 0\ 000+\quad -\sqrt{3.11.17}/5\sqrt{2.7.13}$

$\frac{3}{2}\ \frac{\tilde{1}}{2}\ \tilde{1}\ 0\ 000-\quad -\sqrt{11}/\sqrt{2.7.17}$

$\frac{5}{2}\ \frac{1}{2}\ 3\ 0\ 000-\quad +3\sqrt{11}/8\sqrt{13.17}$

$\frac{5}{2}\ \frac{1}{2}\ 3\ 0\ 100-\quad -5\sqrt{3.5}/8\sqrt{7.17}$

$\frac{5}{2}\ \frac{1}{2}\ \tilde{1}\ 0\ 000-\quad -3\sqrt{3.11}/4\sqrt{13.17}$

$\frac{5}{2}\ \frac{1}{2}\ \tilde{1}\ 0\ 100-\quad -\sqrt{5}/4\sqrt{7.17}$

$\frac{5}{2}\ \frac{3}{2}\ 3\ 0\ 000+\quad -3.19\sqrt{11}/8.2\sqrt{2.5.13.17}$

$\frac{5}{2}\ \frac{3}{2}\ 3\ 0\ 100+\quad +31\sqrt{3}/8.2\sqrt{2.7.17}$

$\frac{5}{2}\ \frac{3}{2}\ 3\ 1\ 000-\quad +11\sqrt{3.11}/8.2.5\sqrt{2.13.17}$

$\frac{5}{2}\ \frac{3}{2}\ 3\ 1\ 100-\quad +73/8.2\sqrt{2.5.7.17}$

$\frac{5}{2}\ \frac{3}{2}\ \tilde{1}\ 0\ 000+\quad +7\sqrt{3.11}/8.5\sqrt{13.17}$

$\frac{5}{2}\ \frac{3}{2}\ \tilde{1}\ 0\ 100+\quad +61/8\sqrt{5.7.17}$

$\frac{5}{2}\ \frac{5}{2}\ 3\ 0\ 000-\quad +7.7\sqrt{3}/8.5\sqrt{2.13.17}$

SO_3–K 3jm Factors (cont.)

$\frac{15}{2}\ \frac{13}{2}\ 3$

$\frac{5}{2}\ \frac{5}{2}\ 3\ 0\ 100-\quad +\sqrt{11.17}/8\sqrt{2.5.7}$

$\frac{5}{2}\ \frac{5}{2}\ 3\ 1\ 000+\quad -19\sqrt{3.5}/8\sqrt{2.13.17}$

$\frac{5}{2}\ \frac{5}{2}\ 3\ 1\ 100+\quad +\sqrt{11}/8\sqrt{2.7.17}$

$\frac{5}{2}\ \frac{5}{2}\ \tilde{1}\ 0\ 000-\quad -7/4.5\sqrt{2.13.17}$

$\frac{5}{2}\ \frac{5}{2}\ \tilde{1}\ 0\ 100-\quad +3\sqrt{3.11}/4\sqrt{2.5.7.17}$

$\frac{5}{2}\ \frac{5}{2}\ \tilde{1}\ 1\ 000-\quad -19/4\sqrt{2.5.13.17}$

$\frac{5}{2}\ \frac{5}{2}\ \tilde{1}\ 1\ 100-\quad -\sqrt{3.11}/4\sqrt{2.7.17}$

$\frac{5}{2}\ \tilde{\frac{1}{2}}\ 3\ 0\ 000+\quad -\sqrt{3}/8\sqrt{17}$

$\frac{5}{2}\ \tilde{\frac{1}{2}}\ 3\ 0\ 100+\quad -\sqrt{5.11.13}/8\sqrt{7.17}$

$\frac{15}{2}\ \frac{13}{2}\ 4$

$\frac{3}{2}\ \frac{1}{2}\ 2\ 0\ 000+\quad +2\sqrt{11}/\sqrt{7.13.17.19}$

$\frac{3}{2}\ \frac{3}{2}\ 2\ 0\ 000-\quad -3\sqrt{5.11}/\sqrt{7.13.17.19}$

$\frac{3}{2}\ \frac{3}{2}\ 3\ 0\ 000+\quad +\sqrt{5.11}/2\sqrt{13.17.19}$

$\frac{3}{2}\ \frac{5}{2}\ 2\ 0\ 000+\quad +\sqrt{5.13}/\sqrt{3.7.17.19}$

$\frac{3}{2}\ \frac{5}{2}\ 2\ 1\ 000-\quad -107\sqrt{5}/3\sqrt{2.7.13.17.19}$

$\frac{3}{2}\ \frac{5}{2}\ 3\ 0\ 000-\quad -\sqrt{19}/\sqrt{3.13.17}$

$\frac{3}{2}\ \frac{5}{2}\ 3\ 1\ 000+\quad -4\sqrt{5}/3\sqrt{13.17.19}$

$\frac{3}{2}\ \tilde{\frac{1}{2}}\ 2\ 0\ 000-\quad -\sqrt{19}/\sqrt{2.7.17}$

$\frac{5}{2}\ \frac{1}{2}\ 2\ 0\ 000-\quad -67/4\sqrt{3.13.17.19}$

$\frac{5}{2}\ \frac{1}{2}\ 2\ 0\ 100-\quad -\sqrt{5.11}/4\sqrt{7.17.19}$

$\frac{5}{2}\ \frac{1}{2}\ 3\ 0\ 000+\quad -5.7\sqrt{5.7}/8\sqrt{3.13.17.19}$

$\frac{5}{2}\ \frac{1}{2}\ 3\ 0\ 100+\quad +11\sqrt{11}/8\sqrt{17.19}$

$\frac{5}{2}\ \frac{3}{2}\ 2\ 0\ 000+\quad -5\sqrt{3.5}/4\sqrt{2.13.17.19}$

$\frac{5}{2}\ \frac{3}{2}\ 2\ 0\ 100+\quad -9\sqrt{11}/4\sqrt{2.7.17.19}$

$\frac{5}{2}\ \frac{3}{2}\ 2\ 1\ 000-\quad +5\sqrt{5}/8.3\sqrt{13.17.19}$

$\frac{5}{2}\ \frac{3}{2}\ 2\ 1\ 100-\quad -61\sqrt{11}/8\sqrt{3.7.17.19}$

$\frac{5}{2}\ \frac{3}{2}\ 3\ 0\ 000-\quad -31\sqrt{7}/8.2\sqrt{2.3.13.17.19}$

$\frac{5}{2}\ \frac{3}{2}\ 3\ 0\ 100-\quad -5\sqrt{5.11}/8.2\sqrt{2.17.19}$

$\frac{5}{2}\ \frac{3}{2}\ 3\ 1\ 000+\quad -277\sqrt{5.7}/8.2.3\sqrt{2.13.17.19}$

$\frac{5}{2}\ \frac{3}{2}\ 3\ 1\ 100+\quad -\sqrt{11.19}/8.2\sqrt{2.3.17}$

$\frac{5}{2}\ \frac{5}{2}\ 2\ 0\ 000-\quad -5.7\sqrt{5}/4\sqrt{2.3.11.13.17.19}$

$\frac{5}{2}\ \frac{5}{2}\ 2\ 0\ 100-\quad +11/4\sqrt{2.7.17.19}$

$\frac{5}{2}\ \frac{5}{2}\ 2\ 1\ 000+\quad +5.7\sqrt{5}/4\sqrt{11.13.17.19}$

$\frac{5}{2}\ \frac{5}{2}\ 2\ 1\ 100+\quad -5.13/4\sqrt{3.7.17.19}$

$\frac{5}{2}\ \frac{5}{2}\ 2\ 2\ 000-\quad +5.101\sqrt{5}/4.3\sqrt{2.11.13.17.19}$

$\frac{5}{2}\ \frac{5}{2}\ 2\ 2\ 100-\quad +79/4\sqrt{2.3.7.17.19}$

$\frac{5}{2}\ \frac{5}{2}\ 3\ 0\ 000+\quad +5.7\sqrt{5.7}/8\sqrt{2.11.13.17.19}$

$\frac{5}{2}\ \frac{5}{2}\ 3\ 0\ 100+\quad +79/8\sqrt{2.3.17.19}$

$\frac{5}{2}\ \frac{5}{2}\ 3\ 1\ 000-\quad -7.37\sqrt{7}/8.3\sqrt{2.11.13.17.19}$

$\frac{5}{2}\ \frac{5}{2}\ 3\ 1\ 100-\quad +31\sqrt{5}/8\sqrt{2.3.17.19}$

$\frac{5}{2}\ \tilde{\frac{1}{2}}\ 2\ 0\ 000+\quad -61/2.3\sqrt{11.17.19}$

$\frac{5}{2}\ \tilde{\frac{1}{2}}\ 2\ 0\ 100+\quad +\sqrt{5.13}/2\sqrt{3.7.17.19}$

$\frac{5}{2}\ \tilde{\frac{1}{2}}\ 3\ 0\ 000-\quad +\sqrt{5.7.19}/8.3\sqrt{11.17}$

$\frac{5}{2}\ \tilde{\frac{1}{2}}\ 3\ 0\ 100-\quad -11\sqrt{13}/8\sqrt{3.17.19}$

$\frac{15}{2}\ \frac{13}{2}\ 5$

$\frac{3}{2}\ \frac{1}{2}\ 1\ 0\ 000+\quad -3\sqrt{5.11}/\sqrt{2.7.13.17.19}$

$\frac{3}{2}\ \frac{1}{2}\ 2\ 0\ 000-\quad +\sqrt{3.11}/\sqrt{2.13.17.19}$

$\frac{3}{2}\ \frac{3}{2}\ 1\ 0\ 000-\quad -3.11\sqrt{11}/\sqrt{2.5.7.13.17.19}$

$\frac{3}{2}\ \frac{3}{2}\ 2\ 0\ 000+\quad -\sqrt{3.5.11}/\sqrt{2.13.17.19}$

$\frac{15}{2}\ \frac{13}{2}\ 5$

$\frac{3}{2}\ \frac{3}{2}\ \tilde{1}\ 0\ 000-\quad -\sqrt{11}/\sqrt{5.13.17.19}$

$\frac{3}{2}\ \frac{5}{2}\ 1\ 0\ 000+\quad +3.109/2\sqrt{2.5.7.13.17.19}$

$\frac{3}{2}\ \frac{5}{2}\ 2\ 0\ 000-\quad -9\sqrt{5}/2\sqrt{2.13.17.19}$

$\frac{3}{2}\ \frac{5}{2}\ 2\ 1\ 000+\quad -\sqrt{3.5}/\sqrt{13.17.19}$

$\frac{3}{2}\ \frac{5}{2}\ \tilde{1}\ 0\ 000+\quad -8\sqrt{2.3}/\sqrt{5.13.17.19}$

$\frac{3}{2}\ \tilde{\frac{1}{2}}\ 2\ 0\ 000+\quad +3\sqrt{3}/2\sqrt{17.19}$

$\frac{3}{2}\ \tilde{\frac{1}{2}}\ \tilde{1}\ 0\ 000-\quad -\sqrt{2.5}/\sqrt{17.19}$

$\frac{5}{2}\ \frac{1}{2}\ 2\ 0\ 000+\quad -3\sqrt{7}/2\sqrt{2.13.17.19}$

$\frac{5}{2}\ \frac{1}{2}\ 2\ 0\ 100+\quad -\sqrt{3.5.11}/2\sqrt{2.17.19}$

$\frac{5}{2}\ \frac{1}{2}\ \tilde{1}\ 0\ 000-\quad +3\sqrt{3.5.7}/2\sqrt{13.17.19}$

$\frac{5}{2}\ \frac{1}{2}\ \tilde{1}\ 0\ 100-\quad -\sqrt{11}/2\sqrt{17.19}$

$\frac{5}{2}\ \frac{3}{2}\ 1\ 0\ 000+\quad +3/\sqrt{5.13.17.19}$

$\frac{5}{2}\ \frac{3}{2}\ 1\ 0\ 100+\quad -\sqrt{3.11}/\sqrt{7.17.19}$

$\frac{5}{2}\ \frac{3}{2}\ 2\ 0\ 000-\quad +3\sqrt{5.7}/2\sqrt{13.17.19}$

$\frac{5}{2}\ \frac{3}{2}\ 2\ 0\ 100-\quad +\sqrt{3.11}/2\sqrt{17.19}$

$\frac{5}{2}\ \frac{3}{2}\ 2\ 1\ 000+\quad -5\sqrt{3.5.7}/2\sqrt{2.13.17.19}$

$\frac{5}{2}\ \frac{3}{2}\ 2\ 1\ 100+\quad -\sqrt{11}/2\sqrt{2.17.19}$

$\frac{5}{2}\ \frac{3}{2}\ \tilde{1}\ 0\ 000+\quad +7\sqrt{3.7}/\sqrt{5.13.17.19}$

$\frac{5}{2}\ \frac{3}{2}\ \tilde{1}\ 0\ 100+\quad -\sqrt{11}/\sqrt{17.19}$

$\frac{5}{2}\ \frac{5}{2}\ 1\ 0\ 000-\quad +3\sqrt{17}/8.4\sqrt{11.13.19}$

$\frac{5}{2}\ \frac{5}{2}\ 1\ 0\ 100-\quad +5.13\sqrt{3.5}/8.4\sqrt{7.17.19}$

$\frac{5}{2}\ \frac{5}{2}\ 1\ 1\ 000-\quad +3.461/8.4\sqrt{5.11.13.17.19}$

$\frac{5}{2}\ \frac{5}{2}\ 1\ 1\ 100-\quad -131\sqrt{3}/8.4\sqrt{7.17.19}$

$\frac{5}{2}\ \frac{5}{2}\ 2\ 0\ 000+\quad 0$

$\frac{5}{2}\ \frac{5}{2}\ 2\ 0\ 100+\quad +3\sqrt{3}/\sqrt{17.19}$

$\frac{5}{2}\ \frac{5}{2}\ 2\ 1\ 000-\quad +7\sqrt{3.5.7.17}/8.2\sqrt{2.11.13.19}$

$\frac{5}{2}\ \frac{5}{2}\ 2\ 1\ 100-\quad +5.7/8.2\sqrt{2.17.19}$

$\frac{5}{2}\ \frac{5}{2}\ 2\ 2\ 000+\quad -41\sqrt{3.5.7}/8.2\sqrt{11.13.17.19}$

$\frac{5}{2}\ \frac{5}{2}\ 2\ 2\ 100+\quad -13/8.2\sqrt{17.19}$

$\frac{5}{2}\ \frac{5}{2}\ \tilde{1}\ 0\ 000-\quad -7.31\sqrt{7}/8\sqrt{2.5.11.13.17.19}$

$\frac{5}{2}\ \frac{5}{2}\ \tilde{1}\ 0\ 100-\quad -3\sqrt{3}/8\sqrt{2.17.19}$

$\frac{5}{2}\ \frac{5}{2}\ \tilde{1}\ 1\ 000-\quad +41\sqrt{7}/8\sqrt{2.11.13.17.19}$

$\frac{5}{2}\ \frac{5}{2}\ \tilde{1}\ 1\ 100-\quad -13\sqrt{3.5}/8\sqrt{2.17.19}$

$\frac{5}{2}\ \tilde{\frac{1}{2}}\ 1\ 0\ 000+\quad +9.3.5\sqrt{5}/8.2\sqrt{2.11.17.19}$

$\frac{5}{2}\ \tilde{\frac{1}{2}}\ 1\ 0\ 100+\quad +\sqrt{3.13.17}/8.2\sqrt{2.7.19}$

$\frac{5}{2}\ \tilde{\frac{1}{2}}\ 2\ 0\ 000-\quad -43\sqrt{3.7}/8.2\sqrt{2.11.17.19}$

$\frac{5}{2}\ \tilde{\frac{1}{2}}\ 2\ 0\ 100-\quad +5\sqrt{5.13}/8.2\sqrt{2.17.19}$

$\frac{15}{2}\ \frac{13}{2}\ 6$

$\frac{3}{2}\ \frac{1}{2}\ 1\ 0\ 000-\quad -11\sqrt{3}/\sqrt{2.13.17.19}$

$\frac{3}{2}\ \frac{1}{2}\ 2\ 0\ 000+\quad +\sqrt{11}/\sqrt{2.5.13.17.19}$

$\frac{3}{2}\ \frac{3}{2}\ 0\ 0\ 000-\quad +11\sqrt{7}/\sqrt{5.13.17.19}$

$\frac{3}{2}\ \frac{3}{2}\ 1\ 0\ 000+\quad -11\sqrt{3}/\sqrt{2.13.17.19}$

$\frac{3}{2}\ \frac{3}{2}\ 2\ 0\ 000-\quad +9\sqrt{11}/5\sqrt{2.13.17.19}$

$\frac{3}{2}\ \frac{3}{2}\ 3\ 0\ 000+\quad -2\sqrt{11}/\sqrt{13.17.19}$

$\frac{3}{2}\ \frac{5}{2}\ 1\ 0\ 000-\quad -3\sqrt{3.11}/2\sqrt{2.13.17.19}$

$\frac{3}{2}\ \frac{5}{2}\ 2\ 0\ 000+\quad +29\sqrt{3}/2.5\sqrt{2.13.17.19}$

$\frac{3}{2}\ \frac{5}{2}\ 2\ 1\ 000-\quad +9.9/5\sqrt{13.17.19}$

$\frac{3}{2}\ \frac{5}{2}\ 3\ 0\ 000-\quad +7\sqrt{3}/\sqrt{5.13.17.19}$

$\frac{3}{2}\ \frac{5}{2}\ 3\ 1\ 000+\quad +9/\sqrt{13.17.19}$

SO_3–K 3jm Factors (cont.)

$\frac{15}{2}$ $\frac{13}{2}$ 6	
$\frac{3}{2}$ $\widetilde{\frac{1}{2}}$ 2 0 000−	$+7/2\sqrt{5.17.19}$
$\frac{5}{2}$ $\frac{1}{2}$ 2 0 000−	$-9\sqrt{3.13}/2\sqrt{2.5.7.17.19}$
$\frac{5}{2}$ $\frac{1}{2}$ 2 0 100−	$-3\sqrt{11}/2\sqrt{2.17.19}$
$\frac{5}{2}$ $\frac{1}{2}$ 3 0 000+	$-8.2\sqrt{3}/\sqrt{7.13.17.19}$
$\frac{5}{2}$ $\frac{1}{2}$ 3 0 100+	0
$\frac{5}{2}$ $\frac{3}{2}$ 1 0 000−	$-2\sqrt{3.11}/\sqrt{7.13.17.19}$
$\frac{5}{2}$ $\frac{3}{2}$ 1 0 100−	0
$\frac{5}{2}$ $\frac{3}{2}$ 2 0 000+	$-191\sqrt{3}/2.5\sqrt{7.13.17.19}$
$\frac{5}{2}$ $\frac{3}{2}$ 2 0 100+	$+3\sqrt{11}/2\sqrt{5.17.19}$
$\frac{5}{2}$ $\frac{3}{2}$ 2 1 000−	$-3\sqrt{13}/2.5\sqrt{2.7.17.19}$
$\frac{5}{2}$ $\frac{3}{2}$ 2 1 100−	$-3\sqrt{3.11}/2\sqrt{2.5.17.19}$
$\frac{5}{2}$ $\frac{3}{2}$ 3 0 000−	$+9.3\sqrt{3}/2\sqrt{2.5.7.13.17.19}$
$\frac{5}{2}$ $\frac{3}{2}$ 3 0 100−	$-3\sqrt{11}/2\sqrt{2.17.19}$
$\frac{5}{2}$ $\frac{3}{2}$ 3 1 000+	$-9.3/2\sqrt{2.7.13.17.19}$
$\frac{5}{2}$ $\frac{3}{2}$ 3 1 100+	$+\sqrt{3.5.11}/2\sqrt{2.17.19}$
$\frac{5}{2}$ $\frac{5}{2}$ 0 0 000−	$+37\sqrt{3}/8\sqrt{5.13.17.19}$
$\frac{5}{2}$ $\frac{5}{2}$ 0 0 100−	$+3\sqrt{7.11}/8\sqrt{17.19}$
$\frac{5}{2}$ $\frac{5}{2}$ 1 0 000+	$-3.139\sqrt{3.5}/8.4\sqrt{7.13.17.19}$
$\frac{5}{2}$ $\frac{5}{2}$ 1 0 100+	$-3\sqrt{11}/8.4\sqrt{17.19}$
$\frac{5}{2}$ $\frac{5}{2}$ 1 1 000+	$+9.3\sqrt{3}/8.4\sqrt{7.13.17.19}$
$\frac{5}{2}$ $\frac{5}{2}$ 1 1 100+	$-3\sqrt{5.11}/8.4\sqrt{17.19}$
$\frac{5}{2}$ $\frac{5}{2}$ 2 0 000−	$+7.7.7\sqrt{3.7}/8.5\sqrt{11.13.17.19}$
$\frac{5}{2}$ $\frac{5}{2}$ 2 0 100−	$-3.11/8\sqrt{5.17.19}$
$\frac{5}{2}$ $\frac{5}{2}$ 2 1 000+	$+3.7.7\sqrt{7}/8.2\sqrt{2.11.13.17.19}$
$\frac{5}{2}$ $\frac{5}{2}$ 2 1 100+	$-7.13\sqrt{3}/8.2\sqrt{2.5.17.19}$
$\frac{5}{2}$ $\frac{5}{2}$ 2 2 000−	$+3.881/8.2.5\sqrt{7.11.13.17.19}$
$\frac{5}{2}$ $\frac{5}{2}$ 2 2 100−	$-5\sqrt{3.5}/8.2\sqrt{17.19}$
$\frac{5}{2}$ $\frac{5}{2}$ 3 0 000+	$-3.7.7\sqrt{7}/4\sqrt{2.11.13.17.19}$
$\frac{5}{2}$ $\frac{5}{2}$ 3 0 100+	$-\sqrt{3.5}/4\sqrt{2.17.19}$
$\frac{5}{2}$ $\frac{5}{2}$ 3 1 000−	$-9.47/4\sqrt{2.5.7.11.13.17.19}$
$\frac{5}{2}$ $\frac{5}{2}$ 3 1 100−	$+5\sqrt{3}/4\sqrt{2.17.19}$
$\frac{5}{2}$ $\widetilde{\frac{1}{2}}$ 1 0 000−	$-37\sqrt{3}/8.2\sqrt{2.7.17.19}$
$\frac{5}{2}$ $\widetilde{\frac{1}{2}}$ 1 0 100−	$-3\sqrt{5.11.13}/8.2\sqrt{2.17.19}$
$\frac{5}{2}$ $\widetilde{\frac{1}{2}}$ 2 0 000+	$-3.103/8.2\sqrt{2.5.7.11.17.19}$
$\frac{5}{2}$ $\widetilde{\frac{1}{2}}$ 2 0 100+	$-\sqrt{3.13.17}/8.2\sqrt{2.19}$
$\frac{5}{2}$ $\widetilde{\frac{1}{2}}$ 3 0 000−	$-8.3/\sqrt{7.11.17.19}$
$\frac{5}{2}$ $\widetilde{\frac{1}{2}}$ 3 0 100−	0

$\frac{15}{2}$ 7 $\frac{1}{2}$	
$\frac{3}{2}$ 1 $\frac{1}{2}$ 0 000+	$+1/\sqrt{5}$
$\frac{3}{2}$ 2 $\frac{1}{2}$ 0 000−	$+1/2\sqrt{5}$
$\frac{5}{2}$ 2 $\frac{1}{2}$ 0 000+	$+\sqrt{7.11}/8\sqrt{3.5}$
$\frac{5}{2}$ 2 $\frac{1}{2}$ 0 100+	$+\sqrt{13}/8$
$\frac{5}{2}$ 3 $\frac{1}{2}$ 0 000−	$-\sqrt{7.13}/8\sqrt{3.5}$
$\frac{5}{2}$ 3 $\frac{1}{2}$ 0 100−	$+\sqrt{11}/8$
$\frac{5}{2}$ $\tilde{1}$ $\frac{1}{2}$ 0 000−	$+1/\sqrt{5}$
$\frac{5}{2}$ $\tilde{1}$ $\frac{1}{2}$ 0 100−	0

$\frac{15}{2}$ 7 $\frac{3}{2}$	
$\frac{3}{2}$ 1 $\frac{3}{2}$ 0 000+	$+\sqrt{17}/5\sqrt{7}$
$\frac{3}{2}$ 2 $\frac{3}{2}$ 0 000−	$-\sqrt{17}/2\sqrt{5.7}$
$\frac{3}{2}$ 3 $\frac{3}{2}$ 0 000+	$-\sqrt{11.13}/2.5\sqrt{7.17}$
$\frac{3}{2}$ $\tilde{1}$ $\frac{3}{2}$ 0 000+	$-\sqrt{3.11}/2.5\sqrt{17}$
$\frac{5}{2}$ 1 $\frac{3}{2}$ 0 000−	$+3\sqrt{11}/2.5\sqrt{2.17}$
$\frac{5}{2}$ 1 $\frac{3}{2}$ 0 100−	$+3\sqrt{3.13}/2\sqrt{2.5.7.17}$
$\frac{5}{2}$ 2 $\frac{3}{2}$ 0 000+	$+\sqrt{11.17}/8\sqrt{2.3.5}$
$\frac{5}{2}$ 2 $\frac{3}{2}$ 0 100+	$-\sqrt{7.13}/8\sqrt{2.17}$
$\frac{5}{2}$ 2 $\frac{3}{2}$ 1 000−	$-11\sqrt{11}/8.2\sqrt{5.17}$
$\frac{5}{2}$ 2 $\frac{3}{2}$ 1 100−	$-3\sqrt{3.13}/8.2\sqrt{7.17}$
$\frac{5}{2}$ 3 $\frac{3}{2}$ 0 000−	$+\sqrt{13.17}/8.2\sqrt{2.3.5}$
$\frac{5}{2}$ 3 $\frac{3}{2}$ 0 100−	$-\sqrt{11.17}/8.2\sqrt{2.7}$
$\frac{5}{2}$ 3 $\frac{3}{2}$ 1 000+	$+3\sqrt{13.17}/8.2.5\sqrt{2}$
$\frac{5}{2}$ 3 $\frac{3}{2}$ 1 100+	$+13\sqrt{3.11}/8.2\sqrt{2.5.7.17}$
$\frac{5}{2}$ $\tilde{1}$ $\frac{3}{2}$ 0 000−	$+7\sqrt{7}/8.2.5\sqrt{17}$
$\frac{5}{2}$ $\tilde{1}$ $\frac{3}{2}$ 0 100−	$-3\sqrt{3.11.13}/8.2\sqrt{5.17}$

$\frac{15}{2}$ 7 $\frac{5}{2}$	
$\frac{3}{2}$ 1 $\frac{5}{2}$ 0 000+	$-\sqrt{3.17}/5\sqrt{7.13}$
$\frac{3}{2}$ 2 $\frac{5}{2}$ 0 000−	$-\sqrt{7}/\sqrt{5.13.17}$
$\frac{3}{2}$ 2 $\frac{5}{2}$ 1 000+	$+139/2\sqrt{2.3.5.7.13.17}$
$\frac{3}{2}$ 3 $\frac{5}{2}$ 0 000+	$+3\sqrt{11}/2\sqrt{5.7.17}$
$\frac{3}{2}$ 3 $\frac{5}{2}$ 1 000−	$-\sqrt{7.11}/2.5\sqrt{3.17}$
$\frac{3}{2}$ $\tilde{1}$ $\frac{5}{2}$ 0 000+	$-9\sqrt{3.11}/2.5\sqrt{2.13.17}$
$\frac{5}{2}$ 1 $\frac{5}{2}$ 0 000−	$+3\sqrt{3.11}/4\sqrt{2.5.13.17}$
$\frac{5}{2}$ 1 $\frac{5}{2}$ 0 100−	$-3.5/4\sqrt{2.7.17}$
$\frac{5}{2}$ 1 $\frac{5}{2}$ 1 000−	$-3.7\sqrt{3.11}/4.5\sqrt{2.13.17}$
$\frac{5}{2}$ 1 $\frac{5}{2}$ 1 100−	$-23/4\sqrt{2.5.7.17}$
$\frac{5}{2}$ 2 $\frac{5}{2}$ 0 000+	$-7\sqrt{11}/4\sqrt{2.5.13.17}$
$\frac{5}{2}$ 2 $\frac{5}{2}$ 0 100+	$+31/4\sqrt{2.3.7.17}$
$\frac{5}{2}$ 2 $\frac{5}{2}$ 1 000−	$+5\sqrt{5.11}/8\sqrt{3.13.17}$
$\frac{5}{2}$ 2 $\frac{5}{2}$ 1 100−	$+11/8.3\sqrt{7.17}$
$\frac{5}{2}$ 2 $\frac{5}{2}$ 2 000+	$+7.7\sqrt{11}/8\sqrt{2.3.5.13.17}$
$\frac{5}{2}$ 2 $\frac{5}{2}$ 2 100+	$-61/8.3\sqrt{2.7.17}$
$\frac{5}{2}$ 3 $\frac{5}{2}$ 0 000−	$-31/8.5\sqrt{2.3.17}$
$\frac{5}{2}$ 3 $\frac{5}{2}$ 0 100−	$+\sqrt{11.13.17}/8.3\sqrt{2.5.7}$
$\frac{5}{2}$ 3 $\frac{5}{2}$ 1 000+	$-11\sqrt{5}/8\sqrt{2.3.17}$
$\frac{5}{2}$ 3 $\frac{5}{2}$ 1 100+	$-\sqrt{7.11.13}/8.3\sqrt{2.17}$
$\frac{5}{2}$ $\tilde{1}$ $\frac{5}{2}$ 0 000−	$-3.7\sqrt{7}/8.5\sqrt{2.13.17}$
$\frac{5}{2}$ $\tilde{1}$ $\frac{5}{2}$ 0 100−	$-13\sqrt{11}/8\sqrt{2.3.5.17}$
$\frac{5}{2}$ $\tilde{1}$ $\frac{5}{2}$ 1 000−	$+19\sqrt{7}/4\sqrt{2.5.13.17}$
$\frac{5}{2}$ $\tilde{1}$ $\frac{5}{2}$ 1 100−	$-\sqrt{11}/4\sqrt{2.3.17}$

$\frac{15}{2}$ 7 $\frac{7}{2}$	
$\frac{3}{2}$ 1 $\frac{5}{2}$ 0 000−	$+11/2\sqrt{7.13.17.19}$
$\frac{3}{2}$ 2 $\frac{5}{2}$ 0 000+	$-\sqrt{7.19}/2\sqrt{3.5.13.17}$
$\frac{3}{2}$ 2 $\frac{5}{2}$ 1 000−	$-191/2\sqrt{2.5.7.13.17.19}$
$\frac{3}{2}$ 2 $\widetilde{\frac{1}{2}}$ 0 000−	$-9.3\sqrt{3}/2\sqrt{2.5.13.17.19}$
$\frac{3}{2}$ 3 $\frac{5}{2}$ 0 000−	$-\sqrt{11.19}/\sqrt{3.5.7.17}$
$\frac{3}{2}$ 3 $\frac{5}{2}$ 1 000+	0
$\frac{3}{2}$ $\tilde{1}$ $\frac{5}{2}$ 0 000−	$+\sqrt{11.13}/2\sqrt{2.17.19}$
$\frac{3}{2}$ $\tilde{1}$ $\widetilde{\frac{1}{2}}$ 0 000+	$-7\sqrt{7.11}/2\sqrt{2.5.13.17.19}$
$\frac{5}{2}$ 1 $\frac{5}{2}$ 0 000+	$-89\sqrt{11}/8.2\sqrt{2.5.13.17.19}$

SO_3–K 3jm Factors (cont.)

$\frac{15}{2}$ 7 $\frac{7}{2}$

$\frac{5}{2}$	1	$\frac{5}{2}$	0 100+	$-\sqrt{3}/8.2\sqrt{2.7.17.19}$
$\frac{5}{2}$	1	$\frac{5}{2}$	1 000+	$+7\sqrt{11.17}/8.2\sqrt{2.13.19}$
$\frac{5}{2}$	1	$\frac{5}{2}$	1 100+	$+3.5\sqrt{3.5}/8.2\sqrt{2.7.17.19}$
$\frac{5}{2}$	1	$\tilde{\frac{1}{2}}$	0 000−	$-29\sqrt{3.7.11}/8.2\sqrt{5.13.17.19}$
$\frac{5}{2}$	1	$\tilde{\frac{1}{2}}$	0 100−	$+9.9/8.2\sqrt{17.19}$
$\frac{5}{2}$	2	$\frac{5}{2}$	0 000−	$-7\sqrt{11.19}/8\sqrt{2.3.5.13.17}$
$\frac{5}{2}$	2	$\frac{5}{2}$	0 100−	$-3.31/8\sqrt{2.7.17.19}$
$\frac{5}{2}$	2	$\frac{5}{2}$	1 000+	$+\sqrt{11.13}/2.3\sqrt{5.17.19}$
$\frac{5}{2}$	2	$\frac{5}{2}$	1 100+	$-47/2\sqrt{3.7.17.19}$
$\frac{5}{2}$	2	$\frac{5}{2}$	2 000−	$+7.23\sqrt{11}/8.3\sqrt{2.5.13.17.19}$
$\frac{5}{2}$	2	$\frac{5}{2}$	2 100−	$-199/8\sqrt{2.3.7.17.19}$
$\frac{5}{2}$	2	$\tilde{\frac{1}{2}}$	0 000+	$-59\sqrt{7.11}/8\sqrt{3.5.13.17.19}$
$\frac{5}{2}$	2	$\tilde{\frac{1}{2}}$	0 100+	$-\sqrt{19}/8\sqrt{17}$
$\frac{5}{2}$	3	$\frac{5}{2}$	0 000+	$-47/8.3\sqrt{2.17.19}$
$\frac{5}{2}$	3	$\frac{5}{2}$	0 100+	$+11\sqrt{5.11.13}/8\sqrt{2.3.7.17.19}$
$\frac{5}{2}$	3	$\frac{5}{2}$	1 000−	$-197/8.3\sqrt{2.5.17.19}$
$\frac{5}{2}$	3	$\frac{5}{2}$	1 100−	$-\sqrt{7.11.13}/8\sqrt{2.3.17.19}$
$\frac{5}{2}$	3	$\tilde{\frac{1}{2}}$	0 000−	$+\sqrt{7.19}/8\sqrt{3.5.17}$
$\frac{5}{2}$	3	$\tilde{\frac{1}{2}}$	0 100−	$+\sqrt{11.13}/8\sqrt{17.19}$
$\frac{5}{2}$	$\tilde{1}$	$\frac{5}{2}$	0 000+	$-5.7\sqrt{7}/8\sqrt{2.3.13.17.19}$
$\frac{5}{2}$	$\tilde{1}$	$\frac{5}{2}$	0 100+	$-\sqrt{5.11}/8\sqrt{2.17.19}$
$\frac{5}{2}$	$\tilde{1}$	$\frac{5}{2}$	1 000+	$+19\sqrt{7.19}/8\sqrt{2.3.5.13.17}$
$\frac{5}{2}$	$\tilde{1}$	$\frac{5}{2}$	1 100+	$+3\sqrt{11}/8\sqrt{2.17.19}$

$\frac{15}{2}$ 7 $\frac{9}{2}$

$\frac{3}{2}$	1	$\frac{3}{2}$	0 000−	$+2.3\sqrt{3.11}/\sqrt{5.7.13.17.19}$
$\frac{3}{2}$	1	$\frac{5}{2}$	0 000+	$+\sqrt{11}/\sqrt{5.13.17.19}$
$\frac{3}{2}$	2	$\frac{3}{2}$	0 000+	$-\sqrt{3.11.13}/2\sqrt{7.17.19}$
$\frac{3}{2}$	2	$\frac{5}{2}$	0 000−	$+2\sqrt{11}/\sqrt{3.13.17.19}$
$\frac{3}{2}$	2	$\frac{5}{2}$	1 000+	$-\sqrt{11}/2\sqrt{2.13.17.19}$
$\frac{3}{2}$	3	$\frac{3}{2}$	0 000−	$-11\sqrt{3}/\sqrt{5.7.17.19}$
$\frac{3}{2}$	3	$\frac{5}{2}$	0 000+	$+\sqrt{19}/2\sqrt{3.17}$
$\frac{3}{2}$	3	$\frac{5}{2}$	1 000−	$+3/2\sqrt{5.17.19}$
$\frac{3}{2}$	$\tilde{1}$	$\frac{3}{2}$	0 000−	$-73/2\sqrt{5.13.17.19}$
$\frac{3}{2}$	$\tilde{1}$	$\frac{5}{2}$	0 000+	$-\sqrt{7}/2\sqrt{2.5.13.17.19}$
$\frac{5}{2}$	1	$\frac{3}{2}$	0 000+	$+31\sqrt{3}/2\sqrt{2.5.13.17.19}$
$\frac{5}{2}$	1	$\frac{3}{2}$	0 100+	$+9.3\sqrt{11}/2.5\sqrt{2.7.17.19}$
$\frac{5}{2}$	1	$\frac{5}{2}$	0 000−	$-5.7\sqrt{7}/4\sqrt{2.13.17.19}$
$\frac{5}{2}$	1	$\frac{5}{2}$	0 100−	$+11\sqrt{3.11}/4\sqrt{2.5.17.19}$
$\frac{5}{2}$	1	$\frac{5}{2}$	1 000−	$+37\sqrt{7}/4\sqrt{2.5.13.17.19}$
$\frac{5}{2}$	1	$\frac{5}{2}$	1 100−	$+3\sqrt{3.11}/4.5\sqrt{2.17.19}$
$\frac{5}{2}$	2	$\frac{3}{2}$	0 000−	$+\sqrt{19}/4\sqrt{2.13.17}$
$\frac{5}{2}$	2	$\frac{3}{2}$	0 100−	$+3\sqrt{3.7.11}/4\sqrt{2.5.17.19}$
$\frac{5}{2}$	2	$\frac{3}{2}$	1 000+	$+5.5/4\sqrt{3.13.17.19}$
$\frac{5}{2}$	2	$\frac{3}{2}$	1 100+	$+23\sqrt{11}/4\sqrt{5.7.17.19}$
$\frac{5}{2}$	2	$\frac{5}{2}$	0 000+	$-7\sqrt{7.13}/8\sqrt{2.3.17.19}$
$\frac{5}{2}$	2	$\frac{5}{2}$	0 100+	$-3.7\sqrt{11}/8\sqrt{2.5.17.19}$
$\frac{5}{2}$	2	$\frac{5}{2}$	1 000−	$+5\sqrt{7.19}/4.3\sqrt{13.17}$
$\frac{5}{2}$	2	$\frac{5}{2}$	1 100−	$+\sqrt{5.11}/4\sqrt{3.17.19}$
$\frac{5}{2}$	2	$\frac{5}{2}$	2 000+	$-5\sqrt{7.13}/4.3\sqrt{2.17.19}$
$\frac{5}{2}$	2	$\frac{5}{2}$	2 100+	$+13\sqrt{11}/4\sqrt{2.3.5.17.19}$
$\frac{5}{2}$	3	$\frac{3}{2}$	0 000+	$-9.3.5/8\sqrt{2.11.17.19}$
$\frac{5}{2}$	3	$\frac{3}{2}$	0 100+	$-\sqrt{3.13.17}/8\sqrt{2.5.7.19}$
$\frac{5}{2}$	3	$\frac{3}{2}$	1 000−	$-571/8\sqrt{2.3.5.11.17.19}$
$\frac{5}{2}$	3	$\frac{3}{2}$	1 100−	$+67\sqrt{13}/8.5\sqrt{2.7.17.19}$
$\frac{5}{2}$	3	$\frac{5}{2}$	0 000−	$-139\sqrt{7}/8.3\sqrt{2.5.11.17.19}$
$\frac{5}{2}$	3	$\frac{5}{2}$	0 100−	$+83\sqrt{13}/8.5\sqrt{2.3.17.19}$
$\frac{5}{2}$	3	$\frac{5}{2}$	1 000+	$+13\sqrt{7}/8.3\sqrt{2.11.17.19}$
$\frac{5}{2}$	3	$\frac{5}{2}$	1 100+	$-\sqrt{5.13}/8\sqrt{2.3.17.19}$
$\frac{5}{2}$	$\tilde{1}$	$\frac{3}{2}$	0 000+	$-7\sqrt{3.5.7}/4\sqrt{11.13.17.19}$
$\frac{5}{2}$	$\tilde{1}$	$\frac{3}{2}$	0 100+	$+3\sqrt{19}/4.5\sqrt{17}$
$\frac{5}{2}$	$\tilde{1}$	$\frac{5}{2}$	0 000−	$-7.7\sqrt{5}/4\sqrt{2.3.11.13.17.19}$
$\frac{5}{2}$	$\tilde{1}$	$\frac{5}{2}$	0 100−	$-41\sqrt{7}/4.5\sqrt{2.17.19}$
$\frac{5}{2}$	$\tilde{1}$	$\frac{5}{2}$	1 000−	$-7.41/8\sqrt{2.3.11.13.17.19}$
$\frac{5}{2}$	$\tilde{1}$	$\frac{5}{2}$	1 100−	$-3.13\sqrt{7}/8\sqrt{2.5.17.19}$

$\frac{15}{2}$ 7 $\frac{11}{2}$

$\frac{3}{2}$	1	$\frac{1}{2}$	0 000−	$-3.11/\sqrt{2.5.13.17.19}$
$\frac{3}{2}$	1	$\frac{3}{2}$	0 000+	$+3\sqrt{7.11}/5\sqrt{13.17.19}$
$\frac{3}{2}$	1	$\frac{5}{2}$	0 000−	$-\sqrt{3.7.11}/5\sqrt{2.13.17.19}$
$\frac{3}{2}$	2	$\frac{1}{2}$	0 000+	$-11\sqrt{5}/\sqrt{2.13.17.19}$
$\frac{3}{2}$	2	$\frac{3}{2}$	0 000−	$-\sqrt{7.11}/2\sqrt{5.13.17.19}$
$\frac{3}{2}$	2	$\frac{5}{2}$	0 000+	$-\sqrt{2.7.11}/\sqrt{5.13.17.19}$
$\frac{3}{2}$	2	$\frac{5}{2}$	1 000−	$+7\sqrt{7.11}/2\sqrt{3.5.13.17.19}$
$\frac{3}{2}$	3	$\frac{3}{2}$	0 000+	$+11\sqrt{7}/2.5\sqrt{17.19}$
$\frac{3}{2}$	3	$\frac{5}{2}$	0 000−	$+3\sqrt{7}/2\sqrt{2.5.17.19}$
$\frac{3}{2}$	3	$\frac{5}{2}$	1 000+	$-31\sqrt{7}/2.5\sqrt{2.3.17.19}$
$\frac{3}{2}$	$\tilde{1}$	$\frac{3}{2}$	0 000+	$+67\sqrt{3}/2.5\sqrt{13.17.19}$
$\frac{3}{2}$	$\tilde{1}$	$\frac{5}{2}$	0 000−	$-3\sqrt{3.19}/2.5\sqrt{13.17}$
$\frac{5}{2}$	1	$\frac{3}{2}$	0 000−	$+3\sqrt{19}/5\sqrt{2.13.17}$
$\frac{5}{2}$	1	$\frac{3}{2}$	0 100−	$+\sqrt{3.7.11}/\sqrt{2.5.17.19}$
$\frac{5}{2}$	1	$\frac{5}{2}$	0 000+	$-4.3\sqrt{3}/\sqrt{5.13.17.19}$
$\frac{5}{2}$	1	$\frac{5}{2}$	0 100+	0
$\frac{5}{2}$	1	$\frac{5}{2}$	1 000+	$+9.9\sqrt{3}/2.5\sqrt{13.17.19}$
$\frac{5}{2}$	1	$\frac{5}{2}$	1 100+	$-\sqrt{7.11}/2\sqrt{5.17.19}$
$\frac{5}{2}$	2	$\frac{1}{2}$	0 000−	$+\sqrt{5.7.11}/8\sqrt{2.3.13.17.19}$
$\frac{5}{2}$	2	$\frac{1}{2}$	0 100−	$+5.11/8.3\sqrt{2.17.19}$
$\frac{5}{2}$	2	$\frac{3}{2}$	0 000+	$+397/8\sqrt{2.3.5.13.17.19}$
$\frac{5}{2}$	2	$\frac{3}{2}$	0 100+	$-5\sqrt{7.11}/8.3\sqrt{2.17.19}$
$\frac{5}{2}$	2	$\frac{3}{2}$	1 000−	$+17\sqrt{17}/8.2\sqrt{5.13.19}$
$\frac{5}{2}$	2	$\frac{3}{2}$	1 100−	$+7\sqrt{7.11}/8.2\sqrt{3.17.19}$
$\frac{5}{2}$	2	$\frac{5}{2}$	0 000−	$+7.7\sqrt{13}/8.2\sqrt{5.17.19}$
$\frac{5}{2}$	2	$\frac{5}{2}$	0 100−	$-5\sqrt{7.11}/8.2\sqrt{3.17.19}$
$\frac{5}{2}$	2	$\frac{5}{2}$	1 000+	$+53\sqrt{5}/8\sqrt{2.3.13.17.19}$
$\frac{5}{2}$	2	$\frac{5}{2}$	1 100+	$-\sqrt{7.11.17}/8.3\sqrt{2.19}$
$\frac{5}{2}$	2	$\frac{5}{2}$	2 000−	$+283/8.2\sqrt{3.5.13.17.19}$
$\frac{5}{2}$	2	$\frac{5}{2}$	2 100−	$+13\sqrt{7.11}/8.2.3\sqrt{17.19}$
$\frac{5}{2}$	3	$\frac{1}{2}$	0 000+	$-47\sqrt{7}/8\sqrt{2.3.5.17.19}$
$\frac{5}{2}$	3	$\frac{1}{2}$	0 100+	$+\sqrt{11.13}/8.3\sqrt{2.17.19}$
$\frac{5}{2}$	3	$\frac{3}{2}$	0 000−	$+991/8.2\sqrt{2.3.5.11.17.19}$

SO_3–K 3jm Factors (cont.)

$\frac{15}{2}$ 7 $\frac{11}{2}$

$\frac{5}{2}$ 3 $\frac{3}{2}$ 0 100− $+5.5\sqrt{7.13}/8.2.3\sqrt{2.17.19}$
$\frac{5}{2}$ 3 $\frac{3}{2}$ 1 000+ $-907/8.2.5\sqrt{2.11.17.19}$
$\frac{5}{2}$ 3 $\frac{3}{2}$ 1 100+ $-13\sqrt{7.13}/8.2\sqrt{2.3.5.17.19}$
$\frac{5}{2}$ 3 $\frac{5}{2}$ 0 000+ $+1259/8.2.5\sqrt{3.11.17.19}$
$\frac{5}{2}$ 3 $\frac{5}{2}$ 0 100+ $-\sqrt{7.13.19}/8.2.3\sqrt{5.17}$
$\frac{5}{2}$ 3 $\frac{5}{2}$ 1 000− $+107/8.2\sqrt{3.5.11.17.19}$
$\frac{5}{2}$ 3 $\frac{5}{2}$ 1 100− $-\sqrt{7.13.19}/8.2.3\sqrt{17}$
$\frac{5}{2}$ $\tilde{1}$ $\frac{1}{2}$ 0 000+ $+3.37/8\sqrt{2.5.13.17.19}$
$\frac{5}{2}$ $\tilde{1}$ $\frac{1}{2}$ 0 100+ $-7\sqrt{7.11}/8\sqrt{2.3.17.19}$
$\frac{5}{2}$ $\tilde{1}$ $\frac{3}{2}$ 0 000− $-3.7.7.7\sqrt{7}/8.2.5\sqrt{11.13.17.19}$
$\frac{5}{2}$ $\tilde{1}$ $\frac{3}{2}$ 0 100− $+43/8.2\sqrt{3.5.17.19}$
$\frac{5}{2}$ $\tilde{1}$ $\frac{5}{2}$ 0 000+ $+3.7.7.7\sqrt{7}/8.2.5\sqrt{11.13.17.19}$
$\frac{5}{2}$ $\tilde{1}$ $\frac{5}{2}$ 0 100+ $+197/8.2\sqrt{3.5.17.19}$
$\frac{5}{2}$ $\tilde{1}$ $\frac{5}{2}$ 1 000+ $+7.41\sqrt{7}/8.2\sqrt{5.11.13.17.19}$
$\frac{5}{2}$ $\tilde{1}$ $\frac{5}{2}$ 1 100+ $-5.13/8.2\sqrt{3.17.19}$

$\frac{15}{2}$ 7 $\frac{13}{2}$

$\frac{3}{2}$ 1 $\frac{1}{2}$ 0 000+ $-11\sqrt{11}/\sqrt{5.13.17.19}$
$\frac{3}{2}$ 1 $\frac{3}{2}$ 0 000− $-2.11\sqrt{11}/5\sqrt{5.13.17.19}$
$\frac{3}{2}$ 1 $\frac{5}{2}$ 0 000+ $-4.11/5\sqrt{5.13.17.19}$
$\frac{3}{2}$ 2 $\frac{1}{2}$ 0 000− $+11\sqrt{11}/2\sqrt{5.13.17.19}$
$\frac{3}{2}$ 2 $\frac{3}{2}$ 0 000+ $+11\sqrt{11}/2\sqrt{13.17.19}$
$\frac{3}{2}$ 2 $\frac{5}{2}$ 0 000− $-\sqrt{17}/2\sqrt{3.13.19}$
$\frac{3}{2}$ 2 $\frac{5}{2}$ 1 000+ $+9/2\sqrt{2.13.17.19}$
$\frac{3}{2}$ 2 $\frac{\tilde{1}}{2}$ 0 000+ $-7/2\sqrt{2.5.17.19}$
$\frac{3}{2}$ 3 $\frac{3}{2}$ 0 000− $+3/5\sqrt{5.17.19}$
$\frac{3}{2}$ 3 $\frac{5}{2}$ 0 000+ $-7\sqrt{11}/5\sqrt{3.17.19}$
$\frac{3}{2}$ 3 $\frac{5}{2}$ 1 000− $+\sqrt{11}/5\sqrt{5.17.19}$
$\frac{3}{2}$ $\tilde{1}$ $\frac{3}{2}$ 0 000− $+\sqrt{3.7}/2.5\sqrt{5.13.17.19}$
$\frac{3}{2}$ $\tilde{1}$ $\frac{5}{2}$ 0 000+ $+37\sqrt{7.11}/2.5\sqrt{2.5.13.17.19}$
$\frac{3}{2}$ $\tilde{1}$ $\frac{\tilde{1}}{2}$ 0 000− $-\sqrt{3.7.11}/2\sqrt{2.5.17.19}$
$\frac{5}{2}$ 1 $\frac{3}{2}$ 0 000+ $+659/2.5\sqrt{2.5.7.13.17.19}$
$\frac{5}{2}$ 1 $\frac{3}{2}$ 0 100+ $+7\sqrt{3.11}/2.5\sqrt{2.17.19}$
$\frac{5}{2}$ 1 $\frac{5}{2}$ 0 000− $-163\sqrt{11}/4.5\sqrt{2.7.13.17.19}$
$\frac{5}{2}$ 1 $\frac{5}{2}$ 0 100− $-13\sqrt{3}/4\sqrt{2.5.17.19}$
$\frac{5}{2}$ 1 $\frac{5}{2}$ 1 000− $-9.3\sqrt{11.17}/4.5\sqrt{2.5.7.13.19}$
$\frac{5}{2}$ 1 $\frac{5}{2}$ 1 100− $-37\sqrt{3}/4.5\sqrt{2.17.19}$
$\frac{5}{2}$ 1 $\frac{\tilde{1}}{2}$ 0 000+ $+\sqrt{11.17}/4\sqrt{5.7.19}$
$\frac{5}{2}$ 1 $\frac{\tilde{1}}{2}$ 0 100+ $-\sqrt{3.13}/4\sqrt{17.19}$
$\frac{5}{2}$ 2 $\frac{1}{2}$ 0 000+ $-47/2\sqrt{3.5.7.13.17.19}$
$\frac{5}{2}$ 2 $\frac{1}{2}$ 0 100+ $-\sqrt{11}/2\sqrt{17.19}$
$\frac{5}{2}$ 2 $\frac{3}{2}$ 0 000− $+83/\sqrt{2.3.7.13.17.19}$
$\frac{5}{2}$ 2 $\frac{3}{2}$ 0 100− $-\sqrt{11}/\sqrt{2.5.17.19}$
$\frac{5}{2}$ 2 $\frac{3}{2}$ 1 000+ $-5.41/8\sqrt{7.13.17.19}$
$\frac{5}{2}$ 2 $\frac{3}{2}$ 1 100+ $+3\sqrt{3.11}/8\sqrt{5.17.19}$
$\frac{5}{2}$ 2 $\frac{5}{2}$ 0 000+ 0
$\frac{5}{2}$ 2 $\frac{5}{2}$ 0 100+ $+3\sqrt{2}/\sqrt{5.17.19}$
$\frac{5}{2}$ 2 $\frac{5}{2}$ 1 000− $+7\sqrt{7.11}/8\sqrt{13.17.19}$
$\frac{5}{2}$ 2 $\frac{5}{2}$ 1 100− $-3.7\sqrt{3}/8\sqrt{5.17.19}$
$\frac{5}{2}$ 2 $\frac{5}{2}$ 2 000+ $-3\sqrt{11}/4\sqrt{2.7.13.17.19}$

$\frac{15}{2}$ 7 $\frac{13}{2}$

$\frac{5}{2}$ 2 $\frac{5}{2}$ 2 100+ $+3.13\sqrt{3}/4\sqrt{2.5.17.19}$
$\frac{5}{2}$ 2 $\frac{\tilde{1}}{2}$ 0 000− $+\sqrt{11.19}/8\sqrt{5.7.17}$
$\frac{5}{2}$ 2 $\frac{\tilde{1}}{2}$ 0 100− $-3\sqrt{3.13}/8\sqrt{17.19}$
$\frac{5}{2}$ 3 $\frac{1}{2}$ 0 000− $-23\sqrt{11}/8\sqrt{3.5.7.17.19}$
$\frac{5}{2}$ 3 $\frac{1}{2}$ 0 100− $+7\sqrt{13}/8\sqrt{17.19}$
$\frac{5}{2}$ 3 $\frac{3}{2}$ 0 000+ $-157\sqrt{11}/8.5\sqrt{2.3.7.17.19}$
$\frac{5}{2}$ 3 $\frac{3}{2}$ 0 100+ $-\sqrt{13.19}/8\sqrt{2.5.17}$
$\frac{5}{2}$ 3 $\frac{3}{2}$ 1 000− $-131\sqrt{11}/8.5\sqrt{2.5.7.17.19}$
$\frac{5}{2}$ 3 $\frac{3}{2}$ 1 100− $-13\sqrt{3.13}/8.5\sqrt{2.17.19}$
$\frac{5}{2}$ 3 $\frac{5}{2}$ 0 000− $-7\sqrt{7}/4.5\sqrt{2.5.17.19}$
$\frac{5}{2}$ 3 $\frac{5}{2}$ 0 100− $+3\sqrt{3.11.13}/4.5\sqrt{2.17.19}$
$\frac{5}{2}$ 3 $\frac{5}{2}$ 1 000+ $+3/4\sqrt{2.7.17.19}$
$\frac{5}{2}$ 3 $\frac{5}{2}$ 1 100+ $+3\sqrt{3.11.13}/4\sqrt{2.5.17.19}$
$\frac{5}{2}$ 3 $\frac{\tilde{1}}{2}$ 0 000+ $+8\sqrt{13}/\sqrt{5.7.17.19}$
$\frac{5}{2}$ 3 $\frac{\tilde{1}}{2}$ 0 100+ 0
$\frac{5}{2}$ $\tilde{1}$ $\frac{1}{2}$ 0 000− $+37\sqrt{11}/8\sqrt{5.13.17.19}$
$\frac{5}{2}$ $\tilde{1}$ $\frac{1}{2}$ 0 100− $+3\sqrt{3.7}/8\sqrt{17.19}$
$\frac{5}{2}$ $\tilde{1}$ $\frac{3}{2}$ 0 000+ $+7.7.7\sqrt{11}/8.5\sqrt{5.13.17.19}$
$\frac{5}{2}$ $\tilde{1}$ $\frac{3}{2}$ 0 100+ $+9\sqrt{3.7}/8.5\sqrt{17.19}$
$\frac{5}{2}$ $\tilde{1}$ $\frac{5}{2}$ 0 000− $+7.7.7\sqrt{3}/2.5\sqrt{2.5.13.17.19}$
$\frac{5}{2}$ $\tilde{1}$ $\frac{5}{2}$ 0 100− $-3\sqrt{7.11}/2.5\sqrt{2.17.19}$
$\frac{5}{2}$ $\tilde{1}$ $\frac{5}{2}$ 1 000− $-\sqrt{2.3}/5\sqrt{13.17.19}$
$\frac{5}{2}$ $\tilde{1}$ $\frac{5}{2}$ 1 100− 0

$\frac{15}{2}$ $\frac{15}{2}$ 0

$\frac{3}{2}$ $\frac{3}{2}$ 0 0 000+ $+1/2$
$\frac{5}{2}$ $\frac{5}{2}$ 0 0 000+ $+\sqrt{3}/2\sqrt{2}$
$\frac{5}{2}$ $\frac{5}{2}$ 0 0 100+ 0
$\frac{5}{2}$ $\frac{5}{2}$ 0 0 110+ $+\sqrt{3}/2\sqrt{2}$

$\frac{15}{2}$ $\frac{15}{2}$ 1

$\frac{3}{2}$ $\frac{3}{2}$ 1 0 000+ $+\sqrt{17}/2.5$
$\frac{5}{2}$ $\frac{3}{2}$ 1 0 000− $+\sqrt{7.11}/2.5\sqrt{2.17}$
$\frac{5}{2}$ $\frac{3}{2}$ 1 0 100− $+\sqrt{3.13}/2\sqrt{2.5.17}$
$\frac{5}{2}$ $\frac{5}{2}$ 1 0 000+ $+29/8.4\sqrt{5.17}$
$\frac{5}{2}$ $\frac{5}{2}$ 1 0 100+ $+\sqrt{3.7.11.13}/8.4\sqrt{17}$
$\frac{5}{2}$ $\frac{5}{2}$ 1 0 110+ $-9\sqrt{5}/8.4\sqrt{17}$
$\frac{5}{2}$ $\frac{5}{2}$ 1 1 000+ $-9.9.3/8.4.5\sqrt{17}$
$\frac{5}{2}$ $\frac{5}{2}$ 1 1 100+ $+\sqrt{3.7.11.13}/8.4\sqrt{5.17}$
$\frac{5}{2}$ $\frac{5}{2}$ 1 1 110+ $+3.13/8.4\sqrt{17}$

$\frac{15}{2}$ $\frac{15}{2}$ 2

$\frac{3}{2}$ $\frac{3}{2}$ 2 0 000+ $-\sqrt{3.17}/2.5\sqrt{7}$
$\frac{5}{2}$ $\frac{3}{2}$ 2 0 000− $+9\sqrt{11}/4.5\sqrt{2.17}$
$\frac{5}{2}$ $\frac{3}{2}$ 2 0 100− $+\sqrt{3.13}/4\sqrt{2.5.7.17}$
$\frac{5}{2}$ $\frac{3}{2}$ 2 1 000+ $-3\sqrt{3.11}/8.5\sqrt{17}$
$\frac{5}{2}$ $\frac{3}{2}$ 2 1 100+ $-\sqrt{13.17}/8\sqrt{5.7}$
$\frac{5}{2}$ $\frac{5}{2}$ 2 0 000+ $-3.7\sqrt{7}/8.4.5\sqrt{17}$
$\frac{5}{2}$ $\frac{5}{2}$ 2 0 100+ $+\sqrt{3.11.13}/8.4\sqrt{5.17}$
$\frac{5}{2}$ $\frac{5}{2}$ 2 0 110+ $+\sqrt{7}/8.4\sqrt{17}$
$\frac{5}{2}$ $\frac{5}{2}$ 2 1 000− 0
$\frac{5}{2}$ $\frac{5}{2}$ 2 1 100− $-\sqrt{11.13}/2\sqrt{2.5.17}$

SO_3–K 3jm Factors (cont.)

$\frac{15}{2}$	$\frac{15}{2}$	**2**			
$\frac{5}{2}$	$\frac{5}{2}$	2	1	110−	0
$\frac{5}{2}$	$\frac{5}{2}$	2	2	000+	$-19\sqrt{3.7}/8.2.5\sqrt{17}$
$\frac{5}{2}$	$\frac{5}{2}$	2	2	100+	$+\sqrt{11.13}/8.2\sqrt{5.17}$
$\frac{5}{2}$	$\frac{5}{2}$	2	2	110+	$-23\sqrt{3}/8.2\sqrt{7.17}$
$\frac{15}{2}$	$\frac{15}{2}$	**3**			
$\frac{3}{2}$	$\frac{3}{2}$	3	0	000+	$-2.9.3\sqrt{3}/5\sqrt{7.13.17.19}$
$\frac{3}{2}$	$\frac{3}{2}$	$\tilde{1}$	0	000+	$-107/2.5\sqrt{7.13.17.19}$
$\frac{5}{2}$	$\frac{3}{2}$	3	0	000−	$+3.29\sqrt{11}/8\sqrt{2.5.13.17.19}$
$\frac{5}{2}$	$\frac{3}{2}$	3	0	100−	$-9.9\sqrt{3}/8\sqrt{2.7.17.19}$
$\frac{5}{2}$	$\frac{3}{2}$	3	1	000+	$+199\sqrt{3.11}/8.5\sqrt{2.13.17.19}$
$\frac{5}{2}$	$\frac{3}{2}$	3	1	100+	$+157/8\sqrt{2.5.7.17.19}$
$\frac{5}{2}$	$\frac{3}{2}$	$\tilde{1}$	0	000−	$+\sqrt{3.11}/8.5\sqrt{13.17.19}$
$\frac{5}{2}$	$\frac{3}{2}$	$\tilde{1}$	0	100−	$+3.41/8\sqrt{5.7.17.19}$
$\frac{5}{2}$	$\frac{5}{2}$	3	0	000+	$-7\sqrt{3.7}/8.2.5\sqrt{13.17.19}$
$\frac{5}{2}$	$\frac{5}{2}$	3	0	100+	$-31\sqrt{11}/8.2\sqrt{5.17.19}$
$\frac{5}{2}$	$\frac{5}{2}$	3	0	110+	$-31\sqrt{3.13}/8.2\sqrt{7.17.19}$
$\frac{5}{2}$	$\frac{5}{2}$	3	1	000−	0
$\frac{5}{2}$	$\frac{5}{2}$	3	1	100−	$-\sqrt{11}/2\sqrt{17.19}$
$\frac{5}{2}$	$\frac{5}{2}$	3	1	110−	0
$\frac{5}{2}$	$\frac{5}{2}$	$\tilde{1}$	0	000+	$-7.53\sqrt{7}/8.4.5\sqrt{13.17.19}$
$\frac{5}{2}$	$\frac{5}{2}$	$\tilde{1}$	0	100+	$-41\sqrt{3.11}/8.4\sqrt{5.17.19}$
$\frac{5}{2}$	$\frac{5}{2}$	$\tilde{1}$	0	110+	$+9.13\sqrt{13}/8.4\sqrt{7.17.19}$
$\frac{5}{2}$	$\frac{5}{2}$	$\tilde{1}$	1	000+	$-19\sqrt{7.19}/8.2\sqrt{5.13.17}$
$\frac{5}{2}$	$\frac{5}{2}$	$\tilde{1}$	1	100+	$+\sqrt{3.11}/8.2\sqrt{17.19}$
$\frac{5}{2}$	$\frac{5}{2}$	$\tilde{1}$	1	110+	$+23\sqrt{5.13}/8.2\sqrt{7.17.19}$
$\frac{15}{2}$	$\frac{15}{2}$	**4**			
$\frac{3}{2}$	$\frac{3}{2}$	2	0	000+	$+11\sqrt{5}/2\sqrt{7.13.17.19}$
$\frac{3}{2}$	$\frac{3}{2}$	3	0	000−	0
$\frac{5}{2}$	$\frac{3}{2}$	2	0	000−	$-\sqrt{5.11.13}/2\sqrt{2.3.17.19}$
$\frac{5}{2}$	$\frac{3}{2}$	2	0	100−	$-1/2\sqrt{2.7.17.19}$
$\frac{5}{2}$	$\frac{3}{2}$	2	1	000+	$+5\sqrt{5.11}/8.3\sqrt{13.17.19}$
$\frac{5}{2}$	$\frac{3}{2}$	2	1	100+	$+97/8\sqrt{3.7.17.19}$
$\frac{5}{2}$	$\frac{3}{2}$	3	0	000+	$+29\sqrt{7.11}/8\sqrt{2.3.13.17.19}$
$\frac{5}{2}$	$\frac{3}{2}$	3	0	100+	$-9.3\sqrt{5}/8\sqrt{2.17.19}$
$\frac{5}{2}$	$\frac{3}{2}$	3	1	000−	$-5\sqrt{5.7.11}/8.3\sqrt{2.13.17.19}$
$\frac{5}{2}$	$\frac{3}{2}$	3	1	100−	$-1/8\sqrt{2.3.17.19}$
$\frac{5}{2}$	$\frac{5}{2}$	2	0	000+	$-5.7\sqrt{5.7}/8.2\sqrt{3.13.17.19}$
$\frac{5}{2}$	$\frac{5}{2}$	2	0	100+	$-\sqrt{11}/8.2\sqrt{17.19}$
$\frac{5}{2}$	$\frac{5}{2}$	2	0	110+	$-3\sqrt{3.5.7.13}/8.2\sqrt{17.19}$
$\frac{5}{2}$	$\frac{5}{2}$	2	1	000−	0
$\frac{5}{2}$	$\frac{5}{2}$	2	1	100−	$-5\sqrt{11}/2\sqrt{2.3.17.19}$
$\frac{5}{2}$	$\frac{5}{2}$	2	1	110−	0
$\frac{5}{2}$	$\frac{5}{2}$	2	2	000+	$-17\sqrt{5.7.17}/8.4.3\sqrt{13.19}$
$\frac{5}{2}$	$\frac{5}{2}$	2	2	100+	$+\sqrt{11.17}/8.4\sqrt{3.19}$
$\frac{5}{2}$	$\frac{5}{2}$	2	2	110+	$-3.5\sqrt{5.13}/8.4\sqrt{7.17.19}$
$\frac{5}{2}$	$\frac{5}{2}$	3	0	000−	0
$\frac{5}{2}$	$\frac{5}{2}$	3	0	100−	$+\sqrt{7.11}/2\sqrt{3.17.19}$
$\frac{5}{2}$	$\frac{5}{2}$	3	0	110−	0
$\frac{5}{2}$	$\frac{5}{2}$	3	1	000+	$+7.103/8.2.3\sqrt{13.17.19}$
$\frac{15}{2}$	$\frac{15}{2}$	**4**			
$\frac{5}{2}$	$\frac{5}{2}$	3	1	100+	$+5\sqrt{5.7.11}/8.2\sqrt{3.17.19}$
$\frac{5}{2}$	$\frac{5}{2}$	3	1	110+	$+9\sqrt{13}/8.2\sqrt{17.19}$
$\frac{15}{2}$	$\frac{15}{2}$	**5**			
$\frac{3}{2}$	$\frac{3}{2}$	1	0	000+	$+9\sqrt{11}/\sqrt{2.5.13.17.19}$
$\frac{3}{2}$	$\frac{3}{2}$	2	0	000−	0
$\frac{3}{2}$	$\frac{3}{2}$	$\tilde{1}$	0	000+	$+\sqrt{7.11}/2\sqrt{5.13.17.19}$
$\frac{5}{2}$	$\frac{3}{2}$	1	0	000−	$+9.7\sqrt{7}/8\sqrt{5.13.17.19}$
$\frac{5}{2}$	$\frac{3}{2}$	1	0	100−	$+5\sqrt{3.11}/8\sqrt{17.19}$
$\frac{5}{2}$	$\frac{3}{2}$	2	0	000+	$-9.7\sqrt{5}/8\sqrt{13.17.19}$
$\frac{5}{2}$	$\frac{3}{2}$	2	0	100+	$+\sqrt{3.7.11}/8\sqrt{17.19}$
$\frac{5}{2}$	$\frac{3}{2}$	2	1	000−	$-5\sqrt{3.5}/2\sqrt{2.13.17.19}$
$\frac{5}{2}$	$\frac{3}{2}$	2	1	100−	$+\sqrt{7.11}/2\sqrt{2.17.19}$
$\frac{5}{2}$	$\frac{3}{2}$	$\tilde{1}$	0	000−	$+59\sqrt{3}/8\sqrt{5.13.17.19}$
$\frac{5}{2}$	$\frac{3}{2}$	$\tilde{1}$	0	100−	$-3\sqrt{7.11}/8\sqrt{17.19}$
$\frac{5}{2}$	$\frac{5}{2}$	1	0	000+	$+3.367/8.4\sqrt{2.11.13.17.19}$
$\frac{5}{2}$	$\frac{5}{2}$	1	0	100+	$-7\sqrt{3.5.7}/8.4\sqrt{2.17.19}$
$\frac{5}{2}$	$\frac{5}{2}$	1	0	110+	$-3\sqrt{11.13}/8.4\sqrt{2.17.19}$
$\frac{5}{2}$	$\frac{5}{2}$	1	1	000+	$+3.811/8.4\sqrt{2.5.11.13.17.19}$
$\frac{5}{2}$	$\frac{5}{2}$	1	1	100+	$-5.7\sqrt{3.7}/8.4\sqrt{2.17.19}$
$\frac{5}{2}$	$\frac{5}{2}$	1	1	110+	$-3\sqrt{5.11.13}/8.4\sqrt{2.17.19}$
$\frac{5}{2}$	$\frac{5}{2}$	2	0	000−	0
$\frac{5}{2}$	$\frac{5}{2}$	2	0	100−	$-3\sqrt{3}/2\sqrt{2.17.19}$
$\frac{5}{2}$	$\frac{5}{2}$	2	0	110−	0
$\frac{5}{2}$	$\frac{5}{2}$	2	1	000+	$+7.7\sqrt{3.5.7}/8.4\sqrt{11.13.17.19}$
$\frac{5}{2}$	$\frac{5}{2}$	2	1	100+	$+5/8.4\sqrt{17.19}$
$\frac{5}{2}$	$\frac{5}{2}$	2	1	110+	$+\sqrt{3.5.7.11.13}/8.4\sqrt{17.19}$
$\frac{5}{2}$	$\frac{5}{2}$	2	2	000−	0
$\frac{5}{2}$	$\frac{5}{2}$	2	2	100−	$+13/2\sqrt{2.17.19}$
$\frac{5}{2}$	$\frac{5}{2}$	2	2	110−	0
$\frac{5}{2}$	$\frac{5}{2}$	$\tilde{1}$	0	000+	$-7.7\sqrt{7}/2\sqrt{5.11.13.17.19}$
$\frac{5}{2}$	$\frac{5}{2}$	$\tilde{1}$	0	100+	$-11\sqrt{3}/4\sqrt{17.19}$
$\frac{5}{2}$	$\frac{5}{2}$	$\tilde{1}$	0	110+	0
$\frac{5}{2}$	$\frac{5}{2}$	$\tilde{1}$	1	000+	$+7.41\sqrt{7}/8.4\sqrt{11.13.17.19}$
$\frac{5}{2}$	$\frac{5}{2}$	$\tilde{1}$	1	100+	$+13\sqrt{3.5}/8.4\sqrt{17.19}$
$\frac{5}{2}$	$\frac{5}{2}$	$\tilde{1}$	1	110+	$-\sqrt{7.11.13}/8.4\sqrt{17.19}$
$\frac{15}{2}$	$\frac{15}{2}$	**6**			
$\frac{3}{2}$	$\frac{3}{2}$	0	0	000+	$-11\sqrt{3.11}/2\sqrt{5.13.17.19}$
$\frac{3}{2}$	$\frac{3}{2}$	1	0	000−	0
$\frac{3}{2}$	$\frac{3}{2}$	2	0	000+	$-11\sqrt{3.7}/5\sqrt{2.13.17.19}$
$\frac{3}{2}$	$\frac{3}{2}$	3	0	000−	0
$\frac{5}{2}$	$\frac{3}{2}$	1	0	000+	$+3.23/8\sqrt{13.17.19}$
$\frac{5}{2}$	$\frac{3}{2}$	1	0	100+	$+\sqrt{3.5.7.11}/8\sqrt{17.19}$
$\frac{5}{2}$	$\frac{3}{2}$	2	0	000−	$-3.29\sqrt{11}/8.5\sqrt{13.17.19}$
$\frac{5}{2}$	$\frac{3}{2}$	2	0	100−	$+7\sqrt{3.7}/8\sqrt{5.17.19}$
$\frac{5}{2}$	$\frac{3}{2}$	2	1	000+	$-3.11\sqrt{3.11}/4.5\sqrt{2.13.17.19}$
$\frac{5}{2}$	$\frac{3}{2}$	2	1	100+	$-13\sqrt{7}/4\sqrt{2.5.17.19}$
$\frac{5}{2}$	$\frac{3}{2}$	3	0	000+	$-3\sqrt{11.17}/8\sqrt{2.5.13.19}$
$\frac{5}{2}$	$\frac{3}{2}$	3	0	100+	$+3\sqrt{3.7}/8\sqrt{2.17.19}$
$\frac{5}{2}$	$\frac{3}{2}$	3	1	000−	$-9.3\sqrt{3.11}/8\sqrt{2.13.17.19}$

SO_3–K 3jm Factors (cont.)

$\frac{15}{2}$ $\frac{15}{2}$ 6

$\frac{5}{2}$ $\frac{3}{2}$ 3 1 100− $+\sqrt{5.7}/8\sqrt{2.17.19}$

$\frac{5}{2}$ $\frac{5}{2}$ 0 0 000+ $+3.37\sqrt{11}/8.2\sqrt{2.5.13.17.19}$

$\frac{5}{2}$ $\frac{5}{2}$ 0 0 100+ $-7\sqrt{3.7}/8.2\sqrt{2.17.19}$

$\frac{5}{2}$ $\frac{5}{2}$ 0 0 110+ $+\sqrt{5.11.13}/8.2\sqrt{2.17.19}$

$\frac{5}{2}$ $\frac{5}{2}$ 1 0 000− 0

$\frac{5}{2}$ $\frac{5}{2}$ 1 0 100− $-\sqrt{2.3}/\sqrt{17.19}$

$\frac{5}{2}$ $\frac{5}{2}$ 1 0 110− 0

$\frac{5}{2}$ $\frac{5}{2}$ 1 1 000− 0

$\frac{5}{2}$ $\frac{5}{2}$ 1 1 100− $+\sqrt{3.5}/\sqrt{2.17.19}$

$\frac{5}{2}$ $\frac{5}{2}$ 1 1 110− 0

$\frac{5}{2}$ $\frac{5}{2}$ 2 0 000+ $+3.7.7.7\sqrt{7}/8.2.5\sqrt{2.13.17.19}$

$\frac{5}{2}$ $\frac{5}{2}$ 2 0 100+ $+7\sqrt{3.11}/8.2\sqrt{2.5.17.19}$

$\frac{5}{2}$ $\frac{5}{2}$ 2 0 110+ $-5\sqrt{7.13}/8.2\sqrt{2.17.19}$

$\frac{5}{2}$ $\frac{5}{2}$ 2 1 000− 0

$\frac{5}{2}$ $\frac{5}{2}$ 2 1 100− $+\sqrt{11}/2\sqrt{5.17.19}$

$\frac{5}{2}$ $\frac{5}{2}$ 2 1 110− 0

$\frac{5}{2}$ $\frac{5}{2}$ 2 2 000+ $-103\sqrt{3.7}/8.2.5\sqrt{2.13.17.19}$

$\frac{5}{2}$ $\frac{5}{2}$ 2 2 100+ $+5\sqrt{5.11}/8.2\sqrt{2.17.19}$

$\frac{5}{2}$ $\frac{5}{2}$ 2 2 110+ $+7\sqrt{3.7.13}/8.2\sqrt{2.17.19}$

$\frac{5}{2}$ $\frac{5}{2}$ 3 0 000− 0

$\frac{5}{2}$ $\frac{5}{2}$ 3 0 100− $+\sqrt{5.11}/2\sqrt{17.19}$

$\frac{5}{2}$ $\frac{5}{2}$ 3 0 110− 0

$\frac{5}{2}$ $\frac{5}{2}$ 3 1 000+ $+23\sqrt{3.7}/8\sqrt{5.13.17.19}$

$\frac{5}{2}$ $\frac{5}{2}$ 3 1 100+ $-5\sqrt{11}/8\sqrt{17.19}$

$\frac{5}{2}$ $\frac{5}{2}$ 3 1 110+ $+\sqrt{3.5.7.13}/8\sqrt{17.19}$

$\frac{15}{2}$ $\frac{15}{2}$ 7

$\frac{3}{2}$ $\frac{3}{2}$ 1 0 000+ $-11\sqrt{7.11.23}/2.5\sqrt{5.13.17.19}$

$\frac{3}{2}$ $\frac{3}{2}$ 2 0 000− 0

$\frac{3}{2}$ $\frac{3}{2}$ 3 0 000+ $-2.7\sqrt{7}/5\sqrt{5.17.19.23}$

$\frac{3}{2}$ $\frac{3}{2}$ $\tilde{1}$ 0 000+ $-4.9.3\sqrt{3}/5\sqrt{5.13.17.19.23}$

$\frac{5}{2}$ $\frac{3}{2}$ 1 0 000− $+659/4.5\sqrt{2.5.13.17.19.23}$

$\frac{5}{2}$ $\frac{3}{2}$ 1 0 100− $+7\sqrt{3.7.11}/4.5\sqrt{2.17.19.23}$

$\frac{5}{2}$ $\frac{3}{2}$ 2 0 000+ $+\sqrt{17.23}/4\sqrt{2.3.13.19}$

$\frac{5}{2}$ $\frac{3}{2}$ 2 0 100+ $+7\sqrt{7.11}/4\sqrt{2.5.17.19.23}$

$\frac{5}{2}$ $\frac{3}{2}$ 2 1 000− $-3.103/8\sqrt{13.17.19.23}$

$\frac{5}{2}$ $\frac{3}{2}$ 2 1 100− $-3.7\sqrt{3.7.11}/8\sqrt{5.17.19.23}$

$\frac{5}{2}$ $\frac{3}{2}$ 3 0 000− $-\sqrt{11.17.23}/8.5\sqrt{2.3.19}$

$\frac{5}{2}$ $\frac{3}{2}$ 3 0 100− $+\sqrt{7.13.23}/8\sqrt{2.5.17.19}$

$\frac{5}{2}$ $\frac{3}{2}$ 3 1 000+ $+11\sqrt{11.23}/8.5\sqrt{2.5.17.19}$

$\frac{5}{2}$ $\frac{3}{2}$ 3 1 100+ $+\sqrt{3.7.13.19}/8.5\sqrt{2.17.23}$

$\frac{5}{2}$ $\frac{3}{2}$ $\tilde{1}$ 0 000− $+7.37\sqrt{7.11}/8.5\sqrt{5.13.17.19.23}$

$\frac{5}{2}$ $\frac{3}{2}$ $\tilde{1}$ 0 100− $+3.73\sqrt{3}/8.5\sqrt{17.19.23}$

$\frac{5}{2}$ $\frac{5}{2}$ 1 0 000+ $+\sqrt{7.11.17}/8.2.5\sqrt{13.19.23}$

$\frac{5}{2}$ $\frac{5}{2}$ 1 0 100+ $+149\sqrt{3}/8.2\sqrt{5.17.19.23}$

$\frac{5}{2}$ $\frac{5}{2}$ 1 0 110+ $-9\sqrt{7.11.13}/8.2\sqrt{17.19.23}$

$\frac{5}{2}$ $\frac{5}{2}$ 1 1 000+ $-3.139\sqrt{7.11}/8.5\sqrt{5.13.17.19.23}$

$\frac{5}{2}$ $\frac{5}{2}$ 1 1 100+ $-167\sqrt{3}/8.5\sqrt{17.19.23}$

$\frac{5}{2}$ $\frac{5}{2}$ 1 1 110+ $-3\sqrt{7.11.13}/8\sqrt{5.17.19.23}$

$\frac{5}{2}$ $\frac{5}{2}$ 2 0 000− 0

$\frac{15}{2}$ $\frac{15}{2}$ 7

$\frac{5}{2}$ $\frac{5}{2}$ 2 0 100− $-3.7/2\sqrt{5.17.19.23}$

$\frac{5}{2}$ $\frac{5}{2}$ 2 0 110− 0

$\frac{5}{2}$ $\frac{5}{2}$ 2 1 000+ $+7\sqrt{7.11.17}/8\sqrt{2.13.19.23}$

$\frac{5}{2}$ $\frac{5}{2}$ 2 1 100+ $+9.11\sqrt{3}/8\sqrt{2.5.17.19.23}$

$\frac{5}{2}$ $\frac{5}{2}$ 2 1 110+ $-3\sqrt{7.11.13}/8\sqrt{2.17.19.23}$

$\frac{5}{2}$ $\frac{5}{2}$ 2 2 000− 0

$\frac{5}{2}$ $\frac{5}{2}$ 2 2 100− $+3.13\sqrt{3}/2\sqrt{5.17.19.23}$

$\frac{5}{2}$ $\frac{5}{2}$ 2 2 110− 0

$\frac{5}{2}$ $\frac{5}{2}$ 3 0 000+ $+7.109\sqrt{7}/8.5\sqrt{5.17.19.23}$

$\frac{5}{2}$ $\frac{5}{2}$ 3 0 100+ $-9.3\sqrt{3.11.13}/8.5\sqrt{17.19.23}$

$\frac{5}{2}$ $\frac{5}{2}$ 3 0 110+ $-3.13\sqrt{7}/8\sqrt{5.17.19.23}$

$\frac{5}{2}$ $\frac{5}{2}$ 3 1 000− 0

$\frac{5}{2}$ $\frac{5}{2}$ 3 1 100− $+3\sqrt{3.11.13}/2\sqrt{5.17.19.23}$

$\frac{5}{2}$ $\frac{5}{2}$ 3 1 110− 0

$\frac{5}{2}$ $\frac{5}{2}$ $\tilde{1}$ 0 000+ $+7.7.7.7\sqrt{3}/8.5\sqrt{5.13.17.19.23}$

$\frac{5}{2}$ $\frac{5}{2}$ $\tilde{1}$ 0 100+ $-3.7\sqrt{7.11}/8.5\sqrt{17.19.23}$

$\frac{5}{2}$ $\frac{5}{2}$ $\tilde{1}$ 0 110+ $+3.13\sqrt{3.13}/8\sqrt{5.17.19.23}$

$\frac{5}{2}$ $\frac{5}{2}$ $\tilde{1}$ 1 000+ $+\sqrt{3.23}/2.5\sqrt{13.17.19}$

$\frac{5}{2}$ $\frac{5}{2}$ $\tilde{1}$ 1 100+ 0

$\frac{5}{2}$ $\frac{5}{2}$ $\tilde{1}$ 1 110+ $-9\sqrt{3.13}/2\sqrt{17.19.23}$

8 4 4

2 2 2 0 000+ $+7\sqrt{2.5.7}/3.17\sqrt{13}$

2 2 2 0 100+ $+7\sqrt{5}/17\sqrt{2.11}$

2 2 2 1 000+ $+11\sqrt{2.7}/3.17\sqrt{13}$

2 2 2 1 100+ $-5.7/3.17\sqrt{2.11}$

2 3 2 0 000− $-\sqrt{13}/2.3.17$

2 3 2 0 100− $+5\sqrt{7}/17\sqrt{11}$

2 3 2 1 000+ $+5.19/2.3.17\sqrt{13}$

2 3 2 1 100+ $-5\sqrt{7}/3.17\sqrt{11}$

2 3 3 0 000+ $-4\sqrt{2.7}/3.17\sqrt{13}$

2 3 3 0 100+ $-8.5\sqrt{2}/3.17\sqrt{11}$

3 2 2 0 000− 0

3 2 2 1 000+ $-2.7/3\sqrt{13.17}$

3 3 2 0 000+ $+2\sqrt{2.7}/3\sqrt{13.17}$

3 3 3 0 000+ $+4\sqrt{2.5}/3\sqrt{13.17}$

$\tilde{1}$ 2 2 0 000− 0

$\tilde{1}$ 3 2 0 000+ $-\sqrt{3}/\sqrt{2.17}$

$\tilde{1}$ 3 3 0 000− 0

8 $\frac{9}{2}$ $\frac{7}{2}$

2 $\frac{3}{2}$ $\frac{5}{2}$ 0 000+ $+3.5\sqrt{5.7}/4.17\sqrt{13}$

2 $\frac{3}{2}$ $\frac{5}{2}$ 0 100+ $+7\sqrt{5}/2.17\sqrt{11}$

2 $\frac{3}{2}$ $\frac{5}{2}$ 1 000− $-9.3\sqrt{3.7}/4.17\sqrt{2.5.13}$

2 $\frac{3}{2}$ $\frac{5}{2}$ 1 100− $+7\sqrt{3.5}/2.17\sqrt{2.11}$

2 $\frac{3}{2}$ $\tilde{\frac{1}{2}}$ 0 000− $-149/4.17\sqrt{2.5.13}$

2 $\frac{3}{2}$ $\tilde{\frac{1}{2}}$ 0 100− $+3\sqrt{5.7}/2.17\sqrt{2.11}$

2 $\frac{5}{2}$ $\frac{5}{2}$ 0 000− $-11\sqrt{3.5}/4.17\sqrt{13}$

2 $\frac{5}{2}$ $\frac{5}{2}$ 0 100− $-3\sqrt{3.5.7}/2.17\sqrt{11}$

2 $\frac{5}{2}$ $\frac{5}{2}$ 1 000+ $+9.9/2.17\sqrt{2.5.13}$

2 $\frac{5}{2}$ $\frac{5}{2}$ 1 100+ $-3\sqrt{5.7}/17\sqrt{2.11}$

SO_3–K 3jm Factors (cont.)

$8\ \tfrac{9}{2}\ \tfrac{7}{2}$

$2\ \tfrac{5}{2}\ \tfrac{5}{2}\ 2\ 000-\quad +9.9/4.17\sqrt{5.13}$
$2\ \tfrac{5}{2}\ \tfrac{5}{2}\ 2\ 100-\quad -3\sqrt{5.7}/2.17\sqrt{11}$
$2\ \tfrac{5}{2}\ \widetilde{\tfrac{1}{2}}\ 0\ 000+\quad +3\sqrt{3.7}/17\sqrt{2.5.13}$
$2\ \tfrac{5}{2}\ \widetilde{\tfrac{1}{2}}\ 0\ 100+\quad +3\sqrt{2.3.5}/17\sqrt{11}$
$3\ \tfrac{3}{2}\ \tfrac{5}{2}\ 0\ 000-\quad +7/2\sqrt{13.17}$
$3\ \tfrac{3}{2}\ \tfrac{5}{2}\ 1\ 000+\quad -7\sqrt{3}/2\sqrt{5.13.17}$
$3\ \tfrac{5}{2}\ \tfrac{5}{2}\ 0\ 000+\quad -3\sqrt{7}/2\sqrt{5.13.17}$
$3\ \tfrac{5}{2}\ \tfrac{5}{2}\ 1\ 000-\quad -3\sqrt{7}/2\sqrt{13.17}$
$3\ \tfrac{5}{2}\ \widetilde{\tfrac{1}{2}}\ 0\ 000-\quad -3\sqrt{3}/\sqrt{2.13.17}$
$\tilde{1}\ \tfrac{3}{2}\ \tfrac{5}{2}\ 0\ 000-\quad -\sqrt{3.7}/4\sqrt{2.5.17}$
$\tilde{1}\ \tfrac{3}{2}\ \widetilde{\tfrac{1}{2}}\ 0\ 000+\quad -3\sqrt{3}/4\sqrt{2.17}$
$\tilde{1}\ \tfrac{5}{2}\ \tfrac{5}{2}\ 0\ 000+\quad +3\sqrt{3}/4\sqrt{5.17}$
$\tilde{1}\ \tfrac{5}{2}\ \tfrac{5}{2}\ 1\ 000+\quad +3\sqrt{3}/4\sqrt{17}$

$8\ \tfrac{9}{2}\ \tfrac{9}{2}$

$2\ \tfrac{3}{2}\ \tfrac{3}{2}\ 0\ 000+\quad +2\sqrt{7}/5.17\sqrt{13}$
$2\ \tfrac{3}{2}\ \tfrac{3}{2}\ 0\ 100+\quad +3.7/17\sqrt{11}$
$2\ \tfrac{5}{2}\ \tfrac{3}{2}\ 0\ 000-\quad +3\sqrt{13}/5.17$
$2\ \tfrac{5}{2}\ \tfrac{3}{2}\ 0\ 100-\quad -\sqrt{7}/17\sqrt{11}$
$2\ \tfrac{5}{2}\ \tfrac{3}{2}\ 1\ 000+\quad -3\sqrt{3}/5.17\sqrt{2.13}$
$2\ \tfrac{5}{2}\ \tfrac{3}{2}\ 1\ 100+\quad +2\sqrt{2.3.7}/17\sqrt{11}$
$2\ \tfrac{5}{2}\ \tfrac{5}{2}\ 0\ 000+\quad +2.7\sqrt{3.7}/5.17\sqrt{13}$
$2\ \tfrac{5}{2}\ \tfrac{5}{2}\ 0\ 100+\quad -2.3\sqrt{3}/17\sqrt{11}$
$2\ \tfrac{5}{2}\ \tfrac{5}{2}\ 1\ 000-\quad 0$
$2\ \tfrac{5}{2}\ \tfrac{5}{2}\ 1\ 100-\quad 0$
$2\ \tfrac{5}{2}\ \tfrac{5}{2}\ 2\ 000+\quad -2.9.3\sqrt{7}/5.17\sqrt{13}$
$2\ \tfrac{5}{2}\ \tfrac{5}{2}\ 2\ 100+\quad -2.3/17\sqrt{11}$
$3\ \tfrac{3}{2}\ \tfrac{3}{2}\ 0\ 000-\quad 0$
$3\ \tfrac{5}{2}\ \tfrac{3}{2}\ 0\ 000+\quad -7\sqrt{7}/2\sqrt{5.13.17}$
$3\ \tfrac{5}{2}\ \tfrac{3}{2}\ 1\ 000-\quad +\sqrt{3.7}/2\sqrt{13.17}$
$3\ \tfrac{5}{2}\ \tfrac{5}{2}\ 0\ 000-\quad 0$
$3\ \tfrac{5}{2}\ \tfrac{5}{2}\ 1\ 000+\quad -2.3/\sqrt{5.13.17}$
$\tilde{1}\ \tfrac{3}{2}\ \tfrac{3}{2}\ 0\ 000-\quad 0$
$\tilde{1}\ \tfrac{5}{2}\ \tfrac{3}{2}\ 0\ 000+\quad -\sqrt{3}/\sqrt{2.17}$
$\tilde{1}\ \tfrac{5}{2}\ \tfrac{5}{2}\ 0\ 000-\quad 0$
$\tilde{1}\ \tfrac{5}{2}\ \tfrac{5}{2}\ 1\ 000-\quad 0$

8 5 3

$2\ 1\ 3\ 0\ 000+\quad +19\sqrt{3.5}/4.17\sqrt{13}$
$2\ 1\ 3\ 0\ 100+\quad -\sqrt{3.5.7}/2.17\sqrt{11}$
$2\ 1\ \tilde{1}\ 0\ 000+\quad +3.5\sqrt{5}/2.17\sqrt{13}$
$2\ 1\ \tilde{1}\ 0\ 100+\quad +\sqrt{5.7}/17\sqrt{11}$
$2\ 2\ 3\ 0\ 000-\quad +89\sqrt{5}/4.17\sqrt{2.7.13}$
$2\ 2\ 3\ 0\ 100-\quad +3.5\sqrt{5}/2.17\sqrt{2.11}$
$2\ 2\ 3\ 1\ 000+\quad -3\sqrt{5.7}/4.17\sqrt{2.13}$
$2\ 2\ 3\ 1\ 100+\quad -3.5\sqrt{5}/2.17\sqrt{2.11}$
$2\ 2\ \tilde{1}\ 0\ 000-\quad +9.3\sqrt{3.5}/2.17\sqrt{2.7.13}$
$2\ 2\ \tilde{1}\ 0\ 100-\quad -5\sqrt{3.5}/17\sqrt{2.11}$
$2\ \tilde{1}\ 3\ 0\ 000+\quad -3\sqrt{2.3.5}/17\sqrt{7.13}$
$2\ \tilde{1}\ 3\ 0\ 100+\quad +3\sqrt{2.3.5}/17\sqrt{11}$
$2\ \tilde{1}\ \tilde{1}\ 0\ 000+\quad -11\sqrt{3.5}/2.17\sqrt{7.13}$

8 5 3

$2\ \tilde{1}\ \tilde{1}\ 0\ 100+\quad -3\sqrt{3.5}/17\sqrt{11}$
$3\ 1\ 3\ 0\ 000-\quad -\sqrt{3.7}/2\sqrt{13.17}$
$3\ 1\ \tilde{1}\ 0\ 000-\quad -\sqrt{7}/\sqrt{13.17}$
$3\ 2\ 3\ 0\ 000+\quad -3\sqrt{5}/2\sqrt{13.17}$
$3\ 2\ \tilde{1}\ 0\ 000+\quad -\sqrt{3.5}/\sqrt{13.17}$
$3\ \tilde{1}\ 3\ 0\ 000-\quad -3\sqrt{3}/\sqrt{2.13.17}$
$\tilde{1}\ 1\ 3\ 0\ 000-\quad -3/4\sqrt{17}$
$\tilde{1}\ 2\ 3\ 0\ 000+\quad -3\sqrt{3.5}/4\sqrt{7.17}$
$\tilde{1}\ 2\ \tilde{1}\ 0\ 000+\quad -\sqrt{3.5}/2\sqrt{2.7.17}$
$\tilde{1}\ \tilde{1}\ \tilde{1}\ 0\ 000-\quad -3\sqrt{3}/2\sqrt{7.17}$

8 5 4

$2\ 1\ 2\ 0\ 000+\quad -\sqrt{7}/17\sqrt{2.3.13}$
$2\ 1\ 2\ 0\ 100+\quad +2.5.7\sqrt{2}/3.17\sqrt{3.11}$
$2\ 1\ 3\ 0\ 000-\quad -23/4.17\sqrt{3.13}$
$2\ 1\ 3\ 0\ 100-\quad -5.5\sqrt{7}/2.3.17\sqrt{3.11}$
$2\ 2\ 2\ 0\ 000-\quad +\sqrt{5.13}/4.17$
$2\ 2\ 2\ 0\ 100-\quad -\sqrt{5.7.11}/2.3.17$
$2\ 2\ 2\ 1\ 000-\quad +3\sqrt{13}/4.17$
$2\ 2\ 2\ 1\ 100-\quad +5.7\sqrt{7}/2.3.17\sqrt{11}$
$2\ 2\ 3\ 0\ 000+\quad -\sqrt{7}/4\sqrt{2.13}$
$2\ 2\ 3\ 0\ 100+\quad +5/2.3\sqrt{2.11}$
$2\ 2\ 3\ 1\ 000-\quad +9.5\sqrt{7}/4.17\sqrt{2.13}$
$2\ 2\ 3\ 1\ 100-\quad -5/2.3.17\sqrt{2.11}$
$2\ \tilde{1}\ 2\ 0\ 000+\quad +9\sqrt{3}/2.17\sqrt{13}$
$2\ \tilde{1}\ 2\ 0\ 100+\quad -5\sqrt{7}/17\sqrt{3.11}$
$2\ \tilde{1}\ 3\ 0\ 000-\quad +3\sqrt{2.3.7}/17\sqrt{13}$
$2\ \tilde{1}\ 3\ 0\ 100-\quad +5\sqrt{2}/17\sqrt{3.11}$
$3\ 1\ 2\ 0\ 000-\quad -7/3\sqrt{3.13.17}$
$3\ 1\ 3\ 0\ 000+\quad +7\sqrt{5.7}/2.3\sqrt{3.13.17}$
$3\ 2\ 2\ 0\ 000+\quad -5\sqrt{7}/3\sqrt{2.13.17}$
$3\ 2\ 2\ 1\ 000-\quad -\sqrt{7}/3\sqrt{2.13.17}$
$3\ 2\ 3\ 0\ 000-\quad -\sqrt{13}/2.3\sqrt{17}$
$3\ \tilde{1}\ 2\ 0\ 000-\quad -2\sqrt{2.7}/\sqrt{3.13.17}$
$3\ \tilde{1}\ 3\ 0\ 000+\quad -\sqrt{5}/\sqrt{2.3.13.17}$
$\tilde{1}\ 1\ 2\ 0\ 000-\quad +\sqrt{7/2.3}\sqrt{17}$
$\tilde{1}\ 1\ 3\ 0\ 000+\quad -5\sqrt{5}/4.3\sqrt{17}$
$\tilde{1}\ 2\ 2\ 0\ 000+\quad +5/2\sqrt{2.3.17}$
$\tilde{1}\ 2\ 3\ 0\ 000-\quad +\sqrt{7}/4\sqrt{3.17}$
$\tilde{1}\ \tilde{1}\ 2\ 0\ 000-\quad +\sqrt{3}/2\sqrt{17}$

8 5 5

$2\ 1\ 1\ 0\ 000+\quad -\sqrt{3.7}/17\sqrt{13.19}$
$2\ 1\ 1\ 0\ 100+\quad -3.5.7\sqrt{3}/2.17\sqrt{11.19}$
$2\ 2\ 1\ 0\ 000-\quad +11\sqrt{3}/17\sqrt{13.19}$
$2\ 2\ 1\ 0\ 100-\quad -5\sqrt{3.7}/2.17\sqrt{11.19}$
$2\ 2\ 2\ 0\ 000+\quad +\sqrt{5.7}/17\sqrt{2.13.19}$
$2\ 2\ 2\ 0\ 100+\quad +3\sqrt{5}/2.17\sqrt{2.11.19}$
$2\ 2\ 2\ 1\ 000+\quad +3.11\sqrt{7}/17\sqrt{2.13.19}$
$2\ 2\ 2\ 1\ 100+\quad -3.5.7/2.17\sqrt{2.11.19}$
$2\ \tilde{1}\ 1\ 0\ 000+\quad -3\sqrt{13}/17\sqrt{19}$

SO_3–K 3jm Factors (cont.)

8 $\frac{11}{2}$ $\frac{5}{2}$ is preceded by:

8 5 5

2	$\tilde{1}$	1	0	100+	$+5\sqrt{7}/17\sqrt{11.19}$
2	$\tilde{1}$	2	0	000−	$+3.7\sqrt{3.7}/17\sqrt{2.13.19}$
2	$\tilde{1}$	2	0	100−	$+4.5\sqrt{2.3}/17\sqrt{11.19}$
2	$\tilde{1}$	$\tilde{1}$	0	000+	$+2.7\sqrt{3.7}/17\sqrt{13.19}$
2	$\tilde{1}$	$\tilde{1}$	0	100+	$-2.3.5\sqrt{3}/17\sqrt{11.19}$
3	2	1	0	000+	$-2\sqrt{2.3.7}/\sqrt{13.17.19}$
3	2	2	0	000−	0
3	2	2	1	000+	$+8.3/\sqrt{13.17.19}$
3	$\tilde{1}$	1	0	000−	$+\sqrt{5.7}/\sqrt{13.17.19}$
3	$\tilde{1}$	2	0	000+	$+\sqrt{3}/\sqrt{13.17.19}$
$\tilde{1}$	2	1	0	000+	$+3\sqrt{2}/\sqrt{17.19}$
$\tilde{1}$	2	2	0	000−	0
$\tilde{1}$	$\tilde{1}$	2	0	000+	$-\sqrt{3.7}/\sqrt{2.17.19}$
$\tilde{1}$	$\tilde{1}$	$\tilde{1}$	0	000−	0

8 $\frac{11}{2}$ $\frac{5}{2}$

2	$\frac{1}{2}$	$\frac{5}{2}$	0	000−	$+\sqrt{5}/\sqrt{3.13}$
2	$\frac{1}{2}$	$\frac{5}{2}$	0	100−	0
2	$\frac{3}{2}$	$\frac{5}{2}$	0	000+	$-11\sqrt{5.11}/2.17\sqrt{3.7.13}$
2	$\frac{3}{2}$	$\frac{5}{2}$	0	100+	$-\sqrt{3.5}/17$
2	$\frac{3}{2}$	$\frac{5}{2}$	1	000−	$-19\sqrt{5.11}/2.17\sqrt{2.7.13}$
2	$\frac{3}{2}$	$\frac{5}{2}$	1	100−	$+\sqrt{5}/17\sqrt{2}$
2	$\frac{5}{2}$	$\frac{5}{2}$	0	000−	$-11\sqrt{5.11}/2.17\sqrt{2.7.13}$
2	$\frac{5}{2}$	$\frac{5}{2}$	0	100−	$-3\sqrt{5}/17\sqrt{2}$
2	$\frac{5}{2}$	$\frac{5}{2}$	1	000+	$-\sqrt{3.5.11}/17\sqrt{7.13}$
2	$\frac{5}{2}$	$\frac{5}{2}$	1	100+	$+\sqrt{3.5}/17$
2	$\frac{5}{2}$	$\frac{5}{2}$	2	000−	$-\sqrt{2.3.5.11}/17\sqrt{7.13}$
2	$\frac{5}{2}$	$\frac{5}{2}$	2	100−	$+\sqrt{2.3.5}/17$
3	$\frac{1}{2}$	$\frac{5}{2}$	0	000+	$+\sqrt{3.7}/2\sqrt{13.17}$
3	$\frac{3}{2}$	$\frac{5}{2}$	0	000−	$+\sqrt{3.11}/2\sqrt{13.17}$
3	$\frac{3}{2}$	$\frac{5}{2}$	1	000+	$+\sqrt{5.11}/2\sqrt{13.17}$
3	$\frac{5}{2}$	$\frac{5}{2}$	0	000+	$+\sqrt{3.5.11}/2\sqrt{2.13.17}$
3	$\frac{5}{2}$	$\frac{5}{2}$	1	000−	$+\sqrt{3.11}/2\sqrt{2.13.17}$
$\tilde{1}$	$\frac{1}{2}$	$\frac{5}{2}$	0	000+	$+1/2\sqrt{17}$
$\tilde{1}$	$\frac{3}{2}$	$\frac{5}{2}$	0	000−	$-\sqrt{5.11}/2\sqrt{2.7.17}$
$\tilde{1}$	$\frac{5}{2}$	$\frac{5}{2}$	0	000+	0
$\tilde{1}$	$\frac{5}{2}$	$\frac{5}{2}$	1	000+	$+3\sqrt{11}/2\sqrt{2.7.17}$

8 $\frac{11}{2}$ $\frac{7}{2}$

2	$\frac{1}{2}$	$\frac{5}{2}$	0	000+	$-\sqrt{5.11}/4.17\sqrt{3.13}$
2	$\frac{1}{2}$	$\frac{5}{2}$	0	100+	$-7\sqrt{5.7}/2.3.17\sqrt{3}$
2	$\frac{3}{2}$	$\frac{5}{2}$	0	000−	$+31\sqrt{5}/17\sqrt{3.7.13}$
2	$\frac{3}{2}$	$\frac{5}{2}$	0	100−	$-5.5\sqrt{5}/3.17\sqrt{3.11}$
2	$\frac{3}{2}$	$\frac{5}{2}$	1	000+	$-5\sqrt{5}/17\sqrt{2.7.13}$
2	$\frac{3}{2}$	$\frac{5}{2}$	1	100+	$-2.3\sqrt{2.5}/17\sqrt{11}$
2	$\frac{3}{2}$	$\frac{\tilde{1}}{2}$	0	000+	$-5\sqrt{3.5}/17\sqrt{2.13}$
2	$\frac{3}{2}$	$\frac{\tilde{1}}{2}$	0	100+	$-\sqrt{2.5.7}/17\sqrt{3.11}$
2	$\frac{5}{2}$	$\frac{5}{2}$	0	000+	$+3.11\sqrt{5}/4.17\sqrt{2.7.13}$
2	$\frac{5}{2}$	$\frac{5}{2}$	0	100+	$+9\sqrt{5}/2.17\sqrt{2.11}$
2	$\frac{5}{2}$	$\frac{5}{2}$	1	000−	$+41\sqrt{3.5}/4.17\sqrt{7.13}$
2	$\frac{5}{2}$	$\frac{5}{2}$	1	100−	$-\sqrt{5}/17\sqrt{3.11}$
2	$\frac{5}{2}$	$\frac{5}{2}$	2	000+	$+19\sqrt{3.5}/4.17\sqrt{2.7.13}$
2	$\frac{5}{2}$	$\frac{5}{2}$	2	100+	$+31\sqrt{5}/2.17\sqrt{2.3.11}$
2	$\frac{5}{2}$	$\frac{\tilde{1}}{2}$	0	000−	$+9\sqrt{5}/17\sqrt{13}$
2	$\frac{5}{2}$	$\frac{\tilde{1}}{2}$	0	100−	$-\sqrt{5.7}/2.17\sqrt{11}$
3	$\frac{1}{2}$	$\frac{5}{2}$	0	000−	$-5\sqrt{7.11}/4.3\sqrt{3.13.17}$
3	$\frac{1}{2}$	$\frac{\tilde{1}}{2}$	0	000+	$-7\sqrt{11}/4\sqrt{3.13.17}$
3	$\frac{3}{2}$	$\frac{5}{2}$	0	000+	$-\sqrt{17}/2.3\sqrt{3.13}$
3	$\frac{3}{2}$	$\frac{5}{2}$	1	000−	$-\sqrt{5}/2.3\sqrt{13.17}$
3	$\frac{5}{2}$	$\frac{5}{2}$	0	000−	$+19\sqrt{5}/4\sqrt{2.3.13.17}$
3	$\frac{5}{2}$	$\frac{5}{2}$	1	000+	$-37/4\sqrt{2.3.13.17}$
3	$\frac{5}{2}$	$\frac{\tilde{1}}{2}$	0	000+	$+\sqrt{7}/4\sqrt{13.17}$
$\tilde{1}$	$\frac{1}{2}$	$\frac{5}{2}$	0	000−	$+\sqrt{11}/4\sqrt{17}$
$\tilde{1}$	$\frac{3}{2}$	$\frac{5}{2}$	0	000+	$-\sqrt{5}/\sqrt{2.7.17}$
$\tilde{1}$	$\frac{3}{2}$	$\frac{\tilde{1}}{2}$	0	000−	$+1/\sqrt{2.17}$
$\tilde{1}$	$\frac{5}{2}$	$\frac{5}{2}$	0	000−	$+\sqrt{5.7}/4\sqrt{2.17}$
$\tilde{1}$	$\frac{5}{2}$	$\frac{5}{2}$	1	000−	$-9/4\sqrt{2.7.17}$

8 $\frac{11}{2}$ $\frac{9}{2}$

2	$\frac{1}{2}$	$\frac{3}{2}$	0	000+	$+\sqrt{3.7.11}/17\sqrt{2.5.13.19}$
2	$\frac{1}{2}$	$\frac{3}{2}$	0	100+	$-2.7\sqrt{2.5}/17\sqrt{3.19}$
2	$\frac{1}{2}$	$\frac{5}{2}$	0	000−	$-43\sqrt{11}/2.17\sqrt{3.5.13.19}$
2	$\frac{1}{2}$	$\frac{5}{2}$	0	100−	$+\sqrt{5.7}/3.17\sqrt{3.19}$
2	$\frac{3}{2}$	$\frac{3}{2}$	0	000−	$-23\sqrt{3}/17\sqrt{5.13.19}$
2	$\frac{3}{2}$	$\frac{3}{2}$	0	100−	$-2.5\sqrt{5.7}/17\sqrt{3.11.19}$
2	$\frac{3}{2}$	$\frac{5}{2}$	0	000+	$+7\sqrt{7.19}/2.17\sqrt{3.5.13}$
2	$\frac{3}{2}$	$\frac{5}{2}$	0	100+	$-43\sqrt{5}/3.17\sqrt{3.11.19}$
2	$\frac{3}{2}$	$\frac{5}{2}$	1	000−	$+61\sqrt{7}/2.17\sqrt{2.5.13.19}$
2	$\frac{3}{2}$	$\frac{5}{2}$	1	100−	$+9.3\sqrt{5}/17\sqrt{2.11.19}$
2	$\frac{5}{2}$	$\frac{3}{2}$	0	000+	$+3\sqrt{2.3.13}/17\sqrt{5.19}$
2	$\frac{5}{2}$	$\frac{3}{2}$	0	100+	$-\sqrt{2.3.5.7}/17\sqrt{11.19}$
2	$\frac{5}{2}$	$\frac{3}{2}$	1	000−	$+9.31/2.17\sqrt{5.13.19}$
2	$\frac{5}{2}$	$\frac{3}{2}$	1	100−	$+\sqrt{5.7}/17\sqrt{11.19}$
2	$\frac{5}{2}$	$\frac{5}{2}$	0	000−	$+2.3.7\sqrt{2.7}/17\sqrt{5.13.19}$
2	$\frac{5}{2}$	$\frac{5}{2}$	0	100−	$-2.9\sqrt{2.5}/17\sqrt{11.19}$
2	$\frac{5}{2}$	$\frac{5}{2}$	1	000+	$+7.11\sqrt{3.7}/2.17\sqrt{5.13.19}$
2	$\frac{5}{2}$	$\frac{5}{2}$	1	100+	$+4\sqrt{5.11}/17\sqrt{3.19}$
2	$\frac{5}{2}$	$\frac{5}{2}$	2	000−	$+37\sqrt{3.7}/2.17\sqrt{2.5.13.19}$
2	$\frac{5}{2}$	$\frac{5}{2}$	2	100−	$-5.5\sqrt{5}/17\sqrt{2.3.11.19}$
3	$\frac{1}{2}$	$\frac{5}{2}$	0	000+	$-7\sqrt{7.11}/2.3\sqrt{3.13.17.19}$
3	$\frac{3}{2}$	$\frac{3}{2}$	0	000+	$+2.11\sqrt{7}/\sqrt{3.5.13.17.19}$
3	$\frac{3}{2}$	$\frac{5}{2}$	0	000−	$-7.23/2.3\sqrt{3.13.17.19}$
3	$\frac{3}{2}$	$\frac{5}{2}$	1	000+	$+193/2.3\sqrt{5.13.17.19}$
3	$\frac{5}{2}$	$\frac{3}{2}$	0	000−	$+2\sqrt{2.3.7}/\sqrt{13.17.19}$
3	$\frac{5}{2}$	$\frac{3}{2}$	1	000+	$+2\sqrt{2.7}/\sqrt{5.13.17.19}$
3	$\frac{5}{2}$	$\frac{5}{2}$	0	000+	$-11/2\sqrt{2.3.5.13.17.19}$
3	$\frac{5}{2}$	$\frac{5}{2}$	1	000−	$+7.7/2\sqrt{2.3.13.17.19}$
$\tilde{1}$	$\frac{1}{2}$	$\frac{5}{2}$	0	000+	$+\sqrt{11}/\sqrt{17.19}$
$\tilde{1}$	$\frac{3}{2}$	$\frac{3}{2}$	0	000+	$-11/\sqrt{5.17.19}$
$\tilde{1}$	$\frac{3}{2}$	$\frac{5}{2}$	0	000−	$-\sqrt{7}/2\sqrt{2.5.17.19}$
$\tilde{1}$	$\frac{5}{2}$	$\frac{3}{2}$	0	000−	$+3/2\sqrt{5.17.19}$

SO_3–K 3jm Factors (cont.)

8 $\frac{11}{2}$ $\frac{9}{2}$	
$\tilde{1}$ $\frac{5}{2}$ $\frac{5}{2}$ 0 000+	$-11\sqrt{7}/2\sqrt{2.5.17.19}$
$\tilde{1}$ $\frac{5}{2}$ $\frac{5}{2}$ 1 000+	0
8 $\frac{11}{2}$ $\frac{11}{2}$	
2 $\frac{3}{2}$ $\frac{1}{2}$ 0 000−	$-4\sqrt{5.11}/17\sqrt{3.13.19}$
2 $\frac{3}{2}$ $\frac{1}{2}$ 0 100−	$-3\sqrt{3.5.7}/2.17\sqrt{19}$
2 $\frac{3}{2}$ $\frac{3}{2}$ 0 000+	$-2.5\sqrt{2.5.7}/17\sqrt{3.13.19}$
2 $\frac{3}{2}$ $\frac{3}{2}$ 0 100+	$+4.3\sqrt{2.3.5}/17\sqrt{11.19}$
2 $\frac{5}{2}$ $\frac{1}{2}$ 0 000+	$-2.5\sqrt{5.11}/17\sqrt{3.13.19}$
2 $\frac{5}{2}$ $\frac{1}{2}$ 0 100+	$+\sqrt{3.5.7}/2.17\sqrt{19}$
2 $\frac{5}{2}$ $\frac{3}{2}$ 0 000−	$-\sqrt{5.7}/17\sqrt{3.13.19}$
2 $\frac{5}{2}$ $\frac{3}{2}$ 0 100−	$-\sqrt{3.5}/2.17\sqrt{11.19}$
2 $\frac{5}{2}$ $\frac{3}{2}$ 1 000+	$+\sqrt{5.7}/\sqrt{2.13.19}$
2 $\frac{5}{2}$ $\frac{3}{2}$ 1 100+	$+\sqrt{5}/\sqrt{2.11.19}$
2 $\frac{5}{2}$ $\frac{5}{2}$ 0 000+	$+7\sqrt{2.5.7}/17\sqrt{13.19}$
2 $\frac{5}{2}$ $\frac{5}{2}$ 0 100+	$-3.5\sqrt{2.5}/17\sqrt{11.19}$
2 $\frac{5}{2}$ $\frac{5}{2}$ 1 000−	0
2 $\frac{5}{2}$ $\frac{5}{2}$ 1 100−	0
2 $\frac{5}{2}$ $\frac{5}{2}$ 2 000+	$+2\sqrt{2.3.5.7}/17\sqrt{13.19}$
2 $\frac{5}{2}$ $\frac{5}{2}$ 2 100+	$+23\sqrt{3.5}/17\sqrt{2.11.19}$
3 $\frac{3}{2}$ $\frac{3}{2}$ 0 000−	0
3 $\frac{5}{2}$ $\frac{1}{2}$ 0 000−	$-\sqrt{3.7.11}/\sqrt{13.17.19}$
3 $\frac{5}{2}$ $\frac{3}{2}$ 0 000+	$+\sqrt{3.13}/2\sqrt{17.19}$
3 $\frac{5}{2}$ $\frac{3}{2}$ 1 000−	$+7\sqrt{5}/2\sqrt{13.17.19}$
3 $\frac{5}{2}$ $\frac{5}{2}$ 0 000−	0
3 $\frac{5}{2}$ $\frac{5}{2}$ 1 000+	$-5\sqrt{2.3}/\sqrt{13.17.19}$
$\tilde{1}$ $\frac{3}{2}$ $\frac{3}{2}$ 0 000−	0
$\tilde{1}$ $\frac{5}{2}$ $\frac{1}{2}$ 0 000−	$+\sqrt{11}/\sqrt{17.19}$
$\tilde{1}$ $\frac{5}{2}$ $\frac{3}{2}$ 0 000+	$-\sqrt{5.7}/\sqrt{2.17.19}$
$\tilde{1}$ $\frac{5}{2}$ $\frac{5}{2}$ 0 000−	0
$\tilde{1}$ $\frac{5}{2}$ $\frac{5}{2}$ 1 000−	0
8 6 2	
2 0 2 0 000+	$+1/\sqrt{13}$
2 0 2 0 100+	0
2 1 2 0 000−	$-2\sqrt{3}/\sqrt{7.13}$
2 1 2 0 100−	0
2 2 2 0 000+	$-11\sqrt{5.11}/2.17\sqrt{2.7.13}$
2 2 2 0 100+	$-3\sqrt{5}/17\sqrt{2}$
2 2 2 1 000+	$+9.3\sqrt{11}/2.17\sqrt{2.7.13}$
2 2 2 1 100+	$-5/17\sqrt{2}$
2 3 2 0 000−	$+\sqrt{11}/2\sqrt{2.7.13}$
2 3 2 0 100−	0
2 3 2 1 000+	$-3\sqrt{11}/2.17\sqrt{2.7.13}$
2 3 2 1 100+	$+5\sqrt{2}/17$
3 1 2 0 000+	$+3\sqrt{3}/\sqrt{2.13.17}$
3 2 2 0 000−	$+3\sqrt{11}/2\sqrt{13.17}$
3 2 2 1 000+	$-\sqrt{11}/2\sqrt{13.17}$
3 3 2 0 000+	$-\sqrt{11}/\sqrt{13.17}$
$\tilde{1}$ 1 2 0 000+	$+3/\sqrt{2.7.17}$
$\tilde{1}$ 2 2 0 000−	$+\sqrt{3.11}/2\sqrt{7.17}$
8 6 2	
$\tilde{1}$ 3 2 0 000+	$-\sqrt{3.11}/2\sqrt{7.17}$
8 6 3	
2 1 3 0 000−	$+9\sqrt{5.7}/4.17\sqrt{13}$
2 1 3 0 100−	$+\sqrt{5.11}/2.17$
2 1 $\tilde{1}$ 0 000−	$+\sqrt{3.5.13}/2.17\sqrt{7}$
2 1 $\tilde{1}$ 0 100−	$-\sqrt{5.11}/17\sqrt{3}$
2 2 3 0 000+	$+23\sqrt{3.5.11}/4.17\sqrt{2.7.13}$
2 2 3 0 100+	$-3\sqrt{3.5}/2.17\sqrt{2}$
2 2 3 1 000−	$+5\sqrt{3.5.11}/4.17\sqrt{2.7.13}$
2 2 3 1 100−	$+\sqrt{5}/2.17\sqrt{2.3}$
2 2 $\tilde{1}$ 0 000+	$+11\sqrt{5.11}/2.17\sqrt{2.7.13}$
2 2 $\tilde{1}$ 0 100+	$+3\sqrt{5}/17\sqrt{2}$
2 3 3 0 000−	$-5\sqrt{3.5.11}/17\sqrt{2.7.13}$
2 3 3 0 100−	$-\sqrt{2.5}/17\sqrt{3}$
2 3 $\tilde{1}$ 0 000−	$-\sqrt{2.5.11}/17\sqrt{7.13}$
2 3 $\tilde{1}$ 0 100−	$+\sqrt{2.5}/17$
3 0 3 0 000−	$-\sqrt{3.7}/\sqrt{2.13.17}$
3 1 3 0 000+	$+5/2\sqrt{13.17}$
3 1 $\tilde{1}$ 0 000+	$+2/\sqrt{3.13.17}$
3 2 3 0 000−	$-\sqrt{5.11}/2\sqrt{3.13.17}$
3 2 $\tilde{1}$ 0 000−	0
3 3 3 0 000−	$-\sqrt{2.11}/\sqrt{3.13.17}$
3 3 $\tilde{1}$ 0 000+	$-\sqrt{2.11}/\sqrt{13.17}$
$\tilde{1}$ 0 $\tilde{1}$ 0 000−	$-1/\sqrt{2.17}$
$\tilde{1}$ 1 3 0 000+	$+3\sqrt{3}/4\sqrt{7.17}$
$\tilde{1}$ 2 3 0 000−	$+\sqrt{5.11}/4\sqrt{7.17}$
$\tilde{1}$ 2 $\tilde{1}$ 0 000−	$-\sqrt{5.11}/2\sqrt{2.7.17}$
$\tilde{1}$ 3 3 0 000+	$-\sqrt{11}/\sqrt{2.7.17}$
8 6 4	
2 0 2 0 000+	$-\sqrt{5.11}/17\sqrt{2.13.19}$
2 0 2 0 100+	$-7\sqrt{2.5.7}/3.17\sqrt{19}$
2 1 2 0 000−	$+\sqrt{5.11.13}/17\sqrt{2.3.7.19}$
2 1 2 0 100−	$-4.7\sqrt{2.5}/3.17\sqrt{3.19}$
2 1 3 0 000+	$+\sqrt{5.11}/4\sqrt{3.13.19}$
2 1 3 0 100+	$-\sqrt{5.7}/2.3\sqrt{3.19}$
2 2 2 0 000+	$+5.157/4.17\sqrt{7.13.19}$
2 2 2 0 100+	$-5.23/2.3.17\sqrt{11.19}$
2 2 2 1 000+	$+5.11\sqrt{5}/4.17\sqrt{7.13.19}$
2 2 2 1 100+	$-\sqrt{5.19}/2.17\sqrt{11}$
2 2 3 0 000−	$+5.23\sqrt{5}/4.17\sqrt{2.13.19}$
2 2 3 0 100−	$+7.13\sqrt{5.7}/2.3.17\sqrt{2.11.19}$
2 2 3 1 000+	$+89\sqrt{5}/4.17\sqrt{2.13.19}$
2 2 3 1 100+	$+3.5\sqrt{5.7}/2.17\sqrt{2.11.19}$
2 3 2 0 000−	$-2.3\sqrt{5.13}/17\sqrt{7.19}$
2 3 2 0 100−	$-7\sqrt{5}/17\sqrt{11.19}$
2 3 2 1 000+	$+8.8\sqrt{5}/17\sqrt{7.13.19}$
2 3 2 1 100+	$+29\sqrt{5}/3.17\sqrt{11.19}$
2 3 3 0 000+	$-41\sqrt{5}/17\sqrt{2.13.19}$
2 3 3 0 100+	$+\sqrt{2.5.7.19}/3.17\sqrt{11}$

SO_3–K 3jm Factors (cont.)

8 6 4		
3 0 3 0	000+	$-7\sqrt{11}/3\sqrt{2.13.17.19}$
3 1 2 0	000+	$-8\sqrt{5.11}/3\sqrt{3.13.17.19}$
3 1 3 0	000−	$+\sqrt{7.11.13}/2.3\sqrt{3.17.19}$
3 2 2 0	000−	$+11\sqrt{2.5}/3\sqrt{13.17.19}$
3 2 2 1	000+	$+5\sqrt{2.5}/\sqrt{13.17.19}$
3 2 3 0	000+	$+5\sqrt{5.7}/2\sqrt{13.17.19}$
3 3 2 0	000+	$-5\sqrt{2.5}/3\sqrt{13.17.19}$
3 3 3 0	000+	$+7\sqrt{2.7}/3\sqrt{13.17.19}$
$\tilde{1}$ 1 2 0	000+	$+11\sqrt{5.11}/2.3\sqrt{7.17.19}$
$\tilde{1}$ 1 3 0	000−	$+7\sqrt{11}/4.3\sqrt{17.19}$
$\tilde{1}$ 2 2 0	000−	$-11\sqrt{5}/2\sqrt{2.3.7.17.19}$
$\tilde{1}$ 2 3 0	000+	$+5\sqrt{5}/4\sqrt{3.17.19}$
$\tilde{1}$ 3 2 0	000+	$+\sqrt{2.5}/\sqrt{3.7.17.19}$
$\tilde{1}$ 3 3 0	000−	$+11/\sqrt{2.3.17.19}$

8 6 5		
2 0 2 0	000−	$-\sqrt{2.3.5.11}/17\sqrt{13.19}$
2 0 2 0	100−	$+\sqrt{2.3.5.7}/17\sqrt{19}$
2 1 1 0	000−	$+3\sqrt{5.11}/17\sqrt{2.13.19}$
2 1 1 0	100−	$-3\sqrt{5.7}/17\sqrt{2.19}$
2 1 2 0	000+	$+3\sqrt{5.7.11}/17\sqrt{2.13.19}$
2 1 2 0	100+	$+13\sqrt{5}/17\sqrt{2.19}$
2 1 $\tilde{1}$ 0	000−	$+\sqrt{3.5.7.11}/17\sqrt{2.13.19}$
2 1 $\tilde{1}$ 0	100−	$-2\sqrt{2.5}/17\sqrt{3.19}$
2 2 1 0	000+	$-9.3\sqrt{5}/17\sqrt{2.13.19}$
2 2 1 0	100+	$-7\sqrt{5.7}/17\sqrt{2.11.19}$
2 2 2 0	000−	$+2.5\sqrt{3.7}/17\sqrt{13.19}$
2 2 2 0	100−	$-5\sqrt{3.11}/2.17\sqrt{19}$
2 2 2 1	000−	$-5\sqrt{3.5.7}/17\sqrt{13.19}$
2 2 2 1	100−	$+23\sqrt{5}/2.17\sqrt{3.11.19}$
2 2 $\tilde{1}$ 0	000+	$+\sqrt{5.7.13}/2.17\sqrt{19}$
2 2 $\tilde{1}$ 0	100+	$-3\sqrt{5}/17\sqrt{11.19}$
2 3 1 0	000−	$+3.29\sqrt{5}/2.17\sqrt{2.13.19}$
2 3 1 0	100−	$-\sqrt{5.7}/17\sqrt{2.11.19}$
2 3 2 0	000+	$-9.3\sqrt{3.5.7}/4.17\sqrt{13.19}$
2 3 2 0	100+	$+\sqrt{3.5}/2.17\sqrt{11.19}$
2 3 2 1	000−	$+\sqrt{3.5.7}/4.17\sqrt{13.19}$
2 3 2 1	100−	$+83\sqrt{5}/2.17\sqrt{3.11.19}$
2 3 $\tilde{1}$ 0	000−	$-4\sqrt{5.7}/17\sqrt{13.19}$
2 3 $\tilde{1}$ 0	100−	$-23\sqrt{5}/17\sqrt{11.19}$
3 1 2 0	000−	$+\sqrt{5.11}/\sqrt{13.17.19}$
3 1 $\tilde{1}$ 0	000+	$+\sqrt{11.13}/\sqrt{2.3.17.19}$
3 2 1 0	000−	$-2\sqrt{5.7}/\sqrt{13.17.19}$
3 2 2 0	000+	$-\sqrt{3.5}/\sqrt{2.13.17.19}$
3 2 2 1	000−	$+5\sqrt{5}/\sqrt{2.3.13.17.19}$
3 2 $\tilde{1}$ 0	000−	$+9\sqrt{5}/\sqrt{2.13.17.19}$
3 3 1 0	000+	$-4\sqrt{2.7}/\sqrt{13.17.19}$
3 3 2 0	000−	$-\sqrt{2.5}/\sqrt{3.13.17.19}$
3 3 $\tilde{1}$ 0	000+	$+5/\sqrt{13.17.19}$
$\tilde{1}$ 0 $\tilde{1}$ 0	000−	$-\sqrt{11}/\sqrt{17.19}$
$\tilde{1}$ 1 2 0	000−	0
$\tilde{1}$ 2 1 0	000−	$+\sqrt{3.5}/\sqrt{17.19}$
$\tilde{1}$ 2 2 0	000+	$+\sqrt{5.7}/\sqrt{2.17.19}$
$\tilde{1}$ 2 $\tilde{1}$ 0	000−	$+\sqrt{5.7}/2\sqrt{17.19}$
$\tilde{1}$ 3 1 0	000+	$+\sqrt{3}/2\sqrt{2.17.19}$
$\tilde{1}$ 3 2 0	000−	$+\sqrt{5.7}/2\sqrt{2.17.19}$

8 6 6		
2 1 1 0	000+	$+11\sqrt{3.5}/17\sqrt{13.19}$
2 1 1 0	100+	$+3\sqrt{3.5.11}/2.17\sqrt{7.19}$
2 2 0 0	000+	$-2\sqrt{5.7.11}/17\sqrt{13.19}$
2 2 0 0	100+	$-3\sqrt{5}/17\sqrt{19}$
2 2 1 0	000−	$-5\sqrt{3.5.11}/17\sqrt{13.19}$
2 2 1 0	100−	$+\sqrt{3.5.19}/2.17\sqrt{7}$
2 2 2 0	000+	$+3.5/17\sqrt{2.13.19}$
2 2 2 0	100+	$+9.5/2.17\sqrt{2.7.11.19}$
2 2 2 1	000+	$-3.11\sqrt{5}/17\sqrt{2.13.19}$
2 2 2 1	100+	$-5.47\sqrt{5}/2.17\sqrt{2.7.11.19}$
2 3 1 0	000+	$+\sqrt{3.5.11}/2.17\sqrt{13.19}$
2 3 1 0	100+	$+5\sqrt{3.5}/17\sqrt{7.19}$
2 3 2 0	000−	$+5.5\sqrt{5}/2.17\sqrt{2.13.19}$
2 3 2 0	100−	$+3\sqrt{5.19}/17\sqrt{2.7.11}$
2 3 2 1	000+	$-9.3\sqrt{5}/2.17\sqrt{2.13.19}$
2 3 2 1	100+	$+3.23\sqrt{5}/17\sqrt{2.7.11.19}$
2 3 3 0	000+	$+2.9\sqrt{2.5}/17\sqrt{13.19}$
2 3 3 0	100+	$-\sqrt{2.5.7}/17\sqrt{11.19}$
3 2 1 0	000+	$+\sqrt{2.3.5.11}/\sqrt{7.13.17.19}$
3 2 2 0	000−	0
3 2 2 1	000+	$-2\sqrt{5.7}/\sqrt{13.17.19}$
3 3 0 0	000+	$-3\sqrt{2.11}/\sqrt{13.17.19}$
3 3 1 0	000−	$-2\sqrt{3.11}/\sqrt{7.13.17.19}$
3 3 2 0	000+	$+4.3\sqrt{5}/\sqrt{7.13.17.19}$
3 3 3 0	000+	$+4.5\sqrt{2}/\sqrt{7.13.17.19}$
$\tilde{1}$ 2 1 0	000+	0
$\tilde{1}$ 2 2 0	000−	0
$\tilde{1}$ 3 1 0	000−	$-3\sqrt{11}/2\sqrt{17.19}$
$\tilde{1}$ 3 2 0	000+	$-\sqrt{3.5}/2\sqrt{17.19}$
$\tilde{1}$ 3 3 0	000−	0

8 $\frac{13}{2}$ $\frac{3}{2}$		
2 $\frac{1}{2}$ $\frac{3}{2}$ 0	000+	$+1/\sqrt{7}$
2 $\frac{1}{2}$ $\frac{3}{2}$ 0	100+	0
2 $\frac{3}{2}$ $\frac{3}{2}$ 0	000−	$-2/\sqrt{5.7}$
2 $\frac{3}{2}$ $\frac{3}{2}$ 0	100−	0
2 $\frac{5}{2}$ $\frac{3}{2}$ 0	000+	$-\sqrt{3.7.11}/4.17\sqrt{5}$
2 $\frac{5}{2}$ $\frac{3}{2}$ 0	100+	$-\sqrt{3.5.13}/2.17$
2 $\frac{5}{2}$ $\frac{3}{2}$ 1	000−	$+9.3\sqrt{11}/4.17\sqrt{2.5.7}$
2 $\frac{5}{2}$ $\frac{3}{2}$ 1	100−	$-\sqrt{5.13}/2.17\sqrt{2}$
2 $\tilde{\frac{1}{2}}$ $\frac{3}{2}$ 0	000−	$-\sqrt{11.13}/4.17\sqrt{2.7}$
2 $\tilde{\frac{1}{2}}$ $\frac{3}{2}$ 0	100−	$+3.5/2.17\sqrt{2}$
3 $\frac{3}{2}$ $\frac{3}{2}$ 0	000+	$-3/\sqrt{5.17}$

SO_3–K 3jm Factors (cont.)

$\mathbf{8\ \tfrac{13}{2}\ \tfrac{3}{2}}$

$3\ \tfrac{5}{2}\ \tfrac{3}{2}\ 0\ 000-\quad 0$

$3\ \tfrac{5}{2}\ \tfrac{3}{2}\ 1\ 000+\quad -\sqrt{11}/\sqrt{5.17}$

$\tilde{1}\ \tfrac{3}{2}\ \tfrac{3}{2}\ 0\ 000+\quad -\sqrt{3.13}/\sqrt{5.7.17}$

$\tilde{1}\ \tfrac{5}{2}\ \tfrac{3}{2}\ 0\ 000-\quad -3\sqrt{11.13}/4\sqrt{2.5.7.17}$

$\tilde{1}\ \tilde{\tfrac{1}{2}}\ \tfrac{3}{2}\ 0\ 000+\quad -\sqrt{3.5.11}/4\sqrt{2.7.17}$

$\mathbf{8\ \tfrac{13}{2}\ \tfrac{5}{2}}$

$2\ \tfrac{1}{2}\ \tfrac{5}{2}\ 0\ 000+\quad +\sqrt{3}/\sqrt{7.13}$

$2\ \tfrac{1}{2}\ \tfrac{5}{2}\ 0\ 100+\quad 0$

$2\ \tfrac{3}{2}\ \tfrac{5}{2}\ 0\ 000-\quad -31\sqrt{3}/17\sqrt{2.5.7.13}$

$2\ \tfrac{3}{2}\ \tfrac{5}{2}\ 0\ 100-\quad +\sqrt{2.5.11}/17\sqrt{3}$

$2\ \tfrac{3}{2}\ \tfrac{5}{2}\ 1\ 000+\quad -241/2.17\sqrt{5.7.13}$

$2\ \tfrac{3}{2}\ \tfrac{5}{2}\ 1\ 100+\quad -\sqrt{5.11}/3.17$

$2\ \tfrac{5}{2}\ \tfrac{5}{2}\ 0\ 000+\quad -43\sqrt{3.11}/2.17\sqrt{2.5.7.13}$

$2\ \tfrac{5}{2}\ \tfrac{5}{2}\ 0\ 100+\quad +\sqrt{5}/17\sqrt{2.3}$

$2\ \tfrac{5}{2}\ \tfrac{5}{2}\ 1\ 000-\quad +5\sqrt{5.11}/2.17\sqrt{7.13}$

$2\ \tfrac{5}{2}\ \tfrac{5}{2}\ 1\ 100-\quad -5\sqrt{5}/2.17$

$2\ \tfrac{5}{2}\ \tfrac{5}{2}\ 2\ 000+\quad -67\sqrt{11}/2.17\sqrt{2.5.7.13}$

$2\ \tfrac{5}{2}\ \tfrac{5}{2}\ 2\ 100+\quad -\sqrt{2.5}/3.17$

$2\ \tilde{\tfrac{1}{2}}\ \tfrac{5}{2}\ 0\ 000-\quad +2\sqrt{11}/17\sqrt{7}$

$2\ \tilde{\tfrac{1}{2}}\ \tfrac{5}{2}\ 0\ 100-\quad +5\sqrt{13}/2.3.17$

$3\ \tfrac{1}{2}\ \tfrac{5}{2}\ 0\ 000-\quad -\sqrt{3.5}/\sqrt{13.17}$

$3\ \tfrac{3}{2}\ \tfrac{5}{2}\ 0\ 000+\quad -3\sqrt{3}/2\sqrt{2.13.17}$

$3\ \tfrac{3}{2}\ \tfrac{5}{2}\ 1\ 000-\quad +7/2.3\sqrt{2.5.13.17}$

$3\ \tfrac{5}{2}\ \tfrac{5}{2}\ 0\ 000-\quad +7\sqrt{11}/2\sqrt{2.5.13.17}$

$3\ \tfrac{5}{2}\ \tfrac{5}{2}\ 1\ 000+\quad -\sqrt{11}/2.3\sqrt{2.13.17}$

$3\ \tilde{\tfrac{1}{2}}\ \tfrac{5}{2}\ 0\ 000+\quad -\sqrt{5.11}/2.3\sqrt{17}$

$\tilde{1}\ \tfrac{1}{2}\ \tfrac{5}{2}\ 0\ 000-\quad -\sqrt{5}/\sqrt{7.17}$

$\tilde{1}\ \tfrac{3}{2}\ \tfrac{5}{2}\ 0\ 000+\quad +1/2\sqrt{5.7.17}$

$\tilde{1}\ \tfrac{5}{2}\ \tfrac{5}{2}\ 0\ 000-\quad -13\sqrt{11}/2\sqrt{2.3.5.7.17}$

$\tilde{1}\ \tfrac{5}{2}\ \tfrac{5}{2}\ 1\ 000-\quad -\sqrt{11}/2\sqrt{2.3.7.17}$

$\mathbf{8\ \tfrac{13}{2}\ \tfrac{7}{2}}$

$2\ \tfrac{1}{2}\ \tfrac{5}{2}\ 0\ 000-\quad +11\sqrt{3.5}/2.17\sqrt{2.7.13.19}$

$2\ \tfrac{1}{2}\ \tfrac{5}{2}\ 0\ 100-\quad +7\sqrt{5.11}/17\sqrt{2.3.19}$

$2\ \tfrac{3}{2}\ \tfrac{5}{2}\ 0\ 000+\quad +269\sqrt{3}/4.17\sqrt{7.13.19}$

$2\ \tfrac{3}{2}\ \tfrac{5}{2}\ 0\ 100+\quad -5\sqrt{11}/2.17\sqrt{3.19}$

$2\ \tfrac{3}{2}\ \tfrac{5}{2}\ 1\ 000-\quad -599/4.17\sqrt{2.7.13.19}$

$2\ \tfrac{3}{2}\ \tfrac{5}{2}\ 1\ 100-\quad +11\sqrt{11}/2.17\sqrt{2.19}$

$2\ \tfrac{3}{2}\ \tilde{\tfrac{1}{2}}\ 0\ 000-\quad -53\sqrt{3}/4.17\sqrt{2.13.19}$

$2\ \tfrac{3}{2}\ \tilde{\tfrac{1}{2}}\ 0\ 100-\quad -3\sqrt{3.7.11}/2.17\sqrt{2.19}$

$2\ \tfrac{5}{2}\ \tfrac{5}{2}\ 0\ 000-\quad -3.43\sqrt{3.11}/4.17\sqrt{7.13.19}$

$2\ \tfrac{5}{2}\ \tfrac{5}{2}\ 0\ 100-\quad +5\sqrt{3}/2.17\sqrt{19}$

$2\ \tfrac{5}{2}\ \tfrac{5}{2}\ 1\ 000+\quad +97\sqrt{11}/17\sqrt{2.7.13.19}$

$2\ \tfrac{5}{2}\ \tfrac{5}{2}\ 1\ 100+\quad +11\sqrt{2}/17\sqrt{19}$

$2\ \tfrac{5}{2}\ \tfrac{5}{2}\ 2\ 000-\quad +23\sqrt{11}/4.17\sqrt{7.13.19}$

$2\ \tfrac{5}{2}\ \tfrac{5}{2}\ 2\ 100-\quad -3/2.17\sqrt{19}$

$2\ \tfrac{5}{2}\ \tilde{\tfrac{1}{2}}\ 0\ 000+\quad +7\sqrt{3.11}/2.17\sqrt{2.13.19}$

$2\ \tfrac{5}{2}\ \tilde{\tfrac{1}{2}}\ 0\ 100+\quad -\sqrt{7.19}/17\sqrt{2.3}$

$2\ \tilde{\tfrac{1}{2}}\ \tfrac{5}{2}\ 0\ 000+\quad +\sqrt{5.11.19}/2.17\sqrt{2.7}$

$2\ \tilde{\tfrac{1}{2}}\ \tfrac{5}{2}\ 0\ 100+\quad -\sqrt{5.13}/17\sqrt{2.19}$

$\mathbf{8\ \tfrac{13}{2}\ \tfrac{7}{2}}$

$3\ \tfrac{1}{2}\ \tfrac{5}{2}\ 0\ 000+\quad -31/\sqrt{2.3.13.17.19}$

$3\ \tfrac{1}{2}\ \tilde{\tfrac{1}{2}}\ 0\ 000-\quad -\sqrt{3.7}/\sqrt{2.13.17.19}$

$3\ \tfrac{3}{2}\ \tfrac{5}{2}\ 0\ 000-\quad -\sqrt{5.17}/2\sqrt{3.13.19}$

$3\ \tfrac{3}{2}\ \tfrac{5}{2}\ 1\ 000+\quad +3.7/2\sqrt{13.17.19}$

$3\ \tfrac{5}{2}\ \tfrac{5}{2}\ 0\ 000+\quad -7\sqrt{11}/2\sqrt{13.17.19}$

$3\ \tfrac{5}{2}\ \tfrac{5}{2}\ 1\ 000-\quad +3\sqrt{5.11}/2\sqrt{13.17.19}$

$3\ \tfrac{5}{2}\ \tilde{\tfrac{1}{2}}\ 0\ 000-\quad -\sqrt{2.5.7.11}/\sqrt{3.13.17.19}$

$3\ \tilde{\tfrac{1}{2}}\ \tfrac{5}{2}\ 0\ 000-\quad +\sqrt{11}/\sqrt{2.17.19}$

$\tilde{1}\ \tfrac{1}{2}\ \tfrac{5}{2}\ 0\ 000+\quad +\sqrt{19}/2\sqrt{2.7.17}$

$\tilde{1}\ \tfrac{3}{2}\ \tfrac{5}{2}\ 0\ 000-\quad +83/4\sqrt{2.7.17.19}$

$\tilde{1}\ \tfrac{3}{2}\ \tilde{\tfrac{1}{2}}\ 0\ 000+\quad -\sqrt{5}/4\sqrt{2.17.19}$

$\tilde{1}\ \tfrac{5}{2}\ \tfrac{5}{2}\ 0\ 000+\quad +\sqrt{3.11}/4\sqrt{7.17.19}$

$\tilde{1}\ \tfrac{5}{2}\ \tfrac{5}{2}\ 1\ 000+\quad -\sqrt{3.5.11}/4\sqrt{7.17.19}$

$\tilde{1}\ \tilde{\tfrac{1}{2}}\ \tilde{\tfrac{1}{2}}\ 0\ 000+\quad -\sqrt{11.13}/2\sqrt{2.17.19}$

$\mathbf{8\ \tfrac{13}{2}\ \tfrac{9}{2}}$

$2\ \tfrac{1}{2}\ \tfrac{3}{2}\ 0\ 000-\quad -2\sqrt{2.3.5.11}/17\sqrt{7.13.19}$

$2\ \tfrac{1}{2}\ \tfrac{3}{2}\ 0\ 100-\quad -7\sqrt{2.3}/17\sqrt{5.19}$

$2\ \tfrac{1}{2}\ \tfrac{5}{2}\ 0\ 000+\quad +\sqrt{3.5.11}/17\sqrt{13.19}$

$2\ \tfrac{1}{2}\ \tfrac{5}{2}\ 0\ 100+\quad -8.4\sqrt{7}/17\sqrt{3.5.19}$

$2\ \tfrac{3}{2}\ \tfrac{3}{2}\ 0\ 000+\quad +11\sqrt{2.3.11}/17\sqrt{7.13.19}$

$2\ \tfrac{3}{2}\ \tfrac{3}{2}\ 0\ 100+\quad -3.7\sqrt{2.3}/5.17\sqrt{19}$

$2\ \tfrac{3}{2}\ \tfrac{5}{2}\ 0\ 000-\quad -5\sqrt{3.11}/17\sqrt{2.13.19}$

$2\ \tfrac{3}{2}\ \tfrac{5}{2}\ 0\ 100-\quad -2.11\sqrt{2.7}/5.17\sqrt{3.19}$

$2\ \tfrac{3}{2}\ \tfrac{5}{2}\ 1\ 000+\quad -\sqrt{11.19}/2.17\sqrt{13}$

$2\ \tfrac{3}{2}\ \tfrac{5}{2}\ 1\ 100+\quad -29\sqrt{7}/5.17\sqrt{19}$

$2\ \tfrac{5}{2}\ \tfrac{3}{2}\ 0\ 000-\quad +3\sqrt{7.13}/17\sqrt{2.19}$

$2\ \tfrac{5}{2}\ \tfrac{3}{2}\ 0\ 100-\quad -8.4.3\sqrt{2}/5.17\sqrt{11.19}$

$2\ \tfrac{5}{2}\ \tfrac{3}{2}\ 1\ 000+\quad -127\sqrt{3}/2.17\sqrt{7.13.19}$

$2\ \tfrac{5}{2}\ \tfrac{3}{2}\ 1\ 100+\quad -229/5.17\sqrt{3.11.19}$

$2\ \tfrac{5}{2}\ \tfrac{5}{2}\ 0\ 000+\quad -3.11\sqrt{3}/2.17\sqrt{2.13.19}$

$2\ \tfrac{5}{2}\ \tfrac{5}{2}\ 0\ 100+\quad +23\sqrt{3.7}/5.17\sqrt{2.11.19}$

$2\ \tfrac{5}{2}\ \tfrac{5}{2}\ 1\ 000-\quad +31/2.17\sqrt{13.19}$

$2\ \tfrac{5}{2}\ \tfrac{5}{2}\ 1\ 100-\quad +37\sqrt{7}/2.17\sqrt{11.19}$

$2\ \tfrac{5}{2}\ \tfrac{5}{2}\ 2\ 000+\quad -5.37/2.17\sqrt{2.13.19}$

$2\ \tfrac{5}{2}\ \tfrac{5}{2}\ 2\ 100+\quad +2.3\sqrt{2.7}/5.17\sqrt{11.19}$

$2\ \tilde{\tfrac{1}{2}}\ \tfrac{3}{2}\ 0\ 000+\quad +9.3\sqrt{3.5}/2.17\sqrt{7.19}$

$2\ \tilde{\tfrac{1}{2}}\ \tfrac{3}{2}\ 0\ 100+\quad +9\sqrt{3.13}/17\sqrt{5.11.19}$

$2\ \tilde{\tfrac{1}{2}}\ \tfrac{5}{2}\ 0\ 000-\quad +4\sqrt{5}/17\sqrt{19}$

$2\ \tilde{\tfrac{1}{2}}\ \tfrac{5}{2}\ 0\ 100-\quad -23\sqrt{7.13}/2.17\sqrt{5.11.19}$

$3\ \tfrac{1}{2}\ \tfrac{5}{2}\ 0\ 000-\quad -\sqrt{7.11}/\sqrt{3.13.17.19}$

$3\ \tfrac{3}{2}\ \tfrac{3}{2}\ 0\ 000-\quad +3\sqrt{2.3.11}/5\sqrt{13.17.19}$

$3\ \tfrac{3}{2}\ \tfrac{5}{2}\ 0\ 000+\quad -11\sqrt{7.11}/2\sqrt{2.3.5.13.17.19}$

$3\ \tfrac{3}{2}\ \tfrac{5}{2}\ 1\ 000-\quad -3.11\sqrt{7.11}/2.5\sqrt{2.13.17.19}$

$3\ \tfrac{5}{2}\ \tfrac{3}{2}\ 0\ 000+\quad +4.7\sqrt{2}/\sqrt{5.13.17.19}$

$3\ \tfrac{5}{2}\ \tfrac{3}{2}\ 1\ 000-\quad -37\sqrt{2}/5\sqrt{3.13.17.19}$

$3\ \tfrac{5}{2}\ \tfrac{5}{2}\ 0\ 000-\quad -29\sqrt{7}/2.5\sqrt{2.13.17.19}$

$3\ \tfrac{5}{2}\ \tfrac{5}{2}\ 1\ 000+\quad -3\sqrt{5.7}/2\sqrt{2.13.17.19}$

$3\ \tilde{\tfrac{1}{2}}\ \tfrac{5}{2}\ 0\ 000+\quad +\sqrt{7}/2\sqrt{17.19}$

$\tilde{1}\ \tfrac{1}{2}\ \tfrac{5}{2}\ 0\ 000-\quad -\sqrt{11}/\sqrt{17.19}$

SO_3–K 3jm Factors (cont.)

8 $\frac{13}{2}$ $\frac{9}{2}$		
$\tilde{1}\ \frac{3}{2}\ \frac{3}{2}$	0 000−	$-11\sqrt{2.11}/5\sqrt{7.17.19}$
$\tilde{1}\ \frac{3}{2}\ \frac{5}{2}$	0 000+	$-\sqrt{11}/2.5\sqrt{17.19}$
$\tilde{1}\ \frac{5}{2}\ \frac{3}{2}$	0 000+	$+9.3\sqrt{3}/2.5\sqrt{7.17.19}$
$\tilde{1}\ \frac{5}{2}\ \frac{5}{2}$	0 000−	$+31\sqrt{3}/2.5\sqrt{2.17.19}$
$\tilde{1}\ \frac{5}{2}\ \frac{5}{2}$	1 000−	$-13\sqrt{3}/2\sqrt{2.5.17.19}$
$\tilde{1}\ \frac{\tilde{1}}{2}\ \frac{3}{2}$	0 000−	$+\sqrt{13}/2\sqrt{7.17.19}$

8 $\frac{13}{2}$ $\frac{11}{2}$		
$2\ \frac{1}{2}\ \frac{3}{2}$	0 000+	$-3\sqrt{3.5.11}/17\sqrt{13.19}$
$2\ \frac{1}{2}\ \frac{3}{2}$	0 100+	$+4\sqrt{3.5}/17\sqrt{7.19}$
$2\ \frac{1}{2}\ \frac{5}{2}$	0 000−	$+\sqrt{3.5.11}/17\sqrt{13.19}$
$2\ \frac{1}{2}\ \frac{5}{2}$	0 100−	$+2.5\sqrt{3.5}/17\sqrt{7.19}$
$2\ \frac{3}{2}\ \frac{1}{2}$	0 000+	$+11\sqrt{3.7}/17\sqrt{2.13.19}$
$2\ \frac{3}{2}\ \frac{1}{2}$	0 100+	$-2\sqrt{2.11}/17\sqrt{3.19}$
$2\ \frac{3}{2}\ \frac{3}{2}$	0 000−	$+11\sqrt{3.11}/17\sqrt{13.19}$
$2\ \frac{3}{2}\ \frac{3}{2}$	0 100−	$+2.29/17\sqrt{3.7.19}$
$2\ \frac{3}{2}\ \frac{5}{2}$	0 000+	$-5\sqrt{3.11}/17\sqrt{2.13.19}$
$2\ \frac{3}{2}\ \frac{5}{2}$	0 100+	$+4.11\sqrt{2}/17\sqrt{3.7.19}$
$2\ \frac{3}{2}\ \frac{5}{2}$	1 000−	$-4\sqrt{11}/17\sqrt{13.19}$
$2\ \frac{3}{2}\ \frac{5}{2}$	1 100−	$+2.59/3.17\sqrt{7.19}$
$2\ \frac{5}{2}\ \frac{1}{2}$	0 000−	$+31\sqrt{7.11}/4.17\sqrt{13.19}$
$2\ \frac{5}{2}\ \frac{1}{2}$	0 100−	$+23/2.3.17\sqrt{19}$
$2\ \frac{5}{2}\ \frac{3}{2}$	0 000+	$+7.7/4.17\sqrt{13.19}$
$2\ \frac{5}{2}\ \frac{3}{2}$	0 100+	$-829/2.3.17\sqrt{7.11.19}$
$2\ \frac{5}{2}\ \frac{3}{2}$	1 000−	$-3.61\sqrt{3}/4.17\sqrt{2.13.19}$
$2\ \frac{5}{2}\ \frac{3}{2}$	1 100−	$+71\sqrt{3}/2.17\sqrt{2.7.11.19}$
$2\ \frac{5}{2}\ \frac{5}{2}$	0 000−	$+5.11\sqrt{3}/4.17\sqrt{2.13.19}$
$2\ \frac{5}{2}\ \frac{5}{2}$	0 100−	$-23\sqrt{7}/2.17\sqrt{2.3.11.19}$
$2\ \frac{5}{2}\ \frac{5}{2}$	1 000+	$-13\sqrt{13}/4.17\sqrt{19}$
$2\ \frac{5}{2}\ \frac{5}{2}$	1 100+	$+3.13/2.17\sqrt{7.11.19}$
$2\ \frac{5}{2}\ \frac{5}{2}$	2 000−	$-29/4.17\sqrt{2.13.19}$
$2\ \frac{5}{2}\ \frac{5}{2}$	2 100−	$-1217/2.3.17\sqrt{2.7.11.19}$
$2\ \frac{\tilde{1}}{2}\ \frac{3}{2}$	0 000−	$-7\sqrt{3.5}/4.17\sqrt{2.19}$
$2\ \frac{\tilde{1}}{2}\ \frac{3}{2}$	0 100−	$+5.13\sqrt{5.13}/2.17\sqrt{2.3.7.11.19}$
$2\ \frac{\tilde{1}}{2}\ \frac{5}{2}$	0 000+	$-7\sqrt{5}/17\sqrt{19}$
$2\ \frac{\tilde{1}}{2}\ \frac{5}{2}$	0 100+	$-23\sqrt{5.13}/3.17\sqrt{7.11.19}$
$3\ \frac{1}{2}\ \frac{5}{2}$	0 000+	$-2.3\sqrt{3.11}/\sqrt{7.13.17.19}$
$3\ \frac{3}{2}\ \frac{3}{2}$	0 000+	$-\sqrt{11.13}/\sqrt{3.7.17.19}$
$3\ \frac{3}{2}\ \frac{5}{2}$	0 000−	0
$3\ \frac{3}{2}\ \frac{5}{2}$	1 000+	$-2.11\sqrt{2.11}/3\sqrt{7.13.17.19}$
$3\ \frac{5}{2}\ \frac{1}{2}$	0 000+	$-2\sqrt{5.11}/3\sqrt{13.17.19}$
$3\ \frac{5}{2}\ \frac{3}{2}$	0 000−	$-2\sqrt{5.13}/3\sqrt{7.17.19}$
$3\ \frac{5}{2}\ \frac{3}{2}$	1 000+	$+29/\sqrt{3.7.13.17.19}$
$3\ \frac{5}{2}\ \frac{5}{2}$	0 000+	$-3\sqrt{13}/\sqrt{2.7.17.19}$
$3\ \frac{5}{2}\ \frac{5}{2}$	1 000−	$-59\sqrt{5}/3\sqrt{2.7.13.17.19}$
$3\ \frac{\tilde{1}}{2}\ \frac{1}{2}$	0 000−	$+2\sqrt{11}/\sqrt{3.17.19}$
$3\ \frac{\tilde{1}}{2}\ \frac{5}{2}$	0 000−	$+5/3\sqrt{7.17.19}$
$\tilde{1}\ \frac{1}{2}\ \frac{5}{2}$	0 000+	$-\sqrt{11}/\sqrt{17.19}$
$\tilde{1}\ \frac{3}{2}\ \frac{3}{2}$	0 000+	0
$\tilde{1}\ \frac{3}{2}\ \frac{5}{2}$	0 000−	$+\sqrt{11}/\sqrt{17.19}$

8 $\frac{13}{2}$ $\frac{11}{2}$		
$\tilde{1}\ \frac{5}{2}\ \frac{1}{2}$	0 000+	$+\sqrt{5.7.11}/4\sqrt{3.17.19}$
$\tilde{1}\ \frac{5}{2}\ \frac{3}{2}$	0 000−	$-5.7/4\sqrt{2.3.17.19}$
$\tilde{1}\ \frac{5}{2}\ \frac{5}{2}$	0 000+	$+13/4\sqrt{2.3.17.19}$
$\tilde{1}\ \frac{5}{2}\ \frac{5}{2}$	1 000+	$+13\sqrt{5}/4\sqrt{2.3.17.19}$
$\tilde{1}\ \frac{\tilde{1}}{2}\ \frac{3}{2}$	0 000+	$+3\sqrt{13}/4\sqrt{2.17.19}$

8 $\frac{13}{2}$ $\frac{13}{2}$		
$2\ \frac{3}{2}\ \frac{1}{2}$	0 000−	$+4.11\sqrt{11}/17\sqrt{5.13.19}$
$2\ \frac{3}{2}\ \frac{1}{2}$	0 100−	$+3\sqrt{5}/17\sqrt{7.19}$
$2\ \frac{3}{2}\ \frac{3}{2}$	0 000+	$+2.3.11\sqrt{11}/5.17\sqrt{13.19}$
$2\ \frac{3}{2}\ \frac{3}{2}$	0 100+	$-3\sqrt{7}/17\sqrt{19}$
$2\ \frac{5}{2}\ \frac{1}{2}$	0 000+	$-11\sqrt{3}/17\sqrt{2.5.13.19}$
$2\ \frac{5}{2}\ \frac{1}{2}$	0 100+	$-\sqrt{2.3.5.11}/17\sqrt{7.19}$
$2\ \frac{5}{2}\ \frac{3}{2}$	0 000−	$-73\sqrt{3}/5.17\sqrt{13.19}$
$2\ \frac{5}{2}\ \frac{3}{2}$	0 100−	$-4\sqrt{3.11}/17\sqrt{7.19}$
$2\ \frac{5}{2}\ \frac{3}{2}$	1 000+	$+3.71/5.17\sqrt{2.13.19}$
$2\ \frac{5}{2}\ \frac{3}{2}$	1 100+	$-2.5\sqrt{2.11}/17\sqrt{7.19}$
$2\ \frac{5}{2}\ \frac{5}{2}$	0 000+	$+37\sqrt{3.11}/5.17\sqrt{13.19}$
$2\ \frac{5}{2}\ \frac{5}{2}$	0 100+	$+9\sqrt{3.7}/2.17\sqrt{19}$
$2\ \frac{5}{2}\ \frac{5}{2}$	1 000−	0
$2\ \frac{5}{2}\ \frac{5}{2}$	1 100−	0
$2\ \frac{5}{2}\ \frac{5}{2}$	2 000+	$+3.31\sqrt{11}/5.17\sqrt{13.19}$
$2\ \frac{5}{2}\ \frac{5}{2}$	2 100+	$+3.5/2.17\sqrt{7.19}$
$2\ \frac{\tilde{1}}{2}\ \frac{3}{2}$	0 000+	$-1/\sqrt{2.5.19}$
$2\ \frac{\tilde{1}}{2}\ \frac{3}{2}$	0 100+	0
$2\ \frac{\tilde{1}}{2}\ \frac{5}{2}$	0 000−	$-3\sqrt{2.11}/17\sqrt{5.19}$
$2\ \frac{\tilde{1}}{2}\ \frac{5}{2}$	0 100−	$+5\sqrt{5.13}/17\sqrt{2.7.19}$
$3\ \frac{3}{2}\ \frac{3}{2}$	0 000−	0
$3\ \frac{5}{2}\ \frac{1}{2}$	0 000−	$+31\sqrt{3}/\sqrt{2.7.13.17.19}$
$3\ \frac{5}{2}\ \frac{3}{2}$	0 000+	$-2\sqrt{3.13}/\sqrt{5.7.17.19}$
$3\ \frac{5}{2}\ \frac{3}{2}$	1 000−	$-4/\sqrt{7.13.17.19}$
$3\ \frac{5}{2}\ \frac{5}{2}$	0 000−	0
$3\ \frac{5}{2}\ \frac{5}{2}$	1 000+	$+4.3\sqrt{11}/\sqrt{5.7.13.17.19}$
$3\ \frac{\tilde{1}}{2}\ \frac{1}{2}$	0 000+	$-3\sqrt{5}/\sqrt{2.7.17.19}$
$3\ \frac{\tilde{1}}{2}\ \frac{5}{2}$	0 000+	$+2\sqrt{2.11}/\sqrt{7.17.19}$
$\tilde{1}\ \frac{3}{2}\ \frac{3}{2}$	0 000−	0
$\tilde{1}\ \frac{5}{2}\ \frac{1}{2}$	0 000−	$+3/\sqrt{2.17.19}$
$\tilde{1}\ \frac{5}{2}\ \frac{3}{2}$	0 000+	$+3/\sqrt{2.17.19}$
$\tilde{1}\ \frac{5}{2}\ \frac{5}{2}$	0 000−	0
$\tilde{1}\ \frac{5}{2}\ \frac{5}{2}$	1 000−	0
$\tilde{1}\ \frac{\tilde{1}}{2}\ \frac{3}{2}$	0 000−	$-\sqrt{3.13}/\sqrt{2.17.19}$
$\tilde{1}\ \frac{\tilde{1}}{2}\ \frac{\tilde{1}}{2}$	0 000−	0

8 7 1		
2 1 1	0 000+	$+1/\sqrt{5}$
2 1 1	0 100+	0
2 2 1	0 000−	$-1/2\sqrt{3}$
2 2 1	0 100−	0
2 3 1	0 000+	$+\sqrt{11.13}/2.17\sqrt{2.3.5}$
2 3 1	0 100+	$-\sqrt{3.5.7}/17\sqrt{2}$
2 $\tilde{1}$ 1	0 000+	$+\sqrt{7.11}/2.17\sqrt{2.5}$

SO_3–K 3jm Factors (cont.)

8 7 1	
2 1̃ 1 0 100+	+√5.13/17√2
3 2 1 0 000+	+√3.7/2√2.17
3 3 1 0 000−	0
3 1̃ 1 0 000−	−√11/2√2.17
1̃ 2 1 0 000+	+√13/2√2.17
1̃ 3 1 0 000−	−√11/2√2.17
8 7 2	
2 1 2 0 000+	−√3/√5.7
2 1 2 0 100+	0
2 2 2 0 000−	−13√3.5/4.17√2.7
2 2 2 0 100−	+√5.11.13/2.17√2.3
2 2 2 1 000−	+139/4.17√2.3.7
2 2 2 1 100−	+5√11.13/2.3.17√2.3
2 3 2 0 000+	+3√3.11.13/4.17√5.7
2 3 2 0 100+	+7√5/2.17√3
2 3 2 1 000−	+√5.11.13/4.17√3.7
2 3 2 1 100−	−11√5/2.3.17√3
2 1̃ 2 0 000+	+9√11/4.17√5
2 1̃ 2 0 100+	−√5.7.13/2.3.17
3 1 2 0 000−	−√2.3/√5.17
3 2 2 0 000+	+√3/4√17
3 2 2 1 000−	+11/4.3√3.17
3 3 2 0 000−	−√2.11.13/3√3.5.17
3 1̃ 2 0 000−	−√7.11/2.3√2.5.17
1̃ 1 2 0 000−	−√2.13/√5.7.17
1̃ 2 2 0 000+	+√13/4√7.17
1̃ 3 2 0 000−	+3√11/2√2.5.7.17
1̃ 1̃ 2 0 000−	−√11.13/4√5.17
8 7 3	
2 1 3 0 000+	+23√3.7/2.17√5.13.19
2 1 3 0 100+	−√3.5.11/17√19
2 1 1̃ 0 000+	+2.9.9/17√5.7.13.19
2 1 1̃ 0 100+	+2√5.11/17√19
2 2 3 0 000−	+107√3/17√2.7.13.19
2 2 3 0 100−	+5√11/2.17√2.3.19
2 2 3 1 000+	−5.11/17√2.3.7.13.19
2 2 3 1 100+	+5.23√11/2.3.17√2.3.19
2 2 1̃ 0 000−	+541/2.17√2.7.13.19
2 2 1̃ 0 100−	−5√11/3.17√2.19
2 3 3 0 000+	−√11.19/17√3.5.7
2 3 3 0 100+	+8√5.13/3.17√3.19
2 3 1̃ 0 000+	−2√11.19/17√5.7
2 3 1̃ 0 100+	−√5.13/3.17√19
2 1̃ 3 0 000+	−3.7√11/17√5.13.19
2 1̃ 3 0 100+	−√5.7.19/2.3.17
2 1̃ 1̃ 0 000+	−3.43√11/2.17√2.5.13.19
2 1̃ 1̃ 0 100+	+√5.7/17√2.19
3 1 3 0 000−	+2√3/√13.17.19
3 1 1̃ 0 000−	+√17/√13.19
8 7 3	
3 2 3 0 000+	+29/3√3.13.17.19
3 2 1̃ 0 000+	+4.7/3√13.17.19
3 3 3 0 000+	+8√11/3√3.17.19
3 3 1̃ 0 000−	+√11/3√17.19
3 1̃ 3 0 000−	−5√7.11/3√13.17.19
1̃ 1 3 0 000−	+√19/2√7.17
1̃ 2 3 0 000+	−2.3/√7.17.19
1̃ 2 1̃ 0 000+	+31/2√2.7.17.19
1̃ 3 3 0 000−	−√11.13/√7.17.19
1̃ 1̃ 1̃ 0 000−	+√11/2√2.17.19
8 7 4	
2 1 2 0 000+	−2.11√2.5/17√3.7.13.19
2 1 2 0 100+	−7√2.11/17√3.5.19
2 1 3 0 000−	−11√5/2.17√3.13.19
2 1 3 0 100−	+11√7.11/17√3.5.19
2 2 2 0 000−	+√3.5.13/4.17√7.19
2 2 2 0 100−	−√5.11/2.17√3.19
2 2 2 1 000−	+5.47√13/4.3.17√3.7.19
2 2 2 1 100−	−11√11/2.17√3.19
2 2 3 0 000+	−5.11/17√2.3.13.19
2 2 3 0 100+	−7√7.11/2.17√2.3.19
2 2 3 1 000−	+5.43/3.17√2.3.13.19
2 2 3 1 100−	+5√7.11/2.17√2.3.19
2 3 2 0 000+	+11√5.11/17√2.3.7.19
2 3 2 0 100+	−7√3.13/17√2.5.19
2 3 2 1 000−	+5.13√5.11/3.17√2.3.7.19
2 3 2 1 100−	+√5.13/17√2.3.19
2 3 3 0 000−	+√5.11/3.17√3.19
2 3 3 0 100−	+4√7.13/17√3.5.19
2 1̃ 2 0 000+	+73√5.11/2.3.17√2.13.19
2 1̃ 2 0 100+	+47√7/3.17√2.5.19
2 1̃ 3 0 000−	+7√5.7.11/3.17√13.19
2 1̃ 3 0 100−	−7√19/2.3.17√5
3 1 2 0 000−	−43/√3.5.13.17.19
3 1 3 0 000+	+2√7/√3.13.17.19
3 2 2 0 000+	−5/√2.3.13.17.19
3 2 2 1 000−	+5.5/√2.3.13.17.19
3 2 3 0 000−	−√7.13/√3.17.19
3 3 2 0 000−	+√3.11/√5.17.19
3 3 3 0 000−	0
3 1̃ 2 0 000−	+11√7.11/3√5.13.17.19
3 1̃ 3 0 000+	−7√11/3√13.17.19
1̃ 1 2 0 000−	+2/√5.7.17.19
1̃ 1 3 0 000+	−7/2√17.19
1̃ 2 2 0 000+	−5.23/2.3√2.7.17.19
1̃ 2 3 0 000−	+4/3√17.19
1̃ 3 2 0 000−	+√11.13/3√5.7.17.19
1̃ 3 3 0 000+	−√11.13/3√17.19
1̃ 1̃ 2 0 000−	+√11/2√2.5.17.19

SO_3–K 3jm Factors (cont.)

8 7 5

2 1 1 0 000+ $-3\sqrt{3.5.11}/17\sqrt{2.13.19}$

2 1 1 0 100+ $-\sqrt{2.3.7}/17\sqrt{5.19}$

2 1 2 0 000− $-\sqrt{3.5.7.11}/17\sqrt{2.13.19}$

2 1 2 0 100− $+9\sqrt{2.3}/17\sqrt{5.19}$

2 1 $\tilde{1}$ 0 000+ 0

2 1 $\tilde{1}$ 0 100+ $-2\sqrt{2}/\sqrt{5.19}$

2 2 1 0 000− $+5.5\sqrt{11}/17\sqrt{2.13.19}$

2 2 1 0 100− $-7\sqrt{7}/3.17\sqrt{2.19}$

2 2 2 0 000+ $-\sqrt{3.5.7.11}/17\sqrt{13.19}$

2 2 2 0 100+ $-3\sqrt{3.5}/2.17\sqrt{19}$

2 2 2 1 000+ $+5\sqrt{7.11}/17\sqrt{3.13.19}$

2 2 2 1 100+ $+113/2.3.17\sqrt{3.19}$

2 2 $\tilde{1}$ 0 000− $+5\sqrt{7.11}/4.17\sqrt{13.19}$

2 2 $\tilde{1}$ 0 100− $+109/2.3.17\sqrt{19}$

2 3 1 0 000+ $-5.5\sqrt{5}/4.17\sqrt{19}$

2 3 1 0 100+ $-59\sqrt{7.13}/2.3.17\sqrt{5.11.19}$

2 3 2 0 000− $+\sqrt{3.5.7}/4.17\sqrt{2.19}$

2 3 2 0 100− $-163\sqrt{13}/2.17\sqrt{2.3.5.11.19}$

2 3 2 1 000+ $-5.5\sqrt{5.7}/4.17\sqrt{2.3.19}$

2 3 2 1 100+ $-\sqrt{5.13}/2.3.17\sqrt{2.3.11.19}$

2 3 $\tilde{1}$ 0 000+ $-2\sqrt{2.5.7}/17\sqrt{19}$

2 3 $\tilde{1}$ 0 100+ $+23\sqrt{2.13}/3.17\sqrt{5.11.19}$

2 $\tilde{1}$ 1 0 000+ $-\sqrt{3.5.7.13}/2.17\sqrt{19}$

2 $\tilde{1}$ 1 0 100+ $+8.4\sqrt{3}/17\sqrt{5.11.19}$

2 $\tilde{1}$ 2 0 000− $+2.3\sqrt{2.5}/17\sqrt{13.19}$

2 $\tilde{1}$ 2 0 100− $+29\sqrt{2.7}/3.17\sqrt{5.11.19}$

2 $\tilde{1}$ $\tilde{1}$ 0 000+ $-3.11\sqrt{5}/4.17\sqrt{13.19}$

2 $\tilde{1}$ $\tilde{1}$ 0 100+ $+23\sqrt{7}/2.17\sqrt{5.11.19}$

3 1 2 0 000+ $-4\sqrt{3.11}/\sqrt{5.13.17.19}$

3 1 $\tilde{1}$ 0 000− $-\sqrt{2.11}/\sqrt{13.17.19}$

3 2 1 0 000+ $+\sqrt{7.11}/4.3\sqrt{13.17.19}$

3 2 2 0 000− $-5\sqrt{11.13}/4\sqrt{2.3.17.19}$

3 2 2 1 000+ $-5.5\sqrt{11}/4.3\sqrt{2.3.13.17.19}$

3 2 $\tilde{1}$ 0 000+ $+4\sqrt{2.11}/3\sqrt{13.17.19}$

3 3 1 0 000− $-4\sqrt{7}/3\sqrt{17.19}$

3 3 2 0 000+ $+8.2/3\sqrt{3.5.17.19}$

3 3 $\tilde{1}$ 0 000− $-\sqrt{2}/3\sqrt{17.19}$

3 $\tilde{1}$ 1 0 000− $-3\sqrt{3}/4\sqrt{13.17.19}$

3 $\tilde{1}$ 2 0 000+ $-7\sqrt{7.17}/4.3\sqrt{5.13.19}$

$\tilde{1}$ 1 2 0 000+ $-\sqrt{7.11}/\sqrt{5.17.19}$

$\tilde{1}$ 2 1 0 000+ $+\sqrt{3.11}/2\sqrt{17.19}$

$\tilde{1}$ 2 2 0 000− 0

$\tilde{1}$ 2 $\tilde{1}$ 0 000+ $-\sqrt{7.11}/4\sqrt{17.19}$

$\tilde{1}$ 3 1 0 000− $+\sqrt{3.13}/4\sqrt{17.19}$

$\tilde{1}$ 3 2 0 000+ $+\sqrt{7.13}/4\sqrt{5.17.19}$

$\tilde{1}$ $\tilde{1}$ 2 0 000+ $+2.3\sqrt{2}/\sqrt{5.17.19}$

$\tilde{1}$ $\tilde{1}$ $\tilde{1}$ 0 000− $+13/4\sqrt{17.19}$

8 7 6

2 1 1 0 000− $-3.11\sqrt{3.11}/17\sqrt{2.5.13.19}$

2 1 1 0 100− $+\sqrt{2.3.5}/17\sqrt{7.19}$

2 1 2 0 000+ $-11\sqrt{3}/17\sqrt{2.5.13.19}$

2 1 2 0 100+ $-\sqrt{2.3.5.11}/17\sqrt{7.19}$

2 1 3 0 000− $+11\sqrt{2.3}/17\sqrt{5.13.19}$

2 1 3 0 100− $+2\sqrt{2.3.5.11}/17\sqrt{7.19}$

2 2 0 0 000− $+11\sqrt{3.7.11}/2.17\sqrt{2.13.19}$

2 2 0 0 100− $-5/17\sqrt{2.3.19}$

2 2 1 0 000+ $+11\sqrt{2.11}/17\sqrt{13.19}$

2 2 1 0 100+ $+5.13/3.17\sqrt{2.7.19}$

2 2 2 0 000− $+3.5\sqrt{3.5}/4.17\sqrt{13.19}$

2 2 2 0 100− $+11\sqrt{5.11}/17\sqrt{3.7.19}$

2 2 2 1 000− $-5.29/4.17\sqrt{3.13.19}$

2 2 2 1 100− $-2.5\sqrt{11}/3.17\sqrt{3.7.19}$

2 2 3 0 000+ $+2.11\sqrt{3}/17\sqrt{13.19}$

2 2 3 0 100+ $-5.5\sqrt{11}/17\sqrt{3.7.19}$

2 2 3 1 000− $+1/17\sqrt{3.13.19}$

2 2 3 1 100− $-5\sqrt{11}/3.17\sqrt{3.7.19}$

2 3 1 0 000− $+61/4.17\sqrt{5.19}$

2 3 1 0 100− $-5\sqrt{5.11.13}/2.3.17\sqrt{7.19}$

2 3 2 0 000+ $-31\sqrt{3.11}/4.17\sqrt{2.5.19}$

2 3 2 0 100+ $+3\sqrt{3.5.13}/2.17\sqrt{2.7.19}$

2 3 2 1 000− $-5\sqrt{5.11}/4.17\sqrt{2.3.19}$

2 3 2 1 100− $-41\sqrt{5.13}/2.3.17\sqrt{2.3.7.19}$

2 3 3 0 000− $-8\sqrt{2.11}/17\sqrt{3.5.19}$

2 3 3 0 100− $-2\sqrt{2.5.7.13}/3.17\sqrt{3.19}$

2 $\tilde{1}$ 1 0 000− $+2.7\sqrt{3.7}/17\sqrt{5.13.19}$

2 $\tilde{1}$ 1 0 100− $+\sqrt{3.5.11}/17\sqrt{19}$

2 $\tilde{1}$ 2 0 000+ $-3\sqrt{7.11.13}/2.17\sqrt{2.5.19}$

2 $\tilde{1}$ 2 0 100+ $+\sqrt{5.19}/3.17\sqrt{2}$

2 $\tilde{1}$ 3 0 000− $-9\sqrt{7.11}/17\sqrt{2.5.13.19}$

2 $\tilde{1}$ 3 0 100− $-2.5\sqrt{2.5}/3.17\sqrt{19}$

3 1 2 0 000− $-2.9.3\sqrt{3}/\sqrt{5.7.13.17.19}$

3 1 3 0 000+ $+2.3\sqrt{2.3}/\sqrt{7.13.17.19}$

3 2 1 0 000− $+5.11\sqrt{11}/4.3\sqrt{7.13.17.19}$

3 2 2 0 000+ $-7\sqrt{7}/4\sqrt{2.3.13.17.19}$

3 2 2 1 000− $+\sqrt{7.17}/4.3\sqrt{2.3.13.19}$

3 2 3 0 000− $+4.5.5\sqrt{2}/3\sqrt{3.7.13.17.19}$

3 3 0 0 000− $+2\sqrt{2}/\sqrt{3.17.19}$

3 3 1 0 000+ $+8.4/3\sqrt{7.17.19}$

3 3 2 0 000− $+2.13\sqrt{11}/3\sqrt{3.5.7.17.19}$

3 3 3 0 000− $+4\sqrt{2.11}/3\sqrt{3.7.17.19}$

3 $\tilde{1}$ 1 0 000+ $+\sqrt{3.13}/4\sqrt{17.19}$

3 $\tilde{1}$ 2 0 000− $+11\sqrt{11}/4.3\sqrt{5.13.17.19}$

3 $\tilde{1}$ 3 0 000+ $+8\sqrt{2.11}/3\sqrt{13.17.19}$

$\tilde{1}$ 1 2 0 000− $-7/\sqrt{5.17.19}$

$\tilde{1}$ 1 3 0 000+ $+\sqrt{2}/\sqrt{17.19}$

$\tilde{1}$ 2 1 0 000− 0

$\tilde{1}$ 2 2 0 000+ $-7/2\sqrt{2.17.19}$

$\tilde{1}$ 2 3 0 000− $+5/\sqrt{2.17.19}$

SO_3–K 3jm Factors (cont.)

8 7 6

$\tilde{1}$ 3 1 0 000+ $-\sqrt{3.13}/4\sqrt{17.19}$
$\tilde{1}$ 3 2 0 000− $-3\sqrt{11.13}/4\sqrt{5.17.19}$
$\tilde{1}$ 3 3 0 000+ 0
$\tilde{1}$ $\tilde{1}$ 0 0 000− $+7/2\sqrt{2.17.19}$
$\tilde{1}$ $\tilde{1}$ 2 0 000− $-\sqrt{7.11}/2\sqrt{2.5.17.19}$

8 7 7

2 1 1 0 000+ $-2.11\sqrt{11.23}/5.17\sqrt{13.19}$
2 1 1 0 100+ $-9/17\sqrt{7.19.23}$
2 2 1 0 000− $-11\sqrt{11.23}/17\sqrt{3.5.13.19}$
2 2 1 0 100− $-2.5\sqrt{3.5}/17\sqrt{7.19.23}$
2 2 2 0 000+ $+83\sqrt{5.11}/17\sqrt{2.3.13.19.23}$
2 2 2 0 100+ $+83\sqrt{3.5}/2.17\sqrt{2.7.19.23}$
2 2 2 1 000+ $-199\sqrt{11}/17\sqrt{2.3.13.19.23}$
2 2 2 1 100+ $+3.5\sqrt{3.23}/2.17\sqrt{2.7.19}$
2 3 1 0 000+ $+53\sqrt{2}/5.17\sqrt{3.19.23}$
2 3 1 0 100+ $-4\sqrt{2.3.11.13}/17\sqrt{7.19.23}$
2 3 2 0 000− $-8/17\sqrt{3.5.19.23}$
2 3 2 0 100− $-\sqrt{3.5.11.13}/17\sqrt{7.19.23}$
2 3 2 1 000+ $+2.11\sqrt{5}/17\sqrt{3.19.23}$
2 3 2 1 100+ $+\sqrt{3.5.11.13}/17\sqrt{7.19.23}$
2 3 3 0 000+ $+2\sqrt{2.11.13}/5.17\sqrt{3.19.23}$
2 3 3 0 100+ $+13\sqrt{2.3.7}/17\sqrt{19.23}$
2 $\tilde{1}$ 1 0 000+ $+7\sqrt{7.13}/5.17\sqrt{2.19.23}$
2 $\tilde{1}$ 1 0 100+ $-\sqrt{2.11}/17\sqrt{19.23}$
2 $\tilde{1}$ 2 0 000− $+\sqrt{7}/\sqrt{5.13.19.23}$
2 $\tilde{1}$ 2 0 100− $-\sqrt{5.11}/2\sqrt{19.23}$
2 $\tilde{1}$ 3 0 000+ $+\sqrt{2.7.11.23}/5.17\sqrt{19}$
2 $\tilde{1}$ 3 0 100+ $-5\sqrt{2.13}/17\sqrt{19.23}$
2 $\tilde{1}$ $\tilde{1}$ 0 000+ $+7.37\sqrt{11}/5.17\sqrt{13.19.23}$
2 $\tilde{1}$ $\tilde{1}$ 0 100+ $+9.7\sqrt{7}/2.17\sqrt{19.23}$
3 2 1 0 000+ $-9.3\sqrt{2.3.11}/\sqrt{5.7.13.17.19.23}$
3 2 2 0 000− 0
3 2 2 1 000+ $+2\sqrt{3.7.11}/\sqrt{13.17.19.23}$
3 3 1 0 000− $-\sqrt{2.3.23}/\sqrt{5.7.17.19}$
3 3 2 0 000+ $+4.3\sqrt{2.3}/\sqrt{5.7.17.19.23}$
3 3 3 0 000+ $-4\sqrt{2.3.11.13}/\sqrt{5.7.17.19.23}$
3 $\tilde{1}$ 1 0 000− $-89\sqrt{2}/\sqrt{5.13.17.19.23}$
3 $\tilde{1}$ 2 0 000+ $+59\sqrt{2}/\sqrt{5.13.17.19.23}$
3 $\tilde{1}$ 3 0 000− $+4\sqrt{2.11}/\sqrt{5.17.19.23}$
$\tilde{1}$ 2 1 0 000+ $-7\sqrt{11}/\sqrt{2.5.17.19.23}$
$\tilde{1}$ 2 2 0 000− 0
$\tilde{1}$ 3 1 0 000− $-\sqrt{2.5.13}/\sqrt{17.19.23}$
$\tilde{1}$ 3 2 0 000+ $+9\sqrt{2.13}/\sqrt{5.17.19.23}$
$\tilde{1}$ 3 3 0 000− 0
$\tilde{1}$ $\tilde{1}$ 2 0 000+ $+2.3\sqrt{7}/\sqrt{5.17.19.23}$
$\tilde{1}$ $\tilde{1}$ $\tilde{1}$ 0 000− 0

8 $\frac{15}{2}$ $\frac{1}{2}$

2 $\frac{3}{2}$ $\frac{1}{2}$ 0 000+ $+1/2$
2 $\frac{3}{2}$ $\frac{1}{2}$ 0 100+ 0
2 $\frac{5}{2}$ $\frac{1}{2}$ 0 000− $+\sqrt{3.7.11}/8.17$
2 $\frac{5}{2}$ $\frac{1}{2}$ 0 010− $+3\sqrt{5.13}/8.17$
2 $\frac{5}{2}$ $\frac{1}{2}$ 0 100− $+5\sqrt{3.13}/4.17$
2 $\frac{5}{2}$ $\frac{1}{2}$ 0 110− $-\sqrt{5.7.11}/4.17$
3 $\frac{5}{2}$ $\frac{1}{2}$ 0 000+ $+\sqrt{3.5.11}/8\sqrt{17}$
3 $\frac{5}{2}$ $\frac{1}{2}$ 0 010+ $+\sqrt{7.13}/8\sqrt{17}$
$\tilde{1}$ $\frac{5}{2}$ $\frac{1}{2}$ 0 000+ 0
$\tilde{1}$ $\frac{5}{2}$ $\frac{1}{2}$ 0 010+ $+\sqrt{3}/\sqrt{17}$

8 $\frac{15}{2}$ $\frac{3}{2}$

2 $\frac{3}{2}$ $\frac{3}{2}$ 0 000+ $-\sqrt{3}/2\sqrt{5}$
2 $\frac{3}{2}$ $\frac{3}{2}$ 0 100+ 0
2 $\frac{5}{2}$ $\frac{3}{2}$ 0 000− $+3\sqrt{7.11}/8.17\sqrt{2.5}$
2 $\frac{5}{2}$ $\frac{3}{2}$ 0 010− $-5\sqrt{3.13}/8.17\sqrt{2}$
2 $\frac{5}{2}$ $\frac{3}{2}$ 0 100− $+3\sqrt{5.13}/4.17\sqrt{2}$
2 $\frac{5}{2}$ $\frac{3}{2}$ 0 110− $+5\sqrt{7.11}/4.17\sqrt{2.3}$
2 $\frac{5}{2}$ $\frac{3}{2}$ 1 000+ $+9\sqrt{3.7.11}/8.2.17\sqrt{5}$
2 $\frac{5}{2}$ $\frac{3}{2}$ 1 010+ $+19\sqrt{13}/8.2.17$
2 $\frac{5}{2}$ $\frac{3}{2}$ 1 100+ $-7\sqrt{5.13}/8.17\sqrt{3}$
2 $\frac{5}{2}$ $\frac{3}{2}$ 1 110+ $+5\sqrt{7.11}/8.17$
3 $\frac{3}{2}$ $\frac{3}{2}$ 0 000− $+\sqrt{3.7}/2\sqrt{5.17}$
3 $\frac{5}{2}$ $\frac{3}{2}$ 0 000+ $+5\sqrt{11}/8.2\sqrt{2.17}$
3 $\frac{5}{2}$ $\frac{3}{2}$ 0 010+ $+\sqrt{5.7.13}/8.2\sqrt{2.3.17}$
3 $\frac{5}{2}$ $\frac{3}{2}$ 1 000− $+19\sqrt{11}/8.2\sqrt{2.3.5.17}$
3 $\frac{5}{2}$ $\frac{3}{2}$ 1 010− $-3\sqrt{7.13}/8.2\sqrt{2.17}$
$\tilde{1}$ $\frac{3}{2}$ $\frac{3}{2}$ 0 000− $+\sqrt{13}/2\sqrt{5.17}$
$\tilde{1}$ $\frac{5}{2}$ $\frac{3}{2}$ 0 000+ $-\sqrt{3.7.11.13}/8.2\sqrt{5.17}$
$\tilde{1}$ $\frac{5}{2}$ $\frac{3}{2}$ 0 010+ $+1/8.2\sqrt{17}$

8 $\frac{15}{2}$ $\frac{5}{2}$

2 $\frac{3}{2}$ $\frac{5}{2}$ 0 000+ $-9.3\sqrt{3}/17\sqrt{5.7.19}$
2 $\frac{3}{2}$ $\frac{5}{2}$ 0 100+ $-\sqrt{5.11.13}/17\sqrt{3.19}$
2 $\frac{3}{2}$ $\frac{5}{2}$ 1 000− $-269/2.17\sqrt{2.5.7.19}$
2 $\frac{3}{2}$ $\frac{5}{2}$ 1 100− $+\sqrt{5.11.13}/3.17\sqrt{2.19}$
2 $\frac{5}{2}$ $\frac{5}{2}$ 0 000− $-43\sqrt{3.11}/4.17\sqrt{2.5.19}$
2 $\frac{5}{2}$ $\frac{5}{2}$ 0 010− $-9.5\sqrt{13}/4.17\sqrt{2.7.19}$
2 $\frac{5}{2}$ $\frac{5}{2}$ 0 100− $+\sqrt{5.7.13}/2.17\sqrt{2.3.19}$
2 $\frac{5}{2}$ $\frac{5}{2}$ 0 110− $+3.5\sqrt{11}/2.17\sqrt{2.19}$
2 $\frac{5}{2}$ $\frac{5}{2}$ 1 000+ $-\sqrt{5.11}/8\sqrt{19}$
2 $\frac{5}{2}$ $\frac{5}{2}$ 1 010+ $+11\sqrt{3.13}/8.17\sqrt{7.19}$
2 $\frac{5}{2}$ $\frac{5}{2}$ 1 100+ 0
2 $\frac{5}{2}$ $\frac{5}{2}$ 1 110+ $+2.5\sqrt{11}/17\sqrt{3.19}$
2 $\frac{5}{2}$ $\frac{5}{2}$ 2 000− $+\sqrt{11}/8.17\sqrt{2.5.19}$
2 $\frac{5}{2}$ $\frac{5}{2}$ 2 010− $-3.37\sqrt{3.13}/8.17\sqrt{2.7.19}$
2 $\frac{5}{2}$ $\frac{5}{2}$ 2 100− $-\sqrt{5.7.13.19}/4.3.17\sqrt{2}$
2 $\frac{5}{2}$ $\frac{5}{2}$ 2 110− $-5.5\sqrt{11}/4.17\sqrt{2.3.19}$
3 $\frac{3}{2}$ $\frac{5}{2}$ 0 000− $-\sqrt{3}/2\sqrt{17.19}$
3 $\frac{3}{2}$ $\frac{5}{2}$ 1 000+ $-71/2.3\sqrt{5.17.19}$
3 $\frac{5}{2}$ $\frac{5}{2}$ 0 000+ $+9\sqrt{7.11}/8\sqrt{2.5.17.19}$
3 $\frac{5}{2}$ $\frac{5}{2}$ 0 010+ $-13\sqrt{13}/8\sqrt{2.3.17.19}$
3 $\frac{5}{2}$ $\frac{5}{2}$ 1 000− $+13\sqrt{7.11}/8.3\sqrt{2.17.19}$

SO_3–K 3jm Factors (cont.)

$8\ \frac{15}{2}\ \frac{5}{2}$

$3\ \frac{5}{2}\ \frac{5}{2}\ 1\ 010-\quad -11\sqrt{5.13}/8\sqrt{2.3.17.19}$

$\tilde{1}\ \frac{3}{2}\ \frac{5}{2}\ 0\ 000-\quad +\sqrt{13.19}/2\sqrt{2.5.7.17}$

$\tilde{1}\ \frac{5}{2}\ \frac{5}{2}\ 0\ 000+\quad +\sqrt{11.13.19}/8\sqrt{2.3.5.17}$

$\tilde{1}\ \frac{5}{2}\ \frac{5}{2}\ 0\ 010+\quad +\sqrt{7}/8\sqrt{2.17.19}$

$\tilde{1}\ \frac{5}{2}\ \frac{5}{2}\ 1\ 000+\quad -\sqrt{11.13}/4\sqrt{2.3.17.19}$

$\tilde{1}\ \frac{5}{2}\ \frac{5}{2}\ 1\ 010+\quad +23\sqrt{5}/4\sqrt{2.7.17.19}$

$8\ \frac{15}{2}\ \frac{7}{2}$

$2\ \frac{3}{2}\ \frac{5}{2}\ 0\ 000-\quad +9.9\sqrt{3}/2.17\sqrt{7.13.19}$

$2\ \frac{3}{2}\ \frac{5}{2}\ 0\ 100-\quad +5\sqrt{3.11}/2.17\sqrt{19}$

$2\ \frac{3}{2}\ \frac{5}{2}\ 1\ 000+\quad +29/2.17\sqrt{2.7.13.19}$

$2\ \frac{3}{2}\ \frac{5}{2}\ 1\ 100+\quad +\sqrt{11}/17\sqrt{2.19}$

$2\ \frac{3}{2}\ \tilde{\frac{1}{2}}\ 0\ 000+\quad -9.3\sqrt{3}/2.17\sqrt{2.13.19}$

$2\ \frac{3}{2}\ \tilde{\frac{1}{2}}\ 0\ 100+\quad +\sqrt{2.3.7.11}/17\sqrt{19}$

$2\ \frac{5}{2}\ \frac{5}{2}\ 0\ 000+\quad +3.43\sqrt{3.11}/8.17\sqrt{2.13.19}$

$2\ \frac{5}{2}\ \frac{5}{2}\ 0\ 010+\quad -9.13\sqrt{5}/8.17\sqrt{2.7.19}$

$2\ \frac{5}{2}\ \frac{5}{2}\ 0\ 100+\quad -5\sqrt{3.7}/4.17\sqrt{2.19}$

$2\ \frac{5}{2}\ \frac{5}{2}\ 0\ 110+\quad +3\sqrt{5.11.13}/4.17\sqrt{2.19}$

$2\ \frac{5}{2}\ \frac{5}{2}\ 1\ 000-\quad +131\sqrt{11}/8.17\sqrt{13.19}$

$2\ \frac{5}{2}\ \frac{5}{2}\ 1\ 010-\quad +43\sqrt{3.5}/8.17\sqrt{7.19}$

$2\ \frac{5}{2}\ \frac{5}{2}\ 1\ 100-\quad -7\sqrt{7}/8.2.17\sqrt{19}$

$2\ \frac{5}{2}\ \frac{5}{2}\ 1\ 110-\quad +3\sqrt{3.5.11.13}/8.2.17\sqrt{19}$

$2\ \frac{5}{2}\ \frac{5}{2}\ 2\ 000+\quad -103\sqrt{11}/8.17\sqrt{2.13.19}$

$2\ \frac{5}{2}\ \frac{5}{2}\ 2\ 010+\quad +3.59\sqrt{3.5}/8.17\sqrt{2.7.19}$

$2\ \frac{5}{2}\ \frac{5}{2}\ 2\ 100+\quad -101\sqrt{7}/8.17\sqrt{2.19}$

$2\ \frac{5}{2}\ \frac{5}{2}\ 2\ 110+\quad -5\sqrt{5.11.13}/8.17\sqrt{2.3.19}$

$2\ \frac{5}{2}\ \tilde{\frac{1}{2}}\ 0\ 000-\quad +7\sqrt{3.7.11}/2.17\sqrt{13.19}$

$2\ \frac{5}{2}\ \tilde{\frac{1}{2}}\ 0\ 010-\quad +5\sqrt{5}/2.17\sqrt{19}$

$2\ \frac{5}{2}\ \tilde{\frac{1}{2}}\ 0\ 100-\quad +7\sqrt{19}/8.2.17\sqrt{3}$

$2\ \frac{5}{2}\ \tilde{\frac{1}{2}}\ 0\ 110-\quad +\sqrt{5.7.11.13}/8.2.17\sqrt{19}$

$3\ \frac{3}{2}\ \frac{5}{2}\ 0\ 000+\quad -\sqrt{3.5}/\sqrt{13.17.19}$

$3\ \frac{3}{2}\ \frac{5}{2}\ 1\ 000-\quad +8.2/\sqrt{13.17.19}$

$3\ \frac{5}{2}\ \frac{5}{2}\ 0\ 000-\quad +11\sqrt{7.11}/8\sqrt{2.13.17.19}$

$3\ \frac{5}{2}\ \frac{5}{2}\ 0\ 010-\quad +3\sqrt{3.5}/8\sqrt{2.17.19}$

$3\ \frac{5}{2}\ \frac{5}{2}\ 1\ 000+\quad +11\sqrt{5.7.11}/8\sqrt{2.13.17.19}$

$3\ \frac{5}{2}\ \frac{5}{2}\ 1\ 010+\quad -\sqrt{19}/8\sqrt{2.3.17}$

$3\ \frac{5}{2}\ \tilde{\frac{1}{2}}\ 0\ 000+\quad +7\sqrt{5.11}/8\sqrt{3.13.17.19}$

$3\ \frac{5}{2}\ \tilde{\frac{1}{2}}\ 0\ 010+\quad +13\sqrt{7}/8\sqrt{17.19}$

$\tilde{1}\ \frac{3}{2}\ \frac{5}{2}\ 0\ 000+\quad +3/2\sqrt{2.7.17.19}$

$\tilde{1}\ \frac{3}{2}\ \tilde{\frac{1}{2}}\ 0\ 000-\quad +5\sqrt{5}/2\sqrt{2.17.19}$

$\tilde{1}\ \frac{5}{2}\ \frac{5}{2}\ 0\ 000-\quad +\sqrt{3.11}/8\sqrt{2.17.19}$

$\tilde{1}\ \frac{5}{2}\ \frac{5}{2}\ 0\ 010-\quad -3\sqrt{5.7.13}/8\sqrt{2.17.19}$

$\tilde{1}\ \frac{5}{2}\ \frac{5}{2}\ 1\ 000-\quad +\sqrt{3.5.11}/8\sqrt{2.17.19}$

$\tilde{1}\ \frac{5}{2}\ \frac{5}{2}\ 1\ 010-\quad +23\sqrt{13}/8\sqrt{2.7.17.19}$

$8\ \frac{15}{2}\ \frac{9}{2}$

$2\ \frac{3}{2}\ \frac{3}{2}\ 0\ 000-\quad -3.11\sqrt{3}/2.17\sqrt{13.19}$

$2\ \frac{3}{2}\ \frac{3}{2}\ 0\ 100-\quad -\sqrt{3.7.11}/5.17\sqrt{19}$

$2\ \frac{3}{2}\ \frac{5}{2}\ 0\ 000+\quad 0$

$2\ \frac{3}{2}\ \frac{5}{2}\ 0\ 100+\quad +\sqrt{3.11}/5\sqrt{19}$

$2\ \frac{3}{2}\ \frac{5}{2}\ 1\ 000-\quad +11\sqrt{7}/2.17\sqrt{2.13.19}$

$8\ \frac{15}{2}\ \frac{9}{2}$

$2\ \frac{3}{2}\ \frac{5}{2}\ 1\ 100-\quad -43\sqrt{11}/5.17\sqrt{2.19}$

$2\ \frac{5}{2}\ \frac{3}{2}\ 0\ 000+\quad +3\sqrt{7.11.13}/4.17\sqrt{2.19}$

$2\ \frac{5}{2}\ \frac{3}{2}\ 0\ 010+\quad -\sqrt{3.5}/4\sqrt{2.19}$

$2\ \frac{5}{2}\ \frac{3}{2}\ 0\ 100+\quad -8.3\sqrt{2}/5.17\sqrt{19}$

$2\ \frac{5}{2}\ \frac{3}{2}\ 0\ 110+\quad 0$

$2\ \frac{5}{2}\ \frac{3}{2}\ 1\ 000-\quad +\sqrt{3.7.11}/4\sqrt{13.19}$

$2\ \frac{5}{2}\ \frac{3}{2}\ 1\ 010-\quad +\sqrt{5.19}/4.17$

$2\ \frac{5}{2}\ \frac{3}{2}\ 1\ 100-\quad +7/5\sqrt{3.19}$

$2\ \frac{5}{2}\ \frac{3}{2}\ 1\ 110-\quad -\sqrt{7.11.13}/17\sqrt{5.19}$

$2\ \frac{5}{2}\ \frac{5}{2}\ 0\ 000-\quad -3.11\sqrt{3.11}/8.17\sqrt{2.13.19}$

$2\ \frac{5}{2}\ \frac{5}{2}\ 0\ 010-\quad +9.3\sqrt{5.7}/8.17\sqrt{2.19}$

$2\ \frac{5}{2}\ \frac{5}{2}\ 0\ 100-\quad +23\sqrt{3.7}/4.5.17\sqrt{2.19}$

$2\ \frac{5}{2}\ \frac{5}{2}\ 0\ 110-\quad +3\sqrt{5.11.13}/4.17\sqrt{2.19}$

$2\ \frac{5}{2}\ \frac{5}{2}\ 1\ 000+\quad +\sqrt{11.13}/4.17\sqrt{19}$

$2\ \frac{5}{2}\ \frac{5}{2}\ 1\ 010+\quad -\sqrt{3.5.7}/4.17\sqrt{19}$

$2\ \frac{5}{2}\ \frac{5}{2}\ 1\ 100+\quad -13\sqrt{7}/4.17\sqrt{19}$

$2\ \frac{5}{2}\ \frac{5}{2}\ 1\ 110+\quad -3\sqrt{3.11.13}/4.17\sqrt{5.19}$

$2\ \frac{5}{2}\ \frac{5}{2}\ 2\ 000-\quad +5\sqrt{11}/4\sqrt{2.13.19}$

$2\ \frac{5}{2}\ \frac{5}{2}\ 2\ 010-\quad -9\sqrt{3.5.7}/4.17\sqrt{2.19}$

$2\ \frac{5}{2}\ \frac{5}{2}\ 2\ 100-\quad -\sqrt{7}/5\sqrt{2.19}$

$2\ \frac{5}{2}\ \frac{5}{2}\ 2\ 110-\quad +\sqrt{11.13}/17\sqrt{2.3.5.19}$

$3\ \frac{3}{2}\ \frac{3}{2}\ 0\ 000+\quad +\sqrt{3.7.13}/5\sqrt{17.19}$

$3\ \frac{3}{2}\ \frac{5}{2}\ 0\ 000-\quad -7\sqrt{3}/2\sqrt{5.13.17.19}$

$3\ \frac{3}{2}\ \frac{5}{2}\ 1\ 000+\quad +37/2.5\sqrt{13.17.19}$

$3\ \frac{5}{2}\ \frac{3}{2}\ 0\ 000-\quad -53\sqrt{11}/8\sqrt{2.5.13.17.19}$

$3\ \frac{5}{2}\ \frac{3}{2}\ 0\ 010-\quad +5\sqrt{3.7}/8\sqrt{2.17.19}$

$3\ \frac{5}{2}\ \frac{3}{2}\ 1\ 000+\quad +\sqrt{11.17}/8.5\sqrt{2.3.13.19}$

$3\ \frac{5}{2}\ \frac{3}{2}\ 1\ 010+\quad +31\sqrt{7}/8\sqrt{2.5.17.19}$

$3\ \frac{5}{2}\ \frac{5}{2}\ 0\ 000+\quad +7\sqrt{7.11.13}/8.5\sqrt{2.17.19}$

$3\ \frac{5}{2}\ \frac{5}{2}\ 0\ 010+\quad +3\sqrt{3.17}/8\sqrt{2.5.19}$

$3\ \frac{5}{2}\ \frac{5}{2}\ 1\ 000-\quad -7\sqrt{5.7.11}/8\sqrt{2.13.17.19}$

$3\ \frac{5}{2}\ \frac{5}{2}\ 1\ 010-\quad -7.7/8\sqrt{2.3.17.19}$

$\tilde{1}\ \frac{3}{2}\ \frac{3}{2}\ 0\ 000+\quad +13/2.5\sqrt{17.19}$

$\tilde{1}\ \frac{3}{2}\ \frac{5}{2}\ 0\ 000-\quad -3.7\sqrt{7}/2.5\sqrt{2.17.19}$

$\tilde{1}\ \frac{5}{2}\ \frac{3}{2}\ 0\ 000-\quad +3\sqrt{3.7.11}/4.5\sqrt{17.19}$

$\tilde{1}\ \frac{5}{2}\ \frac{3}{2}\ 0\ 010-\quad +9\sqrt{13}/4\sqrt{5.17.19}$

$\tilde{1}\ \frac{5}{2}\ \frac{5}{2}\ 0\ 000+\quad +13\sqrt{3.11}/4.5\sqrt{2.17.19}$

$\tilde{1}\ \frac{5}{2}\ \frac{5}{2}\ 0\ 010+\quad -3\sqrt{7.13}/4\sqrt{2.5.17.19}$

$\tilde{1}\ \frac{5}{2}\ \frac{5}{2}\ 1\ 000+\quad -13\sqrt{3.11}/8\sqrt{2.5.17.19}$

$\tilde{1}\ \frac{5}{2}\ \frac{5}{2}\ 1\ 010+\quad -\sqrt{7.13}/8\sqrt{2.17.19}$

$8\ \frac{15}{2}\ \frac{11}{2}$

$2\ \frac{3}{2}\ \frac{1}{2}\ 0\ 000-\quad -11\sqrt{3.11}/17\sqrt{2.13.19}$

$2\ \frac{3}{2}\ \frac{1}{2}\ 0\ 100-\quad -\sqrt{7}/17\sqrt{2.3.19}$

$2\ \frac{3}{2}\ \frac{3}{2}\ 0\ 000+\quad -11\sqrt{3.7}/2.17\sqrt{13.19}$

$2\ \frac{3}{2}\ \frac{3}{2}\ 0\ 100+\quad +2\sqrt{11}/17\sqrt{3.19}$

$2\ \frac{3}{2}\ \frac{5}{2}\ 0\ 000-\quad 0$

$2\ \frac{3}{2}\ \frac{5}{2}\ 0\ 100-\quad +\sqrt{11}/\sqrt{2.3.19}$

$2\ \frac{3}{2}\ \frac{5}{2}\ 1\ 000+\quad -11\sqrt{7}/2.17\sqrt{13.19}$

$2\ \frac{3}{2}\ \frac{5}{2}\ 1\ 100+\quad +2\sqrt{11}/3.17\sqrt{19}$

SO_3–K 3jm Factors (cont.)

8 $\frac{15}{2}$ $\frac{11}{2}$

2 $\frac{5}{2}$ $\frac{1}{2}$ 0 000+ $-173\sqrt{7}/8.17\sqrt{2.13.19}$
2 $\frac{5}{2}$ $\frac{1}{2}$ 0 010+ $-3.5\sqrt{3.5.11}/8.17\sqrt{2.19}$
2 $\frac{5}{2}$ $\frac{1}{2}$ 0 100+ $+67\sqrt{11}/4.3.17\sqrt{2.19}$
2 $\frac{5}{2}$ $\frac{1}{2}$ 0 110+ $-\sqrt{3.5.7.13}/4.17\sqrt{2.19}$
2 $\frac{5}{2}$ $\frac{3}{2}$ 0 000− $-7.31\sqrt{11}/8.17\sqrt{2.13.19}$
2 $\frac{5}{2}$ $\frac{3}{2}$ 0 010− $+9\sqrt{3.5.7}/8.17\sqrt{2.19}$
2 $\frac{5}{2}$ $\frac{3}{2}$ 0 100− $-23\sqrt{7}/4.3.17\sqrt{2.19}$
2 $\frac{5}{2}$ $\frac{3}{2}$ 0 110− $+\sqrt{3.5.11.13}/4.17\sqrt{2.19}$
2 $\frac{5}{2}$ $\frac{3}{2}$ 1 000+ $-3\sqrt{3.11}/8.2\sqrt{13.19}$
2 $\frac{5}{2}$ $\frac{3}{2}$ 1 010+ $+9\sqrt{5.7}/8.2.17\sqrt{19}$
2 $\frac{5}{2}$ $\frac{3}{2}$ 1 100+ $+\sqrt{3.7}/8\sqrt{19}$
2 $\frac{5}{2}$ $\frac{3}{2}$ 1 110+ $+\sqrt{5.11.13}/8.17\sqrt{19}$
2 $\frac{5}{2}$ $\frac{5}{2}$ 0 000+ $-5.11\sqrt{3.11}/8.2.17\sqrt{13.19}$
2 $\frac{5}{2}$ $\frac{5}{2}$ 0 010+ $-9.3\sqrt{5.7}/8.2.17\sqrt{19}$
2 $\frac{5}{2}$ $\frac{5}{2}$ 0 100+ $+23\sqrt{7}/8.17\sqrt{3.19}$
2 $\frac{5}{2}$ $\frac{5}{2}$ 0 110+ $-3\sqrt{5.11.13}/8.17\sqrt{19}$
2 $\frac{5}{2}$ $\frac{5}{2}$ 1 000− $-173\sqrt{11}/8.17\sqrt{2.13.19}$
2 $\frac{5}{2}$ $\frac{5}{2}$ 1 010− $+13\sqrt{3.5.7}/8.17\sqrt{2.19}$
2 $\frac{5}{2}$ $\frac{5}{2}$ 1 100− $-7\sqrt{7}/4.17\sqrt{2.19}$
2 $\frac{5}{2}$ $\frac{5}{2}$ 1 110− $-7\sqrt{5.11.13}/4.17\sqrt{2.3.19}$
2 $\frac{5}{2}$ $\frac{5}{2}$ 2 000+ $+47\sqrt{11}/8.2.17\sqrt{13.19}$
2 $\frac{5}{2}$ $\frac{5}{2}$ 2 010+ $-3.5\sqrt{3.5.7}/8.2.17\sqrt{19}$
2 $\frac{5}{2}$ $\frac{5}{2}$ 2 100+ $-79\sqrt{7}/8.3.17\sqrt{19}$
2 $\frac{5}{2}$ $\frac{5}{2}$ 2 110+ $-5\sqrt{5.11.13}/8.17\sqrt{3.19}$
3 $\frac{3}{2}$ $\frac{3}{2}$ 0 000− $+7.7/2\sqrt{3.13.17.19}$
3 $\frac{3}{2}$ $\frac{5}{2}$ 0 000+ $-3\sqrt{3.5}/2\sqrt{2.13.17.19}$
3 $\frac{3}{2}$ $\frac{5}{2}$ 1 000− $+5.7/2.3\sqrt{2.13.17.19}$
3 $\frac{5}{2}$ $\frac{1}{2}$ 0 000− $-29\sqrt{5}/8.3\sqrt{2.13.17.19}$
3 $\frac{5}{2}$ $\frac{1}{2}$ 0 010− $+\sqrt{3.7.11}/8\sqrt{2.17.19}$
3 $\frac{5}{2}$ $\frac{3}{2}$ 0 000+ $+31\sqrt{5.7.11}/8.2.3\sqrt{2.13.17.19}$
3 $\frac{5}{2}$ $\frac{3}{2}$ 0 010+ $-47\sqrt{3}/8.2\sqrt{2.17.19}$
3 $\frac{5}{2}$ $\frac{3}{2}$ 1 000− $+11\sqrt{7.11}/8.2\sqrt{2.3.13.17.19}$
3 $\frac{5}{2}$ $\frac{3}{2}$ 1 010− $+5.7\sqrt{5}/8.2\sqrt{2.17.19}$
3 $\frac{5}{2}$ $\frac{5}{2}$ 0 000− $-5.5\sqrt{7.11}/8.2\sqrt{13.17.19}$
3 $\frac{5}{2}$ $\frac{5}{2}$ 0 010− $-5.7\sqrt{5}/8.2\sqrt{3.17.19}$
3 $\frac{5}{2}$ $\frac{5}{2}$ 1 000+ $+11\sqrt{5.7.11}/8.2.3\sqrt{13.17.19}$
3 $\frac{5}{2}$ $\frac{5}{2}$ 1 010+ $-\sqrt{17}/8.2\sqrt{3.19}$
$\tilde{1}$ $\frac{3}{2}$ $\frac{3}{2}$ 0 000− $+3\sqrt{7}/2\sqrt{17.19}$
$\tilde{1}$ $\frac{3}{2}$ $\frac{5}{2}$ 0 000+ $+\sqrt{7}/2\sqrt{17.19}$
$\tilde{1}$ $\frac{5}{2}$ $\frac{1}{2}$ 0 000− $+7\sqrt{5.7}/8\sqrt{2.3.17.19}$
$\tilde{1}$ $\frac{5}{2}$ $\frac{1}{2}$ 0 010− $+\sqrt{11.13}/8\sqrt{2.17.19}$
$\tilde{1}$ $\frac{5}{2}$ $\frac{3}{2}$ 0 000+ $+\sqrt{11.17}/8.2\sqrt{3.19}$
$\tilde{1}$ $\frac{5}{2}$ $\frac{3}{2}$ 0 010+ $-\sqrt{5.7.13}/8.2\sqrt{17.19}$
$\tilde{1}$ $\frac{5}{2}$ $\frac{5}{2}$ 0 000− $-31\sqrt{11}/8.2\sqrt{3.17.19}$
$\tilde{1}$ $\frac{5}{2}$ $\frac{5}{2}$ 0 010− $-\sqrt{5.7.13}/8.2\sqrt{17.19}$
$\tilde{1}$ $\frac{5}{2}$ $\frac{5}{2}$ 1 000− $-13\sqrt{5.11}/8.2\sqrt{3.17.19}$
$\tilde{1}$ $\frac{5}{2}$ $\frac{5}{2}$ 1 010− $+\sqrt{7.13}/8.2\sqrt{17.19}$

8 $\frac{15}{2}$ $\frac{13}{2}$

2 $\frac{3}{2}$ $\frac{1}{2}$ 0 000+ $-11\sqrt{3.11.23}/2.17\sqrt{5.13.19}$

8 $\frac{15}{2}$ $\frac{13}{2}$

2 $\frac{3}{2}$ $\frac{1}{2}$ 0 100+ $+2\sqrt{3.5}/17\sqrt{7.19.23}$
2 $\frac{3}{2}$ $\frac{3}{2}$ 0 000− $+11\sqrt{3.11.23}/2.5.17\sqrt{13.19}$
2 $\frac{3}{2}$ $\frac{3}{2}$ 0 100− $-\sqrt{3.19}/17\sqrt{7.23}$
2 $\frac{3}{2}$ $\frac{5}{2}$ 0 000+ $-3.7\sqrt{13}/2.5.17\sqrt{19.23}$
2 $\frac{3}{2}$ $\frac{5}{2}$ 0 100+ $+3\sqrt{11}/17\sqrt{7.19.23}$
2 $\frac{3}{2}$ $\frac{5}{2}$ 1 000− $-3.307\sqrt{3}/2.5.17\sqrt{2.13.19.23}$
2 $\frac{3}{2}$ $\frac{5}{2}$ 1 100− $-5.5\sqrt{3.11}/17\sqrt{2.7.19.23}$
2 $\frac{3}{2}$ $\tilde{\frac{1}{2}}$ 0 000− $-3.7\sqrt{3}/2.17\sqrt{2.5.19.23}$
2 $\frac{3}{2}$ $\tilde{\frac{1}{2}}$ 0 100− $-5\sqrt{3.5.11.13}/17\sqrt{2.7.19.23}$
2 $\frac{5}{2}$ $\frac{1}{2}$ 0 000− $+3.59\sqrt{7}/2.17\sqrt{5.13.19.23}$
2 $\frac{5}{2}$ $\frac{1}{2}$ 0 010− $+\sqrt{3.11}/2\sqrt{19.23}$
2 $\frac{5}{2}$ $\frac{1}{2}$ 0 100− $+9\sqrt{5.11}/2.17\sqrt{19.23}$
2 $\frac{5}{2}$ $\frac{1}{2}$ 0 110− $-5\sqrt{13}/2\sqrt{3.7.19.23}$
2 $\frac{5}{2}$ $\frac{3}{2}$ 0 000+ $+9\sqrt{2.7}/5\sqrt{13.19.23}$
2 $\frac{5}{2}$ $\frac{3}{2}$ 0 010+ $+9\sqrt{2.3.11}/17\sqrt{5.19.23}$
2 $\frac{5}{2}$ $\frac{3}{2}$ 0 100+ $+3\sqrt{11}/2\sqrt{2.19.23}$
2 $\frac{5}{2}$ $\frac{3}{2}$ 0 110+ $+97\sqrt{5.13}/2.17\sqrt{2.3.7.19.23}$
2 $\frac{5}{2}$ $\frac{3}{2}$ 1 000− $-3.431\sqrt{3.7}/8.5.17\sqrt{13.19.23}$
2 $\frac{5}{2}$ $\frac{3}{2}$ 1 010− $-461\sqrt{11}/8.17\sqrt{5.19.23}$
2 $\frac{5}{2}$ $\frac{3}{2}$ 1 100− $-5.5\sqrt{11}/4.17\sqrt{3.19.23}$
2 $\frac{5}{2}$ $\frac{3}{2}$ 1 110− $+3.5\sqrt{5.13}/4.17\sqrt{7.19.23}$
2 $\frac{5}{2}$ $\frac{5}{2}$ 0 000− $+3.37\sqrt{7.11}/5.17\sqrt{2.13.19.23}$
2 $\frac{5}{2}$ $\frac{5}{2}$ 0 010− $-3.29\sqrt{3}/17\sqrt{2.5.19.23}$
2 $\frac{5}{2}$ $\frac{5}{2}$ 0 100− $+9.3.7/2.17\sqrt{2.19.23}$
2 $\frac{5}{2}$ $\frac{5}{2}$ 0 110− $+\sqrt{3.5.7.11.13}/2.17\sqrt{2.19.23}$
2 $\frac{5}{2}$ $\frac{5}{2}$ 1 000+ $+3.37\sqrt{3.7.11}/8.17\sqrt{13.19.23}$
2 $\frac{5}{2}$ $\frac{5}{2}$ 1 010+ $-3.13/8.17\sqrt{5.19.23}$
2 $\frac{5}{2}$ $\frac{5}{2}$ 1 100+ $-3.5.7\sqrt{3}/2.17\sqrt{19.23}$
2 $\frac{5}{2}$ $\frac{5}{2}$ 1 110+ $-\sqrt{5.11.13}/2.17\sqrt{7.19.23}$
2 $\frac{5}{2}$ $\frac{5}{2}$ 2 000− $-9.9.3\sqrt{3.7.11}/4.5.17\sqrt{2.13.19.23}$
2 $\frac{5}{2}$ $\frac{5}{2}$ 2 010− $+9.3\sqrt{5}/4.17\sqrt{2.19.23}$
2 $\frac{5}{2}$ $\frac{5}{2}$ 2 100− $-5\sqrt{3}/2.17\sqrt{2.19.23}$
2 $\frac{5}{2}$ $\frac{5}{2}$ 2 110− $+3\sqrt{5.11.13}/2.17\sqrt{2.7.19.23}$
2 $\frac{5}{2}$ $\tilde{\frac{1}{2}}$ 0 000+ $+3\sqrt{3.7.11.23}/8.17\sqrt{5.19}$
2 $\frac{5}{2}$ $\tilde{\frac{1}{2}}$ 0 010+ $+7\sqrt{13}/8.17\sqrt{19.23}$
2 $\frac{5}{2}$ $\tilde{\frac{1}{2}}$ 0 100+ $+5\sqrt{5.13}/17\sqrt{3.19.23}$
2 $\frac{5}{2}$ $\tilde{\frac{1}{2}}$ 0 110+ $-5.13\sqrt{11}/17\sqrt{7.19.23}$
3 $\frac{3}{2}$ $\frac{3}{2}$ 0 000+ $+9.3\sqrt{3.11}/\sqrt{7.13.17.19.23}$
3 $\frac{3}{2}$ $\frac{5}{2}$ 0 000− $-9\sqrt{23}/\sqrt{5.7.13.17.19}$
3 $\frac{3}{2}$ $\frac{5}{2}$ 1 000+ $-5\sqrt{3.17}/\sqrt{7.13.19.23}$
3 $\frac{5}{2}$ $\frac{1}{2}$ 0 000+ $+3.59/8\sqrt{13.17.19.23}$
3 $\frac{5}{2}$ $\frac{1}{2}$ 0 010+ $-31\sqrt{5.11}/8\sqrt{3.7.17.19.23}$
3 $\frac{5}{2}$ $\frac{3}{2}$ 0 000− $-1019/8\sqrt{2.5.13.17.19.23}$
3 $\frac{5}{2}$ $\frac{3}{2}$ 0 010− $+5.7\sqrt{7.11}/8\sqrt{2.3.17.19.23}$
3 $\frac{5}{2}$ $\frac{3}{2}$ 1 000+ $-797/8\sqrt{2.3.13.17.19.23}$
3 $\frac{5}{2}$ $\frac{3}{2}$ 1 010+ $+\sqrt{5.11.17}/8\sqrt{2.7.19.23}$
3 $\frac{5}{2}$ $\frac{5}{2}$ 0 000+ $-9.7\sqrt{3.11}/4\sqrt{2.13.17.19.23}$
3 $\frac{5}{2}$ $\frac{5}{2}$ 0 010+ $-5\sqrt{5}/4\sqrt{2.7.17.19.23}$
3 $\frac{5}{2}$ $\frac{5}{2}$ 1 000− $-37\sqrt{3.11}/4\sqrt{2.5.13.17.19.23}$

SO_3–K 3jm Factors (cont.)

8 $\frac{15}{2}$ $\frac{13}{2}$	
3 $\frac{5}{2}$ $\frac{5}{2}$ 1 010−	$+3\sqrt{19}/4\sqrt{2.7.17.23}$
3 $\frac{5}{2}$ $\tilde{\frac{1}{2}}$ 0 000−	$-4\sqrt{11}/\sqrt{3.17.19.23}$
3 $\frac{5}{2}$ $\tilde{\frac{1}{2}}$ 0 010−	$+4\sqrt{5.13}/\sqrt{7.17.19.23}$
$\tilde{1}$ $\frac{3}{2}$ $\frac{3}{2}$ 0 000+	$+7\sqrt{11}/2\sqrt{17.19.23}$
$\tilde{1}$ $\frac{3}{2}$ $\frac{5}{2}$ 0 000−	$+3.7\sqrt{3}/2\sqrt{2.17.19.23}$
$\tilde{1}$ $\frac{3}{2}$ $\tilde{\frac{1}{2}}$ 0 000+	$-\sqrt{13}/2\sqrt{2.17.19.23}$
$\tilde{1}$ $\frac{5}{2}$ $\frac{1}{2}$ 0 000+	$-\sqrt{3.7.23}/8\sqrt{17.19}$
$\tilde{1}$ $\frac{5}{2}$ $\frac{1}{2}$ 0 010+	$+\sqrt{5.11.13}/8\sqrt{17.19.23}$
$\tilde{1}$ $\frac{5}{2}$ $\frac{3}{2}$ 0 000−	$+\sqrt{3.7}/8\sqrt{17.19.23}$
$\tilde{1}$ $\frac{5}{2}$ $\frac{3}{2}$ 0 010−	$+\sqrt{5.11.13}/8\sqrt{17.19.23}$
$\tilde{1}$ $\frac{5}{2}$ $\frac{5}{2}$ 0 000+	$+3\sqrt{7.11}/2\sqrt{2.17.19.23}$
$\tilde{1}$ $\frac{5}{2}$ $\frac{5}{2}$ 0 010+	$-\sqrt{3.5.13}/2\sqrt{2.17.19.23}$
$\tilde{1}$ $\frac{5}{2}$ $\frac{5}{2}$ 1 000+	0
$\tilde{1}$ $\frac{5}{2}$ $\frac{5}{2}$ 1 010+	$-3\sqrt{2.3.13}/\sqrt{17.19.23}$
8 $\frac{15}{2}$ $\frac{15}{2}$	
2 $\frac{3}{2}$ $\frac{3}{2}$ 0 000+	$+11\sqrt{7.11.23}/5.17\sqrt{13.19}$
2 $\frac{3}{2}$ $\frac{3}{2}$ 0 100+	$+9/2.17\sqrt{19.23}$
2 $\frac{5}{2}$ $\frac{3}{2}$ 0 000−	$+7.7\sqrt{3.13}/4.5.17\sqrt{2.19.23}$
2 $\frac{5}{2}$ $\frac{3}{2}$ 0 010−	$-3.7\sqrt{7.11}/4.17\sqrt{2.5.19.23}$
2 $\frac{5}{2}$ $\frac{3}{2}$ 0 100−	$-\sqrt{3.7.11}/2.17\sqrt{2.19.23}$
2 $\frac{5}{2}$ $\frac{3}{2}$ 0 110−	$+13\sqrt{5.13}/2.17\sqrt{2.19.23}$
2 $\frac{5}{2}$ $\frac{3}{2}$ 1 000+	$+3.1091/8.5.17\sqrt{13.19.23}$
2 $\frac{5}{2}$ $\frac{3}{2}$ 1 010+	$+3.7\sqrt{3.7.11}/8.17\sqrt{5.19.23}$
2 $\frac{5}{2}$ $\frac{3}{2}$ 1 100+	$-3.5\sqrt{7.11}/8.17\sqrt{19.23}$
2 $\frac{5}{2}$ $\frac{3}{2}$ 1 110+	$+5.5\sqrt{3.5.13}/8.17\sqrt{19.23}$
2 $\frac{5}{2}$ $\frac{5}{2}$ 0 000+	$+7.37\sqrt{3.7.11}/4.5.17\sqrt{13.19.23}$
2 $\frac{5}{2}$ $\frac{5}{2}$ 0 010+	$+9.29/4.17\sqrt{5.19.23}$
2 $\frac{5}{2}$ $\frac{5}{2}$ 0 011+	$+5\sqrt{3.7.11.13}/4.17\sqrt{19.23}$
2 $\frac{5}{2}$ $\frac{5}{2}$ 0 100+	$+9.7.7\sqrt{3}/8.17\sqrt{19.23}$
2 $\frac{5}{2}$ $\frac{5}{2}$ 0 110+	$-3\sqrt{5.7.11.13}/8.17\sqrt{19.23}$
2 $\frac{5}{2}$ $\frac{5}{2}$ 0 111+	$+3.5.13\sqrt{3}/8.17\sqrt{19.23}$
2 $\frac{5}{2}$ $\frac{5}{2}$ 1 000−	0
2 $\frac{5}{2}$ $\frac{5}{2}$ 1 010−	$-3.11\sqrt{2.3}/17\sqrt{5.19.23}$
2 $\frac{5}{2}$ $\frac{5}{2}$ 1 011−	0
2 $\frac{5}{2}$ $\frac{5}{2}$ 1 100−	0
2 $\frac{5}{2}$ $\frac{5}{2}$ 1 110−	$-\sqrt{3.5.7.11.13}/2.17\sqrt{2.19.23}$
2 $\frac{5}{2}$ $\frac{5}{2}$ 1 111−	0
2 $\frac{5}{2}$ $\frac{5}{2}$ 2 000+	$-3.227\sqrt{7.11}/8.2.5.17\sqrt{13.19.23}$
2 $\frac{5}{2}$ $\frac{5}{2}$ 2 010+	$+9.9.3\sqrt{3.5}/8.2.17\sqrt{19.23}$
2 $\frac{5}{2}$ $\frac{5}{2}$ 2 011+	$+9\sqrt{7.11.13}/8.2.17\sqrt{19.23}$
2 $\frac{5}{2}$ $\frac{5}{2}$ 2 100+	$-9.5.5/8.4.17\sqrt{19.23}$
2 $\frac{5}{2}$ $\frac{5}{2}$ 2 110+	$-3.7\sqrt{3.5.7.11.13}/8.4.17\sqrt{19.23}$
2 $\frac{5}{2}$ $\frac{5}{2}$ 2 111+	$-3.5.5.13/8.4.17\sqrt{19.23}$
3 $\frac{3}{2}$ $\frac{3}{2}$ 0 000−	0
3 $\frac{5}{2}$ $\frac{3}{2}$ 0 000+	$+7\sqrt{3.7.17}/8\sqrt{2.5.13.19.23}$
3 $\frac{5}{2}$ $\frac{3}{2}$ 0 010+	$-5.5\sqrt{11}/8\sqrt{2.17.19.23}$
3 $\frac{5}{2}$ $\frac{3}{2}$ 1 000−	$+9.29\sqrt{7}/8\sqrt{2.13.17.19.23}$
3 $\frac{5}{2}$ $\frac{3}{2}$ 1 010−	$+7\sqrt{3.5.11}/8\sqrt{2.17.19.23}$
3 $\frac{5}{2}$ $\frac{5}{2}$ 0 000−	0
8 $\frac{15}{2}$ $\frac{15}{2}$	
3 $\frac{5}{2}$ $\frac{5}{2}$ 0 010−	$+\sqrt{3.5.7}/2\sqrt{17.19.23}$
3 $\frac{5}{2}$ $\frac{5}{2}$ 0 011−	0
3 $\frac{5}{2}$ $\frac{5}{2}$ 1 000+	$-9.3\sqrt{11}/8\sqrt{5.13.17.19.23}$
3 $\frac{5}{2}$ $\frac{5}{2}$ 1 010+	$-3.11\sqrt{3.7}/8\sqrt{17.19.23}$
3 $\frac{5}{2}$ $\frac{5}{2}$ 1 011+	$+3\sqrt{5.11.13}/8\sqrt{17.19.23}$
$\tilde{1}$ $\frac{3}{2}$ $\frac{3}{2}$ 0 000−	0
$\tilde{1}$ $\frac{5}{2}$ $\frac{3}{2}$ 0 000+	$-9.3/8\sqrt{17.19.23}$
$\tilde{1}$ $\frac{5}{2}$ $\frac{3}{2}$ 0 010+	$+\sqrt{3.5.7.11.13}/8\sqrt{17.19.23}$
$\tilde{1}$ $\frac{5}{2}$ $\frac{5}{2}$ 0 000−	0
$\tilde{1}$ $\frac{5}{2}$ $\frac{5}{2}$ 0 010−	$+3\sqrt{5.13}/2\sqrt{17.19.23}$
$\tilde{1}$ $\frac{5}{2}$ $\frac{5}{2}$ 0 011−	0
$\tilde{1}$ $\frac{5}{2}$ $\frac{5}{2}$ 1 000−	0
$\tilde{1}$ $\frac{5}{2}$ $\frac{5}{2}$ 1 010−	$+9\sqrt{13}/2\sqrt{17.19.23}$
$\tilde{1}$ $\frac{5}{2}$ $\frac{5}{2}$ 1 011−	0
8 8 0	
2 2 0 0 000+	$+\sqrt{5}/\sqrt{17}$
2 2 0 0 100+	0
2 2 0 0 110+	$+\sqrt{5}/\sqrt{17}$
3 3 0 0 000+	$+2/\sqrt{17}$
$\tilde{1}$ $\tilde{1}$ 0 0 000+	$+\sqrt{3}/\sqrt{17}$
8 8 1	
2 2 1 0 000+	$-\sqrt{3.5}/2\sqrt{17}$
2 2 1 0 100+	0
2 2 1 0 110+	$+4\sqrt{5}/3\sqrt{3.17}$
3 2 1 0 000−	$-\sqrt{3.5.7}/2.17\sqrt{2}$
3 2 1 0 010−	$-\sqrt{5.11.13}/3.17\sqrt{2.3}$
3 3 1 0 000+	$-4\sqrt{2}/3\sqrt{3.17}$
$\tilde{1}$ 2 1 0 000−	$-\sqrt{5.13}/2.17\sqrt{2}$
$\tilde{1}$ 2 1 0 010−	$+\sqrt{5.7.11}/3.17\sqrt{2}$
$\tilde{1}$ 3 1 0 000+	$-\sqrt{7.13}/2.3\sqrt{2.17}$
8 8 2	
2 2 2 0 000+	$-97\sqrt{3.5}/4.17\sqrt{2.17.19}$
2 2 2 0 100+	$-\sqrt{3.5.7.11.13}/2.17\sqrt{2.17.19}$
2 2 2 0 110+	$-5.5\sqrt{5}/17\sqrt{2.3.17.19}$
2 2 2 1 000+	$+269\sqrt{3}/4.17\sqrt{2.17.19}$
2 2 2 1 100+	$-5\sqrt{7.11.13}/2.17\sqrt{2.3.17.19}$
2 2 2 1 110+	$+5.37/17\sqrt{2.3.17.19}$
3 2 2 0 000−	$-\sqrt{3.7.19}/4.17$
3 2 2 0 010−	$+5\sqrt{11.13}/2.17\sqrt{3.19}$
3 2 2 1 000+	$-11\sqrt{7}/4.17\sqrt{3.19}$
3 2 2 1 010+	$-5\sqrt{11.13}/2.17\sqrt{3.19}$
3 3 2 0 000+	$-\sqrt{2}/\sqrt{3.17.19}$
$\tilde{1}$ 2 2 0 000−	$-\sqrt{13.19}/4.17$
$\tilde{1}$ 2 2 0 010−	$-5\sqrt{7.11}/2.17\sqrt{19}$
$\tilde{1}$ 3 2 0 000+	$-\sqrt{7.13}/2\sqrt{2.17.19}$
$\tilde{1}$ $\tilde{1}$ 2 0 000+	$+1/4\sqrt{17.19}$
8 8 3	
2 2 3 0 000+	$+9.3\sqrt{2.3.5}/17\sqrt{7.17.19}$
2 2 3 0 100+	$+\sqrt{3.5.11.13}/2.17\sqrt{2.17.19}$

SO_3–K 3jm Factors (cont.)

8 8 3	
2 2 3 0 110+	$+\sqrt{2.3.5.7}/17\sqrt{17.19}$
2 2 3 1 000−	0
2 2 3 1 100−	$-\sqrt{5.11.13}/2\sqrt{2.3.17.19}$
2 2 3 1 110−	0
2 2 $\tilde{1}$ 0 000+	$+107\sqrt{5}/2.17\sqrt{2.7.17.19}$
2 2 $\tilde{1}$ 0 100+	$-\sqrt{5.11.13}/17\sqrt{2.17.19}$
2 2 $\tilde{1}$ 0 110+	$-23\sqrt{2.5.7}/3.17\sqrt{17.19}$
3 2 3 0 000−	$+\sqrt{5}/\sqrt{3.19}$
3 2 3 0 010−	0
3 2 $\tilde{1}$ 0 000−	$+2\sqrt{5}/17\sqrt{19}$
3 2 $\tilde{1}$ 0 010−	$+\sqrt{5.7.11.13}/3.17\sqrt{19}$
3 3 3 0 000−	0
3 3 $\tilde{1}$ 0 000+	$-\sqrt{7}/3\sqrt{17.19}$
$\tilde{1}$ 2 3 0 000−	$+5\sqrt{5.13}/17\sqrt{7.19}$
$\tilde{1}$ 2 3 0 010−	$-\sqrt{5.11}/2.17\sqrt{19}$
$\tilde{1}$ 2 $\tilde{1}$ 0 000−	$+9\sqrt{5.13}/2.17\sqrt{2.7.19}$
$\tilde{1}$ 2 $\tilde{1}$ 0 010−	$-3\sqrt{5.11}/17\sqrt{2.19}$
$\tilde{1}$ 3 3 0 000+	$+\sqrt{13}/\sqrt{17.19}$
$\tilde{1}$ $\tilde{1}$ $\tilde{1}$ 0 000+	$-23/2\sqrt{2.7.17.19}$
8 8 4	
2 2 2 0 000+	$+3.5.29/4.17\sqrt{13.17.19}$
2 2 2 0 100+	$+3.5\sqrt{7.11}/2.17\sqrt{17.19}$
2 2 2 0 110+	$-2.5\sqrt{13}/17\sqrt{17.19}$
2 2 2 1 000+	$-3.5.5\sqrt{5}/4.17\sqrt{13.17.19}$
2 2 2 1 100+	$+11\sqrt{5.7.11}/2.3.17\sqrt{17.19}$
2 2 2 1 110+	$+8.2.5\sqrt{5.13}/9.17\sqrt{17.19}$
2 2 3 0 000−	0
2 2 3 0 100−	$-\sqrt{5.11}/2\sqrt{2.17.19}$
2 2 3 0 110−	0
2 2 3 1 000+	$+9\sqrt{2.5.7}/17\sqrt{13.17.19}$
2 2 3 1 100+	$-5\sqrt{5.11.19}/2.3.17\sqrt{2.17}$
2 2 3 1 110+	$-\sqrt{2.5.7.13}/9.17\sqrt{17.19}$
3 2 2 0 000−	$+\sqrt{5.7.13}/17\sqrt{2.19}$
3 2 2 0 010−	$+\sqrt{5.11}/17\sqrt{2.19}$
3 2 2 1 000+	$+5.5\sqrt{5.7}/3.17\sqrt{2.13.19}$
3 2 2 1 010+	$-23\sqrt{5.11}/9.17\sqrt{2.19}$
3 2 3 0 000+	$-5.5\sqrt{5}/3.17\sqrt{13.19}$
3 2 3 0 010+	$-4\sqrt{5.7.11}/9.17\sqrt{19}$
3 3 2 0 000+	$-47\sqrt{5}/9\sqrt{13.17.19}$
3 3 3 0 000+	$-8.7\sqrt{7}/9\sqrt{13.17.19}$
$\tilde{1}$ 2 2 0 000−	$+13\sqrt{5}/2.17\sqrt{2.3.19}$
$\tilde{1}$ 2 2 0 010−	$-\sqrt{5.7.11.13}/3.17\sqrt{2.3.19}$
$\tilde{1}$ 2 3 0 000+	$+5\sqrt{5.7}/17\sqrt{3.19}$
$\tilde{1}$ 2 3 0 010+	$+\sqrt{5.11.13}/2.3.17\sqrt{3.19}$
$\tilde{1}$ 3 2 0 000+	$-\sqrt{5.7}/3\sqrt{3.17.19}$
$\tilde{1}$ 3 3 0 000−	$-13/3\sqrt{3.17.19}$
$\tilde{1}$ $\tilde{1}$ 2 0 000+	$-\sqrt{3.5.13}/2\sqrt{2.17.19}$
8 8 5	
2 2 1 0 000+	$-3.11\sqrt{5.11}/17\sqrt{2.13.17.19}$
2 2 1 0 100+	$-\sqrt{5.7}/17\sqrt{2.17.19}$
2 2 1 0 110+	$-2\sqrt{2.5.11.13}/17\sqrt{17.19}$
2 2 2 0 000−	0
2 2 2 0 100−	$+5\sqrt{3}/2\sqrt{17.19}$
2 2 2 0 110−	0
2 2 2 1 000−	0
2 2 2 1 100−	$+7\sqrt{5}/2\sqrt{3.17.19}$
2 2 2 1 110−	0
2 2 $\tilde{1}$ 0 000+	$-11\sqrt{5.7.11}/4.17\sqrt{13.17.19}$
2 2 $\tilde{1}$ 0 100+	$+47\sqrt{5}/2.17\sqrt{17.19}$
2 2 $\tilde{1}$ 0 110+	$-\sqrt{5.7.11.13}/3.17\sqrt{17.19}$
3 2 1 0 000−	$-127\sqrt{5.7}/4.17\sqrt{11.13.19}$
3 2 1 0 010−	$+\sqrt{5}/2.17\sqrt{19}$
3 2 2 0 000+	$+9.7\sqrt{3.5}/4.17\sqrt{2.11.13.19}$
3 2 2 0 010+	$+\sqrt{3.5.7}/2.17\sqrt{2.19}$
3 2 2 1 000−	$+5.59\sqrt{5}/4.17\sqrt{2.3.11.13.19}$
3 2 2 1 010−	$+3\sqrt{3.5.7}/2.17\sqrt{2.19}$
3 2 $\tilde{1}$ 0 000−	$-2\sqrt{2.5}/17\sqrt{11.13.19}$
3 2 $\tilde{1}$ 0 010−	$-\sqrt{2.5.7}/3.17\sqrt{19}$
3 3 1 0 000+	$+4/\sqrt{11.13.17.19}$
3 3 2 0 000−	0
3 3 $\tilde{1}$ 0 000+	$-53\sqrt{2.7}/3\sqrt{11.13.17.19}$
$\tilde{1}$ 2 1 0 000−	$-\sqrt{3.5}/2\sqrt{11.19}$
$\tilde{1}$ 2 1 0 010−	0
$\tilde{1}$ 2 2 0 000+	$+4\sqrt{2.5.7}/17\sqrt{11.19}$
$\tilde{1}$ 2 2 0 010+	$-\sqrt{2.5.13}/17\sqrt{19}$
$\tilde{1}$ 2 $\tilde{1}$ 0 000−	$-9.3\sqrt{5.7}/4.17\sqrt{11.19}$
$\tilde{1}$ 2 $\tilde{1}$ 0 010−	$-3\sqrt{5.13}/2.17\sqrt{19}$
$\tilde{1}$ 3 1 0 000+	$+9\sqrt{3.7}/4\sqrt{11.17.19}$
$\tilde{1}$ 3 2 0 000−	$+\sqrt{5.19}/4\sqrt{11.17}$
$\tilde{1}$ $\tilde{1}$ 2 0 000−	0
$\tilde{1}$ $\tilde{1}$ $\tilde{1}$ 0 000+	$+\sqrt{7.13}/4\sqrt{11.17.19}$
8 8 6	
2 2 0 0 000+	$-11\sqrt{3.5.11.23}/2.17\sqrt{2.13.17.19}$
2 2 0 0 100+	$-\sqrt{3.5.7}/17\sqrt{2.17.19.23}$
2 2 0 0 110+	$+11\sqrt{2.5.11.13}/17\sqrt{3.17.19.23}$
2 2 1 0 000−	0
2 2 1 0 100−	$+\sqrt{5}/\sqrt{2.17.19.23}$
2 2 1 0 110−	0
2 2 2 0 000+	$-5.59\sqrt{3.7}/4.17\sqrt{13.17.19.23}$
2 2 2 0 100+	$+9.5\sqrt{3.11}/17\sqrt{17.19.23}$
2 2 2 0 110+	$+5.7\sqrt{7.13}/17\sqrt{3.17.19.23}$
2 2 2 1 000+	$+239\sqrt{3.5.7}/4.17\sqrt{13.17.19.23}$
2 2 2 1 100+	$+4.5\sqrt{5.11}/17\sqrt{3.17.19.23}$
2 2 2 1 110+	$-\sqrt{3.5.7.13}/17\sqrt{17.19.23}$
2 2 3 0 000−	0
2 2 3 0 100−	$-\sqrt{3.5.11}/\sqrt{17.19.23}$
2 2 3 0 110−	0
2 2 3 1 000+	$+7\sqrt{3.5.7}/17\sqrt{13.17.19.23}$

SO_3–K 3jm Factors (cont.)

8 8 6

2 2 3 1 100+ $+67\sqrt{5.11}/17\sqrt{3.17.19.23}$
2 2 3 1 110+ $-8.2\sqrt{5.7.13}/17\sqrt{3.17.19.23}$
3 2 1 0 000+ $-149\sqrt{5.11}/4.17\sqrt{13.19.23}$
3 2 1 0 010+ $+3.5\sqrt{5.7}/2.17\sqrt{19.23}$
3 2 2 0 000− $-5.41\sqrt{3.5}/4.17\sqrt{2.13.19.23}$
3 2 2 0 010− $-13\sqrt{5.7.11}/2.17\sqrt{2.3.19.23}$
3 2 2 1 000+ $+347\sqrt{5}/4.17\sqrt{2.3.13.19.23}$
3 2 2 1 010+ $-5\sqrt{3.5.7.11}/2.17\sqrt{2.19.23}$
3 2 3 0 000+ $-8.2.5\sqrt{2.5}/17\sqrt{3.13.19.23}$
3 2 3 0 010+ $-2\sqrt{2.5.7.11}/17\sqrt{3.19.23}$
3 3 0 0 000+ $+2\sqrt{2.11}/\sqrt{3.13.17.19.23}$
3 3 1 0 000− 0
3 3 2 0 000+ $+2.3\sqrt{3.5.7}/\sqrt{13.17.19.23}$
3 3 3 0 000+ $-4\sqrt{2.7}/\sqrt{3.13.17.19.23}$
$\tilde{1}$ 2 1 0 000+ $-2\sqrt{3.5.7.11}/17\sqrt{19.23}$
$\tilde{1}$ 2 1 0 010+ $-3\sqrt{3.5.13}/17\sqrt{19.23}$
$\tilde{1}$ 2 2 0 000− $-5.5\sqrt{5.7}/2.17\sqrt{2.19.23}$
$\tilde{1}$ 2 2 0 010− $+\sqrt{5.11.13}/17\sqrt{2.19.23}$
$\tilde{1}$ 2 3 0 000+ $-\sqrt{5.7}/17\sqrt{2.19.23}$
$\tilde{1}$ 2 3 0 010+ $-2\sqrt{2.5.11.13}/17\sqrt{19.23}$
$\tilde{1}$ 3 1 0 000− $-3\sqrt{3.11}/4\sqrt{17.19.23}$
$\tilde{1}$ 3 2 0 000+ $+5.5\sqrt{5}/4\sqrt{17.19.23}$
$\tilde{1}$ 3 3 0 000− $-8\sqrt{2}/\sqrt{17.19.23}$
$\tilde{1}$ $\tilde{1}$ 0 0 000+ $+\sqrt{11.13}/2\sqrt{2.17.19.23}$
$\tilde{1}$ $\tilde{1}$ 2 0 000+ $-\sqrt{5.7.13}/2\sqrt{2.17.19.23}$

8 8 7

2 2 1 0 000+ $+11\sqrt{3.7.11.23}/17\sqrt{5.13.17.19}$
2 2 1 0 100+ $-4\sqrt{3.5}/17\sqrt{17.19.23}$
2 2 1 0 110+ $+11\sqrt{5.7.11.13}/3.17\sqrt{3.17.19.23}$
2 2 2 0 000− 0
2 2 2 0 100− $-3\sqrt{3.5}/2\sqrt{2.17.19.23}$
2 2 2 0 110− 0
2 2 2 1 000− 0
2 2 2 1 100− $+5/2\sqrt{2.3.17.19.23}$
2 2 2 1 110− 0
2 2 3 0 000+ $+2.3.7\sqrt{3.7}/17\sqrt{5.17.19.23}$
2 2 3 0 100+ $+3\sqrt{3.5.11.13}/17\sqrt{17.19.23}$
2 2 3 0 110+ $+8.13\sqrt{5.7}/17\sqrt{3.17.19.23}$
2 2 3 1 000− 0
2 2 3 1 100− $+\sqrt{5.11.13}/\sqrt{3.17.19.23}$
2 2 3 1 110− 0
2 2 $\tilde{1}$ 0 000+ $+4.9.9.3/17\sqrt{5.13.17.19.23}$
2 2 $\tilde{1}$ 0 100+ $+7\sqrt{5.7.11}/2.17\sqrt{17.19.23}$
2 2 $\tilde{1}$ 0 110+ $-2.41\sqrt{5.13}/3.17\sqrt{17.19.23}$
3 2 1 0 000− $-9.3\sqrt{2.3.11}/17\sqrt{5.13.19.23}$
3 2 1 0 010− $-8.2\sqrt{2.5.7}/3.17\sqrt{3.19.23}$
3 2 2 0 000+ $-2.9.3\sqrt{3.11}/17\sqrt{13.19.23}$
3 2 2 0 010+ $-5.5\sqrt{7}/17\sqrt{3.19.23}$
3 2 2 1 000− $-4\sqrt{11}/17\sqrt{3.13.19.23}$
3 2 2 1 010− $+5.11\sqrt{7}/17\sqrt{3.19.23}$
3 2 3 0 000− $+2.53\sqrt{2}/17\sqrt{3.5.19.23}$
3 2 3 0 010− $-\sqrt{2.5.7.11.13}/17\sqrt{3.19.23}$
3 2 $\tilde{1}$ 0 000− $+79\sqrt{2.7}/17\sqrt{5.13.19.23}$
3 2 $\tilde{1}$ 0 010− $-11\sqrt{2.5.11}/3.17\sqrt{19.23}$
3 3 1 0 000+ $-\sqrt{2.7.11}/3\sqrt{3.13.17.19.23}$
3 3 2 0 000− 0
3 3 3 0 000− 0
3 3 $\tilde{1}$ 0 000+ $+8.2.5\sqrt{2}/3\sqrt{13.17.19.23}$
$\tilde{1}$ 2 1 0 000− $-7\sqrt{7.11}/17\sqrt{2.5.19.23}$
$\tilde{1}$ 2 1 0 010− $+13\sqrt{2.5.13}/3.17\sqrt{19.23}$
$\tilde{1}$ 2 2 0 000+ $-7\sqrt{7.11}/17\sqrt{19.23}$
$\tilde{1}$ 2 2 0 010+ $+5\sqrt{13}/2.3.17\sqrt{19.23}$
$\tilde{1}$ 2 3 0 000− $-\sqrt{2.7.13}/17\sqrt{5.19.23}$
$\tilde{1}$ 2 3 0 010− $+13\sqrt{2.5.11}/3.17\sqrt{19.23}$
$\tilde{1}$ 2 $\tilde{1}$ 0 000− $+3.29/17\sqrt{5.19.23}$
$\tilde{1}$ 2 $\tilde{1}$ 0 010− $-\sqrt{5.7.11.13}/2.17\sqrt{19.23}$
$\tilde{1}$ 3 1 0 000+ $+5\sqrt{2.11}/3\sqrt{17.19.23}$
$\tilde{1}$ 3 2 0 000− $+5\sqrt{2.11}/3\sqrt{17.19.23}$
$\tilde{1}$ 3 3 0 000+ $+4\sqrt{2.13}/3\sqrt{17.19.23}$
$\tilde{1}$ $\tilde{1}$ 2 0 000− 0
$\tilde{1}$ $\tilde{1}$ $\tilde{1}$ 0 000+ $+2.3\sqrt{13}/\sqrt{17.19.23}$

8 8 8

2 2 2 0 000+ $+5.71\sqrt{5.7.11}/17.17\sqrt{2.13.19.23}$
2 2 2 0 100+ $-9.29\sqrt{5}/2.17.17\sqrt{2.19.23}$
2 2 2 0 110+ $+4\sqrt{2.5.7.11.13}/17.17\sqrt{19.23}$
2 2 2 0 111+ $+9.7.13\sqrt{5}/17.17\sqrt{2.19.23}$
2 2 2 1 000+ $-3.5.73\sqrt{7.11}/17.17\sqrt{2.13.19.23}$
2 2 2 1 100+ $-3.5.7.7/2.17.17\sqrt{2.19.23}$
2 2 2 1 110+ $+2.7\sqrt{2.7.11.13}/17.17\sqrt{19.23}$
2 2 2 1 111+ $-5.13.29/17.17\sqrt{2.19.23}$
3 2 2 0 000− 0
3 2 2 0 010− $+\sqrt{7}/\sqrt{17.19.23}$
3 2 2 0 011− 0
3 2 2 1 000+ $-2.3.5\sqrt{11}/17\sqrt{13.17.19.23}$
3 2 2 1 010+ $-5.11\sqrt{7}/17\sqrt{17.19.23}$
3 2 2 1 011+ $-8.2\sqrt{11.13}/17\sqrt{17.19.23}$
3 3 2 0 000+ $+2.11\sqrt{2.7.11}/17\sqrt{13.19.23}$
3 3 2 0 001+ $-\sqrt{2}/17\sqrt{19.23}$
3 3 3 0 000+ $+4\sqrt{2.5.11}/\sqrt{13.17.19.23}$
$\tilde{1}$ 2 2 0 000− 0
$\tilde{1}$ 2 2 0 010− $-3\sqrt{3.13}/2\sqrt{17.19.23}$
$\tilde{1}$ 2 2 0 011− 0
$\tilde{1}$ 3 2 0 000+ $+3\sqrt{2.3.11}/17\sqrt{19.23}$
$\tilde{1}$ 3 2 0 001+ $-3\sqrt{2.3.7.13}/17\sqrt{19.23}$
$\tilde{1}$ 3 3 0 000− 0
$\tilde{1}$ $\tilde{1}$ 2 0 000+ $+\sqrt{3.7.11.13}/17\sqrt{19.23}$
$\tilde{1}$ $\tilde{1}$ 2 0 001+ $+3.13\sqrt{3}/2.17\sqrt{19.23}$
$\tilde{1}$ $\tilde{1}$ $\tilde{1}$ 0 000− 0

14

3j and 6j Symbol Tables

Introduction

This chapter contains the $3j$ phases and $6j$ symbols for the groups $\mathbf{D}_n$ ($n \leqslant 6$), **T**, **O**, and **K**. The $3j$ phases for $\mathbf{C}_n$ ($n \leqslant 6$) are also given, but their $6j$ symbols are omitted as they are unity if the triad conditions are satisfied. The j symbols for the rotation–inversion and rotation–reflection point groups may be read off the table for the pure-rotation group to which they correspond. It is to be noted that the multiplicity separation for the octahedral group is new, and has three advantages over that used previously: more symmetries occur, more zeros occur, and smaller prime factors occur in the $6j$ table; see Section 3.5.

The $3j$ phase is the generalization of the reordering symmetry of coupling coefficients: in $\mathbf{SO}_3$ one has $\{J_1 J_2 J_3\} = (-1)^{J_1+J_2+J_3}$. The $3j$ phase tables also show the allowed triads of the group. A triad is a set of three irreps which couple to give the identity irrep, together with, if required, a product (coupling) multiplicity index. For example, at the bottom of the second column of the O $3j$ phase table on page 439, one has $\{1\,\frac{3}{2}\,\frac{3}{2}\,0\} = -1$ and $\{1\,\frac{3}{2}\,\frac{3}{2}\,1\} = +1$. Because $3j$ phases have modulus 1, only the sign is given, and because the irrep labels may be reordered without introducing a change in value, only one ordering is given.

A $6j$ symbol of a group contains six irreps each of which is involved in two triads. The four triads of a $6j$ appear as shown in (3.3.17). The generalization from the SO_3 $6j$ involves two points: one irrep appearing in the triads 1, 2, and 3 involves the complex conjugate irrep to the one in the $6j$; and a product (coupling) multiplicity index may occur for each triad.

The $6j$ symbols are tabulated by giving the irrep labels of the upper line, in bold, as headers. Under each such set of three labels, we give the irrep labels of the lower line followed first by the four multiplicity indices (if required), the interchange sign in smaller type, and then the value of the $6j$ (note that, as in our other tables, the dots in the $6j$ values are multiplication dots, not decimal points). For example, at the top of the center column of page 440, one has the following O (and T_d) $6j$ symbols:

$$\begin{Bmatrix} \tilde{1} & \frac{3}{2} & \frac{1}{2} \\ 1 & \frac{1}{2} & \frac{3}{2} \end{Bmatrix}_{0100} = \frac{1}{3\sqrt{2}}$$

$$\begin{Bmatrix} \tilde{1} & \frac{3}{2} & \frac{1}{2} \\ 1 & \frac{3}{2} & \frac{1}{2} \end{Bmatrix}_{0000} = \frac{1}{4}$$

while in the third column one can find that

$$\begin{Bmatrix} \tilde{1} & 2 & 1 \\ \frac{1}{2} & \frac{3}{2} & \frac{3}{2} \end{Bmatrix}_{1000} = \frac{1}{2\sqrt{6}}$$

The symmetries of the $6j$ are given by (3.3.18)–(3.3.20). Inspection of the tables shows that column interchanges do not change the sign for most $6j$'s, and that most irreps are real. (Both restrictions always apply to $\mathbf{SO_3}$.) For complex irreps the symmetries have to be applied with some care, but the tables are kept small by the use of the symmetries in the following manner:

1. The largest label (ignoring complex conjugation but otherwise in the character table ordering) is moved to the top left of the $6j$ using interchanges (3.3.18) and using flips (3.3.19). This also fixes the label in the bottom left.
2. The next largest is placed top center by the same means. This fixes the other labels.
3. If appropriate equalities occur, the lower row is ordered, largest on the left, the top right is made larger than the bottom right, and the multiplicity indices are ordered.

Because changes of label can be introduced by the various complex conjugations involved for $\mathbf{D_3}$, $\mathbf{D_5}$, and $\mathbf{T}$, the tabulated form may be elusive, and the complex conjugate of the entire symbol, (3.3.20), may be the tabulated form. On the other hand the tables contain some duplicated entries.

C_1 3j Phases ($C_i = S_2$)

0 0 0 +	½ ½ 0 +

C_2, C_{2y}, and C_s 3j Phases ($C_{2i} = C_{2h}$)

0 0 0 +	-½ ½ 0 +	1 ½ ½ +	1 1 0 +

C_3 3j Phases ($C_{3i} = S_6$)

0 0 0 +	1 -½ -½ +	-1 1 0 +	3/2 3/2 0 +
-½ ½ 0 +	1 1 1 +	3/2 1 ½ +	

C_4 and S_4 3j Phases ($C_{4i} = C_{4h}$)

0 0 0 +	-1 1 0 +	-3/2 3/2 0 +	2 2 0 +
-½ ½ 0 +	3/2 -1 -½ +	2 1 1 +	
1 -½ -½ +	3/2 3/2 1 +	2 3/2 ½ +	

C_5 3j Phases ($C_{5i} = S_{10}$)

0 0 0 +	3/2 -1 -½ +	2 -3/2 -½ +	5/2 2 ½ +
-½ ½ 0 +	-3/2 3/2 0 +	2 2 1 +	5/2 5/2 0 +
1 -½ -½ +	2 -1 -1 +	-2 2 0 +	
-1 1 0 +	2 3/2 3/2 +	5/2 3/2 1 +	

C_6 and C_{3h} 3j Phases ($C_{6i} = C_{6h}$)

0 0 0 +	-3/2 3/2 0 +	5/2 -3/2 -1 +	3 3/2 3/2 +
-½ ½ 0 +	2 -1 -1 +	5/2 2 3/2 +	3 2 1 +
1 -½ -½ +	2 -3/2 -½ +	5/2 -2 -½ +	3 5/2 ½ +
-1 1 0 +	2 2 2 +	5/2 5/2 1 +	3 3 0 +
3/2 -1 -½ +	-2 2 0 +	-5/2 5/2 0 +	

D_2 and C_{2v} 3j Phases ($D_{2i} = D_{2h}$)

$0\ 0\ 0\ +$	$\tilde{0}\ \tilde{0}\ 0\ +$	$\tilde{1}\ \tfrac{1}{2}\ \tfrac{1}{2}\ +$
$\tfrac{1}{2}\ \tfrac{1}{2}\ 0\ -$	$1\ \tfrac{1}{2}\ \tfrac{1}{2}\ +$	$\tilde{1}\ 1\ \tilde{0}\ -$
$\tilde{0}\ \tfrac{1}{2}\ \tfrac{1}{2}\ +$	$1\ 1\ 0\ +$	$\tilde{1}\ \tilde{1}\ 0\ +$

D_2 and C_{2v} 6j Symbols

$0\ 0\ 0$	**$1\ \tfrac{1}{2}\ \tfrac{1}{2}$**	**$\tilde{1}\ \tfrac{1}{2}\ \tfrac{1}{2}$**
$0\ 0\ 0\ +\ \ +1$	$0\ \tfrac{1}{2}\ \tfrac{1}{2}\ +\ \ +1/2$	$\tilde{1}\ \tfrac{1}{2}\ \tfrac{1}{2}\ +\ \ -1/2$
$\tfrac{1}{2}\ \tfrac{1}{2}\ 0$	$\tilde{0}\ \tfrac{1}{2}\ \tfrac{1}{2}\ +\ \ +1/2$	**$\tilde{1}\ 1\ \tilde{0}$**
$0\ 0\ \tfrac{1}{2}\ +\ \ -1/\sqrt{2}$	$1\ \tfrac{1}{2}\ \tfrac{1}{2}\ +\ \ -1/2$	$0\ \tilde{0}\ 1\ +\ \ -1$
$\tfrac{1}{2}\ \tfrac{1}{2}\ 0\ +\ \ -1/2$	**$1\ 1\ 0$**	$\tfrac{1}{2}\ \tfrac{1}{2}\ \tfrac{1}{2}\ +\ \ -1/\sqrt{2}$
$\tilde{0}\ \tfrac{1}{2}\ \tfrac{1}{2}$	$0\ 0\ 1\ +\ \ +1$	$\tilde{1}\ 1\ \tilde{0}\ +\ \ +1$
$0\ \tfrac{1}{2}\ \tfrac{1}{2}\ +\ \ +1/2$	$\tfrac{1}{2}\ \tfrac{1}{2}\ \tfrac{1}{2}\ +\ \ +1/\sqrt{2}$	**$\tilde{1}\ \tilde{1}\ 0$**
$\tilde{0}\ \tfrac{1}{2}\ \tfrac{1}{2}\ +\ \ -1/2$	$1\ 1\ 0\ +\ \ +1$	$0\ 0\ \tilde{1}\ +\ \ +1$
$\tilde{0}\ \tilde{0}\ 0$	**$\tilde{1}\ \tfrac{1}{2}\ \tfrac{1}{2}$**	$\tfrac{1}{2}\ \tfrac{1}{2}\ \tfrac{1}{2}\ +\ \ +1/\sqrt{2}$
$0\ 0\ \tilde{0}\ +\ \ +1$	$0\ \tfrac{1}{2}\ \tfrac{1}{2}\ +\ \ +1/2$	$\tilde{0}\ \tilde{0}\ 1\ +\ \ -1$
$\tfrac{1}{2}\ \tfrac{1}{2}\ \tfrac{1}{2}\ +\ \ +1/\sqrt{2}$	$\tilde{0}\ \tfrac{1}{2}\ \tfrac{1}{2}\ +\ \ +1/2$	$1\ 1\ \tilde{0}\ +\ \ -1$
$\tilde{0}\ \tilde{0}\ 0\ +\ \ +1$	$1\ \tfrac{1}{2}\ \tfrac{1}{2}\ +\ \ +1/2$	$\tilde{1}\ \tilde{1}\ 0\ +\ \ +1$

D_3 and C_{3v} 3j Phases ($D_{3i} = D_{3d}$)

$0\ 0\ 0\ +$	$\tilde{0}\ \tilde{0}\ 0\ +$	$1\ 1\ \tilde{0}\ -$	$\tfrac{3}{2}\ \tfrac{3}{2}\ \tilde{0}\ +$
$\tfrac{1}{2}\ \tfrac{1}{2}\ 0\ -$	$1\ \tfrac{1}{2}\ \tfrac{1}{2}\ +$	$1\ 1\ 1\ +$	$-\tfrac{3}{2}\ \tfrac{3}{2}\ 0\ -$
$\tilde{0}\ \tfrac{1}{2}\ \tfrac{1}{2}\ +$	$1\ 1\ 0\ +$	$\tfrac{3}{2}\ 1\ \tfrac{1}{2}\ -$	

D_3 and C_{3v} 6j Symbols

$0\ 0\ 0$	**$1\ \tfrac{1}{2}\ \tfrac{1}{2}$**	**$1\ 1\ 1$**
$0\ 0\ 0\ +\ \ +1$	$\tilde{0}\ \tfrac{1}{2}\ \tfrac{1}{2}\ +\ \ +1/2$	$\tfrac{1}{2}\ \tfrac{1}{2}\ \tfrac{1}{2}\ -\ \ \ 0$
$\tfrac{1}{2}\ \tfrac{1}{2}\ 0$	$1\ \tfrac{1}{2}\ \tfrac{1}{2}\ +\ \ \ 0$	$1\ 1\ 0\ +\ \ +1/2$
$0\ 0\ \tfrac{1}{2}\ +\ \ -1/\sqrt{2}$	**$1\ 1\ 0$**	$1\ 1\ \tilde{0}\ +\ \ +1/2$
$\tfrac{1}{2}\ \tfrac{1}{2}\ 0\ +\ \ -1/2$	$0\ 0\ 1\ +\ \ +1/\sqrt{2}$	$1\ 1\ 1\ +\ \ \ 0$
$\tilde{0}\ \tfrac{1}{2}\ \tfrac{1}{2}$	$\tfrac{1}{2}\ \tfrac{1}{2}\ \tfrac{1}{2}\ +\ \ +1/2$	**$\tfrac{3}{2}\ 1\ \tfrac{1}{2}$**
$0\ \tfrac{1}{2}\ \tfrac{1}{2}\ +\ \ +1/2$	$\tilde{0}\ \tilde{0}\ 1\ +\ \ -1/\sqrt{2}$	$0\ \tfrac{1}{2}\ 1\ +\ \ -1/2$
$\tilde{0}\ \tfrac{1}{2}\ \tfrac{1}{2}\ +\ \ -1/2$	$1\ 1\ 0\ +\ \ +1/2$	$\tfrac{1}{2}\ 1\ \tfrac{1}{2}\ +\ \ -1/2$
$\tilde{0}\ \tilde{0}\ 0$	**$1\ 1\ \tilde{0}$**	$\tilde{0}\ \tfrac{1}{2}\ 1\ +\ \ -1/2$
$0\ 0\ \tilde{0}\ +\ \ +1$	$\tfrac{1}{2}\ \tfrac{1}{2}\ \tfrac{1}{2}\ +\ \ -1/2$	$1\ \tfrac{1}{2}\ 1\ -\ \ -i/2$
$\tfrac{1}{2}\ \tfrac{1}{2}\ \tfrac{1}{2}\ +\ \ +1/\sqrt{2}$	$\tilde{0}\ 0\ 1\ +\ \ -1/\sqrt{2}$	$\tfrac{3}{2}\ 1\ \tfrac{1}{2}\ +\ \ +1/2$
$\tilde{0}\ \tilde{0}\ 0\ +\ \ +1$	$1\ 1\ 0\ +\ \ -1/2$	$-\tfrac{3}{2}\ 1\ \tfrac{1}{2}\ +\ \ -1/2$
$1\ \tfrac{1}{2}\ \tfrac{1}{2}$	$1\ 1\ \tilde{0}\ +\ \ +1/2$	**$\tfrac{3}{2}\ \tfrac{3}{2}\ \tilde{0}$**
$0\ \tfrac{1}{2}\ \tfrac{1}{2}\ +\ \ +1/2$		$\tfrac{1}{2}\ \tfrac{1}{2}\ 1\ +\ \ +1/\sqrt{2}$

D_3 and C_{3v} 6j Symbols (cont.)

$\frac{3}{2}\ \frac{3}{2}\ \tilde{0}$

$\tilde{0}\ 0\ \text{-}\frac{3}{2}\ +\ \ +1$

$1\ 1\ \frac{1}{2}\ +\ \ +1/\sqrt{2}$

$\frac{3}{2}\ \text{-}\frac{3}{2}\ \tilde{0}\ +\ \ -1$

$\text{-}\frac{3}{2}\ \frac{3}{2}\ 0\ +\ \ +1$

$\text{-}\frac{3}{2}\ \frac{3}{2}\ 0$

$0\ 0\ \frac{3}{2}\ +\ \ -1$

$\frac{1}{2}\ \frac{1}{2}\ 1\ +\ \ -1/\sqrt{2}$

$\tilde{0}\ \tilde{0}\ \text{-}\frac{3}{2}\ +\ \ +1$

$1\ 1\ \frac{1}{2}\ +\ \ -1/\sqrt{2}$

$\text{-}\frac{3}{2}\ \frac{3}{2}\ \mathbf{0}$

$\frac{3}{2}\ \frac{3}{2}\ \tilde{0}\ +\ \ +1$

$\text{-}\frac{3}{2}\ \text{-}\frac{3}{2}\ 0\ +\ \ -1$

D_4, C_{4v}, and D_{2d} 3j Phases ($D_{4i} = D_{4h}$)

$0\ 0\ 0$ +	$1\ 1\ 0$ +	$\frac{3}{2}\ \frac{3}{2}\ 1$ +	$\tilde{2}\ \frac{3}{2}\ \frac{1}{2}$ +
$\frac{1}{2}\ \frac{1}{2}\ 0$ −	$1\ 1\ \tilde{0}$ −	$2\ 1\ 1$ +	$\tilde{2}\ 2\ \tilde{0}$ −
$\tilde{0}\ \frac{1}{2}\ \frac{1}{2}$ +	$\frac{3}{2}\ 1\ \frac{1}{2}$ −	$2\ \frac{3}{2}\ \frac{1}{2}$ +	$\tilde{2}\ \tilde{2}\ 0$ +
$\tilde{0}\ \tilde{0}\ 0$ +	$\frac{3}{2}\ \frac{3}{2}\ 0$ −	$2\ 2\ 0$ +	
$1\ \frac{1}{2}\ \frac{1}{2}$ +	$\frac{3}{2}\ \frac{3}{2}\ \tilde{0}$ +	$\tilde{2}\ 1\ 1$ +	

D_4, C_{4v}, and D_{2d} 6j Symbols

$0\ 0\ 0$

$0\ 0\ 0\ +\ \ +1$

$\frac{1}{2}\ \frac{1}{2}\ 0$

$0\ 0\ \frac{1}{2}\ +\ \ -1/\sqrt{2}$

$\frac{1}{2}\ \frac{1}{2}\ 0\ +\ \ -1/2$

$\tilde{0}\ \frac{1}{2}\ \frac{1}{2}$

$0\ \frac{1}{2}\ \frac{1}{2}\ +\ \ +1/2$

$\tilde{0}\ \frac{1}{2}\ \frac{1}{2}\ +\ \ -1/2$

$\tilde{0}\ \tilde{0}\ 0$

$0\ 0\ \tilde{0}\ +\ \ +1$

$\frac{1}{2}\ \frac{1}{2}\ \frac{1}{2}\ +\ \ +1/\sqrt{2}$

$\tilde{0}\ \tilde{0}\ 0\ +\ \ +1$

$1\ \frac{1}{2}\ \frac{1}{2}$

$0\ \frac{1}{2}\ \frac{1}{2}\ +\ \ +1/2$

$\tilde{0}\ \frac{1}{2}\ \frac{1}{2}\ +\ \ +1/2$

$1\ \frac{1}{2}\ \frac{1}{2}\ +\ \ 0$

$1\ 1\ 0$

$0\ 0\ 1\ +\ \ +1/\sqrt{2}$

$\frac{1}{2}\ \frac{1}{2}\ \frac{1}{2}\ +\ \ +1/2$

$\tilde{0}\ \tilde{0}\ 1\ +\ \ -1/\sqrt{2}$

$1\ 1\ 0\ +\ \ +1/2$

$1\ 1\ \tilde{0}$

$\frac{1}{2}\ \frac{1}{2}\ \frac{1}{2}\ +\ \ -1/2$

$\tilde{0}\ 0\ 1\ +\ \ -1/\sqrt{2}$

$1\ 1\ 0\ +\ \ -1/2$

$1\ 1\ \tilde{0}$

$1\ 1\ \tilde{0}\ +\ \ +1/2$

$\frac{3}{2}\ 1\ \frac{1}{2}$

$0\ \frac{1}{2}\ 1\ +\ \ -1/2$

$\frac{1}{2}\ 1\ \frac{1}{2}\ +\ \ -1/2$

$\tilde{0}\ \frac{1}{2}\ 1\ +\ \ -1/2$

$\frac{3}{2}\ 1\ \frac{1}{2}\ +\ \ 0$

$\frac{3}{2}\ \frac{3}{2}\ 0$

$0\ 0\ \frac{3}{2}\ +\ \ -1/\sqrt{2}$

$\frac{1}{2}\ \frac{1}{2}\ 1\ +\ \ -1/2$

$\tilde{0}\ \tilde{0}\ \frac{3}{2}\ +\ \ +1/\sqrt{2}$

$1\ 1\ \frac{1}{2}\ +\ \ -1/2$

$1\ 1\ \frac{3}{2}\ +\ \ +1/2$

$\frac{3}{2}\ \frac{3}{2}\ 0\ +\ \ -1/2$

$\frac{3}{2}\ \frac{3}{2}\ \tilde{0}$

$\frac{1}{2}\ \frac{1}{2}\ 1\ +\ \ +1/2$

$\tilde{0}\ 0\ \frac{3}{2}\ +\ \ +1/\sqrt{2}$

$1\ 1\ \frac{1}{2}\ +\ \ +1/2$

$1\ 1\ \frac{3}{2}\ +\ \ +1/2$

$\frac{3}{2}\ \frac{3}{2}\ 0\ +\ \ +1/2$

$\frac{3}{2}\ \frac{3}{2}\ \tilde{0}\ +\ \ -1/2$

$\frac{3}{2}\ \frac{3}{2}\ 1$

$\frac{1}{2}\ \frac{1}{2}\ 1\ +\ \ 0$

$1\ 0\ \frac{3}{2}\ +\ \ +1/2$

$1\ \tilde{0}\ \frac{3}{2}\ +\ \ +1/2$

$\frac{3}{2}\ \frac{3}{2}\ 1$

$\frac{3}{2}\ \frac{1}{2}\ 1\ +\ \ -1/2$

$\frac{3}{2}\ \frac{3}{2}\ 0\ +\ \ +1/2$

$\frac{3}{2}\ \frac{3}{2}\ \tilde{0}\ +\ \ +1/2$

$\frac{3}{2}\ \frac{3}{2}\ 1\ +\ \ 0$

$2\ 1\ 1$

$0\ 1\ 1\ +\ \ +1/2$

$\tilde{0}\ 1\ 1\ +\ \ +1/2$

$2\ 1\ 1\ +\ \ +1/2$

$2\ \frac{3}{2}\ \frac{1}{2}$

$0\ \frac{1}{2}\ \frac{3}{2}\ +\ \ +1/2$

$\frac{1}{2}\ 1\ 1\ +\ \ +1/2$

$\tilde{0}\ \frac{1}{2}\ \frac{3}{2}\ +\ \ +1/2$

$1\ \frac{1}{2}\ \frac{3}{2}\ +\ \ -1/2$

$1\ \frac{3}{2}\ \frac{1}{2}\ +\ \ +1/2$

$\frac{3}{2}\ 1\ 1\ +\ \ +1/2$

$2\ \frac{3}{2}\ \frac{1}{2}\ +\ \ -1/2$

$2\ 2\ 0$

$0\ 0\ 2\ +\ \ +1$

$\frac{1}{2}\ \frac{1}{2}\ \frac{3}{2}\ +\ \ +1/\sqrt{2}$

$1\ 1\ 1\ +\ \ +1/\sqrt{2}$

$\frac{3}{2}\ \frac{3}{2}\ \frac{1}{2}\ +\ \ +1/\sqrt{2}$

$2\ 2\ 0\ +\ \ +1$

$\tilde{2}\ 1\ 1$

$0\ 1\ 1\ +\ \ +1/2$

D_4, C_{4v}, and D_{2d} 6j Symbols (cont.)

$\tilde{2}$ 1 1
$\tilde{0}$ 1 1 + +1/2
2 1 1 + −1/2
$\tilde{2}$ 1 1 + +1/2
$\tilde{2}$ $\frac{3}{2}$ $\frac{1}{2}$
0 $\frac{1}{2}$ $\frac{3}{2}$ + +1/2
$\frac{1}{2}$ 1 1 + +1/2
$\tilde{0}$ $\frac{1}{2}$ $\frac{3}{2}$ + +1/2
1 $\frac{1}{2}$ $\frac{3}{2}$ + +1/2
1 $\frac{3}{2}$ $\frac{1}{2}$ + +1/2

$\tilde{2}$ $\frac{3}{2}$ $\frac{1}{2}$
$\frac{3}{2}$ 1 1 + −1/2
2 $\frac{3}{2}$ $\frac{1}{2}$ + +1/2
$\tilde{2}$ $\frac{3}{2}$ $\frac{1}{2}$ + −1/2
$\tilde{2}$ 2 $\tilde{0}$
0 $\tilde{0}$ 2 + −1
$\frac{1}{2}$ $\frac{1}{2}$ $\frac{3}{2}$ + $-1/\sqrt{2}$
1 1 1 + $-1/\sqrt{2}$
$\frac{3}{2}$ $\frac{3}{2}$ $\frac{1}{2}$ + $-1/\sqrt{2}$
$\tilde{2}$ 2 $\tilde{0}$ + +1

$\tilde{2}$ $\tilde{2}$ 0
0 0 $\tilde{2}$ + +1
$\frac{1}{2}$ $\frac{1}{2}$ $\frac{3}{2}$ + $+1/\sqrt{2}$
$\tilde{0}$ $\tilde{0}$ 2 + −1
1 1 1 + $+1/\sqrt{2}$
$\frac{3}{2}$ $\frac{3}{2}$ $\frac{1}{2}$ + $+1/\sqrt{2}$
2 2 $\tilde{0}$ + −1
$\tilde{2}$ $\tilde{2}$ 0 + +1

D_5 and C_{5v} 3j Phases ($D_{5i} = D_{5d}$)

0 0 0 +	1 1 0 +	2 1 1 +	2 2 1 +
$\frac{1}{2}$ $\frac{1}{2}$ 0 −	1 1 $\tilde{0}$ −	2 $\frac{3}{2}$ $\frac{1}{2}$ +	$\frac{5}{2}$ $\frac{3}{2}$ 1 −
$\tilde{0}$ $\frac{1}{2}$ $\frac{1}{2}$ +	$\frac{3}{2}$ 1 $\frac{1}{2}$ −	2 $\frac{3}{2}$ $\frac{3}{2}$ +	$\frac{5}{2}$ 2 $\frac{1}{2}$ −
$\tilde{0}$ $\tilde{0}$ 0 +	$\frac{3}{2}$ $\frac{3}{2}$ 0 −	2 2 0 +	$\frac{5}{2}$ $\frac{5}{2}$ $\tilde{0}$ +
1 $\frac{1}{2}$ $\frac{1}{2}$ +	$\frac{3}{2}$ $\frac{3}{2}$ $\tilde{0}$ +	2 2 $\tilde{0}$ −	-$\frac{5}{2}$ $\frac{5}{2}$ 0 −

D_5 and C_{5v} 6j Symbols

0 0 0
0 0 0 + +1
$\frac{1}{2}$ $\frac{1}{2}$ 0
0 0 $\frac{1}{2}$ + $-1/\sqrt{2}$
$\frac{1}{2}$ $\frac{1}{2}$ 0 + −1/2
$\tilde{0}$ $\frac{1}{2}$ $\frac{1}{2}$
0 $\frac{1}{2}$ $\frac{1}{2}$ + +1/2
$\tilde{0}$ $\frac{1}{2}$ $\frac{1}{2}$ + −1/2
$\tilde{0}$ $\tilde{0}$ 0
0 0 $\tilde{0}$ + +1
$\frac{1}{2}$ $\frac{1}{2}$ $\frac{1}{2}$ + $+1/\sqrt{2}$
$\tilde{0}$ $\tilde{0}$ 0 + +1
1 $\frac{1}{2}$ $\frac{1}{2}$
0 $\frac{1}{2}$ $\frac{1}{2}$ + +1/2
$\tilde{0}$ $\frac{1}{2}$ $\frac{1}{2}$ + +1/2
1 $\frac{1}{2}$ $\frac{1}{2}$ + 0
1 1 0
0 0 1 + $+1/\sqrt{2}$
$\frac{1}{2}$ $\frac{1}{2}$ $\frac{1}{2}$ + +1/2

1 1 0
$\tilde{0}$ $\tilde{0}$ 1 + $-1/\sqrt{2}$
1 1 0 + +1/2
1 1 $\tilde{0}$
$\frac{1}{2}$ $\frac{1}{2}$ $\frac{1}{2}$ + −1/2
$\tilde{0}$ 0 1 + $-1/\sqrt{2}$
1 1 0 + −1/2
1 1 $\tilde{0}$ + +1/2
$\frac{3}{2}$ 1 $\frac{1}{2}$
0 $\frac{1}{2}$ 1 + −1/2
$\frac{1}{2}$ 1 $\frac{1}{2}$ + −1/2
$\tilde{0}$ $\frac{1}{2}$ 1 + −1/2
$\frac{3}{2}$ 1 $\frac{1}{2}$ + 0
$\frac{3}{2}$ $\frac{3}{2}$ 0
0 0 $\frac{3}{2}$ + $-1/\sqrt{2}$
$\frac{1}{2}$ $\frac{1}{2}$ 1 + −1/2
$\tilde{0}$ $\tilde{0}$ $\frac{3}{2}$ + $+1/\sqrt{2}$
1 1 $\frac{1}{2}$ + −1/2
$\frac{3}{2}$ $\frac{3}{2}$ 0 + −1/2

$\frac{3}{2}$ $\frac{3}{2}$ $\tilde{0}$
$\frac{1}{2}$ $\frac{1}{2}$ 1 + +1/2
$\tilde{0}$ 0 $\frac{3}{2}$ + $+1/\sqrt{2}$
1 1 $\frac{1}{2}$ + +1/2
$\frac{3}{2}$ $\frac{3}{2}$ 0 + +1/2
$\frac{3}{2}$ $\frac{3}{2}$ $\tilde{0}$ + −1/2
2 1 1
0 1 1 + +1/2
$\tilde{0}$ 1 1 + +1/2
2 1 1 + 0
2 $\frac{3}{2}$ $\frac{1}{2}$
0 $\frac{1}{2}$ $\frac{3}{2}$ + +1/2
$\frac{1}{2}$ 1 1 + +1/2
$\tilde{0}$ $\frac{1}{2}$ $\frac{3}{2}$ + +1/2
1 $\frac{3}{2}$ $\frac{1}{2}$ + +1/2
2 $\frac{3}{2}$ $\frac{1}{2}$ + 0
2 $\frac{3}{2}$ $\frac{3}{2}$
0 $\frac{3}{2}$ $\frac{3}{2}$ + +1/2
$\frac{1}{2}$ 1 1 − 0

D_5 and C_{5v} 6j Symbols (cont.)

$2\ \frac{3}{2}\ \frac{3}{2}$

$\tilde{0}\ \frac{3}{2}\ \frac{3}{2}\ +\ +1/2$
$2\ \frac{3}{2}\ \frac{1}{2}\ +\ -1/2$
$2\ \frac{3}{2}\ \frac{3}{2}\ +\ 0$

$2\ 2\ 0$

$0\ 0\ 2\ +\ +1/\sqrt{2}$
$\frac{1}{2}\ \frac{1}{2}\ \frac{3}{2}\ +\ +1/2$
$\tilde{0}\ \tilde{0}\ 2\ +\ -1/\sqrt{2}$
$1\ 1\ 1\ +\ +1/2$
$1\ 1\ 2\ +\ +1/2$
$\frac{3}{2}\ \frac{3}{2}\ \frac{1}{2}\ +\ +1/2$
$\frac{3}{2}\ \frac{3}{2}\ \frac{3}{2}\ +\ +1/2$
$2\ 2\ 0\ +\ +1/2$

$2\ 2\ \tilde{0}$

$\frac{1}{2}\ \frac{1}{2}\ \frac{3}{2}\ +\ -1/2$
$\tilde{0}\ 0\ 2\ +\ -1/\sqrt{2}$
$1\ 1\ 1\ +\ -1/2$
$1\ 1\ 2\ +\ +1/2$
$\frac{3}{2}\ \frac{3}{2}\ \frac{1}{2}\ +\ -1/2$
$\frac{3}{2}\ \frac{3}{2}\ \frac{3}{2}\ +\ +1/2$
$2\ 2\ 0\ +\ -1/2$
$2\ 2\ \tilde{0}\ +\ +1/2$

$2\ 2\ 1$

$\frac{1}{2}\ \frac{1}{2}\ \frac{3}{2}\ -\ 0$
$1\ 0\ 2\ +\ +1/2$
$1\ \tilde{0}\ 2\ +\ +1/2$
$\frac{3}{2}\ \frac{1}{2}\ \frac{3}{2}\ +\ +1/2$
$2\ 1\ 1\ +\ +1/2$
$2\ 2\ 0\ +\ +1/2$
$2\ 2\ \tilde{0}\ +\ +1/2$
$2\ 2\ 1\ +\ 0$

$\frac{5}{2}\ \frac{3}{2}\ 1$

$0\ 1\ \frac{3}{2}\ +\ -1/2$
$\frac{1}{2}\ \frac{3}{2}\ 1\ +\ -1/2$
$\tilde{0}\ 1\ \frac{3}{2}\ +\ -1/2$
$2\ 1\ \frac{3}{2}\ -\ +i/2$
$\frac{5}{2}\ \frac{3}{2}\ 1\ +\ -1/2$
$-\frac{5}{2}\ \frac{3}{2}\ 1\ +\ +1/2$

$\frac{5}{2}\ 2\ \frac{1}{2}$

$0\ \frac{1}{2}\ 2\ +\ -1/2$
$\frac{1}{2}\ 1\ \frac{3}{2}\ +\ -1/2$
$\tilde{0}\ \frac{1}{2}\ 2\ +\ -1/2$
$1\ \frac{1}{2}\ 2\ -\ -i/2$
$1\ \frac{3}{2}\ 1\ +\ -1/2$
$\frac{3}{2}\ 1\ \frac{3}{2}\ -\ -i/2$

$\frac{5}{2}\ 2\ \frac{1}{2}$

$\frac{3}{2}\ 2\ \frac{1}{2}\ +\ -1/2$
$2\ \frac{3}{2}\ 1\ -\ -i/2$
$\frac{5}{2}\ 2\ \frac{1}{2}\ +\ +1/2$
$-\frac{5}{2}\ 2\ \frac{1}{2}\ +\ -1/2$

$\frac{5}{2}\ \frac{5}{2}\ \tilde{0}$

$\frac{1}{2}\ \frac{1}{2}\ 2\ +\ +1/\sqrt{2}$
$\tilde{0}\ 0\ -\frac{5}{2}\ +\ +1$
$1\ 1\ \frac{3}{2}\ +\ +1/\sqrt{2}$
$\frac{3}{2}\ \frac{3}{2}\ 1\ +\ +1/\sqrt{2}$
$2\ 2\ \frac{1}{2}\ +\ +1/\sqrt{2}$
$\frac{5}{2}\ -\frac{5}{2}\ \tilde{0}\ +\ -1$
$-\frac{5}{2}\ \frac{5}{2}\ 0\ +\ +1$

$-\frac{5}{2}\ \frac{5}{2}\ 0$

$0\ 0\ \frac{5}{2}\ +\ -1$
$\frac{1}{2}\ \frac{1}{2}\ 2\ +\ -1/\sqrt{2}$
$\tilde{0}\ \tilde{0}\ -\frac{5}{2}\ +\ +1$
$1\ 1\ \frac{3}{2}\ +\ -1/\sqrt{2}$
$\frac{3}{2}\ \frac{3}{2}\ 1\ +\ -1/\sqrt{2}$
$2\ 2\ \frac{1}{2}\ +\ -1/\sqrt{2}$
$\frac{5}{2}\ \frac{5}{2}\ \tilde{0}\ +\ +1$
$-\frac{5}{2}\ -\frac{5}{2}\ 0\ +\ -1$

D_6, C_{6v}, and D_{3h} 3j Phases ($D_{6i} = D_{6h}$)

$0\ 0\ 0\ +$	$\frac{3}{2}\ \frac{3}{2}\ 0\ -$	$\frac{5}{2}\ 2\ \frac{1}{2}\ -$	$3\ 3\ 0\ +$
$\frac{1}{2}\ \frac{1}{2}\ 0\ -$	$\frac{3}{2}\ \frac{3}{2}\ \tilde{0}\ +$	$\frac{5}{2}\ 2\ \frac{3}{2}\ +$	$\tilde{3}\ \frac{3}{2}\ \frac{3}{2}\ +$
$\tilde{0}\ \frac{1}{2}\ \frac{1}{2}\ +$	$2\ 1\ 1\ +$	$\frac{5}{2}\ \frac{5}{2}\ 0\ -$	$\tilde{3}\ 2\ 1\ +$
$\tilde{0}\ \tilde{0}\ 0\ +$	$2\ \frac{3}{2}\ \frac{1}{2}\ +$	$\frac{5}{2}\ \frac{5}{2}\ \tilde{0}\ +$	$\tilde{3}\ \frac{5}{2}\ \frac{1}{2}\ +$
$1\ \frac{1}{2}\ \frac{1}{2}\ +$	$2\ 2\ 0\ +$	$\frac{5}{2}\ \frac{5}{2}\ 1\ +$	$\tilde{3}\ 3\ \tilde{0}\ -$
$1\ 1\ 0\ +$	$2\ 2\ \tilde{0}\ -$	$3\ \frac{3}{2}\ \frac{3}{2}\ +$	$\tilde{3}\ \tilde{3}\ 0\ +$
$1\ 1\ \tilde{0}\ -$	$2\ 2\ 2\ +$	$3\ 2\ 1\ +$	
$\frac{3}{2}\ 1\ \frac{1}{2}\ -$	$\frac{5}{2}\ \frac{3}{2}\ 1\ -$	$3\ \frac{5}{2}\ \frac{1}{2}\ +$	

D_6, C_{6v}, and D_{3h} 6j Symbols

$0\ 0\ 0$
$0\ 0\ 0\ +\ +1$
$\frac{1}{2}\ \frac{1}{2}\ 0$
$0\ 0\ \frac{1}{2}\ +\ -1/\sqrt{2}$
$\frac{1}{2}\ \frac{1}{2}\ 0\ +\ -1/2$
$\tilde{0}\ \frac{1}{2}\ \frac{1}{2}$
$0\ \frac{1}{2}\ \frac{1}{2}\ +\ +1/2$
$\tilde{0}\ \frac{1}{2}\ \frac{1}{2}\ +\ -1/2$
$\tilde{0}\ \tilde{0}\ 0$
$0\ 0\ \tilde{0}\ +\ +1$
$\frac{1}{2}\ \frac{1}{2}\ \frac{1}{2}\ +\ +1/\sqrt{2}$
$\tilde{0}\ \tilde{0}\ 0\ +\ +1$
$1\ \frac{1}{2}\ \frac{1}{2}$
$0\ \frac{1}{2}\ \frac{1}{2}\ +\ +1/2$
$\tilde{0}\ \frac{1}{2}\ \frac{1}{2}\ +\ +1/2$
$1\ \frac{1}{2}\ \frac{1}{2}\ +\ 0$
$1\ 1\ 0$
$0\ 0\ 1\ +\ +1/\sqrt{2}$
$\frac{1}{2}\ \frac{1}{2}\ \frac{1}{2}\ +\ +1/2$
$\tilde{0}\ \tilde{0}\ 1\ +\ -1/\sqrt{2}$
$1\ 1\ 0\ +\ +1/2$
$1\ 1\ \tilde{0}$
$\frac{1}{2}\ \frac{1}{2}\ \frac{1}{2}\ +\ -1/2$
$\tilde{0}\ 0\ 1\ +\ -1/\sqrt{2}$
$1\ 1\ 0\ +\ -1/2$
$1\ 1\ \tilde{0}\ +\ +1/2$
$\frac{3}{2}\ 1\ \frac{1}{2}$
$0\ \frac{1}{2}\ 1\ +\ -1/2$
$\frac{1}{2}\ 1\ \frac{1}{2}\ +\ -1/2$
$\tilde{0}\ \frac{1}{2}\ 1\ +\ -1/2$
$\frac{3}{2}\ 1\ \frac{1}{2}\ +\ 0$
$\frac{3}{2}\ \frac{3}{2}\ 0$
$0\ 0\ \frac{3}{2}\ +\ -1/\sqrt{2}$
$\frac{1}{2}\ \frac{1}{2}\ 1\ +\ -1/2$
$\tilde{0}\ \tilde{0}\ \frac{3}{2}\ +\ +1/\sqrt{2}$
$1\ 1\ \frac{1}{2}\ +\ -1/2$
$\frac{3}{2}\ \frac{3}{2}\ 0\ +\ -1/2$
$\frac{3}{2}\ \frac{3}{2}\ \tilde{0}$
$\frac{1}{2}\ \frac{1}{2}\ 1\ +\ +1/2$
$\tilde{0}\ 0\ \frac{3}{2}\ +\ +1/\sqrt{2}$
$1\ 1\ \frac{1}{2}\ +\ +1/2$
$\frac{3}{2}\ \frac{3}{2}\ \tilde{0}$
$\frac{3}{2}\ \frac{3}{2}\ 0\ +\ +1/2$
$\frac{3}{2}\ \frac{3}{2}\ \tilde{0}\ +\ -1/2$
$2\ 1\ 1$
$0\ 1\ 1\ +\ +1/2$
$\tilde{0}\ 1\ 1\ +\ +1/2$
$2\ 1\ 1\ +\ 0$
$2\ \frac{3}{2}\ \frac{1}{2}$
$0\ \frac{1}{2}\ \frac{3}{2}\ +\ +1/2$
$\frac{1}{2}\ 1\ 1\ +\ +1/2$
$\tilde{0}\ \frac{1}{2}\ \frac{3}{2}\ +\ +1/2$
$1\ \frac{3}{2}\ \frac{1}{2}\ +\ +1/2$
$2\ \frac{3}{2}\ \frac{1}{2}\ +\ 0$
$2\ 2\ 0$
$0\ 0\ 2\ +\ +1/\sqrt{2}$
$\frac{1}{2}\ \frac{1}{2}\ \frac{3}{2}\ +\ +1/2$
$\tilde{0}\ \tilde{0}\ 2\ +\ -1/\sqrt{2}$
$1\ 1\ 1\ +\ +1/2$
$\frac{3}{2}\ \frac{3}{2}\ \frac{1}{2}\ +\ +1/2$
$2\ 2\ 0\ +\ +1/2$
$2\ 2\ \tilde{0}$
$\frac{1}{2}\ \frac{1}{2}\ \frac{3}{2}\ +\ -1/2$
$\tilde{0}\ 0\ 2\ +\ -1/\sqrt{2}$
$1\ 1\ 1\ +\ -1/2$
$\frac{3}{2}\ \frac{3}{2}\ \frac{1}{2}\ +\ -1/2$
$2\ 2\ 0\ +\ -1/2$
$2\ 2\ \tilde{0}\ +\ +1/2$
$2\ 2\ 2$
$1\ 1\ 1\ +\ 0$
$2\ 2\ 0\ +\ +1/2$
$2\ 2\ \tilde{0}\ +\ +1/2$
$2\ 2\ 2\ +\ 0$
$\frac{5}{2}\ \frac{3}{2}\ 1$
$0\ 1\ \frac{3}{2}\ +\ -1/2$
$\frac{1}{2}\ \frac{3}{2}\ 1\ +\ -1/2$
$\tilde{0}\ 1\ \frac{3}{2}\ +\ -1/2$
$\frac{5}{2}\ \frac{3}{2}\ 1\ +\ 0$
$\frac{5}{2}\ 2\ \frac{1}{2}$
$0\ \frac{1}{2}\ 2\ +\ -1/2$
$\frac{1}{2}\ 1\ \frac{3}{2}\ +\ -1/2$
$\tilde{0}\ \frac{1}{2}\ 2\ +\ -1/2$
$\frac{5}{2}\ 2\ \frac{1}{2}$
$1\ \frac{3}{2}\ 1\ +\ -1/2$
$\frac{3}{2}\ 2\ \frac{1}{2}\ +\ -1/2$
$2\ \frac{3}{2}\ 2\ +\ +1/2$
$\frac{5}{2}\ 2\ \frac{1}{2}\ +\ 0$
$\frac{5}{2}\ 2\ \frac{3}{2}$
$0\ \frac{3}{2}\ 2\ +\ +1/2$
$\frac{1}{2}\ 1\ \frac{3}{2}\ +\ 0$
$\frac{1}{2}\ 2\ \frac{3}{2}\ +\ -1/2$
$\tilde{0}\ \frac{3}{2}\ 2\ +\ +1/2$
$2\ \frac{1}{2}\ 2\ +\ +1/2$
$\frac{5}{2}\ 1\ \frac{3}{2}\ +\ +1/2$
$\frac{5}{2}\ 2\ \frac{1}{2}\ +\ -1/2$
$\frac{5}{2}\ 2\ \frac{3}{2}\ +\ 0$
$\frac{5}{2}\ \frac{5}{2}\ 0$
$0\ 0\ \frac{5}{2}\ +\ -1/\sqrt{2}$
$\frac{1}{2}\ \frac{1}{2}\ 2\ +\ -1/2$
$\tilde{0}\ \tilde{0}\ \frac{5}{2}\ +\ +1/\sqrt{2}$
$1\ 1\ \frac{3}{2}\ +\ -1/2$
$1\ 1\ \frac{5}{2}\ +\ +1/2$
$\frac{3}{2}\ \frac{3}{2}\ 1\ +\ -1/2$
$\frac{3}{2}\ \frac{3}{2}\ 2\ +\ +1/2$
$2\ 2\ \frac{1}{2}\ +\ -1/2$
$2\ 2\ \frac{3}{2}\ +\ +1/2$
$\frac{5}{2}\ \frac{5}{2}\ 0\ +\ -1/2$
$\frac{5}{2}\ \frac{5}{2}\ \tilde{0}$
$\frac{1}{2}\ \frac{1}{2}\ 2\ +\ +1/2$
$\tilde{0}\ 0\ \frac{5}{2}\ +\ +1/\sqrt{2}$
$1\ 1\ \frac{3}{2}\ +\ +1/2$
$1\ 1\ \frac{5}{2}\ +\ +1/2$
$\frac{3}{2}\ \frac{3}{2}\ 1\ +\ +1/2$
$\frac{3}{2}\ \frac{3}{2}\ 2\ +\ +1/2$
$2\ 2\ \frac{1}{2}\ +\ +1/2$
$2\ 2\ \frac{3}{2}\ +\ +1/2$
$\frac{5}{2}\ \frac{5}{2}\ 0\ +\ +1/2$
$\frac{5}{2}\ \frac{5}{2}\ \tilde{0}\ +\ -1/2$
$\frac{5}{2}\ \frac{5}{2}\ 1$
$\frac{1}{2}\ \frac{1}{2}\ 2\ +\ 0$
$1\ 0\ \frac{5}{2}\ +\ +1/2$
$1\ \tilde{0}\ \frac{5}{2}\ +\ +1/2$
$\frac{3}{2}\ \frac{1}{2}\ 2\ +\ +1/2$

D_6, C_{6v}, and D_{3h} 6j Symbols (cont.)

$\frac{5}{2}\ \frac{5}{2}\ 1$

$2\ 1\ \frac{3}{2}\ +\ -1/2$

$\frac{5}{2}\ \frac{3}{2}\ 1\ +\ -1/2$

$\frac{5}{2}\ \frac{5}{2}\ 0\ +\ +1/2$

$\frac{5}{2}\ \frac{5}{2}\ \tilde{0}\ +\ +1/2$

$\frac{5}{2}\ \frac{5}{2}\ 1\ +\ 0$

$3\ \frac{3}{2}\ \frac{3}{2}$

$0\ \frac{3}{2}\ \frac{3}{2}\ +\ +1/2$

$\tilde{0}\ \frac{3}{2}\ \frac{3}{2}\ +\ +1/2$

$3\ \frac{3}{2}\ \frac{3}{2}\ +\ -1/2$

$3\ 2\ 1$

$0\ 1\ 2\ +\ +1/2$

$\frac{1}{2}\ \frac{3}{2}\ \frac{3}{2}\ +\ +1/2$

$\tilde{0}\ 1\ 2\ +\ +1/2$

$1\ 2\ 1\ +\ +1/2$

$2\ 1\ 2\ +\ -1/2$

$\frac{5}{2}\ \frac{3}{2}\ \frac{3}{2}\ +\ -1/2$

$3\ 2\ 1\ +\ +1/2$

$3\ \frac{5}{2}\ \frac{1}{2}$

$0\ \frac{1}{2}\ \frac{5}{2}\ +\ +1/2$

$\frac{1}{2}\ 1\ 2\ +\ +1/2$

$\tilde{0}\ \frac{1}{2}\ \frac{5}{2}\ +\ +1/2$

$1\ \frac{1}{2}\ \frac{5}{2}\ +\ -1/2$

$1\ \frac{3}{2}\ \frac{3}{2}\ +\ +1/2$

$\frac{3}{2}\ 1\ 2\ +\ -1/2$

$\frac{3}{2}\ 2\ 1\ +\ +1/2$

$2\ \frac{3}{2}\ \frac{3}{2}\ +\ +1/2$

$2\ \frac{5}{2}\ \frac{1}{2}\ +\ +1/2$

$\frac{5}{2}\ 2\ 1\ +\ +1/2$

$3\ \frac{5}{2}\ \frac{1}{2}$

$3\ \frac{5}{2}\ \frac{1}{2}\ +\ -1/2$

$3\ 3\ 0$

$0\ 0\ 3\ +\ +1$

$\frac{1}{2}\ \frac{1}{2}\ \frac{5}{2}\ +\ +1/\sqrt{2}$

$1\ 1\ 2\ +\ +1/\sqrt{2}$

$\frac{3}{2}\ \frac{3}{2}\ \frac{3}{2}\ +\ +1/\sqrt{2}$

$2\ 2\ 1\ +\ +1/\sqrt{2}$

$\frac{5}{2}\ \frac{5}{2}\ \frac{1}{2}\ +\ +1/\sqrt{2}$

$3\ 3\ 0\ +\ +1$

$\tilde{3}\ \frac{3}{2}\ \frac{3}{2}$

$0\ \frac{3}{2}\ \frac{3}{2}\ +\ +1/2$

$\tilde{0}\ \frac{3}{2}\ \frac{3}{2}\ +\ +1/2$

$3\ \frac{3}{2}\ \frac{3}{2}\ +\ +1/2$

$\tilde{3}\ \frac{3}{2}\ \frac{3}{2}\ +\ -1/2$

$\tilde{3}\ 2\ 1$

$0\ 1\ 2\ +\ +1/2$

$\frac{1}{2}\ \frac{3}{2}\ \frac{3}{2}\ +\ +1/2$

$\tilde{0}\ 1\ 2\ +\ +1/2$

$1\ 2\ 1\ +\ +1/2$

$2\ 1\ 2\ +\ +1/2$

$\frac{5}{2}\ \frac{3}{2}\ \frac{3}{2}\ +\ +1/2$

$3\ 2\ 1\ +\ -1/2$

$\tilde{3}\ 2\ 1\ +\ +1/2$

$\tilde{3}\ \frac{5}{2}\ \frac{1}{2}$

$0\ \frac{1}{2}\ \frac{5}{2}\ +\ +1/2$

$\frac{1}{2}\ 1\ 2\ +\ +1/2$

$\tilde{0}\ \frac{1}{2}\ \frac{5}{2}\ +\ +1/2$

$1\ \frac{1}{2}\ \frac{5}{2}\ +\ +1/2$

$\tilde{3}\ \frac{5}{2}\ \frac{1}{2}$

$1\ \frac{3}{2}\ \frac{3}{2}\ +\ +1/2$

$\frac{3}{2}\ 1\ 2\ +\ +1/2$

$\frac{3}{2}\ 2\ 1\ +\ +1/2$

$2\ \frac{3}{2}\ \frac{3}{2}\ +\ -1/2$

$2\ \frac{5}{2}\ \frac{1}{2}\ +\ +1/2$

$\frac{5}{2}\ 2\ 1\ +\ -1/2$

$3\ \frac{5}{2}\ \frac{1}{2}\ +\ +1/2$

$\tilde{3}\ \frac{5}{2}\ \frac{1}{2}\ +\ -1/2$

$\tilde{3}\ 3\ \tilde{0}$

$0\ \tilde{0}\ 3\ +\ -1$

$\frac{1}{2}\ \frac{1}{2}\ \frac{5}{2}\ +\ -1/\sqrt{2}$

$1\ 1\ 2\ +\ -1/\sqrt{2}$

$\frac{3}{2}\ \frac{3}{2}\ \frac{3}{2}\ +\ -1/\sqrt{2}$

$2\ 2\ 1\ +\ -1/\sqrt{2}$

$\frac{5}{2}\ \frac{5}{2}\ \frac{1}{2}\ +\ -1/\sqrt{2}$

$\tilde{3}\ 3\ \tilde{0}\ +\ +1$

$\tilde{3}\ \tilde{3}\ 0$

$0\ 0\ \tilde{3}\ +\ +1$

$\frac{1}{2}\ \frac{1}{2}\ \frac{5}{2}\ +\ +1/\sqrt{2}$

$\tilde{0}\ \tilde{0}\ 3\ +\ -1$

$1\ 1\ 2\ +\ +1/\sqrt{2}$

$\frac{3}{2}\ \frac{3}{2}\ \frac{3}{2}\ +\ +1/\sqrt{2}$

$2\ 2\ 1\ +\ +1/\sqrt{2}$

$\frac{5}{2}\ \frac{5}{2}\ \frac{1}{2}\ +\ +1/\sqrt{2}$

$3\ 3\ \tilde{0}\ +\ -1$

$\tilde{3}\ \tilde{3}\ 0\ +\ +1$

T 3j Phases ($T_i = T_h$)

$0\ 0\ 0\ 0+$	$1\ 1\ 1\ 0-$	$-\frac{3}{2}\ \frac{3}{2}\ 0\ 0-$	$2\ -\frac{3}{2}\ \frac{1}{2}\ 0-$
$\frac{1}{2}\ \frac{1}{2}\ 0\ 0-$	$1\ 1\ 1\ 1+$	$-\frac{3}{2}\ \frac{3}{2}\ 1\ 0+$	$2\ 2\ 2\ 0+$
$1\ \frac{1}{2}\ \frac{1}{2}\ 0+$	$\frac{3}{2}\ 1\ \frac{1}{2}\ 0+$	$2\ 1\ 1\ 0+$	$-2\ 2\ 0\ 0+$
$1\ 1\ 0\ 0+$	$\frac{3}{2}\ \frac{3}{2}\ 1\ 0+$	$2\ \frac{3}{2}\ \frac{3}{2}\ 0-$	

T 6j Symbols

0 0 0
0 0 0 0000+ $+1$
$\frac{1}{2}$ $\frac{1}{2}$ 0
0 0 $\frac{1}{2}$ 0000+ $-1/\sqrt{2}$
$\frac{1}{2}$ $\frac{1}{2}$ 0 0000+ $-1/2$
1 $\frac{1}{2}$ $\frac{1}{2}$
0 $\frac{1}{2}$ $\frac{1}{2}$ 0000+ $+1/2$
1 $\frac{1}{2}$ $\frac{1}{2}$ 0000+ $+1/2.3$
1 1 0
0 0 1 0000+ $+1/\sqrt{3}$
$\frac{1}{2}$ $\frac{1}{2}$ $\frac{1}{2}$ 0000+ $+1/\sqrt{2.3}$
1 1 0 0000+ $+1/3$
1 1 1
$\frac{1}{2}$ $\frac{1}{2}$ $\frac{1}{2}$ 0000+ $+1/3$
$\frac{1}{2}$ $\frac{1}{2}$ $\frac{1}{2}$ 0001− 0
1 1 0 0000+ $-1/3$
1 1 0 0001− 0
1 1 0 0010− 0
1 1 0 0011+ $+1/3$
1 1 1 0000+ $+1/2.3$
1 1 1 0001− 0
1 1 1 1000− 0
1 1 1 1001+ $+1/2.3$
1 1 1 1100+ $+1/2.3$
1 1 1 1101− 0
1 1 1 1110− 0
1 1 1 1111+ $+1/2.3$
$\frac{3}{2}$ 1 $\frac{1}{2}$
0 $\frac{1}{2}$ 1 0000+ $+1/\sqrt{2.3}$
$\frac{1}{2}$ 1 $\frac{1}{2}$ 0000+ $-1/3$
1 $\frac{1}{2}$ 1 0000+ $-1/2.3$
1 $\frac{1}{2}$ 1 0100− $+i/2\sqrt{3}$
$\frac{3}{2}$ 1 $\frac{1}{2}$ 0000+ $+1/2.3$
-$\frac{3}{2}$ 1 $\frac{1}{2}$ 0000+ $-1/3$
$\frac{3}{2}$ $\frac{3}{2}$ 1
$\frac{1}{2}$ $\frac{1}{2}$ 1 0000+ $+1/3$
1 0 -$\frac{3}{2}$ 0000+ $+1/\sqrt{2.3}$
1 1 $\frac{1}{2}$ 0000+ $-1/3$

$\frac{3}{2}$ $\frac{3}{2}$ 1
1 1 $\frac{1}{2}$ 0010− 0
1 1 $\frac{3}{2}$ 0000+ $-1/2.3$
1 1 $\frac{3}{2}$ 0010− $+i/2\sqrt{3}$
1 1 -$\frac{3}{2}$ 0000+ $-1/2.3$
1 1 -$\frac{3}{2}$ 0010− $-i/2\sqrt{3}$
$\frac{3}{2}$ $\frac{1}{2}$ 1 0000+ $+1/2.3$
$\frac{3}{2}$ $\frac{3}{2}$ 1 0000+ $-1/3$
$\frac{3}{2}$ -$\frac{3}{2}$ 1 0000+ $-1/3$
-$\frac{3}{2}$ $\frac{1}{2}$ 1 0000+ $+1/3$
-$\frac{3}{2}$ $\frac{3}{2}$ 0 0000+ $+1/2$
-$\frac{3}{2}$ $\frac{3}{2}$ 1 0000+ $+1/2.3$
-$\frac{3}{2}$ -$\frac{3}{2}$ 1 0000+ $-1/3$
-$\frac{3}{2}$ $\frac{3}{2}$ 0
0 0 $\frac{3}{2}$ 0000+ $-1/\sqrt{2}$
$\frac{1}{2}$ $\frac{1}{2}$ 1 0000+ $+1/2$
1 1 $\frac{1}{2}$ 0000+ $+1/\sqrt{2.3}$
1 1 $\frac{3}{2}$ 0000+ $+1/\sqrt{2.3}$
1 1 -$\frac{3}{2}$ 0000+ $+1/\sqrt{2.3}$
$\frac{3}{2}$ $\frac{3}{2}$ 1 0000+ $+1/2$
-$\frac{3}{2}$ -$\frac{3}{2}$ 0 0000+ $-1/2$
-$\frac{3}{2}$ $\frac{3}{2}$ 1
$\frac{1}{2}$ $\frac{1}{2}$ 1 0000+ $+1/2.3$
1 0 $\frac{3}{2}$ 0000+ $+1/\sqrt{2.3}$
1 1 $\frac{1}{2}$ 0000+ $-1/2.3$
1 1 $\frac{1}{2}$ 0010− $-i/2\sqrt{3}$
1 1 $\frac{3}{2}$ 0000+ $+1/3$
1 1 $\frac{3}{2}$ 0010− 0
1 1 -$\frac{3}{2}$ 0000+ $-1/2.3$
1 1 -$\frac{3}{2}$ 0010− $+i/2\sqrt{3}$
$\frac{3}{2}$ $\frac{1}{2}$ 1 0000+ $+1/3$
$\frac{3}{2}$ $\frac{3}{2}$ 1 0000+ $+1/2.3$
$\frac{3}{2}$ -$\frac{3}{2}$ 1 0000+ $-1/3$
-$\frac{3}{2}$ $\frac{1}{2}$ 1 0000+ $-1/3$
-$\frac{3}{2}$ $\frac{3}{2}$ 1 0000+ $-1/3$
-$\frac{3}{2}$ -$\frac{3}{2}$ 0 0000+ $+1/2$
-$\frac{3}{2}$ -$\frac{3}{2}$ 1 0000+ $+1/2.3$

2 1 1
0 1 1 0000+ $+1/3$
1 1 1 0000+ $+1/2.3$
1 1 1 0100− $-i/2\sqrt{3}$
1 1 1 0110+ $-1/2.3$
2 1 1 0000+ $+1/3$
-2 1 1 0000+ $+1/3$
2 $\frac{3}{2}$ $\frac{3}{2}$
0 -$\frac{3}{2}$ $\frac{3}{2}$ 0000+ $-1/2$
$\frac{1}{2}$ 1 1 0000+ $-1/\sqrt{2.3}$
1 $\frac{3}{2}$ $\frac{1}{2}$ 0000+ $+1/2$
1 -$\frac{3}{2}$ $\frac{3}{2}$ 0000+ $+1/2$
$\frac{3}{2}$ 1 1 0000+ $+1/\sqrt{2.3}$
-$\frac{3}{2}$ 1 1 0000+ $+1/\sqrt{2.3}$
-2 $\frac{3}{2}$ $\frac{1}{2}$ 0000+ $-1/2$
2 -$\frac{3}{2}$ $\frac{1}{2}$
0 $\frac{1}{2}$ -$\frac{3}{2}$ 0000+ $-1/2$
$\frac{1}{2}$ 1 1 0000+ $+1/\sqrt{2.3}$
1 $\frac{1}{2}$ -$\frac{3}{2}$ 0000+ $+1/2$
1 $\frac{3}{2}$ $\frac{1}{2}$ 0000+ $+1/2$
1 -$\frac{3}{2}$ $\frac{3}{2}$ 0000+ $+1/2$
$\frac{3}{2}$ 1 1 0000+ $+1/\sqrt{2.3}$
-$\frac{3}{2}$ 1 1 0000+ $-1/\sqrt{2.3}$
2 $\frac{3}{2}$ $\frac{1}{2}$ 0000+ $-1/2$
2 2 2
1 1 1 0000+ $+1/\sqrt{3}$
-$\frac{3}{2}$ $\frac{3}{2}$ $\frac{1}{2}$ 0000+ $+1/\sqrt{2}$
-2 2 0 0000+ $+1$
-2 2 0
0 0 2 0000+ $+1$
$\frac{1}{2}$ $\frac{1}{2}$ $\frac{3}{2}$ 0000+ $-1/\sqrt{2}$
1 1 1 0000+ $+1/\sqrt{3}$
$\frac{3}{2}$ $\frac{3}{2}$ -$\frac{3}{2}$ 0000+ $-1/\sqrt{2}$
-$\frac{3}{2}$ -$\frac{3}{2}$ $\frac{1}{2}$ 0000+ $-1/\sqrt{2}$
2 2 -2 0000+ $+1$
-2 -2 0 0000+ $+1$

O and T_d 3j Phases ($O_i = O_h$)

0 0 0 0+	2 1 1 0+	$\tilde{1}$ 2 1 0−	$\tilde{\frac{1}{2}}$ $\tilde{\frac{1}{2}}$ 0 0−
$\frac{1}{2}$ $\frac{1}{2}$ 0 0−	2 $\frac{3}{2}$ $\frac{1}{2}$ 0+	$\tilde{1}$ $\tilde{1}$ 0 0+	$\tilde{\frac{1}{2}}$ $\tilde{\frac{1}{2}}$ 1 0+
1 $\frac{1}{2}$ $\frac{1}{2}$ 0+	2 $\frac{3}{2}$ $\frac{3}{2}$ 0−	$\tilde{1}$ $\tilde{1}$ 1 0−	$\tilde{0}$ $\frac{3}{2}$ $\frac{3}{2}$ 0+
1 1 0 0+	2 2 0 0+	$\tilde{1}$ $\tilde{1}$ 2 0+	$\tilde{0}$ 2 2 0−
1 1 1 0−	2 2 2 0+	$\tilde{1}$ $\tilde{1}$ $\tilde{1}$ 0+	$\tilde{0}$ $\tilde{1}$ 1 0+
$\frac{3}{2}$ 1 $\frac{1}{2}$ 0−	$\tilde{1}$ 1 1 0+	$\tilde{\frac{1}{2}}$ $\frac{3}{2}$ 1 0−	$\tilde{0}$ $\tilde{\frac{1}{2}}$ $\frac{1}{2}$ 0+
$\frac{3}{2}$ $\frac{3}{2}$ 0 0−	$\tilde{1}$ $\frac{3}{2}$ $\frac{1}{2}$ 0+	$\tilde{\frac{1}{2}}$ 2 $\frac{3}{2}$ 0+	$\tilde{0}$ $\tilde{0}$ 0 0+
$\frac{3}{2}$ $\frac{3}{2}$ 1 0+	$\tilde{1}$ $\frac{3}{2}$ $\frac{3}{2}$ 0−	$\tilde{\frac{1}{2}}$ $\tilde{1}$ $\frac{1}{2}$ 0−	
$\frac{3}{2}$ $\frac{3}{2}$ 1 1+	$\tilde{1}$ $\frac{3}{2}$ $\frac{3}{2}$ 1+	$\tilde{\frac{1}{2}}$ $\tilde{1}$ $\frac{3}{2}$ 0+	

O and T_d 6j Symbols

0 0 0
0 0 0 0000+ $+1$
$\frac{1}{2}$ $\frac{1}{2}$ 0
0 0 $\frac{1}{2}$ 0000+ $-1/\sqrt{2}$
$\frac{1}{2}$ $\frac{1}{2}$ 0 0000+ $-1/2$
1 $\frac{1}{2}$ $\frac{1}{2}$
0 $\frac{1}{2}$ $\frac{1}{2}$ 0000+ $+1/2$
1 $\frac{1}{2}$ $\frac{1}{2}$ 0000+ $+1/2.3$
1 1 0
0 0 1 0000+ $+1/\sqrt{3}$
$\frac{1}{2}$ $\frac{1}{2}$ $\frac{1}{2}$ 0000+ $+1/\sqrt{2.3}$
1 1 0 0000+ $+1/3$
1 1 1
$\frac{1}{2}$ $\frac{1}{2}$ $\frac{1}{2}$ 0000+ $+1/3$
1 1 0 0000+ $-1/3$
1 1 1 0000+ $+1/2.3$
$\frac{3}{2}$ 1 $\frac{1}{2}$
0 $\frac{1}{2}$ 1 0000+ $-1/\sqrt{2.3}$
$\frac{1}{2}$ 1 $\frac{1}{2}$ 0000+ $-1/3$
1 $\frac{1}{2}$ 1 0000+ $+1/2.3$
$\frac{3}{2}$ 1 $\frac{1}{2}$ 0000+ $-1/4.3$
$\frac{3}{2}$ $\frac{3}{2}$ 0
0 0 $\frac{3}{2}$ 0000+ $-1/2$
$\frac{1}{2}$ $\frac{1}{2}$ 1 0000+ $-1/2\sqrt{2}$
1 1 $\frac{1}{2}$ 0000+ $-1/2\sqrt{3}$

$\frac{3}{2}$ $\frac{3}{2}$ 0
1 1 $\frac{3}{2}$ 0000+ $+1/2\sqrt{3}$
1 1 $\frac{3}{2}$ 1000+ 0
1 1 $\frac{3}{2}$ 1100+ $+1/2\sqrt{3}$
$\frac{3}{2}$ $\frac{3}{2}$ 0 0000+ $-1/4$
$\frac{3}{2}$ $\frac{3}{2}$ 1
$\frac{1}{2}$ $\frac{1}{2}$ 1 0000+ $+1/2.3\sqrt{2}$
$\frac{1}{2}$ $\frac{1}{2}$ 1 0001+ $+1/3\sqrt{2}$
1 0 $\frac{3}{2}$ 0000+ $+1/2\sqrt{3}$
1 0 $\frac{3}{2}$ 0001+ 0
1 0 $\frac{3}{2}$ 0100+ 0
1 0 $\frac{3}{2}$ 0101+ $+1/2\sqrt{3}$
1 1 $\frac{1}{2}$ 0000+ $-1/2.3\sqrt{2}$
1 1 $\frac{1}{2}$ 0001+ $-1/3\sqrt{2}$
1 1 $\frac{3}{2}$ 0000+ $-1/3\sqrt{2}$
1 1 $\frac{3}{2}$ 0001+ 0
1 1 $\frac{3}{2}$ 1000+ 0
1 1 $\frac{3}{2}$ 1001+ $+1/2.3\sqrt{2}$
1 1 $\frac{3}{2}$ 1100+ $+1/2.3\sqrt{2}$
1 1 $\frac{3}{2}$ 1101+ 0
$\frac{3}{2}$ $\frac{1}{2}$ 1 0000+ $-1/2.3$
$\frac{3}{2}$ $\frac{1}{2}$ 1 0001+ $+1/2.3$
$\frac{3}{2}$ $\frac{1}{2}$ 1 0100+ $+1/2.3$
$\frac{3}{2}$ $\frac{1}{2}$ 1 0101+ $+1/4.3$
$\frac{3}{2}$ $\frac{3}{2}$ 0 0000+ $+1/4$

$\frac{3}{2}$ $\frac{3}{2}$ 1
$\frac{3}{2}$ $\frac{3}{2}$ 0 0001+ 0
$\frac{3}{2}$ $\frac{3}{2}$ 0 0010+ 0
$\frac{3}{2}$ $\frac{3}{2}$ 0 0011+ $+1/4$
$\frac{3}{2}$ $\frac{3}{2}$ 1 0000+ $+1/4.3$
$\frac{3}{2}$ $\frac{3}{2}$ 1 0001+ 0
$\frac{3}{2}$ $\frac{3}{2}$ 1 0010+ 0
$\frac{3}{2}$ $\frac{3}{2}$ 1 0011+ $+1/4.3$
$\frac{3}{2}$ $\frac{3}{2}$ 1 1000+ 0
$\frac{3}{2}$ $\frac{3}{2}$ 1 1001+ $-1/2.3$
$\frac{3}{2}$ $\frac{3}{2}$ 1 1010+ $-1/2.3$
$\frac{3}{2}$ $\frac{3}{2}$ 1 1011+ 0
$\frac{3}{2}$ $\frac{3}{2}$ 1 1100+ $+1/4.3$
$\frac{3}{2}$ $\frac{3}{2}$ 1 1101+ 0
$\frac{3}{2}$ $\frac{3}{2}$ 1 1110+ 0
$\frac{3}{2}$ $\frac{3}{2}$ 1 1111+ $-1/2.3$
2 1 1
0 1 1 0000+ $+1/3$
1 1 1 0000+ $+1/2.3$
2 1 1 0000+ $+1/3$
2 $\frac{3}{2}$ $\frac{1}{2}$
0 $\frac{1}{2}$ $\frac{3}{2}$ 0000+ $+1/2\sqrt{2}$
$\frac{1}{2}$ 1 1 0000+ $+1/2\sqrt{3}$
1 $\frac{1}{2}$ $\frac{3}{2}$ 0000+ $+1/2\sqrt{2}$
1 $\frac{1}{2}$ $\frac{3}{2}$ 0100+ 0

O and T_d 3j Symbols (cont.)

$2\ \frac{3}{2}\ \frac{1}{2}$		
$1\ \frac{3}{2}\ \frac{1}{2}$	0000+	$+1/4$
$1\ \frac{3}{2}\ \frac{3}{2}$	0000+	0
$1\ \frac{3}{2}\ \frac{3}{2}$	0100+	$+1/4$
$\frac{3}{2}\ 1\ 1$	0000+	$-1/2\sqrt{2.3}$
$\frac{3}{2}\ 1\ 1$	0100+	$+1/2\sqrt{2.3}$
$2\ \frac{3}{2}\ \frac{1}{2}$	0000+	$-1/4$
$2\ \frac{3}{2}\ \frac{3}{2}$		
$0\ \frac{3}{2}\ \frac{3}{2}$	0000+	$-1/4$
$\frac{1}{2}\ 1\ 1$	0000+	$+1/2\sqrt{2.3}$
$1\ \frac{3}{2}\ \frac{1}{2}$	0000+	0
$1\ \frac{3}{2}\ \frac{1}{2}$	0010+	$+1/4$
$1\ \frac{3}{2}\ \frac{3}{2}$	0000+	$+1/4$
$1\ \frac{3}{2}\ \frac{3}{2}$	0100+	0
$1\ \frac{3}{2}\ \frac{3}{2}$	0110+	0
$\frac{3}{2}\ 1\ 1$	0000+	0
$\frac{3}{2}\ 1\ 1$	0100+	$+1/2\sqrt{2.3}$
$\frac{3}{2}\ 1\ 1$	0110+	0
$2\ \frac{3}{2}\ \frac{1}{2}$	0000+	$-1/4$
$2\ \frac{3}{2}\ \frac{3}{2}$	0000+	0
$2\ 2\ 0$		
$0\ 0\ 2$	0000+	$+1/\sqrt{2}$
$\frac{1}{2}\ \frac{1}{2}\ \frac{3}{2}$	0000+	$+1/2$
$1\ 1\ 1$	0000+	$+1/\sqrt{2.3}$
$\frac{3}{2}\ \frac{3}{2}\ \frac{1}{2}$	0000+	$+1/2\sqrt{2}$
$\frac{3}{2}\ \frac{3}{2}\ \frac{3}{2}$	0000+	$-1/2\sqrt{2}$
$2\ 2\ 0$	0000+	$+1/2$
$2\ 2\ 2$		
$1\ 1\ 1$	0000+	$-1/2\sqrt{3}$
$\frac{3}{2}\ \frac{3}{2}\ \frac{1}{2}$	0000+	$+1/2\sqrt{2}$
$\frac{3}{2}\ \frac{3}{2}\ \frac{3}{2}$	0000+	0
$2\ 2\ 0$	0000+	$+1/2$
$2\ 2\ 2$	0000+	0
$\tilde{1}\ 1\ 1$		
$0\ 1\ 1$	0000+	$+1/3$
$1\ 1\ 1$	0000+	$+1/2.3$
$2\ 1\ 1$	0000+	$-1/2.3$
$\tilde{1}\ 1\ 1$	0000+	$+1/2.3$
$\tilde{1}\ \frac{3}{2}\ \frac{1}{2}$		
$0\ \frac{1}{2}\ \frac{3}{2}$	0000+	$+1/2\sqrt{2}$
$\frac{1}{2}\ 1\ 1$	0000+	$+1/2\sqrt{3}$
$1\ \frac{1}{2}\ \frac{3}{2}$	0000+	$-1/2.3\sqrt{2}$
$1\ \frac{1}{2}\ \frac{3}{2}$	0100+	$+1/3\sqrt{2}$
$1\ \frac{3}{2}\ \frac{1}{2}$	0000+	$+1/4$
$1\ \frac{3}{2}\ \frac{3}{2}$	0000+	$+1/2.3$
$1\ \frac{3}{2}\ \frac{3}{2}$	0100+	$+1/2.3$
$1\ \frac{3}{2}\ \frac{3}{2}$	1000−	$-1/2.3$
$1\ \frac{3}{2}\ \frac{3}{2}$	1100−	$+1/4.3$
$\frac{3}{2}\ 1\ 1$	0000+	$+1/2\sqrt{2.3}$
$\frac{3}{2}\ 1\ 1$	0100+	0
$2\ \frac{3}{2}\ \frac{1}{2}$	0000+	$+1/4$
$2\ \frac{3}{2}\ \frac{3}{2}$	0000+	0
$2\ \frac{3}{2}\ \frac{3}{2}$	1000−	$+1/4$
$\tilde{1}\ \frac{3}{2}\ \frac{1}{2}$	0000+	$-1/4.3$
$\tilde{1}\ \frac{3}{2}\ \frac{3}{2}$		
$0\ \frac{3}{2}\ \frac{3}{2}$	0000+	$-1/4$
$0\ \frac{3}{2}\ \frac{3}{2}$	0001−	0
$0\ \frac{3}{2}\ \frac{3}{2}$	1000−	0
$0\ \frac{3}{2}\ \frac{3}{2}$	1001+	$+1/4$
$\frac{1}{2}\ 1\ 1$	0000+	$+1/2\sqrt{2.3}$
$\frac{1}{2}\ 1\ 1$	0001−	0
$1\ \frac{3}{2}\ \frac{1}{2}$	0000+	$+1/2.3$
$1\ \frac{3}{2}\ \frac{1}{2}$	0001−	$-1/2.3$
$1\ \frac{3}{2}\ \frac{1}{2}$	0010+	$+1/2.3$
$1\ \frac{3}{2}\ \frac{1}{2}$	0011−	$+1/4.3$
$1\ \frac{3}{2}\ \frac{3}{2}$	0000+	$-1/4.3$
$1\ \frac{3}{2}\ \frac{3}{2}$	0001−	0
$1\ \frac{3}{2}\ \frac{3}{2}$	0100+	0
$1\ \frac{3}{2}\ \frac{3}{2}$	0101−	$+1/2.3$
$1\ \frac{3}{2}\ \frac{3}{2}$	0110+	$+1/4.3$
$1\ \frac{3}{2}\ \frac{3}{2}$	0111−	0
$1\ \frac{3}{2}\ \frac{3}{2}$	1000−	0
$1\ \frac{3}{2}\ \frac{3}{2}$	1001+	$+1/4.3$
$1\ \frac{3}{2}\ \frac{3}{2}$	1100−	$-1/2.3$
$1\ \frac{3}{2}\ \frac{3}{2}$	1101+	0
$1\ \frac{3}{2}\ \frac{3}{2}$	1110−	0
$1\ \frac{3}{2}\ \frac{3}{2}$	1111+	$+1/2.3$
$\frac{3}{2}\ 1\ 1$	0000+	0
$\frac{3}{2}\ 1\ 1$	0001−	0
$\frac{3}{2}\ 1\ 1$	0100+	0
$\frac{3}{2}\ 1\ 1$	0101−	$+1/2\sqrt{2.3}$
$\frac{3}{2}\ 1\ 1$	0110+	$+1/2\sqrt{2.3}$
$\frac{3}{2}\ 1\ 1$	0111−	0
$2\ \frac{3}{2}\ \frac{1}{2}$	0000+	0
$2\ \frac{3}{2}\ \frac{1}{2}$	0001−	$+1/4$
$2\ \frac{3}{2}\ \frac{3}{2}$	0000+	$+1/4$
$2\ \frac{3}{2}\ \frac{3}{2}$	0001−	0
$2\ \frac{3}{2}\ \frac{3}{2}$	1000−	0
$2\ \frac{3}{2}\ \frac{3}{2}$	1001+	0
$\tilde{1}\ \frac{3}{2}\ \frac{1}{2}$	0000+	$-1/2.3$
$\tilde{1}\ \frac{3}{2}\ \frac{1}{2}$	0001−	$-1/2.3$
$\tilde{1}\ \frac{3}{2}\ \frac{1}{2}$	0010−	$-1/2.3$
$\tilde{1}\ \frac{3}{2}\ \frac{1}{2}$	0011+	$+1/4.3$
$\tilde{1}\ \frac{3}{2}\ \frac{3}{2}$	0000+	$+1/4.3$
$\tilde{1}\ \frac{3}{2}\ \frac{3}{2}$	0001−	0
$\tilde{1}\ \frac{3}{2}\ \frac{3}{2}$	0100−	0
$\tilde{1}\ \frac{3}{2}\ \frac{3}{2}$	0101+	$-1/2.3$
$\tilde{1}\ \frac{3}{2}\ \frac{3}{2}$	0110+	$+1/4.3$
$\tilde{1}\ \frac{3}{2}\ \frac{3}{2}$	0111−	0
$\tilde{1}\ \frac{3}{2}\ \frac{3}{2}$	1000−	0
$\tilde{1}\ \frac{3}{2}\ \frac{3}{2}$	1001+	$+1/4.3$
$\tilde{1}\ \frac{3}{2}\ \frac{3}{2}$	1100+	$-1/2.3$
$\tilde{1}\ \frac{3}{2}\ \frac{3}{2}$	1101−	0
$\tilde{1}\ \frac{3}{2}\ \frac{3}{2}$	1110−	0
$\tilde{1}\ \frac{3}{2}\ \frac{3}{2}$	1111+	$-1/2.3$
$\tilde{1}\ 2\ 1$		
$0\ 1\ 2$	0000+	$-1/\sqrt{2.3}$
$\frac{1}{2}\ \frac{1}{2}\ \frac{3}{2}$	0000+	$+1/2\sqrt{3}$
$\frac{1}{2}\ \frac{3}{2}\ \frac{3}{2}$	0000+	$-1/2\sqrt{2.3}$
$\frac{1}{2}\ \frac{3}{2}\ \frac{3}{2}$	1000−	$+1/2\sqrt{2.3}$
$1\ 1\ 1$	0000+	$-1/2\sqrt{3}$
$1\ 2\ 1$	0000+	0
$\frac{3}{2}\ \frac{1}{2}\ \frac{3}{2}$	0000+	$-1/2\sqrt{2.3}$
$\frac{3}{2}\ \frac{3}{2}\ \frac{1}{2}$	0000+	$+1/2\sqrt{2.3}$
$\frac{3}{2}\ \frac{3}{2}\ \frac{1}{2}$	0010+	$+1/2\sqrt{2.3}$
$\frac{3}{2}\ \frac{3}{2}\ \frac{3}{2}$	0000+	0
$\frac{3}{2}\ \frac{3}{2}\ \frac{3}{2}$	0010+	$-1/2\sqrt{2.3}$
$\frac{3}{2}\ \frac{3}{2}\ \frac{3}{2}$	1000−	$-1/2\sqrt{2.3}$
$\frac{3}{2}\ \frac{3}{2}\ \frac{3}{2}$	1010−	0
$2\ 1\ 2$	0000+	$-1/2\sqrt{3}$
$\tilde{1}\ 1\ 1$	0000+	$-1/2.3$
$\tilde{1}\ 2\ 1$	0000+	$+1/3$
$\tilde{1}\ \tilde{1}\ 0$		
$0\ 0\ \tilde{1}$	0000+	$+1/\sqrt{3}$

O and T_d 6j Symbols (cont.)

$\tilde{1}$	**$\tilde{1}$**	**0**		
$\frac{1}{2}$	$\frac{1}{2}$	$\frac{3}{2}$	0000+	$+1/\sqrt{2.3}$
1	1	1	0000+	$+1/3$
1	1	2	0000+	$-1/3$
1	1	$\tilde{1}$	0000+	$-1/3$
$\frac{3}{2}$	$\frac{3}{2}$	$\frac{1}{2}$	0000+	$+1/2\sqrt{3}$
$\frac{3}{2}$	$\frac{3}{2}$	$\frac{3}{2}$	0000+	$-1/2\sqrt{3}$
$\frac{3}{2}$	$\frac{3}{2}$	$\frac{3}{2}$	1000−	0
$\frac{3}{2}$	$\frac{3}{2}$	$\frac{3}{2}$	1100+	$+1/2\sqrt{3}$
2	2	1	0000+	$-1/\sqrt{2.3}$
2	2	$\tilde{1}$	0000+	$+1/\sqrt{2.3}$
$\tilde{1}$	$\tilde{1}$	0	0000+	$+1/3$
$\tilde{1}$	**$\tilde{1}$**	**1**		
$\frac{1}{2}$	$\frac{1}{2}$	$\frac{3}{2}$	0000+	$+1/2.3$
1	0	$\tilde{1}$	0000+	$-1/3$
1	1	1	0000+	$-1/2.3$
1	1	2	0000+	$+1/2.3$
1	1	$\tilde{1}$	0000+	$-1/2.3$
$\frac{3}{2}$	$\frac{1}{2}$	$\frac{3}{2}$	0000+	$-1/2.3\sqrt{2}$
$\frac{3}{2}$	$\frac{1}{2}$	$\frac{3}{2}$	0100−	$+1/3\sqrt{2}$
$\frac{3}{2}$	$\frac{3}{2}$	$\frac{1}{2}$	0000+	$-1/2.3\sqrt{2}$
$\frac{3}{2}$	$\frac{3}{2}$	$\frac{1}{2}$	0010+	$+1/3\sqrt{2}$
$\frac{3}{2}$	$\frac{3}{2}$	$\frac{3}{2}$	0000+	$-1/3\sqrt{2}$
$\frac{3}{2}$	$\frac{3}{2}$	$\frac{3}{2}$	0010+	0
$\frac{3}{2}$	$\frac{3}{2}$	$\frac{3}{2}$	1000−	0
$\frac{3}{2}$	$\frac{3}{2}$	$\frac{3}{2}$	1010−	$+1/2.3\sqrt{2}$
$\frac{3}{2}$	$\frac{3}{2}$	$\frac{3}{2}$	1100+	$-1/2.3\sqrt{2}$
$\frac{3}{2}$	$\frac{3}{2}$	$\frac{3}{2}$	1110+	0
2	1	1	0000+	$-1/2\sqrt{3}$
2	1	$\tilde{1}$	0000+	$+1/2.3$
$\tilde{1}$	1	1	0000+	$+1/2.3$
$\tilde{1}$	1	2	0000+	$-1/2\sqrt{3}$
$\tilde{1}$	1	$\tilde{1}$	0000+	$-1/2.3$
$\tilde{1}$	2	1	0000+	$+1/2.3$
$\tilde{1}$	2	$\tilde{1}$	0000+	$-1/2\sqrt{3}$
$\tilde{1}$	$\tilde{1}$	0	0000+	$-1/3$
$\tilde{1}$	$\tilde{1}$	1	0000+	$+1/2.3$
$\tilde{1}$	**$\tilde{1}$**	**2**		
1	1	1	0000+	$-1/2.3$
1	1	2	0000+	$-1/3$
1	1	$\tilde{1}$	0000+	$+1/2.3$
$\frac{3}{2}$	$\frac{1}{2}$	$\frac{3}{2}$	0000+	$+1/2\sqrt{2.3}$
$\frac{3}{2}$	$\frac{1}{2}$	$\frac{3}{2}$	0100−	$+1/2\sqrt{2.3}$
$\frac{3}{2}$	$\frac{3}{2}$	$\frac{1}{2}$	0000+	$+1/2\sqrt{2.3}$
$\frac{3}{2}$	$\frac{3}{2}$	$\frac{3}{2}$	0000+	0
$\frac{3}{2}$	$\frac{3}{2}$	$\frac{3}{2}$	1000−	$-1/2\sqrt{2.3}$
$\frac{3}{2}$	$\frac{3}{2}$	$\frac{3}{2}$	1100+	0
2	0	$\tilde{1}$	0000+	$+1/\sqrt{2.3}$
2	2	1	0000+	$-1/2\sqrt{3}$
2	2	$\tilde{1}$	0000+	$-1/2\sqrt{3}$
$\tilde{1}$	1	1	0000+	$-1/2\sqrt{3}$
$\tilde{1}$	1	2	0000+	0
$\tilde{1}$	1	$\tilde{1}$	0000+	$-1/2\sqrt{3}$
$\tilde{1}$	$\tilde{1}$	0	0000+	$+1/3$
$\tilde{1}$	$\tilde{1}$	1	0000+	$+1/2.3$
$\tilde{1}$	$\tilde{1}$	2	0000+	$+1/3$
$\tilde{1}$	**$\tilde{1}$**	**$\tilde{1}$**		
1	1	1	0000+	$-1/2.3$
$\frac{3}{2}$	$\frac{3}{2}$	$\frac{1}{2}$	0000+	$+1/2\sqrt{2.3}$
$\frac{3}{2}$	$\frac{3}{2}$	$\frac{1}{2}$	0010−	0
$\frac{3}{2}$	$\frac{3}{2}$	$\frac{3}{2}$	0000+	0
$\frac{3}{2}$	$\frac{3}{2}$	$\frac{3}{2}$	1000−	0
$\frac{3}{2}$	$\frac{3}{2}$	$\frac{3}{2}$	1100+	$-1/2\sqrt{2.3}$
$\frac{3}{2}$	$\frac{3}{2}$	$\frac{3}{2}$	1110−	0
2	1	1	0000+	$-1/2.3$
$\tilde{1}$	1	1	0000+	$-1/2.3$
$\tilde{1}$	2	1	0000+	$-1/2\sqrt{3}$
$\tilde{1}$	$\tilde{1}$	0	0000+	$+1/3$
$\tilde{1}$	$\tilde{1}$	1	0000+	$+1/2.3$
$\tilde{1}$	$\tilde{1}$	2	0000+	$-1/2.3$
$\tilde{1}$	$\tilde{1}$	$\tilde{1}$	0000+	$+1/2.3$
$\tilde{\frac{1}{2}}$	**$\frac{3}{2}$**	**1**		
0	1	$\frac{3}{2}$	0000+	$-1/2\sqrt{3}$
$\frac{1}{2}$	$\frac{3}{2}$	1	0000+	$-1/4$
1	1	$\frac{3}{2}$	0000+	$-1/2.3\sqrt{2}$
1	1	$\frac{3}{2}$	0100+	$+1/3\sqrt{2}$
$\frac{3}{2}$	$\frac{3}{2}$	1	0000+	$-1/2.3$
$\frac{3}{2}$	$\frac{3}{2}$	1	0010+	$-1/2.3$
$\frac{3}{2}$	$\frac{3}{2}$	1	0100+	$-1/2.3$
$\frac{3}{2}$	$\frac{3}{2}$	1	0110+	$+1/4.3$
2	1	$\frac{3}{2}$	0000+	$-1/2\sqrt{2.3}$
$\tilde{1}$	1	$\frac{3}{2}$	0000+	$+1/2\sqrt{2.3}$
$\tilde{1}$	1	$\frac{3}{2}$	0100−	0
$\tilde{\frac{1}{2}}$	**$\frac{3}{2}$**	**1**		
$\tilde{\frac{1}{2}}$	$\frac{3}{2}$	1	0000+	$-1/4.3$
$\tilde{\frac{1}{2}}$	**2**	**$\frac{3}{2}$**		
0	$\frac{3}{2}$	2	0000+	$+1/2\sqrt{2}$
$\frac{1}{2}$	1	$\frac{3}{2}$	0000+	$+1/4$
$\frac{1}{2}$	2	$\frac{3}{2}$	0000+	$+1/4$
1	$\frac{3}{2}$	1	0000+	$+1/2\sqrt{2.3}$
1	$\frac{3}{2}$	1	0010+	$+1/2\sqrt{2.3}$
$\frac{3}{2}$	1	$\frac{3}{2}$	0000+	0
$\frac{3}{2}$	1	$\frac{3}{2}$	0010+	$-1/4$
$\frac{3}{2}$	2	$\frac{3}{2}$	0000+	$-1/4$
2	$\frac{3}{2}$	2	0000+	$-1/2\sqrt{2}$
$\tilde{1}$	$\frac{3}{2}$	1	0000+	$+1/2\sqrt{2.3}$
$\tilde{1}$	$\frac{3}{2}$	1	0010−	$+1/2\sqrt{2.3}$
$\tilde{\frac{1}{2}}$	1	$\frac{3}{2}$	0000+	$+1/4$
$\tilde{\frac{1}{2}}$	2	$\frac{3}{2}$	0000+	$-1/4$
$\tilde{\frac{1}{2}}$	**$\tilde{1}$**	**$\frac{1}{2}$**		
0	$\frac{1}{2}$	$\tilde{1}$	0000+	$-1/\sqrt{2.3}$
$\frac{1}{2}$	1	$\frac{3}{2}$	0000+	$+1/3$
1	$\frac{1}{2}$	$\tilde{1}$	0000+	$+1/3$
1	$\frac{3}{2}$	1	0000+	$+1/2\sqrt{3}$
1	$\frac{3}{2}$	2	0000+	$+1/2\sqrt{3}$
1	$\frac{3}{2}$	$\tilde{1}$	0000+	$+1/2.3$
$\frac{3}{2}$	1	$\frac{3}{2}$	0000+	$-1/2.3\sqrt{2}$
$\frac{3}{2}$	1	$\frac{3}{2}$	0100−	$+1/3\sqrt{2}$
$\frac{3}{2}$	2	$\frac{3}{2}$	0000+	$+1/2\sqrt{2}$
$\frac{3}{2}$	2	$\frac{3}{2}$	0100−	0
$\frac{3}{2}$	$\tilde{1}$	$\frac{1}{2}$	0000+	$-1/3$
$\frac{3}{2}$	$\tilde{1}$	$\frac{3}{2}$	0000+	$+1/2.3\sqrt{2}$
$\frac{3}{2}$	$\tilde{1}$	$\frac{3}{2}$	0100−	$+1/3\sqrt{2}$
2	$\frac{3}{2}$	1	0000+	$-1/2\sqrt{3}$
2	$\frac{3}{2}$	$\tilde{1}$	0000+	$-1/2\sqrt{3}$
$\tilde{1}$	$\frac{3}{2}$	1	0000+	$+1/2.3$
$\tilde{1}$	$\frac{3}{2}$	2	0000+	$+1/2\sqrt{3}$
$\tilde{1}$	$\frac{3}{2}$	$\tilde{1}$	0000+	$+1/2\sqrt{3}$
$\tilde{\frac{1}{2}}$	$\tilde{1}$	$\frac{1}{2}$	0000+	$+1/2.3$
$\tilde{\frac{1}{2}}$	**$\tilde{1}$**	**$\frac{3}{2}$**		
0	$\frac{3}{2}$	$\tilde{1}$	0000+	$+1/2\sqrt{3}$
$\frac{1}{2}$	1	$\frac{3}{2}$	0000+	$+1/4.3$
$\frac{1}{2}$	2	$\frac{3}{2}$	0000+	$+1/4$
$\frac{1}{2}$	$\tilde{1}$	$\frac{3}{2}$	0000+	$-1/4$
1	$\frac{1}{2}$	$\tilde{1}$	0000+	$+1/2.3$

O and T_d 3j Symbols (cont.)

$\tilde{\tfrac{1}{2}}$	$\tilde{1}$	$\tfrac{3}{2}$		
1	$\tfrac{3}{2}$	1	0000+	$+1/2\sqrt{2.3}$
1	$\tfrac{3}{2}$	1	0010+	0
1	$\tfrac{3}{2}$	2	0000+	$-1/2\sqrt{2.3}$
1	$\tfrac{3}{2}$	2	0010+	$+1/2\sqrt{2.3}$
1	$\tfrac{3}{2}$	$\tilde{1}$	0000+	$-1/2.3\sqrt{2}$
1	$\tfrac{3}{2}$	$\tilde{1}$	0010+	$-1/3\sqrt{2}$
$\tfrac{3}{2}$	1	$\tfrac{3}{2}$	0000+	$+1/2.3$
$\tfrac{3}{2}$	1	$\tfrac{3}{2}$	0010+	$-1/2.3$
$\tfrac{3}{2}$	1	$\tfrac{3}{2}$	0100−	$+1/2.3$
$\tfrac{3}{2}$	1	$\tfrac{3}{2}$	0110−	$+1/4.3$
$\tfrac{3}{2}$	2	$\tfrac{3}{2}$	0000+	0
$\tfrac{3}{2}$	2	$\tfrac{3}{2}$	0100−	$-1/4$
$\tfrac{3}{2}$	$\tilde{1}$	$\tfrac{1}{2}$	0000+	$+1/2.3\sqrt{2}$
$\tfrac{3}{2}$	$\tilde{1}$	$\tfrac{1}{2}$	0010−	$+1/3\sqrt{2}$
$\tfrac{3}{2}$	$\tilde{1}$	$\tfrac{3}{2}$	0000+	$-1/2.3$
$\tfrac{3}{2}$	$\tilde{1}$	$\tfrac{3}{2}$	0010−	$+1/2.3$
$\tfrac{3}{2}$	$\tilde{1}$	$\tfrac{3}{2}$	0100−	$+1/2.3$
$\tfrac{3}{2}$	$\tilde{1}$	$\tfrac{3}{2}$	0110+	$+1/4.3$
2	$\tfrac{1}{2}$	$\tilde{1}$	0000+	$-1/2\sqrt{3}$
2	$\tfrac{3}{2}$	1	0000+	$+1/2\sqrt{2.3}$
2	$\tfrac{3}{2}$	$\tilde{1}$	0000+	$-1/2\sqrt{2.3}$
$\tilde{1}$	$\tfrac{1}{2}$	$\tilde{1}$	0000+	$+1/2\sqrt{3}$
$\tilde{1}$	$\tfrac{3}{2}$	1	0000+	$-1/2.3\sqrt{2}$
$\tilde{1}$	$\tfrac{3}{2}$	1	0010−	$+1/3\sqrt{2}$
$\tilde{1}$	$\tfrac{3}{2}$	2	0000+	$-1/2\sqrt{2.3}$
$\tilde{1}$	$\tfrac{3}{2}$	2	0010−	$-1/2\sqrt{2.3}$
$\tilde{1}$	$\tfrac{3}{2}$	$\tilde{1}$	0000+	$+1/2\sqrt{2.3}$
$\tilde{1}$	$\tfrac{3}{2}$	$\tilde{1}$	0010−	0
$\tilde{\tfrac{1}{2}}$	1	$\tfrac{3}{2}$	0000+	$+1/4$
$\tilde{\tfrac{1}{2}}$	2	$\tfrac{3}{2}$	0000+	$+1/4$
$\tilde{\tfrac{1}{2}}$	$\tilde{1}$	$\tfrac{1}{2}$	0000+	$-1/3$
$\tilde{\tfrac{1}{2}}$	$\tilde{1}$	$\tfrac{3}{2}$	0000+	$-1/4.3$

$\tilde{\tfrac{1}{2}}$	$\tilde{\tfrac{1}{2}}$	0		
0	0	$\tilde{\tfrac{1}{2}}$	0000+	$-1/\sqrt{2}$
$\tfrac{1}{2}$	$\tfrac{1}{2}$	$\tilde{1}$	0000+	$-1/2$
1	1	$\tfrac{3}{2}$	0000+	$-1/\sqrt{2.3}$
1	1	$\tilde{\tfrac{1}{2}}$	0000+	$+1/\sqrt{2.3}$
$\tfrac{3}{2}$	$\tfrac{3}{2}$	1	0000+	$-1/2\sqrt{2}$
$\tfrac{3}{2}$	$\tfrac{3}{2}$	2	0000+	$+1/2\sqrt{2}$
$\tfrac{3}{2}$	$\tfrac{3}{2}$	$\tilde{1}$	0000+	$+1/2\sqrt{2}$
2	2	$\tfrac{3}{2}$	0000+	$+1/2$
$\tilde{1}$	$\tilde{1}$	$\tfrac{1}{2}$	0000+	$-1/\sqrt{2.3}$
$\tilde{1}$	$\tilde{1}$	$\tfrac{3}{2}$	0000+	$+1/\sqrt{2.3}$
$\tilde{\tfrac{1}{2}}$	$\tilde{\tfrac{1}{2}}$	0	0000+	$-1/2$

$\tilde{\tfrac{1}{2}}$	$\tilde{\tfrac{1}{2}}$	1		
$\tfrac{1}{2}$	$\tfrac{1}{2}$	$\tilde{1}$	0000+	$+1/2.3$
1	0	$\tilde{\tfrac{1}{2}}$	0000+	$+1/\sqrt{2.3}$
1	1	$\tfrac{3}{2}$	0000+	$-1/2.3$
1	1	$\tilde{\tfrac{1}{2}}$	0000+	$-1/3$
$\tfrac{3}{2}$	$\tfrac{1}{2}$	$\tilde{1}$	0000+	$-1/3$
$\tfrac{3}{2}$	$\tfrac{3}{2}$	1	0000+	$-1/2.3\sqrt{2}$
$\tfrac{3}{2}$	$\tfrac{3}{2}$	1	0010+	$+1/3\sqrt{2}$
$\tfrac{3}{2}$	$\tfrac{3}{2}$	2	0000+	$-1/2\sqrt{2}$
$\tfrac{3}{2}$	$\tfrac{3}{2}$	2	0010+	0
$\tfrac{3}{2}$	$\tfrac{3}{2}$	$\tilde{1}$	0000+	$+1/2.3\sqrt{2}$
$\tfrac{3}{2}$	$\tfrac{3}{2}$	$\tilde{1}$	0010+	$+1/3\sqrt{2}$
2	1	$\tfrac{3}{2}$	0000+	$+1/2\sqrt{3}$
$\tilde{1}$	1	$\tfrac{3}{2}$	0000+	$-1/2\sqrt{3}$
$\tilde{1}$	2	$\tfrac{3}{2}$	0000+	$+1/2\sqrt{3}$
$\tilde{1}$	$\tilde{1}$	$\tfrac{1}{2}$	0000+	$-1/3$
$\tilde{1}$	$\tilde{1}$	$\tfrac{3}{2}$	0000+	$-1/2.3$
$\tilde{\tfrac{1}{2}}$	$\tfrac{3}{2}$	1	0000+	$-1/3$
$\tilde{\tfrac{1}{2}}$	$\tilde{\tfrac{1}{2}}$	0	0000+	$+1/2$
$\tilde{\tfrac{1}{2}}$	$\tilde{\tfrac{1}{2}}$	1	0000+	$+1/2.3$

$\tilde{0}$	$\tfrac{3}{2}$	$\tfrac{3}{2}$		
0	$\tfrac{3}{2}$	$\tfrac{3}{2}$	0000+	$+1/4$
1	$\tfrac{3}{2}$	$\tfrac{3}{2}$	0000+	$-1/4$
1	$\tfrac{3}{2}$	$\tfrac{3}{2}$	0100+	0
1	$\tfrac{3}{2}$	$\tfrac{3}{2}$	0110+	$+1/4$
2	$\tfrac{3}{2}$	$\tfrac{3}{2}$	0000+	$-1/4$
$\tilde{1}$	$\tfrac{3}{2}$	$\tfrac{3}{2}$	0000+	$+1/4$
$\tilde{1}$	$\tfrac{3}{2}$	$\tfrac{3}{2}$	0100−	0
$\tilde{1}$	$\tfrac{3}{2}$	$\tfrac{3}{2}$	0110+	$+1/4$
$\tilde{0}$	$\tfrac{3}{2}$	$\tfrac{3}{2}$	0000+	$-1/4$

$\tilde{0}$	2	2		
0	2	2	0000+	$-1/2$
$\tfrac{1}{2}$	$\tfrac{3}{2}$	$\tfrac{3}{2}$	0000+	$+1/2\sqrt{2}$
$\tfrac{3}{2}$	$\tfrac{3}{2}$	$\tfrac{3}{2}$	0000+	$+1/2\sqrt{2}$
2	2	2	0000+	$+1/2$
$\tilde{\tfrac{1}{2}}$	$\tfrac{3}{2}$	$\tfrac{3}{2}$	0000+	$+1/2\sqrt{2}$
$\tilde{0}$	2	2	0000+	$+1/2$

$\tilde{0}$	$\tilde{1}$	1		
0	1	$\tilde{1}$	0000+	$+1/3$
$\tfrac{1}{2}$	$\tfrac{3}{2}$	$\tfrac{3}{2}$	0000+	$-1/2\sqrt{3}$
1	1	$\tilde{1}$	0000+	$+1/3$
1	2	2	0000+	$+1/\sqrt{2.3}$
1	$\tilde{1}$	1	0000+	$+1/3$
$\tfrac{3}{2}$	$\tfrac{3}{2}$	$\tfrac{3}{2}$	0000+	$+1/2\sqrt{3}$
$\tfrac{3}{2}$	$\tfrac{3}{2}$	$\tfrac{3}{2}$	0010+	0
$\tfrac{3}{2}$	$\tfrac{3}{2}$	$\tfrac{3}{2}$	0100−	0
$\tfrac{3}{2}$	$\tfrac{3}{2}$	$\tfrac{3}{2}$	0110−	$-1/2\sqrt{3}$
2	1	$\tilde{1}$	0000+	$+1/3$
2	$\tilde{1}$	1	0000+	$+1/3$
$\tilde{1}$	1	$\tilde{1}$	0000+	$-1/3$
$\tilde{1}$	2	2	0000+	$-1/\sqrt{2.3}$
$\tilde{1}$	$\tilde{1}$	1	0000+	$-1/3$
$\tilde{\tfrac{1}{2}}$	$\tfrac{3}{2}$	$\tfrac{3}{2}$	0000+	$-1/2\sqrt{3}$
$\tilde{0}$	$\tilde{1}$	1	0000+	$+1/3$

$\tilde{0}$	$\tilde{\tfrac{1}{2}}$	$\tfrac{1}{2}$		
0	$\tfrac{1}{2}$	$\tilde{\tfrac{1}{2}}$	0000+	$+1/2$
$\tfrac{1}{2}$	1	$\tilde{1}$	0000+	$+1/\sqrt{2.3}$
1	$\tfrac{1}{2}$	$\tilde{\tfrac{1}{2}}$	0000+	$+1/2$
1	$\tfrac{3}{2}$	$\tfrac{3}{2}$	0000+	$+1/2\sqrt{2}$
$\tfrac{3}{2}$	1	$\tilde{1}$	0000+	$-1/\sqrt{2.3}$
$\tfrac{3}{2}$	2	2	0000+	$+1/2$
$\tfrac{3}{2}$	$\tilde{1}$	1	0000+	$-1/\sqrt{2.3}$
2	$\tfrac{3}{2}$	$\tfrac{3}{2}$	0000+	$+1/2\sqrt{2}$
$\tilde{1}$	$\tfrac{3}{2}$	$\tfrac{3}{2}$	0000+	$-1/2\sqrt{2}$
$\tilde{1}$	$\tilde{\tfrac{1}{2}}$	$\tfrac{1}{2}$	0000+	$+1/2$
$\tilde{\tfrac{1}{2}}$	$\tilde{1}$	1	0000+	$-1/\sqrt{2.3}$
$\tilde{0}$	$\tilde{\tfrac{1}{2}}$	$\tfrac{1}{2}$	0000+	$-1/2$

$\tilde{0}$	$\tilde{0}$	0		
0	0	$\tilde{0}$	0000+	$+1$
$\tfrac{1}{2}$	$\tfrac{1}{2}$	$\tilde{\tfrac{1}{2}}$	0000+	$+1/\sqrt{2}$
1	1	$\tilde{1}$	0000+	$+1/\sqrt{3}$
$\tfrac{3}{2}$	$\tfrac{3}{2}$	$\tfrac{3}{2}$	0000+	$+1/2$
2	2	2	0000+	$-1/\sqrt{2}$
$\tilde{1}$	$\tilde{1}$	1	0000+	$+1/\sqrt{3}$
$\tilde{\tfrac{1}{2}}$	$\tilde{\tfrac{1}{2}}$	$\tfrac{1}{2}$	0000+	$+1/\sqrt{2}$
$\tilde{0}$	$\tilde{0}$	0	0000+	$+1$

K 3j Phases ($K_i = K_h$)

$0\ 0\ 0\ 0+$	$\frac{5}{2}\ \frac{3}{2}\ 1\ 0-$	$3\ \frac{5}{2}\ \frac{3}{2}\ 0-$	$\tilde{1}\ 3\ 1\ 0-$
$\frac{1}{2}\ \frac{1}{2}\ 0\ 0-$	$\frac{5}{2}\ 2\ \frac{1}{2}\ 0-$	$3\ \frac{5}{2}\ \frac{3}{2}\ 1+$	$\tilde{1}\ 3\ 2\ 0+$
$1\ \frac{1}{2}\ \frac{1}{2}\ 0+$	$\frac{5}{2}\ 2\ \frac{3}{2}\ 0+$	$3\ \frac{5}{2}\ \frac{5}{2}\ 0+$	$\tilde{1}\ 3\ 3\ 0-$
$1\ 1\ 0\ 0+$	$\frac{5}{2}\ 2\ \frac{3}{2}\ 1-$	$3\ \frac{5}{2}\ \frac{5}{2}\ 1-$	$\tilde{1}\ \tilde{1}\ 0\ 0+$
$1\ 1\ 1\ 0-$	$\frac{5}{2}\ \frac{5}{2}\ 0\ 0-$	$3\ 3\ 0\ 0+$	$\tilde{1}\ \tilde{1}\ 2\ 0+$
$\frac{3}{2}\ 1\ \frac{1}{2}\ 0-$	$\frac{5}{2}\ \frac{5}{2}\ 1\ 0+$	$3\ 3\ 1\ 0-$	$\tilde{1}\ \tilde{1}\ \tilde{1}\ 0-$
$\frac{3}{2}\ \frac{3}{2}\ 0\ 0-$	$\frac{5}{2}\ \frac{5}{2}\ 1\ 1+$	$3\ 3\ 2\ 0+$	$\tilde{\frac{1}{2}}\ 2\ \frac{3}{2}\ 0-$
$\frac{3}{2}\ \frac{3}{2}\ 1\ 0+$	$\frac{5}{2}\ \frac{5}{2}\ 2\ 0-$	$3\ 3\ 3\ 0+$	$\tilde{\frac{1}{2}}\ \frac{5}{2}\ 1\ 0-$
$2\ 1\ 1\ 0+$	$\frac{5}{2}\ \frac{5}{2}\ 2\ 1+$	$\tilde{1}\ \frac{3}{2}\ \frac{3}{2}\ 0+$	$\tilde{\frac{1}{2}}\ \frac{5}{2}\ 2\ 0+$
$2\ \frac{3}{2}\ \frac{1}{2}\ 0+$	$\frac{5}{2}\ \frac{5}{2}\ 2\ 2-$	$\tilde{1}\ 2\ 1\ 0+$	$\tilde{\frac{1}{2}}\ 3\ \frac{1}{2}\ 0-$
$2\ \frac{3}{2}\ \frac{3}{2}\ 0-$	$3\ \frac{3}{2}\ \frac{3}{2}\ 0+$	$\tilde{1}\ 2\ 2\ 0-$	$\tilde{\frac{1}{2}}\ 3\ \frac{5}{2}\ 0-$
$2\ 2\ 0\ 0+$	$3\ 2\ 1\ 0+$	$\tilde{1}\ \frac{5}{2}\ \frac{1}{2}\ 0+$	$\tilde{\frac{1}{2}}\ \tilde{1}\ \frac{3}{2}\ 0+$
$2\ 2\ 1\ 0-$	$3\ 2\ 2\ 0-$	$\tilde{1}\ \frac{5}{2}\ \frac{3}{2}\ 0-$	$\tilde{\frac{1}{2}}\ \tilde{\frac{1}{2}}\ 0\ 0-$
$2\ 2\ 2\ 0+$	$3\ 2\ 2\ 1+$	$\tilde{1}\ \frac{5}{2}\ \frac{5}{2}\ 0+$	$\tilde{\frac{1}{2}}\ \tilde{\frac{1}{2}}\ \tilde{1}\ 0+$
$2\ 2\ 2\ 1+$	$3\ \frac{5}{2}\ \frac{1}{2}\ 0+$	$\tilde{1}\ \frac{5}{2}\ \frac{5}{2}\ 1+$	

K 6j Symbols

0 0 0
0 0 0 0000+ $+1$
$\frac{1}{2}$ $\frac{1}{2}$ 0
0 0 $\frac{1}{2}$ 0000+ $-1/\sqrt{2}$
$\frac{1}{2}$ $\frac{1}{2}$ 0 0000+ $-1/2$
1 $\frac{1}{2}$ $\frac{1}{2}$
0 $\frac{1}{2}$ $\frac{1}{2}$ 0000+ $+1/2$
1 $\frac{1}{2}$ $\frac{1}{2}$ 0000+ $+1/2.3$
1 1 0
0 0 1 0000+ $+1/\sqrt{3}$
$\frac{1}{2}$ $\frac{1}{2}$ $\frac{1}{2}$ 0000+ $+1/\sqrt{2.3}$
1 1 0 0000+ $+1/3$
1 1 1
$\frac{1}{2}$ $\frac{1}{2}$ $\frac{1}{2}$ 0000+ $+1/3$
1 1 0 0000+ $-1/3$
1 1 1 0000+ $+1/2.3$
$\frac{3}{2}$ 1 $\frac{1}{2}$
0 $\frac{1}{2}$ 1 0000+ $-1/\sqrt{2.3}$
$\frac{1}{2}$ 1 $\frac{1}{2}$ 0000+ $-1/3$
1 $\frac{1}{2}$ 1 0000+ $+1/2.3$
$\frac{3}{2}$ 1 $\frac{1}{2}$ 0000+ $-1/4.3$
$\frac{3}{2}$ $\frac{3}{2}$ 0
0 0 $\frac{3}{2}$ 0000+ $-1/2$
$\frac{1}{2}$ $\frac{1}{2}$ 1 0000+ $-1/2\sqrt{2}$
1 1 $\frac{1}{2}$ 0000+ $-1/2\sqrt{3}$
1 1 $\frac{3}{2}$ 0000+ $+1/2\sqrt{3}$
$\frac{3}{2}$ $\frac{3}{2}$ 0 0000+ $-1/4$
$\frac{3}{2}$ $\frac{3}{2}$ 1
$\frac{1}{2}$ $\frac{1}{2}$ 1 0000+ $+\sqrt{5}/2.3\sqrt{2}$
1 0 $\frac{3}{2}$ 0000+ $+1/2\sqrt{3}$
1 1 $\frac{1}{2}$ 0000+ $-\sqrt{5}/2.3\sqrt{2}$
1 1 $\frac{3}{2}$ 0000+ $+1/3\sqrt{2.5}$
$\frac{3}{2}$ $\frac{1}{2}$ 1 0000+ $+1/2.3$
$\frac{3}{2}$ $\frac{3}{2}$ 0 0000+ $+1/4$
$\frac{3}{2}$ $\frac{3}{2}$ 1 0000+ $-11/4.3.5$
2 1 1
0 1 1 0000+ $+1/3$
1 1 1 0000+ $+1/2.3$
2 1 1 0000+ $+1/2.3.5$
2 $\frac{3}{2}$ $\frac{1}{2}$
0 $\frac{1}{2}$ $\frac{3}{2}$ 0000+ $+1/2\sqrt{2}$
$\frac{1}{2}$ 1 1 0000+ $+1/2\sqrt{3}$
1 $\frac{1}{2}$ $\frac{3}{2}$ 0000+ $+1/2\sqrt{2.5}$
1 $\frac{3}{2}$ $\frac{1}{2}$ 0000+ $+1/4$
1 $\frac{3}{2}$ $\frac{3}{2}$ 0000+ $+1/2\sqrt{5}$
$\frac{3}{2}$ 1 1 0000+ $+1/2\sqrt{2.3.5}$
2 $\frac{3}{2}$ $\frac{1}{2}$ 0000+ $+1/4.5$
2 $\frac{3}{2}$ $\frac{3}{2}$
0 $\frac{3}{2}$ $\frac{3}{2}$ 0000+ $-1/4$
$\frac{1}{2}$ 1 1 0000+ $+1/2\sqrt{2.3}$
1 $\frac{3}{2}$ $\frac{1}{2}$ 0000+ $+1/2\sqrt{5}$
1 $\frac{3}{2}$ $\frac{3}{2}$ 0000+ $+1/4.5$
$\frac{3}{2}$ 1 1 0000+ $+\sqrt{2}/5\sqrt{3}$
2 $\frac{3}{2}$ $\frac{1}{2}$ 0000+ $-1/2.5$
2 $\frac{3}{2}$ $\frac{3}{2}$ 0000+ $+3/4.5$
2 2 0
0 0 2 0000+ $+1/\sqrt{5}$
$\frac{1}{2}$ $\frac{1}{2}$ $\frac{3}{2}$ 0000+ $+1/\sqrt{2.5}$
1 1 1 0000+ $+1/\sqrt{3.5}$
1 1 2 0000+ $-1/\sqrt{3.5}$
$\frac{3}{2}$ $\frac{3}{2}$ $\frac{1}{2}$ 0000+ $+1/2\sqrt{5}$
$\frac{3}{2}$ $\frac{3}{2}$ $\frac{3}{2}$ 0000+ $-1/2\sqrt{5}$
2 2 0 0000+ $+1/5$
2 2 1
$\frac{1}{2}$ $\frac{1}{2}$ $\frac{3}{2}$ 0000+ $+1/2\sqrt{5}$
1 0 2 0000+ $-1/\sqrt{3.5}$
1 1 1 0000+ $-1/2\sqrt{5}$
1 1 2 0000+ $+1/2.3\sqrt{5}$
$\frac{3}{2}$ $\frac{1}{2}$ $\frac{3}{2}$ 0000+ $-1/2\sqrt{2.5}$
$\frac{3}{2}$ $\frac{3}{2}$ $\frac{1}{2}$ 0000+ $+3/2.5\sqrt{2}$
$\frac{3}{2}$ $\frac{3}{2}$ $\frac{3}{2}$ 0000+ $-1/5\sqrt{2}$
2 1 1 0000+ $-1/2.5$
2 1 2 0000+ $-1/2\sqrt{2.3.5}$
2 1 2 0100+ $+\sqrt{3}/2.5\sqrt{2}$
2 2 0 0000+ $-1/5$
2 2 1 0000+ $+1/2.3$
2 2 2
1 1 1 0000+ $-1/2\sqrt{2.3.5}$
1 1 1 0001+ $+\sqrt{3}/2.5\sqrt{2}$
$\frac{3}{2}$ $\frac{3}{2}$ $\frac{1}{2}$ 0000+ $+1/4\sqrt{5}$
$\frac{3}{2}$ $\frac{3}{2}$ $\frac{1}{2}$ 0001+ $-3/4.5$
$\frac{3}{2}$ $\frac{3}{2}$ $\frac{3}{2}$ 0000+ 0
$\frac{3}{2}$ $\frac{3}{2}$ $\frac{3}{2}$ 0001+ 0
2 2 2
2 1 1 0000+ $-1/2\sqrt{2.3.5}$
2 1 1 0001+ $+\sqrt{3}/2.5\sqrt{2}$
2 2 0 0000+ $+1/5$
2 2 0 0001+ 0
2 2 0 0010+ 0
2 2 0 0011+ $+1/5$
2 2 1 0000+ $+1/4.5$
2 2 1 0001+ $+1/4\sqrt{5}$
2 2 1 0010+ $+1/4\sqrt{5}$
2 2 1 0011+ $-1/4.3.5$
2 2 2 0000+ $+3/4.5$
2 2 2 0001+ 0
2 2 2 1000+ 0
2 2 2 1001+ $-1/4.5$
2 2 2 1100+ $-1/4.5$
2 2 2 1101+ 0
2 2 2 1110+ 0
2 2 2 1111+ $+1/4.3.5$
$\frac{5}{2}$ $\frac{3}{2}$ 1
0 1 $\frac{3}{2}$ 0000+ $-1/2\sqrt{3}$
$\frac{1}{2}$ $\frac{3}{2}$ 1 0000+ $-1/4$
1 1 $\frac{3}{2}$ 0000+ $+1/2\sqrt{2.5}$
$\frac{3}{2}$ $\frac{3}{2}$ 1 0000+ $-1/2.5$
2 1 $\frac{3}{2}$ 0000+ $+1/2.5\sqrt{2.3}$
$\frac{5}{2}$ $\frac{3}{2}$ 1 0000+ $-1/4.3.5$
$\frac{5}{2}$ 2 $\frac{1}{2}$
0 $\frac{1}{2}$ 2 0000+ $-1/\sqrt{2.5}$
$\frac{1}{2}$ 1 $\frac{3}{2}$ 0000+ $+1/\sqrt{3.5}$
1 $\frac{1}{2}$ 2 0000+ $+1/3\sqrt{5}$
1 $\frac{3}{2}$ 1 0000+ $+1/2\sqrt{5}$
1 $\frac{3}{2}$ 2 0000+ $+1/2.3\sqrt{5}$
1 $\frac{3}{2}$ 2 1000− $-1/\sqrt{2.3.5}$
$\frac{3}{2}$ 1 $\frac{3}{2}$ 0000+ $-1/2\sqrt{2.3.5}$
$\frac{3}{2}$ 2 $\frac{1}{2}$ 0000+ $-1/5$
$\frac{3}{2}$ 2 $\frac{3}{2}$ 0000+ $+1/2.5\sqrt{2}$
$\frac{3}{2}$ 2 $\frac{3}{2}$ 1000− $+\sqrt{3}/2.5$
2 $\frac{3}{2}$ 1 0000+ $-1/2.5\sqrt{3}$
2 $\frac{3}{2}$ 2 0000+ $+1/2\sqrt{2.5}$
2 $\frac{3}{2}$ 2 0100+ $+1/2.5\sqrt{2}$
2 $\frac{3}{2}$ 2 1000− 0
2 $\frac{3}{2}$ 2 1100− $+1/5\sqrt{3}$

K 6j Symbols (cont.)

$\frac{5}{2}$	**2**	**$\frac{1}{2}$**		
$\frac{5}{2}$	2	$\frac{1}{2}$	0000+	$-1/2.3.5$
$\frac{5}{2}$	**2**	**$\frac{3}{2}$**		
0	$\frac{3}{2}$	2	0000+	$+1/2\sqrt{5}$
0	$\frac{3}{2}$	2	0001−	0
0	$\frac{3}{2}$	2	1000−	0
0	$\frac{3}{2}$	2	1001+	$-1/2\sqrt{5}$
$\frac{1}{2}$	1	$\frac{3}{2}$	0000+	$+1/4\sqrt{3.5}$
$\frac{1}{2}$	1	$\frac{3}{2}$	0001−	$+1/2\sqrt{2.5}$
$\frac{1}{2}$	2	$\frac{3}{2}$	0000+	$-3/4.5$
$\frac{1}{2}$	2	$\frac{3}{2}$	0001−	$+\sqrt{3}/2.5\sqrt{2}$
$\frac{1}{2}$	2	$\frac{3}{2}$	1000−	$+\sqrt{3}/2.5\sqrt{2}$
$\frac{1}{2}$	2	$\frac{3}{2}$	1001+	$+1/2.5$
1	$\frac{1}{2}$	2	0000+	$+1/2.3\sqrt{5}$
1	$\frac{1}{2}$	2	0001−	$+1/\sqrt{2.3.5}$
1	$\frac{3}{2}$	1	0000+	$+1/2.5\sqrt{2}$
1	$\frac{3}{2}$	1	0001−	$+\sqrt{3/2.5}$
1	$\frac{3}{2}$	2	0000+	$-1/2.3\sqrt{2}$
1	$\frac{3}{2}$	2	0001−	$+1/2.5\sqrt{3}$
1	$\frac{3}{2}$	2	1000−	$-1/2.5\sqrt{3}$
1	$\frac{3}{2}$	2	1001+	0
$\frac{3}{2}$	1	$\frac{3}{2}$	0000+	$-1/2.5\sqrt{3}$
$\frac{3}{2}$	1	$\frac{3}{2}$	0001−	$-1/5\sqrt{2}$
$\frac{3}{2}$	2	$\frac{1}{2}$	0000+	$+1/2.5\sqrt{2}$
$\frac{3}{2}$	2	$\frac{1}{2}$	0001−	$+\sqrt{3/2.5}$
$\frac{3}{2}$	2	$\frac{3}{2}$	0000+	$-1/2.5$
$\frac{3}{2}$	2	$\frac{3}{2}$	0001−	0
$\frac{3}{2}$	2	$\frac{3}{2}$	1000−	0
$\frac{3}{2}$	2	$\frac{3}{2}$	1001+	$-1/2.5$
2	$\frac{1}{2}$	2	0000+	$+1/2\sqrt{2.5}$
2	$\frac{1}{2}$	2	0001−	0
2	$\frac{1}{2}$	2	0100+	$+1/2.5\sqrt{2}$
2	$\frac{1}{2}$	2	0101−	$-1/5\sqrt{3}$
2	$\frac{3}{2}$	1	0000+	$+1/2.5\sqrt{2.3}$
2	$\frac{3}{2}$	1	0001−	$+1/2.5$
2	$\frac{3}{2}$	2	0000+	$+1/4\sqrt{5}$
2	$\frac{3}{2}$	2	0001−	0
2	$\frac{3}{2}$	2	0100+	$+1/4.5$
2	$\frac{3}{2}$	2	0101−	$+1/5\sqrt{2.3}$
2	$\frac{3}{2}$	2	1000−	0
2	$\frac{3}{2}$	2	1001+	$+1/4\sqrt{5}$
2	$\frac{3}{2}$	2	1100−	$-1/5\sqrt{2.3}$
2	$\frac{3}{2}$	2	1101+	$-1/4.5$
$\frac{5}{2}$	1	$\frac{3}{2}$	0000+	$+3/4.5$
$\frac{5}{2}$	1	$\frac{3}{2}$	0001−	$-1/2.5\sqrt{2.3}$
$\frac{5}{2}$	1	$\frac{3}{2}$	0100−	$-1/2.5\sqrt{2.3}$
$\frac{5}{2}$	1	$\frac{3}{2}$	0101+	$+1/3.5$
$\frac{5}{2}$	2	$\frac{1}{2}$	0000+	$-2/3.5$
$\frac{5}{2}$	2	$\frac{1}{2}$	0001−	$+1/5\sqrt{2.3}$
$\frac{5}{2}$	2	$\frac{1}{2}$	0010−	$+1/5\sqrt{2.3}$
$\frac{5}{2}$	2	$\frac{1}{2}$	0011+	$+1/2.3.5$
$\frac{5}{2}$	2	$\frac{3}{2}$	0000+	$-1/4.3$
$\frac{5}{2}$	2	$\frac{3}{2}$	0001−	$-1/2.5\sqrt{2.3}$
$\frac{5}{2}$	2	$\frac{3}{2}$	0010−	$-1/2.5\sqrt{2.3}$
$\frac{5}{2}$	2	$\frac{3}{2}$	0011+	$-1/3.5$
$\frac{5}{2}$	2	$\frac{3}{2}$	0100−	$-1/2.5\sqrt{2.3}$
$\frac{5}{2}$	2	$\frac{3}{2}$	0101+	$+1/2.5$
$\frac{5}{2}$	2	$\frac{3}{2}$	0110+	0
$\frac{5}{2}$	2	$\frac{3}{2}$	0111−	$-1/2.5\sqrt{2.3}$
$\frac{5}{2}$	2	$\frac{3}{2}$	1000−	$-1/2.5\sqrt{2.3}$
$\frac{5}{2}$	2	$\frac{3}{2}$	1001+	0
$\frac{5}{2}$	2	$\frac{3}{2}$	1010+	$+1/2.5$
$\frac{5}{2}$	2	$\frac{3}{2}$	1011−	$-1/2.5\sqrt{2.3}$
$\frac{5}{2}$	2	$\frac{3}{2}$	1100+	$-1/3.5$
$\frac{5}{2}$	2	$\frac{3}{2}$	1101−	$-1/2.5\sqrt{2.3}$
$\frac{5}{2}$	2	$\frac{3}{2}$	1110−	$-1/2.5\sqrt{2.3}$
$\frac{5}{2}$	2	$\frac{3}{2}$	1111+	$-1/4.3$
$\frac{5}{2}$	**$\frac{5}{2}$**	**0**		
0	0	$\frac{5}{2}$	0000+	$-1/\sqrt{2.3}$
$\frac{1}{2}$	$\frac{1}{2}$	2	0000+	$-1/2\sqrt{3}$
1	1	$\frac{3}{2}$	0000+	$-1/3\sqrt{2}$
1	1	$\frac{5}{2}$	0000+	$+1/3\sqrt{2}$
1	1	$\frac{5}{2}$	1000+	0
1	1	$\frac{5}{2}$	1100+	$+1/3\sqrt{2}$
$\frac{3}{2}$	$\frac{3}{2}$	1	0000+	$-1/2\sqrt{2.3}$
$\frac{3}{2}$	$\frac{3}{2}$	2	0000+	$+1/2\sqrt{2.3}$
$\frac{3}{2}$	$\frac{3}{2}$	2	1000−	0
$\frac{3}{2}$	$\frac{3}{2}$	2	1100+	$-1/2\sqrt{2.3}$
2	2	$\frac{1}{2}$	0000+	$-1/\sqrt{2.3.5}$
2	2	$\frac{3}{2}$	0000+	$+1/\sqrt{2.3.5}$
2	2	$\frac{3}{2}$	1000−	0
2	2	$\frac{3}{2}$	1100+	$-1/\sqrt{2.3.5}$
2	2	$\frac{5}{2}$	0000+	$-1/\sqrt{2.3.5}$
$\frac{5}{2}$	**$\frac{5}{2}$**	**0**		
2	2	$\frac{5}{2}$	1000−	0
2	2	$\frac{5}{2}$	1100+	$+1/\sqrt{2.3.5}$
2	2	$\frac{5}{2}$	2000+	0
2	2	$\frac{5}{2}$	2100−	0
2	2	$\frac{5}{2}$	2200+	$-1/\sqrt{2.3.5}$
$\frac{5}{2}$	$\frac{5}{2}$	0	0000+	$-1/2.3$
$\frac{5}{2}$	**$\frac{5}{2}$**	**1**		
$\frac{1}{2}$	$\frac{1}{2}$	2	0000+	$+1/2.3\sqrt{2}$
$\frac{1}{2}$	$\frac{1}{2}$	2	0001+	$+1/2\sqrt{2.5}$
1	0	$\frac{5}{2}$	0000+	$+1/3\sqrt{2}$
1	0	$\frac{5}{2}$	0001+	0
1	0	$\frac{5}{2}$	0100+	0
1	0	$\frac{5}{2}$	0101+	$+1/3\sqrt{2}$
1	1	$\frac{3}{2}$	0000+	$-1/2.3\sqrt{2}$
1	1	$\frac{3}{2}$	0001+	$-1/2\sqrt{2.5}$
1	1	$\frac{5}{2}$	0000+	$-1/2.3\sqrt{2}$
1	1	$\frac{5}{2}$	0001+	0
1	1	$\frac{5}{2}$	1000+	0
1	1	$\frac{5}{2}$	1001+	$+1/2.3\sqrt{2}$
1	1	$\frac{5}{2}$	1100+	$+1/2.3\sqrt{2}$
1	1	$\frac{5}{2}$	1101+	$-1/3\sqrt{2.5}$
$\frac{3}{2}$	$\frac{1}{2}$	2	0000+	$+1/2.3\sqrt{2}$
$\frac{3}{2}$	$\frac{1}{2}$	2	0001+	$-1/2\sqrt{2.5}$
$\frac{3}{2}$	$\frac{1}{2}$	2	0100−	$-1/4\sqrt{3}$
$\frac{3}{2}$	$\frac{1}{2}$	2	0101−	$-1/4\sqrt{3.5}$
$\frac{3}{2}$	$\frac{3}{2}$	1	0000+	$+1/4\sqrt{5}$
$\frac{3}{2}$	$\frac{3}{2}$	1	0001+	$+3/4.5$
$\frac{3}{2}$	$\frac{3}{2}$	2	0000+	$-1/4.3\sqrt{5}$
$\frac{3}{2}$	$\frac{3}{2}$	2	0001+	$+3/4.5$
$\frac{3}{2}$	$\frac{3}{2}$	2	1000−	$+1/2\sqrt{2.3.5}$
$\frac{3}{2}$	$\frac{3}{2}$	2	1001−	$+1/2.5\sqrt{2.3}$
$\frac{3}{2}$	$\frac{3}{2}$	2	1100+	0
$\frac{3}{2}$	$\frac{3}{2}$	2	1101+	$+1/2.5$
2	1	$\frac{3}{2}$	0000+	$+1/2\sqrt{2.5}$
2	1	$\frac{3}{2}$	0001+	$-1/2.5\sqrt{2}$
2	1	$\frac{3}{2}$	0100−	0
2	1	$\frac{3}{2}$	0101−	$-1/5\sqrt{3}$
2	1	$\frac{5}{2}$	0000+	$-1/4.3\sqrt{2}$
2	1	$\frac{5}{2}$	0001+	$+1/4\sqrt{2.5}$
2	1	$\frac{5}{2}$	0100−	0
2	1	$\frac{5}{2}$	0101−	$+1/2\sqrt{3.5}$

K 6j Symbols (cont.)

$\frac{5}{2}$	**$\frac{5}{2}$**	**1**		
2	1	$\frac{5}{2}$	0200+	$+1/4\sqrt{2.3}$
2	1	$\frac{5}{2}$	0201+	$+1/4\sqrt{2.3.5}$
2	1	$\frac{5}{2}$	1000+	$+1/4\sqrt{2.5}$
2	1	$\frac{5}{2}$	1001+	$+7/4.3.5\sqrt{2}$
2	1	$\frac{5}{2}$	1100−	$-1/2\sqrt{3.5}$
2	1	$\frac{5}{2}$	1101−	0
2	1	$\frac{5}{2}$	1200+	$+1/4\sqrt{2.3.5}$
2	1	$\frac{5}{2}$	1201+	$+1/4.5\sqrt{2.3}$
2	2	$\frac{1}{2}$	0000+	$-1/3\sqrt{2.5}$
2	2	$\frac{1}{2}$	0001+	$-1/5\sqrt{2}$
2	2	$\frac{3}{2}$	0000+	$-1/2.3\sqrt{2.5}$
2	2	$\frac{3}{2}$	0001+	$-1/2.5\sqrt{2}$
2	2	$\frac{3}{2}$	1000−	0
2	2	$\frac{3}{2}$	1001−	$+1/5\sqrt{3}$
2	2	$\frac{3}{2}$	1100+	$-1/3\sqrt{2.5}$
2	2	$\frac{3}{2}$	1101+	$-1/3.5\sqrt{2}$
2	2	$\frac{5}{2}$	0000+	0
2	2	$\frac{5}{2}$	0001+	0
2	2	$\frac{5}{2}$	1000−	$+1/4\sqrt{3.5}$
2	2	$\frac{5}{2}$	1001−	$-\sqrt{3}/4.5$
2	2	$\frac{5}{2}$	1100+	0
2	2	$\frac{5}{2}$	1101+	$-1/3.5\sqrt{2}$
2	2	$\frac{5}{2}$	2000+	$-1/2\sqrt{2.3.5}$
2	2	$\frac{5}{2}$	2001+	$-1/2.5\sqrt{2.3}$
2	2	$\frac{5}{2}$	2100−	$-1/4.3\sqrt{5}$
2	2	$\frac{5}{2}$	2101−	$-1/4.3.5$
2	2	$\frac{5}{2}$	2200+	$+1/3\sqrt{2.5}$
2	2	$\frac{5}{2}$	2201+	$-1/3.5\sqrt{2}$
$\frac{5}{2}$	$\frac{3}{2}$	1	0000+	$-1/4.3$
$\frac{5}{2}$	$\frac{3}{2}$	1	0001+	$+1/4\sqrt{5}$
$\frac{5}{2}$	$\frac{3}{2}$	1	0100+	$+1/4\sqrt{5}$
$\frac{5}{2}$	$\frac{3}{2}$	1	0101+	$-1/4.3.5$
$\frac{5}{2}$	$\frac{3}{2}$	2	0000+	$+1/4\sqrt{5}$
$\frac{5}{2}$	$\frac{3}{2}$	2	0001+	$+1/4.5$
$\frac{5}{2}$	$\frac{3}{2}$	2	0100−	0
$\frac{5}{2}$	$\frac{3}{2}$	2	0101−	0
$\frac{5}{2}$	$\frac{3}{2}$	2	0200+	$-1/4\sqrt{3.5}$
$\frac{5}{2}$	$\frac{3}{2}$	2	0201+	$-1/4.5\sqrt{3}$
$\frac{5}{2}$	$\frac{3}{2}$	2	1000−	$-1/4\sqrt{2.3.5}$
$\frac{5}{2}$	$\frac{3}{2}$	2	1001−	$-1/4.5\sqrt{2.3}$
$\frac{5}{2}$	$\frac{3}{2}$	2	1100+	$+1/4\sqrt{5}$

$\frac{5}{2}$	**$\frac{5}{2}$**	**1**		
$\frac{5}{2}$	$\frac{3}{2}$	2	1101+	$-1/4.3$
$\frac{5}{2}$	$\frac{3}{2}$	2	1200−	$-1/4.3\sqrt{2.5}$
$\frac{5}{2}$	$\frac{3}{2}$	2	1201−	$-3/4.5\sqrt{2}$
$\frac{5}{2}$	$\frac{5}{2}$	0	0000+	$+1/2.3$
$\frac{5}{2}$	$\frac{5}{2}$	0	0001+	0
$\frac{5}{2}$	$\frac{5}{2}$	0	0010+	0
$\frac{5}{2}$	$\frac{5}{2}$	0	0011+	$+1/2.3$
$\frac{5}{2}$	$\frac{5}{2}$	1	0000+	$-1/4.3$
$\frac{5}{2}$	$\frac{5}{2}$	1	0001+	0
$\frac{5}{2}$	$\frac{5}{2}$	1	0010+	0
$\frac{5}{2}$	$\frac{5}{2}$	1	0011+	$+1/4.3$
$\frac{5}{2}$	$\frac{5}{2}$	1	1000+	0
$\frac{5}{2}$	$\frac{5}{2}$	1	1001+	$-1/4.3$
$\frac{5}{2}$	$\frac{5}{2}$	1	1010+	$-1/4.3$
$\frac{5}{2}$	$\frac{5}{2}$	1	1011+	$-1/2.3\sqrt{5}$
$\frac{5}{2}$	$\frac{5}{2}$	1	1100+	$+1/4.3$
$\frac{5}{2}$	$\frac{5}{2}$	1	1101+	$-1/2.3\sqrt{5}$
$\frac{5}{2}$	$\frac{5}{2}$	1	1110+	$-1/2.3\sqrt{5}$
$\frac{5}{2}$	$\frac{5}{2}$	1	1111+	$-1/4.3.5$
$\frac{5}{2}$	**$\frac{5}{2}$**	**2**		
1	1	$\frac{3}{2}$	0000+	$+\sqrt{2}/3.5$
1	1	$\frac{3}{2}$	0001−	0
1	1	$\frac{3}{2}$	0002+	$+1/5\sqrt{2.3}$
1	1	$\frac{5}{2}$	0000+	$-1/4.3\sqrt{2}$
1	1	$\frac{5}{2}$	0001−	0
1	1	$\frac{5}{2}$	0002+	$+1/4\sqrt{2.3}$
1	1	$\frac{5}{2}$	1000+	$+1/4\sqrt{2.5}$
1	1	$\frac{5}{2}$	1001−	$+1/2\sqrt{3.5}$
1	1	$\frac{5}{2}$	1002+	$+1/4\sqrt{2.3.5}$
1	1	$\frac{5}{2}$	1100+	$+7/4.3.5\sqrt{2}$
1	1	$\frac{5}{2}$	1101−	0
1	1	$\frac{5}{2}$	1102+	$+1/4.5\sqrt{2.3}$
$\frac{3}{2}$	$\frac{1}{2}$	2	0000+	$-1/5\sqrt{2.3}$
$\frac{3}{2}$	$\frac{1}{2}$	2	0001−	$+1/2.5$
$\frac{3}{2}$	$\frac{1}{2}$	2	0002+	0
$\frac{3}{2}$	$\frac{1}{2}$	2	0100−	$-1/2.5$
$\frac{3}{2}$	$\frac{1}{2}$	2	0101+	$-1/2.5\sqrt{2.3}$
$\frac{3}{2}$	$\frac{1}{2}$	2	0102−	$-1/5\sqrt{3}$
$\frac{3}{2}$	$\frac{3}{2}$	1	0000+	$-1/5\sqrt{3}$
$\frac{3}{2}$	$\frac{3}{2}$	1	0001−	0
$\frac{3}{2}$	$\frac{3}{2}$	1	0002+	$-1/2.5$

$\frac{5}{2}$	**$\frac{5}{2}$**	**2**		
$\frac{3}{2}$	$\frac{3}{2}$	2	0000+	$+1/2.5\sqrt{3}$
$\frac{3}{2}$	$\frac{3}{2}$	2	0001−	0
$\frac{3}{2}$	$\frac{3}{2}$	2	0002+	0
$\frac{3}{2}$	$\frac{3}{2}$	2	1000−	$+1/2.5\sqrt{2}$
$\frac{3}{2}$	$\frac{3}{2}$	2	1001+	$-1/5\sqrt{3}$
$\frac{3}{2}$	$\frac{3}{2}$	2	1002−	$-1/2.5\sqrt{2.3}$
$\frac{3}{2}$	$\frac{3}{2}$	2	1100+	$-1/4.5\sqrt{3}$
$\frac{3}{2}$	$\frac{3}{2}$	2	1101−	0
$\frac{3}{2}$	$\frac{3}{2}$	2	1102+	$+1/4.5$
2	0	$\frac{5}{2}$	0000+	$-1/\sqrt{2.3.5}$
2	0	$\frac{5}{2}$	0001−	0
2	0	$\frac{5}{2}$	0002+	0
2	0	$\frac{5}{2}$	0100−	0
2	0	$\frac{5}{2}$	0101+	$+1/\sqrt{2.3.5}$
2	0	$\frac{5}{2}$	0102−	0
2	0	$\frac{5}{2}$	0200+	0
2	0	$\frac{5}{2}$	0201−	0
2	0	$\frac{5}{2}$	0202+	$-1/\sqrt{2.3.5}$
2	1	$\frac{3}{2}$	0000+	0
2	1	$\frac{3}{2}$	0001−	$-1/5\sqrt{3}$
2	1	$\frac{3}{2}$	0002+	$-1/5\sqrt{2.3}$
2	1	$\frac{3}{2}$	0100−	$+1/5\sqrt{3}$
2	1	$\frac{3}{2}$	0101+	$-1/3.5\sqrt{2}$
2	1	$\frac{3}{2}$	0102−	$+1/3.5$
2	1	$\frac{5}{2}$	0000+	0
2	1	$\frac{5}{2}$	0001−	$-1/4\sqrt{3.5}$
2	1	$\frac{5}{2}$	0002+	$-1/2\sqrt{2.3.5}$
2	1	$\frac{5}{2}$	0100−	$+1/4\sqrt{3.5}$
2	1	$\frac{5}{2}$	0101+	0
2	1	$\frac{5}{2}$	0102−	$+1/4.3\sqrt{5}$
2	1	$\frac{5}{2}$	0200+	$-1/2\sqrt{2.3.5}$
2	1	$\frac{5}{2}$	0201−	$-1/4.3\sqrt{5}$
2	1	$\frac{5}{2}$	0202+	$+1/3\sqrt{2.5}$
2	1	$\frac{5}{2}$	1000+	0
2	1	$\frac{5}{2}$	1001−	$+\sqrt{3}/4.5$
2	1	$\frac{5}{2}$	1002+	$-1/2.5\sqrt{2.3}$
2	1	$\frac{5}{2}$	1100−	$-\sqrt{3}/4.5$
2	1	$\frac{5}{2}$	1101+	$-1/3.5\sqrt{2}$
2	1	$\frac{5}{2}$	1102−	$+1/4.3.5$
2	1	$\frac{5}{2}$	1200+	$-1/2.5\sqrt{2.3}$
2	1	$\frac{5}{2}$	1201−	$-1/4.3.5$

K 6j Symbols (cont.)

$\frac{5}{2}$ $\frac{5}{2}$ 2		
2 1 $\frac{5}{2}$	1202+	$-1/3.5\sqrt{2}$
2 2 $\frac{1}{2}$	0000+	$-1/2\sqrt{3.5}$
2 2 $\frac{1}{2}$	0001−	0
2 2 $\frac{1}{2}$	0002+	0
2 2 $\frac{1}{2}$	0010+	$+1/2.5\sqrt{3}$
2 2 $\frac{1}{2}$	0011−	0
2 2 $\frac{1}{2}$	0012+	$+2/3.5$
2 2 $\frac{3}{2}$	0000+	$+1/2\sqrt{3.5}$
2 2 $\frac{3}{2}$	0001−	0
2 2 $\frac{3}{2}$	0002+	0
2 2 $\frac{3}{2}$	0010+	$-1/2.5\sqrt{3}$
2 2 $\frac{3}{2}$	0011−	0
2 2 $\frac{3}{2}$	0012+	$+1/3.5$
2 2 $\frac{3}{2}$	1000−	0
2 2 $\frac{3}{2}$	1001+	0
2 2 $\frac{3}{2}$	1002−	0
2 2 $\frac{3}{2}$	1010−	0
2 2 $\frac{3}{2}$	1011+	$+1/5\sqrt{3}$
2 2 $\frac{3}{2}$	1012−	0
2 2 $\frac{3}{2}$	1100+	$+1/4\sqrt{3.5}$
2 2 $\frac{3}{2}$	1101−	0
2 2 $\frac{3}{2}$	1102+	$-1/4\sqrt{5}$
2 2 $\frac{3}{2}$	1110+	$+\sqrt{3}/4.5$
2 2 $\frac{3}{2}$	1111−	0
2 2 $\frac{3}{2}$	1112+	$-1/4.3.5$
2 2 $\frac{5}{2}$	0000+	$-1/4\sqrt{3.5}$
2 2 $\frac{5}{2}$	0001−	0
2 2 $\frac{5}{2}$	0002+	0
2 2 $\frac{5}{2}$	0010+	$-\sqrt{3}/4.5$
2 2 $\frac{5}{2}$	0011−	0
2 2 $\frac{5}{2}$	0012+	0
2 2 $\frac{5}{2}$	1000−	0
2 2 $\frac{5}{2}$	1001+	$-1/4\sqrt{3.5}$
2 2 $\frac{5}{2}$	1002−	0
2 2 $\frac{5}{2}$	1010−	0
2 2 $\frac{5}{2}$	1011+	$-1/4.5\sqrt{3}$
2 2 $\frac{5}{2}$	1012−	$-1/5\sqrt{2.3}$
2 2 $\frac{5}{2}$	1100+	$-1/4\sqrt{3.5}$
2 2 $\frac{5}{2}$	1101−	0
2 2 $\frac{5}{2}$	1102+	$-1/4\sqrt{5}$
2 2 $\frac{5}{2}$	1110+	$-1/4.5\sqrt{3}$

$\frac{5}{2}$ $\frac{5}{2}$ 2		
2 2 $\frac{5}{2}$	1111−	0
2 2 $\frac{5}{2}$	1112+	$+1/4.3.5$
2 2 $\frac{5}{2}$	2000+	0
2 2 $\frac{5}{2}$	2001−	0
2 2 $\frac{5}{2}$	2002+	$+1/4\sqrt{3.5}$
2 2 $\frac{5}{2}$	2010+	0
2 2 $\frac{5}{2}$	2011−	$+1/5\sqrt{2.3}$
2 2 $\frac{5}{2}$	2012+	$-1/4.5\sqrt{3}$
2 2 $\frac{5}{2}$	2100−	0
2 2 $\frac{5}{2}$	2101+	$-1/4\sqrt{5}$
2 2 $\frac{5}{2}$	2102−	0
2 2 $\frac{5}{2}$	2110−	$-1/5\sqrt{2.3}$
2 2 $\frac{5}{2}$	2111+	$+1/4.3.5$
2 2 $\frac{5}{2}$	2112−	0
2 2 $\frac{5}{2}$	2200+	$+1/4\sqrt{3.5}$
2 2 $\frac{5}{2}$	2201−	0
2 2 $\frac{5}{2}$	2202+	0
2 2 $\frac{5}{2}$	2210+	$-1/4.5\sqrt{3}$
2 2 $\frac{5}{2}$	2211−	0
2 2 $\frac{5}{2}$	2212+	$+1/3.5$
$\frac{5}{2}$ $\frac{1}{2}$ 2	0000+	$-1/2.5$
$\frac{5}{2}$ $\frac{1}{2}$ 2	0001−	$-1/5\sqrt{2.3}$
$\frac{5}{2}$ $\frac{1}{2}$ 2	0002+	0
$\frac{5}{2}$ $\frac{1}{2}$ 2	0100−	$-1/5\sqrt{2.3}$
$\frac{5}{2}$ $\frac{1}{2}$ 2	0101+	$+1/3.5$
$\frac{5}{2}$ $\frac{1}{2}$ 2	0102−	$+\sqrt{2}/3.5$
$\frac{5}{2}$ $\frac{1}{2}$ 2	0200+	0
$\frac{5}{2}$ $\frac{1}{2}$ 2	0201−	$+\sqrt{2}/3.5$
$\frac{5}{2}$ $\frac{1}{2}$ 2	0202+	$-1/2.5$
$\frac{5}{2}$ $\frac{3}{2}$ 1	0000+	$+1/4\sqrt{5}$
$\frac{5}{2}$ $\frac{3}{2}$ 1	0001−	0
$\frac{5}{2}$ $\frac{3}{2}$ 1	0002+	$-1/4\sqrt{3.5}$
$\frac{5}{2}$ $\frac{3}{2}$ 1	0010−	$-1/4\sqrt{2.3.5}$
$\frac{5}{2}$ $\frac{3}{2}$ 1	0011+	$+1/4\sqrt{5}$
$\frac{5}{2}$ $\frac{3}{2}$ 1	0012−	$-1/4.3\sqrt{2.5}$
$\frac{5}{2}$ $\frac{3}{2}$ 1	0100+	$+1/4.5$
$\frac{5}{2}$ $\frac{3}{2}$ 1	0101−	0
$\frac{5}{2}$ $\frac{3}{2}$ 1	0102+	$-1/4.5\sqrt{3}$
$\frac{5}{2}$ $\frac{3}{2}$ 1	0110−	$-1/4.5\sqrt{2.3}$
$\frac{5}{2}$ $\frac{3}{2}$ 1	0111+	$-1/4.3$
$\frac{5}{2}$ $\frac{3}{2}$ 1	0112−	$-3/4.5\sqrt{2}$

$\frac{5}{2}$ $\frac{5}{2}$ 2		
$\frac{5}{2}$ $\frac{3}{2}$ 2	0000+	$-1/2.5$
$\frac{5}{2}$ $\frac{3}{2}$ 2	0001−	$+1/2.5\sqrt{2.3}$
$\frac{5}{2}$ $\frac{3}{2}$ 2	0002+	0
$\frac{5}{2}$ $\frac{3}{2}$ 2	0010−	0
$\frac{5}{2}$ $\frac{3}{2}$ 2	0011+	$+1/2.5$
$\frac{5}{2}$ $\frac{3}{2}$ 2	0012−	0
$\frac{5}{2}$ $\frac{3}{2}$ 2	0100−	$+1/2.5\sqrt{2.3}$
$\frac{5}{2}$ $\frac{3}{2}$ 2	0101+	$-1/2.3.5$
$\frac{5}{2}$ $\frac{3}{2}$ 2	0102−	$+1/2.3.5\sqrt{2}$
$\frac{5}{2}$ $\frac{3}{2}$ 2	0110+	$-1/4.5$
$\frac{5}{2}$ $\frac{3}{2}$ 2	0111−	$+1/2.5\sqrt{2.3}$
$\frac{5}{2}$ $\frac{3}{2}$ 2	0112+	$+\sqrt{3}/4.5$
$\frac{5}{2}$ $\frac{3}{2}$ 2	0200+	0
$\frac{5}{2}$ $\frac{3}{2}$ 2	0201−	$+1/2.3.5\sqrt{2}$
$\frac{5}{2}$ $\frac{3}{2}$ 2	0202+	$+1/2.5$
$\frac{5}{2}$ $\frac{3}{2}$ 2	0210−	$-1/2.5\sqrt{2}$
$\frac{5}{2}$ $\frac{3}{2}$ 2	0211+	0
$\frac{5}{2}$ $\frac{3}{2}$ 2	0212−	$-1/2.5\sqrt{2.3}$
$\frac{5}{2}$ $\frac{3}{2}$ 2	1000−	0
$\frac{5}{2}$ $\frac{3}{2}$ 2	1001+	$-1/4.5$
$\frac{5}{2}$ $\frac{3}{2}$ 2	1002−	$-1/2.5\sqrt{2}$
$\frac{5}{2}$ $\frac{3}{2}$ 2	1010+	$+1/4.5$
$\frac{5}{2}$ $\frac{3}{2}$ 2	1011−	0
$\frac{5}{2}$ $\frac{3}{2}$ 2	1012+	$+\sqrt{3}/4.5$
$\frac{5}{2}$ $\frac{3}{2}$ 2	1100+	$+1/2.5$
$\frac{5}{2}$ $\frac{3}{2}$ 2	1101−	$+1/2.5\sqrt{2.3}$
$\frac{5}{2}$ $\frac{3}{2}$ 2	1102+	0
$\frac{5}{2}$ $\frac{3}{2}$ 2	1110−	0
$\frac{5}{2}$ $\frac{3}{2}$ 2	1111+	$-1/2.3.5$
$\frac{5}{2}$ $\frac{3}{2}$ 2	1112−	$+1/3.5\sqrt{2}$
$\frac{5}{2}$ $\frac{3}{2}$ 2	1200−	0
$\frac{5}{2}$ $\frac{3}{2}$ 2	1201+	$+\sqrt{3}/4.5$
$\frac{5}{2}$ $\frac{3}{2}$ 2	1202−	$-1/2.5\sqrt{2.3}$
$\frac{5}{2}$ $\frac{3}{2}$ 2	1210+	$+\sqrt{3}/4.5$
$\frac{5}{2}$ $\frac{3}{2}$ 2	1211−	$+1/3.5\sqrt{2}$
$\frac{5}{2}$ $\frac{3}{2}$ 2	1212+	$-1/4.5$
$\frac{5}{2}$ $\frac{5}{2}$ 0	0000+	$-1/2.3$
$\frac{5}{2}$ $\frac{5}{2}$ 0	0001−	0
$\frac{5}{2}$ $\frac{5}{2}$ 0	0002+	0
$\frac{5}{2}$ $\frac{5}{2}$ 0	0010−	0
$\frac{5}{2}$ $\frac{5}{2}$ 0	0011+	$+1/2.3$

K 6j Symbols (cont.)

$\frac{5}{2}$ $\frac{5}{2}$ 2		
$\frac{5}{2}$ $\frac{5}{2}$ 0	0012−	0
$\frac{5}{2}$ $\frac{5}{2}$ 0	0020+	0
$\frac{5}{2}$ $\frac{5}{2}$ 0	0021−	0
$\frac{5}{2}$ $\frac{5}{2}$ 0	0022+	−1/2.3
$\frac{5}{2}$ $\frac{5}{2}$ 1	0000+	+1/8.3
$\frac{5}{2}$ $\frac{5}{2}$ 1	0001−	0
$\frac{5}{2}$ $\frac{5}{2}$ 1	0002+	$+1/8\sqrt{3}$
$\frac{5}{2}$ $\frac{5}{2}$ 1	0010−	0
$\frac{5}{2}$ $\frac{5}{2}$ 1	0011+	+1/4.3
$\frac{5}{2}$ $\frac{5}{2}$ 1	0012−	0
$\frac{5}{2}$ $\frac{5}{2}$ 1	0020+	$+1/8\sqrt{3}$
$\frac{5}{2}$ $\frac{5}{2}$ 1	0021−	0
$\frac{5}{2}$ $\frac{5}{2}$ 1	0022+	−1/8.3
$\frac{5}{2}$ $\frac{5}{2}$ 1	1000+	$-1/8\sqrt{5}$
$\frac{5}{2}$ $\frac{5}{2}$ 1	1001−	0
$\frac{5}{2}$ $\frac{5}{2}$ 1	1002+	$+1/8\sqrt{3.5}$
$\frac{5}{2}$ $\frac{5}{2}$ 1	1010−	0
$\frac{5}{2}$ $\frac{5}{2}$ 1	1011+	$+1/4.3\sqrt{5}$
$\frac{5}{2}$ $\frac{5}{2}$ 1	1012−	$-1/3\sqrt{2.5}$
$\frac{5}{2}$ $\frac{5}{2}$ 1	1020+	$+1/8\sqrt{3.5}$
$\frac{5}{2}$ $\frac{5}{2}$ 1	1021−	$+1/3\sqrt{2.5}$
$\frac{5}{2}$ $\frac{5}{2}$ 1	1022+	$+1/8\sqrt{5}$
$\frac{5}{2}$ $\frac{5}{2}$ 1	1100+	+17/8.3.5
$\frac{5}{2}$ $\frac{5}{2}$ 1	1101−	0
$\frac{5}{2}$ $\frac{5}{2}$ 1	1102+	$+1/8.5\sqrt{3}$
$\frac{5}{2}$ $\frac{5}{2}$ 1	1110−	0
$\frac{5}{2}$ $\frac{5}{2}$ 1	1111+	+1/4.3
$\frac{5}{2}$ $\frac{5}{2}$ 1	1112−	0
$\frac{5}{2}$ $\frac{5}{2}$ 1	1120+	$+1/8.5\sqrt{3}$
$\frac{5}{2}$ $\frac{5}{2}$ 1	1121−	0
$\frac{5}{2}$ $\frac{5}{2}$ 1	1122+	−1/8.3.5
$\frac{5}{2}$ $\frac{5}{2}$ 2	0000+	−1/4.3.5
$\frac{5}{2}$ $\frac{5}{2}$ 2	0001−	0
$\frac{5}{2}$ $\frac{5}{2}$ 2	0002+	0
$\frac{5}{2}$ $\frac{5}{2}$ 2	0010−	0
$\frac{5}{2}$ $\frac{5}{2}$ 2	0011+	+1/4.3.5
$\frac{5}{2}$ $\frac{5}{2}$ 2	0012−	0
$\frac{5}{2}$ $\frac{5}{2}$ 2	0020+	0
$\frac{5}{2}$ $\frac{5}{2}$ 2	0021−	0
$\frac{5}{2}$ $\frac{5}{2}$ 2	0022+	+1/4.3
$\frac{5}{2}$ $\frac{5}{2}$ 2	1000−	0
$\frac{5}{2}$ $\frac{5}{2}$ 2		
$\frac{5}{2}$ $\frac{5}{2}$ 2	1001+	0
$\frac{5}{2}$ $\frac{5}{2}$ 2	1002−	$+1/2.5\sqrt{2}$
$\frac{5}{2}$ $\frac{5}{2}$ 2	1010+	0
$\frac{5}{2}$ $\frac{5}{2}$ 2	1011−	0
$\frac{5}{2}$ $\frac{5}{2}$ 2	1012+	$-1/2.5\sqrt{3}$
$\frac{5}{2}$ $\frac{5}{2}$ 2	1020−	$-1/2.5\sqrt{2}$
$\frac{5}{2}$ $\frac{5}{2}$ 2	1021+	$-1/2.5\sqrt{3}$
$\frac{5}{2}$ $\frac{5}{2}$ 2	1022−	0
$\frac{5}{2}$ $\frac{5}{2}$ 2	1100+	+1/4.3.5
$\frac{5}{2}$ $\frac{5}{2}$ 2	1101−	0
$\frac{5}{2}$ $\frac{5}{2}$ 2	1102+	$+1/4.5\sqrt{3}$
$\frac{5}{2}$ $\frac{5}{2}$ 2	1110−	0
$\frac{5}{2}$ $\frac{5}{2}$ 2	1111+	−1/2.5
$\frac{5}{2}$ $\frac{5}{2}$ 2	1112−	0
$\frac{5}{2}$ $\frac{5}{2}$ 2	1120+	$+1/4.5\sqrt{3}$
$\frac{5}{2}$ $\frac{5}{2}$ 2	1121−	0
$\frac{5}{2}$ $\frac{5}{2}$ 2	1122+	+1/4.5
$\frac{5}{2}$ $\frac{5}{2}$ 2	2000+	0
$\frac{5}{2}$ $\frac{5}{2}$ 2	2001−	$+1/2.5\sqrt{2}$
$\frac{5}{2}$ $\frac{5}{2}$ 2	2002+	−1/4.5
$\frac{5}{2}$ $\frac{5}{2}$ 2	2010−	$-1/2.5\sqrt{2}$
$\frac{5}{2}$ $\frac{5}{2}$ 2	2011+	$+1/4.5\sqrt{3}$
$\frac{5}{2}$ $\frac{5}{2}$ 2	2012−	$+1/2.5\sqrt{2.3}$
$\frac{5}{2}$ $\frac{5}{2}$ 2	2020+	−1/4.5
$\frac{5}{2}$ $\frac{5}{2}$ 2	2021−	$-1/2.5\sqrt{2.3}$
$\frac{5}{2}$ $\frac{5}{2}$ 2	2022+	0
$\frac{5}{2}$ $\frac{5}{2}$ 2	2100−	0
$\frac{5}{2}$ $\frac{5}{2}$ 2	2101+	$-1/2.5\sqrt{3}$
$\frac{5}{2}$ $\frac{5}{2}$ 2	2102−	$+1/2.5\sqrt{2.3}$
$\frac{5}{2}$ $\frac{5}{2}$ 2	2110+	$-1/2.5\sqrt{3}$
$\frac{5}{2}$ $\frac{5}{2}$ 2	2111−	0
$\frac{5}{2}$ $\frac{5}{2}$ 2	2112+	−1/3.5
$\frac{5}{2}$ $\frac{5}{2}$ 2	2120−	$-1/2.5\sqrt{2.3}$
$\frac{5}{2}$ $\frac{5}{2}$ 2	2121+	−1/3.5
$\frac{5}{2}$ $\frac{5}{2}$ 2	2122−	0
$\frac{5}{2}$ $\frac{5}{2}$ 2	2200+	+1/4.3
$\frac{5}{2}$ $\frac{5}{2}$ 2	2201−	0
$\frac{5}{2}$ $\frac{5}{2}$ 2	2202+	0
$\frac{5}{2}$ $\frac{5}{2}$ 2	2210−	0
$\frac{5}{2}$ $\frac{5}{2}$ 2	2211+	+1/4.5
$\frac{5}{2}$ $\frac{5}{2}$ 2	2212−	0
$\frac{5}{2}$ $\frac{5}{2}$ 2		
$\frac{5}{2}$ $\frac{5}{2}$ 2	2220+	0
$\frac{5}{2}$ $\frac{5}{2}$ 2	2221−	0
$\frac{5}{2}$ $\frac{5}{2}$ 2	2222+	−1/4.3.5
3 $\frac{3}{2}$ $\frac{3}{2}$		
0 $\frac{3}{2}$ $\frac{3}{2}$	0000+	+1/4
1 $\frac{3}{2}$ $\frac{3}{2}$	0000+	+3/4.5
2 $\frac{3}{2}$ $\frac{3}{2}$	0000+	+1/4.5
3 $\frac{3}{2}$ $\frac{3}{2}$	0000+	−1/2.5
3 2 1		
0 1 2	0000+	$+1/\sqrt{3.5}$
$\frac{1}{2}$ $\frac{3}{2}$ $\frac{3}{2}$	0000+	$-1/2\sqrt{5}$
1 1 2	0000+	$+1/3\sqrt{5}$
1 2 1	0000+	+1/5
1 2 2	0000+	$+\sqrt{2}/5\sqrt{3}$
1 2 2	1000−	0
$\frac{3}{2}$ $\frac{3}{2}$ $\frac{3}{2}$	0000+	+1/2.5
2 1 2	0000+	$+1/2\sqrt{2.3.5}$
2 1 2	0100+	$+\sqrt{3}/2.5\sqrt{2}$
2 2 1	0000+	+1/3.5
2 2 2	0000+	$-1/4\sqrt{5}$
2 2 2	0100+	+1/4.5
2 2 2	1000−	$-1/4\sqrt{5}$
2 2 2	1100−	−1/4.3
$\frac{5}{2}$ $\frac{3}{2}$ $\frac{3}{2}$	0000+	$+1/5\sqrt{3}$
$\frac{5}{2}$ $\frac{3}{2}$ $\frac{3}{2}$	0100−	$-1/2.5\sqrt{2}$
3 2 1	0000+	+7/4.3.5
3 2 2		
0 2 2	0000+	−1/5
0 2 2	0001−	0
0 2 2	1000−	0
0 2 2	1001+	+1/5
$\frac{1}{2}$ $\frac{3}{2}$ $\frac{3}{2}$	0000+	$+1/5\sqrt{2}$
$\frac{1}{2}$ $\frac{3}{2}$ $\frac{3}{2}$	0001−	0
1 2 1	0000+	$+\sqrt{2}/5\sqrt{3}$
1 2 1	0001−	0
1 2 2	0000+	0
1 2 2	0001−	0
1 2 2	1000−	0
1 2 2	1001+	+2/3.5
$\frac{3}{2}$ $\frac{3}{2}$ $\frac{3}{2}$	0000+	$+1/5\sqrt{2}$
$\frac{3}{2}$ $\frac{3}{2}$ $\frac{3}{2}$	0001−	0

K 6j Symbols (cont.)

3	**2**	**2**		
2	2	1	0000+	$-1/4\sqrt{5}$
2	2	1	0001−	$-1/4\sqrt{5}$
2	2	1	0010+	$+1/4.5$
2	2	1	0011−	$-1/4.3$
2	2	2	0000+	$+1/4.5$
2	2	2	0001−	0
2	2	2	0100+	0
2	2	2	0101−	$+1/4\sqrt{5}$
2	2	2	0110+	$+3/4.5$
2	2	2	0111−	0
2	2	2	1000−	0
2	2	2	1001+	$-1/4.5$
2	2	2	1100−	$-1/4\sqrt{5}$
2	2	2	1101+	0
2	2	2	1110−	0
2	2	2	1111+	$+7/4.3.5$
$\frac{5}{2}$	$\frac{3}{2}$	$\frac{3}{2}$	0000+	0
$\frac{5}{2}$	$\frac{3}{2}$	$\frac{3}{2}$	0001−	0
$\frac{5}{2}$	$\frac{3}{2}$	$\frac{3}{2}$	0100−	$+\sqrt{3}/4.5$
$\frac{5}{2}$	$\frac{3}{2}$	$\frac{3}{2}$	0101+	$+1/4\sqrt{3}$
$\frac{5}{2}$	$\frac{3}{2}$	$\frac{3}{2}$	0110+	0
$\frac{5}{2}$	$\frac{3}{2}$	$\frac{3}{2}$	0111−	0
3	2	1	0000+	$+1/8.5$
3	2	1	0001−	$-1/8$
3	2	1	0010−	$-1/8$
3	2	1	0011+	$-1/8.3$
3	2	2	0000+	$+1/8$
3	2	2	0001−	0
3	2	2	0100−	0
3	2	2	0101+	$+1/8$
3	2	2	0110+	$-3/8.5$
3	2	2	0111−	0
3	2	2	1000−	0
3	2	2	1001+	$-3/8.5$
3	2	2	1100+	$+1/8$
3	2	2	1101−	0
3	2	2	1110−	0
3	2	2	1111+	$-1/8.3.5$

3	$\mathbf{\frac{5}{2}}$	$\mathbf{\frac{1}{2}}$		
0	$\frac{1}{2}$	$\frac{5}{2}$	0000+	$+1/2\sqrt{3}$
$\frac{1}{2}$	1	2	0000+	$+1/3\sqrt{2}$
1	$\frac{1}{2}$	$\frac{5}{2}$	0000+	$+1/3\sqrt{2}$
1	$\frac{1}{2}$	$\frac{5}{2}$	0100+	0
1	$\frac{3}{2}$	$\frac{3}{2}$	0000+	$+1/2\sqrt{2.3}$
1	$\frac{3}{2}$	$\frac{5}{2}$	0000+	$-1/2.3\sqrt{2}$
1	$\frac{3}{2}$	$\frac{5}{2}$	0100+	0
1	$\frac{3}{2}$	$\frac{5}{2}$	1000−	0
1	$\frac{3}{2}$	$\frac{5}{2}$	1100−	$-1/2\sqrt{2.3}$
$\frac{3}{2}$	1	2	0000+	$+1/2.3\sqrt{2}$
$\frac{3}{2}$	1	2	0100−	$-1/4\sqrt{3}$
$\frac{3}{2}$	2	1	0000+	$-1/\sqrt{2.3.5}$
$\frac{3}{2}$	2	2	0000+	$+1/4\sqrt{5}$
$\frac{3}{2}$	2	2	0100−	$+\sqrt{3}/4\sqrt{2.5}$
$\frac{3}{2}$	2	2	1000−	$+1/4\sqrt{5}$
$\frac{3}{2}$	2	2	1100+	$-1/4\sqrt{2.3.5}$
2	$\frac{3}{2}$	$\frac{3}{2}$	0000+	$+1/2\sqrt{2.5}$
2	$\frac{3}{2}$	$\frac{3}{2}$	0100−	0
2	$\frac{3}{2}$	$\frac{5}{2}$	0000+	$-1/4\sqrt{2.3}$
2	$\frac{3}{2}$	$\frac{5}{2}$	0100−	0
2	$\frac{3}{2}$	$\frac{5}{2}$	0200+	$+1/4\sqrt{2}$
2	$\frac{3}{2}$	$\frac{5}{2}$	1000−	$+1/4\sqrt{2.5}$
2	$\frac{3}{2}$	$\frac{5}{2}$	1100+	$-1/2\sqrt{3.5}$
2	$\frac{3}{2}$	$\frac{5}{2}$	1200−	$+1/4\sqrt{2.3.5}$
2	$\frac{5}{2}$	$\frac{1}{2}$	0000+	$+1/2.3$
2	$\frac{5}{2}$	$\frac{3}{2}$	0000+	$-1/4.3$
2	$\frac{5}{2}$	$\frac{3}{2}$	0100−	$-1/4\sqrt{2.3}$
2	$\frac{5}{2}$	$\frac{3}{2}$	1000−	$-1/4\sqrt{3.5}$
2	$\frac{5}{2}$	$\frac{3}{2}$	1100+	$+\sqrt{5}/4.3\sqrt{2}$
2	$\frac{5}{2}$	$\frac{5}{2}$	0000+	$-1/2\sqrt{3.5}$
2	$\frac{5}{2}$	$\frac{5}{2}$	0100−	$-1/2.3\sqrt{2.5}$
2	$\frac{5}{2}$	$\frac{5}{2}$	0200+	$-1/2.3\sqrt{5}$
2	$\frac{5}{2}$	$\frac{5}{2}$	1000−	0
2	$\frac{5}{2}$	$\frac{5}{2}$	1100+	$+1/2.3\sqrt{2}$
2	$\frac{5}{2}$	$\frac{5}{2}$	1200−	0
$\frac{5}{2}$	2	1	0000+	$-1/4.3$
$\frac{5}{2}$	2	1	0100+	$+1/4\sqrt{5}$
$\frac{5}{2}$	2	2	0000+	$+1/2\sqrt{2.3.5}$
$\frac{5}{2}$	2	2	0100−	$-1/4\sqrt{5}$
$\frac{5}{2}$	2	2	0200+	0
$\frac{5}{2}$	2	2	1000−	$+1/2\sqrt{2.3.5}$
$\frac{5}{2}$	2	2	1100+	$+1/4\sqrt{5}$
$\frac{5}{2}$	2	2	1200−	$-1/3\sqrt{2.5}$
3	$\frac{5}{2}$	$\frac{1}{2}$	0000+	$-1/4.3$

3	$\mathbf{\frac{5}{2}}$	$\mathbf{\frac{3}{2}}$		
0	$\frac{3}{2}$	$\frac{5}{2}$	0000+	$-1/2\sqrt{2.3}$
0	$\frac{3}{2}$	$\frac{5}{2}$	0001−	0
0	$\frac{3}{2}$	$\frac{5}{2}$	1000−	0
0	$\frac{3}{2}$	$\frac{5}{2}$	1001+	$+1/2\sqrt{2.3}$
$\frac{1}{2}$	1	2	0000+	$-1/4.3\sqrt{2}$
$\frac{1}{2}$	1	2	0001−	$+\sqrt{3}/4\sqrt{2.5}$
$\frac{1}{2}$	2	2	0000+	$-1/8\sqrt{5}$
$\frac{1}{2}$	2	2	0001−	$+3\sqrt{3/8.5}$
$\frac{1}{2}$	2	2	1000−	$+3/8\sqrt{5}$
$\frac{1}{2}$	2	2	1001+	$+1/8\sqrt{3}$
1	$\frac{1}{2}$	$\frac{5}{2}$	0000+	$-1/2.3\sqrt{2}$
1	$\frac{1}{2}$	$\frac{5}{2}$	0001−	0
1	$\frac{1}{2}$	$\frac{5}{2}$	0100+	0
1	$\frac{1}{2}$	$\frac{5}{2}$	0101−	$+1/2\sqrt{2.3}$
1	$\frac{3}{2}$	$\frac{3}{2}$	0000+	$-1/4\sqrt{3.5}$
1	$\frac{3}{2}$	$\frac{3}{2}$	0001−	$+3/4.5$
1	$\frac{3}{2}$	$\frac{5}{2}$	0000+	$-\sqrt{5}/4.3$
1	$\frac{3}{2}$	$\frac{5}{2}$	0001−	0
1	$\frac{3}{2}$	$\frac{5}{2}$	0100+	0
1	$\frac{3}{2}$	$\frac{5}{2}$	0101−	$-1/4\sqrt{3.5}$
1	$\frac{3}{2}$	$\frac{5}{2}$	1000−	0
1	$\frac{3}{2}$	$\frac{5}{2}$	1001+	$-1/4\sqrt{5}$
1	$\frac{3}{2}$	$\frac{5}{2}$	1100−	$+1/4\sqrt{3.5}$
1	$\frac{3}{2}$	$\frac{5}{2}$	1101+	$+1/2.5$
1	$\frac{5}{2}$	$\frac{3}{2}$	0000+	$+1/8$
1	$\frac{5}{2}$	$\frac{3}{2}$	0001−	$+\sqrt{3}/8\sqrt{5}$
1	$\frac{5}{2}$	$\frac{3}{2}$	1000−	$+\sqrt{3}/8\sqrt{5}$
1	$\frac{5}{2}$	$\frac{3}{2}$	1001+	$-7/8.3.5$
1	$\frac{5}{2}$	$\frac{5}{2}$	0000+	0
1	$\frac{5}{2}$	$\frac{5}{2}$	0001−	$+1/4\sqrt{5}$
1	$\frac{5}{2}$	$\frac{5}{2}$	0100+	$-1/4\sqrt{3.5}$
1	$\frac{5}{2}$	$\frac{5}{2}$	0101−	$+1/3.5$
1	$\frac{5}{2}$	$\frac{5}{2}$	1000−	0
1	$\frac{5}{2}$	$\frac{5}{2}$	1001+	$-1/4.3$
1	$\frac{5}{2}$	$\frac{5}{2}$	1100−	$-1/4\sqrt{3}$
1	$\frac{5}{2}$	$\frac{5}{2}$	1101+	0
$\frac{3}{2}$	1	2	0000+	$-1/2.3\sqrt{5}$
$\frac{3}{2}$	1	2	0001−	$-\sqrt{3}/2.5$
$\frac{3}{2}$	1	2	0100−	$+1/2\sqrt{2.3.5}$

K 6j Symbols (cont.)

$3\ \frac{5}{2}\ \frac{3}{2}$		
$\frac{3}{2}$ 1 2	0101+	$-1/2.5\sqrt{2}$
$\frac{3}{2}$ 2 1	0000+	$-1/4\sqrt{3.5}$
$\frac{3}{2}$ 2 1	0001−	$+3/4.5$
$\frac{3}{2}$ 2 2	0000+	$+1/4\sqrt{2.5}$
$\frac{3}{2}$ 2 2	0001−	$+\sqrt{3}/4.5\sqrt{2}$
$\frac{3}{2}$ 2 2	0100−	0
$\frac{3}{2}$ 2 2	0101+	$-1/2.5$
$\frac{3}{2}$ 2 2	1000−	$-1/4\sqrt{2.5}$
$\frac{3}{2}$ 2 2	1001+	$+1/4\sqrt{2.3}$
$\frac{3}{2}$ 2 2	1100+	$+1/2\sqrt{3.5}$
$\frac{3}{2}$ 2 2	1101−	0
2 $\frac{1}{2}$ $\frac{5}{2}$	0000+	$-1/4\sqrt{2.3}$
2 $\frac{1}{2}$ $\frac{5}{2}$	0001−	$-1/4\sqrt{2.5}$
2 $\frac{1}{2}$ $\frac{5}{2}$	0100−	0
2 $\frac{1}{2}$ $\frac{5}{2}$	0101+	$-1/2\sqrt{3.5}$
2 $\frac{1}{2}$ $\frac{5}{2}$	0200+	$+1/4\sqrt{2}$
2 $\frac{1}{2}$ $\frac{5}{2}$	0201−	$-1/4\sqrt{2.3.5}$
2 $\frac{3}{2}$ $\frac{3}{2}$	0000+	$+1/4\sqrt{5}$
2 $\frac{3}{2}$ $\frac{3}{2}$	0001−	$+\sqrt{3}/4.5$
2 $\frac{3}{2}$ $\frac{3}{2}$	0100−	0
2 $\frac{3}{2}$ $\frac{3}{2}$	0101+	$+1/5\sqrt{2}$
2 $\frac{3}{2}$ $\frac{5}{2}$	0000+	$+1/8\sqrt{3}$
2 $\frac{3}{2}$ $\frac{5}{2}$	0001−	$-1/8\sqrt{5}$
2 $\frac{3}{2}$ $\frac{5}{2}$	0100−	0
2 $\frac{3}{2}$ $\frac{5}{2}$	0101+	$-1/2\sqrt{2.3.5}$
2 $\frac{3}{2}$ $\frac{5}{2}$	0200+	$-1/8$
2 $\frac{3}{2}$ $\frac{5}{2}$	0201−	$-1/8\sqrt{3.5}$
2 $\frac{3}{2}$ $\frac{5}{2}$	1000−	$+1/8\sqrt{5}$
2 $\frac{3}{2}$ $\frac{5}{2}$	1001+	$-7/8.5\sqrt{3}$
2 $\frac{3}{2}$ $\frac{5}{2}$	1100+	$-1/2\sqrt{2.3.5}$
2 $\frac{3}{2}$ $\frac{5}{2}$	1101−	0
2 $\frac{3}{2}$ $\frac{5}{2}$	1200−	$+1/8\sqrt{3.5}$
2 $\frac{3}{2}$ $\frac{5}{2}$	1201+	$-1/8.5$
2 $\frac{5}{2}$ $\frac{1}{2}$	0000+	$-1/4.3$
2 $\frac{5}{2}$ $\frac{1}{2}$	0001−	$-1/4\sqrt{3.5}$
2 $\frac{5}{2}$ $\frac{1}{2}$	0010−	$-1/4\sqrt{2.3}$
2 $\frac{5}{2}$ $\frac{1}{2}$	0011+	$+\sqrt{5}/4.3\sqrt{2}$
2 $\frac{5}{2}$ $\frac{3}{2}$	0000+	$+1/8.3$
2 $\frac{5}{2}$ $\frac{3}{2}$	0001−	$+1/8\sqrt{3.5}$
2 $\frac{5}{2}$ $\frac{3}{2}$	0010−	$-1/4\sqrt{2.3}$
2 $\frac{5}{2}$ $\frac{3}{2}$	0011+	$+\sqrt{5}/4.3\sqrt{2}$

$3\ \frac{5}{2}\ \frac{3}{2}$		
2 $\frac{5}{2}$ $\frac{3}{2}$	0100−	$-1/4\sqrt{2.3}$
2 $\frac{5}{2}$ $\frac{3}{2}$	0101+	$-1/4\sqrt{2.5}$
2 $\frac{5}{2}$ $\frac{3}{2}$	0110+	0
2 $\frac{5}{2}$ $\frac{3}{2}$	0111−	0
2 $\frac{5}{2}$ $\frac{3}{2}$	1000−	$+1/8\sqrt{3.5}$
2 $\frac{5}{2}$ $\frac{3}{2}$	1001+	$+1/8.5$
2 $\frac{5}{2}$ $\frac{3}{2}$	1010+	$-1/4\sqrt{2.5}$
2 $\frac{5}{2}$ $\frac{3}{2}$	1011−	$-\sqrt{3}/4.5\sqrt{2}$
2 $\frac{5}{2}$ $\frac{3}{2}$	1100+	$+\sqrt{5}/4.3\sqrt{2}$
2 $\frac{5}{2}$ $\frac{3}{2}$	1101−	$-\sqrt{3}/4.5\sqrt{2}$
2 $\frac{5}{2}$ $\frac{3}{2}$	1110−	0
2 $\frac{5}{2}$ $\frac{3}{2}$	1111+	$+1/3.5$
2 $\frac{5}{2}$ $\frac{5}{2}$	0000+	$+1/4\sqrt{3.5}$
2 $\frac{5}{2}$ $\frac{5}{2}$	0001−	$+1/4.5$
2 $\frac{5}{2}$ $\frac{5}{2}$	0010−	$+1/8\sqrt{2.5}$
2 $\frac{5}{2}$ $\frac{5}{2}$	0011+	$+\sqrt{3}/8.5\sqrt{2}$
2 $\frac{5}{2}$ $\frac{5}{2}$	0100−	$+1/4.3\sqrt{2.5}$
2 $\frac{5}{2}$ $\frac{5}{2}$	0101+	$-1/4\sqrt{2.3}$
2 $\frac{5}{2}$ $\frac{5}{2}$	0110+	$+\sqrt{3}/8\sqrt{5}$
2 $\frac{5}{2}$ $\frac{5}{2}$	0111−	$+1/8.3$
2 $\frac{5}{2}$ $\frac{5}{2}$	0200+	$+1/4.3\sqrt{5}$
2 $\frac{5}{2}$ $\frac{5}{2}$	0201−	$-1/4.5\sqrt{3}$
2 $\frac{5}{2}$ $\frac{5}{2}$	0210−	$+1/8\sqrt{2.3.5}$
2 $\frac{5}{2}$ $\frac{5}{2}$	0211+	$-13/8.3.5\sqrt{2}$
2 $\frac{5}{2}$ $\frac{5}{2}$	1000−	0
2 $\frac{5}{2}$ $\frac{5}{2}$	1001+	0
2 $\frac{5}{2}$ $\frac{5}{2}$	1010+	$-1/8\sqrt{2}$
2 $\frac{5}{2}$ $\frac{5}{2}$	1011−	$-\sqrt{3}/8\sqrt{2.5}$
2 $\frac{5}{2}$ $\frac{5}{2}$	1100+	$-1/4.3\sqrt{2}$
2 $\frac{5}{2}$ $\frac{5}{2}$	1101−	$+1/4\sqrt{2.3.5}$
2 $\frac{5}{2}$ $\frac{5}{2}$	1110−	$+1/8\sqrt{3}$
2 $\frac{5}{2}$ $\frac{5}{2}$	1111+	$-1/8.3\sqrt{5}$
2 $\frac{5}{2}$ $\frac{5}{2}$	1200−	0
2 $\frac{5}{2}$ $\frac{5}{2}$	1201+	$-1/2\sqrt{3.5}$
2 $\frac{5}{2}$ $\frac{5}{2}$	1210+	$-1/8\sqrt{2.3}$
2 $\frac{5}{2}$ $\frac{5}{2}$	1211−	$-1/8\sqrt{2.5}$
$\frac{5}{2}$ 1 2	0000+	$+1/8\sqrt{5}$
$\frac{5}{2}$ 1 2	0001−	$-1/8.5\sqrt{3}$
$\frac{5}{2}$ 1 2	0100−	$+\sqrt{3}/4\sqrt{2.5}$
$\frac{5}{2}$ 1 2	0101+	$+1/4.3\sqrt{2}$
$\frac{5}{2}$ 1 2	0200+	$-\sqrt{3}/8\sqrt{5}$

$3\ \frac{5}{2}\ \frac{3}{2}$		
$\frac{5}{2}$ 1 2	0201−	$+11/8.3.5$
$\frac{5}{2}$ 2 1	0000+	$+1/4.3$
$\frac{5}{2}$ 2 1	0001−	$-1/4\sqrt{3.5}$
$\frac{5}{2}$ 2 1	0010−	$-1/4\sqrt{2.3}$
$\frac{5}{2}$ 2 1	0011+	$-\sqrt{5}/4.3\sqrt{2}$
$\frac{5}{2}$ 2 1	0100+	$+1/4\sqrt{5}$
$\frac{5}{2}$ 2 1	0101−	$+1/4.5\sqrt{3}$
$\frac{5}{2}$ 2 1	0110−	$+1/4\sqrt{2.3.5}$
$\frac{5}{2}$ 2 1	0111+	$+1/4.3.5\sqrt{2}$
$\frac{5}{2}$ 2 2	0000+	$-1/8\sqrt{2.3.5}$
$\frac{5}{2}$ 2 2	0001−	$-1/8\sqrt{2}$
$\frac{5}{2}$ 2 2	0010−	$+1/4\sqrt{5}$
$\frac{5}{2}$ 2 2	0011+	$-1/4.5\sqrt{3}$
$\frac{5}{2}$ 2 2	0100−	$-1/8\sqrt{5}$
$\frac{5}{2}$ 2 2	0101+	$-\sqrt{3}/8.5$
$\frac{5}{2}$ 2 2	0110+	0
$\frac{5}{2}$ 2 2	0111−	$-1/2.5\sqrt{2}$
$\frac{5}{2}$ 2 2	0200+	$-3/8\sqrt{2.5}$
$\frac{5}{2}$ 2 2	0201−	$+\sqrt{3}/8.5\sqrt{2}$
$\frac{5}{2}$ 2 2	0210−	0
$\frac{5}{2}$ 2 2	0211+	$-1/2.5$
$\frac{5}{2}$ 2 2	1000−	$-1/8\sqrt{2.3.5}$
$\frac{5}{2}$ 2 2	1001+	$-1/8\sqrt{2}$
$\frac{5}{2}$ 2 2	1010+	0
$\frac{5}{2}$ 2 2	1011−	0
$\frac{5}{2}$ 2 2	1100+	$-1/8\sqrt{5}$
$\frac{5}{2}$ 2 2	1101−	$+1/8\sqrt{3}$
$\frac{5}{2}$ 2 2	1110−	$-1/2\sqrt{2.3.5}$
$\frac{5}{2}$ 2 2	1111+	0
$\frac{5}{2}$ 2 2	1200−	$-1/8.3\sqrt{2.5}$
$\frac{5}{2}$ 2 2	1201+	$-1/8\sqrt{2.3}$
$\frac{5}{2}$ 2 2	1210+	$+1/4\sqrt{3.5}$
$\frac{5}{2}$ 2 2	1211−	$+1/4.3$
3 $\frac{3}{2}$ $\frac{3}{2}$	0000+	$-1/8$
3 $\frac{3}{2}$ $\frac{3}{2}$	0001−	$+\sqrt{3}/8\sqrt{5}$
3 $\frac{3}{2}$ $\frac{3}{2}$	0100−	$+\sqrt{3}/8\sqrt{5}$
3 $\frac{3}{2}$ $\frac{3}{2}$	0101+	$+1/8.5$
3 $\frac{3}{2}$ $\frac{5}{2}$	0000+	0
3 $\frac{3}{2}$ $\frac{5}{2}$	0001−	$-1/4\sqrt{3.5}$
3 $\frac{3}{2}$ $\frac{5}{2}$	0100−	0
3 $\frac{3}{2}$ $\frac{5}{2}$	0101+	$+1/4\sqrt{3}$

K 6j Symbols (cont.)

3	**$\frac{5}{2}$**	**$\frac{3}{2}$**		
3	$\frac{3}{2}$	$\frac{5}{2}$	1000−	$+1/4\sqrt{3.5}$
3	$\frac{3}{2}$	$\frac{5}{2}$	1001+	$+1/2.5$
3	$\frac{3}{2}$	$\frac{5}{2}$	1100+	$+1/4\sqrt{3}$
3	$\frac{3}{2}$	$\frac{5}{2}$	1101−	0
3	$\frac{5}{2}$	$\frac{1}{2}$	0000+	$+1/2.3$
3	$\frac{5}{2}$	$\frac{1}{2}$	0001−	0
3	$\frac{5}{2}$	$\frac{1}{2}$	0010−	0
3	$\frac{5}{2}$	$\frac{1}{2}$	0011+	$-1/2.3$
3	$\frac{5}{2}$	$\frac{3}{2}$	0000+	$+1/8.3$
3	$\frac{5}{2}$	$\frac{3}{2}$	0001−	0
3	$\frac{5}{2}$	$\frac{3}{2}$	0010−	0
3	$\frac{5}{2}$	$\frac{3}{2}$	0011+	$-1/8.3$
3	$\frac{5}{2}$	$\frac{3}{2}$	0100−	0
3	$\frac{5}{2}$	$\frac{3}{2}$	0101+	$-1/8$
3	$\frac{5}{2}$	$\frac{3}{2}$	0110+	$+1/8$
3	$\frac{5}{2}$	$\frac{3}{2}$	0111−	$-1/4\sqrt{3.5}$
3	$\frac{5}{2}$	$\frac{3}{2}$	1000−	0
3	$\frac{5}{2}$	$\frac{3}{2}$	1001+	$+1/8$
3	$\frac{5}{2}$	$\frac{3}{2}$	1010+	$-1/8$
3	$\frac{5}{2}$	$\frac{3}{2}$	1011−	$-1/4\sqrt{3.5}$
3	$\frac{5}{2}$	$\frac{3}{2}$	1100+	$-1/8.3$
3	$\frac{5}{2}$	$\frac{3}{2}$	1101−	$-1/4\sqrt{3.5}$
3	$\frac{5}{2}$	$\frac{3}{2}$	1110−	$-1/4\sqrt{3.5}$
3	$\frac{5}{2}$	$\frac{3}{2}$	1111+	$-7/8.3.5$

3	**$\frac{5}{2}$**	**$\frac{5}{2}$**		
0	$\frac{5}{2}$	$\frac{5}{2}$	0000+	$+1/2.3$
0	$\frac{5}{2}$	$\frac{5}{2}$	0001−	0
0	$\frac{5}{2}$	$\frac{5}{2}$	1000−	0
0	$\frac{5}{2}$	$\frac{5}{2}$	1001+	$-1/2.3$
$\frac{1}{2}$	2	2	0000+	$+1/5\sqrt{2}$
$\frac{1}{2}$	2	2	0001−	0
$\frac{1}{2}$	2	2	1000−	0
$\frac{1}{2}$	2	2	1001+	$+1/3\sqrt{2.5}$
1	$\frac{3}{2}$	$\frac{3}{2}$	0000+	$-1/2.5$
1	$\frac{3}{2}$	$\frac{3}{2}$	0001−	0
1	$\frac{5}{2}$	$\frac{3}{2}$	0000+	0
1	$\frac{5}{2}$	$\frac{3}{2}$	0001−	0
1	$\frac{5}{2}$	$\frac{3}{2}$	0010+	$-1/4\sqrt{3.5}$
1	$\frac{5}{2}$	$\frac{3}{2}$	0011−	$-1/4\sqrt{3}$
1	$\frac{5}{2}$	$\frac{3}{2}$	1000−	$+1/4\sqrt{5}$
1	$\frac{5}{2}$	$\frac{3}{2}$	1001+	$-1/4.3$
1	$\frac{5}{2}$	$\frac{3}{2}$	1010−	$+1/3.5$
1	$\frac{5}{2}$	$\frac{3}{2}$	1011+	0
1	$\frac{5}{2}$	$\frac{5}{2}$	0000+	$+1/4.3$
1	$\frac{5}{2}$	$\frac{5}{2}$	0001−	0
1	$\frac{5}{2}$	$\frac{5}{2}$	0100+	$-1/2.3\sqrt{5}$
1	$\frac{5}{2}$	$\frac{5}{2}$	0101−	$-1/4.3$
1	$\frac{5}{2}$	$\frac{5}{2}$	0110+	$-1/4.3.5$
1	$\frac{5}{2}$	$\frac{5}{2}$	0111−	0
1	$\frac{5}{2}$	$\frac{5}{2}$	1000−	0
1	$\frac{5}{2}$	$\frac{5}{2}$	1001+	$+1/4.3$
1	$\frac{5}{2}$	$\frac{5}{2}$	1100−	$+1/4.3$
1	$\frac{5}{2}$	$\frac{5}{2}$	1101+	0
1	$\frac{5}{2}$	$\frac{5}{2}$	1110−	0
1	$\frac{5}{2}$	$\frac{5}{2}$	1111+	$-1/4.3$
$\frac{3}{2}$	2	1	0000+	$+1/4.5\sqrt{3}$
$\frac{3}{2}$	2	1	0001−	$+1/4\sqrt{3.5}$
$\frac{3}{2}$	2	1	0010−	$-11/4.3.5\sqrt{2}$
$\frac{3}{2}$	2	1	0011+	$+1/4.3\sqrt{2.5}$
$\frac{3}{2}$	2	2	0000+	$+1/2.5\sqrt{2}$
$\frac{3}{2}$	2	2	0001−	0
$\frac{3}{2}$	2	2	0100−	$-\sqrt{3}/8.5$
$\frac{3}{2}$	2	2	0101+	$+\sqrt{3}/8\sqrt{5}$
$\frac{3}{2}$	2	2	0110+	$-1/2.5\sqrt{2}$
$\frac{3}{2}$	2	2	0111−	0
$\frac{3}{2}$	2	2	1000−	0
$\frac{3}{2}$	2	2	1001+	$+1/2.3\sqrt{2.5}$
$\frac{3}{2}$	2	2	1100+	$+1/8\sqrt{3}$
$\frac{3}{2}$	2	2	1101−	$+\sqrt{3}/8\sqrt{5}$
$\frac{3}{2}$	2	2	1110−	0
$\frac{3}{2}$	2	2	1111+	$-1/2.3\sqrt{2.5}$
2	$\frac{3}{2}$	$\frac{3}{2}$	0000+	$-1/2.5$
2	$\frac{3}{2}$	$\frac{3}{2}$	0001−	0
2	$\frac{3}{2}$	$\frac{3}{2}$	0100−	$+1/2.5\sqrt{2.3}$
2	$\frac{3}{2}$	$\frac{3}{2}$	0101+	$-1/2\sqrt{2.3.5}$
2	$\frac{3}{2}$	$\frac{3}{2}$	0110+	$+1/2.5$
2	$\frac{3}{2}$	$\frac{3}{2}$	0111−	0
2	$\frac{5}{2}$	$\frac{1}{2}$	0000+	$-1/2\sqrt{3.5}$
2	$\frac{5}{2}$	$\frac{1}{2}$	0001−	0
2	$\frac{5}{2}$	$\frac{1}{2}$	0010−	$-1/2.3\sqrt{2.5}$
2	$\frac{5}{2}$	$\frac{1}{2}$	0011+	$+1/2.3\sqrt{2}$
2	$\frac{5}{2}$	$\frac{1}{2}$	0020+	$-1/2.3\sqrt{5}$
2	$\frac{5}{2}$	$\frac{1}{2}$	0021−	0
2	$\frac{5}{2}$	$\frac{3}{2}$	0000+	$+1/4\sqrt{3.5}$
2	$\frac{5}{2}$	$\frac{3}{2}$	0001−	0
2	$\frac{5}{2}$	$\frac{3}{2}$	0010−	$+1/4.3\sqrt{2.5}$
2	$\frac{5}{2}$	$\frac{3}{2}$	0011+	$-1/4.3\sqrt{2}$
2	$\frac{5}{2}$	$\frac{3}{2}$	0020+	$+1/4.3\sqrt{5}$
2	$\frac{5}{2}$	$\frac{3}{2}$	0021−	0
2	$\frac{5}{2}$	$\frac{3}{2}$	0100−	$+1/8\sqrt{2.5}$
2	$\frac{5}{2}$	$\frac{3}{2}$	0101+	$-1/8\sqrt{2}$
2	$\frac{5}{2}$	$\frac{3}{2}$	0110+	$+\sqrt{3}/8\sqrt{5}$
2	$\frac{5}{2}$	$\frac{3}{2}$	0111−	$+1/8\sqrt{3}$
2	$\frac{5}{2}$	$\frac{3}{2}$	0120−	$+1/8\sqrt{2.3.5}$
2	$\frac{5}{2}$	$\frac{3}{2}$	0121+	$-1/8\sqrt{2.3}$
2	$\frac{5}{2}$	$\frac{3}{2}$	1000−	$+1/4.5$
2	$\frac{5}{2}$	$\frac{3}{2}$	1001+	0
2	$\frac{5}{2}$	$\frac{3}{2}$	1010+	$-1/4\sqrt{2.3}$
2	$\frac{5}{2}$	$\frac{3}{2}$	1011−	$+1/4\sqrt{2.3.5}$
2	$\frac{5}{2}$	$\frac{3}{2}$	1020−	$-1/4.5\sqrt{3}$
2	$\frac{5}{2}$	$\frac{3}{2}$	1021+	$-1/2\sqrt{3.5}$
2	$\frac{5}{2}$	$\frac{3}{2}$	1100+	$+\sqrt{3}/8.5\sqrt{2}$
2	$\frac{5}{2}$	$\frac{3}{2}$	1101−	$-\sqrt{3}/8\sqrt{2.5}$
2	$\frac{5}{2}$	$\frac{3}{2}$	1110−	$+1/8.3$
2	$\frac{5}{2}$	$\frac{3}{2}$	1111+	$-1/8.3\sqrt{5}$
2	$\frac{5}{2}$	$\frac{3}{2}$	1120+	$-13/8.3.5\sqrt{2}$
2	$\frac{5}{2}$	$\frac{3}{2}$	1121−	$-1/8\sqrt{2.5}$
2	$\frac{5}{2}$	$\frac{5}{2}$	0000+	$+1/4.3.5$
2	$\frac{5}{2}$	$\frac{5}{2}$	0001−	0
2	$\frac{5}{2}$	$\frac{5}{2}$	0100−	0
2	$\frac{5}{2}$	$\frac{5}{2}$	0101+	$+1/2\sqrt{2.3.5}$
2	$\frac{5}{2}$	$\frac{5}{2}$	0110+	0
2	$\frac{5}{2}$	$\frac{5}{2}$	0111−	0
2	$\frac{5}{2}$	$\frac{5}{2}$	0200+	$-1/2.5\sqrt{3}$
2	$\frac{5}{2}$	$\frac{5}{2}$	0201−	$-1/4\sqrt{3.5}$
2	$\frac{5}{2}$	$\frac{5}{2}$	0210−	0
2	$\frac{5}{2}$	$\frac{5}{2}$	0211+	$-1/2.3\sqrt{2.5}$
2	$\frac{5}{2}$	$\frac{5}{2}$	0220+	$-1/4.5$
2	$\frac{5}{2}$	$\frac{5}{2}$	0221−	0
2	$\frac{5}{2}$	$\frac{5}{2}$	1000−	0
2	$\frac{5}{2}$	$\frac{5}{2}$	1001+	$+1/4.3$
2	$\frac{5}{2}$	$\frac{5}{2}$	1100+	$+1/2\sqrt{2.3.5}$
2	$\frac{5}{2}$	$\frac{5}{2}$	1101−	0

K 6j Symbols (cont.)

3 $\frac{5}{2}$ $\frac{5}{2}$		
2 $\frac{5}{2}$ $\frac{5}{2}$	1110−	0
2 $\frac{5}{2}$ $\frac{5}{2}$	1111+	0
2 $\frac{5}{2}$ $\frac{5}{2}$	1200−	$+1/4\sqrt{3.5}$
2 $\frac{5}{2}$ $\frac{5}{2}$	1201+	0
2 $\frac{5}{2}$ $\frac{5}{2}$	1210+	$-1/2.3\sqrt{2.5}$
2 $\frac{5}{2}$ $\frac{5}{2}$	1211−	0
2 $\frac{5}{2}$ $\frac{5}{2}$	1220−	0
2 $\frac{5}{2}$ $\frac{5}{2}$	1221+	$+1/4.3$
$\frac{5}{2}$ 2 1	0000+	$-1/8\sqrt{3.5}$
$\frac{5}{2}$ 2 1	0001−	$+1/8\sqrt{3}$
$\frac{5}{2}$ 2 1	0010−	$+1/4\sqrt{2.5}$
$\frac{5}{2}$ 2 1	0011+	$+1/4.3\sqrt{2}$
$\frac{5}{2}$ 2 1	0020+	$-1/8.3\sqrt{5}$
$\frac{5}{2}$ 2 1	0021−	$+1/8.3$
$\frac{5}{2}$ 2 1	0100+	$-3\sqrt{3}/8.5$
$\frac{5}{2}$ 2 1	0101−	$+1/8\sqrt{3.5}$
$\frac{5}{2}$ 2 1	0110−	$-1/4.3\sqrt{2}$
$\frac{5}{2}$ 2 1	0111+	$+1/4.3\sqrt{2.5}$
$\frac{5}{2}$ 2 1	0120+	$-1/8.3.5$
$\frac{5}{2}$ 2 1	0121−	$-7/8.3\sqrt{5}$
$\frac{5}{2}$ 2 2	0000+	0
$\frac{5}{2}$ 2 2	0001−	0
$\frac{5}{2}$ 2 2	0100−	$+1/8.5\sqrt{3}$
$\frac{5}{2}$ 2 2	0101+	$+\sqrt{3}/8\sqrt{5}$
$\frac{5}{2}$ 2 2	0110+	$-1/2.5\sqrt{2}$
$\frac{5}{2}$ 2 2	0111−	0
$\frac{5}{2}$ 2 2	0200+	$-\sqrt{3}/4.5\sqrt{2}$
$\frac{5}{2}$ 2 2	0201−	$-1/4\sqrt{2.3.5}$
$\frac{5}{2}$ 2 2	0210−	$-1/8.5$
$\frac{5}{2}$ 2 2	0211+	$-1/8\sqrt{5}$
$\frac{5}{2}$ 2 2	0220+	$-1/2.5\sqrt{2}$
$\frac{5}{2}$ 2 2	0221−	0
$\frac{5}{2}$ 2 2	1000−	0
$\frac{5}{2}$ 2 2	1001+	0
$\frac{5}{2}$ 2 2	1100+	$-1/8\sqrt{3}$
$\frac{5}{2}$ 2 2	1101−	$+1/8\sqrt{3.5}$
$\frac{5}{2}$ 2 2	1110−	0
$\frac{5}{2}$ 2 2	1111+	$+1/2.3\sqrt{2.5}$
$\frac{5}{2}$ 2 2	1200−	$-1/4\sqrt{2.3}$
$\frac{5}{2}$ 2 2	1201+	$+1/4\sqrt{2.3.5}$
$\frac{5}{2}$ 2 2	1210+	$+1/8.3$

3 $\frac{5}{2}$ $\frac{5}{2}$		
$\frac{5}{2}$ 2 2	1211−	$-1/8\sqrt{5}$
$\frac{5}{2}$ 2 2	1220−	0
$\frac{5}{2}$ 2 2	1221+	$+1/2.3\sqrt{2.5}$
3 $\frac{3}{2}$ $\frac{3}{2}$	0000+	0
3 $\frac{3}{2}$ $\frac{3}{2}$	0001−	0
3 $\frac{3}{2}$ $\frac{3}{2}$	0100−	$-1/4\sqrt{3.5}$
3 $\frac{3}{2}$ $\frac{3}{2}$	0101+	$+1/4\sqrt{3}$
3 $\frac{3}{2}$ $\frac{3}{2}$	0110+	$+1/2.5$
3 $\frac{3}{2}$ $\frac{3}{2}$	0111−	0
3 $\frac{5}{2}$ $\frac{1}{2}$	0000+	$+1/8.3$
3 $\frac{5}{2}$ $\frac{1}{2}$	0001−	$-\sqrt{5}/8.3$
3 $\frac{5}{2}$ $\frac{1}{2}$	0010−	$-\sqrt{5}/8.3$
3 $\frac{5}{2}$ $\frac{1}{2}$	0011+	$-1/8$
3 $\frac{5}{2}$ $\frac{3}{2}$	0000+	$-7/8.2.3$
3 $\frac{5}{2}$ $\frac{3}{2}$	0001−	$+\sqrt{5}/8.2.3$
3 $\frac{5}{2}$ $\frac{3}{2}$	0010−	$+\sqrt{5}/8.2.3$
3 $\frac{5}{2}$ $\frac{3}{2}$	0011+	$-1/8.2$
3 $\frac{5}{2}$ $\frac{3}{2}$	0100−	$+1/8.2\sqrt{3.5}$
3 $\frac{5}{2}$ $\frac{3}{2}$	0101+	$+1/8.2\sqrt{3}$
3 $\frac{5}{2}$ $\frac{3}{2}$	0110+	$+1/8.2\sqrt{3}$
3 $\frac{5}{2}$ $\frac{3}{2}$	0111−	$+\sqrt{5}/8.2\sqrt{3}$
3 $\frac{5}{2}$ $\frac{3}{2}$	1000−	$+1/8.2\sqrt{3.5}$
3 $\frac{5}{2}$ $\frac{3}{2}$	1001+	$+1/8.2\sqrt{3}$
3 $\frac{5}{2}$ $\frac{3}{2}$	1010+	$+1/8.2\sqrt{3}$
3 $\frac{5}{2}$ $\frac{3}{2}$	1011−	$+\sqrt{5}/8.2\sqrt{3}$
3 $\frac{5}{2}$ $\frac{3}{2}$	1100+	$-29/8.2.3.5$
3 $\frac{5}{2}$ $\frac{3}{2}$	1101−	$-\sqrt{5}/8.2.3$
3 $\frac{5}{2}$ $\frac{3}{2}$	1110−	$-\sqrt{5}/8.2.3$
3 $\frac{5}{2}$ $\frac{3}{2}$	1111+	$+1/8.2$
3 $\frac{5}{2}$ $\frac{5}{2}$	0000+	$+1/8.5$
3 $\frac{5}{2}$ $\frac{5}{2}$	0001−	0
3 $\frac{5}{2}$ $\frac{5}{2}$	0100−	0
3 $\frac{5}{2}$ $\frac{5}{2}$	0101+	$-1/8.3$
3 $\frac{5}{2}$ $\frac{5}{2}$	0110+	$+1/8$
3 $\frac{5}{2}$ $\frac{5}{2}$	0111−	0
3 $\frac{5}{2}$ $\frac{5}{2}$	1000−	0
3 $\frac{5}{2}$ $\frac{5}{2}$	1001+	$+1/8$
3 $\frac{5}{2}$ $\frac{5}{2}$	1100+	$-1/8.3$
3 $\frac{5}{2}$ $\frac{5}{2}$	1101−	0
3 $\frac{5}{2}$ $\frac{5}{2}$	1110−	0
3 $\frac{5}{2}$ $\frac{5}{2}$	1111+	$-1/8.3$

3 3 0		
0 0 3	0000+	$+1/2$
$\frac{1}{2}$ $\frac{1}{2}$ $\frac{5}{2}$	0000+	$+1/2\sqrt{2}$
1 1 2	0000+	$+1/2\sqrt{3}$
1 1 3	0000+	$-1/2\sqrt{3}$
$\frac{3}{2}$ $\frac{3}{2}$ $\frac{3}{2}$	0000+	$+1/4$
$\frac{3}{2}$ $\frac{3}{2}$ $\frac{5}{2}$	0000+	$-1/4$
$\frac{3}{2}$ $\frac{3}{2}$ $\frac{5}{2}$	1000−	0
$\frac{3}{2}$ $\frac{3}{2}$ $\frac{5}{2}$	1100+	$+1/4$
2 2 1	0000+	$+1/2\sqrt{5}$
2 2 2	0000+	$-1/2\sqrt{5}$
2 2 2	1000−	0
2 2 2	1100+	$+1/2\sqrt{5}$
2 2 3	0000+	$+1/2\sqrt{5}$
$\frac{5}{2}$ $\frac{5}{2}$ $\frac{1}{2}$	0000+	$+1/2\sqrt{2.3}$
$\frac{5}{2}$ $\frac{5}{2}$ $\frac{3}{2}$	0000+	$-1/2\sqrt{2.3}$
$\frac{5}{2}$ $\frac{5}{2}$ $\frac{3}{2}$	1000−	0
$\frac{5}{2}$ $\frac{5}{2}$ $\frac{3}{2}$	1100+	$+1/2\sqrt{2.3}$
$\frac{5}{2}$ $\frac{5}{2}$ $\frac{5}{2}$	0000+	$+1/2\sqrt{2.3}$
$\frac{5}{2}$ $\frac{5}{2}$ $\frac{5}{2}$	1000−	0
$\frac{5}{2}$ $\frac{5}{2}$ $\frac{5}{2}$	1100+	$-1/2\sqrt{2.3}$
3 3 0	0000+	$+1/4$

3 3 1		
$\frac{1}{2}$ $\frac{1}{2}$ $\frac{5}{2}$	0000+	$+1/2.3\sqrt{2}$
1 0 3	0000+	$-1/2\sqrt{3}$
1 1 2	0000+	$-1/2.3\sqrt{2}$
1 1 3	0000+	$-1/3\sqrt{2}$
$\frac{3}{2}$ $\frac{1}{2}$ $\frac{5}{2}$	0000+	$+1/3\sqrt{2}$
$\frac{3}{2}$ $\frac{1}{2}$ $\frac{5}{2}$	0100−	0
$\frac{3}{2}$ $\frac{3}{2}$ $\frac{3}{2}$	0000+	$+1/4\sqrt{5}$
$\frac{3}{2}$ $\frac{3}{2}$ $\frac{5}{2}$	0000+	$+\sqrt{5}/4.3$
$\frac{3}{2}$ $\frac{3}{2}$ $\frac{5}{2}$	1000−	0
$\frac{3}{2}$ $\frac{3}{2}$ $\frac{5}{2}$	1100+	$+1/4\sqrt{5}$
2 1 2	0000+	$+1/4\sqrt{3.5}$
2 1 2	0100−	$+\sqrt{3}/4\sqrt{5}$
2 1 3	0000+	0
2 2 1	0000+	$-1/3\sqrt{2.5}$
2 2 2	0000+	$+1/4\sqrt{2.5}$
2 2 2	1000−	$+1/4\sqrt{2.5}$
2 2 2	1100+	$+\sqrt{5}/4.3\sqrt{2}$
2 2 3	0000+	$-1/3\sqrt{2.5}$
$\frac{5}{2}$ $\frac{3}{2}$ $\frac{3}{2}$	0000+	$-1/4\sqrt{3}$

K 6j Symbols (cont.)

3 3 1		
$\frac{5}{2}$ $\frac{3}{2}$ $\frac{3}{2}$	0100−	$+1/4\sqrt{5}$
$\frac{5}{2}$ $\frac{3}{2}$ $\frac{5}{2}$	0000+	0
$\frac{5}{2}$ $\frac{3}{2}$ $\frac{5}{2}$	0100−	0
$\frac{5}{2}$ $\frac{3}{2}$ $\frac{5}{2}$	1000−	$+1/2.3\sqrt{5}$
$\frac{5}{2}$ $\frac{3}{2}$ $\frac{5}{2}$	1100+	$-1/2.3$
$\frac{5}{2}$ $\frac{5}{2}$ $\frac{1}{2}$	0000+	$+1/2.3$
$\frac{5}{2}$ $\frac{5}{2}$ $\frac{1}{2}$	0010+	0
$\frac{5}{2}$ $\frac{5}{2}$ $\frac{3}{2}$	0000+	$+1/4.3$
$\frac{5}{2}$ $\frac{5}{2}$ $\frac{3}{2}$	0010+	0
$\frac{5}{2}$ $\frac{5}{2}$ $\frac{3}{2}$	1000−	0
$\frac{5}{2}$ $\frac{5}{2}$ $\frac{3}{2}$	1010−	$+1/4\sqrt{3}$
$\frac{5}{2}$ $\frac{5}{2}$ $\frac{3}{2}$	1100+	$-1/4.3$
$\frac{5}{2}$ $\frac{5}{2}$ $\frac{3}{2}$	1110+	$+1/2.3\sqrt{5}$
$\frac{5}{2}$ $\frac{5}{2}$ $\frac{5}{2}$	0000+	$+1/8$
$\frac{5}{2}$ $\frac{5}{2}$ $\frac{5}{2}$	0010+	$-1/8.3\sqrt{5}$
$\frac{5}{2}$ $\frac{5}{2}$ $\frac{5}{2}$	1000−	$-\sqrt{5}/8.3$
$\frac{5}{2}$ $\frac{5}{2}$ $\frac{5}{2}$	1010−	$-1/8.3$
$\frac{5}{2}$ $\frac{5}{2}$ $\frac{5}{2}$	1100+	$+1/8.3$
$\frac{5}{2}$ $\frac{5}{2}$ $\frac{5}{2}$	1110+	$+\sqrt{5}/8.3$
3 2 1	0000+	$+1/2.3$
3 2 2	0000+	$+1/2\sqrt{2.5}$
3 2 2	1000−	$+1/2.3\sqrt{2.5}$
3 2 3	0000−	$-1/2.3$
3 3 0	0000+	$-1/4$
3 3 1	0000+	$-1/4.3$
3 3 2		
1 1 2	0000+	$-\sqrt{3}/2.5\sqrt{2}$
1 1 3	0000+	0
$\frac{3}{2}$ $\frac{1}{2}$ $\frac{5}{2}$	0000+	0
$\frac{3}{2}$ $\frac{1}{2}$ $\frac{5}{2}$	0100−	$+1/\sqrt{2.3.5}$
$\frac{3}{2}$ $\frac{3}{2}$ $\frac{3}{2}$	0000+	$+3/4.5$
$\frac{3}{2}$ $\frac{3}{2}$ $\frac{5}{2}$	0000+	0
$\frac{3}{2}$ $\frac{3}{2}$ $\frac{5}{2}$	1000−	$-1/4\sqrt{3.5}$
$\frac{3}{2}$ $\frac{3}{2}$ $\frac{5}{2}$	1100+	$-1/2.5$
2 0 3	0000+	$+1/2\sqrt{5}$
2 1 2	0000+	$+3/4.5$
2 1 2	0100−	$-1/4.3$
2 1 3	0000+	$-1/3\sqrt{2.5}$
2 2 1	0000+	$+1/4\sqrt{5}$
2 2 1	0010+	$-7/4.3.5$
2 2 2	0000+	0
3 3 2		
2 2 2	0010+	$+1/4.5$
2 2 2	1000−	$+1/4\sqrt{5}$
2 2 2	1010−	0
2 2 2	1100+	0
2 2 2	1110+	$+1/4.3$
2 2 3	0000+	0
2 2 3	0010+	$+2/3.5$
$\frac{5}{2}$ $\frac{1}{2}$ $\frac{5}{2}$	0000+	$-1/4\sqrt{5}$
$\frac{5}{2}$ $\frac{1}{2}$ $\frac{5}{2}$	0100−	$-1/4.3$
$\frac{5}{2}$ $\frac{3}{2}$ $\frac{3}{2}$	0000+	$-1/4\sqrt{5}$
$\frac{5}{2}$ $\frac{3}{2}$ $\frac{3}{2}$	0010−	$+1/2\sqrt{2.3.5}$
$\frac{5}{2}$ $\frac{3}{2}$ $\frac{3}{2}$	0100−	$-1/4.5\sqrt{3}$
$\frac{5}{2}$ $\frac{3}{2}$ $\frac{3}{2}$	0110+	$+1/2.5\sqrt{2}$
$\frac{5}{2}$ $\frac{3}{2}$ $\frac{5}{2}$	0000+	$-1/4\sqrt{5}$
$\frac{5}{2}$ $\frac{3}{2}$ $\frac{5}{2}$	0010−	$+1/4\sqrt{2.3.5}$
$\frac{5}{2}$ $\frac{3}{2}$ $\frac{5}{2}$	0100−	$+1/4.3$
$\frac{5}{2}$ $\frac{3}{2}$ $\frac{5}{2}$	0110+	$+1/4\sqrt{2.3}$
$\frac{5}{2}$ $\frac{3}{2}$ $\frac{5}{2}$	1000−	$+1/4.5\sqrt{3}$
$\frac{5}{2}$ $\frac{3}{2}$ $\frac{5}{2}$	1010+	$+3/4.5\sqrt{2}$
$\frac{5}{2}$ $\frac{3}{2}$ $\frac{5}{2}$	1100+	$+1/4\sqrt{3.5}$
$\frac{5}{2}$ $\frac{3}{2}$ $\frac{5}{2}$	1110−	$+1/4.3\sqrt{2.5}$
$\frac{5}{2}$ $\frac{5}{2}$ $\frac{1}{2}$	0000+	$+1/4\sqrt{3}$
$\frac{5}{2}$ $\frac{5}{2}$ $\frac{1}{2}$	0010−	0
$\frac{5}{2}$ $\frac{5}{2}$ $\frac{1}{2}$	0020+	$+1/4.3$
$\frac{5}{2}$ $\frac{5}{2}$ $\frac{3}{2}$	0000+	$+1/8\sqrt{3}$
$\frac{5}{2}$ $\frac{5}{2}$ $\frac{3}{2}$	0010−	0
$\frac{5}{2}$ $\frac{5}{2}$ $\frac{3}{2}$	0020+	$+1/8.3$
$\frac{5}{2}$ $\frac{5}{2}$ $\frac{3}{2}$	1000−	$-1/8\sqrt{5}$
$\frac{5}{2}$ $\frac{5}{2}$ $\frac{3}{2}$	1010+	$-1/2\sqrt{2.3.5}$
$\frac{5}{2}$ $\frac{5}{2}$ $\frac{3}{2}$	1020−	$+\sqrt{3}/8\sqrt{5}$
$\frac{5}{2}$ $\frac{5}{2}$ $\frac{3}{2}$	1100+	$+3\sqrt{3}/8.5$
$\frac{5}{2}$ $\frac{5}{2}$ $\frac{3}{2}$	1110−	0
$\frac{5}{2}$ $\frac{5}{2}$ $\frac{3}{2}$	1120+	$+1/8.3.5$
$\frac{5}{2}$ $\frac{5}{2}$ $\frac{5}{2}$	0000+	$-1/8.5\sqrt{3}$
$\frac{5}{2}$ $\frac{5}{2}$ $\frac{5}{2}$	0010−	0
$\frac{5}{2}$ $\frac{5}{2}$ $\frac{5}{2}$	0020+	$+1/8.3.5$
$\frac{5}{2}$ $\frac{5}{2}$ $\frac{5}{2}$	1000−	$+1/8\sqrt{3.5}$
$\frac{5}{2}$ $\frac{5}{2}$ $\frac{5}{2}$	1010+	$+1/3\sqrt{2.5}$
$\frac{5}{2}$ $\frac{5}{2}$ $\frac{5}{2}$	1020−	$+1/8\sqrt{5}$
$\frac{5}{2}$ $\frac{5}{2}$ $\frac{5}{2}$	1100+	$+1/8\sqrt{3}$
$\frac{5}{2}$ $\frac{5}{2}$ $\frac{5}{2}$	1110−	0
3 3 2		
$\frac{5}{2}$ $\frac{5}{2}$ $\frac{5}{2}$	1120+	$-1/8.3$
3 1 2	0000+	$+1/2.3.5$
3 1 3	0000−	$+1/2.3$
3 2 1	0000+	$+1/2\sqrt{2.5}$
3 2 1	0010−	$+1/2.3\sqrt{2.5}$
3 2 2	0000+	$-1/2.5$
3 2 2	0010−	0
3 2 2	1000−	0
3 2 2	1010+	$+1/2.3$
3 2 3	0000−	0
3 2 3	0010+	$+1/3\sqrt{2.5}$
3 3 0	0000+	$+1/4$
3 3 1	0000+	$+1/4.3$
3 3 2	0000+	$+7/4.3.5$
3 3 3		
$\frac{3}{2}$ $\frac{3}{2}$ $\frac{3}{2}$	0000−	0
2 2 1	0000−	0
2 2 1	0010+	$+1/3\sqrt{2.5}$
2 2 2	0000−	0
2 2 2	1000+	$+1/2\sqrt{2.5}$
2 2 2	1100−	0
2 2 2	1110+	$-1/2.3\sqrt{2.5}$
$\frac{5}{2}$ $\frac{3}{2}$ $\frac{3}{2}$	0000−	0
$\frac{5}{2}$ $\frac{3}{2}$ $\frac{3}{2}$	0100+	$-1/4\sqrt{3}$
$\frac{5}{2}$ $\frac{3}{2}$ $\frac{3}{2}$	0110−	0
$\frac{5}{2}$ $\frac{5}{2}$ $\frac{1}{2}$	0000−	0
$\frac{5}{2}$ $\frac{5}{2}$ $\frac{1}{2}$	0010+	$-1/2.3$
$\frac{5}{2}$ $\frac{5}{2}$ $\frac{3}{2}$	0000−	0
$\frac{5}{2}$ $\frac{5}{2}$ $\frac{3}{2}$	0010+	$-1/4.3$
$\frac{5}{2}$ $\frac{5}{2}$ $\frac{3}{2}$	1000+	$+1/4\sqrt{3}$
$\frac{5}{2}$ $\frac{5}{2}$ $\frac{3}{2}$	1010−	0
$\frac{5}{2}$ $\frac{5}{2}$ $\frac{3}{2}$	1100−	0
$\frac{5}{2}$ $\frac{5}{2}$ $\frac{3}{2}$	1110+	$+1/4.3$
$\frac{5}{2}$ $\frac{5}{2}$ $\frac{5}{2}$	0000−	0
$\frac{5}{2}$ $\frac{5}{2}$ $\frac{5}{2}$	1000+	$+1/4.3$
$\frac{5}{2}$ $\frac{5}{2}$ $\frac{5}{2}$	1100−	0
$\frac{5}{2}$ $\frac{5}{2}$ $\frac{5}{2}$	1110+	$+1/4.3$
3 2 1	0000−	$-1/2.3$
3 2 2	0000−	0
3 2 2	1000+	$+1/3\sqrt{2.5}$
3 3 0	0000+	$+1/4$

K 6j Symbols (cont.)

3	**3**	**3**		
3	3	1	0000+	$+1/4.3$
3	3	2	0000+	$-1/4.3$
3	3	3	0000+	$+1/2.3$
$\tilde{1}$	**$\frac{3}{2}$**	**$\frac{3}{2}$**		
0	$\frac{3}{2}$	$\frac{3}{2}$	0000+	$+1/4$
1	$\frac{3}{2}$	$\frac{3}{2}$	0000+	$+3/4.5$
2	$\frac{3}{2}$	$\frac{3}{2}$	0000+	$+1/4.5$
3	$\frac{3}{2}$	$\frac{3}{2}$	0000+	$+3/4.5$
$\tilde{1}$	$\frac{3}{2}$	$\frac{3}{2}$	0000+	$-11/4.3.5$
$\tilde{1}$	**2**	**1**		
0	1	2	0000+	$+1/\sqrt{3.5}$
$\frac{1}{2}$	$\frac{3}{2}$	$\frac{3}{2}$	0000+	$-1/2\sqrt{5}$
1	1	2	0000+	$+1/3\sqrt{5}$
1	2	1	0000+	$+1/5$
1	2	2	0000+	$+\sqrt{2}/5\sqrt{3}$
$\frac{3}{2}$	$\frac{3}{2}$	$\frac{3}{2}$	0000+	$+1/2.5$
2	1	2	0000+	$-1/\sqrt{2.3.5}$
2	1	2	0100+	$-1/5\sqrt{2.3}$
2	2	1	0000+	$+1/3.5$
2	2	2	0000+	0
2	2	2	0100+	$+2/3.5$
$\frac{5}{2}$	$\frac{3}{2}$	$\frac{3}{2}$	0000+	$-\sqrt{3}/2.5$
$\frac{5}{2}$	$\frac{3}{2}$	$\frac{3}{2}$	0100−	$+1/3.5\sqrt{2}$
3	2	1	0000+	$-2/3.5$
3	2	2	0000+	$-1/2.5$
3	2	2	0100−	$+1/2.3$
$\tilde{1}$	2	1	0000+	$+1/5$
$\tilde{1}$	**2**	**2**		
0	2	2	0000+	$-1/5$
$\frac{1}{2}$	$\frac{3}{2}$	$\frac{3}{2}$	0000+	$+1/5\sqrt{2}$
1	2	1	0000+	$+\sqrt{2}/5\sqrt{3}$
1	2	2	0000+	0
$\frac{3}{2}$	$\frac{3}{2}$	$\frac{3}{2}$	0000+	$+1/5\sqrt{2}$
2	2	1	0000+	0
2	2	1	0010+	$+2/3.5$
2	2	2	0000+	$+1/4.5$
2	2	2	0100+	$-1/4\sqrt{5}$
2	2	2	0110+	$-1/4.3.5$
$\frac{5}{2}$	$\frac{3}{2}$	$\frac{3}{2}$	0000+	0
$\frac{5}{2}$	$\frac{3}{2}$	$\frac{3}{2}$	0100−	$-1/2.5\sqrt{3}$
$\frac{5}{2}$	$\frac{3}{2}$	$\frac{3}{2}$	0110+	$-1/2.3\sqrt{2}$

$\tilde{1}$	**2**	**2**		
3	2	1	0000+	$-1/2.5$
3	2	1	0010−	$+1/2.3$
3	2	2	0000+	0
3	2	2	0100−	0
3	2	2	0110+	$+2/3.5$
$\tilde{1}$	2	1	0000+	$+1/3.5$
$\tilde{1}$	2	2	0000+	$+1/2.3$
$\tilde{1}$	**$\frac{5}{2}$**	**$\frac{1}{2}$**		
0	$\frac{1}{2}$	$\frac{5}{2}$	0000+	$+1/2\sqrt{3}$
$\frac{1}{2}$	1	2	0000+	$+1/3\sqrt{2}$
1	$\frac{1}{2}$	$\frac{5}{2}$	0000+	$-1/2.3\sqrt{2}$
1	$\frac{1}{2}$	$\frac{5}{2}$	0100+	$+\sqrt{5}/2.3\sqrt{2}$
1	$\frac{3}{2}$	$\frac{3}{2}$	0000+	$+1/2\sqrt{2.3}$
1	$\frac{3}{2}$	$\frac{5}{2}$	0000+	$+\sqrt{5}/4.3$
1	$\frac{3}{2}$	$\frac{5}{2}$	0100+	$+1/4.3$
$\frac{3}{2}$	1	2	0000+	$-1/3\sqrt{2}$
$\frac{3}{2}$	1	2	0100−	0
$\frac{3}{2}$	2	1	0000+	$-1/\sqrt{2.3.5}$
$\frac{3}{2}$	2	2	0000+	0
$\frac{3}{2}$	2	2	0100−	$+1/\sqrt{2.3.5}$
2	$\frac{3}{2}$	$\frac{3}{2}$	0000+	$-1/2\sqrt{2.5}$
2	$\frac{3}{2}$	$\frac{3}{2}$	0100−	$-1/2\sqrt{3.5}$
2	$\frac{3}{2}$	$\frac{5}{2}$	0000+	$-1/2\sqrt{3.5}$
2	$\frac{3}{2}$	$\frac{5}{2}$	0100−	$-1/2\sqrt{2.5}$
2	$\frac{3}{2}$	$\frac{5}{2}$	0200+	0
2	$\frac{5}{2}$	$\frac{1}{2}$	0000+	$+1/2.3$
2	$\frac{5}{2}$	$\frac{3}{2}$	0000+	$-1/3\sqrt{2.5}$
2	$\frac{5}{2}$	$\frac{3}{2}$	0100−	$+1/2\sqrt{3.5}$
2	$\frac{5}{2}$	$\frac{5}{2}$	0000+	$+1/2\sqrt{3.5}$
2	$\frac{5}{2}$	$\frac{5}{2}$	0100−	$-1/3\sqrt{2.5}$
2	$\frac{5}{2}$	$\frac{5}{2}$	0200+	0
2	$\frac{5}{2}$	$\frac{5}{2}$	1000+	0
2	$\frac{5}{2}$	$\frac{5}{2}$	1100−	0
2	$\frac{5}{2}$	$\frac{5}{2}$	1200+	$-1/2.3$
$\frac{5}{2}$	2	1	0000+	$+1/2.3$
$\frac{5}{2}$	2	1	0100+	$-1/2.3\sqrt{5}$
$\frac{5}{2}$	2	2	0000+	0
$\frac{5}{2}$	2	2	0100−	$+1/3\sqrt{5}$
$\frac{5}{2}$	2	2	0200+	$+1/3\sqrt{2.5}$
3	$\frac{5}{2}$	$\frac{1}{2}$	0000+	$+1/2.3$
3	$\frac{5}{2}$	$\frac{3}{2}$	0000+	$+\sqrt{5}/4.3\sqrt{2}$

$\tilde{1}$	**$\frac{5}{2}$**	**$\frac{1}{2}$**		
3	$\frac{5}{2}$	$\frac{3}{2}$	0100−	$+1/4\sqrt{2.3}$
3	$\frac{5}{2}$	$\frac{5}{2}$	0000+	$-1/2.3$
3	$\frac{5}{2}$	$\frac{5}{2}$	0100−	0
3	$\frac{5}{2}$	$\frac{5}{2}$	1000+	0
3	$\frac{5}{2}$	$\frac{5}{2}$	1100−	$+1/2.3$
$\tilde{1}$	$\frac{5}{2}$	$\frac{1}{2}$	0000+	$-1/2.3$
$\tilde{1}$	**$\frac{5}{2}$**	**$\frac{3}{2}$**		
0	$\frac{3}{2}$	$\frac{5}{2}$	0000+	$-1/2\sqrt{2.3}$
$\frac{1}{2}$	1	2	0000+	$+1/3\sqrt{5}$
$\frac{1}{2}$	2	2	0000+	$+1/5\sqrt{2}$
1	$\frac{1}{2}$	$\frac{5}{2}$	0000+	$+\sqrt{5}/4.3$
1	$\frac{1}{2}$	$\frac{5}{2}$	0100+	$+1/4.3$
1	$\frac{3}{2}$	$\frac{3}{2}$	0000+	$+\sqrt{2}/5\sqrt{3}$
1	$\frac{3}{2}$	$\frac{5}{2}$	0000+	$-1/2.3\sqrt{5}$
1	$\frac{3}{2}$	$\frac{5}{2}$	0100+	$+1/3.5$
1	$\frac{5}{2}$	$\frac{3}{2}$	0000+	$-1/2.5$
1	$\frac{5}{2}$	$\frac{5}{2}$	0000+	$+1/4\sqrt{2.3.5}$
1	$\frac{5}{2}$	$\frac{5}{2}$	0100+	$-7/4.5\sqrt{2.3}$
1	$\frac{5}{2}$	$\frac{5}{2}$	1000+	$+1/4\sqrt{2.3}$
1	$\frac{5}{2}$	$\frac{5}{2}$	1100+	$+1/4\sqrt{2.3.5}$
$\frac{3}{2}$	1	2	0000+	$+1/3.5\sqrt{2}$
$\frac{3}{2}$	1	2	0100−	$-\sqrt{3}/2.5$
$\frac{3}{2}$	2	1	0000+	$+\sqrt{2}/5\sqrt{3}$
$\frac{3}{2}$	2	2	0000+	$-1/2.5$
$\frac{3}{2}$	2	2	0100−	$-1/2.5\sqrt{2.3}$
2	$\frac{1}{2}$	$\frac{5}{2}$	0000+	$-1/2\sqrt{3.5}$
2	$\frac{1}{2}$	$\frac{5}{2}$	0100−	$+1/2\sqrt{2.5}$
2	$\frac{1}{2}$	$\frac{5}{2}$	0200+	0
2	$\frac{3}{2}$	$\frac{3}{2}$	0000+	$-1/5\sqrt{2}$
2	$\frac{3}{2}$	$\frac{3}{2}$	0100−	$+1/2.5\sqrt{3}$
2	$\frac{3}{2}$	$\frac{5}{2}$	0000+	$+1/4.5\sqrt{3}$
2	$\frac{3}{2}$	$\frac{5}{2}$	0100−	0
2	$\frac{3}{2}$	$\frac{5}{2}$	0200+	$+3/4.5$
2	$\frac{5}{2}$	$\frac{1}{2}$	0000+	$-1/3\sqrt{2.5}$
2	$\frac{5}{2}$	$\frac{1}{2}$	0010−	$+1/2\sqrt{3.5}$
2	$\frac{5}{2}$	$\frac{3}{2}$	0000+	$+1/3.5$
2	$\frac{5}{2}$	$\frac{3}{2}$	0010−	$+1/2.5\sqrt{2.3}$
2	$\frac{5}{2}$	$\frac{3}{2}$	0100−	$+1/2.5\sqrt{2.3}$
2	$\frac{5}{2}$	$\frac{3}{2}$	0110+	$+3/4.5$
2	$\frac{5}{2}$	$\frac{5}{2}$	0000+	$-1/5\sqrt{2.3}$
2	$\frac{5}{2}$	$\frac{5}{2}$	0010−	$-1/4.5$

K 6j Symbols (cont.)

$\tilde{1}$	$\frac{5}{2}$	$\frac{3}{2}$		
2	$\frac{5}{2}$	$\frac{5}{2}$	0100−	$-1/4.3$
2	$\frac{5}{2}$	$\frac{5}{2}$	0110+	0
2	$\frac{5}{2}$	$\frac{5}{2}$	0200+	$-1/2.5\sqrt{2}$
2	$\frac{5}{2}$	$\frac{5}{2}$	0210−	$+1/4.5\sqrt{3}$
2	$\frac{5}{2}$	$\frac{5}{2}$	1000+	0
2	$\frac{5}{2}$	$\frac{5}{2}$	1010−	$-1/4\sqrt{5}$
2	$\frac{5}{2}$	$\frac{5}{2}$	1100−	$+1/4\sqrt{5}$
2	$\frac{5}{2}$	$\frac{5}{2}$	1110+	0
2	$\frac{5}{2}$	$\frac{5}{2}$	1200+	$-1/2.3\sqrt{2.5}$
2	$\frac{5}{2}$	$\frac{5}{2}$	1210−	$+1/4\sqrt{3.5}$
$\frac{5}{2}$	1	2	0000+	$+1/5\sqrt{2}$
$\frac{5}{2}$	1	2	0100−	0
$\frac{5}{2}$	1	2	0200+	$-1/5\sqrt{2.3}$
$\frac{5}{2}$	2	1	0000+	$+1/3\sqrt{2.5}$
$\frac{5}{2}$	2	1	0010−	0
$\frac{5}{2}$	2	1	0100+	$-1/3.5\sqrt{2}$
$\frac{5}{2}$	2	1	0110−	$-1/5\sqrt{3}$
$\frac{5}{2}$	2	2	0000+	0
$\frac{5}{2}$	2	2	0010−	$-1/2.5\sqrt{2}$
$\frac{5}{2}$	2	2	0100−	$-1/3.5\sqrt{2}$
$\frac{5}{2}$	2	2	0110+	$+1/5\sqrt{3}$
$\frac{5}{2}$	2	2	0200+	$-2/3.5$
$\frac{5}{2}$	2	2	0210−	$-1/2.5\sqrt{2.3}$
3	$\frac{3}{2}$	$\frac{3}{2}$	0000+	$-1/2\sqrt{2.5}$
3	$\frac{3}{2}$	$\frac{3}{2}$	0100−	$-1/2.5\sqrt{2.3}$
3	$\frac{3}{2}$	$\frac{5}{2}$	0000+	$-1/2.5$
3	$\frac{3}{2}$	$\frac{5}{2}$	0100−	0
3	$\frac{5}{2}$	$\frac{1}{2}$	0000+	$+\sqrt{5}/4.3\sqrt{2}$
3	$\frac{5}{2}$	$\frac{1}{2}$	0010−	$+1/4\sqrt{2.3}$
3	$\frac{5}{2}$	$\frac{3}{2}$	0000+	$-1/4.3$
3	$\frac{5}{2}$	$\frac{3}{2}$	0010−	$-1/4\sqrt{3.5}$
3	$\frac{5}{2}$	$\frac{3}{2}$	0100−	$-1/4\sqrt{3.5}$
3	$\frac{5}{2}$	$\frac{3}{2}$	0110+	$+3/4.5$
3	$\frac{5}{2}$	$\frac{5}{2}$	0000+	$-7/8.3\sqrt{2.5}$
3	$\frac{5}{2}$	$\frac{5}{2}$	0010−	$+1/8.5\sqrt{2.3}$
3	$\frac{5}{2}$	$\frac{5}{2}$	0100−	$+1/8\sqrt{2}$
3	$\frac{5}{2}$	$\frac{5}{2}$	0110+	$-\sqrt{5}/8\sqrt{2.3}$
3	$\frac{5}{2}$	$\frac{5}{2}$	1000+	$-1/8\sqrt{2}$
3	$\frac{5}{2}$	$\frac{5}{2}$	1010−	$-\sqrt{3}/8\sqrt{2.5}$
3	$\frac{5}{2}$	$\frac{5}{2}$	1100−	$-\sqrt{5}/8.3\sqrt{2}$
3	$\frac{5}{2}$	$\frac{5}{2}$	1110+	$-1/8\sqrt{2.3}$

$\tilde{1}$	$\frac{5}{2}$	$\frac{3}{2}$		
$\tilde{1}$	$\frac{3}{2}$	$\frac{3}{2}$	0000+	$-1/2.5$
$\tilde{1}$	$\frac{3}{2}$	$\frac{5}{2}$	0000+	$+3/4.5$
$\tilde{1}$	$\frac{3}{2}$	$\frac{5}{2}$	0100+	$+1/4\sqrt{5}$
$\tilde{1}$	$\frac{5}{2}$	$\frac{1}{2}$	0000+	$-1/2.3$
$\tilde{1}$	$\frac{5}{2}$	$\frac{3}{2}$	0000+	$-1/4.3.5$

$\tilde{1}$	$\frac{5}{2}$	$\frac{5}{2}$		
0	$\frac{5}{2}$	$\frac{5}{2}$	0000+	$+1/2.3$
0	$\frac{5}{2}$	$\frac{5}{2}$	0001+	0
0	$\frac{5}{2}$	$\frac{5}{2}$	1000+	0
0	$\frac{5}{2}$	$\frac{5}{2}$	1001+	$+1/2.3$
$\frac{1}{2}$	2	2	0000+	$-\sqrt{2}/3.5$
$\frac{1}{2}$	2	2	0001+	$+1/3\sqrt{2.5}$
1	$\frac{3}{2}$	$\frac{3}{2}$	0000+	$+1/3.5$
1	$\frac{3}{2}$	$\frac{3}{2}$	0001+	$-1/2.3\sqrt{5}$
1	$\frac{5}{2}$	$\frac{3}{2}$	0000+	$+1/4\sqrt{2.3.5}$
1	$\frac{5}{2}$	$\frac{3}{2}$	0001+	$+1/4\sqrt{2.3}$
1	$\frac{5}{2}$	$\frac{3}{2}$	0010+	$-7/4.5\sqrt{2.3}$
1	$\frac{5}{2}$	$\frac{3}{2}$	0011+	$+1/4\sqrt{2.3.5}$
1	$\frac{5}{2}$	$\frac{5}{2}$	0000+	$-1/8$
1	$\frac{5}{2}$	$\frac{5}{2}$	0001+	$+\sqrt{5}/8.3$
1	$\frac{5}{2}$	$\frac{5}{2}$	0100+	$+1/8.3\sqrt{5}$
1	$\frac{5}{2}$	$\frac{5}{2}$	0101+	$+1/8.3$
1	$\frac{5}{2}$	$\frac{5}{2}$	0110+	$+13/8.3.5$
1	$\frac{5}{2}$	$\frac{5}{2}$	0111+	$+1/8.3\sqrt{5}$
1	$\frac{5}{2}$	$\frac{5}{2}$	1000+	$+\sqrt{5}/8.3$
1	$\frac{5}{2}$	$\frac{5}{2}$	1001+	$+1/8.3$
1	$\frac{5}{2}$	$\frac{5}{2}$	1100+	$+1/8.3$
1	$\frac{5}{2}$	$\frac{5}{2}$	1101+	$+\sqrt{5}/8.3$
1	$\frac{5}{2}$	$\frac{5}{2}$	1110+	$+1/8.3\sqrt{5}$
1	$\frac{5}{2}$	$\frac{5}{2}$	1111+	$-1/8$
$\frac{3}{2}$	2	1	0000+	$-1/5\sqrt{3}$
$\frac{3}{2}$	2	1	0001+	0
$\frac{3}{2}$	2	1	0010−	$+1/3.5\sqrt{2}$
$\frac{3}{2}$	2	1	0011−	$-1/3\sqrt{2.5}$
$\frac{3}{2}$	2	2	0000+	$-1/3.5\sqrt{2}$
$\frac{3}{2}$	2	2	0001+	$-1/3\sqrt{2.5}$
$\frac{3}{2}$	2	2	0100−	$-1/5\sqrt{3}$
$\frac{3}{2}$	2	2	0101−	0
$\frac{3}{2}$	2	2	0110+	$-1/2.5\sqrt{2}$
$\frac{3}{2}$	2	2	0111+	$-1/2.3\sqrt{2.5}$
2	$\frac{3}{2}$	$\frac{3}{2}$	0000+	$-1/2.5$

$\tilde{1}$	$\frac{5}{2}$	$\frac{5}{2}$		
2	$\frac{3}{2}$	$\frac{3}{2}$	0001+	0
2	$\frac{3}{2}$	$\frac{3}{2}$	0100−	$+1/2.5\sqrt{2.3}$
2	$\frac{3}{2}$	$\frac{3}{2}$	0101−	$+1/2\sqrt{2.3.5}$
2	$\frac{3}{2}$	$\frac{3}{2}$	0110+	$-3/4.5$
2	$\frac{3}{2}$	$\frac{3}{2}$	0111+	$+1/4.3\sqrt{5}$
2	$\frac{5}{2}$	$\frac{1}{2}$	0000+	$+1/2\sqrt{3.5}$
2	$\frac{5}{2}$	$\frac{1}{2}$	0001+	0
2	$\frac{5}{2}$	$\frac{1}{2}$	0010−	$-1/3\sqrt{2.5}$
2	$\frac{5}{2}$	$\frac{1}{2}$	0011−	0
2	$\frac{5}{2}$	$\frac{1}{2}$	0020+	0
2	$\frac{5}{2}$	$\frac{1}{2}$	0021+	$-1/2.3$
2	$\frac{5}{2}$	$\frac{3}{2}$	0000+	$-1/5\sqrt{2.3}$
2	$\frac{5}{2}$	$\frac{3}{2}$	0001+	0
2	$\frac{5}{2}$	$\frac{3}{2}$	0010−	$-1/4.3$
2	$\frac{5}{2}$	$\frac{3}{2}$	0011−	$+1/4\sqrt{5}$
2	$\frac{5}{2}$	$\frac{3}{2}$	0020+	$-1/2.5\sqrt{2}$
2	$\frac{5}{2}$	$\frac{3}{2}$	0021+	$-1/2.3\sqrt{2.5}$
2	$\frac{5}{2}$	$\frac{3}{2}$	0100−	$-1/4.5$
2	$\frac{5}{2}$	$\frac{3}{2}$	0101−	$-1/4\sqrt{5}$
2	$\frac{5}{2}$	$\frac{3}{2}$	0110+	0
2	$\frac{5}{2}$	$\frac{3}{2}$	0111+	0
2	$\frac{5}{2}$	$\frac{3}{2}$	0120−	$+1/4.5\sqrt{3}$
2	$\frac{5}{2}$	$\frac{3}{2}$	0121−	$+1/4\sqrt{3.5}$
2	$\frac{5}{2}$	$\frac{5}{2}$	0000+	$+1/4.3.5$
2	$\frac{5}{2}$	$\frac{5}{2}$	0001+	0
2	$\frac{5}{2}$	$\frac{5}{2}$	0100−	0
2	$\frac{5}{2}$	$\frac{5}{2}$	0101−	$-1/2\sqrt{2.3.5}$
2	$\frac{5}{2}$	$\frac{5}{2}$	0110+	$+1/4.3$
2	$\frac{5}{2}$	$\frac{5}{2}$	0111+	$+1/4.3\sqrt{5}$
2	$\frac{5}{2}$	$\frac{5}{2}$	0200+	$-1/2.5\sqrt{3}$
2	$\frac{5}{2}$	$\frac{5}{2}$	0201+	$+1/4\sqrt{3.5}$
2	$\frac{5}{2}$	$\frac{5}{2}$	0210−	0
2	$\frac{5}{2}$	$\frac{5}{2}$	0211−	$+1/2.3\sqrt{2.5}$
2	$\frac{5}{2}$	$\frac{5}{2}$	0220+	$+7/4.3.5$
2	$\frac{5}{2}$	$\frac{5}{2}$	0221+	0
2	$\frac{5}{2}$	$\frac{5}{2}$	1000+	0
2	$\frac{5}{2}$	$\frac{5}{2}$	1001+	$-1/4.3$
2	$\frac{5}{2}$	$\frac{5}{2}$	1100−	$+1/2\sqrt{2.3.5}$
2	$\frac{5}{2}$	$\frac{5}{2}$	1101−	0
2	$\frac{5}{2}$	$\frac{5}{2}$	1110+	$+1/4.3\sqrt{5}$
2	$\frac{5}{2}$	$\frac{5}{2}$	1111+	$+1/4.3$

K 6j Symbols (cont.)

$\tilde{1}\ \frac{5}{2}\ \frac{5}{2}$		
$2\ \frac{5}{2}\ \frac{5}{2}$	1200+	$+1/4\sqrt{3.5}$
$2\ \frac{5}{2}\ \frac{5}{2}$	1201+	0
$2\ \frac{5}{2}\ \frac{5}{2}$	1210−	$-1/2.3\sqrt{2.5}$
$2\ \frac{5}{2}\ \frac{5}{2}$	1211−	0
$2\ \frac{5}{2}\ \frac{5}{2}$	1220+	0
$2\ \frac{5}{2}\ \frac{5}{2}$	1221+	$+1/4.3$
$\frac{5}{2}\ 2\ 1$	0000+	$+1/2\sqrt{3.5}$
$\frac{5}{2}\ 2\ 1$	0001+	0
$\frac{5}{2}\ 2\ 1$	0010−	$-1/2.3\sqrt{2.5}$
$\frac{5}{2}\ 2\ 1$	0011−	$-1/2.3\sqrt{2}$
$\frac{5}{2}\ 2\ 1$	0020+	$+1/2.3\sqrt{5}$
$\frac{5}{2}\ 2\ 1$	0021+	0
$\frac{5}{2}\ 2\ 1$	0100+	$-1/2.5\sqrt{3}$
$\frac{5}{2}\ 2\ 1$	0101+	0
$\frac{5}{2}\ 2\ 1$	0110−	$-1/2.3\sqrt{2}$
$\frac{5}{2}\ 2\ 1$	0111−	$-1/2.3\sqrt{2.5}$
$\frac{5}{2}\ 2\ 1$	0120+	$+1/2.3.5$
$\frac{5}{2}\ 2\ 1$	0121+	$-1/3\sqrt{5}$
$\frac{5}{2}\ 2\ 2$	0000+	0
$\frac{5}{2}\ 2\ 2$	0001+	$-1/2\sqrt{2.5}$
$\frac{5}{2}\ 2\ 2$	0100−	$-1/2.5\sqrt{3}$
$\frac{5}{2}\ 2\ 2$	0101−	0
$\frac{5}{2}\ 2\ 2$	0110+	$+1/3.5\sqrt{2}$
$\frac{5}{2}\ 2\ 2$	0111+	0
$\frac{5}{2}\ 2\ 2$	0200+	$+1/2.5\sqrt{2.3}$
$\frac{5}{2}\ 2\ 2$	0201+	0
$\frac{5}{2}\ 2\ 2$	0210−	$-1/3.5$
$\frac{5}{2}\ 2\ 2$	0211−	$+1/2.3\sqrt{5}$
$\frac{5}{2}\ 2\ 2$	0220+	$+1/3.5\sqrt{2}$
$\frac{5}{2}\ 2\ 2$	0221+	$+1/2.3\sqrt{2.5}$
$3\ \frac{3}{2}\ \frac{3}{2}$	0000+	0
$3\ \frac{3}{2}\ \frac{3}{2}$	0001+	0
$3\ \frac{3}{2}\ \frac{3}{2}$	0100−	$-1/4\sqrt{3.5}$
$3\ \frac{3}{2}\ \frac{3}{2}$	0101−	$+1/4\sqrt{3}$
$3\ \frac{3}{2}\ \frac{3}{2}$	0110+	$+1/2.5$
$3\ \frac{3}{2}\ \frac{3}{2}$	0111+	$+1/2.3\sqrt{5}$
$3\ \frac{5}{2}\ \frac{1}{2}$	0000+	$-1/2.3$
$3\ \frac{5}{2}\ \frac{1}{2}$	0001+	0
$3\ \frac{5}{2}\ \frac{1}{2}$	0010−	0
$3\ \frac{5}{2}\ \frac{1}{2}$	0011−	$+1/2.3$
$3\ \frac{5}{2}\ \frac{3}{2}$	0000+	$-7/8.3\sqrt{2.5}$

$\tilde{1}\ \frac{5}{2}\ \frac{5}{2}$		
$3\ \frac{5}{2}\ \frac{3}{2}$	0001+	$-1/8\sqrt{2}$
$3\ \frac{5}{2}\ \frac{3}{2}$	0010−	$+1/8\sqrt{2}$
$3\ \frac{5}{2}\ \frac{3}{2}$	0011−	$-\sqrt{5}/8.3\sqrt{2}$
$3\ \frac{5}{2}\ \frac{3}{2}$	0100−	$+1/8.5\sqrt{2.3}$
$3\ \frac{5}{2}\ \frac{3}{2}$	0101−	$-\sqrt{3}/8\sqrt{2.5}$
$3\ \frac{5}{2}\ \frac{3}{2}$	0110+	$-\sqrt{5}/8\sqrt{2.3}$
$3\ \frac{5}{2}\ \frac{3}{2}$	0111+	$-1/8\sqrt{2.3}$
$3\ \frac{5}{2}\ \frac{5}{2}$	0000+	$-1/4.3.5$
$3\ \frac{5}{2}\ \frac{5}{2}$	0001+	$-1/2.3\sqrt{5}$
$3\ \frac{5}{2}\ \frac{5}{2}$	0100−	0
$3\ \frac{5}{2}\ \frac{5}{2}$	0101−	$-1/4.3$
$3\ \frac{5}{2}\ \frac{5}{2}$	0110+	$-1/4.3$
$3\ \frac{5}{2}\ \frac{5}{2}$	0111+	0
$3\ \frac{5}{2}\ \frac{5}{2}$	1000+	$-1/2.3\sqrt{5}$
$3\ \frac{5}{2}\ \frac{5}{2}$	1001+	$+1/4.3$
$3\ \frac{5}{2}\ \frac{5}{2}$	1100−	$+1/4.3$
$3\ \frac{5}{2}\ \frac{5}{2}$	1101−	0
$3\ \frac{5}{2}\ \frac{5}{2}$	1110+	0
$3\ \frac{5}{2}\ \frac{5}{2}$	1111+	$+1/4.3$
$\tilde{1}\ \frac{3}{2}\ \frac{3}{2}$	0000+	$+3/4.5$
$\tilde{1}\ \frac{3}{2}\ \frac{3}{2}$	0001+	$+1/4\sqrt{5}$
$\tilde{1}\ \frac{5}{2}\ \frac{1}{2}$	0000+	$-1/2.3$
$\tilde{1}\ \frac{5}{2}\ \frac{1}{2}$	0001+	0
$\tilde{1}\ \frac{5}{2}\ \frac{1}{2}$	0010+	0
$\tilde{1}\ \frac{5}{2}\ \frac{1}{2}$	0011+	$+1/2.3$
$\tilde{1}\ \frac{5}{2}\ \frac{3}{2}$	0000+	$-1/4.3.5$
$\tilde{1}\ \frac{5}{2}\ \frac{3}{2}$	0001+	$+1/4\sqrt{5}$
$\tilde{1}\ \frac{5}{2}\ \frac{3}{2}$	0010+	$+1/4\sqrt{5}$
$\tilde{1}\ \frac{5}{2}\ \frac{3}{2}$	0011+	$-1/4.3$
$\tilde{1}\ \frac{5}{2}\ \frac{5}{2}$	0000+	$-1/4.3.5$
$\tilde{1}\ \frac{5}{2}\ \frac{5}{2}$	0001+	$-1/2.3\sqrt{5}$
$\tilde{1}\ \frac{5}{2}\ \frac{5}{2}$	0100+	$-1/2.3\sqrt{5}$
$\tilde{1}\ \frac{5}{2}\ \frac{5}{2}$	0101+	$-1/4.3$
$\tilde{1}\ \frac{5}{2}\ \frac{5}{2}$	0110+	$+1/4.3$
$\tilde{1}\ \frac{5}{2}\ \frac{5}{2}$	0111+	0
$\tilde{1}\ \frac{5}{2}\ \frac{5}{2}$	1000+	$-1/2.3\sqrt{5}$
$\tilde{1}\ \frac{5}{2}\ \frac{5}{2}$	1001+	$+1/4.3$
$\tilde{1}\ \frac{5}{2}\ \frac{5}{2}$	1100+	$-1/4.3$
$\tilde{1}\ \frac{5}{2}\ \frac{5}{2}$	1101+	0
$\tilde{1}\ \frac{5}{2}\ \frac{5}{2}$	1110+	0
$\tilde{1}\ \frac{5}{2}\ \frac{5}{2}$	1111+	$-1/4.3$

$\tilde{1}\ 3\ 1$		
$0\ 1\ 3$	0000+	$-1/2\sqrt{3}$
$\frac{1}{2}\ \frac{1}{2}\ \frac{5}{2}$	0000+	$+1/3\sqrt{2}$
$\frac{1}{2}\ \frac{3}{2}\ \frac{5}{2}$	0000+	$-\sqrt{5}/4.3$
$1\ 1\ 2$	0000+	$-1/3\sqrt{2}$
$1\ 1\ 3$	0000+	$+1/2.3\sqrt{2}$
$1\ 2\ 2$	0000+	$+1/2\sqrt{3.5}$
$1\ 2\ 3$	0000+	$-1/2\sqrt{2.3}$
$\frac{3}{2}\ \frac{1}{2}\ \frac{5}{2}$	0000+	$-1/4.3\sqrt{2}$
$\frac{3}{2}\ \frac{1}{2}\ \frac{5}{2}$	0100−	$-\sqrt{5}/4\sqrt{2.3}$
$\frac{3}{2}\ \frac{3}{2}\ \frac{3}{2}$	0000+	$+1/2\sqrt{5}$
$\frac{3}{2}\ \frac{3}{2}\ \frac{5}{2}$	0000+	$+1/2.3\sqrt{2}$
$\frac{3}{2}\ \frac{3}{2}\ \frac{5}{2}$	0100−	$-1/2\sqrt{2.3.5}$
$\frac{3}{2}\ \frac{5}{2}\ \frac{3}{2}$	0000+	$-1/2\sqrt{2.3.5}$
$\frac{3}{2}\ \frac{5}{2}\ \frac{5}{2}$	0000+	$+1/8\sqrt{3}$
$\frac{3}{2}\ \frac{5}{2}\ \frac{5}{2}$	0100−	$+7/8.3\sqrt{5}$
$\frac{3}{2}\ \frac{5}{2}\ \frac{5}{2}$	1000+	$+\sqrt{5}/8\sqrt{3}$
$\frac{3}{2}\ \frac{5}{2}\ \frac{5}{2}$	1100−	$-1/8.3$
$2\ 1\ 2$	0000+	$+1/2\sqrt{3.5}$
$2\ 1\ 2$	0100−	$-1/2\sqrt{3.5}$
$2\ 1\ 3$	0000+	$-1/2\sqrt{2.3}$
$2\ 2\ 1$	0000+	$-\sqrt{2}/3\sqrt{5}$
$2\ 2\ 2$	0000+	$+1/2\sqrt{2.5}$
$2\ 2\ 2$	0100−	$+1/2.3\sqrt{2.5}$
$2\ 2\ 3$	0000+	$-1/2.3\sqrt{2.5}$
$2\ 3\ 1$	0000+	$-1/4.3$
$2\ 3\ 2$	0000+	$+1/4\sqrt{2.5}$
$2\ 3\ 2$	0100−	$+7/4.3\sqrt{2.5}$
$2\ 3\ 3$	0000+	$+1/2.3$
$\frac{5}{2}\ \frac{3}{2}\ \frac{3}{2}$	0000+	$+1/4\sqrt{3}$
$\frac{5}{2}\ \frac{3}{2}\ \frac{3}{2}$	0100−	$+1/4.3\sqrt{5}$
$\frac{5}{2}\ \frac{3}{2}\ \frac{5}{2}$	0000+	$-\sqrt{3}/4\sqrt{2.5}$
$\frac{5}{2}\ \frac{3}{2}\ \frac{5}{2}$	0100−	$-1/4\sqrt{2.3}$
$\frac{5}{2}\ \frac{5}{2}\ \frac{1}{2}$	0000+	$+1/4.3$
$\frac{5}{2}\ \frac{5}{2}\ \frac{1}{2}$	0010+	$+\sqrt{5}/4.3$
$\frac{5}{2}\ \frac{5}{2}\ \frac{3}{2}$	0000+	$+\sqrt{5}/4.3\sqrt{2}$
$\frac{5}{2}\ \frac{5}{2}\ \frac{3}{2}$	0010+	$-1/4.3\sqrt{2}$
$\frac{5}{2}\ \frac{5}{2}\ \frac{3}{2}$	0100−	$+1/4\sqrt{2.3}$
$\frac{5}{2}\ \frac{5}{2}\ \frac{3}{2}$	0110−	$+\sqrt{3}/4\sqrt{2.5}$
$\frac{5}{2}\ \frac{5}{2}\ \frac{5}{2}$	0000+	$+1/8.3$
$\frac{5}{2}\ \frac{5}{2}\ \frac{5}{2}$	0010+	$-7/8.3\sqrt{5}$
$\frac{5}{2}\ \frac{5}{2}\ \frac{5}{2}$	0100−	$+\sqrt{5}/8.3$

K 6j Symbols (cont.)

Ĩ 3 1

$\frac{5}{2}$	$\frac{5}{2}$	$\frac{5}{2}$	0110−	$+1/8.3$
$\frac{5}{2}$	$\frac{5}{2}$	$\frac{5}{2}$	1000+	$+\sqrt{5/8.3}$
$\frac{5}{2}$	$\frac{5}{2}$	$\frac{5}{2}$	1010+	$+1/8.3$
$\frac{5}{2}$	$\frac{5}{2}$	$\frac{5}{2}$	1100−	$-1/8$
$\frac{5}{2}$	$\frac{5}{2}$	$\frac{5}{2}$	1110−	$+\sqrt{5/8.3}$
3	2	1	0000+	$-1/2.3$
3	2	2	0000+	$+1/2.3\sqrt{2.5}$
3	2	3	0000−	$+1/2.3$
3	3	1	0000+	$+1/2.3$
3	3	2	0000+	$-1/2.3$
3	3	3	0000−	$-1/2.3$
Ĩ	2	1	0000+	0
Ĩ	2	2	0000+	$+\sqrt{2}/3\sqrt{5}$
Ĩ	2	3	0000+	$+1/2.3$
Ĩ	3	1	0000+	$+1/4$

Ĩ 3 2

0	2	3	0000+	$+1/2\sqrt{5}$
$\frac{1}{2}$	$\frac{3}{2}$	$\frac{5}{2}$	0000+	$+1/4\sqrt{5}$
$\frac{1}{2}$	$\frac{5}{2}$	$\frac{5}{2}$	0000+	$-1/2.3\sqrt{5}$
$\frac{1}{2}$	$\frac{5}{2}$	$\frac{5}{2}$	1000+	$-1/2.3$
1	1	2	0000+	$-1/5\sqrt{2.3}$
1	1	3	0000+	$-1/2\sqrt{2.3}$
1	2	2	0000+	$+1/2.5$
1	2	3	0000+	$+1/2\sqrt{2.5}$
1	3	2	0000+	$+1/5$
1	3	3	0000+	0
$\frac{3}{2}$	$\frac{1}{2}$	$\frac{5}{2}$	0000+	$-1/4\sqrt{2}$
$\frac{3}{2}$	$\frac{1}{2}$	$\frac{5}{2}$	0100−	$+1/4\sqrt{2.3.5}$
$\frac{3}{2}$	$\frac{3}{2}$	$\frac{3}{2}$	0000+	$+1/2.5$
$\frac{3}{2}$	$\frac{3}{2}$	$\frac{5}{2}$	0000+	$-1/2\sqrt{2.5}$
$\frac{3}{2}$	$\frac{3}{2}$	$\frac{5}{2}$	0100−	$-1/2.5\sqrt{2.3}$
$\frac{3}{2}$	$\frac{5}{2}$	$\frac{3}{2}$	0000+	$-1/2.5\sqrt{2}$
$\frac{3}{2}$	$\frac{5}{2}$	$\frac{3}{2}$	0010−	$+1/5\sqrt{3}$
$\frac{3}{2}$	$\frac{5}{2}$	$\frac{5}{2}$	0000+	$+1/8.3\sqrt{5}$
$\frac{3}{2}$	$\frac{5}{2}$	$\frac{5}{2}$	0010−	$+1/4\sqrt{2.3.5}$
$\frac{3}{2}$	$\frac{5}{2}$	$\frac{5}{2}$	0100−	$-\sqrt{3}/8.5$
$\frac{3}{2}$	$\frac{5}{2}$	$\frac{5}{2}$	0110+	$-3/4.5\sqrt{2}$
$\frac{3}{2}$	$\frac{5}{2}$	$\frac{5}{2}$	1000+	$+1/8.3$
$\frac{3}{2}$	$\frac{5}{2}$	$\frac{5}{2}$	1010−	$+1/4\sqrt{2.3}$
$\frac{3}{2}$	$\frac{5}{2}$	$\frac{5}{2}$	1100−	$+\sqrt{5}/8\sqrt{3}$
$\frac{3}{2}$	$\frac{5}{2}$	$\frac{5}{2}$	1110+	$-1/4.3\sqrt{2.5}$
2	1	2	0000+	$+1/2.5$
2	1	2	0100−	$+1/2.3$
2	1	3	0000+	$-1/2.3\sqrt{2.5}$
2	2	1	0000+	0
2	2	1	0010+	$-2/3.5$
2	2	2	0000+	$-1/4\sqrt{5}$
2	2	2	0010+	$-1/4.5$
2	2	2	0100−	$+1/4\sqrt{5}$
2	2	2	0110−	$-1/4.3$
2	2	3	0000+	$-1/4\sqrt{5}$
2	2	3	0010+	$-7/4.3.5$
2	3	1	0000+	$+1/4\sqrt{2.5}$
2	3	1	0010−	$+7/4.3\sqrt{2.5}$
2	3	2	0000+	$+1/8.5$
2	3	2	0010−	$+1/8$
2	3	2	0100−	$+1/8$
2	3	2	0110+	$-1/8.3$
2	3	3	0000+	$-1/2\sqrt{2.5}$
2	3	3	0010−	$+1/2.3\sqrt{2.5}$
$\frac{5}{2}$	$\frac{1}{2}$	$\frac{5}{2}$	0000+	$-1/2.3\sqrt{5}$
$\frac{5}{2}$	$\frac{1}{2}$	$\frac{5}{2}$	0100−	$+1/2.3$
$\frac{5}{2}$	$\frac{3}{2}$	$\frac{3}{2}$	0000+	$+1/4\sqrt{5}$
$\frac{5}{2}$	$\frac{3}{2}$	$\frac{3}{2}$	0010−	$-1/2\sqrt{2.3.5}$
$\frac{5}{2}$	$\frac{3}{2}$	$\frac{3}{2}$	0100−	$+1/4.5\sqrt{3}$
$\frac{5}{2}$	$\frac{3}{2}$	$\frac{3}{2}$	0110+	$+7/2.3.5\sqrt{2}$
$\frac{5}{2}$	$\frac{3}{2}$	$\frac{5}{2}$	0000+	$+11/4.3.5\sqrt{2}$
$\frac{5}{2}$	$\frac{3}{2}$	$\frac{5}{2}$	0010−	$-1/4.5\sqrt{3}$
$\frac{5}{2}$	$\frac{3}{2}$	$\frac{5}{2}$	0100−	$+1/4.3\sqrt{2.5}$
$\frac{5}{2}$	$\frac{3}{2}$	$\frac{5}{2}$	0110+	$+1/4\sqrt{3.5}$
$\frac{5}{2}$	$\frac{5}{2}$	$\frac{1}{2}$	0000+	0
$\frac{5}{2}$	$\frac{5}{2}$	$\frac{1}{2}$	0010−	$-1/2.3\sqrt{2}$
$\frac{5}{2}$	$\frac{5}{2}$	$\frac{1}{2}$	0020+	$+1/2.3$
$\frac{5}{2}$	$\frac{5}{2}$	$\frac{3}{2}$	0000+	$+\sqrt{3}/4\sqrt{2.5}$
$\frac{5}{2}$	$\frac{5}{2}$	$\frac{3}{2}$	0010−	$+1/4.3\sqrt{5}$
$\frac{5}{2}$	$\frac{5}{2}$	$\frac{3}{2}$	0020+	$+1/4.3\sqrt{2.5}$
$\frac{5}{2}$	$\frac{5}{2}$	$\frac{3}{2}$	0100−	$+1/4.5\sqrt{2}$
$\frac{5}{2}$	$\frac{5}{2}$	$\frac{3}{2}$	0110+	$+1/4\sqrt{3}$
$\frac{5}{2}$	$\frac{5}{2}$	$\frac{3}{2}$	0120−	$-1/4.5\sqrt{2.3}$
$\frac{5}{2}$	$\frac{5}{2}$	$\frac{5}{2}$	0000+	$-1/2.5\sqrt{3}$
$\frac{5}{2}$	$\frac{5}{2}$	$\frac{5}{2}$	0010−	$+1/4.3\sqrt{2}$
$\frac{5}{2}$	$\frac{5}{2}$	$\frac{5}{2}$	0020+	$+7/4.3.5$

Ĩ 3 2

$\frac{5}{2}$	$\frac{5}{2}$	$\frac{5}{2}$	0100−	$-1/2\sqrt{3.5}$
$\frac{5}{2}$	$\frac{5}{2}$	$\frac{5}{2}$	0110+	$+1/4.3\sqrt{2.5}$
$\frac{5}{2}$	$\frac{5}{2}$	$\frac{5}{2}$	0120−	$-1/4.3\sqrt{5}$
$\frac{5}{2}$	$\frac{5}{2}$	$\frac{5}{2}$	1000+	0
$\frac{5}{2}$	$\frac{5}{2}$	$\frac{5}{2}$	1010−	$-1/4\sqrt{2.5}$
$\frac{5}{2}$	$\frac{5}{2}$	$\frac{5}{2}$	1020+	$+1/4.3\sqrt{5}$
$\frac{5}{2}$	$\frac{5}{2}$	$\frac{5}{2}$	1100−	0
$\frac{5}{2}$	$\frac{5}{2}$	$\frac{5}{2}$	1110+	$+1/4.3\sqrt{2}$
$\frac{5}{2}$	$\frac{5}{2}$	$\frac{5}{2}$	1120−	$+1/4.3$
3	1	2	0000+	$+2/3.5$
3	1	3	0000−	$-1/2.3$
3	2	1	0000+	$-1/2\sqrt{2.5}$
3	2	1	0010−	$-1/2.3\sqrt{2.5}$
3	2	2	0000+	$-3/4.5$
3	2	2	0010−	$+1/4.3$
3	2	3	0000−	0
3	2	3	0010+	$+1/3\sqrt{2.5}$
3	3	1	0000+	$-1/2.3$
3	3	2	0000+	$+1/2.3.5$
3	3	3	0000−	$-1/2.3$
Ĩ	1	2	0000+	$-2/3.5$
Ĩ	1	3	0000+	$+1/2.3$
Ĩ	2	1	0000+	$+\sqrt{2}/3\sqrt{5}$
Ĩ	2	2	0000+	$+1/3.5$
Ĩ	2	3	0000+	$-1/3\sqrt{2.5}$
Ĩ	3	1	0000+	$-1/4.3$
Ĩ	3	2	0000+	$+7/4.3.5$

Ĩ 3 3

0	3	3	0000+	$-1/4$
$\frac{1}{2}$	$\frac{5}{2}$	$\frac{5}{2}$	0000+	$-\sqrt{5}/4.3$
$\frac{1}{2}$	$\frac{5}{2}$	$\frac{5}{2}$	1000+	$+1/4.3$
1	2	2	0000+	$+1/2\sqrt{2.5}$
1	3	2	0000+	0
1	3	3	0000+	$+1/4$
$\frac{3}{2}$	$\frac{3}{2}$	$\frac{3}{2}$	0000+	$-1/4\sqrt{5}$
$\frac{3}{2}$	$\frac{5}{2}$	$\frac{3}{2}$	0000+	0
$\frac{3}{2}$	$\frac{5}{2}$	$\frac{3}{2}$	0010−	$+1/\sqrt{2.3.5}$
$\frac{3}{2}$	$\frac{5}{2}$	$\frac{5}{2}$	0000+	$+\sqrt{5}/4.3$
$\frac{3}{2}$	$\frac{5}{2}$	$\frac{5}{2}$	0100−	0
$\frac{3}{2}$	$\frac{5}{2}$	$\frac{5}{2}$	0110+	$+1/4\sqrt{5}$
$\frac{3}{2}$	$\frac{5}{2}$	$\frac{5}{2}$	1000+	$-1/4.3$

K 6j Symbols (cont.)

$\tilde{1}$ 3 3

$\frac{3}{2}$	$\frac{5}{2}$	$\frac{5}{2}$	1100−	0
$\frac{3}{2}$	$\frac{5}{2}$	$\frac{5}{2}$	1110+	$+1/4.3$
2	2	1	0000+	$+1/2\sqrt{2.5}$
2	2	1	0010−	$+1/2.3\sqrt{2.5}$
2	2	2	0000+	$+1/4\sqrt{2.5}$
2	2	2	0100−	$-1/4\sqrt{2.5}$
2	2	2	0110+	$+\sqrt{5}/4.3\sqrt{2}$
2	3	1	0000+	$+1/2.3$
2	3	2	0000+	$-1/2\sqrt{2.5}$
2	3	2	0100−	$+1/2.3\sqrt{2.5}$
2	3	3	0000+	$+1/4.3$
$\frac{5}{2}$	$\frac{3}{2}$	$\frac{3}{2}$	0000+	0
$\frac{5}{2}$	$\frac{3}{2}$	$\frac{3}{2}$	0100−	$-1/4\sqrt{3}$
$\frac{5}{2}$	$\frac{3}{2}$	$\frac{3}{2}$	0110+	$+1/2.3\sqrt{5}$
$\frac{5}{2}$	$\frac{5}{2}$	$\frac{1}{2}$	0000+	$+\sqrt{5}/4.3$
$\frac{5}{2}$	$\frac{5}{2}$	$\frac{1}{2}$	0010−	$-1/4.3$
$\frac{5}{2}$	$\frac{5}{2}$	$\frac{3}{2}$	0000+	$-1/4.3\sqrt{2}$
$\frac{5}{2}$	$\frac{5}{2}$	$\frac{3}{2}$	0010−	$-\sqrt{5}/4.3\sqrt{2}$
$\frac{5}{2}$	$\frac{5}{2}$	$\frac{3}{2}$	0100−	$+1/4\sqrt{2.3.5}$
$\frac{5}{2}$	$\frac{5}{2}$	$\frac{3}{2}$	0110+	$+1/4\sqrt{2.3}$
$\frac{5}{2}$	$\frac{5}{2}$	$\frac{5}{2}$	0000+	$+1/8.3\sqrt{5}$
$\frac{5}{2}$	$\frac{5}{2}$	$\frac{5}{2}$	0100−	$-1/8.3$
$\frac{5}{2}$	$\frac{5}{2}$	$\frac{5}{2}$	0110+	$-\sqrt{5}/8.3$
$\frac{5}{2}$	$\frac{5}{2}$	$\frac{5}{2}$	1000+	$-1/8$
$\frac{5}{2}$	$\frac{5}{2}$	$\frac{5}{2}$	1100−	$-\sqrt{5}/8.3$
$\frac{5}{2}$	$\frac{5}{2}$	$\frac{5}{2}$	1110+	$-1/8.3$
3	2	1	0000+	$+1/2.3$
3	2	2	0000+	$-1/3\sqrt{2.5}$
3	3	1	0000−	$-1/2.3$
3	3	2	0000−	$-1/2.3$
3	3	3	0000+	$+1/4.3$
$\tilde{1}$	2	1	0000+	$+1/2.3$
$\tilde{1}$	2	2	0000+	$-1/3\sqrt{2.5}$
$\tilde{1}$	3	1	0000+	$+1/2.3$
$\tilde{1}$	3	2	0000+	$+1/2.3$
$\tilde{1}$	3	3	0000+	$-1/4.3$

$\tilde{1}$ $\tilde{1}$ 0

0	0	$\tilde{1}$	0000+	$+1/\sqrt{3}$
$\frac{1}{2}$	$\frac{1}{2}$	$\frac{5}{2}$	0000+	$+1/\sqrt{2.3}$
1	1	2	0000+	$+1/3$
1	1	3	0000+	$-1/3$
$\frac{3}{2}$	$\frac{3}{2}$	$\frac{3}{2}$	0000+	$+1/2\sqrt{3}$
$\frac{3}{2}$	$\frac{3}{2}$	$\frac{5}{2}$	0000+	$-1/2\sqrt{3}$
2	2	1	0000+	$+1/\sqrt{3.5}$
2	2	2	0000+	$-1/\sqrt{3.5}$
2	2	3	0000+	$+1/\sqrt{3.5}$
2	2	$\tilde{1}$	0000+	$+1/\sqrt{3.5}$
$\frac{5}{2}$	$\frac{5}{2}$	$\frac{1}{2}$	0000+	$+1/3\sqrt{2}$
$\frac{5}{2}$	$\frac{5}{2}$	$\frac{3}{2}$	0000+	$-1/3\sqrt{2}$
$\frac{5}{2}$	$\frac{5}{2}$	$\frac{5}{2}$	0000+	$+1/3\sqrt{2}$
$\frac{5}{2}$	$\frac{5}{2}$	$\frac{5}{2}$	1000+	0
$\frac{5}{2}$	$\frac{5}{2}$	$\frac{5}{2}$	1100+	$+1/3\sqrt{2}$
3	3	1	0000+	$-1/2\sqrt{3}$
3	3	2	0000+	$+1/2\sqrt{3}$
3	3	3	0000+	$-1/2\sqrt{3}$
$\tilde{1}$	$\tilde{1}$	0	0000+	$+1/3$

$\tilde{1}$ $\tilde{1}$ 2

1	1	2	0000+	$-2/3.5$
1	1	3	0000+	$-1/2.3$
$\frac{3}{2}$	$\frac{1}{2}$	$\frac{5}{2}$	0000+	$+1/\sqrt{2.3.5}$
$\frac{3}{2}$	$\frac{3}{2}$	$\frac{3}{2}$	0000+	$+\sqrt{2}/5\sqrt{3}$
$\frac{3}{2}$	$\frac{3}{2}$	$\frac{5}{2}$	0000+	$+1/2.5\sqrt{2.3}$
2	0	$\tilde{1}$	0000+	$+1/\sqrt{3.5}$
2	1	2	0000+	$+\sqrt{2}/5\sqrt{3}$
2	1	3	0000+	$-1/2\sqrt{3.5}$
2	2	1	0000+	$+1/\sqrt{2.3.5}$
2	2	1	0010+	$-1/5\sqrt{2.3}$
2	2	2	0000+	$+1/2\sqrt{2.3.5}$
2	2	2	0010+	$+\sqrt{3}/2.5\sqrt{2}$
2	2	3	0000+	$-1/2\sqrt{2.3.5}$
2	2	3	0010+	$+\sqrt{3}/2.5\sqrt{2}$
2	2	$\tilde{1}$	0000+	$+1/2\sqrt{2.3.5}$
2	2	$\tilde{1}$	0010+	$+\sqrt{3}/2.5\sqrt{2}$
$\frac{5}{2}$	$\frac{1}{2}$	$\frac{5}{2}$	0000+	$+1/\sqrt{2.3.5}$
$\frac{5}{2}$	$\frac{1}{2}$	$\frac{5}{2}$	0100+	0
$\frac{5}{2}$	$\frac{3}{2}$	$\frac{3}{2}$	0000+	$+\sqrt{3}/2.5$
$\frac{5}{2}$	$\frac{3}{2}$	$\frac{3}{2}$	0010−	$-1/2.5\sqrt{2}$
$\frac{5}{2}$	$\frac{3}{2}$	$\frac{5}{2}$	0000+	$+1/5\sqrt{3}$
$\frac{5}{2}$	$\frac{3}{2}$	$\frac{5}{2}$	0010−	$+1/2.5\sqrt{2}$
$\frac{5}{2}$	$\frac{3}{2}$	$\frac{5}{2}$	0100+	0
$\frac{5}{2}$	$\frac{3}{2}$	$\frac{5}{2}$	0110−	$-1/2\sqrt{2.5}$
$\frac{5}{2}$	$\frac{5}{2}$	$\frac{1}{2}$	0000+	$+1/3\sqrt{2}$
$\frac{5}{2}$	$\frac{5}{2}$	$\frac{1}{2}$	0010−	0
$\frac{5}{2}$	$\frac{5}{2}$	$\frac{1}{2}$	0020+	0
$\frac{5}{2}$	$\frac{5}{2}$	$\frac{3}{2}$	0000+	$-1/2.3.5\sqrt{2}$
$\frac{5}{2}$	$\frac{5}{2}$	$\frac{3}{2}$	0010−	0
$\frac{5}{2}$	$\frac{5}{2}$	$\frac{3}{2}$	0020+	$-\sqrt{3}/2.5\sqrt{2}$
$\frac{5}{2}$	$\frac{5}{2}$	$\frac{5}{2}$	0000+	$+1/2.3.5\sqrt{2}$
$\frac{5}{2}$	$\frac{5}{2}$	$\frac{5}{2}$	0010−	0
$\frac{5}{2}$	$\frac{5}{2}$	$\frac{5}{2}$	0020+	$-1/5\sqrt{2.3}$
$\frac{5}{2}$	$\frac{5}{2}$	$\frac{5}{2}$	1000+	0
$\frac{5}{2}$	$\frac{5}{2}$	$\frac{5}{2}$	1010−	$-1/2\sqrt{3.5}$
$\frac{5}{2}$	$\frac{5}{2}$	$\frac{5}{2}$	1020+	$-1/2\sqrt{2.3.5}$
$\frac{5}{2}$	$\frac{5}{2}$	$\frac{5}{2}$	1100+	$-1/2.3\sqrt{2}$
$\frac{5}{2}$	$\frac{5}{2}$	$\frac{5}{2}$	1110−	0
$\frac{5}{2}$	$\frac{5}{2}$	$\frac{5}{2}$	1120+	0
3	1	2	0000+	$-1/5\sqrt{2.3}$
3	1	3	0000+	$+1/2\sqrt{2.3}$
3	2	1	0000+	$+1/2\sqrt{3.5}$
3	2	1	0010−	$+1/2\sqrt{3.5}$
3	2	2	0000+	$-\sqrt{2}/5\sqrt{3}$
3	2	2	0010−	0
3	2	3	0000+	$-1/4\sqrt{3.5}$
3	2	3	0010−	$+\sqrt{3}/4\sqrt{5}$
3	3	1	0000+	$-1/2\sqrt{2.3}$
3	3	2	0000+	$-\sqrt{3}/2.5\sqrt{2}$
3	3	3	0000+	0
$\tilde{1}$	1	2	0000+	$+1/5$
$\tilde{1}$	2	2	0000+	$-1/2.5$
$\tilde{1}$	2	$\tilde{1}$	0000+	$-1/2\sqrt{5}$
$\tilde{1}$	3	2	0000+	$+1/5$
$\tilde{1}$	$\tilde{1}$	0	0000+	$+1/3$
$\tilde{1}$	$\tilde{1}$	2	0000+	$+1/2.3.5$

$\tilde{1}$ $\tilde{1}$ $\tilde{1}$

$\frac{3}{2}$	$\frac{3}{2}$	$\frac{3}{2}$	0000+	$-1/3\sqrt{2.5}$
2	2	1	0000+	$+1/3\sqrt{5}$
2	2	2	0000+	$+1/2.3\sqrt{5}$
$\frac{5}{2}$	$\frac{3}{2}$	$\frac{3}{2}$	0000+	$-1/2\sqrt{2.5}$
$\frac{5}{2}$	$\frac{5}{2}$	$\frac{1}{2}$	0000+	0
$\frac{5}{2}$	$\frac{5}{2}$	$\frac{1}{2}$	0010+	$+1/3\sqrt{2}$
$\frac{5}{2}$	$\frac{5}{2}$	$\frac{3}{2}$	0000+	$+1/2\sqrt{2.5}$
$\frac{5}{2}$	$\frac{5}{2}$	$\frac{3}{2}$	0010+	$+1/2.3\sqrt{2}$
$\frac{5}{2}$	$\frac{5}{2}$	$\frac{5}{2}$	0000+	$+1/3\sqrt{2.5}$

K 6j Symbols (cont.)

$\tilde{1}$	$\tilde{1}$	$\tilde{1}$		
$\frac{5}{2}$	$\frac{5}{2}$	$\frac{5}{2}$	1000+	$-1/2.3\sqrt{2}$
$\frac{5}{2}$	$\frac{5}{2}$	$\frac{5}{2}$	1100+	0
$\frac{5}{2}$	$\frac{5}{2}$	$\frac{5}{2}$	1110+	$+1/2.3\sqrt{2}$
3	2	1	0000+	$+1/3\sqrt{2}$
3	2	2	0000+	$+1/3\sqrt{5}$
3	3	1	0000+	$+1/2.3\sqrt{2}$
3	3	2	0000+	$-1/2.3\sqrt{2}$
3	3	3	0000+	$-1/3\sqrt{2}$
$\tilde{1}$	2	2	0000+	$-1/2\sqrt{5}$
$\tilde{1}$	$\tilde{1}$	0	0000+	$-1/3$
$\tilde{1}$	$\tilde{1}$	2	0000+	$+1/2.3$
$\tilde{1}$	$\tilde{1}$	$\tilde{1}$	0000+	$+1/2.3$

$\tilde{\frac{1}{2}}$	2	$\frac{3}{2}$		
0	$\frac{3}{2}$	2	0000+	$-1/2\sqrt{5}$
$\frac{1}{2}$	2	$\frac{3}{2}$	0000+	$-1/5$
1	$\frac{3}{2}$	2	0000+	$+1/5\sqrt{2}$
$\frac{3}{2}$	2	$\frac{3}{2}$	0000+	$-1/2.5$
2	$\frac{3}{2}$	2	0000+	$+1/4\sqrt{5}$
2	$\frac{3}{2}$	2	0100+	$+3/4.5$
$\frac{5}{2}$	2	$\frac{3}{2}$	0000+	$+1/2.5$
$\frac{5}{2}$	2	$\frac{3}{2}$	0010−	$+\sqrt{3}/2.5\sqrt{2}$
$\frac{5}{2}$	2	$\frac{3}{2}$	0100−	$+\sqrt{3}/2.5\sqrt{2}$
$\frac{5}{2}$	2	$\frac{3}{2}$	0110+	$-3/4.5$
3	$\frac{3}{2}$	2	0000+	$-1/5\sqrt{2}$
3	$\frac{3}{2}$	2	0100−	0
$\tilde{1}$	$\frac{3}{2}$	2	0000+	$+3/2.5\sqrt{2}$
$\tilde{\frac{1}{2}}$	2	$\frac{3}{2}$	0000+	$+1/4.5$

$\tilde{\frac{1}{2}}$	$\frac{5}{2}$	1		
0	1	$\frac{5}{2}$	0000+	$-1/3\sqrt{2}$
$\frac{1}{2}$	$\frac{3}{2}$	2	0000+	$-1/2\sqrt{2.3}$
1	1	$\frac{5}{2}$	0000+	$+1/3\sqrt{2}$
1	1	$\frac{5}{2}$	0100+	0
1	2	$\frac{3}{2}$	0000+	$+1/\sqrt{2.3.5}$
1	2	$\frac{5}{2}$	0000+	0
1	2	$\frac{5}{2}$	0100+	$-1/\sqrt{2.3.5}$
$\frac{3}{2}$	$\frac{3}{2}$	2	0000+	$-1/2\sqrt{3.5}$
$\frac{3}{2}$	$\frac{3}{2}$	2	0100−	$+1/2\sqrt{2.5}$
$\frac{3}{2}$	$\frac{5}{2}$	1	0000+	$-1/2.3$
$\frac{3}{2}$	$\frac{5}{2}$	2	0000+	$-1/2\sqrt{3.5}$
$\frac{3}{2}$	$\frac{5}{2}$	2	0100−	$-1/3\sqrt{2.5}$
2	1	$\frac{5}{2}$	0000+	$-1/2.3\sqrt{2}$
2	1	$\frac{5}{2}$	0100−	0
2	1	$\frac{5}{2}$	0200+	$+1/2\sqrt{2.3}$
2	2	$\frac{3}{2}$	0000+	$-1/\sqrt{2.3.5}$
2	2	$\frac{3}{2}$	0100−	0
2	2	$\frac{5}{2}$	0000+	$-1/2\sqrt{2.3.5}$
2	2	$\frac{5}{2}$	0100−	$+1/3\sqrt{5}$
2	2	$\frac{5}{2}$	0200+	$-1/2.3\sqrt{2.5}$
$\frac{5}{2}$	$\frac{3}{2}$	2	0000+	$+1/4\sqrt{3.5}$
$\frac{5}{2}$	$\frac{3}{2}$	2	0100−	$+1/2\sqrt{2.5}$
$\frac{5}{2}$	$\frac{3}{2}$	2	0200+	$-1/4\sqrt{5}$
$\frac{5}{2}$	$\frac{5}{2}$	1	0000+	$+1/2.3$
$\frac{5}{2}$	$\frac{5}{2}$	1	0010+	0
$\frac{5}{2}$	$\frac{5}{2}$	1	0100+	0
$\frac{5}{2}$	$\frac{5}{2}$	1	0110+	$-1/2.3$
$\frac{5}{2}$	$\frac{5}{2}$	2	0000+	$+1/4\sqrt{3}$
$\frac{5}{2}$	$\frac{5}{2}$	2	0010+	$+1/4\sqrt{3.5}$
$\frac{5}{2}$	$\frac{5}{2}$	2	0100−	0
$\frac{5}{2}$	$\frac{5}{2}$	2	0110−	$-1/3\sqrt{2.5}$
$\frac{5}{2}$	$\frac{5}{2}$	2	0200+	$+1/4.3$
$\frac{5}{2}$	$\frac{5}{2}$	2	0210+	$-1/4\sqrt{5}$
3	2	$\frac{3}{2}$	0000+	$-1/4\sqrt{3}$
3	2	$\frac{3}{2}$	0100−	$+1/4\sqrt{5}$
3	2	$\frac{5}{2}$	0000+	$+1/2.3\sqrt{5}$
3	2	$\frac{5}{2}$	0100−	$+1/2.3$
$\tilde{1}$	2	$\frac{3}{2}$	0000+	$-1/\sqrt{2.3.5}$
$\tilde{1}$	2	$\frac{5}{2}$	0000+	$+1/2.3\sqrt{5}$
$\tilde{1}$	2	$\frac{5}{2}$	0100+	$-1/2.3$
$\tilde{\frac{1}{2}}$	$\frac{5}{2}$	1	0000+	$-1/2.3$

$\tilde{\frac{1}{2}}$	$\frac{5}{2}$	2		
0	2	$\frac{5}{2}$	0000+	$+1/\sqrt{2.3.5}$
$\frac{1}{2}$	$\frac{3}{2}$	2	0000+	$-\sqrt{3}/2.5\sqrt{2}$
$\frac{1}{2}$	$\frac{5}{2}$	2	0000+	$+2/3.5$
1	1	$\frac{5}{2}$	0000+	0
1	1	$\frac{5}{2}$	0100+	$-1/\sqrt{2.3.5}$
1	2	$\frac{3}{2}$	0000+	$-1/5\sqrt{2}$
1	2	$\frac{5}{2}$	0000+	$+1/3\sqrt{2.5}$
1	2	$\frac{5}{2}$	0100+	$-\sqrt{2}/3.5$
$\frac{3}{2}$	$\frac{3}{2}$	2	0000+	$-\sqrt{3}/2.5$
$\frac{3}{2}$	$\frac{3}{2}$	2	0100−	$-1/2.5\sqrt{2}$
$\frac{3}{2}$	$\frac{5}{2}$	1	0000+	$-1/2\sqrt{3.5}$
$\frac{3}{2}$	$\frac{5}{2}$	1	0010−	$-1/3\sqrt{2.5}$
$\frac{3}{2}$	$\frac{5}{2}$	2	0000+	$+1/2.3.5$
$\frac{3}{2}$	$\frac{5}{2}$	2	0010−	$+1/5\sqrt{2.3}$
$\frac{3}{2}$	$\frac{5}{2}$	2	0100−	$+1/5\sqrt{2.3}$
$\frac{3}{2}$	$\frac{5}{2}$	2	0110+	$-2/3.5$
2	1	$\frac{5}{2}$	0000+	$-1/2\sqrt{2.3.5}$
2	1	$\frac{5}{2}$	0100−	$-1/3\sqrt{5}$
2	1	$\frac{5}{2}$	0200+	$-1/2.3\sqrt{2.5}$
2	2	$\frac{3}{2}$	0000+	0
2	2	$\frac{3}{2}$	0010+	$-1/5\sqrt{3}$
2	2	$\frac{3}{2}$	0100−	$+1/2\sqrt{2.5}$
2	2	$\frac{3}{2}$	0110−	$-1/2.5\sqrt{2}$
2	2	$\frac{5}{2}$	0000+	$-1/4\sqrt{3.5}$
2	2	$\frac{5}{2}$	0010+	$+\sqrt{3}/4.5$
2	2	$\frac{5}{2}$	0100−	0
2	2	$\frac{5}{2}$	0110−	0
2	2	$\frac{5}{2}$	0200+	$+1/4\sqrt{5}$
2	2	$\frac{5}{2}$	0210+	$+7/4.3.5$
$\frac{5}{2}$	$\frac{3}{2}$	2	0000+	$-1/4.5$
$\frac{5}{2}$	$\frac{3}{2}$	2	0010−	$-1/2.5\sqrt{2.3}$
$\frac{5}{2}$	$\frac{3}{2}$	2	0100−	$+1/2.5\sqrt{2.3}$
$\frac{5}{2}$	$\frac{3}{2}$	2	0110+	$+1/2.5$
$\frac{5}{2}$	$\frac{3}{2}$	2	0200+	$-1/4\sqrt{3}$
$\frac{5}{2}$	$\frac{3}{2}$	2	0210−	$+1/2.5\sqrt{2}$
$\frac{5}{2}$	$\frac{5}{2}$	1	0000+	$+1/4\sqrt{3}$
$\frac{5}{2}$	$\frac{5}{2}$	1	0010−	0
$\frac{5}{2}$	$\frac{5}{2}$	1	0020+	$+1/4.3$
$\frac{5}{2}$	$\frac{5}{2}$	1	0100+	$+1/4\sqrt{3.5}$
$\frac{5}{2}$	$\frac{5}{2}$	1	0110−	$-1/3\sqrt{2.5}$
$\frac{5}{2}$	$\frac{5}{2}$	1	0120+	$-1/4\sqrt{5}$
$\frac{5}{2}$	$\frac{5}{2}$	2	0000+	$-1/2.5$
$\frac{5}{2}$	$\frac{5}{2}$	2	0010−	$-\sqrt{3}/2.5\sqrt{2}$
$\frac{5}{2}$	$\frac{5}{2}$	2	0020+	0
$\frac{5}{2}$	$\frac{5}{2}$	2	0100−	$-\sqrt{3}/2.5\sqrt{2}$
$\frac{5}{2}$	$\frac{5}{2}$	2	0110+	$+1/3.5$
$\frac{5}{2}$	$\frac{5}{2}$	2	0120−	$+1/2.3.5\sqrt{2}$
$\frac{5}{2}$	$\frac{5}{2}$	2	0200+	0
$\frac{5}{2}$	$\frac{5}{2}$	2	0210−	$+1/2.3.5\sqrt{2}$
$\frac{5}{2}$	$\frac{5}{2}$	2	0220+	$-1/2.5$
3	1	$\frac{5}{2}$	0000+	$+1/2.3\sqrt{5}$
3	1	$\frac{5}{2}$	0100−	$-1/2.3$
3	2	$\frac{3}{2}$	0000+	$-\sqrt{3}/4\sqrt{2.5}$

K 6j Symbols (cont.)

$\tilde{\frac{1}{2}}$	**$\frac{5}{2}$**	**2**		
3	2	$\frac{3}{2}$	0010−	$-1/4\sqrt{2.3.5}$
3	2	$\frac{3}{2}$	0100−	$-1/4.5\sqrt{2}$
3	2	$\frac{3}{2}$	0110+	$+1/4\sqrt{2}$
3	2	$\frac{5}{2}$	0000+	$-1/5\sqrt{2}$
3	2	$\frac{5}{2}$	0010−	0
3	2	$\frac{5}{2}$	0100−	0
3	2	$\frac{5}{2}$	0110+	$-1/3\sqrt{2.5}$
$\tilde{1}$	1	$\frac{5}{2}$	0000+	$+1/2.3\sqrt{5}$
$\tilde{1}$	1	$\frac{5}{2}$	0100+	$-1/2.3$
$\tilde{1}$	2	$\frac{3}{2}$	0000+	$+1/2.5\sqrt{3}$
$\tilde{1}$	2	$\frac{5}{2}$	0000+	$-1/5\sqrt{2}$
$\tilde{1}$	2	$\frac{5}{2}$	0100+	$-1/3\sqrt{2.5}$
$\tilde{\frac{1}{2}}$	$\frac{3}{2}$	2	0000+	$-1/5$
$\tilde{\frac{1}{2}}$	$\frac{5}{2}$	1	0000+	$+1/2.3$
$\tilde{\frac{1}{2}}$	$\frac{5}{2}$	2	0000+	$-1/2.3.5$

$\tilde{\frac{1}{2}}$	**3**	**$\frac{1}{2}$**		
0	$\frac{1}{2}$	3	0000+	$-1/2\sqrt{2}$
$\frac{1}{2}$	1	$\frac{5}{2}$	0000+	$+1/2\sqrt{3}$
1	$\frac{1}{2}$	3	0000+	$+1/2\sqrt{2}$
1	$\frac{3}{2}$	2	0000+	$+1/4$
$\frac{3}{2}$	1	$\frac{5}{2}$	0000+	$+1/2\sqrt{3}$
$\frac{3}{2}$	1	$\frac{5}{2}$	0100−	0
$\frac{3}{2}$	2	$\frac{3}{2}$	0000+	$-1/2\sqrt{5}$
$\frac{3}{2}$	2	$\frac{5}{2}$	0000+	0
$\frac{3}{2}$	2	$\frac{5}{2}$	0100−	$-1/2\sqrt{5}$
2	$\frac{3}{2}$	2	0000+	$+1/4\sqrt{2.5}$
2	$\frac{3}{2}$	2	0100−	$+3/4\sqrt{2.5}$
2	$\frac{5}{2}$	1	0000+	$+1/2\sqrt{2.3}$
2	$\frac{5}{2}$	2	0000+	$+\sqrt{3}/4\sqrt{5}$
2	$\frac{5}{2}$	2	0100−	$-1/4\sqrt{3.5}$
2	$\frac{5}{2}$	3	0000+	$-1/2\sqrt{2.3}$
$\frac{5}{2}$	2	$\frac{3}{2}$	0000+	$-1/4\sqrt{2}$
$\frac{5}{2}$	2	$\frac{3}{2}$	0100−	$+\sqrt{3}/4\sqrt{2.5}$
$\frac{5}{2}$	2	$\frac{5}{2}$	0000+	$+1/2\sqrt{2.3.5}$
$\frac{5}{2}$	2	$\frac{5}{2}$	0100−	$-1/2\sqrt{2.3}$
$\frac{5}{2}$	3	$\frac{1}{2}$	0000+	$-1/4$
$\frac{5}{2}$	3	$\frac{5}{2}$	0000+	$+\sqrt{5}/4\sqrt{2.3}$
$\frac{5}{2}$	3	$\frac{5}{2}$	0100−	$+1/4\sqrt{2.3}$
3	$\frac{5}{2}$	1	0000+	$+1/2\sqrt{2.3}$
3	$\frac{5}{2}$	2	0000+	$-1/2\sqrt{2.3}$
3	$\frac{5}{2}$	3	0000−	$+1/2\sqrt{2.3}$
$\tilde{1}$	$\frac{5}{2}$	1	0000+	$-1/2\sqrt{2.3}$
$\tilde{1}$	$\frac{5}{2}$	2	0000+	$+1/2\sqrt{2.3}$
$\tilde{1}$	$\frac{5}{2}$	3	0000+	$-1/2\sqrt{2.3}$
$\tilde{\frac{1}{2}}$	3	$\frac{1}{2}$	0000+	$+1/4$

$\tilde{\frac{1}{2}}$	**3**	**$\frac{5}{2}$**		
0	$\frac{5}{2}$	3	0000+	$-1/2\sqrt{2.3}$
$\frac{1}{2}$	2	$\frac{5}{2}$	0000+	$+1/2.3$
$\frac{1}{2}$	3	$\frac{5}{2}$	0000+	$-1/2.3$
1	$\frac{3}{2}$	2	0000+	$+1/4\sqrt{5}$
1	$\frac{5}{2}$	2	0000+	$+1/2.3$
1	$\frac{5}{2}$	2	0010+	$+1/2.3\sqrt{5}$
1	$\frac{5}{2}$	3	0000+	$+1/4.3$
1	$\frac{5}{2}$	3	0010+	$-\sqrt{5}/4.3$
$\frac{3}{2}$	1	$\frac{5}{2}$	0000+	0
$\frac{3}{2}$	1	$\frac{5}{2}$	0100−	$-1/2.3$
$\frac{3}{2}$	2	$\frac{3}{2}$	0000+	0
$\frac{3}{2}$	2	$\frac{3}{2}$	0010−	$+1/2\sqrt{2.5}$
$\frac{3}{2}$	2	$\frac{5}{2}$	0000+	$+1/2.3$
$\frac{3}{2}$	2	$\frac{5}{2}$	0010−	0
$\frac{3}{2}$	2	$\frac{5}{2}$	0100−	0
$\frac{3}{2}$	2	$\frac{5}{2}$	0110+	$-1/3\sqrt{2.5}$
$\frac{3}{2}$	3	$\frac{5}{2}$	0000+	$-1/8.3$
$\frac{3}{2}$	3	$\frac{5}{2}$	0010−	$-\sqrt{5}/8\sqrt{3}$
$\frac{3}{2}$	3	$\frac{5}{2}$	0100−	$-\sqrt{5}/8\sqrt{3}$
$\frac{3}{2}$	3	$\frac{5}{2}$	0110+	$+1/8.3$
2	$\frac{1}{2}$	3	0000+	$-1/2\sqrt{2.3}$
2	$\frac{3}{2}$	2	0000+	$+\sqrt{3}/4\sqrt{2.5}$
2	$\frac{3}{2}$	2	0010−	$+1/4\sqrt{5}$
2	$\frac{3}{2}$	2	0100−	$+1/4\sqrt{2.3.5}$
2	$\frac{3}{2}$	2	0110+	$-1/4\sqrt{5}$
2	$\frac{5}{2}$	1	0000+	$+1/4\sqrt{3}$
2	$\frac{5}{2}$	1	0010−	$+1/2.3\sqrt{2}$
2	$\frac{5}{2}$	1	0020+	$+1/4.3$
2	$\frac{5}{2}$	2	0000+	$-1/4\sqrt{2.3.5}$
2	$\frac{5}{2}$	2	0010−	$-1/4\sqrt{5}$
2	$\frac{5}{2}$	2	0020+	$+1/4\sqrt{2.5}$
2	$\frac{5}{2}$	2	0100−	$+\sqrt{3}/4\sqrt{2.5}$
2	$\frac{5}{2}$	2	0110+	$-1/4\sqrt{5}$
2	$\frac{5}{2}$	2	0120−	$-1/4.3\sqrt{2.5}$
2	$\frac{5}{2}$	3	0000+	0
2	$\frac{5}{2}$	3	0010−	0

$\tilde{\frac{1}{2}}$	**3**	**$\frac{5}{2}$**		
2	$\frac{5}{2}$	3	0020+	$+1/2.3$
$\frac{5}{2}$	1	$\frac{5}{2}$	0000+	0
$\frac{5}{2}$	1	$\frac{5}{2}$	0010+	$-1/2.3$
$\frac{5}{2}$	1	$\frac{5}{2}$	0100−	$-1/2.3$
$\frac{5}{2}$	1	$\frac{5}{2}$	0110−	0
$\frac{5}{2}$	2	$\frac{3}{2}$	0000+	$+1/8$
$\frac{5}{2}$	2	$\frac{3}{2}$	0010−	$-1/4\sqrt{2.3}$
$\frac{5}{2}$	2	$\frac{3}{2}$	0020+	$+1/8\sqrt{3}$
$\frac{5}{2}$	2	$\frac{3}{2}$	0100−	$+\sqrt{5}/8\sqrt{3}$
$\frac{5}{2}$	2	$\frac{3}{2}$	0110+	$+1/4\sqrt{2.5}$
$\frac{5}{2}$	2	$\frac{3}{2}$	0120−	$-1/8\sqrt{5}$
$\frac{5}{2}$	2	$\frac{5}{2}$	0000+	0
$\frac{5}{2}$	2	$\frac{5}{2}$	0010−	$+1/2.3\sqrt{2.5}$
$\frac{5}{2}$	2	$\frac{5}{2}$	0020+	$+1/3\sqrt{5}$
$\frac{5}{2}$	2	$\frac{5}{2}$	0100−	0
$\frac{5}{2}$	2	$\frac{5}{2}$	0110+	$+1/2.3\sqrt{2}$
$\frac{5}{2}$	2	$\frac{5}{2}$	0120−	0
$\frac{5}{2}$	3	$\frac{1}{2}$	0000+	$+\sqrt{5}/4\sqrt{2.3}$
$\frac{5}{2}$	3	$\frac{1}{2}$	0010−	$+1/4\sqrt{2.3}$
$\frac{5}{2}$	3	$\frac{5}{2}$	0000+	$+1/8.3$
$\frac{5}{2}$	3	$\frac{5}{2}$	0010−	$+\sqrt{5}/8.3$
$\frac{5}{2}$	3	$\frac{5}{2}$	0100−	$+\sqrt{5}/8.3$
$\frac{5}{2}$	3	$\frac{5}{2}$	0110+	$-1/8$
3	$\frac{1}{2}$	3	0000−	$-1/2\sqrt{2.3}$
3	$\frac{3}{2}$	2	0000+	$-1/4\sqrt{3}$
3	$\frac{3}{2}$	2	0010−	$+1/4\sqrt{5}$
3	$\frac{5}{2}$	1	0000+	$+\sqrt{5}/4.3$
3	$\frac{5}{2}$	1	0010−	$-1/4.3$
3	$\frac{5}{2}$	2	0000+	$+1/4\sqrt{5}$
3	$\frac{5}{2}$	2	0010−	$+1/4.3$
3	$\frac{5}{2}$	3	0000−	0
3	$\frac{5}{2}$	3	0010+	$+1/2.3$
$\tilde{1}$	$\frac{1}{2}$	3	0000+	$-1/2\sqrt{2.3}$
$\tilde{1}$	$\frac{3}{2}$	2	0000+	$-1/\sqrt{2.3.5}$
$\tilde{1}$	$\frac{5}{2}$	1	0000+	$-\sqrt{5}/4.3$
$\tilde{1}$	$\frac{5}{2}$	1	0010+	$-1/4.3$
$\tilde{1}$	$\frac{5}{2}$	2	0000+	$-1/4\sqrt{5}$
$\tilde{1}$	$\frac{5}{2}$	2	0010+	$+1/4.3$
$\tilde{1}$	$\frac{5}{2}$	3	0000+	0
$\tilde{1}$	$\frac{5}{2}$	3	0010+	$+1/2.3$
$\tilde{\frac{1}{2}}$	1	$\frac{5}{2}$	0000+	$+1/2.3$

K 6j Symbols (cont.)

$\tilde{\frac{1}{2}}$	3	$\frac{5}{2}$		
$\tilde{\frac{1}{2}}$	2	$\frac{5}{2}$	0000+	$+1/2.3$
$\tilde{\frac{1}{2}}$	3	$\frac{1}{2}$	0000+	$-1/4$
$\tilde{\frac{1}{2}}$	3	$\frac{5}{2}$	0000+	$-1/4.3$
$\tilde{\frac{1}{2}}$	**$\tilde{1}$**	**$\frac{3}{2}$**		
0	$\frac{3}{2}$	$\tilde{1}$	0000+	$+1/2\sqrt{3}$
$\frac{1}{2}$	1	$\frac{5}{2}$	0000+	$+1/2\sqrt{2.3}$
$\frac{1}{2}$	2	$\frac{5}{2}$	0000+	$-1/2\sqrt{2.3}$
1	$\frac{1}{2}$	3	0000+	$+1/4$
1	$\frac{3}{2}$	2	0000+	$+1/2\sqrt{5}$
1	$\frac{5}{2}$	2	0000+	$+1/3\sqrt{5}$
1	$\frac{5}{2}$	3	0000+	$-\sqrt{5}/4.3$
$\frac{3}{2}$	1	$\frac{5}{2}$	0000+	$-1/2\sqrt{2.3}$
$\frac{3}{2}$	2	$\frac{3}{2}$	0000+	$+1/2\sqrt{5}$
$\frac{3}{2}$	2	$\frac{5}{2}$	0000+	$+1/2\sqrt{2.3.5}$
$\frac{3}{2}$	3	$\frac{5}{2}$	0000+	$-1/2\sqrt{2.3}$
$\frac{3}{2}$	$\tilde{1}$	$\frac{3}{2}$	0000+	$+1/2.3$
2	$\frac{1}{2}$	3	0000+	$+1/4$
2	$\frac{3}{2}$	2	0000+	$-1/2\sqrt{2.5}$
2	$\frac{3}{2}$	$\tilde{1}$	0000+	$-1/2\sqrt{2.3}$
2	$\frac{5}{2}$	1	0000+	0
2	$\frac{5}{2}$	1	0010−	$+1/3\sqrt{2}$
2	$\frac{5}{2}$	2	0000+	$-1/\sqrt{2.3.5}$
2	$\frac{5}{2}$	2	0010−	$+1/2.3\sqrt{5}$
2	$\frac{5}{2}$	3	0000+	$-1/4\sqrt{3}$
2	$\frac{5}{2}$	3	0010−	$-1/2.3\sqrt{2}$
$\frac{5}{2}$	1	$\frac{5}{2}$	0000+	$-1/4.3$
$\frac{5}{2}$	1	$\frac{5}{2}$	0100+	$-\sqrt{5}/4.3$
$\frac{5}{2}$	2	$\frac{3}{2}$	0000+	$+1/2\sqrt{2.5}$
$\frac{5}{2}$	2	$\frac{3}{2}$	0010−	$+1/4\sqrt{3.5}$
$\frac{5}{2}$	2	$\frac{5}{2}$	0000+	$+1/4\sqrt{3.5}$
$\frac{5}{2}$	2	$\frac{5}{2}$	0010−	$+1/2\sqrt{2.5}$
$\frac{5}{2}$	2	$\frac{5}{2}$	0100+	$+1/4\sqrt{3}$
$\frac{5}{2}$	2	$\frac{5}{2}$	0110−	$-1/2.3\sqrt{2}$
$\frac{5}{2}$	3	$\frac{1}{2}$	0000+	$-1/4\sqrt{2}$
$\frac{5}{2}$	3	$\frac{1}{2}$	0010−	$-\sqrt{5}/4\sqrt{2.3}$
$\frac{5}{2}$	3	$\frac{5}{2}$	0000+	$+\sqrt{5}/8\sqrt{3}$
$\frac{5}{2}$	3	$\frac{5}{2}$	0010−	$-1/8$
$\frac{5}{2}$	3	$\frac{5}{2}$	0100+	$+1/8\sqrt{3}$
$\frac{5}{2}$	3	$\frac{5}{2}$	0110−	$+\sqrt{5}/8.3$
$\frac{5}{2}$	$\tilde{1}$	$\frac{3}{2}$	0000+	$-1/4$
3	$\frac{3}{2}$	2	0000+	$+1/2\sqrt{5}$
3	$\frac{5}{2}$	1	0000+	$+1/4\sqrt{3}$
3	$\frac{5}{2}$	1	0010−	$+\sqrt{5}/4.3$
3	$\frac{5}{2}$	2	0000+	$-1/4\sqrt{3}$
3	$\frac{5}{2}$	2	0010−	$-1/4.3\sqrt{5}$
3	$\frac{5}{2}$	3	0000+	$-1/4\sqrt{3}$
3	$\frac{5}{2}$	3	0010−	$+\sqrt{5}/4.3$
$\tilde{1}$	$\frac{3}{2}$	2	0000+	$-1/2\sqrt{2.3.5}$
$\tilde{1}$	$\frac{3}{2}$	$\tilde{1}$	0000+	$-\sqrt{5}/2.3\sqrt{2}$
$\tilde{1}$	$\frac{5}{2}$	2	0000+	$+1/2\sqrt{5}$
$\tilde{\frac{1}{2}}$	2	$\frac{3}{2}$	0000+	$+1/4$
$\tilde{\frac{1}{2}}$	$\tilde{1}$	$\frac{3}{2}$	0000+	$-1/4.3$
$\tilde{\frac{1}{2}}$	**$\tilde{\frac{1}{2}}$**	**0**		
0	0	$\tilde{\frac{1}{2}}$	0000+	$-1/\sqrt{2}$
$\frac{1}{2}$	$\frac{1}{2}$	3	0000+	$-1/2$
1	1	$\frac{5}{2}$	0000+	$-1/\sqrt{2.3}$
$\frac{3}{2}$	$\frac{3}{2}$	2	0000+	$-1/2\sqrt{2}$
$\frac{3}{2}$	$\frac{3}{2}$	$\tilde{1}$	0000+	$+1/2\sqrt{2}$
2	2	$\frac{3}{2}$	0000+	$-1/\sqrt{2.5}$
2	2	$\frac{5}{2}$	0000+	$+1/\sqrt{2.5}$
$\frac{5}{2}$	$\frac{5}{2}$	1	0000+	$-1/2\sqrt{3}$
$\frac{5}{2}$	$\frac{5}{2}$	2	0000+	$+1/2\sqrt{3}$
$\frac{5}{2}$	$\frac{5}{2}$	3	0000+	$-1/2\sqrt{3}$
3	3	$\frac{1}{2}$	0000+	$-1/2\sqrt{2}$
3	3	$\frac{5}{2}$	0000+	$-1/2\sqrt{2}$
$\tilde{1}$	$\tilde{1}$	$\frac{3}{2}$	0000+	$+1/\sqrt{2.3}$
$\tilde{1}$	$\tilde{1}$	$\tilde{\frac{1}{2}}$	0000+	$+1/\sqrt{2.3}$
$\tilde{\frac{1}{2}}$	$\tilde{\frac{1}{2}}$	0	0000+	$-1/2$
$\tilde{\frac{1}{2}}$	**$\tilde{\frac{1}{2}}$**	**$\tilde{1}$**		
$\frac{3}{2}$	$\frac{3}{2}$	2	0000+	$+1/2\sqrt{2.5}$
$\frac{3}{2}$	$\frac{3}{2}$	$\tilde{1}$	0000+	$+\sqrt{5}/2.3\sqrt{2}$
2	1	$\frac{5}{2}$	0000+	$+1/3\sqrt{2}$
2	2	$\frac{3}{2}$	0000+	$-1/2\sqrt{5}$
2	2	$\frac{5}{2}$	0000+	$-1/3\sqrt{5}$
$\frac{5}{2}$	$\frac{1}{2}$	3	0000+	$+1/2\sqrt{3}$
$\frac{5}{2}$	$\frac{3}{2}$	2	0000+	$-1/\sqrt{3.5}$
$\frac{5}{2}$	$\frac{5}{2}$	1	0000+	$+\sqrt{5}/2.3\sqrt{2}$
$\frac{5}{2}$	$\frac{5}{2}$	1	0010+	$-1/2.3\sqrt{2}$
$\frac{5}{2}$	$\frac{5}{2}$	2	0000+	$+1/2\sqrt{2.5}$
$\frac{5}{2}$	$\frac{5}{2}$	2	0010+	$+1/2.3\sqrt{2}$
$\frac{5}{2}$	$\frac{5}{2}$	3	0000+	0
$\frac{5}{2}$	$\frac{5}{2}$	3	0010+	$+1/3\sqrt{2}$
3	1	$\frac{5}{2}$	0000+	$+1/3\sqrt{2}$
3	2	$\frac{5}{2}$	0000+	$+1/3\sqrt{2}$
3	3	$\frac{1}{2}$	0000+	$+1/2\sqrt{2}$
3	3	$\frac{5}{2}$	0000+	$-1/2.3\sqrt{2}$
$\tilde{1}$	0	$\tilde{\frac{1}{2}}$	0000+	$+1/\sqrt{2.3}$
$\tilde{1}$	2	$\frac{3}{2}$	0000+	$+1/2\sqrt{3}$
$\tilde{1}$	$\tilde{1}$	$\frac{3}{2}$	0000+	$-1/2.3$
$\tilde{1}$	$\tilde{1}$	$\tilde{\frac{1}{2}}$	0000+	$+1/3$
$\tilde{\frac{1}{2}}$	$\frac{3}{2}$	$\tilde{1}$	0000+	$-1/3$
$\tilde{\frac{1}{2}}$	$\tilde{\frac{1}{2}}$	0	0000+	$+1/2$
$\tilde{\frac{1}{2}}$	$\tilde{\frac{1}{2}}$	$\tilde{1}$	0000+	$+1/2.3$

15

9 j Symbols

Introduction

The values of the T and O (also T_d) $9j$ symbols are tabulated by using the upper and middle rows of the $9j$ as bold headers. Under each header we give the third row, the three row multiplicity indices, the column interchange sign, the column multiplicity indices, the row interchange sign, and the value. For example, the third T $9j$ appearing in the third column of page 464 is

$$\left\{\begin{matrix} \frac{3}{2} & \frac{3}{2} & 1 \\ \frac{3}{2} & \frac{1}{2} & 1 \\ 1 & 1 & 1 \end{matrix}\right\}\begin{matrix} 0 \\ 0 \\ 0 \end{matrix} \atop \begin{matrix} 0 & 0 & 1 \end{matrix} = \frac{i}{12\sqrt{3}}$$

By column [(3.3.32)] and row [(3.3.33)] interchanges we also have

$$-\left\{\begin{matrix} 1 & \frac{3}{2} & \frac{3}{2} \\ 1 & \frac{1}{2} & \frac{3}{2} \\ 1 & 1 & 1 \end{matrix}\right\}\begin{matrix} 0 \\ 0 \\ 0 \end{matrix} \atop \begin{matrix} 1 & 0 & 0 \end{matrix} = \frac{i}{12\sqrt{3}} = \left\{\begin{matrix} \frac{3}{2} & \frac{3}{2} & 1 \\ 1 & 1 & 1 \\ \frac{3}{2} & \frac{1}{2} & 1 \end{matrix}\right\}\begin{matrix} 0 \\ 0 \\ 0 \end{matrix} \atop \begin{matrix} 0 & 0 & 1 \end{matrix}$$

The $9j$ symbols tabulated have the largest entry in the top left corner, the next largest at top-center (presuming it occurs in a triad including the first), and so on. The interchange symmetries, the transposition symmetry (3.3.35), and (for T $9j$'s) the complex conjugation symmetry (3.3.34) may be required to put an arbitrary $9j$ in this arrangement.

T 9j symbols

0 0 0 0 0 0
0 0 0 000+ 000+ $+1$
$\frac{1}{2}$ $\frac{1}{2}$ 0 $\frac{1}{2}$ 0 $\frac{1}{2}$
0 $\frac{1}{2}$ $\frac{1}{2}$ 000− 000− $+1/4$
$\frac{1}{2}$ $\frac{1}{2}$ 0 $\frac{1}{2}$ $\frac{1}{2}$ 0
0 0 0 000+ 000+ $+1/2$
1 $\frac{1}{2}$ $\frac{1}{2}$ $\frac{1}{2}$ $\frac{1}{2}$ 0
$\frac{1}{2}$ 0 $\frac{1}{2}$ 000+ 000+ $+1/4$
1 $\frac{1}{2}$ $\frac{1}{2}$ $\frac{1}{2}$ 1 $\frac{1}{2}$
$\frac{1}{2}$ $\frac{1}{2}$ 0 000− 000− $+1/4.3$
$\frac{1}{2}$ $\frac{1}{2}$ 1 000+ 000+ $+5/4.9$
1 1 0 $\frac{1}{2}$ $\frac{1}{2}$ 0
$\frac{1}{2}$ $\frac{1}{2}$ 0 000+ 000+ $+1/2\sqrt{3}$
1 1 0 $\frac{1}{2}$ $\frac{1}{2}$ 1
$\frac{1}{2}$ $\frac{1}{2}$ 1 000+ 000+ $-1/2.9$
1 1 0 1 0 1
0 1 1 000+ 000+ $+1/9$
1 1 0 1 $\frac{1}{2}$ $\frac{1}{2}$
0 $\frac{1}{2}$ $\frac{1}{2}$ 000− 000− $+1/2.3$
1 1 0 1 1 0
0 0 0 000+ 000+ $+1/3$
1 1 1 1 $\frac{1}{2}$ $\frac{1}{2}$
0 $\frac{1}{2}$ $\frac{1}{2}$ 000+ 000+ $-1/3\sqrt{2.3}$
0 $\frac{1}{2}$ $\frac{1}{2}$ 100− 000+ 0
1 $\frac{1}{2}$ $\frac{1}{2}$ 000− 000− 0
1 $\frac{1}{2}$ $\frac{1}{2}$ 100+ 000− 0
1 $\frac{1}{2}$ $\frac{1}{2}$ 100+ 100+ $+1/9$
1 1 1 1 1 0
1 0 1 000− 000− $+1/9$
1 0 1 100+ 000− 0
1 0 1 100+ 100+ $+1/9$
1 1 1 1 1 1
0 0 0 000+ 000+ $+1/3\sqrt{3}$
0 0 0 100− 000+ 0
0 0 0 110+ 000+ $+1/3\sqrt{3}$
1 1 0 000+ 000+ $+1/2.9$
1 1 0 100− 000+ 0
1 1 0 100− 100− $+1/2.9$
1 1 0 100− 110+ 0
1 1 0 110+ 000+ $-1/2.9$
1 1 0 110+ 100− 0
1 1 0 110+ 110+ $+1/2.9$
1 1 1 000− 000− 0
1 1 1 100+ 000− 0
1 1 1 100+ 100+ $+1/2.9$
1 1 1 100+ 110− 0
1 1 1 100+ 111+ $+1/2.9$
1 1 1 110− 000− 0
1 1 1 110− 100+ 0
1 1 1 110− 110− 0

1 1 1 1 1 1
1 1 1 110− 111+ 0
1 1 1 111+ 000− 0
1 1 1 111+ 100+ $+1/2.9$
1 1 1 111+ 110− 0
1 1 1 111+ 111+ $+1/2.9$
$\frac{3}{2}$ 1 $\frac{1}{2}$ 1 0 1
$\frac{1}{2}$ 1 $\frac{1}{2}$ 000+ 000+ $+1/9$
$\frac{1}{2}$ 1 $\frac{3}{2}$ 000+ 000+ $+1/9$
$\frac{3}{2}$ 1 $\frac{1}{2}$ 1 $\frac{1}{2}$ $\frac{1}{2}$
$\frac{1}{2}$ $\frac{1}{2}$ 0 000− 000− $+1/2.3$
$\frac{1}{2}$ $\frac{1}{2}$ 1 000+ 000+ $+1/2.9$
$\frac{3}{2}$ 1 $\frac{1}{2}$ 1 $\frac{1}{2}$ $\frac{3}{2}$
$\frac{1}{2}$ $\frac{1}{2}$ 1 000+ 000+ $+1/2.9$
$\frac{1}{2}$ $\frac{3}{2}$ 1 000+ 000+ $+5/4.9$
$\frac{3}{2}$ 1 $\frac{1}{2}$ 1 1 0
$\frac{1}{2}$ 0 $\frac{1}{2}$ 000− 000− $+1/2.3$
$\frac{3}{2}$ 1 $\frac{1}{2}$ 1 1 1
$\frac{1}{2}$ 0 $\frac{1}{2}$ 000+ 000+ $+1/2.3\sqrt{2.3}$
$\frac{1}{2}$ 0 $\frac{1}{2}$ 010− 000+ $-i/2.3\sqrt{2}$
$\frac{1}{2}$ 1 $\frac{1}{2}$ 000− 000− $+1/4.3$
$\frac{1}{2}$ 1 $\frac{1}{2}$ 000− 010+ $-i/4.3\sqrt{3}$
$\frac{1}{2}$ 1 $\frac{1}{2}$ 010+ 000− $+i/4.3\sqrt{3}$
$\frac{1}{2}$ 1 $\frac{1}{2}$ 010+ 010+ $+1/4.9$
$\frac{1}{2}$ 1 $\frac{3}{2}$ 000− 000− 0
$\frac{1}{2}$ 1 $\frac{3}{2}$ 000− 010+ 0
$\frac{1}{2}$ 1 $\frac{3}{2}$ 010+ 000− 0
$\frac{1}{2}$ 1 $\frac{3}{2}$ 010+ 010+ $+1/9$
$\frac{3}{2}$ 1 $\frac{1}{2}$ 1 $\frac{3}{2}$ $\frac{1}{2}$
$\frac{1}{2}$ $\frac{1}{2}$ 0 000− 000− $-1/4.3$
$\frac{1}{2}$ $\frac{1}{2}$ 1 000+ 000+ $+5/4.9$
$\frac{3}{2}$ 1 $\frac{1}{2}$ 1 $\frac{3}{2}$ $\frac{3}{2}$
$\frac{1}{2}$ $\frac{1}{2}$ 1 000+ 000+ $-1/2.9$
$\frac{1}{2}$ $\frac{3}{2}$ 1 000+ 000+ $+1/2.9$
$\frac{3}{2}$ $\frac{3}{2}$ 1 1 $\frac{1}{2}$ $\frac{1}{2}$
$\frac{1}{2}$ 1 $\frac{1}{2}$ 000+ 000+ $+1/9$
$\frac{1}{2}$ 1 $\frac{3}{2}$ 000+ 000+ $-1/2.9$
$\frac{3}{2}$ $\frac{3}{2}$ 1 1 $\frac{1}{2}$ $\frac{3}{2}$
$\frac{1}{2}$ 1 $\frac{1}{2}$ 000+ 000+ $-1/2.9$
$\frac{1}{2}$ 1 $\frac{3}{2}$ 000+ 000+ $+5/4.9$
$\frac{3}{2}$ $\frac{3}{2}$ 1 1 1 0
$\frac{1}{2}$ $\frac{1}{2}$ 1 000+ 000+ $-1/9$
$\frac{3}{2}$ $\frac{3}{2}$ 1 1 1 1
$\frac{1}{2}$ $\frac{1}{2}$ 0 000+ 000+ $+1/3\sqrt{2.3}$
$\frac{1}{2}$ $\frac{1}{2}$ 0 010− 000+ 0
$\frac{1}{2}$ $\frac{1}{2}$ 1 000− 000− 0
$\frac{1}{2}$ $\frac{1}{2}$ 1 000− 001+ 0
$\frac{1}{2}$ $\frac{1}{2}$ 1 010+ 000− $-i/2.3\sqrt{3}$
$\frac{1}{2}$ $\frac{1}{2}$ 1 010+ 001+ $+1/2.9$

$\frac{3}{2}$ $\frac{3}{2}$ 1 $\frac{3}{2}$ $\frac{1}{2}$ 1
1 1 0 000+ 000+ $-1/2.9$
1 1 1 000− 000− $+1/4.3$
1 1 1 000− 001+ $+i/4.3\sqrt{3}$
1 1 1 001+ 000− $-i/4.3\sqrt{3}$
1 1 1 001+ 001+ $+1/4.9$
$\frac{3}{2}$ $\frac{3}{2}$ 1 $\frac{3}{2}$ 1 $\frac{1}{2}$
1 $\frac{1}{2}$ $\frac{1}{2}$ 000+ 000+ $+1/2.9$
1 $\frac{1}{2}$ $\frac{3}{2}$ 000+ 000+ $+1/2.9$
$\frac{3}{2}$ $\frac{3}{2}$ 1 $\frac{3}{2}$ 1 $\frac{3}{2}$
1 $\frac{1}{2}$ $\frac{1}{2}$ 000+ 000+ $+1/9$
1 $\frac{1}{2}$ $\frac{3}{2}$ 000+ 000+ $+1/2.9$
1 $\frac{3}{2}$ $\frac{1}{2}$ 000+ 000+ $+1/2.9$
1 $\frac{3}{2}$ $\frac{3}{2}$ 000+ 000+ $+5/4.9$
$\frac{3}{2}$ $\frac{3}{2}$ 1 $\frac{3}{2}$ $\frac{3}{2}$ 1
1 1 0 000+ 000+ $+1/9$
1 1 1 000− 000− 0
1 1 1 000− 001+ 0
1 1 1 001+ 000− 0
1 1 1 001+ 001+ $+1/9$
-$\frac{3}{2}$ 1 $\frac{1}{2}$ 1 0 1
$\frac{1}{2}$ 1 $\frac{1}{2}$ 000+ 000+ $+1/9$
$\frac{1}{2}$ 1 $\frac{3}{2}$ 000+ 000+ $-1/2.9$
$\frac{1}{2}$ 1 -$\frac{3}{2}$ 000+ 000+ $+1/9$
-$\frac{3}{2}$ 1 $\frac{1}{2}$ 1 $\frac{1}{2}$ $\frac{1}{2}$
$\frac{1}{2}$ $\frac{1}{2}$ 0 000− 000− $+1/2.3$
$\frac{1}{2}$ $\frac{1}{2}$ 1 000+ 000+ $+1/2.9$
-$\frac{3}{2}$ 1 $\frac{1}{2}$ 1 $\frac{1}{2}$ $\frac{3}{2}$
$\frac{1}{2}$ $\frac{1}{2}$ 1 000+ 000+ $-1/9$
$\frac{1}{2}$ $\frac{3}{2}$ 1 000+ 000+ $+1/2.9$
-$\frac{3}{2}$ 1 $\frac{1}{2}$ 1 $\frac{1}{2}$ -$\frac{3}{2}$
$\frac{1}{2}$ $\frac{1}{2}$ 1 000+ 000+ $+1/2.9$
$\frac{1}{2}$ $\frac{3}{2}$ 1 000+ 000+ $+1/2.9$
$\frac{1}{2}$ -$\frac{3}{2}$ 1 000+ 000+ $+5/4.9$
-$\frac{3}{2}$ 1 $\frac{1}{2}$ 1 1 0
$\frac{1}{2}$ 0 $\frac{1}{2}$ 000− 000− $+1/2.3$
-$\frac{3}{2}$ 1 $\frac{1}{2}$ 1 1 1
$\frac{1}{2}$ 0 $\frac{1}{2}$ 000+ 000+ $+1/2.3\sqrt{2.3}$
$\frac{1}{2}$ 0 $\frac{1}{2}$ 010− 000+ $+i/2.3\sqrt{2}$
$\frac{1}{2}$ 1 $\frac{1}{2}$ 000− 000− $+1/4.3$
$\frac{1}{2}$ 1 $\frac{1}{2}$ 000− 010+ $+i/4.3\sqrt{3}$
$\frac{1}{2}$ 1 $\frac{1}{2}$ 010+ 000− $-i/4.3\sqrt{3}$
$\frac{1}{2}$ 1 $\frac{1}{2}$ 010+ 010+ $+1/4.9$
$\frac{1}{2}$ 1 $\frac{3}{2}$ 000− 000− $+1/4.3$
$\frac{1}{2}$ 1 $\frac{3}{2}$ 000− 010+ $-i/4.3\sqrt{3}$
$\frac{1}{2}$ 1 $\frac{3}{2}$ 010+ 000− $+i/4.3\sqrt{3}$
$\frac{1}{2}$ 1 $\frac{3}{2}$ 010+ 010+ $+1/4.9$
$\frac{1}{2}$ 1 -$\frac{3}{2}$ 000− 000− 0
$\frac{1}{2}$ 1 -$\frac{3}{2}$ 000− 010+ 0
$\frac{1}{2}$ 1 -$\frac{3}{2}$ 010+ 000− 0

T 9j Symbols (cont.)

-3/2 1 1/2 1 1 1

1/2 1 -3/2 010+ 010+ +1/9

-3/2 1 1/2 1 3/2 1/2

1/2 1/2 0 000− 000− +1/2.3

1/2 1/2 1 000+ 000+ +1/2.9

-3/2 1 1/2 1 3/2 3/2

1/2 1/2 1 000+ 000+ −1/2.9

1/2 3/2 1 000+ 000+ +5/4.9

-3/2 1 1/2 1 3/2 -3/2

1/2 1/2 1 000+ 000+ −1/9

1/2 3/2 1 000+ 000+ −1/2.9

1/2 -3/2 1 000+ 000+ +1/2.9

-3/2 1 1/2 1 -3/2 1/2

1/2 1/2 0 000− 000− −1/4.3

1/2 1/2 1 000+ 000+ +5/4.9

-3/2 1 1/2 1 -3/2 3/2

1/2 1/2 1 000+ 000+ +1/2.9

1/2 3/2 1 000+ 000+ +1/2.9

-3/2 1 1/2 1 -3/2 -3/2

1/2 1/2 1 000+ 000+ −1/2.9

1/2 3/2 1 000+ 000+ +1/9

1/2 -3/2 1 000+ 000+ +1/2.9

-3/2 3/2 0 1 0 1

1/2 -3/2 1 000− 000− +1/2.3

-3/2 3/2 0 1 1/2 1/2

1/2 1 1/2 000− 000− −1/2.3

-3/2 3/2 0 1 1/2 3/2

1/2 1 -3/2 000− 000− −1/2.3

-3/2 3/2 0 1 1/2 -3/2

1/2 1 3/2 000− 000− +1/4.3

-3/2 3/2 0 1 1 0

1/2 1/2 0 000+ 000+ +1/2√3

-3/2 3/2 0 1 1 1

1/2 1/2 1 000+ 000+ +1/2.3√2.3

1/2 1/2 1 010− 000+ + i/2.3√2

1/2 3/2 1 000+ 000+ +1/3√2.3

1/2 3/2 1 010− 000+ 0

1/2 -3/2 1 000+ 000+ +1/2.3√2.3

1/2 -3/2 1 010− 000+ − i/2.3√2

-3/2 3/2 0 1 3/2 1/2

1/2 1 1/2 000− 000− +1/2.3

-3/2 3/2 0 1 3/2 3/2

1/2 1 -3/2 000− 000− +1/4.3

-3/2 3/2 0 1 3/2 -3/2

1/2 1 3/2 000− 000− +1/2.3

-3/2 3/2 0 1 -3/2 1/2

1/2 0 1/2 000+ 000+ +1/4

1/2 1 1/2 000− 000− +1/4.3

-3/2 3/2 0 1 -3/2 3/2

1/2 1 -3/2 000− 000− −1/2.3

-3/2 3/2 0 1 -3/2 -3/2

1/2 1 3/2 000− 000− +1/2.3

-3/2 3/2 0 3/2 0 -3/2

0 -3/2 3/2 000− 000− +1/4

-3/2 3/2 0 3/2 1/2 1

0 1 1 000− 000− +1/2.3

-3/2 3/2 0 3/2 1 1/2

0 1/2 1/2 000+ 000+ +1/4

-3/2 3/2 0 3/2 1 3/2

0 3/2 -3/2 000+ 000+ +1/4

-3/2 3/2 0 3/2 1 -3/2

0 -3/2 3/2 000+ 000+ +1/4

-3/2 3/2 0 3/2 3/2 1

0 1 1 000− 000− +1/2.3

-3/2 3/2 0 3/2 -3/2 0

0 0 0 000+ 000+ +1/2

-3/2 3/2 0 3/2 -3/2 1

0 1 1 000− 000− +1/2.3

-3/2 3/2 1 1 0 1

1/2 -3/2 1 000+ 000+ +1/2.3√2.3

1/2 -3/2 1 000+ 001− + i/2.3√2

-3/2 3/2 1 1 1/2 1/2

1/2 1 1/2 000+ 000+ +1/2.9

1/2 1 3/2 000+ 000+ +1/2.9

1/2 1 -3/2 000+ 000+ −1/9

-3/2 3/2 1 1 1/2 3/2

1/2 1 1/2 000+ 000+ −1/9

1/2 1 3/2 000+ 000+ −1/2.9

1/2 1 -3/2 000+ 000+ +1/2.9

-3/2 3/2 1 1 1/2 -3/2

1/2 1 1/2 000+ 000+ +1/2.9

1/2 1 3/2 000+ 000+ +5/4.9

1/2 1 -3/2 000+ 000+ −1/2.9

-3/2 3/2 1 1 1 0

1/2 1/2 1 000+ 000+ −1/2.9

1/2 3/2 1 000+ 000+ −1/9

1/2 -3/2 1 000+ 000+ +1/9

-3/2 3/2 1 1 1 1

1/2 1/2 0 000+ 000+ +1/2.3√2.3

1/2 1/2 0 010− 000+ + i/2.3√2

1/2 1/2 1 000− 000− 0

1/2 1/2 1 000− 001+ − i/2.3√3

1/2 1/2 1 010+ 000− 0

1/2 1/2 1 010+ 001+ −1/2.9

1/2 3/2 1 000− 000− 0

1/2 3/2 1 000− 001+ 0

1/2 3/2 1 010+ 000− − i/2.3√3

1/2 3/2 1 010+ 001+ +1/2.9

1/2 -3/2 1 000− 000− +1/4.3

1/2 -3/2 1 000− 001+ − i/4.3√3

1/2 -3/2 1 010+ 000− + i/4.3√3

1/2 -3/2 1 010+ 001+ +1/4.9

-3/2 3/2 1 1 3/2 1/2

1/2 1 1/2 000+ 000+ −1/2.9

1/2 1 3/2 000+ 000+ +1/9

1/2 1 -3/2 000+ 000+ −1/2.9

-3/2 3/2 1 1 3/2 3/2

1/2 1 1/2 000+ 000+ +1/2.9

1/2 1 3/2 000+ 000+ −1/2.9

1/2 1 -3/2 000+ 000+ +5/4.9

-3/2 3/2 1 1 3/2 -3/2

1/2 1 1/2 000+ 000+ +1/9

1/2 1 3/2 000+ 000+ −1/2.9

1/2 1 -3/2 000+ 000+ +1/2.9

-3/2 3/2 1 1 -3/2 1/2

1/2 0 1/2 000− 000− +1/4.3

1/2 1 1/2 000+ 000+ +5/4.9

1/2 1 3/2 000+ 000+ +1/2.9

1/2 1 -3/2 000+ 000+ +1/2.9

-3/2 3/2 1 1 -3/2 3/2

1/2 0 1/2 000− 000− −1/2.3

1/2 1 1/2 000+ 000+ +1/2.9

1/2 1 3/2 000+ 000+ +1/9

1/2 1 -3/2 000+ 000+ +1/2.9

-3/2 3/2 1 1 -3/2 -3/2

1/2 0 1/2 000− 000− +1/2.3

1/2 1 1/2 000+ 000+ −1/2.9

1/2 1 3/2 000+ 000+ −1/2.9

1/2 1 -3/2 000+ 000+ −1/9

-3/2 3/2 1 3/2 0 -3/2

0 -3/2 3/2 000+ 000+ +1/4

1 -3/2 1/2 000− 000− +1/2.3

1 -3/2 3/2 000− 000− −1/4.3

1 -3/2 -3/2 000− 000− +1/2.3

-3/2 3/2 1 3/2 1/2 1

0 1 1 000+ 000+ +1/2.3√2.3

0 1 1 000+ 001− + i/2.3√2

1 1 0 000+ 000+ +1/9

1 1 1 000− 000− +1/4.3

1 1 1 000− 001+ − i/4.3√3

1 1 1 001+ 000− + i/4.3√3

1 1 1 001+ 001+ +1/4.9

-3/2 3/2 1 3/2 1 1/2

0 1/2 1/2 000− 000− −1/4.3

0 3/2 -3/2 000− 000− −1/2.3

0 -3/2 3/2 000− 000− +1/2.3

1 1/2 1/2 000+ 000+ +5/4.9

1 1/2 3/2 000+ 000+ +1/2.9

1 1/2 -3/2 000+ 000+ +1/2.9

T 9j Symbols (cont.)

$-\frac{3}{2}\ \frac{3}{2}\ 1\quad \frac{3}{2}\ 1\ \frac{3}{2}$
$0\ \frac{1}{2}\ \frac{1}{2}$ 000− 000− $-1/2.3$
$0\ \frac{3}{2}\ -\frac{3}{2}$ 000− 000− $-1/4.3$
$0\ -\frac{3}{2}\ \frac{3}{2}$ 000− 000− $+1/2.3$
$1\ \frac{1}{2}\ \frac{1}{2}$ 000+ 000+ $-1/2.9$
$1\ \frac{1}{2}\ \frac{3}{2}$ 000+ 000+ $-1/9$
$1\ \frac{1}{2}\ -\frac{3}{2}$ 000+ 000+ $-1/2.9$
$1\ \frac{3}{2}\ \frac{1}{2}$ 000+ 000+ $+1/2.9$
$1\ \frac{3}{2}\ \frac{3}{2}$ 000+ 000+ $+1/2.9$
$1\ \frac{3}{2}\ -\frac{3}{2}$ 000+ 000+ $+5/4.9$
$-\frac{3}{2}\ \frac{3}{2}\ 1\quad \frac{3}{2}\ 1\ -\frac{3}{2}$
$0\ \frac{1}{2}\ \frac{1}{2}$ 000− 000− $+1/2.3$
$0\ \frac{3}{2}\ -\frac{3}{2}$ 000− 000− $+1/2.3$
$0\ -\frac{3}{2}\ \frac{3}{2}$ 000− 000− $-1/4.3$
$1\ \frac{1}{2}\ \frac{1}{2}$ 000+ 000+ $+1/2.9$
$1\ \frac{1}{2}\ \frac{3}{2}$ 000+ 000+ $+1/2.9$
$1\ \frac{1}{2}\ -\frac{3}{2}$ 000+ 000+ $+1/9$
$1\ \frac{3}{2}\ \frac{1}{2}$ 000+ 000+ $+1/9$
$1\ \frac{3}{2}\ \frac{3}{2}$ 000+ 000+ $+1/2.9$
$1\ \frac{3}{2}\ -\frac{3}{2}$ 000+ 000+ $+1/2.9$
$1\ -\frac{3}{2}\ \frac{1}{2}$ 000+ 000+ $+1/2.9$
$1\ -\frac{3}{2}\ \frac{3}{2}$ 000+ 000+ $+5/4.9$
$1\ -\frac{3}{2}\ -\frac{3}{2}$ 000+ 000+ $+1/2.9$
$-\frac{3}{2}\ \frac{3}{2}\ 1\quad \frac{3}{2}\ \frac{3}{2}\ 1$
$0\ 1\ 1$ 000+ 000+ $+1/2.3\sqrt{2.3}$
$0\ 1\ 1$ 000+ 001− $-i/2.3\sqrt{2}$
$1\ 1\ 0$ 000+ 000+ $+1/9$
$1\ 1\ 1$ 000− 000− $+1/4.3$
$1\ 1\ 1$ 000− 001+ $+i/4.3\sqrt{3}$
$1\ 1\ 1$ 001+ 000− $-i/4.3\sqrt{3}$
$1\ 1\ 1$ 001+ 001+ $+1/4.9$
$-\frac{3}{2}\ \frac{3}{2}\ 1\quad \frac{3}{2}\ -\frac{3}{2}\ 0$
$0\ 1\ 1$ 000− 000− $-1/2.3$
$1\ 0\ 1$ 000− 000− $+1/2.3$
$-\frac{3}{2}\ \frac{3}{2}\ 1\quad \frac{3}{2}\ -\frac{3}{2}\ 1$
$0\ 0\ 0$ 000+ 000+ $+1/2\sqrt{3}$
$0\ 1\ 1$ 000+ 000+ $-1/3\sqrt{2.3}$
$0\ 1\ 1$ 000+ 001− 0
$1\ 0\ 1$ 000+ 000+ $-1/3\sqrt{2.3}$
$1\ 0\ 1$ 000+ 001− 0
$1\ 1\ 0$ 000+ 000+ $-1/2.9$
$1\ 1\ 1$ 000− 000− 0
$1\ 1\ 1$ 000− 001+ 0
$1\ 1\ 1$ 001+ 000− 0
$1\ 1\ 1$ 001+ 001+ $+1/9$
$-\frac{3}{2}\ -\frac{3}{2}\ 1\quad 1\ 0\ 1$
$\frac{1}{2}\ \frac{3}{2}\ 1$ 000+ 000+ $+1/3\sqrt{2.3}$
$\frac{1}{2}\ \frac{3}{2}\ 1$ 000+ 001− 0
$-\frac{3}{2}\ -\frac{3}{2}\ 1\quad 1\ \frac{1}{2}\ \frac{1}{2}$
$\frac{1}{2}\ 1\ \frac{1}{2}$ 000+ 000+ $+1/9$
$\frac{1}{2}\ 1\ \frac{3}{2}$ 000+ 000+ $-1/2.9$
$\frac{1}{2}\ 1\ -\frac{3}{2}$ 000+ 000+ $-1/2.9$
$-\frac{3}{2}\ -\frac{3}{2}\ 1\quad 1\ \frac{1}{2}\ \frac{3}{2}$
$\frac{1}{2}\ 1\ \frac{1}{2}$ 000+ 000+ $-1/2.9$
$\frac{1}{2}\ 1\ \frac{3}{2}$ 000+ 000+ $-1/9$
$\frac{1}{2}\ 1\ -\frac{3}{2}$ 000+ 000+ $-1/2.9$
$-\frac{3}{2}\ -\frac{3}{2}\ 1\quad 1\ \frac{1}{2}\ -\frac{3}{2}$
$\frac{1}{2}\ 1\ \frac{1}{2}$ 000+ 000+ $-1/2.9$
$\frac{1}{2}\ 1\ \frac{3}{2}$ 000+ 000+ $-1/2.9$
$\frac{1}{2}\ 1\ -\frac{3}{2}$ 000+ 000+ $+5/4.9$
$-\frac{3}{2}\ -\frac{3}{2}\ 1\quad 1\ 1\ 0$
$\frac{1}{2}\ \frac{1}{2}\ 1$ 000+ 000+ $-1/9$
$-\frac{3}{2}\ -\frac{3}{2}\ 1\quad 1\ 1\ 1$
$\frac{1}{2}\ \frac{1}{2}\ 0$ 000+ 000+ $+1/3\sqrt{2.3}$
$\frac{1}{2}\ \frac{1}{2}\ 0$ 010− 000+ 0
$\frac{1}{2}\ \frac{1}{2}\ 1$ 000− 000− 0
$\frac{1}{2}\ \frac{1}{2}\ 1$ 000− 001+ 0
$\frac{1}{2}\ \frac{1}{2}\ 1$ 010+ 000− $+i/2.3\sqrt{3}$
$\frac{1}{2}\ \frac{1}{2}\ 1$ 010+ 001+ $+1/2.9$
$-\frac{3}{2}\ -\frac{3}{2}\ 1\quad \frac{3}{2}\ 0\ -\frac{3}{2}$
$0\ \frac{3}{2}\ -\frac{3}{2}$ 000+ 000+ $+1/4$
$1\ \frac{3}{2}\ \frac{1}{2}$ 000− 000− $-1/2.3$
$1\ \frac{3}{2}\ \frac{3}{2}$ 000− 000− $+1/2.3$
$1\ \frac{3}{2}\ -\frac{3}{2}$ 000− 000− $-1/4.3$
$-\frac{3}{2}\ -\frac{3}{2}\ 1\quad \frac{3}{2}\ \frac{1}{2}\ 1$
$0\ 1\ 1$ 000+ 000+ $+1/3\sqrt{2.3}$
$0\ 1\ 1$ 000+ 001− 0
$1\ 1\ 0$ 000+ 000+ $-1/9$
$1\ 1\ 1$ 000− 000− 0
$1\ 1\ 1$ 000− 001+ $+i/2.3\sqrt{3}$
$1\ 1\ 1$ 001+ 000− 0
$1\ 1\ 1$ 001+ 001+ $+1/2.9$
$-\frac{3}{2}\ -\frac{3}{2}\ 1\quad \frac{3}{2}\ 1\ \frac{1}{2}$
$0\ \frac{1}{2}\ \frac{1}{2}$ 000− 000− $-1/2.3$
$0\ \frac{3}{2}\ -\frac{3}{2}$ 000− 000− $-1/2.3$
$0\ -\frac{3}{2}\ \frac{3}{2}$ 000− 000− $-1/4.3$
$1\ \frac{1}{2}\ \frac{1}{2}$ 000+ 000+ $-1/2.9$
$1\ \frac{1}{2}\ \frac{3}{2}$ 000+ 000+ $-1/2.9$
$1\ \frac{1}{2}\ -\frac{3}{2}$ 000+ 000+ $+1/9$
$1\ \frac{3}{2}\ \frac{1}{2}$ 000+ 000+ $+1/9$
$1\ \frac{3}{2}\ \frac{3}{2}$ 000+ 000+ $+1/2.9$
$1\ \frac{3}{2}\ -\frac{3}{2}$ 000+ 000+ $-1/2.9$
$1\ -\frac{3}{2}\ \frac{1}{2}$ 000+ 000+ $+1/2.9$
$1\ -\frac{3}{2}\ \frac{3}{2}$ 000+ 000+ $+5/4.9$
$1\ -\frac{3}{2}\ -\frac{3}{2}$ 000+ 000+ $-1/2.9$
$-\frac{3}{2}\ -\frac{3}{2}\ 1\quad \frac{3}{2}\ 1\ \frac{3}{2}$
$0\ \frac{1}{2}\ \frac{1}{2}$ 000− 000− $-1/4.3$
$0\ \frac{3}{2}\ -\frac{3}{2}$ 000− 000− $+1/2.3$
$0\ -\frac{3}{2}\ \frac{3}{2}$ 000− 000− $+1/2.3$
$1\ \frac{1}{2}\ \frac{1}{2}$ 000+ 000+ $+5/4.9$
$1\ \frac{1}{2}\ \frac{3}{2}$ 000+ 000+ $-1/2.9$
$1\ \frac{1}{2}\ -\frac{3}{2}$ 000+ 000+ $+1/2.9$
$1\ \frac{3}{2}\ \frac{1}{2}$ 000+ 000+ $+1/2.9$
$1\ \frac{3}{2}\ \frac{3}{2}$ 000+ 000+ $-1/9$
$1\ \frac{3}{2}\ -\frac{3}{2}$ 000+ 000+ $+1/2.9$
$1\ -\frac{3}{2}\ \frac{1}{2}$ 000+ 000+ $-1/2.9$
$1\ -\frac{3}{2}\ \frac{3}{2}$ 000+ 000+ $+1/2.9$
$1\ -\frac{3}{2}\ -\frac{3}{2}$ 000+ 000+ $-1/9$
$-\frac{3}{2}\ -\frac{3}{2}\ 1\quad \frac{3}{2}\ 1\ -\frac{3}{2}$
$0\ \frac{1}{2}\ \frac{1}{2}$ 000− 000− $-1/2.3$
$0\ \frac{3}{2}\ -\frac{3}{2}$ 000− 000− $-1/4.3$
$0\ -\frac{3}{2}\ \frac{3}{2}$ 000− 000− $+1/2.3$
$1\ \frac{1}{2}\ \frac{1}{2}$ 000+ 000+ $-1/2.9$
$1\ \frac{1}{2}\ \frac{3}{2}$ 000+ 000+ $+1/9$
$1\ \frac{1}{2}\ -\frac{3}{2}$ 000+ 000+ $+1/2.9$
$1\ \frac{3}{2}\ \frac{1}{2}$ 000+ 000+ $-1/2.9$
$1\ \frac{3}{2}\ \frac{3}{2}$ 000+ 000+ $+1/2.9$
$1\ \frac{3}{2}\ -\frac{3}{2}$ 000+ 000+ $+5/4.9$
$1\ -\frac{3}{2}\ \frac{1}{2}$ 000+ 000+ $-1/9$
$1\ -\frac{3}{2}\ \frac{3}{2}$ 000+ 000+ $+1/2.9$
$1\ -\frac{3}{2}\ -\frac{3}{2}$ 000+ 000+ $+1/2.9$
$-\frac{3}{2}\ -\frac{3}{2}\ 1\quad \frac{3}{2}\ \frac{3}{2}\ 1$
$0\ 0\ 0$ 000+ 000+ $+1/2\sqrt{3}$
$1\ 0\ 1$ 000+ 000+ $+1/2.3\sqrt{2.3}$
$1\ 0\ 1$ 000+ 001− $+i/2.3\sqrt{2}$
$1\ 1\ 0$ 000+ 000+ $-1/2.9$
$1\ 1\ 1$ 000− 000− 0
$1\ 1\ 1$ 000− 001+ 0
$1\ 1\ 1$ 001+ 000− $-i/2.3\sqrt{3}$
$1\ 1\ 1$ 001+ 001+ $-1/2.9$
$-\frac{3}{2}\ -\frac{3}{2}\ 1\quad -\frac{3}{2}\ 0\ \frac{3}{2}$
$1\ \frac{3}{2}\ \frac{1}{2}$ 000− 000− $-1/4.3$
$1\ \frac{3}{2}\ \frac{3}{2}$ 000− 000− $+1/2.3$
$1\ \frac{3}{2}\ -\frac{3}{2}$ 000− 000− $+1/2.3$
$-\frac{3}{2}\ -\frac{3}{2}\ 1\quad -\frac{3}{2}\ \frac{1}{2}\ 1$
$1\ 1\ 0$ 000+ 000+ $-1/2.9$
$1\ 1\ 1$ 000− 000− $+1/4.3$
$1\ 1\ 1$ 000− 001+ $-i/4.3\sqrt{3}$
$1\ 1\ 1$ 001+ 000− $+i/4.3\sqrt{3}$
$1\ 1\ 1$ 001+ 001+ $+1/4.9$
$-\frac{3}{2}\ -\frac{3}{2}\ 1\quad -\frac{3}{2}\ 1\ \frac{1}{2}$
$1\ \frac{1}{2}\ \frac{1}{2}$ 000+ 000+ $+1/2.9$
$1\ \frac{1}{2}\ \frac{3}{2}$ 000+ 000+ $+5/4.9$
$1\ \frac{1}{2}\ -\frac{3}{2}$ 000+ 000+ $+1/2.9$
$-\frac{3}{2}\ -\frac{3}{2}\ 1\quad -\frac{3}{2}\ 1\ \frac{3}{2}$
$1\ \frac{1}{2}\ \frac{1}{2}$ 000+ 000+ $+1/2.9$
$1\ \frac{1}{2}\ \frac{3}{2}$ 000+ 000+ $-1/2.9$
$1\ \frac{1}{2}\ -\frac{3}{2}$ 000+ 000+ $-1/9$

T 9j Symbols (cont.)

$-\frac{3}{2}\ -\frac{3}{2}\ 1\ -\frac{3}{2}\ 1\ \frac{3}{2}$
$1\ \frac{3}{2}\ \frac{1}{2}$ 000+ 000+ $+5/4.9$
$1\ \frac{3}{2}\ \frac{3}{2}$ 000+ 000+ $+1/2.9$
$1\ \frac{3}{2}\ -\frac{3}{2}$ 000+ 000+ $+1/2.9$
$-\frac{3}{2}\ -\frac{3}{2}\ 1\ -\frac{3}{2}\ 1\ -\frac{3}{2}$
$1\ \frac{1}{2}\ \frac{1}{2}$ 000+ 000+ $+1/9$
$1\ \frac{1}{2}\ \frac{3}{2}$ 000+ 000+ $-1/2.9$
$1\ \frac{1}{2}\ -\frac{3}{2}$ 000+ 000+ $+1/2.9$
$1\ \frac{3}{2}\ \frac{1}{2}$ 000+ 000+ $-1/2.9$
$1\ \frac{3}{2}\ \frac{3}{2}$ 000+ 000+ $-1/9$
$1\ \frac{3}{2}\ -\frac{3}{2}$ 000+ 000+ $+1/2.9$
$1\ -\frac{3}{2}\ \frac{1}{2}$ 000+ 000+ $+1/2.9$
$1\ -\frac{3}{2}\ \frac{3}{2}$ 000+ 000+ $+1/2.9$
$1\ -\frac{3}{2}\ -\frac{3}{2}$ 000+ 000+ $+5/4.9$
$-\frac{3}{2}\ -\frac{3}{2}\ 1\ -\frac{3}{2}\ \frac{3}{2}\ 0$
$1\ 0\ 1$ 000− 000− $+1/2.3$
$-\frac{3}{2}\ -\frac{3}{2}\ 1\ -\frac{3}{2}\ \frac{3}{2}\ 1$
$1\ 0\ 1$ 000+ 000+ $+1/2.3\sqrt{2.3}$
$1\ 0\ 1$ 000+ 001− $-i/2.3\sqrt{2}$
$1\ 1\ 0$ 000+ 000+ $+1/9$
$1\ 1\ 1$ 000− 000− $+1/4.3$
$1\ 1\ 1$ 000− 001+ $+i/4.3\sqrt{3}$
$1\ 1\ 1$ 001+ 000− $-i/4.3\sqrt{3}$
$1\ 1\ 1$ 001+ 001+ $+1/4.9$
$-\frac{3}{2}\ -\frac{3}{2}\ 1\ -\frac{3}{2}\ -\frac{3}{2}\ 1$
$1\ 1\ 0$ 000+ 000+ $+1/9$
$1\ 1\ 1$ 000− 000− 0
$1\ 1\ 1$ 000− 001+ 0
$1\ 1\ 1$ 001+ 000− 0
$1\ 1\ 1$ 001+ 001+ $+1/9$
$2\ 1\ 1\ \ 1\ \frac{1}{2}\ \frac{1}{2}$
$1\ \frac{1}{2}\ \frac{1}{2}$ 000+ 000+ $+1/9$
$2\ 1\ 1\ \ 1\ 1\ 0$
$1\ 0\ 1$ 000+ 000+ $+1/9$
$2\ 1\ 1\ \ 1\ 1\ 1$
$1\ 1\ 0$ 000− 000− $+1/2.9$
$1\ 1\ 0$ 000− 010+ $-i/2.3\sqrt{3}$
$1\ 1\ 0$ 010+ 000− $+i/2.3\sqrt{3}$
$1\ 1\ 0$ 010+ 010+ $-1/2.9$
$1\ 1\ 1$ 000+ 000+ $+1/2.9$
$1\ 1\ 1$ 000+ 010− 0
$1\ 1\ 1$ 000+ 011+ $-1/2.9$
$1\ 1\ 1$ 010− 000+ 0
$1\ 1\ 1$ 010− 010− $+1/2.9$
$1\ 1\ 1$ 010− 011+ 0
$1\ 1\ 1$ 011+ 000+ $-1/2.9$
$1\ 1\ 1$ 011+ 010− 0
$1\ 1\ 1$ 011+ 011+ $+1/2.9$
$2\ 1\ 1\ \ 1\ \frac{3}{2}\ \frac{1}{2}$
$1\ \frac{1}{2}\ \frac{1}{2}$ 000+ 000+ $-1/2.9$
$1\ \frac{1}{2}\ \frac{3}{2}$ 000+ 000+ $+1/9$
$1\ \frac{1}{2}\ -\frac{3}{2}$ 000+ 000+ $+1/9$
$1\ \frac{3}{2}\ \frac{1}{2}$ 000+ 000+ $-1/9$
$2\ 1\ 1\ \ 1\ \frac{3}{2}\ \frac{3}{2}$
$1\ \frac{1}{2}\ \frac{1}{2}$ 000+ 000+ $-1/9$
$1\ \frac{3}{2}\ \frac{1}{2}$ 000+ 000+ $+1/9$
$1\ \frac{3}{2}\ \frac{3}{2}$ 000+ 000+ $-1/2.9$
$2\ 1\ 1\ \ 1\ -\frac{3}{2}\ \frac{1}{2}$
$1\ \frac{1}{2}\ \frac{1}{2}$ 000+ 000+ $+1/9$
$1\ \frac{1}{2}\ \frac{3}{2}$ 000+ 000+ $+1/9$
$1\ \frac{1}{2}\ -\frac{3}{2}$ 000+ 000+ $-1/2.9$
$1\ \frac{3}{2}\ \frac{1}{2}$ 000+ 000+ $+1/9$
$1\ \frac{3}{2}\ \frac{3}{2}$ 000+ 000+ $+1/2.9$
$1\ \frac{3}{2}\ -\frac{3}{2}$ 000+ 000+ $+1/9$
$1\ -\frac{3}{2}\ \frac{1}{2}$ 000+ 000+ $+1/2.9$
$2\ 1\ 1\ \ 1\ -\frac{3}{2}\ \frac{3}{2}$
$1\ \frac{1}{2}\ \frac{1}{2}$ 000+ 000+ $+1/9$
$1\ \frac{1}{2}\ \frac{3}{2}$ 000+ 000+ $+1/2.9$
$1\ \frac{1}{2}\ -\frac{3}{2}$ 000+ 000+ $+1/9$
$1\ \frac{3}{2}\ \frac{1}{2}$ 000+ 000+ $-1/2.9$
$1\ \frac{3}{2}\ \frac{3}{2}$ 000+ 000+ $+1/9$
$1\ \frac{3}{2}\ -\frac{3}{2}$ 000+ 000+ $+1/9$
$1\ -\frac{3}{2}\ \frac{1}{2}$ 000+ 000+ $-1/9$
$1\ -\frac{3}{2}\ \frac{3}{2}$ 000+ 000+ $+1/9$
$2\ 1\ 1\ \ 1\ -\frac{3}{2}\ -\frac{3}{2}$
$1\ \frac{1}{2}\ \frac{1}{2}$ 000+ 000+ $+1/2.9$
$1\ \frac{3}{2}\ \frac{1}{2}$ 000+ 000+ $-1/9$
$1\ \frac{3}{2}\ \frac{3}{2}$ 000+ 000+ $+1/9$
$1\ -\frac{3}{2}\ \frac{1}{2}$ 000+ 000+ $+1/9$
$1\ -\frac{3}{2}\ \frac{3}{2}$ 000+ 000+ $-1/2.9$
$1\ -\frac{3}{2}\ -\frac{3}{2}$ 000+ 000+ $+1/9$
$2\ 1\ 1\ \ 1\ 2\ 1$
$1\ 1\ 0$ 000+ 000+ $+1/9$
$1\ 1\ 1$ 000− 000− $-1/9$
$1\ 1\ 1$ 000− 001+ 0
$1\ 1\ 1$ 001+ 000− 0
$1\ 1\ 1$ 001+ 001+ $+1/9$
$1\ 1\ 2$ 000+ 000+ $+1/9$
$2\ \frac{3}{2}\ \frac{3}{2}\ \ 1\ 1\ 0$
$1\ \frac{1}{2}\ -\frac{3}{2}$ 000− 000− $-1/2.3$
$2\ \frac{3}{2}\ \frac{3}{2}\ \ 1\ 1\ 1$
$1\ \frac{1}{2}\ \frac{1}{2}$ 000+ 000+ $+1/3\sqrt{2.3}$
$1\ \frac{1}{2}\ \frac{1}{2}$ 010− 000+ 0
$2\ \frac{3}{2}\ \frac{3}{2}\ \ 1\ \frac{3}{2}\ \frac{1}{2}$
$1\ 1\ 1$ 000+ 000+ $+1/2.3\sqrt{2.3}$
$1\ 1\ 1$ 001− 000+ $+i/2.3\sqrt{2}$
$2\ \frac{3}{2}\ \frac{3}{2}\ \ 1\ \frac{3}{2}\ \frac{3}{2}$
$1\ 1\ 1$ 000+ 000+ $-1/3\sqrt{2.3}$
$1\ 1\ 1$ 001− 000+ 0
$2\ \frac{3}{2}\ \frac{3}{2}\ \ 1\ -\frac{3}{2}\ \frac{1}{2}$
$1\ 0\ 1$ 000− 000− $-1/2.3$
$1\ 1\ 1$ 000+ 000+ $-1/2.3\sqrt{2.3}$
$1\ 1\ 1$ 001− 000+ $+i/2.3\sqrt{2}$
$2\ \frac{3}{2}\ \frac{3}{2}\ \ 1\ -\frac{3}{2}\ \frac{3}{2}$
$1\ 0\ 1$ 000− 000− $+1/2.3$
$1\ 1\ 1$ 000+ 000+ $+1/2.3\sqrt{2.3}$
$1\ 1\ 1$ 001− 000+ $+i/2.3\sqrt{2}$
$1\ 1\ 2$ 000− 000− $-1/2.3$
$2\ \frac{3}{2}\ \frac{3}{2}\ \ 1\ -\frac{3}{2}\ -\frac{3}{2}$
$1\ 1\ 0$ 000− 000− $-1/2.3$
$1\ 1\ 1$ 000+ 000+ $-1/3\sqrt{2.3}$
$1\ 1\ 1$ 001− 000+ 0
$2\ \frac{3}{2}\ \frac{3}{2}\ \ 1\ 2\ 1$
$1\ \frac{3}{2}\ \frac{1}{2}$ 000− 000− $-1/2.3$
$1\ \frac{3}{2}\ \frac{3}{2}$ 000− 000− $-1/2.3$
$1\ \frac{3}{2}\ -\frac{3}{2}$ 000− 000− $-1/2.3$
$2\ \frac{3}{2}\ \frac{3}{2}\ \ \frac{3}{2}\ 1\ \frac{1}{2}$
$\frac{3}{2}\ \frac{1}{2}\ 1$ 000− 000− $+1/4.3$
$2\ \frac{3}{2}\ \frac{3}{2}\ \ \frac{3}{2}\ \frac{3}{2}\ 1$
$\frac{3}{2}\ 1\ \frac{1}{2}$ 000− 000− $+1/2.3$
$\frac{3}{2}\ 1\ \frac{3}{2}$ 000− 000− $-1/4.3$
$\frac{3}{2}\ 1\ -\frac{3}{2}$ 000− 000− $+1/2.3$
$2\ \frac{3}{2}\ \frac{3}{2}\ \ \frac{3}{2}\ -\frac{3}{2}\ 0$
$\frac{3}{2}\ 0\ -\frac{3}{2}$ 000− 000− $+1/4$
$\frac{3}{2}\ 1\ -\frac{3}{2}$ 000+ 000+ $+1/4$
$2\ \frac{3}{2}\ \frac{3}{2}\ \ \frac{3}{2}\ -\frac{3}{2}\ 1$
$\frac{3}{2}\ 0\ -\frac{3}{2}$ 000+ 000+ $+1/4$
$\frac{3}{2}\ 1\ \frac{1}{2}$ 000− 000− $+1/2.3$
$\frac{3}{2}\ 1\ \frac{3}{2}$ 000− 000− $+1/2.3$
$\frac{3}{2}\ 1\ -\frac{3}{2}$ 000− 000− $-1/4.3$
$2\ \frac{3}{2}\ \frac{3}{2}\ \ \frac{3}{2}\ 2\ \frac{3}{2}$
$\frac{3}{2}\ \frac{3}{2}\ 1$ 000+ 000+ $+1/4$
$\frac{3}{2}\ \frac{3}{2}\ 2$ 000− 000− $-1/4$
$2\ -\frac{3}{2}\ \frac{1}{2}\ \ 1\ 1\ 0$
$1\ \frac{1}{2}\ \frac{1}{2}$ 000− 000− $+1/2.3$
$2\ -\frac{3}{2}\ \frac{1}{2}\ \ 1\ 1\ 1$
$1\ \frac{1}{2}\ \frac{1}{2}$ 000+ 000+ $+1/2.3\sqrt{2.3}$
$1\ \frac{1}{2}\ \frac{1}{2}$ 010− 000+ $-i/2.3\sqrt{2}$
$1\ \frac{1}{2}\ \frac{3}{2}$ 000+ 000+ $+1/2.3\sqrt{2.3}$
$1\ \frac{1}{2}\ \frac{3}{2}$ 010− 000+ $+i/2.3\sqrt{2}$
$1\ \frac{1}{2}\ -\frac{3}{2}$ 000+ 000+ $-1/3\sqrt{2.3}$
$1\ \frac{1}{2}\ -\frac{3}{2}$ 010− 000+ 0
$2\ -\frac{3}{2}\ \frac{1}{2}\ \ 1\ 1\ 2$
$1\ \frac{1}{2}\ -\frac{3}{2}$ 000− 000− $+1/2.3$
$2\ -\frac{3}{2}\ \frac{1}{2}\ \ 1\ \frac{3}{2}\ \frac{1}{2}$
$1\ 0\ 1$ 000− 000− $+1/2.3$
$1\ 1\ 0$ 000− 000− $-1/2.3$
$1\ 1\ 1$ 000+ 000+ $-1/3\sqrt{2.3}$
$1\ 1\ 1$ 001− 000+ 0

T 9j Symbols (cont.)

2 -3/2 1/2 1 3/2 3/2
1 0 1 000− 000− −1/2.3
1 1 1 000+ 000+ $-1/2.3\sqrt{2.3}$
1 1 1 001− 000+ $+i/2.3\sqrt{2}$
2 -3/2 1/2 1 3/2 -3/2
1 0 1 000− 000− +1/2.3
1 1 1 000+ 000+ $+1/2.3\sqrt{2.3}$
1 1 1 001− 000+ $+i/2.3\sqrt{2}$
1 1 2 000− 000− −1/2.3
2 -3/2 1/2 1 -3/2 1/2
1 1 0 000− 000− +1/2.3
1 1 1 000+ 000+ $-1/2.3\sqrt{2.3}$
1 1 1 001− 000+ $-i/2.3\sqrt{2}$
2 -3/2 1/2 1 -3/2 3/2
1 1 1 000+ 000+ $+1/3\sqrt{2.3}$
1 1 1 001− 000+ 0
2 -3/2 1/2 1 -3/2 -3/2
1 1 1 000+ 000+ $+1/2.3\sqrt{2.3}$
1 1 1 001− 000+ $-i/2.3\sqrt{2}$
1 1 2 000− 000− −1/2.3
2 -3/2 1/2 1 2 1
1 1/2 1/2 000− 000− −1/2.3
1 1/2 3/2 000− 000− −1/2.3
1 1/2 -3/2 000− 000− −1/2.3
2 -3/2 1/2 3/2 1 1/2
3/2 1/2 1 000− 000− +1/2.3
2 -3/2 1/2 3/2 1 3/2
3/2 1/2 1 000− 000− +1/4.3
2 -3/2 1/2 3/2 1 -3/2
3/2 1/2 1 000− 000− +1/2.3
2 -3/2 1/2 3/2 3/2 1
3/2 0 -3/2 000+ 000+ +1/4
3/2 1 1/2 000− 000− +1/2.3
3/2 1 3/2 000− 000− −1/2.3
3/2 1 -3/2 000− 000− −1/4.3
2 -3/2 1/2 3/2 3/2 2
3/2 0 -3/2 000− 000− +1/4
3/2 1 -3/2 000+ 000+ +1/4
2 -3/2 1/2 3/2 -3/2 0
3/2 1 1/2 000+ 000+ +1/4
2 -3/2 1/2 3/2 -3/2 1
3/2 1 1/2 000− 000− −1/4.3
3/2 1 3/2 000− 000− −1/2.3
3/2 1 -3/2 000− 000− +1/2.3
2 -3/2 1/2 3/2 2 3/2
3/2 1/2 1 000+ 000+ +1/4
2 -3/2 1/2 -3/2 0 3/2
1/2 3/2 1 000+ 000+ +1/4
2 -3/2 1/2 -3/2 1/2 1
1/2 1 1/2 000− 000− +1/2.3

2 -3/2 1/2 -3/2 1/2 1
1/2 1 3/2 000− 000− +1/2.3
1/2 1 -3/2 000− 000− −1/4.3
2 -3/2 1/2 -3/2 1/2 2
1/2 1 -3/2 000+ 000+ +1/4
1/2 2 -3/2 000− 000− +1/4
2 -3/2 1/2 -3/2 1 1/2
1/2 1/2 0 000+ 000+ +1/4
1/2 1/2 1 000− 000− +1/4.3
2 -3/2 1/2 -3/2 1 3/2
1/2 1/2 1 000− 000− −1/2.3
1/2 3/2 1 000− 000− +1/4.3
2 -3/2 1/2 -3/2 1 -3/2
1/2 1/2 1 000− 000− +1/2.3
1/2 3/2 1 000− 000− +1/2.3
1/2 -3/2 1 000− 000− +1/4.3
1/2 -3/2 2 000+ 000+ +1/4
2 -3/2 1/2 -3/2 3/2 0
1/2 0 1/2 000− 000− +1/4
2 -3/2 1/2 -3/2 3/2 1
1/2 0 1/2 000+ 000+ +1/4
1/2 1 1/2 000− 000− −1/4.3
1/2 1 3/2 000− 000− +1/2.3
1/2 1 -3/2 000− 000− +1/2.3
2 -3/2 1/2 -3/2 -3/2 1
1/2 1 1/2 000− 000− +1/2.3
1/2 1 3/2 000− 000− −1/4.3
1/2 1 -3/2 000− 000− +1/2.3
2 -3/2 1/2 -3/2 2 1/2
1/2 1/2 0 000− 000− −1/4
1/2 1/2 1 000+ 000+ +1/4
2 2 2 1 1 1
1 1 1 000+ 000+ $+1/3\sqrt{3}$
1 1 1 010− 000+ 0
1 1 1 011+ 000+ $+1/3\sqrt{3}$
2 2 2 3/2 3/2 1
3/2 3/2 1 000+ 000+ $+1/2\sqrt{3}$
2 2 2 -3/2 1 1/2
1/2 1 -3/2 000+ 000+ $+1/2\sqrt{3}$
2 2 2 -3/2 3/2 1
1/2 3/2 1 000+ 000+ $-1/2\sqrt{3}$
2 2 2 -3/2 -3/2 1
1/2 1/2 1 000+ 000+ $-1/2\sqrt{3}$
2 2 2 2 1 1
2 1 1 000+ 000+ +1/3
2 2 2 2 3/2 3/2
2 3/2 3/2 000+ 000+ +1/2
2 2 2 2 -3/2 1/2
2 1/2 -3/2 000+ 000+ +1/2

2 2 2 2 2 2
2 2 2 000+ 000+ +1
-2 1 1 1 1/2 1/2
1 1/2 1/2 000+ 000+ +1/9
-2 1 1 1 1 0
1 0 1 000+ 000+ +1/9
-2 1 1 1 1 1
1 1 0 000− 000− +1/2.9
1 1 0 000− 010+ $+i/2.3\sqrt{3}$
1 1 0 010+ 000− $-i/2.3\sqrt{3}$
1 1 0 010+ 010+ −1/2.9
1 1 1 000+ 000+ +1/2.9
1 1 1 000+ 010− 0
1 1 1 000+ 011+ −1/2.9
1 1 1 010− 000+ 0
1 1 1 010− 010− +1/2.9
1 1 1 010− 011+ 0
1 1 1 011+ 000+ −1/2.9
1 1 1 011+ 010− 0
1 1 1 011+ 011+ +1/2.9
-2 1 1 1 3/2 1/2
1 1/2 1/2 000+ 000+ +1/9
1 1/2 3/2 000+ 000+ −1/2.9
1 1/2 -3/2 000+ 000+ +1/9
1 3/2 1/2 000+ 000+ +1/2.9
-2 1 1 1 3/2 3/2
1 1/2 1/2 000+ 000+ +1/2.9
1 3/2 1/2 000+ 000+ +1/9
1 3/2 3/2 000+ 000+ +1/9
-2 1 1 1 -3/2 1/2
1 1/2 1/2 000+ 000+ −1/2.9
1 1/2 3/2 000+ 000+ +1/9
1 1/2 -3/2 000+ 000+ +1/9
1 3/2 1/2 000+ 000+ +1/9
1 3/2 3/2 000+ 000+ −1/9
1 3/2 -3/2 000+ 000+ −1/2.9
1 -3/2 1/2 000+ 000+ −1/9
-2 1 1 1 -3/2 3/2
1 1/2 1/2 000+ 000+ +1/9
1 1/2 3/2 000+ 000+ −1/9
1 1/2 -3/2 000+ 000+ −1/2.9
1 3/2 1/2 000+ 000+ +1/9
1 3/2 3/2 000+ 000+ −1/2.9
1 3/2 -3/2 000+ 000+ +1/9
1 -3/2 1/2 000+ 000+ +1/2.9
1 -3/2 3/2 000+ 000+ +1/9
-2 1 1 1 -3/2 -3/2
1 1/2 1/2 000+ 000+ −1/9
1 3/2 1/2 000+ 000+ +1/2.9
1 3/2 3/2 000+ 000+ +1/9

T 9j Symbols (cont.)

-2 1 1 1 $-\frac{3}{2}$ $-\frac{3}{2}$
1 $-\frac{3}{2}$ $\frac{1}{2}$ 000+ 000+ $+1/9$
1 $-\frac{3}{2}$ $\frac{3}{2}$ 000+ 000+ $+1/9$
1 $-\frac{3}{2}$ $-\frac{3}{2}$ 000+ 000+ $-1/2.9$
-2 1 1 1 2 1
1 1 0 000+ 000+ $+1/9$
1 1 1 000− 000− $+1/2.9$
1 1 1 000− 001+ $+i/2.3\sqrt{3}$
1 1 1 001+ 000− $-i/2.3\sqrt{3}$
1 1 1 001+ 001+ $-1/2.9$
1 1 2 000+ 000+ $+1/9$
1 1 -2 000+ 000+ $+1/9$
-2 1 1 1 -2 1
1 1 0 000+ 000+ $+1/9$
1 1 1 000− 000− $-1/9$
1 1 1 000− 001+ 0
1 1 1 001+ 000− 0
1 1 1 001+ 001+ $+1/9$
1 1 2 000+ 000+ $+1/9$
1 1 -2 000+ 000+ $+1/9$
-2 $\frac{3}{2}$ $\frac{1}{2}$ 1 1 0
1 $\frac{1}{2}$ $\frac{1}{2}$ 000− 000− $+1/2.3$
-2 $\frac{3}{2}$ $\frac{1}{2}$ 1 1 1
1 $\frac{1}{2}$ $\frac{1}{2}$ 000+ 000+ $+1/2.3\sqrt{2.3}$
1 $\frac{1}{2}$ $\frac{1}{2}$ 010− 000+ $+i/2.3\sqrt{2}$
1 $\frac{1}{2}$ $\frac{3}{2}$ 000+ 000+ $-1/3\sqrt{2.3}$
1 $\frac{1}{2}$ $\frac{3}{2}$ 010− 000+ 0
1 $\frac{1}{2}$ $-\frac{3}{2}$ 000+ 000+ $+1/2.3\sqrt{2.3}$
1 $\frac{1}{2}$ $-\frac{3}{2}$ 010− 000+ $-i/2.3\sqrt{2}$
-2 $\frac{3}{2}$ $\frac{1}{2}$ 1 1 2
1 $\frac{1}{2}$ $-\frac{3}{2}$ 000− 000− $+1/2.3$
-2 $\frac{3}{2}$ $\frac{1}{2}$ 1 1 -2
1 $\frac{1}{2}$ $\frac{3}{2}$ 000− 000− $+1/2.3$
-2 $\frac{3}{2}$ $\frac{1}{2}$ 1 $\frac{3}{2}$ $\frac{1}{2}$
1 1 0 000− 000− $+1/2.3$
1 1 1 000+ 000+ $-1/2.3\sqrt{2.3}$
1 1 1 001− 000+ $+i/2.3\sqrt{2}$
-2 $\frac{3}{2}$ $\frac{1}{2}$ 1 $\frac{3}{2}$ $\frac{3}{2}$
1 1 1 000+ 000+ $+1/2.3\sqrt{2.3}$
1 1 1 001− 000+ $+i/2.3\sqrt{2}$
1 1 -2 000− 000− $-1/2.3$
-2 $\frac{3}{2}$ $\frac{1}{2}$ 1 $\frac{3}{2}$ $-\frac{3}{2}$
1 1 1 000+ 000+ $+1/3\sqrt{2.3}$
1 1 1 001− 000+ 0
1 1 2 000− 000− $+1/2.3$
-2 $\frac{3}{2}$ $\frac{1}{2}$ 1 $-\frac{3}{2}$ $\frac{1}{2}$
1 0 1 000− 000− $+1/2.3$
1 1 0 000− 000− $-1/2.3$
1 1 1 000+ 000+ $-1/3\sqrt{2.3}$
1 1 1 001− 000+ 0
-2 $\frac{3}{2}$ $\frac{1}{2}$ 1 $-\frac{3}{2}$ $\frac{3}{2}$
1 0 1 000− 000− $+1/2.3$
1 1 1 000+ 000+ $+1/2.3\sqrt{2.3}$
1 1 1 001− 000+ $-i/2.3\sqrt{2}$
1 1 -2 000− 000− $-1/2.3$
-2 $\frac{3}{2}$ $\frac{1}{2}$ 1 $-\frac{3}{2}$ $-\frac{3}{2}$
1 0 1 000− 000− $-1/2.3$
1 1 1 000+ 000+ $-1/2.3\sqrt{2.3}$
1 1 1 001− 000+ $-i/2.3\sqrt{2}$
1 1 2 000− 000− $+1/2.3$
-2 $\frac{3}{2}$ $\frac{1}{2}$ 1 2 1
1 $\frac{3}{2}$ $\frac{1}{2}$ 000− 000− $+1/2.3$
1 $\frac{3}{2}$ $\frac{3}{2}$ 000− 000− $-1/2.3$
1 $\frac{3}{2}$ $-\frac{3}{2}$ 000− 000− $+1/2.3$
-2 $\frac{3}{2}$ $\frac{1}{2}$ 1 -2 1
1 $\frac{1}{2}$ $\frac{1}{2}$ 000− 000− $-1/2.3$
1 $\frac{1}{2}$ $\frac{3}{2}$ 000− 000− $-1/2.3$
1 $\frac{1}{2}$ $-\frac{3}{2}$ 000− 000− $-1/2.3$
-2 $\frac{3}{2}$ $\frac{1}{2}$ $\frac{3}{2}$ 0 $-\frac{3}{2}$
$\frac{1}{2}$ $-\frac{3}{2}$ 1 000+ 000+ $+1/4$
$\frac{1}{2}$ $-\frac{3}{2}$ 2 000− 000− $-1/4$
-2 $\frac{3}{2}$ $\frac{1}{2}$ $\frac{3}{2}$ $\frac{1}{2}$ 1
$\frac{1}{2}$ 1 $\frac{1}{2}$ 000− 000− $+1/2.3$
$\frac{1}{2}$ 1 $\frac{3}{2}$ 000− 000− $-1/4.3$
$\frac{1}{2}$ 1 $-\frac{3}{2}$ 000− 000− $+1/2.3$
-2 $\frac{3}{2}$ $\frac{1}{2}$ $\frac{3}{2}$ $\frac{1}{2}$ -2
$\frac{1}{2}$ 1 $\frac{3}{2}$ 000+ 000+ $+1/4$
$\frac{1}{2}$ -2 $\frac{3}{2}$ 000− 000− $+1/4$
-2 $\frac{3}{2}$ $\frac{1}{2}$ $\frac{3}{2}$ 1 $\frac{1}{2}$
$\frac{1}{2}$ $\frac{1}{2}$ 0 000+ 000+ $+1/4$
$\frac{1}{2}$ $\frac{1}{2}$ 1 000− 000− $+1/4.3$
-2 $\frac{3}{2}$ $\frac{1}{2}$ $\frac{3}{2}$ 1 $\frac{3}{2}$
$\frac{1}{2}$ $\frac{1}{2}$ 1 000− 000− $+1/2.3$
$\frac{1}{2}$ $\frac{3}{2}$ 1 000− 000− $+1/4.3$
$\frac{1}{2}$ $\frac{3}{2}$ -2 000+ 000+ $+1/4$
-2 $\frac{3}{2}$ $\frac{1}{2}$ $\frac{3}{2}$ 1 $-\frac{3}{2}$
$\frac{1}{2}$ $\frac{1}{2}$ 1 000− 000− $-1/2.3$
$\frac{1}{2}$ $\frac{3}{2}$ 1 000− 000− $+1/2.3$
$\frac{1}{2}$ $-\frac{3}{2}$ 1 000− 000− $+1/4.3$
$\frac{1}{2}$ $-\frac{3}{2}$ 2 000+ 000+ $+1/4$
-2 $\frac{3}{2}$ $\frac{1}{2}$ $\frac{3}{2}$ $\frac{3}{2}$ 1
$\frac{1}{2}$ 1 $\frac{1}{2}$ 000− 000− $+1/2.3$
$\frac{1}{2}$ 1 $\frac{3}{2}$ 000− 000− $+1/2.3$
$\frac{1}{2}$ 1 $-\frac{3}{2}$ 000− 000− $-1/4.3$
-2 $\frac{3}{2}$ $\frac{1}{2}$ $\frac{3}{2}$ $\frac{3}{2}$ 2
$\frac{1}{2}$ 1 $-\frac{3}{2}$ 000+ 000+ $+1/4$
$\frac{1}{2}$ 2 $-\frac{3}{2}$ 000− 000− $+1/4$
-2 $\frac{3}{2}$ $\frac{1}{2}$ $\frac{3}{2}$ $-\frac{3}{2}$ 0
$\frac{1}{2}$ 0 $\frac{1}{2}$ 000− 000− $+1/4$
-2 $\frac{3}{2}$ $\frac{1}{2}$ $\frac{3}{2}$ $-\frac{3}{2}$ 1
$\frac{1}{2}$ 0 $\frac{1}{2}$ 000+ 000+ $+1/4$
$\frac{1}{2}$ 1 $\frac{1}{2}$ 000− 000− $-1/4.3$
$\frac{1}{2}$ 1 $\frac{3}{2}$ 000− 000− $+1/2.3$
$\frac{1}{2}$ 1 $-\frac{3}{2}$ 000− 000− $+1/2.3$
-2 $\frac{3}{2}$ $\frac{1}{2}$ $\frac{3}{2}$ 2 $\frac{3}{2}$
$\frac{1}{2}$ $\frac{3}{2}$ 1 000+ 000+ $+1/4$
$\frac{1}{2}$ $\frac{3}{2}$ -2 000− 000− $-1/4$
-2 $\frac{3}{2}$ $\frac{1}{2}$ $\frac{3}{2}$ -2 $\frac{1}{2}$
$\frac{1}{2}$ $\frac{1}{2}$ 0 000− 000− $-1/4$
$\frac{1}{2}$ $\frac{1}{2}$ 1 000+ 000+ $+1/4$
-2 $-\frac{3}{2}$ $-\frac{3}{2}$ 1 1 0
1 $\frac{1}{2}$ $\frac{3}{2}$ 000− 000− $-1/2.3$
-2 $-\frac{3}{2}$ $-\frac{3}{2}$ 1 1 1
1 $\frac{1}{2}$ $\frac{1}{2}$ 000+ 000+ $+1/3\sqrt{2.3}$
1 $\frac{1}{2}$ $\frac{1}{2}$ 010− 000+ 0
-2 $-\frac{3}{2}$ $-\frac{3}{2}$ 1 $\frac{3}{2}$ $\frac{1}{2}$
1 0 1 000− 000− $-1/2.3$
1 1 1 000+ 000+ $-1/2.3\sqrt{2.3}$
1 1 1 001− 000+ $-i/2.3\sqrt{2}$
1 1 2 000− 000− $+1/2.3$
-2 $-\frac{3}{2}$ $-\frac{3}{2}$ 1 $\frac{3}{2}$ $\frac{3}{2}$
1 1 0 000− 000− $-1/2.3$
1 1 1 000+ 000+ $-1/3\sqrt{2.3}$
1 1 1 001− 000+ 0
-2 $-\frac{3}{2}$ $-\frac{3}{2}$ 1 $-\frac{3}{2}$ $\frac{1}{2}$
1 1 1 000+ 000+ $+1/2.3\sqrt{2.3}$
1 1 1 001− 000+ $-i/2.3\sqrt{2}$
1 1 2 000− 000− $-1/2.3$
-2 $-\frac{3}{2}$ $-\frac{3}{2}$ 1 $-\frac{3}{2}$ $\frac{3}{2}$
1 1 0 000− 000− $-1/2.3$
1 1 1 000+ 000+ $+1/2.3\sqrt{2.3}$
1 1 1 001− 000+ $+i/2.3\sqrt{2}$
-2 $-\frac{3}{2}$ $-\frac{3}{2}$ 1 $-\frac{3}{2}$ $-\frac{3}{2}$
1 1 1 000+ 000+ $-1/3\sqrt{2.3}$
1 1 1 001− 000+ 0
-2 $-\frac{3}{2}$ $-\frac{3}{2}$ 1 2 1
1 $\frac{1}{2}$ $\frac{1}{2}$ 000− 000− $+1/2.3$
1 $\frac{1}{2}$ $\frac{3}{2}$ 000− 000− $+1/2.3$
1 $\frac{1}{2}$ $-\frac{3}{2}$ 000− 000− $-1/2.3$
-2 $-\frac{3}{2}$ $-\frac{3}{2}$ 1 -2 1
1 $-\frac{3}{2}$ $\frac{1}{2}$ 000− 000− $-1/2.3$
1 $-\frac{3}{2}$ $\frac{3}{2}$ 000− 000− $-1/2.3$
1 $-\frac{3}{2}$ $-\frac{3}{2}$ 000− 000− $-1/2.3$
-2 $-\frac{3}{2}$ $-\frac{3}{2}$ $\frac{3}{2}$ 1 $\frac{1}{2}$
$\frac{1}{2}$ $\frac{1}{2}$ 1 000− 000− $+1/2.3$
$\frac{1}{2}$ $\frac{3}{2}$ 1 000− 000− $+1/2.3$
$\frac{1}{2}$ $-\frac{3}{2}$ 1 000− 000− $+1/4.3$
$\frac{1}{2}$ $-\frac{3}{2}$ 2 000+ 000+ $+1/4$

T 9j Symbols (cont.)

$-2\ -\frac{3}{2}\ -\frac{3}{2}\quad \frac{3}{2}\ \frac{3}{2}\ 1$
$\frac{1}{2}\ 0\ \frac{1}{2}$ 000+ 000+ $+1/4$
$\frac{1}{2}\ 1\ \frac{1}{2}$ 000− 000− $-1/4.3$
$\frac{1}{2}\ 1\ \frac{3}{2}$ 000− 000− $+1/2.3$
$\frac{1}{2}\ 1\ -\frac{3}{2}$ 000− 000− $-1/2.3$
$-2\ -\frac{3}{2}\ -\frac{3}{2}\quad \frac{3}{2}\ -\frac{3}{2}\ 0$
$\frac{1}{2}\ 1\ \frac{3}{2}$ 000+ 000+ $+1/4$
$\frac{1}{2}\ -2\ \frac{3}{2}$ 000− 000− $+1/4$
$-2\ -\frac{3}{2}\ -\frac{3}{2}\quad \frac{3}{2}\ -\frac{3}{2}\ 1$
$\frac{1}{2}\ 1\ \frac{1}{2}$ 000− 000− $+1/2.3$
$\frac{1}{2}\ 1\ \frac{3}{2}$ 000− 000− $-1/4.3$
$\frac{1}{2}\ 1\ -\frac{3}{2}$ 000− 000− $-1/2.3$
$\frac{1}{2}\ -2\ \frac{3}{2}$ 000+ 000+ $+1/4$
$-2\ -\frac{3}{2}\ -\frac{3}{2}\quad \frac{3}{2}\ 2\ \frac{3}{2}$
$\frac{1}{2}\ \frac{1}{2}\ 0$ 000− 000− $-1/4$
$\frac{1}{2}\ \frac{1}{2}\ 1$ 000+ 000+ $+1/4$
$-2\ -\frac{3}{2}\ -\frac{3}{2}\quad \frac{3}{2}\ -2\ \frac{1}{2}$
$\frac{1}{2}\ -\frac{3}{2}\ 1$ 000+ 000+ $+1/4$
$\frac{1}{2}\ -\frac{3}{2}\ 2$ 000− 000− $-1/4$
$-2\ -\frac{3}{2}\ -\frac{3}{2}\quad -\frac{3}{2}\ 1\ \frac{1}{2}$
$-\frac{3}{2}\ \frac{1}{2}\ 1$ 000− 000− $+1/4.3$
$-\frac{3}{2}\ \frac{1}{2}\ 2$ 000+ 000+ $+1/4$
$-2\ -\frac{3}{2}\ -\frac{3}{2}\quad -\frac{3}{2}\ \frac{3}{2}\ 0$
$-\frac{3}{2}\ 0\ \frac{3}{2}$ 000− 000− $+1/4$
$-\frac{3}{2}\ 1\ \frac{3}{2}$ 000+ 000+ $+1/4$
$-2\ -\frac{3}{2}\ -\frac{3}{2}\quad -\frac{3}{2}\ \frac{3}{2}\ 1$
$-\frac{3}{2}\ 0\ \frac{3}{2}$ 000+ 000+ $+1/4$
$-\frac{3}{2}\ 1\ \frac{1}{2}$ 000− 000− $+1/2.3$
$-\frac{3}{2}\ 1\ \frac{3}{2}$ 000− 000− $-1/4.3$
$-\frac{3}{2}\ 1\ -\frac{3}{2}$ 000− 000− $+1/2.3$
$-2\ -\frac{3}{2}\ -\frac{3}{2}\quad -\frac{3}{2}\ -\frac{3}{2}\ 1$
$-\frac{3}{2}\ 1\ \frac{1}{2}$ 000− 000− $+1/2.3$
$-\frac{3}{2}\ 1\ \frac{3}{2}$ 000− 000− $+1/2.3$
$-\frac{3}{2}\ 1\ -\frac{3}{2}$ 000− 000− $-1/4.3$
$-2\ -\frac{3}{2}\ -\frac{3}{2}\quad -\frac{3}{2}\ 2\ \frac{1}{2}$
$-\frac{3}{2}\ \frac{1}{2}\ 1$ 000+ 000+ $+1/4$
$-\frac{3}{2}\ \frac{1}{2}\ 2$ 000− 000− $-1/4$
$-2\ -\frac{3}{2}\ -\frac{3}{2}\quad -\frac{3}{2}\ -2\ -\frac{3}{2}$
$-\frac{3}{2}\ -\frac{3}{2}\ 1$ 000+ 000+ $+1/4$
$-\frac{3}{2}\ -\frac{3}{2}\ -2$ 000− 000− $-1/4$
$-2\ 2\ 0\quad 1\ 1\ 0$
$1\ 1\ 0$ 000+ 000+ $+1/3$
$-2\ 2\ 0\quad 1\ 1\ 1$
$1\ 1\ 1$ 000+ 000+ $-1/2.3\sqrt{3}$
$1\ 1\ 1$ 010− 000+ $-i/2.3$
$1\ 1\ 1$ 011+ 000+ $-1/2.3\sqrt{3}$
$-2\ 2\ 0\quad 1\ 1\ -2$
$1\ 1\ 2$ 000+ 000+ $+1/3$
$-2\ 2\ 0\quad 1\ \frac{3}{2}\ \frac{1}{2}$
$1\ \frac{3}{2}\ \frac{1}{2}$ 000+ 000+ $-1/2\sqrt{3}$

$-2\ 2\ 0\quad 1\ \frac{3}{2}\ -\frac{3}{2}$
$1\ \frac{3}{2}\ \frac{3}{2}$ 000+ 000+ $+1/2\sqrt{3}$
$-2\ 2\ 0\quad 1\ -\frac{3}{2}\ \frac{1}{2}$
$1\ \frac{1}{2}\ \frac{1}{2}$ 000+ 000+ $+1/2\sqrt{3}$
$-2\ 2\ 0\quad 1\ -\frac{3}{2}\ \frac{3}{2}$
$1\ \frac{1}{2}\ -\frac{3}{2}$ 000+ 000+ $+1/2\sqrt{3}$
$-2\ 2\ 0\quad 1\ -\frac{3}{2}\ -\frac{3}{2}$
$1\ \frac{1}{2}\ \frac{3}{2}$ 000+ 000+ $-1/2\sqrt{3}$
$-2\ 2\ 0\quad 1\ 2\ 1$
$1\ 2\ 1$ 000+ 000+ $+1/3$
$-2\ 2\ 0\quad 1\ -2\ 1$
$1\ 0\ 1$ 000+ 000+ $+1/3$
$-2\ 2\ 0\quad \frac{3}{2}\ 0\ -\frac{3}{2}$
$\frac{1}{2}\ -2\ \frac{3}{2}$ 000+ 000+ $+1/2$
$-2\ 2\ 0\quad \frac{3}{2}\ \frac{1}{2}\ 1$
$\frac{1}{2}\ -\frac{3}{2}\ 1$ 000+ 000+ $+1/2\sqrt{3}$
$-2\ 2\ 0\quad \frac{3}{2}\ \frac{1}{2}\ -2$
$\frac{1}{2}\ -\frac{3}{2}\ 2$ 000+ 000+ $+1/2$
$-2\ 2\ 0\quad \frac{3}{2}\ 1\ \frac{1}{2}$
$\frac{1}{2}\ 1\ \frac{1}{2}$ 000+ 000+ $+1/2\sqrt{3}$
$-2\ 2\ 0\quad \frac{3}{2}\ 1\ \frac{3}{2}$
$\frac{1}{2}\ 1\ -\frac{3}{2}$ 000+ 000+ $-1/2\sqrt{3}$
$-2\ 2\ 0\quad \frac{3}{2}\ 1\ -\frac{3}{2}$
$\frac{1}{2}\ 1\ \frac{3}{2}$ 000+ 000+ $+1/2\sqrt{3}$
$-2\ 2\ 0\quad \frac{3}{2}\ \frac{3}{2}\ 1$
$\frac{1}{2}\ \frac{3}{2}\ 1$ 000+ 000+ $+1/2\sqrt{3}$
$-2\ 2\ 0\quad \frac{3}{2}\ \frac{3}{2}\ 2$
$\frac{1}{2}\ \frac{3}{2}\ -2$ 000+ 000+ $+1/2$
$-2\ 2\ 0\quad \frac{3}{2}\ -\frac{3}{2}\ 0$
$\frac{1}{2}\ \frac{1}{2}\ 0$ 000+ 000+ $+1/2$
$-2\ 2\ 0\quad \frac{3}{2}\ -\frac{3}{2}\ 1$
$\frac{1}{2}\ \frac{1}{2}\ 1$ 000+ 000+ $+1/2\sqrt{3}$
$-2\ 2\ 0\quad \frac{3}{2}\ 2\ \frac{3}{2}$
$\frac{1}{2}\ 2\ -\frac{3}{2}$ 000+ 000+ $-1/2$
$-2\ 2\ 0\quad \frac{3}{2}\ -2\ \frac{1}{2}$
$\frac{1}{2}\ 0\ \frac{1}{2}$ 000+ 000+ $+1/2$
$-2\ 2\ 0\quad -\frac{3}{2}\ 1\ \frac{1}{2}$
$-\frac{3}{2}\ 1\ \frac{1}{2}$ 000+ 000+ $-1/2\sqrt{3}$
$-2\ 2\ 0\quad -\frac{3}{2}\ 1\ -\frac{3}{2}$
$-\frac{3}{2}\ 1\ \frac{3}{2}$ 000+ 000+ $+1/2\sqrt{3}$
$-2\ 2\ 0\quad -\frac{3}{2}\ \frac{3}{2}\ 0$
$-\frac{3}{2}\ \frac{3}{2}\ 0$ 000+ 000+ $+1/2$
$-2\ 2\ 0\quad -\frac{3}{2}\ \frac{3}{2}\ 1$
$-\frac{3}{2}\ \frac{3}{2}\ 1$ 000+ 000+ $+1/2\sqrt{3}$
$-2\ 2\ 0\quad -\frac{3}{2}\ -\frac{3}{2}\ 1$
$-\frac{3}{2}\ \frac{1}{2}\ 1$ 000+ 000+ $+1/2\sqrt{3}$
$-2\ 2\ 0\quad -\frac{3}{2}\ -\frac{3}{2}\ -2$
$-\frac{3}{2}\ \frac{1}{2}\ 2$ 000+ 000+ $+1/2$
$-2\ 2\ 0\quad -\frac{3}{2}\ 2\ \frac{1}{2}$
$-\frac{3}{2}\ 2\ \frac{1}{2}$ 000+ 000+ $-1/2$

$-2\ 2\ 0\quad -\frac{3}{2}\ -2\ -\frac{3}{2}$
$-\frac{3}{2}\ 0\ \frac{3}{2}$ 000+ 000+ $+1/2$
$-2\ 2\ 0\quad 2\ 0\ -2$
$0\ -2\ 2$ 000+ 000+ $+1$
$-2\ 2\ 0\quad 2\ \frac{1}{2}\ -\frac{3}{2}$
$0\ -\frac{3}{2}\ \frac{3}{2}$ 000+ 000+ $+1/2$
$-2\ 2\ 0\quad 2\ 1\ 1$
$0\ 1\ 1$ 000+ 000+ $+1/3$
$-2\ 2\ 0\quad 2\ \frac{3}{2}\ \frac{3}{2}$
$0\ \frac{3}{2}\ -\frac{3}{2}$ 000+ 000+ $+1/2$
$-2\ 2\ 0\quad 2\ -\frac{3}{2}\ \frac{1}{2}$
$0\ \frac{1}{2}\ \frac{1}{2}$ 000+ 000+ $+1/2$
$-2\ 2\ 0\quad 2\ 2\ 2$
$0\ 2\ -2$ 000+ 000+ $+1$
$-2\ 2\ 0\quad 2\ -2\ 0$
$0\ 0\ 0$ 000+ 000+ $+1$
$-2\ -2\ -2\quad 1\ 1\ 1$
$1\ 1\ 1$ 000+ 000+ $+1/3\sqrt{3}$
$1\ 1\ 1$ 010− 000+ 0
$1\ 1\ 1$ 011+ 000+ $+1/3\sqrt{3}$
$-2\ -2\ -2\quad \frac{3}{2}\ 1\ \frac{1}{2}$
$\frac{1}{2}\ 1\ \frac{3}{2}$ 000+ 000+ $+1/2\sqrt{3}$
$-2\ -2\ -2\quad \frac{3}{2}\ \frac{3}{2}\ 1$
$\frac{1}{2}\ \frac{1}{2}\ 1$ 000+ 000+ $-1/2\sqrt{3}$
$-2\ -2\ -2\quad -\frac{3}{2}\ \frac{3}{2}\ 0$
$-\frac{3}{2}\ \frac{1}{2}\ 2$ 000+ 000+ $-1/2$
$-2\ -2\ -2\quad -\frac{3}{2}\ \frac{3}{2}\ 1$
$-\frac{3}{2}\ \frac{1}{2}\ 1$ 000+ 000+ $-1/2\sqrt{3}$
$-2\ -2\ -2\quad -\frac{3}{2}\ -\frac{3}{2}\ 1$
$-\frac{3}{2}\ -\frac{3}{2}\ 1$ 000+ 000+ $+1/2\sqrt{3}$
$-2\ -2\ -2\quad 2\ 1\ 1$
$0\ 1\ 1$ 000+ 000+ $+1/3$
$-2\ -2\ -2\quad 2\ \frac{3}{2}\ \frac{3}{2}$
$0\ \frac{1}{2}\ \frac{1}{2}$ 000+ 000+ $-1/2$
$-2\ -2\ -2\quad 2\ -\frac{3}{2}\ \frac{1}{2}$
$0\ -\frac{3}{2}\ \frac{3}{2}$ 000+ 000+ $-1/2$
$-2\ -2\ -2\quad 2\ 2\ 2$
$0\ 0\ 0$ 000+ 000+ $+1$
$-2\ -2\ -2\quad -2\ 1\ 1$
$-2\ 1\ 1$ 000+ 000+ $+1/3$
$-2\ -2\ -2\quad -2\ \frac{3}{2}\ \frac{1}{2}$
$-2\ \frac{1}{2}\ \frac{3}{2}$ 000+ 000+ $+1/2$
$-2\ -2\ -2\quad -2\ -\frac{3}{2}\ -\frac{3}{2}$
$-2\ -\frac{3}{2}\ -\frac{3}{2}$ 000+ 000+ $+1/2$
$-2\ -2\ -2\quad -2\ 2\ 0$
$-2\ 0\ 2$ 000+ 000+ $+1$
$-2\ -2\ -2\quad -2\ -2\ -2$
$-2\ -2\ -2$ 000+ 000+ $+1$

O and T_d 9j Symbols

$0\ 0\ 0\quad 0\ 0\ 0$
$0\ 0\ 0$ 000+ 000+ +1
$\frac{1}{2}\ \frac{1}{2}\ 0\quad \frac{1}{2}\ 0\ \frac{1}{2}$
$0\ \frac{1}{2}\ \frac{1}{2}$ 000− 000− +1/4
$\frac{1}{2}\ \frac{1}{2}\ 0\quad \frac{1}{2}\ \frac{1}{2}\ 0$
$0\ 0\ 0$ 000+ 000+ +1/2
$1\ \frac{1}{2}\ \frac{1}{2}\quad \frac{1}{2}\ \frac{1}{2}\ 0$
$\frac{1}{2}\ 0\ \frac{1}{2}$ 000+ 000+ +1/4
$1\ \frac{1}{2}\ \frac{1}{2}\quad \frac{1}{2}\ 1\ \frac{1}{2}$
$\frac{1}{2}\ \frac{1}{2}\ 0$ 000− 000− +1/4.3
$\frac{1}{2}\ \frac{1}{2}\ 1$ 000+ 000+ +5/4.9
$1\ 1\ 0\quad \frac{1}{2}\ \frac{1}{2}\ 0$
$\frac{1}{2}\ \frac{1}{2}\ 0$ 000+ 000+ $+1/2\sqrt{3}$
$1\ 1\ 0\quad \frac{1}{2}\ \frac{1}{2}\ 1$
$\frac{1}{2}\ \frac{1}{2}\ 1$ 000+ 000+ −1/2.9
$1\ 1\ 0\quad 1\ 0\ 1$
$0\ 1\ 1$ 000+ 000+ +1/9
$1\ 1\ 0\quad 1\ \frac{1}{2}\ \frac{1}{2}$
$0\ \frac{1}{2}\ \frac{1}{2}$ 000− 000− +1/2.3
$1\ 1\ 0\quad 1\ 1\ 0$
$0\ 0\ 0$ 000+ 000+ +1/3
$1\ 1\ 1\quad 1\ \frac{1}{2}\ \frac{1}{2}$
$0\ \frac{1}{2}\ \frac{1}{2}$ 000+ 000+ $-1/3\sqrt{2.3}$
$1\ \frac{1}{2}\ \frac{1}{2}$ 000− 000− 0
$1\ 1\ 1\quad 1\ 1\ 0$
$1\ 0\ 1$ 000− 000− +1/9
$1\ 1\ 1\quad 1\ 1\ 1$
$0\ 0\ 0$ 000+ 000+ $+1/3\sqrt{3}$
$1\ 1\ 0$ 000+ 000+ +1/2.9
$1\ 1\ 1$ 000− 000− 0
$\frac{3}{2}\ 1\ \frac{1}{2}\quad 1\ 0\ 1$
$\frac{1}{2}\ 1\ \frac{1}{2}$ 000− 000− +1/9
$\frac{1}{2}\ 1\ \frac{3}{2}$ 000+ 000+ +1/4.9
$\frac{3}{2}\ 1\ \frac{1}{2}\quad 1\ \frac{1}{2}\ \frac{1}{2}$
$\frac{1}{2}\ \frac{1}{2}\ 0$ 000+ 000+ +1/2.3
$\frac{1}{2}\ \frac{1}{2}\ 1$ 000− 000− +1/2.9
$\frac{3}{2}\ 1\ \frac{1}{2}\quad 1\ \frac{1}{2}\ \frac{3}{2}$
$\frac{1}{2}\ \frac{1}{2}\ 1$ 000+ 000+ −1/4.9
$\frac{1}{2}\ \frac{3}{2}\ 1$ 000− 000− +11/8.2.9
$\frac{3}{2}\ 1\ \frac{1}{2}\quad 1\ 1\ 0$
$\frac{1}{2}\ 0\ \frac{1}{2}$ 000+ 000+ +1/2.3
$\frac{3}{2}\ 1\ \frac{1}{2}\quad 1\ 1\ 1$
$\frac{1}{2}\ 0\ \frac{1}{2}$ 000− 000− $+1/2.3\sqrt{2.3}$
$\frac{1}{2}\ 1\ \frac{1}{2}$ 000+ 000+ +1/4.3
$\frac{1}{2}\ 1\ \frac{3}{2}$ 000− 000− +1/8.3
$\frac{3}{2}\ 1\ \frac{1}{2}\quad 1\ \frac{3}{2}\ \frac{1}{2}$
$\frac{1}{2}\ \frac{1}{2}\ 0$ 000− 000− +1/8.3
$\frac{1}{2}\ \frac{1}{2}\ 1$ 000+ 000+ +7/8.9
$\frac{3}{2}\ 1\ \frac{1}{2}\quad 1\ \frac{3}{2}\ \frac{3}{2}$
$\frac{1}{2}\ \frac{1}{2}\ 1$ 000− 000− $-1/4.9\sqrt{2}$

$\frac{3}{2}\ 1\ \frac{1}{2}\quad 1\ \frac{3}{2}\ \frac{3}{2}$
$\frac{1}{2}\ \frac{1}{2}\ 1$ 010− 000− $-1/2.9\sqrt{2}$
$\frac{1}{2}\ \frac{3}{2}\ 1$ 000+ 000+ +1/4.9
$\frac{1}{2}\ \frac{3}{2}\ 1$ 000+ 010+ +1/8.9
$\frac{1}{2}\ \frac{3}{2}\ 1$ 010+ 000+ +1/8.9
$\frac{1}{2}\ \frac{3}{2}\ 1$ 010+ 010+ +7/8.2.9
$\frac{3}{2}\ \frac{3}{2}\ 0\quad 1\ 0\ 1$
$\frac{1}{2}\ \frac{3}{2}\ 1$ 000+ 000+ +1/4.3
$\frac{3}{2}\ \frac{3}{2}\ 0\quad 1\ \frac{1}{2}\ \frac{1}{2}$
$\frac{1}{2}\ 1\ \frac{1}{2}$ 000− 000− $+1/2.3\sqrt{2}$
$\frac{3}{2}\ \frac{3}{2}\ 0\quad 1\ \frac{1}{2}\ \frac{3}{2}$
$\frac{1}{2}\ 1\ \frac{3}{2}$ 000− 000− −1/8.2.3
$\frac{3}{2}\ \frac{3}{2}\ 0\quad 1\ 1\ 0$
$\frac{1}{2}\ \frac{1}{2}\ 0$ 000+ 000+ $+1/2\sqrt{2.3}$
$\frac{3}{2}\ \frac{3}{2}\ 0\quad 1\ 1\ 1$
$\frac{1}{2}\ \frac{1}{2}\ 1$ 000+ 000+ $+1/4.3\sqrt{3}$
$\frac{3}{2}\ \frac{3}{2}\ 0\quad \frac{3}{2}\ 0\ \frac{3}{2}$
$0\ \frac{3}{2}\ \frac{3}{2}$ 000− 000− +1/8.2
$\frac{3}{2}\ \frac{3}{2}\ 0\quad \frac{3}{2}\ \frac{1}{2}\ 1$
$0\ 1\ 1$ 000+ 000+ +1/4.3
$\frac{3}{2}\ \frac{3}{2}\ 0\quad \frac{3}{2}\ 1\ \frac{1}{2}$
$0\ \frac{1}{2}\ \frac{1}{2}$ 000− 000− +1/8
$\frac{3}{2}\ \frac{3}{2}\ 0\quad \frac{3}{2}\ 1\ \frac{3}{2}$
$0\ \frac{3}{2}\ \frac{3}{2}$ 000+ 000+ +1/8.2
$0\ \frac{3}{2}\ \frac{3}{2}$ 000+ 010+ 0
$0\ \frac{3}{2}\ \frac{3}{2}$ 010+ 000+ 0
$0\ \frac{3}{2}\ \frac{3}{2}$ 010+ 010+ +1/8.2
$\frac{3}{2}\ \frac{3}{2}\ 0\quad \frac{3}{2}\ \frac{3}{2}\ 0$
$0\ 0\ 0$ 000+ 000+ +1/4
$\frac{3}{2}\ \frac{3}{2}\ 1\quad 1\ 0\ 1$
$\frac{1}{2}\ \frac{3}{2}\ 1$ 000− 000− $-1/4.3\sqrt{2.3}$
$\frac{1}{2}\ \frac{3}{2}\ 1$ 100− 000− $-1/2.3\sqrt{2.3}$
$\frac{3}{2}\ \frac{3}{2}\ 1\quad 1\ \frac{1}{2}\ \frac{1}{2}$
$\frac{1}{2}\ 1\ \frac{1}{2}$ 000+ 000+ $+1/2.9\sqrt{2}$
$\frac{1}{2}\ 1\ \frac{1}{2}$ 100+ 000+ $+1/9\sqrt{2}$
$\frac{1}{2}\ 1\ \frac{3}{2}$ 000− 000− $+1/4.9\sqrt{2}$
$\frac{1}{2}\ 1\ \frac{3}{2}$ 100− 000− $+1/2.9\sqrt{2}$
$\frac{3}{2}\ \frac{3}{2}\ 1\quad 1\ \frac{1}{2}\ \frac{3}{2}$
$\frac{1}{2}\ 1\ \frac{1}{2}$ 000− 000− $+1/4.9\sqrt{2}$
$\frac{1}{2}\ 1\ \frac{1}{2}$ 100− 000− $+1/2.9\sqrt{2}$
$\frac{1}{2}\ 1\ \frac{3}{2}$ 000+ 000+ +7/8.2.9
$\frac{1}{2}\ 1\ \frac{3}{2}$ 000+ 001+ −1/4.9
$\frac{1}{2}\ 1\ \frac{3}{2}$ 100+ 000+ −1/4.9
$\frac{1}{2}\ 1\ \frac{3}{2}$ 100+ 001+ +1/8.2.9
$\frac{3}{2}\ \frac{3}{2}\ 1\quad 1\ 1\ 0$
$\frac{1}{2}\ \frac{1}{2}\ 1$ 000+ 000+ $+1/2.9\sqrt{2}$
$\frac{1}{2}\ \frac{1}{2}\ 1$ 100+ 000+ $+1/9\sqrt{2}$
$\frac{3}{2}\ \frac{3}{2}\ 1\quad 1\ 1\ 1$
$\frac{1}{2}\ \frac{1}{2}\ 0$ 000+ 000+ $-1/4.3\sqrt{3}$
$\frac{1}{2}\ \frac{1}{2}\ 0$ 100+ 000+ $-1/2.3\sqrt{3}$

$\frac{3}{2}\ \frac{3}{2}\ 1\quad 1\ 1\ 1$
$\frac{1}{2}\ \frac{1}{2}\ 1$ 000− 000− 0
$\frac{1}{2}\ \frac{1}{2}\ 1$ 100− 000− 0
$\frac{3}{2}\ \frac{3}{2}\ 1\quad \frac{3}{2}\ 0\ \frac{3}{2}$
$0\ \frac{3}{2}\ \frac{3}{2}$ 000+ 000+ +1/8.2
$0\ \frac{3}{2}\ \frac{3}{2}$ 000+ 001+ 0
$0\ \frac{3}{2}\ \frac{3}{2}$ 100+ 000+ 0
$0\ \frac{3}{2}\ \frac{3}{2}$ 100+ 001+ +1/8.2
$1\ \frac{3}{2}\ \frac{1}{2}$ 000+ 000+ +1/8.3
$1\ \frac{3}{2}\ \frac{1}{2}$ 100+ 000+ −1/8.3
$1\ \frac{3}{2}\ \frac{1}{2}$ 100+ 100+ −1/8.2.3
$1\ \frac{3}{2}\ \frac{3}{2}$ 000− 000− −1/8.2.3
$1\ \frac{3}{2}\ \frac{3}{2}$ 000− 001− 0
$1\ \frac{3}{2}\ \frac{3}{2}$ 001− 000− 0
$1\ \frac{3}{2}\ \frac{3}{2}$ 001− 001− +1/8.3
$1\ \frac{3}{2}\ \frac{3}{2}$ 100− 000− 0
$1\ \frac{3}{2}\ \frac{3}{2}$ 100− 001− −1/8.2.3
$1\ \frac{3}{2}\ \frac{3}{2}$ 100− 100− +1/8.3
$1\ \frac{3}{2}\ \frac{3}{2}$ 100− 101− 0
$1\ \frac{3}{2}\ \frac{3}{2}$ 101− 000− +1/8.3
$1\ \frac{3}{2}\ \frac{3}{2}$ 101− 001− 0
$1\ \frac{3}{2}\ \frac{3}{2}$ 101− 100− 0
$1\ \frac{3}{2}\ \frac{3}{2}$ 101− 101− +1/8.3
$\frac{3}{2}\ \frac{3}{2}\ 1\quad \frac{3}{2}\ \frac{1}{2}\ 1$
$0\ 1\ 1$ 000− 000− $-1/4.3\sqrt{2.3}$
$0\ 1\ 1$ 100− 000− $-1/2.3\sqrt{2.3}$
$1\ 1\ 0$ 000− 000− +1/2.9
$1\ 1\ 0$ 100− 000− −1/2.9
$1\ 1\ 0$ 100− 100− −1/4.9
$1\ 1\ 1$ 000+ 000+ +1/8.3
$1\ 1\ 1$ 100+ 000+ 0
$1\ 1\ 1$ 100+ 100+ +1/8.3
$\frac{3}{2}\ \frac{3}{2}\ 1\quad \frac{3}{2}\ 1\ \frac{1}{2}$
$0\ \frac{1}{2}\ \frac{1}{2}$ 000+ 000+ +1/8.3
$0\ \frac{1}{2}\ \frac{1}{2}$ 100+ 000+ +1/4.3
$0\ \frac{3}{2}\ \frac{3}{2}$ 000+ 000+ +1/8.3
$0\ \frac{3}{2}\ \frac{3}{2}$ 000+ 010+ −1/8.3
$0\ \frac{3}{2}\ \frac{3}{2}$ 100+ 000+ −1/8.3
$0\ \frac{3}{2}\ \frac{3}{2}$ 100+ 010+ −1/8.2.3
$1\ \frac{1}{2}\ \frac{1}{2}$ 000− 000− +5/8.9
$1\ \frac{1}{2}\ \frac{1}{2}$ 100− 000− −1/4.9
$1\ \frac{1}{2}\ \frac{1}{2}$ 100− 100− +1/4.9
$1\ \frac{1}{2}\ \frac{3}{2}$ 000+ 000+ +1/4.9
$1\ \frac{1}{2}\ \frac{3}{2}$ 100+ 000+ +1/8.9
$1\ \frac{1}{2}\ \frac{3}{2}$ 100+ 100+ +7/8.2.9
$\frac{3}{2}\ \frac{3}{2}\ 1\quad \frac{3}{2}\ 1\ \frac{3}{2}$
$0\ \frac{1}{2}\ \frac{1}{2}$ 000− 000− $-1/4.3\sqrt{2}$
$0\ \frac{1}{2}\ \frac{1}{2}$ 010− 000− $+1/4.3\sqrt{2}$
$0\ \frac{1}{2}\ \frac{1}{2}$ 100− 000− $+1/4.3\sqrt{2}$
$0\ \frac{1}{2}\ \frac{1}{2}$ 110− 000− $+1/8.3\sqrt{2}$

O and T_d 9j Symbols (cont.)

$\frac{3}{2}$	**$\frac{3}{2}$**	**1**	**$\frac{3}{2}$ 1 $\frac{3}{2}$**		
0	$\frac{3}{2}$	$\frac{3}{2}$	000−	000−	−1/8.2.3
0	$\frac{3}{2}$	$\frac{3}{2}$	000−	001−	0
0	$\frac{3}{2}$	$\frac{3}{2}$	000−	010−	0
0	$\frac{3}{2}$	$\frac{3}{2}$	000−	011−	+1/8.3
0	$\frac{3}{2}$	$\frac{3}{2}$	010−	000−	0
0	$\frac{3}{2}$	$\frac{3}{2}$	010−	001−	+1/8.3
0	$\frac{3}{2}$	$\frac{3}{2}$	010−	010−	−1/8.2.3
0	$\frac{3}{2}$	$\frac{3}{2}$	010−	011−	0
0	$\frac{3}{2}$	$\frac{3}{2}$	100−	000−	0
0	$\frac{3}{2}$	$\frac{3}{2}$	100−	001−	−1/8.2.3
0	$\frac{3}{2}$	$\frac{3}{2}$	100−	010−	+1/8.3
0	$\frac{3}{2}$	$\frac{3}{2}$	100−	011−	0
0	$\frac{3}{2}$	$\frac{3}{2}$	110−	000−	+1/8.3
0	$\frac{3}{2}$	$\frac{3}{2}$	110−	001−	0
0	$\frac{3}{2}$	$\frac{3}{2}$	110−	010−	0
0	$\frac{3}{2}$	$\frac{3}{2}$	110−	011−	+1/8.3
1	$\frac{1}{2}$	$\frac{1}{2}$	000+	000+	$+1/4.9\sqrt{2}$
1	$\frac{1}{2}$	$\frac{1}{2}$	000+	100+	$+1/2.9\sqrt{2}$
1	$\frac{1}{2}$	$\frac{1}{2}$	010+	000+	$-1/4.9\sqrt{2}$
1	$\frac{1}{2}$	$\frac{1}{2}$	010+	100+	$+1/4.9\sqrt{2}$
1	$\frac{1}{2}$	$\frac{1}{2}$	100+	000+	$-1/4.9\sqrt{2}$
1	$\frac{1}{2}$	$\frac{1}{2}$	100+	100+	$+1/4.9\sqrt{2}$
1	$\frac{1}{2}$	$\frac{1}{2}$	110+	000+	$+5/8.9\sqrt{2}$
1	$\frac{1}{2}$	$\frac{1}{2}$	110+	100+	$+1/2.9\sqrt{2}$
1	$\frac{1}{2}$	$\frac{3}{2}$	000−	000−	+1/8.9
1	$\frac{1}{2}$	$\frac{3}{2}$	000−	001−	+1/4.9
1	$\frac{1}{2}$	$\frac{3}{2}$	000−	100−	−1/8.9
1	$\frac{1}{2}$	$\frac{3}{2}$	000−	101−	+1/8.9
1	$\frac{1}{2}$	$\frac{3}{2}$	010−	000−	−1/8.9
1	$\frac{1}{2}$	$\frac{3}{2}$	010−	001−	−1/4.9
1	$\frac{1}{2}$	$\frac{3}{2}$	010−	100−	+5/8.2.9
1	$\frac{1}{2}$	$\frac{3}{2}$	010−	101−	−1/8.9
1	$\frac{1}{2}$	$\frac{3}{2}$	100−	000−	+1/4.9
1	$\frac{1}{2}$	$\frac{3}{2}$	100−	001−	+1/8.9
1	$\frac{1}{2}$	$\frac{3}{2}$	100−	100−	−1/4.9
1	$\frac{1}{2}$	$\frac{3}{2}$	100−	101−	−1/8.9
1	$\frac{1}{2}$	$\frac{3}{2}$	110−	000−	+1/8.9
1	$\frac{1}{2}$	$\frac{3}{2}$	110−	001−	−1/8.9
1	$\frac{1}{2}$	$\frac{3}{2}$	110−	100−	−1/8.9
1	$\frac{1}{2}$	$\frac{3}{2}$	110−	101−	+1/8.9
1	$\frac{3}{2}$	$\frac{1}{2}$	000−	000−	+1/8.9
1	$\frac{3}{2}$	$\frac{1}{2}$	000−	010−	+1/4.9
1	$\frac{3}{2}$	$\frac{1}{2}$	010−	000−	+1/4.9
1	$\frac{3}{2}$	$\frac{1}{2}$	010−	010−	+1/8.9
1	$\frac{3}{2}$	$\frac{1}{2}$	100−	000−	−1/8.9
1	$\frac{3}{2}$	$\frac{1}{2}$	100−	010−	−1/4.9
1	$\frac{3}{2}$	$\frac{1}{2}$	100−	100−	+5/8.2.9
1	$\frac{3}{2}$	$\frac{1}{2}$	100−	110−	−1/8.9
1	$\frac{3}{2}$	$\frac{1}{2}$	110−	000−	+1/8.9
$\frac{3}{2}$	**$\frac{3}{2}$**	**1**	**$\frac{3}{2}$ 1 $\frac{3}{2}$**		
1	$\frac{3}{2}$	$\frac{1}{2}$	110−	010−	−1/8.9
1	$\frac{3}{2}$	$\frac{1}{2}$	110−	100−	−1/8.9
1	$\frac{3}{2}$	$\frac{1}{2}$	110−	110−	+1/8.9
1	$\frac{3}{2}$	$\frac{3}{2}$	000+	000+	+5/8.2.9
1	$\frac{3}{2}$	$\frac{3}{2}$	000+	001+	0
1	$\frac{3}{2}$	$\frac{3}{2}$	000+	010+	0
1	$\frac{3}{2}$	$\frac{3}{2}$	000+	011+	+1/8.9
1	$\frac{3}{2}$	$\frac{3}{2}$	001+	000+	0
1	$\frac{3}{2}$	$\frac{3}{2}$	001+	001+	+1/8.9
1	$\frac{3}{2}$	$\frac{3}{2}$	001+	010+	+1/8.9
1	$\frac{3}{2}$	$\frac{3}{2}$	001+	011+	0
1	$\frac{3}{2}$	$\frac{3}{2}$	010+	000+	0
1	$\frac{3}{2}$	$\frac{3}{2}$	010+	001+	+1/8.9
1	$\frac{3}{2}$	$\frac{3}{2}$	010+	010+	+5/8.2.9
1	$\frac{3}{2}$	$\frac{3}{2}$	010+	011+	0
1	$\frac{3}{2}$	$\frac{3}{2}$	011+	000+	+1/8.9
1	$\frac{3}{2}$	$\frac{3}{2}$	011+	001+	0
1	$\frac{3}{2}$	$\frac{3}{2}$	011+	010+	0
1	$\frac{3}{2}$	$\frac{3}{2}$	011+	011+	+1/8.9
1	$\frac{3}{2}$	$\frac{3}{2}$	100+	000+	0
1	$\frac{3}{2}$	$\frac{3}{2}$	100+	001+	+5/8.2.9
1	$\frac{3}{2}$	$\frac{3}{2}$	100+	010+	+1/8.9
1	$\frac{3}{2}$	$\frac{3}{2}$	100+	011+	0
1	$\frac{3}{2}$	$\frac{3}{2}$	100+	100+	+1/8.9
1	$\frac{3}{2}$	$\frac{3}{2}$	100+	101+	0
1	$\frac{3}{2}$	$\frac{3}{2}$	100+	110+	0
1	$\frac{3}{2}$	$\frac{3}{2}$	100+	111+	−1/4.9
1	$\frac{3}{2}$	$\frac{3}{2}$	101+	000+	+1/8.9
1	$\frac{3}{2}$	$\frac{3}{2}$	101+	001+	0
1	$\frac{3}{2}$	$\frac{3}{2}$	101+	010+	0
1	$\frac{3}{2}$	$\frac{3}{2}$	101+	011+	+1/8.9
1	$\frac{3}{2}$	$\frac{3}{2}$	101+	100+	0
1	$\frac{3}{2}$	$\frac{3}{2}$	101+	101+	+1/8.9
1	$\frac{3}{2}$	$\frac{3}{2}$	101+	110+	+1/8.9
1	$\frac{3}{2}$	$\frac{3}{2}$	101+	111+	0
1	$\frac{3}{2}$	$\frac{3}{2}$	110+	000+	+1/8.9
1	$\frac{3}{2}$	$\frac{3}{2}$	110+	001+	0
1	$\frac{3}{2}$	$\frac{3}{2}$	110+	010+	0
1	$\frac{3}{2}$	$\frac{3}{2}$	110+	011+	+1/8.9
1	$\frac{3}{2}$	$\frac{3}{2}$	110+	100+	0
1	$\frac{3}{2}$	$\frac{3}{2}$	110+	101+	+1/8.9
1	$\frac{3}{2}$	$\frac{3}{2}$	110+	110+	+1/8.9
1	$\frac{3}{2}$	$\frac{3}{2}$	110+	111+	0
1	$\frac{3}{2}$	$\frac{3}{2}$	111+	000+	0
1	$\frac{3}{2}$	$\frac{3}{2}$	111+	001+	−1/4.9
1	$\frac{3}{2}$	$\frac{3}{2}$	111+	010+	−1/4.9
1	$\frac{3}{2}$	$\frac{3}{2}$	111+	011+	0
1	$\frac{3}{2}$	$\frac{3}{2}$	111+	100+	−1/4.9
1	$\frac{3}{2}$	$\frac{3}{2}$	111+	101+	0
$\frac{3}{2}$	**$\frac{3}{2}$**	**1**	**$\frac{3}{2}$ 1 $\frac{3}{2}$**		
1	$\frac{3}{2}$	$\frac{3}{2}$	111+	110+	0
1	$\frac{3}{2}$	$\frac{3}{2}$	111+	111+	+5/8.2.9
$\frac{3}{2}$	**$\frac{3}{2}$**	**1**	**$\frac{3}{2}$ $\frac{3}{2}$ 0**		
1	0	1	000−	000−	+1/4.3
1	0	1	100−	000−	0
1	0	1	100−	100−	+1/4.3
1	1	1	000+	000+	$+1/2.3\sqrt{2.3}$
1	1	1	000+	100+	0
1	1	1	000+	110+	$-1/4.3\sqrt{2.3}$
1	1	1	100+	000+	0
1	1	1	100+	100+	$-1/4.3\sqrt{2.3}$
1	1	1	100+	110+	0
$\frac{3}{2}$	**$\frac{3}{2}$**	**1**	**$\frac{3}{2}$ $\frac{3}{2}$ 1**		
0	0	0	000+	000+	$+1/4\sqrt{3}$
0	0	0	100+	000+	0
0	0	0	110+	000+	$+1/4\sqrt{3}$
1	0	1	000+	000+	$+1/2.3\sqrt{2.3}$
1	0	1	000+	100+	0
1	0	1	100+	000+	0
1	0	1	100+	100+	$-1/4.3\sqrt{2.3}$
1	0	1	110+	000+	$-1/4.3\sqrt{2.3}$
1	0	1	110+	100+	0
1	1	0	000+	000+	−1/4.9
1	1	0	100+	000+	0
1	1	0	100+	100+	+1/2.9
1	1	0	100+	110+	0
1	1	0	110+	000+	−1/4.9
1	1	0	110+	100+	0
1	1	0	110+	110+	+1/2.9
1	1	1	000−	000−	0
1	1	1	100−	000−	0
1	1	1	100−	100−	+1/8.3
1	1	1	100−	110−	0
1	1	1	110−	000−	0
1	1	1	110−	100−	0
1	1	1	110−	110−	0
2	**1**	**1**	**1 $\frac{1}{2}$ $\frac{1}{2}$**		
1	$\frac{1}{2}$	$\frac{1}{2}$	000+	000+	+1/9
2	**1**	**1**	**1 1 0**		
1	0	1	000+	000+	+1/9
2	**1**	**1**	**1 1 1**		
1	1	0	000−	000−	+1/2.9
1	1	1	000+	000+	+1/2.9
2	**1**	**1**	**1 $\frac{3}{2}$ $\frac{1}{2}$**		
1	$\frac{1}{2}$	$\frac{1}{2}$	000−	000−	+1/4.9
1	$\frac{1}{2}$	$\frac{3}{2}$	000+	000+	+5/8.9
1	$\frac{3}{2}$	$\frac{1}{2}$	000+	000+	$-1/9\sqrt{2}$
1	$\frac{3}{2}$	$\frac{1}{2}$	000+	010+	$+1/4.9\sqrt{2}$

O and T_d 9j Symbols (cont.)

2 1 1 1 3/2 3/2
1 ½ ½ 000+ 000+ $-1/9\sqrt{2}$
1 ½ ½ 010+ 000+ $+1/4.9\sqrt{2}$
1 3/2 ½ 000− 000− +1/8.9
1 3/2 ½ 000− 010− −1/8.9
1 3/2 ½ 010− 000− −1/8.9
1 3/2 ½ 010− 010− +1/2.9
1 3/2 3/2 000+ 000+ +1/2.9
1 3/2 3/2 000+ 010+ 0
1 3/2 3/2 000+ 011+ +1/2.9
1 3/2 3/2 010+ 000+ 0
1 3/2 3/2 010+ 010+ +1/8.9
1 3/2 3/2 010+ 011+ 0
1 3/2 3/2 011+ 000+ +1/2.9
1 3/2 3/2 011+ 010+ 0
1 3/2 3/2 011+ 011+ +1/8.9
2 1 1 1 2 1
1 1 0 000+ 000+ +1/9
1 1 1 000− 000− −1/4.9
1 1 2 000+ 000+ +1/9
2 3/2 ½ 1 1 0
1 ½ ½ 000+ 000+ $+1/2.3\sqrt{2}$
2 3/2 ½ 1 1 1
1 ½ ½ 000− 000− $+1/4.3\sqrt{3}$
1 ½ 3/2 000+ 000+ $+1/8.3\sqrt{3}$
2 3/2 ½ 1 1 2
1 ½ 3/2 000− 000− +1/4.3
2 3/2 ½ 1 3/2 ½
1 0 1 000− 000− −1/4.3
1 1 0 000− 000− −1/4.3
1 1 0 000− 010− +1/4.3
1 1 1 000+ 000+ $-1/2.3\sqrt{2.3}$
1 1 1 000+ 010+ $-1/4.3\sqrt{2.3}$
2 3/2 ½ 1 3/2 3/2
1 0 1 000+ 000+ $-1/4.3\sqrt{2}$
1 0 1 010+ 000+ $+1/4.3\sqrt{2}$
1 1 1 000− 000− $+1/8.3\sqrt{3}$
1 1 1 000− 010− $+1/4.3\sqrt{3}$
1 1 1 010− 000− $-1/8.3\sqrt{3}$
1 1 1 010− 010− $+1/8.3\sqrt{3}$
1 1 2 000+ 000+ +1/8.3
1 1 2 000+ 010+ −1/8.3
1 1 2 010+ 000+ −1/8.3
1 1 2 010+ 010+ +1/8.3
2 3/2 ½ 1 2 1
1 ½ ½ 000+ 000+ +1/4.3
1 ½ 3/2 000− 000− −1/4.3
1 3/2 ½ 000− 000− $-1/4.3\sqrt{2}$
1 3/2 3/2 000+ 000+ +1/8.3
1 3/2 3/2 001+ 000+ −1/8.3

2 3/2 ½ 3/2 0 3/2
½ 3/2 1 000+ 000+ +1/8.2
½ 3/2 2 000− 000− −1/8.2
2 3/2 ½ 3/2 ½ 1
½ 1 ½ 000− 000− +1/4.3
½ 1 3/2 000+ 000+ +1/8.2.3
2 3/2 ½ 3/2 ½ 2
½ 1 3/2 000− 000− +1/8.2
½ 2 3/2 000+ 000+ +1/8.2
2 3/2 ½ 3/2 1 ½
½ ½ 0 000+ 000+ +1/8
½ ½ 1 000− 000− +1/8.3
2 3/2 ½ 3/2 1 3/2
½ ½ 1 000+ 000+ $+1/4.3\sqrt{2}$
½ ½ 1 010+ 000+ $-1/4.3\sqrt{2}$
½ 3/2 1 000− 000− +1/8.2.3
½ 3/2 1 000− 010− +1/8.3
½ 3/2 1 010− 000− +1/8.3
½ 3/2 1 010− 010− +1/8.2.3
½ 3/2 2 000+ 000+ +1/8.2
½ 3/2 2 000+ 010+ 0
½ 3/2 2 010+ 000+ 0
½ 3/2 2 010+ 010+ +1/8.2
2 3/2 ½ 3/2 3/2 0
½ 0 ½ 000+ 000+ +1/8
2 3/2 ½ 3/2 3/2 1
½ 0 ½ 000− 000− −1/8
½ 0 ½ 010− 000− 0
½ 1 ½ 000+ 000+ −1/8.3
½ 1 ½ 000+ 010+ 0
½ 1 ½ 010+ 000+ 0
½ 1 ½ 010+ 010+ +1/4.3
½ 1 3/2 000− 000− +1/4.3
½ 1 3/2 000− 010− 0
½ 1 3/2 010− 000− 0
½ 1 3/2 010− 010− +1/8.2.3
2 3/2 ½ 3/2 3/2 2
½ 1 3/2 000+ 000+ 0
½ 1 3/2 000+ 010+ +1/8.2
½ 2 3/2 000− 000− +1/8.2
2 3/2 ½ 3/2 2 ½
½ ½ 0 000− 000− −1/8
½ ½ 1 000+ 000+ +1/8
2 3/2 ½ 3/2 2 3/2
½ ½ 1 000− 000− 0
½ 3/2 1 000+ 000+ +1/8.2
½ 3/2 2 000− 000− −1/8.2
2 3/2 3/2 1 1 0
1 ½ 3/2 000+ 000+ $-1/4.3\sqrt{2}$

2 3/2 3/2 1 1 1
1 ½ ½ 000+ 000+ $+1/2.3\sqrt{2.3}$
2 3/2 3/2 1 3/2 ½
1 0 1 000+ 000+ $-1/4.3\sqrt{2}$
1 1 1 000− 000− $+1/8.3\sqrt{3}$
1 1 1 000− 010− $-1/8.3\sqrt{3}$
1 1 2 000+ 000+ +1/8.3
1 1 2 000+ 010+ −1/8.3
2 3/2 3/2 1 3/2 3/2
1 1 0 000− 000− 0
1 1 0 000− 010− $-1/4.3\sqrt{2}$
1 1 0 010− 000− $-1/4.3\sqrt{2}$
1 1 0 010− 010− 0
1 1 1 000+ 000+ 0
1 1 1 000+ 010+ $-1/8.3\sqrt{3}$
1 1 1 000+ 011+ 0
1 1 1 010+ 000+ $+1/4.3\sqrt{3}$
1 1 1 010+ 010+ 0
1 1 1 010+ 011+ $+1/4.3\sqrt{3}$
2 3/2 3/2 1 2 1
1 ½ ½ 000− 000− $-1/4.3\sqrt{2}$
1 ½ 3/2 000+ 000+ +1/8.3
1 ½ 3/2 000+ 001+ −1/8.3
1 3/2 ½ 000+ 000+ +1/8.3
1 3/2 3/2 000− 000− −1/8.3
1 3/2 3/2 000− 001− 0
1 3/2 3/2 001− 000− 0
1 3/2 3/2 001− 001− −1/8.3
2 3/2 3/2 3/2 1 ½
½ ½ 1 000+ 000+ $-1/4.3\sqrt{2}$
½ 3/2 1 000− 000− +1/8.3
½ 3/2 1 000− 010− +1/8.2.3
½ 3/2 2 000+ 000+ 0
½ 3/2 2 000+ 010+ +1/8.2
3/2 ½ 1 000− 000− +1/8.2.3
3/2 ½ 2 000+ 000+ +1/8.2
2 3/2 3/2 3/2 3/2 0
½ 1 3/2 000− 000− 0
½ 1 3/2 000− 010− +1/8.2
½ 2 3/2 000+ 000+ +1/8.2
3/2 0 3/2 000− 000− +1/8.2
3/2 1 3/2 000+ 000+ +1/8.2
3/2 1 3/2 000+ 010+ 0
3/2 1 3/2 001+ 000+ 0
3/2 1 3/2 001+ 010+ 0
2 3/2 3/2 3/2 3/2 1
½ 0 ½ 000+ 000+ 0
½ 0 ½ 010+ 000+ $-1/8\sqrt{2}$
½ 1 ½ 000− 000− 0
½ 1 ½ 000− 010− $+1/4.3\sqrt{2}$

O and T_d 9j Symbols (cont.)

2 $\tfrac{3}{2}$ $\tfrac{3}{2}$ $\tfrac{3}{2}$ $\tfrac{3}{2}$ 1

$\tfrac{1}{2}$ 1 $\tfrac{1}{2}$ 010− 000− $-1/8.3\sqrt{2}$

$\tfrac{1}{2}$ 1 $\tfrac{1}{2}$ 010− 010− 0

$\tfrac{1}{2}$ 1 $\tfrac{3}{2}$ 000+ 000+ 0

$\tfrac{1}{2}$ 1 $\tfrac{3}{2}$ 000+ 001+ 0

$\tfrac{1}{2}$ 1 $\tfrac{3}{2}$ 000+ 010+ $-1/8.2.3$

$\tfrac{1}{2}$ 1 $\tfrac{3}{2}$ 000+ 011+ $-1/8.3$

$\tfrac{1}{2}$ 1 $\tfrac{3}{2}$ 010+ 000+ $+1/8.3$

$\tfrac{1}{2}$ 1 $\tfrac{3}{2}$ 010+ 001+ $-1/8.3$

$\tfrac{1}{2}$ 1 $\tfrac{3}{2}$ 010+ 010+ 0

$\tfrac{1}{2}$ 1 $\tfrac{3}{2}$ 010+ 011+ 0

$\tfrac{1}{2}$ 2 $\tfrac{3}{2}$ 000− 000− $+1/8.2$

$\tfrac{1}{2}$ 2 $\tfrac{3}{2}$ 000− 001− 0

$\tfrac{1}{2}$ 2 $\tfrac{3}{2}$ 010− 000− 0

$\tfrac{1}{2}$ 2 $\tfrac{3}{2}$ 010− 001− 0

$\tfrac{3}{2}$ 0 $\tfrac{3}{2}$ 000+ 000+ $+1/8.2$

$\tfrac{3}{2}$ 0 $\tfrac{3}{2}$ 000+ 001+ 0

$\tfrac{3}{2}$ 0 $\tfrac{3}{2}$ 010+ 000+ 0

$\tfrac{3}{2}$ 0 $\tfrac{3}{2}$ 010+ 001+ 0

$\tfrac{3}{2}$ 1 $\tfrac{1}{2}$ 000+ 000+ $+1/8.3$

$\tfrac{3}{2}$ 1 $\tfrac{1}{2}$ 000+ 010+ 0

$\tfrac{3}{2}$ 1 $\tfrac{1}{2}$ 010+ 000+ 0

$\tfrac{3}{2}$ 1 $\tfrac{1}{2}$ 010+ 010+ $+1/8.3$

$\tfrac{3}{2}$ 1 $\tfrac{3}{2}$ 000− 000− $-1/8.2.3$

$\tfrac{3}{2}$ 1 $\tfrac{3}{2}$ 000− 001− 0

$\tfrac{3}{2}$ 1 $\tfrac{3}{2}$ 000− 010− 0

$\tfrac{3}{2}$ 1 $\tfrac{3}{2}$ 000− 011− 0

$\tfrac{3}{2}$ 1 $\tfrac{3}{2}$ 001− 000− 0

$\tfrac{3}{2}$ 1 $\tfrac{3}{2}$ 001− 001− $+1/8.3$

$\tfrac{3}{2}$ 1 $\tfrac{3}{2}$ 001− 010− 0

$\tfrac{3}{2}$ 1 $\tfrac{3}{2}$ 001− 011− 0

$\tfrac{3}{2}$ 1 $\tfrac{3}{2}$ 010− 000− 0

$\tfrac{3}{2}$ 1 $\tfrac{3}{2}$ 010− 001− 0

$\tfrac{3}{2}$ 1 $\tfrac{3}{2}$ 010− 010− $+1/8.3$

$\tfrac{3}{2}$ 1 $\tfrac{3}{2}$ 010− 011− 0

$\tfrac{3}{2}$ 1 $\tfrac{3}{2}$ 011− 000− 0

$\tfrac{3}{2}$ 1 $\tfrac{3}{2}$ 011− 001− 0

$\tfrac{3}{2}$ 1 $\tfrac{3}{2}$ 011− 010− 0

$\tfrac{3}{2}$ 1 $\tfrac{3}{2}$ 011− 011− $-1/8.2.3$

2 $\tfrac{3}{2}$ $\tfrac{3}{2}$ $\tfrac{3}{2}$ 2 $\tfrac{1}{2}$

$\tfrac{1}{2}$ $\tfrac{1}{2}$ 1 000− 000− 0

$\tfrac{1}{2}$ $\tfrac{3}{2}$ 1 000+ 000+ $+1/8.2$

$\tfrac{1}{2}$ $\tfrac{3}{2}$ 2 000− 000− $-1/8.2$

$\tfrac{3}{2}$ $\tfrac{1}{2}$ 1 000+ 000+ $+1/8.2$

$\tfrac{3}{2}$ $\tfrac{1}{2}$ 2 000− 000− $-1/8.2$

$\tfrac{3}{2}$ $\tfrac{3}{2}$ 1 000− 000− 0

$\tfrac{3}{2}$ $\tfrac{3}{2}$ 1 001− 000− 0

$\tfrac{3}{2}$ $\tfrac{3}{2}$ 2 000+ 000+ 0

2 $\tfrac{3}{2}$ $\tfrac{3}{2}$ $\tfrac{3}{2}$ 2 $\tfrac{3}{2}$

$\tfrac{1}{2}$ $\tfrac{1}{2}$ 0 000− 000− $+1/8\sqrt{2}$

$\tfrac{1}{2}$ $\tfrac{1}{2}$ 1 000+ 000+ $+1/8\sqrt{2}$

$\tfrac{1}{2}$ $\tfrac{1}{2}$ 1 000+ 001+ 0

$\tfrac{1}{2}$ $\tfrac{3}{2}$ 1 000− 000− 0

$\tfrac{1}{2}$ $\tfrac{3}{2}$ 1 000− 001− 0

$\tfrac{1}{2}$ $\tfrac{3}{2}$ 2 000+ 000+ 0

$\tfrac{3}{2}$ $\tfrac{1}{2}$ 1 000− 000− 0

$\tfrac{3}{2}$ $\tfrac{1}{2}$ 1 000− 001− 0

$\tfrac{3}{2}$ $\tfrac{1}{2}$ 2 000+ 000+ 0

$\tfrac{3}{2}$ $\tfrac{3}{2}$ 0 000− 000− 0

$\tfrac{3}{2}$ $\tfrac{3}{2}$ 1 000+ 000+ 0

$\tfrac{3}{2}$ $\tfrac{3}{2}$ 1 000+ 001+ 0

$\tfrac{3}{2}$ $\tfrac{3}{2}$ 1 001+ 000+ 0

$\tfrac{3}{2}$ $\tfrac{3}{2}$ 1 001+ 001+ $+1/8.2$

$\tfrac{3}{2}$ $\tfrac{3}{2}$ 2 000− 000− $-1/8.2$

2 2 0 1 1 0

1 1 0 000+ 000+ $+1/3\sqrt{2}$

2 2 0 1 1 1

1 1 1 000+ 000+ $-1/2.3\sqrt{2.3}$

2 2 0 1 1 2

1 1 2 000+ 000+ $+1/2.3$

2 2 0 $\tfrac{3}{2}$ 0 $\tfrac{3}{2}$

$\tfrac{1}{2}$ 2 $\tfrac{3}{2}$ 000− 000− $+1/8$

2 2 0 $\tfrac{3}{2}$ $\tfrac{1}{2}$ 1

$\tfrac{1}{2}$ $\tfrac{3}{2}$ 1 000+ 000+ $+1/4\sqrt{2.3}$

2 2 0 $\tfrac{3}{2}$ $\tfrac{1}{2}$ 2

$\tfrac{1}{2}$ $\tfrac{3}{2}$ 2 000+ 000+ $+1/8$

2 2 0 $\tfrac{3}{2}$ 1 $\tfrac{1}{2}$

$\tfrac{1}{2}$ 1 $\tfrac{1}{2}$ 000− 000− $+1/4\sqrt{3}$

$\tfrac{3}{2}$ 1 $\tfrac{1}{2}$ 000+ 000+ $-1/4\sqrt{2.3}$

2 2 0 $\tfrac{3}{2}$ 1 $\tfrac{3}{2}$

$\tfrac{1}{2}$ 1 $\tfrac{3}{2}$ 000− 000− $+1/8\sqrt{3}$

$\tfrac{1}{2}$ 1 $\tfrac{3}{2}$ 010− 000− $-1/8\sqrt{3}$

$\tfrac{3}{2}$ 1 $\tfrac{3}{2}$ 000+ 000+ 0

$\tfrac{3}{2}$ 1 $\tfrac{3}{2}$ 010+ 000+ $+1/8\sqrt{3}$

$\tfrac{3}{2}$ 1 $\tfrac{3}{2}$ 011+ 000+ 0

2 2 0 $\tfrac{3}{2}$ $\tfrac{3}{2}$ 0

$\tfrac{1}{2}$ $\tfrac{1}{2}$ 0 000+ 000+ $+1/4$

$\tfrac{3}{2}$ $\tfrac{3}{2}$ 0 000+ 000+ $+1/4\sqrt{2}$

2 2 0 $\tfrac{3}{2}$ $\tfrac{3}{2}$ 1

$\tfrac{1}{2}$ $\tfrac{1}{2}$ 1 000+ 000+ $-1/4\sqrt{3}$

$\tfrac{1}{2}$ $\tfrac{1}{2}$ 1 010+ 000+ 0

$\tfrac{3}{2}$ $\tfrac{1}{2}$ 1 000− 000− 0

$\tfrac{3}{2}$ $\tfrac{1}{2}$ 1 010− 000− $+1/4\sqrt{2.3}$

$\tfrac{3}{2}$ $\tfrac{3}{2}$ 1 000+ 000+ $+1/4\sqrt{2.3}$

$\tfrac{3}{2}$ $\tfrac{3}{2}$ 1 010+ 000+ 0

$\tfrac{3}{2}$ $\tfrac{3}{2}$ 1 011+ 000+ 0

2 2 0 $\tfrac{3}{2}$ $\tfrac{3}{2}$ 2

$\tfrac{3}{2}$ $\tfrac{1}{2}$ 2 000− 000− $+1/8$

$\tfrac{3}{2}$ $\tfrac{3}{2}$ 2 000+ 000+ 0

2 2 0 2 0 2

0 2 2 000+ 000+ $+1/4$

2 2 0 2 $\tfrac{1}{2}$ $\tfrac{3}{2}$

0 $\tfrac{3}{2}$ $\tfrac{3}{2}$ 000− 000− $+1/8$

2 2 0 2 1 1

0 1 1 000+ 000+ $+1/2.3$

2 2 0 2 $\tfrac{3}{2}$ $\tfrac{1}{2}$

0 $\tfrac{1}{2}$ $\tfrac{1}{2}$ 000− 000− $+1/4$

2 2 0 2 $\tfrac{3}{2}$ $\tfrac{3}{2}$

0 $\tfrac{3}{2}$ $\tfrac{3}{2}$ 000+ 000+ $+1/8$

2 2 0 2 2 0

0 0 0 000+ 000+ $+1/2$

2 2 2 1 1 1

1 1 1 000+ 000+ $-1/2.3\sqrt{3}$

2 2 2 $\tfrac{3}{2}$ 1 $\tfrac{1}{2}$

$\tfrac{1}{2}$ 1 $\tfrac{3}{2}$ 000+ 000+ $-1/8\sqrt{3}$

2 2 2 $\tfrac{3}{2}$ $\tfrac{3}{2}$ 0

$\tfrac{3}{2}$ $\tfrac{1}{2}$ 2 000− 000− $+1/8$

2 2 2 $\tfrac{3}{2}$ $\tfrac{3}{2}$ 1

$\tfrac{1}{2}$ $\tfrac{1}{2}$ 1 000+ 000+ 0

$\tfrac{1}{2}$ $\tfrac{1}{2}$ 1 010+ 000+ $-1/4\sqrt{2.3}$

$\tfrac{3}{2}$ $\tfrac{1}{2}$ 1 000− 000− $+1/8\sqrt{3}$

$\tfrac{3}{2}$ $\tfrac{1}{2}$ 1 010− 000− 0

$\tfrac{3}{2}$ $\tfrac{3}{2}$ 1 000+ 000+ 0

$\tfrac{3}{2}$ $\tfrac{3}{2}$ 1 010+ 000+ 0

$\tfrac{3}{2}$ $\tfrac{3}{2}$ 1 011+ 000+ $-1/8\sqrt{3}$

2 2 2 2 1 1

0 1 1 000+ 000+ $-1/2.3\sqrt{2}$

2 1 1 000+ 000+ $+1/4.3$

2 2 2 2 $\tfrac{3}{2}$ $\tfrac{1}{2}$

0 $\tfrac{3}{2}$ $\tfrac{3}{2}$ 000− 000− $-1/8$

2 $\tfrac{1}{2}$ $\tfrac{3}{2}$ 000+ 000+ $+1/8$

2 2 2 2 $\tfrac{3}{2}$ $\tfrac{3}{2}$

0 $\tfrac{1}{2}$ $\tfrac{1}{2}$ 000+ 000+ $+1/4\sqrt{2}$

0 $\tfrac{3}{2}$ $\tfrac{3}{2}$ 000+ 000+ 0

2 $\tfrac{3}{2}$ $\tfrac{1}{2}$ 000− 000− 0

2 $\tfrac{3}{2}$ $\tfrac{3}{2}$ 000+ 000+ $+1/8$

2 2 2 2 2 0

2 0 2 000+ 000+ $+1/4$

2 2 2 2 2 2

0 0 0 000+ 000+ $+1/2\sqrt{2}$

2 2 0 000+ 000+ 0

2 2 2 000+ 000+ $+1/4$

$\tilde{1}$ 1 1 1 $\tfrac{1}{2}$ $\tfrac{1}{2}$

1 $\tfrac{1}{2}$ $\tfrac{1}{2}$ 000+ 000+ $+1/9$

$\tilde{1}$ 1 1 1 1 0

1 0 1 000+ 000+ $+1/9$

$\tilde{1}$ 1 1 1 1 1

1 1 0 000− 000− $+1/2.9$

1 1 1 000+ 000+ $+1/2.9$

O and T_d 9j Symbols (cont.)

$\tilde{1}$ 1 1 1 $\frac{3}{2}$ $\frac{1}{2}$

1 $\frac{1}{2}$ $\frac{1}{2}$ 000− 000− +1/4.9

1 $\frac{1}{2}$ $\frac{3}{2}$ 000+ 000+ +5/8.9

1 $\frac{3}{2}$ $\frac{1}{2}$ 000+ 000+ $+1/2.9\sqrt{2}$

1 $\frac{3}{2}$ $\frac{1}{2}$ 000+ 010+ $-1/2.9\sqrt{2}$

$\tilde{1}$ 1 1 1 $\frac{3}{2}$ $\frac{3}{2}$

1 $\frac{1}{2}$ $\frac{1}{2}$ 000+ 000+ $+1/2.9\sqrt{2}$

1 $\frac{1}{2}$ $\frac{1}{2}$ 010+ 000+ $-1/2.9\sqrt{2}$

1 $\frac{3}{2}$ $\frac{1}{2}$ 000− 000− +1/8.9

1 $\frac{3}{2}$ $\frac{1}{2}$ 000− 010− +1/4.9

1 $\frac{3}{2}$ $\frac{1}{2}$ 010− 000− +1/4.9

1 $\frac{3}{2}$ $\frac{1}{2}$ 010− 010− +1/8.9

1 $\frac{3}{2}$ $\frac{3}{2}$ 000+ 000+ +1/2.9

1 $\frac{3}{2}$ $\frac{3}{2}$ 000+ 010+ 0

1 $\frac{3}{2}$ $\frac{3}{2}$ 000+ 011+ −1/4.9

1 $\frac{3}{2}$ $\frac{3}{2}$ 010+ 000+ 0

1 $\frac{3}{2}$ $\frac{3}{2}$ 010+ 010+ +1/8.9

1 $\frac{3}{2}$ $\frac{3}{2}$ 010+ 011+ 0

1 $\frac{3}{2}$ $\frac{3}{2}$ 011+ 000+ −1/4.9

1 $\frac{3}{2}$ $\frac{3}{2}$ 011+ 010+ 0

1 $\frac{3}{2}$ $\frac{3}{2}$ 011+ 011+ +1/2.9

$\tilde{1}$ 1 1 1 2 1

1 1 0 000+ 000+ −1/2.9

1 1 1 000− 000− +1/2.9

1 1 2 000+ 000+ +1/4.9

1 1 $\tilde{1}$ 000+ 000+ +1/2.9

$\tilde{1}$ 1 1 1 $\tilde{1}$ 1

1 1 0 000+ 000+ +1/2.9

1 1 1 000− 000− 0

1 1 2 000+ 000+ +1/2.9

1 1 $\tilde{1}$ 000+ 000+ +1/2.9

1 2 1 000+ 000+ 0

1 2 $\tilde{1}$ 000− 000− −1/2.9

1 $\tilde{1}$ 1 000+ 000+ −1/2.9

$\tilde{1}$ 1 1 1 $\tilde{1}$ 2

1 1 1 000+ 000+ 0

1 1 $\tilde{1}$ 000− 000− −1/2.9

1 2 1 000− 000− +1/4.3

1 2 $\tilde{1}$ 000+ 000+ +1/4.9

1 $\tilde{1}$ 1 000− 000− 0

$\tilde{1}$ 1 1 1 $\tilde{1}$ $\tilde{1}$

1 1 1 000+ 000+ −1/2.9

1 2 1 000− 000− 0

1 $\tilde{1}$ 1 000− 000− 0

1 $\tilde{1}$ 2 000+ 000+ +1/2.9

1 $\tilde{1}$ $\tilde{1}$ 000+ 000+ +1/2.9

$\tilde{1}$ $\frac{3}{2}$ $\frac{1}{2}$ 1 1 0

1 $\frac{1}{2}$ $\frac{1}{2}$ 000+ 000+ $+1/2.3\sqrt{2}$

$\tilde{1}$ $\frac{3}{2}$ $\frac{1}{2}$ 1 1 1

1 $\frac{1}{2}$ $\frac{1}{2}$ 000− 000− $+1/4.3\sqrt{3}$

$\tilde{1}$ $\frac{3}{2}$ $\frac{1}{2}$ 1 1 1

1 $\frac{1}{2}$ $\frac{3}{2}$ 000+ 000+ $+1/8.3\sqrt{3}$

$\tilde{1}$ $\frac{3}{2}$ $\frac{1}{2}$ 1 1 2

1 $\frac{1}{2}$ $\frac{3}{2}$ 000− 000− −1/8.3

$\tilde{1}$ $\frac{3}{2}$ $\frac{1}{2}$ 1 1 $\tilde{1}$

1 $\frac{1}{2}$ $\frac{3}{2}$ 000− 000− +1/8.3

$\tilde{1}$ $\frac{3}{2}$ $\frac{1}{2}$ 1 $\frac{3}{2}$ $\frac{1}{2}$

1 0 1 000− 000− −1/4.3

1 1 0 000− 000− +1/4.3

1 1 0 000− 010− 0

1 1 1 000+ 000+ 0

1 1 1 000+ 010+ $-1/2.3\sqrt{2.3}$

$\tilde{1}$ $\frac{3}{2}$ $\frac{1}{2}$ 1 $\frac{3}{2}$ $\frac{3}{2}$

1 0 1 000+ 000+ $+1/4.3\sqrt{2}$

1 0 1 010+ 000+ 0

1 1 1 000− 000− $-1/8.3\sqrt{3}$

1 1 1 000− 010− 0

1 1 1 010− 000− $+1/4.3\sqrt{3}$

1 1 1 010− 010− $+1/8.3\sqrt{3}$

1 1 2 000+ 000+ +1/8.3

1 1 2 000+ 010+ −1/8.3

1 1 2 010+ 000+ 0

1 1 2 010+ 010+ −1/8.3

1 1 $\tilde{1}$ 000+ 000+ +1/8.9

1 1 $\tilde{1}$ 000+ 010+ −1/4.9

1 1 $\tilde{1}$ 010+ 000+ −1/4.9

1 1 $\tilde{1}$ 010+ 010+ +1/8.9

$\tilde{1}$ $\frac{3}{2}$ $\frac{1}{2}$ 1 2 1

1 $\frac{1}{2}$ $\frac{1}{2}$ 000+ 000+ −1/4.3

1 $\frac{1}{2}$ $\frac{3}{2}$ 000− 000− −1/8.3

1 $\frac{3}{2}$ $\frac{1}{2}$ 000− 000− 0

1 $\frac{3}{2}$ $\frac{3}{2}$ 000+ 000+ −1/8.3

1 $\frac{3}{2}$ $\frac{3}{2}$ 001+ 000+ −1/8.3

$\tilde{1}$ $\frac{3}{2}$ $\frac{1}{2}$ 1 2 $\tilde{1}$

1 $\frac{1}{2}$ $\frac{3}{2}$ 000+ 000+ −1/8.3

1 $\frac{3}{2}$ $\frac{3}{2}$ 000− 000− +1/8.3

1 $\frac{3}{2}$ $\frac{3}{2}$ 001− 000− −1/8.3

$\tilde{1}$ $\frac{3}{2}$ $\frac{1}{2}$ 1 $\tilde{1}$ 1

1 $\frac{1}{2}$ $\frac{1}{2}$ 000+ 000+ +1/4.9

1 $\frac{1}{2}$ $\frac{3}{2}$ 000− 000− −5/8.9

1 $\frac{3}{2}$ $\frac{1}{2}$ 000− 000− $-1/2.9\sqrt{2}$

1 $\frac{3}{2}$ $\frac{1}{2}$ 000− 010+ $+1/2.9\sqrt{2}$

1 $\frac{3}{2}$ $\frac{3}{2}$ 000+ 000+ −1/8.9

1 $\frac{3}{2}$ $\frac{3}{2}$ 000+ 010− −1/4.9

1 $\frac{3}{2}$ $\frac{3}{2}$ 001+ 000+ −1/4.9

1 $\frac{3}{2}$ $\frac{3}{2}$ 001+ 010− −1/8.9

$\tilde{1}$ $\frac{3}{2}$ $\frac{1}{2}$ 1 $\tilde{1}$ 2

1 $\frac{1}{2}$ $\frac{3}{2}$ 000+ 000+ −1/8.3

1 $\frac{3}{2}$ $\frac{3}{2}$ 000− 000− −1/8.3

1 $\frac{3}{2}$ $\frac{3}{2}$ 000− 010+ −1/8.3

$\tilde{1}$ $\frac{3}{2}$ $\frac{1}{2}$ 1 $\tilde{1}$ 2

1 $\frac{3}{2}$ $\frac{3}{2}$ 001− 000− 0

1 $\frac{3}{2}$ $\frac{3}{2}$ 001− 010+ +1/8.3

$\tilde{1}$ $\frac{3}{2}$ $\frac{1}{2}$ 1 $\tilde{1}$ $\tilde{1}$

1 $\frac{1}{2}$ $\frac{3}{2}$ 000+ 000+ $+1/8.3\sqrt{3}$

1 $\frac{3}{2}$ $\frac{3}{2}$ 000− 000− $-1/8.3\sqrt{3}$

1 $\frac{3}{2}$ $\frac{3}{2}$ 000− 010+ 0

1 $\frac{3}{2}$ $\frac{3}{2}$ 001− 000− $-1/4.3\sqrt{3}$

1 $\frac{3}{2}$ $\frac{3}{2}$ 001− 010+ $+1/8.3\sqrt{3}$

$\tilde{1}$ $\frac{3}{2}$ $\frac{1}{2}$ $\frac{3}{2}$ 0 $\frac{3}{2}$

$\frac{1}{2}$ $\frac{3}{2}$ 1 000+ 000+ +1/8.2

$\frac{1}{2}$ $\frac{3}{2}$ 2 000− 000− +1/8.2

$\frac{1}{2}$ $\frac{3}{2}$ $\tilde{1}$ 000− 000− −1/8.2.3

$\tilde{1}$ $\frac{3}{2}$ $\frac{1}{2}$ $\frac{3}{2}$ $\frac{1}{2}$ 1

$\frac{1}{2}$ 1 $\frac{1}{2}$ 000− 000− +1/4.3

$\frac{1}{2}$ 1 $\frac{3}{2}$ 000+ 000+ +1/8.2.3

$\tilde{1}$ $\frac{3}{2}$ $\frac{1}{2}$ $\frac{3}{2}$ $\frac{1}{2}$ 2

$\frac{1}{2}$ 1 $\frac{3}{2}$ 000− 000− −1/8.2

$\frac{1}{2}$ 2 $\frac{3}{2}$ 000+ 000+ +1/8.2

$\tilde{1}$ $\frac{3}{2}$ $\frac{1}{2}$ $\frac{3}{2}$ $\frac{1}{2}$ $\tilde{1}$

$\frac{1}{2}$ 1 $\frac{3}{2}$ 000− 000− +1/8.2.3

$\frac{1}{2}$ 2 $\frac{3}{2}$ 000+ 000+ +1/8.2.3

$\frac{1}{2}$ $\tilde{1}$ $\frac{3}{2}$ 000+ 000+ +11/8.2.9

$\tilde{1}$ $\frac{3}{2}$ $\frac{1}{2}$ $\frac{3}{2}$ 1 $\frac{1}{2}$

$\frac{1}{2}$ $\frac{1}{2}$ 0 000+ 000+ +1/8

$\frac{1}{2}$ $\frac{1}{2}$ 1 000− 000− +1/8.3

$\tilde{1}$ $\frac{3}{2}$ $\frac{1}{2}$ $\frac{3}{2}$ 1 $\frac{3}{2}$

$\frac{1}{2}$ $\frac{1}{2}$ 1 000+ 000+ $-1/4.3\sqrt{2}$

$\frac{1}{2}$ $\frac{1}{2}$ 1 010+ 000+ 0

$\frac{1}{2}$ $\frac{3}{2}$ 1 000− 000− +1/8.2.3

$\frac{1}{2}$ $\frac{3}{2}$ 1 000− 010− 0

$\frac{1}{2}$ $\frac{3}{2}$ 1 010− 000− 0

$\frac{1}{2}$ $\frac{3}{2}$ 1 010− 010− +1/8.2

$\frac{1}{2}$ $\frac{3}{2}$ 2 000+ 000+ +1/8.2.3

$\frac{1}{2}$ $\frac{3}{2}$ 2 000+ 010+ −1/8.3

$\frac{1}{2}$ $\frac{3}{2}$ 2 010+ 000+ −1/8.3

$\frac{1}{2}$ $\frac{3}{2}$ 2 010+ 010+ +1/8.2.3

$\frac{1}{2}$ $\frac{3}{2}$ $\tilde{1}$ 000+ 000+ +7/8.2.9

$\frac{1}{2}$ $\frac{3}{2}$ $\tilde{1}$ 000+ 010+ +1/4.9

$\frac{1}{2}$ $\frac{3}{2}$ $\tilde{1}$ 010+ 000+ +1/4.9

$\frac{1}{2}$ $\frac{3}{2}$ $\tilde{1}$ 010+ 010+ +1/8.2.9

$\tilde{1}$ $\frac{3}{2}$ $\frac{1}{2}$ $\frac{3}{2}$ $\frac{3}{2}$ 0

$\frac{1}{2}$ 0 $\frac{1}{2}$ 000+ 000+ +1/8

$\tilde{1}$ $\frac{3}{2}$ $\frac{1}{2}$ $\frac{3}{2}$ $\frac{3}{2}$ 1

$\frac{1}{2}$ 0 $\frac{1}{2}$ 000− 000− +1/8.3

$\frac{1}{2}$ 0 $\frac{1}{2}$ 010− 000− −1/4.3

$\frac{1}{2}$ 1 $\frac{1}{2}$ 000+ 000+ +5/8.9

$\frac{1}{2}$ 1 $\frac{1}{2}$ 000+ 010+ +1/4.9

$\frac{1}{2}$ 1 $\frac{1}{2}$ 010+ 000+ +1/4.9

$\frac{1}{2}$ 1 $\frac{1}{2}$ 010+ 010+ +1/4.9

O and T_d 9j Symbols (cont.)

$\tilde{1}$	**$\frac{3}{2}$**	**$\frac{1}{2}$**	**$\frac{3}{2}$ $\frac{3}{2}$ 1**		
$\frac{1}{2}$	1	$\frac{3}{2}$	000−	000−	$+1/4.9$
$\frac{1}{2}$	1	$\frac{3}{2}$	000−	010−	$-1/8.9$
$\frac{1}{2}$	1	$\frac{3}{2}$	010−	000−	$-1/8.9$
$\frac{1}{2}$	1	$\frac{3}{2}$	010−	010−	$+7/8.2.9$
$\tilde{1}$	**$\frac{3}{2}$**	**$\frac{1}{2}$**	**$\frac{3}{2}$ $\frac{3}{2}$ 2**		
$\frac{1}{2}$	1	$\frac{3}{2}$	000+	000+	$+1/8.3$
$\frac{1}{2}$	1	$\frac{3}{2}$	000+	010+	$-1/8.2.3$
$\frac{1}{2}$	2	$\frac{3}{2}$	000−	000−	$+1/8.2$
$\tilde{1}$	**$\frac{3}{2}$**	**$\frac{1}{2}$**	**$\frac{3}{2}$ $\frac{3}{2}$ $\tilde{1}$**		
$\frac{1}{2}$	1	$\frac{3}{2}$	000+	000+	0
$\frac{1}{2}$	1	$\frac{3}{2}$	000+	010+	$+1/8.3$
$\frac{1}{2}$	1	$\frac{3}{2}$	010−	000+	$+1/8.3$
$\frac{1}{2}$	1	$\frac{3}{2}$	010−	010+	$+1/8.2.3$
$\frac{1}{2}$	2	$\frac{3}{2}$	000−	000−	$+1/8.3$
$\frac{1}{2}$	2	$\frac{3}{2}$	010+	000−	$-1/8.2.3$
$\frac{1}{2}$	$\tilde{1}$	$\frac{3}{2}$	000−	000−	$+1/4.9$
$\frac{1}{2}$	$\tilde{1}$	$\frac{3}{2}$	000−	010+	$-1/8.9$
$\frac{1}{2}$	$\tilde{1}$	$\frac{3}{2}$	010+	000−	$-1/8.9$
$\frac{1}{2}$	$\tilde{1}$	$\frac{3}{2}$	010+	010+	$+7/8.2.9$
$\tilde{1}$	**$\frac{3}{2}$**	**$\frac{1}{2}$**	**$\frac{3}{2}$ 2 $\frac{1}{2}$**		
$\frac{1}{2}$	$\frac{1}{2}$	0	000−	000−	$+1/8$
$\frac{1}{2}$	$\frac{1}{2}$	1	000+	000+	$+1/8.3$
$\tilde{1}$	**$\frac{3}{2}$**	**$\frac{1}{2}$**	**$\frac{3}{2}$ 2 $\frac{3}{2}$**		
$\frac{1}{2}$	$\frac{1}{2}$	1	000−	000−	$+1/4.3\sqrt{2}$
$\frac{1}{2}$	$\frac{3}{2}$	1	000+	000+	$+1/8.2.3$
$\frac{1}{2}$	$\frac{3}{2}$	2	000−	000−	$+1/8.2$
$\frac{1}{2}$	$\frac{3}{2}$	$\tilde{1}$	000−	000−	$+1/8.2.3$
$\tilde{1}$	**$\frac{3}{2}$**	**$\frac{1}{2}$**	**$\frac{3}{2}$ $\tilde{1}$ $\frac{1}{2}$**		
$\frac{1}{2}$	$\frac{1}{2}$	0	000−	000−	$-1/8.3$
$\frac{1}{2}$	$\frac{1}{2}$	1	000+	000+	$+7/8.9$
$\tilde{1}$	**$\frac{3}{2}$**	**$\frac{1}{2}$**	**$\frac{3}{2}$ $\tilde{1}$ $\frac{3}{2}$**		
$\frac{1}{2}$	$\frac{1}{2}$	1	000−	000−	$+1/4.9\sqrt{2}$
$\frac{1}{2}$	$\frac{1}{2}$	1	010+	000−	$+1/2.9\sqrt{2}$
$\frac{1}{2}$	$\frac{3}{2}$	1	000+	000+	$+7/8.2.9$
$\frac{1}{2}$	$\frac{3}{2}$	1	000+	010−	$-1/4.9$
$\frac{1}{2}$	$\frac{3}{2}$	1	010−	000+	$-1/4.9$
$\frac{1}{2}$	$\frac{3}{2}$	1	010−	010−	$+1/8.2.9$
$\frac{1}{2}$	$\frac{3}{2}$	2	000−	000−	$+1/8.2.3$
$\frac{1}{2}$	$\frac{3}{2}$	2	000−	010+	$+1/8.3$
$\frac{1}{2}$	$\frac{3}{2}$	2	010+	000−	$+1/8.3$
$\frac{1}{2}$	$\frac{3}{2}$	2	010+	010+	$+1/8.2.3$
$\frac{1}{2}$	$\frac{3}{2}$	$\tilde{1}$	000−	000−	$+1/8.2.3$
$\frac{1}{2}$	$\frac{3}{2}$	$\tilde{1}$	000−	010+	0
$\frac{1}{2}$	$\frac{3}{2}$	$\tilde{1}$	010+	000−	0
$\frac{1}{2}$	$\frac{3}{2}$	$\tilde{1}$	010+	010+	$+1/8.2$
$\tilde{1}$	**$\frac{3}{2}$**	**$\frac{3}{2}$**	**1 1 0**		
1	$\frac{1}{2}$	$\frac{3}{2}$	000+	000+	$-1/4.3\sqrt{2}$
1	$\frac{1}{2}$	$\frac{3}{2}$	100−	000+	0
$\tilde{1}$	**$\frac{3}{2}$**	**$\frac{3}{2}$**	**1 1 1**		
1	$\frac{1}{2}$	$\frac{1}{2}$	000+	000+	$+1/2.3\sqrt{2.3}$
1	$\frac{1}{2}$	$\frac{1}{2}$	100−	000+	0
$\tilde{1}$	**$\frac{3}{2}$**	**$\frac{3}{2}$**	**1 $\frac{3}{2}$ $\frac{1}{2}$**		
1	0	1	000+	000+	$-1/4.3\sqrt{2}$
1	0	1	100−	000+	0
1	1	1	000−	000−	$-1/8.3\sqrt{3}$
1	1	1	000−	010−	0
1	1	1	100+	000−	$-1/4.3\sqrt{3}$
1	1	1	100+	010−	$+1/8.3\sqrt{3}$
1	1	2	000+	000+	$-1/8.3$
1	1	2	000+	010+	$-1/8.3$
1	1	2	100−	000+	0
1	1	2	100−	010+	$+1/8.3$
1	1	$\tilde{1}$	000+	000+	$-1/8.9$
1	1	$\tilde{1}$	000+	010+	$-1/4.9$
1	1	$\tilde{1}$	100−	000+	$-1/4.9$
1	1	$\tilde{1}$	100−	010+	$-1/8.9$
$\tilde{1}$	**$\frac{3}{2}$**	**$\frac{3}{2}$**	**1 $\frac{3}{2}$ $\frac{3}{2}$**		
1	1	0	000−	000−	0
1	1	0	000−	010−	0
1	1	0	010−	000−	0
1	1	0	010−	010−	$-1/4.3\sqrt{2}$
1	1	0	100+	000−	0
1	1	0	100+	010−	$-1/4.3\sqrt{2}$
1	1	0	110+	000−	$+1/4.3\sqrt{2}$
1	1	0	110+	010−	0
1	1	1	000+	000+	0
1	1	1	000+	010+	0
1	1	1	000+	011+	$+1/4.3\sqrt{3}$
1	1	1	010+	000+	0
1	1	1	010+	010+	$+1/8.3\sqrt{3}$
1	1	1	010+	011+	0
1	1	1	100−	000+	0
1	1	1	100−	010+	$-1/8.3\sqrt{3}$
1	1	1	100−	011+	0
1	1	1	110−	000+	0
1	1	1	110−	010+	0
1	1	1	110−	011+	0
$\tilde{1}$	**$\frac{3}{2}$**	**$\frac{3}{2}$**	**1 2 1**		
1	$\frac{1}{2}$	$\frac{1}{2}$	000−	000−	0
1	$\frac{1}{2}$	$\frac{1}{2}$	100+	000−	$+1/4.3\sqrt{2}$
1	$\frac{1}{2}$	$\frac{3}{2}$	000+	000+	$-1/8.3$
1	$\frac{1}{2}$	$\frac{3}{2}$	000+	001+	$-1/8.3$
1	$\frac{1}{2}$	$\frac{3}{2}$	100−	000+	0
1	$\frac{1}{2}$	$\frac{3}{2}$	100−	001+	$-1/8.3$
1	$\frac{3}{2}$	$\frac{1}{2}$	000+	000+	$-1/8.3$
1	$\frac{3}{2}$	$\frac{1}{2}$	100−	000+	$+1/8.3$
1	$\frac{3}{2}$	$\frac{3}{2}$	000−	000−	0
1	$\frac{3}{2}$	$\frac{3}{2}$	000−	001−	0
$\tilde{1}$	**$\frac{3}{2}$**	**$\frac{3}{2}$**	**1 2 1**		
1	$\frac{3}{2}$	$\frac{3}{2}$	001−	000−	$+1/8.3$
1	$\frac{3}{2}$	$\frac{3}{2}$	001−	001−	0
1	$\frac{3}{2}$	$\frac{3}{2}$	100+	000−	$+1/8.3$
1	$\frac{3}{2}$	$\frac{3}{2}$	100+	001−	0
1	$\frac{3}{2}$	$\frac{3}{2}$	101+	000−	0
1	$\frac{3}{2}$	$\frac{3}{2}$	101+	001−	0
$\tilde{1}$	**$\frac{3}{2}$**	**$\frac{3}{2}$**	**1 $\tilde{1}$ 1**		
1	$\frac{1}{2}$	$\frac{1}{2}$	000−	000−	$-1/2.9\sqrt{2}$
1	$\frac{1}{2}$	$\frac{1}{2}$	100+	000−	$-1/2.9\sqrt{2}$
1	$\frac{1}{2}$	$\frac{3}{2}$	000+	000+	$-1/8.9$
1	$\frac{1}{2}$	$\frac{3}{2}$	000+	001+	$-1/4.9$
1	$\frac{1}{2}$	$\frac{3}{2}$	100−	000+	$+1/4.9$
1	$\frac{1}{2}$	$\frac{3}{2}$	100−	001+	$+1/8.9$
1	$\frac{3}{2}$	$\frac{1}{2}$	000+	000+	$+1/8.9$
1	$\frac{3}{2}$	$\frac{1}{2}$	000+	010−	$+1/4.9$
1	$\frac{3}{2}$	$\frac{1}{2}$	100−	000+	$-1/4.9$
1	$\frac{3}{2}$	$\frac{1}{2}$	100−	010−	$-1/8.9$
1	$\frac{3}{2}$	$\frac{3}{2}$	000−	000−	$-1/2.9$
1	$\frac{3}{2}$	$\frac{3}{2}$	000−	001−	0
1	$\frac{3}{2}$	$\frac{3}{2}$	000−	010+	0
1	$\frac{3}{2}$	$\frac{3}{2}$	000−	011+	$+1/4.9$
1	$\frac{3}{2}$	$\frac{3}{2}$	001−	000−	0
1	$\frac{3}{2}$	$\frac{3}{2}$	001−	001−	$-1/8.9$
1	$\frac{3}{2}$	$\frac{3}{2}$	001−	010+	$-1/8.9$
1	$\frac{3}{2}$	$\frac{3}{2}$	001−	011+	0
1	$\frac{3}{2}$	$\frac{3}{2}$	100+	000−	0
1	$\frac{3}{2}$	$\frac{3}{2}$	100+	001−	$+1/8.9$
1	$\frac{3}{2}$	$\frac{3}{2}$	100+	010+	$+1/8.9$
1	$\frac{3}{2}$	$\frac{3}{2}$	100+	011+	0
1	$\frac{3}{2}$	$\frac{3}{2}$	101+	000−	$-1/4.9$
1	$\frac{3}{2}$	$\frac{3}{2}$	101+	001−	0
1	$\frac{3}{2}$	$\frac{3}{2}$	101+	010+	0
1	$\frac{3}{2}$	$\frac{3}{2}$	101+	011+	$+1/2.9$
$\tilde{1}$	**$\frac{3}{2}$**	**$\frac{3}{2}$**	**1 $\tilde{1}$ 2**		
1	$\frac{1}{2}$	$\frac{1}{2}$	000+	000+	0
1	$\frac{1}{2}$	$\frac{1}{2}$	100−	000+	$-1/4.3\sqrt{2}$
1	$\frac{1}{2}$	$\frac{3}{2}$	000−	000−	$+1/8.3$
1	$\frac{1}{2}$	$\frac{3}{2}$	100+	000−	$-1/8.3$
1	$\frac{3}{2}$	$\frac{1}{2}$	000−	000−	$-1/8.3$
1	$\frac{3}{2}$	$\frac{1}{2}$	000−	010+	$-1/8.3$
1	$\frac{3}{2}$	$\frac{1}{2}$	100+	000−	0
1	$\frac{3}{2}$	$\frac{1}{2}$	100+	010+	$-1/8.3$
1	$\frac{3}{2}$	$\frac{3}{2}$	000+	000+	0
1	$\frac{3}{2}$	$\frac{3}{2}$	000+	010−	0
1	$\frac{3}{2}$	$\frac{3}{2}$	001+	000+	$+1/8.3$
1	$\frac{3}{2}$	$\frac{3}{2}$	001+	010−	0
1	$\frac{3}{2}$	$\frac{3}{2}$	100−	000+	$+1/8.3$
1	$\frac{3}{2}$	$\frac{3}{2}$	100−	010−	0
1	$\frac{3}{2}$	$\frac{3}{2}$	101−	000+	0

O and T_d 9j Symbols (cont.)

$\tilde{1}\ \frac{3}{2}\ \frac{3}{2}\quad 1\ \tilde{1}\ 2$

$1\ \frac{3}{2}\ \frac{3}{2}$	101−	010−	0

$\tilde{1}\ \frac{3}{2}\ \frac{3}{2}\quad 1\ \tilde{1}\ \tilde{1}$

$1\ \frac{1}{2}\ \frac{1}{2}$	000+	000+	$-1/2.3\sqrt{2.3}$
$1\ \frac{1}{2}\ \frac{1}{2}$	100−	000+	0
$1\ \frac{3}{2}\ \frac{1}{2}$	000−	000−	$-1/8.3\sqrt{3}$
$1\ \frac{3}{2}\ \frac{1}{2}$	000−	010+	0
$1\ \frac{3}{2}\ \frac{1}{2}$	100+	000−	$+1/4.3\sqrt{3}$
$1\ \frac{3}{2}\ \frac{1}{2}$	100+	010+	$-1/8.3\sqrt{3}$
$1\ \frac{3}{2}\ \frac{3}{2}$	000+	000+	0
$1\ \frac{3}{2}\ \frac{3}{2}$	000+	010−	0
$1\ \frac{3}{2}\ \frac{3}{2}$	000+	011+	$-1/4.3\sqrt{3}$
$1\ \frac{3}{2}\ \frac{3}{2}$	001+	000+	0
$1\ \frac{3}{2}\ \frac{3}{2}$	001+	010−	$-1/8.3\sqrt{3}$
$1\ \frac{3}{2}\ \frac{3}{2}$	001+	011+	0
$1\ \frac{3}{2}\ \frac{3}{2}$	100−	000+	0
$1\ \frac{3}{2}\ \frac{3}{2}$	100−	010−	$-1/8.3\sqrt{3}$
$1\ \frac{3}{2}\ \frac{3}{2}$	100−	011+	0
$1\ \frac{3}{2}\ \frac{3}{2}$	101−	000+	0
$1\ \frac{3}{2}\ \frac{3}{2}$	101−	010−	0
$1\ \frac{3}{2}\ \frac{3}{2}$	101−	011+	0

$\tilde{1}\ \frac{3}{2}\ \frac{3}{2}\quad \frac{3}{2}\ 1\ \frac{1}{2}$

$\frac{1}{2}\ \frac{1}{2}\ 1$	000+	000+	$-1/4.3\sqrt{2}$
$\frac{1}{2}\ \frac{1}{2}\ 1$	100−	000+	0
$\frac{1}{2}\ \frac{3}{2}\ 1$	000−	000−	0
$\frac{1}{2}\ \frac{3}{2}\ 1$	000−	010−	$+1/8.3$
$\frac{1}{2}\ \frac{3}{2}\ 1$	100+	000−	$-1/8.3$
$\frac{1}{2}\ \frac{3}{2}\ 1$	100+	010−	$+1/8.2.3$
$\frac{1}{2}\ \frac{3}{2}\ 2$	000+	000+	0
$\frac{1}{2}\ \frac{3}{2}\ 2$	000+	010+	0
$\frac{1}{2}\ \frac{3}{2}\ 2$	100−	000+	0
$\frac{1}{2}\ \frac{3}{2}\ 2$	100−	010+	$+1/8.2$
$\frac{1}{2}\ \frac{3}{2}\ \tilde{1}$	000+	000+	0
$\frac{1}{2}\ \frac{3}{2}\ \tilde{1}$	000+	010+	$+1/8.3$
$\frac{1}{2}\ \frac{3}{2}\ \tilde{1}$	100−	000+	$-1/8.3$
$\frac{1}{2}\ \frac{3}{2}\ \tilde{1}$	100−	010+	$-1/8.2.3$
$\frac{3}{2}\ \frac{1}{2}\ 1$	000−	000−	$+1/8.2.3$
$\frac{3}{2}\ \frac{1}{2}\ 1$	100+	000−	0
$\frac{3}{2}\ \frac{1}{2}\ 1$	100+	100+	$+1/8.2$
$\frac{3}{2}\ \frac{1}{2}\ 2$	000+	000+	$+1/8.2.3$
$\frac{3}{2}\ \frac{1}{2}\ 2$	100−	000+	$+1/8.3$
$\frac{3}{2}\ \frac{1}{2}\ 2$	100−	100−	$+1/8.2.3$
$\frac{3}{2}\ \frac{1}{2}\ \tilde{1}$	000+	000+	$+7/8.2.9$
$\frac{3}{2}\ \frac{1}{2}\ \tilde{1}$	100−	000+	$-1/4.9$
$\frac{3}{2}\ \frac{1}{2}\ \tilde{1}$	100−	100−	$+1/8.2.9$

$\tilde{1}\ \frac{3}{2}\ \frac{3}{2}\quad \frac{3}{2}\ \frac{3}{2}\ 0$

$\frac{1}{2}\ 1\ \frac{3}{2}$	000−	000−	$+1/8.3$
$\frac{1}{2}\ 1\ \frac{3}{2}$	000−	010−	$+1/8.3$
$\frac{1}{2}\ 1\ \frac{3}{2}$	100+	000−	$+1/8.3$
$\frac{1}{2}\ 1\ \frac{3}{2}$	100+	010−	$-1/8.2.3$
$\frac{1}{2}\ 2\ \frac{3}{2}$	000+	000+	0
$\frac{1}{2}\ 2\ \frac{3}{2}$	100−	000+	$+1/8.2$
$\frac{1}{2}\ \tilde{1}\ \frac{3}{2}$	000+	000+	$+1/8.3$
$\frac{1}{2}\ \tilde{1}\ \frac{3}{2}$	000+	010−	$-1/8.3$
$\frac{1}{2}\ \tilde{1}\ \frac{3}{2}$	100−	000+	$-1/8.3$
$\frac{1}{2}\ \tilde{1}\ \frac{3}{2}$	100−	010−	$-1/8.2.3$
$\frac{3}{2}\ 0\ \frac{3}{2}$	000−	000−	$+1/8.2$
$\frac{3}{2}\ 0\ \frac{3}{2}$	100+	000−	0
$\frac{3}{2}\ 0\ \frac{3}{2}$	100+	100+	$+1/8.2$
$\frac{3}{2}\ 1\ \frac{3}{2}$	000+	000+	$-1/8.2.3$
$\frac{3}{2}\ 1\ \frac{3}{2}$	000+	010+	0
$\frac{3}{2}\ 1\ \frac{3}{2}$	000+	100−	0
$\frac{3}{2}\ 1\ \frac{3}{2}$	000+	110−	$+1/8.3$
$\frac{3}{2}\ 1\ \frac{3}{2}$	001+	000+	0
$\frac{3}{2}\ 1\ \frac{3}{2}$	001+	010+	$+1/8.2.3$
$\frac{3}{2}\ 1\ \frac{3}{2}$	001+	100−	$-1/8.3$
$\frac{3}{2}\ 1\ \frac{3}{2}$	001+	110−	0
$\frac{3}{2}\ 1\ \frac{3}{2}$	100−	000+	0
$\frac{3}{2}\ 1\ \frac{3}{2}$	100−	010+	$+1/8.3$
$\frac{3}{2}\ 1\ \frac{3}{2}$	100−	100−	$-1/8.2.3$
$\frac{3}{2}\ 1\ \frac{3}{2}$	100−	110−	0
$\frac{3}{2}\ 1\ \frac{3}{2}$	101−	000+	$-1/8.3$
$\frac{3}{2}\ 1\ \frac{3}{2}$	101−	010+	0
$\frac{3}{2}\ 1\ \frac{3}{2}$	101−	100−	0
$\frac{3}{2}\ 1\ \frac{3}{2}$	101−	110−	$-1/8.3$

$\tilde{1}\ \frac{3}{2}\ \frac{3}{2}\quad \frac{3}{2}\ \frac{3}{2}\ 1$

$\frac{1}{2}\ 0\ \frac{1}{2}$	000+	000+	$-1/4.3\sqrt{2}$
$\frac{1}{2}\ 0\ \frac{1}{2}$	010+	000+	$-1/4.3\sqrt{2}$
$\frac{1}{2}\ 0\ \frac{1}{2}$	100−	000+	$-1/4.3\sqrt{2}$
$\frac{1}{2}\ 0\ \frac{1}{2}$	110−	000+	$+1/8.3\sqrt{2}$
$\frac{1}{2}\ 1\ \frac{1}{2}$	000−	000−	$+1/4.9\sqrt{2}$
$\frac{1}{2}\ 1\ \frac{1}{2}$	000−	010−	$-1/2.9\sqrt{2}$
$\frac{1}{2}\ 1\ \frac{1}{2}$	010−	000−	$+1/4.9\sqrt{2}$
$\frac{1}{2}\ 1\ \frac{1}{2}$	010−	010−	$+1/4.9\sqrt{2}$
$\frac{1}{2}\ 1\ \frac{1}{2}$	100+	000−	$+1/4.9\sqrt{2}$
$\frac{1}{2}\ 1\ \frac{1}{2}$	100+	010−	$+1/4.9\sqrt{2}$
$\frac{1}{2}\ 1\ \frac{1}{2}$	110+	000−	$+5/8.9\sqrt{2}$
$\frac{1}{2}\ 1\ \frac{1}{2}$	110+	010−	$-1/2.9\sqrt{2}$
$\frac{1}{2}\ 1\ \frac{3}{2}$	000+	000+	$+1/8.9$
$\frac{1}{2}\ 1\ \frac{3}{2}$	000+	001+	$+1/4.9$
$\frac{1}{2}\ 1\ \frac{3}{2}$	000+	010+	$+1/8.9$
$\frac{1}{2}\ 1\ \frac{3}{2}$	000+	011+	$-1/8.9$
$\frac{1}{2}\ 1\ \frac{3}{2}$	010+	000+	$-1/4.9$
$\frac{1}{2}\ 1\ \frac{3}{2}$	010+	001+	$-1/8.9$
$\frac{1}{2}\ 1\ \frac{3}{2}$	010+	010+	$-1/4.9$
$\frac{1}{2}\ 1\ \frac{3}{2}$	010+	011+	$-1/8.9$
$\frac{1}{2}\ 1\ \frac{3}{2}$	100−	000+	$+1/8.9$
$\frac{1}{2}\ 1\ \frac{3}{2}$	100−	001+	$+1/4.9$
$\frac{1}{2}\ 1\ \frac{3}{2}$	100−	010+	$+5/8.2.9$
$\frac{1}{2}\ 1\ \frac{3}{2}$	100−	011+	$-1/8.9$
$\frac{1}{2}\ 1\ \frac{3}{2}$	110−	000+	$+1/8.9$
$\frac{1}{2}\ 1\ \frac{3}{2}$	110−	001+	$-1/8.9$
$\frac{1}{2}\ 1\ \frac{3}{2}$	110−	010+	$+1/8.9$
$\frac{1}{2}\ 1\ \frac{3}{2}$	110−	011+	$-1/8.9$
$\frac{1}{2}\ 2\ \frac{3}{2}$	000−	000−	0
$\frac{1}{2}\ 2\ \frac{3}{2}$	000−	001−	$-1/8.3$
$\frac{1}{2}\ 2\ \frac{3}{2}$	010−	000−	0
$\frac{1}{2}\ 2\ \frac{3}{2}$	010−	001−	$-1/8.3$
$\frac{1}{2}\ 2\ \frac{3}{2}$	100+	000−	$-1/8.2.3$
$\frac{1}{2}\ 2\ \frac{3}{2}$	100+	001−	0
$\frac{1}{2}\ 2\ \frac{3}{2}$	110+	000−	$+1/8.3$
$\frac{1}{2}\ 2\ \frac{3}{2}$	110+	001−	0
$\frac{1}{2}\ \tilde{1}\ \frac{3}{2}$	000−	000−	$+1/8.9$
$\frac{1}{2}\ \tilde{1}\ \frac{3}{2}$	000−	001−	$-1/4.9$
$\frac{1}{2}\ \tilde{1}\ \frac{3}{2}$	000−	010+	$-1/8.9$
$\frac{1}{2}\ \tilde{1}\ \frac{3}{2}$	000−	011+	$-1/8.9$
$\frac{1}{2}\ \tilde{1}\ \frac{3}{2}$	010−	000−	$-1/4.9$
$\frac{1}{2}\ \tilde{1}\ \frac{3}{2}$	010−	001−	$+1/8.9$
$\frac{1}{2}\ \tilde{1}\ \frac{3}{2}$	010−	010+	$+1/4.9$
$\frac{1}{2}\ \tilde{1}\ \frac{3}{2}$	010−	011+	$-1/8.9$
$\frac{1}{2}\ \tilde{1}\ \frac{3}{2}$	100+	000−	$-1/8.9$
$\frac{1}{2}\ \tilde{1}\ \frac{3}{2}$	100+	001−	$+1/4.9$
$\frac{1}{2}\ \tilde{1}\ \frac{3}{2}$	100+	010+	$+5/8.2.9$
$\frac{1}{2}\ \tilde{1}\ \frac{3}{2}$	100+	011+	$+1/8.9$
$\frac{1}{2}\ \tilde{1}\ \frac{3}{2}$	110+	000−	$-1/8.9$
$\frac{1}{2}\ \tilde{1}\ \frac{3}{2}$	110+	001−	$-1/8.9$
$\frac{1}{2}\ \tilde{1}\ \frac{3}{2}$	110+	010+	$+1/8.9$
$\frac{1}{2}\ \tilde{1}\ \frac{3}{2}$	110+	011+	$+1/8.9$
$\frac{3}{2}\ 0\ \frac{3}{2}$	000+	000+	$-1/8.2.3$
$\frac{3}{2}\ 0\ \frac{3}{2}$	000+	001+	0
$\frac{3}{2}\ 0\ \frac{3}{2}$	000+	100−	0
$\frac{3}{2}\ 0\ \frac{3}{2}$	000+	101−	$-1/8.3$
$\frac{3}{2}\ 0\ \frac{3}{2}$	010+	000+	0
$\frac{3}{2}\ 0\ \frac{3}{2}$	010+	001+	$+1/8.2.3$
$\frac{3}{2}\ 0\ \frac{3}{2}$	010+	100−	$+1/8.3$
$\frac{3}{2}\ 0\ \frac{3}{2}$	010+	101−	0
$\frac{3}{2}\ 0\ \frac{3}{2}$	100−	000+	0
$\frac{3}{2}\ 0\ \frac{3}{2}$	100−	001+	$-1/8.3$
$\frac{3}{2}\ 0\ \frac{3}{2}$	100−	100−	$-1/8.2.3$
$\frac{3}{2}\ 0\ \frac{3}{2}$	100−	101−	0
$\frac{3}{2}\ 0\ \frac{3}{2}$	110−	000+	$+1/8.3$
$\frac{3}{2}\ 0\ \frac{3}{2}$	110−	001+	0
$\frac{3}{2}\ 0\ \frac{3}{2}$	110−	100−	0
$\frac{3}{2}\ 0\ \frac{3}{2}$	110−	101−	$-1/8.3$
$\frac{3}{2}\ 1\ \frac{1}{2}$	000+	000+	$+1/8.9$
$\frac{3}{2}\ 1\ \frac{1}{2}$	000+	010+	$+1/8.9$
$\frac{3}{2}\ 1\ \frac{1}{2}$	010+	000+	$+1/8.9$
$\frac{3}{2}\ 1\ \frac{1}{2}$	010+	010+	$+5/8.2.9$

O and T_d 9j Symbols (cont.)

$\tilde{1}$ $\frac{3}{2}$ $\frac{3}{2}$	$\frac{3}{2}$ $\frac{3}{2}$ 1				
$\frac{3}{2}$	1	$\frac{1}{2}$	100−	000+	$-1/4.9$
$\frac{3}{2}$	1	$\frac{1}{2}$	100−	010+	$-1/4.9$
$\frac{3}{2}$	1	$\frac{1}{2}$	100−	100−	$+1/8.9$
$\frac{3}{2}$	1	$\frac{1}{2}$	100−	110−	$+1/8.9$
$\frac{3}{2}$	1	$\frac{1}{2}$	110−	000+	$+1/8.9$
$\frac{3}{2}$	1	$\frac{1}{2}$	110−	010+	$+1/8.9$
$\frac{3}{2}$	1	$\frac{1}{2}$	110−	100−	$+1/8.9$
$\frac{3}{2}$	1	$\frac{1}{2}$	110−	110−	$+1/8.9$
$\frac{3}{2}$	1	$\frac{3}{2}$	000−	000−	$+5/8.2.9$
$\frac{3}{2}$	1	$\frac{3}{2}$	000−	001−	0
$\frac{3}{2}$	1	$\frac{3}{2}$	000−	010−	0
$\frac{3}{2}$	1	$\frac{3}{2}$	000−	011−	$-1/8.9$
$\frac{3}{2}$	1	$\frac{3}{2}$	001−	000−	0
$\frac{3}{2}$	1	$\frac{3}{2}$	001−	001−	$+1/8.9$
$\frac{3}{2}$	1	$\frac{3}{2}$	001−	010−	$-5/8.2.9$
$\frac{3}{2}$	1	$\frac{3}{2}$	001−	011−	0
$\frac{3}{2}$	1	$\frac{3}{2}$	010−	000−	0
$\frac{3}{2}$	1	$\frac{3}{2}$	010−	001−	$-5/8.2.9$
$\frac{3}{2}$	1	$\frac{3}{2}$	010−	010−	$+1/8.9$
$\frac{3}{2}$	1	$\frac{3}{2}$	010−	011−	0
$\frac{3}{2}$	1	$\frac{3}{2}$	011−	000−	$-1/8.9$
$\frac{3}{2}$	1	$\frac{3}{2}$	011−	001−	0
$\frac{3}{2}$	1	$\frac{3}{2}$	011−	010−	0
$\frac{3}{2}$	1	$\frac{3}{2}$	011−	011−	$+1/8.9$
$\frac{3}{2}$	1	$\frac{3}{2}$	100+	000−	0
$\frac{3}{2}$	1	$\frac{3}{2}$	100+	001−	$-1/8.9$
$\frac{3}{2}$	1	$\frac{3}{2}$	100+	010−	$+1/8.9$
$\frac{3}{2}$	1	$\frac{3}{2}$	100+	011−	0
$\frac{3}{2}$	1	$\frac{3}{2}$	100+	100+	$+5/8.2.9$
$\frac{3}{2}$	1	$\frac{3}{2}$	100+	101+	0
$\frac{3}{2}$	1	$\frac{3}{2}$	100+	110+	0
$\frac{3}{2}$	1	$\frac{3}{2}$	100+	111+	$+1/4.9$
$\frac{3}{2}$	1	$\frac{3}{2}$	101+	000−	$-1/8.9$
$\frac{3}{2}$	1	$\frac{3}{2}$	101+	001−	0
$\frac{3}{2}$	1	$\frac{3}{2}$	101+	010−	0
$\frac{3}{2}$	1	$\frac{3}{2}$	101+	011−	$+1/8.9$
$\frac{3}{2}$	1	$\frac{3}{2}$	101+	100+	0
$\frac{3}{2}$	1	$\frac{3}{2}$	101+	101+	$+1/8.9$
$\frac{3}{2}$	1	$\frac{3}{2}$	101+	110+	$-1/8.9$
$\frac{3}{2}$	1	$\frac{3}{2}$	101+	111+	0
$\frac{3}{2}$	1	$\frac{3}{2}$	110+	000−	$+1/8.9$
$\frac{3}{2}$	1	$\frac{3}{2}$	110+	001−	0
$\frac{3}{2}$	1	$\frac{3}{2}$	110+	010−	0
$\frac{3}{2}$	1	$\frac{3}{2}$	110+	011−	$-1/8.9$
$\frac{3}{2}$	1	$\frac{3}{2}$	110+	100+	0
$\frac{3}{2}$	1	$\frac{3}{2}$	110+	101+	$-1/8.9$
$\frac{3}{2}$	1	$\frac{3}{2}$	110+	110+	$+1/8.9$
$\frac{3}{2}$	1	$\frac{3}{2}$	110+	111+	0
$\frac{3}{2}$	1	$\frac{3}{2}$	111+	000−	0

$\tilde{1}$ $\frac{3}{2}$ $\frac{3}{2}$	$\frac{3}{2}$ $\frac{3}{2}$ 1				
$\frac{3}{2}$	1	$\frac{3}{2}$	111+	001−	$-1/4.9$
$\frac{3}{2}$	1	$\frac{3}{2}$	111+	010−	$+1/4.9$
$\frac{3}{2}$	1	$\frac{3}{2}$	111+	011−	0
$\frac{3}{2}$	1	$\frac{3}{2}$	111+	100+	$+1/4.9$
$\frac{3}{2}$	1	$\frac{3}{2}$	111+	101+	0
$\frac{3}{2}$	1	$\frac{3}{2}$	111+	110+	0
$\frac{3}{2}$	1	$\frac{3}{2}$	111+	111+	$+5/8.2.9$

$\tilde{1}$ $\frac{3}{2}$ $\frac{3}{2}$	$\frac{3}{2}$ 2 $\frac{1}{2}$				
$\frac{1}{2}$	$\frac{1}{2}$	1	000−	000−	$+1/4.3\sqrt{2}$
$\frac{1}{2}$	$\frac{1}{2}$	1	100+	000−	$+1/4.3\sqrt{2}$
$\frac{1}{2}$	$\frac{3}{2}$	1	000+	000+	$-1/8.3$
$\frac{1}{2}$	$\frac{3}{2}$	1	100−	000+	$+1/8.2.3$
$\frac{1}{2}$	$\frac{3}{2}$	2	000−	000−	0
$\frac{1}{2}$	$\frac{3}{2}$	2	100+	000−	$+1/8.2$
$\frac{1}{2}$	$\frac{3}{2}$	$\tilde{1}$	000−	000−	$-1/8.3$
$\frac{1}{2}$	$\frac{3}{2}$	$\tilde{1}$	100+	000−	$-1/8.2.3$
$\frac{3}{2}$	$\frac{1}{2}$	1	000+	000+	$+1/8.2.3$
$\frac{3}{2}$	$\frac{1}{2}$	1	100−	000+	$-1/8.3$
$\frac{3}{2}$	$\frac{1}{2}$	1	100−	100−	$+1/8.2.3$
$\frac{3}{2}$	$\frac{1}{2}$	2	000−	000−	$+1/8.2$
$\frac{3}{2}$	$\frac{1}{2}$	2	100+	000−	0
$\frac{3}{2}$	$\frac{1}{2}$	2	100+	100+	$+1/8.2$
$\frac{3}{2}$	$\frac{1}{2}$	$\tilde{1}$	000−	000−	$+1/8.2.3$
$\frac{3}{2}$	$\frac{1}{2}$	$\tilde{1}$	100+	000−	$+1/8.3$
$\frac{3}{2}$	$\frac{1}{2}$	$\tilde{1}$	100+	100+	$+1/8.2.3$
$\frac{3}{2}$	$\frac{3}{2}$	1	000−	000−	0
$\frac{3}{2}$	$\frac{3}{2}$	1	000−	100+	$+1/8.3$
$\frac{3}{2}$	$\frac{3}{2}$	1	001−	000−	$+1/8.2.3$
$\frac{3}{2}$	$\frac{3}{2}$	1	001−	100+	0
$\frac{3}{2}$	$\frac{3}{2}$	1	100+	000−	0
$\frac{3}{2}$	$\frac{3}{2}$	1	100+	100+	$+1/8.3$
$\frac{3}{2}$	$\frac{3}{2}$	1	101+	000−	$-1/8.3$
$\frac{3}{2}$	$\frac{3}{2}$	1	101+	100+	0
$\frac{3}{2}$	$\frac{3}{2}$	2	000+	000+	$+1/8.2$
$\frac{3}{2}$	$\frac{3}{2}$	2	000+	100−	0
$\frac{3}{2}$	$\frac{3}{2}$	2	100−	000+	0
$\frac{3}{2}$	$\frac{3}{2}$	2	100−	100−	0
$\frac{3}{2}$	$\frac{3}{2}$	$\tilde{1}$	000+	000+	0
$\frac{3}{2}$	$\frac{3}{2}$	$\tilde{1}$	000+	100−	$+1/8.3$
$\frac{3}{2}$	$\frac{3}{2}$	$\tilde{1}$	001−	000+	$+1/8.2.3$
$\frac{3}{2}$	$\frac{3}{2}$	$\tilde{1}$	001−	100−	0
$\frac{3}{2}$	$\frac{3}{2}$	$\tilde{1}$	100−	000+	0
$\frac{3}{2}$	$\frac{3}{2}$	$\tilde{1}$	100−	100−	$-1/8.3$
$\frac{3}{2}$	$\frac{3}{2}$	$\tilde{1}$	101+	000+	$+1/8.3$
$\frac{3}{2}$	$\frac{3}{2}$	$\tilde{1}$	101+	100−	0

$\tilde{1}$ $\frac{3}{2}$ $\frac{3}{2}$	$\frac{3}{2}$ 2 $\frac{3}{2}$				
$\frac{1}{2}$	$\frac{1}{2}$	0	000−	000−	0
$\frac{1}{2}$	$\frac{1}{2}$	0	100+	000−	$-1/8\sqrt{2}$
$\frac{1}{2}$	$\frac{1}{2}$	1	000+	000+	0

$\tilde{1}$ $\frac{3}{2}$ $\frac{3}{2}$	$\frac{3}{2}$ 2 $\frac{3}{2}$				
$\frac{1}{2}$	$\frac{1}{2}$	1	000+	001+	$+1/4.3\sqrt{2}$
$\frac{1}{2}$	$\frac{1}{2}$	1	100−	000+	$+1/8.3\sqrt{2}$
$\frac{1}{2}$	$\frac{1}{2}$	1	100−	001+	0
$\frac{1}{2}$	$\frac{3}{2}$	1	000−	000−	$+1/8.3$
$\frac{1}{2}$	$\frac{3}{2}$	1	000−	001−	0
$\frac{1}{2}$	$\frac{3}{2}$	1	100+	000−	0
$\frac{1}{2}$	$\frac{3}{2}$	1	100+	001−	$-1/8.3$
$\frac{1}{2}$	$\frac{3}{2}$	2	000+	000+	0
$\frac{1}{2}$	$\frac{3}{2}$	2	100−	000+	0
$\frac{1}{2}$	$\frac{3}{2}$	$\tilde{1}$	000+	000+	$+1/8.3$
$\frac{1}{2}$	$\frac{3}{2}$	$\tilde{1}$	000+	001−	0
$\frac{1}{2}$	$\frac{3}{2}$	$\tilde{1}$	100−	000+	0
$\frac{1}{2}$	$\frac{3}{2}$	$\tilde{1}$	100−	001−	$-1/8.3$
$\frac{3}{2}$	$\frac{1}{2}$	1	000−	000−	0
$\frac{3}{2}$	$\frac{1}{2}$	1	000−	001−	$+1/8.2.3$
$\frac{3}{2}$	$\frac{1}{2}$	1	000−	100+	0
$\frac{3}{2}$	$\frac{1}{2}$	1	000−	101+	$-1/8.3$
$\frac{3}{2}$	$\frac{1}{2}$	1	100+	000−	$+1/8.3$
$\frac{3}{2}$	$\frac{1}{2}$	1	100+	001−	0
$\frac{3}{2}$	$\frac{1}{2}$	1	100+	100+	$+1/8.3$
$\frac{3}{2}$	$\frac{1}{2}$	1	100+	101+	0
$\frac{3}{2}$	$\frac{1}{2}$	2	000+	000+	$+1/8.2$
$\frac{3}{2}$	$\frac{1}{2}$	2	000+	100−	0
$\frac{3}{2}$	$\frac{1}{2}$	2	100−	000+	0
$\frac{3}{2}$	$\frac{1}{2}$	2	100−	100−	0
$\frac{3}{2}$	$\frac{1}{2}$	$\tilde{1}$	000+	000+	0
$\frac{3}{2}$	$\frac{1}{2}$	$\tilde{1}$	000+	001−	$+1/8.2.3$
$\frac{3}{2}$	$\frac{1}{2}$	$\tilde{1}$	000+	100−	0
$\frac{3}{2}$	$\frac{1}{2}$	$\tilde{1}$	000+	101+	$+1/8.3$
$\frac{3}{2}$	$\frac{1}{2}$	$\tilde{1}$	100−	000+	$+1/8.3$
$\frac{3}{2}$	$\frac{1}{2}$	$\tilde{1}$	100−	001−	0
$\frac{3}{2}$	$\frac{1}{2}$	$\tilde{1}$	100−	100−	$-1/8.3$
$\frac{3}{2}$	$\frac{1}{2}$	$\tilde{1}$	100−	101+	0
$\frac{3}{2}$	$\frac{3}{2}$	0	000−	000−	$+1/8.2$
$\frac{3}{2}$	$\frac{3}{2}$	0	100+	000−	0
$\frac{3}{2}$	$\frac{3}{2}$	0	100+	100+	0
$\frac{3}{2}$	$\frac{3}{2}$	1	000+	000+	$+1/8.2.3$
$\frac{3}{2}$	$\frac{3}{2}$	1	000+	001+	0
$\frac{3}{2}$	$\frac{3}{2}$	1	001+	000+	0
$\frac{3}{2}$	$\frac{3}{2}$	1	001+	001+	0
$\frac{3}{2}$	$\frac{3}{2}$	1	100−	000+	0
$\frac{3}{2}$	$\frac{3}{2}$	1	100−	001+	$+1/8.3$
$\frac{3}{2}$	$\frac{3}{2}$	1	100−	100−	0
$\frac{3}{2}$	$\frac{3}{2}$	1	100−	101−	0
$\frac{3}{2}$	$\frac{3}{2}$	1	101−	000+	0
$\frac{3}{2}$	$\frac{3}{2}$	1	101−	001+	0
$\frac{3}{2}$	$\frac{3}{2}$	1	101−	100−	0
$\frac{3}{2}$	$\frac{3}{2}$	1	101−	101−	$+1/8.2.3$
$\frac{3}{2}$	$\frac{3}{2}$	2	000−	000−	0

O and T_d 9j Symbols (cont.)

$\tilde{1}$ 3/2 3/2	3/2 2 3/2				
3/2	3/2	2	100+	000−	0
3/2	3/2	2	100+	100+	+1/8.2
3/2	3/2	$\tilde{1}$	000−	000−	+1/8.2.3
3/2	3/2	$\tilde{1}$	000−	001+	0
3/2	3/2	$\tilde{1}$	001+	000−	0
3/2	3/2	$\tilde{1}$	001+	001+	0
3/2	3/2	$\tilde{1}$	100+	000−	0
3/2	3/2	$\tilde{1}$	100+	001+	+1/8.3
3/2	3/2	$\tilde{1}$	100+	100+	0
3/2	3/2	$\tilde{1}$	100+	101−	0
3/2	3/2	$\tilde{1}$	101−	000−	0
3/2	3/2	$\tilde{1}$	101−	001+	0
3/2	3/2	$\tilde{1}$	101−	100+	0
3/2	3/2	$\tilde{1}$	101−	101−	+1/8.2.3

$\tilde{1}$ 3/2 3/2	3/2 $\tilde{1}$ 1/2				
1/2	1/2	1	000−	000−	$+1/4.9\sqrt{2}$
1/2	1/2	1	100+	000−	$-1/2.9\sqrt{2}$
1/2	3/2	1	000+	000+	+1/4.9
1/2	3/2	1	000+	010−	+1/8.9
1/2	3/2	1	100−	000+	−1/8.9
1/2	3/2	1	100−	010−	−7/8.2.9
1/2	3/2	2	000−	000−	−1/4.3
1/2	3/2	2	000−	010+	0
1/2	3/2	2	100+	000−	0
1/2	3/2	2	100+	010+	+1/8.2.3
1/2	3/2	$\tilde{1}$	000−	000−	−1/4.9
1/2	3/2	$\tilde{1}$	000−	010+	+1/8.9
1/2	3/2	$\tilde{1}$	100+	000−	−1/8.9
1/2	3/2	$\tilde{1}$	100+	010+	+7/8.2.9
3/2	1/2	1	000+	000+	+7/8.2.9
3/2	1/2	1	100−	000+	+1/4.9
3/2	1/2	1	100−	100−	+1/8.2.9
3/2	1/2	2	000−	000−	+1/8.2.3
3/2	1/2	2	100+	000−	−1/8.3
3/2	1/2	2	100+	100+	+1/8.2.3
3/2	1/2	$\tilde{1}$	000−	000−	+1/8.2.3
3/2	1/2	$\tilde{1}$	100+	000−	0
3/2	1/2	$\tilde{1}$	100+	100+	+1/8.2
3/2	3/2	1	000−	000−	−1/8.9
3/2	3/2	1	000−	010+	+1/8.9
3/2	3/2	1	000−	100+	+1/4.9
3/2	3/2	1	000−	110−	+1/8.9
3/2	3/2	1	001−	000−	+1/8.9
3/2	3/2	1	001−	010+	−5/8.2.9
3/2	3/2	1	001−	100+	−1/4.9
3/2	3/2	1	001−	110−	−1/8.9
3/2	3/2	1	100+	000−	+1/4.9
3/2	3/2	1	100+	010+	−1/4.9
3/2	3/2	1	100+	100+	−1/8.9

$\tilde{1}$ 3/2 3/2	3/2 $\tilde{1}$ 1/2				
3/2	3/2	1	100+	110−	+1/8.9
3/2	3/2	1	101+	000−	+1/8.9
3/2	3/2	1	101+	010+	−1/8.9
3/2	3/2	1	101+	100+	+1/8.9
3/2	3/2	1	101+	110−	−1/8.9
3/2	3/2	2	000+	000+	0
3/2	3/2	2	000+	010−	−1/8.2.3
3/2	3/2	2	000+	100−	0
3/2	3/2	2	000+	110+	+1/8.3
3/2	3/2	2	100−	000+	−1/8.3
3/2	3/2	2	100−	010−	0
3/2	3/2	2	100−	100−	−1/8.3
3/2	3/2	2	100−	110+	0
3/2	3/2	$\tilde{1}$	000+	000+	+1/8.9
3/2	3/2	$\tilde{1}$	000+	010−	+1/8.9
3/2	3/2	$\tilde{1}$	000+	100−	−1/4.9
3/2	3/2	$\tilde{1}$	000+	110+	+1/8.9
3/2	3/2	$\tilde{1}$	001−	000+	−1/8.9
3/2	3/2	$\tilde{1}$	001−	010−	−5/8.2.9
3/2	3/2	$\tilde{1}$	001−	100−	+1/4.9
3/2	3/2	$\tilde{1}$	001−	110+	−1/8.9
3/2	3/2	$\tilde{1}$	100−	000+	+1/4.9
3/2	3/2	$\tilde{1}$	100−	010−	+1/4.9
3/2	3/2	$\tilde{1}$	100−	100−	−1/8.9
3/2	3/2	$\tilde{1}$	100−	110+	−1/8.9
3/2	3/2	$\tilde{1}$	101+	000+	+1/8.9
3/2	3/2	$\tilde{1}$	101+	010−	+1/8.9
3/2	3/2	$\tilde{1}$	101+	100−	+1/8.9
3/2	3/2	$\tilde{1}$	101+	110+	+1/8.9

$\tilde{1}$ 3/2 3/2	3/2 $\tilde{1}$ 3/2				
1/2	1/2	0	000−	000−	$+1/4.3\sqrt{2}$
1/2	1/2	0	010+	000−	$-1/4.3\sqrt{2}$
1/2	1/2	0	100+	000−	$+1/4.3\sqrt{2}$
1/2	1/2	0	110−	000−	$+1/8.3\sqrt{2}$
1/2	1/2	1	000+	000+	$+1/4.9\sqrt{2}$
1/2	1/2	1	000+	001+	$+1/2.9\sqrt{2}$
1/2	1/2	1	010−	000+	$-1/4.9\sqrt{2}$
1/2	1/2	1	010−	001+	$+1/4.9\sqrt{2}$
1/2	1/2	1	100−	000+	$+1/4.9\sqrt{2}$
1/2	1/2	1	100−	001+	$-1/4.9\sqrt{2}$
1/2	1/2	1	110+	000+	$-5/8.9\sqrt{2}$
1/2	1/2	1	110+	001+	$-1/2.9\sqrt{2}$
1/2	3/2	1	000−	000−	−1/8.9
1/2	3/2	1	000−	001−	+1/8.9
1/2	3/2	1	000−	010+	−1/4.9
1/2	3/2	1	000−	011+	−1/8.9
1/2	3/2	1	010+	000−	−1/4.9
1/2	3/2	1	010+	001−	+1/4.9
1/2	3/2	1	010+	010+	−1/8.9

$\tilde{1}$ 3/2 3/2	3/2 $\tilde{1}$ 3/2				
1/2	3/2	1	010+	011+	+1/8.9
1/2	3/2	1	100+	000−	−1/8.9
1/2	3/2	1	100+	001−	+5/8.2.9
1/2	3/2	1	100+	010+	−1/4.9
1/2	3/2	1	100+	011+	−1/8.9
1/2	3/2	1	110−	000−	+1/8.9
1/2	3/2	1	110−	001−	−1/8.9
1/2	3/2	1	110−	010+	−1/8.9
1/2	3/2	1	110−	011+	+1/8.9
1/2	3/2	2	000+	000+	0
1/2	3/2	2	000+	010−	+1/8.3
1/2	3/2	2	010−	000+	0
1/2	3/2	2	010−	010−	−1/8.3
1/2	3/2	2	100−	000+	+1/8.2.3
1/2	3/2	2	100−	010−	0
1/2	3/2	2	110+	000+	+1/8.3
1/2	3/2	2	110+	010−	0
1/2	3/2	$\tilde{1}$	000+	000+	+1/8.9
1/2	3/2	$\tilde{1}$	000+	001−	+1/8.9
1/2	3/2	$\tilde{1}$	000+	010−	−1/4.9
1/2	3/2	$\tilde{1}$	000+	011+	+1/8.9
1/2	3/2	$\tilde{1}$	010−	000+	+1/4.9
1/2	3/2	$\tilde{1}$	010−	001−	+1/4.9
1/2	3/2	$\tilde{1}$	010−	010−	−1/8.9
1/2	3/2	$\tilde{1}$	010−	011+	−1/8.9
1/2	3/2	$\tilde{1}$	100−	000+	−1/8.9
1/2	3/2	$\tilde{1}$	100−	001−	−5/8.2.9
1/2	3/2	$\tilde{1}$	100−	010−	+1/4.9
1/2	3/2	$\tilde{1}$	100−	011+	−1/8.9
1/2	3/2	$\tilde{1}$	110+	000+	+1/8.9
1/2	3/2	$\tilde{1}$	110+	001−	+1/8.9
1/2	3/2	$\tilde{1}$	110+	010−	+1/8.9
1/2	3/2	$\tilde{1}$	110+	011+	+1/8.9
3/2	1/2	1	000−	000−	−1/8.9
3/2	1/2	1	000−	001−	+1/8.9
3/2	1/2	1	000−	100+	+1/4.9
3/2	1/2	1	000−	101+	+1/8.9
3/2	1/2	1	010+	000−	+1/8.9
3/2	1/2	1	010+	001−	−5/8.2.9
3/2	1/2	1	010+	100+	−1/4.9
3/2	1/2	1	010+	101+	−1/8.9
3/2	1/2	1	100+	000−	+1/4.9
3/2	1/2	1	100+	001−	−1/4.9
3/2	1/2	1	100+	100+	−1/8.9
3/2	1/2	1	100+	101+	+1/8.9
3/2	1/2	1	110−	000−	+1/8.9
3/2	1/2	1	110−	001−	−1/8.9
3/2	1/2	1	110−	100+	+1/8.9
3/2	1/2	1	110−	101+	−1/8.9

O and T_d 9j Symbols (cont.)

1̃	3/2	3/2	3/2 1̃ 3/2		
3/2	1/2	2	000+	000+	0
3/2	1/2	2	000+	100−	−1/8.3
3/2	1/2	2	010−	000+	−1/8.2.3
3/2	1/2	2	010−	100−	0
3/2	1/2	2	100−	000+	0
3/2	1/2	2	100−	100−	−1/8.3
3/2	1/2	2	110+	000+	+1/8.3
3/2	1/2	2	110+	100−	0
3/2	1/2	1̃	000+	000+	+1/8.9
3/2	1/2	1̃	000+	001−	−1/8.9
3/2	1/2	1̃	000+	100−	+1/4.9
3/2	1/2	1̃	000+	101+	+1/8.9
3/2	1/2	1̃	010−	000+	+1/8.9
3/2	1/2	1̃	010−	001−	−5/8.2.9
3/2	1/2	1̃	010−	100−	+1/4.9
3/2	1/2	1̃	010−	101+	+1/8.9
3/2	1/2	1̃	100−	000+	−1/4.9
3/2	1/2	1̃	100−	001−	+1/4.9
3/2	1/2	1̃	100−	100−	−1/8.9
3/2	1/2	1̃	100−	101+	+1/8.9
3/2	1/2	1̃	110+	000+	+1/8.9
3/2	1/2	1̃	110+	001−	−1/8.9
3/2	1/2	1̃	110+	100−	−1/8.9
3/2	1/2	1̃	110+	101+	+1/8.9
3/2	3/2	0	000−	000−	+1/8.2.3
3/2	3/2	0	000−	010+	0
3/2	3/2	0	010+	000−	0
3/2	3/2	0	010+	010+	+1/8.2.3
3/2	3/2	0	100+	000−	0
3/2	3/2	0	100+	010+	+1/8.3
3/2	3/2	0	100+	100+	+1/8.2.3
3/2	3/2	0	100+	110−	0
3/2	3/2	0	110−	000−	+1/8.3
3/2	3/2	0	110−	010+	0
3/2	3/2	0	110−	100+	0
3/2	3/2	0	110−	110−	−1/8.3
3/2	3/2	1	000+	000+	+5/8.2.9
3/2	3/2	1	000+	001+	0
3/2	3/2	1	000+	010−	0
3/2	3/2	1	000+	011−	+1/8.9
3/2	3/2	1	001+	000+	0
3/2	3/2	1	001+	001+	+5/8.2.9
3/2	3/2	1	001+	010−	+1/8.9
3/2	3/2	1	001+	011−	0
3/2	3/2	1	010−	000+	0
3/2	3/2	1	010−	001+	+1/8.9
3/2	3/2	1	010−	010−	+5/8.2.9
3/2	3/2	1	010−	011−	0
3/2	3/2	1	011−	000+	+1/8.9

1̃	3/2	3/2	3/2 1̃ 3/2		
3/2	3/2	1	011−	001+	0
3/2	3/2	1	011−	010−	0
3/2	3/2	1	011−	011−	+1/8.9
3/2	3/2	1	100−	000+	0
3/2	3/2	1	100−	001+	−1/8.9
3/2	3/2	1	100−	010−	−1/8.9
3/2	3/2	1	100−	011−	0
3/2	3/2	1	100−	100−	+5/8.2.9
3/2	3/2	1	100−	101−	0
3/2	3/2	1	100−	110+	0
3/2	3/2	1	100−	111+	−1/4.9
3/2	3/2	1	101−	000+	−1/8.9
3/2	3/2	1	101−	001+	0
3/2	3/2	1	101−	010−	0
3/2	3/2	1	101−	011−	−1/8.9
3/2	3/2	1	101−	100−	0
3/2	3/2	1	101−	101−	+1/8.9
3/2	3/2	1	101−	110+	+1/8.9
3/2	3/2	1	101−	111+	0
3/2	3/2	1	110+	000+	−1/8.9
3/2	3/2	1	110+	001+	0
3/2	3/2	1	110+	010−	0
3/2	3/2	1	110+	011−	−1/8.9
3/2	3/2	1	110+	100−	0
3/2	3/2	1	110+	101−	+1/8.9
3/2	3/2	1	110+	110+	+1/8.9
3/2	3/2	1	110+	111+	0
3/2	3/2	1	111+	000+	0
3/2	3/2	1	111+	001+	+1/4.9
3/2	3/2	1	111+	010−	+1/4.9
3/2	3/2	1	111+	011−	0
3/2	3/2	1	111+	100−	−1/4.9
3/2	3/2	1	111+	101−	0
3/2	3/2	1	111+	110+	0
3/2	3/2	1	111+	111+	+5/8.2.9
3/2	3/2	2	000−	000−	+1/8.2.3
3/2	3/2	2	000−	010+	0
3/2	3/2	2	010+	000−	0
3/2	3/2	2	010+	010+	0
3/2	3/2	2	100+	000−	0
3/2	3/2	2	100+	010+	+1/8.3
3/2	3/2	2	100+	100+	0
3/2	3/2	2	100+	110−	0
3/2	3/2	2	110−	000−	0
3/2	3/2	2	110−	010+	0
3/2	3/2	2	110−	100+	0
3/2	3/2	2	110−	110−	+1/8.2.3
3/2	3/2	1̃	000−	000−	−5/8.2.9
3/2	3/2	1̃	000−	001+	0

1̃	3/2	3/2	3/2 1̃ 3/2		
3/2	3/2	1̃	000−	010+	0
3/2	3/2	1̃	000−	011−	−1/8.9
3/2	3/2	1̃	001+	000−	0
3/2	3/2	1̃	001+	001+	+5/8.2.9
3/2	3/2	1̃	001+	010+	+1/8.9
3/2	3/2	1̃	001+	011−	0
3/2	3/2	1̃	010+	000−	0
3/2	3/2	1̃	010+	001+	+1/8.9
3/2	3/2	1̃	010+	010+	+5/8.2.9
3/2	3/2	1̃	010+	011−	0
3/2	3/2	1̃	011−	000−	−1/8.9
3/2	3/2	1̃	011−	001+	0
3/2	3/2	1̃	011−	010+	0
3/2	3/2	1̃	011−	011−	−1/8.9
3/2	3/2	1̃	100+	000−	0
3/2	3/2	1̃	100+	001+	+1/8.9
3/2	3/2	1̃	100+	010+	+1/8.9
3/2	3/2	1̃	100+	011−	0
3/2	3/2	1̃	100+	100+	+5/8.2.9
3/2	3/2	1̃	100+	101−	0
3/2	3/2	1̃	100+	110−	0
3/2	3/2	1̃	100+	111+	−1/4.9
3/2	3/2	1̃	101−	000−	−1/8.9
3/2	3/2	1̃	101−	001+	0
3/2	3/2	1̃	101−	010+	0
3/2	3/2	1̃	101−	011−	−1/8.9
3/2	3/2	1̃	101−	100+	0
3/2	3/2	1̃	101−	101−	−1/8.9
3/2	3/2	1̃	101−	110−	−1/8.9
3/2	3/2	1̃	101−	111+	0
3/2	3/2	1̃	110−	000−	−1/8.9
3/2	3/2	1̃	110−	001+	0
3/2	3/2	1̃	110−	010+	0
3/2	3/2	1̃	110−	011−	−1/8.9
3/2	3/2	1̃	110−	100+	0
3/2	3/2	1̃	110−	101−	−1/8.9
3/2	3/2	1̃	110−	110−	−1/8.9
3/2	3/2	1̃	110−	111+	0
3/2	3/2	1̃	111+	000−	0
3/2	3/2	1̃	111+	001+	−1/4.9
3/2	3/2	1̃	111+	010+	−1/4.9
3/2	3/2	1̃	111+	011−	0
3/2	3/2	1̃	111+	100+	−1/4.9
3/2	3/2	1̃	111+	101−	0
3/2	3/2	1̃	111+	110−	0
3/2	3/2	1̃	111+	111+	+5/8.2.9
1̃	**2**	**1**	**1 1 1**		
1	1	0	000+	000+	$+1/2.3\sqrt{3}$
1	1	1	000−	000−	0

O and T_d 9j Symbols (cont.)

$\tilde{1}$ 2 1 1 1 2
1 1 1 000+ 000+ $-1/4.3\sqrt{3}$
$\tilde{1}$ 2 1 1 1 $\tilde{1}$
1 1 1 000+ 000+ 0
1 1 2 000- 000- $+1/4.3$
1 1 $\tilde{1}$ 000- 000- 0
$\tilde{1}$ 2 1 1 $\frac{3}{2}$ $\frac{1}{2}$
1 $\frac{1}{2}$ $\frac{1}{2}$ 000+ 000+ $-1/4.3$
1 $\frac{1}{2}$ $\frac{3}{2}$ 000- 000- $+1/8.3$
1 $\frac{3}{2}$ $\frac{1}{2}$ 000- 000- 0
$\tilde{1}$ 2 1 1 $\frac{3}{2}$ $\frac{3}{2}$
1 $\frac{1}{2}$ $\frac{1}{2}$ 000- 000- 0
1 $\frac{1}{2}$ $\frac{1}{2}$ 010- 000- $+1/4.3\sqrt{2}$
1 $\frac{1}{2}$ $\frac{3}{2}$ 000+ 000+ $-1/8.3$
1 $\frac{1}{2}$ $\frac{3}{2}$ 000+ 001+ $-1/8.3$
1 $\frac{1}{2}$ $\frac{3}{2}$ 010+ 000+ 0
1 $\frac{1}{2}$ $\frac{3}{2}$ 010+ 001+ $+1/8.3$
1 $\frac{3}{2}$ $\frac{1}{2}$ 000+ 000+ $+1/8.3$
1 $\frac{3}{2}$ $\frac{1}{2}$ 010+ 000+ $+1/8.3$
1 $\frac{3}{2}$ $\frac{3}{2}$ 000- 000- 0
1 $\frac{3}{2}$ $\frac{3}{2}$ 000- 001- 0
1 $\frac{3}{2}$ $\frac{3}{2}$ 010 000- $-1/8.3$
1 $\frac{3}{2}$ $\frac{3}{2}$ 010- 001- 0
1 $\frac{3}{2}$ $\frac{3}{2}$ 011- 000- 0
1 $\frac{3}{2}$ $\frac{3}{2}$ 011- 001- 0
$\tilde{1}$ 2 1 1 2 1
1 0 1 000- 000- $-1/2.3\sqrt{2}$
1 2 1 000- 000- 0
$\tilde{1}$ 2 1 1 2 $\tilde{1}$
1 0 1 000+ 000+ $+1/2.3\sqrt{2.3}$
1 2 1 000+ 000+ $-1/2.3\sqrt{3}$
1 2 $\tilde{1}$ 000- 000- 0
$\tilde{1}$ 2 1 1 $\tilde{1}$ 1
1 1 0 000- 000- $+1/2.9$
1 1 1 000+ 000+ $+1/2.9$
1 1 2 000- 000- $-1/4.9$
1 1 $\tilde{1}$ 000- 000- $-1/2.9$
1 $\tilde{1}$ 1 000- 000- 0
$\tilde{1}$ 2 1 1 $\tilde{1}$ 2
1 1 1 000- 000- $-1/4.3\sqrt{3}$
1 1 $\tilde{1}$ 000+ 000+ $+1/4.9$
1 $\tilde{1}$ 1 000+ 000+ $-1/4.3$
$\tilde{1}$ 2 1 1 $\tilde{1}$ $\tilde{1}$
1 1 1 000- 000- $-1/2.9$
1 1 2 000+ 000+ $-1/4.3\sqrt{3}$
1 1 $\tilde{1}$ 000+ 000+ $-1/2.9$
1 $\tilde{1}$ 1 000+ 000+ 0
1 $\tilde{1}$ 2 000- 000- $+1/4.3\sqrt{3}$
1 $\tilde{1}$ $\tilde{1}$ 000- 000- 0

$\tilde{1}$ 2 1 $\frac{3}{2}$ 0 $\frac{3}{2}$
$\frac{1}{2}$ 2 $\frac{3}{2}$ 000+ 000+ $-1/8\sqrt{3}$
$\frac{1}{2}$ 2 $\frac{3}{2}$ 000+ 001+ $-1/8\sqrt{3}$
$\tilde{1}$ 2 1 $\frac{3}{2}$ $\frac{1}{2}$ 1
$\frac{1}{2}$ $\frac{3}{2}$ 1 000- 000- $+1/8\sqrt{3}$
$\frac{1}{2}$ $\frac{3}{2}$ 2 000+ 000+ 0
$\frac{1}{2}$ $\frac{3}{2}$ $\tilde{1}$ 000+ 000+ $-1/8.3$
$\tilde{1}$ 2 1 $\frac{3}{2}$ $\frac{1}{2}$ 2
$\frac{1}{2}$ $\frac{3}{2}$ 1 000+ 000+ 0
$\frac{1}{2}$ $\frac{3}{2}$ $\tilde{1}$ 000- 000- $-1/4.3$
$\tilde{1}$ 2 1 $\frac{3}{2}$ $\frac{1}{2}$ $\tilde{1}$
$\frac{1}{2}$ $\frac{3}{2}$ 1 000+ 000+ $-1/8.3$
$\frac{1}{2}$ $\frac{3}{2}$ 2 000- 000- $+1/4.3$
$\frac{1}{2}$ $\frac{3}{2}$ $\tilde{1}$ 000- 000- $-1/8.3\sqrt{3}$
$\tilde{1}$ 2 1 $\frac{3}{2}$ 1 $\frac{1}{2}$
$\frac{1}{2}$ 1 $\frac{1}{2}$ 000+ 000+ $-1/4.3$
$\frac{1}{2}$ 1 $\frac{3}{2}$ 000- 000- $-1/8.3$
$\frac{1}{2}$ $\tilde{1}$ $\frac{3}{2}$ 000+ 000+ $-1/8.3$
$\frac{3}{2}$ 1 $\frac{1}{2}$ 000- 000- 0
$\frac{3}{2}$ 1 $\frac{1}{2}$ 000- 100+ $+1/4.3\sqrt{2}$
$\tilde{1}$ 2 1 $\frac{3}{2}$ 1 $\frac{3}{2}$
$\frac{1}{2}$ 1 $\frac{1}{2}$ 000- 000- 0
$\frac{1}{2}$ 1 $\frac{1}{2}$ 010- 000- $-1/4.3\sqrt{2}$
$\frac{1}{2}$ 1 $\frac{3}{2}$ 000+ 000+ $+1/8.3$
$\frac{1}{2}$ 1 $\frac{3}{2}$ 000+ 001+ $+1/8.3$
$\frac{1}{2}$ 1 $\frac{3}{2}$ 010+ 000+ $-1/8.3$
$\frac{1}{2}$ 1 $\frac{3}{2}$ 010+ 001+ 0
$\frac{1}{2}$ $\tilde{1}$ $\frac{3}{2}$ 000- 000- $+1/8.9$
$\frac{1}{2}$ $\tilde{1}$ $\frac{3}{2}$ 000- 001- $+1/8.9$
$\frac{1}{2}$ $\tilde{1}$ $\frac{3}{2}$ 010- 000- $+1/8.9$
$\frac{1}{2}$ $\tilde{1}$ $\frac{3}{2}$ 010- 001- $+1/2.9$
$\frac{3}{2}$ 1 $\frac{1}{2}$ 000+ 000+ $+1/8.3$
$\frac{3}{2}$ 1 $\frac{1}{2}$ 000+ 100- $-1/8.3$
$\frac{3}{2}$ 1 $\frac{1}{2}$ 010+ 000+ $+1/8.3$
$\frac{3}{2}$ 1 $\frac{1}{2}$ 010+ 100- 0
$\frac{3}{2}$ 1 $\frac{3}{2}$ 000- 000- 0
$\frac{3}{2}$ 1 $\frac{3}{2}$ 000- 001- 0
$\frac{3}{2}$ 1 $\frac{3}{2}$ 000- 100+ 0
$\frac{3}{2}$ 1 $\frac{3}{2}$ 000- 101+ 0
$\frac{3}{2}$ 1 $\frac{3}{2}$ 010- 000- 0
$\frac{3}{2}$ 1 $\frac{3}{2}$ 010- 001- $-1/8.3$
$\frac{3}{2}$ 1 $\frac{3}{2}$ 010- 100+ $+1/8.3$
$\frac{3}{2}$ 1 $\frac{3}{2}$ 010- 101+ 0
$\frac{3}{2}$ 1 $\frac{3}{2}$ 011- 000- 0
$\frac{3}{2}$ 1 $\frac{3}{2}$ 011- 001- 0
$\frac{3}{2}$ 1 $\frac{3}{2}$ 011- 100+ 0
$\frac{3}{2}$ 1 $\frac{3}{2}$ 011- 101+ $-1/8.3$
$\tilde{1}$ 2 1 $\frac{3}{2}$ $\frac{3}{2}$ 0
$\frac{1}{2}$ $\frac{1}{2}$ 1 000+ 000+ $-1/4.3$
$\frac{1}{2}$ $\frac{3}{2}$ 1 000- 000- $-1/4.3\sqrt{2}$

$\tilde{1}$ 2 1 $\frac{3}{2}$ $\frac{3}{2}$ 0
$\frac{3}{2}$ $\frac{1}{2}$ 1 000- 000- $+1/4.3\sqrt{2}$
$\frac{3}{2}$ $\frac{1}{2}$ 1 000- 100+ $+1/4.3\sqrt{2}$
$\tilde{1}$ 2 1 $\frac{3}{2}$ $\frac{3}{2}$ 1
$\frac{1}{2}$ $\frac{1}{2}$ 0 000+ 000+ $-1/4.3$
$\frac{1}{2}$ $\frac{1}{2}$ 0 010+ 000+ $-1/4.3$
$\frac{1}{2}$ $\frac{1}{2}$ 1 000- 000- $+1/2.3\sqrt{2.3}$
$\frac{1}{2}$ $\frac{1}{2}$ 1 010- 000- $-1/4.3\sqrt{2.3}$
$\frac{1}{2}$ $\frac{3}{2}$ 1 000+ 000+ $-1/8.3\sqrt{3}$
$\frac{1}{2}$ $\frac{3}{2}$ 1 010+ 000+ $-1/8.3\sqrt{3}$
$\frac{1}{2}$ $\frac{3}{2}$ 2 000- 000- $+1/8.3$
$\frac{1}{2}$ $\frac{3}{2}$ 2 010- 000- $+1/8.3$
$\frac{1}{2}$ $\frac{3}{2}$ $\tilde{1}$ 000- 000- $-1/8.3$
$\frac{1}{2}$ $\frac{3}{2}$ $\tilde{1}$ 010- 000- $+1/8.3$
$\frac{3}{2}$ $\frac{1}{2}$ 1 000+ 000+ $+1/8.3\sqrt{3}$
$\frac{3}{2}$ $\frac{1}{2}$ 1 000+ 100- $+1/8.3\sqrt{3}$
$\frac{3}{2}$ $\frac{1}{2}$ 1 010+ 000+ $-1/4.3\sqrt{3}$
$\frac{3}{2}$ $\frac{1}{2}$ 1 010+ 100- $+1/8.3\sqrt{3}$
$\frac{3}{2}$ $\frac{1}{2}$ 2 000- 000- $-1/8.3$
$\frac{3}{2}$ $\frac{1}{2}$ 2 000- 100+ $-1/8.3$
$\frac{3}{2}$ $\frac{1}{2}$ 2 010- 000- $-1/8.3$
$\frac{3}{2}$ $\frac{1}{2}$ 2 010- 100+ $-1/8.3$
$\frac{3}{2}$ $\frac{1}{2}$ $\tilde{1}$ 000- 000- $-1/8.3$
$\frac{3}{2}$ $\frac{1}{2}$ $\tilde{1}$ 000- 100+ $-1/8.3$
$\frac{3}{2}$ $\frac{1}{2}$ $\tilde{1}$ 010- 000- 0
$\frac{3}{2}$ $\frac{1}{2}$ $\tilde{1}$ 010- 100+ $+1/8.3$
$\frac{3}{2}$ $\frac{3}{2}$ 0 000+ 000+ 0
$\frac{3}{2}$ $\frac{3}{2}$ 0 000+ 100- $-1/4.3\sqrt{2}$
$\frac{3}{2}$ $\frac{3}{2}$ 0 010+ 000+ $-1/4.3\sqrt{2}$
$\frac{3}{2}$ $\frac{3}{2}$ 0 010+ 100- 0
$\frac{3}{2}$ $\frac{3}{2}$ 1 000- 000- 0
$\frac{3}{2}$ $\frac{3}{2}$ 1 000- 100+ $-1/4.3\sqrt{3}$
$\frac{3}{2}$ $\frac{3}{2}$ 1 010- 000- $+1/8.3\sqrt{3}$
$\frac{3}{2}$ $\frac{3}{2}$ 1 010- 100+ 0
$\frac{3}{2}$ $\frac{3}{2}$ 1 011- 000- 0
$\frac{3}{2}$ $\frac{3}{2}$ 1 011- 100+ $+1/4.3\sqrt{3}$
$\tilde{1}$ 2 1 $\frac{3}{2}$ $\frac{3}{2}$ 2
$\frac{1}{2}$ $\frac{1}{2}$ 1 000+ 000+ $-1/4.3\sqrt{2}$
$\frac{1}{2}$ $\frac{3}{2}$ 1 000- 000- $-1/8.3$
$\frac{1}{2}$ $\frac{3}{2}$ $\tilde{1}$ 000+ 000+ $+1/8.3$
$\frac{3}{2}$ $\frac{1}{2}$ 1 000- 000- $+1/8.3$
$\frac{3}{2}$ $\frac{1}{2}$ 1 000- 100+ $+1/8.3$
$\frac{3}{2}$ $\frac{1}{2}$ $\tilde{1}$ 000+ 000+ $-1/8.3$
$\frac{3}{2}$ $\frac{1}{2}$ $\tilde{1}$ 000+ 100- $+1/8.3$
$\frac{3}{2}$ $\frac{3}{2}$ 1 000+ 000+ $+1/8.3$
$\frac{3}{2}$ $\frac{3}{2}$ 1 000+ 100- 0
$\frac{3}{2}$ $\frac{3}{2}$ 1 001+ 000+ 0
$\frac{3}{2}$ $\frac{3}{2}$ 1 001+ 100- $-1/8.3$
$\tilde{1}$ 2 1 $\frac{3}{2}$ $\frac{3}{2}$ $\tilde{1}$
$\frac{1}{2}$ $\frac{1}{2}$ 1 000+ 000+ 0

O and T_d 9j Symbols (cont.)

$\tilde{1}$	2	1	$\frac{3}{2}$	$\frac{3}{2}$	$\tilde{1}$
$\frac{1}{2}$	$\frac{1}{2}$	1	010−	000+	$-1/4.3\sqrt{2}$
$\frac{1}{2}$	$\frac{3}{2}$	1	000−	000−	$-1/8.3$
$\frac{1}{2}$	$\frac{3}{2}$	1	010+	000−	$+1/8.3$
$\frac{1}{2}$	$\frac{3}{2}$	2	000+	000+	$-1/8.3$
$\frac{1}{2}$	$\frac{3}{2}$	2	010−	000+	$-1/8.3$
$\frac{1}{2}$	$\frac{3}{2}$	$\tilde{1}$	000+	000+	$-1/8.3\sqrt{3}$
$\frac{1}{2}$	$\frac{3}{2}$	$\tilde{1}$	010−	000+	$-1/8.3\sqrt{3}$
$\frac{3}{2}$	$\frac{1}{2}$	1	000−	000−	$+1/8.3$
$\frac{3}{2}$	$\frac{1}{2}$	1	000−	100+	$-1/8.3$
$\frac{3}{2}$	$\frac{1}{2}$	1	010+	000−	0
$\frac{3}{2}$	$\frac{1}{2}$	1	010+	100+	$-1/8.3$
$\frac{3}{2}$	$\frac{1}{2}$	2	000+	000+	$+1/8.3$
$\frac{3}{2}$	$\frac{1}{2}$	2	000+	100−	$-1/8.3$
$\frac{3}{2}$	$\frac{1}{2}$	2	010−	000+	$-1/8.3$
$\frac{3}{2}$	$\frac{1}{2}$	2	010−	100−	$+1/8.3$
$\frac{3}{2}$	$\frac{1}{2}$	$\tilde{1}$	000+	000+	$-1/8.3\sqrt{3}$
$\frac{3}{2}$	$\frac{1}{2}$	$\tilde{1}$	000+	100−	$+1/8.3\sqrt{3}$
$\frac{3}{2}$	$\frac{1}{2}$	$\tilde{1}$	010−	000+	$-1/4.3\sqrt{3}$
$\frac{3}{2}$	$\frac{1}{2}$	$\tilde{1}$	010−	100−	$-1/8.3\sqrt{3}$
$\frac{3}{2}$	$\frac{3}{2}$	1	000+	000+	0
$\frac{3}{2}$	$\frac{3}{2}$	1	000+	100−	0
$\frac{3}{2}$	$\frac{3}{2}$	1	001+	000+	$+1/8.3$
$\frac{3}{2}$	$\frac{3}{2}$	1	001+	100−	0
$\frac{3}{2}$	$\frac{3}{2}$	1	010−	000+	$+1/8.3$
$\frac{3}{2}$	$\frac{3}{2}$	1	010−	100−	0
$\frac{3}{2}$	$\frac{3}{2}$	1	011−	000+	0
$\frac{3}{2}$	$\frac{3}{2}$	1	011−	100−	0
$\frac{3}{2}$	$\frac{3}{2}$	2	000−	000−	$+1/8.3$
$\frac{3}{2}$	$\frac{3}{2}$	2	000−	100+	0
$\frac{3}{2}$	$\frac{3}{2}$	2	010+	000−	0
$\frac{3}{2}$	$\frac{3}{2}$	2	010+	100+	$+1/8.3$
$\frac{3}{2}$	$\frac{3}{2}$	$\tilde{1}$	000−	000−	0
$\frac{3}{2}$	$\frac{3}{2}$	$\tilde{1}$	000−	100+	$-1/4.3\sqrt{3}$
$\frac{3}{2}$	$\frac{3}{2}$	$\tilde{1}$	010+	000−	$+1/8.3\sqrt{3}$
$\frac{3}{2}$	$\frac{3}{2}$	$\tilde{1}$	010+	100+	0
$\frac{3}{2}$	$\frac{3}{2}$	$\tilde{1}$	011−	000−	0
$\frac{3}{2}$	$\frac{3}{2}$	$\tilde{1}$	011−	100+	$+1/4.3\sqrt{3}$

$\tilde{1}$	2	1	$\frac{3}{2}$	2	$\frac{1}{2}$
$\frac{1}{2}$	0	$\frac{1}{2}$	000+	000+	$-1/4\sqrt{3}$
$\frac{1}{2}$	2	$\frac{3}{2}$	000−	000−	$-1/8\sqrt{3}$
$\frac{3}{2}$	0	$\frac{3}{2}$	000+	000+	$-1/8\sqrt{3}$
$\frac{3}{2}$	0	$\frac{3}{2}$	000+	100−	$-1/8\sqrt{3}$
$\frac{3}{2}$	2	$\frac{1}{2}$	000−	000−	0
$\frac{3}{2}$	2	$\frac{1}{2}$	000−	100+	$+1/4\sqrt{2.3}$

$\tilde{1}$	2	1	$\frac{3}{2}$	2	$\frac{3}{2}$
$\frac{1}{2}$	0	$\frac{1}{2}$	000−	000−	$-1/4\sqrt{2.3}$
$\frac{1}{2}$	2	$\frac{3}{2}$	000+	000+	$-1/8\sqrt{3}$
$\frac{1}{2}$	2	$\frac{3}{2}$	000+	001+	0
$\frac{3}{2}$	0	$\frac{3}{2}$	000−	000−	0
$\frac{3}{2}$	0	$\frac{3}{2}$	000−	001−	$+1/8\sqrt{3}$
$\frac{3}{2}$	0	$\frac{3}{2}$	000−	100+	$-1/8\sqrt{3}$
$\frac{3}{2}$	0	$\frac{3}{2}$	000−	101+	0
$\frac{3}{2}$	2	$\frac{1}{2}$	000+	000+	$+1/8\sqrt{3}$
$\frac{3}{2}$	2	$\frac{1}{2}$	000+	100−	0
$\frac{3}{2}$	2	$\frac{3}{2}$	000−	000−	0
$\frac{3}{2}$	2	$\frac{3}{2}$	000−	001−	0
$\frac{3}{2}$	2	$\frac{3}{2}$	000−	100+	0
$\frac{3}{2}$	2	$\frac{3}{2}$	000−	101+	$-1/8\sqrt{3}$

$\tilde{1}$	2	1	$\frac{3}{2}$	$\tilde{1}$	$\frac{1}{2}$
$\frac{1}{2}$	1	$\frac{1}{2}$	000−	000−	$-1/4.9$
$\frac{1}{2}$	1	$\frac{3}{2}$	000+	000+	$+5/8.9$
$\frac{1}{2}$	$\tilde{1}$	$\frac{3}{2}$	000−	000−	$-1/8.3$
$\frac{3}{2}$	1	$\frac{1}{2}$	000+	000+	$-1/9\sqrt{2}$
$\frac{3}{2}$	1	$\frac{1}{2}$	000+	100−	$-1/4.9\sqrt{2}$
$\frac{3}{2}$	1	$\frac{3}{2}$	000−	000−	$+1/8.9$
$\frac{3}{2}$	1	$\frac{3}{2}$	000−	100+	$+1/8.9$
$\frac{3}{2}$	1	$\frac{3}{2}$	001−	000−	$-1/8.9$
$\frac{3}{2}$	1	$\frac{3}{2}$	001−	100+	$-1/2.9$
$\frac{3}{2}$	$\tilde{1}$	$\frac{1}{2}$	000−	000−	0
$\frac{3}{2}$	$\tilde{1}$	$\frac{1}{2}$	000−	100+	$-1/4.3\sqrt{2}$

$\tilde{1}$	2	1	$\frac{3}{2}$	$\tilde{1}$	$\frac{3}{2}$
$\frac{1}{2}$	1	$\frac{1}{2}$	000+	000+	$-1/9\sqrt{2}$
$\frac{1}{2}$	1	$\frac{1}{2}$	010−	000+	$+1/4.9\sqrt{2}$
$\frac{1}{2}$	1	$\frac{3}{2}$	000−	000−	$+1/8.9$
$\frac{1}{2}$	1	$\frac{3}{2}$	000−	001−	$-1/8.9$
$\frac{1}{2}$	1	$\frac{3}{2}$	010+	000−	$-1/8.9$
$\frac{1}{2}$	1	$\frac{3}{2}$	010+	001−	$+1/2.9$
$\frac{1}{2}$	$\tilde{1}$	$\frac{3}{2}$	000+	000+	$+1/8.3$
$\frac{1}{2}$	$\tilde{1}$	$\frac{3}{2}$	000+	001+	$-1/8.3$
$\frac{1}{2}$	$\tilde{1}$	$\frac{3}{2}$	010−	000+	$+1/8.3$
$\frac{1}{2}$	$\tilde{1}$	$\frac{3}{2}$	010−	001+	0
$\frac{3}{2}$	1	$\frac{1}{2}$	000−	000−	$-1/8.9$
$\frac{3}{2}$	1	$\frac{1}{2}$	000−	100+	$-1/8.9$
$\frac{3}{2}$	1	$\frac{1}{2}$	010+	000−	$+1/8.9$
$\frac{3}{2}$	1	$\frac{1}{2}$	010+	100+	$+1/2.9$
$\frac{3}{2}$	1	$\frac{3}{2}$	000+	000+	$+1/2.9$
$\frac{3}{2}$	1	$\frac{3}{2}$	000+	001+	0
$\frac{3}{2}$	1	$\frac{3}{2}$	000+	100−	0
$\frac{3}{2}$	1	$\frac{3}{2}$	000+	101−	$-1/2.9$
$\frac{3}{2}$	1	$\frac{3}{2}$	001+	000+	0
$\frac{3}{2}$	1	$\frac{3}{2}$	001+	001+	$+1/8.9$
$\frac{3}{2}$	1	$\frac{3}{2}$	001+	100−	$-1/8.9$
$\frac{3}{2}$	1	$\frac{3}{2}$	001+	101−	0
$\frac{3}{2}$	1	$\frac{3}{2}$	010−	000+	0
$\frac{3}{2}$	1	$\frac{3}{2}$	010−	001+	$+1/8.9$
$\frac{3}{2}$	1	$\frac{3}{2}$	010−	100−	$-1/8.9$
$\frac{3}{2}$	1	$\frac{3}{2}$	010−	101−	0
$\frac{3}{2}$	1	$\frac{3}{2}$	011−	000+	$+1/2.9$

$\tilde{1}$	2	1	$\frac{3}{2}$	$\tilde{1}$	$\frac{3}{2}$
$\frac{3}{2}$	1	$\frac{3}{2}$	011−	001+	0
$\frac{3}{2}$	1	$\frac{3}{2}$	011−	100−	0
$\frac{3}{2}$	1	$\frac{3}{2}$	011−	101−	$-1/8.9$
$\frac{3}{2}$	$\tilde{1}$	$\frac{1}{2}$	000+	000+	$+1/8.3$
$\frac{3}{2}$	$\tilde{1}$	$\frac{1}{2}$	000+	100−	$+1/8.3$
$\frac{3}{2}$	$\tilde{1}$	$\frac{1}{2}$	010−	000+	$+1/8.3$
$\frac{3}{2}$	$\tilde{1}$	$\frac{1}{2}$	010−	100−	0
$\frac{3}{2}$	$\tilde{1}$	$\frac{3}{2}$	000−	000−	0
$\frac{3}{2}$	$\tilde{1}$	$\frac{3}{2}$	000−	001−	0
$\frac{3}{2}$	$\tilde{1}$	$\frac{3}{2}$	000−	100+	0
$\frac{3}{2}$	$\tilde{1}$	$\frac{3}{2}$	000−	101+	0
$\frac{3}{2}$	$\tilde{1}$	$\frac{3}{2}$	010+	000−	0
$\frac{3}{2}$	$\tilde{1}$	$\frac{3}{2}$	010+	001−	$-1/8.3$
$\frac{3}{2}$	$\tilde{1}$	$\frac{3}{2}$	010+	100+	$-1/8.3$
$\frac{3}{2}$	$\tilde{1}$	$\frac{3}{2}$	010+	101+	0
$\frac{3}{2}$	$\tilde{1}$	$\frac{3}{2}$	011−	000−	0
$\frac{3}{2}$	$\tilde{1}$	$\frac{3}{2}$	011−	001−	0
$\frac{3}{2}$	$\tilde{1}$	$\frac{3}{2}$	011−	100+	0
$\frac{3}{2}$	$\tilde{1}$	$\frac{3}{2}$	011−	101+	$+1/8.3$

$\tilde{1}$	2	1	2	0	2
1	2	1	000−	000−	0
1	2	$\tilde{1}$	000+	000+	$+1/2.3$

$\tilde{1}$	2	1	2	$\frac{1}{2}$	$\frac{3}{2}$
1	$\frac{3}{2}$	$\frac{1}{2}$	000+	000+	0
1	$\frac{3}{2}$	$\frac{3}{2}$	000−	000−	$+1/8.3$
1	$\frac{3}{2}$	$\frac{3}{2}$	000−	001−	$+1/8.3$
1	$\frac{3}{2}$	$\frac{3}{2}$	001−	000−	$+1/8.3$
1	$\frac{3}{2}$	$\frac{3}{2}$	001−	001−	$+1/8.3$

$\tilde{1}$	2	1	2	1	1
1	1	0	000−	000−	0
1	1	1	000+	000+	$+1/4.3$
1	1	2	000−	000−	0
1	1	$\tilde{1}$	000−	000−	$+1/4.3$

$\tilde{1}$	2	1	2	1	$\tilde{1}$
1	1	1	000−	000−	$+1/4.3\sqrt{3}$
1	1	2	000+	000+	0
1	1	$\tilde{1}$	000+	000+	$-1/4.3\sqrt{3}$
1	$\tilde{1}$	1	000+	000+	$+1/4.9$
1	$\tilde{1}$	2	000−	000−	$+1/9$
1	$\tilde{1}$	$\tilde{1}$	000−	000−	$+1/4.9$

$\tilde{1}$	2	1	2	$\frac{3}{2}$	$\frac{1}{2}$
1	$\frac{1}{2}$	$\frac{1}{2}$	000−	000−	$+1/4.3$
1	$\frac{1}{2}$	$\frac{3}{2}$	000+	000+	$+1/4.3$

$\tilde{1}$	2	1	2	$\frac{3}{2}$	$\frac{3}{2}$
1	$\frac{1}{2}$	$\frac{1}{2}$	000+	000+	$-1/4.3\sqrt{2}$
1	$\frac{1}{2}$	$\frac{3}{2}$	000−	000−	$-1/8.3$
1	$\frac{1}{2}$	$\frac{3}{2}$	000−	001−	$+1/8.3$
1	$\frac{3}{2}$	$\frac{1}{2}$	000−	000−	$+1/8.3$
1	$\frac{3}{2}$	$\frac{3}{2}$	000+	000+	$+1/8.3$

O and T_d 9j Symbols (cont.)

$\tilde{1}$ 2 1 2 $\frac{3}{2}$ $\frac{3}{2}$
1 $\frac{3}{2}$ $\frac{3}{2}$ 000+ 001+ 0
1 $\frac{3}{2}$ $\frac{3}{2}$ 001+ 000+ 0
1 $\frac{3}{2}$ $\frac{3}{2}$ 001+ 001+ $+1/8.3$
$\tilde{1}$ 2 1 2 2 0
1 0 1 000− 000− $+1/2.3$
$\tilde{1}$ 2 1 2 2 2
1 0 1 000− 000− $+1/2.3\sqrt{2}$
1 2 1 000− 000− $+1/4.3$
1 2 $\tilde{1}$ 000+ 000+ $+1/4.3$
$\tilde{1}$ 2 1 2 $\tilde{1}$ 1
1 1 0 000+ 000+ $+1/9$
1 1 1 000− 000− $+1/4.9$
1 1 2 000+ 000+ $+1/9$
1 1 $\tilde{1}$ 000+ 000+ $+1/4.9$
$\tilde{1}$ 2 1 2 $\tilde{1}$ $\tilde{1}$
1 1 1 000+ 000+ $-1/4.3\sqrt{3}$
1 1 2 000− 000− 0
1 1 $\tilde{1}$ 000− 000− $+1/4.3\sqrt{3}$
1 $\tilde{1}$ 1 000− 000− $+1/4.3$
1 $\tilde{1}$ 2 000+ 000+ 0
1 $\tilde{1}$ $\tilde{1}$ 000+ 000+ $+1/4.3$
$\tilde{1}$ $\tilde{1}$ 0 1 1 0
1 1 0 000+ 000+ $+1/3\sqrt{3}$
$\tilde{1}$ $\tilde{1}$ 0 1 1 1
1 1 1 000+ 000+ $-1/2.9$
$\tilde{1}$ $\tilde{1}$ 0 1 1 2
1 1 2 000+ 000+ $-1/2.3\sqrt{2.3}$
$\tilde{1}$ $\tilde{1}$ 0 1 1 $\tilde{1}$
1 1 $\tilde{1}$ 000+ 000+ $+1/2.9$
$\tilde{1}$ $\tilde{1}$ 0 $\frac{3}{2}$ 0 $\frac{3}{2}$
$\frac{1}{2}$ $\tilde{1}$ $\frac{3}{2}$ 000− 000− $+1/4.3$
$\tilde{1}$ $\tilde{1}$ 0 $\frac{3}{2}$ $\frac{1}{2}$ 1
$\frac{1}{2}$ $\frac{3}{2}$ 1 000+ 000+ $+1/4.3$
$\tilde{1}$ $\tilde{1}$ 0 $\frac{3}{2}$ $\frac{1}{2}$ 2
$\frac{1}{2}$ $\frac{3}{2}$ 2 000+ 000+ $-1/4\sqrt{2.3}$
$\tilde{1}$ $\tilde{1}$ 0 $\frac{3}{2}$ $\frac{1}{2}$ $\tilde{1}$
$\frac{1}{2}$ $\frac{3}{2}$ $\tilde{1}$ 000+ 000+ $+1/4.9$
$\tilde{1}$ $\tilde{1}$ 0 $\frac{3}{2}$ 1 $\frac{1}{2}$
$\frac{1}{2}$ 1 $\frac{1}{2}$ 000− 000− $+1/2.3\sqrt{2}$
$\frac{3}{2}$ 1 $\frac{1}{2}$ 000+ 000+ $-1/4.3$
$\frac{3}{2}$ 1 $\frac{1}{2}$ 000+ 100− 0
$\tilde{1}$ $\tilde{1}$ 0 $\frac{3}{2}$ 1 $\frac{3}{2}$
$\frac{1}{2}$ 1 $\frac{3}{2}$ 000− 000− $-1/4.3\sqrt{2}$
$\frac{1}{2}$ 1 $\frac{3}{2}$ 010− 000− 0
$\frac{1}{2}$ 2 $\frac{3}{2}$ 000+ 000+ $-1/4.3\sqrt{2}$
$\frac{1}{2}$ 2 $\frac{3}{2}$ 010+ 000+ $-1/4.3\sqrt{2}$
$\frac{1}{2}$ $\tilde{1}$ $\frac{3}{2}$ 000+ 000+ $+1/4.3\sqrt{2.3}$
$\frac{1}{2}$ $\tilde{1}$ $\frac{3}{2}$ 010+ 000+ $-1/2.3\sqrt{2.3}$
$\frac{3}{2}$ 1 $\frac{3}{2}$ 000+ 000+ 0

$\tilde{1}$ $\tilde{1}$ 0 $\frac{3}{2}$ 1 $\frac{3}{2}$
$\frac{3}{2}$ 1 $\frac{3}{2}$ 000+ 100− 0
$\frac{3}{2}$ 1 $\frac{3}{2}$ 010+ 000+ 0
$\frac{3}{2}$ 1 $\frac{3}{2}$ 010+ 100− $+1/4.3\sqrt{2}$
$\frac{3}{2}$ 1 $\frac{3}{2}$ 011+ 000+ $+1/4.3\sqrt{2}$
$\frac{3}{2}$ 1 $\frac{3}{2}$ 011+ 100− 0
$\tilde{1}$ $\tilde{1}$ 0 $\frac{3}{2}$ $\frac{3}{2}$ 0
$\frac{1}{2}$ $\frac{1}{2}$ 0 000+ 000+ $+1/2\sqrt{2.3}$
$\frac{3}{2}$ $\frac{3}{2}$ 0 000+ 000+ $+1/4\sqrt{3}$
$\frac{3}{2}$ $\frac{3}{2}$ 0 000+ 100− 0
$\frac{3}{2}$ $\frac{3}{2}$ 0 000+ 110+ $+1/4\sqrt{3}$
$\tilde{1}$ $\tilde{1}$ 0 $\frac{3}{2}$ $\frac{3}{2}$ 1
$\frac{1}{2}$ $\frac{1}{2}$ 1 000+ 000+ $+1/2.9\sqrt{2}$
$\frac{1}{2}$ $\frac{1}{2}$ 1 010+ 000+ $-1/9\sqrt{2}$
$\frac{3}{2}$ $\frac{1}{2}$ 1 000− 000− $+1/2.9$
$\frac{3}{2}$ $\frac{1}{2}$ 1 000− 100+ $+1/2.9$
$\frac{3}{2}$ $\frac{1}{2}$ 1 010− 000− $+1/2.9$
$\frac{3}{2}$ $\frac{1}{2}$ 1 010− 100+ $-1/4.9$
$\frac{3}{2}$ $\frac{3}{2}$ 1 000+ 000+ $-1/4.9$
$\frac{3}{2}$ $\frac{3}{2}$ 1 000+ 100− 0
$\frac{3}{2}$ $\frac{3}{2}$ 1 000+ 110+ $-1/4.9$
$\frac{3}{2}$ $\frac{3}{2}$ 1 010+ 000+ 0
$\frac{3}{2}$ $\frac{3}{2}$ 1 010+ 100− $+1/2.9$
$\frac{3}{2}$ $\frac{3}{2}$ 1 010+ 110+ 0
$\frac{3}{2}$ $\frac{3}{2}$ 1 011+ 000+ $+1/4.9$
$\frac{3}{2}$ $\frac{3}{2}$ 1 011+ 100− 0
$\frac{3}{2}$ $\frac{3}{2}$ 1 011+ 110+ $-1/2.9$
$\tilde{1}$ $\tilde{1}$ 0 $\frac{3}{2}$ $\frac{3}{2}$ 2
$\frac{3}{2}$ $\frac{1}{2}$ 2 000− 000− 0
$\frac{3}{2}$ $\frac{1}{2}$ 2 000− 100+ $+1/4\sqrt{2.3}$
$\frac{3}{2}$ $\frac{3}{2}$ 2 000+ 000+ $-1/4\sqrt{2.3}$
$\frac{3}{2}$ $\frac{3}{2}$ 2 000+ 100− 0
$\frac{3}{2}$ $\frac{3}{2}$ 2 000+ 110+ 0
$\tilde{1}$ $\tilde{1}$ 0 $\frac{3}{2}$ $\frac{3}{2}$ $\tilde{1}$
$\frac{3}{2}$ $\frac{1}{2}$ $\tilde{1}$ 000− 000− $+1/2.9$
$\frac{3}{2}$ $\frac{1}{2}$ $\tilde{1}$ 000− 100+ $-1/2.9$
$\frac{3}{2}$ $\frac{1}{2}$ $\tilde{1}$ 010+ 000− $-1/2.9$
$\frac{3}{2}$ $\frac{1}{2}$ $\tilde{1}$ 010+ 100+ $-1/4.9$
$\frac{3}{2}$ $\frac{3}{2}$ $\tilde{1}$ 000+ 000+ $-1/4.9$
$\frac{3}{2}$ $\frac{3}{2}$ $\tilde{1}$ 000+ 100− 0
$\frac{3}{2}$ $\frac{3}{2}$ $\tilde{1}$ 000+ 110+ $+1/4.9$
$\frac{3}{2}$ $\frac{3}{2}$ $\tilde{1}$ 010− 000+ 0
$\frac{3}{2}$ $\frac{3}{2}$ $\tilde{1}$ 010− 100− $+1/2.9$
$\frac{3}{2}$ $\frac{3}{2}$ $\tilde{1}$ 010− 110+ 0
$\frac{3}{2}$ $\frac{3}{2}$ $\tilde{1}$ 011+ 000+ $+1/4.9$
$\frac{3}{2}$ $\frac{3}{2}$ $\tilde{1}$ 011+ 100− 0
$\frac{3}{2}$ $\frac{3}{2}$ $\tilde{1}$ 011+ 110+ $+1/2.9$
$\tilde{1}$ $\tilde{1}$ 0 2 0 2
1 $\tilde{1}$ 2 000− 000− $+1/2.3$

$\tilde{1}$ $\tilde{1}$ 0 2 $\frac{1}{2}$ $\frac{3}{2}$
1 $\frac{3}{2}$ $\frac{3}{2}$ 000+ 000+ $-1/4.3\sqrt{2}$
1 $\frac{3}{2}$ $\frac{3}{2}$ 001+ 000+ $-1/4.3\sqrt{2}$
$\tilde{1}$ $\tilde{1}$ 0 2 1 1
1 1 1 000− 000− $+1/2.3\sqrt{3}$
1 2 1 000+ 000+ 0
1 $\tilde{1}$ 1 000+ 000+ $+1/2.3\sqrt{3}$
$\tilde{1}$ $\tilde{1}$ 0 2 1 $\tilde{1}$
1 1 $\tilde{1}$ 000− 000− $-1/2.9$
1 2 $\tilde{1}$ 000+ 000+ $+1/9$
1 $\tilde{1}$ $\tilde{1}$ 000+ 000+ $+1/2.9$
$\tilde{1}$ $\tilde{1}$ 0 2 $\frac{3}{2}$ $\frac{1}{2}$
1 $\frac{1}{2}$ $\frac{1}{2}$ 000+ 000+ $-1/2.3\sqrt{2}$
1 $\frac{3}{2}$ $\frac{1}{2}$ 000− 000− $+1/4.3$
1 $\frac{3}{2}$ $\frac{1}{2}$ 000− 010+ $+1/4.3$
$\tilde{1}$ $\tilde{1}$ 0 2 $\frac{3}{2}$ $\frac{3}{2}$
1 $\frac{1}{2}$ $\frac{3}{2}$ 000+ 000+ $-1/4.3\sqrt{2}$
1 $\frac{3}{2}$ $\frac{3}{2}$ 000− 000− 0
1 $\frac{3}{2}$ $\frac{3}{2}$ 000− 010+ $+1/4.3\sqrt{2}$
1 $\frac{3}{2}$ $\frac{3}{2}$ 001− 000− $-1/4.3\sqrt{2}$
1 $\frac{3}{2}$ $\frac{3}{2}$ 001− 010+ 0
$\tilde{1}$ $\tilde{1}$ 0 2 2 0
1 1 0 000+ 000+ $+1/3\sqrt{2}$
$\tilde{1}$ $\tilde{1}$ 0 2 2 2
1 1 2 000+ 000+ $+1/2.3\sqrt{2}$
$\tilde{1}$ $\tilde{1}$ 0 $\tilde{1}$ 0 $\tilde{1}$
0 $\tilde{1}$ $\tilde{1}$ 000+ 000+ $+1/9$
$\tilde{1}$ $\tilde{1}$ 0 $\tilde{1}$ $\frac{1}{2}$ $\frac{3}{2}$
0 $\frac{3}{2}$ $\frac{3}{2}$ 000− 000− $+1/4.3$
$\tilde{1}$ $\tilde{1}$ 0 $\tilde{1}$ 1 1
0 1 1 000+ 000+ $+1/9$
$\tilde{1}$ $\tilde{1}$ 0 $\tilde{1}$ 1 2
0 2 2 000− 000− $+1/2.3$
$\tilde{1}$ $\tilde{1}$ 0 $\tilde{1}$ 1 $\tilde{1}$
0 $\tilde{1}$ $\tilde{1}$ 000− 000− $+1/9$
$\tilde{1}$ $\tilde{1}$ 0 $\tilde{1}$ $\frac{3}{2}$ $\frac{1}{2}$
0 $\frac{1}{2}$ $\frac{1}{2}$ 000− 000− $+1/2.3$
$\tilde{1}$ $\tilde{1}$ 0 $\tilde{1}$ $\frac{3}{2}$ $\frac{3}{2}$
0 $\frac{3}{2}$ $\frac{3}{2}$ 000+ 000+ $+1/4.3$
0 $\frac{3}{2}$ $\frac{3}{2}$ 000+ 010− 0
0 $\frac{3}{2}$ $\frac{3}{2}$ 010− 000+ 0
0 $\frac{3}{2}$ $\frac{3}{2}$ 010− 010− $+1/4.3$
$\tilde{1}$ $\tilde{1}$ 0 $\tilde{1}$ 2 1
0 1 1 000− 000− $+1/9$
$\tilde{1}$ $\tilde{1}$ 0 $\tilde{1}$ 2 $\tilde{1}$
0 $\tilde{1}$ $\tilde{1}$ 000+ 000+ $+1/9$
$\tilde{1}$ $\tilde{1}$ 0 $\tilde{1}$ $\tilde{1}$ 0
0 0 0 000+ 000+ $+1/3$
$\tilde{1}$ $\tilde{1}$ 1 1 1 1
1 1 0 000+ 000+ $+1/2.9$

O and T_d 9j Symbols (cont.)

$\tilde{1}$	**$\tilde{1}$**	**1**	**1 1 1**		
1	1	1	000−	000−	0
$\tilde{1}$	**$\tilde{1}$**	**1**	**1 1 2**		
1	1	1	000+	000+	$+1/2.9$
$\tilde{1}$	**$\tilde{1}$**	**1**	**1 1 $\tilde{1}$**		
1	1	1	000+	000+	$-1/2.9$
1	1	2	000−	000−	0
1	1	$\tilde{1}$	000−	000−	0
$\tilde{1}$	**$\tilde{1}$**	**1**	**$\frac{3}{2}$ 0 $\frac{3}{2}$**		
$\frac{1}{2}$	$\tilde{1}$	$\frac{3}{2}$	000+	000+	$+1/4.3\sqrt{2.3}$
$\frac{1}{2}$	$\tilde{1}$	$\frac{3}{2}$	000+	001+	$-1/2.3\sqrt{2.3}$
$\tilde{1}$	**$\tilde{1}$**	**1**	**$\frac{3}{2}$ $\frac{1}{2}$ 1**		
$\frac{1}{2}$	$\frac{3}{2}$	1	000−	000−	$+1/8.3$
$\frac{1}{2}$	$\frac{3}{2}$	2	000+	000+	$-1/8\sqrt{3}$
$\frac{1}{2}$	$\frac{3}{2}$	$\tilde{1}$	000+	000+	$+1/8.3\sqrt{3}$
$\tilde{1}$	**$\tilde{1}$**	**1**	**$\frac{3}{2}$ $\frac{1}{2}$ 2**		
$\frac{1}{2}$	$\frac{3}{2}$	1	000+	000+	$-1/8\sqrt{3}$
$\frac{1}{2}$	$\frac{3}{2}$	$\tilde{1}$	000−	000−	$-1/8.3\sqrt{3}$
$\tilde{1}$	**$\tilde{1}$**	**1**	**$\frac{3}{2}$ $\frac{1}{2}$ $\tilde{1}$**		
$\frac{1}{2}$	$\frac{3}{2}$	1	000+	000+	$+1/8.3\sqrt{3}$
$\frac{1}{2}$	$\frac{3}{2}$	2	000−	000−	$-1/8.3\sqrt{3}$
$\frac{1}{2}$	$\frac{3}{2}$	$\tilde{1}$	000−	000−	$+1/8.3$
$\tilde{1}$	**$\tilde{1}$**	**1**	**$\frac{3}{2}$ 1 $\frac{1}{2}$**		
$\frac{1}{2}$	1	$\frac{1}{2}$	000+	000+	$-1/4.3\sqrt{3}$
$\frac{1}{2}$	1	$\frac{3}{2}$	000−	000−	$-1/8.3\sqrt{3}$
$\frac{1}{2}$	2	$\frac{3}{2}$	000+	000+	$-1/8\sqrt{3}$
$\frac{1}{2}$	$\tilde{1}$	$\frac{3}{2}$	000+	000+	$+1/8.3$
$\frac{3}{2}$	1	$\frac{1}{2}$	000−	000−	0
$\frac{3}{2}$	1	$\frac{1}{2}$	000−	100+	$-1/2.3\sqrt{2.3}$
$\tilde{1}$	**$\tilde{1}$**	**1**	**$\frac{3}{2}$ 1 $\frac{3}{2}$**		
$\frac{1}{2}$	1	$\frac{1}{2}$	000−	000−	$-1/2.3\sqrt{2.3}$
$\frac{1}{2}$	1	$\frac{1}{2}$	010−	000−	0
$\frac{1}{2}$	1	$\frac{3}{2}$	000+	000+	$+1/8.3\sqrt{3}$
$\frac{1}{2}$	1	$\frac{3}{2}$	000+	001+	0
$\frac{1}{2}$	1	$\frac{3}{2}$	010+	000+	$+1/4.3\sqrt{3}$
$\frac{1}{2}$	1	$\frac{3}{2}$	010+	001+	$-1/8.3\sqrt{3}$
$\frac{1}{2}$	2	$\frac{3}{2}$	000−	000−	$+1/8.3\sqrt{3}$
$\frac{1}{2}$	2	$\frac{3}{2}$	000−	001−	$-1/4.3\sqrt{3}$
$\frac{1}{2}$	2	$\frac{3}{2}$	010−	000−	$+1/8.3\sqrt{3}$
$\frac{1}{2}$	2	$\frac{3}{2}$	010−	001−	$+1/8.3\sqrt{3}$
$\frac{1}{2}$	$\tilde{1}$	$\frac{3}{2}$	000−	000−	$+1/8.3$
$\frac{1}{2}$	$\tilde{1}$	$\frac{3}{2}$	000−	001−	0
$\frac{1}{2}$	$\tilde{1}$	$\frac{3}{2}$	010−	000−	0
$\frac{1}{2}$	$\tilde{1}$	$\frac{3}{2}$	010−	001−	$+1/8.3$
$\frac{3}{2}$	1	$\frac{1}{2}$	000+	000+	$-1/8.3\sqrt{3}$
$\frac{3}{2}$	1	$\frac{1}{2}$	000+	100−	0
$\frac{3}{2}$	1	$\frac{1}{2}$	010+	000+	$+1/4.3\sqrt{3}$
$\frac{3}{2}$	1	$\frac{1}{2}$	010+	100−	$+1/8.3\sqrt{3}$
$\frac{3}{2}$	1	$\frac{3}{2}$	000−	000−	0
$\frac{3}{2}$	1	$\frac{3}{2}$	000−	001−	0
$\tilde{1}$	**$\tilde{1}$**	**1**	**$\frac{3}{2}$ 1 $\frac{3}{2}$**		
$\frac{3}{2}$	1	$\frac{3}{2}$	000−	100+	0
$\frac{3}{2}$	1	$\frac{3}{2}$	000−	101+	$+1/4.3\sqrt{3}$
$\frac{3}{2}$	1	$\frac{3}{2}$	010−	000−	0
$\frac{3}{2}$	1	$\frac{3}{2}$	010−	001−	$+1/8.3\sqrt{3}$
$\frac{3}{2}$	1	$\frac{3}{2}$	010−	100+	$+1/8.3\sqrt{3}$
$\frac{3}{2}$	1	$\frac{3}{2}$	010−	101+	0
$\frac{3}{2}$	1	$\frac{3}{2}$	011−	000−	0
$\frac{3}{2}$	1	$\frac{3}{2}$	011−	001−	0
$\frac{3}{2}$	1	$\frac{3}{2}$	011−	100+	0
$\frac{3}{2}$	1	$\frac{3}{2}$	011−	101+	0
$\tilde{1}$	**$\tilde{1}$**	**1**	**$\frac{3}{2}$ $\frac{3}{2}$ 0**		
$\frac{1}{2}$	$\frac{1}{2}$	1	000+	000+	$-1/4.3\sqrt{3}$
$\frac{3}{2}$	$\frac{1}{2}$	1	000−	000−	$+1/4.3\sqrt{2.3}$
$\frac{3}{2}$	$\frac{1}{2}$	1	000−	100+	$-1/2.3\sqrt{2.3}$
$\tilde{1}$	**$\tilde{1}$**	**1**	**$\frac{3}{2}$ $\frac{3}{2}$ 1**		
$\frac{1}{2}$	$\frac{1}{2}$	0	000+	000+	$+1/4.3\sqrt{3}$
$\frac{1}{2}$	$\frac{1}{2}$	0	010+	000+	$-1/2.3\sqrt{3}$
$\frac{1}{2}$	$\frac{1}{2}$	1	000−	000−	0
$\frac{1}{2}$	$\frac{1}{2}$	1	010−	000−	0
$\frac{3}{2}$	$\frac{1}{2}$	1	000+	000+	$-1/8.3$
$\frac{3}{2}$	$\frac{1}{2}$	1	000+	100−	0
$\frac{3}{2}$	$\frac{1}{2}$	1	010+	000+	0
$\frac{3}{2}$	$\frac{1}{2}$	1	010+	100−	$-1/8.3$
$\frac{3}{2}$	$\frac{1}{2}$	2	000−	000−	$+1/8.3\sqrt{3}$
$\frac{3}{2}$	$\frac{1}{2}$	2	000−	100+	$-1/4.3\sqrt{3}$
$\frac{3}{2}$	$\frac{1}{2}$	2	010−	000−	$+1/8.3\sqrt{3}$
$\frac{3}{2}$	$\frac{1}{2}$	2	010−	100+	$+1/8.3\sqrt{3}$
$\frac{3}{2}$	$\frac{1}{2}$	$\tilde{1}$	000−	000−	$-1/8.3\sqrt{3}$
$\frac{3}{2}$	$\frac{1}{2}$	$\tilde{1}$	000−	100+	0
$\frac{3}{2}$	$\frac{1}{2}$	$\tilde{1}$	010−	000−	$-1/4.3\sqrt{3}$
$\frac{3}{2}$	$\frac{1}{2}$	$\tilde{1}$	010−	100+	$+1/8.3\sqrt{3}$
$\frac{3}{2}$	$\frac{3}{2}$	0	000+	000+	$-1/2.3\sqrt{2.3}$
$\frac{3}{2}$	$\frac{3}{2}$	0	000+	100−	0
$\frac{3}{2}$	$\frac{3}{2}$	0	000+	110+	$+1/4.3\sqrt{2.3}$
$\frac{3}{2}$	$\frac{3}{2}$	0	010+	000+	0
$\frac{3}{2}$	$\frac{3}{2}$	0	010+	100−	$+1/4.3\sqrt{2.3}$
$\frac{3}{2}$	$\frac{3}{2}$	0	010+	110+	0
$\frac{3}{2}$	$\frac{3}{2}$	1	000−	000−	0
$\frac{3}{2}$	$\frac{3}{2}$	1	000−	100+	0
$\frac{3}{2}$	$\frac{3}{2}$	1	000−	110−	0
$\frac{3}{2}$	$\frac{3}{2}$	1	010−	000−	0
$\frac{3}{2}$	$\frac{3}{2}$	1	010−	100+	$-1/8.3$
$\frac{3}{2}$	$\frac{3}{2}$	1	010−	110−	0
$\frac{3}{2}$	$\frac{3}{2}$	1	011−	000−	0
$\frac{3}{2}$	$\frac{3}{2}$	1	011−	100+	0
$\frac{3}{2}$	$\frac{3}{2}$	1	011−	110−	0
$\tilde{1}$	**$\tilde{1}$**	**1**	**$\frac{3}{2}$ $\frac{3}{2}$ 2**		
$\frac{1}{2}$	$\frac{1}{2}$	1	000+	000+	$+1/2.3\sqrt{2.3}$
$\frac{3}{2}$	$\frac{1}{2}$	1	000−	000−	$+1/8.3\sqrt{3}$
$\tilde{1}$	**$\tilde{1}$**	**1**	**$\frac{3}{2}$ $\frac{3}{2}$ 2**		
$\frac{3}{2}$	$\frac{1}{2}$	1	000−	100+	$+1/8.3\sqrt{3}$
$\frac{3}{2}$	$\frac{1}{2}$	$\tilde{1}$	000+	000+	$-1/8.3\sqrt{3}$
$\frac{3}{2}$	$\frac{1}{2}$	$\tilde{1}$	000+	100−	$+1/8.3\sqrt{3}$
$\frac{3}{2}$	$\frac{3}{2}$	1	000+	000+	0
$\frac{3}{2}$	$\frac{3}{2}$	1	000+	100−	$+1/8.3\sqrt{3}$
$\frac{3}{2}$	$\frac{3}{2}$	1	000+	110+	0
$\frac{3}{2}$	$\frac{3}{2}$	1	001+	000+	$+1/4.3\sqrt{3}$
$\frac{3}{2}$	$\frac{3}{2}$	1	001+	100−	0
$\frac{3}{2}$	$\frac{3}{2}$	1	001+	110+	$+1/4.3\sqrt{3}$
$\tilde{1}$	**$\tilde{1}$**	**1**	**$\frac{3}{2}$ $\frac{3}{2}$ $\tilde{1}$**		
$\frac{1}{2}$	$\frac{1}{2}$	1	000+	000+	$-1/2.3\sqrt{2.3}$
$\frac{1}{2}$	$\frac{1}{2}$	1	010−	000+	0
$\frac{3}{2}$	$\frac{1}{2}$	1	000−	000−	$+1/8.3\sqrt{3}$
$\frac{3}{2}$	$\frac{1}{2}$	1	000−	100+	0
$\frac{3}{2}$	$\frac{1}{2}$	1	010+	000−	$-1/4.3\sqrt{3}$
$\frac{3}{2}$	$\frac{1}{2}$	1	010+	100+	$-1/8.3\sqrt{3}$
$\frac{3}{2}$	$\frac{1}{2}$	2	000+	000+	$-1/8.3\sqrt{3}$
$\frac{3}{2}$	$\frac{1}{2}$	2	000+	100−	$-1/4.3\sqrt{3}$
$\frac{3}{2}$	$\frac{1}{2}$	2	010−	000+	$+1/8.3\sqrt{3}$
$\frac{3}{2}$	$\frac{1}{2}$	2	010−	100−	$-1/8.3\sqrt{3}$
$\frac{3}{2}$	$\frac{1}{2}$	$\tilde{1}$	000+	000+	$+1/8.3$
$\frac{3}{2}$	$\frac{1}{2}$	$\tilde{1}$	000+	100−	0
$\frac{3}{2}$	$\frac{1}{2}$	$\tilde{1}$	010−	000+	0
$\frac{3}{2}$	$\frac{1}{2}$	$\tilde{1}$	010−	100−	$+1/8.3$
$\frac{3}{2}$	$\frac{3}{2}$	1	000+	000+	0
$\frac{3}{2}$	$\frac{3}{2}$	1	000+	100−	0
$\frac{3}{2}$	$\frac{3}{2}$	1	000+	110+	$-1/4.3\sqrt{3}$
$\frac{3}{2}$	$\frac{3}{2}$	1	001+	000+	0
$\frac{3}{2}$	$\frac{3}{2}$	1	001+	100−	$+1/8.3\sqrt{3}$
$\frac{3}{2}$	$\frac{3}{2}$	1	001+	110+	0
$\frac{3}{2}$	$\frac{3}{2}$	1	010−	000+	0
$\frac{3}{2}$	$\frac{3}{2}$	1	010−	100−	$+1/8.3\sqrt{3}$
$\frac{3}{2}$	$\frac{3}{2}$	1	010−	110+	0
$\frac{3}{2}$	$\frac{3}{2}$	1	011−	000+	0
$\frac{3}{2}$	$\frac{3}{2}$	1	011−	100−	0
$\frac{3}{2}$	$\frac{3}{2}$	1	011−	110+	0
$\frac{3}{2}$	$\frac{3}{2}$	2	000−	000−	0
$\frac{3}{2}$	$\frac{3}{2}$	2	000−	100+	$+1/8.3\sqrt{3}$
$\frac{3}{2}$	$\frac{3}{2}$	2	000−	110−	0
$\frac{3}{2}$	$\frac{3}{2}$	2	010+	000−	$-1/4.3\sqrt{3}$
$\frac{3}{2}$	$\frac{3}{2}$	2	010+	100+	0
$\frac{3}{2}$	$\frac{3}{2}$	2	010+	110−	$+1/4.3\sqrt{3}$
$\frac{3}{2}$	$\frac{3}{2}$	$\tilde{1}$	000−	000−	0
$\frac{3}{2}$	$\frac{3}{2}$	$\tilde{1}$	000−	100+	0
$\frac{3}{2}$	$\frac{3}{2}$	$\tilde{1}$	000−	110−	0
$\frac{3}{2}$	$\frac{3}{2}$	$\tilde{1}$	010+	000−	0
$\frac{3}{2}$	$\frac{3}{2}$	$\tilde{1}$	010+	100+	$+1/8.3$
$\frac{3}{2}$	$\frac{3}{2}$	$\tilde{1}$	010+	110−	0
$\frac{3}{2}$	$\frac{3}{2}$	$\tilde{1}$	011−	000−	0

O and T_d 9j Symbols (cont.)

1̃ 1̃ 1 3/2 3/2 1̃
3/2 3/2 1̃ 011− 100+ 0
3/2 3/2 1̃ 011− 110− 0
1̃ 1̃ 1 2 0 2
1 1̃ 1 000− 000− −1/2.3√2
1 1̃ 1̃ 000+ 000+ +1/2.3√2.3
1̃ 1̃ 1 2 1/2 3/2
1 3/2 1/2 000+ 000+ −1/8√3
1 3/2 3/2 000− 000− +1/8.3√3
1 3/2 3/2 000− 001− −1/4.3√3
1 3/2 3/2 001− 000− +1/8.3√3
1 3/2 3/2 001− 001− +1/8.3√3
1̃ 1̃ 1 2 1 1
1 1 0 000− 000− −1/2.3√3
1 1 1 000+ 000+ 0
1 1 2 000− 000− +1/4.3√3
1 1 1̃ 000− 000− 0
1 2 1 000− 000− +1/4.3
1 2 1̃ 000+ 000+ −1/4.3√3
1 1̃ 1 000− 000− 0
1 1̃ 2 000+ 000+ +1/4.3
1 1̃ 1̃ 000+ 000+ 0
1̃ 1̃ 1 2 1 1̃
1 1 1 000− 000− +1/2.9
1 1 2 000+ 000+ −1/4.3√3
1 1 1̃ 000+ 000+ +1/2.9
1 2 1 000+ 000+ −1/4.3√3
1 2 1̃ 000− 000− −1/4.9
1 1̃ 1 000+ 000+ −1/2.9
1 1̃ 2 000− 000− +1/4.9
1 1̃ 1̃ 000− 000− +1/2.9
1̃ 1̃ 1 2 3/2 1/2
1 1/2 1/2 000− 000− −1/4.3√3
1 1/2 3/2 000+ 000+ +1/8.3√3
1 3/2 1/2 000+ 000+ −1/2.3√2.3
1 3/2 1/2 000+ 010− +1/4.3√2.3
1 3/2 3/2 000− 000− +1/8.3√3
1 3/2 3/2 000− 010+ −1/4.3√3
1 3/2 3/2 001− 000− −1/8.3√3
1 3/2 3/2 001− 010+ −1/8.3√3
1̃ 1̃ 1 2 3/2 3/2
1 1/2 1/2 000+ 000+ −1/2.3√2.3
1 1/2 3/2 000− 000− +1/8.3√3
1 1/2 3/2 000− 001− −1/8.3√3
1 3/2 1/2 000− 000− −1/8.3√3
1 3/2 1/2 000− 010+ −1/8.3√3
1 3/2 3/2 000+ 000+ 0
1 3/2 3/2 000+ 001+ +1/8.3√3
1 3/2 3/2 000+ 010− −1/8.3√3
1 3/2 3/2 000+ 011− 0

1̃ 1̃ 1 2 3/2 3/2
1 3/2 3/2 001+ 000+ −1/4.3√3
1 3/2 3/2 001+ 001+ 0
1 3/2 3/2 001+ 010− 0
1 3/2 3/2 001+ 011− +1/4.3√3
1̃ 1̃ 1 2 2 0
1 1 1 000+ 000+ +1/2.3√2.3
1̃ 1̃ 1 2 2 2
1 1 1 000+ 000+ −1/2.3√3
1 1 1̃ 000− 000− 0
1̃ 1̃ 1 1̃ 0 1̃
0 1̃ 1̃ 000− 000− +1/9
1 1̃ 1 000− 000− +1/2.9
1 1̃ 2 000+ 000+ +1/2.9
1 1̃ 1̃ 000+ 000+ +1/2.9
1̃ 1̃ 1 1̃ 1/2 3/2
0 3/2 3/2 000+ 000+ +1/4.3√2.3
0 3/2 3/2 000+ 001+ −1/2.3√2.3
1 3/2 1/2 000+ 000+ +1/8.3
1 3/2 3/2 000− 000− +1/8.3
1 3/2 3/2 000− 001− 0
1 3/2 3/2 001− 000− 0
1 3/2 3/2 001− 001− +1/8.3
1̃ 1̃ 1 1̃ 1 1
0 1 1 000− 000− +1/2.9
0 2 2 000− 000− −1/2.3√2
0 1̃ 1̃ 000− 000− +1/2.9
1 1 0 000− 000− +1/2.9
1 1 1 000+ 000+ +1/2.9
1 1 2 000− 000− +1/2.9
1 1 1̃ 000− 000− 0
1̃ 1̃ 1 1̃ 1 2
0 1 1 000+ 000+ +1/2.3√3
0 1̃ 1̃ 000+ 000+ +1/2.9
1 1 1 000− 000− 0
1 1 1̃ 000+ 000+ −1/2.9
1 2 1 000+ 000+ +1/4.3
1 2 1̃ 000− 000− +1/4.9
1̃ 1̃ 1 1̃ 1 1̃
0 1 1 000+ 000+ −1/2.9
0 2 2 000+ 000+ +1/2.3√2.3
0 1̃ 1̃ 000+ 000+ +1/2.9
1 1 1 000− 000− 0
1 1 2 000+ 000+ 0
1 1 1̃ 000+ 000+ +1/2.9
1 2 1 000+ 000+ 0
1 2 1̃ 000− 000− +1/2.9
1 1̃ 1 000+ 000+ +1/2.9
1 1̃ 2 000− 000− +1/2.9
1 1̃ 1̃ 000− 000− 0

1̃ 1̃ 1 1̃ 3/2 1/2
0 1/2 1/2 000+ 000+ −1/2.3√2.3
0 3/2 3/2 000+ 000+ −1/4.3√2.3
0 3/2 3/2 000+ 010− +1/2.3√2.3
1 1/2 1/2 000− 000− +1/4.3
1 1/2 3/2 000+ 000+ +1/8.3
1̃ 1̃ 1 1̃ 3/2 3/2
0 1/2 1/2 000− 000− +1/4.3√3
0 1/2 1/2 010+ 000− −1/2.3√3
0 3/2 3/2 000− 000− −1/2.3√2.3
0 3/2 3/2 000− 001− 0
0 3/2 3/2 000− 010+ 0
0 3/2 3/2 000− 011+ −1/4.3√2.3
0 3/2 3/2 010+ 000− 0
0 3/2 3/2 010+ 001− −1/4.3√2.3
0 3/2 3/2 010+ 010+ +1/4.3√2.3
0 3/2 3/2 010+ 011+ 0
1 1/2 1/2 000+ 000+ 0
1 1/2 1/2 010− 000+ 0
1 1/2 3/2 000− 000− +1/8.3
1 1/2 3/2 000− 001− 0
1 1/2 3/2 010+ 000− 0
1 1/2 3/2 010+ 001− −1/8.3
1 3/2 1/2 000− 000− +1/8.3
1 3/2 1/2 000− 010+ 0
1 3/2 1/2 010+ 000− 0
1 3/2 1/2 010+ 010+ +1/8.3
1 3/2 3/2 000+ 000+ 0
1 3/2 3/2 000+ 001+ 0
1 3/2 3/2 000+ 010− 0
1 3/2 3/2 000+ 011− 0
1 3/2 3/2 001+ 000+ 0
1 3/2 3/2 001+ 001+ +1/8.3
1 3/2 3/2 001+ 010− +1/8.3
1 3/2 3/2 001+ 011− 0
1 3/2 3/2 010− 000+ 0
1 3/2 3/2 010− 001+ +1/8.3
1 3/2 3/2 010− 010− +1/8.3
1 3/2 3/2 010− 011− 0
1 3/2 3/2 011− 000+ 0
1 3/2 3/2 011− 001+ 0
1 3/2 3/2 011− 010− 0
1 3/2 3/2 011− 011− 0
1̃ 1̃ 1 1̃ 2 1
0 1 1 000+ 000+ +1/2.9
0 1̃ 1̃ 000+ 000+ +1/2.3√3
1 1 0 000+ 000+ +1/2.9
1 1 1 000− 000− +1/2.9
1 1 2 000+ 000+ −1/4.9
1 1 1̃ 000+ 000+ +1/2.9

O and T_d 9j Symbols (cont.)

$\tilde{1}\ \tilde{1}\ 1\quad \tilde{1}\ 2\ \tilde{1}$
0 1 1 000− 000− $-1/2.3\sqrt{3}$
0 $\tilde{1}$ $\tilde{1}$ 000− 000− $-1/2.9$
1 1 1 000+ 000+ 0
1 1 2 000− 000− $+1/4.3$
1 1 $\tilde{1}$ 000− 000− 0
1 $\tilde{1}$ 1 000− 000− $+1/2.9$
1 $\tilde{1}$ 2 000+ 000+ $-1/4.9$
1 $\tilde{1}$ $\tilde{1}$ 000+ 000+ $+1/2.9$

$\tilde{1}\ \tilde{1}\ 1\quad \tilde{1}\ \tilde{1}\ 0$
1 0 1 000− 000− $+1/9$
1 1 1 000+ 000+ $-1/2.9$

$\tilde{1}\ \tilde{1}\ 1\quad \tilde{1}\ \tilde{1}\ 1$
0 0 0 000+ 000+ $+1/3\sqrt{3}$
1 0 1 000+ 000+ $-1/2.9$
1 1 0 000+ 000+ $+1/2.9$
1 1 1 000− 000− 0
1 1 2 000+ 000+ $+1/2.9$
1 1 $\tilde{1}$ 000+ 000+ $+1/2.9$

$\tilde{1}\ \tilde{1}\ 2\quad 1\ 1\ 1$
1 1 1 000+ 000+ $-1/2.9$

$\tilde{1}\ \tilde{1}\ 2\quad 1\ 1\ 2$
1 1 0 000+ 000+ $-1/2.3\sqrt{2.3}$
1 1 2 000+ 000+ $-1/2.3\sqrt{3}$

$\tilde{1}\ \tilde{1}\ 2\quad 1\ 1\ \tilde{1}$
1 1 1 000− 000− 0
1 1 $\tilde{1}$ 000+ 000+ $+1/2.9$

$\tilde{1}\ \tilde{1}\ 2\quad \frac{3}{2}\ 0\ \frac{3}{2}$
$\frac{1}{2}$ $\tilde{1}$ $\frac{3}{2}$ 000− 000− $+1/4.3\sqrt{2}$

$\tilde{1}\ \tilde{1}\ 2\quad \frac{3}{2}\ \frac{1}{2}\ 1$
$\frac{1}{2}$ $\frac{3}{2}$ 1 000+ 000+ $-1/8.3$
$\frac{1}{2}$ $\frac{3}{2}$ $\tilde{1}$ 000− 000− $+1/8.3$

$\tilde{1}\ \tilde{1}\ 2\quad \frac{3}{2}\ \frac{1}{2}\ 2$
$\frac{1}{2}$ $\frac{3}{2}$ 2 000+ 000+ $-1/8\sqrt{3}$

$\tilde{1}\ \tilde{1}\ 2\quad \frac{3}{2}\ \frac{1}{2}\ \tilde{1}$
$\frac{1}{2}$ $\frac{3}{2}$ 1 000− 000− $+1/8.3$
$\frac{1}{2}$ $\frac{3}{2}$ $\tilde{1}$ 000+ 000+ $+5/8.9$

$\tilde{1}\ \tilde{1}\ 2\quad \frac{3}{2}\ 1\ \frac{1}{2}$
$\frac{1}{2}$ 1 $\frac{3}{2}$ 000+ 000+ $-1/8.3$
$\frac{1}{2}$ 2 $\frac{3}{2}$ 000− 000− 0
$\frac{1}{2}$ $\tilde{1}$ $\frac{3}{2}$ 000− 000− $-1/8\sqrt{3}$

$\tilde{1}\ \tilde{1}\ 2\quad \frac{3}{2}\ 1\ \frac{3}{2}$
$\frac{1}{2}$ 1 $\frac{1}{2}$ 000+ 000+ 0
$\frac{1}{2}$ 1 $\frac{1}{2}$ 010+ 000+ $-1/4.3\sqrt{2}$
$\frac{1}{2}$ 1 $\frac{3}{2}$ 000− 000− $-1/8.3$
$\frac{1}{2}$ 1 $\frac{3}{2}$ 010− 000− $+1/8.3$
$\frac{1}{2}$ 2 $\frac{3}{2}$ 000+ 000+ $-1/8.3$
$\frac{1}{2}$ 2 $\frac{3}{2}$ 010+ 000+ $-1/8.3$
$\frac{1}{2}$ $\tilde{1}$ $\frac{3}{2}$ 000+ 000+ $+1/8.3\sqrt{3}$
$\frac{1}{2}$ $\tilde{1}$ $\frac{3}{2}$ 010+ 000+ $+1/8.3\sqrt{3}$

$\tilde{1}\ \tilde{1}\ 2\quad \frac{3}{2}\ 1\ \frac{3}{2}$
$\frac{3}{2}$ 1 $\frac{1}{2}$ 000− 000− $-1/8.3$
$\frac{3}{2}$ 1 $\frac{1}{2}$ 000− 100+ $+1/8.3$
$\frac{3}{2}$ 1 $\frac{1}{2}$ 010− 000− 0
$\frac{3}{2}$ 1 $\frac{1}{2}$ 010− 100+ $+1/8.3$
$\frac{3}{2}$ 1 $\frac{3}{2}$ 000+ 000+ 0
$\frac{3}{2}$ 1 $\frac{3}{2}$ 000+ 100− 0
$\frac{3}{2}$ 1 $\frac{3}{2}$ 010+ 000+ $-1/8.3$
$\frac{3}{2}$ 1 $\frac{3}{2}$ 010+ 100− 0
$\frac{3}{2}$ 1 $\frac{3}{2}$ 011+ 000+ 0
$\frac{3}{2}$ 1 $\frac{3}{2}$ 011+ 100− 0

$\tilde{1}\ \tilde{1}\ 2\quad \frac{3}{2}\ \frac{3}{2}\ 0$
$\frac{3}{2}$ $\frac{1}{2}$ 2 000− 000− $+1/8\sqrt{3}$
$\frac{3}{2}$ $\frac{1}{2}$ 2 000− 100+ $+1/8\sqrt{3}$

$\tilde{1}\ \tilde{1}\ 2\quad \frac{3}{2}\ \frac{3}{2}\ 1$
$\frac{1}{2}$ $\frac{1}{2}$ 1 000+ 000+ $-1/9\sqrt{2}$
$\frac{1}{2}$ $\frac{1}{2}$ 1 010+ 000+ $-1/4.9\sqrt{2}$
$\frac{3}{2}$ $\frac{1}{2}$ 1 000− 000− $+1/8.9$
$\frac{3}{2}$ $\frac{1}{2}$ 1 000− 100+ $+1/8.9$
$\frac{3}{2}$ $\frac{1}{2}$ 1 010− 000− $+1/8.9$
$\frac{3}{2}$ $\frac{1}{2}$ 1 010− 100+ $+1/2.9$
$\frac{3}{2}$ $\frac{1}{2}$ $\tilde{1}$ 000+ 000+ $+1/8.3$
$\frac{3}{2}$ $\frac{1}{2}$ $\tilde{1}$ 000+ 100− $+1/8.3$
$\frac{3}{2}$ $\frac{1}{2}$ $\tilde{1}$ 010+ 000+ $-1/8.3$
$\frac{3}{2}$ $\frac{1}{2}$ $\tilde{1}$ 010+ 100− 0
$\frac{3}{2}$ $\frac{3}{2}$ 1 000+ 000+ $+1/2.9$
$\frac{3}{2}$ $\frac{3}{2}$ 1 000+ 100− 0
$\frac{3}{2}$ $\frac{3}{2}$ 1 000+ 110+ $+1/2.9$
$\frac{3}{2}$ $\frac{3}{2}$ 1 010+ 000+ 0
$\frac{3}{2}$ $\frac{3}{2}$ 1 010+ 100− $+1/8.9$
$\frac{3}{2}$ $\frac{3}{2}$ 1 010+ 110+ 0
$\frac{3}{2}$ $\frac{3}{2}$ 1 011+ 000+ $-1/2.9$
$\frac{3}{2}$ $\frac{3}{2}$ 1 011+ 100− 0
$\frac{3}{2}$ $\frac{3}{2}$ 1 011+ 110+ $-1/8.9$

$\tilde{1}\ \tilde{1}\ 2\quad \frac{3}{2}\ \frac{3}{2}\ 2$
$\frac{1}{2}$ $\frac{1}{2}$ 0 000+ 000+ $+1/4\sqrt{2.3}$
$\frac{3}{2}$ $\frac{1}{2}$ 2 000− 000− $-1/8\sqrt{3}$
$\frac{3}{2}$ $\frac{1}{2}$ 2 000− 100+ 0
$\frac{3}{2}$ $\frac{3}{2}$ 0 000+ 000+ 0
$\frac{3}{2}$ $\frac{3}{2}$ 0 000+ 100− $+1/8\sqrt{3}$
$\frac{3}{2}$ $\frac{3}{2}$ 0 000+ 110+ 0
$\frac{3}{2}$ $\frac{3}{2}$ 2 000+ 000+ 0
$\frac{3}{2}$ $\frac{3}{2}$ 2 000+ 100− 0
$\frac{3}{2}$ $\frac{3}{2}$ 2 000+ 110+ $+1/8\sqrt{3}$

$\tilde{1}\ \tilde{1}\ 2\quad \frac{3}{2}\ \frac{3}{2}\ \tilde{1}$
$\frac{1}{2}$ $\frac{1}{2}$ 1 000− 000− 0
$\frac{1}{2}$ $\frac{1}{2}$ 1 010+ 000− $+1/4.3\sqrt{2}$
$\frac{3}{2}$ $\frac{1}{2}$ 1 000+ 000+ $+1/8.3$
$\frac{3}{2}$ $\frac{1}{2}$ 1 000+ 100− $-1/8.3$
$\frac{3}{2}$ $\frac{1}{2}$ 1 010− 000+ $+1/8.3$

$\tilde{1}\ \tilde{1}\ 2\quad \frac{3}{2}\ \frac{3}{2}\ \tilde{1}$
$\frac{3}{2}$ $\frac{1}{2}$ 1 010− 100− 0
$\frac{3}{2}$ $\frac{1}{2}$ $\tilde{1}$ 000− 000− $+1/8.9$
$\frac{3}{2}$ $\frac{1}{2}$ $\tilde{1}$ 000− 100+ $-1/8.9$
$\frac{3}{2}$ $\frac{1}{2}$ $\tilde{1}$ 010+ 000− $-1/8.9$
$\frac{3}{2}$ $\frac{1}{2}$ $\tilde{1}$ 010+ 100+ $+1/2.9$
$\frac{3}{2}$ $\frac{3}{2}$ 1 000− 000− 0
$\frac{3}{2}$ $\frac{3}{2}$ 1 000− 100+ 0
$\frac{3}{2}$ $\frac{3}{2}$ 1 000− 110− 0
$\frac{3}{2}$ $\frac{3}{2}$ 1 001− 000− 0
$\frac{3}{2}$ $\frac{3}{2}$ 1 001− 100+ $-1/8.3$
$\frac{3}{2}$ $\frac{3}{2}$ 1 001− 110− 0
$\frac{3}{2}$ $\frac{3}{2}$ 1 010+ 000− 0
$\frac{3}{2}$ $\frac{3}{2}$ 1 010+ 100+ $-1/8.3$
$\frac{3}{2}$ $\frac{3}{2}$ 1 010+ 110− 0
$\frac{3}{2}$ $\frac{3}{2}$ 1 011+ 000− 0
$\frac{3}{2}$ $\frac{3}{2}$ 1 011+ 100+ 0
$\frac{3}{2}$ $\frac{3}{2}$ 1 011+ 110− $-1/8.3$
$\frac{3}{2}$ $\frac{3}{2}$ $\tilde{1}$ 000+ 000+ $+1/2.9$
$\frac{3}{2}$ $\frac{3}{2}$ $\tilde{1}$ 000+ 100− 0
$\frac{3}{2}$ $\frac{3}{2}$ $\tilde{1}$ 000+ 110+ $-1/2.9$
$\frac{3}{2}$ $\frac{3}{2}$ $\tilde{1}$ 010− 000+ 0
$\frac{3}{2}$ $\frac{3}{2}$ $\tilde{1}$ 010− 100− $+1/8.9$
$\frac{3}{2}$ $\frac{3}{2}$ $\tilde{1}$ 010− 110+ 0
$\frac{3}{2}$ $\frac{3}{2}$ $\tilde{1}$ 011+ 000+ $-1/2.9$
$\frac{3}{2}$ $\frac{3}{2}$ $\tilde{1}$ 011+ 100− 0
$\frac{3}{2}$ $\frac{3}{2}$ $\tilde{1}$ 011+ 110+ $+1/8.9$

$\tilde{1}\ \tilde{1}\ 2\quad 2\ 0\ 2$
1 $\tilde{1}$ 2 000− 000− $+1/2.3\sqrt{2}$

$\tilde{1}\ \tilde{1}\ 2\quad 2\ \frac{1}{2}\ \frac{3}{2}$
1 $\frac{3}{2}$ $\frac{1}{2}$ 000− 000− 0
1 $\frac{3}{2}$ $\frac{3}{2}$ 000+ 000+ $-1/8.3$
1 $\frac{3}{2}$ $\frac{3}{2}$ 001+ 000+ $-1/8.3$

$\tilde{1}\ \tilde{1}\ 2\quad 2\ 1\ 1$
1 1 1 000− 000− $-1/4.3\sqrt{3}$
1 1 $\tilde{1}$ 000+ 000+ $-1/4.3$
1 2 1 000+ 000+ 0
1 2 $\tilde{1}$ 000− 000− 0
1 $\tilde{1}$ 1 000+ 000+ $-1/4.3\sqrt{3}$
1 $\tilde{1}$ $\tilde{1}$ 000− 000− $+1/4.3$

$\tilde{1}\ \tilde{1}\ 2\quad 2\ 1\ \tilde{1}$
1 1 1 000+ 000+ $-1/4.3\sqrt{3}$
1 1 $\tilde{1}$ 000− 000− $+1/4.9$
1 2 1 000− 000− 0
1 2 $\tilde{1}$ 000+ 000+ $+1/9$
1 $\tilde{1}$ 1 000− 000− $-1/4.3\sqrt{3}$
1 $\tilde{1}$ $\tilde{1}$ 000+ 000+ $-1/4.9$

$\tilde{1}\ \tilde{1}\ 2\quad 2\ \frac{3}{2}\ \frac{1}{2}$
1 $\frac{1}{2}$ $\frac{3}{2}$ 000− 000− $+1/4.3$
1 $\frac{3}{2}$ $\frac{3}{2}$ 000+ 000+ $-1/8.3$

O and T_d 9j Symbols (cont.)

$\tilde{1}$ $\tilde{1}$ 2 2 $\frac{3}{2}$ $\frac{1}{2}$
1 $\frac{3}{2}$ $\frac{3}{2}$ 000+ 010− $-1/8.3$
1 $\frac{3}{2}$ $\frac{3}{2}$ 001+ 000+ $+1/8.3$
1 $\frac{3}{2}$ $\frac{3}{2}$ 001+ 010− $+1/8.3$
$\tilde{1}$ $\tilde{1}$ 2 2 $\frac{3}{2}$ $\frac{3}{2}$
1 $\frac{1}{2}$ $\frac{1}{2}$ 000− 000− $-1/4.3\sqrt{2}$
1 $\frac{1}{2}$ $\frac{3}{2}$ 000+ 000+ $+1/8.3$
1 $\frac{3}{2}$ $\frac{1}{2}$ 000+ 000+ $+1/8.3$
1 $\frac{3}{2}$ $\frac{1}{2}$ 000+ 010− $+1/8.3$
1 $\frac{3}{2}$ $\frac{3}{2}$ 000− 000− $-1/8.3$
1 $\frac{3}{2}$ $\frac{3}{2}$ 000− 010+ 0
1 $\frac{3}{2}$ $\frac{3}{2}$ 001− 000− 0
1 $\frac{3}{2}$ $\frac{3}{2}$ 001− 010+ $+1/8.3$
$\tilde{1}$ $\tilde{1}$ 2 2 2 0
1 1 2 000+ 000+ $+1/2.3$
$\tilde{1}$ $\tilde{1}$ 2 2 2 2
1 1 0 000+ 000+ $+1/2.3\sqrt{2}$
1 1 2 000+ 000+ $-1/4.3$
$\tilde{1}$ $\tilde{1}$ 2 $\tilde{1}$ 0 $\tilde{1}$
0 $\tilde{1}$ $\tilde{1}$ 000+ 000+ $+1/9$
1 $\tilde{1}$ 1 000+ 000+ $+1/2.3\sqrt{3}$
1 $\tilde{1}$ $\tilde{1}$ 000− 000− $-1/2.9$
2 $\tilde{1}$ 1 000− 000− 0
2 $\tilde{1}$ $\tilde{1}$ 000+ 000+ $+1/9$
$\tilde{1}$ $\tilde{1}$ 2 $\tilde{1}$ $\frac{1}{2}$ $\frac{3}{2}$
0 $\frac{3}{2}$ $\frac{3}{2}$ 000− 000− $+1/4.3\sqrt{2}$
1 $\frac{3}{2}$ $\frac{1}{2}$ 000− 000− $-1/8\sqrt{3}$
1 $\frac{3}{2}$ $\frac{3}{2}$ 000+ 000+ $+1/8.3\sqrt{3}$
1 $\frac{3}{2}$ $\frac{3}{2}$ 001+ 000+ $+1/8.3\sqrt{3}$
2 $\frac{3}{2}$ $\frac{1}{2}$ 000+ 000+ 0
2 $\frac{3}{2}$ $\frac{3}{2}$ 000− 000− $+1/8.3$
$\tilde{1}$ $\tilde{1}$ 2 $\tilde{1}$ 1 1
0 1 1 000+ 000+ $-1/2.9$
0 $\tilde{1}$ $\tilde{1}$ 000+ 000+ $+1/2.3\sqrt{3}$
1 1 1 000− 000− $+1/2.9$
1 1 $\tilde{1}$ 000+ 000+ 0
1 2 1 000+ 000+ $-1/4.3\sqrt{3}$
1 2 $\tilde{1}$ 000− 000 $-1/4.3\sqrt{3}$
1 $\tilde{1}$ 1 000+ 000+ $-1/2.9$
1 $\tilde{1}$ $\tilde{1}$ 000− 000− 0
2 1 1 000+ 000+ $+1/4.9$
2 1 $\tilde{1}$ 000− 000− $+1/4.3$
$\tilde{1}$ $\tilde{1}$ 2 $\tilde{1}$ 1 2
0 2 2 000− 000− $+1/2.3\sqrt{2}$
1 1 0 000− 000− $+1/2.3\sqrt{2}$
1 1 2 000− 000− 0
1 $\tilde{1}$ 2 000+ 000+ $-1/2.3\sqrt{3}$
2 2 0 000− 000− 0
2 2 2 000− 000− $+1/4.3$

$\tilde{1}$ $\tilde{1}$ 2 $\tilde{1}$ 1 $\tilde{1}$
0 1 1 000− 000− $-1/2.3\sqrt{3}$
0 $\tilde{1}$ $\tilde{1}$ 000− 000− $-1/2.9$
1 1 1 000+ 000+ 0
1 1 $\tilde{1}$ 000− 000− $+1/2.9$
1 2 1 000− 000− $+1/4.3$
1 2 $\tilde{1}$ 000+ 000+ $-1/4.9$
1 $\tilde{1}$ 1 000− 000− 0
1 $\tilde{1}$ $\tilde{1}$ 000+ 000+ $+1/2.9$
2 1 1 000− 000− $+1/4.3\sqrt{3}$
2 1 $\tilde{1}$ 000+ 000+ $-1/4.3\sqrt{3}$
2 $\tilde{1}$ 1 000+ 000+ $+1/4.3$
2 $\tilde{1}$ $\tilde{1}$ 000− 000− $+1/4.9$
$\tilde{1}$ $\tilde{1}$ 2 $\tilde{1}$ $\frac{3}{2}$ $\frac{1}{2}$
0 $\frac{3}{2}$ $\frac{3}{2}$ 000− 000− $-1/4.3\sqrt{2}$
0 $\frac{3}{2}$ $\frac{3}{2}$ 000− 010+ $-1/4.3\sqrt{2}$
1 $\frac{1}{2}$ $\frac{3}{2}$ 000− 000− $-1/8.3\sqrt{3}$
1 $\frac{3}{2}$ $\frac{3}{2}$ 000+ 000+ $-1/8.3\sqrt{3}$
1 $\frac{3}{2}$ $\frac{3}{2}$ 000+ 010− $-1/8.3\sqrt{3}$
1 $\frac{3}{2}$ $\frac{3}{2}$ 001+ 000+ $-1/4.3\sqrt{3}$
1 $\frac{3}{2}$ $\frac{3}{2}$ 001+ 010− $+1/8.3\sqrt{3}$
2 $\frac{1}{2}$ $\frac{3}{2}$ 000+ 000+ $+1/4.3$
$\tilde{1}$ $\tilde{1}$ 2 $\tilde{1}$ $\frac{3}{2}$ $\frac{3}{2}$
0 $\frac{1}{2}$ $\frac{1}{2}$ 000+ 000+ $+1/4.3$
0 $\frac{1}{2}$ $\frac{1}{2}$ 010− 000+ $+1/4.3$
0 $\frac{3}{2}$ $\frac{3}{2}$ 000+ 000+ 0
0 $\frac{3}{2}$ $\frac{3}{2}$ 000+ 010− $-1/4.3\sqrt{2}$
0 $\frac{3}{2}$ $\frac{3}{2}$ 010− 000+ $-1/4.3\sqrt{2}$
0 $\frac{3}{2}$ $\frac{3}{2}$ 010− 010− 0
1 $\frac{1}{2}$ $\frac{1}{2}$ 000− 000− $+1/2.3\sqrt{2.3}$
1 $\frac{1}{2}$ $\frac{1}{2}$ 010+ 000− $-1/4.3\sqrt{2.3}$
1 $\frac{1}{2}$ $\frac{3}{2}$ 000+ 000+ $+1/8.3\sqrt{3}$
1 $\frac{1}{2}$ $\frac{3}{2}$ 010− 000+ $+1/8.3\sqrt{3}$
1 $\frac{3}{2}$ $\frac{1}{2}$ 000+ 000+ $+1/8.3\sqrt{3}$
1 $\frac{3}{2}$ $\frac{1}{2}$ 000+ 010− $+1/8.3\sqrt{3}$
1 $\frac{3}{2}$ $\frac{1}{2}$ 010− 000+ $-1/4.3\sqrt{3}$
1 $\frac{3}{2}$ $\frac{1}{2}$ 010− 010− $+1/8.3\sqrt{3}$
1 $\frac{1}{2}$ $\frac{3}{2}$ 000− 000− 0
1 $\frac{3}{2}$ $\frac{3}{2}$ 000− 010+ $+1/4.3\sqrt{3}$
1 $\frac{3}{2}$ $\frac{3}{2}$ 001− 000− $+1/8.3\sqrt{3}$
1 $\frac{3}{2}$ $\frac{3}{2}$ 001− 010+ 0
1 $\frac{3}{2}$ $\frac{3}{2}$ 010+ 000− $-1/8.3\sqrt{3}$
1 $\frac{3}{2}$ $\frac{3}{2}$ 010+ 010+ 0
1 $\frac{3}{2}$ $\frac{3}{2}$ 011+ 000− 0
1 $\frac{3}{2}$ $\frac{3}{2}$ 011+ 010+ $-1/4.3\sqrt{3}$
2 $\frac{1}{2}$ $\frac{3}{2}$ 000− 000− $+1/8.3$
2 $\frac{1}{2}$ $\frac{3}{2}$ 010+ 000− $+1/8.3$
2 $\frac{3}{2}$ $\frac{1}{2}$ 000− 000− $+1/8.3$
2 $\frac{3}{2}$ $\frac{1}{2}$ 000− 010+ $+1/8.3$
2 $\frac{3}{2}$ $\frac{1}{2}$ 010+ 000− $+1/8.3$

$\tilde{1}$ $\tilde{1}$ 2 $\tilde{1}$ $\frac{3}{2}$ $\frac{3}{2}$
2 $\frac{3}{2}$ $\frac{1}{2}$ 010+ 010+ $+1/8.3$
2 $\frac{3}{2}$ $\frac{3}{2}$ 000+ 000+ $+1/8.3$
2 $\frac{3}{2}$ $\frac{3}{2}$ 000+ 010− 0
2 $\frac{3}{2}$ $\frac{3}{2}$ 010− 000+ 0
2 $\frac{3}{2}$ $\frac{3}{2}$ 010− 010− $+1/8.3$
$\tilde{1}$ $\tilde{1}$ 2 $\tilde{1}$ 2 1
0 1 1 000− 000− $+1/9$
0 $\tilde{1}$ $\tilde{1}$ 000− 000− 0
1 1 1 000+ 000+ $-1/4.9$
1 1 $\tilde{1}$ 000− 000− $+1/4.3\sqrt{3}$
1 $\tilde{1}$ 1 000− 000− $+1/4.3\sqrt{3}$
1 $\tilde{1}$ $\tilde{1}$ 000+ 000+ $+1/4.3$
2 1 1 000− 000− $+1/9$
2 1 $\tilde{1}$ 000+ 000+ 0
$\tilde{1}$ $\tilde{1}$ 2 $\tilde{1}$ 2 $\tilde{1}$
0 1 1 000+ 000+ 0
0 $\tilde{1}$ $\tilde{1}$ 000+ 000+ $+1/9$
1 1 1 000− 000− $-1/4.3$
1 1 $\tilde{1}$ 000+ 000+ $-1/4.3\sqrt{3}$
1 $\tilde{1}$ 1 000+ 000+ $-1/4.3\sqrt{3}$
1 $\tilde{1}$ $\tilde{1}$ 000− 000− $+1/4.9$
2 1 1 000+ 000+ 0
2 1 $\tilde{1}$ 000− 000− 0
2 $\tilde{1}$ 1 000− 000− 0
2 $\tilde{1}$ $\tilde{1}$ 000+ 000+ $+1/9$
$\tilde{1}$ $\tilde{1}$ 2 $\tilde{1}$ $\tilde{1}$ 0
1 1 2 000+ 000+ $-1/2.3\sqrt{2.3}$
2 0 2 000+ 000+ $+1/2.3$
2 2 2 000+ 000+ $-1/2.3\sqrt{2}$
$\tilde{1}$ $\tilde{1}$ 2 $\tilde{1}$ $\tilde{1}$ 1
1 0 1 000− 000− $+1/2.9$
1 1 1 000+ 000+ $-1/2.9$
1 1 $\tilde{1}$ 000− 000− 0
2 1 1 000− 000− $+1/4.9$
2 1 $\tilde{1}$ 000+ 000+ $+1/4.3$
$\tilde{1}$ $\tilde{1}$ 2 $\tilde{1}$ $\tilde{1}$ 2
0 0 0 000+ 000+ $+1/3\sqrt{2}$
1 1 0 000+ 000+ $-1/2.3\sqrt{2.3}$
1 1 2 000+ 000+ $-1/2.3\sqrt{3}$
2 0 2 000+ 000+ $-1/2.3\sqrt{2}$
2 2 0 000+ 000+ $+1/2.3$
2 2 2 000+ 000+ $+1/4.3$
$\tilde{1}$ $\tilde{1}$ $\tilde{1}$ 1 1 1
1 1 1 000+ 000+ $-1/2.9$
$\tilde{1}$ $\tilde{1}$ $\tilde{1}$ $\frac{3}{2}$ 1 $\frac{1}{2}$
$\frac{1}{2}$ 1 $\frac{3}{2}$ 000+ 000+ $-1/8.3$
$\frac{1}{2}$ 2 $\frac{3}{2}$ 000− 000− $-1/8.3$
$\frac{1}{2}$ $\tilde{1}$ $\frac{3}{2}$ 000− 000− $-1/8.3\sqrt{3}$

O and T_d 9j Symbols (cont.)

$\tilde{1}\ \tilde{1}\ \tilde{1}\quad \frac{3}{2}\ \frac{3}{2}\ 0$

$\frac{3}{2}\ \frac{1}{2}\ \tilde{1}$ 000− 000− $+1/4.3\sqrt{2}$

$\frac{3}{2}\ \frac{1}{2}\ \tilde{1}$ 000− 100+ 0

$\tilde{1}\ \tilde{1}\ \tilde{1}\quad \frac{3}{2}\ \frac{3}{2}\ 1$

$\frac{1}{2}\ \frac{1}{2}\ 1$ 000+ 000+ $-1/2.9\sqrt{2}$

$\frac{1}{2}\ \frac{1}{2}\ 1$ 010+ 000+ $-1/2.9\sqrt{2}$

$\frac{3}{2}\ \frac{1}{2}\ 1$ 000− 000− $-1/8.9$

$\frac{3}{2}\ \frac{1}{2}\ 1$ 000− 100+ $+1/4.9$

$\frac{3}{2}\ \frac{1}{2}\ 1$ 010− 000− $+1/4.9$

$\frac{3}{2}\ \frac{1}{2}\ 1$ 010− 100+ $-1/8.9$

$\frac{3}{2}\ \frac{1}{2}\ 2$ 000+ 000+ $+1/8.3$

$\frac{3}{2}\ \frac{1}{2}\ 2$ 000+ 100− 0

$\frac{3}{2}\ \frac{1}{2}\ 2$ 010+ 000+ $-1/8.3$

$\frac{3}{2}\ \frac{1}{2}\ 2$ 010+ 100− $-1/8.3$

$\frac{3}{2}\ \frac{1}{2}\ \tilde{1}$ 000+ 000+ $-1/8.3\sqrt{3}$

$\frac{3}{2}\ \frac{1}{2}\ \tilde{1}$ 000+ 100− $+1/4.3\sqrt{3}$

$\frac{3}{2}\ \frac{1}{2}\ \tilde{1}$ 010+ 000+ 0

$\frac{3}{2}\ \frac{1}{2}\ \tilde{1}$ 010+ 100− $+1/8.3\sqrt{3}$

$\frac{3}{2}\ \frac{3}{2}\ 1$ 000+ 000+ $-1/2.9$

$\frac{3}{2}\ \frac{3}{2}\ 1$ 000+ 100− 0

$\frac{3}{2}\ \frac{3}{2}\ 1$ 000+ 110+ $+1/4.9$

$\frac{3}{2}\ \frac{3}{2}\ 1$ 010+ 000+ 0

$\frac{3}{2}\ \frac{3}{2}\ 1$ 010+ 100− $-1/8.9$

$\frac{3}{2}\ \frac{3}{2}\ 1$ 010+ 110+ 0

$\frac{3}{2}\ \frac{3}{2}\ 1$ 011+ 000+ $-1/4.9$

$\frac{3}{2}\ \frac{3}{2}\ 1$ 011+ 100− 0

$\frac{3}{2}\ \frac{3}{2}\ 1$ 011+ 110+ $+1/2.9$

$\tilde{1}\ \tilde{1}\ \tilde{1}\quad 2\ 1\ 1$

$1\ 1\ 1$ 000− 000− 0

$1\ 2\ 1$ 000+ 000+ $-1/4.3$

$1\ \tilde{1}\ 1$ 000+ 000+ 0

$1\ \tilde{1}\ 2$ 000− 000− $+1/4.3\sqrt{3}$

$1\ \tilde{1}\ \tilde{1}$ 000− 000− 0

$\tilde{1}\ \tilde{1}\ \tilde{1}\quad 2\ \frac{3}{2}\ \frac{1}{2}$

$1\ \frac{1}{2}\ \frac{3}{2}$ 000− 000− $+1/8.3$

$1\ \frac{3}{2}\ \frac{3}{2}$ 000+ 000+ $+1/8.3$

$1\ \frac{3}{2}\ \frac{3}{2}$ 000+ 010− 0

$1\ \frac{3}{2}\ \frac{3}{2}$ 001+ 000+ $-1/8.3$

$1\ \frac{3}{2}\ \frac{3}{2}$ 001+ 010− $+1/8.3$

$\tilde{1}\ \tilde{1}\ \tilde{1}\quad 2\ \frac{3}{2}\ \frac{3}{2}$

$1\ \frac{1}{2}\ \frac{1}{2}$ 000− 000− 0

$1\ \frac{3}{2}\ \frac{1}{2}$ 000+ 000+ $+1/8.3$

$1\ \frac{3}{2}\ \frac{1}{2}$ 000+ 010− $-1/8.3$

$1\ \frac{3}{2}\ \frac{3}{2}$ 000− 000− 0

$1\ \frac{3}{2}\ \frac{3}{2}$ 000− 010+ $+1/8.3$

$1\ \frac{3}{2}\ \frac{3}{2}$ 000− 011− 0

$1\ \frac{3}{2}\ \frac{3}{2}$ 001− 000− 0

$1\ \frac{3}{2}\ \frac{3}{2}$ 001− 010+ 0

$1\ \frac{3}{2}\ \frac{3}{2}$ 001− 011− 0

$\tilde{1}\ \tilde{1}\ \tilde{1}\quad 2\ 2\ 0$

$1\ 1\ \tilde{1}$ 000+ 000+ $+1/2.3\sqrt{2.3}$

$\tilde{1}\ \tilde{1}\ \tilde{1}\quad 2\ 2\ 2$

$1\ 1\ 1$ 000− 000− 0

$\tilde{1}\ \tilde{1}\ \tilde{1}\quad \tilde{1}\ 1\ 1$

$0\ 1\ 1$ 000+ 000+ $-1/2.9$

$0\ 2\ 2$ 000+ 000+ $+1/2.3\sqrt{2.3}$

$0\ \tilde{1}\ \tilde{1}$ 000+ 000+ $+1/2.9$

$1\ 1\ 1$ 000− 000− 0

$1\ 2\ 1$ 000+ 000+ 0

$1\ \tilde{1}\ 1$ 000+ 000+ $-1/2.9$

$1\ \tilde{1}\ 2$ 000− 000− $-1/2.9$

$1\ \tilde{1}\ \tilde{1}$ 000− 000− 0

$2\ 1\ 1$ 000+ 000+ $-1/2.9$

$2\ 2\ 2$ 000+ 000+ $-1/2.3\sqrt{3}$

$2\ \tilde{1}\ 1$ 000− 000− 0

$2\ \tilde{1}\ \tilde{1}$ 000+ 000+ $+1/2.9$

$\tilde{1}\ 1\ 1$ 000+ 000+ $+1/2.9$

$\tilde{1}\ \tilde{1}\ \tilde{1}\quad \tilde{1}\ \frac{3}{2}\ \frac{1}{2}$

$0\ \frac{3}{2}\ \frac{3}{2}$ 000− 000− $-1/4.3\sqrt{2}$

$0\ \frac{3}{2}\ \frac{3}{2}$ 000− 010+ 0

$1\ \frac{1}{2}\ \frac{3}{2}$ 000− 000− $+1/8.3\sqrt{3}$

$1\ \frac{3}{2}\ \frac{3}{2}$ 000+ 000+ $-1/8.3\sqrt{3}$

$1\ \frac{3}{2}\ \frac{3}{2}$ 000+ 010− $-1/4.3\sqrt{3}$

$1\ \frac{3}{2}\ \frac{3}{2}$ 001+ 000+ 0

$1\ \frac{3}{2}\ \frac{3}{2}$ 001+ 010− $-1/8.3\sqrt{3}$

$2\ \frac{1}{2}\ \frac{3}{2}$ 000+ 000+ $-1/8.3$

$2\ \frac{3}{2}\ \frac{3}{2}$ 000− 000− $+1/8.3$

$2\ \frac{3}{2}\ \frac{3}{2}$ 000− 010+ $-1/8.3$

$\tilde{1}\ \frac{1}{2}\ \frac{3}{2}$ 000+ 000+ $+5/8.9$

$\tilde{1}\ \tilde{1}\ \tilde{1}\quad \tilde{1}\ \frac{3}{2}\ \frac{3}{2}$

$0\ \frac{1}{2}\ \frac{1}{2}$ 000+ 000+ $+1/4.3$

$0\ \frac{1}{2}\ \frac{1}{2}$ 010− 000+ 0

$0\ \frac{3}{2}\ \frac{3}{2}$ 000+ 000+ 0

$0\ \frac{3}{2}\ \frac{3}{2}$ 000+ 010− 0

$0\ \frac{3}{2}\ \frac{3}{2}$ 000+ 011+ $-1/4.3\sqrt{2}$

$0\ \frac{3}{2}\ \frac{3}{2}$ 010− 000+ 0

$0\ \frac{3}{2}\ \frac{3}{2}$ 010− 010− $-1/4.3\sqrt{2}$

$0\ \frac{3}{2}\ \frac{3}{2}$ 010− 011+ 0

$1\ \frac{1}{2}\ \frac{1}{2}$ 000− 000− 0

$1\ \frac{1}{2}\ \frac{1}{2}$ 010+ 000− $+1/2.3\sqrt{2.3}$

$1\ \frac{3}{2}\ \frac{1}{2}$ 000+ 000+ $+1/8.3\sqrt{3}$

$1\ \frac{3}{2}\ \frac{1}{2}$ 000+ 010− $+1/4.3\sqrt{3}$

$1\ \frac{3}{2}\ \frac{1}{2}$ 010− 000+ 0

$1\ \frac{3}{2}\ \frac{1}{2}$ 010− 010− $-1/8.3\sqrt{3}$

$1\ \frac{3}{2}\ \frac{3}{2}$ 000− 000− 0

$1\ \frac{3}{2}\ \frac{3}{2}$ 000− 010+ 0

$1\ \frac{3}{2}\ \frac{3}{2}$ 000− 011− 0

$1\ \frac{3}{2}\ \frac{3}{2}$ 001− 000− 0

$1\ \frac{3}{2}\ \frac{3}{2}$ 001− 010+ $-1/8.3\sqrt{3}$

$\tilde{1}\ \tilde{1}\ \tilde{1}\quad \tilde{1}\ \frac{3}{2}\ \frac{3}{2}$

$1\ \frac{3}{2}\ \frac{3}{2}$ 001− 011− 0

$1\ \frac{3}{2}\ \frac{3}{2}$ 010+ 000− 0

$1\ \frac{3}{2}\ \frac{3}{2}$ 010+ 010+ $+1/8.3\sqrt{3}$

$1\ \frac{3}{2}\ \frac{3}{2}$ 010+ 011− 0

$1\ \frac{3}{2}\ \frac{3}{2}$ 011+ 000− $-1/4.3\sqrt{3}$

$1\ \frac{3}{2}\ \frac{3}{2}$ 011+ 010+ 0

$1\ \frac{3}{2}\ \frac{3}{2}$ 011+ 011− 0

$2\ \frac{3}{2}\ \frac{1}{2}$ 000− 000− $-1/8.3$

$2\ \frac{3}{2}\ \frac{1}{2}$ 000− 010+ 0

$2\ \frac{3}{2}\ \frac{1}{2}$ 010+ 000− $-1/8.3$

$2\ \frac{3}{2}\ \frac{1}{2}$ 010+ 010+ $+1/8.3$

$2\ \frac{3}{2}\ \frac{3}{2}$ 000+ 000+ 0

$2\ \frac{3}{2}\ \frac{3}{2}$ 000+ 010− $-1/8.3$

$2\ \frac{3}{2}\ \frac{3}{2}$ 000+ 011+ 0

$2\ \frac{3}{2}\ \frac{3}{2}$ 010− 000+ 0

$2\ \frac{3}{2}\ \frac{3}{2}$ 010− 010− 0

$2\ \frac{3}{2}\ \frac{3}{2}$ 010− 011+ 0

$\tilde{1}\ \frac{3}{2}\ \frac{1}{2}$ 000− 000− $+1/8.9$

$\tilde{1}\ \frac{3}{2}\ \frac{1}{2}$ 000− 010+ $+1/4.9$

$\tilde{1}\ \frac{3}{2}\ \frac{1}{2}$ 010+ 000− $+1/4.9$

$\tilde{1}\ \frac{3}{2}\ \frac{1}{2}$ 010+ 010+ $+1/8.9$

$\tilde{1}\ \frac{3}{2}\ \frac{3}{2}$ 000+ 000+ $+1/2.9$

$\tilde{1}\ \frac{3}{2}\ \frac{3}{2}$ 000+ 010− 0

$\tilde{1}\ \frac{3}{2}\ \frac{3}{2}$ 000+ 011+ $+1/4.9$

$\tilde{1}\ \frac{3}{2}\ \frac{3}{2}$ 010− 000+ 0

$\tilde{1}\ \frac{3}{2}\ \frac{3}{2}$ 010− 010− $+1/8.9$

$\tilde{1}\ \frac{3}{2}\ \frac{3}{2}$ 010− 011+ 0

$\tilde{1}\ \frac{3}{2}\ \frac{3}{2}$ 011+ 000+ $+1/4.9$

$\tilde{1}\ \frac{3}{2}\ \frac{3}{2}$ 011+ 010− 0

$\tilde{1}\ \frac{3}{2}\ \frac{3}{2}$ 011+ 011+ $+1/2.9$

$\tilde{1}\ \tilde{1}\ \tilde{1}\quad \tilde{1}\ 2\ 1$

$0\ 1\ 1$ 000− 000− $+1/2.9$

$0\ \tilde{1}\ \tilde{1}$ 000− 000− $-1/2.3\sqrt{3}$

$1\ 1\ 1$ 000+ 000+ $-1/2.9$

$1\ 1\ 2$ 000− 000− $-1/4.3\sqrt{3}$

$1\ 1\ \tilde{1}$ 000− 000− $+1/2.9$

$1\ \tilde{1}\ 1$ 000− 000− 0

$1\ \tilde{1}\ 2$ 000+ 000+ $-1/4.3\sqrt{3}$

$1\ \tilde{1}\ \tilde{1}$ 000+ 000+ 0

$2\ 1\ 1$ 000− 000− $-1/4.9$

$2\ 1\ \tilde{1}$ 000+ 000+ $-1/4.3\sqrt{3}$

$2\ \tilde{1}\ 1$ 000+ 000+ $-1/4.3$

$2\ \tilde{1}\ \tilde{1}$ 000− 000− $+1/4.3\sqrt{3}$

$\tilde{1}\ 1\ 1$ 000− 000− $+1/2.9$

$\tilde{1}\ 1\ 2$ 000+ 000+ $+1/4.9$

$\tilde{1}\ 1\ \tilde{1}$ 000+ 000+ $+1/2.9$

$\tilde{1}\ \tilde{1}\ \tilde{1}\quad \tilde{1}\ \tilde{1}\ 0$

$1\ 1\ \tilde{1}$ 000+ 000+ $+1/2.9$

$2\ 1\ \tilde{1}$ 000− 000− $+1/2.3\sqrt{3}$

O and T_d 9j Symbols (cont.)

$\tilde{1}\ \tilde{1}\ \tilde{1}\quad \tilde{1}\ \tilde{1}\ 0$			
$\tilde{1}\ 0\ \tilde{1}$	000+	000+	$+1/9$
$\tilde{1}\ 1\ \tilde{1}$	000−	000−	$-1/2.9$
$\tilde{1}\ 2\ \tilde{1}$	000+	000+	$-1/2.9$
$\tilde{1}\ \tilde{1}\ \tilde{1}\quad \tilde{1}\ \tilde{1}\ 1$			
$1\ 0\ 1$	000−	000−	$-1/2.9$
$1\ 1\ 1$	000+	000+	$+1/2.9$
$1\ 1\ 2$	000−	000−	0
$1\ 1\ \tilde{1}$	000−	000−	0
$2\ 0\ 2$	000−	000−	$+1/2.3\sqrt{2}$
$2\ 1\ 1$	000−	000−	$+1/2.9$
$2\ 1\ \tilde{1}$	000+	000+	0
$2\ 2\ 2$	000−	000−	0
$\tilde{1}\ 0\ \tilde{1}$	000−	000−	$-1/2.9$
$\tilde{1}\ 1\ 1$	000−	000−	0
$\tilde{1}\ 1\ 2$	000+	000+	$+1/2.9$
$\tilde{1}\ 1\ \tilde{1}$	000+	000+	$+1/2.9$
$\tilde{1}\ 2\ 1$	000+	000+	0
$\tilde{1}\ 2\ \tilde{1}$	000−	000−	$-1/2.9$
$\tilde{1}\ \tilde{1}\ 1$	000+	000+	$-1/2.9$
$\tilde{1}\ \tilde{1}\ \tilde{1}\quad \tilde{1}\ \tilde{1}\ 2$			
$1\ 0\ 1$	000+	000+	$+1/2.3\sqrt{3}$
$1\ 1\ 1$	000−	000−	0
$1\ 1\ \tilde{1}$	000+	000+	$+1/2.9$
$2\ 1\ 1$	000+	000+	$-1/4.3\sqrt{3}$
$2\ 1\ \tilde{1}$	000−	000−	$-1/4.3\sqrt{3}$
$\tilde{1}\ 0\ \tilde{1}$	000+	000+	$-1/2.9$
$\tilde{1}\ 1\ 1$	000+	000+	0
$\tilde{1}\ 1\ \tilde{1}$	000−	000−	$-1/2.9$
$\tilde{1}\ 2\ 1$	000−	000−	$+1/4.3$
$\tilde{1}\ 2\ \tilde{1}$	000+	000+	$+1/4.9$
$\tilde{1}\ \tilde{1}\ 1$	000−	000−	0
$\tilde{1}\ \tilde{1}\ \tilde{1}\quad \tilde{1}\ \tilde{1}\ \tilde{1}$			
$0\ 0\ 0$	000+	000+	$+1/3\sqrt{3}$
$1\ 1\ 0$	000+	000+	$-1/2.9$
$1\ 1\ 1$	000−	000−	0
$2\ 1\ 1$	000+	000+	$-1/2.9$
$2\ 2\ 0$	000+	000+	$-1/2.3\sqrt{2.3}$
$2\ 2\ 2$	000+	000+	$-1/2.3\sqrt{3}$
$\tilde{1}\ 1\ 1$	000+	000+	$-1/2.9$
$\tilde{1}\ 2\ 1$	000−	000−	0
$\tilde{1}\ \tilde{1}\ 0$	000+	000+	$+1/2.9$
$\tilde{1}\ \tilde{1}\ 1$	000−	000−	0
$\tilde{1}\ \tilde{1}\ 2$	000+	000+	$+1/2.9$
$\tilde{1}\ \tilde{1}\ \tilde{1}$	000+	000+	$+1/2.9$
$\tilde{\frac{1}{2}}\ \frac{3}{2}\ 1\quad \frac{3}{2}\ 0\ \frac{3}{2}$			
$1\ \frac{3}{2}\ \frac{1}{2}$	000−	000−	$+1/8.2$
$1\ \frac{3}{2}\ \frac{3}{2}$	000+	000+	$+1/8.3$
$1\ \frac{3}{2}\ \frac{3}{2}$	000+	001+	$+1/8.3$
$1\ \frac{3}{2}\ \frac{3}{2}$	001+	000+	$+1/8.3$
$1\ \frac{3}{2}\ \frac{3}{2}$	001+	001+	$-1/8.2.3$
$1\ \frac{3}{2}\ \tilde{\frac{1}{2}}$	000−	000−	$+1/8.2.3$
$\tilde{\frac{1}{2}}\ \frac{3}{2}\ 1\quad \frac{3}{2}\ \frac{1}{2}\ 1$			
$1\ 1\ 0$	000+	000+	$+1/4.3$
$1\ 1\ 1$	000−	000−	$+1/8.3$
$1\ 1\ 2$	000+	000+	$-1/8.3$
$1\ 1\ \tilde{1}$	000+	000+	$+1/8.3$
$\tilde{\frac{1}{2}}\ \frac{3}{2}\ 1\quad \frac{3}{2}\ \frac{1}{2}\ 2$			
$1\ 1\ 1$	000+	000+	$+1/8\sqrt{3}$
$1\ 1\ \tilde{1}$	000−	000−	$+1/8.3$
$1\ 2\ 1$	000−	000−	0
$1\ 2\ \tilde{1}$	000+	000+	$+1/4.3$
$\tilde{\frac{1}{2}}\ \frac{3}{2}\ 1\quad \frac{3}{2}\ \frac{1}{2}\ \tilde{1}$			
$1\ 1\ 1$	000+	000+	$-1/8.3\sqrt{3}$
$1\ 1\ 2$	000−	000−	$-1/8.3$
$1\ 1\ \tilde{1}$	000−	000−	$+1/8.3\sqrt{3}$
$1\ 2\ 1$	000−	000−	$+1/8.3$
$1\ 2\ \tilde{1}$	000+	000+	$-1/8.3\sqrt{3}$
$1\ \tilde{1}\ 1$	000−	000−	$+5/8.9$
$1\ \tilde{1}\ 2$	000+	000+	$+5/8.9$
$1\ \tilde{1}\ \tilde{1}$	000+	000+	$-1/8.3$
$\tilde{\frac{1}{2}}\ \frac{3}{2}\ 1\quad \frac{3}{2}\ 1\ \frac{1}{2}$			
$1\ \frac{1}{2}\ \frac{1}{2}$	000+	000+	$+1/4.3$
$1\ \frac{1}{2}\ \frac{3}{2}$	000−	000−	$+1/8.2.3$
$\tilde{\frac{1}{2}}\ \frac{3}{2}\ 1\quad \frac{3}{2}\ 1\ \frac{3}{2}$			
$1\ \frac{1}{2}\ \frac{1}{2}$	000−	000−	$-1/4.3\sqrt{2}$
$1\ \frac{1}{2}\ \frac{1}{2}$	010−	000−	0
$1\ \frac{1}{2}\ \frac{3}{2}$	000+	000+	0
$1\ \frac{1}{2}\ \frac{3}{2}$	000+	001+	$-1/8.3$
$1\ \frac{1}{2}\ \frac{3}{2}$	010+	000+	$+1/8.3$
$1\ \frac{1}{2}\ \frac{3}{2}$	010+	001+	$-1/8.2.3$
$1\ \frac{3}{2}\ \frac{1}{2}$	000+	000+	$-1/8.2.3$
$1\ \frac{3}{2}\ \frac{1}{2}$	000+	010+	0
$1\ \frac{3}{2}\ \frac{1}{2}$	010+	000+	0
$1\ \frac{3}{2}\ \frac{1}{2}$	010+	010+	$+1/8.2$
$1\ \frac{3}{2}\ \frac{3}{2}$	000−	000−	$+1/8.9$
$1\ \frac{3}{2}\ \frac{3}{2}$	000−	001−	$+1/8.9$
$1\ \frac{3}{2}\ \frac{3}{2}$	000−	010−	$-1/4.9$
$1\ \frac{3}{2}\ \frac{3}{2}$	000−	011−	$+1/8.9$
$1\ \frac{3}{2}\ \frac{3}{2}$	001−	000−	$+1/8.9$
$1\ \frac{3}{2}\ \frac{3}{2}$	001−	001−	$+5/8.2.9$
$1\ \frac{3}{2}\ \frac{3}{2}$	001−	010−	$-1/4.9$
$1\ \frac{3}{2}\ \frac{3}{2}$	001−	011−	$+1/8.9$
$1\ \frac{3}{2}\ \frac{3}{2}$	010−	000−	$-1/4.9$
$1\ \frac{3}{2}\ \frac{3}{2}$	010−	001−	$-1/4.9$
$1\ \frac{3}{2}\ \frac{3}{2}$	010−	010−	$+1/8.9$
$1\ \frac{3}{2}\ \frac{3}{2}$	010−	011−	$+1/8.9$
$1\ \frac{3}{2}\ \frac{3}{2}$	011−	000−	$+1/8.9$
$1\ \frac{3}{2}\ \frac{3}{2}$	011−	001−	$+1/8.9$
$\tilde{\frac{1}{2}}\ \frac{3}{2}\ 1\quad \frac{3}{2}\ 1\ \frac{3}{2}$			
$1\ \frac{3}{2}\ \frac{3}{2}$	011−	010−	$+1/8.9$
$1\ \frac{3}{2}\ \frac{3}{2}$	011−	011−	$+1/8.9$
$1\ \frac{3}{2}\ \tilde{\frac{1}{2}}$	000+	000+	$+7/8.2.9$
$1\ \frac{3}{2}\ \tilde{\frac{1}{2}}$	000+	010+	$+1/4.9$
$1\ \frac{3}{2}\ \tilde{\frac{1}{2}}$	010+	000+	$+1/4.9$
$1\ \frac{3}{2}\ \tilde{\frac{1}{2}}$	010+	010+	$+1/8.2.9$
$\tilde{\frac{1}{2}}\ \frac{3}{2}\ 1\quad \frac{3}{2}\ 1\ \tilde{\frac{1}{2}}$			
$1\ \frac{1}{2}\ \frac{3}{2}$	000−	000−	$+1/8.2.3$
$1\ \frac{3}{2}\ \frac{3}{2}$	000+	000+	$+1/4.9$
$1\ \frac{3}{2}\ \frac{3}{2}$	000+	010+	$-1/8.9$
$1\ \frac{3}{2}\ \frac{3}{2}$	001+	000+	$-1/8.9$
$1\ \frac{3}{2}\ \frac{3}{2}$	001+	010+	$+7/8.2.9$
$1\ \frac{3}{2}\ \tilde{\frac{1}{2}}$	000−	000−	$+1/4.9\sqrt{2}$
$1\ \frac{3}{2}\ \tilde{\frac{1}{2}}$	000−	010−	$-1/2.9\sqrt{2}$
$1\ \tilde{\frac{1}{2}}\ \frac{3}{2}$	000−	000−	$+11/8.2.9$
$1\ \tilde{\frac{1}{2}}\ \tilde{\frac{1}{2}}$	000+	000+	$-1/4.9$
$\tilde{\frac{1}{2}}\ \frac{3}{2}\ 1\quad \frac{3}{2}\ \frac{3}{2}\ 0$			
$1\ 0\ 1$	000+	000+	$+1/4.3$
$\tilde{\frac{1}{2}}\ \frac{3}{2}\ 1\quad \frac{3}{2}\ \frac{3}{2}\ 1$			
$1\ 0\ 1$	000−	000−	$-1/4.3\sqrt{2.3}$
$1\ 0\ 1$	010−	000−	$+1/2.3\sqrt{2.3}$
$1\ 1\ 0$	000−	000−	$+1/2.9$
$1\ 1\ 0$	000−	010−	$+1/2.9$
$1\ 1\ 0$	010−	000−	$+1/2.9$
$1\ 1\ 0$	010−	010−	$-1/4.9$
$1\ 1\ 1$	000+	000+	$+1/8.3$
$1\ 1\ 1$	000+	010+	0
$1\ 1\ 1$	010+	000+	0
$1\ 1\ 1$	010+	010+	$+1/8.3$
$1\ 1\ 2$	000−	000−	$+1/8.9$
$1\ 1\ 2$	000−	010−	$+1/8.9$
$1\ 1\ 2$	010−	000−	$+1/8.9$
$1\ 1\ 2$	010−	010−	$+1/2.9$
$1\ 1\ \tilde{1}$	000−	000−	$+1/8.9$
$1\ 1\ \tilde{1}$	000−	010−	$-1/4.9$
$1\ 1\ \tilde{1}$	010−	000−	$-1/4.9$
$1\ 1\ \tilde{1}$	010−	010−	$+1/8.9$
$\tilde{\frac{1}{2}}\ \frac{3}{2}\ 1\quad \frac{3}{2}\ \frac{3}{2}\ 2$			
$1\ 0\ 1$	000+	000+	$+1/4.3\sqrt{2}$
$1\ 1\ 1$	000−	000−	$-1/8.3\sqrt{3}$
$1\ 1\ 1$	000−	010−	$-1/8.3\sqrt{3}$
$1\ 1\ \tilde{1}$	000+	000+	$-1/8.3$
$1\ 1\ \tilde{1}$	000+	010+	$+1/8.3$
$1\ 2\ 1$	000+	000+	$+1/8.3$
$1\ 2\ \tilde{1}$	000−	000−	$+1/8.3$
$\tilde{\frac{1}{2}}\ \frac{3}{2}\ 1\quad \frac{3}{2}\ \frac{3}{2}\ \tilde{1}$			
$1\ 0\ 1$	000+	000+	$-1/4.3\sqrt{2}$
$1\ 0\ 1$	010−	000+	0
$1\ 1\ 1$	000−	000−	$-1/8.3\sqrt{3}$

O and T_d 9j Symbols (cont.)

$\tilde{\frac{1}{2}}$	$\frac{3}{2}$	1	$\frac{3}{2}$	$\frac{3}{2}$	$\tilde{1}$
1	1	1	000−	010−	0
1	1	1	010+	000−	$-1/4.3\sqrt{3}$
1	1	1	010+	010−	$-1/8.3\sqrt{3}$
1	1	2	000+	000+	+1/8.3
1	1	2	000+	010+	−1/8.3
1	1	2	010−	000+	−1/8.3
1	1	2	010−	010+	0
1	1	$\tilde{1}$	000+	000+	$-1/8.3\sqrt{3}$
1	1	$\tilde{1}$	000+	010+	$-1/4.3\sqrt{3}$
1	1	$\tilde{1}$	010−	000+	0
1	1	$\tilde{1}$	010−	010+	$-1/8.3\sqrt{3}$
1	2	1	000+	000+	+1/8.3
1	2	1	010−	000+	−1/8.3
1	2	$\tilde{1}$	000−	000−	$+1/8.3\sqrt{3}$
1	2	$\tilde{1}$	010+	000−	$+1/8.3\sqrt{3}$
1	$\tilde{1}$	1	000+	000+	+1/8.9
1	$\tilde{1}$	1	000+	010−	−1/4.9
1	$\tilde{1}$	1	010−	000+	−1/4.9
1	$\tilde{1}$	1	010−	010−	+1/8.9
1	$\tilde{1}$	2	000−	000−	+1/8.9
1	$\tilde{1}$	2	000−	010+	+1/8.9
1	$\tilde{1}$	2	010+	000−	+1/8.9
1	$\tilde{1}$	2	010+	010+	+1/2.9
1	$\tilde{1}$	$\tilde{1}$	000−	000−	+1/8.3
1	$\tilde{1}$	$\tilde{1}$	000−	010+	0
1	$\tilde{1}$	$\tilde{1}$	010+	000−	0
1	$\tilde{1}$	$\tilde{1}$	010+	010+	+1/8.3

$\tilde{\frac{1}{2}}$	$\frac{3}{2}$	1	$\frac{3}{2}$	2	$\frac{1}{2}$
1	$\frac{1}{2}$	$\frac{1}{2}$	000−	000−	+1/4.3
1	$\frac{1}{2}$	$\frac{3}{2}$	000+	000+	+1/8.2.3

$\tilde{\frac{1}{2}}$	$\frac{3}{2}$	1	$\frac{3}{2}$	2	$\frac{3}{2}$
1	$\frac{1}{2}$	$\frac{1}{2}$	000+	000+	$+1/4.3\sqrt{2}$
1	$\frac{1}{2}$	$\frac{3}{2}$	000−	000−	+1/8.3
1	$\frac{1}{2}$	$\frac{3}{2}$	000−	001−	+1/8.2.3
1	$\frac{3}{2}$	$\frac{1}{2}$	000−	000−	−1/8.2.3
1	$\frac{3}{2}$	$\frac{3}{2}$	000+	000+	+1/8.3
1	$\frac{3}{2}$	$\frac{3}{2}$	000+	001+	0
1	$\frac{3}{2}$	$\frac{3}{2}$	001+	000+	0
1	$\frac{3}{2}$	$\frac{3}{2}$	001+	001+	+1/8.3
1	$\frac{3}{2}$	$\tilde{\frac{1}{2}}$	000−	000−	−1/8.2.3

$\tilde{\frac{1}{2}}$	$\frac{3}{2}$	1	$\frac{3}{2}$	2	$\tilde{\frac{1}{2}}$
1	$\frac{1}{2}$	$\frac{3}{2}$	000+	000+	+1/8.2
1	$\frac{3}{2}$	$\frac{3}{2}$	000−	000−	−1/8.3
1	$\frac{3}{2}$	$\frac{3}{2}$	001−	000−	+1/8.2.3
1	$\frac{3}{2}$	$\tilde{\frac{1}{2}}$	000+	000+	$+1/4.3\sqrt{2}$
1	$\tilde{\frac{1}{2}}$	$\frac{3}{2}$	000+	000+	+1/8.2.3
1	$\tilde{\frac{1}{2}}$	$\tilde{\frac{1}{2}}$	000−	000−	+1/4.3

$\tilde{\frac{1}{2}}$	$\frac{3}{2}$	1	$\frac{3}{2}$	$\tilde{1}$	$\frac{1}{2}$
1	$\frac{1}{2}$	$\frac{1}{2}$	000−	000−	−1/4.9
1	$\frac{1}{2}$	$\frac{3}{2}$	000+	000+	+11/8.2.9

$\tilde{\frac{1}{2}}$	$\frac{3}{2}$	1	$\frac{3}{2}$	$\tilde{1}$	$\frac{3}{2}$
1	$\frac{1}{2}$	$\frac{1}{2}$	000+	000+	$-1/4.9\sqrt{2}$
1	$\frac{1}{2}$	$\frac{1}{2}$	010−	000+	$-1/2.9\sqrt{2}$
1	$\frac{1}{2}$	$\frac{3}{2}$	000−	000−	+1/4.9
1	$\frac{1}{2}$	$\frac{3}{2}$	000−	001−	+1/8.9
1	$\frac{1}{2}$	$\frac{3}{2}$	010+	000−	+1/8.9
1	$\frac{1}{2}$	$\frac{3}{2}$	010+	001−	+7/8.2.9
1	$\frac{3}{2}$	$\frac{1}{2}$	000−	000−	+7/8.2.9
1	$\frac{3}{2}$	$\frac{1}{2}$	000−	010+	−1/4.9
1	$\frac{3}{2}$	$\frac{1}{2}$	010+	000−	−1/4.9
1	$\frac{3}{2}$	$\frac{1}{2}$	010+	010+	+1/8.2.9
1	$\frac{3}{2}$	$\frac{3}{2}$	000+	000+	+1/8.9
1	$\frac{3}{2}$	$\frac{3}{2}$	000+	001+	−1/8.9
1	$\frac{3}{2}$	$\frac{3}{2}$	000+	010−	+1/4.9
1	$\frac{3}{2}$	$\frac{3}{2}$	000+	011−	+1/8.9
1	$\frac{3}{2}$	$\frac{3}{2}$	001+	000+	−1/8.9
1	$\frac{3}{2}$	$\frac{3}{2}$	001+	001+	+5/8.2.9
1	$\frac{3}{2}$	$\frac{3}{2}$	001+	010−	−1/4.9
1	$\frac{3}{2}$	$\frac{3}{2}$	001+	011−	−1/8.9
1	$\frac{3}{2}$	$\frac{3}{2}$	010−	000+	+1/4.9
1	$\frac{3}{2}$	$\frac{3}{2}$	010−	001+	−1/4.9
1	$\frac{3}{2}$	$\frac{3}{2}$	010−	010−	+1/8.9
1	$\frac{3}{2}$	$\frac{3}{2}$	010−	011−	−1/8.9
1	$\frac{3}{2}$	$\frac{3}{2}$	011−	000+	+1/8.9
1	$\frac{3}{2}$	$\frac{3}{2}$	011−	001+	−1/8.9
1	$\frac{3}{2}$	$\frac{3}{2}$	011−	010−	−1/8.9
1	$\frac{3}{2}$	$\frac{3}{2}$	011−	011−	+1/8.9
1	$\frac{3}{2}$	$\tilde{\frac{1}{2}}$	000−	000−	−1/8.2.3
1	$\frac{3}{2}$	$\tilde{\frac{1}{2}}$	000−	010+	0
1	$\frac{3}{2}$	$\tilde{\frac{1}{2}}$	010+	000−	0
1	$\frac{3}{2}$	$\tilde{\frac{1}{2}}$	010+	010+	+1/8.2

$\tilde{\frac{1}{2}}$	$\frac{3}{2}$	1	$\frac{3}{2}$	$\tilde{1}$	$\tilde{\frac{1}{2}}$
1	$\frac{1}{2}$	$\frac{3}{2}$	000+	000+	+1/8.2.3
1	$\frac{3}{2}$	$\frac{3}{2}$	000−	000−	0
1	$\frac{3}{2}$	$\frac{3}{2}$	000−	010+	−1/8.3
1	$\frac{3}{2}$	$\frac{3}{2}$	001−	000−	+1/8.3
1	$\frac{3}{2}$	$\frac{3}{2}$	001−	010+	−1/8.2.3
1	$\frac{3}{2}$	$\tilde{\frac{1}{2}}$	000+	000+	$+1/4.3\sqrt{2}$
1	$\frac{3}{2}$	$\tilde{\frac{1}{2}}$	000+	010−	0
1	$\tilde{\frac{1}{2}}$	$\frac{3}{2}$	000+	000+	+1/8.2.3
1	$\tilde{\frac{1}{2}}$	$\tilde{\frac{1}{2}}$	000−	000−	+1/4.3

$\tilde{\frac{1}{2}}$	$\frac{3}{2}$	1	$\frac{3}{2}$	$\tilde{\frac{1}{2}}$	1
1	1	0	000+	000+	+1/4.9
1	1	1	000−	000−	−1/8.3
1	1	2	000+	000+	+5/8.9
1	1	$\tilde{1}$	000+	000+	+5/8.9

$\tilde{\frac{1}{2}}$	$\frac{3}{2}$	1	$\frac{3}{2}$	$\tilde{\frac{1}{2}}$	2
1	1	1	000+	000+	$-1/8.3\sqrt{3}$
1	1	$\tilde{1}$	000−	000−	+1/8.3
1	2	1	000−	000−	+1/4.3
1	2	$\tilde{1}$	000+	000+	0

$\tilde{\frac{1}{2}}$	$\frac{3}{2}$	1	$\frac{3}{2}$	$\tilde{\frac{1}{2}}$	$\tilde{1}$
1	1	1	000+	000+	$+1/8.3\sqrt{3}$
1	1	2	000−	000−	−1/8.3
1	1	$\tilde{1}$	000−	000−	$-1/8.3\sqrt{3}$
1	2	1	000−	000−	+1/8.3
1	2	$\tilde{1}$	000+	000+	$+1/8\sqrt{3}$
1	$\tilde{1}$	1	000−	000−	+1/8.3
1	$\tilde{1}$	2	000+	000+	−1/8.3
1	$\tilde{1}$	$\tilde{1}$	000+	000+	+1/8.3

$\tilde{\frac{1}{2}}$	2	$\frac{3}{2}$	$\frac{3}{2}$	0	$\frac{3}{2}$
1	2	1	000−	000−	$+1/8\sqrt{3}$
1	2	1	000−	001−	$+1/8\sqrt{3}$
1	2	$\tilde{1}$	000+	000+	$+1/8\sqrt{3}$
1	2	$\tilde{1}$	000+	001−	$-1/8\sqrt{3}$

$\tilde{\frac{1}{2}}$	2	$\frac{3}{2}$	$\frac{3}{2}$	$\frac{1}{2}$	1
1	$\frac{3}{2}$	$\frac{1}{2}$	000+	000+	+1/8.2
1	$\frac{3}{2}$	$\frac{3}{2}$	000−	000−	0
1	$\frac{3}{2}$	$\frac{3}{2}$	000−	001−	0
1	$\frac{3}{2}$	$\frac{3}{2}$	001−	000−	0
1	$\frac{3}{2}$	$\frac{3}{2}$	001−	001−	+1/8.2
1	$\frac{3}{2}$	$\tilde{\frac{1}{2}}$	000+	000+	+1/8.2

$\tilde{\frac{1}{2}}$	2	$\frac{3}{2}$	$\frac{3}{2}$	$\frac{1}{2}$	2
1	$\frac{3}{2}$	$\frac{1}{2}$	000−	000−	+1/8.2
1	$\frac{3}{2}$	$\frac{3}{2}$	000+	000+	0
1	$\frac{3}{2}$	$\frac{3}{2}$	001+	000+	+1/8.2
1	$\frac{3}{2}$	$\tilde{\frac{1}{2}}$	000−	000−	−1/8.2

$\tilde{\frac{1}{2}}$	2	$\frac{3}{2}$	$\frac{3}{2}$	$\frac{1}{2}$	$\tilde{1}$
1	$\frac{3}{2}$	$\frac{1}{2}$	000−	000−	+1/8.2.3
1	$\frac{3}{2}$	$\frac{3}{2}$	000+	000+	−1/4.3
1	$\frac{3}{2}$	$\frac{3}{2}$	000+	001−	0
1	$\frac{3}{2}$	$\frac{3}{2}$	001+	000+	0
1	$\frac{3}{2}$	$\frac{3}{2}$	001+	001−	−1/8.2.3
1	$\frac{3}{2}$	$\tilde{\frac{1}{2}}$	000−	000−	+1/8.2.3

$\tilde{\frac{1}{2}}$	2	$\frac{3}{2}$	$\frac{3}{2}$	1	$\frac{1}{2}$
1	1	1	000+	000+	$-1/8\sqrt{3}$
1	1	2	000−	000−	0
1	1	$\tilde{1}$	000−	000−	+1/8.3
1	$\tilde{1}$	1	000−	000−	+1/8.3
1	$\tilde{1}$	2	000+	000+	0
1	$\tilde{1}$	$\tilde{1}$	000+	000+	$-1/8\sqrt{3}$

$\tilde{\frac{1}{2}}$	2	$\frac{3}{2}$	$\frac{3}{2}$	1	$\frac{3}{2}$
1	1	0	000+	000+	$+1/4.3\sqrt{2}$
1	1	0	010+	000+	$+1/4.3\sqrt{2}$
1	1	1	000−	000−	$-1/8.3\sqrt{3}$
1	1	1	000−	001−	$+1/4.3\sqrt{3}$
1	1	1	010−	000−	$-1/8.3\sqrt{3}$

O and T_d 9j Symbols (cont.)

$\tilde{\frac{1}{2}}$ 2 $\frac{3}{2}$ $\frac{3}{2}$ 1 $\frac{3}{2}$
1 1 1 010− 001− $-1/8.3\sqrt{3}$
1 1 2 000+ 000+ $+1/8.3$
1 1 2 010+ 000+ $+1/8.3$
1 1 $\tilde{1}$ 000+ 000+ $+1/8.3$
1 1 $\tilde{1}$ 000+ 001− 0
1 1 $\tilde{1}$ 010+ 000+ $-1/8.3$
1 1 $\tilde{1}$ 010+ 001− $+1/8.3$
1 $\tilde{1}$ 1 000+ 000+ $+1/8.3$
1 $\tilde{1}$ 1 000+ 001+ 0
1 $\tilde{1}$ 1 010+ 000+ $-1/8.3$
1 $\tilde{1}$ 1 010+ 001+ $-1/8.3$
1 $\tilde{1}$ 2 000− 000− $-1/8.3$
1 $\tilde{1}$ 2 010− 000− $-1/8.3$
1 $\tilde{1}$ $\tilde{1}$ 000− 000− $-1/8.3\sqrt{3}$
1 $\tilde{1}$ $\tilde{1}$ 000− 001+ $-1/4.3\sqrt{3}$
1 $\tilde{1}$ $\tilde{1}$ 010− 000− $-1/8.3\sqrt{3}$
1 $\tilde{1}$ $\tilde{1}$ 010− 001+ $+1/8.3\sqrt{3}$
$\tilde{\frac{1}{2}}$ 2 $\frac{3}{2}$ $\frac{3}{2}$ 1 $\tilde{\frac{1}{2}}$
1 1 1 000+ 000+ $-1/8.3\sqrt{3}$
1 1 2 000− 000− $+1/4.3$
1 1 $\tilde{1}$ 000− 000− $-1/8.3$
1 $\tilde{1}$ 1 000− 000− $-1/8.3$
1 $\tilde{1}$ 2 000+ 000+ $+1/4.3$
1 $\tilde{1}$ $\tilde{1}$ 000+ 000+ $-1/8.3\sqrt{3}$
$\tilde{\frac{1}{2}}$ 2 $\frac{3}{2}$ $\frac{3}{2}$ $\frac{3}{2}$ 0
1 $\frac{1}{2}$ $\frac{3}{2}$ 000+ 000+ $+1/8.2$
1 $\frac{3}{2}$ $\frac{3}{2}$ 000− 000− 0
1 $\frac{3}{2}$ $\frac{3}{2}$ 001− 000− $+1/8.2$
1 $\tilde{\frac{1}{2}}$ $\frac{3}{2}$ 000+ 000+ $+1/8.2$
$\tilde{\frac{1}{2}}$ 2 $\frac{3}{2}$ $\frac{3}{2}$ $\frac{3}{2}$ 1
1 $\frac{1}{2}$ $\frac{1}{2}$ 000+ 000+ $+1/4.3\sqrt{2}$
1 $\frac{1}{2}$ $\frac{1}{2}$ 010+ 000+ $+1/4.3\sqrt{2}$
1 $\frac{1}{2}$ $\frac{3}{2}$ 000− 000− $+1/8.2.3$
1 $\frac{1}{2}$ $\frac{3}{2}$ 000− 001− $+1/8.3$
1 $\frac{1}{2}$ $\frac{3}{2}$ 010− 000− $-1/8.3$
1 $\frac{1}{2}$ $\frac{3}{2}$ 010− 001− $-1/8.2.3$
1 $\frac{3}{2}$ $\frac{1}{2}$ 000− 000− $-1/8.3$
1 $\frac{3}{2}$ $\frac{1}{2}$ 010− 000− $+1/8.2.3$
1 $\frac{3}{2}$ $\frac{3}{2}$ 000+ 000+ 0
1 $\frac{3}{2}$ $\frac{3}{2}$ 000+ 001+ $-1/8.3$
1 $\frac{3}{2}$ $\frac{3}{2}$ 001+ 000+ $+1/8.2.3$
1 $\frac{3}{2}$ $\frac{3}{2}$ 001+ 001+ 0
1 $\frac{3}{2}$ $\frac{3}{2}$ 010+ 000+ 0
1 $\frac{3}{2}$ $\frac{3}{2}$ 010+ 001+ $-1/8.3$
1 $\frac{3}{2}$ $\frac{3}{2}$ 011+ 000+ $-1/8.3$
1 $\frac{3}{2}$ $\frac{3}{2}$ 011+ 001+ 0
1 $\frac{3}{2}$ $\tilde{\frac{1}{2}}$ 000− 000− $+1/8.3$
1 $\frac{3}{2}$ $\tilde{\frac{1}{2}}$ 010− 000− $-1/8.2.3$
1 $\tilde{\frac{1}{2}}$ $\frac{3}{2}$ 000− 000− $+1/8.2.3$

$\tilde{\frac{1}{2}}$ 2 $\frac{3}{2}$ $\frac{3}{2}$ $\frac{3}{2}$ 1
1 $\tilde{\frac{1}{2}}$ $\frac{3}{2}$ 000− 001− $-1/8.3$
1 $\tilde{\frac{1}{2}}$ $\frac{3}{2}$ 010− 000− $-1/8.3$
1 $\tilde{\frac{1}{2}}$ $\frac{3}{2}$ 010− 001− $+1/8.2.3$
1 $\tilde{\frac{1}{2}}$ $\tilde{\frac{1}{2}}$ 000+ 000+ $-1/4.3\sqrt{2}$
1 $\tilde{\frac{1}{2}}$ $\tilde{\frac{1}{2}}$ 010+ 000+ $-1/4.3\sqrt{2}$
$\tilde{\frac{1}{2}}$ 2 $\frac{3}{2}$ $\frac{3}{2}$ $\frac{3}{2}$ 2
1 $\frac{1}{2}$ $\frac{1}{2}$ 000− 000− 0
1 $\frac{1}{2}$ $\frac{3}{2}$ 000+ 000+ $-1/8.2$
1 $\frac{3}{2}$ $\frac{1}{2}$ 000+ 000+ $+1/8.2$
1 $\frac{3}{2}$ $\frac{3}{2}$ 000− 000− 0
1 $\frac{3}{2}$ $\frac{3}{2}$ 001− 000− 0
1 $\frac{3}{2}$ $\tilde{\frac{1}{2}}$ 000+ 000+ $+1/8.2$
1 $\tilde{\frac{1}{2}}$ $\frac{3}{2}$ 000+ 000+ $+1/8.2$
1 $\tilde{\frac{1}{2}}$ $\tilde{\frac{1}{2}}$ 000− 000− 0
$\tilde{\frac{1}{2}}$ 2 $\frac{3}{2}$ $\frac{3}{2}$ $\frac{3}{2}$ $\tilde{1}$
1 $\frac{1}{2}$ $\frac{1}{2}$ 000− 000− $+1/4.3\sqrt{2}$
1 $\frac{1}{2}$ $\frac{1}{2}$ 010+ 000− $+1/4.3\sqrt{2}$
1 $\frac{1}{2}$ $\frac{3}{2}$ 000+ 000+ $+1/8.2.3$
1 $\frac{1}{2}$ $\frac{3}{2}$ 000+ 001− $+1/8.3$
1 $\frac{1}{2}$ $\frac{3}{2}$ 010− 000+ $-1/8.3$
1 $\frac{1}{2}$ $\frac{3}{2}$ 010− 001− $-1/8.2.3$
1 $\frac{3}{2}$ $\frac{1}{2}$ 000+ 000+ $+1/8.3$
1 $\frac{3}{2}$ $\frac{1}{2}$ 010− 000+ $-1/8.2.3$
1 $\frac{3}{2}$ $\frac{3}{2}$ 000− 000− 0
1 $\frac{3}{2}$ $\frac{3}{2}$ 000− 001+ $+1/8.3$
1 $\frac{3}{2}$ $\frac{3}{2}$ 001− 000− $-1/8.2.3$
1 $\frac{3}{2}$ $\frac{3}{2}$ 001− 001+ 0
1 $\frac{3}{2}$ $\frac{3}{2}$ 010+ 000− 0
1 $\frac{3}{2}$ $\frac{3}{2}$ 010+ 001+ $+1/8.3$
1 $\frac{3}{2}$ $\frac{3}{2}$ 011+ 000 $+1/8.3$
1 $\frac{3}{2}$ $\frac{3}{2}$ 011+ 001+ 0
1 $\frac{3}{2}$ $\tilde{\frac{1}{2}}$ 000+ 000+ $-1/8.3$
1 $\frac{3}{2}$ $\tilde{\frac{1}{2}}$ 010− 000+ $+1/8.2.3$
1 $\tilde{\frac{1}{2}}$ $\frac{3}{2}$ 000+ 000+ $+1/8.2.3$
1 $\tilde{\frac{1}{2}}$ $\frac{3}{2}$ 000+ 001− $-1/8.3$
1 $\tilde{\frac{1}{2}}$ $\frac{3}{2}$ 010− 000+ $-1/8.3$
1 $\tilde{\frac{1}{2}}$ $\frac{3}{2}$ 010− 001− $+1/8.2.3$
1 $\tilde{\frac{1}{2}}$ $\tilde{\frac{1}{2}}$ 000− 000− $-1/4.3\sqrt{2}$
1 $\tilde{\frac{1}{2}}$ $\tilde{\frac{1}{2}}$ 010+ 000− $-1/4.3\sqrt{2}$
$\tilde{\frac{1}{2}}$ 2 $\frac{3}{2}$ $\frac{3}{2}$ 2 $\frac{1}{2}$
1 0 1 000+ 000+ $+1/4\sqrt{2.3}$
1 2 1 000+ 000+ $+1/8\sqrt{3}$
1 2 $\tilde{1}$ 000− 000− $-1/8\sqrt{3}$
$\tilde{\frac{1}{2}}$ 2 $\frac{3}{2}$ $\frac{3}{2}$ 2 $\frac{3}{2}$
1 0 1 000− 000− 0
1 0 1 000− 001− $+1/4\sqrt{2.3}$
1 2 1 000− 000− $-1/8\sqrt{3}$
1 2 1 000− 001− 0
1 2 $\tilde{1}$ 000+ 000+ $+1/8\sqrt{3}$

$\tilde{\frac{1}{2}}$ 2 $\frac{3}{2}$ $\frac{3}{2}$ 2 $\frac{3}{2}$
1 2 $\tilde{1}$ 000+ 001− 0
$\tilde{\frac{1}{2}}$ 2 $\frac{3}{2}$ $\frac{3}{2}$ 2 $\tilde{\frac{1}{2}}$
1 0 1 000+ 000+ $+1/4\sqrt{2.3}$
1 2 1 000+ 000+ $-1/8\sqrt{3}$
1 2 $\tilde{1}$ 000− 000− $+1/8\sqrt{3}$
$\tilde{\frac{1}{2}}$ 2 $\frac{3}{2}$ $\frac{3}{2}$ $\tilde{1}$ $\frac{1}{2}$
1 1 1 000− 000− $-1/8.3\sqrt{3}$
1 1 2 000+ 000+ $-1/4.3$
1 1 $\tilde{1}$ 000+ 000+ $-1/8.3$
1 $\tilde{1}$ 1 000+ 000+ $-1/8.3$
1 $\tilde{1}$ 2 000− 000− $-1/4.3$
1 $\tilde{1}$ $\tilde{1}$ 000− 000− $-1/8.3\sqrt{3}$
$\tilde{\frac{1}{2}}$ 2 $\frac{3}{2}$ $\frac{3}{2}$ $\tilde{1}$ $\frac{3}{2}$
1 1 0 000− 000− $+1/4.3\sqrt{2}$
1 1 0 010+ 000− $-1/4.3\sqrt{2}$
1 1 1 000+ 000+ $-1/8.3\sqrt{3}$
1 1 1 000+ 001+ $-1/4.3\sqrt{3}$
1 1 1 010− 000+ $+1/8.3\sqrt{3}$
1 1 1 010− 001+ $-1/8.3\sqrt{3}$
1 1 2 000− 000− $-1/8.3$
1 1 2 010+ 000− $+1/8.3$
1 1 $\tilde{1}$ 000− 000− $+1/8.3$
1 1 $\tilde{1}$ 000− 001+ 0
1 1 $\tilde{1}$ 010+ 000− $+1/8.3$
1 1 $\tilde{1}$ 010+ 001+ $+1/8.3$
1 $\tilde{1}$ 1 000− 000− $+1/8.3$
1 $\tilde{1}$ 1 000− 001− 0
1 $\tilde{1}$ 1 010+ 000− $+1/8.3$
1 $\tilde{1}$ 1 010+ 001− $-1/8.3$
1 $\tilde{1}$ 2 000+ 000+ $+1/8.3$
1 $\tilde{1}$ 2 010− 000+ $-1/8.3$
1 $\tilde{1}$ $\tilde{1}$ 000+ 000+ $-1/8.3\sqrt{3}$
1 $\tilde{1}$ $\tilde{1}$ 000+ 001− $+1/4.3\sqrt{3}$
1 $\tilde{1}$ $\tilde{1}$ 010− 000+ $+1/8.3\sqrt{3}$
1 $\tilde{1}$ $\tilde{1}$ 010− 001− $+1/8.3\sqrt{3}$
$\tilde{\frac{1}{2}}$ 2 $\frac{3}{2}$ $\frac{3}{2}$ $\tilde{1}$ $\tilde{\frac{1}{2}}$
1 1 1 000− 000− $-1/8\sqrt{3}$
1 1 2 000+ 000+ 0
1 1 $\tilde{1}$ 000+ 000+ $+1/8.3$
1 $\tilde{1}$ 1 000+ 000+ $+1/8.3$
1 $\tilde{1}$ 2 000− 000− 0
1 $\tilde{1}$ $\tilde{1}$ 000− 000− $-1/8\sqrt{3}$
$\tilde{\frac{1}{2}}$ 2 $\frac{3}{2}$ $\frac{3}{2}$ $\tilde{\frac{1}{2}}$ 1
1 $\frac{3}{2}$ $\frac{1}{2}$ 000+ 000+ $+1/8.2.3$
1 $\frac{3}{2}$ $\frac{3}{2}$ 000− 000− $-1/4.3$
1 $\frac{3}{2}$ $\frac{3}{2}$ 000− 001− 0
1 $\frac{3}{2}$ $\frac{3}{2}$ 001− 000− 0
1 $\frac{3}{2}$ $\frac{3}{2}$ 001− 001− $-1/8.2.3$
1 $\frac{3}{2}$ $\tilde{\frac{1}{2}}$ 000+ 000+ $+1/8.2.3$

O and T_d 9j Symbols (cont.)

$\widetilde{\tfrac{1}{2}}\ 2\ \tfrac{3}{2}\quad \tfrac{3}{2}\ \widetilde{\tfrac{1}{2}}\ 2$
$1\ \tfrac{3}{2}\ \tfrac{1}{2}$ 000− 000− +1/8.2
$1\ \tfrac{3}{2}\ \tfrac{3}{2}$ 000+ 000+ 0
$1\ \tfrac{3}{2}\ \tfrac{3}{2}$ 001+ 000+ −1/8.2
$1\ \tfrac{3}{2}\ \widetilde{\tfrac{1}{2}}$ 000− 000− −1/8.2
$\widetilde{\tfrac{1}{2}}\ 2\ \tfrac{3}{2}\quad \tfrac{3}{2}\ \widetilde{\tfrac{1}{2}}\ \tilde{1}$
$1\ \tfrac{3}{2}\ \tfrac{1}{2}$ 000− 000− +1/8.2
$1\ \tfrac{3}{2}\ \tfrac{3}{2}$ 000+ 000+ 0
$1\ \tfrac{3}{2}\ \tfrac{3}{2}$ 000+ 001− 0
$1\ \tfrac{3}{2}\ \tfrac{3}{2}$ 001+ 000+ 0
$1\ \tfrac{3}{2}\ \tfrac{3}{2}$ 001+ 001− +1/8.2
$1\ \tfrac{3}{2}\ \widetilde{\tfrac{1}{2}}$ 000− 000− +1/8.2
$\widetilde{\tfrac{1}{2}}\ 2\ \tfrac{3}{2}\quad 2\ 0\ 2$
$\tfrac{3}{2}\ 2\ \tfrac{1}{2}$ 000+ 000+ −1/8
$\tfrac{3}{2}\ 2\ \tfrac{3}{2}$ 000− 000− +1/8
$\tfrac{3}{2}\ 2\ \widetilde{\tfrac{1}{2}}$ 000+ 000+ +1/8
$\widetilde{\tfrac{1}{2}}\ 2\ \tfrac{3}{2}\quad 2\ \tfrac{1}{2}\ \tfrac{3}{2}$
$\tfrac{3}{2}\ \tfrac{3}{2}\ 0$ 000− 000− −1/8.2
$\tfrac{3}{2}\ \tfrac{3}{2}\ 1$ 000+ 000+ −1/8.2
$\tfrac{3}{2}\ \tfrac{3}{2}\ 1$ 000+ 001+ 0
$\tfrac{3}{2}\ \tfrac{3}{2}\ 1$ 001+ 000+ 0
$\tfrac{3}{2}\ \tfrac{3}{2}\ 1$ 001+ 001+ +1/8.2
$\tfrac{3}{2}\ \tfrac{3}{2}\ 2$ 000− 000− +1/8.2
$\tfrac{3}{2}\ \tfrac{3}{2}\ \tilde{1}$ 000− 000− +1/8.2
$\tfrac{3}{2}\ \tfrac{3}{2}\ \tilde{1}$ 000− 001+ 0
$\tfrac{3}{2}\ \tfrac{3}{2}\ \tilde{1}$ 001+ 000− 0
$\tfrac{3}{2}\ \tfrac{3}{2}\ \tilde{1}$ 001+ 001+ +1/8.2
$\widetilde{\tfrac{1}{2}}\ 2\ \tfrac{3}{2}\quad 2\ 1\ 1$
$\tfrac{3}{2}\ 1\ \tfrac{1}{2}$ 000− 000− 0
$\tfrac{3}{2}\ 1\ \tfrac{3}{2}$ 000+ 000+ +1/8.3
$\tfrac{3}{2}\ 1\ \tfrac{3}{2}$ 000+ 001+ +1/8.3
$\tfrac{3}{2}\ 1\ \tfrac{3}{2}$ 001+ 000+ +1/8.3
$\tfrac{3}{2}\ 1\ \tfrac{3}{2}$ 001+ 001+ +1/8.3
$\tfrac{3}{2}\ 1\ \widetilde{\tfrac{1}{2}}$ 000− 000− +1/4.3
$\widetilde{\tfrac{1}{2}}\ 2\ \tfrac{3}{2}\quad 2\ 1\ \tilde{1}$
$\tfrac{3}{2}\ 1\ \tfrac{1}{2}$ 000+ 000+ −1/4.3
$\tfrac{3}{2}\ 1\ \tfrac{3}{2}$ 000− 000− −1/8.3
$\tfrac{3}{2}\ 1\ \tfrac{3}{2}$ 000− 001+ +1/8.3
$\tfrac{3}{2}\ 1\ \tfrac{3}{2}$ 001− 000− +1/8.3
$\tfrac{3}{2}\ 1\ \tfrac{3}{2}$ 001− 001+ −1/8.3
$\tfrac{3}{2}\ 1\ \widetilde{\tfrac{1}{2}}$ 000+ 000+ 0
$\tfrac{3}{2}\ \tilde{1}\ \tfrac{1}{2}$ 000− 000− 0
$\tfrac{3}{2}\ \tilde{1}\ \tfrac{3}{2}$ 000+ 000+ +1/8.3
$\tfrac{3}{2}\ \tilde{1}\ \tfrac{3}{2}$ 000+ 001− +1/8.3
$\tfrac{3}{2}\ \tilde{1}\ \tfrac{3}{2}$ 001− 000+ +1/8.3
$\tfrac{3}{2}\ \tilde{1}\ \tfrac{3}{2}$ 001− 001− +1/8.3
$\tfrac{3}{2}\ \tilde{1}\ \widetilde{\tfrac{1}{2}}$ 000− 000− +1/4.3
$\widetilde{\tfrac{1}{2}}\ 2\ \tfrac{3}{2}\quad 2\ \tfrac{3}{2}\ \tfrac{1}{2}$
$\tfrac{3}{2}\ \tfrac{1}{2}\ 1$ 000− 000− +1/8.2
$\tfrac{3}{2}\ \tfrac{1}{2}\ 2$ 000+ 000+ +1/8.2
$\tfrac{3}{2}\ \tfrac{1}{2}\ \tilde{1}$ 000+ 000+ +1/8.2
$\widetilde{\tfrac{1}{2}}\ 2\ \tfrac{3}{2}\quad 2\ \tfrac{3}{2}\ \tfrac{3}{2}$
$\tfrac{3}{2}\ \tfrac{1}{2}\ 1$ 000+ 000+ 0
$\tfrac{3}{2}\ \tfrac{1}{2}\ 1$ 000+ 001+ −1/8.2
$\tfrac{3}{2}\ \tfrac{1}{2}\ 2$ 000− 000− −1/8.2
$\tfrac{3}{2}\ \tfrac{1}{2}\ \tilde{1}$ 000− 000− 0
$\tfrac{3}{2}\ \tfrac{1}{2}\ \tilde{1}$ 000− 001+ −1/8.2
$\tfrac{3}{2}\ \tfrac{3}{2}\ 0$ 000+ 000+ +1/8.2
$\tfrac{3}{2}\ \tfrac{3}{2}\ 1$ 000− 000− +1/8.2
$\tfrac{3}{2}\ \tfrac{3}{2}\ 1$ 000− 001− 0
$\tfrac{3}{2}\ \tfrac{3}{2}\ 1$ 001− 000− 0
$\tfrac{3}{2}\ \tfrac{3}{2}\ 1$ 001− 001− 0
$\tfrac{3}{2}\ \tfrac{3}{2}\ 2$ 000+ 000+ 0
$\tfrac{3}{2}\ \tfrac{3}{2}\ \tilde{1}$ 000+ 000+ +1/8.2
$\tfrac{3}{2}\ \tfrac{3}{2}\ \tilde{1}$ 000+ 001− 0
$\tfrac{3}{2}\ \tfrac{3}{2}\ \tilde{1}$ 001− 000+ 0
$\tfrac{3}{2}\ \tfrac{3}{2}\ \tilde{1}$ 001− 001− 0
$\widetilde{\tfrac{1}{2}}\ 2\ \tfrac{3}{2}\quad 2\ \tfrac{3}{2}\ \widetilde{\tfrac{1}{2}}$
$\tfrac{3}{2}\ \tfrac{1}{2}\ 1$ 000− 000− −1/8.2
$\tfrac{3}{2}\ \tfrac{1}{2}\ 2$ 000+ 000+ +1/8.2
$\tfrac{3}{2}\ \tfrac{1}{2}\ \tilde{1}$ 000+ 000+ −1/8.2
$\tfrac{3}{2}\ \tfrac{3}{2}\ 1$ 000+ 000+ 0
$\tfrac{3}{2}\ \tfrac{3}{2}\ 1$ 001+ 000+ −1/8.2
$\tfrac{3}{2}\ \tfrac{3}{2}\ 2$ 000− 000− +1/8.2
$\tfrac{3}{2}\ \tfrac{3}{2}\ \tilde{1}$ 000− 000− 0
$\tfrac{3}{2}\ \tfrac{3}{2}\ \tilde{1}$ 001+ 000− −1/8.2
$\tfrac{3}{2}\ \widetilde{\tfrac{1}{2}}\ 1$ 000− 000− +1/8.2
$\tfrac{3}{2}\ \widetilde{\tfrac{1}{2}}\ 2$ 000+ 000+ +1/8.2
$\tfrac{3}{2}\ \widetilde{\tfrac{1}{2}}\ \tilde{1}$ 000+ 000+ +1/8.2
$\widetilde{\tfrac{1}{2}}\ 2\ \tfrac{3}{2}\quad 2\ 2\ 0$
$\tfrac{3}{2}\ 0\ \tfrac{3}{2}$ 000− 000− +1/8
$\widetilde{\tfrac{1}{2}}\ 2\ \tfrac{3}{2}\quad 2\ 2\ 2$
$\tfrac{3}{2}\ 0\ \tfrac{3}{2}$ 000− 000− −1/8
$\tfrac{3}{2}\ 2\ \tfrac{1}{2}$ 000+ 000+ +1/8
$\tfrac{3}{2}\ 2\ \tfrac{3}{2}$ 000− 000− 0
$\tfrac{3}{2}\ 2\ \widetilde{\tfrac{1}{2}}$ 000+ 000+ +1/8
$\widetilde{\tfrac{1}{2}}\ 2\ \tfrac{3}{2}\quad 2\ \tilde{1}\ 1$
$\tfrac{3}{2}\ 1\ \tfrac{1}{2}$ 000+ 000+ +1/4.3
$\tfrac{3}{2}\ 1\ \tfrac{3}{2}$ 000− 000− +1/8.3
$\tfrac{3}{2}\ 1\ \tfrac{3}{2}$ 000− 001− −1/8.3
$\tfrac{3}{2}\ 1\ \tfrac{3}{2}$ 001− 000− −1/8.3
$\tfrac{3}{2}\ 1\ \tfrac{3}{2}$ 001− 001− +1/8.3
$\tfrac{3}{2}\ 1\ \widetilde{\tfrac{1}{2}}$ 000+ 000+ 0
$\widetilde{\tfrac{1}{2}}\ 2\ \tfrac{3}{2}\quad 2\ \tilde{1}\ \tilde{1}$
$\tfrac{3}{2}\ 1\ \tfrac{1}{2}$ 000− 000− 0
$\tfrac{3}{2}\ 1\ \tfrac{3}{2}$ 000+ 000+ −1/8.3
$\tfrac{3}{2}\ 1\ \tfrac{3}{2}$ 000+ 001− −1/8.3
$\tfrac{3}{2}\ 1\ \tfrac{3}{2}$ 001+ 000+ −1/8.3
$\tfrac{3}{2}\ 1\ \tfrac{3}{2}$ 001+ 001− −1/8.3
$\widetilde{\tfrac{1}{2}}\ 2\ \tfrac{3}{2}\quad 2\ \tilde{1}\ \tilde{1}$
$\tfrac{3}{2}\ 1\ \widetilde{\tfrac{1}{2}}$ 000− 000− −1/4.3
$\tfrac{3}{2}\ \tilde{1}\ \tfrac{1}{2}$ 000+ 000+ +1/4.3
$\tfrac{3}{2}\ \tilde{1}\ \tfrac{3}{2}$ 000− 000− +1/8.3
$\tfrac{3}{2}\ \tilde{1}\ \tfrac{3}{2}$ 000− 001+ −1/8.3
$\tfrac{3}{2}\ \tilde{1}\ \tfrac{3}{2}$ 001+ 000− −1/8.3
$\tfrac{3}{2}\ \tilde{1}\ \tfrac{3}{2}$ 001+ 001+ +1/8.3
$\tfrac{3}{2}\ \tilde{1}\ \widetilde{\tfrac{1}{2}}$ 000+ 000+ 0
$\widetilde{\tfrac{1}{2}}\ 2\ \tfrac{3}{2}\quad 2\ \widetilde{\tfrac{1}{2}}\ \tfrac{3}{2}$
$\tfrac{3}{2}\ \tfrac{3}{2}\ 0$ 000− 000− +1/8.2
$\tfrac{3}{2}\ \tfrac{3}{2}\ 1$ 000+ 000+ +1/8.2
$\tfrac{3}{2}\ \tfrac{3}{2}\ 1$ 000+ 001+ 0
$\tfrac{3}{2}\ \tfrac{3}{2}\ 1$ 001+ 000+ 0
$\tfrac{3}{2}\ \tfrac{3}{2}\ 1$ 001+ 001+ +1/8.2
$\tfrac{3}{2}\ \tfrac{3}{2}\ 2$ 000− 000− +1/8.2
$\tfrac{3}{2}\ \tfrac{3}{2}\ \tilde{1}$ 000− 000− −1/8.2
$\tfrac{3}{2}\ \tfrac{3}{2}\ \tilde{1}$ 000− 001+ 0
$\tfrac{3}{2}\ \tfrac{3}{2}\ \tilde{1}$ 001+ 000− 0
$\tfrac{3}{2}\ \tfrac{3}{2}\ \tilde{1}$ 001+ 001+ +1/8.2
$\widetilde{\tfrac{1}{2}}\ \tilde{1}\ \tfrac{1}{2}\quad \tfrac{3}{2}\ 0\ \tfrac{3}{2}$
$1\ \tilde{1}\ 1$ 000+ 000+ −1/4.3
$1\ \tilde{1}\ 2$ 000− 000− +1/4.3
$1\ \tilde{1}\ \tilde{1}$ 000− 000− $-1/4.3\sqrt{3}$
$\widetilde{\tfrac{1}{2}}\ \tilde{1}\ \tfrac{1}{2}\quad \tfrac{3}{2}\ \tfrac{1}{2}\ 1$
$1\ \tfrac{3}{2}\ \tfrac{1}{2}$ 000− 000− −1/4.3
$1\ \tfrac{3}{2}\ \tfrac{3}{2}$ 000+ 000+ $-1/4.3\sqrt{2}$
$1\ \tfrac{3}{2}\ \tfrac{3}{2}$ 001+ 000+ 0
$1\ \widetilde{\tfrac{1}{2}}\ \tfrac{3}{2}$ 000− 000− +1/8.3
$\widetilde{\tfrac{1}{2}}\ \tilde{1}\ \tfrac{1}{2}\quad \tfrac{3}{2}\ \tfrac{1}{2}\ 2$
$1\ \tfrac{3}{2}\ \tfrac{3}{2}$ 000− 000− $+1/4.3\sqrt{2}$
$1\ \tfrac{3}{2}\ \tfrac{3}{2}$ 001− 000− $+1/4.3\sqrt{2}$
$1\ \widetilde{\tfrac{1}{2}}\ \tfrac{3}{2}$ 000+ 000+ +1/8.3
$\widetilde{\tfrac{1}{2}}\ \tilde{1}\ \tfrac{1}{2}\quad \tfrac{3}{2}\ \tfrac{1}{2}\ \tilde{1}$
$1\ \tfrac{3}{2}\ \tfrac{3}{2}$ 000− 000− $+1/4.9\sqrt{2}$
$1\ \tfrac{3}{2}\ \tfrac{3}{2}$ 001− 000− $-1/2.9\sqrt{2}$
$1\ \tfrac{3}{2}\ \widetilde{\tfrac{1}{2}}$ 000+ 000+ −1/4.9
$1\ \widetilde{\tfrac{1}{2}}\ \tfrac{3}{2}$ 000+ 000+ +7/8.9
$1\ \widetilde{\tfrac{1}{2}}\ \widetilde{\tfrac{1}{2}}$ 000− 000− +1/2.9
$\widetilde{\tfrac{1}{2}}\ \tilde{1}\ \tfrac{1}{2}\quad \tfrac{3}{2}\ 1\ \tfrac{1}{2}$
$1\ 1\ 0$ 000+ 000+ $-1/2.3\sqrt{2}$
$1\ 1\ 1$ 000− 000− $+1/4.3\sqrt{3}$
$1\ 2\ 1$ 000+ 000+ −1/4.3
$1\ \tilde{1}\ 1$ 000+ 000+ $+1/4.3\sqrt{3}$
$\widetilde{\tfrac{1}{2}}\ \tilde{1}\ \tfrac{1}{2}\quad \tfrac{3}{2}\ 1\ \tfrac{3}{2}$
$1\ 1\ 1$ 000+ 000+ $-1/2.3\sqrt{2.3}$
$1\ 1\ 1$ 010+ 000+ 0
$1\ 1\ 2$ 000− 000− 0
$1\ 1\ 2$ 010− 000− $-1/4.3\sqrt{2}$
$1\ 1\ \tilde{1}$ 000− 000− $+1/2.9\sqrt{2}$
$1\ 1\ \tilde{1}$ 010− 000− $+1/2.9\sqrt{2}$

O and T_d 9j Symbols (cont.)

$\tilde{\frac{1}{2}}\ \tilde{1}\ \frac{1}{2}\quad \frac{3}{2}\ 1\ \frac{3}{2}$			
$1\ 2\ 1$	000−	000−	0
$1\ 2\ 1$	010−	000−	$+1/4.3\sqrt{2}$
$1\ 2\ \tilde{1}$	000+	000+	$-1/9\sqrt{2}$
$1\ 2\ \tilde{1}$	010+	000+	$-1/4.9\sqrt{2}$
$1\ \tilde{1}\ 1$	000−	000−	0
$1\ \tilde{1}\ 1$	010−	000−	$+1/2.3\sqrt{2.3}$
$1\ \tilde{1}\ 2$	000+	000+	$+1/2.3\sqrt{2.3}$
$1\ \tilde{1}\ 2$	010+	000+	$-1/4.3\sqrt{2.3}$
$1\ \tilde{1}\ \tilde{1}$	000+	000+	0
$1\ \tilde{1}\ \tilde{1}$	010+	000+	0

$\tilde{\frac{1}{2}}\ \tilde{1}\ \frac{1}{2}\quad \frac{3}{2}\ 1\ \tilde{\frac{1}{2}}$			
$1\ 1\ \tilde{1}$	000+	000+	$+1/4.9$
$1\ 2\ \tilde{1}$	000−	000−	$+1/4.9$
$1\ \tilde{1}\ \tilde{1}$	000−	000−	$+1/4.3$

$\tilde{\frac{1}{2}}\ \tilde{1}\ \frac{1}{2}\quad \frac{3}{2}\ \frac{3}{2}\ 0$			
$1\ \frac{1}{2}\ \frac{1}{2}$	000+	000+	$-1/2.3\sqrt{2}$
$1\ \frac{3}{2}\ \frac{1}{2}$	000−	000−	$-1/8.3$
$1\ \frac{3}{2}\ \frac{1}{2}$	000−	010+	$+1/4.3$

$\tilde{\frac{1}{2}}\ \tilde{1}\ \frac{1}{2}\quad \frac{3}{2}\ \frac{3}{2}\ 1$			
$1\ \frac{1}{2}\ \frac{1}{2}$	000−	000−	$-1/2.9\sqrt{2}$
$1\ \frac{1}{2}\ \frac{1}{2}$	010−	000−	$+1/9\sqrt{2}$
$1\ \frac{1}{2}\ \frac{3}{2}$	000+	000+	$-1/4.9\sqrt{2}$
$1\ \frac{1}{2}\ \frac{3}{2}$	010+	000+	$+1/2.9\sqrt{2}$
$1\ \frac{3}{2}\ \frac{1}{2}$	000+	000+	$+5/8.9$
$1\ \frac{3}{2}\ \frac{1}{2}$	000+	010−	$+1/4.9$
$1\ \frac{3}{2}\ \frac{1}{2}$	010+	000+	$+1/4.9$
$1\ \frac{3}{2}\ \frac{1}{2}$	010+	010−	$+1/4.9$
$1\ \frac{3}{2}\ \frac{3}{2}$	000−	000−	$-1/4.9\sqrt{2}$
$1\ \frac{3}{2}\ \frac{3}{2}$	000−	010+	$+1/2.9\sqrt{2}$
$1\ \frac{3}{2}\ \frac{3}{2}$	001−	000−	$+1/4.9\sqrt{2}$
$1\ \frac{3}{2}\ \frac{3}{2}$	001−	010+	$+1/4.9\sqrt{2}$
$1\ \frac{3}{2}\ \frac{3}{2}$	010−	000−	$-1/4.9\sqrt{2}$
$1\ \frac{3}{2}\ \frac{3}{2}$	010−	010+	$-1/4.9\sqrt{2}$
$1\ \frac{3}{2}\ \frac{3}{2}$	011−	000−	$+5/8.9\sqrt{2}$
$1\ \frac{3}{2}\ \frac{3}{2}$	011−	010+	$-1/2.9\sqrt{2}$
$1\ \tilde{\frac{1}{2}}\ \frac{3}{2}$	000+	000+	$-1/4.3\sqrt{2}$
$1\ \tilde{\frac{1}{2}}\ \frac{3}{2}$	010+	000+	0

$\tilde{\frac{1}{2}}\ \tilde{1}\ \frac{1}{2}\quad \frac{3}{2}\ \frac{3}{2}\ 2$			
$1\ \frac{1}{2}\ \frac{3}{2}$	000−	000−	$-1/4.3\sqrt{2}$
$1\ \frac{3}{2}\ \frac{3}{2}$	000+	000+	0
$1\ \frac{3}{2}\ \frac{3}{2}$	000+	010−	$+1/4.3\sqrt{2}$
$1\ \frac{3}{2}\ \frac{3}{2}$	001+	000+	$+1/8.3\sqrt{2}$
$1\ \frac{3}{2}\ \frac{3}{2}$	001+	010−	0
$1\ \tilde{\frac{1}{2}}\ \frac{3}{2}$	000−	000−	$+1/4.3\sqrt{2}$

$\tilde{\frac{1}{2}}\ \tilde{1}\ \frac{1}{2}\quad \frac{3}{2}\ \frac{3}{2}\ \tilde{1}$			
$1\ \frac{1}{2}\ \frac{3}{2}$	000−	000−	$+1/4.3\sqrt{2}$
$1\ \frac{1}{2}\ \frac{3}{2}$	010+	000−	0
$1\ \frac{3}{2}\ \frac{3}{2}$	000+	000+	$+1/4.9\sqrt{2}$
$1\ \frac{3}{2}\ \frac{3}{2}$	000+	010−	$+1/2.9\sqrt{2}$
$1\ \frac{3}{2}\ \frac{3}{2}$	001+	000+	$+1/4.9\sqrt{2}$
$1\ \frac{3}{2}\ \frac{3}{2}$	001+	010−	$-1/4.9\sqrt{2}$
$1\ \frac{3}{2}\ \frac{3}{2}$	010−	000+	$+1/4.9\sqrt{2}$
$1\ \frac{3}{2}\ \frac{3}{2}$	010−	010−	$-1/4.9\sqrt{2}$
$1\ \frac{3}{2}\ \frac{3}{2}$	011−	000+	$+5/8.9\sqrt{2}$
$1\ \frac{3}{2}\ \frac{3}{2}$	011−	010−	$+1/2.9\sqrt{2}$
$1\ \frac{3}{2}\ \tilde{\frac{1}{2}}$	000−	000−	$-5/8.9$
$1\ \frac{3}{2}\ \tilde{\frac{1}{2}}$	000−	010+	$+1/4.9$
$1\ \frac{3}{2}\ \tilde{\frac{1}{2}}$	010+	000−	$-1/4.9$
$1\ \frac{3}{2}\ \tilde{\frac{1}{2}}$	010+	010+	$+1/4.9$
$1\ \tilde{\frac{1}{2}}\ \frac{3}{2}$	000−	000−	$+1/4.9\sqrt{2}$
$1\ \tilde{\frac{1}{2}}\ \frac{3}{2}$	010+	000−	$-1/2.9\sqrt{2}$
$1\ \tilde{\frac{1}{2}}\ \tilde{\frac{1}{2}}$	000+	000+	$-1/2.9\sqrt{2}$
$1\ \tilde{\frac{1}{2}}\ \tilde{\frac{1}{2}}$	010−	000+	$+1/9\sqrt{2}$

$\tilde{\frac{1}{2}}\ \tilde{1}\ \frac{1}{2}\quad \frac{3}{2}\ 2\ \frac{1}{2}$			
$1\ 1\ 0$	000−	000−	$+1/2.3\sqrt{2}$
$1\ 1\ 1$	000+	000+	$-1/4.3\sqrt{3}$
$1\ \tilde{1}\ 1$	000−	000−	$-1/4.3$

$\tilde{\frac{1}{2}}\ \tilde{1}\ \frac{1}{2}\quad \frac{3}{2}\ 2\ \frac{3}{2}$			
$1\ 1\ 1$	000−	000−	$-1/2.3\sqrt{2.3}$
$1\ 1\ 2$	000+	000+	$+1/4.3\sqrt{2}$
$1\ 1\ \tilde{1}$	000+	000+	0
$1\ \tilde{1}\ 1$	000+	000+	0
$1\ \tilde{1}\ 2$	000−	000−	$+1/4.3\sqrt{2}$
$1\ \tilde{1}\ \tilde{1}$	000−	000−	$+1/2.3\sqrt{2.3}$

$\tilde{\frac{1}{2}}\ \tilde{1}\ \frac{1}{2}\quad \frac{3}{2}\ 2\ \tilde{\frac{1}{2}}$			
$1\ 1\ \tilde{1}$	000−	000−	$-1/4.3$
$1\ \tilde{1}\ \tilde{1}$	000+	000+	$-1/4.3\sqrt{3}$

$\tilde{\frac{1}{2}}\ \tilde{1}\ \frac{1}{2}\quad \frac{3}{2}\ \tilde{1}\ \frac{1}{2}$			
$1\ 0\ 1$	000−	000−	$+1/9$
$1\ 1\ 0$	000−	000−	$-1/2.3\sqrt{2.3}$
$1\ 1\ 1$	000+	000+	$-1/4.3$
$1\ 2\ 1$	000−	000−	$+1/4.9$
$1\ \tilde{1}\ 1$	000−	000−	$-1/4.9$

$\tilde{\frac{1}{2}}\ \tilde{1}\ \frac{1}{2}\quad \frac{3}{2}\ \tilde{1}\ \frac{3}{2}$			
$1\ 0\ 1$	000+	000+	$+1/2.9\sqrt{2}$
$1\ 0\ 1$	010−	000+	$+1/9\sqrt{2}$
$1\ 1\ 1$	000−	000−	0
$1\ 1\ 1$	010+	000−	0
$1\ 1\ 2$	000+	000+	$+1/2.3\sqrt{2.3}$
$1\ 1\ 2$	010−	000+	$+1/4.3\sqrt{2.3}$
$1\ 1\ \tilde{1}$	000+	000+	0
$1\ 1\ \tilde{1}$	010−	000+	$+1/2.3\sqrt{2.3}$
$1\ 2\ 1$	000+	000+	$-1/9\sqrt{2}$
$1\ 2\ 1$	010−	000+	$+1/4.9\sqrt{2}$
$1\ 2\ \tilde{1}$	000−	000−	0
$1\ 2\ \tilde{1}$	010+	000−	$-1/4.3\sqrt{2}$
$1\ \tilde{1}\ 1$	000+	000+	$-1/2.9\sqrt{2}$
$1\ \tilde{1}\ 1$	010−	000+	$+1/2.9\sqrt{2}$
$1\ \tilde{1}\ 2$	000−	000−	0
$1\ \tilde{1}\ 2$	010+	000−	$+1/4.3\sqrt{2}$
$1\ \tilde{1}\ \tilde{1}$	000−	000−	$+1/2.3\sqrt{2.3}$
$1\ \tilde{1}\ \tilde{1}$	010+	000−	0

$\tilde{\frac{1}{2}}\ \tilde{1}\ \frac{1}{2}\quad \frac{3}{2}\ \tilde{1}\ \tilde{\frac{1}{2}}$			
$1\ 1\ \tilde{1}$	000−	000−	$-1/4.3\sqrt{3}$
$1\ 2\ \tilde{1}$	000+	000+	$-1/4.3$
$1\ \tilde{1}\ \tilde{1}$	000+	000+	$-1/4.3\sqrt{3}$

$\tilde{\frac{1}{2}}\ \tilde{1}\ \frac{1}{2}\quad \frac{3}{2}\ \tilde{\frac{1}{2}}\ 1$			
$1\ \frac{1}{2}\ \frac{1}{2}$	000+	000+	$+1/2.9$
$1\ \frac{1}{2}\ \frac{3}{2}$	000−	000−	$-7/8.9$
$1\ \frac{3}{2}\ \frac{1}{2}$	000−	000−	$+1/4.9$
$1\ \frac{3}{2}\ \frac{3}{2}$	000+	000+	$-1/4.9\sqrt{2}$
$1\ \frac{3}{2}\ \frac{3}{2}$	001+	000+	$-1/2.9\sqrt{2}$

$\tilde{\frac{1}{2}}\ \tilde{1}\ \frac{1}{2}\quad \frac{3}{2}\ \tilde{\frac{1}{2}}\ 2$			
$1\ \frac{1}{2}\ \frac{3}{2}$	000+	000+	$+1/8.3$
$1\ \frac{3}{2}\ \frac{3}{2}$	000−	000−	$+1/4.3\sqrt{2}$
$1\ \frac{3}{2}\ \frac{3}{2}$	001−	000−	$-1/4.3\sqrt{2}$

$\tilde{\frac{1}{2}}\ \tilde{1}\ \frac{1}{2}\quad \frac{3}{2}\ \tilde{\frac{1}{2}}\ \tilde{1}$			
$1\ \frac{1}{2}\ \frac{3}{2}$	000+	000+	$-1/8.3$
$1\ \frac{3}{2}\ \frac{3}{2}$	000−	000−	$+1/4.3\sqrt{2}$
$1\ \frac{3}{2}\ \frac{3}{2}$	001−	000−	0
$1\ \frac{3}{2}\ \tilde{\frac{1}{2}}$	000+	000+	$+1/4.3$

$\tilde{\frac{1}{2}}\ \tilde{1}\ \frac{1}{2}\quad 2\ 0\ 2$			
$\frac{3}{2}\ \tilde{1}\ \frac{3}{2}$	000+	000+	$+1/4\sqrt{3}$
$\frac{3}{2}\ \tilde{1}\ \frac{3}{2}$	001−	000+	0

$\tilde{\frac{1}{2}}\ \tilde{1}\ \frac{1}{2}\quad 2\ \frac{1}{2}\ \frac{3}{2}$			
$\frac{3}{2}\ \frac{3}{2}\ 1$	000−	000−	$-1/4.3\sqrt{2}$
$\frac{3}{2}\ \frac{3}{2}\ 1$	001−	000−	$-1/4.3\sqrt{2}$
$\frac{3}{2}\ \frac{3}{2}\ 2$	000+	000+	0
$\frac{3}{2}\ \frac{3}{2}\ \tilde{1}$	000+	000+	$+1/4.3\sqrt{2}$
$\frac{3}{2}\ \frac{3}{2}\ \tilde{1}$	001−	000+	$-1/4.3\sqrt{2}$
$\frac{3}{2}\ \tilde{\frac{1}{2}}\ 1$	000+	000+	$+1/8.3$
$\frac{3}{2}\ \tilde{\frac{1}{2}}\ 2$	000−	000−	$+1/8$
$\frac{3}{2}\ \tilde{\frac{1}{2}}\ \tilde{1}$	000−	000−	$-1/8.3$

$\tilde{\frac{1}{2}}\ \tilde{1}\ \frac{1}{2}\quad 2\ 1\ 1$			
$\frac{3}{2}\ 1\ \frac{1}{2}$	000+	000+	$-1/4.3$
$\frac{3}{2}\ 1\ \frac{3}{2}$	000−	000−	0
$\frac{3}{2}\ 1\ \frac{3}{2}$	001−	000−	$-1/4.3\sqrt{2}$
$\frac{3}{2}\ 2\ \frac{1}{2}$	000−	000−	$+1/4.3$
$\frac{3}{2}\ 2\ \frac{3}{2}$	000+	000+	$+1/4.3\sqrt{2}$
$\frac{3}{2}\ \tilde{1}\ \frac{1}{2}$	000−	000−	$+1/4.3\sqrt{3}$
$\frac{3}{2}\ \tilde{1}\ \frac{3}{2}$	000+	000+	$-1/2.3\sqrt{2.3}$
$\frac{3}{2}\ \tilde{1}\ \frac{3}{2}$	001−	000+	$-1/4.3\sqrt{2.3}$

$\tilde{\frac{1}{2}}\ \tilde{1}\ \frac{1}{2}\quad 2\ 1\ \tilde{1}$			
$\frac{3}{2}\ 1\ \frac{3}{2}$	000+	000+	0
$\frac{3}{2}\ 1\ \frac{3}{2}$	001+	000+	$-1/4.3\sqrt{2}$
$\frac{3}{2}\ 1\ \tilde{\frac{1}{2}}$	000−	000−	$+1/4.3$
$\frac{3}{2}\ 2\ \frac{3}{2}$	000−	000−	$+1/4.3\sqrt{2}$

O and T_d 9j Symbols (cont.)

$\tilde{\frac{1}{2}}\ \tilde{1}\ \frac{1}{2}\quad 2\ 1\ \tilde{1}$

$\frac{3}{2}\ 2\ \tilde{\frac{1}{2}}$ 000+ 000+ $+1/4.3$

$\frac{3}{2}\ \tilde{1}\ \frac{3}{2}$ 000− 000− $+1/2.3\sqrt{2.3}$

$\frac{3}{2}\ \tilde{1}\ \frac{3}{2}$ 001+ 000− $-1/4.3\sqrt{2.3}$

$\frac{3}{2}\ \tilde{1}\ \tilde{\frac{1}{2}}$ 000+ 000+ $-1/4.3\sqrt{3}$

$\tilde{\frac{1}{2}}\ \tilde{1}\ \frac{1}{2}\quad 2\ \frac{3}{2}\ \frac{1}{2}$

$\frac{3}{2}\ \frac{1}{2}\ 1$ 000+ 000+ $-1/4.3$

$\frac{3}{2}\ \frac{3}{2}\ 0$ 000+ 000+ $+1/8$

$\frac{3}{2}\ \frac{3}{2}\ 0$ 000+ 010− 0

$\frac{3}{2}\ \frac{3}{2}\ 1$ 000− 000− $+1/8.3$

$\frac{3}{2}\ \frac{3}{2}\ 1$ 000− 010+ 0

$\frac{3}{2}\ \frac{3}{2}\ 1$ 001− 000− 0

$\frac{3}{2}\ \frac{3}{2}\ 1$ 001− 010+ $+1/4.3$

$\frac{3}{2}\ \tilde{\frac{1}{2}}\ 1$ 000+ 000+ $-1/4.3$

$\tilde{\frac{1}{2}}\ \tilde{1}\ \frac{1}{2}\quad 2\ \frac{3}{2}\ \frac{3}{2}$

$\frac{3}{2}\ \frac{1}{2}\ 1$ 000− 000− $-1/4.3\sqrt{2}$

$\frac{3}{2}\ \frac{1}{2}\ 2$ 000+ 000+ 0

$\frac{3}{2}\ \frac{1}{2}\ \tilde{1}$ 000+ 000+ $-1/4.3\sqrt{2}$

$\frac{3}{2}\ \frac{3}{2}\ 1$ 000+ 000+ 0

$\frac{3}{2}\ \frac{3}{2}\ 1$ 000+ 010− $-1/4.3\sqrt{2}$

$\frac{3}{2}\ \frac{3}{2}\ 1$ 001+ 000+ $-1/8.3\sqrt{2}$

$\frac{3}{2}\ \frac{3}{2}\ 1$ 001+ 010− 0

$\frac{3}{2}\ \frac{3}{2}\ 2$ 000− 000− $+1/8\sqrt{2}$

$\frac{3}{2}\ \frac{3}{2}\ 2$ 000− 010+ 0

$\frac{3}{2}\ \frac{3}{2}\ \tilde{1}$ 000− 000− 0

$\frac{3}{2}\ \frac{3}{2}\ \tilde{1}$ 000− 010+ $-1/4.3\sqrt{2}$

$\frac{3}{2}\ \frac{3}{2}\ \tilde{1}$ 001+ 000− $+1/8.3\sqrt{2}$

$\frac{3}{2}\ \frac{3}{2}\ \tilde{1}$ 001+ 010+ 0

$\frac{3}{2}\ \tilde{\frac{1}{2}}\ 1$ 000− 000− $+1/4.3\sqrt{2}$

$\frac{3}{2}\ \tilde{\frac{1}{2}}\ 2$ 000+ 000+ 0

$\frac{3}{2}\ \tilde{\frac{1}{2}}\ \tilde{1}$ 000+ 000+ $+1/4.3\sqrt{2}$

$\tilde{\frac{1}{2}}\ \tilde{1}\ \frac{1}{2}\quad 2\ \frac{3}{2}\ \tilde{\frac{1}{2}}$

$\frac{3}{2}\ \frac{1}{2}\ \tilde{1}$ 000− 000− $+1/4.3$

$\frac{3}{2}\ \frac{3}{2}\ \tilde{1}$ 000+ 000+ $-1/8.3$

$\frac{3}{2}\ \frac{3}{2}\ \tilde{1}$ 000+ 010− 0

$\frac{3}{2}\ \frac{3}{2}\ \tilde{1}$ 001− 000+ 0

$\frac{3}{2}\ \frac{3}{2}\ \tilde{1}$ 001− 010− $+1/4.3$

$\frac{3}{2}\ \tilde{\frac{1}{2}}\ \tilde{1}$ 000− 000− $+1/4.3$

$\tilde{\frac{1}{2}}\ \tilde{1}\ \frac{1}{2}\quad 2\ 2\ 0$

$\frac{3}{2}\ 1\ \frac{1}{2}$ 000+ 000+ $+1/4\sqrt{3}$

$\frac{3}{2}\ \tilde{1}\ \frac{1}{2}$ 000− 000− $-1/4\sqrt{3}$

$\tilde{\frac{1}{2}}\ \tilde{1}\ \frac{1}{2}\quad 2\ 2\ 2$

$\frac{3}{2}\ 1\ \frac{3}{2}$ 000− 000− 0

$\frac{3}{2}\ 1\ \frac{3}{2}$ 001− 000− $+1/4\sqrt{2.3}$

$\frac{3}{2}\ \tilde{1}\ \frac{3}{2}$ 000+ 000+ 0

$\frac{3}{2}\ \tilde{1}\ \frac{3}{2}$ 001− 000+ $+1/4\sqrt{2.3}$

$\tilde{\frac{1}{2}}\ \tilde{1}\ \frac{1}{2}\quad 2\ \tilde{1}\ 1$

$\frac{3}{2}\ 0\ \frac{3}{2}$ 000− 000− $-1/4.3$

$\frac{3}{2}\ 1\ \frac{1}{2}$ 000− 000− $+1/4.3\sqrt{3}$

$\frac{3}{2}\ 1\ \frac{3}{2}$ 000+ 000+ $-1/2.3\sqrt{2.3}$

$\tilde{\frac{1}{2}}\ \tilde{1}\ \frac{1}{2}\quad 2\ \tilde{1}\ 1$

$\frac{3}{2}\ 1\ \frac{3}{2}$ 001+ 000+ $-1/4.3\sqrt{2.3}$

$\frac{3}{2}\ 2\ \frac{1}{2}$ 000+ 000+ $-1/4.3$

$\frac{3}{2}\ 2\ \frac{3}{2}$ 000− 000− $+1/4.3\sqrt{2}$

$\frac{3}{2}\ \tilde{1}\ \frac{1}{2}$ 000+ 000+ $-1/4.3$

$\frac{3}{2}\ \tilde{1}\ \frac{3}{2}$ 000− 000− 0

$\frac{3}{2}\ \tilde{1}\ \frac{3}{2}$ 001+ 000− $-1/4.3\sqrt{2}$

$\tilde{\frac{1}{2}}\ \tilde{1}\ \frac{1}{2}\quad 2\ \tilde{1}\ \tilde{1}$

$\frac{3}{2}\ 0\ \frac{3}{2}$ 000+ 000+ $+1/4.3$

$\frac{3}{2}\ 1\ \frac{3}{2}$ 000− 000− $+1/2.3\sqrt{2.3}$

$\frac{3}{2}\ 1\ \frac{3}{2}$ 001− 000− $-1/4.3\sqrt{2.3}$

$\frac{3}{2}\ 1\ \tilde{\frac{1}{2}}$ 000+ 000+ $-1/4.3\sqrt{3}$

$\frac{3}{2}\ 2\ \frac{3}{2}$ 000+ 000+ $+1/4.3\sqrt{2}$

$\frac{3}{2}\ 2\ \tilde{\frac{1}{2}}$ 000− 000− $-1/4.3$

$\frac{3}{2}\ \tilde{1}\ \frac{3}{2}$ 000+ 000+ 0

$\frac{3}{2}\ \tilde{1}\ \frac{3}{2}$ 001− 000+ $-1/4.3\sqrt{2}$

$\frac{3}{2}\ \tilde{1}\ \tilde{\frac{1}{2}}$ 000− 000− $+1/4.3$

$\tilde{\frac{1}{2}}\ \tilde{1}\ \frac{1}{2}\quad 2\ \tilde{\frac{1}{2}}\ \frac{3}{2}$

$\frac{3}{2}\ \frac{1}{2}\ 1$ 000+ 000+ $+1/8.3$

$\frac{3}{2}\ \frac{1}{2}\ 2$ 000− 000− $-1/8$

$\frac{3}{2}\ \frac{1}{2}\ \tilde{1}$ 000− 000− $-1/8.3$

$\frac{3}{2}\ \frac{3}{2}\ 1$ 000− 000− $-1/4.3\sqrt{2}$

$\frac{3}{2}\ \frac{3}{2}\ 1$ 001− 000− $+1/4.3\sqrt{2}$

$\frac{3}{2}\ \frac{3}{2}\ 2$ 000+ 000+ 0

$\frac{3}{2}\ \frac{3}{2}\ \tilde{1}$ 000+ 000+ $+1/4.3\sqrt{2}$

$\frac{3}{2}\ \frac{3}{2}\ \tilde{1}$ 001− 000+ $+1/4.3\sqrt{2}$

$\tilde{\frac{1}{2}}\ \tilde{1}\ \frac{1}{2}\quad \tilde{1}\ 0\ \tilde{1}$

$\frac{1}{2}\ \tilde{1}\ \frac{3}{2}$ 000− 000− $+1/9$

$\frac{1}{2}\ \tilde{1}\ \tilde{\frac{1}{2}}$ 000+ 000+ $-1/2.9$

$\tilde{\frac{1}{2}}\ \tilde{1}\ \frac{1}{2}\quad \tilde{1}\ \frac{1}{2}\ \frac{3}{2}$

$\frac{1}{2}\ \frac{3}{2}\ 1$ 000+ 000+ $+1/4.3$

$\frac{1}{2}\ \frac{3}{2}\ 2$ 000− 000− $+1/4.3$

$\frac{1}{2}\ \frac{3}{2}\ \tilde{1}$ 000− 000− $-1/4.9$

$\tilde{\frac{1}{2}}\ \tilde{1}\ \frac{1}{2}\quad \tilde{1}\ \frac{1}{2}\ \tilde{\frac{1}{2}}$

$\frac{1}{2}\ \frac{3}{2}\ \tilde{1}$ 000+ 000+ $+1/2.9$

$\frac{1}{2}\ \tilde{\frac{1}{2}}\ \tilde{1}$ 000− 000− $+5/4.9$

$\tilde{\frac{1}{2}}\ \tilde{1}\ \frac{1}{2}\quad \tilde{1}\ 1\ 1$

$\frac{1}{2}\ 1\ \frac{1}{2}$ 000− 000− $+1/9$

$\frac{1}{2}\ 1\ \frac{3}{2}$ 000+ 000+ $+1/4.9$

$\tilde{\frac{1}{2}}\ \tilde{1}\ \frac{1}{2}\quad \tilde{1}\ 1\ 2$

$\frac{1}{2}\ 1\ \frac{3}{2}$ 000− 000− $+1/4.3$

$\frac{1}{2}\ 2\ \frac{3}{2}$ 000+ 000+ $+1/4.3$

$\tilde{\frac{1}{2}}\ \tilde{1}\ \frac{1}{2}\quad \tilde{1}\ 1\ \tilde{1}$

$\frac{1}{2}\ 1\ \frac{3}{2}$ 000− 000− $-1/4.3\sqrt{3}$

$\frac{1}{2}\ 2\ \frac{3}{2}$ 000+ 000+ $+1/4.3\sqrt{3}$

$\frac{1}{2}\ \tilde{1}\ \frac{3}{2}$ 000+ 000+ $+1/4.3$

$\frac{1}{2}\ \tilde{1}\ \tilde{\frac{1}{2}}$ 000− 000− 0

$\tilde{\frac{1}{2}}\ \tilde{1}\ \frac{1}{2}\quad \tilde{1}\ \frac{3}{2}\ \frac{1}{2}$

$\frac{1}{2}\ \frac{1}{2}\ 0$ 000+ 000+ $+1/2.3$

$\frac{1}{2}\ \frac{1}{2}\ 1$ 000− 000− $+1/2.9$

$\tilde{\frac{1}{2}}\ \tilde{1}\ \frac{1}{2}\quad \tilde{1}\ \frac{3}{2}\ \frac{3}{2}$

$\frac{1}{2}\ \frac{1}{2}\ 1$ 000+ 000+ $+1/2.9\sqrt{2}$

$\frac{1}{2}\ \frac{1}{2}\ 1$ 010− 000+ $-1/9\sqrt{2}$

$\frac{1}{2}\ \frac{3}{2}\ 1$ 000− 000− $+5/8.9$

$\frac{1}{2}\ \frac{3}{2}\ 1$ 000− 010+ $+1/4.9$

$\frac{1}{2}\ \frac{3}{2}\ 1$ 010+ 000− $+1/4.9$

$\frac{1}{2}\ \frac{3}{2}\ 1$ 010+ 010+ $+1/4.9$

$\frac{1}{2}\ \frac{3}{2}\ 2$ 000+ 000+ $-1/8.3$

$\frac{1}{2}\ \frac{3}{2}\ 2$ 000+ 010− 0

$\frac{1}{2}\ \frac{3}{2}\ 2$ 010− 000+ 0

$\frac{1}{2}\ \frac{3}{2}\ 2$ 010− 010− $+1/4.3$

$\frac{1}{2}\ \frac{3}{2}\ \tilde{1}$ 000+ 000+ $+5/8.9$

$\frac{1}{2}\ \frac{3}{2}\ \tilde{1}$ 000+ 010− $-1/4.9$

$\frac{1}{2}\ \frac{3}{2}\ \tilde{1}$ 010− 000+ $-1/4.9$

$\frac{1}{2}\ \frac{3}{2}\ \tilde{1}$ 010− 010− $+1/4.9$

$\tilde{\frac{1}{2}}\ \tilde{1}\ \frac{1}{2}\quad \tilde{1}\ \frac{3}{2}\ \tilde{\frac{1}{2}}$

$\frac{1}{2}\ \frac{3}{2}\ \tilde{1}$ 000− 000− $+1/2.9\sqrt{2}$

$\frac{1}{2}\ \frac{3}{2}\ \tilde{1}$ 000− 010+ $+1/9\sqrt{2}$

$\frac{1}{2}\ \tilde{\frac{1}{2}}\ \tilde{1}$ 000+ 000+ $+1/2.9$

$\tilde{\frac{1}{2}}\ \tilde{1}\ \frac{1}{2}\quad \tilde{1}\ 2\ 1$

$\frac{1}{2}\ 1\ \frac{1}{2}$ 000+ 000+ $+1/9$

$\frac{1}{2}\ 1\ \frac{3}{2}$ 000− 000− $+1/4.9$

$\tilde{\frac{1}{2}}\ \tilde{1}\ \frac{1}{2}\quad \tilde{1}\ 2\ \tilde{1}$

$\frac{1}{2}\ 1\ \frac{3}{2}$ 000+ 000+ $-1/4.3$

$\frac{1}{2}\ \tilde{1}\ \frac{3}{2}$ 000− 000− $+1/4.9$

$\frac{1}{2}\ \tilde{1}\ \tilde{\frac{1}{2}}$ 000+ 000+ $+1/9$

$\tilde{\frac{1}{2}}\ \tilde{1}\ \frac{1}{2}\quad \tilde{1}\ \tilde{1}\ 0$

$\frac{1}{2}\ 0\ \frac{1}{2}$ 000+ 000+ $+1/2.3$

$\tilde{\frac{1}{2}}\ \tilde{1}\ \frac{1}{2}\quad \tilde{1}\ \tilde{1}\ 1$

$\frac{1}{2}\ 0\ \frac{1}{2}$ 000− 000− $+1/3\sqrt{2.3}$

$\frac{1}{2}\ 1\ \frac{1}{2}$ 000+ 000+ 0

$\frac{1}{2}\ 1\ \frac{3}{2}$ 000− 000− $+1/4.3$

$\tilde{\frac{1}{2}}\ \tilde{1}\ \frac{1}{2}\quad \tilde{1}\ \tilde{1}\ 2$

$\frac{1}{2}\ 1\ \frac{3}{2}$ 000+ 000+ $+1/4.3\sqrt{3}$

$\frac{1}{2}\ 2\ \frac{3}{2}$ 000− 000− $+1/4.3$

$\tilde{\frac{1}{2}}\ \tilde{1}\ \frac{1}{2}\quad \tilde{1}\ \tilde{1}\ \tilde{1}$

$\frac{1}{2}\ 1\ \frac{3}{2}$ 000+ 000+ $-1/4.3\sqrt{3}$

$\frac{1}{2}\ 2\ \frac{3}{2}$ 000− 000− $+1/4.3$

$\frac{1}{2}\ \tilde{1}\ \frac{3}{2}$ 000− 000− $+1/4.9$

$\frac{1}{2}\ \tilde{1}\ \tilde{\frac{1}{2}}$ 000+ 000+ $+1/9$

$\tilde{\frac{1}{2}}\ \tilde{1}\ \frac{1}{2}\quad \tilde{1}\ \tilde{\frac{1}{2}}\ \frac{1}{2}$

$\frac{1}{2}\ \frac{1}{2}\ 0$ 000− 000− $-1/4.3$

$\frac{1}{2}\ \frac{1}{2}\ 1$ 000+ 000+ $+5/4.9$

$\tilde{\frac{1}{2}}\ \tilde{1}\ \frac{1}{2}\quad \tilde{1}\ \tilde{\frac{1}{2}}\ \frac{3}{2}$

$\frac{1}{2}\ \frac{1}{2}\ 1$ 000− 000− $+1/2.9$

$\frac{1}{2}\ \frac{3}{2}\ 1$ 000+ 000+ $-1/4.9$

$\frac{1}{2}\ \frac{3}{2}\ 2$ 000− 000− $+1/4.3$

$\frac{1}{2}\ \frac{3}{2}\ \tilde{1}$ 000− 000− $+1/4.3$

$\tilde{\frac{1}{2}}\ \tilde{1}\ \frac{3}{2}\quad \frac{3}{2}\ 0\ \frac{3}{2}$

$1\ \tilde{1}\ 1$ 000− 000− $+1/4.3\sqrt{2}$

O and T_d 9j Symbols (cont.)

$\tilde{\frac{1}{2}}$	$\tilde{1}$	$\frac{3}{2}$	$\frac{3}{2}$	0	$\frac{3}{2}$
1	$\tilde{1}$	1	000−	001−	0
1	$\tilde{1}$	2	000+	000+	$+1/4.3\sqrt{2}$
1	$\tilde{1}$	$\tilde{1}$	000+	000+	$-1/4.3\sqrt{2.3}$
1	$\tilde{1}$	$\tilde{1}$	000+	001−	$-1/2.3\sqrt{2.3}$

$\tilde{\frac{1}{2}}$	$\tilde{1}$	$\frac{3}{2}$	$\frac{3}{2}$	$\frac{1}{2}$	1
1	$\frac{3}{2}$	$\frac{1}{2}$	000+	000+	+1/8.2.3
1	$\frac{3}{2}$	$\frac{3}{2}$	000−	000−	0
1	$\frac{3}{2}$	$\frac{3}{2}$	000−	001−	+1/8.3
1	$\frac{3}{2}$	$\frac{3}{2}$	001−	000−	−1/8.3
1	$\frac{3}{2}$	$\frac{3}{2}$	001−	001−	+1/8.2.3
1	$\frac{3}{2}$	$\tilde{\frac{1}{2}}$	000+	000+	+1/8.2.3
1	$\tilde{\frac{1}{2}}$	$\frac{3}{2}$	000+	000+	$-1/4.3\sqrt{2}$
1	$\tilde{\frac{1}{2}}$	$\frac{3}{2}$	000+	001+	0
1	$\tilde{\frac{1}{2}}$	$\tilde{\frac{1}{2}}$	000−	000−	+1/4.3

$\tilde{\frac{1}{2}}$	$\tilde{1}$	$\frac{3}{2}$	$\frac{3}{2}$	$\frac{1}{2}$	2
1	$\frac{3}{2}$	$\frac{1}{2}$	000−	000−	+1/8.2
1	$\frac{3}{2}$	$\frac{3}{2}$	000+	000+	−1/8.3
1	$\frac{3}{2}$	$\frac{3}{2}$	001+	000+	+1/8.2.3
1	$\frac{3}{2}$	$\tilde{\frac{1}{2}}$	000−	000−	+1/8.2.3
1	$\tilde{\frac{1}{2}}$	$\frac{3}{2}$	000−	000−	$+1/4.3\sqrt{2}$
1	$\tilde{\frac{1}{2}}$	$\tilde{\frac{1}{2}}$	000+	000+	+1/4.3

$\tilde{\frac{1}{2}}$	$\tilde{1}$	$\frac{3}{2}$	$\frac{3}{2}$	$\frac{1}{2}$	$\tilde{1}$
1	$\frac{3}{2}$	$\frac{1}{2}$	000−	000−	−1/8.2.3
1	$\frac{3}{2}$	$\frac{3}{2}$	000+	000+	+1/4.9
1	$\frac{3}{2}$	$\frac{3}{2}$	000+	001−	−1/8.9
1	$\frac{3}{2}$	$\frac{3}{2}$	001+	000+	−1/8.9
1	$\frac{3}{2}$	$\frac{3}{2}$	001+	001−	+7/8.2.9
1	$\frac{3}{2}$	$\tilde{\frac{1}{2}}$	000−	000−	−11/8.2.9
1	$\tilde{\frac{1}{2}}$	$\frac{3}{2}$	000−	000−	$+1/4.9\sqrt{2}$
1	$\tilde{\frac{1}{2}}$	$\frac{3}{2}$	000−	001+	$-1/2.9\sqrt{2}$
1	$\tilde{\frac{1}{2}}$	$\tilde{\frac{1}{2}}$	000+	000+	+1/4.9

$\tilde{\frac{1}{2}}$	$\tilde{1}$	$\frac{3}{2}$	$\frac{3}{2}$	1	$\frac{1}{2}$
1	1	1	000+	000+	$-1/8.3\sqrt{3}$
1	1	2	000−	000−	−1/8.3
1	1	$\tilde{1}$	000−	000−	+1/8.3
1	2	1	000−	000−	+1/8.3
1	2	$\tilde{1}$	000+	000+	−1/8.3
1	$\tilde{1}$	1	000−	000−	$-1/8.3\sqrt{3}$
1	$\tilde{1}$	2	000+	000+	$+1/8\sqrt{3}$
1	$\tilde{1}$	$\tilde{1}$	000+	000+	−1/8.3

$\tilde{\frac{1}{2}}$	$\tilde{1}$	$\frac{3}{2}$	$\frac{3}{2}$	1	$\frac{3}{2}$
1	1	0	000+	000+	$+1/4.3\sqrt{2}$
1	1	0	010+	000+	0
1	1	1	000−	000−	$-1/8.3\sqrt{3}$
1	1	1	000−	001−	0
1	1	1	010−	000−	$-1/4.3\sqrt{3}$
1	1	1	010−	001−	$+1/8.3\sqrt{3}$
1	1	2	000+	000+	+1/8.3
1	1	2	010+	000+	−1/8.3
1	1	$\tilde{1}$	000+	000+	−1/8.9
1	1	$\tilde{1}$	000+	001−	−1/4.9
1	1	$\tilde{1}$	010+	000+	+1/4.9
1	1	$\tilde{1}$	010+	001−	+1/8.9
1	2	1	000+	000+	+1/8.3
1	2	1	000+	001+	+1/8.3
1	2	1	010+	000+	−1/8.3
1	2	1	010+	001+	0
1	2	$\tilde{1}$	000−	000−	−1/8.9
1	2	$\tilde{1}$	000−	001+	+1/8.9
1	2	$\tilde{1}$	010−	000−	−1/8.9
1	2	$\tilde{1}$	010−	001+	+1/2.9
1	$\tilde{1}$	1	000+	000+	$+1/8.3\sqrt{3}$
1	$\tilde{1}$	1	000+	001+	$-1/4.3\sqrt{3}$
1	$\tilde{1}$	1	010+	000+	0
1	$\tilde{1}$	1	010+	001+	$-1/8.3\sqrt{3}$
1	$\tilde{1}$	2	000−	000−	$-1/8.3\sqrt{3}$
1	$\tilde{1}$	2	010−	000−	$-1/8.3\sqrt{3}$
1	$\tilde{1}$	$\tilde{1}$	000−	000−	+1/8.3
1	$\tilde{1}$	$\tilde{1}$	000−	001+	0
1	$\tilde{1}$	$\tilde{1}$	010−	000−	0
1	$\tilde{1}$	$\tilde{1}$	010−	001+	−1/8.3

$\tilde{\frac{1}{2}}$	$\tilde{1}$	$\frac{3}{2}$	$\frac{3}{2}$	1	$\tilde{\frac{1}{2}}$
1	1	1	000+	000+	$+1/8.3\sqrt{3}$
1	1	2	000−	000−	−1/8.3
1	1	$\tilde{1}$	000−	000−	+5/8.9
1	2	1	000−	000−	+1/8.3
1	2	$\tilde{1}$	000+	000+	+5/8.9
1	$\tilde{1}$	1	000−	000−	$+1/8.3\sqrt{3}$
1	$\tilde{1}$	2	000+	000+	$-1/8.3\sqrt{3}$
1	$\tilde{1}$	$\tilde{1}$	000+	000+	+1/8.3

$\tilde{\frac{1}{2}}$	$\tilde{1}$	$\frac{3}{2}$	$\frac{3}{2}$	$\frac{3}{2}$	0
1	$\frac{1}{2}$	$\frac{3}{2}$	000+	000+	+1/8.2.3
1	$\frac{3}{2}$	$\frac{3}{2}$	000−	000−	−1/8.3
1	$\frac{3}{2}$	$\frac{3}{2}$	000−	010+	−1/8.3
1	$\frac{3}{2}$	$\frac{3}{2}$	001−	000−	+1/8.3
1	$\frac{3}{2}$	$\frac{3}{2}$	001−	010+	−1/8.2.3
1	$\tilde{\frac{1}{2}}$	$\frac{3}{2}$	000+	000+	+1/8.2

$\tilde{\frac{1}{2}}$	$\tilde{1}$	$\frac{3}{2}$	$\frac{3}{2}$	$\frac{3}{2}$	1
1	$\frac{1}{2}$	$\frac{1}{2}$	000+	000+	$-1/4.9\sqrt{2}$
1	$\frac{1}{2}$	$\frac{1}{2}$	010+	000+	$+1/2.9\sqrt{2}$
1	$\frac{1}{2}$	$\frac{3}{2}$	000−	000−	−7/8.2.9
1	$\frac{1}{2}$	$\frac{3}{2}$	000−	001−	+1/4.9
1	$\frac{1}{2}$	$\frac{3}{2}$	010−	000−	−1/4.9
1	$\frac{1}{2}$	$\frac{3}{2}$	010−	001−	+1/8.2.9
1	$\frac{3}{2}$	$\frac{1}{2}$	000−	000−	−1/4.9
1	$\frac{3}{2}$	$\frac{1}{2}$	000−	010+	+1/8.9
1	$\frac{3}{2}$	$\frac{1}{2}$	010−	000−	+1/8.9
1	$\frac{3}{2}$	$\frac{1}{2}$	010−	010+	−7/8.2.9
1	$\frac{3}{2}$	$\frac{3}{2}$	000+	000+	+1/8.9
1	$\frac{3}{2}$	$\frac{3}{2}$	000+	001+	+1/4.9
1	$\frac{3}{2}$	$\frac{3}{2}$	000+	010−	+1/8.9
1	$\frac{3}{2}$	$\frac{3}{2}$	000+	011−	−1/8.9
1	$\frac{3}{2}$	$\frac{3}{2}$	001+	000+	−1/8.9
1	$\frac{3}{2}$	$\frac{3}{2}$	001+	001+	−1/4.9
1	$\frac{3}{2}$	$\frac{3}{2}$	001+	010−	−5/8.2.9
1	$\frac{3}{2}$	$\frac{3}{2}$	001+	011−	+1/8.9
1	$\frac{3}{2}$	$\frac{3}{2}$	010+	000+	−1/4.9
1	$\frac{3}{2}$	$\frac{3}{2}$	010+	001+	−1/8.9
1	$\frac{3}{2}$	$\frac{3}{2}$	010+	010−	−1/4.9
1	$\frac{3}{2}$	$\frac{3}{2}$	010+	011−	−1/8.9
1	$\frac{3}{2}$	$\frac{3}{2}$	011+	000+	−1/8.9
1	$\frac{3}{2}$	$\frac{3}{2}$	011+	001+	+1/8.9
1	$\frac{3}{2}$	$\frac{3}{2}$	011+	010−	−1/8.9
1	$\frac{3}{2}$	$\frac{3}{2}$	011+	011−	+1/8.9
1	$\frac{3}{2}$	$\tilde{\frac{1}{2}}$	000−	000−	0
1	$\frac{3}{2}$	$\tilde{\frac{1}{2}}$	000−	010+	+1/8.3
1	$\frac{3}{2}$	$\tilde{\frac{1}{2}}$	010−	000−	−1/8.3
1	$\frac{3}{2}$	$\tilde{\frac{1}{2}}$	010−	010+	+1/8.2.3
1	$\tilde{\frac{1}{2}}$	$\frac{3}{2}$	000−	000−	+1/8.2.3
1	$\tilde{\frac{1}{2}}$	$\frac{3}{2}$	000−	001−	0
1	$\tilde{\frac{1}{2}}$	$\frac{3}{2}$	010−	000−	0
1	$\tilde{\frac{1}{2}}$	$\frac{3}{2}$	010−	001−	+1/8.2
1	$\tilde{\frac{1}{2}}$	$\tilde{\frac{1}{2}}$	000+	000+	$+1/4.3\sqrt{2}$
1	$\tilde{\frac{1}{2}}$	$\tilde{\frac{1}{2}}$	010+	000+	0

$\tilde{\frac{1}{2}}$	$\tilde{1}$	$\frac{3}{2}$	$\frac{3}{2}$	$\frac{3}{2}$	2
1	$\frac{1}{2}$	$\frac{1}{2}$	000−	000−	$-1/4.3\sqrt{2}$
1	$\frac{1}{2}$	$\frac{3}{2}$	000+	000+	+1/8.2.3
1	$\frac{3}{2}$	$\frac{1}{2}$	000+	000+	+1/8.3
1	$\frac{3}{2}$	$\frac{1}{2}$	000+	010−	−1/8.2.3
1	$\frac{3}{2}$	$\frac{3}{2}$	000−	000−	+1/8.3
1	$\frac{3}{2}$	$\frac{3}{2}$	000−	010+	0
1	$\frac{3}{2}$	$\frac{3}{2}$	001−	000−	0
1	$\frac{3}{2}$	$\frac{3}{2}$	001−	010+	−1/8.3
1	$\frac{3}{2}$	$\tilde{\frac{1}{2}}$	000+	000+	−1/8.3
1	$\frac{3}{2}$	$\tilde{\frac{1}{2}}$	000+	010−	−1/8.2.3
1	$\tilde{\frac{1}{2}}$	$\frac{3}{2}$	000+	000+	+1/8.2.3
1	$\tilde{\frac{1}{2}}$	$\tilde{\frac{1}{2}}$	000−	000−	$-1/4.3\sqrt{2}$

$\tilde{\frac{1}{2}}$	$\tilde{1}$	$\frac{3}{2}$	$\frac{3}{2}$	$\frac{3}{2}$	$\tilde{1}$
1	$\frac{1}{2}$	$\frac{1}{2}$	000−	000−	$+1/4.3\sqrt{2}$
1	$\frac{1}{2}$	$\frac{1}{2}$	010+	000−	0
1	$\frac{1}{2}$	$\frac{3}{2}$	000+	000+	−1/8.2.3
1	$\frac{1}{2}$	$\frac{3}{2}$	000+	001−	0
1	$\frac{1}{2}$	$\frac{3}{2}$	010−	000+	0
1	$\frac{1}{2}$	$\frac{3}{2}$	010−	001−	+1/8.2
1	$\frac{3}{2}$	$\frac{1}{2}$	000+	000+	0
1	$\frac{3}{2}$	$\frac{1}{2}$	000+	010−	+1/8.3
1	$\frac{3}{2}$	$\frac{1}{2}$	010−	000+	+1/8.3

O and T_d 9j Symbols (cont.)

$\tilde{\frac{1}{2}}$	$\tilde{1}$	$\frac{3}{2}$	$\frac{3}{2}$	$\frac{3}{2}$	$\tilde{1}$
1	$\frac{3}{2}$	$\frac{1}{2}$	010−	010−	$+1/8.2.3$
1	$\frac{3}{2}$	$\frac{3}{2}$	000−	000−	$-1/8.9$
1	$\frac{3}{2}$	$\frac{3}{2}$	000−	001+	$+1/4.9$
1	$\frac{3}{2}$	$\frac{3}{2}$	000−	010+	$+1/8.9$
1	$\frac{3}{2}$	$\frac{3}{2}$	000−	011−	$+1/8.9$
1	$\frac{3}{2}$	$\frac{3}{2}$	001−	000−	$-1/8.9$
1	$\frac{3}{2}$	$\frac{3}{2}$	001−	001+	$+1/4.9$
1	$\frac{3}{2}$	$\frac{3}{2}$	001−	010+	$+5/8.2.9$
1	$\frac{3}{2}$	$\frac{3}{2}$	001−	011−	$+1/8.9$
1	$\frac{3}{2}$	$\frac{3}{2}$	010+	000−	$+1/4.9$
1	$\frac{3}{2}$	$\frac{3}{2}$	010+	001+	$-1/8.9$
1	$\frac{3}{2}$	$\frac{3}{2}$	010+	010+	$-1/4.9$
1	$\frac{3}{2}$	$\frac{3}{2}$	010+	011−	$+1/8.9$
1	$\frac{3}{2}$	$\frac{3}{2}$	011+	000−	$-1/8.9$
1	$\frac{3}{2}$	$\frac{3}{2}$	011+	001+	$-1/8.9$
1	$\frac{3}{2}$	$\frac{3}{2}$	011+	010+	$+1/8.9$
1	$\frac{3}{2}$	$\frac{3}{2}$	011+	011−	$+1/8.9$
1	$\frac{3}{2}$	$\tilde{\frac{1}{2}}$	000+	000+	$+1/4.9$
1	$\frac{3}{2}$	$\tilde{\frac{1}{2}}$	000+	010−	$+1/8.9$
1	$\frac{3}{2}$	$\tilde{\frac{1}{2}}$	010−	000+	$-1/8.9$
1	$\frac{3}{2}$	$\tilde{\frac{1}{2}}$	010−	010−	$-7/8.2.9$
1	$\tilde{\frac{1}{2}}$	$\frac{3}{2}$	000+	000+	$+7/8.2.9$
1	$\tilde{\frac{1}{2}}$	$\frac{3}{2}$	000+	001−	$+1/4.9$
1	$\tilde{\frac{1}{2}}$	$\frac{3}{2}$	010−	000+	$+1/4.9$
1	$\tilde{\frac{1}{2}}$	$\frac{3}{2}$	010−	001−	$+1/8.2.9$
1	$\tilde{\frac{1}{2}}$	$\tilde{\frac{1}{2}}$	000−	000−	$-1/4.9\sqrt{2}$
1	$\tilde{\frac{1}{2}}$	$\tilde{\frac{1}{2}}$	010+	000−	$+1/2.9\sqrt{2}$

$\tilde{\frac{1}{2}}$	$\tilde{1}$	$\frac{3}{2}$	$\frac{3}{2}$	2	$\frac{1}{2}$
1	1	1	000−	000−	$+1/8.3\sqrt{3}$
1	1	2	000+	000+	$-1/4.3$
1	1	$\tilde{1}$	000+	000+	$+1/8.3$
1	$\tilde{1}$	1	000+	000+	$+1/8.3$
1	$\tilde{1}$	2	000−	000−	0
1	$\tilde{1}$	$\tilde{1}$	000−	000−	$-1/8\sqrt{3}$

$\tilde{\frac{1}{2}}$	$\tilde{1}$	$\frac{3}{2}$	$\frac{3}{2}$	2	$\frac{3}{2}$
1	1	0	000−	000−	$+1/4.3\sqrt{2}$
1	1	1	000+	000+	$+1/8.3\sqrt{3}$
1	1	1	000+	001+	$-1/8.3\sqrt{3}$
1	1	2	000−	000−	$-1/8.3$
1	1	$\tilde{1}$	000−	000−	$+1/8.3$
1	1	$\tilde{1}$	000−	001+	$+1/8.3$
1	$\tilde{1}$	1	000−	000−	$-1/8.3$
1	$\tilde{1}$	1	000−	001−	$-1/8.3$
1	$\tilde{1}$	2	000+	000+	$+1/8.3$
1	$\tilde{1}$	$\tilde{1}$	000+	000+	$-1/8.3\sqrt{3}$
1	$\tilde{1}$	$\tilde{1}$	000+	001−	$+1/8.3\sqrt{3}$

$\tilde{\frac{1}{2}}$	$\tilde{1}$	$\frac{3}{2}$	$\frac{3}{2}$	2	$\tilde{\frac{1}{2}}$
1	1	1	000−	000−	$-1/8\sqrt{3}$
1	1	2	000+	000+	0
1	1	$\tilde{1}$	000+	000+	$+1/8.3$
1	$\tilde{1}$	1	000+	000+	$+1/8.3$
1	$\tilde{1}$	2	000−	000−	$+1/4.3$
1	$\tilde{1}$	$\tilde{1}$	000−	000−	$+1/8.3\sqrt{3}$

$\tilde{\frac{1}{2}}$	$\tilde{1}$	$\frac{3}{2}$	$\frac{3}{2}$	$\tilde{1}$	$\frac{1}{2}$
1	0	1	000+	000+	$+1/4.9$
1	1	1	000−	000−	$-1/8.3$
1	1	2	000+	000+	$-1/8.3\sqrt{3}$
1	1	$\tilde{1}$	000+	000+	$-1/8.3\sqrt{3}$
1	2	1	000+	000+	$+5/8.9$
1	2	$\tilde{1}$	000−	000−	$+1/8.3$
1	$\tilde{1}$	1	000+	000+	$-5/8.9$
1	$\tilde{1}$	2	000−	000−	$-1/8.3$
1	$\tilde{1}$	$\tilde{1}$	000−	000−	$-1/8.3\sqrt{3}$

$\tilde{\frac{1}{2}}$	$\tilde{1}$	$\frac{3}{2}$	$\frac{3}{2}$	$\tilde{1}$	$\frac{3}{2}$
1	0	1	000−	000−	$-1/2.9$
1	0	1	000−	001−	$+1/2.9$
1	0	1	010+	000−	$+1/2.9$
1	0	1	010+	001−	$+1/4.9$
1	1	0	000−	000−	$-1/4.3\sqrt{2.3}$
1	1	0	010+	000−	$-1/2.3\sqrt{2.3}$
1	1	1	000+	000+	$-1/8.3$
1	1	1	000+	001+	0
1	1	1	010−	000+	0
1	1	1	010−	001+	$-1/8.3$
1	1	2	000−	000−	$+1/8.3\sqrt{3}$
1	1	2	010+	000−	$-1/8.3\sqrt{3}$
1	1	$\tilde{1}$	000−	000−	$-1/8.3\sqrt{3}$
1	1	$\tilde{1}$	000−	001+	$+1/4.3\sqrt{3}$
1	1	$\tilde{1}$	010+	000−	0
1	1	$\tilde{1}$	010+	001+	$-1/8.3\sqrt{3}$
1	2	1	000−	000−	$-1/8.9$
1	2	1	000−	001−	$+1/8.9$
1	2	1	010+	000−	$+1/8.9$
1	2	1	010+	001−	$-1/2.9$
1	2	$\tilde{1}$	000+	000+	$+1/8.3$
1	2	$\tilde{1}$	000+	001−	$+1/8.3$
1	2	$\tilde{1}$	010−	000+	$+1/8.3$
1	2	$\tilde{1}$	010−	001−	0
1	$\tilde{1}$	1	000−	000−	$+1/8.9$
1	$\tilde{1}$	1	000−	001−	$+1/4.9$
1	$\tilde{1}$	1	010+	000−	$+1/4.9$
1	$\tilde{1}$	1	010+	001−	$+1/8.9$
1	$\tilde{1}$	2	000+	000+	$-1/8.3$
1	$\tilde{1}$	2	010−	000+	$-1/8.3$
1	$\tilde{1}$	$\tilde{1}$	000+	000+	$+1/8.3\sqrt{3}$
1	$\tilde{1}$	$\tilde{1}$	000+	001−	0
1	$\tilde{1}$	$\tilde{1}$	010−	000+	$-1/4.3\sqrt{3}$
1	$\tilde{1}$	$\tilde{1}$	010−	001−	$+1/8.3\sqrt{3}$

$\tilde{\frac{1}{2}}$	$\tilde{1}$	$\frac{3}{2}$	$\frac{3}{2}$	$\tilde{1}$	$\tilde{\frac{1}{2}}$
1	0	1	000+	000+	$+1/4.3$
1	1	1	000−	000−	$+1/8.3$
1	1	2	000+	000+	$+1/8\sqrt{3}$
1	1	$\tilde{1}$	000+	000+	$+1/8.3\sqrt{3}$
1	2	1	000+	000+	$-1/8.3$
1	2	$\tilde{1}$	000−	000−	$+1/8.3$
1	$\tilde{1}$	1	000+	000+	$-1/8.3$
1	$\tilde{1}$	2	000−	000−	$-1/8.3$
1	$\tilde{1}$	$\tilde{1}$	000−	000−	$+1/8.3\sqrt{3}$

$\tilde{\frac{1}{2}}$	$\tilde{1}$	$\frac{3}{2}$	$\frac{3}{2}$	$\tilde{\frac{1}{2}}$	1
1	$\frac{1}{2}$	$\frac{1}{2}$	000−	000−	$+1/4.9$
1	$\frac{1}{2}$	$\frac{3}{2}$	000+	000+	$-1/4.9\sqrt{2}$
1	$\frac{1}{2}$	$\frac{3}{2}$	000+	001+	$-1/2.9\sqrt{2}$
1	$\frac{3}{2}$	$\frac{1}{2}$	000+	000+	$+11/8.2.9$
1	$\frac{3}{2}$	$\frac{3}{2}$	000−	000−	$-1/4.9$
1	$\frac{3}{2}$	$\frac{3}{2}$	000−	001−	$-1/8.9$
1	$\frac{3}{2}$	$\frac{3}{2}$	001−	000−	$-1/8.9$
1	$\frac{3}{2}$	$\frac{3}{2}$	001−	001−	$-7/8.2.9$
1	$\frac{3}{2}$	$\tilde{\frac{1}{2}}$	000+	000+	$+1/8.2.3$

$\tilde{\frac{1}{2}}$	$\tilde{1}$	$\frac{3}{2}$	$\frac{3}{2}$	$\tilde{\frac{1}{2}}$	2
1	$\frac{1}{2}$	$\frac{1}{2}$	000+	000+	$-1/4.3$
1	$\frac{1}{2}$	$\frac{3}{2}$	000−	000−	$-1/4.3\sqrt{2}$
1	$\frac{3}{2}$	$\frac{1}{2}$	000−	000−	$+1/8.2.3$
1	$\frac{3}{2}$	$\frac{3}{2}$	000+	000+	$+1/8.3$
1	$\frac{3}{2}$	$\frac{3}{2}$	001+	000+	$+1/8.2.3$
1	$\frac{3}{2}$	$\tilde{\frac{1}{2}}$	000−	000−	$+1/8.2$

$\tilde{\frac{1}{2}}$	$\tilde{1}$	$\frac{3}{2}$	$\frac{3}{2}$	$\tilde{\frac{1}{2}}$	$\tilde{1}$
1	$\frac{1}{2}$	$\frac{1}{2}$	000+	000+	$+1/4.3$
1	$\frac{1}{2}$	$\frac{3}{2}$	000−	000−	$+1/4.3\sqrt{2}$
1	$\frac{1}{2}$	$\frac{3}{2}$	000−	001+	0
1	$\frac{3}{2}$	$\frac{1}{2}$	000−	000−	$-1/8.2.3$
1	$\frac{3}{2}$	$\frac{3}{2}$	000+	000+	0
1	$\frac{3}{2}$	$\frac{3}{2}$	000+	001−	$+1/8.3$
1	$\frac{3}{2}$	$\frac{3}{2}$	001+	000+	$-1/8.3$
1	$\frac{3}{2}$	$\frac{3}{2}$	001+	001−	$-1/8.2.3$
1	$\frac{3}{2}$	$\tilde{\frac{1}{2}}$	000−	000−	$-1/8.2.3$

$\tilde{\frac{1}{2}}$	$\tilde{1}$	$\frac{3}{2}$	2	0	2
$\frac{3}{2}$	$\tilde{1}$	$\frac{1}{2}$	000+	000+	$-1/4\sqrt{2.3}$
$\frac{3}{2}$	$\tilde{1}$	$\frac{3}{2}$	000−	000−	0
$\frac{3}{2}$	$\tilde{1}$	$\frac{3}{2}$	001+	000−	$-1/4\sqrt{2.3}$
$\frac{3}{2}$	$\tilde{1}$	$\tilde{\frac{1}{2}}$	000+	000+	$-1/4\sqrt{2.3}$

$\tilde{\frac{1}{2}}$	$\tilde{1}$	$\frac{3}{2}$	2	$\frac{1}{2}$	$\frac{3}{2}$
$\frac{3}{2}$	$\frac{3}{2}$	0	000−	000−	$-1/8.2$
$\frac{3}{2}$	$\frac{3}{2}$	1	000+	000+	$+1/8.2.3$
$\frac{3}{2}$	$\frac{3}{2}$	1	000+	001+	$+1/8.3$
$\frac{3}{2}$	$\frac{3}{2}$	1	001+	000+	$-1/8.3$
$\frac{3}{2}$	$\frac{3}{2}$	1	001+	001+	$-1/8.2.3$
$\frac{3}{2}$	$\frac{3}{2}$	2	000−	000−	$+1/8.2$
$\frac{3}{2}$	$\frac{3}{2}$	$\tilde{1}$	000−	000−	$-1/8.2.3$

O and T_d 9j Symbols (cont.)

$\tilde{\frac{1}{2}}\ \tilde{1}\ \frac{3}{2}\quad 2\ \frac{1}{2}\ \frac{3}{2}$			
$\frac{3}{2}\ \frac{3}{2}\ \tilde{1}$	000−	001+	$-1/8.3$
$\frac{3}{2}\ \frac{3}{2}\ \tilde{1}$	001+	000−	$-1/8.3$
$\frac{3}{2}\ \frac{3}{2}\ \tilde{1}$	001+	001+	$-1/8.2.3$
$\frac{3}{2}\ \tilde{\frac{1}{2}}\ 1$	000−	000−	$-1/4.3\sqrt{2}$
$\frac{3}{2}\ \tilde{\frac{1}{2}}\ 1$	000−	001−	$+1/4.3\sqrt{2}$
$\frac{3}{2}\ \tilde{\frac{1}{2}}\ 2$	000+	000+	0
$\frac{3}{2}\ \tilde{\frac{1}{2}}\ \tilde{1}$	000+	000+	$+1/4.3\sqrt{2}$
$\frac{3}{2}\ \tilde{\frac{1}{2}}\ \tilde{1}$	000+	001−	$-1/4.3\sqrt{2}$
$\tilde{\frac{1}{2}}\ \tilde{1}\ \frac{3}{2}\quad 2\ 1\ 1$			
$\frac{3}{2}\ 1\ \frac{1}{2}$	000−	000−	$-1/8.3$
$\frac{3}{2}\ 1\ \frac{3}{2}$	000+	000+	$-1/8.3$
$\frac{3}{2}\ 1\ \frac{3}{2}$	000+	001+	$-1/8.3$
$\frac{3}{2}\ 1\ \frac{3}{2}$	001+	000+	0
$\frac{3}{2}\ 1\ \frac{3}{2}$	001+	001+	$+1/8.3$
$\frac{3}{2}\ 1\ \tilde{\frac{1}{2}}$	000−	000−	$+1/8.3$
$\frac{3}{2}\ 2\ \frac{1}{2}$	000+	000+	$-1/4.3$
$\frac{3}{2}\ 2\ \frac{3}{2}$	000−	000−	$-1/8.3$
$\frac{3}{2}\ 2\ \frac{3}{2}$	000−	001−	$+1/8.3$
$\frac{3}{2}\ 2\ \tilde{\frac{1}{2}}$	000+	000+	0
$\frac{3}{2}\ \tilde{1}\ \frac{1}{2}$	000+	000+	$+1/8.3\sqrt{3}$
$\frac{3}{2}\ \tilde{1}\ \frac{3}{2}$	000−	000−	$-1/8.3\sqrt{3}$
$\frac{3}{2}\ \tilde{1}\ \frac{3}{2}$	000−	001−	$+1/8.3\sqrt{3}$
$\frac{3}{2}\ \tilde{1}\ \frac{3}{2}$	001+	000−	$-1/4.3\sqrt{3}$
$\frac{3}{2}\ \tilde{1}\ \frac{3}{2}$	001+	001−	$-1/8.3\sqrt{3}$
$\frac{3}{2}\ \tilde{1}\ \tilde{\frac{1}{2}}$	000+	000+	$+1/8\sqrt{3}$
$\tilde{\frac{1}{2}}\ \tilde{1}\ \frac{3}{2}\quad 2\ 1\ \tilde{1}$			
$\frac{3}{2}\ 1\ \frac{1}{2}$	000+	000+	$-1/8.3$
$\frac{3}{2}\ 1\ \frac{3}{2}$	000−	000−	$+1/8.3$
$\frac{3}{2}\ 1\ \frac{3}{2}$	000−	001+	$-1/8.3$
$\frac{3}{2}\ 1\ \frac{3}{2}$	001−	000−	0
$\frac{3}{2}\ 1\ \frac{3}{2}$	001−	001+	$-1/8.3$
$\frac{3}{2}\ 1\ \tilde{\frac{1}{2}}$	000+	000+	$+1/8.3$
$\frac{3}{2}\ 2\ \frac{1}{2}$	000−	000−	0
$\frac{3}{2}\ 2\ \frac{3}{2}$	000+	000+	$-1/8.3$
$\frac{3}{2}\ 2\ \frac{3}{2}$	000+	001−	$-1/8.3$
$\frac{3}{2}\ 2\ \tilde{\frac{1}{2}}$	000−	000−	$-1/4.3$
$\frac{3}{2}\ \tilde{1}\ \frac{1}{2}$	000−	000−	$-1/8\sqrt{3}$
$\frac{3}{2}\ \tilde{1}\ \frac{3}{2}$	000+	000+	$+1/8.3\sqrt{3}$
$\frac{3}{2}\ \tilde{1}\ \frac{3}{2}$	000+	001−	$+1/8.3\sqrt{3}$
$\frac{3}{2}\ \tilde{1}\ \frac{3}{2}$	001−	000+	$-1/4.3\sqrt{3}$
$\frac{3}{2}\ \tilde{1}\ \frac{3}{2}$	001−	001−	$+1/8.3\sqrt{3}$
$\frac{3}{2}\ \tilde{1}\ \tilde{\frac{1}{2}}$	000−	000−	$-1/8.3\sqrt{3}$
$\tilde{\frac{1}{2}}\ \tilde{1}\ \frac{3}{2}\quad 2\ \frac{3}{2}\ \frac{1}{2}$			
$\frac{3}{2}\ \frac{1}{2}\ 1$	000−	000−	$-1/8.2.3$
$\frac{3}{2}\ \frac{1}{2}\ 2$	000+	000+	$-1/8.2$
$\frac{3}{2}\ \frac{1}{2}\ \tilde{1}$	000+	000+	$+1/8.2$
$\frac{3}{2}\ \frac{3}{2}\ 1$	000+	000+	$-1/4.3$
$\frac{3}{2}\ \frac{3}{2}\ 1$	000+	010−	0
$\frac{3}{2}\ \frac{3}{2}\ 1$	001+	000+	0

$\tilde{\frac{1}{2}}\ \tilde{1}\ \frac{3}{2}\quad 2\ \frac{3}{2}\ \frac{1}{2}$			
$\frac{3}{2}\ \frac{3}{2}\ 1$	001+	010−	$+1/8.2.3$
$\frac{3}{2}\ \frac{3}{2}\ 2$	000−	000−	0
$\frac{3}{2}\ \frac{3}{2}\ 2$	000−	010+	$+1/8.2$
$\frac{3}{2}\ \frac{3}{2}\ \tilde{1}$	000−	000−	0
$\frac{3}{2}\ \frac{3}{2}\ \tilde{1}$	000−	010+	0
$\frac{3}{2}\ \frac{3}{2}\ \tilde{1}$	001+	000−	0
$\frac{3}{2}\ \frac{3}{2}\ \tilde{1}$	001+	010+	$-1/8.2$
$\frac{3}{2}\ \tilde{\frac{1}{2}}\ 1$	000−	000−	$-1/8.2.3$
$\frac{3}{2}\ \tilde{\frac{1}{2}}\ 2$	000+	000+	$+1/8.2$
$\frac{3}{2}\ \tilde{\frac{1}{2}}\ \tilde{1}$	000+	000+	$+1/8.2$
$\tilde{\frac{1}{2}}\ \tilde{1}\ \frac{3}{2}\quad 2\ \frac{3}{2}\ \frac{3}{2}$			
$\frac{3}{2}\ \frac{1}{2}\ 1$	000+	000+	$-1/8.3$
$\frac{3}{2}\ \frac{1}{2}\ 1$	000+	001+	$-1/8.2.3$
$\frac{3}{2}\ \frac{1}{2}\ 2$	000−	000−	$+1/8.2$
$\frac{3}{2}\ \frac{1}{2}\ \tilde{1}$	000−	000−	$-1/8.3$
$\frac{3}{2}\ \frac{1}{2}\ \tilde{1}$	000−	001+	$-1/8.2.3$
$\frac{3}{2}\ \frac{3}{2}\ 0$	000+	000+	0
$\frac{3}{2}\ \frac{3}{2}\ 0$	000+	010−	$-1/8.2$
$\frac{3}{2}\ \frac{3}{2}\ 1$	000−	000−	0
$\frac{3}{2}\ \frac{3}{2}\ 1$	000−	001−	0
$\frac{3}{2}\ \frac{3}{2}\ 1$	000−	010+	$+1/8.2.3$
$\frac{3}{2}\ \frac{3}{2}\ 1$	000−	011+	$+1/8.3$
$\frac{3}{2}\ \frac{3}{2}\ 1$	001−	000−	$+1/8.3$
$\frac{3}{2}\ \frac{3}{2}\ 1$	001−	001−	$-1/8.3$
$\frac{3}{2}\ \frac{3}{2}\ 1$	001−	010+	0
$\frac{3}{2}\ \frac{3}{2}\ 1$	001−	011+	0
$\frac{3}{2}\ \frac{3}{2}\ 2$	000+	000+	0
$\frac{3}{2}\ \frac{3}{2}\ 2$	000+	010−	0
$\frac{3}{2}\ \frac{3}{2}\ \tilde{1}$	000+	000+	0
$\frac{3}{2}\ \frac{3}{2}\ \tilde{1}$	000+	001	0
$\frac{3}{2}\ \frac{3}{2}\ \tilde{1}$	000+	010−	$+1/8.2.3$
$\frac{3}{2}\ \frac{3}{2}\ \tilde{1}$	000+	011+	$+1/8.3$
$\frac{3}{2}\ \frac{3}{2}\ \tilde{1}$	001−	000+	$-1/8.3$
$\frac{3}{2}\ \frac{3}{2}\ \tilde{1}$	001−	001−	$+1/8.3$
$\frac{3}{2}\ \frac{3}{2}\ \tilde{1}$	001−	010−	0
$\frac{3}{2}\ \frac{3}{2}\ \tilde{1}$	001−	011+	0
$\frac{3}{2}\ \tilde{\frac{1}{2}}\ 1$	000+	000+	$+1/8.3$
$\frac{3}{2}\ \tilde{\frac{1}{2}}\ 1$	000+	001+	$+1/8.2.3$
$\frac{3}{2}\ \tilde{\frac{1}{2}}\ 2$	000−	000−	$+1/8.2$
$\frac{3}{2}\ \tilde{\frac{1}{2}}\ \tilde{1}$	000−	000−	$+1/8.3$
$\frac{3}{2}\ \tilde{\frac{1}{2}}\ \tilde{1}$	000−	001+	$+1/8.2.3$
$\tilde{\frac{1}{2}}\ \tilde{1}\ \frac{3}{2}\quad 2\ \frac{3}{2}\ \tilde{\frac{1}{2}}$			
$\frac{3}{2}\ \frac{1}{2}\ 1$	000−	000−	$-1/8.2$
$\frac{3}{2}\ \frac{1}{2}\ 2$	000+	000+	$-1/8.2$
$\frac{3}{2}\ \frac{1}{2}\ \tilde{1}$	000+	000+	$+1/8.2.3$
$\frac{3}{2}\ \frac{3}{2}\ 1$	000+	000+	0
$\frac{3}{2}\ \frac{3}{2}\ 1$	000+	010−	0
$\frac{3}{2}\ \frac{3}{2}\ 1$	001+	000+	0
$\frac{3}{2}\ \frac{3}{2}\ 1$	001+	010−	$-1/8.2$

$\tilde{\frac{1}{2}}\ \tilde{1}\ \frac{3}{2}\quad 2\ \frac{3}{2}\ \tilde{\frac{1}{2}}$			
$\frac{3}{2}\ \frac{3}{2}\ 2$	000−	000−	0
$\frac{3}{2}\ \frac{3}{2}\ 2$	000−	010+	$-1/8.2$
$\frac{3}{2}\ \frac{3}{2}\ \tilde{1}$	000−	000−	$+1/4.3$
$\frac{3}{2}\ \frac{3}{2}\ \tilde{1}$	000−	010+	0
$\frac{3}{2}\ \frac{3}{2}\ \tilde{1}$	001+	000−	0
$\frac{3}{2}\ \frac{3}{2}\ \tilde{1}$	001+	010+	$+1/8.2.3$
$\frac{3}{2}\ \tilde{\frac{1}{2}}\ 1$	000−	000−	$-1/8.2$
$\frac{3}{2}\ \tilde{\frac{1}{2}}\ 2$	000+	000+	$+1/8.2$
$\frac{3}{2}\ \tilde{\frac{1}{2}}\ \tilde{1}$	000+	000+	$+1/8.2.3$
$\tilde{\frac{1}{2}}\ \tilde{1}\ \frac{3}{2}\quad 2\ 2\ 0$			
$\frac{3}{2}\ 1\ \frac{3}{2}$	000+	000+	$+1/8\sqrt{3}$
$\frac{3}{2}\ 1\ \frac{3}{2}$	001+	000+	$-1/8\sqrt{3}$
$\frac{3}{2}\ \tilde{1}\ \frac{3}{2}$	000−	000−	$-1/8\sqrt{3}$
$\frac{3}{2}\ \tilde{1}\ \frac{3}{2}$	001+	000−	$+1/8\sqrt{3}$
$\tilde{\frac{1}{2}}\ \tilde{1}\ \frac{3}{2}\quad 2\ 2\ 2$			
$\frac{3}{2}\ 1\ \frac{1}{2}$	000−	000−	$+1/8\sqrt{3}$
$\frac{3}{2}\ 1\ \frac{3}{2}$	000+	000+	$-1/8\sqrt{3}$
$\frac{3}{2}\ 1\ \frac{3}{2}$	001+	000+	0
$\frac{3}{2}\ 1\ \tilde{\frac{1}{2}}$	000−	000−	$-1/8\sqrt{3}$
$\frac{3}{2}\ \tilde{1}\ \frac{1}{2}$	000+	000+	$+1/8\sqrt{3}$
$\frac{3}{2}\ \tilde{1}\ \frac{3}{2}$	000−	000−	$-1/8\sqrt{3}$
$\frac{3}{2}\ \tilde{1}\ \frac{3}{2}$	001+	000−	0
$\frac{3}{2}\ \tilde{1}\ \tilde{\frac{1}{2}}$	000+	000+	$-1/8\sqrt{3}$
$\tilde{\frac{1}{2}}\ \tilde{1}\ \frac{3}{2}\quad 2\ \tilde{1}\ 1$			
$\frac{3}{2}\ 0\ \frac{3}{2}$	000+	000+	$+1/4.3\sqrt{2}$
$\frac{3}{2}\ 0\ \frac{3}{2}$	000+	001+	$-1/4.3\sqrt{2}$
$\frac{3}{2}\ 1\ \frac{1}{2}$	000+	000+	$+1/8.3\sqrt{3}$
$\frac{3}{2}\ 1\ \frac{3}{2}$	000−	000−	$-1/8.3\sqrt{3}$
$\frac{3}{2}\ 1\ \frac{3}{2}$	000−	001−	$+1/8.3\sqrt{3}$
$\frac{3}{2}\ 1\ \frac{3}{2}$	001−	000−	$-1/4.3\sqrt{3}$
$\frac{3}{2}\ 1\ \frac{3}{2}$	001−	001−	$-1/8.3\sqrt{3}$
$\frac{3}{2}\ 1\ \tilde{\frac{1}{2}}$	000+	000+	$+1/8\sqrt{3}$
$\frac{3}{2}\ 2\ \frac{1}{2}$	000−	000−	$+1/4.3$
$\frac{3}{2}\ 2\ \frac{3}{2}$	000+	000+	$-1/8.3$
$\frac{3}{2}\ 2\ \frac{3}{2}$	000+	001+	$+1/8.3$
$\frac{3}{2}\ 2\ \tilde{\frac{1}{2}}$	000−	000−	0
$\frac{3}{2}\ \tilde{1}\ \frac{1}{2}$	000−	000−	$-1/8.3$
$\frac{3}{2}\ \tilde{1}\ \frac{3}{2}$	000+	000+	$-1/8.3$
$\frac{3}{2}\ \tilde{1}\ \frac{3}{2}$	000+	001+	$-1/8.3$
$\frac{3}{2}\ \tilde{1}\ \frac{3}{2}$	001−	000+	0
$\frac{3}{2}\ \tilde{1}\ \frac{3}{2}$	001−	001+	$+1/8.3$
$\frac{3}{2}\ \tilde{1}\ \tilde{\frac{1}{2}}$	000−	000−	$+1/8.3$
$\tilde{\frac{1}{2}}\ \tilde{1}\ \frac{3}{2}\quad 2\ \tilde{1}\ \tilde{1}$			
$\frac{3}{2}\ 0\ \frac{3}{2}$	000−	000−	$-1/4.3\sqrt{2}$
$\frac{3}{2}\ 0\ \frac{3}{2}$	000−	001+	$-1/4.3\sqrt{2}$
$\frac{3}{2}\ 1\ \frac{1}{2}$	000−	000−	$-1/8\sqrt{3}$
$\frac{3}{2}\ 1\ \frac{3}{2}$	000+	000+	$+1/8.3\sqrt{3}$
$\frac{3}{2}\ 1\ \frac{3}{2}$	000+	001−	$+1/8.3\sqrt{3}$
$\frac{3}{2}\ 1\ \frac{3}{2}$	001+	000+	$-1/4.3\sqrt{3}$

O and T_d 9j Symbols (cont.)

$\tilde{\frac{1}{2}}$	**$\tilde{1}$**	**$\frac{3}{2}$**	**2**	**$\tilde{1}$**	**$\tilde{1}$**
$\frac{3}{2}$ 1 $\frac{3}{2}$	001+	001−	$+1/8.3\sqrt{3}$		
$\frac{3}{2}$ 1 $\tilde{\frac{1}{2}}$	000−	000−	$-1/8.3\sqrt{3}$		
$\frac{3}{2}$ 2 $\frac{1}{2}$	000+	000+	0		
$\frac{3}{2}$ 2 $\frac{3}{2}$	000−	000−	$-1/8.3$		
$\frac{3}{2}$ 2 $\frac{3}{2}$	000−	001+	$-1/8.3$		
$\frac{3}{2}$ 2 $\tilde{\frac{1}{2}}$	000+	000+	$+1/4.3$		
$\frac{3}{2}$ $\tilde{1}$ $\frac{1}{2}$	000+	000+	$-1/8.3$		
$\frac{3}{2}$ $\tilde{1}$ $\frac{3}{2}$	000−	000−	$+1/8.3$		
$\frac{3}{2}$ $\tilde{1}$ $\frac{3}{2}$	000−	001+	$-1/8.3$		
$\frac{3}{2}$ $\tilde{1}$ $\frac{3}{2}$	001+	000−	0		
$\frac{3}{2}$ $\tilde{1}$ $\frac{3}{2}$	001+	001+	$-1/8.3$		
$\frac{3}{2}$ $\tilde{1}$ $\tilde{\frac{1}{2}}$	000+	000+	$+1/8.3$		
$\tilde{\frac{1}{2}}$	**$\tilde{1}$**	**$\frac{3}{2}$**	**2**	**$\tilde{\frac{1}{2}}$**	**$\frac{3}{2}$**
$\frac{3}{2}$ $\frac{1}{2}$ 1	000−	000−	$-1/4.3\sqrt{2}$		
$\frac{3}{2}$ $\frac{1}{2}$ 1	000−	001−	$+1/4.3\sqrt{2}$		
$\frac{3}{2}$ $\frac{1}{2}$ 2	000+	000+	0		
$\frac{3}{2}$ $\frac{1}{2}$ $\tilde{1}$	000+	000+	$+1/4.3\sqrt{2}$		
$\frac{3}{2}$ $\frac{1}{2}$ $\tilde{1}$	000+	001−	$-1/4.3\sqrt{2}$		
$\frac{3}{2}$ $\frac{3}{2}$ 0	000−	000−	$-1/8.2$		
$\frac{3}{2}$ $\frac{3}{2}$ 1	000+	000+	$+1/8.2.3$		
$\frac{3}{2}$ $\frac{3}{2}$ 1	000+	001+	$+1/8.3$		
$\frac{3}{2}$ $\frac{3}{2}$ 1	001+	000+	$+1/8.3$		
$\frac{3}{2}$ $\frac{3}{2}$ 1	001+	001+	$+1/8.2.3$		
$\frac{3}{2}$ $\frac{3}{2}$ 2	000−	000−	$-1/8.2$		
$\frac{3}{2}$ $\frac{3}{2}$ $\tilde{1}$	000−	000−	$-1/8.2.3$		
$\frac{3}{2}$ $\frac{3}{2}$ $\tilde{1}$	000−	001+	$-1/8.3$		
$\frac{3}{2}$ $\frac{3}{2}$ $\tilde{1}$	001+	000−	$+1/8.3$		
$\frac{3}{2}$ $\frac{3}{2}$ $\tilde{1}$	001+	001+	$+1/8.2.3$		
$\tilde{\frac{1}{2}}$	**$\tilde{1}$**	**$\frac{3}{2}$**	**$\tilde{1}$**	**0**	**$\tilde{1}$**
$\frac{1}{2}$ $\tilde{1}$ $\frac{3}{2}$	000+	000+	$+1/2.9\sqrt{2}$		
$\frac{1}{2}$ $\tilde{1}$ $\frac{3}{2}$	000+	001−	$-1/9\sqrt{2}$		
$\frac{1}{2}$ $\tilde{1}$ $\tilde{\frac{1}{2}}$	000−	000−	$-1/9$		
$\frac{3}{2}$ $\tilde{1}$ $\frac{1}{2}$	000+	000+	$+1/4.3$		
$\frac{3}{2}$ $\tilde{1}$ $\frac{3}{2}$	000−	000−	$+1/2.9$		
$\frac{3}{2}$ $\tilde{1}$ $\frac{3}{2}$	000−	001+	$+1/2.9$		
$\frac{3}{2}$ $\tilde{1}$ $\frac{3}{2}$	001+	000−	$+1/2.9$		
$\frac{3}{2}$ $\tilde{1}$ $\frac{3}{2}$	001+	001+	$-1/4.9$		
$\frac{3}{2}$ $\tilde{1}$ $\tilde{\frac{1}{2}}$	000+	000+	$+1/4.9$		
$\tilde{\frac{1}{2}}$	**$\tilde{1}$**	**$\frac{3}{2}$**	**$\tilde{1}$**	**$\frac{1}{2}$**	**$\frac{3}{2}$**
$\frac{1}{2}$ $\frac{3}{2}$ 1	000−	000−	$-1/4.3\sqrt{2}$		
$\frac{1}{2}$ $\frac{3}{2}$ 1	000−	001−	0		
$\frac{1}{2}$ $\frac{3}{2}$ 2	000+	000+	$+1/4.3\sqrt{2}$		
$\frac{1}{2}$ $\frac{3}{2}$ $\tilde{1}$	000+	000+	$-1/4.9\sqrt{2}$		
$\frac{1}{2}$ $\frac{3}{2}$ $\tilde{1}$	000+	001−	$-1/2.9\sqrt{2}$		
$\frac{1}{2}$ $\tilde{\frac{1}{2}}$ $\tilde{1}$	000−	000−	$+1/2.9\sqrt{2}$		
$\frac{1}{2}$ $\tilde{\frac{1}{2}}$ $\tilde{1}$	000−	001+	$+1/9\sqrt{2}$		
$\frac{3}{2}$ $\frac{3}{2}$ 0	000−	000−	$+1/8.2$		
$\frac{3}{2}$ $\frac{3}{2}$ 1	000+	000+	$-1/8.2.3$		
$\frac{3}{2}$ $\frac{3}{2}$ 1	000+	001+	0		

$\tilde{\frac{1}{2}}$	**$\tilde{1}$**	**$\frac{3}{2}$**	**$\tilde{1}$**	**$\frac{1}{2}$**	**$\frac{3}{2}$**
$\frac{3}{2}$ $\frac{3}{2}$ 1	001+	000+	0		
$\frac{3}{2}$ $\frac{3}{2}$ 1	001+	001+	$+1/8.2$		
$\frac{3}{2}$ $\frac{3}{2}$ 2	000−	000−	$-1/8.2.3$		
$\frac{3}{2}$ $\frac{3}{2}$ $\tilde{1}$	000−	000−	$+7/8.2.9$		
$\frac{3}{2}$ $\frac{3}{2}$ $\tilde{1}$	000−	001+	$-1/4.9$		
$\frac{3}{2}$ $\frac{3}{2}$ $\tilde{1}$	001+	000−	$-1/4.9$		
$\frac{3}{2}$ $\frac{3}{2}$ $\tilde{1}$	001+	001+	$+1/8.2.9$		
$\tilde{\frac{1}{2}}$	**$\tilde{1}$**	**$\frac{3}{2}$**	**$\tilde{1}$**	**$\frac{1}{2}$**	**$\tilde{\frac{1}{2}}$**
$\frac{1}{2}$ $\frac{3}{2}$ 1	000+	000+	$-1/8.3$		
$\frac{1}{2}$ $\frac{3}{2}$ 2	000−	000−	$-1/8.3$		
$\frac{1}{2}$ $\frac{3}{2}$ $\tilde{1}$	000−	000−	$+7/8.9$		
$\frac{1}{2}$ $\tilde{\frac{1}{2}}$ $\tilde{1}$	000+	000+	$+1/2.9$		
$\frac{3}{2}$ $\frac{3}{2}$ 1	000−	000−	$-1/4.3\sqrt{2}$		
$\frac{3}{2}$ $\frac{3}{2}$ 1	001−	000−	0		
$\frac{3}{2}$ $\frac{3}{2}$ 2	000+	000+	$+1/4.3\sqrt{2}$		
$\frac{3}{2}$ $\frac{3}{2}$ $\tilde{1}$	000+	000+	$-1/4.9\sqrt{2}$		
$\frac{3}{2}$ $\frac{3}{2}$ $\tilde{1}$	001−	000+	$-1/2.9\sqrt{2}$		
$\frac{3}{2}$ $\tilde{\frac{1}{2}}$ 1	000+	000+	$+1/4.3$		
$\frac{3}{2}$ $\tilde{\frac{1}{2}}$ 2	000−	000−	$+1/4.3$		
$\frac{3}{2}$ $\tilde{\frac{1}{2}}$ $\tilde{1}$	000−	000−	$-1/4.9$		
$\tilde{\frac{1}{2}}$	**$\tilde{1}$**	**$\frac{3}{2}$**	**$\tilde{1}$**	**1**	**1**
$\frac{1}{2}$ 1 $\frac{1}{2}$	000+	000+	$-1/4.9$		
$\frac{1}{2}$ 1 $\frac{3}{2}$	000−	000−	$+1/2.9\sqrt{2}$		
$\frac{1}{2}$ 1 $\frac{3}{2}$	000−	001−	$-1/2.9\sqrt{2}$		
$\frac{1}{2}$ 2 $\frac{3}{2}$	000+	000+	0		
$\frac{1}{2}$ 2 $\frac{3}{2}$	000+	001+	$-1/4.3\sqrt{2}$		
$\frac{1}{2}$ $\tilde{1}$ $\frac{3}{2}$	000+	000+	$+1/2.3\sqrt{2.3}$		
$\frac{1}{2}$ $\tilde{1}$ $\frac{3}{2}$	000+	001+	0		
$\frac{1}{2}$ $\tilde{1}$ $\tilde{\frac{1}{2}}$	000−	000−	$+1/4.3\sqrt{3}$		
$\frac{3}{2}$ 1 $\frac{1}{2}$	000−	000−	$+5/8.9$		
$\frac{3}{2}$ 1 $\frac{3}{2}$	000+	000+	$+1/8.9$		
$\frac{3}{2}$ 1 $\frac{3}{2}$	000+	001+	$+1/4.9$		
$\frac{3}{2}$ 1 $\frac{3}{2}$	001+	000+	$+1/4.9$		
$\frac{3}{2}$ 1 $\frac{3}{2}$	001+	001+	$+1/8.9$		
$\frac{3}{2}$ 1 $\tilde{\frac{1}{2}}$	000−	000−	$+1/8.3$		
$\tilde{\frac{1}{2}}$	**$\tilde{1}$**	**$\frac{3}{2}$**	**$\tilde{1}$**	**1**	**2**
$\frac{1}{2}$ 1 $\frac{1}{2}$	000−	000−	$+1/4.3$		
$\frac{1}{2}$ 1 $\frac{3}{2}$	000+	000+	0		
$\frac{1}{2}$ 2 $\frac{3}{2}$	000−	000−	$-1/4.3\sqrt{2}$		
$\frac{1}{2}$ $\tilde{1}$ $\frac{3}{2}$	000−	000−	$+1/2.3\sqrt{2.3}$		
$\frac{1}{2}$ $\tilde{1}$ $\tilde{\frac{1}{2}}$	000+	000+	$-1/4.3\sqrt{3}$		
$\frac{3}{2}$ 1 $\frac{1}{2}$	000+	000+	$+1/8.3$		
$\frac{3}{2}$ 1 $\frac{3}{2}$	000−	000−	$-1/8.3$		
$\frac{3}{2}$ 1 $\frac{3}{2}$	001−	000−	$-1/8.3$		
$\frac{3}{2}$ 1 $\tilde{\frac{1}{2}}$	000+	000+	$+1/8.3$		
$\frac{3}{2}$ 2 $\frac{1}{2}$	000−	000−	0		
$\frac{3}{2}$ 2 $\frac{3}{2}$	000+	000+	$+1/8.3$		
$\frac{3}{2}$ 2 $\tilde{\frac{1}{2}}$	000−	000−	$+1/4.3$		

$\tilde{\frac{1}{2}}$	**$\tilde{1}$**	**$\frac{3}{2}$**	**$\tilde{1}$**	**1**	**$\tilde{1}$**
$\frac{1}{2}$ 1 $\frac{1}{2}$	000−	000−	$-1/4.3\sqrt{3}$		
$\frac{1}{2}$ 1 $\frac{3}{2}$	000+	000+	0		
$\frac{1}{2}$ 1 $\frac{3}{2}$	000+	001−	$+1/2.3\sqrt{2.3}$		
$\frac{1}{2}$ 2 $\frac{3}{2}$	000−	000−	$+1/2.3\sqrt{2.3}$		
$\frac{1}{2}$ 2 $\frac{3}{2}$	000−	001+	$-1/4.3\sqrt{2.3}$		
$\frac{1}{2}$ $\tilde{1}$ $\frac{3}{2}$	000−	000−	0		
$\frac{1}{2}$ $\tilde{1}$ $\frac{3}{2}$	000−	001+	0		
$\frac{1}{2}$ $\tilde{1}$ $\tilde{\frac{1}{2}}$	000+	000+	$+1/4.3$		
$\frac{3}{2}$ 1 $\frac{1}{2}$	000+	000+	$-1/8.3\sqrt{3}$		
$\frac{3}{2}$ 1 $\frac{3}{2}$	000−	000−	$-1/8.3\sqrt{3}$		
$\frac{3}{2}$ 1 $\frac{3}{2}$	000−	001+	0		
$\frac{3}{2}$ 1 $\frac{3}{2}$	001−	000−	$+1/4.3\sqrt{3}$		
$\frac{3}{2}$ 1 $\frac{3}{2}$	001−	001+	$+1/8.3\sqrt{3}$		
$\frac{3}{2}$ 1 $\tilde{\frac{1}{2}}$	000+	000+	$+1/8.3\sqrt{3}$		
$\frac{3}{2}$ 2 $\frac{1}{2}$	000−	000−	$+1/8\sqrt{3}$		
$\frac{3}{2}$ 2 $\frac{3}{2}$	000+	000+	$+1/8.3\sqrt{3}$		
$\frac{3}{2}$ 2 $\frac{3}{2}$	000+	001−	$+1/8.3\sqrt{3}$		
$\frac{3}{2}$ 2 $\tilde{\frac{1}{2}}$	000−	000−	$-1/8.3\sqrt{3}$		
$\frac{3}{2}$ $\tilde{1}$ $\frac{1}{2}$	000−	000−	$+1/8.3$		
$\frac{3}{2}$ $\tilde{1}$ $\frac{3}{2}$	000+	000+	$+1/8.3$		
$\frac{3}{2}$ $\tilde{1}$ $\frac{3}{2}$	000+	001−	0		
$\frac{3}{2}$ $\tilde{1}$ $\frac{3}{2}$	001−	000+	0		
$\frac{3}{2}$ $\tilde{1}$ $\frac{3}{2}$	001−	001−	$+1/8.3$		
$\frac{3}{2}$ $\tilde{1}$ $\tilde{\frac{1}{2}}$	000−	000−	$-1/8.3$		
$\tilde{\frac{1}{2}}$	**$\tilde{1}$**	**$\frac{3}{2}$**	**$\tilde{1}$**	**$\frac{3}{2}$**	**$\frac{1}{2}$**
$\frac{1}{2}$ $\frac{1}{2}$ 1	000+	000+	$-1/4.9$		
$\frac{1}{2}$ $\frac{3}{2}$ 1	000−	000−	$-1/4.9\sqrt{2}$		
$\frac{1}{2}$ $\frac{3}{2}$ 1	000−	010+	$+1/2.9\sqrt{2}$		
$\frac{1}{2}$ $\frac{3}{2}$ 2	000+	000+	$+1/4.3\sqrt{2}$		
$\frac{1}{2}$ $\frac{3}{2}$ 2	000+	010−	$+1/4.3\sqrt{2}$		
$\frac{1}{2}$ $\frac{3}{2}$ $\tilde{1}$	000+	000+	$-1/4.3\sqrt{2}$		
$\frac{1}{2}$ $\frac{3}{2}$ $\tilde{1}$	000+	010−	0		
$\frac{1}{2}$ $\tilde{\frac{1}{2}}$ $\tilde{1}$	000−	000−	$+1/4.3$		
$\frac{3}{2}$ $\frac{1}{2}$ 1	000−	000−	$+11/8.2.9$		
$\frac{3}{2}$ $\frac{1}{2}$ 2	000+	000+	$+1/8.2.3$		
$\frac{3}{2}$ $\frac{1}{2}$ $\tilde{1}$	000+	000+	$+1/8.2.3$		
$\tilde{\frac{1}{2}}$	**$\tilde{1}$**	**$\frac{3}{2}$**	**$\tilde{1}$**	**$\frac{3}{2}$**	**$\frac{3}{2}$**
$\frac{1}{2}$ $\frac{1}{2}$ 0	000+	000+	$+1/8.3$		
$\frac{1}{2}$ $\frac{1}{2}$ 0	010−	000+	$-1/4.3$		
$\frac{1}{2}$ $\frac{1}{2}$ 1	000−	000−	$+5/8.9$		
$\frac{1}{2}$ $\frac{1}{2}$ 1	000−	001−	$-1/4.9$		
$\frac{1}{2}$ $\frac{1}{2}$ 1	010+	000−	$+1/4.9$		
$\frac{1}{2}$ $\frac{1}{2}$ 1	010+	001−	$-1/4.9$		
$\frac{1}{2}$ $\frac{3}{2}$ 1	000+	000+	$+1/4.9\sqrt{2}$		
$\frac{1}{2}$ $\frac{3}{2}$ 1	000+	001+	$-1/4.9\sqrt{2}$		
$\frac{1}{2}$ $\frac{3}{2}$ 1	000+	010−	$+1/4.9\sqrt{2}$		
$\frac{1}{2}$ $\frac{3}{2}$ 1	000+	011−	$-5/8.9\sqrt{2}$		
$\frac{1}{2}$ $\frac{3}{2}$ 1	010−	000+	$-1/2.9\sqrt{2}$		
$\frac{1}{2}$ $\frac{3}{2}$ 1	010−	001+	$-1/4.9\sqrt{2}$		

O and T_d 9j Symbols (cont.)

$\tilde{\frac{1}{2}}$	$\tilde{1}$	$\frac{3}{2}$	$\tilde{1}$	$\frac{3}{2}$	$\frac{3}{2}$
$\frac{1}{2}$	$\frac{3}{2}$	1	010−	010−	$+1/4.9\sqrt{2}$
$\frac{1}{2}$	$\frac{3}{2}$	1	010−	011−	$+1/2.9\sqrt{2}$
$\frac{1}{2}$	$\frac{3}{2}$	2	000−	000−	0
$\frac{1}{2}$	$\frac{3}{2}$	2	000−	010+	$+1/8.3\sqrt{2}$
$\frac{1}{2}$	$\frac{3}{2}$	2	010+	000−	$-1/4.3\sqrt{2}$
$\frac{1}{2}$	$\frac{3}{2}$	2	010+	010+	0
$\frac{1}{2}$	$\frac{3}{2}$	$\tilde{1}$	000−	000−	$+1/4.9\sqrt{2}$
$\frac{1}{2}$	$\frac{3}{2}$	$\tilde{1}$	000−	001+	$-1/4.9\sqrt{2}$
$\frac{1}{2}$	$\frac{3}{2}$	$\tilde{1}$	000−	010+	$-1/4.9\sqrt{2}$
$\frac{1}{2}$	$\frac{3}{2}$	$\tilde{1}$	000−	011−	$+5/8.9\sqrt{2}$
$\frac{1}{2}$	$\frac{3}{2}$	$\tilde{1}$	010+	000−	$+1/2.9\sqrt{2}$
$\frac{1}{2}$	$\frac{3}{2}$	$\tilde{1}$	010+	001+	$+1/4.9\sqrt{2}$
$\frac{1}{2}$	$\frac{3}{2}$	$\tilde{1}$	010+	010+	$+1/4.9\sqrt{2}$
$\frac{1}{2}$	$\frac{3}{2}$	$\tilde{1}$	010+	011−	$+1/2.9\sqrt{2}$
$\frac{1}{2}$	$\tilde{\frac{1}{2}}$	$\tilde{1}$	000+	000+	$+5/8.9$
$\frac{1}{2}$	$\tilde{\frac{1}{2}}$	$\tilde{1}$	000+	001−	$-1/4.9$
$\frac{1}{2}$	$\tilde{\frac{1}{2}}$	$\tilde{1}$	010−	000+	$-1/4.9$
$\frac{1}{2}$	$\tilde{\frac{1}{2}}$	$\tilde{1}$	010−	001−	$+1/4.9$
$\frac{3}{2}$	$\frac{1}{2}$	1	000+	000+	$+1/4.9$
$\frac{3}{2}$	$\frac{1}{2}$	1	000+	001+	$+1/8.9$
$\frac{3}{2}$	$\frac{1}{2}$	1	010−	000+	$-1/8.9$
$\frac{3}{2}$	$\frac{1}{2}$	1	010−	001+	$-7/8.2.9$
$\frac{3}{2}$	$\frac{1}{2}$	2	000−	000−	$-1/8.3$
$\frac{3}{2}$	$\frac{1}{2}$	2	010+	000−	$+1/8.2.3$
$\frac{3}{2}$	$\frac{1}{2}$	$\tilde{1}$	000−	000−	0
$\frac{3}{2}$	$\frac{1}{2}$	$\tilde{1}$	000−	001+	$-1/8.3$
$\frac{3}{2}$	$\frac{1}{2}$	$\tilde{1}$	010+	000−	$+1/8.3$
$\frac{3}{2}$	$\frac{1}{2}$	$\tilde{1}$	010+	001+	$-1/8.2.3$
$\frac{3}{2}$	$\frac{3}{2}$	0	000+	000+	$+1/8.3$
$\frac{3}{2}$	$\frac{3}{2}$	0	000+	010−	$+1/8.3$
$\frac{3}{2}$	$\frac{3}{2}$	0	010−	000+	$+1/8.3$
$\frac{3}{2}$	$\frac{3}{2}$	0	010−	010−	$-1/8.2.3$
$\frac{3}{2}$	$\frac{3}{2}$	1	000−	000−	$+1/8.9$
$\frac{3}{2}$	$\frac{3}{2}$	1	000−	001−	$+1/4.9$
$\frac{3}{2}$	$\frac{3}{2}$	1	000−	010+	$+1/8.9$
$\frac{3}{2}$	$\frac{3}{2}$	1	000−	011+	$-1/8.9$
$\frac{3}{2}$	$\frac{3}{2}$	1	001−	000−	$+1/4.9$
$\frac{3}{2}$	$\frac{3}{2}$	1	001−	001−	$+1/8.9$
$\frac{3}{2}$	$\frac{3}{2}$	1	001−	010+	$+1/4.9$
$\frac{3}{2}$	$\frac{3}{2}$	1	001−	011+	$+1/8.9$
$\frac{3}{2}$	$\frac{3}{2}$	1	010+	000−	$+1/8.9$
$\frac{3}{2}$	$\frac{3}{2}$	1	010+	001−	$+1/4.9$
$\frac{3}{2}$	$\frac{3}{2}$	1	010+	010+	$+5/8.2.9$
$\frac{3}{2}$	$\frac{3}{2}$	1	010+	011+	$-1/8.9$
$\frac{3}{2}$	$\frac{3}{2}$	1	011+	000−	$-1/8.9$
$\frac{3}{2}$	$\frac{3}{2}$	1	011+	001−	$+1/8.9$
$\frac{3}{2}$	$\frac{3}{2}$	1	011+	010+	$-1/8.9$
$\frac{3}{2}$	$\frac{3}{2}$	1	011+	011+	$+1/8.9$
$\frac{3}{2}$	$\frac{3}{2}$	2	000+	000+	$+1/8.3$

$\tilde{\frac{1}{2}}$	$\tilde{1}$	$\frac{3}{2}$	$\tilde{1}$	$\frac{3}{2}$	$\frac{3}{2}$
$\frac{3}{2}$	$\frac{3}{2}$	2	000+	010−	0
$\frac{3}{2}$	$\frac{3}{2}$	2	010−	000+	0
$\frac{3}{2}$	$\frac{3}{2}$	2	010−	010−	$+1/8.3$
$\frac{3}{2}$	$\frac{3}{2}$	$\tilde{1}$	000+	000+	$+1/8.9$
$\frac{3}{2}$	$\frac{3}{2}$	$\tilde{1}$	000+	001−	$+1/4.9$
$\frac{3}{2}$	$\frac{3}{2}$	$\tilde{1}$	000+	010−	$-1/8.9$
$\frac{3}{2}$	$\frac{3}{2}$	$\tilde{1}$	000+	011+	$+1/8.9$
$\frac{3}{2}$	$\frac{3}{2}$	$\tilde{1}$	001−	000+	$+1/4.9$
$\frac{3}{2}$	$\frac{3}{2}$	$\tilde{1}$	001−	001−	$+1/8.9$
$\frac{3}{2}$	$\frac{3}{2}$	$\tilde{1}$	001−	010−	$-1/4.9$
$\frac{3}{2}$	$\frac{3}{2}$	$\tilde{1}$	001−	011+	$-1/8.9$
$\frac{3}{2}$	$\frac{3}{2}$	$\tilde{1}$	010−	000+	$-1/8.9$
$\frac{3}{2}$	$\frac{3}{2}$	$\tilde{1}$	010−	001−	$-1/4.9$
$\frac{3}{2}$	$\frac{3}{2}$	$\tilde{1}$	010−	010−	$+5/8.2.9$
$\frac{3}{2}$	$\frac{3}{2}$	$\tilde{1}$	010−	011+	$-1/8.9$
$\frac{3}{2}$	$\frac{3}{2}$	$\tilde{1}$	011+	000+	$+1/8.9$
$\frac{3}{2}$	$\frac{3}{2}$	$\tilde{1}$	011+	001−	$-1/8.9$
$\frac{3}{2}$	$\frac{3}{2}$	$\tilde{1}$	011+	010−	$-1/8.9$
$\frac{3}{2}$	$\frac{3}{2}$	$\tilde{1}$	011+	011+	$+1/8.9$

$\tilde{\frac{1}{2}}$	$\tilde{1}$	$\frac{3}{2}$	$\tilde{1}$	$\frac{3}{2}$	$\tilde{\frac{1}{2}}$
$\frac{1}{2}$	$\frac{1}{2}$	1	000+	000+	$+1/4.3$
$\frac{1}{2}$	$\frac{3}{2}$	1	000−	000−	$-1/4.3\sqrt{2}$
$\frac{1}{2}$	$\frac{3}{2}$	1	000−	010+	0
$\frac{1}{2}$	$\frac{3}{2}$	2	000+	000+	$+1/4.3\sqrt{2}$
$\frac{1}{2}$	$\frac{3}{2}$	2	000+	010−	$-1/4.3\sqrt{2}$
$\frac{1}{2}$	$\frac{3}{2}$	$\tilde{1}$	000+	000+	$-1/4.9\sqrt{2}$
$\frac{1}{2}$	$\frac{3}{2}$	$\tilde{1}$	000+	010−	$-1/2.9\sqrt{2}$
$\frac{1}{2}$	$\tilde{\frac{1}{2}}$	$\tilde{1}$	000−	000−	$-1/4.9$
$\frac{3}{2}$	$\frac{1}{2}$	1	000−	000−	$+1/8.2.3$
$\frac{3}{2}$	$\frac{1}{2}$	2	000+	000+	$+1/8.2$
$\frac{3}{2}$	$\frac{1}{2}$	$\tilde{1}$	000+	000+	$+1/8.2.3$
$\frac{3}{2}$	$\frac{3}{2}$	1	000+	000+	0
$\frac{3}{2}$	$\frac{3}{2}$	1	000+	010−	$-1/8.3$
$\frac{3}{2}$	$\frac{3}{2}$	1	001+	000+	$-1/8.3$
$\frac{3}{2}$	$\frac{3}{2}$	1	001+	010−	$+1/8.2.3$
$\frac{3}{2}$	$\frac{3}{2}$	2	000−	000−	$+1/8.3$
$\frac{3}{2}$	$\frac{3}{2}$	2	000−	010+	$+1/8.2.3$
$\frac{3}{2}$	$\frac{3}{2}$	$\tilde{1}$	000−	000−	$+1/4.9$
$\frac{3}{2}$	$\frac{3}{2}$	$\tilde{1}$	000−	010+	$+1/8.9$
$\frac{3}{2}$	$\frac{3}{2}$	$\tilde{1}$	001+	000−	$+1/8.9$
$\frac{3}{2}$	$\frac{3}{2}$	$\tilde{1}$	001+	010+	$+7/8.2.9$
$\frac{3}{2}$	$\tilde{\frac{1}{2}}$	1	000−	000−	$+1/8.2.3$
$\frac{3}{2}$	$\tilde{\frac{1}{2}}$	2	000+	000+	$+1/8.2.3$
$\frac{3}{2}$	$\tilde{\frac{1}{2}}$	$\tilde{1}$	000+	000+	$+11/8.2.9$

$\tilde{\frac{1}{2}}$	$\tilde{1}$	$\frac{3}{2}$	$\tilde{1}$	2	1
$\frac{1}{2}$	1	$\frac{1}{2}$	000−	000−	$-1/4.9$
$\frac{1}{2}$	1	$\frac{3}{2}$	000+	000+	$-1/9\sqrt{2}$
$\frac{1}{2}$	1	$\frac{3}{2}$	000+	001+	$+1/4.9\sqrt{2}$
$\frac{1}{2}$	$\tilde{1}$	$\frac{3}{2}$	000−	000−	0
$\frac{1}{2}$	$\tilde{1}$	$\frac{3}{2}$	000−	001−	$+1/4.3\sqrt{2}$
$\frac{1}{2}$	$\tilde{1}$	$\tilde{\frac{1}{2}}$	000+	000+	$-1/4.3$
$\frac{3}{2}$	1	$\frac{1}{2}$	000+	000+	$+5/8.9$
$\frac{3}{2}$	1	$\frac{3}{2}$	000−	000−	$+1/8.9$
$\frac{3}{2}$	1	$\frac{3}{2}$	000−	001−	$-1/8.9$
$\frac{3}{2}$	1	$\frac{3}{2}$	001−	000−	$-1/8.9$
$\frac{3}{2}$	1	$\frac{3}{2}$	001−	001−	$+1/2.9$
$\frac{3}{2}$	1	$\tilde{\frac{1}{2}}$	000+	000+	$-1/8.3$

$\tilde{\frac{1}{2}}$	$\tilde{1}$	$\frac{3}{2}$	$\tilde{1}$	2	$\tilde{1}$
$\frac{1}{2}$	1	$\frac{1}{2}$	000+	000+	$-1/4.3$
$\frac{1}{2}$	1	$\frac{3}{2}$	000−	000−	0
$\frac{1}{2}$	1	$\frac{3}{2}$	000−	001+	$-1/4.3\sqrt{2}$
$\frac{1}{2}$	$\tilde{1}$	$\frac{3}{2}$	000+	000+	$-1/9\sqrt{2}$
$\frac{1}{2}$	$\tilde{1}$	$\frac{3}{2}$	000+	001−	$-1/4.9\sqrt{2}$
$\frac{1}{2}$	$\tilde{1}$	$\tilde{\frac{1}{2}}$	000−	000−	$-1/4.9$
$\frac{3}{2}$	1	$\frac{1}{2}$	000−	000−	$-1/8.3$
$\frac{3}{2}$	1	$\frac{3}{2}$	000+	000+	$+1/8.3$
$\frac{3}{2}$	1	$\frac{3}{2}$	000+	001−	$-1/8.3$
$\frac{3}{2}$	1	$\frac{3}{2}$	001+	000+	$+1/8.3$
$\frac{3}{2}$	1	$\frac{3}{2}$	001+	001−	0
$\frac{3}{2}$	1	$\tilde{\frac{1}{2}}$	000−	000−	$-1/8.3$
$\frac{3}{2}$	$\tilde{1}$	$\frac{1}{2}$	000+	000+	$-1/8.3$
$\frac{3}{2}$	$\tilde{1}$	$\frac{3}{2}$	000−	000−	$+1/8.9$
$\frac{3}{2}$	$\tilde{1}$	$\frac{3}{2}$	000−	001+	$+1/8.9$
$\frac{3}{2}$	$\tilde{1}$	$\frac{3}{2}$	001+	000−	$+1/8.9$
$\frac{3}{2}$	$\tilde{1}$	$\frac{3}{2}$	001+	001+	$+1/2.9$
$\frac{3}{2}$	$\tilde{1}$	$\tilde{\frac{1}{2}}$	000+	000+	$+5/8.9$

$\tilde{\frac{1}{2}}$	$\tilde{1}$	$\frac{3}{2}$	$\tilde{1}$	$\tilde{1}$	0
$\frac{1}{2}$	1	$\frac{3}{2}$	000−	000−	$-1/4.3\sqrt{3}$
$\frac{1}{2}$	2	$\frac{3}{2}$	000+	000+	$-1/4.3$
$\frac{1}{2}$	$\tilde{1}$	$\frac{3}{2}$	000+	000+	$+1/4.3$
$\frac{3}{2}$	0	$\frac{3}{2}$	000−	000−	$+1/4.3$

$\tilde{\frac{1}{2}}$	$\tilde{1}$	$\frac{3}{2}$	$\tilde{1}$	$\tilde{1}$	1
$\frac{1}{2}$	0	$\frac{1}{2}$	000+	000+	$+1/2.3\sqrt{2.3}$
$\frac{1}{2}$	1	$\frac{1}{2}$	000−	000−	$+1/4.3$
$\frac{1}{2}$	1	$\frac{3}{2}$	000+	000+	0
$\frac{1}{2}$	1	$\frac{3}{2}$	000+	001+	0
$\frac{1}{2}$	2	$\frac{3}{2}$	000−	000−	$+1/2.3\sqrt{2.3}$
$\frac{1}{2}$	2	$\frac{3}{2}$	000−	001−	$+1/4.3\sqrt{2.3}$
$\frac{1}{2}$	$\tilde{1}$	$\frac{3}{2}$	000−	000−	0
$\frac{1}{2}$	$\tilde{1}$	$\frac{3}{2}$	000−	001−	$-1/2.3\sqrt{2.3}$
$\frac{1}{2}$	$\tilde{1}$	$\tilde{\frac{1}{2}}$	000+	000+	$-1/4.3\sqrt{3}$
$\frac{3}{2}$	0	$\frac{3}{2}$	000+	000+	$+1/4.3\sqrt{2.3}$
$\frac{3}{2}$	0	$\frac{3}{2}$	000+	001+	$+1/2.3\sqrt{2.3}$
$\frac{3}{2}$	1	$\frac{1}{2}$	000+	000+	$-1/8.3$
$\frac{3}{2}$	1	$\frac{3}{2}$	000−	000−	$+1/8.3$
$\frac{3}{2}$	1	$\frac{3}{2}$	000−	001−	0
$\frac{3}{2}$	1	$\frac{3}{2}$	001−	000−	0
$\frac{3}{2}$	1	$\frac{3}{2}$	001−	001−	$+1/8.3$

O and T_d 9j Symbols (cont.)

$\tilde{\frac{1}{2}}\ \tilde{1}\ \frac{3}{2}\quad \tilde{1}\ \tilde{1}\ 1$

$\frac{3}{2}\ 1\ \tilde{\frac{1}{2}}$ 000+ 000+ $+1/8.3$

$\tilde{\frac{1}{2}}\ \tilde{1}\ \frac{3}{2}\quad \tilde{1}\ \tilde{1}\ 2$

$\frac{1}{2}\ 0\ \frac{1}{2}$ 000− 000− $+1/2.3\sqrt{2}$

$\frac{1}{2}\ 1\ \frac{1}{2}$ 000+ 000+ $-1/4.3\sqrt{3}$

$\frac{1}{2}\ 1\ \frac{3}{2}$ 000− 000− $-1/2.3\sqrt{2.3}$

$\frac{1}{2}\ 2\ \frac{3}{2}$ 000+ 000+ $+1/4.3\sqrt{2}$

$\frac{1}{2}\ \tilde{1}\ \frac{3}{2}$ 000+ 000+ 0

$\frac{1}{2}\ \tilde{1}\ \tilde{\frac{1}{2}}$ 000− 000− $+1/4.3$

$\frac{3}{2}\ 0\ \frac{3}{2}$ 000− 000− $-1/4.3\sqrt{2}$

$\frac{3}{2}\ 1\ \frac{1}{2}$ 000− 000− $-1/8.3\sqrt{3}$

$\frac{3}{2}\ 1\ \frac{3}{2}$ 000+ 000+ $-1/8.3\sqrt{3}$

$\frac{3}{2}\ 1\ \frac{3}{2}$ 001+ 000+ $+1/8.3\sqrt{3}$

$\frac{3}{2}\ 1\ \tilde{\frac{1}{2}}$ 000− 000− $+1/8\sqrt{3}$

$\frac{3}{2}\ 2\ \frac{1}{2}$ 000+ 000+ $+1/4.3$

$\frac{3}{2}\ 2\ \frac{3}{2}$ 000− 000− $+1/8.3$

$\frac{3}{2}\ 2\ \tilde{\frac{1}{2}}$ 000+ 000+ 0

$\tilde{\frac{1}{2}}\ \tilde{1}\ \frac{3}{2}\quad \tilde{1}\ \tilde{1}\ \tilde{1}$

$\frac{1}{2}\ 0\ \frac{1}{2}$ 000− 000− $-1/2.3\sqrt{2}$

$\frac{1}{2}\ 1\ \frac{1}{2}$ 000+ 000+ $+1/4.3\sqrt{3}$

$\frac{1}{2}\ 1\ \frac{3}{2}$ 000− 000− $+1/2.3\sqrt{2.3}$

$\frac{1}{2}\ 1\ \frac{3}{2}$ 000− 001+ 0

$\frac{1}{2}\ 2\ \frac{3}{2}$ 000+ 000+ 0

$\frac{1}{2}\ 2\ \frac{3}{2}$ 000+ 001− $+1/4.3\sqrt{2}$

$\frac{1}{2}\ \tilde{1}\ \frac{3}{2}$ 000+ 000+ $+1/2.9\sqrt{2}$

$\frac{1}{2}\ \tilde{1}\ \frac{3}{2}$ 000+ 001− $+1/2.9\sqrt{2}$

$\frac{1}{2}\ \tilde{1}\ \tilde{\frac{1}{2}}$ 000− 000− $-1/4.9$

$\frac{3}{2}\ 0\ \frac{3}{2}$ 000− 000− $+1/4.3\sqrt{2}$

$\frac{3}{2}\ 0\ \frac{3}{2}$ 000− 001+ 0

$\frac{3}{2}\ 1\ \frac{1}{2}$ 000− 000− $+1/8.3\sqrt{3}$

$\frac{3}{2}\ 1\ \frac{3}{2}$ 000+ 000+ $-1/8.3\sqrt{3}$

$\frac{3}{2}\ 1\ \frac{3}{2}$ 000+ 001− $-1/4.3\sqrt{3}$

$\frac{3}{2}\ 1\ \frac{3}{2}$ 001+ 000+ 0

$\frac{3}{2}\ 1\ \frac{3}{2}$ 001+ 001− $+1/8.3\sqrt{3}$

$\frac{3}{2}\ 1\ \tilde{\frac{1}{2}}$ 000− 000− $-1/8.3\sqrt{3}$

$\frac{3}{2}\ 2\ \frac{1}{2}$ 000+ 000+ $+1/8.3$

$\frac{3}{2}\ 2\ \frac{3}{2}$ 000− 000− $+1/8.3$

$\frac{3}{2}\ 2\ \frac{3}{2}$ 000− 001+ $-1/8.3$

$\frac{3}{2}\ 2\ \tilde{\frac{1}{2}}$ 000+ 000+ $+1/8.3$

$\frac{3}{2}\ \tilde{1}\ \frac{1}{2}$ 000+ 000+ $+1/8.3$

$\frac{3}{2}\ \tilde{1}\ \frac{3}{2}$ 000− 000− $+1/8.9$

$\frac{3}{2}\ \tilde{1}\ \frac{3}{2}$ 000− 001+ $-1/4.9$

$\frac{3}{2}\ \tilde{1}\ \frac{3}{2}$ 001+ 000− $-1/4.9$

$\frac{3}{2}\ \tilde{1}\ \frac{3}{2}$ 001+ 001+ $+1/8.9$

$\frac{3}{2}\ \tilde{1}\ \tilde{\frac{1}{2}}$ 000+ 000+ $+5/8.9$

$\tilde{\frac{1}{2}}\ \tilde{1}\ \frac{3}{2}\quad \tilde{1}\ \tilde{\frac{1}{2}}\ \frac{1}{2}$

$\frac{1}{2}\ \frac{1}{2}\ 1$ 000− 000− $+1/2.9$

$\frac{1}{2}\ \frac{3}{2}\ 1$ 000+ 000+ $+7/8.9$

$\frac{1}{2}\ \frac{3}{2}\ 2$ 000− 000− $-1/8.3$

$\frac{1}{2}\ \frac{3}{2}\ \tilde{1}$ 000− 000− $-1/8.3$

$\frac{3}{2}\ \frac{1}{2}\ 1$ 000+ 000+ $-1/4.9$

$\frac{3}{2}\ \frac{1}{2}\ 2$ 000− 000− $+1/4.3$

$\frac{3}{2}\ \frac{1}{2}\ \tilde{1}$ 000− 000− $+1/4.3$

$\tilde{\frac{1}{2}}\ \tilde{1}\ \frac{3}{2}\quad \tilde{1}\ \tilde{\frac{1}{2}}\ \frac{3}{2}$

$\frac{1}{2}\ \frac{1}{2}\ 0$ 000− 000− $-1/2.3\sqrt{2}$

$\frac{1}{2}\ \frac{1}{2}\ 1$ 000+ 000+ $+1/2.9\sqrt{2}$

$\frac{1}{2}\ \frac{1}{2}\ 1$ 000+ 001+ $+1/9\sqrt{2}$

$\frac{1}{2}\ \frac{3}{2}\ 1$ 000− 000− $-1/4.9\sqrt{2}$

$\frac{1}{2}\ \frac{3}{2}\ 1$ 000− 001− $-1/2.9\sqrt{2}$

$\frac{1}{2}\ \frac{3}{2}\ 2$ 000+ 000+ $-1/4.3\sqrt{2}$

$\frac{1}{2}\ \frac{3}{2}\ \tilde{1}$ 000+ 000+ $-1/4.3\sqrt{2}$

$\frac{1}{2}\ \frac{3}{2}\ \tilde{1}$ 000+ 001− 0

$\frac{3}{2}\ \frac{1}{2}\ 1$ 000− 000− $-1/4.9\sqrt{2}$

$\frac{3}{2}\ \frac{1}{2}\ 1$ 000− 001− $-1/2.9\sqrt{2}$

$\frac{3}{2}\ \frac{1}{2}\ 2$ 000+ 000+ $-1/4.3\sqrt{2}$

$\frac{3}{2}\ \frac{1}{2}\ \tilde{1}$ 000+ 000+ $-1/4.3\sqrt{2}$

$\frac{3}{2}\ \frac{1}{2}\ \tilde{1}$ 000+ 001− 0

$\frac{3}{2}\ \frac{3}{2}\ 0$ 000− 000− $+1/8.2.3$

$\frac{3}{2}\ \frac{3}{2}\ 1$ 000+ 000+ $+7/8.2.9$

$\frac{3}{2}\ \frac{3}{2}\ 1$ 000+ 001+ $-1/4.9$

$\frac{3}{2}\ \frac{3}{2}\ 1$ 001+ 000+ $-1/4.9$

$\frac{3}{2}\ \frac{3}{2}\ 1$ 001+ 001+ $+1/8.2.9$

$\frac{3}{2}\ \frac{3}{2}\ 2$ 000− 000− $-1/8.2.3$

$\frac{3}{2}\ \frac{3}{2}\ \tilde{1}$ 000− 000− $-1/8.2.3$

$\frac{3}{2}\ \frac{3}{2}\ \tilde{1}$ 000− 001+ 0

$\frac{3}{2}\ \frac{3}{2}\ \tilde{1}$ 001+ 000− 0

$\frac{3}{2}\ \frac{3}{2}\ \tilde{1}$ 001+ 001+ $+1/8.2$

$\tilde{\frac{1}{2}}\ \tilde{\frac{1}{2}}\ 0\quad \frac{3}{2}\ 0\ \frac{3}{2}$

$1\ \tilde{\frac{1}{2}}\ \frac{3}{2}$ 000− 000− $+1/8$

$\tilde{\frac{1}{2}}\ \tilde{\frac{1}{2}}\ 0\quad \frac{3}{2}\ \frac{1}{2}\ 1$

$1\ \tilde{1}\ 1$ 000+ 000+ $-1/2.3\sqrt{2}$

$\tilde{\frac{1}{2}}\ \tilde{\frac{1}{2}}\ 0\quad \frac{3}{2}\ \frac{1}{2}\ 2$

$1\ \tilde{1}\ 2$ 000+ 000+ $-1/4\sqrt{3}$

$\tilde{\frac{1}{2}}\ \tilde{\frac{1}{2}}\ 0\quad \frac{3}{2}\ \frac{1}{2}\ \tilde{1}$

$1\ \tilde{1}\ \tilde{1}$ 000+ 000+ $+1/2.3\sqrt{2.3}$

$\tilde{\frac{1}{2}}\ \tilde{\frac{1}{2}}\ 0\quad \frac{3}{2}\ 1\ \frac{1}{2}$

$1\ \frac{3}{2}\ \frac{1}{2}$ 000− 000− $+1/8$

$\tilde{\frac{1}{2}}\ \tilde{\frac{1}{2}}\ 0\quad \frac{3}{2}\ 1\ \frac{3}{2}$

$1\ \frac{3}{2}\ \frac{3}{2}$ 000− 000− $-1/4.3\sqrt{2}$

$1\ \frac{3}{2}\ \frac{3}{2}$ 001− 000− $-1/4.3\sqrt{2}$

$1\ \frac{3}{2}\ \frac{3}{2}$ 010− 000− $-1/4.3\sqrt{2}$

$1\ \frac{3}{2}\ \frac{3}{2}$ 011− 000− $+1/8.3\sqrt{2}$

$1\ \tilde{\frac{1}{2}}\ \frac{3}{2}$ 000+ 000+ $-1/8.3$

$1\ \tilde{\frac{1}{2}}\ \frac{3}{2}$ 010+ 000+ $+1/4.3$

$\tilde{\frac{1}{2}}\ \tilde{\frac{1}{2}}\ 0\quad \frac{3}{2}\ 1\ \tilde{\frac{1}{2}}$

$1\ \frac{3}{2}\ \tilde{\frac{1}{2}}$ 000− 000− $+1/8.3$

$1\ \tilde{\frac{1}{2}}\ \tilde{\frac{1}{2}}$ 000+ 000+ $+1/2.3$

$\tilde{\frac{1}{2}}\ \tilde{\frac{1}{2}}\ 0\quad \frac{3}{2}\ \frac{3}{2}\ 0$

$1\ 1\ 0$ 000+ 000+ $+1/2\sqrt{2.3}$

$\tilde{\frac{1}{2}}\ \tilde{\frac{1}{2}}\ 0\quad \frac{3}{2}\ \frac{3}{2}\ 1$

$1\ 1\ 1$ 000+ 000+ $-1/4.3\sqrt{3}$

$1\ 1\ 1$ 010+ 000+ $+1/2.3\sqrt{3}$

$\tilde{\frac{1}{2}}\ \tilde{\frac{1}{2}}\ 0\quad \frac{3}{2}\ \frac{3}{2}\ 2$

$1\ 1\ 2$ 000+ 000+ $+1/4\sqrt{2.3}$

$\tilde{\frac{1}{2}}\ \tilde{\frac{1}{2}}\ 0\quad \frac{3}{2}\ \frac{3}{2}\ \tilde{1}$

$1\ 1\ \tilde{1}$ 000+ 000+ $-1/4.3$

$1\ 1\ \tilde{1}$ 010− 000+ 0

$\tilde{\frac{1}{2}}\ \tilde{\frac{1}{2}}\ 0\quad 2\ 0\ 2$

$\frac{3}{2}\ \tilde{\frac{1}{2}}\ 2$ 000− 000− $+1/4$

$\tilde{\frac{1}{2}}\ \tilde{\frac{1}{2}}\ 0\quad 2\ \frac{1}{2}\ \frac{3}{2}$

$\frac{3}{2}\ \tilde{1}\ \frac{3}{2}$ 000+ 000+ $+1/8$

$\frac{3}{2}\ \tilde{1}\ \frac{3}{2}$ 001− 000+ 0

$\tilde{\frac{1}{2}}\ \tilde{\frac{1}{2}}\ 0\quad 2\ 1\ 1$

$\frac{3}{2}\ \frac{3}{2}\ 1$ 000− 000− $+1/4.3$

$\frac{3}{2}\ \frac{3}{2}\ 1$ 001− 000− $+1/4.3$

$\frac{3}{2}\ \tilde{\frac{1}{2}}\ 1$ 000+ 000+ $+1/2.3\sqrt{2}$

$\tilde{\frac{1}{2}}\ \tilde{\frac{1}{2}}\ 0\quad 2\ 1\ \tilde{1}$

$\frac{3}{2}\ \frac{3}{2}\ \tilde{1}$ 000− 000− $-1/4.3$

$\frac{3}{2}\ \frac{3}{2}\ \tilde{1}$ 001+ 000− $+1/4.3$

$\frac{3}{2}\ \tilde{\frac{1}{2}}\ \tilde{1}$ 000+ 000+ $-1/2.3\sqrt{2}$

$\tilde{\frac{1}{2}}\ \tilde{\frac{1}{2}}\ 0\quad 2\ \frac{3}{2}\ \frac{1}{2}$

$\frac{3}{2}\ 1\ \frac{1}{2}$ 000+ 000+ $+1/8$

$\frac{3}{2}\ 2\ \frac{1}{2}$ 000− 000− $+1/8$

$\frac{3}{2}\ \tilde{1}\ \frac{1}{2}$ 000− 000− $+1/8$

$\tilde{\frac{1}{2}}\ \tilde{\frac{1}{2}}\ 0\quad 2\ \frac{3}{2}\ \frac{3}{2}$

$\frac{3}{2}\ 1\ \frac{3}{2}$ 000+ 000+ 0

$\frac{3}{2}\ 1\ \frac{3}{2}$ 001+ 000+ $+1/8\sqrt{2}$

$\frac{3}{2}\ 2\ \frac{3}{2}$ 000− 000− $+1/8\sqrt{2}$

$\frac{3}{2}\ \tilde{1}\ \frac{3}{2}$ 000− 000− 0

$\frac{3}{2}\ \tilde{1}\ \frac{3}{2}$ 001+ 000− $-1/8\sqrt{2}$

$\tilde{\frac{1}{2}}\ \tilde{\frac{1}{2}}\ 0\quad 2\ \frac{3}{2}\ \tilde{\frac{1}{2}}$

$\frac{3}{2}\ 1\ \tilde{\frac{1}{2}}$ 000+ 000+ $+1/8$

$\frac{3}{2}\ 2\ \tilde{\frac{1}{2}}$ 000− 000− $-1/8$

$\frac{3}{2}\ \tilde{1}\ \tilde{\frac{1}{2}}$ 000− 000− $+1/8$

$\tilde{\frac{1}{2}}\ \tilde{\frac{1}{2}}\ 0\quad 2\ 2\ 0$

$\frac{3}{2}\ \frac{3}{2}\ 0$ 000+ 000+ $+1/4$

$\tilde{\frac{1}{2}}\ \tilde{\frac{1}{2}}\ 0\quad 2\ 2\ 2$

$\frac{3}{2}\ \frac{3}{2}\ 2$ 000+ 000+ $-1/4\sqrt{2}$

$\tilde{\frac{1}{2}}\ \tilde{\frac{1}{2}}\ 0\quad \tilde{1}\ 0\ \tilde{1}$

$\frac{1}{2}\ \tilde{\frac{1}{2}}\ \tilde{1}$ 000+ 000+ $+1/2.3$

$\frac{3}{2}\ \tilde{\frac{1}{2}}\ \tilde{1}$ 000− 000− $+1/2.3$

$\tilde{\frac{1}{2}}\ \tilde{\frac{1}{2}}\ 0\quad \tilde{1}\ \frac{1}{2}\ \frac{3}{2}$

$\frac{1}{2}\ \tilde{1}\ \frac{3}{2}$ 000− 000− $+1/2.3\sqrt{2}$

$\frac{3}{2}\ \tilde{1}\ \frac{3}{2}$ 000+ 000+ $+1/8.3$

$\frac{3}{2}\ \tilde{1}\ \frac{3}{2}$ 001− 000+ $-1/4.3$

$\tilde{\frac{1}{2}}\ \tilde{\frac{1}{2}}\ 0\quad \tilde{1}\ \frac{1}{2}\ \tilde{\frac{1}{2}}$

$\frac{1}{2}\ \tilde{1}\ \tilde{\frac{1}{2}}$ 000− 000− $+1/4.3$

$\frac{3}{2}\ \tilde{1}\ \tilde{\frac{1}{2}}$ 000+ 000+ $+1/2.3$

O and T_d 9j Symbols (cont.)

$\tilde{\frac{1}{2}}\ \tilde{\frac{1}{2}}\ 0\quad \tilde{1}\ 1\ 1$
$\frac{1}{2}\ \frac{3}{2}\ 1$ 000+ 000+ $-1/2.3\sqrt{2}$
$\frac{3}{2}\ \frac{3}{2}\ 1$ 000− 000− $+1/4.3$
$\frac{3}{2}\ \frac{3}{2}\ 1$ 001− 000− 0
$\frac{3}{2}\ \tilde{\frac{1}{2}}\ 1$ 000+ 000+ $-1/2.3\sqrt{2}$
$\tilde{\frac{1}{2}}\ \tilde{\frac{1}{2}}\ 0\quad \tilde{1}\ 1\ 2$
$\frac{1}{2}\ \frac{3}{2}\ 2$ 000+ 000+ $-1/4\sqrt{3}$
$\frac{3}{2}\ \frac{3}{2}\ 2$ 000− 000− $-1/4\sqrt{2.3}$
$\frac{3}{2}\ \tilde{\frac{1}{2}}\ 2$ 000+ 000+ $-1/4\sqrt{3}$
$\tilde{\frac{1}{2}}\ \tilde{\frac{1}{2}}\ 0\quad \tilde{1}\ 1\ \tilde{1}$
$\frac{1}{2}\ \frac{3}{2}\ \tilde{1}$ 000+ 000+ $+1/2.3\sqrt{2.3}$
$\frac{1}{2}\ \tilde{\frac{1}{2}}\ \tilde{1}$ 000− 000− $-1/3\sqrt{2.3}$
$\frac{3}{2}\ \frac{3}{2}\ \tilde{1}$ 000− 000− $+1/4.3\sqrt{3}$
$\frac{3}{2}\ \frac{3}{2}\ \tilde{1}$ 001+ 000− $+1/2.3\sqrt{3}$
$\frac{3}{2}\ \tilde{\frac{1}{2}}\ \tilde{1}$ 000+ 000+ $+1/2.3\sqrt{2.3}$
$\tilde{\frac{1}{2}}\ \tilde{\frac{1}{2}}\ 0\quad \tilde{1}\ \frac{3}{2}\ \frac{1}{2}$
$\frac{1}{2}\ 1\ \frac{1}{2}$ 000− 000− $-1/2.3$
$\frac{3}{2}\ 1\ \frac{1}{2}$ 000+ 000+ $+1/8.3$
$\frac{3}{2}\ 2\ \frac{1}{2}$ 000− 000− $+1/8$
$\frac{3}{2}\ \tilde{1}\ \frac{1}{2}$ 000− 000− $-1/8$
$\tilde{\frac{1}{2}}\ \tilde{\frac{1}{2}}\ 0\quad \tilde{1}\ \frac{3}{2}\ \frac{3}{2}$
$\frac{1}{2}\ 1\ \frac{3}{2}$ 000− 000− $-1/8.3$
$\frac{1}{2}\ 1\ \frac{3}{2}$ 010+ 000− $+1/4.3$
$\frac{1}{2}\ 2\ \frac{3}{2}$ 000+ 000+ $+1/8$
$\frac{1}{2}\ 2\ \frac{3}{2}$ 010− 000+ 0
$\frac{1}{2}\ \tilde{1}\ \frac{3}{2}$ 000+ 000+ $+1/8.3$
$\frac{1}{2}\ \tilde{1}\ \frac{3}{2}$ 010− 000+ $+1/4.3$
$\frac{3}{2}\ 1\ \frac{3}{2}$ 000+ 000+ $-1/4.3\sqrt{2}$
$\frac{3}{2}\ 1\ \frac{3}{2}$ 001+ 000+ $+1/4.3\sqrt{2}$
$\frac{3}{2}\ 1\ \frac{3}{2}$ 010− 000+ $-1/4.3\sqrt{2}$
$\frac{3}{2}\ 1\ \frac{3}{2}$ 011− 000+ $-1/8.3\sqrt{2}$
$\frac{3}{2}\ 2\ \frac{3}{2}$ 000− 000− 0
$\frac{3}{2}\ 2\ \frac{3}{2}$ 010+ 000− $+1/8\sqrt{2}$
$\frac{3}{2}\ \tilde{1}\ \frac{3}{2}$ 000− 000− $+1/4.3\sqrt{2}$
$\frac{3}{2}\ \tilde{1}\ \frac{3}{2}$ 001+ 000− $+1/4.3\sqrt{2}$
$\frac{3}{2}\ \tilde{1}\ \frac{3}{2}$ 010+ 000− $-1/4.3\sqrt{2}$
$\frac{3}{2}\ \tilde{1}\ \frac{3}{2}$ 011− 000− $+1/8.3\sqrt{2}$
$\tilde{\frac{1}{2}}\ \tilde{\frac{1}{2}}\ 0\quad \tilde{1}\ \frac{3}{2}\ \tilde{\frac{1}{2}}$
$\frac{1}{2}\ \tilde{1}\ \tilde{\frac{1}{2}}$ 000+ 000+ $+1/2.3$
$\frac{3}{2}\ 1\ \tilde{\frac{1}{2}}$ 000+ 000+ $+1/8$
$\frac{3}{2}\ 2\ \tilde{\frac{1}{2}}$ 000− 000− $+1/8$
$\frac{3}{2}\ \tilde{1}\ \tilde{\frac{1}{2}}$ 000− 000− $-1/8.3$
$\tilde{\frac{1}{2}}\ \tilde{\frac{1}{2}}\ 0\quad \tilde{1}\ 2\ 1$
$\frac{1}{2}\ \frac{3}{2}\ 1$ 000− 000− $+1/2.3\sqrt{2}$
$\frac{3}{2}\ \frac{3}{2}\ 1$ 000+ 000+ $+1/4.3$
$\frac{3}{2}\ \frac{3}{2}\ 1$ 001+ 000+ $-1/4.3$
$\tilde{\frac{1}{2}}\ \tilde{\frac{1}{2}}\ 0\quad \tilde{1}\ 2\ \tilde{1}$
$\frac{1}{2}\ \frac{3}{2}\ \tilde{1}$ 000− 000− $-1/2.3\sqrt{2}$
$\frac{3}{2}\ \frac{3}{2}\ \tilde{1}$ 000+ 000+ $-1/4.3$
$\frac{3}{2}\ \frac{3}{2}\ \tilde{1}$ 001− 000+ $+1/4.3$
$\tilde{\frac{1}{2}}\ \tilde{\frac{1}{2}}\ 0\quad \tilde{1}\ \tilde{1}\ 0$
$\frac{1}{2}\ \frac{1}{2}\ 0$ 000+ 000+ $+1/2\sqrt{3}$
$\frac{3}{2}\ \frac{3}{2}\ 0$ 000+ 000+ $+1/2\sqrt{2.3}$
$\tilde{\frac{1}{2}}\ \tilde{\frac{1}{2}}\ 0\quad \tilde{1}\ \tilde{1}\ 1$
$\frac{1}{2}\ \frac{1}{2}\ 1$ 000+ 000+ $+1/3\sqrt{2.3}$
$\frac{3}{2}\ \frac{1}{2}\ 1$ 000− 000− $-1/2.3\sqrt{2.3}$
$\frac{3}{2}\ \frac{3}{2}\ 1$ 000+ 000+ $+1/4.3\sqrt{3}$
$\frac{3}{2}\ \frac{3}{2}\ 1$ 001+ 000+ $+1/2.3\sqrt{3}$
$\tilde{\frac{1}{2}}\ \tilde{\frac{1}{2}}\ 0\quad \tilde{1}\ \tilde{1}\ 2$
$\frac{3}{2}\ \frac{1}{2}\ 2$ 000− 000− $-1/4\sqrt{3}$
$\frac{3}{2}\ \frac{3}{2}\ 2$ 000+ 000+ $-1/4\sqrt{2.3}$
$\tilde{\frac{1}{2}}\ \tilde{\frac{1}{2}}\ 0\quad \tilde{1}\ \tilde{1}\ \tilde{1}$
$\frac{3}{2}\ \frac{1}{2}\ \tilde{1}$ 000− 000− $+1/2.3\sqrt{2}$
$\frac{3}{2}\ \frac{3}{2}\ \tilde{1}$ 000+ 000+ $+1/4.3$
$\frac{3}{2}\ \frac{3}{2}\ \tilde{1}$ 001− 000+ 0
$\tilde{\frac{1}{2}}\ \tilde{\frac{1}{2}}\ 0\quad \tilde{\frac{1}{2}}\ 0\ \tilde{\frac{1}{2}}$
$0\ \tilde{\frac{1}{2}}\ \tilde{\frac{1}{2}}$ 000− 000− $+1/4$
$\tilde{\frac{1}{2}}\ \tilde{\frac{1}{2}}\ 0\quad \tilde{\frac{1}{2}}\ \frac{1}{2}\ \tilde{1}$
$0\ \tilde{1}\ \tilde{1}$ 000+ 000+ $+1/2.3$
$\tilde{\frac{1}{2}}\ \tilde{\frac{1}{2}}\ 0\quad \tilde{\frac{1}{2}}\ 1\ \frac{3}{2}$
$0\ \frac{3}{2}\ \frac{3}{2}$ 000− 000− $+1/8$
$\tilde{\frac{1}{2}}\ \tilde{\frac{1}{2}}\ 0\quad \tilde{\frac{1}{2}}\ 1\ \tilde{\frac{1}{2}}$
$0\ \tilde{\frac{1}{2}}\ \tilde{\frac{1}{2}}$ 000+ 000+ $+1/4$
$\tilde{\frac{1}{2}}\ \tilde{\frac{1}{2}}\ 0\quad \tilde{\frac{1}{2}}\ \frac{3}{2}\ 1$
$0\ 1\ 1$ 000+ 000+ $+1/2.3$
$\tilde{\frac{1}{2}}\ \tilde{\frac{1}{2}}\ 0\quad \tilde{\frac{1}{2}}\ \frac{3}{2}\ 2$
$0\ 2\ 2$ 000− 000− $+1/4$
$\tilde{\frac{1}{2}}\ \tilde{\frac{1}{2}}\ 0\quad \tilde{\frac{1}{2}}\ \frac{3}{2}\ \tilde{1}$
$0\ \tilde{1}\ \tilde{1}$ 000− 000− $+1/2.3$
$\tilde{\frac{1}{2}}\ \tilde{\frac{1}{2}}\ 0\quad \tilde{\frac{1}{2}}\ 2\ \frac{3}{2}$
$0\ \frac{3}{2}\ \frac{3}{2}$ 000+ 000+ $+1/8$
$\tilde{\frac{1}{2}}\ \tilde{\frac{1}{2}}\ 0\quad \tilde{\frac{1}{2}}\ \tilde{1}\ \frac{1}{2}$
$0\ \frac{1}{2}\ \frac{1}{2}$ 000− 000− $+1/4$
$\tilde{\frac{1}{2}}\ \tilde{\frac{1}{2}}\ 0\quad \tilde{\frac{1}{2}}\ \tilde{1}\ \frac{3}{2}$
$0\ \frac{3}{2}\ \frac{3}{2}$ 000+ 000+ $+1/8$
$\tilde{\frac{1}{2}}\ \tilde{\frac{1}{2}}\ 0\quad \tilde{\frac{1}{2}}\ \tilde{\frac{1}{2}}\ 0$
$0\ 0\ 0$ 000+ 000+ $+1/2$
$\tilde{\frac{1}{2}}\ \tilde{\frac{1}{2}}\ 1\quad \frac{3}{2}\ 0\ \frac{3}{2}$
$1\ \tilde{\frac{1}{2}}\ \frac{3}{2}$ 000+ 000+ $-1/8.3$
$1\ \tilde{\frac{1}{2}}\ \frac{3}{2}$ 000+ 001+ $+1/4.3$
$1\ \tilde{\frac{1}{2}}\ \tilde{\frac{1}{2}}$ 000− 000− $-1/2.3\sqrt{2}$
$\tilde{\frac{1}{2}}\ \tilde{\frac{1}{2}}\ 1\quad \frac{3}{2}\ \frac{1}{2}\ 1$
$1\ \tilde{1}\ 1$ 000− 000− $+1/4.3\sqrt{3}$
$1\ \tilde{1}\ 2$ 000+ 000+ $+1/4.3$
$1\ \tilde{1}\ \tilde{1}$ 000+ 000+ $-1/4.3\sqrt{3}$
$\tilde{\frac{1}{2}}\ \tilde{\frac{1}{2}}\ 1\quad \frac{3}{2}\ \frac{1}{2}\ 2$
$1\ \tilde{1}\ 1$ 000+ 000+ $-1/4.3$
$1\ \tilde{1}\ \tilde{1}$ 000− 000− $+1/4.3\sqrt{3}$
$\tilde{\frac{1}{2}}\ \tilde{\frac{1}{2}}\ 1\quad \frac{3}{2}\ \frac{1}{2}\ \tilde{1}$
$1\ \tilde{1}\ 1$ 000+ 000+ $+1/4.9$
$1\ \tilde{1}\ 2$ 000− 000− $+1/4.9$
$1\ \tilde{1}\ \tilde{1}$ 000− 000− $-1/4.3$
$\tilde{\frac{1}{2}}\ \tilde{\frac{1}{2}}\ 1\quad \frac{3}{2}\ 1\ \frac{1}{2}$
$1\ \frac{3}{2}\ \frac{1}{2}$ 000+ 000+ $+1/8.3$
$1\ \frac{3}{2}\ \frac{3}{2}$ 000− 000− $+1/4.3\sqrt{2}$
$1\ \frac{3}{2}\ \frac{3}{2}$ 001− 000− 0
$1\ \tilde{\frac{1}{2}}\ \frac{3}{2}$ 000+ 000+ $+1/4.3$
$\tilde{\frac{1}{2}}\ \tilde{\frac{1}{2}}\ 1\quad \frac{3}{2}\ 1\ \frac{3}{2}$
$1\ \frac{3}{2}\ \frac{1}{2}$ 000− 000− $+1/4.3\sqrt{2}$
$1\ \frac{3}{2}\ \frac{1}{2}$ 010− 000− 0
$1\ \frac{3}{2}\ \frac{3}{2}$ 000+ 000+ $-1/4.9\sqrt{2}$
$1\ \frac{3}{2}\ \frac{3}{2}$ 000+ 001+ $+1/2.9\sqrt{2}$
$1\ \frac{3}{2}\ \frac{3}{2}$ 001+ 000+ $-1/4.9\sqrt{2}$
$1\ \frac{3}{2}\ \frac{3}{2}$ 001+ 001+ $-1/4.9\sqrt{2}$
$1\ \frac{3}{2}\ \frac{3}{2}$ 010+ 000+ $-1/4.9\sqrt{2}$
$1\ \frac{3}{2}\ \frac{3}{2}$ 010+ 001+ $-1/4.9\sqrt{2}$
$1\ \frac{3}{2}\ \frac{3}{2}$ 011+ 000+ $-5/8.9\sqrt{2}$
$1\ \frac{3}{2}\ \frac{3}{2}$ 011+ 001+ $+1/2.9\sqrt{2}$
$1\ \frac{3}{2}\ \tilde{\frac{1}{2}}$ 000− 000− $+1/4.9\sqrt{2}$
$1\ \frac{3}{2}\ \tilde{\frac{1}{2}}$ 010− 000− $-1/2.9\sqrt{2}$
$1\ \tilde{\frac{1}{2}}\ \frac{3}{2}$ 000− 000− $+5/8.9$
$1\ \tilde{\frac{1}{2}}\ \frac{3}{2}$ 000− 001− $+1/4.9$
$1\ \tilde{\frac{1}{2}}\ \frac{3}{2}$ 010− 000− $+1/4.9$
$1\ \tilde{\frac{1}{2}}\ \frac{3}{2}$ 010− 001− $+1/4.9$
$1\ \tilde{\frac{1}{2}}\ \tilde{\frac{1}{2}}$ 000+ 000+ $-1/2.9\sqrt{2}$
$1\ \tilde{\frac{1}{2}}\ \tilde{\frac{1}{2}}$ 010+ 000+ $+1/9\sqrt{2}$
$\tilde{\frac{1}{2}}\ \tilde{\frac{1}{2}}\ 1\quad \frac{3}{2}\ 1\ \tilde{\frac{1}{2}}$
$1\ \frac{3}{2}\ \frac{3}{2}$ 000− 000− $+1/4.9\sqrt{2}$
$1\ \frac{3}{2}\ \frac{3}{2}$ 001− 000− $-1/2.9\sqrt{2}$
$1\ \frac{3}{2}\ \tilde{\frac{1}{2}}$ 000+ 000+ $+7/8.9$
$1\ \tilde{\frac{1}{2}}\ \frac{3}{2}$ 000+ 000+ $-1/4.9$
$1\ \tilde{\frac{1}{2}}\ \tilde{\frac{1}{2}}$ 000− 000− $+1/2.9$
$\tilde{\frac{1}{2}}\ \tilde{\frac{1}{2}}\ 1\quad \frac{3}{2}\ \frac{3}{2}\ 0$
$1\ 1\ 1$ 000+ 000+ $-1/4.3\sqrt{3}$
$\tilde{\frac{1}{2}}\ \tilde{\frac{1}{2}}\ 1\quad \frac{3}{2}\ \frac{3}{2}\ 1$
$1\ 1\ 0$ 000+ 000+ $-1/2.9\sqrt{2}$
$1\ 1\ 0$ 010+ 000+ $+1/9\sqrt{2}$
$1\ 1\ 1$ 000− 000− 0
$1\ 1\ 1$ 010− 000− 0
$1\ 1\ 2$ 000+ 000+ $+1/9\sqrt{2}$
$1\ 1\ 2$ 010+ 000+ $+1/4.9\sqrt{2}$
$1\ 1\ \tilde{1}$ 000+ 000+ $-1/2.9\sqrt{2}$
$1\ 1\ \tilde{1}$ 010+ 000+ $-1/2.9\sqrt{2}$
$\tilde{\frac{1}{2}}\ \tilde{\frac{1}{2}}\ 1\quad \frac{3}{2}\ \frac{3}{2}\ 2$
$1\ 1\ 1$ 000+ 000+ $+1/2.3\sqrt{2.3}$
$1\ 1\ \tilde{1}$ 000− 000− 0
$\tilde{\frac{1}{2}}\ \tilde{\frac{1}{2}}\ 1\quad \frac{3}{2}\ \frac{3}{2}\ \tilde{1}$
$1\ 1\ 1$ 000+ 000+ $-1/2.3\sqrt{2.3}$
$1\ 1\ 1$ 010− 000+ 0

O and T_d 9j Symbols (cont.)

$\tilde{\frac{1}{2}}$	$\tilde{\frac{1}{2}}$	**1**	$\frac{3}{2}$	$\frac{3}{2}$	$\tilde{1}$
1	1	2	000−	000−	0
1	1	2	010+	000−	$+1/4.3\sqrt{2}$
1	1	$\tilde{1}$	000−	000−	0
1	1	$\tilde{1}$	010+	000−	$-1/2.3\sqrt{2.3}$
$\tilde{\frac{1}{2}}$	$\tilde{\frac{1}{2}}$	**1**	**2**	**0**	**2**
$\frac{3}{2}$	$\tilde{\frac{1}{2}}$	1	000−	000−	$-1/4\sqrt{3}$
$\frac{3}{2}$	$\tilde{\frac{1}{2}}$	$\tilde{1}$	000+	000+	$-1/4\sqrt{3}$
$\tilde{\frac{1}{2}}$	$\tilde{\frac{1}{2}}$	**1**	**2**	$\frac{1}{2}$	$\frac{3}{2}$
$\frac{3}{2}$	$\tilde{1}$	$\frac{1}{2}$	000+	000+	$+1/4.3$
$\frac{3}{2}$	$\tilde{1}$	$\frac{3}{2}$	000−	000−	$+1/8.3$
$\frac{3}{2}$	$\tilde{1}$	$\frac{3}{2}$	000−	001−	0
$\frac{3}{2}$	$\tilde{1}$	$\frac{3}{2}$	001+	000−	0
$\frac{3}{2}$	$\tilde{1}$	$\frac{3}{2}$	001+	001−	$-1/4.3$
$\frac{3}{2}$	$\tilde{1}$	$\tilde{\frac{1}{2}}$	000+	000+	$+1/4.3$
$\tilde{\frac{1}{2}}$	$\tilde{\frac{1}{2}}$	**1**	**2**	**1**	**1**
$\frac{3}{2}$	$\frac{3}{2}$	0	000−	000−	$-1/4.3$
$\frac{3}{2}$	$\frac{3}{2}$	1	000+	000+	$-1/2.3\sqrt{2.3}$
$\frac{3}{2}$	$\frac{3}{2}$	1	001+	000+	$+1/4.3\sqrt{2.3}$
$\frac{3}{2}$	$\frac{3}{2}$	2	000−	000−	$-1/4.3\sqrt{2}$
$\frac{3}{2}$	$\frac{3}{2}$	$\tilde{1}$	000−	000−	0
$\frac{3}{2}$	$\frac{3}{2}$	$\tilde{1}$	001+	000−	$+1/4.3\sqrt{2}$
$\frac{3}{2}$	$\tilde{\frac{1}{2}}$	1	000−	000−	$+1/4.3\sqrt{3}$
$\frac{3}{2}$	$\tilde{\frac{1}{2}}$	2	000+	000+	$+1/4.3$
$\frac{3}{2}$	$\tilde{\frac{1}{2}}$	$\tilde{1}$	000+	000+	$+1/4.3$
$\tilde{\frac{1}{2}}$	$\tilde{\frac{1}{2}}$	**1**	**2**	**1**	$\tilde{1}$
$\frac{3}{2}$	$\frac{3}{2}$	1	000−	000−	0
$\frac{3}{2}$	$\frac{3}{2}$	1	001−	000−	$-1/4.3\sqrt{2}$
$\frac{3}{2}$	$\frac{3}{2}$	2	000+	000+	$+1/4.3\sqrt{2}$
$\frac{3}{2}$	$\frac{3}{2}$	$\tilde{1}$	000+	000+	$-1/2.3\sqrt{2.3}$
$\frac{3}{2}$	$\frac{3}{2}$	$\tilde{1}$	001−	000+	$-1/4.3\sqrt{2.3}$
$\frac{3}{2}$	$\tilde{\frac{1}{2}}$	1	000+	000+	$+1/4.3$
$\frac{3}{2}$	$\tilde{\frac{1}{2}}$	2	000−	000−	$+1/4.3$
$\frac{3}{2}$	$\tilde{\frac{1}{2}}$	$\tilde{1}$	000−	000−	$+1/4.3\sqrt{3}$
$\tilde{\frac{1}{2}}$	$\tilde{\frac{1}{2}}$	**1**	**2**	$\frac{3}{2}$	$\frac{1}{2}$
$\frac{3}{2}$	1	$\frac{1}{2}$	000−	000−	$-1/8.3$
$\frac{3}{2}$	1	$\frac{3}{2}$	000+	000+	$-1/4.3\sqrt{2}$
$\frac{3}{2}$	1	$\frac{3}{2}$	001+	000+	$+1/4.3\sqrt{2}$
$\frac{3}{2}$	2	$\frac{1}{2}$	000+	000+	$+1/8$
$\frac{3}{2}$	2	$\frac{3}{2}$	000−	000−	0
$\frac{3}{2}$	$\tilde{1}$	$\frac{1}{2}$	000+	000+	$-1/8.3$
$\frac{3}{2}$	$\tilde{1}$	$\frac{3}{2}$	000−	000−	$-1/4.3\sqrt{2}$
$\frac{3}{2}$	$\tilde{1}$	$\frac{3}{2}$	001+	000−	$+1/4.3\sqrt{2}$
$\tilde{\frac{1}{2}}$	$\tilde{\frac{1}{2}}$	**1**	**2**	$\frac{3}{2}$	$\frac{3}{2}$
$\frac{3}{2}$	1	$\frac{1}{2}$	000+	000+	$-1/4.3\sqrt{2}$
$\frac{3}{2}$	1	$\frac{3}{2}$	000−	000−	0
$\frac{3}{2}$	1	$\frac{3}{2}$	000−	001−	$-1/4.3\sqrt{2}$
$\frac{3}{2}$	1	$\frac{3}{2}$	001−	000−	$+1/8.3\sqrt{2}$
$\frac{3}{2}$	1	$\frac{3}{2}$	001−	001−	0
$\frac{3}{2}$	1	$\tilde{\frac{1}{2}}$	000+	000+	$+1/4.3\sqrt{2}$

$\tilde{\frac{1}{2}}$	$\tilde{\frac{1}{2}}$	**1**	**2**	$\frac{3}{2}$	$\frac{3}{2}$
$\frac{3}{2}$	2	$\frac{1}{2}$	000−	000−	0
$\frac{3}{2}$	2	$\frac{3}{2}$	000+	000+	$-1/8\sqrt{2}$
$\frac{3}{2}$	2	$\frac{3}{2}$	000+	001+	0
$\frac{3}{2}$	2	$\tilde{\frac{1}{2}}$	000−	000−	0
$\frac{3}{2}$	$\tilde{1}$	$\frac{1}{2}$	000−	000−	$+1/4.3\sqrt{2}$
$\frac{3}{2}$	$\tilde{1}$	$\frac{3}{2}$	000+	000+	0
$\frac{3}{2}$	$\tilde{1}$	$\frac{3}{2}$	000+	001+	$+1/4.3\sqrt{2}$
$\frac{3}{2}$	$\tilde{1}$	$\frac{3}{2}$	001−	000+	$-1/8.3\sqrt{2}$
$\frac{3}{2}$	$\tilde{1}$	$\frac{3}{2}$	001−	001+	0
$\frac{3}{2}$	$\tilde{1}$	$\tilde{\frac{1}{2}}$	000−	000−	$-1/4.3\sqrt{2}$
$\tilde{\frac{1}{2}}$	$\tilde{\frac{1}{2}}$	**1**	**2**	$\frac{3}{2}$	$\tilde{\frac{1}{2}}$
$\frac{3}{2}$	1	$\frac{3}{2}$	000+	000+	$-1/4.3\sqrt{2}$
$\frac{3}{2}$	1	$\frac{3}{2}$	001+	000+	$-1/4.3\sqrt{2}$
$\frac{3}{2}$	1	$\tilde{\frac{1}{2}}$	000−	000−	$+1/8.3$
$\frac{3}{2}$	2	$\frac{3}{2}$	000−	000−	0
$\frac{3}{2}$	2	$\tilde{\frac{1}{2}}$	000+	000+	$+1/8$
$\frac{3}{2}$	$\tilde{1}$	$\frac{3}{2}$	000−	000−	$-1/4.3\sqrt{2}$
$\frac{3}{2}$	$\tilde{1}$	$\frac{3}{2}$	001+	000−	$-1/4.3\sqrt{2}$
$\frac{3}{2}$	$\tilde{1}$	$\tilde{\frac{1}{2}}$	000+	000+	$+1/8.3$
$\tilde{\frac{1}{2}}$	$\tilde{\frac{1}{2}}$	**1**	**2**	**2**	**0**
$\frac{3}{2}$	$\frac{3}{2}$	1	000+	000+	$+1/4\sqrt{3}$
$\frac{3}{2}$	$\frac{3}{2}$	1	001+	000+	0
$\tilde{\frac{1}{2}}$	$\tilde{\frac{1}{2}}$	**1**	**2**	**2**	**2**
$\frac{3}{2}$	$\frac{3}{2}$	1	000+	000+	0
$\frac{3}{2}$	$\frac{3}{2}$	1	001+	000+	$-1/4\sqrt{2.3}$
$\frac{3}{2}$	$\frac{3}{2}$	$\tilde{1}$	000−	000−	0
$\frac{3}{2}$	$\frac{3}{2}$	$\tilde{1}$	001+	000−	$-1/4\sqrt{2.3}$
$\tilde{\frac{1}{2}}$	$\tilde{\frac{1}{2}}$	**1**	$\tilde{1}$	**0**	$\tilde{1}$
$\frac{1}{2}$	$\tilde{\frac{1}{2}}$	$\tilde{1}$	000−	000−	$-1/3\sqrt{2.3}$
$\frac{3}{2}$	$\tilde{\frac{1}{2}}$	1	000−	000−	$+1/2.3\sqrt{2}$
$\frac{3}{2}$	$\tilde{\frac{1}{2}}$	2	000+	000+	$-1/2.3\sqrt{2}$
$\frac{3}{2}$	$\tilde{\frac{1}{2}}$	$\tilde{1}$	000+	000+	$+1/2.3\sqrt{2.3}$
$\tilde{\frac{1}{2}}$	$\tilde{\frac{1}{2}}$	**1**	$\tilde{1}$	$\frac{1}{2}$	$\frac{3}{2}$
$\frac{1}{2}$	$\tilde{1}$	$\frac{3}{2}$	000+	000+	$-1/2.9\sqrt{2}$
$\frac{1}{2}$	$\tilde{1}$	$\frac{3}{2}$	000+	001+	$+1/9\sqrt{2}$
$\frac{1}{2}$	$\tilde{1}$	$\tilde{\frac{1}{2}}$	000−	000−	$+1/2.9$
$\frac{3}{2}$	$\tilde{1}$	$\frac{1}{2}$	000+	000+	$-1/4.3$
$\frac{3}{2}$	$\tilde{1}$	$\frac{3}{2}$	000−	000−	$+5/8.9$
$\frac{3}{2}$	$\tilde{1}$	$\frac{3}{2}$	000−	001−	$+1/4.9$
$\frac{3}{2}$	$\tilde{1}$	$\frac{3}{2}$	001+	000−	$+1/4.9$
$\frac{3}{2}$	$\tilde{1}$	$\frac{3}{2}$	001+	001−	$+1/4.9$
$\frac{3}{2}$	$\tilde{1}$	$\tilde{\frac{1}{2}}$	000+	000+	$+1/4.9$
$\tilde{\frac{1}{2}}$	$\tilde{\frac{1}{2}}$	**1**	$\tilde{1}$	$\frac{1}{2}$	$\tilde{\frac{1}{2}}$
$\frac{1}{2}$	$\tilde{1}$	$\frac{3}{2}$	000−	000−	$+1/2.9$
$\frac{1}{2}$	$\tilde{1}$	$\tilde{\frac{1}{2}}$	000+	000+	$+5/4.9$
$\frac{3}{2}$	$\tilde{1}$	$\frac{3}{2}$	000+	000+	$-1/2.9\sqrt{2}$
$\frac{3}{2}$	$\tilde{1}$	$\frac{3}{2}$	001−	000+	$+1/9\sqrt{2}$
$\frac{3}{2}$	$\tilde{1}$	$\tilde{\frac{1}{2}}$	000−	000−	$-1/2.9$

$\tilde{\frac{1}{2}}$	$\tilde{\frac{1}{2}}$	**1**	$\tilde{1}$	**1**	**1**
$\frac{1}{2}$	$\frac{3}{2}$	1	000−	000−	$+1/4.3\sqrt{3}$
$\frac{1}{2}$	$\frac{3}{2}$	2	000+	000+	$-1/4.3$
$\frac{1}{2}$	$\frac{3}{2}$	$\tilde{1}$	000+	000+	$+1/4.9$
$\frac{1}{2}$	$\tilde{\frac{1}{2}}$	$\tilde{1}$	000−	000−	$+1/9$
$\frac{3}{2}$	$\frac{3}{2}$	0	000−	000−	$+1/4.3$
$\frac{3}{2}$	$\frac{3}{2}$	1	000+	000+	0
$\frac{3}{2}$	$\frac{3}{2}$	1	001+	000+	$-1/2.3\sqrt{2.3}$
$\frac{3}{2}$	$\frac{3}{2}$	2	000−	000−	0
$\frac{3}{2}$	$\frac{3}{2}$	$\tilde{1}$	000−	000−	$-1/2.9\sqrt{2}$
$\frac{3}{2}$	$\frac{3}{2}$	$\tilde{1}$	001+	000−	$+1/2.9\sqrt{2}$
$\frac{3}{2}$	$\tilde{\frac{1}{2}}$	1	000−	000−	$-1/4.3\sqrt{3}$
$\frac{3}{2}$	$\tilde{\frac{1}{2}}$	2	000+	000+	$+1/4.3$
$\frac{3}{2}$	$\tilde{\frac{1}{2}}$	$\tilde{1}$	000+	000+	$+1/4.9$
$\tilde{\frac{1}{2}}$	$\tilde{\frac{1}{2}}$	**1**	$\tilde{1}$	**1**	**2**
$\frac{1}{2}$	$\frac{3}{2}$	1	000+	000+	$+1/4.3$
$\frac{1}{2}$	$\frac{3}{2}$	$\tilde{1}$	000−	000−	$+1/4.9$
$\frac{1}{2}$	$\tilde{\frac{1}{2}}$	$\tilde{1}$	000+	000+	$+1/9$
$\frac{3}{2}$	$\frac{3}{2}$	1	000−	000−	0
$\frac{3}{2}$	$\frac{3}{2}$	1	001−	000−	$+1/4.3\sqrt{2}$
$\frac{3}{2}$	$\frac{3}{2}$	$\tilde{1}$	000+	000+	$+1/9\sqrt{2}$
$\frac{3}{2}$	$\frac{3}{2}$	$\tilde{1}$	001−	000+	$-1/4.9\sqrt{2}$
$\frac{3}{2}$	$\tilde{\frac{1}{2}}$	1	000+	000+	$-1/4.3$
$\frac{3}{2}$	$\tilde{\frac{1}{2}}$	$\tilde{1}$	000−	000−	$+1/4.9$
$\tilde{\frac{1}{2}}$	$\tilde{\frac{1}{2}}$	**1**	$\tilde{1}$	**1**	$\tilde{1}$
$\frac{1}{2}$	$\frac{3}{2}$	1	000+	000+	$-1/4.3\sqrt{3}$
$\frac{1}{2}$	$\frac{3}{2}$	2	000−	000−	$+1/4.3\sqrt{3}$
$\frac{1}{2}$	$\frac{3}{2}$	$\tilde{1}$	000−	000−	$-1/4.3$
$\frac{1}{2}$	$\tilde{\frac{1}{2}}$	$\tilde{1}$	000+	000+	0
$\frac{3}{2}$	$\frac{3}{2}$	1	000−	000−	$-1/2.3\sqrt{2.3}$
$\frac{3}{2}$	$\frac{3}{2}$	1	001−	000−	0
$\frac{3}{2}$	$\frac{3}{2}$	2	000+	000+	$-1/2.3\sqrt{2.3}$
$\frac{3}{2}$	$\frac{3}{2}$	$\tilde{1}$	000+	000+	0
$\frac{3}{2}$	$\frac{3}{2}$	$\tilde{1}$	001−	000+	0
$\frac{3}{2}$	$\tilde{\frac{1}{2}}$	1	000+	000+	$-1/4.3\sqrt{3}$
$\frac{3}{2}$	$\tilde{\frac{1}{2}}$	2	000−	000−	$+1/4.3\sqrt{3}$
$\frac{3}{2}$	$\tilde{\frac{1}{2}}$	$\tilde{1}$	000−	000−	$+1/4.3$
$\tilde{\frac{1}{2}}$	$\tilde{\frac{1}{2}}$	**1**	$\tilde{1}$	$\frac{3}{2}$	$\frac{1}{2}$
$\frac{1}{2}$	1	$\frac{1}{2}$	000+	000+	$-1/2.9$
$\frac{1}{2}$	1	$\frac{3}{2}$	000−	000−	$-1/4.9$
$\frac{1}{2}$	2	$\frac{3}{2}$	000+	000+	$+1/4.3$
$\frac{1}{2}$	$\tilde{1}$	$\frac{3}{2}$	000+	000+	$-1/4.3$
$\frac{3}{2}$	1	$\frac{1}{2}$	000−	000−	$+7/8.9$
$\frac{3}{2}$	1	$\frac{3}{2}$	000+	000+	$+1/4.9\sqrt{2}$
$\frac{3}{2}$	1	$\frac{3}{2}$	001+	000+	$+1/2.9\sqrt{2}$
$\frac{3}{2}$	2	$\frac{1}{2}$	000+	000+	$-1/8.3$
$\frac{3}{2}$	2	$\frac{3}{2}$	000−	000−	$+1/4.3\sqrt{2}$
$\frac{3}{2}$	$\tilde{1}$	$\frac{1}{2}$	000+	000+	$+1/8.3$
$\frac{3}{2}$	$\tilde{1}$	$\frac{3}{2}$	000−	000−	$-1/4.3\sqrt{2}$
$\frac{3}{2}$	$\tilde{1}$	$\frac{3}{2}$	001+	000−	0

O and T_d 9j Symbols (cont.)

$\tilde{\frac{1}{2}}$	$\tilde{\frac{1}{2}}$	1	$\tilde{1}$	$\frac{3}{2}$	$\frac{3}{2}$
$\frac{1}{2}$	1	$\frac{1}{2}$	000−	000−	$+1/2.9\sqrt{2}$
$\frac{1}{2}$	1	$\frac{1}{2}$	010+	000−	$-1/9\sqrt{2}$
$\frac{1}{2}$	1	$\frac{3}{2}$	000+	000+	−5/8.9
$\frac{1}{2}$	1	$\frac{3}{2}$	000+	001+	+1/4.9
$\frac{1}{2}$	1	$\frac{3}{2}$	010−	000+	−1/4.9
$\frac{1}{2}$	1	$\frac{3}{2}$	010−	001+	+1/4.9
$\frac{1}{2}$	2	$\frac{3}{2}$	000−	000−	+1/8.3
$\frac{1}{2}$	2	$\frac{3}{2}$	000−	001−	0
$\frac{1}{2}$	2	$\frac{3}{2}$	010+	000−	0
$\frac{1}{2}$	2	$\frac{3}{2}$	010+	001−	+1/4.3
$\frac{1}{2}$	$\tilde{1}$	$\frac{3}{2}$	000−	000−	+5/8.9
$\frac{1}{2}$	$\tilde{1}$	$\frac{3}{2}$	000−	001−	+1/4.9
$\frac{1}{2}$	$\tilde{1}$	$\frac{3}{2}$	010+	000−	−1/4.9
$\frac{1}{2}$	$\tilde{1}$	$\frac{3}{2}$	010+	001−	−1/4.9
$\frac{1}{2}$	$\tilde{1}$	$\tilde{\frac{1}{2}}$	000+	000+	$-1/2.9\sqrt{2}$
$\frac{1}{2}$	$\tilde{1}$	$\tilde{\frac{1}{2}}$	010−	000+	$-1/9\sqrt{2}$
$\frac{3}{2}$	1	$\frac{1}{2}$	000+	000+	$+1/4.9\sqrt{2}$
$\frac{3}{2}$	1	$\frac{1}{2}$	010−	000+	$-1/2.9\sqrt{2}$
$\frac{3}{2}$	1	$\frac{3}{2}$	000−	000−	$+1/4.9\sqrt{2}$
$\frac{3}{2}$	1	$\frac{3}{2}$	000−	001−	$+1/2.9\sqrt{2}$
$\frac{3}{2}$	1	$\frac{3}{2}$	001−	000−	$-1/4.9\sqrt{2}$
$\frac{3}{2}$	1	$\frac{3}{2}$	001−	001−	$+1/4.9\sqrt{2}$
$\frac{3}{2}$	1	$\frac{3}{2}$	010+	000−	$+1/4.9\sqrt{2}$
$\frac{3}{2}$	1	$\frac{3}{2}$	010+	001−	$-1/4.9\sqrt{2}$
$\frac{3}{2}$	1	$\frac{3}{2}$	011+	000−	$-5/8.9\sqrt{2}$
$\frac{3}{2}$	1	$\frac{3}{2}$	011+	001−	$-1/2.9\sqrt{2}$
$\frac{3}{2}$	1	$\tilde{\frac{1}{2}}$	000+	000+	$+1/4.3\sqrt{2}$
$\frac{3}{2}$	1	$\tilde{\frac{1}{2}}$	010−	000+	0
$\frac{3}{2}$	2	$\frac{1}{2}$	000−	000−	$-1/4.3\sqrt{2}$
$\frac{3}{2}$	2	$\frac{1}{2}$	010+	000−	$-1/4.3\sqrt{2}$
$\frac{3}{2}$	2	$\frac{3}{2}$	000+	000+	0
$\frac{3}{2}$	2	$\frac{3}{2}$	000+	001+	$+1/4.3\sqrt{2}$
$\frac{3}{2}$	2	$\frac{3}{2}$	010−	000+	$+1/8.3\sqrt{2}$
$\frac{3}{2}$	2	$\frac{3}{2}$	010−	001+	0
$\frac{3}{2}$	2	$\tilde{\frac{1}{2}}$	000−	000−	$-1/4.3\sqrt{2}$
$\frac{3}{2}$	2	$\tilde{\frac{1}{2}}$	010+	000−	$+1/4.3\sqrt{2}$
$\frac{3}{2}$	$\tilde{1}$	$\frac{1}{2}$	000−	000−	$-1/4.3\sqrt{2}$
$\frac{3}{2}$	$\tilde{1}$	$\frac{1}{2}$	010+	000−	0
$\frac{3}{2}$	$\tilde{1}$	$\frac{3}{2}$	000+	000+	$-1/4.9\sqrt{2}$
$\frac{3}{2}$	$\tilde{1}$	$\frac{3}{2}$	000+	001+	$+1/2.9\sqrt{2}$
$\frac{3}{2}$	$\tilde{1}$	$\frac{3}{2}$	001−	000+	$-1/4.9\sqrt{2}$
$\frac{3}{2}$	$\tilde{1}$	$\frac{3}{2}$	001−	001+	$-1/4.9\sqrt{2}$
$\frac{3}{2}$	$\tilde{1}$	$\frac{3}{2}$	010−	000+	$+1/4.9\sqrt{2}$
$\frac{3}{2}$	$\tilde{1}$	$\frac{3}{2}$	010−	001+	$+1/4.9\sqrt{2}$
$\frac{3}{2}$	$\tilde{1}$	$\frac{3}{2}$	011+	000+	$+5/8.9\sqrt{2}$
$\frac{3}{2}$	$\tilde{1}$	$\frac{3}{2}$	011+	001+	$-1/2.9\sqrt{2}$
$\frac{3}{2}$	$\tilde{1}$	$\tilde{\frac{1}{2}}$	000−	000−	$-1/4.9\sqrt{2}$
$\frac{3}{2}$	$\tilde{1}$	$\tilde{\frac{1}{2}}$	010+	000−	$-1/2.9\sqrt{2}$

$\tilde{\frac{1}{2}}$	$\tilde{\frac{1}{2}}$	1	$\tilde{1}$	$\frac{3}{2}$	$\tilde{\frac{1}{2}}$
$\frac{1}{2}$	1	$\frac{3}{2}$	000−	000−	+1/4.3
$\frac{1}{2}$	2	$\frac{3}{2}$	000+	000+	+1/4.3
$\frac{1}{2}$	$\tilde{1}$	$\frac{3}{2}$	000+	000+	+1/4.9
$\frac{1}{2}$	$\tilde{1}$	$\tilde{\frac{1}{2}}$	000−	000−	−1/2.9
$\frac{3}{2}$	1	$\frac{1}{2}$	000+	000+	$+1/4.3\sqrt{2}$
$\frac{3}{2}$	1	$\frac{1}{2}$	001+	000+	0
$\frac{3}{2}$	1	$\tilde{\frac{1}{2}}$	000−	000−	+1/8.3
$\frac{3}{2}$	2	$\frac{1}{2}$	000−	000−	$-1/4.3\sqrt{2}$
$\frac{3}{2}$	2	$\tilde{\frac{1}{2}}$	000+	000+	+1/8.3
$\frac{3}{2}$	$\tilde{1}$	$\frac{1}{2}$	000−	000−	$-1/4.9\sqrt{2}$
$\frac{3}{2}$	$\tilde{1}$	$\frac{1}{2}$	001+	000−	$+1/2.9\sqrt{2}$
$\frac{3}{2}$	$\tilde{1}$	$\tilde{\frac{1}{2}}$	000+	000+	+7/8.9

$\tilde{\frac{1}{2}}$	$\tilde{\frac{1}{2}}$	1	$\tilde{1}$	2	1
$\frac{1}{2}$	$\frac{3}{2}$	1	000+	000+	$-1/4.3\sqrt{3}$
$\frac{1}{2}$	$\frac{3}{2}$	2	000−	000−	−1/4.3
$\frac{1}{2}$	$\frac{3}{2}$	$\tilde{1}$	000−	000−	+1/4.3
$\frac{3}{2}$	$\frac{3}{2}$	0	000+	000+	−1/4.3
$\frac{3}{2}$	$\frac{3}{2}$	1	000−	000−	$+1/2.3\sqrt{2.3}$
$\frac{3}{2}$	$\frac{3}{2}$	1	001−	000−	$+1/4.3\sqrt{2.3}$
$\frac{3}{2}$	$\frac{3}{2}$	2	000+	000+	$+1/4.3\sqrt{2}$
$\frac{3}{2}$	$\frac{3}{2}$	$\tilde{1}$	000+	000+	0
$\frac{3}{2}$	$\frac{3}{2}$	$\tilde{1}$	001−	000+	$+1/4.3\sqrt{2}$

$\tilde{\frac{1}{2}}$	$\tilde{\frac{1}{2}}$	1	$\tilde{1}$	2	$\tilde{1}$
$\frac{1}{2}$	$\frac{3}{2}$	1	000−	000−	+1/4.3
$\frac{1}{2}$	$\frac{3}{2}$	2	000+	000+	+1/4.3
$\frac{1}{2}$	$\frac{3}{2}$	$\tilde{1}$	000+	000+	$-1/4.3\sqrt{3}$
$\frac{3}{2}$	$\frac{3}{2}$	1	000+	000+	0
$\frac{3}{2}$	$\frac{3}{2}$	1	001+	000+	$+1/4.3\sqrt{2}$
$\frac{3}{2}$	$\frac{3}{2}$	2	000−	000−	$+1/4.3\sqrt{2}$
$\frac{3}{2}$	$\frac{3}{2}$	$\tilde{1}$	000−	000−	$+1/2.3\sqrt{2.3}$
$\frac{3}{2}$	$\frac{3}{2}$	$\tilde{1}$	001+	000−	$+1/4.3\sqrt{2.3}$

$\tilde{\frac{1}{2}}$	$\tilde{\frac{1}{2}}$	1	$\tilde{1}$	$\tilde{1}$	0
$\frac{1}{2}$	$\frac{1}{2}$	1	000+	000+	+1/2.9
$\frac{3}{2}$	$\frac{1}{2}$	1	000−	000−	−1/9
$\frac{3}{2}$	$\frac{3}{2}$	1	000+	000+	$-1/2.9\sqrt{2}$
$\frac{3}{2}$	$\frac{3}{2}$	1	001+	000+	$-1/9\sqrt{2}$

$\tilde{\frac{1}{2}}$	$\tilde{\frac{1}{2}}$	1	$\tilde{1}$	$\tilde{1}$	1
$\frac{1}{2}$	$\frac{1}{2}$	0	000+	000+	$-1/3\sqrt{2.3}$
$\frac{1}{2}$	$\frac{1}{2}$	1	000−	000−	0
$\frac{3}{2}$	$\frac{1}{2}$	1	000+	000+	+1/4.3
$\frac{3}{2}$	$\frac{1}{2}$	2	000−	000−	$-1/4.3\sqrt{3}$
$\frac{3}{2}$	$\frac{1}{2}$	$\tilde{1}$	000−	000−	$-1/4.3\sqrt{3}$
$\frac{3}{2}$	$\frac{3}{2}$	0	000+	000+	$+1/4.3\sqrt{3}$
$\frac{3}{2}$	$\frac{3}{2}$	1	000−	000−	0
$\frac{3}{2}$	$\frac{3}{2}$	1	001−	000−	0
$\frac{3}{2}$	$\frac{3}{2}$	2	000+	000+	$+1/2.3\sqrt{2.3}$
$\frac{3}{2}$	$\frac{3}{2}$	$\tilde{1}$	000+	000+	$+1/2.3\sqrt{2.3}$
$\frac{3}{2}$	$\frac{3}{2}$	$\tilde{1}$	001−	000+	0

$\tilde{\frac{1}{2}}$	$\tilde{\frac{1}{2}}$	1	$\tilde{1}$	$\tilde{1}$	2
$\frac{1}{2}$	$\frac{1}{2}$	1	000+	000+	−1/9
$\frac{3}{2}$	$\frac{1}{2}$	1	000−	000−	−1/4.9
$\frac{3}{2}$	$\frac{1}{2}$	$\tilde{1}$	000+	000+	+1/4.3
$\frac{3}{2}$	$\frac{3}{2}$	1	000+	000+	$+1/9\sqrt{2}$
$\frac{3}{2}$	$\frac{3}{2}$	1	001+	000+	$-1/4.9\sqrt{2}$
$\frac{3}{2}$	$\frac{3}{2}$	$\tilde{1}$	000−	000−	0
$\frac{3}{2}$	$\frac{3}{2}$	$\tilde{1}$	001+	000−	$+1/4.3\sqrt{2}$

$\tilde{\frac{1}{2}}$	$\tilde{\frac{1}{2}}$	1	$\tilde{1}$	$\tilde{1}$	$\tilde{1}$
$\frac{1}{2}$	$\frac{1}{2}$	1	000+	000+	+1/9
$\frac{3}{2}$	$\frac{1}{2}$	1	000−	000−	+1/4.9
$\frac{3}{2}$	$\frac{1}{2}$	2	000+	000+	−1/4.3
$\frac{3}{2}$	$\frac{1}{2}$	$\tilde{1}$	000+	000+	$-1/4.3\sqrt{3}$
$\frac{3}{2}$	$\frac{3}{2}$	1	000+	000+	$+1/2.9\sqrt{2}$
$\frac{3}{2}$	$\frac{3}{2}$	1	001+	000+	$-1/2.9\sqrt{2}$
$\frac{3}{2}$	$\frac{3}{2}$	2	000−	000−	0
$\frac{3}{2}$	$\frac{3}{2}$	$\tilde{1}$	000−	000−	0
$\frac{3}{2}$	$\frac{3}{2}$	$\tilde{1}$	001+	000−	$+1/2.3\sqrt{2.3}$

$\tilde{\frac{1}{2}}$	$\tilde{\frac{1}{2}}$	1	$\tilde{\frac{1}{2}}$	0	$\tilde{\frac{1}{2}}$
0	$\tilde{\frac{1}{2}}$	$\tilde{\frac{1}{2}}$	000+	000+	+1/4
1	$\tilde{\frac{1}{2}}$	$\frac{3}{2}$	000+	000+	+1/2.3
1	$\tilde{\frac{1}{2}}$	$\tilde{\frac{1}{2}}$	000−	000−	−1/4.3

$\tilde{\frac{1}{2}}$	$\tilde{\frac{1}{2}}$	1	$\tilde{\frac{1}{2}}$	$\frac{1}{2}$	$\tilde{1}$
0	$\tilde{1}$	$\tilde{1}$	000−	000−	$-1/3\sqrt{2.3}$
1	$\tilde{1}$	1	000−	000−	+1/9
1	$\tilde{1}$	2	000+	000+	+1/9
1	$\tilde{1}$	$\tilde{1}$	000+	000+	0

$\tilde{\frac{1}{2}}$	$\tilde{\frac{1}{2}}$	1	$\tilde{\frac{1}{2}}$	1	$\frac{3}{2}$
0	$\frac{3}{2}$	$\frac{3}{2}$	000+	000+	−1/8.3
0	$\frac{3}{2}$	$\frac{3}{2}$	000+	001+	+1/4.3
0	$\tilde{\frac{1}{2}}$	$\tilde{\frac{1}{2}}$	000+	000+	+1/2.3
1	$\frac{3}{2}$	$\frac{1}{2}$	000+	000+	+1/4.3
1	$\frac{3}{2}$	$\frac{3}{2}$	000−	000−	+5/8.9
1	$\frac{3}{2}$	$\frac{3}{2}$	000−	001−	+1/4.9
1	$\frac{3}{2}$	$\frac{3}{2}$	001−	000−	+1/4.9
1	$\frac{3}{2}$	$\frac{3}{2}$	001−	001−	+1/4.9
1	$\frac{3}{2}$	$\tilde{\frac{1}{2}}$	000+	000+	−1/4.9

$\tilde{\frac{1}{2}}$	$\tilde{\frac{1}{2}}$	1	$\tilde{\frac{1}{2}}$	1	$\tilde{\frac{1}{2}}$
0	$\frac{3}{2}$	$\frac{3}{2}$	000−	000−	$-1/2.3\sqrt{2}$
0	$\tilde{\frac{1}{2}}$	$\tilde{\frac{1}{2}}$	000−	000−	−1/4.3
1	$\frac{3}{2}$	$\frac{3}{2}$	000+	000+	$-1/2.9\sqrt{2}$
1	$\frac{3}{2}$	$\frac{3}{2}$	001+	000+	$+1/9\sqrt{2}$
1	$\frac{3}{2}$	$\tilde{\frac{1}{2}}$	000−	000−	+1/2.9
1	$\tilde{\frac{1}{2}}$	$\frac{3}{2}$	000−	000−	+1/2.9
1	$\tilde{\frac{1}{2}}$	$\tilde{\frac{1}{2}}$	000+	000+	+5/4.9

$\tilde{\frac{1}{2}}$	$\tilde{\frac{1}{2}}$	1	$\tilde{\frac{1}{2}}$	$\frac{3}{2}$	1
0	1	1	000−	000−	$-1/2.3\sqrt{2.3}$
0	2	2	000−	000−	$-1/4\sqrt{3}$
0	$\tilde{1}$	$\tilde{1}$	000−	000−	$+1/2.3\sqrt{2}$
1	1	0	000−	000−	+1/9

O and T_d 9j Symbols (cont.)

$\tilde{\frac{1}{2}}$	$\tilde{\frac{1}{2}}$	1	$\tilde{\frac{1}{2}}$	$\frac{3}{2}$	1
1	1	1	000+	000+	+1/4.3
1	1	2	000−	000−	+1/4.9
1	1	$\tilde{1}$	000−	000−	+1/4.9
$\tilde{\frac{1}{2}}$	**$\tilde{\frac{1}{2}}$**	**1**	**$\tilde{\frac{1}{2}}$**	**$\frac{3}{2}$**	**2**
0	1	1	000+	000+	$+1/2.3\sqrt{2}$
0	$\tilde{1}$	$\tilde{1}$	000+	000+	$-1/2.3\sqrt{2}$
1	1	1	000−	000−	$+1/4.3\sqrt{3}$
1	1	$\tilde{1}$	000+	000+	+1/4.3
1	2	1	000+	000+	+1/4.3
1	2	$\tilde{1}$	000−	000−	+1/4.3
$\tilde{\frac{1}{2}}$	**$\tilde{\frac{1}{2}}$**	**1**	**$\tilde{\frac{1}{2}}$**	**$\frac{3}{2}$**	**$\tilde{1}$**
0	1	1	000+	000+	$-1/2.3\sqrt{2}$
0	2	2	000+	000+	$-1/4\sqrt{3}$
0	$\tilde{1}$	$\tilde{1}$	000+	000+	$+1/2.3\sqrt{2.3}$
1	1	1	000−	000−	$-1/4.3\sqrt{3}$
1	1	2	000+	000+	−1/4.3
1	1	$\tilde{1}$	000+	000+	$-1/4.3\sqrt{3}$
1	2	1	000+	000+	+1/4.3
1	2	$\tilde{1}$	000−	000−	$+1/4.3\sqrt{3}$
1	$\tilde{1}$	1	000+	000+	+1/4.9
1	$\tilde{1}$	2	000−	000−	+1/4.9
1	$\tilde{1}$	$\tilde{1}$	000−	000−	+1/4.3
$\tilde{\frac{1}{2}}$	**$\tilde{\frac{1}{2}}$**	**1**	**$\tilde{\frac{1}{2}}$**	**2**	**$\frac{3}{2}$**
0	$\frac{3}{2}$	$\frac{3}{2}$	000−	000−	+1/8
0	$\frac{3}{2}$	$\frac{3}{2}$	000−	001−	0
1	$\frac{3}{2}$	$\frac{1}{2}$	000−	000−	+1/4.3
1	$\frac{3}{2}$	$\frac{3}{2}$	000+	000+	−1/8.3
1	$\frac{3}{2}$	$\frac{3}{2}$	000+	001+	0
1	$\frac{3}{2}$	$\frac{3}{2}$	001+	000+	0
1	$\frac{3}{2}$	$\frac{3}{2}$	001+	001+	+1/4.3
1	$\frac{3}{2}$	$\tilde{\frac{1}{2}}$	000−	000−	+1/4.3
$\tilde{\frac{1}{2}}$	**$\tilde{\frac{1}{2}}$**	**1**	**$\tilde{\frac{1}{2}}$**	**$\tilde{1}$**	**$\frac{1}{2}$**
0	$\frac{1}{2}$	$\frac{1}{2}$	000+	000+	+1/4.3
0	$\frac{3}{2}$	$\frac{3}{2}$	000+	000+	$+1/2.3\sqrt{2}$
1	$\frac{1}{2}$	$\frac{1}{2}$	000−	000−	+5/4.9
1	$\frac{1}{2}$	$\frac{3}{2}$	000+	000+	+1/2.9
$\tilde{\frac{1}{2}}$	**$\tilde{\frac{1}{2}}$**	**1**	**$\tilde{\frac{1}{2}}$**	**$\tilde{1}$**	**$\frac{3}{2}$**
0	$\frac{1}{2}$	$\frac{1}{2}$	000−	000−	−1/2.3
0	$\frac{3}{2}$	$\frac{3}{2}$	000−	000−	−1/8.3
0	$\frac{3}{2}$	$\frac{3}{2}$	000−	001−	−1/4.3
1	$\frac{1}{2}$	$\frac{1}{2}$	000+	000+	+1/2.9
1	$\frac{1}{2}$	$\frac{3}{2}$	000−	000−	$+1/2.9\sqrt{2}$
1	$\frac{1}{2}$	$\frac{3}{2}$	000−	001−	$+1/9\sqrt{2}$
1	$\frac{3}{2}$	$\frac{1}{2}$	000−	000−	−1/4.9
1	$\frac{3}{2}$	$\frac{3}{2}$	000+	000+	+5/8.9
1	$\frac{3}{2}$	$\frac{3}{2}$	000+	001+	−1/4.9
1	$\frac{3}{2}$	$\frac{3}{2}$	001+	000+	−1/4.9
1	$\frac{3}{2}$	$\frac{3}{2}$	001+	001+	+1/4.9
1	$\frac{3}{2}$	$\tilde{\frac{1}{2}}$	000−	000−	+1/4.3

$\tilde{\frac{1}{2}}$	$\tilde{\frac{1}{2}}$	1	$\tilde{\frac{1}{2}}$	$\tilde{\frac{1}{2}}$	0
1	0	1	000−	000−	+1/2.3
1	1	1	000+	000+	$+1/3\sqrt{2.3}$
$\tilde{\frac{1}{2}}$	**$\tilde{\frac{1}{2}}$**	**1**	**$\tilde{\frac{1}{2}}$**	**$\tilde{\frac{1}{2}}$**	**1**
0	0	0	000+	000+	$+1/2\sqrt{3}$
1	0	1	000+	000+	$+1/3\sqrt{2.3}$
1	1	0	000+	000+	−1/2.9
1	1	1	000−	000−	0
1	1	2	000+	000+	+1/9
1	1	$\tilde{1}$	000+	000+	+1/9
$\tilde{0}$	**$\frac{3}{2}$**	**$\frac{3}{2}$**	**$\frac{3}{2}$**	**1**	**$\frac{1}{2}$**
$\frac{3}{2}$	$\frac{1}{2}$	1	000+	000+	+1/8.2
$\frac{3}{2}$	$\frac{1}{2}$	2	000−	000−	+1/8.2
$\frac{3}{2}$	$\frac{1}{2}$	$\tilde{1}$	000−	000−	−1/8.2.3
$\tilde{0}$	**$\frac{3}{2}$**	**$\frac{3}{2}$**	**$\frac{3}{2}$**	**$\frac{3}{2}$**	**0**
$\frac{3}{2}$	0	$\frac{3}{2}$	000+	000+	+1/8.2
$\frac{3}{2}$	1	$\frac{3}{2}$	000−	000−	+1/8.2
$\frac{3}{2}$	1	$\frac{3}{2}$	000−	010−	0
$\frac{3}{2}$	1	$\frac{3}{2}$	001−	000−	0
$\frac{3}{2}$	1	$\frac{3}{2}$	001−	010−	−1/8.2
$\tilde{0}$	**$\frac{3}{2}$**	**$\frac{3}{2}$**	**$\frac{3}{2}$**	**$\frac{3}{2}$**	**1**
$\frac{3}{2}$	0	$\frac{3}{2}$	000−	000−	+1/8.2
$\frac{3}{2}$	0	$\frac{3}{2}$	000−	001−	0
$\frac{3}{2}$	0	$\frac{3}{2}$	010−	000−	0
$\frac{3}{2}$	0	$\frac{3}{2}$	010−	001−	−1/8.2
$\frac{3}{2}$	1	$\frac{1}{2}$	000−	000−	+1/8.3
$\frac{3}{2}$	1	$\frac{1}{2}$	000−	010−	+1/8.3
$\frac{3}{2}$	1	$\frac{1}{2}$	010−	000−	+1/8.3
$\frac{3}{2}$	1	$\frac{1}{2}$	010−	010−	−1/8.2.3
$\frac{3}{2}$	1	$\frac{3}{2}$	000+	000+	−1/8.2.3
$\frac{3}{2}$	1	$\frac{3}{2}$	000+	001+	0
$\frac{3}{2}$	1	$\frac{3}{2}$	000+	010+	0
$\frac{3}{2}$	1	$\frac{3}{2}$	000+	011+	−1/8.3
$\frac{3}{2}$	1	$\frac{3}{2}$	001+	000+	0
$\frac{3}{2}$	1	$\frac{3}{2}$	001+	001+	+1/8.3
$\frac{3}{2}$	1	$\frac{3}{2}$	001+	010+	+1/8.2.3
$\frac{3}{2}$	1	$\frac{3}{2}$	001+	011+	0
$\frac{3}{2}$	1	$\frac{3}{2}$	010+	000+	0
$\frac{3}{2}$	1	$\frac{3}{2}$	010+	001+	+1/8.2.3
$\frac{3}{2}$	1	$\frac{3}{2}$	010+	010+	+1/8.3
$\frac{3}{2}$	1	$\frac{3}{2}$	010+	011+	0
$\frac{3}{2}$	1	$\frac{3}{2}$	011+	000+	−1/8.3
$\frac{3}{2}$	1	$\frac{3}{2}$	011+	001+	0
$\frac{3}{2}$	1	$\frac{3}{2}$	011+	010+	0
$\frac{3}{2}$	1	$\frac{3}{2}$	011+	011+	+1/8.3
$\frac{3}{2}$	1	$\tilde{\frac{1}{2}}$	000−	000−	+1/8.3
$\frac{3}{2}$	1	$\tilde{\frac{1}{2}}$	000−	010−	−1/8.3
$\frac{3}{2}$	1	$\tilde{\frac{1}{2}}$	010−	000−	−1/8.3
$\frac{3}{2}$	1	$\tilde{\frac{1}{2}}$	010−	010−	−1/8.2.3

$\tilde{0}$	$\frac{3}{2}$	$\frac{3}{2}$	$\frac{3}{2}$	2	$\frac{1}{2}$
$\frac{3}{2}$	$\frac{1}{2}$	1	000−	000−	+1/8.2
$\frac{3}{2}$	$\frac{1}{2}$	2	000+	000+	−1/8.2
$\frac{3}{2}$	$\frac{1}{2}$	$\tilde{1}$	000+	000+	+1/8.2
$\frac{3}{2}$	$\frac{3}{2}$	1	000+	000+	0
$\frac{3}{2}$	$\frac{3}{2}$	1	001+	000+	+1/8.2
$\frac{3}{2}$	$\frac{3}{2}$	2	000−	000−	−1/8.2
$\frac{3}{2}$	$\frac{3}{2}$	$\tilde{1}$	000−	000−	0
$\frac{3}{2}$	$\frac{3}{2}$	$\tilde{1}$	001+	000−	+1/8.2
$\tilde{0}$	**$\frac{3}{2}$**	**$\frac{3}{2}$**	**$\frac{3}{2}$**	**2**	**$\frac{3}{2}$**
$\frac{3}{2}$	$\frac{1}{2}$	1	000+	000+	0
$\frac{3}{2}$	$\frac{1}{2}$	1	000+	001+	+1/8.2
$\frac{3}{2}$	$\frac{1}{2}$	2	000−	000−	−1/8.2
$\frac{3}{2}$	$\frac{1}{2}$	$\tilde{1}$	000−	000−	0
$\frac{3}{2}$	$\frac{1}{2}$	$\tilde{1}$	000−	001+	+1/8.2
$\frac{3}{2}$	$\frac{3}{2}$	0	000+	000+	−1/8.2
$\frac{3}{2}$	$\frac{3}{2}$	1	000−	000−	+1/8.2
$\frac{3}{2}$	$\frac{3}{2}$	1	000−	001−	0
$\frac{3}{2}$	$\frac{3}{2}$	1	001−	000−	0
$\frac{3}{2}$	$\frac{3}{2}$	1	001−	001−	0
$\frac{3}{2}$	$\frac{3}{2}$	2	000+	000+	0
$\frac{3}{2}$	$\frac{3}{2}$	$\tilde{1}$	000+	000+	+1/8.2
$\frac{3}{2}$	$\frac{3}{2}$	$\tilde{1}$	000+	001−	0
$\frac{3}{2}$	$\frac{3}{2}$	$\tilde{1}$	001−	000+	0
$\frac{3}{2}$	$\frac{3}{2}$	$\tilde{1}$	001−	001−	0
$\frac{3}{2}$	$\frac{3}{2}$	$\tilde{0}$	000−	000−	−1/8.2
$\tilde{0}$	**$\frac{3}{2}$**	**$\frac{3}{2}$**	**$\frac{3}{2}$**	**$\tilde{1}$**	**$\frac{1}{2}$**
$\frac{3}{2}$	$\frac{1}{2}$	1	000−	000−	−1/8.2.3
$\frac{3}{2}$	$\frac{1}{2}$	2	000+	000+	+1/8.2
$\frac{3}{2}$	$\frac{1}{2}$	$\tilde{1}$	000+	000+	+1/8.2
$\frac{3}{2}$	$\frac{3}{2}$	1	000+	000+	+1/8.3
$\frac{3}{2}$	$\frac{3}{2}$	1	000+	010−	−1/8.3
$\frac{3}{2}$	$\frac{3}{2}$	1	001+	000+	−1/8.3
$\frac{3}{2}$	$\frac{3}{2}$	1	001+	010−	−1/8.2.3
$\frac{3}{2}$	$\frac{3}{2}$	2	000−	000−	0
$\frac{3}{2}$	$\frac{3}{2}$	2	000−	010+	−1/8.2
$\frac{3}{2}$	$\frac{3}{2}$	$\tilde{1}$	000−	000−	−1/8.3
$\frac{3}{2}$	$\frac{3}{2}$	$\tilde{1}$	000−	010+	−1/8.3
$\frac{3}{2}$	$\frac{3}{2}$	$\tilde{1}$	001+	000−	+1/8.3
$\frac{3}{2}$	$\frac{3}{2}$	$\tilde{1}$	001+	010+	−1/8.2.3
$\tilde{0}$	**$\frac{3}{2}$**	**$\frac{3}{2}$**	**$\frac{3}{2}$**	**$\tilde{1}$**	**$\frac{3}{2}$**
$\frac{3}{2}$	$\frac{1}{2}$	1	000+	000+	+1/8.3
$\frac{3}{2}$	$\frac{1}{2}$	1	000+	001+	−1/8.3
$\frac{3}{2}$	$\frac{1}{2}$	1	010−	000+	−1/8.3
$\frac{3}{2}$	$\frac{1}{2}$	1	010−	001+	−1/8.2.3
$\frac{3}{2}$	$\frac{1}{2}$	2	000−	000−	0
$\frac{3}{2}$	$\frac{1}{2}$	2	010+	000−	−1/8.2
$\frac{3}{2}$	$\frac{1}{2}$	$\tilde{1}$	000−	000−	−1/8.3
$\frac{3}{2}$	$\frac{1}{2}$	$\tilde{1}$	000−	001+	+1/8.3
$\frac{3}{2}$	$\frac{1}{2}$	$\tilde{1}$	010+	000−	−1/8.3

O and T_d 9j Symbols (cont.)

$\tilde{0}\ \frac{3}{2}\ \frac{3}{2}\ \ \frac{3}{2}\ \tilde{1}\ \frac{3}{2}$			
$\frac{3}{2}\ \frac{1}{2}\ \tilde{1}$	010+	001+	$-1/8.2.3$
$\frac{3}{2}\ \frac{3}{2}\ 0$	000+	000+	$+1/8.2$
$\frac{3}{2}\ \frac{3}{2}\ 0$	000+	010−	0
$\frac{3}{2}\ \frac{3}{2}\ 0$	010−	000+	0
$\frac{3}{2}\ \frac{3}{2}\ 0$	010−	010−	$+1/8.2$
$\frac{3}{2}\ \frac{3}{2}\ 1$	000−	000−	$+1/8.2.3$
$\frac{3}{2}\ \frac{3}{2}\ 1$	000−	001−	0
$\frac{3}{2}\ \frac{3}{2}\ 1$	000−	010+	0
$\frac{3}{2}\ \frac{3}{2}\ 1$	000−	011+	$-1/8.3$
$\frac{3}{2}\ \frac{3}{2}\ 1$	001−	000−	0
$\frac{3}{2}\ \frac{3}{2}\ 1$	001−	001−	$+1/8.2.3$
$\frac{3}{2}\ \frac{3}{2}\ 1$	001−	010+	$-1/8.3$
$\frac{3}{2}\ \frac{3}{2}\ 1$	001−	011+	0
$\frac{3}{2}\ \frac{3}{2}\ 1$	010+	000−	0
$\frac{3}{2}\ \frac{3}{2}\ 1$	010+	001−	$-1/8.3$
$\frac{3}{2}\ \frac{3}{2}\ 1$	010+	010+	$+1/8.2.3$
$\frac{3}{2}\ \frac{3}{2}\ 1$	010+	011+	0
$\frac{3}{2}\ \frac{3}{2}\ 1$	011+	000−	$-1/8.3$
$\frac{3}{2}\ \frac{3}{2}\ 1$	011+	001−	0
$\frac{3}{2}\ \frac{3}{2}\ 1$	011+	010+	0
$\frac{3}{2}\ \frac{3}{2}\ 1$	011+	011+	$-1/8.3$
$\frac{3}{2}\ \frac{1}{2}\ 2$	000+	000+	$+1/8.2$
$\frac{3}{2}\ \frac{3}{2}\ 2$	000+	010−	0
$\frac{3}{2}\ \frac{3}{2}\ 2$	010−	000+	0
$\frac{3}{2}\ \frac{3}{2}\ 2$	010−	010−	0
$\frac{3}{2}\ \frac{3}{2}\ \tilde{1}$	000+	000+	$-1/8.2.3$
$\frac{3}{2}\ \frac{3}{2}\ \tilde{1}$	000+	001−	0
$\frac{3}{2}\ \frac{3}{2}\ \tilde{1}$	000+	010−	0
$\frac{3}{2}\ \frac{3}{2}\ \tilde{1}$	000+	011+	$+1/8.3$
$\frac{3}{2}\ \frac{3}{2}\ \tilde{1}$	001−	000+	0
$\frac{3}{2}\ \frac{3}{2}\ \tilde{1}$	001−	001−	$+1/8.2.3$
$\frac{3}{2}\ \frac{3}{2}\ \tilde{1}$	001−	010−	$-1/8.3$
$\frac{3}{2}\ \frac{3}{2}\ \tilde{1}$	001−	011+	0
$\frac{3}{2}\ \frac{3}{2}\ \tilde{1}$	010−	000+	0
$\frac{3}{2}\ \frac{3}{2}\ \tilde{1}$	010−	001−	$-1/8.3$
$\frac{3}{2}\ \frac{3}{2}\ \tilde{1}$	010−	010−	$+1/8.2.3$
$\frac{3}{2}\ \frac{3}{2}\ \tilde{1}$	010−	011+	0
$\frac{3}{2}\ \frac{3}{2}\ \tilde{1}$	011+	000+	$+1/8.3$
$\frac{3}{2}\ \frac{3}{2}\ \tilde{1}$	011+	001−	0
$\frac{3}{2}\ \frac{3}{2}\ \tilde{1}$	011+	010−	0
$\frac{3}{2}\ \frac{3}{2}\ \tilde{1}$	011+	011+	$+1/8.3$
$\frac{3}{2}\ \frac{3}{2}\ \tilde{0}$	000−	000−	$-1/8.2$
$\frac{3}{2}\ \frac{3}{2}\ \tilde{0}$	000−	010+	0
$\frac{3}{2}\ \frac{3}{2}\ \tilde{0}$	010+	000−	0
$\frac{3}{2}\ \frac{3}{2}\ \tilde{0}$	010+	010+	$+1/8.2$
$\tilde{0}\ \frac{3}{2}\ \frac{3}{2}\ \ \frac{3}{2}\ \tilde{\frac{3}{2}}\ 1$			
$\frac{3}{2}\ 1\ \frac{1}{2}$	000+	000+	$+1/8.2.3$
$\frac{3}{2}\ 1\ \frac{3}{2}$	000−	000−	$+1/8.3$
$\frac{3}{2}\ 1\ \frac{3}{2}$	000−	001−	$-1/8.3$

$\tilde{0}\ \frac{3}{2}\ \frac{3}{2}\ \ \frac{3}{2}\ \tilde{\frac{3}{2}}\ 1$			
$\frac{3}{2}\ 1\ \frac{3}{2}$	001−	000−	$-1/8.3$
$\frac{3}{2}\ 1\ \frac{3}{2}$	001−	001−	$-1/8.2.3$
$\frac{3}{2}\ 1\ \tilde{\frac{3}{2}}$	000+	000+	$+1/8.2$
$\frac{3}{2}\ 2\ \frac{1}{2}$	000−	000−	$-1/8.2$
$\frac{3}{2}\ 2\ \frac{3}{2}$	000+	000+	0
$\frac{3}{2}\ 2\ \frac{3}{2}$	000+	001+	$-1/8.2$
$\frac{3}{2}\ 2\ \tilde{\frac{3}{2}}$	000−	000−	$-1/8.2$
$\frac{3}{2}\ \tilde{1}\ \frac{1}{2}$	000−	000−	$+1/8.2$
$\frac{3}{2}\ \tilde{1}\ \frac{3}{2}$	000+	000+	$+1/8.3$
$\frac{3}{2}\ \tilde{1}\ \frac{3}{2}$	000+	001+	$+1/8.3$
$\frac{3}{2}\ \tilde{1}\ \frac{3}{2}$	001−	000+	$+1/8.3$
$\frac{3}{2}\ \tilde{1}\ \frac{3}{2}$	001−	001+	$-1/8.2.3$
$\frac{3}{2}\ \tilde{1}\ \tilde{\frac{3}{2}}$	000−	000−	$+1/8.2.3$
$\tilde{0}\ \frac{3}{2}\ \frac{3}{2}\ \ \frac{3}{2}\ \tilde{\frac{3}{2}}\ 2$			
$\frac{3}{2}\ 1\ \frac{1}{2}$	000−	000−	$-1/8.2$
$\frac{3}{2}\ 1\ \frac{3}{2}$	000+	000+	0
$\frac{3}{2}\ 1\ \frac{3}{2}$	001+	000+	$-1/8.2$
$\frac{3}{2}\ 1\ \tilde{\frac{3}{2}}$	000−	000−	$-1/8.2$
$\frac{3}{2}\ 2\ \frac{1}{2}$	000+	000+	$+1/8.2$
$\frac{3}{2}\ 2\ \frac{3}{2}$	000−	000−	$+1/8.2$
$\frac{3}{2}\ 2\ \tilde{\frac{3}{2}}$	000+	000+	$-1/8.2$
$\frac{3}{2}\ \tilde{1}\ \frac{1}{2}$	000+	000+	$+1/8.2$
$\frac{3}{2}\ \tilde{1}\ \frac{3}{2}$	000−	000−	0
$\frac{3}{2}\ \tilde{1}\ \frac{3}{2}$	001+	000−	$-1/8.2$
$\frac{3}{2}\ \tilde{1}\ \tilde{\frac{3}{2}}$	000+	000+	$+1/8.2$
$\tilde{0}\ \frac{3}{2}\ \frac{3}{2}\ \ \frac{3}{2}\ \tilde{\frac{3}{2}}\ \tilde{1}$			
$\frac{3}{2}\ 1\ \frac{1}{2}$	000−	000−	$+1/8.2$
$\frac{3}{2}\ 1\ \frac{3}{2}$	000+	000+	$+1/8.3$
$\frac{3}{2}\ 1\ \frac{3}{2}$	000+	001−	$+1/8.3$
$\frac{3}{2}\ 1\ \frac{3}{2}$	001+	000+	$+1/8.3$
$\frac{3}{2}\ 1\ \frac{3}{2}$	001+	001−	$-1/8.2.3$
$\frac{3}{2}\ 1\ \tilde{\frac{3}{2}}$	000−	000−	$+1/8.2.3$
$\frac{3}{2}\ 2\ \frac{1}{2}$	000+	000+	$+1/8.2$
$\frac{3}{2}\ 2\ \frac{3}{2}$	000−	000−	0
$\frac{3}{2}\ 2\ \frac{3}{2}$	000−	001+	$-1/8.2$
$\frac{3}{2}\ 2\ \tilde{\frac{3}{2}}$	000+	000+	$+1/8.2$
$\frac{3}{2}\ \tilde{1}\ \frac{1}{2}$	000+	000+	$+1/8.2.3$
$\frac{3}{2}\ \tilde{1}\ \frac{3}{2}$	000−	000−	$+1/8.3$
$\frac{3}{2}\ \tilde{1}\ \frac{3}{2}$	000−	001+	$-1/8.3$
$\frac{3}{2}\ \tilde{1}\ \frac{3}{2}$	001+	000−	$-1/8.3$
$\frac{3}{2}\ \tilde{1}\ \frac{3}{2}$	001+	001+	$-1/8.2.3$
$\frac{3}{2}\ \tilde{1}\ \tilde{\frac{3}{2}}$	000+	000+	$+1/8.2$
$\tilde{0}\ \frac{3}{2}\ \frac{3}{2}\ \ \frac{3}{2}\ \tilde{0}\ \frac{3}{2}$			
$\frac{3}{2}\ \frac{3}{2}\ 0$	000−	000−	$-1/8.2$
$\frac{3}{2}\ \frac{3}{2}\ 1$	000+	000+	$+1/8.2$
$\frac{3}{2}\ \frac{3}{2}\ 1$	000+	001+	0
$\frac{3}{2}\ \frac{3}{2}\ 1$	001+	000+	0
$\frac{3}{2}\ \frac{3}{2}\ 1$	001+	001+	$+1/8.2$
$\frac{3}{2}\ \frac{3}{2}\ 2$	000−	000−	$-1/8.2$

$\tilde{0}\ \frac{3}{2}\ \frac{3}{2}\ \ \frac{3}{2}\ \tilde{0}\ \frac{3}{2}$			
$\frac{3}{2}\ \frac{3}{2}\ \tilde{1}$	000−	000−	$-1/8.2$
$\frac{3}{2}\ \frac{3}{2}\ \tilde{1}$	000−	001+	0
$\frac{3}{2}\ \frac{3}{2}\ \tilde{1}$	001+	000−	0
$\frac{3}{2}\ \frac{3}{2}\ \tilde{1}$	001+	001+	$+1/8.2$
$\frac{3}{2}\ \frac{3}{2}\ \tilde{0}$	000+	000+	$+1/8.2$
$\tilde{0}\ 2\ 2\ \ \frac{3}{2}\ \frac{3}{2}\ 0$			
$\frac{3}{2}\ \frac{1}{2}\ 2$	000+	000+	$-1/8$
$\tilde{0}\ 2\ 2\ \ \frac{3}{2}\ \frac{3}{2}\ 1$			
$\frac{3}{2}\ \frac{1}{2}\ 1$	000+	000+	$-1/8\sqrt{3}$
$\frac{3}{2}\ \frac{1}{2}\ 1$	010+	000+	$-1/8\sqrt{3}$
$\frac{3}{2}\ \frac{1}{2}\ \tilde{1}$	000−	000−	$+1/8\sqrt{3}$
$\frac{3}{2}\ \frac{1}{2}\ \tilde{1}$	010−	000−	$-1/8\sqrt{3}$
$\frac{3}{2}\ \frac{3}{2}\ 1$	000−	000−	0
$\frac{3}{2}\ \frac{3}{2}\ 1$	010−	000−	$-1/8\sqrt{3}$
$\frac{3}{2}\ \frac{3}{2}\ 1$	011−	000−	0
$\tilde{0}\ 2\ 2\ \ \frac{3}{2}\ 2\ \frac{1}{2}$			
$\frac{3}{2}\ 0\ \frac{3}{2}$	000+	000+	$-1/8$
$\tilde{0}\ 2\ 2\ \ \frac{3}{2}\ 2\ \frac{3}{2}$			
$\frac{3}{2}\ 0\ \frac{3}{2}$	000−	000−	$+1/8$
$\frac{3}{2}\ 2\ \frac{1}{2}$	000+	000+	$+1/8$
$\frac{3}{2}\ 2\ \frac{3}{2}$	000−	000−	0
$\tilde{0}\ 2\ 2\ \ \frac{3}{2}\ \tilde{1}\ \frac{1}{2}$			
$\frac{3}{2}\ 1\ \frac{3}{2}$	000−	000−	$+1/8\sqrt{3}$
$\frac{3}{2}\ 1\ \frac{3}{2}$	001−	000−	$-1/8\sqrt{3}$
$\tilde{0}\ 2\ 2\ \ \frac{3}{2}\ \tilde{1}\ \frac{3}{2}$			
$\frac{3}{2}\ 1\ \frac{1}{2}$	000−	000−	$-1/8\sqrt{3}$
$\frac{3}{2}\ 1\ \frac{1}{2}$	010+	000−	$+1/8\sqrt{3}$
$\frac{3}{2}\ 1\ \frac{3}{2}$	000+	000+	0
$\frac{3}{2}\ 1\ \frac{3}{2}$	001+	000+	$+1/8\sqrt{3}$
$\frac{3}{2}\ 1\ \frac{3}{2}$	010−	000+	$+1/8\sqrt{3}$
$\frac{3}{2}\ 1\ \frac{3}{2}$	011−	000+	0
$\frac{3}{2}\ 1\ \tilde{\frac{3}{2}}$	000−	000−	$+1/8\sqrt{3}$
$\frac{3}{2}\ 1\ \tilde{\frac{3}{2}}$	010+	000−	$+1/8\sqrt{3}$
$\frac{3}{2}\ \tilde{1}\ \frac{1}{2}$	000+	000+	$-1/8\sqrt{3}$
$\frac{3}{2}\ \tilde{1}\ \frac{1}{2}$	010−	000+	$-1/8\sqrt{3}$
$\frac{3}{2}\ \tilde{1}\ \frac{3}{2}$	000−	000−	0
$\frac{3}{2}\ \tilde{1}\ \frac{3}{2}$	010+	000−	$+1/8\sqrt{3}$
$\frac{3}{2}\ \tilde{1}\ \frac{3}{2}$	011−	000−	0
$\tilde{0}\ 2\ 2\ \ \frac{3}{2}\ \tilde{\frac{3}{2}}\ 1$			
$\frac{3}{2}\ \frac{3}{2}\ 1$	000+	000+	$+1/8\sqrt{3}$
$\frac{3}{2}\ \frac{3}{2}\ 1$	001+	000+	$-1/8\sqrt{3}$
$\frac{3}{2}\ \frac{3}{2}\ \tilde{1}$	000−	000−	$+1/8\sqrt{3}$
$\frac{3}{2}\ \frac{3}{2}\ \tilde{1}$	001+	000−	$-1/8\sqrt{3}$
$\tilde{0}\ 2\ 2\ \ \frac{3}{2}\ \tilde{\frac{3}{2}}\ 2$			
$\frac{3}{2}\ \frac{3}{2}\ 0$	000+	000+	$-1/8$
$\frac{3}{2}\ \frac{3}{2}\ 2$	000+	000+	$-1/8$
$\frac{3}{2}\ \frac{3}{2}\ \tilde{0}$	000−	000−	$-1/8$
$\tilde{0}\ 2\ 2\ \ \frac{3}{2}\ \tilde{\frac{3}{2}}\ \tilde{1}$			
$\frac{3}{2}\ \frac{3}{2}\ 1$	000−	000−	$+1/8\sqrt{3}$

O and T_d 9j Symbols (cont.)

$\tilde{0}\ 2\ 2\ \ \frac{3}{2}\ \tilde{\frac{1}{2}}\ \tilde{1}$
$\frac{3}{2}\ \frac{3}{2}\ 1$ 001− 000− $+1/8\sqrt{3}$
$\frac{3}{2}\ \frac{3}{2}\ \tilde{1}$ 000+ 000+ $+1/8\sqrt{3}$
$\frac{3}{2}\ \frac{3}{2}\ \tilde{1}$ 001− 000+ $+1/8\sqrt{3}$
$\tilde{0}\ 2\ 2\ \ \frac{3}{2}\ \tilde{0}\ \frac{3}{2}$
$\frac{3}{2}\ 2\ \frac{1}{2}$ 000− 000− $-1/8$
$\frac{3}{2}\ 2\ \frac{3}{2}$ 000+ 000+ $+1/8$
$\frac{3}{2}\ 2\ \tilde{\frac{1}{2}}$ 000− 000− $-1/8$
$\tilde{0}\ 2\ 2\ \ 2\ 1\ 1$
$2\ 1\ 1$ 000− 000− 0
$\tilde{0}\ 2\ 2\ \ 2\ \frac{3}{2}\ \frac{1}{2}$
$2\ \frac{1}{2}\ \frac{3}{2}$ 000− 000− $+1/8$
$\tilde{0}\ 2\ 2\ \ 2\ \frac{3}{2}\ \frac{3}{2}$
$2\ \frac{3}{2}\ \frac{1}{2}$ 000+ 000+ $+1/8$
$2\ \frac{3}{2}\ \frac{3}{2}$ 000− 000− 0
$\tilde{0}\ 2\ 2\ \ 2\ 2\ 0$
$2\ 0\ 2$ 000− 000− $+1/4$
$\tilde{0}\ 2\ 2\ \ 2\ 2\ 2$
$2\ 2\ 0$ 000− 000− $+1/4$
$2\ 2\ 2$ 000− 000− 0
$\tilde{0}\ 2\ 2\ \ 2\ \tilde{1}\ 1$
$2\ 1\ 1$ 000+ 000+ $+1/2.3$
$2\ 1\ \tilde{1}$ 000− 000− 0
$2\ \tilde{1}\ 1$ 000− 000− 0
$\tilde{0}\ 2\ 2\ \ 2\ \tilde{1}\ \tilde{1}$
$2\ 1\ 1$ 000− 000− 0
$2\ \tilde{1}\ 1$ 000+ 000+ $+1/2.3$
$2\ \tilde{1}\ \tilde{1}$ 000− 000− 0
$\tilde{0}\ 2\ 2\ \ 2\ \tilde{\frac{1}{2}}\ \frac{3}{2}$
$2\ \frac{3}{2}\ \frac{1}{2}$ 000− 000− $+1/8$
$2\ \frac{3}{2}\ \frac{3}{2}$ 000+ 000+ $+1/8$
$2\ \frac{3}{2}\ \tilde{\frac{1}{2}}$ 000− 000− $-1/8$
$\tilde{0}\ 2\ 2\ \ 2\ \tilde{0}\ 2$
$2\ 2\ 0$ 000+ 000+ $+1/4$
$2\ 2\ 2$ 000+ 000+ $+1/4$
$2\ 2\ \tilde{0}$ 000− 000− $-1/4$
$\tilde{0}\ \tilde{1}\ 1\ \ \frac{3}{2}\ 1\ \frac{1}{2}$
$\frac{3}{2}\ 1\ \frac{1}{2}$ 000+ 000+ $-1/4.3$
$\tilde{0}\ \tilde{1}\ 1\ \ \frac{3}{2}\ 1\ \frac{3}{2}$
$\frac{3}{2}\ 1\ \frac{1}{2}$ 000− 000− $+1/4.3\sqrt{2}$
$\frac{3}{2}\ 1\ \frac{1}{2}$ 010− 000− 0
$\frac{3}{2}\ 1\ \frac{3}{2}$ 000+ 000+ 0
$\frac{3}{2}\ 1\ \frac{3}{2}$ 000+ 001+ 0
$\frac{3}{2}\ 1\ \frac{3}{2}$ 010+ 000+ 0
$\frac{3}{2}\ 1\ \frac{3}{2}$ 010+ 001+ $+1/4.3\sqrt{2}$
$\frac{3}{2}\ 1\ \frac{3}{2}$ 011+ 000+ $-1/4.3\sqrt{2}$
$\frac{3}{2}\ 1\ \frac{3}{2}$ 011+ 001+ 0
$\tilde{0}\ \tilde{1}\ 1\ \ \frac{3}{2}\ 1\ \tilde{\frac{1}{2}}$
$\frac{3}{2}\ 1\ \frac{3}{2}$ 000− 000− $-1/4.3\sqrt{2}$
$\frac{3}{2}\ 1\ \frac{3}{2}$ 001− 000− 0

$\tilde{0}\ \tilde{1}\ 1\ \ \frac{3}{2}\ 1\ \tilde{\frac{1}{2}}$
$\frac{3}{2}\ 1\ \tilde{\frac{1}{2}}$ 000+ 000+ $+1/4.3$
$\tilde{0}\ \tilde{1}\ 1\ \ \frac{3}{2}\ \frac{3}{2}\ 0$
$\frac{3}{2}\ \frac{1}{2}\ 1$ 000+ 000+ $-1/4.3$
$\tilde{0}\ \tilde{1}\ 1\ \ \frac{3}{2}\ \frac{3}{2}\ 1$
$\frac{3}{2}\ \frac{1}{2}\ 1$ 000− 000− $+1/4.3\sqrt{2.3}$
$\frac{3}{2}\ \frac{1}{2}\ 1$ 010− 000− $-1/2.3\sqrt{2.3}$
$\frac{3}{2}\ \frac{1}{2}\ 2$ 000+ 000+ $+1/4.3\sqrt{2}$
$\frac{3}{2}\ \frac{1}{2}\ 2$ 010+ 000+ $+1/4.3\sqrt{2}$
$\frac{3}{2}\ \frac{1}{2}\ \tilde{1}$ 000+ 000+ $-1/4.3\sqrt{2}$
$\frac{3}{2}\ \frac{1}{2}\ \tilde{1}$ 010+ 000+ 0
$\frac{3}{2}\ \frac{3}{2}\ 0$ 000− 000− $+1/4.3$
$\frac{3}{2}\ \frac{3}{2}\ 0$ 000− 010+ 0
$\frac{3}{2}\ \frac{3}{2}\ 0$ 010− 000− 0
$\frac{3}{2}\ \frac{3}{2}\ 0$ 010− 010+ $+1/4.3$
$\frac{3}{2}\ \frac{3}{2}\ 1$ 000+ 000+ $-1/2.3\sqrt{2.3}$
$\frac{3}{2}\ \frac{3}{2}\ 1$ 000+ 010− 0
$\frac{3}{2}\ \frac{3}{2}\ 1$ 010+ 000+ 0
$\frac{3}{2}\ \frac{3}{2}\ 1$ 010+ 010− $+1/4.3\sqrt{2.3}$
$\frac{3}{2}\ \frac{3}{2}\ 1$ 011+ 000+ $-1/4.3\sqrt{2.3}$
$\frac{3}{2}\ \frac{3}{2}\ 1$ 011+ 010− 0
$\tilde{0}\ \tilde{1}\ 1\ \ \frac{3}{2}\ \frac{3}{2}\ 2$
$\frac{3}{2}\ \frac{1}{2}\ 1$ 000+ 000+ $-1/4.3\sqrt{2}$
$\frac{3}{2}\ \frac{1}{2}\ \tilde{1}$ 000− 000− $+1/4.3\sqrt{2}$
$\frac{3}{2}\ \frac{3}{2}\ 1$ 000− 000− 0
$\frac{3}{2}\ \frac{3}{2}\ 1$ 000− 010+ $-1/4.3\sqrt{2}$
$\frac{3}{2}\ \frac{3}{2}\ 1$ 001− 000− $+1/4.3\sqrt{2}$
$\frac{3}{2}\ \frac{3}{2}\ 1$ 001− 010+ 0
$\tilde{0}\ \tilde{1}\ 1\ \ \frac{3}{2}\ \frac{3}{2}\ \tilde{1}$
$\frac{3}{2}\ \frac{1}{2}\ 1$ 000+ 000+ $+1/4.3\sqrt{2}$
$\frac{3}{2}\ \frac{1}{2}\ 1$ 010− 000+ 0
$\frac{3}{2}\ \frac{1}{2}\ 2$ 000− 000− $-1/4.3\sqrt{2}$
$\frac{3}{2}\ \frac{1}{2}\ 2$ 010+ 000− $+1/4.3\sqrt{2}$
$\frac{3}{2}\ \frac{1}{2}\ \tilde{1}$ 000− 000− $-1/4.3\sqrt{2.3}$
$\frac{3}{2}\ \frac{1}{2}\ \tilde{1}$ 010+ 000− $-1/2.3\sqrt{2.3}$
$\frac{3}{2}\ \frac{3}{2}\ 1$ 000− 000− 0
$\frac{3}{2}\ \frac{3}{2}\ 1$ 000− 010+ 0
$\frac{3}{2}\ \frac{3}{2}\ 1$ 001− 000− 0
$\frac{3}{2}\ \frac{3}{2}\ 1$ 001− 010+ $+1/4.3\sqrt{2}$
$\frac{3}{2}\ \frac{3}{2}\ 1$ 010+ 000− 0
$\frac{3}{2}\ \frac{3}{2}\ 1$ 010+ 010+ $+1/4.3\sqrt{2}$
$\frac{3}{2}\ \frac{3}{2}\ 1$ 011+ 000− $+1/4.3\sqrt{2}$
$\frac{3}{2}\ \frac{3}{2}\ 1$ 011+ 010+ 0
$\frac{3}{2}\ \frac{3}{2}\ 2$ 000+ 000+ 0
$\frac{3}{2}\ \frac{3}{2}\ 2$ 000+ 010− $-1/4.3\sqrt{2}$
$\frac{3}{2}\ \frac{3}{2}\ 2$ 010− 000+ $+1/4.3\sqrt{2}$
$\frac{3}{2}\ \frac{3}{2}\ 2$ 010− 010− 0
$\frac{3}{2}\ \frac{3}{2}\ \tilde{1}$ 000+ 000+ $+1/2.3\sqrt{2.3}$
$\frac{3}{2}\ \frac{3}{2}\ \tilde{1}$ 000+ 010− 0
$\frac{3}{2}\ \frac{3}{2}\ \tilde{1}$ 010− 000+ 0

$\tilde{0}\ \tilde{1}\ 1\ \ \frac{3}{2}\ \frac{3}{2}\ \tilde{1}$
$\frac{3}{2}\ \frac{3}{2}\ \tilde{1}$ 010− 010− $+1/4.3\sqrt{2.3}$
$\frac{3}{2}\ \frac{3}{2}\ \tilde{1}$ 011+ 000+ $+1/4.3\sqrt{2.3}$
$\frac{3}{2}\ \frac{3}{2}\ \tilde{1}$ 011+ 010− 0
$\tilde{0}\ \tilde{1}\ 1\ \ \frac{3}{2}\ \frac{3}{2}\ \tilde{0}$
$\frac{3}{2}\ \frac{1}{2}\ \tilde{1}$ 000+ 000+ $+1/4.3$
$\frac{3}{2}\ \frac{3}{2}\ \tilde{1}$ 000− 000− $+1/4.3$
$\frac{3}{2}\ \frac{3}{2}\ \tilde{1}$ 000− 010+ 0
$\frac{3}{2}\ \frac{3}{2}\ \tilde{1}$ 001+ 000− 0
$\frac{3}{2}\ \frac{3}{2}\ \tilde{1}$ 001+ 010+ $+1/4.3$
$\tilde{0}\ \tilde{1}\ 1\ \ \frac{3}{2}\ 2\ \frac{1}{2}$
$\frac{3}{2}\ 1\ \frac{1}{2}$ 000− 000− $+1/4.3$
$\frac{3}{2}\ 1\ \frac{3}{2}$ 000+ 000+ $-1/4.3\sqrt{2}$
$\frac{3}{2}\ 1\ \frac{3}{2}$ 001+ 000+ $+1/4.3\sqrt{2}$
$\tilde{0}\ \tilde{1}\ 1\ \ \frac{3}{2}\ 2\ \frac{3}{2}$
$\frac{3}{2}\ 1\ \frac{1}{2}$ 000+ 000+ $+1/4.3\sqrt{2}$
$\frac{3}{2}\ 1\ \frac{3}{2}$ 000− 000− 0
$\frac{3}{2}\ 1\ \frac{3}{2}$ 000− 001− $+1/4.3\sqrt{2}$
$\frac{3}{2}\ 1\ \frac{3}{2}$ 001− 000− $+1/4.3\sqrt{2}$
$\frac{3}{2}\ 1\ \frac{3}{2}$ 001− 001− 0
$\frac{3}{2}\ 1\ \tilde{\frac{1}{2}}$ 000+ 000+ $-1/4.3\sqrt{2}$
$\tilde{0}\ \tilde{1}\ 1\ \ \frac{3}{2}\ 2\ \tilde{\frac{1}{2}}$
$\frac{3}{2}\ 1\ \frac{3}{2}$ 000+ 000+ $+1/4.3\sqrt{2}$
$\frac{3}{2}\ 1\ \frac{3}{2}$ 001+ 000+ $+1/4.3\sqrt{2}$
$\frac{3}{2}\ 1\ \tilde{\frac{1}{2}}$ 000− 000− $+1/4.3$
$\tilde{0}\ \tilde{1}\ 1\ \ \frac{3}{2}\ \tilde{1}\ \frac{1}{2}$
$\frac{3}{2}\ 0\ \frac{3}{2}$ 000− 000− $+1/4.3$
$\frac{3}{2}\ 1\ \frac{1}{2}$ 000− 000− $-1/4.3\sqrt{3}$
$\frac{3}{2}\ 1\ \frac{3}{2}$ 000+ 000+ $+1/4.3\sqrt{2.3}$
$\frac{3}{2}\ 1\ \frac{3}{2}$ 001+ 000+ $+1/2.3\sqrt{2.3}$
$\frac{3}{2}\ 2\ \frac{1}{2}$ 000+ 000+ $-1/4.3$
$\frac{3}{2}\ 2\ \frac{3}{2}$ 000− 000− $-1/4.3\sqrt{2}$
$\frac{3}{2}\ \tilde{1}\ \frac{1}{2}$ 000+ 000+ $+1/4.3$
$\tilde{0}\ \tilde{1}\ 1\ \ \frac{3}{2}\ \tilde{1}\ \frac{3}{2}$
$\frac{3}{2}\ 0\ \frac{3}{2}$ 000+ 000+ $+1/4.3$
$\frac{3}{2}\ 0\ \frac{3}{2}$ 000+ 001+ 0
$\frac{3}{2}\ 0\ \frac{3}{2}$ 010− 000+ 0
$\frac{3}{2}\ 0\ \frac{3}{2}$ 010− 001+ $+1/4.3$
$\frac{3}{2}\ 1\ \frac{1}{2}$ 000+ 000+ $-1/4.3\sqrt{2.3}$
$\frac{3}{2}\ 1\ \frac{1}{2}$ 010− 000+ $-1/2.3\sqrt{2.3}$
$\frac{3}{2}\ 1\ \frac{3}{2}$ 000− 000− $-1/2.3\sqrt{2.3}$
$\frac{3}{2}\ 1\ \frac{3}{2}$ 000− 001− 0
$\frac{3}{2}\ 1\ \frac{3}{2}$ 001− 000− 0
$\frac{3}{2}\ 1\ \frac{3}{2}$ 001− 001− $+1/4.3\sqrt{2.3}$
$\frac{3}{2}\ 1\ \frac{3}{2}$ 010+ 000− 0
$\frac{3}{2}\ 1\ \frac{3}{2}$ 010+ 001− $+1/4.3\sqrt{2.3}$
$\frac{3}{2}\ 1\ \frac{3}{2}$ 011+ 000− $+1/4.3\sqrt{2.3}$
$\frac{3}{2}\ 1\ \frac{3}{2}$ 011+ 001− 0
$\frac{3}{2}\ 1\ \tilde{\frac{1}{2}}$ 000+ 000+ $-1/4.3\sqrt{2.3}$
$\frac{3}{2}\ 1\ \tilde{\frac{1}{2}}$ 010− 000+ $+1/2.3\sqrt{2.3}$

O and T_d 9j Symbols (cont.)

$\tilde{0}\ \tilde{1}\ 1\quad \frac{3}{2}\ \tilde{1}\ \frac{3}{2}$

$\frac{3}{2}$ 2 $\frac{1}{2}$ 000− 000− $-1/4.3\sqrt{2}$
$\frac{3}{2}$ 2 $\frac{1}{2}$ 010+ 000− $+1/4.3\sqrt{2}$
$\frac{3}{2}$ 2 $\frac{3}{2}$ 000+ 000+ 0
$\frac{3}{2}$ 2 $\frac{3}{2}$ 000+ 001+ $+1/4.3\sqrt{2}$
$\frac{3}{2}$ 2 $\frac{3}{2}$ 010− 000+ $+1/4.3\sqrt{2}$
$\frac{3}{2}$ 2 $\frac{3}{2}$ 010− 001+ 0
$\frac{3}{2}$ 2 $\tilde{\frac{1}{2}}$ 000− 000− $+1/4.3\sqrt{2}$
$\frac{3}{2}$ 2 $\tilde{\frac{1}{2}}$ 010+ 000− $+1/4.3\sqrt{2}$
$\frac{3}{2}$ $\tilde{1}$ $\frac{1}{2}$ 000− 000− $-1/4.3\sqrt{2}$
$\frac{3}{2}$ $\tilde{1}$ $\frac{1}{2}$ 010+ 000− 0
$\frac{3}{2}$ $\tilde{1}$ $\frac{3}{2}$ 000+ 000+ 0
$\frac{3}{2}$ $\tilde{1}$ $\frac{3}{2}$ 000+ 001+ 0
$\frac{3}{2}$ $\tilde{1}$ $\frac{3}{2}$ 010− 000+ 0
$\frac{3}{2}$ $\tilde{1}$ $\frac{3}{2}$ 010− 001+ $-1/4.3\sqrt{2}$
$\frac{3}{2}$ $\tilde{1}$ $\frac{3}{2}$ 011+ 000+ $+1/4.3\sqrt{2}$
$\frac{3}{2}$ $\tilde{1}$ $\frac{3}{2}$ 011+ 001+ 0

$\tilde{0}\ \tilde{1}\ 1\quad \frac{3}{2}\ \tilde{1}\ \tilde{\frac{1}{2}}$

$\frac{3}{2}$ 0 $\frac{3}{2}$ 000− 000− $+1/4.3$
$\frac{3}{2}$ 1 $\frac{3}{2}$ 000+ 000+ $+1/4.3\sqrt{2.3}$
$\frac{3}{2}$ 1 $\frac{3}{2}$ 001+ 000+ $-1/2.3\sqrt{2.3}$
$\frac{3}{2}$ 1 $\tilde{\frac{1}{2}}$ 000− 000− $+1/4.3\sqrt{3}$
$\frac{3}{2}$ 2 $\frac{3}{2}$ 000− 000− $+1/4.3\sqrt{2}$
$\frac{3}{2}$ 2 $\tilde{\frac{1}{2}}$ 000+ 000+ $-1/4.3$
$\frac{3}{2}$ $\tilde{1}$ $\frac{3}{2}$ 000− 000− $+1/4.3\sqrt{2}$
$\frac{3}{2}$ $\tilde{1}$ $\frac{3}{2}$ 001+ 000− 0
$\frac{3}{2}$ $\tilde{1}$ $\tilde{\frac{1}{2}}$ 000+ 000+ $-1/4.3$

$\tilde{0}\ \tilde{1}\ 1\quad \frac{3}{2}\ \tilde{\frac{1}{2}}\ 1$

$\frac{3}{2}$ $\frac{1}{2}$ 1 000+ 000+ $-1/4.3\sqrt{3}$
$\frac{3}{2}$ $\frac{1}{2}$ 2 000− 000− $-1/4.3$
$\frac{3}{2}$ $\frac{1}{2}$ $\tilde{1}$ 000− 000− $-1/4.3$
$\frac{3}{2}$ $\frac{3}{2}$ 0 000+ 000+ $-1/4.3$
$\frac{3}{2}$ $\frac{3}{2}$ 1 000− 000− $-1/4.3\sqrt{2.3}$
$\frac{3}{2}$ $\frac{3}{2}$ 1 001− 000− $-1/2.3\sqrt{2.3}$
$\frac{3}{2}$ $\frac{3}{2}$ 2 000+ 000+ $+1/4.3\sqrt{2}$
$\frac{3}{2}$ $\frac{3}{2}$ $\tilde{1}$ 000+ 000+ $+1/4.3\sqrt{2}$
$\frac{3}{2}$ $\frac{3}{2}$ $\tilde{1}$ 001− 000+ 0

$\tilde{0}\ \tilde{1}\ 1\quad \frac{3}{2}\ \tilde{\frac{1}{2}}\ 2$

$\frac{3}{2}$ $\frac{1}{2}$ 1 000− 000− $-1/4.3$
$\frac{3}{2}$ $\frac{1}{2}$ $\tilde{1}$ 000+ 000+ $-1/4.3$
$\frac{3}{2}$ $\frac{3}{2}$ 1 000+ 000+ $-1/4.3\sqrt{2}$
$\frac{3}{2}$ $\frac{3}{2}$ 1 001+ 000+ $+1/4.3\sqrt{2}$
$\frac{3}{2}$ $\frac{3}{2}$ $\tilde{1}$ 000− 000− $-1/4.3\sqrt{2}$
$\frac{3}{2}$ $\frac{3}{2}$ $\tilde{1}$ 001+ 000− $-1/4.3\sqrt{2}$

$\tilde{0}\ \tilde{1}\ 1\quad \frac{3}{2}\ \tilde{\frac{1}{2}}\ \tilde{1}$

$\frac{3}{2}$ $\frac{1}{2}$ 1 000− 000− $+1/4.3$
$\frac{3}{2}$ $\frac{1}{2}$ 2 000+ 000+ $+1/4.3$
$\frac{3}{2}$ $\frac{1}{2}$ $\tilde{1}$ 000+ 000+ $+1/4.3\sqrt{3}$
$\frac{3}{2}$ $\frac{3}{2}$ 1 000+ 000+ $-1/4.3\sqrt{2}$
$\frac{3}{2}$ $\frac{3}{2}$ 1 001+ 000+ 0
$\frac{3}{2}$ $\frac{3}{2}$ 2 000− 000− $+1/4.3\sqrt{2}$
$\frac{3}{2}$ $\frac{3}{2}$ $\tilde{1}$ 000− 000− $+1/4.3\sqrt{2.3}$
$\frac{3}{2}$ $\frac{3}{2}$ $\tilde{1}$ 001+ 000− $-1/2.3\sqrt{2.3}$
$\frac{3}{2}$ $\frac{3}{2}$ $\tilde{0}$ 000+ 000+ $+1/4.3$

$\tilde{0}\ \tilde{1}\ 1\quad \frac{3}{2}\ \tilde{0}\ \frac{3}{2}$

$\frac{3}{2}$ 1 $\frac{1}{2}$ 000− 000− $-1/4.3$
$\frac{3}{2}$ 1 $\frac{3}{2}$ 000+ 000+ $+1/4.3$
$\frac{3}{2}$ 1 $\frac{3}{2}$ 000+ 001+ 0
$\frac{3}{2}$ 1 $\frac{3}{2}$ 001+ 000+ 0
$\frac{3}{2}$ 1 $\frac{3}{2}$ 001+ 001+ $+1/4.3$
$\frac{3}{2}$ 1 $\tilde{\frac{1}{2}}$ 000− 000− $-1/4.3$

$\tilde{0}\ \tilde{1}\ 1\quad 2\ 1\ 1$

2 1 1 000+ 000+ $-1/2.3\sqrt{2}$

$\tilde{0}\ \tilde{1}\ 1\quad 2\ 1\ \tilde{1}$

2 1 1 000− 000− $+1/2.3\sqrt{2.3}$
2 1 $\tilde{1}$ 000+ 000+ $+1/2.3\sqrt{2}$

$\tilde{0}\ \tilde{1}\ 1\quad 2\ \frac{3}{2}\ \frac{1}{2}$

2 $\frac{1}{2}$ $\frac{3}{2}$ 000+ 000+ $+1/4\sqrt{2.3}$
2 $\frac{3}{2}$ $\frac{1}{2}$ 000+ 000+ $-1/4\sqrt{3}$
2 $\frac{3}{2}$ $\frac{1}{2}$ 000+ 010− 0

$\tilde{0}\ \tilde{1}\ 1\quad 2\ \frac{3}{2}\ \frac{3}{2}$

2 $\frac{1}{2}$ $\frac{3}{2}$ 000− 000− 0
2 $\frac{1}{2}$ $\frac{3}{2}$ 000− 001− $+1/4\sqrt{2.3}$
2 $\frac{3}{2}$ $\frac{1}{2}$ 000− 000− 0
2 $\frac{3}{2}$ $\frac{1}{2}$ 000− 010+ $+1/4\sqrt{2.3}$
2 $\frac{3}{2}$ $\frac{3}{2}$ 000+ 000+ $-1/4\sqrt{2.3}$
2 $\frac{3}{2}$ $\frac{3}{2}$ 000+ 001+ 0
2 $\frac{3}{2}$ $\frac{3}{2}$ 000+ 010− 0
2 $\frac{3}{2}$ $\frac{3}{2}$ 000+ 011− 0

$\tilde{0}\ \tilde{1}\ 1\quad 2\ \frac{3}{2}\ \tilde{\frac{1}{2}}$

2 $\frac{1}{2}$ $\frac{3}{2}$ 000+ 000+ $+1/4\sqrt{2.3}$
2 $\frac{3}{2}$ $\frac{3}{2}$ 000− 000− 0
2 $\frac{3}{2}$ $\frac{3}{2}$ 000− 010+ $-1/4\sqrt{2.3}$
2 $\frac{3}{2}$ $\tilde{\frac{1}{2}}$ 000+ 000+ $+1/4\sqrt{3}$
2 $\frac{3}{2}$ $\tilde{\frac{1}{2}}$ 000+ 010− 0

$\tilde{0}\ \tilde{1}\ 1\quad 2\ 2\ 0$

2 1 1 000+ 000+ $-1/2.3$

$\tilde{0}\ \tilde{1}\ 1\quad 2\ 2\ 2$

2 1 1 000+ 000+ $-1/2.3\sqrt{2}$
2 1 $\tilde{1}$ 000− 000− $-1/2.3\sqrt{2}$

$\tilde{0}\ \tilde{1}\ 1\quad 2\ 2\ \tilde{0}$

2 1 $\tilde{1}$ 000+ 000+ $+1/2.3$

$\tilde{0}\ \tilde{1}\ 1\quad 2\ \tilde{1}\ 1$

2 0 2 000− 000− $+1/2.3$
2 1 1 000− 000− $+1/2.3\sqrt{2.3}$
2 1 $\tilde{1}$ 000+ 000+ $-1/2.3\sqrt{2}$
2 2 0 000− 000− $+1/2.3$
2 2 2 000− 000− $-1/2.3\sqrt{2}$
2 $\tilde{1}$ 1 000+ 000+ $+1/2.3\sqrt{2}$

$\tilde{0}\ \tilde{1}\ 1\quad 2\ \tilde{1}\ \tilde{1}$

2 0 2 000+ 000+ $+1/2.3$
2 1 1 000+ 000+ $+1/2.3\sqrt{2}$
2 1 $\tilde{1}$ 000− 000− $-1/2.3\sqrt{2.3}$
2 2 2 000+ 000+ $+1/2.3\sqrt{2}$
2 2 $\tilde{0}$ 000− 000− $-1/2.3$
2 $\tilde{1}$ 1 000− 000− $-1/2.3\sqrt{2.3}$
2 $\tilde{1}$ $\tilde{1}$ 000+ 000+ $-1/2.3\sqrt{2}$

$\tilde{0}\ \tilde{1}\ 1\quad 2\ \tilde{\frac{1}{2}}\ \frac{3}{2}$

2 $\frac{1}{2}$ $\frac{3}{2}$ 000+ 000+ $-1/4\sqrt{3}$
2 $\frac{1}{2}$ $\frac{3}{2}$ 000+ 001+ 0
2 $\frac{3}{2}$ $\frac{1}{2}$ 000+ 000+ $+1/4\sqrt{2.3}$
2 $\frac{3}{2}$ $\frac{3}{2}$ 000− 000− 0
2 $\frac{3}{2}$ $\frac{3}{2}$ 000− 001− $+1/4\sqrt{2.3}$
2 $\frac{3}{2}$ $\tilde{\frac{1}{2}}$ 000+ 000+ $+1/4\sqrt{2.3}$

$\tilde{0}\ \tilde{1}\ 1\quad 2\ \tilde{0}\ 2$

2 1 1 000− 000− $-1/2.3$
2 1 $\tilde{1}$ 000+ 000+ $+1/2.3$

$\tilde{0}\ \tilde{1}\ 1\quad \tilde{1}\ 0\ \tilde{1}$

1 $\tilde{1}$ 1 000+ 000+ $+1/9$
1 $\tilde{1}$ 2 000− 000− $+1/9$
1 $\tilde{1}$ $\tilde{1}$ 000− 000− $-1/9$
1 $\tilde{1}$ $\tilde{0}$ 000+ 000+ $+1/9$

$\tilde{0}\ \tilde{1}\ 1\quad \tilde{1}\ \frac{1}{2}\ \frac{3}{2}$

1 $\frac{3}{2}$ $\frac{1}{2}$ 000− 000− $+1/4.3$
1 $\frac{3}{2}$ $\frac{3}{2}$ 000+ 000+ $+1/2.9$
1 $\frac{3}{2}$ $\frac{3}{2}$ 000+ 001+ $+1/2.9$
1 $\frac{3}{2}$ $\frac{3}{2}$ 001+ 000+ $+1/2.9$
1 $\frac{3}{2}$ $\frac{3}{2}$ 001+ 001+ $-1/4.9$
1 $\frac{3}{2}$ $\tilde{\frac{1}{2}}$ 000− 000− $+1/4.9$

$\tilde{0}\ \tilde{1}\ 1\quad \tilde{1}\ \frac{1}{2}\ \tilde{\frac{1}{2}}$

1 $\frac{3}{2}$ $\frac{3}{2}$ 000− 000− $+1/2.9\sqrt{2}$
1 $\frac{3}{2}$ $\frac{3}{2}$ 001− 000− $-1/9\sqrt{2}$
1 $\frac{3}{2}$ $\tilde{\frac{1}{2}}$ 000+ 000+ $+1/9$
1 $\tilde{\frac{1}{2}}$ $\frac{3}{2}$ 000+ 000+ $+1/9$
1 $\tilde{\frac{1}{2}}$ $\tilde{\frac{1}{2}}$ 000− 000− $-1/2.9$

$\tilde{0}\ \tilde{1}\ 1\quad \tilde{1}\ 1\ 1$

1 1 0 000+ 000+ $+1/9$
1 1 1 000− 000− $+1/2.9$
1 1 2 000+ 000+ $-1/2.9$
1 1 $\tilde{1}$ 000+ 000+ $+1/2.9$

$\tilde{0}\ \tilde{1}\ 1\quad \tilde{1}\ 1\ 2$

1 1 1 000+ 000+ $-1/2.3\sqrt{3}$
1 1 $\tilde{1}$ 000− 000− $-1/2.9$
1 2 1 000− 000− 0
1 2 $\tilde{1}$ 000+ 000+ $+1/9$

$\tilde{0}\ \tilde{1}\ 1\quad \tilde{1}\ 1\ \tilde{1}$

1 1 1 000+ 000+ $+1/2.9$
1 1 2 000− 000− $+1/2.3\sqrt{3}$
1 1 $\tilde{1}$ 000− 000− $-1/2.9$

O and T_d 9j Symbols (cont.)

$\tilde{0}\ \tilde{1}\ 1\quad \tilde{1}\ 1\ \tilde{1}$
$1\ 2\ 1$ 000− 000− $+1/2.3\sqrt{3}$
$1\ 2\ \tilde{1}$ 000+ 000+ $-1/2.9$
$1\ \tilde{1}\ 1$ 000− 000− $+1/2.9$
$1\ \tilde{1}\ 2$ 000+ 000+ $+1/2.9$
$1\ \tilde{1}\ \tilde{1}$ 000+ 000+ $+1/2.9$
$1\ \tilde{1}\ \tilde{0}$ 000− 000− $-1/9$
$\tilde{0}\ \tilde{1}\ 1\quad \tilde{1}\ 1\ \tilde{0}$
$1\ 1\ \tilde{1}$ 000+ 000+ $+1/9$
$1\ 2\ \tilde{1}$ 000− 000− $+1/9$
$1\ \tilde{1}\ \tilde{1}$ 000− 000− $+1/9$
$1\ \tilde{0}\ \tilde{1}$ 000+ 000+ $+1/9$
$\tilde{0}\ \tilde{1}\ 1\quad \tilde{1}\ \frac{3}{2}\ \frac{1}{2}$
$1\ \frac{1}{2}\ \frac{1}{2}$ 000+ 000+ $+1/9$
$1\ \frac{1}{2}\ \frac{3}{2}$ 000− 000− $+1/4.9$
$\tilde{0}\ \tilde{1}\ 1\quad \tilde{1}\ \frac{3}{2}\ \frac{3}{2}$
$1\ \frac{1}{2}\ \frac{1}{2}$ 000− 000− $+1/2.9\sqrt{2}$
$1\ \frac{1}{2}\ \frac{1}{2}$ 010+ 000− $-1/9\sqrt{2}$
$1\ \frac{1}{2}\ \frac{3}{2}$ 000+ 000+ $-1/2.9$
$1\ \frac{1}{2}\ \frac{3}{2}$ 000+ 001+ $+1/2.9$
$1\ \frac{1}{2}\ \frac{3}{2}$ 010− 000+ $-1/2.9$
$1\ \frac{1}{2}\ \frac{3}{2}$ 010− 001+ $-1/4.9$
$1\ \frac{3}{2}\ \frac{1}{2}$ 000+ 000+ $+1/2.9$
$1\ \frac{3}{2}\ \frac{1}{2}$ 000+ 010− $+1/2.9$
$1\ \frac{3}{2}\ \frac{1}{2}$ 010− 000+ $+1/2.9$
$1\ \frac{3}{2}\ \frac{1}{2}$ 010− 010− $-1/4.9$
$1\ \frac{3}{2}\ \frac{3}{2}$ 000− 000− $-1/4.9$
$1\ \frac{3}{2}\ \frac{3}{2}$ 000− 001− 0
$1\ \frac{3}{2}\ \frac{3}{2}$ 000− 010+ 0
$1\ \frac{3}{2}\ \frac{3}{2}$ 000− 011+ $+1/4.9$
$1\ \frac{3}{2}\ \frac{3}{2}$ 001− 000− 0
$1\ \frac{3}{2}\ \frac{3}{2}$ 001− 001− $+1/2.9$
$1\ \frac{3}{2}\ \frac{3}{2}$ 001− 010+ $-1/2.9$
$1\ \frac{3}{2}\ \frac{3}{2}$ 001− 011+ 0
$1\ \frac{3}{2}\ \frac{3}{2}$ 010+ 000− 0
$1\ \frac{3}{2}\ \frac{3}{2}$ 010+ 001− $-1/2.9$
$1\ \frac{3}{2}\ \frac{3}{2}$ 010+ 010+ $+1/2.9$
$1\ \frac{3}{2}\ \frac{3}{2}$ 010+ 011+ 0
$1\ \frac{3}{2}\ \frac{3}{2}$ 011+ 000− $+1/4.9$
$1\ \frac{3}{2}\ \frac{3}{2}$ 011+ 001− 0
$1\ \frac{3}{2}\ \frac{3}{2}$ 011+ 010+ 0
$1\ \frac{3}{2}\ \frac{3}{2}$ 011+ 011+ $+1/2.9$
$1\ \frac{3}{2}\ \tilde{\frac{1}{2}}$ 000+ 000+ $+1/2.9$
$1\ \frac{3}{2}\ \tilde{\frac{1}{2}}$ 000+ 010− $-1/2.9$
$1\ \frac{3}{2}\ \tilde{\frac{1}{2}}$ 010− 000+ $-1/2.9$
$1\ \frac{3}{2}\ \tilde{\frac{1}{2}}$ 010− 010− $-1/4.9$
$\tilde{0}\ \tilde{1}\ 1\quad \tilde{1}\ \frac{3}{2}\ \tilde{\frac{1}{2}}$
$1\ \frac{1}{2}\ \frac{3}{2}$ 000− 000− $+1/4.3$
$1\ \frac{3}{2}\ \frac{3}{2}$ 000+ 000+ $-1/2.9$
$1\ \frac{3}{2}\ \frac{3}{2}$ 000+ 010− $+1/2.9$
$1\ \frac{3}{2}\ \frac{3}{2}$ 001+ 000+ $-1/2.9$
$1\ \frac{3}{2}\ \frac{3}{2}$ 001+ 010− $-1/4.9$
$1\ \frac{3}{2}\ \tilde{\frac{1}{2}}$ 000− 000− $-1/2.9\sqrt{2}$
$1\ \frac{3}{2}\ \tilde{\frac{1}{2}}$ 000− 010+ $-1/9\sqrt{2}$
$1\ \tilde{\frac{1}{2}}\ \frac{3}{2}$ 000− 000− $+1/4.9$
$1\ \tilde{\frac{1}{2}}\ \tilde{\frac{1}{2}}$ 000+ 000+ $+1/9$
$\tilde{0}\ \tilde{1}\ 1\quad \tilde{1}\ 2\ 1$
$1\ 1\ 0$ 000− 000− $+1/9$
$1\ 1\ 1$ 000+ 000+ $+1/2.9$
$1\ 1\ 2$ 000− 000− $+1/9$
$1\ 1\ \tilde{1}$ 000− 000− $-1/2.9$
$\tilde{0}\ \tilde{1}\ 1\quad \tilde{1}\ 2\ \tilde{1}$
$1\ 1\ 1$ 000− 000− $+1/2.3\sqrt{3}$
$1\ 1\ 2$ 000+ 000+ 0
$1\ 1\ \tilde{1}$ 000+ 000+ $+1/2.3\sqrt{3}$
$1\ \tilde{1}\ 1$ 000+ 000+ $-1/2.9$
$1\ \tilde{1}\ 2$ 000− 000− $+1/9$
$1\ \tilde{1}\ \tilde{1}$ 000− 000− $+1/2.9$
$1\ \tilde{1}\ \tilde{0}$ 000+ 000+ $+1/9$
$\tilde{0}\ \tilde{1}\ 1\quad \tilde{1}\ \tilde{1}\ 0$
$1\ 0\ 1$ 000+ 000+ $+1/9$
$\tilde{0}\ \tilde{1}\ 1\quad \tilde{1}\ \tilde{1}\ 1$
$1\ 0\ 1$ 000− 000− $-1/9$
$1\ 1\ 0$ 000− 000− $-1/9$
$1\ 1\ 1$ 000+ 000+ $+1/2.9$
$1\ 1\ 2$ 000− 000− $+1/2.9$
$1\ 1\ \tilde{1}$ 000− 000− $+1/2.9$
$\tilde{0}\ \tilde{1}\ 1\quad \tilde{1}\ \tilde{1}\ 2$
$1\ 0\ 1$ 000+ 000+ $+1/9$
$1\ 1\ 1$ 000− 000− $+1/2.9$
$1\ 1\ \tilde{1}$ 000+ 000+ $-1/2.3\sqrt{3}$
$1\ 2\ 1$ 000+ 000+ $+1/9$
$1\ 2\ \tilde{1}$ 000− 000− 0
$\tilde{0}\ \tilde{1}\ 1\quad \tilde{1}\ \tilde{1}\ \tilde{1}$
$1\ 0\ 1$ 000+ 000+ $-1/9$
$1\ 1\ 1$ 000− 000− $-1/2.9$
$1\ 1\ 2$ 000+ 000+ $+1/2.3\sqrt{3}$
$1\ 1\ \tilde{1}$ 000+ 000+ $+1/2.9$
$1\ 2\ 1$ 000+ 000+ $+1/2.9$
$1\ 2\ \tilde{1}$ 000− 000− $+1/2.3\sqrt{3}$
$1\ \tilde{1}\ 1$ 000+ 000+ $+1/2.9$
$1\ \tilde{1}\ 2$ 000− 000− $-1/2.9$
$1\ \tilde{1}\ \tilde{1}$ 000− 000− $+1/2.9$
$1\ \tilde{1}\ \tilde{0}$ 000+ 000+ $+1/9$
$\tilde{0}\ \tilde{1}\ 1\quad \tilde{1}\ \tilde{\frac{1}{2}}\ \frac{1}{2}$
$1\ \frac{1}{2}\ \frac{1}{2}$ 000− 000− $-1/2.9$
$1\ \frac{1}{2}\ \frac{3}{2}$ 000+ 000+ $+1/9$
$\tilde{0}\ \tilde{1}\ 1\quad \tilde{1}\ \tilde{\frac{1}{2}}\ \frac{3}{2}$
$1\ \frac{1}{2}\ \frac{1}{2}$ 000+ 000+ $-1/9$
$1\ \frac{1}{2}\ \frac{3}{2}$ 000− 000− $+1/2.9\sqrt{2}$
$1\ \frac{1}{2}\ \frac{3}{2}$ 000− 001− $+1/9\sqrt{2}$
$1\ \frac{3}{2}\ \frac{1}{2}$ 000− 000− $+1/4.9$
$1\ \frac{3}{2}\ \frac{3}{2}$ 000+ 000+ $+1/2.9$
$1\ \frac{3}{2}\ \frac{3}{2}$ 000+ 001+ $-1/2.9$
$1\ \frac{3}{2}\ \frac{3}{2}$ 001+ 000+ $-1/2.9$
$1\ \frac{3}{2}\ \frac{3}{2}$ 001+ 001+ $-1/4.9$
$1\ \frac{3}{2}\ \tilde{\frac{1}{2}}$ 000− 000− $+1/4.3$
$\tilde{0}\ \tilde{1}\ 1\quad \tilde{1}\ \tilde{0}\ 1$
$1\ 1\ 0$ 000+ 000+ $+1/9$
$1\ 1\ 1$ 000− 000− $-1/9$
$1\ 1\ 2$ 000+ 000+ $+1/9$
$1\ 1\ \tilde{1}$ 000+ 000+ $+1/9$
$\tilde{0}\ \tilde{\frac{1}{2}}\ \frac{1}{2}\quad \frac{3}{2}\ \frac{3}{2}\ 0$
$\frac{3}{2}\ 1\ \frac{1}{2}$ 000+ 000+ $+1/8$
$\tilde{0}\ \tilde{\frac{1}{2}}\ \frac{1}{2}\quad \frac{3}{2}\ \frac{3}{2}\ 1$
$\frac{3}{2}\ 1\ \frac{1}{2}$ 000− 000− $+1/8.3$
$\frac{3}{2}\ 1\ \frac{1}{2}$ 010− 000− $-1/4.3$
$\frac{3}{2}\ 1\ \frac{3}{2}$ 000+ 000+ $-1/4.3\sqrt{2}$
$\frac{3}{2}\ 1\ \frac{3}{2}$ 001+ 000+ $+1/4.3\sqrt{2}$
$\frac{3}{2}\ 1\ \frac{3}{2}$ 010+ 000+ $-1/4.3\sqrt{2}$
$\frac{3}{2}\ 1\ \frac{3}{2}$ 011+ 000+ $-1/8.3\sqrt{2}$
$\tilde{0}\ \tilde{\frac{1}{2}}\ \frac{1}{2}\quad \frac{3}{2}\ \frac{3}{2}\ 2$
$\frac{3}{2}\ 1\ \frac{3}{2}$ 000− 000− 0
$\frac{3}{2}\ 1\ \frac{3}{2}$ 001− 000− $+1/8\sqrt{2}$
$\tilde{0}\ \tilde{\frac{1}{2}}\ \frac{1}{2}\quad \frac{3}{2}\ \frac{3}{2}\ \tilde{1}$
$\frac{3}{2}\ 1\ \frac{3}{2}$ 000− 000− $-1/4.3\sqrt{2}$
$\frac{3}{2}\ 1\ \frac{3}{2}$ 001− 000− $-1/4.3\sqrt{2}$
$\frac{3}{2}\ 1\ \frac{3}{2}$ 010+ 000− $+1/4.3\sqrt{2}$
$\frac{3}{2}\ 1\ \frac{3}{2}$ 011+ 000− $-1/8.3\sqrt{2}$
$\frac{3}{2}\ 1\ \tilde{\frac{1}{2}}$ 000+ 000+ $+1/8.3$
$\frac{3}{2}\ 1\ \tilde{\frac{1}{2}}$ 010− 000+ $+1/4.3$
$\tilde{0}\ \tilde{\frac{1}{2}}\ \frac{1}{2}\quad \frac{3}{2}\ \frac{3}{2}\ \tilde{0}$
$\frac{3}{2}\ 1\ \tilde{\frac{1}{2}}$ 000− 000− $+1/8$
$\tilde{0}\ \tilde{\frac{1}{2}}\ \frac{1}{2}\quad \frac{3}{2}\ 2\ \frac{1}{2}$
$\frac{3}{2}\ \frac{3}{2}\ 0$ 000− 000− $+1/8$
$\frac{3}{2}\ \frac{3}{2}\ 1$ 000+ 000+ $+1/8$
$\frac{3}{2}\ \frac{3}{2}\ 1$ 001+ 000+ 0
$\tilde{0}\ \tilde{\frac{1}{2}}\ \frac{1}{2}\quad \frac{3}{2}\ 2\ \frac{3}{2}$
$\frac{3}{2}\ \frac{3}{2}\ 1$ 000− 000− 0
$\frac{3}{2}\ \frac{3}{2}\ 1$ 001− 000− $+1/8\sqrt{2}$
$\frac{3}{2}\ \frac{3}{2}\ 2$ 000+ 000+ $-1/8\sqrt{2}$
$\frac{3}{2}\ \frac{3}{2}\ \tilde{1}$ 000+ 000+ 0
$\frac{3}{2}\ \frac{3}{2}\ \tilde{1}$ 001− 000+ $-1/8\sqrt{2}$
$\tilde{0}\ \tilde{\frac{1}{2}}\ \frac{1}{2}\quad \frac{3}{2}\ 2\ \tilde{\frac{1}{2}}$
$\frac{3}{2}\ \frac{3}{2}\ \tilde{1}$ 000− 000− $+1/8$
$\frac{3}{2}\ \frac{3}{2}\ \tilde{1}$ 001+ 000− 0
$\frac{3}{2}\ \frac{3}{2}\ \tilde{0}$ 000+ 000+ $+1/8$

O and T_d 9j Symbols (cont.)

$\tilde{0}\ \tilde{\frac{1}{2}}\ \frac{1}{2}\quad \frac{3}{2}\ \tilde{1}\ \frac{1}{2}$

$\frac{3}{2}\ \frac{1}{2}\ 1$ 000− 000− $+1/2.3\sqrt{2}$

$\frac{3}{2}\ \frac{3}{2}\ 0$ 000− 000− $-1/8$

$\frac{3}{2}\ \frac{3}{2}\ 1$ 000+ 000+ $+1/8.3$

$\frac{3}{2}\ \frac{3}{2}\ 1$ 001+ 000+ $+1/4.3$

$\tilde{0}\ \tilde{\frac{1}{2}}\ \frac{1}{2}\quad \frac{3}{2}\ \tilde{1}\ \frac{3}{2}$

$\frac{3}{2}\ \frac{1}{2}\ 1$ 000+ 000+ $+1/8.3$

$\frac{3}{2}\ \frac{1}{2}\ 1$ 010− 000+ $+1/4.3$

$\frac{3}{2}\ \frac{1}{2}\ 2$ 000− 000− $+1/8$

$\frac{3}{2}\ \frac{1}{2}\ 2$ 010+ 000− 0

$\frac{3}{2}\ \frac{1}{2}\ \tilde{1}$ 000− 000− $-1/8.3$

$\frac{3}{2}\ \frac{1}{2}\ \tilde{1}$ 010+ 000− $+1/4.3$

$\frac{3}{2}\ \frac{3}{2}\ 1$ 000− 000− $+1/4.3\sqrt{2}$

$\frac{3}{2}\ \frac{3}{2}\ 1$ 001− 000− $-1/4.3\sqrt{2}$

$\frac{3}{2}\ \frac{3}{2}\ 1$ 010+ 000− $-1/4.3\sqrt{2}$

$\frac{3}{2}\ \frac{3}{2}\ 1$ 011+ 000− $-1/8.3\sqrt{2}$

$\frac{3}{2}\ \frac{3}{2}\ 2$ 000+ 000+ 0

$\frac{3}{2}\ \frac{3}{2}\ 2$ 010− 000+ $-1/8\sqrt{2}$

$\frac{3}{2}\ \frac{3}{2}\ \tilde{1}$ 000+ 000+ $-1/4.3\sqrt{2}$

$\frac{3}{2}\ \frac{3}{2}\ \tilde{1}$ 001− 000+ $-1/4.3\sqrt{2}$

$\frac{3}{2}\ \frac{3}{2}\ \tilde{1}$ 010− 000+ $-1/4.3\sqrt{2}$

$\frac{3}{2}\ \frac{3}{2}\ \tilde{1}$ 011+ 000+ $+1/8.3\sqrt{2}$

$\tilde{0}\ \tilde{\frac{1}{2}}\ \frac{1}{2}\quad \frac{3}{2}\ \tilde{1}\ \tilde{\frac{1}{2}}$

$\frac{3}{2}\ \frac{1}{2}\ \tilde{1}$ 000+ 000+ $-1/2.3\sqrt{2}$

$\frac{3}{2}\ \frac{3}{2}\ \tilde{1}$ 000− 000− $-1/8.3$

$\frac{3}{2}\ \frac{3}{2}\ \tilde{1}$ 001+ 000− $+1/4.3$

$\frac{3}{2}\ \frac{3}{2}\ \tilde{0}$ 000+ 000+ $+1/8$

$\tilde{0}\ \tilde{\frac{1}{2}}\ \frac{1}{2}\quad \frac{3}{2}\ \tilde{\frac{1}{2}}\ 1$

$\frac{3}{2}\ 0\ \frac{3}{2}$ 000+ 000+ $+1/8$

$\frac{3}{2}\ 1\ \frac{1}{2}$ 000+ 000+ $+1/2.3\sqrt{2}$

$\frac{3}{2}\ 1\ \frac{3}{2}$ 000 000− $-1/8.3$

$\frac{3}{2}\ 1\ \frac{3}{2}$ 001− 000− $-1/4.3$

$\tilde{0}\ \tilde{\frac{1}{2}}\ \frac{1}{2}\quad \frac{3}{2}\ \tilde{\frac{1}{2}}\ 2$

$\frac{3}{2}\ 0\ \frac{3}{2}$ 000− 000− $-1/8$

$\frac{3}{2}\ 1\ \frac{3}{2}$ 000+ 000+ $-1/8$

$\frac{3}{2}\ 1\ \frac{3}{2}$ 001+ 000+ 0

$\tilde{0}\ \tilde{\frac{1}{2}}\ \frac{1}{2}\quad \frac{3}{2}\ \tilde{\frac{1}{2}}\ \tilde{1}$

$\frac{3}{2}\ 0\ \frac{3}{2}$ 000− 000− $+1/8$

$\frac{3}{2}\ 1\ \frac{3}{2}$ 000+ 000+ $-1/8.3$

$\frac{3}{2}\ 1\ \frac{3}{2}$ 001+ 000+ $+1/4.3$

$\frac{3}{2}\ 1\ \tilde{\frac{1}{2}}$ 000− 000− $+1/2.3\sqrt{2}$

$\tilde{0}\ \tilde{\frac{1}{2}}\ \frac{1}{2}\quad \frac{3}{2}\ \tilde{0}\ \frac{3}{2}$

$\frac{3}{2}\ \frac{1}{2}\ 1$ 000− 000− $-1/8$

$\frac{3}{2}\ \frac{1}{2}\ 2$ 000+ 000+ $+1/8$

$\frac{3}{2}\ \frac{1}{2}\ \tilde{1}$ 000+ 000+ $+1/8$

$\tilde{0}\ \tilde{\frac{1}{2}}\ \frac{1}{2}\quad 2\ \frac{3}{2}\ \frac{1}{2}$

$2\ 1\ 1$ 000+ 000+ $-1/4\sqrt{3}$

$\tilde{0}\ \tilde{\frac{1}{2}}\ \frac{1}{2}\quad 2\ \frac{3}{2}\ \frac{3}{2}$

$2\ 1\ 1$ 000− 000− $-1/4\sqrt{2.3}$

$2\ 1\ \tilde{1}$ 000+ 000+ $+1/4\sqrt{2.3}$

$\tilde{0}\ \tilde{\frac{1}{2}}\ \frac{1}{2}\quad 2\ \frac{3}{2}\ \tilde{\frac{1}{2}}$

$2\ 1\ \tilde{1}$ 000− 000− $-1/4\sqrt{3}$

$\tilde{0}\ \tilde{\frac{1}{2}}\ \frac{1}{2}\quad 2\ 2\ 0$

$2\ \frac{3}{2}\ \frac{1}{2}$ 000+ 000+ $-1/4$

$\tilde{0}\ \tilde{\frac{1}{2}}\ \frac{1}{2}\quad 2\ 2\ 2$

$2\ \frac{3}{2}\ \frac{3}{2}$ 000− 000− $+1/4\sqrt{2}$

$\tilde{0}\ \tilde{\frac{1}{2}}\ \frac{1}{2}\quad 2\ 2\ \tilde{0}$

$2\ \frac{3}{2}\ \tilde{\frac{1}{2}}$ 000− 000− $+1/4$

$\tilde{0}\ \tilde{\frac{1}{2}}\ \frac{1}{2}\quad 2\ \tilde{1}\ 1$

$2\ \frac{1}{2}\ \frac{3}{2}$ 000− 000− $+1/4\sqrt{3}$

$2\ \frac{3}{2}\ \frac{1}{2}$ 000− 000− $-1/4\sqrt{3}$

$2\ \frac{3}{2}\ \frac{3}{2}$ 000+ 000+ $-1/4\sqrt{2.3}$

$\tilde{0}\ \tilde{\frac{1}{2}}\ \frac{1}{2}\quad 2\ \tilde{1}\ \tilde{1}$

$2\ \frac{1}{2}\ \frac{3}{2}$ 000+ 000+ $-1/4\sqrt{3}$

$2\ \frac{3}{2}\ \frac{3}{2}$ 000− 000− $-1/4\sqrt{2.3}$

$2\ \frac{3}{2}\ \tilde{\frac{1}{2}}$ 000+ 000+ $+1/4\sqrt{3}$

$\tilde{0}\ \tilde{\frac{1}{2}}\ \frac{1}{2}\quad 2\ \tilde{\frac{1}{2}}\ \frac{3}{2}$

$2\ 0\ 2$ 000+ 000+ $-1/4$

$2\ 1\ 1$ 000+ 000+ $-1/4\sqrt{3}$

$2\ 1\ \tilde{1}$ 000− 000− $-1/4\sqrt{3}$

$\tilde{0}\ \tilde{\frac{1}{2}}\ \frac{1}{2}\quad 2\ \tilde{0}\ 2$

$2\ \frac{1}{2}\ \frac{3}{2}$ 000− 000− $-1/4$

$\tilde{0}\ \tilde{\frac{1}{2}}\ \frac{1}{2}\quad \tilde{1}\ 0\ \tilde{1}$

$1\ \tilde{\frac{1}{2}}\ \frac{3}{2}$ 000− 000− $-1/2.3$

$1\ \tilde{\frac{1}{2}}\ \tilde{\frac{1}{2}}$ 000+ 000+ $-1/2.3$

$\tilde{0}\ \tilde{\frac{1}{2}}\ \frac{1}{2}\quad \tilde{1}\ \frac{1}{2}\ \frac{3}{2}$

$1\ \tilde{1}\ 1$ 000+ 000+ $+1/2.3\sqrt{2}$

$1\ \tilde{1}\ 2$ 000− 000− $-1/2.3\sqrt{2}$

$1\ \tilde{1}\ \tilde{1}$ 000− 000− $+1/2.3\sqrt{2.3}$

$1\ \tilde{0}\ \tilde{1}$ 000+ 000+ $+1/2.3$

$\tilde{0}\ \tilde{\frac{1}{2}}\ \frac{1}{2}\quad \tilde{1}\ \frac{1}{2}\ \tilde{\frac{1}{2}}$

$1\ \tilde{1}\ \tilde{1}$ 000+ 000+ $-1/3\sqrt{2.3}$

$1\ \tilde{1}\ \tilde{0}$ 000− 000− $-1/2.3$

$1\ \tilde{0}\ \tilde{1}$ 000− 000− $+1/2.3$

$\tilde{0}\ \tilde{\frac{1}{2}}\ \frac{1}{2}\quad \tilde{1}\ 1\ 1$

$1\ \frac{3}{2}\ \frac{1}{2}$ 000− 000− $-1/2.3\sqrt{2}$

$1\ \frac{3}{2}\ \frac{3}{2}$ 000+ 000+ $-1/4.3$

$1\ \frac{3}{2}\ \frac{3}{2}$ 001+ 000+ 0

$1\ \tilde{\frac{1}{2}}\ \frac{3}{2}$ 000− 000− $-1/2.3\sqrt{2}$

$\tilde{0}\ \tilde{\frac{1}{2}}\ \frac{1}{2}\quad \tilde{1}\ 1\ 2$

$1\ \frac{3}{2}\ \frac{3}{2}$ 000− 000− $-1/4.3$

$1\ \frac{3}{2}\ \frac{3}{2}$ 001− 000− $-1/4.3$

$1\ \tilde{\frac{1}{2}}\ \frac{3}{2}$ 000+ 000+ $+1/2.3\sqrt{2}$

$\tilde{0}\ \tilde{\frac{1}{2}}\ \frac{1}{2}\quad \tilde{1}\ 1\ \tilde{1}$

$1\ \frac{3}{2}\ \frac{3}{2}$ 000− 000− $-1/4.3\sqrt{3}$

$1\ \frac{3}{2}\ \frac{3}{2}$ 001− 000− $+1/2.3\sqrt{3}$

$1\ \frac{3}{2}\ \tilde{\frac{1}{2}}$ 000+ 000+ $+1/2.3\sqrt{2.3}$

$1\ \tilde{\frac{1}{2}}\ \frac{3}{2}$ 000+ 000+ $+1/2.3\sqrt{2.3}$

$1\ \tilde{\frac{1}{2}}\ \tilde{\frac{1}{2}}$ 000− 000− $-1/3\sqrt{2.3}$

$\tilde{0}\ \tilde{\frac{1}{2}}\ \frac{1}{2}\quad \tilde{1}\ 1\ \tilde{0}$

$1\ \frac{3}{2}\ \tilde{\frac{1}{2}}$ 000− 000− $+1/2.3$

$1\ \tilde{\frac{1}{2}}\ \tilde{\frac{1}{2}}$ 000+ 000+ $+1/2.3$

$\tilde{0}\ \tilde{\frac{1}{2}}\ \frac{1}{2}\quad \tilde{1}\ \frac{3}{2}\ \frac{1}{2}$

$1\ 1\ 0$ 000+ 000+ $-1/2.3$

$1\ 1\ 1$ 000− 000− $+1/2.3\sqrt{2.3}$

$1\ 2\ 1$ 000+ 000+ $+1/2.3\sqrt{2}$

$1\ \tilde{1}\ 1$ 000+ 000+ $-1/2.3\sqrt{2}$

$\tilde{0}\ \tilde{\frac{1}{2}}\ \frac{1}{2}\quad \tilde{1}\ \frac{3}{2}\ \frac{3}{2}$

$1\ 1\ 1$ 000+ 000+ $+1/4.3\sqrt{3}$

$1\ 1\ 1$ 010− 000+ $-1/2.3\sqrt{3}$

$1\ 1\ 2$ 000− 000− $+1/4.3$

$1\ 1\ 2$ 010+ 000− $+1/4.3$

$1\ 1\ \tilde{1}$ 000− 000− $-1/4.3$

$1\ 1\ \tilde{1}$ 010+ 000− 0

$1\ 2\ 1$ 000− 000− $+1/4.3$

$1\ 2\ 1$ 010+ 000− $+1/4.3$

$1\ 2\ \tilde{1}$ 000+ 000+ $-1/4.3$

$1\ 2\ \tilde{1}$ 010− 000+ $+1/4.3$

$1\ \tilde{1}\ 1$ 000− 000− $+1/4.3$

$1\ \tilde{1}\ 1$ 010+ 000− 0

$1\ \tilde{1}\ 2$ 000+ 000+ $+1/4.3$

$1\ \tilde{1}\ 2$ 010− 000+ $-1/4.3$

$1\ \tilde{1}\ \tilde{1}$ 000+ 000+ $-1/4.3\sqrt{3}$

$1\ \tilde{1}\ \tilde{1}$ 010− 000+ $-1/2.3\sqrt{3}$

$\tilde{0}\ \tilde{\frac{1}{2}}\ \frac{1}{2}\quad \tilde{1}\ \frac{3}{2}\ \tilde{\frac{1}{2}}$

$1\ 1\ \tilde{1}$ 000+ 000+ $+1/2.3\sqrt{2}$

$1\ 2\ \tilde{1}$ 000− 000− $-1/2.3\sqrt{2}$

$1\ \tilde{1}\ \tilde{1}$ 000− 000− $-1/2.3\sqrt{2.3}$

$1\ \tilde{1}\ \tilde{0}$ 000+ 000+ $+1/2.3$

$\tilde{0}\ \tilde{\frac{1}{2}}\ \frac{1}{2}\quad \tilde{1}\ 2\ 1$

$1\ \frac{3}{2}\ \frac{1}{2}$ 000+ 000+ $+1/2.3\sqrt{2}$

$1\ \frac{3}{2}\ \frac{3}{2}$ 000− 000− $-1/4.3$

$1\ \frac{3}{2}\ \frac{3}{2}$ 001− 000− $+1/4.3$

$\tilde{0}\ \tilde{\frac{1}{2}}\ \frac{1}{2}\quad \tilde{1}\ 2\ \tilde{1}$

$1\ \frac{3}{2}\ \frac{3}{2}$ 000+ 000+ $-1/4.3$

$1\ \frac{3}{2}\ \frac{3}{2}$ 001+ 000+ $-1/4.3$

$1\ \frac{3}{2}\ \tilde{\frac{1}{2}}$ 000− 000− $+1/2.3\sqrt{2}$

$\tilde{0}\ \tilde{\frac{1}{2}}\ \frac{1}{2}\quad \tilde{1}\ \tilde{1}\ 0$

$1\ \frac{1}{2}\ \frac{1}{2}$ 000+ 000+ $+1/2.3$

$1\ \frac{3}{2}\ \frac{1}{2}$ 000− 000− $+1/2.3$

$\tilde{0}\ \tilde{\frac{1}{2}}\ \frac{1}{2}\quad \tilde{1}\ \tilde{1}\ 1$

$1\ \frac{1}{2}\ \frac{1}{2}$ 000− 000− $+1/3\sqrt{2.3}$

$1\ \frac{1}{2}\ \frac{3}{2}$ 000+ 000+ $+1/2.3\sqrt{2.3}$

$1\ \frac{3}{2}\ \frac{1}{2}$ 000+ 000+ $-1/2.3\sqrt{2.3}$

$1\ \frac{3}{2}\ \frac{3}{2}$ 000− 000− $+1/4.3\sqrt{3}$

$1\ \frac{3}{2}\ \frac{3}{2}$ 001− 000− $+1/2.3\sqrt{3}$

$\tilde{0}\ \tilde{\frac{1}{2}}\ \frac{1}{2}\quad \tilde{1}\ \tilde{1}\ 2$

$1\ \frac{1}{2}\ \frac{3}{2}$ 000− 000− $+1/2.3\sqrt{2}$

$1\ \frac{3}{2}\ \frac{3}{2}$ 000+ 000+ $+1/4.3$

O and T_d 9j Symbols (cont.)

$\tilde{0}\ \tilde{\tfrac12}\ \tfrac12\quad \tilde{1}\ \tilde{1}\ 2$

$1\ \tfrac32\ \tfrac32$ 001+ 000+ $-1/4.3$

$\tilde{0}\ \tilde{\tfrac12}\ \tfrac12\quad \tilde{1}\ \tilde{1}\ \tilde{1}$

$1\ \tfrac12\ \tfrac32$ 000− 000− $-1/2.3\sqrt{2}$

$1\ \tfrac32\ \tfrac32$ 000+ 000+ $+1/4.3$

$1\ \tfrac32\ \tfrac32$ 001+ 000+ 0

$1\ \tfrac32\ \tilde{\tfrac12}$ 000− 000− $+1/2.3\sqrt{2}$

$\tilde{0}\ \tilde{\tfrac12}\ \tfrac12\quad \tilde{1}\ \tilde{\tfrac12}\ \tfrac12$

$1\ 0\ 1$ 000− 000− $-1/2.3$

$1\ 1\ 0$ 000− 000− $-1/2.3$

$1\ 1\ 1$ 000+ 000+ $-1/3\sqrt{2.3}$

$\tilde{0}\ \tilde{\tfrac12}\ \tfrac12\quad \tilde{1}\ \tilde{\tfrac12}\ \tfrac32$

$1\ 0\ 1$ 000+ 000+ $-1/2.3$

$1\ 1\ 1$ 000− 000− $+1/2.3\sqrt{2.3}$

$1\ 1\ 2$ 000+ 000+ $+1/2.3\sqrt{2}$

$1\ 1\ \tilde{1}$ 000+ 000+ $+i/2.3\sqrt{2}$

$\tilde{0}\ \tilde{\tfrac12}\ \tfrac12\quad \tilde{1}\ \tilde{0}\ 1$

$1\ \tfrac12\ \tfrac12$ 000+ 000+ $+1/2.3$

$1\ \tfrac12\ \tfrac32$ 000− 000− $-1/2.3$

$\tilde{0}\ \tilde{\tfrac12}\ \tfrac12\quad \tilde{\tfrac12}\ 0\ \tilde{\tfrac12}$

$\tfrac12\ \tilde{\tfrac12}\ \tilde{1}$ 000+ 000+ $+1/4$

$\tfrac12\ \tilde{\tfrac12}\ \tilde{0}$ 000− 000− $-1/4$

$\tilde{0}\ \tilde{\tfrac12}\ \tfrac12\quad \tilde{\tfrac12}\ \tfrac12\ \tilde{1}$

$\tfrac12\ \tilde{1}\ \tfrac32$ 000− 000− $+1/2.3$

$\tfrac12\ \tilde{1}\ \tilde{\tfrac12}$ 000+ 000+ $-1/4.3$

$\tilde{0}\ \tilde{\tfrac12}\ \tfrac12\quad \tilde{\tfrac12}\ \tfrac12\ \tilde{0}$

$\tfrac12\ \tilde{1}\ \tilde{\tfrac12}$ 000− 000− $+1/4$

$\tfrac12\ \tilde{0}\ \tilde{\tfrac12}$ 000+ 000+ $+1/4$

$\tilde{0}\ \tilde{\tfrac12}\ \tfrac12\quad \tilde{\tfrac12}\ 1\ \tfrac32$

$\tfrac12\ \tfrac32\ 1$ 000+ 000+ $+1/8$

$\tfrac12\ \tfrac32\ 2$ 000− 000− $+1/8$

$\tfrac12\ \tfrac32\ \tilde{1}$ 000− 000− $-1/8.3$

$\tilde{0}\ \tilde{\tfrac12}\ \tfrac12\quad \tilde{\tfrac12}\ 1\ \tilde{\tfrac12}$

$\tfrac12\ \tfrac32\ \tilde{1}$ 000+ 000+ $-1/2.3$

$\tfrac12\ \tilde{\tfrac12}\ \tilde{1}$ 000− 000− $+1/4.3$

$\tfrac12\ \tilde{\tfrac12}\ \tilde{0}$ 000+ 000+ $+1/4$

$\tilde{0}\ \tilde{\tfrac12}\ \tfrac12\quad \tilde{\tfrac12}\ \tfrac32\ 1$

$\tfrac12\ 1\ \tfrac12$ 000− 000− $+1/2.3$

$\tfrac12\ 1\ \tfrac32$ 000+ 000+ $+1/8.3$

$\tilde{0}\ \tilde{\tfrac12}\ \tfrac12\quad \tilde{\tfrac12}\ \tfrac32\ 2$

$\tfrac12\ 1\ \tfrac32$ 000− 000− $-1/8$

$\tfrac12\ 2\ \tfrac32$ 000+ 000+ $+1/8$

$\tilde{0}\ \tilde{\tfrac12}\ \tfrac12\quad \tilde{\tfrac12}\ \tfrac32\ \tilde{1}$

$\tfrac12\ 1\ \tfrac32$ 000− 000− $+1/8$

$\tfrac12\ 2\ \tfrac32$ 000+ 000+ $+1/8$

$\tfrac12\ \tilde{1}\ \tfrac32$ 000+ 000+ $+1/8.3$

$\tfrac12\ \tilde{1}\ \tilde{\tfrac12}$ 000− 000− $+1/2.3$

$\tilde{0}\ \tilde{\tfrac12}\ \tfrac12\quad \tilde{\tfrac12}\ 2\ \tfrac32$

$\tfrac12\ \tfrac32\ 1$ 000− 000− $+1/8$

$\tfrac12\ \tfrac32\ 2$ 000+ 000+ $-1/8$

$\tilde{0}\ \tilde{\tfrac12}\ \tfrac12\quad \tilde{\tfrac12}\ 2\ \tfrac32$

$\tfrac12\ \tfrac32\ \tilde{1}$ 000+ 000+ $+1/8$

$\tilde{0}\ \tilde{\tfrac12}\ \tfrac12\quad \tilde{\tfrac12}\ \tilde{1}\ \tfrac12$

$\tfrac12\ \tfrac12\ 0$ 000+ 000+ $+1/4$

$\tfrac12\ \tfrac12\ 1$ 000− 000− $+1/4.3$

$\tilde{0}\ \tilde{\tfrac12}\ \tfrac12\quad \tilde{\tfrac12}\ \tilde{1}\ \tfrac32$

$\tfrac12\ \tfrac12\ 1$ 000+ 000+ $+1/2.3$

$\tfrac12\ \tfrac32\ 1$ 000− 000− $-1/8.3$

$\tfrac12\ \tfrac32\ 2$ 000+ 000+ $+1/8$

$\tfrac12\ \tfrac32\ \tilde{1}$ 000+ 000+ $+1/8$

$\tilde{0}\ \tilde{\tfrac12}\ \tfrac12\quad \tilde{\tfrac12}\ \tilde{\tfrac12}\ 0$

$\tfrac12\ 0\ \tfrac12$ 000+ 000+ $+1/4$

$\tilde{0}\ \tilde{\tfrac12}\ \tfrac12\quad \tilde{\tfrac12}\ \tilde{\tfrac12}\ 1$

$\tfrac12\ 0\ \tfrac12$ 000− 000− $-1/4$

$\tfrac12\ 1\ \tfrac12$ 000+ 000+ $-1/4.3$

$\tfrac12\ 1\ \tfrac32$ 000− 000− $+1/2.3$

$\tilde{0}\ \tilde{\tfrac12}\ \tfrac12\quad \tilde{\tfrac12}\ \tilde{0}\ \tfrac12$

$\tfrac12\ \tfrac12\ 0$ 000− 000− $-1/4$

$\tfrac12\ \tfrac12\ 1$ 000+ 000+ $+1/4$

$\tilde{0}\ \tilde{0}\ 0\quad \tfrac32\ \tfrac32\ 0$

$\tfrac32\ \tfrac32\ 0$ 000+ 000+ $+1/4$

$\tilde{0}\ \tilde{0}\ 0\quad \tfrac32\ \tfrac32\ 1$

$\tfrac32\ \tfrac32\ 1$ 000+ 000+ $+1/4\sqrt{3}$

$\tfrac32\ \tfrac32\ 1$ 010+ 000+ 0

$\tfrac32\ \tfrac32\ 1$ 011+ 000+ $-1/4\sqrt{3}$

$\tilde{0}\ \tilde{0}\ 0\quad \tfrac32\ \tfrac32\ 2$

$\tfrac32\ \tfrac32\ 2$ 000+ 000+ $-1/4\sqrt{2}$

$\tilde{0}\ \tilde{0}\ 0\quad \tfrac32\ \tfrac32\ \tilde{1}$

$\tfrac32\ \tfrac32\ \tilde{1}$ 000+ 000+ $+1/4\sqrt{3}$

$\tfrac32\ \tfrac32\ \tilde{1}$ 010− 000+ 0

$\tfrac32\ \tfrac32\ \tilde{1}$ 011+ 000+ $-1/4\sqrt{3}$

$\tilde{0}\ \tilde{0}\ 0\quad \tfrac32\ \tfrac32\ \tilde{0}$

$\tfrac32\ \tfrac32\ \tilde{0}$ 000+ 000+ $+1/4$

$\tilde{0}\ \tilde{0}\ 0\quad 2\ \tfrac32\ \tfrac12$

$2\ \tfrac32\ \tfrac12$ 000+ 000+ $-1/4$

$\tilde{0}\ \tilde{0}\ 0\quad 2\ \tfrac32\ \tfrac32$

$2\ \tfrac32\ \tfrac32$ 000+ 000+ $+1/4\sqrt{2}$

$\tilde{0}\ \tilde{0}\ 0\quad 2\ \tfrac32\ \tilde{\tfrac12}$

$2\ \tfrac32\ \tilde{\tfrac12}$ 000+ 000+ $-1/4$

$\tilde{0}\ \tilde{0}\ 0\quad 2\ 2\ 0$

$2\ 2\ 0$ 000+ 000+ $+1/2$

$\tilde{0}\ \tilde{0}\ 0\quad 2\ 2\ 2$

$2\ 2\ 2$ 000+ 000+ $-1/2\sqrt{2}$

$\tilde{0}\ \tilde{0}\ 0\quad 2\ 2\ \tilde{0}$

$2\ 2\ \tilde{0}$ 000+ 000+ $+1/2$

$\tilde{0}\ \tilde{0}\ 0\quad \tilde{1}\ 0\ \tilde{1}$

$1\ \tilde{0}\ \tilde{1}$ 000+ 000+ $+1/3$

$\tilde{0}\ \tilde{0}\ 0\quad \tilde{1}\ \tfrac12\ \tfrac32$

$1\ \tilde{\tfrac12}\ \tfrac32$ 000− 000− $-1/2\sqrt{2.3}$

$\tilde{0}\ \tilde{0}\ 0\quad \tilde{1}\ \tfrac12\ \tilde{\tfrac12}$

$1\ \tilde{\tfrac12}\ \tilde{\tfrac12}$ 000− 000− $+1/2\sqrt{3}$

$\tilde{0}\ \tilde{0}\ 0\quad \tilde{1}\ 1\ 1$

$1\ \tilde{1}\ 1$ 000+ 000+ $+1/3\sqrt{3}$

$\tilde{0}\ \tilde{0}\ 0\quad \tilde{1}\ 1\ 2$

$1\ \tilde{1}\ 2$ 000+ 000+ $-1/3\sqrt{2}$

$\tilde{0}\ \tilde{0}\ 0\quad \tilde{1}\ 1\ \tilde{1}$

$1\ \tilde{1}\ \tilde{1}$ 000+ 000+ $+1/3\sqrt{3}$

$\tilde{0}\ \tilde{0}\ 0\quad \tilde{1}\ 1\ \tilde{0}$

$1\ \tilde{1}\ \tilde{0}$ 000+ 000+ $+1/3$

$\tilde{0}\ \tilde{0}\ 0\quad \tilde{1}\ \tfrac32\ \tfrac12$

$1\ \tfrac32\ \tfrac12$ 000− 000− $-1/2\sqrt{2.3}$

$\tilde{0}\ \tilde{0}\ 0\quad \tilde{1}\ \tfrac32\ \tfrac32$

$1\ \tfrac32\ \tfrac32$ 000− 000− $-1/4\sqrt{3}$

$1\ \tfrac32\ \tfrac32$ 001− 000− 0

$1\ \tfrac32\ \tfrac32$ 010+ 000− 0

$1\ \tfrac32\ \tfrac32$ 011+ 000− $+1/4\sqrt{3}$

$\tilde{0}\ \tilde{0}\ 0\quad \tilde{1}\ \tfrac32\ \tilde{\tfrac12}$

$1\ \tfrac32\ \tilde{\tfrac12}$ 000− 000− $-1/2\sqrt{2.3}$

$\tilde{0}\ \tilde{0}\ 0\quad \tilde{1}\ 2\ 1$

$1\ 2\ 1$ 000− 000− $-1/3\sqrt{2}$

$\tilde{0}\ \tilde{0}\ 0\quad \tilde{1}\ 2\ \tilde{1}$

$1\ 2\ \tilde{1}$ 000− 000− $-1/3\sqrt{2}$

$\tilde{0}\ \tilde{0}\ 0\quad \tilde{1}\ \tilde{1}\ 0$

$1\ 1\ 0$ 000+ 000+ $+1/3$

$\tilde{0}\ \tilde{0}\ 0\quad \tilde{1}\ \tilde{1}\ 1$

$1\ 1\ 1$ 000+ 000+ $-1/3\sqrt{3}$

$\tilde{0}\ \tilde{0}\ 0\quad \tilde{1}\ \tilde{1}\ 2$

$1\ 1\ 2$ 000+ 000+ $+1/3\sqrt{2}$

$\tilde{0}\ \tilde{0}\ 0\quad \tilde{1}\ \tilde{1}\ \tilde{1}$

$1\ 1\ \tilde{1}$ 000+ 000+ $-1/3\sqrt{3}$

$\tilde{0}\ \tilde{0}\ 0\quad \tilde{\tfrac12}\ 0\ \tilde{\tfrac12}$

$\tfrac12\ \tilde{0}\ \tilde{\tfrac12}$ 000− 000− $+1/2$

$\tilde{0}\ \tilde{0}\ 0\quad \tilde{\tfrac12}\ \tfrac12\ \tilde{1}$

$\tfrac12\ \tilde{\tfrac12}\ \tilde{1}$ 000+ 000+ $+1/2\sqrt{3}$

$\tilde{0}\ \tilde{0}\ 0\quad \tilde{\tfrac12}\ \tfrac12\ \tilde{0}$

$\tfrac12\ \tilde{\tfrac12}\ \tilde{0}$ 000+ 000+ $+1/2$

$\tilde{0}\ \tilde{0}\ 0\quad \tilde{\tfrac12}\ 1\ \tfrac32$

$\tfrac12\ \tilde{1}\ \tfrac32$ 000− 000− $-1/2\sqrt{2.3}$

$\tilde{0}\ \tilde{0}\ 0\quad \tilde{\tfrac12}\ 1\ \tilde{\tfrac12}$

$\tfrac12\ \tilde{1}\ \tilde{\tfrac12}$ 000− 000− $+1/2\sqrt{3}$

$\tilde{0}\ \tilde{0}\ 0\quad \tilde{\tfrac12}\ \tfrac32\ 1$

$\tfrac12\ \tfrac32\ 1$ 000+ 000+ $+1/2\sqrt{2.3}$

$\tilde{0}\ \tilde{0}\ 0\quad \tilde{\tfrac12}\ \tfrac32\ 2$

$\tfrac12\ \tfrac32\ 2$ 000+ 000+ $-1/4$

$\tilde{0}\ \tilde{0}\ 0\quad \tilde{\tfrac12}\ \tfrac32\ \tilde{1}$

$\tfrac12\ \tfrac32\ \tilde{1}$ 000+ 000+ $+1/2\sqrt{2.3}$

$\tilde{0}\ \tilde{0}\ 0\quad \tilde{\tfrac12}\ 2\ \tfrac32$

$\tfrac12\ 2\ \tfrac32$ 000+ 000+ $-1/4$

O and T_d 9j Symbols (cont.)

$\tilde{0}\ \tilde{0}\ 0\quad \tilde{\frac{1}{2}}\ \tilde{1}\ \frac{1}{2}$

$\frac{1}{2}\ 1\ \frac{1}{2}$ 000− 000− $+1/2\sqrt{3}$

$\tilde{0}\ \tilde{0}\ 0\quad \tilde{\frac{1}{2}}\ \tilde{1}\ \frac{3}{2}$

$\frac{1}{2}\ 1\ \frac{3}{2}$ 000− 000− $+1/2\sqrt{2.3}$

$\tilde{0}\ \tilde{0}\ 0\quad \tilde{\frac{1}{2}}\ \tilde{\frac{1}{2}}\ 0$

$\frac{1}{2}\ \frac{1}{2}\ 0$ 000+ 000+ $+1/2$

$\tilde{0}\ \tilde{0}\ 0\quad \tilde{\frac{1}{2}}\ \tilde{\frac{1}{2}}\ 1$

$\frac{1}{2}\ \frac{1}{2}\ 1$ 000+ 000+ $-1/2\sqrt{3}$

$\tilde{0}\ \tilde{0}\ 0\quad \tilde{0}\ 0\ \tilde{0}$

$0\ \tilde{0}\ \tilde{0}$ 000+ 000+ $+1$

$\tilde{0}\ \tilde{0}\ 0\quad \tilde{0}\ \frac{1}{2}\ \tilde{\frac{1}{2}}$

$0\ \tilde{\frac{1}{2}}\ \tilde{\frac{1}{2}}$ 000− 000− $+1/2$

$\tilde{0}\ \tilde{0}\ 0\quad \tilde{0}\ 1\ \tilde{1}$

$0\ \tilde{1}\ \tilde{1}$ 000+ 000+ $+1/3$

$\tilde{0}\ \tilde{0}\ 0\quad \tilde{0}\ \frac{3}{2}\ \frac{3}{2}$

$0\ \frac{3}{2}\ \frac{3}{2}$ 000− 000− $+1/4$

$\tilde{0}\ \tilde{0}\ 0\quad \tilde{0}\ 2\ 2$

$0\ 2\ 2$ 000− 000− $+1/2$

$\tilde{0}\ \tilde{0}\ 0\quad \tilde{0}\ \tilde{1}\ 1$

$0\ 1\ 1$ 000+ 000+ $+1/3$

$\tilde{0}\ \tilde{0}\ 0\quad \tilde{0}\ \tilde{\frac{1}{2}}\ \frac{1}{2}$

$0\ \frac{1}{2}\ \frac{1}{2}$ 000− 000− $+1/2$

$\tilde{0}\ \tilde{0}\ 0\quad \tilde{0}\ \tilde{0}\ 0$

$0\ 0\ 0$ 000+ 000+ $+1$

16

Bases in Terms of Spherical Harmonics

Properties of spherical harmonic bases are to be found in most books on quantum theory. A summary of several of the properties of the JM rotation matrices in terms of Euler angles is given in Section 16.1, and a table of these matrices up to $J=3$ is included. The other parts of this chapter describe the partners of irreps of the point groups in various bases. The partners of $\mathbf{SO}_3$ irreps in the $\mathbf{SO}_2$ (that is the JM) basis are given in terms of rectangular coordinates in Section 16.2, while transformations between several point group bases of $\mathbf{SO}_3$ are given at the end of the chapter. In these tables the layout is such that nonzero transformation coefficients are given following each partner. For example, on line 18 of page 522 we find that a partner of the $\frac{5}{2}(\mathbf{SO}_3)$ irrep in the SO_3–O–D_3–C_3 basis is written in the JM (SO_3–SO_2) basis as

$$|\tfrac{5}{2}(SO_3)\tfrac{3}{2}(O)\tfrac{3}{2}(D_3)\tfrac{3}{2}(C_3)\rangle = |\tfrac{5}{2}(SO_3)\tfrac{3}{2}(SO_2)\rangle\left(\frac{5}{3\sqrt{6}}+\frac{i}{3\sqrt{3}}\right)$$

$$+|\tfrac{5}{2}-\tfrac{3}{2}\rangle\left(\frac{1}{3\sqrt{3}}-\frac{5i}{3\sqrt{6}}\right)$$

The group labels and the parentheses around the coefficients are omitted in the tables.

In several of the tables at the end of the chapter, the transformation from one point group basis to another requires a rotation, and in some cases this rotation cannot be expressed in simple surds for the spin functions. In these cases every spinor transformation coefficient must be divided by the square root of the number appearing at the head of the table.

Tables 16.1 and 16.2 were not computer-typeset. They were computer-generated, however, using the algebraic language REDUCE (Hearn 1973).

16.1 Rotation Matrices in the JM Basis

The explicit form for the rotation matrices depends on the choice of several conventions. We use Condon and Shortley's (1935) convention for the JM partners (the SO_3–SO_2 spherical harmonics). Most quantum mechanics texts do the same. It is customary in quantum mechanics to use Euler angles defined about z, y, z axes, but classical mechanics texts usually use z, x, z axes. Wolf (1969) discusses the equivalence of the Euler angle definitions of Messiah (1965), Rose (1957), and Brink and Satchler (1962), and notes some discrepancies in Edmonds (1957). We follow Messiah's conventions, but note that his diagram (pp. 524 and 1068) has the α direction mislabeled; it should be in positive sense.

With one exception (in Figure 5.1) we have taken an active view of rotations. This view may be illustrated by considering the function $\psi(\theta)$, θ being the angle from the x axis in the x–y plane. Let $R_z(\alpha)$ be the rotation through angle α about the z axis, taking the system ψ to a primed system ψ' where

$$\psi'(\theta+\alpha)=\psi(\theta) \tag{16.1.1}$$

The passive view has one rotating the axis system by angle $-\alpha$ about z to give primed axes and primed parameter θ'. A passive rotation is equivalent to the inverse of an active rotation.

The rotation matrices are parametrized by three angles, the Euler angles α, β, and γ. Our angles are defined as follows:

1. Rotate the system by angle γ about the laboratory z axis. This is achieved by the operator $R_z(\gamma)=e^{-i\gamma J_z}$.

2. Rotate the system by angle β about the laboratory y axis. The operator is $R_y(\beta)=e^{-i\beta J_y}$.

3. Rotate the system by angle α about the laboratory z axis. The operator is $R_z(\alpha)=e^{-i\alpha J_z}$.

The combined operator $R(\alpha,\beta,\gamma)$ is therefore

$$\begin{aligned} R(\alpha,\beta,\gamma)&=R_z(\alpha)R_y(\beta)R_z(\gamma) \\ &=e^{-i\alpha J_z}e^{-i\beta J_y}e^{-i\gamma J_z} \end{aligned} \tag{16.1.2}$$

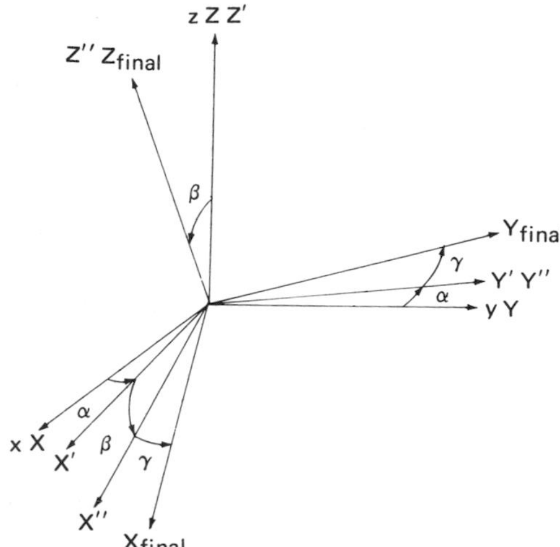

Figure 16.1. The Euler angles in terms of body-fixed XYZ axes as seen in the laboratory space xyz. [Each successive position of the body axes is indicated by attaching a prime; see (16.1.3).]

Note that operators act on their right, the γ rotation is performed first, and the parametrization is not unique. An equivalent definition in terms of rotations about body-fixed XYZ axes rather than space-fixed xyz axes is provided by the equation

$$R(\alpha,\beta,\gamma)=R_{Z''}(\gamma)R_{Y'}(\beta)R_{Z}(\alpha) \tag{16.1.3}$$

where the later positions of the body axes are denoted by attaching primes. The proof of the equivalence is as follows:

The initial axes coincide, so

$$R_{Z}(\alpha)=R_{z}(\alpha) \tag{16.1.4}$$

The body axes $X'Y'Z'$ at the time of the β rotation are rotated by $R_z(\alpha)$ from the XYZ axes,

$$R_{Y'}(\beta)=R_z(\alpha)R_y(\beta)R_z(\alpha)^{-1} \tag{16.1.5}$$

and the body axes for the γ rotation are further rotated,

$$R_{Z''}(\gamma)=R_{Y'}(\beta)R_z(\gamma)R_{Y'}(\beta)^{-1} \tag{16.1.6}$$

as can be seen from Figure 16.1. Combining (16.1.4)–(16.1.6) shows the equivalence of (16.1.2) and (16.1.3). The inverse follows from the definition (16.1.2),

$$R(\alpha,\beta,\gamma)^{-1}=R(-\gamma,-\beta,-\alpha) \tag{16.1.7}$$

The matrix elements of the Euler rotation operator are as follows in the JM basis:

$$D^J(\alpha\beta\gamma)_{MM'}=J(\alpha\beta\gamma)_{MM'}$$

$$=\langle xJM|R(\alpha\beta\gamma)|xJM'\rangle$$

$$=\langle xJM|e^{-i\alpha J_z}e^{-i\beta J_y}e^{-i\gamma J_z}|xJM'\rangle$$

$$=e^{-i\alpha M}r^J(\beta)_{KK'}e^{-i\alpha M'} \quad (16.1.8)$$

where

$$r^J(\beta)_{MM'}=\langle JM|e^{-i\beta J_y}|JM'\rangle \quad (16.1.9)$$

The values of this matrix are given in Table 16.1 for the values of $J \leqslant 3$. They have been computed from the Wigner formula,

$$r^J_{MM'}=\sum_t(-1)^t\frac{[(J+M)!(J-M)!(J+M')!(J-M')!]^{1/2}}{(J+M-t)!(J-M'-t)!t!(t-M+M')!}$$

$$\times\left(\cos\tfrac{1}{2}\beta\right)^{2J+M-M'-2t}\left(\sin\tfrac{1}{2}\beta\right)^{2t-M+M'} \quad (16.1.10)$$

The sum is over all t for which the factorials have meaning. The special cases of rotations about one axis through an angle π are

$$R_x(\pi)|JM\rangle=|J-M\rangle e^{-i\pi J}$$

$$R_y(\pi)|JM\rangle=|J-M\rangle(-1)^{J-M} \quad (16.1.11)$$

$$R_z(\pi)|JM\rangle=|JM\rangle e^{-i\pi M}$$

A rotation by 2π about any axis u is

$$R_u(2\pi)|JM\rangle=|JM\rangle(-1)^{2J} \quad (16.1.12)$$

Table 16.1 The Matrices $r^J(\beta)_{MM'}$ Describing a Rotation About the y Axis in the JM Basis

Only the values for M positive and $M > M'$ are given; the others may be obtained from $r^J(\beta)_{MM'} = r^J(\beta)_{-M'-M}$ and $r^J(\beta)_{MM'} = (-1)^{M-M'} r^J(\beta)_{-M-M'}$.

For convenience, the superscript J and the argument β are omitted in the entries below.

$J=0$ .... $r_{00}=1$

$J=1/2$ .... $r_{\frac{1}{2}\frac{1}{2}}=\cos\frac{1}{2}\beta$

$r_{\frac{1}{2}-\frac{1}{2}}=-\sin\frac{1}{2}\beta$

$J=1$ .... $r_{11}=(1/2)(1+\cos\beta)$

$r_{10}=-(1/\sqrt{2})\sin\beta$

$r_{1-1}=(1/2)(1-\cos\beta)$

$r_{00}=\cos\beta$

$J=3/2$ .... $r_{\frac{3}{2}\frac{3}{2}}=(1/4)(3\cos\frac{1}{2}\beta+\cos\frac{3}{2}\beta)$

$r_{\frac{3}{2}\frac{1}{2}}=(\sqrt{3}/2)(-\sin\frac{1}{2}\beta-\sin\frac{3}{2}\beta)$

$r_{\frac{3}{2}-\frac{1}{2}}=(\sqrt{3}/2)(\cos\frac{1}{2}\beta-\cos\frac{3}{2}\beta)$

$r_{\frac{3}{2}-\frac{3}{2}}=(1/4)(-3\sin\frac{1}{2}\beta+\sin\frac{3}{2}\beta)$

$r_{\frac{1}{2}\frac{1}{2}}=(1/4)(\cos\frac{1}{2}\beta+3\cos\frac{3}{2}\beta)$

$r_{\frac{1}{2}-\frac{1}{2}}=(1/4)(\sin\frac{1}{2}\beta-3\sin\frac{3}{2}\beta)$

$J=2$ .... $r_{22}=(1/8)(3+4\cos\beta+\cos 2\beta)$

$r_{21}=-(1/4)(2\sin\beta+\sin 2\beta)$

$r_{20}=(\sqrt{3}/4\sqrt{2})(1-\cos 2\beta)$

$r_{2-1}=(1/4)(-2\sin\beta+\sin 2\beta)$

$r_{2-2}=(1/8)(3-4\cos\beta+\cos 2\beta)$

$r_{11}=(1/2)(\cos\beta+\cos 2\beta)$

$r_{10}=-(\sqrt{3}/2\sqrt{2})\sin 2\beta$

$r_{1-1}=(1/2)(\cos\beta-\cos 2\beta)$

$r_{00}=(1/4)(1+3\cos 2\beta)$

$J=5/2$ .... $r_{\frac{5}{2}\frac{5}{2}}=(1/16)(10\cos\frac{1}{2}\beta+\cos\frac{5}{2}\beta+5\cos\frac{3}{2}\beta)$

$r_{\frac{5}{2}\frac{3}{2}}=(\sqrt{5}/16)(-2\sin\frac{1}{2}\beta-\sin\frac{5}{2}\beta-3\sin\frac{3}{2}\beta)$

$r_{\frac{5}{2}\frac{1}{2}}=(\sqrt{5}/8\sqrt{2})(2\cos\frac{1}{2}\beta-\cos\frac{5}{2}\beta-\cos\frac{3}{2}\beta)$

$r_{\frac{5}{2}-\frac{1}{2}}=(\sqrt{5}/8\sqrt{2})(-2\sin\frac{1}{2}\beta+\sin\frac{5}{2}\beta-\sin\frac{3}{2}\beta)$

$r_{\frac{5}{2}-\frac{3}{2}}=(\sqrt{5}/16)(2\cos\frac{1}{2}\beta+\cos\frac{5}{2}\beta-3\cos\frac{3}{2}\beta)$

$r_{\frac{5}{2}-\frac{5}{2}}=(1/16)(-10\sin\frac{1}{2}\beta-\sin\frac{5}{2}\beta+5\sin\frac{3}{2}\beta)$

$r_{\frac{3}{2}\frac{3}{2}}=(1/16)(2\cos\frac{1}{2}\beta+5\cos\frac{5}{2}\beta+9\cos\frac{3}{2}\beta)$

$r_{\frac{3}{2}\frac{1}{2}}=(1/8\sqrt{2})(2\sin\frac{1}{2}\beta-5\sin\frac{5}{2}\beta-3\sin\frac{3}{2}\beta)$

$r_{\frac{3}{2}-\frac{1}{2}}=(1/8\sqrt{2})(2\cos\frac{1}{2}\beta-5\cos\frac{5}{2}\beta+3\cos\frac{3}{2}\beta)$

Table 16.1 *Continued*

Only the values for M positive and $M>M'$ are given; the others may be obtained from $r^J(\beta)_{MM'}=r^J(\beta)_{-M'-M}$ and $r^J(\beta)_{MM'}=(-1)^{M-M'}r^J(\beta)_{-M-M'}$.
For convenience, the superscript J and the argument β are omitted in the entries below.

$$r_{\frac{3}{2}-\frac{3}{2}}=(1/16)(2\sin\tfrac{1}{2}\beta+5\sin\tfrac{5}{2}\beta-9\sin\tfrac{3}{2}\beta)$$
$$r_{\frac{1}{2}\frac{1}{2}}=(1/8)(2\cos\tfrac{1}{2}\beta+5\cos\tfrac{5}{2}\mathrm{B}+\cos\tfrac{3}{2}\beta)$$
$$r_{\frac{1}{2}-\frac{1}{2}}=(1/8)(-2\sin\tfrac{1}{2}\beta-5\sin\tfrac{5}{2}\beta+\sin\tfrac{3}{2}\beta)$$

$J=3$ ....

$$r_{33}=(1/32)(10+15\cos\beta+\cos 3\beta+6\cos 2\beta)$$
$$r_{32}=(\sqrt{3}/16\sqrt{2})(-5\sin\beta-\sin 3\beta-4\sin 2\beta)$$
$$r_{31}=(\sqrt{15}/32)(2+\cos\beta-\cos 3\beta-2\cos 2\beta)$$
$$r_{30}=(\sqrt{5}/16)(-3\sin\beta+\sin 3\beta)$$
$$r_{3-1}=(\sqrt{15}/32)(2-\cos\beta+\cos 3\beta-2\cos 2\beta)$$
$$r_{3-2}=(\sqrt{3}/16\sqrt{2})(-5\sin\beta-\sin 3\beta+4\sin 2\beta)$$
$$r_{3-3}=(1/32)(10-15\cos\beta-\cos 3\beta+6\cos 2\beta)$$
$$r_{22}=(1/16)(5\cos\beta+3\cos 3\beta+8\cos 2\beta)$$
$$r_{21}=(\sqrt{5}/16\sqrt{2})(\sin\beta-3\sin 3\beta-4\sin 2\beta)$$
$$r_{20}=(\sqrt{15}/8\sqrt{2})(\cos\beta-\cos 3\beta)$$
$$r_{2-1}=(\sqrt{5}/16\sqrt{2})(-\sin\beta+3\sin 3\beta-4\sin 2\beta)$$
$$r_{2-2}=(1/16)(5\cos\beta+3\cos 3\beta-8\cos 2\beta)$$
$$r_{11}=(1/32)(6+\cos\beta+15\cos 3\beta+10\cos 2\beta)$$
$$r_{10}=(\sqrt{3}/16)(-\sin\beta-5\sin 3\beta)$$
$$r_{1-1}=(1/32)(6-\cos\beta-15\cos 3\beta+10\cos 2\beta)$$
$$r_{00}=(1/8)(3\cos\beta+5\cos 3\beta)$$

16.2 Spherical Harmonics in Rectangular Coordinates

A formula was given previously, (4.4.7)–(4.4.8), for the spherical harmonics with the Condon and Shortley phase convention and in terms of spherical polar coordinates. The expression may be converted to rectangular coordinates by use of the transformation

$$\begin{aligned} x&=r\sin\theta\cos\phi \\ y&=r\sin\theta\sin\phi \\ z&=r\cos\theta \end{aligned} \tag{16.2.1}$$

Note that because

$$x+iy=r\sin\theta e^{i\phi} \tag{16.2.2}$$

the transformation of spherical harmonics between polar and rectangular coordinates is trivial.

Table 16.2 gives the spherical harmonics up to $L=8$. For $L>2$ only those with positive M are tabulated; the relationship

$$Y_{LM}^{*}=(-1)^{M}Y_{L-M} \tag{16.2.3}$$

may be used to obtain the others.

Two normalizations are given. The tabulated $|LM\rangle$ have the normalization that was used in several examples of this book. Namely the $|LM\rangle$ functions transform as the partners of $\mathbf{SO_3}$ irreps in the SO_3–SO_2 basis, and have for $L=1$

$$\begin{aligned}&\langle x|x\rangle=\langle y|y\rangle=\langle z|z\rangle=1\\&\langle x|y\rangle=\langle y|z\rangle=\langle z|x\rangle=0\end{aligned} \tag{16.2.4}$$

The normalization of products of such functions arises from the replacement of x_1, x_2, etc. by x. If one wishes to use (16.2.4), one must reverse the procedure, e.g., $(x+iy)^2\to(x_1+iy_1)(x_2+iy_2)$, while $(x+iy)z\to\frac{1}{2}(x_1+iy_1)z_2+\frac{1}{2}z_1(x_2+iy_2)$. See also the comments after (5.3.12a). Also given for each L value is a normalizing factor N_L by which the xyz functions are to be multiplied to obtain the spherical harmonic normalization of unity,

$$\int_{\theta=0}^{\pi}\int_{\phi=0}^{2\pi}Y_{LM}(\theta,\phi)^{*}Y_{L'M'}(\theta,\phi)\sin\theta d\theta d\phi=\delta_{LL'}\delta_{MM'} \tag{16.2.5}$$

In particular,

$$Y_{LM}=|LM\rangle N_L \tag{16.2.6}$$

Table 16.2 Spherical Harmonics in Rectangular Coordinates, $Y_{LM} = |LM\rangle N_L$

L	N_L	$\|LM\rangle$
0	$\left(\frac{1}{4\pi}\right)^{1/2}$	$\|00\rangle = 1$
1	$\left(\frac{3}{4\pi}\right)^{1/2} r^{-1}$	$\|11\rangle = -(1/\sqrt{2})(x+iy)$
		$\|10\rangle = z$
		$\|1-1\rangle = (1/\sqrt{2})(x-iy)$
2	$\left(\frac{5}{4\pi}\right)^{1/2}\left(\frac{\sqrt{3}}{\sqrt{2}}\right) r^{-2}$	$\|22\rangle = (1/2)(x+iy)^2$
		$\|21\rangle = -(x+iy)z$
		$\|20\rangle = (1/\sqrt{6})(3z^2 - r^2)$
		$\|2-1\rangle = (x-iy)z$
		$\|2-2\rangle = (1/2)(x-iy)^2$
3	$\left(\frac{7}{4\pi}\right)^{1/2}\left(\frac{\sqrt{5}}{\sqrt{2}}\right) r^{-3}$	$\|33\rangle = -(1/2\sqrt{2})(x+iy)^3$
		$\|32\rangle = (\sqrt{3}/2)(x+iy)^2 z$
		$\|31\rangle = -(\sqrt{3}/2\sqrt{2.5})(x+iy)(5z^2 - r^2)$
		$\|30\rangle = (1/\sqrt{2.5})(5z^3 - 3zr^2)$
4	$\left(\frac{9}{4\pi}\right)^{1/2} \frac{\sqrt{5.7}}{2\sqrt{2}} r^{-4}$	$\|44\rangle = (1/4)(x+iy)^4$
		$\|43\rangle = -(1/\sqrt{2})(x+iy)^3 z$
		$\|42\rangle = (1/2\sqrt{7})(x+iy)^2(7z^2 - r^2)$
		$\|41\rangle = -(1/\sqrt{2.7})(x+iy)(7z^3 - 3zr^2)$
		$\|40\rangle = (1/2\sqrt{2.5.7})(35z^4 - 30z^2r^2 + 3r^4)$
5	$\left(\frac{11}{4\pi}\right)^{1/2}\left(\frac{3\sqrt{7}}{2\sqrt{2}}\right) r^{-5}$	$\|55\rangle = -(1/4\sqrt{2})(x+iy)^5$
		$\|54\rangle = (\sqrt{5}/4)(x+iy)^4 z$
		$\|53\rangle = -(\sqrt{5}/4.3\sqrt{2})(x+iy)^3(9z^2 - r^2)$
		$\|52\rangle = (\sqrt{5}/2\sqrt{3})(x+iy)^2(3z^3 - zr^2)$
		$\|51\rangle = -(\sqrt{5}/4\sqrt{3.7})(x+iy)(21z^4 - 14z^2r^2 + r^4)$
		$\|50\rangle = (1/2.3\sqrt{2.7})(63z^5 - 70z^3r^2 + 15zr^4)$

Table 16.2 *Continued*

L	N_L	$\|LM\rangle$
6	$\left(\frac{13}{4\pi}\right)^{1/2}\frac{\sqrt{3.7.11}}{4}r^{-6}$	$\|66\rangle=(1/8)(x+iy)^6$
		$\|65\rangle=-(\sqrt{3}\ /4)(x+iy)^5z$
		$\|64\rangle=(\sqrt{3}\ /4\sqrt{2.11}\ )(x+iy)^4(11z^2-r^2)$
		$\|63\rangle=-(\sqrt{5}/4\ \sqrt{11}\ )(x+iy)^3(11z^3-3zr^2)$
		$\|62\rangle=(\sqrt{5}\ /8\sqrt{11}\ )(x+iy)^2(33z^4-18z^2r^2+r^4)$
		$\|61\rangle=-(1/2\sqrt{2.11}\ )(x+iy)(33z^5-30z^3r^2+5zr^4)$
		$\|60\rangle=(1/4\sqrt{3.7.11}\ )(231z^6-315z^4r^2+105z^2r^4-5r^6)$
7	$\left(\frac{15}{4\pi}\right)^{1/2}\frac{\sqrt{3.11.13}}{4}r^{-7}$	$\|77\rangle=-(1/8\sqrt{2}\ )(x+iy)^7$
		$\|76\rangle=(\sqrt{7}\ /8)(x+iy)^6z$
		$\|75\rangle=-(\sqrt{7}\ /8\sqrt{2.13}\ )(x+iy)^5(13z^2-r^2)$
		$\|74\rangle=(\sqrt{7}\ /4\sqrt{2.13}\ )(x+iy)^4(13z^3-3zr^2)$
		$\|73\rangle=-(\sqrt{7}\ /8\sqrt{2.11.13}\ )(x+iy)^3(143z^4-66z^2r^2+3r^4)$
		$\|72\rangle=(\sqrt{7}\ /8\sqrt{11.13}\ )(x+iy)^2$ $\times(143z^5-110z^3r^2+15zr^4)$
		$\|71\rangle=-(\sqrt{7}\ /8\sqrt{2.3.11.13}\ )(x+iy)$ $\times(429z^6-495z^4r^2+135z^2r^4-5r^6)$
		$\|70\rangle=(1/4\sqrt{3.11.13}\ )$ $\times(429z^7-693z^5r^2+315z^3r^4-35zr^6)$
8	$\left(\frac{17}{4\pi}\right)^{1/2}\frac{3\sqrt{5.11.13}}{8\sqrt{2}}r^{-8}$	$\|88\rangle=(1/16)(x+iy)^8$
		$\|87\rangle=-(1/4\sqrt{3}\ )(x+iy)^7z$
		$\|86\rangle=(1/4\sqrt{2.3.5}\ )(x+iy)^6(15z^2-r^2)$
		$\|85\rangle=-(7/4\sqrt{5}\ )(x+iy)^5(5z^3-zr^2)$
		$\|84\rangle=\sqrt{7}\ /8\sqrt{5.13}\ )(x+iy)^4(65z^4-26z^2r^2+r^4)$
		$\|83\rangle=-(\sqrt{7}\ /4\sqrt{3.13}\ )(x+iy)^3(39z^5-26z^3r^2+3zr^4)$
		$\|82\rangle=(\sqrt{7}\ /4\sqrt{2.11.13}\ )(x+iy)^2$ $\times(143z^6-143z^4r^2+33z^2r^4-r^6)$
		$\|81\rangle=-(1/4\sqrt{5.11.13}\ )(x+iy)$ $\times(715z^7-1001z^5r^2+385z^3r^4-35zr^6)$
		$\|80\rangle=(1/8.3\sqrt{2.5.11.13}\ )$ $\times(6435z^8-12012z^6r^2+6930z^4r^4-1260z^2r^6+35r^8)$

SO_3–O–D_3–C_3 Partners as JM Partners

$|0\;0\;0\;0\rangle = |0\;0\rangle + 1$

$|\tfrac12\;\tfrac12\;\tfrac12\;\tfrac12\rangle = |\tfrac12\;\tfrac12\rangle + 1$
$|\tfrac12\;\tfrac12\;\tfrac12\,\text{-}\tfrac12\rangle = |\tfrac12\,\text{-}\tfrac12\rangle + 1$

$|1\;1\;\tilde{0}\;0\rangle = |1\;0\rangle + 1$
$|1\;1\;1\;1\rangle = |1\;1\rangle - 1$
$|1\;1\;1\,\text{-}1\rangle = |1\,\text{-}1\rangle - 1$

$|\tfrac32\;\tfrac32\;\tfrac12\;\tfrac12\rangle = |\tfrac32\;\tfrac12\rangle + 1$
$|\tfrac32\;\tfrac32\;\tfrac12\,\text{-}\tfrac12\rangle = |\tfrac32\,\text{-}\tfrac12\rangle - 1$
$|\tfrac32\;\tfrac32\;\tfrac32\;\tfrac32\rangle = |\tfrac32\;\tfrac32\rangle + i/\sqrt{2} + |\tfrac32\,\text{-}\tfrac32\rangle - 1/\sqrt{2}$
$|\tfrac32\;\tfrac32\,\text{-}\tfrac32\;\tfrac32\rangle = |\tfrac32\;\tfrac32\rangle + 1/\sqrt{2} + |\tfrac32\,\text{-}\tfrac32\rangle - i/\sqrt{2}$

$|2\;2\;1\;1\rangle = |2\;1\rangle - \sqrt{2}/\sqrt{3} + |2\,\text{-}2\rangle + 1/\sqrt{3}$
$|2\;2\;1\,\text{-}1\rangle = |2\;2\rangle + 1/\sqrt{3} + |2\,\text{-}1\rangle + \sqrt{2}/\sqrt{3}$
$|2\;\tilde{1}\;0\;0\rangle = |2\;0\rangle - 1$
$|2\;\tilde{1}\;1\;1\rangle = |2\;1\rangle - 1/\sqrt{3} + |2\,\text{-}2\rangle - \sqrt{2}/\sqrt{3}$
$|2\;\tilde{1}\;1\,\text{-}1\rangle = |2\;2\rangle - \sqrt{2}/\sqrt{3} + |2\,\text{-}1\rangle + 1/\sqrt{3}$

$|\tfrac52\;\tfrac32\;\tfrac12\;\tfrac12\rangle = |\tfrac52\;\tfrac12\rangle - 2/3 + |\tfrac52\,\text{-}\tfrac52\rangle + \sqrt{5}/3$
$|\tfrac52\;\tfrac32\;\tfrac12\,\text{-}\tfrac12\rangle = |\tfrac52\;\tfrac52\rangle - \sqrt{5}/3 + |\tfrac52\,\text{-}\tfrac12\rangle - 2/3$
$|\tfrac52\;\tfrac32\;\tfrac32\;\tfrac32\rangle = |\tfrac52\;\tfrac32\rangle + 5/3\sqrt{2.3} + i/3\sqrt{3} + |\tfrac52\,\text{-}\tfrac32\rangle + 1/3\sqrt{3} - 5\,i/3\sqrt{2.3}$
$|\tfrac52\;\tfrac32\,\text{-}\tfrac32\;\tfrac32\rangle = |\tfrac52\;\tfrac32\rangle + 1/3\sqrt{3} + 5\,i/3\sqrt{2.3} + |\tfrac52\,\text{-}\tfrac32\rangle - 5/3\sqrt{2.3} + i/3\sqrt{3}$
$|\tfrac52\;\tilde{\tfrac12}\;\tfrac12\;\tfrac12\rangle = |\tfrac52\;\tfrac12\rangle - \sqrt{5}/3 + |\tfrac52\,\text{-}\tfrac52\rangle - 2/3$
$|\tfrac52\;\tilde{\tfrac12}\;\tfrac12\,\text{-}\tfrac12\rangle = |\tfrac52\;\tfrac52\rangle + 2/3 + |\tfrac52\,\text{-}\tfrac12\rangle - \sqrt{5}/3$

$|3\;1\;\tilde{0}\;0\rangle = |3\;3\rangle + \sqrt{5}/3\sqrt{2} + |3\;0\rangle + 2/3 + |3\,\text{-}3\rangle - \sqrt{5}/3\sqrt{2}$
$|3\;1\;1\;1\rangle = |3\;1\rangle + 1/\sqrt{2.3} + |3\,\text{-}2\rangle - \sqrt{5}/\sqrt{2.3}$
$|3\;1\;1\,\text{-}1\rangle = |3\;2\rangle + \sqrt{5}/\sqrt{2.3} + |3\,\text{-}1\rangle + 1/\sqrt{2.3}$
$|3\;\tilde{1}\;0\;0\rangle = |3\;3\rangle - 1/\sqrt{2} + |3\,\text{-}3\rangle - 1/\sqrt{2}$
$|3\;\tilde{1}\;1\;1\rangle = |3\;1\rangle + \sqrt{5}/\sqrt{2.3} + |3\,\text{-}2\rangle + 1/\sqrt{2.3}$
$|3\;\tilde{1}\;1\,\text{-}1\rangle = |3\;2\rangle - 1/\sqrt{2.3} + |3\,\text{-}1\rangle + \sqrt{5}/\sqrt{2.3}$
$|3\;\tilde{0}\;\tilde{0}\;0\rangle = |3\;3\rangle + \sqrt{2}/3 + |3\;0\rangle - \sqrt{5}/3 + |3\,\text{-}3\rangle - \sqrt{2}/3$

$|\tfrac72\;\tfrac12\;\tfrac12\;\tfrac12\rangle = |\tfrac72\;\tfrac72\rangle + \sqrt{5}/3\sqrt{2.3} + |\tfrac72\;\tfrac12\rangle + \sqrt{7}/3\sqrt{3} + |\tfrac72\,\text{-}\tfrac52\rangle - \sqrt{5.7}/3\sqrt{2.3}$
$|\tfrac72\;\tfrac12\;\tfrac12\,\text{-}\tfrac12\rangle = |\tfrac72\;\tfrac52\rangle - \sqrt{5.7}/3\sqrt{2.3} + |\tfrac72\,\text{-}\tfrac12\rangle - \sqrt{7}/3\sqrt{3} + |\tfrac72\,\text{-}\tfrac72\rangle + \sqrt{5}/3\sqrt{2.3}$
$|\tfrac72\;\tfrac32\;\tfrac12\;\tfrac12\rangle = |\tfrac72\;\tfrac72\rangle + \sqrt{2.7}/3\sqrt{3} + |\tfrac72\;\tfrac12\rangle + \sqrt{5}/3\sqrt{3} + |\tfrac72\,\text{-}\tfrac52\rangle + 2\sqrt{2}/3\sqrt{3}$
$|\tfrac72\;\tfrac32\;\tfrac12\,\text{-}\tfrac12\rangle = |\tfrac72\;\tfrac52\rangle + 2\sqrt{2}/3\sqrt{3} + |\tfrac72\,\text{-}\tfrac12\rangle - \sqrt{5}/3\sqrt{3} + |\tfrac72\,\text{-}\tfrac72\rangle + \sqrt{2.7}/3\sqrt{3}$
$|\tfrac72\;\tfrac32\;\tfrac32\;\tfrac32\rangle = |\tfrac72\;\tfrac32\rangle + 1/\sqrt{3} - i/\sqrt{2.3} + |\tfrac72\,\text{-}\tfrac32\rangle + 1/\sqrt{2.3} + i/\sqrt{3}$
$|\tfrac72\;\tfrac32\,\text{-}\tfrac32\;\tfrac32\rangle = |\tfrac72\;\tfrac32\rangle - 1/\sqrt{2.3} + i/\sqrt{3} + |\tfrac72\,\text{-}\tfrac32\rangle + 1/\sqrt{3} + i/\sqrt{2.3}$
$|\tfrac72\;\tilde{\tfrac12}\;\tfrac12\;\tfrac12\rangle = |\tfrac72\;\tfrac72\rangle - \sqrt{7}/3\sqrt{2} + |\tfrac72\;\tfrac12\rangle + \sqrt{5}/3 + |\tfrac72\,\text{-}\tfrac52\rangle + 1/3\sqrt{2}$
$|\tfrac72\;\tilde{\tfrac12}\;\tfrac12\,\text{-}\tfrac12\rangle = |\tfrac72\;\tfrac52\rangle + 1/3\sqrt{2} + |\tfrac72\,\text{-}\tfrac12\rangle - \sqrt{5}/3 + |\tfrac72\,\text{-}\tfrac72\rangle - \sqrt{7}/3\sqrt{2}$

$|4\;0\;0\;0\rangle = |4\;3\rangle - \sqrt{2.5}/3\sqrt{3} + |4\;0\rangle - \sqrt{7}/3\sqrt{3} + |4\,\text{-}3\rangle + \sqrt{2.5}/3\sqrt{3}$
$|4\;1\;\tilde{0}\;0\rangle = |4\;3\rangle - 1/\sqrt{2} + |4\,\text{-}3\rangle - 1/\sqrt{2}$
$|4\;1\;1\;1\rangle = |4\;4\rangle - \sqrt{2}/3 + |4\;1\rangle - \sqrt{7}/3\sqrt{2} + |4\,\text{-}2\rangle + \sqrt{7}/3\sqrt{2}$
$|4\;1\;1\,\text{-}1\rangle = |4\;2\rangle + \sqrt{7}/3\sqrt{2} + |4\,\text{-}1\rangle + \sqrt{7}/3\sqrt{2} + |4\,\text{-}4\rangle - \sqrt{2}/3$
$|4\;2\;1\;1\rangle = |4\;4\rangle - \sqrt{7}/3\sqrt{3} + |4\;1\rangle - 2/3\sqrt{3} + |4\,\text{-}2\rangle - 4/3\sqrt{3}$
$|4\;2\;1\,\text{-}1\rangle = |4\;2\rangle - 4/3\sqrt{3} + |4\,\text{-}1\rangle + 2/3\sqrt{3} + |4\,\text{-}4\rangle - \sqrt{7}/3\sqrt{3}$

SO_3–O–D_3–C_3 Partners (cont.)

$|4\,\tilde{1}\,0\,0\rangle = |4\,3\rangle+\sqrt{7}/3\sqrt{2.3} + |4\,0\rangle-2\sqrt{5}/3\sqrt{3} + |4\,{-3}\rangle-\sqrt{7}/3\sqrt{2.3}$

$|4\,\tilde{1}\,1\,1\rangle = |4\,4\rangle-\sqrt{2.7}/3\sqrt{3} + |4\,1\rangle+5/3\sqrt{2.3} + |4\,{-2}\rangle+1/3\sqrt{2.3}$

$|4\,\tilde{1}\,1\,{-1}\rangle = |4\,2\rangle+1/3\sqrt{2.3} + |4\,{-1}\rangle-5/3\sqrt{2.3} + |4\,{-4}\rangle-\sqrt{2.7}/3\sqrt{3}$

$|\tfrac{9}{2}\,\tfrac{1}{2}\,\tfrac{1}{2}\,\tfrac{1}{2}\rangle = |\tfrac{9}{2}\,\tfrac{7}{2}\rangle-4/3\sqrt{3} + |\tfrac{9}{2}\,\tfrac{1}{2}\rangle-\sqrt{7}/3\sqrt{3} + |\tfrac{9}{2}\,{-\tfrac{5}{2}}\rangle+2/3\sqrt{3}$

$|\tfrac{9}{2}\,\tfrac{1}{2}\,\tfrac{1}{2}\,{-\tfrac{1}{2}}\rangle = |\tfrac{9}{2}\,\tfrac{5}{2}\rangle-2/3\sqrt{3} + |\tfrac{9}{2}\,{-\tfrac{1}{2}}\rangle-\sqrt{7}/3\sqrt{3} + |\tfrac{9}{2}\,{-\tfrac{7}{2}}\rangle+4/3\sqrt{3}$

$|\tfrac{9}{2}{}_{0}\,\tfrac{3}{2}\,\tfrac{1}{2}\,\tfrac{1}{2}\rangle = |\tfrac{9}{2}\,\tfrac{7}{2}\rangle-4/3\sqrt{5} + |\tfrac{9}{2}\,\tfrac{1}{2}\rangle+2\sqrt{7}/3\sqrt{5} + |\tfrac{9}{2}\,{-\tfrac{5}{2}}\rangle-1/3\sqrt{5}$

$|\tfrac{9}{2}{}_{0}\,\tfrac{3}{2}\,\tfrac{1}{2}\,{-\tfrac{1}{2}}\rangle = |\tfrac{9}{2}\,\tfrac{5}{2}\rangle+1/3\sqrt{5} + |\tfrac{9}{2}\,{-\tfrac{1}{2}}\rangle+2\sqrt{7}/3\sqrt{5} + |\tfrac{9}{2}\,{-\tfrac{7}{2}}\rangle+4/3\sqrt{5}$

$|\tfrac{9}{2}{}_{0}\,\tfrac{3}{2}\,\tfrac{3}{2}\,\tfrac{3}{2}\rangle = |\tfrac{9}{2}\,\tfrac{9}{2}\rangle-2.7/9\sqrt{3.5}+8i\sqrt{2}/9\sqrt{3.5} + |\tfrac{9}{2}\,\tfrac{3}{2}\rangle+\sqrt{5.7}/9\sqrt{2}+i\sqrt{7}/9\sqrt{5}$

$+ |\tfrac{9}{2}\,{-\tfrac{3}{2}}\rangle+\sqrt{7}/9\sqrt{5}-i\sqrt{5.7}/9\sqrt{2} + |\tfrac{9}{2}\,{-\tfrac{9}{2}}\rangle-8\sqrt{2}/9\sqrt{3.5}-2.7i/9\sqrt{3.5}$

$|\tfrac{9}{2}{}_{0}\,\tfrac{3}{2}\,{-\tfrac{3}{2}}\,\tfrac{3}{2}\rangle = |\tfrac{9}{2}\,\tfrac{9}{2}\rangle+8\sqrt{2}/9\sqrt{3.5}-2.7i/9\sqrt{3.5} + |\tfrac{9}{2}\,\tfrac{3}{2}\rangle+\sqrt{7}/9\sqrt{5}+i\sqrt{5.7}/9\sqrt{2}$

$+ |\tfrac{9}{2}\,{-\tfrac{3}{2}}\rangle-\sqrt{5.7}/9\sqrt{2}+i\sqrt{7}/9\sqrt{5} + |\tfrac{9}{2}\,{-\tfrac{9}{2}}\rangle-2.7/9\sqrt{3.5}-8i\sqrt{2}/9\sqrt{3.5}$

$|\tfrac{9}{2}{}_{1}\,\tfrac{3}{2}\,\tfrac{1}{2}\,\tfrac{1}{2}\rangle = |\tfrac{9}{2}\,\tfrac{7}{2}\rangle-\sqrt{7}/3\sqrt{3.5} + |\tfrac{9}{2}\,\tfrac{1}{2}\rangle-4/3\sqrt{3.5} + |\tfrac{9}{2}\,{-\tfrac{5}{2}}\rangle-4\sqrt{7}/3\sqrt{3.5}$

$|\tfrac{9}{2}{}_{1}\,\tfrac{3}{2}\,\tfrac{1}{2}\,{-\tfrac{1}{2}}\rangle = |\tfrac{9}{2}\,\tfrac{5}{2}\rangle+4\sqrt{7}/3\sqrt{3.5} + |\tfrac{9}{2}\,{-\tfrac{1}{2}}\rangle-4/3\sqrt{3.5} + |\tfrac{9}{2}\,{-\tfrac{7}{2}}\rangle+\sqrt{7}/3\sqrt{3.5}$

$|\tfrac{9}{2}{}_{1}\,\tfrac{3}{2}\,\tfrac{3}{2}\,\tfrac{3}{2}\rangle = |\tfrac{9}{2}\,\tfrac{9}{2}\rangle+\sqrt{7}/3\sqrt{5}+i\sqrt{7}/3\sqrt{2.5} + |\tfrac{9}{2}\,\tfrac{3}{2}\rangle+2i/\sqrt{3.5} + |\tfrac{9}{2}\,{-\tfrac{3}{2}}\rangle+2/\sqrt{3.5}$

$+ |\tfrac{9}{2}\,{-\tfrac{9}{2}}\rangle-\sqrt{7}/3\sqrt{2.5}+i\sqrt{7}/3\sqrt{5}$

$|\tfrac{9}{2}{}_{1}\,\tfrac{3}{2}\,{-\tfrac{3}{2}}\,\tfrac{3}{2}\rangle = |\tfrac{9}{2}\,\tfrac{9}{2}\rangle+\sqrt{7}/3\sqrt{2.5}+i\sqrt{7}/3\sqrt{5} + |\tfrac{9}{2}\,\tfrac{3}{2}\rangle+2/\sqrt{3.5} + |\tfrac{9}{2}\,{-\tfrac{3}{2}}\rangle+2i/\sqrt{3.5}$

$+ |\tfrac{9}{2}\,{-\tfrac{9}{2}}\rangle+\sqrt{7}/3\sqrt{5}-i\sqrt{7}/3\sqrt{2.5}$

$|5{}_{0}\,1\,\tilde{0}\,0\rangle = |5\,3\rangle-\sqrt{5.7}/3\sqrt{2.11} + |5\,0\rangle+8/3\sqrt{11} + |5\,{-3}\rangle+\sqrt{5.7}/3\sqrt{2.11}$

$|5{}_{0}\,1\,1\,1\rangle = |5\,4\rangle-\sqrt{5.7}/9\sqrt{2.11} + |5\,1\rangle+4\sqrt{5}/3\sqrt{3.11} + |5\,{-2}\rangle+\sqrt{2.5.7}/3\sqrt{3.11}$

$+ |5\,{-5}\rangle+\sqrt{7.11}/9\sqrt{2}$

$|5{}_{0}\,1\,1\,{-1}\rangle = |5\,5\rangle+\sqrt{7.11}/9\sqrt{2} + |5\,2\rangle-\sqrt{2.5.7}/3\sqrt{3.11} + |5\,{-1}\rangle+4\sqrt{5}/3\sqrt{3.11}$

$+ |5\,{-4}\rangle+\sqrt{5.7}/9\sqrt{2.11}$

$|5{}_{1}\,1\,\tilde{0}\,0\rangle = |5\,3\rangle-4\sqrt{2}/3\sqrt{11} + |5\,0\rangle-\sqrt{5.7}/3\sqrt{11} + |5\,{-3}\rangle+4\sqrt{2}/3\sqrt{11}$

$|5{}_{1}\,1\,1\,1\rangle = |5\,4\rangle+2\sqrt{2}/\sqrt{11} + |5\,1\rangle+\sqrt{7}/\sqrt{3.11} + |5\,{-2}\rangle-\sqrt{2}/\sqrt{3.11}$

$|5{}_{1}\,1\,1\,{-1}\rangle = |5\,2\rangle+\sqrt{2}/\sqrt{3.11} + |5\,{-1}\rangle+\sqrt{7}/\sqrt{3.11} + |5\,{-4}\rangle-2\sqrt{2}/\sqrt{11}$

$|5\,2\,1\,1\rangle = |5\,4\rangle+4/9 + |5\,1\rangle-\sqrt{2.7}/3\sqrt{3} + |5\,{-2}\rangle+1/3\sqrt{3} + |5\,{-5}\rangle+2\sqrt{5}/9$

$|5\,2\,1\,{-1}\rangle = |5\,5\rangle+2\sqrt{5}/9 + |5\,2\rangle-1/3\sqrt{3} + |5\,{-1}\rangle-\sqrt{2.7}/3\sqrt{3} + |5\,{-4}\rangle-4/9$

$|5\,\tilde{1}\,0\,0\rangle = |5\,3\rangle+1/\sqrt{2} + |5\,{-3}\rangle+1/\sqrt{2}$

$|5\,\tilde{1}\,1\,1\rangle = |5\,4\rangle+1/3\sqrt{2} + |5\,{-2}\rangle+\sqrt{2}/\sqrt{3} + |5\,{-5}\rangle-\sqrt{5}/3\sqrt{2}$

$|5\,\tilde{1}\,1\,{-1}\rangle = |5\,5\rangle-\sqrt{5}/3\sqrt{2} + |5\,2\rangle-\sqrt{2}/\sqrt{3} + |5\,{-4}\rangle-1/3\sqrt{2}$

$|\tfrac{11}{2}\,\tfrac{1}{2}\,\tfrac{1}{2}\,\tfrac{1}{2}\rangle = |\tfrac{11}{2}\,\tfrac{7}{2}\rangle-\sqrt{5.7}/9\sqrt{2.3} + |\tfrac{11}{2}\,\tfrac{1}{2}\rangle+4\sqrt{2}/9 + |\tfrac{11}{2}\,{-\tfrac{5}{2}}\rangle+\sqrt{5.7}/9\sqrt{2}$

$+ |\tfrac{11}{2}\,{-\tfrac{11}{2}}\rangle+\sqrt{7.11}/9\sqrt{3}$

$|\tfrac{11}{2}\,\tfrac{1}{2}\,\tfrac{1}{2}\,{-\tfrac{1}{2}}\rangle = |\tfrac{11}{2}\,\tfrac{11}{2}\rangle-\sqrt{7.11}/9\sqrt{3} + |\tfrac{11}{2}\,\tfrac{5}{2}\rangle+\sqrt{5.7}/9\sqrt{2} + |\tfrac{11}{2}\,{-\tfrac{1}{2}}\rangle-4\sqrt{2}/9$

$+ |\tfrac{11}{2}\,{-\tfrac{7}{2}}\rangle-\sqrt{5.7}/9\sqrt{2.3}$

$|\tfrac{11}{2}{}_{0}\,\tfrac{3}{2}\,\tfrac{1}{2}\,\tfrac{1}{2}\rangle = |\tfrac{11}{2}\,\tfrac{7}{2}\rangle+8\sqrt{2.5}/9\sqrt{3.11} + |\tfrac{11}{2}\,\tfrac{1}{2}\rangle-4\sqrt{2.7}/9\sqrt{11} + |\tfrac{11}{2}\,{-\tfrac{5}{2}}\rangle+\sqrt{2.5}/9\sqrt{11}$

$+ |\tfrac{11}{2}\,{-\tfrac{11}{2}}\rangle+11/9\sqrt{3}$

$|\tfrac{11}{2}{}_{0}\,\tfrac{3}{2}\,\tfrac{1}{2}\,{-\tfrac{1}{2}}\rangle = |\tfrac{11}{2}\,\tfrac{11}{2}\rangle-11/9\sqrt{3} + |\tfrac{11}{2}\,\tfrac{5}{2}\rangle+\sqrt{2.5}/9\sqrt{11} + |\tfrac{11}{2}\,{-\tfrac{1}{2}}\rangle+4\sqrt{2.7}/9\sqrt{11}$

$+ |\tfrac{11}{2}\,{-\tfrac{7}{2}}\rangle+8\sqrt{2.5}/9\sqrt{3.11}$

$|\tfrac{11}{2}{}_{0}\,\tfrac{3}{2}\,\tfrac{3}{2}\,\tfrac{3}{2}\rangle = |\tfrac{11}{2}\,\tfrac{9}{2}\rangle-\sqrt{11}/9\sqrt{2}-5i/9\sqrt{11} + |\tfrac{11}{2}\,\tfrac{3}{2}\rangle+2\sqrt{2.5}/3\sqrt{3.11}+4i\sqrt{5}/3\sqrt{3.11}$

$+ |\tfrac{11}{2}\,{-\tfrac{3}{2}}\rangle-4\sqrt{5}/3\sqrt{3.11}+2i\sqrt{2.5}/3\sqrt{3.11} + |\tfrac{11}{2}\,{-\tfrac{9}{2}}\rangle-5/9\sqrt{11}+i\sqrt{11}/9\sqrt{2}$

$|\tfrac{11}{2}{}_{0}\,\tfrac{3}{2}\,{-\tfrac{3}{2}}\,\tfrac{3}{2}\rangle = |\tfrac{11}{2}\,\tfrac{9}{2}\rangle-5/9\sqrt{11}-i\sqrt{11}/9\sqrt{2} + |\tfrac{11}{2}\,\tfrac{3}{2}\rangle+4\sqrt{5}/3\sqrt{3.11}+2i\sqrt{2.5}/3\sqrt{3.11}$

$+ |\tfrac{11}{2}\,{-\tfrac{3}{2}}\rangle+2\sqrt{2.5}/3\sqrt{3.11}-4i\sqrt{5}/3\sqrt{3.11} + |\tfrac{11}{2}\,{-\tfrac{9}{2}}\rangle+\sqrt{11}/9\sqrt{2}-5i/9\sqrt{11}$

$|\tfrac{11}{2}{}_{1}\,\tfrac{3}{2}\,\tfrac{1}{2}\,\tfrac{1}{2}\rangle = |\tfrac{11}{2}\,\tfrac{7}{2}\rangle-4/\sqrt{3.11} + |\tfrac{11}{2}\,\tfrac{1}{2}\rangle-\sqrt{5.7}/3\sqrt{11} + |\tfrac{11}{2}\,{-\tfrac{5}{2}}\rangle+4/3\sqrt{11}$

$|\tfrac{11}{2}{}_{1}\,\tfrac{3}{2}\,\tfrac{1}{2}\,{-\tfrac{1}{2}}\rangle = |\tfrac{11}{2}\,\tfrac{5}{2}\rangle+4/3\sqrt{11} + |\tfrac{11}{2}\,{-\tfrac{1}{2}}\rangle+\sqrt{5.7}/3\sqrt{11} + |\tfrac{11}{2}\,{-\tfrac{7}{2}}\rangle-4/\sqrt{3.11}$

$|\tfrac{11}{2}{}_{1}\,\tfrac{3}{2}\,\tfrac{3}{2}\,\tfrac{3}{2}\rangle = |\tfrac{11}{2}\,\tfrac{9}{2}\rangle-2i\sqrt{2.5}/3\sqrt{11} + |\tfrac{11}{2}\,\tfrac{3}{2}\rangle+2/3\sqrt{3.11}-7i/3\sqrt{2.3.11}$

$+ |\tfrac{11}{2}\,{-\tfrac{3}{2}}\rangle+7/3\sqrt{2.3.11}+2i/3\sqrt{3.11} + |\tfrac{11}{2}\,{-\tfrac{9}{2}}\rangle-2\sqrt{2.5}/3\sqrt{11}$

$|\tfrac{11}{2}{}_{1}\,\tfrac{3}{2}\,{-\tfrac{3}{2}}\,\tfrac{3}{2}\rangle = |\tfrac{11}{2}\,\tfrac{9}{2}\rangle-2\sqrt{2.5}/3\sqrt{11} + |\tfrac{11}{2}\,\tfrac{3}{2}\rangle-7/3\sqrt{2.3.11}+2i/3\sqrt{3.11}$

$+ |\tfrac{11}{2}\,{-\tfrac{3}{2}}\rangle+2/3\sqrt{3.11}+7i/3\sqrt{2.3.11} + |\tfrac{11}{2}\,{-\tfrac{9}{2}}\rangle-2i\sqrt{2.5}/3\sqrt{11}$

SO_3–O–D_3–C_3 Partners (cont.)

$|\tfrac{11}{2}\ \tilde{\tfrac{1}{2}}\ \tfrac{1}{2}\ \tfrac{1}{2}> \ = \ |\tfrac{11}{2}\ \tfrac{7}{2}>+\sqrt{11}/3\sqrt{2.3} \ + \ |\tfrac{11}{2}\,\text{-}\tfrac{5}{2}>+\sqrt{11}/3\sqrt{2} \ + \ |\tfrac{11}{2}\,\text{-}\tfrac{11}{2}>-\sqrt{5}/3\sqrt{3}$

$|\tfrac{11}{2}\ \tilde{\tfrac{1}{2}}\ \tfrac{1}{2}\,\text{-}\tfrac{1}{2}> \ = \ |\tfrac{11}{2}\ \tfrac{11}{2}>+\sqrt{5}/3\sqrt{3} \ + \ |\tfrac{11}{2}\ \tfrac{5}{2}>+\sqrt{11}/3\sqrt{2} \ + \ |\tfrac{11}{2}\,\text{-}\tfrac{7}{2}>+\sqrt{11}/3\sqrt{2.3}$

$|\,6\ 0\ 0\ 0> \ = \ |\,6\ 6>-\sqrt{7.11}/9\sqrt{2.3} \ + \ |\,6\ 3>+\sqrt{5.7}/9\sqrt{3} \ + \ |\,6\ 0>-4\sqrt{2}/9$

$\qquad + \ |\,6\,\text{-}3>-\sqrt{5.7}/9\sqrt{3} \ + \ |\,6\,\text{-}6>-\sqrt{7.11}/9\sqrt{2.3}$

$|\,6\ 1\ \tilde{0}\ 0> \ = \ |\,6\ 6>-\sqrt{11}/3\sqrt{3} \ + \ |\,6\ 3>+\sqrt{5}/3\sqrt{2.3} \ + \ |\,6\,\text{-}3>+\sqrt{5}/3\sqrt{2.3}$

$\qquad + \ |\,6\,\text{-}6>+\sqrt{11}/3\sqrt{3}$

$|\,6\ 1\ 1\ 1> \ = \ |\,6\ 4>+5/9\sqrt{2} \ + \ |\,6\ 1>-4/3\sqrt{3} \ + \ |\,6\,\text{-}2>-\sqrt{5}/3\sqrt{3} \ + \ |\,6\,\text{-}5>-\sqrt{11}/9\sqrt{2}$

$|\,6\ 1\ 1\,\text{-}1> \ = \ |\,6\ 5>+\sqrt{11}/9\sqrt{2} \ + \ |\,6\ 2>-\sqrt{5}/3\sqrt{3} \ + \ |\,6\,\text{-}1>+4/3\sqrt{3} \ + \ |\,6\,\text{-}4>+5/9\sqrt{2}$

$|\,6\ 2\ 1\ 1> \ = \ |\,6\ 4>-4\sqrt{2}/9 \ + \ |\,6\ 1>-2/3\sqrt{3} \ + \ |\,6\,\text{-}2>+\sqrt{5}/3\sqrt{3} \ + \ |\,6\,\text{-}5>-\sqrt{2.11}/9$

$|\,6\ 2\ 1\,\text{-}1> \ = \ |\,6\ 5>+\sqrt{2.11}/9 \ + \ |\,6\ 2>+\sqrt{5}/3\sqrt{3} \ + \ |\,6\,\text{-}1>+2/3\sqrt{3} \ + \ |\,6\,\text{-}4>-4\sqrt{2}/9$

$|\,6\ _0\ \tilde{1}\ 0\ 0> \ = \ |\,6\ 6>+\sqrt{5}/\sqrt{3.7} \ + \ |\,6\ 3>+\sqrt{11}/\sqrt{2.3.7} \ + \ |\,6\,\text{-}3>-\sqrt{11}/\sqrt{2.3.7}$

$\qquad + \ |\,6\,\text{-}6>+\sqrt{5}/\sqrt{3.7}$

$|\,6\ _0\ \tilde{1}\ 1\ 1> \ = \ |\,6\ 4>+\sqrt{5.11}/3\sqrt{2.7} \ + \ |\,6\,\text{-}2>+\sqrt{11}/\sqrt{3.7} \ + \ |\,6\,\text{-}5>-\sqrt{5}/3\sqrt{2.7}$

$|\,6\ _0\ \tilde{1}\ 1\,\text{-}1> \ = \ |\,6\ 5>+\sqrt{5}/3\sqrt{2.7} \ + \ |\,6\ 2>+\sqrt{11}/\sqrt{3.7} \ + \ |\,6\,\text{-}4>+\sqrt{5.11}/3\sqrt{2.7}$

$|\,6\ _1\ \tilde{1}\ 0\ 0> \ = \ |\,6\ 6>-4\sqrt{11}/9\sqrt{3.7} \ + \ |\,6\ 3>+4\sqrt{2.5}/9\sqrt{3.7} \ + \ |\,6\ 0>+7/9$

$\qquad + \ |\,6\,\text{-}3>-4\sqrt{2.5}/9\sqrt{3.7} \ + \ |\,6\,\text{-}6>-4\sqrt{11}/9\sqrt{3.7}$

$|\,6\ _1\ \tilde{1}\ 1\ 1> \ = \ |\,6\ 4>+2\sqrt{2}/9\sqrt{7} \ + \ |\,6\ 1>+\sqrt{7}/3\sqrt{3} \ + \ |\,6\,\text{-}2>-2\sqrt{5}/3\sqrt{3.7}$

$\qquad + \ |\,6\,\text{-}5>-4\sqrt{2.11}/9\sqrt{7}$

$|\,6\ _1\ \tilde{1}\ 1\,\text{-}1> \ = \ |\,6\ 5>+4\sqrt{2.11}/9\sqrt{7} \ + \ |\,6\ 2>-2\sqrt{5}/3\sqrt{3.7} \ + \ |\,6\,\text{-}1>-\sqrt{7}/3\sqrt{3}$

$\qquad + \ |\,6\,\text{-}4>+2\sqrt{2}/9\sqrt{7}$

$|\,6\ \tilde{0}\ \tilde{0}\ 0> \ = \ |\,6\ 6>+\sqrt{5}/3\sqrt{2.3} \ + \ |\,6\ 3>+\sqrt{11}/3\sqrt{3} \ + \ |\,6\,\text{-}3>+\sqrt{11}/3\sqrt{3}$

$\qquad + \ |\,6\,\text{-}6>-\sqrt{5}/3\sqrt{2.3}$

$|\tfrac{13}{2}\ \tfrac{1}{2}\ \tfrac{1}{2}\ \tfrac{1}{2}> \ = \ |\tfrac{13}{2}\ \tfrac{13}{2}>-\sqrt{11.13}/9\sqrt{2.3} \ + \ |\tfrac{13}{2}\ \tfrac{7}{2}>+5\sqrt{2}/9\sqrt{3} \ + \ |\tfrac{13}{2}\ \tfrac{1}{2}>-4\sqrt{2}/9$

$\qquad + \ |\tfrac{13}{2}\,\text{-}\tfrac{5}{2}>-2\sqrt{5}/9\sqrt{3} \ + \ |\tfrac{13}{2}\,\text{-}\tfrac{11}{2}>-\sqrt{11}/9\sqrt{2.3}$

$|\tfrac{13}{2}\ \tfrac{1}{2}\ \tfrac{1}{2}\,\text{-}\tfrac{1}{2}> \ = \ |\tfrac{13}{2}\ \tfrac{11}{2}>-\sqrt{11}/9\sqrt{2.3} \ + \ |\tfrac{13}{2}\ \tfrac{5}{2}>+2\sqrt{5}/9\sqrt{3} \ + \ |\tfrac{13}{2}\,\text{-}\tfrac{1}{2}>-4\sqrt{2}/9$

$\qquad + \ |\tfrac{13}{2}\,\text{-}\tfrac{7}{2}>-5\sqrt{2}/9\sqrt{3} \ + \ |\tfrac{13}{2}\,\text{-}\tfrac{13}{2}>-\sqrt{11.13}/9\sqrt{2.3}$

$|\tfrac{13}{2}\,0\ \tfrac{3}{2}\ \tfrac{1}{2}\ \tfrac{1}{2}> \ = \ |\tfrac{13}{2}\ \tfrac{13}{2}>-2\sqrt{13}/3\sqrt{3.7} \ + \ |\tfrac{13}{2}\ \tfrac{7}{2}>-\sqrt{11}/3\sqrt{3.7} \ + \ |\tfrac{13}{2}\,\text{-}\tfrac{5}{2}>+\sqrt{2.5.11}/3\sqrt{3.7}$

$\qquad + \ |\tfrac{13}{2}\,\text{-}\tfrac{11}{2}>-4/3\sqrt{3.7}$

$|\tfrac{13}{2}\,0\ \tfrac{3}{2}\ \tfrac{1}{2}\,\text{-}\tfrac{1}{2}> \ = \ |\tfrac{13}{2}\ \tfrac{11}{2}>-4/3\sqrt{3.7} \ + \ |\tfrac{13}{2}\ \tfrac{5}{2}>-\sqrt{2.5.11}/3\sqrt{3.7} \ + \ |\tfrac{13}{2}\,\text{-}\tfrac{7}{2}>+\sqrt{11}/3\sqrt{3.7}$

$\qquad + \ |\tfrac{13}{2}\,\text{-}\tfrac{13}{2}>-2\sqrt{13}/3\sqrt{3.7}$

$|\tfrac{13}{2}\,0\ \tfrac{3}{2}\ \tfrac{3}{2}\ \tfrac{3}{2}> \ = \ |\tfrac{13}{2}\ \tfrac{9}{2}>+23/9\sqrt{3.7}-i\sqrt{7}/9\sqrt{2.3} \ + \ |\tfrac{13}{2}\ \tfrac{3}{2}>+\sqrt{11}/9\sqrt{7}+2i\sqrt{2.11}/9\sqrt{7}$

$\qquad + \ |\tfrac{13}{2}\,\text{-}\tfrac{3}{2}>+2\sqrt{2.11}/9\sqrt{7}-i\sqrt{11}/9\sqrt{7} \ + \ |\tfrac{13}{2}\,\text{-}\tfrac{9}{2}>+\sqrt{7}/9\sqrt{2.3}+23i/9\sqrt{3.7}$

$|\tfrac{13}{2}\,0\ \tfrac{3}{2}\,\text{-}\tfrac{3}{2}\ \tfrac{3}{2}> \ = \ |\tfrac{13}{2}\ \tfrac{9}{2}>-\sqrt{7}/9\sqrt{2.3}+23i/9\sqrt{3.7} \ + \ |\tfrac{13}{2}\ \tfrac{3}{2}>+2\sqrt{2.11}/9\sqrt{7}+i\sqrt{11}/9\sqrt{7}$

$\qquad + \ |\tfrac{13}{2}\,\text{-}\tfrac{3}{2}>-\sqrt{11}/9\sqrt{7}+2i\sqrt{2.11}/9\sqrt{7} \ + \ |\tfrac{13}{2}\,\text{-}\tfrac{9}{2}>+23/9\sqrt{3.7}+i\sqrt{7}/9\sqrt{2.3}$

$|\tfrac{13}{2}\,1\ \tfrac{3}{2}\ \tfrac{1}{2}\ \tfrac{1}{2}> \ = \ |\tfrac{13}{2}\ \tfrac{13}{2}>+2\sqrt{2.11.13}/9.3\sqrt{7} \ + \ |\tfrac{13}{2}\ \tfrac{7}{2}>-17\sqrt{2}/9.3\sqrt{7} \ + \ |\tfrac{13}{2}\ \tfrac{1}{2}>-2\sqrt{2.7}/9\sqrt{3}$

$\qquad + \ |\tfrac{13}{2}\,\text{-}\tfrac{5}{2}>-\sqrt{5}/9.3\sqrt{7} \ + \ |\tfrac{13}{2}\,\text{-}\tfrac{11}{2}>-2.5\sqrt{2.11}/9.3\sqrt{7}$

$|\tfrac{13}{2}\,1\ \tfrac{3}{2}\ \tfrac{1}{2}\,\text{-}\tfrac{1}{2}> \ = \ |\tfrac{13}{2}\ \tfrac{11}{2}>-2.5\sqrt{2.11}/9.3\sqrt{7} \ + \ |\tfrac{13}{2}\ \tfrac{5}{2}>+\sqrt{5}/9.3\sqrt{7} \ + \ |\tfrac{13}{2}\,\text{-}\tfrac{1}{2}>-2\sqrt{2.7}/9\sqrt{3}$

$\qquad + \ |\tfrac{13}{2}\,\text{-}\tfrac{7}{2}>+17\sqrt{2}/9.3\sqrt{7} \ + \ |\tfrac{13}{2}\,\text{-}\tfrac{13}{2}>+2\sqrt{2.11.13}/9.3\sqrt{7}$

$|\tfrac{13}{2}\,1\ \tfrac{3}{2}\ \tfrac{3}{2}\ \tfrac{3}{2}> \ = \ |\tfrac{13}{2}\ \tfrac{9}{2}>+4\sqrt{2.11}/9.3\sqrt{7}+i\sqrt{7.11}/9.3 \ + \ |\tfrac{13}{2}\ \tfrac{3}{2}>+23/9\sqrt{2.3.7}-17i/9\sqrt{3.7}$

$\qquad + \ |\tfrac{13}{2}\,\text{-}\tfrac{3}{2}>-17/9\sqrt{3.7}-23i/9\sqrt{2.3.7} \ + \ |\tfrac{13}{2}\,\text{-}\tfrac{9}{2}>-\sqrt{7.11}/9.3+4i\sqrt{2.11}/9.3\sqrt{7}$

$|\tfrac{13}{2}\,1\ \tfrac{3}{2}\,\text{-}\tfrac{3}{2}\ \tfrac{3}{2}> \ = \ |\tfrac{13}{2}\ \tfrac{9}{2}>+\sqrt{7.11}/9.3+4i\sqrt{2.11}/9.3\sqrt{7} \ + \ |\tfrac{13}{2}\ \tfrac{3}{2}>-17/9\sqrt{3.7}+23i/9\sqrt{2.3.7}$

$\qquad + \ |\tfrac{13}{2}\,\text{-}\tfrac{3}{2}>-23/9\sqrt{2.3.7}-17i/9\sqrt{3.7} \ + \ |\tfrac{13}{2}\,\text{-}\tfrac{9}{2}>+4\sqrt{2.11}/9.3\sqrt{7}-i\sqrt{7.11}/9.3$

$|\tfrac{13}{2}\,0\ \tilde{\tfrac{1}{2}}\ \tfrac{1}{2}\ \tfrac{1}{2}> \ = \ |\tfrac{13}{2}\ \tfrac{13}{2}>+\sqrt{5.13}/3\sqrt{2.3.7} \ + \ |\tfrac{13}{2}\ \tfrac{7}{2}>+\sqrt{2.5.11}/3\sqrt{3.7} \ + \ |\tfrac{13}{2}\,\text{-}\tfrac{5}{2}>+2\sqrt{11}/3\sqrt{3.7}$

$\qquad + \ |\tfrac{13}{2}\,\text{-}\tfrac{11}{2}>-\sqrt{5}/3\sqrt{2.3.7}$

$|\tfrac{13}{2}\,0\ \tilde{\tfrac{1}{2}}\ \tfrac{1}{2}\,\text{-}\tfrac{1}{2}> \ = \ |\tfrac{13}{2}\ \tfrac{11}{2}>-\sqrt{5}/3\sqrt{2.3.7} \ + \ |\tfrac{13}{2}\ \tfrac{5}{2}>-2\sqrt{11}/3\sqrt{3.7} \ + \ |\tfrac{13}{2}\,\text{-}\tfrac{7}{2}>-\sqrt{2.5.11}/3\sqrt{3.7}$

$\qquad + \ |\tfrac{13}{2}\,\text{-}\tfrac{13}{2}>+\sqrt{5.13}/3\sqrt{2.3.7}$

$|\tfrac{13}{2}\,1\ \tilde{\tfrac{1}{2}}\ \tfrac{1}{2}\ \tfrac{1}{2}> \ = \ |\tfrac{13}{2}\ \tfrac{13}{2}>-4\sqrt{11}/9.3\sqrt{7} \ + \ |\tfrac{13}{2}\ \tfrac{7}{2}>+4\sqrt{13}/9.3\sqrt{7} \ + \ |\tfrac{13}{2}\ \tfrac{1}{2}>+\sqrt{7.13}/9\sqrt{3}$

$\qquad + \ |\tfrac{13}{2}\,\text{-}\tfrac{5}{2}>-2\sqrt{2.5.13}/9.3\sqrt{7} \ + \ |\tfrac{13}{2}\,\text{-}\tfrac{11}{2}>-4\sqrt{11.13}/9.3\sqrt{7}$

SO_3–O–D_3–C_3 Partners (cont.)

$|\tfrac{13}{2}\,{}_1\,\tilde{\tfrac{1}{2}}\;\tfrac{1}{2}\,\text{-}\tfrac{1}{2}\rangle = |\tfrac{13}{2}\,\tfrac{11}{2}\rangle-4\sqrt{11.13}/9.3\sqrt{7} + |\tfrac{13}{2}\,\tfrac{5}{2}\rangle+2\sqrt{2.5.13}/9.3\sqrt{7} + |\tfrac{13}{2}\,\text{-}\tfrac{1}{2}\rangle+\sqrt{7.13}/9\sqrt{3}$
$\quad + |\tfrac{13}{2}\,\text{-}\tfrac{7}{2}\rangle-4\sqrt{13}/9.3\sqrt{7} + |\tfrac{13}{2}\,\text{-}\tfrac{13}{2}\rangle-4\sqrt{11}/9.3\sqrt{7}$

$|7\,{}_0\,1\,\tilde{0}\;0\rangle = |7\;6\rangle-\sqrt{11.13}/9\sqrt{2.3.5} + |7\;3\rangle+2\sqrt{2.5}/9\sqrt{3} + |7\;0\rangle-4\sqrt{2.7}/9\sqrt{5}$
$\quad + |7\,\text{-}3\rangle-2\sqrt{2.5}/9\sqrt{3} + |7\,\text{-}6\rangle-\sqrt{11.13}/9\sqrt{2.3.5}$

$|7\,{}_0\,1\;1\;1\rangle = |7\;7\rangle+\sqrt{7.11.13}/9\sqrt{2.3.5} + |7\;4\rangle-\sqrt{5.11}/9\sqrt{3} + |7\;1\rangle+8\sqrt{2}/9\sqrt{5}$
$\quad + |7\,\text{-}2\rangle+\sqrt{2.5}/9\sqrt{3} + |7\,\text{-}5\rangle+\sqrt{11}/9\sqrt{2.3.5}$

$|7\,{}_0\,1\;1\,\text{-}1\rangle = |7\;5\rangle+\sqrt{11}/9\sqrt{2.3.5} + |7\;2\rangle-\sqrt{2.5}/9\sqrt{3} + |7\,\text{-}1\rangle+8\sqrt{2}/9\sqrt{5}$
$\quad + |7\,\text{-}4\rangle+\sqrt{5.11}/9\sqrt{3} + |7\,\text{-}7\rangle+\sqrt{7.11.13}/9\sqrt{2.3.5}$

$|7\,{}_1\,1\,\tilde{0}\;0\rangle = |7\;6\rangle+2\sqrt{2.7.13}/9.3\sqrt{5} + |7\;3\rangle+\sqrt{5.7.11}/9.3\sqrt{2}$
$\quad + |7\;0\rangle+2\sqrt{2.11}/9\sqrt{3.5} + |7\,\text{-}3\rangle-\sqrt{5.7.11}/9.3\sqrt{2}$
$\quad + |7\,\text{-}6\rangle+2\sqrt{2.7.13}/9.3\sqrt{5}$

$|7\,{}_1\,1\;1\;1\rangle = |7\;7\rangle+\sqrt{2.13}/9.3\sqrt{5} + |7\;4\rangle+\sqrt{5.7}/9.3 + |7\;1\rangle+\sqrt{7.11}/9\sqrt{2.3.5}$
$\quad + |7\,\text{-}2\rangle-\sqrt{5.7.11}/9.3\sqrt{2} + |7\,\text{-}5\rangle+13\sqrt{2.7}/9.3\sqrt{5}$

$|7\,{}_1\,1\;1\,\text{-}1\rangle = |7\;5\rangle+13\sqrt{2.7}/9.3\sqrt{5} + |7\;2\rangle+\sqrt{5.7.11}/9.3\sqrt{2} + |7\,\text{-}1\rangle+\sqrt{7.11}/9\sqrt{2.3.5}$
$\quad + |7\,\text{-}4\rangle-\sqrt{5.7}/9.3 + |7\,\text{-}7\rangle+\sqrt{2.13}/9.3\sqrt{5}$

$|7\;2\;1\;1\rangle = |7\;7\rangle+\sqrt{2.7.13}/9.3 + |7\;4\rangle-1/9.3 + |7\;1\rangle-\sqrt{2.11}/9\sqrt{3}$
$\quad + |7\,\text{-}2\rangle-4\sqrt{2.11}/9.3 + |7\,\text{-}5\rangle-8\sqrt{2}/9.3$

$|7\;2\;1\,\text{-}1\rangle = |7\;5\rangle-8\sqrt{2}/9.3 + |7\;2\rangle+4\sqrt{2.11}/9.3 + |7\,\text{-}1\rangle-\sqrt{2.11}/9\sqrt{3}$
$\quad + |7\,\text{-}4\rangle+1/9.3 + |7\,\text{-}7\rangle+\sqrt{2.7.13}/9.3$

$|7\,{}_0\,\tilde{1}\;0\;0\rangle = |7\;6\rangle-\sqrt{13}/3\sqrt{2.3.7} + |7\;3\rangle-2\sqrt{2.11}/3\sqrt{3.7} + |7\,\text{-}3\rangle-2\sqrt{2.11}/3\sqrt{3.7}$
$\quad + |7\,\text{-}6\rangle+\sqrt{13}/3\sqrt{2.3.7}$

$|7\,{}_0\,\tilde{1}\;1\;1\rangle = |7\;7\rangle+\sqrt{13}/3\sqrt{2.3} + |7\;4\rangle+11/3\sqrt{3.7} + |7\,\text{-}2\rangle+\sqrt{2.11}/3\sqrt{3.7}$
$\quad + |7\,\text{-}5\rangle-1/3\sqrt{2.3.7}$

$|7\,{}_0\,\tilde{1}\;1\,\text{-}1\rangle = |7\;5\rangle-1/3\sqrt{2.3.7} + |7\;2\rangle-\sqrt{2.11}/3\sqrt{3.7} + |7\,\text{-}4\rangle-11/3\sqrt{3.7}$
$\quad + |7\,\text{-}7\rangle+\sqrt{13}/3\sqrt{2.3}$

$|7\,{}_1\,\tilde{1}\;0\;0\rangle = |7\;6\rangle-2\sqrt{2.11}/3\sqrt{3.7} + |7\;3\rangle+\sqrt{13}/3\sqrt{2.3.7} + |7\,\text{-}3\rangle+\sqrt{13}/3\sqrt{2.3.7}$
$\quad + |7\,\text{-}6\rangle+2\sqrt{2.11}/3\sqrt{3.7}$

$|7\,{}_1\,\tilde{1}\;1\;1\rangle = |7\;7\rangle-\sqrt{2.11}/9\sqrt{3} + |7\;4\rangle+\sqrt{11.13}/9\sqrt{3.7} + |7\;1\rangle+\sqrt{7.13}/9\sqrt{2}$
$\quad + |7\,\text{-}2\rangle-5\sqrt{13}/9\sqrt{2.3.7} + |7\,\text{-}5\rangle-\sqrt{2.11.13}/9\sqrt{3.7}$

$|7\,{}_1\,\tilde{1}\;1\,\text{-}1\rangle = |7\;5\rangle-\sqrt{2.11.13}/9\sqrt{3.7} + |7\;2\rangle+5\sqrt{13}/9\sqrt{2.3.7} + |7\,\text{-}1\rangle+\sqrt{7.13}/9\sqrt{2}$
$\quad + |7\,\text{-}4\rangle-\sqrt{11.13}/9\sqrt{3.7} + |7\,\text{-}7\rangle-\sqrt{2.11}/9\sqrt{3}$

$|7\,\tilde{0}\,\tilde{0}\;0\rangle = |7\;6\rangle-4\sqrt{11}/9.3 + |7\;3\rangle+2\sqrt{13}/9.3 + |7\;0\rangle+\sqrt{7.13}/9\sqrt{3}$
$\quad + |7\,\text{-}3\rangle-2\sqrt{13}/9.3 + |7\,\text{-}6\rangle-4\sqrt{11}/9.3$

$|\tfrac{15}{2}\,\tfrac{1}{2}\;\tfrac{1}{2}\;\tfrac{1}{2}\rangle = |\tfrac{15}{2}\,\tfrac{13}{2}\rangle+2\sqrt{13}/9.3 + |\tfrac{15}{2}\,\tfrac{7}{2}\rangle+5\sqrt{7}/9.3\sqrt{2} + |\tfrac{15}{2}\,\tfrac{1}{2}\rangle+\sqrt{11}/9\sqrt{3}$
$\quad + |\tfrac{15}{2}\,\text{-}\tfrac{5}{2}\rangle-\sqrt{5.7.11}/9.3\sqrt{2} + |\tfrac{15}{2}\,\text{-}\tfrac{11}{2}\rangle+2\sqrt{7.13}/9.3$

$|\tfrac{15}{2}\,\tfrac{1}{2}\;\tfrac{1}{2}\,\text{-}\tfrac{1}{2}\rangle = |\tfrac{15}{2}\,\tfrac{11}{2}\rangle-2\sqrt{7.13}/9.3 + |\tfrac{15}{2}\,\tfrac{5}{2}\rangle-\sqrt{5.7.11}/9.3\sqrt{2} + |\tfrac{15}{2}\,\text{-}\tfrac{1}{2}\rangle-\sqrt{11}/9\sqrt{3}$
$\quad + |\tfrac{15}{2}\,\text{-}\tfrac{7}{2}\rangle+5\sqrt{7}/9.3\sqrt{2} + |\tfrac{15}{2}\,\text{-}\tfrac{13}{2}\rangle-2\sqrt{13}/9.3$

$|\tfrac{15}{2}\,{}_0\,\tfrac{3}{2}\;\tfrac{1}{2}\;\tfrac{1}{2}\rangle = |\tfrac{15}{2}\,\tfrac{13}{2}\rangle-\sqrt{7.11.13}/2.9\sqrt{2.3.5} + |\tfrac{15}{2}\,\tfrac{7}{2}\rangle+\sqrt{5.11}/9\sqrt{3} + |\tfrac{15}{2}\,\tfrac{1}{2}\rangle-4\sqrt{2.7}/9\sqrt{5}$
$\quad + |\tfrac{15}{2}\,\text{-}\tfrac{5}{2}\rangle-5/9\sqrt{3} + |\tfrac{15}{2}\,\text{-}\tfrac{11}{2}\rangle-\sqrt{11.13}/2.9\sqrt{2.3.5}$

$|\tfrac{15}{2}\,{}_0\,\tfrac{3}{2}\;\tfrac{1}{2}\,\text{-}\tfrac{1}{2}\rangle = |\tfrac{15}{2}\,\tfrac{11}{2}\rangle+\sqrt{11.13}/2.9\sqrt{2.3.5} + |\tfrac{15}{2}\,\tfrac{5}{2}\rangle-5/9\sqrt{3} + |\tfrac{15}{2}\,\text{-}\tfrac{1}{2}\rangle+4\sqrt{2.7}/9\sqrt{5}$
$\quad + |\tfrac{15}{2}\,\text{-}\tfrac{7}{2}\rangle+\sqrt{5.11}/9\sqrt{3} + |\tfrac{15}{2}\,\text{-}\tfrac{13}{2}\rangle+\sqrt{7.11.13}/2.9\sqrt{2.3.5}$

$|\tfrac{15}{2}\,{}_0\,\tfrac{3}{2}\;\tfrac{3}{2}\;\tfrac{3}{2}\rangle = |\tfrac{15}{2}\,\tfrac{15}{2}\rangle-i\sqrt{7.11.13}/4.9\sqrt{3} + |\tfrac{15}{2}\,\tfrac{9}{2}\rangle+\sqrt{11}/4.9\sqrt{3.5}+i\sqrt{5.11}/9\sqrt{2.3}$
$\quad + |\tfrac{15}{2}\,\tfrac{3}{2}\rangle-\sqrt{5}/9\sqrt{2.3}-4i/3\sqrt{3.5} + |\tfrac{15}{2}\,\text{-}\tfrac{3}{2}\rangle+4/3\sqrt{3.5}-i\sqrt{5}/9\sqrt{2.3}$
$\quad + |\tfrac{15}{2}\,\text{-}\tfrac{9}{2}\rangle+\sqrt{5.11}/9\sqrt{2.3}-i\sqrt{11}/4.9\sqrt{3.5} + |\tfrac{15}{2}\,\text{-}\tfrac{15}{2}\rangle+\sqrt{7.11.13}/4.9\sqrt{3}$

$|\tfrac{15}{2}\,{}_0\,\tfrac{3}{2}\,\text{-}\tfrac{3}{2}\;\tfrac{3}{2}\rangle = |\tfrac{15}{2}\,\tfrac{15}{2}\rangle-\sqrt{7.11.13}/4.9\sqrt{3} + |\tfrac{15}{2}\,\tfrac{9}{2}\rangle+\sqrt{5.11}/9\sqrt{2.3}+i\sqrt{11}/4.9\sqrt{3.5}$
$\quad + |\tfrac{15}{2}\,\tfrac{3}{2}\rangle-4/3\sqrt{3.5}-i\sqrt{5}/9\sqrt{2.3} + |\tfrac{15}{2}\,\text{-}\tfrac{3}{2}\rangle-\sqrt{5}/9\sqrt{2.3}+4i/3\sqrt{3.5}$
$\quad + |\tfrac{15}{2}\,\text{-}\tfrac{9}{2}\rangle-\sqrt{11}/4.9\sqrt{3.5}+i\sqrt{5.11}/9\sqrt{2.3} + |\tfrac{15}{2}\,\text{-}\tfrac{15}{2}\rangle+i\sqrt{7.11.13}/4.9\sqrt{3}$

SO_3–O–D_3–C_3 Partners (cont.)

$|\frac{15}{2}\,{}_{1}\,\frac{3}{2}\,\frac{1}{2}\,\frac{1}{2}> \;=\; |\frac{15}{2}\,\frac{13}{2}> +2\sqrt{13}/3\sqrt{3.5} \;+\; |\frac{15}{2}\,\frac{7}{2}> +\sqrt{2.5.7}/9\sqrt{3} \;+\; |\frac{15}{2}\,\frac{1}{2}> +\sqrt{11}/9\sqrt{5}$
$\quad + \; |\frac{15}{2}\,\text{-}\frac{11}{2}> -2\sqrt{7.13}/9\sqrt{3.5}$
$|\frac{15}{2}\,{}_{1}\,\frac{3}{2}\,\frac{1}{2}\,\text{-}\frac{1}{2}> \;=\; |\frac{15}{2}\,\frac{11}{2}> +2\sqrt{7.13}/9\sqrt{3.5} \;+\; |\frac{15}{2}\,\text{-}\frac{1}{2}> -\sqrt{11}/9\sqrt{5} \;+\; |\frac{15}{2}\,\text{-}\frac{7}{2}> +\sqrt{2.5.7}/9\sqrt{3}$
$\quad + \; |\frac{15}{2}\,\text{-}\frac{13}{2}> -2\sqrt{13}/3\sqrt{3.5}$
$|\frac{15}{2}\,{}_{1}\,\frac{3}{2}\,\frac{3}{2}\,\frac{3}{2}> \;=\; |\frac{15}{2}\,\frac{15}{2}> -i\sqrt{2.13}/9.3\sqrt{3} \;+\; |\frac{15}{2}\,\frac{9}{2}> +13\sqrt{2.7}/9.3\sqrt{3.5} -2i\sqrt{5.7}/9.3\sqrt{3}$
$\quad + \; |\frac{15}{2}\,\frac{3}{2}> +\sqrt{5.7.11}/9.3\sqrt{3} -i\sqrt{7.11}/9\sqrt{2.3.5}$
$\quad + \; |\frac{15}{2}\,\text{-}\frac{3}{2}> +\sqrt{7.11}/9\sqrt{2.3.5} +i\sqrt{5.7.11}/9.3\sqrt{3}$
$\quad + \; |\frac{15}{2}\,\text{-}\frac{9}{2}> -2\sqrt{5.7}/9.3\sqrt{3} -13i\sqrt{2.7}/9.3\sqrt{3.5} \;+\; |\frac{15}{2}\,\text{-}\frac{15}{2}> +\sqrt{2.13}/9.3\sqrt{3}$
$|\frac{15}{2}\,{}_{1}\,\frac{3}{2}\,\text{-}\frac{3}{2}\,\frac{3}{2}> \;=\; |\frac{15}{2}\,\frac{15}{2}> -\sqrt{2.13}/9.3\sqrt{3} \;+\; |\frac{15}{2}\,\frac{9}{2}> -2\sqrt{5.7}/9.3\sqrt{3} +13i\sqrt{2.7}/9.3\sqrt{3.5}$
$\quad + \; |\frac{15}{2}\,\frac{3}{2}> -\sqrt{7.11}/9\sqrt{2.3.5} +i\sqrt{5.7.11}/9.3\sqrt{3}$
$\quad + \; |\frac{15}{2}\,\text{-}\frac{3}{2}> +\sqrt{5.7.11}/9.3\sqrt{3} +i\sqrt{7.11}/9\sqrt{2.3.5}$
$\quad + \; |\frac{15}{2}\,\text{-}\frac{9}{2}> -13\sqrt{2.7}/9.3\sqrt{3.5} -2i\sqrt{5.7}/9.3\sqrt{3} \;+\; |\frac{15}{2}\,\text{-}\frac{15}{2}> +i\sqrt{2.13}/9.3\sqrt{3}$
$|\frac{15}{2}\,{}_{2}\,\frac{3}{2}\,\frac{1}{2}\,\frac{1}{2}> \;=\; |\frac{15}{2}\,\frac{13}{2}> +\sqrt{5.7}/2.9\sqrt{2.3} \;+\; |\frac{15}{2}\,\frac{7}{2}> -\sqrt{5.13}/9\sqrt{3} \;+\; |\frac{15}{2}\,\text{-}\frac{5}{2}> -\sqrt{11.13}/9\sqrt{3}$
$\quad + \; |\frac{15}{2}\,\text{-}\frac{11}{2}> -7\sqrt{5}/2.9\sqrt{2.3}$
$|\frac{15}{2}\,{}_{2}\,\frac{3}{2}\,\frac{1}{2}\,\text{-}\frac{1}{2}> \;=\; |\frac{15}{2}\,\frac{11}{2}> +7\sqrt{5}/2.9\sqrt{2.3} \;+\; |\frac{15}{2}\,\frac{5}{2}> -\sqrt{11.13}/9\sqrt{3} \;+\; |\frac{15}{2}\,\text{-}\frac{7}{2}> -\sqrt{5.13}/9\sqrt{3}$
$\quad + \; |\frac{15}{2}\,\text{-}\frac{13}{2}> -\sqrt{5.7}/2.9\sqrt{2.3}$
$|\frac{15}{2}\,{}_{2}\,\frac{3}{2}\,\frac{3}{2}\,\frac{3}{2}> \;=\; |\frac{15}{2}\,\frac{15}{2}> -2\sqrt{2.7}/9\sqrt{3} -i\sqrt{7}/4.9.3\sqrt{3}$
$\quad + \; |\frac{15}{2}\,\frac{9}{2}> -7\sqrt{5.13}/4.9.3\sqrt{3} +i\sqrt{5.13}/9.3\sqrt{2.3} \;+\; |\frac{15}{2}\,\frac{3}{2}> +\sqrt{5.11.13}/9.3\sqrt{2.3}$
$\quad + \; |\frac{15}{2}\,\text{-}\frac{3}{2}> +i\sqrt{5.11.13}/9.3\sqrt{2.3} \;+\; |\frac{15}{2}\,\text{-}\frac{9}{2}> +\sqrt{5.13}/9.3\sqrt{2.3} +7i\sqrt{5.13}/4.9.3\sqrt{3}$
$\quad + \; |\frac{15}{2}\,\text{-}\frac{15}{2}> +\sqrt{7}/4.9.3\sqrt{3} -2i\sqrt{2.7}/9\sqrt{3}$
$|\frac{15}{2}\,{}_{2}\,\frac{3}{2}\,\text{-}\frac{3}{2}\,\frac{3}{2}> \;=\; |\frac{15}{2}\,\frac{15}{2}> -\sqrt{7}/4.9.3\sqrt{3} -2i\sqrt{2.7}/9\sqrt{3}$
$\quad + \; |\frac{15}{2}\,\frac{9}{2}> +\sqrt{5.13}/9.3\sqrt{2.3} -7i\sqrt{5.13}/4.9.3\sqrt{3} \;+\; |\frac{15}{2}\,\frac{3}{2}> +i\sqrt{5.11.13}/9.3\sqrt{2.3}$
$\quad + \; |\frac{15}{2}\,\text{-}\frac{3}{2}> +\sqrt{5.11.13}/9.3\sqrt{2.3} \;+\; |\frac{15}{2}\,\text{-}\frac{9}{2}> +7\sqrt{5.13}/4.9.3\sqrt{3} +i\sqrt{5.13}/9.3\sqrt{2.3}$
$\quad + \; |\frac{15}{2}\,\text{-}\frac{15}{2}> -2\sqrt{2.7}/9\sqrt{3} +i\sqrt{7}/4.9.3\sqrt{3}$
$|\frac{15}{2}\,\tilde{\frac{1}{2}}\,\frac{1}{2}\,\frac{1}{2}> \;=\; |\frac{15}{2}\,\frac{13}{2}> -2\sqrt{7.11}/9.3 \;+\; |\frac{15}{2}\,\frac{7}{2}> +\sqrt{11.13}/9.3\sqrt{2} \;+\; |\frac{15}{2}\,\frac{1}{2}> +\sqrt{7.13}/9\sqrt{3}$
$\quad + \; |\frac{15}{2}\,\text{-}\frac{5}{2}> -\sqrt{5.13}/9.3\sqrt{2} \;+\; |\frac{15}{2}\,\text{-}\frac{11}{2}> -2\sqrt{11}/9.3$
$|\frac{15}{2}\,\tilde{\frac{1}{2}}\,\frac{1}{2}\,\text{-}\frac{1}{2}> \;=\; |\frac{15}{2}\,\frac{11}{2}> +2\sqrt{11}/9.3 \;+\; |\frac{15}{2}\,\frac{5}{2}> -\sqrt{5.13}/9.3\sqrt{2} \;+\; |\frac{15}{2}\,\text{-}\frac{1}{2}> -\sqrt{7.13}/9\sqrt{3}$
$\quad + \; |\frac{15}{2}\,\text{-}\frac{7}{2}> +\sqrt{11.13}/9.3\sqrt{2} \;+\; |\frac{15}{2}\,\text{-}\frac{13}{2}> +2\sqrt{7.11}/9.3$

$|8\;0\;0\;0> \;=\; |8\;6> -4\sqrt{13}/9.3 \;+\; |8\;3> -2\sqrt{5.7}/9.3 \;+\; |8\;0> -\sqrt{11}/9\sqrt{3}$
$\quad + \; |8\,\text{-}3> +2\sqrt{5.7}/9.3 \;+\; |8\,\text{-}6> -4\sqrt{13}/9.3$
$|8\;{}_{0}\;1\;\tilde{0}\;0> \;=\; |8\;6> -2\sqrt{2.13}/9\sqrt{3} \;+\; |8\;3> -\sqrt{5.7}/9\sqrt{2.3} \;+\; |8\,\text{-}3> -\sqrt{5.7}/9\sqrt{2.3}$
$\quad + \; |8\,\text{-}6> +2\sqrt{2.13}/9\sqrt{3}$
$|8\;{}_{0}\;1\;1\;1> \;=\; |8\;7> -\sqrt{2.5.13}/9.3 \;+\; |8\;4> -5\sqrt{7}/9.3 \;+\; |8\;1> -\sqrt{11}/9\sqrt{2}$
$\quad + \; |8\,\text{-}2> +\sqrt{5.7.11}/9.3\sqrt{2} \;+\; |8\,\text{-}5> -\sqrt{2.7.13}/9.3$
$|8\;{}_{0}\;1\;1\,\text{-}1> \;=\; |8\;5> +\sqrt{2.7.13}/9.3 \;+\; |8\;2> +\sqrt{5.7.11}/9.3\sqrt{2} \;+\; |8\,\text{-}1> +\sqrt{11}/9\sqrt{2}$
$\quad + \; |8\,\text{-}4> -5\sqrt{7}/9.3 \;+\; |8\,\text{-}7> +\sqrt{2.5.13}/9.3$
$|8\;{}_{1}\;1\;\tilde{0}\;0> \;=\; |8\;6> -\sqrt{5.7}/9\sqrt{2.3} \;+\; |8\;3> +2\sqrt{2.13}/9\sqrt{3} \;+\; |8\,\text{-}3> +2\sqrt{2.13}/9\sqrt{3}$
$\quad + \; |8\,\text{-}6> +\sqrt{5.7}/9\sqrt{2.3}$
$|8\;{}_{1}\;1\;1\;1> \;=\; |8\;7> -\sqrt{7}/9.3\sqrt{2} \;+\; |8\;4> +\sqrt{5.13}/9.3 \;+\; |8\,\text{-}2> +\sqrt{2.11.13}/9.3$
$\quad + \; |8\,\text{-}5> +7\sqrt{5}/9.3\sqrt{2} \;+\; |8\,\text{-}8> -2\sqrt{7}/9$
$|8\;{}_{1}\;1\;1\,\text{-}1> \;=\; |8\;8> -2\sqrt{7}/9 \;+\; |8\;5> -7\sqrt{5}/9.3\sqrt{2} \;+\; |8\;2> +\sqrt{2.11.13}/9.3$
$\quad + \; |8\,\text{-}4> +\sqrt{5.13}/9.3 \;+\; |8\,\text{-}7> +\sqrt{7}/9.3\sqrt{2}$
$|8\;{}_{0}\;2\;1\;1> \;=\; |8\;7> -\sqrt{7.11.13}/2.9.3\sqrt{3} \;+\; |8\;4> -11\sqrt{5.11}/2.9.3\sqrt{2.3}$
$\quad + \; |8\;1> -\sqrt{5.7}/2.9.3\sqrt{3} \;+\; |8\,\text{-}2> -8.2/9.3\sqrt{3} \;+\; |8\,\text{-}5> +\sqrt{5.11.13}/9.3\sqrt{3}$
$\quad + \; |8\,\text{-}8> -\sqrt{7.11.13}/2.9.3\sqrt{2.3}$
$|8\;{}_{0}\;2\;1\,\text{-}1> \;=\; |8\;8> -\sqrt{7.11.13}/2.9.3\sqrt{2.3} \;+\; |8\;5> -\sqrt{5.11.13}/9.3\sqrt{3}$
$\quad + \; |8\;2> -8.2/9.3\sqrt{3} \;+\; |8\,\text{-}1> +\sqrt{5.7}/2.9.3\sqrt{3} \;+\; |8\,\text{-}4> -11\sqrt{5.11}/2.9.3\sqrt{2.3}$
$\quad + \; |8\,\text{-}7> +\sqrt{7.11.13}/2.9.3\sqrt{3}$

SO_3–O–D_3–C_3 Partners (cont.)

$$
\begin{aligned}
|8_1\,2\,1\,1\rangle &= |8\;7\rangle+\sqrt{5}/2.3 \;+\; |8\;4\rangle-\sqrt{7.13}/2.9\sqrt{2} \;+\; |8\;1\rangle+\sqrt{11.13}/2.9 \\
&+ |8\,\text{-}5\rangle-\sqrt{7}/9 \;+\; |8\,\text{-}8\rangle-5\sqrt{5}/2.9\sqrt{2} \\
|8_1\,2\,1\,\text{-}1\rangle &= |8\;8\rangle-5\sqrt{5}/2.9\sqrt{2} \;+\; |8\;5\rangle+\sqrt{7}/9 \;+\; |8\,\text{-}1\rangle-\sqrt{11.13}/2.9 \\
&+ |8\,\text{-}4\rangle-\sqrt{7.13}/2.9\sqrt{2} \;+\; |8\,\text{-}7\rangle-\sqrt{5}/2.3 \\
|8_0\,\tilde{1}\,0\,0\rangle &= |8\;6\rangle+2\sqrt{2.7.11}/9\sqrt{3.17} \;+\; |8\;3\rangle-\sqrt{5.11.13}/9\sqrt{2.3.17} \\
&+ |8\;0\rangle-2\sqrt{2.7.13}/9\sqrt{17} \;+\; |8\,\text{-}3\rangle+\sqrt{5.11.13}/9\sqrt{2.3.17} \\
&+ |8\,\text{-}6\rangle+2\sqrt{2.7.11}/9\sqrt{3.17} \\
|8_0\,\tilde{1}\,1\,1\rangle &= |8\;7\rangle-\sqrt{2.5.7.11}/9\sqrt{17} \;+\; |8\;4\rangle+\sqrt{11.13}/9\sqrt{17} \;+\; |8\;1\rangle+\sqrt{7.13}/3\sqrt{2.17} \\
&+ |8\,\text{-}2\rangle-\sqrt{5.13}/9\sqrt{2.17} \;+\; |8\,\text{-}5\rangle-\sqrt{2.11}/9\sqrt{17} \\
|8_0\,\tilde{1}\,1\,\text{-}1\rangle &= |8\;5\rangle+\sqrt{2.11}/9\sqrt{17} \;+\; |8\;2\rangle-\sqrt{5.13}/9\sqrt{2.17} \;+\; |8\,\text{-}1\rangle-\sqrt{7.13}/3\sqrt{2.17} \\
&+ |8\,\text{-}4\rangle+\sqrt{11.13}/9\sqrt{17} \;+\; |8\,\text{-}7\rangle+\sqrt{2.5.7.11}/9\sqrt{17} \\
|8_1\,\tilde{1}\,0\,0\rangle &= |8\;6\rangle+5\sqrt{5.13}/9.3\sqrt{2.17} \;+\; |8\;3\rangle-2.7\sqrt{2.7}/9.3\sqrt{17} \\
&+ |8\;0\rangle+4\sqrt{2.5.11}/9\sqrt{3.17} \;+\; |8\,\text{-}3\rangle+2.7\sqrt{2.7}/9.3\sqrt{17} \\
&+ |8\,\text{-}6\rangle+5\sqrt{5.13}/9.3\sqrt{2.17} \\
|8_1\,\tilde{1}\,1\,1\rangle &= |8\;7\rangle+5\sqrt{13}/9.3\sqrt{2.3.17} \;+\; |8\;4\rangle-7\sqrt{5.7}/9.3\sqrt{3.17} \\
&+ |8\;1\rangle+8\sqrt{2.5.11}/9.3\sqrt{3.17} \;+\; |8\,\text{-}2\rangle+7\sqrt{2.7.11}/9.3\sqrt{3.17} \\
&+ |8\,\text{-}5\rangle+5\sqrt{5.7.13}/9.3\sqrt{2.3.17} \;+\; |8\,\text{-}8\rangle+2\sqrt{13.17}/9.3\sqrt{3} \\
|8_1\,\tilde{1}\,1\,\text{-}1\rangle &= |8\;8\rangle+2\sqrt{13.17}/9.3\sqrt{3} \;+\; |8\;5\rangle-5\sqrt{5.7.13}/9.3\sqrt{2.3.17} \\
&+ |8\;2\rangle+7\sqrt{2.7.11}/9.3\sqrt{3.17} \;+\; |8\,\text{-}1\rangle-8\sqrt{2.5.11}/9.3\sqrt{3.17} \\
&+ |8\,\text{-}4\rangle-7\sqrt{5.7}/9.3\sqrt{3.17} \;+\; |8\,\text{-}7\rangle-5\sqrt{13}/9.3\sqrt{2.3.17}
\end{aligned}
$$

SO_3–O–D_4–C_4 Partners as JM Partners

$$
\begin{aligned}
|0\;0\;0\;0\rangle &= |0\;0\rangle+1
\end{aligned}
$$

$$
\begin{aligned}
|\tfrac12\;\tfrac12\;\tfrac12\;\tfrac12\rangle &= |\tfrac12\;\tfrac12\rangle+1 \\
|\tfrac12\;\tfrac12\;\tfrac12\,\text{-}\tfrac12\rangle &= |\tfrac12\,\text{-}\tfrac12\rangle+1
\end{aligned}
$$

$$
\begin{aligned}
|1\;1\;\tilde{0}\;0\rangle &= |1\;0\rangle+1 \\
|1\;1\;1\;1\rangle &= |1\;1\rangle-1 \\
|1\;1\;1\,\text{-}1\rangle &= |1\,\text{-}1\rangle-1
\end{aligned}
$$

$$
\begin{aligned}
|\tfrac32\;\tfrac32\;\tfrac12\;\tfrac12\rangle &= |\tfrac32\;\tfrac12\rangle+1 \\
|\tfrac32\;\tfrac32\;\tfrac12\,\text{-}\tfrac12\rangle &= |\tfrac32\,\text{-}\tfrac12\rangle-1 \\
|\tfrac32\;\tfrac32\;\tfrac32\;\tfrac32\rangle &= |\tfrac32\;\tfrac32\rangle+1 \\
|\tfrac32\;\tfrac32\;\tfrac32\,\text{-}\tfrac32\rangle &= |\tfrac32\,\text{-}\tfrac32\rangle+1
\end{aligned}
$$

$$
\begin{aligned}
|2\;2\;0\;0\rangle &= |2\;0\rangle-1 \\
|2\;2\;2\;2\rangle &= |2\;2\rangle-1/\sqrt{2} \;+\; |2\,\text{-}2\rangle-1/\sqrt{2} \\
|2\;\tilde{1}\;1\;1\rangle &= |2\;1\rangle-1 \\
|2\;\tilde{1}\;1\,\text{-}1\rangle &= |2\,\text{-}1\rangle+1 \\
|2\;\tilde{1}\;\tilde{2}\;2\rangle &= |2\;2\rangle+1/\sqrt{2} \;+\; |2\,\text{-}2\rangle-1/\sqrt{2}
\end{aligned}
$$

$$
\begin{aligned}
|\tfrac52\;\tfrac32\;\tfrac12\;\tfrac12\rangle &= |\tfrac52\;\tfrac12\rangle+1 \\
|\tfrac52\;\tfrac32\;\tfrac12\,\text{-}\tfrac12\rangle &= |\tfrac52\,\text{-}\tfrac12\rangle+1 \\
|\tfrac52\;\tfrac32\;\tfrac32\;\tfrac32\rangle &= |\tfrac52\;\tfrac32\rangle-1/\sqrt{2.3} \;+\; |\tfrac52\,\text{-}\tfrac52\rangle-\sqrt{5}/\sqrt{2.3} \\
|\tfrac52\;\tfrac32\;\tfrac32\,\text{-}\tfrac32\rangle &= |\tfrac52\;\tfrac52\rangle+\sqrt{5}/\sqrt{2.3} \;+\; |\tfrac52\,\text{-}\tfrac32\rangle+1/\sqrt{2.3} \\
|\tfrac52\;\tilde{\tfrac12}\;\tfrac32\;\tfrac32\rangle &= |\tfrac52\;\tfrac32\rangle+\sqrt{5}/\sqrt{2.3} \;+\; |\tfrac52\,\text{-}\tfrac52\rangle-1/\sqrt{2.3} \\
|\tfrac52\;\tilde{\tfrac12}\;\tfrac32\,\text{-}\tfrac32\rangle &= |\tfrac52\;\tfrac52\rangle+1/\sqrt{2.3} \;+\; |\tfrac52\,\text{-}\tfrac32\rangle-\sqrt{5}/\sqrt{2.3}
\end{aligned}
$$

SO_3–O–D_4–C_4 Partners (cont.)

$|3\ 1\ \tilde{0}\ 0\rangle = |3\ 0\rangle -1$
$|3\ 1\ 1\ 1\rangle = |3\ 1\rangle -\sqrt{3}/2\sqrt{2} + |3\,{-3}\rangle -\sqrt{5}/2\sqrt{2}$
$|3\ 1\ 1\,{-1}\rangle = |3\ 3\rangle -\sqrt{5}/2\sqrt{2} + |3\,{-1}\rangle -\sqrt{3}/2\sqrt{2}$
$|3\ \tilde{1}\ 1\ 1\rangle = |3\ 1\rangle -\sqrt{5}/2\sqrt{2} + |3\,{-3}\rangle +\sqrt{3}/2\sqrt{2}$
$|3\ \tilde{1}\ 1\,{-1}\rangle = |3\ 3\rangle +\sqrt{3}/2\sqrt{2} + |3\,{-1}\rangle -\sqrt{5}/2\sqrt{2}$
$|3\ \tilde{1}\ \tilde{2}\ 2\rangle = |3\ 2\rangle -1/\sqrt{2} + |3\,{-2}\rangle -1/\sqrt{2}$
$|3\ \tilde{0}\ 2\ 2\rangle = |3\ 2\rangle -1/\sqrt{2} + |3\,{-2}\rangle +1/\sqrt{2}$

$|\tfrac{7}{2}\ \tfrac{1}{2}\ \tfrac{1}{2}\ \tfrac{1}{2}\rangle = |\tfrac{7}{2}\ \tfrac{1}{2}\rangle -\sqrt{7}/2\sqrt{3} + |\tfrac{7}{2}\,{-\tfrac{7}{2}}\rangle -\sqrt{5}/2\sqrt{3}$
$|\tfrac{7}{2}\ \tfrac{1}{2}\ \tfrac{1}{2}\,{-\tfrac{1}{2}}\rangle = |\tfrac{7}{2}\ \tfrac{7}{2}\rangle +\sqrt{5}/2\sqrt{3} + |\tfrac{7}{2}\,{-\tfrac{1}{2}}\rangle +\sqrt{7}/2\sqrt{3}$
$|\tfrac{7}{2}\ \tfrac{3}{2}\ \tfrac{1}{2}\ \tfrac{1}{2}\rangle = |\tfrac{7}{2}\ \tfrac{1}{2}\rangle -\sqrt{5}/2\sqrt{3} + |\tfrac{7}{2}\,{-\tfrac{7}{2}}\rangle +\sqrt{7}/2\sqrt{3}$
$|\tfrac{7}{2}\ \tfrac{3}{2}\ \tfrac{1}{2}\,{-\tfrac{1}{2}}\rangle = |\tfrac{7}{2}\ \tfrac{7}{2}\rangle -\sqrt{7}/2\sqrt{3} + |\tfrac{7}{2}\,{-\tfrac{1}{2}}\rangle +\sqrt{5}/2\sqrt{3}$
$|\tfrac{7}{2}\ \tfrac{3}{2}\ \tfrac{3}{2}\ \tfrac{3}{2}\rangle = |\tfrac{7}{2}\ \tfrac{3}{2}\rangle +\sqrt{3}/2 + |\tfrac{7}{2}\,{-\tfrac{5}{2}}\rangle +1/2$
$|\tfrac{7}{2}\ \tfrac{3}{2}\ \tfrac{3}{2}\,{-\tfrac{3}{2}}\rangle = |\tfrac{7}{2}\ \tfrac{5}{2}\rangle +1/2 + |\tfrac{7}{2}\,{-\tfrac{3}{2}}\rangle +\sqrt{3}/2$
$|\tfrac{7}{2}\ \tilde{\tfrac{1}{2}}\ \tfrac{3}{2}\ \tfrac{3}{2}\rangle = |\tfrac{7}{2}\ \tfrac{3}{2}\rangle +1/2 + |\tfrac{7}{2}\,{-\tfrac{5}{2}}\rangle -\sqrt{3}/2$
$|\tfrac{7}{2}\ \tilde{\tfrac{1}{2}}\ \tfrac{3}{2}\,{-\tfrac{3}{2}}\rangle = |\tfrac{7}{2}\ \tfrac{5}{2}\rangle -\sqrt{3}/2 + |\tfrac{7}{2}\,{-\tfrac{3}{2}}\rangle +1/2$

$|4\ 0\ 0\ 0\rangle = |4\ 4\rangle +\sqrt{5}/2\sqrt{2.3} + |4\ 0\rangle +\sqrt{7}/2\sqrt{3} + |4\,{-4}\rangle +\sqrt{5}/2\sqrt{2.3}$
$|4\ 1\ \tilde{0}\ 0\rangle = |4\ 4\rangle +1/\sqrt{2} + |4\,{-4}\rangle -1/\sqrt{2}$
$|4\ 1\ 1\ 1\rangle = |4\ 1\rangle +\sqrt{7}/2\sqrt{2} + |4\,{-3}\rangle +1/2\sqrt{2}$
$|4\ 1\ 1\,{-1}\rangle = |4\ 3\rangle -1/2\sqrt{2} + |4\,{-1}\rangle -\sqrt{7}/2\sqrt{2}$
$|4\ 2\ 0\ 0\rangle = |4\ 4\rangle -\sqrt{7}/2\sqrt{2.3} + |4\ 0\rangle +\sqrt{5}/2\sqrt{3} + |4\,{-4}\rangle -\sqrt{7}/2\sqrt{2.3}$
$|4\ 2\ 2\ 2\rangle = |4\ 2\rangle -1/\sqrt{2} + |4\,{-2}\rangle -1/\sqrt{2}$
$|4\ \tilde{1}\ 1\ 1\rangle = |4\ 1\rangle +1/2\sqrt{2} + |4\,{-3}\rangle -\sqrt{7}/2\sqrt{2}$
$|4\ \tilde{1}\ 1\,{-1}\rangle = |4\ 3\rangle +\sqrt{7}/2\sqrt{2} + |4\,{-1}\rangle -1/2\sqrt{2}$
$|4\ \tilde{1}\ \tilde{2}\ 2\rangle = |4\ 2\rangle +1/\sqrt{2} + |4\,{-2}\rangle -1/\sqrt{2}$

$|\tfrac{9}{2}\ \tfrac{1}{2}\ \tfrac{1}{2}\ \tfrac{1}{2}\rangle = |\tfrac{9}{2}\ \tfrac{9}{2}\rangle +\sqrt{3}/2\sqrt{2} + |\tfrac{9}{2}\ \tfrac{1}{2}\rangle +\sqrt{7}/2\sqrt{3} + |\tfrac{9}{2}\,{-\tfrac{7}{2}}\rangle +1/2\sqrt{2.3}$
$|\tfrac{9}{2}\ \tfrac{1}{2}\ \tfrac{1}{2}\,{-\tfrac{1}{2}}\rangle = |\tfrac{9}{2}\ \tfrac{7}{2}\rangle +1/2\sqrt{2.3} + |\tfrac{9}{2}\,{-\tfrac{1}{2}}\rangle +\sqrt{7}/2\sqrt{3} + |\tfrac{9}{2}\,{-\tfrac{9}{2}}\rangle +\sqrt{3}/2\sqrt{2}$
$|\tfrac{9}{2}\ 0\ \tfrac{3}{2}\ \tfrac{1}{2}\ \tfrac{1}{2}\rangle = |\tfrac{9}{2}\ \tfrac{9}{2}\rangle +1/\sqrt{2.5} + |\tfrac{9}{2}\,{-\tfrac{7}{2}}\rangle -3/\sqrt{2.5}$
$|\tfrac{9}{2}\ 0\ \tfrac{3}{2}\ \tfrac{1}{2}\,{-\tfrac{1}{2}}\rangle = |\tfrac{9}{2}\ \tfrac{7}{2}\rangle -3/\sqrt{2.5} + |\tfrac{9}{2}\,{-\tfrac{9}{2}}\rangle +1/\sqrt{2.5}$
$|\tfrac{9}{2}\ 0\ \tfrac{3}{2}\ \tfrac{3}{2}\ \tfrac{3}{2}\rangle = |\tfrac{9}{2}\ \tfrac{3}{2}\rangle -\sqrt{7}/\sqrt{2.5} + |\tfrac{9}{2}\,{-\tfrac{5}{2}}\rangle +\sqrt{3}/\sqrt{2.5}$
$|\tfrac{9}{2}\ 0\ \tfrac{3}{2}\ \tfrac{3}{2}\,{-\tfrac{3}{2}}\rangle = |\tfrac{9}{2}\ \tfrac{5}{2}\rangle -\sqrt{3}/\sqrt{2.5} + |\tfrac{9}{2}\,{-\tfrac{3}{2}}\rangle +\sqrt{7}/\sqrt{2.5}$
$|\tfrac{9}{2}\ 1\ \tfrac{3}{2}\ \tfrac{1}{2}\ \tfrac{1}{2}\rangle = |\tfrac{9}{2}\ \tfrac{9}{2}\rangle +\sqrt{3.7}/2\sqrt{2.5} + |\tfrac{9}{2}\ \tfrac{1}{2}\rangle -\sqrt{5}/2\sqrt{3} + |\tfrac{9}{2}\,{-\tfrac{7}{2}}\rangle +\sqrt{7}/2\sqrt{2.3.5}$
$|\tfrac{9}{2}\ 1\ \tfrac{3}{2}\ \tfrac{1}{2}\,{-\tfrac{1}{2}}\rangle = |\tfrac{9}{2}\ \tfrac{7}{2}\rangle +\sqrt{7}/2\sqrt{2.3.5} + |\tfrac{9}{2}\,{-\tfrac{1}{2}}\rangle -\sqrt{5}/2\sqrt{3} + |\tfrac{9}{2}\,{-\tfrac{9}{2}}\rangle +\sqrt{3.7}/2\sqrt{2.5}$
$|\tfrac{9}{2}\ 1\ \tfrac{3}{2}\ \tfrac{3}{2}\ \tfrac{3}{2}\rangle = |\tfrac{9}{2}\ \tfrac{3}{2}\rangle -\sqrt{3}/\sqrt{2.5} + |\tfrac{9}{2}\,{-\tfrac{5}{2}}\rangle -\sqrt{7}/\sqrt{2.5}$
$|\tfrac{9}{2}\ 1\ \tfrac{3}{2}\ \tfrac{3}{2}\,{-\tfrac{3}{2}}\rangle = |\tfrac{9}{2}\ \tfrac{5}{2}\rangle +\sqrt{7}/\sqrt{2.5} + |\tfrac{9}{2}\,{-\tfrac{3}{2}}\rangle +\sqrt{3}/\sqrt{2.5}$

$|5_0\ 1\ \tilde{0}\ 0\rangle = |5\ 4\rangle -\sqrt{5.7}/2\sqrt{2.11} + |5\ 0\rangle +3/2\sqrt{11} + |5\,{-4}\rangle -\sqrt{5.7}/2\sqrt{2.11}$
$|5_0\ 1\ 1\ 1\rangle = |5\ 5\rangle -\sqrt{7}/4\sqrt{2.11} + |5\ 1\rangle +\sqrt{3.5}/4\sqrt{11} + |5\,{-3}\rangle -3\sqrt{5.7}/4\sqrt{2.11}$
$|5_0\ 1\ 1\,{-1}\rangle = |5\ 3\rangle -3\sqrt{5.7}/4\sqrt{2.11} + |5\,{-1}\rangle +\sqrt{3.5}/4\sqrt{11} + |5\,{-5}\rangle -\sqrt{7}/4\sqrt{2.11}$
$|5_1\ 1\ \tilde{0}\ 0\rangle = |5\ 4\rangle +3/2\sqrt{2.11} + |5\ 0\rangle +\sqrt{5.7}/2\sqrt{11} + |5\,{-4}\rangle +3/2\sqrt{2.11}$
$|5_1\ 1\ 1\ 1\rangle = |5\ 5\rangle -3\sqrt{5}/2\sqrt{2.11} + |5\ 1\rangle -\sqrt{3.7}/2\sqrt{11} + |5\,{-3}\rangle -1/2\sqrt{2.11}$
$|5_1\ 1\ 1\,{-1}\rangle = |5\ 3\rangle -1/2\sqrt{2.11} + |5\,{-1}\rangle -\sqrt{3.7}/2\sqrt{11} + |5\,{-5}\rangle -3\sqrt{5}/2\sqrt{2.11}$
$|5\ 2\ 0\ 0\rangle = |5\ 4\rangle -1/\sqrt{2} + |5\,{-4}\rangle +1/\sqrt{2}$
$|5\ 2\ 2\ 2\rangle = |5\ 2\rangle +1/\sqrt{2} + |5\,{-2}\rangle -1/\sqrt{2}$
$|5\ \tilde{1}\ 1\ 1\rangle = |5\ 5\rangle -\sqrt{3.5}/4\sqrt{2} + |5\ 1\rangle +\sqrt{7}/4 + |5\,{-3}\rangle +\sqrt{3}/4\sqrt{2}$
$|5\ \tilde{1}\ 1\,{-1}\rangle = |5\ 3\rangle +\sqrt{3}/4\sqrt{2} + |5\,{-1}\rangle +\sqrt{7}/4 + |5\,{-5}\rangle -\sqrt{3.5}/4\sqrt{2}$
$|5\ \tilde{1}\ \tilde{2}\ 2\rangle = |5\ 2\rangle -1/\sqrt{2} + |5\,{-2}\rangle -1/\sqrt{2}$

SO_3–O–D_4–C_4 Partners (cont.)

$|\tfrac{11}{2}\,\tfrac{1}{2}\,\tfrac{1}{2}\,\tfrac{1}{2}\rangle = |\tfrac{11}{2}\,\tfrac{9}{2}\rangle - \sqrt{7}/4\sqrt{3} + |\tfrac{11}{2}\,\tfrac{1}{2}\rangle + 1/2\sqrt{2} + |\tfrac{11}{2}\,{-\tfrac{7}{2}}\rangle - \sqrt{5.7}/4\sqrt{3}$

$|\tfrac{11}{2}\,\tfrac{1}{2}\,\tfrac{1}{2}\,{-\tfrac{1}{2}}\rangle = |\tfrac{11}{2}\,\tfrac{7}{2}\rangle + \sqrt{5.7}/4\sqrt{3} + |\tfrac{11}{2}\,{-\tfrac{1}{2}}\rangle - 1/2\sqrt{2} + |\tfrac{11}{2}\,{-\tfrac{9}{2}}\rangle + \sqrt{7}/4\sqrt{3}$

$|\tfrac{11}{2}\,0\,\tfrac{3}{2}\,\tfrac{1}{2}\,\tfrac{1}{2}\rangle = |\tfrac{11}{2}\,\tfrac{9}{2}\rangle + 19/4\sqrt{3.11} + |\tfrac{11}{2}\,\tfrac{1}{2}\rangle - \sqrt{7}/2\sqrt{2.11} + |\tfrac{11}{2}\,{-\tfrac{7}{2}}\rangle - 5\sqrt{5}/4\sqrt{3.11}$

$|\tfrac{11}{2}\,0\,\tfrac{3}{2}\,\tfrac{1}{2}\,{-\tfrac{1}{2}}\rangle = |\tfrac{11}{2}\,\tfrac{7}{2}\rangle + 5\sqrt{5}/4\sqrt{3.11} + |\tfrac{11}{2}\,{-\tfrac{1}{2}}\rangle + \sqrt{7}/2\sqrt{2.11} + |\tfrac{11}{2}\,{-\tfrac{9}{2}}\rangle - 19/4\sqrt{3.11}$

$|\tfrac{11}{2}\,0\,\tfrac{3}{2}\,\tfrac{3}{2}\,\tfrac{3}{2}\rangle = |\tfrac{11}{2}\,\tfrac{11}{2}\rangle - 1/4 + |\tfrac{11}{2}\,\tfrac{3}{2}\rangle + \sqrt{3.5}/2\sqrt{2.11} + |\tfrac{11}{2}\,{-\tfrac{5}{2}}\rangle - 3\sqrt{3.5}/4\sqrt{11}$

$|\tfrac{11}{2}\,0\,\tfrac{3}{2}\,\tfrac{3}{2}\,{-\tfrac{3}{2}}\rangle = |\tfrac{11}{2}\,\tfrac{5}{2}\rangle - 3\sqrt{3.5}/4\sqrt{11} + |\tfrac{11}{2}\,{-\tfrac{3}{2}}\rangle + \sqrt{3.5}/2\sqrt{2.11} + |\tfrac{11}{2}\,{-\tfrac{11}{2}}\rangle - 1/4$

$|\tfrac{11}{2}\,1\,\tfrac{3}{2}\,\tfrac{1}{2}\,\tfrac{1}{2}\rangle = |\tfrac{11}{2}\,\tfrac{9}{2}\rangle + \sqrt{3.5}/2\sqrt{2.11} + |\tfrac{11}{2}\,\tfrac{1}{2}\rangle + \sqrt{5.7}/2\sqrt{11} + |\tfrac{11}{2}\,{-\tfrac{7}{2}}\rangle + \sqrt{3}/2\sqrt{2.11}$

$|\tfrac{11}{2}\,1\,\tfrac{3}{2}\,\tfrac{1}{2}\,{-\tfrac{1}{2}}\rangle = |\tfrac{11}{2}\,\tfrac{7}{2}\rangle - \sqrt{3}/2\sqrt{2.11} + |\tfrac{11}{2}\,{-\tfrac{1}{2}}\rangle - \sqrt{5.7}/2\sqrt{11} + |\tfrac{11}{2}\,{-\tfrac{9}{2}}\rangle - \sqrt{3.5}/2\sqrt{2.11}$

$|\tfrac{11}{2}\,1\,\tfrac{3}{2}\,\tfrac{3}{2}\,\tfrac{3}{2}\rangle = |\tfrac{11}{2}\,\tfrac{11}{2}\rangle + \sqrt{5}/2\sqrt{2} + |\tfrac{11}{2}\,\tfrac{3}{2}\rangle + 7/2\sqrt{3.11} + |\tfrac{11}{2}\,{-\tfrac{5}{2}}\rangle + 1/2\sqrt{2.3.11}$

$|\tfrac{11}{2}\,1\,\tfrac{3}{2}\,\tfrac{3}{2}\,{-\tfrac{3}{2}}\rangle = |\tfrac{11}{2}\,\tfrac{5}{2}\rangle + 1/2\sqrt{2.3.11} + |\tfrac{11}{2}\,{-\tfrac{3}{2}}\rangle + 7/2\sqrt{3.11} + |\tfrac{11}{2}\,{-\tfrac{11}{2}}\rangle + \sqrt{5}/2\sqrt{2}$

$|\tfrac{11}{2}\,\tilde{\tfrac{1}{2}}\,\tfrac{3}{2}\,\tfrac{3}{2}\rangle = |\tfrac{11}{2}\,\tfrac{11}{2}\rangle + \sqrt{5}/4 + |\tfrac{11}{2}\,\tfrac{3}{2}\rangle - \sqrt{11}/2\sqrt{2.3} + |\tfrac{11}{2}\,{-\tfrac{5}{2}}\rangle - \sqrt{11}/4\sqrt{3}$

$|\tfrac{11}{2}\,\tilde{\tfrac{1}{2}}\,\tfrac{3}{2}\,{-\tfrac{3}{2}}\rangle = |\tfrac{11}{2}\,\tfrac{5}{2}\rangle - \sqrt{11}/4\sqrt{3} + |\tfrac{11}{2}\,{-\tfrac{3}{2}}\rangle - \sqrt{11}/2\sqrt{2.3} + |\tfrac{11}{2}\,{-\tfrac{11}{2}}\rangle + \sqrt{5}/4$

$|6\,0\,0\,0\rangle = |6\,4\rangle + \sqrt{7}/4 + |6\,0\rangle - 1/2\sqrt{2} + |6\,{-4}\rangle + \sqrt{7}/4$

$|6\,1\,\tilde{0}\,0\rangle = |6\,4\rangle + 1/\sqrt{2} + |6\,{-4}\rangle - 1/\sqrt{2}$

$|6\,1\,1\,1\rangle = |6\,5\rangle + \sqrt{11}/4\sqrt{2} + |6\,1\rangle - \sqrt{3}/4 + |6\,{-3}\rangle + \sqrt{3.5}/4\sqrt{2}$

$|6\,1\,1\,{-1}\rangle = |6\,3\rangle - \sqrt{3.5}/4\sqrt{2} + |6\,{-1}\rangle + \sqrt{3}/4 + |6\,{-5}\rangle - \sqrt{11}/4\sqrt{2}$

$|6\,2\,0\,0\rangle = |6\,4\rangle + 1/4 + |6\,0\rangle + \sqrt{7}/2\sqrt{2} + |6\,{-4}\rangle + 1/4$

$|6\,2\,2\,2\rangle = |6\,6\rangle + \sqrt{11}/4\sqrt{2} + |6\,2\rangle + \sqrt{5}/4\sqrt{2} + |6\,{-2}\rangle + \sqrt{5}/4\sqrt{2}$
$\quad + |6\,{-6}\rangle + \sqrt{11}/4\sqrt{2}$

$|6\,0\,\tilde{1}\,1\,1\rangle = |6\,5\rangle + \sqrt{3.5}/4\sqrt{2.7} + |6\,1\rangle - \sqrt{5.11}/4\sqrt{7} + |6\,{-3}\rangle - 3\sqrt{11}/4\sqrt{2.7}$

$|6\,0\,\tilde{1}\,1\,{-1}\rangle = |6\,3\rangle + 3\sqrt{11}/4\sqrt{2.7} + |6\,{-1}\rangle + \sqrt{5.11}/4\sqrt{7} + |6\,{-5}\rangle - \sqrt{3.5}/4\sqrt{2.7}$

$|6\,0\,\tilde{1}\,\tilde{2}\,2\rangle = |6\,6\rangle + 3\sqrt{5}/4\sqrt{7} + |6\,2\rangle - \sqrt{11}/4\sqrt{7} + |6\,{-2}\rangle + \sqrt{11}/4\sqrt{7}$
$\quad + |6\,{-6}\rangle - 3\sqrt{5}/4\sqrt{7}$

$|6\,1\,\tilde{1}\,1\,1\rangle = |6\,5\rangle - \sqrt{3.11}/2\sqrt{2.7} + |6\,1\rangle - 3/2\sqrt{7} + |6\,{-3}\rangle + \sqrt{5}/2\sqrt{2.7}$

$|6\,1\,\tilde{1}\,1\,{-1}\rangle = |6\,3\rangle - \sqrt{5}/2\sqrt{2.7} + |6\,{-1}\rangle + 3/2\sqrt{7} + |6\,{-5}\rangle + \sqrt{3.11}/2\sqrt{2.7}$

$|6\,1\,\tilde{1}\,\tilde{2}\,2\rangle = |6\,6\rangle + \sqrt{11}/4\sqrt{7} + |6\,2\rangle + 3\sqrt{5}/4\sqrt{7} + |6\,{-2}\rangle - 3\sqrt{5}/4\sqrt{7}$
$\quad + |6\,{-6}\rangle - \sqrt{11}/4\sqrt{7}$

$|6\,\tilde{0}\,2\,2\rangle = |6\,6\rangle - \sqrt{5}/4\sqrt{2} + |6\,2\rangle + \sqrt{11}/4\sqrt{2} + |6\,{-2}\rangle + \sqrt{11}/4\sqrt{2}$
$\quad + |6\,{-6}\rangle - \sqrt{5}/4\sqrt{2}$

$|\tfrac{13}{2}\,\tfrac{1}{2}\,\tfrac{1}{2}\,\tfrac{1}{2}\rangle = |\tfrac{13}{2}\,\tfrac{9}{2}\rangle + \sqrt{11}/4 + |\tfrac{13}{2}\,\tfrac{1}{2}\rangle - 1/2\sqrt{2} + |\tfrac{13}{2}\,{-\tfrac{7}{2}}\rangle + \sqrt{3}/4$

$|\tfrac{13}{2}\,\tfrac{1}{2}\,\tfrac{1}{2}\,{-\tfrac{1}{2}}\rangle = |\tfrac{13}{2}\,\tfrac{7}{2}\rangle + \sqrt{3}/4 + |\tfrac{13}{2}\,{-\tfrac{1}{2}}\rangle - 1/2\sqrt{2} + |\tfrac{13}{2}\,{-\tfrac{9}{2}}\rangle + \sqrt{11}/4$

$|\tfrac{13}{2}\,0\,\tfrac{3}{2}\,\tfrac{1}{2}\,\tfrac{1}{2}\rangle = |\tfrac{13}{2}\,\tfrac{9}{2}\rangle + 1/2\sqrt{2.7} + |\tfrac{13}{2}\,\tfrac{1}{2}\rangle - \sqrt{11}/2\sqrt{7} + |\tfrac{13}{2}\,{-\tfrac{7}{2}}\rangle - \sqrt{3.11}/2\sqrt{2.7}$

$|\tfrac{13}{2}\,0\,\tfrac{3}{2}\,\tfrac{1}{2}\,{-\tfrac{1}{2}}\rangle = |\tfrac{13}{2}\,\tfrac{7}{2}\rangle - \sqrt{3.11}/2\sqrt{2.7} + |\tfrac{13}{2}\,{-\tfrac{1}{2}}\rangle - \sqrt{11}/2\sqrt{7} + |\tfrac{13}{2}\,{-\tfrac{9}{2}}\rangle + 1/2\sqrt{2.7}$

$|\tfrac{13}{2}\,0\,\tfrac{3}{2}\,\tfrac{3}{2}\,\tfrac{3}{2}\rangle = |\tfrac{13}{2}\,\tfrac{11}{2}\rangle - 1/\sqrt{7} + |\tfrac{13}{2}\,\tfrac{3}{2}\rangle + \sqrt{11}/2\sqrt{7} + |\tfrac{13}{2}\,{-\tfrac{13}{2}}\rangle + \sqrt{13}/2\sqrt{7}$

$|\tfrac{13}{2}\,0\,\tfrac{3}{2}\,\tfrac{3}{2}\,{-\tfrac{3}{2}}\rangle = |\tfrac{13}{2}\,\tfrac{13}{2}\rangle - \sqrt{13}/2\sqrt{7} + |\tfrac{13}{2}\,{-\tfrac{3}{2}}\rangle - \sqrt{11}/2\sqrt{7} + |\tfrac{13}{2}\,{-\tfrac{11}{2}}\rangle + 1/\sqrt{7}$

$|\tfrac{13}{2}\,1\,\tfrac{3}{2}\,\tfrac{1}{2}\,\tfrac{1}{2}\rangle = |\tfrac{13}{2}\,\tfrac{9}{2}\rangle - \sqrt{3.11}/4\sqrt{7} + |\tfrac{13}{2}\,\tfrac{1}{2}\rangle - 3\sqrt{3}/2\sqrt{2.7} + |\tfrac{13}{2}\,{-\tfrac{7}{2}}\rangle + 5/4\sqrt{7}$

$|\tfrac{13}{2}\,1\,\tfrac{3}{2}\,\tfrac{1}{2}\,{-\tfrac{1}{2}}\rangle = |\tfrac{13}{2}\,\tfrac{7}{2}\rangle + 5/4\sqrt{7} + |\tfrac{13}{2}\,{-\tfrac{1}{2}}\rangle - 3\sqrt{3}/2\sqrt{2.7} + |\tfrac{13}{2}\,{-\tfrac{9}{2}}\rangle - \sqrt{3.11}/4\sqrt{7}$

$|\tfrac{13}{2}\,1\,\tfrac{3}{2}\,\tfrac{3}{2}\,\tfrac{3}{2}\rangle = |\tfrac{13}{2}\,\tfrac{11}{2}\rangle + 5\sqrt{11}/4\sqrt{2.3.7} + |\tfrac{13}{2}\,\tfrac{3}{2}\rangle - \sqrt{3}/4\sqrt{2.7} + |\tfrac{13}{2}\,{-\tfrac{5}{2}}\rangle + \sqrt{5.7}/4\sqrt{2.3}$
$\quad + |\tfrac{13}{2}\,{-\tfrac{13}{2}}\rangle + \sqrt{11.13}/4\sqrt{2.3.7}$

$|\tfrac{13}{2}\,1\,\tfrac{3}{2}\,\tfrac{3}{2}\,{-\tfrac{3}{2}}\rangle = |\tfrac{13}{2}\,\tfrac{13}{2}\rangle - \sqrt{11.13}/4\sqrt{2.3.7} + |\tfrac{13}{2}\,\tfrac{5}{2}\rangle - \sqrt{5.7}/4\sqrt{2.3} + |\tfrac{13}{2}\,{-\tfrac{3}{2}}\rangle + \sqrt{3}/4\sqrt{2.7}$
$\quad + |\tfrac{13}{2}\,{-\tfrac{11}{2}}\rangle - 5\sqrt{11}/4\sqrt{2.3.7}$

$|\tfrac{13}{2}\,0\,\tilde{\tfrac{1}{2}}\,\tfrac{3}{2}\,\tfrac{3}{2}\rangle = |\tfrac{13}{2}\,\tfrac{11}{2}\rangle - \sqrt{5}/4\sqrt{2.7} + |\tfrac{13}{2}\,\tfrac{3}{2}\rangle + \sqrt{5.11}/4\sqrt{2.7} + |\tfrac{13}{2}\,{-\tfrac{5}{2}}\rangle + 3\sqrt{11}/4\sqrt{2.7}$
$\quad + |\tfrac{13}{2}\,{-\tfrac{13}{2}}\rangle - \sqrt{5.13}/4\sqrt{2.7}$

$|\tfrac{13}{2}\,0\,\tilde{\tfrac{1}{2}}\,\tfrac{3}{2}\,{-\tfrac{3}{2}}\rangle = |\tfrac{13}{2}\,\tfrac{13}{2}\rangle + \sqrt{5.13}/4\sqrt{2.7} + |\tfrac{13}{2}\,\tfrac{5}{2}\rangle - 3\sqrt{11}/4\sqrt{2.7} + |\tfrac{13}{2}\,{-\tfrac{3}{2}}\rangle - \sqrt{5.11}/4\sqrt{2.7}$
$\quad + |\tfrac{13}{2}\,{-\tfrac{11}{2}}\rangle + \sqrt{5}/4\sqrt{2.7}$

$|\tfrac{13}{2}\,1\,\tilde{\tfrac{1}{2}}\,\tfrac{3}{2}\,\tfrac{3}{2}\rangle = |\tfrac{13}{2}\,\tfrac{11}{2}\rangle + \sqrt{11.13}/4\sqrt{3.7} + |\tfrac{13}{2}\,\tfrac{3}{2}\rangle + \sqrt{3.13}/4\sqrt{7} + |\tfrac{13}{2}\,{-\tfrac{5}{2}}\rangle - \sqrt{5.13}/4\sqrt{3.7}$
$\quad + |\tfrac{13}{2}\,{-\tfrac{13}{2}}\rangle - \sqrt{11}/4\sqrt{3.7}$

$|\tfrac{13}{2}\,1\,\tilde{\tfrac{1}{2}}\,\tfrac{3}{2}\,{-\tfrac{3}{2}}\rangle = |\tfrac{13}{2}\,\tfrac{13}{2}\rangle + \sqrt{11}/4\sqrt{3.7} + |\tfrac{13}{2}\,\tfrac{5}{2}\rangle + \sqrt{5.13}/4\sqrt{3.7} + |\tfrac{13}{2}\,{-\tfrac{3}{2}}\rangle - \sqrt{3.13}/4\sqrt{7}$
$\quad + |\tfrac{13}{2}\,{-\tfrac{11}{2}}\rangle - \sqrt{11.13}/4\sqrt{3.7}$

SO_3–O–D_4–C_4 Partners (cont.)

$|7\,{\scriptstyle 0}\,1\,\tilde{0}\,0\rangle = |7\,4\rangle + \sqrt{3.11}/4\sqrt{5} + |7\,0\rangle - \sqrt{7}/2\sqrt{2.5} + |7\,{-4}\rangle + \sqrt{3.11}/4\sqrt{5}$

$|7\,{\scriptstyle 0}\,1\,1\,1\rangle = |7\,5\rangle - \sqrt{3.11}/2\sqrt{2.5} + |7\,1\rangle + 1/\sqrt{2.5} + |7\,{-3}\rangle - \sqrt{3}/2\sqrt{2.5}$

$|7\,{\scriptstyle 0}\,1\,1\,{-1}\rangle = |7\,3\rangle - \sqrt{3}/2\sqrt{2.5} + |7\,{-1}\rangle + 1/\sqrt{2.5} + |7\,{-5}\rangle - \sqrt{3.11}/2\sqrt{2.5}$

$|7\,{\scriptstyle 1}\,1\,\tilde{0}\,0\rangle = |7\,4\rangle + \sqrt{7}/4\sqrt{5} + |7\,0\rangle + \sqrt{3.11}/2\sqrt{2.5} + |7\,{-4}\rangle + \sqrt{7}/4\sqrt{5}$

$|7\,{\scriptstyle 1}\,1\,1\,1\rangle = |7\,5\rangle + \sqrt{7}/8\sqrt{2.5} + |7\,1\rangle + \sqrt{3.7.11}/8\sqrt{2.5} + |7\,{-3}\rangle + \sqrt{7.11}/8\sqrt{2.5}$
$\quad + |7\,{-7}\rangle + \sqrt{5.13}/8\sqrt{2}$

$|7\,{\scriptstyle 1}\,1\,1\,{-1}\rangle = |7\,7\rangle + \sqrt{5.13}/8\sqrt{2} + |7\,3\rangle + \sqrt{7.11}/8\sqrt{2.5} + |7\,{-1}\rangle + \sqrt{3.7.11}/8\sqrt{2.5}$
$\quad + |7\,{-5}\rangle + \sqrt{7}/8\sqrt{2.5}$

$|7\,2\,0\,0\rangle = |7\,4\rangle - 1/\sqrt{2} + |7\,{-4}\rangle + 1/\sqrt{2}$

$|7\,2\,2\,2\rangle = |7\,6\rangle + \sqrt{13}/4\sqrt{3} + |7\,2\rangle - \sqrt{11}/4\sqrt{3} + |7\,{-2}\rangle + \sqrt{11}/4\sqrt{3}$
$\quad + |7\,{-6}\rangle - \sqrt{13}/4\sqrt{3}$

$|7\,{\scriptstyle 0}\,\tilde{1}\,1\,1\rangle = |7\,5\rangle - 1/4\sqrt{2.7} + |7\,1\rangle + \sqrt{3.11}/4\sqrt{2.7} + |7\,{-3}\rangle + 3\sqrt{11}/4\sqrt{2.7}$
$\quad + |7\,{-7}\rangle - \sqrt{13}/4\sqrt{2}$

$|7\,{\scriptstyle 0}\,\tilde{1}\,1\,{-1}\rangle = |7\,7\rangle - \sqrt{13}/4\sqrt{2} + |7\,3\rangle + 3\sqrt{11}/4\sqrt{2.7} + |7\,{-1}\rangle + \sqrt{3.11}/4\sqrt{2.7}$
$\quad + |7\,{-5}\rangle - 1/4\sqrt{2.7}$

$|7\,{\scriptstyle 0}\,\tilde{1}\,\tilde{2}\,2\rangle = |7\,6\rangle - \sqrt{13}/4\sqrt{2.7} + |7\,2\rangle + 3\sqrt{11}/4\sqrt{2.7} + |7\,{-2}\rangle + 3\sqrt{11}/4\sqrt{2.7}$
$\quad + |7\,{-6}\rangle - \sqrt{13}/4\sqrt{2.7}$

$|7\,{\scriptstyle 1}\,\tilde{1}\,1\,1\rangle = |7\,5\rangle + \sqrt{11.13}/8\sqrt{2.7} + |7\,1\rangle + 3\sqrt{3.13}/8\sqrt{2.7} + |7\,{-3}\rangle - 5\sqrt{13}/8\sqrt{2.7}$
$\quad + |7\,{-7}\rangle - \sqrt{11}/8\sqrt{2}$

$|7\,{\scriptstyle 1}\,\tilde{1}\,1\,{-1}\rangle = |7\,7\rangle - \sqrt{11}/8\sqrt{2} + |7\,3\rangle - 5\sqrt{13}/8\sqrt{2.7} + |7\,{-1}\rangle + 3\sqrt{3.13}/8\sqrt{2.7}$
$\quad + |7\,{-5}\rangle + \sqrt{11.13}/8\sqrt{2.7}$

$|7\,{\scriptstyle 1}\,\tilde{1}\,\tilde{2}\,2\rangle = |7\,6\rangle + 3\sqrt{11}/4\sqrt{2.7} + |7\,2\rangle + \sqrt{13}/4\sqrt{2.7} + |7\,{-2}\rangle + \sqrt{13}/4\sqrt{2.7}$
$\quad + |7\,{-6}\rangle + 3\sqrt{11}/4\sqrt{2.7}$

$|7\,\tilde{0}\,2\,2\rangle = |7\,6\rangle - \sqrt{11}/4\sqrt{3} + |7\,2\rangle - \sqrt{13}/4\sqrt{3} + |7\,{-2}\rangle + \sqrt{13}/4\sqrt{3}$
$\quad + |7\,{-6}\rangle + \sqrt{11}/4\sqrt{3}$

$|\tfrac{15}{2}\,\tfrac{1}{2}\,\tfrac{1}{2}\,\tfrac{1}{2}\rangle = |\tfrac{15}{2}\,\tfrac{9}{2}\rangle + \sqrt{7}/8\sqrt{3} + |\tfrac{15}{2}\,\tfrac{1}{2}\rangle + \sqrt{3.11}/8 + |\tfrac{15}{2}\,{-\tfrac{7}{2}}\rangle + \sqrt{7}/8$
$\quad + |\tfrac{15}{2}\,{-\tfrac{15}{2}}\rangle + \sqrt{5.13}/8\sqrt{3}$

$|\tfrac{15}{2}\,\tfrac{1}{2}\,\tfrac{1}{2}\,{-\tfrac{1}{2}}\rangle = |\tfrac{15}{2}\,\tfrac{15}{2}\rangle - \sqrt{5.13}/8\sqrt{3} + |\tfrac{15}{2}\,\tfrac{7}{2}\rangle - \sqrt{7}/8 + |\tfrac{15}{2}\,{-\tfrac{1}{2}}\rangle - \sqrt{3.11}/8$
$\quad + |\tfrac{15}{2}\,{-\tfrac{9}{2}}\rangle - \sqrt{7}/8\sqrt{3}$

$|\tfrac{15}{2}\,0\,\tfrac{3}{2}\,\tfrac{1}{2}\,\tfrac{1}{2}\rangle = |\tfrac{15}{2}\,\tfrac{9}{2}\rangle + 3\sqrt{11}/4\sqrt{2.5} + |\tfrac{15}{2}\,\tfrac{1}{2}\rangle - \sqrt{7}/2\sqrt{2.5} + |\tfrac{15}{2}\,{-\tfrac{7}{2}}\rangle + \sqrt{3.11}/4\sqrt{2.5}$

$|\tfrac{15}{2}\,0\,\tfrac{3}{2}\,\tfrac{1}{2}\,{-\tfrac{1}{2}}\rangle = |\tfrac{15}{2}\,\tfrac{7}{2}\rangle - \sqrt{3.11}/4\sqrt{2.5} + |\tfrac{15}{2}\,{-\tfrac{1}{2}}\rangle + \sqrt{7}/2\sqrt{2.5} + |\tfrac{15}{2}\,{-\tfrac{9}{2}}\rangle - 3\sqrt{11}/4\sqrt{2.5}$

$|\tfrac{15}{2}\,0\,\tfrac{3}{2}\,\tfrac{3}{2}\,\tfrac{3}{2}\rangle = |\tfrac{15}{2}\,\tfrac{11}{2}\rangle + \sqrt{11.13}/4\sqrt{2.5} + |\tfrac{15}{2}\,\tfrac{3}{2}\rangle - \sqrt{3}/2\sqrt{2.5} + |\tfrac{15}{2}\,{-\tfrac{5}{2}}\rangle + 1/4\sqrt{2}$

$|\tfrac{15}{2}\,0\,\tfrac{3}{2}\,\tfrac{3}{2}\,{-\tfrac{3}{2}}\rangle = |\tfrac{15}{2}\,\tfrac{5}{2}\rangle + 1/4\sqrt{2} + |\tfrac{15}{2}\,{-\tfrac{3}{2}}\rangle - \sqrt{3}/2\sqrt{2.5} + |\tfrac{15}{2}\,{-\tfrac{11}{2}}\rangle + \sqrt{11.13}/4\sqrt{2.5}$

$|\tfrac{15}{2}\,1\,\tfrac{3}{2}\,\tfrac{1}{2}\,\tfrac{1}{2}\rangle = |\tfrac{15}{2}\,\tfrac{9}{2}\rangle + 7\sqrt{7}/8.3\sqrt{5} + |\tfrac{15}{2}\,\tfrac{1}{2}\rangle + 3\sqrt{11}/8\sqrt{5} + |\tfrac{15}{2}\,{-\tfrac{7}{2}}\rangle - \sqrt{7}/8\sqrt{3.5}$
$\quad + |\tfrac{15}{2}\,{-\tfrac{15}{2}}\rangle - 5\sqrt{13}/8.3$

$|\tfrac{15}{2}\,1\,\tfrac{3}{2}\,\tfrac{1}{2}\,{-\tfrac{1}{2}}\rangle = |\tfrac{15}{2}\,\tfrac{15}{2}\rangle + 5\sqrt{13}/8.3 + |\tfrac{15}{2}\,\tfrac{7}{2}\rangle + \sqrt{7}/8\sqrt{3.5} + |\tfrac{15}{2}\,{-\tfrac{1}{2}}\rangle - 3\sqrt{11}/8\sqrt{5}$
$\quad + |\tfrac{15}{2}\,{-\tfrac{9}{2}}\rangle - 7\sqrt{7}/8.3\sqrt{5}$

$|\tfrac{15}{2}\,1\,\tfrac{3}{2}\,\tfrac{3}{2}\,\tfrac{3}{2}\rangle = |\tfrac{15}{2}\,\tfrac{11}{2}\rangle - \sqrt{7.13}/8.3\sqrt{5} + |\tfrac{15}{2}\,\tfrac{3}{2}\rangle - \sqrt{3.7.11}/8\sqrt{5} + |\tfrac{15}{2}\,{-\tfrac{5}{2}}\rangle - \sqrt{7.11}/8.3$
$\quad + |\tfrac{15}{2}\,{-\tfrac{13}{2}}\rangle - \sqrt{5.13}/8.3$

$|\tfrac{15}{2}\,1\,\tfrac{3}{2}\,\tfrac{3}{2}\,{-\tfrac{3}{2}}\rangle = |\tfrac{15}{2}\,\tfrac{13}{2}\rangle - \sqrt{5.13}/8.3 + |\tfrac{15}{2}\,\tfrac{5}{2}\rangle - \sqrt{7.11}/8.3 + |\tfrac{15}{2}\,{-\tfrac{3}{2}}\rangle - \sqrt{3.7.11}/8\sqrt{5}$
$\quad + |\tfrac{15}{2}\,{-\tfrac{11}{2}}\rangle - \sqrt{7.13}/8.3\sqrt{5}$

$|\tfrac{15}{2}\,2\,\tfrac{3}{2}\,\tfrac{1}{2}\,\tfrac{1}{2}\rangle = |\tfrac{15}{2}\,\tfrac{9}{2}\rangle + \sqrt{5.13}/4.3\sqrt{2} + |\tfrac{15}{2}\,{-\tfrac{7}{2}}\rangle - \sqrt{5.13}/4\sqrt{2.3} + |\tfrac{15}{2}\,{-\tfrac{15}{2}}\rangle + \sqrt{7}/2.3\sqrt{2}$

$|\tfrac{15}{2}\,2\,\tfrac{3}{2}\,\tfrac{1}{2}\,{-\tfrac{1}{2}}\rangle = |\tfrac{15}{2}\,\tfrac{15}{2}\rangle - \sqrt{7}/2.3\sqrt{2} + |\tfrac{15}{2}\,\tfrac{7}{2}\rangle + \sqrt{5.13}/4\sqrt{2.3} + |\tfrac{15}{2}\,{-\tfrac{9}{2}}\rangle - \sqrt{5.13}/4.3\sqrt{2}$

$|\tfrac{15}{2}\,2\,\tfrac{3}{2}\,\tfrac{3}{2}\,\tfrac{3}{2}\rangle = |\tfrac{15}{2}\,\tfrac{11}{2}\rangle - \sqrt{5}/4.3\sqrt{2} + |\tfrac{15}{2}\,{-\tfrac{5}{2}}\rangle + \sqrt{11.13}/4.3\sqrt{2} + |\tfrac{15}{2}\,{-\tfrac{13}{2}}\rangle - \sqrt{5.7}/2.3\sqrt{2}$

$|\tfrac{15}{2}\,2\,\tfrac{3}{2}\,\tfrac{3}{2}\,{-\tfrac{3}{2}}\rangle = |\tfrac{15}{2}\,\tfrac{13}{2}\rangle - \sqrt{5.7}/2.3\sqrt{2} + |\tfrac{15}{2}\,\tfrac{5}{2}\rangle + \sqrt{11.13}/4.3\sqrt{2} + |\tfrac{15}{2}\,{-\tfrac{11}{2}}\rangle - \sqrt{5}/4.3\sqrt{2}$

$|\tfrac{15}{2}\,\tilde{\tfrac{1}{2}}\,\tfrac{3}{2}\,\tfrac{3}{2}\rangle = |\tfrac{15}{2}\,\tfrac{11}{2}\rangle - \sqrt{11}/8\sqrt{3} + |\tfrac{15}{2}\,\tfrac{3}{2}\rangle - \sqrt{13}/8 + |\tfrac{15}{2}\,{-\tfrac{5}{2}}\rangle + \sqrt{5.13}/8\sqrt{3}$
$\quad + |\tfrac{15}{2}\,{-\tfrac{13}{2}}\rangle + \sqrt{7.11}/8\sqrt{3}$

$|\tfrac{15}{2}\,\tilde{\tfrac{1}{2}}\,\tfrac{3}{2}\,{-\tfrac{3}{2}}\rangle = |\tfrac{15}{2}\,\tfrac{13}{2}\rangle + \sqrt{7.11}/8\sqrt{3} + |\tfrac{15}{2}\,\tfrac{5}{2}\rangle + \sqrt{5.13}/8\sqrt{3} + |\tfrac{15}{2}\,{-\tfrac{3}{2}}\rangle - \sqrt{13}/8$
$\quad + |\tfrac{15}{2}\,{-\tfrac{11}{2}}\rangle - \sqrt{11}/8\sqrt{3}$

SO_3–O–D_4–C_4 Partners (cont.)

$|\,8\ 0\ 0\ 0\rangle = |\,8\ 8\rangle - \sqrt{5.13}/8\sqrt{2.3} + |\,8\ 4\rangle - \sqrt{7}/4\sqrt{2.3} + |\,8\ 0\rangle - \sqrt{3.11}/8$
$+ |\,8\,\text{-}4\rangle - \sqrt{7}/4\sqrt{2.3} + |\,8\,\text{-}8\rangle - \sqrt{5.13}/8\sqrt{2.3}$
$|\,8\ {}_0\ 1\ \tilde{0}\ 0\rangle = |\,8\ 8\rangle - \sqrt{5.13}/4.3 + |\,8\ 4\rangle - \sqrt{7}/4.3 + |\,8\,\text{-}4\rangle + \sqrt{7}/4.3$
$+ |\,8\,\text{-}8\rangle + \sqrt{5.13}/4.3$
$|\,8\ {}_0\ 1\ 1\ 1\rangle = |\,8\ 5\rangle - \sqrt{7.13}/8.3\sqrt{2} + |\,8\ 1\rangle - 3\sqrt{11}/8\sqrt{2} + |\,8\,\text{-}3\rangle - \sqrt{5.7}/8\sqrt{2.3}$
$+ |\,8\,\text{-}7\rangle - \sqrt{5.13}/8.3\sqrt{2}$
$|\,8\ {}_0\ 1\ 1\,\text{-}1\rangle = |\,8\ 7\rangle + \sqrt{5.13}/8.3\sqrt{2} + |\,8\ 3\rangle + \sqrt{5.7}/8\sqrt{2.3} + |\,8\,\text{-}1\rangle + 3\sqrt{11}/8\sqrt{2}$
$+ |\,8\,\text{-}5\rangle + \sqrt{7.13}/8.3\sqrt{2}$
$|\,8\ {}_1\ 1\ \tilde{0}\ 0\rangle = |\,8\ 8\rangle + \sqrt{7}/4.3 + |\,8\ 4\rangle - \sqrt{5.13}/4.3 + |\,8\,\text{-}4\rangle + \sqrt{5.13}/4.3$
$+ |\,8\,\text{-}8\rangle - \sqrt{7}/4.3$
$|\,8\ {}_1\ 1\ 1\ 1\rangle = |\,8\ 5\rangle - \sqrt{5}/2.3\sqrt{2} + |\,8\,\text{-}3\rangle + \sqrt{13}/2\sqrt{2.3} + |\,8\,\text{-}7\rangle - \sqrt{7}/3\sqrt{2}$
$|\,8\ {}_1\ 1\ 1\,\text{-}1\rangle = |\,8\ 7\rangle + \sqrt{7}/3\sqrt{2} + |\,8\ 3\rangle - \sqrt{13}/2\sqrt{2.3} + |\,8\,\text{-}5\rangle + \sqrt{5}/2.3\sqrt{2}$
$|\,8\ {}_0\ 2\ 0\ 0\rangle = |\,8\ 8\rangle + \sqrt{7.11.13}/8.4\sqrt{3} + |\,8\ 4\rangle - \sqrt{5.11}/8.2\sqrt{3} + |\,8\ 0\rangle - \sqrt{3.5.7}/8.2\sqrt{2}$
$+ |\,8\,\text{-}4\rangle - \sqrt{5.11}/8.2\sqrt{3} + |\,8\,\text{-}8\rangle + \sqrt{7.11.13}/8.4\sqrt{3}$
$|\,8\ {}_0\ 2\ 2\ 2\rangle = |\,8\ 2\rangle + 1/\sqrt{2} + |\,8\,\text{-}2\rangle + 1/\sqrt{2}$
$|\,8\ {}_1\ 2\ 0\ 0\rangle = |\,8\ 8\rangle - \sqrt{5}/8.4 + |\,8\ 4\rangle - \sqrt{7.13}/8.2 + |\,8\ 0\rangle + \sqrt{11.13}/8.2\sqrt{2}$
$+ |\,8\,\text{-}4\rangle - \sqrt{7.13}/8.2 + |\,8\,\text{-}8\rangle - \sqrt{5}/8.4$
$|\,8\ {}_1\ 2\ 2\ 2\rangle = |\,8\ 6\rangle - 1/\sqrt{2} + |\,8\,\text{-}6\rangle - 1/\sqrt{2}$
$|\,8\ {}_0\ \tilde{1}\ 1\ 1\rangle = |\,8\ 5\rangle - \sqrt{3.11}/8\sqrt{2.17} + |\,8\ 1\rangle - \sqrt{3.7.13}/8\sqrt{2.17}$
$+ |\,8\,\text{-}3\rangle + \sqrt{5.11.13}/8\sqrt{2.17} + |\,8\,\text{-}7\rangle + \sqrt{3.5.7.11}/8\sqrt{2.17}$
$|\,8\ {}_0\ \tilde{1}\ 1\,\text{-}1\rangle = |\,8\ 7\rangle - \sqrt{3.5.7.11}/8\sqrt{2.17} + |\,8\ 3\rangle - \sqrt{5.11.13}/8\sqrt{2.17}$
$+ |\,8\,\text{-}1\rangle + \sqrt{3.7.13}/8\sqrt{2.17} + |\,8\,\text{-}5\rangle + \sqrt{3.11}/8\sqrt{2.17}$
$|\,8\ {}_0\ \tilde{1}\ \tilde{2}\ 2\rangle = |\,8\ 6\rangle - \sqrt{7.11}/4\sqrt{2.17} + |\,8\ 2\rangle - \sqrt{3.5.13}/4\sqrt{2.17}$
$+ |\,8\,\text{-}2\rangle + \sqrt{3.5.13}/4\sqrt{2.17} + |\,8\,\text{-}6\rangle + \sqrt{7.11}/4\sqrt{2.17}$
$|\,8\ {}_1\ \tilde{1}\ 1\ 1\rangle = |\,8\ 5\rangle - \sqrt{5.7.13}/4\sqrt{2.17} + |\,8\ 1\rangle + \sqrt{5.11}/4\sqrt{2.17} + |\,8\,\text{-}3\rangle - \sqrt{3.7}/4\sqrt{2.17}$
$+ |\,8\,\text{-}7\rangle + \sqrt{13}/4\sqrt{2.17}$
$|\,8\ {}_1\ \tilde{1}\ 1\,\text{-}1\rangle = |\,8\ 7\rangle - \sqrt{13}/4\sqrt{2.17} + |\,8\ 3\rangle + \sqrt{3.7}/4\sqrt{2.17} + |\,8\,\text{-}1\rangle - \sqrt{5.11}/4\sqrt{2.17}$
$+ |\,8\,\text{-}5\rangle + \sqrt{5.7.13}/4\sqrt{2.17}$
$|\,8\ {}_1\ \tilde{1}\ \tilde{2}\ 2\rangle = |\,8\ 6\rangle + \sqrt{3.5.13}/4\sqrt{2.17} + |\,8\ 2\rangle - \sqrt{7.11}/4\sqrt{2.17} + |\,8\,\text{-}2\rangle + \sqrt{7.11}/4\sqrt{2.17}$
$+ |\,8\,\text{-}6\rangle - \sqrt{3.5.13}/4\sqrt{2.17}$

SO_3–K–D_3–C_3 Partners as JM Partners

$|\,0\ 0\ 0\ 0\rangle = |\,0\ 0\rangle + 1$

$|\,\tfrac{1}{2}\ \tfrac{1}{2}\ \tfrac{1}{2}\ \tfrac{1}{2}\rangle = |\,\tfrac{1}{2}\ \tfrac{1}{2}\rangle + 1$
$|\,\tfrac{1}{2}\ \tfrac{1}{2}\ \tfrac{1}{2}\,\text{-}\tfrac{1}{2}\rangle = |\,\tfrac{1}{2}\,\text{-}\tfrac{1}{2}\rangle + 1$

$|\,1\ 1\ \tilde{0}\ 0\rangle = |\,1\ 0\rangle + 1$
$|\,1\ 1\ 1\ 1\rangle = |\,1\ 1\rangle - 1$
$|\,1\ 1\ 1\,\text{-}1\rangle = |\,1\,\text{-}1\rangle - 1$

$|\,\tfrac{3}{2}\ \tfrac{3}{2}\ \tfrac{1}{2}\ \tfrac{1}{2}\rangle = |\,\tfrac{3}{2}\ \tfrac{1}{2}\rangle + 1$
$|\,\tfrac{3}{2}\ \tfrac{3}{2}\ \tfrac{1}{2}\,\text{-}\tfrac{1}{2}\rangle = |\,\tfrac{3}{2}\,\text{-}\tfrac{1}{2}\rangle - 1$
$|\,\tfrac{3}{2}\ \tfrac{3}{2}\ \tfrac{3}{2}\ \tfrac{3}{2}\rangle = |\,\tfrac{3}{2}\ \tfrac{3}{2}\rangle + i/\sqrt{2} + |\,\tfrac{3}{2}\,\text{-}\tfrac{3}{2}\rangle - 1/\sqrt{2}$
$|\,\tfrac{3}{2}\ \tfrac{3}{2}\,\text{-}\tfrac{3}{2}\ \tfrac{3}{2}\rangle = |\,\tfrac{3}{2}\ \tfrac{3}{2}\rangle + 1/\sqrt{2} + |\,\tfrac{3}{2}\,\text{-}\tfrac{3}{2}\rangle - i/\sqrt{2}$

$|\,2\ 2\ 0\ 0\rangle = |\,2\ 0\rangle - 1$
$|\,2\ 2\ {}_0\ 1\ 1\rangle = |\,2\ 1\rangle - 1$

SO_3–K–D_3–C_3 Partners (cont.)

$|\,2\ 2\ 0\ 1\,{-1}\rangle = |\,2\,{-1}\rangle{+1}$
$|\,2\ 2\ 1\ 1\ 1\rangle = |\,2\,{-2}\rangle{+1}$
$|\,2\ 2\ 1\ 1\,{-1}\rangle = |\,2\ 2\rangle{+1}$

$|\,\tfrac{5}{2}\ \tfrac{5}{2}\ 0\ \tfrac{1}{2}\ \tfrac{1}{2}\rangle = |\,\tfrac{5}{2}\ \tfrac{1}{2}\rangle{-1}$
$|\,\tfrac{5}{2}\ \tfrac{5}{2}\ 0\ \tfrac{1}{2}\,{-\tfrac{1}{2}}\rangle = |\,\tfrac{5}{2}\,{-\tfrac{1}{2}}\rangle{-1}$
$|\,\tfrac{5}{2}\ \tfrac{5}{2}\ 1\ \tfrac{1}{2}\ \tfrac{1}{2}\rangle = |\,\tfrac{5}{2}\,{-\tfrac{5}{2}}\rangle{+1}$
$|\,\tfrac{5}{2}\ \tfrac{5}{2}\ 1\ \tfrac{1}{2}\,{-\tfrac{1}{2}}\rangle = |\,\tfrac{5}{2}\ \tfrac{5}{2}\rangle{-1}$
$|\,\tfrac{5}{2}\ \tfrac{5}{2}\ \tfrac{3}{2}\ \tfrac{3}{2}\rangle = |\,\tfrac{5}{2}\ \tfrac{3}{2}\rangle{+i/\sqrt{2}} + |\,\tfrac{5}{2}\,{-\tfrac{3}{2}}\rangle{+1/\sqrt{2}}$
$|\,\tfrac{5}{2}\ \tfrac{5}{2}\,{-\tfrac{3}{2}}\ \tfrac{3}{2}\rangle = |\,\tfrac{5}{2}\ \tfrac{3}{2}\rangle{+1/\sqrt{2}} + |\,\tfrac{5}{2}\,{-\tfrac{3}{2}}\rangle{+i/\sqrt{2}}$

$|\,3\ 3\ 0\ 0\rangle = |\,3\ 3\rangle{-1/\sqrt{2}} + |\,3\,{-3}\rangle{-1/\sqrt{2}}$
$|\,3\ 3\ \tilde{0}\ 0\rangle = |\,3\ 3\rangle{-1/3\sqrt{2}} + |\,3\ 0\rangle{+2\sqrt{2}/3} + |\,3\,{-3}\rangle{+1/3\sqrt{2}}$
$|\,3\ 3\ 1\ 1\rangle = |\,3\ 1\rangle{+1/\sqrt{3}} + |\,3\,{-2}\rangle{-\sqrt{2}/\sqrt{3}}$
$|\,3\ 3\ 1\,{-1}\rangle = |\,3\ 2\rangle{+\sqrt{2}/\sqrt{3}} + |\,3\,{-1}\rangle{+1/\sqrt{3}}$
$|\,3\ \tilde{1}\ \tilde{0}\ 0\rangle = |\,3\ 3\rangle{-2/3} + |\,3\ 0\rangle{-1/3} + |\,3\,{-3}\rangle{+2/3}$
$|\,3\ \tilde{1}\ 1\ 1\rangle = |\,3\ 1\rangle{-\sqrt{2}/\sqrt{3}} + |\,3\,{-2}\rangle{-1/\sqrt{3}}$
$|\,3\ \tilde{1}\ 1\,{-1}\rangle = |\,3\ 2\rangle{+1/\sqrt{3}} + |\,3\,{-1}\rangle{-\sqrt{2}/\sqrt{3}}$

$|\,\tfrac{7}{2}\ \tfrac{5}{2}\ 0\ \tfrac{1}{2}\ \tfrac{1}{2}\rangle = |\,\tfrac{7}{2}\ \tfrac{7}{2}\rangle{+\sqrt{7}/9} + |\,\tfrac{7}{2}\ \tfrac{1}{2}\rangle{-5/9} + |\,\tfrac{7}{2}\,{-\tfrac{5}{2}}\rangle{-7/9}$
$|\,\tfrac{7}{2}\ \tfrac{5}{2}\ 0\ \tfrac{1}{2}\,{-\tfrac{1}{2}}\rangle = |\,\tfrac{7}{2}\ \tfrac{5}{2}\rangle{-7/9} + |\,\tfrac{7}{2}\,{-\tfrac{1}{2}}\rangle{+5/9} + |\,\tfrac{7}{2}\,{-\tfrac{7}{2}}\rangle{+\sqrt{7}/9}$
$|\,\tfrac{7}{2}\ \tfrac{5}{2}\ 1\ \tfrac{1}{2}\ \tfrac{1}{2}\rangle = |\,\tfrac{7}{2}\ \tfrac{7}{2}\rangle{+2\sqrt{2.7}/9} + |\,\tfrac{7}{2}\ \tfrac{1}{2}\rangle{+7/9\sqrt{2}} + |\,\tfrac{7}{2}\,{-\tfrac{5}{2}}\rangle{-1/9\sqrt{2}}$
$|\,\tfrac{7}{2}\ \tfrac{5}{2}\ 1\ \tfrac{1}{2}\,{-\tfrac{1}{2}}\rangle = |\,\tfrac{7}{2}\ \tfrac{5}{2}\rangle{-1/9\sqrt{2}} + |\,\tfrac{7}{2}\,{-\tfrac{1}{2}}\rangle{-7/9\sqrt{2}} + |\,\tfrac{7}{2}\,{-\tfrac{7}{2}}\rangle{+2\sqrt{2.7}/9}$
$|\,\tfrac{7}{2}\ \tfrac{5}{2}\ \tfrac{3}{2}\ \tfrac{3}{2}\rangle = |\,\tfrac{7}{2}\ \tfrac{3}{2}\rangle{-7/2.3\sqrt{3}-i\sqrt{5}/2.3\sqrt{3}} + |\,\tfrac{7}{2}\,{-\tfrac{3}{2}}\rangle{+\sqrt{5}/2.3\sqrt{3}-7i/2.3\sqrt{3}}$
$|\,\tfrac{7}{2}\ \tfrac{5}{2}\,{-\tfrac{3}{2}}\ \tfrac{3}{2}\rangle = |\,\tfrac{7}{2}\ \tfrac{3}{2}\rangle{-\sqrt{5}/2.3\sqrt{3}-7i/2.3\sqrt{3}} + |\,\tfrac{7}{2}\,{-\tfrac{3}{2}}\rangle{-7/2.3\sqrt{3}+i\sqrt{5}/2.3\sqrt{3}}$
$|\,\tfrac{7}{2}\ \tilde{\tfrac{7}{2}}\ \tfrac{1}{2}\ \tfrac{1}{2}\rangle = |\,\tfrac{7}{2}\ \tfrac{7}{2}\rangle{-\sqrt{2}/3} + |\,\tfrac{7}{2}\ \tfrac{1}{2}\rangle{+\sqrt{7}/3\sqrt{2}} + |\,\tfrac{7}{2}\,{-\tfrac{5}{2}}\rangle{-\sqrt{7}/3\sqrt{2}}$
$|\,\tfrac{7}{2}\ \tilde{\tfrac{7}{2}}\ \tfrac{1}{2}\,{-\tfrac{1}{2}}\rangle = |\,\tfrac{7}{2}\ \tfrac{5}{2}\rangle{-\sqrt{7}/3\sqrt{2}} + |\,\tfrac{7}{2}\,{-\tfrac{1}{2}}\rangle{-\sqrt{7}/3\sqrt{2}} + |\,\tfrac{7}{2}\,{-\tfrac{7}{2}}\rangle{-\sqrt{2}/3}$

$|\,4\ 2\ 0\ 0\rangle = |\,4\ 3\rangle{-2\sqrt{7}/9} + |\,4\ 0\rangle{+5/9} + |\,4\,{-3}\rangle{+2\sqrt{7}/9}$
$|\,4\ 2\ 0\ 1\ 1\rangle = |\,4\ 4\rangle{+\sqrt{7}/9\sqrt{3}} + |\,4\ 1\rangle{-2\sqrt{2.5}/9\sqrt{3}} + |\,4\,{-2}\rangle{-2.7/9\sqrt{3}}$
$|\,4\ 2\ 0\ 1\,{-1}\rangle = |\,4\ 2\rangle{-2.7/9\sqrt{3}} + |\,4\,{-1}\rangle{+2\sqrt{2.5}/9\sqrt{3}} + |\,4\,{-4}\rangle{+\sqrt{7}/9\sqrt{3}}$
$|\,4\ 2\ 1\ 1\ 1\rangle = |\,4\ 4\rangle{+2\sqrt{5.7}/9\sqrt{3}} + |\,4\ 1\rangle{+7\sqrt{2}/9\sqrt{3}} + |\,4\,{-2}\rangle{-\sqrt{5}/9\sqrt{3}}$
$|\,4\ 2\ 1\ 1\,{-1}\rangle = |\,4\ 2\rangle{-\sqrt{5}/9\sqrt{3}} + |\,4\,{-1}\rangle{-7\sqrt{2}/9\sqrt{3}} + |\,4\,{-4}\rangle{+2\sqrt{5.7}/9\sqrt{3}}$
$|\,4\ 3\ 0\ 0\rangle = |\,4\ 3\rangle{-5/9\sqrt{2}} + |\,4\ 0\rangle{-2\sqrt{2.7}/9} + |\,4\,{-3}\rangle{+5/9\sqrt{2}}$
$|\,4\ 3\ \tilde{0}\ 0\rangle = |\,4\ 3\rangle{+1/\sqrt{2}} + |\,4\,{-3}\rangle{+1/\sqrt{2}}$
$|\,4\ 3\ 1\ 1\rangle = |\,4\ 4\rangle{-4\sqrt{2}/9} + |\,4\ 1\rangle{+\sqrt{5.7}/9} + |\,4\,{-2}\rangle{-\sqrt{2.7}/9}$
$|\,4\ 3\ 1\,{-1}\rangle = |\,4\ 2\rangle{-\sqrt{2.7}/9} + |\,4\,{-1}\rangle{-\sqrt{5.7}/9} + |\,4\,{-4}\rangle{-4\sqrt{2}/9}$

$|\,\tfrac{9}{2}\ \tfrac{3}{2}\ \tfrac{1}{2}\ \tfrac{1}{2}\rangle = |\,\tfrac{9}{2}\ \tfrac{7}{2}\rangle{+\sqrt{7}/3\sqrt{5}} + |\,\tfrac{9}{2}\ \tfrac{1}{2}\rangle{-\sqrt{2}/3} + |\,\tfrac{9}{2}\,{-\tfrac{5}{2}}\rangle{-2\sqrt{7}/3\sqrt{5}}$
$|\,\tfrac{9}{2}\ \tfrac{3}{2}\ \tfrac{1}{2}\,{-\tfrac{1}{2}}\rangle = |\,\tfrac{9}{2}\ \tfrac{5}{2}\rangle{+2\sqrt{7}/3\sqrt{5}} + |\,\tfrac{9}{2}\,{-\tfrac{1}{2}}\rangle{-\sqrt{2}/3} + |\,\tfrac{9}{2}\,{-\tfrac{7}{2}}\rangle{-\sqrt{7}/3\sqrt{5}}$
$|\,\tfrac{9}{2}\ \tfrac{3}{2}\ \tfrac{3}{2}\ \tfrac{3}{2}\rangle = |\,\tfrac{9}{2}\ \tfrac{9}{2}\rangle{-2\sqrt{2.7}/9\sqrt{3}-i\sqrt{7}/9\sqrt{2.3.5}} + |\,\tfrac{9}{2}\ \tfrac{3}{2}\rangle{-7\sqrt{2}/9\sqrt{5}+i\sqrt{2}/9}$
$\quad + |\,\tfrac{9}{2}\,{-\tfrac{3}{2}}\rangle{+\sqrt{2}/9+7i\sqrt{2}/9\sqrt{5}} + |\,\tfrac{9}{2}\,{-\tfrac{9}{2}}\rangle{+\sqrt{7}/9\sqrt{2.3.5}-2i\sqrt{2.7}/9\sqrt{3}}$
$|\,\tfrac{9}{2}\ \tfrac{3}{2}\,{-\tfrac{3}{2}}\ \tfrac{3}{2}\rangle = |\,\tfrac{9}{2}\ \tfrac{9}{2}\rangle{-\sqrt{7}/9\sqrt{2.3.5}-2i\sqrt{2.7}/9\sqrt{3}} + |\,\tfrac{9}{2}\ \tfrac{3}{2}\rangle{+\sqrt{2}/9-7i\sqrt{2}/9\sqrt{5}}$
$\quad + |\,\tfrac{9}{2}\,{-\tfrac{3}{2}}\rangle{+7\sqrt{2}/9\sqrt{5}+i\sqrt{2}/9} + |\,\tfrac{9}{2}\,{-\tfrac{9}{2}}\rangle{-2\sqrt{2.7}/9\sqrt{3}+i\sqrt{7}/9\sqrt{2.3.5}}$
$|\,\tfrac{9}{2}\ \tfrac{5}{2}\ 0\ \tfrac{1}{2}\ \tfrac{1}{2}\rangle = |\,\tfrac{9}{2}\ \tfrac{7}{2}\rangle{-11\sqrt{2}/9\sqrt{5}} + |\,\tfrac{9}{2}\ \tfrac{1}{2}\rangle{+\sqrt{7}/9} + |\,\tfrac{9}{2}\,{-\tfrac{5}{2}}\rangle{-8\sqrt{2}/9\sqrt{5}}$
$|\,\tfrac{9}{2}\ \tfrac{5}{2}\ 0\ \tfrac{1}{2}\,{-\tfrac{1}{2}}\rangle = |\,\tfrac{9}{2}\ \tfrac{5}{2}\rangle{+8\sqrt{2}/9\sqrt{5}} + |\,\tfrac{9}{2}\,{-\tfrac{1}{2}}\rangle{+\sqrt{7}/9} + |\,\tfrac{9}{2}\,{-\tfrac{7}{2}}\rangle{+11\sqrt{2}/9\sqrt{5}}$
$|\,\tfrac{9}{2}\ \tfrac{5}{2}\ 1\ \tfrac{1}{2}\ \tfrac{1}{2}\rangle = |\,\tfrac{9}{2}\ \tfrac{7}{2}\rangle{+2\sqrt{5}/9} + |\,\tfrac{9}{2}\ \tfrac{1}{2}\rangle{+2\sqrt{2.7}/9} + |\,\tfrac{9}{2}\,{-\tfrac{5}{2}}\rangle{-\sqrt{5}/9}$
$|\,\tfrac{9}{2}\ \tfrac{5}{2}\ 1\ \tfrac{1}{2}\,{-\tfrac{1}{2}}\rangle = |\,\tfrac{9}{2}\ \tfrac{5}{2}\rangle{+\sqrt{5}/9} + |\,\tfrac{9}{2}\,{-\tfrac{1}{2}}\rangle{+2\sqrt{2.7}/9} + |\,\tfrac{9}{2}\,{-\tfrac{7}{2}}\rangle{-2\sqrt{5}/9}$
$|\,\tfrac{9}{2}\ \tfrac{5}{2}\ \tfrac{3}{2}\ \tfrac{3}{2}\rangle = |\,\tfrac{9}{2}\ \tfrac{9}{2}\rangle{+\sqrt{2}/3-i\sqrt{2}/3\sqrt{5}} + |\,\tfrac{9}{2}\ \tfrac{3}{2}\rangle{-\sqrt{2.7}/3\sqrt{3.5}+i\sqrt{7}/3\sqrt{2.3}}$
$\quad + |\,\tfrac{9}{2}\,{-\tfrac{3}{2}}\rangle{+\sqrt{7}/3\sqrt{2.3}+i\sqrt{2.7}/3\sqrt{3.5}} + |\,\tfrac{9}{2}\,{-\tfrac{9}{2}}\rangle{+\sqrt{2}/3\sqrt{5}+i\sqrt{2}/3}$
$|\,\tfrac{9}{2}\ \tfrac{5}{2}\,{-\tfrac{3}{2}}\ \tfrac{3}{2}\rangle = |\,\tfrac{9}{2}\ \tfrac{9}{2}\rangle{-\sqrt{2}/3\sqrt{5}+i\sqrt{2}/3} + |\,\tfrac{9}{2}\ \tfrac{3}{2}\rangle{+\sqrt{7}/3\sqrt{2.3}-i\sqrt{2.7}/3\sqrt{3.5}}$
$\quad + |\,\tfrac{9}{2}\,{-\tfrac{3}{2}}\rangle{+\sqrt{2.7}/3\sqrt{3.5}+i\sqrt{7}/3\sqrt{2.3}} + |\,\tfrac{9}{2}\,{-\tfrac{9}{2}}\rangle{+\sqrt{2}/3+i\sqrt{2}/3\sqrt{5}}$

SO_3–K–D_3–C_3 Partners (cont.)

$|5\;1\;\tilde{0}\;0\rangle = |5\;3\rangle - \sqrt{7}/3\sqrt{2} + |5\;0\rangle + \sqrt{2}/3 + |5\,\text{-}3\rangle + \sqrt{7}/3\sqrt{2}$

$|5\;1\;1\;1\rangle = |5\;4\rangle - \sqrt{7}/9\sqrt{2} + |5\;1\rangle + \sqrt{5}/3\sqrt{2.3} + |5\,\text{-}2\rangle + \sqrt{2.7}/3\sqrt{3}$

$\quad + |5\,\text{-}5\rangle - 2\sqrt{7}/9$

$|5\;1\;1\,\text{-}1\rangle = |5\;5\rangle - 2\sqrt{7}/9 + |5\;2\rangle - \sqrt{2.7}/3\sqrt{3} + |5\,\text{-}1\rangle + \sqrt{5}/3\sqrt{2.3}$

$\quad + |5\,\text{-}4\rangle + \sqrt{7}/9\sqrt{2}$

$|5\;2\;0\;0\rangle = |5\;3\rangle + 1/\sqrt{2} + |5\,\text{-}3\rangle + 1/\sqrt{2}$

$|5\;2\;0\;1\;1\rangle = |5\;4\rangle - 13/9\sqrt{2.3} + |5\;1\rangle + \sqrt{5.7}/9\sqrt{2} + |5\,\text{-}2\rangle + \sqrt{2}/9$

$\quad + |5\,\text{-}5\rangle + 2.5/9\sqrt{3}$

$|5\;2\;0\;1\,\text{-}1\rangle = |5\;5\rangle + 2.5/9\sqrt{3} + |5\;2\rangle - \sqrt{2}/9 + |5\,\text{-}1\rangle + \sqrt{5.7}/9\sqrt{2}$

$\quad + |5\,\text{-}4\rangle + 13/9\sqrt{2.3}$

$|5\;2\;1\;1\;1\rangle = |5\;4\rangle + 2\sqrt{2.5}/9\sqrt{3} + |5\;1\rangle + 2\sqrt{2.7}/9 + |5\,\text{-}2\rangle - \sqrt{2.5}/9$

$\quad + |5\,\text{-}5\rangle - \sqrt{5}/9\sqrt{3}$

$|5\;2\;1\;1\,\text{-}1\rangle = |5\;5\rangle - \sqrt{5}/9\sqrt{3} + |5\;2\rangle + \sqrt{2.5}/9 + |5\,\text{-}1\rangle + 2\sqrt{2.7}/9$

$\quad + |5\,\text{-}4\rangle - 2\sqrt{2.5}/9\sqrt{3}$

$|5\;\tilde{1}\;\tilde{0}\;0\rangle = |5\;3\rangle + 1/3 + |5\;0\rangle + \sqrt{7}/3 + |5\,\text{-}3\rangle - 1/3$

$|5\;\tilde{1}\;1\;1\rangle = |5\;4\rangle - 2/3 + |5\,\text{-}2\rangle - 1/\sqrt{3} + |5\,\text{-}5\rangle - \sqrt{2}/3$

$|5\;\tilde{1}\;1\,\text{-}1\rangle = |5\;5\rangle - \sqrt{2}/3 + |5\;2\rangle + 1/\sqrt{3} + |5\,\text{-}4\rangle + 2/3$

$|\tfrac{11}{2}\;\tfrac{1}{2}\;\tfrac{1}{2}\;\tfrac{1}{2}\rangle = |\tfrac{11}{2}\;\tfrac{7}{2}\rangle - \sqrt{7.11}/9\sqrt{2.3} + |\tfrac{11}{2}\;\tfrac{1}{2}\rangle + \sqrt{11}/9 + |\tfrac{11}{2}\,\text{-}\tfrac{5}{2}\rangle + \sqrt{7.11}/9\sqrt{2}$

$\quad + |\tfrac{11}{2}\,\text{-}\tfrac{11}{2}\rangle - 2\sqrt{2.7}/9\sqrt{3}$

$|\tfrac{11}{2}\;\tfrac{1}{2}\;\tfrac{1}{2}\,\text{-}\tfrac{1}{2}\rangle = |\tfrac{11}{2}\;\tfrac{11}{2}\rangle + 2\sqrt{2.7}/9\sqrt{3} + |\tfrac{11}{2}\;\tfrac{5}{2}\rangle + \sqrt{7.11}/9\sqrt{2} + |\tfrac{11}{2}\,\text{-}\tfrac{1}{2}\rangle - \sqrt{11}/9$

$\quad + |\tfrac{11}{2}\,\text{-}\tfrac{7}{2}\rangle - \sqrt{7.11}/9\sqrt{2.3}$

$|\tfrac{11}{2}\;\tfrac{3}{2}\;\tfrac{1}{2}\;\tfrac{1}{2}\rangle = |\tfrac{11}{2}\;\tfrac{7}{2}\rangle - 8\sqrt{2}/9\sqrt{3} + |\tfrac{11}{2}\;\tfrac{1}{2}\rangle + \sqrt{7}/9 + |\tfrac{11}{2}\,\text{-}\tfrac{5}{2}\rangle - \sqrt{2}/9$

$\quad + |\tfrac{11}{2}\,\text{-}\tfrac{11}{2}\rangle + 2\sqrt{2.11}/9\sqrt{3}$

$|\tfrac{11}{2}\;\tfrac{3}{2}\;\tfrac{1}{2}\,\text{-}\tfrac{1}{2}\rangle = |\tfrac{11}{2}\;\tfrac{11}{2}\rangle - 2\sqrt{2.11}/9\sqrt{3} + |\tfrac{11}{2}\;\tfrac{5}{2}\rangle - \sqrt{2}/9 + |\tfrac{11}{2}\,\text{-}\tfrac{1}{2}\rangle - \sqrt{7}/9$

$\quad + |\tfrac{11}{2}\,\text{-}\tfrac{7}{2}\rangle - 8\sqrt{2}/9\sqrt{3}$

$|\tfrac{11}{2}\;\tfrac{3}{2}\;\tfrac{3}{2}\;\tfrac{3}{2}\rangle = |\tfrac{11}{2}\;\tfrac{9}{2}\rangle - 2/9 + i\sqrt{5}/9 + |\tfrac{11}{2}\;\tfrac{3}{2}\rangle - 2\sqrt{2}/3\sqrt{3} - i\sqrt{5}/3\sqrt{2.3}$

$\quad + |\tfrac{11}{2}\,\text{-}\tfrac{3}{2}\rangle + \sqrt{5}/3\sqrt{2.3} - 2i\sqrt{2}/3\sqrt{3} + |\tfrac{11}{2}\,\text{-}\tfrac{9}{2}\rangle + \sqrt{5}/9 + 2i/9$

$|\tfrac{11}{2}\;\tfrac{3}{2}\,\text{-}\tfrac{3}{2}\;\tfrac{3}{2}\rangle = |\tfrac{11}{2}\;\tfrac{9}{2}\rangle + \sqrt{5}/9 - 2i/9 + |\tfrac{11}{2}\;\tfrac{3}{2}\rangle - \sqrt{5}/3\sqrt{2.3} - 2i\sqrt{2}/3\sqrt{3}$

$\quad + |\tfrac{11}{2}\,\text{-}\tfrac{3}{2}\rangle - 2\sqrt{2}/3\sqrt{3} + i\sqrt{5}/3\sqrt{2.3} + |\tfrac{11}{2}\,\text{-}\tfrac{9}{2}\rangle + 2/9 + i\sqrt{5}/9$

$|\tfrac{11}{2}\;\tfrac{5}{2}\;0\;\tfrac{1}{2}\;\tfrac{1}{2}\rangle = |\tfrac{11}{2}\;\tfrac{7}{2}\rangle + 11/9\sqrt{2.3} + |\tfrac{11}{2}\;\tfrac{1}{2}\rangle + \sqrt{7}/9 + |\tfrac{11}{2}\,\text{-}\tfrac{5}{2}\rangle + 7/9\sqrt{2}$

$\quad + |\tfrac{11}{2}\,\text{-}\tfrac{11}{2}\rangle + 2\sqrt{2.11}/9\sqrt{3}$

$|\tfrac{11}{2}\;\tfrac{5}{2}\;0\;\tfrac{1}{2}\,\text{-}\tfrac{1}{2}\rangle = |\tfrac{11}{2}\;\tfrac{11}{2}\rangle - 2\sqrt{2.11}/9\sqrt{3} + |\tfrac{11}{2}\;\tfrac{5}{2}\rangle + 7/9\sqrt{2} + |\tfrac{11}{2}\,\text{-}\tfrac{1}{2}\rangle - \sqrt{7}/9$

$\quad + |\tfrac{11}{2}\,\text{-}\tfrac{7}{2}\rangle + 11/9\sqrt{2.3}$

$|\tfrac{11}{2}\;\tfrac{5}{2}\;1\;\tfrac{1}{2}\;\tfrac{1}{2}\rangle = |\tfrac{11}{2}\;\tfrac{7}{2}\rangle + 4/9\sqrt{3} + |\tfrac{11}{2}\;\tfrac{1}{2}\rangle + 2\sqrt{2.7}/9 + |\tfrac{11}{2}\,\text{-}\tfrac{5}{2}\rangle - 4/9 + |\tfrac{11}{2}\,\text{-}\tfrac{11}{2}\rangle - \sqrt{11}/9\sqrt{3}$

$|\tfrac{11}{2}\;\tfrac{5}{2}\;1\;\tfrac{1}{2}\,\text{-}\tfrac{1}{2}\rangle = |\tfrac{11}{2}\;\tfrac{11}{2}\rangle + \sqrt{11}/9\sqrt{3} + |\tfrac{11}{2}\;\tfrac{5}{2}\rangle - 4/9 + |\tfrac{11}{2}\,\text{-}\tfrac{1}{2}\rangle - 2\sqrt{2.7}/9 + |\tfrac{11}{2}\,\text{-}\tfrac{7}{2}\rangle + 4/9\sqrt{3}$

$|\tfrac{11}{2}\;\tfrac{5}{2}\;\tfrac{3}{2}\;\tfrac{3}{2}\rangle = |\tfrac{11}{2}\;\tfrac{9}{2}\rangle + 4\sqrt{2}/9\sqrt{3} + 5i\sqrt{5}/9\sqrt{2.3} + |\tfrac{11}{2}\;\tfrac{3}{2}\rangle + 2/9 - i\sqrt{5}/9$

$\quad + |\tfrac{11}{2}\,\text{-}\tfrac{3}{2}\rangle + \sqrt{5}/9 + 2i/9 + |\tfrac{11}{2}\,\text{-}\tfrac{9}{2}\rangle + 5\sqrt{5}/9\sqrt{2.3} - 4i\sqrt{2}/9\sqrt{3}$

$|\tfrac{11}{2}\;\tfrac{5}{2}\,\text{-}\tfrac{3}{2}\;\tfrac{3}{2}\rangle = |\tfrac{11}{2}\;\tfrac{9}{2}\rangle + 5\sqrt{5}/9\sqrt{2.3} + 4i\sqrt{2}/9\sqrt{3} + |\tfrac{11}{2}\;\tfrac{3}{2}\rangle - \sqrt{5}/9 + 2i/9$

$\quad + |\tfrac{11}{2}\,\text{-}\tfrac{3}{2}\rangle + 2/9 + i\sqrt{5}/9 + |\tfrac{11}{2}\,\text{-}\tfrac{9}{2}\rangle - 4\sqrt{2}/9\sqrt{3} + 5i\sqrt{5}/9\sqrt{2.3}$

$|6\;0\;0\;0\rangle = |6\;6\rangle + 2\sqrt{7}/9\sqrt{3} + |6\;3\rangle + \sqrt{7.11}/9\sqrt{3} + |6\;0\rangle - \sqrt{11}/9$

$\quad + |6\,\text{-}3\rangle - \sqrt{7.11}/9\sqrt{3} + |6\,\text{-}6\rangle + 2\sqrt{7}/9\sqrt{3}$

$|6\;1\;\tilde{0}\;0\rangle = |6\;6\rangle + 2\sqrt{2}/3\sqrt{3} + |6\;3\rangle + \sqrt{11}/3\sqrt{2.3} + |6\,\text{-}3\rangle + \sqrt{11}/3\sqrt{2.3}$

$\quad + |6\,\text{-}6\rangle - 2\sqrt{2}/3\sqrt{3}$

$|6\;1\;1\;1\rangle = |6\;4\rangle + \sqrt{5.11}/9\sqrt{2} + |6\;1\rangle - \sqrt{11}/3\sqrt{2.3} + |6\,\text{-}2\rangle - \sqrt{11}/3\sqrt{3}$

$\quad + |6\,\text{-}5\rangle + 2/9$

$|6\;1\;1\,\text{-}1\rangle = |6\;5\rangle - 2/9 + |6\;2\rangle - \sqrt{11}/3\sqrt{3} + |6\,\text{-}1\rangle + \sqrt{11}/3\sqrt{2.3}$

$\quad + |6\,\text{-}4\rangle + \sqrt{5.11}/9\sqrt{2}$

$|6\;2\;0\;0\rangle = |6\;6\rangle - 2\sqrt{2.11}/9\sqrt{3} + |6\;3\rangle + 5/9\sqrt{2.3} + |6\;0\rangle - \sqrt{2.7}/9$

$\quad + |6\,\text{-}3\rangle - 5/9\sqrt{2.3} + |6\,\text{-}6\rangle - 2\sqrt{2.11}/9\sqrt{3}$

SO_3–K–D_3–C_3 Partners (cont.)

$|\,6\ 2_0\ 1\ 1> \;=\; |\,6\ 4>+7\sqrt{5}/9\sqrt{2.3}\;+\;|\,6\ 1>-1/9\sqrt{2}\;+\;|\,6\text{-}2>+5/9$

$\qquad +\;|\,6\text{-}5>-2\sqrt{11}/9\sqrt{3}$

$|\,6\ 2_0\ 1\text{-}1> \;=\; |\,6\ 5>+2\sqrt{11}/9\sqrt{3}\;+\;|\,6\ 2>+5/9\;+\;|\,6\text{-}1>+1/9\sqrt{2}$

$\qquad +\;|\,6\text{-}4>+7\sqrt{5}/9\sqrt{2.3}$

$|\,6\ 2_1\ 1\ 1> \;=\; |\,6\ 4>+2\sqrt{2}/9\sqrt{3}\;+\;|\,6\ 1>+2\sqrt{2.5}/9\;+\;|\,6\text{-}2>-2\sqrt{5}/9$

$\qquad +\;|\,6\text{-}5>-\sqrt{5.11}/9\sqrt{3}$

$|\,6\ 2_1\ 1\text{-}1> \;=\; |\,6\ 5>+\sqrt{5.11}/9\sqrt{3}\;+\;|\,6\ 2>-2\sqrt{5}/9\;+\;|\,6\text{-}1>-2\sqrt{2.5}/9$

$\qquad +\;|\,6\text{-}4>+2\sqrt{2}/9\sqrt{3}$

$|\,6\ 3\ 0\ 0> \;=\; |\,6\ 6>+\sqrt{11}/9\sqrt{2.3}\;+\;|\,6\ 3>-4\sqrt{2}/9\sqrt{3}\;+\;|\,6\ 0>-2\sqrt{2.7}/9$

$\qquad +\;|\,6\text{-}3>+4\sqrt{2}/9\sqrt{3}\;+\;|\,6\text{-}6>+\sqrt{11}/9\sqrt{2.3}$

$|\,6\ 3\ \tilde{0}\ 0> \;=\; |\,6\ 6>+\sqrt{11}/3\sqrt{2.3}\;+\;|\,6\ 3>-2\sqrt{2}/3\sqrt{3}\;+\;|\,6\text{-}3>-2\sqrt{2}/3\sqrt{3}$

$\qquad +\;|\,6\text{-}6>-\sqrt{11}/3\sqrt{2.3}$

$|\,6\ 3\ 1\ 1> \;=\; |\,6\ 4>-\sqrt{2.5}/9\;+\;|\,6\ 1>-2\sqrt{2}/3\sqrt{3}\;+\;|\,6\text{-}2>-1/3\sqrt{3}\;+\;|\,6\text{-}5>-2\sqrt{11}/9$

$|\,6\ 3\ 1\text{-}1> \;=\; |\,6\ 5>+2\sqrt{11}/9\;+\;|\,6\ 2>-1/3\sqrt{3}\;+\;|\,6\text{-}1>+2\sqrt{2}/3\sqrt{3}\;+\;|\,6\text{-}4>-\sqrt{2.5}/9$

$|\,\tfrac{13}{2}\ \tfrac{1}{2}\ \tfrac{1}{2}\ \tfrac{1}{2}> \;=\; |\,\tfrac{13}{2}\,\tfrac{13}{2}>+2\sqrt{13}/9\sqrt{3}\;+\;|\,\tfrac{13}{2}\,\tfrac{7}{2}>+\sqrt{2.5.11}/9\sqrt{3}\;+\;|\,\tfrac{13}{2}\,\tfrac{1}{2}>-\sqrt{11}/9$

$\qquad +\;|\,\tfrac{13}{2}\text{-}\tfrac{5}{2}>-2\sqrt{11}/9\sqrt{3}\;+\;|\,\tfrac{13}{2}\text{-}\tfrac{11}{2}>+2/9\sqrt{3}$

$|\,\tfrac{13}{2}\ \tfrac{1}{2}\ \tfrac{1}{2}\text{-}\tfrac{1}{2}> \;=\; |\,\tfrac{13}{2}\,\tfrac{11}{2}>+2/9\sqrt{3}\;+\;|\,\tfrac{13}{2}\,\tfrac{5}{2}>+2\sqrt{11}/9\sqrt{3}\;+\;|\,\tfrac{13}{2}\text{-}\tfrac{1}{2}>-\sqrt{11}/9$

$\qquad +\;|\,\tfrac{13}{2}\text{-}\tfrac{7}{2}>-\sqrt{2.5.11}/9\sqrt{3}\;+\;|\,\tfrac{13}{2}\text{-}\tfrac{13}{2}>+2\sqrt{13}/9\sqrt{3}$

$|\,\tfrac{13}{2}\ \tfrac{3}{2}\ \tfrac{1}{2}\ \tfrac{1}{2}> \;=\; |\,\tfrac{13}{2}\,\tfrac{13}{2}>+4\sqrt{2.13}/9\sqrt{3.5}\;+\;|\,\tfrac{13}{2}\,\tfrac{7}{2}>+\sqrt{11}/9\sqrt{3}\;+\;|\,\tfrac{13}{2}\,\tfrac{1}{2}>+\sqrt{2.11}/9\sqrt{5}$

$\qquad +\;|\,\tfrac{13}{2}\text{-}\tfrac{5}{2}>+\sqrt{2.5.11}/9\sqrt{3}\;+\;|\,\tfrac{13}{2}\text{-}\tfrac{11}{2}>-8\sqrt{2}/9\sqrt{3.5}$

$|\,\tfrac{13}{2}\ \tfrac{3}{2}\ \tfrac{1}{2}\text{-}\tfrac{1}{2}> \;=\; |\,\tfrac{13}{2}\,\tfrac{11}{2}>-8\sqrt{2}/9\sqrt{3.5}\;+\;|\,\tfrac{13}{2}\,\tfrac{5}{2}>-\sqrt{2.5.11}/9\sqrt{3}\;+\;|\,\tfrac{13}{2}\text{-}\tfrac{1}{2}>+\sqrt{2.11}/9\sqrt{5}$

$\qquad +\;|\,\tfrac{13}{2}\text{-}\tfrac{7}{2}>-\sqrt{11}/9\sqrt{3}\;+\;|\,\tfrac{13}{2}\text{-}\tfrac{13}{2}>+4\sqrt{2.13}/9\sqrt{3.5}$

$|\,\tfrac{13}{2}\ \tfrac{3}{2}\ \tfrac{3}{2}\ \tfrac{3}{2}> \;=\; |\,\tfrac{13}{2}\,\tfrac{9}{2}>-2\sqrt{2}/9\sqrt{3.5}-11\,i/9\sqrt{2.3}\;+\;|\,\tfrac{13}{2}\,\tfrac{3}{2}>-\sqrt{11}/9+2\,i\sqrt{11}/9\sqrt{5}$

$\qquad +\;|\,\tfrac{13}{2}\text{-}\tfrac{3}{2}>+2\sqrt{11}/9\sqrt{5}+i\sqrt{11}/9\;+\;|\,\tfrac{13}{2}\text{-}\tfrac{9}{2}>+11/9\sqrt{2.3}-2\,i\sqrt{2}/9\sqrt{3.5}$

$|\,\tfrac{13}{2}\ \tfrac{3}{2}\text{-}\tfrac{3}{2}\ \tfrac{3}{2}> \;=\; |\,\tfrac{13}{2}\,\tfrac{9}{2}>-11/9\sqrt{2.3}-2\,i\sqrt{2}/9\sqrt{3.5}\;+\;|\,\tfrac{13}{2}\,\tfrac{3}{2}>+2\sqrt{11}/9\sqrt{5}-i\sqrt{11}/9$

$\qquad +\;|\,\tfrac{13}{2}\text{-}\tfrac{3}{2}>+\sqrt{11}/9+2\,i\sqrt{11}/9\sqrt{5}\;+\;|\,\tfrac{13}{2}\text{-}\tfrac{9}{2}>-2\sqrt{2}/9\sqrt{3.5}+11\,i/9\sqrt{2.3}$

$|\,\tfrac{13}{2}\ \tfrac{5}{2}\ 0\ \tfrac{1}{2}\ \tfrac{1}{2}> \;=\; |\,\tfrac{13}{2}\,\tfrac{13}{2}>-\sqrt{11.13}/9\sqrt{5}\;+\;|\,\tfrac{13}{2}\,\tfrac{7}{2}>+\sqrt{2}/3\;+\;|\,\tfrac{13}{2}\,\tfrac{1}{2}>-4/3\sqrt{3.5}$

$\qquad +\;|\,\tfrac{13}{2}\text{-}\tfrac{5}{2}>+\sqrt{5}/9\;+\;|\,\tfrac{13}{2}\text{-}\tfrac{11}{2}>-\sqrt{11}/3\sqrt{5}$

$|\,\tfrac{13}{2}\ \tfrac{5}{2}\ 0\ \tfrac{1}{2}\text{-}\tfrac{1}{2}> \;=\; |\,\tfrac{13}{2}\,\tfrac{11}{2}>-\sqrt{11}/3\sqrt{5}\;+\;|\,\tfrac{13}{2}\,\tfrac{5}{2}>-\sqrt{5}/9\;+\;|\,\tfrac{13}{2}\text{-}\tfrac{1}{2}>-4/3\sqrt{3.5}\;+\;|\,\tfrac{13}{2}\text{-}\tfrac{7}{2}>-\sqrt{2}/3$

$\qquad +\;|\,\tfrac{13}{2}\text{-}\tfrac{13}{2}>-\sqrt{11.13}/9\sqrt{5}$

$|\,\tfrac{13}{2}\ \tfrac{5}{2}\ 1\ \tfrac{1}{2}\ \tfrac{1}{2}> \;=\; |\,\tfrac{13}{2}\,\tfrac{7}{2}>+1/9\;+\;|\,\tfrac{13}{2}\,\tfrac{1}{2}>+\sqrt{2.5}/3\sqrt{3}\;+\;|\,\tfrac{13}{2}\text{-}\tfrac{5}{2}>-\sqrt{5}/3\sqrt{2}$

$\qquad +\;|\,\tfrac{13}{2}\text{-}\tfrac{11}{2}>-\sqrt{5.11}/9\sqrt{2}$

$|\,\tfrac{13}{2}\ \tfrac{5}{2}\ 1\ \tfrac{1}{2}\text{-}\tfrac{1}{2}> \;=\; |\,\tfrac{13}{2}\,\tfrac{11}{2}>-\sqrt{5.11}/9\sqrt{2}\;+\;|\,\tfrac{13}{2}\,\tfrac{5}{2}>+\sqrt{5}/3\sqrt{2}\;+\;|\,\tfrac{13}{2}\text{-}\tfrac{1}{2}>+\sqrt{2.5}/3\sqrt{3}$

$\qquad +\;|\,\tfrac{13}{2}\text{-}\tfrac{7}{2}>-1/9$

$|\,\tfrac{13}{2}\ \tfrac{5}{2}\ \tfrac{3}{2}\ \tfrac{3}{2}> \;=\; |\,\tfrac{13}{2}\,\tfrac{9}{2}>+\sqrt{11}/3\sqrt{2.3.5}-i\sqrt{11}/3\sqrt{2.3}\;+\;|\,\tfrac{13}{2}\,\tfrac{3}{2}>+1/2-i/2.3\sqrt{5}$

$\qquad +\;|\,\tfrac{13}{2}\text{-}\tfrac{3}{2}>-1/2.3\sqrt{5}-i/2\;+\;|\,\tfrac{13}{2}\text{-}\tfrac{9}{2}>+\sqrt{11}/3\sqrt{2.3}+i\sqrt{11}/3\sqrt{2.3.5}$

$|\,\tfrac{13}{2}\ \tfrac{5}{2}\text{-}\tfrac{3}{2}\ \tfrac{3}{2}> \;=\; |\,\tfrac{13}{2}\,\tfrac{9}{2}>-\sqrt{11}/3\sqrt{2.3}+i\sqrt{11}/3\sqrt{2.3.5}\;+\;|\,\tfrac{13}{2}\,\tfrac{3}{2}>-1/2.3\sqrt{5}+i/2$

$\qquad +\;|\,\tfrac{13}{2}\text{-}\tfrac{3}{2}>-1/2-i/2.3\sqrt{5}\;+\;|\,\tfrac{13}{2}\text{-}\tfrac{9}{2}>+\sqrt{11}/3\sqrt{2.3.5}+i\sqrt{11}/3\sqrt{2.3}$

$|\,\tfrac{13}{2}\ \tilde{\tfrac{1}{2}}\ \tfrac{1}{2}\ \tfrac{1}{2}> \;=\; |\,\tfrac{13}{2}\,\tfrac{13}{2}>+\sqrt{2.11}/9\sqrt{3}\;+\;|\,\tfrac{13}{2}\,\tfrac{7}{2}>-\sqrt{5.13}/9\sqrt{3}\;+\;|\,\tfrac{13}{2}\,\tfrac{1}{2}>-\sqrt{2.13}/9$

$\qquad +\;|\,\tfrac{13}{2}\text{-}\tfrac{5}{2}>-\sqrt{13}/9\sqrt{2.3}\;+\;|\,\tfrac{13}{2}\text{-}\tfrac{11}{2}>-\sqrt{11.13}/9\sqrt{2.3}$

$|\,\tfrac{13}{2}\ \tilde{\tfrac{1}{2}}\ \tfrac{1}{2}\text{-}\tfrac{1}{2}> \;=\; |\,\tfrac{13}{2}\,\tfrac{11}{2}>-\sqrt{11.13}/9\sqrt{2.3}\;+\;|\,\tfrac{13}{2}\,\tfrac{5}{2}>+\sqrt{13}/9\sqrt{2.3}\;+\;|\,\tfrac{13}{2}\text{-}\tfrac{1}{2}>-\sqrt{2.13}/9$

$\qquad +\;|\,\tfrac{13}{2}\text{-}\tfrac{7}{2}>+\sqrt{5.13}/9\sqrt{3}\;+\;|\,\tfrac{13}{2}\text{-}\tfrac{13}{2}>+\sqrt{2.11}/9\sqrt{3}$

$|\,7\ 1\ \tilde{0}\ 0> \;=\; |\,7\ 6>+2\sqrt{13}/9\sqrt{3.5}\;+\;|\,7\ 3>+2\sqrt{2.11}/9\sqrt{3}\;+\;|\,7\ 0>-\sqrt{7.11}/9\sqrt{5}$

$\qquad +\;|\,7\text{-}3>-2\sqrt{2.11}/9\sqrt{3}\;+\;|\,7\text{-}6>+2\sqrt{13}/9\sqrt{3.5}$

$|\,7\ 1\ 1\ 1> \;=\; |\,7\ 7>-2\sqrt{7.13}/9\sqrt{3.5}\;+\;|\,7\ 4>-11/9\sqrt{3}\;+\;|\,7\ 1>+2\sqrt{11}/9\sqrt{5}$

$\qquad +\;|\,7\text{-}2>+\sqrt{2.11}/9\sqrt{3}\;+\;|\,7\text{-}5>-2/9\sqrt{3.5}$

$|\,7\ 1\ 1\text{-}1> \;=\; |\,7\ 5>-2/9\sqrt{3.5}\;+\;|\,7\ 2>-\sqrt{2.11}/9\sqrt{3}\;+\;|\,7\text{-}1>+2\sqrt{11}/9\sqrt{5}$

$\qquad +\;|\,7\text{-}4>+11/9\sqrt{3}\;+\;|\,7\text{-}7>-2\sqrt{7.13}/9\sqrt{3.5}$

SO_3–K–D_3–C_3 Partners (cont.)

$|\,7\;2\;0\;0\rangle \;=\; |\,7\;6\rangle - \surd 13/9 \;+\; |\,7\;3\rangle - \surd 5.11/9\surd 2 \;+\; |\,7\text{-}3\rangle - \surd 5.11/9\surd 2$
$\quad + \;|\,7\text{-}6\rangle + \surd 13/9$

$|\,7\;2\;{\scriptstyle 0}\;1\;1\rangle \;=\; |\,7\;7\rangle - \surd 7.13/9\surd 3 \;+\; |\,7\;1\rangle - \surd 11/9 \;+\; |\,7\text{-}2\rangle - \surd 2.5.11/9\surd 3$
$\quad + \;|\,7\text{-}5\rangle + 1/3\surd 3$

$|\,7\;2\;{\scriptstyle 0}\;1\text{-}1\rangle \;=\; |\,7\;5\rangle + 1/3\surd 3 \;+\; |\,7\;2\rangle + \surd 2.5.11/9\surd 3 \;+\; |\,7\text{-}1\rangle - \surd 11/9$
$\quad + \;|\,7\text{-}7\rangle - \surd 7.13/9\surd 3$

$|\,7\;2\;{\scriptstyle 1}\;1\;1\rangle \;=\; |\,7\;4\rangle + 1/9\surd 3 \;+\; |\,7\;1\rangle + \surd 5.11/2.9 \;+\; |\,7\text{-}2\rangle - \surd 11/3\surd 2.3$
$\quad + \;|\,7\text{-}5\rangle - 11\surd 5/2.9\surd 3$

$|\,7\;2\;{\scriptstyle 1}\;1\text{-}1\rangle \;=\; |\,7\;5\rangle - 11\surd 5/2.9\surd 3 \;+\; |\,7\;2\rangle + \surd 11/3\surd 2.3 \;+\; |\,7\text{-}1\rangle + \surd 5.11/2.9$
$\quad + \;|\,7\text{-}4\rangle - 1/9\surd 3$

$|\,7\;3\;0\;0\rangle \;=\; |\,7\;6\rangle - \surd 5.11/9\surd 2 \;+\; |\,7\;3\rangle + \surd 13/9 \;+\; |\,7\text{-}3\rangle + \surd 13/9$
$\quad + \;|\,7\text{-}6\rangle + \surd 5.11/9\surd 2$

$|\,7\;3\;\tilde{0}\;0\rangle \;=\; |\,7\;6\rangle + 11\surd 11/9.3\surd 2.5 \;+\; |\,7\;3\rangle + \surd 13/9.3 \;+\; |\,7\;0\rangle + 2\surd 2.7.13/9\surd 3.5$
$\quad + \;|\,7\text{-}3\rangle - \surd 13/9.3 \;+\; |\,7\text{-}6\rangle + 11\surd 11/9.3\surd 2.5$

$|\,7\;3\;1\;1\rangle \;=\; |\,7\;7\rangle + 2\surd 2.7.11/9.3\surd 5 \;+\; |\,7\;4\rangle - \surd 2.11.13/9.3 \;+\; |\,7\;1\rangle - 4\surd 2.13/9\surd 3.5$
$\quad + \;|\,7\text{-}2\rangle - \surd 13/9.3 \;+\; |\,7\text{-}5\rangle - \surd 2.11.13/9.3\surd 5$

$|\,7\;3\;1\text{-}1\rangle \;=\; |\,7\;5\rangle - \surd 2.11.13/9.3\surd 5 \;+\; |\,7\;2\rangle + \surd 13/9.3 \;+\; |\,7\text{-}1\rangle - 4\surd 2.13/9\surd 3.5$
$\quad + \;|\,7\text{-}4\rangle + \surd 2.11.13/9.3 \;+\; |\,7\text{-}7\rangle + 2\surd 2.7.11/9.3\surd 5$

$|\,7\;\tilde{1}\;\tilde{0}\;0\rangle \;=\; |\,7\;6\rangle - \surd 7.11.13/9.3\surd 5 \;+\; |\,7\;3\rangle + 5\surd 7/9.3\surd 2 \;+\; |\,7\;0\rangle + 8.2/9\surd 3.5$
$\quad + \;|\,7\text{-}3\rangle - 5\surd 7/9.3\surd 2 \;+\; |\,7\text{-}6\rangle - \surd 7.11.13/9.3\surd 5$

$|\,7\;\tilde{1}\;1\;1\rangle \;=\; |\,7\;7\rangle - 2\surd 11.13/9.3\surd 5 \;+\; |\,7\;4\rangle + \surd 7.11/9.3 \;+\; |\,7\;1\rangle - 13\surd 7/2.9\surd 3.5$
$\quad + \;|\,7\text{-}2\rangle + 7\surd 7/9.3\surd 2 \;+\; |\,7\text{-}5\rangle - 7\surd 7.11/2.9.3\surd 5$

$|\,7\;\tilde{1}\;1\text{-}1\rangle \;=\; |\,7\;5\rangle - 7\surd 7.11/2.9.3\surd 5 \;+\; |\,7\;2\rangle - 7\surd 7/9.3\surd 2 \;+\; |\,7\text{-}1\rangle - 13\surd 7/2.9\surd 3.5$
$\quad + \;|\,7\text{-}4\rangle - \surd 7.11/9.3 \;+\; |\,7\text{-}7\rangle - 2\surd 11.13/9.3\surd 5$

$|\,\tfrac{15}{2}\;\tfrac{3}{2}\;\tfrac{1}{2}\;\tfrac{1}{2}\rangle \;=\; |\,\tfrac{15}{2}\;\tfrac{13}{2}\rangle + \surd 7.13/9\surd 3.5 \;+\; |\,\tfrac{15}{2}\;\tfrac{7}{2}\rangle + 11/9\surd 3 \;+\; |\,\tfrac{15}{2}\;\tfrac{1}{2}\rangle - \surd 7.11/9\surd 5$
$\quad + \;|\,\tfrac{15}{2}\text{-}\tfrac{5}{2}\rangle - \surd 5.11/9\surd 3 \;+\; |\,\tfrac{15}{2}\text{-}\tfrac{11}{2}\rangle + \surd 13/9\surd 3.5$

$|\,\tfrac{15}{2}\;\tfrac{3}{2}\;\tfrac{1}{2}\text{-}\tfrac{1}{2}\rangle \;=\; |\,\tfrac{15}{2}\;\tfrac{11}{2}\rangle - \surd 13/9\surd 3.5 \;+\; |\,\tfrac{15}{2}\;\tfrac{5}{2}\rangle - \surd 5.11/9\surd 3 \;+\; |\,\tfrac{15}{2}\text{-}\tfrac{1}{2}\rangle + \surd 7.11/9\surd 5$
$\quad + \;|\,\tfrac{15}{2}\text{-}\tfrac{7}{2}\rangle + 11/9\surd 3 \;+\; |\,\tfrac{15}{2}\text{-}\tfrac{13}{2}\rangle - \surd 7.13/9\surd 3.5$

$|\,\tfrac{15}{2}\;\tfrac{3}{2}\;\tfrac{3}{2}\;\tfrac{3}{2}\rangle \;=\; |\,\tfrac{15}{2}\;\tfrac{15}{2}\rangle + i\surd 7.13/9\surd 2.3 \;+\; |\,\tfrac{15}{2}\;\tfrac{9}{2}\rangle - 1/9\surd 2.3.5 + 11\,i/9\surd 2.3$
$\quad + \;|\,\tfrac{15}{2}\;\tfrac{3}{2}\rangle - \surd 11/9\surd 2.3 - i\surd 11/3\surd 2.3.5 \;+\; |\,\tfrac{15}{2}\text{-}\tfrac{3}{2}\rangle + \surd 11/3\surd 2.3.5 - i\surd 11/9\surd 2.3$
$\quad + \;|\,\tfrac{15}{2}\text{-}\tfrac{9}{2}\rangle + 11/9\surd 2.3 + i/9\surd 2.3.5 \;+\; |\,\tfrac{15}{2}\text{-}\tfrac{15}{2}\rangle - \surd 7.13/9\surd 2.3$

$|\,\tfrac{15}{2}\;\tfrac{3}{2}\text{-}\tfrac{3}{2}\;\tfrac{3}{2}\rangle \;=\; |\,\tfrac{15}{2}\;\tfrac{15}{2}\rangle + \surd 7.13/9\surd 2.3 \;+\; |\,\tfrac{15}{2}\;\tfrac{9}{2}\rangle + 11/9\surd 2.3 - i/9\surd 2.3.5$
$\quad + \;|\,\tfrac{15}{2}\;\tfrac{3}{2}\rangle - \surd 11/3\surd 2.3.5 - i\surd 11/9\surd 2.3 \;+\; |\,\tfrac{15}{2}\text{-}\tfrac{3}{2}\rangle - \surd 11/9\surd 2.3 + i\surd 11/3\surd 2.3.5$
$\quad + \;|\,\tfrac{15}{2}\text{-}\tfrac{9}{2}\rangle + 1/9\surd 2.3.5 + 11\,i/9\surd 2.3 \;+\; |\,\tfrac{15}{2}\text{-}\tfrac{15}{2}\rangle - i\surd 7.13/9\surd 2.3$

$|\,\tfrac{15}{2}\,0\;\tfrac{5}{2}\;0\;\tfrac{1}{2}\;\tfrac{1}{2}\rangle \;=\; |\,\tfrac{15}{2}\;\tfrac{13}{2}\rangle - \surd 11.13/2.9.3\surd 5 \;+\; |\,\tfrac{15}{2}\;\tfrac{7}{2}\rangle - \surd 7.11/4.9.3 \;+\; |\,\tfrac{15}{2}\;\tfrac{1}{2}\rangle + 41/2.9\surd 3.5$
$\quad + \;|\,\tfrac{15}{2}\text{-}\tfrac{5}{2}\rangle - 11\surd 5.7/4.9.3 \;+\; |\,\tfrac{15}{2}\text{-}\tfrac{11}{2}\rangle + \surd 7.11.13/9.3\surd 5$

$|\,\tfrac{15}{2}\,0\;\tfrac{5}{2}\;0\;\tfrac{1}{2}\text{-}\tfrac{1}{2}\rangle \;=\; |\,\tfrac{15}{2}\;\tfrac{11}{2}\rangle - \surd 7.11.13/9.3\surd 5 \;+\; |\,\tfrac{15}{2}\;\tfrac{5}{2}\rangle - 11\surd 5.7/4.9.3 \;+\; |\,\tfrac{15}{2}\text{-}\tfrac{1}{2}\rangle - 41/2.9\surd 3.5$
$\quad + \;|\,\tfrac{15}{2}\text{-}\tfrac{7}{2}\rangle - \surd 7.11/4.9.3 \;+\; |\,\tfrac{15}{2}\text{-}\tfrac{13}{2}\rangle + \surd 11.13/2.9.3\surd 5$

$|\,\tfrac{15}{2}\,0\;\tfrac{5}{2}\;1\;\tfrac{1}{2}\;\tfrac{1}{2}\rangle \;=\; |\,\tfrac{15}{2}\;\tfrac{13}{2}\rangle - \surd 5.11.13/9.3\surd 2 \;+\; |\,\tfrac{15}{2}\;\tfrac{7}{2}\rangle + \surd 2.7.11/9.3 \;+\; |\,\tfrac{15}{2}\;\tfrac{1}{2}\rangle + 11\surd 5/4.9\surd 2.3$
$\quad + \;|\,\tfrac{15}{2}\text{-}\tfrac{5}{2}\rangle - \surd 5.7/2.9.3\surd 2 \;+\; |\,\tfrac{15}{2}\text{-}\tfrac{11}{2}\rangle - \surd 5.7.11.13/4.9.3\surd 2$

$|\,\tfrac{15}{2}\,0\;\tfrac{5}{2}\;1\;\tfrac{1}{2}\text{-}\tfrac{1}{2}\rangle \;=\; |\,\tfrac{15}{2}\;\tfrac{11}{2}\rangle + \surd 5.7.11.13/4.9.3\surd 2 \;+\; |\,\tfrac{15}{2}\;\tfrac{5}{2}\rangle - \surd 5.7/2.9.3\surd 2$
$\quad + \;|\,\tfrac{15}{2}\text{-}\tfrac{1}{2}\rangle - 11\surd 5/4.9\surd 2.3 \;+\; |\,\tfrac{15}{2}\text{-}\tfrac{7}{2}\rangle + \surd 2.7.11/9.3 \;+\; |\,\tfrac{15}{2}\text{-}\tfrac{13}{2}\rangle + \surd 5.11.13/9.3\surd 2$

$|\,\tfrac{15}{2}\,0\;\tfrac{5}{2}\;\tfrac{3}{2}\;\tfrac{3}{2}\rangle \;=\; |\,\tfrac{15}{2}\;\tfrac{15}{2}\rangle + \surd 5.11.13/2.9.3\surd 3 + i\surd 11.13/9.3\surd 3$
$\quad + \;|\,\tfrac{15}{2}\;\tfrac{9}{2}\rangle - 17\surd 7.11/4.9.3\surd 3.5 - i\surd 7.11/8.9.3\surd 3$
$\quad + \;|\,\tfrac{15}{2}\;\tfrac{3}{2}\rangle + 11\surd 7/8.9.3\surd 3 + 37\,i\surd 7/2.9.3\surd 3.5$
$\quad + \;|\,\tfrac{15}{2}\text{-}\tfrac{3}{2}\rangle - 37\surd 7/2.9.3\surd 3.5 + 11\,i\surd 7/8.9.3\surd 3$
$\quad + \;|\,\tfrac{15}{2}\text{-}\tfrac{9}{2}\rangle - \surd 7.11/8.9.3\surd 3 + 17\,i\surd 7.11/4.9.3\surd 3.5$
$\quad + \;|\,\tfrac{15}{2}\text{-}\tfrac{15}{2}\rangle - \surd 11.13/9.3\surd 3 + i\surd 5.11.13/2.9.3\surd 3$

SO_3–K–D_3–C_3 Partners (cont.)

$|\tfrac{15}{2}\,0\,\tfrac{5}{2}\,-\tfrac{3}{2}\,\tfrac{3}{2}> \;=\; |\tfrac{15}{2}\,\tfrac{15}{2}> + \surd 11.13/9.3\surd 3 + i\surd 5.11.13/2.9.3\surd 3$
$+\; |\tfrac{15}{2}\,\tfrac{9}{2}> - \surd 7.11/8.9.3\surd 3 - 17\,i\surd 7.11/4.9.3\surd 3.5$
$+\; |\tfrac{15}{2}\,\tfrac{3}{2}> + 37\surd 7/2.9.3\surd 3.5 + 11\,i\surd 7/8.9.3\surd 3$
$+\; |\tfrac{15}{2}\,-\tfrac{3}{2}> + 11\surd 7/8.9.3\surd 3 - 37\,i\surd 7/2.9.3\surd 3.5$
$+\; |\tfrac{15}{2}\,-\tfrac{9}{2}> + 17\surd 7.11/4.9.3\surd 3.5 - i\surd 7.11/8.9.3\surd 3$
$+\; |\tfrac{15}{2}\,-\tfrac{15}{2}> + \surd 5.11.13/2.9.3\surd 3 - i\surd 11.13/9.3\surd 3$
$|\tfrac{15}{2}\,1\,\tfrac{5}{2}\,0\,\tfrac{1}{2}\,\tfrac{1}{2}> \;=\; |\tfrac{15}{2}\,\tfrac{13}{2}> - 17\surd 7/2.9.3\surd 3 \;+\; |\tfrac{15}{2}\,\tfrac{7}{2}> - 11\surd 5.13/4.9.3\surd 3$
$+\; |\tfrac{15}{2}\,\tfrac{1}{2}> - \surd 7.11.13/2.9.3 \;+\; |\tfrac{15}{2}\,-\tfrac{5}{2}> - 7\surd 11.13/4.9.3\surd 3 \;+\; |\tfrac{15}{2}\,-\tfrac{11}{2}> - 1/9.3\surd 3$
$|\tfrac{15}{2}\,1\,\tfrac{5}{2}\,0\,\tfrac{1}{2}\,-\tfrac{1}{2}> \;=\; |\tfrac{15}{2}\,\tfrac{11}{2}> + 1/9.3\surd 3 \;+\; |\tfrac{15}{2}\,\tfrac{5}{2}> - 7\surd 11.13/4.9.3\surd 3 \;+\; |\tfrac{15}{2}\,-\tfrac{1}{2}> + \surd 7.11.13/2.9.3$
$+\; |\tfrac{15}{2}\,-\tfrac{7}{2}> - 11\surd 5.13/4.9.3\surd 3 \;+\; |\tfrac{15}{2}\,-\tfrac{13}{2}> + 17\surd 7/2.9.3\surd 3$
$|\tfrac{15}{2}\,1\,\tfrac{5}{2}\,1\,\tfrac{1}{2}\,\tfrac{1}{2}> \;=\; |\tfrac{15}{2}\,\tfrac{13}{2}> + 11\surd 7/9.3\surd 2.3 \;+\; |\tfrac{15}{2}\,\tfrac{7}{2}> - \surd 2.5.13/9.3\surd 3$
$+\; |\tfrac{15}{2}\,\tfrac{1}{2}> + \surd 7.11.13/4.9.3\surd 2 \;+\; |\tfrac{15}{2}\,-\tfrac{5}{2}> - 5\surd 11.13/2.9.3\surd 2.3$
$+\; |\tfrac{15}{2}\,-\tfrac{11}{2}> - 11.17/4.9.3\surd 2.3$
$|\tfrac{15}{2}\,1\,\tfrac{5}{2}\,1\,\tfrac{1}{2}\,-\tfrac{1}{2}> \;=\; |\tfrac{15}{2}\,\tfrac{11}{2}> + 11.17/4.9.3\surd 2.3 \;+\; |\tfrac{15}{2}\,\tfrac{5}{2}> - 5\surd 11.13/2.9.3\surd 2.3$
$+\; |\tfrac{15}{2}\,-\tfrac{1}{2}> - \surd 7.11.13/4.9.3\surd 2 \;+\; |\tfrac{15}{2}\,-\tfrac{7}{2}> - \surd 2.5.13/9.3\surd 3$
$+\; |\tfrac{15}{2}\,-\tfrac{13}{2}> - 11\surd 7/9.3\surd 2.3$
$|\tfrac{15}{2}\,1\,\tfrac{5}{2}\,\tfrac{3}{2}\,\tfrac{3}{2}> \;=\; |\tfrac{15}{2}\,\tfrac{15}{2}> - 11\surd 7/2.9.9 + 5\,i\surd 5.7/9.9 \;+\; |\tfrac{15}{2}\,\tfrac{9}{2}> + 29\surd 13/4.9.9 - 11\,i\surd 5.13/8.9.9$
$+\; |\tfrac{15}{2}\,\tfrac{3}{2}> + 11\surd 5.11.13/8.9.9 + i\surd 11.13/2.9.9$
$+\; |\tfrac{15}{2}\,-\tfrac{3}{2}> - \surd 11.13/2.9.9 + 11\,i\surd 5.11.13/8.9.9$
$+\; |\tfrac{15}{2}\,-\tfrac{9}{2}> - 11\surd 5.13/8.9.9 - 29\,i\surd 13/4.9.9 \;+\; |\tfrac{15}{2}\,-\tfrac{15}{2}> - 5\surd 5.7/9.9 - 11\,i\surd 7/2.9.9$
$|\tfrac{15}{2}\,1\,\tfrac{5}{2}\,-\tfrac{3}{2}\,\tfrac{3}{2}> \;=\; |\tfrac{15}{2}\,\tfrac{15}{2}> + 5\surd 5.7/9.9 - 11\,i\surd 7/2.9.9 \;+\; |\tfrac{15}{2}\,\tfrac{9}{2}> - 11\surd 5.13/8.9.9 + 29\,i\surd 13/4.9.9$
$+\; |\tfrac{15}{2}\,\tfrac{3}{2}> + \surd 11.13/2.9.9 + 11\,i\surd 5.11.13/8.9.9$
$+\; |\tfrac{15}{2}\,-\tfrac{3}{2}> + 11\surd 5.11.13/8.9.9 - i\surd 11.13/2.9.9$
$+\; |\tfrac{15}{2}\,-\tfrac{9}{2}> - 29\surd 13/4.9.9 - 11\,i\surd 5.13/8.9.9 \;+\; |\tfrac{15}{2}\,-\tfrac{15}{2}> - 11\surd 7/2.9.9 - 5\,i\surd 5.7/9.9$

$|8\,0\,2\,0\,0> \;=\; |8\,6> - \surd 7.13/9\surd 3.17 \;+\; |8\,3> - 5.11/9\surd 2.3.17 \;+\; |8\,0> + 2\surd 7.11/9\surd 17$
$+\; |8\,-3> + 5.11/9\surd 2.3.17 \;+\; |8\,-6> - \surd 7.13/9\surd 3.17$
$|8\,0\,2\,0\,1\,1> \;=\; |8\,7> - \surd 5.7.13/9\surd 3.17 \;+\; |8\,4> - 2.11\surd 5/9\surd 3.17$
$+\; |8\,1> + \surd 7.11/3\surd 3.17 \;+\; |8\,-2> + 5\surd 2.11/9\surd 3.17 \;+\; |8\,-5> - \surd 13/9\surd 3.17$
$|8\,0\,2\,0\,1\,-1> \;=\; |8\,5> + \surd 13/9\surd 3.17 \;+\; |8\,2> + 5\surd 2.11/9\surd 3.17 \;+\; |8\,-1> - \surd 7.11/3\surd 3.17$
$+\; |8\,-4> - 2.11\surd 5/9\surd 3.17 \;+\; |8\,-7> + \surd 5.7.13/9\surd 3.17$
$|8\,0\,2\,1\,1\,1> \;=\; |8\,4> + 1/9\surd 3.17 \;+\; |8\,1> + \surd 5.7.11/2.9\surd 3.17 \;+\; |8\,-2> - \surd 5.11/3\surd 2.3.17$
$+\; |8\,-5> - 11\surd 5.13/2.9\surd 3.17 \;+\; |8\,-8> + 2\surd 5.7.13/9\surd 3.17$
$|8\,0\,2\,1\,1\,-1> \;=\; |8\,8> + 2\surd 5.7.13/9\surd 3.17 \;+\; |8\,5> + 11\surd 5.13/2.9\surd 3.17$
$+\; |8\,2> - \surd 5.11/3\surd 2.3.17 \;+\; |8\,-1> - \surd 5.7.11/2.9\surd 3.17 \;+\; |8\,-4> + 1/9\surd 3.17$
$|8\,1\,2\,0\,0> \;=\; |8\,6> - 4.5\surd 11/9.3\surd 3.17 \;+\; |8\,3> - \surd 2.7.11.13/9.3\surd 3.17$
$+\; |8\,0> - 5.5\surd 13/9.3\surd 17 \;+\; |8\,-3> + \surd 2.7.11.13/9.3\surd 3.17$
$+\; |8\,-6> - 4.5\surd 11/9.3\surd 3.17$
$|8\,1\,2\,0\,1\,1> \;=\; |8\,7> + 8\surd 5.11/9.9\surd 3.17 \;+\; |8\,4> + \surd 5.7.11.13/9.9\surd 3.17$
$+\; |8\,1> + 4.5.5\surd 13/9.9\surd 3.17 \;+\; |8\,-2> - 11\surd 2.7.13/9.9\surd 3.17$
$+\; |8\,-5> + 8.5\surd 7.11/9.9\surd 3.17 \;+\; |8\,-8> + 4\surd 11.17/9.9\surd 3$
$|8\,1\,2\,0\,1\,-1> \;=\; |8\,8> + 4\surd 11.17/9.9\surd 3 \;+\; |8\,5> - 8.5\surd 7.11/9.9\surd 3.17$
$+\; |8\,2> - 11\surd 2.7.13/9.9\surd 3.17 \;+\; |8\,-1> - 4.5.5\surd 13/9.9\surd 3.17$
$+\; |8\,-4> + \surd 5.7.11.13/9.9\surd 3.17 \;+\; |8\,-7> - 8\surd 5.11/9.9\surd 3.17$
$|8\,1\,2\,1\,1\,1> \;=\; |8\,7> - 8\surd 11.17/9.9\surd 3 \;+\; |8\,4> + 2.5\surd 7.11.13/9.9\surd 3.17$
$+\; |8\,1> + 11\surd 5.13/9.9\surd 3.17 \;+\; |8\,-2> + 5\surd 2.5.7.13/9.9\surd 3.17$
$+\; |8\,-5> - \surd 5.7.11/9.9\surd 3.17 \;+\; |8\,-8> + \surd 5.11/9.9\surd 3.17$
$|8\,1\,2\,1\,1\,-1> \;=\; |8\,8> + \surd 5.11/9.9\surd 3.17 \;+\; |8\,5> + \surd 5.7.11/9.9\surd 3.17$
$+\; |8\,2> + 5\surd 2.5.7.13/9.9\surd 3.17 \;+\; |8\,-1> - 11\surd 5.13/9.9\surd 3.17$
$+\; |8\,-4> + 2.5\surd 7.11.13/9.9\surd 3.17 \;+\; |8\,-7> + 8\surd 11.17/9.9\surd 3$

SO_3–K–D_3–C_3 Partners (cont.)

$|8\;3\;0\;0\rangle = |8\;6\rangle + 11\sqrt{13}/9.3\sqrt{2.3} + |8\;3\rangle - 5\sqrt{7}/9.3\sqrt{3} + |8\;0\rangle - 2\sqrt{2.11}/9.3$
$\quad + |8\,{-3}\rangle + 5\sqrt{7}/9.3\sqrt{3} + |8\,{-6}\rangle + 11\sqrt{13}/9.3\sqrt{2.3}$

$|8\;3\;\tilde{0}\;0\rangle = |8\;6\rangle + \sqrt{13}/3\sqrt{2.3} + |8\;3\rangle + \sqrt{7}/3\sqrt{3} + |8\,{-3}\rangle + \sqrt{7}/3\sqrt{3}$
$\quad + |8\,{-6}\rangle - \sqrt{13}/3\sqrt{2.3}$

$|8\;3\;1\;1\rangle = |8\;7\rangle + 2\sqrt{2.5.13}/9.9 + |8\;4\rangle + \sqrt{2.5.7}/9.9 + |8\;1\rangle - 4\sqrt{2.11}/9.9$
$\quad + |8\,{-2}\rangle + 7\sqrt{7.11}/9.9 + |8\,{-5}\rangle + \sqrt{2.7.13}/9.9 + |8\,{-8}\rangle + 8\sqrt{2.13}/9.9$

$|8\;3\;1\,{-1}\rangle = |8\;8\rangle + 8\sqrt{2.13}/9.9 + |8\;5\rangle - \sqrt{2.7.13}/9.9 + |8\;2\rangle + 7\sqrt{7.11}/9.9$
$\quad + |8\,{-1}\rangle + 4\sqrt{2.11}/9.9 + |8\,{-4}\rangle + \sqrt{2.5.7}/9.9 + |8\,{-7}\rangle - 2\sqrt{2.5.13}/9.9$

$|8\;\tilde{1}\;\tilde{0}\;0\rangle = |8\;6\rangle + \sqrt{7}/3\sqrt{3} + |8\;3\rangle - \sqrt{13}/3\sqrt{2.3} + |8\,{-3}\rangle - \sqrt{13}/3\sqrt{2.3}$
$\quad + |8\,{-6}\rangle - \sqrt{7}/3\sqrt{3}$

$|8\;\tilde{1}\;1\;1\rangle = |8\;7\rangle - 2\sqrt{5.7}/9.3 + |8\;4\rangle - \sqrt{5.13}/9.3 + |8\;1\rangle - \sqrt{7.11.13}/2.9.3$
$\quad + |8\,{-2}\rangle - \sqrt{11.13}/9.3\sqrt{2} + |8\,{-5}\rangle + 19/2.9.3 + |8\,{-8}\rangle + 4\sqrt{7}/9.3$

$|8\;\tilde{1}\;1\,{-1}\rangle = |8\;8\rangle + 4\sqrt{7}/9.3 + |8\;5\rangle - 19/2.9.3 + |8\;2\rangle - \sqrt{11.13}/9.3\sqrt{2}$
$\quad + |8\,{-1}\rangle + \sqrt{7.11.13}/2.9.3 + |8\,{-4}\rangle - \sqrt{5.13}/9.3 + |8\,{-7}\rangle + 2\sqrt{5.7}/9.3$

SO_3–K–D_5–C_5 Partners as JM Partners

$|0\;0\;0\;0\rangle = |0\;0\rangle + 1$

$|\tfrac12\;\tfrac12\;\tfrac12\;\tfrac12\rangle = |\tfrac12\;\tfrac12\rangle + 1$
$|\tfrac12\;\tfrac12\;\tfrac12\,{-\tfrac12}\rangle = |\tfrac12\,{-\tfrac12}\rangle + 1$

$|1\;1\;\tilde{0}\;0\rangle = |1\;0\rangle + 1$
$|1\;1\;1\;1\rangle = |1\;1\rangle - 1$
$|1\;1\;1\,{-1}\rangle = |1\,{-1}\rangle - 1$

$|\tfrac32\;\tfrac32\;\tfrac12\;\tfrac12\rangle = |\tfrac32\;\tfrac12\rangle + 1$
$|\tfrac32\;\tfrac32\;\tfrac12\,{-\tfrac12}\rangle = |\tfrac32\,{-\tfrac12}\rangle - 1$
$|\tfrac32\;\tfrac32\;\tfrac32\;\tfrac32\rangle = |\tfrac32\;\tfrac32\rangle + 1$
$|\tfrac32\;\tfrac32\;\tfrac32\,{-\tfrac32}\rangle = |\tfrac32\,{-\tfrac32}\rangle + 1$

$|2\;2\;0\;0\rangle = |2\;0\rangle - 1$
$|2\;2\;1\;1\rangle = |2\;1\rangle - 1$
$|2\;2\;1\,{-1}\rangle = |2\,{-1}\rangle + 1$
$|2\;2\;2\;2\rangle = |2\;2\rangle - 1$
$|2\;2\;2\,{-2}\rangle = |2\,{-2}\rangle - 1$

$|\tfrac52\;\tfrac52\;\tfrac12\;\tfrac12\rangle = |\tfrac52\;\tfrac12\rangle - 1$
$|\tfrac52\;\tfrac52\;\tfrac12\,{-\tfrac12}\rangle = |\tfrac52\,{-\tfrac12}\rangle - 1$
$|\tfrac52\;\tfrac52\;\tfrac32\;\tfrac32\rangle = |\tfrac52\;\tfrac32\rangle + 1$
$|\tfrac52\;\tfrac52\;\tfrac32\,{-\tfrac32}\rangle = |\tfrac52\,{-\tfrac32}\rangle - 1$
$|\tfrac52\;\tfrac52\;\tfrac52\;\tfrac52\rangle = |\tfrac52\;\tfrac52\rangle + i/\sqrt{2} + |\tfrac52\,{-\tfrac52}\rangle - 1/\sqrt{2}$
$|\tfrac52\;\tfrac52\,{-\tfrac52}\;\tfrac52\rangle = |\tfrac52\;\tfrac52\rangle + 1/\sqrt{2} + |\tfrac52\,{-\tfrac52}\rangle - i/\sqrt{2}$

$|3\;3\;1\;1\rangle = |3\;1\rangle + 1$
$|3\;3\;1\,{-1}\rangle = |3\,{-1}\rangle + 1$
$|3\;3\;2\;2\rangle = |3\;2\rangle - \sqrt{2}/\sqrt{5} + |3\,{-3}\rangle + \sqrt{3}/\sqrt{5}$
$|3\;3\;2\,{-2}\rangle = |3\;3\rangle + \sqrt{3}/\sqrt{5} + |3\,{-2}\rangle + \sqrt{2}/\sqrt{5}$

SO_3–K–D_5–C_5 Partners (cont.)

$|3\ \tilde{1}\ \tilde{0}\ 0\rangle = |3\ 0\rangle -1$

$|3\ \tilde{1}\ 2\ 2\rangle = |3\ 2\rangle -\sqrt{3}/\sqrt{5} + |3\ {-3}\rangle -\sqrt{2}/\sqrt{5}$

$|3\ \tilde{1}\ 2\ {-2}\rangle = |3\ 3\rangle -\sqrt{2}/\sqrt{5} + |3\ {-2}\rangle +\sqrt{3}/\sqrt{5}$

$|\tfrac{7}{2}\ \tfrac{5}{2}\ \tfrac{1}{2}\ \tfrac{1}{2}\rangle = |\tfrac{7}{2}\ \tfrac{1}{2}\rangle +1$

$|\tfrac{7}{2}\ \tfrac{5}{2}\ \tfrac{1}{2}\ {-\tfrac{1}{2}}\rangle = |\tfrac{7}{2}\ {-\tfrac{1}{2}}\rangle -1$

$|\tfrac{7}{2}\ \tfrac{5}{2}\ \tfrac{3}{2}\ \tfrac{3}{2}\rangle = |\tfrac{7}{2}\ \tfrac{3}{2}\rangle +\sqrt{3}/\sqrt{2.5} + |\tfrac{7}{2}\ {-\tfrac{7}{2}}\rangle +\sqrt{7}/\sqrt{2.5}$

$|\tfrac{7}{2}\ \tfrac{5}{2}\ \tfrac{3}{2}\ {-\tfrac{3}{2}}\rangle = |\tfrac{7}{2}\ \tfrac{7}{2}\rangle -\sqrt{7}/\sqrt{2.5} + |\tfrac{7}{2}\ {-\tfrac{3}{2}}\rangle +\sqrt{3}/\sqrt{2.5}$

$|\tfrac{7}{2}\ \tfrac{5}{2}\ \tfrac{5}{2}\ \tfrac{5}{2}\rangle = |\tfrac{7}{2}\ \tfrac{5}{2}\rangle +7/2.5- i/2.5 + |\tfrac{7}{2}\ {-\tfrac{5}{2}}\rangle -1/2.5-7i/2.5$

$|\tfrac{7}{2}\ \tfrac{5}{2}\ {-\tfrac{5}{2}}\ \tfrac{5}{2}\rangle = |\tfrac{7}{2}\ \tfrac{5}{2}\rangle -1/2.5+7i/2.5 + |\tfrac{7}{2}\ {-\tfrac{5}{2}}\rangle -7/2.5- i/2.5$

$|\tfrac{7}{2}\ \tfrac{\tilde{1}}{2}\ \tfrac{3}{2}\ \tfrac{3}{2}\rangle = |\tfrac{7}{2}\ \tfrac{3}{2}\rangle -\sqrt{7}/\sqrt{2.5} + |\tfrac{7}{2}\ {-\tfrac{7}{2}}\rangle +\sqrt{3}/\sqrt{2.5}$

$|\tfrac{7}{2}\ \tfrac{\tilde{1}}{2}\ \tfrac{3}{2}\ {-\tfrac{3}{2}}\rangle = |\tfrac{7}{2}\ \tfrac{7}{2}\rangle -\sqrt{3}/\sqrt{2.5} + |\tfrac{7}{2}\ {-\tfrac{3}{2}}\rangle -\sqrt{7}/\sqrt{2.5}$

$|4\ 2\ 0\ 0\rangle = |4\ 0\rangle -1$

$|4\ 2\ 1\ 1\rangle = |4\ 1\rangle +2\sqrt{2}/\sqrt{3.5} + |4\ {-4}\rangle +\sqrt{7}/\sqrt{3.5}$

$|4\ 2\ 1\ {-1}\rangle = |4\ 4\rangle +\sqrt{7}/\sqrt{3.5} + |4\ {-1}\rangle -2\sqrt{2}/\sqrt{3.5}$

$|4\ 2\ 2\ 2\rangle = |4\ 2\rangle -1/\sqrt{3.5} + |4\ {-3}\rangle -\sqrt{2.7}/\sqrt{3.5}$

$|4\ 2\ 2\ {-2}\rangle = |4\ 3\rangle +\sqrt{2.7}/\sqrt{3.5} + |4\ {-2}\rangle -1/\sqrt{3.5}$

$|4\ 3\ 1\ 1\rangle = |4\ 1\rangle -\sqrt{7}/\sqrt{3.5} + |4\ {-4}\rangle +2\sqrt{2}/\sqrt{3.5}$

$|4\ 3\ 1\ {-1}\rangle = |4\ 4\rangle +2\sqrt{2}/\sqrt{3.5} + |4\ {-1}\rangle +\sqrt{7}/\sqrt{3.5}$

$|4\ 3\ 2\ 2\rangle = |4\ 2\rangle -\sqrt{2.7}/\sqrt{3.5} + |4\ {-3}\rangle +1/\sqrt{3.5}$

$|4\ 3\ 2\ {-2}\rangle = |4\ 3\rangle -1/\sqrt{3.5} + |4\ {-2}\rangle -\sqrt{2.7}/\sqrt{3.5}$

$|\tfrac{9}{2}\ \tfrac{3}{2}\ \tfrac{1}{2}\ \tfrac{1}{2}\rangle = |\tfrac{9}{2}\ \tfrac{1}{2}\rangle +3\sqrt{2/5} + |\tfrac{9}{2}\ {-\tfrac{9}{2}}\rangle +\sqrt{7/5}$

$|\tfrac{9}{2}\ \tfrac{3}{2}\ \tfrac{1}{2}\ {-\tfrac{1}{2}}\rangle = |\tfrac{9}{2}\ \tfrac{9}{2}\rangle -\sqrt{7/5} + |\tfrac{9}{2}\ {-\tfrac{1}{2}}\rangle +3\sqrt{2/5}$

$|\tfrac{9}{2}\ \tfrac{3}{2}\ \tfrac{3}{2}\ \tfrac{3}{2}\rangle = |\tfrac{9}{2}\ \tfrac{3}{2}\rangle -2/5 + |\tfrac{9}{2}\ {-\tfrac{7}{2}}\rangle -\sqrt{3.7/5}$

$|\tfrac{9}{2}\ \tfrac{3}{2}\ \tfrac{3}{2}\ {-\tfrac{3}{2}}\rangle = |\tfrac{9}{2}\ \tfrac{7}{2}\rangle -\sqrt{3.7/5} + |\tfrac{9}{2}\ {-\tfrac{3}{2}}\rangle +2/5$

$|\tfrac{9}{2}\ \tfrac{5}{2}\ \tfrac{1}{2}\ \tfrac{1}{2}\rangle = |\tfrac{9}{2}\ \tfrac{1}{2}\rangle -\sqrt{7/5} + |\tfrac{9}{2}\ {-\tfrac{9}{2}}\rangle +3\sqrt{2/5}$

$|\tfrac{9}{2}\ \tfrac{5}{2}\ \tfrac{1}{2}\ {-\tfrac{1}{2}}\rangle = |\tfrac{9}{2}\ \tfrac{9}{2}\rangle -3\sqrt{2/5} + |\tfrac{9}{2}\ {-\tfrac{1}{2}}\rangle -\sqrt{7/5}$

$|\tfrac{9}{2}\ \tfrac{5}{2}\ \tfrac{3}{2}\ \tfrac{3}{2}\rangle = |\tfrac{9}{2}\ \tfrac{3}{2}\rangle -\sqrt{3.7/5} + |\tfrac{9}{2}\ {-\tfrac{7}{2}}\rangle +2/5$

$|\tfrac{9}{2}\ \tfrac{5}{2}\ \tfrac{3}{2}\ {-\tfrac{3}{2}}\rangle = |\tfrac{9}{2}\ \tfrac{7}{2}\rangle +2/5 + |\tfrac{9}{2}\ {-\tfrac{3}{2}}\rangle +\sqrt{3.7/5}$

$|\tfrac{9}{2}\ \tfrac{5}{2}\ \tfrac{5}{2}\ \tfrac{5}{2}\rangle = |\tfrac{9}{2}\ \tfrac{5}{2}\rangle +\sqrt{2}/\sqrt{5}+ i/\sqrt{2.5} + |\tfrac{9}{2}\ {-\tfrac{5}{2}}\rangle -1/\sqrt{2.5}+ i\sqrt{2}/\sqrt{5}$

$|\tfrac{9}{2}\ \tfrac{5}{2}\ {-\tfrac{5}{2}}\ \tfrac{5}{2}\rangle = |\tfrac{9}{2}\ \tfrac{5}{2}\rangle +1/\sqrt{2.5}+ i\sqrt{2}/\sqrt{5} + |\tfrac{9}{2}\ {-\tfrac{5}{2}}\rangle +\sqrt{2}/\sqrt{5}- i/\sqrt{2.5}$

$|5\ 1\ \tilde{0}\ 0\rangle = |5\ 5\rangle +\sqrt{7}/5\sqrt{2} + |5\ 0\rangle -3\sqrt{2}/5 + |5\ {-5}\rangle -\sqrt{7}/5\sqrt{2}$

$|5\ 1\ 1\ 1\rangle = |5\ 1\rangle -\sqrt{3}/\sqrt{2.5} + |5\ {-4}\rangle -\sqrt{7}/\sqrt{2.5}$

$|5\ 1\ 1\ {-1}\rangle = |5\ 4\rangle +\sqrt{7}/\sqrt{2.5} + |5\ {-1}\rangle -\sqrt{3}/\sqrt{2.5}$

$|5\ 2\ 0\ 0\rangle = |5\ 5\rangle -1/\sqrt{2} + |5\ {-5}\rangle -1/\sqrt{2}$

$|5\ 2\ 1\ 1\rangle = |5\ 1\rangle -\sqrt{7}/\sqrt{2.5} + |5\ {-4}\rangle +\sqrt{3}/\sqrt{2.5}$

$|5\ 2\ 1\ {-1}\rangle = |5\ 4\rangle -\sqrt{3}/\sqrt{2.5} + |5\ {-1}\rangle -\sqrt{7}/\sqrt{2.5}$

$|5\ 2\ 2\ 2\rangle = |5\ 2\rangle +\sqrt{2}/\sqrt{5} + |5\ {-3}\rangle +\sqrt{3}/\sqrt{5}$

$|5\ 2\ 2\ {-2}\rangle = |5\ 3\rangle +\sqrt{3}/\sqrt{5} + |5\ {-2}\rangle -\sqrt{2}/\sqrt{5}$

$|5\ \tilde{1}\ \tilde{0}\ 0\rangle = |5\ 5\rangle -3/5 + |5\ 0\rangle -\sqrt{7/5} + |5\ {-5}\rangle +3/5$

$|5\ \tilde{1}\ 2\ 2\rangle = |5\ 2\rangle +\sqrt{3}/\sqrt{5} + |5\ {-3}\rangle -\sqrt{2}/\sqrt{5}$

$|5\ \tilde{1}\ 2\ {-2}\rangle = |5\ 3\rangle -\sqrt{2}/\sqrt{5} + |5\ {-2}\rangle -\sqrt{3}/\sqrt{5}$

$|\tfrac{11}{2}\ \tfrac{1}{2}\ \tfrac{1}{2}\ \tfrac{1}{2}\rangle = |\tfrac{11}{2}\ \tfrac{11}{2}\rangle +\sqrt{7}/5\sqrt{2.3} + |\tfrac{11}{2}\ \tfrac{1}{2}\rangle -\sqrt{11}/5 + |\tfrac{11}{2}\ {-\tfrac{9}{2}}\rangle -\sqrt{7.11}/5\sqrt{2.3}$

$|\tfrac{11}{2}\ \tfrac{1}{2}\ \tfrac{1}{2}\ {-\tfrac{1}{2}}\rangle = |\tfrac{11}{2}\ \tfrac{9}{2}\rangle -\sqrt{7.11}/5\sqrt{2.3} + |\tfrac{11}{2}\ {-\tfrac{1}{2}}\rangle +\sqrt{11}/5 + |\tfrac{11}{2}\ {-\tfrac{11}{2}}\rangle +\sqrt{7}/5\sqrt{2.3}$

$|\tfrac{11}{2}\ \tfrac{3}{2}\ \tfrac{1}{2}\ \tfrac{1}{2}\rangle = |\tfrac{11}{2}\ \tfrac{11}{2}\rangle +\sqrt{2.11}/5\sqrt{3} + |\tfrac{11}{2}\ \tfrac{1}{2}\rangle -\sqrt{7}/5 + |\tfrac{11}{2}\ {-\tfrac{9}{2}}\rangle +4\sqrt{2}/5\sqrt{3}$

$|\tfrac{11}{2}\ \tfrac{3}{2}\ \tfrac{1}{2}\ {-\tfrac{1}{2}}\rangle = |\tfrac{11}{2}\ \tfrac{9}{2}\rangle +4\sqrt{2}/5\sqrt{3} + |\tfrac{11}{2}\ {-\tfrac{1}{2}}\rangle +\sqrt{7}/5 + |\tfrac{11}{2}\ {-\tfrac{11}{2}}\rangle +\sqrt{2.11}/5\sqrt{3}$

$|\tfrac{11}{2}\ \tfrac{3}{2}\ \tfrac{3}{2}\ \tfrac{3}{2}\rangle = |\tfrac{11}{2}\ \tfrac{3}{2}\rangle +\sqrt{3}/\sqrt{5} + |\tfrac{11}{2}\ {-\tfrac{7}{2}}\rangle +\sqrt{2}/\sqrt{5}$

SO_3–K–D_5–C_5 Partners (cont.)

$$|\tfrac{11}{2}\,\tfrac{3}{2}\,\tfrac{3}{2}\,{-\tfrac{3}{2}}> = |\tfrac{11}{2}\,\tfrac{7}{2}>-\sqrt{2}/\sqrt{5} + |\tfrac{11}{2}\,{-\tfrac{3}{2}}>+\sqrt{3}/\sqrt{5}$$
$$|\tfrac{11}{2}\,\tfrac{5}{2}\,\tfrac{1}{2}\,\tfrac{1}{2}> = |\tfrac{11}{2}\,\tfrac{11}{2}>-\sqrt{3.11}/5\sqrt{2} + |\tfrac{11}{2}\,\tfrac{1}{2}>-\sqrt{7}/5 + |\tfrac{11}{2}\,{-\tfrac{9}{2}}>+\sqrt{3}/5\sqrt{2}$$
$$|\tfrac{11}{2}\,\tfrac{5}{2}\,\tfrac{1}{2}\,{-\tfrac{1}{2}}> = |\tfrac{11}{2}\,\tfrac{9}{2}>+\sqrt{3}/5\sqrt{2} + |\tfrac{11}{2}\,{-\tfrac{1}{2}}>+\sqrt{7}/5 + |\tfrac{11}{2}\,{-\tfrac{11}{2}}>-\sqrt{3.11}/5\sqrt{2}$$
$$|\tfrac{11}{2}\,\tfrac{5}{2}\,\tfrac{3}{2}\,\tfrac{3}{2}> = |\tfrac{11}{2}\,\tfrac{3}{2}>+\sqrt{2}/\sqrt{5} + |\tfrac{11}{2}\,{-\tfrac{7}{2}}>-\sqrt{3}/\sqrt{5}$$
$$|\tfrac{11}{2}\,\tfrac{5}{2}\,\tfrac{3}{2}\,{-\tfrac{3}{2}}> = |\tfrac{11}{2}\,\tfrac{7}{2}>+\sqrt{3}/\sqrt{5} + |\tfrac{11}{2}\,{-\tfrac{3}{2}}>+\sqrt{2}/\sqrt{5}$$
$$|\tfrac{11}{2}\,\tfrac{5}{2}\,\tfrac{5}{2}\,\tfrac{5}{2}> = |\tfrac{11}{2}\,\tfrac{5}{2}>+3/5\sqrt{2}-2i\sqrt{2/5} + |\tfrac{11}{2}\,{-\tfrac{5}{2}}>-2\sqrt{2/5}-3i/5\sqrt{2}$$
$$|\tfrac{11}{2}\,\tfrac{5}{2}\,{-\tfrac{5}{2}}\,\tfrac{5}{2}> = |\tfrac{11}{2}\,\tfrac{5}{2}>-2\sqrt{2/5}+3i/5\sqrt{2} + |\tfrac{11}{2}\,{-\tfrac{5}{2}}>-3/5\sqrt{2}-2i\sqrt{2/5}$$

$$|6\,0\,0\,0> = |6\,5>-\sqrt{7}/5 + |6\,0>+\sqrt{11}/5 + |6\,{-5}>+\sqrt{7}/5$$
$$|6\,1\,\tilde{0}\,0> = |6\,5>-1/\sqrt{2} + |6\,{-5}>-1/\sqrt{2}$$
$$|6\,1\,1\,1> = |6\,6>-\sqrt{3}/5 + |6\,1>+\sqrt{3.11}/5\sqrt{2} + |6\,{-4}>+\sqrt{11}/5\sqrt{2}$$
$$|6\,1\,1\,{-1}> = |6\,4>+\sqrt{11}/5\sqrt{2} + |6\,{-1}>-\sqrt{3.11}/5\sqrt{2} + |6\,{-6}>-\sqrt{3}/5$$
$$|6\,2\,0\,0> = |6\,5>+\sqrt{11}/5\sqrt{2} + |6\,0>+\sqrt{2.7}/5 + |6\,{-5}>-\sqrt{11}/5\sqrt{2}$$
$$|6\,2\,1\,1> = |6\,6>-\sqrt{11}/5 + |6\,1>+1/5\sqrt{2} + |6\,{-4}>-3\sqrt{3}/5\sqrt{2}$$
$$|6\,2\,1\,{-1}> = |6\,4>-3\sqrt{3}/5\sqrt{2} + |6\,{-1}>-1/5\sqrt{2} + |6\,{-6}>-\sqrt{11}/5$$
$$|6\,2\,2\,2> = |6\,2>-2/\sqrt{5} + |6\,{-3}>-1/\sqrt{5}$$
$$|6\,2\,2\,{-2}> = |6\,3>+1/\sqrt{5} + |6\,{-2}>-2/\sqrt{5}$$
$$|6\,3\,1\,1> = |6\,6>+\sqrt{11}/5 + |6\,1>+2\sqrt{2}/5 + |6\,{-4}>-\sqrt{2.3}/5$$
$$|6\,3\,1\,{-1}> = |6\,4>-\sqrt{2.3}/5 + |6\,{-1}>-2\sqrt{2}/5 + |6\,{-6}>+\sqrt{11}/5$$
$$|6\,3\,2\,2> = |6\,2>-1/\sqrt{5} + |6\,{-3}>+2/\sqrt{5}$$
$$|6\,3\,2\,{-2}> = |6\,3>-2/\sqrt{5} + |6\,{-2}>-1/\sqrt{5}$$

$$|\tfrac{13}{2}\,\tfrac{1}{2}\,\tfrac{1}{2}\,\tfrac{1}{2}> = |\tfrac{13}{2}\,\tfrac{11}{2}>-2\sqrt{3}/5 + |\tfrac{13}{2}\,\tfrac{1}{2}>+\sqrt{11}/5 + |\tfrac{13}{2}\,{-\tfrac{9}{2}}>+\sqrt{2}/5$$
$$|\tfrac{13}{2}\,\tfrac{1}{2}\,\tfrac{1}{2}\,{-\tfrac{1}{2}}> = |\tfrac{13}{2}\,\tfrac{9}{2}>-\sqrt{2}/5 + |\tfrac{13}{2}\,{-\tfrac{1}{2}}>+\sqrt{11}/5 + |\tfrac{13}{2}\,{-\tfrac{11}{2}}>+2\sqrt{3}/5$$
$$|\tfrac{13}{2}\,\tfrac{3}{2}\,\tfrac{1}{2}\,\tfrac{1}{2}> = |\tfrac{13}{2}\,\tfrac{11}{2}>-3\sqrt{2.3}/5\sqrt{5} + |\tfrac{13}{2}\,\tfrac{1}{2}>-\sqrt{2.11}/5\sqrt{5} + |\tfrac{13}{2}\,{-\tfrac{9}{2}}>-7/5\sqrt{5}$$
$$|\tfrac{13}{2}\,\tfrac{3}{2}\,\tfrac{1}{2}\,{-\tfrac{1}{2}}> = |\tfrac{13}{2}\,\tfrac{9}{2}>+7/5\sqrt{5} + |\tfrac{13}{2}\,{-\tfrac{1}{2}}>-\sqrt{2.11}/5\sqrt{5} + |\tfrac{13}{2}\,{-\tfrac{11}{2}}>+3\sqrt{2.3}/5\sqrt{5}$$
$$|\tfrac{13}{2}\,\tfrac{3}{2}\,\tfrac{3}{2}\,\tfrac{3}{2}> = |\tfrac{13}{2}\,\tfrac{13}{2}>+\sqrt{2.13}/5\sqrt{5} + |\tfrac{13}{2}\,\tfrac{3}{2}>-2\sqrt{2.11}/5\sqrt{5} + |\tfrac{13}{2}\,{-\tfrac{7}{2}}>-\sqrt{11}/5\sqrt{5}$$
$$|\tfrac{13}{2}\,\tfrac{3}{2}\,\tfrac{3}{2}\,{-\tfrac{3}{2}}> = |\tfrac{13}{2}\,\tfrac{7}{2}>-\sqrt{11}/5\sqrt{5} + |\tfrac{13}{2}\,{-\tfrac{3}{2}}>+2\sqrt{2.11}/5\sqrt{5} + |\tfrac{13}{2}\,{-\tfrac{13}{2}}>+\sqrt{2.13}/5\sqrt{5}$$
$$|\tfrac{13}{2}\,\tfrac{5}{2}\,\tfrac{1}{2}\,\tfrac{1}{2}> = |\tfrac{13}{2}\,\tfrac{11}{2}>+\sqrt{11}/5\sqrt{5} + |\tfrac{13}{2}\,\tfrac{1}{2}>+4\sqrt{3}/5\sqrt{5} + |\tfrac{13}{2}\,{-\tfrac{9}{2}}>-\sqrt{2.3.11}/5\sqrt{5}$$
$$|\tfrac{13}{2}\,\tfrac{5}{2}\,\tfrac{1}{2}\,{-\tfrac{1}{2}}> = |\tfrac{13}{2}\,\tfrac{9}{2}>+\sqrt{2.3.11}/5\sqrt{5} + |\tfrac{13}{2}\,{-\tfrac{1}{2}}>+4\sqrt{3}/5\sqrt{5} + |\tfrac{13}{2}\,{-\tfrac{11}{2}}>-\sqrt{11}/5\sqrt{5}$$
$$|\tfrac{13}{2}\,\tfrac{5}{2}\,\tfrac{3}{2}\,\tfrac{3}{2}> = |\tfrac{13}{2}\,\tfrac{13}{2}>+\sqrt{11.13}/5\sqrt{2.5} + |\tfrac{13}{2}\,\tfrac{3}{2}>+3/5\sqrt{2.5} + |\tfrac{13}{2}\,{-\tfrac{7}{2}}>+7/5\sqrt{5}$$
$$|\tfrac{13}{2}\,\tfrac{5}{2}\,\tfrac{3}{2}\,{-\tfrac{3}{2}}> = |\tfrac{13}{2}\,\tfrac{7}{2}>+7/5\sqrt{5} + |\tfrac{13}{2}\,{-\tfrac{3}{2}}>-3/5\sqrt{2.5} + |\tfrac{13}{2}\,{-\tfrac{13}{2}}>+\sqrt{11.13}/5\sqrt{2.5}$$
$$|\tfrac{13}{2}\,\tfrac{5}{2}\,\tfrac{5}{2}\,\tfrac{5}{2}> = |\tfrac{13}{2}\,\tfrac{5}{2}>+1/2\sqrt{5}+3i/2\sqrt{5} + |\tfrac{13}{2}\,{-\tfrac{5}{2}}>-3/2\sqrt{5}+i/2\sqrt{5}$$
$$|\tfrac{13}{2}\,\tfrac{5}{2}\,{-\tfrac{5}{2}}\,\tfrac{5}{2}> = |\tfrac{13}{2}\,\tfrac{5}{2}>+3/2\sqrt{5}+i/2\sqrt{5} + |\tfrac{13}{2}\,{-\tfrac{5}{2}}>+1/2\sqrt{5}-3i/2\sqrt{5}$$
$$|\tfrac{13}{2}\,\tilde{\tfrac{1}{2}}\,\tfrac{3}{2}\,\tfrac{3}{2}> = |\tfrac{13}{2}\,\tfrac{13}{2}>-\sqrt{11}/5\sqrt{2} + |\tfrac{13}{2}\,\tfrac{3}{2}>-\sqrt{13}/5\sqrt{2} + |\tfrac{13}{2}\,{-\tfrac{7}{2}}>+\sqrt{13}/5$$
$$|\tfrac{13}{2}\,\tilde{\tfrac{1}{2}}\,\tfrac{3}{2}\,{-\tfrac{3}{2}}> = |\tfrac{13}{2}\,\tfrac{7}{2}>+\sqrt{13}/5 + |\tfrac{13}{2}\,{-\tfrac{3}{2}}>+\sqrt{13}/5\sqrt{2} + |\tfrac{13}{2}\,{-\tfrac{13}{2}}>-\sqrt{11}/5\sqrt{2}$$

$$|7\,1\,\tilde{0}\,0> = |7\,5>-2\sqrt{2.3}/5\sqrt{5} + |7\,0>+\sqrt{7.11}/5\sqrt{5} + |7\,{-5}>+2\sqrt{2.3}/5\sqrt{5}$$
$$|7\,1\,1\,1> = |7\,6>+\sqrt{2.3.13}/5\sqrt{5} + |7\,1>-2\sqrt{11}/5\sqrt{5} + |7\,{-4}>-\sqrt{3}/5\sqrt{5}$$
$$|7\,1\,1\,{-1}> = |7\,4>+\sqrt{3}/5\sqrt{5} + |7\,{-1}>-2\sqrt{11}/5\sqrt{5} + |7\,{-6}>-\sqrt{2.3.13}/5\sqrt{5}$$
$$|7\,2\,0\,0> = |7\,5>+1/\sqrt{2} + |7\,{-5}>+1/\sqrt{2}$$
$$|7\,2\,1\,1> = |7\,6>+\sqrt{2.13}/5\sqrt{3} + |7\,1>+\sqrt{11}/5 + |7\,{-4}>+4/5\sqrt{3}$$
$$|7\,2\,1\,{-1}> = |7\,4>-4/5\sqrt{3} + |7\,{-1}>+\sqrt{11}/5 + |7\,{-6}>-\sqrt{2.13}/5\sqrt{3}$$
$$|7\,2\,2\,2> = |7\,7>-\sqrt{7.13}/2.5\sqrt{3} + |7\,2>+\sqrt{3.11}/5\sqrt{2} + |7\,{-3}>+\sqrt{11}/2.5\sqrt{3}$$
$$|7\,2\,2\,{-2}> = |7\,3>+\sqrt{11}/2.5\sqrt{3} + |7\,{-2}>-\sqrt{3.11}/5\sqrt{2} + |7\,{-7}>-\sqrt{7.13}/2.5\sqrt{3}$$
$$|7\,3\,1\,1> = |7\,6>-\sqrt{11}/5\sqrt{3.5} + |7\,1>-\sqrt{2.13}/5\sqrt{5} + |7\,{-4}>+\sqrt{2.11.13}/5\sqrt{3.5}$$
$$|7\,3\,1\,{-1}> = |7\,4>-\sqrt{2.11.13}/5\sqrt{3.5} + |7\,{-1}>-\sqrt{2.13}/5\sqrt{5} + |7\,{-6}>+\sqrt{11}/5\sqrt{3.5}$$
$$|7\,3\,2\,2> = |7\,7>-\sqrt{2.7.11}/5\sqrt{3.5} + |7\,2>-\sqrt{3.13}/5\sqrt{5} + |7\,{-3}>+2\sqrt{2.13}/5\sqrt{3.5}$$
$$|7\,3\,2\,{-2}> = |7\,3>+2\sqrt{2.13}/5\sqrt{3.5} + |7\,{-2}>+\sqrt{3.13}/5\sqrt{5} + |7\,{-7}>-\sqrt{2.7.11}/5\sqrt{3.5}$$
$$|7\,\tilde{1}\,\tilde{0}\,0> = |7\,5>+\sqrt{7.11}/5\sqrt{2.5} + |7\,0>+4\sqrt{3}/5\sqrt{5} + |7\,{-5}>-\sqrt{7.11}/5\sqrt{2.5}$$

SO_3–K–D_5–C_5 Partners (cont.)

$|7\ \tilde{1}\ 2\ 2> = |7\ 7> -\sqrt{11.13}/2.5\sqrt{5} + |7\ 2> -\sqrt{7}/5\sqrt{2.5} + |7\text{-}3> -7\sqrt{7}/2.5\sqrt{5}$

$|7\ \tilde{1}\ 2\text{-}2> = |7\ 3> -7\sqrt{7}/2.5\sqrt{5} + |7\text{-}2> +\sqrt{7}/5\sqrt{2.5} + |7\text{-}7> -\sqrt{11.13}/2.5\sqrt{5}$

$|\tfrac{15}{2}\ \tfrac{3}{2}\ \tfrac{1}{2}\ \tfrac{1}{2}> = |\tfrac{15}{2}\ \tfrac{11}{2}> -\sqrt{3.13}/5\sqrt{5} + |\tfrac{15}{2}\ \tfrac{1}{2}> +\sqrt{7.11}/5\sqrt{5} + |\tfrac{15}{2}\text{-}\tfrac{9}{2}> +3/5\sqrt{5}$

$|\tfrac{15}{2}\ \tfrac{3}{2}\ \tfrac{1}{2}\text{-}\tfrac{1}{2}> = |\tfrac{15}{2}\ \tfrac{9}{2}> +3/5\sqrt{5} + |\tfrac{15}{2}\text{-}\tfrac{1}{2}> -\sqrt{7.11}/5\sqrt{5} + |\tfrac{15}{2}\text{-}\tfrac{11}{2}> -\sqrt{3.13}/5\sqrt{5}$

$|\tfrac{15}{2}\ \tfrac{3}{2}\ \tfrac{3}{2}\ \tfrac{3}{2}> = |\tfrac{15}{2}\ \tfrac{13}{2}> -\sqrt{7.13}/5\sqrt{5} + |\tfrac{15}{2}\ \tfrac{3}{2}> +\sqrt{3.11}/5\sqrt{5} + |\tfrac{15}{2}\text{-}\tfrac{7}{2}> +1/5\sqrt{5}$

$|\tfrac{15}{2}\ \tfrac{3}{2}\ \tfrac{3}{2}\text{-}\tfrac{3}{2}> = |\tfrac{15}{2}\ \tfrac{7}{2}> -1/5\sqrt{5} + |\tfrac{15}{2}\text{-}\tfrac{3}{2}> +\sqrt{3.11}/5\sqrt{5} + |\tfrac{15}{2}\text{-}\tfrac{13}{2}> +\sqrt{7.13}/5\sqrt{5}$

$|\tfrac{15}{2}\ 0\ \tfrac{5}{2}\ \tfrac{1}{2}\ \tfrac{1}{2}> = |\tfrac{15}{2}\ \tfrac{11}{2}> +\sqrt{7.11.13}/4.5\sqrt{5} + |\tfrac{15}{2}\ \tfrac{1}{2}> +4\sqrt{3}/5\sqrt{5} + |\tfrac{15}{2}\text{-}\tfrac{9}{2}> -\sqrt{3.7.11}/4.5\sqrt{5}$

$|\tfrac{15}{2}\ 0\ \tfrac{5}{2}\ \tfrac{1}{2}\text{-}\tfrac{1}{2}> = |\tfrac{15}{2}\ \tfrac{9}{2}> -\sqrt{3.7.11}/4.5\sqrt{5} + |\tfrac{15}{2}\text{-}\tfrac{1}{2}> -4\sqrt{3}/5\sqrt{5} + |\tfrac{15}{2}\text{-}\tfrac{11}{2}> +\sqrt{7.11.13}/4.5\sqrt{5}$

$|\tfrac{15}{2}\ 0\ \tfrac{5}{2}\ \tfrac{3}{2}\ \tfrac{3}{2}> = |\tfrac{15}{2}\ \tfrac{13}{2}> -\sqrt{11.13}/4.5\sqrt{2.5} + |\tfrac{15}{2}\ \tfrac{3}{2}> -\sqrt{3.7}/2.5\sqrt{2.5}$

$\qquad + |\tfrac{15}{2}\text{-}\tfrac{7}{2}> -7\sqrt{7.11}/4.5\sqrt{2.5}$

$|\tfrac{15}{2}\ 0\ \tfrac{5}{2}\ \tfrac{3}{2}\text{-}\tfrac{3}{2}> = |\tfrac{15}{2}\ \tfrac{7}{2}> +7\sqrt{7.11}/4.5\sqrt{2.5} + |\tfrac{15}{2}\text{-}\tfrac{3}{2}> -\sqrt{3.7}/2.5\sqrt{2.5}$

$\qquad + |\tfrac{15}{2}\text{-}\tfrac{13}{2}> +\sqrt{11.13}/4.5\sqrt{2.5}$

$|\tfrac{15}{2}\ 0\ \tfrac{5}{2}\ \tfrac{5}{2}\ \tfrac{5}{2}> = |\tfrac{15}{2}\ \tfrac{15}{2}> -\sqrt{3.11.13}/4.5\sqrt{5} + i\sqrt{3.11.13}/8.5\sqrt{5}$

$\qquad + |\tfrac{15}{2}\ \tfrac{5}{2}> -11\sqrt{7}/8.5\sqrt{5} -3i\sqrt{7}/2.5\sqrt{5} + |\tfrac{15}{2}\text{-}\tfrac{5}{2}> -3\sqrt{7}/2.5\sqrt{5} +11i\sqrt{7}/8.5\sqrt{5}$

$\qquad + |\tfrac{15}{2}\text{-}\tfrac{15}{2}> -\sqrt{3.11.13}/8.5\sqrt{5} - i\sqrt{3.11.13}/4.5\sqrt{5}$

$|\tfrac{15}{2}\ 0\ \tfrac{5}{2}\text{-}\tfrac{5}{2}\ \tfrac{5}{2}> = |\tfrac{15}{2}\ \tfrac{15}{2}> +\sqrt{3.11.13}/8.5\sqrt{5} - i\sqrt{3.11.13}/4.5\sqrt{5}$

$\qquad + |\tfrac{15}{2}\ \tfrac{5}{2}> -3\sqrt{7}/2.5\sqrt{5} -11i\sqrt{7}/8.5\sqrt{5} + |\tfrac{15}{2}\text{-}\tfrac{5}{2}> +11\sqrt{7}/8.5\sqrt{5} -3i\sqrt{7}/2.5\sqrt{5}$

$\qquad + |\tfrac{15}{2}\text{-}\tfrac{15}{2}> -\sqrt{3.11.13}/4.5\sqrt{5} - i\sqrt{3.11.13}/8.5\sqrt{5}$

$|\tfrac{15}{2}\ 1\ \tfrac{5}{2}\ \tfrac{1}{2}\ \tfrac{1}{2}> = |\tfrac{15}{2}\ \tfrac{11}{2}> +\sqrt{3}/4 + |\tfrac{15}{2}\text{-}\tfrac{9}{2}> +\sqrt{13}/4$

$|\tfrac{15}{2}\ 1\ \tfrac{5}{2}\ \tfrac{1}{2}\text{-}\tfrac{1}{2}> = |\tfrac{15}{2}\ \tfrac{9}{2}> +\sqrt{13}/4 + |\tfrac{15}{2}\text{-}\tfrac{11}{2}> +\sqrt{3}/4$

$|\tfrac{15}{2}\ 1\ \tfrac{5}{2}\ \tfrac{3}{2}\ \tfrac{3}{2}> = |\tfrac{15}{2}\ \tfrac{13}{2}> -3\sqrt{3.7}/4.5\sqrt{2} + |\tfrac{15}{2}\ \tfrac{3}{2}> -\sqrt{11.13}/2.5\sqrt{2} + |\tfrac{15}{2}\text{-}\tfrac{7}{2}> +\sqrt{3.13}/4.5\sqrt{2}$

$|\tfrac{15}{2}\ 1\ \tfrac{5}{2}\ \tfrac{3}{2}\text{-}\tfrac{3}{2}> = |\tfrac{15}{2}\ \tfrac{7}{2}> -\sqrt{3.13}/4.5\sqrt{2} + |\tfrac{15}{2}\text{-}\tfrac{3}{2}> -\sqrt{11.13}/2.5\sqrt{2} + |\tfrac{15}{2}\text{-}\tfrac{13}{2}> +3\sqrt{3.7}/4.5\sqrt{2}$

$|\tfrac{15}{2}\ 1\ \tfrac{5}{2}\ \tfrac{5}{2}\ \tfrac{5}{2}> = |\tfrac{15}{2}\ \tfrac{15}{2}> +11\sqrt{7}/4.5.5 +29i\sqrt{7}/8.5.5 + |\tfrac{15}{2}\ \tfrac{5}{2}> +3\sqrt{3.11.13}/8.5.5 - i\sqrt{3.11.13}/2.5.5$

$\qquad + |\tfrac{15}{2}\text{-}\tfrac{5}{2}> -\sqrt{3.11.13}/2.5.5 -3i\sqrt{3.11.13}/8.5.5 + |\tfrac{15}{2}\text{-}\tfrac{15}{2}> -29\sqrt{7}/8.5.5 +11i\sqrt{7}/4.5.5$

$|\tfrac{15}{2}\ 1\ \tfrac{5}{2}\text{-}\tfrac{5}{2}\ \tfrac{5}{2}> = |\tfrac{15}{2}\ \tfrac{15}{2}> +29\sqrt{7}/8.5.5 +11i\sqrt{7}/4.5.5 + |\tfrac{15}{2}\ \tfrac{5}{2}> -\sqrt{3.11.13}/2.5.5 +3i\sqrt{3.11.13}/8.5.5$

$\qquad + |\tfrac{15}{2}\text{-}\tfrac{5}{2}> -3\sqrt{3.11.13}/8.5.5 - i\sqrt{3.11.13}/2.5.5 + |\tfrac{15}{2}\text{-}\tfrac{15}{2}> +11\sqrt{7}/4.5.5 -29i\sqrt{7}/8.5.5$

$|8\ 0\ 2\ 0\ 0> = |8\ 5> +3\sqrt{13}/5\sqrt{2.17} + |8\ 0> -2\sqrt{7.11}/5\sqrt{17} + |8\text{-}5> -3\sqrt{13}/5\sqrt{2.17}$

$|8\ 0\ 2\ 1\ 1> = |8\ 6> +\sqrt{2.7.13}/5\sqrt{17} + |8\ 1> -\sqrt{3.7.11}/5\sqrt{17} + |8\text{-}4> -2\sqrt{3}/5\sqrt{17}$

$|8\ 0\ 2\ 1\text{-}1> = |8\ 4> -2\sqrt{3}/5\sqrt{17} + |8\text{-}1> +\sqrt{3.7.11}/5\sqrt{17} + |8\text{-}6> +\sqrt{2.7.13}/5\sqrt{17}$

$|8\ 0\ 2\ 2\ 2> = |8\ 7> +\sqrt{3.7.13}/2\sqrt{5.17} + |8\ 2> -\sqrt{3.11}/\sqrt{2.5.17} + |8\text{-}3> -1/2\sqrt{5.17}$

$|8\ 0\ 2\ 2\text{-}2> = |8\ 3> +1/2\sqrt{5.17} + |8\text{-}2> -\sqrt{3.11}/\sqrt{2.5.17} + |8\text{-}7> -\sqrt{3.7.13}/2\sqrt{5.17}$

$|8\ 1\ 2\ 0\ 0> = |8\ 5> -\sqrt{2.7.11}/5\sqrt{17} + |8\ 0> -3\sqrt{13}/5\sqrt{17} + |8\text{-}5> +\sqrt{2.7.11}/5\sqrt{17}$

$|8\ 1\ 2\ 1\ 1> = |8\ 6> +\sqrt{2.11}/5\sqrt{17} + |8\ 1> +4\sqrt{13}/5\sqrt{3.17} + |8\text{-}4> -\sqrt{7.11.13}/5\sqrt{3.17}$

$|8\ 1\ 2\ 1\text{-}1> = |8\ 4> -\sqrt{7.11.13}/5\sqrt{3.17} + |8\text{-}1> -4\sqrt{13}/5\sqrt{3.17} + |8\text{-}6> +\sqrt{2.11}/5\sqrt{17}$

$|8\ 1\ 2\ 2\ 2> = |8\ 7> -\sqrt{11}/5\sqrt{3.5.17} + |8\ 2> -\sqrt{2.7.13}/5\sqrt{3.5.17}$

$\qquad + |8\text{-}3> +\sqrt{7.11.13}/5\sqrt{5.17} + |8\text{-}8> -\sqrt{11.17}/5\sqrt{3.5}$

$|8\ 1\ 2\ 2\text{-}2> = |8\ 8> -\sqrt{11.17}/5\sqrt{3.5} + |8\ 3> -\sqrt{7.11.13}/5\sqrt{5.17}$

$\qquad + |8\text{-}2> -\sqrt{2.7.13}/5\sqrt{3.5.17} + |8\text{-}7> +\sqrt{11}/5\sqrt{3.5.17}$

$|8\ 3\ 1\ 1> = |8\ 6> -\sqrt{13}/5 + |8\ 1> -\sqrt{2.11}/5\sqrt{3} + |8\text{-}4> -\sqrt{2.7}/5\sqrt{3}$

$|8\ 3\ 1\text{-}1> = |8\ 4> -\sqrt{2.7}/5\sqrt{3} + |8\text{-}1> +\sqrt{2.11}/5\sqrt{3} + |8\text{-}6> -\sqrt{13}/5$

$|8\ 3\ 2\ 2> = |8\ 7> +\sqrt{2.13}/5\sqrt{3.5} + |8\ 2> +\sqrt{7.11}/5\sqrt{3.5} + |8\text{-}3> +2\sqrt{2.7}/5\sqrt{5}$

$\qquad + |8\text{-}8> +2\sqrt{2.13}/5\sqrt{3.5}$

$|8\ 3\ 2\text{-}2> = |8\ 8> +2\sqrt{2.13}/5\sqrt{3.5} + |8\ 3> -2\sqrt{2.7}/5\sqrt{5} + |8\text{-}2> +\sqrt{7.11}/5\sqrt{3.5}$

$\qquad + |8\text{-}7> -\sqrt{2.13}/5\sqrt{3.5}$

$|8\ \tilde{1}\ \tilde{0}\ 0> = |8\ 5> +1/\sqrt{2} + |8\text{-}5> +1/\sqrt{2}$

$|8\ \tilde{1}\ 2\ 2> = |8\ 7> +3\sqrt{7}/2.5\sqrt{5} + |8\ 2> +\sqrt{11.13}/5\sqrt{2.5} + |8\text{-}3> -\sqrt{3.13}/2.5\sqrt{5}$

$\qquad + |8\text{-}8> -2\sqrt{7}/5\sqrt{5}$

$|8\ \tilde{1}\ 2\text{-}2> = |8\ 8> -2\sqrt{7}/5\sqrt{5} + |8\ 3> +\sqrt{3.13}/2.5\sqrt{5} + |8\text{-}2> +\sqrt{11.13}/5\sqrt{2.5}$

$\qquad + |8\text{-}7> -3\sqrt{7}/2.5\sqrt{5}$

SO_3–D_∞–D_6–C_6 Partners as JM Partners

$|\,0\ 0\ 0\ 0\rangle = |\,0\ 0\rangle{+1}$

$|\,\tfrac{1}{2}\ \tfrac{1}{2}\ \tfrac{1}{2}\ \tfrac{1}{2}\rangle = |\,\tfrac{1}{2}\ \tfrac{1}{2}\rangle{+1}$
$|\,\tfrac{1}{2}\ \tfrac{1}{2}\ \tfrac{1}{2}\text{-}\tfrac{1}{2}\rangle = |\,\tfrac{1}{2}\text{-}\tfrac{1}{2}\rangle{+1}$

$|\,1\ \tilde{0}\ \tilde{0}\ 0\rangle = |\,1\ 0\rangle{+1}$
$|\,1\ 1\ 1\ 1\rangle = |\,1\ 1\rangle{-1}$
$|\,1\ 1\ 1\text{-}1\rangle = |\,1\text{-}1\rangle{-1}$

$|\,\tfrac{3}{2}\ \tfrac{1}{2}\ \tfrac{1}{2}\ \tfrac{1}{2}\rangle = |\,\tfrac{3}{2}\ \tfrac{1}{2}\rangle{-1}$
$|\,\tfrac{3}{2}\ \tfrac{1}{2}\ \tfrac{1}{2}\text{-}\tfrac{1}{2}\rangle = |\,\tfrac{3}{2}\text{-}\tfrac{1}{2}\rangle{+1}$
$|\,\tfrac{3}{2}\ \tfrac{3}{2}\ \tfrac{3}{2}\ \tfrac{3}{2}\rangle = |\,\tfrac{3}{2}\ \tfrac{3}{2}\rangle{+1}$
$|\,\tfrac{3}{2}\ \tfrac{3}{2}\ \tfrac{3}{2}\text{-}\tfrac{3}{2}\rangle = |\,\tfrac{3}{2}\text{-}\tfrac{3}{2}\rangle{+1}$

$|\,2\ 0\ 0\ 0\rangle = |\,2\ 0\rangle{+1}$
$|\,2\ 1\ 1\ 1\rangle = |\,2\ 1\rangle{+1}$
$|\,2\ 1\ 1\text{-}1\rangle = |\,2\text{-}1\rangle{-1}$
$|\,2\ 2\ 2\ 2\rangle = |\,2\ 2\rangle{-1}$
$|\,2\ 2\ 2\text{-}2\rangle = |\,2\text{-}2\rangle{-1}$

$|\,\tfrac{5}{2}\ \tfrac{1}{2}\ \tfrac{1}{2}\ \tfrac{1}{2}\rangle = |\,\tfrac{5}{2}\ \tfrac{1}{2}\rangle{+1}$
$|\,\tfrac{5}{2}\ \tfrac{1}{2}\ \tfrac{1}{2}\text{-}\tfrac{1}{2}\rangle = |\,\tfrac{5}{2}\text{-}\tfrac{1}{2}\rangle{+1}$
$|\,\tfrac{5}{2}\ \tfrac{3}{2}\ \tfrac{3}{2}\ \tfrac{3}{2}\rangle = |\,\tfrac{5}{2}\ \tfrac{3}{2}\rangle{-1}$
$|\,\tfrac{5}{2}\ \tfrac{3}{2}\ \tfrac{3}{2}\text{-}\tfrac{3}{2}\rangle = |\,\tfrac{5}{2}\text{-}\tfrac{3}{2}\rangle{+1}$
$|\,\tfrac{5}{2}\ \tfrac{5}{2}\ \tfrac{5}{2}\ \tfrac{5}{2}\rangle = |\,\tfrac{5}{2}\ \tfrac{5}{2}\rangle{+1}$
$|\,\tfrac{5}{2}\ \tfrac{5}{2}\ \tfrac{5}{2}\text{-}\tfrac{3}{2}\rangle = |\,\tfrac{5}{2}\text{-}\tfrac{5}{2}\rangle{+1}$

$|\,3\ \tilde{0}\ \tilde{0}\ 0\rangle = |\,3\ 0\rangle{+1}$
$|\,3\ 1\ 1\ 1\rangle = |\,3\ 1\rangle{-1}$
$|\,3\ 1\ 1\text{-}1\rangle = |\,3\text{-}1\rangle{-1}$
$|\,3\ 2\ 2\ 2\rangle = |\,3\ 2\rangle{+1}$
$|\,3\ 2\ 2\text{-}2\rangle = |\,3\text{-}2\rangle{-1}$
$|\,3\ 3\ 3\ 3\rangle = |\,3\ 3\rangle{-1/\sqrt{2}} + |\,3\text{-}3\rangle{-1/\sqrt{2}}$
$|\,3\ 3\ \tilde{3}\ 3\rangle = |\,3\ 3\rangle{+1/\sqrt{2}} + |\,3\text{-}3\rangle{-1/\sqrt{2}}$

$|\,\tfrac{7}{2}\ \tfrac{1}{2}\ \tfrac{1}{2}\ \tfrac{1}{2}\rangle = |\,\tfrac{7}{2}\ \tfrac{1}{2}\rangle{-1}$
$|\,\tfrac{7}{2}\ \tfrac{1}{2}\ \tfrac{1}{2}\text{-}\tfrac{1}{2}\rangle = |\,\tfrac{7}{2}\text{-}\tfrac{1}{2}\rangle{+1}$
$|\,\tfrac{7}{2}\ \tfrac{3}{2}\ \tfrac{3}{2}\ \tfrac{3}{2}\rangle = |\,\tfrac{7}{2}\ \tfrac{3}{2}\rangle{+1}$
$|\,\tfrac{7}{2}\ \tfrac{3}{2}\ \tfrac{3}{2}\text{-}\tfrac{3}{2}\rangle = |\,\tfrac{7}{2}\text{-}\tfrac{3}{2}\rangle{+1}$
$|\,\tfrac{7}{2}\ \tfrac{5}{2}\ \tfrac{5}{2}\ \tfrac{5}{2}\rangle = |\,\tfrac{7}{2}\ \tfrac{5}{2}\rangle{-1}$
$|\,\tfrac{7}{2}\ \tfrac{5}{2}\ \tfrac{5}{2}\text{-}\tfrac{5}{2}\rangle = |\,\tfrac{7}{2}\text{-}\tfrac{5}{2}\rangle{+1}$
$|\,\tfrac{7}{2}\ \tfrac{7}{2}\ \tfrac{5}{2}\ \tfrac{5}{2}\rangle = |\,\tfrac{7}{2}\text{-}\tfrac{7}{2}\rangle{-1}$
$|\,\tfrac{7}{2}\ \tfrac{7}{2}\ \tfrac{5}{2}\text{-}\tfrac{5}{2}\rangle = |\,\tfrac{7}{2}\ \tfrac{7}{2}\rangle{+1}$

$|\,4\ 0\ 0\ 0\rangle = |\,4\ 0\rangle{+1}$
$|\,4\ 1\ 1\ 1\rangle = |\,4\ 1\rangle{+1}$
$|\,4\ 1\ 1\text{-}1\rangle = |\,4\text{-}1\rangle{-1}$
$|\,4\ 2\ 2\ 2\rangle = |\,4\ 2\rangle{-1}$
$|\,4\ 2\ 2\text{-}2\rangle = |\,4\text{-}2\rangle{-1}$
$|\,4\ 3\ 3\ 3\rangle = |\,4\ 3\rangle{+1/\sqrt{2}} + |\,4\text{-}3\rangle{-1/\sqrt{2}}$
$|\,4\ 3\ \tilde{3}\ 3\rangle = |\,4\ 3\rangle{-1/\sqrt{2}} + |\,4\text{-}3\rangle{-1/\sqrt{2}}$

SO_3–D_∞–D_6–C_6 Partners (cont.)

$| 4\ 4\ 2\ 2> \ = \ | 4\,\text{-}4> -1$
$| 4\ 4\ 2\,\text{-}2> \ = \ | 4\ 4> -1$

$| \frac{9}{2}\ \frac{1}{2}\ \frac{1}{2}\ \frac{1}{2}> \ = \ | \frac{9}{2}\ \frac{1}{2}> +1$
$| \frac{9}{2}\ \frac{1}{2}\ \frac{1}{2}\,\text{-}\frac{1}{2}> \ = \ | \frac{9}{2}\,\text{-}\frac{1}{2}> +1$
$| \frac{9}{2}\ \frac{3}{2}\ \frac{3}{2}\ \frac{3}{2}> \ = \ | \frac{9}{2}\ \frac{3}{2}> -1$
$| \frac{9}{2}\ \frac{3}{2}\ \frac{3}{2}\,\text{-}\frac{3}{2}> \ = \ | \frac{9}{2}\,\text{-}\frac{3}{2}> +1$
$| \frac{9}{2}\ \frac{5}{2}\ \frac{5}{2}\ \frac{5}{2}> \ = \ | \frac{9}{2}\ \frac{5}{2}> +1$
$| \frac{9}{2}\ \frac{5}{2}\ \frac{5}{2}\,\text{-}\frac{5}{2}> \ = \ | \frac{9}{2}\,\text{-}\frac{5}{2}> +1$
$| \frac{9}{2}\ \frac{7}{2}\ \frac{5}{2}\ \frac{5}{2}> \ = \ | \frac{9}{2}\,\text{-}\frac{7}{2}> -1$
$| \frac{9}{2}\ \frac{7}{2}\ \frac{5}{2}\,\text{-}\frac{5}{2}> \ = \ | \frac{9}{2}\ \frac{7}{2}> -1$
$| \frac{9}{2}\ \frac{9}{2}\ \frac{3}{2}\ \frac{3}{2}> \ = \ | \frac{9}{2}\,\text{-}\frac{9}{2}> -1$
$| \frac{9}{2}\ \frac{9}{2}\ \frac{3}{2}\,\text{-}\frac{3}{2}> \ = \ | \frac{9}{2}\ \frac{9}{2}> +1$

$| 5\ \tilde{0}\ \tilde{0}\ 0> \ = \ | 5\ 0> +1$
$| 5\ 1\ 1\ 1> \ = \ | 5\ 1> -1$
$| 5\ 1\ 1\,\text{-}1> \ = \ | 5\,\text{-}1> -1$
$| 5\ 2\ 2\ 2> \ = \ | 5\ 2> +1$
$| 5\ 2\ 2\,\text{-}2> \ = \ | 5\,\text{-}2> -1$
$| 5\ 3\ 3\ 3> \ = \ | 5\ 3> -1/\sqrt{2} \ + \ | 5\,\text{-}3> -1/\sqrt{2}$
$| 5\ 3\ \tilde{3}\ 3> \ = \ | 5\ 3> +1/\sqrt{2} \ + \ | 5\,\text{-}3> -1/\sqrt{2}$
$| 5\ 4\ 2\ 2> \ = \ | 5\,\text{-}4> -1$
$| 5\ 4\ 2\,\text{-}2> \ = \ | 5\ 4> +1$
$| 5\ 5\ 1\ 1> \ = \ | 5\,\text{-}5> -1$
$| 5\ 5\ 1\,\text{-}1> \ = \ | 5\ 5> -1$

$| \frac{11}{2}\ \frac{1}{2}\ \frac{1}{2}\ \frac{1}{2}> \ = \ | \frac{11}{2}\ \frac{1}{2}> -1$
$| \frac{11}{2}\ \frac{1}{2}\ \frac{1}{2}\,\text{-}\frac{1}{2}> \ = \ | \frac{11}{2}\,\text{-}\frac{1}{2}> +1$
$| \frac{11}{2}\ \frac{3}{2}\ \frac{3}{2}\ \frac{3}{2}> \ = \ | \frac{11}{2}\ \frac{3}{2}> +1$
$| \frac{11}{2}\ \frac{3}{2}\ \frac{3}{2}\,\text{-}\frac{3}{2}> \ = \ | \frac{11}{2}\,\text{-}\frac{3}{2}> +1$
$| \frac{11}{2}\ \frac{5}{2}\ \frac{5}{2}\ \frac{5}{2}> \ = \ | \frac{11}{2}\ \frac{5}{2}> -1$
$| \frac{11}{2}\ \frac{5}{2}\ \frac{5}{2}\,\text{-}\frac{5}{2}> \ = \ | \frac{11}{2}\,\text{-}\frac{5}{2}> +1$
$| \frac{11}{2}\ \frac{7}{2}\ \frac{5}{2}\ \frac{5}{2}> \ = \ | \frac{11}{2}\,\text{-}\frac{7}{2}> -1$
$| \frac{11}{2}\ \frac{7}{2}\ \frac{5}{2}\,\text{-}\frac{5}{2}> \ = \ | \frac{11}{2}\ \frac{7}{2}> +1$
$| \frac{11}{2}\ \frac{9}{2}\ \frac{3}{2}\ \frac{3}{2}> \ = \ | \frac{11}{2}\,\text{-}\frac{9}{2}> -1$
$| \frac{11}{2}\ \frac{9}{2}\ \frac{3}{2}\,\text{-}\frac{3}{2}> \ = \ | \frac{11}{2}\ \frac{9}{2}> -1$
$| \frac{11}{2}\ \frac{11}{2}\ \frac{1}{2}\ \frac{1}{2}> \ = \ | \frac{11}{2}\,\text{-}\frac{11}{2}> -1$
$| \frac{11}{2}\ \frac{11}{2}\ \frac{1}{2}\,\text{-}\frac{1}{2}> \ = \ | \frac{11}{2}\ \frac{11}{2}> +1$

$| 6\ 0\ 0\ 0> \ = \ | 6\ 0> +1$
$| 6\ 1\ 1\ 1> \ = \ | 6\ 1> +1$
$| 6\ 1\ 1\,\text{-}1> \ = \ | 6\,\text{-}1> -1$
$| 6\ 2\ 2\ 2> \ = \ | 6\ 2> -1$
$| 6\ 2\ 2\,\text{-}2> \ = \ | 6\,\text{-}2> -1$
$| 6\ 3\ 3\ 3> \ = \ | 6\ 3> +1/\sqrt{2} \ + \ | 6\,\text{-}3> -1/\sqrt{2}$
$| 6\ 3\ \tilde{3}\ 3> \ = \ | 6\ 3> -1/\sqrt{2} \ + \ | 6\,\text{-}3> -1/\sqrt{2}$
$| 6\ 4\ 2\ 2> \ = \ | 6\,\text{-}4> -1$
$| 6\ 4\ 2\,\text{-}2> \ = \ | 6\ 4> -1$
$| 6\ 5\ 1\ 1> \ = \ | 6\,\text{-}5> -1$
$| 6\ 5\ 1\,\text{-}1> \ = \ | 6\ 5> +1$
$| 6\ 6\ 0\ 0> \ = \ | 6\ 6> +1/\sqrt{2} \ + \ | 6\,\text{-}6> +1/\sqrt{2}$
$| 6\ 6\ \tilde{0}\ 0> \ = \ | 6\ 6> +1/\sqrt{2} \ + \ | 6\,\text{-}6> -1/\sqrt{2}$

SO_3–D_∞–D_6–C_6 Partners (cont.)

$|\tfrac{13}{2}\ \tfrac{1}{2}\ \tfrac{1}{2}\ \tfrac{1}{2}\rangle = |\tfrac{13}{2}\ \tfrac{1}{2}\rangle + 1$

$|\tfrac{13}{2}\ \tfrac{1}{2}\ \tfrac{1}{2}\,\text{-}\tfrac{1}{2}\rangle = |\tfrac{13}{2}\,\text{-}\tfrac{1}{2}\rangle + 1$

$|\tfrac{13}{2}\ \tfrac{3}{2}\ \tfrac{3}{2}\ \tfrac{3}{2}\rangle = |\tfrac{13}{2}\ \tfrac{3}{2}\rangle - 1$

$|\tfrac{13}{2}\ \tfrac{3}{2}\ \tfrac{3}{2}\,\text{-}\tfrac{3}{2}\rangle = |\tfrac{13}{2}\,\text{-}\tfrac{3}{2}\rangle + 1$

$|\tfrac{13}{2}\ \tfrac{5}{2}\ \tfrac{5}{2}\ \tfrac{5}{2}\rangle = |\tfrac{13}{2}\ \tfrac{5}{2}\rangle + 1$

$|\tfrac{13}{2}\ \tfrac{5}{2}\ \tfrac{5}{2}\,\text{-}\tfrac{5}{2}\rangle = |\tfrac{13}{2}\,\text{-}\tfrac{5}{2}\rangle + 1$

$|\tfrac{13}{2}\ \tfrac{7}{2}\ \tfrac{5}{2}\ \tfrac{5}{2}\rangle = |\tfrac{13}{2}\,\text{-}\tfrac{7}{2}\rangle - 1$

$|\tfrac{13}{2}\ \tfrac{7}{2}\ \tfrac{5}{2}\,\text{-}\tfrac{5}{2}\rangle = |\tfrac{13}{2}\ \tfrac{7}{2}\rangle - 1$

$|\tfrac{13}{2}\ \tfrac{9}{2}\ \tfrac{3}{2}\ \tfrac{3}{2}\rangle = |\tfrac{13}{2}\,\text{-}\tfrac{9}{2}\rangle - 1$

$|\tfrac{13}{2}\ \tfrac{9}{2}\ \tfrac{3}{2}\,\text{-}\tfrac{3}{2}\rangle = |\tfrac{13}{2}\ \tfrac{9}{2}\rangle + 1$

$|\tfrac{13}{2}\ \tfrac{11}{2}\ \tfrac{1}{2}\ \tfrac{1}{2}\rangle = |\tfrac{13}{2}\,\text{-}\tfrac{11}{2}\rangle - 1$

$|\tfrac{13}{2}\ \tfrac{11}{2}\ \tfrac{1}{2}\,\text{-}\tfrac{1}{2}\rangle = |\tfrac{13}{2}\ \tfrac{11}{2}\rangle - 1$

$|\tfrac{13}{2}\ \tfrac{13}{2}\ \tfrac{1}{2}\ \tfrac{1}{2}\rangle = |\tfrac{13}{2}\ \tfrac{13}{2}\rangle + 1$

$|\tfrac{13}{2}\ \tfrac{13}{2}\ \tfrac{1}{2}\,\text{-}\tfrac{1}{2}\rangle = |\tfrac{13}{2}\,\text{-}\tfrac{13}{2}\rangle + 1$

$|7\ \tilde{0}\ \tilde{0}\ 0\rangle = |7\ 0\rangle + 1$

$|7\ 1\ 1\ 1\rangle = |7\ 1\rangle - 1$

$|7\ 1\ 1\,\text{-}1\rangle = |7\,\text{-}1\rangle - 1$

$|7\ 2\ 2\ 2\rangle = |7\ 2\rangle + 1$

$|7\ 2\ 2\,\text{-}2\rangle = |7\,\text{-}2\rangle - 1$

$|7\ 3\ 3\ 3\rangle = |7\ 3\rangle - 1/\sqrt{2} + |7\,\text{-}3\rangle - 1/\sqrt{2}$

$|7\ 3\ \tilde{3}\ 3\rangle = |7\ 3\rangle + 1/\sqrt{2} + |7\,\text{-}3\rangle - 1/\sqrt{2}$

$|7\ 4\ 2\ 2\rangle = |7\,\text{-}4\rangle - 1$

$|7\ 4\ 2\,\text{-}2\rangle = |7\ 4\rangle + 1$

$|7\ 5\ 1\ 1\rangle = |7\,\text{-}5\rangle - 1$

$|7\ 5\ 1\,\text{-}1\rangle = |7\ 5\rangle - 1$

$|7\ 6\ 0\ 0\rangle = |7\ 6\rangle - 1/\sqrt{2} + |7\,\text{-}6\rangle + 1/\sqrt{2}$

$|7\ 6\ \tilde{0}\ 0\rangle = |7\ 6\rangle - 1/\sqrt{2} + |7\,\text{-}6\rangle - 1/\sqrt{2}$

$|7\ 7\ 1\ 1\rangle = |7\ 7\rangle - 1$

$|7\ 7\ 1\,\text{-}1\rangle = |7\,\text{-}7\rangle - 1$

$|\tfrac{15}{2}\ \tfrac{1}{2}\ \tfrac{1}{2}\ \tfrac{1}{2}\rangle = |\tfrac{15}{2}\ \tfrac{1}{2}\rangle - 1$

$|\tfrac{15}{2}\ \tfrac{1}{2}\ \tfrac{1}{2}\,\text{-}\tfrac{1}{2}\rangle = |\tfrac{15}{2}\,\text{-}\tfrac{1}{2}\rangle + 1$

$|\tfrac{15}{2}\ \tfrac{3}{2}\ \tfrac{3}{2}\ \tfrac{3}{2}\rangle = |\tfrac{15}{2}\ \tfrac{3}{2}\rangle + 1$

$|\tfrac{15}{2}\ \tfrac{3}{2}\ \tfrac{3}{2}\,\text{-}\tfrac{3}{2}\rangle = |\tfrac{15}{2}\,\text{-}\tfrac{3}{2}\rangle + 1$

$|\tfrac{15}{2}\ \tfrac{5}{2}\ \tfrac{5}{2}\ \tfrac{5}{2}\rangle = |\tfrac{15}{2}\ \tfrac{5}{2}\rangle - 1$

$|\tfrac{15}{2}\ \tfrac{5}{2}\ \tfrac{5}{2}\,\text{-}\tfrac{5}{2}\rangle = |\tfrac{15}{2}\,\text{-}\tfrac{5}{2}\rangle + 1$

$|\tfrac{15}{2}\ \tfrac{7}{2}\ \tfrac{5}{2}\ \tfrac{5}{2}\rangle = |\tfrac{15}{2}\,\text{-}\tfrac{7}{2}\rangle - 1$

$|\tfrac{15}{2}\ \tfrac{7}{2}\ \tfrac{5}{2}\,\text{-}\tfrac{5}{2}\rangle = |\tfrac{15}{2}\ \tfrac{7}{2}\rangle + 1$

$|\tfrac{15}{2}\ \tfrac{9}{2}\ \tfrac{3}{2}\ \tfrac{3}{2}\rangle = |\tfrac{15}{2}\,\text{-}\tfrac{9}{2}\rangle - 1$

$|\tfrac{15}{2}\ \tfrac{9}{2}\ \tfrac{3}{2}\,\text{-}\tfrac{3}{2}\rangle = |\tfrac{15}{2}\ \tfrac{9}{2}\rangle - 1$

$|\tfrac{15}{2}\ \tfrac{11}{2}\ \tfrac{1}{2}\ \tfrac{1}{2}\rangle = |\tfrac{15}{2}\,\text{-}\tfrac{11}{2}\rangle - 1$

$|\tfrac{15}{2}\ \tfrac{11}{2}\ \tfrac{1}{2}\,\text{-}\tfrac{1}{2}\rangle = |\tfrac{15}{2}\ \tfrac{11}{2}\rangle + 1$

$|\tfrac{15}{2}\ \tfrac{13}{2}\ \tfrac{1}{2}\ \tfrac{1}{2}\rangle = |\tfrac{15}{2}\ \tfrac{13}{2}\rangle - 1$

$|\tfrac{15}{2}\ \tfrac{13}{2}\ \tfrac{1}{2}\,\text{-}\tfrac{1}{2}\rangle = |\tfrac{15}{2}\,\text{-}\tfrac{13}{2}\rangle + 1$

$|\tfrac{15}{2}\ \tfrac{15}{2}\ \tfrac{3}{2}\ \tfrac{3}{2}\rangle = |\tfrac{15}{2}\ \tfrac{15}{2}\rangle + 1$

$|\tfrac{15}{2}\ \tfrac{15}{2}\ \tfrac{3}{2}\,\text{-}\tfrac{3}{2}\rangle = |\tfrac{15}{2}\,\text{-}\tfrac{15}{2}\rangle + 1$

$|8\ 0\ 0\ 0\rangle = |8\ 0\rangle + 1$

$|8\ 1\ 1\ 1\rangle = |8\ 1\rangle + 1$

$|8\ 1\ 1\,\text{-}1\rangle = |8\,\text{-}1\rangle - 1$

$|8\ 2\ 2\ 2\rangle = |8\ 2\rangle - 1$

$|8\ 2\ 2\,\text{-}2\rangle = |8\,\text{-}2\rangle - 1$

SO_3–D_∞–D_6–C_6 Partners (cont.)

$|8\;3\;3\;3> = |8\;3>+1/\sqrt{2} + |8\;{-3}>-1/\sqrt{2}$
$|8\;3\;\tilde{3}\;3> = |8\;3>-1/\sqrt{2} + |8\;{-3}>-1/\sqrt{2}$
$|8\;4\;2\;2> = |8\;{-4}>-1$
$|8\;4\;2\;{-2}> = |8\;4>-1$
$|8\;5\;1\;1> = |8\;{-5}>-1$
$|8\;5\;1\;{-1}> = |8\;5>+1$
$|8\;6\;0\;0> = |8\;6>+1/\sqrt{2} + |8\;{-6}>+1/\sqrt{2}$
$|8\;6\;\tilde{0}\;0> = |8\;6>+1/\sqrt{2} + |8\;{-6}>-1/\sqrt{2}$
$|8\;7\;1\;1> = |8\;7>+1$
$|8\;7\;1\;{-1}> = |8\;{-7}>-1$
$|8\;8\;2\;2> = |8\;8>-1$
$|8\;8\;2\;{-2}> = |8\;{-8}>-1$

D_2–C_{2y} Partners as D_2–C_2 Partners

$|0\;0> = |0\;0>+1$

$|\tfrac{1}{2}\;\tfrac{1}{2}> = |\tfrac{1}{2}\;\tfrac{1}{2}>+1/\sqrt{2} + |\tfrac{1}{2}\;{-\tfrac{1}{2}}>+i/\sqrt{2}$
$|\tfrac{1}{2}\;{-\tfrac{1}{2}}> = |\tfrac{1}{2}\;\tfrac{1}{2}>+i/\sqrt{2} + |\tfrac{1}{2}\;{-\tfrac{1}{2}}>+1/\sqrt{2}$

$|\tilde{0}\;1> = |\tilde{0}\;0>-i$

$|1\;0> = |1\;1>-i$

$|\tilde{1}\;1> = |\tilde{1}\;1>+1$

D_3–C_2 Partners as D_3–C_3 Partners

$|0\;0> = |0\;0>+1$

$|\tfrac{1}{2}\;\tfrac{1}{2}> = |\tfrac{1}{2}\;\tfrac{1}{2}>+1/\sqrt{2} + |\tfrac{1}{2}\;{-\tfrac{1}{2}}>+i/\sqrt{2}$
$|\tfrac{1}{2}\;{-\tfrac{1}{2}}> = |\tfrac{1}{2}\;\tfrac{1}{2}>+i/\sqrt{2} + |\tfrac{1}{2}\;{-\tfrac{1}{2}}>+1/\sqrt{2}$

$|\tilde{0}\;1> = |\tilde{0}\;0>-i$

$|1\;0> = |1\;1>-i/\sqrt{2} + |1\;{-1}>-i/\sqrt{2}$
$|1\;1> = |1\;1>-1/\sqrt{2} + |1\;{-1}>+1/\sqrt{2}$

$|\tfrac{3}{2}\;\tfrac{1}{2}> = |\tfrac{3}{2}\;\tfrac{3}{2}>+1$

$|{-\tfrac{3}{2}}\;{-\tfrac{1}{2}}> = |{-\tfrac{3}{2}}\;\tfrac{3}{2}>+1$

D_4–D_2–C_2 Partners as D_4–C_4 Partners

$|\,0\;0\;0\rangle = |\,0\;0\rangle{+1}$

$|\,\tfrac{1}{2}\;\tfrac{1}{2}\;\tfrac{1}{2}\rangle = |\,\tfrac{1}{2}\;\tfrac{1}{2}\rangle{+1}$
$|\,\tfrac{1}{2}\;\tfrac{1}{2}\,\text{-}\tfrac{1}{2}\rangle = |\,\tfrac{1}{2}\,\text{-}\tfrac{1}{2}\rangle{+1}$

$|\,\tilde{0}\;\tilde{0}\;0\rangle = |\,\tilde{0}\;0\rangle{+1}$

$|\,1\;1\;1\rangle = |\,1\;1\rangle{+1/\sqrt{2}} + |\,1\,\text{-}1\rangle{+1/\sqrt{2}}$
$|\,1\;\tilde{1}\;1\rangle = |\,1\;1\rangle{-1/\sqrt{2}} + |\,1\,\text{-}1\rangle{+1/\sqrt{2}}$

$|\,\tfrac{3}{2}\;\tfrac{1}{2}\;\tfrac{1}{2}\rangle = |\,\tfrac{3}{2}\,\text{-}\tfrac{3}{2}\rangle{-1}$
$|\,\tfrac{3}{2}\;\tfrac{1}{2}\,\text{-}\tfrac{1}{2}\rangle = |\,\tfrac{3}{2}\;\tfrac{3}{2}\rangle{+1}$

$|\,2\;0\;0\rangle = |\,2\;2\rangle{-1}$

$|\,\tilde{2}\;\tilde{0}\;0\rangle = |\,\tilde{2}\;2\rangle{+1}$

D_5–C_2 Partners as D_5–C_5 Partners

$|\,0\;0\rangle = |\,0\;0\rangle{+1}$

$|\,\tfrac{1}{2}\;\tfrac{1}{2}\rangle = |\,\tfrac{1}{2}\;\tfrac{1}{2}\rangle{+1/\sqrt{2}} + |\,\tfrac{1}{2}\,\text{-}\tfrac{1}{2}\rangle{+\,i/\sqrt{2}}$
$|\,\tfrac{1}{2}\,\text{-}\tfrac{1}{2}\rangle = |\,\tfrac{1}{2}\;\tfrac{1}{2}\rangle{+\,i/\sqrt{2}} + |\,\tfrac{1}{2}\,\text{-}\tfrac{1}{2}\rangle{+1/\sqrt{2}}$

$|\,\tilde{0}\;1\rangle = |\,\tilde{0}\;0\rangle{-\,i}$

$|\,1\;0\rangle = |\,1\;1\rangle{-\,i/\sqrt{2}} + |\,1\,\text{-}1\rangle{-\,i/\sqrt{2}}$
$|\,1\;1\rangle = |\,1\;1\rangle{-1/\sqrt{2}} + |\,1\,\text{-}1\rangle{+1/\sqrt{2}}$

$|\,\tfrac{3}{2}\;\tfrac{1}{2}\rangle = |\,\tfrac{3}{2}\;\tfrac{3}{2}\rangle{+\,i/\sqrt{2}} + |\,\tfrac{3}{2}\,\text{-}\tfrac{3}{2}\rangle{-1/\sqrt{2}}$
$|\,\tfrac{3}{2}\,\text{-}\tfrac{1}{2}\rangle = |\,\tfrac{3}{2}\;\tfrac{3}{2}\rangle{+1/\sqrt{2}} + |\,\tfrac{3}{2}\,\text{-}\tfrac{3}{2}\rangle{-\,i/\sqrt{2}}$

$|\,2\;0\rangle = |\,2\;2\rangle{-1/\sqrt{2}} + |\,2\,\text{-}2\rangle{-1/\sqrt{2}}$
$|\,2\;1\rangle = |\,2\;2\rangle{+\,i/\sqrt{2}} + |\,2\,\text{-}2\rangle{-\,i/\sqrt{2}}$

$|\,\tfrac{5}{2}\;\tfrac{1}{2}\rangle = |\,\tfrac{5}{2}\;\tfrac{5}{2}\rangle{+\,i}$

$|\text{-}\tfrac{5}{2}\,\text{-}\tfrac{1}{2}\rangle = |\text{-}\tfrac{5}{2}\;\tfrac{5}{2}\rangle{-\,i}$

D_6–D_3–C_3 Partners as D_6–C_6 Partners

$| 0\ 0\ 0\rangle = | 0\ 0\rangle + 1$

$| \tfrac{1}{2}\ \tfrac{1}{2}\ \tfrac{1}{2}\rangle = | \tfrac{1}{2}\ \tfrac{1}{2}\rangle + 1$
$| \tfrac{1}{2}\ \tfrac{1}{2}\, \text{-}\tfrac{1}{2}\rangle = | \tfrac{1}{2}\, \text{-}\tfrac{1}{2}\rangle + 1$

$| \tilde{0}\ \tilde{0}\ 0\rangle = | \tilde{0}\ 0\rangle + 1$

$| 1\ 1\ 1\rangle = | 1\ 1\rangle + 1$
$| 1\ 1\, \text{-}1\rangle = | 1\, \text{-}1\rangle + 1$

$| \tfrac{3}{2}\ \tfrac{3}{2}\ \tfrac{3}{2}\rangle = | \tfrac{3}{2}\ \tfrac{3}{2}\rangle + i/\sqrt{2}\ +\ | \tfrac{3}{2}\, \text{-}\tfrac{3}{2}\rangle - 1/\sqrt{2}$
$| \tfrac{3}{2}\, \text{-}\tfrac{3}{2}\ \tfrac{3}{2}\rangle = | \tfrac{3}{2}\ \tfrac{3}{2}\rangle + 1/\sqrt{2}\ +\ | \tfrac{3}{2}\, \text{-}\tfrac{3}{2}\rangle - i/\sqrt{2}$

$| 2\ 1\ 1\rangle = | 2\, \text{-}2\rangle + i$
$| 2\ 1\, \text{-}1\rangle = | 2\ 2\rangle + i$

$| \tfrac{5}{2}\ \tfrac{1}{2}\ \tfrac{1}{2}\rangle = | \tfrac{5}{2}\, \text{-}\tfrac{5}{2}\rangle + i$
$| \tfrac{5}{2}\ \tfrac{1}{2}\, \text{-}\tfrac{1}{2}\rangle = | \tfrac{5}{2}\ \tfrac{5}{2}\rangle - i$

$| 3\ 0\ 0\rangle = | 3\ 3\rangle - i$

$| \tilde{3}\ \tilde{0}\ 0\rangle = | \tilde{3}\ 3\rangle + i$

T–D_2–C_2 Partners as T–C_3 Partners

The square of the norm of each spinor is $+8.3 - 4.3\sqrt{2} - 8\sqrt{3} + 4\sqrt{2.3}$

$| 0\ 0\ 0\rangle = | 0\ 0\rangle + 1$

$| \tfrac{1}{2}\ \tfrac{1}{2}\ \tfrac{1}{2}\rangle = | \tfrac{1}{2}\ \tfrac{1}{2}\rangle + \sqrt{2} - 2i + i\sqrt{2}\ +\ | \tfrac{1}{2}\, \text{-}\tfrac{1}{2}\rangle - 1 + \sqrt{3} - i + i\sqrt{2} + i\sqrt{3} - i\sqrt{2.3}$
$| \tfrac{1}{2}\ \tfrac{1}{2}\, \text{-}\tfrac{1}{2}\rangle = | \tfrac{1}{2}\ \tfrac{1}{2}\rangle + 1 - \sqrt{3} - i + i\sqrt{2} + i\sqrt{3} - i\sqrt{2.3}\ +\ | \tfrac{1}{2}\, \text{-}\tfrac{1}{2}\rangle + \sqrt{2} + 2i - i\sqrt{2}$

$| 1\ \tilde{0}\ 0\rangle = | 1\ 0\rangle + 1/\sqrt{3}\ +\ | 1\ 1\rangle + 1/\sqrt{3}\ +\ | 1\, \text{-}1\rangle - 1/\sqrt{3}$
$| 1\ 1\ 1\rangle = | 1\ 0\rangle + i/\sqrt{3}\ +\ | 1\ 1\rangle + 1/2 - i/2\sqrt{3}\ +\ | 1\, \text{-}1\rangle + 1/2 + i/2\sqrt{3}$
$| 1\ \tilde{1}\ 1\rangle = | 1\ 0\rangle + 1/\sqrt{3}\ +\ | 1\ 1\rangle - 1/2\sqrt{3} + i/2\ +\ | 1\, \text{-}1\rangle + 1/2\sqrt{3} + i/2$

$| \tfrac{3}{2}\ \tfrac{1}{2}\ \tfrac{1}{2}\rangle = | \tfrac{3}{2}\ \tfrac{1}{2}\rangle + \sqrt{3} - 1 - i\sqrt{2.3} + i\sqrt{3} + i\sqrt{2} - i\ +\ | \tfrac{3}{2}\ \tfrac{3}{2}\rangle - \sqrt{2} + 2i - i\sqrt{2}$
$| \tfrac{3}{2}\ \tfrac{1}{2}\, \text{-}\tfrac{1}{2}\rangle = | \tfrac{3}{2}\ \tfrac{1}{2}\rangle + \sqrt{2} - i\sqrt{2} + 2i\ +\ | \tfrac{3}{2}\ \tfrac{3}{2}\rangle + \sqrt{3} - 1 - i\sqrt{3} + i\sqrt{2.3} + i - i\sqrt{2}$

$|\text{-}\tfrac{3}{2}\ \tfrac{1}{2}\ \tfrac{1}{2}\rangle = |\text{-}\tfrac{3}{2}\, \text{-}\tfrac{1}{2}\rangle - \sqrt{2} + 2i - i\sqrt{2}\ +\ |\text{-}\tfrac{3}{2}\ \tfrac{3}{2}\rangle - \sqrt{3} + 1 + i\sqrt{2.3} - i\sqrt{3} - i\sqrt{2} + i$
$|\text{-}\tfrac{3}{2}\ \tfrac{1}{2}\, \text{-}\tfrac{1}{2}\rangle = |\text{-}\tfrac{3}{2}\, \text{-}\tfrac{1}{2}\rangle + \sqrt{3} - 1 - i\sqrt{3} + i\sqrt{2.3} + i - i\sqrt{2}\ +\ |\text{-}\tfrac{3}{2}\ \tfrac{3}{2}\rangle - \sqrt{2} + i\sqrt{2} - 2i$

$| 2\ 0\ 0\rangle = | 2\ 1\rangle - 1$

$|\text{-}2\ 0\ 0\rangle = |\text{-}2\, \text{-}1\rangle - 1$

O–T–D_2–C_2 Partners as O–D_4–D_2–C_2 Partners

$| 0\ 0\ 0\ 0> \;=\; | 0\ 0\ 0\ 0> +1$

$| \tfrac{1}{2}\ \tfrac{1}{2}\ \tfrac{1}{2}\ \tfrac{1}{2}> \;=\; | \tfrac{1}{2}\ \tfrac{1}{2}\ \tfrac{1}{2}\ \tfrac{1}{2}> +1$

$| \tfrac{1}{2}\ \tfrac{1}{2}\ \tfrac{1}{2}\ {-\tfrac{1}{2}}> \;=\; | \tfrac{1}{2}\ \tfrac{1}{2}\ \tfrac{1}{2}\ {-\tfrac{1}{2}}> +1$

$| 1\ 1\ \tilde{0}\ 0> \;=\; | 1\ \tilde{0}\ \tilde{0}\ 0> +1$

$| 1\ 1\ 1\ 1> \;=\; | 1\ 1\ 1\ 1> +1$

$| 1\ 1\ \tilde{1}\ 1> \;=\; | 1\ 1\ \tilde{1}\ 1> +1$

$| \tfrac{3}{2}\ \tfrac{3}{2}\ \tfrac{1}{2}\ \tfrac{1}{2}> \;=\; | \tfrac{3}{2}\ \tfrac{1}{2}\ \tfrac{1}{2}\ \tfrac{1}{2}> +1/\sqrt{2} \;+\; | \tfrac{3}{2}\ \tfrac{3}{2}\ \tfrac{1}{2}\ \tfrac{1}{2}> + i/\sqrt{2}$

$| \tfrac{3}{2}\ \tfrac{3}{2}\ \tfrac{1}{2}\ {-\tfrac{1}{2}}> \;=\; | \tfrac{3}{2}\ \tfrac{1}{2}\ \tfrac{1}{2}\ {-\tfrac{1}{2}}> +1/\sqrt{2} \;+\; | \tfrac{3}{2}\ \tfrac{3}{2}\ \tfrac{1}{2}\ {-\tfrac{1}{2}}> + i/\sqrt{2}$

$| \tfrac{3}{2}\ {-\tfrac{3}{2}}\ \tfrac{1}{2}\ \tfrac{1}{2}> \;=\; | \tfrac{3}{2}\ \tfrac{1}{2}\ \tfrac{1}{2}\ \tfrac{1}{2}> +1/\sqrt{2} \;+\; | \tfrac{3}{2}\ \tfrac{3}{2}\ \tfrac{1}{2}\ \tfrac{1}{2}> - i/\sqrt{2}$

$| \tfrac{3}{2}\ {-\tfrac{3}{2}}\ \tfrac{1}{2}\ {-\tfrac{1}{2}}> \;=\; | \tfrac{3}{2}\ \tfrac{1}{2}\ \tfrac{1}{2}\ {-\tfrac{1}{2}}> +1/\sqrt{2} \;+\; | \tfrac{3}{2}\ \tfrac{3}{2}\ \tfrac{1}{2}\ {-\tfrac{1}{2}}> - i/\sqrt{2}$

$| 2\ 2\ 0\ 0> \;=\; | 2\ 0\ 0\ 0> +1/\sqrt{2} \;+\; | 2\ 2\ 0\ 0> + i/\sqrt{2}$

$| 2\ {-2}\ 0\ 0> \;=\; | 2\ 0\ 0\ 0> +1/\sqrt{2} \;+\; | 2\ 2\ 0\ 0> - i/\sqrt{2}$

$| \tilde{1}\ 1\ \tilde{0}\ 0> \;=\; | \tilde{1}\ \tilde{2}\ \tilde{0}\ 0> - i$

$| \tilde{1}\ 1\ 1\ 1> \;=\; | \tilde{1}\ 1\ 1\ 1> + i$

$| \tilde{1}\ 1\ \tilde{1}\ 1> \;=\; | \tilde{1}\ 1\ \tilde{1}\ 1> - i$

$| \tilde{\tfrac{1}{2}}\ \tfrac{1}{2}\ \tfrac{1}{2}\ \tfrac{1}{2}> \;=\; | \tilde{\tfrac{1}{2}}\ \tfrac{3}{2}\ \tfrac{1}{2}\ \tfrac{1}{2}> - i$

$| \tilde{\tfrac{1}{2}}\ \tfrac{1}{2}\ \tfrac{1}{2}\ {-\tfrac{1}{2}}> \;=\; | \tilde{\tfrac{1}{2}}\ \tfrac{3}{2}\ \tfrac{1}{2}\ {-\tfrac{1}{2}}> - i$

$| \tilde{0}\ 0\ 0\ 0> \;=\; | \tilde{0}\ 2\ 0\ 0> + i$

O–T–C_3 Partners as O–D_3–C_3 Partners

$| 0\ 0\ 0> \ = \ | 0\ 0\ 0> +1$

$| \frac{1}{2}\ \frac{1}{2}\ \frac{1}{2}> \ = \ | \frac{1}{2}\ \frac{1}{2}\ \frac{1}{2}> +1$
$| \frac{1}{2}\ \frac{1}{2}\text{-}\frac{1}{2}> \ = \ | \frac{1}{2}\ \frac{1}{2}\text{-}\frac{1}{2}> +1$

$| 1\ 1\ 0> \ = \ | 1\ \tilde{0}\ 0> +1$
$| 1\ 1\ 1> \ = \ | 1\ 1\ 1> +1$
$| 1\ 1\text{-}1> \ = \ | 1\ 1\text{-}1> +1$

$| \frac{3}{2}\ \frac{3}{2}\ \frac{1}{2}> \ = \ | \frac{3}{2}\ \frac{1}{2}\ \frac{1}{2}> +1$
$| \frac{3}{2}\ \frac{3}{2}\ \frac{3}{2}> \ = \ | \frac{3}{2}\ \frac{3}{2}\ \frac{3}{2}> +1/\sqrt{3} - i/\sqrt{2.3} \ + \ | \frac{3}{2}\text{-}\frac{3}{2}\ \frac{3}{2}> +1/\sqrt{2.3} - i/\sqrt{3}$
$| \frac{3}{2}\text{-}\frac{3}{2}\text{-}\frac{1}{2}> \ = \ | \frac{3}{2}\ \frac{1}{2}\text{-}\frac{1}{2}> +1$
$| \frac{3}{2}\text{-}\frac{3}{2}\ \frac{3}{2}> \ = \ | \frac{3}{2}\ \frac{3}{2}\ \frac{3}{2}> -1/\sqrt{2.3} - i/\sqrt{3} \ + \ | \frac{3}{2}\text{-}\frac{3}{2}\ \frac{3}{2}> +1/\sqrt{3} + i/\sqrt{2.3}$

$| 2\ 2\ 1> \ = \ | 2\ 1\ 1> +1$
$| 2\text{-}2\text{-}1> \ = \ | 2\ 1\text{-}1> +1$

$| \tilde{1}\ 1\ 0> \ = \ | \tilde{1}\ 0\ 0> +1$
$| \tilde{1}\ 1\ 1> \ = \ | \tilde{1}\ 1\ 1> +1$
$| \tilde{1}\ 1\text{-}1> \ = \ | \tilde{1}\ 1\text{-}1> -1$

$| \tilde{\frac{1}{2}}\ \frac{1}{2}\ \frac{1}{2}> \ = \ | \tilde{\frac{1}{2}}\ \frac{1}{2}\ \frac{1}{2}> -1$
$| \tilde{\frac{1}{2}}\ \frac{1}{2}\text{-}\frac{1}{2}> \ = \ | \tilde{\frac{1}{2}}\ \frac{1}{2}\text{-}\frac{1}{2}> +1$

$| \tilde{0}\ 0\ 0> \ = \ | \tilde{0}\ \tilde{0}\ 0> -1$

O–D$_3$–C$_3$ Partners as O–D$_4$–C$_4$ Partners

The square of the norm of each spinor is $+8.3-4.3\sqrt{2}-8\sqrt{3}+4\sqrt{2.3}$

$$|\,0\;0\;0\rangle = |\,0\;0\;0\rangle +1$$

$$\begin{aligned}
|\,\tfrac12\;\tfrac12\;\tfrac12\rangle &= |\,\tfrac12\;\tfrac12\;\tfrac12\rangle+\sqrt{2}+2i-i\sqrt{2} \;+\; |\,\tfrac12\;\tfrac12\,\text{-}\tfrac12\rangle+1-\sqrt{3}+i-i\sqrt{2}-i\sqrt{3}+i\sqrt{2.3}\\
|\,\tfrac12\;\tfrac12\,\text{-}\tfrac12\rangle &= |\,\tfrac12\;\tfrac12\;\tfrac12\rangle-1+\sqrt{3}+i-i\sqrt{2}-i\sqrt{3}+i\sqrt{2.3} \;+\; |\,\tfrac12\;\tfrac12\,\text{-}\tfrac12\rangle+\sqrt{2}-2i+i\sqrt{2}
\end{aligned}$$

$$\begin{aligned}
|\,1\;\tilde{0}\;0\rangle &= |\,1\;\tilde{0}\;0\rangle+1/\sqrt{3} \;+\; |\,1\;1\;1\rangle-1/\sqrt{2.3}-i/\sqrt{2.3} \;+\; |\,1\;1\,\text{-}1\rangle+1/\sqrt{2.3}-i/\sqrt{2.3}\\
|\,1\;1\;1\rangle &= |\,1\;\tilde{0}\;0\rangle+1/\sqrt{3} \;+\; |\,1\;1\;1\rangle+1/2\sqrt{2}+1/2\sqrt{2.3}+i/2\sqrt{2}+i/2\sqrt{2.3}\\
&\;+\; |\,1\;1\,\text{-}1\rangle+1/2\sqrt{2}-1/2\sqrt{2.3}-i/2\sqrt{2}+i/2\sqrt{2.3}\\
|\,1\;1\,\text{-}1\rangle &= |\,1\;\tilde{0}\;0\rangle-1/\sqrt{3} \;+\; |\,1\;1\;1\rangle+1/2\sqrt{2}-1/2\sqrt{2.3}+i/2\sqrt{2}-i/2\sqrt{2.3}\\
&\;+\; |\,1\;1\,\text{-}1\rangle+1/2\sqrt{2}+1/2\sqrt{2.3}-i/2\sqrt{2}-i/2\sqrt{2.3}
\end{aligned}$$

$$\begin{aligned}
|\,\tfrac32\;\tfrac12\;\tfrac12\rangle &= |\,\tfrac32\;\tfrac12\;\tfrac12\rangle-1/\sqrt{2}+\sqrt{3}/\sqrt{2}-i+i\sqrt{3}+i/\sqrt{2}-i\sqrt{3}/\sqrt{2} \;+\; |\,\tfrac32\;\tfrac12\,\text{-}\tfrac12\rangle+1-i\sqrt{2}+i\\
&\;+\; |\,\tfrac32\;\tfrac32\;\tfrac32\rangle+\sqrt{2}-1+i \;+\; |\,\tfrac32\;\tfrac32\,\text{-}\tfrac32\rangle+\sqrt{3}-1-\sqrt{3}/\sqrt{2}+1/\sqrt{2}-i\sqrt{3}/\sqrt{2}+i/\sqrt{2}\\
|\,\tfrac32\;\tfrac12\,\text{-}\tfrac12\rangle &= |\,\tfrac32\;\tfrac12\;\tfrac12\rangle-1+i-i\sqrt{2} \;+\; |\,\tfrac32\;\tfrac12\,\text{-}\tfrac12\rangle+\sqrt{3}/\sqrt{2}-1/\sqrt{2}+i\sqrt{3}/\sqrt{2}-i\sqrt{3}+i-i/\sqrt{2}\\
&\;+\; |\,\tfrac32\;\tfrac32\;\tfrac32\rangle+1-\sqrt{3}-1/\sqrt{2}+\sqrt{3}/\sqrt{2}+i/\sqrt{2}-i\sqrt{3}/\sqrt{2} \;+\; |\,\tfrac32\;\tfrac32\,\text{-}\tfrac32\rangle+\sqrt{2}-1-i\\
|\,\tfrac32\;\tfrac32\;\tfrac32\rangle &= |\,\tfrac32\;\tfrac12\;\tfrac12\rangle-1/\sqrt{2}-\sqrt{3}/2+3/2-i\sqrt{3}/\sqrt{2}+i\sqrt{3}/2-i/2\\
&\;+\; |\,\tfrac32\;\tfrac12\,\text{-}\tfrac12\rangle+\sqrt{2}-1/2+\sqrt{3}/2-\sqrt{3}/\sqrt{2}+i/\sqrt{2}-i/2-i\sqrt{3}/2\\
&\;+\; |\,\tfrac32\;\tfrac32\;\tfrac32\rangle+1/2-5/2\sqrt{3}-1/\sqrt{2}+\sqrt{2}/\sqrt{3}-i/2-i/2\sqrt{3}+i\sqrt{2}-i/\sqrt{2.3}\\
&\;+\; |\,\tfrac32\;\tfrac32\,\text{-}\tfrac32\rangle-\sqrt{3}/\sqrt{2}+1/2+1/2\sqrt{3}-i/\sqrt{2}+i\sqrt{2}/\sqrt{3}+3i/2-i\sqrt{3}/2\\
|\,\tfrac32\,\text{-}\tfrac32\;\tfrac32\rangle &= |\,\tfrac32\;\tfrac12\;\tfrac12\rangle+\sqrt{3}/\sqrt{2}-\sqrt{2}-\sqrt{3}/2+1/2+i/\sqrt{2}-i\sqrt{3}/2-i/2\\
&\;+\; |\,\tfrac32\;\tfrac12\,\text{-}\tfrac12\rangle-1/\sqrt{2}+3/2-\sqrt{3}/2+i/2-i\sqrt{3}/2+i\sqrt{3}/\sqrt{2}\\
&\;+\; |\,\tfrac32\;\tfrac32\;\tfrac32\rangle-1/2-1/2\sqrt{3}+\sqrt{3}/\sqrt{2}+3i/2-i\sqrt{3}/2-i/\sqrt{2}+i\sqrt{2}/\sqrt{3}\\
&\;+\; |\,\tfrac32\;\tfrac32\,\text{-}\tfrac32\rangle-1/\sqrt{2}+\sqrt{2}/\sqrt{3}+1/2-5/2\sqrt{3}-i\sqrt{2}+i/\sqrt{2.3}+i/2+i/2\sqrt{3}
\end{aligned}$$

$$\begin{aligned}
|\,2\;1\;1\rangle &= |\,2\;0\;0\rangle-1/\sqrt{2} \;+\; |\,2\;2\;2\rangle+i/\sqrt{2}\\
|\,2\;1\,\text{-}1\rangle &= |\,2\;0\;0\rangle-1/\sqrt{2} \;+\; |\,2\;2\;2\rangle-i/\sqrt{2}
\end{aligned}$$

$$\begin{aligned}
|\,\tilde{1}\;0\;0\rangle &= |\,\tilde{1}\;1\;1\rangle+1/\sqrt{2.3}+i/\sqrt{2.3} \;+\; |\,\tilde{1}\;1\,\text{-}1\rangle+1/\sqrt{2.3}-i/\sqrt{2.3} \;+\; |\,\tilde{1}\;\tilde{2}\;2\rangle-i/\sqrt{3}\\
|\,\tilde{1}\;1\;1\rangle &= |\,\tilde{1}\;1\;1\rangle-1/2\sqrt{2.3}+1/2\sqrt{2}-i/2\sqrt{2.3}+i/2\sqrt{2}\\
&\;+\; |\,\tilde{1}\;1\,\text{-}1\rangle-1/2\sqrt{2}-1/2\sqrt{2.3}+i/2\sqrt{2}+i/2\sqrt{2.3} \;+\; |\,\tilde{1}\;\tilde{2}\;2\rangle-i/\sqrt{3}\\
|\,\tilde{1}\;1\,\text{-}1\rangle &= |\,\tilde{1}\;1\;1\rangle-1/2\sqrt{2}-1/2\sqrt{2.3}-i/2\sqrt{2}-i/2\sqrt{2.3}\\
&\;+\; |\,\tilde{1}\;1\,\text{-}1\rangle-1/2\sqrt{2.3}+1/2\sqrt{2}+i/2\sqrt{2.3}-i/2\sqrt{2} \;+\; |\,\tilde{1}\;\tilde{2}\;2\rangle-i/\sqrt{3}
\end{aligned}$$

$$\begin{aligned}
|\,\tilde{\tfrac12}\;\tfrac12\;\tfrac12\rangle &= |\,\tilde{\tfrac12}\;\tfrac32\;\tfrac32\rangle+\sqrt{3}-\sqrt{2.3}+\sqrt{2}-1-i\sqrt{3}+i \;+\; |\,\tilde{\tfrac12}\;\tfrac32\,\text{-}\tfrac32\rangle+2-\sqrt{2}-i\sqrt{2}\\
|\,\tilde{\tfrac12}\;\tfrac12\,\text{-}\tfrac12\rangle &= |\,\tilde{\tfrac12}\;\tfrac32\;\tfrac32\rangle-2+\sqrt{2}-i\sqrt{2} \;+\; |\,\tilde{\tfrac12}\;\tfrac32\,\text{-}\tfrac32\rangle-\sqrt{2.3}+\sqrt{3}-1+\sqrt{2}+i\sqrt{3}-i
\end{aligned}$$

$$|\,\tilde{0}\;\tilde{0}\;0\rangle = |\,\tilde{0}\;2\;2\rangle+i$$

O–D_3–C_2 Partners as O–D_4–C_4 Partners

The square of the norm of each spinor is $+8.2.3+8\sqrt{2.3}-8.3\sqrt{2}-8.2\sqrt{3}$

$$|\,0\ 0\ 0\rangle = |\,0\ 0\ 0\rangle+1$$

$$\begin{aligned}
|\,\tfrac{1}{2}\ \tfrac{1}{2}\ \tfrac{1}{2}\rangle &= |\,\tfrac{1}{2}\ \tfrac{1}{2}\ \tfrac{1}{2}\rangle+2\sqrt{2}+\sqrt{3}-1-\sqrt{2.3}+i-i\sqrt{2}+i\sqrt{3}\\
&+|\,\tfrac{1}{2}\ \tfrac{1}{2}\text{-}\tfrac{1}{2}\rangle+3-\sqrt{2}-\sqrt{3}+i-i\sqrt{3}+i\sqrt{2.3}\\
|\,\tfrac{1}{2}\ \tfrac{1}{2}\text{-}\tfrac{1}{2}\rangle &= |\,\tfrac{1}{2}\ \tfrac{1}{2}\ \tfrac{1}{2}\rangle-3+\sqrt{2}+\sqrt{3}+i-i\sqrt{3}+i\sqrt{2.3}\\
&+|\,\tfrac{1}{2}\ \tfrac{1}{2}\text{-}\tfrac{1}{2}\rangle+2\sqrt{2}+\sqrt{3}-1-\sqrt{2.3}-i+i\sqrt{2}-i\sqrt{3}
\end{aligned}$$

$$\begin{aligned}
|\,1\ \tilde{0}\ 1\rangle &= |\,1\ \tilde{0}\ 0\rangle-i/\sqrt{3}\ +\ |\,1\ 1\ 1\rangle-1/\sqrt{2.3}+i/\sqrt{2.3}\ +\ |\,1\ 1\text{-}1\rangle-1/\sqrt{2.3}-i/\sqrt{2.3}\\
|\,1\ 1\ 0\rangle &= |\,1\ 1\ 1\rangle+1/2-i/2\ +\ |\,1\ 1\text{-}1\rangle-1/2-i/2\\
|\,1\ 1\ 1\rangle &= |\,1\ \tilde{0}\ 0\rangle-\sqrt{2}/\sqrt{3}\ +\ |\,1\ 1\ 1\rangle-1/2\sqrt{3}-i/2\sqrt{3}\ +\ |\,1\ 1\text{-}1\rangle+1/2\sqrt{3}-i/2\sqrt{3}
\end{aligned}$$

$$\begin{aligned}
|\,\tfrac{3}{2}\ \tfrac{1}{2}\ \tfrac{1}{2}\rangle &= |\,\tfrac{3}{2}\ \tfrac{1}{2}\ \tfrac{1}{2}\rangle+\sqrt{3}/\sqrt{2}-1+1/\sqrt{2}-i\sqrt{3}/\sqrt{2}-2i+i\sqrt{3}+i/\sqrt{2}\\
&+|\,\tfrac{3}{2}\ \tfrac{1}{2}\text{-}\tfrac{1}{2}\rangle-\sqrt{3}/\sqrt{2}+\sqrt{3}+1/\sqrt{2}+i\sqrt{3}/\sqrt{2}+i-3i/\sqrt{2}\\
&+|\,\tfrac{3}{2}\ \tfrac{3}{2}\ \tfrac{3}{2}\rangle-1+1/\sqrt{2}+\sqrt{3}/\sqrt{2}+2i-i/\sqrt{2}-i\sqrt{3}+i\sqrt{3}/\sqrt{2}\\
&+|\,\tfrac{3}{2}\ \tfrac{3}{2}\text{-}\tfrac{3}{2}\rangle+1/\sqrt{2}+\sqrt{3}-\sqrt{3}/\sqrt{2}+3i/\sqrt{2}-i-i\sqrt{3}/\sqrt{2}\\
|\,\tfrac{3}{2}\ \tfrac{1}{2}\text{-}\tfrac{1}{2}\rangle &= |\,\tfrac{3}{2}\ \tfrac{1}{2}\ \tfrac{1}{2}\rangle+\sqrt{3}/\sqrt{2}-1/\sqrt{2}-\sqrt{3}+i+i\sqrt{3}/\sqrt{2}-3i/\sqrt{2}\\
&+|\,\tfrac{3}{2}\ \tfrac{1}{2}\text{-}\tfrac{1}{2}\rangle+1/\sqrt{2}-1+\sqrt{3}/\sqrt{2}-i/\sqrt{2}+2i+i\sqrt{3}/\sqrt{2}-i\sqrt{3}\\
&+|\,\tfrac{3}{2}\ \tfrac{3}{2}\ \tfrac{3}{2}\rangle+\sqrt{3}/\sqrt{2}-\sqrt{3}-1/\sqrt{2}-i\sqrt{3}/\sqrt{2}+3i/\sqrt{2}-i\\
&+|\,\tfrac{3}{2}\ \tfrac{3}{2}\text{-}\tfrac{3}{2}\rangle+1/\sqrt{2}+\sqrt{3}/\sqrt{2}-1+i\sqrt{3}+i/\sqrt{2}-i\sqrt{3}/\sqrt{2}-2i\\
|\,\tfrac{3}{2}\ \tfrac{3}{2}\ \tfrac{1}{2}\rangle &= |\,\tfrac{3}{2}\ \tfrac{1}{2}\ \tfrac{1}{2}\rangle-1-\sqrt{3}/\sqrt{2}+3/\sqrt{2}+i\sqrt{3}/\sqrt{2}-i/\sqrt{2}-i\sqrt{3}\\
&+|\,\tfrac{3}{2}\ \tfrac{1}{2}\text{-}\tfrac{1}{2}\rangle+2+\sqrt{3}/\sqrt{2}-1/\sqrt{2}-\sqrt{3}+i-i\sqrt{3}/\sqrt{2}-i/\sqrt{2}\\
&+|\,\tfrac{3}{2}\ \tfrac{3}{2}\ \tfrac{3}{2}\rangle-5/\sqrt{2.3}+2/\sqrt{3}+1/\sqrt{2}-1-i/\sqrt{2.3}-i/\sqrt{3}-i/\sqrt{2}+2i\\
&+|\,\tfrac{3}{2}\ \tfrac{3}{2}\text{-}\tfrac{3}{2}\rangle-\sqrt{3}+1/\sqrt{2}+1/\sqrt{2.3}+2i/\sqrt{3}+3i/\sqrt{2}-i\sqrt{3}/\sqrt{2}-i\\
|\,\tfrac{3}{2}\text{-}\tfrac{3}{2}\text{-}\tfrac{1}{2}\rangle &= |\,\tfrac{3}{2}\ \tfrac{1}{2}\ \tfrac{1}{2}\rangle+\sqrt{3}+1/\sqrt{2}-\sqrt{3}/\sqrt{2}-2-i/\sqrt{2}-i\sqrt{3}/\sqrt{2}+i\\
&+|\,\tfrac{3}{2}\ \tfrac{1}{2}\text{-}\tfrac{1}{2}\rangle-\sqrt{3}/\sqrt{2}+3/\sqrt{2}-1-i\sqrt{3}/\sqrt{2}+i\sqrt{3}+i/\sqrt{2}\\
&+|\,\tfrac{3}{2}\ \tfrac{3}{2}\ \tfrac{3}{2}\rangle-1/\sqrt{2}-1/\sqrt{2.3}+\sqrt{3}+3i/\sqrt{2}-i-i\sqrt{3}/\sqrt{2}+2i/\sqrt{3}\\
&+|\,\tfrac{3}{2}\ \tfrac{3}{2}\text{-}\tfrac{3}{2}\rangle-1-5/\sqrt{2.3}+1/\sqrt{2}+2/\sqrt{3}-2i+i/\sqrt{2.3}+i/\sqrt{2}+i/\sqrt{3}
\end{aligned}$$

$$\begin{aligned}
|\,2\ 1\ 0\rangle &= |\,2\ 0\ 0\rangle+i\\
|\,2\ 1\ 1\rangle &= |\,2\ 2\ 2\rangle-i
\end{aligned}$$

$$\begin{aligned}
|\,\tilde{1}\ 0\ 0\rangle &= |\,\tilde{1}\ 1\ 1\rangle+1/\sqrt{2.3}+i/\sqrt{2.3}\ +\ |\,\tilde{1}\ 1\text{-}1\rangle+1/\sqrt{2.3}-i/\sqrt{2.3}\ +\ |\,\tilde{1}\ \tilde{2}\ 2\rangle-i/\sqrt{3}\\
|\,\tilde{1}\ 1\ 0\rangle &= |\,\tilde{1}\ 1\ 1\rangle-1/2\sqrt{3}+i/2\sqrt{3}\ +\ |\,\tilde{1}\ 1\text{-}1\rangle+1/2\sqrt{3}+i/2\sqrt{3}\ +\ |\,\tilde{1}\ \tilde{2}\ 2\rangle-\sqrt{2}/\sqrt{3}\\
|\,\tilde{1}\ 1\ 1\rangle &= |\,\tilde{1}\ 1\ 1\rangle-1/2-i/2\ +\ |\,\tilde{1}\ 1\text{-}1\rangle+1/2-i/2
\end{aligned}$$

$$\begin{aligned}
|\,\tilde{\tfrac{1}{2}}\ \tfrac{1}{2}\ \tfrac{1}{2}\rangle &= |\,\tilde{\tfrac{1}{2}}\ \tfrac{3}{2}\ \tfrac{3}{2}\rangle+2\sqrt{2}-1-\sqrt{2.3}+\sqrt{3}+i\sqrt{2}-i-i\sqrt{3}\\
&+|\,\tilde{\tfrac{1}{2}}\ \tfrac{3}{2}\text{-}\tfrac{3}{2}\rangle+3-\sqrt{2}-\sqrt{3}-i-i\sqrt{2.3}+i\sqrt{3}\\
|\,\tilde{\tfrac{1}{2}}\ \tfrac{1}{2}\text{-}\tfrac{1}{2}\rangle &= |\,\tilde{\tfrac{1}{2}}\ \tfrac{3}{2}\ \tfrac{3}{2}\rangle-3+\sqrt{3}+\sqrt{2}-i\sqrt{2.3}-i+i\sqrt{3}\\
&+|\,\tilde{\tfrac{1}{2}}\ \tfrac{3}{2}\text{-}\tfrac{3}{2}\rangle-\sqrt{2.3}-1+\sqrt{3}+2\sqrt{2}+i+i\sqrt{3}-i\sqrt{2}
\end{aligned}$$

$$|\,\tilde{0}\ \tilde{0}\ 1\rangle = |\,\tilde{0}\ 2\ 2\rangle+1$$

K–T–C_3 Partners as K–D_3–C_3 Partners

The square of the norm of each spinor is $+8.2+4\sqrt{2.5}$

$|\ 0\ 0\ 0> \ = \ |\ 0\ 0\ 0>+1$

$|\ \tfrac12\ \tfrac12\ \tfrac12> \ = \ |\ \tfrac12\ \tfrac12\ \tfrac12>+\sqrt3+2i\sqrt2+i\sqrt5$
$|\ \tfrac12\ \tfrac12\text{-}\tfrac12> \ = \ |\ \tfrac12\ \tfrac12\text{-}\tfrac12>+\sqrt3-2i\sqrt2-i\sqrt5$

$|\ 1\ 1\ 0> \ = \ |\ 1\ \tilde{0}\ 0>+1$
$|\ 1\ 1\ 1> \ = \ |\ 1\ 1\ 1>-\sqrt5/2\sqrt2+i\sqrt3/2\sqrt2$
$|\ 1\ 1\text{-}1> \ = \ |\ 1\ 1\text{-}1>-\sqrt5/2\sqrt2-i\sqrt3/2\sqrt2$

$|\ \tfrac32\ \tfrac32\ \tfrac12> \ = \ |\ \tfrac32\ \tfrac12\ \tfrac12>+\sqrt3+2i\sqrt2+i\sqrt5$
$|\ \tfrac32\ \tfrac32\ \tfrac32> \ = \ |\ \tfrac32\ \tfrac32\ \tfrac32>-\sqrt5/\sqrt2-1-1/2\sqrt3-\sqrt5/\sqrt{2.3}+i/\sqrt{2.3}+i\sqrt5/\sqrt3+i\sqrt5/2+i/\sqrt2$
$\qquad + \ |\ \tfrac32\text{-}\tfrac32\ \tfrac32>+1/\sqrt{2.3}+\sqrt5/\sqrt3-\sqrt5/2-1/\sqrt2+i\sqrt5/\sqrt2+i-i/2\sqrt3-i\sqrt5/\sqrt{2.3}$
$|\ \tfrac32\text{-}\tfrac32\text{-}\tfrac12> \ = \ |\ \tfrac32\ \tfrac12\text{-}\tfrac12>+\sqrt3-2i\sqrt2-i\sqrt5$
$|\ \tfrac32\text{-}\tfrac32\ \tfrac32> \ = \ |\ \tfrac32\ \tfrac32\ \tfrac32>+\sqrt5/2+1/\sqrt2-1/\sqrt{2.3}-\sqrt5/\sqrt3-i/2\sqrt3-i\sqrt5/\sqrt{2.3}+i\sqrt5/\sqrt2+i$
$\qquad + \ |\ \tfrac32\text{-}\tfrac32\ \tfrac32>-1/2\sqrt3-\sqrt5/\sqrt{2.3}-\sqrt5/\sqrt2-1-i\sqrt5/2-i/\sqrt2-i/\sqrt{2.3}-i\sqrt5/\sqrt3$

$|\ 2\ 1\ 0> \ = \ |\ 2\ 0\ 0>+i$
$|\ 2\ 1\ 1> \ = \ |\ 2\,{\scriptstyle 0}\ 1\ 1>-1/2\sqrt2-i\sqrt5/2\sqrt{2.3} \ + \ |\ 2\,{\scriptstyle 1}\ 1\ 1>+\sqrt5/2\sqrt2-i/2\sqrt{2.3}$
$|\ 2\ 1\text{-}1> \ = \ |\ 2\,{\scriptstyle 0}\ 1\text{-}1>-1/2\sqrt2+i\sqrt5/2\sqrt{2.3} \ + \ |\ 2\,{\scriptstyle 1}\ 1\text{-}1>+\sqrt5/2\sqrt2+i/2\sqrt{2.3}$
$|\ 2\ 2\ 1> \ = \ |\ 2\,{\scriptstyle 0}\ 1\ 1>-\sqrt5/2\sqrt3+i/2 \ + \ |\ 2\,{\scriptstyle 1}\ 1\ 1>+1/4\sqrt3+i\sqrt5/4$
$|\ 2\text{-}2\text{-}1> \ = \ |\ 2\,{\scriptstyle 0}\ 1\text{-}1>-\sqrt5/2\sqrt3-i/2 \ + \ |\ 2\,{\scriptstyle 1}\ 1\text{-}1>+1/4\sqrt3-i\sqrt5/4$

$|\ \tfrac52\ \tfrac12\ \tfrac12> \ = \ |\ \tfrac52\,{\scriptstyle 0}\ \tfrac12\ \tfrac12>-2\sqrt{2.5/3}-5/3+i\sqrt5/\sqrt3 \ + \ |\ \tfrac52\,{\scriptstyle 1}\ \tfrac12\ \tfrac12>+\sqrt{5/3}-\sqrt{2/3}-i\sqrt3-i\sqrt{2.5}/\sqrt3$
$|\ \tfrac52\ \tfrac12\text{-}\tfrac12> \ = \ |\ \tfrac52\,{\scriptstyle 0}\ \tfrac12\text{-}\tfrac12>-2\sqrt{2.5/3}-5/3-i\sqrt5/\sqrt3 \ + \ |\ \tfrac52\,{\scriptstyle 1}\ \tfrac12\text{-}\tfrac12>+\sqrt{5/3}-\sqrt{2/3}+i\sqrt3+i\sqrt{2.5}/\sqrt3$
$|\ \tfrac52\ \tfrac32\ \tfrac12> \ = \ |\ \tfrac52\,{\scriptstyle 0}\ \tfrac12\ \tfrac12>-4\sqrt{2/3}-2\sqrt{5/3}+2i/\sqrt3 \ + \ |\ \tfrac52\,{\scriptstyle 1}\ \tfrac12\ \tfrac12>-5/2.3+\sqrt5/3\sqrt2+i\sqrt{3.5/2}+5i/\sqrt{2.3}$
$|\ \tfrac52\ \tfrac32\ \tfrac32> \ = \ |\ \tfrac52\ \tfrac32\ \tfrac32>+1/2.3+\sqrt5/3\sqrt2-2/\sqrt3-\sqrt{2.5}/\sqrt3+i\sqrt5/2\sqrt3+i/\sqrt{2.3}-2i\sqrt{5/3}-i\sqrt{2/3}$
$\qquad + \ |\ \tfrac52\text{-}\tfrac32\ \tfrac32>+\sqrt5/2\sqrt3+1/\sqrt{2.3}+2\sqrt{5/3}+\sqrt{2/3}-i/2.3-i\sqrt5/3\sqrt2-2i/\sqrt3-i\sqrt{2.5}/\sqrt3$
$|\ \tfrac52\text{-}\tfrac32\text{-}\tfrac12> \ = \ |\ \tfrac52\,{\scriptstyle 0}\ \tfrac12\text{-}\tfrac12>-4\sqrt{2/3}-2\sqrt{5/3}-2i/\sqrt3 \ + \ |\ \tfrac52\,{\scriptstyle 1}\ \tfrac12\text{-}\tfrac12>-5/2.3+\sqrt5/3\sqrt2-i\sqrt{3.5/2}-5i/\sqrt{2.3}$
$|\ \tfrac52\text{-}\tfrac32\ \tfrac32> \ = \ |\ \tfrac52\ \tfrac32\ \tfrac32>-\sqrt{2/3}-2\sqrt{5/3}-1/\sqrt{2.3}-\sqrt5/2\sqrt3-i\sqrt{2.5}/\sqrt3-2i/\sqrt3-i\sqrt5/3\sqrt2-i/2.3$
$\qquad + \ |\ \tfrac52\text{-}\tfrac32\ \tfrac32>-\sqrt{2.5}/\sqrt3-2/\sqrt3+\sqrt5/3\sqrt2+1/2.3+i\sqrt{2/3}+2i\sqrt{5/3}-i/\sqrt{2.3}-i\sqrt5/2\sqrt3$

$|\ 3\ 0\ 0> \ = \ |\ 3\ 0\ 0>-\sqrt3/2\sqrt2 \ + \ |\ 3\ \tilde{0}\ 0>+i\sqrt5/2\sqrt2$
$|\ 3\ 1\ 0> \ = \ |\ 3\ 0\ 0>+i\sqrt5/2\sqrt2 \ + \ |\ 3\ \tilde{0}\ 0>-\sqrt3/2\sqrt2$
$|\ 3\ 1\ 1> \ = \ |\ 3\ 1\ 1>-i$
$|\ 3\ 1\text{-}1> \ = \ |\ 3\ 1\text{-}1>+i$

$|\ \tilde{1}\ 1\ 0> \ = \ |\ \tilde{1}\ \tilde{0}\ 0>-1$
$|\ \tilde{1}\ 1\ 1> \ = \ |\ \tilde{1}\ 1\ 1>+\sqrt5/2\sqrt2+i\sqrt3/2\sqrt2$
$|\ \tilde{1}\ 1\text{-}1> \ = \ |\ \tilde{1}\ 1\text{-}1>+\sqrt5/2\sqrt2-i\sqrt3/2\sqrt2$

$|\ \tilde{\tfrac12}\ \tfrac12\ \tfrac12> \ = \ |\ \tilde{\tfrac12}\ \tfrac12\ \tfrac12>-2\sqrt2-\sqrt5-i\sqrt3$
$|\ \tilde{\tfrac12}\ \tfrac12\text{-}\tfrac12> \ = \ |\ \tilde{\tfrac12}\ \tfrac12\text{-}\tfrac12>-2\sqrt2-\sqrt5+i\sqrt3$

References

Note: [BvD] indicates that the paper is reprinted in Biedenharn and van Dam (1965).

Abragam, A., and Bleaney, B. (1970). *Electron Paramagnetic Resonance of Transition Ions.* Oxford University Press, Oxford.

Altmann, S. L. (1979). Double groups as projective representatives. I: General theory, *Mol. Phys.* **38** 489–511.

Arfken G. (1970). *Mathematical Methods for Physicists*, 2nd ed. Academic Press, New York.

Bacher, R. F., and Goudsmit, S. (1934). Atomic energy relations. I, *Phys. Rev.* **46**, 948–69.

Bethe, H. (1929). Term splitting in crystals, *Ann. Phys. (Leipzig)* **3**, 133–208.

Bickerstaff, R. P., and Wybourne, B. G. (1976). Integrity bases, invariant operators and the state labelling problem for finite subgroups of SO_3, *J. Phys. A* **9**, 1051–68.

Biedenharn, L. C. (1953). An identity satisfied by Racah coefficients, *J. Math. Phys.* **31**, 287–293 [BvD].

Biedenharn, L. C. (1961). Wigner coefficients for the R_4 group and some applications, *J. Math. Phys.* **2**, 433–441 [BvD].

Biedenharn, L. C., and van Dam, H. (1965) (Eds). *Quantum Theory of Angular Momentum. A Collection of Reprints and Original Papers*. Academic Press, New York.

Biedenharn, L. C., Blatt, J.M., and Rose, M. E. (1952). Some properties of the Racah and associated coefficients, *Rev. Mod. Phys.* **24**, 249–257 [BvD].

Brink, D. M., and Satchler, G. R. (1962). *Angular Momentum.* Oxford University Press, Oxford.

Burnside, W. (1911). *The Theory of Groups of Finite Order*, 2nd Ed. Cambridge (Dover reprint, 1955).

Butler, P. H. (1975). Coupling coefficients and tensor operators for chains of groups, *Phil. Trans. R. Soc. London* **277**, 545–585.

Butler, P. H. (1976). Calculation of *j* and *jm* symbols for arbitrary compact groups. II. An alternate procedure for angular momentum, *Int. J. Quantum Chem.* **10**, 599–613.

Butler, P. H., and Ford, A. M. (1979). Special symmetries of *jm* and *j* symbols, *J. Phys. A.* **12**, 1357–1365.

Butler, P. H., and King, R. C. (1974). Symmetrized Kronecker products of group representations, *Can. J. Math.* **23**, 329–39.

Butler, P. H., and Reid, M. F. (1979). *j* symbols and *jm* factors for all dihedral and cyclic groups, *J. Phys. A.* **12**, 1655–1666.

Butler, P. H., and Wybourne, B. G. (1970a). Generalised Racah tensors and the structure of mixed configurations, *J. Math. Phys.* **11**, 2512–2518.

Butler, P. H., and Wybourne, B. G. (1970b). Is the group R_4 an approximate symmetry for many electron theory?, *J. Math. Phys.* **11**, 2517–24.

Butler, P. H., and Wybourne, B. G. (1976a). Calculation of j and jm symbols for arbitrary compact groups. I. Methodology, *Int. J. Quantum Chem.* **10**, 581–598.

Butler, P. H., and Wybourne, B. G. (1976b). Calculation of j and jm symbols for arbitrary compact groups. III. Application to $SO_3 \supset T \supset C_3 \supset C_1$, *Int. J. Quantum Chem.* **10**, 615–628.

Butler, P. H., Haase, R. A., and Wybourne, B. G. (1978). Calculation of $6j$ symbols for the exceptional group E_7, *Aust. J. Phys.* **31**, 131–135.

Butler, P. H., Haase, R. W., and Wybourne, B. G. (1979). Calculation of $3jm$ factors and the matrix elements of E_7 group generators, *Aust. J. Phys.* **32**, 137–154.

Butler, P. H., Minchin, P. E., and Wybourne, B. G. (1971). Tables of hydrogenic Slater radial integrals, *Atomic Data* **3**, 153–168.

Condon, E. U., and Shortley, G. H. (1935). *The Theory of Atomic Spectra*. Cambridge University Press, Cambridge.

Derome, J. R. (1966). Symmetry properties of the $3j$-symbols for an arbitrary group, *J. Math. Phys.* **7**, 612–615.

Derome, J. R., and Sharp, W. T. (1965). Racah algebra for an arbitrary group, *J. Math. Phys.* **6**, 1584–1590.

Dirac, P. A. M. (1930). *The Principles of Quantum Mechanics*. Clarendon Press, Oxford (4th ed., 1958).

Dobosh, P. A. (1972). Irreducible-tensor theory for the group O^*. I. V and W coefficients, *Phys. Rev. A* **5**, 2376–86.

Edmonds, A. R. (1957). *Angular Momentum in Quantum Mechanics*. Princeton University Press, Princeton, New Jersey (2nd ed., 1963).

Elliott, J. P. (1953). Theoretical studies in nuclear structure. V. The matrix elements of non-central forces with an application to the $2p$-shell, *Proc. R. Soc. London, Ser. A* **218**, 345–370 [BvD].

Elliott, J. P. (1958). Collective motion in the Nuclear Shell model. I. Classification schemes for states of mixed configurations, *Proc. R. Soc. London, Ser. A* **245**, 128–145.

Fano, U., and Racah, G. (1959). *Irreducible Tensorial Sets.* Academic Press, New York.

Feneuille, S. (1967a). Application de la théorie de groupes de Lie aux configurations mélangées, *J. Phys. (Paris)* **28**, 61–6.

Feneuille, S. (1967b). Symmetrie des operateurs de l'interaction Coulombienne pour les configuration $(d+s)^n$, *J. Phys. (Paris)* **28**, 215–27.

Golding, R. M., and Newmarch, J. D. (1977). Symmetry coupling coefficients for double groups *CN*-star, *DN*-star, and *T*-star, *Mol. Phys.* **33**, 1301–1318.

Griffith, J. S. (1961). *The Theory of Transition-Metal Ions.* Cambridge.

Griffith, J. S. (1962). *The Irreducible Tensor Method for Molecular Symmetry Groups*. Prentice-Hall, Englewood Cliffs, New Jersey.

Hamermesh, M. (1962). *Group Theory and Its Application to Physical Problems*. Addison-Wesley, Reading, Massachusetts.

Harnung, S. E. (1973). Irreducible tensors in the octahedral spinor group, *Mol. Phys.* **26**, 473–502.

Harnung, S. E., and Schäffer, C. E. (1972). Phase-fixed 3-Γ Symbols and coupling coefficients for the point groups, *Struct. Bond.* **12**, 201–255.

Hearn, A. C. (1973). *Reduce II Users Manual*. University of Utah.

Jansen, L., and Boon, M. (1967). *Theory of Finite Groups. Applications in Physics.* North-Holland, Amsterdam.

Jucys, A. P. (1968). *Tables of 9j Coefficients for Integral Values of the Parameters with One Parameter Equal to Unity* (in Russian and English). Computing Centre of USSR, Moscow.

Jucys, A. P., Levinson, I. B., and Vanagas, V. V. (1960). *Mathematical Apparatus of the Theory of Angular Momentum* (in Russian). Mintis, Vilnius. [A translation by Sen and Sen has been published, with Jucys transliterated as Yutsis, by the Israel Program for Scientific Translations, Jerusalem (1962).]

Judd, B. R. (1963). *Operator Techniques in Atomic Spectroscopy*. McGraw-Hill, New York.

Judd, B. R. (1967). *Second Quantisation and Atomic Spectroscopy*. Johns Hopkins Press, Baltimore, Maryland.

Judd, B. R. (1968). Group theory in atomic spectroscopy, in *Group Theory and Its Applications*, E. M. Loebl, ed. Academic Press, New York.

Judd, B. R. (1974). Lie groups and the Jahn–Teller effect, *Can. J. Phys.* **52**, 999–1044.

Judd, B. R., and Elliott, J. P. (1970). *Topics in Atomic and Nuclear Theory*. University of Canterbury Press, Christchurch, New Zealand.

Judd, B. R., and Runciman, W. A. (1976). Transverse Zeeman effect for ions in uniaxial crystals, *Proc. R. Soc. London, Ser. A* **352**, 91–108.

Karasev, V. P., and Shelepin, L. A. (1973). Finite Differences, Clebsch–Gordan coefficients, and hypergeometric functions, *Theor. Math. Phys.* 991–8 [*Teor. Mat. Fiz.* **17**, 67–78 (1973)].

Kibler, M. R., and Grenet, G. (1977). Clebsch–Gordan coefficients for chains of groups of interest in quantum chemistry. 2. Chain $SU(2)\supset D_{\infty}\supset D_4\supset D_2$, *Int. J. Quantum Chem.* **11**, 359–379.

Kibler, M. R., and Guichon, P. A. M. (1976). Clebsch–Gordan coefficients for chains of groups of interest in quantum chemistry, *Int. J. Quantum Chem.* **10**, 87–111.

King, R. C. (1975). Branching rules for classical Lie groups using tensor spinor methods, *J. Phys. A* **8**, 429–449.

Konig, E., and Kremer, S. (1973). Symmetry coupling coefficients for point groups and the importance of Racah's lemma for the standardization of phase, *Theor. Chim. Acta.* **32**, 27–40.

Koster, G. F., Dimmock, J. O., Wheeler, R. G., and Statz, H. (1963). *Properties of the Thirty-Two Point Groups*. MIT Press, Cambridge, Massachusetts.

Kramer, P., and Seligman, T. H. (1969). Origin of Regge symmetry for Wigner coefficients of $SU(2)$, *Z. Physik* **219**, 105–113.

Kustov, E. F. (1977). Reduction symmetry method in crystal field theory (III), *Phys. Stat. Sol. (b)* **81**, 421–432.

Lax, M. (1974). *Symmetry Principles in Solid State and Molecular Physics*. Wiley, New York.

Littlewood, D. E. (1950). *The Theory of Group Characters*, 2nd ed. Oxford University Press, Oxford.

Messiah, A. (1965). *Quantum Mechanics*. North-Holland, Amsterdam. (First printing, 1961; original French ed. Dunod, Paris, 1959.)

Mulliken, R. S. (1933). Electronic structures of polyatomic molecules and valence. IV. Electronic states, quantum theory of the double bond, *Phys. Rev.* **43**, 279–302.

Nielson, C. W., and Koster, G. F. (1963). *Spectroscopic Coefficients for the p^N, d^N and f^N Configurations*. MIT Press, Cambridge, Massachusetts.

O'Brien, M. C. M. (1976). Some fractional parentage coefficients for R_5 and their application to $T\otimes(E+\tau_2)$ Jahn–Teller systems such as F and F^+ centres, *J. Phys. C* **9**, 3153–3163.

Piepho, S. B., and Schatz, P. N. (1981). *Group Theory in Spectroscopy with Applications to Magnetic Circular Dichroism*. Wiley, New York. (In press.)

Pooler, D. R., and O'Brien, M. C. M. (1977). The Jahn–Teller effect in a Γ_8 quartet: Equal coupling to ϵ and τ_2 vibrations, *J. Phys. C* **10**, 3769–3791.

Racah, G. (1942). Theory of complex spectra II, *Phys. Rev.* **62**, 438–462 [BvD].

Racah, G. (1943). Theory of complex spectra III, *Phys. Rev.* **63**, 367–382 [BvD].

Racah, G. (1949). Theory of complex spectra IV, *Phys. Rev.* **76**, 1352–1365 [BvD].

Reid, M. F. (1979). Coupling coefficients and crystal field theory, M.Sc. Thesis, University of Canterbury, Christchurch, New Zealand.

Reid, M. F., and Butler, P. H. (1980). Orientations of point groups: Phase choices in the Racah–Wigner algebra, *J. Phys.* **A13**, 2887–2902.

Rose, M. E. (1957). *Elementary Theory of Angular Momentum.* Wiley, New York.

Rotenberg, M., Bivins, R., Metropolis, N., and Wooten, J. K. (1959). *The 3j and 6j symbols.* Technology Press—MIT, Cambridge, Massachusetts.

Smorodinskii, Y. A., and Shelepin, L. A. (1972). Clebsch–Gordan coefficients, viewed from different sides, *Sov. Phys.—Uspekhi* **15**, 1–24 [*Usp. Fiz. Nauk* **106**, 3–45 (1972)].

Stedman, G. E. (1975). A diagram technique for coupling calculations in compact groups, *J. Phys. A* **8**, 1021–1037.

Stedman, G. E. (1976). A diagram technique for basis functions and their transformation, with application to group–subgroup bases and crystal tensors, *J. Phys. A* **9**, 1999–2019.

Stedman, G. E., and Butler, P. H. (1980), Time reversal symmetry in applications of point group theory, *J. Phys. A* **13**, 3125–3140.

Sullivan, J. J. (1978). Recoupling coefficients of the general linear group in bases adapted to shell theories, *J. Math. Phys.* **19**, 1681–7.

Vanagas, V. V. (1972). *Algebraic Methods in Nuclear Theory* (in Russian). Mintis, Vilnius.

Wigner, E. P. (1940). On the matrices which reduce the Kronecker products of representations of S.R. groups (unpublished) [BvD].

Wigner, E. P. (1959). *Group Theory*. Academic Press, New York. [A version of *Gruppen Theorie* (Viereg and Sohn, Brunswick, Germany, 1931), translated by J. J. Griffin and revised and updated by Wigner.]

Wolf, A. A. (1969). Rotation operators, *Am. J. Phys.* **37**, 531–536.

Wybourne, B. G. (1965). *Spectroscopic Properties of the Rare Earths*. Wiley, New York.

Wybourne, B. G. (1970). *Symmetry Principles and Atomic Spectroscopy* (including an appendix of tables by P. H. Butler). Wiley, New York.

Wybourne, B. G. (1973). Lie algebras in quantum chemistry: symmetrized orbitals, *Int. J. Quantum Chem.* **7**, 1117–37.

Subject Index

Abelian, 9–11, 48, 49, 54, 58
Abragam, A., 153, 553
Abstract property, 149
Accuracy of tables, 46, 187
Active rotations, 6, 122, 190, 514
ALGOL, 187
Alternating group
  $\mathbf{A}_4$, 200
  $\mathbf{A}_5$, 205
Alternative bases, 21
Altmann, S. L., 39, 40, 553
Ambivalent, 54, 99
Angular momentum: *see J or JM*
Antilinear operator, 154
Antilinear transformation, 57
Antisymmetric function, 174
Antisymmetric irrep of $\mathbf{U}_n$, 145
Antiunitary, 154
Approximations, 87
Arfken, G., 96, 553
Arithmetic routines, 188
Asterisk, 5
Atomic orbitals, 158
Axial vectors, 103, 108, 113–115
Axioms
  of group, 8
  of vector space, 9
Axis
  direction, 190, 195
  specification, 107–127
  system, 121–122
$\mathbf{A}_4$, 200
$\mathbf{A}_5$, 205

Bacher, R. F., 139, 140, 553
Basis, 7, 9
  functions, 190
  group–subgroup, 47
  independent of, 69
  information, 68
  operator, 84–86
  specification, 107–127
  transformations of, 7, 67
  triads, 75
Bethe, H., 182, 190, 553
Bickerstaff, R. P., 22, 553
Biedenharn, L. C., 1, 4, 5, 61, 94, 156, 553
Biedenharn–Elliott sum, 61–63, 70, 188
Bivins, R., 2, 3, 4, 111, 129, 132, 185, 217, 556
Blatt, J. M., 156, 553
Bleaney, G., 153, 553
Block diagonal, 18, 32
Body axes, 121
Bohr magneton, 176
Boldface, 6
Boon, M., 100, 153, 554
Bra, 8, 154
Branching
  definition, 19, 22
  examples, 108–109
  freedom, 71
  labels, 56
  multiplicity, 19, 22, 45, 217, 218
  for parentage, 148, 150
  resolution, 80
  rotation–reflection, 103
  giving scalars, 87
  $\mathbf{SO}_3$–$\mathbf{D}_\infty$, 131
  tables, 207–216

Brink, D. M., 156, 514, 553
Burnside, W., 29, 40, 553
Butler, P. H., 28, 40, 41, 43, 45, 51, 52, 55, 69, 76, 95, 105, 129, 133, 137, 145, 148, 151, 153, 173, 188, 195, 553–556

Calculations
  computer, 188
  $\mathbf{D}_n$, etc., 130, 136
  rotation–reflections, 104–105
  $\mathbf{SO}_3$–$\mathbf{D}_\infty$, 134
  tetrahedral, 200
  see also $3jm$, $6j$, Transformations, *etc.*
Canonical basis, 11
Cartan, E., 39, 143
Central field, 172
Cfp: *see* Parentage
Chains of groups, 18–22, 195
  $\mathbf{SO}_3$–$\mathbf{D}_\infty$–$\mathbf{SO}_2$, 132–136
Character, 13, 44
  of continuous groups, 144–145
  definition, 30
  orthogonality, 33
  showing determinant, 149
  of similar representations, 30
  table, 35, 189–206
  theory, 7, 14, 22, 28–36, 69
  theory of $\mathbf{D}_n$, 130–131
Characteristic, 7, 29, 33
Characters and symmetries, 51
Choices, 71–81
  JM case, 133
  orientation, 190, 208
  partners, 48
  roots, 79–80, 190
  $2j$, 49
  $2jm$, 45, 78
  $3j$, 50–52, 72–73
  $3jm$, 78–81
  $6j$, 73–77
  see also Phases, Partners, Multiplicity
$\mathbf{C}_i$, inversion group, 99, 191
C_q^k, tensor, 96
Class
  characteristic of, 33
  conjugacy, 33
  $\mathbf{D}_n$, 130
  double groups, 100
  labels, 34, 190, 195
Coincident axes, 120
Combinatorics, 29, 144–145
Commuting groups, 101; *see also* Abelian
Compact, 10; *see also* Continuous
Completeness of recursion, 70
Completeness theorems, 14
Complex conjugate character, 35–36
Complex conjugate labels, 56
Complex conjugate sign, 5
Complex conjugation, 36–38, 153, 219
  operator, K_Q, 155
Complex irreps, 37–38, 54, 200, 430
  and time reversal, 168
Complex numbers, 14
  need for $\mathbf{D}_n$, 132
  need for $3jm$'s, 79
Computation methods, 21, 68–71
Computational simplifications, 142
Computer program, 46, 187–188, 514
Condon, E. U., 1, 172, 176, 514, 554
Condon and Shortley phases, 45, 46, 518
Conjugacy class, 33, 54
Conjugate linear, 154
Conjugate unitary, 154
Continuous groups, 22, 54, 69, 75, 142–145, 188
  characters of, 34, 142–145
  generalizations needed, 52, 54–55, 99
Coordinate functions, 8
Coordinate vectors, 108
Corepresentation, 159, 160, 167
Coulomb interaction, 142, 171–175
Coupling
  coefficients, 1–7, 23–28
  definition, 25
  examples, 107–127
  factor, 28
  freedoms, 25–26, 74
  JM example, 111
  multiplicity, 25, 56, 76, 217–218
  phases, 28, 72, 76–79
  polar vectors, 116
  $\mathbf{SO}_3$–$\mathbf{O}$–$\mathbf{D}_3$–$\mathbf{C}_3$ example, 112–114
  tensors, 89–94; *see also* Reduced matrix elements
Covering groups, 147, 195
  double, 39; *see also* Spin
Crystal field, 87–89
Crystallographic groups, 29, 100, 191, 193, 205
Cube, 121–122, 202, 204
Cubic site, 184–186
Cyclic groups, 48, 191–194
  j, jm symbols, 101–102, 217; *see also* *Table Index*
Cyclic permutations, 51, 52, 54, 59, 219
  see also Symmetries
$\mathbf{C}_1$, omission of, 20
$\mathbf{C}_3$, example of, 10, 11

Degeneracy, 151, 174
under time reversal, 153, 159
Derome, J. R., 1, 4, 43, 49, 51, 57, 554
Derome–Sharp lemma, 55, 57
Derome–Sharp matrix choice 73, 74
Descent in symmetry, 3
Determinant, 142
operator, 149
Determinantal state, 140
Diagram techniques, 2, 67
Dihedral groups, 129–137, 195–199
orientation, 137
structure, 130–132
Dimension
of irreps, 23
of operator space, 84
of space, 9; *see also* Characters *in Table Index*
Dimmock, J. O., 41, 100, 103, 189, 190, 555
Dirac, P. A. M., 8, 554
Dirac notation, 8–10, 84
Dirac δ-function, 154
Direct product
groups, 101–102
matrices, 24, 35
subgroups, 144
Discrete groups, 34
$\mathbf{D}_n$: *see* Dihedral
Dobosh, P. A., 77, 554
Dodecahedron, 205
Double group, 39, 100, 190
Double tensor, 95, 175
Double-valued, 39, 130
Doubling of degeneracy, 158–160
Dual operator, 154
Dual space, 8, 154
Duality
cube–octahedron, 202
icosahedron–dodecahedron, 205

Edmonds, A. R., 156, 514, 554
e_g, parentage, 150
Eigenfunction, 151
Eigenvalue
of matrix, 29
of operator, 149
Electron spin, 38–41
Electrostatic field, 182
Elementary particles, 75
Elements of a group, 6, 8, 30, 130
Elliott, J. P., 61, 95, 146, 149, 151, 554, 555
Elliott model, 140
Environment, 183
Equivalent irreps, 13
Equivalent spaces, 17
Euler angles, 6, 119, 121, 122, 190, 513–518
Even–odd, 191; *see also* Axial vectors
Even permutations, groups of, 200, 205
Exceptional groups, 143, 145
Expansion in unit tensors, 165, 183
Explicit functions, 107
$\mathbf{E}_6, \mathbf{E}_7, \mathbf{E}_8$, 143, 145

Factorization
of cfp, 142
of coupling coefficients, 28
of *jm*'s, 43–48, 87–89
Racah's, 26, 87
restrictions for dihedral chains, 136–137
see also Wigner–Eckart theorem
Faithful, 40, 150
Fano, U., 69, 156, 554
Feneuille, S., 95, 151, 554
Fermions, 161
Feynman diagrams, 2
Finite groups, 10, 22, 99
Fivefold axes, 205
Ford, A. M., 76, 105, 533
Fractional linear, 38
Fractional parentage: *see* Parentage
Free ion, 171–175
Free phase, 25; *see also* Choice
Freedoms, 71–81
Frobenius–Schur invariant, 38, 49
f-shell parentage, 143
Functionals, 8
$\mathbf{F}_4$, 143, 145

$\tilde{\mathbf{G}}$, definition, 103
General equations, 52
General results, 4
Generating functions, 145
Generation of tables, 187–188
Generators of group, 148
Gerade, 101, 149
$\mathbf{G}_i$, definition, 101
GL(n,**C**), definition, 142
Golding, R. M., 129, 554
Goudsmit, S., 139, 140, 553
Gramm–Schmidt orthogonalization, 69
Great orthogonality theorem, 30
Greek letters, 6
Grenet, G., 69, 129, 555
Griffin, J. J., 556
Griffith, J. S., 1, 46, 554
Group
abstract structure, 29
arbitrary, 1

Group (*cont.*)
axioms, 8, 142
double, 39, 100, 190
integration, 57
labels, 101, 141–145
multiplication law, 38
operators, 9
order of, 32
parentage, 141
possible point, 100
simply reducible, 5
see also Continuous
Group–subgroup basis, 18–22, 47
"Group theory," 7
g–u labels, 101, 149–151, 191
Guichon, P. A. M., 69, 555
Gyromagnetic ratio, 176
$\mathbf{G}_2$, 143, 145–148

$H(J_1J_2J)$, 46, 90, 111
Haar integral, 34
Haase, R. A., 40, 145, 188, 554
Hamermesh, M., 29, 38, 39, 76, 100, 104, 145, 190, 554
Hamiltonian
approximate, 151
distortion, 87
free-ion, 171–175
full, 188
perturbation, 22, 81–89
phenomenological, 184, 186
time reversal of, 156
Harnung, S. E., 69, 77, 129, 554
Hearn, A. C., 514, 554
Heisenberg representation, 155
Hermitian, 154, 161
Hermiticity, 153
Hilbert space, 83; *see also* Vector space
Historical phase, 46, 90, 111
Hole–particle, 149, 160
Homomorphic, 9
Horizontal plane, 101, 191
Hydrogenic solutions, 173
Hyperfine, 175

Icosahedral group, 101, 205–206
Icosahedral symmetry, 183
Icosahedron, 205–206
Identity group, $\mathbf{C}_1$, 101
Identity operation, 32
Image ket, 9
Imbeddings, 207–216
abstract, 117
equivalent, 195
nonfaithful, 150
Infinitesimal operator, 3
Inner product, group, 24, 93–94
Inner product, vector, 8
Integer label, 40
Integer representation, 39
Interactions, 83
Interchanges, 51, 52, 57, 219, 430
Invariance group, 3
Invariant space, 7, 14
Invariant subgroup, 151
Invariants, 44
Inversion, 99, 100, 189, 207
Irreducible, 11, 32
Irreducible tensorial set, 86
Irreducibility under $\mathbf{D}_n$, 130
Irrep
complex conjugate, 36, 200
content, 35
definition, 11
faithful, 40
labels, 12, 189–206
power of, 41, 70
primitive, 40
simple phase, 43
space, 7–18
structure, 29
uniqueness, 33
Isomorphic groups, 101
Isomorphic imbeddings, 208
Isomorphic parentage groups, 149
Isomorphic subgroups, 68, 102–107
Isoscalar factor, 2, 5, 28

$\mathbf{J(SO_3)}$, 6, 18
j: *see* $2j$, $3j$, $6j$, etc.
Jahn–Teller effect, 149
Jansen, L., 100, 153, 554
jj coupled, 54
JM basis, 18, 43
reality properties, 37
JM $3jm$ factor, not factorizable, 132
jm: *see* $2jm$, $3jm$, etc.
jm factors, for parentage, 146
Jucys, A. P., 2, 3, 67, 555
Judd, B. R., 5, 6, 69, 86, 87, 95, 140, 144, 146, 147, 148, 151, 156, 173, 175, 184, 555

K, icosahedral group, 101, 205–206
Karasev, V. P., 4, 555
K_{CS}, 156
Ket, 7
Ket labels, 12, 17–18, 107–127, 141–145
Ket space axioms, 8
Kibler, M. R., 69, 129, 555

King, R. C., 43, 51, 54, 145, 553, 555
Konig, E., 69, 555
Koster, G. F., 41, 86, 95, 96, 100, 103, 139, 148, 149, 175, 184, 185, 189, 190, 555
Kramer, P., 94, 555
Kramers degeneracy, 154, 158, 160, 168–169
Kremer, S., 69, 555
Kronecker powers, 40
Kronecker product, 24, 35, 200,
  as inner, 93
  multiplicity, 25
K_t, 155
Kustov, E. F., 151, 555

Labeling
  classes, 34, 190, 195
  groups, 101, 141–145
  irreps, 38–41, 141–145
  kets, 108
  mirror planes, 101, 191
  parentage, 142, 149–150
Labeling operator, 22
Labeling schemes, 189
Labels, 12–13, 17–18
  need for, 22, 48, 217–218
Laboratory axes, 121, 514–515
Ladder operators, 69, 133
Landé g-factor, 177
Latin letters, 5, 6
Lax, M., 38, 153, 190, 555
Legendre polynomials, 96, 173
Levinson, I. B., 2, 67, 555
Lie groups, 10, 142, 144
Ligand field, 168–169, 182–186
  parameters A, B, 182
  parameter X, 183–186
Linear algebra, 7, 84
Linear operators, 153
Linearity, 8
Littlewood, D. E., 39, 144, 145, 555
LS coupling: *see SL*

$\mathbf{M(SO_2)}$, 6, 18
Magnetic: *see* Zeeman
Many-electron: *see* Reduced matrix element
Many-particle states, 23
Many-valued, 39
Mathematical structure, 146
Matrix diagonalization, 188
Matrix elements of group operator, 12–14
Matrix groups, 142–145
Matrix irrep, 11
Matrix representation of operator, 13, 17, 155
Maximal imbeddings (chains), 136–137, 151
Messiah, A., 5, 6, 86, 154, 155, 162, 514, 555
Metropolis, N., 2, 3, 4, 111, 129, 132, 185, 217, 556
Minchin, P.E., 173, 553
Mirror planes, 101, 191
Mixed configurations, 95, 140, 149, 151
Mixed symmetry, 51
Molecular function, 176
Mulliken, R. S., 190, 555
Mulliken notation, 20, 41, 190
Multiconfigurations, 95
Multiplicity, 5
  in **O**, 76–77, 105, 429
  in product, 25, 200
  separation 71–81, 105, 136, 160, 200, 429

N-electron state, 95, 148
Natural labels, 40
Newmarch, J. D., 129, 554
Nielson, C. W., 86, 95, 96, 139, 148, 149, 175, 184, 185, 555
Nielson and Koster tensor, 96
Noncompact groups, 34
Nonprimitive *j*'s and *jm*'s, 68
Normal subgroups, 151
Normalization, 86–87, 94–97, 183–184, 519
Notation conventions, 4–6
Nuclear model, 140, 149

O'Brien, M. C. M., 151, 555
Occupation numbers, 148, 151
Occupied states, 146
Octahedron, 202
Odd-electron systems, 159
$\mathbf{O}_n$ definition, 143
Operator
  basis, 85–86
  group, 6, 8–11
  Hermitian conjugate, 161
  time-reversal conjugate, 162
  see also Euler, Group, Reflection, Rotation, *etc.*
Order of group, 32
Ordering of irreps, 209
Orders within tables, 220, 430, 463
Ordinary representation, 39
Orientation, 69, 71, 100, 107–127
  in branching, 79, 208
  in character tables 190, 195
  choice, 77–80, 117, 195
  of cube, 121, 122, 202, 204
  default, 127
  dihedral groups, 137

Orientation (*cont.*)
  effect of choice, 125–127
  of field, 179–181, 184
  of Hamiltonian, 87
  relationship to tilde, 103
  of tetrahedron, 201
Orthogonal group, 93, 143
Orthogonal irreps, 37–38
Orthogonal matrix, 37, 143
Orthogonal subspaces, 10
Orthogonality
  of characters, 32–34
  of coupling factors, 28
  of $3jm$, 57–58
  of $6j$, 60, 74, 188
Orthogonally irreducible, 14
$\mathbf{O}_3$
  definition, 99, 143
  subgroups of, 99–127, 208
  see also $\mathbf{SO}_3$
$\mathbf{O}_3^p$, 147, 151

p orbitals, 8, 149
Parameters, ligand field, 183–186
Parametrization of $\mathbf{D}_n$, 130
Parentage, 1, 69, 96, 139–151, 174
  cfp definition, 141
  complete schemes, 146–148
  group of, 87, 147
  irrep labels, 141–145, 148
  strong field, 149–151
  under K_Q, 160
Parents of a state, 140
Parity
  of irrep, 208
  of orbitals, 156
  selection rules, 97, 166
  of state, 106
  in time reversal, 161; *see also* Axial, Polar
Particle–hole, 149, 160
Partition, 144
Partners, 11, 190, 513–521
  choice, 18–22, 48
  of complex-conjugate irreps, 36
  freedom, 13, 16, 68–69, 77–81
  in JM basis 107–127, 522–544
  in other bases, 544–551
Passive rotations, 6, 121, 190, 514
Permutation group, 43, 145
Permutations, 29, 200; *see also* Symmetries
Perturbation fields, 171
Phase choices, 71–81
  for $\mathbf{D}_n$, 136–137
  for JM, 129, 133, 514
  effects, 119
Phase conventions, 1, 4–6
Phase freedom, 44
Phase information
  in cfp, 140
  in primitives, 68–69
Phase matrix, 55
Physical structure, 146
Piepho, S. B., 5, 555
Point groups, possible, 100
Polar coordinates, 518–519
Polar vectors, 103, 108, 116–118
Polyboranes, 183
Pooler, D. R., 151, 556
Power of irrep, 41, 70
Power order, 209
Prime numbers, 77, 81, 429
Primitive characters, 33
Primitive irrep, 40, 63, 70
  for rotation–reflections, 104
Primitive triad, 74
Primitive $3jm$, 68, 77–81
  for $\mathbf{D}_n$, 136–137
Primitive $6j$, 63, 68–77
Product labels, 56
Product resolution, 76
Projective representation, 39
Projective transformation, 38
Pseudoreal, 38
Pseudoscalar, 116–117
Pure rotation, 189

Quasi-ambivalent, 54, 57
Quasi-orthogonal, 54–55, 73
Quasiparticle, 147
Quasisymplectic, 54–55

Racah, G., 1, 4, 26, 68, 69, 87, 139, 146, 156, 554, 556
Racah backcoupling, 60, 68, 188
Racah factorization, 44–45, 87, 113
  proof, 26–28
  $\mathbf{SO}_3$–$\mathbf{D}_\infty$–$\mathbf{C}_\infty$, 129
Racah's parentage, 144
Racah tensors, 94–97
  $V^{(11)}$ definition, 97
Radial information, 182
Radial integrals, 173
Rare earth ions, 146
Real irreps, 37–38, 75, 430
Reality or otherwise
  of JM $3jm$, 133
  of $6j$ of $\mathbf{D}_n$, 132
  see also Complex
Rearrangement theorem, 30

Rectangular coordinates, 513, 518–519
Recursion relations, 63, 64, 70–71
for $\mathbf{SO}_3$–$\mathbf{D}_\infty$, 134
REDUCE, 514
Reduced matrix element, 83, 86
angular momentum operator, 97, 177
conjugation symmetries, 165–169
many-electron, 87
single-electron, 94
SL basis, 175
SO_3 scalar, 93
$\mathbf{SO}_3$, *H* phase, 90
spherical tensor, 97
tensor
on one part, $=6j$, 92
of two, $=6j$, 90; of two, $=9j$, 91
scalar product, 91, 93
$\mathbf{SO}_3$, 97
under time reversal, 163–169
unit operator, 87
Zeeman, 175–178
Reducible representation, 10
Reduction into irreps, 25, 35
Reflection, 100, 195, 207
Reflection operators σ_v, σ_h, 106–107
Reflection–rotation, 102–107, 116–118, 189–191, 195, 200, 202
Regge symmetry, 94
Reid, M. F., 129, 137, 188, 195, 553, 556
Relative phases, 46, 72
Relativistic corrections, 174
Reordering information, 68
Reordering symmetry, 429
"Representation," 155
Representation matrices, 10, 11
Representation property for parentage, 141
Representation space, 9
Representations, spin or true, 39
Restrictions on tables, 217
Rose, M. E., 156, 514, 553, 556
Rotation
into coincidence, 121–127
definition, 514–515
full group of, 99
and inversion, 102, 191, 207
matrix, 6, 513–518
point groups, 100
pure, 207
reflection, 108–120, 195, 207
spin 1/2 matrices, 110
2π, 516
Rotenberg, M., 2, 3, 4, 5, 111, 129, 132, 185, 217, 556
Runciman, W. A., 6, 156, 555
Russell–Saunders coupling, 140
Satchler, G. R., 156, 514, 553
Scalar irrep, 44
Scalar operators, 87
Scalar partners, 119
Scalar product, 8, 92, 173
Schäffer, C. E., 69, 129, 554
Schönflies notation, 189
Schrödinger quantum mechanics, 155, 160
Schur, 38–39
function, 144–145
lemmas, 1, 7, 12–14, 30–31, 49, 53, 57, 83, 85
proof, 15–16
summary, 17
Schwinger g_s, 176
Second quantization, 146
Selection rules, 44, 69, 217
and parity, 97
and time reversal, 160–167
Seligman, T. H., 94, 555
Seniority, 139
Sharp, W. T., 1, 4, 43, 49, 57, 554; *see also* Derome–Sharp
Shatz, P. N., 5, 555
Shelepin, L. A., 4, 555, 556
Shell of states, 148
Shortley, G. H., 1, 172, 176, 514, 554
Similar irreps, 12, 13
Similarity transformation, 29, 33
Simple characters, 33
Simple group, 143
Simple phase
generalization, 99
group, 51–52
irrep, 43
Simply reducible, 43, 129
Simultaneous equations, 40
Single configuration, 175
Single-particle state, 148, 172
SL coupled, 54, 92, 140–142
S_l: *see* Symmetric group
Smorodinskii, Y. A., 4, 556
$\mathbf{S}_n$: *see* Symmetric group
$\mathbf{SO}_n$, 143
$\mathbf{SO}_2$, 10
$\mathbf{SO}_3$, 33, 99
$\mathbf{SO}_3$–$\mathbf{D}_\infty$, $3jm$ values, 135
$\mathbf{SO}_3$ tensors, 86–87, 94–97
$\mathbf{SO}_4$, 94
Space groups, 39
Space–time, 158
Spatial functions, 139, 146
Spatial group, 151
Special unitary group, 142
Specification of axes, 107–127

Spectroscopic label, 141
Spherical harmonics, 96, 513–521
  expansion, 519
  i^l convention, 156
  under reflection, 106
Spherical polar, 182
Spherical tensors, 86, 173, 182
Spin, 37, 38–41, 54, 99, 189
  of $\mathbf{D}_n$, 130
  of $\mathbf{O}_n$, 145
  under reflections, 106
  rotations, 123
  and time reversal, 156, 161
  $2j$ for, 49
Spin–orbit, 174
Spin–spin, 175
$\mathbf{Sp}_n$, 143
Spur of matrix, 29
Standard basis, 11–14
Standard transformation, 24
Stark effect, 166
Statz, H., 41, 100, 103, 189, 190, 555
Stedman, G. E., 2, 67, 153, 556
Stern–Gerlach experiment, 38
Strictly complex, 36
Strong field, 139, 149–151
Structural information, 69
Structural properties, 29
Subgroup labels, 19
Subgroups
  of $\mathbf{D}_\infty$, 136–137
  of $\mathbf{O}_3$, 99–127, 208
Subspace, proper, 15
Sullivan, J. J., 68, 556
$\mathbf{SU}_n$, 142
Superfluous labels, 48
Symmetric groups, 29, 68, 76, 145
  $\mathbf{S}_4$, 202
Symmetric square, 51
Symmetries
  approximate, 146, 205
  $\mathbf{D}_\infty$ examples, 132–134
  of Hamiltonian, 172
  of j, 56, 59–60, 65, 429–430, 463
  of jm, 48, 57, 218–220
  of Kronecker product, 36
  rules, 165–169
  tilde, 76–77
  Wigner's, 26
Symmetrized bases, 7
Symmetrized products, 51–52
  of $\mathbf{D}_n$, 131
Symmetry chains, 107–127
Symplectic group, 143
Symplectic irreps, 37–38
Symplectic matrix, 38, 143

Tables, other, 1–6
Tensor operators, 1–6, 83, 89–94
  definition, 86
  for mixed configurations, 95
  parentage labels of, 148
  for $\mathbf{SO}_3$, 94–97
  see also Reduced matrix element
Tensor product, 23–25, 35
Tetrahedral group, 200–201
Tetrahedron, 200–202
Tilde labels, 41, 76–77, 103, 197, 209
  choice of, 103
Tilde symmetry, 76–77
Time reversal, 6, 107, 153–169
  conjugate, 162
  invariant, 158–159
  operator, 155, 160
  parity, 161
  signature, 162
Time-reversed state, 156, 158, 160
Trace, 7, 29, 130
Transformation,
  between bases, 119–127, 513
  brackets, 21
  coefficients, 21, 119, 125, 188
  JM to $\mathbf{SO}_3$-$\mathbf{D}_\infty$-$\mathbf{SO}_2$, 135
  of operator, 83–85
  to symmetrized basis, 140
Transition metal, 184, 202
Triad, 47, 217–220, 429
  basis, 75
  primitive, 74
  for $\mathbf{SO}_3$, 59
  in $6j$, 58–59
Trivial j's and jm's, 68
True irreps, 39–41, 49, 54, 99, 159
Typesetting, 187
t_{2g} parentage, 150

u–g labels, 191
$\mathbf{U}_n$, 141–151
Uniqueness
  of basis, 12
  of orientation, 125
Unit matrix, 32
Unit operator, 9, 87
Unit tensors, 95
Unitary group, 139, 141–151
Unitary matrix, 10, 142

Unitary operator, 154
Unitary representation, 10

V coefficients, 4
Van Dam, H., 1, 4, 5, 553
Vanagas, V. V., 1, 2, 67, 555, 556
Vector space, 7, 84
  axioms, 8
  representation, 39
Vertex-face duality, 202, 205
Vibrational, 149
$V^{(11)}$, 96
V_p, 8
V_r, 8, 10

W coefficients, 4
Wave function, 8
Weak field, 139
Wheeler, R. G., 41, 100, 103, 189, 190, 555
Wigner, E. P., 1, 5, 26, 43, 54, 129, 154, 155, 158, 217, 556
Wigner–Eckart theorem, 43, 69, 83–97
  evaluation example, 87
  factorization, 87
  and parentage, 146–151
  proof, 86
  and time reversal, 160–169
  see also Reduced matrix element
Wigner notation, 5, 43, 217
Wigner relation, 64,
Wigner rotation formula, 516
Wigner's phase, 46
Wolf, A. A., 6, 514, 556
Wooten, J. K., 2, 3, 4, 111, 129, 132, 185, 217, 556
Wybourne, B. G., 5, 22, 29, 40, 41, 45, 52, 55, 69, 95, 144, 145, 146, 147, 148, 151, 173, 177, 182, 188, 553, 554, 556

X coefficients, 4

Y_{lm}, 96, 519–521; *see also* Spherical harmonic

Zeeman interaction, 88, 116–118, 176–181

1*j* of Wigner, 5, 43
2*j*, 5, 48–49
  choice, 49, 55
  for $\mathbf{SO}_3$, 52
  as special case of 3*j*, 56
2*jm*, 5, 44
  choices, 45, 78
2*jm* (*cont.*)
  factorization, 47, 56–57
  reality, 45
  $\mathbf{SO}_3$–$\mathbf{SO}_2$, 45, 57
  as special case of 3*jm*, 56
2*jm* operator *T*, 36–37, 156
3*j*, 49, 217
  choices, 50–52, 56, 72–73
  for $\mathbf{D}_n$, 131
  for $\mathbf{SO}_3$, 52
  as special case of 6*j*, 60
  symmetries, 56
3*j* of Rotenberg, 5, 217
3*j* of Wigner, 5, 43, 217
3*jm*, 2–5
  and basis changes, 67–68
  choices, 52, 78
  complex, 79–80
  coupling factor, 48, 56
  definition, 45
  factorization, 3, 47–48
  $\mathbf{G}\times\mathbf{G}\to\mathbf{G}$ is 9*j*, 94
  labels, 48, 217–219
  multiplicity resolution in, 80–81
  orthogonality, 57–58
  as overlap, 46
  recursion, 64, 71
  for rotation–reflection, 104–105
  for $\mathbf{SO}_3$–$\mathbf{SO}_2$, 2–5, 58, 217
  symmetries, 57, 218–219
  with 6*j* in sums 58, 63, 64
  values: *see Table Index*
4*jm*, 52
6*j*, 4, 52–54
  Biedenharn–Elliott sum rule, 61–63, 70
  multiplicity resolution in, 76
  orthogonality, 60
  Racah backcoupling, 60
  for $\mathbf{SO}_3$, 5, 59
  symmetries, 59–60, 429–430
  triad positions, 58, 429
  for rotation–reflection groups, 103
  3*j* as special case, 60
  as *jm*'s, 58
  values: *see Table Index*
9*j*, 54–55
  symmetries, 65, 463
  as 3*jm* sum, 66
  as 3*jm* of product group, 93
  values: *see Table Index, or use 6j sum*, 66
  6*j* as special case, 66
  as 6*j* sum, 66
Γ symbols, 4
σ_h and σ_v, 106–107, 191

Table Index

Branching rules
$C_2 \to C_1$ to $C_6 \to C_3$, 210
$C_{6h} \to C_{3h}$ to $D_4 \to D_2$, 211
$D_{4h} \to C_{4v}$ to $D_6 \to D_3$, 212
$D_{6h} \to C_{6v}$ to $D_\infty \to D_6$, 213
$T \to C_3$ to $O \to T$, 214
$K \to D_3$ to $SO_3 \to D_\infty$, 215
$SO_3 \to O$ to $SO_3 \to K$, 216
Characters and general information
C_i, 191
$C_1, S_2, C_2, C_s, C_{2h}, C_3, S_6$, 192
$C_4, S_4, C_{4h}, C_5, S_{10}$, 193
C_6, C_{3h}, C_{6h}, 194
$D_2, C_{2v}, D_{2h}, D_3, C_{3v}, D_{3d}$, 196
$D_4, C_{4v}, D_{2d}, D_{4h}$, 197
D_5, C_{5v}, D_{5d}, 198
$D_6, C_{6v}, D_{3h}, D_{6h}$, 199
T, T_h, 200–201
O, T_d, O_h, 202–204
K, K_h, 205–206
Parentage t_{2g}^N, e_g^N, 150
Partners
JM as xyz, 520
SO_3–O–D_3–C_3 as JM, 522
SO_3–O–D_4–C_4 as JM, 527
SO_3–K–D_3–C_3 as JM, 531
SO_3–K–D_5–C_5 as JM, 537
SO_3–D_∞–SO_2 as JM, 135
SO_3–D_∞–D_6–C_6 as JM, 541
D_2–C_{2y} as D_2–C_2, 544
D_3–C_2 as D_3–C_3, 544
D_4–D_2–C_2 as D_4–C_4, 545
D_5–C_2 as D_5–C_5, 545
D_6–D_3–C_3 as D_6–C_6, 546
T–D_2–C_2 as T–C_3, 546
O–T–D_2–C_2 as O–D_4–D_2–C_2, 547
O–T–C_3 as O–D_3–C_3, 548
O–D_3–C_3 as O–D_4–C_4, 549
O–D_3–C_2 as O–D_4–C_4, 550
K–T–C_3 as K–D_3–C_3, 551
Rotation matrices, 517
Subgroup chains
crystallographic groups, 209
pure rotation groups, 208
$2jm$ and $3jm$ tables
D_2–C_2, C_{2v}–C_2, D_{2h}–C_{2h}, 220
D_2–C_{2y}, C_{2v}–C_s, 220
D_{2h}–C_{2v}, 221
D_3–C_2, C_{3v}–C_s, D_{3d}–C_{2h}, 221
D_3–C_3, C_{3v}–C_3, D_{3d}–C_{3i}, 222
D_{3d}–C_{3v}, 222
D_4–C_4, C_{4v}–C_4, D_{2d}–S_4, D_{4h}–C_{4h}, 223
D_4–D_2, C_{4v}–C_{2v}, D_{2d}–D_2, D_{2d}–C_{2v}, D_{4h}–D_{2h}, 223
D_{4h}–C_{4v}, 224
D_{4h}–D_{2d}, 225
D_5–C_2, C_{5v}–C_s, D_{5d}–C_{2h}, 226
D_5–C_5, C_{5v}–C_5, D_{5d}–C_{5i}, 227
D_{5d}–C_{5v}, 227
D_6–C_6, C_{6v}–C_6, D_{3h}–C_{3h}, D_{6h}–C_{6h}, 228
D_6–D_2, C_{6v}–C_{2v}, D_{3h}–C_{2v}, D_{6h}–D_{2h}, 229
D_6–D_3, C_{6v}–C_{3v}, D_{3h}–D_3, D_{3h}–C_{3v}, D_{6h}–D_{3d}, 230
D_{6h}–C_{6v}, 231

D_{6h}–D_{3h}, 232
D_∞–D_4, $C_{\infty v}$–C_{4v}, $D_{\infty h}$–D_{4h}, 234
D_∞–D_5, $C_{\infty v}$–C_{5v}, $D_{\infty h}$–D_{5d}, 236
D_∞–D_6, $C_{\infty v}$–C_{6v}, $D_{\infty h}$–D_{6h}, 238
T–C_3, T_h–C_{3i}, 240
T–D_2, T_h–D_{2h}, 241
O–D_3, T_d–C_{3v}, O_h–D_{3d}, 242
O–D_4, T_d–D_{2d}, O_h–D_{4h}, 243
O–T, T_d–T, O_h–T_h, 245
O_h–T_d, 245
K–D_3, K_h–D_{3d}, 248
K–D_5, K_h–D_{5d}, 254
K–T, K_h–T_h, 258
D_∞–SO_2, $D_{\infty h}$–$C_{\infty h}$, 133
SO_3–O, O_3–O_h, 262
SO_3–K, O_3–K_h, 378
SO_3–D_∞, O_3–$D_{\infty h}$, 135
3j and 6j tables
C_1, S_2, 431
C_2, C_{2y}, C_s, C_{2h}, 431
C_3, S_6, 431
C_4, S_4, C_{4h}, 431
C_5, S_{10}, 431
C_6, C_{3h}, C_{6h}, 431
D_2, C_{2v}, D_{2h}, 432
D_3, C_{3v}, D_{3d}, 432
D_4, C_{4v}, D_{2d}, D_{4h}, 433
D_5, C_{5v}, D_{5d}, 434
D_6, C_{6v}, D_{3h}, D_{6h}, 435
T, T_h, 437
O, T_d, O_h, 439
K, K_h, 443
9j tables
T, T_h, 464
O, T_d, O_h, 471
for other groups see the formula on page 66